The MEMS Handbook

The Mechanical Engineering Handbook Series

Series Editor
Frank Kreith
Consulting Engineer

Published Titles

Handbook of Heating, Ventilation, and Air Conditioning
    Jan F. Kreider
Computational Intelligence in Manufacturing Handbook
    Jun Wang and Andrew Kusiak
The CRC Handbook of Mechanical Engineering
    Frank Kreith
The CRC Handbook of Thermal Engineering
    Frank Kreith
The Handbook of Fluid Dynamics
    Richard W. Johnson
Hazardous and Radioactive Waste Treatment Technologies Handbook
    Chang Ho Oh
The MEMS Handbook
    Mohamed Gad-el-Hak

Forthcoming Titles

Biomedical Technology and Devices Handbook
    James Moore and George Zouridakis
Fuel Cell Technology Handbook
    Gregor Hoogers
Air Pollution Control Technology Handbook
    Karl B. Schnelle and Charles A. Brown
Handbook of Mechanical Engineering, Second Edition
    Frank Kreith and Massimo Capobianchi
Handbook of Non-Destructive Testing and Evaluation Engineering
    James Tulenko and David Hintenlang
Inverse Engineering Handbook
    Keith A. Woodbury

The
MEMS Handbook

Edited by
Mohamed Gad-el-Hak
University of Notre Dame

CRC PRESS

Boca Raton   London   New York   Washington, D.C.

Cover Photographs

Foreground: The first walking microrobot with a Swedish wasp relishing a ride on its back. The out-of-plane rotation of the eight legs is obtained by thermal shrinkage of polyimide in V-grooves (PVG). Leg movements are effected by sending heating pulses via integrated heaters causing the polyimide joints to expand. The size of the silicon legs is $1000 \times 600 \times 30$ μm, and the overall chip size of the robot is $15 \times 5 \times 0.5$ mm. The walking speed is 6 mm/s and the robot can carry 50 times its own weight. (Photograph by Per Westergård, Vetenskapsjournalisterna, Sweden, courtesy of Thorbjörn Ebefors, Royal Institute of Technology, Sweden.)

Background: A 12-layer microchain fabricated in nickel using the Electrochemical Fabrication (EFAB) technology. Overall height of the chain is approximately 100 μm and the width of a chain link is about 290 μm. All horizontal links are free to move, while the vertical links are attached to the substrate. By simply including a sacrificial layer beneath the links, the entire chain can be released from the substrate. The microchain is fabricated in a pre-assembled state, without the need for actual assembly. A humble, picnic-loving ant would tower over the microchain shown here. (The scanning electron micrograph courtesy of Adam L. Cohen, MEMGen Corporation, Torrance, California.)

Library of Congress Cataloging-in-Publication Data

The MEMS handbook / edited by Mohamed Gad-el-Hak.
       p. cm.— (Mechanical engineering handbook series)
    Includes bibliographical references and index.
    ISBN 0-8493-0077-0 (alk. paper)
    1. Mechanical engineering—Handbooks, manuals, etc. I. Gad-el-Hak, M. II. Series.

TJ151 .M46 2001
621—dc21   2001037402

This book contains information obtained from authentic and highly regarded sources. Reprinted material is quoted with permission, and sources are indicated. A wide variety of references are listed. Reasonable efforts have been made to publish reliable data and information, but the authors and the publisher cannot assume responsibility for the validity of all materials or for the consequences of their use.

Neither this book nor any part may be reproduced or transmitted in any form or by any means, electronic or mechanical, including photocopying, microfilming, and recording, or by any information storage or retrieval system, without prior permission in writing from the publisher.

All rights reserved. Authorization to photocopy items for internal or personal use, or the personal or internal use of specific clients, may be granted by CRC Press LLC, provided that $1.50 per page photocopied is paid directly to Copyright Clearance Center, 222 Rosewood Drive, Danvers, MA 01923 USA The fee code for users of the Transactional Reporting Service is ISBN 0-8493-0077-0/02/$0.00+$1.50. The fee is subject to change without notice. For organizations that have been granted a photocopy license by the CCC, a separate system of payment has been arranged.

The consent of CRC Press LLC does not extend to copying for general distribution, for promotion, for creating new works, or for resale. Specific permission must be obtained in writing from CRC Press LLC for such copying.

Direct all inquiries to CRC Press LLC, 2000 N.W. Corporate Blvd., Boca Raton, Florida 33431.

Trademark Notice: Product or corporate names may be trademarks or registered trademarks, and are used only for identification and explanation, without intent to infringe.

Visit the CRC Press Web site at www.crcpress.com

© 2002 by CRC Press LLC

No claim to original U.S. Government works
International Standard Book Number 0-8493-0077-0
Library of Congress Card Number 2001037402
Printed in the United States of America   3 4 5 6 7 8 9 0
Printed on acid-free paper

Preface

In a little time I felt something alive moving on my left leg, which advancing gently forward over my breast, came almost up to my chin; when bending my eyes downward as much as I could, I perceived it to be a human creature not six inches high, with a bow and arrow in his hands, and a quiver at his back. ...I had the fortune to break the strings, and wrench out the pegs that fastened my left arm to the ground; for, by lifting it up to my face, I discovered the methods they had taken to bind me, and at the same time with a violent pull, which gave me excessive pain, I a little loosened the strings that tied down my hair on the left side, so that I was just able to turn my head about two inches. ...These people are most excellent mathematicians, and arrived to a great perfection in mechanics by the countenance and encouragement of the emperor, who is a renowned patron of learning. This prince has several machines fixed on wheels, for the carriage of trees and other great weights.

(From *Gulliver's Travels—A Voyage to Lilliput*, by Jonathan Swift, 1726.)

The length-scale of man, at slightly more than 10^0 m, amazingly fits right in the middle of the smallest subatomic particle, which is approximately 10^{-26} m, and the extent of the observable universe, which is of the order of 10^{26} m. Toolmaking has always differentiated our species from all others on Earth. Aerodynamically correct wooden spears were carved by *archaic Homo sapiens* close to 400,000 years ago. Man builds things consistent with his size, typically in the range of two orders of magnitude larger or smaller than himself. But humans have always strived to explore, build and control the extremes of length and time scales. In the *Voyages to Lilliput* and *Brobdingnag of Gulliver's Travels*, Jonathan Swift speculates on the remarkable possibilities that diminution or magnification of physical dimensions provides. The Great Pyramid of Khufu was originally 147 m high when completed around 2600 B.C., while the Empire State Building, constructed in 1931, is 449 m tall. At the other end of the spectrum of manmade artifacts, a dime is slightly less than 2 cm in diameter. Watchmakers have practiced the art of miniaturization since the 13th century. The invention of the microscope in the 17th century opened the way for direct observation of microbes and plant and animal cells. Smaller things were manmade in the latter half of the 20th century. The transistor in today's integrated circuits has a size of 0.18 μm in production and approaches 10 nm in research laboratories.

Microelectromechanical systems (MEMS) refer to devices that have a characteristic length of less than 1 mm but more than 1 μm, that combine electrical and mechanical components and that are fabricated using integrated circuit batch-processing technologies. Current manufacturing techniques for MEMS include surface silicon micromachining; bulk silicon micromachining; lithography, electrodeposition and plastic molding; and electrodischarge machining. The multidisciplinary field has witnessed explosive growth during the last decade, and the technology is progressing at a rate that far exceeds that of our understanding of the physics involved. Electrostatic, magnetic, electromagnetic, pneumatic and thermal actuators, motors, valves, gears, cantilevers, diaphragms and tweezers of less than 100-μm size have been fabricated. These have been used as sensors for pressure, temperature, mass flow, velocity, sound and chemical composition; as actuators for linear and angular motions; and as simple components for complex systems such as robots, micro heat engines and micro heat pumps. Worldwide market projections for MEMS devices tend to be optimistic, reaching $30 billion by the year 2004.

This handbook covers several aspects of microelectromechanical systems or, more broadly, the art and science of electromechanical miniaturization. MEMS design, fabrication and application as well as the physical modeling of their materials, transport phenomena and operations are all discussed. Chapters on the electrical, structural, fluidic, transport and control aspects of MEMS are included. Other chapters cover existing and potential applications of microdevices in a variety of fields including instrumentation and distributed control. The book is divided into four parts: Part I provides background and physical considerations, Part II discusses the design and fabrication of microdevices, Part III reviews a few of the applications of microsensors and microactuators, and Part IV ponders the future of the field. There are 36 chapters written by the world's foremost authorities on this mutidisciplinary subject. The 54 contributing authors come from the U.S., China (Hong Kong), Israel, Korea, Sweden and Taiwan and are affiliated with academia, government and industry. Without compromising rigorousness, the text is designed for maximum readability by a broad audience having an engineering or science background. As expected when several authors are involved, and despite the editor's best effort, the different chapters vary in length, depth, breadth and writing style.

The MEMS Handbook should be useful as a reference book to scientists and engineers already experienced in the field or as a primer to researchers and graduate students just getting started in the art and science of electromechanical miniaturization. The editor-in-chief is very grateful to all the contributing authors for their dedication to this endeavor and selfless, generous giving of their time with no material reward other than the knowledge that their hard work may one day make the difference in someone else's life. Ms. Cindy Renee Carelli has been our acquisition editor and lifeline to CRC Press. Cindy's talent, enthusiasm and indefatigability were highly contagious and percolated throughout the entire endeavor.

Mohamed Gad-el-Hak
Notre Dame, Indiana
1 January 2001

Editor-in-Chief

Mohamed Gad-el-Hak received his B.Sc. (*summa cum laude*) in mechanical engineering from Ain Shams University in 1966 and his Ph.D. in fluid mechanics from the Johns Hopkins University in 1973. Dr. Gad-el-Hak has since taught and conducted research at the University of Southern California, University of Virginia, Institut National Polytechnique de Grenoble, Université de Poitiers and Friedrich-Alexander-Universität Erlangen-Nürnberg and has lectured extensively at seminars in the U.S. and overseas. Dr. Gad-el-Hak is currently Professor of Aerospace and Mechanical Engineering at the University of Notre Dame. Prior to that, he was a Senior Research Scientist and Program Manager at Flow Research Company in Seattle, WA, where he managed a variety of aerodynamic and hydrodynamic research projects.

Dr. Gad-el-Hak is world renowned for advancing several novel diagnostic tools for turbulent flows, including the laser-induced fluorescence (LIF) technique for flow visualization; for discovering the efficient mechanism via which a turbulent region rapidly grows by destabilizing a surrounding laminar flow; for introducing the concept of targeted control to achieve drag reduction, lift enhancement and mixing augmentation in boundary-layer flows; and for developing a novel viscous pump suited for microelectromechanical systems (MEMS) applications. Gad-el-Hak's work on Reynolds number effects in turbulent boundary layers, published in 1994, marked a significant paradigm shift in the subject. He holds two patents: one for a drag-reducing method for airplanes and underwater vehicles and the other for a lift-control device for delta wings. Dr. Gad-el-Hak has published over 340 articles, authored/edited eight books and conference proceedings, and presented 195 invited lectures. He is the author of the book *Flow Control: Passive, Active, and Reactive Flow Management*, and editor of the books *Frontiers in Experimental Fluid Mechanics*, *Advances in Fluid Mechanics Measurements*, and *Flow Control: Fundamentals and Practices*.

Dr. Gad-el-Hak is a fellow and life member of the American Physical Society, a fellow of the American Society of Mechanical Engineers, an associate fellow of the American Institute of Aeronautics and Astronautics, a member of the American Academy of Mechanics, a research fellow of the American Biographical Institute and a member of the European Mechanics Society. From 1988 to 1991, Dr. Gad-el-Hak served as Associate Editor for *AIAA Journal*. He is currently serving as Associate Editor for *Applied Mechanics Reviews* as well as Contributing Editor for Springer-Verlag's *Lecture Notes in Engineering* and *Lecture Notes in Physics*, for McGraw-Hill's *Year Book of Science and Technology*, and for CRC Press' *Mechanical Engineering Series*.

Dr. Gad-el-Hak serves as consultant to the governments of Egypt, France, Germany, Sweden and the U.S., the United Nations, and numerous industrial organizations. During the 1991/1992 academic year, he was a visiting professor at Institut de Mécanique de Grenoble, France. During the summers of 1993, 1994 and 1997, Dr. Gad-el-Hak was, respectively, a distinguished faculty fellow at Naval Undersea Warfare Center, Newport, RI; a visiting exceptional professor at Université de Poitiers, France; and a Gastwissenschaftler (guest scientist) at Forschungszentrum Rossendorf, Dresden, Germany. In 1998, Professor Gad-el-Hak was named the Fourteenth ASME Freeman Scholar. In 1999, Gad-el-Hak was awarded the prestigious Alexander von Humboldt Prize—Germany's highest research award for senior U.S. scientists and scholars in all disciplines.

Contributors

Professor Ronald J. Adrian
Department of Theoretical and
  Applied Mechanics
University of Illinois at
  Urbana–Champaign
Urbana, Illinois
E-mail: r-adrian@uiuc.edu

Professor Ramesh K. Agarwal
National Institute for Aviation
  Research
Wichita State University
Wichita, Kansas
E-mail: agarwal@niar.twsu.edu

Dr. Glenn M. Beheim
NASA Glenn Research Center
Cleveland, Ohio
E-mail:
  Glenn.M.Beheim@grc.nasa.gov

Professor Paul L. Bergstrom
Department of Electrical and
  Computer Engineering
Michigan Technological University
Houghton, Michigan
E-mail: paulb@mtu.edu

Professor Gary H. Bernstein
Department of Electrical
  Engineering
University of Notre Dame
Notre Dame, Indiana
E-mail: bernstein.1@nd.edu

Professor Ali Beskok
Department of Mechanical
  Engineering
Texas A&M University
College Station, Texas
E-mail: ABeskok@mengr.tamu.edu

Professor Thomas R. Bewley
Department of Mechanical and
  Aerospace Engineering
University of California, San Diego
La Jolla, California
E-mail: bewley@ucsd.edu

Professor Kenneth S. Breuer
Division of Engineering
Brown University
Providence, Rhode Island
E-mail:
  Kenneth_Breuer@brown.edu

Professor Hsueh-Chia Chang
Department of Chemical
  Engineering
University of Notre Dame
Notre Dame, Indiana
E-mail: hchang@nd.edu

Dr. Liang-Yu Chen
NASA Glenn Research Center
Cleveland, Ohio
E-mail: Liangyu.Chen@grc.nasa.gov

Professor Haecheon Choi
School of Mechanical and Aerospace
  Engineering
Seoul National University
Seoul, Republic of Korea
E-mail: choi@socrates.snu.ac.kr

Dr. Todd Christenson
Photonics and Microfabrication
Sandia National Laboratories
Albuquerque, New Mexico
E-mail: trchris@sandia.gov

Mr. Adam L. Cohen
MEMGen Corporation
Torrance, California
E-mail: acohen@memgen.com

Dr. Thorbjörn Ebefors
Department of Signals, Sensors and
  Systems
Royal Institute of Technology
Stockholm, Sweden
E-mail: thorbjorn.ebefors@s3.kth.se

Professor Mohamed Gad-el-Hak
Department of Aerospace and
  Mechanical Engineering
University of Notre Dame
Notre Dame, Indiana
E-mail: gadelhak@nd.edu

Professor Yogesh Gianchandani
Department of Electrical and
  Computer Engineering
University of Wisconsin–Madison
Madison, Wisconsin
E-mail: yogesh@engr.wisc.edu

Professor Holly V. Goodson
Department of Chemistry and
  Biochemistry
University of Notre Dame
Notre Dame, Indiana
E-mail: hgoodson@nd.edu

Professor Bill Goodwine
Department of Aerospace and
  Mechanical Engineering
University of Notre Dame
Notre Dame, Indiana
E-mail: jgoodwin@nd.edu

Dr. Gary W. Hunter
NASA Glenn Research Center
Cleveland, Ohio
E-mail:
 Gary.W.Hunter@grc.nasa.gov

Professor George Em Karniadakis
Center for Fluid Mechanics
Brown University
Providence, Rhode Island
E-mail: gk@cfm.brown.edu

Mr. Robert M. Kirby
Division of Applied Mathematics
Brown University
Providence, Rhode Island
E-mail: kirby@cfm.brown.edu

Dr. Jih-Fen Lei
NASA Glenn Research Center
Cleveland, Ohio
E-mail: Jih-Fen.Lei@grc.nasa.gov

Dr. Gary G. Li
OMM, Inc.
San Diego, California
E-mail: gli@omminc.com

Professor Chung-Chiun Liu
Electronics Design Center
Case Western Reserve University
Cleveland, Ohio
E-mail: cxl9@po.cwru.edu

Professor Lennart Löfdahl
Thermo and Fluid Dynamics
Chalmers University of Technology
Göteborg, Sweden
E-mail: lelo@tfd.chalmers.se

Dr. Marc J. Madou
Nanogen, Inc.
San Diego, California
E-mail: mmadou@nanogen.com

Dr. Darby B. Makel
Makel Engineering, Inc.
Chico, California
E-mail:
 dmakel@makelengineering.com

Professor Kartikeya Mayaram
Department of Electrical and
 Computer Engineering
Oregon State University
Corvallis, Oregon
E-mail: karti@ece.orst.edu

Mr. J. Jay McMahon
Electrcial Engineering and
 Computer Science Department
Case Western Reserve University
Cleveland, Ohio
E-mail: jjm@cwru.edu

Professor Mehran Mehregany
Electrical Engineering and
 Computer Science Department
Case Western Reserve University
Cleveland, Ohio
E-mail: mxm31@cwru.edu

Dr. Oleg Mikulchenko
Department of Electrical and
 Computer Engineering
Oregon State University
Corvallis, Oregon
E-mail: mikul@ece.orst.edu

Mr. Joshua I. Molho
Department of Mechanical
 Engineering
Stanford University
Stanford, California
E-mail: jmolho@stanford.edu

Professor E. Phillip Muntz
Department of Aerospace and
 Mechanical Engineering
University of Southern California
Los Angeles, California
E-mail: muntz@spock.usc.edu

Professor Ahmed Naguib
Department of Mechanical
 Engineering
Michigan State University
East Lansing, Michigan
E-mail: naguib@egr.msu.edu

Dr. Robert S. Okojie
NASA Glenn Research Center
Cleveland, Ohio
E-mail:
 Robert.S.Okojie@grc.nasa.gov

Dr. Andrew D. Oliver
Electromechanical Engineering
 Department
Sandia National Laboratories
Albuquerque, New Mexico
E-mail: adolive@sandia.gov

Professor Alexander Oron
Department of Mechanical
 Engineering
Technion–Israel Institute of
 Technology
Haifa, Israel
E-mail:
 meroron@bsf97.technion.ac.il

Mr. Jae-Sung Park
Department of Electrical and
 Computer Engineering
University of Wisconsin–Madison
Madison, Wisconsin
E-mail: jae-sung@cae.wisc.edu

Professor G. P. "Bud" Peterson
Provost Office
Rensselaer Polytechnic Institute
Troy, New York
E-mail: peterson@rpi.edu

Mr. David W. Plummer
Sandia National Laboratories
Albuquerque, New Mexico
E-mail: dwplumm@sandia.gov

Professor Juan G. Santiago
Department of Mechanical
 Engineering
Stanford University
Stanford, California
E-mail:
 santiago@vonkarman.stanford.edu

Professor Mihir Sen
Department of Aerospace and
 Mechanical Engineering
University of Notre Dame
Notre Dame, Indiana
E-mail: msen@nd.edu

Dr. Kendra V. Sharp
Department of Theoretical and
 Applied Mechanics
University of Illinois at
 Urbana–Champaign
Urbana, Illinois
E-mail: kvsharp@students.uiuc.edu

Professor William N. Sharpe, Jr.
Department of Mechanical
 Engineering
Johns Hopkins University
Baltimore, Maryland
E-mail: sharpe@jhu.edu

Professor Gregory L. Snider
Department of Electrical
 Engineering
University of Notre Dame
Notre Dame, Indiana
E-mail: gsnider@nd.edu

Professor Göran Stemme
Department of Signals, Sensors and
 Systems
Royal Institute of Technology
Stockholm, Sweden
E-mail: goran.stemme@s3.kth.se

Mr. Robert H. Stroud
The Aerospace Corporation
Chantilly, Virginia
E-mail: robertstroud@earthlink.net

Dr. William Trimmer
Belle Mead Research, Inc.
Hillsborough, New Jersey
E-mail: william@trimmer.net

Professor Fan-Gang Tseng
Department of Engineering and
 System Science
National Tsing Hua University
Hsinchu, Taiwan
E-mail: fangang@ess.nthu.edu.tw

Dr. Stephen E. Vargo
SiWave, Inc.
Arcadia, California
E-mail: s.vargo@siwaveinc.com

Mr. Chester Wilson
Department of Electrical and
 Computer Engineering
University of Wisconsin–Madison
Madison, Wisconsin
E-mail:
 cgwilso1@students.wisc.edu

Dr. Keon-Young Yun
National Institute for Aviation
 Research
Wichita State University
Wichita, Kansas
E-mail: kyun@twsu.edu

Professor Yitshak Zohar
Department of Mechanical
 Engineering
Hong Kong University of Science
 and Technology
Kowloon, Hong Kong
E-mail: mezohar@ust.hk

Dr. Christian A. Zorman
Electrical Engineering and
 Computer Science Department
Case Western Reserve University
Cleveland, Ohio
E-mail: caz@po.cwru.edu

Contents

Preface v

Editor-in-Chief vii

Contributors ix

I Background and Fundamentals

1 Introduction   *Mohamed Gad-el-Hak* 1-1

2 Scaling of Micromechanical Devices   *William Trimmer and Robert H. Stroud* 2-1

3 Mechanical Properties of MEMS Materials   *William N. Sharpe, Jr.* 3-1

4 Flow Physics   *Mohamed Gad-el-Hak* 4-1

5 Integrated Simulation for MEMS: Coupling Flow-Structure-Thermal-Electrical Domains   *Robert M. Kirby, George Em Karniadakis, Oleg Mikulchenko and Kartikeya Mayaram* 5-1

6 Liquid Flows in Microchannels   *Kendra V. Sharp, Ronald J. Adrian, Juan G. Santiago and Joshua I. Molho* 6-1

7 Burnett Simulations of Flows in Microdevices   *Ramesh K. Agarwal and Keon-Young Yun* 7-1

8 Molecular-Based Microfluidic Simulation Models   *Ali Beskok* 8-1

9 Lubrication in MEMS   *Kenneth S. Breuer* 9-1

10 Physics of Thin Liquid Films *Alexander Oron* 10-1

11 Bubble/Drop Transport in Microchannels *Hsueh-Chia Chang* 11-1

12 Fundamentals of Control Theory *Bill Goodwine* 12-1

13 Model-Based Flow Control for Distributed Architectures *Thomas R. Bewley* 13-1

14 Soft Computing in Control *Mihir Sen and Bill Goodwine* 14-1

II Design and Fabrication

15 Materials for Microelectromechanical Systems *Christian A. Zorman and Mehran Mehregany* 15-1

16 MEMS Fabrication *Marc J. Madou* 16-1

17 LIGA and Other Replication Techniques *Marc J. Madou* 17-1

18 X-Ray-Based Fabrication *Todd Christenson* 18-1

19 Electrochemical Fabrication (EFAB™) *Adam L. Cohen* 19-1

20 Fabrication and Characterization of Single-Crystal Silicon Carbide MEMS *Robert S. Okojie* 20-1

21 Deep Reactive Ion Etching for Bulk Micromachining of Silicon Carbide *Glenn M. Beheim* 21-1

22 Microfabricated Chemical Sensors for Aerospace Applications *Gary W. Hunter, Chung-Chiun Liu and Darby B. Makel* 22-1

23 Packaging of Harsh-Environment MEMS Devices *Liang-Yu Chen and Jih-Fen Lei* 23-1

III Applications of MEMS

24 Inertial Sensors   *Paul L. Bergstrom and Gary G. Li* **24**-1

25 Micromachined Pressure Sensors   *Jae-Sung Park,
Chester Wilson and Yogesh B. Gianchandani* **25**-1

26 Sensors and Actuators for Turbulent Flows   *Lennart Löfdahl
and Mohamed Gad-el-Hak* **26**-1

27 Surface-Micromachined Mechanisms   *Andrew D. Oliver
and David W. Plummer* **27**-1

28 Microrobotics   *Thorbjörn Ebefors and Göran Stemme* **28**-1

29 Microscale Vacuum Pumps   *E. Phillip Muntz and Stephen E. Vargo* **29**-1

30 Microdroplet Generators   *Fan-Gang Tseng* **30**-1

31 Micro Heat Pipes and Micro Heat Spreaders   *G.P. "Bud" Peterson* **31**-1

32 Microchannel Heat Sinks   *Yitshak Zohar* **32**-1

33 Flow Control   *Mohamed Gad-el-Hak* **33**-1

IV The Future

34 Reactive Control for Skin-Friction Reduction   *Haecheon Choi* **34**-1

35 Towards MEMS Autonomous Control of Free-Shear Flows
Ahmed Naguib **35**-1

36 Fabrication Technologies for Nanoelectromechanical Systems
Gary H. Bernstein, Holly V. Goodson and Gregory L. Snider **36**-1

Index I-1

*The farther backward you can look,
the farther forward you are likely to see.*

(Sir Winston Leonard Spencer Churchill, 1874–1965)

Janus, Roman god of gates, doorways and all beginnings, gazing both forward and backward.

As for the future, your task is not to foresee, but to enable it.

(Antoine-Marie-Roger de Saint-Exupéry, 1900–1944,
in Citadelle [The Wisdom of the Sands])

I

Background and Fundamentals

1 **Introduction** *Mohamed Gad-el-Hak* 1-1

2 **Scaling of Micromechanical Devices** *William Trimmer, Robert H. Stroud* 2-1
 Introduction • The Log Plot • Scaling of Mechanical Systems

3 **Mechanical Properties of MEMS Materials** *William N. Sharpe, Jr.* 3-1
 Introduction • Mechanical Property Definitions • Test Methods • Mechanical
 Properties • Initial Design Values

4 **Flow Physics** *Mohamed Gad-el-Hak* 4-1
 Introduction • Flow Physics • Fluid Modeling • Continuum
 Model • Compressibility • Boundary Conditions • Molecular-Based
 Models • Liquid Flows • Surface Phenomena • Parting Remarks

5 **Integrated Simulation for MEMS: Coupling Flow-Structure-Thermal-Electrical
 Domains** *Robert M. Kirby, George Em Karniadakis, Oleg Mikulchenko,
 Kartikeya Mayaram* 5-1
 Abstract • Introduction • Coupled Circuit-Device Simulation • Overview of
 Simulators • Circuit-Microfluidic Device Simulation • Demonstrations of the Integrated
 Simulation Approach • Summary and Discussion

6 **Liquid Flows in Microchannels** *Kendra V. Sharp, Ronald J. Adrian,
 Juan G. Santiago, Joshua I. Molho* 6-1
 Introduction • Experimental Studies of Flow Through Microchannels • Electrokinetics
 Background • Summary and Conclusions

7 **Burnett Simulations of Flows in Microdevices** *Ramesh K. Agarwal,
 Keon-Young Yun* 7-1
 Abstract • Introduction • History of Burnett Equations • Governing Equations •
 Wall-Boundary Conditions • Linearized Stability Analysis of Burnett Equations •
 Numerical Method • Numerical Simulations • Conclusions

8 **Molecular-Based Microfluidic Simulation Models** *Ali Beskok* 8-1
 Abstract • Introduction • Gas Flows • Liquid and Dense Gas Flows • Summary and
 Conclusions

9 **Lubrication in MEMS** *Kenneth S. Breuer* 9-1
   Introduction • Fundamental Scaling Issues • Governing Equations for
   Lubrication • Couette-Flow Damping • Squeeze-Film Damping • Lubrication
   in Rotating Devices • Constraints on MEMS Bearing Geometries • Thrust
   Bearings • Journal Bearings • Fabrication Issues • Tribology
   and Wear • Conclusions

10 **Physics of Thin Liquid Films** *Alexander Oron* 10-1
   Introduction • The Evolution Equation for a Liquid Film on a Solid Surface •
   Isothermal Films • Thermal Effects • Change of Phase: Evaporation
   and Condensation • Closing Remarks

11 **Bubble/Drop Transport in Microchannels** *Hsueh-Chia Chang* 11-1
   Introduction • Fundamentals • The Bretherton Problem for Pressure-Driven
   Bubble/Drop Transport • Bubble Transport by Electrokinetic Flow • Future
   Directions

12 **Fundamentals of Control Theory** *Bill Goodwine* 12-1
   Introduction • Classical Linear Control • "Modern" Control • Nonlinear
   Control • Parting Remarks

13 **Model-Based Flow Control for Distributed Architectures** *Thomas R. Bewley* 13-1
   Introduction • Linearization: Life in a Small Neighborhood • Linear Stabilization:
   Leveraging Modern Linear Control Theory • Decentralization: Designing for Massive
   Arrays • Localization: Relaxing Nonphysical Assumptions • Compensator Reduction:
   Eliminating Unnecessary Complexity • Extrapolation: Linear Control of Nonlinear
   Systems • Generalization: Extending to Spatially Developing Flows • Nonlinear
   Optimization: Local Solutions for Full Navier–Stokes • Robustification: Appealing to
   Murphy's Law • Unification: Synthesizing a General Framework • Decomposition:
   Simulation-Based System Modeling • Global Stabilization: Conservatively Enhancing
   Stability • Adaptation: Accounting for a Changing Environment • Performance
   Limitation: Identifying Ideal Control Targets • Implementation: Evaluating
   Engineering Trade-Offs • Discussion: A Common Language for Dialog •
   The Future: A Renaissance

14 **Soft Computing in Control** *Mihir Sen, Bill Goodwine* 14-1
   Introduction • Artificial Neural Networks • Genetic Algorithms • Fuzzy Logic
   and Fuzzy Control • Conclusions

1
Introduction

Mohamed Gad-el-Hak
University of Notre Dame

How many times when you are working on something frustratingly tiny, like your wife's wrist watch, have you said to yourself, "If I could only train an ant to do this!" What I would like to suggest is the possibility of training an ant to train a mite to do this. What are the possibilities of small but movable machines? They may or may not be useful, but they surely would be fun to make.

(From the talk "There's Plenty of Room at the Bottom," delivered by Richard P. Feynman at the annual meeting of the American Physical Society, Pasadena, CA, December 29, 1959.)

Tool making has always differentiated our species from all others on Earth. Aerodynamically correct wooden spears were carved by *archaic Homo sapiens* close to 400,000 years ago. Man builds things consistent with his size, typically in the range of two orders of magnitude larger or smaller than himself, as indicated in Figure 1.1. Though the extremes of the length scale are outside the range of this figure, man, at slightly more than 10^0 m, amazingly fits right in the middle of the smallest subatomic particle, which is approximately 10^{-26} m, and the extent of the observable universe, which is of the order of 10^{26} m (15 billion light years)—neither geocentric nor heliocentric but rather an egocentric universe! But humans have always striven to explore, build and control the extremes of length and time scales. In the voyages to Lilliput and Brobdingnag of *Gulliver's Travels*, Jonathan Swift (1726) speculates on the remarkable possibilities which diminution or magnification of physical dimensions provides.[1] The Great Pyramid of Khufu was originally 147 m high when completed around 2600 B.C., while the Empire State Building constructed in 1931 is currently—after the addition of a television antenna mast in 1950—449 m high. At the other end of the spectrum of man-made artifacts, a dime is slightly less than 2 cm in diameter. Watchmakers have practiced the art of miniaturization since the 13th century. The invention of the microscope in the 17th century opened the way for direct observation of microbes and plant and animal cells. Smaller things were man-made in the latter half of the 20th century. The transistor—invented in 1947—in today's integrated circuits has a size[2] of 0.18 μm (180 nm) in production and approaches 10 nm in research laboratories using electron beams. But what about the miniaturization of mechanical parts—machines—envisioned by Feynman (1961) in his legendary speech quoted above?

---

[1] *Gulliver's Travels* was originally designed to form part of a satire on the abuse of human learning. At the heart of the story is a radical critique of human nature in which subtle ironic techniques work to part the reader from any comfortable preconceptions and challenge him to rethink from first principles his notions of man.

[2] The smallest feature on a microchip is defined by its smallest linewidth, which in turn is related to the wavelength of light employed in the basic lithographic process used to create the chip.

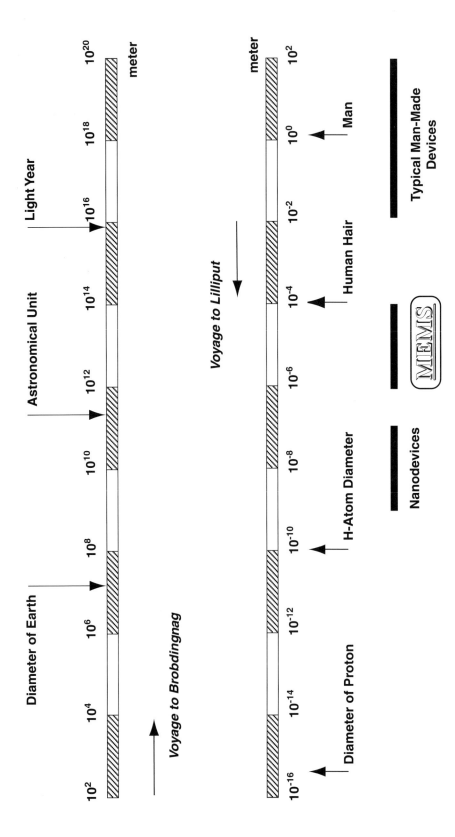

FIGURE 1.1 Scale of things, in meters. Lower scale continues in the upper bar from left to right. One meter is 10^6 μm, 10^9 nm or 10^{10} Å.

Introduction

Manufacturing processes that can create extremely small machines have been developed in recent years [Angell et al., 1983; Gabriel et al., 1988; 1992; O'Connor, 1992; Gravesen et al., 1993; Bryzek et al., 1994; Gabriel, 1995; Ashley, 1996; Ho and Tai, 1996; 1998; Hogan, 1996; Ouellette, 1996; Paula, 1996; Robinson et al., 1996a; 1996b; Madou, 1997; Tien, 1997; Amato, 1998; Busch-Vishniac, 1998; Kovacs, 1998; Knight, 1999; Epstein, 2000; Goldin et al., 2000; O'Connor and Hutchinson, 2000; Chalmers, 2001; Tang and Lee, 2001]. Electrostatic, magnetic, electromagnetic, pneumatic and thermal actuators, motors, valves, gears, cantilevers, diaphragms and tweezers less than 100 µm in size have been fabricated. These have been used as sensors for pressure, temperature, mass flow, velocity, sound and chemical composition; as actuators for linear and angular motions; and as simple components for complex systems such as robots, micro-heat-engines and micro-heat-pumps [Lipkin, 1993; Garcia and Sniegowski, 1993; 1995; Sniegowski and Garcia, 1996; Epstein and Senturia, 1997; Epstein et al., 1997].

Microelectromechanical systems (MEMS) refer to devices that have a characteristic length of less than 1 mm but more than 1 µm, that combine electrical and mechanical components and that are fabricated using integrated circuit batch-processing technologies. The books by Madou (1997) and Kovacs (1998) provide excellent sources for microfabrication technology. Current manufacturing techniques for MEMS include surface silicon micromachining; bulk silicon micromachining; lithography, electrodeposition and plastic molding (or, in its original German, *Lithographie Galvanoformung Abformung*, LIGA); and electrodischarge machining (EDM). As indicated in Figure 1.1, MEMS are more than four orders of magnitude larger than the diameter of the hydrogen atom, but about four orders of magnitude smaller than the traditional man-made artifacts. Microdevices can have characteristic lengths smaller than the diameter of a human hair. Nanodevices (some say NEMS) further push the envelope of electromechanical miniaturization [Roco, 2001].

The famed physicist Richard P. Feynman delivered a mere two, but profound, lectures[3] on electromechanical miniaturization: "There's Plenty of Room at the Bottom," quoted above, and "Infinitesimal Machinery," presented at the Jet Propulsion Laboratory on February 23, 1983. He could not see a lot of use for micromachines, lamenting in 1959: "[Small but movable machines] may or may not be useful, but they surely would be fun to make," and, 24 years later, "There is no use for these machines, so I still don't understand why I'm fascinated by the question of making small machines with movable and controllable parts." Despite Feynman's demurring regarding the usefulness of small machines, MEMS are finding increased applications in a variety of industrial and medical fields, with a potential worldwide market in the billions of dollars ($30 billion by 2004). Accelerometers for automobile airbags, keyless entry systems, dense arrays of micromirrors for high-definition optical displays, scanning electron microscope tips to image single atoms, micro-heat-exchangers for cooling of electronic circuits, reactors for separating biological cells, blood analyzers and pressure sensors for catheter tips are but a few in current use. Microducts are used in infrared detectors, diode lasers, miniature gas chromatographs and high-frequency fluidic control systems. Micropumps are used for ink-jet printing, environmental testing and electronic cooling. Potential medical applications for small pumps include controlled delivery and monitoring of minute amounts of medication, manufacturing of nanoliters of chemicals and development of an artificial pancreas.

This multidisciplinary field has witnessed explosive growth during the last decade. Several new journals are dedicated to the science and technology of MEMS—for example, *Journal of Microelectromechanical Systems*, *Journal of Micromechanics and Microengineering* and *Microscale Thermophysical Engineering*. Numerous professional meetings are devoted to micromachines—for example, Solid-State Sensor and Actuator Workshop, International Conference on Solid-State Sensors and Actuators (Transducers), Micro Electro Mechanical Systems Workshop, Micro Total Analysis Systems, Eurosensors, etc.

This handbook covers several aspects of microelectromechanical systems, or more broadly the art and science of electromechanical miniaturization. MEMS design, fabrication and application as well as the physical modeling of their materials, transport phenomena and operations are discussed. Chapters on

---

[3] Both talks have been reprinted in the *Journal of Microelectromechanical Systems* **1**(1), pp. 60–66, 1992; **2**(1), pp. 4–14, 1993.

the electrical, structural, fluidic, transport and control aspects of MEMS are included. Other chapters cover existing and potential applications of microdevices in a variety of fields including instrumentation and distributed control. Physical understanding of the different phenomena unique to micromachines is emphasized throughout this book. The handbook is divided into four parts: Part I provides background and physical considerations, Part II discusses the design and fabrication of microdevices, Part III reviews a few of the applications of microsensors and microactuators, and Part IV ponders the future of the field. The 36 chapters are written by 54 of the world's foremost authorities on this multidisciplinary subject. The contributing authors come from academia, government and industry. Without compromising rigorousness, the text is designed for maximum readability by a broad audience having an engineering or science background. The nature of the book—being a handbook and not an encyclopedia—and its size limitation dictate the exclusion of several important topics in the MEMS area of research and development.

Our objective is to provide a current overview of the fledgling discipline and its future developments for the benefit of working professionals and researchers. The handbook will be a useful guide and reference to the explosive literature on MEMS and should provide the definitive word for the fundamentals and applications of microfabrication and microdevices. Glancing at the table of contents, the reader may rightly sense an overemphasis on the physics of microdevices. This is consistent with the strong conviction of the editor-in-chief that the MEMS technology is moving too fast relative to our understanding of the unconventional physics involved. This technology can certainly benefit from a solid foundation of the underlying fundamentals. If the physics is better understood, better, less expensive and more efficient microdevices can be designed, built and operated for a variety of existing and yet-to-be-dreamed applications. Consistent with this philosophy, chapters on control theory, distributed control and soft computing are included as the backbone of the futuristic idea of using colossal numbers of microsensors and microactuators in reactive control strategies aimed at taming turbulent flows to achieve substantial energy savings and performance improvements of vehicles and other man-made devices.

I shall leave you now for the many wonders of the small world you are about to encounter when navigating through the various chapters that follow. May your voyage to Lilliput be as exhilarating, enchanting and enlightening as *Lemuel Gulliver's Travels into Several Remote Nations of the World. Hekinah degul!* Jonathan Swift may not have been a good biologist and his scaling laws were not as good as those of William Trimmer (see Chapter 2 of this book), but Swift most certainly was a magnificent storyteller. *Hnuy illa nyha majah Yahoo!*

References

Amato, I. (1998) "Formenting a Revolution, in Miniature," *Science* **282**(5388), 16 October, pp. 402–405.
Angell, J.B., Terry, S.C., and Barth, P.W. (1983) "Silicon Micromechanical Devices," *Faraday Trans. I* **68**, pp. 744–748.
Ashley, S. (1996) "Getting a Microgrip in the Operating Room," *Mech. Eng.* **118**, September, pp. 91–93.
Bryzek, J., Peterson, K., and McCulley, W. (1994) "Micromachines on the March," *IEEE Spectrum* **31**, May, pp. 20–31.
Busch-Vishniac, I.J. (1998) "Trends in Electromechanical Transduction," *Phys. Today* **51**, July, pp. 28–34.
Chalmers, P. (2001) "Relay Races," *Mech. Eng.* **123**, January, pp. 66–68.
Epstein, A.H. (2000) "The Inevitability of Small," *Aerosp. Am.* **38**, March, pp. 30–37.
Epstein, A.H., and Senturia, S.D. (1997) "Macro Power from Micro Machinery," *Science* **276**, 23 May, p. 1211.
Epstein, A.H., Senturia, S.D., Al-Midani, O., Anathasuresh, G., Ayon, A., Breuer, K., Chen, K.-S., Ehrich, F.F., Esteve, E., Frechette, L., Gauba, G., Ghodssi, R., Groshenry, C., Jacobson, S.A., Kerrebrock, J.L., Lang, J.H., Lin, C.-C., London, A., Lopata, J., Mehra, A., Mur Miranda, J.O., Nagle, S., Orr, D.J., Piekos, E., Schmidt, M.A., Shirley, G., Spearing, S.M., Tan, C.S., Tzeng, Y.-S., and Waitz, I.A. (1997) "Micro-Heat Engines, Gas Turbines, and Rocket Engines—The MIT Microengine Project," AIAA Paper No. 97-1773, American Institute of Aeronautics and Astronautics, Reston, VA.

Feynman, R.P. (1961) "There's Plenty of Room at the Bottom," in *Miniaturization,* ed. H.D. Gilbert, pp. 282–296, Reinhold Publishing, New York.

Gabriel, K.J. (1995) "Engineering Microscopic Machines," *Sci. Am.* **260**, September, pp. 150–153.

Garcia, E.J., and Sniegowski, J.J. (1993) "The Design and Modelling of a Comb-Drive-Based Microengine for Mechanism Drive Applications," in Proc. Seventh Int. Conf. on Solid-State Sensors and Actuators (Transducers '93), pp. 763–766, 7–10 June, Yokohama, Japan.

Garcia, E.J., and Sniegowski, J.J. (1995) "Surface Micromachined Microengine," *Sensors and Actuators A* **48**, pp. 203–214.

Gabriel, K.J., Jarvis, J., and Trimmer, W., eds. (1988) *Small Machines, Large Opportunities: A Report on the Emerging Field of Microdynamics,* National Science Foundation, AT&T Bell Laboratories, Murray Hill, NJ.

Gabriel, K.J., Tabata, O., Shimaoka, K., Sugiyama, S., and Fujita, H. (1992) "Surface-Normal Electrostatic/Pneumatic Actuator," in Proc. IEEE Micro Electro Mechanical Systems '92, pp. 128–131, 4–7 February, Travemünde, Germany.

Goldin, D.S., Venneri, S.L., and Noor, A.K. (2000) "The Great out of the Small," *Mech. Eng.* **122**, November, pp. 70–79.

Gravesen, P., Branebjerg, J., and Jensen, O.S. (1993) "Microfluidics—A Review," *J. Micromech. Microeng.* **3**, pp. 168–182.

Ho, C.-M., and Tai, Y.-C. (1996) "Review: MEMS and Its Applications for Flow Control," *J. Fluids Eng.* **118**, pp. 437–447.

Ho, C.-M., and Tai, Y.-C. (1998) "Micro-Electro-Mechanical Systems (MEMS) and Fluid Flows," *Annu. Rev. Fluid Mech.* **30**, pp. 579–612.

Hogan, H. (1996) "Invasion of the Micromachines," *New Sci.* **29**, June, pp. 28–33.

Knight, J. (1999) "Dust Mite's Dilemma," *New Sci.* **162**(2180), 29 May, pp. 40–43.

Kovacs, G.T.A. (1998) *Micromachined Transducers Sourcebook,* McGraw-Hill, New York.

Lipkin, R. (1993) "Micro Steam Engine Makes Forceful Debut," *Sci. News* **144**, September, p. 197.

Madou, M. (1997) *Fundamentals of Microfabrication,* CRC Press, Boca Raton, FL.

O'Connor, L. (1992) "MEMS: Micromechanical Systems," *Mech. Eng.* **114**, February, pp. 40–47.

O'Connor, L., and Hutchinson, H. (2000) "Skyscrapers in a Microworld," *Mech. Eng.* **122**, March, pp. 64–67.

Ouellette, J. (1996) "MEMS: Mega Promise for Micro Devices," *Mech. Eng.* **118**, October, pp. 64–68.

Paula, G. (1996) "MEMS Sensors Branch Out," *Aerosp. Am.* **34**, September, pp. 26–32.

Robinson, E.Y., Helvajian, H., and Jansen, S.W. (1996a) "Small and Smaller: The World of MNT," *Aerosp. Am.* **34**, September, pp. 26–32.

Robinson, E.Y., Helvajian, H., and Jansen, S.W. (1996b) "Big Benefits from Tiny Technologies," *Aerosp. Am.* **34**, October, pp. 38–43.

Roco, M.C. (2001) "A Frontier for Engineering," *Mech. Eng.* **123**, January, pp. 52–55.

Sniegowski, J.J., and Garcia, E.J. (1996) "Surface Micromachined Gear Trains Driven by an On-Chip Electrostatic Microengine," *IEEE Electron Device Lett.* **17**, July, p. 366.

Swift, J. (1726) *Gulliver's Travels,* 1840 reprinting of *Lemuel Gulliver's Travels into Several Remote Nations of the World,* Hayward & Moore, London.

Tang, W.C., and Lee, A.P. (2001) "Military Applications of Microsystems," *Ind. Physicist* **7**, February, pp. 26–29.

Tien, N.C. (1997) "Silicon Micromachined Thermal Sensors and Actuators," *Microscale Thermophys. Eng.* **1**, pp. 275–292.

2
Scaling of Micromechanical Devices

William Trimmer
Belle Mead Research, Inc.

Robert H. Stroud
The Aerospace Corporation

2.1 Introduction 2-1
2.2 The Log Plot 2-2
2.3 Scaling of Mechanical Systems 2-4

2.1 Introduction

A revolution in understanding and utilizing micromechanical devices is starting. The utility of these devices will be enormous, and with time these microdevices will fill the niches of our lives as pervasively as electronics. What form will these microdevices take? What will actuate them, and how will they interact with their environment? We cannot foresee where the developing technology will take us.

How, then, do we start to design this world of the micro? As you will discover in this book, there are a large number of ways to fabricate microdevices and a vast number of designs. The number of possible things we could try is beyond possibility. Should we just start trying approaches until something works? Perhaps there is a better way.

Scaling theory is a valuable guide to what may work and what will not. By understanding how phenomena behave and change, as the scale size changes, we can gain some insight and better understand the profitable approaches. This chapter examines how things change with size and will develop mathematics that helps find the profitable approaches.

Three general scale sizes will be discussed: astronomical objects; the normal objects we deal with, called *macro-objects*; and very small objects, called *micro-objects*. Things effective at one of these scale sizes often are insignificant at another scale size. As an example, gravitational forces dominate on an astronomical scale. The motion of our planet around the sun and our sun around the galaxy is driven mostly by gravitational forces. Yet, on the macroscale of my desk top, the gravitational force between two objects, such as my tape dispenser and stapler, is insignificant. A few simple scaling calculations later in this chapter will tell us this: On astronomical scales, be concerned with gravity; on smaller scales, look to other forces to move objects.

What is obvious on an astronomical-scale size or on a macroscale size is often not obvious on the microscale. For example, take the case of an electric motor. It is really a magnetic motor, and almost all macrosized electric actuators use magnetic fields to generate forces. Hence, one's first intuition would be to use magnetic motors in designing microdevices. However, most of the common micromotor designs use electrostatic fields instead of magnetic. The reasons for this will become obvious in the following discussion of how forces scale.

This field of micromechanical devices is extremely broad. It encompasses all of the traditional science and engineering disciplines, only on a smaller scale. Try to think of a traditional science or engineering discipline that does not have a microequivalent. What we are about in our new discipline is replicating the macroscience and engineering on a microscale. As a result, technical people from all science and engineering disciplines can make important contributions to this field.

The time scale from conception to utilization has been collapsing. Alessandro Volta and Andre Marie Ampere developed the basic concepts of electricity. About a hundred years later, Nikola Tesla and Thomas Alva Edison developed practical electric generators and motors. In contrast, the microcomb drive motor was described in 1989 and currently is being used in automobiles as an airbag sensor [Tang et al., 1989]. Volta and Ampere's ideas took a hundred years to culminate in practical implementation, but the microcomb drive motor took less than a dozen years from conception to full-scale implementation.

One of the marvelous things about nature is the widely varying scale sizes available to us. Section 2.2 will discuss this broad range of scales. Section 2.3 will show how scaling theory can be used as a guide to understand how phenomena change with scale. We hope the material that follows encourages you to explore the broad scope of this new field.

2.2 The Log Plot

As the scale, or size, of a system changes by several orders of magnitude, the system tends to behave differently. Consider, for example, a glass of water that is about 5 cm on a side. Pour the glass of water onto a table, and notice how the water flows and runs off the edge of the table. If the size of the glass is decreased by two orders of magnitude, or a factor of 100, the glass is now 0.05 cm (or half a millimeter) on a side. Pour this glass onto the table, and see that the surface tension pulls the water into a drop that sticks to the table. Turn the table on its side, and observe that it is difficult to make the drop flow to the edge of the table. In each case, the substance is the same, water, and it is the same table, but changing the scale size makes the water behave very differently.

By decreasing the size of the glass again by two orders of magnitude, the glass is now 0.0005 cm, or 5 μm, on a side. If you pour a drop this size onto the table, it most likely will not even reach the table. Some air current will entrain the drop and carry it away like mist flowing through the city at night. Again, the behavior of the water is dramatically different because of its size. Even the act of pouring the glass over the table is different. The 5-cm glass pours, whereas water in the 0.05- and 0.0005-cm glasses is constrained by surface tension. Different physical effects manifest themselves differently because of the system size.

Figure 2.1 shows the full range of sizes available to us, from atoms to the Universe. Atoms are the smallest mechanical system we will manipulate in the near future; their size is several Angstroms (10^{-10} of a meter). The Universe is the largest mechanical system we can observe. Depending upon the particular astronomical theory, the Universe is about 10^{37} m in diameter. Hence, the full range available for us to investigate and use is about 10^{47} m or 47 orders of magnitude.

Along the horizontal axis in Figure 2.1 is plotted the size of the system. The short vertical lines in the center of the plot represent a factor of 10 change in the system size. The long vertical lines represent a change of 100,000, or five orders of magnitude. Along the top, the size of the system is given in meters, and in the central band the size of the system is given in Angstroms. Figure 2.1 is plotted as a log plot for two reasons: (1) to enable everything to be depicted on the same piece of paper, and (2) to easily visualize the different size domains.

One can get a sense of the size of things by looking at the Ant, Man and Whale. These familiar objects span about five orders of magnitude. Several orders of magnitude smaller than the Ant are Bacteria and Virus. Going to larger systems, the U.S. road system is about five orders or magnitude larger than the Whale, and the Earth's orbit is about five orders of magnitude larger than the U.S. road system. Increasing another five or six orders of magnitude brings us to interstellar distances.

Scaling of Micromechanical Devices

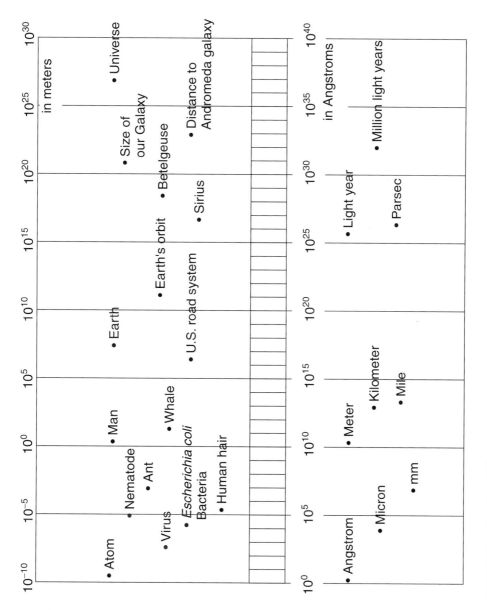

FIGURE 2.1 Log plot of all mechanical systems available for exploration.

The bottom portion of Figure 2.1 shows the units we use to measure things. The Angstrom, micron, millimeter, meter, kilometer, and mile are familiar units, but then we see a gap of about a dozen orders of magnitude before the astronomical units of light year and parsec are reached.

The microregion of interest to this chapter ranges from about a millimeter to an Angstrom (from about 10^{-3} to 10^{-10} meters). This region comprises roughly a fifth of the full range of domains available for us to explore and may seem like a small portion, but consider that the U.S. roadway system is one of the largest mechanical systems we will build for quite a while. Buildings and ships are probably the largest self-contained mechanical systems we will construct in the near future. Most of the larger domains are so large, they simply are not accessible to us. Thus, the microregion represents the majority of the new domains available for exploration!

This microdomain is enticing. Part of the charm is that conventional designs do not work well, and ingenuity is needed to make new designs. For example, macro- and microdevices that transfer water tend to use different physical principles. An open ditch works at one scale, and a capillary works at a smaller scale. Because microdesigners are left without the conventional solutions, they have the opportunity to find their own solutions.

2.3 Scaling of Mechanical Systems

As the size of a system changes, the physical parameters change, often in a dramatic way [Trimmer et al., 1989; Madou, 1997]. To understand how these parameters change, consider the scale factor S. This scale factor is similar to the small notation on the corner of a mechanical drawing which might say the scale of the drawing is 1:10. The actual object to be made is 10 times the size of the drawing. A scale of 1:100 means the actual object is 100 times larger. In the microdomain, the scale might be 100:1, meaning the object is 100 times smaller than the drawing. When the scale size changes, all the dimensions of the object change by exactly the same amount S such that 1:S.

This scale factor S can be used to describe how physical phenomena change. All the lengths of the drawing scale by the factor S, but other parameters such as the volume scale differently. Volume V is length L times width W times height H, or

$$V = L \cdot W \cdot H \tag{2.1}$$

When the scale changes by 1/100 (that is, decreases by a factor of 100), the length and width and height all change by 1/100, and the volume decreases by $(1/100)^3$ or 1/1,000,000. The volume decreases by a factor of a million when the scale size decreases by a factor of a hundred. Volume is an example of a parameter that scales as S^3. The force due to surface tension scales as S^1, the force due to electrostatics scales as S^2, the force due to certain magnetic forces scales as S^3, and gravitational forces scale as S^4. Now, if the size of the system decreases from a meter to a millimeter, this is a decrease in size of a factor of a thousand, $S = 1/1000$. The surface tension force decreases by a factor of a thousand, $S^1 = (1/1000)^1$; the electrostatic force decreases by a factor of a million, $S^2 = (1/1000)^2 = 1/1,000,000$; the magnetic force decreases by a factor of a billion, $S^3 = (1/1000)^3 = 1/1,000,000,000$; and the gravitational force decreases by a factor of a trillion, $S^4 = (1/1000)^4 = 1/1,000,000,000,000$. Indeed, changing the size of a mechanical system changes what forces are important.

Knowing how a physical phenomenon scales, whether it scales as S^1 or S^2 or S^3 or S^4 or some other power of S, guides our understanding of how to design small mechanical systems. As an example, consider a water bug. The weight of the water bug scales as the volume, or S^3, while the force used to support the bug scales as the surface tension (S^1) times the distance around the bug's foot (S^1), and the force on the bug's foot scales as $S^1 \times S^1 = S^2$. When the scale size, S, decreases, the weight decreases more rapidly than the surface tension forces. Changing from a 2-m-sized man to a 2-mm-sized bug decreases the weight by a factor of a billion, while the surface tension force decreases by only a factor of a million. Hence, the bug can walk on water.

Scaling provides a good guide to how things behave and helps us gain insight into small systems, but scaling is just that, a good guide. It usually does not provide exact solutions. For example, the scaling above does not take into account the difference between the water bug's foot and a person's foot. Water bug's feet are designed for water, and we expect superior performance. Creativity and intuition are what make an excellent design; scaling is a guide to understanding what is important.

A mathematical notation is now discussed that captures the different scaling laws into a convenient form. This notation shows many different scaling laws at once and can be used to easily understand what happens to the different terms and parameters of an equation as the scale size is changed.

Consider four different forces laws, $F = S^1$, $F = S^2$, $F = S^3$, $F = S^4$, and collect these different cases into a vertical Trimmer bracket:

$$F = \begin{bmatrix} S^1 \\ S^2 \\ S^3 \\ S^4 \end{bmatrix} \quad (2.2)$$

The top-most element of this bracket refers to the case where the force scales as S^1, the next element down refers to the case where the force scales as S^2, etc.

To continue, let us do something with this bracket. Work W is force F times distance D, or

$$W = F \cdot D \quad (2.3)$$

and, extending our notation,

$$W = F \cdot D = \begin{bmatrix} S^1 \\ S^2 \\ S^3 \\ S^4 \end{bmatrix} \begin{bmatrix} S^1 \\ S^1 \\ S^1 \\ S^1 \end{bmatrix} = \begin{bmatrix} S^2 \\ S^3 \\ S^4 \\ S^5 \end{bmatrix} \quad (2.4)$$

or

$$W = \begin{bmatrix} S^2 \\ S^3 \\ S^4 \\ S^5 \end{bmatrix} \quad (2.5)$$

Note that distance D, shown above, always scales as S^1, and its bracket consists of all S^1's. In the top case where the force scales as S^1, the distance scales as S^1, and their product scales as S^2. In the second element down, the force scales as S^2, the distance scales as S^1, and their product scales as S^3. Here in one notation we have shown how the work scales for the four different force laws. For example, the gravitational force between an object and the Earth scales as S^3 (the Earth's mass remains constant in this example, and the mass of the object scales as its volume, S^3), and looking at the third element down we see that a force scaling of S^3 gives us a work, or energy, scaling of S^4. If the size of a system decreases by a factor of a thousand (say, from 10 cm to 0.10 mm), the gravitational energy required to move an object from the bottom to the top of the machine under consideration decreases by $(1/1000)^4 = 1/1,000,000,000,000$. The gravitational work decreases by a factor of a trillion. We know this intuitively: Drop an ant from ten times his height, and he walks away. Please do not try this with a horse.

How do the acceleration and transit times change for the different force scaling laws? Acceleration a is equal to force F divided by the mass m:

$$a = \frac{F}{m} = F \cdot m^{-1} \tag{2.6}$$

and we know the mass scales as S^3, and m^{-1} scales as S^{-3}, giving:

$$a = \begin{bmatrix} S^1 \\ S^2 \\ S^3 \\ S^4 \end{bmatrix} \begin{bmatrix} S^3 \\ S^3 \\ S^3 \\ S^3 \end{bmatrix}^{-1} = \begin{bmatrix} S^1 \\ S^2 \\ S^3 \\ S^4 \end{bmatrix} \begin{bmatrix} S^{-3} \\ S^{-3} \\ S^{-3} \\ S^{-3} \end{bmatrix} = \begin{bmatrix} S^{-2} \\ S^{-1} \\ S^0 \\ S^1 \end{bmatrix} \tag{2.7}$$

This is an interesting result. When the force scales as S^1, the acceleration scales as S^{-2}. If the size of the system decreases by a factor of 100, the acceleration increases by $(1/100)^{-2} = 10,000$. And, as the system becomes smaller, the acceleration increases. A predominance of the forces we use in the microdomain scale as S^2. For these forces, the acceleration scales as S^{-1}, and decreasing the size by a factor of 100 increases the acceleration by a factor of 100, still a nice increase in acceleration. In general, small systems tend to accelerate very rapidly. For the case where the force scales as S^3, the acceleration remains constant, $(1/100)^0 = 1$, and the acceleration decreases for forces that scale as S^4.

The transit time t to move from point A to B in our scalable drawing can be calculated as:

$$x = \frac{1}{2}at^2 \tag{2.8}$$

$$t = \sqrt{\frac{2x}{a}} = \sqrt{2} \cdot (x)^{0.5} \cdot (a)^{-0.5} \tag{2.9}$$

and

$$t = \begin{bmatrix} S^0 \\ S^0 \\ S^0 \\ S^0 \end{bmatrix} \begin{bmatrix} S^1 \\ S^1 \\ S^1 \\ S^1 \end{bmatrix}^{0.5} \begin{bmatrix} S^{-2} \\ S^{-1} \\ S^0 \\ S^1 \end{bmatrix}^{-0.5} = \begin{bmatrix} S^0 \\ S^0 \\ S^0 \\ S^0 \end{bmatrix} \begin{bmatrix} S^{0.5} \\ S^{0.5} \\ S^{0.5} \\ S^{0.5} \end{bmatrix} \begin{bmatrix} S^1 \\ S^{0.5} \\ S^0 \\ S^{-0.5} \end{bmatrix} = \begin{bmatrix} S^{1.5} \\ S^1 \\ S^{0.5} \\ S^0 \end{bmatrix} \tag{2.10}$$

$$t = \begin{bmatrix} S^{1.5} \\ S^1 \\ S^{0.5} \\ S^0 \end{bmatrix} \tag{2.11}$$

For the case where the force scales as S^2, transit time t scales as S^1. If the system decreases by a factor of 100, the transit time decreases by a factor of 100. Again, we know this intuitively; small things tend to be fast.

Scaling of Micromechanical Devices

Depending upon the equation and variables of interest, the Trimmer brackets can be configured differently. To continue the above example, one might be interested in how things behave if the acceleration, instead of the force, scales in different ways. One could write:

$$a = \begin{bmatrix} S^1 \\ S^2 \\ S^3 \\ S^4 \end{bmatrix} \quad (2.12)$$

From above:

$$t = \sqrt{\frac{2x}{a}} = \sqrt{2} \cdot (x)^{0.5} \cdot (a)^{-0.5} \quad (2.13)$$

and

$$t = \begin{bmatrix} S^0 \\ S^0 \\ S^0 \\ S^0 \end{bmatrix} \begin{bmatrix} S^1 \\ S^1 \\ S^1 \\ S^1 \end{bmatrix}^{0.5} \begin{bmatrix} S^1 \\ S^2 \\ S^3 \\ S^4 \end{bmatrix}^{-0.5} = \begin{bmatrix} S^0 \\ S^0 \\ S^0 \\ S^0 \end{bmatrix} \begin{bmatrix} S^{0.5} \\ S^{0.5} \\ S^{0.5} \\ S^{0.5} \end{bmatrix} \begin{bmatrix} S^{-0.5} \\ S^{-1} \\ S^{-1.5} \\ S^{-2} \end{bmatrix} = \begin{bmatrix} S^0 \\ S^{-0.5} \\ S^{-1} \\ S^{-1.5} \end{bmatrix} \quad (2.14)$$

$$t = \begin{bmatrix} S^0 \\ S^{-0.5} \\ S^{-1} \\ S^{-1.5} \end{bmatrix} \quad (2.15)$$

The top element in this bracket describes how time scales when the acceleration scales as S^1. (In the earlier discussion, the top element describes how time scales when the *force* scales as S^1.) These brackets can be arranged for the convenience of the problem at hand. We need not even use integer exponents. For example, we could have defined the acceleration as:

$$a = \begin{bmatrix} S^{0.1} \\ S^{0.2} \\ S^{0.3} \\ S^2 \\ S^4 \end{bmatrix} \quad (2.16)$$

and then calculated the transit times for these five new scaling functions.

Let us examine the gravitational example in the introduction. As you will see in a moment, gravitational forces scale as S^4 and are a dominant force in large systems, but not small systems. The force between two objects is

$$F = G\frac{M_1 \cdot M_2}{r^2} \quad (2.17)$$

where F is the force; G is the gravitational constant (= 6.670×10^{-11} N m^2 kg^{-2}), which does not change with scale size; M_1 and M_2 are the masses of the two objects; and r is the separation. The masses scale as:

$$M = \rho \cdot V = S^0 \cdot S^3 = S^3 \qquad (2.18)$$

where the density, ρ, is assumed constant (S^0), and V is the volume (S^3). Now, force F scales as:

$$F = S^0 \frac{S^3 \cdot S^3}{S^2} = S^4 \qquad (2.19)$$

Now, let's make a different assumption and suppose the density is not constant with scale size. The density could be represented as:

$$\rho = \begin{bmatrix} S^0 \\ S^{-1} \\ S^{-2} \\ S^{-3} \end{bmatrix} \qquad (2.20)$$

and force F becomes:

$$F = G \frac{M_1 \cdot M_2}{r^2} = G \frac{\rho \cdot V_1 \cdot \rho V_2}{r^2} = G \cdot \rho^2 \cdot V_1 \cdot V_2 \cdot R^{-2} \qquad (2.21)$$

$$F = S^0 \begin{bmatrix} S^0 \\ S^{-1} \\ S^{-2} \\ S^{-3} \end{bmatrix}^2 S^3 S^3 S^{-2} = S^0 \begin{bmatrix} S^0 \\ S^{-2} \\ S^{-4} \\ S^{-6} \end{bmatrix} S^3 S^3 S^{-2} = \begin{bmatrix} S^4 \\ S^2 \\ S^0 \\ S^{-2} \end{bmatrix} \qquad (2.22)$$

From the top element, where the density does not change with size, the force scales as S^4. From the third element down, when the density scales as S^{-2}, the gravitational force remains constant as the scale size changes. That is, if astronomical objects become less dense as they become larger (as $\rho = S^{-2}$), then the gravitational force between objects remains constant ($F = S^0$) as one considers different-sized astronomical systems.

It is useful to understand how different forces scale. A more complete listing of forces and their scaling is given below,

$$F = \begin{bmatrix} S^1 \\ S^2 \\ S^3 \\ S^4 \end{bmatrix} = \begin{bmatrix} \text{Surface tension} \\ \text{Electrostatic, Pressure, Biological, Magnetic } (J = S^{-1}) \\ \text{Magnetic } (J = S^{-0.5}) \\ \text{Gravitational, Magnetic } (J = S^0) \end{bmatrix} \qquad (2.23)$$

Surface tension has the delightful scaling of S^1 and increases rapidly relative to other forces as a system becomes smaller. However, changing the surface tension usually requires changing the temperature, adding a surfactant, or changing some other parameter that is usually difficult to control. Most forces currently used by microdesigners scale as S^2. These include electrostatic forces, forces generated by pressures, and biological forces (the force an animal can exert generally depends upon as the cross section of the muscle). How magnetic forces scale depends upon how the current density (current per unit area of the coils) scales. If the current density, J, in the coil remains constant (S^0), the magnetic force between two coils scales as S^4, and in this case the magnetic forces become weak in the microdomain. However, one can remove heat much more efficiently from a small volume, and the current density of a microcoil can be much higher than in a large coil. If the current density scales as S^{-1} when the system decreases by a factor of ten, the current density increases by a factor of ten. In this case, the coil has much higher resistive losses, but the force scales much more advantageously as S^2.

References

Madou, M. (1997) *Fundamentals of Microfabrication,* CRC Press, Boca Raton, FL, pp. 405–412.

Tang, W.C., Nguyen, T.-C.H., and Howe, R.T. (1989) "Laterally Driven Polysilicon Resonant Microstructures," Proceedings of the IEEE Micro Electro Mechanical Systems Workshop, February 1989; reprinted in *Micromechanics and MEMS: Classic and Seminal Papers to 1990,* ed. W. Trimmer, Institute of Electrical and Electronics Engineers, New York, 1997, pp. 187–193.

Trimmer, W.S.N.T. (1989) "Microrobots and Micromechanical Systems," *Sensors Actuators,* September; reprinted in *Micromechanics and MEMS: Classic and Seminal Papers to 1990,* ed. W. Trimmer, Institute of Electrical and Electronics Engineers, New York, 1997, pp. 96–116.

3
Mechanical Properties of MEMS Materials

William N. Sharpe, Jr.
Johns Hopkins University

3.1 Introduction 3-1
3.2 Mechanical Property Definitions 3-2
3.3 Test Methods.. 3-3
   Specimen and Test Structure Preparation • Dimension Measurement • Force and Displacement Measurement • Strain Measurement • Tensile Tests • Bend Tests • Resonant Structure Tests • Membrane Tests • Indentation Tests • Other Test Methods • Fracture Tests • Fatigue Tests • Creep Tests • Round Robin Tests
3.4 Mechanical Properties... 3-17
3.5 Initial Design Values 3-24
   Acknowledgments 3-25

3.1 Introduction

New technologies tend to originate with new materials and manufacturing processes, which are used for new products. In the early stages, the emphasis is on novel devices and systems as well as ways of making them. Studies of fundamental issues such as mechanical properties and design procedures come later. For example, in 1830 there were 23 miles of railroad track in the U.S., but by 1870 there were 53,000 miles of track. However, the Bessemer steelmaking process did not originate until 1856, and the American Society for Testing and Materials (ASTM) was not organized until 1898.

The same is true for microelectromechanical systems (MEMS). The emphasis over the past dozen or so years has been on new materials, new manufacturing processes, and new microdevices—and rightfully so. These technological advances have been paralleled by an increasing interest in mechanical testing of materials used in MEMS. More researchers are becoming involved, with the topic appearing in symposia sponsored by the Society for Experimental Mechanics, the American Society of Mechanical Engineers, and the Materials Research Society. Further, the November 2000 ASTM symposium, "Mechanical Testing of Structural Films," was an important first step toward standardization of test methods. This increase in MEMS material testing has occurred over the past ten or so years, and this chapter is a review of the current status of the field.

Mechanical properties of interest fall into three general categories: elastic, inelastic, and strength. The designer of a microdevice needs to know the elastic properties in order to predict the amount of deflection from an applied force, or vice versa. If the material is ductile and the deformed structure does not need to return to its initial state, then the inelastic material behavior is necessary. The strength of the material must be known so that the allowable operating limits can be set. The manufacturer of a MEMS device needs to understand the relation between the processing and the properties of the material.

The importance of mechanical properties was recognized early on by a leader in the MEMS field, Richard Muller, who wrote in 1990, "Research on the mechanical properties of the electrical materials forming microdynamic structures (which previously had exclusively electrical uses), on the scaling of mechanical design, and on the effective uses of computer aids is needed to provide the engineering base that will make it possible to exploit fully this technology" [Muller, 1990]. Later, and expanded, conclusions and recommendations were made in a 1997 report of a National Research Council committee, of which Muller was chair [Muller, 1997]. One conclusion was, "Test-and-characterization methods and metrologies are required to (1) help fabrication facilities define MEMS materials for potential users, (2) facilitate consistent evaluations of material and process properties at the required scales, and (3) provide a basis for comparisons among materials fabricated at different facilities." One recommendation was, "Studies that address fundamental mechanical properties (e.g., Young's modulus, fatigue strength, residual stress, internal friction) and the engineering physics of long-term reliability, friction, and wear are vitally needed." These rather obvious statements are part of the development of a new technology.

There have been other reviews of the topic. The first on freestanding thin films by Menter and Pashley (1959) is interesting to read from a historical point of view. This author reviewed existing techniques and introduced new ones in 1996 [Sharpe et al., 1996] and looked at the variation in the mechanical properties of polysilicon as tested by several researchers in 1997 [Sharpe et al., 1997b]. Obermeier (1996) reviewed test methods for mechanical and thermophysical properties. Ballarini (1998) prepared a report for the Air Force Research Laboratory, which reviewed pertinent experimental and theoretical work up until then. Yi and Kim (1999a) published a review article, "Measurement of Mechanical Properties for MEMS Materials" on just this topic. Schweitz and Ericson (1999) reviewed the state of the art and offered some interesting conclusions and advice. Chang and Sharpe (1999) wrote an introductory chapter on the subject, and Spearing (2000) wrote a comprehensive exposition from a materials aspect. This chapter is intended to be a comprehensive survey focusing on both the test methods and the properties that have been measured.

After brief definitions of the mechanical properties of interest, the current test methods for MEMS materials are reviewed. Then, the properties of the various materials are summarized in a comprehensive set of tables. In almost all cases, these properties are not yet firmly established with the same confidence one is accustomed to when looking in a handbook, so a final table of initial design values completes this chapter as an aid to initial consideration and design of MEMS.

If the reader is interested in the experimental methods, then the review of test methods will guide him or her to the appropriate references. If details about mechanical properties of specific materials are desired, then the tables and the references will prove useful. Finally, if one wants only typical properties for an initial design concept, the last section provides a succinct answer.

3.2 Mechanical Property Definitions

The properties of interest here are *material properties*; that is, the measured value is independent of the test method. Implicit is the understanding that the property is also independent of the size of the specimen, but that may not necessarily be the case for MEMS materials. The fabrication process for, say, thin-film silicon carbide is completely different from that of bulk silicon carbide, and it is reasonable to expect different mechanical behavior. The question of specimen size effect needs to be considered at the appropriate length scale—in this case, whether a 200×200-μm cross-section tensile specimen behaves the same way as a 2×2-μm one. That question is not very easy to answer until test methods exist with sufficient sensitivity and reproducibility to differentiate the material behavior.

The American Society for Testing Materials defines standard test procedures through a lengthy process of draft and review. Many of the common standards for structural materials were set in the early part of the twentieth century. However, new ones appear to meet the demands of new technologies, and a complete set of standards is issued each year. For example, the field of fracture mechanics as a usable measure of material and structural response emerged in the early 1950s. The first draft of a standard measure of fracture toughness did not appear until 1965, with the first complete standard appearing in 1970 [ASTM, 1970]. It will be some time before standards for measuring the mechanical properties of MEMS materials are established, but it is useful to be guided by the accepted definitions of mechanical properties. The pertinent standards for testing the mechanical properties of metals appear in [ASTM, 2000a] and those for ceramics in [ASTM, 2000b]. ASTM Standard E-8 gives directions for tension testing of metals, while E-9 covers compression testing. ASTM Standards C-1273 and C-1161 cover the tension and creep testing of ceramics. Once the stress–strain curve is obtained, various approximations or curve-fits can be used to insert the material behavior into the design process.

Young's modulus is the slope of the linear part of the stress–strain curve of a material; it is a measure of its stiffness. ASTM E-111 specifies that, "The test specimen is loaded uniaxially and load and strain are measured, either incrementally or continuously." It then goes on to prescribe how the slope is determined along with a myriad of other details. *Poisson's ratio* is a measure of the lateral contraction or expansion of a material when subjected to an axial stress within the elastic region. ASTM E-132 requires that, "In this test method, the value of Poisson's ratio is obtained from strains resulting from uniaxial stress only." Note that these elastic properties are defined for isotropic materials only. Neither of these is easy to measure at the MEMS material size scale as will be seen in the next section. When a material is inelastic (and nonlinear), one needs the complete stress–strain curve to specify the material behavior.

The strength of a material enables one to determine how much force can be applied to a component or structure. ASTM E-6 defines *fracture strength* as "the normal stress at the beginning of fracture"; it is the useful measure for brittle materials. ASTM C-1161 defines *flexural strength* as "a measure of the ultimate strength of a specified beam in bending"; note the linking of the strength measure to a particular size and shape of specimen. If the material is inelastic, then *yield strength* (defined by a prescribed deviation from initial linearity) is the one that defines the departure from elastic response, and *tensile strength* denotes the maximum stress the material will support before complete failure. *Compressive strength* is more difficult to establish unless the material is brittle.

Fracture toughness is a generic term for various measures of resistance to extension of a crack. The most familiar version is plane-strain fracture toughness, ASTM E-399, which requires that the test specimen be thick enough to produce a state of plane strain at the tip of the crack. In this case, the value measured is indeed a material property that is independent of specimen size. Perhaps a more appropriate measure for MEMS is plane-stress fracture toughness, ASTM E-561, but it requires that one either measure or infer the actual crack extension. Implicit in all fracture testing is the condition that the radius at the tip of the crack be very small relative to other dimensions; this is a difficult requirement at the MEMS size scale.

The response of a material to cyclic loading is presented as the *S–N curve*, which is a plot of the applied stress, S, on the ordinate vs. the number of cycles to failure, N, on the abscissa. One obtains such a plot by testing many samples at various levels of applied stress and recording the number of cycles until the specimen breaks in two. ASTM E-466 gives the detailed procedures for metals; this is obviously an expensive kind of test.

Creep is the time-dependent increase in strain under applied stress. Although important in systems operating at high temperature, there is no ASTM standard for creep testing of metals. ASTM C-1291 defines procedures for testing ceramics. As in fatigue testing, results are usually presented in the form of plots.

3.3 Test Methods

Measuring mechanical properties of materials manufactured by processes used in MEMS is not easy. One must be able to: (1) obtain and mount a specimen, (2) measure its dimensions, (3) apply force or displacement to deform it, (4) measure the force, and (5) measure the displacement, or preferably, measure the strain. All of these steps are fully developed and standardized for common structural materials where the minimum dimension of the gauge section of a tensile specimen is 2 mm or so. ASTM E-345 does describe procedures for testing metallic foils that are less than 150 µm thick, but the rest of the dimensions are large. ASTM E-8 includes wires and even describes special grips, but does not state a minimum diameter. One can get guidance from these two standards, but neither is completely appropriate for small MEMS specimens.

One would prefer to determine mechanical properties by *direct methods* similar to the approaches of ASTM. To obtain Young's modulus, a uniform stress is applied, which is calculated from direct measurements of the applied force and the dimensions of the specimen. Strain is measured directly as the force is applied. The specimen is designed to have a uniform gauge section, which is long enough to assure that the stress field is not affected by the grip ends and to permit strain measurement.

This is not always possible for MEMS materials; in fact, it is most often neither possible nor practical. It is then necessary to resort to *inverse methods* in which a model (simple or complex) is constructed of the test structure. Force is applied to the test structure and displacement is measured with the elastic, inelastic, or strength properties then extracted from the model. A simple example, and one that has been widely used in MEMS material testing, is a cantilever beam. If it is sufficiently long and thin, then only the Young's modulus enters as a material property into the formula relating force and displacement. Other examples are resonant structures and bulge tests with pressurized membranes; these are described later. If more than one material property appears in the model, then different geometries must be tested.

The formulas for determining Young's modulus, E, by various methods are

Static Beam	Resonant Beam	Bulge Test	Tensile Test
$\dfrac{4PL^3}{\delta bh^3}$	$\dfrac{ML^3\omega^2}{2bh^3}$	$\dfrac{p(1-\nu)a^4}{\delta^3 h\, c(\nu)}$	$\dfrac{P}{bh\varepsilon}$

where h, b and L are the thickness, width and length of the specimen; P and p are the applied force and pressure; M is the effective mass; ω is the resonant frequency; a is the dimension of a square membrane; and δ and ε are the measured deflection and strain, respectively. The function of Poisson's ratio, $c(\nu)$, depends upon the geometry and is often approximated. The simplicity of the tensile test is an obvious advantage.

Johnson et al. (1999) have compared the uniaxial and bending tests and point out that uncertainty in specimen dimensions are more of a problem in bending tests, while overall elongation is difficult to measure in a tension test. However, if strain can be measured directly, the overall elongation does not need to be measured. They also point out that strength due to misalignment is more of a problem in tension than in bending.

As will be seen later in this chapter, there is an alarming variability among measured values of even such a basic property as Young's modulus for the most widely studied MEMS material—polysilicon. Senturia (1998) attributes this to two primary reasons: "insufficiently precise models used to interpret the data and metrology errors in establishing the geometry of the test devices." He is referring to inverse methods in the first point; whether the boundary conditions of the actual structure actually match the model is a real question. The point on metrology applies to all test methods—direct or inverse.

It is useful to distinguish between *on-chip* test methods and *property tests*. It is very important in this technology to be able to obtain a measure of mechanical properties from *test structures* that are on the same chip (or die) as the manufactured MEMS. That usually precludes a direct property measurement on a *specimen*, which must be larger to allow gripping and pulling even though the size of the gauge section is the same size scale as the microdevice. This is not an issue in mechanical and civil engineering fields where the required specimen size is smaller than the system or structure. One may regard property tests as basic or baseline and on-chip tests as practical. Obviously, direct comparisons of the two approaches with specimens and test structures on the same die are required for completeness.

3.3.1 Specimen and Test Structure Preparation

One does not take a billet of bulk material and shape it into the final MEMS product as is common for most manufacturing processes. Rather, the microdevice is produced by deposition and etching processes. This means that the specimen or test structure cannot be cut from the bulk material but must be produced by the same processes as the product. A tensile specimen must be designed so that one end remains fixed to the die and the other end accommodates some sort of gripping mechanism. A test structure must be designed so that the boundaries are indeed fixed and must incorporate some sort of actuating mechanism to produce force and a sensor to determine displacement or perhaps strain.

An early and interesting approach to producing tensile specimens of thin foils was conceived by Neugebauer (1960), who deposited gold films onto oriented rocksalt crystals. The gold film was glued to the grips of a test machine and the test section covered with sealing wax while the salt was dissolved away. These specimens ranged in thickness from 0.05 to 1.5 μm and were 1 to 2 mm wide and approximately 1 cm long. He found the tensile strength to be 2 to 4 times higher than for annealed bulk material but observed no dependence on film thickness. The Young's modulus values agreed with those of the bulk material.

This is a simple example of specimen preparation, but it is illustrative of the methods used in the mechanical test methods for MEMS materials. One deposits the material of interest and removes the unwanted portions of the supporting substrate. An additional step is the patterning of the test material through photolithography.

3.3.2 Dimension Measurement

Minimum features in MEMS are usually on the order of 1 μm—a bit larger than in microelectronics. One would think that measuring the 2 × 2-μm cross section of a tensile specimen or the equivalent dimensions of a test structure is easy, but it is not. The thickness of a layer is well controlled and measured by the manufacturer. Lengths are large enough to measure with sufficient accuracy in an optical microscope. It is the width of the specimens or test structure components that is difficult to determine when it is small.

A major problem is the fact that the cross section is not sharply defined or even rectangular as expected. Figure 3.1 is a scanning electron microscope (SEM) photograph of the end of a polysilicon tensile specimen after testing. This specimen is from the MUMPs process at Cronos, which deposits a first layer of polysilicon that is 2.0 μm thick and then a second layer that is 1.5 μm thick. The interface between these two layers is visible in the photograph. The designed width is 2.0 μm, which is approximately the case at the bottom. The fact that the cross section is not a perfect rectangle contributes to the uncertainty in the area. One can also see that the corners are somewhat rounded, which makes it difficult to establish the edges when making a plan-view measurement.

FIGURE 3.1 Scanning electron micrograph of the end of a broken tensile specimen [Sharpe et al., 1999a]. The specimen is 3.5 μm thick and 2 μm wide at the bottom.

The dimensions of a specimen or test structure are normally established before the experiment, but a more accurate measurement may be made after the specimen is broken. Optical or scanning electron microscopy, mechanical or optical profilometry, and interferometry are possible measurement techniques, but some of these can be quite time consuming and expensive. Johnson et al. (1999) state that it is typical to assign an uncertainty of between ±0.05 and ±0.10 μm to width measurements. A 2 μm-wide specimen would therefore have at most ±5% relative uncertainty in its width, which is actually quite reasonable for such small specimens. This short discussion reinforces the statement by Senturia (1998) that metrology is a major problem in determining mechanical properties.

3.3.3 Force and Displacement Measurement

Johnson et al. (1999) explain that tensile tests require the measurement of larger forces and smaller overall displacements, while the opposite is true for bending tests. A polysilicon tensile specimen 2×2 μm square with a fracture strength of 2 GPa requires a force of 8 mN to break it. If that same specimen is 50 μm long, fixed at one end, and with a transverse point load at the other end, a force of only 0.05 mN is required to break it. If that material has a modulus of 160 GPa, the elongation of a tensile specimen is 0.62 μm, while the deflection at the end of a bending specimen is 10.4 μm.

Commercial force transducers are readily available with a range of ±5 g (50 mN) and a resolution of 0.001 g. This author prefers to use the lower range of a ±100-g-load cell because it is stiffer relative to a tensile specimen and to calibrate it with weights. A resolution of 0.01 g with a full-scale uncertainty of ±1% is achieved [Sharpe et al., 1999a]. Howard and Fu (1997) review suitable force transducers, and others such as Greek et al. (1995) and Saif and MacDonald (1996) construct their own.

Commercial capacitance-based displacement transducers can be used to measure the overall displacement of a test system to a resolution of 0.01 μm and full-scale uncertainty of ±1% [Sharpe et al., 1999a]. Schemes to measure mechanical deflections at the optical microscope level are attractive, and Pan and Hsu (1999) present a vernier gauge approach to measure residual stress. This approach can be electrically instrumented with differential capacitance measurement as shown by Que et al. (1999).

3.3.4 Strain Measurement

It is, of course, preferable to measure strain directly whether the test arrangement is bending or tension; this is difficult to do on such small specimens. The author and his colleagues have developed a laser-based strain measurement system in which two reflective lines are deposited on the gauge section of a tensile specimen during manufacture. These lines are perpendicular to the loading axis, and when they are illuminated with a low-power laser beam interference fringe patterns are formed. These fringes move as the lines separate when the specimen is strained; the motion is tracked with diode arrays and a computer system to enable real-time strain measurement on specimens as narrow as 20 μm. A set of four lines on wider specimens permits one to measure Poisson's ratio. Details are given in Sharpe et al. (1997c; 1997d), and the resolution is approximately ±5 microstrain with a relative uncertainty of 5% at 0.5% strain.

Detailed full-field strain measurements at the MEMS size scale are desirable but difficult. Micro-Raman spectroscopy can probe very small areas, on the order of 1 μm in diameter, on thin films. Analysis of frequency shifts as force is applied to a specimen leads to local strain measurements [Benrakkad et al., 1995; Pinardi et al., 1997; Zhang et al., 1997; Amimoto et al., 1998]. The Moiré method using e-beam lithography to write high-frequency line and dot gratings at a small scale has been demonstrated by Dally and Read (1993), but this is a very challenging process. Chasiotis and Knauss (1998) are developing digital image correlation methods to measure strains in tensile specimens; the resolution is 300 to 500 microstrain. Mazza et al. (1996a) have demonstrated this to be a viable technique on single-crystal silicon specimens. Laser speckle methods can give full-field results and have been demonstrated by Anwander et al. (2000) and Chang et al. (2000). None of these techniques has been applied to extensive studies of mechanical properties of MEMS materials.

3.3.5 Tensile Tests

There are three arrangements used in tensile tests of MEMS materials: specimen in a supporting frame, specimen fixed to die at one end, and separate specimen. A fourth clever approach was introduced early on by Koskinen et al. (1993) but has not been continued. They deposited a grid of long, thin tensile specimens, which were all fastened to larger portions at each end; the appearance was similar to a foil resistance strain gage. One end of the arrangement was fixed and the other attached to a movable grip that could be rotated about an axis perpendicular to the grid. This caused all of the specimens to buckle, each a different amount than its neighbor. When the grip moved, each specimen in turn was straightened and pulled. The recorded force-displacement record enabled measurement of modulus and strength.

3.3.5.1 Specimen in Frame

Read and Dally (1992) introduced a very effective way of handling thin-film specimens in 1992. The tensile specimen is patterned onto the surface of a wafer and then a window is etched in the back of the wafer to expose the gauge section. The result is a specimen suspended across a rectangular frame, which can be handled easily and placed into a test machine. The two larger ends of the frame are fastened to grips, and the two narrower sides are cut to completely free the specimen. This is an extension of the much earlier approach by Neugebauer (1960) and has been adopted by others [Cunningham et al., 1995; Emery et al., 1997; Ogawa et al., 1997; Sharpe et al., 1997c; Cornella et al., 1998; Yi and Kim, 1999b]. A SEM photograph of such a specimen while still in the frame is shown in Figure 3.2.

3.3.5.2 Specimen Fixed at One End

Tsuchiya introduced the concept of a tensile specimen fixed to the die at one end and gripped with an electrostatic probe at the other end [Tsuchiya et al., 1998]. This approach has been adopted by this author and his students [Sharpe et al., 1998a], and Figure 3.3 is a photograph of this type of specimen. The gauge section is 3.5 µm thick, 50 µm wide, and 2 mm long. The fixed end is topped with a gold layer for electrical contact. The grip end is filled with etch holes, as are the two curved transition regions from the grips to the gauge section. The large grip end is held in place during the etch release process by four anchor straps, which are broken before testing.

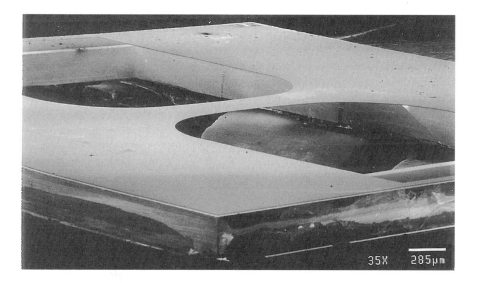

FIGURE 3.2 Scanning electron micrograph of a polysilicon tensile specimen in a supporting single-crystal silicon frame. (From Sharpe, W.N., Jr., Yuan, B., Vaidyanathan, R., and Edwards, R.L. (1996) *Proc. SPIE* **2880**, pp. 78–91. With permission.)

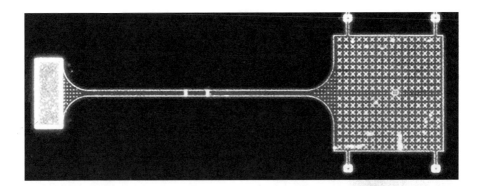

FIGURE 3.3 A tensile specimen fixed at the left end with a free grip end at the right end. (From Sharpe, W.N., Jr., and Jackson, K. (2000) *Microscale Systems: Mechanics and Measurements Symposium*, Society for Experimental Mechanics, pp. ix–xiv. With permission.)

Chasiotis and Knauss (2000) have developed procedures for gluing the grip end of a similar specimen to a force/displacement transducer, which enables application of larger forces. A different approach is to fabricate the grip end in the shape of a ring and insert a pin into it to make the connection to the test system. Greek et al. (1995) originated this with a custom-made setup, and LaVan et al. (2000a) use the probe of a nanoindenter for the same purpose.

It is possible to build the deforming mechanism onto or into the wafer, although it can be difficult to get an accurate measure of the forces and deflections. Biebl and von Philipsborn (1995) stretched polysilicon specimens in tension with residual stresses in the structure. Yoshioka et al. (1996) etched a hinged paddle in the silicon wafer, which could be deflected to pull a thin single crystal specimen. Nieva et al. (1998) produced a framed specimen and heated the frame to pull the specimen, as did Kapels et al. (2000).

3.3.5.3 Separate Specimen

The challenge of picking up a tensile specimen only a few microns thick and placing it into a test machine is formidable. However, if the specimens are on the order of tens or hundreds of microns thick, as they are for LIGA-deposited materials, it is perfectly possible. This author and students developed techniques to test steel microspecimens having submillimeter dimensions [Sharpe et al., 1998b]. The steel dog biscuit-shaped specimens were obtained by cutting thin slices from the bulk material and cutting them out with a small CNC mill. Electroplated nickel specimens can be patterned into a similar shape in LIGA molds as shown in Figure 3.4. These specimens are released from the substrate by etching, picked up, and put into grips with inserts that match the wedge-shaped ends [Sharpe et al., 1997e].

McAleavey et al. (1998) used the same sort of specimen to test SU-8 polymer specimens. Mazza et al. (1996b) prepared nickel specimens of similar size in the gauge section, but with much larger grip ends. Christenson et al. (1998) fabricated LIGA nickel specimens of a more conventional shape that were approximately 2 cm long with flat grip ends; these were large enough to test in a commercial table-top electrohydraulic test machine.

3.3.5.4 Smaller Specimens

All of the above may appear impressive to the materials test engineer accustomed to common structural materials, but there is a continuing push toward smaller structural components at the nanoscale. Yu et al. (2000) have successfully attached the ends of carbon nanotubes as small as 20 nm in diameter and a few microns long to atomic force microscopy (AFM) probes. As the probes are moved apart inside a SEM, their deflections are measured and used to extract both the force in the tube and its overall elongation. They report strengths up to 63 GPa and modulus values up to 950 GPa.

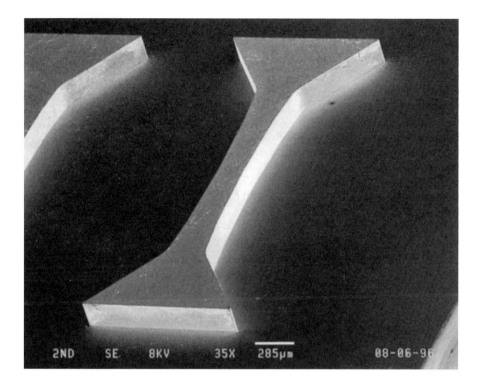

FIGURE 3.4 Nickel microspecimen produced by the LIGA method. The overall length is 3.1 mm, and the width of the specimen at the center is 200 μm. (From Sharpe, W.N., Jr., et al. (1997) *Proc. Int. Solid State Sensors and Actuators Conf.—Transducers '97*, pp. 607–610. © 1997 IEEE. With permission.)

3.3.6 Bend Tests

Three arrangements are also used in bend tests of structural films: out-of-plane bending of cantilever beams, beams fastened at both ends, and in-plane bending of beams. Larger specimens, which can be individually handled, can also be tested in bending fixtures similar to those used for ceramics.

3.3.6.1 Out-of-Plane Bending

The approach here is simple. One patterns long, narrow, and thin beams of the test material onto a substrate and then etches away the material underneath to leave a cantilever beam hanging over the edge. By measuring the force vs. deflection at or near the end of the beam, one can extract Young's modulus via the formula in Section 3.3. However, this is not so easy because if the beams are long and thin, the deflections can be large, but the forces are small. The converse is true if the beam is short and thick, but then the applicability of simple beam theory comes into question. If the beam is narrow enough, Poisson's ratio does not enter the formula; otherwise, beams of different geometries must be tested to determine it.

Weihs et al. (1988) introduced this method in 1988 by measuring the force and deflection with a nanoindenter having a force resolution of 0.25 μN and a displacement resolution of 0.3 nm. Typical specimens had a thickness, width and length of 1.0, 20 and 30 μm, respectively. Figure 3.5 shows a cantilever beam deflected by a nanoindenter tip in a later investigation [Hollman et al., 1995].

Biebl et al. (1995a) attracted the end of a cantilever down to the substrate with electrostatic forces and recorded the capacitance change as the voltage was increased to pull more of the beam into contact. Fitting these measurements to an analytical model permitted a determination of Young's modulus.

Krulevitch (1996) proposed a technique for measuring Poisson's ratio of thin films fabricated in the shapes of beams and plates by comparing the measured curvatures. These were two-layer composite

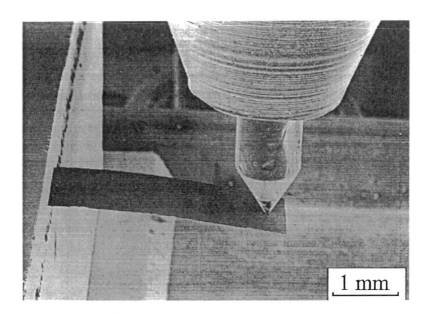

FIGURE 3.5 A cantilever microbeam deflected out of plane by a diamond stylus. The beam was cut from a free-standing diamond film. (Reprinted from Hollman, P., et al. (1995) *Thin Solid Films* **270**, pp. 137–142. With permission from Elsevier Science.)

structures, so the properties of the substrate must be known. Kraft et al. (1998) also tested composite beams by measuring the force-deflection response with a nanoindenter. Bi-layer cantilever beams have been tested by Tada et al. (1998) who heated the substrate and measured the curvature.

More sensitive measurements of force and displacement on smaller cantilever beams can be made by using an AFM probe, as shown by Serre et al. (1998), Namazu et al. (2000), Comella and Scanlon (2000) and Kazinczi et al. (2000). A specially designed test machine using an electromagnetic actuator has been developed by Komai et al. (1998).

3.3.6.2 Beams with Fixed Ends

It is somewhat easier to work with a beam that is fixed at both ends; it is stiffer and more robust. Tai and Muller (1990) used a surface profilometer to trace the shapes of fixed–fixed beams at various load settings. By comparison of measured traces and using a finite element analysis of the structure, they were able to determine Young's modulus.

A promising on-chip test structure has been developed over the years by Senturia and his students; it is shown schematically in Figure 3.6. A voltage is applied between the conductive polysilicon beam and the substrate to pull the beam down, and the voltage that causes the beam to make contact is a measure of its stiffness. This concept was introduced early on by Petersen and Guarnieri (1979) and further developed by Gupta et al. (1996). A similar approach and analysis were described by Zou et al. (1995). The considerable advantage here is that the measurements can be made entirely with electrical probing in a manner similar to that used to check microelectronic circuits. This opens the opportunity for process monitoring and quality control.

It is clear that the "fixed" ends exert a major influence on the stiffness of the test structure. Kobrinsky et al. (1999) have thoroughly examined this effect and shown its importance. The problem is that a particular manufacturing process, or even variations within the same process, may etch the substrate slightly differently and change the rigidity of the ends. Nevertheless, this is a potentially very useful method for monitoring the consistency of MEMS materials and processes.

Zhang et al. (2000) recently conducted a thorough study of silicon nitride in which microbridges (fixed–fixed beams) were deflected with a nanoindenter with a wedge-shaped indenter. By fitting the

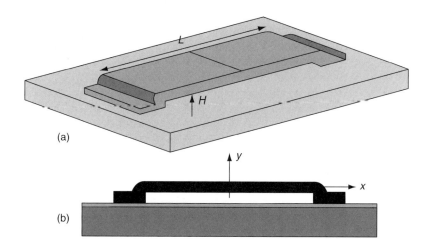

FIGURE 3.6 Schematic of a fixed-fixed beam. (From Kobrinsky, M., et al. (1999) *MEMS Microelectromechanical System* **1**, pp. 3–10, ASME, New York. With permission.)

measured force-deflection records to their analytical model, they extracted both Young's modulus and residual stress.

3.3.6.3 In-Plane Bending

In-plane bending may be a more appropriate test method in that the structural supports of MEMS accelerometers are subjected to that mode of deformation. Jaecklin et al. (1994) pushed long, thin cantilever beams with a probe until they broke; optical micrographs gave the maximum deflections from which the fracture strain was determined. Jones et al. (1996) constructed a test structure consisting of cantilever beams of different lengths fastened to a movable shuttle. As the shuttle was pushed, the beams contacted fixed stops on the substrate; the deformed shape was videotaped and the fracture strain determined. Figure 3.7 is a photograph of one of their deformed specimens.

Kahn et al. (1996) developed a double cantilever beam arrangement to measure the fracture toughness of polysilicon and used the measured displacement between the two beams to determine Young's modulus via a finite element model. The beams were separated by forcing a mechanical probe between them and pushing it toward the notched end. A similar approach has been taken by Fitzgerald et al. (1998) to measure crack growth and fracture toughness in single crystal silicon, but they use a clever structure that permits opening the beams by compression of cantilever extensions.

3.3.6.4 Bending of Larger Specimens

Microelectromechanical technology is not restricted to only thin-film structures, although they are far-and-away the predominant type. Materials fabricated with thicknesses on the order of tens or hundreds of microns are of current interest and likely to become more important in the future.

Ruther et al. (1995) manufactured a microtesting system by the LIGA process to test electroplated copper. The interesting feature is that the in-plane cantilever beam and the test system are fabricated together on the die; however, this requires a rather complex assembly. Stephens et al. (1998) fabricated rows of LIGA nickel beams sticking up from the substrate and then measured the force applied near the upper tip of the beam while displacing the substrate. The resulting force-displacement curve permitted extraction of Young's modulus, and the recorded maximum force gave a modulus of rupture.

Larger structures, such as the microengine under development at MIT, have thicknesses on the order of several millimeters. It is then necessary to test specimens of similar sizes in what is sometimes called the "mesocale" region—dimensions from 0.1 mm to 1 cm for a general definition. Single-crystal silicon is the material of interest for initial versions and Chen et al. (1998) have developed a method for bend

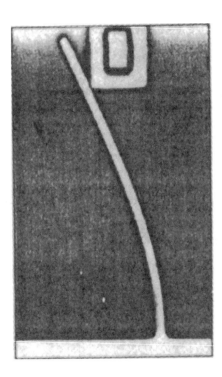

FIGURE 3.7 A polysilicon cantilever beam subjected to in-plane bending. The beam is 2.8 μm wide, and the vertical distance between the fixed end at the bottom and the deflected end at the top is 70 μm. (From Sharpe, W.N., Jr., et al. (1998) *Microelectromechanical Structures for Materials Research Symposium*, pp. 56–65. With permission.)

testing square plates simply supported over a circular hole and recording the force as a small steel ball is pushed into the center of the plate. Fracture strengths are obtained, and this efficient arrangement permits study of the effects of various manufacturing processes on the load-carrying capability of the material.

3.3.7 Resonant Structure Tests

Frequency and changes in frequency can be measured precisely, and elastic properties of modeled structures can be determined. The microstructures can be very small and excited by capacitive comb drives, which require only electrical contact. This makes this approach suitable for on-chip testing; in fact, the MUMPs process at Cronos includes a resonant structure on each die. That microstructure moves parallel to the substrate, but others vibrate perpendicularly.

Petersen and Guarnieri (1979) introduced the resonant structure concept in 1979 by fabricating arrays of thin, narrow cantilever beams of various lengths extending over an anisotropically etched pit in the substrate. The die containing the beams was excited by variable frequency electrostatic attraction between the substrate and the beams, and the vibration perpendicular to the substrate was measured by reflection from an incident laser beam, as shown by the schematic in Figure 3.8. Yang and Fujita (1997) took a similar approach to study the effect of resistive heating on U-shaped beams. Commercial AFM cantilevers were tested in a similar manner by Hoummady et al. (1997), who measured the higher resonant modes of a cantilever beam with a mass on the end. Zhang et al. (1991) measured vibrations of a beam fixed at both ends by using laser interferometry. An elaborate and carefully modeled micromirror that was excited by electrostatic attraction was developed by Michalicek et al. (1995). Deflection was also measured by laser interferometry, and experiments determined Young's modulus over a range of temperatures as well as validated the model.

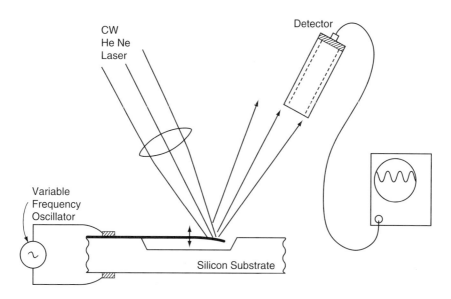

FIGURE 3.8 Schematic of the resonant structure system of Petersen and Guarnieri (1979). (Reprinted from Petersen, K.E., and Guarnieri, C.R. (1979) *J. Appl. Phys.* **50**, pp. 6761–6766. With permission from American Institute of Physics.)

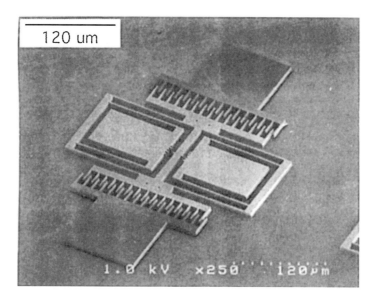

FIGURE 3.9 Scanning electron micrograph of the in-plane resonant structure of Kahn et al. (1998). (From Kahn, H., et al. (1998) *Microelectromechanical Structures for Materials Research Symposium*, pp. 33–38. With permission.)

Microstructures that vibrate parallel to the plane of the substrate require less processing; the substrate does not have to be removed. Biebl et al. (1995b) introduced this concept, and Kahn et al. (1998) have used a more recent version to study the effects of heating on the Young's modulus of films sputtered onto the structure. Figure 3.9 is a SEM image of their structure, which is easy to model. Pads A, B, C and D are fixed to the substrate; the rest of the structure is free. Electrostatic comb drives excite the two symmetrical substructures, which consist of four flexural springs and a rigid mass. The resonant frequency of this device is around 47 kHz. A different approach in which a small notched specimen is fabricated as part of a large resonant fan-shaped component has been developed by Brown et al. (1997).

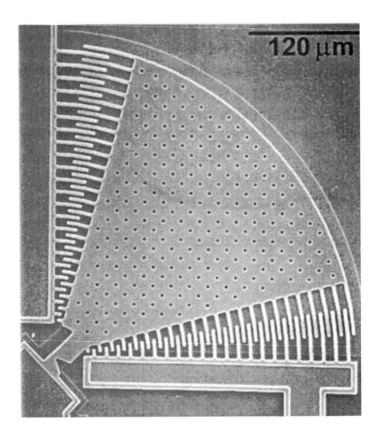

FIGURE 3.10 Scanning electron micrograph of the in-plane resonant structure of Brown et al. (1997). (From Brown, S.B., et al. (1997) *Proc. Int. Solid-State Sensors and Actuators Conf.—Transducers '97*, pp. 591–593. © 1997 IEEE. With permission.)

This resonant structure, shown in Figure 3.10, has been primarily used for fatigue and crack growth studies, but Young's modulus of polysilicon has been extracted from its finite element model [Sharpe et al., 1998c].

3.3.8 Membrane Tests

It is relatively easy to fabricate a thin membrane of test material by etching away the substrate; the membrane is then pressurized and the measured deflection can be used to determine the biaxial modulus. An advantage of this approach is that tensile residual stress in the membrane can be measured, but the value of Poisson's ratio must be assumed. This method, often called bulge testing, was first introduced by Beams (1959), who tested thin films of gold and silver and measured the center deflection of the circular membrane as a function of applied pressure. Jacodine and Schlegel (1966) used this approach to measure Young's modulus of silicon oxide. Tabata et al. (1989) tested rectangular membranes whose deflections were measured by observations of Newton's rings, as did Maier-Schneider et al. (1995). The variation of Hong et al. (1990) used circular membranes with force deflection measured at the center with a nanoindenter. Pressurized square membranes with the deflection measured by a stage-mounted microscope were tested by Walker et al. (1990) to study the effect of hydrofluoric acid exposure on polysilicon; a similar approach to determine biaxial modulus, residual stress and strength was used by Cardinale and Tustison (1992). Vlassak and Nix (1992) eliminated the need to assume a value of Poisson's ratio by testing rectangular silicon nitride films with different aspect ratios. More recently, Jayaraman et al. (1998) took this same approach to measure Young's modulus and Poisson's ratio of polysilicon.

3.3.9 Indentation Tests

A nanoindenter is, in the fewest words, simply a miniature and highly sensitive hardness tester. Both force and displacement are measured, and modulus and strength can be obtained from the resulting plot. Penetration depths can be very small (a few nanometers), and automated machines permit multiple measurements to enhance confidence in the results and also to scan small areas for variations in properties.

Weihs et al. (1989) measured the Young's modulus of an amorphous silicon oxide film and a nontextured gold film with a nanoindenter and obtained only limited agreement with their microbeam deflection experiments. The modulus measured by indentation was consistently higher and the large pressure of the indenter tip was the probable cause. Taylor (1991) used nanoindenter measurements restricted to penetrations of 200 nm into 1-μm-thick silicon nitride films to study the effects of processing on mechanical properties. Young's modulus decreased with decreasing density of the films.

Bhushan and Li (1997) and Li and Bhusan (1999) have studied the tribological properties of MEMS materials and used a nanoindenter to measure the modulus and a microhardness tester to measure the fracture toughness of thin films. Measurements of Young's modulus of polysilicon showed a wide scatter. Bucheit et al. (1999) examined the mechanical properties of LIGA-fabricated nickel and copper by using a nanoindenter as one of the tools. In most cases, Young's modulus from nanoindenter measurements were higher than from tension tests, but the nanoindenter does allow one to look at both sides of the thin film as well as sectioned areas.

3.3.10 Other Test Methods

The readily observed buckling of a column-like structure under compression can be used to measure forces in specimens; if the specimen breaks, an estimate of the fracture strength can be made. Tai and Muller (1988) fabricated long, thin polysilicon specimens with one end fixed and the other enclosed in slides. The movable end was pushed with a micromanipulator, and its displacement when the structure buckled was used to determine the strain (not stress) at fracture. Ziebart and colleagues have analyzed thin films with various boundary conditions ranging from fixed along two sides [Ziebart et al., 1997] to fixed on all four sides [Ziebart et al., 1999]. The first arrangement permitted the measurement of Poisson's ratio when the side supports were compressed, and the second determined prestrains induced by processing. Beautiful patterns are obtained, but the analysis and the specimen preparation can be time consuming.

Another clever approach based on buckling is described by Cho et al. (1997). They etched away the silicon substrate under an overhanging strip of diamond-like carbon film and used the buckled pattern to determine the residual stress in the film. A more traditional creep test was used by Teh et al. (1999) to study creep in $2 \times 2 \times 100$-μm polysilicon strips fixed at each end. As current passed through the specimens, they heated up and their buckled deflection over time at a constant current was used to extract a strain vs. time creep curve. This approach is complicated by the nonuniformity of the strain in the specimen.

Although torsion is an important mode of deformation in certain MEMS such as digital mirrors, few test methods have been developed. Saif and MacDonald (1996) introduced a system to twist very small (10 μm long and 1 μm on a side) pillars of single-crystal silicon and measure both the force and deflection. Larger (300 μm long with side dimensions varying from 30 to 180 μm) of both silicon and LIGA nickel were tested by Schiltges et al. (1998). Emphasis was on the elastic properties only with the shear modulus values agreeing with expected bulk values.

Nondestructive measurements of elastic properties of thin films can be accomplished with laser-induced ultrasonic surface waves. An impulse is generated in the film by a laser pulse, and the surface wave is sensed by a piezoelectric transducer. In principle, Young's modulus, density and thickness can be determined, but this cannot be achieved for all combinations of film and substrate materials. Schneider and Tucker (1996) describe this test method and present results for a wide range of films; the Young's modulus values generally agree with other thin-film measurements. A drawback here is the planar size of the film; the input and output must be several millimeters apart. A related technique uses Brillouin scattering as described in Monteiro et al. (1996).

3.3.11 Fracture Tests

Single-crystal silicon and polysilicon are both brittle materials, and it is therefore natural to want to measure their fracture toughness. This is even more difficult than measuring their fracture strength because of the need for a crack with a tip radius small relative to the specimen dimensions.

Photolithography processes for typical thin films have a minimum feature radius of approximately 1 μm. Fan et al. (1990), Sharpe et al. (1997f) and Tsuchiya et al. (1998) have tested polysilicon films in tension using edge cracks, center cracks and edge cracks, respectively. Kahn et al. (1999) modeled a double-cantilever specimen with a long crack and wedged it open with an electrostatic actuator.

Fitzgerald et al. (1999) prepared sharp cracks in double-cantilever silicon crystal specimens by etching, and Suwito et al. (1997) modeled the sharp corner of a tensile specimen to measure the fracture toughness. Van Arsdell and Brown (1999) introduced cracks at notches in polysilicon with a diamond indenter. A promising new approach using a focused ion beam (FIB) can prepare cracks with tip radii of 30 nm according to K. Jackson (pers. comm.).

3.3.12 Fatigue Tests

Many MEMS operate for billions of cycles, but that kind of testing is conducted on microdevices such as digital mirrors instead of the more basic reversed bending or push–pull tests so familiar to the metal fatigue community. Brown and his colleagues have developed a fan-shaped, electrostatically driven notched specimen that has been used for fatigue and crack growth studies [Brown et al., 1993; 1997; Van Arsdell and Brown, 1999]. Minoshima et al. (1999) have tested single crystal silicon in bending fatigue, and Sharpe et al. (1999) reported some preliminary tension–tension tests on polysilicon. As noted earlier, fatigue data are reported as stress vs. life plots, and Kapels et al. (2000) present a plot that looks much like one would expect for a metal; the allowable applied stress decreases from 2.9 GPa for a monotonic test to 2.2 GPa at one million cycles.

3.3.13 Creep Tests

Some MEMS are thermally actuated, so the possibility of creep failure exists. No techniques similar to the familiar dead-weight loading to produce strain vs. time curves exist. Teh et al. (1999) have observed the buckling of heated fixed-end polysilicon strips.

3.3.14 Round Robin Tests

Mechanical testing of MEMS materials presents unique challenges as the above review shows. Convergence of test methods into a standard is still far in the future, but progress in that direction usually begins with a "round robin" program in which a common material is tested by the method-of-choice in participating laboratories. That first step was taken in 1997/1998 with the results reported at the Spring 1998 meeting of the Materials Research Society [Sharpe et al., 1998c]. Polysilicon from the MUMPs 19 and 21 runs of Cronos were tested in bending (Figure 3.7), resonance (Figure 3.10) and tension (Figure 3.3). Young's modulus was measured as 174 ± 20 GPa in bending, 137 ± 5 GPa in resonance and 139 ± 20 GPa in tension. Strengths in bending were 2.8 ± 0.5 GPa, in resonance 2.7 ± 0.2 GPa and in tension 1.3 ± 0.2 GPa. These variations were alarming, but in retrospect perhaps not too surprising given the newness of the test methods at that time.

A more recent interlaboratory study of the fracture strength of polysilicon manufactured at Sandia has been arranged by LaVan et al. (2000b). Strengths measured on similar tensile specimens by Tsuchiya in Japan and at Johns Hopkins were 3.23 ± 0.25 and 2.85 ± 0.40 GPa, respectively. LaVan tested in tension with a different approach and obtained 4.27 ± 0.61 GPa. It seems clear that more effort needs to be devoted to the development of test methods that can be used in a standardized manner by anyone who is interested.

3.4 Mechanical Properties

This section lists in tabular form the results of measurements of mechanical properties of materials used in MEMS structural components. The intent of this section is to provide the reader not only values of mechanical properties but also access to references on materials and test methods of interest. Because there is as yet no standard test method and such a wide variety in the values is obtained for supposedly identical materials, the reader with a strong interest in the mechanical behavior of a particular material can use the tables to identify pertinent references.

Almost all the data listed comes from experiments directly related to free-standing structural films. The only exceptions are the results from ultrasonic measurements by Schneider and Tucker (1996) because they tested a number of materials of interest. It proved too cumbersome to include information on the processing conditions for each reference, but it is hoped the short comments in the tables will be useful. Many of the results are average values of multiple replications, and the standard deviations are included when they are available. Most of the materials used in MEMS are ceramics and show linear and brittle behavior, in which case only the fracture strength is listed. The tables for ductile materials show both yield and ultimate strengths. It should also be noted that the values in the tables are edited from a larger list. Some of the same values had been presented in two different venues (e.g., a conference publication and a journal paper), in which case the more archival version was referenced. A limited number of studies have been conducted on the effects of environment (temperature, hydrofluoric acid, saltwater, etc.) on MEMS materials, but that area of research is in its infancy and is not included.

First, typical stress–strain curves are plotted in Figure 3.11 to compare the mechanical behavior of MEMS materials with a common structural steel, A533-B, which is moderately strong (yield strength of 440 MPa) but ductile and tough. Polysilicon is linear and brittle and much stronger. LIGA nickel is ductile and considerably stronger than bulk pure nickel; one must test materials as they are produced for MEMS instead of relying on bulk material values.

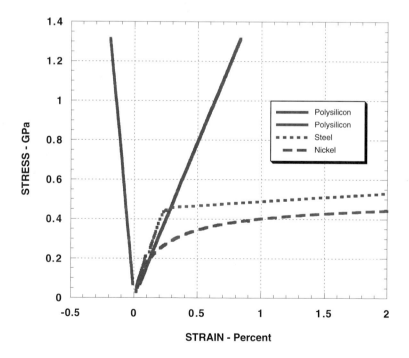

FIGURE 3.11 Representative stress-strain curves of polysilicon, electroplated nickel and A-533B steel. These are from microspecimens tested in the author's laboratory.

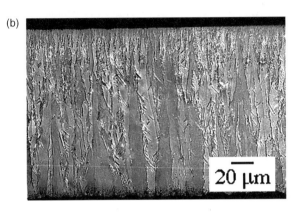

FIGURE 3.12 Microstructure of two common MEMS materials. Note the columnar grain structure perpendicular to the plane of the film. (a) Polysilicon deposited in two layers; the bottom layer is 2.0 µm thick and the top one is 1.5 µm thick. (Reprinted from Sharpe et al. (1998c). © 1998 IEEE. With permission.) (b) Nickel electroplated into LIGA molds. (Reprinted from Sharpe et al. (1997d). © 1998 IEEE. With permission.)

The microstructure of these MEMS materials is also different from bulk materials. The physics of the thin-film deposition process cause the grains to be columnar in a direction perpendicular to the film as shown in Figure 3.12. The result is similar to the cross section of a piece of bamboo or wood, and the material is transversely isotropic. Test methods are not sensitive enough to measure the anisotropic constants.

Table 3.1 lists metal films tested in a free-standing manner such that they would be appropriate for use in MEMS. Only aluminum is currently used in that fashion, but these are materials commonly used in the electronics industry and may be of interest. Note that all of these are ductile materials; the complete stress–strain curves are included in many of the references. The values of Young's modulus as measured for pure bulk materials are listed for reference.

Carbon can be deposited to form an amorphous or crystalline structure, which is often referred to as diamond-like carbon, or DLC. Diamond itself has a very high stiffness and strength as well as a low coefficient of friction, and for these reasons DLC offers exciting possibilities in MEMS. The very limited results to date, shown in Table 3.2, support this line of reasoning although they are far too sparse to be conclusive.

Electroplated nickel and nickel–iron MEMS, usually manufactured via the LIGA process, offer the possibility of larger and stronger actuators and connectors. The microstructure and mechanical properties of an electroplated material are highly dependent upon the composition of the plating bath along with the current and temperature. Similarly, the composition of a nickel–iron alloy makes a big difference. Young's modulus and strength values are listed in Tables 3.3 and 3.4 for nickel and nickel–iron, respectively. The modulus of bulk nickel is around 200 GPa, and the yield strength of pure fine-grained nickel is approximately 60 MPa [ASM, 1990]. Table 3.3 shows that the modulus of nickel is generally somewhat lower and the strength considerably higher. Nickel–iron has a smaller modulus as expected, but can be a very strong material as seen from the few results in Table 3.4.

TABLE 3.1 Metals

	Young's Modulus (GPa)	Yield Strength (GPa)	Ultimate Strength (GPa)	Ref.	Method	Comments
Aluminum;	8–38	—	0.04–0.31	Hoffman (1989)	Tension	110–160 μm thick
modulus of bulk material = 69 GPa	40	—	0.15	Ogawa et al. (1996)	Tension	1.0 μm thick
	69–85	—	—	Comella and Scanlon (2000)	Bending	Various lengths
Copper;	86–137	0.12–0.24	0.33–0.38	Buchheit et al. (1999)	Tension	Plated; annealed
modulus of bulk material = 117 GPa	108–145	—	—	Buchheit et al. (1999)	Indentation	Various locations
	98 ± 4	—	—	Anwander et al. (2000)	Tension	Laser speckle
Gold;	40–80	—	0.2–0.4	Neugebauer (1960)	Tension	0.06–16 μm thick
modulus of bulk material = 74 GPa	57	0.26	—	Weihs et al. (1988)	Bending	~1 μm thick
	74	—	—	Weihs et al. (1988)	Indentation	~1 μm thick
	82	—	0.33–0.36	Emery et al. (1997)	Tension	0.8 μm thick
	—	—	0.22–0.27	Kraft et al. (1998)	Bending	Composite beam
Titanium; modulus of bulk material = 110 GPa	96 ± 12	—	0.95 ± 0.15	Ogawa et al. (1997)	Tension	0.5 μm thick
Ti-Al-Ti	—	0.07–0.12	0.14–0.19	Read and Dally (1992)	Tension	Composite film

TABLE 3.2 Diamond-Like Carbon

Young's Modulus (GPa)	Fracture Strength (GPa)	Ref.	Method	Comments
600–1100	0.8–1.8	Hollman et al. (1995)	Bending	Hot flame deposited
800–1140	—	Schneider and Tucker (1996)	Ultrasonic	CVD diamond
150–800	—	Schneider and Tucker (1996)	Ultrasonic	Laser arc deposited
580	—	Monteiro et al. (1996)	Brillouin	CVD diamond
94–128	—	Cho et al. (1998)	Buckling	Poisson's ratio = 0.22
—	8.5 ± 1.4	LaVan et al. (2000a)	Tension	Amorphous diamond

The most common MEMS material is polysilicon, and it is also the most tested, as Table 3.5 demonstrates. The stiffness coefficients of single-crystal silicon are well established, and the modulus in different directions can vary from 125 to 180 GPa [Sato et al., 1997]. Aggregate theories predict that randomly oriented polycrystalline silicon should have a Young's modulus between 163 and 166 GPa [Guo et al., 1992; Jayaraman et al., 1999]. Most of the modulus values in Table 3.5 are near or within this range, but some vary widely, especially when a test method is first used. An estimate of what the fracture strength should be is more difficult as it depends on the flaws in the material. Even though strength is easier to measure than modulus (one only needs to measure force), there are fewer entries. This is because many of the bending, resonance and bulge tests do not lead to failure in the specimen.

Single-crystal silicon has also been studied extensively, as Table 3.6 shows. The modulus values there are measured along particular crystallographic directions, so they should not be expected to compare with the polysilicon values.

Silicon carbide holds promise for MEMS because of its expected high stiffness, strength and chemical and temperature stability, and Sarro (2000) provides a thorough overview of its potential. Bulk silicon carbide is commonly available, but manufacturing processes for thin, free-standing films are still in development. Table 3.7 lists results from the few tests to date; note that no strength values appear.

Silicon nitride commonly appears in both MEMS and in microelectronics as an insulating layer. There is growing interest in it as a structural material, and Table 3.8 lists its properties. Silicon oxide is also

TABLE 3.3 Nickel

Young's Modulus (GPa)	Yield Strength (GPa)	Ultimate Strength (GPa)	Ref.	Method	Comments
202	0.40	0.78	Mazza et al. (1996b)	Tension	Vibration for modulus
~200	—	—	Schneider and Tucker (1996)	Ultrasonic	3–75 μm thick
168–182	0.1 ± 0.01	—	Basrour et al. (1997)	FE Model	Microgrippers
205	—	—	Dual et al. (1997)	Resonance	Also fatigue
68*	—	—	Dual et al. (1997)	Torsion	*Shear modulus
176 ± 30	0.32 ± 0.03	0.55	Sharpe et al. (1997e)	Tension	~200 μm thick
131–160	0.28–0.44	0.46–0.76	Christenson et al. (1998)	Tension	Varied current
231 ± 12	1.55 ± 05	2.47 ± 0.07	Greek and Ericson (1998)	Tension	6 μm thick
180 ± 12	—	—	Kahn et al. (1998)	Resonance	Film on resonator
181 ± 36	0.33 ± 0.03	0.44 ± 0.04	Sharpe and McAleavey (1998)	Tension	LIGA 3 films
158 ± 22	0.32 ± 0.02	0.52 ± 0.02	Sharpe and McAleavey (1998)	Tension	LIGA 4 films
182 ± 22	0.42 ± 0.02	0.60 ± 0.01	Sharpe and McAleavey (1998)	Tension	HI-MEMS films
153 ± 14	—	1.28 ± 0.24*	Stephens et al. (1998)	Bending	*Modulus of rupture
156 ± 9	0.44 ± 0.03	—	Buchheit et al. (1999)	Tension	Current = 20 ma/cm^2
92	0.06/0.16*	—	Buchheit et al. (1999)	*Tension/compression	Annealed
160 ± 1	0.28/0.27*	—	Buchheit et al. (1999)	*Tension/compression	Current = 50 ma/cm^2
146–184	—	—	Buchheit et al. (1999)	Indentation	Various locations
194	—	—	Anwander et al. (2000)	Tension	Laser speckle

TABLE 3.4 Nickel–Iron

Young's Modulus (GPa)	Yield Strength (GPa)	Ultimate Strength (GPa)	Ref.	Method	Comments
65	—	—	Chung and Allen (1996)	Fixed ends	80% Ni–20% Fe
119	0.73	1.62	Dual et al. (1997)	Tension	50% Ni–50% Fe
115	—	—	Dual et al. (1997)	Resonance	50% Ni–50% Fe
15–54*	—	—	Dual et al. (1997)	Torsion	*Shear modulus
155	—	2.26	Greek and Ericson (1998)	Tension	Electroplated
—	1.83–2.20	2.26–2.49	Sharpe and McAleavey (1998)	Tension	HI-MEMS films

typically included in a MEMS or microelectronics process but is less likely to be used as a structural component because of its low stiffness and strength, as shown in Table 3.9.

The main application to date of the polymer SU-8 is as a mask material for thicker electroplated metal MEMS. The use of it as a structural component is possible, but the values of stiffness and strength in Table 3.10 are very low.

Fracture toughness values have been measured for polysilicon, and Table 3.11 lists the results. Note that this is not the plane-strain fracture toughness that is a materials property; one must be careful, as some authors list this value as K_{Ic}.

Poisson's ratio is an important materials property when the stress state is bi-axial, but only a very limited number of measurements have been made. Those are listed in the "Comments" column of the tables.

The question is often asked about the effect of size on the strength of MEMS materials. It is well known that fine single-crystal "whiskers" of materials can have very high strengths, the premise being that they

TABLE 3.5 Polysilicon

Young's Modulus (GPa)	Fracture Strength (GPa)	Ref.	Method	Comments
160	—	Tabata et al. (1989)	Bulge	Obtains residual stress
123	—	Tai and Muller (1990)	Fixed ends	Heavily doped
190–240	—	Walker et al. (1990)	Bulge	Various etches
164–176	2.86–3.37	Koskinen et al. (1993)	Tension	Varied grain size
—	2.11–2.77	Biebl et al. (1995a)	Bending	CMOS process
147 ± 6	—	Biebl et al. (1995b)	Resonance	Temperature effects
170	—	Biebl and Philipsborn (1995)	Bending	Varied doping
—	0.57–0.77	Greek et al. (1995)	Tension	Weibull analysis
151–162	—	Maier-Schneider et al. (1995)	Bulge	Various anneals
163	—	Michalicek et al. (1995)	Resonance	Temperature effects
171–176	—	Zou et al. (1995)	Fixed ends	Pull-in voltage
149 ± 10	—	Gupta et al. (1996)	Fixed ends	Pull-in voltage
150 ± 30	—	Kahn et al. (1996)	Resonance	10 μm thick
140*	0.70	Read and Marshall (1996)	Tension	*Approximate
152–171	—	Schneider and Tucker (1996)	Ultrasonic	0.4 μm thick
176–201	—	Bhushan and Li (1997)	Indentation	Different depths
160–167	1.08–1.25	Greek and Johansson (1997)	Tension	Weibull analysis
178 ± 3	—	Gupta (1997)	Fixed ends	Ph.D. thesis
169 ± 6	1.20 ± 0.15	Sharpe et al. (1997d)	Tension	Poisson's ratio = 0.22 ± .01
174 ± 20	2.8 ± 0.5	Sharpe et al. (1998c)	Bending	Tested by Jones et al.
132	—	Sharpe et al. (1998c)	Tension	Tested by Chasiotis et al.
137 ± 5	2.7 ± 0.2	Sharpe et al. (1998c)	Resonance	Tested by Brown et al.
140 ± 14	1.3 ± 0.1	Sharpe et al. (1998c)	Tension	Tested by Sharpe et al.
172 ± 7	1.76	Greek and Ericson (1998)	Tension	10 μm thick
162 ± 4	—	Jayaraman et al. (1998)	Bulge	Poisson's ratio = 0.19 ± .03
168 ± 4	—	Kahn et al. (1998)	Resonance	0.45–0.9 μm thick
135 ± 10	—	Serre et al. (1998)	Bending	AFM
95–167	—	Sundararajan and Bhushan (1998)	Indentation	Also wear tests
167	2.0–2.7	Tsuchiya et al. (1998a)	Tension	Modulus from bulge; P-doped
163	2.0–2.8	Tsuchiya et al. (1998a)	Tension	Modulus from bulge; undoped
—	1.8–3.7	Tsuchiya et al. (1998b)	Tension	Different sizes and anneals
95/175	—	Li and Bhushan (1998)	Indentation	Doped and undoped
198	—	Que et al. (1999)	Bending	Capacitive device
166 ± 5	1.0 ± 0.1	Chasiotis and Knauss (2000)	Tension	Force-displacement
—	4.27 ± 0.61	LaVan et al. (2000b)	Tension	By LaVan et al.
—	2.85 ± 0.40	LaVan et al. (2000b)	Tension	By Sharpe et al.
—	3.23 ± 0.25	LaVan et al. (2000b)	Tension	By Tsuchiya et al.
158 ± 8	1.56 ± 0.25	Sharpe and Jackson (2000)	Tension	Size effects
159 and 169	—	Yi (pers. comm.)	Tension	Two specimens from Sharpe
—	3.2 ± 0.3	Jones et al. (2000)	Bending	Assumed E = 160 GPa
—	2.9 ± 0.5	Kapels et al. (2000)	Tension	4 μm thick
—	3.4 ± 0.5	Kapels et al. (2000)	Bending	4 μm thick

have fewer imperfections. MEMS structural components can be on the same size scale, which leads to the question being asked. However, there are no dramatic increases in strength because the materials still have fine grains relative to the specimen size. Tsuchiya et al. (1998) found an increase in the tensile strength of 2.0-μm-thick polysilicon specimens as their length increased from 30 to 300 μm, but that was only a 30% increase. Recent results show that the modulus of polysilicon does not vary with specimen size, but the strength increases from 1.21 to 1.65 GPa with decreasing specimen size [Sharpe et al., 2001].

TABLE 3.6 Silicon Crystals

Young's Modulus (GPa)	Fracture Strength (GPa)	Ref.	Method	Comments
177 ± 18	2.0–4.3	Johansson et al. (1988)	Bending	⟨110⟩
188	—	Weihs et al. (1989)	Indentation	
163	>3.4	Weihs et al. (1989)	Bending	⟨110⟩
122 ± 2	—	Ding et al. (1989)	Bending	⟨110⟩
125 ± 1	—	Ding et al. (1989)	Resonance	⟨110⟩
131	—	Zhang et al. (1991)	Resonance	
173 ± 13	—	Osterberg et al. (1994)	Bending	⟨110⟩
147	0.26–0.82	Cunningham et al. (1995)	Tension	⟨110⟩
—	8.5–20	Saif and MacDonald (1996)	Torsion	Shear and normal
60–200	—	Bhushan and Li (1997)	Indentation	Various doping
130	—	Dual et al. (1997)	Resonance	⟨100⟩
75	—	Dual et al. (1997)	Torsion	Shear modulus
125–180	1.3–2.1	Sato et al. (1997)	Tension	Three orientations
—	9.5–26.4	Chen et al. (1998)	Bending	Various etches
—	0.7–3.0	Chen et al. (1999)	Bending	Measured roughness
142 ± 9	1.73	Greek and Ericson (1998)	Tension	⟨100⟩
165 ± 20	2–8	Komai et al. (1998)	Bending	Fatigue tests also
168	—	Li and Bhushan (1999)	Indentation	⟨100⟩
—	0.59 ± 0.02	Mazza and Dual (1999)	Tension	⟨100⟩
—	2–6	Minoshima et al. (1999)	Bending	Fatigue also
169.2 ± 3.5	0.6–1.2	Yi and Kim (1999b)	Tension	Various etches
115–191	—	Yi and Kim (1999c)	Tension	Three orientations
164.9 ± 4	—	Anwander et al. (2000)	Tension	Laser speckle
169.9	0.5–17	Namazu et al. (2000)	Bending	Various sizes

TABLE 3.7 Silicon Carbide

Young's Modulus (GPa)	Fracture Strength (GPa)	Ref.	Method	Comments
394	—	Tong and Mehregany (1992)	Bulge	3C–SiC
88 ± 10 to 242 ± 30	—	El Khakani et al. (1993)	Bulge + indentation	Amorphous SiC
694	—	Su and Wettig (1995)	Resonance	3C–SiC
100–150	—	Schneider and Tucker (1996)	Ultrasonic	0.2–0.3 μm thick
331	—	Mehregany et al. (1997)	Bulge	3C–SiC; assumed ν = 0.25
196 and 273	—	Cros et al. (1997)	Acoustic microscopy	Amorphous SiC
395	—	Sundararajan and Bhushan (1998)	Indentation	3C–SiC
470 ± 10	—	Serre et al. (1999)	Bending	3C–SiC

From a practical point of view, one does not need to worry about the effect of size on strength for common MEMS structural components.

On the other hand, Namazu et al. (2000) tested silicon crystal beams ranging in size from widths, thicknesses and lengths of 0.2, 0.25 and 6 μm to 1.04, 0.52 and 9.85 mm, respectively. These were prepared by anisotropic etching; the smallest ones were tested using an atomic force microscope, and the largest ones with a microhardness tester. The mean bending strengths covered an astonishing range from 0.47 to 17.5 GPa—a factor of 37!

TABLE 3.8 Silicon Nitride

Young's Modulus (GPa)	Fracture Strength (GPa)	Ref.	Method	Comments
130–146 ± 20%	—	Petersen and Guarnieri (1979)	Resonance	~0.3 μm thick
230 and 330	—	Hong et al. (1990)	Bulge	Different processing
373	—	Tai and Muller (1990)	Fixed ends	Low stress
101–251*	—	Taylor (1991)	Indentation	*Assume Poisson's ratio = 0.27
110 and 160*	0.39–0.42	Cardinale and Tustison (1991)	Bulge	*Biaxial modulus
222 ± 3	—	Vlassak and Nix (1992)	Bulge	Poisson's ratio = 0.28 ± 0.05
216 ± 10	—	Vlassak and Nix (1992)	Indentation	
230–265	—	Schneider and Tucker (1996)	Ultrasonic	0.2–0.3 μm thick
192	—	Buchaillot et al. (1997)	Resonance	
194.25 ± 1%	—	Hoummady et al. (1997)	Resonance	
130	—	Ziebart et al. (1999)	Buckling	
290	7.0 ± 0.9	Kuhn et al. (2000)	Bending	
202.57 ± 15.80	12.26 ± 1.69*	Zhang et al. (2000)	Fixed ends	*Bending strength
255 ± 3	6.4 ± 1.1	G. Coles (pers. comm.)	Tension	Poisson's ratio = 0.23 ± 0.01

TABLE 3.9 Silicon Oxide

Young's Modulus (GPa)	Fracture Strength (GPa)	Ref.	Method	Comments
66*	—	Jaccodine and Schlegel (1966)	Bulge	*Assumed ν = 0.18
57–92 ± 20%	—	Petersen and Guarnieri (1979)	Resonance	Various depositions
64	>0.6	Weihs et al. (1988)	Indentation	
83	—	Weihs et al. (1988)	Bending	
—	0.6–1.9	Tsushiya et al. (1999)	Tension	In vacuum and in air

TABLE 3.10 SU-8

Young's Modulus (GPa)	Yield Strength (GPa)	Ultimate Strength (GPa)	Ref.	Method	Comments
~3	—	0.12–0.13	McAleavey et al. (1998)	Tension	
1.5–3.1	0.03–0.05	0.05–0.08	Chang et al. (2000)	Tension	Strain by SIEM

TABLE 3.11 Fracture Toughness Values

Fracture Toughness (MPa-m$^{1/2}$)	Ref.	Test Method	Material
1.8 ± 0.3	Fan et al. (1990)	Tension; edge crack	Silicon nitride; two kinds
1.2	DeBoer et al. (1993)	Indentation	Silicon crystal
0.96–1.65	Fitzgerald et al. (1999)	Double cantilever	Silicon crystal
1.4 ± 0.6	Sharpe et al. (1997f)	Tension; center crack	Polysilicon
1.9–4.5	Tsuchiya et al. (1997)	Tension; edge crack	Polysilicon
3.5–5.0	Ballarini et al. (1998)	Notched specimen	Polysilicon; various dopings
1.1–2.7	Kahn et al. (1999)	Notched specimen	Polysilicon; various dopings
1.2 ± 0.3	Kahn et al. (2000)	Sharp pre-crack	Polysilicon
1.6 ± 0.3	K. Jackson (pers. comm.)	Tension; corner	Polysilicon
1.0 ± 0.1	J. Bagdahn (pers. comm.)	Surface crack	Polysilicon

3.5 Initial Design Values

If the manufacturing and testing technology for materials used in MEMS were as fully developed as those associated with common structural materials, such as aluminum, for example, then this entire chapter could have been reduced to a one-page table. However, that is not the case; the materials themselves are actually new, and the test methods are still in their infancy.

It may be useful to list "best guesses" at the material properties of MEMS materials to be used in an initial design, and that is done in Table 3.12. The reader is cautioned that these are only estimates and that the actual properties resulting from a particular manufacturing process may be quite different from these nominal values.

Aluminum, copper and gold have essentially the same modulus values as the bulk materials, but the ultimate strengths are slightly higher than those found for commercially pure materials. Young's modulus for thin-film nickel can vary depending upon the deposition parameters, but it is conservative to assume that it will be lower (at 180 GPa) than the 200 GPa expected for bulk pure nickel. There are fewer results for nickel–iron, so the modulus of 120 GPa is only a rough estimate. However, it is clear that thin-film nickel and nickel–iron alloys are quite a bit stronger than one would expect from knowledge of bulk behavior.

The values listed for diamond-like carbon are only an optimistic guide. There are many variations of this material, and very few tests. These properties are included because such a material would be very attractive if it could be realized.

Polysilicon has certainly been thoroughly tested and is widely used, so one must wonder why there is no "standard" value—at least for Young's modulus—by now. The explanation is, of course, the difficulty in testing at this size scale, but there is a clear trend toward a modulus in the neighborhood of 160 GPa. An assumption of that number ±10 GPa can be used with confidence in the initial design of a microdevice. It is also clear that the strength can vary depending upon the manufacturing process, but will fall in the range of 1.2 to 3.0 GPa.

Single-crystal silicon has been thoroughly characterized, to the point that it has been used as a "standard material" to validate test systems. The modulus depends on orientation, and the strength range is enormous, with some extremely high values being reported.

Silicon carbide is widely promoted as a MEMS material, but conclusive measurements of its modulus have yet to be made and there are no measurements of strength. One should use the modulus value with caution. The situation is better for silicon nitride, as it has been more widely used and tested.

Although numbers to three significant figures are listed in Table 3.12, the reader will surely appreciate their unreliability and wonder as to their value. But, there are many other uncertainties that occur between the initial design and the product. Dimensions may not come out as specified, and that can have a profound effect on the stiffnesses of small components. Boundary conditions may not be as specified

TABLE 3.12 Initial Design Values

Material	Young's Modulus (GPa)	Poisson's Ratio	Yield Strength (GPa)	Ultimate or Fracture Strength (GPa)
Aluminum	70	—	—	0.15
Copper	120	—	0.15	0.35
Gold	70	—	—	0.30
Nickel	180	—	0.30	0.50
Nickel–Iron	120	—	0.70	1.60
Diamond-like carbon	800	0.22	—	8.0
Polysilicon	160	0.22	—	1.2–3.0
Silicon crystal	125–180	—	—	>1.0
Silicon carbide	400	0.25	—	—
Silicon nitride	250	0.23	—	6.0
Silicon oxide	70	—	—	1.0

either, due to variations in etch release processes. Nevertheless, one must start somewhere, and the values in Table 3.12 enable that. A user should certainly look at the more detailed information in the other tables and probably should consult some of the appropriate references.

Acknowledgments

The author is grateful for the interactions with students and colleagues over the past five years that has provided the background for this chapter. The assistance of Vanessa Coleman with the preparation is appreciated. Effort sponsored by the Defense Advanced Research Projects Agency (DARPA) and Air Force Research Laboratory, Air Force Materiel Command, USAF, under agreement number F30602-99-2-0553. The U.S. government is authorized to reproduce and distribute reprints for governmental purposes notwithstanding any copyright annotation thereon.

References

Amimoto, S.T., Chang, D.J., and Birkitt, A. (1998) "Stress Measurements in MEMS using Raman Spectroscopy," *Proc. SPIE* **3512**, pp. 123–129.

Anwander, M., Kaindl, G., Klein, M., and Weiss, B. (2000) "Noncontacting Laser Based Techniques for the Determination of Elastic Constants of Thin Foils," Micromat 2000, pp. 1100–1103, 17–19 April, Berlin, Germany.

ASM (1990) *Metals Handbook,* tenth edition, Vol. 2, ASM International, Materials Park, OH, p. 1143.

ASTM (1970) *Review of Developments in Plane Strain Fracture Toughness Testing,* STP 463, American Society for Testing and Materials, New York.

ASTM (2000a) *Metals—Mechanical Testing; Elevated and Low-Temperature Tests; Metallography,* Vol. 03.01, Annual Book of ASTM Standards, American Society for Testing and Materials, New York.

ASTM (2000b) *Refractories; Carbon and Graphite Products; Activated Carbon,* Vol. 15.01, Annual Book of ASTM Standards, American Society for Testing and Materials, New York.

Ballarini, R. (1998) "The Role of Mechanics in Microelectromechanical Systems (MEMS) Technology," AFRL-ML-WP-TR-1998-4209, Air Force Research Laboratory, Wright-Patterson Air Force Base, OH.

Ballarini, R., Mullen, R.L., Kahn, H., and Heuer, A.H. (1998) "The Fracture Toughness of Polysilicon Microdevices," in *Microelectromechanical Structures for Materials Research,* Materials Research Society Symposium **518**, pp. 137–142, 15–16 April, San Francisco, CA.

Basrour, S., Robert, L., Ballandras, S., and Hauden D. (1997) "Mechanical Characterization of Microgrippers Realized by LIGA Technique," in *Proc. Int. Solid-State Sensors and Actuators Conf.— Transducers '97,* pp. 599–602, 16–19 June, Chicago, IL.

Beams, J.W. (1959) "Mechanical Properties of Thin Films of Gold and Silver," in Proc. Int. Conf. Sponsored by Air Force Office of Scientific Research, Air Research and Development Command and The General Electric Research Laboratory '59, pp. 183–192, 9–11 September, Bolton Landing, NY.

Benrakkad, M.S., Benitex, M.A., Esteve, J., Lopez-Villegas, J.M., Samitier, J., and Morante, J.R. (1995) "Stress Measurement by Microraman Spectroscopy of Polycrystalline Silicon Structures," *J. Micromech. Microeng.* **5**, pp. 132–135.

Bhushan, B., and Li, X. (1997) "Micromechanical and Tribological Characterization of Doped Single-Crystal Silicon and Polysilicon Films for Microelectromechanical Systems Devices," *J. Mater. Res.* **12**, pp. 54–63.

Biebl, M., and von Philipsborn, H. (1995) "Fracture Strength of Doped and Undoped Polysilicon," in *Proc. Int. Solid-State Sensors and Actuators Conf.—Transducers '95,* pp. 72–75, 25–29 June, Stockholm, Sweden.

Biebl, M., Brandl, G., and Howe, R.T. (1995a) "Youngs Modulus of In-Situ Phosphorus-Doped Silicon," in *Proc. Int. Solid-State Sensors and Actuators Conf.—Transducers '95,* pp. 80–83, 25–29 June, Stockholm, Sweden.

Biebl, M., Scheiter, T., Hierold, C., Philipsborn, H., and Klose, H. (1995b) "Micromechanics Compatible with an 0.8 mu m CMOS Process," *Sensors and Actuators A (Phys.)* **46–47**, pp. 593–597.

Brown, S.B., Povirk, G., and Connally, J. (1993) "Measurement of Slow Crack Growth in Silicon and Nickel Mechanical Devices," in *Proc. IEEE Micro Electro Mechanical Systems*, pp. 99–104, 7–10 February, Fort Lauderdale, FL.

Brown, S.B., Van Arsdell, W., and Muhlstein, C.L. (1997) "Materials Reliability in MEMS Devices," in *Proc. Int. Solid-State Sensors and Actuators Conf.—Transducers '97*, pp. 591–593, 16–19 June, Chicago, IL.

Buchaillot, L., Farnault, E., Hoummady, M., and Fujita, H. (1997) "Silcon Nitride Thin Films Young's Modulus Determination by an Optical Non Destructive Method," *Jpn. J. Appl. Phys., Part 2 (Lett.)* **36**, pp. L794–L797.

Buchheit, T.E., Christenson, T.R., Schmale, D.T., and Lavan, D.A. (1999) "Understanding and Tailoring the Mechanical Properties of LIGA Fabricated Materials," in *Materials Science of Microelectromechanical Systems (MEMS) Devices*, Materials Research Society Symposium **546**, pp. 121–126, 1–2 December, Boston, MA.

Cardinale, G.F., and Tustison, R.W. (1992) "Fracture Strength and Biaxial Modulus Measurement of Plasma Silicon Nitride Films," *Thin Solid Films* **207**, pp. 126–130.

Chang, D.J., and Sharpe, W.N., Jr. (1999) "Mechanical Analysis and Properties of MEMS Materials," *Microengineering for Aerospace Systems*, The Aerospace Press of the Aerospace Corporation, El Segundo, CA, pp. 73–118.

Chang, S., Warren, J., and Chiang, F.P. (2000) "Mechanical Testing of EPON SU-8 with SIEM," in *Microscale Systems: Mechanics and Measurements Symposium*," Society for Experimental Mechanics, pp. 46–49, 8 June, Orlando, FL.

Chasiotis, I., and Knauss, W.G. (1998) "Mechanical Properties of Thin Polysilicon Films by Means of Probe Microscopy," *Proc. SPIE* **3512**, pp. 66–75.

Chasiotis, I., and Knauss, W.G. (2000) "Instrumentation Requirements in Mechanical testing of MEMS Materials," in *Microscale Systems: Mechanics and Measurements Symposium*, Society for Experimental Mechanics, pp. 56–61, 8 June, Orlando, FL.

Chen, K.S., Ayon, A., and Spearing, M. (1998) "Silicon Strength Testing for Mesoscale Structural Applications," in *Microelectromechanical Structures for Materials Research*, Materials Research Society Symposium **518**, pp. 123–130, 15–16 April, San Francisco, CA.

Chen, K.S., Ayon, A., Lohner, K.A., Kepets, M.A., Melconian, T.K., and Spearing, S.M. (1999) "Dependence of Silicon Fracture Strength and Surface Morphology on Deep Reactive Ion Etching Parameters," in *Materials Science of Microelectromechanical Systems (MEMS) Devices*, Materials Research Society Symposium **546**, pp. 21–26, 1–2 December, Boston, MA.

Cho, S.J., Lee, K.R., Eun, K.Y., Han, J.H., and Ko, D.H. (1997) "A Method For Independent Measurement of Elastic Modulus and Poisson's Ratio of Diamond-Like Carbon Films," in *Thin-Films-Stresses and Mechanical Properties VII*, Materials Research Society Symposium **505** pp. 33–38, 1–5 December, Boston, MA.

Cho, S.J., Kwang-Ryeol, L., Eun, K.Y., and Ko, D.H. (1998) "Measurement of Elastic Modulus and Poisson's Ratio of Diamond-Like Carbon Films," in *Microelectromechanical Structures for Materials Research*, Materials Research Society Symposium **518**, pp. 203–208, 15–16 April, San Francisco, CA.

Christenson, T.R., Buchheit, T.E., Schmale, D.T., and Bourcier, R.J. (1998) "Mechanical and Metallographic Characterization of LIGA Fabricated Nickel and 80%Ni–20%Fe Permalloy," in *Microelectromechanical Structures for Materials Research*, Materials Research Society Symposium **518**, pp. 185–190, 15–16 April, San Francisco, CA.

Chung, C.C., and Allen, M.G. (1996) "Measurement of Mechanical Properties of Electroplated Nickel–Iron Alloys," *ASME Dynamic Syst. Contr. Div. DSC* **59**, pp. 453–457.

Comella, B.T., and Scanlon, M.R. (2000) "The Determination of the Elastic Modulus of Microcantilever Beams Using Atomic Force Microscopy," *J. Mater. Sci.* **35**, pp. 567–572.

Connally, J.A., and Brown, S.B. (1993) "Micromechanical Fatigue Testing," *Exp. Mech.* **33**, pp. 81–90.

Cornella, G., Vinci, R.P., Iyer, R.S., and Dauskardt, R.H. (1998) "Observations of Low-Cycle Fatigue of Al Thin Films for MEMS Applications," in *Microelectromechanical Structures for Materials Research*, Materials Research Society Symposium **518**, pp. 81–86, 15–16 April, San Francisco, CA.

Cros, B., Gat, E., and Saurel, J. (1997) "Characterization of the Elastic Properties of Amorphous Silicon Carbide Thin Films by Acoustic Microscopy," *J. Non-Crystalline Solids* **209**, pp. 273–282.

Cunningham, S.J., Wan, S., and Read, D.T. (1995) "Tensile Testing of Epitaxial Silicon Films," in *Proc. Int. Solid-State Sensors and Actuators Conf.—Tranducers '95*, pp. 96–9, 25–29 June, Stockholm, Sweden.

Dally, J.W., and Read, D.T. (1993) "Electron Beam Moire," *Exp. Mech.* **33**, pp. 270–277.

DeBoer, M.P., Huang, H., Nelson, J.C., Jiang, Z.P., and Gerberich, W. (1993) "Fracture Toughness of Silicon and Thin Film Micro Structures by Wedge Indentation," in *Microelectromechanical Structures for Materials Research*, Materials Research Society Symposium **308**, pp. 647–652, 30 November–2 December, Boston, MA.

Ding, X., Ko, W.H., and Mansour, J.M. (1989) "Residual Stress and Mechanical Properties of Boron-Doped p$^+$-Silicon Films," in *5th Int. Conf. on Solid-State Sensors and Actuators and Eurosensors III*, pp. 866–871, 25–30 June, Montreux, Switzerland.

Dual, J., Mazza, E., Schiltges, G., and Schlums, D. (1997) "Mechanical Properties of Microstructures: Experiments and Theory," *Proc. SPIE* **3225**, pp. 12–22.

El Khakani, M.A., Chaker, M., Jean, A., Boily, S., and Kieffer, J.C. (1993) "Hardness and Young's Modulus of Amorphous *a*-SiC Thin Films Determined by Nanoindentation and Bulge Tests," *J. Mater. Res.* **9**, pp. 96–103.

Emery, R.D., Lenshek, D.X., Behin, B., Gherasimova, M., and Povrik, G.L. (1997) "Tensile Behavior of Free-Standing Gold Films," in *Thin-Films-Stresses and Mechanical Properties VII*, Materials Research Society Symposium **505**, pp. 361–366, New Haven, CT.

Fan, L.S., Howe, R.T., and Muller, R.S. (1990) "Fracture Toughness Characterization of Brittle Thin Films," *Sensors and Actuators A* **21–A23**, pp. 872–874.

Fitzgerald, A.M., Iyer, R.S., Dauskart, R.H., and Kenny, T.W. (1998) "Fracture and Sub-Critical Crack Growth Behavior of Micromachined Single Crystal Silicon Structures," *ASME Dynamic System Contr. Div. DSC* **66**, pp. 395–399.

Fitzgerald, A.M., Iyer, R.S., Dauskardt, R.H., and Kenny, T.W. (1999) "Fracture Toughness and Crack Growth Phenomena of Plasma-Etched Single Crystal Silicon," in *Transducers '99: 10th Int. Conf. on Solid-State Sensors and Actuators*, pp. 194–199, 7–10 June, Sendai, Japan.

Greek, S., and Ericson, F. (1998) "Young's Modulus, Yield Strength and Fracture Strength of Microelements Determined by Tensile Testing," in *Microelectromechanical Structures for Materials Research*, Materials Research Society Symposium **518**, pp. 51–56, 15–16 April, San Francisco, CA.

Greek, S., and Johansson, S. (1997) "Tensile Testing of Thin Film Microstructures," *Proc. SPIE* **3224**, pp. 344–351.

Greek, S., Ericson, F., Johansson, S., and Schweitz, J. (1995) "In Situ Tensile Strength Measurement of Thick-Film and Thin-Film Micromachined Structures," in *Proc. Int. Solid-State Sensors and Actuators Conf.—Transducers '95*, pp. 56–59, 25–29 June, Stockholm, Sweden.

Guo, S., Daowen, Z., and Wang, W. (1992) "Theoretical Calculation for the Young's Modulus of Poly-Si and a-Si films," in *Smart Materials Fabrication*, Materials Research Society Symposium **276**, pp. 233–238, 28–30 April, San Francisco, CA.

Gupta, R. (1997) "Electrosatic Pull-In Test Structure Design for In-Situ Mechanical Property Measurements of Microelectromechanical Systems (MEMS)," Ph.D. thesis, Massachusetts Institute of Technology, Cambridge, MA.

Gupta, R.K., Osterberg, P.M., and Senturia, S.D. (1996) "Material Property Measurements of Micromechanical Polysilicon Beams," *Proc. SPIE* **2880**, pp. 39–45.

Hoffman, R.W. (1989) "Nanomechanics of Thin Films: Emphasis: Tensile Properties," in *Thin Films: Stresses and Mechanical Properties Symposium*, Materials Research Society Symposium **130**, pp. 295–307, 28–30 November, Boston, MA.

Hollman, P., Alahelisten, A., Olsson, M., and Hogmark, S. (1995) "Residual Stress, Young's Modulus and Fracture Stress of Hot Flame Deposited Diamond," *Thin Solid Films* **270**, pp. 137–142.

Hong, S., Weihs, T.P., Bravman, J.C., and Nix, W.D. (1990) "Measuring Stiffnesses and Residual Stresses of Silicon Nitride Thin Films," *J. Electron. Mater.* **19**, pp. 903–909.

Hoummady, M., Farnault, E., Kawakatsu, H., and Masuzawa, T. (1997) "Applications of Dynamic Techniques for Accurate Determination of Silicon Nitride Young's Moduli," in *Proc. Int. Solid-State Sensors and Actuators Conf.—Transducers '97*, pp. 615–618, 16–19 June, Chicago, IL.

Howard, L.P., and Fu, J. (1997) "Accurate Force Measurements for Miniature Mechanical Systems: A Review of Progress," *Proc. SPIE* **3225**, pp. 2–11.

Jaccodine, R.J., and Schlegel, W.A. (1966) "Measurement of strains at Si-O_2 Interface," *J. Appl. Phys.* **37**, pp. 2429–2434.

Jaecklin, V.P., Linder, C., Brugger, J., and deRooij, N.F. (1994) "Mechanical and Optical Properties of Surface Micromachined Torsional Mirrors in Silicon, Polysilicon and Aluminum," *Sensors and Actuators A (Phys.)* **43**, pp. 269–275.

Jayaraman, S., Edwards, R.L., and Hemker, K. (1998) "Determination of The Mechanical Properties of Polysilicon Thin Films Using Bulge Testing. Thin-Films-Stresses and Mechanical Properties VII," in *Microelectromechanical Structures for Materials Research*, Materials Research Society Symposium **505**, pp. 623–628, 1–5 December, Boston, MA.

Jayaraman, S., Edwards, R.L., and Hemker, K. (1999) "Relating Mechanical Testing and Microstructural Features of Polysilicon Thin Films," *J. Mater. Res.* **14**, pp. 688–697.

Johansson, S., Schweitz, J.A., Tenerz, L., and Tiren, J. (1988) "Fracture Testing of Silicon Microelements In Situ in a Scanning Electron Microscope" *J. Appl. Phys.* **63**, pp. 4799–4803.

Johnson, G.C., Jones, P.T., and Howe, R.T. (1999) "Materials Characterization for MEMS—A Comparison of Uniaxial and Bending Tests," *Proc. SPIE* **3874**, pp. 94–101.

Jones, P.T., Johnson, G.C., and Howe, R.T. (1996) "Micromechanical Structures for Fracture Testing of Brittle Thin Films," *ASME Dynamic System Contr. Div. DSC* **59**, pp. 325–330.

Jones, P.T., Johnson, G.C., and Howe, R.T. (2000) "Statistical Characterization of Fracture of Brittle MEMS Materials," in *Proc. SPIE* **3880**, pp. 20–29.

Kahn, H., Stemmer, S., Nandakumar, K., Heuer, A.H., Mullen, R.L., Ballarini, R., and Huff, M.A. (1996) "Mechanical Properties of Thick, Surface Micromachined Polysilicon Films," in *Proc. Ninth Int. Workshop on Micro Electromechanical Systems*, pp. 343–348, 11–15 February, San Diego, CA.

Kahn, H., Huff, M.A., and Heuer, A.H. (1998) "Heating Effects on the Young's Modulus of Films Sputtered onto Micromachined Resonators," in *Microelectromechanical Structures for Materials Research*, Materials Research Society Symposium **518**, pp. 33–8, 15–16 April, San Francisco, CA.

Kahn, H., Ballarini, R., Mullen, R.L., and Heuer, A.H. (1999) "Electrostatically Actuated Failure of Microfabricated Polysilicon Fracture Mechanics Specimens," in *Proc. R. Soc. London, Ser. A* **455**, pp. 3807–3823.

Kahn, H., Tayebi, N., Ballarina, R., Mullen, R.L., and Heur, A.H. (2000) "Fracture Toughness of Polysilicon MEMS Devices," *Sensors and Actuators A (Phys.)* **82**, pp. 274–280.

Kapels, H., Aigner, R., and Binder, J. (2000) "Fracture Strength and Fatigue of Polysilicon Determined by a Novel Thermal Actuator," *IEEE Trans. Electron Devices* **47**, pp. 1522–1528.

Kazinczi, R., Mollinger, J.R., and Bossche, A. (2000) "Versatile Tool for Characterising Long-Term Stability and Reliability of Micromechanical Structures," *Sensors and Actuators A (Phys.)* **85**, pp. 84–89.

Kobrinsky, M., Deutsch, E., and Senturia S. (1999) "Influence of Support Compliance and Residual Stress on the Shape of Doubly-Supported Surface Micromachined Beams," *MEMS Microelectromech. Syst.* **1**, pp. 3–10.

Komai, K., Minoshima, K., and Inoue, S. (1998) "Fracture and Fatigue Behavior of Single Crystal Silicon Microelements and Nanoscopic AFM Damage Evaluation," *Microsystem Technol.* **5**, pp. 30–7.

Koskinen, J., Steinwall, J.E., Soave, R., and Johnson, H.H. (1993) "Microtensile Testing of Free-Standing Polysilicon Fibers of Various Grain Sizes." *J. Micromech. Microeng.* **3**, pp. 13–17.

Kraft, O., Schwaiger, R., and Nix, W.D. (1998) "Measurement of Mechanical Properties in Small Dimensions by Microbeam Deflection," in *Microelectromechanical Structures for Materials Research*, Materials Research Society Symposium **518**, pp. 39–44, 15–16 April, San Francisco, CA.

Krulevitch, P. (1996) "Technique for Determining the Poisson's Ratio of Thin Films," *ASME Dynamic Syst. Contr. Div. DSC* **59**, pp. 319–323.

Kuhn, J., Fettig, R.K., Moseley, S.H., Kutyrev, A.S., and Orloff, J. (2000) "Fracture Tests of Etched Components Using a Focused Ion Beam Machine," NASA/Goddard Space Flight Center, Greenbelt, MD (pre-publication report).

LaVan, D.A., Bucheit, T.E., and Kotula, P.G. (2000a) "Mechanical and Microstructural Characterization of Critical Features of MEMS Materials," in *Microscale Systems: Mechanics and Measurements Symposium*, Society for Experimental Mechanics, pp. 41–45.

LaVan, D.A., Tsuchiya, T., and Coles, G. (2000b) "Cross Comparison of Direct Tensile Testing Techniques on Summit Polysilicon Films," in *Mechanical Properties of Structural Films*, ASTM STP 1413, American Society for Testing and Materials, submitted for publication.

Li, X., and Bhushan, B. (1998) "Measurement of Fracture Toughness of Ultra-Thin Amorphous Carbon Films," *Thin Solid Films* **315**, pp. 214–221.

Li, X., and Bhushan, B. (1999) "Micro/Nanomechanical Characterization of Ceramic Films for Microdevices," *Thin Solid Films* **340**, pp. 210–217.

Maier-Schneider, D., Maibach, J., Obemeier, E., and Schneider, D. (1995) "Variations in Young's Modulus and Intrinsic Stress of LPCVD-Polysilicon Due to High-Temperature Annealing," *J. Micromech. Microeng.* **5**, pp. 121–124.

Mazza, E., and Dual, J. (1999) "Mechanical Behaviour of a μm-Sized Single Crystal Silicon Structure with Sharp Notches," *J. Mech. Phys. Solids* **47**, pp. 1795–1821.

Mazza, E., Danuser, G., and Dual, J. (1996a) "Light Optical Deformation Measurements in Microbars with Nanometer Resolution," *Microsystem Technol.* **2**, pp. 83–91.

Mazza, E., Abel, S., and Dual, J. (1996b) "Experimental Determination of Mechanical Properties of Ni and Ni-Fe Microbars." *Microsystem Technol.* **2**, pp. 197–202.

McAleavey, A., Coles, G., Edwards, R.L., and Sharpe, W.N. (1998) "Mechanical Properties of SU-8," in *Microelectromechanical Structures for Microelectromechanical Structures for Materials Research*, Materials Research Society Symposium **546**, pp. 213–218, 1–2 December, Boston, MA.

Mehregany, M., Tong, L., Matus, L.G., and Larkin, D.J. (1997) "Internal Stress and Elastic Modulus Measurements on Micromachined 3C-SiC Thin Films," *IEEE Trans. Electron. Devices* **44**, pp. 74–79.

Menter, J.W., and Pashley, D.W. (1959) "The Microstructure and Mechanical Properties of Thin Films," in Proc. Int. Conf. Sponsored by Air Force Office of Scientific Research, Air Research and Development Command and The General Electric Research Laboratory '59, pp. 111–150, 9–11 September, Bolton Landing, NY.

Michalicek, A.M., Sene, D.E., and Bright, V.M. (1995) "Advanced Modeling of Micromirror Devices," in *Proc. Int. Conf. Integrated Micro/Nanotechnology for Space Applications*, pp. 214–229, 30 October–3 November, Houston, TX.

Minoshima, K., Inoue, S., Terada, T., and Komai, K. (1999) "Influence of Specimen Size and Sub-Micron Notch on the Fracture Behavior of Single Crystal Silicon Microelements and Nanoscopic AFM Damage Evaluation," in *Microelectromechanical Structures for Materials Research*, Materials Research Society Symposium **546**, pp. 15–20, 1–2 December, Boston, MA.

Monteiro, O.R., Brown, I.G., Sooryakumar, R., and Chirita, M. (1996) "Elastic Properties of Diamond Like Carbon Thin Films: A Brillouin Scattering Study," in *Microelectromechanical Structures for Materials Research*, Materials Research Society Symposium **444**, pp. 93–98, 4–5 December, Boston, MA.

Muller, R.S. (1990) "Microdynamics," *Sensors and Actuators A (Phys.)* **21–A23**, pp. 1–8.

Muller, R.S. (1997) *Microelectromechanical Systems*, National Academy of Sciences, Washington, D.C., 59 pp.

Namazu, T., Isono, Y., and Tanaka, T. (2000) "Nano-Scale Bending Test of Si Beam for MEMS," in *Proc. IEEE Thirteenth Annual Int. Conf. on Micro Electro Mechanical Systems,* pp. 205–210, 23–27 January, Miyazaki, Japan.

Neugebauer, G. (1960). "Tensile Properties of Thin, Evaporated Gold Films," *J. Appl. Phys.* **31**, pp. 1096–1101.

Nieva, P., Tada, H., Zavracky, P., Adams, G., Miaoulis, I., and Wong, P. (1998) "Mechanical and Thermophysical Properties of Silicon Nitride Thin Films at High Temperatures Using *in situ* MEMS Temperature Sensors," in *Microelectromechanical Structures for Materials Research,* Materials Research Society Symposium **546**, pp. 97–102, 1–2 December, Boston, MA.

Obermeier, E. (1996) "Mechanical and Thermophysical Properties of Thin Film Materials for MEMS: Techniques and Devices," in *Microelectromechanical Structures for Materials Research,* Materials Research Society Symposium **444**, pp. 39–57, 4–5 December, Boston, MA.

Ogawa, H., Ishikawa, Y., and Kitahara, T. (1996) "Measurements of Stress–Strain Diagrams of Thin Films by a Developed Tensile Machine." *Proc. SPIE* **2880**, pp. 272–279.

Ogawa, H., and Suzuki, K. et al. (1997) "Measurements of Mechanical Properties of Microfabricated Thin Films," in *Proc. IEEE Tenth Annual Int. Workshop on Micro Electro Mechanical Syst.,* pp. 430–435, 26–30 January, Nagoya, Japan.

Osterberg, P.M., R.K. Gupta, Gilbert, J.R., and Senturia, S.D. (1994) "Quantitative Models for the Measurement of Residual Stress, Poisson Ratio and Young's Modulus Using Electrostatic Pull-In of Beams and Diaphragms," in *Technical Digest Solid-State Sensor and Actuator Workshop,* pp. 184–188, 13–16 June, Hilton Head Island, SC.

Pan, C.S., and Hsu, W. (1999) "A Microstructure For In Situ Determination of Residual Strain," *J. Microelectromechanical Syst.* **8**, pp. 200–207.

Petersen, K.E., and Guarnieri, C.R. (1979) "Young's Modulus Measurements of Thin Films Using Micromechanics," *J. Appl. Phys.* **50**, pp. 6761–6766.

Pinardi, K., Jain, S.C., Maes, H.E., Van Overstraeten, R., Willander, M., and Atkinson, A. (1997) "Measurement of Nonuniform Stresses in Semiconductors by the Micro-Raman Method," in *Microelectromechanical Structures for Materials Research,* Materials Research Society Symposium **505**, pp. 507–512, 1–5 December, Boston, MA.

Que, L., Li, L., Chu, L., and Gianchandani, Y.B. (1999) "A Micromachined Strain Sensor with Differential Capacitive Readout," in *Proc. 12th Int. Workshop on Micro Electro Mechanical Syst.* pp. 552–557, 17–21 January, Orlando, FL.

Read, D.T., and Dally J.W. (1992). "A New Method for Measuring the Constitutive Properties of Thin Films." *J. Mater. Res.* **8**, pp. 1542–1549.

Read, D.T., and Marshall, R.C. (1996) "Measurements of Fracture Strength and Young's Modulus of Surface-Micromachined Polysilicon." *Proc. SPIE* **2880**, pp. 56–63.

Ruther, P., Bacher, W., Feit, K., and Menz, W. (1995) "Microtesting System Made by the LIGA Process To Measure the Young's Modulus in Cantilever Microbeams," *ASME Dynamic Syst. Contr. Div. DSC* **57-2**, pp. 963–967.

Saif, M.T.A., and MacDonald, N.C. (1996) "Micro Mechanical Single Crystal Silicon Fracture Studies—Torsion and Bending," in *Ninth Annual Int. Workshop on Micro Electro Mechanical Syst.,* pp. 105–109, 11–15 February, San Diego, CA.

Saif, T., and MacDonald, N.C. (1998) "Failure of Micron Scale Single Crystal Silicon Bars Due to Torsion Developed by MEMS Micro Instruments," in *Microelectromechanical Structures for Materials Research,* Materials Research Society Symposium **518**, pp. 45–49, San Francisco, CA.

Sarro, P. (2000) "Silicon Carbide as a New MEMS Technology," *Sensors and Actuators A (Phys.)* **82**, pp. 210–218.

Sato, K., Shikida, M., Yoshioka, T., Ando, T., and Kawbata, T. (1997) "Micro Tensile-Test of Silicon Film Having Different Crystallographic Orientations," in *Proc. Int. Solid-State Sensors and Actuators Conf.—Transducers '97,* pp. 595–598, 16–19 June 1997, Chicago, IL.

Schiltges, G., Gsell, D., and Jual, J. (1998) "Torsional Tests on Microstructures: Two Methods To Determine Shear-Moduli," *Microsystem Technol.* **5**, pp. 22–29.

Schneider, D., and Tucker, M.D. (1996) "Non-Destructive Characterization and Evaluation of Thin Films by Laser Induced Ultrasonic Surface Waves," *Thin Solid Films* **209–291**, pp. 305–311.

Schweitz, J.A., and Ericson, F. (1999) "Evaluation of Mechanical Materials Properites by Means of Surface Micromachined Structures," *Sensors and Actuators A (Phys.)* **74**, pp. 126–133.

Senturia, S.D. (1998) "CAD Challengers for Microsensors, Microactuators, and Microsystems." *Proc. IEEE* **86**, pp. 1611–1626.

Serre, C., Gorostiza, P., Perez-Rodriquez, A., Sanz, F., and Morante, J.R. (1998) "Measurement of Micromechanical Properties of Polysilicon Microstructures with an Atomic Force Microscope," *Sensors and Actuators A (Phys.)* **67**, pp. 215–219.

Serre, C., Perez-Rodriguez, A., Romano-Rodriguez, A., Morante, J.R., Esteve, J., and Acero, M.C. (1999) "Test Microstructures for Measurement of SiC Thin Film Mechanical Properties," *J. Micromech. Microeng.* **9**, pp. 190–193.

Sharpe, W.N., Jr. (1999) "Fatigue Testing of Materials Used in Microelectromechanical Systems," in *FATIGUE '99, Proc. Seventh Int. Fatigue Congress,* pp. 1837–1844, 8–12 June, Beijing, China.

Sharpe, W.N., Jr., and Jackson, K. (2000) "Tensile Testing of MEMS Materials," in *Microscale Systems: Mechanics and Measurements Symposium,* Society for Experimental Mechanics, pp. ix–xiv, 8 June, Orlando, FL.

Sharpe, W.N., Jr., and McAleavey, A. (1998) "Tensile Properties of LIGA Nickel," *Proc. SPIE* **3512**, pp. 30–137.

Sharpe, W.N., Jr., Yuan, B., Vaidyanathan, R., and Edwards, R.L. (1996) "New Test Structures and Techniques for Measurement of Mechanical Properties of MEMS Materials," *Proc. SPIE* **2880**, pp. 78–91.

Sharpe, W.N., Jr., LaVan, D.A., and McAleavey, A. (1997a) "Mechanical Testing of Thicker MEMS Materials," *ASME Dynamic Syst. Contr. Div. DSC* **62**, pp. 93–97.

Sharpe, W.N., Jr., Yuan, B., and Edwards, R.L. (1997b) "Variations in Mechanical Properties of Polysilicon," in *43rd Int. Symp. Instrumentation Society of America,* pp. 179–188, 4–8 May, Orlando, FL.

Sharpe, W.N., Jr., Yuan, B., and Edwards, R. L. (1997c) "A New Technique for Measuring the Mechanical Properties of Thin Films," *J. Microelectromech. Syst.* **6**, pp. 193–199.

Sharpe, W.N., Jr., Yuan, B., and Vaidyanathan, R. (1997d) "Measurements of Young's Modulus, Poisson's Ratio, and Tensile Strength of Polysilicon," *Proc. IEEE Tenth Annual Int. Workshop on Micro Electro Mechanical Systems,* pp. 424–429, 26–30 January, Nagoya, Japan.

Sharpe, W.N., Jr., LaVan, D.A., and Edwards, R.L. (1997e) "Mechanical Properties of LIGA-Deposited Nickel for MEMS Transducers," *Proc. Int. Solid-State Sensors and Actuators Conf.—Transducers '97,* pp. 607–610, 16–19 June, Chicago, IL.

Sharpe, W.N., Jr., Yuan, B., and Edwards, R.L. (1997f) "Fracture Tests of Polysilicon Film," in *Microelectromechanical Structures for Materials Research,* Materials Research Society Symposium **505**, pp. 51–56, 1–5 December, Boston, MA.

Sharpe, W.N., Jr., Turner, K., and Edwards, R.L. (1998a) "Polysilicon Tensile Testing with Electrostatic Gripping," in *Microelectromechanical Structures for Materials Research,* Materials Research Society Symposium **518**, pp. 191–196, 15–16 April, San Francisco, CA.

Sharpe, W.N., Jr., Danley, D., and LaVan, D.A. (1998b) "Microspecimen Tensile Tests of A533-B Steel," in *Small Specimen Test Techniques, ASTM STP 1329,* American Society of Testing and Materials (ASTM), pp. 497–512, 13–14 January, 1997, New Orleans, LA.

Sharpe, W.N., Jr., Brown, S., Johnson, G.C., and Knauss, W. (1998c) "Round-Robin Tests of Modulus and Strength of Polysilicon," in *Microelectromechanical Structures for Materials Research,* Materials Research Society Symposium **518**, pp. 56–65, 15–16 April, Francisco, CA.

Sharpe, W.N., Jr., Turner, K.T., and Edwards, R.L. (1999) "Tensile Testing of Polysilicon," *Exp. Mech.* **39**, pp. 162–170.

Sharpe, W.N., Jr., Jackson, K., Hemker, K.J., and Zie, Z. (2001) "Effect of Specimen Size on Young's Modulus and Strength of Polysilicon," *J. Microelectromech. Syst.,* September.

Spearing, S.M. (2000) "Materials Issues in Microelectromechanical Systems (MEMS)," *Acta Materialia* **48**, pp. 179–196.

Stephens, L.S., Kelly, K.W., Meletis, E.I., and Simhadri, S. (1998) "Mechanical Property Evaluation of Electroplated High Aspect Ratio Microstructures," in *Microelectromechanical Structures for Materials Research*, Materials Research Society Symposium **518**, pp. 173–178, 15–16 April, San Francisco, CA.

Su, C.M., and Wuttig, M. (1995) "Elastic and Anelastic Properties of Chemical Vapor Deposited Epitaxial 3C-SiC," *J. Appl. Phys.* **77**, pp. 5611–5615.

Sundararajan, S., and B. Bhushan (1998) "Micro/Nanotribological Studies of Polysilicon and SiC films for MEMS Applications," *Wear* **217**, pp. 251–261.

Suwito, W., Dunn, M. L., and Cunningham, S. (1997) "Strength and Fracture of Micromachined Silicon Structures," *ASME Dynamic Syst. Contr. Div. DSC* **62**, pp. 99–104.

Tabata, O., Kawahata, K., Sugiyama, S., and Igarashi, I. (1989) "Mechanical Property Measurements of Thin Films Using Load-Deflection of Composite Rectangular Membranes," *Sensors Actuators* **20**, pp. 135–141.

Tada, H., Nieva, P., Zavracky, P., Miaoulis, N., and Wong, P.Y. (1998) "Determining the High-Temperature Properties of Thin Films Using Bilayered Cantilevers," in *Microelectromechanical Structures for Materials Research*, Materials Research Society Symposium **546**, pp. 39–44, 1–2 December, Boston, MA.

Tai, Y.C., and R.S. Muller (1988) "Fracture Strain of LPCVD Polysilicon," in *1988 Sensor and Actuator Workshop*, pp. 88–91, 6–9 June, Hilton Head Island, SC.

Tai, Y.C., and R.S. Muller (1990) "Measurement of Young's Modulus on Microfabricated Structures Using a Surface Profiler," in *IEEE Micro Electro Mechanical Systems*, pp. 147–152, 11–14 February, Napa Valley, CA.

Taylor, J.A. (1991). "The Mechanical Properties and Microstructure of Plasma Enhanced Chemical Vapor Deposited Silicon Nitride Thin Films," *J. Vacuum Sci. Technol. A* **9**, pp. 2464–2468.

Teh, K.S., Lin, L., and Chiao M. (1999) "The Creep Behavior of Polysilicon Microstructures," in *Transducers '99*, pp. 508–511, 7–10 June, Sendai, Japan.

Tong, L., and Mehregany, M. (1992) "Mechanical Properties of 3C Silicon Carbide," *Appl. Phys. Lett.* **60**, pp. 2992–2994.

Tsuchiya, T., Tabata, O., Sakata, J., and Taga, Y. (1998a) "Specimen Size Effect on Tensile Strength of Surface Micromachined Polycrystalline Silicon Thin Films," *J. Microelectromech. Syst.* pp. 106–113.

Tsuchiya, T., Sakata, J., and Taga, Y. (1998b) "Tensile Strength and Fracture Toughness of Surface Micromachined Polycrystalline Silicon Thin Films Prepared Under Various Conditions," in *Microelectromechanical Structures for Materials Research*, Materials Research Society Symposium **505**, pp. 285–290, 1–5 December, Boston, MA.

Tsuchiya, T., Inoue, A., and Sakata, J. (1999) "Tensile Testing of Insulating Thin Films; Humidity Effect on Tensile Strength of SiO_2 Films," in *Proc. 10th Int. Conf. on Solid-State Sensors and Actuators—Transducers '99*, pp. 488–491, 7–10 June, Sendai, Japan.

Van Arsdell, W.W., and Brown, S.B. (1999) "Subcritical Crack Growth in Silicon MEMS," *J. Microelectromech. Syst.* **8**, pp. 319–327.

Vlassak, J.J., and Nix, W.D. (1992). "A New Bulge Test Technique for the Determination of Young's Modulus and Poisson's Ratio of Thin Films," *J. Mater. Res.* **7**, pp. 3242–3249.

Walker, J.A., Gabriel, K.J., and Mehregany, M. (1990) "Mechanical Integrity of Polysilicon Films Exposed to Hydrofluoric Acid Solutions," *J. Electron. Mater.* **20**, pp. 665–670.

Weihs, T.P., Hong, S., Bravman, J.C., and Nix, W.D. (1988) "Mechanical Deflection of Cantilever Microbeams: A New Technique for Testing the Mechanical Properties of Thin Films," *J. Mater. Res.* **3**, pp. 931–942.

Weihs, T.P., Hong, S., Bravman, J.C., and Nix, W.D. (1989) "Measuring the Strength and Stiffness of Thin Film Materials by Mechanically Deflecting Cantilever Microbeams," in *Thin Films: Stresses and Mechanical Properties Symposium, Microelectromechanical Structures for Materials Research*, Materials Research Society Symposium **402**, pp. 87–92, 28–30 November, Boston, MA.

Yang, E.H., and Fujita, H. (1997) "Fabrication and Characterization of U-Shaped Beams for the Determination of Young's Modulus Modification Due to Joule Heating of Polysilicon Microstructures," in *Proc. Int. Solid-State Sensors and Actuators Conf.—Transducers '97*, pp. 603–606, 16–19 June, Chicago, IL.

Yi, T., and Kim, C.J. (1999a) "Measurement of Mechanical Properties for MEMS Materials," *Measurement Sci. Technol.* **10**, pp. 706–716.

Yi, T., and Kim, C.J. (1999b) "Microscale Material Testing: Etchant Effect on the Tensile Strength," in *Transducers '99,* pp. 518–521, 7–10 June, Sendai, Japan.

Yi, T., and Kim, C.J. (1999c) "Tension Test with Single Crystalline Silicon Microspecimen," *MEMS Microelectromech. Syst.* **1**, pp. 81–86.

Yoshioka, T., Yamasaki, M., Shikida, M., and Sato, K. (1996) "Tensile Testing of Thin-Film Materials on a Silicon Chip," in *MHS '96 Proc. Seventh Int. Symp. on Micro Machine and Human Science,* pp. 111–117, 2–4 October, Nagoya, Japan.

Yu, M.F., Lourie, O., Dyer, M.J., Moloni, K., Kelly, T.F., and Ruoff, R.S. (2000) "Strength and Breaking Mechanism of Multiwalled Carbon Nanotubes Under Tensile Load," *Science* **287**, pp. 637–640.

Zhang, L.M., Uttamchandani, D., and Culshaw, B. (1991) "Measurement of the Mechanical Properties of Silicon Microresonators," *Sensors and Actuators A (Phys.)* **29**, pp. 79–84.

Zhang, X., Zhang, T.Y., Wong, M., and Zohar, Y. (1997) "Effects of High-Temperature Rapid Thermal Annealing on the Residual Stress of LPCVD-Polysilicon Thin Films," in *Proc. IEEE Tenth Annual Int. Workshop on Micro Electro Mechanical Systems,* pp. 535–540, 26–30 January, Nagoya, Japan.

Zhang, T.Y., Su, Y.J., Qian, C.F., Zhao, M.H., and Chen, L.Q. (2000) "Microbridge Testing of Silicon Nitride Thin Films Deposited on Silicon Wafers," *Acta Materialia* **48**, pp. 2843–2857.

Ziebart, V., Paul, O., Munch, U., and Baltes, H. (1997) "A Novel Method To Measure Poisson's Ratio of Thin Films," in *Microelectromechanical Structures for Materials Research,* Materials Research Society Symposium **505**, pp. 103–108, 1–2, December, Boston, MA.

Ziebart, V., Paul, O., Munch, U., and Baltes, H. (1999) "Strongly Buckled Square Micromachined Membranes," *J. Microelectromech. Syst.* **8**, pp. 423–432.

Zou, Q., Li, Z. et al. (1995). "New Methods for Measuring Mechanical Properties of Thin Films in Micromachining: Beam Pull-In Voltage (VPI) Method and Long Beam Deflection (LBD) Method," *Sensors and Actuators A (Phys.)* **48**, pp. 137–143.

4
Flow Physics

Mohamed Gad-el-Hak
University of Notre Dame

4.1 Introduction 4-1
4.2 Flow Physics.. 4-2
4.3 Fluid Modeling.. 4-3
4.4 Continuum Model 4-6
4.5 Compressibility... 4-9
4.6 Boundary Conditions 4-11
4.7 Molecular-Based Models.. 4-18
4.8 Liquid Flows 4-24
4.9 Surface Phenomena.. 4-30
4.10 Parting Remarks 4-33

One of the first men who speculated on the remarkable possibilities which magnification or diminution of physical dimensions provides was Jonathan Swift, who, in Gulliver's Travels, *drew some conclusions as to what dwarfs and giants would really look like, and what sociological consequences size would have. Some time ago Florence Moog (Scientific American, November 1948) showed that Swift was a "bad biologist," or Gulliver a "poor liar." She showed that a linear reduction in size would carry with it a reduction in the number of brain cells, and hence a reduction in intellectual capacity in Lilliputians, whereas the enormous Brobdingnagians were physically impossible; they could have had physical reality only if their necks and legs had been short and thick. These 90-ton monsters could never have walked on dry land, nor could their tremendous weight have been carried on proportionately-sized feet.*

Even though Swift, in his phantasy, committed a number of physical errors, because he was not sufficiently aware of the fact that some physical properties of a body are proportional to the linear dimensions (height), whereas others vary with the third power of linear size (such as weight and cell number), yet he surpassed his medieval predecessors in many respects and drew a number of excellent conclusions, bringing both giants and dwarfs close to physical reality.

(From "The Size of Man," by F.W. Went, *American Scientist* **56**(4), pp. 400–413, 1968.)

4.1 Introduction

In this chapter, we review the status of our understanding of fluid flow physics particular to microdevices. The chapter is an update of an earlier publication by the same author [Gad-el-Hak, 1999]. The coverage here is broad, leaving the details to other chapters in the handbook that treat specialized problems in microscale fluid mechanics. Not all microelectromechanical systems (MEMS) devices involve fluid flows, of course, but this chapter will focus on the ones that do. Microducts, micronozzles, micropumps, microturbines and microvalves are examples of small devices involving the flow of liquids and gases. MEMS can also be related to fluid flows in an indirect way. The availability of inexpensive, batch-processing-produced microsensors and microactuators provides opportunities for targeting small-scale coherent structures in macroscopic turbulent shear flows. Flow control using MEMS promises a quantum leap in control system performance [Gad-el-Hak, 2000]. Additionally, the extremely small sensors made possible by microfabrication technology allow measurements with spatial and temporal resolutions not achievable before. For example, high-Reynolds-number turbulent flow diagnoses are now feasible down to the Kolmogorov scales [Löfdahl and Gad-el-Hak, 1999]. Those indirect topics are also left to other chapters in the book.

4.2 Flow Physics

The rapid progress in fabricating and utilizing microelectromechanical systems during the last decade has not been matched by corresponding advances in our understanding of the unconventional physics involved in the manufacture and operation of small devices [Madou, 1997; Kovacs, 1998; Gad-el-Hak, 1999; Knight, 1999]. Providing such understanding is crucial to designing, optimizing, fabricating and utilizing improved MEMS devices. This chapter focuses on the physics of fluid flows in microdevices.

Fluid flows in small devices differ from those in macroscopic machines. The operation of MEMS-based ducts, nozzles, valves, bearings, turbomachines, etc. cannot always be predicted from conventional flow models such as the Navier–Stokes equations with a no-slip-boundary condition at a fluid–solid interface, as is routinely and successfully applied for larger flow devices. Many questions have been raised when the results of experiments with microdevices could not be explained via traditional flow modeling. The pressure gradient in a long microduct was observed to be nonconstant and the measured flowrate was higher than that predicted from the conventional continuum flow model. Load capacities of microbearings were diminished and electric currents necessary to move micromotors were extraordinarily high. The dynamic response of micromachined accelerometers operating at atmospheric conditions was observed to be over-damped.

In the early stages of development of this exciting new field, the objective was to build MEMS devices as productively as possible. Microsensors were reading something, but not many researchers seemed to know exactly what. Microactuators were moving, but conventional modeling could not precisely predict their motion. After a decade of unprecedented progress in MEMS technology, perhaps the time is now ripe to take stock, slow down a bit and answer the many questions that have arisen. The ultimate aim of this long-term exercise is to achieve rational-design capability for useful microdevices and to be able to characterize definitively and with as little empiricism as possible the operations of microsensors and microactuators.

In dealing with fluid flow-through microdevices, one is faced with the question of which model to use, which boundary condition to apply and how to proceed to obtain solutions to the problem at hand. Obviously surface effects dominate in small devices. The surface-to-volume ratio for a machine with a characteristic length of 1 m is 1 m^{-1}, while that for a MEMS device having a size of 1 µm is 10^6 m^{-1}. The millionfold increase in surface area relative to the mass of the minute device substantially affects the transport of mass, momentum and energy through the surface. The small length scale of microdevices may invalidate the continuum approximation altogether. Slip flow, thermal creep, rarefaction, viscous dissipation, compressibility, intermolecular forces and other unconventional effects may have to be taken into account, preferably using only first principles such as conservation of mass, Newton's second law, conservation of energy, etc.

In this chapter, we discuss continuum as well as molecular-based flow models and the choices to be made. Computing typical Reynolds, Mach and Knudsen numbers for the flow through a particular device is a good start to characterize the flow. For gases, microfluid mechanics has been studied by incorporating slip-boundary conditions, thermal creep and viscous dissipation, as well as compressibility effects, into the continuum equations of motion. Molecular-based models have also been attempted for certain ranges of the operating parameters. Use is made of the well-developed kinetic theory of gases, embodied in the Boltzmann equation, and direct simulation methods such as Monte Carlo. Microfluid mechanics of liquids is more complicated. The molecules are much more closely packed at normal pressures and temperatures, and the attractive or cohesive potential between the liquid molecules as well as between the liquid and solid ones plays a dominant role if the characteristic length of the flow is sufficiently small. In cases when the traditional continuum model fails to provide accurate predictions or postdictions, expensive molecular dynamics simulations seem to be the only first-principle approach available to rationally characterize liquid flows in microdevices. Such simulations are not yet feasible for realistic flow extent or number of molecules. As a consequence, the microfluid mechanics of liquids is much less developed than that for gases.

4.3 Fluid Modeling

A flowfield can be modeled either as the fluid really is—a collection of molecules—or as a continuum where the matter is assumed continuous and indefinitely divisible. The former modeling is subdivided into deterministic methods and probabilistic ones, while in the latter approach the velocity, density, pressure, etc. are defined at every point in space and time, and conservation of mass, energy and momentum lead to a set of nonlinear partial differential equations (Euler, Navier–Stokes, Burnett, etc.). Fluid modeling classification is depicted schematically in Figure 4.1.

The continuum model, embodied in the Navier–Stokes equations, is applicable to numerous flow situations. The model ignores the molecular nature of gases and liquids and regards the fluid as a continuous medium describable in terms of the spatial and temporal variations of density, velocity, pressure, temperature and other macroscopic flow quantities. For dilute gas flows near equilibrium, the Navier–Stokes equations are derivable from the molecularly based Boltzmann equation but can also be derived independently of that for both liquids and gases. In the case of direct derivation, some empiricism is necessary to close the resulting indeterminate set of equations. The continuum model is easier to handle mathematically (and is also more familiar to most fluid dynamicists) than the alternative molecular models. Continuum models should therefore be used as long as they are applicable. Thus, careful considerations of the validity of the Navier–Stokes equations and the like are in order.

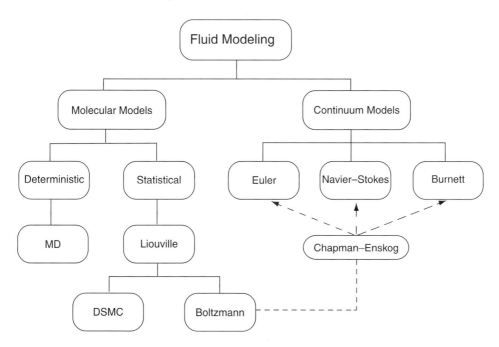

FIGURE 4.1 Molecular and continuum flow models. (From Gad-el-Hak, M. (1999) *J. Fluids Eng.* **121**, pp. 5–33, ASME, New York. With permission.)

Basically, the continuum model leads to fairly accurate predictions as long as local properties such as density and velocity can be defined as averages over elements large compared with the microscopic structure of the fluid but small enough in comparison with the scale of the macroscopic phenomena to permit the use of differential calculus to describe them. Additionally, the flow must not be too far from thermodynamic equilibrium. The former condition is almost always satisfied, but it is the latter which usually restricts the validity of the continuum equations. As will be seen in the following section, the continuum flow equations do not form a determinate set. The shear stress and heat flux must be expressed in terms of lower order macroscopic quantities such as velocity and temperature, and the simplest (i.e., linear) relations are valid only when the flow is near thermodynamic equilibrium. Worse yet, the traditional no-slip-boundary condition at a solid–fluid interface breaks down even before the linear stress–strain relation becomes invalid.

To be more specific, we temporarily restrict the discussion to gases where the concept of mean free path is well defined. Liquids are more problematic and we defer their discussion to a later section. For gases, the mean free path $\mathcal{L}$ is the average distance traveled by molecules between collisions. For an ideal gas modeled as rigid spheres, the mean free path is related to temperature T and pressure p as follows:

$$\mathcal{L} = \frac{1}{\sqrt{2}\pi n \sigma^2} = \frac{kT}{\sqrt{2}\pi p \sigma^2} \tag{4.1}$$

where n is the number density (number of molecules per unit volume), σ is the molecular diameter, and k is the Boltzmann constant (1.38×10^{-23} J/K · molecule).

The continuum model is valid when $\mathcal{L}$ is much smaller than a characteristic flow dimension L. As this condition is violated, the flow is no longer near equilibrium and the linear relation between stress and rate of strain and the no-slip-velocity condition are no longer valid. Similarly, the linear relation between heat flux and temperature gradient and the no-jump temperature condition at a solid–fluid interface are no longer accurate when $\mathcal{L}$ is not much smaller than L.

The length scale L can be some overall dimension of the flow, but a more precise choice is the scale of the gradient of a macroscopic quantity, as, for example, the density ρ,

$$L = \frac{\rho}{\left|\frac{\partial \rho}{\partial y}\right|} \tag{4.2}$$

The ratio between the mean free path and the characteristic length is known as the Knudsen number:

$$Kn = \frac{\mathcal{L}}{L} \tag{4.3}$$

and generally the traditional continuum approach is valid, albeit with modified boundary conditions, as long as $Kn < 0.1$.

There are two more important dimensionless parameters in fluid mechanics, and the Knudsen number can be expressed in terms of those two. The Reynolds number is the ratio of inertial forces to viscous ones:

$$Re = \frac{v_o L}{\nu} \tag{4.4}$$

where v_o is a characteristic velocity, and ν is the kinematic viscosity of the fluid. The Mach number is the ratio of flow velocity to the speed of sound:

$$Ma = \frac{v_o}{a_o} \tag{4.5}$$

The Mach number is a dynamic measure of fluid compressibility and may be considered as the ratio of inertial forces to elastic ones. From the kinetic theory of gases, the mean free path is related to the viscosity as follows:

$$\nu = \frac{\mu}{\rho} = \frac{1}{2}\mathcal{L}\bar{v}_m \tag{4.6}$$

where μ is the dynamic viscosity, and $\bar{v}_m$ is the mean molecular speed which is somewhat higher than the sound speed a_o,

$$\bar{v}_m = \sqrt{\frac{8}{\pi\gamma}}\, a_o \tag{4.7}$$

where γ is the specific heat ratio (i.e., the isentropic exponent). Combining Eqs. (4.3) to (4.7), we reach the required relation:

$$Kn = \sqrt{\frac{\pi\gamma}{2}}\frac{Ma}{Re} \tag{4.8}$$

In boundary layers, the relevant length scale is the shear-layer thickness δ, and for laminar flows:

$$\frac{\delta}{L} \sim \frac{1}{\sqrt{Re}} \tag{4.9}$$

$$Kn \sim \frac{Ma}{Re_\delta} \sim \frac{Ma}{\sqrt{Re}} \tag{4.10}$$

where Re_δ is the Reynolds number based on the freestream velocity v_o and the boundary layer thickness δ, and Re is based on v_o and the streamwise length scale L.

Rarefied gas flows are in general encountered in flows in small geometries such as MEMS devices and in low-pressure applications such as high-altitude flying and high-vacuum gadgets. The local value of Knudsen number in a particular flow determines the degree of rarefaction and the degree of validity of the continuum model. The different Knudsen number regimes are determined empirically and are therefore only approximate for a particular flow geometry. The pioneering experiments in rarefied gas dynamics were conducted by Knudsen in 1909. In the limit of zero Knudsen number, the transport terms in the continuum momentum and energy equations are negligible and the Navier–Stokes equations then reduce to the inviscid Euler equations. Both heat conduction and viscous diffusion and dissipation are negligible, and the flow is then approximately isentropic (i.e., adiabatic and reversible) from the continuum viewpoint, while the equivalent molecular viewpoint is that the velocity distribution function is everywhere of the local equilibrium or Maxwellian form. As Kn increases, rarefaction effects become more important, and eventually the continuum approach breaks down altogether. The different Knudsen number regimes are depicted in Figure 4.2, and can be summarized as follows:

Euler equations (neglect molecular diffusion):	$Kn \to 0$ ($Re \to \infty$)
Navier–Stokes equations with no-slip boundary conditions:	$Kn < 10^{-3}$
Navier–Stokes equations with slip boundary conditions:	$10^{-3} \leq Kn < 10^{-1}$
Transition regime:	$10^{-1} \leq Kn < 10$
Free-molecule flow:	$Kn \geq 10$

As an example, consider air at standard temperature ($T = 288$ K) and pressure ($p = 1.01 \times 10^5$ N/m^2). A cube 1 μm on a side contains 2.54×10^7 molecules separated by an average distance of 0.0034 μm. The gas is considered dilute if the ratio of this distance to the molecular diameter exceeds 7, and in the current example this ratio is 9, barely satisfying the dilute gas assumption. The mean free path computed from Eq. (1) is $\mathcal{L} = 0.065$ μm. A microdevice with characteristic length of 1 μm would have $Kn = 0.065$, which is in the slip-flow regime. At lower pressures, the Knudsen number increases. For example, if the

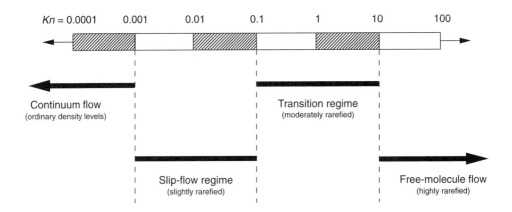

FIGURE 4.2 Knudsen number regimes. (From Gad-el-Hak, M. (1999) *J. Fluids Eng.* **121**, pp. 5–33, ASME, New York. With permission.)

pressure is 0.1 atm and the temperature remains the same, $Kn = 0.65$ for the same 1-µm device, and the flow is then in the transition regime. There would still be over 2 million molecules in the same 1-µm cube, and the average distance between them would be 0.0074 µm. The same device at 100 km altitude would have $Kn = 3 \times 10^4$, well into the free-molecule flow regime. The Knudsen number for the flow of a light gas such as helium is about 3 times larger than that for air flow at otherwise the same conditions.

Consider a long microchannel where the entrance pressure is atmospheric and the exit conditions are near vacuum. As air goes down the duct, the pressure and density decrease while the velocity, Mach number and Knudsen number increase. The pressure drops to overcome viscous forces in the channel. If isothermal conditions prevail,[1] density also drops and conservation of mass requires the flow to accelerate down the constant-area tube. The fluid acceleration in turn affects the pressure gradient, resulting in a nonlinear pressure drop along the channel. The Mach number increases down the tube, limited only by choked-flow condition $Ma = 1$. Additionally, the normal component of velocity is no longer zero. With lower density, the mean free path increases and Kn correspondingly increases. All flow regimes depicted in Figure 4.2 may occur in the same tube: continuum with no-slip-boundary conditions, slip-flow regime, transition regime and free-molecule flow. The air flow may also change from incompressible to compressible as it moves down the microduct. A similar scenario may take place if the entrance pressure is, say, 5 atm, while the exit is atmospheric. This deceivingly simple duct flow may in fact manifest every single complexity discussed in this section. In the following six sections, we discuss in turn the Navier–Stokes equations, compressibility effects, boundary conditions, molecular-based models, liquid flows and surface phenomena.

4.4 Continuum Model

We recall in this section the traditional conservation relations in fluid mechanics. A concise derivation of these equations can be found in Gad-el-Hak (2000). In here, we re-emphasize the precise assumptions needed to obtain a particular form of the equations. A continuum fluid implies that the derivatives of all the dependent variables exist in some reasonable sense. In other words, local properties such as density and velocity are defined as averages over elements large compared with the microscopic structure of the fluid but small enough in comparison with the scale of the macroscopic phenomena to permit the use of differential calculus to describe them. As mentioned earlier, such conditions are almost always met. For such fluids, and assuming the laws of nonrelativistic mechanics hold, the conservation of mass,

---

[1] More likely, the flow will be somewhere in between isothermal and adiabatic, Fanno flow. In that case, both density and temperature decrease downstream, the former not as fast as in the isothermal case, None of that changes the qualitative arguments made in the example.

momentum and energy can be expressed at every point in space and time as a set of partial differential equations as follows:

$$\frac{\partial \rho}{\partial t} + \frac{\partial}{\partial x_k}(\rho u_k) = 0 \tag{4.11}$$

$$\rho\left(\frac{\partial u_i}{\partial t} + u_k \frac{\partial u_i}{\partial x_k}\right) = \frac{\partial \Sigma_{ki}}{\partial x_k} + \rho g_i \tag{4.12}$$

$$\rho\left(\frac{\partial e}{\partial t} + u_k \frac{\partial e}{\partial x_k}\right) = -\frac{\partial q_k}{\partial x_k} + \Sigma_{ki}\frac{\partial u_i}{\partial x_k} \tag{4.13}$$

where ρ is the fluid density, u_k is an instantaneous velocity component (u, v, w), Σ_{ki} is the second-order stress tensor (surface force per unit area), g_i is the body force per unit mass, e is the internal energy and q_k is the sum of heat flux vectors due to conduction and radiation. The independent variables are time t and the three spatial coordinates x_1, x_2 and x_3 or (x, y, z).

Equations (4.11) to (4.13) constitute five differential equations for the 17 unknowns: ρ, u_i, Σ_{ki}, e and q_k. Absent any body couples, the stress tensor is symmetric, having only six independent components, which reduces the number of unknowns to 14. Obviously, the continuum flow equations do not form a determinate set. To close the conservation equations, the relation between the stress tensor and deformation rate, the relation between the heat flux vector and the temperature field and appropriate equations of state relating the different thermodynamic properties are needed. The stress–rate-of-strain relation and the heat-flux–temperature-gradient relation are approximately linear if the flow is not too far from thermodynamic equilibrium. This is a phenomenological result but can be rigorously derived from the Boltzmann equation for a dilute gas assuming the flow is near equilibrium. For a Newtonian, isotropic, Fourier, ideal gas, for example, those relations read:

$$\Sigma_{ki} = -p\,\delta_{ki} + \mu\left(\frac{\partial u_i}{\partial x_k} + \frac{\partial u_k}{\partial x_i}\right) + \lambda\left(\frac{\partial u_j}{\partial x_j}\right)\delta_{ki} \tag{4.14}$$

$$q_i = -\kappa\frac{\partial T}{\partial x_i} + \text{heat flux due to radiation} \tag{4.15}$$

$$de = c_v\,dT \quad \text{and} \quad p = \rho \mathcal{R} T \tag{4.16}$$

where p is the thermodynamic pressure, μ and λ are the first and second coefficients of viscosity, respectively, δ_{ki} is the unit second-order tensor (Kronecker delta), κ is the thermal conductivity, T is the temperature field, c_v is the specific heat at constant volume and $\mathcal{R}$ is the gas constant given by the Boltzmann constant divided by the mass of an individual molecule $k = m\mathcal{R}$. The Stokes' hypothesis relates the first and second coefficients of viscosity, thus $\lambda + \frac{2}{3}\mu = 0$, although the validity of this assumption for other than dilute, monatomic gases has occasionally been questioned [Gad-el-Hak, 1995]. With the above constitutive relations and neglecting radiative heat transfer, Eqs. (4.11), (4.12) and (4.13), respectively, read:

$$\frac{\partial \rho}{\partial t} + \frac{\partial}{\partial x_k}(\rho u_k) = 0 \tag{4.17}$$

$$\rho\left(\frac{\partial u_i}{\partial t} + u_k \frac{\partial u_i}{\partial x_k}\right) = -\frac{\partial p}{\partial x_i} + \rho g_i + \frac{\partial}{\partial x_k}\left[\mu\left(\frac{\partial u_i}{\partial x_k} + \frac{\partial u_k}{\partial x_i}\right) + \delta_{ki}\lambda\frac{\partial u_j}{\partial x_j}\right] \tag{4.18}$$

$$\rho\left(\frac{\partial e}{\partial t} + u_k \frac{\partial e}{\partial x_k}\right) = \frac{\partial}{\partial x_k}\left(\kappa\frac{\partial T}{\partial x_k}\right) - p\frac{\partial u_k}{\partial x_k} + \phi \tag{4.19}$$

The three components of the vector Eq. (4.18) are the Navier–Stokes equations expressing the conservation of momentum for a Newtonian fluid. In the thermal energy Eq. (4.19), ϕ is the always positive dissipation function expressing the irreversible conversion of mechanical energy to internal energy as a result of the deformation of a fluid element. The second term on the right-hand side of Eq. (4.19) is the reversible work done (per unit time) by the pressure as the volume of a fluid material element changes. For a Newtonian, isotropic fluid, the viscous dissipation rate is given by:

$$\phi = \frac{1}{2}\mu\left(\frac{\partial u_i}{\partial x_k} + \frac{\partial u_k}{\partial x_i}\right)^2 + \lambda\left(\frac{\partial u_j}{\partial x_j}\right)^2 \tag{4.20}$$

There are now six unknowns: ρ, u_i, p and T; the five coupled Eqs. (4.17) to (4.19); plus the equation of state relating pressure, density and temperature. These six equations together with a sufficient number of initial and boundary conditions constitute a well-posed, albeit formidable, problem. The system of Eqs. (4.17) to (4.19) is an excellent model for the laminar or turbulent flow of most fluids such as air and water under many circumstances, including high-speed gas flows for which the shock waves are thick relative to the mean free path of the molecules.

Considerable simplification is achieved if the flow is assumed incompressible, usually a reasonable assumption provided that the characteristic flow speed is less than 0.3 of the speed of sound. The incompressibility assumption is readily satisfied for almost all liquid flows and many gas flows. In such cases, the density is assumed either a constant or a given function of temperature (or species concentration). The governing equations for such flow are

$$\frac{\partial u_k}{\partial x_k} = 0 \tag{4.21}$$

$$\rho\left(\frac{\partial u_i}{\partial t} + u_k\frac{\partial u_i}{\partial x_k}\right) = -\frac{\partial p}{\partial x_i} + \frac{\partial}{\partial x_k}\left[\mu\left(\frac{\partial u_i}{\partial x_k} + \frac{\partial u_k}{\partial x_i}\right)\right] + \rho g_i \tag{4.22}$$

$$\rho c_p\left(\frac{\partial T}{\partial t} + u_k\frac{\partial T}{\partial x_k}\right) = \frac{\partial}{\partial x_k}\left(\kappa\frac{\partial T}{\partial x_k}\right) + \phi_{\text{incomp}} \tag{4.23}$$

where ϕ_{incomp} is the incompressible limit of Eq. (4.20). These are now five equations for the five dependent variables u_i, p and T. Note that the left-hand side of Eq. (4.23) has the specific heat at constant pressure c_p and not c_v. It is the convection of enthalpy—and not internal energy—that is balanced by heat conduction and viscous dissipation. This is the correct incompressible-flow limit—of a compressible fluid—as discussed in detail in Section 10.9 of Panton (1996); a subtle point perhaps but one that is frequently missed in textbooks.

For both the compressible and the incompressible equations of motion, the transport terms are neglected away from solid walls in the limit of infinite Reynolds number ($Kn \to 0$). The fluid is then approximated as inviscid and nonconducting, and the corresponding equations read (for the compressible case):

$$\frac{\partial \rho}{\partial t} + \frac{\partial}{\partial x_k}(\rho u_k) = 0 \tag{4.24}$$

$$\rho\left(\frac{\partial u_i}{\partial t} + u_k\frac{\partial u_i}{\partial x_k}\right) = -\frac{\partial p}{\partial x_i} + \rho g_i \tag{4.25}$$

$$\rho c_v\left(\frac{\partial T}{\partial t} + u_k\frac{\partial T}{\partial x_k}\right) = -p\frac{\partial u_k}{\partial x_k} \tag{4.26}$$

The Euler Eq. (4.25) can be integrated along a stream line, and the resulting Bernoulli's equation provides a direct relation between the velocity and pressure.

4.5 Compressibility

The issue of whether to consider the continuum flow compressible or incompressible seems to be rather straightforward but is in fact full of potential pitfalls. If the local Mach number is less than 0.3, then the flow of a compressible fluid such as air can—according to the conventional wisdom—be treated as incompressible. But the well-known $Ma < 0.3$ criterion is only a necessary not a sufficient one to allow a treatment of the flow as approximately incompressible. In other words, situations exist where the Mach number can be exceedingly small while the flow is compressible. As is well documented in heat transfer textbooks, strong wall heating or cooling may cause the density to change sufficiently and the incompressible approximation to break down, even at low speeds. Less known is the situation encountered in some microdevices where the pressure may strongly change due to viscous effects even though the speeds may not be high enough for the Mach number to go above the traditional threshold of 0.3. Corresponding to the pressure changes would be strong density changes that must be taken into account when writing the continuum equations of motion. In this section, we systematically explain all situations where compressibility effects must be considered. Let us rewrite the full continuity Eq. (4.11) as follows:

$$\frac{D\rho}{Dt} + \rho \frac{\partial u_k}{\partial x_k} = 0 \qquad (4.27)$$

where $\frac{D}{Dt}$ is the substantial derivative $\left(\frac{\partial}{\partial t} + u_k \frac{\partial}{\partial x_k}\right)$, expressing changes following a fluid element. The proper criterion for the incompressible approximation to hold is that $\left(\frac{1}{\rho}\frac{D\rho}{Dt}\right)$ is vanishingly small. In other words, if density changes following a fluid particle are small, the flow is approximately incompressible. Density may change arbitrarily from one particle to another without violating the incompressible flow assumption. This is the case for example in the stratified atmosphere and ocean, where the variable-density/temperature/salinity flow is often treated as incompressible.

From the state principle of thermodynamics, we can express the density changes of a simple system in terms of changes in pressure and temperature:

$$\rho = \rho(p, T) \qquad (4.28)$$

Using the chain rule of calculus,

$$\frac{1}{\rho}\frac{D\rho}{Dt} = \alpha \frac{Dp}{Dt} - \beta \frac{DT}{Dt} \qquad (4.29)$$

where α and β are, respectively, the isothermal compressibility coefficient and the bulk expansion coefficient—two thermodynamic variables that characterize the fluid susceptibility to change of volume—which are defined by the following relations:

$$\alpha(p, T) \equiv \frac{1}{\rho}\frac{\partial \rho}{\partial p}\bigg|_T \qquad (4.30)$$

$$\beta(p, T) \equiv -\frac{1}{\rho}\frac{\partial \rho}{\partial T}\bigg|_p \qquad (4.31)$$

For ideal gases, $\alpha = 1/p$ and $\beta = 1/T$. Note, however, that in the following arguments it will not be necessary to invoke the ideal gas assumption. The flow must be treated as compressible if pressure and/or temperature changes—following a fluid element—are sufficiently strong. Equation (4.29) must of course be properly nondimensionalized before deciding whether a term is large or small. In here, we follow closely the procedure detailed in Panton (1996).

Consider first the case of adiabatic walls. Density is normalized with a reference value ρ_o; velocities, with a reference speed v_o; spatial coordinates and time, with, respectively, L and L/v_o, the isothermal compressibility coefficient and bulk expansion coefficient with reference values α_o and β_o. The pressure is nondimensionalized with the inertial pressure scale $\rho_o v_o^2$. This scale is twice the dynamic pressure (i.e., the pressure change as an inviscid fluid moving at the reference speed is brought to rest).

Temperature changes for the case of adiabatic walls can only result from the irreversible conversion of mechanical energy into internal energy via viscous dissipation. Temperature is therefore nondimensionalized as follows:

$$T^* = \frac{T - T_o}{\left(\frac{\mu_o v_o^2}{c_{\kappa_o}}\right)} = \frac{T - T_o}{Pr\left(\frac{v_o^2}{c_{p_o}}\right)} \quad (4.32)$$

where T_o is a reference temperature, μ_o, κ_o and c_{p_o} are, respectively, reference viscosity, thermal conductivity and specific heat at constant pressure, and Pr is the reference Prandtl number, $(\mu_o c_{p_o})/\kappa_o$.

In the present formulation, the scaling used for pressure is based on the Bernoulli's equation and therefore neglects viscous effects. This particular scaling guarantees that the pressure term in the momentum equation will be of the same order as the inertia term. The temperature scaling assumes that the conduction, convection and dissipation terms in the energy equation have the same order of magnitude. The resulting dimensionless form of Eq. (4.29) reads:

$$\frac{1}{\rho^*}\frac{D\rho^*}{Dt^*} = \gamma_o Ma^2 \left\{ \alpha^* \frac{Dp^*}{Dt^*} - \frac{Pr\, B\beta^*}{A}\frac{DT^*}{Dt^*} \right\} \quad (4.33)$$

where the superscript $*$ indicates a nondimensional quantity, Ma is the reference Mach number $(v_o/a_o$, where a_o is the reference speed of sound) and A and B are dimensionless constants defined by $A \equiv \alpha_o \rho_o c_{p_o} T_o$ and $B \equiv \beta_o T_o$. If the scaling is properly chosen, the terms having the $*$ superscript in the right-hand side should be of order one, and the relative importance of such terms in the equations of motion is determined by the magnitude of the dimensionless parameter(s) appearing to their left (e.g., Ma, Pr, etc.). Therefore, as $Ma^2 \rightarrow 0$, temperature changes due to viscous dissipation are neglected (unless Pr is very large, as, for example, in the case of highly viscous polymers and oils). Within the same order of approximation, all thermodynamic properties of the fluid are assumed constant.

Pressure changes are also neglected in the limit of zero Mach number. Hence, for $Ma < 0.3$ (i.e., $Ma^2 < 0.09$), density changes following a fluid particle can be neglected and the flow can then be approximated as incompressible;[2] however, there is a caveat in this argument. Pressure changes due to inertia can indeed be neglected at small Mach numbers and this is consistent with the way we nondimensionalized the pressure term above. If, on the other hand, pressure changes are mostly due to viscous effects, as is the case in a long microduct or a micro-gas-bearing, pressure changes may be significant even at low speeds (low Ma). In that case, the term $\frac{Dp^*}{Dt}$ in Eq. (4.33) is no longer of order one and may be large regardless of the value of Ma. Density then may change significantly and the flow must be treated as compressible. Had pressure been nondimensionalized using the viscous scale $\left(\frac{\mu_o v_o}{L}\right)$ instead of the inertial one $(\rho_o v_o^2)$, the revised Eq. (4.33) would have Re^{-1} appearing explicitly in the first term in the right-hand side, accentuating the importance of this term when viscous forces dominate.

---

[2] With an error of about 10% at $Ma = 0.3$, 4% at $Ma = 0.2$, 1% at $Ma = 0.1$ and so on.

A similar result can be gleaned when the Mach number is interpreted as follows:

$$Ma^2 = \frac{v_o^2}{a_o^2} = v_o^2 \left.\frac{\partial \rho}{\partial p}\right|_s = \frac{\rho_o v_o^2}{\rho_o} \left.\frac{\partial \rho}{\partial p}\right|_s \sim \frac{\Delta p}{\rho_o}\frac{\Delta \rho}{\Delta p} = \frac{\Delta \rho}{\rho_o} \quad (4.34)$$

where s is the entropy. Again, the above equation assumes that pressure changes are inviscid; therefore, a small Mach number means negligible pressure and density changes. In a flow dominated by viscous effects—such as that inside a microduct—density changes may be significant even in the limit of zero Mach number.

Identical arguments can be made in the case of isothermal walls. Here, strong temperature changes may be the result of wall heating or cooling, even if viscous dissipation is negligible. The proper temperature scale in this case is given in terms of the wall temperature T_w and the reference temperature T_o as follows:

$$\widehat{T} = \frac{T - T_o}{T_w - T_o} \quad (4.35)$$

where $\widehat{T}$ is the new dimensionless temperature. The nondimensional form of Eq. (4.29) now reads:

$$\frac{1}{\rho^*}\frac{D\rho^*}{Dt^*} = \gamma_o Ma^2 \alpha^* \frac{Dp^*}{Dt^*} - \beta^* B\left(\frac{T_w - T_o}{T_o}\right)\frac{D\widehat{T}}{Dt^*} \quad (4.36)$$

Here, we notice that the temperature term is different from that in Eq. (4.33). Ma is no longer appearing in this term, and strong temperature changes, i.e., large $(T_w - T_o)/T_o$, may cause strong density changes regardless of the value of the Mach number. Additionally, the thermodynamic properties of the fluid are not constant but depend on temperature, and as a result the continuity, momentum and energy equations all couple. The pressure term in Eq. (4.36), on the other hand, is exactly as it was in the adiabatic case and the same arguments made before apply: The flow should be considered compressible if $Ma > 0.3$, or if pressure changes due to viscous forces are sufficiently large.

Experiments in gaseous microducts confirm the above arguments. For both low- and high-Mach-number flows, pressure gradients in long microchannels are nonconstant, consistent with the compressible flow equations. Such experiments were conducted by, among others, Prud'homme et al. (1986), Pfahler et al. (1991), van den Berg et al. (1993), Liu et al. (1993; 1995), Pong et al. (1994), Harley et al. (1995), Piekos and Breuer (1996), Arkilic (1997) and Arkilic et al. (1995; 1997a; 1997b). Sample results will be presented in the following section.

Three additional scenarios exist in which significant pressure and density changes may take place without inertial, viscous or thermal effects. First is the case of quasistatic compression/expansion of a gas in, for example, a piston–cylinder arrangement. The resulting compressibility effects are, however, compressibility of the fluid and not of the flow. Two other situations where compressibility effects must also be considered are problems with length scales comparable to the scale height of the atmosphere and rapidly varying flows as in sound propagation [Lighthill, 1963].

4.6 Boundary Conditions

The continuum equations of motion described earlier require a certain number of initial and boundary conditions for proper mathematical formulation of flow problems. In this section, we describe the boundary conditions at a fluid–solid interface. Boundary conditions in the inviscid flow theory pertain only to the velocity component normal to a solid surface. The highest spatial derivative of velocity in the inviscid equations of motion is first order, and only one velocity boundary condition at the surface is admissible. The normal velocity component at a fluid–solid interface is specified, and no statement can be made regarding the tangential velocity component. The normal-velocity condition simply states that a fluid-particle path cannot go through an impermeable wall. Real fluids are of course viscous, and the corresponding momentum equation has second-order derivatives of velocity, thus requiring an additional boundary condition on the velocity component tangential to a solid surface.

Traditionally, the no-slip condition at a fluid–solid interface is enforced in the momentum equation and an analogous no-temperature-jump condition is applied in the energy equation. The notion underlying the no-slip/no-jump condition is that within the fluid there cannot be any finite discontinuities of velocity/temperature. Those would involve infinite velocity/temperature gradients and so produce infinite viscous stress/heat flux that would destroy the discontinuity in infinitesimal time. The interaction between a fluid particle and a wall is similar to that between neighboring fluid particles; therefore, no discontinuities are allowed at the fluid–solid interface either. In other words, the fluid velocity must be zero relative to the surface and the fluid temperature must equal to that of the surface. But, strictly speaking, those two boundary conditions are valid only if the fluid flow adjacent to the surface is in thermodynamic equilibrium. This requires an infinitely high frequency of collisions between the fluid and the solid surface. In practice, the no-slip/no-jump condition leads to fairly accurate predictions as long as $Kn < 0.001$ (for gases). Beyond that, the collision frequency is simply not high enough to ensure equilibrium and a certain degree of tangential-velocity slip and temperature jump must be allowed. This is a case frequently encountered in MEMS flows, and we develop the appropriate relations in this section.

For both liquids and gases, the linear Navier boundary condition empirically relates the tangential velocity slip at the wall $\Delta u|_w$ to the local shear:

$$\Delta u|_w = u_{\text{fluid}} - u_{\text{wall}} = L_s \frac{\partial u}{\partial y}\bigg|_w \tag{4.37}$$

where L_s is the constant slip length, and $\frac{\partial u}{\partial y}\big|_w$ is the strain rate computed at the wall. In most practical situations, the slip length is so small that the no-slip condition holds. In MEMS applications, however, that may not be the case. Once again we defer the discussion of liquids to a later section and focus for now on gases.

Assuming isothermal conditions prevail, the above slip relation has been rigorously derived by Maxwell (1879) from considerations of the kinetic theory of dilute, monatomic gases. Gas molecules, modeled as rigid spheres, continuously strike and reflect from a solid surface, just as they continuously collide with each other. For an idealized perfectly smooth wall (at the molecular scale), the incident angle exactly equals the reflected angle and the molecules conserve their tangential momentum and thus exert no shear on the wall. This is termed *specular reflection* and results in perfect slip at the wall. For an extremely rough wall, on the other hand, the molecules reflect at some random angle uncorrelated with their entry angle. This perfectly diffuse reflection results in zero tangential momentum for the reflected fluid molecules to be balanced by a finite slip velocity in order to account for the shear stress transmitted to the wall. A force balance near the wall leads to the following expression for the slip velocity:

$$u_{\text{gas}} - u_{\text{wall}} = \mathcal{L} \frac{\partial u}{\partial y}\bigg|_w \tag{4.38}$$

where $\mathcal{L}$ is the mean free path. The right-hand side can be considered as the first term in an infinite Taylor series, sufficient if the mean free path is relatively small enough. The equation above states that significant slip occurs only if the mean velocity of the molecules varies appreciably over a distance of one mean free path. This is the case, for example, in vacuum applications and/or flow in microdevices. The number of collisions between the fluid molecules and the solid in those cases is not large enough for even an approximate flow equilibrium to be established. Furthermore, additional (nonlinear) terms in the Taylor series would be needed as $\mathcal{L}$ increases and the flow is further removed from the equilibrium state.

For real walls some molecules reflect diffusively and some reflect specularly. In other words, a portion of the momentum of the incident molecules is lost to the wall and a (typically smaller) portion is retained by the reflected molecules. The tangential-momentum-accommodation coefficient σ_v is defined as the fraction of molecules reflected diffusively. This coefficient depends on the fluid, the solid and the surface finish and has been determined experimentally to be between 0.2 and 0.8 [Thomas and Lord, 1974; Seidl and Steiheil, 1974; Porodnov et al., 1974; Arkilic et al., 1997b; Arkilic, 1997], the lower limit being for exceptionally smooth surfaces while the upper limit is typical of most practical surfaces. The final expression derived by Maxwell for an isothermal wall reads:

$$u_{gas} - u_{wall} = \frac{2 - \sigma_v}{\sigma_v} \mathcal{L} \frac{\partial u}{\partial y}\bigg|_w \quad (4.39)$$

For $\sigma_v = 0$, the slip velocity is unbounded, while for $\sigma_v = 1$, Eq. (4.39) reverts to Eq. (4.38).

Similar arguments were made for the temperature-jump-boundary condition by von Smoluchowski (1898). For an ideal gas flow in the presence of wall-normal and tangential temperature gradients, the complete (first-order) slip-flow and temperature-jump-boundary conditions read:

$$u_{gas} - u_{wall} = \frac{2 - \sigma_v}{\sigma_v} \frac{1}{\rho \sqrt{\frac{2 \mathcal{R} T_{gas}}{\pi}}} \tau_w + \frac{3}{4} \frac{Pr(\gamma - 1)}{\gamma \rho \mathcal{R} T_{gas}} (-q_x)_w = \frac{2 - \sigma_v}{\sigma_v} \mathcal{L} \left(\frac{\partial u}{\partial y}\right)_w + \frac{3}{4} \frac{\mu}{\rho T_{gas}} \left(\frac{\partial u}{\partial x}\right)_w \quad (4.40)$$

$$T_{gas} - T_{wall} = \frac{2 - \sigma_T}{\sigma_T} \left[\frac{2(\gamma - 1)}{(\gamma + 1)}\right] \frac{1}{\rho \mathcal{R} \sqrt{\frac{2 \mathcal{R} T_{gas}}{\pi}}} (-q_y)_w = \frac{2 - \sigma_T}{\sigma_T} \left[\frac{2\gamma}{(\gamma + 1)}\right] \frac{\mathcal{L}}{Pr} \left(\frac{\partial T}{\partial y}\right)_w \quad (4.41)$$

where x and y are the streamwise and normal coordinates; ρ and μ are, respectively, the fluid density and viscosity; $\mathcal{R}$ is the gas constant; T_{gas} is the temperature of the gas adjacent to the wall; T_{wall} is the wall temperature; τ_w is the shear stress at the wall; Pr is the Prandtl number; γ is the specific heat ratio; q_x and q_y are, respectively, the tangential and normal heat flux at the wall.

The tangential-momentum-accommodation coefficient σ_v and the thermal-accommodation coefficient σ_T are given, respectively, by:

$$\sigma_v = \frac{\tau_i - \tau_r}{\tau_i - \tau_w} \quad (4.42)$$

$$\sigma_T = \frac{dE_i - dE_r}{dE_i - dE_w} \quad (4.43)$$

where the subscripts i, r and w stand for, respectively, incident, reflected and solid wall conditions; τ is a tangential momentum flux; dE is an energy flux.

The second term in the right-hand side of Eq. (4.40) is the *thermal creep*, which generates slip velocity in the fluid opposite to the direction of the tangential heat flux (i.e., flow in the direction of increasing temperature). At sufficiently high Knudsen numbers, a streamwise temperature gradient in a conduit leads to a measurable pressure gradient along the tube. This may be the case in vacuum applications and MEMS devices. Thermal creep is the basis for the so-called Knudsen pump—a device with no moving parts—in which rarefied gas is hauled from one cold chamber to a hot one.[3] Clearly, such a pump

---

[3] The terminology *Knudsen pump* has been used by, for example, Vargo and Muntz (1996), but, according to Loeb (1961), the original experiments demonstrating such a pump were carried out by Osborne Reynolds.

performs best at high Knudsen numbers and is typically designed to operate in the free-molecule flow regime.

In dimensionless form, Eqs. (4.40) and (4.41), respectively, read:

$$u^*_{\text{gas}} - u^*_{\text{wall}} = \frac{2-\sigma_v}{\sigma_v} Kn\left(\frac{\partial u^*}{\partial y^*}\right)_w + \frac{3}{2\pi}\frac{(\gamma-1)}{\gamma}\frac{Kn^2 Re}{Ec}\left(\frac{\partial T^*}{\partial x^*}\right)_w \qquad (4.44)$$

$$T^*_{\text{gas}} - T^*_{\text{wall}} = \frac{2-\sigma_T}{\sigma_T}\left[\frac{2\gamma}{(\gamma+1)}\right]\frac{Kn}{Pr}\left(\frac{\partial T^*}{\partial y^*}\right)_w \qquad (4.45)$$

where the superscript * indicates dimensionless quantity, Kn is the Knudsen number, Re is the Reynolds number and Ec is the Eckert number defined by:

$$Ec = \frac{v_o^2}{c_p \Delta T} = (\gamma-1)\frac{T_o}{\Delta T} Ma^2 \qquad (4.46)$$

where v_o is a reference velocity, $\Delta T = (T_{\text{gas}} - T_o)$ and T_o is a reference temperature. Note that very low values of σ_v and σ_T lead to substantial velocity slip and temperature jump even for flows with small Knudsen number.

The first term in the right-hand side of Eq. (4.44) is first order in Knudsen number, while the thermal creep term is second order, meaning that the creep phenomenon is potentially significant at large values of the Knudsen number. Equation (4.45) is first order in Kn. Using Eqs. (4.8) and (4.46), the thermal creep term in Eq. (4.44) can be rewritten in terms of ΔT and Reynolds number. Thus,

$$u^*_{\text{gas}} - u^*_{\text{wall}} = \frac{2-\sigma_v}{\sigma_v} Kn\left(\frac{\partial u^*}{\partial y^*}\right)_w + \frac{3}{4}\frac{\Delta T}{T_o}\frac{1}{Re}\left(\frac{\partial T^*}{\partial x^*}\right)_w \qquad (4.47)$$

It is clear that large temperature changes along the surface or low Reynolds numbers lead to significant thermal creep.

The continuum Navier–Stokes equations with no-slip-/no-temperature-jump-boundary conditions are valid as long as the Knudsen number does not exceed 0.001. First-order slip-/temperature-jump-boundary conditions should be applied to the Navier–Stokes equations in the range of $0.001 < Kn < 0.1$. The transition regime spans the range of $0.1 < Kn < 10$, and second-order or higher slip-/temperature-jump-boundary conditions are applicable there. Note, however, that the Navier–Stokes equations are first-order accurate in Kn as will be shown later and are themselves not valid in the transition regime. Either higher order continuum equations (e.g., Burnett equations) should be used there or molecular modeling should be invoked, abandoning the continuum approach altogether.

For isothermal walls, Beskok (1994) derived a higher order, slip-velocity condition as follows:

$$u_{\text{gas}} - u_{\text{wall}} = \frac{2-\sigma_v}{\sigma_v}\left[\mathcal{L}\left(\frac{\partial u}{\partial y}\right)_w + \frac{\mathcal{L}^2}{2!}\left(\frac{\partial^2 u}{\partial y^2}\right)_w + \frac{\mathcal{L}^3}{3!}\left(\frac{\partial^3 u}{\partial y^3}\right)_w + \cdots\right] \qquad (4.48)$$

Attempts to implement the above slip condition in numerical simulations are rather difficult. Second-order and higher derivatives of velocity cannot be computed accurately near the wall. Based on asymptotic analysis, Beskok (1996) and Beskok and Karniadakis (1994; 1999) proposed the following alternative higher order boundary condition for the tangential velocity, including the thermal creep term:

$$u^*_{\text{gas}} - u^*_{\text{wall}} = \frac{2-\sigma_v}{\sigma_v}\frac{Kn}{1-bKn}\left(\frac{\partial u^*}{\partial y^*}\right)_w + \frac{3}{2\pi}\frac{(\gamma-1)}{\gamma}\frac{Kn^2 Re}{Ec}\left(\frac{\partial T^*}{\partial x^*}\right)_w \qquad (4.49)$$

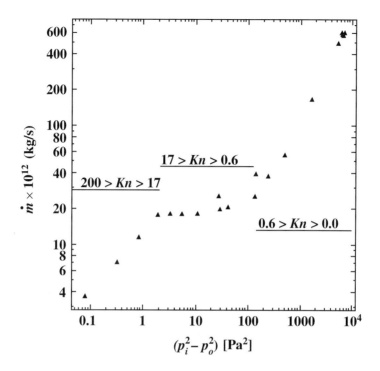

FIGURE 4.3 Variation of mass flowrate as a function of $(p_i^2 - p_o^2)$. (Original data acquired by S.A. Tison and plotted by Beskok et al. (1996). Figure from Gad-el-Hak, M. (1999) *J. Fluids Eng.* **121**, pp. 5–33, ASME, New York. With permission.)

where b is a high-order slip coefficient determined from the presumably known no-slip solution, thus avoiding the computational difficulties mentioned above. If this high-order slip coefficient is chosen as $b = u_w'''/u_w'$, where the prime denotes derivative with respect to y and the velocity is computed from the no-slip Navier–Stokes equations, Eq. (4.49) becomes second-order accurate in Knudsen number. Beskok's procedure can be extended to third and higher orders for both the slip-velocity and thermal creep terms.

Similar arguments can be applied to the temperature-jump-boundary condition, and the resulting Taylor series reads in dimensionless form (Beskok, 1996):

$$T_{\text{gas}}^* - T_{\text{wall}}^* = \frac{2-\sigma_T}{\sigma_T}\left[\frac{2\gamma}{(\gamma+1)}\right]\frac{1}{Pr}\left[Kn\left(\frac{\partial T^*}{\partial y^*}\right)_w + \frac{Kn^2}{2!}\left(\frac{\partial^2 T^*}{\partial y^{*2}}\right)_w + \cdots\right] \quad (4.50)$$

Again, the difficulties associated with computing second- and higher order derivatives of temperature are alleviated using a procedure identical to that utilized for the tangential velocity boundary condition.

Several experiments in low-pressure macroducts or in microducts confirm the necessity of applying the slip-boundary condition at sufficiently large Knudsen numbers. Among them are those conducted by Knudsen (1909), Pfahler et al. (1991), Tison (1993), Liu et al. (1993; 1995), Pong et al. (1994), Arkilic et al. (1995), Harley et al. (1995) and Shih et al. (1995; 1996). The experiments are complemented by the numerical simulations carried out by Beskok (1994; 1996), Beskok and Karniadakis (1994; 1999), and Beskok et al. (1996). Here we present selected examples of the experimental and numerical results.

Tison (1993) conducted pipe flow experiments at very low pressures. His pipe has a diameter of 2 mm and a length-to-diameter ratio of 200. Both inlet and outlet pressures were varied to yield Knudsen number in the range of $Kn = 0$ to 200. Figure 4.3 shows the variation of mass flowrate as a function of $(p_i^2 - p_o^2)$, where p_i is the inlet pressure and p_o is the outlet pressure. The pressure drop in this rarefied pipe flow is nonlinear, characteristic of low-Reynolds-number, compressible flows. Three distinct flow

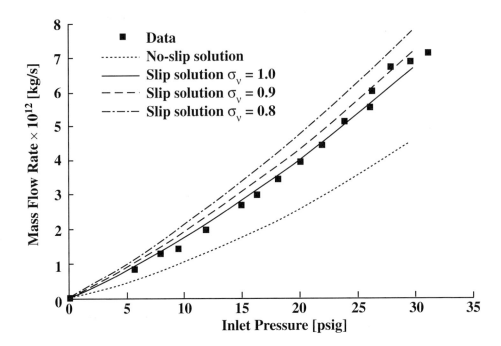

FIGURE 4.4 Mass flowrate vs. inlet pressure in a microchannel. (From Shih, J.C., et al. (1995) "Non-linear Pressure Distribution in Uniform Microchannels," ASME AMD-MD-Vol. 238, American Society of Mechanical Engineers, New York. With permission.)

regimes are identified: (1) slip-flow regime, $0 < Kn < 0.6$; (2) transition regime, $0.6 < Kn < 17$, where the mass flowrate is almost constant as the pressure changes; and (3) free-molecule flow, $Kn > 17$. Note that the demarkation between these three regimes is slightly different from that mentioned earlier. As stated, the different Knudsen number regimes are determined empirically and are therefore only approximate for a particular flow geometry.

Shih et al. (1995) conducted their experiments in a microchannel using helium as a fluid. The inlet pressure varied but the duct exit was atmospheric. Microsensors were fabricated *in situ* along their MEMS channel to measure the pressure. Figure 4.4 shows their measured mass flowrate vs. the inlet pressure. The data are compared to the no-slip solution and the slip solution using three different values of the tangential-momentum-accommodation coefficient: 0.8, 0.9 and 1.0. The agreement is reasonable with the case $\sigma_v = 1$, indicating perhaps that the channel used by Shih et al. was quite rough on the molecular scale. In a second experiment [Shih et al., 1996], nitrous oxide was used as the fluid. The square of the pressure distribution along the channel is plotted in Figure 4.5 for five different inlet pressures. The experimental data (symbols) compare well with the theoretical predictions (solid lines). Again, the nonlinear pressure drop shown indicates that the gas flow is compressible.

Arkilic (1997) provided an elegant analysis of the compressible, rarefied flow in a microchannel. The results of his theory are compared to the experiments of Pong et al. (1994) in Figure 4.6. The dotted line is the incompressible flow solution, where the pressure is predicted to drop linearly with streamwise distance. The dashed line is the compressible flow solution that neglects rarefaction effects (assumes $Kn = 0$). Finally, the solid line is the theoretical result that takes into account both compressibility and rarefaction via the slip-flow-boundary condition computed at the exit Knudsen number of $Kn = 0.06$. That theory compares most favorably with the experimental data. In the compressible flow through the constant-area duct, density decreases and thus velocity increases in the streamwise direction. As a result, the pressure distribution is nonlinear with negative curvature. A moderate Knudsen number (i.e., moderate slip) actually diminishes, albeit rather weakly, this curvature. Thus, compressibility and rarefaction effects lead to opposing trends, as pointed out by Beskok et al. (1996).

Flow Physics

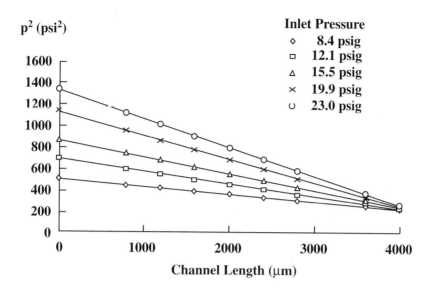

FIGURE 4.5 Pressure distribution of nitrous oxide in a microduct. Solid lines are theoretical predictions. (From Shih, J.C., et al. (1996) in *Applications of Microfabrication to Fluid Mechanics*, ed. K. Breuer et al., American Society of Mechanical Engineers, New York, pp. 197–203. With permission.)

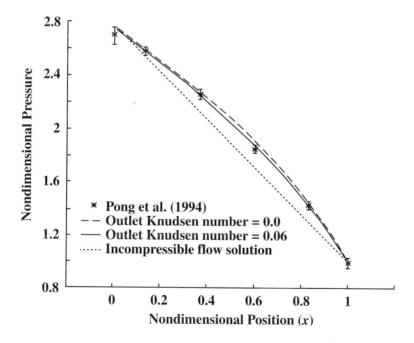

FIGURE 4.6 Pressure distribution in a long microchannel. The symbols are experimental data while the lines are different theoretical predictions. (From Arkilic, E.B. (1997) "Measurement of the Mass Flow and Tangential Momentum Accommodation Coefficient in Silicon Micromachines Channels," Ph.D. thesis, Massachusettes Institute of Technology, Cambridge, MA. With permission.)

4.7 Molecular-Based Models

In the continuum models discussed thus far, the macroscopic fluid properties are the dependent variables, while the independent variables are the three spatial coordinates and time. The molecular models recognize the fluid as a myriad of discrete particles: molecules, atoms, ions and electrons. The goal here is to determine the position, velocity and state of all particles at all times. The molecular approach is either deterministic or probabilistic (refer to Figure 4.1). Provided that there is a sufficient number of microscopic particles within the smallest significant volume of a flow, the macroscopic properties at any location in the flow can then be computed from the discrete-particle information by a suitable averaging or weighted averaging process. This section discusses molecular-based models and their relation to the continuum models previously considered.

The most fundamental of the molecular models is a deterministic one. The motion of the molecules is governed by the laws of classical mechanics, although, at the expense of greatly complicating the problem, the laws of quantum mechanics can also be considered in special circumstances. The modern molecular dynamics (MD) computer simulations have been pioneered by Alder and Wainwright (1957; 1958; 1970) and reviewed by Ciccotti and Hoover (1986), Allen and Tildesley (1987), Haile (1993) and Koplik and Banavar (1995). The simulation begins with a set of N molecules in a region of space, each assigned a random velocity corresponding to a Boltzmann distribution at the temperature of interest. The interaction between the particles is prescribed typically in the form of a two-body potential energy, and the time evolution of the molecular positions is determined by integrating Newton's equations of motion. Because MD is based on the most basic set of equations, it is valid in principle for any flow extent and any range of parameters. The method is straightforward in principle but has two hurdles: choosing a proper and convenient potential for particular fluid and solid combinations and the colossal computer resources required to simulate a reasonable flowfield extent.

For purists, the former difficulty is a sticky one. There is no totally rational methodology by which a convenient potential can be chosen. Part of the art of MD is to pick an appropriate potential and validate the simulation results with experiments or other analytical/computational results. A commonly used potential between two molecules is the generalized Lennard–Jones 6–12 potential, to be used in the following section and further discussed in the section following that.

The second difficulty, and by far the most serious limitation of molecular dynamics simulations, is the number of molecules N that can realistically be modeled on a digital computer. Because the computation of an element of trajectory for any particular molecule requires consideration of *all* other molecules as potential collision partners, the amount of computation required by the MD method is proportional to N^2. Some saving in computer time can be achieved by cutting off the weak tail of the potential (see Figure 4.11) at, say, $r_c = 2.5\ \sigma$ and shifting the potential by a linear term in r so that the force goes smoothly to zero at the cutoff. As a result, only nearby molecules are treated as potential collision partners, and the computation time for N molecules no longer scales with N^2.

The state of the art of molecular dynamics simulations in the early 2000s is such that with a few hours of CPU time general-purpose supercomputers can handle around 100,000 molecules. At enormous expense, the fastest parallel machine available can simulate around 10 million particles. Because of the extreme diminution of molecular scales, the above translates into regions of liquid flow of about 0.02 μm (200 Å) in linear size, over time intervals of around 0.001 μs, enough for continuum behavior to set in for simple molecules. To simulate 1 s of real time for complex molecular interactions (e.g., including vibration modes, reorientation of polymer molecules, collision of colloidal particles) requires unrealistic CPU time measured in hundreds of years.

Molecular dynamics simulations are highly inefficient for dilute gases where the molecular interactions are infrequent. The simulations are more suited for dense gases and liquids. Clearly, molecular dynamics simulations are reserved for situations where the continuum approach or the statistical methods are inadequate to compute from first principles important flow quantities. Slip-boundary conditions for liquid flows in extremely small devices are such a case, as will be discussed in the following section.

An alternative to the deterministic molecular dynamics is the statistical approach where the goal is to compute the probability of finding a molecule at a particular position and state. If the appropriate conservation equation can be solved for the probability distribution, important statistical properties such as the mean number, momentum or energy of the molecules within an element of volume can be computed from a simple weighted averaging. In a practical problem, it is such average quantities that concern us rather than the detail for every single molecule. Clearly, however, the accuracy of computing average quantities, via the statistical approach, improves as the number of molecules in the sampled volume increases. The kinetic theory of dilute gases is well advanced, but that for dense gases and liquids is much less so due to the extreme complexity of having to include multiple collisions and intermolecular forces in the theoretical formulation. The statistical approach is well covered in books such as those by Kennard (1938), Hirschfelder et al. (1954), Schaaf and Chambré (1961), Vincenti and Kruger (1965), Kogan (1969), Chapman and Cowling (1970), Cercignani (1988; 2000) and Bird (1994), and review articles such as those by Kogan (1973), Muntz (1989) and Oran et al. (1998).

In the statistical approach, the fraction of molecules in a given location and state is the sole dependent variable. The independent variables for monatomic molecules are time, the three spatial coordinates and the three components of molecular velocity. Those describe a six-dimensional phase space.[4] For diatomic or polyatomic molecules, the dimension of phase space is increased by the number of internal degrees of freedom. Orientation adds an extra dimension for molecules that are not spherically symmetric. Finally, for mixtures of gases, separate probability distribution functions are required for each species. Clearly, the complexity of the approach increases dramatically as the dimension of phase space increases. The simplest problems are, for example, those for steady, one-dimensional flow of a simple monatomic gas.

To simplify the problem we restrict the discussion here to monatomic gases having no internal degrees of freedom. Furthermore, the fluid is restricted to dilute gases, and molecular chaos is assumed. The former restriction requires the average distance between molecules, δ, to be an order of magnitude larger than their diameter, σ. That will almost guarantee that all collisions between molecules are binary collisions, avoiding the complexity of modeling multiple encounters.[5] The molecular chaos restriction improves the accuracy of computing the macroscopic quantities from the microscopic information. In essence, the volume over which averages are computed has to have sufficient number of molecules to reduce statistical errors. It can be shown that computing macroscopic flow properties by averaging over a number of molecules will result in statistical fluctuations with a standard deviation of approximately 0.1% if one million molecules are used and around 3% if one thousand molecules are used. The molecular chaos limit requires the length scale L for the averaging process to be at least 100 times the average distance between molecules (i.e., typically averaging over at least one million molecules).

Figure 4.7, adapted from Bird (1994), shows the limits of validity of the dilute gas approximation ($\delta/\sigma > 7$), the continuum approach ($Kn < 0.1$, as discussed previously) and the neglect of statistical fluctuations ($L/\delta > 100$). Using a molecular diameter of $\sigma = 4 \times 10^{-10}$ m as an example, the three limits are conveniently expressed as functions of the normalized gas density ρ/ρ_o or number density n/n_o, where the reference densities ρ_o and n_o are computed at standard conditions. All three limits are straight lines in the log–log plot of L vs. ρ/ρ_o, as depicted in Figure 4.7. Note the shaded triangular wedge inside which both the Boltzmann and Navier–Stokes equations are valid. Additionally, the lines describing the three limits very nearly intersect at a single point. As a consequence, the continuum breakdown limit always lies between the dilute gas limit and the limit for molecular chaos. As density or characteristic dimension is reduced in a dilute gas, the Navier–Stokes model breaks down before the level of statistical fluctuations becomes significant. In a dense gas, on the other hand, significant fluctuations may be present even when the Navier–Stokes model is still valid.

The starting point in statistical mechanics is the Liouville equation, which expresses the conservation of the N-particle distribution function in $6N$-dimensional phase space,[6] where N is the number of particles

---

[4] The evolution equation of the probability distribution is considered; hence, time is the seventh independent variable.
[5] Dissociation and ionization phenomena involve triple collisions and therefore require separate treatment.
[6] Three positions and three velocities for *each* molecule of a monatomic gas with no internal degrees of freedom.

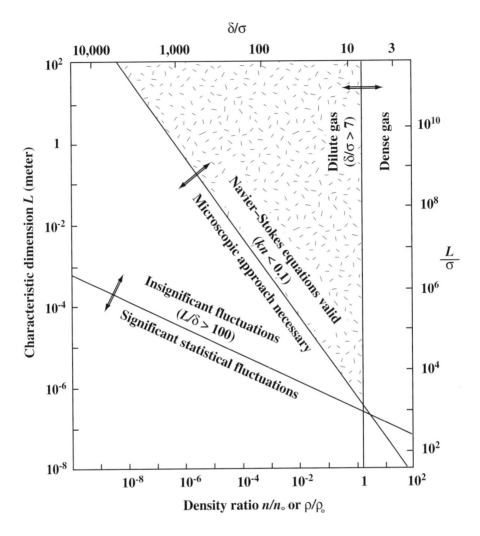

FIGURE 4.7 Effective limits of different flow models. (From Bird, G.A. (1994) *Molecular Gas Dynamics and the Direct Stimulation of Gas Flows*, Clarendon Press, Oxford. With permission.)

under consideration. Considering only external forces that do not depend on the velocity of the molecules,[7] the Liouville equation for a system of N mass points reads:

$$\frac{\partial \mathscr{F}}{\partial t} + \sum_{k=1}^{N} \vec{\xi}_k \cdot \frac{\partial \mathscr{F}}{\partial \vec{x}_k} + \sum_{k=1}^{N} \vec{F}_k \cdot \frac{\partial \mathscr{F}}{\partial \vec{\xi}_k} = 0 \qquad (4.51)$$

where $\mathscr{F}$ is the probability of finding a molecule at a particular point in phase space, t is time, $\vec{\xi}_k$ is the three-dimensional velocity vector for the kth molecule, $\vec{x}_k$ is the three-dimensional position vector for the kth molecule and $\vec{F}$ is the external force vector. Note that the dot product in the above equation is carried out over each of the three components of the vectors $\vec{\xi}$, $\vec{x}$ and $\vec{F}$, and that the summation is over all molecules. Obviously such an equation is not tractable for realistic number of particles.

A hierarchy of reduced distribution functions may be obtained by repeated integration of the Liouville equation above. The final equation in the hierarchy is for the single-particle distribution, which also

---

[7] This excludes Lorentz forces, for example.

Flow Physics

involves the two-particle distribution function. Assuming molecular chaos, that final equation becomes a closed one (i.e., one equation in one unknown) and is known as the Boltzmann equation, the fundamental relation of the kinetic theory of gases. That final equation in the hierarchy is the only one that carries any hope of obtaining analytical solutions.

A simpler direct derivation of the Boltzmann equation is provided by Bird (1994). For monatomic gas molecules in binary collisions, the integro-differential Boltzmann equation reads:

$$\frac{\partial(nf)}{\partial t} + \xi_j \frac{\partial(nf)}{\partial x_j} + F_j \frac{\partial(nf)}{\partial \xi_j} = J(f, f^*), \quad j = 1, 2, 3 \qquad (4.52)$$

where nf is the product of the number density and the normalized velocity distribution function ($dn/n = f d\vec{\xi}$); x_j and ξ_j are, respectively, the coordinates and speeds of a molecule;[8] F_j is a known external force; $J(f, f^*)$ is the nonlinear collision integral that describes the net effect of populating and depopulating collisions on the distribution function. The collision integral is the source of difficulty in obtaining analytical solutions to the Boltzmann equation, and is given by:

$$J(f, f^*) = \int_{-\infty}^{\infty} \int_0^{4\pi} n^2 (f f_1^* - f f_1) \vec{\xi}_r \sigma d\Omega (d\vec{\xi})_1 \qquad (4.53)$$

where the superscript $*$ indicates post-collision values, f and f_1 represent two different molecules, $\vec{\xi}_r$ is the relative speed between two molecules, σ is the molecular cross-section, Ω is the solid angle, and $d\vec{\xi} = d\xi_1 d\xi_2 d\xi_3$.

Once a solution for f is obtained, macroscopic quantities, such as density, velocity, temperature, etc., can be computed from the appropriate weighted integral of the distribution function. For example,

$$\rho = mn = m \int (nf) d\vec{\xi} \qquad (4.54)$$

$$u_i = \int \xi_i f d\vec{\xi} \qquad (4.55)$$

$$\frac{3}{2} kT = \int \frac{1}{2} m \xi_i \xi_i f d\vec{\xi} \qquad (4.56)$$

If the Boltzmann equation is nondimensionalized with a characteristic length L and characteristic speed $[2(k/m)T]^{1/2}$, where k is the Boltzmann constant, m is the molecular mass, and T is temperature, the inverse Knudsen number appears explicitly in the right-hand side of the equation as follows:

$$\frac{\partial \hat{f}}{\partial \hat{t}} + \hat{\xi}_j \frac{\partial \hat{f}}{\partial \hat{x}_j} + \hat{F}_j \frac{\partial \hat{f}}{\partial \hat{\xi}_j} = \frac{1}{Kn} \hat{J}(\hat{f}, \hat{f}^*), \quad j = 1, 2, 3 \qquad (4.57)$$

where the topping symbol $\hat{}$ represents a dimensionless variable, and $\hat{f}$ is nondimensionalized using a reference number density n_o.

The five conservation equations for the transport of mass, momentum and energy can be derived by multiplying the Boltzmann equation above by, respectively, the molecular mass, momentum and energy, then integrating over all possible molecular velocities. Subject to the restrictions of dilute gas and molecular chaos stated earlier, the Boltzmann equation is valid for all ranges of Knudsen number from 0 to ∞. Analytical solutions to this equation for arbitrary geometries are difficult mostly because of the

---

[8] Constituting, together with time, the seven independent variables of the single-dependent-variable equation.

nonlinearity of the collision integral. Simple models of this integral have been proposed to facilitate analytical solutions; see, for example, Bhatnagar et al. (1954).

There are two important asymptotes to Eq. (4.57). First, as $Kn \to \infty$, molecular collisions become unimportant. This is the free-molecule-flow regime depicted in Figure 4.2 for $Kn > 10$, where the only important collision is that between a gas molecule and the solid surface of an obstacle or a conduit. Analytical solutions are then possible for simple geometries, and numerical simulations for complicated geometries are straightforward once the surface-reflection characteristics are accurately modeled. Second, as $Kn \to 0$, collisions become important and the flow approaches the continuum regime of conventional fluid dynamics. The Second Law specifies a tendency for thermodynamic systems to revert to equilibrium state, smoothing out any discontinuities in macroscopic flow quantities. The number of molecular collisions in the limit $Kn \to 0$ is so large that the flow approaches the equilibrium state in a time short compared to the macroscopic time scale. For example, for air at standard conditions ($T = 288$ K; $p = 1$ atm), each molecule experiences, on the average, 10 collisions per nanosecond and travels 1 μm in the same time period. Such a molecule has already *forgotten* its previous state after 1 ns. In a particular flowfield, if the macroscopic quantities vary little over a distance of 1 μm or over a time interval of 1 ns, the flow of standard temperature and pressure (STP) air is near equilibrium.

At $Kn = 0$, the velocity distribution function is everywhere of the local equilibrium or Maxwellian form:

$$\widehat{f}^{(0)} = \frac{n}{n_o} \pi^{-3/2} \exp[-(\widehat{\xi} - \widehat{u})^2] \qquad (4.58)$$

where $\widehat{\xi}$ and $\widehat{u}$ are, respectively, the dimensionless speeds of a molecule and the flow. In this Knudsen number limit, the velocity distribution of each element of the fluid instantaneously adjusts to the equilibrium thermodynamic state appropriate to the local macroscopic properties as this molecule moves through the flowfield. From the continuum viewpoint, the flow is isentropic and heat conduction and viscous diffusion and dissipation vanish from the continuum conservation relations.

The Chapman–Enskog theory attempts to solve the Boltzmann equation by considering a small perturbation of $\widehat{f}$ from the equilibrium Maxwellian form. For small Knudsen numbers, the distribution function can be expanded in terms of Kn in the form of a power series:

$$\widehat{f} = \widehat{f}^{(0)} + Kn \widehat{f}^{(1)} + Kn^2 \widehat{f}^{(2)} + \cdots \qquad (4.59)$$

By substituting the above series in the Boltzmann Eq. (4.57) and equating terms of equal order, the following recurrent set of integral equations result:

$$\begin{aligned}\widehat{J}(\widehat{f}^{(0)}, \widehat{f}^{(0)}) &= 0 \\ \widehat{J}(\widehat{f}^{(0)}, \widehat{f}^{(0)}) &= \frac{\partial \widehat{f}^{(0)}}{\partial \widehat{t}} + \widehat{\xi}_j \frac{\partial \widehat{f}^{(0)}}{\partial \widehat{x}_j} + \widehat{F}_j \frac{\partial \widehat{f}^{(0)}}{\partial \widehat{\xi}_j}, \cdots\end{aligned} \qquad (4.60)$$

The first integral is nonlinear and its solution is the local Maxwellian distribution, Eq. (4.58). The distribution functions $\widehat{f}^{(1)}, \widehat{f}^{(2)}$, etc. each satisfies an inhomogeneous linear equation whose solution leads to the transport terms needed to close the continuum equations appropriate to the particular level of approximation. The continuum stress tensor and heat flux vector can be written in terms of the distribution function, which in turn can be specified in terms of the macroscopic velocity and temperature and their derivatives [Kogan, 1973]. The zeroth-order equation yields the Euler equations, the first-order equation results in the linear transport terms of the Navier–Stokes equations, the second-order equation gives the nonlinear transport terms of the Burnett equations, and so on. Keep in mind, however, that the Boltzmann equation as developed in this section is for a monatomic gas. This excludes the all important air which is composed largely of diatomic nitrogen and oxygen.

As discussed earlier, the Navier–Stokes equations can and should be used up to a Knudsen number of 0.1. Beyond that, the transition flow regime commences ($0.1 < Kn < 10$). In this flow regime, the molecular mean free path for a gas becomes significant relative to a characteristic distance for important flow-property changes to take place. The Burnett equations can be used to obtain analytical/numerical solutions for at least a portion of the transition regime for a monatomic gas, although their complexity has precluded much results for realistic geometries [Agarwal et al., 1999]. A certain degree of uncertainty also exists about the proper boundary conditions to use with the continuum Burnett equations, and experimental validation of the results has been very scarce. Additionally, as the gas flow further departs from equilibrium, the bulk viscosity ($= \lambda + \frac{2}{3}\mu$, where λ is the second coefficient of viscosity) is no longer zero, and Stokes' hypothesis no longer holds (see Gad-el-Hak, 1995, for an interesting summary of the issue of bulk viscosity).

In the transition regime, the molecularly based Boltzmann equation cannot easily be solved either, unless the nonlinear collision integral is simplified. So, clearly the transition regime is one in dire need of alternative methods of solution. MD simulations as mentioned earlier are not suited for dilute gases. The best approach for the transition regime right now is the direct simulation Monte Carlo (DSMC) method developed by Bird (1963; 1965; 1976; 1978; 1994) and briefly described below. Some recent reviews of DSMC include those by Muntz (1989), Cheng (1993), Cheng and Emmanuel (1995) and Oran et al. (1998). The mechanics as well as the history of the DSMC approach and its ancestors are well described in the book by Bird (1994).

Unlike molecular dynamics simulations, DSMC is a statistical computational approach to solving rarefied gas problems. Both approaches treat the gas as discrete particles. Subject to the dilute gas and molecular chaos assumptions, the direct simulation Monte Carlo method is valid for all ranges of Knudsen number, although it becomes quite expensive for $Kn < 0.1$. Fortunately, this is the continuum regime where the Navier–Stokes equations can be used analytically or computationally. DSMC is therefore ideal for the transition regime ($0.1 < Kn < 10$), where the Boltzmann equation is difficult to solve. The Monte Carlo method is, like its namesake, a random number strategy based directly on the physics of the individual molecular interactions. The idea is to track a large number of randomly selected, statistically representative particles and to use their motions and interactions to modify their positions and states. The primary approximation of the direct simulation Monte Carlo method is to uncouple the molecular motions and the intermolecular collisions over small time intervals. A significant advantage of this approximation is that the amount of computation required is proportional to N, in contrast to N^2 for molecular dynamics simulations. In essence, particle motions are modeled deterministically while collisions are treated probabilistically, each simulated molecule representing a large number of actual molecules. Typical computer runs of DSMC in the 1990s involve tens of millions of intermolecular collisions and fluid–solid interactions.

The DSMC computation is started from some initial condition and followed in small time steps that can be related to physical time. Colliding pairs of molecules in a small geometric cell in physical space are randomly selected after each computational time step. Complex physics such as radiation, chemical reactions and species concentrations can be included in the simulations without the necessity of non-equilibrium thermodynamic assumptions that commonly afflict nonequilibrium continuum-flow calculations. DSMC is more computationally intensive than classical continuum simulations and should therefore be used only when the continuum approach is not feasible.

The DSMC technique is explicit and time marching, and therefore always produces unsteady flow simulations. For macroscopically steady flows, Monte Carlo simulation proceeds until a steady flow is established, within a desired accuracy, at sufficiently large time. The macroscopic flow quantities are then the time average of all values calculated after reaching the steady state. For macroscopically unsteady flows, ensemble averaging of many independent Monte Carlo simulations is carried out to obtain the final results within a prescribed statistical accuracy.

4.8 Liquid Flows

From the continuum point of view, liquids and gases are both fluids obeying the same equations of motion. For incompressible flows, for example, the Reynolds number is the primary dimensionless parameter that determines the nature of the flowfield. True, water, for example, has density and viscosity that are, respectively, three and two orders of magnitude higher than those for air, but if the Reynolds number and geometry are matched, liquid and gas flows should be identical.[9] For MEMS applications, however, we anticipate the possibility of nonequilibrium flow conditions and the consequent invalidity of the Navier–Stokes equations and the no-slip-boundary conditions. Such circumstances can best be researched using the molecular approach. This was discussed for gases earlier, and the corresponding arguments for liquids will be given in the present section. The literature on non-Newtonian fluids in general and polymers in particular is vast (for example, the bibliographic survey by Nadolink and Haigh, 1995, cites over 4900 references on polymer drag reduction alone) and provides a rich source of information on the molecular approach for liquid flows.

Solids, liquids and gases are distinguished merely by the degree of proximity and the intensity of motions of their constituent molecules. In solids, the molecules are packed closely and confined, each hemmed in by its neighbors [Chapman and Cowling, 1970]. Only rarely would one solid molecule slip from its neighbors to join a new set. As the solid is heated, molecular motion becomes more violent and a slight thermal expansion takes place. At a certain temperature that depends on ambient pressure, sufficiently intense motion of the molecules enables them to pass freely from one set of neighbors to another. The molecules are no longer confined but are nevertheless still closely packed, and the substance is now considered a liquid. Further heating of the matter eventually releases the molecules altogether, allowing them to break the bonds of their mutual attractions. Unlike solids and liquids, the resulting gas expands to fill any volume available to it.

Unlike solids, both liquids and gases cannot resist finite shear force without continuous deformation—that is, the definition of a fluid medium. In contrast to the reversible, elastic, static deformation of a solid, the continuous deformation of a fluid resulting from the application of a shear stress results in an irreversible work that eventually becomes random thermal motion of the molecules—viscous dissipation. Around 25 million molecules of STP air are in a 1-μm cube. The same cube would contain around 34 billion molecules of water. So, liquid flows are a continuum even in extremely small devices through which gas flows would not be. The average distance between molecules in the gas example is one order of magnitude higher than the diameter of its molecules, while that for the liquid phase approaches the molecular diameter. As a result, liquids are almost incompressible. Their isothermal compressibility coefficient α and bulk expansion coefficient β are much smaller compared to those for gases. For water, for example, a hundredfold increase in pressure leads to less than 0.5% decrease in volume. Sound speeds through liquids are also high relative to those for gases, and as a result most liquid flows are incompressible,[10] the exception being propagation of ultra-high-frequency sound waves and cavitation phenomena.

The mechanism by which liquids transport mass, momentum and energy must be very different from that for gases. In dilute gases, intermolecular forces play no role, and the molecules spend most of their time in free flight between brief collisions at which instances the direction and speed of the molecules abruptly change. The random molecular motions are responsible for gaseous transport processes. In liquids, on the other hand, the molecules are closely packed though not fixed in one position. In essence, the liquid molecules are always in a collision state. Applying a shear force must create a velocity gradient so that the molecules move relative to one another, *ad infinitum* as long as the stress is applied. For liquids, momentum transport due to the random molecular motion is negligible compared to that due to the intermolecular forces. The straining between liquid molecules causes some to separate from their original neighbors, bringing them into the force field of new molecules. Across the plane of the shear stress, the

---

[9] Barring phenomena unique to liquids such as cavitation, free surface flows, etc.

[10] Note that we distinguish between a fluid and a flow as being compressible or incompressible. For example, the *flow* of the highly compressible air can be either compressible or incompressible.

sum of all intermolecular forces must, on the average, balance the imposed shear. Liquids at rest transmit only normal force, but when a velocity gradient occurs the net intermolecular force would have a tangential component.

The incompressible Navier–Stokes equations describe liquid flows under most circumstances. Liquids, however, do not have a well-advanced molecular-based theory as dilute gases do. The concept of mean free path is not very useful for liquids, and the conditions under which a liquid flow fails to be in quasi-equilibrium state are not well defined. There is no Knudsen number for liquid flows to guide us through the maze. We do not know, from first principles, the conditions under which the no-slip-boundary condition becomes inaccurate, or the point at which the stress–rate-of-strain relation or the heat-flux–temperature-gradient relation fails to be linear. Certain empirical observations indicate that those simple relations that we take for granted occasionally fail to accurately model liquid flows. For example, it has been shown in rheological studies [Loose and Hess, 1989] that non-Newtonian behavior commences when the strain rate approximately exceeds twice the molecular frequency scale:

$$\dot{\gamma} = \frac{\partial u}{\partial y} \geq 2\mathcal{T}^{-1} \quad (4.61)$$

where the molecular time-scale $\mathcal{T}$ is given by:

$$\mathcal{T} = \left[\frac{m\sigma^2}{\varepsilon}\right]^{\frac{1}{2}} \quad (4.62)$$

where m is the molecular mass, and σ and ε are, respectively, the characteristic length and energy scale for the molecules. For ordinary liquids such as water, this time scale is extremely small, and the threshold shear rate for the onset of non-Newtonian behavior is therefore extraordinarily high. For high-molecular-weight polymers, on the other hand, m and σ are both many orders of magnitude higher than their respective values for water, and the linear stress–strain relation breaks down at realistic values of the shear rate.

The moving contact line when a liquid spreads on a solid substrate is an example where slip flow must be allowed to avoid singular or unrealistic behavior in the Navier–Stokes solutions [Dussan and Davis, 1974; Dussan, 1976; 1979; Thompson and Robbins, 1989]. Other examples where slip flow must be admitted include corner flows [Moffatt, 1964; Koplik and Banavar, 1995] and extrusion of polymer melts from capillary tubes [Pearson and Petrie, 1968; Richardson, 1973; Den, 1990].

Existing experimental results of liquid flow in microdevices are contradictory. This is not surprising given the difficulty of such experiments and the lack of a guiding rational theory. Pfahler et al. (1990; 1991), Pfahler (1992) and Bau (1994) summarize the relevant literature. For small-length-scale flows, a phenomenological approach for analyzing the data is to define an *apparent* viscosity μ_a calculated so that if it were used in the traditional no-slip Navier–Stokes equations instead of the fluid viscosity μ, the results would be in agreement with experimental observations. Israelachvili (1986) and Gee et al. (1990) found that $\mu_a = \mu$ for thin-film flows as long as the film thickness exceeds 10 molecular layers (≈ 5 nm). For thinner films, μ_a depends on the number of molecular layers and can be as much as 10^5 times larger than μ. Chan and Horn's (1985) results are somewhat different: The apparent viscosity deviates from the fluid viscosity for films thinner than 50 nm.

In polar-liquid flows through capillaries, Migun and Prokhorenko (1987) report that μ_a increases for tubes smaller than 1 μm in diameter. In contrast, Debye and Cleland (1959) report μ_a smaller than μ for paraffin flow in porous glass with average pore size several times larger than the molecular length scale. Experimenting with microchannels ranging in depths from 0.5 to 50 μm, Pfahler et al. (1991) found that μ_a is consistently smaller than μ for both liquid (isopropyl alcohol, silicone oil) and gas (nitrogen, helium) flows in microchannels. For liquids, the apparent viscosity decreases with decreasing channel depth. Other researchers using small capillaries report that μ_a is about the same as μ [Anderson and Quinn, 1972; Tuckermann and Pease, 1981; 1982; Tuckermann, 1984; Guvenc, 1985; Nakagawa et al., 1990].

Very recently, Sharp (2001) and Sharp et al. (2001) asserted that, despite the significant inconsistencies in the literature regarding liquid flows in microchannels, such flows are well predicted by macroscale continuum theory. A case can be made to the contrary, however, as will be seen at the end of this section, and the final verdict on this controversy is yet to come.

The above contradictory results point to the need for replacing phenomenological models by first-principles ones. The lack of a molecular-based theory of liquids—despite extensive research by the rheology and polymer communities—leaves molecular dynamics simulations as the best weapon in the first-principles arsenal. MD simulations offer a unique approach to checking the validity of the traditional continuum assumptions. However, as was pointed out earlier, such simulations are limited to exceedingly minute flow extent.

Thompson and Troian (1997) provide molecular dynamics simulations to quantify the dependence of the slip-flow-boundary condition on shear rate. Recall the linear Navier boundary condition introduced earlier:

$$\Delta u|_w = u_{\text{fluid}} - u_{\text{wall}} = L_s \frac{\partial u}{\partial y}\bigg|_w \quad (4.63)$$

where L_s is the constant slip length, and $\frac{\partial u}{\partial y}\big|_w$ is the strain rate computed at the wall. The goal of Thompson and Troian's simulations was to determine the degree of slip at a solid–liquid interface as the interfacial parameters and the shear rate change. In their simulations, a simple liquid underwent planar shear in a Couette cell, as shown in Figure 4.8. The typical cell measured $12.51 \times 7.22 \times h$, in units of molecular length scale σ, where the channel depth h varied in the range of $16.71\,\sigma$ to $24.57\,\sigma$, and the corresponding number of molecules simulated ranged from 1152 to 1728. The liquid is treated as an isothermal ensemble of spherical molecules. A shifted Lennard–Jones 6–12 potential is used to model intermolecular interactions, with energy and length scales ε and σ and cut-off distance $r_c = 2.2\sigma$:

$$V(r) = 4\varepsilon\left[\left(\frac{r}{\sigma}\right)^{-12} - \left(\frac{r}{\sigma}\right)^{-6} - \left(\frac{r_c}{\sigma}\right)^{-12} + \left(\frac{r_c}{\sigma}\right)^{-6}\right] \quad (4.64)$$

The truncated potential is set to zero for $r > r_c$.

The fluid–solid interaction is also modeled with a truncated Lennard–Jones potential, with energy and length scales ε^{wf} and σ^{wf} and cut-off distance r_c. The equilibrium state of the fluid is a well-defined liquid phase characterized by number density $n = 0.81\sigma^{-3}$ and temperature $T = 1.1\varepsilon/k$, where k is the Boltzmann constant.

The steady-state velocity profiles resulting from Thompson and Troian's (1997) MD simulations are depicted in Figure 4.8 for different values of the interfacial parameters $\varepsilon^{wf}, \sigma^{wf}$ and n^w. Those parameters, shown in units of the corresponding fluid parameters ε, σ and n, characterize, respectively, the strength of the liquid–solid coupling, the thermal roughness of the interface and the commensurability of wall and liquid densities. The macroscopic velocity profiles recover the expected flow behavior from continuum hydrodynamics with boundary conditions involving varying degrees of slip. Note that when slip exists, the shear rate $\dot{\gamma}$ no longer equals U/h. The degree of slip increases (i.e., the amount of momentum transfer at the wall–fluid interface decreases) as the relative wall density n^w increases or the strength of the wall–fluid coupling σ^{wf} decreases—in other words, when the relative surface energy corrugation of the wall decreases. Conversely, the corrugation is maximized when the wall and fluid densities are commensurate and the strength of the wall–fluid coupling is large. In this case, the liquid *feels* the corrugations in the surface energy of the solid owing to the atomic close packing. Consequently,

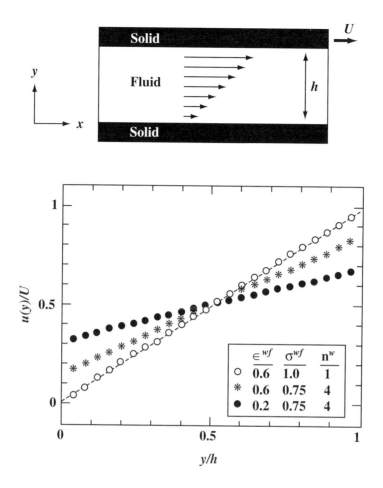

FIGURE 4.8 Velocity profiles in a Couette flow geometry at different interfacial parameters. All three profiles are for $U = \sigma \mathcal{T}^{-1}$, and $h = 24.57\sigma$. The dashed line is the no-slip Couette-flow solution. (From Thompson, P.A., and Troian, S.M. (1997) *Phys. Rev. Lett.* **63**, pp. 766–769. With permission.)

there is efficient momentum transfer and the no-slip condition applies, or in extreme cases, a "stick" boundary condition takes hold.

Variations of the slip length L_s and viscosity μ as functions of shear rate $\dot{\gamma}$ are shown in parts (a) and (b) of Figure 4.9, for five different sets of interfacial parameters. For Couette flow, the slip length is computed from its definition, $\Delta u|_w / \dot{\gamma} = (U/\dot{\gamma} - h)/2$. The slip length, viscosity and shear rate are normalized in the figure using the respective molecular scales for length σ, viscosity $\varepsilon \mathcal{T} \sigma^{-3}$ and inverse time $\mathcal{T}^{-1}$. The viscosity of the fluid is constant over the entire range of shear rates (Figure 4.9b), indicating Newtonian behavior. As indicated earlier, non-Newtonian behavior is expected for $\dot{\gamma} \geq 2\mathcal{T}^{-1}$, well above the shear rates used in Thompson and Troian's simulations.

At low shear rates, the slip length behavior is consistent with the Navier model (i.e., is independent of the shear rate). Its limiting value L_s^0 ranges from 0 to ~17 σ for the range of interfacial parameters chosen (Figure 4.9a). In general, the amount of slip increases with decreasing surface-energy corrugation. Most interestingly, at high shear rates, the Navier condition breaks down as the slip length increases rapidly with $\dot{\gamma}$. The critical shear-rate value for the slip length to diverge, $\dot{\gamma}_c$, decreases as the surface energy corrugation decreases. Surprisingly, the boundary condition is nonlinear even though the liquid is still Newtonian. In dilute gases, the linear slip condition and the Navier–Stokes equations, with their linear stress–strain relation, are both valid to the same order of approximation in Knudsen number. In other

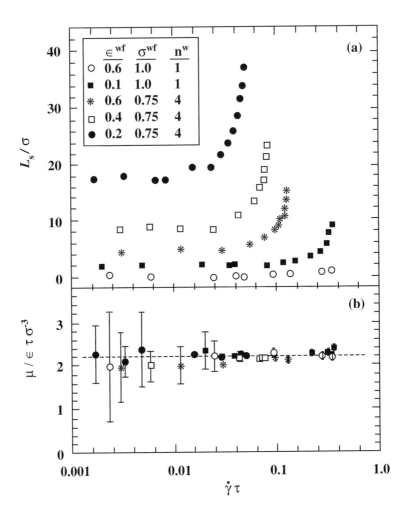

FIGURE 4.9 Variation of slip length and viscosity as functions of shear rate. (From Thompson, P.A., and Troian, S.M. (1997) *Phys. Rev. Lett.* **63**, pp. 766–769. With permission.)

words, deviation from linearity is expected to take place at the same value of $Kn = 0.1$. In liquids, in contrast, the slip length appears to become nonlinear and to diverge at a critical value of shear rate well below the shear rate at which the linear stress–strain relation fails. Moreover, the boundary condition deviation from linearity is not gradual but is rather catastrophic. The critical value of shear rate $\dot{\gamma}_c$ signals the point at which the solid can no longer impart momentum to the liquid. This means that the same liquid molecules sheared against different substrates will experience varying amounts of slip and vice versa.

Based on the above results, Thompson and Troian (1997) suggest a universal boundary condition at a solid–liquid interface. Scaling the slip length L_s by its asymptotic limiting value L_s^o and the shear rate $\dot{\gamma}$ by its critical value $\dot{\gamma}_c$ collapses the data in the single curve shown in Figure 4.10. The data points are well described by the relation:

$$L_s = L_s^o \left[1 - \frac{\dot{\gamma}}{\dot{\gamma}_c} \right]^{-\frac{1}{2}} \quad (4.65)$$

The nonlinear behavior close to a critical shear rate suggests that the boundary condition can significantly affect flow behavior at macroscopic distances from the wall. Experiments with polymers confirm this

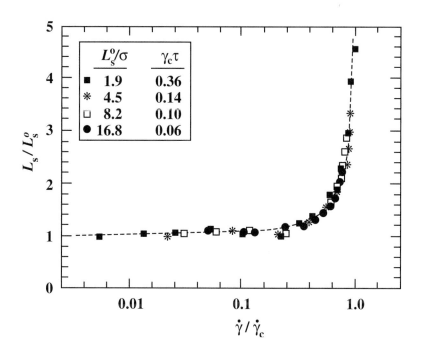

FIGURE 4.10 Universal relation of slip length as a function of shear rate. (From Thompson, P.A., and Troian, S.M. (1997) *Phys. Rev. Lett.* **63**, pp. 766–769. With permission.)

observation [Atwood and Schowalter, 1989]. The rapid change in the slip length suggests that, for flows in the vicinity of $\dot{\gamma}_c$, small changes in surface properties can lead to large fluctuations in the apparent boundary condition. Thompson and Troian (1997) conclude that the Navier slip condition is but the low-shear-rate limit of a more generalized universal relationship which is nonlinear and divergent. Their relation provides a mechanism for relieving the stress singularity in spreading contact lines and corner flows, as it naturally allows for varying degrees of slip on approach to regions of higher rate of strain.

To place the above results in physical terms, consider water[11] at a temperature of $T = 288$ K. The energy scale in the Lennard–Jones potential is then $\varepsilon = 3.62 \times 10^{-21}$ J. For water, $m = 2.99 \times 10^{-26}$ kg, $\sigma = 2.99 \times 10^{-10}$ m, and at standard temperature $n = 3.35 \times 10^{28}$ molecules/m^3. The molecular time scale can thus be computed as $\mathcal{T} = [m\sigma^2/\varepsilon]^{1/2} = 8.31 \times 10^{-13}$ s. For the third case depicted in Figure 4.10 (the open squares), $\dot{\gamma}_c \mathcal{T} = 0.1$, and the critical shear rate at which the slip condition diverges is thus $\dot{\gamma}_c = 1.2 \times 10^{11}$ s^{-1}. Such an enormous rate of strain[12] may be found in extremely small devices having extremely high speeds. On the other hand, the conditions to achieve a measurable slip of 17σ (the solid circles in Figure 4.9) are not difficult to encounter in microdevices: a density of the solid four times that of the liquid and an energy scale for the wall–fluid interaction that is one fifth the energy scale for the liquid.

The limiting value of slip length is independent of the shear rate and can be computed for water as $L_s^o = 17\sigma = 4.91 \times 10^{-9}$ m. Consider a water microbearing having a shaft diameter of 100 μm, rotation rate of 20,000 rpm and minimum gap of $h = 1$ μm. In this case, $U = 0.1$ m/s and the no-slip shear rate is $U/h = 10^5$ s^{-1}. When slip occurs at the limiting value just computed, the shear rate and the wall slip

---

[11] Water molecules are complex ones, forming directional, short-range covalent bonds, thus requiring a more complex potential than the Lennard–Jones (L–J) to describe the intermolecular interactions. For the purpose of the qualitative example described here, however, we use the computational results of Thompson and Troian (1997) who employed the L–J potential.

[12] Note, however, that $\dot{\gamma}_c$ for high-molecular-weight polymers would be many orders of magnitude smaller than the value developed here for water.

velocity are computed as follows:

$$\dot{\gamma} = \frac{U}{h + 2L_s^o} = 9.90 \times 10^4 \text{ s}^{-1} \tag{4.66}$$

$$\Delta u|_w = \dot{\gamma} L_s = 4.87 \times 10^{-4} \text{ m/s} \tag{4.67}$$

As a result of the Navier slip, the shear rate is reduced by 1% from its no-slip value, and the slip velocity at the wall is about 0.5% of U, small but not insignificant.

4.9 Surface Phenomena

The surface-to-volume ratio for a machine with a characteristic length of 1 m is 1 m^{-1}, while that for a MEMS device having a size of 1 µm is 10^6 m^{-1}. The million-fold increase in surface area relative to the mass of the minute device substantially affects the transport of mass, momentum and energy through the surface. Obviously surface effects dominate in small devices. The surface boundary conditions in MEMS flows have already been discussed earlier. In microdevices, it has been shown that it is possible to have measurable slip-velocity and temperature jump at a solid–fluid interface. In this section, we illustrate other ramifications of the large surface-to-volume ratio unique to MEMS, and provide a molecular viewpoint to surface forces.

In microdevices, both radiative and convective heat loss/gain are enhanced by the huge surface-to-volume ratio. Consider a device having a characteristic length L_s. Use of the lumped capacitance method to compute the rate of convective heat transfer, for example, is justified if the Biot number ($\equiv hL_s/\kappa_s$, where h is the convective heat transfer coefficient of the fluid and κ_s is the thermal conductivity of the solid) is less than 0.1. Small L_s implies small Biot number and a nearly uniform temperature within the solid. Within this approximation, the rate at which heat is lost to the surrounding fluid is given by:

$$\rho_s L_s^3 c_s \frac{dT}{dt} = -hL_s^2 (T_s - T_\infty) \tag{4.68}$$

where ρ_s and c_s are, respectively, the density and specific heat of the solid; T_s is its (uniform) temperature; T_∞ is the ambient fluid temperature. Solution of the above equation is trivial, and the temperature of a hot surface drops exponentially with time from an initial temperature T_i,

$$\frac{T_s(t) - T_\infty}{T_i - T_\infty} = \exp\left[-\frac{t}{\mathcal{T}}\right] \tag{4.69}$$

where the time constant $\mathcal{T}$ is given by:

$$\mathcal{T} = \frac{\rho_s L_s^3 c_s}{hL_s^2} \tag{4.70}$$

For small devices, the time it takes the solid to cool down is proportionally small. Clearly, the million-fold increase in surface-to-volume ratio implies a proportional increase in the rate at which heat escapes. Identical scaling arguments can be made regarding mass transfer.

Another effect of the diminished scale is the increased importance of surface forces and the waning importance of body forces. Based on biological studies, Went (1968) concludes that the demarkation length scale is around 1 mm. Below that, surface forces dominate over gravitational forces. A 10-mm piece of paper will fall down when gently placed on a smooth, vertical wall, while a 0.1-mm piece will stick. Try it! *Stiction* is a major problem in MEMS applications. Certain structures such as long, thin polysilicon beams and large, thin comb drives have a propensity to stick to their substrates and thus fail to perform as designed [Mastrangelo and Hsu, 1992; Tang et al., 1989].

Conventional dry friction between two solids in relative motion is proportional to the normal force, which is usually a component of the moving device weight. The friction is independent of the contact-surface area because the van der Waals cohesive forces are negligible relative to the weight of the macroscopic device. In MEMS applications, the cohesive intermolecular forces between two surfaces are significant and the stiction is independent of the device mass but is proportional to its surface area. The first micromotor did not move—despite large electric current through it—until the contact area between the 100-μm rotor and the substrate was reduced significantly by placing dimples on the rotor's surface [Fan et al., 1988; 1989; Tai and Muller, 1989].

One last example of surface effects that, to my knowledge, has not been investigated for microflows is the adsorbed layer in gaseous wall-bounded flows. It is well known [Brunauer, 1944; Lighthill, 1963] that when a gas flows in a duct, the gas molecules are attracted to the solid surface by the van der Waals and other forces of cohesion. The potential energy of the gas molecules drops on reaching the surface. The adsorbed layer partakes the thermal vibrations of the solid, and the gas molecules can only escape when their energy exceeds the potential energy minimum. In equilibrium, at least part of the solid would be covered by a monomolecular layer of adsorbed gas molecules. Molecular species with significant partial pressure—relative to their vapor pressure—may locally form layers two or more molecules thick. Consider, for example, the flow of a mixture of dry air and water vapor at STP. The energy of adsorption of water is much larger than that for nitrogen and oxygen, making it more difficult for water molecules to escape the potential energy trap. It follows that the lifetime of water molecules in the adsorbed layer significantly exceeds that for the air molecules (by 60,000-fold, in fact) and, as a result, the thin surface layer would be mostly water. For example, if the proportion of water vapor in the ambient air is 1:1,000 (i.e., very low humidity level), the ratio of water to air in the adsorbed layer would be 60:1. Microscopic roughness of the solid surface causes partial condensation of the water along portions having sufficiently strong concave curvature. So, surfaces exposed to non-dry-air flows are mainly liquid water surfaces. In most applications, this thin adsorbed layer has little effect on the flow dynamics, despite the fact that the density and viscosity of liquid water are far greater than those for air. In MEMS applications, however, the layer thickness may not be an insignificant portion of the characteristic flow dimension and the water layer may have a measurable effect on the gas flow. A hybrid approach of molecular dynamics and continuum flow simulations or MD–Monte Carlo simulations may be used to investigate this issue.

It should be noted that recently Majumdar and Mezic (1998; 1999) have studied the stability and rupture into droplets of thin liquid films on solid surfaces. They point out that the free energy of a liquid film consists of a surface tension component as well as highly nonlinear volumetric intermolecular forces resulting from van der Waals, electrostatic, hydration and elastic strain interactions. For water films on hydrophilic surfaces such as silica and mica, Majumdar and Mezic (1998) estimate the equilibrium film thickness to be about 0.5 nm (2 monolayers) for a wide range of ambient-air relative humidities. The equilibrium thickness grows very sharply, however, as the relative humidity approaches 100%.

Majumdar and Mezic's (1998; 1999) results open many questions. What are the stability characteristics of their water film in the presence of air flow above it? Would this water film affect the accommodation coefficient for microduct air flow? In a modern Winchester-type hard disk, the drive mechanism has a read/write head that floats 50 nm above the surface of the spinning platter. The head and platter together with the air layer in between form a slider bearing. Would the computer performance be affected adversely by the high relative humidity on a particular day when the adsorbed water film is no longer "thin"? If a microduct hauls liquid water, would the water film adsorbed by the solid walls influence the effective viscosity of the water flow? Electrostatic forces can extend to almost 1 μm (the Debye length), and that length is known to be highly pH dependent. Would the water flow be influenced by the surface and liquid chemistry? Would this explain the contradictory experimental results of liquid flows in microducts discussed earlier?

The few examples above illustrate the importance of surface effects in small devices. From the continuum viewpoint, forces at a solid–fluid interface are the limit of pressure and viscous forces acting on a parallel elementary area displaced into the fluid, when the displacement distance is allowed to tend to zero. From the molecular point of view, all macroscopic surface forces are ultimately traced to intermolecular forces, a subject extensively covered in the book by Israelachvilli (1991) and references therein. Here we provide a very brief introduction to the molecular viewpoint. The four forces in nature are (1) the strong and (2) weak forces describing the interactions among neutrons, protons, electrons, etc.; (3) the electromagnetic forces between atoms and molecules; and (4) gravitational forces between masses. The range of action of the first two forces is around 10^{-5} nm; hence, neither concerns us overly in MEMS applications. The electromagnetic forces are effective over a much larger though still small distance on the order of the interatomic separations (0.1 to 0.2 nm). Effects over longer range—several orders of magnitude longer—can and do rise from the short-range intermolecular forces. For example, the rise of liquid column in capillaries and the action of detergent molecules in removing oily dirt from fabric are the result of intermolecular interactions. Gravitational forces decay with the distance to second power, while intermolecular forces decay much quicker, typically with the seventh power. Cohesive forces are therefore negligible once the distance between molecules exceeds a few molecular diameters, while massive bodies such as stars and planets are still strongly interacting, via gravity, over astronomical distances.

Electromagnetic forces are the source of all intermolecular interactions and the cohesive forces holding atoms and molecules together in solids and liquids. They can be classified into (1) purely electrostatic arising from the Coulomb force between charges, interactions between charges, permanent dipoles, quadrupoles, etc.; (2) polarization forces arising from the dipole moments induced in atoms and molecules by the electric field of nearby charges and permanent dipoles; and (3) quantum mechanical forces that give rise to covalent or chemical bonding and to repulsive steric or exchange interactions that balance the attractive forces at very short distances. The Hellman–Feynman theorem of quantum mechanics states that once the spatial distribution of the electron clouds has been determined by solving the appropriate Schrödinger equation, intermolecular forces may be calculated on the basis of classical electrostatics, in effect reducing all intermolecular forces to Coulombic forces. Note, however, that intermolecular forces exist even when the molecules are totally neutral. Solutions of the Schrödinger equation for general atoms and molecules are not easy, of course, and alternative modeling are sought to represent intermolecular forces. The van der Waals attractive forces are usually represented with a potential that varies as the inverse sixth power of distance, while the repulsive forces are represented with either a power or an exponential potential.

A commonly used potential between two molecules is the generalized Lennard–Jones (L–J 6–12) pair potential given by:

$$V_{ij}(r) = 4\varepsilon \left[c_{ij}\left(\frac{r}{\sigma}\right)^{-12} - d_{ij}\left(\frac{r}{\sigma}\right)^{-6} \right] \qquad (4.71)$$

where V_{ij} is the potential energy between two particles i and j; r is the distance between the two molecules; ε and σ are, respectively, characteristic energy and length scales; and c_{ij} and d_{ij} are parameters to be chosen for the particular fluid and solid combinations under consideration. The first term in the right-hand side is the strong repulsive force that is felt when two molecules are at extremely close range comparable to the molecular length scale. That short-range repulsion prevents overlap of the molecules in physical space. The second term is the weaker, van der Waals attractive force that commences when the molecules are sufficiently close (several times σ). That negative part of the potential represents the attractive polarization interaction of neutral, spherically symmetric particles. The power of 6 associated with this term is derivable from quantum mechanics considerations, while the power of the repulsive part of the potential is found empirically. The Lennard–Jones potential is zero at very large distances, has a weak negative peak at r slightly larger than σ, is zero at $r = \sigma$ and is infinite as $r \to 0$.

The force field resulting from this potential is given by:

$$F_{ij}(r) = -\frac{\partial V_{ij}}{\partial r} = \frac{48\varepsilon}{\sigma}\left[c_{ij}\left(\frac{r}{\sigma}\right)^{-13} - \frac{d_{ij}}{2}\left(\frac{r}{\sigma}\right)^{-7} \right] \qquad (4.72)$$

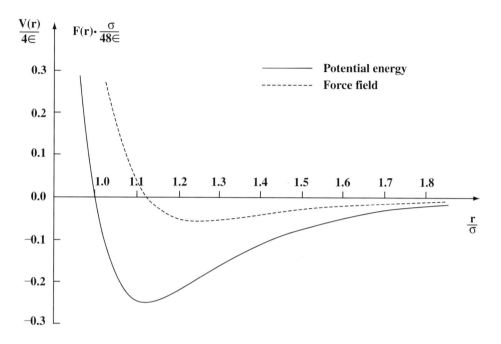

FIGURE 4.11 Typical Lennard–Jones 6–12 potential and the intermolecular force field resulting from it. Only a small portion of the potential function is shown for clarity. (From Gad-el-Hak, M. (1999) *J. Fluids Eng.* **121**, pp. 5–33, ASME, New York. With permission.)

A typical L–J 6–12 potential and force field are shown in Figure 4.11, for $c = d = 1$. The minimum potential $V_{min} = -\varepsilon$ corresponds to the equilibrium position (zero force) and occurs at $r = 1.12\sigma$. The attractive van der Waals contribution to the minimum potential is -2ε, while the repulsive energy contribution is $+\varepsilon$. Thus, the inverse 12th-power repulsive force term decreases the strength of the binding energy at equilibrium by 50%.

The L–J potential is commonly used in molecular dynamics simulations to model intermolecular interactions between dense gas or liquid molecules and between fluid and solid molecules. As mentioned earlier, such potential is not accurate for complex substances such as water whose molecules form directional covalent bonds. As a result, MD simulations for water are much more involved.

4.10 Parting Remarks

The 40-year-old vision of Richard Feynman of building minute machines is now a reality. Microelectromechanical systems have witnessed explosive growth during the last decade and are finding increased applications in a variety of industrial and medical fields. The physics of fluid flows in microdevices has been explored in this chapter. While we now know considerably more than we did just few years ago, much physics remains to be explored so that rational tools can be developed for the design, fabrication and operation of MEMS devices.

The traditional Navier–Stokes model of fluid flows with no-slip-boundary conditions works only for a certain range of the governing parameters. This model basically demands two conditions: (1) The fluid is a continuum, which is almost always satisfied as there are usually more than 1 million molecules in the smallest volume in which appreciable macroscopic changes take place; this is the molecular chaos restriction. (2) The flow is not too far from thermodynamic equilibrium, which is satisfied if there is a sufficient number of molecular encounters during a time period that is small compared to the smallest time scale for flow changes. During this time period the average molecule would have moved a distance small compared to the smallest flow length scale.

For gases, the Knudsen number determines the degree of rarefaction and the applicability of traditional flow models. As $Kn \to 0$, the time and length scales of molecular encounters are vanishingly small compared to those for the flow, and the velocity distribution of each element of the fluid instantaneously adjusts to the equilibrium thermodynamic state appropriate to the local macroscopic properties as this molecule moves through the flowfield. From the continuum viewpoint, the flow is isentropic and heat conduction and viscous diffusion and dissipation vanish from the continuum conservation relations, leading to the Euler equations of motion. At small but finite Kn, the Navier–Stokes equations describe near-equilibrium, continuum flows.

Slip flow must be taken into account for $Kn > 0.001$. The slip-boundary condition is at first linear in Knudsen number, then nonlinear effects take over beyond a Knudsen number of 0.1. At the same transition regime ($0.1 < Kn < 10$), the linear-stress–rate-of-strain and heat-flux–temperature-gradient relations—needed to close the Navier–Stokes equations—also break down, and alternative continuum equations (e.g., Burnett or higher order equations) or molecular-based models must be invoked. In the transition regime, provided that the dilute gas and molecular chaos assumptions hold, solutions to the difficult Boltzmann equation are sought, but physical simulations such as Monte Carlo methods are more readily executed in this range of Knudsen number. In the free-molecule-flow regime ($Kn > 10$), the nonlinear collision integral is negligible and the Boltzmann equation is drastically simplified. Analytical solutions are possible in this case for simple geometries, and numerical integration of the Boltzmann equation is straightforward for arbitrary geometries, provided that the surface-reflection characteristics are accurately modeled.

Gaseous flows are often compressible in microdevices even at low Mach numbers. Viscous effects can cause sufficient pressure drop and density changes for the flow to be treated as compressible. In a long, constant-area microduct, all Knudsen number regimes may be encountered, and the degree of rarefaction increases along the tube. The pressure drop is nonlinear and the Mach number increases downstream, limited only by choked-flow condition.

Similar deviation and breakdown of the traditional Navier–Stokes equations occur for liquids as well, but there the situation is more murky. Existing experiments are contradictory. There is no kinetic theory of liquids, and first-principles prediction methods are scarce. Molecular dynamics simulations can be used, but they are limited to extremely small flow extents. Nevertheless, measurable slip is predicted from MD simulations at realistic shear rates in microdevices.

Much nontraditional physics is still to be learned and many exciting applications of microdevices are yet to be discovered. The future is bright for this emerging field of science and technology. Richard Feynman was right about the possibility of building mite-size machines but was somewhat cautious in forecasting that such machines, while they "would be fun to make," may or may not be useful.

References

Agarwal, R., Yun, K., and Balakrishnan, R. (1999) "Beyond Navier–Stokes: Burnett Equations for Flow Simulations in Continuum–Transition Regime," AIAA Paper No. 99-3580, American Institute of Aeronautics and Astronautics, Reston, VA.

Alder, B.J., and Wainwright, T.E. (1957) "Studies in Molecular Dynamics," *J. Chem. Phys.* **27**, pp. 1208–1209.

Alder, B.J., and Wainwright, T.E. (1958) "Molecular Dynamics by Electronic Computers," in *Transport Processes in Statistical Mechanics*, ed. I. Prigogine, pp. 97–131, Interscience, New York.

Alder, B.J., and Wainwright, T.E. (1970) "Decay of the Velocity Auto-Correlation Function," *Phy. Rev. A* **1**, pp. 18–21.

Allen, M.P., and Tildesley, D.J. (1987) *Computer Simulation of Liquids*, Clarendon Press, Oxford.

Anderson, J.L., and Quinn, J.A. (1972) "Ionic Mobility in Microcapillaries," *J. Chem. Phys.* **27**, pp. 1208–1209.

Arkilic, E.B. (1997) "Measurement of the Mass Flow and Tangential Momentum Accommodation Coefficient in Silicon Micromachined Channels," Ph.D. thesis, Massachusetts Institute of Technology, Cambridge, MA.

Arkilic, E.B., Schmidt, M.A., and Breuer, K.S. (1995) "Slip Flow in Microchannels," in *Rarefied Gas Dynamics*, Vol. 19, ed. J. Harvey and G. Lord, Oxford University Press, Oxford.

Arkilic, E.B., Schmidt, M.A., and Breuer, K.S. (1997a) "Gaseous Slip Flow in Long Microchannels," *J. MEMS* **6**, pp. 167–178.

Arkilic, E.B., Schmidt, M.A., and Breuer, K.S. (1997b) "TMAC Measurement in Silicon Micromachined Channels," in *Rarefied Gas Dynamics*, Vol. 20, ed. C. Shen, Beijing University Press, Beijing, China, 6 pp.

Atwood, B.T., and Schowalter, W.R. (1989) "Measurements of Slip at the Wall during Flow of High-Density Polyethylene through a Rectangular Conduit," *Rheologica Acta* **28**, pp. 134–146.

Bau, H.H. (1994) "Transport Processes Associated with Micro-Devices," *Thermal Sci. Eng.* **2**, pp. 172–178.

Beskok, A. (1994) "Simulation of Heat and Momentum Transfer in Complex Micro-Geometries," M.Sc. thesis, Princeton University, Princeton, NJ.

Beskok, A. (1996) "Simulations and Models of Gas Flows in Microgeometries," Ph.D. thesis, Princeton University, Princeton, NJ.

Beskok, A., and Karniadakis, G.E. (1994) "Simulation of Heat and Momentum Transfer in Complex Micro-Geometries," *J. Thermophys. Heat Transfer* **8**, pp. 355–370.

Beskok, A., and Karniadakis, G.E. (1999) "A Model for Flows in Channels, Pipes and Ducts at Micro and Nano Scales," *Microscale Thermophys. Eng.* **3**, pp. 43–77.

Beskok, A., Karniadakis, G.E., and Trimmer, W. (1996) "Rarefaction and Compressibility Effects in Gas Microflows," *J. Fluids Eng.* **118**, pp. 448–456.

Bhatnagar, P.L., Gross, E.P., and Krook, M. (1954) "A Model for Collision Processes in Gases. I. Small Amplitude Processes in Charged and Neutral One-Component Systems," *Phys. Rev.* **94**, pp. 511–524.

Bird, G.A. (1963) "Approach to Translational Equilibrium in a Rigid Sphere Gas," *Phys. Fluids* **6**, pp. 1518–1519.

Bird, G.A. (1965) "The Velocity Distribution Function within a Shock Wave," *J. Fluid Mech.* **30**, pp. 479–487.

Bird, G.A. (1976) *Molecular Gas Dynamics*, Clarendon Press, Oxford.

Bird, G.A. (1978) "Monte Carlo Simulation of Gas Flows," *Annu. Rev. Fluid Mech.* **10**, pp. 11–31.

Bird, G.A. (1994) *Molecular Gas Dynamics and the Direct Simulation of Gas Flows*, Clarendon Press, Oxford.

Brunauer, S. (1944) *Physical Adsorption of Gases and Vapours*, Oxford University Press, Oxford.

Cercignani, C. (1988) *The Boltzmann Equation and Its Applications*, Springer-Verlag, Berlin.

Cercignani, C. (2000) *Rarefied Gas Dynamics: From Basic Concepts to Actual Calculations*, Cambridge University Press, Cambridge, U.K.

Chan, D.Y.C., and Horn, R.G. (1985) "Drainage of Thin Liquid Films," *J. Chem. Phys.* **83**, pp. 5311–5324.

Chapman, S., and Cowling, T.G. (1970) *The Mathematical Theory of Non-Uniform Gases*, third edition, Cambridge University Press, Cambridge, U.K.

Cheng, H.K. (1993) "Perspectives on Hypersonic Viscous Flow Research," *Annu. Rev. Fluid Mech.* **25**, pp. 455–484.

Cheng, H.K., and Emmanuel, G. (1995) "Perspectives on Hypersonic Nonequilibrium Flow," *AIAA J.* **33**, pp. 385–400.

Ciccotti, G., and Hoover, W.G., eds. (1986) *Molecular Dynamics Simulation of Statistical Mechanics Systems*, North-Holland, Amsterdam.

Debye, P., and Cleland, R.L. (1959) "Flow of Liquid Hydrocarbons in Porous Vycor," *J. Appl. Phys.* **30**, pp. 843–849.

Den, L.M. (1990) "Issues in Viscoelastic Fluid Mechanics," *Annu. Rev. Fluid Mech.* **22**, pp. 13–34.

Dussan, E.B. (1976) "The Moving Contact Line: the Slip Boundary Condition," *J. Fluid Mech.* **77**, pp. 665–684.

Dussan, E.B. (1979) "On the Spreading of Liquids on Solid Surfaces: Static and Dynamic Contact Lines," *Annu. Rev. Fluid Mech.* **11**, pp. 371–400.

Dussan, E.B., and Davis, S.H. (1974) "On the Motion of Fluid-Fluid Interface Along a Solid Surface," *J. Fluid Mech.* **65**, pp. 71–95.

Fan, L.-S., Tai, Y.-C., and Muller, R.S. (1988) "Integrated Movable Micromechanical Structures for Sensors and Actuators," *IEEE Trans. Electron. Devices* **35**, pp. 724–730.

Fan, L.-S., Tai, Y.-C., and Muller, R.S. (1989) "IC-Processed Electrostatic Micromotors," *Sensors and Actuators A* **20**, pp. 41–47.

Gad-el-Hak, M. (1995) "Questions in Fluid Mechanics: Stokes' Hypothesis for a Newtonian, Isotropic Fluid," *J. Fluids Eng.* **117**, pp. 3–5.

Gad-el-Hak, M. (1999) "The Fluid Mechanics of Microdevices—The Freeman Scholar Lecture," *J. Fluids Eng.* **121**, pp. 5–33.

Gad-el-Hak, M. (2000) *Flow Control: Passive, Active, and Reactive Flow Management,* Cambridge University Press, Cambridge, U.K.

Gee, M.L., McGuiggan, P.M., Israelachvili, J.N., and Homola, A.M. (1990) "Liquid to Solidlike Transitions of Molecularly Thin Films under Shear," *J. Chem. Phys.* **93**, pp. 1895–1906.

Guvenc, M.G. (1985) "V-Groove Capillary for Low Flow Control and Measurement," in *Micromachining and Micropackaging of Transducers,* eds. C.D. Fung, P.W. Cheung, W.H. Ko, and D.G. Fleming, pp. 215–223, Elsevier, Amsterdam.

Haile, J.M. (1993) *Molecular Dynamics Simulation: Elementary Methods,* Wiley, New York.

Harley, J.C., Huang, Y., Bau, H.H., and Zemel, J.N. (1995) "Gas Flow in Micro-Channels," *J. Fluid Mech.* **284**, pp. 257–274.

Hirschfelder, J.O., Curtiss, C.F., and Bird, R.B. (1954) *Molecular Theory of Gases and Liquids,* Wiley, New York.

Israelachvili, J.N. (1986) "Measurement of the Viscosity of Liquids in Very Thin Films," *J. Colloid Interface Sci.* **110**, pp. 263–271.

Israelachvili, J.N. (1991) *Intermolecular and Surface Forces,* second edition, Academic Press, New York.

Kennard, E.H. (1938) *Kinetic Theory of Gases,* McGraw-Hill, New York.

Knight, J. (1999) "Dust Mite's Dilemma," *New Scientist* **162**(2180), 29 May, pp. 40–43.

Knudsen, M. (1909) "Die Gesetze der Molekularströmung und der inneren Reibungsströmung der Gase durch Röhren," *Annalen der Physik* **28**, pp. 75–130.

Kogan, M.N. (1969) *Rarefied Gas Dynamics,* Nauka, Moscow (translated from Russian, ed. L. Trilling, Plenum, New York).

Kogan, M.N. (1973) "Molecular Gas Dynamics," *Annu. Rev. Fluid Mech.* **5**, pp. 383–404.

Koplik, J., and Banavar, J.R. (1995) "Continuum Deductions from Molecular Hydroynamics," *Annu. Rev. Fluid Mech.* **27**, pp. 257–292.

Kovacs, G.T.A. (1998) *Micromachined Transducers Sourcebook,* McGraw-Hill, New York.

Lighthill, M.J. (1963) "Introduction. Real and Ideal Fluids," in *Laminar Boundary Layers,* ed. L. Rosenhead, pp. 1–45, Clarendon Press, Oxford.

Liu, J., Tai, Y.C., Lee, J., Pong, K.C., Zohar, Y., and Ho, C.M. (1993) "In-Situ Monitoring and Universal Modeling of Sacrificial PSG Etching Using Hydrofluoric Acid," in *Proc. IEEE Micro Electro Mechanical Systems '93,* pp. 71–76, Institute of Electrical and Electronics Engineers, New York.

Liu, J., Tai, Y.C., Pong, K., and Ho, C.M. (1995) "MEMS for Pressure Distribution Studies of Gaseous Flows in Microchannels," in *Proc. IEEE Micro Electro Mechanical Systems '95,* pp. 209–215, IEEE, New York.

Loeb, L.B. (1961) *The Kinetic Theory of Gases,* third edition, Dover, New York.

Löfdahl, L., and Gad-el-Hak, M. (1999) "MEMS Applications in Turbulence and Flow Control," *Prog. Aero. Sci.* **35**, pp. 101–203.

Loose, W., and Hess, S. (1989) "Rheology of Dense Fluids via Nonequilibrium Molecular Hydrodynamics: Shear Thinning and Ordering Transition," *Rheologica Acta* **28**, pp. 91–101.

Madou, M. (1997) *Fundamentals of Microfabrication,* CRC Press, Boca Raton, FL.

Majumdar, A., and Mezic, I. (1998) "Stability Regimes of Thin Liquid Films," *Microscale Thermophys. Eng.* **2**, pp. 203–213.

Majumdar, A., and Mezic, I. (1999) "Instability of Ultra-Thin Water Films and the Mechanism of Droplet Formation on Hydrophilic Surfaces," in Proc. ASME–JSME Thermal Engineering and Solar Energy Joint Conference, San Diego, CA, 15–19 March (also published in *J. Heat Transfer* **121**, pp. 964–971).

Mastrangelo, C., and Hsu, C.H. (1992) "A Simple Experimental Technique for the Measurement of the Work of Adhesion of Microstructures," in *Technical Digest IEEE Solid-State Sensors and Actuators Workshop,* pp. 208–212, Institute of Electrical and Electronics Engineers, New York.

Maxwell, J.C. (1879) "On Stresses in Rarefied Gases Arising from Inequalities of Temperature," *Phil. Trans. R. Soc. Part 1* **170**, pp. 231–256.

Migun, N.P., and Prokhorenko, P.P. (1987) "Measurement of the Viscosity of Polar Liquids in Microcapillaries," *Colloid J. USSR* **49**, pp. 894–897.

Moffatt, H.K. (1964) "Viscous and Resistive Eddies Near a Sharp Corner," *J. Fluid Mech.* **18**, pp. 1–18.

Muntz, E.P. (1989) "Rarefied Gas Dynamics," *Annu. Rev. Fluid Mech.* **21**, pp. 387–417.

Nadolink, R.H., and Haigh, W.W. (1995) "Bibliography on Skin Friction Reduction with Polymers and Other Boundary-Layer Additives," *Appl. Mech. Rev.* **48**, pp. 351–459.

Nakagawa, S., Shoji, S., and Esashi, M. (1990) "A Micro-Chemical Analyzing System Integrated on Silicon Chip," in *Proc. IEEE: Micro Electro Mechanical Systems,* IEEE 90CH2832-4, Institute of Electrical and Electronics Engineers, New York.

Oran, E.S., Oh, C.K., and Cybyk, B.Z. (1998) "Direct Simulation Monte Carlo: Recent Advances and Applications," *Annu. Rev. Fluid Mech.* **30**, pp. 403–441.

Panton, R.L. (1996) *Incompressible Flow,* second edition, Wiley-Interscience, New York.

Pearson, J.R.A., and Petrie, C.J.S. (1968) "On Melt Flow Instability of Extruded Polymers," in *Polymer Systems: Deformation and Flow,* eds. R.E. Wetton and R.W. Whorlow, pp. 163–187, Macmillian, London.

Pfahler, J. (1992) "Liquid Transport in Micron and Submicron Size Channels," Ph.D. thesis, University of Pennsylvania, Philadelphia.

Pfahler, J., Harley, J., Bau, H., and Zemel, J.N. (1990) "Liquid Transport in Micron and Submicron Channels," *Sensors and Actuators A* **21–23**, pp. 431–434.

Pfahler, J., Harley, J., Bau, H., and Zemel, J.N. (1991) "Gas and Liquid Flow in Small Channels," in *Symp. on Micromechanical Sensors, Actuators, and Systems,* eds. D. Cho et al., ASME DSC-Vol. 32, pp. 49–60, American Society of Mechanical Engineers, New York.

Piekos, E.S., and Breuer, K.S. (1996) "Numerical Modeling of Micromechanical Devices Using the Direct Simulation Monte Carlo Method," *J. Fluids Eng.* **118**, pp. 464–469.

Pong, K.-C., Ho, C.-M., Liu, J., and Tai, Y.-C. (1994) "Non-Linear Pressure Distribution in Uniform Microchannels," in *Application of Microfabrication to Fluid Mechanics,* eds. P.R. Bandyopadhyay, K.S. Breuer, and C.J. Belchinger, ASME FED-Vol. 197, pp. 47–52, American Society of Mechanical Engineers, New York.

Porodnov, B.T., Suetin, P.E., Borisov, S.F., and Akinshin, V.D. (1974) "Experimental Investigation of Rarefied Gas Flow in Different Channels," *J. Fluid Mech.* **64**, pp. 417–437.

Prud'homme, R.K., Chapman, T.W., and Bowen, J.R. (1986) "Laminar Compressible Flow in a Tube," *Appl. Sci. Res.* **43**, pp. 67–74.

Richardson, S. (1973) "On the No-Slip Boundary Condition," *J. Fluid Mech.* **59**, pp. 707–719.

Schaaf, S.A., and Chambré, P.L. (1961) *Flow of Rarefied Gases,* Princeton University Press, Princeton, NJ.

Seidl, M., and Steinheil, E. (1974) "Measurement of Momentum Accommodation Coefficients on Surfaces Characterized by Auger Spectroscopy, SIMS and LEED," in *Rarefied Gas Dynamics,* Vol. 9, eds. M. Becker and M. Fiebig, pp. E9.1–E9.2, DFVLR-Press, Porz-Wahn, Germany.

Sharp, K.V. (2001) "Experimental Investigation of Liquid and Particle-Laden Flows in Microtubes," Ph.D. thesis, University of Illinois at Urbana–Champaign.

Sharp, K.V., Adrian, R.J., Santiago, J.G., and Molho, J.I. (2001) "Liquid Flow in Microchannels," in *The MEMS Handbook*, ed. M. Gad-el-Hak, pp. 4-1–4-38, CRC Press, Boca Raton, FL.

Shih, J.C., Ho, C.-M., Liu, J., and Tai, Y.-C. (1995) "Non-Linear Pressure Distribution in Uniform Microchannels," ASME AMD-MD-Vol. 238, American Society of Mechanical Engineers, New York.

Shih, J.C., Ho, C.-M., Liu, J., and Tai, Y.-C. (1996) "Monatomic and Polyatomic Gas Flow through Uniform Microchannels," in *Applications of Microfabrication to Fluid Mechanics*, eds. K. Breuer, P. Bandyopadhyay, and M. Gad-el-Hak, ASME DSC-Vol. 59, pp. 197–203, American Society of Mechanical Engineers, New York.

Tai, Y.-C., and Muller, R.S. (1989) "IC-Processed Electrostatic Synchronous Micromotors," *Sensors and Actuators A* **20**, pp. 49–55.

Tang, W.C., Nguyen, T.-C., and Howe, R.T. (1989) "Laterally Driven Polysilicon Resonant Microstructures," *Sensors and Actuators A* **20**, pp. 25–32.

Thomas, L.B., and Lord, R.G. (1974) "Comparative Measurements of Tangential Momentum and Thermal Accommodations on Polished and on Roughened Steel Spheres," in *Rarefied Gas Dynamics*, Vol. 8, ed. K. Karamcheti, Academic Press, New York.

Thompson, P.A., and Robbins, M.O. (1989) "Simulations of Contact Line Motion: Slip and the Dynamic Contact Line," *Nature* **389**, 25 September, pp. 360–362.

Thompson, P.A., and Troian, S.M. (1997) "A General Boundary Condition for Liquid Flow at Solid Surfaces," *Phys. Rev. Lett.* **63**, pp. 766–769.

Tison, S.A. (1993) "Experimental Data and Theoretical Modeling of Gas Flows through Metal Capillary Leaks," *Vacuum* **44**, pp. 1171–1175.

Tuckermann, D.B. (1984) "Heat Transfer Microstructures for Integrated Circuits," Ph.D. thesis, Stanford University, Stanford, CA.

Tuckermann, D.B., and Pease, R.F.W. (1981) "High-Performance Heat Sinking for VLSI," *IEEE Electron. Device Lett. EDL-2*, No. 5, May.

Tuckermann, D.B., and Pease, R.F.W. (1982) "Optimized Convective Cooling Using Micromachined Structures," *J. Electrochem. Soc.* **129**(3), March, C98.

Van den Berg, H.R., Seldam, C.A., and Gulik, P.S. (1993) "Compressible Laminar Flow in a Capillary," *J. Fluid Mech.* **246**, pp. 1–20.

Vargo, S.E., and Muntz, E.P. (1996) "A Simple Micromechanical Compressor and Vacuum Pump for Flow Control and Other Distributed Applications," AIAA Paper No. 96-0310, American Institute of Aeronautics and Astronautics, Washington, D.C.

Vincenti, W.G., and Kruger, C.H., Jr. (1965) *Introduction to Physical Gas Dynamics*, Wiley, New York.

Von Smoluchowski, M. (1898) "Ueber Wärmeleitung in verdünnten Gasen," *Annalen der Physik und Chemie* **64**, pp. 101–130.

Went, F.W. (1968) "The Size of Man," *Am. Sci.* **56**, pp. 400–413.

5
Integrated Simulation for MEMS: Coupling Flow-Structure-Thermal-Electrical Domains

Robert M. Kirby
Brown University

George Em Karniadakis
Brown University

Oleg Mikulchenko
Oregon State University

Kartikeya Mayaram
Oregon State University

5.1 Abstract 5-1
5.2 Introduction 5-2
   Full-System Simulation • Computational Complexity of MEMS Flows • Coupled-Domain Problems • A Prototype Problem
5.3 Coupled Circuit-Device Simulation 5-7
5.4 Overview of Simulators 5-8
   The Circuit Simulator: SPICE3 • The Fluid Simulator: NεκTαr • Formulation for Flow-Structure Interactions • Grid Velocity Algorithm • The Structural Simulator • Differences between Circuit, Fluid and Solid Simulators
5.5 Circuit-Microfluidic Device Simulation 5-15
   Software Integration • Lumped Element and Compact Models for Devices
5.6 Demonstrations of the Integrated Simulation Approach 5-20
   Microfluidic System Description • SPICE3–NεκTαr Integration • Simulation Results
5.7 Summary and Discussion 5-22
   Acknowledgments 5-23

5.1 Abstract

Full-system simulation of microelectromechanical systems (MEMS) involves coupling of many diverse processes with disparate spatial and temporal scales. In its simplest form, all elements in a MEMS device are represented as equivalent analog circuits so that robust simulators such as SPICE can solve for the entire system. However, devices, and especially fluidic devices, do not usually have such equivalent analogs and may require full simulation of individual components. This is particularly true for new designs, which often involve unfamiliar physics; such lumped models and continuum approximations are inappropriate in this case. In this chapter, we address such issues for an integrated simulation approach for MEMS with the ultimate objective of deriving low-dimensional models that can be used for MEMS design in full-system simulations. In particular, we present an integrated circuit and microfluidic simulator, the SPICE–$N\varepsilon\kappa T\alpha r$ simulator, which allows coupled simulation of flow, structure, thermal and electrical domains. The overall architecture and various algorithms, including the coupling of the circuit and microfluidic simulators, are described. An application of the simulator is demonstrated for a controlled microliquid dosing system using detailed numerical models for the fluid field, a low-dimensional model for the flow sensor and circuit elements for the electronic control. Unlike other simulators, which employ mostly lumped models for the various components, we formulate here a methodology for distributed systems.

5.2 Introduction

5.2.1 Full-System Simulation

Microelectromechanical systems (MEMS) involve complex functions governed by diverse transient physical and electrical processes for each of their many components. The design complexity and functionality complexity of MEMS exceed by far the complexity of VLSI systems. Today, however, very large-scale integration (VLSI) systems are simulated routinely thanks to the many advances in CAD and simulation tools achieved over the last two decades. It is clear that similar and even greater advances are required in the MEMS field in order to make full-system simulation of MEMS a reality in the near future. This will enable the MEMS community to explore new pathways of discovery and expand the role and influence of MEMS at a rapid rate.

To develop a systems-level simulation framework that is sufficiently accurate and robust, all processes involved must be simulated at a comparable degree of accuracy and integrated seamlessly. That is, circuits, semiconductors, springs and masses, beams and membranes as well as the flow field should be simulated in a consistent and compatible way and in reasonable computational time! This coupling of diverse domains has already been addressed by the electrical engineering community, primarily for mixed circuit-device simulation.

The combination of circuits and devices necessitates the use of different levels of description. At a first level for analog circuits represented by lumped continuum models, the use of ordinary differential equations (ODEs) and algebraic equations (AEs) is sufficient. However, some other devices and circuits can be described as digital automata, and Boolean equations of mathematical logic should be employed in the description; they correspond to digital circuit simulation on the digital level. Finally, some semiconductor devices of the kind encountered in MEMS have to be described as linear and nonlinear partial differential equations (PDEs), and they are usually employed on the device simulation level. Mixed-level simulation is implemented for digital–analog (or analog-mixed) circuit simulation and for analog circuit-device simulation. In the following, we briefly review the common practice in simulating circuits with some nonfluidic devices.

The code SPICE (Simulation Program with Integrated Circuit Emphasis) was developed in the 1970s at the University of California, Berkeley [Nagel and Pederson, 1973] and since then has become the unofficial industrial standard for integrated circuit (IC) designers. SPICE is a general-purpose simulation program for circuits that may contain resistors, capacitors, inductors, switches, transmission lines, etc., as well as the five most common semiconductor devices: diodes, bipolar junction transistors (BJTs), junction field effect transistors (JFETs), metal semiconductor field effect transistors (MESFETs) and metal oxide semiconductor field effect transistors (MOSFETs). SPICE has built-in models for the semiconductor devices, and the user specifies only the pertinent model parameter values. However, these devices are typically simple and can be described by lumped models—combinations of ordinary differential equations and algebraic equations (ODEs/AEs). In some cases, such as in submicron devices, even for usual semiconductor devices (i.e., MOSFET), simple modeling is not straightforward and it is more art than science to transfer from basic PDEs to approximated ODEs and algebraic equations. Mechanical systems can be recast into electrical systems, which can be handled by SPICE. This can be understood more clearly by considering the analogy of a mass–spring–damper system driven by an external force with a parallel-connected RLC (resistance, inductance, capacitance) circuit with a current source, taking mass to correspond to capacitance, dampers to resistors, springs to inductive elements, and forces to currents.

However, there are other devices that cannot be represented by lumped models and such an analogy may not be valid. While SPICE is essentially an ODE solver (i.e., an analog circuit simulator only), another successful code, CODECS (Coupled Device and Circuit Simulator), provides a truly mixed level description of both circuits and devices. This code, too, was developed at the University of California, Berkeley [Mayaram and Pederson, 1992] and employs combinations of both ODEs and PDEs with algebraic equations. CODECS incorporates SPICE3, the latest version of SPICE written in C [Quarles, 1989], for the circuit simulation capability. The multirate dynamics introduced by combinations of devices and circuits is handled efficiently by a multilevel Newton method or a full-Newton method for transient analysis, unlike the standard Newton method employed in SPICE. CODECS is appropriate for one- and two-dimensional devices, but recent developments have produced efficient algorithms for three-dimensional devices as well [Mayaram et al., 1993].

The aforementioned simulation tools for IC design can be used for MEMS simulations, and, in fact, SPICE has been use to model electrostatic lateral resonators [Lo et al., 1996]. The assumption here is that all device components can be recast as equivalent analog circuit elements that SPICE recognizes. Clearly, this approach can be used in some well-studied structures, such as membranes or simple microbeams, but very rarely for microflows. However, in the last decade there has been an intense effort to produce such models and corresponding software, such as MEMCAD [Senturia et al., 1992], which has become a commercial package [Gilbert et al., 1993] for electrostatic and mechanical analysis of microstructures. Other such packages are the SOLIDIS and IntelliCAD (IntelliSense and ISE). In these simulation approaches, the flowfield is not simulated but its effect is typically expressed by the equivalent of a drag coefficient that provides damping. In some cases, as in the squeezed gas film in silicon accelerometers, an equivalent RLC circuit can also be obtained [Veijola et al., 1995]; however, this is the exception rather than the rule. Even the structural components are often modeled analytically, and significant effort has been devoted to constructing reduced-order macromodels [Hung et al., 1997; Gabbay, 1998]. These are typically nonlinear, low-dimensional models obtained from projections of full three-dimensional simulations to a few representative modal shapes. Nonlinear function fitting is then employed so that analytical forms can be written and these structural models are then imported to SPICE as analog circuit equivalent elements.

This reduced-order macromodeling approach has been used with success in a variety of applications, including, for example, the electrostatic actuation of a suspended beam and elastically suspended plates [Gabbay, 1998]. Their great advantage is computational speed, but they are limited to small displacements and small deformations (i.e., mostly in the linear regime) and are appropriate for familiar designs only. Unfortunately, most of the MEMS devices are operating in nonlinear regimes, including electrostatic actuators, flow fields and structures. More importantly, the real impact of MEMS and anticipated benefits will come from new designs, yet unknown, which may be pre-tested using full simulations and all processes are simulated accurately without sacrificing important details of the physics. MEMS simulation based on full-physics models may then be more appropriate for exploring new concepts, whereas macromodeling may be employed efficiently for familiar designs and in known operating regimes.

In the following, we address some of the specific issues encountered in each of the coupled domains, (i.e., fluid, electric, structure, thermal), analyze their corresponding computational complexity and propose algorithms for their integration.

5.2.2 Computational Complexity of MEMS Flows

Liquid and gas flows in microdevices are characterized by a low Reynolds number, typically of order one or less in channels with heights in the submillimeter range [Ho and Tai, 1998; Gad-el-Hak, 1999]. They are unsteady due to external excitation from a moving boundary or an electric field, often driven by high-frequency oscillators—for example, 50 KHz for the MIT electrostatic comb-drive [Freeman et al., 1998]. The domain of microflows is three dimensional and geometrically complex, consisting of large-aspect-ratio components, abrupt expansions and rough boundaries. In addition, microdevices interact with larger devices, resulting in fluid flow going through disparate regimes.

Accurate and efficient simulation of microflows should take into account the above factors. For example, the significant geometric complexity of MEMS flows suggests that finite elements and boundary elements are more suitable than finite differences for efficient discretization [Ye et al., 1999]. However, because of the nonlinear effects, either through the convection or boundary conditions, boundary element methods are also limited in their application range despite their efficiency for linear flows [Aluru and White, 1996]. A particularly promising approach developed recently for MEMS flows makes use of meshless and mesh-free approaches [Aluru, 1999], where particles are "sprinkled" almost randomly into the flow and boundary. This approach handles the geometric complexity of MEMS flows effectively but the issues of accuracy and efficiency have not been fully resolved yet. In regard to nonlinearities, one may argue that at such a low Reynolds number the convection effects should be neglected, but in complex geometries with abrupt turns the convective acceleration terms may be substantial, thus they need to be taken into account.

The computational difficulties for liquid and gas flows are of different types. Gas microflows are compressible and can experience large density variations. In addition, for channels below 10 μm or at subatmospheric conditions, serious *rarefaction effects* may be present; that is, the walls "move" (see Beskok et al., 1996; Beskok, this volume). In this case, either modified Navier–Stokes equations with appropriate *slip-boundary conditions* or higher order approximations are necessary to describe the governing flow dynamics. To this end, a nondimensional number, the *Knudsen number*, defined as the ratio of the mean free path to the characteristic domain size, defines which model (and correspondingly which numerical method) is appropriate for simulating gas microflows [Bird, 1994]. For submicron devices, atomistic or molecular simulations are necessary as the familiar concept of continuum description breaks down. The direct simulation Monte Carlo (DSMC) method (described elsewhere in this volume) is one efficient method of simulating highly rarefied flows.

On the other hand, liquid flows in microscales are "granular;" that is, they form a layering structure very close to the wall over a distance of a few molecule diameters [Koplik and Banavar, 1995]. This is accompanied by large-density fluctuations very close to the wall that lead to anomalous heat and momentum transport. Liquid flows, in particular, are very sensitive to the wall type and, although such an issue may not be important for averaged heat and momentum transport rates in flow domains of 100 μm or greater, it is extremely important in smaller domains. This distinction suggests two possible approaches in simulating liquid flows in microscales: a phenomenological one, using the Navier–Stokes equation and similar to macrodomain flows, and a molecular approach based on the molecular dynamics (MD) approach [Allen and Tildesley, 1994; Koplik and Banavar, 1995]. The MD approach is deterministic, following the trajectories of all molecules involved, unlike the DSMC approach, which is stochastic and represents collisions as a random process. The drawback of the Navier–Stokes approach is that events at the molecular level are modeled via continuum-like parameters. For example, consider the problem of routing microdroplets on a silicon surface, effectively altering dynamically the contact line of the microdrop. This is a molecular-level process, but in the continuum approach it is determined via a macrodomain-type formulation (e.g., via gradients), which may lead to erroneous results. Accurate MD modeling of the contact line will be truly predictive, as it will take into account the wall–fluid interaction at the molecular level. The wall type and the specific fluid type are taken into account by different potentials that describe intermolecular structure and force. However, such a detailed simulation requires an enormous amount of molecules (e.g., hundreds of millions of molecules) and is limited to a very small region, probably around the contact line region only. It is thus important to develop new hybrid approaches that combine the best features of both methods [Hadjiconstantinou, 1999].

In summary, geometry and surface effects, compressibility and rarefaction, unsteadiness and unfamiliar physics make simulation of microflows a challenging task. The true challenge, however, comes from the interaction of the fluidic system with other system components (e.g., the structure, the electric field and the thermal field). In the following sections, we discuss this interaction.

5.2.3 Coupled-Domain Problems

In coupled-domain problems, such as flow–structure or structure–electric, or a combination of both, there are significant disparities in temporal and spatial scales. These disparities, in turn, imply that multiple grids and heterogeneous time-stepping algorithms may be needed for discretization, leading to very complicated and consequently computational prohibitive simulation algorithms. Simplifications are typically made, with one of the fields represented at a reduced resolution level or by low-dimensional systems or even by equivalent lumped dynamical models. For example, consider the electric activation of a cantilever microbeam made of piezoelectric material. The emphasis may be on modeling the electronic circuit and the motion; thus, a simple model for the motion-induced hydrodynamic damping may be constructed avoiding full simulation of the flow around the beam.

Possible construction of low-order dynamical models could be achieved by projecting the results of detailed numerical simulations onto spaces spanned by a very small number of degrees of freedom—the so-called *nonlinear macromodeling* approach (see Gabbay [1998] and Senturia et al. [1997]). To clarify the concept of a macromodel we give a specific example (taken from Senturia et al. [1997]) for a suspended membrane of thickness t and deflected at its center by an amplitude d under the action of uniform pressure force P. Let $2a$ be the length of the membrane; E, the Young's module; ν, the Poisson ratio; and σ, the residual stress. One can use analytical methods to obtain the resulting form of the pressure-deflection relation (e.g., power series, assuming a circular thin membrane). This can be extended to more general shapes and nonlinear responses, for example:

$$P = \frac{C_1 t}{a^2} + \frac{C_2 f(\nu)}{a^4} \frac{E}{1-\nu} d^3 \tag{5.1}$$

where C_1 and C_2 are dimensionlesss constants that depend on the shape of the membrane, and $f(\nu)$ is a slowly varying function of the Poisson ratio. This function is determined from detailed finite element simulations over a range of length a, thickness t, and material properties ν and E. Such "best-fits" are tabulated and are used in the simulation according to the specific structure considered, without the need for solving the partial differential equations governing the dynamics of the structure. They can also be built automatically as has been demonstrated by Gabbay (1998). Another type of a macromodel based on neural networks training will be presented later for a flow sensor.

Unfortunately, construction of such macromodels is not always possible, and this lack of simplified models for the many and diverse components of microsystems makes system-level simulation a challenging task. On the other hand, model development for electronic components (transistors, resistors, capacitors, etc.) has reached a state of maturity. Therefore, considerable attention should be focused on models for nonelectronic components. This is necessary for the design and verification of complete microsystems. In this chapter, we describe an integrated approach for simulation of microsystems with the emphasis being on microfluidic systems. To this end, we resort to full simulation of the fluidic system, which involves interactions with moving structures. To illustrate the formulation more clearly, we next present a target simulation problem that represents the aforementioned challenges.

5.2.4 A Prototype Problem

An example of a microfluidic system is a microliquid dosing system shown schematically in Figure 5.1. This system is made up of a micropump, a microflow sensor and an electronic control circuit. The electronic circuit adjusts the pump flow rate so that a constant flow is maintained in the microchannel. A realization of this system is shown in Figure 5.2 along with the details of the control circuit. Simulation of the complete system requires models for the micropump, the microflow sensor and the electronic components shown in Figure 5.2. When models, low-order or full-physics, are available for all components,

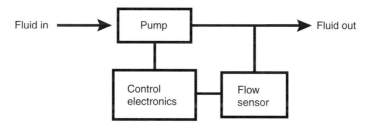

FIGURE 5.1 Block diagram of a generic microfluidic system. The flow sensor senses the flow rate, which is controlled by the electronic circuit controlling the pump.

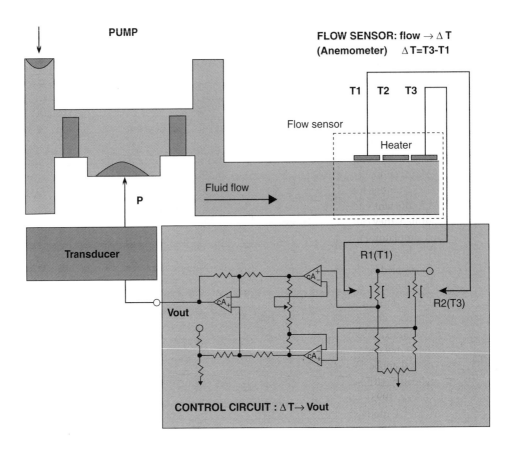

FIGURE 5.2 Realization of the microfluidic system showing the electronic control circuit. The fluid flow determines the temperature ΔT of the flow sensor. This temperature is transformed by the control electronics into the voltage V_{out}, which in turn controls the pump pressure P by a transformation of the voltage to a proportional pressure.

including the fluid flow, the complete system can be simulated using a standard circuit simulator such as SPICE [Nagel, 1975; Quarles, 1989].

In the absence of macromodels for the micropump and the microflow sensor, the typical approach for microsystem simulation makes use of lumped-element equivalent circuit descriptions for these devices [Tilmans, 1996]. However, such an approach has two main limitations:

1. It is suitable only for open-loop systems, where there is no feedback from the output to the input.
2. It is applicable only for small-signal conditions.

These two limitations arise in the model development process where several assumptions are made in order to construct the lumped-element equivalent circuits; therefore, this approach would not be suitable when the large-signal behavior of a closed-loop system is of interest.

To address the above problem, we present a coupled circuit/microfluidic device simulator that efficiently couples the discretized Navier–Stokes equations describing a microfluidic device (numerical model) to the solution of circuit equations. Such a capability is unique in that it allows direct and efficient simulation of microfluidic systems without the need for mapping finite element descriptions into equivalent networks [Tilmans, 1996] or analog hardware description languages (AHDLs) [Bielefeld et al., 1997].

The remainder of this chapter is organized as follows: An overview of coupled circuit and device simulation is given in Section 5.3, followed by a description of the circuit and fluidic simulators in Section 5.4. The details of the coupled circuit/fluidic simulator are presented in Section 5.5, and an illustrative example is described in Section 5.6. Conclusions are provided in Section 5.7.

5.3 Coupled Circuit-Device Simulation

Coupled simulation techniques have been used for simulation of a sensor system [Schroth et al., 1995]. In this approach, the finite element program ANSYS [Moaveni, 1999] is coupled to an electrical simulator PSPICE [Keown, 1997]. Although such an approach has been demonstrated to work for system simulations, the coupling is not efficient. Special coupling algorithms and time-stepping schemes are required to enable fast simulation of microsystems; therefore, a tight coupling between the circuit and device simulators is necessary for simulation efficiency [Mayaram and Pederson, 1992; Mayaram et al., 1993].

The coupled circuit-device simulator allows verification of microfluidic systems. It provides accurate large- and small-signal simulation of systems even in the absence of proper macromodels for the microfluidic devices. On the other hand, the coupled simulator is important for constructing and validating macromodels. As important effects (such as highly nonlinear or distributed behavior, compressibility or slip flow) are identified, they can be implemented in the macromodels and verified for system simulation using the coupled simulator. Furthermore, critical devices can be simulated using the full physics-based numerical models when there are stringent accuracy requirements on the simulated results.

The concept of a coupled circuit and device simulator has proved to be extremely beneficial in the domain of integrated circuits. Since the first of such simulators, MEDUSA [Engl et al., 1982], became available in the early 1980s, significant work has addressed coupled simulation. These activities have focused on improved algorithms and faster execution speeds and applications. Commercial TCAD vendors also support a mixed circuit-device simulation capability [Silvaco International, 1995; Technology Modeling Associates, 1997]. Because the computational costs of these simulators are high, they are not used on a routine basis. However, there are several critical applications in which these simulators are extremely valuable. These include simulation of radio frequency (RF) circuits [Rotella et al., 1997], single-event-upset simulation of memories [Woodruff and Rudeck, 1993], simulation of power devices [Ravanelli and Hu, 1991], and validation of nonquasistatic MOSFET models [Park et al., 1991].

The coupled circuit-device simulator for microfluidic applications is illustrated in Figure 5.3. This simulator supports compact models for the electronic components and available macromodels for microfluidic devices. In addition, numerical models are available for the microfluidic components which can be utilized when detailed and accurate modeling is required. As an example, specific components such as

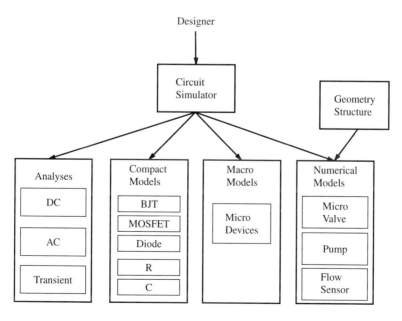

FIGURE 5.3 The coupled circuit/fluidic device simulator. Microfluidic systems, including the control electronics, can be simulated using accurate numerical models for all components.

microvalves, pumps and flow sensors are shown in Figure 5.3. The coupling of the circuit and microfluidic components is handled by imposing suitable boundary conditions on the fluid solver. This simulator allows the simulation of a complete microfluidic system including the associated control electronics. The details of the various simulators and coupling methods are described in the sections below.

One of the biggest disadvantages of such an approach is the high computational cost involved. The major cost comes from solving the three-dimensional, time-dependent Navier–Stokes equations in complex geometric domains. Thus, efficient flow solvers are critical to the success of a coupled circuit–microfluidic device simulator. Any performance improvements in the solution of the Navier–Stokes equations directly translate into a significant performance gain for the coupled simulator.

5.4 Overview of Simulators

The circuit simulator employed here is based on the circuit simulator SPICE3f5 [Quarles, 1989] and the microfluidic simulator on the code $N\varepsilon\kappa T\alpha r$ [Karniadakis and Sherwin, 1999; Kirby et al., 1999]. A brief description of the algorithms and software structure of each of these simulators is provided in this section.

5.4.1 The Circuit Simulator: SPICE3

Electrical circuits consist of many components (resistors, capacitors, inductors, transistors, diodes and independent sources) which are described by algebraic and/or differential relations between the component's currents and voltages. These relationships are called the *branch constitutive relations* [Sangiovanni-Vincentelli, 1981]. The circuits also satisfy conservation laws known as the Kirchhoff's laws; these laws result in algebraic equations. Therefore, a circuit is described by a set of coupled, nonlinear, differential algebraic equations, which are both highly nonlinear and stiff, which imposes certain limitations on the solution methods. One of the most commonly used analyses is the *time-domain transient analysis*. We briefly describe below the solution approach used for this analysis.

5.4.1.1 Time Discretization

At each time step of the transient analysis, the time derivatives are replaced by an algebraic equation using an integration method. Typically, an implicit linear multistep method of the backward-differentiation type suitable for stiff ODEs is used [Sangiovanni-Vincentelli, 1981]:

$$\dot{v} \approx \alpha_0 v_{t_n} + \sum_{k=1}^{n} \alpha_k v_{t_{n-k}} \tag{5.2}$$

5.4.1.2 Linearization

Time discretization yields a system of nonlinear algebraic equations, which are typically solved by a Newton–Raphson method. The nonlinear components are replaced by linear equivalent models for each iteration of the Newton's method:

$$f(v_{t_n}^{j+1}) \approx f(v_{t_n}^{j}) + \partial f(v)/\partial v \Big|_{v_{t_n}^{j}} \cdot (v_{t_n}^{j+1} - v_{t_n}^{j}) \tag{5.3}$$

5.4.1.3 Equation Solution

After time discretization and application of Newton's method a linear system of equations is obtained at each iteration of the Newton method. These equations are described by:

$$Av^{j+1} = b \tag{5.4}$$

where $A \in \Re^{n \times n}$, $v^{j+1} \in \Re^n$ and $b \in \Re^n$. They can be solved by sparse matrix techniques [Kundert, 1987].

Integrated Simulation for MEMS

The time-domain simulation algorithm can be summarized in the following steps [Sangiovanni-Vincentelli, 1981]:

1. Read circuit description and initialize data structures.
2. Increment time, $t_n = t_{n-1} + h$.
3. Update values of independent sources at t_n.
4. Predict values of unknown variables at t_n.
5. Apply integration Eq. (1) to capacitors and inductors.
6. Apply linearization Eq. (2) to nonlinear circuit elements.
7. Assemble linear circuit equations.
8. Solve linear circuit equations.
9. Check convergence; if not converged, go to step 6.
10. Estimate local truncation error.
11. Select new time step h; rollback time if truncation error is unacceptable.
12. If $t_n < t_{stop}$, go to step 3.

5.4.2 The Fluid Simulator: NεκTαr

The flow solver corresponds to a particular version of the code *NεκTαr*, which is a general-purpose computational fluid dynamics (CFD) code for simulating incompressible, compressible and plasma flows in unsteady three-dimensional geometries. The major algorithmic developments are described in Sherwin (1995) and Warburton (1999), and the capabilities are summarized in Figure 5.4. The code uses meshes similar to those for standard finite element and finite volume and consist of structured or unstructured grids, or a combination of both. The formulation is also similar to those methods, corresponding to Galerkin and discontinuous Galerkin projections for the incompressible and compressible Navier–Stokes

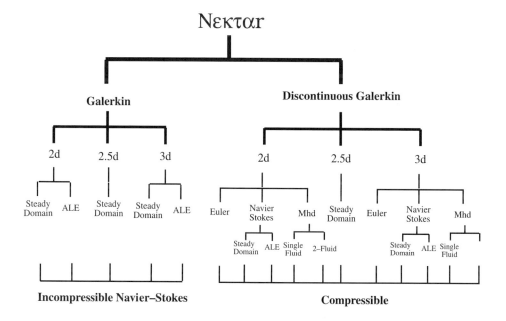

FIGURE 5.4 Hierarchy of the *NεκTαr* code. Note that "2.5d" refers to a three-dimensional capability with one of the directions being homogeneous in the geometry. Also, ALE refers to moving computational domains required in dynamic flow–structure interactions. Gaseous microflows can be simulated by either the compressible or incompressible version, depending on the pressure/density variations.

equations, respectively. Field variables, data and geometry are represented in terms of hierarchical (Jacobi) polynomial expansions [Karniadakis and Sherwin, 1999]; both iso-parametric and super-parametric representations are employed. These expansions are ordered in vertex, edge, face and interior (or bubble) modes. For the Galerkin formulation, the required C^0 continuity across elements is imposed by choosing the appropriate edge (and face, in three-dimensional) modes; at low order expansions, this formulation reduces to the standard finite element formulation. The discontinuous Galerkin is a flux-based formulation, and all field variables have L^2 continuity; at low order, this formulation reduces to the standard finite-volume formulation.

This new generation of Galerkin and discontinuous Galerkin spectral/h–p element methods implemented in the code NɛκTαr does not replace but rather extends the classical finite element and finite volumes with which the CFD practitioners are familiar [Karniadakis and Sherwin, 1999]. The additional advantages are that convergence of the discretization and thus solution verification can be obtained without remeshing (h-refinement) and that the quality of the solution does not depend on the quality of the original discretization. In Figure 5.4 we summarize the major current capabilities of the general code NɛκTαr for incompressible, compressible and even plasma flows. In particular for microflows, both the compressible and incompressible versions are used. For gas microflows we account for rarefaction by using velocity-slip and temperature-jump boundary conditions [Beskok et al., 1996; Beskok and Karniadakis, 1999; Beskok, this volume]. An extension of the classical Maxwell's boundary condition is employed in the code in the form:

$$U_g - U_w = \frac{Kn}{1 - bKn}(\nabla U)_w \cdot \hat{n} \tag{5.5}$$

Here we define the Knudsen number as $Kn = \lambda/L$, with λ the mean free path of the gas molecules and L the characteristic length scale in the flow. Also, U_g is the velocity (tangential component) of the gas at the wall, U_w is the wall velocity, and $\hat{n}$ is the unit normal vector. The constant b is adjusted to reflect the physics of the problem as we go from the slightly rarefied regime (slip flow) to the transition regime ($Kn \approx 1$) or free molecular regime ($Kn > 5$ to 10). For $b = 0$, we recover the classical linear relationship between velocity slip and shear stress, first proposed by Maxwell. However, for $b = -1$, we obtain a second-order accuracy [Beskok and Karniadakis, 1999], and in general for $b \neq 0$, Eq. (5.5) leads to finite slip at the wall, unlike the linear boundary condition (for $b = 0$) used in most codes. The boundary condition in Eq. (5.5) has been used with success in the entire Knudsen number regime, $Kn \approx 0$ to 200 (see several examples in Beskok and Karniadakis, 1999).

One of the key points in obtaining *efficiency* in simulations of moving domains is the type of discretization employed in the flow solver. In NɛκTαr we employ the so-called h–p version of the finite element method with spectral Jacobi polynomials as basic functions. Convergence is obtained via a dual path in this approach, either by increasing the number of elements (h-refinement) or by increasing the order of the spectral polynomial (p-refinement). In the latter case, a faster convergence is obtained without the need for remeshing. Instead, the number of degrees of freedom is increased in the *modal space* by increasing the polynomial order (p) while keeping the mesh unchanged. It is, of course, the cost of reconstructing the mesh that is orders of magnitude higher in time-dependent simulations both in terms of computer and human time.

Regarding the type of elements (subdomains), NɛκTαr uses hybrid meshes (both structured and unstructured). For example, in three-dimensional simulations, a hybrid grid may consist of tetrahedra, hexahedra, triangular prisms, and even pyramids. In Figure 5.5 we plot the mesh used in the simulation of the pump, and in Figure 5.6 we plot the flow field at three different time instances.

In the following, we briefly describe how we formulate the algorithm for a compatible and efficient flow-structure coupling.

Integrated Simulation for MEMS

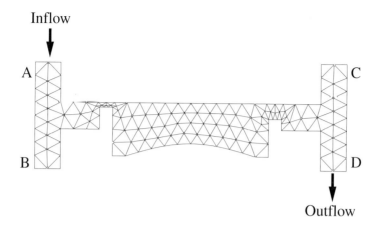

FIGURE 5.5 Mesh of the pump used in the flow simulator $N\epsilon\kappa T\alpha r$. This device was first introduced by Beskok and Warburton (1998) as a mixing device between two microchannels. Here, B and C are blocked so the device is operating as a pump from A to D.

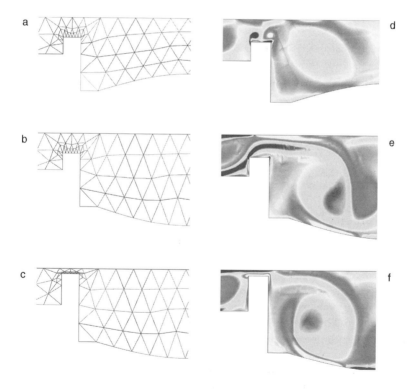

FIGURE 5.6 (Color figure follows p. **12**-26.) Close up of the vorticity contours for Re = 30 simulation at the left valve (meshes shown on right side). Top: $\tau\omega = 0.28$, corresponding to the beginning of the suction stage; start-up vortices due to the motion of the inlet valve can be identified. Middle: $\tau\omega = 0.72$, corresponding to the end of the suction stage; a vortex jet pair is visible in the pump cavity. Bottom: $\tau\omega = 84$, corresponding to early ejection stage; further evolution of the vortex jet and the start-up vortex of the exit valve can be identified. (Courtesy of A. Beskok.)

5.4.3 Formulation for Flow-Structure Interactions

We consider the incompressible Navier–Stokes equations in a time-dependent domain, $\Omega(t)$:

$$u_{i,t} + u_j u_{i,j} = -(p\delta_{ij})_j + \nu u_{i,jj} + f_i \text{ in } \Omega(t) \tag{5.6}$$

$$u_{j,j} = 0 \text{ in } \Omega(t) \tag{5.7}$$

where ν is the viscosity and f_i is a body force. We assume for clarity homogeneous boundary conditions; velocity-slip-boundary conditions can be included relatively easily in the Galerkin framework as mixed (Robin) boundary conditions. Multiplying Eq. (5.6) by test functions and integrating by parts, we obtain:

$$\int_{\Omega(t)} v_i(u_{i,t} + u_j u_{i,j}) dx = \int_{\Omega(t)} v_{i,j}(p\delta_{ij} - \nu u_{i,j} + v_i f_i) dx \tag{5.8}$$

The next step is to define the reference system on which time differentiation takes place. This was accomplished by Ho (1989) by use of the Reynold's transport theorem and by using the fact that the test function v_i is following the material points; therefore, its time derivative in that reference frame is zero:

$$\frac{dv_i}{dt}\bigg|_{x_p} = v_{i,t} + w_j v_{i,j} = 0$$

where w_j is a velocity that describes the motion of the time-dependent domain $\Omega(t)$; x_p denotes the material point. The final variational statement then becomes:

$$\frac{d}{dt}\int_{\Omega(t)} v_i u_i \, dx + \int_{\Omega(t)} [v_i(u_j - w_j)u_{i,j} - v_i u_i w_{j,j}] dx = \int_{\Omega(t)} [v_{i,j} p\delta_{ij} - \nu v_{i,j} u_{i,j} + v_i f_i] dx \tag{5.9}$$

This is the ALE formulation of the momentum equation. It reduces to the familiar Eulerian and Lagrangian forms by setting $w_j = 0$ and $w_j = u_j$, respectively. However, w_j can be chosen arbitrarily to minimize the mesh deformation. We discuss this algorithm next.

5.4.4 Grid Velocity Algorithm

The grid velocity is arbitrary in the ALE formulation; therefore, great latitude exists in the choice of technique for updating it. Mesh constraints such as smoothness, consistency, and lack of edge crossover, combined with computational constraints such as memory use and efficiency, dictate the update algorithm used. In the current work, we address the problem of solving for the mesh velocity in terms of its graph theory equivalent problem. Mesh positions are obtained using methods based on a graph theory analogy to the spring problem. Vertices are treated as *nodes*, while edges are treated as *springs* of varying length and tension. At each time step, the mesh coordinate positions are updated by equilibration of the spring network. Once the new vertex positions are calculated, the mesh velocity is obtained through differences between the original and equilibrated mesh vertex positions.

Specifically, we incorporate the idea of variable diffusivity while maintaining computational efficiency by avoiding solving full Laplacian equations. The method we use for updating the mesh velocity is a variation of the barycenter method [Battista et al., 1998] and relies on graph theory. Given the graph $G = (V, E)$ of element vertices V and connecting edges E, we define a partition $V = V_0 \cup V_1 \cup V_2$ of V such that V_0 contains all vertices affixed to the moving boundary, V_1 contains all vertices on the outer boundary of the computational domain, and V_2 contains all remaining interior vertices. To create the effect of variable diffusivity, we use the concept of layers. As discussed in Lohner and Yang (1996), it is desirable for the

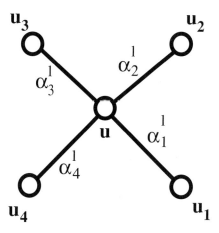

FIGURE 5.7 Graph showing vertices with associated velocities and edges with associated weights.

vertices very close to the moving boundary to have a grid velocity almost equivalent to that of the boundary. Hence, locally the mesh appears to move with solid movement, whereas far away from the moving boundary the velocity must gradually go to zero. To accomplish this in our formulation, we use the concept of local tension within layers to allow us to prescribe the rigidity of our system. Each vertex is assigned to a layer value which heuristically denotes its distance from the moving boundary. Weights are chosen such that vertices closer to the moving boundary have a higher influence on the updated velocity value. To find the updated grid velocity u^g at a vertex $v \in V_2$, we use a force-directed method. Given a configuration as shown in Figure 5.7, the grid velocity at the center vertex is given by:

$$u^g = \sum_{i=1}^{\deg(v)} \alpha_i u^l u_i, \quad \sum_{i=1}^{\deg(v)} \alpha_i^l = 1$$

where $\deg(v)$ is the number of edges meeting at the vertex v and α_i^l is the lth layer weight associated with the ith edge. This is subjected to the following constraints: $u^g = 0 (\forall v \in V_1)$, and $u^g (\forall v \in V_0)$ is prescribed to be the wall velocity. This procedure is repeated for a few cycles following an incomplete iteration algorithm, over all $v \in V_2$. (Here, by "incomplete" we mean that only a few sweeps are performed and full convergence is not sought.) Once the grid velocity is known at every vertex, the updated vertex positions are determined using explicit time integration of the newly found grid velocities.

An example of the improved computational speed gained by following the graph-theory approach vs. the classical approach of employing Poisson solvers to update the grid velocity is shown in Figure 5.8. We have computed the portion of CPU time devoted exclusively to the solver as a function of the spectral order employed in the discretization. The problem we considered involved the motion of a piezoelectric membrane induced by vortex shedding caused by a bluff body in front of the membrane. We see that a two- to three-orders of magnitude speed-up can be obtained using the graph-based algorithm.

5.4.5 The Structural Simulator

The membrane of the micropump is modeled using the linear string-beam equation:

$$\frac{d^2 y}{dt^2} + \frac{R}{m}\frac{dy}{dt} + \frac{EI}{m}\frac{d^4 y}{dx^4} - \frac{T}{m}\frac{d^2 y}{dx^2} = \frac{F}{m} \qquad (5.10)$$

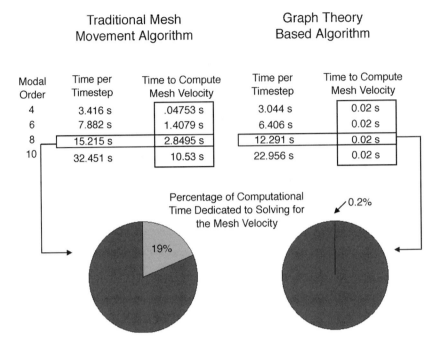

FIGURE 5.8 Comparison of CPU time for the grid velocity algorithm between the classical approach (Poisson solver) and the new approach (graph algorithm). On the left-most column is the order of spectral polynomial approximation.

where

    E is the Young's modulus of elasticity.
    I is the second moment of inertia.
    T is the axial tension.
    F is the hydrodynamic forcing.
    R is the coefficient of structural damping.
    m is the structural mass per unit length.

In this model, the coefficients are given by the physical parameters of the membrane used within the pump, and the hydrodynamic forcing on the membrane is provided by N$\varepsilon\kappa T\alpha r$.

Assume that the membrane lies in the interval $[0, L]$. For the micropump configuration, we have chosen the boundary conditions $y(0) = y(L) = 0$, $y''(0) = y''(L) = 0$ which correspond to a fixed-hinged membrane. Equation (5.10) combined with these boundary conditions lends itself to the use of eigenfunction decomposition for the efficient solution of the membrane motion. We begin by transforming the problem to lie on the interval $[0, 1]$ using the linear mapping $x = L\xi$, $\xi \in [0, 1]$. The eigenfunctions of this system are given by $\phi_n = \frac{1}{2}\sin\sqrt{\lambda_n}\xi; \sqrt{\lambda_n} = (n-1)\pi n = 1,2,\ldots,\infty$. If we assume a solution of the form $y(\xi, t) = \Sigma_{n=1}^{N} A_n(t)\phi_n(\xi)$, then by employing the Galerkin method we obtain the following evolution equation for the coefficients $A_n(t)$:

$$\frac{d^2 A_n}{dt^2} + \frac{R}{m}\frac{dA_n}{dt} + \left(\frac{EI}{mL^4}\lambda_n - \frac{T}{m^2}\right)\lambda_n A_n = \frac{1}{m}\int_0^1 F d\xi \tag{5.11}$$

We then solve this evolution equation using the Newmark scheme [Hughes, 1987], which returns the coefficients for the displacement, velocity and acceleration of the membrane. This information is then returned to N$\varepsilon\kappa T\alpha r$ as demonstrated in Figure 5.9.

Integrated Simulation for MEMS

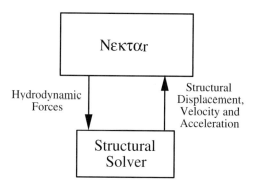

FIGURE 5.9 Coupling between NεκTαr and the structural solver. NεκTαr provides the hydrodynamic force information on the membrane. With this information the structural solver calculates the membrane's response. Structural displacement, velocity and acceleration are then returned to NεκTαr for determining the influence of the structure's motion on the fluid.

5.4.6 Differences between Circuit Fluid and Solid Simulators

The above descriptions suggest some differences between the various simulators. The key distinguishing features are

- The fluid simulator is computationally more expensive than the structure and circuit simulators.
- SPICE3 has a reliable error estimation for time discretization; therefore, a rollback in time can be done if the truncation error is unacceptable. As a result, SPICE3 automatically controls the simulation time step to ensure an acceptable user-specified error. NεκTαr is a much more complex code and does not have an automatic time-step control scheme for coupled fluid–structure simulation.
- SPICE3 uses implicit numerical integration methods for time-domain simulation. These methods are efficient for circuit simulation, because the circuit equations are stiff. For the fluid solver, however, explicit methods are simpler to implement and reasonably efficient. For this reason, NεκTαr uses semi-implicit methods for the time-domain integration (explicit for the advection terms and implicit for the diffusion terms of the Navier–Stokes equations), which suffer from the standard CFL (Courant–Friedrichs–Levy condition for the time step) restrictions. However, the flow time step is much higher than the electronics time step due to the relevant physical time scales. Also, the Newmark scheme for the structure is unconditionally stable.

5.5 Circuit-Microfluidic Device Simulation

For coupled circuit/microfluidic device simulation, four different physical domains (electrical, structure mechanical, fluid mechanical and thermal) must be considered, as shown in Figure 5.10. These domains are coupled to one another as described below.

In Figure 5.2 four types of coupling can be identified:

1. Electromechanical coupling for a piezoelectric actuation of the pump membrane
2. Fluid–structure coupling due to volume displacement of the pump membrane
3. Fluid–thermal coupling because of the thermoresistor cooling in the fluid when an anemometer type of microflow sensor is used
4. Electrothermal thermoresistor heating due to current flow in the microflow sensor

The overall system can be simulated using different approaches. One could use detailed physical simulation for each coupled domain. Another way is the use of lumped-element equivalent circuits,

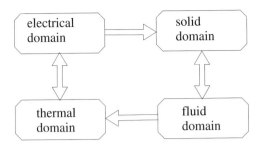

FIGURE 5.10 Coupling between the various physical domains.

compact or macromodels and/or analog hardware description languages. A third possibility is to use a combination of coupled solvers, compact models and lumped elements. In this work, we will demonstrate this third approach.

5.5.1 Software Integration

The interaction of the full system is based on different abstraction levels, using lumped circuit elements, compact/macromodels and a direct interconnection of solvers for various domains. The circuit simulator SPICE3 is chosen as the controlling solver for the following reasons:

- SPICE3 has advanced time-step control.
- Models for different abstraction levels can be easily implemented in SPICE3.
- Lumped-element equivalent circuits can be readily simulated.

Relatively simple elements are implemented as lumped elements or compact models. These elements are electromechanical transducers (piezoelectric actuator) and thermoresistors. Flow sensors are much more complicated but often the fluid flow around sensors is relatively simple. For example, if the fluid flow in a channel is fully developed then it has a parabolic profile for the velocity, thus this profile can be used (compact model) for the flow sensors as well. It is important to note that these compact models are parameterized and can be highly nonlinear. These models are obtained by insight gained from detailed physical-level simulations, such as Navier–Stokes simulations, DSMC, and linearized solutions of the Boltzmann equation [Beskok and Karniadakis, 1999]. The pump can also be described as a lumped element [Klein et al., 1998]. However, these lumped-element descriptions are applicable only for small variations in the fluid flow. Usually pumps operate in a nonlinear and non-smooth mode of fluid flow with a strong fluid–structure interaction; therefore, a detailed physical-level simulation of the pump is required. A simplification can be made by employing a macromodel of the form described in Eq. (1) but here we employ full Navier–Stokes simulations with full dynamics. For this reason, the following options are used:

- Electromechanical actuators, thermoresistors and flow sensors are described as lumped elements and/or compact models.
- The pump is modeled at the detailed physical level.
- All lumped elements and models are implemented in SPICE3.
- The pump is implemented as a direct SPICE3–NεκTαr interconnection (Figure 5.11). SPICE3 transfers the time t_{spice} and pressure P for the membrane activation to NεκTαr and receives the flowrate Q and the time t_{call} for the next call to NεκTαr.

A detailed description of this coupling is provided later.

Integrated Simulation for MEMS

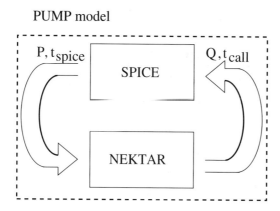

FIGURE 5.11 The SPICE3–NɛκTαr interaction for the pump microsystem of Figure 5.2. SPICE3 provides the time t_{spice} and pressure P for the membrane actuation to NɛκTαr. NɛκTαr transfers the flowrate Q at time t_{call} for the next call of NɛκTαr by SPICE3.

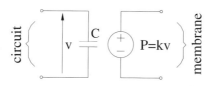

FIGURE 5.12 Lumped model for piezoelectric actuation. The voltage V is transformed into a pressure P that is used to activate the membrane of the pump.

5.5.1 Lumped Element and Compact Models for Devices

5.5.1.1 Model for Piezoelectric Transducers

The model for electromechanical coupling with a piezoelectric actuation of the membrane is shown in Figure 5.12. This model forms the interface between the electrical and mechanical networks. The electrical characteristics of the piezoelectric actuator are described by the capacitor C. The input voltage V translates into an output pressure P by virtue of the piezoelectric effect with coefficient k. This pressure is an input argument to NɛκTαr. The mechanical characteristics of the piezoelectric actuator are coupled with the mechanical characteristics of the substrate [Timoshenko and Woinowsky-Krieger, 1970; Klein, 1997].

5.5.1.2 Compact Model for Flow Sensor

For the anemometer type of flow sensor [Rasmussen and Zaghloul, 1999] shown in Figure 5.13, a macromodel has been developed by Mikulchenko et al. (2000). This macromodel (Figure 5.14) is based on neural networks trained using data from detailed physical simulations.

The inputs to the neural network are the flow velocity U and the vector of geometrical and physical parameters Θ. The results from this model are in good agreement with the simulated data for a large range of parameters [Mikulchenko et al., 2000].

The dynamic macromodel is incorporated in SPICE3 by coupling it with a sensor circuit and a model for thermoresistors for the heater and sensors as shown in Figure 5.15. Based on the fluid flow rate, the thermoresistor temperatures (T_1, T_2 and T_3) change, which in turn alter the resistance values and the sensing circuit currents and voltages.

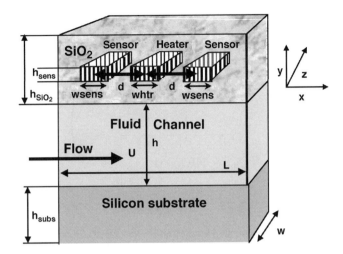

FIGURE 5.13 Structure of an anemometer-type flow sensor (thermocouple). This sensor is made up of a heating element and two sensing elements. The temperature difference between the sensors is used to measure the flow.

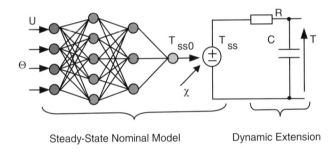

FIGURE 5.14 Dynamic macromodel for the flow sensor. The steady-state solution T_{SS0} corresponds to a nominal power for the heat source χ. The neural network output T_{SS0} is a multivariate function of the flow velocity U and the vector of geometrical and physical parameters Θ. T_{SS} is a linear function of the heat source χ and T_{SS0}.

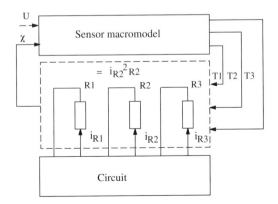

FIGURE 5.15 Macromodel implementation in SPICE3. Based on the fluid flow rate the thermoresistor temperatures (T_1, T_2 and T_3) change, which in turn alter the resistance values and the sensing circuit currents and voltages.

5.5.1.3 Effective Time-Stepping Algorithms

In general, the flow solver NεκTαr can be implemented as one big model in SPICE3. This is accomplished by calling NεκTαr from SPICE3 for each Newton iteration. However, such a coupling is extremely inefficient, because a call to NεκTαr is computationally very expensive. Furthermore, the time scales and nonlinearities are extremely different for the circuit and fluidic devices. If one considers only the circuit element, then a SPICE3 simulation results in nonuniform time steps and several Newton iterations for each time step. Typical time constants for circuits are of the order of 10^{-12} to 10^{-6} seconds. On the other hand, fluidic devices have a typical time constant of the order 10^{-4} to 10^{-1} seconds.

This property can be exploited to improve simulation performance by calling NεκTαr only at some of the circuit time points following a subcycling type of algorithm. Between these time points, the NεκTαr outputs can be modeled as constant values. Further improvement in performance is possible by taking into account the usage of semi-explicit methods for fluid simulation. In this case, the flowrate Q_n for time point t_n is calculated by the explicit scheme: $Q_n = F(\mathbf{P}_{n-1}, \mathbf{V}_{n-1}, t_n)$, where $\mathbf{P}$ is the vector of the pressure at mesh points, and $\mathbf{V}$ is the vector of velocities at mesh points. For the SPICE3–NεκTαr interaction described earlier, the important quantities are the distributed pressure P for the pump membrane and the flowrate Q_n. This functional relationship can be expressed as follows: $Q_n = f(P_{n-1}, Q_{n-1}, t_n)$.

Based on this observation an efficient time-stepping scheme is obtained as shown in Figure 5.16. Here, time is plotted on the horizontal axis and the SPICE3 iterations are plotted on the vertical axis; $t_{S,k}$ and $t_{N,k}$ are the SPICE3 and NεκTαr time points, respectively. NεκTαr selects a time step $h_{N,i} = t_{N,i} - t_{N,i-1}$ independent of SPICE3, based on the Courant number (CFL) constraint for convection. The NεκTαr time points $t_{N,i}$ are used as synchronization time points with SPICE3, whereby $t_{N,i} = t_{S,k}$. The flowrate Q has a constant value between these synchronization time points. The membrane pressure $p_{j,k}$ is calculated

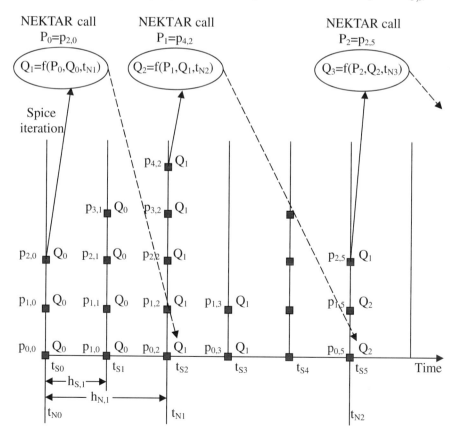

FIGURE 5.16 Time-stepping scheme for SPICE3–NεκTαr coupling.

as a function of the circuit behavior for each SPICE3 call at time $t_{S,k}$ and iteration j. The pressure $p_i = p_{M,k}$ at the final SPICE3 iteration M, for a synchronization time point $t_{S,k} = t_{N,i}$, is an input. A NɛκTαr call is made at $t_{N,i}$ and a new value of Q is computed using the relation $Q_{i+1} = f(P_i, Q_i, t_{N,i+1})$. This value is then used for the next NɛκTαr time point, $t_{N,i+1}$.

The main features of this time-stepping scheme can be summarized as follows:

- NɛκTαr is called from SPICE3.
- The time step for SPICE3 is much smaller than the time step for NɛκTαr.
- NɛκTαr specifies the next synchronization time point.

From this, it can be concluded that the number of NɛκTαr calls are the same as that of stand-alone NɛκTαr. This is the best possible situation in terms of efficiency for the coupled SPICE3–NɛκTαr simulation.

5.6 Demonstrations of the Integrated Simulation Approach

5.6.1 Microfluidic System Description

A microliquid dosing system is used as an illustrative example. This system is made up of a micropump, a flow sensor and an electronic control circuit. The electronic circuit adjusts the pump flow rate. A simplified simulation circuit is shown in Figure 5.17.

In this system, the flow rate Q determines the flow sensor velocity U for a given set of geometry parameters (h, d, wsens). Based on the fluid flow rate, the thermoresistor temperatures (T_1, T_2 and T_3) change, which in turn alter the resistance values $R_1(T_1)$, $R_2(T_2)$ and $R_3(T_3)$. The resistances $R_1(T_1)$ and $R_3(T_3)$

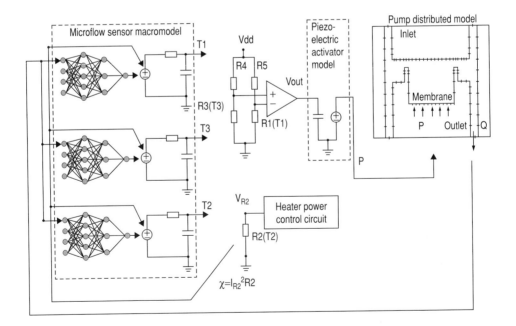

FIGURE 5.17 Description of the complete system for simulation. The pump flow rate Q determines the flow sensor velocity U. This yields the temperatures for the sensor thermoresistors. The difference between the resistance values $R_1(T_1)$ and $R_3(T_3)$ is transformed into the voltage V_{out} by the control electronics, which is used to control the pressure P for the pump membrane. This, in turn, determines the flow rate Q.

Integrated Simulation for MEMS

are included in a Wheatstone-bridge arrangement with two fixed resistors R_4 and R_5. The voltage difference $V_{R_3(T_3)} - V_{R_1(T_1)}$ is directly proportional to the temperature difference $T_3 - T_1$. This voltage difference is linearly transformed to the output voltage V_{out} by an operational amplifier with a controlled gain. This output voltage determines the pressure P, which activates the pump membrane and changes the flow rate Q. The thermoresistor of the heater (R_2) is activated by the control electronics that maintain a constant heater temperature.

5.6.2 SPICE3–NεκTαr Integration

As mentioned earlier, NεκTαr is embedded as a subroutine in SPICE3. The interaction with SPICE3 is by means of the model code and the simulation engine. Synchronization time points are determined by NεκTαr and used by the SPICE3 transient analysis engine. The pump is modeled as a SPICE3 element with NεκTαr being the underlying simulation engine. The other elements in the circuit are described by lumped element descriptions and/or compact models.

5.6.3 Simulation Results

The simulation results from the coupled simulator are presented in Figure 5.18. In this simulation, one can determine the pressure on the pump membrane, the flow velocity and the output control voltage as a function of time for various component parameters. As an example, consider the microflow sensor, the characteristics for which are shown in Figure 5.19. It is seen that for the given range of flow velocity the temperature difference between the upstream and downstream sensor temperatures is in the range 12 to 7 K. This simulation required approximately 5 minutes of CPU time on a 300-MHz Pentium II processor. Thus, the coupled simulator is reasonably efficient and provides valuable information to the system or device developers.

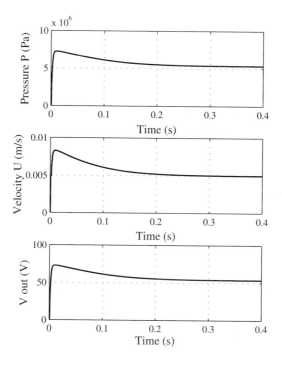

FIGURE 5.18 External pressure for the pump membrane, inlet velocity for the microflow sensor, and amplifier output voltage for the simulation of the microfluidic system as a function of time.

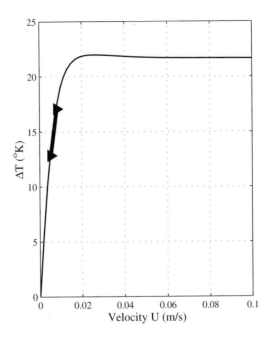

FIGURE 5.19 Flow-sensor characteristics and region of operation. A small change in velocity results in a large change in ΔT, the difference of the upstream and downstream sensor temperatures.

5.7 Summary and Discussion

Coupled-domain simulation is necessary in MEMS applications, as many different physical phenomena are present and different processes are taking place simultaneously. Depending on the specific application (e.g., a microsensor vs. a microactuator or a more complex system), some aspects of the device must be simulated in detail at high resolution while others must be accounted for by a low-dimensional description. Nonlinear macromodels are a possibility, but this is inadequate for the microfluidic system, which is typically highly unsteady and nonlinear. In addition, in the microdomain certain nonstandard flow features have to be modeled accurately such as velocity slip or temperature jump in gas flows, viscous electrokinetic effects in liquid flows and particle trajectories in particulate flows. To this end, we have developed a code that can simulate flows in the micro- and macrodomains both for liquids and gases. In addition, it includes a library of linear and nonlinear structures, such as beams, membranes, cables, etc.

For the coupled-domain simulation, the main driver program is SPICE3, a popular code for circuit simulation. In this chapter, a coupled circuit and microfluidic device simulator was presented. The simulator allows simulation of a complete microfluidic system in which thermal, flow, structural and electrical domains are integrated. The coupling of these simulators was described and demonstrated for a microliquid dosing system. The integrated simulator can be utilized for parametric studies and optimal design of microfluidic systems.

The integration of different simulators required for complete MEMS simulations is a difficult problem with challenges well beyond software integration. It involves disparate temporal and spatial scales leading to great stiffness and inefficiencies, new physical assumptions and approximations for some of the components, issues of numerical stability, staggered time-marching procedures, new fast solvers for coupled problems and optimization and control algorithms. Most of the mature algorithms are inefficient in this context so new methods are required in order to produce a new generation of simulation algorithms for MEMS devices. In this chapter, we have demonstrated that this is possible by coupling two accurate codes and resolving, at least at some level, some of these coupling issues. However, significant improvements can be made for specific devices. For example, for the membrane-driven micropump presented here, convergence of the coupling algorithm could be accelerated by inspecting the time-dependent, mass-conservation equation for every SPICE time step and obtaining a new estimate of flow from:

$$Q_{new} = Q_{old} + \frac{\Delta V}{\Delta t}$$

where ΔV is the change in volume due to the change in the membrane position, and Δt is the time between two consecutive SPICE calls. This requires solving for the structure only but not necessarily for the entire flow field, which is the most computationally intensive part. The structure solver is very fast and can be called as often as necessary without a serious computational overhead.

Acknowledgments

This work is supported in part by DARPA under agreement number F30602-98-2-0178. We would like to thank Prof. A. Beskok for many useful suggestions regarding this work.

References

Allen, M., and Tildesley, D. (1994) *Computer Simulation of Liquids,* Clarendon Press, Oxford.

Aluru, N., (1999) "A Reproducing Kernel Particle Method for Meshless Analysis," *Computational Mech.* **23**, pp. 324–338.

Aluru, N., and White, J. (1996) "Direct-Newton Finite-Element/Boundary-Element Technique for Micro-Electro-Mechanical Analysis," in *Tech. Digest, Solid-State Sensor and Actuator Workshop,* Hilton Head Island, S.C., pp. 54–57.

Battista, G.D., Eades, P., Tamassia, R., and Tollis, I., (1998) *Graph Drawing,* Prentice-Hall, Englewood Cliffs, NJ.

Beskok, A., and Karniadakis, G.E., (1999) "A Model for Flows in Channels, Pipes and Ducts at Micro- and Nano-Scales," *J. Microscale Thermophys. Eng.* **3**, pp. 43–77.

Beskok, A., Karniadakis, G.E., and Trimmer, W. (1996) "Rarefaction and Compressibility Effects in Gas Microflows," *J. Fluids Eng.* **118**, p. 448.

Bielefeld, J., Pelz, G., and Zimmer, G. (1997) "AHDL Model of a 2D Mechanical Finite-Element Usable for Microelectro-Mechanical Systems," *BMAS '97,* pp. 177–181.

Bird, G. (1994) *Molecular Gas Dynamic and the Direct Simulation of Gas Flows,* Oxford Science Publications, New York.

Engl, W.L., Laur, R., and Dirks, H.K. (1982) "MEDUSA—A Simulator for Modular Circuits," *IEEE Trans. Computer-Aided Design* **1**, pp. 85–93.

Freeman, D., Aranyosi, A., Gordon, M., and Hong, S. (1998) "Multidimensional Motion Analysis of MEMS Using Computer Microvision," in *Solid-State Sensor and Actuator Workshop,* Hilton Head Island, S.C., pp. 150–155.

Gabbay, L.D. (1998) "Computer Aided Macromodeling for MEMS," Ph.D. thesis, Department of Electrical Engineering and Computer Science, Massachusetts Institute of Technology, Cambridge, MA.

Gad-El-Hak, M. (1999) "The Fluid Mechanics of Microdevices," *J. Fluids Eng.* **12**(1), pp. 5–33.

Gilbert, J.R. et al. (1993) "Implementation of a MEMCAD System for Electrostatic and Mechanical Analysis of Complex Structures from Mask Descriptions," in *Proc. IEEE Workshop on Microelectromechanical Systems, MEMS'93,* Ft. Lauderdale, FL, pp. 207–212.

Hadjiconstantinou, N. (1999) "Hybrid Atomistic-Continuum Formulations and the Moving Contact-Line Problem," *J. Comp. Phys.* **154**, pp. 245–265.

Ho, C.-M., and Tai, Y.-C. (1998) "Microelectromechanical Systems (MEMS) and Fluid Flows," *Ann. Rev. Fluid Mech.* **30**, pp. 579–612.

Ho, L.-W. (1989) "A Legendre Spectral Element Method for Simulation of Incompressible Unsteady Free-Surface Flows," Ph.D. thesis, Department of Mechanical Engineering, Massachusetts Institute of Technology, Cambridge, MA.

Hughes, T.J.R. (1987) *The Finite Element Method,* Prentice-Hall, Englewood Cliffs, NJ.

Hung, E.S., Yang, Y.-J., and Senturia, S. (1997) "Low-Order Models for Fast Dynamical Simulations of MEMS Microstructures," *Transducers '97,* pp. 1101–1104.

Karniadakis, G.E., and Sherwin, S.J., (1999) *Spectral/hp Element Methods for CFD,* Oxford University Press, London.

Keown, J., (1997) *Microsim PSPCE and Circuit Analysis,* third edition, Prentice-Hall, Englewood Cliffs, NJ.

Kirby, R.M., Warburton, T.C., Sherwin, S.J., Beskok, A., and Karniadakis, G.E. (1999) "The NεκTαr Code: Dynamic Simulations with Remeshing," in *Second Int. Conf. on Computational Technologies for Fluid/Thermal/Chemical Systems with Industrial Applications,* Boston, MA, August 1–5.

Klein, A. (1997) "Modelling of Piezoelectric Bimorph Structures Using an Analog Hardware Description Language," in *Second Int. Conf. on the Simulation and Design of Microsystems and Microstructures (MICROSIM '97),* Lausanne, Switzerland, September.

Klein, A., Matsumoto, S., and Gerlach, G. (1998) "Modeling and Design Optimization of a Novel Micropump," in *First Int. Conf. on Modeling and Simulation of Microsystems, Semiconductors, Sensors and Actuators,* Santa Clara, CA, April 6–8.

Koplik, J., and Banavar, J., (1995) "Continuum Deductions from Molecular Hydrodynamics," *Annu. Rev. Fluid Mech.* **27**, pp. 257–292.

Kundert, K.S. (1987) "Sparse-Matrix Techniques and Their Application to Circuit Simulation," in *Circuit Analysis, Simulation and Design,* ed. A.E. Ruehli, North-Holland, Amsterdam.

Lo, N.R. et al. (1996) "Parameterized Layout, Synthesis, Extraction, and SPICE Simulation for MEMS," *ICASE '96,* pp. 481–484.

Lohner, R., and Yang, C. (1996) "Improved ALE Mesh Velocities for Moving Bodies," *Commun. Num. Meth. Eng., Phys.* **12**, pp. 599–608.

Mayaram, K., and Pederson, D.O. (1987) "Codecs: An Object-Oriented Mixed-Level Circuit and Device Simulator," in *1987 IEEE Int. Symp. on Circuits and Systems,* pp. 604–607, Philadelphia, May 4–7.

Mayaram, K., and Pederson, D.O. (1992) "Coupling Algorithms for Mixed-Level Circuit and Device Simulation," *IEEE Trans. Computer-Aided Design* **11**, pp. 1003–1012.

Mayaram, K., Chern, J., and Yang, P. (1993) "Algorithms for Transient Three-Dimensional Mixed-Level Circuit and Device Simulation," *IEEE Trans. Computer-Aided Design* **12**, pp. 1726–1733.

Mikulchenko, O., Rasmussen, A., and Mayaram, K. (2000) "A Neural Network-Based Macromodel of Microflow Sensors," *Int. Conf. on Modeling and Simulation of Microsystems,* pp. 540–543, March.

Moaveni, S. (1999) *Finite Element Analysis: Theory and Application with ANSYS,* Prentice-Hall, Englewood Cliffs, NJ.

Nagel, L.W. (1975) "SPICE2: A Computer Program To Simulate Semiconductor Circuits," Tech. Rep. No. UCB/ERL M520, Electronics Research Lab., University of California, Berkeley.

Nagel, L.W., and Pederson, D.O. (1973) "SPICE—Simulation Program with Integrated Circuit Emphasis," in *Proc. 16th Midwest Symp. Circuit Theory,* Waterloo, Canada; Tech. Rep. No. UCB/ERL M382, Electronics Research Laboratory, University of California, Berkeley.

Park, H.J., Ko, P.K., and Hu, C. (1991) "A Charge Conserving Non-Quasi-Static (NQS) Model for SPICE Transient Analysis," *IEEE Trans. Computer-Aided Design* **10**, pp. 629–642.

Quarles, T.L. (1989) "The SPICE3 Implementation Guide," Tech. Rep. No. UCB/ERL M89/44, Electronics Research Lab., University of California, Berkeley.

Rasmussen, A., and Zaghloul, M.E. (1999) "The Design and Fabrication of Microfluidic Flow Sensors," in *Proc. ISCAS-99,* pp. 136–139.

Ravanelli, E., and Hu, C. (1991) "Device-Circuit Mixed Simulation of VDMOS Charge Transients," *Solid State Electron.* **34**, pp. 1353–1360.

Rotella, F.M., Troyanovsky, B., Yu, Z., Dutton, R., and Ma, G. (1997) "Harmonic Balance Device Analysis of an LDMOS RF Power Amplifier with Parasitics and Matching Network," in *SISPAD-97*, pp. 157–159.

Sangiovanni-Vincentelli, A.L. (1981) "Circuit Simulation," in *Computer Design Aids for VLSI Circuits*, ed. P. Antognetti, D.O. Pederson, and H. De Man, pp. 19–113, Sijthoff & Noordhoff, Rockville, MD.

Schroth, A., Blochwitz, T., and Gerlach, G. (1995) "Simulation of a Complex Sensor System Using Coupled Simulation Programs," in *Transducers '95*, pp. 33–35.

Senturia, S. et al. (1992) "A Computer-Aided Design System for Microelectromechanical Systems (MEMCAD)," *J. Microelectromech. Syst.* **1**, pp. 3–13.

Senturia, S., Aluru, N., and White, J. (1997) "Simulating the Behavior of MEMS Devices: Computational Methods and Needs," *IEEE Computational Sci. Eng.* January–March, pp. 30–43.

Sherwin, S.J. (1995) "Triangular and Tetrahedral Spectral/hp Finite Element Methods for Fluid Dynamics," Ph.D. thesis, Princeton University, Princeton, NJ.

Silvaco International (1995) *ATLAS User's Manual: Mixed Mode*, Santa Clara, CA, June.

Technology Modeling Associates (1997), *MEDICI User's Manual: Circuit Analysis*, Sunnyvale, CA, February.

Tilmans, H.A.C. (1996) "Equivalent Circuit Representation of Electromechanical Transducers. I. Lumped-Parameter Systems," *J. Micromech. Microeng.* **6**, pp. 157–176.

Timoshenko, S.P., and Woinowsky-Krieger, S. (1970) *Theory of Plates and Shells*, second edition, McGraw-Hill, New York.

Veijola T. et al. (1995) "Equivalent-Circuit Model of the Squeezed Gas Film in a Silicon Accelerometer," *Sensors and Actuators A* **48** pp. 239–248.

Warburton, T.C. (1999) "Spectral/hp Methods on Polymorphic Multi-Domains: Algorithms and Applications," Ph.D. thesis, Division of Applied Mathematics, Brown University, Providence, RI.

Woodruff, R.L., and Rudeck, P.J. (1993) "Three-Dimensional Numerical Simulation of Single Event Upset of an SRAM Cell," *IEEE Trans. Nucl. Sci.* **40**, pp. 1795–1803.

Ye, W., Kanapka, J., and White, J. (1999) "A Fast 3D Solver for Unsteady Stokes Flow with Applications to Micro-Electro-Mechanical Systems," in *Proc. Second Int. Conf. on Modeling and Simulation of Microsystems*, pp. 518–521.

6
Liquid Flows in Microchannels

Kendra V. Sharp
University of Illinois at Urbana–Champaign

Ronald J. Adrian
University of Illinois at Urbana–Champaign

Juan G. Santiago
Stanford University

Joshua I. Molho
Stanford University

6.1 Introduction 6-1
   Unique Aspects of Liquids in Microchannels • Continuum Hydrodynamics of Pressure Driven Flow in Channels • Hydraulic Diameter • Flow in Round Capillaries • Entrance Length Development • Transition to Turbulent Flow • Noncircular Channels
6.2 Experimental Studies of Flow Through Microchannels 6-10
   Proposed Explanations for Measured Behavior • Measurements of Velocity in Microchannels • Nonlinear Channels • Capacitive Effects
6.3 Electrokinetics Background 6-17
   Electrical Double Layers • EOF with Finite EDL • Thin EDL Electro-Osmotic Flow • Electrophoresis • Similarity between Electric and Velocity Fields for Electro-Osmosis and Electrophoresis • Electrokinetic Microchips • Engineering Considerations: Flowrate and Pressure of Electro-Osmotic Flow • Electrical Analogy and Microfluidic Networks • Practical Considerations
6.4 Summary and Conclusions 6-33
   Nomenclature 6-33

6.1 Introduction

Nominally, microchannels can be defined as channels whose dimensions are less than 1 mm and greater than 1 μm. Above 1 mm the flow exhibits behavior that is the same as most macroscopic flows. Below 1 μm the flow is better characterized as nanoscopic. Currently, most microchannels fall into the range of 30 to 300 μm. Microchannels can be fabricated in many materials—glass, polymers, silicon, metals—using various processes including surface micromachining, bulk micromachining, molding, embossing and conventional machining with microcutters. These methods and the characteristics of the resulting flow channels are discussed elsewhere in this handbook.

Microchannels offer advantages due to their high surface-to-volume ratio and their small volumes. The large surface-to-volume ratio leads to a high rate of heat and mass transfer, making microdevices excellent tools for compact heat exchangers. For example, the device in Figure 6.1 is a cross-flow heat exchanger constructed from a stack of 50 14-mm × 14-mm foils, each containing 34 200-μm-wide × 100-μm-deep channels machined into the 200-μm-thick stainless steel foils by the process of direct, high-precision mechanical micromachining [Brandner et al., 2000; Schaller et al., 1999]. The direction of flow in adjacent foils is alternated 90°, and the foils are attached by means of diffusion bonding to create a stack of cross-flow heat exchangers capable of transferring 10 kW at a temperature difference of

FIGURE 6.1 (Color figure follows p. 12-26.) Micro heat exchanger constructed from rectangular channels machined in metal. (Courtesy of K. Schubert and D. Cacuci, Forschungszentrum, Karlsruhe.)

80 K using water flowing at 750 kg/hr. The impressively large rate of heat transfer is accomplished mainly by the large surface area covered by the interior of the microchannel: approximately 3600 mm^2 packed into a 14-mm cube.

A second example of the application of microchannels is in the area of microelectromechanical systems (MEMS) devices for biological and chemical analyses. The primary advantages of microscale devices in these applications are the good match with the scale of biological structures and the potential for placing multiple functions for chemical analysis on a small area—the concept of a "chemistry laboratory on a chip."

Microchannels are used to transport biological materials such as (in order of size) proteins, DNA, cells and embryos or to transport chemical samples and analytes. Typical of such devices is the i-STAT blood sample analysis cartridge shown in Figure 6.2. The sample is taken on board the chip through a port and moved through the microchannels by pressure to various sites where it is mixed with analyte and moved to a different site where the output is read. Flows in biological devices and chemical analysis microdevices are usually much slower than those in heat transfer and chemical reactor microdevices.

6.1.1 Unique Aspects of Liquids in Microchannels

Flows in microscale devices differ from their macroscopic counterparts for two reasons: the small scale makes molecular effects such as wall slip more important, and it amplifies the magnitudes of certain ordinary continuum effects to extreme levels. Consider, for example, strain rate and shear rate which

FIGURE 6.2 Blood sample cartridge using microfluidic channels. (Courtesy of i-Stat, East Windsor, NJ.)

scale in proportion to the velocity scale, U_s, and inverse proportion to the length scale, L_s. Thus, 100-mm/s flow in a 10-μm channel experiences a shear rate of the order of $10^4 \, s^{-1}$. Acceleration scales as U_s^2/L_s and is similarly enhanced. The effect is even more dramatic if one tries to maintain the same volume flux while scaling down. The flux scales as $Q \sim U_s L_s^2$, so at constant flux $U_s \sim L_s^{-2}$ and both shear and acceleration go as L_s^{-3}. Fluids that are Newtonian at ordinary rates of shear and extension can become non-Newtonian at very high rates. The pressure gradient becomes especially large in small cross-section channels. For fixed volume flux, the pressure gradient increases as L_s^{-4}.

Electrokinetic effects occur at the interface between liquids and solids such as glass due to chemical interaction. The result is an electrically charged double layer that induces a charge distribution in a very thin layer of fluid close to the wall. Application of an electric field to this layer creates a body force capable of moving the fluid as if it were slipping over the wall. The electro-osmotic effect and the electrophoretic effect (charges around particles) will be discussed in detail in a later section. Neither occurs in gases.

The effects of molecular structure are quite different in gases and liquids. If the Knudsen number (defined as $Kn = \lambda/L_s$, where λ is the mean free path in a gas and L_s is the characteristic channel dimension) is greater than 10^{-3} [Gad-el-Hak, 1999; Janson et al., 1999], nonequilibrium effects may start to occur. Modified slip boundary conditions can be used in continuum models for Knudsen numbers between 10^{-1} and 10^{-3} [Gad-el-Hak, 1999]. As the Knudsen number continues to increase, continuum assumptions and fluid theory are no longer applicable. Analysis of such flow requires consideration of different physical phenomena [see Chapter 8 in this book on Analytical and Computational Models for Microscale Flows; Arkilic et al., 1997; Gad-el-Hak, 1999; Harley et al., 1995; Janson et al., 1999].

Because the density of liquids is about 1000 times the density of gases, the spacing between molecules in liquids is approximately ten times less than the spacing in gases. Liquid molecules do not have a mean free path, but following Bridgman (1923), the lattice spacing, δ, may be used as a similar measure. The lattice spacing, δ, is defined as [Probstein, 1994]:

$$\delta \sim \left(\frac{\bar{V}_1}{N_A}\right)^{1/3} \tag{6.1}$$

where $\bar{V}_1$ is the molar volume and N_A is Avogadro's number. For water, this spacing is 0.3 nm. In a 1-μm gap and a 50-μm-diameter channel, the equivalent Knudsen numbers are 3×10^{-4} and 6×10^{-6}, respectively, well within the range of obeying continuum flow. In gases, effects such as slip at the wall occur when the mean free path length of the molecules is more than about one tenth of the flow dimension (i.e., flow dimensions of order less than 650 nm in air at STP). (Note that the mean free path length of a gas is longer than the mean spacing between its molecules; see Chapter 4 for a detailed discussion.) In liquids, this condition will not occur unless the channels are smaller than approximately 3 nm, and continuum hydrodynamics may provide a useful description at scales even smaller than this because the forces of interaction between molecules in liquids are long range. For example, Stokes' classical result for drag on a sphere is routinely applied to particles whose diameters are well below 100 nm. Thus, liquid flow in microdevices should be described adequately by continuum hydrodynamics well below dimensions of 1 μm.

Molecular effects in liquids are difficult to predict because the transport theory is less well developed than the kinetic theory of gases. For this reason, studies of liquid microflows in which molecular effects may play a role are much more convincing if done experimentally.

Liquids are essentially incompressible. Consequently, the density of a liquid in microchannel flow remains very nearly constant as a function of distance along the channel, despite the very large pressure gradients that characterize microscale flow. This behavior greatly simplifies the analysis of liquid flows relative to gas flows wherein the large pressure drop in a channel leads to large expansion and large changes of thermal heat capacity.

The large heat capacity of liquids relative to gases implies that the effects of internal heating due to viscous dissipation are much less significant in liquid flows. The pressure drop in microchannel flow can be very large, and because all of the work of the pressure difference against the mean flow ultimately goes into viscous dissipation, effects due to internal heating by viscous dissipation may be significant. However, they will be substantially less in liquids than in gases, and they can often be ignored, allowing one to approximate the liquid as a constant density, constant property fluid.

The dynamic viscosity, μ, of a liquid is larger than that of a gas by a factor of about 100 (see Table 6.1). This implies much higher resistance to flow through the channels. The kinematic viscosity of a liquid is typically much less than the kinematic viscosity of a gas, owing to the much higher density of liquids (Table 6.1). This implies much more rapid diffusion of momentum in gases. Similar statements pertain qualitatively to the thermal conductivity and the thermal diffusivity.

Liquids in contact with solids or gases have surface tension in the interface. At the microscale, the surface tension force becomes one of the most important forces, far exceeding body forces such as gravity and electrostatic fields.

Bubbles can occur in liquids, for good or ill. Unwanted bubbles can block channels, or substantially alter the flow. But, bubbles can also be used to apply pressure and to perform pumping by heating and cooling the gases inside the bubble.

Particulates and droplets suspended in liquids have densities that match liquid more closely. Settling is much less rapid in liquids, and suspensions have the ability to follow the accelerations of the flow. This effect can also keep suspended impurities in suspension for much longer, thereby increasing the probability that an impurity will introduce unwanted behavior.

Liquids can interact with solids to form an electric double layer at the interface. This is the basis for the phenomena of electro-osmosis and electrophoresis, both of which can be used to move fluid and particles in channels. These topics will be discussed in detail in a later section. Liquids can be non-Newtonian, especially at the high shear rates encountered in microchannels.

TABLE 6.1 Dynamic and Kinematic Viscosities of Typical Liquids Compared to Air at 1 atm

Fluid	Dynamic Viscosity μ (gm cm^{-1} s^{-1})	Kinematic Viscosity ν (cm^2 s^{-1})	Thermal Conductivity k (J K^{-1} s^{-1} cm^{-1})	Thermal Diffusivity κ (cm^2 s^{-1})
Water @15°C	0.0114	0.0114	0.0059	0.00140
Ethyl alcohol @ 15°C	0.0134	0.0170	0.00183	0.00099
Glycerin @15°C	23.3	18.50	0.0029	0.00098
Air @15°C	0.000178	0.145	0.000253	0.202

6.1.2 Continuum Hydrodynamics of Pressure Driven Flow in Channels

The general continuum description of the flow of an incompressible, Newtonian fluid flow with variable properties and no body forces other than gravity (i.e., no electrical forces) consists of the incompressible continuity equation:

$$\frac{\partial u_j}{\partial x_j} = 0 \qquad (6.2)$$

and the momentum equation:

$$\rho\left(\frac{\partial u_i}{\partial t} + u_j \frac{\partial u_i}{\partial x_j}\right) = \frac{\partial \tau_{ij}}{\partial x_j} + \rho b_i \qquad (6.3)$$

where the fluid stress is given by Stokes' law of viscosity:

$$\tau_{ij} = -p\delta_{ij} + \mu\left(\frac{\partial u_i}{\partial x_j} + \frac{\partial u_j}{\partial x_i}\right) \qquad (6.4)$$

Here, u_i is the ith component of the velocity vector $\mathbf{u}(\mathbf{x}, t)$, ρ is the mass density (kg/m^3), b_i is the body force per unit mass (m/s^2) (often, $b_i = g_i$, the gravitational acceleration), and τ_{ij} is the stress tensor (N/m^2). The corresponding enthalpy equation is

$$\rho c_p\left(\frac{\partial T}{\partial t} + u_j \frac{\partial T}{\partial x_j}\right) = -\frac{\partial q_j}{\partial x_j} + \Phi \qquad (6.5)$$

where T is the temperature, and q is the heat flux (J/s m^2) given by Fourier's law of heat conduction by molecular diffusion k,

$$q_i = -k\frac{\partial T}{\partial x_i} \qquad (6.6)$$

The rate of conversion of mechanical energy into heat due to internal viscous heating is

$$\Phi = \mu\left(\frac{\partial u_i}{\partial x_j} + \frac{\partial u_j}{\partial x_i}\right)\frac{\partial u_i}{\partial x_j} \qquad (6.7)$$

Consider a long parallel duct or channel with the x-direction along the axis of the channel and the coordinates y and z in the plane perpendicular to the axis of the channel (Figure 6.3). The entering flow

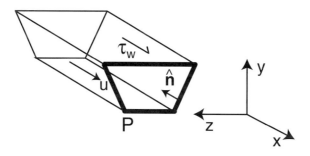

FIGURE 6.3 Flow in a duct of arbitrary cross section A. P is the perimeter and τ_w is the wall shear stress.

undergoes a transient response in which the velocity and temperature profiles change in the streamwise direction. This process continues until the flow properties become independent of the streamwise position. In this state of fully developed velocity profile, the velocity field is unidirectional, $\mathbf{u}(x) = (u(y, z), 0, 0)$, and there is no acceleration of the fluid. Thus, for fully developed flow with gravitational body force $\mathbf{g}$ the equations become, very simply:

$$\rho \frac{\partial u}{\partial t} = -\frac{dp}{dx} + \rho g_x + \frac{\partial}{\partial y}\left(\mu \frac{\partial u}{\partial y}\right) + \frac{\partial}{\partial z}\left(\mu \frac{\partial u}{\partial z}\right) \tag{6.8}$$

$$\rho c_p \frac{\partial T}{\partial t} = \frac{\partial}{\partial y}\left(k \frac{\partial T}{\partial y}\right) + \frac{\partial}{\partial z}\left(k \frac{\partial T}{\partial z}\right) + \Phi \tag{6.9}$$

Finally, if the flow is steady and the temperature and properties are constant, then the equation for streamwise velocity profiles becomes a simple Poisson equation:

$$\frac{\partial^2 u}{\partial y^2} + \frac{\partial^2 u}{\partial z^2} = \frac{1}{\mu} \frac{d}{dx}(p - \rho g_x x) \tag{6.10}$$

In the absence of electrokinetic effects and for shear rates less than about 10^{12} s^{-1}, the appropriate boundary condition is the no-slip condition:

$$u = 0 \quad \text{on the boundary } P \tag{6.11}$$

6.1.3 Hydraulic Diameter

Control volume analysis of fully developed flow leads naturally to the concept of the *hydraulic diameter*. Figure 6.3 shows flow in a duct of arbitrary cross section. Because the flow is fully developed and unidirectional (assuming a straight duct), the acceleration is zero and control volume analysis of the momentum reduces to a simple force balance in the streamwise direction,

$$-\frac{dp}{dx} A = \bar{\tau}_w P \tag{6.12}$$

where

$$\bar{\tau}_w = \frac{1}{P} \oint_P \tau_w \, dl \tag{6.13}$$

is the wall shear stress averaged around the perimeter, and the local wall shear stress is given by:

$$\tau_w = \mu \frac{\partial u}{\partial n}\bigg|_{n=0} \tag{6.14}$$

Equation (6.12) displays the relevance of the ratio of the area A to the perimeter P. In practice, the hydraulic diameter is defined to be

$$D_h = \frac{4A}{P} \tag{6.15}$$

Liquid Flows in Microchannels

so that when the cross section is a circle, D_h equals the diameter of the circle. The hydraulic diameter provides a convenient way to characterize a duct with a single length scale and a basis for comparison between ducts of different shapes. A common approximation is to also estimate the flow resistance in a duct or channel as the resistance of a round duct whose diameter is equal to the hydraulic diameter. This approximation is useful but subject to errors of order 10 to 20%. Because solution of Poisson's equation to obtain the exact wall shear stress is accomplished readily by numerical means, it is not necessary.

6.1.4 Flow in Round Capillaries

Flow in a round tube is the archetype for all duct and channel flows. While microfabrication characteristically yields channels of noncircular cross section, the round cross section is a useful and familiar point of reference, and microcapillaries are not uncommon. Extensive macroscale research on pipe flows dates back to Hagen's (1839), Poiseuille's (1841) and Reynolds' (1883) original studies in the 19th century. Independently, both Hagen (1839) and Poiseuille (1841) observed the relation between pressure head and velocity and its inverse proportionality to the fourth power of tube diameter.

In a round capillary of radius $a = D/2$ and radial coordinate r, it is well known that the velocity profile across a diameter is parabolic:

$$u = u_{max}\left(1 - \frac{r^2}{a^2}\right) \qquad (6.16)$$

where the maximum velocity is given by:

$$u_{max} = \frac{a^2}{4\mu}\left(-\frac{dp}{dx}\right) \qquad (6.17)$$

The volume flow rate, Q, is given by:

$$Q = \bar{U}A \qquad (6.18)$$

where the average velocity, $\bar{U}$, defined by:

$$\bar{U} = \frac{1}{\pi a^2}\int_0^a u(r)2\pi r\, dr \qquad (6.19)$$

is numerically equal to

$$\bar{U} = \frac{1}{2}u_{max} \qquad (6.20)$$

Using these relations it is easily shown that the pressure drop in a length L, $\Delta p = (-dp/dx)L$, is given by:

$$\Delta p = \frac{8\mu L Q}{\pi a^4} \qquad (6.21)$$

The Darcy friction factor, f, is defined so that:

$$\Delta p = f\frac{L}{D}\rho\frac{\bar{U}^2}{2} \qquad (6.22)$$

(the Fanning friction factor is one fourth the Darcy friction factor). The Reynolds number is defined, in terms of a characteristic length scale,[1] L_s, by:

$$Re = \frac{\rho \bar{U} L_s}{\mu} \tag{6.23}$$

For a round pipe, the characteristic length scale is the diameter of the pipe, D. The friction factor for laminar flow in a round capillary is given by:

$$f = \frac{64}{Re} \tag{6.24}$$

The Poiseuille number is sometimes used to describe flow resistance in ducts of arbitrary cross section. It is defined by:

$$Po = f\,Re/4$$
$$= -\frac{1}{\mu}\frac{dp}{dx}\frac{D_h^2}{2\bar{U}} \tag{6.25}$$

where L_s used in the calculation of Re is D_h. The Poiseuille number has a value of 16 for a round capillary.

The inverse relationship between friction factor and Reynolds number has been well documented on the macroscale. It means that the pressure drop is linearly proportional to the flow rate, Q. In the laminar region there is no dependence on surface roughness.

As the Reynolds number increases above 2000 in a circular duct, the flow begins to transition to turbulence. At this point, the friction factor increases dramatically, and the flow resistance ultimately becomes proportional to Q^2 rather than Q.

6.1.5 Entrance Length Development

Before the flow reaches the state of a fully developed velocity profile, it must transition from the profile of the velocity at the entrance to the microduct, whatever that is, to the fully developed limit. This transition occurs in the *entrance length* of the duct. In this region, the flow looks like a boundary layer that grows as it progresses downstream. Ultimately, the viscously retarded layers meet in the center of the duct at the end of the entrance length.

The pressure drop from the beginning of the duct to a location x is given by:

$$p_0 - p(x) = \left(f\frac{x}{D_h} + K(x)\right)\frac{\rho \bar{U}^2}{2} \tag{6.26}$$

where $K(x)$ is the pressure drop parameter given in Figure 6.4 for a circular duct and for parallel plates [White, 1991]. One sees that the flow development is largely completed by $x/D = 0.065\,Re$.

6.1.6 Transition to Turbulent Flow

In 1883, Reynolds found a critical value of velocity, u_{crit}, above which the form of the flow resistance changes. The corresponding dimensionless parameter is the critical Reynolds number, Re_{crit}, below which disturbances in the flow are not maintained. Such disturbances may be caused by inlet conditions such as a sharp edge or unsteadiness in the flow source. Depending on the Reynolds number, disturbances may also be introduced by natural transition to turbulent flow.

---

[1] In the remainder of this chapter, the characteristic length scale used in calculating Re is to be inferred from context, e.g., generally D_h for a rectangular channel and D for a circular tube.

Liquid Flows in Microchannels

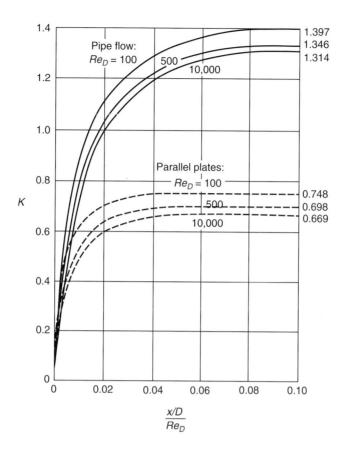

FIGURE 6.4 Entrance length parameter K for laminar flow in the inlet of a duct. (From White, F.M. (1991) *Viscous Flow*, second edition, McGraw-Hill, New York, p. 292. With permission.)

Reynolds found Re_{crit} to be approximately 2000, and this value has been generally accepted. Once the flow is fully turbulent, the empirical relationship often used to correlate friction factor and the Reynolds number for smooth pipes and initially proposed by Blasius is

$$f \approx 0.3164 \, Re^{-0.25} \tag{6.27}$$

For rough pipes, the friction factor departs from the Blasius relation in the turbulent region. This departure occurs at different values of Re depending on the magnitude of the surface roughness. The Moody chart summarizes the traditional friction factor curves and is readily available in any basic fluids textbook (e.g., White, 1994, p. 318).

6.1.7 Noncircular Channels

Fully developed flow in noncircular ducts is found by solving the Poisson equation (Eq. 6.10). Frequently, analytical solutions can also be found, but the numerical approach is so reliable that there is little need for exact solutions. Developing flow in the entrance region is more difficult, but here again numerical approaches are relatively straightforward. Table 6.2 summarizes the flow resistance for various laminar flows. One sees that the effect of the shape of the channel is relatively weak.

As mentioned earlier, a common approximation made in analyzing flow in ducts of noncircular cross section is to use the results for circular ducts, but with the hydraulic diameter of the noncircular duct replacing that of the round duct. For example, this can be done to estimate the flow resistance of fully developed flow and the resistance in the entrance region.

TABLE 6.2 Resistance to Flow in Fully Developed Flow Through Straight Microchannels of Various Cross-Sectional Geometries

Cross Section	$f\,Re$	u_{max}/u_B
Circle, $D=2a$	64	2.000
Square, $2a \times 2a$	56.92	2.0962
Rectangle, $2a \times 2b$, $\alpha = b/a$	$96[1 - 1.3553\alpha + 1.9467\alpha^2$ $-1.7012\alpha^3 + 0.9564\alpha^4$ $-0.2537\alpha^5]$	—
Parallel plates, $\alpha \to 0$	96	1.5000
Hexagon, $60°$	60	—
Trapezoid, $60°$, $2a \times 2b$	$\underline{2b/2a}$ 4.000 55.66 2.000 55.22 1.000 56.60 0.500 62.77 0.250 72.20	2.181 2.162 2.119 1.969 1.766
Trapezoid, $54.72°$, $2a \times 2b$	1.000 56.15	2.137

Source: Data from Shah, R.K., and London, A.L. (1978) *Advances in Heat Transfer,* Suppl. 1, Academic Press, New York.

6.2 Experimental Studies of Flow Through Microchannels

Despite the fundamental simplicity of laminar flow in straight ducts, experimental studies of microscale flow have often failed to reveal the expected relationship between the friction factor and Reynolds number. Further, flow discrepancies are neither consistently higher nor lower than macroscale predictions. A summary of the experiments that have been conducted to investigate the behavior of fluid flow in microchannels, over a large range of Reynolds numbers, geometries and experimental conditions, is presented in Table 6.3. In reviewing these results, they will be grouped according to the results of friction factor measurements (follow macroscale predictions, higher than predictions, and lower than predictions).

The first experimental investigations of flow through microchannels in the early 1980s were motivated by the interest in high-performance heat sinking. The large surface-to-volume ratios of microchannels make them excellent candidates for efficient heat transfer devices, as discussed in the introduction to this chapter. Tuckerman and Pease (1981) studied flow through an array of microchannels with approximately rectangular cross sections (height range 50 to 56 μm, width range 287 to 320 μm). Although this study was focused primarily on heat transfer characteristics, they "confirmed that the flow rate obeyed

TABLE 6.3 Experimental Conditions for Flow Resistance Experiments

Study	Working Fluid	Channel Description	Reynolds Numbers
Choi et al. (1991)	Nitrogen gas	Circular: diameter 3–81 μm	~30–20,000
Flockhart and Dhariwal (1998)	Water	Trapezoidal: depth 27–63 μm, width 100–1000 μm, length 12–36 mm	<600
Jiang et al. (1995)	Water	Circular: diameter 8–42 μm; rectangular, trapezoidal and triangular: depth 13.4–46 μm, width 35–110 μm, length 2.5–10 mm	Circular: < ~1.2; unclear for other geometries
Mala and Li (1999)	Water	Circular: diameter 50–254 μm	Up to 2500
Papautsky et al. (1999a)	Water	Array of rectangular channels, each: depth 30 μm, width 600 μm, length 3 mm	1–18
Papautsky et al. (1999b)	Water	Rectangular array of channels: height 22.71–26.35 μm, width 150–600 μm, length 7.75 mm	0.001–10
Peng et al. (1994)	Water	Rectangular: height 100–300 μm, width 200–400 μm, length 50 mm	50–4,000
Pfahler et al. (1991)	n-Propanol, silicone oil, nitrogen gas, helium gas	Rectangular and trapezoidal, depending on channel: depth 0.48–38.7 μm, width 55–115 μm, length 10.2–10.9 mm	Liquids: <<1 to approx. 80
Qu et al. (2000)	Water	Trapezoidal: depth 28–114 μm, width 148–523 μm, length 28 mm	~10–1,450
Sharp et al. (2000)	Water	Circular: diameter 75–242 μm	50–2500
Tuckerman and Pease (1981)	Water	Array of etched rectangular/trapezoidal, each: depth 50–56 μm, width 287–320 μm, length 1 cm	~200–600
Wilding et al. (1994)	Water, biological fluids	Trapezoidal: depth 20–40 μm, width 40–150 μm, length 11.7 mm	Water: 17–126
Wu and Little (1983)	Nitrogen, helium and argon	Trapezoidal or U-shaped: depth 28–65 μm, width 133–200 μm, length 7.6–40.3 mm	~200–15,000
Yu et al. (1995)	Nitrogen gas, water	Circular: diameter 19–102 μm	250–20,000

Note: When a range of values for channel dimension is given, it indicates that several channels were used, with different individual dimensions.

Poiseuille's equation." Shortly thereafter, a study of microchannels for use in small Joule–Thomson refrigerators was performed [Wu and Little, 1983]. Significant roughnesses were present in some of these etched silicon or glass channels, but friction factors measured in the smoothest channel showed reasonable agreement with theoretical macroscale predictions. Several other experiments have also shown general agreement with macroscale theoretical predictions for friction factor, in at least certain parameter ranges [Flockhart and Dhariwal, 1998; Jiang et al., 1995; Sharp et al., 2000; Wilding et al., 1994].

Geometrical differences were emphasized in Jiang et al.'s (1995) experiments, which employed circular glass microchannels and silicon microchannels with rectangular, trapezoidal and triangular cross sections. Although linear relationships between flow rate and pressure drop were observed in all the microchannels, a direct analysis of the friction factor was not performed except in the case of the circular microchannels ($D = 8$–42 μm, $Re < 1.2$). In the circular case, the friction factor matched theoretical predictions to within 10 to 20%.

Focusing primarily on biological fluids, but also using water and saline as working fluids, Wilding et al. (1994) found that results for water flowing in silicon micromachined channels (Reynolds numbers ~17 to 126) agreed well with theory for at least the lower Reynolds numbers tested. The disagreement at higher Reynolds numbers was noted to be a result of "entrance effects and inertial losses" not included in the theory.

Saline was also used as the working fluid without much change in the results, suggesting that surface charge effects are minimal. Biological fluids, as expected, exhibited shear thinning and non-Newtonian behavior.

Comparison of numerical calculations for flow in trapezoidal channels finds good agreement between numerics and experiment for $Re < 600$, although some entrance effects were observed in the shortest channels [Flockhart and Dhariwal, 1998]. The numerical calculations were motivated by the fact that this particular trapezoidal geometry is very common in microfluidic applications as a result of anisotropic etching techniques employed in fabrication, particularly in <100> silicon.

Microscale measurements of the friction factor by Sharp et al. (2000) generally agree with the macroscale laminar theory to within ±2% experimental error over all Reynolds numbers up to transition ($\sim 50 < Re < 2000$), for water flowing through circular fused silica microchannels with $D \sim 75$ to $242\ \mu m$. Similar agreement is also obtained using a 20% solution of glycerol and 1-propanol. Occasional discrepancies that fall outside of the error bars tend to occur in the Reynolds number range $1200 < Re < 2000$, but in no case is the discrepancy greater than +4%.

Studies finding an increase in friction factor on the microscale under certain conditions include Wu and Little (1983), Peng et al. (1994), Mala and Li (1999), Qu et al. (2000) and Papautsky et al. (1999a; 1999b). In some cases, the departure from agreement with theoretical predictions is reasonably linked with a change in roughness or geometrical parameter. For example, Wu and Little's (1983) friction factor measurements appear to correlate with surface roughness, as the results agreed well with theory for smooth channels, but the agreement decreased as the roughness increased. In the rougher channels (absolute roughness up to $40\ \mu m$), the measured friction factors were higher than theoretical predictions. A linear relationship between f and Re was maintained in the laminar region, and the authors suggested that a transition to turbulence appeared at Re lower than expected (namely, 400) for the roughest glass channel.

In an effort to understand the influence of geometrical parameters (specifically, hydraulic diameter and aspect ratio) on flow resistance, Peng et al. (1994) considered water flows in rectangular machined steel grooves enclosed with a fiberglass cover. A large range of Re were obtained (50 to 4000), and a geometrical dependence was observed. For the most part, f increased with increasing H/W and also with increasing D_h (holding H/W constant).

The friction factor was observed to be higher than macroscale predictions for five of seven geometrical cases, all except $H/W = 0.5$. Because the relationship between f and Re no longer appeared inversely proportional, new power laws were suggested for this correlation in the laminar and turbulent regions, namely, $f \propto Re^{-1.98}$ and $f \propto Re^{-1.72}$, respectively.

In addition to proposing new correlations in the laminar and fully turbulent regions, Peng et al. (1994) suggested that the regions themselves must be redefined depending upon the geometry of the microchannel. As in macroscale flows (*cf.* standard Moody chart, White, 1994, p. 318), two distinct relationships between f and Re were observed, one in the lower Re region and one in the higher Re region. The area between the two trends represents the transition region. The critical Reynolds number, Re_{crit} (the Re above which the low Re linear trend is no longer appropriate), exhibits a dependence on the hydraulic diameter. Re_{crit} was estimated at 200 for $D_h < 220\ \mu m$, 400 for $D_h = 240\ \mu m$ and 700 for $260 < D_h < 360\ \mu m$, implying, according to Peng et al. (1994), the occurrence of "early transition to turbulence" in these microchannels.

A second study of the dependence on geometrical parameters, but for much smaller Re (~ 0.001 to 10), measured friction factors generally higher than macroscale theory [Papautsky et al., 1999b], more so for the smallest Re (<0.01). As the Reynolds number increased to 10, the measured friction factor neared the theoretical macroscale predictions, though it still remained up to 20% high, even considering experimental error. A possible dependence of friction characteristics due to size was observed at a small aspect ratios (<0.2), although the authors were not able to attribute this independently to either the small dimension required for small aspect ratio (O, 20 to $40\ \mu m$) or very low Reynolds number.

Nonlinear trends between pressure drop and flowrate were observed for Re as low as 300 by Mala and Li (1999), specifically for water flowing through a 130-μm-diameter stainless steel microtube. At small Reynolds numbers ($Re < 100$) the measured friction factors were in "rough agreement" with conventional theory, but for all other Re (up to 2500), the friction factors were consistently higher in stainless steel and fused silica microtubes (circular cross section). Measured flow friction for trapezoidal channels was 8 to 38% higher than macroscale predictions for the range of parameters studied by Qu et al. (2000), and a dependence on D_h and Re was also observed.

Another group of studies found the flow resistance to be less than theoretical macroscale predictions for certain conditions [Choi et al., 1991; Peng et al., 1994; Pfahler et al., 1990a; 1990b; 1991; Yu et al., 1995]. In the study of geometrical effects by Peng et al. (1994), the friction factor was found to be lower than the macroscale theoretical prediction for two out of seven cases, namely, when $H/W = 0.5$. A clear trend between Re_{crit} and aspect ratio was observable. Although previously described, the trends between friction factor behavior and aspect ratio or hydraulic diameter were less consistent.

Comparing the friction factors for flow through channels with several fluids (2-propanol, n-propanol and silicone oil), Pfahler et al. (1990a; 1990b; 1991) found the measured friction factors to be lower than theoretical values for all but two cases: a small depth and very low Re (0.8 μm, $Re \ll 1$) case and the case with largest depth (~40 μm). The authors acknowledged that their results depend heavily on channel size measurement, which was extremely difficult to measure accurately for the shallowest channels (depth <1 μm). The Reynolds numbers in their liquid flow experiments varied from much less than 1 to approximately 80 [Pfahler et al., 1991]. A possible size effect was noted for the silicone oil experiments; namely, for a channel with 3.0-μm depth, $f Re$ did not remain constant as Re varied from 0.01 to 0.3.

In addition to measuring friction factors, Choi et al. (1991) measured surface roughness and heat transfer coefficients for nitrogen gas flowing through microtubes with diameters in the range of 3 to 81 μm and L/D ratios of 640 to 8100. The measured friction factors in both the laminar and turbulent region were lower than predicted by macroscale theory. The measured absolute root mean square (rms) roughness was in the range of 10 to 80 nm, such that the normalized rms roughnesss, ε/D, was less than 0.006 for all but one case. No variations were observed that appeared to depend on roughness.

A similar study of flows through microtubes with diameters of 19 to 102 μm obtained extremely high Reynolds numbers for microscale conditions, up to 20,000 [Yu et al., 1995]. The normalized roughness of the tubes, ε/D, was estimated at 0.0003. For both water and gas, the measured friction factors were lower than macroscale predictions, by approximately 19% in laminar flow and 5% in the turbulent flow regime.

To aid in comparing the results of these studies, a normalized friction factor, C^*, is defined as:

$$C^* = \frac{(f\,Re)_{experimental}}{(f\,Re)_{theoretical}} \tag{6.28}$$

The wide variability of results is illustrated in Figure 6.5. There is also wide variability in experimental conditions, microchannel geometries and methodology. However, the inconsistencies demonstrate the need for detailed velocity measurements in order to elucidate potential microscale effects and mechanisms in these channels.

6.2.1 Proposed Explanations for Measured Behavior

Thus far, the explanations offered in the literature for anomalous behavior of friction factor and flow resistance in microchannels include surface/roughness effects and electrical charge, variations in viscosity, microrotational effects of individual fluid molecules, "early" transition to turbulence, entrance effects and inaccuracies in measuring channel dimensions.

In macroscale theory, small to moderate surface roughness does not affect the flow resistance relationship in the laminar region [White, 1994]. Microscale results are inconclusive regarding the effects of surface roughness on resistance in the laminar region. Resistance results have shown both a strong increase due to roughness [Wu and Little, 1983] and no effect due to roughness [Choi et al., 1991].

In terms of surface effects on viscosity, a roughness viscosity model (RVM) has been introduced by Mala (1999), based on work by Merkle et al. (1974). Assuming that surface roughness increases the momentum transfer near the wall, Mala (1999) proposed that the roughness viscosity, μ_r, as a function of r, is higher near the wall and proportional to the Reynolds number. Implementing this roughness-viscosity model for water flowing through trapezoidal channels, Qu et al. (2000) found reasonable agreement with model prediction and experimental results in most cases. However, the model did not

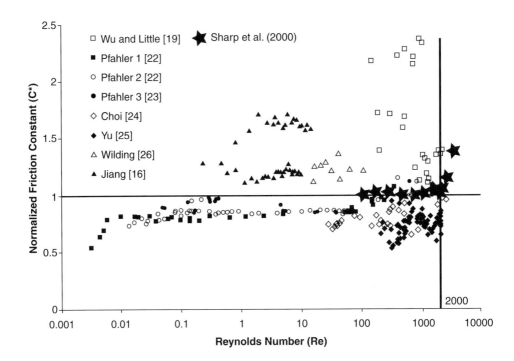

FIGURE 6.5 Comparison of C^* vs. Reynolds number for several previous studies. Data from Papautsky et al. (1999a), with additional data from Sharp et al. (2000). References are [19] Wu and Little (1983), [22] Pfahler et al. (1990a), [23] Pfahler et al. (1991), [24] Choi et al. (1991), [25] Yu et al. (1995), [26] Wilding et al. (1994) and [16] Jiang et al. (1995).

accurately depict the increased slope in the relationship between pressure drop and Reynolds number observed in the same experiments for $Re \geq 500$.

It is possible that unexpected electrokinetic effects occur at the interface between the channel and the fluid due to chemical interactions. These effects have not been adequately addressed to date in the flow resistance studies.

Direct measurement of viscosity in very thin layers, or thin films, has been performed by Israelachvili (1986). The viscosity of water was found to retain its bulk viscosity value to within 10% even in a film as thin as 5 nm. Concentrated and dilute NaCl/KCl solutions were also tested to assess the impact of double-layer forces on the value of viscosity near a surface. The viscosity of these solutions also remained only minimally affected until the last molecular layer near the wall. Based on these measurements, the viscosity of fluid in the wall region is not expected to vary significantly from the bulk value, even in the presence of possible charging effects, bringing the explanations of Mala and Li (1999) and Qu et al. (2000) into question.

As previously mentioned, it is possible that the very high shear rates in these microchannels cause normally Newtonian fluids to behave in a non-Newtonian fashion. The shear rates in Sharp et al. (2000) were as high as $7.2 \times 10^5 \, s^{-1}$. Measuring the rheology of fluids at very high shear rates is challenging. Novotny and Eckert (1974), using a flat plate rheometer, determined that the relationship between shear stress and shear rate is still linear for water at a shear rate of $10{,}000 \, s^{-1}$. The possibility that anomalous effects are caused by non-Newtonian behavior above shear rates of $10^4 \, s^{-1}$ needs to be explored.

A model based upon micropolar fluid theory has been implemented by Papautsky et al. (1999a) in order to interpret their experimental results (Reynolds numbers 1 to 20). Eringen (1964) first proposed this type of theory, which takes into account microvolume elements capable of independently translating, rotating and deforming. The equations for a "micropolar fluid" include additional terms accounting for the independent behavior of these elements. The equations have been solved numerically in Papautsky et al. (1999a) for flow in rectangular channels. The model predicts an increase in flow resistance, in better agreement with their data than conventional predictions.

One explanation often offered to explain the increase of flow resistance in microchannels is that scaling considerations somehow trigger the transition to turbulence at a critical Reynolds number lower than the value of 1800 to 2000 commonly accepted in macroscale pipe flows. If the critical Reynolds number is lower, then disturbances introduced into the flow by unsteadiness in the flow source, sharp inlet conditions or any other source would begin to grow and eventually the flow would become fully turbulent. The majority of flow behavior studies thus far have relied upon bulk flow measurements and have not been able to quantitatively assess the spatial or temporal instabilities in the microflows. Thus, the evidence suggesting "early transition to turbulence" has been based solely on the trends in the resistance data (see, for example, the discussion of the experiment of Peng et al. [1994] presented previously). "Early transition to turbulence" has also been suggested as an explanation for anomalous behavior of the relationship between the pressure gradient, $\Delta p/L$, and Q by Mala and Li (1999), who reported a transition to turbulence at $Re > 300$ to 900 and fully developed turbulent flows at $Re \geq 1000$ to 1500.

In Sharp et al.'s (2000) resistance experiments, there is no evidence that transition to turbulence occurs for $Re < 2000$. The magnitude of spatial and temporal variations measured using micro-particle image velocimetry (PIV) is also inconsistent with a transition to turbulence for $Re < 2000$.

Certainly, the inclusion or exclusion of entrance effects could affect the magnitude of the measured friction factor. For example, the L/D ratio in Peng et al.'s (1994) experiments ranged from 145 to 376. The Reynolds number range was 50 to 4000. The entrance length can be approximated as $L_e/D = 0.06 \times Re$ [White, 1994], which brings into question the validity of certain regions of the results. In particular, in certain experiments ($D_h = 267$ µm) of Peng et al. (1994), the entrance length is estimated at half the length of the test section for the higher Reynolds number data by the above criterion.

Regardless of the geometry, it is extremely difficult to accurately measure the dimensions of these microchannels, particularly when one of the dimensions is on the order of a couple of microns. The pressure drop in a round capillary is inversely proportional to D^4 (Eq. (21), where $D = 2a$), so an inaccuracy of 5% in measuring D can bias resistance results by 20%, enough to explain the majority of the discrepancies between the conventional macroscopic resistance predictions and the observed values in Figure 6.5.

6.2.2 Measurements of Velocity in Microchannels

Along with the growth of research in microdevices, rapid development of experimental techniques for investigating flows in such devices is also underway, including modification of experimental techniques commonly applied at the macroscale and development of new techniques. Measurements of velocities in microchannels have been obtained using bulk flow, point-wise and field measurements. Each technique has certain advantages making it more suitable to provide a specific type of flowfield information.

The majority of flow resistance data to date has been obtained through the use of bulk flow measurements. This type of measurement generally includes a method for measuring flowrate, such as an in-line flowmeter, the timed collection of fluid at the outlet and pressure taps located at the inlet and outlet or simply at the inlet if the pressure at the outlet is known. Bulk flow measurements require no optical access to the microchannel, there are no restrictions on the geometrical parameters of the channel, and no seeding is required. However, given the disagreement in results regarding microscale effects on flow resistance in particular, bulk flow measurements are lacking in detail sufficient to discern potential mechanisms causing deviation from macroscale theory. Detailed measurements of flow velocity are also useful for optimizing the design of complex microdevices for mixing, separation, reaction and thermal control.

The first micro-PIV measurements were made by Santiago et al. (1998) in a Hele–Shaw cell. These velocity field measurements were resolved to 6.9 µm in the lateral directions and 1.5 µm in the depth direction and demonstrated the applicability of the well-established PIV technique for microflows. Micro-PIV measurements in a rectangular glass microchannel with 200-nm fluorescent tracer particles ($Re < 1$) have been described in Meinhart et al. (1999). With improved acquisition and analysis, the lateral resolution was 13.6 µm in the streamwise direction and significantly better, 0.9 µm, in the cross-stream direction, the direction of highest velocity variation. The first application of micro-PIV to a circular capillary was performed by Koutsiaris et al. (1999) in a 236-μm-diameter channel with $Re \ll 1$. The seeding particles were 10-μm glass spheres, and the resolution of the measurements was 26.2 μm in the cross-stream direction. The measured velocity profiles agreed well with the predicted laminar parabolic profiles. More recently, micro-PIV has been used to study the velocity profiles and turbulence statistics of water flows in circular channels with $D \sim 100$ to 250 μm and Re up to 3000 by Sharp et al. (2000), using 2-μm fluorescent particles.

Alternate visual methods applied to microchannel velocity measurements have been demonstrated by numerous researchers [Brody et al., 1996; Ovryn, 1999; Paul et al., 1998b; Taylor and Yeung, 1993; Webb and Maynes, 1999]. Molecular tagging velocimetry (MTV) was adapted to the microscale by Webb and Maynes (1999), where velocity profiles were obtained in circular tubes with $D = 705$ μm for $Re = 800$ to 2200. The spatial resolution of these measurements was 2.1 μm. The measured velocity profiles were consistent with macroscale laminar predictions for $Re \leq 1600$. A departure from laminar flow theory was noted at $Re = 2168$. Relevant development issues for microscale MTV are similar to those for PIV, namely, optical access and index of refraction compensation, particularly for curved surfaces, and optimized detection of the tracking particles (PIV) or beams (MTV).

Particle tracking, streak quantification or dye visualization can be relatively simple to implement given optical access and the ability to illuminate the flow. Care must be exercised in the extraction of quantitative data, particularly if the depth of field of the imaging device is large, if optical complications exist due to complex microchannel geometries, or if the particles or molecules are not accurately following the flow due to charge, size or density effects. Implementations of such techniques may be found in Taylor and Yeung (1993) and Brody et al. (1996).

Novel three-dimensional measurement techniques for microchannel flows are currently in development [Hitt and Lowe, 1999; Ovryn, 1999]. Building upon a technique already developed for the study of microscale structures, Hitt and Lowe (1999) used confocal imaging to build up a three-dimensional map of the "separation surface" following a bifurcation, where the "separation surface" describes the interfacial boundary between two components from different branches of the bifurcation. Using two laser scanning confocal microscopes, a series of thin (4.5 or 7.1 μm) horizontal slices were acquired and reconstruction software was used to combine these slices into a three-dimensional map. Again, optical access and effects are primary issues in the implementation of this method, and it is not suitable for unsteady flows. Ovryn (1999) sought to resolve and interpret the scattering pattern of a particle to determine its three-dimensional position, and he has applied this technique to laminar flow.

X-ray imaging techniques do not require optical access in the channel, though a contrast medium detectable by X-rays must be used as the working fluid. Lanzillotto et al. (1996) obtained flow displacement information from microradiograph images of emulsion flow through a 640-μm-diameter tube and iodododecane flow through a silicon V-groove chip.

The level of complexity increases when electrokinetic flows are considered. Paul et al. (1998b), Cummings et al. (1999) and Taylor and Yeung (1993) performed visual measurements in this environment. Paul et al. (1998b) seeded the flow with an uncaged fluorescent dye. Once the dye was uncaged by an initial ultraviolet (UV) laser pulse, the flow was illuminated by succeeding pulses of blue light for charge-coupled device (CCD) image acquisition, causing only the uncaged dye molecules to be excited. This technique was applied to both pressure-driven and electrokinetic flows in circular capillaries with diameters of the order 100 μm. Since the dye transport represented both convection and diffusion, requisite care was necessary to separate the effects, as in Paul et al. (1998b). This method can also be used to acquire quantitative information regarding diffusion effects. Micro-PIV and nuclear magnetic resonance (NMR) techniques have also been applied to the measurement of velocities in electro-osmotic flow (Cummings et al., 1999; Manz et al., 1995).

Point-wise velocity measurements techniques have been applied at the microscale by Chen et al. (1997), Yazdanfar et al. (1997) and Tieu et al. (1995). Optical Doppler tomography, a point-wise measurement technique demonstrated by Chen et al. (1997), combines elements of Doppler velocimetry with optical coherence tomography in an effort to develop a system that can quantify the flow in biological tissues.

Chen et al. (1997) applied the technique to a 580-μm-diameter conduit seeded with 1.7-μm particles. An approximate parabolic profile was measured in the first test, and in the second it was shown that fluid particle velocities could be measured even with the conduit submerged in a highly scattering medium, as would be the case for particles in biological tissues. A similar measurement technique has been used for *in vivo* measurements in Yazdanfar et al. (1997). An adaptation of laser Doppler anemometry (LDA) techniques to microscale flows was demonstrated by Tieu et al. (1995), and point-wise data were obtained in a 175-μm channel.

6.2.3 Nonlinear Channels

For practical MEMS applications, it is often useful to consider mixing or separation of components in microchannels. Numerous designs have been proposed, including T- and H-shaped channels, zigzag-shaped channels, two- and three-dimensional serpentine channels and multilaminators.

For example, Weigl and Yager (1999) have designed a T-sensor for implementation of assays in microchannels, as shown in Figure 6.6. A reference stream, a detection stream and a sample stream have been introduced through multiple T-junctions into a common channel. The design relied upon the differential diffusion of different-sized molecules to separate components in the sample stream. Differential diffusion rates are also fundamental to the design of the H-filter, used to separate components [Schulte et al., 2000]. Application of a slightly different T-channel design has been demonstrated for measurement of diffusion coefficients of a species in a complex fluid [Galambos and Forster, 1998].

A layering approach has been implemented by Branebjerg et al. (1996), splitting the streams and re-layering to increase interfacial area, thus promoting mixing. Adding complexity to the flowfield also has potential to increase the amount of mixing between streams, as demonstrated by Branebjerg et al.'s (1995) zigzag channel and the serpentine channels introduced by Liu et al. (2000). The three-dimensional serpentine channel in Liu et al. (2000) was designed to introduce chaotic advection into the system and further enhance mixing over a two-dimensional serpentine channel. A schematic of the three-dimensional serpentine channel is shown in Figure 6.7.

6.2.4 Capacitive Effects

While liquids are incompressible, the systems through which they flow may expand or contract in response to pressure in the liquid. This behavior can be described by analogy to flow in electrical circuits. In this analogy, fluid pressure corresponds to electrical voltage, $p \sim V$; the volume flow rate corresponds to electrical current, $Q \sim I$; and the flow resistance through a fluid element corresponds to an electrical resistor, $R_{flow} \sim R_{elec}$. Thus, for capillary flow, $\Delta p = R_{flow} Q$, where $R_{flow} = 8\mu L/\pi a^4$ (*cf.* Eq. (21)), whereas in the electrical analogy $\Delta V = R_{elec} I$. If a fluid element is able to change its volume (expansion of plastic tubing, flexing in pressure transducer diaphragm, etc.), fluid continuity implies that:

$$\Delta Q = C_{flow} \frac{dp}{dt} \qquad (6.29)$$

where C_{flow} is the capacitance of the fluid element. The corresponding electrical law is $I = C_{elec} dV/dt$, where C_{elec} is the electrical capacitance.

It is well known, in the context of electrical circuits, that a resistor and capacitor in combination cause transients whose time constant, τ, is proportional to $R_{elec} C_{elec}$. In a microfluidic circuit, any capacitive element in combination with a flow resistance leads to analogous transients whose time constant is proportional to $R_{flow} C_{flow}$. Since R_{flow} can be very large in microchannels, the time constant can be surprisingly large (i.e., 10^3 s). Consequently, capacitive effects can cause significant and inconveniently long transients.

6.3 Electrokinetics Background

The first demonstration of electrokinetic phenomena is attributed to F.F. Reuss, who demonstrated electro-osmotic flow through a sand column in a paper published in the *Proceedings of the Imperial Society*

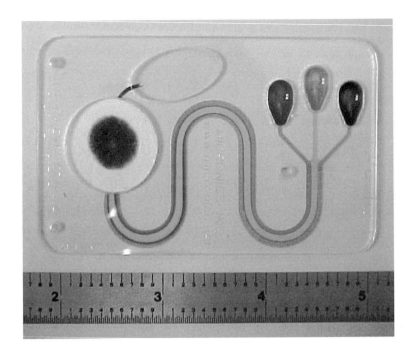

FIGURE 6.6 T-Sensor™ – Self-calibrating microchemical reactor and sensor. This design allows for self-calibration through the simultaneous flow of a reference solution on the opposite flank of the indicator stream from the sample to be analyzed. (Courtesy of Micronics, Inc., Redmond, WA.)

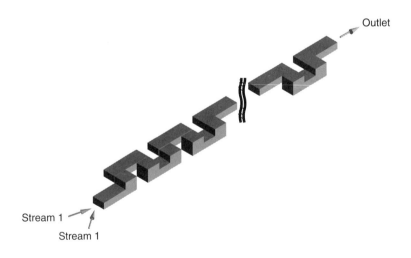

FIGURE 6.7 Three-dimensional serpentine channel. (From Liu, R.H., et al. (2000) *J. MEMS* **9**, 190–197, ASME, New York. With permission.)

of Naturalists of Moscow in 1809 [Probstein, 1994]. In the latter part of the 20th century, the main applications of electrokinetic phenomena have been fairly wide ranging, from the dewatering of soils and waste sludges using electric fields [Hiemenz and Rajagopalan, 1997] to the study of the stability of colloidal suspensions for household paint to devices that use electrophoretic mass transfer of colloidal suspensions to produce images on a planar substrate [Kitahara and Watanabe, 1984]. A community that has paid particular attention to the study of mass and momentum transport using electrokinetic effects is the developers of capillary electrophoresis (CE) devices [Manz et al., 1994]. CE devices are used to

Liquid Flows in Microchannels

separate biological and chemical species by their electrophoretic mobility (roughly speaking, by their mass-to-charge ratio). These traditional CE systems incorporate on-line detection schemes such as ultraviolet radiation scatter/absorption and laser-induced fluorescence [Baker, 1995].

Electrokinetics is the general term describing phenomena that involve the interaction between solid surfaces, ionic solutions and macroscopic electric fields. Two important classes of electrokinetics are electrophoresis and electro-osmosis, where the motions of solid bodies and fluids, respectively, occur when an external electric field is applied to the system. Electrophoresis is the induced motion of colloidal particles or molecules suspended in liquids that results from the application of an electric field. Electro-osmosis describes the motion of electrolyte liquids with respect to a fixed solid that results when an electric field is applied parallel to the surface. An example of electro-osmosis is the fluid pumping that occurs in a microcapillary when an electric field is applied along the axis of the capillary [Probstein, 1994]. Two other phenomena also classified under electrokinetics are flows with a finite streaming potential and sedimentation potential. These phenomena are counter-examples of electro-osmosis and electrophoresis, respectively. Streaming potential is the spontaneous generation of an electric potential from a pressure-driven flow in a charged microchannel [Hunter, 1981; Scales et al., 1992]. Sedimentation potential is the generation of an electric potential that results from the sedimentation (e.g., due to gravity) of a charged particle [Russel et al., 1999]. All of the phenomena classified under the term electrokinetics are manifestations of the electrostatic component of the Lorentz force (on ions and surface charges) and Newton's second law of motion. These interactions between charged particles and electric fields often involve electric double layers formed at liquid/solid interfaces and an introduction to this phenomenon is presented below.

6.3.1 Electrical Double Layers

Most solid surfaces acquire a surface electric charge when brought into contact with an electrolyte (liquid). Mechanisms for the spontaneous charging of surface layers include the differential adsorption of ions from an electrolyte onto solid surfaces (e.g., ionic surfactants), the differential solution of ions from the surface to the electrolyte, and the deprotonation/ionization of surface groups [Hunter, 1981]. The most common of these in microfluidic electrokinetic systems is the deprotonation of surface groups on the surface of materials such as silica, glass, acrylic and polyester. In the case of glass and silica, the deprotonation of surface silanol groups (SiOH) determines the generated surface charge density. The magnitude of the net surface charge density at the liquid–solid interface is a function of the local pH. The equilibrium reaction associated with this deprotonation can be represented as:

$$SiOH \Leftrightarrow SiO^- + H^+ \tag{6.30}$$

Models describing this reaction have been proposed for several types of glass and silica [Hayes et al., 1993; Huang et al., 1993; Scales et al., 1992]. In practice, the full deprotonation of the glass surface, and therefore the maximum electro-osmotic flow mobility, is achieved for pH values greater than about 9.

In response to the surface charge generated at a liquid–solid interface, nearby ions of opposite charge in the electrolyte are attracted by the electric field produced by the surface charge, and ions of like charge are repelled. The spontaneously formed surface charge therefore forms a region near the surface called an electrical double layer (EDL) that supports a net excess of mobile ions with a polarity opposite to that of the wall. Figure 6.8 shows a schematic of the EDL for a negatively charged wall (e.g., as in the case of a glass surface). The region of excess charge formed by the counterions shielding the electric field of the wall can be used to impart a force (through ion drag) on the bulk fluid.

As shown in Figure 6.8, counterions reside in two regions divided into the Stern and Gouy–Chapman diffuse layers [Adamson and Gast, 1997]. The Stern layer counterions are adsorbed onto the wall, while the ions of the Gouy–Chapman diffuse layer are free to diffuse into the bulk fluid and therefore are available to impart work on the fluid. The plane separating the Stern and Gouy–Chapman layers is called the *shear plane*. The bulk liquid far from the wall is assumed to have net neutral charge. Also shown in Figure 6.8 is a sketch of the potential associated with the EDL. The magnitude of this potential is a maximum at the wall and drops rapidly through the Stern layer. The potential at the shear plane, which is also the

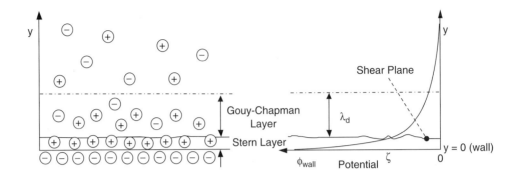

FIGURE 6.8 Schematic of the electrical double layer (EDL): (Left) Distribution of co- and counterions near a charged wall. The Stern and Gouy–Chapman layers are shown with the Gouy–Chapman thickness roughly approximated as the Debye length of the solution. (Right) A plot of the negative potential distribution near a glass wall indicating the zeta potential, wall potential and location of the shear plane.

boundary of the fluid flow problem, is called the *zeta potential*, ζ. Because of the difficulties associated with predicting the properties of the Stern layer [Hunter, 1981], the zeta potential is typically determined empirically from electro-osmotic or streaming potential flow measurements.

The simple treatment of the physics of the diffuse portion of the EDL presented here assumes a liquid with constant properties (i.e., constant viscosity and electrical permittivity). A more detailed model of the diffuse portion of the electrical double layer should include noncontinuum effects such as finite ion size effects and gradients in the dielectric strength and viscosity of the fluid [Hunter, 1981]. The width of the diffuse portion of the EDL is determined by the opposing forces of electrostatic attraction and thermal diffusion. This balance between electromigration and diffusive fluxes, together with the Nernst–Einstein equation relating ion diffusivity and mobility [Hiemenz and Rajagopalan, 1997], can be used to show that the concentration profile is described by the Boltzmann distribution. For an EDL on a flat plate, the Boltzmann distribution of ions of species i, c_i, is

$$c_i(y) = c_{\infty,i} \exp\left(-\frac{ze\phi(y)}{kT}\right) \qquad (6.31)$$

where $c_{\infty,i}$ is the molar concentration of ion i in the bulk, z is the valance number of the ion, ϕ is the local potential, T is temperature, e is the charge of an electron and k is Boltzmann's constant. The coordinate y is perpendicular to the wall, and the origin is at the shear plane of the EDL. The net charge density in the EDL, ρ_E, is related to the molar concentrations of N species using the relation:

$$\rho_E = F \sum_{i=1}^{N} z_i c_i \qquad (6.32)$$

where F is Faraday's constant. The net charge density can also be related to the local potential in the diffuse EDL by the Poisson equation:

$$\nabla^2 \phi = \frac{-\rho_E}{\varepsilon} \qquad (6.33)$$

where ε is the permittivity of the liquid. Substituting Eqs. (6.31) and (6.32) into Eq. (6.33), we find that:

$$\frac{d^2\phi}{dy^2} = \frac{-F}{\varepsilon} \sum_{i=1}^{N} z_i c_{\infty,i} \exp\left(-\frac{ze\phi(y)}{kT}\right) \qquad (6.34)$$

For the simple case of a symmetric electrolyte with (two) monovalent ions, this relation becomes:

$$\frac{d^2\phi}{dy^2} = \frac{2Fz_ic_\infty}{\varepsilon}\sinh\left(\frac{ze\phi(y)}{kT}\right) \tag{6.35}$$

where c_∞ is the molar concentration of each of the two ion species in the bulk. This relation is the nonlinear Poisson–Boltzmann equation. A closed form, analytical solution of this equation for the EDL on a flat wall is given by Adamson and Gast (1997).

A well-known approximation to the Poisson–Boltzmann solution, known as the Debye–Huckle limit, is to consider the case where the potential energy of ions in the EDL is small compared to their thermal energy so that the argument of the hyperbolic sine function in Eq. (6.35) is small. Applying this approximation, Eq. (6.35) becomes:

$$\frac{d^2\phi}{dy^2} = \frac{\phi(y)}{\lambda_D^2} \tag{6.36}$$

where λ_D is the Debye length of the electrolyte defined as:

$$\lambda_D \equiv \left(\frac{\varepsilon kT}{2z^2F^2c_\infty}\right)^{\frac{1}{2}} \tag{6.37}$$

for a symmetric, monovalent electrolyte. The Debye length describes the characteristic thickness of the EDL which varies inversely with the square root of ion molar concentration. At typical biochemical, singly ionized buffer concentrations of 10 mM, the thickness of the EDL is therefore on the order of a few nanometers [Hiemenz and Rajagopalan, 1997]. In analyzing electrokinetic flow in microchannels, the Debye length should be compared to the characteristic dimension of the microchannel in order to classify the pertinent flow regime. Overbeek (1952a) points out that the Debye–Huckle approximation of the potential of the EDL holds remarkably well for values of the ratio $ze\phi/(kT)$ up to approximately 2. This value is equivalent to a zeta potential of about 50 mV which is within the typical range of microfluidic applications.

Models of the physics of the EDLs can be used to extrapolate the zeta potential of particles and microchannels across a significant range of buffer concentration, fluid viscosity, electrical permittivity of electrolytes, and field strengths given only a few measurements. One of the most difficult zeta potential extrapolations to make is across different values of pH, as pH changes the equilibrium reactions associated with the charge at the liquid/solid interface.

A full formulation of the coupled system of equations describing electro-osmotic and electrokinetic flow includes the convective diffusion equations for each of the charged species in the system, the Poisson equation for both the applied electric field and the potential of the EDL and the equations of fluid motion. A few solutions to this transport problem relevant to microfluidic systems are presented below.

6.3.2 EOF with Finite EDL

Electro-osmotic flow (EOF) results when an electric field is applied through a liquid-filled microchannel having an EDL at the channel surfaces, as described above. This applied electric field introduces an electrostatic Lorentz body force:

$$\rho\mathbf{b} = \rho_E\mathbf{E} \tag{6.38}$$

into the equation of motion for the fluid, Eq. (6.3). Within the EDL, the electric field exerts a net force on the liquid, causing the fluid near the walls to move. Alternatively, one can describe the effect as simply the ion drag on the fluid associated with the electrophoresis of the ions in the EDL. The fluid in the EDL exerts a viscous force on the rest of the (net zero charge) liquid in the bulk of the channel. For EDLs

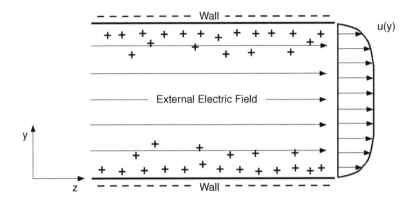

FIGURE 6.9 Schematic of an electro-osmotic flow channel with a finite EDL. The charges drawn in the figure indicate net charge. The boundary layers on either wall have a thickness on the order the Debye length of the solution. For nonoverlapping EDLs, the region near the center of the channel is net neutral.

much smaller than the channel dimension D, the fluid velocity reaches steady state in a short time t that is on the order of D^2/ν, where ν is the kinematic viscosity of the fluid. The resulting bulk electro-osmotic flow is depicted schematically in Figure 6.9.

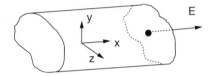

FIGURE 6.10 Illustration of a section of a long, straight channel having an arbitrary cross section.

The equation of motion for steady, low Reynolds number flow in the microchannel is given by:

$$\nabla p = \mu \nabla^2 \mathbf{u} + \rho_E \mathbf{E} \tag{6.39}$$

Substituting Eq. (6.33) for the charge density results in:

$$\nabla^2 \left(\mathbf{u} - \frac{\varepsilon \mathbf{E}}{\mu} \phi \right) = \frac{\nabla p}{\mu} \tag{6.40}$$

In Eq. (6.40), the electric field, $\mathbf{E}$, can be brought into the Laplace operator because $\nabla \cdot \mathbf{E} = \nabla \times \mathbf{E} = 0$. Equation (6.40) is linear so the velocities caused by the pressure gradient and the electric field can be considered separately and then superposed as follows:

$$\nabla^2 \left(\mathbf{u}_{\text{EOF}} - \frac{\varepsilon \mathbf{E}}{\mu} \phi \right) = 0 \tag{6.41}$$

$$\nabla^2 \mathbf{u}_{\text{pressure}} = \frac{\nabla p}{\mu} \tag{6.42}$$

Together with Eq. (6.2), these are the general equations for electro-osmotic flow in a microchannel. Evaluation of the pressure-driven flow component of velocity in a microchannel can leverage analytical solutions available for channels of various cross sections [White, 1991]. The pressure gradient can be

Liquid Flows in Microchannels

applied externally or may arise internally because of variations in the zeta-potential at the channel walls [Herr et al., 2000].

Now consider electro-osmosis in a long, straight microchannel with a finite width electrical double layer and an arbitrary cross section that remains constant along the flow direction (x-axis), as shown in Figure 6.10. The applied electric field is assumed to be uniform and along the x-axis of the microchannel. For the case where the potential at the wall is uniform, the solution to Eq. (6.41) is

$$u_{EOF} - \frac{\varepsilon E \zeta}{\mu} \phi = \frac{-\varepsilon E \zeta}{\mu} \qquad (6.43)$$

with the zeta potential, ζ, being the value of ϕ at the top of the double layer. In Eq. (6.43), u_{EOF} and E are the unidirectional velocity and unidirectional applied electric field, respectively. The general expression for the electro-osmotic velocity, implicit in the potential, is then

$$u_{EOF}(y,z) = \frac{-\varepsilon E \zeta}{\mu}\left(1 - \frac{\phi(y,z)}{\zeta}\right) \qquad (6.44)$$

To compute values for the velocity given in Eq. (6.44), an expression for the potential $\phi(y,z)$ is required. In general, $\phi(y,z)$ can be computed numerically from Eq. (6.34), but analytical solutions exist for several geometries. Using the Boltzmann equation for a symmetric analyte and the Debye–Huckel approximation discussed in the previous section, Rice and Whitehead (1965) give the solution for electro-osmosis in a long cylindrical capillary:

$$u_{EOF}(r) = \frac{-\varepsilon E \zeta}{\mu}\left(1 - \frac{I_0(r/\lambda_D)}{I_0(r/a)}\right) \qquad (6.45)$$

In Eq. (6.45), I_0 is the zero-order modified Bessel function of the first kind, r is the radial direction and a is the radius of the cylindrical capillary. This solution can be superposed with the solution of Eq. (6.42) for a constant pressure gradient. The resulting composite solution is

$$u(r) = \frac{-\varepsilon E \zeta}{\mu}\left(1 - \frac{I_0(r/\lambda_D)}{I_0(r/a)}\right) - \frac{dp}{dx}\frac{a^2}{4\mu}\left(1 - \frac{r^2}{a^2}\right) \qquad (6.46)$$

Burgeen and Nakache (1964) give a solution for electro-osmotic flow between two long, parallel plates for a finite EDL thickness. For other more complex geometries and many unsteady problems, numerical solutions for the electro-osmotic flow are required. However, when the Debye length is finite, but much smaller than other dimensions (e.g., the width of the microchannel), the disparate length scales can make numerical solutions difficult [Bianchi et al., 2000; Patankar and Hu, 1998]. In many cases, EOF in complex geometries can be determined numerically using a thin double-layer assumption described in the next section.

6.3.3 Thin EDL Electro-Osmotic Flow

This section presents a brief analysis of electro-osmotic flow in microchannels with thin EDLs. Figure 6.11 shows a schematic of an electro-osmotic flow in a microchannel with zero pressure gradient. As shown in the figure, the Debye length of typical electrolytes used in microfabricated electrokinetic systems is much smaller than the hydraulic diameter of the channels. Typical ratios of Debye length-to-channel diameter are less than 10^{-4}. For low Reynolds number, electro-osmotic flow in a cylindrical channel in the presence of a constant axial pressure gradient and a Debye length much smaller than the capillary radius, the solution of the velocity field is simply:

$$u(r) = -\frac{\varepsilon \zeta E}{\mu} - \frac{dp}{dx}\frac{(a^2 - r^2)}{4\mu} \qquad (6.47)$$

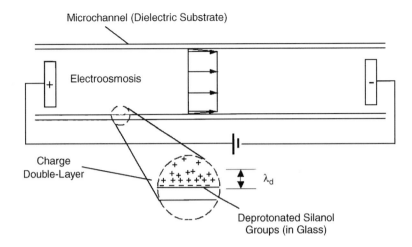

FIGURE 6.11 Schematic of electro-osmotic flow in a glass microchannel with a thin EDL. A zero pressure gradient "plug flow" is shown. The electrodes on the ends of the channel indicate the polarity of the electric field.

This equation can be derived by evaluating Eq. (6.46) in the limit of a thin EDL (i.e., a small value of λ_D/a).

The zeta potential typically determines flow velocities and flow rates in common thin EDL systems. As mentioned above, this quantity can often be interpreted as an empirically measured mobility parameter that determines the local velocity of the flow at the top of the electrical double layer. The zeta potential

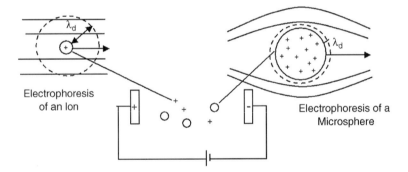

FIGURE 6.12 Two limiting limits of electrophoresis in an electrolyte. Shown are electrophoretic particles in the electric field generated between two electrodes. On the left is the detail of a charged ion with a characteristic dimension much smaller than the Debye length of the electrolyte. On the right is a charged microsphere with a diameter much larger than the Debye length.

can be approximately related to the local surface charge density on the wall and the bulk fluid properties by applying continuum field and flow theory. Theoretically, the zeta potential is defined as the value of the electrostatic potential at the plane that separates double-layer counterions that are mobile from those that are fixed. For the case of zero applied pressure gradients, Eq. (6.47) reduces to the well-known Helmholtz–Smoulochowski relation for electro-osmotic flow: $u = \varepsilon\zeta E/\mu$ [Probstein, 1994].

6.3.4 Electrophoresis

Many electrokinetic microfluidic systems leverage the combination of electro-osmotic and electrophoresis to achieve biological separations and to transport charged particles (e.g., biological assay microbeads)

Liquid Flows in Microchannels

and ions. Because of this, we present here a short introduction to electrophoresis. Electrophoresis is the induced motion of colloidal particles or molecules suspended in polar solutions that results from application of an electric field. Two important regimes of electrophoresis depicted in Figure 6.12 are for the electromigration of species that are either large or small compared to the Debye length of the ionic solution in which they are suspended.

Electrophoresis of ionic molecules and macromolecules can be described as a simple balance between the electrostatic force on the molecule and the viscous drag associated with its resulting motion. As a result, the electrophoretic mobility (velocity-to-electric-field ratio) of molecules is a function of the size/molecular weight of the molecule and directly proportional to their valence number, or

$$u = \frac{qE}{3\pi\mu d} \quad (d \ll \lambda_d) \tag{6.48}$$

where q is the total molecule charge and d is the Stokes diameter of the particle (the diameter of a sphere of equal drag). In comparison, the electrophoresis of relatively large solid particles, such as 100- to 10,000-nm-diameter polystyrene spheres and single-celled organisms is a function of the electrostatic forces on the surface charge, the electrostatic forces on their charge double layers, and the viscous drag associated with both the motion of the body and the motion of the ionic cloud. For a wide range of cases where the ratio of particle diameter to Debye length is large so that, locally, the ionic cloud near the particle surface can be approximated by the EDL relations for a flat plate, the velocity of an electrophoretic particle reduces simply to:

$$u = \frac{\varepsilon \zeta E}{\mu} \quad (d \gg \lambda_d) \tag{6.49}$$

where the dimension d in the inequality condition is a characteristic dimension of the particle (e.g., its Stokes diameter). This equation was shown by Smoluchowski (1903) to be independent of particle shape. The reader will note that this, again, is the Helmholtz–Smoluchowski equation introduced earlier (with a change of sign).

The two expressions above describing the electrophoresis of particles can be expressed in terms of a mobility v_{eph} equal to $q/(3\pi\mu d)$ and $\varepsilon\zeta/\mu$ for characteristic particle dimensions much smaller and much larger than the Debye length, respectively. Note also that for the simple case of a fluid with uniform properties, the solution of the drift velocity of electrophoretic particles with respect to the bulk liquid are similar (i.e., parallel and directly proportional) to lines of electric field.

Several solutions of the particle velocity and velocity field in the region of an electrophoretic particle with a finite EDL exist [Hunter, 1981; Russel et al., 1999]. A well-known solution is that of Henry (1948) for the flow around an electrophoretic sphere in the Debye–Huckle limit. The $d \gg \lambda_d$ limit of Henry's solution results in Eq. (6.49).

6.3.5 Similarity between Electric and Velocity Fields for Electro-Osmosis and Electrophoresis

The previous sections have described the solution for electro-osmotic velocity field in straight, uniform cross-section channels. In general, solving for the electro-osmotic velocity field in more complex geometries requires a solution of the electric field and charge density in the microchannel, together with a solution to the Navier–Stokes equations. For simple electro-osmotic flows, however, Overbeek (1952b) has argued that the fluid velocity is parallel to the electric field everywhere. This concept has also been discussed by Cummings et al. (2000) and Santiago (2001). A set of sufficient conditions for which there exists a velocity field solution that is similar to the electric field is

- Uniform zeta potential

- Electric double-layers thin compared to channel dimension
- Electrically insulating channel walls
- Low Reynolds number
- Low product of Reynolds and Strouhal numbers
- Parallel flow at inlets and outlets
- Uniform fluid properties

When these conditions are met, the electro-osmotic streamlines exactly correspond to the electric field-lines. The approximation is applicable to systems with a microchannel length scale less than 100 µm, a Debye length less than 10 nm, a velocity scale less than 1 mm/s and a characteristic forcing function time scale greater than 10 ms [Santiago, 2001]. An important part of this similarity proof is to show the applicability of the Helmholtz–Smoulochowski equation in describing the local velocity field at the slip surface that bounds the "internal" flow of the microchannel which excludes the EDL. The Helmholtz–Smoulochowski equation can be shown to hold for most microfluidic systems where the motion of the EDL is dominated by the Lorentz and viscous forces. In such systems, we can consider the velocity field of the fluid outside of the EDL as a three-dimensional, unsteady flow of a viscous fluid of zero net charge that is bounded by the following slip velocity condition:

$$u_{slip} = -\frac{\varepsilon \zeta}{\mu} E_{slip} \tag{6.50}$$

where the subscript "slip" indicates a quantity evaluated at the slip surface at the top of the EDL (in practice, a few Debye lengths from the wall). The velocity along this slip surface is, for thin EDLs, similar to the electric field. This equation and the condition of similarity also hold for inlets and outlets of the flow domain which have zero imposed pressure gradients.

The complete velocity field of the flow bounded by the slip surface (and inlets and outlets) can be shown to be similar to the electric field [Santiago, 2001]. We nondimensionalize the Navier–Stokes equations by a characteristic velocity and length scale U_s and L_s, respectively. The pressure p is nondimensionalized by the viscous pressure $\mu U_s/L_s$. The Reynolds and Strouhal numbers are $Re = \rho L_s U_s/\mu$ and $St = L_s/\tau U_s$, respectively, where τ is the characteristic time scale of a forcing function. The equation of motion is

$$Re\, St\, \frac{\partial \mathbf{u}'}{\partial t'} + (Re\mathbf{u}' \cdot \nabla \mathbf{u}') = -\nabla p' + \nabla^2 \mathbf{u}' \tag{6.51}$$

Note that the right-most term in Eq. (6.51) can be expanded using a well-known vector identity:

$$\nabla^2 \mathbf{u}' = \nabla(\nabla \cdot \mathbf{u}') - \nabla \times \nabla \times \mathbf{u}' \tag{6.52}$$

We can now propose a solution to Eq. (6.52) that is proportional to the electric field and of the form:

$$\mathbf{u}' = \frac{c_o}{U} \mathbf{E} \tag{6.53}$$

where c_o is a proportionality constant and $\mathbf{E}$ is the electric field driving the fluid. Because we have assumed that the EDL is thin, the electric field at the slip surface can be approximated by the electric field at the wall. The electric field bounded by the slip surface satisfies Faraday's and Gauss' laws:

$$\nabla \cdot \mathbf{E} = \nabla \times \mathbf{E} = 0 \tag{6.54}$$

Substituting Eqs. (6.53) and (6.54) into Eq. (6.51) yields:

$$Re\, St \frac{\partial \mathbf{u}'}{\partial t'} + Re(\mathbf{u}' \cdot \nabla \mathbf{u}') = -\nabla p' \qquad (6.55)$$

This is the condition that must hold for Eq. (6.53) to be a solution to Eq. (6.51). One limiting case where this holds is for very high Reynolds number flows where inertial and pressure forces are much larger than viscous forces. Such flows are not applicable to microfluidics. Another limiting case applicable here is when Re and $Re\,St$ are both small, so that the condition for Eq. (6.53) to hold becomes:

$$\nabla p' = 0 \qquad (6.56)$$

Therefore, we see that for small Re and $Re\,St$ and the pressure gradient at the inlets and outlets equal to zero, Eq. (6.53) is a valid solution to the flow bounded by the slip surface, inlets and outlets (note that these arguments do not show the uniqueness of this solution). We can now consider the boundary conditions required to determine the value of the proportionality constant c_o. Setting Eq. (6.50) equal to Eq. (6.53) we see that $c_o = \varepsilon\zeta/\eta$, so that if the simple flow conditions are met, then the velocity everywhere in the fluid bounded by the slip surface is given by:

$$\mathbf{u}(x, y, z, t) = -\frac{\varepsilon\zeta}{\mu} \mathbf{E}(x, y, z, t) \qquad (6.57)$$

Equation (6.57) is the Helmholtz–Smoluchowski equation shown to be a valid solution to the quasi-steady velocity field in electro-osmotic flow with ζ the value of the zeta potential at the slip surface. This result greatly simplifies the modeling of simple electro-osmotic flows, as simple Laplace equation solvers can be used to solve for the electric potential and then, using Eq. (6.57), for the velocity field. This approach has been applied to the optimization of microchannel geometries and verified experimentally [Molho et al., 2000]. Figure 6.13 shows the superposition of particle pathlines/streamlines and predicted electric fieldlines [Devasenathipathy and Santiago, 2000] in a steady flow that meets the simple electro-osmotic flow conditions summarized above. As shown in the figure, the electro-osmotic flow field streamlines are very well approximated by electric fieldlines.

Note that for the simple electro-osmotic flow conditions analyzed here, the electrophoretic drift velocities (with respect to the bulk fluid) are also similar to the electric field, as mentioned above. Therefore, the time-averaged, total (drift plus fluid) velocity field of electrophoretic particles can be shown to be

$$\mathbf{u}_{particle} = \left(v_{eph} - \frac{\varepsilon\zeta}{\mu}\right)\mathbf{E} \qquad (6.58)$$

Here, we use the electrophoretic mobility v_{eph} that was defined earlier and $-\varepsilon\zeta/\mu$ is the electro-osmotic flow mobility of the microchannel walls.

6.3.6 Electrokinetic Microchips

The advent of microfabrication and microelectromechanical systems (MEMS) technology has seen an application of electrokinetics as a method for pumping fluids on microchips [Jacobson et al., 1994; Manz et al., 1994]. On-chip electro-osmotic pumping is easily incorporated into electrophoretic and chromatographic separations and these "laboratories on a chip" offer distinct advantages over the traditional, free-standing capillary systems. Advantages include reduced reagent use, tight control of geometry, the ability to network and control multiple channels on a chip, the possibility of massively parallel analytical

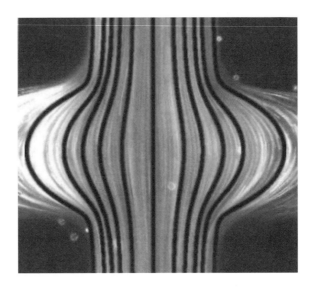

FIGURE 6.13 Comparison between experimentally determined electrokinetic particle pathlines at a microchannel intersection and predicted electric field lines. The light streaks show the path lines of 0- to 5-μm-diameter particles advecting through an intersection of two microchannels. The electrophoretic drift velocities and electro-osmotic flow velocities of the particles are approximately equal. The channels have a trapezoidal cross section having a hydraulic diameter of 18 μm (130 μm wide at the top, 60 μm wide at the base and 50 μm deep). The superposed, heavy black lines correspond to a prediction of electric field lines in the same geometry. The predicted electric field lines very closely approximate the experimentally determined pathlines of the flow. (From Devasenathipathy, S. and Santiago, J.G., unpublished results, Stanford University, Stanford, CA. With permission.)

process on a single chip, the use of chip substrate as a heat sink (for high field separations) and the many advantages that follow the realization of a portable device [Khaledi, 1998]. Electrokinetic effects significantly extend the current design space of microsystems technology by offering unique methods of sample handling, mixing, separation and detection of biological species including cells, microparticles and molecules.

This section presents typical characteristics of a microchannel network fabricated using microlithographic techniques (see description of fabrication in the next section). Figure 6.14 shows a top view schematic of a typical microchannel fluidic chip used for capillary electrophoresis [Manz et al., 1994]. In this simple example, the channels are etched on a dielectric substrate and bonded to a plate (coverslip) of the same material. The circles in the schematic represent liquid reservoirs that connect with the channels through holes drilled through the coverslip. The parameters V_1 through V_4 are time-dependent voltages applied at each reservoir "well." A typical voltage switching system may apply voltages with on/off ramp profiles of approximately 10,000 V/s or less so that the flow can often be approximated as quasi-steady.

The four-well system shown in Figure 6.14 can be used to perform an electrophoretic separation by injecting a sample from well #3 to well #2 by applying a potential difference between these wells. During this injection phase, the sample is confined or "pinched" to a small region within the separation channel by flowing solution from well #1 to #2 and from well #4 to well #2. The amount of desirable pinching is generally a trade-off between separation efficiency and sensitivity. Alarie et al. (2000) have found that the pinching was optimized (implying a minimization of the injected sample plug) when the flow rate of sample from well #3 was 44% of the total amount of flow entering the intersection. Next, the injection phase potential is deactivated and a potential is applied between well #1 and well #4 to dispense the injection plug into the separation channel and begin the electrophoretic separation. The potential between wells #1 and #2 is referred to as the *separation potential*. During the separation phase, potentials are applied at wells #2 and #3 which "retract" or "pull back" the solution-filled streams on either side of the separation channel. As with the pinching described above, the amount of "pull back" is a trade-off between

Liquid Flows in Microchannels

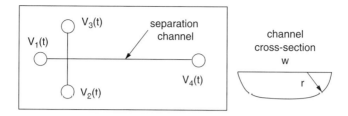

FIGURE 6.14 Schematic of a typical electrokinetic microchannel chip. V_1 through V_5 represent time-dependent voltages applied to each microchannel. The channel cross section shown is for the [common case] of an isotropically etched glass substrate with a mask line width of $(w - 2r)$.

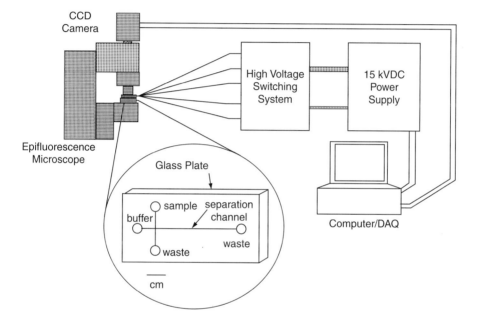

FIGURE 6.15 Schematic of microfabricated capillary electrophoresis system, flow imaging system, high voltage control box and data acquisition computer.

separation efficiency and sensitivity. Alarie et al. (2000) have found that dispensing the sample plug into the separation channel was optimized when the flow rate of sample toward well #4 was 48% of the total amount of flow entering the intersection.

Figure 6.15 shows a schematic of a system that was used to perform and image an electrophoretic separation in a microfluidic chip. The microchip depicted schematically in Figure 6.15 is commercially available from Micralyne, Inc., Alberta, Canada. The width and depth of the channels are 50 and 20 μm, respectively. The separation channel is 80 mm from the intersection to the waste well (well #4 in Figure 6.14). A high-voltage switching system allows for rapid switching between the injection and separation voltages and a computer, epifluorescent microscope and CCD camera are used to image the electrophoretic separation. The system depicted in Figure 6.15 is used to design and characterize electrokinetic injections; in a typical electrophoresis application, the CCD camera would be replaced with a point detector (e.g., a photomultiplier tube) near well #4.

Figure 6.16 shows an injection and separation sequence of 200 μM solutions of fluorescein and Bodipy dyes (Molecular Probes, Inc., Eugene, OR). Images (a) through (d) are each 20-ms exposures separated by 250 ms. In Figure 6.16a, the sample is injected applying 0.5 kV and ground to well #3 and well #2, respectively. The sample volume at the intersection is "pinched" by flowing buffer from well #1 and well #4. Once a

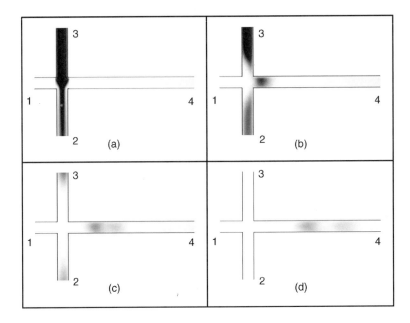

FIGURE 6.16 Separation sequence of Bodipy and fluorescein in a microfabricated capillary electrophoresis system. The channels shown are 50 μm wide and 20 μm deep. The fluorescence images are 20-ms exposures, and consecutive images are separated by 250 ms. A background image has been subtracted from each of the images, and the channel walls were drawn in for clarity. (From Bharadwaj, R., and Santiago, J.G., unpublished results, Stanford University, Stanford, CA. With permission.)

steady flow condition is achieved, the voltages are switched to inject a small sample plug into the separation channel. During this separation phase, the voltages applied at well #1 and well #4 are 2.4 kV and ground, respectively. The sample remaining in the injection channel is "retracted" from the intersection by applying 1.4 kV to both well #2 and well #3. During the separation, the electric field strength in the separation channel is about 200 V/cm. The electrokinetic injection introduces an approximately 400-pL volume of the homogenous sample mixture into the separation channel, as seen in Figure 6.16b. The Bodipy dye is neutral so its species velocity is identical to that of the electro-osmotic flow velocity. The relatively high electro-osmotic flow velocity in the capillary carries both the neutral Bodipy and negatively charged fluorescein toward well #4. The negative electrophoretic mobility of the fluorescein moves it against the electro-osmotic bulk flow, so it travels more slowly than the Bodipy dye. This difference in electrophoretic mobilities results in a separation of the two dyes into distinct analyte bands, as seen in Figures 6.16c and 6.16d. The zeta potential of the microchannel walls for the system used in this experiment was estimated at −50 mV from the velocity of the neutral Bodipy dye [Bharadwaj and Santiago, 2000].

6.3.7 Engineering Considerations: Flowrate and Pressure of Electro-Osmotic Flow

As we have seen, the velocity field of systems with thin EDLs is approximately independent of the location in the microchannel and is therefore a "plug flow" profile for any cross section of the channel. The volume flow rate of such a flow is well approximated by the product of the electro-osmotic flow velocity and the cross-sectional area of the inner capillary:

$$Q = -\frac{\varepsilon \zeta E A}{\mu} \tag{6.59}$$

Liquid Flows in Microchannels

For the typical case of electrokinetic systems with a bulk ion concentration in excess of about 100 µM and characteristic dimension greater than about 10 µm, the vast majority of the current carried within the microchannel is the electromigration current of the bulk liquid. For such typical flows, we can rewrite the fluid flow rate in terms of the net conductivity of the solution, σ:

$$Q = -\frac{\varepsilon \zeta I}{\mu \sigma} \quad (6.60)$$

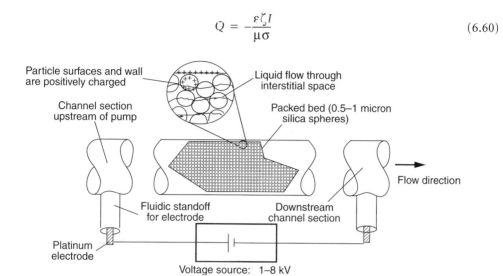

FIGURE 6.17 Schematic of electrokinetic pump fabricated using a glass microchannel packed with silica spheres. The interstitial spaces of the packed bead structure create a network of submicron microchannels which can be used to generate pressures in excess of 5000 psi.

where I is the current consumed, and we have made the reasonable assumption that the electromigration component of the current flux dominates. The flow rate of a microchannel is therefore a function of the current carried by the channel and otherwise independent of geometry.

Another interesting case is that of an electrokinetic capillary with an imposed axial pressure gradient. For this case, we can use Eq. (6.47) to show the magnitude of the pressure that an electrokinetic microchannel can achieve. To this end, we solve Eq. (6.47) for the maximum pressure generated by a capillary with a sealed end and an applied voltage ΔV, noting that the electric field and the pressure gradient can be expressed as $\Delta V/L$ and $\Delta p/L$, respectively. Such a microchannel will produce zero net flow but will provide a significant pressure gradient in the direction of the electric field (in the case of a negatively charged wall). Imposing a zero net flow condition $Q = \int_A \mathbf{u} \cdot d\mathbf{A} = 0$, the solution for pressure generated in a thin EDL microchannel is then:

$$\Delta p = -\frac{8\varepsilon \zeta \Delta V}{a^2} \quad (6.61)$$

which shows that the generated pressure will be directly proportional to voltage and inversely proportional to the square of the capillary radius. Eq. (6.61) dictates that decreasing the characteristic radius of the microchannel will result in higher pressure generation. This principle has, for example, been demonstrated effectively by electrokinetic pumps that have been built recently [Paul et al., 1998a; Zeng et al., 2000]. Pressures in excess of 20 atm have been demonstrated. Figure 6.17 shows a schematic of one of these pumps. This structure achieves a network of submicron-diameter microchannels by packing 0.5- to 1-µm spheres in fused silica capillaries, using the interstitial spaces in these packed beds as flow passages.

6.3.8 Electrical Analogy and Microfluidic Networks

There is a strong analogy between electro-osmotic and electrophoretic transport and resistive electrical networks of microchannels with long axial-to-radial-dimension ratios. As described above, the electro-osmotic flowrate is directly proportional to the current. This analogy holds provided that the previously described conditions for electric/velocity field similarity also hold. Therefore, Kirkoff's current and voltage laws can be used to predict flow rates in a network of electro-osmotic channels given voltage at endpoint "nodes" of the system. In this one-dimensional analogy, all of the current, and hence all of the flow entering a node, must also leave that node. The resistance of each segment of the network can be determined by knowing the cross-sectional area, the conductivity of the liquid buffer and the length of the segment. Once the resistances and applied voltages are known, the current and electro-osmotic flow rate in every part of the network can be determined using Eq. (6.60).

6.3.9 Practical Considerations

A few practical considerations should be considered in the design, fabrication and operation of electrokinetic microfluidic systems. These considerations include the dimensions of the system, the choice of liquid and buffer ions, the field strengths used and the characteristics of the flow reservoirs and interconnects. A few examples of these design issues are given here.

In the case of microchannels used to generate pressure, Eq. (6.60) shows that a low liquid conductivity is essential for increasing thermodynamic efficiency of an electrokinetic pump, as Joule heating is an important contributor to dissipation [Zeng et al., 2000]. In electrokinetic systems for chemical analysis, on the other hand, the need for a stable pH requires a finite buffer strength, and typical buffer strengths are in the 1- to 100-mM range. The need for a stable pH, therefore, often conflicts with a need for high fields to achieve high-efficiency separations because of the effects of Joule heating of the liquid.

Joule heating of the liquid in electrokinetic systems can be detrimental in two ways. First, temperature gradients within the microchannel cause a nonuniformity in the local mobility of electrophoretic particles because the local viscosity is a function of temperature. This nonuniformity in mobility results in a dispersion associated with the transport of electrophoretic species [Bosse and Arce, 2000; Grushka et al., 1989; Knox, 1988]. The second effect of Joule heating is the rise in the absolute temperature of the buffer. This temperature rise results in higher electro-osmotic mobilities and higher sample diffusivities. In microchip electrophoretic separations, the effect of increased diffusivity on separation efficiency is somewhat offset by the associated decrease in separation time. In addition, the authors have found that an important limitation to the electric field magnitude in microchannel electrokinetics is that the associated elevated temperatures and decreases in gas solubility of the solution often result in the nucleation of gas bubbles in the channel. This effect of driving gas out of solution typically occurs well before the onset of boiling and can be catastrophic to the electrokinetic system, as gas bubbles grow and eventually break the electrical circuit required to drive the flow. This effect can be reduced by outgassing of the solution and is, of course, a strong function of the channel geometry, buffer conductivity and the thermal properties of the substrate material.

Finally, an important consideration in electrokinetic experiments is the inadvertent application and/or generation of pressure gradients in the microchannel. Probably the most common cause of this is a mismatch in the height of the fluid level at the reservoirs. Note that, although there may not be a mismatch of fluid level at the start of an experiment, the flow rates created by electro-osmotic flow may eventually create a fluid level mismatch. Also, the fluid level in each reservoir, particularly in reservoirs of 1 mm diameter or less, may be affected by electrolytic gas generation at each electrode. Because electro-osmotic flowrate scales as channel diameter squared and pressure-driven flow scales as channel diameter to the fourth power, this effect is greatly reduced by decreasing the characteristic channel diameter. Another common method of reducing this pressure head effect is to increase the length of the channel for a given cross section. This length increase, of course, implies an increase in operating voltages to achieve the same flowrate. A second source of pressure gradients is a nonuniformity in the surface charge in the channel. An elegant, closed-form solution for the flow in a microchannel with arbitrary axial zeta potential distribution is presented by Anderson and Idol (1985). Recently, Herr et al. (2000) have visualized this effect and offered a simple analytical expression to the pressure-driven flow components associated with zeta potential gradients in fully developed channel flows.

6.4 Summary and Conclusions

In microchannels, the flow of a liquid differs fundamentally from that of a gas, primarily due to the effects of compressibility and potential rarefaction in gases. Although significant differences from continuum macroscale theory have been reported, Sharp et al. (2000) have shown that in the range of accepted laminar flow behavior, the friction factor agrees with classical continuum hydrodynamic theory to within small or negligible differences. Transition to turbulence occurs at $Re = 2000$ [Sharp et al., 2000].

However, large differences in observed pressure drop in microchannels have been reported by many investigators. These differences may be due to imperfections in the flow system of the experiment. Because imperfections may well occur in real engineering systems, it is essential to understand the sources of the observed discrepancies in order to avoid them, control them or factor them into the designs.

The development of measurement techniques for liquid flows is advancing quickly, both as macroscale methods are adapted to these smaller scales and as novel techniques are being developed. Further insight into phenomena present in the microscale flows, including those due to imperfections in channels or flow systems, is likely to occur rapidly given the evolving nature of the measurement techniques. Complex, nonlinear channels can be used efficiently to design for functionality.

Electrokinetics is a convenient and easily controlled method of achieving sample handling and separations on a microchip. Because the body force exerted on the liquid is limited to a region within a few nanometers from the wall, the resulting profiles, in the absence of imposed pressure gradients, are typically plug-like for channel dimensions greater than about 10 µm and ion concentrations greater than about 10 µM. For simple electro-osmotic flows with thin EDLs, low Reynolds number, uniform surface charge and zero imposed pressure gradients, the velocity field of these systems is well approximated by potential flow theory. This significant simplification can, in many cases, be used to predict and optimize the performance of electrokinetic systems. Further, electrokinetics can be used to generate large pressures (>20 atm) on a microfabricated device. In principle, the handling and rapid mixing of less than 1 pL of sample volumes should be possible using electrokinetic systems built with current microfabrication technologies.

Nomenclature

δ	lattice spacing
ε	surface roughness or permittivity of liquid
ϕ	local potential
κ	thermal diffusivity
λ	mean free path
λ_D	Debye length
μ	dynamic viscosity
μ_r	roughness viscosity
ν	kinematic viscosity
ν_{eph}	electrophoretic mobility
ρ	density
ρ_E	net charge density in EDL
σ	net conductivity of solution
τ	time constant or characteristic time scale of a forcing function
τ_{ij}	stress tensor
τ_w	local wall shear stress
$\bar{\tau}_w$	wall shear stress, averaged around the perimeter
ζ	zeta potential

Φ	rate of conversation of mechanical energy to thermal energy viscous dissipation function)
a	radius of cylindrical capillary, $=D/2$
b_i	body force per unit mass in the ith direction
b	body force per unit mass
c_i	Boltzmann distribution of ions of species i
c_o	proportionality constant
c_p	specific heat at constant pressure
$c_{\infty,i}$	molar concentration of ion i in the bulk
d	Stokes diameter of the particle
e	charge of an electron
f	Darcy friction factor
g_i	gravitational acceleration in the ith direction
g	gravitational acceleration
k	thermal conductivity or Boltzmann's constant
n	local unit vector
p	pressure
q	heat flux
q	total molecule charge
r	radial coordinate
t	time
u_{crit}	critical velocity
u_{EOF}	electro-osmostic velocity
u_i	ith component of velocity
$u_{\max}$	maximum u velocity
u_{particle}	total velocity of electrophoretic particles
u_{pressure}	pressure-driven velocity
u_{slip}	velocity evaluated at slip surface at the top of the EDL
u	velocity vector
x	streamwise coordinate
x_i	coordinate in the ith direction
y	coordinate perpendicular to the streamwise direction
z	coordinate perpendicular to the streamwise direction or valance number of the ion
A	cross-sectional area
C_{elec}	electrical capacitance
C_{flow}	capacitance of a fluid element
C^*	normalized friction coefficient
D	characteristic channel dimension (usually diameter)
D_h	hydraulic diameter
E	electric field driving the fluid
E_{slip}	electric field evaluated at slip surface at the top of the EDL
F	Faraday's constant
H	channel height
I	electrical current
I_0	zero-order modified Bessel function of the first kind
K	pressure drop parameter
Kn	Knudsen number
L	channel length
L_s	characteristic length scale
L_e	entrance length

N_A	Avogadro's number
P	perimeter
Q	flux
PO	Poiseuille number
R_{flow}	flow resistance
R_{elec}	electrical resistance
Re	Reynolds number
Re_{crit}	critical Reynolds number
St	Strouhal number
T	temperature
U_s	characteristic velocity scale
$\bar{U}$	area-averaged velocity
V	voltage
V_1	molar volume
V_i	time-dependent voltage applied to a reservoir "well"
W	channel width
STP	standard temperature and pressure

References

Adamson, A.W., and Gast, A.P. (1997) *Physical Chemistry of Surfaces,* sixth edition, John Wiley & Sons, New York.

Alarie, J.P., Jacobson, S.C., Culbertson, C.T. et al. (2000) "Effects of the Electric Field Distribution on Microchip Valving Performance," *Electrophoresis* **21**, pp. 100–106.

Anderson, J.L., and Idol, W.K. (1985) "Electroosmosis Through Pores with Nonuniformly Charged Walls," *Chem. Eng. Commun.* **38**, pp. 93–106.

Arkilic, E.B., Schmidt, M.A., and Breuer, K.S. (1997) "Gaseous Slip Flow in Long Microchannels," *J. MEMS* **6**, pp. 167–178.

Baker, D.R. (1995) *Capillary Electrophoresis,* Techniques in Analytical Chemistry Series, John Wiley & Sons, New York.

Bharadwaj, R., and Santiago, J.G. (2000) unpublished results, Stanford University, Stanford, CA.

Bianchi, F., Ferrigno, R., and Girault, H.H. (2000) "Finite Element Simulation of an Electroosmotic-Driven Flow Division at a T-Junction of Microscale Dimensions," *Anal. Chem.* **72**, pp. 1987–1993.

Bosse, M.A., and Arce, P. (2000) "Role of Joule Heating in Dispersive Mixing Effects in Electrophoretic Cells: Convective-Diffusive Transport Aspects," *Electrophoresis* **21**, pp. 1026–1033.

Brandner, J., Fichtner, M., Schygulla, U., and Schubert, K. (2000) "Improving the Efficiency of Micro Heat Exchangers and Reactors," in *Proc. 4th Int. Conf. Microreaction Technology, AIChE,* March 5–9, Atlanta, GA, pp. 244–249.

Branebjerg, J., Fabius, B., and Gravesen, P. (1995) "Application of Miniature Analyzers: From Microfluidic Components to μTAS," in *Micro Total Analysis Systems,* eds. A. van den Berg and P. Bergveld, Kluwer Academic, Dordrecht, pp. 141–151.

Branebjerg, J., Gravesen, P., Krog, J.P., and Nielsen, C.R. (1996) "Fast Mixing by Lamination," in *Proc. 9th Annual Workshop of Micro Electro Mechanical Systems,* February 11–15, San Diego, CA, pp. 441–446.

Bridgman, P.W. (1923) "The Thermal Conductivity of Liquids Under Pressure," *Proc. Am. Acad. Arts Sci.* **59**, pp. 141–169.

Brody, J.P., Yager, P., Goldstein, R.E., and Austin, R.H. (1996) "Biotechnology at Low Reynolds Numbers," *Biophys. J.* **71**, pp. 3430–3441.

Burgreen, D., and Nakache, F.R. (1964) "ElectroKinetic Flow in Ultrafine Capillary Slits," *J. Phys. Chem.* **68**, pp. 1084–1091.

Chen, Z., Milner, T.E., Dave, D., and Nelson, J.S. (1997) "Optical Doppler Tomographic Imaging of Fluid Flow Velocity in Highly Scattering Media," *Opt. Lett.* **22**, pp. 64–66.

Choi, S.B., Barron, R.F., and Warrington, R.O. (1991) "Fluid Flow and Heat Transfer in Microtubes," in *DSC-Vol. 32, Micromechanical Sensors, Actuators and Systems,* ASME Winter Annual Meeting, Atlanta, GA, pp. 123–134.

Cummings, E.B., Griffiths, S.K., and Nilson, R.H. (1999) "Irrotationality of Uniform Electroosmosis," in *SPIE Conf. on Microfluidic Devices and Systems II,* Santa Clara, CA, **3877**, pp. 180–189.

Cummings, E.B., Griffiths, S.K., Nilson, R.H. et al. (2000) "Conditions for Similitude between the Fluid Velocity and Electric Field in Electroosmotic Flow," *Anal. Chem.* **72**, pp. 2526–2532.

Devasenathipathy, S., and Santiago, J.G. (2000) unpublished results, Stanford University, Stanford, CA.

Eringen, A.C. (1964) "Simple Microfluids," *Int. J. Eng. Sci.* **2**, pp. 205–217.

Flockhart, S.M., and Dhariwal, R.S. (1998) "Experimental and Numerical Investigation into the Flow Characteristics of Channels Etched in <100> Silicon," *J. Fluids Eng.* **120**, pp. 291–295.

Gad-el-Hak, M. (1999) "The Fluid Mechanics of Microdevices—The Freeman Scholar Lecture," *J. Fluids. Eng.* **121**, pp. 5–33.

Galambos, P., and Forster, F.K. (1998) "Micro-Fluidic Diffusion Coefficient Measurement," in *Micro Total Analysis Systems,* eds. D.J. Harrison and A. van den Berg, Kluwer Academic, Dordrecht, pp. 189–192.

Grushka, E., McCormick, R.M., and Kirkland, J.J. (1989) "Effect of Temperature Gradients on the Efficiency of Capillary Zone Electrophoresis Separations," *Anal. Chem.* **61**, pp. 241–246.

Hagen, G. (1839) "On the Motion of Water in Narrow Cylindrical Tubes [German]," *Pogg. Ann.* **46**, p. 423.

Harley, J.C., Huang, Y., Bau, H.H., and Zemel, J.N. (1995) "Gas Flow in Micro-Channels," *J. Fluid Mech.* **284**, pp. 257–274.

Hayes, M., Kheterpal, I., and Ewing, A. (1993) "Effects of Buffer pH on Electoosmotic Flow Control by an Applied Radial Voltage for Capillary Zone Electrophoresis," *Anal. Chem.* **65**, pp. 27–31.

Henry, D.C. (1948) "The Electrophoresis of Suspended Particles. IV. The Surface Conductivity Effect," *Trans. Faraday Soc.* **44**, pp. 1021–1026.

Herr, A.E., Molho, J.I., Santiago, J.G. et al. (2000) "Electroosmotic Capillary Flow with Nonuniform Zeta Potential," *Anal. Chem.* **72**, pp. 1053–1057.

Hiemenz, P.C., and Rajagopalan, R. (1997) *Principles of Colloid and Surface Chemistry,* third edition, Marcel Dekker, New York.

Hitt, D.L., and Lowe, M.L. (1999) "Confocal Imaging of Flows in Artificial Venular Bifurcations," *Trans. ASME J. Biomech. Eng.* **121**, pp. 170–177.

Huang, T., Tsai, P., Wu, Ch., and Lee, C. (1993) "Mechanistic Studies of Electroosmotic Control at the Capillary-Solution Interface," *Anal. Chem.* **65**, pp. 2887–2893.

Hunter, R.J. (1981) *Zeta Potential in Colloid Science,* Academic Press, London.

Israelachvili, J.N. (1986) "Measurement of the Viscosity of Liquids in Very Thin Films," *J. Coll. Interface Sci.* **110**, pp. 263–271.

Jacobson, S.C., Hergenroder, R., Moore, A.W., Jr., and Ramsey, J.M. (1994) "Precolumn Reactions with Electrophoretic Analysis Integrated on a Microchip," *Anal. Chem.* **66**, pp. 4127–4132.

Janson, S.W., Helvajian, H., and Breuer, K. (1999) "MEMS, Microengineering and Aerospace Systems," in *30th AIAA Fluid Dyn. Conf.,* June 28–July 1, Norfolk, VA, AIAA 99-3802.

Jiang, X.N., Zhou, Z.Y., Yao, J., Li, Y., and Ye, X.Y. (1995) "Micro-Fluid Flow in Microchannel," in *Transducers '95: Eurosensor IX,* 8th Int. Conf. on Solid-State Sensors and Actuators and Eurosensors IX, Sweden, pp. 317–320.

Khaledi, M.G. (1998) "High-Performance Capillary Electrophoresis," in *Chemical Analysis: A Series of Monographs on Analytical Chemistry and its Applications,* Vol. 146, ed. J.D. Winefordner, John Wiley & Sons, New York.

Kitahara, A., and Watanabe, A. (1984) *Electrical Phenomena at Interfaces: Fundamentals, Measurements, and Applications,* Vol. 15, Surfactant Science Series, Marcel Dekker, New York.

Knox, J.H., (1988) "Thermal Effects and Band Spreading in Capillary Electro-Separation," *Chromatographia* **26**, pp. 329–337.

Koutsiaris, A.G., Mathiouslakis, D.S., and Tsangaris, S. (1999) "Microscope PIV for Velocity-Field Measurement of Particle Suspensions Flowing Inside Glass Capillaries," *Meas. Sci. Technol.* **10**, pp. 1037–1046.

Lanzillotto, A.-M., Leu, T.-S., Amabile, M., and Wildes, R. (1996) "An Investigation of Microstructure and Microdynamics of Fluid Flow in MEMS," in *AD-Vol. 52, Proc. of ASME Aerospace Division*, Atlanta, GA, pp. 789–796.

Liu, R.H., Stremler, M.A., Sharp, K.V., Olsen, M.G., Santiago, J.G., Adrian, R.J., Aref, H., and Beebe, D.J. (2000) "Passive Mixing in a Three-Dimensional Serpentine Microchannel," *J. MEMS* **9**, pp. 190–197.

Mala, G.M. (1999) "Heat Transfer and Fluid Flow in Microchannels," Ph.D. thesis, University of Alberta.

Mala, G.M., and Li, D. (1999) "Flow Characteristics of Water in Microtubes," *Int. J. Heat Fluid Flow* **20**, pp. 142–148.

Manz, A., Effenhauser, C.S., Burggraf, N. et al. (1994) "Electroosmotic Pumping and Electrophoretic Separations for Miniaturized Chemical Analysis Systems," *J. Micromech. Microeng.* **4**, pp. 257–265.

Manz, B., Stilbs, P., Jönsson, B., Söderman, O., and Callaghan, P.T. (1995) "NMR Imaging of the Time Evolution of Electroosmotic Flow in a Capillary," *J. Phys. Chem.* **99**, pp. 11297–11301.

Meinhart, C.D., Wereley, S.T., and Santiago, J.G. (1999) "PIV Measurements of a Microchannel Flow," *Exp. Fluids* **27**, pp. 414–419.

Merkle, C.L., Kubota, T., and Ko, D.R.S. (1974) "An Analytical Study of the Effects of Surface Roughness on Boundary-Layer Transition," AF Office of Scientific Research, Space and Missile Sys. Org., AD/A004786.

Molho, J.I., Herr, A.E., Mosier, B.P., Santiago, J.G., Kenny, T.W., Brennen, R.A., and Gordon, G. (2000) "Designing Corner Compensation for Electrophoresis in Compact Geometries," in *Micro Total Analysis Systems*, eds. A. van den Berg, W. Olthuis, and P. Bergveld, Kluwer Academic, Dordrecht, pp. 287–290.

Novotny, E.J., and Eckert, R.E. (1974) "Rheological Properties of Viscoelastic Fluids from Continuous Flow Through a Channel Approximating Infinite Parallel Plates," *Trans. Soc. Rheol.* **18**, pp. 1–26.

Overbeek, J.T.G. (1952a) "Electrochemistry of the Double Layer," in *Colloid Science*, ed. H.R. Kruyt, Elsevier, Amsterdam, pp. 115–277.

Overbeek, J.T.G. (1952b) "Electrokinetic Phenomena," in *Colloid Science*, Vol. 1, ed. H.R. Kruyt, Elsevier, Amsterdam.

Ovryn, B. (1999) "Three-Dimensional Forward Scattering Particle Image Velocimetry in a Microscopic Field-of-View," in *Proc. 3rd Int. Workshop PIV*, Sept. 16–18, Santa Barbara, CA, pp. 385–393.

Papautsky, I., Brazzle, J., Ameel, T., and Frazier, A.B. (1999a) "Laminar Fluid Behavior in Microchannels Using Micropolar Fluid Theory," *Sensors and Actuators A* **73**, pp. 101–108.

Papautsky, I., Gale, B.K., Mohanty, S., Ameel, T.A., and Frazier, A.B. (1999b) "Effects of Rectangular Microchannel Aspect Ratio on Laminar Friction Constant," in *SPIE Conf. on Microfluidic Devices and Systems II*, Santa Clara, CA, **3877**, pp. 147–158.

Patankar, N.A., and Hu, H.H. (1998) "Numerical Simulation of Electroosmotic Flow," *Anal. Chem.* **70**, pp. 1870–1881.

Paul, P.H., Arnold, D.W., and Rakestraw, D.J. (1998a) "Electrokinetic generation of high pressures using porous microstructures," in *Micro Total Analysis Systems*, ed. D.J. Harrison and A. van den Berg, Kluwer Academic, Dordrecht, pp. 49–52.

Paul, P.H., Garguilo, M.G., and Rakestraw, D.J. (1998b) "Imaging of Pressure- and Electrokinetically Driven Flows Through Open Capillaries," *Anal. Chem.* **70**, pp. 2459–2467.

Peng, X.F., Peterson, G.P., and Wang, B.X. (1994) "Frictional Flow Characteristics of Water Flowing Through Rectangular Microchannels," *Exp. Heat Transfer* **7**, pp. 249–264.

Pfahler, J., Harley, J., Bau, H., and Zemel, J. (1990a) "Liquid Transport in Micron and Submicron Channels," *Sensors and Actuators A* **21–A23**, pp. 431–434.

Pfahler, J., Harley, J., Bau, H.H., and Zemel, J. (1990b) "Liquid and Gas Transport in Small Channels," in *DSC-Vol. 19, Microstructures, Sensors and Actuators*, ASME Winter Annual Meeting, Dallas, TX, pp. 149–157.

Pfahler, J., Harley, J., Bau, H., and Zemel, J.N. (1991) "Gas and Liquid Flow in Small Channels," in *DSC-Vol. 32, Micromechanical Sensors, Actuators and Systems*, ASME Winter Annual Meeting, Atlanta, GA, pp. 49–59.

Poiseuille, M. (1840, 1841) "Recherches Expérimentales Sur le Mouvement des Liquides dans les Tubes de Très Petits Diamètres," *CR Hebdomaires des Séances Acad. Sci.* **11**.

Prandtl, L., and Tietjens, O.G. (1934) *Applied Hydro- and Aeromechanics*, McGraw-Hill, New York.

Probstein, R.F. (1994) *Physicochemical Hydrodynamics: An Introduction,* second edition, John Wiley & Sons, New York.

Qu, W., Mala, G.M., and Li, D. (2000) "Pressure-Driven Water Flows in Trapezoidal Silicon Microchannels," *Int. J. Heat Mass Transfer* **43**, pp. 353–364.

Reynolds, O. (1883) "An Experimental Investigation of the Circumstances Which Determine Whether the Motion of Water Will Be Direct or Sinuous, and the Law of Resistance in Parallel Channels," *Phil. Trans. R. Soc. London* **2**, p. 51.

Rice, C.L., and Whitehead, R. (1965) "Electrokinetic Flow in a Narrow Cylindrical Capillary," *J. Phys. Chem.* **69**, pp. 4017–4024.

Russel, W.B., Saville, D.A., and Schowalter, W.R. (1999) *Colloidal Dispersions,* ed. G.K. Batchelor, Cambridge Mongraphs on Mechanics and Applied Mathematics, Cambridge University Press, Cambridge, U.K.

Santiago, J.G. (2001) "Electroosmotic Flows in Microchannels with Finite Inertial and Pressure Forces," *Anal. Chem.* **73**, pp. 2353–2365.

Santiago, J.G., Wereley, S.T., Meinhart, C.D., Beebe, D.J., and Adrian, R.J. (1998) "A Particle Image Velocimetry System for Microfluidics," *Exp. Fluids* **25**, pp. 316–319.

Scales, P., Grieser, F., and Healy, T. (1992) "Electrokinetics of the Silica-Solution Interface: A Flat Plate Streaming Potential Study," *ACS J. Langmuir Surf. Colloids* **8**, pp. 965–974.

Schaller, Th., Bolin, L., Mayer, J., and Schubert, K. (1999) "Microstructure Grooves with a Width Less Than 50 µm Cut with Ground Hard Metal Micro End Mills," *Precision Eng.* **23**, pp. 229–235.

Schulte, T.H., Bardell, R.L., and Weigl, B.H. (2000) "On-Chip Microfluidic Sample Preparation," *J. Lab. Autom.* **5**, p. 83.

Shah, R.K., and London, A.L. (1978) "Laminar Flow Forced Convection in Ducts," in *Advances in Heat Transfer,* Suppl. 1, Academic Press, New York.

Sharp, K.V., Adrian, R.J., and Beebe, D.J. (2000) "Anomalous Transition to Turbulence in Microtubes," in *Proc. Int. Mech. Eng. Cong. Expo., 5th Micro-Fluidic Symp.,* Nov. 5–10, Orlando, FL, in press.

Smoluchowski, M.V. (1903) "Contribution à la théorie de l'endosmose électrique et de quelques phenomènes corrélatifs," *Bull. Int. Acad. Sci. Cracovie* **8**, pp. 182–200.

Taylor, J.A., and Yeung, E.S. (1993) "Imaging of Hydrodynamic and Electrokinetic Flow Profiles in Capillaries," *Anal. Chem.* **65**, pp. 2928–2932.

Tieu, A.K., Mackenzie, M.R., and Li, E.B. (1995) "Measurements in Microscopic Flow with a Solid-State LDA," *Exp. Fluids* **19**, pp. 293–294.

Tuckerman, D.B., and Pease, R.F.W. (1981) "High-Performance Heat Sinking for VLSI," *IEEE Electron Device Lett.* **EDL-2**, pp. 126–129.

Webb, A., and Maynes, D. (1999) "Velocity Profile Measurements in Microtubes," in *30th AIAA Fluid Dyn. Conf.,* June 28–July 1, Norfolk, VA, AIAA 99-3803.

Weigl, B.H., and Yager, P. (1999) "Microfluidic Diffusion-Based Separation and Detection," *Science* **283**, pp. 346–347.

White, F.M. (1991) *Viscous Fluid Flow,* second edition, eds. J.P. Holman and J.R. Lloyd, McGraw-Hill, New York.

White, F. M. (1994) *Fluid Mechanics,* third edition, McGraw-Hill, New York.

Wilding, P., Pfahler, J., Bau, H.H., Zemel, J.N., and Kricka, L.J. (1994) "Manipulation and Flow of Biological Fluids in Straight Channels Micromachined in Silicon," *Clin. Chem.* **40**, pp. 43–47.

Wu, P., and Little, W.A. (1983) "Measurement of Friction Factors for the Flow of Gases in Very Fine Channels Used for Microminiature Joule-Thomson Refrigerators," *Cryogenics* **23**, pp. 273–277.

Yazdanfar, S., Kulkarni, M.D., and Izatt, J.A. (1997) "High Resolution Imaging of In Vivo Cardiac Dynamics Using Color Doppler Optical Coherence Tomography," *Opt. Exp.* **1**, pp. 424–431.

Yu, D., Warrington, R., Barron, R., and Ameel, T. (1995) "An Experimental and Theoretical Investigation of Fluid Flow and Heat Transfer in Microtubes," in *Proc. of ASME/JSME Thermal Engineering Joint Conf.,* March 19–24, Maui, HI, pp. 523–530.

Zeng, S., Chen, C., Mikkelsen, J.C. et al. (2000) "Fabrication and Characterization of Electrokinetic Micro Pumps," in *7th Intersoc. Conf. on Thermal and Thermomechanical Phenomena in Electronic Systems,* May 23–26, Las Vegas, NV.

7
Burnett Simulations of Flows in Microdevices

Ramesh K. Agarwal
Wichita State University

Keon-Young Yun
Wichita State University

7.1 Abstract 7-1
7.2 Introduction 7-2
7.3 History of Burnett Equations.................................... 7-5
7.4 Governing Equations 7-7
7.5 Wall-Boundary Conditions 7-11
7.6 Linearized Stability Analysis of Burnett Equations 7-12
7.7 Numerical Method... 7-15
7.8 Numerical Simulations 7-15
   Application to Hypersonic Shock Structure • Application to Two-Dimensional Hypersonic Blunt Body Flow • Application to Axisymmetric Hypersonic Blunt Body Flow • Application to NACA 0012 Airfoil • Subsonic Flow in a Microchannel • Supersonic Flow in a Microchannel
7.9 Conclusions 7-33
   Nomenclature 7-33

7.1 Abstract

In recent years, interest in computing gas flows at high Knudsen numbers in microdevices has been considerable. At low Knudsen numbers, models based on the solution of compressible Navier–Stokes equations with slip-boundary conditions are adequate. At high Knudsen numbers, either higher order (beyond Navier–Stokes) continuum equations or the particle methods such as direct simulation Monte Carlo (DSMC) are employed to compute the flows. Higher order continuum approximations are based on the Chapman–Enskog expansion of the Boltzmann equation (leading to Burnett and super-Burnett equations), or moment methods based on taking the moments of the Boltzmann equation with flow variables (leading to Grad's 13-moment equations or Levermore' moment equations, for example). In this chapter, the history of the Burnett equations and a variety of Burnett approximations (conventional Burnett equations, augmented Burnett equations and BGK–Burnett equations) are presented. The physical and numerical issues related to these approximations are discussed. Traditionally, Burnett equations have been employed to compute the high-altitude, low-density hypersonic flows in the continuum–transition regime. Therefore, some Navier–Stokes and Burnett solutions are presented for hypersonic shock structure and blunt-body flows and rarefied subsonic airfoil flow to provide some perspective on the applicability and suitability of Burnett equations for computing flows at high Knudsen numbers. Calculations are then presented for both subsonic and supersonic flows in microchannels. These computations are compared with Navier–Stokes solutions with slip-boundary conditions and DSMC solutions.

The computations provide some assessment of Burnett equations for computing microchannel flows at high Knudsen numbers.

7.2 Introduction

Microelectromechanical systems (MEMS) are currently attracting a great deal of interest and attention because of their great potential in many industrial and medical applications. As a result, considerable effort is being devoted to the design and fabrication of MEMS. MEMS refer to devices that have characteristic length of less than 1 mm but more than 1 µm, that combine electrical and mechanical components and that are fabricated using integrated-circuit, batch-processing technologies. A few examples of MEMS are microsensors, microactuators, micromotors, microvalves, micropumps, microducts etc. Fluid flows in the microdevices such as microvalves, micropumps and microducts are significantly different from those in macroscopic devices due to their small characteristic sizes. Hence, understanding of the physics of the flows in the microdevices is very important in their development and design.

Various regimes of fluid flows can be broadly classified into the continuum, continuum–transition, transition and free molecular regimes as shown in Table 7.1. For a large class of flows, Navier–Stokes equations based on the continuum approximation are adequate to model the fluid behavior. Continuum approximation implies that the mean free path of the molecules λ in a gas is much smaller than the characteristic length L of interest (say, the body dimension); that is, the Knudsen number Kn ($=\lambda/L$) is very small ($\ll 1$). However, for a variety of flows, this assumption is not valid; Knudsen number is of $O(1)$. In these flows, the gas is neither completely in the continuum regime nor in the rarefied (free molecular flow) regime. Therefore, such flows have been categorized as continuum–transition or transitional flows. Examples of such flows include the hypersonic flows about space vehicles in low Earth orbit [Ivanov and Gimelshein, 1998] or flows in microchannels of MEMS [Gad-el-Hak, 1999].

In high-altitude hypersonic flows, low density gives rise to high Knudsen number effects, while in microscale flows, which usually occur at atmospheric conditions, small length scales create regions of high Knudsen numbers. In the case of high-altitude hypersonic flows, the shock layer thickness at the nose of the space vehicle (shuttle) is much thicker than that predicted from the Navier–Stokes equations. In a long microchannel, the pressure gradient is observed to be nonconstant and the experimentally measured mass flow rate is higher than that predicted from the conventional continuum flow model [Arkilic et al., 1997; Harley et al., 1995; Liu et al., 1993; Pong et al., 1994]. In such a microscale flow, the mean free path of the molecules can be of the same order of magnitude as the characteristic length of the microchannel: $Kn \sim O(1)$. For a microchannel defined by ratio $\varepsilon = H/L$, where H and L are width

TABLE 7.1 Flow Regimes and Fluid Models

Knudsen Number	Fluid Model
$Kn \to 0$ (continuum, no molecular diffusion)	Euler equations
$Kn \leq 10^{-3}$ (continuum with molecular diffusion)	Navier–Stokes equations with no-slip-boundary conditions
$10^{-3} \leq Kn \leq 10^{-1}$ (continuum–transition)	Navier–Stokes equations with slip-boundary conditions
$10^{-1} \leq Kn \leq 10$ (transition)	Burnett equations with slip-boundary conditions Moment equations DSMC Lattice Boltzmann
$Kn > 10$ (free molecular flow)	Collisionless Boltzmann DSMC Lattice Boltzmann

TABLE 7.2 Flow Regimes in a Microchannel for Different Knudsen Numbers

	Re		
M	$O(\varepsilon)$	$O(1)$	$O(1/\varepsilon)$
$O(\varepsilon)$	$Kn = O(1)$; creeping microflow	$Kn = O(\varepsilon)$; moderate microflow	$Kn = O(\varepsilon^2)$; low M Fanno flow
$O(1)$	$Kn = O(1/\varepsilon)$; transonic free molecular flow	$Kn = O(1)$; transonic microflow	$Kn = O(\varepsilon)$; transonic Fanno flow
$O(1/\varepsilon)$	$Kn = O(1/\varepsilon^2)$; hypersonic free molecular flow	$Kn = O(1/\varepsilon)$; hypersonic free molecular flow	$Kn = O(1)$; hypersonic Fanno (transitional) flow

Source: From Arkilic, E.B. et al. (1997) *J. MEMS* **6**, 167–178. With permission.

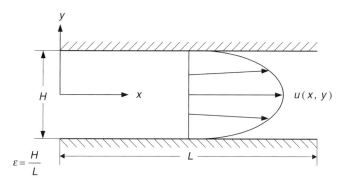

FIGURE 7.1 Flow in a microchannel. Relevant flow parameters: Mach number, Reynolds number and Knudsen number are $M = \bar{u}/c$, $Re = \bar{\rho}\bar{u}H/\mu$, and $Kn = (\pi\gamma/2)^{0.5} M/Re$, respectively. "–" denotes the average outlet conditions.

and length of the channel, respectively, as shown in Figure 7.1, Arkilic et al. (1997) have characterized various flow regimes depending upon the Reynolds number Re and Mach number M of the flow as shown in Table 7.2. Tables 7.1 and 7.2 together now can be used to select an appropriate fluid model for simulation of the flowfield in a microchannel. It should be noted here that both low-density and microscale effects can be local in a flow so that the entire flow is in both the continuum and transition regimes. (Readers may refer to Chapter 4 in this handbook for further understanding of the physics of the flows in microdevices and appropriate models for various flow regimes.)

As shown in Table 7.1, Navier–Stokes equations are not adequate to model the flows in the continuum–transition regime; the Boltzmann equation describes the flow in all the regimes—continuum, continuum–transition and free molecular. The techniques available for solving the Boltzmann equation can be classified into particulate methods and moment methods. The direct simulation Monte Carlo (DSMC) method falls in the category of particulate methods [Bird, 1994]. Moment methods derive the higher order fluid dynamics approximations beyond Navier–Stokes equations to account for departures from thermal equilibrium. The higher order fluid dynamic models are known as the extended hydrodynamic equations (EHE) or generalized hydrodynamic equations (GHE). However, both classes of methods have significant limitations, either in describing the physics or in terms of the computational resources needed for accurate simulation, for modeling flows in continuum–transition regime. Currently, the DSMC method, perhaps, can be considered as the most accurate and widely used technique for computation of low-density flows. However, in the continuum–transition regime, where the densities are not "low enough," the DSMC method requires a large number of particles for accurate simulation, making the technique prohibitively expensive both in terms of computational time and memory requirements. For example, it has been shown by Koppenwallner (1987) that the nose-up pitching moment of the space shuttle was predicted inaccurately by the DSMC method in the continuum–transition regime due to the inadequate number of particles used in modeling. The nose-up pitching moment could be corrected by deflecting the body flap to 15 degrees—twice the amount predicted by DSMC. A similar situation may occur in microscale flows due to the relatively high density and low velocity requiring enormous computational power and resulting in large statistical scatter in the DSMC simulations [Nance et al., 1998].

At present, the majority of the computations with the DSMC method, especially in two and three dimensions for $Kn = O(1)$, are beyond the currently available computing power. As an alternative, higher order extended or generalized hydrodynamic equations have been proposed that can perform equally well in both the continuum and continuum–transition regimes. The extended hydrodynamic equations have been derived from the Boltzmann equation using either one of the following approaches. In one approach, higher order constitutive relations (beyond Navier–Stokes) for stress and heat transfer terms are obtained using the Chapman–Enskog expansion of the Boltzmann equation with the Knudsen number as a parameter. In the Chapman–Enskog expansion, the first-term represents the Maxwellian (equilibrium) distribution function f_0. The first moment of the Boltzmann equation with the collision invariant vector, with f_0 as the approximation for the distribution function, results in the Euler equations. The first two terms in the Chapman–Enskog expansion—$(f_0 + Kn\, f_1)$—give a distribution function corresponding to the Navier–Stokes equations, representing a first-order departure from thermal equilibrium. The first three terms $(f_0 + Kn\, f_1 + Kn^2 f_2)$ in the expansion give a distribution function, which results in the so-called "Burnett equations," representing a second-order departure from the equilibrium.

Burnett equations have been a subject of considerable investigation in recent years and are the main subject of this chapter. Higher order approximations beyond Burnett equations, the so-called "super-Burnett equations" etc., can be derived by continuing the Chapman–Enskog expansion to higher orders. However, at the present time, the complexity of the highly nonlinear Burnett stress and heat transfer terms itself is enormously challenging both computationally and in terms of understanding the physics, so the consideration of super-Burnett equations and beyond is meaningless.

Burnett stress and heat transfer terms contain higher than second-order derivatives. Therefore, an additional boundary condition is necessary for the solution to the Burnett equations to be uniquely determined; different solutions can result based on the choice of boundary values [Lee, 1994]. Furthermore, it has also been shown that the conventional Burnett equations can violate the second law of thermodynamics at high Knudsen numbers [Comeaux et al., 1995]. Because the focus of this chapter is on Burnett equations, these are described in detail in Section 7.3.

In another approach, the extended hydrodynamic equations are derived using the moment method, which employs the equations of transfer instead of dealing with the distribution function. In the moment method, the distribution function f is expanded in moments of physical variables (density, velocity, pressure, temperature etc.) and the evolution equations for moments are derived from the Boltzmann equation. In principle, this approach should result in a set of macroscopic equations consistent with the second law of thermodynamics, but many of the methods (for example, Grad's 13-moment method [Grad, 1949]) result in the entropy equation violating the Gibb's relation [Holway, 1964; Weiss, 1996]. This problem was addressed in the recent work of Levermore (1996) and Levermore and Morokoff (1998) on the so-called "Gaussian closure." The Gaussian closure is based on a more elegant choice of a finite-dimensional linear subspace and yields a hyperbolic system of moment equations. Because the hyperbolic equations are easier to solve numerically, Groth et al. (1995) have developed some computational models based on this closure. However, the Gaussian closure is of limited practical interest, as the primary system with ten variables admits no heat flux. Other moment systems (for example, the 35-moment system of Brown [1996]) do not yield numerical solutions above Mach numbers of approximately two. Furthermore, it should be noted here that the application of the 13- or 35-moment systems to three-dimensional problems remains computationally prohibitive at present.

Because of the physical and numerical difficulties associated with the Burnett equations and moment equations, Myong (1999) has recently suggested yet another set of generalized hydrodynamic equations based on the work of Eu (1992). Eu's equations are based on a nonequilibrium canonical distribution function and a cumulant expansion of the collision integral in Boltzmann equation. These equations can be considered as the most thermodynamically consistent macroscopic equations, as the second law of thermodynamics is satisfied to every order of approximation. It also turns out that they recover the correct behavior at both the continuum and free molecular limits. Myong (1999) has recently developed a computational model based on Eu's evolution equations within the framework of 13 moments. This model so far has been applied to some one-dimensional problems, but the full potential of this set of equations for calculating two- and three-dimensional flows in the continuum–transition regime remains to be determined and will require several years of intensive computational effort. Furthermore, the solution of Eu's equations for a three-dimensional problem will remain computationally prohibitive in the near future.

Because of the limitations of the DSMC method due to the requirements of computing power (currently unavailable) and the difficulties associated with higher order fluid dynamics models as described above, a hybrid approach has been suggested by many investigators in recent years [Oran et al., 1998; Roveda et al., 1998]. The hybrid method couples a Euler or Navier–Stokes solver with DSMC. The hybrid codes have been developed for problems that contain disconnected nonequilibrium regions embedded in a continuum flow [Roveda et al., 1998]. However, the development of a hybrid code is not simple, as two issues need to be resolved before implementation: (1) when to switch between the two methods, and (2) how to pass information from one method to the other [Boyd et al., 1995]. Furthermore, a conceptual inconsistency remains, as the hybrid method must recover both the continuum and free molecular limits. (For molecular-based microfluidic simulation models, the reader is referred to Chapter 8 in this handbook.)

In recent years, several modifications to the original Burnett equations that have been proposed are discussed in Section 7.3. Sections 7.4 and 7.5 describe the governing equations and the wall-boundary conditions, respectively. Section 7.6 deals with the linearized stability analysis of one-dimensional Burnett equations. Section 7.7 briefly describes the numerical scheme and other computational aspects of the three-dimensional Burnett solver. In Section 7.8, computational results are presented for one- and two-dimensional problems. They include computations for hypersonic shock structures, blunt body flows, subsonic flow past an airfoil, and subsonic and supersonic flow in a microchannel. Although the focus of this chapter is on flows in microdevices, the hypersonic flow computations for blunt body flows etc. are presented here because traditionally the Burnett equations have been applied to compute this class of flows over the past decade and computational results from Navier–Stokes and DSMC simulations can be used for the purpose of comparison. These solutions are instructive in providing some assessment of the accuracy and applicability of Burnett equations for computing flows in the continuum–transition regime.

7.3 History of Burnett Equations

Table 7.3 briefly traces the history of Burnett equations. In 1935, Burnett (1935) developed constitutive relationships for the stress and heat transfer terms by applying the Chapman–Enskog expansion to the Boltzmann equation for second-order departures from collisional equilibrium. These equations are referred to as the *original* Burnett equations. In 1939, Chapman and Cowling (1970) replaced the material derivatives in the original Burnett equations by spatial derivatives obtained from inviscid Euler equations. This alternative form of the original Burnett equations is referred to as the *conventional* Burnett equations. Expressing the material derivatives in terms of the spatial derivatives was considered acceptable as the Navier–Stokes and Burnett equations were considered to be first- and second-order corrections to the Euler equations. The use of Euler equations to express the material derivatives retained the second-order accuracy of the Burnett equations. For reasons unknown, the conventional Burnett equations and not the original Burnett equations became the set of higher order constitutive relations studied during the past six decades.

In recent years, Fiscko and Chapman (1988) and Zhong (1991) have employed the conventional Burnett equations to extend the numerical methods for continuum flow into the continuum–transition regime by incorporating the additional linear and nonlinear stress and heat transfer terms in the standard Navier–Stokes solvers. In one of the earliest attempts to numerically solve the conventional Burnett equations, Fiscko and Chapman (1988) solved the hypersonic shock structure problem by relaxing an initial solution to steady state. They obtained solutions for a variety of Mach numbers and concluded

TABLE 7.3 Brief History of Burnett Set of Equations

Equations	Ref.	Comments
Burnett equations	Burnett (1935)	Derived from Boltzmann equation by considering the first three terms of the Chapman–Enskog expansion; appearance of material derivatives, $D(\ )/Dt$, in the second-order (Burnett) flux vectors.
Conventional Burnett equations	Chapman and Cowling (1970)	Euler equations were used to express the material derivatives in terms of the spatial derivatives.
Conventional Burnett equations	Fiscko and Chapman (1988)	Encountered problem of small wavelength instability as the grids were refined.
Augmented Burnett equations	Zhong (1991)	Linearized third-order terms were added to stabilize the Burnett equations; not entirely successful for computing blunt body wakes and flat plate boundary layers.
Conventional Burnett equations	Welder et al. (1993)	Due to the nonlinear terms in the Burnett equations, linear stability analysis alone is not sufficient to explain the instability at high Knudsen numbers.
Conventional Burnett equations	Comeaux et al. (1995)	Burnett equations can violate the second law of thermodynamics at high Knudsen numbers.
BGK–Burnett equations	Balakrishnan and Agarwal (1996)	Nonlinear collision integral in the Boltzmann equation was simplified by representing it with the Bhatnagar–Gross–Krook (BGK) model; material derivatives expressed in terms of the spatial derivatives using Navier–Stokes equations; linear stability analysis shows unconditional stability for all Knudsen numbers; when Euler equations are used to express the material derivatives, they guarantee unconditional stability for monatomic gases; entropy consistent (satisfies the Boltzmann's H-theorem) for a wide range of Knudsen numbers.

that the conventional Burnett equations do indeed describe the normal shock structure better than the Navier–Stokes equations at high Mach numbers. They, however, experienced stability problems when the computational grids were made progressively finer.

In a subsequent attempt, Zhong (1991) showed that the conventional Burnett equations could be stabilized by adding linear third-order terms from the super-Burnett equations, in order to maintain second-order accuracy, to the stress and heat transfer terms in the Burnett equations. This set of equations was termed the augmented Burnett equations. The coefficients (weights) of these linear third-order terms were determined by carrying out a linearized stability analysis of the augmented Burnett equations. The augmented Burnett equations did not present any stability problems when they were used to compute the hypersonic shock structure and hypersonic blunt body flows. However, attempts at computing the flowfields for blunt body wakes and flat-plate boundary layers with the augmented Burnett equations have not been entirely successful. Furthermore, the *ad hoc* addition of the linear super-Burnett terms and their necessity raises the question of whether the approximation used to create the *conventional* Burnett equations from the original Burnett equations introduces the small wavelength instabilities. It has been noted by Welder et al. (1993) that linear stability analysis alone is not sufficient to explain the instability of Burnett equations with increasing Knudsen numbers as this analysis does not take into account many nonlinear terms, products of first- and higher-order derivatives, that are present in the conventional Burnett equations. Comeaux et al. (1995) have recently surmised that this instability may also be attributed to the fact that the *conventional* Burnett equations can violate second law of thermodynamics at high Knudsen numbers.

In order to overcome the difficulties associated with the conventional Burnett equations (violation of the second law of thermodynamics and instability at high Knudsen numbers), Balakrishnan and Agarwal (1996; 1997) have recently derived a new set of Burnett equations designated as "BGK–Burnett" equations, which are entropy consistent and satisfy the Boltzmann H-theorem. The highly nonlinear nature of the collision integral in the Boltzmann equation presents the biggest hurdle in devising a higher order distribution function. This problem can be circumvented by representing the collision integral in the

Bhatnagar–Gross–Krook (BGK) form [Bhatnagar et al., 1954]. This approximation assumes that any slight departure from the equilibrium distribution will eventually settle down to the equilibrium distribution exponentially. This approximation also assumes that the gas is dilute, hence the collision processes are predominantly binary in nature. Because only binary collisions are considered, the time taken for the nonequilibrium distribution to settle down to the equilibrium level is equal to the reciprocal of the collision frequency. With the BGK approximation to the collision integral, the exact closed-form analytical expression for the distribution function to any order can be obtained.

Balakrishnan and Agarwal (1997) have derived the BGK–Burnett equations by considering the first three terms in the Chapman–Enskog expansion. In this derivation, Euler equations were used to approximate the material derivatives in the first-order distribution function. Moments of the first-order distribution function with the collision invariant vector yield the Navier–Stokes equations. In order to keep in step with the iterative refinement technique, it was conjectured that the Navier–Stokes equations could be used to approximate the material derivatives in the second-order distribution function. It has been shown that this formulation ensures a positive entropy change. The BGK–Burnett equations are obtained by taking moments of this second-order distribution function with the collision invariant vector. This set of equations contains all the stress and heat transfer terms reported by Fiscko and Chapman (1988) and has additional terms which are similar to the super-Burnett terms. Using linearized stability analysis, it has been shown that these additional terms make the BGK–Burnett equations unconditionally stable for monatomic as well as polyatomic gases. In order to check if the entropy production is positive throughout the flowfield, the Boltzmann H-theorem was applied to the second-order distribution function. It was shown that "H" is a monotonically decreasing function, thereby ensuring that the equations do not violate the second law of thermodynamics [Balakrishnan et al., 1997]. Thus the BGK–Burnett equations overcome the problems associated with the conventional Burnett equations—namely, violation of the second law of thermodynamics and instability at high Knudsen numbers.

7.4 Governing Equations

In this section, the augmented Burnett and BGK–Burnett equations are presented. For original and conventional Burnett equations, the reader is referred to papers by Burnett (1935) and Chapman and Cowling (1970), respectively. Here, we present only the two-dimensional augmented and BGK–Burnett equations for the sake of brevity; three-dimensional augmented Burnett equations and BGK–Burnett equations are given in Yun et al. (1998).

The governing equations for two-dimensional, unsteady, compressible, viscous flow can be written in Cartesian coordinates as:

$$\frac{\partial \mathbf{Q}}{\partial t} + \frac{\partial \mathbf{E}}{\partial x} + \frac{\partial \mathbf{F}}{\partial y} = 0 \tag{7.1}$$

where

$$\mathbf{Q} = \begin{bmatrix} \rho \\ \rho u \\ \rho v \\ e_t \end{bmatrix} \tag{7.2}$$

In Eq. (7.1), $\mathbf{E}$ and $\mathbf{F}$ are the flux vectors of the flow variables $\mathbf{Q}$ in the x and y directions, respectively. These flux vectors can be written as:

$$\mathbf{E} = \mathbf{E}_I + \mathbf{E}_V$$
$$\mathbf{F} = \mathbf{F}_I + \mathbf{F}_V \tag{7.3}$$

where $\mathbf{E}_I$ and $\mathbf{F}_I$ are the inviscid-flux terms and $\mathbf{E}_V$ and $\mathbf{F}_V$ are the viscous-flux terms given as follows:

$$\mathbf{E}_I = \begin{bmatrix} \rho u \\ \rho u^2 + p \\ \rho u v \\ (e_t + p) u \end{bmatrix}, \quad \mathbf{E}_V = \begin{bmatrix} 0 \\ \sigma_{11} \\ \sigma_{12} \\ \sigma_{11} u + \sigma_{12} v + q_1 \end{bmatrix} \tag{7.4}$$

$$\mathbf{F}_I = \begin{bmatrix} \rho v \\ \rho u v \\ \rho v^2 + p \\ (e_t + p) v \end{bmatrix}, \quad \mathbf{F}_V = \begin{bmatrix} 0 \\ \sigma_{21} \\ \sigma_{22} \\ \sigma_{21} u + \sigma_{22} v + q_2 \end{bmatrix} \tag{7.5}$$

The constitutive equations for a gas flow near thermodynamic equilibrium can be derived as approximate solutions of the Boltzmann equation using the Chapman–Enskog expansion. This method yields the general constitutive relations for the stress tensor σ_{ij} and the heat-flux vector q_i as follows:

$$\begin{aligned} \sigma_{ij} &= \sigma_{ij}^{(0)} + \sigma_{ij}^{(1)} + \sigma_{ij}^{(2)} + \sigma_{ij}^{(3)} + \cdots + \sigma_{ij}^{(n)} + O(Kn^{n+1}) \\ q_i &= q_i^{(0)} + q_i^{(1)} + q_i^{(2)} + q_i^{(3)} + \cdots + q_i^{(n)} + O(Kn^{n+1}) \end{aligned} \tag{7.6}$$

where n represents the order of accuracy with respect to Kn. Kn is defined as:

$$Kn = \lambda/L \tag{7.7}$$

where L is the macroscopic characteristic length, and the mean free path λ is given by:

$$\lambda = \frac{16\mu}{5\rho\sqrt{2\pi RT}} \tag{7.8}$$

In the case of $Kn \approx 0$, only the first terms in Eq. (7.6) are important. The zeroth-order approximation ($n = 0$) results in the Euler equations with:

$$\sigma_{ij}^{(0)} = 0 \quad \text{and} \quad q_i^{(0)} = 0 \tag{7.9}$$

When $Kn < 0.1$, the first two terms in Eq. (7.6) become important for the accurate representation of the stress and heat transfer properties of the gas flow. This first-order approximation represents the Navier–Stokes equations. The stress tensor and the heat-flux terms ($n = 1$) are given as:

$$\begin{aligned} \sigma_{11}^{(1)} &= -\mu(\delta_1 u_x + \delta_2 v_y) \\ \sigma_{12}^{(1)} &= \sigma_{21}^{(1)} = -\mu(u_y + v_x) \\ \sigma_{22}^{(1)} &= -\mu(\delta_1 v_y + \delta_2 u_x) \\ q_1^{(1)} &= -\kappa T_x \\ q_2^{(1)} &= -\kappa T_y \end{aligned} \tag{7.10}$$

where $(\ )_x = \partial/\partial x$ and $(\ )_y = \partial/\partial y$. The coefficients (δ_1, δ_2) are (1.333, −0.666) and (1.6, −0.4) for the augmented Burnett equations and the BGK–Burnett equations (for $\gamma = 1.4$), respectively.

As Kn becomes larger (>0.1), additional higher order terms in Eq. (7.6) are required. The second-order approximation yields the Burnett equations that retain the first three terms in Eq. (7.6). The expression for stress and heat-flux terms ($n = 2$) are obtained as [Yun et al., 1998]:

$$\sigma_{11}^{(2)} = \frac{\mu^2}{p}\left(\alpha_1 u_x^2 + \alpha_2 u_x v_y + \alpha_3 v_y^2 + \alpha_4 u_x v_y + \alpha_5 u_y^2 + \alpha_6 v_x^2 + \alpha_7 RT_{xx} + \alpha_8 RT_{yy} + \alpha_9 \frac{RT}{\rho}\rho_{xx}\right.$$
$$\left. + \alpha_{10}\frac{RT}{\rho}\rho_{yy} + \alpha_{11}\frac{RT}{\rho^2}\rho_x^2 + \alpha_{12}\frac{R}{\rho}T_x\rho_x + \alpha_{13}\frac{R}{T}T_x^2 + \alpha_{14}\frac{RT}{\rho^2}\rho_y^2 + \alpha_{15}\frac{R}{\rho}T_y\rho_y + \alpha_{16}\frac{R}{T}T_y^2\right) \quad (7.11)$$

$$\sigma_{22}^{(2)} = \frac{\mu^2}{p}\left(\alpha_1 v_y^2 + \alpha_2 u_x v_y + \alpha_3 u_x^2 + \alpha_4 u_y v_x + \alpha_5 v_x^2 + \alpha_6 u_y^2 + \alpha_7 RT_{yy} + \alpha_8 RT_{xx} + \alpha_9 \frac{RT}{\rho}\rho_{yy}\right.$$
$$\left. + \alpha_{10}\frac{RT}{\rho}\rho_{xx} + \alpha_{11}\frac{RT}{\rho^2}\rho_y^2 + \alpha_{12}\frac{R}{\rho}T_y\rho_y + \alpha_{13}\frac{R}{T}T_y^2 + \alpha_{14}\frac{RT}{\rho^2}\rho_x^2 + \alpha_{15}\frac{R}{\rho}T_x\rho_x + \alpha_{16}\frac{R}{T}T_x^2\right) \quad (7.12)$$

$$\sigma_{12}^{(2)} = \sigma_{21}^{(2)} = \frac{\mu^2}{p}\left(\beta_1 u_x u_y + \beta_2 u_y v_y + \beta_2 u_x v_x + \beta_1 v_x v_y + \beta_3 RT_{xy} + \beta_4 \frac{RT}{\rho}\rho_{xy}\right.$$
$$\left. + \beta_5 \frac{R}{T}T_x T_y + \beta_6 \frac{RT}{\rho^2}\rho_x\rho_y + \beta_7 \frac{R}{\rho}\rho_x T_y + \beta_7 \frac{R}{\rho}T_x\rho_y\right) \quad (7.13)$$

$$q_1^{(2)} = \frac{\mu^2}{p}\left(\gamma_1 \frac{1}{T}T_x u_x + \gamma_2 \frac{1}{T}T_x v_y + \gamma_3 u_{xx} + \gamma_4 u_{yy} + \gamma_5 v_{xy} + \gamma_6 \frac{1}{T}T_y v_x\right.$$
$$\left. + \gamma_7 \frac{1}{T}T_y u_y + \gamma_8 \frac{1}{\rho}\rho_x u_x + \gamma_9 \frac{1}{\rho}\rho_x v_y + \gamma_{10} \frac{1}{\rho}\rho_y u_y + \gamma_{11}\frac{1}{\rho}\rho_y v_x\right) \quad (7.14)$$

$$q_2^{(2)} = \frac{\mu^2}{p}\left(\gamma_1 \frac{1}{T}T_y u_y + \gamma_2 \frac{1}{T}T_y v_x + \gamma_3 v_{yy} + \gamma_4 v_{xx} + \gamma_5 u_{xy} + \gamma_6 \frac{1}{T}T_x u_y\right.$$
$$\left. + \gamma_7 \frac{1}{T}T_x v_x + \gamma_8 \frac{1}{\rho}\rho_y v_y + \gamma_9 \frac{1}{\rho}\rho_y u_x + \gamma_{10}\frac{1}{\rho}\rho_x v_x + \gamma_{11}\frac{1}{\rho}\rho_x u_y\right) \quad (7.15)$$

Both the augmented Burnett and BGK–Burnett equations have the same forms of the stress tensor and heat-flux terms in the second-order approximation; however, the two sets of equations have different values of the coefficients. The coefficients for the augmented Burnett equations (for a hard sphere gas) are $\alpha_1 = 1.199$, $\alpha_2 = 0.153$, $\alpha_3 = -0.600$, $\alpha_4 = -0.115$, $\alpha_5 = 1.295$, $\alpha_6 = -0.733$, $\alpha_7 = 0.260$, $\alpha_8 = -0.130$, $\alpha_9 = -1.352$, $\alpha_{10} = 0.676$, $\alpha_{11} = 1.352$, $\alpha_{12} = -0.898$, $\alpha_{13} = 0.600$, $\alpha_{14} = -0.676$, $\alpha_{15} = 0.449$, $\alpha_{16} = -0.300$, $\beta_1 = -0.115$, $\beta_2 = 1.913$, $\beta_3 = 0.390$, $\beta_4 = -2.028$, $\beta_5 = 0.900$, $\beta_6 = 2.028$, $\beta_7 = -0.676$, $\gamma_1 = 10.830$, $\gamma_2 = 0.407$, $\gamma_3 = -2.269$, $\gamma_4 = 1.209$, $\gamma_5 = -3.478$, $\gamma_6 = -0.611$, $\gamma_7 = 11.033$, $\gamma_8 = -2.060$, $\gamma_9 = 1.030$, $\gamma_{10} = -1.545$, and $\gamma_{11} = -1.545$.

The coefficients for the BGK–Burnett equations (for $\gamma = 1.4$) are $\alpha_1 = -2.24$, $\alpha_2 = -0.48$, $\alpha_3 = 0.56$, $\alpha_4 = -1.20$, $\alpha_5 = 0.0$, $\alpha_6 = 0.0$, $\alpha_7 = -19.6$, $\alpha_8 = -5.6$, $\alpha_9 = -1.6$, $\alpha_{10} = 0.4$, $\alpha_{11} = 1.6$, $\alpha_{12} = -19.6$, $\alpha_{13} = -18.0$, $\alpha_{14} = -0.4$, $\alpha_{15} = -5.6$, $\alpha_{16} = -6.9$, $\beta_1 = -1.4$, $\beta_2 = -1.4$, $\beta_3 = 0.0$, $\beta_4 = -2.0$, $\beta_5 = 2.0$, $\beta_6 = 2.0$, $\beta_7 = 0.0$, $\gamma_1 = -25.241$, $\gamma_2 = -0.2$, $\gamma_3 = -1.071$, $\gamma_4 = -2.0$, $\gamma_5 = -2.8$, $\gamma_6 = -7.5$, $\gamma_7 = -11.0$, $\gamma_8 = -1.271$, $\gamma_9 = 1.0$, $\gamma_{10} = -3.0$, and $\gamma_{11} = -3.0$.

The third-order approximation ($n = 3$) represents the super-Burnett equations; however, not all of the third-order terms of the super-Burnett equations are used in the augmented Burnett and the BGK–Burnett equations. In the augmented Burnett equations, the third-order terms are added on an *ad hoc* basis to obtain stable numerical solutions while maintaining second-order accuracy of the solutions. The third-order terms in the augmented Burnett equations are given as [Yun et al., 1998]:

$$\sigma_{11}^{(a)} = \frac{\mu^3}{p^2} RT(\alpha_{17} u_{xxx} + \alpha_{17} u_{xyy} + \alpha_{18} v_{xxy} + \alpha_{18} v_{yyy}) \tag{7.16}$$

$$\sigma_{22}^{(a)} = \frac{\mu^3}{p^2} RT(\alpha_{17} v_{yyy} + \alpha_{17} v_{xxy} + \alpha_{18} u_{xyy} + \alpha_{18} u_{xxx}) \tag{7.17}$$

$$\sigma_{12}^{(a)} = \sigma_{21}^{(a)} = \frac{\mu^3}{p^2} RT(\beta_8 u_{xxy} + \beta_8 u_{yyy} + \beta_8 v_{xyy} + \beta_8 v_{xxx}) \tag{7.18}$$

$$q_1^{(a)} = \frac{\mu^3}{p\rho} R\left(\gamma_{12} T_{xxx} + \gamma_{12} T_{xyy} + \gamma_{13} \frac{T}{\rho} \rho_{xxx} + \gamma_{13} \frac{T}{\rho} \rho_{xyy}\right) \tag{7.19}$$

$$q_2^{(a)} = \frac{\mu^3}{p\rho} R\left(\gamma_{12} T_{yyy} + \gamma_{12} T_{xxy} + \gamma_{13} \frac{T}{\rho} \rho_{yyy} + \gamma_{13} \frac{T}{\rho} \rho_{xxy}\right) \tag{7.20}$$

The superscript (a) denotes the augmented Burnett terms. The coefficients in stress and heat-flux terms are $\alpha_{17} = 0.2222$, $\alpha_{18} = -0.1111$, $\beta_8 = 0.1667$, $\gamma_{12} = 0.6875$, and $\gamma_{13} = -0.625$.

The BGK–Burnett equations have more additional third-order terms than the augmented Burnett equations. These are not added on an *ad hoc* basis but are derived from the second-order Chapman–Enskog expansion of the BGK–Boltzmann equation. The third-order terms in the BGK–Burnett equations are obtained as [Yun et al., 1998]:

$$\begin{aligned}\sigma_{11}^B &= \frac{\mu^3}{p^2} RT(\theta_1 u_{xxx} + \theta_2 u_{xyy} + \theta_3 v_{xxy} + \theta_4 v_{yyy}) \\ &\quad - \frac{\mu^3}{p^2} \frac{RT}{\rho} (\theta_1 \rho_x u_{xx} + \theta_5 \rho_x v_{xy} + \theta_6 \rho_x u_{yy} + \theta_7 \rho_y v_{xx} + \theta_8 \rho_y u_{xy} + \theta_4 \rho_y v_{yy}) \\ &\quad + \frac{\mu^3}{p^2} (\theta_9 u_x^3 + 3\theta_{10} u_x^2 v_y + \theta_{11} u_x v_y^2 - \theta_4 u_x u_y^2 - 2\theta_4 u_x u_y v_x \\ &\quad - \theta_4 u_x v_x^2 + \theta_{10} v_y^3 - \theta_{12} v_y u_y^2 - 2\theta_{12} u_y v_x v_y - \theta_{12} v_x^2 v_y) \\ &\quad + \frac{\mu^3}{p^2} R(\theta_{13} u_x T_{xx} + \theta_{13} u_x T_{yy} + \theta_{14} v_y T_{xx} + \theta_{14} v_y T_{yy})\end{aligned} \tag{7.21}$$

$$\begin{aligned}\sigma_{22}^B &= \frac{\mu^3}{p^2} RT(\theta_1 v_{yyy} + \theta_2 v_{xxy} + \theta_3 u_{xyy} + \theta_4 u_{xxx}) \\ &\quad - \frac{\mu^3}{p^2} \frac{RT}{\rho} (\theta_1 \rho_y v_{yy} + \theta_5 \rho_y u_{xy} + \theta_6 \rho_y v_{xx} + \theta_7 \rho_x u_{yy} + \theta_8 \rho_x v_{xy} + \theta_4 \rho_x u_{xx}) \\ &\quad + \frac{\mu^3}{p^2} (\theta_9 v_y^3 + 3\theta_{10} v_y^2 u_x + \theta_{11} v_y u_x^2 - \theta_4 v_y v_x^2 - 2\theta_4 v_y v_x u_y \\ &\quad - \theta_4 v_y u_y^2 + \theta_{10} u_x^3 - \theta_{12} u_x v_x^2 - 2\theta_{12} v_x u_y u_x - \theta_{12} u_y^2 u_x) \\ &\quad + \frac{\mu^3}{p^2} R(\theta_{13} v_y T_{yy} + \theta_{13} v_y T_{xx} + \theta_{14} u_x T_{yy} + \theta_{14} u_x T_{xx})\end{aligned} \tag{7.22}$$

$$\sigma_{12}^{(B)} = \frac{\mu^3}{p^2}RT(\theta_{15}u_{xxy} + u_{yyy} + \theta_{15}v_{xyy} + v_{xxx})$$

$$-\frac{\mu^3}{p^2}\frac{RT}{\rho}(\theta_6\rho_y u_{xx} + \theta_{16}\rho_y v_{xy} + \rho_y u_{yy} + \rho_x v_{xx} + \theta_{16}\rho_x u_{xy} + \theta_6\rho_x v_{yy})$$

$$-\frac{\mu^3}{p^2}(u_y + v_x)(\theta_4 u_x^2 + 2\theta_{12}u_x v_y + 2\theta_7 u_y v_x + \theta_7 u_y^2 + \theta_7 v_x^2 + \theta_4 v_y^2)$$

$$+\frac{\mu^3}{p^2}R(\theta_{17}u_y T_{xx} + \theta_{17}u_y T_{yy} + \theta_{17}v_x T_{xx} + \theta_{17}v_x T_{yy}) \quad (7.23)$$

$$q_1^{(B)} = \frac{\mu^3}{p\rho}R\left(\theta_{18}T_{xxx} + \theta_{18}T_{xyy} - \theta_{18}\frac{1}{\rho}\rho_x T_{xx} - \theta_{18}\frac{1}{\rho}\rho_x T_{yy}\right)$$

$$+\frac{\mu^3}{p\rho}(\theta_{19}u_x u_{xx} + \theta_{20}u_x v_{xy} + \theta_6 u_x u_{yy} + \theta_{21}v_y u_{xx} + \theta_{22}v_y v_{xy} + \theta_7 v_y u_{yy}$$

$$+\theta_{23}u_y v_{xx} + \theta_{24}u_y u_{xy} + \theta_6 u_y v_{yy} + \theta_{23}v_x v_{xx} + \theta_{24}v_x u_{xy} + \theta_6 v_x v_{yy})$$

$$-\frac{\mu^3}{p\rho}\left(\frac{1}{\rho}\rho_x + \frac{1}{T}T_x\right)(\theta_{13}u_x^2 + 2\theta_{14}u_x v_y + 2\theta_{17}u_y v_x + \theta_{17}u_y^2 + \theta_{17}v_x^2 + \theta_{13}v_y^2)$$

$$+\frac{\mu^3}{p\rho}\frac{R}{T}(\theta_{18}T_x T_{xx} + \theta_{18}T_x T_{yy}) \quad (7.24)$$

$$q_2^{(B)} = \frac{\mu^3}{p\rho}R\left(\theta_{18}T_{yyy} + \theta_{18}T_{xxy} - \theta_{18}\frac{1}{\rho}\rho_y T_{yy} - \theta_{18}\frac{1}{\rho}\rho_y T_{xx}\right)$$

$$+\frac{\mu^3}{p\rho}(\theta_{19}v_y v_{yy} + \theta_{20}v_y u_{xy} + \theta_6 v_y v_{xx} + \theta_{21}u_x v_{yy} + \theta_{22}u_x u_{xy} + \theta_7 u_x v_{xx}$$

$$+\theta_{23}v_x u_{yy} + \theta_{24}v_x v_{xy} + \theta_6 v_x u_{xx} + \theta_{23}u_y u_{yy} + \theta_{24}u_y v_{xy} + \theta_6 u_y u_{xx})$$

$$-\frac{\mu^3}{p\rho}\left(\frac{1}{\rho}\rho_y + \frac{1}{T}T_y\right)(\theta_{13}u_x^2 + 2\theta_{14}u_x v_y + 2\theta_{17}u_y v_x + \theta_{17}u_y^2 + \theta_{17}v_x^2 + \theta_{13}v_y^2)$$

$$+\frac{\mu^3}{p\rho}\frac{R}{T}(\theta_{18}T_y T_{xx} + \theta_{18}T_y T_{yy}) \quad (7.25)$$

The superscript (B) denotes third-order stress and heat-flux terms in the BGK–Burnett equations. The θ_i are given as follows for $\gamma = 1.4$: $\theta_1 = 2.56$, $\theta_2 = 1.36$, $\theta_3 = 0.56$, $\theta_4 = -0.64$, $\theta_5 = 0.96$, $\theta_6 = 1.6$, $\theta_7 = -0.4$, $\theta_8 = -0.24$, $\theta_9 = 1.024$, $\theta_{10} = -0.256$, $\theta_{11} = 1.152$, $\theta_{12} = 0.16$, $\theta_{13} = 2.24$, $\theta_{14} = -0.56$, $\theta_{15} = 3.6$, $\theta_{16} = 0.6$, $\theta_{17} = 1.4$, $\theta_{18} = 4.9$, $\theta_{19} = 7.04$, $\theta_{20} = -0.16$, $\theta_{21} = -1.76$, $\theta_{22} = 4.24$, $\theta_{23} = 3.8$, and $\theta_{24} = 3.4$.

Finally, governing Eq. (7.1) is nondimensionalized by a reference length and freestream variables written in a curvilinear coordinate system (ξ, η) by employing a coordinate transformation:

$$\tau = t, \quad \xi = \xi(x, y), \quad \eta = \eta(x, y) \quad (7.26)$$

7.5 Wall-Boundary Conditions

The no-slip-/no-temperature-jump boundary conditions are employed at the wall when solving the continuum Navier–Stokes equations for $Kn < 0.001$. In the continuum–transition regimes, the non-slip-boundary conditions are no longer correct. First-order slip/temperature-jump boundary conditions should be applied to both the Navier–Stokes equations and Burnett equations in the range $0.001 < Kn < 0.1$.

The transition regime spans the range 0.01 < Kn < 10; the second-order slip/temperature-jump conditions should be used in this regime with the Navier–Stokes as well as the Burnett equations. It should be noted here that the Navier–Stokes equations are first-order accurate in Kn while the Burnett equations are second-order accurate in Kn. Both first- and second-order Maxwell–Smoluchowski slip/temperature-jump boundary conditions are generally employed on the body surface when solving the Burnett equations.

The first-order Maxwell–Smoluchowski slip-boundary conditions in Cartesian coordinates are [Smoluchowski, 1998]:

$$U_s = \frac{2-\bar{\sigma}}{\bar{\sigma}} \frac{2\mu}{\rho} \sqrt{\frac{\pi}{8RT}} \left(\frac{\partial U}{\partial y}\right)_s + \frac{3}{4} \frac{\mu}{\rho T} \left(\frac{\partial T}{\partial x}\right)_s \tag{7.27}$$

and

$$T_s - T_w = \frac{2-\bar{\alpha}}{\bar{\alpha}} \frac{2\gamma}{\gamma+1} \frac{2\mu}{\rho} \sqrt{\frac{\pi}{8RT}} \frac{1}{Pr} \left(\frac{\partial T}{\partial y}\right)_s \tag{7.28}$$

The subscript s denotes the flow variables on the solid surface of the body. First-order Maxwell–Smoluchowski slip-boundary conditions can be derived by considering the momentum and energy-flux balance on the wall surface. The reflection coefficient $\bar{\sigma}$ and the accommodation coefficient $\bar{\alpha}$ are assumed to be equal to unity (for complete accommodation) in the calculations presented in this chapter.

Beskok's slip-boundary condition [Beskok et al, 1996] is the second-order extension of the Maxwell's slip-velocity-boundary condition excluding the thermal creep terms, given as:

$$U_s = \frac{2-\bar{\sigma}}{\bar{\sigma}} \left[\frac{Kn}{1-bKn}\left(\frac{\partial U}{\partial y}\right)_s\right] \tag{7.29}$$

where b is the slip coefficient determined analytically in the slip flow regime and empirically in transitional and free molecular regimes.

Langmuir's slip-boundary condition has also been employed in the literature [Myong, 1999]. Langmuir's slip-boundary condition is based on the theory of adsorption phenomena at the solid wall. Gas molecules do not in general rebound elastically but condense on the surface, being held by the field of force of the surface atoms. These molecules may subsequently evaporate from the surface, resulting in some time lag. Slip is the direct result of this time lag. The slip velocity at the wall is given as:

$$U_s = \frac{1}{1+\beta p} \tag{7.30}$$

where β is the adsorption coefficient determined empirically or by theoretical prediction.

In this chapter, these slip boundary conditions are applied and compared to determine their influence on the solution.

7.6 Linearized Stability Analysis of Burnett Equations

It was shown by Bobylev (1982) that the conventional Burnett equations are not stable to small wavelength disturbances; hence, the solutions to conventional Burnett equations tend to blow up when the mesh size is made progressively finer. Balakrishnan and Agarwal (1999) recently performed the linearized stability of one-dimensional original Burnett equations, conventional Burnett equations, augmented equations and the BGK–Burnett equations. They considered the response of a uniform gas subjected to small one-dimensional periodic perturbations. Burnett equations were linearized by neglecting products and powers of small perturbations, and a linearized set of equations for small perturbation variables $V' = [\rho', u', T']^T$ was obtained. It was assumed that the solution is of the form:

$$V' = \bar{V} e^{i\omega x} e^{\phi t} \quad (7.31)$$

where $\phi = \alpha + i\beta$, and α and β denote the attenuation and dispersion coefficients, respectively. For stability, $\alpha \leq 0$ as the Knudsen number increases. Substitution of Eq. (31) in the equations for small perturbation quantities V' results in a characteristic equation, $|F(\phi, \omega)| = 0$. The trajectory of the roots of this characteristic equation is plotted in a complex plane on which the real axis denotes the attenuation coefficient and the imaginary axis denotes the dispersion coefficient. For stability, the roots must lie to the left of the imaginary axis as the Knudsen number increases. Figures 7.2 to 7.5 show the trajectory of

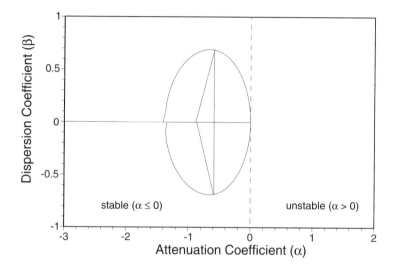

FIGURE 7.2 Characteristic trajectories of the one-dimensional Navier–Stokes equations.

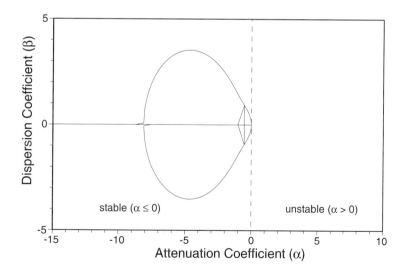

FIGURE 7.3 Characteristic trajectories of the one-dimensional augmented Burnett equations ($\gamma = 1.667$); Euler equations are used to express the material derivatives $D(\ )/Dt$ in terms of spatial derivatives.

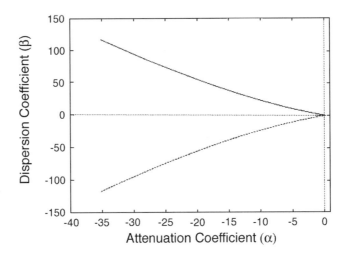

FIGURE 7.4 Characteristic trajectories of the one-dimensional BGK–Burnett equations ($\gamma = 1.667$); Euler equations are used to express the material derivatives $D(\ )/Dt$ in terms of spatial derivatives.

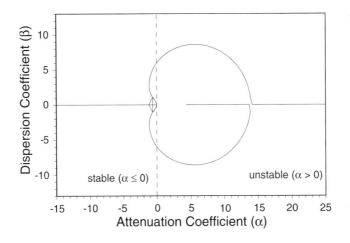

FIGURE 7.5 Characteristic trajectories of the one-dimensional conventional Burnett equations; Euler equations are used to express the material derivatives $D(\ )/Dt$ in terms of spatial derivatives.

the three roots of the characteristic equations as the Knudsen number increases. From the plots, it can be observed that the Navier–Stokes equations, the augmented Burnett equations and the BGK–Burnett equations (with $\gamma = 1.667$) are stable, but the conventional Burnett equations are unstable. Euler equations are employed to approximate the material derivatives in all three types of Burnett equations. It should be noted here that the BGK–Burnett equations, however, become unstable for $\gamma = 1.4$. On the other hand, if the material derivatives are approximated using the Navier–Stokes equations, the conventional, augmented and BGK–Burnett equations are all stable to small wavelength disturbances.

Based on these observations, we have employed the Navier–Stokes equations to approximate the material derivatives in the conventional, augmented and BGK–Burnett equations presented in Section 7.4. For the detailed analysis behind Figures 7.2 to 7.5, the reader is referred to Balakrishnan and Agarwal (1999). The linearized stability analysis of conventional, augmented and super-Burnett equations has also been performed in three dimensions with similar conclusions [Yun and Agarwal, 2000].

7.7 Numerical Method

An explicit finite-difference scheme is employed to solve the governing equations of Section 7.4. The Steger–Warming flux-vector splitting method [Steger and Warming, 1981] is applied to the inviscid-flux terms. The second-order, central-differencing scheme is applied to discretize the stress tensor and heat-flux terms. Converged solutions were obtained with a reduction in residuals of six orders of magnitude. All the calculations were performed on a sequence of successively refined grids to assure grid independence of the solutions.

7.8 Numerical Simulations

Numerical simulations have been performed for both the hypersonic flows and microscale flows in the continuum–transition regime. Hypersonic flow calculations include one-dimensional shock structure, two-dimensional and axisymmetric blunt bodies, and a space shuttle re-entry condition. Microscale flows include the subsonic flow and supersonic flow in a microchannel.

7.8.1 Application to Hypersonic Shock Structure

The hypersonic shock for argon was computed using the BGK–Burnett equations. The upstream flow conditions were specified and the downstream conditions were determined from the Rankine–Hugoniot relations. For purposes of comparison, the same flow conditions as in Fiscko and Chapman (1988) were used in the computations. The parameters used were

$$T_\infty = 300 \text{ K}, \quad P_\infty = 1.01323 \times 10^5 \text{ N/m}^2, \quad \gamma_{argon} = 1.667, \quad \mu_{argon} = 22.7 \times 10^{-6} \text{ kg/sec} \cdot \text{m}$$

The Navier–Stokes solution was taken as the initial value. This initial Navier–Stokes spatial distribution of variables was imposed on a mesh that encloses the shock. The length of the control volume enclosing the shock was chosen to be $1000 \times \lambda_\infty$ where the mean free path based on the freestream parameters is given by the expression $\lambda_\infty = 16\mu/(5\rho_\infty\sqrt{2\pi R T_\infty})$. This is the mean free path that would exist in the unshocked region if the gas were composed of hard elastic spheres and had the same viscosity, density and temperature as of the gas being considered. The solution was marched in time until the observed deviations were smaller than a preset convergence criterion.

A set of computational experiments was carried out to compare the BGK–Burnett solutions with the Burnett solutions of Fiscko and Chapman (1988). Tests were conducted at Mach 20 and Mach 35. In order to test for instabilities to small wavelength disturbances, the grid points were increased from 101 to 501 points. Figures 7.6 and 7.7 show variations of specific entropy across the shock wave. It is observed that the BGK–Burnett equations show a positive entropy change throughout the flowfield, while the conventional Burnett equations give rise to a negative entropy spike just ahead of the shock as the number of grid points is increased. This spike increases in magnitude until the conventional Burnett equations break down completely. The BGK–Burnett equations did not exhibit any instabilities for the range of grid points considered. Figure 7.8 shows the variation of reciprocal density thickness with Mach number. BGK–Burnett calculations compare well to those of Woods and simplified Woods equations [Reese et al., 1995] and the experimental data of Alsmeyer (1976). Extensive calculations for one-dimensional hypersonic shock structure using various higher order kinetic formulations are given in the Ph.D. thesis of Balakrishnan (1999).

7.8.2 Application to Two-Dimensional Hypersonic Blunt Body Flow

The two-dimensional augmented Burnett code was employed to compute the hypersonic flow over a cylindrical leading edge with a nose radius of 0.02 m in the continuum–transition regime. The grid system (50×82 mesh) used in the computations is shown in Figure 7.9. The results were compared with those of Zhong (1991).

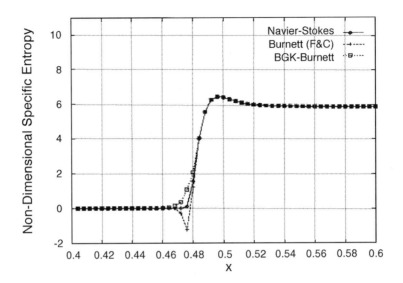

FIGURE 7.6 Specific entropy variation across a Mach 20 normal shock in a monatomic gas (argon), $\Delta x/\lambda_\infty = 4.0$ and $\gamma = 1.667$.

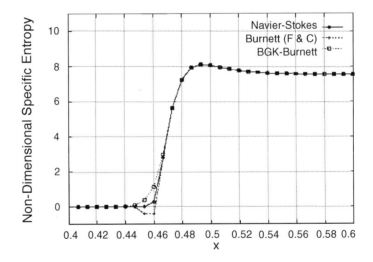

FIGURE 7.7 Specific entropy variation across a Mach 35 normal shock in a monatomic gas (argon), $\Delta x/\lambda_\infty = 4.0$ and $\gamma = 1.667$.

The flow conditions for this case are as follows:

$$M_\infty = 10, \quad Kn_\infty = 0.1, \quad Re_\infty = 167.9,$$

$$P_\infty = 2.3881 \text{ N/m}^2, \quad T_\infty = 208.4 \text{ K}, \quad T_w = 1000.0 \text{ K}$$

The viscosity is calculated by Sutherland's law, $\mu = c_1 T^{1.5}/(T + c_2)$. The coefficients c_1 and c_2 for air are 1.458×10^{-6} kg/(sec m $K^{1/2}$) and 110.4 K, respectively. Other constants used in this computation for air are $\gamma = 1.4$, $Pr = 0.72$ and $R = 287.04$ m^2/(sec^2 K).

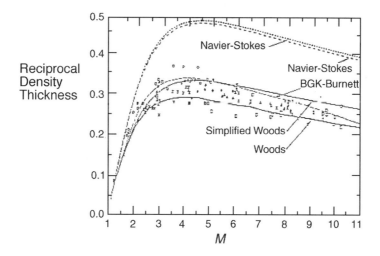

FIGURE 7.8 Plot showing the variation of reciprocal density thickness with Mach number, obtained with the Navier–Stokes, Wood and Simplified Woods [Reese et al., 1995] and BGK–Burnett equations for a monatomic gas (argon). Experimental data were obtained from Alsmeyer (1976). (From Alsmeyer, H. (1976) *J. Fluid Mech.* **74**, pp. 497–513, ASME, New York. With permission.)

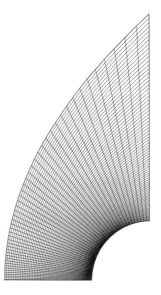

FIGURE 7.9 Two-dimensional computational grid (50 × 82 mesh) around a blunt body, $r_n = 0.02$ m.

The comparisons of density, velocity and temperature distributions along the stagnation streamline are shown in Figures 7.10, 7.11 and 7.12, respectively. The results agree well with those of Zhong for both the Navier–Stokes and the augmented Burnett computations. Because the flow is in the continuum–transition regime ($Kn = 0.1$), the Navier–Stokes equations become inaccurate, and the differences between the Navier–Stokes and the augmented Burnett solutions are obvious. In particular, the difference between the Navier–Stokes and Burnett solutions for the temperature distribution is significant across the shock. Temperature and Mach number contours of the Navier–Stokes solutions and the augmented Burnett solutions are compared in Figures 7.13 and 7.14, respectively. The shock structure of the augmented Burnett solutions agrees well with that of Zhong. The shock layer of the augmented Burnett solutions is thicker and the shock location starts upstream of that of the Navier–Stokes solutions. However, because the local Knudsen number decreases and the flow tends toward equilibrium as it approaches the wall

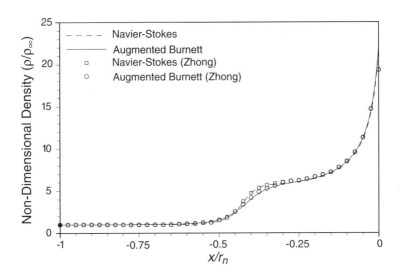

FIGURE 7.10 Density distributions along stagnation streamline for blunt body flow: Air, $M_\infty = 10$ and $Kn_\infty = 0.1$.

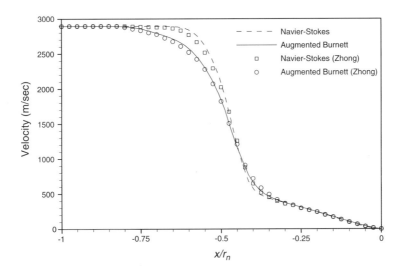

FIGURE 7.11 Velocity distributions along stagnation streamline for blunt body flow: Air, $M_\infty = 10$ and $Kn_\infty = 0.1$.

surface, the differences between the Navier–Stokes and augmented Burnett solutions become negligible near the wall, especially near the stagnation point. Thus, the Maxwell–Smoluchowski slip boundary conditions can be applied for both the Navier–Stokes and the augmented Burnett calculations for the hypersonic blunt body.

7.8.3 Application to Axisymmetric Hypersonic Blunt Body Flow

The results of the axisymmetric augmented Burnett computations are compared with the DSMC results obtained by Vogenitz and Takara (1971) for the axisymmetric hemispherical nose. The computed results are also compared with Zhong and Furumoto's (1995) axisymmetric augmented Burnett solutions. The

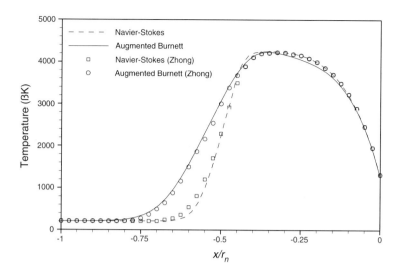

FIGURE 7.12 Temperature distributions along stagnation streamline for blunt body flow: Air, $M_\infty = 10$ and $Kn_\infty = 0.1$.

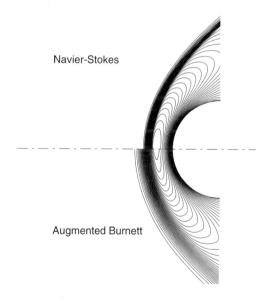

FIGURE 7.13 Comparison of temperature contours for blunt body flow: Air, $M_\infty = 10$ and $Kn_\infty = 0.1$.

flow conditions for this case are

$$M_\infty = 10, \quad Kn_\infty = 0.1$$

$$\frac{T_w}{T_\infty} = 1.0, \quad \frac{T_w}{T_0} = 0.029$$

T_0 is the stagnation temperature. The gas is assumed to be a monatomic gas with a hard-sphere model. The viscosity coefficient is calculated by the power law $\mu = \mu_r (T/T_r)^{0.5}$. The reference viscosity μ_r and the reference temperature T_r used in this case are 2.2695×10^{-5} kg/(sec m) and 300 K, respectively. Other constants used in this computation are $\gamma = 1.67$ and $Pr = 0.67$.

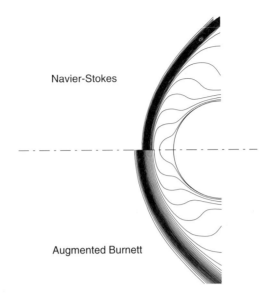

FIGURE 7.14 Comparison of Mach number contours for blunt body flow: Air, $M_\infty = 10$ and $Kn_\infty = 0.1$.

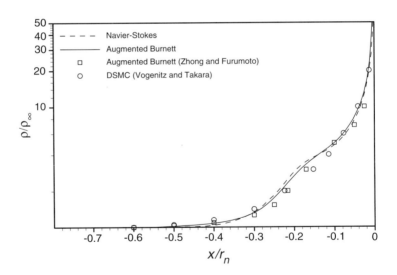

FIGURE 7.15 Density distributions along stagnation streamline for a hemispherical nose: Hard-sphere gas, $M_\infty = 10$ and $Kn_\infty = 0.1$.

The comparisons of density and temperature distributions along the stagnation streamline among the current axisymmetric augmented Burnett solutions, the axisymmetric augmented Burnett solutions of Zhong and Furumoto and the DSMC results are shown in Figures 7.15 and 7.16, respectively. The corresponding Navier–Stokes solutions are also compared in these figures. The axisymmetric augmented Burnett solutions agree well with Zhong and Furumoto's axisymmetric augmented Burnett solutions in both density and temperature. The density distributions for both the Navier–Stokes and augmented Burnett equations show little differences from the DSMC results. The temperature distributions, however, show that the DSMC method predicts a thicker shock than the augmented Burnett equations. The maximum

Burnett Simulations of Flows in Microdevices 7-21

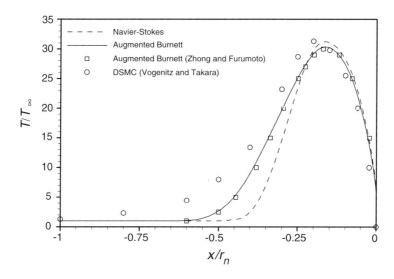

FIGURE 7.16 Temperature distribution along stagnation streamline for a hemispherical nose: Hard-sphere gas, $M_\infty = 10$ and $Kn_\infty = 0.1$.

temperature of the DSMC results is slightly higher than those of the augmented Burnett solutions. However, the augmented Burnett solutions show much closer agreement with the DSMC results than the Navier–Stokes solutions.

In conclusion, the axisymmetric augmented Burnett solutions presented here agree well with Zhong's axisymmetric augmented Burnett solutions and describe the shock structure closer to the DSMC results than the Navier–Stokes solutions.

As another application to the hypersonic blunt body, the augmented Burnett equations are applied to compute the hypersonic flowfield at re-entry condition encountered by the nose region of the space shuttle. The computations are compared with the DSMC results of Moss and Bird (1984). The DSMC method accounts for the translational, rotational, vibrational and chemical nonequilibrium effects.

An "equivalent axisymmetric body" concept [Moss and Bird, 1984] is applied to model the windward centerline of the space shuttle at a given angle of attack. A hyperboloid with nose radius of 1.362 m and asymptotic half angle of 42.5 degrees is employed as the equivalent axisymmetric body to simulate the nose of the shuttle. Figure 7.17 shows the side view of the grid (61×100 mesh) around the hyperboloid. The freestream conditions at an altitude of 104.93 km as given in Moss and Bird are

$$M_\infty = 25.3, \quad Kn_\infty = 0.227, \quad Re_\infty = 163.8,$$

$$\rho_\infty = 2.475 \times 10^{-7} \text{ kg/m}^3, \quad T_\infty = 223 \text{ K}, \quad T_w = 560 \text{ K}$$

The viscosity is calculated by the power law. The reference viscosity μ_r and the reference temperature T_r are taken as 1.47×10^{-5} kg/(sec m) and 223 K, respectively.

Figures 7.18 and 7.19 show comparisons of the density and temperature distributions along the stagnation streamline between the Navier–Stokes solutions, the augmented Burnett solutions and the DSMC results. The differences between the augmented Burnett solutions and the DSMC results are significant in both density and temperature distributions. In Figure 7.18, the density distribution of the DSMC results is lower and smoother than that of the augmented Burnett solutions. In Figure 7.19, the DSMC method predicts about 30% thicker shock layer and 9% lower maximum temperature than the augmented Burnett equations. The DSMC results can be considered to be more accurate than the augmented Burnett

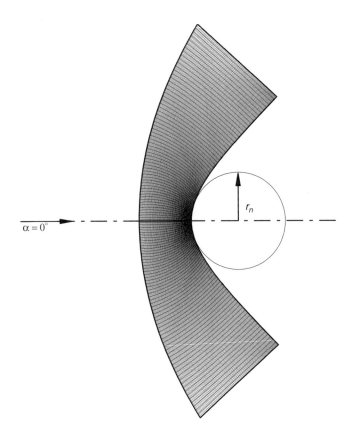

FIGURE 7.17 Side view of the grid (61 × 100 mesh) around a hyperboloid nose of radius $r_n = 1.362$ m.

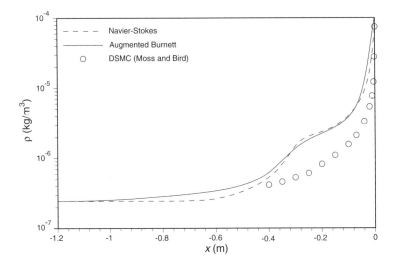

FIGURE 7.18 Density distributions along stagnation streamline for a hyperboloid nose: Air, $M_\infty = 25.3$ and $Kn_\infty = 0.227$.

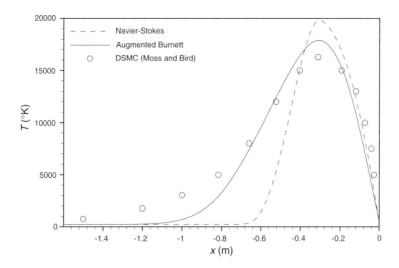

FIGURE 7.19 Temperature distributions along stagnation streamline for a hyperboloid nose: Air, $M_\infty = 25.3$ and $Kn_\infty = 0.227$.

solutions, as the DSMC method accounts for all the effects of translational, rotational, vibrational and chemical nonequilibrium, while the augmented Burnett equations do not. However, the augmented Burnett solutions agree much better with the DSMC results than the Navier–Stokes computations. The difference between the Navier–Stokes and the augmented Burnett solutions in temperature distributions is very significant. The shock layer of the augmented Burnett solutions is almost two times thicker than the Navier–Stokes solutions. The augmented Burnett solutions predict about 11% less maximum temperature than the Navier–Stokes solutions.

7.8.4 Application to NACA 0012 Airfoil

The Navier–Stokes equations are applied to compute the rarefied subsonic flow over a NACA 0012 airfoil with chord length of 0.04 m. The grid system in the physical domain is shown in Figure 7.20. The flow conditions are

$$M_\infty = 0.8, \quad Re_\infty = 73, \quad \rho_\infty = 1.116 \times 10^{-4} \text{ kg/m}^3, \quad T_\infty = 257 \text{ K}, \quad \text{and} \quad Kn_\infty = 0.014$$

Various constants used in the calculation for air are $\gamma = 1.4$, $Pr = 0.72$, and $R = 287.04 \text{ m}^2/(\text{sec}^2 \text{ K})$.

Figure 7.21 shows the density contours of the Navier–Stokes solution with the first-order Maxwell–Smoluchowski slip-boundary conditions. These contours using the continuum approach agree well with those of Sun et al. (2000) using the information preservation (IP) particle method. At these Mach and Knudsen numbers, the contours from the DSMC calculations are not smooth due to the statistical scatter. The comparison of pressure distribution along the surface between our Navier–Stokes solution with a slip-boundary condition and the DSMC calculation [Sun et al., 2000] is shown in Figure 7.22; the agreement between the solutions is good. Figure 7.23 compares the surface slip velocity from the DSMC, IP and Navier–Stokes methods as calculated by Sun et al. and by our Navier–Stokes calculation. The slip velocity distribution from our Navier–Stokes calculation shows good agreement with that obtained from the DSMC and IP methods except near the trailing edge. However, there is considerable disagreement between our Navier–Stokes results and those reported in Sun et al. (2000). This calculation again demonstrates that Navier–Stokes equations with slip-boundary conditions can provide accurate flow simulation $0.001 < Kn < 0.1$.

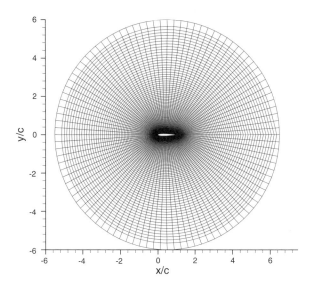

FIGURE 7.20 Grid (101 × 91 mesh) around a NACA 0012 airfoil, $c = 0.04$ m.

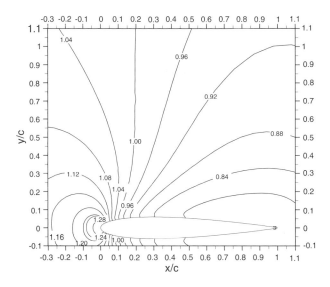

FIGURE 7.21 Density contours for NACA 0012 airfoil: Air, $M_\infty = 0.8$ and $Kn_\infty = 0.014$.

7.8.5 Subsonic Flow in a Microchannel

The augmented Burnett equations are employed for computation of subsonic flow in a microchannel with a ratio of channel length to height of 20 ($L/H = 20$). For the wall boundary conditions, Beskok's and Langmuir's boundary conditions are employed and compared. The augmented Burnett solutions are also compared with the Navier–Stokes solutions. Flow conditions at the entrance and exit of the channel are $Kn_{in} = 0.088$, $Kn_{out} = 0.2$ and $P_{in}/P_{out} = 2.28$.

Figure 7.24 compares the velocity profiles at various streamwise locations. Both Navier–Stokes and augmented Burnett equations using either Beskok's or Langmuir's boundary conditions show almost identical velocity profiles. These velocity profiles agree well with the velocity profiles from the μ flow calculation by Beskok and Karniadakis (1999). Nondimensional mass flowrates along the microchannel are shown in Figure 7.25. All the mass flowrates from both equations and both slip-boundary conditions

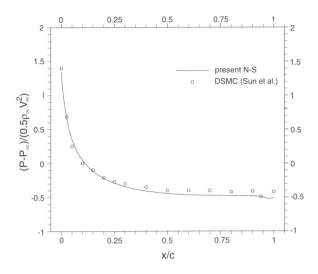

FIGURE 7.22 Pressure distributions along NACA 0012 airfoil surface: Air, $M_\infty = 0.8$ and $Kn_\infty = 0.014$.

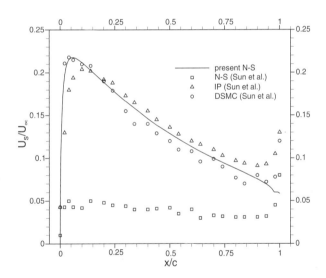

FIGURE 7.23 Slip velocity distributions along NACA 0012 airfoil surface: Air, $M_\infty = 0.8$ and $Kn_\infty = 0.014$.

are about 0.76 and almost constant along the channel, as should be the case. This mass flowrate is 13% higher than that predicted with a no-slip-boundary condition, which is 0.667. Figure 7.26 shows comparison of pressure distribution along the centerline. Both the Navier–Stokes and the augmented Burnett equations show a nonconstant pressure gradient along the channel. The solutions using Beskok's slip-boundary condition show less change in pressure gradient than those from the Langmuir's boundary condition. Figure 7.27 compares the streamwise velocity distributions along the centerline. The streamwise velocity distributions are almost identical except near the exit. Figure 7.28 compares the slip velocity distributions along the wall. The slip velocity profiles obtained from both Navier–Stokes and augmented Burnett equations are identical when the same wall-boundary conditions are employed. However, the Beskok's slip-boundary condition and Langmuir's slip-boundary condition show a large difference. The Beskok's slip-boundary condition predicts lower slip velocity near the entrance and higher slip velocity near the exit. As the figures show, there is very little difference between the Navier–Stokes and augmented Burnett

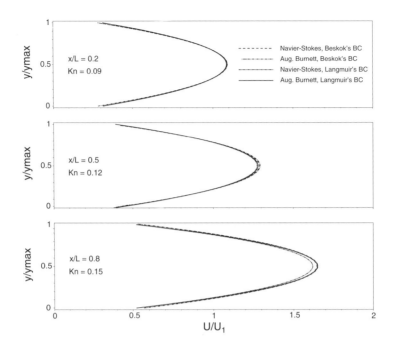

FIGURE 7.24 Comparison of velocity profiles at various streamwise locations: $Kn_{in} = 0.088$, $Kn_{out} = 0.2$ and $P_{in}/P_{out} = 2.28$.

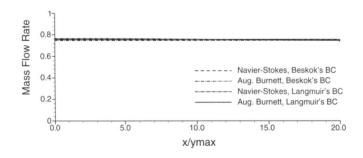

FIGURE 7.25 Comparison of mass flow rates along the microchannel: $Kn_{in} = 0.088$, $Kn_{out} = 0.2$ and $P_{in}/P_{out} = 2.28$.

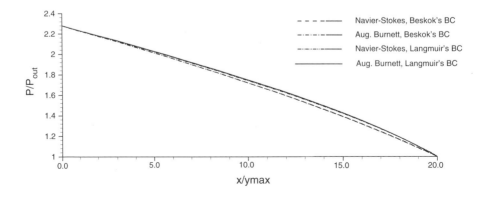

FIGURE 7.26 Comparison of pressure distribution along the centerline: $Kn_{in} = 0.088$, $Kn_{out} = 0.2$ and $P_{in}/P_{out} = 2.28$.

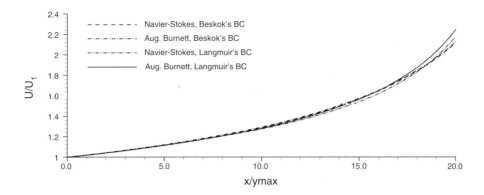

FIGURE 7.27 Comparison of streamwise velocity distributions along the centerline: $Kn_{in} = 0.088$, $Kn_{out} = 0.2$ and $P_{in}/P_{out} = 2.28$.

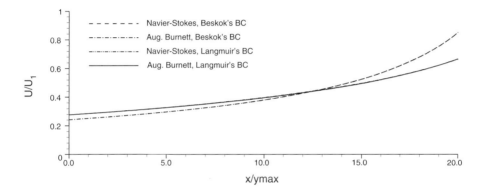

FIGURE 7.28 Comparison of slip velocity distributions along the wall: $Kn_{in} = 0.088$, $Kn_{out} = 0.2$ and $P_{in}/P_{out} = 2.28$.

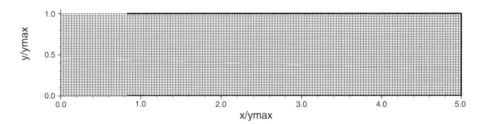

FIGURE 7.29 Microchannel geometry and grid.

solutions at the entrance, but, as the local Knudsen number increases toward the exit of the channel, the difference between the Navier–Stokes and augmented Burnett solutions increases as expected.

7.8.6 Supersonic Flow in a Microchannel

The Navier–Stokes equations and the augmented Burnett equations are applied to compute the supersonic flow in a microchannel. The geometry and grid of the microchannel are shown in Figure 7.29. As the flow enters the channel, the tangential velocity component to the wall is retained, while the velocity component normal to the wall is neglected at wall boundaries in the region $0 \leq x \leq 1$ μm. The first-order Maxwell–Smoluchowski slip-boundary conditions are employed at the rest of the wall boundaries.

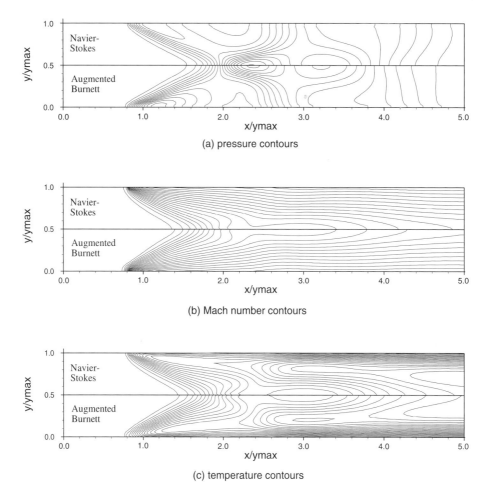

FIGURE 7.30 Comparisons of contours between Navier–Stokes and augmented Burnett equations: Helium, $M_\infty = 5$ and $Kn_\infty = 0.7$.

The channel height and length are 2.4 and 12 μm, respectively. The flow conditions at the entrance for the helium flow are

$$M_\infty = 5.0, \quad P_\infty = 1.01 \times 10^6 \text{ dyne/cm}^2,$$

$$Kn_\infty = 0.07, \quad T_\infty = 298 \text{ K}$$

Figures 7.30a-c compare the pressure, Mach number and temperature contours obtained from the Navier–Stokes and augmented Burnett equations. Solutions from the Navier–Stokes and augmented Burnett equations do not show significant differences. These flow property contours also agree well with the DSMC solutions obtained by Oh et al. (1997). Figure 7.31 compares the density, temperature, pressure and Mach number profiles along the centerline of the channel using the Navier–Stokes, augmented Burnett and DSMC formulations [Oh et al., 1997]. The profiles generally agree well with the DSMC results. The temperature and Mach number profiles especially show very close agreement with the DSMC results. The augmented Burnett solutions are closer to the DSMC solutions than the Navier–Stokes solutions in the temperature and Mach number profiles. Figure 7.32 compares the density, temperature, pressure and Mach number profiles along the channel wall using the Navier–Stokes, augmented Burnett and DSMC formulations. Both the Navier–Stokes and augmented Burnett solutions show some difference with the DSMC solutions. Figures 7.33 and 7.34 compare the temperature and velocity profiles across

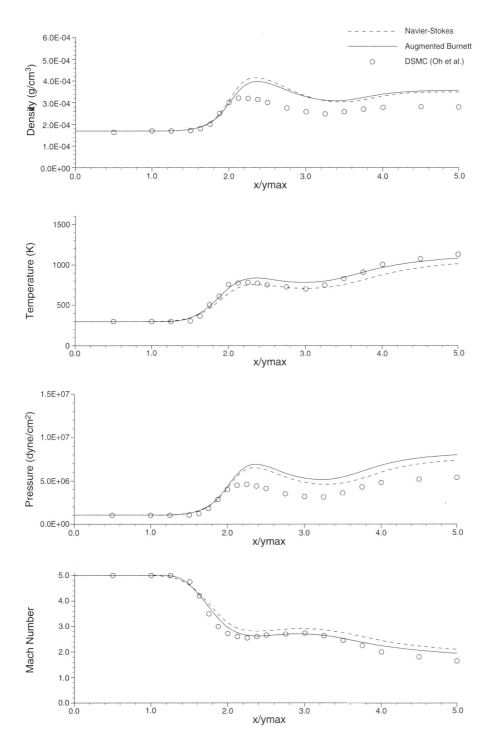

FIGURE 7.31 Comparisons of density, temperature, pressure and Mach number profiles along the centerline of the channel: Helium, $M_\infty = 5$ and $Kn_\infty = 0.7$.

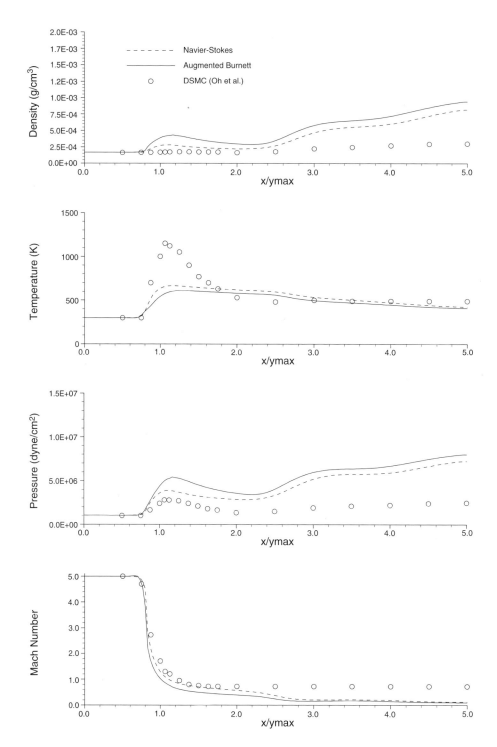

FIGURE 7.32 Comparisons of density, temperature, pressure and Mach number profiles along the wall of the channel: Helium, $M_\infty = 5$ and $Kn_\infty = 0.7$.

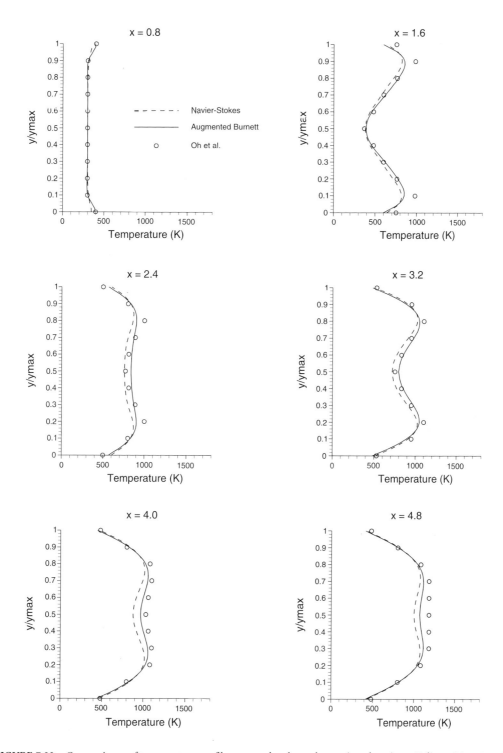

FIGURE 7.33 Comparisons of temperature profiles across the channel at various locations: Helium, $M_\infty = 5$ and $Kn_\infty = 0.7$.

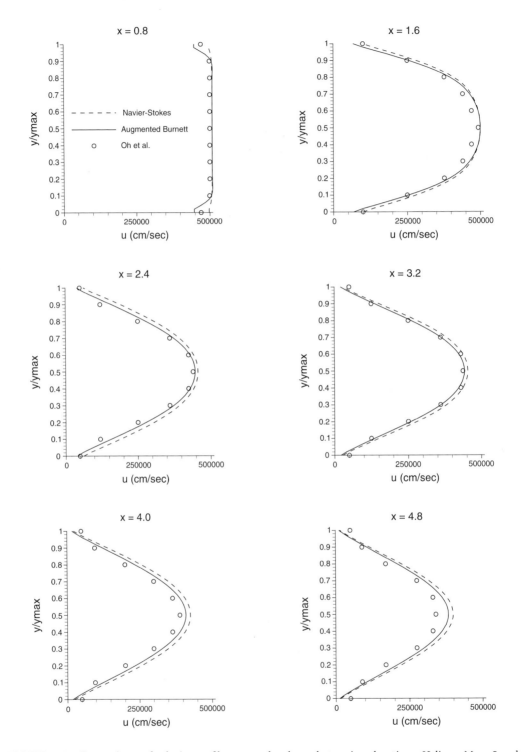

FIGURE 7.34 Comparisons of velocity profiles across the channel at various locations: Helium, $M_\infty = 5$ and $Kn_\infty = 0.7$.

the channel at various streamwise locations using the Navier–Stokes, augmented Burnett and DSMC formulations, respectively. The profiles obtained from the augmented Burnett solutions are closer to the DSMC results than the Navier–Stokes solutions.

Unfortunately, experimental data are not available to assess the accuracy of the Navier–Stokes, Burnett and DSMC models for computing the microchannel flows. A substantial amount of both experimental and computational simulation work is needed to determine the applicability and accuracy of various fluid models for computing high Knudsen number flow in microchannels.

7.9 Conclusions

For computing flows in the continuum–transition regime, higher order fluid dynamics models beyond Navier–Stokes are needed. These models are known as extended or generalized hydrodynamics models in the literature. Some of these models are the Burnett equations, 13-moment Grad's equations, Gaussian closure or Levermore moment equations, and Eu's equations. An alternative to generalized hydrodynamic models is the hybrid approach, which combines a Euler or Navier–Stokes solver with the DSMC method. Most of the generalized hydrodynamic models proposed to date suffer from the following drawbacks: They do not capture the physics properly or they are too computationally intensive, or both. In this chapter, the history of a set of extended hydrodynamics equations based on the Chapman–Enskog expansion of Boltzmann equations to $O(Kn^2)$, known as the Burnett equations, has been reviewed. The various sets known in the literature as conventional, augmented and BGK–Burnett equations have been considered and critically examined. Computations for hypersonic flows and microscale flows show that both the augmented and BGK–Burnett equations can be effectively applied to accurately compute flows in the continuum–transition regime. However, a great deal of additional work is needed, both experimentally and computationally, to assess the range of applicability and accuracy of Navier–Stokes, Burnett and DSMC approximations for simulating the flows in transition regime.

Nomenclature

c	=	speed of sound
e_t	=	total energy
Kn	=	Knudsen number
L	=	characteristic length
M	=	Mach number
Pr	=	Prandtl number
p	=	pressure
q_i	=	heat flux
R	=	gas constant
Re	=	Reynolds number
T	=	temperature
T_w	=	wall temperature
t	=	time
x, y	=	Cartesian coordinates
u, v	=	velocity components in x and y direction
α_i, β_i	=	coefficients of stress terms in Burnett equations
γ	=	specific heat ratio
γ_i	=	coefficients of heat-flux terms in Burnett equations
δ_i	=	coefficients of stress terms in Navier–Stokes equations
θ_i	=	coefficients of third-order terms in BGK–Burnett equations
κ	=	thermal conductivity

λ = mean free path of gas molecules
μ = coefficient of viscosity
ρ = density
σ_{ij} = stress tensor

References

Alsmeyer, H. (1976) "Density Profiles in Argon and Nitrogen Shock Waves Measured by the Absorption of an Electron Beam," *J. Fluid Mech.* **74**, pp. 497–513.

Arkilic, E.B., Schmidt, M.A., and Breuer, K.S. (1997) "Gaseous Flow in Long Microchannel," *J. MEMS* **6**, pp. 167–178.

Balakrishnan, R. (1999) "Entropy Consistent Formulation and Numerical Simulation of the BGK–Burnett Equations for Hypersonic Flows in the Continuum–Transition Regime," Ph.D. thesis, Wichita State University, Wichita, KS.

Balakrishnan, R., and Agarwal, R.K. (1996) "Entropy Consistent Formulation and Numerical Simulation of the BGK–Burnett Equations for Hypersonic Flows in the Continuum–Transition Regime," in *Proc. of the Int. Conf. on Numerical Methods in Fluid Dynamics*, Springer–Verlag, New York.

Balakrishnan, R., and Agarwal, R.K. (1997) "Numerical Simulation of the BGK–Burnett for Hypersonic Flows," *J. Thermophys. Heat Transfer* **11**, pp. 391–399.

Balakrishnan, R., and Agarwal, R.K. (1999) "A Comparative Study of Several Higher-Order Kinetic Formulations Beyond Navier–Stokes for Computing the Shock Structure," AIAA Paper 99-0224, American Institute of Aeronautics and Astronomics, Reno, NV.

Balakrishnan, R., Agarwal, R.K., and Yun, K.Y. (1997) "Higher-Order Distribution Functions, BGK–Burnett Equations and Boltzmann's H-Theorem," AIAA Paper 97-2552, American Institute of Aeronautics and Astronomics, Atlanta, GA.

Beskok, A., and Karniadakis, G. (1999) "A Model for Flows in Channels, Pipes, and Ducts at Micro and Nano Scales," *Microscale Thermophys. Eng.* **8**, pp. 43–77.

Beskok, A., Karniadakis, G., and Trimmer, W. (1996) "Rarefaction and Compressibility Effects in Gas Microflows," *J. Fluid Eng.* **118**, pp. 448–456.

Bhatnagar, P.L., Gross, E.P., and Krook, M. (1954) "A Model for Collision Process in Gas," *Phys. Rev.* **94**, pp. 511–525.

Bird, G.A. (1994) *Molecular Gas Dynamics and the Direct Simulation of Gas Flows*, Oxford Science Publications, New York.

Bobylev, A.V. (1982) "The Chapman–Enskog and Grad Methods for Solving the Boltzmann Equation," *Soviet Physics Doklady* **27**, pp. 29–31.

Boyd, I., Chen, G., and Candler, G. (1995) "Predicting Failure of the Continuum Fluid Equations in Transitional Hypersonic Flows," *Phys. Fluids* **7**, pp. 210–219.

Brown, S. (1996) "Approximate Riemann Solvers for Moment Models of Dilute Gases," Ph.D. thesis, University of Michigan, Ann Arbor.

Burnett, D. (1935) "The Distribution of Velocities and Mean Motion in a Slight Non-Uniform Gas," *Proc. London Math. Soc.* **39**, pp. 385–430.

Chapman, S., and Cowling, T.G. (1970) *The Mathematical Theory of Non-Uniform Gases*, Cambridge University Press, New York.

Comeaux, K.A., Chapman, D.R., and MacCormack, R.W. (1995) "An Analysis of the Burnett Equations Based in the Second Law of Thermodynamics," AIAA Paper 95-0415, American Institute of Aeronautics and Astronautics, Reno, NV.

Eu, B.C. (1992) *Kinetic Theory and Irreversible Thermodynamics*, John Wiley & Sons, New York.

Fiscko, K.A., and Chapman, D.R. (1988) "Comparison of Burnett, Super-Burnett and Monte Carlo Solutions for Hypersonic Shock Structure," in *Proc. 16th Int. Symp. on Rarefied Gas Dynamics*, Pasadena, CA, pp. 374–395.

Gad-el-Hak, M. (1999) "The Fluid Mechanics of Microdevices," *J. Fluids Eng.* **121**, pp. 5–33.

Grad, H. (1949) "On the Kinetic Theory of Rarefied Gases," *Comm. Pure Appl. Math.* **2**, pp. 325–331.

Groth, C.P.T., Roe, P.L., Gombosi, T.I., and Brown, S.L. (1995) "On the Nonstationary Wave Structure of 35-Moment Closure for Rarefied Gas Dynamics," AIAA Paper 95-2312, American Institute of Aeronautics and Astronautics, San Diego, CA.

Harley, J.C., Huang, Y., Bau, H.H., and Zemel, J.N. (1995) "Gas Flow in Microchannels," *J. Fluid Mech.* **284**, pp. 257–274.

Holway, L.H. (1964) "Existence of Kinetic Theory Solutions to the Shock Structure Problem," *Phys. Fluids* **7**, pp. 911–913.

Ivanov, M.S., and Gimelshein, S.F. (1998) "Computational Hypersonic Rarefied Flows," *Annu. Rev. Fluid Mech.* **30**, pp. 469–505.

Koppenwallner, G. (1987) "Low Reynolds Number Influence on the Aerodynamic Performance of Hypersonic Lifting Vehicles," *Aerodyn. Hypersonic Lifting Vehicles* (AGARD, CP-428) **11**, pp. 1–14.

Lee, C. J. (1994) "Unique Determination of Solutions to the Burnett Equations," *AIAA J.* **32**, pp. 985–990.

Levermore, C.D. (1996) "Moment Closure Hierarchies for Kinetic theory," *J. Stat. Phys.* **83**, pp. 1021–1065.

Levermore C.D., and Morokoff, W.J. (1998) "The Gaussian Moment Closure for Gas Dynamics," *SIAM J. Appl. Math.* **59**, pp. 72–96.

Liu, J.Q., Tai, Y.C., Pong, K.C., and Ho, C.M. (1993) "Micromachined Channel/Pressure Sensor Systems for Micro Flow Studies," in *Proc. of the 7th Int. Conf. on Solid-State Sensors and Actuators—Transducers '93*, pp. 995–998.

Moss, J.N., and Bird, G.A. (1984) "Direct Simulation of Transitional Flow for Hypersonic Reentry Conditions," AIAA Paper 84-0223, American Institute of Aeronautics and Astronautics, Reno, NV.

Myong, R. (1999) "A New Hydrodynamic Approach to Computational Hypersonic Rarefied Gas Dynamics," AIAA Paper 99-3578, American Institute of Aeronautics and Astronautics, Norfolk, VA.

Nance, R.P., Hash, D.B., and Hassan, H.A. (1998) "Role of Boundary Conditions in Monte Carlo Simulation of Microelectromechanical Systems," *J. Spacecraft Rockets* **12**, pp. 447–449.

Oh, C.K., Oran, E.S., and Sinkovits, R.S. (1997) "Computations of High-Speed, High Knudsen Number Microchannel Flows," *J. Thermophys. Heat Transfer* **11**, pp. 497–505.

Oran, E.S., Oh, C.K., and Cybyk, B.Z. (1998) "Direct Simulation Monte Carlo: Recent Advances and Application," *Annu. Rev. Fluid Mech.* **30**, pp. 403–441.

Pong, K.C., Ho, C.M., Liu, J.Q., and Tai, Y.C. (1994) "Non-linear Pressure Distribution in Uniform Microchannels," *ASME FED* **197**, pp. 51–56.

Reese, J.M., Woods, L.C., Thivet, F.J.P., and Candel, S.M. (1995) "A Second-Order Description of Shock Structure," *J. Computational Phys.* **117**, pp. 240–250.

Roveda, R., Goldstein, D.B., and Varghese, P.L. (1998) "Hybrid Euler/Particle Approach for Continuum/Rarefied Flows," *J. Spacecraft Rockets* **35**, pp. 258–265.

Smoluchowski, von M. (1998) "Veder Warmeleitung in Verdumteu Gasen," *Ann. Phys. Chem.* **64**, pp. 101–130.

Steger, J.L., and Warming, R.F. (1981) "Flux Vector Splitting of the Inviscid Gas Dynamics Equations with Application to Finite-Difference Methods," *J. Computational Phys.* **40**, pp. 263–293.

Sun, Q., Boyd, I.D., and Fan, J. (2000) "Development of Particle Methods for Computing MEMS Gas Flows," *J. MEMS* **2**, pp. 563–569.

Vogenitz, F.W., and Takara, G.Y. (1971) "Monte Carlo Study of Blunt Body Hypersonic Viscous Shock Layers," *Rarefied Gas Dynamics* **2**, pp. 911–918.

Weiss, W. (1996) "Comments on Existence of Kinetic Theory Solutions to the Shock Structure Problem," *Phys. Fluids* **8**, pp. 1689–1690.

Welder, W.T., Chapman, D.R., and MacCormack, R.W. (1993) "Evaluation of Various Forms of the Burnett Equations," AIAA Paper 93-3094, American Institute of Aeronautics and Astronautics, Orlando, FL.

Yun, K.Y., and Agarwal, R.K. (2000) "Numerical Simulation of 3-D Augmented Burnett Equations for Hypersonic Flow in Continuum–Transition Regime," AIAA Paper 2000-0339, American Institute of Aeronautics and Astronautics, Reno, NV.

Yun, K.Y., Agarwal, R.K., and Balakrishnan, R. (1998) "Three-Dimensional Augmented and BGK–Burnett Equations," unpublished report, Wichita State University, Wichita, KS.

Yun, K.Y., Agarwal, R.K., and Balakrishnan, R. (1998) "Augmented Burnett and Bhatnagar–Gross–Krook–Burnett for Hypersonic Flow," *J. Thermophys. Heat Transfer* **12**, pp. 328–335.

Zhong, X. (1991) "Development and Computation of Continuum Higher Order Constitutive Relations for High-Altitude Hypersonic Flow," Ph.D. thesis, Stanford University, Stanford, CA.

Zhong, X., and Furumoto, G. (1995) "Augmented Burnett Equation Solutions over Axisymmetric Blunt Bodies in Hypersonic Flow," *J. Spacecraft Rockets* **32**, pp. 588–595.

8
Molecular-Based Microfluidic Simulation Models

Ali Beskok
Texas A&M University

8.1 Abstract 8-1
8.2 Introduction 8-1
8.3 Gas Flows 8-2
   Molecular Magnitudes • An Overview of the Direct Simulation Monte Carlo Method • Limitations, Error Sources and Disadvantages of the DSMC Approach • Some DSMC-Based Gas Microflow Results • Recent Advances in the DSMC Method • DSMC Coupling with Continuum Equations • Boltzmann Equation Research • Hybrid Boltzmann/Continuum Simulation Methods • Lattice Boltzmann Methods
8.4 Liquid and Dense Gas Flows 8-20
   Electric Double Layer and Electrokinetic Effects • The Electro-Osmotic Flow • Molecular Dynamics Method • Treatment of Surfaces
8.5 Summary and Conclusions 8-23

8.1 Abstract

This chapter concentrates on molecular-based numerical simulation methods for liquid and gas microflows. After a brief introduction of molecular magnitudes, a detailed coverage of the direct simulation Monte Carlo method for gas flows is presented. Brief descriptions of other simulation methodologies, such as the Boltzmann and lattice Boltzmann methods and the molecular dynamics method, are also given. Throughout the chapter, extensive references to books, research and review articles are supplied with the perspective of guiding the reader toward recent literature.

8.2 Introduction

Simulation of microscale thermal fluidic transport is gaining importance due to the emerging technologies of the 21st century, such as microelectromechanical systems (MEMS) and nanotechnologies. Miniaturization of device scales, for the first time, has enabled the possibility of integration of sensing, computation, actuation, control, communication and power generation within the same microchip. The small size, light weight and high-durability of MEMS, combined with the mass-fabrication ability, result in low-cost systems with a wide variety of applications spanning from control systems to advanced energy systems to biological, medical and chemical applications. Despite these diverse prospects and fast growth of MEMS, further miniaturization of device scales is faced with the challenge of "better understanding" of micron and submicron scale physics.

The microscale thermal/fluidic transport phenomenon differs from its larger scale counterparts mainly due to the size, surface and interface effects [Ho and Tai, 1998; Gad-el-Hak, 1999]. Reduction of the characteristic device dimensions to micrometer scale drastically decreases the volume-to-surface area ratio. Hence, the surface forces are more dominant than the body forces in such small scales. The origin of the surface forces is atomistic and based on the short-ranged van der Waals forces and longer ranged electrostatic or Coulombic forces. Although a molecular-simulation-based approach for understanding fluid forces on surfaces is fundamental in nature, it is very difficult to apply to engineering problems due to the vast number of molecules involved in the analysis; however, direct application of the well-known continuum equations is not appropriate, either. For example, the Navier–Stokes level of constitutive relations that model the shear stress, being linearly proportional to the strain rate, is not valid for gases when the Knudsen number $Kn > 0.1$, or for liquids when the strain rate exceeds twice the molecular frequency scale [Gad-el-Hak, 1999]. Significant differences between the thermal/fluidic transport of gas and liquid states also exist. For example, dilute gases spend most of their time in free flight with abrupt changes in their direction and speed caused by binary intermolecular collisions. However, the liquid molecules are closely packed, and they experience multiple collisions with large intermolecular forces. The fundamental simulation approaches, from a microscopic point of view, differ for liquid and gas flows. In this chapter, we will address separately the numerical simulation methods relevant for dilute gases and liquids. However, the main emphasis of this chapter is microscale gas transport modeling with the direct simulation Monte Carlo (DSMC) algorithm. Other microscopic simulation methods, such as the Boltzmann equation approach, lattice Boltzmann method and molecular dynamics (MD), are briefly introduced to guide the reader to the appropriate resources in these areas.

8.3 Gas Flows

The ratio of the gas mean free path (λ) to a characteristic microfluidic length scale (h) is known as the Knudsen number, $Kn = \lambda/h$. Because the momentum and energy transfers happen with intermolecular and gas/wall collisions, the mean free path indicates an intrinsic length scale of thermal/fluidic transport for gases. In standard pressure and temperature (STP), the mean free path for air is about 65 nm. For macroscopic devices, the Knudsen number is very small, so the surrounding air can be treated as a continuous medium. However, in microscales, the Knudsen number can be fairly large due to the small length scales. Momentum and energy transport in micron and submicron scales show significant deviations from their larger scale counterparts. For example, recent microchannel experiments show increased mass flowrates compared to the Navier–Stokes-based continuum estimates [Arkilic et al., 1997; Harley et al., 1995; Liu et al., 1993; Pong et al., 1994]. Similarly, in the case of magnetic disk storage units, the head floating about 50 nm above the media exhibits an order of magnitude reduction of load capacity compared to predictions by the continuum Reynolds equations [Fukui and Kaneko, 1990]. These deviations are explained as a function of the Knudsen number by dividing the flow into four regimes: continuum ($Kn \leq 0.01$), slip ($0.01 \leq Kn \leq 0.1$), transitional ($0.1 \leq Kn \leq 10$) and free-molecular ($Kn > 10$). Operation regimes of typical MEMS devices at standard temperature and pressure are shown in Figure 8.1. MEMS operate in a wide variety of flow regimes covering the continuum, slip and early transitional flow regimes. Further miniaturization of MEMS device components and nanotechnology applications [Drexler, 1990] correspond to higher Knudsen numbers, making it necessary to study the mass, momentum and energy transport in the entire Knudsen regime.

It may be misleading to identify the flow regimes as "slip" and "continuum." Here, we must clarify that within this text and in most of the microscale transport literature "continuum" refers to the Navier–Stokes equations subject to the "no-slip"-boundary conditions. This identification leads to two common misconceptions. First, if the Navier–Stokes equations cannot be applied, then the continuum approximation should break down. This is misleading, as we will see shortly that it is possible to derive

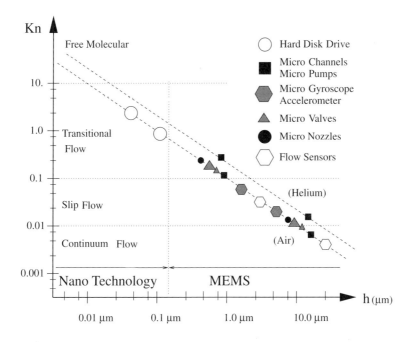

FIGURE 8.1 The Operation range for typical MEMS and nanotechnology applications under standard conditions spans the entire Knudsen regime (continuum, slip, transition and free molecular flow regimes).

conservation equations with more advanced constitutive laws than the Navier–Stokes equations. One example of this is the Burnett equations. The second misconception is that, in the "slip flow" regime, the boundary conditions suddenly change from no slip to slip. This is also misleading, as the no-slip-boundary condition is just an empirical finding and the Navier–Stokes equations are valid both for "slip" and "continuum" flow regimes. Hence, the slip effects become important gradually with increased Kn. Nevertheless, the identification of flow regimes was made for rarefied gas flows almost a century ago. For $Kn \leq 0.1$ flows, the Navier–Stokes equations subject to the velocity-slip and temperature-jump boundary conditions should be used. The slip conditions are [Kennard, 1938; Schaaf and Chambre, 1961]:

$$u_s - u_w = \frac{2-\sigma_v}{\sigma_v}\frac{1}{\rho(2RT_w/\pi)^{1/2}}\tau_s + \frac{3}{4}\frac{Pr(\gamma-1)}{\gamma\rho RT_w}(-q_s) \tag{8.1}$$

$$T_s - T_w = \frac{2-\sigma_T}{\sigma_T}\left[\frac{2(\gamma-1)}{\gamma+1}\right]\frac{1}{R\rho(2RT_w/\pi)^{1/2}}(-q_n) \tag{8.2}$$

where q_n, q_s are the normal and tangential heat-flux components; τ_s is the viscous stress component corresponding to the skin friction; R is the specific gas constant; γ is the ratio of specific heats, ρ is the density; Pr is the Prandtl number; and T_w and u_w are the wall temperature and velocity, respectively. The gas slip velocity and temperature near the wall (jump) are given by u_s and T_s, respectively. The term in the above equation proportional to $(-q_s)$ is associated with the phenomenon of *thermal creep*, which can cause variations of pressure along tubes in the presence of tangential temperature gradients [Beskok et al., 1995; Sone, 2000; Vargo and Muntz, 1996; Vargo et al., 1998].

In a recent work, a Padé approximation of (1) was developed, resulting in a velocity slip condition valid in the entire Knudsen regime. Excluding the thermal creep terms, this new slip condition is given in the following form [Beskok et al., 1996; Beskok and Karniadakis, 1999]:

$$U_s - U_w = \frac{2-\sigma_v}{\sigma_v}\left[\frac{Kn}{1-bKn}\right]\frac{\partial U}{\partial n} \qquad (8.3)$$

where U_w and U_s are the wall and gas-slip velocity nondimensionalized with a reference velocity, respectively. Here, b is the general slip coefficient, determined by the following procedures:

- A perturbation expansion in Kn for $Kn < 1$, such that Eq. (8.3) is equivalent to a second-order slip condition [Beskok et al., 1996].
- Matching the velocity profiles with the direct simulation Monte Carlo (DSMC) results in the transitional and free molecular flow regimes [Beskok and Karniadakis, 1999].

Hence the value of b is defined analytically in the slip and early transition flow regime, but in the transitional and free molecular flow regimes it is an empirical parameter.

In Eqs. (8.1) and (8.3), σ_v, and σ_T are the tangential momentum and thermal accommodation coefficients, respectively. The accommodation coefficients model the momentum and energy exchange of gas molecules impinging on the walls. Hence, they characterize the surface effects. For example, $\sigma_v = 0.2$ enhances the apparent slip by almost an order of magnitude! The accommodation coefficients are usually determined experimentally. Due to the difficulties of experimentation in microscales, the accommodation coefficients are obtained by assuming slip flow and matching the value of the accommodation coefficients to maintain the measured mass flow rate. This has resulted in $\sigma_v = 0.80$ for nitrogen, argon and carbon dioxide in contact with prime silicon crystal [Arkilic et al., 1997]. Lower accommodation coefficients are expected due to the low surface roughness of the prime silicon crystal. However, for a general micromachined surface and gas pair, the values of the accommodation coefficients are not known *a priori*. For low-pressure, rarefied gas flows, the values of the accommodation coefficients are tabulated as a function of the specific gas and surface quality [Seidl and Steinheil, 1974]; under laboratory conditions, values as low as 0.2 have been observed [Lord, 1976]. Very low values of σ_v will increase the slip on the walls considerably, even for small Knudsen number flows.

In the transitional flow regime, the constitutive laws defining the stress tensor and the heat-flux vector must be updated for increased rarefaction effects, resulting in Wood's, Burnett's or Grad's equations. It is also possible to use the Boltzmann transport equation in this regime (see Section 8.3.7). In a recent work, Myong has developed thermodynamically consistent hydrodynamic computational models for high-Knudsen-number gas flows, uniformly valid in all Mach numbers and satisfying the second law of thermodynamics [Myong, 1999].

In the free molecular flow regime ($Kn \geq 10$), the molecule–wall interactions dominate the transport with significantly reduced intermolecular collisions. Hence, the collisionless Boltzmann equation is commonly used in this flow regime.

8.3.1 Molecular Magnitudes

Before studying the molecular-based numerical simulation algorithms, it is crucial to understand the complexity of the molecular simulation problem. In this section, we present relationships for the number density of molecules n, mean molecular spacing δ, molecular diameter d_m, mean free path λ, mean collision time t_c and mean square molecular speed $\sqrt{\overline{C^2}}$.

The number of molecules in one mole of gas is a constant known as the Avogadro's number, 6.02252×10^{23}/mole, and the volume occupied by one mole of gas at a given temperature and pressure is a constant, regardless of the composition of the gas [Vincenti and Kruger, 1977]. This leads to the perfect gas relationship given by:

$$P = nk_b T \tag{8.4}$$

where P is the pressure, T is the temperature, n is the number density of the gas, and k_b is the Boltzmann constant ($k_b = 1.3805 \times 10^{-23}$ J/K). This ideal gas law is valid for dilute gases at any pressure (above the saturation pressure). Hence, for most microscale gas flow applications we can predict the number density of the molecules at a given temperature and pressure using Eq. (8.4). At atmospheric pressure and 0°C (standard conditions) the number density is about $n \cong 2.69 \times 10^{25}$ m^{-3}. If all of these molecules are placed in a 1-m cube in an equidistant fashion, the mean molecular spacing will be

$$\delta = n^{-1/3} \tag{8.5}$$

Under standard conditions the mean molecular spacing is $\delta \cong 3.3 \times 10^{-9}$ m.

The mean molecular diameter (d_m) of typical gases, based on the measured coefficient of viscosity and the Chapman-Enskong theory of transport properties for hard-sphere molecules, is on the order of 10^{-10} m. For air under standard conditions, $d_m \cong 3.7 \times 10^{-10}$ m, as tabulated in Bird (1994). Comparison of the mean molecular spacing δ and the typical molecular diameter d_m shows an order of magnitude difference. This leads to the concept of "dilute gas," where $\delta/d_m \gg 1$. For dilute gases, binary intermolecular collisions are more likely than the simultaneous multiple collisions. On the other hand, dense gases and liquids go through multiple collisions at a given instant, making the treatment of the intermolecular collision process more difficult. The dilute gas approximations, along with molecular chaos and equipartition of energy principles, lead us to the well-established kinetic theory of gases and formulation of the Boltzmann transport equation starting from the Liouville equation. The assumptions and simplifications of this derivation are given in Vincenti and Kruger (1977) and Bird (1994).

Momentum and energy transport in the bulk of the fluid and settling to a thermodynamic equilibrium state happen with intermolecular collisions. Hence, the time and length scales associated with the intermolecular collisions are important parameters for many applications. The distance traveled by the molecules between the intermolecular collisions is known as the mean free path. For a simple gas of hard-sphere molecules in thermodynamic equilibrium, the mean free path is given in the following form [Bird, 1994]:

$$\lambda = (2^{1/2} \pi d_m^2 n)^{-1} \tag{8.6}$$

The gas molecules are traveling with high speeds proportional to the speed of sound. By simple considerations, the mean-square molecular speed of the gas molecules is given by [Vincenti and Kruger, 1977]:

$$\sqrt{\overline{C^2}} = \sqrt{\frac{3P}{\rho}} = \sqrt{3RT} \tag{8.7}$$

where R is the specific gas constant. For air under standard conditions, this corresponds to 486 m/s. This value is about 3 to 5 orders of magnitude larger than the typical average speeds obtained in gas microflows. (The importance of this discrepancy will be discussed in Section 8.3.3.) In regard to the time scales of intermolecular collisions, we can obtain an average value by taking the ratio of the mean free path to the mean-square molecular speed. This results in $t_c \cong 10^{-10}$ for air under standard conditions. This time scale should be compared to a typical microscale process time scale to determine the validity of the thermodynamic equilibrium assumption.

So far we have identified the vast number of molecules and the associated time and length scales for gas flows. It is our common experience that it is possible to lump all of the microscopic quantities into time- and/or space-averaged macroscopic quantities, such as fluid density, temperature and velocity. It is crucial to determine the limitations of these continuum-based descriptions; in other words:

- How small should a sample size be so that we can still talk about the macroscopic properties and their spatial variations?
- At what length scales do the statistical fluctuations become significant?

It turns out that a sampling volume that contains 10,000 molecules typically results in 1% statistical fluctuations in the averaged quantities [Bird, 1994]. This corresponds to a volume of 3.7×10^{-22} m^3 for air at standard conditions. If we try to measure an "instantaneous" macroscopic quantity such as velocity in a three-dimensional space, one side of our sampling cube will typically be about 72 nm. This length scale is slightly larger than the mean free path of air (λ) under standard conditions. Therefore, in complex microgeometries, where three-dimensional spatial gradients are expected, the definition of "instantaneous" macroscopic values may become problematic for $Kn > 1$. If we would like to subdivide this domain further to obtain an instantaneous velocity distribution, the statistical fluctuations will be increased significantly as the sample volume is decreased. Hence, we may not be able to define instantaneous velocity distribution in a 72-nm^3 volume. On the other hand, it is always possible to perform time or ensemble averaging of the data at such small scales. Hence, we can still talk about a velocity profile in an averaged sense.

To describe the statistical fluctuation issues further, we present in Figure 8.2 the flow regimes and the limit of the onset of statistical fluctuations as a function of the characteristic dimension L and the

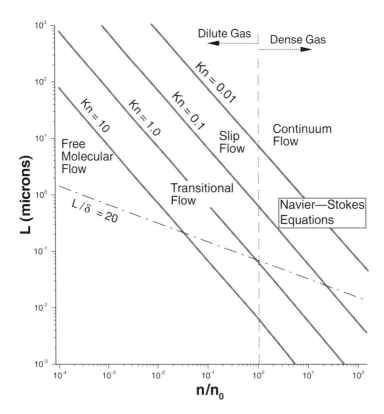

FIGURE 8.2 Limit of approximations in modeling gas microflows. L is the characteristic length, n/n_o is the number density normalized with the corresponding standard conditions. The lines that define the various Knudsen regimes are based on air at isothermal conditions ($T = 273$ K). The $L/\delta = 20$ line corresponds to the 1% statistical scatter in the macroscopic properties. The area below this line experiences increased statistical fluctuations.

normalized number density n/n_o. The 1% statistical scatterline is defined in a cubic volume of side L which contains approximately 10,000 molecules. Using Eq. (8.5) one can find that $L/\delta \approx 20$ satisfies this condition approximately, and the 1% fluctuation line varies as $(n/n_o)^{-1/3}$. Under standard conditions, 1% fluctuation is observed at $L = 72$ nm, and the Knudsen number based on this value is $Kn \approx 1$. Figure 8.2 also shows the continuum, slip, transitional and free molecular flow regimes for air at 273 K and at various pressures. The mean free path varies inversely with the pressure. Hence, at isothermal conditions, the Knudsen number varies as $(n/n_o)^{-1}$. The fundamental question of dynamic similarity of low-pressure gas flows to gas microflows under geometrically similar and identical Knudsen, Mach and Reynolds number conditions can be answered, to some degree, by inspection of Figure 8.2. Provided that there are no unforeseen microscale-specific effects, the two flow cases should be dynamically similar. However, a distinction between the low-pressure and gas microflows is the difference in the length scales for which the statistical fluctuations become important.

It is interesting to note that for low-pressure rarefied gas flows the length scales for the onset of significant statistical scatter correspond to much larger Knudsen values compared to the gas microflows. For example, $Kn = 1.0$ flow obtained at standard conditions in a 72-nm cube volume permits us to perform one instantaneous measurement in the entire volume with 1% scatter. However, at 100 Pascal pressure and 273 K temperature, $Kn = 1.0$ flow corresponds to a length scale of 65 μm. For this case, 1% statistical scatter in the macroscopic quantities is observed in a cubic volume of side 0.72 μm, allowing about 90 pointwise instantaneous measurements. The discussion presented above is valid for instantaneous measurements of macroscopic properties in complex three-dimensional conduits. In large-aspect-ratio microdevices one can always perform spanwise averaging to define an averaged velocity profile. Also, for practical reasons one can also define averaged macroscopic properties either by time or ensemble averaging (such examples are presented in Section 8.3.4).

8.3.2 An Overview of the Direct Simulation Monte Carlo Method

In this section, we present the algorithmic details, advantages and disadvantages of using the direct simulation Monte Carlo algorithm for microfluidic applications. The DSMC method was invented by Professor Graeme A. Bird (1976, 1994). Several review articles about the DSMC method are currently available [Bird 1978; 1998; Muntz, 1989; Oran et al., 1998]. Most of these articles present an extended review of the DSMC method for low-pressure rarefied gas flow applications, with the exception of Oran et al. (1998), who also address microfluidic applications.

In the previous section, we described molecular magnitudes and associated time and length scales. Under standard conditions in a volume of 10 μm^3, there are about 2.69×10^{10} molecules. A molecular-based simulation model that can compute the motion and interactions of all these molecules is not possible. The typical DSMC method uses hundreds of thousands or even millions of "simulated" molecules/particles that mimic the motion of real molecules.

The DSMC method is based on splitting the molecular motion and intermolecular collisions by choosing a time step less than the mean collision time and tracking the evolution of this molecular process in space and time. For efficient numerical implementation, the space is divided into cells similar to the finite-volume method. The DSMC cells are chosen proportional to the mean free path λ. In order to resolve large gradients in flow with realistic (physical) viscosity values, the average cell size should be $\Delta x_c \cong \lambda/3$ [Oran et al., 1998]. The time- and cell-averaged molecular quantities are obtained as the macroscopic values at the cell centers. The DSMC involves four main processes: motion of the particles, indexing and cross-referencing of particles, simulation of collisions and sampling of the flow field. The basic steps of a DSMC algorithm are given in Figure 8.3 and summarized below.

The first process involves motion of the simulated molecules during a time interval of Δt. Because the molecules will go through intermolecular collisions, the time step for simulation is chosen smaller than the mean collision time Δt_c. Once the molecules are advanced in space, some of them will have gone through wall collisions or will have left the computational domain through the inflow–outflow boundaries.

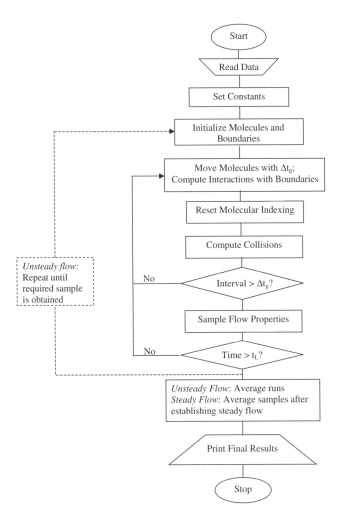

FIGURE 8.3 Typical steps for a DSMC method. (From Oran, E.S., et al. (1998) *Ann. Rev. Fluid Mech.* **30**, 403–441. With permission.)

Hence, the boundary conditions must be enforced at this level, and the macroscopic properties along the solid surfaces must be sampled. This is done by modeling the surface molecule interactions by application of the conservation laws on individual molecules, rather than using a velocity distribution function that is commonly utilized in the Boltzmann algorithms. This approach allows inclusion of many other physical processes, such as chemical reactions, radiation effects, three-body collisions and ionized flow effects, without major modifications to the basic DSMC procedure [Oran et al., 1998]. However, *a priori* knowledge of the accommodation coefficients must be used in this process. Hence, this constitutes a weakness of the DSMC method, similar to the Navier–Stokes-based slip and even Boltzmann equation-based simulation models. This issue is discussed in detail in the following section.

The second process is the indexing and tracking of the particles. This is necessary because, during the first stage, the molecules might have moved to new cell locations. The new cell location of the molecules is indexed, thus the intermolecular collisions and flowfield sampling can be handled accurately. This is a crucial step for an efficient DSMC algorithm. The indexing, molecule tracking and data structuring algorithms should be carefully designed for the specific computing platforms, such as vector super computers and workstation architectures.

The third step is simulation of collisions via a probabilistic process. Because only a small portion of the molecules is simulated, and the motion and collision processes are decoupled, probabilistic treatment becomes necessary. A common collision model is the no-time-counter technique of Bird (1994) that is used in conjunction with the subcell technique, where the collision rates are calculated within the cells and the collision pairs are selected within the subcells. This improves the accuracy of the method by maintaining the collisions of the molecules with their closest neighbors [Oran et al., 1998].

The last process is calculation of appropriate macroscopic properties by the sampling of molecular (microscopic) properties within a cell. The macroscopic properties for unsteady flow conditions are obtained by the ensemble average of many independent calculations. For steady flows, time averaging is also used.

8.3.3 Limitations, Error Sources and Disadvantages of the DSMC Approach

Following the work of Oran et al., (1998), we identify several possible limitations and error sources within a DSMC method.

1. *Finite cell size*: The typical DSMC cell should be about one third of the local mean free path. Values of cell sizes larger than this may result in enhanced diffusion coefficients. In DSMC, one cannot directly specify the dynamic viscosity and thermal conductivity of the fluid. The dynamic viscosity is calculated via diffusion of linear momentum. Breuer et al. (1995), performed one-dimensional Rayleigh flow problems in the continuum flow regime and showed that, for cell sizes larger than one mean free path, the apparent viscosity of the fluid was increased. Some numerical experimentation details for this finding are also given in Beskok (1996). More recently, the viscosity and thermal conductivity dependence on cell size have been obtained more systematically by using the Green–Kubo theory [Alexander et al., 1998; Hadjiconstaninou, 2000].

2. *Finite time step*: Due to the time splitting of the molecular motion and collisions, the maximum allowable time step is smaller than the local collision time t_c. Values of time steps larger than t_c result in molecules traveling through several cells prior to a cell-based collision calculation.

    The time-step and cell-size restrictions presented in items 1 and 2 are not a Courant–Friederichs–Lewy (CFL) stability condition. The DSMC method is always stable. However, overlooking the physical restrictions stated in items 1 and 2 may result in highly diffusive numerical results.

3. *Ratio of the simulated particles to the real molecules*: Due to the vast number of molecules and limited computational resources, one always has to choose a sample of molecules to simulate. If the ratio of the actual to the simulated molecules gets too high the statistical scatter of the solution is increased. The details for the statistical error sources and the corresponding remedies can be found in Oran et al. (1998), Bird (1994) and Chen and Boyd (1996). A relatively well-resolved DSMC calculation requires a minimum of 20 simulated particles per cell.

4. *Boundary condition treatment*: The inflow–outflow boundary conditions can become particularly important in a microfluidic simulation. A subsonic microchannel flow simulation may require specification of inlet and exit pressures. The flow will develop under this pressure gradient and result in a certain mass flowrate. During such simulations, specification of back pressure for subsonic flows is challenging. In the DSMC studies, one can simulate the entry problem to the channels by specifying the number density, temperature and average macroscopic velocity of the molecules at the inlet of the channel. At the outflow, the number density and temperature corresponding to the desired back pressure can be specified. By this and similar treatments it is possible to significantly reduce the spurious numerical boundary layers at inflow and outflow regions. For high Knudsen number flows (i.e., $Kn > 1$) in a channel with blockage (such as a sphere in a pipe), the location of the inflow and outflow boundaries is important. For example, the molecules reflected from the front of the body may reach the inflow region with very few intermolecular collisions, creating a diffusing flow at the front of the bluff body [Liu et al., 1998]. (The details of this case are presented in Section 8.3.4.)

5. *Uncertainties in the physical input parameters*: These typically include the input for molecular collision cross-section models, such as the hard sphere (HS), variable hard sphere (VHS) and variable soft sphere (VSS) models [Oran et al., 1998; Vijayakumar et al., 1999]. The HS model is usually sufficient for monatomic gases or for cases with negligible vibrational and rotational non-equilibrium effects, such as in the case of nearly isothermal flow conditions.

Along with these possible error sources and limitations are some particular disadvantages of the DSMC method for simulation of gas microflows:

1. *Slow convergence*: The error in the DSMC method is inversely proportional to the square root of the number of simulated molecules. Reducing the error by a factor of two requires increasing the number of simulated molecules by a factor of four! This is a very slow convergence rate compared to the continuum-based simulations with spatial accuracy of second or higher order. Therefore, continuum-based simulation models should be preferred over the DSMC method, whenever it is possible.

2. *Large statistical noise*: Gas microflows are usually low subsonic flows with typical speeds of 1 mm/s to 1 m/s (exceptions to this are the micronozzles utilized in synthetic jets and satellite thruster control applications). The macroscopic fluid velocity is obtained by time or ensemble averaging of the molecular velocities. This difference of five to two orders of magnitude between the molecular and average speeds results in large statistical noise and requires a very long time averaging for gas microflow simulations. The statistical fluctuations decrease with the square root of the sample size. Time or ensemble averages of low-speed microflows on the order of 0.1 m/s require about 10^8 samples in order to be able to distinguish such small macroscopic velocities. In a recent paper, Fan and Shen (1999) introduced the information preservation (IP) technique for the DSMC method, which enables efficient DSMC simulations for low-speed flows (the IP scheme is briefly covered in Section 8.3.5).

3. *Extensive number of simulated molecules*: If we discretize a rectangular domain of 1 mm × 100 μm × 1 μm under standard conditions for $Kn = 0.065$ flow, we will need at least 20 cells per micrometer length scale. This results in a total of 8×10^8 cells. Each of these cells should contain at least 20 simulated molecules, resulting in a total of 1.6×10^{10} particles. Combined with the number of time-step restrictions, simulation of low-speed microflows with DSMC easily exceeds the capabilities of current computers. An alternative treatment to overcome the extensive number of simulated molecules and long integration times is utilization of the dynamic similarity of low-pressure rarefied gas flows to gas microflows under atmospheric conditions. The key parameters for the dynamic similarity are the geometric similarity and matching of the flow Knudsen, Mach and Reynolds numbers. Performing actual experiments under dynamically similar conditions may be very difficult; however, parametric studies via numerical simulations are possible. The fundamental question to answer for such an approach is whether or not a specific, unforeseen microscale phenomenon is missed with the dynamic similarity approach. In response to this question, we must note that all numerical simulations are inherently model based. Unless microscale-specific models are implemented in the algorithm, we will not be able to obtain more physical information from a microscopic simulation than from a dynamically similar low-pressure simulation. In regard to utilization of the dynamic similarity concept, one limitation is the onset of statistical scatter in the instantaneous macroscopic flow quantities for gas microflows for $Kn > 1$ (see Section 8.3.1 and Figure 8.2 for details). Here, we must also remember that DSMC utilizes time or ensemble averages to sample the macroscopic properties from the microscopic variables. Hence, DSMC already determines the macroscopic properties in an averaged sense.

4. *Lack of deterministic surface effects*: Molecule wall interactions are specified by the accommodation coefficients (σ_v, σ_T). For diffuse reflection $\sigma = 1$, and the reflected molecules lose their incoming tangential velocity while being reflected with the tangential wall velocity. For $\sigma = 0$, the tangential velocity of the impinging molecules are not changed during the molecule/wall collisions. For any other value of σ, a combination of these procedures can be applied. The molecule–wall interaction treatment implemented in DSMC is more flexible than the slip conditions given by Eqs. (8.1) and (8.2).

However, it still requires specification of the accommodation coefficients, which are not known for any gas surface pair with a specified surface root mean square (rms) roughness. Recently, the tangential momentum accommodation coefficients for helium, nitrogen, argon and carbon dioxide on single-crystal silicon were measured by careful microchannel experiments [Arkilic, 1997].

8.3.4 Some DSMC-Based Gas Microflow Results

In this section, some DSMC results applied to gas microflows are presented.

8.3.4.1 Microchannel Flows

In this section, the DSMC simulation results for subsonic gas flows in microchannels are presented. Due to the computational difficulties explained in the previous sections, a low-aspect-ratio, two-dimensional channel with relatively high inlet velocities is studied. The results presented in the figures are for microchannels with a length-to-height ratio (L/h) of 20 under various inlet-to-exit-pressure ratios. The DSMC results are performed with 24,000 cells, of which 400 cells were in the flow direction and 60 cells across the channel. A total of 480,000 molecules are simulated. The results are sampled (time averaged) for 10^5 times and the sampling is performed every 10 time steps.

In the following simulations, diffuse reflection ($\sigma_V = 1.0$) is assumed for interaction of gas molecules with the surfaces. Because the slip amount can be affected significantly by small variations in σ_V (see Eq. (8.1)), the apparent value of the accommodation coefficient σ_V is monitored throughout the simulations by recording the tangential momentum of the impinging (τ_i) and reflected (τ_r) gas molecules. Based on these values, the apparent tangential momentum accommodation coefficient, $\sigma_V = (\tau_i - \tau_r)/(\tau_i - \tau_w) = 0.99912$ with standard deviation of $\sigma_{rms} = 0.01603$, is obtained.

The velocity profiles normalized with the corresponding average speed are presented in Figure 8.4 for pressure-driven microchannel flows at $Kn = 0.1$ and $Kn = 2.0$. The figure also presents the molecule/cell refinement studies as well as predictions of the VHS and VSS models. The DSMC results are compared against the linearized Boltzmann solutions [Ohwada et al., 1989], and excellent agreements of the VHS and VSS models with the linearized Boltzmann solutions are observed for these nearly isothermal flows. In regard to the molecule/cell refinement study, the number of cells and the number of simulated molecules are identified for each case. The first VHS case utilized only 6000 cells with 80,000 simulated molecules, and the results are sampled about 5×10^5 times. Sampling is performed every 20 time steps. The refined VHS and VSS cases utilized 24,000 cells and a total of 480,000 molecules. The results for these are sampled 10^5 times, every 10 time steps. Although the velocity profiles for the low-resolution case (6000 cells) seem acceptable, the density and pressure profiles show large fluctuations.

The DSMC and μFlow (spectral-element-based, continuum computational fluid dynamics (CFD) solver) predictions of density and pressure variations along a pressure-driven microchannel flow are shown in Figure 8.5. For this case, the ratio of inlet to exit pressure is $\Pi = 2.28$, and the Knudsen number at the channel outlet is 0.2. Deviations of the slip flow pressure distribution from the no-slip solution are also presented in the figure. Good agreements between the DSMC and μFlow simulations are achieved. The curvature in the pressure distribution is due to the compressibility effect, and the rarefaction negates this curvature, as seen in Figure 8.5. The slip velocity variation on the channel wall is shown in Figure 8.6. Overall good agreements between both methods are observed. In a recent article, Pan et al. (1999) used the DSMC simulations to determine the slip distance as a function of various physical conditions such as the number density, wall temperature and the gas mass. They determined that an appropriate slip distance is 1.125 λ_{gw}, where the subscript gw indicates the gas–wall conditions [Pan et al., 1999].

In the transitional flow regime, Beskok and Karniadakis (1999) studied the Burnett equations for low-speed isothermal flows. This analysis has shown that the velocity profiles remain parabolic even for large Kn flows. In order to verify this hypothesis, they performed several DSMC simulations; the velocity distribution nondimensionalized with the local average speed is shown in Figure 8.7. They also obtained

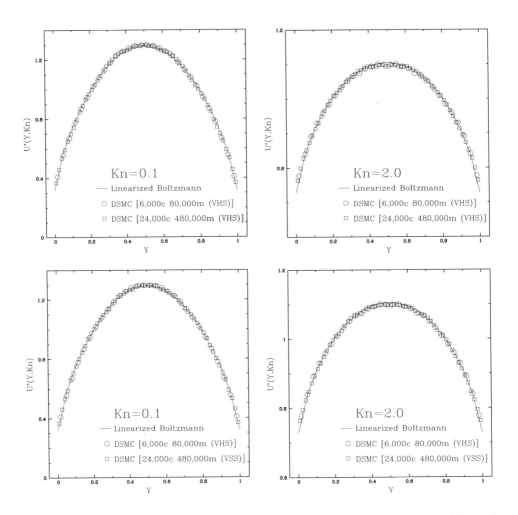

FIGURE 8.4 Velocity profiles normalized with the local average velocity in the slip and transitional flow regimes. The DSMC predictions with the VHS and VSS models agree well with the linearized Boltzmann solutions of Ohwada et al. (1989). The number of cells and simulated molecules are identified on each figure.

an approximation to this nondimensionalized velocity distribution in the following form:

$$U^*(y, Kn) \equiv U(x,y)/\overline{U}(x) = \left[\frac{-\left(\frac{y}{h}\right)^2 + \frac{y}{h} + \frac{Kn}{1-bKn}}{\frac{1}{6} + \frac{Kn}{1-bKn}} \right] \quad (8.8)$$

where the extended slip condition given in Eq. (8.3) is used. In the above relation, the value of $b = -1$ is determined analytically for channel and pipe flows [Beskok and Karniadakis, 1999]. In Figure 8.7, the nondimensional velocity variation obtained in a series of DSMC simulations for $Kn = 0.1$, $Kn = 1.0$, $Kn = 5.0$ and $Kn = 10.0$ flows are presented along with the corresponding linearized Boltzmann solutions [Ohwada et al., 1989]. It is seen that the DSMC velocity distribution and the linearized Boltzmann solutions agree quite well. One can use Eq. (8.8) to compare the results with the DSMC/linearized Boltzmann data by varying the parameter b. The case $b = 0$ corresponds to Maxwell's first-order slip model and $b = -1$ corresponds to Beskok's second-order slip boundary condition. It is clear from Figure 8.7 that Eq. (8.8) with $b = -1$ results in a uniformly valid representation of the velocity distribution in the entire Knudsen regime.

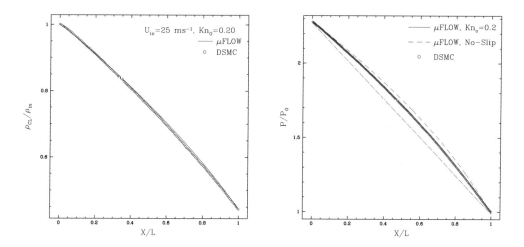

FIGURE 8.5 Density (left) and pressure (right) variation along a microchannel. Comparisons of the Navier–Stokes and DSMC predictions for ratio of inlet to exit pressure of $\Pi = 2.28$ and $Kn_o = 0.20$. (From Beskok, A. (1996) Ph.D. thesis, Princeton University, Princeton, NJ.)

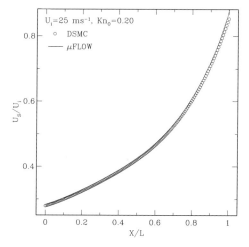

FIGURE 8.6 Wall slip velocity variation along a microchannel predicted by Navier–Stokes and DSMC simulations. (From Beskok, A. (1996) Ph.D. thesis, Princeton University, Princeton, NJ.)

The nondimensionalized centerline and wall velocities for $0.01 \leq Kn \leq 30$ flows are shown in Figure 8.8. Included in the figure are the data for the slip velocity and centerline velocity from 20 different DSMC runs, of which 15 are for nitrogen (diatomic molecules) and 5 for helium (monatomic molecules). The differences between the nitrogen and helium simulations are negligible, thus this velocity scaling is independent of the gas type. The linearized Boltzmann solutions [Ohwada et al., 1989] for a monatomic gas are also shown in Figure 8.8 by triangles. The Boltzmann solutions closely match the DSMC predictions. Maxwell's first-order boundary condition ($b = 0$) (shown by solid lines) predicts, erroneously, a uniform nondimensional velocity profile for large Knudsen numbers. The breakdown of the slip flow theory based on the first-order slip-boundary conditions is realized around $Kn = 0.1$ and $Kn = 0.4$ for wall and centerline velocities, respectively. This finding is consistent with the commonly accepted limits of the slip flow regime [Schaaf and Chambre, 1961]. The prediction using $b = -1$ is shown by small dashed lines. The corresponding centerline velocity closely follows the DSMC results, while the slip velocity of the model with $b = -1$ deviates from DSMC in the intermediate range for $0.1 < Kn < 5$.

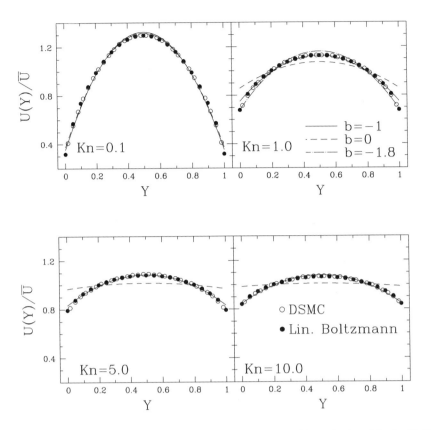

FIGURE 8.7 Comparison of the velocity profiles obtained by new slip model Eqs. (8.3) and (8.8) with DSMC and linearized Boltzmann solutions [Ohwada et al., 1989]. Maxwell's first-order boundary condition is shown by the dashed lines ($b=0$), and the general slip boundary condition ($b=-1$) is shown by solid lines. (From Beskok, A., and Karniadakis, G.E. (1999) *Microscale Thermophys. Eng.* **3**, pp. 43–77. Reproduced with permission of Taylor & Francis, Inc.)

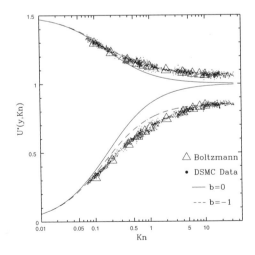

FIGURE 8.8 Centerline and wall slip velocity variations in the entire Knudsen regime. The linearized Boltzmann solutions of Ohwada et al. (1989) are shown by triangles, and the DSMC simulations are shown by closed circles. Theoretical predictions of the velocity scaling obtained by Eq. (8.8) are shown for different values of b. The $b=0$ case corresponds to Maxwell's first-order boundary condition, and $b=-1$ corresponds to the general slip-boundary condition.

One possible reason for this is the effect of the Knudsen layer. For small Kn flows, the Knudsen layer is thin and does not affect the overall velocity distribution too much. For very large Kn flows, the Knudsen layer covers the channel entirely. However, for intermediate Kn values, both the fully developed viscous flow and the Knudsen layer co-exist in the channel. At this intermediate range, approximating the velocity profile as parabolic ignores the Knudsen layers. For this reason, the model with $b = -1$ results in 10% error in the slip velocity at $Kn = 1$. However, the velocity distribution in the rest of the channel is described accurately for the entire flow regime. Based on these results, Beskok and Karniadakis (1999) developed a unified flow model that can predict the velocity profiles, pressure distribution and mass flowrate in channels, pipes and arbitrary aspect-ratio rectangular ducts in the entire Knudsen regime, including Knudsen's minimum effects [Beskok and Karniadakis, 1999; Kennard, 1938; Tison, 1993].

8.3.4.2 Separated Rarefied Gas Flows

Gas flow through complex microgeometries are prone to flow separation and recirculation. Most of the DSMC-based microflow analyses were performed in straight channels [Mavriplis et al., 1997; Oh et al., 1997] and for smooth microdiffusers [Piekos and Breuer, 1996]. A discussion of the Monte Carlo simulation for MEMS devices is given by Nance et al. (1997). The mainstream approach for gas flow modeling in MEMS is solution of the Navier–Stokes equations with slip models. This is more practical and numerically efficient than utilization of the DSMC method. However, rarefied separated gas flows are not studied extensively. In order to investigate the validity of slip-boundary conditions under severe adverse pressure gradients and separation, Beskok and Karniadakis (1997) performed a series of numerical simulations using the classical backwards-facing step geometry with a step-to-channel-height ratio of 0.467. The variations of pressure and streamwise velocity along a step microchannel, obtained at five different cross-flow locations (y/h), are presented in Figure 8.9. The values of pressure and velocity are nondimensionalized with the

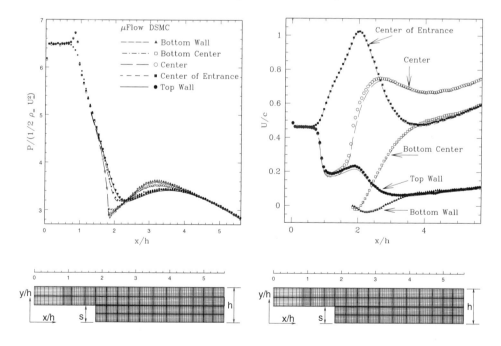

FIGURE 8.9 Pressure (left) and streamwise velocity (right) distribution along a backward-facing step channel at five selected locations. Predictions of both DSMC (symbols) and continuum-based spectral element CFD code (lines) are presented. A 10th-order spectral element grid is also shown. The top wall is at $y/h = 0.98325$; the center of the entrance is at $y/h = 0.75$; the center is at $y/h = 0.48$; the bottom center is at $y/h = 0.25$; and the bottom wall is at $y/h = 0.017$. The simulation conditions are for $Re = 80$, $Kn_{out} = 0.04$, $M_{in} = 0.45$ and $Pr = 0.7$ (From Beskok, A., and Karniadakis, G.E. (1997) 28th AIAA Fluid Dynamics Conf., AIAA 97-1883, June 29–July 2. Copyright © 1997 by the American Institute of Aeronautics and Astronautics, Inc. Reprinted with permission.)

FIGURE 8.10 Velocity contours for a sphere in a pipe in the transitional flow regime ($Kn = 3.5$). Molecules reflected from the sphere create a diffusive layer at the entrance of the pipe [Liu et al., 1998].

corresponding freestream dynamic head and the local sound speed, respectively. The specific y/h locations are selected to coincide with the DSMC cell centers to avoid interpolations or extrapolations of the DSMC method. The results show reasonable agreements of the slip-based Navier–Stokes simulations with the DSMC data. The flow recirculation and re-attachment location at the bottom wall are predicted equally well with both methods. The DSMC simulations utilized 28,000 cells (700 × 40) with 420,000 simulated molecules. The solution is sampled 10^5 times. The continuum-based simulations are performed by 52 spectral elements with 10th-order polynomial expansions for each direction.

8.3.4.3 Transitional Flow Past a Sphere in a Pipe

The DSMC simulations of high Kn rarefied flows at the entry of channels or pipes show diffusion of the molecules from the entry toward the free stream region. In order to demonstrate this counterintuitive effect, Liu et al. (1998) simulated flow past a sphere in a pipe problem with diffuse reflection from the surfaces. In order to incorporate the molecules diffusing out from the entry of the pipe, the computational domain for the freestream region had to be extended more than expected. In high Knudsen number subsonic flows, the molecules reflected from the sphere can travel toward the pipe inlet with very few intermolecular collisions and then diffuse out. Figure 8.10 presents the velocity contours for $Kn = 3.5$ flow. Diffusion of the molecules toward the inflow can easily be identified from the velocity contours. This effect was studied earlier by Kannenberg and Boyd (1996) for transitional flow entering a channel. For $Kn = 3.5$ results presented in Figure 8.10, the length of the freestream region is equal to the length of the pipe; hence, the computational cost is increased significantly.

8.3.5 Recent Advances in the DSMC Method

In this section, recent developments in the application and implementation of the DSMC method are presented.

8.3.5.1 Information Preservation DSMC Scheme

Fan and Shen (1999) developed an information preservation (IP) DSMC scheme for low-speed rarefied gas flows. Their method uses the molecular velocities of the DSMC method as well as an information velocity that records the collective velocity of an enormous number of molecules that a simulated particle represents. The information velocity evolves with inelastic molecular collisions, and the results presented for Couette, Poiseuille and Rayleigh flows in the slip, transition and free molecular regimes show very good agreements with the corresponding analytical solutions. This approach seems to cut down the sample size and correspondingly the CPU time required by a regular DSMC method for low-speed flows by orders of magnitude. This is a tremendous gain in computational time that can lead to the effective use of IP DSMC schemes for microfluidic and MEMS simulations. Currently, the IP DSMC schemes are being validated in two-dimensional, complex-geometry flows, and extensions of the IP technique for three-dimensional flows are also being developed [Cai et al., 2000].

8.3.5.2 DSMC with Moving Boundaries

Some of the microflow applications require numerical simulation of moving surfaces. In continuum-based approaches, arbitrary Lagrangian Eulerian (ALE) algorithms are successfully utilized for such applications [Beskok and Warburton, 2000a; 2000b]. A similar effort to expand the DSMC method for grid adaptation, including the moving external and internal boundaries, combined the DSMC method with a monotonic Lagrangian grid (MLG) method [Cybyk et al., 1995; Oran et al., 1998].

8.3.5.3 Parallel DSMC Algorithms

Because the DSMC calculations involve vast numbers of molecules, utilization of parallel algorithms with efficient interprocessor communications and load balancing can have a significant impact on the effectiveness of simulations. Recent developments in parallel DSMC algorithms are addressed by Oran et al. (1998). For example, Dietrich and Boyd (1996) were able to obtain 90% parallel efficiency with 400 processors, simulating over 100 million molecules on a 400-node IBM SP2 computer. The computing power of their code is comparable to 75 single-processor Cray C90 vector computers. Good parallel efficiencies for DSMC algorithms can be achieved with effective load-balancing methods based on the number of molecules. This is because the computational work of the DSMC method is proportional to the number of simulated molecules.

8.3.6 DSMC Coupling with Continuum Equations

In this section, an overview of the DSMC/Navier–Stokes and DSMC/Euler equation coupling strategies is presented. These equations are particularly important for simulation of gas flows in MEMS components. If we consider a micromotor or a microcomb drive mechanism, gas flow in most of the device can be simulated using the slip-based continuum models. The DSMC method should be utilized only when the gap between the surfaces becomes submicron, or when the local $Kn > 0.1$. Hence, it is necessary to implement multidomain DSMC/continuum solvers. Depending on the specific application, hybrid Euler/DSMC [Roveda et al., 1998] or DSMC/Navier-Stokes algorithms [Hash and Hassan, 1995] can be used. Such hybrid methods require compatible kinetic-split fluxes for the Navier–Stokes portion of the scheme [Chou and Baganoff, 1997; Lou et al., 1998] so that an efficient coupling can be achieved. Recently, an adaptive mesh and algorithm refinement (AMAR) procedure, which embeds a DSMC-based particle method within a continuum grid, has been developed that enables molecular-based treatments even within the continuum region [Garcia et al., 1999]. Hence, the AMAR procedure can be used to deliver microscopic and macroscopic information within the same flow region.

Simulation results for a Navier-Stokes/DSMC coupling procedure obtained by Liu (2001) is shown in Figure 8.11. A structured spectral element algorithm, μFlow [Beskok, 1996], is coupled with an unstructured DSMC method, UDSMC 2-D, with an overlapping domain. Both the grid and the streamwise velocity contours are shown in the figure; smooth transition of the velocity contours from the continuum-based slip region to the DSMC region can be observed. The details of the coupling procedure are given in Liu (2001).

8.3.7 Boltzmann Equation Research

Microscale thermal/fluidic transport in the entire Knudsen regime ($0 \leq Kn < \infty$) is governed by the Boltzmann equation (BE). The Boltzmann equation describes the evolution of a velocity distribution function by molecular transport and binary intermolecular collisions. The assumption of binary intermolecular collisions is a key limitation in the Boltzmann formulation, making it applicable for dilute gases only. The Boltzmann equation for a simple dilute gas is given in the following form [Bird, 1994]:

$$\frac{\partial nf}{\partial t} + \vec{c} \cdot \frac{\partial nf}{\partial \vec{x}} + \vec{F} \cdot \frac{\partial nf}{\partial \vec{c}} = \int_{-\infty}^{\infty} \int_{0}^{4\pi} n^2 (f^* f_1^* - f f_1) c_r \sigma \, d\Omega \, d\vec{c}_1 \qquad (8.9)$$

where f is the velocity distribution function, n is the number density, $\vec{c}$ is the molecular velocity, $\vec{F}$ is the external force per unit mass, c_r is the relative speed of class $\vec{c}$ molecules with respect to class $\vec{c}_1$ molecules, and σ is the differential collision cross section. The definition of each term in Eq. (8.9) is as follows: The first term is the rate of change of the number of class $\vec{c}$ molecules in the phase space.

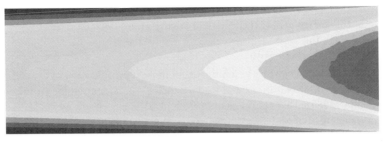

NS (Spectral Element - Slip Model)   DSMC (Unstructured Mesh)

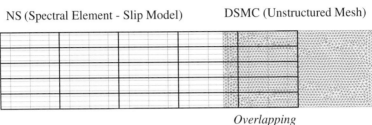

Overlapping

FIGURE 8.11 Streamwise velocity contours for rarefied gas flow in mixed slip/transitional regimes, obtained by a coupled DSMC/continuum solution method. Most of the channel is in the slip flow regime, and a spectral element method μFlow is utilized to solve the compressible Navier–Stokes equations with slip. The rest of the channel is in the transitional flow regime, where a DSMC method with unstructured cells is utilized. (From Liu, H.F. (2001) Ph.D. thesis, Brown University, Providence, RI.)

The second term shows convection of molecules across a fluid volume by molecular velocity $\vec{c}$. The third term is convection of molecules across the velocity space as a result of the external force $\vec{F}$. The fourth term is the binary collision integral. The term $(-ff_1)$ describes the collision of molecules of class $\vec{c}$ with molecules of class $\vec{c}_1$ (resulting in creation of molecules of class $\vec{c}^*$ and $\vec{c}_1^*$, respectively), and it is known as the loss term. Similarly, there are inverse collisions where class $\vec{c}^*$ molecules collide with class $\vec{c}_1^*$ molecules, creating class $\vec{c}$ and $\vec{c}_1$ molecules. This is shown by $f^* f_1^*$, known as the gain term. Assuming binary elastic collisions enables us to determine class $\vec{c}^*$ and $\vec{c}_1^*$ conditions [Bird, 1994]. The difficulty of the Boltzmann equation arises due to the nonlinearity and complexity of the collision integral terms and the multidimensionality of the equation (x, c, t).

Current numerical methods are usually very expensive, and they are applied for simple geometries, such as pipes and channels. In particular, a number of investigators have considered semi-analytical and numerical solutions of the linearized Boltzmann equation to be valid for flows with small pressure and temperature gradients [Aoki, 1989; Huang et al., 1966; Loyalka and Hamoodi, 1990; Ohwada et al., 1989; Sone, 1989]. These studies used HS and Maxwellian molecular models. Simplifications for the collision integral based on the BGK model [Bhatnagar et al., 1954] are used in the Boltzmann equation studies. The BGK model for a rarefied gas with no external forcing is given as:

$$\frac{\partial nf}{\partial t} + \vec{c} \cdot \frac{\partial nf}{\partial \vec{x}} = \nu n(f_o - f) \tag{8.10}$$

where ν is the collision frequency and f_o is the local Maxwellian (equilibrium) distribution. The right-hand side of Eq. (8.10) becomes zero when the flow is in local equilibrium (continuum flow) or when the collision frequency goes to zero (corresponding to the free molecular flow). The BGK model captures both limits correctly. However, there are justified concerns about the validity of the BGK model in the transition flow regime. A model's ability to capture the two asymptotic limits ($Kn \to 0$ and $Kn \to \infty$) is not necessarily sufficient for its accuracy in the intermediate regimes [Bird, 1994].

Recently, Veijola et al. (1995; 1998) presented a Boltzmann equation analysis of silicon accelerometer motion and squeeze-film damping as a function of the Knudsen number and the time-periodic motion of the surfaces. Although the squeeze-film damping analysis is very challenging due to the mixed compressibility and rarefaction effects, it has many practical applications, including computer disk hard drives, micro-accelerometers and noncontact gas buffer seals [Fukui and Kaneko, 1988; 1990]. A comprehensive theory of internal rarefied gas flows, including the numerical simulation data, is given by Saripov and Seleznev (1998). The reader is referred to this article for further theoretical and numerical details on the Boltzmann equation.

The wall-boundary conditions for Boltzmann solutions typically use diffuse and mixed diffuse/specular reflections. For diffuse reflection, the molecules reflected from a solid surface are assumed to have reached thermodynamic equilibrium with the surface. Thus, they are reflected with a Maxwellian distribution corresponding to the temperature and velocity of the surface.

8.3.8 Hybrid Boltzmann/Continuum Simulation Methods

Solution of the Navier–Stokes equation is numerically more efficient than solution of the Boltzmann equation; therefore it is desirable to develop coupled multidomain Boltzmann/Navier–Stokes models for simulation of mixed regime flows in MEMS and microfluidic applications. Because the typical DSMC method for this coupling results in large statistical noise, as discussed before, solution of the Boltzmann equation may be preferred. The hybrid Boltzmann/Navier–Stokes simulation approach can be achieved by calculating the macroscopic fluid properties from the Boltzmann solutions by moment methods [Bird, 1994], and using the kinetic flux-vector splitting procedure of Chou and Baganoff (1997). Another continuum to Boltzmann coupling can be obtained by using local Chapman–Enskog expansions to the BGK equation [Chapman and Cowling, 1970] and evaluating the distribution function for the kinetic region [Jamamato and Sanryo, 1990].

8.3.9 Lattice Boltzmann Methods

Another approach for simulating flows in microscales is the lattice Boltzmann method (LBM), which is based on the solution of the Boltzmann equation on a previously defined lattice structure with simplistic molecular collision rules. Details of the lattice Boltzmann method are given in a review article by Chen and Doolen (1998). The LBM can be viewed as a special finite differencing scheme for the kinetic equation of the discrete-velocity distribution function, and it is possible to recover the Navier–Stokes equations from the discrete lattice Boltzmann equation with sufficient lattice symmetry [Frisch et al., 1986].

The main advantages of the LBM compared to other continuum-based numerical methods include [Chen and Doolen, 1998]:

- The convection operator is linear in the phase space.
- The LBM is able to obtain both compressible and incompressible Navier–Stokes limits.
- The LBM utilizes a minimal set of velocities in the phase space compared to the continuous velocity distribution function of the Boltzmann algorithms.

With these advantages, the LBM has developed significantly within the last decade. The molecular motions for LBM are allowed on a previously defined lattice structure with restriction on molecular velocities to a few values. Particles move to a neighboring lattice location in every time step. Rules of molecular interactions conserve mass and momentum. Successful thermal and hydrodynamic analysis of multiphase flows including real gas effects can also be obtained [He, et al., 1998; Lou, 1998; Qian, 1993; Shan and Chen, 1994]. Another useful application of the LBM is for granular flows, which can be expanded to include flow-through microfiltering systems [Angelopoulos et al., 1998; Bernsdorf et al., 1999; Michael et al., 1997; Spaid and Phelan, 1997; Vangenabeek and Rothman, 1996].

Lattice Boltzmann methods have relatively simple algorithms, and they are introduced as an alternative to the solution of the Navier–Stokes equations [Frisch et al., 1986; Qian et al., 1992]. In contrast to the continuum algorithms, which have difficulties in simulating rarefied flows with consistent slip-boundary conditions, the lattice Boltzmann method initially had difficulties in imposing the no-slip-boundary condition accurately. However, this problem has been successfully resolved [Inamuro et al., 1997; Lavallée et al., 1991; Noble et al., 1995; Zou and He, 1997].

Rapid development of the lattice Boltzmann method with relatively simpler algorithms that can handle both the rarefied and continuum gas flows from a kinetic theory point of view and the ability of the method to capture the incompressible flow limit make the LBM a great candidate for microfluidic simulations. At this time, the author is not familiar with applications of the lattice Boltzmann method specifically for microfluidic problems.

8.4 Liquid and Dense Gas Flows

Liquids do not have a well-advanced molecular-based theory. Similar limitations also exist for dense gases where simultaneous intermolecular collisions can exist. The stand-alone Navier–Stokes simulations cannot describe the liquid and dense gas flows in submicron-scale conduits. The effects of van der Waals forces between the fluid and the wall molecules and the presence of longer range Coulombic forces and an electrical double layer (EDL) can significantly affect the microscale transport [Ho and Tai, 1998; Gad-el-Hak, 1999]. For example, the streaming potential effect present in pressure-driven flows under the influence of EDL can explain deviations in the Poiseuille number reported in the seminal liquid microflow experiments of Pfahler et al. (1991).

In recent years, there has been increased interest in the development of micropumps and valves with nonmoving components for medical, pharmaceutical, defense and environmental-monitoring applications. Electrokinetic body forces can be used to develop microfluidic flow control and manipulation systems with nonmoving components. In this section, a brief review of continuum equations for electrokinetic phenomena and the electric double layer is presented.

8.4.1 Electric Double Layer and Electrokinetic Effects

The electric double layer is formed due to the presence of static charges on the surfaces. Generally, a dielectric surface acquires charges when it is in contact with a polar medium or by chemical reaction, ionization or ion absorption. For example, when a glass surface is immersed in water, it undergoes a chemical reaction that results in a net negative surface potential [Cummings et al., 1999; Probstein, 1994]. This influences the distribution of ions in the buffer solution. Figure 8.12 shows the schematic view of a solid surface in contact with a polar medium. Here, a net negative electric potential is generated on the surface. Due to this surface electric potential, positive ions in the liquid are attracted to the wall; on the other hand, the negative ions are repelled from the wall. This results in redistribution of the ions close to the wall, keeping the bulk of the liquid far away from the wall electrically neutral. The distance from the wall, where the electric potential energy is equal to the thermal energy, is known as the Debye length (λ), and this zone is known as the electric double layer. The electric potential distribution within the fluid is described by the Poisson–Boltzmann equation:

$$\nabla^2(\psi/\zeta) = \frac{-4\pi h^2 \rho_e}{D\zeta} = \beta \sinh(\alpha \psi/\zeta) \qquad (8.11)$$

where the ψ is the electric potential field, ζ is the zeta potential, ρ_e is the net charge density, D is the dielectric constant, and α is the ionic energy parameter given as:

$$\alpha = ez\zeta/k_b T \qquad (8.12)$$

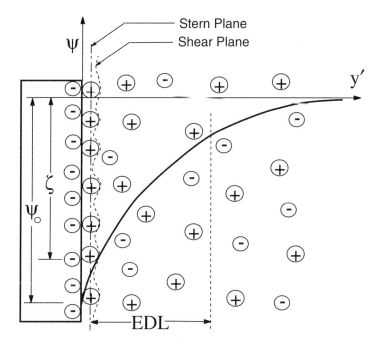

FIGURE 8.12 Schematic diagram of the electric double layer next to a negatively charged solid surface. Here, Ψ is the electric potential, Ψ_s is the surface electric potential, ζ is the zeta potential, and y' is the distance measured from the wall.

where e is the electron charge, z is the valence, k_b is the Boltzmann constant, and T is the temperature. In Eq. (8.11), the spatial gradients are nondimensionalized with a characteristic length h. Parameter β relates the ionic energy parameter α and the characteristic length h to the Debye–Hückel parameter $\omega = 1/\lambda$ as shown below:

$$\beta = (\omega h)^2/\alpha$$

where

$$\omega = \frac{1}{\lambda} = \sqrt{\frac{2n_0 e^2 z^2}{\varepsilon k_b T}}$$

The electrokinetic phenomenon can be divided into four parts [Probstein, 1994]:

1. *Electro-osmosis*: Motion of ionized liquid relative to the stationary charged surface by an applied electric field
2. *Streaming potential*: Electric field created by the motion of ionized fluid along stationary charged surfaces (opposite of electro-osmosis)
3. *Electrophoresis*: Motion of the charged surface relative to the stationary liquid, by an applied electric field
4. *Sedimentation potential*: Electric field created by the motion of charged particles relative to a stationary liquid (opposite of electrophoresis).

8.4.2 The Electro-Osmotic Flow

The electro-osmotic flow is created by applying an electric field in the streamwise direction, where this electric field ($\vec{E}$) interacts with the electric charge distribution in the channel (ρ_e) and creates an electrokinetic body force on the fluid. The ionized incompressible fluid flow with electro-osmotic body forces is governed by the incompressible Navier–Stokes equation:

$$\rho_f \left(\frac{\partial \vec{V}}{\partial t} + (\vec{V} \cdot \nabla) \vec{V} \right) = -\nabla P + \mu \nabla^2 \vec{V} + \rho_e \vec{E}. \tag{8.13}$$

The main simplifying assumptions and approximations are

- The fluid viscosity is independent of the shear rate; hence, the Newtonian fluid is assumed.
- The fluid viscosity is independent of the local electric field strength. This condition is an approximation. Because the ion concentration within the EDL is increased, the viscosity of the fluid may be affected; however, such effects are usually neglected for dilute solutions.
- The Poisson–Boltzmann equation, Eq. (8.11), is valid; hence, the ion convection effects are negligible.
- The solvent is continuous and its permittivity is not affected by the overall and local electric field strength.

Based on these continuum-based equations, various researchers have developed numerical models to simulate electrokinetic effects in microdevices [Yang et al., 1998; Ermakov et al., 1998; Dutta et al., 1999; 2000]. Here we must add that the EDL thickness can be as small as a few nanometers, and in such small scales the continuum description given by the Poisson–Boltzmann equation may break down [Dutta and Beskok, 2001]. Hence, the molecular dynamics method can be used to study the EDL effects in such small scales [Lyklema et al., 1998].

8.4.3 Molecular Dynamics Method

The molecular dynamics method requires simulation of motion and interactions of all molecules in a given volume. The intermolecular interactions are described by a potential energy function, typically the Lennard–Jones 12–6 potential given as [Allen and Tildesley, 1994]:

$$V^{LJ}(r) = 4\varepsilon \left[c_{ij} \left(\frac{\sigma}{r} \right)^{12} - d_{ij} \left(\frac{\sigma}{r} \right)^6 \right] \tag{8.14}$$

where r is the molecular separation, and σ and ε are the length and energy parameters in the pair potential, respectively. The coefficients c_{ij} and d_{ij} are parameters chosen for particular fluid–fluid and fluid–surface combinations (see Allen and Tildesley (1994) for details). The first term on the right-hand side shows the strong repulsive force felt when the two molecules are extremely close to each other, and the second term represents the van der Waals forces. The force field is found by differentiation of this potential for each molecule pair, and the molecular motions are obtained by numerical integration of Newton's equations of motion. Because the motion and interactions of all molecules are simulated, MD simulations are expensive. The computational work scales like the square of the number of the simulated molecules $O(N^2)$. Reduction in the computational intensity can be achieved by fast multipole algorithms or by implementation of a simple cutoff distance [Gad-el-Hak, 1999]. The MD simulations are usually performed for simple liquid molecules and for dense gases. Potential functions, other than the Lennard–Jones 12–6 potential, are also available. However, an intrinsic limitation of the molecular dynamics method, besides the prohibitively large number of molecules involved in the simulation, is the insight required to select the appropriate potential functions. For example, the electrokinetic transport simulations require inclusion of electrostatic forces in the potential function.

8.4.4 Treatment of Surfaces

In the molecular dynamics method, the fluid–surface interactions can be handled more realistically by including solid atoms attached to the lattice sites via a confining potential and letting them interact with gas–liquid molecules through a Lennard–Jones potential, Eq. (8.14). The solid atoms exhibit random thermal motions corresponding to the surface temperature T_{wall}. The desired temperature of the simulation is maintained by controlling the outer parts of the solid-lattice structure [Koplik and Banavar, 1995a]. Using realistic atomistic surface discretization increases the number of molecules even further, but this may become necessary to determine the surface roughness effects in microtransport. Also considering that the microfabricated surfaces can have rms surface roughness on the order of a few nanometers (depending on the fabrication process), realistic molecular-based surface treatments for liquid flows in nanoscales can be achieved using the molecular dynamics method [Tehver et al., 1998].

The molecular dynamics method is used to determine the validity range of the Newtonian fluid approximation and the no-slip-boundary conditions for simple liquids in submicron and nanoscale channels. Koplik et al. (1989) investigated dense gas and liquid Poiseuille flows with MD simulations. The molecular structure of the wall is also included in these simulations, resulting in fluid–wall interactions for smooth surfaces. Their findings for liquid flows show insignificant velocity slip effects. However, considerable slip effects with decreasing density are reported for gas molecules, consistent with Maxwell's slip-boundary conditions given in Eq. (8.1). There are some conflicting findings in the literature in regard to the validity of the no-slip conditions, and the appropriate viscosity of liquids in nanoscale channels (see Section 2.7 in Gad-el-Hak, 1999). In a recent study by Granick (1999), the behavior of complex liquids with chain-molecule structures in nanoscales has been reported. Confined fluid behavior, solidification, melting and rapid deformation of liquid thin films can also be studied by the molecular dynamics method.

One must keep in mind that MD is restricted to very small (nanoscale) volumes, and the maximum integration time is also limited to a few thousand mean collision times. Hence, molecular simulations should be used whenever the corresponding continuum equations are suspected of failing or are expected to fail, such as in the case of fast time-scale processes, thin films or interfaces and in the presence of geometric singularities [Koplik and Banavar, 1995]. In order to apply MD to larger scale, thermal/fluidic transport problems, Hadjiconstantinou and Patera (1997) developed coupled atomistic/continuum simulation methods and recently extended this work to include multifluid interfaces [Hadjiconstantinou, 1999]. Due to the computational and practical restrictions of MD, it is crucial to develop hybrid atomistic/continuum simulation methodologies for further studies of liquid transport in micron and nanoscales.

8.5 Summary and Conclusions

In this chapter molecular-based simulation methodologies for liquid and gas flows in micron and submicron scales were presented. For simulation of gas flows, the main emphasis was given to the direct simulation Monte Carlo (DSMC) method. Its algorithmic details, limitations, advantages and disadvantages were presented. Although the DSMC is quite popular for analysis of high-speed rarefied gas flows, it is not as effective for simulation of gas microflows. It suffers from slow convergence and large statistical noise, and it requires an extensive number of simulated molecules. These disadvantages can be eliminated to some degree by using the newly developed information preservation (IP) technique. However, the IP-DSMC is still undergoing development and validation. An alternative to the DSMC method is solution of the Boltzmann transport equation, which is an integro-differential equation with seven independent variables. It is clear that the Boltzmann equation algorithms are very complicated to implement for general engineering applications, but they can be used for simple geometry cases such as in microchannels. A final alternative for simulation of gas microflows is the lattice Boltzmann method (LBM), which has been developed extensively within the past decade. The LBM has relatively simpler algorithms that can handle both the rarefied and continuum gas flows from a kinetic theory point of view, and the ability of the LBM to capture the incompressible flow limit can make this method a great candidate for microfluidic simulations.

The molecular dynamics (MD) method was introduced for liquid flows. Because MD requires modeling of every molecule, it is computationally expensive and is usually applied to very small volumes in order to verify the onset of the continuum behavior in liquids. MD is general enough to handle the interactions of long-chain molecules with each other and the surfaces in very thin gaps. The wall surface roughness and its molecular structure can also be included in the simulations. Thus, realistic molecule/surface interactions can be obtained using the MD method. The main drawback of MD is its prohibitively large computational cost.

The DSMC, MD and Boltzmann equation models are numerically more expensive than solution of the Navier-Stokes equations. Considering that the microtransport applications cover a wide range of length scales from submillimeter to tens of nanometers, it is numerically more efficient to implement hybrid continuum/atomistic models, where the atomistic simulations take place only at a small section of the entire computational domain. References to recent developments of hybrid schemes were given for each model.

All numerical methods are inherently model based, including the constitutive laws and the boundary conditions of the Navier–Stokes equations, as well as the molecular interaction models of the MD and the DSMC methods. Although it may seem as though the molecular simulation methods are more fundamental in nature, they require assumptions and models of more fundamental levels. For example, the molecular dynamics method requires specification of the Lennard–Jones potentials and their coefficients. Physical insight about a problem is of utmost importance for any model. After all, numerical models can only deliver the physics implemented within them.

References

Alexander, F.J., Garcia, A.L., and Alder, B.J., 1998, "Cell Size Dependence of Transport Coefficients in Stochastic Particle Algorithms," *Phys. Fluids,* **10**(6), pp. 1540–1542.

Allen, M.P., and Tildesley D.J. (1994) *Computer Simulation of Liquids,* Oxford Science Publications, New York.

Angelopoulos, A.D., Paunov, V.N., Burganos, V.N., and Payatakes, A.C. (1998) "Lattice Boltzmann Simulation of Non-Ideal Vapor–Liquid Flow in Porous Media," *Phys. Rev. E 57* **3**, pp. 3237–3245.

Aoki, K. (1989) "Numerical Analysis of Rarefied Gas Flows by Finite-Difference Method," in *Rarefied Gas Dynamics: Theoretical and Computational Techniques,* ed. E.P. Muntz, D.P. Weaver, and D.H. Campbell, AIAA, New York, pp. 297–322.

Arkilic, E.B. (1997) "Measurement of the Mass Flow and Tangential Momentum Accommodation Coefficient in Silicon Micro-Machined Channels," Ph.D. thesis, MIT, Cambridge, MA.

Arkilic, E., Schmidt, M.A., and Breuer, K.S. (1997) "Gaseous Flow in Long Microchannels," *J. MEMS,* **6**, pp. 2–7.

Bernsdorf, J., Durst, F., and Schafer, M. (1999) "Comparison of Cellular Automata and Finite Volume Techniques for Simulation of Incompressible Flows in Complex Geometries," *Int. J. Numerical Methods Fluids* **29**, pp. 251–264.

Beskok, A. (1996) "Simulations and Models for Gas Flows in Microgeometries," Ph.D thesis, Department of Mechanical and Aerospace Engineering, Princeton University, Princeton, NJ.

Beskok, A., Trimmer, W., and Karniadakis, G.E. (1995) "Rarefaction Compressibility and Thermal Creep Effects in Gas Microflows," in *IMECE 95, Proc. ASME Dynamic Systems and Control Division,* DSC-Vol. 57-2, pp. 877–892.

Beskok, A., Karniadakis, G.E., and Trimmer, W. (1996) "Rarefaction and Compressibility Effects in Gas Microflows," *J. Fluids Eng.* **118**, pp. 448–456.

Beskok, A., and Karniadakis, G.E. (1997) "Modeling Separation in Rarefied Gas Flows," 28th AIAA Fluid Dynamics Conference, AIAA 97-1883, June 29–July 2.

Beskok, A., and Karniadakis, G.E. (1999) "A Model for Flows in Channels, Pipes, and Ducts at Micro and Nano scales," *Microscale Thermophys. Eng.* **3**, pp. 43–77.

Beskok, A., and Warburton, T.C. (2000a) "CFD Analysis of a Bi-Directional Micro-Pump," *Int. J. Computational Sci. Eng.* **2**(1), March.

Beskok, A., and Warburton, T.C. (2000b) "An Unstructured H/P Finite Element Scheme for Fluid Flow and Heat Transfer in Moving Domains," *J. Computational Phys.*

Bhatnagar, P.L., Gross, E.P., and Krook, K. (1954) "A Model for Collision Processes in Gases," *Phys. Rev.* **94**, pp. 511–524.

Bird, G. (1976) *Molecular Gas Dynamics,* Clarendon Press, Oxford, U.K.

Bird, G. (1978) "Monte Carlo Simulation of Gas Flows," *Ann. Rev. Fluid Mech.* **10**, pp. 11–31.

Bird, G. (1994) *Molecular Gas Dynamics and the Direct Simulation of Gas Flows,* Oxford Science Publications, New York.

Bird, G.A. (1998) "Recent Advances and Current Challenges for DSMC," *Computers Math. Applic.* **35**(1), pp. 1–14.

Breuer, K.S., Piekos, E.S., and Gonzales, D.A. (1995) "DSMC Simulations of Continuum Flows," AIAA Paper 95-2088, Thermophysics Conf., June 19–22, American Institute of Aeronautics and Astronautics, San Diego, CA.

Cai, C.P., Boyd, I.D., Fan, J., and Candler G.V. (2000) "Direct Simulation Methods for Low Speed Microchannel Flows," *AIAA J. Thermophys. Heat Transfer* **14**(3), pp. 368–378.

Chapman, S., and Cowling, T.G. (1970) *The Mathematical Theory of Non-Uniform Gases,* Cambridge University Press, New York.

Chen, G., and Boyd, I.D. (1996) "Statistical error analysis for the direct simulation Monte Carlo Technique," *J. Computational Phys.* **126**, pp. 434–448.

Chen, S., and Doolen, G.D. (1998) "Lattice Boltzmann Method for Fluid Flows," *Ann. Rev. Fluid Mech.* **30**, pp. 329–364.

Chou, S.Y., and Baganoff, D. (1997) "Kinetic Flux-Vector Splitting for the Navier–Stokes Equations," *J. Computational Phys.* **130**, pp. 217–230.

Cummings, E.B., Griffiths, S.K., and Nilson, R.H. (1999) "Irrotationality of uniform electroosmosis," *Proc. of SPIE Microfluidic Devices and Systems II,* pp. 180–189.

Cybyk, B.Z., Oran, E.S., Boris, J.P., and Anderson, J.D., Jr. (1995) "Combining the Monotonic Lagrangian Grid with a Direct Simulation Monte Carlo Model," *J. Computational Phys.* **122**, pp. 323–334.

Deitrich, S., and Boyd, I.D. (1996) "Scalar and Parallel Optimized Implementation of the Direct Simulation Monte Carlo Method," *J. Computational Phys.* **126**, pp. 328–342.

Drexler, K.E. (1990) *Engines of Creation: The Coming Era of Nanotechnology,* Anchor Books/Doubleday, New York.

Dutta, P., and Beskok, A. (2001) "Analytical Solution of Combined Electro-osmotic/Pressure Driven Flows in Two-Dimensional Straight Channels: Finite Debye Layer Effects" *Analyt. Chem.* **73**(9), pp. 1979–1986.

Dutta, P., Warburton, T.C., and Beskok, A. (1999) "Numerical Modeling of Electroosmotically Driven Micro Flows," *Proc. ASME IMECE Meeting, J. MEMS* **1**, pp. 467–474.

Dutta P., Beskok, A., and Warburton, T.C. (2000) "Electroosmotic Flow Control in Complex Micro-Geometries," *J. MEMS.*

Ermakov, S.V., Jacobson, S.C., and Ramsey, J.M. (1998) "Computer Simulations of Electrokinetic Transport in Microfabricated Channel Structure," *Analyt. Chem.* **70**, pp. 4494–4505.

Fan, J., and Shen, C. (1999) "Statistical Simulation of Low-Speed Unidirectional Flows in Transition Regime," in *Rarefied Gas Dynamics,* Vol. 2, ed. R. Brun et al., Cepadues-Editions, Toulouse, France, pp. 245–252.

Frisch, U., Hasslacher, B., and Pomeau, Y. (1986) "Lattice-Gas Automata for the Navier–Stokes Equation," *Phys. Rev. Lett.* **56**(14), pp. 1505–1508.

Fukui, S., and Kaneko, R. (1988) "Analysis of Ultra Thin Gas Film Lubrication Based on Linearized Boltzmann Equation: First Report—Derivation of a Generalized Lubrication Equation Including Thermal Creep Flow," *J. Tribol.* **110**, pp. 253–262.

Fukui, S., and Kaneko, R. (1990) "A database for interpolation of Poiseuille flow rates for high Knudsen number lubrication problems," *J. Tribol.* **112**, pp. 78–83.

Gad-el-Hak, M. (1999) "The Fluid Mechanics of Microdevices—The Freeman Scholar Lecture," *J. Fluids Eng.* **121**, pp. 5–33.

Garcia, A.L., Bell, J.B., Crutchfield, W.Y., and Alder, B.J. (1999) "Adaptive Mesh and Algorithm Refinement Using Direct Simulation Monte Carlo," *J. Computational Phys.* **154**, pp. 134–155.

Granick, S. (1999) "Soft Matter in a Tight Spot," *Phys. Today* **52**(7), pp. 26–31.

Hadjiconstantinou, N.G. (1999) "Hybrid Atomistic-Continuum Formulations and the Moving Contact-Line Problem," *J. Computational Phys.* **154**(2), pp. 245–265.

Hadjiconstantinou, N.G. (2000) "Analysis of Discretization in the Direct Simulation Monte Carlo," *Phys. Fluids* **12**(10), pp. 2634–2638.

Hadjiconstantinou, N.G., and Patera, A.T. (1997) "Heterogeneous Atomistic-Continuum Representations for Dense Fluid Systems," *Int. J. Mod. Phys.* **C8**(4), pp. 967–976.

Harley, J.C., Huang, Y., Bau, H.H., and Zemel, J.N. (1995) "Gas Flow in Micro-Channels," *J. Fluid Mech.* **284**, pp. 257–274.

Hash, D.B., and Hassan, H.A. (1995) "A Hybrid DSMC/Navier–Stokes Solver," AIAA Paper No. 95-0410, American Institute of Aeronautics and Astronautics, Reston, VA.

He, X.Y., Shan, X.W., and Doolen, G.D. (1998) "Discrete Boltzmann Equation Model for Nonideal Gases," *Phys. Rev. E* **57**(1), pp. R13–R16.

Ho, C.M., and Tai, Y.C. (1998) "Micro-Electro-Mechanical Systems (MEMS) and Fluid Flows," *Ann. Rev. Fluid Mech.* **30**, pp. 579–612.

Huang, A.B., Giddens, D.P., and Bagnal, C.W. (1966) "Rarefied Gas Flow Between Parallel Plates Based on the Discrete Ordinate Method," *Phys. Fluids* **10**(3), pp. 498–502.

Inamuro, T., Yoshino, M., and Ogino, F. (1997) "Accuracy of Lattice Boltzmann Method for Small Knudsen Number with Finite Reynolds Number," *Phys. Fluids,* **9**(11), pp. 3535–3542.

Jamamoto, K., and Sanryo, T. (1991) "Flow of Rarefied Gas Past 2-D Body of an Arbitrary Shape at Small Mach Numbers," in *17th Rarefied Gas Dynamics Symp.,* ed. A.E. Beylich, VCH Verlag, Weinheim, p. 273.

Kannenberg, K.C., and Boyd, I.D. (1996) "Monte Carlo Computation of Rarefied Supersonic Flow into a Pitot Probe," *AIAA J.* **34**(1).

Kennard, E.H. (1938) *Kinetic Theory of Gases,* McGraw-Hill, New York.

Koplik, J., and Banavar, J.R. (1995a) "Continuum Deductions from Molecular Hydrodynamics," *Ann. Rev. Fluid Mech.* **27**, pp. 257–292.

Koplik, J., and Banavar, J.R. (1995b) "Corner Flow in the Sliding Plate Problem," *Phys. Fluids* **7**(12), pp. 3118–3125.

Koplik, J., Banavar, J.R., and Willemsen, J.F. (1989) "Molecular Dynamics of Fluid Flow at Solid Surfaces," *Phys. Fluids A* **1**, pp. 781–794.

Lavallée, P., Boon, J.P., and Noullez, A. (1991) "Boundaries in Lattice Gas Flows," *Physica D* **47**, pp. 233–240.

Liu, H.F. (2001) "2D and 3D Unstructured Grid Simulation and Coupling Techniques for Micro-Geometries and Rarefied Gas Flows," Ph.D. thesis, Brown University, Providence, RI.

Liu, J.Q., Tai, Y.C., Pong, K.C., and Ho, C.M. (1993) "Micromachined Channel/Pressure Sensor Systems for Micro Flow Studies," in *7th Int. Conf. on Solid-State Sensors and Actuators—Transducers '93,* pp. 995–998.

Liu, H.F., Gatsonis, N., Beskok, A., and Karniadakis, G.E. (1998) "Simulation Models for Rarefied Flow Past a Sphere in a Pipe," 21st International Rarefied Gas Dynamics Conference, France.

Lord, R.G. (1976) "Tangential Momentum Coefficients of Rare Gases on Polycrystalline Surfaces," in *Proc. 10th Int. Symp. on Rarefied Gas Dynamics,* pp. 531–538.

Lou, T., Dahlby, D.C., and Baganoff, D. (1998) "A Numerical Study Comparing Kinetic Flux Vector Splitting for the Navier–Stokes Equations with a Particle Method," *J. Computational Phys.* **145**, pp. 489–510.

Loyalka, S.K., and Hamoodi, S.A. (1990) "Poiseuille Flow of a Rarefied Gas in a Cylindrical Tube: Solution of Linearized Boltzmann Equation," *Phys. Fluids A* **2**(11), pp. 2061–2065.

Luo, L.S. (1998) "Unified Theory of Lattice Boltzmann Models for Non-Ideal Gases," *Phys. Rev. Lett.* **81**(8), pp. 1618–1621.

Lyklema, J., Rovillard, S., and Coninck, J. D. (1998) "Electrokinetics: The Properties of the Stagnant Layer Unraveled," *J. Surfaces Colloids* **14**(20), pp. 5659–5663.

Mavriplis, C., Ahn, J.C., and Goulard, R. (1997) "Heat Transfer and Flow Fields in Short Microchannels Using Direct Simulation Monte Carlo," *J. Thermophys. Heat Transfer* **11**(4), pp. 489–496.

Michael, A.A., Spaid, M.A.A., and Phelan, F.R. (1997) "Lattice Boltzmann Methods for Modeling Microscale Flow in Fibrous Porous Media," *Phys. Fluids* **9**(9), pp. 2468–2474.

Muntz, E.P. (1989) "Rarefied Gas Dynamics," *Ann. Rev. Fluid Mech.* **21**, pp. 387–417.

Myong, R.S. (1999) "Thermodynamically Consistent Hydrodynamic Computational Models for High-Knudsen-Number Gas Flows," *Phys. Fluids* **11**(9), pp. 2788–2802.

Nance, R.P., Hash, D., and Hassan, H.A. (1997) "Role of Boundary Conditions in Monte Carlo Simulation of MEMS Devices," *J. Thermophys. Heat Transfer* **11**, pp. 497–505.

Noble, D.R., Chen, S., Georgiadis, J.G., and Buckius, R.O. (1995) "A Consistent Hydrodynamic Boundary Condition for the Lattice Boltzmann Method," *Phys. Fluids* **7**, pp. 203–209.

Oh, C.K., Oran, E.S., and Sinkovits, R.S. (1997) "Computations of High-Speed, High Knudsen Number Microchannel flows," *J. Thermophys. Heat Transfer,* **11**(4), pp. 497–505.

Ohwada, T., Sone, Y., and Aoki, K. (1989) "Numerical Analysis of the Poiseuille and Thermal Transpiration Flows between Two Parallel Plates on the Basis of the Boltzmann Equation for Hard Sphere Molecules," *Phys. Fluids A* **1**(12), pp. 2042–2049.

Oran, E.S., Oh, C.K., and Cybyk, B.Z. (1998) "Direct Simulation Monte Carlo: Recent Advances and Applications," *Ann. Rev. Fluid Mech.* **30**, pp. 403–441.

Pan, L.S., Liu, G.R., and Lam, K.Y. (1999) "Determination of Slip Coefficient for Rarefied Gas Flows Using Direct Simulation Monte Carlo," *J. Micromech. Microeng.* **9**, pp. 89–96.

Pfahler, J., Harley, J., Bau, H., and Zemel, J. (1991) "Gas and Liquid Flow in Small Channels," *Proc. of ASME Winter Ann. Mtg. (DSC)* **32**, pp. 49–59.

Piekos, E.S., and Breuer, K.S. (1996) "Numerical Modeling of Micromechanical Devices Using the Direct Simulation Monte Carlo Method," *J. Fluids Eng.* **118**(3), pp. 464–469.

Pong, K.C., Ho, C.M., Liu, J., and Tai, Y.C. (1994) "Non-linear Pressure Distribution in Uniform Microchannels" *Appl. Microfabrication Fluid Mech.,* pp. 51–56.

Probstein, R.F. (1994) *Physiochemical Hydrodynamics: An Introduction,* Second edition, John Wiley & Sons, New York.

Qian, Y.H. (1993) "Simulating Thermohydrodynamics with Lattice BGK Models," *J. Sci. Comp.* **8**, pp. 231–241.

Qian, Y.H., D'Humiéres, D., and Lallemand, P. (1992) "Lattice BGK Models for Navier–Stokes Equation," *Europhys. Lett.* **17**, pp. 479–482.

Roveda, R., Goldstein, D.B., and Varghese, P.L. (1998) "Hybrid Euler/Particle Approach for Continuum/Rarefied Flows," *J. Spacecraft Rockets* **35**(3), pp. 258–265.

Saripov, F., and Seleznev V. (1998) "Data on Internal Rarefied Gas Flows," *J. Phys. Chem. Ref. Data* **27**(3), pp. 657–706.

Schaaf, S.A., and Chambre, P.L. (1961) *Flow of Rarefied Gases,* Princeton University Press, Princeton, NJ.

Seidl, M., and Steinheil, E. (1974) "Measurement of Momentum Accommodation Coefficients on Surfaces Characterized by Auger Spectroscopy, SIMS and LEED," in *9th Int. Symp. on Rarefied Gas Dynamics,* pp. E9.1–E9.2.

Shan, X., and Chen, H. (1994) "Simulation of Non-ideal Gases and Liquid Gas Phase Transitions by the Lattice Boltzmann Equation," *Phys. Rev. E* **49**, pp. 2941–2948.

Sone, Y. (1989) "Analytical and Numerical Studies of Rarefied Gas Flows on the Basis of the Boltzmann Equation for Hard Sphere Molecules," *Phys. Fluids A* **1**(12), pp. 2042–2049.

Sone, Y. (2000) "Flows Induced by Temperature Fields in a Rarefied Gas and Their Ghost Effect on the Behavior of a Gas in the Continuum Limit," *Ann. Rev. Fluid Mech.* **32**, pp. 779–811.

Spaid, M.A.A., and Phelan F.R. Jr. (1997) "Lattice Boltzmann Methods for Modeling Microscale Flow in Fibrous Porous Media," *Phys. Fluids* **9**, pp. 2468–2474.

Tehver, R., Toigo, F., Koplik, J., and Banavar, J.R. (1998) "Thermal Walls in Computer Simulations," *Phys. Rev. E* **57**(1), pp. R17–R20.

Tison, S.A. (1993) "Experimental Data and Theoretical Modeling of Gas Flows through Metal Capillary Leaks," *Vacuum* **44**, pp. 1171–1175.

Vangenabeek, O., and Rothman, D.H. (1996) "Macroscopic Manifestations of Microscopic Flows through Porous Media Phenomenology from Simulation," *Ann. Rev. Earth Planet Sci.* **24**, pp. 63–87.

Vargo, S.E., and Muntz, E.P. (1996) "A Simple Micromechanical Compressor and Vacuum Pump for Flow Control and Other Distributed Applications," 34th Aerospace Sciences Meeting and Exhibit, January 15–18, Reno, NV.

Vargo, S.E., Muntz, E.P., Shiflett, G.R., and Tang, W.C. (1998) "Knudsen Compressor as a Micro- and Macroscale Vacuum Pump without Moving Parts or Fluids," *J. Vacuum Sci. Technol. A Vacuum Surfaces Films* **17**(4), pp. 2308–2313.

Veijola, T., Kuisma, H., and Lahdenperä, J. (1995) "Equivalent Circuit Model of the Squeezed Gas Film in a Silicon Accelerometer," *Sensors and Actuators A* **48**, pp. 239–248.

Veijola, T., Kuisma, H., and Lahdenperä, J. (1998) "The Influence of Gas Surface on Gas Film Damping in a Silicon Accelerometer," *Sensors and Actuators A* **66**, pp. 83–92.

Vergeles, M., Keblinski, P., Koplik, J., and Banavar, J.R. (1996) "Stokes Drag and Lubrication Flows: A Molecular Dynamics Study," *Phys. Rev. E* **53**(5), pp. 4852–4864.

Vijayakumar, P., Sun, Q., and Boyd, L.D. (1999) "Vibrational-Translational Energy Exchange Models for the Direct Simulation Monte Carlo Method," *Phys. Fluids* **11**(8), pp. 2117–2126.

Vincenti, W.G., and Kruger, C.H. (1977) *Introduction to Physical Gas Dynamics*, Robert E. Krieger Publishing, Huntington, NY.

Yang, C., Li, D., and Masliyah, J.H. (1998) "Modeling Forced Liquid Convection in Rectangular Microchannels with Electrokinetic Effects," *Int. J. Heat Mass Transfer* **41**, pp. 4229–4249.

Zou, Q., and He, X. (1997) "On Pressure and Velocity Boundary Conditions for the Lattice Boltzmann BGK Model," *Phys. Fluids* **9**, pp. 1591–1598.

9
Lubrication in MEMS

Kenneth S. Breuer
Brown University

9.1 Introduction 9-1
    Objectives and Outline
9.2 Fundamental Scaling Issues... 9-2
    The Cube-Square Law • Applicability of the Continuum
    Hypothesis • Surface Roughness
9.3 Governing Equations for Lubrication 9-4
9.4 Couette-Flow Damping 9-5
    Limit of Molecular Flow
9.5 Squeeze-Film Damping... 9-7
    Derivation of Governing Equations • Effects of Vent
    Holes • Reduced-Order Models for Complex Geometries
9.6 Lubrication in Rotating Devices 9-10
9.7 Constraints on MEMS Bearing Geometries.................. 9-11
    Device Aspect Ratio • Minimum Etchable Clearance
9.8 Thrust Bearings 9-12
9.9 Journal Bearings.. 9-14
    Hydrodynamic Operation • Static Journal Bearing
    Behavior • Journal Bearing Stability • Advanced Journal
    Bearing Designs • Side Pressurization • Hydrostatic
    Operation
9.10 Fabrication Issues.. 9-22
    Cross-Wafer Uniformity • Deep Etch Uniformity • Material
    Properties
9.11 Tribology and Wear... 9-25
    Stiction • The Tribology of Silicon
9.12 Conclusions 9-25
    Acknowledgments 9-26

9.1 Introduction

As microengineering technology continues to advance, driven by increasingly complex and capable microfabrication and materials technologies, the need for greater sophistication in microelectromechanical systems (MEMS) design will increase. Fluid film lubrication has been a critical issue from the outset of MEMS development, particularly in the prediction and control of viscous damping in vibrating devices such as accelerometers and gyros. Much attention has been showered on the development of models for accurate prediction of viscous damping and of fabrication techniques for minimizing the damping (which destroys the quality, or Q-factor, of a resonant system). In addition to the development and optimization of these oscillatory devices, rotating devices—micromotors, microengines, etc.—have also captured the attention of MEMS researchers since the early days of MEMS development, and there have been several demonstrations of micromotors driven by electrostatic forces [Bart et al., 1988; Mehregany et al., 1992; Nagle and Lang, 1999; Sniegowski and Garcia, 1996]. For the most part these motors were very small, with rotors of the order of 100 μm in diameter, and, although they had high rotational speeds (hundreds of thousands of revolutions per minute), their tip speeds and rotational energy (which scales with tip speed) were quite small (tip speeds of 1 m/s are typical). In addition, the focus of these projects was the considerable challenge associated with the fabrication of a freely moving part and the integration of the drive electrodes. Lubrication and protection against wear were low down on the "to-do" list and the demonstrated engines relied on dry-rubbing bearings in which the rotor was held in place by a bushing, but there was no design or integration of a lubrication system. The low surface speeds of these engines meant that they could have operated for long times using this primitive bearing; however, failure was observed frequently due to rotor/bearing wear.

As MEMS devices become more sophisticated and have more stringent design and longevity requirements, the need for more accurate and extensive design tools for lubrication has increased. In addition, the energy density of MEMS is increasing. Devices for power generation, propulsion and so forth are actively under development. In such devices the temperatures and stresses are stretched to the material limits. Hence, the requirement for protection of moving surfaces becomes more than a casual interest—it is critical for the success of a "power MEMS" device (imagine, for comparison, an aircraft engine without lubrication—it would melt and fail within seconds of operation!).

9.1.1 Objectives and Outline

The objective of this chapter is to briefly summarize some the issues associated with lubrication in MEMS. Lubrication is a vast topic, and clearly there is no way in which these few pages can adequately introduce and treat an entire field as diverse and complex as this. For this reason, some hard choices have been made. Our focus is to review the essential features of lubrication theory and design practice and to highlight the difficulties that arise in the design of a lubrication system for MEMS devices. A key feature of MEMS is that the fabrication, material properties and mechanical and electrical design are all tightly interwoven and cannot be separated. For this reason, some attention is devoted to the important issue of how a successful lubrication system is influenced by manufacturing constraints and material properties.

Finally, one should always remember that MEMS is a rapidly developing and maturing manufacturing technology. The range of geometric options, available materials and dimensional control is continually expanding and improving. This chapter necessarily focuses on the current state of the art in MEMS fabrication, and for this reason it tends to favor silicon-based fabrication processes and the constraints that lithographic-based batch fabrication techniques place on lubrication system design and performance, but all of this may be rendered obsolete in 5 or 10 years!

Examples are drawn from several sources. In the sections on translational and squeeze-film damping examples are drawn from the extensive literature associated with accelerometer and gyro design. The rotating system lubrication discussion draws heavily on the MIT Microengine project [Epstein et al., 1997], which, to our knowledge, is the only MEMS device to date with rotating elements using a fluid-film lubrication system. This device is described in some detail further on in the chapter, and the analysis and examples of thrust bearings and journal bearings are drawn from that device.

9.2 Fundamental Scaling Issues

9.2.1 The Cube-Square Law

The most dominant effect that changes our intuitive appreciation of the behavior of microsystems is the so-called *cube-square law*, which simply reflects the fact that volumes scale with the cube of the typical length scale, while areas (including surface areas) scale with the square of the length scale. Thus, as a device shrinks, surface phenomena become relatively more important than volumetric phenomena. The most important consequence of this is the observation that the device mass (and inertia) becomes negligibly small at the micro- and nanoscales. For lubrication, the implication of this is that the volumetric loads that require support (such as the weight of a rotor) quickly become negligible. As an example, consider the ratio of the weight of a microfabricated rotor (a cylinder of density ρ, diameter D and length L) compared to the pressure p acting on its projected surface area. This can be expressed as a nondimensional load parameter:

$$\zeta = \frac{\rho \pi L D^2 / 4}{pLD} \propto \frac{D}{p}$$

from which it can clearly be seen that the load parameter decreases linearly as the device shrinks. For example, the load parameter (due to the rotor mass) for the MIT Microengine (which is a relatively large MEMS device, measuring 4 mm in diameter and 300 μm deep) is approximately 10^{-3}. The benefits of this scaling are that orientation of the freely suspended part becomes effectively irrelevant and that unloaded operation is easy to accomplish, should one desire to do so. In addition, because the gravity loading is negligible, the primary forces to be supported are pressure-induced loads and (in a rotating device) loads due to rotational imbalance. This last load is, in fact, *very* important and will be discussed in more detail in connection to rotating lubrication requirements. The chief disadvantage of the low natural loading is that unloaded operation is often undesirable (in hydrodynamic lubrication, where a minimum eccentricity is required for journal bearing stability); in practice, gravity loading is often used to advantage. Thus, a scheme for applying an artificial load must be developed. This is discussed in more detail later in the chapter.

9.2.2 Applicability of the Continuum Hypothesis

A common concern in microfluidic devices is the appropriateness of the continuum hypothesis as the device scale continues to fall. At some scale, the typical intermolecular distances will be comparable to the device scales, and the use of continuum fluid equations becomes suspect. For gases, the Knudsen number (Kn) is the ratio of the mean free path to the typical device scale. Numerous experiments (e.g., see Arkilic et al., 1993; 1997; Arkilic et al., 2001) have determined that noncontinuum effects become observable when Kn reaches approximately 0.1, and continuum equations become meaningless (the "transition flow regime") at Kn of approximately 0.3. For atmospheric temperature and pressure, the mean free path (of air) is approximately 70 nm. Thus, atmospheric devices with features smaller than approximately 0.2 μm will be subject to nonnegligible, noncontinuum effects. In many cases, such small dimensions are not present, thus the fluidic analysis can safely use the standard Navier–Stokes equations (as is the case for the microengine).

Nevertheless, in applications where viscous damping is to be avoided (for example, in high-Q resonating devices, such as accelerometers or gyroscopes) the operating gaps are typically quite small (perhaps a few microns) and the gaps serve double duty as both a physical standoff and a sense gap where capacitive sensing is accomplished. In such examples, one is "stuck" with the small dimension, so, in order to minimize viscous effects, the device is packaged at low pressures where noncontinuum effects are clearly evident. For small Knudsen numbers, the Navier–Stokes equations can be used with a single modification—the boundary condition is relaxed from the standard non-slip condition to that of a slip-flow condition, where the velocity at the wall is related to the Knudsen number and the gradient of velocity at the wall:

$$u_w = \lambda \frac{2-\sigma}{\sigma} \frac{\partial u}{\partial y}\bigg|_w$$

where σ is the tangential momentum accommodation coefficient (TMAC) which varies between 0 and 1. Experimental measurements [Arkilic et al., 2001] indicate that smooth native silicon has a TMAC of approximately 0.7 in contact with several commonly used gases.

Despite the fact that the slip-flow theory is only valid for low Kn, it is often used (incorrectly) with great success at much higher Knudsen numbers. Its adoption beyond its range of applicability stems primarily from the lack of any better approach (short of solving the Bolzmann equation or using direct simulation Monte Carlo [DSMC] computations [Beskok and Karniadakis, 1994; Cai et al., 2000]).

However, for many simple geometries, the "extended" slip-flow theory works much better than it should and provides quite adequate results. This is demonstrated in the sections on Couette and squeeze-film damping later in the chapter.

9.2.3 Surface Roughness

Another peculiar feature of MEMS devices is that the surface roughness of the material used can become a significant factor in the overall device geometry. MEMS surface finishes are quite varied, ranging from atomically smooth surfaces found on polished single-crystal silicon substrates to the rough surfaces left by various etching processes. The effects of these topologies can be important in several areas for microdevice performance. Probably the most important effect is the way in which the roughness can affect structural characteristics (crack initiation, yield strength, etc.), although this will not be explored in this chapter. Second, the surface finish can affect fluidic phenomena such as the energy and momentum accommodation coefficient, as described above, and consequently the momentum and heat transfer. Finally, the surface characteristics (although not only the roughness, but also the surface chemistry and affinity) can strongly affect its adhesive force. This again is not treated in any detail in this discussion, although it is mentioned briefly at the end of the chapter in connection with tribology issues in MEMS.

9.3 Governing Equations for Lubrication

With the proviso that the continuum hypothesis holds for micron-scale devices (perhaps with a modified boundary condition), the equations for microlubrication are identical to those used in conventional lubrication analysis and can be found in any standard lubrication textbook (for example, see Hamrock, 1984). We present the essential results here for convenience and completeness, but the reader is referred to more complete treatments for full derivations and a detailed discussion of the appropriate limitations.

Starting with the Navier–Stokes equations, we can make a number of simplifying assumptions appropriate for lubrication problems:

1. *Inertia*: The terms representing transfer of momentum due to inertia may be neglected. This arises due to the small dimensions that characterize lubrication geometries and MEMS, in particular. In very high-speed devices (such as the MIT Microengine) inertial terms may not be as small as one might like, and corrections for inertia may be applied. However, preliminary studies suggest that these corrections are small [Piekos, 2000].
2. *Curvature*: Lubrication geometries are typically characterized by a thin fluid film with a slowly varying film thickness. The critical dimension in such systems is the film thickness, which is assumed to be much smaller than any radius of curvature associated with the overall system. This assumption is particularly important in a rotating system where a circular journal bearing is used. However, assuming that the radius of the bearing, R, is much larger than the typical film thickness c (i.e., $c/R \ll 1$) greatly simplifies the governing equations.
3. *Isothermal*: Because volumes are small and surface areas are large, thermal contact between the fluid and the surrounding solid is very good in a MEMS device. In addition, common MEMS materials are good thermal conductors. For both these reasons, it is very safe to assume that the lubrication film is isothermal.

With these restrictions, the Navier–Stokes equations, the equation for the conservation of mass and the equation of state for a perfect gas may be combined to yield the Reynolds equation (named after Osborne Reynolds, 1886, who first derived it), written here for two-dimensional films:

$$0 = \frac{\partial}{\partial x}\left(-\frac{\rho h^3}{12\mu}\frac{\partial p}{\partial x}\right) + \frac{\partial}{\partial y}\left(-\frac{\rho h^3}{12\mu}\frac{\partial p}{\partial y}\right) + \frac{\partial}{\partial x}\left(\frac{\rho h(u_a + u_b)}{2}\right) + \frac{\partial}{\partial y}\left(\frac{\rho h(v_a + v_b)}{2}\right)$$
$$+ \rho(w_a - w_b) - \rho u_a \frac{\partial h}{\partial x} - \rho v_a \frac{\partial h}{\partial y} + h\frac{\partial \rho}{\partial t}$$

where x and y are the coordinates in the lubrication plane, and u_a etc. are the velocities of the upper and lower surfaces. An alternate (and more general) version may be derived [Burgdorfer, 1959] by non-dimensionalization with the film length and width l and b, the minimum clearance h_{min}, characteristic shearing velocity u_b and characteristic unsteady frequency ω. In addition, gas rarefaction can be incorporated for low Knudsen numbers by assuming a slip-flow wall boundary condition:

$$\frac{\partial}{\partial X}\left[(1+6K)PH^3\frac{\partial P}{\partial X}\right] + A^2\frac{\partial}{\partial Y}\left[(1+6K)PH^3\frac{\partial P}{\partial Y}\right] = \Lambda\frac{\partial(PH)}{\partial X} + \sigma\frac{\partial(PH)}{\partial T}$$

where

$$A = \frac{l}{b}; \quad \Lambda = \frac{6\mu u_b l^2}{p_a h_{min}^2}; \quad \sigma = \frac{12\mu \omega l^2}{p_a h_{min}^2}$$

Here, A is the film aspect ratio, Λ is the bearing number and σ is the squeeze number, representing unsteady effects.

Solution of the Reynolds equation is straightforward but not trivial. A chief difficulty comes from the fact that gas films are notoriously unstable if they operate in the wrong parameter space. In order to determine the stability or instability of the numerically generated solution, both the steady Reynolds equation and its unsteady counterpart need to be addressed and with some accuracy. These issues are discussed more by Piekos and Breuer (1998), as well as others.

9.4 Couette-Flow Damping

The viscous damping of a plate oscillating in parallel motion to a substrate has been a problem of tremendous importance in MEMS devices, particularly in the development of resonating structures such as accelerometers and gyros. The problem arises due to the fact that the proof mass, which may be hundreds of microns in the lateral dimension, is typically suspended above the substrate with a separation of only a few microns. A simple analysis of Couette-flow damping for rarefied flows is easy to demonstrate by choosing a model problem of a one-dimensional proof mass (i.e., ignoring the dimension perpendicular to the plate motion). This is shown schematically in Figure 9.1.

The Navier–Stokes equations for this simple geometry reduce simply to:

$$\frac{\partial u}{\partial t} = \mu\frac{\partial^2 u}{\partial y^2}$$

in which only viscous stresses due to the velocity gradient and the unsteady terms survive. This can be solved using separation of variables and employing a slip-flow boundary condition [Arkilic and Breuer, 1993],

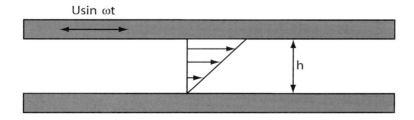

FIGURE 9.1 Schematic of Couette-flow damping geometry. The upper plate vibrates with a proscribed amplitude and frequency; however, for most MEMS geometries and frequencies, the unsteadiness can usually be neglected.

yielding the solution the drag force experienced by the moving plate:

$$D = \frac{4\pi U^2}{\beta}\left[\frac{\sinh\beta + \sin\beta}{(\cosh\beta - \cos\beta) + D_R}\right]$$

where

$$\beta = \sqrt{\frac{\omega h^2}{\mu}}$$

is a Stokes number, representing the balance between unsteady and viscous effects, and D_R is a correction due to slip-flow at the wall:

$$D_R = 2Kn\beta(\sinh\beta + \sin\beta) + 2Kn^2\beta^2(\cosh\beta + \cos\beta)$$

A typical MEMS geometry might have a plate separation of 1 μm and an operating frequency of 10 kHz. With these parameters, we see that the Stokes number, as defined, is very small (approximately 0.1), so that the flow may be considered quasi-steady to a high degree of approximation. In addition, the rarefaction effects indicated by D_R are also vanishingly small at atmospheric conditions.

9.4.1 Limit of Molecular Flow

Although the slip-flow solution is limited to low Knudsen numbers, the damping due to a gas at high degrees of rarefaction can be easily computed using a free molecular-flow approximation. In such cases,

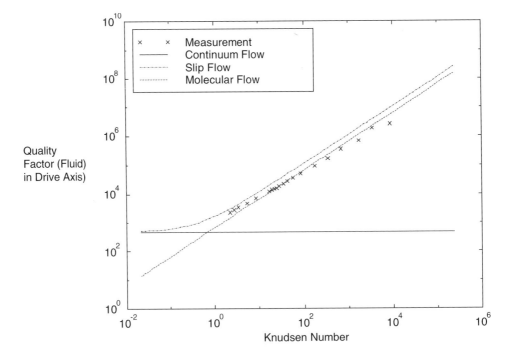

FIGURE 9.2 Theory and measurements of Couette damping in a tuning fork gyro [Kwok et al., 2001]. Note that in the high Knudsen number limit, the free molecular approximation predicts the damping more closely, but that the slip-flow model, though totally inappropriate at this high Kn level, is not too far from the experimental measurements.

the friction factor on a flat plate is given by Rohsenow and Choi (1961):

$$C_f = \sqrt{\frac{2}{\pi \gamma}} \frac{1}{M}$$

where γ is the ratio of specific heats and M is the Mach number. It is important to recognize that the damping (and Q) in this case is provided not only by the flow in the gap, but also by the flow above the vibrating plate. However, it is unlikely that the fluid damping provides the dominant source of damping at such extremely low pressures. More likely, damping derived from the structure (for example, flexing of the support tethers, nonelastic strain at material interfaces etc.) will take over as the dominant energy-loss mechanism. Kwok et al. (2001) compared the continuum, slip-flow and free molecular-flow models for Couette damping with data obtained by measuring the "ring down" of a tuning fork gyroscope fabricated by Draper Laboratories. The measurements and theory are shown in Figure 9.2, confirming the functional behavior of the damping as the pressure drops (Kn increases) and the unexpected accuracy of these rather simple models. Although the trends are well predicted, the absolute value of the Q-factor is still in error by a factor of two, suggesting that more detailed computations are still of interest.

9.5 Squeeze-Film Damping

Squeeze-film damping arises when the gap size changes in an oscillatory manner, squeezing the trapped fluid (Figure 9.3). Fluid (usually air) is trapped between the vibrating proof mass and the substrate, resulting in a squeeze film that can significantly reduce the quality factor of the resonator. In some cases, this damping is desirable, but, as with the case of Couette-flow damping, it is often parasitic, and the MEMS designer tries to minimize its effects and thus maximize the resonant Q-factor of the device. Common methods for alleviating squeeze-film effects are to fabricate breathing holes ("chimneys") throughout the proof mass which relieve the buildup of pressure and/or by packaging the device at low pressure. Both of these solutions have drawbacks. The introduction of vent holes reduces the vibrating mass, necessitating an even larger structure, while the low-pressure packaging adds considerable complexity to the overall device development and cost. An example of this is shown in Figure 9.4, which illustrates a high-performance tuning fork gyroscope fabricated by Draper Laboratories.

9.5.1 Derivation of Governing Equations

An analysis of squeeze-film damping is presented in the following section. The Reynolds equations may be used as the starting point. However, a particularly elegant and complete solution was published by Blech (1983) for the case of the continuum flow and extended by Kwok et al. (2001) to the case of slip-flow and flows films with vent holes. This analysis is summarized here.

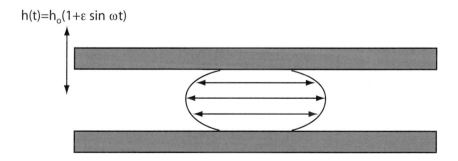

FIGURE 9.3 Schematic of squeeze-film damping between parallel plates. As with Couette damping, for most practical embodiments of MEMS, the damping is quasi-steady.

FIGURE 9.4 Photograph of a typical microfabricated vibrating proof mass used in a high-performance tuning fork gyroscope. (Courtesy of M. Weinberg, Draper Laboratories, Cambridge, MA.)

The Navier–Stokes equations are written for the case of a parallel plate, vibrating sinusoidally in a proscribed manner in the vertical direction. If we assume that the motion and subsequent pressure and velocity perturbations are small, a perturbation analysis yields the classical squeeze-film equation derived by Blech, with an additional term due to the rarefaction:

$$\frac{\partial}{\partial X}\left(\Psi H^3 \frac{\partial \Psi}{\partial X}\right) + \frac{\partial}{\partial X}\left(6KH^2 \frac{\partial \Psi}{\partial X}\right) = \sigma \frac{\partial(\Psi H)}{\partial T}$$

where the variables have been nondimensionalized, so that H represents the film gap, normalized by the nominal film gap: $H = h(x, y, t)/h_0$; Ψ is the pressure, normalized by ambient pressure: $\Psi = P(x, y, t)/P_0$; X and Y are the coordinates, normalized by the characteristic plate geometry: $X = x/L_x$, $Y = y/L_y$; T is time, normalized by the vibration frequency: $T = \omega t$; and, finally, the squeeze number σ is defined as before:

$$\sigma = \frac{12\mu\omega L_x^2}{P_0 h_0^2}$$

Assuming small-amplitude, harmonic forcing of the gap, $H = 1 + \varepsilon \sin T$, and by assuming a harmonic response of the pressure we can derive a pair of coupled equations describing the in-phase (Ψ_0) and out-of-phase (Ψ_1) pressure distributions in the gap (representing stiffness and damping coefficients, respectively):

$$\frac{\partial^2 \Psi_0}{\partial x^2} + \frac{\sigma}{1+6K}\Psi_1 + \frac{\sigma}{1+6K} = 0$$

$$\frac{\partial^2 \Psi_1}{\partial x^2} + \frac{\sigma}{1+6K}\Psi_0 = 0$$

Note that these equations represent the standard conditions with the adoption of a modified squeeze number, $\sigma_m \equiv \sigma/(1 + 6K)$. The solutions are easy to achieve, either by manual substitution of Fourier sine and cosine series or by direct numerical solution. The results, for rectangular plates (with no vent holes) are shown in Figure 9.5.

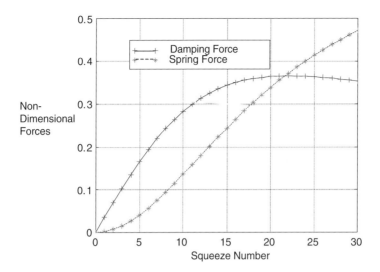

FIGURE 9.5 Solutions to the squeeze-film equation for a rectangular plate. The stiffness and damping coefficients are presented as functions of the modified squeeze number, which includes a correction due to first-order rarefaction effects. (Adapted from Blech, 1983; Kwok et al., 2001.)

9.5.2 Effects of Vent Holes

The equations as derived above can be made more useful by extending them to account for the presence of vent holes in the vibrating proof mass. In such cases, the boundary condition at each vent hole is no longer atmospheric pressure ($\Psi_0 = \Psi_1 = 0$), but rather an elevated pressure proscribed by the pressure drop through the "chimney" that vents the squeeze film to the ambient. Kwok et al. (2001) demonstrate that this can be incorporated into the above model (in the limit of low squeeze number) by a modified boundary condition for the squeeze film equations for Ψ_0:

$$\Psi = \left[\frac{32 \frac{t}{L_x} \left(\frac{h_0}{L_x}\right)^3 \left(1 - \left(\frac{L_h}{L_x}\right)^2\right)}{12 \left(\frac{L_h}{L_x}\right)^4} \right] \left[\frac{1}{1 + 8\frac{\lambda}{L_h}} \right] \sigma$$

This boundary condition has three components: The first is a geometric component dependent on the plate thickness t, length L_x, hole size L_h and nominal gap size h_0. The second component is a rarefaction component (here based on the hole size), and the third is a time-dependent component—the squeeze number σ. Note that as the thickness of the plate decreases and the chimney pressure drop falls, the boundary condition approaches zero. Similarly, as the open area fraction of the plate increases (more venting), the boundary condition approaches that of the ambient (as one would expect). This boundary condition can be applied at the chimney locations and can thus accurately simulate the squeeze-film damping of perforated micromachined plates.

9.5.3 Reduced-Order Models for Complex Geometries

Unfortunately, most devices of practical interest have geometries that are usually too complex to enable full numerical simulation of the kind described above. Reduced-order models are of great value in such cases. Many such models have been developed, including those based on acousto-electric models (see, for example, Veijola et al. [1995]). In the case of squeeze-film damping in the limit of low squeeze numbers, such models reduce to solution of a resistor network that models the pressure drops associated with each segment of the squeeze film. This is effectively a finite-element approach to the problem; however, instead of modeling a large number of elements, as is generally the case in a "brute-force" numerical solution, a relatively small number of discrete elements can be used if higher order solutions can be employed to connect each element together. Kwok et al. (2001) demonstrate this approach and model the damping associated with an inclined plate with vent holes—in this case, the pressure at the center of each vent hole. More complex numerical solution techniques based on boundary integral techniques have also been presented [Aluru and White, 1998; Ye et al., 1999], providing a good balance between solution fidelity and required computing power.

9.6 Lubrication in Rotating Devices

Rotating MEMS devices bring a new level of complexity to both MEMS fabrication as well as the lubrication considerations. As discussed in the introduction, many rotors and motors have been demonstrated with dry-rubbing bearings, and the success of these devices is due to the low surface speeds of the rotors. However, as the surface speed increases (in order to get high power densities), the dry-rubbing bearings are no longer an option and true lubrication systems need to be considered. An example of such "power-MEMS" development is provided by a project initiated at MIT in 1995 to demonstrate a fully functional microfabricated gas turbine engine [Epstein et al., 1997]. The baseline engine, illustrated in Figure 9.6, consists of a centrifugal compressor, fuel injectors (hydrogen is the initial fuel, although hydrocarbons are planned for later configurations), a combustor operating at 1600 K and a radial inflow turbine. The device is constructed from single-crystal silicon and is fabricated by extensive and complex fabrication of multiple silicon wafers fusion-bonded in a stack to form the complete device. An electrostatic induction generator may also be mounted on a shroud above the compressor to produce electric power instead of thrust [Nagle and Lang, 1999]. The baseline MIT Microengine has at its core a "stepped" rotor consisting of a compressor with an 8-mm diameter and a journal bearing and turbine with a diameter of 6 mm. The rotor spins at 1.2 million rpm.

Key to the successful realization of such a device is clearly the ability to spin a silicon rotor at high speed in a controlled and sustained manner, and the key to spinning a rotor at such high speeds is the demonstration of efficient lubrication between the rotating and stationary structures. The lubrication system needs to be simple enough to be fabricated and yet have sufficient performance and robustness to be of practical use both in the development program and in future devices. To develop this technology, a microbearing rig was fabricated and is illustrated in Figure 9.7. Here, the core of the rotating machinery has been implemented, but without the substantial complications of the thermal environment that the full engine brings. The rig consists of a radial inflow turbine mounted on a rotor and embedded inside two thrust bearings, which provide axial support. A journal bearing located around the disk periphery provides radial support for the disk as it rotates.

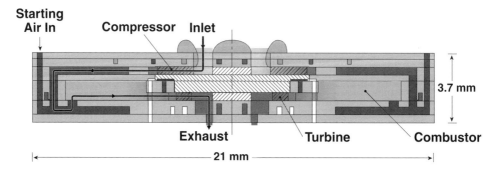

FIGURE 9.6 Schematic of the MIT Microengine, showing the air path through the compressor, combustor and turbine. Forward and aft thrust bearings located on the centerline hold the rotor in axial equilibrium, while a journal bearing around the rotor periphery holds the rotor in radial equilibrium [Protz, 2000].

Lubrication in MEMS   9-11

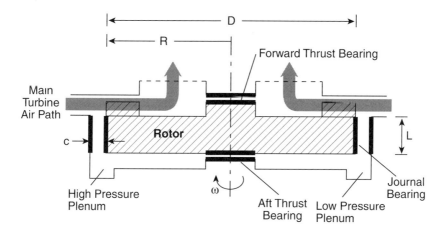

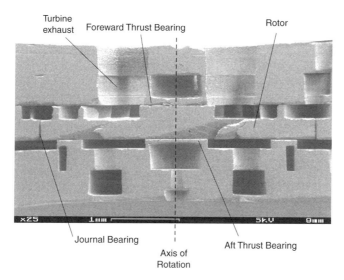

FIGURE 9.7 Illustrating schematic and corresponding SEM of a typical microfabricated rotor, supported by axial thrust bearings and a radial journal bearing.

9.7 Constraints on MEMS Bearing Geometries

9.7.1 Device Aspect Ratio

Perhaps the most restrictive aspect of microbearing design is the fact that MEMS devices are limited to rather shallow etches, resulting in devices with a low aspect ratio. Even with the advent of deep reactive ion etchers (DRIE), in which the ion etching cycle is interleaved with a polymer passivation step, the maximum practical etch depth that can be achieved while maintaining dimensional control is about 500 μm (even this has an etch time of about 9 hours which makes its adoption a very costly decision). This is in comparison with typical rotor dimensions of a few millimeters. The result is that microbearings are characterized by very low aspect ratios (length/depth, or L/D). In the case of the MIT microturbine test rig, the journal bearing is nominally 300 μm deep, while the rotor is 4 mm in diameter, yielding an aspect ratio of 0.075. To put this in perspective, commonly available design charts [Wilcox, 1972] present data for values of L/D as low as 0.5 or perhaps 0.25. Prior to this work, there were no data for lower L/D. The implications of the low-aspect-ratio bearings are that the task becomes one of supporting a *disk* rather than a *shaft*.

In fact, the low-aspect-ratio bearings do not have terrible performance by any standard. The key features of the low L/D bearings are

- The load capacity is reduced compared to "conventional" designs. This is because the fluid leaking out of the ends relieves any tendency for the bearing to build up a pressure distribution. Thus, for a given geometry and speed, a short bearing supports a lower load (per unit length) than its longer counterpart.
- The bearing acts as an incompressible bearing over a wide range of operation. The reason for this is as above—pressure rises, which might lead to gas compressibility, are minimized by the flow leaking out of the short bearing. The implication of this is that incompressible behavior (without, of course, the usual fluid cavitation commonly assumed in incompressible liquid bearings) can be observed to relatively high speeds and eccentricities.

9.7.2 Minimum Etchable Clearance

It is reasonable to question why one could not fabricate a 300-µm-long "shaft" but with a much smaller diameter and thus greatly enhance the L/D. For example, a shaft with a diameter of 300 µm would result in a reasonable value for L/D; however, this leads us to the second major constraint on bearing design with current microfabrication technologies—that of minimum etchable clearance.

In the current microengine manufacturing process, the bearing/rotor combination is defined by a single deep and narrow etch, currently 300 µm deep and about 12 µm in width. No foreseeable advance in fabrication technology will make it possible to significantly reduce the minimum etchable clearance, which has considerable implications for bearing design. In particular, if one were to fabricate a bearing with a diameter of 300 µm (in an attempt to improve the L/D ratio), it would result in a bearing with a ratio of clearance to radius (c/R) of 12/300, or 0.04. For a fluid bearing, this is *huge* (two orders of magnitude above conventional bearings) and has several detrimental implications. The most severe of these is the impact on the dynamic stability of the bearing. The nondimensional mass of the rotor depends on c/R raised to the *fifth* power [Piekos, 2000]. Thus, bringing the bearing into the center of the disk and raising the c/R by a factor of 10 result in a mass parameter increasing by a factor of 10^5. This increase in effective mass has severe implications for the stability of the bearing that are difficult to accommodate.

All of these factors, and others not enumerated here (such as the issue of where to locate the thrust bearings), make the implementation of an inner-radius bearing less attractive; thus, the constraint of small L/D is, in our opinion, unassailable as long as one requires that the microdevice be fabricated *in situ*. If one were to imagine a change in the fabrication process such that the rotor and bearing could be fabricated separately and subsequently assembled (*reliably*), this situation would be quite different. In such an event, the bearing gap is not constrained by the minimum etch dimension of the fabrication process, and almost any "conventional" bearing geometry could be considered and would probably be superior in performance to the bearings discussed here. Such fabrication could be considered for a "one-off" device but does not appear feasible for mass production, which relies on the monolithic fabrication of the parts. Finally, the risk of contamination during assembly—a common concern for all precision-machined MEMS—effectively rules out piece-by-piece manufacture and assembly and again constrains the bearing geometry as described.

9.8 Thrust Bearings

Thrust bearings are required to support any axial loads generated by rotating devices, such as turbines, engines or motors. Current fabrication techniques require that the axis of rotation in MEMS devices lie normal to the lithographic plane. This immediately lends a significant advantage to the design and operation of thrust bearings, as the area available for the thrust bearing is relatively large (defined by lithography), while the weight of the rotating elements will be typically small (due to the cube-square law and the low thicknesses of microfabricated parts). For these reasons, thrust bearings are one (rare) area of microlubrication where solutions abound and problems are dealt with relatively easily.

Two options exist: (1) hydrostatic (externally pressured) thrust bearings, in which the fluid is fed from a high-pressure source to a lubrication film, and (2) hydrodynamic, where the supporting pressure is generated by a viscous pump fabricated on the surface of the thrust bearing itself (see Figure 9.9). Hydrostatic bearings are easy to operate and relatively easy to fabricate. These have been successfully demonstrated in the MIT Microengine program [Fréchette et al., 2000]; the thrust bearing is shown in Figure 9.8, which shows an SEM of the fabricated device cut though the middle to reveal the plenum, restrictor holes and bearing lubrication gap, which is approximately 1 μm wide. Key to the successful operation of hydrostatic

FIGURE 9.8 Close-up cutaway view of microthrust bearing showing the pressure plenum (on top), feed holes and bearing gap (faintly visible). (SEM courtesy of C.-C. Lin.)

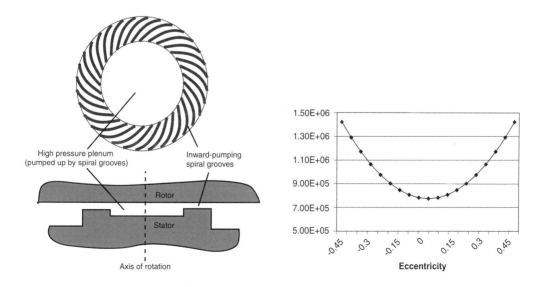

FIGURE 9.9 Schematic of hydrodynamic thrust bearings and predicted performance (stiffness, in N/m, vs. axial eccentricity) for a typical spiral-groove thrust bearing for use in a high-speed MEMS rotor.

thrust bearings is the accurate manufacture of the restrictor holes, maintenance of sharp edges at the restrictor exit and carefully controlling the dimension of the lubrication film. In an initial fabrication run, the restrictor holes were fabricated 2 μm larger than specified. While the bearing operated, its performance was well below its design peak, due to the off-design restrictor size. Current specifications of the fabrication protocols control the restrictor size carefully, ensuring close to optimal operation.

Hydrodynamic, or spiral groove, bearings (SGB; illustrated in Figure 9.9) were first analyzed in some detail some 40 years ago [Muijderman, 1966] but have not received much attention due to their low load capacity (compared to hydrostatic thrust bearings) and complex manufacturing requirements.

Spiral groove bearings operate by using the rotor motion against a series of spiral grooves (etched in the bearing) to viscously pump fluid into the lubrication gap and thus create a high-pressure cushion on which the rotor can ride. The devices typically have relatively low load capacity, which has limited their use in macroscopic applications. However, the load capacity becomes more than adequate at microscales due to favorable cube-square scaling. Thus, they gain considerable advantage when compared with conventional hydrostatic thrust bearings as the scale (and Reynolds number) decreases. In addition, the fabrication of the multitude of shallow spiral features (which is an expensive task for a traditional SGB) is ideally suited for lithographic fabrication technologies such as MEMS. Figure 9.9 illustrates the bearing stiffness for a single-point design for the MIT microrotor rotating at design speed (2.4 million rpm) and supported by matched forward and aft spiral groove bearings. The stiffness at full speed is quite impressive (superior to comparable hydrostatic bearings), but the SGB do suffer at lower speeds as the bearing stiffness is roughly proportional to rotational speed. However, for this design, the lift-off speed (the speed at which the film can support the weight of the rotor and the pressure distribution associated with the turbine flow) is only a few thousand rpm, so the dry rubbing endured during startup will be minimal. They also have the strong advantage that the two matched spiral groove bearings (forward and aft) naturally balance each other (no supply pressures to maintain or adjust). Also, and most importantly, removal of the thrust bearing plena and restrictor holes considerably simplifies the overall device fabrication and can allow for the use of two fewer wafers in the wafer-bonded stack—a considerable advantage from the perspective of manufacturing process cost and yield.

9.9 Journal Bearings

Journal bearings, which are used to support radial loads in a rotating machine, have somewhat unusual requirements in MEMS deriving from the extremely shallow structures that are currently fabricated. Thus, rotating devices tend to be disk shaped, and their corresponding journal bearings are characterized by very low aspect ratios, defined as the ratio of the bearing height to its radius. In addition, the minimum etchable gap allowed by current (and forseeable) fabrication techniques results in a paradoxically large bearing clearance—a 10-μm gap over a 2-mm-radius rotor, or a c/R of 1/200, which is large in comparison to conventional journal bearings, which typically have c/R ratios lower by a factor of 10. Journal bearings can operate in two distinct modes: hydrodynamic and hydrostatic, although typically any operating condition will contain aspects of both modes. Each of these modes is discussed in the following sections.

9.9.1 Hydrodynamic Operation

Hydrodynamic operation occurs when the rotor is forced to operate at an eccentric position in the bearing housing. As a consequence of this, a pressure distribution develops in the gap to balance the viscous stresses that arise due to the rotor motion. This pressure distribution is used to support the rotor both statically (against the applied force) and dynamically (to suppress random excursions of the rotor due to vibration etc.). Hydrodynamic operation has the advantage that it requires no external supply of lubrication fluid; however, it has two distinct drawbacks: First, it requires a means to load the rotor to an eccentric position, and, second, insufficient eccentricity results in instability (the so-called "fractional speed whirl") and likely failure. Both of these issues are particularly difficult to resolve in the case of MEMS bearings.

9.9.2 Static Journal Bearing Behavior

The static behavior of a MEMS journal bearing is shown in Figure 9.10, which presents the load capacity ζ and the accompanying attitude angle (the angle between the applied load and the eccentricity vector) as functions of the bearing number and the operating eccentricity. The geometry considered here is for a low-aspect-ratio bearing ($L/D = 0.075$) typical of a deep reactive ion-etched rotor such as the MIT Microengine. The bearing number is defined as:

$$\Lambda = \frac{6\mu\omega}{p}\left[\frac{R}{c}\right]^2$$

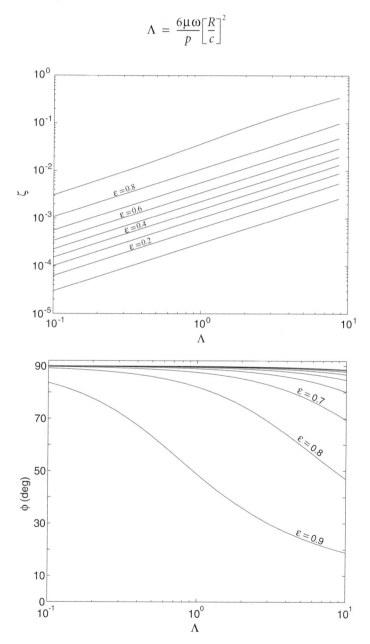

FIGURE 9.10 Static performance (eccentricity and attitude angle vs. bearing number) for a journal bearing with $L/D = 0.075$. Notice that the load lines are almost constant (linear), indicating the absence of compressibility effects. This is also indicated by the attitude angle, which remains close to 90° except at very high eccentricities. (From Piekos, E.S., and Breuer, K.S. (1999) *J. Tribol.* **121**, pp. 604–609, ASME, New York. With permission.)

where μ is the fluid viscosity, ω the rotation rate, p the ambient pressure and R/c the ratio of the radius to clearance. For a given bearing geometry, Λ can be interpreted as operating speed.

Several aspects of these results should be noted. First, the load capacity is quite small when compared with bearings of higher L/D. As mentioned before, this is due to the fact that, for very short bearings, the applied load simply squeezes the fluid out of the bearing ends, which has difficulty developing significant restoring force. The same mechanism is responsible for the fact that the load lines are straight, indicating that little compressibility of the fluid is taking place (which usually results in a "saturation" of the load parameter at higher values of the bearing number). Again, this is because any tendency to compress the gas is alleviated by the fluid venting at the bearing edge. This point is driven home by the behavior of the attitude angle, which maintains a high angle (close to $\pi/2$) over a wide range of bearing numbers and eccentricities. This value of the attitude angle corresponds to the analytic behavior of a Full–Sommerfeld incompressible short bearing [Orr, 1999], which clearly is a good approximation for such short bearings at low to moderate eccentricities, so long as the eccentricity remains below approximately 0.6, at which point compressibility finally becomes important. This incompressible behavior is much more extensive than "conventional" gas bearings of higher aspect ratio and has profound ramifications, particularly with respect to the dynamic properties of the system.

9.9.3 Journal Bearing Stability

The stability of a hydrodynamic journal bearing has long been recognized as troublesome and is foreshadowed by the static behavior shown in Figure 9.10. The high attitude angle suggests that the bearing spring stiffness is dominated by cross stiffness as opposed to direct stiffness. Thus, any perturbation to the rotor will result in its motion perpendicular to the applied force. If this reaction is not damped, the rotor will enter a whirling motion. This is precisely what is observed, and gas bearings are notorious for their susceptibility to fractional-speed whirl. The instability is suppressed by the generation of more damping and/or increased direct stiffness, both of which are obtained by increasing the loading and the static eccentricity of the rotor.

Figure 9.11 shows a somewhat unusual presentation of the stability boundaries for a low-aspect-ratio MEMS journal bearing. The vertical axis shows the nondimensional mass of the rotor, $\overline{M}$, which is

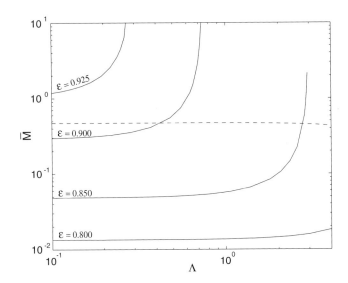

FIGURE 9.11 Stability boundaries for a typical microbearing, plotted vs. bearing number (speed for a fixed geometry). The dotted line represents an operating line for a microbearing that has almost constant $\overline{M}$ (varying only due to centrifugal expansion of the rotor at high speeds) [Piekos, 2000].

defined as:

$$\overline{M} = \frac{mp}{72L\mu^2}\left[\frac{c}{R}\right]^5$$

This is the "mass" that appears in the nondimensionalized equations of motion for the rotor, and it is fixed for a given geometry (close inspection of Figure 9.11 indicates that $\overline{M}$ does change very slightly with speed due to elastic expansion of the rotor due to centrifugal forces, the variation in the ambient pressure at different speeds and temperature effects on viscosity). The horizontal axis of Figure 9.11 shows the bearing number, which can be interpreted as speed, for a fixed-bearing geometry. The contours on the graph represent the stability boundary at fixed eccentricity. Stable operation lies above each line. Thus, for a fixed $\overline{M}$ at low bearing number (i.e., speed), a minimum eccentricity must be obtained to ensure stability. As the speed increases, this minimum eccentricity remains almost constant (the lines are horizontal) until a particular speed, at which point the lines break upward and the minimum eccentricity required for stability starts to drop (note that as Λ increases the load required to maintain a fixed eccentricity increases linearly due to the stiffening of the hydrodynamic bearing, as indicated by Figure 9.10). The key feature of this chart is that the minimum eccentricities are *very* high and suggest that stable operation requires running very close to the wall, which is troublesome. The high eccentricities are driven by high values of the mass parameter $\overline{M}$, which is due to the relatively high value of the clearance-to-radius ratio (c/R) and the short length L. The low aspect ratio (L/D) also contributes to high minimum eccentricities. At high speeds, the problem becomes less severe, as the high speed has allowed the bearing to generate sufficient direct stiffness. However, even at these points, the eccentricity is very high and may not be manageable in practical operation.

Orr (1999) has demonstrated, on a scaled-up experimental rig that matches the microengine geometry, that stable high-eccentricity operation is possible for extended periods of time. His experiments achieved 46,000 rpm, which when translated to the equivalent speed at the microscale corresponds to approximately 1.6 million rpm. However, in order to accomplish this high eccentricity operation, he noted that the rotor system must (1) have very good axial thrust bearings to control axial and tipping modes of the rotor system, and (2) be well balanced—a rotor with imbalance of more than a few percent could not be started from rest. Piekos (2000) also explored the tolerance of the microbearing system to imbalance and found that it was surprisingly robust to imbalance of several percent. However, his computations were achieved assuming that the rotor was at full speed and then carefully subjected to imbalance. In practice, the imbalance will exist at rest and so it might well be the case that the rotor is stable at full speed but unable to accelerate to that point. This "operating line" issue is discussed in some more detail by Savoulides et al. (2001), who explored several options for accelerating microbearings from rest under both hydrodynamic and hydrostatic modes of operation.

A convenient summary of the trade-offs for design of a hydrodynamic MEMS bearing is shown in Figure 9.12. This presents the variation of the low-speed minimum eccentricity asymptote (or worst-case eccentricity, as demonstrated in Figure 9.11) as functions of the mass parameter $\overline{M}$ and other geometric factors (L/D, clearance c etc.). Notice that the worst-case eccentricity improves as the L/D increases and the $\overline{M}$ decreases. However, it is interesting to note that the physical running distance from the wall is actually increased (slightly) by running at a higher eccentricity, but with a larger bearing gap. In all cases, however, the stable eccentricity is alarmingly high and other alternatives must be sought for simpler stable operation.

9.9.4 Advanced Journal Bearing Designs

One prospect for further improvement in the journal bearing performance is the incorporation of "wave bearings" [Dimofte, 1995], as illustrated in Figure 9.13. These bearing geometries have been demonstrated to suppress the subsynchronous whirl due to the excitation of multisynchronous pressure perturbations imposed by the bearing geometry. As an added attraction, the geometric complexity of the wave bearing is no problem for lithographic manufacturing processes such as are used for MEMS, thus alleviating many of the reservations and cost that might inhibit their adoption. However, because the MEMS constraint is the minimum gap dimension, the wave bearing in a MEMS machine can only be implemented by

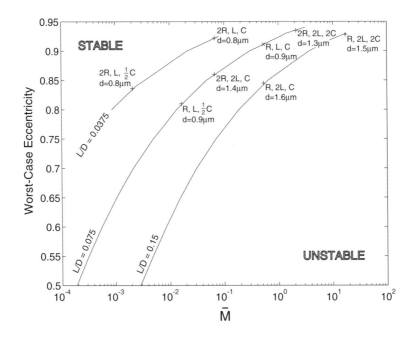

FIGURE 9.12 Trade-off chart for microbearing design. For a given length-to-diameter ratio (L/D) and a given $\overline{M}$, the worst-case (i.e., low-speed) eccentricity is shown for a variety of geometric perturbations. In general, lower eccentricities are preferred [Piekos, 2000].

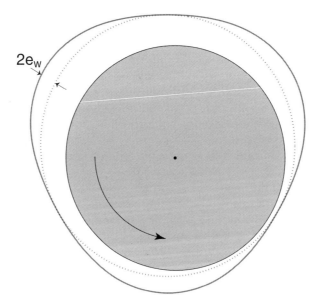

FIGURE 9.13 Geometry of a wave bearing, with the clearance greatly exaggerated for clarity [Piekos, 2000].

selectively enlarging the bearing gap. Piekos (2000) has analyzed the performance of the wave bearing for the microengine geometry and has found (Figure 9.14) that, while the load capacity is diminished, the stability is enhanced (as expected) and the load required to maintain stable operation (i.e., to achieve the minimum stable eccentricity) is reduced considerably with the introduction of a wave geometry. In microbearings, the load capacity is usually quite sufficient (the cube-square law to our rescue!), thus the wave bearing looks very attractive as a stabilizing mechanism.

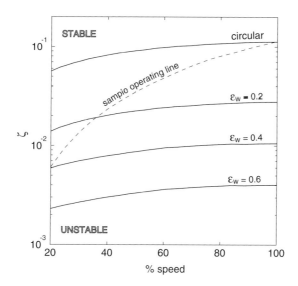

FIGURE 9.14 Effect of wave-bearing amplitude on journal-bearing stability as a function of rotor speed. The dotted line shows a typical operating line for a microengine [Piekos, 2000].

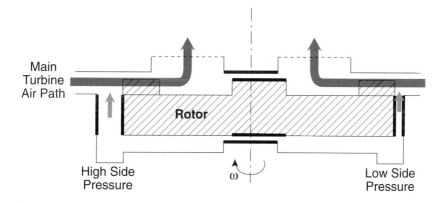

FIGURE 9.15 Schematic of the pressure-loading scheme used in the microengine to provide a side load to the rotor during hydrodynamic operation. The side load is developed by applying a differential pressure to the two plena located on the aft side of the rotor.

Having said that, rotor imbalance, which is increasingly becoming a first-order issue, can only be contained with excess load capacity, so this trade-off is by no means clear. The adoption and testing of wave-bearing geometries are scheduled to be explored in the near future as part of our development program.

9.9.5 Side Pressurization

Due to the small mass of the rotor in a MEMS device, any eccentricity required to enable stable hydrodynamic operation *must* be applied using some other means. Typically, this requires the use of a pressure distribution introduced around the circumference of the bearing, which serves to load the bearing preferentially to one side. Such a scheme is illustrated in Figure 9.15 for the MIT Microengine. Here, the aft side of the rotor is divided into two plena, isolated by seals. Each plenum can be separately pressurized. The pressure in each plenum forces an axial flow through the journal bearing to the forward side (which is assumed to be at a uniform pressure) and establishes two differing pressure distributions on the high- and low-pressure sides of the rotor. Of course, nothing comes without its consequences, and the axial flow through the bearing has consequences of its own—the generation of a hydrostatic stiffness mechanism and an associated hydrostatic critical frequency, discussed in the following section.

9.9.6 Hydrostatic Operation

Although hydrodynamically lubricated bearings with low aspect ratio are predicted to operate successfully and have now been demonstrated on a scaled-up level [Orr, 1999], a number of issues make them undesirable in a practical MEMS rotor system. The primary difficulty is that in order to satisfy the requirements of subsynchronous stability, the rotor needs to operate at very high eccentricity (made unavoidable due to the low aspect ratio of the journal). For a MEMS device, this means operating 1 to 2 μm from the wall. This is hard to control, particularly with the limited available instrumentation that plagues MEMS devices. An alternative mode of operation is to use a hydrostatic lubrication system. In this mode, fluid is forced from a high-pressure source through a series of restrictors, each of which imparts a fixed resistance. The fluid then flows through the lubrication passage (the bearing gap). If the rotor moves to one side, the restrictor/lubrication film acts as a pressure divider such that the pressure in the lubrication film rises, forcing the rotor back toward the center of the bearing. The advantages of using hydrostatic lubrication in MEMS devices include:

- The rotor tends to operate near the center of the housing, thus small clearances are avoided. This is "safer," being far more tolerant of any motion induced by rotor imbalance, and results in lower viscous resistance.
- Because the hydrostatic system is a zero-eccentricity-based system, no position information about the rotor is needed. This greatly simplifies instrumentation requirements.

Significant disadvantages to a hydrostatic system include:

- Pressurized air must be supplied to the bearing. This requires supply channels, which complicate the fabrication process. This requirement also comes with a system cost—the high-pressure air must come from somewhere (in a turbomachinery application, bleed air from the compressor could be used).
- Because the bearing gaps are relatively large (due to minimum etchable dimensions discussed above), the mass flow through hydrostatic systems can be substantial and might be impractical in anything but demonstration experiments.
- Fabrication constraints make the manufacture of effective flow restrictors very difficult. Flow restrictors must have very well-controlled dimensions and sharp edges, as well as other specific geometric features. Only the simplest restrictors can be implemented without undue cost and effort, severely limiting the hydrostatic design.

For journal bearings with low aspect ratio, a novel method for achieving hydrostatic lubrication was demonstrated by Orr et al. (2000). The mechanism relies on the fact that small pressure differences will inevitably exist between the forward and aft sides of the rotor and that, for a short bearing of the kind seen in MEMS devices, the flow resistance to that pressure differences is small enough that an axial flow will ensue.

As the flow enters the bearing channel, boundary layers develop along the wall, eventually merging to form the fully developed lubrication film. This boundary layer development (Figure 9.16) acts as an inherent restrictor so that if the rotor moves off the centerline and disturbs the axisymmetric symmetry of the flow, a restoring force is generated. This source of hydrostatic stiffness acts to support the bearing at zero eccentricity and is effective even when the rotor is not moving. The stiffness is also enhanced by the conventional inherent restriction of the flow entering the lubrication channel. The stiffness coupled with the rotor mass defines a natural frequency which was measured by Orr (1999) (indeed the presence of this frequency led to the discovery of the axial-through-flow mechanism). Simple theory [Orr et al., 2000]

Lubrication in MEMS 9-21

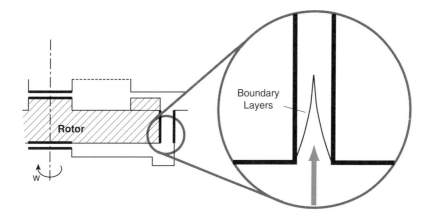

FIGURE 9.16 Schematic illustrating the origin of the axial-through-flow hydrostatic mechanism.

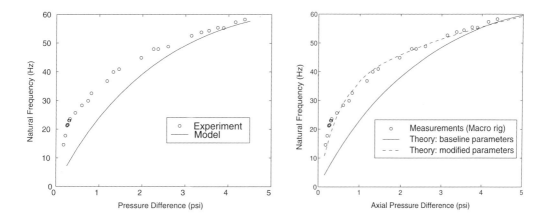

FIGURE 9.17 Prediction and measurement of natural frequency associated with axial through-flow in a low-aspect-ratio microbearing. The left-hand figure shows the measurements (from a 26:1 scaled-up experimental rig) along with the theoretical predictions based on the assumed geometry, while the right-hand figure shows the same measurements compared with the same model but using slightly modified geometric parameters [Orr et al., 2000].

was also able to predict the frequency (in a scaled-up experimental facility) with reasonable accuracy (Figure 9.17). However, at this writing, there is a severe gap between our ability to accurately predict and account for all the hydrostatic lubrication phenomena in a real microrotor. Nevertheless, experiments conducted at the microscale [Fréchette et al., 2000] demonstrate successful operation at high speeds (1.4 million rpm at the time of writing), despite theoretical predictions of failure! Indeed, experimental measurements suggest that both the natural frequency is higher than predicted by the simple axial-through-flow theory of Piekos (2000) and that the damping is sufficient to operate at critical speed ratios (rotor frequency, scaled by the natural frequency of the hydrostatic system) greater than 10. Conventional analysis [Savoulides, 2001] suggests that the instability occurs at critical speed ratios of 2, and this discrepancy suggests that the real bearing exhibits significantly higher damping than is accounted for by the theory, perhaps deriving from the turbine which drives the rotor or some other source of fluid damping, not yet considered. Resolution of these discrepancies clearly requires more attention and will be aided greatly by improved models as well as more detailed measurements of the microrotor in operation.

9.10 Fabrication Issues

A key challenge to the successful operation of a high-speed microbearing is the accurate fabrication of bearing geometries. Two aspects of this need to be considered: the need to hold design tolerances in any given fabrication process and the ability to manufacture multiple devices with good uniformity in a single fabrication run.

The first issue, that of achieving design tolerances, is a matter of process maturity, and the attention paid to the maintenance of tight tolerances and small details is the hallmark of a well-established fabrication process. The microengine process is *very* complex and continually advances the state of the art in micromachining complexity. Almost any fabrication run that results in a freely rotating turbine should deservedly be considered a manufacturing triumph. However from the standpoint of the success of the *system*, we must have much more stringent manufacturing requirements. The bearing designs are very sensitive to critical dimensions such as the bearing-rotor gap and the size of restrictor holes for hydrostatic injectors, and the failure to hold these dimensions within a specified tolerance can make the difference between a device that operates with a lubricated film and one that grinds the rotor and stator surface until failure. The very first version of the microbearing rig ran in just this mode, with very occasional demonstrations of lubricated operation. However, subsequent designs and builds have paid extreme attention to dimensional accuracy, and the fabrication protocols at this stage are quite mature so that this precision is ensured from one build to the next.

9.10.1 Cross-Wafer Uniformity

Typically, multiple microengines are fabricated in parallel on a single silicon wafer. Thus, in addition to the accurate manufacture of critical dimensions on a single microengine die, manufacturing uniformity from one die to the next on a single wafer is vital and has revealed itself to be a major obstacle to device yield. It is very common for a given process to exhibit cross-wafer variations. For example, a shallow plasma etch into a silicon substrate might show a variance of as much as 10% from one side of the wafer to the other due to variations in the plasma that are intrinsic to the fabrication tool. All fabrication processes will exhibit such variations to one degree or another, and any microfabrication process should identify and accommodate these variations. Should they be unacceptably large, either the fabrication tool must be improved or a different processing path needs to be considered. This need is a common driver throughout both the MEMS and microelectronics industry (which also desires greater process uniformity as feature size diminishes and processing moves to larger and larger wafers).

As discussed above, for microbearing design several critical etches must be controlled to a high degree of precision. The difficulty in maintaining cross-wafer uniformity thus results in some operational devices on the wafer (typically from the center of the wafer, where the process was initially honed to precision), while many devices (from the wafer edge) are out of spec and will not operate satisfactorily. At this stage, most of the uniformity issues have been addressed; however, two items are still troublesome. First, the DRIE, being a relatively new tool (perhaps 5 years in maturity), exhibits a fairly significant variation in etch rate between the center and edge of a wafer. This variation results in a gradient in etch depth that is particularly severe on devices lying on the wafer periphery (perhaps 3-μm variation across a 4-mm rotor wheel). This gradient may not seem important, but it contributes to a mass imbalance of as much as 25% of the bearing gap, rendering the bearing inoperable at high speed (the imbalance force increases with the square of the rotational speed).

The second continuing difficulty is that of front-to-back mask alignment during fabrication. It is common during the fabrication process for a single silicon wafer to be patterned on both the front and back surfaces. For example, the rotor has the turbine blades patterned from one side and the bearing gap patterned from the other side. Any slight misalignment between the lithography on the front and back surfaces of the wafer will result in an offset of front and back features which, as with the etch-depth gradient, leads to a rotor imbalance. At the present, mask alignment of critical features (primarily the rotor blades and the bearing gap) must be maintained to within 0.5 μm or better in order to ensure

Lubrication in MEMS

operable rotors from every die on the wafer. This is an extremely tight, but achievable, tolerance and work continues to improve the alignment even further and to improve process design to minimize imbalance.

9.10.2 Deep Etch Uniformity

The last issue for fabrication precision is that of deep-etch uniformity. Any high-speed bearing depends critically on the straightness and parallelism of the sidewalls that constitute the bearing and rotor surfaces. This is particularly true for hydrodynamic operation at high eccentricity. Unfortunately, in the drive to generate deep trenches so that the bearing aspect ratio is minimized, the quality of the bearing etch is often compromised. These two issues—the etch depth and the etch quality—constantly pull against each other, and their relative advantages need to be weighed against each other in any final design.

For DRIE, typical nonuniform etch profiles shown in Figure 9.18 illustrate two common phenomena—etch bow (where the trench widens in the middle) and etch taper (where the trench widens, usually at the bottom). The effects of these nonuniformities have been analyzed computationally [Piekos, 2000]. As one might expect, the static performance of the bearing (load capacity) is degraded by the blow-out, particularly in the case of the tapered bearing where fluid pressure cannot accumulate in the gap but rather leaks out the enlarged end. The bowed bearing, due to its concave curvature, tends to "hold" the pressure a bit more successfully, and the loss in load capacity is typically less severe. However, as mentioned earlier, load capacity is less of an issue in microbearings, and it is the effect on hydrodynamic stability that is of most interest. The effects of bow and taper on hydrodynamic operation are summarized in Figure 9.19, which shows the minimum stable eccentricity as a function of bearing number (i.e., speed, for a fixed bearing geometry) for different levels of both bow and taper. Clearly, the effects of taper are most severe, and considerable effort has been spent in the fabrication process design to minimize bearing taper.

9.10.3 Material Properties

One of the key benefits that can be realized at the microscale is the improvement in strength-related material properties. This is particularly true in silicon-based MEMS where the baseline structural material is single-crystal silicon, which can be fabricated to have very good mechanical properties. The strength of brittle materials is controlled primarily by flaws and to some extent by grain boundaries, both of which

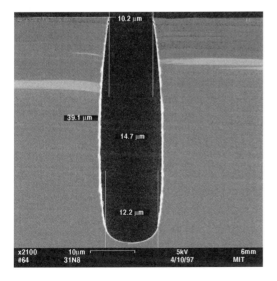
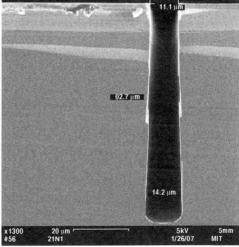

FIGURE 9.18 SEMS of bearing etches, illustrating typical manufacturing nonuniformities. The left-hand SEM shows an etch with a bow in the center, while the right-hand SEM shows an etch with a taper. (SEMs courtesy of A. Ayon.)

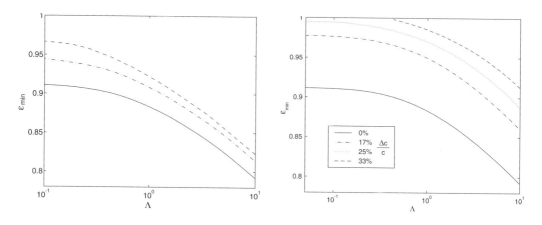

FIGURE 9.19 Degradation of hydrodynamic stability due to bow (left) and taper (right), as indicated by the minimum eccentricity required for hydrodynamic stability at a given bearing number (speed) [Piekos, 2001].

FIGURE 9.20 Photograph of the remains of a silicon rotor after experiencing a high-speed crash. The instant fracture of the rotor (largely along crystallographic planes) is clearly visible.

become smaller or nonexistent in a single-crystal material with surfaces defined by microfabrication processes. In addition, the device size becomes comparable with the flaw distribution, such that the incidence of "super-strong" devices increases in microscale systems. Finally, silicon is a light material with a density (2330 kg/m^3) lower than that of aluminum (2700 kg/m^3). For these reasons, the strength-to-weight ratio of silicon micromachined structures is unparalleled, which is a key for high-speed rotating machinery. However, notwithstanding its high specific strength, silicon is still a very brittle material. For a high-speed rotating system, such as a turbine, this can be problematic, as an impact or touchdown at any appreciable speed is likely to result in a catastrophic failure rather than an elastic rebound or more benign plastic deformation. This is illustrated in Figure 9.20, which shows a photograph of a microturbine rotor that crashed during a high-speed test run. The importance of robust bearings is emphasized, as the material is extremely unforgiving.

Lubrication in MEMS 9-25

9.11 Tribology and Wear

When lubrication fails, tribology and wear become important as our focus shifts from the prevention of contact to mitigation of its effects. Clearly, tribology has been a subject of technical and industrial importance since the industrial revolution, and a certain level of accomplishment has allowed the design and operation of complex machinery in difficult environments. With the advent of microengineering (and the development of the atomic force microscope), the field has defined a new set of problems and has witnessed a rebirth with a focus on the micro- and nanoscale processes associated with friction and wear. This chapter will not address in any complete manner the progress in micro- and nanotribology. However, a few general comments can be made that are of value in practical MEMS devices.

9.11.1 Stiction

A common problem associated with the failure of MEMS devices is that of stiction, in which two surfaces touch and stick together due to the high surface energies, high areas in close contact and small separating gaps. The problem is exacerbated by the use of wet etching during fabrication and by the relatively smooth surfaces (and thus high contact areas) associated with MEMS materials. Many lubricants have been used to mitigate the problem. Self-assembling monolayers (SAMs), such as perfluoro-decyl-trichlorosilane, coat the surface with a monolayer of a long molecule, which adheres to the surface at one end and is hydrophobic at the other end (thus preventing stiction). Other lubricants are also under investigation, such as Fomblin-Zdol, which is used in the hard-disk industry to protect the disk surface during head crashes and can be vapor deposited during manufacture. Other remedies include the intentional design of textured surfaces or standoffs to prevent large areas from coming into direct contact.

9.11.2 The Tribology of Silicon

Because most MEMS devices are silicon based, the tribology of silicon has received considerable attention in the past several years. By itself, silicon is not a very desirable bearing material (see, for example, Gardos, 2001) and exhibits high wear rates, high coefficients of friction and poor stiction characteristics. Surface treatments (e.g., silicon carbide, fluorocarbon or Teflon^{TM}-like materials) have been demonstrated to improve matters considerably, as well as appropriate design. However, most MEMS have not *yet* been designed with significant surface motions that require either supporting lubrication films or protective coatings. This is likely to change as fabrication processes enable more complex devices with higher power densities and more challenging lubrication and wear requirements. In addition, the lifetime and reliability requirements of MEMS are becoming more stringent as devices are being developed for space, medical and national defense applications. Each of these has unique requirements for predicting lifetime, wear and failure, so this oft-neglected corner of the industry will receive the attention it deserves.

9.12 Conclusions

In this chapter, we have tried to focus on some of the key issues that face us in the design, manufacture and operation of microdevices that must operate with minimal friction or wear. Many of the issues in Couette and squeeze-film lubrication have been successfully addressed, although fundamental surface models (such as gas accommodation coefficients) are still unknown and reliable prediction methods are only just becoming available. Stiction issues remain problematic, although they are dealt with in an engineering manner. As before, increased understanding of surface science is again needed to solve this problem.

Rotating MEMS devices still require much more development before they can be reliably manufactured and used. Even with dramatic advances in lubricant and surface treatment technologies, high-speed operation will likely require gas bearings, which have always offered high-speed, low-wear operation with the attendant cost of a narrow and treacherous window of stable operation. Many of the commonly held assumptions and design rules that have guided previous fluid-film bearings in conventional machinery have had to be revisited due to the consequences of scaling and the current (and foreseeable) limitations in microfabrication technology.

Future research will focus on many areas. On the manufacturing side, the single largest obstacle to trouble-free production of gas bearings for high-speed rotors and shafts is the issue of precision microfabrication. Macroscopic systems, with typical scale of 1 m, require precision manufacturing, in some places with submillimeter tolerances. So, too, microdevices with typical dimensions of 1 mm will require tolerances of 1 µm or less. The ability to manufacture with such precision will require much improved understanding of micromachining technologies such as etching and deposition so that cross-device and cross-wafer uniformities can be improved.

At the level of lubrication technologies, the new parameter regimes that are exposed by microfabricated systems (very low aspect ratios, relatively large clearances, insignificant inertial properties etc.) must be further explored and understood. Despite the low Reynolds numbers, inertial losses are critically important for hydrostatic lubrication mechanisms and must be better understood and predicted. Similarly, the coupled fluid–structure interactions at high eccentricities and the interactions between hydrodynamic and hydrostatic mechanisms should be more fully explored. Finally, fundamental issues of fluid and solid physics need to be addressed as the scale continues to shrink. Gas surface interactions, momentum and energy accommodation phenomena and the effects of surface contamination (whether deliberate or accidental) all must be rigorously studied so that the macroscopic behavior can be predicted with some certainty. These issues are only going to become more important as manufacturing scales decrease further and as continuum assumptions become more and more problematic.

Acknowledgments

I am indebted to my several friends and collaborators over the years with whom I have been fortunate to work in this exciting area. In particular, the contributions of Errol Arkilic, Stu Jacobson, Fred Ehrich, Alan Epstein, Luc Fréchette, Peter Kwok, C.-C. Lin, D.J. Orr, Ed Piekos, Nick Savoulides, Marc Weinberg and C.-W. Wong are gratefully acknowledged. Some of the contents of this paper are drawn from a similar chapter, presented at a NIST-sponsored workshop [Hsu, 2001].

References

Aluru, N.R., and White, J. (1988) "A Fast Integral Equation Technique for Analysis of Microflow Sensors Based on Drag Force Calculations," Int. Conf. on Modeling and Simulation of Microsystems, Semiconductors, Sensors and Actuators, April 1998, Santa Clara, CA, pp. 283–286.

Arkilic, E., and Breuer, K.S. (1993) "Gaseous Flow in Small Channels," AIAA 93-3270, AIAA Shear Flow Conference, July 1993.

Arkilic, E.B., Schmidt, M.A., and Breuer, K.S. (1997) "Gaseous Slip Flow in Long Microchannels," J. MEMS 6(2).

Arkilic, E.B., Breuer, K.S., and Schmidt, M.A. (2001) "Slip Flow and Tangential Momentum Accommodation in Silicon Micromachined Channels," J. Fluid Mech. 437, pp. 29–43.

Bart, S.F., Lorber, T.A., Howe, R.T., Lang, J.H., and Schlecht, M.F. (1988) "Design Considerations for Micromachined Electric Actuators," Sensors Actuators 14, pp. 269–292.

Beskok, A., and Karniadakis, G.E. (1994) "Simulation of Heat and Momentum Transfer in Complex Micro-Geometries," J. Thermophys. Heat Transfer 8(4), pp. 647–655.

Blech, J.J. (1983) "On Isothermal Squeeze Films," J. Lubrication Technol. 105, pp. 615–620.

Burgdorfer, A. (1959) "The Influence of the Molecular Mean Free Path on the Performance of Hydrodynamics Gas Lubricated Bearing," J. Basic Eng. 81, pp. 94–99.

Cai, C.-P., Boyd, I.D., Fan, J., and Candler, G.V. (2000) "Direct Simulation Methods for Low-Speed Microchannel Flows," J. Thermophys. Heat Transfer 14(3).

Dimofte, F. (1995) "Wave Journal Bearing with Compressible Lubricant. Part 1: The Wave Bearing Concept and a Comparison to the Plain Circular Bearing," *Tribology Trans.* **38**(1), pp. 153–160.

Epstein, A.H., Senturia, S.D., Al-Midani, O. et al. (1997) "Micro-Heat Engines, Gas Turbines and Rocket Engines—the MIT Microengine Project," AIAA Paper 97-1773, Snowmass, CO.

Fréchette, L.G., Jacobson, S.A., Breuer, K.S., et al. (2000) "Demonstration of a Microfabricated High-Speed Turbine Supported on Gas Bearings," IEEE Solid State Sensors and Actuators Workshop, Hilton Head, SC.

Fukui, S., and Kaneko, R. (1988) "Analysis of Ultra-Thin Gas Film Lubrication Based on Linearized Boltzmann Equation: First Report—Derivation of a Generalized Lubrication Equation Including Thermal Creep Flow," *J. Tribol.* **110**, pp. 253–262.

Gardos, M. (2001) In *Nanotribology*, ed. S. Hsu, Kluwer Academic, Dordrecht/Norwell, MA.

Hamrock, B.J. (1984) *Fundamentals of Fluid Film Lubrication*, McGraw-Hill, New York.

Hsu, S., ed. (2001) *Nanotribology*, Kluwer Academic, Dordrecht/Norwell, MA.

Kwok, P., Breuer, K.S., and Weinberg, M. (2001) "A Model for Proof Mass Damping in MEMS Devices with Venting Geometries," manuscript in preparation.

Lin, C.-C., Ghodssi, R., Ayon, A.A., Chen, D.-Z., Jacobson, S., Breuer, K.S., Epstein, A.H., and Schmidt, M.A. (1999) "Fabrication and Characterization of Micro Turbine/Bearing Rig," *Proc. MEMS '99*, IEEE, Orlando, FL, January.

Mehregany, M., Senturia, S.D., and Lang, J.H. (1992) "Measurement of Wear in Polysilicon Micromotors," *IEEE Trans. Electron. Devices* **39**(5), pp. 1136–1143.

Muijderman, E.A. (1966) *Spiral Groove Bearings*, Springer Verlag, Berlin, 1966.

Nagle, S.F., and Lang, J.H. (1999) "A Micro-Scale Electric-Induction Machine for a Micro Gas-Turbine Generator," presented at the 27th Meeting of the Electrostatics Society of America, Boston, MA.

Orr, D.J. (1999) "Macro-Scale Investigation of High Speed Gas Bearings for MEMS Devices," Ph.D. thesis, Department of Aeronautics and Astronautics, Massachusetts Institute of Technology, Cambridge.

Orr, D.J., Piekos, E.S., and Breuer, K.S. (2000) "A Novel Mechanism for Hydrostatic Lubrication in Short Journal Bearings," manuscript in preparation.

Piekos, E.S. (2000) "Numerical Simulations of Gas-Lubricated Journal Bearings for Microfabricated Machines," Ph.D. thesis, Department of Aeronautics and Astronautics, Massachusetts Institute of Technology, Cambridge.

Piekos, E.S., and Breuer, K.S. (1999) "Pseudo-Spectral Orbit Simulation of Non-Ideal Gas-Lubricated Journal Bearings for Microfabricated Turbomachines," *J. Tribol.* **121**, pp. 604–609.

Protz, J. (2000) "An Assessment of the Aerodynamic, Thermodynamic, and Manufacturing Issues for the Design, Development and Microfabrication of a Demonstration Micro Engine," Ph.D. thesis, Department of Aeronautics and Astronautics, Massachusetts Institute of Technology, Cambridge.

Reynolds, O. (1886) "On the Theory of Lubrication and Its Application to Mr. Beauchamp Tower's Experiments, Including an Experimental Determination of the Viscosity of Olive Oil," *R. Soc. Phil. Trans., Pt. 1*, 1886.

Rohsenow, W.M., and Choi, H.Y. (1961) *Heat, Mass, and Momentum Transfer*, Prentice-Hall, Englewood Cliffs, NJ.

Savoulides, N. (2001) "Low Order Models for Hybrid Gas Bearings," *J. Tribol.* **123**, pp. 368–375.

Sniegowski, J., and Garcia, E. (1996) "Surface-Micromachined Geartrains Driven by an On-Chip Electrostatic Microengine," *IEEE Electron. Device Lett.* **17**(7), p. 366.

Sun, Q., Boyd, I.D, and Fan, J. (2000) "Development of an Information Preservation Method for Subsonic, Micro-Scale Gas Flows," Presented at 22nd Rarefied Gas Dynamics Symp., July 2000, Sydney, Australia.

Veijola, T., Kuisma, H., Lahdenpera, J., and Ryhanen, T. (1995) "Equivalent-Circuit Model of the Squeezed Gas Film in a Silicon Accelerometer," *Sensors Actuators A* **A48**, pp. 239–248.

Wilcox, D.F., ed. (1972) *Design of Gas Bearings*, M.T.I., Inc., Latham, NY.

Ye, W., Kanapka, J., and White, J. (1999) "A Fast 3D Solver for Unsteady Stokes Flow with Applications to Micro-Electro-Mechanical Systems," Int. Conf. on Modeling and Simulation of Microsystems, Semiconductors, Sensors and Actuators, April 1999, San Juan, PR.

10
Physics of Thin Liquid Films

Alexander Oron
Technion–Israel Institute of Technology

10.1 Introduction 10-1
10.2 The Evolution Equation for a Liquid Film
 on a Solid Surface 10-9
10.3 Isothermal Films... 10-15
 Constant Surface Tension and Gravity • van der Waals
 Forces and Constant Surface Tension • Homogeneous
 Substrates • Heterogeneous Substrates • Flow on a
 Rotating Disc
10.4 Thermal Effects 10-29
 Thermocapillarity, Surface Tension and Gravity • Liquid Film
 on a Thick Substrate
10.5 Change of Phase: Evaporation and Condensation 10-36
 Interfacial Conditions • Evaporation/Condensation
 Only • Evaporation/Condensation, Vapor Recoil, Capillarity
 and Thermocapillarity • Flow on a Rotating Disc
10.6 Closing Remarks... 10-44
 Acknowledgments 10-44

10.1 Introduction

Various aspects of fluid mechanics in microelectromechanical systems (MEMS), such as flows in microconfigurations, flow transducers and flow control by microsystems, were reviewed by Ho and Tai (1998). However, the issue of thin liquid films and their dynamics in the context of microelectromechanical systems was left out of the scope of that important work. This chapter is intended to fill this gap.

Thin liquid films are encountered in a variety of phenomena and technological applications [Myers, 1998]. On a large scale, they emerge in geophysics as gravity currents under water or as lava flows [Huppert and Simpson, 1980; Huppert, 1982]. On the engineering scale, liquid films serve in heat and mass transfer processes to control fluxes and protect surfaces, and their various applications arise in paints, coatings and adhesives. They also occur in foams [Schramm, 1994; Prud'homme and Khan, 1996], emulsions [Ivanov, 1988; Edwards et al., 1991] and detergents [Adamson, 1990]. In biological applications, they appear as membranes, as linings of mammalian lungs [Grotberg, 1994] or as tear films in the eye [Sharma and Ruckenstein, 1986]. On the microscale in MEMS, thin liquid films are used to produce an insulating coating of solid surfaces, to form stable liquid bridges at specified locations, to create networks of microchannels on patterned microchips [Herminghaus et al., 1999; 2000] and to design fluid microreactors [Ichimura et al., 2000].

The presence of the deformable interface between the liquid and the ambient (normally gaseous, but possibly also another liquid) phases engenders various kinds of dynamics driven by one or usually several physical factors simultaneously. Liquid films may spontaneously or under the influence of external factors undergo diverse profound changes in their shapes. These changes are related to different kinds of instability, which may interact. Such interactions may lead either to a mutual enhancement or quenching of each other, so that the overall film dynamics may be rather complex. The film can rupture when its local depth vanishes and dewet the solid leading to holes in the liquid that expose the substrate to the ambient gas. The continuous character of the film changes in this case, if droplets of liquid are detached from the film. Changes in structure may occur in flows with contact lines leading to wavy fronts, fingered patterns or rivulets. Liquid films may be isothermal or subjected to the influence of the temperature field, which normally alters their dynamics. Liquid films may also undergo phase changes, such as mass loss by evaporation, mass gain by condensation or solidification. Liquid films exhibit many fascinating examples of behavior, some of which are presented below.

Sharma and Reiter (1996) studied the process of spontaneous dewetting of thin (less than 60 nm thick) polystyrene films on various coated silicon wafers and found a wealth of types of pattern formation. Different stages of dewetting identified in their experiments were (1) rupture of the film and emergence of holes; (2) expansion of the holes, their coalescence and formation of polygonal cellular pattern, where most of the liquid gathers in the ridges (see also Reiter [1998]); and (3) disintegration of liquid ridges into isolated, ultimately spherical drops. Also, fingering instability of the hole rim during hole expansion was observed on low wettability coatings, which resulted in the emergence of separate drops (see Elbaum and Lipson [1994]). It was found that the growth rate of the initial disturbance, the time of rupture, the number density of holes and the size of the polygons depend *only* on the solid substrate and is independent of the coating. The contact angle, which strongly depends on the choice of the coating layer, was found to affect the generation of droplets via the fingering instability, which is faster for larger contact angles. Also, the size of spherical drops forming as a result of the breakup of the liquid ridges was found to depend on the contact angle. Figure 10.1 displays the final pattern established by a 45-nm-thick liquid polysterene film on a coated silicon wafer, where the average contact angle for this combination of solid and liquid was about 22°. The pattern presented in Figure 10.1 consists of spherical droplets of various sizes aligned along the polygonal structure obtained as a result of the evolution of an initially uniform film that went through all the stages mentioned above. Problems associated with dewetting of solid surfaces by liquid films are discussed below in Section 10.3.

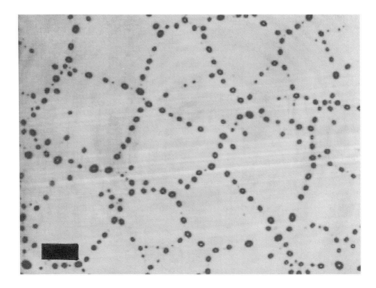

FIGURE 10.1 Photograph of the final polygonal pattern of spherical droplets for an initially flat polysterene film of mean thickness 45 nm on a coated silicon wafer. For reference, the length of the bar is 70 μm. (From Sharma, A., and Reiter, G. (1996) *J. Coll. Interface Sci.* **178**, 383–399. With permission.)

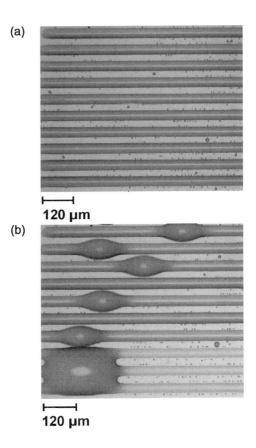

FIGURE 10.2 Water condensation on hydrophilic (magnesium fluoride) stripes of an elsewhere hydrophobic (silicone rubber) substrate. (a) Low-condensate-volume regime: The parallel channels of condensed water have a constant cross section and small contact angle; several droplets are also seen on the hydrophobic domains of the substrate. (b) High-condensate-volume regime: Some of the liquid channels develop a single drop when the contact angle exceeds a certain critical value; if two drops are in a close proximity, they merge to form a microbridge between the two neighboring microchannels. (From Herminghaus, S. et al. (2000) *J. Phys. Condens. Matter* **12**, A57–A74. With permission.)

Thiele et al. (1998) carried out experiments with thin (around 10 nm thick) volatile films consisting of collagen solution in acetic acid in various conditions of ambient humidity. For low ambient humidity, and thus a high evaporation rate, the pattern of very small holes along with several large ones was observed, suggesting that the small holes emerged due to polar interactions with the short residence time, while the large holes nucleated because of defects, such as dust particles or imperfections of the substrate. However, when the humidity was high and therefore the evaporation rate was low, the pattern of homogeneous polygonal network with large spacing was found. The large size of the polygons was explained by the long residence time, during which holes created by nucleation were able to open up. The effects associated with evaporation are discussed below in the section dedicated to phase changes.

Figure 10.2 displays the patterns generated by condensation of water on a spatially heterogeneous, solid, poorly wetted (hydrophobic) silicon rubber substrate with well-wetted (hydrophilic) magnesium fluoride stripes. Figure 10.2a shows the intermediate stage of water condensation, in which the liquid phase forms parallel microchannels with a certain contact angle θ at the edges of the stripes where the liquid appears to be pinned. The cross section of these microchannels is position independent and represents a circular cap. Several drops of the condensate between the hydrophilic stripes are also observed on Figure 10.2a. When the process of water condensation proceeds further, the contact angle θ increases beyond a certain value, and the microchannels undergo morphological change. As a result of this, droplets

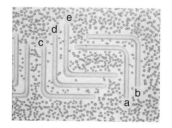
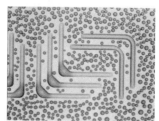
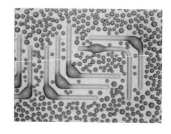

FIGURE 10.3 Deposition of water onto a patterned surface with hydrophilic microchannels with corners. The width of the channel in the corner region increases from channel (a) to channel (e). Time and the volume of the condensate increase from top to bottom. When a microchannel undergoes a morphological change of its shape, the drop moves to the corner to maximize the contact area with the hydrophilic part of the substrate. (From Herminghaus, S. et al. (2000) *J. Phys. Condens. Matter* **12**, A57–A74. With permission.)

emerge on some of the microchannels, one per channel, as seen in Figure 10.2b. When such droplets develop near each other, they merge to create a bridge, as shown in the left corner in the bottom of Figure 10.2b. These fundamental phenomena can be used to guide the liquid into the desired location(s) on the substrate with a specially designed wettability pattern [Herminghaus et al., 2000].

Figure 10.3 shows the time evolution (from top to bottom) of deposition of water condensing onto curved wettable patches of different width in the corners. This width increases from channel (a) to channel (e). When water condenses and gradually fills the channels, the behavior of the liquid depends solely on the width of the corner. If the latter is large, as for case (e), the drop develops in the corner. However, if it is small, as for case (a), the uniform channel configuration becomes unstable and a droplet develops in the straight part before it occurs in the corner. In the intermediate range of (b) to (d), the corner first develops a structure similar to case (a), but then it suddenly and discontinuously moves into the corner when a critical value of capillary pressure is attained. In such a configuration, the contact area of the liquid with the hydrophilic patch of the substrate is maximized. If the corners are sufficiently close each to another, two droplets will merge to produce a microbridge between the two neighboring microchannels, similar to what is shown in Figure 10.2.

Van Hook et al. (1997) carried out experiments aiming to study the thermocapillary convection produced by variations of surface tension in bilayer systems containing silicon oil films 0.007 to 0.027 cm thick and an overlying gas gap. Figure 10.4 shows some of their representative results. The dominant feature for thinner films was found to be the emergence of a drained spot, Figure 10.4a, or of a localized elevated structure with a peak touching the upper lid, Figure 10.4b. The drained spot may contain in certain circumstances isolated droplets trapped inside it. All these are manifestations of the long-wave instability of the film. These and other effects will be discussed below in the section dedicated to thermal effects. However, along with long-wave features of the system, other phenomena were observed for generally thicker films. These short-wave phenomena (whose length scale is comparable with the mean thickness of the layer) displayed the formation of hexagons or a combination of hexagons with the emergence of a dry spot, as shown in Figures 10.4c,d. Mathematical treatment of such short-wave phenomena will not

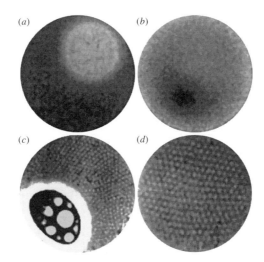

FIGURE 10.4 Infrared images of various states as seen in the experiments. The temperature increases with increasing brightness, so warm depression regions are white, except in (c), and cool elevated regions are dark. Each image has its own brightness, so temperatures in different images cannot be compared. (a) A localized depression (dry spot) with a helium gas layer and $d = 0.025$ cm; (b) a localized elevation (high spot) with an air gas layer and $d = 0.037$ cm; (c) a dry spot with hexagons in the surrounding region and $d = 0.025$ cm; (d) hexagons with an air gas layer and $d = 0.045$ cm. For further details, the reader is referred to Van Hook et al. (1997). (From Van Hook, S.J. et al. (1997) *J. Fluid Mech.* **345**, 45–78. Reprinted with permission of Cambridge University Press.)

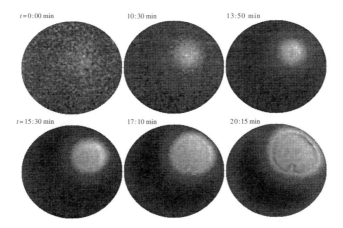

FIGURE 10.5 Evolution of a localized depression and formation of a dry spot in silicon oil of depth $d = 0.0267 \pm 0.0008$ cm and helium in the gas layer. At $t = 0$ (an arbitrary starting point), deformation of the interface is negligible. The liquid layer begins to form a localized depression (the white circle) and in 15 minutes the interface has ruptured ($h_{min} \to 0$) and formed a dry spot. The dry spot continues to grow for several more minutes before saturating. Bright (dark) regions are hot (cool) because they are closer (farther) to (from) the heater. All images have the same intensity scaling. (From Van Hook, S.J. et al. (1997) *J. Fluid Mech.* **345**, 45–78. Reprinted with permission of Cambridge University Press.)

be considered here. It turns out that rupture of thin liquid films and, following that, growth of the dry spot are frequent features associated with various physical mechanisms. Such an evolution of a local depression to the stage of film rupture and further to the growth and saturation of the dry spot driven by the thermocapillary effect (see Section 10.4) is shown in Figure 10.5.

A low-cost, high-yield passive alignment method known as controlled collapse chip connection (C4) process was designed [Goldmann, 1969] and used in opto-electronic packaging, where alignment accuracies at the submicron level are required [Wale and Edge, 1990; Lin et al., 1995]. Such precision alignment techniques, as illustrated in Figure 10.6, are employed for coupling fibers and waveguides to devices such as lasers and photo detectors and are being actively developed and improved. They use molten solder and are based on the restoring force arising from surface tension that drives the misaligned solder joint to become well aligned and to minimize by this the total interfacial energy of the system. The final well-aligned configuration is then fixed by cooling down and further solidifying the solder. Figure 10.6a presents such a misaligned layout of the two chips with four solder joints seen as dark circles. The misalignment is illustrated by the position of the center of the cross with respect to the point between the four squares. Figure 10.6b is a zoom of the area designated by the large circle on Figure 10.6a and presents the initial position of the misaligned system, which moves through the intermediate state, Figure 10.6c, to its final well-aligned position, Figure 10.6d. Figure 10.6e displays the final cross section of the solder between the two chips. It is interesting to note that the alignment process presented here takes place in the time range of a second. A similar mechanism, by which a hard contact lens centers itself over the cornea in a human eye, was discussed by Moriarty and Terrill (1996).

The centrifugal spinning of volatile solutions has been found to be a convenient and efficient means of coating planar solids with thin films. This process, known as spin coating, has been widely used in many technological processes, such as deposition of dielectric layers onto silicon wafers in the microelectronic industry, formation of ultrathin antireflective coatings for deep ultraviolet lithography and others. Two important stages of the process are usually considered. The first one occurs shortly after the liquid volume is delivered to the disk surface, rotating usually at the speed of 1000 to 10,000 rpm. At the beginning of this stage, the liquid film is relatively thick (usually greater than 500 µm). The film thins mainly due to radial drainage under the influence of centrifugal forces. Inertial forces are important and may lead to the appearance of instabilities of the spinning film. The second stage occurs when the film has thinned to the point where inertia is no longer important (film thickness usually less than 100 µm), and the flow slows down considerably, but deformations of the fluid interface may still be present due to the instabilities that appeared during the first stage. The film continues to thin mainly due to solvent evaporation until the solvent becomes depleted and the film solidifies and ceases to flow. Such problems will be discussed further below.

Numerous applications relevant for MEMS involve the dynamics of liquid films or drops. This area is experiencing constant progress, and new, exciting developments are often reported in the literature. Knight et al. (1998) described a new method of enhancement and control of nanoscale fluid jets. They demonstrated their method with the design of a continuous-flow mixer capable of mixing flowrates of nanoliters per second within the time scale of 10 µs. Such a mixer can be useful in nanofabrication techniques and serve as an essential part of a microreactor built on a chip.

Spatially controlled changes in the chemical structure of a solid substrate can guide a deposited liquid along the substrate. Ichimura et al. (2000) reported their experimental results showing the possibility of reversible guidance of liquid motion by light irradiation of a photoresponsive solid substrate. Asymmetric irradiation of the solid surface with blue light led to movement of a 2-µl olive oil droplet with a typical speed of 35 µm/s. Similar irradiation with a homogeneous blue light stopped the movement of the droplet completely. The speed of the droplet and the direction of its movement were found adjustable to the conditions of such irradiation. The phenomenon described has potential applicability in the design of microreactors and microchips.

A new technique of creation and replication of lateral structures in films of submicron-length scales was proposed by Schaeffer et al. (2000). It is based on the fact that lateral gradients of the electric field applied in the vicinity of the film interface induce variations of surface tension and thus lead to the electrocapillary effect. The latter is similar to the thermocapillary effect already mentioned above and is addressed more thoroughly in Section 10.4. The electrocapillary effect triggering the electrocapillary instability of the film results in formation of ordered patterns on the film interface and focusing of the interfacial troughs and peaks in the desired locations following the master pattern of the electrodes. Schaeffer et al. (2000) reported the replication of patterns of lateral dimensions on the order of 140 nm while employing this technique. A complete investigation of the electrocapillary instability of thin liquid

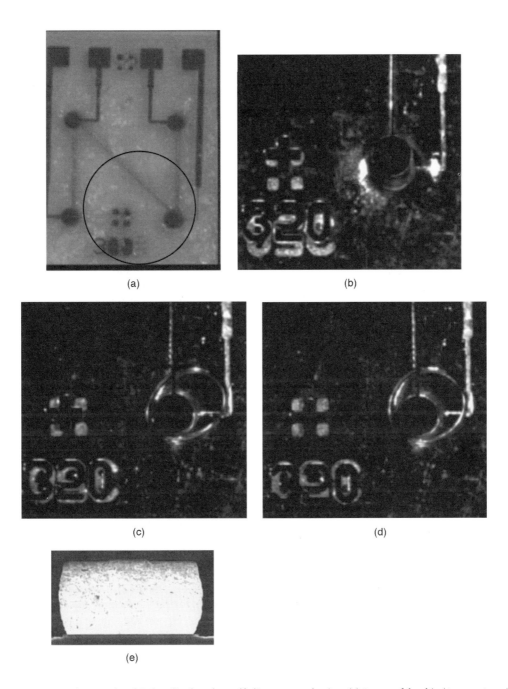

FIGURE 10.6 Photographs of C4 bonding based on self-alignment mechanism. (a) Layout of the chip (4 mm × 4 mm) which consists of four solder joints made of 63Sn37Pb. The upper chip is not aligned with the lower one, as can be seen from the position of the upper cross relative to the four squares at the lower chip. Initial misalignment is 150 μm. (b) Enlarged picture of one of the solder joints at the initial moment. (c) Intermediate stage. (d) Final position. (e) Side view showing the cross section of the solder joint at the final stage. (From Salalha, S.S. (2000) "Prediction of Yield for Flip-Chip Packaging," presented at International CIRP Design Seminar, Haifa, Israel. With permission.)

films has not yet appeared in the literature. Lee and Kim (2000) presented a liquid micromotor and liquid-metal droplets rotating along a microchannel loop being driven by a continuous electrowetting (CEW) phenomenon based on the electrocapillary effect. They identified and developed key technologies to design, manufacture and test the first MEMS devices employing CEW.

An attempt to provide a mathematical treatment of such phenomena must confront the fact that the interface of the film lying or flowing on a solid surface is partially or entirely a free boundary whose configuration evolving both temporally and spatially must be determined as an integral part of the solution of the governing equations. This renders the problem too difficult and very often almost intractable analytically, which may lead researchers to rely on computing only. The latter also becomes very complicated due to the free-boundary character of the problem which requires a careful design of adequate numerical methods.

Another property of such mathematical problems is their strong inherent nonlinearity, which is present in both governing equations and boundary conditions. Such nonlinearity of the problem presents another complexity that has to be faced. Consideration of coupled phenomena, such as those mentioned above, requires compact description of simultaneous instabilities that interact in intricate ways. This compact form must be tractable and at the same time still complex enough to retain the main features of the problem at hand.

The most appropriate analytical method of dealing with the above complexities is to analyze long-scale phenomena only, in which the characteristic lateral length scales are much larger than the average film thickness, the flowfield and temperature variations along the film are much more gradual than those normal to it, and the time variations are slow. Similar theories arise in a variety of areas of classical physics: shallow-water theory for water waves, lubrication theory in viscous flows and slender-body theory in aerodynamics and in dynamics of jets [e.g., Yarin, 1993]. In all these, a geometrical disparity is utilized in order to practically separate the variables and simplify the analysis. In thin viscous films, it turns out that most rupture and instability phenomena do occur on long scales, and a long-wave approach explained below is, in fact, very useful.

The long-wave theory approach is based on the asymptotic reduction of the governing equations and boundary conditions of a simplified system, which consists often, but not always, of a single, nonlinear, partial differential equation formulated in terms of the local thickness of the film varying in time and space. The rest of the unknowns (i.e., the fluid velocity, pressure, temperature etc.) are then determined via functionals of the solution of this differential equation, usually called an *evolution* equation. The notorious complexity of a free-boundary problem is thus removed. The corresponding penalty is, however, the presence of strong nonlinearity in the evolution equation(s) and the higher order spatial derivatives (usually up to the fourth) appearing there. A simplified linear stability analysis of the problem can be carried out based on the resulting evolution equation. A weakly nonlinear analysis of the problem is also possible through that equation. However, the fully nonlinear analysis that allows one to study finite-amplitude deformations of the film interface must be performed numerically. Still, numerical solution of the evolution equation is incomparably less difficult than that of the original, free-boundary problem.

Several encouraging verifications of the long-wave theory vs. the experimental results have appeared in the literature. Burelbach et al. (1990) carried out a series of experiments attempting to check the long-wave theory of Tan et al. (1990) for steady thermocapillary flows induced by nonuniform heating of the solid substrate. The measured steady shapes were favorably tested against theoretical predictions for layers less than 1 mm thick under moderate heating conditions. However, the relative error was large for conditions near rupture, where the long-wave theory is *formally* invalid [Burelbach et al., 1988], but in all other cases the predicted and measured values of the minimal film thickness agreed to within 20%. The theory also predicts rupture when the parameter L (see Eq. (3.6) in Tan et al. [1990]) exceeds a certain critical value, but steady patterns otherwise. Experimental results [Burelbach et al., 1990, Figure 1] showed that L is an excellent qualitative indicator of whether or not the film ruptures.

Van Hook et al. (1995; 1997) performed experiments on the onset of the long-wavelength instability in thin layers of silicon oil of varying thickness, aspect ratios and transverse temperature gradients across the layer. Formation of "dry spots" at randomly varying locations was found above the critical temperature difference across the layer, in qualitative agreement with corresponding numerical simulations. More about the experimental support for theoretical results can be found below in various sections of this chapter.

Another test for the validity of an asymptotic theory, such as the long-wave theory presented here, is achieved via comparison between the numerical solutions for the *full* free-boundary problem in its original form and the solutions obtained for the corresponding long-wave evolution equations. Due to the difficulty of carrying out the direct numerical simulations already discussed above, the number of such comparative studies is quite limited. Krishnamoorthy et al. (1995) performed a full-scale direct numerical simulation of the governing equations to study the rupture of thin liquid films due to thermocapillarity and found very good qualitative agreement with the results arising from solution of the corresponding evolution equation, except for times prior to rupture. Oron (2000b) found even better agreement at rupture between his results and direct simulations of the Navier–Stokes equations of Krishnamoorthy et al. (1995). There has been a long debate in the literature about the validity of fingered structures of the film interface arising from solution of the evolution equations and whether they are artifacts of the asymptotic reduction applied. Direct solution of the Navier–Stokes equations [Krishnamoorthy et al., 1995] provides convincing evidence supporting the validity of the evolution equations even in the domain where some assumptions leading to their derivation are violated.

There has been major progress in the analysis of thin liquid films in recent years. In the review article by Oron et al. (1997), such analyses were unified into a simple framework in which the special cases naturally emerged. In this chapter, the physics of thin liquid films is reviewed with an emphasis on the phenomena of considerable interest for MEMS. The theory of drop spreading, in spite of its importance, is not included here and the reader is referred to other reviews [de Gennes, 1985; Leger and Joanny, 1992; Oron et al., 1997] and the references therein.

The general evolution equation describing the general dynamics of thin liquid films will be derived following Oron et al. (1997) and discussed in Section 10.2. The topic to be addressed in Section 10.3 is isothermal films, for which the physical effects discussed are viscous, surface tension, gravity and centrifugal forces along with van der Waals interactions. Section 10.4 examines the influence of thermal effects on the dynamics of liquid films. Section 10.5 considers the dynamics of liquid films undergoing phase changes, such as evaporation and condensation.

10.2 The Evolution Equation for a Liquid Film on a Solid Surface

We now describe the long-wave approach and apply it to a flow of a viscous liquid in a film supported below by a solid horizontal plate and bounded above by an interface separating the liquid and a passive gas, slowly evolving in space and time, as given by its equation $z = h(x, y, t)$. We assume the possibility of the existence of external interfacial forces Π with the components $\{\Pi_3, \Pi_1, \Pi_2\}$ in the normal and both of the tangential to the film surface directions, respectively, determined by the vectors:

$$\mathbf{n} = \frac{\{-h_x, -h_y, 1\}}{\sqrt{1 + h_x^2 + h_y^2}}, \quad \mathbf{t}_1 = \frac{\{1, 0, h_x\}}{\sqrt{1 + h_x^2}}, \quad \mathbf{t}_2 = \frac{\{0, 1, h_y\}}{\sqrt{1 + h_y^2}}. \tag{10.1}$$

The components of the vectors $\mathbf{n}, \mathbf{t}_1, \mathbf{t}_2$ in Eq. (10.1) are specified in the order of the x, y and z directions, where x and y are the spatial coordinates in the given solid plane, while z is normal to the latter and directed across the film. The presence of a conservative body force determined by the potential ϕ acting on the liquid phase, such as gravity, centrifugal or van der Waals force, is accounted for as well. We note that the vectors $\mathbf{t}_1, \mathbf{t}_2$ are not orthogonal, but it is sufficient for our application below that $(\mathbf{n}, \mathbf{t}_1)$ and $(\mathbf{n}, \mathbf{t}_2)$ constitute pairs of orthogonal vectors. From here on, the letter subscripts denote the partial derivatives with respect to the corresponding variable.

The liquid considered in this work is assumed to be a simple Newtonian incompressible viscous fluid whose dynamics are well described by the Navier–Stokes and mass conservation equations, provided that the length scales characteristic for the flow domain are within the continuum range exceeding several molecular spacings. The mass conservation and Navier–Stokes equations for such a liquid in three dimensions have the form:

$$u_x + v_y + w_z = 0 \tag{10.2a}$$

$$\rho(u_t + uu_x + vu_y + wu_z) = -p_x + \mu(u_{xx} + u_{yy} + u_{zz}) - \phi_x \tag{10.2b}$$

$$\rho(v_t + uv_x + vv_y + wv_z) = -p_y + \mu(v_{xx} + v_{yy} + v_{zz}) - \phi_y \tag{10.2c}$$

$$\rho(w_t + uw_x + vw_y + ww_z) = -p_z + \mu(w_{xx} + w_{yy} + w_{zz}) - \phi_z \tag{10.2d}$$

where ρ and μ are, respectively, the density and kinematic viscosity of the liquid; u, v, w are the respective components of the fluid velocity vector $\mathbf{v}$ in the directions x, y, z; t is time; and p is pressure.

The classical boundary conditions between the liquid and the solid surface supporting it are no penetration, $w = 0$, and no slip, $u = 0$, $v = 0$. These conditions are appropriate for the continuous films to be considered here. Problems related to the case where a contact line is a junction of the three phases (solid, liquid, gas) exist, and the liquid on a solid surface spreading or receding will not be examined in this chapter. The reader interested in this topic is referred to the review papers by de Gennes (1985), Leger and Joanny (1992), Oron et al. (1997) and references therein.

The boundary conditions at the solid surface are, therefore:

$$w = 0, \; u = 0, \; v = 0 \quad \text{at} \; z = 0 \tag{10.3}$$

At the film surface $z = h(x, y, t)$, the boundary conditions are formulated in the vector form [e.g., Wehausen and Laitone, 1960]:

$$h_t + \mathbf{v} \cdot \nabla^* h - w = 0 \tag{10.4a}$$

$$\mathbf{T} \cdot \mathbf{n} = -2\tilde{H}\sigma\mathbf{n} + \nabla_s \sigma + \Pi \tag{10.4b}$$

where $\mathbf{T}$ is the stress tensor of the liquid, Π is the prescribed forcing at the interface, $\tilde{H}$ is the mean curvature of the interface determined from:

$$2\tilde{H} = \nabla^* \cdot \mathbf{n} = -\frac{h_{xx}(1 + h_y^2) + h_{yy}(1 + h_x^2) - 2h_x h_y h_{xy}}{(1 + h_x^2 + h_y^2)^{3/2}} \tag{10.5}$$

$\nabla^* = (\partial/\partial x, \partial/\partial y, \partial/\partial z)$ is the gradient operator, and ∇_s is the surface gradient with respect to the interface $z = h(x, y, t)$. Note that in Eq. (10.4) the dot represents both the inner product of two vectors and the product of a tensor and a vector, respectively.

Equation (10.4a) is the kinematic boundary condition formulated in the absence of interfacial mass transfer that represents the balance between the normal component of the liquid velocity at the interface and the velocity of the interface itself. An appropriate change should be made in Eq. (10.4a) to accommodate the phenomena of evaporation or condensation (see Section 10.5). Equation (10.4b), which constitutes the balance of interfacial stresses in the absence of interfacial mass transfer, has three components. The physical meaning of its two tangential components is that the shear stress at the interface is balanced by the sum of the respective Π_i, $i = 1, 2$, and the surface gradient of surface tension σ, whereas the normal component of Eq. (10.4b) states that the difference between the normal interfacial stress and Π_3 exhibits a jump equal to the product of twice the mean curvature of the film interface and surface tension. This jump is known in the literature as the capillary pressure. When the external force Π is zero, and the fluid has zero viscosity or the fluid is static, $\mathbf{v} = 0$, then $\mathbf{T} \cdot \mathbf{n} \cdot \mathbf{n} = -p$, and Eq. (10.4b) reduces to the well-known Young–Laplace equation which describes, for instance, the excess pressure in an air bubble gauged to the external pressure as twice the surface tension divided by the bubble radius (see, e.g., Landau and Lifshitz [1987]). The derivations below closely follow those made by Oron et al. (1997), being explicitly extended into three dimensions.

Projecting Eq. (10.4b) onto the directions $\mathbf{n}, \mathbf{t}_1, \mathbf{t}_2$, respectively, yields:

$$-p + \frac{2\mu[u_x(h_x^2-1) + v_y(h_y^2-1) + h_x h_y(u_y+v_x) - h_x(u_z+w_x) - h_y(v_z+w_y)]}{1+h_x^2+h_y^2} = 2H\sigma + \Pi_3$$

$$\mu[(u_z+w_x)(1-h_x^2) - (v_z+w_y)h_x h_y - (u_y+v_x)h_y - 2(u_x-w_z)h_x] = \left(\Pi_1 + \frac{\partial \sigma}{\partial x}\right)(1+h_x^2+h_y^2)^{1/2}$$

$$\mu[-(u_z+w_x)h_x h_y + (v_z+w_y)(1-h_y^2) - (u_y+v_x)h_x - 2(v_y-w_z)h_y] = \left(\Pi_2 + \frac{\partial \sigma}{\partial y}\right)(1+h_x^2+h_y^2)^{1/2}$$

(10.6)

Let us now introduce scales appropriate for thin films where the transverse length scale is much smaller than the lateral scales. Assume length scales in the lateral directions x and y to be defined by wavelength λ of the interfacial disturbance on a film of mean thickness d. The film is referred to as a *thin* film if the interfacial distortions are much longer than the mean film thickness; that is,

$$\varepsilon = \frac{d}{\lambda} \ll 1 \tag{10.7}$$

The z coordinate (normal to the solid substrate) is normalized with respect to d, while the coordinates x and y are scaled with λ or equivalently d/ε. Thus, the dimensionless z coordinate is defined as:

$$\varsigma = \frac{z}{d} \tag{10.8a}$$

while the dimensionless x- and y coordinates are given by:

$$\xi = \frac{\varepsilon x}{d}, \quad \eta = \frac{\varepsilon y}{d} \tag{10.8b}$$

It is assumed that in the new spatial variables no rapid variations occur as $\varepsilon \to 0$; then,

$$\frac{\partial}{\partial \xi}, \frac{\partial}{\partial \eta}, \frac{\partial}{\partial \varsigma} = O(1) \tag{10.8c}$$

If the lateral components of the velocity field u and v are assumed to be of order one and U_0 denotes the characteristic velocity of the problem, the dimensionless fluid velocities in the x and y directions are defined as:

$$U = \frac{u}{U_0}, \quad V = \frac{v}{U_0} \quad (10.8d)$$

Then, the continuity equation, Eq. (10.2a), requires that the z component of the velocity field w be small, and the dimensionless fluid velocity in the z direction is defined as:

$$W = \frac{w}{\varepsilon U_0} \quad (10.8e)$$

We stress that the characteristic velocity U_0 is not specified here for the sake of generality. The freedom of choosing this value is thus given to the user. We just note one of the possible choices but not the unique one, $U_0 = \mu/\rho d$, which is known in the literature as a viscous velocity.

Time is scaled in the units of λ/U_0, so the asymptotically long-time behavior of the film is to be considered. The dimensionless time is therefore defined via:

$$\tau = \frac{\varepsilon U_0 t}{d} \quad (10.8f)$$

Finally, because of the assumed slow lateral variation of the film interface, one expects locally parallel flow in the liquid so that the pressure gradient is balanced with the viscous stress $p_x \propto \mu u_{zz}$, and the dimensionless interfacial stresses, body-force potential and pressure are defined, respectively, as:

$$(\Pi_1, \Pi_2, \Pi_3) = \frac{d}{\mu U_0}(\hat{\Pi}_1, \hat{\Pi}_2, \varepsilon\hat{\Pi}_3), \quad (\Phi, P) = \frac{\varepsilon d}{\mu U_0}(\phi, p) \quad (10.8g)$$

Notice that pressure is asymptotically large similar to the situation arising in the lubrication effect [Schlichting, 1968].

If all these dimensionless variables are substituted into the governing system, Eqs. (10.2) to (10.5), the following scaled system is obtained:

$$U_\xi + V_\eta + W_\varsigma = 0 \quad (10.9a)$$

$$\varepsilon R(U_\tau + UU_\xi + VU_\eta + WU_\varsigma) = -P_\xi + U_{\varsigma\varsigma} + \varepsilon^2(U_{\xi\xi} + U_{\eta\eta}) - \Phi_\xi \quad (10.9b)$$

$$\varepsilon R(V_\tau + UV_\xi + VV_\eta + WV_\varsigma) = -P_\eta + V_{\varsigma\varsigma} + \varepsilon^2(V_{\xi\xi} + V_{\eta\eta}) - \Phi_\eta \quad (10.9c)$$

$$\varepsilon^3 R(W_\tau + UW_\xi + VW_\eta + WW_\varsigma) = -P_\varsigma + \varepsilon^2 W_{\varsigma\varsigma} + \varepsilon^4(W_{\xi\xi} + W_{\eta\eta}) - \Phi_\varsigma \quad (10.9d)$$

At $\varsigma = 0$:

$$W = 0, \quad U = 0, \quad V = 0 \quad (10.10)$$

At $\varsigma = H$:

$$W = H_\tau + UH_\xi + VH_\eta \quad (10.11a)$$

$$\frac{2\varepsilon^2[U_\xi(\varepsilon^2 H_\xi^2 - 1) + V_\eta(\varepsilon^2 H_\eta^2 - 1) + \varepsilon^2 H_\xi H_\eta(U_\eta + V_\xi) - H_\xi(U_\varsigma + W_\xi) - H_\eta(V_\varsigma + W_\eta)]}{1 + \varepsilon^2(H_\xi^2 + H_\eta^2)}$$

$$= P + \hat{\Pi}_3 + \frac{\bar{S}\varepsilon^3[H_{\xi\xi}(1 + \varepsilon^2 H_\eta^2) + H_{\eta\eta}(1 + \varepsilon^2 H_\xi^2) - 2\varepsilon^2 H_\xi H_\eta H_{\xi\eta}]}{[1 + \varepsilon^2(H_\xi^2 + H_\eta^2)]^{3/2}} \quad (10.11b)$$

$$(U_\varsigma + \varepsilon^2 W_\xi)(1 - \varepsilon^2 H_\xi^2) - \varepsilon^2(V_\varsigma + \varepsilon^2 W_\eta)H_\xi H_\eta - \varepsilon^2(U_\eta + V_\xi)H_\eta - 2\varepsilon^2(U_\xi - W_\varsigma)H_\xi$$
$$= (\hat{\Pi}_1 + \Sigma_\xi)[1 + \varepsilon^2(H_\xi^2 + H_\eta^2)]^{1/2} \qquad (10.11c)$$

$$(V_\varsigma + \varepsilon^2 W_\eta)(1 - \varepsilon^2 H_\eta^2) - \varepsilon^2(U_\varsigma + \varepsilon^2 W_\xi)H_\xi H_\eta - \varepsilon^2(U_\eta + V_\xi)H_\xi - 2\varepsilon^2(V_\eta - W_\varsigma)H_\eta$$
$$= (\hat{\Pi}_2 + \Sigma_\eta)[1 + \varepsilon^2(H_\xi^2 + H_\eta^2)]^{1/2} \qquad (10.11d)$$

Here, $H = h/d$ is the dimensionless thickness of the film and $\Sigma = \varepsilon\sigma/\mu U_0$ is the dimensionless surface tension normalized with respect to its characteristic value. The Reynolds number R and the inverse capillary number $\bar{S}$ are defined by:

$$R = \frac{U_0 d\rho}{\mu}, \quad \bar{S} = \frac{\sigma}{U_0\mu} \qquad (10.12)$$

The continuity equation, Eq. (10.9a), is now integrated in ς across the film from 0 to $H(\xi, \eta, \tau)$, and Eqs. (10.10) and (10.11a) are used along with integration by parts to obtain:

$$H_\tau + \frac{\partial}{\partial \xi}\int_0^H U\,d\varsigma + \frac{\partial}{\partial \eta}\int_0^H V\,d\varsigma = 0 \qquad (10.13)$$

Equation (10.13) is a more convenient form of the kinematic condition in the sense that only two of three components of the fluid velocity field appear there explicitly. It also warrants conservation of mass in a domain with a deflecting upper boundary.

The solution of the governing equations, Eqs. (10.2) to (10.5), is sought in the form of expansion of the dependent variables into asymptotic series in powers of the small parameter ε:

$$\begin{aligned} U &= U^{(0)} + \varepsilon U^{(1)} + \varepsilon^2 U^{(2)} + \cdots, & V &= V^{(0)} + \varepsilon V^{(1)} + \varepsilon^2 V^{(2)} + \cdots \\ W &= W^{(0)} + \varepsilon W^{(1)} + \varepsilon^2 W^{(2)} + \cdots, & P &= P^{(0)} + \varepsilon P^{(1)} + \varepsilon^2 P^{(2)} + \cdots \end{aligned} \qquad (10.14)$$

One way to approximate the solution of the governing system is to assume that $R, \bar{S} = O(1)$ as $\varepsilon \to 0$. Under this assumption, the inertial terms measured by εR are one order of magnitude smaller than the dominant viscous terms, consistent with the local-parallel-flow assumption, while the surface tension terms measured by $\bar{S}\varepsilon^3$ are two orders of magnitude smaller and would be lost. However, it is essential to retain surface-tension effects at the leading order, so it is assumed that capillary effects are strong relative to those of viscosity and

$$\bar{S} = S\varepsilon^{-3} \qquad (10.15)$$

It is then assumed that $R, S = O(1)$, as $\varepsilon \to 0$.

Equations (10.14) and (10.15) are substituted into Eqs. (10.9) to (10.11) and (10.13), and the resulting equations are sorted with respect to the powers of ε. At the leading order in ε the governing system becomes, after omitting the superscript zero in $U^{(0)}$, $V^{(0)}$, $W^{(0)}$ and $P^{(0)}$:

$$\begin{aligned} U_{\varsigma\varsigma} &= (P + \Phi)_\xi \\ V_{\varsigma\varsigma} &= (P + \Phi)_\eta \\ (P + \Phi)_\varsigma &= 0 \\ H_\tau + UH_\xi + VH_\eta - W &= 0 \\ U_\xi + V_\eta + W_\varsigma &= 0 \end{aligned} \qquad (10.16)$$

with the boundary conditions at $\varsigma = 0$:

$$W = 0, \quad U = 0, \quad V = 0 \tag{10.17}$$

and at $\varsigma = H$:

$$\begin{aligned}
P &= -\hat{\Pi}_3 - S(H_{\xi\xi} + H_{\eta\eta}) \\
U_\varsigma &= \hat{\Pi}_1 + \Sigma_\xi \\
V_\varsigma &= \hat{\Pi}_2 + \Sigma_\eta
\end{aligned} \tag{10.18}$$

We note here that Eqs. (10.16) to (10.18) are linear with respect to the variables U, V, W, P, and the only nonlinearity of this problem is associated, as seen from Eq. (10.19) in conjunction with the kinematic condition Eq. (10.16d), with the local film thickness $H(\xi, \eta, \tau)$. Solving Eqs. (10.16) to (10.18) yields:

$$\begin{aligned}
U &= \left[\tfrac{1}{2}\varsigma^2 - H\varsigma\right](\Phi - \hat{\Pi}_3|_{\varsigma=H} - S\nabla^2 H)_\xi + \varsigma(\hat{\Pi}_1 + \Sigma_\xi) \\
V &= \left[\tfrac{1}{2}\varsigma^2 - H\varsigma\right](\Phi - \hat{\Pi}_3|_{\varsigma=H} - S\nabla^2 H)_\eta + \varsigma(\hat{\Pi}_2 + \Sigma_\eta) \\
W &= -\int_0^\varsigma (U_\xi + V_\eta)\, d\varsigma, \quad P = -\hat{\Pi}_3|_{\varsigma=H} - S\nabla^2 H
\end{aligned} \tag{10.19}$$

If Eq. (10.19) is substituted into the mass conservation Eq. (10.13), one obtains the appropriate evolution equation for the interface,

$$H_\tau + \tfrac{1}{2}\nabla \cdot [H^2(\hat{\Pi}^* + \nabla\Sigma)] + \tfrac{1}{3}\nabla \cdot \{H^3[\nabla(\hat{\Pi}_3 - \Phi|_{\varsigma=H}) + S\nabla\nabla^2 H]\} = 0 \tag{10.20}$$

where $\hat{\Pi}^* = (\hat{\Pi}_1, \hat{\Pi}_2)$ is the tangential projection of the dimensionless vector $\hat{\Pi}$, $\nabla \equiv (\partial/\partial\xi, \partial/\partial\eta)$ and $\nabla^2 \equiv \partial^2/\partial\xi^2 + \partial^2/\partial\eta^2$.

In two dimensions ($\partial/\partial\eta = 0$), this evolution equation reduces to:

$$H_\tau + \tfrac{1}{2}[H^2(\hat{\Pi}_1 + \Sigma_\xi)]_\xi + \tfrac{1}{3}\{H^3[(\hat{\Pi}_3 - \Phi|_{\varsigma=H})_\xi + SH_{\xi\xi\xi}]\}_\xi = 0 \tag{10.21}$$

In these equations, the location of the film interface $H = H(\xi, \eta, \tau)$ is unknown and has to be determined from the solution of the corresponding partial differential equation. When such a solution is obtained, the components of the velocity and the pressure fields can be determined from Eq. (10.19).

The physical significance of the terms becomes apparent when Eqs. (10.20) and (10.21) are written in the original dimensional variables:

$$\mu h_t + \tfrac{1}{2}\overline{\nabla} \cdot [h^2(\Pi^* + \overline{\nabla}\sigma)] + \tfrac{1}{3}\overline{\nabla} \cdot \{h^3[\overline{\nabla}(\Pi_3 - \phi|_{z=h}) + \sigma\overline{\nabla}\overline{\nabla}^2 h]\} = 0 \tag{10.22}$$

with $\overline{\nabla} \equiv (\partial/\partial x, \partial/\partial y)$, $\overline{\nabla}^2 \equiv (\partial^2/\partial x^2 + \partial^2/\partial y^2)$ and

$$\mu h_t + \tfrac{1}{2}[h^2(\Pi_1 + \sigma_x)]_x + \tfrac{1}{3}\{h^3[(\Pi_3 - \phi|_{z=h})_x + \sigma h_{xxx}]\}_x = 0 \tag{10.23}$$

Physics of Thin Liquid Films

The first terms in Eqs. (10.22) and (10.23) represent the effect of viscous damping, while the next ones account, respectively, for the effects of the imposed tangential interfacial stress, nonuniformity of surface tension, imposed normal interfacial stress, body forces and surface tension on the dynamics of the film.

In the examples discussed below, both two- and three-dimensional cases are examined. Further, unless specified, only disturbances periodic in x and y are discussed. Thus, λ is the wavelength of these disturbances and $2\pi d/\lambda$ is the corresponding dimensionless wave number. In accordance with this, Eqs. (10.20) to (10.23) are normally solved with periodic boundary conditions. These equations, whether in two or three dimensions, are of fourth order in each of the spatial variables; therefore, four boundary conditions are needed to define a well-posed mathematical problem. These four boundary conditions imply periodicity of the solution H and its first, second and third derivatives with respect to the corresponding spatial variable. On the other hand, Eqs. (10.20) to (10.23) are of first order in time, thus one initial condition is needed to complete the well-posed statement of the problem. Such an initial condition representing the location of the film interface at $t = 0$ or $\tau = 0$ is usually taken as a small-amplitude random or sinusoidal disturbance on top of the uniform state given by $H = 1$. In two dimensions it can be written by:

$$H(\tau = 0, \xi) = 1 + \delta \sin(\xi + \varphi) \quad \text{or} \quad H(\tau = 0, \xi) = 1 + \delta \, \text{rand}(\xi) \quad (10.24)$$

where $\delta \ll 1$, φ is a phase, and $\text{rand}(\xi)$ is a random function uniformly distributed in the interval $(-1, 1)$. An obvious extension of Eq. (10.24) can be obtained in the three-dimensional case.

10.3 Isothermal Films

We now examine the dynamics of films whose temperature remains unchanged and phase changes do not occur.

10.3.1 Constant Surface Tension and Gravity

Consider the simplest case in which the film is supported from below by a solid surface and subjected to the influence of gravity and constant surface tension. In this case, one has $\Sigma_\xi = \Sigma_\eta = \hat{\Pi}_1 = \hat{\Pi}_2 = \hat{\Pi}_3 = 0$ and $\Phi = G\varsigma$, so that in two dimensions Eq. (10.21) becomes:

$$H_\tau - \frac{1}{3}G(H^3 H_\xi)_\xi + \frac{1}{3}S(H^3 H_{\xi\xi\xi})_\xi = 0 \quad (10.25a)$$

where G is the unit-order positive gravity number,

$$G = \frac{\rho g d^2}{\mu U_0}$$

The second term of Eq. (10.25a) accounts for the influence of gravity, while the third one describes the effect of the capillary forces. The dimensional version of Eq. (10.25a) is obtained from Eq. (10.23) as:

$$\mu h_t - \frac{1}{3}\rho g(h^3 h_x)_x + \frac{1}{3}\sigma(h^3 h_{xxx})_x = 0 \quad (10.25b)$$

In the absence of surface tension, Eq. (10.25b) is a nonlinear (forward) diffusion equation so that one can envision that no disturbance to $h = d$ experiences growth in time. Surface tension acts through a fourth-order (forward) dissipation term only, enhancing stabilization of the interface, so that no instabilities would occur in the case described by Eq. (10.25b) for $G > 0$.

To formally assess these intuitive observations one can investigate the stability properties of the uniform film, $h = d$, perturbing it by a small disturbance, h', periodic in x (i.e., $h = d + h'$, with $h' \ll d$). Substituting this into Eq. (10.25b) and linearizing it with respect to h', one obtains the linear-stability equation for the uniform state $h = d$. Since this equation has coefficients independent of t and x, one can seek separable solutions of the form:

$$h' = h'_0 \exp(ikx + \omega t), \quad h'_0 = \text{const}$$

which constitute a complete set of "normal modes" and can be used to represent any disturbance by means of the Fourier series. Here, k is the wave number of the disturbance in the x direction. If these normal modes are substituted into the linear-stability equation, one obtains the following characteristic equation for ω:

$$\mu\omega = -\frac{1}{3d}(\rho g d^2 + \sigma a^2)a^2 \qquad (10.26)$$

where $a = kd$ is the nondimensional wave number and ω is the growth rate of the perturbation. In general, the amplitude of the perturbation will decay if the real part of the growth rate, $Re(\omega)$, is negative and will grow if $Re(\omega)$ is positive. Purely imaginary values of ω will correspond to translation along the x-axis and give rise to traveling-wave solutions. Finally, zero values of $Re(\omega)$ will correspond to neutral, stationary-in-time perturbations.

Two remarks are now in order. First, the linear stability analysis is carried out here in the dimensional form, but it could be done also in the same way in the dimensionless form when its starting point would be Eq. (10.25a). Second, the linear stability analysis is carried out here in the two-dimensional case. The same can be done in the three-dimensional case with respect to the normal modes:

$$h' = h'_0 \exp(ik_x x + ik_y y + \omega t), \quad h'_0 = \text{const}$$

where k_x and k_y are, respectively, the wave numbers in the x and y directions. As in the physical problem at hand, the symmetry is such that the spatial variables x and y are interchangeable; the characteristic equation for ω will be identical to Eq. (10.26), but now $k = (k_x^2 + k_y^2)^{1/2}$ is the total wave number of the disturbance.

Equation (10.26) describes the rate of film leveling, as $\omega < 0$ for any value of the dimensionless wave number a and the rest of parameters. This implies that if at time $t = 0$ a small bump is imposed on the interface, Eq. (10.26) describes how it will relax to zero and the interface will return to $h = d$.

The overall rate of film leveling can be estimated by the maximal value of the growth (decay, in the case at hand) rate ω, as given by Eq. (10.26). If the lateral size of the film is L, the fastest decaying mode is the longest available one, so that its wave number is $k = 2\pi/L$. Thus, the rate of disturbance decay is given by:

$$\omega_m = -\frac{4\pi^2 d^3}{3\mu L^2}\left(\rho g + \frac{4\pi^2 \sigma}{L^2}\right)$$

so the amplitude of the disturbance will reach the value of, say a thousandth of the initial amplitude at the time of $t = (\ln 0.001)/\omega_m$. However, this is only an estimate based on the linear stability analysis, and the effect of nonlinearities on the rate of film leveling can be found only from the solution of Eq. (10.25).

Equations (10.25a,b), with the obvious change in the sign of the gravity term in each of these, also apply to the case of a film on the underside of a plate. This case is known in the literature as the Rayleigh–Taylor instability [Chandrasekhar, 1961] of a thin viscous layer. To study the stability properties of such a system, one replaces g by $-g$ in Eq. (10.26) and finds that:

$$\mu\omega = \frac{1}{3d}(\rho|g|d^2 - \sigma a^2)a^2 \qquad (10.27)$$

The film is linearly unstable if

$$a^2 < a_c^2 \equiv \frac{\rho|g|d^2}{\sigma} \equiv Bo$$

That is, if the perturbations are so long that the nondimensional wave number is smaller than the square root of the Bond number Bo, which measures the relative importance of gravity and capillary effects. The value of a_c is often called the (dimensionless) *cutoff wave number* for neutral stability. The cutoff wave number is defined in a way that all perturbations with the wave number larger than a_c are damped, while those with the wave number smaller than a_c are amplified.

We should point out that Eq. (10.25) constitutes the valid limit to the governing set of equations and boundary conditions when the Bond number Bo is asymptotically small. This follows from the relationships $G = Bo\bar{S}$ and $G = O(1)$ and the large value of $\bar{S}$, as assumed in Eq. (10.15).

The case of Rayleigh–Taylor instability was studied by Yiantsios and Higgins (1989; 1991) for a thin film of a light fluid atop the plate, overlain by a large body of a heavy fluid, and by Oron and Rosenau (1992) for a thin liquid film on the underside of a plane. It was found that evolution of an interfacial disturbance of small amplitude leads to rupture of the film; that is, at certain location(s) the local thickness of the film is driven to zero.

The dimensionless wave number of the fastest growing mode is determined for a film of an infinite lateral extent from Eq. (10.27) as $a = \sqrt{Bo/2}$, and its growth rate is determined from Eq. (10.27) as:

$$\omega_m = \frac{\rho^2 g^2 d^3}{12\mu\sigma}$$

Thus, the time of film rupture can be estimated by $t = \ln(d/h_0')/\omega_m$.

Yiantsios and Higgins (1989) showed that Eq. (10.25a) with $G < 0$ admits several steady solutions. These consist of various numbers of sinusoidal drops separated by "dry" spots of zero film thickness, as shown in Figure 8 in Yiantsios and Higgins (1989). The examination of an appropriate free energy functional suggests that multidrop states are energetically less preferred than a one-drop state. These analytical results were partially confirmed by numerical simulations. As found in the long time limit, the solutions can asymptotically approach multihumped states with different amplitudes and spacings, as well. This suggests that terminal states depend upon the choice of initial data [Yiantsios and Higgins, 1989]. It was also found that if the overlying semi-infinite fluid phase is more viscous than the thin liquid film, the process of the film rupture slows down in comparison with the single-fluid case.

It is interesting to note that Eq. (10.25a) with $G < 0$ was also derived and studied by Hammond (1983) in the context of capillary instability of a thin liquid film on the inner side of a cylindrical surface, when gravity was neglected. The gravitational term there was due to the destabilizing effect of the capillary forces arising from longitudinal (along the axis of the cylinder) disturbances. Hammond (1983) also showed that the film ruptures but the process of rupture is infinitely long.

The three-dimensional version of the problem of the Rayleigh–Taylor instability was considered by Fermigier et al. (1992) using the weakly nonlinear analysis. Formation of patterns of different symmetries and transition between these patterns were experimentally studied. Axially symmetric cells and hexagons were found to be preferred. Droplet detachment was observed at the final stage of the experiment as a manifestation of a film rupture. The growth of an axisymmetric drop was shown in Figure 5 in Fermigier et al. (1992). A theoretical study of the Rayleigh–Taylor instability in an extended geometry [Fermigier et al., 1992] on the basis of the long-wave equation showed the tendency of the hexagonal structures to emerge as a preferred pattern in agreement with their own experimental observations.

Saturation of the Rayleigh–Taylor instability of a thin liquid film and therefore prevention of its rupture by an imposed advection in the longitudinal (parallel to the interface) direction was discussed by Babchin et al. (1983). Similarly, capillary instability of an annular film saturates due to a through flow [Frenkel et al., 1987].

Stillwagon and Larson (1988) considered the problem of a film leveling under the action of capillary force on a substrate with topography given by $z = \ell(x)$. Using the approach described above, they derived the evolution equation that for the case of zero gravity reads:

$$\mu h_t + \frac{1}{3}\sigma[h^3(h+\ell)_{xxx}]_x = 0 \qquad (10.28)$$

Numerical solutions of Eq. (10.28) showed a good agreement with their own experimental data. At short times there is film deplanarization due to the emergence of capillary humps, but these relax at longer times.

10.3.2 van der Waals Forces and Constant Surface Tension

Because of very small typical length scales of MEMS applications in general and of liquid film thickness in particular, which go down into the range of fractions of a micrometer, a new physics related mainly to intermolecular forces has to be brought into consideration. These fundamental types of forces acting on interatomic or intermolecular distances can affect the dynamics of macroscopic thin liquid films. Although some of them, such as weak and strong interactions, are short-range (i.e., much beyond the validity limits of continuum theory considered here), others, such as electromagnetic and gravitational forces, are of a long range and will be thus of a great importance for the subject of the current review.

Israelachvili (1992) presents a classification of electromagnetic forces into three categories. The first category includes purely electrostatic forces that stem from the Coulomb interaction. These forces include interactions between charges, dipoles etc. The second category includes polarization forces that stem from the dipole moments induced in totally neutral particles by the electric fields associated with other neighboring particles and permanent dipoles. These forces include interactions in a solvent medium. The third category includes forces of quantum mechanics origin. Such forces lead to chemical bonding and to repulsive steric interactions. Among these forces one can also find the force that acts, similar to the gravitational force, between all kinds of particles whether charged or neutral. It is called *dispersion force* or *London force*. The origin of the dispersion force is explained by the following consideration: In an electrically neutral particle whose time-averaged dipole moment vanishes, an instantaneous dipole moment does not vanish according to a time-varying relative distribution of negative and positive charges. Such an instantaneous dipole moment gives rise to a dipole moment in the neighboring neutral particles, and the interaction between these dipoles induces the force with a nonvanishing time-averaged value. These dispersion forces are long-range forces acting at distances of several Angstroms to several hundreds of Angstroms. They play, as we shall see later, a very important role in the dynamics of ultrathin liquid films whose average thickness is in the above range and in various phenomena such as wetting and adhesion. The dispersion forces can be either attractive or repulsive affecting by this the properties of good or poor wetting of solids by liquids. The presence of other bodies alters the dispersion interaction between the molecules, thus the latter is nonadditive. As shown in Table 6.3 in Israelachvili (1992), the dispersion force constitutes in many cases, except for highly polar water molecules, the main contribution to the total intermolecular force called *van der Waals force*. Various types of potentials describing the forces acting between molecules were reviewed by Israelachvili (1992), and the reader is referred to pages 112 to 114 therein.

Dzyaloshinskii et al. (1959) developed a theory for van der Waals interactions in which an integral representation is given for the excess Helmholtz free energy of the layer as functions of the frequency-dependent dielectric properties of the materials in the layered system.

The potential ϕ of the van der Waals forces is frequently specified in terms of the excess intermolecular free energy ΔG. These two values are related each to other via:

$$\phi = \frac{\partial \Delta G}{\partial h} \qquad (10.29)$$

It follows in this case from Eq. (10.22) in the three-dimensional case and Eq. (10.23) in the two-dimensional case that the film is unstable to infinitesimal disturbances only if:

$$\frac{\partial \phi}{\partial h} < 0 \quad \text{or equivalently} \quad \frac{\partial^2 \Delta G}{\partial h^2} < 0 \tag{10.30}$$

It follows from Eq. (10.30) that the film is unstable only if the potential ϕ has a decreasing branch or ΔG displays a negative curvature, both as functions of the film thickness h.

In the special case of an apolar film with parallel boundaries and nonretarded forces,

$$\phi = \phi_r + A' h^{-3}/6\pi \tag{10.31a}$$

where ϕ_r is an additive reference value for the body-force potential omitted hereafter and A' is the dimensional Hamaker constant [Dzyaloshinskii et al., 1959]. When $A' > 0$, there is negative disjoining pressure (referred to sometimes as conjoining pressure), and a corresponding attraction of the two interfaces (solid–liquid and liquid–gas) toward each other causes the instability of the flat state of the film surface and eventually its breakup. When the disjoining pressure is positive, $A' < 0$, the interfaces repel each other, and the flat state of the film surface is energetically preferred.

Various forms for the potential ϕ accounting for more complex physical situations are encountered in the literature. Mitlin (1993), Mitlin and Petviashvili (1994), Khanna and Sharma (1998) and others used the Lennard–Jones 6–12 potential for van der Waals interactions between the solid and the apolar liquid:

$$\phi = A'_3 h^{-3} - A'_9 h^{-9} \tag{10.31b}$$

with positive dimensional Hamaker coefficients A'_j. In this case, the two interfaces of the film are mutually attracting when the separation distance is relatively large, which drives the instability of the flat state of the film surface, and mutually repelling when the separation distance is relatively short, which finally saturates the amplitude of the interfacial undulation.

If the solid substrate is coated with a layer of thickness δ, the potential of the intermolecular pairwise interactions among the solid, coating, passive air and apolar liquid phases is given by [Bankoff, 1990; Hirasaki, 1991; Sharma and Reiter, 1996; Khanna et al., 1996; Oron and Bankoff, 1999]:

$$\phi = A'_3 h^{-3} + \hat{A}'_3 (h+\delta)^{-3} \tag{10.31c}$$

where $A'_3 = (A'_{\ell\ell} - A'_{c\ell})/6\pi$, $\hat{A}'_3 = (A'_{s\ell} - A'_{c\ell})/6\pi$ with A'_{ij} being the Hamaker constant related to the interaction between the phases i and j, $A'_{ij} = A'^{1/2}_{ii} A'^{1/2}_{jj}$ [Israelachvili, 1992], and subscripts s, c and ℓ correspond, respectively, to solid, coating and liquid phases.

Oron and Bankoff (1999) derived the potential topologically similar to the Lennard–Jones potential (10.31a) but with different exponents:

$$\phi = A'_3 h^{-3} - A'_4 h^{-4} \tag{10.31d}$$

to model the simultaneous action of the attractive $(A'_3 > 0)$ long-range and repulsive $(A'_4 > 0)$ (relatively) short-range van der Waals interactions and their influence on the dynamics of the film. To obtain the potential of Eq. (10.31d), Eq. (10.31a) was expanded into the Taylor series in h under the assumption of $\delta \ll d$ with $\hat{A}'_3 > 0$, $A'_3 + \hat{A}'_3 > 0$, and only two leading terms of this expansion being kept. Thus, the coefficients A'_3, A'_4 are specified by the properties of the three phases. The potential of the form of Eq. (10.31d) is also appropriate for liquid films on a rough solid substrate [Teletzke et al., 1987; Mitlin, 2000].

A combination of long-range apolar (van der Waals) and shorter range polar intermolecular interactions gives rise to the generalized disjoining pressure expressed by the potential:

$$\phi = A'_3 h^{-3} - S^p \exp(-h/\ell)/\ell \qquad (10.31e)$$

where S^p and ℓ are dimensional constants [Williams, 1981; Sharma and Jameel, 1993; Jameel and Sharma, 1994; Paulsen et al., 1996; Sharma and Khanna, 1998; and others], being, respectively, the strength of the polar interaction and its decay length ℓ, called the *correlation length* for polar interaction. The polar component of the potential is repulsive, if $S^p > 0$, and attractive, if $S^p < 0$. Sharma and Jameel (1993) classified films with polar and apolar components into four groups: type I systems, with both polar and apolar attractive forces ($A'_3 > 0$, $S^p < 0$); type II systems, with apolar attractions and polar repulsions ($A'_3 > 0$, $S^p > 0$); type III systems, with both polar and apolar repulsions ($A'_3 < 0$, $S^p > 0$); and type IV systems, with apolar repulsions and polar attractions ($A'_3 < 0$, $S^p < 0$). Type I films are always unstable and their dynamics in many ways are similar to those of apolar films described by the potential, Eq. (10.31a), while those of type III are always stable. Type II and IV films display ranges of stability and instability according to the sign of the derivative $\partial \phi / \partial h$; see the instability criterion Eq. (10.30c).

10.3.3 Homogeneous Substrates

Scheludko (1967) observed experimentally the spontaneous breakup of ultrathin static films and proposed that negative disjoining pressure is responsible. He also used linear stability analysis in order to calculate a critical thickness of the film below which breakup occurs, while neglecting the presence of electric double layers. Since then a great deal of scientific activity has been focused around the phenomenon.

The dynamics of ultrathin liquid films and the process of dewetting of solid surfaces have attracted special interest during the last decade. Progress and development of experimental techniques such as ellipsometry, X-ray reflectometry, atomic force microscopy (AFM) and others, on the one hand, and computational techniques along with the availability and affordability of fast computers, on the other hand, have helped to advance the study of the pertinent phenomena. The main interest was and still is centered around pattern formation and the dominant mechanisms driving film evolution. In the context of the latter issue, the polemics is ongoing between the two candidates, namely, thin-film instability arising from the interaction between the intermolecular and capillary forces (called sometimes in the literature *spinodal dewetting* or *spinodal mode*) and nucleation of holes from impurities or defects. It should be noted that most if not all of the experiments with dewetting recorded in the literature were carried out on liquid polymer films, while the theory is currently available for simple Newtonian liquids. The reasons for using polymer films in terms of controllability of the experiments were discussed by Sharma and Reiter (1996) and Reiter et al. (1999b).

Bischof et al. (1996) performed experiments on ultrathin ($\approx$40-nm-thick) metal (gold, copper and nickel) films on a fused silica substrate irradiated by a laser and turned into the liquid phase. Both isolated and coalesced holes along with typical rims surrounding them were observed. Little humps were found in the center of many holes, and the mechanism of heterogeneous hole nucleation was suggested to be responsible for formation of these. However, along with this mechanism, growing film surface deformations were detected, thus the mechanism of spinodal dewetting is also in effect. The characteristic size of film surface deformations was found to be well correlated with the wavelength of the most amplified linear mode proportional to d^2. Similar conclusions about the dominance of the nucleation mechanism were drawn later by Jacobs et al. (1998). Experimental evidences of spinodal dewetting were given by Brochard-Wyart and Daillant (1990), Reiter (1992), Sharma and Reiter (1996), Xie et al. (1998), Reiter et al. (1999b; 2000) and others. Reiter et al. (2000) showed for the first time that the spinodal length and time scales are consistent with the results of their experiments. Independent molecular dynamics simulations [Koplik and Banavar, 2000] support the spinodal character of dewetting.

Khanna et al. (2000) presented the first real-time experimental observation of the pattern formation in thin unstable polydimethylsiloxane (PDMS) films placed on coated silicon wafers and bounded by aqueous surfactant solutions. The process of film disintegration ("self-destruction") was described as the sequence of the stages of self-organization of the pattern and selective amplification of the interfacial disturbance, breakup of the film and formation of isolated circular holes, lateral expansion of the holes and emergence of long liquid ridges and, finally, breakup of the ridges into droplets standing on an equilibrium film plateau and ripening of the droplet structure.

Muller-Buschbaum et al. (1997) studied the process of dewetting of thin polystyrene films on silicon wafers covered with an oxide layer of different thicknesses and observed the emergence of "nano-dewetting structures" inside the dewetted areas. These structures in the form of troughs about 70 nm in diameter indicated that the dewetted areas were neither completely dry nor covered with a flat ultrathin layer of the liquid. Such patterns were detected along with micrometer-size drops usually observed in similar situations on top of oxide layers 24 Å thick but were not present on thinner oxide layers where only drops emerged. The dependence of the mean drop size as well as the trough diameter on the initial thickness of the film was in agreement with theoretical predictions based on the assumption of spinodal dewetting [Muller-Buschbaum et al., 1997].

Consider now a film under the influence of van der Waals forces and constant surface tension only, so that $\Pi_1 = \Pi_2 = \Pi_3 = \sigma_x = \sigma_y = 0$. As we shall see shortly, the planar film is unstable when $A' > 0$ and stable when $A' < 0$. In two dimensions, Eq. (10.23) in the case at hand becomes [Williams and Davis, 1982]:

$$\mu h_t + \frac{1}{6\pi} A'(h^{-1} h_x)_x + \frac{1}{3}\sigma(h^3 h_{xxx})_x = 0 \tag{10.32a}$$

Its dimensionless version reads:

$$H_\tau + A(H^{-1} H_\xi)_\xi + \frac{1}{3} S (H^3 H_{\xi\xi\xi})_\xi = 0 \tag{10.32b}$$

where

$$A = \frac{\varepsilon A'}{6\pi\rho\nu^2 d}$$

is the scaled dimensionless Hamaker constant and $\nu = \mu/\rho$ is the kinematic viscosity of the liquid. Here, the characteristic velocity was chosen as $U_0 = \nu/d$.

If Eq. (10.32a) is linearized around $h = d$, the following characteristic equation for ω,

$$\mu\omega = \left(\frac{a}{d}\right)^2 \left(\frac{A'}{6\pi d} - \frac{1}{3}\sigma d a^2\right) \tag{10.33a}$$

is obtained. It follows from Eq. (10.33a) that there is instability for $A' > 0$, driven by the long-range molecular forces, and stabilization is due to surface tension. The cutoff wave number a_c is given then by:

$$a_c = \frac{1}{d}\left(\frac{A'}{2\pi\sigma}\right)^{1/2} \tag{10.33b}$$

which reflects the fact that an initially corrugated interface has its thin regions thinned further by van der Waals forces while surface tension cuts off the small scales. Instability is possible only if $0 < a < a_c$, as seen by combining Eqs. (10.33a) and (10.33b):

$$\mu\omega = \frac{\sigma a^2}{3d}(a_c^2 - a^2) \tag{10.34}$$

Similar results were obtained in the linear stability analysis presented by Jain and Ruckenstein (1974). On the periodic infinite domain of wavelength $\lambda = 2\pi/k$, the linearized theory predicts that the film is always unstable as all wave numbers are available to the system. In an experimental situation, the film resides in a container of finite width, say, L. The solution obtained from the linear stability theory for $0 \leq \xi \leq L$ would show that only perturbations of the nondimensional wave number lower than a_c (see Eq. (10.34)) and those of small enough wavelength that "fit" in the box (i.e., $\lambda < L$) are unstable. Hence, no instability would occur by this estimate if $2\pi d/L > a_c$. One sees from this theory that it is inappropriate to seek a "global" critical thickness from the theory but only a critical thickness for a given experiment, as the condition depends on the system size L.

The evolution of the film interface as described by Eq. (10.32) with periodic boundary conditions and an initial linearly unstable perturbation of the uniform state leads to the rupture of the film in a finite (nondimensional) time τ_R [Williams and Davis, 1982]. This rupture manifests itself by the fact that at a certain time the local thickness of the film becomes zero. The time of rupture of the film of an infinite lateral extent can be estimated from the linear stability theory by:

$$t_R = \frac{48\pi^2 d^5 \sigma}{A'} \ln\left(\frac{d}{h'_0}\right)$$

However, the rate of film thinning, measured as the rate of decrease of the minimal thickness of the film, explosively increases with time and becomes much larger than the disturbance growth rate given by Eq. (10.33a) according to the linear theory. This fact could be foreseen by observation of Eq. (10.32b), noticing that the "effective" diffusion coefficient, proportional to H^{-1}, in the backward diffusion A term increases indefinitely as the film becomes thinner, $H \to 0$, while the local stabilization effect provided by surface tension weakens proportionally to H^3. This observation was confirmed numerically from the solution of Eq. (10.32b) [Williams and Davis, 1982] and analytically by weakly nonlinear theory [Sharma and Ruckenstein, 1986; Hwang et al., 1993]. Hwang et al. (1997) studied the three-dimensional version of this problem using the natural extension of Eq. (10.32b). They confirmed the fact of film rupture and found that it occurs pointwise and not along a line. Moreover, the rupture time in the three-dimensional case is shorter than in the two-dimensional one.

Burelbach et al. (1988) used numerical analysis to show that in a certain time range near the rupture point surface tension has a minor effect and therefore the local behavior of the interface there is governed by the backward diffusion equation,

$$H_\tau + A(H^{-1}H_\xi)_\xi = 0 \tag{10.35}$$

Looking for separable solutions for Eq. (10.35) in the form $H(\xi,\tau) = T(\tau)X(\xi)$, Oron et al. (1997) used the known temporal asymptotics [Burelbach et al., 1988] and found that [see also Rosenau, 1995]:

$$H(\xi,\tau) = A\frac{b^2}{2}(\tau_R - \tau)\sec^2\left(\frac{b\xi}{2}\right) \tag{10.36}$$

where τ_R is the time of rupture and b is the constant which should be determined from the matching with the far-from-rupture solution. The minimal thickness of the film close to the rupture point is therefore expected to decrease linearly with time. This allows the long-wave analysis to be extrapolated closer to the point where adsorbed layers and/or moving contact lines appear. However, the solution to Eq. (10.35) is not expected to be valid very close to the rupture point, as the film progresses toward rupture and the fluid velocities diverge. Recently, the existence of infinite sets of similarity solutions in which both van der Waals and surface tension forces are equally important near rupture was shown [Zhang and Lister, 1999; Witelski and Bernoff, 1999; 2000]. These solutions have the same form in both two-dimensional and axisymmetric cases

$$H(\xi, \tau) = (\tau_R - \tau)^{1/5} g[\xi(\tau_R - \tau)^{-2/5}] \qquad (10.37)$$

where g is a function to be further determined. Among this infinite set of self-similar solutions the fundamental solution stable to linear perturbations was identified as the only asymptotic behavior observed in the direct numerical solution of Eq. (10.32b) [Zhang and Lister, 1999; Witelski and Bernoff, 1999]. It is described by the function g being the least oscillatory one among the possible solutions of the corresponding ordinary differential equation. The point rupture was found to be the preferred mode of film rupture in three dimensions [Witelski and Bernoff, 2000].

Several authors [Kheshgi and Scriven, 1991; Mitlin, 1993; Sharma and Jameel, 1993; Jameel and Sharma, 1994; Mitlin and Petviashvili, 1994; Oron and Bankoff, 1999] have considered the dynamics of thin liquid films in the process of dewetting of a solid surface. The effects important for a meaningful description of the process are gravity, capillarity and, if necessary, the use of a generalized disjoining pressure, which contains a sum of intermolecular attractive and repulsive potentials. The generalized disjoining pressure of the Mie type is destabilizing (attractive) or stabilizing (repulsive) for the film of a larger (smaller) thickness, still within the range of several hundreds of Angstroms [Israelachvili, 1992] where van der Waals interactions are effective. Equations (10.21) and (10.23) can be rewritten in the situation considered, respectively, in the form:

$$H_\tau - \frac{1}{3}[H^3(GH - SH_{\xi\xi} + \Phi)_\xi]_\xi = 0 \qquad (10.38a)$$

$$\mu h_t - \frac{1}{3}[h^3(\rho g h - \sigma h_{xx} + \phi)_x]_x = 0 \qquad (10.38b)$$

Linearizing Eq. (10.38b) around $h = d$, one obtains:

$$\mu \omega = -\frac{1}{3}a^2 d\left(\rho g + \frac{\partial \phi}{\partial h}d + \frac{\sigma a^2}{d^2}\right) \qquad (10.39)$$

It follows from Eq. (39) that the necessary condition for linear instability is

$$\frac{\partial \phi}{\partial h}d < -\rho g \qquad (10.40)$$

i.e., the destabilizing effect of the van der Waals force has to be stronger than the leveling effect of gravity.

Kheshgi and Scriven (1991) studied the evolution of the film using Eq. (10.38a) with the potential (10.31a) and found that smaller disturbances decay due to the presence of gravity leveling, while larger ones grow and lead to film rupture propelled by van der Waals force. Mitlin (1993) and Mitlin and Petviashvili (1994) discussed possible stationary states for the late stage of solid-surface dewetting with the potential (10.31b) and drew the formal analogy between the latter and the Cahn theory of spinodal decomposition [Cahn, 1961]. Sharma and Jameel (1993; 1994) followed the film evolution as described by Eqs. (10.38) and (10.31e) with no gravity ($G = 0$) and concluded that thicker films break up, while thinner ones undergo "morphological phase separation" that manifests itself in creation of steady structures of drops separated by ultrathin, practically flat liquid films (holes). Similar patterns of morphological phase separation were also observed by Oron and Bankoff (1999) in their study of the dynamics of thin spots near film breakup. Figure 2 in Oron and Bankoff (1999) shows typical steady-state solutions for Eq. (10.38a) with the potential (10.31d) and $G = 0$ for different sets of parameters.

Khanna and Sharma (1998) used the Lennard–Jones potential, Eq. (10.31b), to study the three-dimensional dynamics of an apolar liquid film on a solid substrate. Their investigation based on the dimensionless evolution equation,

$$H_\tau - \frac{1}{3}\nabla \cdot (H^3 \nabla \Phi) + \frac{1}{3} S \nabla \cdot (H^3 \nabla \nabla^2 H) = 0 \qquad (10.41)$$

showed that in the case of $A'_9 d^6 \ll A'_3$ the corresponding film evolution displays the formation of steep holes. These holes are axisymmetric when the size of the periodic domain slightly exceeds the critical wavelength. However, they are nonaxisymmetric with uneven rims surrounding the holes for larger domains.

Sharma and Khanna (1998) studied the film dynamics governed by Eq. (10.41) with the potential (10.31e) that engenders short-range polar repulsion, intermediate-range van der Waals attraction and long-range polar repulsion. The linear and weakly nonlinear analyses fail to predict the structure of the emerging patterns. The former, however, can successfully predict the length scale of the resulting pattern. Two characteristic, morphologically different patterns were found and in both of them true dewetting does not occur. A microfilm covering the solid surface emerges and persists instead. The first pattern is typical for films with a thickness closer to the upper critical thickness. In this case, the film undergoes the stages of reorganization into a pattern of a length scale corresponding to the fastest growing linear mode, emergence of circular holes with rims uneven in height, coalescence of the holes and slow evolution into circular drops standing on top of a flat microfilm. The second one typical for relatively thin films of initial thickness near the lower critical thickness does not exhibit formation of circular holes and produces droplets that tend to be circular subject to the capillary forces. This type of a film evolution seems to be less frequent but has also been observed in experiments [Xie et al., 1998]. The flat microfilm covering the substrate emerges after the formation of isolated drops. Finally, a stable state that consists of a single circular drop standing on a flat equilibrium film is reached. In the intermediate range of the initial film thickness, the patterns consisting of holes, ridges and drops coexist when the number of each of these depends on the initial film thickness. As will be discussed below, all kinds of structures that contain holes, drops and ridges may coexist on heterogeneous substrates [Konnur et al., 2000].

Sharma et al. (2000) attributed the type of film dewetting to the relative position of the average thickness of the film d and the location of the minimum of the function $\partial \phi / \partial h$: When the film is thicker than the thickness corresponding to the minimum of $\partial \phi / \partial h$, the film dewets by formation of holes, while in the opposite case dewetting sets in by formation of liquid ridges which further break up into droplets. In either case, ripening of the droplet structure takes place and larger droplets grow at the expense of smaller ones.

Oron (2000c) studied the evolution of a film on a coated solid substrate as described by Eq. (10.41) with the potential Eq. (10.31d) given in dimensionless form as $\Phi = A_3 H^{-3} - A_4 H^{-4}$, where A_3 and A_4 are positive nondimensional Hamaker constants. As noted above, this potential acts as long-range van der Waals attraction and short-range repulsion, both apolar. The evolution of a small-amplitude disturbance of a flat initial state $H = 1$ leads to self-organization of the surface, emergence of holes, their expansion, coalescence and formation of polygonal network of liquid ridges on top of the essentially flat microlayer. Later on the liquid ridges break up into isolated drops and ridges that pump their liquid by means of the capillary forces into the largest drop, making the latter bigger and more circular. The existence of a "thick" microlayer facilitates a relatively free liquid flow along the coated substrate and the accumulation of the liquid in an isolated drop standing on a plateau, minimizing by that the free energy of the system. Finally, a steady state is reached, where a circular drop persists standing on a flat equilibrium film (Figure 10.7). The film evolution described was found to follow the typical sequence of events as described in the experiments [Khanna et al., 2000].

Reiter et al. (1999a) carried out theoretical and experimental studies of the dynamics of films on wettable solid surfaces and being in contact with the ambient phase of various physicochemical identity. By exchanging the ambient phase it is possible to vary the total Hamaker constant of the system and even to change its sign, thus turning the initially stable configuration into the unstable one. Experiments with PDMS films on a silicon wafer with alternating air and water ambient phases provided an example

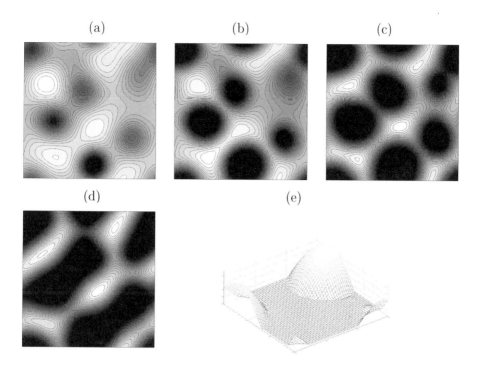

FIGURE 10.7 Stages of evolution of a nonevaporating film as described by Eq. (10.41) with the potential (10.31d). The first four images are given in the form of a contour plot, while the last one is in the form of a surface plot. Each image has its own brightness, so the film thickness in different images cannot be compared. A polygonal network of liquid ridges qualitatively similar to the experimental observations made by Sharma and Reiter (1996) is seen in parts (b) to (d). Bright and dark shades correspond to elevations and depressions, respectively. (From Oron, A. (2000) *Phys. Rev. Lett.* **85**, pp. 2108–2111. Copyright 2000 by the American Physical Society. With permission.)

of such a system [Reiter et al., 1999a; 1999b]. When in contact with air, the film remained flat and did not exhibit any evidence of instability; however, while in contact with water, instability set in and the film, whose initial thickness ranged between 30 and 110 nm, finally reached the state in which small droplets stood on top of a thin wetting layer. This phenomenon was studied theoretically [Reiter et al., 1999a] using a three-dimensional evolution equation (10.41) with potential topologically similar to that of Eq. (10.31b). Qualitative agreement between theory and experiments was found to be quite good. However, as noted by Reiter et al. (1999), even quantitative agreement between the two could be achieved but for an "unexpectedly high effective Hamaker constant." The reason for this is still unclear.

10.3.4 Heterogeneous Substrates

A study of the dynamics of thin liquid films on a heterogeneous substrate can be motivated by the presence of dust particles or other types of impurities, oxidized or rough patches, or varying chemical composition leading to nonuniform wettability properties of the solid surface underlying the film. These and other types of heterogeneity of the substrate may be present unintentionally or may be created deliberately to achieve a certain goal.

The governing equation to be studied in this context is again Eq. (10.41) but, in contrast to the case of the homogeneous substrate where the potential of the intermolecular forces depends solely on the film thickness, $\Phi = \Phi(H)$, in the current case the potential explicitly depends on the lateral spatial coordinates as well. This dependence enters the equations via spatial variation of the Hamaker coefficients.

A series of papers [Lenz and Lipowsky, 1998; Herminghaus et al., 1999; 2000; Gau et al, 1999; Lenz, 1999; Lipowsky et al., 2000] examined the morphological transitions of liquid layers on heterogeneous

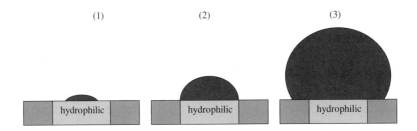

FIGURE 10.8 Equilibrium states of the droplets on a heterogeneous substrate that consists of alternating hydrophilic and hydrophobic patches. These equilibria depend on the droplet volume.

structured substrates. Lenz and Lipowsky (1998) showed by minimization of the total interfacial free energy that for a domain containing a hydrophilic patch confined between the hydrophobic ones three different regimes, depending on the volume of the droplet, are possible, as shown in Figure 10.8. In regimes 1 and 3, the respective contact angles are prescribed *a priori* by the phases chosen and satisfy the Young equation. Regime 2 is characterized by the droplet volume and the contact angle spanning the range between the respective values of regimes 1 and 3. In the limiting case of perfectly wettable hydrophilic and nonwettable hydrophobic patches, only regime 2 is possible. In the case of a two-dimensional, square, periodic lattice of N circular hydrophilic patches surrounded by hydrophobic domains, the equilibrium state for a low total liquid volume consists of N identical droplets, each of them covering its own hydrophilic patch similar to regime 1 for the case of an isolated patch. As the total volume of the liquid increases, the droplets grow and the system undergoes transition to a heterogeneous equilibrium state that consists of one large drop and $N-1$ small identical drops. More complex heterogeneous states were found to be unstable [Gau et al., 1999]. When the total volume of the liquid increases further beyond a certain value, a third equilibrium state that represents a single completely wetting layer covering the entire system becomes possible. The transition to this equilibrium state is possible from either of the aforementioned states. For striped periodic domains, all of the three equilibrium states found in the previous case persist. However, a new kind of transition from the homogeneous state to the film state exists here. This transition consists of the stages where identical droplets span over several hydrophilic patches and the hydrophobic ones in between.

Gau et al. (1999) performed a series of experiments with liquid microchannels created by hydrophilic stripes about 40 μm wide and further condensation of water onto the substrate. When the total amount of condensed water was low, the microchannels had a shape of cylindrical caps of a constant cross section with a small contact angle θ between the liquid and the solid. However, when the total volume of water exceeded a certain value, the straight channels underwent instability, which led to formation of a single bulge on each of the stripes. Moreover, when the bulges on two neighboring channels were in close proximity, they merged to form a big drop or microbridge between the channels. Gau et al. (1999) found theoretically that the cylindrical cap configuration with contact angle θ is linearly stable for $\theta < 90°$ and unstable to long-wave disturbances for $\theta > 90°$, provided that the wavelength of the disturbance is sufficiently large:

$$\lambda > \lambda_c = \left[\frac{\pi/2}{\theta^2 - (\pi/2)^2} \right]^{1/2} \frac{\theta}{\sin\theta} a_\gamma$$

where a_γ is the width of the hydrophilic stripe. The presence of this instability disallows the emergence of long homogeneous liquid channels with a contact angle larger than 90°. The onset of the instability occurs at $\theta = 90°$, and the wavelength of the critical disturbance is infinite. This explains the formation of a single bulge on the microchannel [Herminghaus et al., 2000]. The precise shape of the configuration of liquid microchannels with bulges was numerically calculated by Gau et al. (1999) using minimization of the total free energy. A very good agreement was found between the experimental and theoretical results.

Konnur et al. (2000) and Kargupta et al. (2000) studied the three-dimensional dynamics of liquid crystal films using Eq. (10.41) with the potential (10.31e) with different sets of fixed positive values—a_3, S^p, ℓ—on the patches of the substrate. They reported a new mechanism of film instability associated with the substrate heterogeneity which is driven by the pressure gradient generated by the spatial variation of ϕ and directed from the less to the more wettable domains on the solid. The potential (10.31e) employed by Konnur et al. (2000) prescribes instability for both relatively thin and thick films, while films in the intermediate range are stable. They found that the presence of heterogeneity is able to destabilize even spinodally stable films, speed up the rupture process of the film and produce spatially complex and locally ordered patterns. Destabilization of spinodally stable films arises even when the heterogeneous patch is much smaller than the spinodal length scale determined as the wavelength of the fastest growing linearly unstable disturbance. It was found that true rupture can occur for spinodally stable films if the local thickness of the film is reduced by the heterogeneous mechanism to the value where the spinodal instability condition is met and both of the mechanisms propel the film to rupture. The evolution of an initially flat film typically exhibits such morphological patterns as a lack of surface deformations prior to the formation of a hole, emergence of a non-growing hole on a perfectly wetted substrate or in a spinodally stable film, formation of a "castle-moat" pattern with a central drop surrounded by a ring-like depression or hole and formation of locally ordered structures with alternating depressions and rims [Konnur et al. 2000; Kargupta et al. 2000]. The heterogeneous mechanism was found to be strong for relatively thick films and its time scale several orders of magnitude lower than that of the spinodal mechanism. Kargupta et al. (2000) also considered the two-dimensional dynamics of film on a substrate with a heterogeneous patch of varying size. They found that the presence of heterogeneity always causes the emergence of local interfacial depression, which can evolve into film rupture when the length of the patch becomes sufficiently large. The rupture time was found to rapidly decrease when the patch length increases beyond the critical length and to become independent of the patch length when the latter is large.

Studies of the two- and three-dimensional film dynamics on heterogeneous substrates with alternating parallel stripes of hydrophilic and hydrophobic patterns are now underway [Oron, 2000d; Sharma, 2000] and will be shortly reported elsewhere in the literature.

10.3.5 Flow on a Rotating Disc

The isothermal, axisymmetric flow of an incompressible viscous liquid was considered on a horizontal rotating disk by Reisfeld et al. (1991). Cylindrical polar coordinates r, θ, z are used in the frame of reference rotating with the disk. The film interface is located at $z = h(r, t)$. In the coordinate system chosen, the outward unit normal vector $\mathbf{n}$ and unit tangent vector $\mathbf{t}$ are

$$\mathbf{n} = \frac{(-h_r, 0, 1)}{(1 + h_r^2)^{1/2}}, \quad \mathbf{t} = \frac{(1, 0, h_r)}{(1 + h_r^2)^{1/2}}$$

The hydrodynamic equations analogous to Eq. (10.2) taking into account both the centrifugal forces and Coriolis acceleration are written in the vector form as:

$$\nabla \cdot \mathbf{v} = 0, \quad \rho[\mathbf{v}_t + (\mathbf{v} \cdot \nabla)\mathbf{v}] = -\nabla p + \mu \nabla^2 \mathbf{v} - \rho[\mathbf{g} + 2\boldsymbol{\omega} \times \mathbf{v} + \boldsymbol{\omega} \times \boldsymbol{\omega} \times \mathbf{v}]$$

where $\boldsymbol{\omega}$ is the angular-velocity vector with the components $(0, 0, \varpi)$. The boundary conditions are given by Eq. (10.4) formulated in cylindrical polar coordinates with $\nabla_s \sigma = 0$ and $\Pi = 0$.

The characteristic length scale in the horizontal direction is chosen as the radius of the rotating disk $\bar{R}$, and the velocity scale is taken as $U_0 = \rho \varpi^2 \bar{R} d^2/\mu$. A small parameter ε is defined in accordance with Eq. (10.7) as $\varepsilon = d/\bar{R}$. The dimensionless parameters of the problem are the Reynolds number R as given in Eq. (10.12), the scaled inverse capillary number S given by Eqs. (10.12) and (10.15) and the

Froude number F,

$$F = \frac{U_0}{(gd)^{1/2}}$$

Using the procedures outlined above, one obtains at the leading order the following evolution equation:

$$H_\tau + \frac{1}{3r}\left\{r^2 H^3 + SrH^3\left[\frac{1}{r}(rH_r)_r\right]_r\right\} = 0 \qquad (10.42)$$

Here, the terms describing the effect of inertia and gravity appeared in the terms of first order in ε and thus were omitted. However, they may be retained to investigate the dynamics of the rotating film in the first phase of the process including inertia and amplification of kinematic waves [Reisleld et al., 1991]. Equation (10.42) models the combined effect of capillary forces and centrifugal drainage, none of which describes any kind of instability.

For most spin-coating applications, S is very small and the corresponding term may be neglected, although it may be very important in planarization studies, where the leveling of liquid films on rough surfaces is investigated. Equation (10.42) can be thus simplified as:

$$H_\tau + \frac{1}{3r}(r^2 H^3)_r = 0 \qquad (10.43)$$

This simplified equation can then be used for further analysis. Looking for flat basic states $H = H(\tau)$, Eq. (10.43) is reduced to the ordinary differential equation which is to be solved with the initial condition $H(0) = 1$. The film thins due to centrifugal drainage according to the solution

$$H(\tau) = \left(1 + \frac{4}{3}\tau\right)^{-1/2}$$

which predicts a decrease of the thickness to zero at the infinite time. The cases where inertia was taken into account were considered by Reisfeld et al. (1991), who also provided linear stability analysis of flat base states.

Stillwagon and Larson (1990) considered the spin-coating process and leveling of a nonvolatile liquid film over an axisymmetric, uneven solid substrate. For a given local dimensionless height of the substrate $\ell(r)$, their equation derived from the Cartesian version valid for capillary leveling of a film in a trench resembles Eq. (10.42) and reads:

$$H_\tau + \frac{1}{3\hat{r}}[\alpha\hat{r}^2 H^3 + \hat{S}\hat{r}H^3(H_{\xi\xi\xi} + \ell_{\xi\xi\xi})]_\xi = 0 \qquad (10.44)$$

where $\hat{r}$ and ξ are, respectively, the radial coordinate and the radial distance from the trench, both scaled with the trench width, and α is the ratio of the width and the location of the trench. Equation (10.44) can be further simplified under the assumption that the width of the trench is small compared to its radial position and can be brought to the form:

$$H_\tau + \frac{1}{3}[H^3 + \Omega^{-2} H^3(H_{\xi\xi\xi} + \ell_{\xi\xi\xi})]_\xi = 0, \qquad (10.45)$$

where Ω^2 is the ratio between the centrifugal and capillary forces. Stillwagon and Larson (1990) calculated quasi-steady-state solutions close to the trench by solving the time-independent version of Eq. (10.44)

$$H^3(H_{\xi\xi\xi} + \ell_{\xi\xi\xi}) + \Omega^2 H^3 = \Omega^2 \qquad (10.46)$$

where the term on the right arises from the condition of uniformity of the film far from the trench. Experiments with liquid films reported by Stillwagon and Larson (1990) demonstrated quantitative agreement between measured film profiles and those obtained from Eq. (10.46).

Wu et al. [Wu and Chou, 1999; Wu et al., 1999] used Eq. (10.46) to study the degree of planarization for periodic uneven substrates expressed as the ratio between the amplitude of the deformed film interface and the average thickness of the film. They showed that this value is independent of Ω and varies slightly with the trench spacing for large Ω, while it decreases with the increase of spacing for small fixed values of Ω.

Chou and Wu (2000) studied the effect of air shear on the process of film planarization. Similar to the case considered in Section IIC of Oron et al. (1997; see Eq. (2.31)), where the term proportional to the imposed shear stress multiplied by hh_x arises in the evolution equation, air shear produces the advective term proportional to H^2, which has to be added to the expressions in the square brackets of the left-hand side of Eqs. (10.44) and (10.45). Corresponding additional terms will appear in Eq. (10.46) as well. Chou and Wu (2000) studied such an extended Eq. (10.46) and found that the shear stress enhances the amplitude of the film interface and thus opposes film planarization during spin coating for both isolated and periodic features of the substrate.

Puerrung and Graves (1993) considered three-dimensional quasi-steady states for spin coating over topography using the natural extension of Eq. (10.45) into three dimensions. Their theoretical and experimental results agree qualitatively, both showing the emergence of wake-like structures at the downstream side of the protrusion with crests extending along each of the corners and the depression near the center.

10.4 Thermal Effects

One of the best known fluid flows under the influence of heat transfer is the buoyancy or Rayleigh convection [Chandrasekhar, 1961] of a stagnant liquid layer lying on a horizontal solid surface triggered by heating from below and subsequent establishment of unstable density stratification. This convection sets in when the temperature difference across the layer exceeds a certain critical value, which is proportional among other physical parameters of the system to the third power of the layer thickness d. Due to the fact that the range of very small values of the film thickness is of a major interest in the context of MEMS, the Rayleigh effect is much weaker than the thermocapillary or Marangoni effect addressed next. The latter scales with the first power of d in contrast with d^3 in the case of the Rayleigh effect.

10.4.1 Thermocapillarity, Surface Tension and Gravity

The thermocapillary or Marangoni effect (e.g., see Davis [1987]) accounts for the emergence of interfacial shear stresses, due to the variation of surface tension with temperature ϑ, $\sigma = \sigma(\vartheta)$, which is in most cases monotonically decreasing. Such a shear stress is mathematically expressed by $\nabla_s \sigma$ [Edwards et al., 1991]. In order to incorporate the thermocapillary effect into the equations, one needs to add an energy equation and the appropriate boundary conditions related to heat transfer to the governing system Eqs. (10.2) to (10.4).

The energy equation in three dimensions and the boundary conditions have the form:

$$\rho c(\vartheta_t + u\vartheta_x + v\vartheta_y + w\vartheta_z) = k_{th}(\vartheta_{xx} + \vartheta_{yy} + \vartheta_{zz}) + \dot{q} \qquad (10.47)$$

$$\vartheta = \vartheta_0 \quad \text{at} \quad z = 0 \qquad (10.48a)$$

$$k_{th}\mathbf{n} \cdot \nabla\vartheta + \alpha_{th}(\vartheta - \vartheta_\infty) = 0 \quad \text{at} \quad z = h(x, y, t) \qquad (10.48b)$$

Here, c is the specific heat of the fluid; k_{th} is its thermal conductivity; ϑ_0 is the temperature of the rigid substrate, assumed to be uniform; and $\dot{q}$ is the rate of internal heat generation. The boundary condition

(Eq. 10.48b) is Newton's cooling law, and α_{th} is the heat-transfer coefficient describing the rate of heat transfer from the liquid to the ambient gas phase held at the constant temperature ϑ_∞.

Turning to the two-dimensional case, scaling the temperature by:

$$\Theta = \frac{\vartheta - \vartheta_\infty}{\vartheta_0 - \vartheta_\infty} \tag{10.49}$$

and substituting scales (10.8) into Eqs. (10.47) and (10.48) yield:

$$\varepsilon RP(\Theta_\tau + U\Theta_\xi + W\Theta_\varsigma) = \varepsilon^2 \Theta_{\xi\xi} + \Theta_{\varsigma\varsigma} + 2Qf(\varsigma) \tag{10.50}$$

$$\Theta = 1 \quad \text{at} \quad \varsigma = 0 \tag{10.51a}$$

$$\Theta_\varsigma - \varepsilon^2 \Theta_\xi H_\xi + B\Theta(1 + \varepsilon^2 H_\xi^2)^{1/2} = 0 \quad \text{at} \quad \varsigma = H \tag{10.51b}$$

where P and B are, respectively, the Prandtl and Biot numbers; Q is the dimensionless measure of the rate of internal energy generation defined by:

$$P = \frac{\rho c v}{k_{th}}, \quad B = \frac{\alpha_{th} d}{k_{th}}, \quad Q = \frac{\dot{q} d^2}{2 k_{th} \vartheta_r}, \tag{10.52}$$

where ϑ_r is the reference temperature chosen as $\vartheta_r = \vartheta_0 - \vartheta_\infty$ if $\vartheta_0 > \vartheta_\infty$, and as $\vartheta_r = \vartheta_0$ if $\vartheta_0 = \vartheta_\infty$. Furthermore, $f(\varsigma)$ expresses the dependence of the rate of internal energy generation on the vertical coordinate ς.

Begin first with the case of no internal heat generation, $\dot{q} = 0$, leading to $Q = 0$. Expand the temperature Θ in a perturbation series in ε, along with the expansions, Eq. (10.14), and substitute these into the system given by Eqs. (10.50) and (10.51). Assume again that $R = O(1)$ and further let $P, B = O(1)$ so that the convective terms in Eq. (10.50) are delayed to the next order; that is, declare that conduction in the liquid is dominant, and the conductive heat flux at the interface balances the heat lost to the environment.

At the leading order in ε the governing system for $\Theta^{(0)}$ consists of condition (10.51a):

$$\Theta_{\varsigma\varsigma} = 0 \tag{10.53}$$

and

$$\Theta_\varsigma + B\Theta = 0 \quad \text{at} \quad \varsigma = H \tag{10.54}$$

where the superscript zero has been dropped. The solution to this system is

$$\Theta = 1 - \frac{B\varsigma}{1 + BH} \quad \text{and} \quad \Theta_i = \frac{1}{1 + BH} \tag{10.55}$$

where $\Theta_i = \Theta(\tau, \xi)$ is the surface temperature.

According to Eq. (10.21) we are now required to determine the thermocapillary stress Σ_ξ. By the chain rule,

$$\Sigma_\xi = M(d\Sigma/d\Theta)(\Theta_\xi + H_\xi \Theta_\varsigma) \equiv -M\frac{\gamma(H)H_\xi}{(1 + BH)^2} \tag{10.56a}$$

where

$$\gamma(H) = -(d\Sigma/d\Theta)_{\Theta=\Theta_i}$$
$$M = \frac{\Delta\sigma}{\mu U_0} \tag{10.56b}$$

is the Marangoni number, and the sign change is inserted because $d\Sigma/d\Theta$ is negative for most common materials. Here, $\Delta\sigma$ is the change of surface tension over the temperature domain between the characteristic temperatures, usually ϑ_∞ and ϑ_0. To be more precise, if $\vartheta_\infty < \vartheta_0$ (heating at the bottom of the layer), then $\Delta\sigma > 0$ for standard fluid pairs with surface tension decreasing with temperature. For heating at the interface side, $\vartheta_\infty > \vartheta_0$ and $\Delta\sigma < 0$. The shear stress condition, Eq. (10.18b), has at the leading order in ε the form:

$$U_\varsigma + M\frac{\gamma(H)H_\xi}{(1+BH)^2} = 0 \quad \text{at } \varsigma = H \tag{10.57}$$

Thus, in this case, $\hat{\Pi}_1 = \hat{\Pi}_3 = \Phi = 0$, and Eq. (10.21) becomes:

$$H_\tau + \frac{1}{2}MB\left[\frac{H^2\gamma(H)H_\xi}{(1+BH)^2}\right]_\xi + \frac{1}{3}S(H^3 H_{\xi\xi\xi})_\xi = 0 \tag{10.58}$$

If gravity forces are to be included, $\Phi = g\varsigma$ and Eq. (10.58) becomes:

$$H_\tau + \left\{\left[\frac{1}{2}MB\frac{H^2\gamma(H)}{(1+BH)^2} - \frac{1}{3}GH^3\right]H_\xi\right\}_\xi + \frac{1}{3}S(H^3 H_{\xi\xi\xi})_\xi = 0. \tag{10.59}$$

For the most ubiquitous case in which surface tension is a linearly decreasing function of temperature $\sigma = \sigma(\vartheta)$, the value $(d\Sigma/d\Theta) = const$ and $\gamma(H) = 1$. Equations (10.58) and (10.59), with $\gamma(H) = 1$, appeared in Davis (1983) for $B \ll 1$, Kopbosynov and Pukhnachev (1986), Bankoff and Davis (1987), Burelbach et al. (1988), Oron and Rosenau (1992), Deissler and Oron (1992), Van Hook et al. (1995) and Oron (2000b).

For $B \ll 1$, Eq. (10.59) in dimensional form becomes:

$$\mu h_t + \frac{\alpha_{th}\Delta\sigma}{2k_{th}}(h^2 h_x)_x - \frac{1}{3}\rho g(h^3 h_x)_x + \frac{1}{3}\sigma(h^3 h_{xxx})_x = 0 \tag{10.60}$$

Linearization of Eq. (10.60) around the state $h = d$ yields the characteristic equation:

$$\mu\omega = \left(\frac{\alpha_{th}\Delta\sigma}{2k_{th}} - \frac{1}{3}\rho g d - \frac{\sigma}{3d}a^2\right)a^2 \tag{10.61}$$

Equation (10.61) shows that if $g > 0$ (gravity acting toward the base of the film), gravity has a stabilizing effect (similar to that described in Section 10.3), while thermocapillarity has a destabilizing effect on the interface. It is clear from Eq. (10.61) that the gravitational stabilization is enhanced with the thickness of the film. The dimensionless cutoff wave number a_c is given in this case by:

$$a_c = \left(\frac{3}{2}B\frac{\Delta\sigma}{\sigma} - Bo\right)^{1/2} \tag{10.62a}$$

Thermocapillary destabilization of a film can be explained by examining the behavior of an initially deformed interface in the linear temperature field produced by the heat transfer at the interface. The depression lies in the region of higher temperature than its neighbors; hence, if surface tension is a decreasing function of temperature, interfacial stresses proportional to the surface gradient of the surface tension [e.g., Levich, 1962; Landau and Lifshitz, 1987] drive the interfacial liquid away from it. Because the liquid is viscous, it is dragged away from the depression, causing it to deepen further. Hydrostatic and capillary forces cannot prevent this deepening, and as shall be seen further the film proceeds to zero thickness at some location (i.e., ruptures).

Studies of Eq. (10.58) with $\gamma(H) = 1$ [Oron and Rosenau, 1992] reveal that evolution of small-amplitude initial data usually results in rupture of the film qualitatively similar to that displayed in Figure 2 in Oron and Rosenau (1992). The three-dimensional version of Eq. (10.58) was studied by Van Hook et al. (1995) and the results were tested against their experiments. The existence of a ruptured region predicted by Eq. (10.58) was qualitatively confirmed by the experiment. However, the theoretical predictions of the instability threshold were about 50% higher than the experimental data. Becerril et al. (1998) recently addressed this discrepancy in terms of side-wall effects and deflected initial interface shapes.

Van Hook et al. (1997) developed a "two-layer" theory modeling the dynamics of systems containing superposed layers of a liquid and a passive gas confined between two horizontal, rigid, differentially heated surfaces. This approach takes into consideration the change in the temperature profile in the air due to deformation of the interface. The two-layer setting leads to the thermal problem containing Eq. (10.53) formulated in each layer with the boundary conditions of temperature and heat flux continuity at the liquid–gas interface. Two new parameters arise from the solution of this thermal problem:

$$\eta = \frac{k_{th,g}d}{k_{th}d_g}, \quad F = \frac{(d/d_g) - \eta}{1 + \eta} \tag{10.63}$$

where d_g is the thickness of the gas layer and $k_{th,g}$ is its thermal conductivity. The parameter η replaces the Biot number in the "one-layer" model described above.

In the case at hand, the dimensionless interfacial temperature is found to be

$$\Theta_i = 1 - \frac{H}{1 + F - FH} \tag{10.64}$$

and the corresponding dimensionless evolution equation in the standard case of $\gamma(H) = 1$ in two dimensions is obtained in the form [Van Hook et al., 1997]:

$$H_\tau + \left\{\left[\frac{1}{2}M(1+F)\frac{H^2}{(1+F-FH)^2} - \frac{1}{3}GH^3\right]H_\xi\right\}_\xi + \frac{1}{3}S(H^3 H_{\xi\xi\xi})_\xi = 0 \tag{10.65}$$

Equation (10.65) reduces to the "one-layer" model, Eq. (10.59) when $F = -\eta/(1 + \eta)$ which corresponds to $d/d_g \to 0$. For one-layer systems, the parameter F is always non-positive, while for two-layer systems F is usually positive. For both of these cases, $F > -1$. The solutions of Eq. (10.65) were found to be of two distinct types, namely, "dry spots" that represent rupture at the bottom solid surface (see Figure 10.4a), and "high spots" that represent rupture at the top solid surface by the elevated film interface (see Figure 10.4b). Dry spots emerge for $F < 1/2$, while high spots form when $F > 1/2$. The transition between these two different kinds of solutions, which was found to depend on the value of the Bond number Bo and the initial condition, occurs in the vicinity of $F = 1/2$ [Van Hook et al., 1997]. As in the "one-layer" theory reviewed above, steady nonruptured states of the system were not found. The experimental results of Van Hook et al. (1997) are found to qualitatively agree for certain liquid depths with a "two-layer" model.

In the case of "negative gravity," $g < 0$; that is, when the film is on the underside of the solid plane, the Rayleigh–Taylor instability (heavy fluid overlying light fluid) enhances the thermocapillary instability and broadens the band of linearly unstable modes:

$$a_c = \left(\frac{3}{2}B\frac{\Delta\sigma}{\sigma} + Bo\right)^{1/2} \quad (10.66)$$

Stabilization of the Rayleigh–Taylor instability by thermocapillarity was investigated by Oron and Rosenau (1992) and Deissler and Oron (1992) for two- and three-dimensional cases, respectively. They found that negative thermocapillarity (i.e., with $\Delta\sigma < 0$), corresponding to heating at the interface side or cooling at the rigid bottom, in conjunction with surface tension can lead to saturation of the Rayleigh–Taylor instability and formation of steady drops. The experimental confirmation of such a saturation was recently obtained by Burgess et al. (2001).

We now turn to the case where the $\dot{q}$ term is present in Eq. (10.47). Its presence stands for the effect of internal energy generation, which might be induced by irradiation of the film and further absorption of the radiation energy within the nonscattering liquid phase. In this context, the solid substrate is assumed to be black (i.e., absorbing all radiation penetrating through the liquid film). Oron and Peles (1998) considered the simplified case of spatially uniform energy absorption, $f(\varsigma) \equiv 1$. In this situation, the solution of the thermal problem of Eqs. (10.50) and (10.51) is given at the leading order by:

$$\Theta = \Theta_0\left(1 - \frac{B\varsigma}{1+BH}\right) + Q\left(-\varsigma^2 + \frac{BH^2 + 2H}{1+BH}\varsigma\right) \quad \text{and} \quad \Theta_i = \frac{1}{1+BH}(\Theta_0 + QH^2) \quad (10.67)$$

where $\Theta_0 = 1$ if $\vartheta_0 > \vartheta_\infty$, and $\Theta_0 = 0$ if $\vartheta_0 = \vartheta_\infty$. The corresponding evolution equation of the film obtained upon calculation of the corresponding value of Σ_ξ using Eq. (10.67) and its substitution into Eq. (10.21) with gravity neglected and $\gamma(H) = 1$ is

$$H_\tau + \frac{1}{2}M\left[H^2\frac{B\Theta_0 - Q(BH^2 + 2H)}{(1+BH)^2}H_\xi\right]_\xi + \frac{1}{3}S(H^3 H_{\xi\xi\xi})_\xi = 0 \quad (10.68)$$

The main result found by Oron and Peles (1998) was that internal heat generation stabilizes the interface via the thermocapillary effect associated with it. In the simplest case, where the temperature of the solid is equal to the saturation temperature, $\vartheta_0 = \vartheta_\infty$ (i.e., $\Theta_0 = 0$), which follows directly from the fact that the interfacial temperature Θ_i is an increasing function of the film thickness H. The effect of stabilization becomes apparent, as at the leading order the heat transfer in a thin liquid film is one dimensional across it, and the energy input from absorption of radiation energy in the thicker part of the film is greater than in its thinner part. Thus, the interfacial temperature at the depression is lower than at the crest of the interface, and the thermocapillary stress drives the liquid into the depression, promoting stabilization of the interface. All this is different from the standard case discussed in the literature, when internal energy generation is absent. In the latter case of $Q = 0$ and $\Theta_w = 1$, the interfacial temperature decreases with H, and instability of the spatially uniform state of the interface is thus triggered. Also, when the internal heat generation is sufficiently large ($Q \geq 1$), the film becomes unconditionally stable. When the film is linearly unstable, the range of unstable modes narrows with the increase of Q for a fixed value of B [Oron and Peles, 1998].

Oron (2000a) considered thin liquid films with an optically smooth, nonreflective, deformed interface irradiated with a monochromatic beam of specified wavelength λ. The intensity of radiation i_λ of such a beam normally impinging on the optically smooth, nonreflective interface was shown in the absence of emission by the irradiated liquid phase to decay exponentially with distance from the flat liquid surface [Siegel and Howell, 1992]:

$$i_\lambda(z) = i_\lambda(z_0)e^{-K_\lambda(z_0-z)} \tag{10.69}$$

where z_0 is the location of the film surface, and K_λ is the extinction coefficient of the given liquid assumed to be constant. The extinction coefficient is a property of the medium and in general varies with its temperature, pressure and the wavelength of the incident radiation.

Equation (10.69) is often referred as to Bouguer's [Siegel and Howell, 1992] or Beer's law. The attenuation of the radiation intensity is associated with the absorption and scattering of energy. The extinction coefficient is, in general, represented as a sum of the absorption and scattering components $K_\lambda = a_\lambda + a_{s,\lambda}$. In the case of a vanishing scattering $a_{s,\lambda} = 0$, $K_\lambda = a_\lambda$, and the optical thickness κ_λ of a liquid film of a uniform thickness d can be defined as:

$$\kappa_\lambda \equiv a_\lambda d = d/\ell_m$$

where ℓ_m is the mean penetration length of the incident radiation by the wave λ. Assuming that the film is nonscattering with a nonreflecting surface, Oron (2000a) used an expression compatible with Eq. (10.69) for the intensity of absorbed radiation and therefore for the intensity of the induced internal heat sources:

$$\dot{q}(z) = \dot{q}\exp[-a_\lambda(h-z)]$$

where h represents the location of the interface and $\dot{q} = i_\lambda(z_0)a_\lambda$ is the *constant* representing the rate of energy absorption at the film interface. Note that the intensity of the heat sources $\dot{q}(z)$ varies with x, y, z, t when $a_\lambda \neq 0$. By comparing the value of the optical thickness of the film with unity, one can examine the following limiting cases: (1) if $\kappa_\lambda \ll 1$, the radiation passes through the film, and such a film is called *optically thin* or *transparent*; or (2) if $\kappa_\lambda \gg 1$, the radiation penetrates only into a very thin boundary layer adjacent to the film interface, and in this case the film is called *optically thick* or *opaque*.

Solving the thermal problem given by Eqs. (10.50) and (10.51) with $f(\varsigma) = \exp[-\beta(H-\varsigma)]$, constant a_λ, and $\beta = a_\lambda d$ being a nondimensional attenuation coefficient yields the interfacial temperature in the form:

$$\Theta_i = \frac{B}{1+BH}\left[\Theta_0 + \frac{2Q}{\beta^2}(e^{-\beta H} + \beta H - 1)\right] \tag{10.70}$$

The corresponding evolution equation of the film obtained upon calculation of the term Σ_ξ based on Eq. (10.70) and its substitution into Eq. (10.21) with gravity neglected and $\gamma(H) = 1$ is

$$H_\tau + \frac{M}{2}\left\{\frac{H^2}{(1+BH)^2}\left[B\Theta_0 - \frac{2Q}{\beta^2}((\beta+BH)(1-e^{-\beta H}) - \beta BHe^{-\beta H})\right]H_\xi\right\}_\xi + \frac{S}{3}(H^3 H_{\xi\xi\xi})_\xi = 0 \tag{10.71}$$

As in the case of uniform heat generation, Oron (2000a) found for irradiated films following the Bouguer's law that an increase of the radiation intensity leads to stabilization of the interface due to the appropriate change in the profile of the interfacial temperature. In the presence of heating across the film, $\Theta_0 = 1$, there exists a critical value of $Q = Q_c$ depending on the Biot number B such that the film becomes

Physics of Thin Liquid Films

linearly stable when $Q > Q_c$ and remains linearly unstable when $Q < Q_c$, although the rate of disturbance growth slows down in comparison to the case of $Q = 0$. This critical value Q_c is obtained from the linear stability analysis as:

$$Q_c = \frac{\beta^2}{2}[(1 + \beta B^{-1})(1 - e^{-\beta}) - \beta e^{-\beta}]^{-1} \tag{10.72}$$

This critical value tends to its limiting value of $Q_c = B/(2 + B)$ for optically thin films $\beta \ll 1$, which is the limit of the spatially uniform absorption, and to $Q_c = \beta B/2$ for optically thick films, $\beta \gg 1$.

It was experimentally discovered that dilute aqueous solutions of long-chain alcohols exhibit non-monotonic dependence of surface tension on temperature [Legros et al., 1984; Legros, 1986]. This dependence can be approximated quite well by the quadratic polynomial,

$$\sigma(\vartheta) = \delta(\vartheta - \vartheta_m)^2 \tag{10.73a}$$

where δ is constant and ϑ_m is the temperature corresponding to the minimal surface tension. In this case,

$$\gamma(H) \propto \left(\frac{\vartheta_m - \vartheta_\infty}{\vartheta_0 - \vartheta_\infty} - \frac{1}{1 + BH} \right) \tag{10.73b}$$

The instability arising from the variation of surface tension given by Eq. (10.73a) was called "quadratic Marangoni (QM)" instability and studied by Oron and Rosenau (1994). In contrast with the case of the standard thermocapillary instability described by Eq. (10.58) with $\gamma(H) = 1$, evolution of QM instability may result in a nonruptured steady state. Figure 4 in Oron and Rosenau (1994) displays such a state along with the streamlines of the flow field. It results from solving Eq. (10.58) with $\gamma(H)$ given by Eq. (10.73b). The intersections of the Θ_0 line with the film interface in Figure 4 in Oron and Rosenau (1994) correspond to the locations of the minimal surface tension. The existence of these creates surface shear stresses acting in opposite directions and leads to film stabilization.

10.4.2 Liquid Film on a Thick Substrate

The methods described in the previous sections can be easily implemented in the case of a liquid film lying on top of a solid slab of thickness small compared to the characteristic wavelength of the interfacial disturbance [Oron et al., 1996]. In this case, the thermal conduction equation in the solid has to be solved simultaneously with the energy equation in the liquid. This coupled thermal problem is written at the leading order in ε as:

$$\begin{aligned} \Theta_{w,\varsigma\varsigma} &= 0, \quad -\frac{d_w}{d} \leq \varsigma \leq 0 \\ \Theta_{\varsigma\varsigma} &= 0, \quad 0 \leq \varsigma \leq H \end{aligned} \tag{10.74}$$

with the boundary conditions:

$$\begin{aligned} \Theta_w &= \Theta, \quad -k_{th,w}\Theta_{w,\varsigma} = -k_{th}\Theta_\varsigma \quad \text{at } \varsigma = 0 \\ \Theta_w &= 1 \quad \text{at } \varsigma = -\frac{d_w}{d} \end{aligned} \tag{10.75}$$

where Θ_w and d_w/d are the dimensionless temperature and scaled with d thickness of the solid slab; $k_{th,w}$ is its thermal conductivity. The upper equations in Eq. (10.75) express the conditions of continuity of both the temperature and heat flux at the solid–liquid boundary, while the lower one defines a uniform temperature at the bottom of the solid substrate. The last boundary condition is taken at the film interface, and at the leading order in ε it is given by Eq. (10.54). Appropriate extension has to be made in the case of a volatile liquid (see Section 10.5).

Solution of Eqs. (10.74) and (10.75) results in:

$$\Theta = 1 - \frac{B(\bar{\kappa}\varsigma + d_w/d)}{\bar{\kappa}(1+BH) + Bd_w/d}, \quad \Theta_w = 1 - \frac{B(\varsigma + d_w/d)}{\bar{\kappa}(1+BH) + Bd_w/d} \quad (10.76a)$$

with $\bar{\kappa} = k_{th,w}/k_{th}$, which implies the interfacial temperature in the form:

$$\Theta_i = \left[1 + B\left(H + \frac{d_w/k_{th,w}}{d/k_{th}}\right)\right]^{-1} \quad (10.76b)$$

Comparing the expressions for the interfacial temperatures Θ_i, as given by Eqs. (10.55) and (10.76b), one notices that, in addition to the thermal resistance due to each conduction and convection at the interface in the former case, the latter contains a thermal resistance owing to conduction in the solid. The evolution equation, analogous to Eq. (10.58), will have the same form, except for the obvious change in the denominator of the second term containing an additional additive term:

$$a \equiv \frac{d_w/k_{th,w}}{d/k_{th}} \quad (10.77)$$

This additional term represents the ratio between the values of the thermal conductive resistance of the solid and the liquid.

Using Eq. (10.76) one can derive the expressions for the temperatures along the gas–liquid (GL) and solid–liquid (SL) interfaces: $\Theta_{GL} \equiv \Theta(\varsigma = H)$ and $\Theta_{SL} \equiv \Theta_w(\varsigma = 0)$. When the film ruptures (i.e., $H = 0$), the values for Θ_{GL} and Θ_{SL} are equal if:

$$a = \frac{d_w}{\bar{\kappa}d} \neq 0 \quad (10.78)$$

However, the temperature singularity, $\Theta_{GL} \neq \Theta_{SL}$, emerges at the rupture point when $a = 0$. Equation (10.78) is the sufficient condition to be satisfied in order to relieve this singularity [Oron et al., 1996a]. Indeed, if it is satisfied,

$$\lim_{H \to 0} \lim_{a \to 0} \Theta_{GL} = \lim_{H \to 0} \lim_{a \to 0} \Theta_{SL} = 0$$

and the singularity is removed. The problematic case of $a \to 0$ materializes if the substrate is of a negligible thickness.

10.5 Change of Phase: Evaporation and Condensation

10.5.1 Interfacial Conditions

We now consider the case of an evaporating (condensing) thin film of a simple liquid lying on a heated (cooled) plane surface held at constant temperature ϑ_0 which is higher (lower) than the saturation temperature at the given vapor pressure. It is assumed that the speed of vapor particles is sufficiently low so that the vapor can be considered an incompressible fluid.

The boundary conditions appropriate for phase transformation at the film interface $z = h$ are now formulated. The mass conservation equation at the interface is given by the balance between the liquid

Physics of Thin Liquid Films

and vapor fluxes through the interface:

$$j = \rho_v(\mathbf{v}_v - \mathbf{v}_i) \cdot \mathbf{n} = \rho_f(\mathbf{v}_f - \mathbf{v}_i) \cdot \mathbf{n} \tag{10.79a}$$

where j is the mass flux due to evaporation; ρ_v and ρ_f are, respectively, the densities of the vapor and the liquid; $\mathbf{v}_v$ and $\mathbf{v}_f$ are the vapor and liquid velocities at $z = h$; and $\mathbf{v}_i$ is the velocity of the interface. Equation (10.79a) provides the relationship between the normal components of the vapor and liquid velocities at the interface. The tangential components of both of the velocity fields are equal at the interface:

$$(\mathbf{v}_f - \mathbf{v}_v) \cdot \mathbf{t}_m = 0, \quad m = 1, 2 \tag{10.79b}$$

The boundary condition that expresses the stress balance and extends Eq. (10.4b) to the case of phase transformation reads [Delhaye, 1974; Burelbach et al., 1988]:

$$j(\mathbf{v}_f - \mathbf{v}_v) - (\mathbf{T} - \mathbf{T}_v) \cdot \mathbf{n} = 2\tilde{H}\sigma(\vartheta)\mathbf{n} - \nabla_s \sigma \tag{10.80a}$$

where $\mathbf{T}_v$ is the stress tensor in the vapor phase and temperature dependence of surface tension is accounted for.

The energy balance at $z = h$ is given by [Delhaye, 1974; Burelbach et al., 1988]:

$$j\left(L + \frac{1}{2}v_{v,n}^2 - \frac{1}{2}v_{f,n}^2\right) + (k_{th}\nabla\vartheta - k_{th,v}\nabla\vartheta_v) \cdot \mathbf{n} + 2\mu(\mathbf{e}_f \cdot \mathbf{n}) \cdot \mathbf{v}_{f,r} - 2\mu_v(\mathbf{e}_v \cdot \mathbf{n}) \cdot \mathbf{v}_{v,r} = 0 \tag{10.80b}$$

where L is the latent heat of vaporization per unit mass; $k_{th,v}$, μ_v, ϑ_v are, respectively, the thermal conductivity, viscosity and the temperature of the vapor; $\mathbf{v}_{v,r} = \mathbf{v}_v - \mathbf{v}_i$ and $\mathbf{v}_{f,r} = \mathbf{v}_f - \mathbf{v}_i$ are the vapor and liquid velocities relative to the interface, respectively; $v_{v,r} = \mathbf{v}_{v,r} \cdot \mathbf{n}$ and $v_{f,n} = \mathbf{v}_{f,r} \cdot \mathbf{n}$ are the normal components of the latter; and $\mathbf{e}_f$ and $\mathbf{e}_v$ are the rate-of-deformation tensors in the liquid and the vapor, respectively. In Eq. (10.80b) the first term represents the contribution of the latent heat, the combination of the second and the third terms represents the interfacial jump in the momentum flux, the combination of the fourth and the fifth terms represents the jump in the conductive heat flux at both sides of the interface, while the combination of the last two terms is associated with the viscous dissipation of energy at both sides of the interface.

Because $\rho_v/\rho_f \ll 1$, typically of order 10^{-3}, it follows from Eq. (10.79a) that the magnitude of the normal velocity of the vapor relative to the interface is much greater than that of the liquid. Hence, the phase transformation causes large accelerations of the vapor at the interface, where the back reaction, called the *vapor recoil*, represents a force exerted on the interface. During evaporation (condensation) the troughs of the deformed interface are closer to the hot (cold) plate than the crests, so they have greater evaporation (condensation) rates j. The dynamic pressure at the vapor side of the interface is much larger than that at the liquid side:

$$\rho_v v_{v,n}^2 = \frac{j^2}{\rho_v} \gg \rho_f v_{f,n}^2 = \frac{j^2}{\rho_f} \tag{10.81}$$

Momentum fluxes are thus greater in the troughs than at the crests of surface waves. Vapor recoil is found to be a destabilizing factor for the interface dynamics for both evaporation ($j > 0$) and condensation ($j < 0$) [Burelbach et al., 1988]. Being scaled with j^2 (see Eq. (10.84) below), the vapor recoil is only important for applications where very high mass fluxes are involved.

Vapor recoil generally exerts a reactive downward pressure on a horizontal evaporating film. Bankoff (1961) introduced the effect of vapor recoil in the analysis of film boiling, where the liquid overlays the vapor layer generated by boiling, thus leading to the Rayleigh–Taylor instability of an evaporating liquid–vapor interface above a hot horizontal wall. In this case, the vapor recoil stabilizes the film boiling, as the reactive force is greater for the wave crests approaching the wall than for the troughs.

To obtain closure for the system of governing equations and boundary conditions, an equation relating the dependence of the interfacial temperature ϑ_i and the local pressure in the vapor phase is added [Plesset and Prosperetti, 1976; Palmer, 1976; Sadhal and Plesset, 1979]. Its linearized form is

$$\tilde{K}j = \vartheta_i - \vartheta_s \equiv \Delta\vartheta_i, \tag{10.82}$$

where

$$\tilde{K} = \frac{\vartheta_s^{3/2}}{\hat{\alpha}\rho_v L}\left(\frac{2\pi R_v}{M_w}\right)^{1/2}$$

ϑ_s is the absolute saturation temperature, $\hat{\alpha}$ is the accommodation coefficient, R_v is the universal gas constant and M_w is the molecular weight of the vapor [Palmer, 1976; Plesset and Prosperetti, 1976; Burelbach et al., 1988]. Note that the absolute saturation temperature ϑ_s serves now as the reference temperature instead of ϑ_∞ in the normalization equation, Eq. (10.49). When $\Delta\vartheta_i = 0$, the phases are in thermal equilibrium with each other, and, in order for net mass transport to take place, a vapor pressure driving force must exist and is given for ideal gases by kinetic theory [Schrage, 1953], which is represented in the linear approximation by the parameter $\tilde{K}$ [Burelbach et al., 1988]. Departure from ideal behavior is addressed in the parameter $\tilde{K}$ by the presence of an accommodation coefficient $\hat{\alpha}$, which depends on interface/molecule orientation and steric effects and represents the probability of a vapor molecule sticking upon hitting the liquid–vapor interface.

The set of the boundary conditions Eq. (10.80) can be simplified to what is known as a "one-sided" model for evaporation or condensation [Burelbach et al., 1988] in which the dynamics of the liquid are decoupled from those of the vapor. This simplification is enabled due to the assumption of smallness of density, viscosity and thermal conductivity of the vapor with respect to the respective properties of the liquid. The vapor dynamics are ignored in the one-sided model and only the mass conservation and the effect of vapor recoil indicate the presence of the vapor phase.

The energy balance (10.80b) becomes:

$$-k_{th}\nabla\vartheta \cdot \mathbf{n} = j\left(L + \frac{1}{2}\frac{j^2}{\rho_v^2}\right) \tag{10.83}$$

suggesting that the heat flux conducted to the interface in the liquid is converted to latent heat of evaporation and the kinetic energy of vapor particles.

The stress balance at the interface equation, Eq. (10.80a), is reduced and now rewritten explicitly for the components of the normal and tangential stresses there as:

$$-\frac{j^2}{\rho_v} - \mathbf{T} \cdot \mathbf{n} \cdot \mathbf{n} = 2\tilde{H}\sigma(\vartheta)$$

$$\mathbf{T} \cdot \mathbf{n} \cdot \mathbf{t} = \nabla_s \sigma \cdot \mathbf{t} \tag{10.84}$$

In Eq. (10.84) the j^2 term stands for the contribution of vapor recoil. Finally, the remaining boundary conditions (10.79) and (10.82) are unchanged.

The procedure of asymptotic expansions outlined in the beginning of the chapter is used again to derive the pertinent evolution equation. The dimensionless mass balance Eq. (10.13) is modified by the presence of the nondimensional evaporative mass flux, $J = jdL/k_{th}(\vartheta_0 - \vartheta_s)$.

$$EJ = (-H_\tau - UH_\xi - VH_\eta + W)(1 + H_\xi^2)^{-1/2} \tag{10.85a}$$

or at the leading order of approximation:

$$H_\tau + Q^{(x)}_\xi + Q^{(y)}_\eta + EJ = 0 \qquad (10.85b)$$

where $Q^{(x)}(\xi, \eta, \tau) = \int_0^H U\, d\varsigma$, $Q^{(y)}(\xi, \eta, \tau) = \int_0^H V\, d\varsigma$ are the components of the scaled volumetric flowrate per unit width parallel to the wall. The parameter E in Eq. (10.85) is an evaporation number:

$$E = \frac{k_{th}(\vartheta_0 - \vartheta_s)}{\rho \nu L}$$

which represents the ratio of the viscous time scale, $t_v = d^2/\nu$, to the evaporative time scale, $t_e = \rho d^2 L/k_{th}(\vartheta_0 - \vartheta_s)$ [Burelbach et al., 1988].

The dimensionless versions of Eqs. (10.82) and (10.83) are

$$\begin{aligned} KJ &= \Theta \quad \text{at } \varsigma = H \\ \Theta_\varsigma &= -J \quad \text{at } \varsigma = H \end{aligned} \qquad (10.86)$$

where

$$K = \tilde{K}\frac{k_{th}}{dL}$$

In the lower equation in Eq. (10.86) the kinetic energy term is neglected. For details, the reader is referred to Burelbach et al. (1988). Equations (10.18), (10.19), (10.53) and (10.86) pose the problem, whose solution is substituted into Eq. (10.85b) to obtain the sought evolution equation. The general dimensionless evolution equation, Eq. (10.21), will then contain an additional term EJ, which arises from the mass flux due to evaporation/condensation now being expressed via the local film thickness H.

A different approach to theoretically describing the rate of evaporative flux j in the *isothermal* case is known in the literature [Sharma, 1998; Padmakar et al., 1999]. This approach is based on the extended Kelvin equation that accounts for the local interfacial curvature and the disjoining/conjoining pressures, both entering the resulting expression for the evaporative mass flux j. It was shown by Padmakar et al. (1999) that their evaporation model admits the emergence of a flat adsorbed layer remaining in equilibrium with the ambient vapor phase, and thus in this state the evaporation rate from the film vanishes. This adsorbed layer, however, is usually several molecular spacings thick, which is beyond the resolution of continuum theory.

10.5.2 Evaporation/Condensation Only

We first consider the case of an evaporating/condensing thin liquid layer lying on a rigid plane held at constant temperature. Mass loss or gain is retained, while all other effects are neglected.

Solving first Eq. (10.53) along with boundary conditions Eqs. (10.51a) and (10.86) and eliminating the mass flux J from the latter yield the dimensionless temperature field and evaporative mass flux through the interface,

$$\Theta = 1 - \frac{\varsigma}{H + K}, \quad J = \frac{1}{H + K} \qquad (10.87)$$

An initially flat interface will remain flat as evaporation or condensation proceeds and if surface tension, thermocapillary and convective thermal effects are negligible; if $M = S = \varepsilon RP = 0$, it will give rise to a scaled evolution equation of the form:

$$H_\tau + \frac{\bar{E}}{H + K} = 0 \qquad (10.88)$$

where $\bar{E} = \varepsilon^{-1}E$, positive in the evaporative case and negative in the condensing one. K, the scaled interfacial thermal resistance, is equivalent to the inverse Biot number, B^{-1}. On the physical grounds, $K \neq 0$ represents a temperature jump from the liquid surface temperature to the uniform temperature of the saturated vapor, ϑ_s. This jump drives the mass transfer. The conductive resistance of the liquid film is proportional to H, and the total thermal resistance, assuming infinite thermal conductivity of the solid, is given by $(H + K)^{-1}$. For a specified temperature difference, $\vartheta_0 - \vartheta_s$, Eq. (10.88) represents a volumetric balance, whose solution, subject to the initial condition $H(\tau = 0) = 1$, is

$$H = -K + [(K+1)^2 - 2\bar{E}\tau]^{1/2} \qquad (10.89)$$

In the case of evaporation $\bar{E} > 0$, and when $K \neq 0$ the film vanishes in a finite time $\tau_e = (2K + 1)/2\bar{E}$, and the rate of disappearance of the film at $\tau = \tau_e$ is finite:

$$\left.\frac{dH}{d\tau}\right|_{\tau = \tau_e} = -\frac{\bar{E}}{K}$$

For $K \neq 0$, the value of $dH/d\tau$ remains finite, because as the film thins the interface temperature ϑ_i, nominally at its saturation value ϑ_s, increases to the wall temperature. If $K = 0$, however, the problem becomes singular. In this case, the thermal resistance vanishes and the mass flux will increase indefinitely if a finite temperature difference $\vartheta_0 - \vartheta_s$ is sustained. The speed of the interface at rupture becomes infinite as well.

Burelbach et al. (1988) showed that the interfacial thermal resistance $K = 10$ for a 10-nm-thick water film. Because K is inversely proportional to the initial film thickness, $K \approx 1$ for $d = 100$ nm, so $H/K \approx 1$ at this point. However, $H/K \approx 10^{-1}$ at $d = 30$ nm, so the conduction resistance becomes small compared to the interfacial transport resistance shortly after van der Waals forces become appreciable.

10.5.3 Evaporation/Condensation, Vapor Recoil, Capillarity and Thermocapillarity

The dimensionless vapor recoil gives an additional normal stress at the interface determined by the j^2 term in Eq. (10.84), $\hat{\Pi}_3 = -\frac{3}{2}\bar{E}^2 D^{-1} J^2$, where D is a unit-order scaled ratio between the vapor and liquid densities:

$$D = \frac{3}{2}\varepsilon^{-3}\frac{\rho_v}{\rho}$$

This stress can be calculated using Eq. (10.87). The resulting scaled evolution equation for an evaporating film on an isothermal horizontal surface neglecting the thermocapillary effect and body forces is obtained using the combination of Eqs. (10.21) and (10.88) with $\Pi_1 = 0$, $\Sigma_\xi = 0$ [Burelbach et al., 1988]:

$$H_\tau + \frac{\bar{E}}{H+K} + \left[\bar{E}^2 D^{-1}\left(\frac{H}{H+K}\right)^3 H_\xi\right]_\xi + \frac{1}{3}S(H^3 H_{\xi\xi\xi})_\xi = 0 \qquad (10.90)$$

Because usually $t_e \gg t_v$, $\bar{E}$, can itself be a small number and then can be used as an expansion parameter for slow evaporation compared to the nonevaporating base state [Burelbach et al., 1988] appropriate to very thin evaporating films.

Taking into account van der Waals forces and thermocapillarity, the complete evolution equation for a thin heated or cooled film on a horizontal plane surface was given by Burelbach et al. (1988) in the form:

$$H_\tau + \frac{\bar{E}}{H+K} + \left\{\left[AH^{-1} + \bar{E}^2 D^{-1}\left(\frac{H}{H+K}\right)^3 + K\bar{M}P^{-1}\left(\frac{H}{H+K}\right)^2\right]H_\xi\right\}_\xi + \frac{1}{3}S(H^3 H_{\xi\xi\xi})_\xi = 0 \qquad (10.91)$$

with $\bar{M} = \varepsilon M$. Here, the first term represents the rate of volumetric change; the second one, the mass loss/gain; the third, fourth and fifth ones, the attractive van der Waals, vapor recoil and thermocapillary terms, respectively, all destabilizing; while the sixth term describes the stabilizing capillary force. This was the first full statement of the possible competition among various stabilizing and destabilizing effects on a horizontal plate, with scaling taken to make them present at the same order. Other effects such as gravity may be also included in Eq. (10.91). Joo et al. (1991) extended the work to an evaporating (condensing) liquid film draining down a heated (cooled) inclined plate.

Oron and Bankoff (1999) studied the two-dimensional dynamics of an evaporating ultrathin film on a coated solid surface when the potential (Eq. (10.31d)) was used. Three different types of the evolution of a volatile film were identified. One of them is related to low evaporation rates associated with relatively small $\bar{E} > 0$, when holes covered by a liquid microlayer emerge and expansion of such holes is governed mainly by the action of the attractive molecular forces. These forces impart the squeeze effect to the film, and, as a result, the liquid flows away from the hole. In this stage the role of evaporation is secondary. Figure 10.9 displays such an evolution of a volatile liquid film. Following the nucleation of the hole and during the process of surface dewetting, one can identify the formation of a large ridge, or drop, on either side of the trough. The former grows during the evolution of the film until the drops at both ends of the periodic domain collide. A further recession of the walls of the dry spot leads to the formation of a single large drop that flattens and ultimately disappears, according to Eq. (10.89). The stages of the film evolution shown in Figure 10.9a are found to be very similar to that sketched in Figure 3 in Elbaum and Lipson (1995). This type of evolution also resembles the results obtained by Padmakar et al. (1998) for the isothermal film subject to hydrophobic interactions and evaporation driven by the difference between the equilibrium vapor pressure and the pressure in the vapor phase. Such films were found to thin uniformly to a critical thickness and then spontaneously to dewet the solid substrate by the formation of growing dry spots, when the solid was partially wetted. In the completely wetted case, thin liquid films evolved to an array of islands that disappeared by evaporation to a thin equilibrium flat film. Two other regimes corresponding to intermediate and high evaporation rates were discussed by Oron and Bankoff (1999).

An important phenomenon was found in the last stage of the evolution of an evaporating film, when the latter finally disappears by evaporation: Prior to that the film equilibrates, so that its disappearance is practically uniform in space. The film equilibration is due to what has been called the *reservoir effect*, which is driven by the difference in disjoining pressures and manifests itself by feeding the liquid from the large drops into the ultrathin film that bridges between them.

Oron and Bankoff (2001) studied the dynamics of condensing thin films on a horizontal coated solid surface. In the case of a relatively fast condensation, where the initial depression of the interface rapidly fills up due to the enhanced mass gain there, the film equilibrates and grows uniformly in space according to Eq. (10.89); note that $\bar{E} < 0$. When condensation is relatively slow, the evolution of the film exhibits several distinct stages. The first stage dominated by attractive van der Waals forces leads to opening of a hole covered by a microlayer, as shown in the first three snapshots of Figure 10.10a. This is accompanied by continuous condensation with the highest rate of mass gain attained in the microlayer region corresponding to the smallest thickness H in Eq. (10.87). However, opposite to the evaporative case [Oron and Bankoff, 1999], where the reservoir effect arising from the difference between the disjoining pressures causes feeding of the liquid from the large drops into the microlayer and film equilibration, in the condensing case the excess liquid is driven from the microlayer into the large drops. This effect has been referred as to the *reversed reservoir effect*. The thickness of the microlayer remains nearly constant due to local mass gain by condensation compensating for the impact of the reverse reservoir effect. The first stage of the film evolution terminates in the situation where the size of the hole is the largest. The receding of the drops stops due to the increase of the drop curvature and buildup of the capillary pressure that comes to balance with the squeeze effect of the attractive van der Waals forces. From this moment on, the hole closes, driven by condensation (as shown in Figures 10.10a, b). Once the hole closes, the depression fills up rapidly, the amplitude of the interfacial disturbance decreases and the film thus tends to flatten out. The film then grows uniformly in space following the solution of Eq. (10.89) with negative $\bar{E}$.

Oron (2000c) studied the three-dimensional evolution of an evaporating film on a coated solid surface subject to the potential [Eq. (10.31d)]. The main stages of the evolution repeat those mentioned above

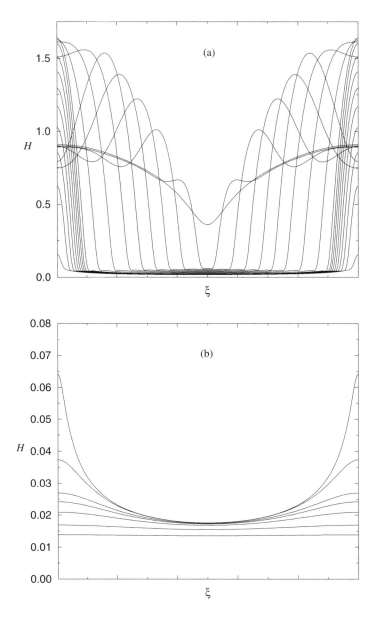

FIGURE 10.9 Evolution of a slowly evaporating film. (a) Initial and intermediate stages of the film evolution. (b) Final stage of the evolution. The curves in both graphs correspond to the interfacial shapes in consecutive times (not necessarily equidistant). (From Oron, A., and Bankoff, S.G. (1999) *J. Coll. Interface Sci.* **218**, 152–166. With permission.)

in the case of a nonvolatile film (Section 10.3), except for the stage of disappearance accompanied by the reservoir effect. Because of the reservoir effect, the minimal film thickness decreases very slowly during the stage of film equilibration.

10.5.4 Flow on a Rotating Disc

Reisfeld et al. (1991) considered the axisymmetric flow of an incompressible viscous volatile liquid on a horizontal, rotating disk. The liquid was assumed to evaporate due to the difference between the vapor pressures of the solvent species at the fluid–vapor interface and in the gas phase. This situation is analogous to phase two of the spin coating process.

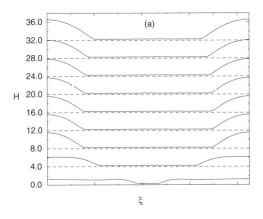
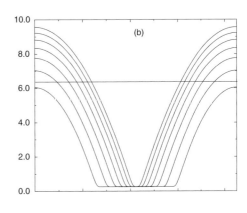

FIGURE 10.10 Evolution of a slowly condensing film on the horizontal plane. (a) Curves from the bottom to the top correspond to consecutive times (not necessarily equidistant). (b) Curves from the left to the right correspond to consecutive times (not necessarily equidistant). The flat curve corresponds to the interface at a certain time after which the film grows uniformly in space according to Eq. (10.89). In part (a), the dashed lines represent the location of the solid substrate $H = 0$. (From Oron, A., and Bankoff, S.G. (2001) *Phys. Fluids* **18**, pp. 1107–1117. With permission.)

The analysis is similar to what was done in Section 10.3, but now it includes an additional parameter describing the process of evaporation, which for a prescribed evaporative mass flux j is defined as:

$$E = \frac{3j}{2\varepsilon\rho U_0}$$

Using the procedures outlined before one obtains at the leading order the following evolution equation:

$$H_\tau + \frac{2}{3}E + \frac{1}{3r}\left\{r^2 H^3 + S r H^3 \left[\frac{1}{r}(rH_r)_r\right]_r\right\}_r = 0 \qquad (10.92)$$

Equation (10.92) models the combined effect of local mass loss, capillary forces and centrifugal drainage, none of which describes any kind of instability.

For most spin coating applications, S is very small and the corresponding term may be neglected, although it may be very important in planarization studies, where the leveling of liquid films on rough surfaces is investigated. Equation (10.92) can be thus simplified:

$$H_\tau + \frac{2}{3}E + \frac{1}{3r}(r^2 H^3)_r = 0 \qquad (10.93)$$

This simplified equation can then be used for further analysis. Looking for flat basic states $H = H(\tau)$, Eq. (10.93) is reduced to the ordinary differential equation which is to be solved with the initial condition $H(0) = 1$. In the case of $E > 0$, both evaporation and drainage cause thinning of the layer. Equation (10.93) describes the evolution in which the films thin monotonically to zero thickness in a finite time in contrast with an infinite thinning time by centrifugal drainage only. Explicit expressions for $H(\tau)$ and for the time of film disappearance are given by Reisfeld et al. (1991). In the condensing case, $E < 0$, drainage competes with condensation to thin the film. Initially, the film thins due to drainage, until the rate of mass gain due to condensation balances the rate of mass loss by drainage. At this point, the film interface reaches its steady location $H = |E|^{1/3}$. The cases where inertia is taken into account are also considered by Reisfeld et al. (1991), who provide a linear stability analysis of flat base states.

Experiments with volatile rotating liquid films [Stillwagon and Larson, 1990] showed that the final stage of film leveling was affected by an evaporative shrinkage of the films. Therefore, they suggested to separate the analysis of the evolution of evaporating spinning films into two stages with fluid flow dominating the first stage and solvent evaporation dominating the second one [Stillwagon and Larson, 1992].

10.6 Closing Remarks

In this chapter, the physics of thin liquid films was reviewed and various examples of their dynamics relevant for MEMS were presented, some of them along with reference to the corresponding experimental results. The examples discussed examined isothermal, nonisothermal with no phase changes and evaporating/condensing films under the influence of surface tension, gravity, van der Waals and centrifugal forces. The long-wave theory has been proven to be a powerful tool of research for the dynamics of thin liquid films.

However, several optional approaches are suitable for a study of the dynamics of thin liquid films. Direct numerical simulation of the hydrodynamic equations (Navier–Stokes and continuity) [Scardovelli and Zaleski, 1999] mentioned briefly in the introduction represents one of these options. A variety of methods have been developed to carry out such simulations: techniques based on finite elements method (FEM) [Ho and Patera, 1990; Salamon et al., 1994; Krishnamoorthy et al., 1995; Tsai and Yue, 1996; Ramaswamy et al., 1997], techniques based on the boundary-integral method [Pozrikidis, 1992; 1997; Newhouse and Pozrikidis, 1992; Boos and Thess, 1999], surface tracking technique [Yiantsios and Higgins, 1989] and others. Another optional approach is that of molecular dynamics (MD) simulations [Allen and Tildesley, 1987; Koplik and Banavar, 1995; 2000]. The details of all these are not given here and the reader is referred directly to these works and the references therein.

Acknowledgments

It is a pleasure to express my gratitude to A. Sharma, G. Reiter and U. Thiele for reading the manuscript and sharing their comments and thoughts with me. A. Sharma, S. Herminghaus, M.F. Schatz and E. Zussman are acknowledged for providing me with their experimental results to be used in this chapter.

References

Adamson, A.W. (1990) *Physical Chemistry of Surfaces,* Wiley, New York.
Allen, M.P., and Tildesley, D. J. (1987) *Computer Simulation of Liquids,* Clarendon, Oxford.
Babchin, A.J., Frenkel, A.L., Levich, B.G., and Sivashinsky, G.I. (1983) "Nonlinear Saturation of Rayleigh-Taylor Instability," *Phys. Fluids* **26**, pp. 3159–3161.
Bankoff, S.G. (1961) "Taylor Instability of an Evaporating Plane Interface," *AIChE J.* **7**, pp. 485–487.
Bankoff, S.G. (1990) "Dynamics and Stability of Thin Heated Films," *J. Heat Transfer, Trans. ASME* **112**, pp. 538–546.
Bankoff, S.G., and Davis, S.H. (1987) "Stability of Thin Films," *Phys. Chem. Hydrodyn.* **9**, pp. 5–7.
Becerril, R., Van Hook, S.J., and Swift, J.B. (1998) "The Influence of Interface Profile on the Onset of Long-Wavelength Marangoni Convection," *Phys. Fluids* **10**, pp. 3230–3232.
Benney, D.J., 1966, "Long Waves on Liquid Films," *J. Math. Phys.* **45**, pp. 150–155.
Bischof, J., Scherer, D., Herminghaus, S., and Leiderer, P. (1996) "Dewetting Modes of Thin Metallic Films: Nucleation of Holes and Spinodal Dewetting," *Phys. Rev. Lett.* **77**, pp. 1536–1539.
Boos, W., and Thess, A. (1999) "Cascade of Structures in Long-Wavelength Marangoni Instability," *Phys. Fluids* **11**, pp. 1484–1494.
Brochard-Wyart, F., and Daillant, J. (1990) "Drying of Solids Wetted by Thin Liquid Films," *Can. J. Phys.* **68**, pp. 1084–1088.

Burelbach, J.P., Bankoff, S.G., and Davis, S.H. (1988) "Nonlinear Stability of Evaporating/Condensing Liquid Films," *J. Fluid Mech.* **195**, pp. 463–494.

Burelbach, J.P., Bankoff, S.G., and Davis, S.H. (1990) "Steady Thermocapillary Flows of Thin Liquid Layers. II. Experiment," *Phys. Fluids A* **2**, pp. 322–333.

Burgess, J.M., Juel, A., McCormick, W.D., Swift, J.B., and Swinney. H.L. (2001) "Suppression of Dripping from a Ceiling," *Phys. Rev. Lett.* **86**, pp. 1203–1206.

Cahn, J.W. (1961) "On Spinodal Decomposition," *Acta Metall.* **9**, pp. 795–801.

Chandrasekhar, S. (1961) *Hydrodynamic and Hydromagnetic Stability*, Clarendon, Oxford.

Chou, F.-C., and Wu, P.-Y. (2000) "Effect of Air Shear on Film Planarization during Spin Coating," *J. Electrochem. Soc.* **147**, pp. 699–705.

Davis, S.H. (1983) "Rupture of Thin Films," in *Waves on Fluid Interfaces*, ed. R.E. Meyer, Academic Press, New York, pp. 291–302.

Davis, S.H. (1987) "Thermocapillary Instabilities," *Ann. Rev. Fluid Mech.* **19**, pp. 403–435.

Deissler, R.J., and Oron, A. (1992) "Stable Localized Patterns in Thin Liquid Films," *Phys. Rev. Lett.* **68**, pp. 2948–2951.

Delhaye, J.M. (1974) "Jump Conditions and Entropy Sources in Two-Phase Systems. Local Instant Formulation," *Int. J. Multiphase Flow* **1**, pp. 395–409.

Dzyaloshinskii, I.E., Lifshitz, E.M., and Pitaevskii, L.P. (1959) *Zh. Eksp. Teor. Fiz.* **37**, pp. 229–241; (1960) "Van der Waals Forces in Liquid Films," *Sov. Phys. JETP* **10**, pp. 161–170.

Edwards, D.A., Brenner, H., and Wasan, D.T. (1991) *Interfacial Transport Processes and Rheology*, Butterworth-Heinemann, Boston.

Elbaum, M., and Lipson, S.G. (1994) "How Does a Thin Wetted Film Dry Up?" *Phys. Rev. Lett.* **72**, pp. 3562–3565.

Elbaum, M., and Lipson, S.G. (1995), "Pattern Formation in the Evaporation of Thin Liquid Films," *Israel J. Chem.* **35**, pp. 27–32.

Fermigier, M., Limat, L., Wesfreid, J.E., Boudinet, P., and Quilliet, C. (1992) "Two-Dimensional Patterns in Rayleigh–Taylor Instability of a Thin Layer," *J. Fluid Mech.* **236**, pp. 349–383.

Frenkel, A.L., Babchin, A.J., Levich, B.G., Shlang, T., and Sivashinsky, G.I. (1987) "Annular Flow Can Keep Unstable Flow from Breakup: Nonlinear Saturation of Capillary Instability," *J. Coll. Interface Sci.* **115**, pp. 225–233.

Gau, H., Herminghaus, S., Lenz, P., and Lipowsky, R. (1999) "Liquid Morphologies on Structured Surfaces: From Microchannels to Microchips," *Science* **283**, pp. 46–49.

de Gennes, P.G. (1985) "Wetting: Statics and Dynamics," *Rev. Mod. Phys.* **57**, pp. 827–863.

Goldmann, L.S. (1969) "Geometric Optimization of Controlled Collapse Interconnections," *IBM J. Res. Develop.* **13**.

Grotberg, J.B. (1994) "Pulmonary Flow and Transport Phenomena," *Ann. Rev. Fluid Mech.* **26**, pp. 529–571.

Hammond, P.S. (1983) "Nonlinear Adjustment of a Thin Annular Film of Viscous Fluid Surrounding a Thread of Another within a Circular Cylindrical Pipe," *J. Fluid Mech.* **137**, pp. 363–384.

Herminghaus, S., Jacobs, K., Mecke, K., Bischof, J., Fery, A., Ibn-Elhaj, M., and Schlagowski, S. (1998) "Spinodal Dewetting in Liquid Crystal and Liquid Metal Films," *Science* **282**, pp. 916–919.

Herminghaus, S., Gau, H., and Moench, W. (1999) "Artificial Liquid Microstructures," *Adv. Mater.* **11**, pp. 1393–1395.

Herminghaus, S., Fery, A., Schlagowski, S., Jacobs, K., Seeman, R., Gau, H., Moench, W., and Pompe, T. (2000) "Liquid Microstructures at Solid Interfaces," *J. Phys. Condens. Matter* **12**, pp. A57–A74.

Hirasaki, G.J. (1991) "Thin Films and Fundamentals of Wetting Phenomena," in *Interfacial Phenomena in Petroleum Recovery*, ed. N.R. Morrow, Marcel Dekker, New York.

Ho, C.-M., and Tai, Y.-C. (1998) "Micro-Electro-Mechanical Systems (MEMS) and Fluid Flows," *Ann. Rev. Fluid Mech.* **30**, pp. 579–612.

Ho, L.-W., and Patera, A.T. (1990) "A Legendre Spectral Element Method for Simulation of Unsteady Incompressible Viscous Free-Surface Flow," *Comp. Meth. Appl. Mech. Eng.* **80**, pp. 355–366.

Huppert, H.E. (1982) "The Propagation of Two-Dimensional and Axisymmetric Viscous Gravity Currents over a Rigid Horizontal Surface," *J. Fluid Mech.* **121**, pp. 43–58.

Huppert, H.E., and Simpson, J.E. (1980) "The Slumping of Gravity Currents," *J. Fluid Mech.* **99**, pp. 785–799.

Hwang, C.-C., Chang, S.-H., and Chen, J.-L. (1993) "On the Rupture Process of Thin Liquid Film," *J. Coll. Interface Sci.* **159**, pp. 184–188.

Hwang, C.-C., Lin, C.-K., and Uen, W.-Y. (1997) "A Nonlinear Three-Dimensional Rupture Theory of Thin Liquid Films," *J. Coll. Interface Sci.* **190**, pp. 250–252.

Ichimura, K., Oh, S.-K., and Nakagawa, M. (2000) "Light-Driven Motion of Liquids on a Photoresistive Surface," *Science* **288**, pp. 1624–1626.

Israelachvili, J.N. (1992) *Intermolecular and Surface Forces,* second edition, Academic Press, London.

Ivanov, I.B. (1988) *Thin Liquid Films: Fundamentals and Applications,* Marcel Dekker, New York.

Jacobs, K., Herminghaus, S., and Mecke, K.R. (1998) "Thin Liquid Polymer Films Rupture via Defects," *Langmuir* **14**, pp. 965–969.

Jain, R.K., and Ruckenstein, E. (1974) "Spontaneous Rupture of Thin Liquid Films," *J. Chem. Soc. Faraday Trans. II* **70**, pp. 132–147.

Jameel, A.T., and Sharma, A. (1994) "Morphological Phase Separation in Thin Liquid Films," *J. Coll. Interface Sci.* **164**, pp. 416–427.

Joo, S.W., Davis, S.H., and Bankoff, S. G. (1991) "Long-Wave Instabilities of Heated Films: Two Dimensional Theory of Uniform Layers," *J. Fluid Mech.* **230**, pp. 117–146.

Kargupta, K., Konnur, R., and Sharma, A. (2000) "Instability and Pattern Formation in Thin Liquid Films on Chemically Heterogeneous Substrates," *Langmuir* **16**, pp. 10243–10253.

Khanna, R., and. Sharma, A. (1998) "Pattern Formation in Spontaneous Dewetting of Thin Apolar Films," *J. Coll. Interface Sci.* **195**, pp. 42–50.

Khanna, R., Jameel, A.T., and Sharma, A. (1996) "Stability and Breakup of Thin Polar Films on Coated Substrates: Relationship to Macroscopic Parameters of Wetting," *Ind. Eng. Chem. Res.* **35**, pp. 3081–3092.

Khanna, R., Sharma, A., and Reiter, G. (2000) "The ABC of Pattern Evolution in Self-Destruction of Thin Polymer Films," *Euro. Phys. J. E* **2**, pp. 1–9.

Kheshgi, H.S., and Scriven, L.E. (1991) "Dewetting: Nucleation and Growth of Dry Regions," *Chem. Eng. Sci.* **46**, pp. 519–526.

Knight, J.B., Vishwanath, A., Brody, J.P., and Austin, R.H. (1998) "Hydrodynamic Focusing on a Silicon Chip: Mixing Nanoliters in Microseconds," *Phys. Rev. Lett.* **80**, pp. 3863–3866.

Konnur, R., Kargupta, K., and Sharma, A. (2000) "Instability and Morphology of Thin Liquid Films on Chemically Heterogeneous Substrates," *Phys. Rev. Lett.* **84**, pp. 931–934.

Kopbosynov, B.K., and Pukhnachev, V.V. (1986) "Thermocapillary Flow in Thin Liquid Films," *Fluid Mech. Sov. Res.* **15**, pp. 95–106.

Koplik, J., and Banavar, J.R. (1995) "Continuum Deductions from Molecular Hydrodynamics," *Ann. Rev. Fluid Mech.* **27**, pp. 257–292.

Koplik, J., and Banavar, J.R. (2000) "Molecular Simulations of Dewetting," *Phys. Rev. Lett.* **84**, pp. 4401–4404.

Krishnamoorthy, S., Ramaswamy, B., and Joo, S.W. (1995) "Spontaneous Rupture of Thin Liquid Films Due to Thermocapillarity: A Full-Scale Direct Numerical Simulation," *Phys. Fluids A* **7**, pp. 2291–2293.

Landau, L.D., and Lifshitz, E.M. (1987) *Fluid Mechanics,* Pergamon, Oxford.

Lee, J., and Kim, C.-J. (2000) "Surface-Tension-Driven Microactuation Based on Continuous Electrowetting," *J. Microelectromech. Sys.* **9**, pp. 171–180.

Leger, L., and Joanny, J.F. (1992) "Liquid Spreading," *Rep. Prog. Phys.* **55**, pp. 431–486.

Legros, J.C. (1986) "Problems Related to Non-Linear Variations of Surface Tension," *Acta Astron.* **13**, pp. 697–703.

Legros, J.C., Limbourg-Fontaine, M.C., and Petre, G. (1984) "Influence of Surface Tension Minimum as a Function of Temperature on the Marangoni Convection," *Acta Astron.* **14**, pp. 143–147.

Lenz, P. (1999) "Wetting Phenomena on Structured Surfaces," *Adv. Mater.* **11**, pp. 1531–1533.

Lenz, P., and Lipowsky, R. (1998) "Morphological Transitions of Wetting Layers on Structured Surfaces," *Phys. Rev. Lett.* **80**, pp. 1920–1923.

Levich, V.G. (1962) *Physicochemical Hydrodynamics*, Prentice-Hall, Englewood Cliffs, NJ.

Lin, W., Patra, S.K., and Lee, Y.C. (1995) "Design of Solder Joints for Self-Aligned Optoelectronic Assemblies," *IEEE Trans. Compon. Packaging. Manuf. Technol. B* **18**, pp. 543–551.

Lipowsky, R., Lenz, P., and Swain, P.S. (2000) "Wetting and Dewetting of Structured and Imprinted Surfaces," *Coll. Surf. A* **161**, pp. 3–22.

Mitlin, V.S. (1993) "Dewetting of Solid Surface: Analogy with Spinodal Decomposition," *J. Coll. Interface Sci.* **156**, pp. 491–497.

Mitlin, V.S. (2000) "Dewetting Revisited: New Asymptotics of the Film Stability Diagram and the Metastable Regime of Nucleation and Growth of Dry Zones," *J. Coll. Interface Sci.* **227**, pp. 371–379.

Mitlin, V.S., and Petviashvili, N.V. (1994) "Nonlinear Dynamics of Dewetting: Kinetically Stable Structures," *Phys. Lett. A* **192**, pp. 323–326.

Moriarty, J.A., and Terrill, E.L. (1996) "Mathematical Modeling of the Motion of Hard Contact Lenses," *Eur. J. Appl. Math.* **7**, pp. 575–594.

Muller-Buschbaum, P., Vanhoorne, P., Scheumann, V., and Stamm, M. (1997) "Observation of Nano-Dewetting Structures," *Europhys. Lett.* **40**, pp. 655–660.

Myers, T.G. (1998) "Thin Films with High Surface Tension," *SIAM Rev.* **40**, pp. 441–462.

Newhouse, L.A., and Pozrikidis, C. (1992) "The Capillary Instability of Annular Layers and Liquid Threads," *J. Fluid Mech.* **242**, pp. 193–209.

Oron, A. (2000a) "Nonlinear Dynamics of Irradiated Thin Volatile Liquid Films," *Phys. Fluids* **12**, pp. 29–41.

Oron, A. (2000b) "Nonlinear Dynamics of Three-Dimensional Long-Wave Marangoni Instability in Thin Liquid Films," *Phys. Fluids* **12**, pp. 1633–1645.

Oron, A. (2000c) "Three-Dimensional Nonlinear Dynamics of Thin Liquid Films," *Phys. Rev. Lett.* **85**, pp. 2108–2111.

Oron, A. (2000d) "Three-Dimensional Nonlinear Dynamics of Thin Liquid Films on Patterned Solid Surfaces," submitted.

Oron, A., and Bankoff, S.G. (1999) "Dewetting of a Heated Surface by an Evaporating Liquid Film under Conjoining/Disjoining Pressures," *J. Coll. Interface Sci.* **218**, pp. 152–166.

Oron, A., and Bankoff, S.G. (2001) "Dynamics of a Condensing Liquid Film under Conjoining/Disjoining Pressures," *Phys. Fluids* **13**, pp. 1107–1117.

Oron, A., and Peles, Y. (1998) "Stabilization of Thin Liquid Films by Internal Heat Generation," *Phys. Fluids A* **10**, pp. 537–539.

Oron, A., and Rosenau, P. (1992) "Formation of Patterns Induced by Thermocapillarity and Gravity," *J. Phys. II France* **2**, pp. 131–146.

Oron, A., and Rosenau, P. (1994) "On a Nonlinear Thermocapillary Effect in Thin Liquid Layers," *J. Fluid Mech.* **273**, pp. 361–374.

Oron, A., Bankoff, S.G., and Davis, S.H. (1996) "Thermal Singularities in Film Rupture," *Phys. Fluids A* **8**, pp. 3433–3435.

Oron, A., Davis, S.H., and Bankoff, S.G. (1997) "Long-Scale Evolution of Thin Liquid Films," *Rev. Modern Phys.* **68**, pp. 931–980.

Padmakar, A.S., Kargupta, K., and Sharma, A. (1999) "Stability and Dewetting of Evaporating Thin Water Films on Partially and Completely Wettable Substrates," *J. Chem. Phys.* **110**, pp. 1735–1744.

Palmer, H.J. (1976) "The Hydrodynamic Stability of Rapidly Evaporating Liquids at Reduced Pressure," *J. Fluid Mech.* **75**, pp. 487–511.

Paulsen, F.G., Pan, R., Bousfield, D.W., and Thompson, E.V. (1996) "The Dynamics of Bubble/Particle Attachment and the Application of Two Disjoining Film Rupture Models to Flotation," *J. Coll. Interface Sci.* **178**, pp. 400–410.

Peurrung, L.M., and Graves, D.B. (1993) "Spin Coating Over Topography," *IEEE Trans. Semiconductor Manuf.* **6**, pp. 72–76.

Plesset, M.S. (1952) "Note on the Flow of Vapor between Liquid Surfaces," *J. Chem. Phys.* **20**, pp. 790–793.

Plesset, M.S., and Prosperetti, A. (1976) "Flow of Vapor in a Liquid Enclosure," *J. Fluid Mech.* **78**, pp. 433–444.

Pozrikidis, C. (1992) *Boundary Integral and Singularity Methods for Linearized Viscous Flow*, Cambridge University, Cambridge, U.K.

Pozrikidis, C. (1997) "Numerical Studies of Singularity Formation at Free Surfaces and Fluid Interfaces in Two-Dimensional Stokes Flow," *J. Fluid Mech.* **331**, pp. 145–167.

Prud'homme, R.K., and Khan, S.A., eds. (1996) *Foams: Theory, Measurements and Applications*, Marcel Dekker, New York.

Ramaswamy, B., Krishnamoorthy, S., and Joo, S.W. (1997) "Three-Dimensional Simulation of Instabilities and Rivulet Formation in Heated Falling Films," *J. Comp. Phys.* **131**, pp. 70–88.

Reisfeld, B., Bankoff, S.G., and Davis, S.H. (1991) "The Dynamics and Stability of Thin Liquid Films During Spin-Coating. I. Films with Constant Rates of Evaporation or Absorption," *J. App. Phys.* **70**, pp. 5258–5266; "II. Films with Unit-Order and Large Peclet Numbers," *J. App. Phys.* **70**, pp. 5267–5277.

Reiter, G. (1992) "Dewetting of Thin Polymer Films," *Phys. Rev. Lett.* **68**, pp. 75–78.

Reiter, G. (1998) "The Artistic Side of Intermolecular Forces," *Science* **282**, pp. 888–889.

Reiter, G., Sharma, A., Casoli, A., David, M.-O., Khanna, R., and Auroy, P. (1999a) "Destabilizing Effect of Long-Range Forces in Thin Liquid Films on Wettable Substrates," *Europhys. Lett.* **46**, pp. 512–518.

Reiter, G., Sharma, A., Casoli, A., David, M.-O., Khanna, R., and Auroy, P. (1999b) "Thin Film Instability Induced by Long-Range Forces," *Langmuir* **15**, pp. 2551–2558.

Reiter, G., Khanna, R., and Sharma, A. (2000) "Enhanced Instability in Thin Liquid Films by Improved Compatibility," *Phys. Rev. Lett.* **85**, pp. 1432–1435.

Rosenau, P. (1995) "Fast and Superfast Diffusion Processes," *Phys. Rev. Lett.* **74**, pp. 1056–1059.

Sadhal, S.S., and Plesset, M.S. (1979) "Effect of Solid Properties and Contact Angle in Dropwise Condensation and Evaporation," *J. Heat Transfer* **101**, pp. 48–54.

Salalha, W., Zussman, E., Meltser, M., and Kaldor, S. (2000) "Prediction of Yield for Flip-Chip Packaging," International CIRP Design Seminar, Haifa, Israel.

Salamon, T.R., Armstrong, R.C., and Brown, R.A. (1994) "Traveling Waves on Vertical Films: Numerical Analysis Using the Finite Elements Method," *Phys. Fluids A* **6**, pp. 2202–2220.

Scardovelli, R., and Zaleski, S. (1999) "Direct Numerical Simulation of Free-Surface and Interfacial Flow," *Ann. Rev. Fluid Mech.* **31**, pp. 567–603.

Schaeffer, E., Thurn-Albrecht, T., Russell, T. P., and Steiner, U. (2000) "Electrically Induced Structure Formation and Pattern Transfer," *Nature* **403**, pp. 874–877.

Scheludko, A.D. (1967) "Thin Liquid Films," *Adv. Coll. Interface Sci.* **1**, pp. 391–464.

Schlichting, H. (1968) *Boundary-Layer Theory*, McGraw-Hill, New York.

Schrage, R.W. (1953) *A Theoretical Study of Interphase Mass Transfer*, Columbia University, New York.

Schramm, L.A., ed. (1994) *Foams: Fundamentals and Applications in the Petroleum Industry*, American Chemical Society, Washington, D.C.

Sharma, A. (1998) "Equilibrium and Dynamics of Evaporating or Condensing Thin Fluid Domains: Thin Film Stability and Heterogeneous Nucleation," *Langmuir* **14**, pp. 4915–4928.

Sharma, A. (2000) private communication.

Sharma, A., and Jameel, A.T. (1993) "Nonlinear Stability, Rupture, and Morphological Phase Separation of Thin Fluid Films on Apolar and Polar Substrates," *J. Coll. Interface Sci.* **161**, pp. 190–208.

Sharma, A., and Khanna, R. (1998) "Pattern Formation in Unstable Thin Liquid Films," *Phys. Rev. Lett.* **81**, pp. 3463–3466.

Sharma, A., and Reiter, G. (1996) "Instability of Thin Polymer Films on Coated Substrates: Rupture, Dewetting, and Drop Formation," *J. Coll. Interface Sci.* **178**, pp. 383–399.

Sharma, A., and Ruckenstein, E. (1986), "An Analytical Nonlinear Theory of Thin Film Rupture and Its Application to Wetting Films," *J. Coll. Interface Sci.* **113**, pp. 456–479.

Sharma, A., Konnur, R., Khanna, R., and Reiter, G. (2000) "Morphological Pathways of Pattern Evolution and Dewetting in Thin Liquid Films," in *Emulsions, Foams and Thin Films,* ed. K.L. Mittal, Marcel Dekker, New York, pp. 211–232.

Siegel, R., and Howell, J.R. (1992) *Thermal Radiation Heat Transfer,* Hemisphere, Washington, D.C.

Stillwagon, L.E., and Larson, R.G. (1988) "Fundamentals of Topographic Substrate Leveling," *J. App. Phys.* **63**, pp. 5251–5258.

Stillwagon, L.E., and Larson, R.G. (1990) "Leveling of Thin Films over Uneven Substrates during Spin Coating," *Phys. Fluids A* **2**, pp. 1937–1944.

Stillwagon, L.E., and Larson, R.G. (1992) "Planarization during Spin Coating," *Phys. Fluids A* **4**, pp. 895–903.

Tan, M.J., Bankoff, S.G., and Davis, S.H. (1990) "Steady Thermocapillary Flows of Thin Liquid Layers," *Phys. Fluids A* **2**, pp. 313–321.

Teletzke, G.F., Davis, H.T., and Scriven, L.E. (1987) "How Liquids Spread on Solids," *Chem. Eng. Comm.* **55**, pp. 41–82.

Thiele, U., Mertig, M., and Pompe, W. (1998) "Dewetting of an Evaporating Thin Liquid Film: Heterogeneous Nucleation and Surface Instability," *Phys. Rev. Lett.* **80**, pp. 2869–2871.

Tsai, W.T., and Yue, D.K.P. (1996) "Computation of Nonlinear Free-Surface Flow," *Ann. Rev. Fluid Mech.* **28**, pp. 249–278.

Van Hook, S.J., Schatz, M.F., McCormick, W.D., Swift, J.B., and Swinney, H.L. (1995) "Long-Wavelength Instability in Surface-Tension-Driven Benard Convection," *Phys. Rev. Lett.* **75**, pp. 4397–4400.

Van Hook, S.J., Schatz, M.F., Swift, J.B., McCormick, W.D., and Swinney, H.L. (1997) "Long-Wavelength Instability in Surface-Tension-Driven Benard Convection: Experiment and Theory," *J. Fluid Mech.* **345**, pp. 45–78.

Wale, M.J., and Edge, C. (1990) "Self-Aligned, Flip-Chip Assembly of Photonic Devices with Electrical and Optical Connections," *IEEE Trans. Compon. Hybrids Manuf. Technol.* **13**, pp. 780–786.

Wehausen, J.V., and Laitone, E.V. (1960) "Surface Waves," in *Encyclopedia of Physics,* ed. S. Flugge, Vol. IX, Fluid Dynamics III, Springer-Verlag, Berlin, pp. 446–778.

Williams, M.B. (1981) "Nonlinear Theory of Film Rupture," Ph.D. thesis, Johns Hopkins University, Baltimore, MD.

Williams, M.B., and Davis, S.H. (1982) "Nonlinear Theory of Film Rupture," *J. Coll. Interface Sci.* **90**, pp. 220–228.

Witelski, T.P., and Bernoff, A.J. (1999) "Stability of Self-Similar Solutions for van der Waals Driven Thin Film Rupture," *Phys. Fluids* **11**, pp. 2443–2445.

Witelski, T.P., and Bernoff, A.J. (2000) "Dynamics of Three-Dimensional Thin Film Rupture," *Physica D* **147**, pp. 155–176.

Wu, P.-Y., and Chou, F.-C. (1999) "Complete Analytical Solutions of Film Planarization during Spin Coating," *J. Electrochem. Soc.* **146**, pp. 3819–3826.

Wu, P.-Y., Chou, F.-C., and Gong, S.-C. (1999) "Analytical Solutions of Film Planarization for Periodic Features," *J. Appl. Phys.* **86**, pp. 4657–4659.

Xie, R., Karim, A., Douglas, J.F., Han, C.C., and Weiss, R.A. (1998) "Spinodal Dewetting of Thin Polymer Films," *Phys. Rev. Lett.* **81**, pp. 1251–1254.

Yarin, A.L. (1993) *Free Liquid Jets and Films,* Longman, New York.

Yiantsios, S.G., and Higgins, B.G. (1989) "Rayleigh–Taylor Instability in Thin Viscous Films," *Phys. Fluids A* **1**, pp. 1484–1501.

Yiantsios, S.G., and Higgins, B.G. (1991) "Rupture of Thin Films: Nonlinear Stability Analysis," *J. Coll. Interface Sci.* **147**, pp. 341–350.

Zhang, W.W., and Lister, J.R. (1999) "Similarity Solutions for van der Waals Rupture of a Thin Film on a Solid Substrate," *Phys. Fluids* **11**, pp. 2454–2462.

11
Bubble/Drop Transport in Microchannels

Hsueh-Chia Chang
University of Notre Dame

11.1 Introduction 11.1
11.2 Fundamentals 11.2
11.3 The Bretherton Problem for Pressure-Driven Bubble/Drop Transport 11.4
  Corrections to the Bretherton Results for Pressure-Driven Flow
11.4 Bubble Transport by Electrokinetic Flow.......................11.8
11.5 Future Directions 11.12
  Acknowledgments 11.12

11.1 Introduction

Many microdevices involve fluid flows. Microducts, micronozzles, micropumps, microturbines and microvalves are examples of small devices with gas or liquid flow. It would be extremely desirable to design similar devices for two-phase flows, and many attractive applications can be envisioned if microreactors and microlaboratories could include immiscible liquid–liquid and gas–liquid systems. Miniature evaporative and distillation units, bubble generators, multiphase extraction/separation units and many other conventional multiphase chemical processes could then be fabricated at microscales. Efficient multiphase heat exchangers could be designed for MEM devices to minimize Joule or frictional heating effects. Even for the current generation of microlaboratories using electrokinetic flow, multiphase flow has many advantages. Drops of organic samples could be transported by flowing electrolytes, thus extending the electrokinetic concept to a broader class of samples. Gas bubbles could be used as spacers for samples in a channel or to act as a piston to produce pressure-driven flow on top of the electrokinetic flow. Flow valves and pumps that employ air bubbles, like those in the ink reservoirs of ink jet printers, are already being tested for microchannels. Drug-delivery and diagnostic devices involving colloids, molecules and biological cells are also active areas of research.

Before multiphase flow in microchannels becomes a reality, however, several fundamental problems that arise from the small dimension of the channels must be solved. Most of these problems originate from the large curvature of the interface between two phases in these small channels. As a result, capillary effects and other related phenomena dominate in multiphase microfluidics. Contact-line resistance, for example, is often negligible in macroscopic flows. The contact-line region, defined by intermolecular and capillary forces, is small compared to the macroscopic length scales. However, in microchannels, the contact-line region is comparable in dimension to the channel size. As a result, the large stress in that region (the classical contact-line logarithm stress singularity) can dominate the total viscous dissipation [Kalliadasis and Chang, 1994; Veretennikov et al., 1998; Indeikina and Chang, 1999]; hence, it is inadvisable to have contact lines in microchannels unless one is prepared to apply enormous pressure or electric potential driving forces. One fluid should wet the channel or capillary walls while the other is dispersed in the form of drops or bubbles. Due to the small channel dimension, though, the drops and bubbles usually have a free radius larger than the channel radius—it is typically difficult to generate colloid-size drops or bubbles smaller than the channel. Biological cells and protein molecules are also comparable in dimension to microchannels and represent a unique class of "drops" in such devices. This chapter addresses several fundamental issues in the transport of these "large" bubbles/drops and suggests the most realistic and attainable conditions for such multiphase microfluidic flows.

11.2 Fundamentals

Schematics of a bubble immersed in a wetting liquid within a capillary of radius R are shown in Figure 11.1. The dimensionless coordinate r is scaled by the capillary radius R. If the bubble is not translating, the capillary pressure drop across the bubble cap is of order σ/R, where σ is the interfacial tension. In contrast, the pressure drop necessary to drive a liquid slug of length l at speed U in the same channel is of order $Ul\,\mu/R^2$. Hence, the slug length l scales as RCa^{-1} where $Ca = \mu U/\sigma$ is the capillary number. In microchannels, Ca ranges from 10^{-8} to 10^{-4} (for aqueous solutions moving at 10^{-4} to 1 mm/sec), thus the equivalent slug length l is many orders of magnitude higher than R. Equivalently, the capillary pressure across the static meniscus can drive a liquid slug of length R at the astronomically large dimensionless speed of $Ca = 1$. For electrokinetic flow, in contrast, such speeds can only be achieved by an electric field of more than 10^4 volts/cm! The capillary pressure across a static meniscus in a capillary is sometimes called the *invasion pressure* and it is the required pressure to insert a meniscus in the capillary. Once the bubble is set into motion, the required pressure to sustain its motion is less than σ/R but, as will be shown below, it is still significant.

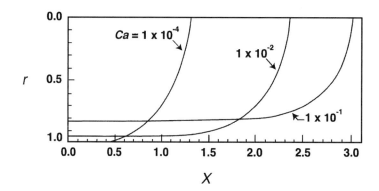

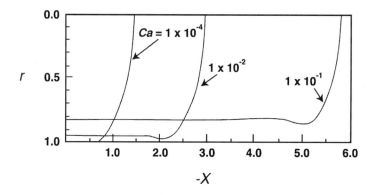

FIGURE 11.1 Front and back profiles for very long bubbles.

The thickness of the wetting film around a moving bubble in a capillary and the pressure drop across it were first studied by Bretherton (1961). For capillary radii R smaller than the capillary length $(\sigma/\Delta\rho g)^{1/2}$, which is about 1 mm for aqueous solutions, buoyancy effects are negligible and the bubble is axisymmetrically placed within the capillary. The flat annular film around the bubble allows only unidirectional longitudinal flow. Moreover, this lubrication limit stipulates that the pressure be constant across the film and determined by the local interfacial curvature, the sum of the axial and azimuthal curvatures of the axisymmetric bubble. Pressure variation is hence only in the longitudinal direction. For pressure-driven mobile bubbles, however, the flat annular film at the middle of the drop indicates that no pressure gradient is present and there is no flow in the film. Liquid flow only occurs at the transition regions near the caps where the film is no longer flat in the longitudinal direction. Near the front cap, the azimuthal curvature decreases behind the tip and the resulting capillary pressure gradient drives fluid into the annular film. The reverse happens near the back cap to pick up the stagnant liquid laid down by the front cap. Unlike the usual symmetric Stokes flow, however, the flow around the two caps are not mirror images of each other in this free-surface problem. If they were reflectively symmetric, the net pressure drop across the bubble would be zero, which is impossible for a translating bubble. In essence, pressure-driven liquid flow before and after the bubble is due to the same negative bulk pressure gradient. However, the capillary pressure gradients at the two caps are in opposite directions relative to this bulk gradient. As such, the two caps are not mirror images of each other and the capillary pressure across the back cap must be smaller than that at the front cap.

Simple scaling arguments can immediately determine the pressure drop across the bubble and the thickness of the surrounding film. The leading order estimate of capillary pressure drops at both caps is indeed identical at $2\sigma/R$ and the axial curvature at the tips is $1/R$. The axial curvature at the surrounding annular film is d^2h/dx^2 where h is the interfacial thickness measured from the capillary wall and x is the longitudinal direction. (The azimuthal curvature gradient scales as h_x and is negligible compared to the axial curvature gradient h_{xxx} in the short transition region.) Balancing the axial curvature d^2h/dx^2 to $1/R$ immediately reveals that the ratio of the length of the transition region scales as the square root of the film thickness, with both lengths small compared to the capillary radius R. However, the pressure gradient in the transition regions provided by the capillary pressure drives a liquid flow at the speed U of the bubble. Balancing the viscous dissipation estimate $\mu U/h^2$ with dp/dx and using the above scalings for each quantity, we immediately conclude that the ratio of h^2 to the transition length x is of order Ca. Reconciling this with the relative scalings imposed by curvature matching, we obtain the classical Bretherton scalings—the film thickness scales as $RCa^{2/3}$ while the transition regions near the cap are of the order of $RCa^{1/3}$ long, with $Ca \ll 1$. The total viscous dissipation due to the flow at the caps is the integral of μ times the normal gradient of the flow field at the wall over the transition length. This is just the capillary pressure required to balance the dissipation and, using the above scaling, this capillary pressure is of the order $(\sigma/R)Ca^{2/3}$. Due to the asymmetry of the two caps, this capillary pressure is different at the two caps. The difference in the pressure drop across the two caps is then of the order $(\sigma/R)Ca^{2/3}$.

Using this new estimate for the pressure drop, we conclude that the equivalent slug length l scales as $RCa^{-1/3}$. Equivalently, in a train of translating bubbles spaced by continuous liquid slugs, the pressure drop across each bubble roughly corresponds to a liquid slug of lengh $RCa^{-1/3}$, or 10 to 1000 times the capillary radius! Hence, the pressure drop required to drive most bubble trains occurs at the bubble caps. Even without contact-line resistance, pressure-driven multiphase transport in microchannels is thus expected to require orders of magnitude higher pressure drops. In the next section, we will estimate this pressure drop with and without Marangoni traction introduced by surfactants and sketch the effects of drop viscosity and noncircular capillaries. It is unlikely, however, that pressure-driven multiphase flow can be achieved under realistic conditions. In the following section, we show that electrokinetically driven multiphase flow is very much achievable and demonstrate that bubble speed can reach as high as the electrokinetic speed of pure liquids. However, such flows occur under very specific conditions, which will be described in some detail. We conclude with some conjectures on other multiphase microfluidics in the final section.

11.3 The Bretherton Problem for Pressure-Driven Bubble/Drop Transport

The above scaling arguments can be made more precise with matched asymptotics. Using a local Cartesian coordinate for the thin-film region, the usual lubrication analysis yields the following longitudinal velocity profile:

$$u(y) = \frac{-\sigma}{\mu} \frac{\partial^3 h}{\partial x^3} \left(\frac{y^2}{2} - yh \right) \quad (11.1)$$

The normal coordinate y is measured from the capillary wall. The pressure p is independent of y and is equal to $-\sigma h_{xx}$, the axial curvature of the film. Hence, integrating over the film thickness, one obtains the flow rate $q = \left(\frac{\sigma h^3}{3\mu} \right) \frac{\partial^3 h}{\partial x^3}$. The cubic power dependence arises from the parabolic profile of $u(y)$ in Eq. (11.1). Mass balance over the entire film cross section yields:

$$\frac{\partial h}{\partial t} = -\frac{\partial q}{\partial x} \quad (11.2)$$

In a frame moving with the bubble speed U, the time derivative is converted into $-Uh_x$ in the moving frame. Integrating from the flat-film region where the third derivative vanishes into the transition region yields:

$$3\left(\frac{\mu U}{\sigma} \right)(h - h_\infty) = h^3 h_{xx} \quad (11.3)$$

Scaling h by the unknown flat-film thickness h_∞ and the x coordinate by $h_\infty/(3Ca)^{1/3}$, respectively, we obtain the Bretherton equation:

$$H_{XXX} = \frac{H-1}{H^3} \quad (11.4)$$

This nonlinear equation for H describes the transition regions of both caps. However, the front one corresponds to $X \to \infty$ while the back one corresponds to $X \to -\infty$. The two asymptotic behaviors are clearly not identical, indicating that the two caps are not mirror images. Nevertheless, as H blows up in both infinities, its third derivative must vanish according to Eq. (11.4) and one expects quadratic blowup in both directions. These quadratic asymptotes must then be matched to the outer cap solutions. As h blows up, viscous effects become negligible and the outer caps are, to the leading order, just static solutions of the Laplace–Young equation. Without gravitational effects, these axisymmetric solutions are just spherical caps of radius R that make quadratic contact with the wall.

Linearizing about $H = 1$, the behavior away from the flat film is seen to be governed by three eigenvalues, 1 and $-\frac{1}{2} \pm \frac{\sqrt{3}}{2} i$. There is, then, only monotonic blowup in the positive X direction due to a lone positive real eigenvalue, and a numerical integration of Eq. (11.4) yields the front cap asymptote:

$$H(X \to \infty) = \alpha^+ X^2 + \gamma^+ X + \beta^+ \quad (11.5)$$

The second coefficient can be changed due to an arbitrary shift of X but the quadratic coefficient is universal.

We then choose the origin of X until γ^+ vanishes. Equivalently, we can vary $H - 1$ with $H_X = H_{XX} = 0$ for the initial condition in our forward integration of Eq. (11.4). This one-parameter iteration yields:

$$\alpha^+ = 0.32171 \quad \beta^+ = 2.898 \quad (11.6)$$

Hence, the asymptotic curvature of the annular film toward the front cap is $H_{XX} = 2\alpha^+$ or, in the original dimensional coordinate:

$$h_{xx} = \frac{(3Ca)^{2/3}}{h_\infty} H_{XX} = \frac{2\alpha^+(3Ca)^{2/3}}{h_\infty} \quad (11.7)$$

This must match with the front spherical cap of radius R that makes the quadratic tangent with the capillary. Hence, matching Eq. (11.7) to this quadratic contact, we obtain the leading order estimate of the film thickness:

$$h_\infty/R = 0.6434(3Ca)^{2/3} \quad (11.8)$$

The back matching is more intricate. We note first that the complex eigenvalues suggest that the back film is undulating. A pronounced dimple due to this undulation is evident in the back profiles of Figure 11.1 computed by Lu and Chang (1988). This film oscillation is indeed confirmed by the photographs of Friz (1965). There is an arbitrary phase between these two complex modes that must be specified. This extra degree of freedom is not present for the positive direction with only one real eigenvalue. Due to the quadratic contact of the back cap, we again iterate on the origin of X to obtain the back asymptote:

$$H(X \to -\infty) = \alpha^- X^2 + \beta^- \quad (11.9)$$

Because of the extra degree of freedom in the phase of the two complex conjugate modes, both α^- and β^- are functions of the phase, thus the pair (α^-, β^-) is a one-parameter family. To the leading order, however, this asymptote must also match a sphere of radius R that makes tangential contact with the capillary. Hence, $\alpha^- = \alpha^+ = 0.32171$. For this value of α^-, the corresponding value of β^- is $\beta^- = -0.8415$. (This is the most accurate estimate obtained by Chang and Demekhin [1999]. It is slightly different from many earlier values, including Bretherton's.)

The capillary pressure drops at the two caps arise from the $\beta^\pm$ terms. Consider the two spherical caps of radius $R' = R(1 + \varepsilon)$ different from the capillary radius R. Then, the expansion of the cap near the contact point, $\frac{dh(0)}{dx} = 0$, is

$$h \sim \frac{x^2}{R} - R\varepsilon \quad (11.10)$$

Matching this expansion of the outer cap solution near the capillary to the two asymptotes of the inner film solutions, Eqs. (11.5) and (11.9), one immediately sees that the front cap has a radius smaller than R and the back has a larger radius. The difference is of order $Ca^{2/3}$, the scalings for H in both equations. Hence, the pressure drop across the entire bubble is then the difference in the two cap capillary pressures σ/R':

$$\Delta p/(\sigma/R) = \frac{2}{1+\varepsilon^-} - \frac{2}{1+\varepsilon^+} \sim 2(\varepsilon^+ - \varepsilon^-)$$
$$= 2(\beta^+ - \beta^-)(h_\infty/R) = 10.0 Ca^{2/3} \quad (11.11)$$

The scaling of this pressure drop is consistent with the earlier order-of-magnitude arguments of the previous section. The unit-order coefficients are now specified by this classical matched asymptotic analysis. We note in passing that there is also an inner $X \ln X$ asymptotic behavior that needs to be matched to similar expansions in the outer solution [Kalliadasis and Chang, 1996]. Such high-order matching becomes important only when contact lines appear.

11.3.1 Corrections to the Bretherton Results for Pressure-Driven Flow

At higher values of Ca, between 0.01 and 0.1, the film thickness and pressure drop across the bubble must be solved numerically instead of by matched asymptotics. This effort was carried out by Reinelt and Saffman (1985) and Lu and Chang (1988). The pressure drop can be correlated up to $Ca = 0.1$ as [Ratulowski and Chang, 1989]:

$$\Delta p/(\sigma/R) = 10.0 Ca^{2/3} - 12.6 Ca^{0.95} \qquad (11.12)$$

However, the capillary number rarely exceeds 10^{-4} in microfluidics and the Bretherton results of the previous section are usually adequate.

Bretherton finds his film thickness prediction to be smaller than the measured values at low Ca, exactly where the matched asymptotic analysis is most valid. This is confirmed by a series of experiments summarized by Ratulowski and Chang (1990), who attribute the deviation to Marangoni effects of surfactant contaminants that are most pronounced at the thin films of low Ca. The film thickness is determined only by how the front cap lays down a thin film by its capillary pressure. In this region, the film interface is stretched considerably and the interfacial surfactant concentration decreases from the cap to the film. The film surface tension is then larger than the cap and this Marangoni traction drags additional liquid into the film to thicken it.

For soluble surfactants, a complex model involving bulk-interface transport must be constructed to account for this new mechanism. For insoluble surfactants, however, a simple correction can be obtained almost trivially. In the limit of very small Ca, this traction approaches infinity and the free surface in the transition region can be treated as a deformable but rigid interface that is laid onto the stagnant film. The velocity at the rigid interface vanishes and the parabolic velocity of Eq. (11.1) simply becomes:

$$u(y) = -\frac{\sigma}{\mu}\frac{\partial^3 h}{\partial x^3}\left(\frac{y^2}{2} - \frac{yh}{2}\right) \qquad (11.13)$$

The flow rate q is then corrected by the factor of 4 due to the interface traction. The same correction yields a factor of 4 to the left-hand side of the Bretherton equation, Eq. (11.3). As a result, one simply scales Ca by 4 and the same dimensionless Eq. (11.4) results. Hence, in the limit of low Ca, soluble surfactants will correct the film thickness by a factor of $4^{2/3}$. This asymptote is clearly approached by the experimental data in Figure 11.2 at low Ca. Ratulowski and Chang show that these asymptotic values at infinite traction are also the maximum values attainable for other more complex surfactant transport at low Ca. The correction to pressure drop is more intricate as it requires the resolution of the entire bubble. Because the surfactants accumulate at the back cap (or near a stagnation point near the back cap), correction requires a model for surfactant accumulation. Such a model was constructed by Park (1992) who then showed that the pressure drop across the bubble now has a $Ca^{1/3}$ scaling due to the accumulation. The pressure drop, then, increases by a factor of $Ca^{1/3}$ in the presence of surfactants.

One particularly interesting phenomenon concerning Marangoni effect is remobilization [Stebe et al., 1991] at high bulk surfactant concentrations when the entire interface can saturate even as it is being stretched. As such, the Marangoni traction vanishes and the mobile limit is again attained. Still, this strategy only reduces the pressure drop by a factor of order unity and does not change the basic scalings.

For bubble trains whose bubbles are separated by thin lamellae instead of spherical caps (see Figure 11.3), Ratulowski and Chang (1989) show that the pressure drop remains constant to the leading order, while the film thickness decreases as adjacent bubbles are compressed (larger contact radius r_c in Figure 11.3). It is clear from geometric considerations that a larger compression between adjacent bubbles will decrease the film thickness. In fact, an expansion of the Laplace–Young equation for the lamellae about zero contact radius shows that the film thickness is related to the free bubble thickness at $r_c = 0$ by:

$$h_\infty(r_c) = (1 - r_c)h_\infty(0) \qquad (11.14)$$

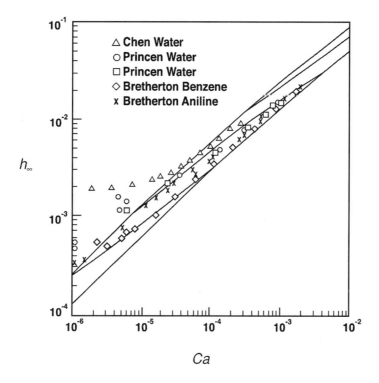

FIGURE 11.2 The film thickness of a bubble translating in various surfactant solutions. The capillary number Ca is a dimensionless speed and the film thickness is scaled by the capillary radius. The theoretical curves correspond to different surfactant equilibrium constants between the interface and the bulk. At low Ca, they all approach the same asymptote derived in the text.

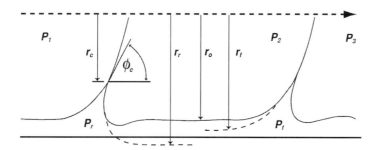

FIGURE 11.3 Schematic of a bubble train. The contact radius r_c represents the degree of compression.

What is less apparent is that, because the lamella is a constant curvature axisymmetric surface, its contribution to the curvatures of both asymptotes of the thin annular film is identical. As a result, the pressure drop across the bubble is independent of the contact radius.

Schwartz et al. (1986) examine drop transport and find that the thickness and pressure drop increase monotonically with respect to the viscosity ratio between the drop and the wetting fluid. The maximum occurs at infinitely large viscosity, corresponding to a solid drop, and it is found to be larger than Bretherton's result by a factor of $2^{2/3}$. The difference in this correction factor from the Marangoni correction is because the latter corresponds to a stationary rigid film while the former is a translating one. The Bretherton scaling results are hence very robust estimates for circular capillaries. They are only slightly corrected by Marangoni tractions due to surfactants, drop viscosity and even bubble spacing.

The Bretherton scaling arguments break down, however, for channels with noncircular channels. Ratulowski and Chang (1989) examined the square channel numerically. Because the bubble caps of isolated bubbles are axisymmetric, contact must be made with the wall at low Ca. This is estimated to be at $Ca = 0.04$, below which contact lines are expected and the liquid does not wet the channel wall. Thus, favorable operating conditions only exist for Ca larger than 0.04 and the numerical results show that the film thickness and pressure drop show peculiar scaling:

$$h_\infty = 0.69 - 0.10 \ln Ca \qquad \Delta p/(\sigma/R) = 3.14 Ca^{0.14} \qquad (11.15)$$

The radius R corresponds to a cylindrical capillary with the same cross-section area.

Estimates for other channel geometries have not been computed but the formulation by Ratulowski and Chang can still be used. In essence, one must solve the following two-dimensional unit cell equation for each dimensionless bubble radius r, scaled with respect to R:

$$\nabla^2 \psi = -1 \qquad (11.16)$$

This fundamental solution is solved within a cross section of the straight capillary with a Dirichlet boundary condition at the capillary wall and Neumann condition at the circular interface with the dimensionless radius r. The flow rate–capillary pressure relation then becomes:

$$q = -K \frac{\partial p}{\partial x} = K(r) \frac{\partial^3 r}{\partial x^3} \qquad (11.17)$$

The permeability constant $K(r)$ is the cross-section average of the fundamental solution above multiplied by the factor $\sigma R^4/\mu$. A higher order version of the curvature can be used in place of the second derivative of r. To avoid contact between bubble and capillary, the capillary cross-section geometry must be nearly axisymmetric. As such, one does not expect the pressure drop to be significantly different from Bretherton's estimate for circular capillaries, despite the difference in Ca scaling.

11.4 Bubble Transport by Electrokinetic Flow

The large pressure drop required to drive multiphase microchannel flow immediately suggests the electrokinetic driving force is more desirable. If the electrokinetic flow behind the bubble or drop is larger than that of the surrounding film, a high-pressure region can build up behind the bubble/drop to drive it with the same capillary pressure mechanism as before. As will be shown below, it is much easier to build pressure with electrokinetic flow than pressure-driven flow behind the bubble. As such, the required driving force is not as large as that of pressure-driven flow. The task is then to reduce the flow around the bubble/drop without shorting the current required to drive the fluid.

This design consideration requires some knowledge of electrokinetic flow [Russel et al., 1989; Probstein, 1994]. Electrokinetic flow occurs when the dielectric channel wall contains some surface charges that attract co-ions of opposite charge in the solution to a thin double layer of thickness λ, the Debye length, which ranges from 10 nm to microns depending on the bulk electrolyte concentration. The counter-ion concentration increases from the bulk value toward the wall within this double layer, while the co-ion decreases from its bulk value. Both bulk values are identical due to charge neutrality so there is a net charge within the thin double layer. The total amount of this charge is determined through ionization equilibrium by the surface charge on the capillary.

Within the double layer, the potential ϕ is governed by the Poisson equation:

$$\frac{\partial^2 \phi}{\partial y^2} = \frac{F\rho}{\varepsilon} \qquad (11.18)$$

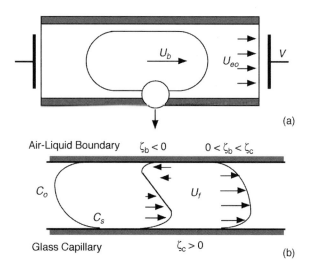

FIGURE 11.4 Electrokinetically driven bubble transport. The electrolyte ion concentration profile C_0 is shown below with the velocity profiles U_f for both negative and positive zeta potentials at the bubble interface. The bubble translates with bubble speed U_b and the liquid with electrokinetic velocity U_{eo}.

The charge density is ρ and the potential is set to zero at the bulk when y approaches infinity. The potential at the surface is called the *zeta potential* ζ. Due to the Boltzmann distributions of the co-ion and counter-ion, the counter-ion concentration increases much faster than the co-ion concentration decreases toward the wall. As a result, the total ion concentration in the double layer exceeds that in the bulk by a factor of $\exp(\zeta/kT)$, as seen in Figure 11.4. Consequently, the charge density ρ also increases from zero at the bulk to a value at the wall equal to the bulk concentration multiplied by $\exp(\zeta/kT)$. Hence, at low ζ/kT, Eq. (11.18) indicates that the scaling for λ is inversely proportional to the square root of the bulk ion concentration. Moreover, by integrating Eq. (11.18) over the double layer, its total charge is seen to scale linearly with respect to the potential gradient at the wall. The latter quantity scales as ζ/λ and one concludes that, for a given capillary–electrolyte pair, the zeta potential ζ scales as λ, inversely as the square root of the bulk electrolyte concentration.

In the presence of a tangential electric field E, there is a net body force on the electrolyte that scales as $E\rho$. This body force vanishes in the neutral bulk but accelerates the ions in the double layer to large speeds. These streaming ions then drag the entire fluid body in the capillary along with them. Because the body force is concentrated in the thin double layer, it acts like a surface force and the entire bulk liquid translates rigidly with a uniform tangential velocity. The momentum transfer in y for the tangential velocity field involves the viscous dissipation term $\mu d^2 u/d^2 y$ balanced by the body force $E\rho$. Because this is in the same form as the Poisson equation, one sees that u scales linearly with respect to the electric potential ϕ but approaches a constant value away from the double layer. This asymptote is called the *electrokinetic velocity*:

$$u_c = -\frac{\varepsilon_0 \varepsilon \zeta_c E}{\mu} \qquad (11.19)$$

The constants ε_0 and ε are the dielectric permittivities that we have omitted in the above scaling arguments.

Because the electrokinetic velocity is flat away from the thin double layer (see Figure 11.4), the flowrate scales as the cross-section area, or R^2. In contrast, for pressure-driven flow the flow rate scales as the square of the area, or R^4. For the small dimensions of microchannels, one can then obtain much larger flow rates with electrokinetic flow. Even in microchannels that are microns in width and depth, electrokinetic velocity of an electrolyte solution can often exceed 0.1 mm/sec. Similarly, it is easier to build up back pressure behind bubbles and drops with electrokinetic flow than with pressure-driven flow, if one can reduce the film flow.

Unfortunately, the same flat electrokinetic velocity profile now serves to prevent cessation of film flow. By simple current-voltage calculation in the longitudinal direction, the local electric field E is easily shown to scale as the inverse of the cross-section area of the electrolyte across the capillary. By Eq. (11.19), the electrokinetic velocity scales the same way. However, the flowrate scales as the electrokinetic velocity times the cross-section area and hence is independent of the cross-section area! The flowrate behind the bubble is the same as that in its surrounding film. As a result, there is no back pressure buildup and the electrolyte simply flows around the bubble. This is indeed observed when air bubbles are driven electrokinetically in a KCl/H_2SO_4 electrolyte (about 10^{-2} and 10^{-6} mol/L each) in a 5-cm-long glass capillary with a 1.0-mm inner diameter and with a voltage drop of 30 to 70 V [Takhistov et al., 2000]. At these conditions, the electrokinetic velocity of the electrolyte is at a healthy 0.1 to 1.0 mm/sec. Nevertheless, the bubble remains stationary as the electrolyte flows past it.

There are several possible means of breaking the above flowrate invariance to cross-section area to reduce the film flow. One can endow the interface with traction, by using finite viscosity drops or interfacial surfactants such that the film profile is no longer flat. The film electric field can also be reduced by reducing the electrolyte concentration such that the high ion concentration within the thickened double layer can increase the film conductivity. More intriguingly, one can use an ionic surfactant to endow a double layer at the interface that has a different charge from the capillary double layer. In this manner, the velocity at the interface is not zero in the moving frame but is negative (see Figure 11.4). This could effectively reduce the film flow to zero.

Because the glass capillary surface of Takhistov et al.'s experiment has a positive charge such that its double layer contains a negative charge, an anionic surfactant, sodium dodecyl sulfate (SDS), is employed to test the above idea. Most glass surfaces are negatively charged but the charge has been reversed through chemical treatment to allow an interfacial double layer of the opposite charge. About 10^{-5} mol/L of such surfactants is added and, after some equilibration time, the bubbles indeed begin to move. In Figure 11.5, the measured bubble speed Ca is recorded as a function of the electrolyte concentration, applied field and the surfactant concentration. The last quantity is presented as a concentration-normalized field obtained from an electrokinetic theory [Takhistov et al., 2000]. Bubble speeds approaching the liquid electrokinetic velocity (without bubble) of $Ca = 10^{-4}$ are observed, indicating a complete cessation of film flow. A more specific set of data is shown in Figure 11.6 in dimensional quantities, showing a robust 0.14-mm/sec bubble speed.

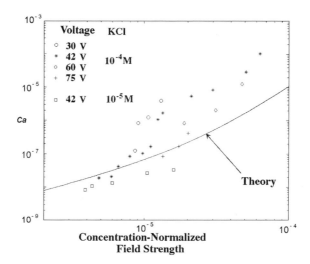

FIGURE 11.5 Electrokinetically driven bubble speed as a function of a concentration-normalized electric field for the KCl electrolyte of indicated concentrations. The unnormalized data scatter over 5 decades and are collapsed by the theory.

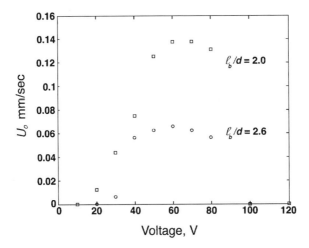

FIGURE 11.6 Raw bubble speed data u_0 as a function of applied voltage and bubble length l_b normalized by the capillary diameter d. The bubble speed approaches that of the electrolyte electrokinetic velocity without bubble before dropping to zero abruptly at a critical voltage of 80 V.

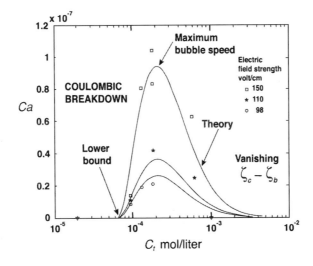

FIGURE 11.7 The window of total ion concentration C_t where bubble motion is possible.

There are, however, limitations to such electrokinetically driven bubble flow in micochannels. The interfacial zeta potential endowed by the surfactants is a strong function of the electrolyte concentration due to the strong screening effects near the anions of the surfactants [Schultz, 1984]. At high concentrations, the interface double layer can become negligibly thin such that again the film velocity profile approaches a flat one and the bubble speed approaches zero. At very low bulk electrolyte concentrations, on the other hand, the Debye thickness approaches the film thickness. As a result, the ion concentration and conductivity in the film can increase by a factor of $\exp(\zeta/kT)$. Because ζ is large at low concentration, by the arguments presented before, a significant increase in the film conductivity can occur at low concentrations. This reduces the field strength E and further reduces the film flow. However, because the interfacial and capillary double layers are oppositely charged, their increased thickness leads to significant Coulombic attraction between the two and will eventually collapse the entire film. These limits are observed in our experimental data for different KCl concentrations shown in Figure 11.7. The fact that the bubble speed approaches zero at high concentrations suggests that interfacial traction provided by the surfactants is negligible. There is, then, a wide but finite window of electrolyte concentration through which multiphase microfluidic flow can be achieved with an electrokinetic driving force.

The Bretherton analysis can be extended to electrokinetic flow [Takhistov et al., 2000]. The theory now includes the electrolyte concentration dependence of the zeta potential and the important interfacial double layer. The resulting theoretical predictions collapse the data in Figure 11.5 and provide accurate estimates of the data in Figures 11.6 and 11.7. Also included in the theory are the transients necessary to establish a steadily translating bubble. Surfactant adsorption equilibration at the interface and capillary double-layer equilibration are just some of the important transients that must be considered for the design of microdevices.

11.5 Future Directions

It appears from the above arguments that electrokinetic flow is the only means of overcoming the large capillary forces involved in transporting bubbles and drops in microchannels. Electrokinetic flow in a circular capillary is possible only within certain windows of operation. For noncircular channels, current and flow leakage at the corners are additional concerns. Such leaks must be prevented to have complete cessation of film flow. Further complications arise at channel junctions or constrictions where bubbles and drops may break up or coalesce. Jet streaming of small drops from the front tip of a bubble being sheared in front of a constriction has also been observed. Most of these phenomena must be carefully avoided in the device design. However, the rapid ion motion in the double layers of electrokinetic flow can have certain desirable applications. We have observed charge separation along the bubble that is sufficient to break up the bubble. This may suggest a means of electrophoretic bubble motion and bubble breakup in microreactors. The same polarization can also induce bubble coalescence. Streaming potentials with opposite flows is another possibility but a rather difficult task because of the capillary pressure again. Evaporation and condensation phenomena are also profoundly different in microchannels due to capillary forces. Biological cells with internal and external charges and of the same dimension as the microchannels exhibit a rich spectrum of electrophoretic, electrokinetic and stress-induced adsorption dynamics in microfluidics. These and other known and new phenomena of ionic flow await future studies for applications in microdevice designs.

Acknowledgments

It has been my privilege to work with a remarkably talented group of students in our studies of bubble transport. They are, in chronological order, J. Ratulowski (Shell), S. Kalliadasis (Leeds), A. Indeikina (Notre Dame) and P. Takhistov (Notre Dame). Our research has been supported by grants from ACS-PRF and NSF (XYZ-on-a-Chip).

References

Bretherton, F.P. (1961) "The Motion of Long Drops and Bubbles in Tubes," *J. Fluid Mech.* **10**, pp. 166–188.

Chang, H.-C., and Demekhin, E.A. (1999) "Mechanism for Drop Formation on a Coated Vertical Fibre," *J. Fluid Mech.* **380**, pp. 233–255.

Friz, V.G. (1965) "Über den dynamischen Randwindel im Fall der vollstädigen Benetzung," *A. für angewandto Phys.* **19**, pp. 374–378.

Indeikina, A., and Chang, H.-C. (1999) "A Molecular Theory for Dynamic Contact Angles," *Proc. IUTAM Symp. on Nonlinear Singularities in Deformation and Flow,* eds. D. Durban and J.R.A. Pearson, Kluwer Academic, Dordrecht/Norwell, MA, pp. 321–368.

Kalliadasis, S., and Chang, H.-C. (1994) "Apparent Dynamic Contact Angle of an Advancing Gas–Liquid Meniscus," *Phys. Fluids* **6**, pp. 12–23.

Kalliadasis, S., and Chang, H.-C. (1996) "Effects of Wettability on Spreading Dynamics," *Ind. Eng. Chem. Fluid* **35**, pp. 2860–2874.

Lu, W.-Q., and Chang, H.-C. (1988) "A Boundary Integral Study of Bubble Formation and Transport in Channels Filled with a Viscous Fluid," *J. Comp. Phys.* **340**, pp. 77–89.

Park, C.-W. (1992) "Influence of Soluble Surfactants on the Motion of a Finite Bubble in a Capillary," *Phys. Fluids A* **4**, pp. 2335–2347.

Probstein, R.F. (1994) *Physiochemical Hydrodynamics*, John Wiley & Sons, New York.

Ratulowski, J., and Chang, H.-C. (1989) "Transport of Gas Bubbles in Capillaries," *Phys. Fluids A* **1**, pp. 1642–1655.

Ratulowski, J., and Chang, H-C. (1990) "Maragoni Effects of Trace Impurities on the Motion of Long Gas Bubbles in Capillaries," *J. Fluid Mech.* **210**, pp. 303–328.

Reinelt, D.A., and Saffman, P.G. (1985) "The Penetration of a Finger into a Viscous Fluid in a Channel and Tube," *SIAM J. Stat. Comp.* **6**, pp. 542–561.

Russel, W.B., Saville, D.A., and Schowalter, W.R. (1989) *Colloidal Dispersion*, Cambridge University Press, Cambridge, U.K.

Schultz, H.J. (1984) *Physico-Chemical Elementary Processes in Flotation*, Elsevier, New York.

Schwartz, L.W., Princen, H.M., and Kiss, A.D. (1986) "On the Motion of Bubbles in Capillary Tubes," *J. Fluid Mech.* **172**, pp. 259–275.

Stebe, K.S., Lin, S.Y., and Maldarelli, C. (1991) "Remobilizing Surfactant Retarded Particle Interfaces," *Phys. Fluids A* **3**, pp. 3–20.

Takhistov, P., Indeikina, A., and Chang, H.-C. (2000) "Electrokinetically Driven Bubbles in Microchannels," *Phys. Fluids*, submitted.

Veretennikov, I., Indeikina, A., and Chang, H.-C. (1998) "Front Dynamics and Fingering of a Driven Contact Line," *J. Fluid Mech.* **373**, pp. 81–110.

12
Fundamentals of Control Theory

Bill Goodwine
University of Notre Dame

12.1 Introduction 12-1
12.2 Classical Linear Control 12-2
   Mathematical Preliminaries • Control System Analysis and Design • Other Topics
12.3 "Modern" Control 12-17
   Pole Placement • The Linear Quadratic Regulator • Basic Robust Control
12.4 Nonlinear Control 12-24
   SISO Feedback Linearization • MIMO Full-State Feedback Linearization • Control Applications of Lyapunov Stability Theory • Hybrid Systems
12.5 Parting Remarks 12-30

12.1 Introduction

This chapter reviews the fundamentals of linear and nonlinear control. This topic is particularly important in microelectromechanical systems (MEMS) applications for two reasons. First, as electromechanical systems, MEMS devices often must be controlled in order to be utilized in an effective manner. Second, important applications of MEMS technology are controls-related because of the utility of MEMS devices in sensor and actuator technologies. Because the area of control is far too vast to be entirely presented in one self-contained chapter, the approach adopted for this chapter is to outline a variety of techniques used for control system synthesis and analysis, provide at least a brief description of their mathematical foundation, discuss the advantages and disadvantages of each of the techniques and provide sufficient references so that the reader can find a starting point in the literature to fully implement any described techniques. The material varies from the extremely basic (e.g., root locus design) to relatively advanced material (e.g., sliding mode control) to cutting-edge research (hybrid systems). Some examples are provided; additionally, many references to the literature are provided to help the reader find further examples of a particular analysis or synthesis technique.

This chapter is divided into three sections, each of which considers both the stability and performance of a control system. The term *performance* includes both the qualitative nature of any transient response of the system, reference signal tracking properties of the system and the long-term or steady-state performance of the system. The first section considers "classical control," which is the study of single-input, single-output (SISO) linear control systems, which relies heavily upon mathematical techniques from complex variable theory. The material in this section outlines what is typically covered in an elementary undergraduate controls course. The second section considers so-called "modern control" which is the study of multi-input, multi-output (MIMO) control systems in state space. Included in this section is what is sometimes called "post-modern control" (Zhou, 1996) which is a study of robust system performance and stability in the presence of unmodeled system dynamics. Finally, the third section considers nonlinear control techniques. Not considered in this chapter are model-free control techniques based upon concepts from soft computing, which are outlined in Chapter 14. Also not considered in this chapter are nonlinear, open-loop control techniques (for recent advances in this area, the reader is referred to Lafferriere and Sussmann [1993]; Bullo et al. [2000]; Goodwine and Burdick [2000]).

12.2 Classical Linear Control

Classical linear control relies heavily upon mathematical techniques from complex variable theory. This is apparently an historical consequence of the importance of frequency analyses of feedback amplifiers, which motivated much of the development of classical control theory, as well as a consequence of the fact that convolution in the time domain is simple multiplication in the frequency domain which greatly simplifies the analysis of the natural input/output and "block diagram" structure of many control systems. This topic is typically thoroughly covered in undergraduate controls courses. Good references include Dorf (1992), Franklin et al. (1994), Gajec and Lelic (1996), Kuo (1995), Ogata (1997), Raven (1995) and Shinners (1992).

12.2.1 Mathematical Preliminaries

The main tool is the Laplace transform, which transforms the linear ordinary differential equation (ODE) into an algebraic equation, thus reducing the task of solving an ODE into simple algebra. The Laplace transform of a function, $f(t)$, is defined as:

$$L[f(t)] = F(s) = \int_0^\infty e^{-st} f(t) dt \quad (12.1)$$

and the inverse Laplace transform of $F(s)$ as:

$$L^{-1}[F(s)] = f(t) = \frac{1}{2\pi j} \int_{c-j\omega}^{c+j\omega} F(s) e^{st} ds, \quad \text{for } t > 0 \quad (12.2)$$

A discussion of extremely important mathematical details concerning convergence and the proper lower limit of integration is found in Ogata (1997). As a practical matter, evaluating the integrals in the definition of the Laplace transform and the inverse Laplace transform is rarely necessary as extensive tables of Laplace transform pairs are readily available. A few Laplace transform pairs for typical functions are listed in Table 12.1. More complete tables can be found in any undergraduate text on classical control theory such as the references listed previously.

Important properties of the Laplace transform are as follows:

1. Real differentiation: $L[\frac{d}{dt} f(t)] = sF(s) - f(0)$.
2. Linearity: $L[\alpha f_1(t) \pm \beta f_2(t)] = \alpha F_1(s) \pm \beta F_2(s)$.
3. Convolution: $L[\int_0^t f_1(t-\tau) f_2(\tau) d\tau] = F_1(s) F_2(s)$.
4. Final value theorem: If all the poles of $sF(s)$ are in the left half of the complex plane, then $\lim_{t \to \infty} f(t) = \lim_{s \to 0} sF(s)$.

A basic result from the first three properties is that to solve a linear ordinary differential equation, one can take the Laplace transform of each side of the equation, which converts the differential equation into an algebraic equation in s, then algebraically solve the expression for the Laplace transform of the dependent variable, and then take the inverse Laplace transform of the resulting function.

TABLE 12.1 Laplace Transform Pairs for Basic Functions

	$F(t)$	$F(s)$
1	Unit impulse, $\delta(t)$	1
2	Unit step, $1(t)$	$\dfrac{1}{s}$
3	t	$\dfrac{1}{s^2}$
4	$t^n, n = 1, 2, 3, \ldots$	$\dfrac{n!}{s^{n+1}}$
5	e^{-at}	$\dfrac{1}{s+a}$
6	$t^n e^{-at}$	$\dfrac{n!}{(s+a)^{n+1}}$
7	$\sin \omega t$	$\dfrac{\omega}{s^2 + \omega^2}$
8	$\cos \omega t$	$\dfrac{s}{s^2 + \omega^2}$
9	$e^{-at} \cos bt$	$\dfrac{s+a}{(s+a)^2 + b^2}$
8	$e^{-at} \sin bt$	$\dfrac{b}{(s+a)^2 + b^2}$

Example

As a simple example, consider the differential equation:

$$\ddot{x} + x = 0$$
$$x(0) = 0 \qquad (12.3)$$
$$\dot{x}(0) = 1$$

Taking the Laplace transform of the equation yields:

$$s^2 X(s) - sx(0) - \dot{x}(0) + X(s) = 0 \qquad (12.4)$$

Algebraic manipulation gives:

$$X(s) = \frac{1}{s^2 + 1} \qquad (12.5)$$

consequently, from the table of Laplace transform pairs:

$$x(t) = \sin(t) \qquad (12.6)$$

For more examples, see Ogata (1997), Raven (1995), Kuo (1995), Franklin et al. (1994) etc.

Due to the convolution property of Laplace transforms, a convenient representation of a linear control system is the block diagram representation illustrated in Figure 12.1. In such a block diagram representation, each block contains the Laplace transform of the differential equation representing that component of the control system that relates the input to the block to its output. Arrows between blocks indicate that the output from the preceding block is transferred to the input of the subsequent block. The output of the preceding block multiplies the contents of the block to which it is an input. Simple algebra will yield the overall transfer function of a block diagram representation for a system.

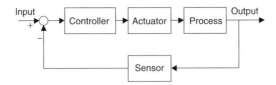

FIGURE 12.1 Typical block diagram representation of a control system.

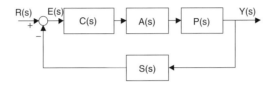

FIGURE 12.2 Generic block diagram including transfer functions.

Example

The transfer function for the system illustrated in Figure 12.2 can be computed by observing that:

$$E(s) = R(s) - Y(s)S(s) \qquad (12.7)$$

and

$$Y(s) = E(s)C(s)A(s)P(s) \qquad (12.8)$$

which can be combined to yield

$$\frac{Y(s)}{R(s)} = \frac{C(s)A(s)P(s)}{1 + C(s)A(s)P(s)S(s)} \qquad (12.9)$$

A more complete exposition on block diagram algebra can be found in any of the previously cited undergraduate texts. Note that the numerator and denominator of the transfer function will typically be polynomials in s. The denominator is called the *characteristic equation* for the system. From the table of Laplace transform pairs, it should be clear that if the characteristic polynomial has any roots with a positive real part, then the system will be unstable, as it will correspond to an exponentially increasing solution (see entry 5 in Table 12.1). Given a reference input, $R(s)$, determining the response of the system is straightforward: Multiply the transfer function by the reference input, perform a partial fraction expansion (i.e., expand):

$$Y(s) = \frac{R(s)C(s)A(s)P(s)}{1 + C(s)A(s)P(s)S(s)} = \frac{C_1}{s - p_1} + \frac{C_2}{s - p_2} + \cdots + \frac{C_n}{s - p_n} \qquad (12.10)$$

where each term in the sum on the right-hand side of the equation is something in the form of one of the entries in Table 12.1. In this manner, the contribution to the response of each individual term can be determined by referring to a Laplace transform table and can be superimposed to determine the overall solution:

$$y(t) = y_1(t) + y_2(t) + \cdots + y_n(t) \qquad (12.11)$$

where each term in the sum is the inverse Laplace transform of the corresponding term in the partial fraction expansion.

Example

For the block diagram in Figure 12.2 if $C(s) = \frac{1}{s}$, $A(s) = 1$, $P(s) = \frac{\omega_n^2}{s + 2\zeta\omega_n}$, $S(s) = 1$ and $R(s) = \frac{1}{s}$ (a unit step input), then:

$$\begin{aligned}
Y(s) &= \frac{\omega_n^2}{s(s^2 + 2\zeta\omega_n s + \omega_n^2)} \\
&= \frac{1}{s} - \frac{s + 2\zeta\omega_n}{s^2 + 2\zeta\omega_n s + \omega_n^2} \\
&= \frac{1}{s} - \frac{s + \zeta\omega_n}{(s + \zeta\omega_n)^2 + \omega_d} - \frac{\zeta\omega_n}{(s + \zeta\omega_n)^2 + \omega_d} \\
&= \frac{1}{s} - \frac{s + \zeta\omega_n}{(s + \zeta\omega_n)^2 + \omega_d} - \frac{\zeta\omega_n}{\omega_d} \frac{\omega_d}{(s + \zeta\omega_n)^2 + \omega_d}
\end{aligned} \quad (12.12)$$

where $\omega_d = \omega_n \sqrt{1 - \zeta^2}$. Referring to the table of Laplace transform pairs (Table 12.1), and assuming that $\zeta < 1$,

$$y(t) = 1 - e^{-\zeta\omega_n t}\left(\cos(\omega_d t) + \frac{\zeta}{\sqrt{1-\zeta^2}} \sin(\omega_d t)\right) \quad (12.13)$$

12.2.2 Control System Analysis and Design

Control system analysis and design considers primarily stability and performance. One approach to the former is briefly dispensed with first. The stability of a system with the closed-loop transfer function (note that in such a case a controller has already been specified):

$$T(s) = \frac{b_0 s^m + b_1 s^{m-1} + \cdots + b_m}{s^n + a_1 s^{n-1} + \cdots + a_n} \quad (12.14)$$

is determined by the roots of the denominator, or characteristic equation. In fact, it is possible to determine whether or not the system is stable without actually computing the roots of the characteristic equation. A *necessary condition* for stability is that each of the coefficients a_i appearing in the characteristic equation be positive. Because this is only a necessary condition, if any of the a_i are negative, then the system is unstable, but the converse is not necessarily true: Even if all the a_i are positive, the system may still be unstable. Routh (1975) devised a method to check *necessary and sufficient* conditions for stability.

The method is to construct the Routh array, defined as follows:

Row n	s^n:	1	a_2	a_4	...
Row $n-1$	s^{n-1}:	a_1	a_3	a_5	...
Row $n-2$	s^{n-2}:	b_1	b_2	b_3	...
Row $n-3$	s^{n-3}:	c_1	c_2	c_3	...
⋮	⋮	⋮	⋮	⋮	⋮
Row 2	s^2:	*	*		
Row 1	s^1:	*			
Row 0	s^0:	*			

in which the a_i are from the denominator of Eq. (12.14); b_i and c_i are defined as:

$$b_1 = -\frac{\det\begin{bmatrix} 1 & a_2 \\ a_1 & a_3 \end{bmatrix}}{a_1} \qquad b_2 = -\frac{\det\begin{bmatrix} 1 & a_4 \\ a_1 & a_5 \end{bmatrix}}{a_1} \qquad b_3 = -\frac{\det\begin{bmatrix} 1 & a_6 \\ a_1 & a_7 \end{bmatrix}}{a_1}$$

$$c_1 = -\frac{\det\begin{bmatrix} a_1 & a_3 \\ b_1 & b_2 \end{bmatrix}}{b_1} \qquad c_2 = -\frac{\det\begin{bmatrix} a_1 & a_5 \\ b_1 & b_3 \end{bmatrix}}{b_1} \qquad c_3 = -\frac{\det\begin{bmatrix} a_1 & a_7 \\ b_1 & b_4 \end{bmatrix}}{b_1}.$$

The basic result is that the number of poles in the right-half plane (i.e., unstable solutions) is equal to the number of sign changes among the elements in the first column of the Routh array. If they are all positive, then the system is stable. When a zero is encountered, it should be replaced with a small positive constant, ε, which will then be propagated to lower rows in the array. Then, the result can be obtained by taking the limit as $\varepsilon \to 0$.

Example

Construct the Routh array and determine the stability of the system described by the transfer function:

$$\frac{Y(s)}{R(s)} = \frac{1}{s^4 + 4s^3 + 9s^2 + 10s + 8} \tag{12.15}$$

The Routh array is

$$\begin{array}{llll}
s^4: & 1 & 9 & 8 \\
s^3: & 4 & 10 & 0 \\
s^2: & \dfrac{-(10-36)}{1} = 26 & \dfrac{-(0-32)}{1} = 32 & 0 \\
s^1: & \dfrac{-(128-260)}{4} = 33 & 0 & 0 \\
s^0: & \dfrac{-(0-1056)}{26} = 40.6 & 0 & 0
\end{array} \tag{12.16}$$

so the system is stable, as there are no sign changes in the elements in the first column of the array.

One aspect of performance concerns the steady-state error exhibited by the system. For example, from the time-domain solution from the example above, it is clear that as $t \to \infty$, $y(t) \to 1$. However, the final value theorem can be utilized to determine this without actually solving for the time domain solution.

Example

Determine the steady-state value for the time-domain function, $y(t)$, if its Laplace tranfrorm is given by $Y(s) = \frac{\omega_n^2}{s(s^2 + 2\zeta\omega_n s + \omega_n^2)}$. Because all the solutions of $s^2 + 2\zeta\omega_n s + \omega_n^2 = 0$ have a negative real part, all the poles of $sY(s)$ lie in the left half of the complex plane. Therefore, the final value theorem can be applied to yield:

$$\lim_{t \to \infty} y(t) = \lim_{s \to 0} sY(s) = \lim_{s \to 0} s\frac{\omega_n^2}{s(s^2 + 2\zeta\omega_n s + \omega_n^2)} = 1 \tag{12.17}$$

which clearly is identical to the limit of the time domain solution as $t \to \infty$.

FIGURE 12.3 Robot arm model.

12.2.2.1 Proportional–Integral–Derivative (PID) Control

Perhaps the most common control implementation is so-called PID control, where the commanded control input (the output of the "controller" box in Figures 12.1 and 12.2) is equal to the sum of three terms: one term proportional to the error signal (the input to the "controller" box in Figures 12.1 and 12.2), the next term proportional to the derivative of the error signal and the third term proportional to the time integral of the error signal. In this case, from Figure 12.2, $D(s) = K_P + \frac{K_I}{s} + K_d s$, where K_P is the *proportional gain*, K_I is the *integral gain* and K_d is the *derivative gain*. A simple analysis of a second-order system shows that increasing K_P and K_I generally increases the speed of the response at the cost of generally reducing stability; whereas, increasing K_d generally increases damping and stability of the response. With $K_I = 0$, there may be a nonzero steady-state error, but when K_I is nonzero the effect of the integral control effort is to typically eliminate steady-state error.

Example—PID Control of a Robot Arm

Consider a robot arm illustrated in Figure 12.3. Linearizing the equations of motion about $\theta = 0$ (the configuration in Figure 12.3) gives:

$$I\ddot{\theta} + mg\theta = u \tag{12.18}$$

where I is the moment of inertia of the arm, m is the mass of the arm, θ is the angle of the arm and u is a torque applied to the arm. For PID control,

$$u = K_p(\theta_{desired} - \theta_{actual}) + K_d(\dot{\theta}_{desired} - \dot{\theta}_{actual}) + K_I \int_0^t (\theta_{desired} - \theta_{actual})\, dt \tag{12.19}$$

If we set $I = 1$ and $m = 1/g$, the block diagram representation for the system is illustrated in Figure 12.4. Thus, the closed-loop transfer function is

$$T(s) = \frac{K_d s^2 + K_p s + K_I}{s^3 + K_d s^2 + (K_p + 1)s + K_I} \tag{12.20}$$

Figure 12.5 illustrates the step response of the system for proportional control ($K_P = 1$, $K_I = 0$, $K_d = 0$), PD control ($K_P = 1$, $K_I = 0$, $K_d = 1$) and PID control ($K_P = 1$, $K_I = 1$, $K_d = 1$). Note that for proportional

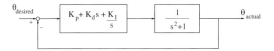

FIGURE 12.4 Robot arm block diagram.

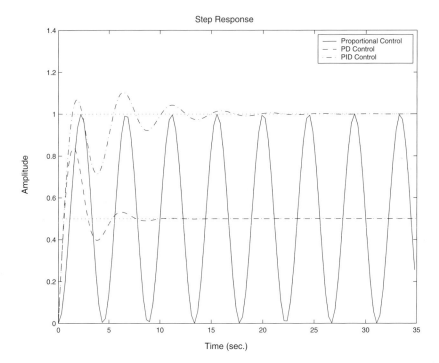

FIGURE 12.5 PID control response.

control and PD controls, there is a final steady-state error, which is eliminated with PI control (also note that both of these facts could be verified analytically using the final value theorem). Finally, note that the system response for pure proportional control is oscillatory, whereas with derivative control the response is much more damped.

The subjects contained in the subsequent sections consider controller synthesis issues. For PID controllers, tuning methods exist and the interested reader is directed to the undergraduate texts cited previously or the papers by Ziegler and Nichols (1942; 1943).

12.2.2.2 The Root Locus Design Method

Although, as mentioned above in the discussion of PID control, various rules of thumb can be determined to relate system performance to changes in gains, a systematic approach is clearly more desirable. Because pole locations determine the characteristics of the response of the system (recall the partial fraction expansion), one natural design technique is to plot how pole locations change as a system parameter or control gain is varied [Evans, 1948; 1950]. Because the real part of the pole corresponds to exponential solutions, if all the poles are in the left-half plane, the poles closest to the $j\omega$-axis will dominate the system response. In particular, if we focus a second-order system of the form:

$$H(s) = \frac{\omega_n^2}{s^2 + 2\zeta\omega_n s + \omega_n^2} \qquad (12.21)$$

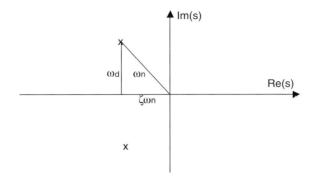

FIGURE 12.6 Complex conjugate poles, natural frequency, damped natural frequency and damping ratio.

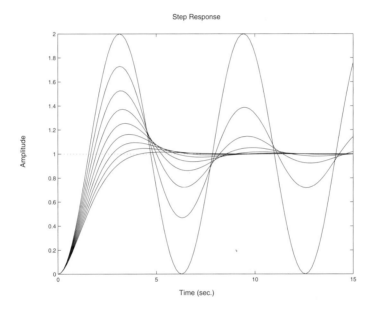

FIGURE 12.7 Step response for various damping factors.

the poles of the system are as illustrated in Figure 12.6. The terms ω_n, ω_d and ζ are the *natural frequency*, the *damped natural frequency* and the *damping ratio*, respectively. Multiplying $H(s)$ by $\frac{1}{s}$ (unit step) and performing a partial fractions expansion give:

$$Y(s) = \frac{1}{s} - \frac{s + \zeta\omega_n}{(s+\zeta\omega_n)^2 + \omega_n^2(1-\zeta^2)} - \frac{\zeta\omega_n}{(s+\zeta\omega_n)^2 + \omega_n^2(1-\zeta^2)} \tag{12.22}$$

so the time response for the system is

$$y(t) = 1 - e^{-\zeta\omega_n t}\left(\cos\omega_d t + \frac{\zeta}{\sqrt{1-\zeta^2}}\sin\omega_d t\right) \tag{12.23}$$

where $\omega_d = \omega_n\sqrt{1-\zeta^2}$ and $0 \leq \zeta < 1$. Plots of the response for various values of ζ are illustrated in Figure 12.7. Clearly, referring to the previous equation and Figure 12.7, if the damping ratio is increased, the oscillatory nature of the response is increasingly damped.

Because the natural frequency and damping are directly related to the location of the poles, one effective approach to designing controllers is to pick control gains based upon desired pole locations. A *root locus*

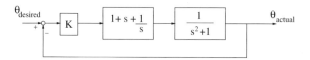

FIGURE 12.8 Robot arm block diagram.

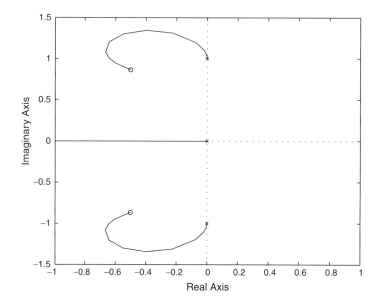

FIGURE 12.9 Root locus for robot arm PID controller.

plot is a plot of pole locations as a system parameter or controller gain is varied. Once the root locus has been plotted, it is straightforward to pick the location on the root locus with the desired pole locations to give the desired system response. There is a systematic procedure to plot the root locus by hand (refer to the cited undergraduate texts), and computer packages such as Matlab (using the rlocus() and rlocfind() functions) make it even easier. Figure 12.9 illustrates a root locus plot for the robot arm above with the block diagram as the single gain K is varied from 0 to ∞ as illustrated in Figure 12.8. Note that for the usual root locus plot, only one gain can be varied at a time. In the above example, the ratio of the proportional, integral and derivative gains was fixed, and a multiplicative scaling factor is what is varied in the root locus plot.

Because the roots of the characteristic equation start at each pole when $K = 0$ and approach each 0 of the characteristic equation as $K \to \infty$, the desired K can be determined from the root locus plot by finding the part of the locus that most closely matches the desired natural frequency ω_n and damping ratio ζ (recall Figure 12.7).

Typically, control system performance is specified in terms of *time-domain specifications*, such as *rise time, maximum overshoot, peak time* and *settling time*, each of which is illustrated in Figure 12.10. Rough estimates of the relationship between the time domain specifications and the natural frequency and damping ratio are given in Table 12.2 [Franklin et al., 1994].

Example

Returning to the robot arm example, assume that the desired system performance is to have the system rise time be less than 1.4 sec, the maximum overshoot less than 30%, and the 1% settling time less than 10 sec. From the first row in Table 12.2, the natural frequency must be greater than 1.29, and from the third and fourth rows the damping ratio should be greater than approximately 0.4. Figure 12.11 illustrates the root locus plot along with the pole locations and corresponding gain, and K (rlocfind() is the

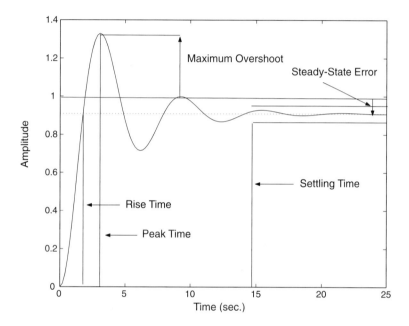

TABLE 12.2 Time-Domain Specifications as a Function of Natural Frequency, Damped Natural Frequency and Damping Ratio

Rise time: $t_r \cong \dfrac{1.8}{\omega_n}$

Peak time: $t_p \cong \dfrac{\pi}{\omega_d}$

Overshoot: $M_p = e^{-\pi\zeta/\sqrt{1-\zeta^2}}$

Settling time (1%): $t_s = \dfrac{4.6}{\zeta\omega_n}$

Note: Results are from Franklin et al., 1994.

FIGURE 12.10 Time domain control specifications.

Matlab command for retrieving the gain value for a particular location on the root locus), which provide a damping ratio of approximately .45 and a natural frequency of approximately 1.38. Figure 12.12 illustrates the step response of the system to a unit step input verifying these system parameters.

12.2.2.3 Frequency Response Design Methods

An alternative approach to controller design and analysis is the so-called *frequency response methods*. Frequency response controller design techniques have two main advantages: They provide good controller design even with uncertainty with respect to high-frequency plant characteristics and using experimental data for controller design purposes is straightforward. The two main tools are *Bode* and *Nyquist plots* (see Bode [1945] and Nyquist [1932] for first-source references) and stability analyses are considered first.

A Bode plot is a plot of two curves. The first curve is the logarithm of the magnitude of the response of the *open-loop* transfer function with respect to unit sinusoidal inputs of frequency ω. The second curve is the phase of the open-loop transfer function response as a function of input frequency ω. The Bode plot for the transfer function:

$$G(s) = \frac{1}{s^3 + 25s^2 + s} \qquad (12.24)$$

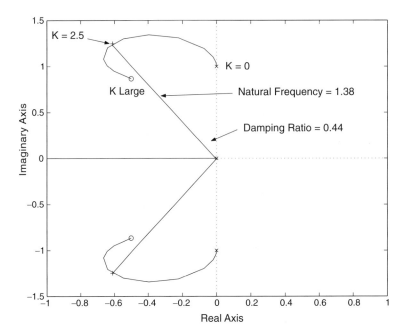

FIGURE 12.11 Selecting pole locations for a desired system response.

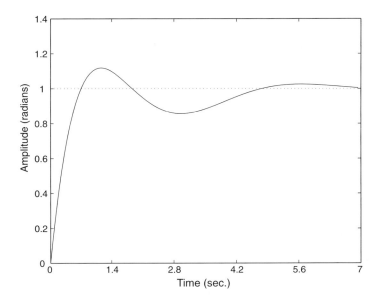

FIGURE 12.12 Robot arm step response.

is illustrated in Figure 12.13. Clearly, as the frequency of the sinusoidal input in increased, the magnitude of the system response decreases. Additionally, the phase difference between the sinusoidal input and system response starts near −90° and approaches −270° as the input frequency becomes large.

An advantage of Bode plots is that they are easy to sketch by hand. Because the magnitude of the system response is plotted on a logarithmic scale, the contributions to the magnitude of the response due to individual factors in the transfer function simply add together. Similarly, due to basic facts related to the polar representation of complex numbers, the phase contributions of each factor simply add as well.

Fundamentals of Control Theory

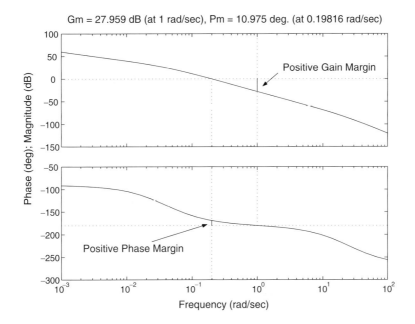

FIGURE 12.13 Bode plot.

Recipes for sketching Bode plots by hand can be found in any undergraduate controls text, such as Franklin et al. (1994), Raven (1995), Ogata (1997) and Kuo (1995).

For systems where the magnitude of the response passes through the value of 1 only one time and for systems where *increasing* the transfer function gain leads to instability (the most common, but not exclusive scenario), the *gain margin* and *phase margin* can be determined directly from the Bode plot to provide a measure of system stability under unity feedback. Figure 12.13 also illustrates the definition of gain and phase margin. Positive gain and phase margins indicate stability under unity feedback; conversely, negative gain and phase margins indicate instability under unity feedback. The class of systems for which Bode plots can be used to determine stability are called *minimum phase systems*. A system is minimum phase if all of its open-loop poles and zeros are in the left-half plane.

Bode plots also provide a means to determine the steady-state error under unity feedback for various types of reference inputs (steps, ramps etc.). In particular, if the low-frequency asymptote of the magnitude plot has a slope of zero and if the value of this asymptote is denoted by K, then the steady-state error of the system under unity feedback to a step input is

$$\lim_{t \to \infty} e = \frac{1}{1+K} \qquad (12.25)$$

If the slope of the magnitude plot at low frequencies is -20 dB/decade and if the value where the asymptote intersects the vertical line $\omega = 1$ is denoted by K, then the steady-state error to a ramp input is

$$\lim_{t \to \infty} e = \frac{1}{K} \qquad (12.26)$$

Example

Consider the system illustrated in Figure 12.2 where $C(s) = A(s) = S(s) = 1$ and $P(s) = \frac{1/2}{s+1}$. The Bode plot for the open-loop transfer function $P(s) = \frac{1/2}{s+1}$ is illustrated in Figure 12.14. The low-frequency asymptote is approximately at -6, so $20\log K = -6 \Rightarrow K \approx 0.5012 \Rightarrow y_{ss} \approx 0.6661$, where $y_{ss} = \lim_{t \to \infty} p(t)$. Figure 12.15 illustrates the unity feedback closed-loop step response of the system, verifying that the steady-state value for $y(t)$ is as computed from the Bode plot.

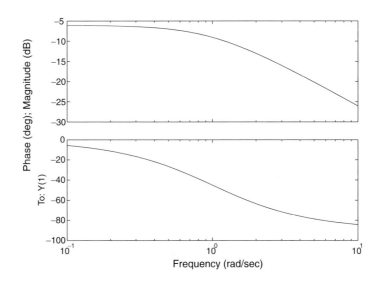

FIGURE 12.14 Bode plot for example problem.

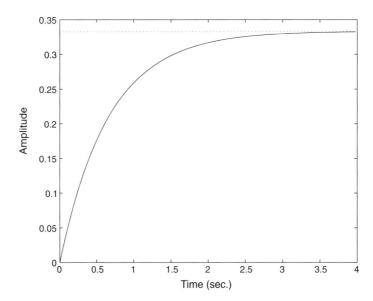

FIGURE 12.15 Step response for example problem.

A Nyquist plot is a more sophisticated means to determine stability and is not limited to cases where only increasing gain leads to system instability. It is based on the well-known result from complex variable theory called the *principle of the argument*. Consider the (factored) transfer function:

$$G(s) = \frac{\prod_i (s + z_i)}{\prod_j (s + p_j)} \qquad (12.27)$$

Fundamentals of Control Theory

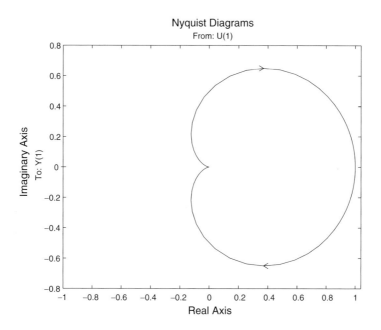

FIGURE 12.16 Nyquist plot for a stable system.

By basic complex variable theory, $\angle G(s) = \Sigma_i \theta_i - \Sigma_j \varphi_j$, where θ_i are the angles between s and the zeros z_i, and the ϕ_j are the angles between s and the poles p_j. Thus, a plot of $G(s)$ as s follows a closed contour (in the clockwise direction) in the complex plane will encircle the origin in the clockwise direction the same number of times that there are zeros of $G(s)$ within the contour minus the number of times that there are poles of $G(s)$ within the contour. Thus, an easy check for stability is to plot the *open loop* $G(s)$ on a contour that encircles the entire left-half complex plane. Assuming that $G(s)$ has no right-half plane poles (poles of $G(s)$ *itself*, in contrast to poles of the closed-loop transfer function), an encirclement of -1 by the plot will indicate a right-half plane zero of $1 + G(s)$, which is an unstable right-half plane pole of the unity feedback closed-loop transfer function:

$$\frac{G(s)}{1 + G(s)} \tag{12.28}$$

The Nyquist plot for a unity feedback system with open-loop transfer function given by:

$$G(s) = \frac{1}{(s+1)(s+1)} \tag{12.29}$$

which is stable under unity feedback is illustrated in Figure 12.16, and a Nyquist plot for a system that is unstable under unity feedback is illustrated in Figure 12.17.

12.2.2.4 Lead-Lag Compensation

Lead-lag controller design is another extremely popular compensation technique. In this case, the compensator (the $C(s)$ block in Figure 12.2) is of the form:

$$C(s) = K\beta \frac{As + 1}{\alpha As + 1} \frac{Bs + 1}{\beta Bs + 1} \tag{12.30}$$

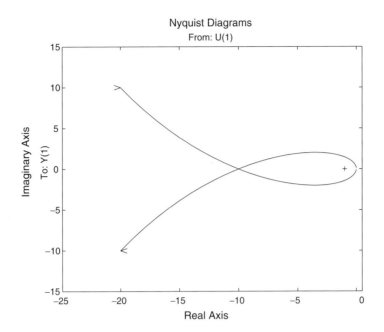

FIGURE 12.17 Nyquist plot of an unstable system.

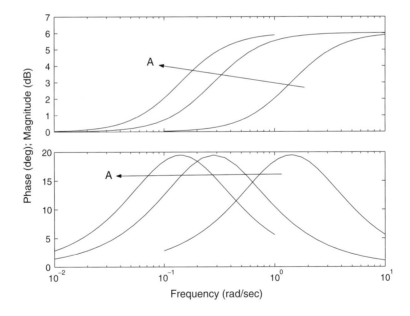

FIGURE 12.18 Bode plots of various lead compensators.

where $\alpha < 1$ and $\beta > 1$. The first fraction is the lead portion of the compensator and can provide increased stability with an appropriate choice for A. The second term is the lag compensator and provides decreased steady-state error. To understand the first of these assertions, Figure 12.18 plots the Bode plot for a lead compensator for various values of the parameter A. Because the lead compensator shifts the phase plot up, by an appropriate choice of the parameter A, the crossover point where the magnitude plot crosses through the value of 0 dB can be shifted to the right, increasing the gain margin.

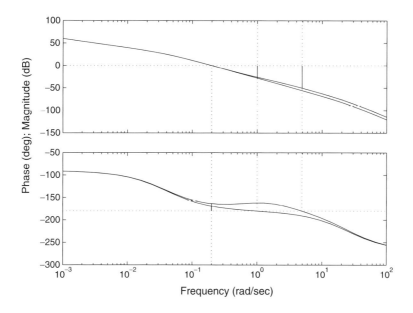

FIGURE 12.19 Lead compensated example system.

Example

Figure 12.19 plots the bode plot for the compensated system:

$$G(s) = \frac{As+1}{\alpha As+1} \frac{1}{s^3 + 25s^2 + s} \qquad (12.31)$$

where $A = 0, 1$ and $\alpha = 0.5$. As can be seen, the magnitude crossover point has been shifted to the left, increasing the gain margin. In a similar manner, unstable systems (which would originally have negative gain and phase margins) can possibly be stabilized.

Lag compensation works in a very similar manner to increase the magnitude plot for low frequencies, which decreases the steady-state error for the system. Lead and lag controllers can be used in series to both increase stability as well as decrease steady-state error. Systematic approaches for determining the parameters, α, β, A and B can be found in the references, particularly Franklin et al. (1994).

12.2.3 Other Topics

Various other topics are typically considered in classical control but will not be outlined here due to space limitations. Such topics include, but are not limited to, systematic methods for tuning PID regulators, lead-lag compensation and techniques for considering and modeling time delay. Interested readers should consult the references, particularly Franklin et al. (1994), Ogata (1997), Kuo (1995) and Raven (1995).

12.3 "Modern" Control

In contrast to classical control, which is essentially a complex-variable, frequency-based approach for SISO systems, modern control is a time-domain approach that is amenable to MIMO systems. The basic tools are from the theory of ordinary differential equations and matrix algebra. The topics outlined in this section are the *pole placement* and *linear quadratic regulator* (LQR) problems. Additionally, the basics of robust control are outlined.

12.3.1 Pole Placement

First, a multistate but single-input control system will be examined. Consider a control system written in state space:

$$\dot{\mathbf{x}} = \mathbf{A}\mathbf{x} + \mathbf{B}u \qquad (12.32)$$

where $\mathbf{x}$ is the $1 \times n$ state vector, u is the scalar input, $\mathbf{A}$ is an $n \times n$ constant matrix and $\mathbf{B}$ is an $n \times 1$ constant matrix. If we assume that the control input u can be expressed as a combination of the current state variables (called full state feedback), we can write:

$$u = -k_1 x_1 - k_2 x_2 - \cdots - k_n x_n = -\mathbf{K}\mathbf{x} \qquad (12.33)$$

where $\mathbf{K}$ is a row vector comprised of each of the gains k_i. Then, the state-space description of the system becomes:

$$\dot{\mathbf{x}} = (\mathbf{A} - \mathbf{B}\mathbf{K})\mathbf{x} \qquad (12.34)$$

so that the solution of this equation is

$$\mathbf{x}(t) = e^{(\mathbf{A} - \mathbf{B}\mathbf{K})t}\mathbf{x}(0) \qquad (12.35)$$

where $e^{(\mathbf{A} - \mathbf{B}\mathbf{K})t}$ is the *matrix exponential* of the matrix $\mathbf{A} - \mathbf{B}\mathbf{K}$ defined by:

$$e^{(\mathbf{A} - \mathbf{B}\mathbf{K})t} = \mathbf{I} + (\mathbf{A} - \mathbf{B}\mathbf{K})t + \frac{(\mathbf{A} - \mathbf{B}\mathbf{K})^2 t^2}{2!} + \frac{(\mathbf{A} - \mathbf{B}\mathbf{K})^3 t^3}{3!} + \cdots \qquad (12.36)$$

Basic theory from linear algebra and ordinary differential equations [Hirsch and Smale, 1974] indicates that the stability and characteristics of the transient response will be determined by the eigenvalues of the matrix $\mathbf{A} - \mathbf{B}\mathbf{K}$. In fact, if

$$\text{rank}[\mathbf{B}|\mathbf{A}\mathbf{B}|\mathbf{A}^2\mathbf{B}|\mathbf{A}^3\mathbf{B}|\ldots|\mathbf{A}^{n-1}\mathbf{B}] = n \qquad (12.37)$$

then it can be shown that the eigenvalues of $\mathbf{A} - \mathbf{B}\mathbf{K}$ can be placed arbitrarily as a function of the elements of $\mathbf{K}$. Techniques to solve the problem by hand by way of a similarity transformation exist (see the standard undergraduate controls books), and Matlab has functions for the computations as well.

Example—Pole Placement for Inverted Pendulum System

Consider the cart/pendulum system illustrated in Figure 12.20. In state-space form, the equations of motion are

$$\frac{d}{dt}\begin{bmatrix} x \\ \dot{x} \\ \theta \\ \dot{\theta} \end{bmatrix} = \begin{bmatrix} 0 & 1 & 0 & 0 \\ 0 & 0 & \frac{-gm}{M} & 0 \\ 0 & 0 & 0 & 1 \\ 0 & 0 & \frac{-(m+M)g}{lM} & 0 \end{bmatrix} \begin{bmatrix} x \\ \dot{x} \\ \theta \\ \dot{\theta} \end{bmatrix} + \begin{bmatrix} 0 \\ \frac{1}{M} \\ 0 \\ \frac{-1}{lM} \end{bmatrix} u \qquad (12.38)$$

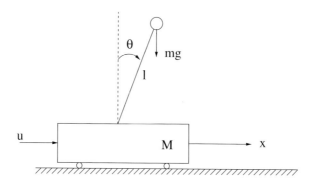

FIGURE 12.20 Cart and pendulum system.

Setting $M = 10$, $m = 1$, $g = 9.81$ and $l = 1$ and letting $u = -k_1 x + k_2 \dot{x} + k_3 \theta + k_4 \dot{\theta}$ if the desired pole locations for the system are at

$$\begin{aligned} \lambda_1 &= -1 - i \\ \lambda_2 &= -1 + i \\ \lambda_3 &= -8 \\ \lambda_4 &= -9 \end{aligned} \quad (12.39)$$

the Matlab function place() can be used to compute the values for the corresponding k_i. For this problem, the gain values are

$$\begin{aligned} k_1 &= 122.32 \\ k_2 &= 151.21 \\ k_3 &= -849.77 \\ k_4 &= -38.79 \end{aligned} \quad (12.40)$$

With initial conditions $x(0) = 0.25$, $\dot{x}(0) = 0$, $\theta(0) = 0.25$ and $\dot{\theta}(0) = 0$, the response of the system is illustrated in Figure 12.21. Note that the cart position, x, initially moves in the "wrong" direction in order to compensate for the pendulum position.

12.3.2 The Linear Quadratic Regulator

The LQR problem is not limited to scalar input problems and seeks to find a control input,

$$\mathbf{u} = -\mathbf{K}\mathbf{x}(t) \quad (12.41)$$

for the system:

$$\dot{\mathbf{x}} = \mathbf{A}\mathbf{x} + \mathbf{B}\mathbf{u} \quad (12.42)$$

that minimizes the *performance index*:

$$J = \int_0^\infty (\mathbf{x}^T \mathbf{Q} \mathbf{x} + \mathbf{u}^T \mathbf{R} \mathbf{u}) \, dt \quad (12.43)$$

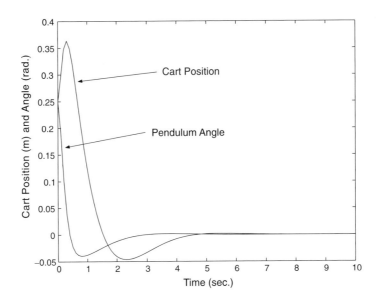

FIGURE 12.21 Cart and pendulum system pole placement response.

where **Q** and **R** are positive definite, real symmetric matrices. By the second method of Lyapunov [Khalil, 1996; Sastry, 2000], the control input that minimizes the performance index is

$$u = -\mathbf{R}^{-1}\mathbf{B}^T\mathbf{P}\mathbf{x}(t) \tag{12.44}$$

where **R** and **B** are from the performance index and equations of motion, respectively, and **P** satisfies the *reduced matrix Riccati equation*:

$$\mathbf{A}^T\mathbf{P} + \mathbf{P}\mathbf{A} - \mathbf{P}\mathbf{B}\mathbf{R}^{-1}\mathbf{B}^T\mathbf{P} + \mathbf{Q} = 0 \tag{12.45}$$

Example—LQR for Inverted Pendulum System

For the same cart and pole system as in the previous example with

$$Q = \begin{bmatrix} 1 & 0 & 0 & 0 \\ 0 & 1 & 0 & 0 \\ 0 & 0 & 1 & 0 \\ 0 & 0 & 0 & 1 \end{bmatrix} \tag{12.46}$$

(which weights all the states equally), and $R = 0.001$, the optimal gains (computed via the Matlab `lqr()` function) are

$$\begin{aligned} k_1 &= 31.62 \\ k_2 &= 145.75 \\ k_3 &= -95.53 \\ k_4 &= -21.65 \end{aligned} \tag{12.47}$$

Fundamentals of Control Theory

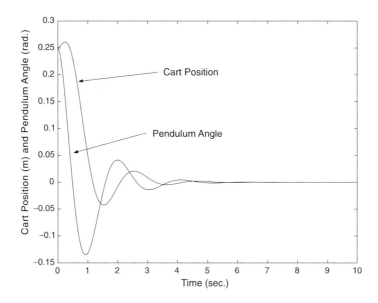

FIGURE 12.22 Cart and pendulum LQR response.

and the response of the system with initial conditions $x(0) = 0.25$, $\dot{x}(0) = 0$, $\theta(0) = 0.25$ and $\dot{\theta}(0) = 0$ is illustrated in Figure 12.22. If the Q matrix is modified to provide a heavy weighting for the θ state,

$$Q = \begin{bmatrix} 1 & 0 & 0 & 0 \\ 0 & 1 & 0 & 0 \\ 0 & 0 & 100 & 0 \\ 0 & 0 & 0 & 1 \end{bmatrix} \quad (12.48)$$

the system response is illustrated in Figure 12.23. Note that the pendulum angle goes to zero very rapidly, but at the "expense" of a slower response and greater deviation for the cart position.

12.3.3 Basic Robust Control

The main idea motivating modern robust control techniques is to incorporate explicitly plant uncertainty representations into system modeling and control synthesis methods. The basic material here outlines the presentation in Doyle et al. (1992), and the more advanced material is from Zhou (1996). Modern robust control is a very involved subject and only the briefest outline can be provided here.

Consider the unity feedback SISO system illustrated in Figure 12.24, where P and C are the plant and controller transfer functions; $R(s)$ is the reference signal; $Y(s)$ is the output; $D(s)$ and $N(s)$ are external disturbances and sensor noise, respectively; $E(s)$ is the error signal; and $U(s)$ is the control input.

Now, define the *loop transfer function*, $L = CP$, and the *sensitivity function*:

$$S = \frac{1}{1+L} \quad (12.49)$$

which is the transfer function from the reference input $R(s)$ to the error $E(s)$ which provides a measure of the sensitivity of the closed loop (or *complementary sensitivity*) transfer function:

$$T = \frac{PC}{1+PC} \quad (12.50)$$

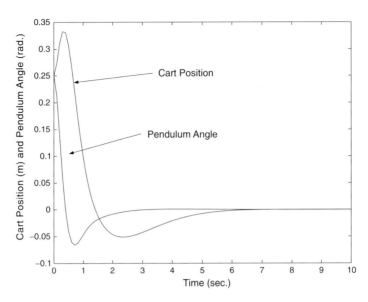

FIGURE 12.23 Cart and pendulum LQR response with large pendulum angle weighting.

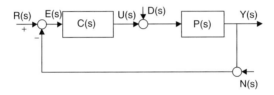

FIGURE 12.24 Robust control feedback block diagram.

to infinitesimal variations in the plant P. Now, given a (frequency-dependent) weighting function $W_1(s)$, a natural *performance specification* (relating tracking error to classes of reference signals) is

$$\|W_1 S\|_\infty < 1 \tag{12.51}$$

where $\|\cdot\|_\infty$ denotes the infinity norm. An easy graphical test for the performance specification is that the Nyquist plot of L must always lie outside a disk of radius $|W_1|$ centered at -1.

To incorporate plant uncertainty into the model, consider a nominal plant P and perturbed plant $\tilde{P}$ where P and $\tilde{P}$ differ by some multiplicative or other type of uncertainty. Let W_2 be a stable transfer function and Δ be a variable stable transfer function satisfying $\|\Delta\|_\infty \leq 1$. Then, common uncertainty models can be constructed by appropriate combinations of P, Δ and W. It can be shown that the system is *internally stable* (this is a stronger definition than simple I/O stability; see Doyle [1992]) for the conditions shown in Table 12.3.

Recall that the nominal performance condition was $\|W_1 S\|_\infty < 1$. Not surprising, then, the *robust performance* condition is a combination of the two (for the $(1 + \Delta W_2)P$ perturbation):

$$\| |W_1 S| + |W_2 T| \|_\infty < 1 \tag{12.52}$$

Other robust performance measures for various types of uncertainty can be found in Doyle et al. (1992) and Zhou (1996).

Recall that W_1 is the performance specification weighting function and W_2 is the plant uncertainty transfer function. Consider the following facts:

TABLE 12.3 Internal Stability Conditions

Perturbation	Condition
$(1 + \Delta W_2)P$	$\|W_2 T\|_\infty < 1$
$P + \Delta W_2$	$\|W_2 CS\|_\infty < 1$
$\dfrac{P}{1 + \Delta W_2 P}$	$\|W_2 PS\|_\infty < 1$
$\dfrac{P}{1 + \Delta W_2}$	$\|W_2 S\|_\infty < 1$

1. Plant uncertainty is greatest for high frequencies.
2. It is only reasonable to demand high performance for low frequencies.

Typically, then,

$$|W_1| > 1 > |W_2| \tag{12.53}$$

for low frequencies, and

$$|W_1| < 1 < |W_2| \tag{12.54}$$

for high frequencies (it can be shown that the magnitude of either W_1 or W_2 must be less than 1). Now, by considering the relationship between L, S and T, the following can be derived:

$$|W_1| \gg 1 > |W_2| \Rightarrow |L| > \frac{|W_1|}{1 - |W_2|} \tag{12.55}$$

and

$$|W_1| < 1 \ll |W_2| \Rightarrow |L| < \frac{1 - |W_1|}{|W_2|} \tag{12.56}$$

Loopshaping [Bower and Schultheiss, 1961; Horowitz, 1963] controller design is the task of determining an L (and hence C) that satisfies the low-frequency performance criterion as well as the high-frequency robustness criterion. Hence, the task is to design C so that the magnitude versus frequency plot of L appears as in Figure 12.25. In the figure, the indicated low-frequency performance bound is a plot of:

$$\frac{|W_1|}{1 - |W_2|} \tag{12.57}$$

for low frequencies, and the high-frequency stability bound is a plot of:

$$\frac{1 - |W_1|}{|W_2|} \tag{12.58}$$

for high frequencies.

Two more aspects of this problem have been developed in recent years. The first concerns optimality and the second concerns multivariable systems. For both aspects of these recent developments, interested readers are referred to the comprehensive book by Zhou (1996).

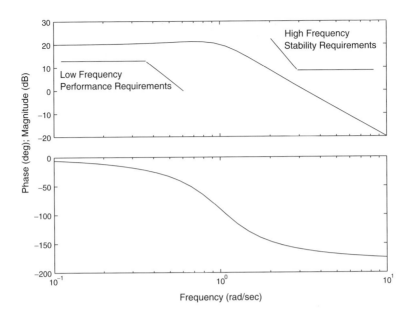

FIGURE 12.25 Loopshaping concepts.

12.4 Nonlinear Control

Aside from the developments of robust optimal control briefly outlined in the previous section, the area of most recent development in control theory has been nonlinear control, where, in contrast to ignoring nonlinear effects via linearization, the nonlinearities in the control system are either expressly recognized or are even exploited for control purposes. Much, but not all, development in nonlinear control has utilized tools from differential geometry. While the control techniques will be outlined here, the basics of differential geometry will not, and the interested reader is referred to Abraham et al. (1988), Boothby (1986), Isidori (1996) and Nijmeijer and van der Schaft (1990) for details.

The general nonlinear model considered here is of the form:

$$\dot{\mathbf{x}} = \mathbf{f}(\mathbf{x}) + \sum_{i=1}^{n} \mathbf{g_i}(\mathbf{x}) u_i \qquad (12.59)$$

where $\mathbf{x}$ is a $1 \times n$ vector, the $\mathbf{f}(\mathbf{x})$ and $\mathbf{g_i}(\mathbf{x})$ are smooth vector fields, and the u_i are scalar control inputs. Note that this is not the most general form for nonlinear systems, as the u_i are assumed to enter the equations in an affine manner (i.e., they simply multiply the $\mathbf{g_i}(\mathbf{x})$ vector fields). For some aerodynamic problems, for example, this assumption may not be true.

12.4.1 SISO Feedback Linearization

In contrast to the standard Jacobian linearization of a nonlinear control system, feedback linearization is a technique to construct a nonlinear change of coordinates which converts a nonlinear system in the original coordinates to a linear system in the new coordinates. Thus, whereas the Jacobian linearization is an approximation of the original system, a feedback linearized system is still exactly the original system. For clarity of presentation, SISO systems will be considered first, followed by MIMO systems. Excellent references for feedback linearization are Isidori (1996), Nijmeijer and van der Schaft (1990), Krener (1987), Khalil (1996) and Sastry (2000). Developmental papers or current research in this area are considered

Fundamentals of Control Theory

in Slotine and Hedrick (1993), Brockett (1978), Dayawansa et al. (1985), Isidori et al. (1981a; 1981b) and Krener (1987).

Consider the nonlinear system:

$$\dot{\mathbf{x}} = \mathbf{f}(\mathbf{x}) + \mathbf{g}(\mathbf{x})u$$
$$y = h(\mathbf{x}) \tag{12.60}$$

where the function $h(\mathbf{x})$ is called the *output function*. Let $L_f h$ denote the *Lie derivative* of the function h with respect to the vector field $\mathbf{f}$, which is defined in coordinates as:

$$L_f h(x) = \sum_{i=1}^{n} \frac{\partial h}{\partial x_i}(\mathbf{x}) f_i(\mathbf{x}) \tag{12.61}$$

where

$$\mathbf{x} = \begin{bmatrix} x_1 \\ x_2 \\ \vdots \\ x_{n-1} \\ x_n \end{bmatrix} \quad \text{and} \quad \mathbf{f} = \begin{bmatrix} f_1(\mathbf{x}) \\ f_2(\mathbf{x}) \\ \vdots \\ f_{n-1}(\mathbf{x}) \\ f_n(\mathbf{x}) \end{bmatrix} \tag{12.62}$$

so it is simply the directional derivative of h along $\mathbf{f}$. Now, because the system evolves according to the state equations, the time derivative of the output function $\dot{y}$ is simply the directional derivative of the output function along the control system:

$$\dot{y} = \dot{h} = \frac{\partial h}{\partial \mathbf{x}} \dot{\mathbf{x}} = \frac{\partial h}{\partial \mathbf{x}}(\mathbf{f}(\mathbf{x}) + \mathbf{g}(\mathbf{x})u) = L_{f+gu} h = L_f h + L_g h u \tag{12.63}$$

The *relative degree* of a system is defined as follows: A SISO nonlinear system is said to have *strict relative degree* γ at the point x if:

1. $L_g L_f^i h(x) \equiv 0 \quad i = 0, 1, 2, \ldots, \gamma - 2$ \hfill (12.64)

2. $L_g L_f^{\gamma-1} h(x) \neq 0$ \hfill (12.65)

In the case where $\gamma = n$, the system is *full state feedback linearizable*, and it is possible to construct the following change of coordinates where the original coordinates x_i are mapped to a new set of coordinates ξ_i as follows:

$$\begin{aligned} \xi_1 &= h(x) \\ \xi_2 &= \dot{\xi}_1 = \dot{h} = L_f h \\ \xi_3 &= \dot{\xi}_2 = \ddot{h} = L_f^2 h \\ &\vdots \\ \xi_n &= \dot{\xi}_{n-1} = L_f^{\gamma-1} h \end{aligned} \tag{12.66}$$

Computing derivatives, the control system becomes:

$$\dot{\xi}_1 = \xi_2$$
$$\dot{\xi}_2 = \xi_3$$
$$\vdots \quad (12.67)$$
$$\dot{\xi}_{n-1} = \xi_n$$
$$\dot{\xi}_n = L_f^\gamma h + L_g L_f^{\gamma-1} h u$$

or, setting

$$u = \frac{1}{L_g L_f^{\gamma-1} h}(-L_f^\gamma h + v) \quad (12.68)$$

the system is

$$\dot{\xi}_1 = \xi_2$$
$$\dot{\xi}_2 = \xi_3$$
$$\vdots \quad (12.69)$$
$$\dot{\xi}_{n-1} = \xi_n$$
$$\dot{\xi}_n = v$$

which is both linear and in controllable canonical form. At this point, it is simple to determine an appropriate v to stabilize the system or track desired values of $h(x)$. One approach is pole placement; that is, $v = -K\xi$. Note that the overall approach was to determine an output function h that could be differentiated n times before the control input appeared. This essentially constructs a system known as a *chain of integrators*, as the derivative of the ith state in the ξ variables is equal to the $(i+1)$th state variable.

There are two main limitations to feedback linearization approaches. The first is that not all systems are feedback linearizable, although analytical tests exist to determine whether a particular system is linearizable. Second, determining the output function, $h(x)$ involves solving a system of partial differential equations (generally, not easy to do).

Example—SISO Full State Feedback Linearization

Consider, as a mathematical example of the computations involved in feedback linearization, the system:

$$\begin{bmatrix} \dot{x}_1 \\ \dot{x}_2 \\ \dot{x}_3 \\ \dot{x}_4 \end{bmatrix} = \begin{bmatrix} x_3 \\ x_4 \\ x_1 + x_2 + x_3 \\ x_1 - x_3 \end{bmatrix} + \begin{bmatrix} 0 \\ 0 \\ 0 \\ 1 \end{bmatrix} u = f(x) + g(x)u \quad (12.70)$$

with output function, $y = h(x) = x_1$. The system has a relative degree equal to 4, so the system is full state feedback linearizable and the coordinate transformation is given by:

$$\xi_1 = h(x) = x_1$$
$$\xi_2 = L_f h(x) = x_3$$
$$\xi_3 = L_f^2 h(x) = x_1 + x_2 + x_3 \quad (12.71)$$
$$\xi_4 = L_f^3 h(x) = x_1 + x_2 + 2x_3 + x_4$$

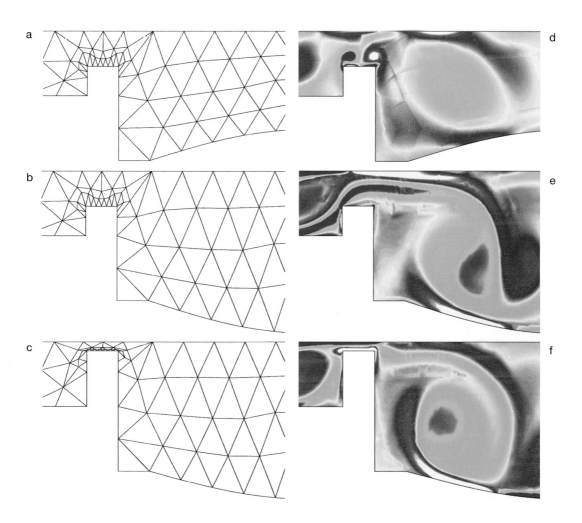

COLOR FIGURE 5.6 Close-up of the vorticity contours for $Re = 30$ simulation at the left valve (meshes shown on right side). Top: $\tau\omega = 0.28$, corresponding to the beginning of the suction stage; start-up vortices due to the motion of the inlet valve can be identified. Middle: $\tau\omega = 0.72$, corresponding to the end of the suction stage; a vortex jet pair is visible in the pump cavity. Bottom: $\tau\omega = 0.84$, corresponding to early ejection stage; further evolution of the vortex jet and the start-up vortex of the exit valve can be identified. (Courtesy of A. Beskok.)

COLOR FIGURE 6.1 Micro heat exchanger constructed from rectangular channels machined in metal. (Courtesy of K. Schubert and D. Cacuci, Forschungszentrum, Karlsruhe.)

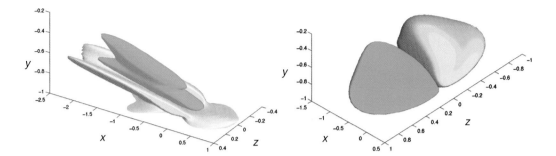

COLOR FIGURE 13.8 Localized controller gains relating the state estimate $\hat{x}$ inside the domain to the control forcing **u** at the point $\{x = 0, y = -1, z = 0\}$ on the wall. Visualized are a positive and negative isosurface of the convolution kernels for (left) the wall-normal component of velocity and (right) the wall-normal component of vorticity. (From Högberg, M., and Bewley, T.R., *Automatica* (submitted). With permission from Elsevier Science.)

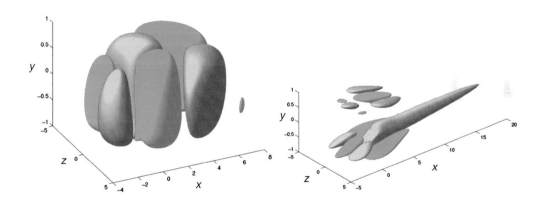

COLOR FIGURE 13.9 Localized estimator gains relating the measurement error $(\mathbf{y} - \hat{\mathbf{y}})$ at the point $\{x = 0, y = -1, z = 0\}$ on the wall to the estimator forcing terms **v** inside the domain. Visualized are a positive and negative isosurface of the convolution kernels for (left) the wall-normal component of velocity and (right) the wall-normal component of vorticity. (From Högberg, M., and Bewley, T.R., *Automatica* (submitted). With permission from Elsevier Science.)

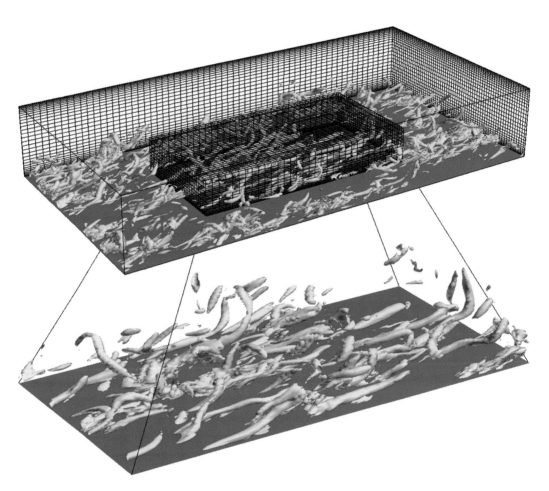

COLOR FIGURE 13.11 Visualization of the coherent structures of uncontrolled near-wall turbulence at $Re_\tau = 180$. Despite the geometric simplicity of this flow (see Figure 13.1), it is phenomenologically rich and is characterized by a large range of length scales and time scales over which energy transport and scalar mixing occur. The relevant spectra characterizing these complex nonlinear phenomena are continuous over this large range of scales, thus such flows have largely eluded accurate description via dynamic models of low state dimension. The nonlinearity, the distributed nature and the inherent complexity of their dynamics make turbulent flow systems particularly challenging for successful application of control theory. (Simulation by Bewley et al. (2001). Reprinted with permission of Cambridge University Press.)

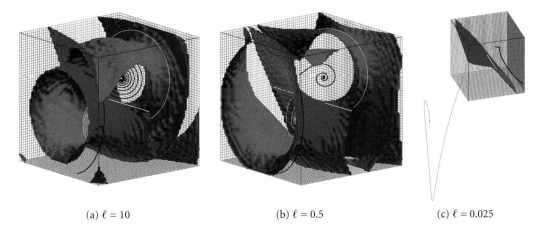

(a) ℓ = 10 (b) ℓ = 0.5 (c) ℓ = 0.025

COLOR FIGURE 13.12 Example of the spectacular failure of linear control theory to stabilize a simple nonlinear chaotic convection system governed by the Lorenz equation. Plotted are the regions of attraction to the desired stationary point (blue) and to an undesired stationary point (red) in the linearly controlled nonlinear system, and typical trajectories in each region (black and green, respectively). The cubical domain illustrated is $\Omega = (-25, 25)^3$ in all subfigures; for clarity, different viewpoints are used in each subfigure. (Reprinted from Bewley, T.R. (1999) *Phys. Fluids* **11**, 1169–1186. Copyright 1999, American Institute of Physics. With permission.)

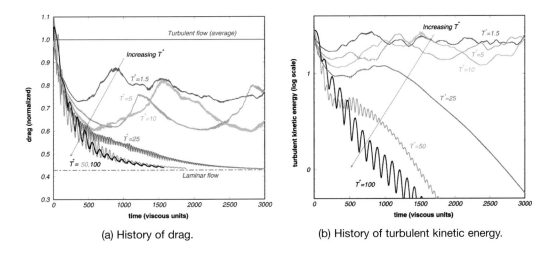

(a) History of drag. (b) History of turbulent kinetic energy.

COLOR FIGURE 13.14 Performance of optimized blowing/suction controls for formulations based on minimizing $T_0(\phi)$, case c (see Section 13.9.1.2), as a function of the optimization horizon T^*. The direct numerical simulations of turbulent channel flow reported here were conducted at $Re_\tau = 100$. For small optimization horizons ($T^* = O(1)$, sometimes called the "suboptimal approximation"), approximately 20% drag reduction is obtained, a result that can be obtained with a variety of other approaches. For sufficiently large optimization horizons ($T^* \geq 25$), the flow is returned to the region of stability of the laminar flow, and the flow relaminarizes with no further control effort required. No other control algorithm tested in this flow to date has achieved this result with this type of flow actuation. (From Bewley, T.R., Moin, P., and Temam, R., *J. Fluid Mech.*, to appear. With permission of Cambridge University Press.)

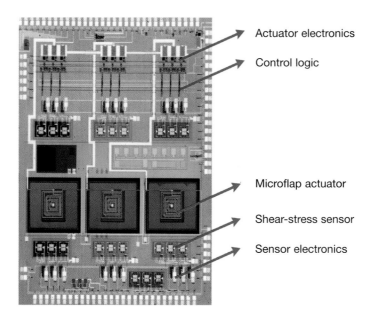

COLOR FIGURE 13.20 A MEMS tile integrating sensors, actuators and control logic for distributed flow control applications. (Developed by Profs. Chih-Ming Ho, UCLA, and Yu-Chong Tai, Caltech.)

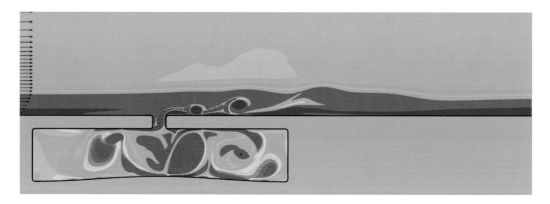

COLOR FIGURE 13.21 Simulation of a proposed driven-cavity actuator design (Prof. Rajat Mittal, University of Florida). The fluid-filled cavity is driven by vertical motions of the membrane along its lower wall. Numerical simulation and reduced-order modeling of the the influence of such flow-control actuators on the system of interest will be essential for the development of feedback control algorithms to coordinate arrays of realistic sensor/actuator configurations.

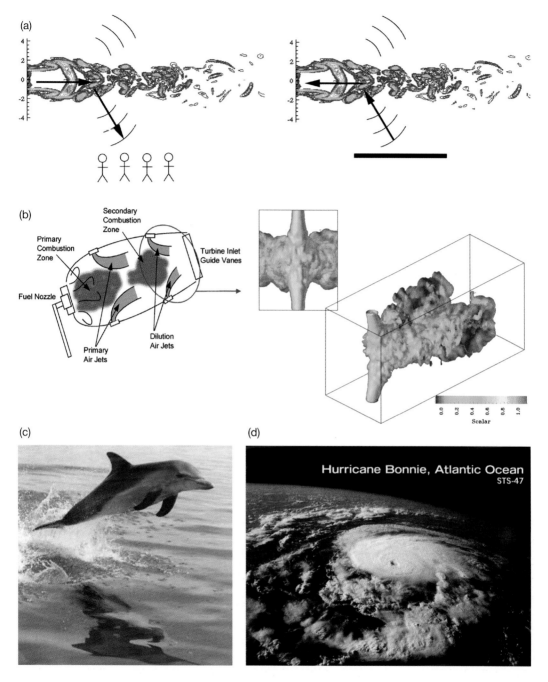

COLOR FIGURE 13.22 Future interdisciplinary problems in flow control amenable to adjoint-based analysis:
(a) Minimization of sound radiating from a turbulent jet (simulation by Prof. Jon Freund, UCLA)
(b) Maximization of mixing in interacting cross-flow jets (simulation by Dr. Peter Blossey, UCSD) [Schematic of jet engine combustor is shown at left. Simulation of interacting cross-flow dilution jets, designed to keep the turbine inlet vanes cool, is visualized at right.]
(c) Optimization of surface compliance properties to minimize turbulent skin friction
(d) Accurate forecasting of inclement weather system

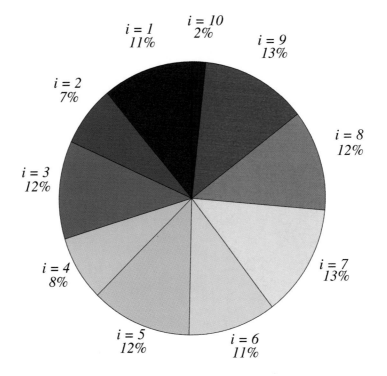

COLOR FIGURE 14.11 "Old" fitness.

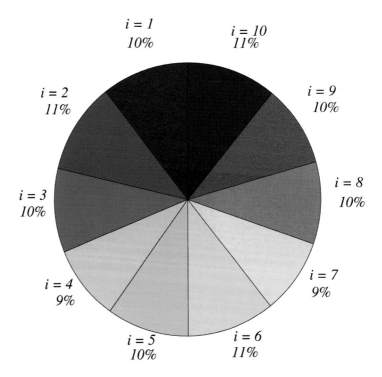

COLOR FIGURE 14.13 "New" fitness.

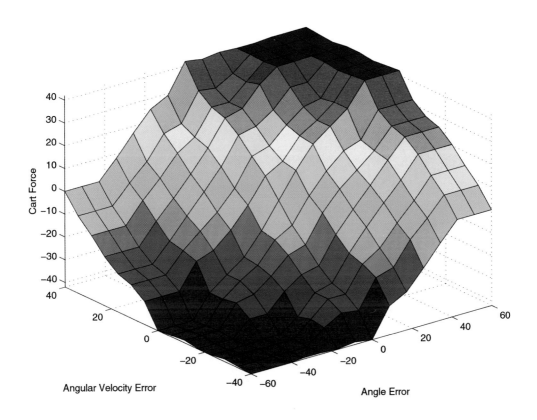

COLOR FIGURE 14.29 Response surface for pendulum fuzzy controller.

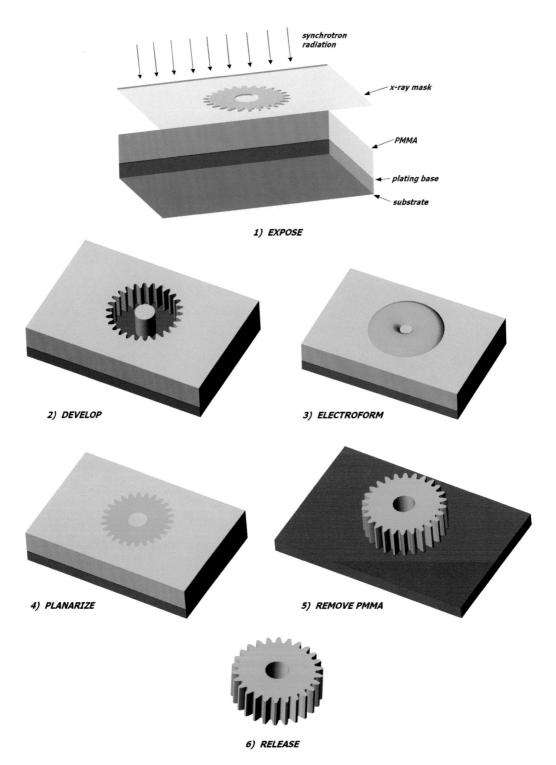

COLOR FIGURE 18.1 DXRL-based (direct LIGA) microfabrication process.

COLOR FIGURE 21.11 Packaged SiC pressure-sensor chip. Through the semitransparent SiC can be seen the edge of the well that has been etched in the backside of the wafer to form the circular diaphragm. The metal-covered, n-type SiC that connects the strain gauges is highly visible, while the n-type SiC strain gauges are more faintly visible. There is a U-shaped strain gauge over the edge of the diaphragm at top and bottom, and there are two vertically oriented linear gauges in the center of the diaphragm.

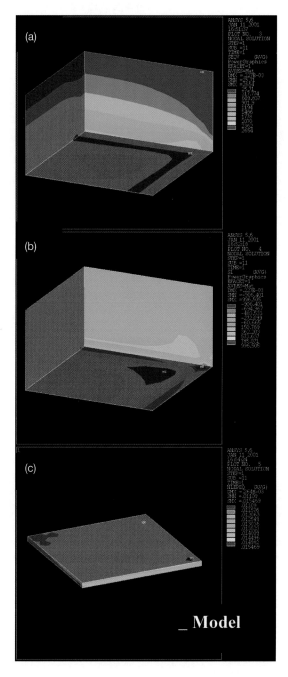

COLOR FIGURE 23.8 Thermal stress and strain distribution in SiC die and Au thick-film layer. (a) Von-Mises stress distribution in SiC die; (b) principal stress distribution in SiC die; (c) equivalent plastic strain in Au thick-film layer.

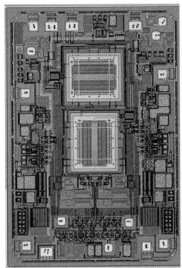

COLOR FIGURE 24.1 Examples of two high-volume accelerometer products. On the left is Analog Devices, Inc. ADXL250 two-axis lateral monolithically integrated accelerometer. On the right is a Motorola, Inc. wafer-scale packaged accelerometer and control chips mounted on a lead frame prior to plastic injection molding. (Photographs courtesy of Analog Devices, Inc. and Motorola, Inc.)

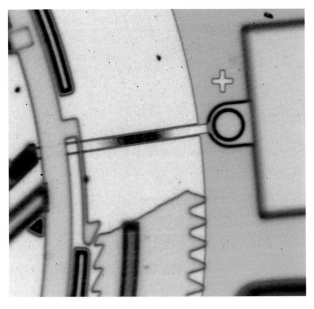

COLOR FIGURE 27.31 The blackened trace near the "+" sign was caused by a dimple on a grounded gear touching a high potential signal line. (Photograph courtesy of S. Barnes, Sandia National Laboratories.)

COLOR FIGURE 28.17 Photograph of the microrobot platform used for walking during a load test. The load of 2500 mg is equivalent to maximum 625 mg/leg (or more than 30 times the weight of the robot itself). The power supply is maintained through three 30-µm-thin and 5- to 10-cm-long bonding wires of gold. The robot walks using the asynchronous ciliary motion principle described in Figure 28.6c. The legs are actuated using the polyimide V-groove joint technology which was described in Figure 28.10. (Photo by P. Westergård; published with permission.) (Note: Videos on different experiments that have been performed using the microrobot shown above are available at: http://www.s3.kth.se/mst/research/gallery/microrobot_video.html.)

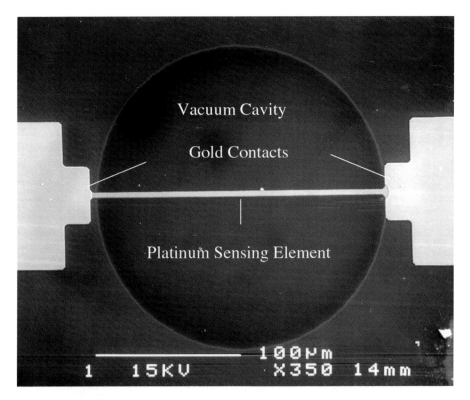

COLOR FIGURE 35.35 SEM view of University of Florida MEMS wall-shear sensor.

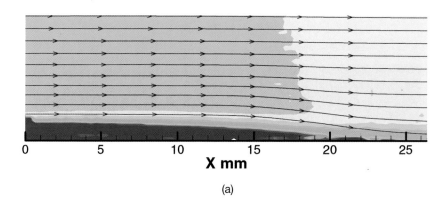

(a)

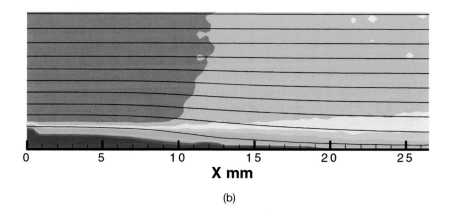

(b)

COLOR FIGURE 35.37 Streamlines of the flow over the step without (a) and with (b) actuation.

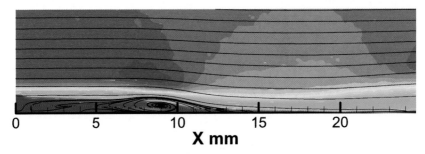

(a) 5.4% of input waveform, flap deflection = 14.1 μm

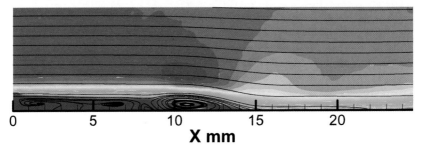

(a) 25.4% of input waveform, flap deflection = 22.3 μm

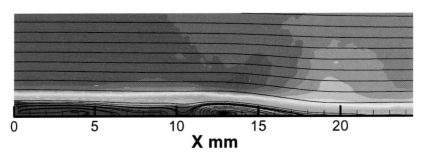

(a) 45.4% of input waveform, flap deflection = −0.3 μm

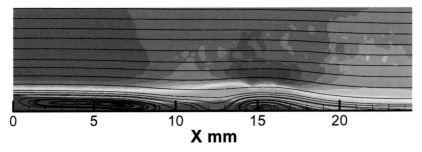

(a) 65.4% of input waveform, flap deflection = −22.5 μm

COLOR FIGURE 35.38 Phase-averaged streamline plots at different phases of the forcing cycle.

Fundamentals of Control Theory

The above equations and the fact that the system has a relative degree of 4 is verified by the following detailed calculations:

$$L_g h(x) = L_g x_1 = \begin{bmatrix} 1 & 0 & 0 & 0 \end{bmatrix} \begin{bmatrix} 0 \\ 0 \\ 0 \\ 1 \end{bmatrix} = 0,$$

$$L_g L_f h(x) = L_g l_f x_1 = L_g \begin{bmatrix} 1 & 0 & 0 & 0 \end{bmatrix} \begin{bmatrix} x_3 \\ x_4 \\ x_1 + x_2 + x_3 \\ x_1 - x_3 \end{bmatrix} = L_g x_3 = \begin{bmatrix} 0 & 0 & 1 & 0 \end{bmatrix} \begin{bmatrix} 0 \\ 0 \\ 0 \\ 1 \end{bmatrix} = 0,$$

$$L_g L_f^2 h(x) = L_g L_f^2 x_1 = L_g L_f x_3 = L_g \begin{bmatrix} 0 & 0 & 1 & 0 \end{bmatrix} \begin{bmatrix} x_3 \\ x_4 \\ x_1 + x_2 + x_3 \\ x_1 - x_3 \end{bmatrix} \quad (12.72)$$

$$= L_g(x_1 + x_2 + x_3) = \begin{bmatrix} 1 & 1 & 1 & 0 \end{bmatrix} \begin{bmatrix} 0 \\ 0 \\ 0 \\ 1 \end{bmatrix} = 0,$$

$$L_g L_f^3 h(x) = L_g L_f^3 x_1 = L_g L_f (x_1 + x_2 + x_3) = L_g \begin{bmatrix} 1 & 1 & 1 & 0 \end{bmatrix} \begin{bmatrix} x_3 \\ x_4 \\ x_1 + x_2 + x_3 \\ x_1 - x_3 \end{bmatrix}$$

$$= L_g(x_1 + x_2 + 2x_3 + x_4) = \begin{bmatrix} 1 & 1 & 2 & 1 \end{bmatrix} \begin{bmatrix} 0 \\ 0 \\ 0 \\ 1 \end{bmatrix} = 1$$

Therefore, a controller of the form:

$$u = \frac{1}{L_g L_f^3 h}(-L_f^4 h + v)$$

$$= (-(3x_1 + 2x_2 + 2x_3 + x_4) + k_1 x_1 + k_2 x_3 + k_3(x_1 + x_2 + x_3) + k_4(x_1 + x_2 + 2x_3 + x_{er4})) \quad (12.73)$$

with the gains k_i picked via pole placement, for example, will allow the system to track trajectories of the output function $h(x) = x_1$.

So far, this section has considered full state feedback linearization where the relative degree of a system is equal to the dimension of its state space. Partial feedback linearization is also possible where the relative degree is less than the dimension of the state space; however, for such systems an analysis of the stability of the *zero dynamics* is necessary. In particular, if the relative degree $\gamma < n$, then the change of coordinates is typically expressed in the form:

$$\begin{aligned}
\xi_1 &= h(x) \\
\xi_2 &= \dot{\xi}_1 = \dot{h} = L_f h \\
\xi_3 &= \dot{\xi}_2 = \ddot{h} = L_f^2 h \\
&\vdots \\
\xi_\gamma &= \dot{\xi}_{\gamma-1} = L_f^{\gamma-1} h \\
\eta_1 &= \eta_1(x) \\
\eta_2 &= \eta_2(x) \\
&\vdots \\
\eta_{n-\gamma} &= \eta_{n-\gamma}(x)
\end{aligned} \tag{12.74}$$

where the η_i are chosen so that the matrix:

$$\begin{bmatrix} dh(x) \\ dL_f h(x) \\ \vdots \\ dL_f^{\gamma-1} h(x) \\ d\eta_1(x) \\ \vdots \\ d\eta_{n-\gamma}(x) \end{bmatrix} \tag{12.75}$$

is full rank. The dynamics of the system in the new coordinates will be of the form:

$$\begin{aligned}
\dot{\xi}_1 &= \xi_2 \\
\dot{\xi}_2 &= \xi_3 \\
&\vdots \\
\dot{\xi}_\gamma &= b(\xi, \eta) + a(\xi, \eta) u \\
\dot{\eta}_1 &= q_1(\xi, \eta) \\
&\vdots \\
\dot{\eta}_{n-\gamma} &= q_{n-\gamma}(\xi, \eta)
\end{aligned} \tag{12.76}$$

The zero dynamics are the dynamics expressed by the η equations, the stability of which must be considered independently of the linearized ξ equations. The interested reader is referrred to texts by Isidori (1996), Khalil (1996), Nijmeijer and van der Schaft (1990) and Sastry (2000) for the relevant details.

12.4.2 MIMO Full-State Feedback Linearization

The MIMO feedback linearization is a slight extension of the SISO feedback linearization by which the SISO linearization construction is repeated for m output functions for a system with m control inputs:

$$\dot{\mathbf{x}} = \mathbf{f}(\mathbf{x}) + \mathbf{g}_1(\mathbf{x})u_1 + \mathbf{g}_2(\mathbf{x})u_2 + \cdots + \mathbf{g}(\mathbf{x})_m u_m$$
$$y_1 = h_1(\mathbf{x})$$
$$y_2 = h_2(\mathbf{x})$$
$$\vdots$$
$$y_m = h_m(\mathbf{x})$$
(12.77)

Now, the vector relative degree is defined as a combination of relative degrees for each of the output functions. Considering the jth output y_j,

$$\dot{y}_j = L_\mathbf{f} h_j + L_{\mathbf{g}_1} h_j u_1 + L_{\mathbf{g}_2} h_j u_2 + \cdots + L_{\mathbf{g}_m} h_j u_m$$
(12.78)

If $L_{\mathbf{g}_i} h_j \equiv 0$ for each i, then the inputs do not appear in the derivative. Now let γ_j be the smallest integer such that $L_\mathbf{g} L_\mathbf{f}^{\gamma_j - 1} h_j \neq 0$ for at least one i. Define the matrix:

$$A(x) = \begin{bmatrix} L_{\mathbf{g}_1} L_\mathbf{f}^{\gamma_1 - 1} h_1 & \cdots & L_{\mathbf{g}_m} L_\mathbf{f}^{\gamma_1 - 1} h_1 \\ \vdots & \ddots & \vdots \\ L_{\mathbf{g}_1} L_\mathbf{f}^{\gamma_m - 1} h_m & \cdots & L_{\mathbf{g}_m} L_\mathbf{f}^{\gamma_m - 1} h_m \end{bmatrix}$$
(12.79)

Now, the system has vector relative degree $\gamma_1, \gamma_2, \ldots, \gamma_m$ at x if $L_{\mathbf{g}_i} L_\mathbf{f}^k h_1 \equiv 0$ $0 \leq k \leq \gamma_i - 2$ for $i = 1, \ldots, m$ and the matrix $A(x)$ is nonsingular.

12.4.3 Control Applications of Lyapunov Stability Theory

Lyapunov theory for autonomous differential equations states that if $x = 0$ is an equilibrium point for a differential equation $\dot{x} = f(x)$, and there exists a continuously differentiable function, $V(x) > 0$ except for $V(0) = 0$ and $\dot{V}(x) < 0$ and in some domain containing zero where:

$$\dot{V}(x) = \sum_{i=1}^{n} \frac{\partial V}{\partial x_i} \dot{x}_i = \sum_{i=1}^{n} \frac{\partial V}{\partial x_i} f_i(x)$$
(12.80)

then the point $x = 0$ is an asymptotically stable equilibrium point for the differential equation. The utility of Lyapunov theory in control is that controller synthesis techniques can be designed to ensure the negative definiteness of a Lyapunov function to ensure stability or boundedness of the system trajectories.

As fully described in Khalil (1996), the main applications of Lyapunov stability theory to control system design are *Lyapunov redesign*, *backstepping*, *sliding mode control* and *adaptive control*. Due to space limitations, the basic concepts of each will only be briefly outlined here.

Lyapunov redesign is an instance of nonlinear robust control design; however, there is a severe restriction upon how the uncertainties are expressed in the equations of motion with a corresponding restriction on the types of systems that are amenable to this technique. In particular, consider the system:

$$\dot{x} = f(t, x) + G(t, x)u + G(t, x)\delta(t, x, u)$$
(12.81)

where f and G are known, δ is unknown but is bounded by a known, but not necessarily small, function. The main restriction here is that the uncertainty enters the system in exactly the same manner as the control input. In order to use Lyapunov redesign, a stabilizing control law exists for the nominal system (ignoring δ) and a Lyapunov function for the nominal system must be known. (Note that one nice aspect of the feedback linearization discussed previously is that if a controller is designed using that technique, a Lyapunov function is straightforward to determine due to the simple form of the equations of motion after the nonlinear coordinate transformation.) Due to the way that the uncertainty enters the system, it is easy to modify the nominal control law to compensate for the uncertainty to ensure that $\dot{V}(x) < 0$. References concerning Lyapunov redesign include Corless (1993), Corless and Heitmann (1981), Barmish et al. (1983) and Spong and Vidyasager (1989).

Backstepping is a recursive controller design procedure where the entire control system is decomposed into smaller, simpler subsystems for which it may be easier to design a stabilizing controller. By considering the appropriate way to modify a Lyapunov function after each smaller subsystem is designed, a stabilizing controller for the full system may be obtained. The main restriction for this technique is a limitation on the structure of the equations of motion (a type of hierarchical structure is required). Extensions of this procedure to account for certain system uncertainties have also been developed. For references, see Krstic et al. (1995), Qu (1993) and Slotine and Hedrick (1993).

The basic idea in sliding mode control is to drive the system in finite time to a certain submanifold of the configuration space, called the *sliding manifold*, upon which the system should indefinitely evolve. Because the sliding manifold has a lower dimension than the full state space for the system, a lower order model can describe the evolution of the system on the sliding manifold. If a stabilizing controller can be designed for the sliding manifold, the problem reduces to designing a controller to drive the system to the sliding manifold. The advantage of sliding mode control is that it is very robust with respect to system uncertainties. One disadvantage is that it is a bit mathematically quirky as there are discontinuities in the control law when switching from the full state of the system to the sliding manifold. Additionally, "chattering," wherein the system constantly alternates between the two sides of the submanifold, is a common problem. See Utkin (1992) and DeCarlo et al. (1988) for overviews of the approach.

Finally, there is a vast literature in the area of adaptive control. In adaptive control, some system performance index is measured and the adaptive controller modifies adjustable parameters in the controller in order to maintain the performance index of the control system close to a desired value (or set of desired values). This is clearly desirable in cases where system parameters are unknown or change with time. Representative references concerning adaptive control include Anderson et al. (1986), Ioannou and Sun (1995), Krstic et al. (1995), Landau et al. (1998), Narendra and Annaswamy (1989) and Sastry and Bodson (1989).

12.4.4 Hybrid Systems

Hybrid systems are systems characterized by both continuous as well as discrete dynamics. Examples of hybrid systems include, but certainly are not limited to, digital computer-controlled systems, distributed control systems governed by a hierarchical logical interaction structure, multi-agent systems (such as the air traffic management system [Tomlin, 1998]) and systems characterized by intermittent physical contact [Goodwine, 2000]. Recent papers considering modeling and control synthesis methods for such complicated systems include Alur and Henzinger (1996), Antsaklis et al. (1995; 1997), Branicky et al. (1998), Henzinger and Sastry (1998) and Lygeros et al. (1999).

12.5 Parting Remarks

This chapter provided a brief overview of the fundamental concepts in analysis and design of control systems. Included was an outline of "classical linear control," including stability concepts (the Routh array) and controller design techniques (root locus and lead-lag synthesis). Additionally, more recent advances in control including pole placement, the linear quadratic regulator and the basic concepts from robust control were outlined and examples were provided. Finally, recent developments in nonlinear control, including feedback linearization (for both single-input, single-output and multi-input, multi-output systems), were outlined along with basic approaches utilizing Lyapunov stability theory.

Defining Terms

Adaptive control: A controller design technique wherein typically some parameter in the controller is varied or changed in response to variations in the controlled system.
Backstepping: A nonlinear controller design technique based upon Lyapunov stability theory.
Block diagram: A graphical representation of a differential equation describing a linear control system.
Bode plot: A plot of the magnitude of the response of a system to sinusoidal inputs vs. the frequency of the inputs and a plot of the phase difference between the sinusoidal input and response of the system.
Characteristic equation: The equation obtained by equating the denominator of a transfer function with zero. Analysis of the characteristic equation yields insight into the stability of a transfer function.
Feedback linearization: A nonlinear controller design technique based upon determining a nonlinear change of coordinates which transform the differential equations describing the system into a simple, canonical and controllable form.
Gain margin: A measure of stability or instability of a control system which can be determined from a Bode plot by considering the magnitude of the system response when the phase difference between the input and system response is $-180°$.
Laplace transform: A mathematical transformation useful to transform ordinary differential equations into algebraic equations.
Loopshaping: A controller design method based upon obtaining a desired "shape" for the Bode plot of a system.
LQR control: An optimal control design technique based upon minimizing a cost function defined as a combination of the magnitude of the system response and the control effort. LQR is an acronym for linear quadratic regulator.
Lyapunov redesign: A nonlinear robust controller design technique based upon Lyapunov stability theory.
Lyapunov stability theory: The analysis of the stability of (nonlinear) differential equations based upon the time derivative of a Lyapunov function.
Nonlinear control: The design and analysis of control systems described by nonlinear differential equations.
Nyquist plot: A contour plot in the complex plane of a transfer function as the dependent variable encircles the entire right-half complex plane.
Phase margin: A measure of stability or instability of a control system which can be determined from a Bode plot by considering the magnitude of the phase difference between the input and system response when the magnitude of the response is equal to 0 dB.
PID control: A very common control law wherein the input to the system is proportional to the error, the derivative of the error and the time integral of the error. PID is an acronym for proportional–integral–derivative control.
Poles: Roots of the denominator of a transfer function.
Pole placement: A state-space controller design technique based upon specifying certain eigenvalues for the system.
Robust control: The design and analysis of control systems which explicitly account for unmodeled system dynamics and disturbances.
Root locus: A controller synthesis technique wherein the poles of the transfer function of a control system are plotted as a parameter (usually a controller gain) is varied.
Routh array: A technique to determine the number of roots of a polynomial that have a positive real part without having to factor the polynomial. This is used to determine the stability of a transfer function based upon its characteristic equation.

Sliding mode control: A nonlinear controller design technique based upon Lyapunov stability theory with the goal of driving the system to a center manifold upon which the dynamics of the system are simpler and easily controlled.

Transfer function: The algebraic expression relating the Laplace transform of the input to the output of a control system.

Zeros: Roots of the numerator of a transfer function.

References

Abraham, R., Marsden, J.E., and Ratiu, T.T. (1988) *Manifolds, Tensor Analysis, and Applications,* Springer-Verlag, New York.

Aeyels, D. (1985) "Stabilization of a Class of Nonlinear Systems by a Smooth Feedback Control," *Syst. Control Lett.* **5**, pp. 289–294.

Alur, R., and Henzinger, T., eds. (1996) *Hybrid Systems III: Verification and Control,* Springer-Verlag, New York.

Anderson, B.D.O., Bitmead, R.R., Johnson, C.R., Kokotovic, P.V., Kosut, R.L., Mareels, I.M.Y., Praly, L., and Riedle, B.D. (1986) *Stability of Adaptive Systems,* MIT Press, Cambridge, MA.

Antsaklis, P., Kohn, W., Nerode, A., and Sastry, S., eds. (1995) *Hybrid Systems II,* Springer-Verlag, New York.

Antsaklis, P., Kohn, W., Nerode, A., and Sastry, S., eds. (1997) *Hybrid Systems IV,* Springer-Verlag, New York.

Barmish, B.R., Corless, M., and Leitmann, G. (1983) "A New Class of Stabilizing Controllers for Uncertain Dynamical Systems," *SIAM J. Control Optimization* **21**, pp. 246–355.

Bestle, D., and Zeitz, M. (1983) "Canonical Form Observer Design for Nonlinear Time-Variable Systems," *Int. J. Control* **38**, pp. 419–431.

Bode, H.W. (1945) *Network Analysis and Feedback Amplifier Design,* D Van Nostrand, Princeton, NJ.

Boothby, W.M. (1986) *An Introduction to Differentiable Manifolds and Reimannian Geometry,* Academic Press, Boston.

Bower, J.L., and Schultheiss, P. (1961) *Introduction to the Design of Servomechanisms,* Wiley, New York.

Branicky, M., Borkar, V., and Mitter, S.K. (1998) "A Unified Framework for Hybrid Control: Model and Optimal Control Theory," *IEEE Trans. Autom. Control* **AC-43**, pp. 31–45.

Brockett, R.W. (1978) "Feedback Invariants for Nonlinear Systems," in *Proceedings of the 1978 IFAC Congress,* Helsinki, Finland, Pergamon Press, Oxford.

Bullo, F., Leonard, N.E., and Lewis, A. (2000) "Controllability and Motion Algorithms for Underactuated Lagrangian Systems on Lie Groups," *IEEE Trans. Autom. Control* **45**, pp. 1437–1454.

Byrnes, C.I., and Isidori, A. (1988) "Local Stabilization of Minimum-Phase Nonlinear Systems," *Syst. Control Lett.* **11**, pp. 9–17.

Chen, M.J., and Desoer, C.A. (1982) "Necessary and Sufficient Condition for Robust Stability of Linear, Distributed Feedback Systems," *Int. J. Control* **35**, pp. 255–267.

Corless, M. (1993) "Control of Uncertain Nonlinear Systems," *J. Dynamical Syst. Meas. Control* **115**, pp. 362–372.

Corless, M., and Leitmann, G. (1981) "Continuous State Feedback Guaranteeing Uniform Ultimate Boundedness for Uncertain Dynamic Systems," *IEEE Trans. Autom. Control* **AC-26**, pp. 1139–1144.

Dayawansa, W.P., Boothby, W.M., and Elliott, D. (1985) "Global State and Feedback Equivalence of Nonlinear Systems," *Syst. Control Lett.* **6**, pp. 517–535.

DeCarlo, R.A., Zak, S.H., and Matthews, G.P. (1988) "Variable Structure Control of Nonlinear Multivariable Systems: A Tutorial," *Proc. IEEE* **76**, pp. 212–232.

Dorf, R.C. (1992) *Modern Control Systems,* Addison-Wesley, Reading, MA.

Doyle, J.C., Francis, B. A., and Tannenbaum, A.R. (1992) *Feedback Control Theory,* Macmillan, New York.

Doyle, J.C., and Stein, G. (1981) "Multivariable Feedback Design: Concepts for a Classical Modern Synthesis," *IEEE Trans. Autom. Control* **AC-26**, pp. 4–16.

Evans, W.R. (1948) "Graphical Analysis of Control Systems," *AIEE Trans. Part II,* pp. 547–551.

Evans, W.R. (1950) "Control System Synthesis by Root Locus Method," *AIEE Trans. Part II*, pp. 66–69.

Franklin, G.F., Powell, D.J., and Emami-Naeini, A. (1994) *Feedback Control of Dynamic Systems*, Addison-Wesley, Reading, MA.

Friedland, B. (1986) *Control System Design*, McGraw-Hill, New York.

Gajec, Z., and Lelic, M.M. (1996) *Modern Control Systems Engineering*, Prentice-Hall, London.

Gibson, J.E. (1963) *Nonlinear Automatic Control*, McGraw-Hill, New York.

Goodwine, B., and Burdick, J. (2000) "Motion Planning for Kinematic Stratified Systems with Application to Quasistatic Legged Locomotion and Finger Gaiting," *IEEE J. Robotics Autom.*, accepted for publication.

Hahn, W. (1963) *Theory and Application of Liapunov's Direct Method*, Prentice-Hall, Upper Saddle River, NJ.

Henzinger, T., and Sastry, S., eds. (1998) *Hybrid Systems: Computation and Control*, HSCC 2000, Pittsburgh, PA.

Hirsch, M., and Smale, S. (1974) *Differential Equations, Dynamical Systems, and Linear Algebra*, Academic Press, Boston.

Horowitz, I.M. (1963) *Synthesis of Feedback Mechanisms*, Academic Press, New York.

Ioannou, P.A., and Sun, J. (1995) *Robust Adaptive Control*, Prentice-Hall, Englewood Cliffs, NJ.

Isidori, A. (1996) *Nonlinear Control Systems*, Springer-Verlag, Berlin.

Isidori, A., Krener, A.J., Gori-Giorgi, C., and Monaco, S. (1981a) "Locally (f, g)-Invariant Distributions," *Syst. Control Lett.* **1**, pp. 12–15.

Isidori, A., Krener, A.J., Gori-Giorgi, C., and Monaco, S. (1981b) "Nonlinear Decoupling Via Feedback: A Differential Geometric Approach," *IEEE Trans. Autom. Control* **AC-26**, pp. 331–345.

Jakubczyk, B., and Respondek, W. (1980) "On Linearization of Control Systems," *Bull. Acad. Polonaise Sci. Ser. Sci. Math* **28**, pp. 517–522.

Junt, L.R., Su, R., and Meyer, G. (1983) "Design for Multi-Input Nonlinear Systems," in *Differential Geometric Control Theory*, eds. R.W. Brockett, R.S. Millman, and H. Sussmann, Birkhäuser, Basel.

Khalil, H.K. (1996) *Nonlinear Systems*, Prentice-Hall, Englewood Cliffs, NJ.

Krener, A.J. (1987) "Normal Forms for Linear and Nonlinear Systems," *Contempor. Math* **68**, pp. 157–189.

Krener, A.J. (1998) "Feedback Linearization," in *Mathematical Control Theory*, eds. J. Baillieul and J. Willems, Springer-Verlag, New York.

Krstic, M., Kanellakopoulos, I., and Kokotovic, P. (1995) *Nonlinear and Adaptive Control Systems Design*, John Wiley & Sons, New York.

Kuo, B.C. (1995) *Automatic Control Systems*, Prentice-Hall, Englewood Cliffs, NJ.

Lafferriere, G., and Sussmann, H. (1993) "A Differential Geometric Approach to Motion Planning," in *Nonholonomic Motion Planning*, eds. Z. Li and J. F. Canny, Academic Press, New York, pp. 235–270.

Landau, Y.D., Lozano, R., and M'Saad, M. (1998) *Adaptive Control*, Springer-Verlag, Berlin.

LaSalle, J.P., and Lefschetz, S. (1961) *Stability by Liapunov's Direct Method with Applications*, Academic Press, New York.

Lygeros, J., Godbole, D., and Sastry, S. (1999) "Controllers for Reachability Specifications for Hybrid Systems," *Automatica* **35**, pp. 349–370.

Marino, R., and Tomei, P. (1995) *Nonlinear Control Design*, Prentice-Hall, New York.

Marino, R. (1988) "Feedback Stabilization of Single-Input Nonlinear Systems," *Syst. Control Lett.* **10**, pp. 201–206.

Narendra, K.S., and Annaswamy, A.M. (1989) *Stable Adaptive Systems*, Prentice-Hall, Englewood Cliffs, NJ.

Nijmeijer, H., and van der Schaft, A.J. (1990) *Nonlinear Dynamical Control Systems*, Springer-Verlag, New York.

Nyquist, H. (1932) "Regeneration Theory," *Bell System Technol. J.* **11**, pp. 126–47.

Ogata, K. (1997) *Modern Control Engineering*, Prentice-Hall, Englewood Cliffs, NJ.

Qu, Z. (1993) "Robust Control of Nonlinear Uncertain Systems under Generalized Matching Conditions," *Automatica* **29**, pp. 985–998.

Raven, F.H. (1995) *Automatic Control Engineering*, McGraw-Hill, New York.
Routh, E.J. (1975) *Stability of Motion*, Taylor & Francis, London.
Sastry, S. (2000) *Nonlinear Systems: Analysis, Stability and Control*, Springer-Verlag, New York.
Sastry, S., and Bodson, M. (1989) *Adaptive Systems: Stability, Convergence and Robustness*, Prentice-Hall, Englewood Cliffs, NJ.
Shinners, S.M. (1992) *Modern Control System Theory and Design*, John Wiley & Sons, New York.
Slotine, J.-J.E., and Hedrick, J.K. (1993) "Robust Input–Output Feedback Linearization," *Int. J. Control* **57**, pp. 1133–1139.
Sontag, E.D. (1998) *Mathematical Control Theory: Deterministic Finite Dimensional Systems*, Second Edition, Springer, New York, 1990.
Spong, M.W., and Vidyasager, M. (1989) *Robot Dynamics and Control*, Wiley, New York.
Su, R. (1982) "On the Linear Equivalents of Nonlinear Systems," *Syst. Control Lett.* **2**, pp. 48–52.
Tomlin, C., Pappas, G., and Sastry, S. (1998) "Conflict Resolution in Air Traffic Management: A Study in Multi-Agent Hybrid Systems," *IEEE Trans. Autom. Control* **43**(4), pp. 509–521.
Truxal, J.G. (1955) *Automatic Feedback Systems Synthesis*, McGraw-Hill, New York.
Utkin, V.I. (1992) *Sliding Modes in Optimization and Control*, Springer-Verlag, New York.
Zames, G., and Francis, B.A. (1983) "Feedback, Minimax Sensitivity, and Optimal Robustness," *IEEE Trans. Autom. Control* **AC-28**, pp. 585–601.
Zames, G. (1981) "Feedback and Optimal Sensitivity: Model Reference Transformations, Multiplicative Seminorms, and Approximate Inverses," *IEEE Trans. Autom. Control* **AC-26**, pp. 301–320.
Zeitz, M. (1983) "Controllability Canonical (Phase-Variable) Forms for Nonlinear Time-Variable Systems," *Int. J. Control* **37**, pp. 1449–1457.
Zhou, K. (1996) *Robust and Optimal Control*, Prentice-Hall, Englewood Cliffs, NJ.
Ziegler, J.G., and Nichols, N.B. (1942) "Optimum Settings for Automatic Controllers," *ASME Trans.* **64**, pp. 759–768.
Ziegler, J.G., and Nichols, N.B. (1943) "Process Lags in Automatic Control Circuits," *ASME Trans.* **65**, pp. 433–444.

For Further Information

Most of the material covered in the chapter is thoroughly covered in textbooks. For the section on classical control, the undergraduate texts by Dorf (1992), Franklin et al. (1994), Kuo (1995), Ogata (1997) and Raven (1995) provide a complete mathematical treatment of root locus design, PID control, lead-lag compensation and basic state-space methods. Textbooks for linear, robust control include Doyle et al. (1992) and Zhou (1996).

The standard textbooks for geometric nonlinear control are Isidori (1996) and Nijmeijer and van der Schaft (1990). Additional material concerning the mathematical basis for differential geometry is found in Abraham et al. (1988) and Boothby (1986). Sastry (2000) provides an overview of differential geometric techniques but also considers Lyapunov-based methods and nonlinear dynamical systems in general. Khalil (1996) focuses primarily on Lyapunov methods and includes a chapter on differential geometric techniques.

13
Model-Based Flow Control for Distributed Architectures

Thomas R. Bewley
University of California, San Diego

13.1 Introduction 13-2
13.2 Linearization: Life in a Small Neighborhood 13-3
13.3 Linear Stabilization: Leveraging Modern Linear
Control Theory 13-6
The $\mathcal{H}_\infty$ Approach to Control Design • Advantages of Modern Control Design for Non-Normal Systems • Effectiveness of Control Feedback at Particular Wavenumber Pairs
13.4 Decentralization: Designing for Massive Arrays 13-14
Centralized Approach • Decentralized Approach
13.5 Localization: Relaxing Nonphysical Assumptions 13-16
Open Questions
13.6 Compensator Reduction: Eliminating
Unnecessary Complexity 13-19
Fourier-Space Compensator Reduction • Physical-Space Compensator Reduction • Nonspatially Invariant Systems
13.7 Extrapolation: Linear Control
of Nonlinear Systems 13-21
13.8 Generalization: Extending to Spatially
Developing Flows 13-25
13.9 Nonlinear Optimization: Local Solutions
for Full Navier–Stokes 13-26
Adjoint-Based Optimization Approach • Continuous Adjoint vs. Discrete Adjoint
13.10 Robustification: Appealing to Murphy's Law 13-34
Well-Posedness • Convergence of Numerical Algorithm
13.11 Unification: Synthesizing a General Framework 13-36
13.12 Decomposition: Simulation-Based
System Modeling 13-37
13.13 Global Stabilization: Conservatively
Enhancing Stability 13-37
13.14 Adaptation: Accounting for a Changing
Environment 13-38
13.15 Performance Limitation: Identifying Ideal
Control Targets 13-38

13.16	Implementation: Evaluating Engineering Trade-Offs	**13**-39
13.17	Discussion: A Common Language for Dialog	**13**-41
13.18	The Future: A Renaissance	**13**-41
	Acknowledgments	**13**-43

As traditional scientific disciplines individually grow towards their maturity, many new opportunities for significant advances lie at their intersection. For example, remarkable developments in control theory in the last few decades have considerably expanded the selection of available tools which may be applied to regulate physical and electrical systems. When combined with microelectromechanical (MEMS) techniques for distributed sensing and actuation, as highlighted elsewhere in this handbook, these techniques hold great promise for several applications in fluid mechanics, including the delay of transition and the regulation of turbulence. Such applications of control theory require a very balanced perspective, in which one considers the relevant flow physics when designing the control algorithms and, conversely, takes into account the requirements and limitations of control algorithms when designing both reduced-order flow models and the fluid-mechanical systems to be controlled themselves. Such a balanced perspective is elusive, however, as both the research establishment in general and universities in particular are accustomed only to the dissemination and teaching of component technologies in isolated fields. To advance, we must not toss substantial new interdisciplinary questions over the fence for fear of them being "outside our area;" rather, we must break down these very fences that limit us and attack these challenging new questions with a Renaissance approach. In this spirit, this chapter surveys a few recent attempts at bridging the gaps between the several scientific disciplines comprising the field of flow control, in an attempt to clarify the author's perspective on how recent advances in these constituent disciplines fit together in a manner that opens up significant new research opportunities.

13.1 Introduction

Flow control is perhaps the most difficult Grand Challenge application area for MEMS technology. Potentially, it is also one of the most rewarding, as a common feature in many fluid systems is the existence of natural instability mechanisms by which a small input, when coordinated correctly, can lead to a large response in the overall system. As one of the key driving application areas for MEMS, it is thus appropriate to survey recent developments in the fundamental framework for flow control in this handbook.

The area of flow control plainly resides at the intersection of disciplines, incorporating essential and nontrivial elements from control theory, fluid mechanics, Navier–Stokes mathematics, numerical methods and, of course, fabrication technology for "small" (millimeter-scale), self-contained, durable devices which can integrate the functions of sensing, actuation and control logic. Recent developments in the integration of these disciplines, while grounding us with appropriate techniques to address some fundamental open questions, hint at the solution of several new questions which are yet to be asked. To follow up on these new directions, it is essential to have a clear vision of how recent advances in these fields fit together and to know where the significant unresolved issues at their intersection lie.

This chapter will attempt to elucidate the utility of an interdisciplinary perspective to this type of problem by focusing on the control of a prototypical and fundamental fluid system: plane channel flow. The control of the flow in this simple geometry embodies a myriad of complex issues and interrelationships whose understanding requires us to draw from a variety of traditional disciplines. Only when these issues and perspectives are combined is a complete understanding of the state of the art achieved and a vision of where to proceed next identified.

Though plane channel flow will be the focus problem we will discuss here, the purpose of this work goes well beyond simply controlling this particular flow with a particular actuator/sensor configuration. At its core, the research effort we will describe is devoted to the development of an integrated, interdisciplinary understanding that will allow us to synthesize the necessary tools to attack a variety of flow control problems in the future. The focus problem of control of channel flow is chosen not simply because of its technological relevance or fundamental character, but because it embodies many of the important unsolved issues to be encountered in the assortment of new flow control problems that will inevitably follow. The primary objective of this work is to lay a solid, integrated footing upon which these future efforts may be based.

To this end, this chapter will describe mostly the efforts with which the author has been directly involved, in an attempt to weave the story that threads these projects together as part of the fabric of a substantial new area of interdisciplinary research. Space does not permit complete development of these projects; rather, the chapter will survey a selection of recent results that bring the relevant issues to light. The reader is referred to the appropriate full journal articles for all of the relevant details and careful placement of these projects in context with the works of others. Space limitations also do not allow this brief chapter to adequately review the various directions all my friends and colleagues are taking in this field. Rather than attempt such a review and fail, I refer the reader to a host of other recent reviews which, taken together, themselves span only a fraction of the current work being done in this active area of research. For an experimental perspective, the reader is referred to several other chapters in this handbook, and to the recent reviews of Ho and Tai (1996; 1998), McMichael (1996), Gad-el-Hak (1996) and Löfdahl and Gad-el-Hak (1999). For a mathematical perspective, the reader is referred to the recent dedicated volumes compiled by Banks (1992), Banks et al. (1993), Gunzburger (1995), Lagnese et al. (1995) and Sritharan (1998) for a sampling of recent results in this area.

13.2 Linearization: Life in a Small Neighborhood

As a starting point for the introduction of control theory into the fluid-mechanical setting, we first consider the linearized system arising from the equation governing small perturbations to a laminar flow. From a physical point of view, such perturbations are quite significant, as they represent the initial stages of the complex process of transition to turbulence, and thus their mitigation or enhancement has a substantial effect on the evolution of the flow.

To be concrete, an enlightening problem that captures the essential physics of many important features of both transition and turbulence in wall-bounded flows is that of plane channel flow, as illustrated in Figure 13.1. Without loss of generality, we assume the walls are located at $y = \pm 1$. We begin our study by

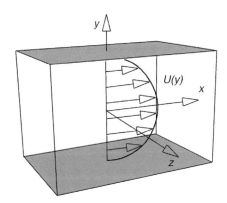

FIGURE 13.1 Geometry of plane channel flow. The flow is sustained by an externally applied pressure gradient in the x direction. This canonical problem provides an excellent testbed for the study of both transition and turbulence in wall-bounded flows. Note that many of the important flow phenomena in this geometry, in both the linear and nonlinear setting, are fundamentally three dimensional. A nonphysical assumption of periodicity of the flow perturbations in the x and z directions is often assumed for numerical convenience, with the box size chosen to be large enough that this nonphysical assumption has minimal effect on the observed flow statistics. It is important to evaluate critically the implications of such assumptions during the process of control design, as discussed in detail in Sections 13.4 and 13.5.

analyzing small perturbations $\{u, v, w, p\}$ to the (parabolic) laminar flow profile $U(y)$ in this geometry, which are governed by the linearized incompressible Navier–Stokes equation:

$$\frac{\partial u}{\partial x} + \frac{\partial v}{\partial y} + \frac{\partial w}{\partial z} = 0, \tag{13.1a}$$

$$\dot{u} + U\frac{\partial}{\partial x}u + U'v = -\frac{\partial p}{\partial x} + \frac{1}{\mathrm{Re}}\Delta u, \tag{13.1b}$$

$$\dot{v} + U\frac{\partial}{\partial x}v = -\frac{\partial p}{\partial y} + \frac{1}{\mathrm{Re}}\Delta v, \tag{13.1c}$$

$$\dot{w} + U\frac{\partial}{\partial x}w = -\frac{\partial p}{\partial z} + \frac{1}{\mathrm{Re}}\Delta w. \tag{13.1d}$$

Equation (13.1a), the continuity equation, constrains the solution of Eqs. (13.1b) to (13.1d), the momentum equations, to be divergence free. This constraint is imposed through the ∇p terms in the momentum equations, which act as Lagrange multipliers to maintain the velocity field on a divergence-free submanifold of the space of square-integrable vector fields. In the discretized setting, such systems are called *descriptor systems* or *differential-algebraic equations* and, defining a state vector $\mathbf{x}$ and a control vector $\mathbf{u}$, may be written in the generalized state-space form:

$$E\dot{\mathbf{x}} = A\mathbf{x} + B\mathbf{u}. \tag{13.2}$$

Note that if the Navier–Stokes equation (13.1) is put directly into this form, E is singular. This is an essential feature of the Navier–Stokes equation that necessitates careful treatment in both simulation and control design in order to avoid spurious numerical artifacts. A variety of techniques exist to express the system of equations (13.1) with a reduced set of variables or spatially distributed functions with only two degrees of freedom per spatial location, referred to as a *divergence-free basis*. In such a basis, the continuity equation is applied implicitly, and the pressure is eliminated from the set of governing equations. All three velocity components and the pressure (up to an arbitrary constant) may be determined from solutions represented in such a basis. When discretized and represented in the form of Eq. (13.2), the Navier–Stokes equation written in such a basis leads to an expression for E that is nonsingular.

For the geometry indicated in Figure 13.1, a suitable choice for this reduced set of variables, which is convenient in terms of the implementation of boundary conditions, is the wall-normal velocity, v, and the wall-normal vorticity, $\omega \triangleq \partial u/\partial z - \partial w/\partial x$. Taking the Fourier transform of Eq. (13.1) in the streamwise and spanwise directions and manipulating these equations and their derivatives leads to the classical Orr–Sommerfeld/Squire formulation of the Navier–Stokes equation at each wavenumber pair $\{k_x, k_z\}$:

$$\hat{\Delta}\dot{\hat{v}} = \{-ik_x U\hat{\Delta} + ik_x U'' + \hat{\Delta}(\hat{\Delta}/\mathrm{Re})\}\hat{v}, \tag{13.3a}$$

$$\dot{\hat{\omega}} = \{-ik_z U'\}\hat{v} + \{-ik_x U + \hat{\Delta}/\mathrm{Re}\}\hat{\omega}, \tag{13.3b}$$

where the hats (^) indicate Fourier coefficients and the Laplacian now takes the form $\hat{\Delta} \triangleq \partial^2/\partial y^2 - k_x^2 - k_z^2$. Note that particular care is needed when solving this system; in order to invert the Laplacian on the LHS of Eq. (13.3a), the boundary conditions on v must be accounted for properly. By manipulation of the governing equations and casting them in a derivative form, we effectively trade one numerical difficulty (singularity of E) for another (a tricky boundary condition inclusion to make the Laplacian on the LHS of Eq. (13.3a) invertible).

Note the spatially invariant structure of the present geometry: every point on each wall is, statistically speaking, identical to every other point on that wall. Canonical problems with this sort of spatially invariant structure in one or more directions form the backbone of much of the literature on flow transition and turbulence. It is this structure that facilitates the use of Fourier transforms to completely decouple the system state $\{\hat{v}, \hat{\omega}\}$ at each wavenumber pair $\{k_x, k_z\}$ from the system state at every other wavenumber pair, as indicated in Eq. (13.3). Such decoupling of the Fourier modes of the unforced linear system in the directions of spatial invariance is a classical result upon which much of the available linear theory for the stability of Navier–Stokes systems is based. As noted by Bewley and Agarwal (1996), taking the Fourier transform of both the control variables and the measurement variables maintains this system decoupling in the control formulation, greatly reducing the complexity of the control design problem to several smaller, completely decoupled control design problems at each wavenumber pair $\{k_x, k_z\}$, each of which requires spatial discretization in the y direction only.

Once a tractable form of the governing equation has been selected, in order to pose the flow control problem completely, several steps remain:

- the state equation must be spatially discretized,
- boundary conditions must be chosen and enforced,
- the variables representing the controls and the available measurements must be identified and extracted,
- the disturbances must be modeled, and
- the "control objective" must be precisely defined.

To identify a fundamental yet physically relevant flow control problem, the decisions made at each of these steps require engineering judgment. Such judgment is based on physical insight concerning the flow system to be controlled and how the essential features of such a system may be accurately modeled. An example of how to accomplish these steps is described in some detail by Bewley and Liu (1998). In short, we may choose:

- a Chebyshev spatial discretization in y,
- no-slip boundary conditions ($u = w = 0$ on the walls) with the distribution of v on the walls (the blowing/suction profile) prescribed as the control,
- skin friction measurements distributed on the walls,
- idealized disturbances exciting the system, and
- an objective of minimizing flow perturbation energy.

As we learn more about the physics of the system to be controlled, there is significant room for improvement in this problem formulation, particularly in modeling the structure of relevant system disturbances and in the precise statement of the control objective.

Once the above-mentioned steps are complete, the present decoupled system at each wavenumber pair $\{k_x, k_z\}$ may finally be manipulated into the standard state-space form:

$$\dot{\mathbf{x}} = A\mathbf{x} + B_1\mathbf{w} + B_2\mathbf{u},$$
$$\mathbf{y} = C_2\mathbf{x} + D_{21}\mathbf{w}, \qquad (13.4)$$

with

$$B_1 \triangleq (G_1 \quad 0), \qquad C_2 \triangleq G_2^{-1} C, \qquad D_{21} \triangleq (0 \quad \alpha I), \qquad \mathbf{w} \triangleq \begin{pmatrix} \mathbf{w}_1 \\ \mathbf{w}_2 \end{pmatrix},$$

where **x** denotes the state, **u** denotes the control, **y** denotes the available measurements (scaled as discussed below), and **w** accounts for the external disturbances (including the state disturbances $\mathbf{w}_1$ and the measurement noise $\mathbf{w}_2$, scaled as discussed below). Note that $C\mathbf{x}$ denotes the raw vector of measured variables, and G_1 and αG_2 represent the square root of any known or expected covariance structure of the state disturbances and measurement noise, respectively. The scalar α^2 is identified as an adjustable parameter that defines the ratio of the maximum singular value of the covariance of the measurement noise divided by the maximum singular value of the covariance of the state disturbances; w.l.o.g., we take $\bar{\sigma}(G_1) = \bar{\sigma}(G_2) = 1$. Effectively, the matrix G_1 reflects which state disturbances are strongest, and the matrix G_2 reflects which measurements are most corrupted by noise. Small α implies relatively high overall confidence in the measurements, whereas large α implies relatively low overall confidence in the measurements.

Not surprisingly, there is a wide body of theory surrounding how to control a linear system in the standard form of Eq. (13.4). The application of one popular technique (to a related two-dimensional problem), called *proportional–integral* (PI) *control* and generally referred to as "classical" control design, is presented in Joshi et al. (1997). The application of another technique, called $\mathcal{H}_\infty$ control and generally referred to as "modern" control design, is laid out in Bewley and Liu (1998), hereafter referred to as BL98. The application of a related modern control strategy (to the two-dimensional problem), called *loop transfer recovery* (LTR), is presented in Cortelezzi and Speyer (1998). More recent publications by these groups further extend these seminal efforts.

It is useful, to some extent, to understand the various theoretical implications of the control design technique chosen. Ultimately, however, flow control boils down to the *design* of a control that achieves the desired engineering objective (transition delay, drag reduction, mixing enhancement, etc.) to the maximum extent possible. The theoretical implications of the particular control technique chosen are useful only to the degree to which they help attain this objective. Engineering judgment, based both on an understanding of the merits of the various control theories and on the suitability of such theories to the structure of the fluid-mechanical problem of interest, guides the selection of an appropriate control design strategy. In the following section, we summarize the $\mathcal{H}_\infty$ control design approach, illustrate why this approach is appropriate for the structure of the problem at hand, and highlight an important distinguishing characteristic of the present system when controls computed via this approach are applied.

13.3 Linear Stabilization: Leveraging Modern Linear Control Theory

As only a limited number of noisy measurements **y** of the state **x** are available in any practical control implementation, it is beneficial to develop a filter that extracts as much useful information as possible from the available flow measurements before using this filtered information to compute a suitable control. In modern control theory, a model of the system itself is used as this filter, and the filtered information extracted from the measurements is simply an estimate of the state of the physical system. This intuitive framework is illustrated schematically in Figure 13.2. By modeling (or neglecting) the influence of the unknown disturbances in Eq. (13.4), the system model takes the form:

$$\dot{\hat{\mathbf{x}}} = A\hat{\mathbf{x}} + B_1\hat{\mathbf{w}} + B_2\mathbf{u} - \mathbf{v}, \qquad (13.5a)$$

$$\hat{\mathbf{y}} = C_2\hat{\mathbf{x}} + D_{21}\hat{\mathbf{w}}, \qquad (13.5b)$$

where $\hat{\mathbf{x}}$ is the state estimate, $\hat{\mathbf{w}}$ is a disturbance estimate, and **v** is a feedback term based on the difference between the measurement of the state, **y**, and the corresponding quantity in the model, $\hat{\mathbf{y}}$, such that:

$$\mathbf{v} = L(\mathbf{y} - \hat{\mathbf{y}}). \qquad (13.5c)$$

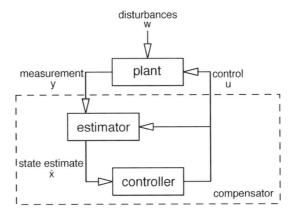

FIGURE 13.2 Flow of information in a modern control realization. The plant, forced by external disturbances, has an internal state **x** which cannot be observed. Instead, a noisy measurement **y** is made, with which a state estimate $\hat{\mathbf{x}}$ is determined. This state estimate is then used to determine the control **u** to be applied to the plant to regulate **x** to zero. Essentially, the full equation for the plant (or a reduced model thereof) is used in the estimator as a filter to extract useful information about the state from the available measurements.

The control **u**, in turn, is based on the state estimate $\hat{\mathbf{x}}$ such that:

$$\mathbf{u} = K\hat{\mathbf{x}}. \tag{13.6}$$

Equation (13.4) is referred to as the "plant," (13.5) is referred to as the "estimator," and (13.6) is referred to as the "controller." The estimator, Eq. (13.5), and the controller, Eq. (13.6), taken together, will be referred to as the "compensator." The problem at hand is to compute linear time-invariant (LTI) matrices K and L and some estimate of the disturbance, $\hat{\mathbf{w}}$, such that:

1. the estimator feedback **v** forces $\hat{\mathbf{x}}$ towards **x**, and
2. the controller feedback **u** forces **x** towards zero,

even as unknown disturbances **w** both disrupt the system evolution and corrupt the available measurements of the system state.

13.3.1 The $\mathcal{H}_\infty$ Approach to Control Design

Several textbooks describe in detail how the $\mathcal{H}_\infty$ technique determines K, L, and $\hat{\mathbf{w}}$ for systems of the form Eqs. (13.4) to (13.6) in the presence of structured and unstructured disturbances **w**. The reader is referred to the seminal paper by Doyle et al. (1989), the more accessible textbook by Green and Limebeer (1995), and the more advanced texts by Zhou et al. (1996) and Zhou and Doyle (1998) for derivation and further discussion of these control theories, and to BL98 for an extended discussion in the context of the present problem. To summarize this approach briefly, a cost function $\mathcal{J}$ describing the control problem at hand is defined that weighs together the state **x**, the control **u**, and the disturbance **w** such that:

$$\mathcal{J} \triangleq E[\mathbf{x}^*Q\mathbf{x} + \ell^2 \mathbf{u}^*\mathbf{u} - \gamma^2 \mathbf{w}^*\mathbf{w}] \triangleq E[\mathbf{z}^*\mathbf{z} - \gamma^2 \mathbf{w}^*\mathbf{w}], \tag{13.7a}$$

where

$$\mathbf{z} \triangleq C_1 \mathbf{x} + D_{12}\mathbf{u}, \quad C_1 \triangleq \begin{pmatrix} Q^{1/2} \\ 0 \end{pmatrix}, \quad D_{12} \triangleq \begin{pmatrix} 0 \\ \ell I \end{pmatrix}. \tag{13.7b}$$

The matrix Q shaping the dependence on the state in the cost function, $\mathbf{x}^*Q\mathbf{x}$, may be selected to numerically approximate any of a variety of physical properties of the flow, such as the flow perturbation energy, its enstrophy, the mean square of the drag measurements, etc. The matrix Q may also be biased to place extra penalty on flow perturbations in a specific region in space of particular physical significance. The choice of Q has a profound effect on the final closed-loop behavior, and it must be selected with care. Based on our numerical tests to date, cost functions related to the energy of the flow perturbations have been the most successful for the purpose of transition delay. To simplify the algebra that follows, we have set the matrices R and S shaping the $\mathbf{u}^*R\mathbf{u}$ and $\mathbf{w}^*S\mathbf{w}$ terms in the cost function equal to I. As shown in Lauga and Bewley (2000), it is straightforward to generalize this result to other positive-definite choices for R and S. Such a generalization is particularly useful when designing controls for a discretization of a partial differential equation (PDE) in a consistent manner such that the feedback kernels converge to continuous functions as the computational grid is refined.

Given the structure of the system defined in Eqs. (13.4) to (13.6) and the control objective defined in Eq. (13.7), the $\mathcal{H}_\infty$ compensator is determined by simultaneously minimizing the cost function $\mathcal{J}$ with respect to the control $\mathbf{u}$ and maximizing $\mathcal{J}$ with respect to the disturbance $\mathbf{w}$. In such a way, a control $\mathbf{u}$ is found that maximally attains the control objective even in the presence of a disturbance $\mathbf{w}$ that maximally disrupts this objective. For *sufficiently large* γ and a system that is both *stabilizable* and *detectable* via the controls and measurements chosen,[1] this results in finite values for $\mathbf{u}$, $\mathbf{v}$, and $\mathbf{w}$, the magnitudes of which may be adjusted by variation of the three scalar parameters ℓ, α, and γ, respectively. Reducing ℓ, modeling the "price of the control" in the engineering design, generally results in increased levels of control feedback $\mathbf{u}$. Reducing α, modeling the "relative level of corruption" of the measurements by noise, generally results in increased levels of estimator feedback $\mathbf{v}$. Reducing γ, modeling the "price" of the disturbance to Nature (in the spirit of a noncooperative game), generally results in increased levels of disturbances $\mathbf{w}$ of maximally disruptive structure to be accounted for during the design of the compensator.

The $\mathcal{H}_\infty$ control solution [Doyle et al., 1989] may be described as follows: a compensator that minimizes $\mathcal{J}$ in the presence of that disturbance which simultaneously maximizes $\mathcal{J}$ is given by:

$$K = -\frac{1}{\ell^2}B_2^*X, \qquad L = -\frac{1}{\alpha^2}ZYC_2^*, \qquad \hat{\mathbf{w}} = \frac{1}{\gamma^2}B_1^*X\hat{\mathbf{x}}, \tag{13.8}$$

where

$$X = \operatorname{Ric}\begin{pmatrix} A & \frac{1}{\gamma^2}B_1B_1^* - \frac{1}{\ell^2}B_2B_2^* \\ -C_1^*C_1 & -A^* \end{pmatrix},$$

$$Y = \operatorname{Ric}\begin{pmatrix} A^* & \frac{1}{\gamma^2}C_1^*C_1 - \frac{1}{\alpha^2}C_2^*C_2 \\ -B_1B_1^* & -A \end{pmatrix},$$

$$Z = \left(1 - \frac{YX}{\gamma^2}\right)^{-1},$$

where $\operatorname{Ric}(\cdot)$ denotes the positive-definite solution of the associated Riccati equation [Laub, 1991].[2] The simple structure of the above solution, and its profound implications in terms of the performance and

---

[1] Further description of these important technical requirements for solvability of the control problem is deferred to the above-mentioned texts. These requirements are easily met in many practical settings.

[2] Note that, for the control problem to be soluble, γ must be sufficiently large so that: (a) X and Y may be found that are positive definite, and (b) $\rho(XY) < \gamma^2$, where $\rho(\cdot)$ denotes the spectral radius. An approximate lower bound on γ that meets these conditions, denoted γ_0, may be determined by trial and error.

robustness of the resulting closed-loop system, is one of the most elegant results of linear control theory. We comment below on a few of the more salient features of this result.

Algebraic manipulation of Eqs. (13.4) to (13.8) leads to the closed-loop form:

$$\dot{\tilde{\mathbf{x}}} = \tilde{A}\tilde{\mathbf{x}} + \tilde{B}\mathbf{w},$$
$$\mathbf{z} = \tilde{C}\tilde{\mathbf{x}},$$
(13.9)

where

$$\tilde{\mathbf{x}} = \begin{pmatrix} \mathbf{x} \\ \mathbf{x} - \hat{\mathbf{x}} \end{pmatrix},$$

$$\tilde{A} = \begin{pmatrix} A + B_2 K & -B_2 K \\ -\gamma^{-2} B_1 B_1^* & A + LC_2 + \gamma^{-2} B_1 B_1^* \end{pmatrix},$$

$$\tilde{B} = \begin{pmatrix} B_1 \\ B_1 + LD_{21} \end{pmatrix},$$

$$\tilde{C} = (C_1 + D_{12}K \quad -D_{12}K).$$

Taking the Laplace transform of Eq. (13.9), it is easy to define the transfer function $T_{zw}(s)$ from $\mathbf{w}(s)$ to $\mathbf{z}(s)$ (the Laplace transforms of $\mathbf{w}$ and $\mathbf{z}$) such that:

$$\mathbf{z}(s) = \tilde{C}(sI - \tilde{A})^{-1}\tilde{B}\mathbf{w}(s) \triangleq T_{zw}(s)\mathbf{w}(s).$$

Norms of the system transfer function $T_{zw}(s)$ quantify how the system output of interest, $\mathbf{z}$, responds to disturbances $\mathbf{w}$ exciting the closed-loop system.

The expected value of the root mean square (rms) of the output $\mathbf{z}$ over the rms of the input $\mathbf{w}$ for disturbances $\mathbf{w}$ of maximally disruptive structure is denoted by the ∞–norm of the system transfer function,

$$\|T_{zw}\|_\infty \triangleq \sup_\omega \bar{\sigma}[T_{zw}(j\omega)].$$

$\mathcal{H}_\infty$ control is often referred to as "robust" control, as $\|T_{zw}\|_\infty$, reflecting the worst-case amplification of disturbances by the system from the input $\mathbf{w}$ to the output $\mathbf{z}$, is in fact bounded from above by the value of γ used in the problem formulation. Subject to this ∞–norm bound, $\mathcal{H}_\infty$ control minimizes the expected value of the rms of the output $\mathbf{z}$ over the rms of the input $\mathbf{w}$ for white Gaussian disturbances $\mathbf{w}$ with identity covariance, denoted by the 2–norm of the system transfer function:

$$\|T_{zw}\|_2 \triangleq \left(\frac{1}{2\pi} \int_{-\infty}^{\infty} \text{trace}[T_{zw}(j\omega)^* T_{zw}(j\omega)] d\omega \right)^{1/2}.$$

Note that $\|T_{zw}\|_2$ is often cited as a measure of performance of the closed-loop system, whereas $\|T_{zw}\|_\infty$ is often cited as a measure of its robustness. Further motivation for consideration of control theories related to these particular norms is elucidated by Skogestad and Postlethwaite (1996). Efficient numerical algorithms to solve the Riccati equations for X and Y in the compensator design and to compute the transfer function norms $\|T_{zw}\|_2$ and $\|T_{zw}\|_\infty$ quantifying the closed-loop system behavior are well developed, and are discussed further in the standard texts.

Note that, for high-dimensional discretizations of infinite dimensional systems, it is not feasible to perform a parametric variation on the individual elements of the matrices defining the control problem. The control design approach taken here represents a balance of engineering judgment in the construction of the matrices defining the structure of the control problem, $\{B_1, B_2, C_1, C_2\}$, and parametric variation of the three scalar parameters involved, $\{\ell, \alpha, \gamma\}$, in order to achieve the desired trade-offs between performance, robustness, and the control effort required. This approach retains a sufficient but not excessive degree of flexibility in the control design process. In general, intermediate values of the three parameters $\{\ell, \alpha, \gamma\}$ are found to lead to the most suitable control designs.

$\mathcal{H}_2$ control (also known as linear quadratic Gaussian control, or LQG) is an important limiting case of $\mathcal{H}_\infty$ control. It is obtained in the present formulation by relaxing the bound γ on the infinity norm of the closed-loop system, taking the limit as $\gamma \to \infty$ in the controller formulation. Such a control formulation focuses solely on performance (i.e., minimizing $\|T_{zw}\|_2$). As LQG does not provide any guarantees about system behavior for disturbances of particularly disruptive structure ($\|T_{zw}\|_\infty$), it is often referred to as "optimal" control. Though one might confirm *a posteriori* that a particular LQG design has favorable robustness properties, such properties are not guaranteed by the LQG control design process. When designing a large number of compensators for an entire array of wavenumber pairs $\{k_x, k_z\}$ via an automated algorithm, as is necessary in the current problem, it is useful to have a control design tool that inherently builds in system robustness, such as $\mathcal{H}_\infty$. For isolated low-dimensional systems, as often encountered in many industrial processes, *a posteriori* robustness checks on hand-tuned LQG designs are often sufficient.

It is also interesting to note that certain favorable robustness properties may be assured by the LQG approach by strategies involving either:

1. setting $B_1 = (B_2 \ \ 0)$ and taking $\alpha \to 0$, or
2. setting $C_1 = \binom{C_2}{0}$ and taking $\ell \to 0$.

These two approaches are referred to as loop transfer recovery (LQG/LTR), and are further explained in Stein and Athans (1987). Such a strategy is explored by Cortelezzi and Speyer (1998) in the two-dimensional setting of the current problem. In the present system, both B_2 and C_2 are very low rank, as there is only a single control variable and a single measurement variable at each wall in the Fourier-space representation of the physical system at each wavenumber pair $\{k_x, k_z\}$. However, the state itself is a high-dimensional approximation of an infinite-dimensional system. It is beneficial in such a problem to allow the modeled state disturbances $\mathbf{w}_1$ to input the system, via the matrix B_1, at more than just the actuator inputs, and to allow the response of the system $\mathbf{x}$ to be weighted in the cost function, via the matrix C_1, at more than just the sensor outputs. The LQG/LTR approach of assuring closed-loop system robustness, however, requires us to sacrifice one of these features in the control formulation, in addition to taking $\alpha \to 0$ or $\ell \to 0$, in order to apply one of the two strategies listed above. It is noted here that the $\mathcal{H}_\infty$ approach, when soluble, allows for the design of compensators with inherent robustness guarantees without such sacrifices of flexibility in the definition of the control problem of interest, thereby giving significantly more latitude in the design of a "robust" compensator.

The names $\mathcal{H}_2$ and $\mathcal{H}_\infty$ are derived from the system norms $\|T_{zw}\|_2$ and $\|T_{zw}\|_\infty$ that these control theories address, with the symbol $\mathcal{H}$ denoting the particular "Hardy space" in which these transfer function norms are well defined. It deserves mention that the difference between $\|T_{zw}\|_2$ and $\|T_{zw}\|_\infty$ might be expected to be increasingly significant as the dimension of the system is increased. Neglecting, for the moment, the dependence on ω in the definition of the system norms, the matrix Frobenius norm, $(\text{trace}[T^*T]^{1/2})$, and the matrix 2–norm, $\bar{\sigma}[T]$, are "equivalent" up to a constant. Indeed, for scalar systems these two matrix norms are identical, and for low-dimensional systems their ratio is bounded by a constant related to the dimension of the system. For high-dimensional discretizations of infinite-dimensional systems, however, this norm equivalence is relaxed, and the differences between these two matrix norms may be substantial. The temporal dependence of the two system norms $\|T_{zw}\|_2$ and $\|T_{zw}\|_\infty$ distinguishes them even for low-dimensional systems; the point here is only that, for high-dimensional systems, the important differences between these two system norms is even more pronounced, and control techniques such as $\mathcal{H}_\infty$ that account for both such norms might prove to be beneficial. Techniques (such as $\mathcal{H}_\infty$) that bound $\|T_{zw}\|_\infty$ are especially appropriate for the present problem, as transition is often associated with the triggering of a "worst-case" phenomenon, which is well characterized by this measure.

13.3.2 Advantages of Modern Control Design for Non-Normal Systems

Matrices A arising from the discretization of systems in fluid mechanics are often highly "non-normal," which means that the eigenvectors of A are highly nonorthogonal. This is especially true for transition in a plane channel. Important characteristics of this system, such as $O(1000)$ transient energy growth and large amplification of external disturbance energy in stable flows at subcritical Reynolds numbers, cannot be explained by examination of its eigenvalues alone. Discretizations of Eq. (13.3), when put into the state-space form of Eq. (13.4), lead to system matrices of the form:

$$A = \begin{pmatrix} L & 0 \\ C & S \end{pmatrix}. \tag{13.10}$$

For certain wavenumber pairs (specifically, those with $k_x \approx 0$ and $k_z = O(1)$), the eigenvalues of A are real and stable, the matrices L and S are quite similar in structure, and $\bar{\sigma}(C)$ is disproportionately large.

In order to illustrate the behavior of a system matrix with such structure, consider a reduced system matrix of the above form but where L, C, and S are scalars. Specifically, compare the two stable closed-loop system matrices:

$$A_1 = \begin{pmatrix} -0.01 & 0 \\ 0 & -0.011 \end{pmatrix},$$

$$A_2 = \begin{pmatrix} -0.01 & 0 \\ 1 & -0.011 \end{pmatrix}.$$

Both matrices have the same eigenvalues. However, the eigenvectors of A_1 are orthogonal, whereas the eigenvectors of A_2 are

$$\xi_1 = \begin{pmatrix} 0.001 \\ 1.000 \end{pmatrix} \quad \text{and} \quad \xi_2 = \begin{pmatrix} 0 \\ 1.000 \end{pmatrix}.$$

Even though its eigenvalues differ by 10%, the eigenvectors of A_2 are less than $0.06°$ from being exactly parallel. It is in this sense that we define this system as being "non-normal" or "nearly defective."[3] This severe nonorthogonality of the system eigenvectors is a direct result of the disproportionately large coupling term C. Compensators that reduce C will make the eigenvectors of A_2 closer to orthogonal without necessarily changing the system eigenvalues.

The consequences of nonorthogonality of the system eigenvectors are significant. Though the "energy" (the Euclidean norm) of the state of the system $\dot{x} = A_1 x$ uniformly decreases in time from all initial conditions, the "energy" of the state of the system $\dot{x} = A_2 x$ from the initial condition $x(0) = \xi_1 - \xi_2$ grows by a factor of over a thousand before eventually decaying due to the stability of the system. This is referred to as the *transient energy growth* of the stable non-normal system, and is a result of the reduced destructive interference exhibited by the two modes of the solution as they decay at different rates. In fluid mechanics, transient energy growth is thought to be an important linear mechanism leading to transition in subcritical flows, which are linearly stable but nonlinearly unstable [Butler and Farrell, 1992].

The excitation of such systems by external disturbances is well described in terms of the system norms $\|T_{zw}\|_2$ and $\|T_{zw}\|_\infty$, which (as described previously) quantify the rms amplification of Gaussian and worst-case disturbances by the system. For example, consider a closed-loop system of the form of Eq. (13.9) with $\tilde{B} = \tilde{C} = I$. Taking the system matrix $\tilde{A} = A_1$, the norms of the system transfer function are $\|T_{zw}\|_2 = 9.8$ and $\|T_{zw}\|_\infty = 100$. On the other hand, taking the system matrix $\tilde{A} = A_2$, the 2-norm of the

---

[3] Though this definition is dependent on the coordinate system and norm chosen to define the orthogonality of the eigenvectors, the physical systems we will consider lend themselves naturally to preferred norm definitions motivated by the energetics of the system. In particular, the nonlinear terms of the Navier–Stokes equation (which are neglected in the present linear analysis) are orthogonal only under certain norms related to the kinetic energy of the system at hand, suggesting a natural, physically motivated choice of norm for the systems we will consider.

system transfer function is 48 times larger and the ∞–norm is 91 times larger, though the two systems have identical closed-loop eigenvalues. Large system-transfer-function norms and large values of maximum transient energy growth are often highly correlated, as they both come about due to nonnormality in a stable system.

Graphical interpretations of $\|T_{zw}\|_2$ and $\|T_{zw}\|_\infty$ for the present channel flow system are given in Figures 13.3 and 13.4 by examining contour plots of the appropriate matrix norms of $T_{zw}(s)$ in the complex plane s. Recall that $T_{zw}(s) \triangleq \tilde{C}\,(sI - \tilde{A})^{-1}\tilde{B}$, so these contours approach infinity in the neighborhood of each eigenvalue of $\tilde{A}$. Contour plots of this type have recently become known as the *pseudospectra* of an input/output system and have become a popular generalization of plots of the eigenvalues of $\tilde{A}$ in recent efforts to study nonnormality in uncontrolled fluid systems [Trefethen et al., 1993]. For the open-loop systems depicted in these figures, we define $\tilde{A} = A$, $\tilde{B} = B_1$, and $\tilde{C} = C_1$. The severe non-normality of the present fluid system for Fourier modes with $k_x \approx 0$ is reflected by the elliptical isolines surrounding each pair of eigenvalues with nearly parallel eigenvectors in these pseudospectra, a feature that is much more pronounced in the system depicted in Figure 13.3 than in that depicted in Figure 13.4. The severe non-normality of the system depicted in Figure 13.3 is also reflected by its much larger value of $\|T_{zw}\|_\infty$.

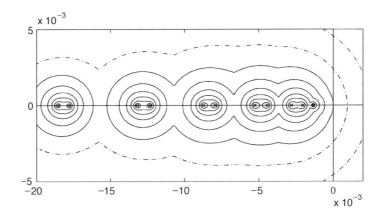

(a) Isocontours of $\bar{\sigma}[T_{zw}(s)]$ in the complex plane s. The peak value of this matrix norm on the $j\omega$ axis is defined as the system norm $\|T_{zw}\|_\infty$ and corresponds to the solid isoline with the smallest value.

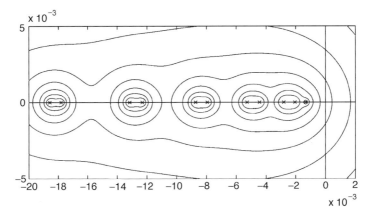

(b) Isocontours of $(\text{trace}[T^*T])^{1/2}$ in the complex plane s. The system norm $\|T_{zw}\|_2$ is related to the integral of the square of this matrix norm over the $j\omega$ axis.

FIGURE 13.3 Graphical interpretations (a.k.a. "pseudospectra") of the transfer function norms $\|T_{zw}\|_\infty$ (a) and $\|T_{zw}\|_2$ (b) for the present system in open loop, obtained at $k_x = 0$, $k_z = 2$, and $Re = 5000$. The eigenvalues of the system matrix A are marked with an ×. All isoline values are separated by a factor of 2, and the isolines with the largest value are those nearest to the eigenvalues. For this system, $\|T_{zw}\|_\infty = 2.6 \times 10^5$.

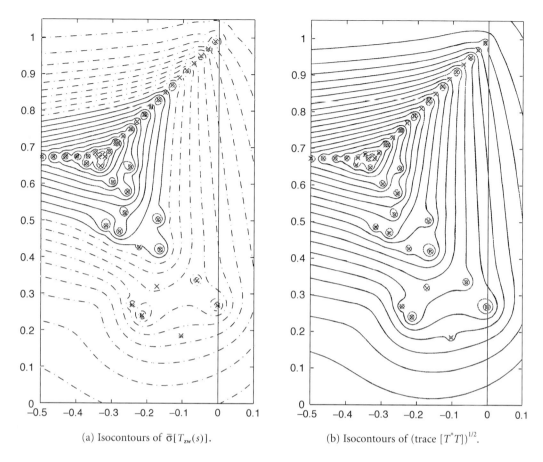

FIGURE 13.4 Pseudospectra interpretations of $\|T_{zw}\|_\infty$ (a) and $\|T_{zw}\|_2$ (b) for the open loop system at $k_x = -1$, $k_z = 0$, and $Re = 5000$. For plotting details, see Figure 13.3. For this system, $\|T_{zw}\|_\infty = 1.9 \times 10^4$.

As $\{\tilde{A}, \tilde{B}, \tilde{C}\}$ may be defined for either the open-loop or the closed-loop case, this technique for analysis of non-normality extends directly to the characterization of controlled fluid systems.

The $\mathcal{H}_\infty$ control technique is in fact based on minimizing the 2–norm of the system transfer function while simultaneously bounding the ∞–norm of the system-transfer function. In the current transition problem, our control objective is to inhibit the (linear) formation of energetic flow perturbations that can lead to nonlinear instability and transition to turbulence. It is natural that control techniques such as $\mathcal{H}_\infty$, which are designed upon the very transfer function norms that quantify the excitation of such flow perturbations by external disturbances, will have a distinct advantage for achieving this objective over control techniques that account for the eigenvalues only, such as those based on the analysis of root-locus plots.

13.3.3 Effectiveness of Control Feedback at Particular Wavenumber Pairs

The application of the modern control design approach described in Section 13.3.1 to the Orr-Sommerfeld/Squire problem laid out in Section 13.2 was explored extensively in BL98 for two particular wavenumber pairs and Reynolds numbers. The control effectiveness was quantified using several different techniques, including eigenmode analysis, transient energy growth, and transfer function norms. The control was remarkably effective and the trends with $\{\ell, \alpha, \gamma\}$ were all as expected; the reader is referred to the journal article for complete tabulation of the results. One of the most notable features of this paper is that the application of the control resulted in the closed-loop eigenvectors becoming significantly closer to orthogonal, as illustrated in Figure 13.5; note especially the high degree of correlation between the second

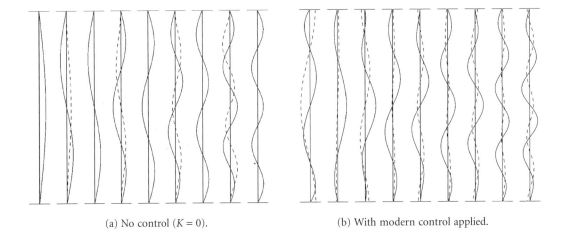

(a) No control ($K = 0$).   (b) With modern control applied.

FIGURE 13.5   The nine least stable eigenmodes of the closed-loop system matrix $A + B_2 K$ for $k_x = 0$, $k_z = 2$, and $Re = 5000$. Plotted are the nonzero part of the $\hat{\omega}$ component of the eigenvectors (solid) and the nonzero part of the $\hat{v}$ component of the eigenvectors (dashed) as a function of y from the lower wall (bottom) to the upper wall (top). In (a), the dashed line is magnified by a factor of 1000 with respect to the solid line; in (b), the dashed line is magnified by a factor of 300. Note that the eigenvectors become significantly closer to orthogonal by the application of the control. (From Bewley, T.R., and Liu, S. (1998) *J. Fluid Mech.* **365**, 305–349. Reprinted with permission of Cambridge University Press.)

and third eigenvectors of Figure 13.5a, and how this correlation is disrupted in Figure 13.5b. This was accompanied by concomitant reductions in both transient energy growth and the system transfer function norms in the controlled system. Note that the nearly parallel nature of the pairs of eigenvectors $\{\xi_2, \xi_3\}$, $\{\xi_4, \xi_5\}$, $\{\xi_6, \xi_7\}$, and $\{\xi_8, \xi_9\}$ in the uncontrolled case (Figure 13.5a) is also reflected by the elliptical isolines surrounding the corresponding eigenvalues illustrated by the pseudospectra of Figure 13.3.

Note the nonzero value of $\hat{v}$ at the walls in Figure 13.5b; this reflects the wall blowing/suction applied as the control. Note also that half of the eigenvectors in Figure 13.5a have zero $\hat{v}$ components. These are commonly referred to as the Squire modes of the system and are decoupled from the perturbations in $\hat{v}$ because of the block of zeros in the upper-right corner of A. Such decoupling is not seen in Figure 13.5b, because the closed-loop system matrix $A + B_2 K$ is full.

13.4   Decentralization: Designing for Massive Arrays

As illustrated in Figures 13.6 and 13.7, there are two possible approaches for experimental implementation of linear compensators for this problem:

1. a centralized approach, applied in Fourier space, or
2. a decentralized approach, applied in physical space.

Both of these approaches may be used to apply boundary control (such as distributions of blowing/suction) based on wall information (such as distributions of skin friction measurements). Both approaches may be used to implement the $\mathcal{H}_\infty$ compensators developed in Section 13.3, LQG/LTR compensators, PID feedback, or a host of other types of control designs. However, there are important differences in terms of the applicability of these two approaches to physical systems. The pros and cons of these approaches are now presented.

13.4.1   Centralized Approach

The centralized approach is simplest in terms of its derivation, as most linear compensators in this geometry are designed in Fourier space, leveraging the spatially invariant structure of this system mentioned previously and the complete decoupling into Fourier modes which this structure provides [Bewley

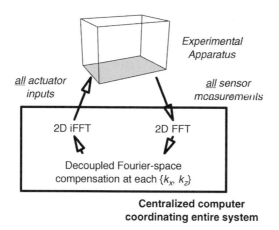

FIGURE 13.6 Centralized approach to the control of plane channel flow in Fourier space.

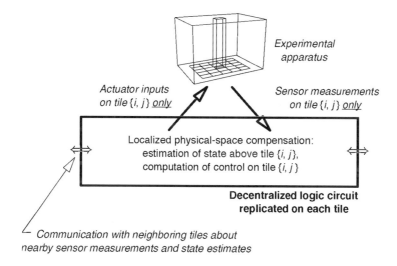

FIGURE 13.7 Decentralized approach to the control of plane channel flow in physical space.

and Agarwal, 1996]. As indicated in Figure 13.6, implementation of this approach is straightforward. This type of experimental realization was recommended by Cortelezzi and Speyer (1998) in related work. There are two major shortcomings of this approach:

1. The approach requires an online two-dimensional fast Fourier transform (FFT) of the entire measurement vector and an online two-dimensional inverse FFT (iFFT) of the entire control vector.
2. The approach assumes spatial periodicity of the flow perturbations.

With regard to point 1, it is important to note that the expense of centralized computations of two-dimensional FFTs and iFFTs will grow rapidly with the size of the array of sensors and actuators; to be specific, the computational expense is proportional to $N_x N_z \log(N_x N_z)$. This will rapidly decrease the bandwidth possible as the array size (and the number of Fourier modes) is increased for a fixed speed of the central processing unit (CPU). Communication of signals to and from the CPU is also an important limiting factor as the array size grows. Thus, this approach does not extend well to massive arrays of sensors and actuators.

With regard to point 2, it is important to note that transition phenomena in physical systems, such as boundary layers and plane channels, are not spatially periodic, though it is often useful to characterize the solutions of such systems with Fourier modes. The application of Fourier-space controllers that assume spatial periodicity in their formulation to physical systems that are not spatially periodic will be corrupted by Gibb's phenomenon, the well-known effect in which a Fourier transform is spoiled across all frequencies when the data one is transforming are not themselves spatially periodic. In order to correct for this phenomenon in formulations based on Fourier-space computations of the control, windowing functions such as the Hanning window are appropriate. Windowing functions filter the signals coming into the compensator such that they are driven to zero near the edges of the physical domain under consideration, thus artificially imposing spatial periodicity on the non-spatially-periodic measurement vector.

13.4.2 Decentralized Approach

The decentralized approach, applied in physical space, is not as convenient to derive. Riccati equations of the size of the entire discretized three-dimensional system pictured in Figure 13.1 and governed by Eq. (13.1), represented in physical space appear to be numerically intractable.

However, if such a problem could be solved, one would expect that the controller feedback kernels relating the state estimate $\hat{\mathbf{x}}$ inside the domain to the control forcing $\mathbf{u}$ at some point on the wall should decay quickly as a function of distance from the control point, as the control authority of any blowing/suction hole drilled into the wall on the surrounding flow decays rapidly with distance in a distributed viscous system.

Similarly, the estimator feedback kernels relating measurement errors $(\mathbf{y} - \hat{\mathbf{y}})$ at some point on the wall to the estimator forcing terms $\mathbf{v}$ on the system model inside the domain should decay quickly as a function of distance from the measurement point, as the correlation of any two flow-perturbation variables is known to decay rapidly with distance in a distributed viscous system.

Finally, due to the spatially invariant structure of the problem at hand, the control and estimation kernels for each sensor and actuator on the wall should be identical, though spatially shifted.

In other words, the physical-space kernels sought to determine the control and estimator feedback are spatially localized convolution kernels. If their spatial decay rate is rapid enough (e.g., exponential), then we will be able to truncate them at a finite distance from each actuator and sensor while maintaining a prescribed degree of accuracy in the feedback computation, resulting in spatially compact convolution kernels with finite support.

With such spatially compact convolution kernels, decentralized control of the present system becomes possible, as illustrated in Figure 13.7. In such an approach, several tiles are fabricated, each with sensors, actuators and an identical logic circuit. The computations on each tile are limited in spatial extent, with the individual logic circuit on each tile responsible for the (physical-space) computation of the state estimate only in the volume immediately above that tile. Each tile communicates its local measurements and state estimates with its immediate neighbors, with the number of tiles over which such information propagates in each direction depending on the tile size and spatial extent of the truncated convolution kernels. By replication, we can extend such an approach to arbitrarily large arrays of sensors and actuators. Though additional truncation of the kernels will disrupt the effectiveness of this control strategy near the edges of the array, such edge effects are limited to the edges in this case (unlike Gibbs' phenomenon) and should become insignificant as the array size is increased.

13.5 Localization: Relaxing Nonphysical Assumptions

As discussed previously, though the physical-space representation of the three-dimensional linear system is intractable in the controls setting, the (completely decoupled) one-dimensional systems at each wavenumber pair $\{k_x, k_z\}$ in the Fourier-space representation of this problem are easily managed. Remarkably, these two representations are completely equivalent. Performing a Fourier transform (which is simply a linear change of variables) of the *entire* three-dimensional system (including the state, the controls, the measurements and the disturbances) block diagonalizes *all* of the matrices involved in the three-dimensional physical-space control problem. With such block-diagonal structure, the constituent $\mathcal{H}_\infty$ control problems

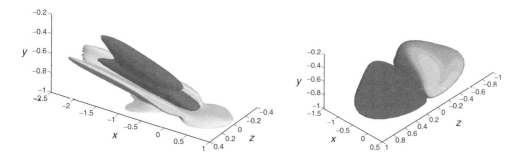

FIGURE 13.8 (Color figure follows p. 12-26.) Localized controller gains relating the state estimate $\hat{x}$ inside the domain to the control forcing **u** at the point $\{x = 0, y = -1, z = 0\}$ on the wall. Visualized are a positive and negative isosurface of the convolution kernels for (left) the wall-normal component of velocity and (right) the wall-normal component of vorticity. (From Högberg, M., and Bewley, T.R., *Automatica* (submitted). With permission from Elsevier Science.)

at each wavenumber pair $\{k_x, k_z\}$ may be solved independently and, once solved, reassembled in physical space with an inverse Fourier transform. If the numerics are handled properly, this approach is equivalent to solving the three-dimensional physical-space control problem directly.

Recent theoretical work on this problem by Bamieh et al. (2000), and related work by D'Andrea and Dullerud (2000), further support the notion that an array of $\mathcal{H}_\infty$ compensators developed at each wavenumber pair, when inverse-transformed back to the physical domain, should in fact result in spatially localized convolution kernels with exponential decay. This exponential decay, in turn, allows truncation of the kernels to any prescribed degree of accuracy. Thus, if the truncated kernels are allowed to be sufficiently large in streamwise and spanwise extent, favorable closed-loop system properties, such as robust stability and reduced system transfer function norms, may be retained. Until very recently, however, it has not been possible to obtain such kernels for Navier–Stokes systems, due to an assortment of numerical challenges.

In Högberg and Bewley (2000), spatially localized convolution kernels for both the control and estimation of plane channel flow have finally been obtained. The technique used was based on that described previously, deriving (in our initial efforts) $\mathcal{H}_2$ compensation at an array of wavenumber pairs $\{k_x, k_z\}$ and then inverse-transforming the lot, with special attention paid to the details of the control formulation and the numerical method. In particular, a numerical discretization technique not plagued by spurious eigenvalues was chosen, and the control formulation was slightly modified such that the time derivative of the blowing/suction velocities is penalized in the cost function. The resulting localized kernels are illustrated in Figures 13.8 and 13.9. Such kernels facilitate the decentralized control implementation discussed in Section 13.4.2 and depicted in Figure 13.7, paving the way for experimental implementation with massive arrays of tiles integrating sensing, actuating and the control logic.

Note that the control convolution kernels shown in Figure 13.8 angle away from the wall in the *upstream* direction. Coupled with the mean flow profile indicated in Figure 13.1, this accounts for the convective delay which requires us to anticipate flow perturbations on the interior of the domain with actuation on the wall somewhere downstream. The estimation convolution kernels shown in Figure 13.9, on the other hand, extend well *downstream* of the measurement point. This accounts for the delay between the motions of the convecting flow structures on the interior of the domain and the eventual influence of these motions on the local drag profile on the wall; during this time delay, the flow structures responsible for these motions convect downstream. Note that the upstream bias of the control kernels and the downstream bias of the estimation kernels, though physically tenable, were not prescribed in the problem formulation. *A posteriori* study of the streamwise, spanwise, and wall-normal extent, the symmetry, and the shape of such control and estimation kernels provides us with a powerful new tool with which the fundamental physics of this distributed fluid-mechanical system may be characterized.

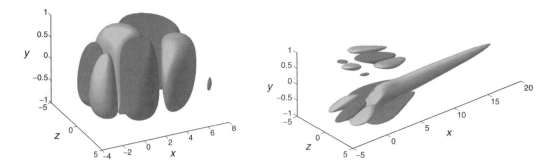

FIGURE 13.9 (Color figure follows p. 12-26.) Localized estimator gains relating the measurement error $(\mathbf{y} - \hat{\mathbf{y}})$ at the point $\{x = 0, y = -1, z = 0\}$ on the wall to the estimator forcing terms $\mathbf{v}$ inside the domain. Visualized are a positive and negative isosurface of the convolution kernels for (left) the wall-normal component of velocity and (right) the wall-normal component of vorticity. (From Högberg, M., and Bewley, T.R., *Automatica* (submitted). With permission from Elsevier Science.)

The localized convolution kernels illustrated in Figures 13.8 and 13.9 are approximately independent of the size of the computational box in which they were computed, so long as this box is sufficiently large. Thus, when implementing these kernels, we may effectively assume that they were derived in an *infinite*-sized box, relaxing the nonphysical assumption of spatial periodicity used in the problem formulation and modeling the physical situation of spatially evolving flow perturbations in a spatially invariant geometry and mean flow.

The localized convolution kernels illustrated in Figures 13.8 and 13.9 are also approximately independent of the computational mesh resolution with which they were computed, so long as this computational mesh is sufficiently fine. Indeed, a computational mesh sufficient to resolve the flow under consideration also adequately resolves these convolution kernels.

13.5.1 Open Questions

As we have shown, the framework for decentralized $\mathcal{H}_\infty$ control of the fully resolved transition problem in the geometry depicted in Figure 13.1 is now established. Obtaining spatial localization of the convolution kernels in physical space was the final remaining conceptual/numerical hurdle to be overcome. This work paves the way for decentralized application of such compensation with massive arrays of identical control tiles integrating sensing, actuation and the control logic (Figure 13.7). Though in some sense "complete," this effort has also exposed several fundamental open questions, which will now be briefly discussed.

For a given choice of the matrices $\{B_1, B_2, C_1, C_2\}$ and design parameters $\{\ell, \alpha, \gamma > \gamma_0\}$ selected, decentralized $\mathcal{H}_\infty$ compensators may be determined using the procedure described above, and performance and robustness benchmarks may be obtained via simulation. As a final step in the control design process, it is of interest to explore how much the computational effort required by the logic on each tile may be reduced without significant degradation in the closed-loop system behavior. This can lead to a significant reduction in the number of floating point operations per second required by the logic circuit on each tile. However, as will be discussed in Section 13.6, compensator reduction in the decentralized setting remains a significant unsolved problem; standard reduction strategies developed for finite, closed systems are not applicable and new research is motivated.

With the decentralized linear control framework established and prototypical numerical examples solved, we are now in a position to explore the effectiveness of compensators computed via this framework to the finite-amplitude perturbations that actually lead to transition, and to the "large" amplitude perturbations of fully developed turbulence, in the nonlinear equations of fluid motion. An extensive analytical and numerical study within this framework is underway. Issues regarding our preliminary efforts in this direction are briefly reviewed in Section 13.7. As emphasized in the introduction, such a study should be guided by an interdisciplinary perspective in order to be maximally successful. Specifically, such a study should fully incorporate the known or postulated linear mechanisms leading to transition or, in the case of turbulence, the linear mechanisms thought to be at least partially responsible for sustaining the turbulent cascade of energy. In addition, this effort motivates the development of new analytical tools that might help clarify the types of state disturbances and flow perturbations that are particularly important in such phenomena. Armed with such an understanding, large benefits might be realized in the compensator design, as the modeling of the structure of the state disturbances exciting the system, G_1, and the weighting on the flow perturbations of interest in the cost function, Q, are important design criteria. In fact, we fully expect that the transfer of information between our physical understanding of fundamental flow phenomena and our knowledge of how to control such phenomena will be a two-way transfer. Such a strategy promises to provide powerful new tools for obtaining fundamental physical understanding of classical problems in fluid mechanics as we gain new insight in how to modify these phenomena by the action of control feedback.

A host of other canonical flow control problems, including the control of spatially developing boundary layers, bluff-body flows, and free shear layers, should also be amenable to linear control application using the framework outlined here. A few such extensions are discussed briefly in Section 13.8.

13.6 Compensator Reduction: Eliminating Unnecessary Complexity

Strategies for the development of reduced-order decentralized compensators of the present form remain a key unsolved issue. With the $\mathcal{H}_2/\mathcal{H}_\infty$ approach, as described previously, a physical-space state estimate in the volume immediately above each tile must be updated online by the logic circuit on each tile as the flow evolves. However, it is not at all necessary for the compensator to compute an accurate state estimate as an intermediate variable; indeed, our only requirement is that, based on whatever filtered information the dynamic compensator does extract from the noisy system measurements, suitable controls may be determined to achieve the desired closed-loop system behavior. It should be possible to reduce substantially the complexity of the dynamic compensator and still achieve this more modest objective.

There are two possible representations in which the complexity of the compensator can be reduced: in Fourier space (where the compensator is designed) or in physical space (where the decentralized compensation is applied).

13.6.1 Fourier-Space Compensator Reduction

At any particular wavenumber pair $\{k_x, k_z\}$, there is one actuator variable at each wall, one sensor variable at each wall, and a spatial discretization in y of the state variables across the domain stretching between these walls. Due to the complete decoupling of the control problem into separate Fourier modes, the system model used in the estimator at each particular wavenumber pair is not referenced by the compensator at any other wavenumber pair. Thus, the compensators at each wavenumber pair are completely decoupled and may be reduced independently. At certain wavenumber pairs, it might be important to retain several degrees of freedom in the dynamic compensator, while at other wavenumber pairs, it might be possible to retain significantly fewer degrees of freedom without significant degradation in the closed-loop system behavior. Several existing compensator reduction strategies are well suited to this problem, and their application in this setting is straightforward. Cortelezzi and Speyer (1998) successfully applied the balanced truncation technique of open-loop model reduction in this Fourier-space framework in order to facilitate the design of a reduced-complexity dynamic compensator.

As mentioned earlier, it is the nonorthogonality of the entire set of system eigenvectors that leads to the peculiar (and important) possibilities for energy amplification in these systems, so compensator reduction techniques mindful of the relevant transfer function norms are necessary. In addition, as eloquently described by Obinata and Anderson (2000), it is most appropriate when designing low-order compensators for high-order plants to reduce the compensator while accounting for how it performs in the closed loop. An assortment of closed-loop compensator reduction techniques are now available and should be tested in future work.

In the setting of designing a decentralized compensator, there is an important shortcoming to performing standard compensator reductions in Fourier space. As the compensator reduction problem is independent at each wavenumber pair, we might be left with a different number of degrees of freedom in the reduced-order compensator at each wavenumber pair, leaving us with a dynamical system model that is impossible to inverse transform back into the physical domain. Even if we restrict the compensator reduction algorithm to reduce to the same number of degrees of freedom at each wavenumber pair (a restrictive assumption that should be unnecessary), there appears to be no appropriate strategy currently available to coordinate this reduction process across all wavenumbers in a consistent manner such that the inverse transform of the reduced dynamic model is spatially localized. Without such coordination, it seems inevitable that the ordering and representation of the various modes of this dynamic model will be scrambled during the process of compensator reduction at each wavenumber pair, resulting in an inverse-transform back in physical space that does not exhibit the spatial localization which is essential to facilitate decentralized control.

13.6.2 Physical-Space Compensator Reduction

As an alternative to Fourier-space compensator reduction, one might consider instead the reduction of the physical-space model and its associated localized convolution kernels. This has several advantages linked to the fact that this is the actual compensation to be computed on each tile. The first advantage is that spatial localization will be retained, as compensator reduction is applied after the localized kernels are obtained. Another important advantage is that this setting allows us to keep more degrees of freedom in the dynamical system model to represent streamwise and spanwise fluctuations of the state near the wall than we retain to represent the behavior of the state on the interior of the domain. This effectively relaxes the restrictive assumption referred to in the previous paragraph. Such an emphasis on resolving the state near the wall is motivated by inspection of the convolution kernels plotted in Figures 13.8 and 13.9, in which it is clear that the details of the flow near the wall are of increased importance when computing the feedback.

However, note that the system model simulated on each individual tile is not self-contained, due to the interconnections with neighboring tiles indicated in Figure 13.7. Thus, if one reduces the system model above a single tile, all neighboring tiles that reference this state estimate will be affected. As the system model is not self-contained, as it was in the Fourier-space case, existing compensator reduction approaches are not applicable.

An important observation, however, is that the structure of the system model carried by each tile is identical. Due to the repeated structure of the model represented on the array, it is sufficient to optimize the system model carried by a single tile. The repeated structure of the distributed physical-space model should make the compensator reduction problem tractable. This fundamental problem of reducing distributed, interconnected dynamic compensators in the decentralized closed-loop setting remains, as yet, unsolved.

13.6.3 Nonspatially Invariant Systems

Finally, it should be stated that the Fourier-space decoupling leveraged at the outset of this problem formulation has been one of the key ingredients that have permitted accurate solution of well-resolved canonical flow control problems to date. The linear control technique we have used to solve these control problems involves the solution of matrix Riccati equations, which are accurately soluble for state dimensions only up to $O(10^3)$. As we move to more applied flow control problems in which such Fourier-space decoupling is either more restrictive or not available, if we continue to use Riccati-based control approaches, creative new compensator reduction strategies will be required. We might need to apply "open-loop" model reduction strategies (in advance of computing the control feedback and closing the loop) in order to make manageable the dimension of the Riccati equations to be solved in the compensator formulation. As mentioned earlier, it is most appropriate when designing low-order compensators for high-order plants to reduce the compensator while accounting for how it performs in the closed loop. Unfortunately, extremely high-order discretizations of nonspatially invariant PDE systems will not likely afford us this luxury, as such systems do not decouple (via Fourier transforms) into constituent lower-order control problems amenable to matrix-based compensator design strategies.

13.7 Extrapolation: Linear Control of Nonlinear Systems

Once a decentralized linear compensator of the present form is developed, a verification of its utility for the transition problem may be obtained by applying it to the laminar flow depicted in Figure 13.1 with either finite-amplitude (but sufficiently small) initial flow perturbations and/or finite-amplitude (but sufficiently small) applied external disturbances. The resulting finite-amplitude flow perturbations are governed by the fully nonlinear Navier–Stokes equation and have been simulated in well resolved direct numerical simulations (DNS) with the code benchmarked in Bewley et al. (2001). Representative simulations are shown in Figure 13.10, indicating that linear compensators can indeed relaminarize perturbed flows that would otherwise proceed rapidly towards transition to turbulence. With the framework presented here, extensive numerical studies promise to significantly extend our fundamental understanding of the process of transition and how this process may be inhibited by control feedback.

It is also of interest to consider the application of decentralized linear compensation to the fully nonlinear problem of a turbulent flow, such as that shown in Figure 13.11. The first reason to try such an approach is simply because we can: linear control theory leads to implementable control algorithms and grants a lot of flexibility in the compensator design. Nonlinear turbulence control strategies, though currently under active development (see Sections 13.9 to 13.13), are much more difficult to design and implement and require substantial further research before they will provide implementable control strategies as flexible and powerful as those which we currently have at our disposal in the linear setting.

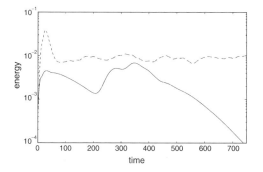
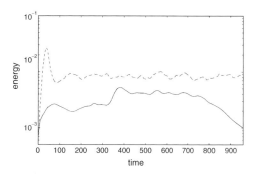

FIGURE 13.10 Evolution of oblique waves (left) and an initially random flow perturbation (right) added to a laminar flow at $Re = 2000$, with and without decentralized linear control feedback. The magnitude of the initial flow perturbations in these simulations greatly exceed the thresholds reported by Reddy et al. (1998) that lead to transition to turbulence in an uncontrolled flow (by a factor of 225 for the oblique waves and by a factor of 15 for the random initial perturbation). Solid lines indicate the energy evolution in the controlled case, dashed lines indicate the energy evolution in the uncontrolled case. Both of the uncontrolled systems lead quickly to transition to turbulence, whereas both of the controlled systems relaminarize. For the controlled cases, initial perturbations with greater energy fail to relaminarize, whereas initial perturbations with less energy relaminarize earlier. (From Högberg, M., and Bewley, T.R., *Automatica* (submitted). With permission from Elsevier Science.)

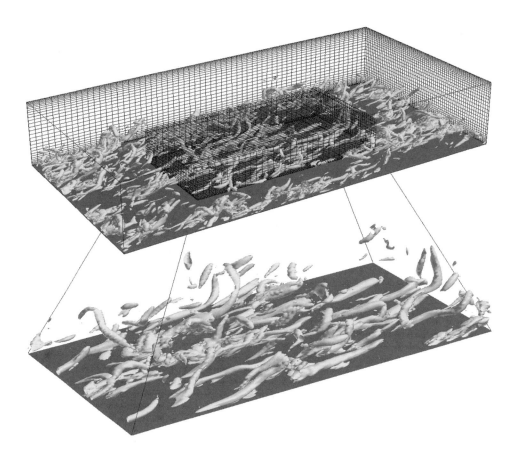

FIGURE 13.11 (Color figure follows p. 12-26.) Visualization of the coherent structures of uncontrolled near-wall turbulence at $Re_\tau = 180$. Despite the geometric simplicity of this flow (see Figure 13.1), it is phenomenologically rich and is characterized by a large range of length scales and time scales over which energy transport and scalar mixing occur. The relevant spectra characterizing these complex nonlinear phenomena are continuous over this large range of scales, thus such flows have largely eluded accurate description via dynamic models of low state dimension. The nonlinearity, the distributed nature, and the inherent complexity of its dynamics make turbulent flow systems particularly challenging for successful application of control theory. (Simulation by Bewley, T.R., Moin, P., and Temam, R. (2001) *J. Fluid Mech*. Reprinted with permission of Cambridge University Press.)

There is some evidence in the fluids literature that applying linear control feedback to turbulence might be at least partially effective. Though the significance of this result has been debated in the fluid mechanics community, Farrell and Ioannou (1993) have clearly shown that linearized Navier–Stokes systems in plane channel flows, when excited with the appropriate stochastic forcing, exhibit behavior reminiscent of the streamwise vortices and streamwise streaks that characterize actual near-wall turbulence. Whatever information the linearized Navier–Stokes equation actually contains about the mechanisms sustaining these turbulence structures, the present linear control framework (perhaps restricted to a finite horizon) should be able to exploit. Though the life cycle of the near-wall coherent structures of turbulence appears to involve important nonlinear phenomena [see, e.g., Hamilton et al., 1995], that in itself does not disqualify the utility of linear control strategies to effectively disrupt critical linear terms of this nonlinear process. Indeed, recent numerical experiments by Kim and Lim (2000) support this idea by conclusively demonstrating the importance of the coupling term C in the linearized system matrix A (see Eq. (13.10)) for maintaining near-wall turbulence in nonlinear simulations.

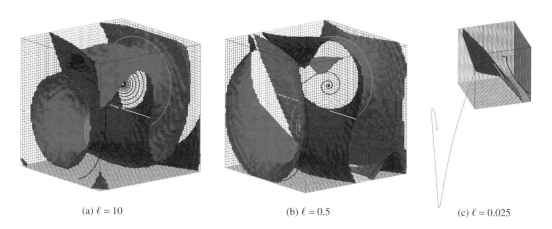

(a) $\ell = 10$ (b) $\ell = 0.5$ (c) $\ell = 0.025$

FIGURE 13.12 (Color figure follows p. 12-26.) Example of the spectacular failure of linear control theory to stabilize a simple nonlinear chaotic convection system governed by the Lorenz equation. Plotted are the regions of attraction to the desired stationary point (blue) and to an undesired stationary point (red) in the linearly controlled nonlinear system, and typical trajectories in each region (black and green, respectively). The cubical domain illustrated is $\Omega = (-25, 25)^3$ in all subfigures; for clarity, different viewpoints are used in each subfigure. (Reprinted from Bewley, T.R. (1999) *Phys. Fluids* **11**, 1169–1186. Copyright 1999, American Institute of Physics. With permission.)

In order to understand the possible pitfalls of applying linear feedback to nonlinear systems, a low-order nonlinear convection problem governed by the Lorenz equation was studied by Bewley (1999). As with the problem of turbulent channel flow, but in a low-order system easily amenable to analysis, control feedback was determined with linear control theory by linearizing the governing equation about a desired fixed point. Once a linear controller was determined by such an approach, it was then applied directly to the fully nonlinear system. The result is depicted in Figure 13.12.

For control feedback determined by linear control theory with a large weighting ℓ on the control effort, direct application of linear feedback to the full nonlinear system stabilizes both the desired state and an undesired state, indicated by the two trajectories marked in Figure 13.12a. An unstable manifold exists between these two states, indicated by the contorted surface shown. Any initial state on one side of this manifold will converge to the desired state, and any initial state on the other side of this manifold will converge to the undesired state.

As seen in Figures 13.12b and c, as the weighting on the control effort ℓ is turned down and the desired stationary state is stabilized more aggressively, the domain of convergence to the undesired stabilized state remains large. This undesired state is "aggravated" by the enhanced control feedback, moving farther from the origin. The undesired state eventually escapes to infinity for sufficiently small ℓ, indicating instability of the nonlinear system from a wide range of initial conditions even though the desired stationary point is endowed with a high degree of linear stability. Implication: strong linear stabilization of a desired system state (such as laminar flow) will not necessarily eliminate undesired nonlinear system behavior (such as turbulence) in a chaotic system.

Some form of nonlinearity in the feedback rule was required to eliminate this undesired behavior. One effective technique is to apply a switch such that the linear control feedback is turned on only when the state $\mathbf{x}(t)$ is within some sufficiently small neighborhood of the desired stabilized state $\bar{\mathbf{x}}$ in the linearly controlled system. The chaotic dynamics of the uncontrolled Lorenz system will bring the system into this neighborhood in finite time, after which control may be applied to "catch" the system at the desired equilibrium state.

Thus, even in this simple model problem, linear feedback can have a destabilizing influence if applied outside the neighborhood for which it was designed. For the full Navier–Stokes problem, though a certain set of linear feedback gains might stabilize the laminar state, on the "other side of the manifold" might lie a turbulent state aggravated by the same linear controls. Application of linear control to nonlinear chaotic systems must therefore be done with vigilance, lest nonlinearities destabilize the closed-loop system, as shown here. The easy fix found for this low-order model problem (that is, simply turn off the control until the chaotic dynamics bring the state into a neighborhood of the desired state) might not be available for the (high-dimensional) problem of turbulence, as fully turbulent flows appear to remain at all times far from the laminar state.

In our preliminary attempts at applying the decentralized compensators developed above to turbulence, we have succeeded in reducing the drag of a fully developed turbulent flow by 25% with state-feedback controllers, as shown in Figure 13.13. Interestingly, for the choice of control parameters selected here, there is no evidence of an aggravated turbulent state. A 25% drag reduction, though significant, is comparable to the drag reductions obtained with a variety of other *ad hoc* control approaches in this flow. We are actively pursuing modification of this linear control feedback to improve upon this result. Interdisciplinary considerations, such as those involved in the design of linear compensation for the problem of transition, are essential in this effort. Specifically, the (unmodeled) nonlinear terms in the Navier–Stokes equation provide insight as to the structure of the disturbances, G_1, to be accounted for in the linear control formulation in order to best compensate for their unmodeled effects. Additionally,

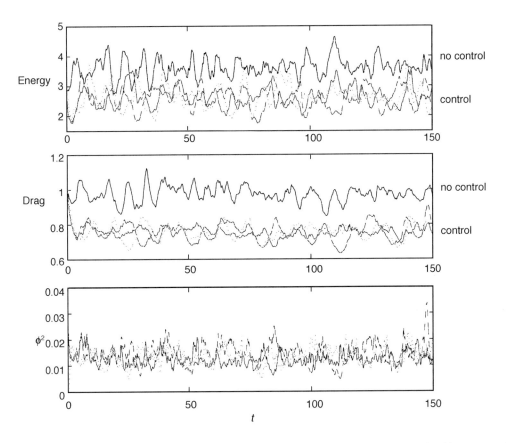

FIGURE 13.13 Evolution of fully developed turbulence at $RE_\tau = 100$ with and without decentralized linear control feedback. Note that this flow has approximately the same mass flux as the laminar flow at $Re = 2000$. (Top) Energy of flow perturbation. (Middle) Drag (note approximately 25% reduction in the controlled cases). (Bottom) Control effort used. The uncontrolled energy and drag are the (upper) solid lines in the top and middle figures. A gain scheduling approach is used to tune the control feedback gains to the instantaneous mean flow profile. (From Högberg, M., and Bewley, T.R., *Automatica* (submitted). With permission from Elsevier Science.)

the coherent structures of fully developed near-wall turbulence, believed to be a major player in the self-sustaining nonlinear process of turbulence generation near the wall, provide a phenomenological target that may be exploited in the selection of the weighting on the flow perturbations Q in the cost function.

13.8 Generalization: Extending to Spatially Developing Flows

Extension of the decentralized linear control framework developed here to a large class of slightly nonparallel flows is heuristic but straightforward. To accomplish this, the parabolic mean flow profile $U(y)$ indicated in Figure 13.1 is replaced with an appropriate "quasi-one-dimensional" profile, such as the Blasius boundary layer profile. As long as the mean flow profile evolves slowly enough in space (as compared to the wavelengths of the significant instabilities in the problem), it may be assumed to be constant in space for the purpose of developing the linear control feedback. Such an assumption of slow spatial divergence forms the foundation of the study of local and global modes used in the characterization of absolute and convective instabilities [Huerre and Monkewitz, 1990], and has proven to be a quite powerful concept. For the appropriate flows, we believe this concept is also appropriate in the context of the development of control feedback.

Implementation of the decentralized control concept in this setting is a heuristic extension of the approach presented in Figure 13.7: gradual variations in the mean flow are accounted for by local extension of the mean flow profile in the compensator derivation for each tile, gradually scaling the compensation rules from one tile to the next as the flow develops downstream. For example, we may consider developing this strategy for the laminar boundary layer (LBL) solutions of the Falkner–Skan–Cooke family, found by solving the ordinary differential equation (ODE)

$$f''' + ff'' + \beta(1 - f'^2) = 0$$

with $f(0) = f'(0) = 0$ and $f'(\infty) \to 1$ and defining

$$U = U_0 f'(\eta) \quad \text{and} \quad V = \sqrt{\frac{\nu U_0}{2x}}[\eta f'(\eta) - f(\eta)].$$

Cases of interest include the Blasius profile, modeling a zero-pressure-gradient, flat-plate LBL with

$$U_0 = U_\infty, \quad \beta = 0, \quad \eta = y\sqrt{\frac{U_0}{2\nu x}},$$

the Falkner–Skan profile, modeling a nonzero-pressure-gradient LBL or wedge flow by taking

$$U_0 = Kx^m, \quad \beta = \frac{2m}{1+m}, \quad \eta = y\sqrt{\frac{(m+1)U_0}{2\nu x}},$$

and the Falkner–Skan–Cooke profile, which models the addition of sweep to the leading edge by solving the supplemental ODE

$$g'' + fg' = 0$$

with $g(0) = 0$ and $g(\infty) \to 1$ and defining $W = W_\infty g(\eta)$. Note that the self-similarity of the LBL profiles might lead to simplified parameterizations of the convolution kernels for the control and estimation problems. Extension of this approach to a variety of other spatially developing flows (self-similar or otherwise) should also be straightforward.

13.9 Nonlinear Optimization: Local Solutions for Full Navier–Stokes

Given an idealized setting of full state information, no disturbances, and extensive computational resources, significant finite-horizon optimization problems may be formulated and (locally) solved for complex nonlinear systems using iterative, adjoint-based, gradient optimization strategies. Such optimization problems can now be solved for high-dimensional discretizations of turbulent flow systems, incorporating the full nonlinear Navier–Stokes equation, locally minimizing cost functionals representing a variety of control problems of physical interest within a given space of feasible control variables. The mathematical framework for such optimizations will be reviewed briefly in Section 13.9.1 and is described in greater detail by Bewley, Moin, and Temam (2001), hereafter referred to as BMT01.

The optimizations obtained via this approach are, strictly speaking, only "local" over the domain of feasible controls (that is, unless restrictive assumptions are made in the formulation of the control problem). Thus, the performance obtained via this approach usually cannot be guaranteed to be "globally optimal." However, the performance obtained with such nonlinear optimizations often far exceeds that possible with other control design approaches (see, for example, Figure 13.14). In addition, this approach is quite flexible, as it can be used to iteratively improve high-dimensional control distributions directly, as will be illustrated below, or, alternatively, to optimize open-loop forcing schedules, shape functions, or the coefficients of practical, implementable, and possibly nonlinear feedback control rules. Thus, interest in adjoint-based optimization strategies for turbulent flow systems goes far beyond that of establishing performance benchmarks via predictive optimizations of the control distribution itself. Establishing such benchmarks is only a first step toward a much wider range of applications for adjoint-based tools in turbulent flow systems.

The general idea of this approach, often referred to as *model predictive control*, is well motivated by comparing and contrasting it to massively parallel brute-force algorithms recently developed to play the game of chess. The goal when playing chess is to capture the other player's king through an alternating

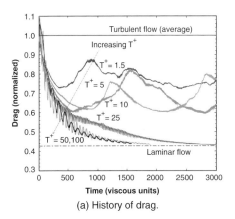
(a) History of drag.

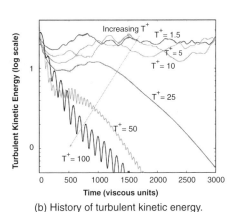
(b) History of turbulent kinetic energy.

FIGURE 13.14 (Color figure follows p. 12-26.) Performance of optimized blowing/suction controls for formulations based on minimizing $\mathcal{J}_o(\phi)$, case *c* (see Section 13.9.1.2), as a function of the optimization horizon T^+. The direct numerical simulations of turbulent channel flow reported here were conducted at $Re_\tau = 100$. For small optimization horizons ($T^+ = O(1)$, sometimes called the "suboptimal approximation"), approximately 20% drag reduction is obtained, a result that can be obtained with a variety of other approaches. For sufficiently large optimization horizons ($T^+ \geq 25$), the flow is returned to the region of stability of the laminar flow, and the flow relaminarizes with no further control effort required. No other control algorithm tested in this flow to date has achieved this result with this type of flow actuation. (From Bewley, T.R., Moin, P., and Temam, R., *J. Fluid Mech.*, to appear. With permission of Cambridge University Press.)

series of discrete moves with the opponent; at any particular turn, a player has to select one move out of at most 20 or 30 legal alternatives.

To accomplish its optimization, a computer program designed to play the comparatively "simple" game of chess, such as *Deep Blue* [Newborn, 1997], must, in the worst case, plan ahead by iteratively examining a tree of possible evolutions of the game several moves into the future [Atkinson, 1993], a strategy based on "function evaluations" alone. At each step, the program selects the move that leads to its best expected outcome, given that the opponent is doing the same in a truly noncooperative competition. The version of *Deep Blue* that defeated Garry Kasparov in 1997 was able to calculate up to 200 billion moves in the three minutes it was allowed to conduct each turn. Even with this extreme number of function evaluations at its disposal on this relatively simple problem, the algorithm was only about an even match with Kasparov's human intuition.

An improved algorithm compared to those based on function evaluations alone, suitable for optimizing the present problem in a reasonable amount of time, is available because (1) we know the equation governing the evolution of the present system, and (2) we can state the problem of interest as a functional to be minimized. Taking these two facts together, we may devise an iterative procedure based on gradient information, derived from an *adjoint field*, to optimize the controls for the desired purpose on the prediction horizon of interest in an efficient manner. Only by exploiting such gradient information can the high-dimensional optimization problem at hand (up to $O(10^7)$ control variables per optimization horizon in some of our simulations) be made tractable.

13.9.1 Adjoint-Based Optimization Approach

13.9.1.1 Governing Equation

The problem we consider here is the control of a fully developed turbulent channel flow with full flowfield information and copious computational resources available to the control algorithm. The flow is governed by the incompressible Navier–Stokes equation inside a three-dimensional rectangular domain (Figure 13.15) with unsteady wall-normal velocity boundary conditions ϕ applied on the walls as the control. Three vector fields are first defined: the flow state $\mathbf{q}$, the flow perturbation state $\mathbf{q}'$, and the adjoint state $\mathbf{q}^*$:

$$\mathbf{q}(\mathbf{x},t) = \begin{pmatrix} p(\mathbf{x},t) \\ \mathbf{u}(\mathbf{x},t) \end{pmatrix}, \qquad \mathbf{q}'(\mathbf{x},t) = \begin{pmatrix} p'(\mathbf{x},t) \\ \mathbf{u}'(\mathbf{x},t) \end{pmatrix}, \qquad \mathbf{q}^*(\mathbf{x},t) = \begin{pmatrix} p^*(\mathbf{x},t) \\ \mathbf{u}^*(\mathbf{x},t) \end{pmatrix}.$$

Each of these vector fields is composed of a pressure component and a velocity component, all of which are continuous functions of space $\mathbf{x}$ and time t. The velocity components themselves are also vectors, with

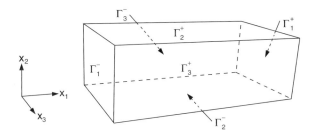

FIGURE 13.15 Channel flow geometry. The interior of the domain is denoted Ω and the boundaries of the domain in the x_i direction are denoted $\Gamma_i^\pm$. Unsteady wall-normal velocity boundary conditions are applied on the walls $\Gamma_2^\pm$ as the control, with periodic boundary conditions applied in the streamwise direction x_1 and spanwise direction x_3. An external pressure gradient is applied to induce a mean flow in the x_1 direction.

components in the streamwise direction x_1, the wall-normal direction x_2, and the spanwise direction x_3. Partial differential equations governing all three of these fields will be derived in due course, and the motivation for introducing $\mathbf{q}'$ and $\mathbf{q}^*$ will be given as the need for these fields arises in the control derivation. Only after the optimization approach has been derived completely in differential form is it discretized in space and time; an alternative strategy, discretizing the state equation in space before determining the adjoint operator, is discussed in Section 13.9.2.

The governing equation is written as

$$\mathcal{N}(\mathbf{q}) = \mathbf{F} \quad \text{in } \Omega,$$
$$\mathbf{u} = -\phi \mathbf{n} \quad \text{on } \Gamma_2^\pm, \tag{13.11}$$
$$\mathbf{u} = \mathbf{u}_0 \quad \text{at } t = 0,$$

where $\mathcal{N}(\mathbf{q})$ is the (nonlinear) Navier–Stokes operator

$$\mathcal{N}(\mathbf{q}) = \begin{pmatrix} \dfrac{\partial u_j}{\partial x_j} \\ \dfrac{\partial u_i}{\partial t} + \dfrac{\partial u_j u_i}{\partial x_j} - \nu \dfrac{\partial^2 u_i}{\partial x_j^2} + \dfrac{\partial p}{\partial x_i} \end{pmatrix},$$

$\mathbf{F}$ is a forcing vector accounting for an externally applied mean pressure gradient driving the flow in the streamwise direction, and $\mathbf{n}$ is the unit outward normal to the boundary $\partial \Omega$. The boundary conditions on the state $\mathbf{q}$ are periodic in the streamwise and spanwise directions. A wall-normal control velocity ϕ is distributed over the walls as indicated, and is constrained to inject zero net mass such that $\forall t$, $\int_{\Gamma_2^+} \phi \, d\mathbf{x} = \int_{\Gamma_2^-} \phi \, d\mathbf{x} = 0$. Initial conditions on the velocity, $\mathbf{u}_0$, of fully developed turbulent channel flow are prescribed.

13.9.1.2 Cost Functional

As in the linear setting, an essential step in the framing of the nonlinear optimization problem is the representation of the control objective as a cost functional to be minimized. Several cases of physical interest may be represented by a cost functional of the generic form

$$\mathcal{J}_0(\phi) = \frac{1}{2}\int_0^T \int_\Omega |C_1 \mathbf{u}|^2 \, d\mathbf{x} \, dt + \frac{1}{2}\int_\Omega |C_2 \mathbf{u}(\mathbf{x}, T)|^2 \, d\mathbf{x} - \int_0^T \int_{\Gamma_2^\pm} C_3 \nu \frac{\partial \mathbf{u}}{\partial n} \cdot \mathbf{r} \, d\mathbf{x} \, dt + \frac{\ell^2}{2}\int_0^T \int_{\Gamma_2^\pm} |\phi|^2 \, d\mathbf{x} \, dt.$$

Four cases of particular interest are:

a. $C_1 = d_1 I$ and $C_2 = C_3 = 0 \Rightarrow$ regulation of turbulent kinetic energy;
b. $C_1 = d_2 \nabla \times$ and $C_2 = C_3 = 0 \Rightarrow$ regulation of the square of the vorticity;
c. $C_2 = d_3 I$ and $C_1 = C_3 = 0 \Rightarrow$ terminal control of turbulent kinetic energy;
d. $C_3 = d_4 I$ and $C_1 = C_2 = 0 \Rightarrow$ minimization of the time-averaged skin friction in the direction $\mathbf{r}$ integrated over the boundary of the domain, where $\mathbf{r}$ is a unit vector in the streamwise direction.

All four of these cases, and many others, may be considered in the current framework, and the extension to other cost functionals is straightforward. The dimensional constants d_i (which are the appropriate functions of the kinematic viscosity, the channel width and the bulk velocity), as well as ℓ, are included to make the cost functional dimensionally consistent and to account for the relative weight of each individual term.

Clearly, in both the chess problem and the turbulence problem, the further into the future one can optimize the problem the better (Figure 13.14); however, both problems get exponentially more difficult to optimize as the prediction horizon is increased. Because only intermediate-term optimization is tractable, it is not always the best approach to represent the final objective in the cost functional. In the chess problem, though the final aim is to capture the other player's king, it is most effective to adopt a mid-game strategy of establishing good board position and achieving material advantage. Similarly, if the turbulence control objective is reducing drag, it was found in BMT01 that it is most effective along the way to minimize a finite-horizon cost functional related to the turbulent kinetic energy of the flow, as the turbulent transport of momentum is responsible for inducing a substantial portion of the drag in a turbulent flow. In a sense, turbulence is the "cause" and high drag is the "effect," and it is most effective to target the "cause" in the cost functional when optimizations on only intermediate prediction horizons are possible.

In addition, a smart optimization algorithm allows for excursions in the short term if it leads to a long-term advantage. For example, in chess, a good player is willing to sacrifice a lesser piece if, by so doing, a commanding board position is attained and/or a restoring exchange is forced a few moves later. Similarly, by allowing a turbulence control scheme to increase (temporarily) the turbulent kinetic energy of a flow, a transient may ensue which, eventually, effectively diminishes the strength of the near-wall coherent structures. It was found in BMT01 that terminal control strategies, aimed at minimizing the turbulence only at the end of each optimization period, have a decided advantage over regulation strategies, which penalize excursions of, for example, the turbulent kinetic energy over the entire prediction horizon.

13.9.1.3 Gradient of Cost Functional

As suggested by Abergel and Temam (1990), a rigorous procedure may be developed to determine the sensitivity of a cost functional $\mathcal{J}$ to small modifications of the control ϕ for nonlinear problems of this sort. To do this, consider the perturbation to the cost functional resulting from a small perturbation to the control ϕ in the direction ϕ'. (Note that this control perturbation direction ϕ' is arbitrary and scaled to have unit norm.) Define $\mathcal{J}'$ as the Fréchet differential [Vainberg, 1964] of a cost functional $\mathcal{J}$ such that

$$\mathcal{J}' \triangleq \lim_{\varepsilon \to 0} \frac{\mathcal{J}(\phi + \varepsilon \phi') - \mathcal{J}(\phi)}{\varepsilon} \triangleq \int_0^T \int_{\Gamma_2^\pm} \frac{\mathcal{D}\mathcal{J}(\phi)}{\mathcal{D}(\phi)} \phi' \, dt \, d\mathbf{x}.$$

The quantity $\mathcal{J}'$ is the cost functional perturbation due to a control perturbation $\varepsilon \phi'$ scaled by the inverse of the control perturbation magnitude ε in the limit that $\varepsilon \to 0$. The above relation, considered for arbitrary ϕ', also defines the gradient of the cost functional $\mathcal{J}$ with respect to the control ϕ, which is written $\mathcal{D}\mathcal{J}(\phi)/\mathcal{D}\phi$.

In the current approach, the cost functional perturbation $\mathcal{J}'$ defined above will be expressed as a simple linear function of the direction of the control perturbation ϕ' through the solution of an adjoint problem. By the above formula, such a representation then reveals the gradient direction $\mathcal{D}\mathcal{J}(\phi)/\mathcal{D}\phi$ directly. With this gradient information, the control ϕ is updated on $(0, T]$ in the direction that, at least locally (i.e., for infinitesimal control updates), most effectively reduces the cost functional. The finite distance the control is updated in this direction is then found by a line search routine, which makes this iteration procedure stable even when controlling nonlinear phenomena. The flow resulting from this modified control is then computed according to the (nonlinear) Navier–Stokes equation (13.11), the sensitivity of this new flow to further control modification is computed, and the process repeated. Upon convergence of this iteration, the flow is advanced over the interval $(0, T_1]$, where $T_1 \leq T$, and an iteration for the optimal control over a new time interval $(T_1, T_1 + T]$ begins anew.

The cost functional perturbation $\mathcal{J}'$ resulting from a control perturbation in the direction ϕ' is given by

$$\mathcal{J}'_0(\phi) = \frac{1}{2}\int_0^T \int_\Omega C_1^* C_1 \mathbf{u} \cdot \mathbf{u}' \, d\mathbf{x} \, dt + \frac{1}{2}\int_\Omega (C_2^* C_2 \mathbf{u} \cdot \mathbf{u}')_{t=T} \, d\mathbf{x} - \int_0^T \int_{\Gamma_2^\pm} \nu C_3^* \mathbf{r} \cdot \frac{\partial \mathbf{u}'}{\partial n} \, d\mathbf{x} \, dt$$

$$+ \ell^2 \int_0^T \int_{\Gamma_2^\pm} \phi \phi' \, d\mathbf{x} \, dt \triangleq \int_0^T \int_{\Gamma_2^\pm} \frac{\mathcal{D}\mathcal{J}_0(\phi)}{\mathcal{D}(\phi)} \phi' \, d\mathbf{x} \, dt,$$

where **u**′ is the Fréchet differential of **u**, as defined in the following subsection. Adjoint calculus is used simply to re-express the integrals involving **u**′ as a linear function of ϕ'. Once this is accomplished, ϕ' is factored out of the integrands and, as the equation holds for arbitrary ϕ', an expression for the gradient $\mathcal{D}\mathcal{J}_o(\phi)/\mathcal{D}\phi$ is identified.

13.9.1.4 Linearized Perturbation Field

Now consider the linearized perturbation **q**′ to the flow **q** resulting from a perturbation ϕ' to the control ϕ. Again, the quantity **q**′ may be defined by the limiting process of a Fréchet differential such that

$$\mathbf{q}' \triangleq \lim_{\varepsilon \to 0} \frac{\mathbf{q}'(\phi + \varepsilon\phi') - \mathbf{q}(\phi)}{\varepsilon}.$$

For the purpose of gaining physical intuition, it is useful to note that the quantity **q**′, described above as a differential quantity, may instead be defined as the small perturbation to the state **q** arising from a small control perturbation ϕ' to the control ϕ. In such derivations, the notations $\delta\phi$ and $\delta\mathbf{q}$, denoting small perturbations to ϕ and **q**, are used instead of the differential quantities ϕ' and **q**′. The two derivations are roughly equivalent, though the present derivation does not assume that primed quantities are small.

The equation governing the dependence of the linearized flow perturbation **q**′ on the control perturbation ϕ' may be found by taking the Fréchet differential of the state equation (13.11). The result is

$$\begin{aligned}
\mathcal{N}'(\mathbf{q})\mathbf{q}' &= 0 && \text{in } \Omega, \\
\mathbf{u}' &= -\phi'\mathbf{n} && \text{on } \Gamma_2^{\pm}, \\
\mathbf{u}' &= 0 && \text{at } t = 0,
\end{aligned} \qquad (13.12)$$

where the linearized Navier–Stokes operation $\mathcal{N}'(\mathbf{q})\mathbf{q}'$ is given by

$$\mathcal{N}'(\mathbf{q})\mathbf{q}' = \begin{pmatrix} \dfrac{\partial u'_j}{\partial x_j} \\ \dfrac{\partial u'_i}{\partial t} + \dfrac{\partial}{\partial x_j}(u_j u'_i + u'_j u_i) - \nu\dfrac{\partial^2 u'_i}{\partial x_j^2} + \dfrac{\partial p'}{\partial x_i} \end{pmatrix}.$$

The operation $\mathcal{N}'(\mathbf{q})\mathbf{q}'$ is a linear operation on the perturbation field **q**′, though the operator $\mathcal{N}'(\mathbf{q})\mathbf{q}'$ is itself a function of the solution **q** of the Navier–Stokes problem. Equation (13.12) thus reflects the linear dependence of the perturbation field **q**′ in the interior of the domain on the control perturbation ϕ' at the boundary. However, the implicit linear relationship $\mathbf{q}' = \mathbf{q}'(\phi')$ given by this equation is not yet tractable for expressing $\mathcal{J}'_0$ in a simple form from which $\mathcal{D}\mathcal{J}_0(\phi)/\mathcal{D}\phi$ may be deduced. For the purpose of determining a more useful relationship with which we may determine $\mathcal{D}\mathcal{J}_0(\phi)/\mathcal{D}\phi$, we now appeal to an adjoint identity.

13.9.1.5 Statement of Adjoint Identity

This subsection derives the adjoint of the linear partial differential operator $\mathcal{N}'(\mathbf{q})\mathbf{q}'$. For readers not familiar with this approach, a review of the derivation of an adjoint operator for a very simple case in the present notation is given in Appendix A of BMT01. The adjoint derivation presented below extends in a straightforward manner to more complex equations, such as the compressible Euler equation, as shown in Appendix B of BMT01 (again, using the same notation). Such generality highlights the versatility of the present approach.

Define an inner product over the domain in space-time under consideration such that

$$\langle \mathbf{q}^*, \mathbf{q}' \rangle = \int_0^T \!\!\int_\Omega \mathbf{q}^* \cdot \mathbf{q}' \, d\mathbf{x}\, dt$$

and consider the identity

$$\langle \mathbf{q}^*, \mathcal{N}'(\mathbf{q})\mathbf{q}' \rangle = \langle \mathcal{N}'(\mathbf{q})^*\mathbf{q}^*, \mathbf{q}' \rangle + b. \tag{13.13}$$

Integration by parts may be used to move all differential operations from $\mathbf{q}'$ on the left-hand side of Eq. (13.13) to $\mathbf{q}^*$ on the right-hand side, resulting in the derivation of the adjoint operator

$$\mathcal{N}'(\mathbf{q})^*\mathbf{q}^* = \begin{pmatrix} -\dfrac{\partial u_j^*}{\partial x_j} \\ -\dfrac{\partial u_i^*}{\partial t} - u_j \left(\dfrac{\partial u_i^*}{\partial x_j} + \dfrac{\partial u_j^*}{\partial x_i} \right) - \nu \dfrac{\partial^2 u_i^*}{\partial x_j^2} - \dfrac{\partial p^*}{\partial x_i} \end{pmatrix},$$

where, again, the operation $\mathcal{N}'(\mathbf{q})^*\mathbf{q}^*$ is a linear operation on the adjoint field $\mathbf{q}^*$, and the operator $\mathcal{N}'(\mathbf{q})^*$ is itself a function of the solution $\mathbf{q}$ of the Navier–Stokes problem. From the integrations by parts, we also get several boundary terms:

$$b = \int_\Omega (u_j^* u_i')_{t=0}^{t=T} d\mathbf{x} + \int_0^T \int_\Omega n_j \left[u_i^* (u_j u_i' + u_j' u_i) + p^* u_j' - \nu \left(u_i^* \dfrac{\partial u_i'}{\partial x_j} - u_i' \dfrac{\partial u_i^*}{\partial x_j} \right) + u_j^* p' \right] d\mathbf{x}\, dt.$$

The identity (13.13) is the key to expressing $\mathcal{J}'$ in the desired form. An adjoint field $\mathbf{q}^*$ is first defined using the operator $\mathcal{N}'(\mathbf{q})^*$ together with appropriate forcing on an interior equation with appropriate boundary conditions and initial conditions. There is here some flexibility which we exploit to obtain a simple expression of $\mathcal{J}'$. Indeed, combining this definition of $\mathbf{q}^*$ with the definitions of $\mathbf{q}$ in Eq. (13.11) and $\mathbf{q}'$ in Eq. (13.12), the identity (13.13) reveals the desired expression, as will now be shown.

13.9.1.6 Definition of Adjoint Field

Consider an adjoint state defined (as yet, arbitrarily) by

$$\begin{aligned} \mathcal{N}'(\mathbf{q})^*\mathbf{q}^* &= \begin{pmatrix} 0 \\ C_1^* C_1 \mathbf{u} \end{pmatrix} \quad \text{in } \Omega, \\ \mathbf{u}^* &= C_3^* \mathbf{r} \quad \text{on } \Gamma_2^\pm, \\ \mathbf{u}^* &= C_2^* C_2 \mathbf{u} \quad \text{at } t = T, \end{aligned} \tag{13.14}$$

where the adjoint operation $\mathcal{N}'(\mathbf{q})^*$ is derived in the previous subsection. Note by Eq. (13.14) that, depending on where the cost functional weighs the flow perturbations (see Section 13.9.1.2), the adjoint problem may be driven by the initial conditions, by the boundary conditions, or by the RHS of the adjoint PDE itself. Note also that the adjoint "initial" conditions are defined at $t = T$ and are thus best referred to as "terminal" conditions. With this definition, the adjoint field must be marched *backward* in time over the optimization horizon; due to the sign of the time derivative and viscous terms in the adjoint operator $\mathcal{N}'(\mathbf{q})^*$, this is the natural direction for this time march. However, as both the adjoint operator $\mathcal{N}'(\mathbf{q})^*$ and the RHS forcing on Eq. (13.14a) are functions of $\mathbf{q}$, computation of the adjoint field $\mathbf{q}^*$ requires storage of the flow field $\mathbf{q}$ on $t \in [0, T]$, which itself must be computed with a *forward* march. This storage issue presents one of the numerical complications that preclude solution of the present optimization problem for large optimization intervals T. However, this storage issue is not insurmountable for intermediate values of $T^+ < O(100)$. The adjoint problem Eq. (13.14), though linear, has complexity similar to that of the Navier–Stokes problem, Eq. (13.11), and may be solved with similar numerical methods.

13.9.1.7 Identification of Gradient

The identity (13.13) is now simplified using the equations defining the state field, Eq. (13.11), the perturbation field, Eq. (13.12), and the adjoint field, Eq. (13.14). Due to the judicious choice of the forcing terms driving the adjoint problem, the identity (13.13) reduces (after some manipulation) to

$$\int_0^T \int_\Omega C_1^* C_1 \mathbf{u} \cdot \mathbf{u}' \, d\mathbf{x} \, dt + \int_\Omega (C_2^* C_2 \mathbf{u} \cdot \mathbf{u}')_{t=T} \, d\mathbf{x} - \int_0^T \int_{\Gamma_2^\pm} \nu C_3^* \mathbf{r} \cdot \frac{\partial \mathbf{u}'}{\partial n} \, d\mathbf{x} \, dt = \int_0^T \int_{\Gamma_2^\pm} p^* \, \phi' \, d\mathbf{x} \, dt.$$

Using this equation, the cost functional perturbation $\mathcal{J}_0'$ may be rewritten as

$$\mathcal{J}_0'(\phi; \phi') = \int_0^T \int_{\Gamma_2^\pm} (p^* + \ell^2 \phi) \phi' \, d\mathbf{x} \, dt \triangleq \int_0^T \int_{\Gamma_2^\pm} \frac{\mathcal{D}\mathcal{J}_0(\phi)}{\mathcal{D}(\phi)} \phi' \, d\mathbf{x} \, dt.$$

As ϕ' is arbitrary, we may identify (weakly) the desired gradient as

$$\frac{\mathcal{D}\mathcal{J}_0(\phi)}{\mathcal{D}(\phi)} = p^* + \ell^2 \phi.$$

The desired gradient $\mathcal{D}\mathcal{J}_0(\phi)/\mathcal{D}\phi$ is thus found to be a simple function of the solution of the adjoint problem proposed in Eq. (13.14); specifically, in the present case of boundary forcing by wall-normal blowing and suction, it is found to be a simple function of the adjoint pressure on the walls.

In fact, this simple result hints at the more fundamental physical interpretation of what the adjoint field actually represents:

> The adjoint field $\mathbf{q}^*$, when properly defined, is a measure of the sensitivity of the terms of the cost functional that appraise the state $\mathbf{q}$ to additional forcing of the state equation.

Note that there are exactly as many components of the adjoint field $\mathbf{q}^*$ as there are components of the state PDE on the interior of the domain, and that the adjoint field may take nontrivial values at the initial time $t = 0$ and on the boundaries $\Gamma_2^\pm$. Depending upon where the control is applied to the state equation (13.11), (i.e., on the RHS of the mass or momentum equations on the interior of the domain, on the boundary conditions, or on the initial conditions), the adjoint field will appear in the resulting expression for the gradient accordingly.

To summarize, the forcing on the adjoint problem is a function of where the flow perturbations are weighed in the cost functional. The dependence of the gradient $\mathcal{D}\mathcal{J}(\phi)/\mathcal{D}\phi$ on the resulting adjoint field, on the other hand, is a function of where the control enters the state equation.

13.9.1.8 Gradient Update to Control

A control optimization strategy using a steepest descent algorithm may now be proposed such that

$$\phi^k = \phi^{k-1} - \alpha^k \frac{\mathcal{D}\mathcal{J}_0(\phi^{k-1})}{\mathcal{D}\phi}$$

over the entire time interval $t \in (0, T]$, where k indicates the iteration number and α^k is a parameter of descent that governs how large an update is made, which is adjusted at each iteration step to be the value that minimizes $\mathcal{J}$. This algorithm updates ϕ at each iteration in the direction of maximum decrease of $\mathcal{J}$. As $k \to \infty$, the algorithm should converge to some local minimum of $\mathcal{J}$ over the domain of the control ϕ on the time interval $t \in (0, T]$. Note that convergence to a global minimum will not in general be attained by such a scheme and that, as time proceeds, $\mathcal{J}$ will not necessarily decrease.

The steepest descent algorithm described above illustrates the essence of the approach, but is usually not very efficient. Even in linear low-dimensional problems, for cases in which the cost functional has a long, narrow "valley," the lack of a momentum term from one iteration to the next tends to cause the steepest descent algorithm to bounce from one side of the valley to the other without turning to proceed along the valley floor. Standard nonlinear conjugate gradient algorithms [e.g., Press et al., 1986] improve this behavior considerably with relatively little added computational cost or algorithmic complexity, as discussed further in BMT01.

As mentioned previously, the dimension of the control in the present problem (once discretized) is quite large, which precludes the use of second-order techniques based on the computation or approximation of the Hessian matrix $\partial^2 \mathcal{J}/\partial \phi_i \partial \phi_j$ or its inverse during the control optimization. The number of elements in such a matrix scales with the square of the number of control variables and is unmanageable in the present case. However, reduced-storage variants of variable metric methods [Vanderplaats, 1984], such as the Davidon–Fletcher–Powell (DFP) method, the Broydon–Fletcher–Goldfarb–Shanno (BFGS) method, and the sequential quadratic programming (SQP) method, approximate the inverse Hessian information by outer products of stored gradient vectors and thus achieve nearly second-order convergence without storage of the Hessian matrix itself. Such techniques should be explored further for very large-scale optimization problems such as the present in future work.

13.9.2 Continuous Adjoint vs. Discrete Adjoint

Direct numerical simulations (DNS) of the current three-dimensional nonlinear system necessitate carefully chosen numerical techniques involving a stretched, staggered grid, an energy-conserving spatial discretization, and a mixture of implicit and multistep explicit schemes for accurate time advancement, with incompressibility enforced by an involved fractional step algorithm. The optimization approach described above, which will be referred to as "optimize then discretize" (OTD), avoids all of these cumbersome numerical details by deriving the gradient of the cost functional in the continuous setting, discretizing in time and space only as the final step before implementation in numerical code. The remarkable similarity of the flow and adjoint systems allows both to be coded with similar numerical techniques. For systems which are well resolved in the numerical discretization, this approach is entirely justifiable, and is found to yield adjoint systems which are easy to derive and implement in numerical code.

Unfortunately, many PDE systems, such as high-Reynolds-number turbulent flows, are difficult or impossible to simulate with sufficient resolution to capture accurately all of the important dynamic phenomena of the continuous system. Such systems are often simulated on coarse grids, usually with some "subgrid-scale model" to account for the unresolved dynamics. This setting is referred to as large eddy simulation (LES), and a variety of techniques are currently under development to model the significant subgrid-scale effects.

There are important unresolved issues concerning how to approach large eddy simulations in the optimization framework. If we continue with the OTD approach, in which the optimization equations are determined before the numerical discretization is applied, it is not yet clear at what point the LES model should be introduced. Prof. Scott Collis' group (Rice University) has modified the numerical code of BMT01 in order to study this issue; Chang and Collis (1999) report on their preliminary findings.

An alternative approach to the OTD setting, in which one spatially discretizes the governing equation before determining the optimization equations, may also be considered.[4] After spatially discretizing the governing equation, this approach, which will be referred to as "discretize then optimize" (DTO), follows an analogous sequence of steps as the OTD approach presented previously, with these steps now applied in the discrete setting. Derivation of the adjoint operator is significantly more cumbersome in this discrete setting. In general, the processes of optimization and discretization do not commute, and

---

[4] Note that we may also consider temporally discretizing the governing equation before determining the optimization equations. Generally, the spatial discretization of a turbulent system is the most restrictive issue, however, so we focus on that problem here.

thus the OTD and DTO approaches are not necessarily equivalent even upon refinement of the space/time grid [Vogel and Wade, 1995]. However, by carefully framing the discrete identity defining the DTO adjoint operator as a discrete approximation of the identity given in Eq. (13.13), these two approaches can be posed in an equivalent fashion for Navier–Stokes systems.

It remains the topic of some debate whether or not the DTO approach is better than the OTD approach for marginally resolved PDE systems. The argument for DTO is that it clearly is the most direct way to optimize the discrete problem actually being solved by the computer. The argument against DTO is that one really wants to optimize the continuous problem, so gradient information that identifies and exploits deficiencies in the numerical discretization that can lead to performance improvements in the discrete problem might be misleading when interpreting the numerical results in terms of the physical system.

13.10 Robustification: Appealing to Murphy's Law

Though optimal control approaches possess an attractive mathematical elegance and are now proven to provide excellent results in terms of drag and turbulent kinetic energy reduction in fully developed turbulent flows, they are often impractical. One of the most significant drawbacks of this nonlinear optimization approach is that it tends to "over-optimize" the system, leaving a high degree of design-point sensitivity. This phenomenon has been encountered frequently in, for example, the adjoint-based optimization of the shape of aircraft wings. Overly optimized wing shapes might work quite well at exactly the flow conditions for which they were designed, but their performance is often abysmal at off-design conditions. In order to abate such system sensitivity, the noncooperative framework of robust control provides a natural means to "detune" the optimized results. This concept can be applied easily to a broad range of related applications.

The noncooperative approach to robust control, one might say, amounts to Murphy's law taken seriously:

If a worst-case disturbance can disrupt a controlled closed-loop system, it will.

When designing a robust controller, therefore, one might plan on a finite component of the worst-case disturbance aggravating the system, and design a controller suited to handle even this extreme situation. A controller designed to work even in the presence of a finite component of the worst-case disturbance will also be robust to a wide class of other possible disturbances which, by definition, are not as detrimental to the control objective as the worst-case disturbance. This concept is exactly that which leads to the $\mathcal{H}_\infty$ control formulation discussed previously in the linear setting, and can easily be extended to the optimization of nonlinear systems.

Based on the ideas of $\mathcal{H}_\infty$ control theory presented in Section 13.3, the extension of the nonlinear optimization approach presented in Section 13.9 to the noncooperative setting is straightforward. A disturbance is first introduced to the governing equation (13.11); as an example, we may consider disturbances that perturb the state PDE itself such that

$$\mathcal{N}(\mathbf{q}) = \mathbf{F} + \mathbf{B}_1(\psi) \quad \text{in } \Omega.$$

(Accounting for disturbances to the boundary conditions and initial conditions of the governing equation is also straightforward.) The cost functional is then extended to penalize these disturbances in the noncooperative framework, as was also done in the linear setting

$$\mathcal{J}_r(\psi, \phi) = \mathcal{J}_0 - \frac{\gamma^2}{2} \int_0^T \int_\Omega |\psi|^2 \, d\mathbf{x} \, dt.$$

This cost functional is simultaneously minimized with respect to the controls ϕ and maximized with respect to the disturbances ψ (Figure 13.16). The parameter γ is used to scale the magnitude of the

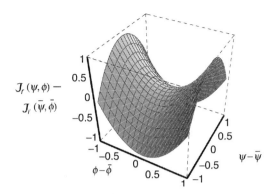

FIGURE 13.16 Schematic of a saddle point representing the neighborhood of a solution to a robust control problem with one scalar disturbance variable ψ and one scalar control variable ϕ. When the robust control problem is solved, the cost function $\mathcal{J}_r$ is simultaneously maximized with respect to ψ and minimized with respect to ϕ, and a saddle point such as $(\bar{\psi}, \bar{\phi})$ is reached. An essentially infinite-dimensional extension of this concept may be formulated to achieve robustness to disturbances and insensitivity to design point in fluid-mechanical systems. In such approaches, the cost $\mathcal{J}_r$ is related to a distributed disturbance ψ and a distributed control ϕ through the solution of the Navier–Stokes equation.

disturbances accounted for in this noncooperative competition, with the limit of large γ recovering the optimal approach discussed in Section 13.9 (i.e., $\psi \to 0$). A gradient-based algorithm may then be devised to march to the saddle point, such as the simple algorithm given by:

$$\phi^k = \phi^{k-1} - \alpha^k \frac{\mathcal{D}\mathcal{J}_r(\psi^{k-1}; \phi^{k-1})}{\mathcal{D}\phi},$$

$$\psi^k = \psi^{k-1} + \beta^k \frac{\mathcal{D}\mathcal{J}_r(\psi^{k-1}; \phi^{k-1})}{\mathcal{D}\psi}.$$

The robust control problem is considered to be solved when a saddle point $(\bar{\psi}, \bar{\phi})$ is reached; note that such a solution, if it exists, is not necessarily unique.

The gradients $\mathcal{D}\mathcal{J}_r(\psi; \phi)/\mathcal{D}\phi$ and $\mathcal{D}\mathcal{J}_r(\psi; \phi)/\mathcal{D}\psi$ may be found in a manner analogous to that leading to $\mathcal{D}\mathcal{J}_0(\phi)/\mathcal{D}\phi$ discussed in Section 13.9. In fact, both gradients may be extracted from the single adjoint field defined by Eq. (13.14). Thus, the additional computational complexity introduced by the noncooperative component of the robust control problem is simply a matter of updating and storing the appropriate disturbance variables.

13.10.1 Well-Posedness

Based on the extensive mathematical literature on the Navier–Stokes equation, Abergel and Temam (1990) established the well-posedness of the mathematical framework for the optimization problem presented in Section 13.9. This characterization was generalized and extended to the noncooperative framework of Section 13.10 in Bewley, Temam, and Ziane (2000).

Due to the fact that the inequalities currently available for estimating the magnitude of the various terms of the Navier–Stokes equation are limited, the mathematical characterizations in both of these articles are quite conservative. In our numerical simulations, we regularly apply numerical optimization techniques to control problems that are well outside the range over which we can mathematically establish well-posedness. However, such mathematical characterizations are still quite important, as they give us confidence that, for example, if ℓ and γ are at least taken to be large enough, a saddle point of the noncooperative optimization problem will exist. Once such mathematical characterizations are derived, numerically determining the values of ℓ and γ for which solutions of the control problem may still be obtained is reduced to a simple matter of implementation.

13.10.2 Convergence of Numerical Algorithm

Saddle points are typically more difficult to find than minimum points, and particular care needs to be taken to craft efficient but stable numerical algorithms for finding them. In the approach described above, sufficiently small values of α^k and β^k must be selected to ensure convergence. Fortunately, the same mathematical inequalities used to characterize well-posedness of the control problem can also be used to characterize convergence of proposed numerical algorithms. Such characterizations lend valuable insight when designing practical numerical algorithms. Preliminary work in the development of such saddle point algorithms is reported by Tachim Medjo (2000).

13.11 Unification: Synthesizing a General Framework

The various cost functionals considered previously led to three possible sources of forcing for the adjoint problem: the right-hand side of the PDE, the boundary conditions, and the initial conditions. Similarly, three different locations of forcing may be identified for the flow problem. As illustrated in Figures 13.17 and 13.18 and discussed further in Bewley, Temam, and Ziane (2000), the various regions of forcing of the flow and adjoint problems together form a general framework that can be applied to a wide variety of problems in fluid mechanics including both flow control (e.g., drag reduction, mixing enhancement, and noise control) and flow forecasting (e.g., weather prediction and storm forecasting). Related techniques, but applied to the time-averaged Navier–Stokes equation, have also been used extensively to optimize the shapes of airfoils [see, e.g., Reuther et al., 1996].

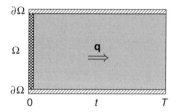

FIGURE 13.17 Schematic of the space–time domain over which the flow field **q** is defined. The possible regions of forcing in the system defining **q** are: (1) the right-hand side of the PDE, indicated with shading, representing flow control by interior volume forcing (e.g., externally applied electromagnetic forcing by wall-mounted magnets and electrodes); (2) the boundary conditions, indicated with diagonal stripes, representing flow control by boundary forcing (e.g., wall transpiration); (3) the initial conditions, indicated with checkerboard, representing optimization of the initial state in a data assimilation framework (e.g., the weather forecasting problem).

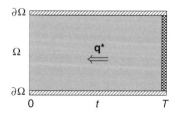

FIGURE 13.18 Schematic of the space–time domain over which the adjoint field **q*** is defined. The possible regions of forcing in the system defining **q***, corresponding exactly to the possible domains in which the cost functional can depend on **q**, are: (1) the right-hand side of the PDE, indicated with shading, representing regulation of an interior quantity (e.g., turbulent kinetic energy); (2) the boundary conditions, indicated with diagonal stripes, representing regulation of a boundary quantity (e.g., wall skin friction); (3) the terminal conditions, indicated with checkerboard, representing terminal control of an interior quantity (e.g., turbulent kinetic energy).

By identifying a range of problems that all fit into the same general framework, we can better understand how to extend, for example, the idea of noncooperative optimizations to a full suite of related problems in fluid mechanics. Though advanced research projects must often be highly focused and specialized in order to obtain solid results, the importance of making connections of such research to a large scope of related problems must be recognized in order to realize fully the potential impact of the techniques developed.

13.12 Decomposition: Simulation-Based System Modeling

For the purpose of developing model-based feedback control strategies for turbulent flows, reduced-order nonlinear models of turbulence that are effective in the closed-loop setting are highly desired. Recent work in this direction, using proper orthogonal decompositions (POD) to obtain these reduced-order representations, is reviewed by Lumley and Blossey (1998).

The POD technique uses analysis of a simulation database to develop an efficient reduced-order basis for the system dynamics represented within the database [Holmes et al., 1996]. One of the primary challenges of this approach is that the dynamics of the system in closed loop (after the control is turned on) is often quite different than the dynamics of the open-loop (uncontrolled) system. Thus, development of simulation-based reduced-order models for turbulent flows should probably be coordinated with the design of the control algorithm itself in order to determine system models that are maximally effective in the closed-loop setting. Such coordination of simulation-based modeling and control design is largely an unsolved problem. A particularly sticky issue is the fact that, as the controls are turned on, the dynamics of the turbulent flow system are nonstationary (they evolve in time). The system eventually relaminarizes if the control is sufficiently effective. In such nonstationary problems, it is not clear which dynamics the POD should represent (of the flow shortly after the control is turned on, of the nearly relaminarized flow, or something in between), or if in fact several PODs should be created and used in a scheduled approach in an attempt to capture several different stages of the nonstationary relaminarization process.

Note that reduced-order models that are effective in the closed-loop setting need not capture the majority of the energetics of the unsteady flow. Rather, the essential feature of a system model for the purpose of control design is that the model capture the important effects of the control on the system dynamics. Future control-oriented modeling efforts might benefit by deviating from the standard POD mindset of simply attempting to capture the energetics of the system dynamics, instead focusing on capturing the significant effects of the control on the system in a reduced-order fashion.

13.13 Global Stabilization: Conservatively Enhancing Stability

Global stabilization approaches based on Lyapunov analysis of the system energetics have recently been explored for two-dimensional channel-flow systems (in the continuous setting) by Balogh et al. (2001). In the setting considered there, localized tangential wall motions are coordinated with local measurements of skin friction via simple proportional feedback strategies. Analysis of the flow at $Re \leq 0.125$ motivates such feedback rules, indicating appropriate values of proportional feedback coefficients that enhance the L^2 stability of the flow. Though such an approach is very conservative, rigorously guaranteeing enhanced stability of the channel-flow system only at extremely low Reynolds numbers, extrapolation of the feedback strategies so determined to much higher Reynolds numbers also indicates effective enhancements of system stability, even for three-dimensional systems up to $Re = 2000$ (A. Balogh, pers. comm.).

An alternative approach for achieving global stabilization of a nonlinear PDE is the application of nonlinear backstepping to the discretized system equation. Boškovic and Krstic (2001) report on recent efforts in this direction (applied to a thermal convection loop). Backstepping is typically an aggressive approach to stabilization. One of the primary difficulties with this approach is that proofs of convergence to a continuous, bounded function upon refinement of the grid are difficult to attain due to increasing controller complexity as the grid is refined. Significant advancements will be necessary before this approach will be practical for turbulent flow systems.

13.14 Adaptation: Accounting for a Changing Environment

Adaptive control algorithms, such as least mean squares (LMS), neural networks (NN), genetic algorithms (GA), simulated annealing, extremum seeking, and the like, play an important role in the control of fluid-mechanical systems when the number of undetermined parameters in the control problem is fairly small ($O(10)$) and individual "function evaluations" (i.e., quantitative characterizations of the effectiveness of the control) can be performed relatively quickly. Many control problems in fluid mechanics are of this type, and are readily approachable by a wide variety of well-established adaptive control strategies. A significant advantage of such approaches over those discussed previously is that they do not require extensive analysis or coding of localized convolution kernels, adjoint fields, etc., but may instead be applied directly "out of the box" to optimize the parameters of interest in a given fluid-mechanical problem. This also poses a bit of a disadvantage, however, because the analysis required during the development of model-based control strategies can sometimes yield significant physical insight that black-box optimizations fail to provide.

In order to apply the adaptive approach, one needs an inexpensive simulation code or an experimental apparatus in which the control parameters of interest can be altered by an automated algorithm. Any of a number of established methodological strategies can then be used to search the parameter space for favorable closed-loop system behavior. Given enough function evaluations and a small enough number of control parameters, such strategies usually converge to effective control solutions. Koumoutsakos et al. (1998) demonstrate this approach (computationally) to determine effective control parameters for exciting instabilities in a round jet. Rathnasingham and Breuer (1998) demonstrate this approach (experimentally) for the feedforward reduction of turbulence intensities in a boundary layer.

Unfortunately, due to an effect known as "the curse of dimensionality," as the number of control parameters to be optimized is increased, the ability of adaptive strategies to converge to effective control solutions based on function evaluations alone is diminished. For example, in a system with 1000 control parameters, it takes 1000 function evaluations to determine the gradient information available in a single adjoint computation. Thus, for problems in which the number of control variables to be optimized is large, the convergence of adaptive strategies based on function evaluations alone is generally quite poor. In such high-dimensional problems, for cases in which the control problem of interest is plagued by multiple minima, a blend of an efficient adjoint-based gradient optimization approach with GA-type management of parameter "mutations" or the simulated annealing approach of varying levels of "noise" added to the optimization process might prove to be beneficial.

Adaptive strategies are also quite valuable for recognizing and responding to changing conditions in the flow system. In the low-dimensional setting, they can be used online to update controller gains directly as the system evolves in time (for instance, as the mean speed or direction of the flow changes or as the sensitivity of a sensor degrades). In the high-dimensional setting, adaptive strategies can be used to identify certain critical aspects of the flow (such as the flow speed), and, based on this identification, an appropriate control strategy may be selected from a look-up table of previously computed controller gains.

The selection of what level of adaptation is appropriate for a particular flow control problem of interest is, again, a consideration that must be guided by physical insight of the particular problem at hand.

13.15 Performance Limitation: Identifying Ideal Control Targets

Another important, but as yet largely unrealized, role for mathematical analysis in the field of flow control is in the identification of fundamental limitations on the performance that can be achieved in certain flow control problems. For example, motivated by the active debate surrounding the proposed physical mechanism for channel-flow drag reduction illustrated in Figure 13.19, we formally state the following, as yet unproven, conjecture:

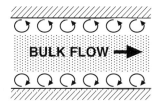

FIGURE 13.19 An enticing picture: fundamental restructuring of the near-wall unsteadiness to insulate the wall from the viscous effects of the bulk flow. It has been argued [Nosenchuck, 1994; Koumoutsakos, 1999] that it might be possible to maintain a series of so-called "fluid rollers" to effectively reduce the drag of a near-wall flow. Such rollers are depicted in the figure above by indicating total velocity vectors in a reference frame convecting with the vortices themselves; in this frame, the generic picture of fluid rollers is similar to a series of stationary Kelvin–Stuart cat's eye vortices. A possible mechanism for drag reduction might be akin to a series of solid cylinders serving as an effective conveyor belt, with the bulk flow moving to the right above the vortices and the wall moving to the left below the vortices. It is still the topic of some debate whether or not a continuous flow can be maintained in such a configuration by an unsteady control in such a way as to sustain the mean skin friction below laminar levels. Such a control might be implemented either by interior electromagnetic forcing (applied with wall-mounted magnets and electrodes) or by boundary controls such as zero-net mass-flux blowing/suction.

Conjecture: The lowest sustainable drag of an incompressible constant mass-flux channel flow, in either two or three dimensions, when controlled via a distribution of zero-net mass-flux blowing/suction over the channel walls, is exactly that of the laminar flow.

Note that by "sustainable drag" we mean the long-time average of the instantaneous drag, given by:

$$D_\infty = \lim_{(T \to \infty)} \frac{-1}{T} \int_0^T \int_{\Gamma_2^\pm} \nu \frac{\partial u_1}{\partial n} d\mathbf{x}\, dt$$

Proof (by mathematical analysis) or disproof (by counterexample) of this conjecture would be quite significant and lead to greatly improved physical understanding of the channel flow problem. If proven to be correct, it would provide rigorous motivation for targeting flow relaminarization when the problem one actually seeks to solve is minimization of drag. If shown to be incorrect, our target trajectories for future flow control strategies might be substantially altered.

Similar fundamental performance limitations may also be sought for exterior flow problems, such as the minimum drag of a circular cylinder subject to a class of zero-net control actions, such as rotation or transverse oscillation (B. Protas, pers. comm.).

13.16 Implementation: Evaluating Engineering Trade-Offs

We are still some years away from applying the distributed control techniques discussed herein to microelectromechanical systems (MEMS) arrays of sensors and actuators, such as that depicted in Figure 13.20. One of the primary hurdles left to be tackled in order to bring us closer to actual implementation is that of accounting for practical designs of sensors and actuators in the control formulations, rather than the idealized distributions of blowing/suction and skin-friction measurements that we have assumed here. Detailed simulations, such as that shown in Figure 13.21, of proposed actuator designs will be essential for developing reduced-order models of the effects of the actuators on the system of interest in order to make control design for realistic arrays of sensors and actuators tractable.

By performing analysis and control design in a high-dimensional, unconstrained setting, as discussed in this chapter, it is believed that we can obtain substantial insight into the physical characteristics of highly effective control strategies. Such insight naturally guides the engineering trade-offs that follow in order to make the design of the turbulence control system practical. Particular traits of the present control solutions in which we are especially interested include the times scales and the streamwise and spanwise

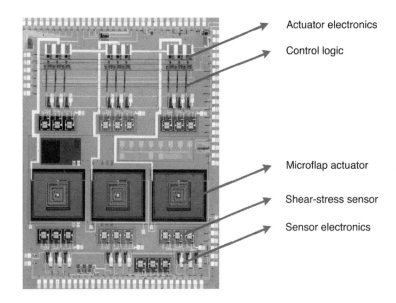

FIGURE 13.20 (Color figure follows p. 12-26.) A MEMS tile integrating sensors, actuators and control logic for distributed flow control applications. (Developed by Profs. Chih-Ming Ho, UCLA, and Yu-Chong Tai, Caltech.)

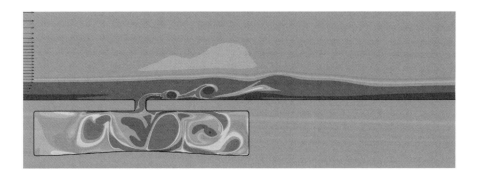

FIGURE 13.21 (Color figure follows p. 12-26.) Simulation of a proposed driven-cavity actuator design (Prof. Rajat Mittal, University of Florida). The fluid-filled cavity is driven by vertical motions of the membrane along its lower wall. Numerical simulation and reduced-order modeling of the the influence of such flow-control actuators on the system of interest will be essential for the development of feedback control algorithms to coordinate arrays of realistic sensor/actuator configurations.

length scales that are dominant in the optimized control computations (which shed insight on suitable actuator bandwidth, dimensions, and spacing) and the extent and structure of the convolution kernels (which indicate the distance and direction over which sensor measurements and state estimates should propagate when designing the communication architecture of the tiled array).

It is recognized that the control algorithm finally to be implemented must be kept fairly simple for its realization in the on-board electronics to be feasible. We believe that an appropriate strategy for determining implementable feedback algorithms that are both effective and simple is to learn how to solve the high-dimensional, fully resolved control problem first, as discussed herein. This results in high-dimensional compensator designs that are highly effective in the closed-loop setting. Compensator reduction strategies combined with engineering judgment may then be used to distill the essential features of such well-resolved control solutions to implementable feedback designs with minimal degradation of the closed-loop system behavior.

13.17 Discussion: A Common Language for Dialog

It is imperative that an accessible language be developed that provides a common ground upon which people from the fields of fluid mechanics, mathematics and controls can meet, communicate and develop new theories and techniques for flow control. Pierre-Simon de Laplace (quoted by Rose, 1998) once said

> Such is the advantage of a well-constructed language that its simplified notation often becomes the source of profound theories.

Similarly, it was recognized by Gottfried Wilhelm Leibniz (quoted by Simmons, 1992) that

> In symbols one observes an advantage in discovery which is greatest when they express the exact nature of a thing briefly…then indeed the labor of thought is wonderfully diminished.

Profound new theories are still possible in this young field. To a large extent, however, we have not yet homed in on a common language in which such profound theories can be framed. Such a language needs to be actively pursued; time spent on identifying, implementing and explaining a clear "compromise" language that is approachable by those from the related "traditional" disciplines is time well spent.

In particular, care should be taken to respect the meaning of certain "loaded" words which imply specific techniques, qualities, or phenomena in some disciplines but only general notions in others. When both writing and reading papers on flow control, one must be especially alert, as these words are sometimes used outside of their more narrow, specialized definitions, creating undue confusion. With time, a common language will develop. In the meantime, avoiding the use of such words outside of their specialized definitions, precisely defining such words when they are used, and identifying and using the existing names for specialized techniques already well established in some disciplines when introducing such techniques into other disciplines, will go a long way towards keeping us focused and in sync as an extended research community.

There are, of course, some significant obstacles to the implementation of a common language. For example, fluid mechanicians have historically used **u** to denote flow velocities and **x** to denote spatial coordinates, whereas the controls community overwhelmingly adopts **x** as the state vector and **u** as the control. The simplified two-dimensional system that fluid mechanicians often study examines the flow in a vertical plane, whereas the simplified two-dimensional system that meteorologists often study examines the flow in a horizontal plane; thus, when studying three-dimensional problems such as turbulence, those with a background in fluid mechanics usually introduce their third coordinate, z, in a horizontal direction, whereas those with a background in meteorology normally have "their zed in the clouds." Writing papers in a manner conscious to such different backgrounds and notations, elucidating, motivating, and distilling the suitable control strategies, the relevant flow physics, the useful mathematical inequalities, and the appropriate numerical methods to a general audience of specialists from other fields is certainly extra work. However, such efforts are necessary to make flow control research accessible to the broad audience of scientists, mathematicians, and engineers whose talents will be instrumental in advancing this field in the years to come.

13.18 The Future: A Renaissance

The field of flow control is now poised for explosive growth and exciting new discoveries. The relative maturity of the constituent traditional scientific disciplines contributing to this field provides us with key elements that future efforts in this field may leverage. The work described herein represents only our first, preliminary steps towards laying an integrated, interdisciplinary footing upon which future efforts in this field may be based. Many technologically significant and fundamentally important problems lie before us, awaiting analysis and new understanding in this setting. With each of these new applications come significant new questions about how best to integrate the constituent disciplines. The answers to these difficult questions will only come about through a broad knowledge of what these disciplines have to offer and how they can best be used in concert. A few problems that might be studied in the near future in the present interdisciplinary framework are highlighted in Figure 13.22.

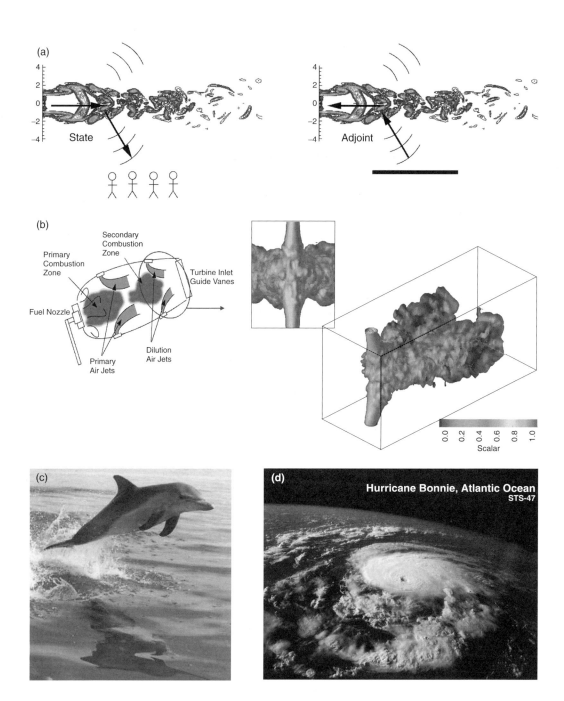

FIGURE 13.22 (Color figure follows p. 12-26.) Future interdisciplinary problems in flow control amenable to adjoint-based analysis:
(a) minimization of sound radiating from a turbulent jet (simulation by Prof. Jon Freund, UCLA),
(b) maximization of mixing in interacting cross-flow jets (simulation by Dr. Peter Blossey, UCSD) [Schematic of jet engine combustor is shown at left. Simulation of interacting cross-flow dilution jets, designed to keep the turbine inlet vanes cool, are visualized at right.],
(c) optimization of surface compliance properties to minimize turbulent skin friction, and
(d) accurate forecasting of inclement weather systems.

Unfortunately, there are particular difficulties in pursuing truly interdisciplinary investigations of fundamental problems in flow control in our current society, as it is impossible to conduct such investigations from the perspective of any particular traditional discipline alone. Though the language of interdisciplinary research is in vogue, many university departments, funding agencies, technical journals and, therefore, college professors fall back on the pervasive tendency of the twentieth-century scientist to categorize and isolate difficult scientific questions, often to the exclusion of addressing the fundamentally interdisciplinary issues. The proliferation and advancement of science in the twentieth century was, in fact, largely due to such an approach; by isolating specific and difficult problems with single-minded focus into narrowly defined scientific disciplines, great advances could once be achieved. To a large extent, however, the opportunities once possible with such a narrow focus have stagnated in many fields, though we are left with the scientific infrastructure in which that approach once flourished. To advance, we must courageously lead our research groups outside of the various neatly defined scientific domains into which this infrastructure injects us, and pursue the significant new opportunities appearing at their intersection. University departments and technical journals can and will follow suit as increasingly successful interdisciplinary efforts, such as those in the field of flow control, gain momentum. The endorsement that professional societies, technical journals and funding agencies might bring to such interdisciplinary efforts holds the potential to significantly accelerate this reformation of the scientific infrastructure.

In order to promote interdisciplinary work in the scientific community at large, describing oneself as working at the intersection of disciplines X and Y (or, where they are still disjoint, the bridge between such disciplines) needs to become more commonplace. People often resort to the philosophy "I do X... oh, and I also sometimes dabble a bit with Y," but the philosophy "I do $X * Y$," where $*$ denotes something of the nature of an integral convolution, has not been in favor since the Renaissance. Perhaps the primary reason for this is that X and Y (and $Z, W, \ldots$) have gotten progressively more and more difficult. By specialization (though often to the point of isolation), we are able to "master" our more and more narrowly defined disciplines. In the experience of the author, not only is it often the case that X and Y are not immiscible, but the solution sought may often not be formulated with the ingredients of X or Y alone. In order to advance, the essential ingredients of X and Y must be crystallized and communicated across the artificial disciplinary boundaries. New research must then be conducted at the intersection of X and Y. To be successful in the years to come, we must prepare ourselves and our students with the training, perspective and resolve to seize the new opportunities appearing at such intersections with a Renaissance approach.

Acknowledgments

This chapter is adapted from the article by the same author, "Flow control: new challenges for a new Renaissance," *Progress in Aerospace Sciences*, **37**, 21–58, 2001, with permission from Elsevier Science.

The author is thankful to Prof. Robert Skelton, for his vision to promote integrated controls research at the intersection of disciplines, to Profs. Roger Temam and Mohammed Ziane, for their patient attention to the new mathematical challenges laying the foundation for this development, and to Profs. Dan Henningson, Patrick Huerre, John Kim and Parviz Moin, for their physical insight and continued support of this sometimes unconventional effort within the fluid mechanics community. The author has also benefited from numerous technical discussions with Profs. Jeff Baggett, Bassam Bamieh, Bob Bitmead, Scott Collis, Brian Farrell, Jon Freund, Petros Ioannou, and Miroslav Krstic, and remains indebted to the several graduate students and post-docs who are making it all happen, including Dr. Peter Blossey, Markus Högberg, Eric Lauga and Scott Miller. We also thank the AFOSR, NSF, and DARPA for the foresight to sponsor several of these investigations. This article is dedicated to the memory of Prof. Mohammed Dahleh, whose charisma and intellect endure as a continual inspiration to all who knew him.

References

Abergel, F., and Teman, R. (1990) "On Some Control Problems in Fluid Mechanics," *Theor. Comput. Fluid Dyn.* **1**, pp. 303–325.

Atkinson, G.W. (1993) *Chess and Machine Intuition,* Ablex, Norwood, NJ.

Balogh, A., Liu, W.-J., and Krstic, M. (2001) "Stability Enhancement by Boundary Control in 2D Channel Flow," *IEEE Trans. Autom. Contr.* (submitted).

Bamieh, B., Paganini, F., and Dahleh, M. (2001) "Distributed Control of Spatially Invariant Systems," *IEEE Trans. Autom. Contr.* (to appear).

Banks, H.T., ed. (1992) "Control and Estimation in Distributed Parameter System," in *Frontiers in Applied Mathematics,* Vol. **11**, SIAM.

Banks, H.T., Fabiano, R.H., and Ito, K., eds. (1993) "Identification and Control in Systems Governed by Partial Differential Equations," in *Proceedings in Applied Mathematics,* Vol. **68**, SIAM.

Bewley, T.R. (1999) "Linear Control and Estimation of Nonlinear Chaotic Convection: Harnessing the Butterfly Effect," *Phys. Fluids* **11**, pp. 1169–1186.

Bewley, T.R., and Agarwal, R. (1996) "Optimal and Robust Control of Transition," *Center for Turbulence Research, Proc. of the Summer Program 1996,* pp. 405–432.

Bewley, T.R., and Liu, S. (1998) "Optimal and Robust control and Estimation of Linear Paths to Transition," *J. Fluid Mech.* **365**, pp. 305–349.

Bewley, T.R., Moin, P., and Temam, R. (2001) "DNS-Based Predictive Control of Turbulence: An Optimal Target for Feedback Algorithms," *J. Fluid Mech.* (to appear).

Bewley, T.R., Temam, R., and Ziane, M. (2000) "A General Framework for Robust Control in Fluid Mechanics," *Physica D* **138**, pp. 360–392.

Boškovic, D.M., and Krstic, M. (2001) "Global Stabilization of a Thermal Convection Loop, *Automatica* (submitted).

Butler, K.M., and Farrell, B.F. (1992) "Three-Dimensional Optimal Perturbations in Viscous Shear Flows," *Phys. Fluids A* **4**(8), pp. 1637–1650.

Chang, Y., and Collis, S.S. (1999) "Active Control of Turbulent Channel Flows Based on Large-Eddy Simulation," *Proc. of the 3rd ASME/JSME Joint Fluids Engineering Conf.,* FEDSM 99-6929, July 18–23, San Francisco, CA.

Cortelezzi, L., and Speyer, J.L. (1998) "Robust Reduced-Order Controller of Laminar Boundary Layer Transitions," *Phys. Rev. E* **58**, pp. 1906–1910.

D'Andrea, R., and Dullerud, G.E. (2000) "Distributed Control of Spatially Interconnected Systems," *IEEE Trans. Autom. Contr.* (submitted).

Doyle, J.C., Glover, K., Khargonekar, P.P., and Francis, B.A. (1989) "State-Space Solutions to Standard and Control Problems," *IEEE Trans. Autom. Contr.* **34**, pp. 831–847.

Farrell, B.F., and Ioannou, P.J. (1993) "Stochastic Forcing of the Linearized Navier–Stokes Equation," *Phys. Fluids A* **5**, pp. 2600–2609.

Gad-el-Hak, M. (1996) "Modern Developments in Flow Control," *Appl. Mech. Rev.* **49**, pp. 365–379.

Green, M., and Limebeer, D.J.N. (1995) *Linear Robust Control,* Prentice-Hall, Englewood Cliffs, NJ.

Gunzburger, M.D., ed. (1995) *Flow Control,* Springer-Verlag, Berlin.

Hamilton, J.M., Kim, J., and Waleffe, F. (1995) "Regeneration Mechanisms of Near-Wall Turbulence Structures," *J. Fluid Mech.* **287**, pp. 317–348.

Ho, C.-M., and Tai, Y.-C. (1996) "Review: MEMS and Its Applications for Flow Control," *ASME J. Fluid Eng.* **118**, 437–447.

Ho, C.-M., and Tai, Y.-C. (1998) "Micro-Electro-Mechanical Systems (MEMS) and Fluid Flows," *Annu. Rev. Fluid Mech.* **30**, pp. 579–612.

Högberg, M., and Bewley, T.R. (2000) "Spatially Compact Convolution Kernels for Decentralized Control and Estimation of Transition in Plane Channel Flow, *Automatica* (submitted).

Holmes, P., Lumley, J.L., and Berkooz, G. (1996) *Turbulence, Coherent Structures, Dynamical Systems and Symmetry,* Cambridge University Press, Cambridge, U.K.

Huerre, P., and Monkewitz, P.A. (1990) "Local and Global Instabilities in Spatially Developing Flows," *Annu. Rev. Fluid Mech.* **22**, pp. 473–537.

Joshi, S.S., Speyer, J.L., and Kim, J. (1997) "A Systems Theory Approach to the Feedback Stabilization of Infinitesimal and Finite-Amplitude Disturbances in Plane Poiseuille Flow," *J. Fluid Mech.* **332**, pp. 157–184.

Kim, J., and Lim, J. (2000) "A Linear Process in Wall-Bounded Turbulent Shear Flows," *Phys. Fluid* **12**(8), pp. 1885–1888.

Koumoutsakos, P. (1999) "Vorticity Flux Control for a Turbulent Channel Flow," *Phys. Fluids* **11**(2), pp. 248–250.

Koumoutsakos, P., Freund, J., and Parekh, D. (1998) "Evolution Strategies for Parameter Optimization in Controlled Jet Flows," *Proc. of the 1998 CTR Summer Program,* Center for Turbulence Research, Stanford University/NASA Ames.

Lagnese, J.E., Russell, D.L., and White, L.W., eds. (1995) *Control and Optimal Design of Distributed Parameter Systems,* Springer-Verlag, Berlin.

Laub, A.J. (1991) "Invariant Subspace Methods for the Numerical Solution of Riccati Equations," in *The Riccati Equation,* eds. Bittaini, Laub, and Willems, Springer-Verlag, Berlin, pp. 163–196.

Lauga, E., and Bewley, T.R. (2000) "Robust Control of Linear Global Instability in Models of Non-Parallel Shear Flows" (under preparation).

Löfdahl, L., and Gad-el-Hak, M. (1999) "MEMS Applications in Turbulence and Flow Control," *Prog. Aerospace Sci.* **35**, pp. 101–203.

Lumley, J., and Blossey, P. (1998) "Control of Turbulence," *Annu. Rev. Fluid Mech.* **30**, pp. 311–327.

McMichael, J.M. (1996) *Progress and Prospects for Active Flow Control Using Microfabricated Electro-Mechanical Systems (MEMS),* AIAA Paper 96-0306.

Newborn, M. (1997) *Kasparov Versus Deep Blue: Computer Chess Comes of Age,* Springer-Verlag, Berlin.

Nosenchuck, D.M. (1994) "Electromagnetic Turbulent Boundary-Layer Control," *Bull. Am. Phys. Soc.,* **39**, p. 1938.

Obinata, G., and Anderson, B.D.O. (2000) *Model Reduction for Control System Design* (in press).

Press, W.H., Flannery, B.P., Teukolsky, S.A., and Vetterling, W.T. (1986) *Numerical Recipes,* Cambridge University Press, Cambridge, U.K.

Rathnasingham, R., and Breuer, K.S. (1997) "System Identification and Control of a Turbulent Boundary Layer," *Phys. Fluids* **9**, pp. 1867–1869.

Reddy, S.C., Schmid, P.J., Baggett, J.S., and Henningson, D.S. (1998) "On Stability of Streamwise Streaks and Transition Thresholds in Plane Channel Flows," *J. Fluid Mech.* **365**, pp. 269–303.

Reuther, J., Jameson, A., Farmer, J., Martinelli, L., and Saunders, D. (1996) *Aerodynamic Shape Optimization of Complex Aircraft Configurations via an Adjoint Formulation,* AIAA Paper 96-0094.

Rose, N., ed. (1998) *Mathematical Maxims and Minims,* Rome Press, Raleigh, NC.

Simmons, G. (1992) *Calculus Gems,* McGraw-Hill, New York.

Skogestad, S., and Postlethwaite, I. (1996) *Multivariable Feedback Control,* John Wiley & Sons, New York.

Sritharan, S.S. (1998) *Optimal Control of Viscous Flows,* SIAM.

Stein, G., and Athans, M. (1987) "The LQG/LTR Procedure for Multivariable Feedback Control Design," *IEEE Trans. Autom. Contr.* **32**, pp. 105–114.

Tachim Medjo, T. (2000) "Iterative Methods for Robust Control Problems" (in preparation).

Trefethen, L.N., Trefethen, A.E., Reddy, S.C., and Driscoll, T.A. (1993) "Hydrodynamic Stability without Eigenvalues," *Science* **261**, pp. 578–584.

Vainberg, M. (1964) *Variational Methods for the Study of Nonlinear Operators,* Holden-Day, Oakland, CA.

Vanderplaats, G. (1984) *Numerical Optimization Techniques for Engineering Design,* McGraw-Hill, New York.

Vogel and Wade (1995) "Analysis of Costate Discretizations in Parameter Estimation for Linear Evolution Equations," *SIAM J. Contr. Optimization* **33**, pp. 227–254.

Zhou, K., and Doyle, J.C. (1998) *Essentials of Robust Control,* Prentice-Hall, Englewood Cliffs, NJ.

Zhou, K., Doyle, J.C., and Glover, K. (1996) *Robust and Optimal Control,* Prentice-Hall, Englewood Cliffs, NJ.

14
Soft Computing in Control

Mihir Sen
University of Notre Dame

Bill Goodwine
University of Notre Dame

14.1 Introduction 14-1
14.2 Artificial Neural Networks 14-2
   Background • Feedforward ANN • Training
   • Implementation Issues • Neurocontrol • Heat Exchanger
   Application • Other Applications • Concluding Remarks
14.3 Genetic Algorithms 14-13
   Procedure • Heat Exchanger Application • Other
   Applications • Final Remarks
14.4 Fuzzy Logic and Fuzzy Control 14-18
   Introduction • Example Implementation of Fuzzy
   Control • Fuzzy Sets and Fuzzy Logic • Fuzzy Logic
   • Alternative Inference Systems • Other Applications
14.5 Conclusions 14-32
   Acknowledgments 14-32

14.1 Introduction

Many important applications of microelectromechanical systems (MEMS) devices involve the control of complex systems, be they fluid, solid or thermal. For example, MEMS have been used for microsensors and microactuators [Subramanian et al., 1997; Nagaoka et al., 1997]. They have also been used in conjunction with optimal closed-loop control to increase the critical buckling load of a pinned-end column and for structural stability [Berlin et al., 1998]. Another example is that of the active control of noise inside an enclosure from exterior noise sources that is accomplished by the use of MEMS sensors placed on the enclosure wall. Varadan et al. (1995) used them for the active vibration and noise control of thin plates. Vandelli et al. (1998) developed a MEMS microvalve array for fluid flow control. Nelson et al. (1998) applied control theory to the microassembly of MEMS devices.

Solving the problem of control of other complex systems, such as fluid flows and structures, using these techniques appears to be promising. Ho and Tai (1996, 1998) have reviewed the applications of MEMS to flow control, Gad-el-Hak (1999) discussed the fluid mechanics of microdevices, and Löfdahl and Gad-el-Hak (1999) provided an overview of the applications of MEMS technology to turbulence and flow control. Sen and Yang (2000) have reviewed applications of artificial neural networks and genetic algorithms to thermal systems.

A previous chapter has outlined the basics of control theory and some of its applications. Apart from the traditional approach, there is another perspective that can be taken towards control, that of artificial intelligence (AI). This is a body of diverse techniques that have been recently developed in the computer science community to solve problems that could not be solved, or were difficult to solve, by other means. Though AI is difficult to define, it is often referred to as doing through means of a computer what a human being would do to solve a given problem. The objective here, however, is not to discuss AI but to point out that some of the techniques developed in the context of AI can be transported to applications that involve MEMS. If the latter is the hardware of the future, the former may be its software. AI encompasses a broad spectrum of computational techniques and methods which are often thought of as being based on heuristics rather than algorithms. Thus, they are often not guaranteed to work, but have a high probability of success. As a general characteristic they imitate nature in some sense, though the difference between manmade computers and nature itself is so vast that the analogy is far from perfect.

There are some specific techniques within AI, collectively known as soft computing (SC), that have matured to the point of being computationally useful for complex engineering problems. SC includes artificial neural networks, genetic algorithms and fuzzy logic and related techniques in probabilistic reasoning. SC techniques are especially useful for complex systems, which, for purposes of this discussion, will be defined to mean those that can be broken into a number of subsystems that are individually simple and may be analytically or numerically computed but together cannot be analyzed in real time for control purposes. The physical phenomena behind these complex systems may or may not be known, or the system may or may not be possible to model mathematically. Often the model equations, as in the case of turbulent fluid flow, are too many or too difficult to permit analytical solutions or rapid numerical computations.

The objective of this chapter is to describe the basic SC techniques that can be applied to control problems relevant to MEMS. The sections below will describe the artificial neural network, genetic algorithm and fuzzy logic methodologies in outline. The description will be at an introductory level so as to let the reader decide whether the technique has some utility to his or her own application, and then will point to references for further details. It is important to emphasize that these are tools, and as such work much better in some circumstances than in others so that caution must always be used in their use. Reference will also be made to some of the applications that have been reported in the literature to give an idea of the kind of problems that can be approached.

Several excellent books and texts can be consulted on the general subject of SC. Aminzadeh and Jamshidi (1994), Yager and Zadeh (1994), Bouchon-Meunier et al. (1995) and Jang et al. (1997) cover broad aspects of SC; Schalkoff (1997) and Haykin (1999) deal with artificial neural networks; Fogel (1999) provides an outline of evolutionary programming; Goldberg (1989) presents an exposition on genetic algorithms; and Mordeson and Nair (1988) introduce the topic of fuzzy logic. Books covering more specific areas include those by Jain and Fukuda (1998) on the application of SC to robotic systems and Buckley and Feuring (1999) and Pal and Mitra (1999) on neuro-fuzzy systems.

14.2 Artificial Neural Networks

One of the most common SC-based techniques is that of artificial neural networks (ANN). Excellent introductory texts are available on the subject, among them those by Schalkoff (1997) and Haykin (1999), in which an account of its history and mathematical background can be found. The technique has been applied to such diverse fields as philosophy, psychology, business and economics, in addition to science and engineering. What all these applications have in common is complexity, for which the ANN is particularly suitable.

14.2.1 Background

Inspiration for the ANN comes from the study of biological neurons in humans and other animals. These neurons learn from experience and are also able to handle and store information that is not precise [Eeckman, 1992]. Each neuron in a biological network of interconnecting neurons receives input signals from other neurons and, if the accumulation of inputs exceeds a certain threshold, puts out a signal that is sent to other neurons to which it is connected. The decision to fire or not represents the ability of the ANN to learn and store information. In spite of the analogy between the biological and computational neurons, there are significant differences that must be kept in mind. Though the biological processes in a neuron are slower, the connections are massively parallel as compared to its computational analogue, which must be limited by the speed of the currently available hardware.

Artificial neural networks are designed to mimic the biological behavior of natural neurons. Each artificial neuron (or node) in this network has connections (or synapses) with other neurons and has an input/output characteristic function. An ANN is composed of a number of artificial neurons, and each interneural connection has associated with it a certain weight and each neuron itself with a certain bias. Central to the use of an ANN is its training, which is also called the *learning process*. Training is the use of existing data to find a suitable set of weights and biases.

Many different ANN structures and configurations have been proposed along with various methodologies of training [Warwick et al., 1992]. The one we shall discuss in some detail is the multilayer ANN operating in the feedforward mode, and we will use the backpropagation algorithm for training. This combination has been found to be useful for many engineering and control purposes [Zeng, 1998]. The ANN, once trained, works as an input–output system with multiple inputs and outputs, and learning is accomplished by adjusting the weights and biases so that the training data are reproduced.

14.2.2 Feedforward ANN

Figure 14.1 shows a feedforward ANN consisting of a series of layers, each with a number of neurons. The first and last layers are for input and output, respectively, while the ones in between are the hidden layers. The ANN is said to be fully connected when any neuron in a given layer is connected to all the neurons in the adjacent layers.

Though notation in this subject is not standard, we will use the following. The jth neuron in the ith layer will be written (i, j). The input of the neuron (i, j), is x_{ij}, its output is y_{ij}, its bias is θ_{ij}, and $w_{i-1,k}^{i,j}$ is the synaptic weight between neurons $(i-1, k)$ and (i, j). A number of parameters determine the configuration: I is the total number of layers, and J_i is the number of neurons in the ith layer. There are thus J_1 input values to the ANN and J_1 output.

Each neuron processes the information between its input and output. The input of a neuron is the sum of all the outputs from the previous neurons modified by the respective internodal synaptic weights and a bias at the neuron. Thus, the relation between the output of the neurons $(i-1, k)$ for $k = 1, \ldots, J_{i-1}$

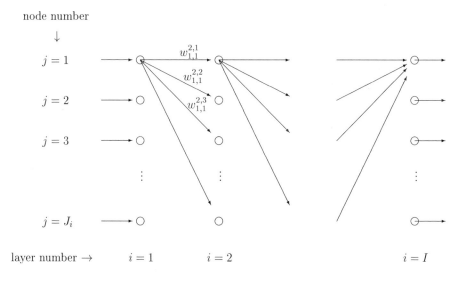

FIGURE 14.1 Schematic of feedforward neural network.

in one layer and the input of a neuron (i, j) in the following layer is

$$x_{i,j} = \theta_{i,j} + \sum_{k=1}^{J_{i-1}} w^{i,j}_{i-1,k} \, y_{i-1,k} \tag{14.1}$$

The input and output of the neuron (i, j) are related by:

$$y_{i,j} = \phi_{i,j}(x_{i,j}) \quad \text{for } i > 1 \tag{14.2}$$

and

$$y_{i,j} = x_{i,j} \quad \text{for } i = 1 \tag{14.3}$$

The function $\phi_{i,j}(x)$, called the *activation function*, plays a central role in the processing of information by the ANN. When the input signal is small, the neuron suppresses the signal altogether, resulting in a small output, and when the input exceeds a certain threshold the neuron fires and sends a signal to all the neurons in the next layer. Several appropriate activation functions have been used, the most popular being the logistic sigmoid function:

$$\phi_{i,j}(x) = \frac{1}{1 + \exp(-x/c)} \tag{14.4}$$

where c is a parameter that determines the steepness of the function. This function has several useful characteristics: It is an approximation to the step function that simulates the operation of firing and not firing but with continuous derivatives, and its output always lies between 0 and 1.

14.2.3 Training

Once the configuration of an ANN is fixed, the weights between the neurons and the bias at each neuron define its input–output characteristics. The weights determine the relative importance of each one of the signals received by a neuron from those of the previous layer, and the bias is the propensity for the combined input to trigger a response from the neuron. The process of training is the adjustment of the weights and biases to reproduce a known set of input–output values that have to be provided.

Though there are many methods in use, the backpropagation technique is a widely used deterministic training algorithm for this type of ANN [Rumelhart et al., 1986]. The method is based on the minimization of an error function by the method of steepest descent. Descriptions of this algorithm may be found in many recent texts on ANN (for instance, Rzempoluck [1998]), and only a brief outline will be given here. First, initial values are assigned to the weights and biases. Then, for a given input to the ANN, the output is determined. The synaptic weights and biases are iteratively modified until the output values differ little from the target outputs.

In the backpropagation method, an error $\delta_{I,j}$ is quantified for the last layer by:

$$\delta_{I,j} = (y^T_{I,j} - y_{I,j}) y_{I,j} (1 - y_{I,j}) \tag{14.5}$$

where $y^T_{I,j}$ is the target output for the jth neuron of the last layer. This equation comes from a finite-difference approximation of the derivative of the sigmoid function. After all the $\delta_{I,j}$ have been calculated, the computation moves back to the layer $I - 1$. There are no target outputs for this layer, so the value,

$$\delta_{I-1,k} = y_{I-1,k}(1 - y_{I-1,k}) \sum_{j=1}^{J_I} \delta_{I,j} w^{I,j}_{I-1,k} \tag{14.6}$$

is used instead. A similar procedure is used for all the inner layers until layer 2 is reached. After all these errors have been determined, changes in the weights and biases are calculated from:

$$\Delta w_{i-1,k}^{I,j} = \lambda \delta_{i,j} y_{i-1,k}$$
$$\Delta \theta_{i,j} = \lambda \delta_{i,j} \tag{14.7}$$

for $i < I$, where λ is the learning rate. From this the new weights and biases are determined.

In one cycle of training, a new set of synaptic weights and biases is determined for all training data after which the error defined by:

$$E = \frac{1}{2} \sum_{j=1}^{J_I} (y_{I,j}^T - y_{I,j})^2 \tag{14.8}$$

is calculated. The error of the ANN at the end of each cycle can be based on either a maximum or averaged value for the output errors. The process is repeated over many cycles with the weights and biases being continuously updated throughout the training runs and cycles. The training is terminated when the error is low enough to be satisfactory in some pre-determined sense.

14.2.4 Implementation Issues

Several choices must be made to construct a suitable ANN for a given problem, and they are fairly important in achieving good results. There is no general theoretical basis for these choices, and experience combined with trial and error are still the best guides.

14.2.4.1 Configuration

The first choice that must be made in using an ANN is its configuration (i.e., the number of layers and the number of neurons in each layer). Though the accuracy of prediction sometimes becomes better (at other times, all it does is pick up additional noise) as the number of layers and neurons becomes larger, the number of cycles to achieve this also increases. It is possible to do some optimization by beginning with one hidden layer as a starting point and then adding more neurons and layers while checking the prediction error [Flood and Kartam, 1994]. Practical considerations dictate a compromise between accuracy and computational speed. It is usual to not go beyond two or three hidden layers, though many users prefer only one.

There are other suggestions for choosing the parameters of the ANN. Karmin (1990) used a relatively large ANN reduced in size by removing neurons that do not significantly affect the results. In the so-called radial-Gaussian system, hidden neurons are added to the ANN in a systematic way during the training process [Gagarin et al., 1994]. It is also possible to use evolutionary programming to optimize the ANN configuration [Angeline et al., 1994]. Some authors, e.g., Thibault and Grandjean (1991), present studies of the effect of varying these parameters.

14.2.4.2 Normalization

The data that the ANN handles are usually dimensional and thus have to be normalized. Furthermore, the slope of the sigmoid function $\phi_{i,j}(x)$ used as the activation function becomes smaller as $x \to \pm \infty$. To use the central part of the function it is desirable to normalize all physical variables. In other words, the range between the minimum and maximum values in the training data is linearly mapped into a restricted range such as [0.15, 0.85] or [0.1, 0.9]. The exact choice is somewhat arbitrary and the operation of the ANN is not very sensitive to these values.

14.2.4.3 Learning Rate

The learning rate, λ, is another parameter that must be arbitrarily assumed. If it is large, the changes in the weights and biases in each step will be large; the ANN will learn quickly, but the training process could be oscillatory or unstable. Small learning rates, on the other hand, lead to a longer training period to achieve the same accuracy. Its value is usually taken to be around 0.4 and determined by trial and error; there are also other possibilities [Kamarthi et al., 1992].

14.2.4.4 Initial Values

Initial values of weights and biases have to be assigned at the beginning of the training process. Its importance lies in the fact that both the final values reached after training and the number of cycles needed to reach a reasonable convergence depend on the initial values. A simple method is to assign the values in a random fashion, though Wessels and Barnard (1992), Drago and Ridella (1992) and Lehtokangas et al. (1995) have suggested other methods for determining the initial assignment. Sometimes, the ANN is trained or upgraded on new data for which the old values can be used as the initial weights and biases.

14.2.4.5 Training Cutoff

Training is repeated until a certain criterion is reached. A simple one is a fixed number of cycles. It is also common to specify the minimum in the error-number of cycles curve as the end of training. This has a possible pitfall in that there may be local minimum, beyond which the error may decrease some more.

14.2.5 Neurocontrol

The ANN described up to now is a static input–out system; that is, given an input vector it is able to predict an output. For purposes of real-time control the procedure must be extended to variables that are changing in time. This means that time t must also be a variable for both training and predictions. There are two ways in which this can be done: Either the time t or a time-step Δt between predictions can be considered to be additional inputs to the ANN. The latter procedure, which is convenient for microprocessor applications, has the advantage in that the initial values not really relevant to a control system after a long time quickly become irrelevant. The time step Δt in this procedure may be constant or may vary according to the needs of the prediction as time goes on.

The ANN can be trained as before. The trained ANN can predict values of variables at an instant $t + \Delta t$ if the values at t are given. The dynamic simulation can be introduced into a prediction-based controller to control the behavior of a system.

14.2.6 Heat Exchanger Application

The general procedures and methodology of ANNs have been discussed up to now. In this section, we will apply the method to the specific problem of heat exchangers. A heat exchanger is merely an example of a system that, under the definition used here, is complex. Though the physical phenomena going on are well known, in the face of problems such as turbulence, secondary flows, developing flows, complicated geometry, property variations, conduction along walls, etc. it becomes impossible to compute the desired operating variables for prediction or, what is more difficult, in real time for control purposes. It will be shown that an ANN can be used for this purpose. Much of the work reported here is in Díaz (2000).

The heat exchanger used in tests is schematically shown in Figure 14.2. The experiments were carried out in a variable-speed, open wind-tunnel facility [Zhao, 1995]. Hot water flows inside the tubes of the heat exchanger, and room air is drawn over the outside of the tubes. The flow rates of air and water are measured, along with the temperatures of the two fluids going in and out.

14.2.6.1 Steady State

For a given heat exchanger, the heat transfer rate $\dot{Q}$ under steady-state conditions depends on the flow rates of air and water, $\dot{m}_a$ and $\dot{m}_w$, respectively, and their inlet temperatures, T_a^{in} and T_w^{in}, respectively. From the heat transfer rate, secondary quantities such as the fluid outlet temperatures, T_a^{out} and T_w^{out}, respectively, can be determined. For the present experiments, a total of 259 runs were made, of which only data for 197 runs were used for training, while the rest were used for testing the predictions.

Soft Computing in Control

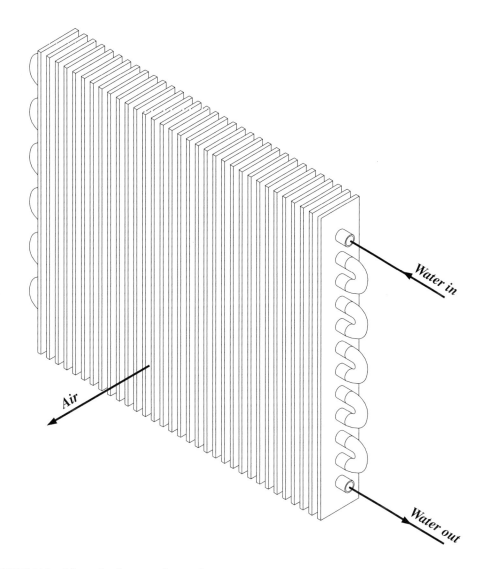

FIGURE 14.2 Schematic of compact heat exchanger.

For each test run the six quantities, $\dot{Q}$, $\dot{m}_a$, $\dot{m}_w$, T_a^{in}, T_w^{in}, T_a^{out} and T_w^{out} were measured or determined from measurements. The work here is described in detail in Díaz et al. (1999).

From the data an ANN was trained. This network had four inputs, $\dot{m}_a$, $\dot{m}_w$, T_a^{in} and T_w^{in}, and a single output, $\dot{Q}$. Many different configurations and numbers of training cycles were tried. Good results were found for a 4–5–5–1 (i.e., four layers with 4, 5, 5 and 1 neurons in each layer, respectively) configuration trained for 200,00 cycles. The trained ANN was tested on the dataset set aside for the purpose. Figure 14.3 shows the results of the ANN prediction, $\dot{Q}_{ANN}^p$, plotted against the actual measurement, $\dot{Q}^e$. The 45° line that is shown would be an exact prediction, while the dotted lines represent errors of ±10%. The prediction of a power-law correlation for the same data, $\dot{Q}_{cor}^p$ [Zhao, 1995], is also shown. The ANN is seen to do a better job than the correlation.

14.2.6.2 Thermal Neurocontrol

Once again the same heat exchanger was used to develop and use the neurocontrol methodology. Dynamic data were obtained by varying the water inlet temperatures by changing the heater settings while keeping the other variables constant. Training data were obtained from experiments in which the water inlet

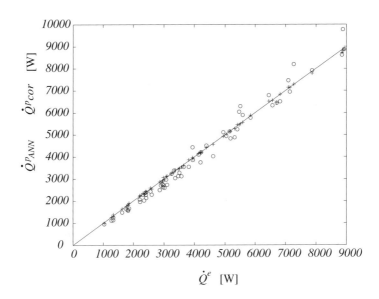

FIGURE 14.3 Predictions of 4–5–5–1 neural network (+) and correlation (○).

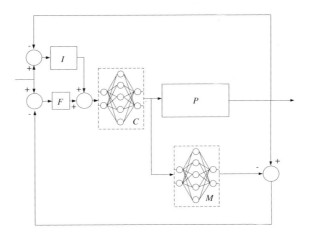

FIGURE 14.4 Closed-loop neurocontrol (P = plant, M = ANN model of plant, C = neorocontroller, F = filter, I = integral controller).

temperature was varied in small increments of 5.56°C from 32.2°C up to 65.6°C. Further details are in Díaz et al. (2001a). A nonlinear system may be controlled in many different ways. The method chosen for neurocontrol testing is shown in Figure 14.4. Control of the heat exchanger is by an inverse-ANN controller, C, while the plant, P, is modeled by a forward ANN, M. The controller is trained as a dynamic ANN. The desired control objective was to keep the outlet air temperature T_a^{out} constant.

In the first test, the system was subjected to a step change in the set temperature. The system was stabilized around $T_a^{out} = 32°C$, after which the set temperature was suddenly changed to 36°C. Figure 14.5 shows how the neurocontroller behaved as compared to conventional PID (proportional–integral–derivative) and PI (proportional–integral) controllers. All the controllers work properly, but the neurocontroller has fewer oscillations. In a second test, we looked at the disturbance rejection ability of the controller. At steady operation, the water flow is completely shut down for a short interval: between $t = 40$ s and $t = 70$ s. Figure 14.6 shows how the system variables respond to PID and neurocontrol during and after the disturbance pulse. In the neurocontroller, oscillations are quickly damped out.

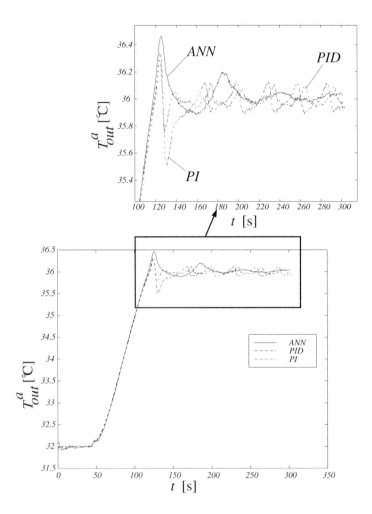

FIGURE 14.5 Response to change in set point.

The procedure outlined above in which the ANN simulates the plant to be controlled is fairly straightforward. Other special aspects should be examined further.

14.2.6.2.1 Stabilization of Feedback Loop
(See Díaz et al. [2001b].) The static ANN, once incorporated into the feedback loop, may lead to a dynamical system that is unstable. In order to avoid such a possibility, the ANN has to be trained not only to make accurate predictions but also to give dynamical stability to the loop. Because the weights are not unique, it is possible to come up with an algorithm that does both. The stabilization algorithm was designed and tested on the heat exchanger facility. Figure 14.7 shows the behavior of two controllers C_1 and C_2, where the former is a stable controller and the latter is unstable. In each case, the air flow rate is being controlled and the air outlet temperature is shown as a function of time. The stable controller works well, while in the unstable case the air flow rate is increased as far it as will go without achieving the desired result.

14.2.6.2.2 Adaptive Neurocontrol
(See Díaz et al. [2001c].) The major advantage of neurocontrollers is that they can be made adaptive. For such a system, the ANN can be made to go through a process of retraining if its predictions are not accurate due to change in system characteristics. The adaptive controller is tested in two different ways. The first is a disturbance on the water side, shown in Figure 14.8; v_a in the figure is the air velocity.

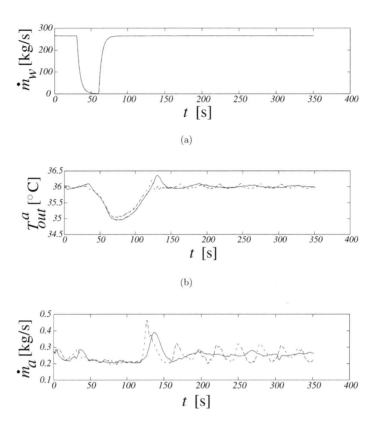

FIGURE 14.6 Disturbance rejection (continuous line is neural network, broken line is PID).

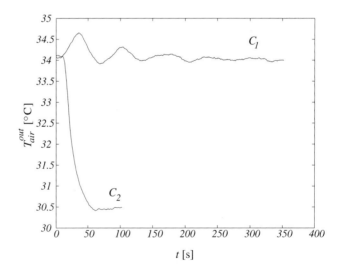

FIGURE 14.7 Performance of stable and unstable neorocontrollers.

Initially, the neurocontroller keeps the system close to $T_a^{out} = 34°C$. From $t = 100$ s to $t = 130$ s, the water is shut off. The neurocontroller tries to control until $t = 110$ s, at which point it hands off to a backup PID controller while it adapts to the new circumstances. Then, once it is able to make reasonable predictions, it resumes control at $t = 170$ s. The second test is a sudden reduction of inlet area of the wind tunnel test facility, shown in Figure 14.9. The controller keeps T_a^{out} at 34°C, and then half of the inlet is

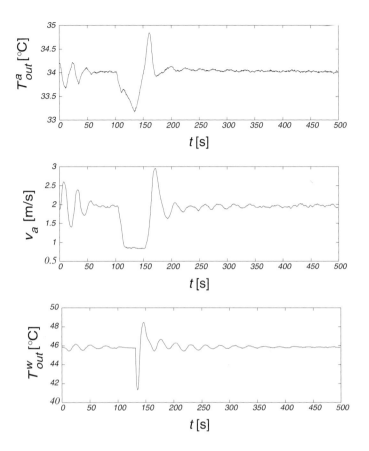

FIGURE 14.8 Response of adaptive neurocontroller to pulsed stoppage in water flow.

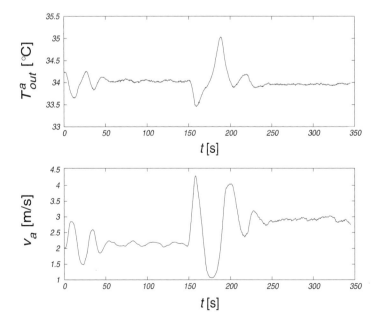

FIGURE 14.9 Response of adaptive neurocontroller to sudden change in wind-tunnel inlet area.

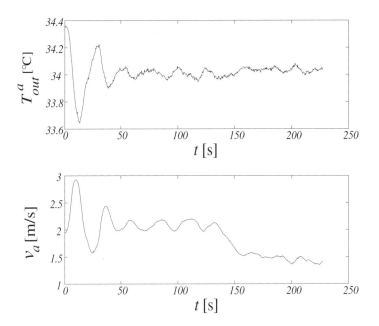

FIGURE 14.10 Application of energy minimization procedure.

suddenly blocked at $t = 150$ s. The neurocontroller adapts until it learns the behavior of the new system, until finally at $t = 240$ s it resumes control of the system.

14.2.6.2.3 Optimal Control
For systems where there is more than one variable to be controlled, it is often required that it be controlled under an optimizing constraint. For thermal systems, for example, it may be possible to require that the system use the least energy at the new setpoint. Figure 14.10 shows the neurocontroller being used in this mode with the water and air flow rates being the control variables. In this case, the ANN has been fed information related to the energy consumption of the energy-consuming components of the facility: the hydraulic pump, the fan and the electric heater. We first let the controller stabilize T_a^{out} at 34°C. Around $t = 130$ s, the energy minimization routine is turned on and the controller adjusts both the water and air flow rates to minimize energy usage while keeping T_a^{out} roughly at its set value.

14.2.7 Other Applications

Many other applications of neurocontrol have been reported in the literature, including inflow-related problems. Gad-el-Hak (1994) has given an overview of the problem of flow control in turbulent boundary layers. Control has been attempted by Jacobson and Reynolds (1993); Lee et al. (1997) used ANNs for turbulence control for drag reduction purposes; Suzuki and Kasagi (1997) were able to use it for the optimal control of vortex shedding behind a square cylinder, minimizing, at the same time, the angular motion of the cylinder; and Chan and Rad (2000) used it for the purpose of real-time flow control.

14.2.8 Concluding Remarks

The implementation of ANN procedures in a complex problem is really very straightforward. It relieves the necessity of having a first-principles model that may not be available for devices, such as those based on MEMS which are relatively new. Predictions can be made and systems can be controlled on the basis of available information without the need for models. In addition, the ANN is also extremely adaptable to changing circumstances.

14.3 Genetic Algorithms

Evolutionary algorithms are those that change or evolve as they do their work. The genetic algorithm (GA) is a specific type of search technique based on the Darwinian evolutionary principles of natural selection to attain its objective of optimization. Optimization by itself is fundamental to many applications in engineering. For example, it is important in the design of systems; although analysis permits the prediction of the behavior of a given system, optimization is the technique that searches among all possible designs of the system to find the one that is the best for the application. Optimization is also intimately related to the control problem where parameters have to be chosen: The configuration of neural networks has to be selected, and the constants of PID control have to be set. The importance of this problem has given rise to a wide variety of techniques that help search for the optimum. In this context, local and global optima must be distinguished. If one visualizes a multivariable function with many peaks, any one of these is a local maximum, while only one is the highest.

Genetic algorithms are described in monographs by Goldberg (1989), Michalewicz (1992), Mitchell (1997) and Man et al. (1999). GAs are generally used for the purpose of global optimization. They are not gradient-based and are an alternative to other global optimization techniques such as simulated annealing. Because local gradient information is not used, a GA usually finds the global optimum as opposed to a local one, a characteristic that is often useful. Furthermore, the fact that gradients are not used may be significant in problems in which the variables to be optimized are functions of discrete quantities and their derivatives are not possible.

Robustness, a central characteristic for survival of natural species, is also a feature of genetic algorithms. Goldberg (1989) compared the genetic algorithm from this perspective with other search techniques. Methods based on the calculus, such as equating to zero the derivatives of a function to obtain the extrema, are indirect. In a direct method, the iteration moves in a direction determined by the local gradient of the function. Both these methods are local and depend on the local existence of derivatives. Another class of methods is based on evaluation of a function at every point on a fine but finite grid. For most practical applications this approach is too time consuming. In yet another class of techniques, randomness is used in some way. Simulated annealing and genetic algorithm are examples of this. They search from a population of points rather than from a single point, they use only the function rather than its derivative, and they use probabilistic rather than deterministic rules. The search using many points makes the method global rather than local.

14.3.1 Procedure

There are many variants of the genetic algorithm procedure. A simple approach is described here, but it is not the only one. We can illustrate the GA procedure by finding the maximum of a function $f(x)$ within a given domain $x \in U \subset \Re^n$. In a gradient-based method, the slope of the tangent plane at a given value of x indicates which way to "go up" within an iterative procedure. The GA does not work this way. For simplicity, in the following we will assume that $U = [a, b] \subset \Re$, though the method can be easily generalized to higher dimensions. Furthermore, we will map the interval $[a, b]$ to $\lfloor 0, 2^c - 1 \rfloor$. This way, the independent variable x can be represented by a binary string of length c running from 000...000 all the way up to 111...111. The example function chosen for maximization is $f(x) = x(1 - x)$.

Step 1: We begin by randomly selecting r candidate numbers within the desired domain (i.e., $x_1,\ldots, x_r$). This is the first generation. An example is shown in Table 14.1, where we have taken $r = 10$ and $c = 6$. The second column shows the numbers in decimal form, the third column the same numbers in binary form, and the fourth the normalized binary version of the same. The function $f(x_i)$ for each number which, in the context of this algorithm, is called the fitness is indicated in the fifth column. The reason for this name is that, the higher the fitness is, the closer x_i is to its value where $f(x_i)$ is a maximum. In other words, we seek the value of x for which $f(x)$ is the fittest. The maximum fitness of the members of the this generation is 0.2469 for $x_i = 0.4444$. The last column is the normalized fitness $s(x_i)$, the values of the previous column divided by the sum of the fitnesses; thus, $s(x_i) = f(x)/\Sigma f(x_i)$.

TABLE 14.1 Binary, Decimal and Normalized Forms of the First Generation of Candidate Numbers and Their Absolute and Normalized Fitnesses

i	$x_i(d)$	$x_i(b)$	$x_i(d, n)$	$f(x_i)(d)$	$s(x_i)(d)$
1	18	010010	0.2857	0.2041	0.1096
2	53	110101	0.8413	0.1335	0.0717
3	43	101011	0.6825	0.2167	0.1164
4	11	001011	0.1746	0.1441	0.0774
5	22	010110	0.3492	0.2273	0.1221
6	46	101110	0.7302	0.1970	0.1058
7	28	011100	0.4444	0.2469	0.1326
8	42	101010	0.6667	0.2222	0.1194
9	25	011001	0.3968	0.2394	0.1286
10	61	111101	0.9683	0.0307	0.0165

Note: b = binary; d = decimal; n = normalized.

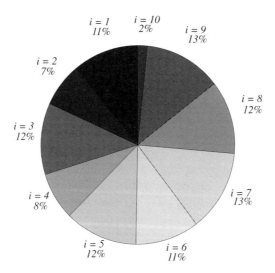

FIGURE 14.11 (Color figure follows p. 12-26.) "Old" fitness.

Step 2: From the first generation of numbers, pairs of parents are chosen which then give rise to offspring that form the next generation. To visualize the process we draw a pie chart, shown in Figure 14.11, with slices that have angles proportional to the normalized fitness $s(x_i)$. To form parents, pairs of numbers are randomly selected from the chart as if it were a roulette wheel. Of course, the numbers with larger normalized fitnesses have the higher probability of being selected. The result of such a selection process is shown in the second column of Table 14.2 as generation $G = 1/4$. These are then shuffled to produce column $G = 1/2$. The first two entries in this column are a set of parents, the next two another set, and so on.

Step 3: Each pair of parents produces a pair of offspring by crossover, which can be done in different ways. Here, we have randomly chosen a point along the two binary strings that form the parents and have interchanged the part of the string beyond this point. This is a single-point crossover. An example is shown in Figure 14.12, where the crossover point is in the middle of the string and the numbers 011100 and 011001 produce the offspring 011001 and 011100. The crossover point in the other pairs may be different. The final result of crossover between the parents is column $G = 3/4$.

Step 4: The column $G = 1$ is obtained from $G = 3/4$ by mutation. This is obtained by selecting a randomly chosen bit and changing it from 0 to 1, or vice versa. One bit in the fourth number can be seen to have been changed, as shown in Figure 14.12. The procedure above gives a new generation with

TABLE 14.2 From One Generation to the Next

$G=0$	$G=1/4$	$G=1/2$	$G=3/4$	$G=1$
010010	011100	011100	011001	011001
110101	010110	011001	011100	011100
101011	011001	010110	010110	010110
001011	101010	010110	010110	010010
010110	010010	101010	101001	101001
101110	101011	011001	011010	011010
011100	011100	010010	010010	010010
101010	010110	101010	101010	101010
011001	011001	101011	101010	101010
111101	010110	011100	011101	011101

FIGURE 14.12 (a) Crossover and (b) mutation in a genetic algorithm.

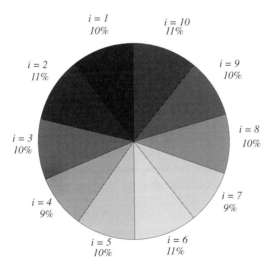

FIGURE 14.13 (Color figure follows p. 12-26.) "New" fitness.

members that are generally fitter (i.e., give a value of the function that is closer to the maximum). The new pie-chart of the normalized fitness, Figure 14.13, shows much more uniformity in fitnesses since the "unfit" have disappeared. The maximum fitness of this generation turns out to be $f(x_i) = 0.2484$ at $x_i = 0.4603$ compared to the value of $f(x_i) = 0.2469$ at $x_i = 0.4444$ for the previous generation. The process of finding a new generation is repeated several times until some criterion is satisfied. Either a desired number of cycles

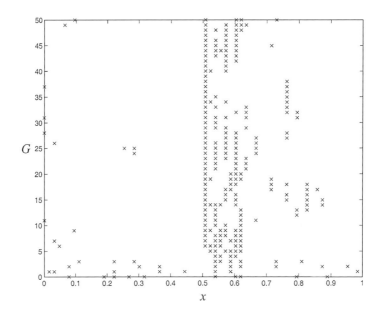

FIGURE 14.14 Change in population as a function of generation number.

have been completed, the maximum fitness of the generation does not change much, or the value of x at which this maximum fitness is obtained does not change significantly.

Some programming details need to be taken care of in actual implementation. Parameters such as the number of members of a generation and the length of a binary string must be decided upon. In addition, a finite probability for crossover and mutation must be prescribed. The crossover does the bulk of the work in selecting the new generation. Although the probability of mutation is generally kept small, it is vital as mutation enables a possible solution to break out of the neighborhood of a local optimum and go somewhere else that may turn out to have a better local optimum. It is also common to keep the best of each generation in the next to make sure that the fitness of the generation is nondecreasing. The algorithm itself is probabilistic so that it cannot be guaranteed to find the global optimum. In fact, every time it is run, the exact results obtained will be different.

Figure 14.14 shows the result of running a GA code (written by A. Pacheco-Vega) to continue the process indicated in Tables 14.1 and 14.2. The probability of crossover is taken to be unity, and that of mutation is 0.03. The abscissa shows the distribution of the population $x_1, \ldots, x_r$ at a generation number G indicated in the ordinate. The initial population at $G = 0$ is indicated by 10 different crosses (the exact values are different from those in Tables 14.1 and 14.2 because of the generation of different random numbers). In the following generations, some of the crosses overlap as numbers may repeat themselves within a population. The code has been terminated after 50 generations. A certain crowding of the population around the correct value $x = 0.5$ is observed as well as the presence of values relatively far from it. This is a consequence of the global nature of the search and is a characteristic of the GA. In addition, even at $G = 50$ the value of x_i that gives the highest value of the function is close to the correct value but is not exact.

14.3.2 Heat Exchanger Application

Application of this SC technique will again be illustrated by its use in the heat exchanger described before. The optimization problem here is to find the best correlation that fits experimental data. A set of $N = 214$ experimental runs provided the database. In each case, the heat rate $\dot{Q}$ is found as a function of the two flow rates m_w and m_a as well as the two inlet fluid temperatures T_a^{in} and T_w^{in}. Details are in Pacheco-Vega et al. (1998).

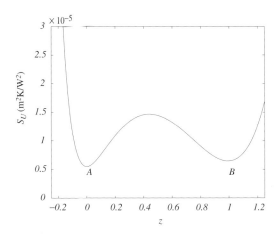

FIGURE 14.15 Global vs. local minima in optimization problem.

There are two resistances to the flow of heat by convection: on the inside with water and on the outside with air. The conventional way of handling data is to determine correlations for the inner and outer heat transfer coefficients. For example, in power-law relations of the form $Nu = a Re^n$ between the Nusselt and Reynolds numbers, Nu and Re, respectively, on both sides of the tube wall are often assumed. There are then four constants to determine: a_1, a_2, n_1 and n_2. One possible procedure is to minimize the root mean square (rms) error $S_U(a_1, a_2, n_1, n_2)$ in total thermal resistance to heat transfer between prediction and data in the least-square sense. The total resistance is the sum of the air-side and water-side resistances.

It is found that this procedure leads to a large number of local minima due to the nonlinearity of the function to be minimized. Figure 14.15 shows a pair of such minima. In the figure, a section of the error surface $S_U(a_1, a_2, n_1, n_2)$ that passes though two local minima A and B is shown. The coordinate z is a linear combination of a_1, a_2, n_1 and n_2 such that it is zero at A and unity at B, and the ordinate is the rms error. Though the values S_U of for the two correlations obtained at A and at B are very similar and the heat rate predictions for the resulting correlations are also almost equally accurate, a_1, a_2, n_1, n_2 and the predictions of the thermal resistances on either side are very different. This shows the importance, at least in this case, of using global minimization techniques for nonlinear regression analysis. If the GA is used to find the global minimum, the point A is in fact seen to be the global minimum. The correlation (not shown) found as a result of the global search is the best that fits the assumed power laws and is closest to the experimental data.

14.3.3 Other Applications

Many other applications of GAs to optimization and control problems include optimization of a control scheme by Seywald et al. (1995), Michalewicz et al. (1992), Perhinschi (1998) and Tang et al. (1996b). Reis et al. (1997) and Kao (1999) have used the GA to find the optimal location of control valves in a piping network. Gaudenzi et al. (1998) optimized the control of a beam using the technique. Several workers have applied the method to the motion of robots [Nakashima et al., 1998; Nordin et al., 1998]. Katisikas et al. (1995) and Tang et al. (1996a) used the genetic algorithm for active noise control. Nagaya and Ryu (1996) controlled the shape of a flexible beam using a shape memory alloy and Keane (1995) optimized the geometry of structures for vibration control. Dimeo and Lee (1995) controlled a boiler and turbine using the genetic algorithm. Sharatchandra et al. (1998) used the GA for shape optimization of a micropump. Kaboudan (1999) have used genetic algorithms for time-series prediction. Luk et al. (1999) developed a GA-based fuzzy logic control of a solar power plant using distributed collector fields. Additional applications of GAs combined with other SC techniques have been used for optimization of the control process [e.g., Matsuura et al., 1995; Trebi-Ollennu and White, 1997; Rahmoun and Benmohamed, 1998; Ranganath et al., 1999; Lin and Lee, 1999].

14.3.4 Final Remarks

There are two main advantages in using a genetic or evolutionary approach to optimization. One is that the methods seek the global optimum. The other advantage is that they can be used in discrete systems, in which derivatives do not exist or are meaningless. Examples of this are piping networks and positioning of electronic components. As with all tools, the reader who is interested in this procedure must evaluate the advantages and disadvantages in terms of specific applications.

14.4 Fuzzy Logic and Fuzzy Control

14.4.1 Introduction

Fuzzy sets and fuzzy logic date back to the early 1960s with the work of Lotfi Zadeh [Zadeh, 1965; 1968a;1968b; 1971] concerning complex systems and have been present in controls applications since the late 1970s [Mamdani, 1974; Mamdani and Assilian, 1975; Mamdani and Baaklini, 1975]. The main utility of fuzzy logic and its application to feedback control is comprised of two components. First, fuzzy logic is not model based so it can be applied to systems for which developing analytical models, either from first principles or from some identification techniques, is impractical or expensive. Second, it provides a convenient mechanism for application to feedback control of human (or expert) intuition regarding how a system should be controlled. This section outlines basic fuzzy set definitions, fuzzy logic concepts and their primary application to control systems. First, an illustrative controls application of fuzzy logic is presented in complete detail. The example is followed by a more complete exposition of the mathematics of fuzzy logic intended to provide the reader with a complete set of tools with which to approach a fuzzy control problem.

14.4.2 Example Implementation of Fuzzy Control

This section first introduces a typical structure of fuzzy controllers by presenting an example of a common fuzzy control application—namely, to stabilize the inverted pendulum system illustrated in Figure 14.16, where the control input is a force of magnitude u. In this problem, only the pendulum angle is stabilized. This is accomplished via linguistic variables and fuzzy if–then rules such as:

1. If the pendulum angle is zero and the angular velocity is zero, then the control force should be zero.
2. If the pendulum angle is positive and small and the angular velocity is zero, then the control force should be positive and small.

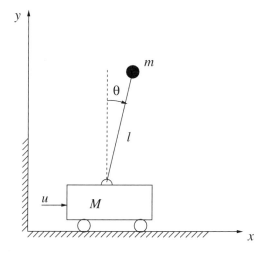

FIGURE 14.16 Pendulum system.

Soft Computing in Control

TABLE 14.3 Fuzzy Logic Rules to Determine Control Force

	Angular Velocity				
Error	Negative Large (1)	Negative Small (2)	Zero (3)	Positive Small (4)	Positive Large (5)
(1) Negative large	Negative large	Negative large	Negative large	Negative small	Zero
(2) Negative small	Negative large	Negative large	Negative small	Zero	Positive small
(3) Zero	Negative large	Negative small	Zero	Positive small	Positive large
(4) Positive small	Negative small	Zero	Positive small	Positive large	Positive large
(5) Positive large	Zero	Positive small	Positive large	Positive large	Positive large

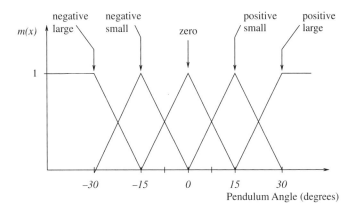

FIGURE 14.17 Pendulum angle fuzzy set.

3. If the pendulum angle is positive and large and the angular velocity is zero, then the control force should be positive and large.
4. If the pendulum angle is positive and small and the pendulum angular velocity is negative and small, then the control force should be zero.

The linguistic variables are the angle error and the angular velocity. These rules are better expressed in tabular form in Table 14.3. The first enumerated rule is expressed in the third column and third row of the table. The second rule is in the third column and fourth row. The third rule is in the third column and fifth row. The fourth rule is in the second column and fourth row. An important point is that these rules were determined by intuition. For example, whether the second column and second row should be "negative small" or "negative large" is determined by experience, guesswork or tuning.

The next basic element of the fuzzy controller is the fuzzy set, which basically encapsulates the notion of to what degree the angle is "zero," "negative small," etc. The fuzzy sets that define the fuzzy state of the angle of the pendulum system are illustrated in Figure 14.17. In the figure, if the pendulum angle is, for example, −7.5°, then the degree of membership in the "negative small" fuzzy set is 0.5 and the degree of membership in the "zero" fuzzy set is also 0.5. The degree of membership in the other fuzzy sets is 0. Similar fuzzy sets can be defined for the angular velocity and the control force, as illustrated in Figures 14.18 and 14.19, respectively.

The overall control structure is illustrated in Figure 14.20. First, the state $(\theta, \dot{\theta})$ is measured by some sensor. Next, the state is "fuzzified" by computing the degree of membership of the state in each of the fuzzy sets, A_j, used in the if–then rules. Then, the if–then rules in the rule base are evaluated in parallel, and the output of each rule is the fuzzy set (control force), which has the shape of the fuzzy set associated with the output of the if–then rule but is "capped" or "cut off" at the degree of membership of the state

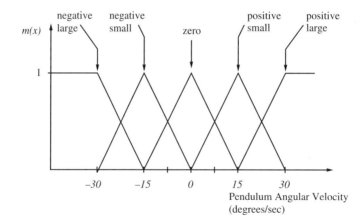

FIGURE 14.18 Pendulum velocity fuzzy set.

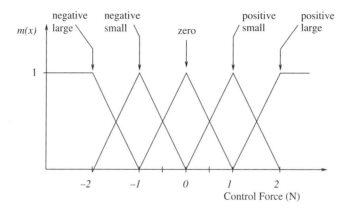

FIGURE 14.19 Pendulum force fuzzy set.

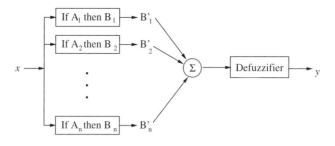

FIGURE 14.20 Fuzzy control structure.

in the associated fuzzy set. If there is a logical operation, such as "and" in the antecedent (the "if" part) of the rule, then the minimum of the degree of membership in each of the fuzzy sets is used.

As a concrete example of this "fuzzy inference," consider the case where the pendulum angle is −20° and the angular velocity is +22.5°/s. Then, the fuzzy state of the angle of the system is determined according to Figure 14.21, where the state of the system is represented by a 0.25 degree of membership

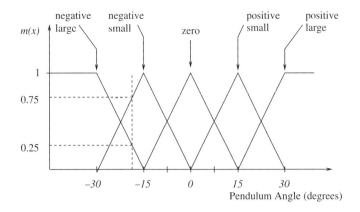

FIGURE 14.21 Fuzzification of pendulum angle.

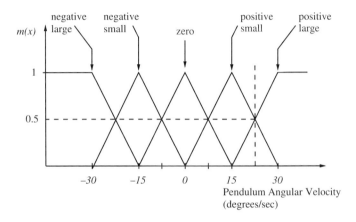

FIGURE 14.22 Fuzzification of pendulum anglular velocity.

in the "negative large" fuzzy set, and a 0.75 degree of membership in the "negative small" fuzzy set. Correspondingly, the velocity is characterized by a 0.5 degree of membership in both the "positive large" fuzzy set and the "positive small" fuzzy set, as is illustrated in Figure 14.22.

Now, the output of each rule will be the corresponding force fuzzy set, but modified so that its maximum value is capped to be the minimum degree of membership of the two elements of the antecedent part of each rule. In particular, only four of the rules listed in the table will evaluate to nonzero values—namely, the top two rows in the last two columns of Table 14.3. Considering the "negative large" position and "positive small" velocity first, the "negative small" force output will be capped at 0.25, which is the degree of membership in the "negative large" position fuzzy set which is less than the 0.5 membership of the angular velocity in the "positive small" fuzzy set. In the "negative large" position and "positive large" velocity, the output will again be capped at 0.25, as similarly, it is less than the 0.5 membership of the angular velocity in the "positive large" fuzzy set. In the cases of "negative small" position and "positive small" velocity, as well as "negative small" position and "positive large" velocity, the output of the "zero" and "positive small" output force fuzzy sets will both be capped at 0.5. Once the outputs from each if–then rule are computed, they are aggregated into one large fuzzy set. In this aggregation, if two of the fuzzy outputs overlap, then (opposite to the "and" combination for the fuzzy rules) the maximum of the two sets is taken. Returning to the example, the aggregation of the four rules for the angle of −20° and angular velocity of +22.5°/s is illustrated in Figure 14.23. "Defuzzification" is necessary

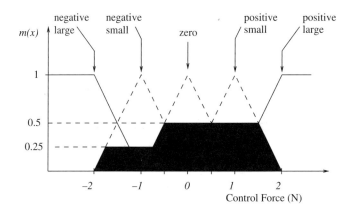

FIGURE 14.23   Aggregation of fuzzy output sets.

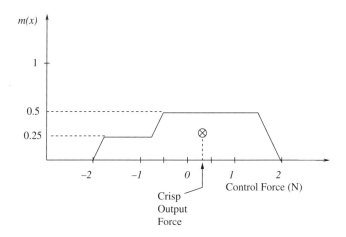

FIGURE 14.24   Defuzzification of output by computing centroid.

to have a crisp output force, and a common technique to do that is to compute the value of the crisp output as the centroid of the aggregated fuzzy output set, as is illustrated Figure 14.24.

Simulating such a system is straightforward using Matlab. If the pendulum mass is 0.1 kg, the cart mass 2.0 kg, and the length of the pendulum 0.5 m, and the values of the membership functions are as illustrated in Figure 14.25, the response of the cart and pendulum system is illustrated in Figures 14.26 and 14.27. Figure 14.26 illustrates the response of the pendulum angle and Figure 14.27 illustrates the velocity of the pendulum. The control effort is illustrated in Figure 14.28. Because the cart position was not controlled, its steady-state response is actually a constant, nonzero velocity. The "response surface" (i.e., the plot of the function defining the control force computed by the fuzzy controller as a function of the two input variables) is illustrated in Figure 14.29.

The remainder of this section will outline the mathematical foundations of fuzzy logic which will allow the reader to adapt this example as necessary for a particular application. In particular, note that in the pendulum example, the "and" conjunction, the aggregation of the outputs and the means to defuzzify the output were all implemented in certain, specific ways. However, these are not necessarily the only or best implementations. The mathematical outline will consider in more general terms fuzzy statements such as, "If A and B, then C" or "If A or B, then C," which will lead to a list of possible alternative implementations of such a fuzzy inference system. Which type of implementation is "best" may be application dependent, although the procedure outlined above is the predominant approach to fuzzy control.

Soft Computing in Control

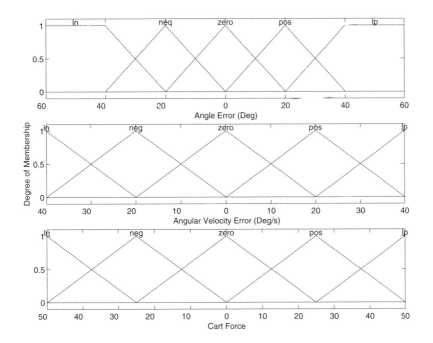

FIGURE 14.25 Membership functions for cart and pendulum simulation.

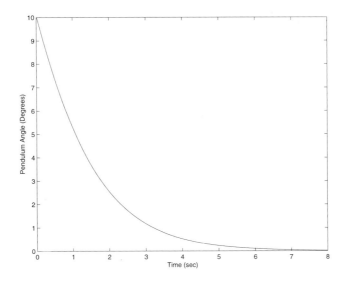

FIGURE 14.26 Pendulum position.

14.4.3 Fuzzy Sets and Fuzzy Logic

14.4.3.1 Introduction

This section introduces fuzzy sets and fuzzy logic and the mathematical foundations thereof. First, this section considers the concept of a membership function which extends the notion of whether or not an element belongs to a set to casting membership in a set to be a matter of degree; that is, instead of either belonging or not belonging to a crisp set, an element can partially belong to a "fuzzy" set. Several examples of fuzzy sets are provided, and the properties of traditional crisp sets are compared with the analogous

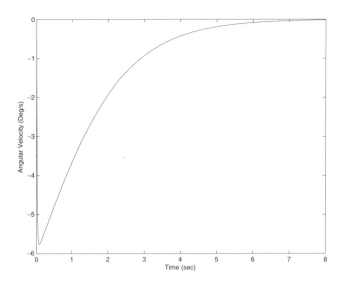

FIGURE 14.27 Pendulum velocity.

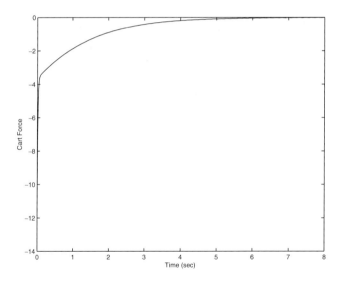

FIGURE 14.28 Control effort required to stabilize inverted pendulum.

properties of fuzzy sets. There is a "crisp" aspect to the normal definition of fuzzy sets in that the membership function returns a crisp value. Fuzzy sets defined as such can be generalized to have fuzzy-valued membership functions. After defining fuzzy sets and outlining their properties, operations on fuzzy sets such as the compliment, intersection, etc. are defined and contrasted with the analogous operations on crisp sets. Finally, fuzzy arithmetic and fuzzy logic are introduced as well as the notion of an additive fuzzy system, which is the basic framework utilized in most fuzzy control (in fact, the pendulum example above utilized this type of inference system).

14.4.3.2 Fuzzy vs. Crisp Sets

The traditional notion of a set will be called a *crisp set*. Examples of crisp sets include:

1. The set of integers $\{\ldots, -2, -1, 0, 1, 2, \ldots\}$
2. The set of all people taller than five feet, eight inches

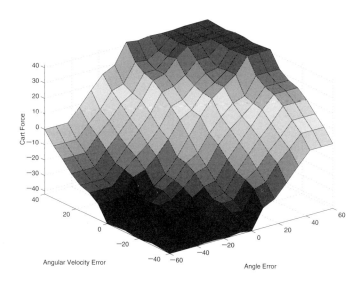

FIGURE 14.29 (Color figure follows p. 12-26.) Response surface for pendulum fuzzy controller.

3. Closed or open intervals of real numbers between a and b: $[a, b]$, (a, b), respectively
4. A set defined by explicitly listing its elements, such as the set containing the letters a, b and c: $\{a, b, c\}$.

Unless otherwise indicated, crisp sets will not be considered ordered. The fundamental characteristic of crisp sets that will later distinguish them from fuzzy sets is that an element either is a member of the set or it is not a member of the set. Mathematically, then, one can define a membership function, m, which maps from a universal set, U, which is the set of all possible elements, to the set $\{0, 1\}$, where for set A and element $x \in U$,

$$m : U \to \{0, 1\} \qquad (14.9)$$

That is, the membership function returns a 1 if x is a member of A, and returns 0 if x is not a member of A.

Crisp sets have a list of standard properties related to concepts in classical logic. In particular, if the following operations are defined:

1. Complement: $\bar{A} = U - A = \{x \in U | x \notin A\}$
2. Union: $A \cup B = \{x \in U | x \in A \text{ or } x \in B\}$
3. Intersection: $A \cap B = \{x \in U | x \in A \text{ and } x \in B\}$

then verifying the following partial list of fundamental properties of crisp sets is straightforward:

1. Involution: $\bar{\bar{A}} = A$
2. Contradiction: $A \cap \bar{A} = \phi$
3. Excluded middle: $A \cup \bar{A} = U$

Having defined the membership function as a mapping from the universal set to the set containing zero and one, it is natural to consider a generalization of the mapping. Instead of considering the membership function as a binary mapping, the membership function for a fuzzy set is a mapping to the interval $[0, 1]$, i.e.:

$$m : U \to [0, 1] \qquad (14.10)$$

Now the mapping returns a value anywhere in the range between and including zero and one which encapsulates the notion that membership can be a matter of degree. This notion of degree enables

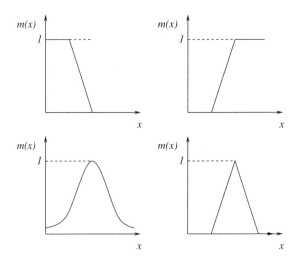

FIGURE 14.30 Examples of membership functions. (Adapted from Klir, G.J., and Yuan, B. (1995) *Fuzzy Sets and Fuzzy Logic: Theory and Application,* Prentice-Hall, Englewood Cliffs, NJ.)

fuzzy sets to express transitions between membership in sets where the transition is gradual (as opposed to crisp).

A prototypical example is temperature and whether the temperature on any given day is hot or cold. There is the set of "hot" days and the set of "cold" days. If these sets were crisp, they would require sharp boundaries. For example, if the temperature is above 80°F, it is hot; otherwise, it is not hot. Similarly, if the temperature is below 45°F, it is cold; otherwise, it is not cold. Such a rigid mathematical treatment of the notions of hot and cold is not appealing, as human inclination is to treat the transition to and from the set of hot and cold temperatures as gradual. A more appealing notion is that a given temperature may have a degree of membership in the set of hot days having a value of zero or one or some value between zero and one. These "in-between" values represent the transition from a day being not hot to the day being hot.

So far, membership functions have been described only as a mapping from the universal set to the interval from zero to one. Several examples of typical membership functions are illustrated in Figure 14.30. The membership function illustrated in the upper left figure is an example of a membership function that may model "cold" where the variable x represents temperature. For low temperatures, the value of the membership function is one, illustrating that the temperature is "cold." In contrast, high temperatures do not belong to the set of cold days, hence the value of the membership function is zero. Between the two extremes is a transition period where the temperature only partially belongs to the set of cold days. The figure in the upper right-hand corner is the analogous membership function for the set of hot days. Other fuzzy sets may require that only values within a certain range have a significant degree of membership in the fuzzy set. Possible examples of such membership functions are illustrated in the bottom two figures, which could, for example, represent warm days.

An interesting feature of all the examples of fuzzy sets presented above is that the membership functions are crisp values; that is, $m(x)$ is a crisp number. Depending on the application, requiring m to return a crisp value may be overly precise. Fuzzy sets can be generalized by defining membership functions to return a range of values instead of a crisp value. In particular,

$$m : U \to I([0, 1]) \tag{14.11}$$

where I represents the family of all closed intervals of real numbers in [0, 1] as is illustrated by the shaded portion in Figure 14.31. Note that further generalization is possible in that interval valued

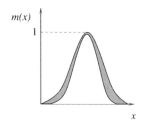

FIGURE 14.31 Fuzzy set defined by a fuzzy membership function.

membership functions can be generalized to have their intervals be fuzzy. Further generalizations are subsequently possible in a recursive fashion. The reader is referred to Klir and Yuan (1995) for complete details.

14.4.3.3 Operations on Fuzzy Sets

Analogous to operations on crisp sets, a variety of operations can be defined on fuzzy sets. Adopting the standard notational shortcut where:

$$A(x) = m(x) \qquad (14.12)$$

where $m(x)$ is the membership function that defines the fuzzy set A, we define the "standard" fuzzy complement, intersection and union as follows:

1. Complement: $\bar{A}(x) = 1 - A(x)$
2. Intersection: $(A \cap B)(x) = \min[A(x), B(x)]$
3. Union: $(A \cup B)(x) = \max[A(x), B(x)]$
4. Subsethood: $A \subseteq B \Leftrightarrow A(x) \leq B(x)$

where each operation holds for all x. It is important to note that these are not the only ways to define these operations, although they are the typical ways. In fact, the intersection can also be defined in three other common ways:

$$\begin{aligned}
(A \cap B)(x) &= A(x) \cdot B(x), \\
(A \cap B)(x) &= \max[0, A(x) + B(x) - 1] \\
(A \cap B)(x) &= \begin{cases} a & \text{if } b = 1 \\ b & \text{if } a = 1 \\ 0 & \text{otherwise} \end{cases}
\end{aligned} \qquad (14.13)$$

Similarly, the union can alternatively be defined by:

$$\begin{aligned}
(A \cup B)(x) &= A(x) + B(x) - A(x) \cdot B(x), \\
(A \cup B)(x) &= \min[1, A(x) + B(x)], \\
(A \cup B)(x) &= \begin{cases} a & \text{if } b = 0 \\ b & \text{if } a = 0 \\ 0 & \text{otherwise} \end{cases}
\end{aligned} \qquad (14.14)$$

For a more complete, axiomatic development, and a list of further possible definitions of intersections and unions of fuzzy sets, see Klir and Yuan (1995). In the more mathematical literature, intersections may be called t-norms, and unions may be called t-conorms. It is straightforward to show that most properties associated with crisp sets still hold for fuzzy sets, except for the properties of contradiction and excluded middle. The equality conditions of contradiction and excluded middle for crisp sets are replaced by subset conditions for fuzzy sets:

1. Contradiction: $A \cap \bar{A} \supset \phi$
2. Excluded Middle: $A \cup \bar{A} \subset U$

14.4.4 Fuzzy Logic

Fuzzy sets and their operations and properties provide the mathematical foundation for fuzzy logic, which in turn is the basis for fuzzy control and other applications of fuzzy logic. Because feedback control is based upon measuring state variables, an important type of fuzzy set for fuzzy control is defined by a membership function whose domain is the set of real numbers:

$$m : \Re \to [0, 1] \tag{14.15}$$

which provides the degree to which a given variable is "close" to a specified value. Arithmetic operations on fuzzy numbers can then be defined as follows:

1. Addition: $(A + B)(z) = \sup_z \min[A(x), B(y)]$,
   $z = x + y$

2. Subtraction: $(A + B)(z) = \sup_z \min[A(x), B(y)]$,
   $z = x - y$

3. Multiplication: $(A + B)(z) = \sup_z \min[A(x), B(y)]$,
   $z = x \cdot y$

4. Division: $(A + B)(z) = \sup_z \min[A(x), B(y)]$,
   $z = x/y$

This arithmetic basis provides the foundation for the application of the use of linguistic variables in fuzzy control algorithms. A linguistic variable is a fuzzy number that represents some sort of linguistic concept such as "very cold," "cold," "chilly," "comfortable," "warm," "hot" or "very hot." An example of a linguistic variable was actually previously illustrated in the pendulum example where the elements of the state of the pendulum ($\theta, \dot{\theta}$) could be described in linguistic terms such as "negative large," "positive small," etc. Linguistic variables allow linguistic terms to represent the approximate condition of the state of the system to be represented by fuzzy numbers. As illustrated in the pendulum example, the utility of linguistic variables is that they are an effective means to "translate" human expertise germane to a controls application into appropriate fuzzy rules utilized in a fuzzy controller.

The inference system typically utilized in fuzzy controllers is best illustrated by developing the so-called standard additive model [Kosko, 1997] utilizing the Mamdani inference system, which is the framework underlying most fuzzy controllers (and was the framework of the pendulum controller example above). The standard additive model is illustrated in Figure 14.20 [Kosko, 1997].

Central to this system is a set of if–then rules, which require some basic fuzzy logic and inference. Considering the linguistic variables that represent the fuzzy numbers representing the state of the pendulum, it is clear that there are the basic, or primary, terms "negative," "zero" and "positive" and two hedges, "small" and "large." Clearly, for other applications, different primary terms can be utilized, as well as different hedges, such as "very," "more," "less," "extremely" etc.

Several operators on fuzzy numbers are useful for implementing a fuzzy inference system. In particular, a fuzzy number can be concentrated or dilated according to

$$A^k(x) = (A(x))^k \qquad (14.16)$$

where it is the concentration operator if $k > 1$ or the dilation operator if $k < 1$ that can be utilized to represent the linguistic hedges "very" and "more or less," respectively. Also, the operator "not" and the relations "and" and "or" are related to the definitions of complement, intersection and union as follows:

1. Not A $\neg A(x) = \bar{A}(x) = 1 - A(x)$
2. A and B $(A \text{ and } B)(x) = (A \cap B)(x)$
3. A or B $(A \text{ or } B)(x) = (A \cup B)(x)$

Note that the definitions of "and" and "or" are not unique, as the definitions of the complement, intersection and union are not unique. Thus, any of the possible definitions of intersection and union can be used to implement the logical "and" or logical "or."

An example of one way to evaluate the multiconditional approximate reasoning inference system in the standard additive model typical for fuzzy controllers is as follows. Given a measured state variable, x, it may be "fuzzified" to account for measurement uncertainty. (Such a fuzzification was not considered in the pendulum example—in that case, the degree of membership of the crisp state value was used). As illustrated in Figure 14.32, if a measurement from a sensor is x, then the fuzzified set, $X(x)$, may be defined to account for sensor uncertainty, where the shape of the membership function defining the fuzzy set $X(x)$ depends upon the type of uncertainty expected from the sensor. Then the degree of consistency between the fuzzified state measurement and a fuzzy set A_i is computed as the height of the intersection between $X(x)$ and $A_i(x)$. This is essentially determining the degree to which "if X is A_i" is satisfied. Because there are various means to compute the intersection of two fuzzy sets, the value of this degree of consistency will depend upon the definition of intersection utilized. In particular, if the standard intersection is utilized, then the degree of consistency is given by:

$$r_i(X) = \sup_x \min[X(x), A_i(x)] \qquad (14.17)$$

where the "min" function computes the standard intersection, and the "sup" function determines its maximum value, as is illustrated for two arbitrary fuzzy sets in Figure 14.33. Note that this is simply a generalization of utilizing the degree of membership of a crisp value, as the degree of membership is the supremum

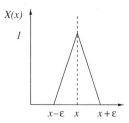

FIGURE 14.32 Fuzzifying a crisp variable.

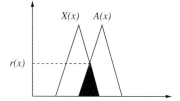

FIGURE 14.33 Degree of consistency between fuzzy sets $X(x)$ and $A(x)$.

of the intersection of the line representing the crisp value of the variable and the fuzzy set, as is illustrated, for example, in Figure 14.21.

Having determined the degree to which "if X is A_i," is satisfied, the result of "then Y is B_i," must be determined. The most common (and apparently most effective) technique was illustrated in the pendulum example. This technique is to let the resulting fuzzy set, B', be determined according to $B' = \min[r_i, B]$ which is simply the "clipping" approach illustrated in the pendulum example.

The overall general formulation to do so is as follows. Given an if–then rule, if "X is A, then Y is B," where X and Y are fuzzy sets representing the state of linguistic variables, the task is to determine the application of this rule to a fuzzy set A' which is not necessarily identical to A to determine the appropriate conclusion, B', as illustrated in the following chart:

Rule: If X is A, then Y is B.
Fact: X is A'.
Conclusion: Y is B'.

It is important to note that the "min" operator utilized to determine the degree of consistency neither satisfies the rules of classical (Boolean) logic when reduced to the crisp case [Terano, 1992] nor does it satisfy all the axioms that may be generated as reasonable extensions of the classical case [Klir and Yuan, 1995]. Possibilities other than the "min" operator as fuzzy implications include $\max[1 - A(x), \min[A(x), B(y)]]$ (due to Zadeh), or $\min[1, 1 - A(x) + B(x)]$ (the Lukasiewicz implication). A list of such fuzzy implications, as well as a full exposition regarding their properties can be found in Klir (1995) or Jang et al. (1997). A more basic presentation is in Terano (1992) or Jang et al. (1997). From a controls perspective, note that "very good results are obtained" from the more general implications, but that Mamdani (1974), attempting to actually control a steam engine, "obtained excellent results from the max–min compositions" illustrated. A complete and rigorous exposition of fuzzy logic is based upon considerations of fuzzy relations and fuzzy implications, which are beyond the scope of this section. The interested reader is referred to the references.

The final step is defuzzification, where, as is typical, there are various alternative approaches to the centroid method presented in the pendulum example. In particular, in addition to the centroid, the following are possible methods for defuzzification:

1. Bisector of area
2. Mean of the maximum
3. Smallest of maximum
4. Largest of maximum

These concepts are illustrated in Figure 14.34.

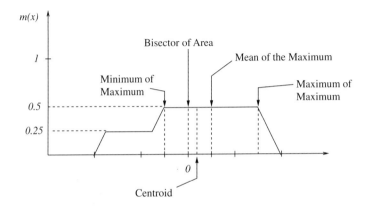

FIGURE 14.34 Defuzzification methods.

14.4.5 Alternative Inference Systems

The Mamdani inference system considered so far in this presentation is not the only inference system utilized in fuzzy control applications. In particular, the so-called TSK fuzzy model (named for Takagi, Sugeno and Kang [Jang et al., 1997]) is an alternative model which has the advantage that it does not require defuzzification of the output, which can possibly be computationally costly.

In particular, in the TSK model, fuzzy rules are of the form "if X is A and Y is B, then $z = f(x, y)$." In contrast to the Mamdani model, the output of the rules is a function, as opposed to a fuzzy set. For the pendulum example, possible TSK rules may include:

1. If the pendulum angle is zero and the angular velocity is zero, then $u = 0$.
2. If the pendulum angle is positive and small and the angular velocity is zero, then $u = 0.5\theta$.
3. If the pendulum angle is positive and large and the angular velocity is zero, then $u = 0.7\theta$.
4. If the pendulum angle is positive and small and the pendulum angular velocity is negative and small, then $u = 0.4\theta + 0.6\dot{\theta}$.

As indicated, defuzzification of the outputs is not required, but the outputs from each of the rules still need to be combined. Two possible alternatives are often employed: weighted average, and weighted sum.

For the weighted average, if z_1 and z_2 are the output functions for two rules, and r_1 and r_2 are the degrees of consistency between the input data and antecedent fuzzy sets, A_1 and A_2, then the output is computed as:

$$u = \frac{r_1 z_1 + r_2 z_2}{r_1 + r_2} \qquad (14.18)$$

If the weighted sum is used, then simply:

$$u = r_1 z_1 + r_2 z_2 \qquad (14.19)$$

A final control paradigm briefly summarized here is model-based fuzzy control, which considers the design of fuzzy rules given the (nonlinear) model of the system to be controlled, which is in contrast with the heuristic approach of the traditional fuzzy logic control paradigm outlined above. The advantage of this approach is that it makes use of analytical model information that may be available but is completely ignored in the standard fuzzy control paradigm.

At least two different forms of model based fuzzy control paradigms exist: the so-called Takagi–Sugeno fuzzy logic controllers (TSFLCs) and sliding-mode fuzzy logic controllers (SMFLCs). For TSFLCs, rules are determined by considering the dynamics of the system in various "fuzzy regimes" of the state space and determining appropriate (linear) control laws at the center of each of these fuzzy regimes. SMFLC rules can be determined by considering the distance between the state vector and a desired "sliding surface." For further details, the reader is referred to Palm et al. (1997).

14.4.6 Other Applications

Finally, although feedback control is the primary application of fuzzy logic, it certainly is not the exclusive application. Other applications include identification and classification techniques such as handwriting recognition [Yen and Langari, 1999], robotics and intelligent agents [Yen and Langari, 1999], database information retrieval [Yen and Langari, 1999], nonlinear system identification and adaptive noise cancellation [Jang et al., 1997], modeling [Babuska, 1998], PID controller tuning [Yen and Langari, 1999], process control and analysis [Ruan, 1997] and traffic control [Dubois, 1980].

14.5 Conclusions

We have reviewed some of the major soft computing (SC) techniques used for complex systems. Due to limitations of space, SC has been described here only in outline. The purpose has been to show the way the methods work and the possible range of applications and to interest the reader in these new technologies. SC techniques are not model based so are most suitable for applications for which first-principles-based approaches are either not possible or too slow. There are many such instances in the control area for which soft computing is especially appropriate. As MEMS devices are in the frontiers of hardware, many of the issues related to their understanding are still not completely clear and the model equations cannot always be computed quickly enough for real-time control purposes. It is possible, then, that SC techniques could lend a hand to the use of these devices in real applications.

Acknowledgments

M.S. wishes to thank his colleagues K.T. Yang and R.L. McClain and his students X. Zhao, G. Díaz, A. Pacheco-Vega and W. Franco for collaboration on artificial neural networks and genetic algorithms. He also thanks D.K. Dorini of BRDG-TNDR for sponsoring the research and CONACyT (Mexico) and the Organization of American States for support of the students. B.G. would like to thank Neil Petroff for his many thoughtful comments and suggestions concerning the section on fuzzy logic and fuzzy control.

Defining Terms

Neural Networks

Activation function: The input–output relation of a single neuron.
Artificial neural network (ANN): A network of neurons used for computational purposes, as opposed to the natural neural networks in humans and animals.
Backpropagation: A specific error correction technique that enables weights and biases to be adjusted during training.
Bias: A characteristic of each neuron that determines the threshold at which it will fire.
Configuration: The arrangement of neurons and layers in an artificial neural network.
Feedforward artificial neural network: A kind of network in which the information flows in one direction from the input to the output.
Hidden layer: A layer of the network in between the input and output.
Layer: A slice of the feedforward artificial neural network with several neurons in it.
Learning rate: A coefficient that controls how fast the weights and biases change during training.
Multilayer network: A network with three or more layers of neurons.
Neurocontrol: Control by means of an artificial neural network.
Neuron: The smallest, single computational unit; also called a node.
Node: See neuron.
Output layer: The layer of the network in which the information goes out.
Synaptic weight: Same as weight.
Training: A procedure by which weights and biases are adjusted.
Weight: A characteristic of every interneuronal connection that defines it.

Genetic Algorithms

Binary string: A series of zeros and ones that may represent a number.
Crossover: An operation between two strings in which part of each string is detached from its original and interchanged with the other.
Evolutionary algorithm: An algorithm that changes in some probablistic way according to the results being obtained.

Fitness: A numerical quantity that characterizes how good a proposed solution is.
Genetic algorithm: An optimization algorithm that uses the principles of evolution in a specific way.
Global optimum: The best in some predefined sense among a possibly infinite set of possibilities.
Gradient-based optimization: Optimization methods that use derivatives of functions.
Local optimum: The best in some predefined sense in an infinitesimally small neighborhood.
Mutation: The operation in which a bit of a binary string changes from zero to one or vice versa.
Offspring: The pair of strings produced from parents belonging to the old generation.
Parents: The pair of strings in the old generation that produces a pair of offspring in the new.
Population size: The number of candidate strings in a given generation.

Fuzzy Logic

Additive fuzzy system: A fuzzy logic inference system used frequently in feedback control applications.
Crisp set: A "standard" set wherein an element must either be a member of the set or not a member of the set.
Defuzzification: The process by which a crisp value is obtained from a fuzzy set.
Fuzzification: The process by which the degree of membership of the state of a system is determined in various fuzzy sets.
Fuzzy control: A feedback control method utilizing concepts from fuzzy logic.
Fuzzy logic: A logical inference system based upon fuzzy set theory.
Fuzzy set: A generalization of "standard" set theory wherein the degree of membership of an element to the set has a continuous range of values.
Linguistic variable: A qualitative variable described quantitatively by a fuzzy set.
Mamdani inference system: A feedback control approach that facilitates the application of human intuition or expertise to feedback control systems by the use of fuzzy set theory and if–then rules.
Membership function: The function describing the degree of membership an element has in a fuzzy set.
T-conorm: An operator used to define in mathematically general terms the union of two fuzzy sets.
T-norm: An operator used to define in mathematically general terms the intersection of two fuzzy sets.

References

Aminzadeh, F., and Jamshidi, M. (1994) *Soft Computing: Fuzzy Logic, Neural Networks, and Distributed Artificial Intelligence,* Prentice-Hall, Englewood Cliffs, NJ.
Angeline, P.J., Saunder, G.M., and Pollack, J.B. (1994) "Complete Induction of Recurrent Neural Networks," in *Proc. of the Third Annual Conf. on Evolutionary Programming,* eds. A.V. Sebald and L.J. Fogel, World Scientific, Singapore, pp. 1–8.
Babuska, R. (1998) *Fuzzy Modeling for Control,* Kluwer Academic, Boston.
Berlin, A.A., Chase, J.G., and Jacobsen, S.C. (1998) "MEMS-Based Control of Structural Dynamic Instability," *J. Intelligent Mater. Syst. Structures* **9**(7), pp. 574–586.
Bouchon-Meunier, B., Yager, R.R., and Zadeh, L.A., eds. (1995) *Fuzzy Logic and Soft Computing,* World Scientific, River Edge, NJ.
Buckley, J.J., and Feuring, T. (1999) *Fuzzy and Neural: Interactions and Applications,* Physica-Verlag, New York.
Chan, H.L., and Rad, A.B. (2000) "Real-Time Flow Control Using Neural Networks," *ISA Trans.* **39**(1), pp. 93–101.
Díaz, G. (2000) "Simulation and Control of Heat Exchangers Using Artificial Neural Networks," Ph.D. dissertation, Department of Aerospace and Mechanical Engineering, University of Notre Dame, Notre Dame, IN.
Díaz, G., Sen, M., Yang, K.T., and McClain, R.L. (1999) "Simulation of Heat Exchanger Performance by Artificial Neural Networks," *Int. J. HVAC&R Res.* **5**(3), pp. 195–208.
Díaz, G., Sen, M., Yang, K.T., and McClain, R.L. (2001a) "Dynamic Prediction and Control of Heat Exchangers Using Artificial Neural Networks," *Int. J. Heat Mass Transfer* **44**(9), pp. 1671–1679.

Díaz, G., Sen, M., Yang, K.T., and McClain, R.L. (2001b) "Stabilization of Thermal Neurocontrollers," submitted for review.
Díaz, G., Sen, M., Yang, K.T., and McClain, R.L. (2001c) "Adaptive Neurocontrol of Heat Exchangers," *ASME J. Heat Transfer* **123**(3), pp. 556–562.
Dimeo, R., and Lee, K.Y. (1995) "Boiler-Turbine Control System Design Using a Genetic Algorithm," *IEEE Trans. Energy Conv.* **10**(4), p. 752.
Drago, G.P., and Ridella, S. (1992) "Statistically Controlled Activation Weight Initialization," *IEEE Neural.* **3**(4), pp. 627–631.
Dubois, D. (1980) *Fuzzy Sets and Systems—Theory and Applications*, Academic Press, New York.
Eeckman, F.H., ed. (1992) *Analysis and Modeling of Neural Systems*, Kluwer Academic, Boston, MA.
Flood, I., and Kartam, N. (1994) "Neural Networks in Civil Engineering. I. Principles and Understanding," *ASCE J. Comp. Civil Eng.* **8**(2), pp. 131–148.
Fogel, L.J. (1999) *Intelligence Through Simulated Evolution*, John Wiley & Sons, New York.
Gad-el-Hak, M. (1994) "Interactive Control of Turbulent Boundary Layers: A Futuristic Overview," *AIAA J.* **32**, pp. 1753–1765.
Gad-el-Hak, M. (1999) "The Fluid Mechanics of Microdevices—The Freeman Scholar Lecture," *J. Fluid Eng. Trans. ASME* **121**(1), pp. 5–33.
Gagarin, N., Flood, I., and Albrecht, P. (1994) "Computing Truck Attributes with Artificial Neural Networks," *ASCE J. Comput. Civil Eng.* **8**(2), pp. 179–200.
Gaudenzi, P., Fantini, E., Gantes, and Charis J. (1998) "Genetic Algorithm Optimization for the Active Control of a Beam by Means of PAT Actuators," *J. Intelligent Mater. Syst. Structures* **9**(4), p. 291.
Goldberg, D.E. (1989) *Genetic Algorithms in Search, Optimization and Machine Learning*, Addison-Wesley, Reading, MA.
Haykin, S. (1999) *Neural Networks, A Comprehensive Foundation*, 2nd ed., Prentice-Hall, Englewood Cliffs, NJ.
Ho, C.-M., and Tai, Y.-C. (1996) "Review: MEMS and Its Applications for Flow Control," *J. Fluid Eng. Trans. ASME* **118**(3), pp. 437–447.
Ho, C.-M., and Tai, Y.-C. (1998) "Micro-Electro-Mechanical-Systems (MEMS) and Fluid Flows," *Annu. Rev. Fluid Mech.* **30**, pp. 579–612.
Jacobson, S.A., and Reynolds, W.C. (1993) "Active Control of Boundary Layer Wall Shear Stress Using Self-Learning Neural Networks," AIAA Paper No. 93-3272, American Institute of Aeronautics and Astronautics, Washington, D.C.
Jain, L.C., and Fukuda, T., eds. (1998) *Soft Computing for Intelligent Robotic Systems*, Physica-Verlag, Heidelberg, Germany.
Jang, J.S.R., Sun, C.T., and Mizutani, E. (1997) *Neuro-Fuzzy and Soft Computing: A Computational Approach to Learning and Machine Intelligence*, Prentice-Hall, Englewood Cliffs, NJ.
Kaboudan, M.A. (1999) "A Measure of Time Series' Predictability Using Genetic Programming Applied to Stock Returns," *J. Forecasting* **18**(5/6), p. 345.
Kamarthi, S., Sanvido, V., and Kumara, R. (1992) "Neuroform—Neural Network System for Vertical Formwork Selection," *ASCE J. Comp. Civ. Eng.* **6**(2), pp. 178–199.
Kao J.J. (1999) "Optimal Location of Control Valves in Pipe Networks by Genetic Algorithm—Closure," *J. Water Resour. Planning Manage. Div. ASCE* **125**(1), pp. 68–69.
Karmin, E.D. (1990) "Simple Procedure for Pruning Back Propagation Trained Neural Networks," *IEEE Trans. Neural Networks* **1**(2), pp. 239–242.
Katisikas, S.K., Tsahalis, D., and Xanthakis, S.A. (1995) "A Genetic Algorithm for Active Noise Control Actuator Positioning," *Mech. Syst. Signal Proc.* **9**(6), p. 697.
Keane, A.J. (1995) "Passive Vibration Control via Unusual Geometries the Application of Genetic Algorithm Optimization to Structural Design," *J. Sound Vib.* **185**(3), p. 441.
Klir, G.J., and Yuan, B. (1995) *Fuzzy Sets and Fuzzy Logic: Theory and Applications*, Prentice-Hall, Englewood Cliffs, NJ.
Kosko, B. (1997) *Fuzzy Engineering*, Prentice-Hall, Englewood Cliffs, NJ.

Lee, C., Kim, J., Babcock, D., and Goodman, R. (1997) "Application of Neural Networks to Turbulence Control for Drag Reduction," *Phys. Fluids* **9**(6), p. 1740.

Lehtokangas, M., Saarinen, J., and Kaski, K. (1995) "Initializing Weights of a Multilayer Perceptron Network by Using the Orthogonal Least Squares Algorithm," *Neural Computation* **7**, pp. 982–999.

Lin, L.C., and Lee, G.Y. (1999) "Hierarchical Fuzzy Control for C-axis of CNC Turning Centers Using Genetic Algorithms," *J. Intelligent Robotic Syst.* **25**(3), pp. 255–275.

Löfdahl, L., and Gad-el-Hak, M. (1999) "MEMS Applications in Turbulence and Flow Control," *Prog. Aerosp. Sci.* **35**(2), pp. 101–203.

Luk, P.C.K., Low, K.C., and Sayiah, A. (1999) "GA-Based Fuzzy Logic Control of a Solar Power Plant Using Distributed Collector Fields," *Renew. Energy* **16**(1–4), pp. 765–768.

Mamdani, E.H. (1974) "Application of Fuzzy Algorithms for Control of Simple Dynamic Plant," *IEEE Proc.* **121**(12).

Mamdani, E.H., and Assilian, S. (1975) "An Experiment in Linguistic Synthesis with a Fuzzy Logic Controller," *Int. J. Machine Stud.* **7**(1).

Mamdani, E.H., and Baaklini, N. (1975) "Perspective Method for Deriving Control Policy in a Fuzzy-Logic Controller," *Electron. Lett.* **11**, pp. 625–626.

Man, K.F., Tang, K.S., and Kwong, S. (1999) *Genetic Algorithms*, Springer-Verlag, Berlin.

Matsuura, K., Shiba, H., and Hamachi, M. (1995) "Optimizing Control of Sensory Evaluation in the Sake Mashing Process by Decentralized Learning of Fuzzy Inference Using a Genetic Algorithm," *J. Ferment. Bioeng.* **80**(3), p. 251.

Michalewicz, Z. (1992) *Genetic Algorithm + Data Structure = Evolution Programs*, Springer-Verlag, Berlin.

Michalewicz, Z., Janikow, C.Z., and Krawczyk, J.B. (1992) "A Modified Genetic Algorithm for Optimal Control Problems," *Comp. Math. Appl.* **23**(12), p. 8; 3.

Mitchell, M. (1997) *An Introduction to Genetic Algorithms*, MIT Press, Cambridge, MA.

Mordeson, J.N., and Nair, P.S. (1988) *Fuzzy Mathematics: An Introduction for Engineers and Scientists*, Physica-Verlag, New York.

Nagaoka, Y., Alexander, H.G., Liu, W., and Ho, C.M. (1997) "Shear Stress Measurements on an Airfoil Surface Using Micro-Machined Sensors," *JSME Int. J. Series B—Fluids Thermal Eng.* **40**(2), pp. 265–272.

Nagaya, K., and Ryu, H. (1996) "Deflection Shape Control of a Flexible Beam by Using Shape Memory Alloy Wires Under the Genetic Algorithm Control," *J. Intelligent Mater. Syst. Structures* **7**(3), p. 336.

Nakashima, M., Maruyama, Y., and Hasegawa, T. (1998) "Basic Experiments on Robot-Based Vibration Control of the Hot-Line Work Robot System Using Genetic Algorithm," *J. Electr. Eng. Jpn.* **123**(2), p. 40.

Nelson, B.J., Zhou, Y., and Vikramaditya, B. (1998) "Sensor-Based Microassembly of Hybrid MEMS Devices," *IEEE Contr. Syst. Mag.* **18**(6), p. 35.

Nordin, P., Banzhaf, W., and Brameier, M. (1998) "Evolution of a World Model for a Miniature Robot Using Genetic Programming," *Robot Autonomous Syst.* **25**(1–2), pp. 105–116.

Pacheco-Vega, A., Sen, M., Yang, K.T., and McClain, R.L. (1998) "Genetic-Algorithm-Based Prediction of a Fin-Tube Heat Exchanger Performance," *Proc. 11th Int. Heat Trans. Conf.* **6**, pp. 137–142.

Pacheco-Vega, A., Diaz, G., Sen, M., Yang, K.T., and McClain, R.L. (2001) "Heat Rate Predictions in Humid Air-Water Heat Exchangers Using Correlations and Neural Networks," *ASME J. Heat Transfer* **123**(2), pp. 348–354.

Pal, S.K., and Mitra, S. (1999) *Neuro-Fuzzy Pattern Recognition*, John Wiley & Sons, New York.

Palm, R., Driankov, D., and Hellendoorn, H. (1997) *Model Based Fuzzy Control*, Springer-Verlag, Berlin.

Passino, K.M., and Yurkovich, S. (1998) *Fuzzy Control*, Addison-Wesley, Menlo Park, CA.

Perhinschi, M.G. (1998) "Optimal Control System Design Using a Genetic Algorithm," *ZAMM* **78**(suppl. 3), p. S1035.

Rahmoun, A., and Benmohamed, M. (1998) "Genetic Algorithm Based Methodology to Generate Automatically Optimal Fuzzy Systems," *IEE Proc. D,* **145**(6), pp. 583–586.

Ranganath, M., Renganathan, S., and Rao, C.S. (1999) "Genetic Algorithm Based Fuzzy Logic Control of a Fed-Batch Fermenter," *Bioprocess. Eng.* **21**(3), pp. 215–218.

Reis, L.F.R., Porto, R.M., and Chaudhry, F.H. (1997) "Optimal Location of Control Valves in Pipe Networks by Genetic Algorithm," *J. Water Resour. Planning Manage.* **123**(6), p. 317.

Ruan, D., Ed. (1997) *Intelligent Hybrid Systems,* Kluwer, Norwell, MA.

Rumelhart, D.E., Hinton, D.E., and Williams, R.J. (1986) "Learning Internal Representations by Error Propagation," in *Parallel Distributed Processing: Exploration in the Microstructure of Cognition,* Vol. 1, eds. D.E. Rumelhart and J.L. McClelland, MIT Press, Cambridge, MA.

Rzempoluck, E.J. (1998) *Neural Network Data and Analysis Using Simulnet,* Springer, New York.

Schalkoff, R.J. (1997) *Artificial Neural Networks,* McGraw-Hill, New York.

Sharatchandra, M.C., Sen, M., and Gad-el-Hak, M. (1998) "New Approach to Constrained Shape Optimization Using Genetic Algorithms," *AIAA J.* **38**(1), pp. 51–61.

Sen, M., and Yang, K.T. (2000) "Applications of Artificial Neural Networks and Genetic Algorithms in Thermal Engineering," in *The CRC Handbook of Thermal Engineering,* F. Kreith, Ed., CRC Press, Boca Raton, FL, pp. 620–661.

Seywald, H., Kumar, R.R., and Deshpande, S.M. (1995) "Genetic Algorithm Approach for Optimal Control Problems with Linearly Appearing Controls," *J. Guid. Contr. Dyn.* **18**(1), p. 177.

Subramanian, H., Varadan, V.K., Varadan, V.V., and Vellekoop, M.J. (1997) "Design and Fabrication of Wireless Remotely Readable MEMS Based Microaccelerometers," *Smart Mater. Struct.* **6**(6), pp. 730–738.

Suzuki, Y., and Kasagi, N. (1997) "Active Flow Control with Neural Network and Its Application to Vortex Shedding," in *Proc. 11th Symp. on Turbulent Shear Flows,* Grenoble, pp. 9.18–9.23.

Takemori, T., Miyasaka, N., and Hirose, S. (1991) "Neural Network Air-Conditioning System for Individual Comfort," *Proc. SPIE Int. Soc. Optical Eng. Appl. Artificial Neural Networks II* **1469**(pt. 1), pp. 157–165.

Tang, K.S., Man, K.F., and Chu, C.Y. (1996a) "Application of the Genetic Algorithm to Real-Time Active Noise Control," *Real-Time Syst.* **11**(3), p. 289.

Tang, K.S., Man, K.F., and Gu, D.W. (1996b) "Structured Genetic Algorithm for Robust H Control Systems Design," *IEEE Trans. Ind. Electron.* **43**(5), p. 575.

Terano, T. (1992) *Fuzzy Systems Theory and Its Applications,* Academic Press, San Diego, CA.

Thibault, J., and Grandjean, B.P.A. (1991) "Neural Network Methodology for Heat Transfer Data Analysis," *Int. J. Heat Mass Transfer* **34**(8), pp. 2063–2070.

Trebi-Ollennu, A., and White, B.A. (1997) "Multiobjective Fuzzy Genetic Algorithm Optimisation Approach to Nonlinear Control System Design," *IEE Proc. D* **144**(2), p. 137.

Vandelli, N., Wroblewski, D., Velonis, M., and Bifano, T. (1998) "Development of a MEMS Microvalve Array for Fluid Flow Control," *J Microelectromechanical Syst.* **7**(4), pp. 395–403.

Varadan, V.K., Varadan, V.V., and Bao, X.Q. (1995) "Comparison of MEMS and PZT Sensor Performance in Active Vibration and Noise Control of Thin Plates," *J. Wave-Mater. Interaction* **10**(4), p. 51.

Warwick, K., Irwin, G.W., and Hunt, K.J. (1992) *Neural Networks for Control and Systems,* Short Run Press, Ltd., Exeter.

Wessels, L., and Barnard, E. (1992) "Avoiding Fake Local Minima by Proper Initialization of Connections," *IEEE Neural.* **3**(6), pp. 899–905.

Yager, R.R., and Zadeh, L.A. (1994) *Fuzzy Sets, Neural Networks, and Soft Computing,* Van Nostrand-Reinhold, New York.

Yen, J., and Langari, R. (1999) *Fuzzy Logic,* Prentice-Hall, Englewood Cliffs, NJ.

Zadeh, L.A. (1965) "Fuzzy Sets," *Inf. Contr.* **8**.

Zadeh, L.A. (1968a) "Probability Measures and Fuzzy Systems," *J. Math. Anal. Appl.* **23**(2), pp. 421–427.

Zadeh, L.A. (1968b) "Fuzzy Algorithm," *Inf. Contr.* **12**, pp. 94–102.

Zadeh, L.A. (1971) "Toward a Theory of Fuzzy Systems," in *Aspects of Network and System Theory,* eds. R.E. Kalman and N. Dellaris, Holt, Rinehart and Winston, New York.

Zeng, P. (1998) "Neural Computing in Mechanics," *AMR* **51**(2), pp. 173–197.

Zhao, X. (1995) "Performance of a Single-Row Heat Exchanger at Low In-Tube Flow Rates," M.S. thesis, Department of Aerospace and Mechanical Engineering, University of Notre Dame, Notre Dame, IN.

For Further Information

For neural networks, the monographs by Schalkoff (1997) and Haykin (1999) are very good for beginners. For more advanced readers, the list of references in this chapter can be consulted for papers pertaining to specific areas. For genetic algorithms, the books by Goldberg (1989), Michalewicz (1992) and Mitchell (1997) are recommended for beginners. There are many excellent fuzzy control texts, however, the authors found Kosko (1997) and Passino and Yurkovich (1998) particularly clear and useful. Texts with indepth and complete developments of fuzzy logic include Klir and Yuan (1995) and Jang et al. (1997).

II

Design and Fabrication

15 Materials for Microelectromechanical Systems *Christian A. Zorman, Mehran Mehregany* 15-1
Introduction • Single-Crystal Silicon • Polysilicon • Silicon Dioxide • Silicon Nitride • Germanium-Based Materials • Metals • Silicon Carbide • Diamond • III–V Materials • Piezoelectric Materials • Conclusions

16 MEMS Fabrication *Marc J. Madou* 16-1
Micromachining: Introduction • Historical Note • Silicon Crystallography • Silicon as a Substrate and Structural Material • Wet Isotropic and Anisotropic Etching • Etching With Bias and/or Illumination of the Semiconductor • Etch-Stop Techniques • Problems with Wet Bulk Micromachining • Wet Bulk Micromachining Examples • Surface Micromachining: Introduction • Historical Note • Mechanical Properties of Thin Films • Surface Micromachining Processes • Poly-Si Surface Micromachining Modifications • Non-Poly-Si Surface Micromachining Modifications • Resists as Structural Elements and Molds in Surface Micromachining • Materials Case Studies • Polysilicon Surface Micromachining Examples

17 LIGA and Other Replication Techniques *Marc J. Madou* 17-1
Introduction • LIGA Processes—Introduction to the Technology • LIGA Applications • Technological Barriers and Competing Technologies

18 X-Ray-Based Fabrication *Todd Christenson* 18-1
Introduction • DXRL Fundamentals • Mold Filling • Material Characterization and Modification • Planarization • Angled and Re-entrant Geometry • Multilayer DXRL Processing • Sacrificial Layers and Assembly • Applications • Conclusions

19 Electrochemical Fabrication (EFAB™) *Adam L. Cohen* 19-1
Introduction • Background • A New SFF Process • Instant Masking • EFAB • Detailed Process Flow • Microfabricated Structures • Automated EFAB Process Tool • Materials for EFAB • EFAP Performance • EFAB Compared with Other SFF Processes • EFAB Compared with Other Microfabrication Processes • EFAB Limitations and Shortcomings • EFAB Applications • The Future of EFAB

20 Fabrication and Characterization of Single-Crystal Silicon Carbide MEMS *Robert S. Okojie* 20-1
Introduction • Photoelectrochemical Fabrication Principles of 6H-SiC • Characterization of 6H-SiC Gauge Factor • High-Temperature Metallization • Sensor Characteristics • Summary

21 Deep Reactive Ion Etching for Bulk Micromachining of Silicon Carbide
Glenn M. Beheim 21-1
Introduction • Fundamentals of High-Density Plasma Etching • Fundamentals of SiC Etching Using Fluorine Plasmas • Applications of SiC DRIE: Review • Applications of SiC DRIE: Experimental Results • Applications of SiC DRIE: Fabrication of a Bulk Micromachined SiC Pressure Sensor • Summary

22 Microfabricated Chemical Sensors for Aerospace Applications *Gary W. Hunter, Chung-Chiun Liu, Darby B. Makel*.. 22-1
Introduction • Aerospace Applications • Sensor Fabrication Technologies • Chemical Sensor Development • Future Directions, Sensor Arrays and Commercialization • Commercial Applications • Summary

23 Packaging of Harsh-Environment MEMS Devices *Liang-Yu Chen, Jih-Fen Lei* 23-1
Introduction • Material Requirements for Packaging Harsh-Environment MEMS • High-Temperature Electrical Interconnection System • Thermomechanical Properties of Die-Attach • Discussion

15
Materials for Microelectromechanical Systems

Christian A. Zorman
Case Western Reserve University

Mehran Mehregany
Case Western Reserve University

15.1	Introduction	**15**-1
15.2	Single-Crystal Silicon	**15**-2
15.3	Polysilicon	**15**-3
15.4	Silicon Dioxide	**15**-9
15.5	Silicon Nitride	**15**-11
15.6	Germanium-Based Materials	**15**-14
15.7	Metals	**15**-16
15.8	Silicon Carbide	**15**-17
15.9	Diamond	**15**-20
15.10	III–V Materials	**15**-21
15.11	Piezoelectric Materials	**15**-22
15.12	Conclusions	**15**-23

15.1 Introduction

Without question, one of the most exciting technological developments during the last decade of the 20th century was the field of microelectromechanical systems (MEMS). MEMS consists of microfabricated mechanical and electrical structures working in concert for perception and control of the local environment. It was no accident that the development of MEMS accelerated rapidly during the 1990s, as the field was able to take advantage of innovations created during the integrated circuit revolution of the 1960s to the 1980s, in terms of processes, equipment and materials. A well-rounded understanding of MEMS requires a mature knowledge of the materials used to construct the devices, as the material properties of each component can influence device performance. Because the fabrication of MEMS structures often depends on the use of structural, sacrificial and masking materials on a common substrate, issues related to etch selectivity, adhesion, microstructure and a host of other properties are important design considerations. A discussion of the materials used in MEMS is really a discussion of the material systems used in MEMS, as the fabrication technologies rarely utilize a single material, but rather a collection of materials, each providing a critical function. It is in this light that this chapter is constructed. This chapter does not attempt to present a comprehensive review of all materials used in MEMS because the list of materials is just too long. It does, however, detail a selection of material systems that illustrate the importance of viewing MEMS in terms of material systems as opposed to individual materials.

15.2 Single-Crystal Silicon

Use of silicon (Si) as a material for microfabricated sensors can be traced to 1954, when the first paper describing the piezoresistive effect in germanium (Ge) and Si was published [Smith, 1954]. The results of this study suggested that strain gauges made from these materials could be 10 to 20 times larger than those for conventional metal strain gauges, which eventually led to the commercial development of Si strain gauges in the late 1950s. Throughout the 1960s and early 1970s, techniques to mechanically and chemically micromachine Si substrates into miniature, flexible mechanical structures on which the strain gauges could be fabricated were developed and ultimately led to commercially viable, high-volume production of Si-based pressure sensors in the mid 1970s. These lesser known developments in Si microfabrication technology happened concurrently with more popular developments in the areas of Si-based solid-state devices and integrated-circuit (IC) technologies that have revolutionized modern life. The conjoining of Si IC processing with Si micromachining techniques during the 1980s marked the advent of MEMS, and positioned Si as the primary material for MEMS.

There is little question that Si is the most widely known semiconducting material in use today. Single-crystal Si has a diamond (cubic) crystal structure. It has an electronic band gap of 1.1 eV, and, like many semiconducting materials, it can be doped with impurities to alter its conductivity. Phosphorus (P) is a common dopant for n-type Si and boron (B) is commonly used to produce p-type Si. A solid-phase oxide (SiO_2) that is chemically stable under most conditions can readily be grown on Si surfaces. Mechanically, Si is a brittle material with a Young's modulus of about 190 GPa, a value that is comparable to steel (210 GPa). Si is among the most abundant elements on Earth that can readily be refined from sand to produce electronic-grade material. Mature industrial processes exist for the low-cost production of single-crystal Si wafered substrates that have large surface areas (>8-in. diameter) and very low defect densities.

For MEMS applications, single-crystal Si serves several key functions. Single-crystal Si is perhaps the most versatile material for bulk micromachining, owing to the availability of well-characterized anisotropic etches and etch-mask materials. For surface micromachining applications, single-crystal Si substrates are used as mechanical platforms on which device structures are fabricated, whether they are made from Si or other materials. In the case of Si-based integrated MEMS devices, single-crystal Si is the primary electronic material from which the IC devices are fabricated.

Bulk micromachining of Si uses wet and dry etching techniques in conjunction with etch masks and etch stops to sculpt micromechanical devices from the Si substrate. From the materials perspective, two key capabilities make bulk micromachining a viable technology: (1) the availability of anisotropic etchants such as ethylene-diamine pyrocatecol (EDP) and potassium hydroxide (KOH), which preferentially etch single-crystal Si along select crystal planes; and (2) the availability of Si-compatible etch-mask and etch-stop materials that can be used in conjunction with the etch chemistries to protect select regions of the substrate from removal.

One of the most important characteristics of etching is the directionality (or profile) of the etching process. If the etch rate in all directions is equal, the process is said to be *isotropic*. By comparison, etch processes that are *anisotropic* generally have etch rates in the direction perpendicular to the wafer surface that are much larger than the lateral etch rates. It should be noted that an "anisotropic" sidewall profile could also be produced in virtually any Si substrate by deep reactive ion etching, ion beam milling or laser drilling.

Isotropic etching of a semiconductor in liquid reagents is commonly used for removal of work-damaged surfaces, creation of structures in single-crystal slices, and patterning single-crystal or polycrystalline semiconductor films. For isotropic etching of Si, the most commonly used etchants are mixtures of hydrofluoric (HF) and nitric (HNO_3) acid in water or acetic acid (CH_3COOH), usually called the HNA etching system.

Anisotropic Si etchants attack the (100) and (110) crystal planes significantly faster than the (111) crystal planes. For example, the (100) to (111) etch-rate ratio is about 400:1 for a typical KOH/water etch solution. Silicon dioxide (SiO_2), silicon nitride (Si_3N_4) and some metallic thin films (e.g., Cr, Au) provide good etch masks for most Si anisotropic etchants. In structures requiring long etching times in KOH, Si_3N_4 is the preferred masking material, due to its chemical durability.

In terms of etch stops, heavily B-doped Si ($>7 \times 10^{19}$/cm^3), commonly referred to as a p+ etch stop, is effective for some etch chemistries. Fundamentally, etching is a charge transfer process, with etch rates dependent on dopant type and concentration. Highly doped material might be expected to exhibit higher etch rates because of the greater availability of mobile carriers. This is true for isotropic etchants such as HNA, where typical etch rates are 1 to 3 µm/min for p- or n-type dopant concentrations greater than 10^{18}/cm^3 and essentially zero for concentrations less than 10^{17}/cm^3. On the other hand, anisotropic etchants such as EDP and KOH exhibit a much different preferential etching behavior. Si that is heavily doped with B ($>7 \times 10^{19}$/cm^3) etches at a rate that is about 5 to 100 times slower than undoped Si when etched in KOH and 250 times slower in EDP. Etch stops formed by the p+ technique are often less than 10 µm thick, as the B doping is often done by diffusion. Using high diffusion temperatures (e.g., 1175°C) and long diffusion times (e.g., 15 to 20 hours), thick (~20 µm) p+ etch stop layers can be created. It is also possible to create a p+ etch stop below the Si surface using ion implantation; however, the implant depth is limited to a few microns and a high-energy/high-current ion accelerator is required for implantation. While techniques are available to grow a B-doped Si epitaxial layer on top of a p+ etch stop to increase the thickness of the final structure, this is seldom utilized due to the expense of the epitaxial process step.

Due to the high concentration of B, p+ Si has a high density of defects. These defects are generated as a result of stresses created in the Si lattice due to the fact that B is a smaller atom than Si. Studies of p+ Si report that stress in the resultant films can either be tensile [Ding et al., 1990] or compressive [Maseeh and Senturia, 1990]. These variations may be due to post-processing steps. For instance, thermal oxidation can significantly modify the residual stress distribution in the near-surface region of p+ Si films, thereby changing the overall stress in the film. In addition to the generation of crystalline defects, the high concentration of dopants in the p+ etch-stops prevents the fabrication of electronic devices in these layers. Despite some of these shortcomings, the p+ etch-stop technique is widely used in Si bulk micromachining due to its effectiveness and simplicity.

A large number of dry etch processes are available to pattern single-crystal Si. The process spectrum ranges from physical etching via sputtering and ion milling to chemical plasma etching. Two processes, reactive ion etching (RIE) and reactive ion beam etching (RIBE) combine aspects of both physical and chemical etching. In general, dry etch processes utilize a plasma of ionized gases, along with neutral particles to remove material from the etch surface. Details regarding the physical processes involved in dry etching can be found elsewhere [Wolfe and Tauber, 1999].

Reactive ion etching is the most commonly used dry etch process to pattern Si. In general, fluorinated compounds such as CF_4, SF_6 and NF_3 or chlorinated compounds such as CCl_4 or Cl_2 sometimes mixed with He, O_2 or H_2 are used. The RIE process is highly directional, thereby enabling direct pattern transfer from the masking material to the etched Si surface. The selection of masking material is dependent on the etch chemistry and the desired etch depth. For MEMS applications, photoresist and SiO_2 thin films are often used. Si etch rates in RIE processes are typically less than 1 µm/min, so dry etching is mostly used to pattern layers on the order of several microns in thickness. The plasmas selectively etch Si relative to Si_3N_4 or SiO_2, so these materials can be used as etch masks or etch-stop layers. Recent development of deep reactive ion etching processes has extended Si etch depths well beyond several hundred microns, thereby enabling a multitude of new designs for bulk micromachined structures.

15.3 Polysilicon

Without doubt the most common material system for the fabrication of surface micromachined MEMS devices utilizes polycrystalline Si (polysilicon) as the primary structural material, SiO_2 as the sacrificial material and Si_3N_4 for electrical isolation of device structures. Heavy reliance on this material system stems in part from the fact these three materials find uses in the fabrication of ICs and, as a result, film deposition and etching technologies are readily and widely available. Like single-crystal Si, polysilicon can be doped during or after film deposition using standard IC processing techniques. SiO_2 can be grown or deposited over a broad temperature range (e.g., 200 to 1150°C) to meet various process and material requirements. SiO_2 is readily dissolvable in hydrofluoric acid (HF), an IC-compatible chemical, without etching the polysilicon structural material [Adams, 1988]. HF does not wet bare Si surfaces; as a result, it is automatically rejected from microscopic cavities in between polysilicon layers after a SiO_2 sacrificial layer is completely dissolved.

For surface micromachined structures, polysilicon is an attractive material because it has mechanical properties that are comparable to single-crystal Si, the required processing technology has been developed for IC applications, and it is resistant to SiO_2 etchants. In other words, polysilicon surface micromachining leverages on the significant capital investment made by the IC industry in the important areas of film deposition, patterning and material characterization.

For MEMS and IC applications, polysilicon thin films are commonly deposited by a process known as low-pressure chemical vapor deposition (LPCVD). This deposition technique was first commercialized in the mid-1970s [Rosler, 1977] and has since been a standard process in the microelectronics industry. The typical polysilicon LPCVD reactor (or furnace) is based on a hot-wall, resistance-heated, horizontal, fused-silica tube design. The temperature of the wafers in the furnace is maintained by heating the tube using resistive heating elements. The furnaces are equipped with quartz boats that have closely spaced, vertically oriented slots that hold the wafers. The close spacing requires that the deposition process be performed in the reaction-limited regime to obtain uniform deposition across each wafer surface. In the reaction-limited deposition regime, the deposition rate is determined by the reaction rate of the reacting species on the substrate surface, as opposed to the arrival rate of the reacting species to the surface (which is the diffusion-controlled regime). The relationship between the deposition rate and the substrate temperature in the reaction-limited regime is exponential; therefore, precise temperature control of the reaction chamber is required. Operating in the reaction-limited regime facilitates conformal deposition of the film over the substrate topography, an important aspect of multilayer surface micromachining. Commercial equipment is available to accommodate furnace loads exceeding 100 wafers.

Typical deposition conditions utilize temperatures from 580 to 650°C and pressures ranging from 100 to 400 mtorr. The most commonly used source gas is silane (SiH_4), which readily decomposes into Si on substrates heated to these temperatures. Gas flow rates depend on the tube diameter and other conditions. For processes performed at 630°C, the polysilicon deposition rate is about 100 Å/min. The gas inlets are typically at the load door end of the tube, with the outlet to the vacuum pump located at the opposite end. For door injection systems, depletion of the source gas occurs along the length of the tube. To keep the deposition rate uniform, a temperature gradient is maintained along the tube so that the increased deposition rate associated with higher substrate temperatures offsets the reduction due to gas depletion. Typical temperature gradients range from 5 to 15°C along the tube length. Some systems incorporate an injector inside the tube to allow for the additional supply of source gas to offset depletion effects. In this case, the temperature gradient along the tube is zero. This is an important modification, as the microstructure and physical properties of the deposited polysilicon are a function of the deposition temperature.

Polysilicon is made up of small single-crystal domains called *grains*, whose orientations and/or alignment vary with respect to each other. The roughness often observed on polysilicon surfaces is due to the granular nature of polysilicon. The microstructure of the as-deposited polysilicon is a function of the deposition conditions [Kamins, 1998]. For typical LPCVD processes (e.g., 100% SiH_4 source gas, 200 mtorr deposition pressure), the amorphous-to-polycrystalline transition temperature is about 570°C, with amorphous films deposited below this temperature (Figure 15.1) and polycrystalline films above this temperature (Figure 15.2). As the deposition temperature increases significantly above 570°C, the grain structure of the as-deposited polysilicon films changes in dramatic fashion. For example, at 600°C, the grains are very fine and equiaxed, while at 625°C, the grains are larger and have a columnar structure that is aligned perpendicular to the plane of the substrate [Kamins, 1998]. In general, the grain size tends to increase with film thickness across the entire range of deposition temperatures. As with grain size, the crystalline orientation of the polysilicon grains is dependent on the deposition temperature. For example, under standard LPCVD conditions (100% SiH_4, 200 mtorr), the crystal orientation of polysilicon is predominantly (110) for substrate temperatures between 600 and 650°C. In contrast, the (100) orientation is dominant for substrate temperatures between 650 and 700°C.

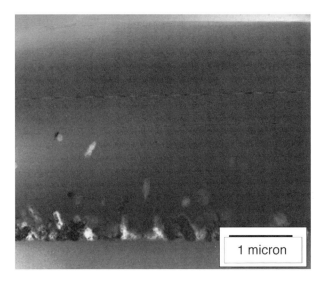

FIGURE 15.1 TEM micrograph of an amorphous Si film deposited at 570°C.

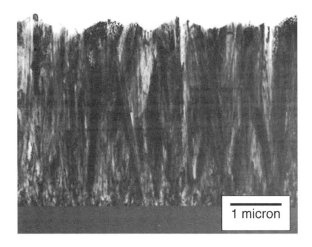

FIGURE 15.2 TEM micrograph of a polysilicon film deposited at 620°C.

During the fabrication of micromechanical devices, polysilicon films typically undergo one or more high-temperature processing steps (e.g., doping, thermal oxidation, annealing) after deposition. These high-temperature steps can cause recrystallization of the polysilicon grains, leading to a reorientation of the film and a significant increase in average grain size. Consequently, the polysilicon surface roughness increases with the increase in grain size, an undesirable outcome from a fabrication point of view because surface roughness limits pattern resolution. Smooth surfaces are desired for many mechanical structures, as defects associated with surface roughness can act as initiating points of structural failure. To address these concerns, chemical–mechanical polishing processes that reduce surface roughness with minimal film removal can be used.

Three phenomena influence the growth of polysilicon grains, namely strain-induced growth, grain-boundary growth, and impurity drag [Kamins, 1998]. If the dominant driving force for grain growth is the release of stored strain energy caused by such things as doping or mechanical deformation (wafer warpage), grain growth will increase linearly with increasing annealing time. To minimize the energy associated with grain boundaries, the gains tend to grow in a way that minimizes the grain boundary area.

This driving force is inversely proportional to the radius of curvature of the grain boundary, and the growth rate is proportional to the square root of the annealing time. Heavy P-doping causes significant grain growth at temperatures as low as 900°C because P increases grain boundary mobility. If other impurities are incorporated in the gain boundaries, they may retard grain growth, which then results in the growth rate being proportional to the cube root of the annealing time.

Thermal oxidation of polysilicon is carried out in a manner essentially identical to that of single-crystal Si. The oxidation rate of undoped polysilicon is typically between that of (100)- and (111)-oriented single-crystal Si. Heavily P-doped polysilicon oxidizes at a rate significantly higher than undoped polysilicon. However, this impurity-enhanced oxidation effect is smaller in polysilicon than in single-crystal Si. The effect is most noticeable at lower oxidation temperatures (<1000°C). Like single-crystal Si, oxidation of polysilicon can be modeled by using process simulation software. For first-order estimates, however, the oxidation rate of (100) Si can be used to estimate the oxidation rate of polysilicon.

The resistivity of polysilicon can be modified by impurity doping using the methods developed for single-crystal doping. Polysilicon doping can be achieved during deposition (called *in situ* doping) or after film deposition either by diffusion or ion implantation. *In situ* doping is achieved by adding reaction gases such as diborane (B_2H_6) and phosphine (PH_3) to the Si-containing source gas. The addition of dopants during the deposition process not only affects the conductivity of the as-deposited films, but also affects the deposition rate. Relative to the deposition of undoped polysilicon, the addition of P reduces the deposition rate while the addition of B increases the deposition rate. *In situ* doping can be used to produce conductive films with uniform doping profiles through the film thickness without the need for high-temperature steps commonly associated with diffusion or ion implantation. Nonuniform doping through the thickness of a polysilicon film can lead to microstructural variations in the thickness direction that can result in stress gradients in the films and subsequent bending of released structural components. In addition, minimizing the maximum required temperature and duration of high-temperature processing steps is important for the fabrication of micromechanical components on wafers that contain temperature-sensitive layers.

The primary disadvantage of *in situ* doping is the complexity of the deposition process. The control of film thickness, deposition rate, and deposition uniformity is more complicated than the process used to deposit undoped polysilicon films, in part because a second gas with a different set of temperature- and pressure-related reaction parameters is included. Additionally, the cleanliness standards of the reactor are more demanding for the doped furnace. Therefore, many MEMS fabrication facilities use diffusion-based doping processes. Diffusion is an effective method for doping polysilicon films, especially for very heavy doping (e.g., resistivities of 10^{-4} Ω-cm) of thick (>2 μm) films. However, diffusion is a high-temperature process, typically from 900 to 1000°C. Therefore, fabrication processes that require long diffusion times to achieve uniform doping at significant depths may not be compatible with pre-MEMS, complementary metal-oxide-semiconductor (CMOS) integration schemes. Like *in situ* doping, diffusion processes must be performed properly to ensure that the dopant distribution through the film thickness is uniform, so that dopant-related variations in the mechanical properties through the film thickness are minimized. As will be discussed below, the use of doped oxide sacrificial layers relaxes some of the concerns associated with doping the film uniformly by diffusion, because the sacrificial doped SiO_2 can also be used as a diffusion source. Phosphorous, which is the most commonly used dopant in polysilicon MEMS, diffuses significantly faster in polysilicon than in single-crystal Si, due primarily to enhanced diffusion rates along grain boundaries. The diffusivity in polysilicon thin films (i.e., small equiaxed grains) is about 1×10^{12} cm^2/s.

Ion implantation is also used to dope polysilicon films. The implantation energy is typically adjusted so that the peak of the concentration profile is near the midpoint of the film. When necessary, several implant steps are performed at various energies in order to distribute the dopant uniformly through the thickness of the film. A high-temperature anneal step is usually required to electrically activate the implanted dopant, as well as to repair implant-related damage in the polysilicon film. In general, the resistivity of implanted polysilicon films is not as low as films doped by diffusion. In addition, the need for specialized implantation equipment limits the use of this method in polysilicon MEMS.

The electrical properties of polysilicon depend strongly on the grain structure of the film. The grain boundaries provide a potential barrier to the moving charge carriers, thus affecting the conductivity of the films. For P-doped polysilicon, the resistivity decreases as the amount of P increases for concentrations up to about $1 \times 10^{21}/cm^3$. Above this value, the resistivity reaches a plateau of about 4×10^4 Ω-cm after a 1000°C anneal. The maximum mobility for such a highly P-doped polysilicon is about 30 cm^2/Vs. Grain boundary and ionized impurity scattering are important factors limiting the mobility [Kamins, 1988].

The thermal conductivity of polysilicon is a strong function of the grain structure of the film [Kamins 1998]. For fine-grain films, the thermal conductivity is about 0.30 to 0.35 W/cm-K, which is about 20 to 25% of the single-crystal value. For thick films with large grains, the thermal conductivity ranges between 50 and 85% of the single-crystal value.

In general, thin films are generally under a state of stress, commonly referred to as *residual stress*, and polysilicon is no exception. In polysilicon micromechanical structures, the residual stress in the films can greatly affect the performance of the device. Like the electrical and thermal properties of polysilicon, the as-deposited residual stress in polysilicon films depends on microstructure. In general, as-deposited polysilicon films have compressive residual stresses, although reports regarding polysilicon films with tensile stress can be found in the literature [Kim et al., 1998]. The highest compressive stresses are found in amorphous Si films and polysilicon films with a strong, columnar (110) texture. For films with fine-grained microstructures, the stress tends to be tensile. For the same deposition conditions, thick polysilicon films tend to have lower residual stress values than thin films, especially true for films with a columnar microstructure. Annealing can be used to reduce the compressive stress in as-deposited polysilicon films. For polysilicon films doped with phosphorus by diffusion, a decrease in the magnitude of compressive stress has been correlated with grain growth [Kamins, 1998]. For polysilicon films deposited at 650°C, the compressive residual stress is typically on the order of 5×10^9 to 10×10^9 dyne/cm^2. However, these stresses can be reduced to less than 10^8 dyne/cm^2 by annealing the films at high temperature (1000°C) in a N$_2$ ambient [Guckel et al., 1985; Howe and Muller, 1983]. Compressive stresses in fine-grained polysilicon films deposited at 580°C (100-Å grain size) can be reduced from 1.5×10^{10} to less than 10^8 dyne/cm^2 by annealing above 1000°C, or even made to be tensile (5×10^9 dynes/cm^2) by annealing at temperatures between 650 and 850°C [Guckel et al., 1988]. Recent advances in the area of rapid thermal annealing (RTA) as applied to polysilicon indicate that RTA is a fast and effective method of stress reduction in polysilicon films. For polysilicon films deposited at 620°C with compressive stresses of about 340 MPa, a 10-s anneal at 1100°C was sufficient to completely relieve the stress [Zhang et al., 1998].

A second approach, called the *multipoly process*, has recently been developed to address issues related to residual stress [Yang et al., 2000]. As the name implies, the multipoly process is a deposition method to produce a polysilicon-based multilayer structure where the composite has a predetermined stress level. The multilayer structure is comprised of alternating tensile and compressive polysilicon layers deposited sequentially. The overall stress of the composite is simply the superposition of the stress in each individual layer. The tensile layers consist of fine-grained polysilicon grown at a temperature of 570°C while the compressive layers are made up of polysilicon deposited at 615°C and having a columnar microstructure. The overall stress in the composite film depends on the number of alternating layers and the thickness of each layer. With the proper set of parameters, a composite polysilicon film can be deposited with a near-zero residual stress. Moreover, despite the fact that the composite has a clearly changing microstructure through the thickness of the film, the stress gradient is also nearly zero. The clear advantage of the multipoly process lies in the fact that stress reduction can be achieved without the need for high-temperature annealing, a considerable advantage for polysilicon MEMS processses with on-chip CMOS integration. A transmission electron microscopy (TEM) micrograph of a multipoly structure is shown in Figure 15.3.

Conventional techniques to deposit polysilicon films for MEMS applications utilize LPCVD systems with deposition rates that limit film the maximum film thickness to roughly 5 μm. Many device designs, however, require thick structural layers that are not readily achievable using LPCVD processes. For these devices, wafer bonding and etchback techniques are often used to produce thick (>10 μm) single-crystal Si films on sacrificial substrate layers. There is, however, a deposition technique to produce thick polysilicon films on sacrificial substrates. These thick polysilicon films are called *epi-poly* films because

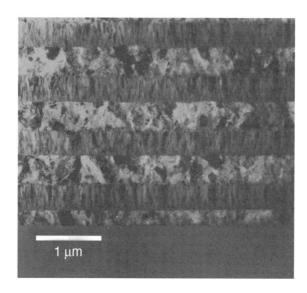

FIGURE 15.3 TEM micrograph of a polysilicon multilayer film created using the multipoly process.

epitaxial Si reactors are used to deposit these films using a high-temperature process. Unlike conventional LPCVD polysilicon deposition processes which have deposition rates of 100 Å/min, epi-poly processes have deposition rates on the order of 1 μm/min [Gennissen et al., 1997]. The high deposition rates are a result of the deposition conditions used—specifically, much higher substrate temperatures (>1000°C) and deposition pressures (>50 torr). The polysilicon films are usually deposited on SiO_2 sacrificial substrate layers and have been used in the fabrication of mechanical properties test structures [Lange et al., 1996; Gennissen et al., 1997; Greek et al., 1999], thermal actuators [Gennissen et al., 1997], electrostatically actuated accelerometers [Gennissen et al., 1997] and gryoscopes [Funk et al., 1999]. An LPCVD polysilicon seed layer is used in order to control nucleation, grain size and surface roughness. In general, the microstructure and residual stress of epi-poly films is related to deposition conditions, with compressive films having a mixture of (110) and (311) grains [Lange et al., 1996; Greek et al., 1999] and tensile films having a random mix of (110), (100), (111) and (311) grains [Lange et al., 1996]. The Young's modulus of epi-poly measured from micromachined test structures is comparable to LPCVD polysilicon [Greek et al., 1999].

Porous Si is a "type" of Si finding applications in MEMS technology. Porous Si is made by room-temperature electrochemical etching of Si in HF. Under normal conditions, Si is not etched by HF, hence its widespread use as an etchant of sacrificial oxide in polysilicon surface micromachining. In an electrochemical circuit using an HF-based solution, however, positive charge carriers (holes) at the Si surface facilitate the exchange of F atoms with the H atoms terminating the Si surface bonds. The exchange continues with the exchange of subsurface bonds, leading to the eventual removal of the fluorinated Si. The quality of the etched surface is related to the density of holes at the surface which is controlled by the applied current density. For high current densities, the density of holes is high and the etched surface is smooth. For low current density, the density of holes is low and they are clustered in highly localized regions associated with surface defects. The surface defects become enlarged by etching, leading to the formation of pores. Pore size and density are related to the type of Si used and the conditions of the electrochemical cell. Both single-crystal and polycrystalline Si can be converted to porous Si, with porosities of up to 80% possible.

The large surface-to-volume ratios make porous Si attractive for many MEMS applications. As one might expect, use of porous Si has been proposed for a number of gaseous and liquid applications, including filter membranes and absorbing layers for chemical and mass sensing [Anderson et al., 1994]. The large surface-to-volume ratio also permits the use of porous Si as the starting material for the formation of thick thermal oxides, as the proper pore size can be selected to account for the volume expansion of the thermal oxide. When single-crystal substrates are used in the formation of porous Si films, the unetched material remains single crystalline, thus providing the appropriate template for epitaxial growth. It has been shown that CVD coatings will not penetrate the porous regions, but rather overcoat the pores at the surface [Lang et al., 1995]. The formation of localized, surface-micromachinable, Si on insulator structures is possible by simply combining electrochemical etching, epitaxial growth, dry etching (to create access holes) and thermal oxidation. A third, MEMS-related application is the direct use of porous Si as a sacrificial layer in polysilicon and single-crystalline Si surface micromachining. The process involves the electrical isolation of the structural Si layer either by the formation of pn-junctions through selective doping or by use of electrically insulating thin films [Lang, 1995]. In essence the formation of pores only occurs on electrically charged surfaces. A weak Si etchant aggressively attacks the porous regions with little damage to the structural Si layers. Porous Si may be an attractive option for micromachining processes that are chemically stable in HF but are tolerant of high-temperature processing steps.

With the possible exception of porous Si, all of the above-mentioned processes to prepare polysilicon for MEMS applications utilize substrate temperatures in excess of 570°C, either during film deposition or in subsequent stress-relieving annealing steps. Such high-temperature processing restricts the use of non-Si derivative materials, such as aluminum for metallization and polymers for sacrificial layers, both of which are relatively straightforward to deposit and pattern and would be of great benefit to polysilicon micromachining if they could be used throughout the process. Work in developing low-temperature deposition processes for polysilicon has focused on sputter deposition techniques [Abe and Reed, 1996; Honer and Kovacs, 2000]. Early work [Abe and Reed, 1996] emphasized the ability to deposit very smooth (25-Å roughness average) films at reasonable deposition rates (191 Å/min) and with low residual compressive stresses. The process involved DC magnitron sputtering from a Si target using an Ar sputtering gas, a chamber pressure of 5 mtorr and a power of 100 W. The substrates consisted of thermally oxidized Si wafers. The authors reported that a post-deposition anneal at 700°C in N_2 for 2 hr was performed to crystallize the deposited film and perhaps lower the stress. A second group [Honer and Kovacs, 2000] sought to develop a polymer-friendly, Si-based, surface-micromachining process. The Si films were sputter-deposited on polyimide sacrificial layers. To improve the conductivity of the micromachined Si structures, the sputtered Si films were sandwiched between two TiW cladding layers. The device structures were released by etching the polyimide in a O_2 plasma. The processing step with the highest temperature was the polyimide cure, which was performed for 1 hr at 350°C. To test the robustness of the process, sputter-deposited Si microstructures were fabricated on substrates containing CMOS devices. As expected from thermal budget considerations, the authors reported no measurable degradation of device performance.

15.4 Silicon Dioxide

SiO_2 can be grown thermally on Si substrates as well as also deposited using a variety of processes to satisfy a wide range of different requirements. In polysilicon surface micromachining, SiO_2 is used as a sacrificial material, as it can be easily dissolved using etchants that do not attack polysilicon. In a less prominent role, SiO_2 is used as an etch mask for dry etching of thick polysilicon films, as it is chemically resistant to dry polysilicon etch chemistries.

The SiO_2 growth and deposition processes most widely used in polysilicon surface micromachining are thermal oxidation and LPCVD. Thermal oxidation of Si is performed at high temperatures (e.g., 900 to 1000°C) in the presence of oxygen or steam. Because thermal oxidation is a self-limiting process (i.e., the oxide growth rate decreases with increasing film thickness), the maximum practical film thickness that can be obtained is about 2 μm, which for many sacrificial applications is sufficient.

SiO_2 films for MEMS applications can also be deposited using an LPCVD process known as low-temperature oxidation (LTO). In general, LPCVD provides a means for depositing thick (>2 μm) SiO_2 films at temperatures much lower than thermal oxidation. Not only are LTO films deposited at low temperatures, but the films also have a higher etch rate in HF than thermal oxides which results in significantly faster releases of polysilicon surface-micromachined devices. An advantage of the LPCVD processes is that dopant gases can be included in the flow of source gases in order to dope the as-deposited SiO_2 films. One such example is the incorporation of P to form phosphosilicate glass (PSG). PSG is formed using the same deposition process as LTO, with PH_3 added to dope the glass with a P content ranging from 2 to 8 wt%. PSG has an even higher etch rate in HF than LTO, further facilitating the release of polysilicon surface-micromachined components. PSG flows at high temperatures (e.g., 1000 to 1100°C) which can be exploited to create a smooth surface topography. Additionally, PSG layers sandwiching a polysilicon film can be used as a P doping source, improving the uniformity of diffusion-based doping.

Phosphosilicate glass and LTO films are deposited in hot-wall, low-pressure, fused-silica reactors in a manner similar to the systems described previously for polysilicon. Typical deposition rates are about 100 Å/min. Precursor gases include SiH_4 as a Si source, O_2 as an oxygen source and, in the case of PSG, PH_3 as a source of phosphorus. Because SiH_4 is pyrophoric (i.e., spontaneously combusts in the presence of O_2), door injection of the deposition gases would result in a large depletion of the gases at deposition temperatures of 400 to 500°C and nonuniform deposition along the tube. Therefore, the gases are introduced in the furnace through injectors distributed along the length of the tube. The wafers are placed vertically in caged boats to ensure uniform gas transport to the wafers. In the caged boats, two wafers are placed back to back in each slot, thus minimizing the deposition of SiO_2 on the back of the wafers. The typical load of an LTO system is over 100 wafers.

Low-temperature oxidation and PSG films are typically deposited at temperatures of 425 to 450°C and pressures ranging from 200 to 400 mtorr. The low deposition temperatures result in LTO and PSG films that are slightly less dense than thermal oxides due to incorporation of hydrogen in the films. LTO films can, however, be densified by an annealing step at high temperature (1000°C). The low density of LTO and PSG films is partially responsible for the increased etch rate in HF, which makes them attractive sacrificial materials for polysilicon surface micromachining. LTO and PSG deposition processes are not typically conformal to nonplanar surfaces because the low substrate temperatures result in low surface migration of reacting species. Step coverage is, however, sufficient for many polysilicon surface-micromachining applications, although deposited films tend to thin at the bottom surfaces of deep trenches and therefore must be thoroughly characterized for each application.

The dissolution of the sacrificial SiO_2 to release free-standing structures is a critical step in polysilicon surface micromachining. Typically, 49% (by weight) HF is used for the release process. To pattern oxide films using wet chemistries, etching in buffered HF (28 ml 49% HF, 170 ml H_2O, 113 g NH_4F), also known as buffered oxide etch (BOE), is common for large structures. A third wet etchant, known as P-etch, is traditionally used to selectively remove PSG over undoped oxide (e.g., to deglaze a wafer straight from a diffusion furnace).

Thermal SiO_2, LTO, and PSG are electrical insulators suitable for many MEMS applications. The dielectric constants of thermal oxide and LTO are 3.9 and 4.3, respectively. The dielectric strength of thermal SiO_2 is 1.1×10^6 V/cm, and for LTO it is about 80% that of thermal SiO_2 [Ghandhi, 1983]. Thermal SiO_2 is in compression with a stress level of about 3×10^9 dyne/cm^2 [Ghandhi, 1983]. For LTO, however, the as-deposited residual stress is tensile, with a magnitude of about 1 to 4×10^9 dyne/cm^2 [Ghandhi, 1983]. The addition of phosphorous to LTO (i.e., PSG) decreases the tensile residual stress to about 10^8 dyne/cm^2 for a phosphorus concentration of 8% [Pilskin, 1977]. These data are representative of oxide films deposited directly on Si substrates under typical conditions; however, the final value of the stress in an oxide film can be a strong function of the process parameters as well as any post-processing steps.

A recent report documents the development of another low-pressure process, known as plasma-enhanced chemical vapor deposition (PECVD), for MEMS applications. The objective was to deposit low-stress, very thick (10 to 20 μm) SiO_2 films for insulating layers in micromachined gas turbine engines [Zhang et al., 2000]. PECVD was selected, in part, because it offers the possibility to deposit films of the desired thickness at a reasonable deposition rate. The process used a conventional parallel plate reactor with tetraethylorthosilicate (TEOS), a commonly used precursor in LPCVD processes, as the source gas.

As expected, the authors found that film stress is related to the concentration of dissolved gases in the film and that annealed films tend to suffer from cracking. By using a thin Si_3N_4 film in conjunction with the thick SiO_2 film, conditions were found where a low-stress, crack-free SiO_2 film could be produced.

Two other materials in the SiO_2 family are receiving increasing attention from MEMS fabricators, especially now that the material systems have expanded beyond conventional Si processing. The first of these is crystalline quartz. The chemical composition of quartz is SiO_2. Quartz is optically transparent and, like its amorphous counterpart, quartz is electrically insulating. However, the crystalline nature of quartz gives it piezoelectric properties that have been exploited for many years in electronic circuitry. Like single-crystal Si, quartz substrates are available as high-quality, large-area wafers. Also like single-crystal Si, quartz can be bulk micromachined using anisotropic etchants based on heated HF and ammonium fluoride (NH_4F) solutions, albeit the structural shapes that can be etched into quartz do not resemble the shapes that can be etched into Si. A short review of the basics of quartz etching and its applications to the fabrication of a micromachined acceleration sensor can be found in Danel et al. (1990).

A second SiO_2-related material that has found utility in MEMS is spin-on-glass (SOG), which is used in thin-film form as a planarization dielectric material in IC processing. As the name implies, SOG is applied to a substrate by spin coating. The material is polymer based with a viscosity suitable for spin-coating, and once dispensed at room temperature on the spinning substrate, it is cured at elevated temperatures to form a solid thin film. Two recent publications illustrate the potential uses of SOG in MEMS. In the first example, SOG was developed as a thick-film sacrificial molding material to pattern thick polysilicon films [Yasseen et al., 1999]. The authors reported a process to produce SOG films that were 20 μm thick, complete with a chemical–mechanical polishing (CMP) procedure and etching techniques. The thick SOG films were patterned into molds that were filled with 10-μm-thick LPCVD polysilicon films, planarized by selective CMP and subsequently dissolved in a $HCl:HF:H_2O$ wet etchant to reveal the patterned polysilicon structures. The cured SOG films were completely compatible with the polysilicon deposition process, indicating that SOG could be used to produce MEMS devices with extremely large gaps between structural layers. In the second example, high-aspect-ratio channel-plate microstrucures were fabricated from SOG [Liu et al., 1999]. The process required the use of molds to create the structures. Electroplated nickel (Ni) was used as the molding material, with Ni channel plate molds fabricated using a conventional LIGA process. The Ni molds are then filled with SOG, and the sacrificial Ni molds are removed in a reverse electroplating process. In this case, the fabricated SOG structures were over 100 μm tall, essentially bulk micromachined structures fabricated using a sacrificial molding material system.

15.5 Silicon Nitride

Si_3N_4 is widely used in MEMS for electrical isolation, surface passivation, etch masking and as a mechanical material. Two deposition methods are commonly used to deposit Si_3N_4 thin films: LPCVD and PECVD. PECVD Si_3N_4 is generally nonstoichiometric and may contain significant concentrations of hydrogen. Use of PECVD Si_3N_4 in micromachining applications is somewhat limited because its etch rate in HF can be high (e.g., often higher than that of thermally grown SiO_2) due to the porosity of the film. However, PECVD offers the potential to deposit nearly stress-free Si_3N_4 films, an attractive property for many MEMS applications, especially in the area of encapsulation and packaging. Unlike its PECVD counterpart, LPCVD Si_3N_4 is extremely resistant to chemical attack, thereby making it the material of choice for many Si bulk and surface micromachining applications. LPCVD Si_3N_4 is commonly used as an insulating layer to isolate device structures from the substrate and from other device structures, because it is a good insulator with a resistivity of 10^{16} Ω-cm and a field breakdown limit of 10^7 V/cm.

The LPCVD Si_3N_4 films are deposited in horizontal furnaces similar to those used for polysilicon deposition. Typical deposition temperatures and pressures range between 700 and 900°C and 200 to 500 mtorr, respectively. A typical deposition rate is about 30 Å/min. The standard source gases are dichlorosilane (SiH_2Cl_2) and ammonia (NH_3). SiH_2Cl_2 is used in place of SiH_4 because it produces films with a higher degree of thickness uniformity at the required deposition temperature and it allows the wafers to be spaced close together, thus increasing the number of wafers per furnace load. To produce stoichiometric Si_3N_4, a NH_3-to-SiH_2Cl_2 ratio of 10:1 is commonly used. The standard furnace configuration uses door injection of the source gases with a temperature gradient along the tube axis to accommodate for the gas depletion effects. LPCVD Si_3N_4 films deposited between 700 and 900°C are amorphous; therefore, the material properties do not vary significantly along the length of tube despite the temperature gradient. As with polysilicon deposition, a typical furnace can accommodate over 100 wafers. Because Si_3N_4 is deposited in the reaction-limited regime, film is deposited on both sides of each wafer with equal thickness.

The residual stress in stochiometric Si_3N_4 is large and tensile, with a magnitude of about 10^{10} dyne/cm^2. Such a large residual stress limits the practical thickness of a deposited Si_3N_4 film to a few thousand angstroms because thicker films tend to crack. Nevertheless, stoichiometric Si_3N_4 films have been used as mechanical support structures and electrical insulating layers in piezoresistive pressure sensors [Folkmer et al., 1995]. To reduce the residual stress, thus enabling the use of thick Si_3N_4 films for applications that require durable, chemically resistant membranes, nonstoichiometric silicon nitride (Si_xN_y) films can be deposited by LPCVD. These films, often referred to as Si-rich or low-stress nitride, are intentionally deposited with an excess of Si by simply decreasing the NH_3 to SiH_2Cl_2 ratio in the reaction furnace. For a NH_3-to-SiH_2Cl_2 ratio of 1:6 at a deposition temperature of 850°C and pressure of 500 mtorr, the as-deposited films are nearly stress free [Sekimoto et al., 1982]. The increase in Si content not only leads to a reduction in tensile stress, but also decreases the etch rate of the film in HF. As a result, low-stress silicon nitride films have replaced stoichiometric Si_3N_4 in many MEMS applications and even have enabled the development of fabrication techniques that would otherwise not be feasible with stoichiometric Si_3N_4. For example, low-stress silicon nitride has been successfully used as a structural material in a surface micromachining process that uses polysilicon as the sacrificial material [Monk et al., 1993]. In this case, Si anisotropic etchants such as KOH and EDP were used for dissolving the sacrificial polysilicon. A second low-stress nitride surface micromachining process used PSG as a sacrificial layer, which was removed using a HF-based solution [French et al., 1997]. Of course, wide use of Si_3N_4 as a MEMS material is restricted by its dielectric properties; however, its Young's modulus (146 GPa) is on par with Si (~190 GPa), making it an attractive material for mechanical components.

The essential interactions between substrate, electrical isolation layer, sacrificial layers, and structural layers are best illustrated by examining the critical steps in a multilevel surface micromachining process. The example used here (shown in Figure 15.4) is the fabrication of a Si micromotor using a technique called the *rapid prototyping process*. The rapid prototyping process utilizes three deposition and three photolithography steps to implement flange-bearing, side-drive micromotors such as in the SEM of Figure 15.5. The device consists of heavily P-doped LPCVD polysilicon structural components deposited on a Si wafer, using LTO both as a sacrificial layer and as an electrical isolation layer. Initially, a 2.4-µm-thick LTO film is deposited on the Si substrate. A 2-µm-thick doped polysilicon layer is then deposited on the LTO film. Photolithography and RIE steps are then performed to define the rotor, stator and rotor/stator gap. To fabricate the flange, a sacrificial mold is created by etching into the LTO film with an isotropic etchant, then partially oxidizing the polysilicon rotor and stator structures to form what is called the *bearing clearance oxide*. This oxidation step also forms the bottom of the bearing flange mold. A 1- to 2-µm-thick, heavily doped polysilicon film is then deposited and patterned by photolithography and RIE to form the bearing. At this point, the structural components of the micromotor are completely formed, and all that remains is to release the rotor by etching the sacrificial oxide in HF and performing an appropriate drying procedure (detailed later in this chapter). In this example, the LTO film serves three purposes: It is the sacrificial underlayer for the free-spinning rotor, it comprises part of the flange mold, and it serves as an insulating anchor for the stators and bearing post. Likewise, the thermal oxide serves as a mold and electrical isolation layer. The material properties of LTO and thermal oxide allow for these films to be used as they are in the rapid prototyping process, thus enabling the fabrication of multilayer structures with a minimum of processing steps.

Without question, SiO_2 is an excellent sacrificial material for polysilicon surface micromachining; however, other materials could also be used. In terms of chemical properties, aluminum (Al) would

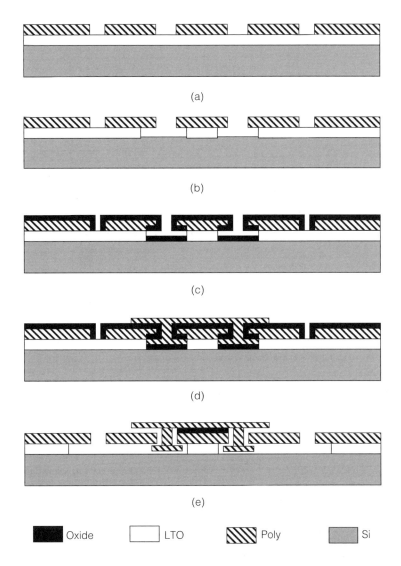

FIGURE 15.4 Cross-sectional schematics of the rapid prototyping process used to fabricate polysilicon micromotors by surface micromachining: (a) after the rotor/stator etch, (b) after the flange mold etch, (c) after the bearing clearance oxidation step, (d) after the bearing etch, and (e) after the release step.

certainly be a satisfactory candidate as a sacrificial layer, as it can be dissolved in acidic-based Al etchants that do not etch polysilicon. However, the thermal properties tell a different story. LPCVD polysilicon is often deposited at temperatures between 580 and 630°C, which are excessively close to the Al melting temperature at the deposition pressure. Independent of the temperature incompatibility, polysilicon is often used as the gate material in MOS processes. As a result, for MEMS and IC processes that share the same LPCVD polysilicon furnace, as might be the case for an integrated MEMS process, it would be inadvisable to put Al-coated wafers in a polysilicon furnace due to cross-contamination considerations.

The release process associated with polysilicon surface micromachining is simple in principle but can be complicated in practice. The objective is to completely dissolve the sacrificial oxide from beneath the freestanding components without etching the polysilicon structural components. The wafers/dies are simply immersed in the appropriate solution for a period of time sufficient to release all desired parts. This is done with various concentrations of electronic-grade HF, including BOE, as the etch rates of SiO_2 and polysilicon are significantly different. It has been observed, however, that during the HF release step,

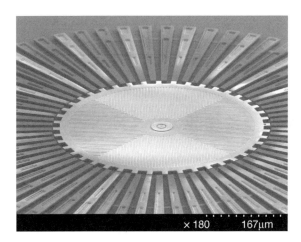

FIGURE 15.5 SEM micrograph of a polysilicon micromotor fabricated using the rapid prototyping process.

the mechanical properties of polysilicon, including residual stress, Young's modulus and fracture strain, can be affected [Walker et al., 1991]. In general, the modulus and fracture strain of polysilicon decreases with increasing time of exposure to HF and with increasing HF concentration. This decrease in the modulus and fracture strain indicates a degradation of the film mechanical integrity. To minimize the HF release time, structures are designed with access holes and cuts of sufficient size to facilitate the flow of HF to the sacrificial oxide. In this manner, polysilicon structures can be released without appreciable degradation to film properties and hence device performance.

Following the HF release step, the devices must be rinsed and dried. A simple process includes rinses in deionized (DI) water then in methanol, followed by a drying step using N_2. The primary difficulty with the wet release process is that surface tension forces, which are related to the surface properties of the material, tend to pull the micromechanical parts toward the substrate as the devices are immersed and pulled out of the solutions. Release processes that avoid the surface tension problem by using frozen alcohols that are sublimated at the final rinse step have been developed [Guckel et al., 1990]. Processes based on the use of supercritical fluids [Mulhern et al., 1993], such as CO_2 at 35°C and 1100 psi, to extinguish surface tension effects vanish are now commonplace in many MEMS facilities.

15.6 Germanium-Based Materials

Germanium (Ge) has a long history in the development of semiconducting materials, dating back to the development of the earliest transistors. The same is true in the development of micromachined transducers and the early work on the piezoresistive effect in semiconducting materials [Smith, 1954]. Development of Ge for microelectronic devices may have continued if only a water-insoluable oxide could be formed on Ge surfaces. Nonetheless, there is a renewed interest in Ge for micromachined devices, especially for devices that require use of low-temperature processes.

Thin polycrystalline Ge (poly-Ge) films can be deposited by LPCVD at temperatures much lower than polysilicon, namely, 325°C at a pressure of 300 mtorr on Si, Ge or SiGe substrates [Li et al., 1999]. Ge does not nucleate on SiO_2 surfaces, which prohibits use of thermal oxides and LTO films as sacrificial substrate layers but does enable use of these films as sacrificial molds, as selective growth using SiO_2 masking films is possible. Residual stress in poly-Ge films deposited on Si substrates is about 125 MPa compressive, which can be reduced to nearly zero after a 30-s anneal at 600°C. Poly-Ge is essentially impervious to KOH, tetramethyl ammonium hydroxide (TMAH) and BOE, making it an ideal masking and etch-stop material in Si micromachining. In fact, the combination of low residual stress and inertness to Si anisotropic etches enables the fabrication of Ge membranes on Si substrates [Li et al., 1999]. The mechanical properties of poly-Ge are comparable with polysilicon, with a Young's modulus measured at

132 GPa and a fracture stress ranging between 1.5 GPa and 3.0 GPa [Franke et al., 1999]. Poly-Ge can also be used as a sacrificial layer. Typical wet etchants are based on mixtures of HNO_3, H_2O and HCl and H_2O, H_2O_2 and HCl, as well as the RCA SC-1 cleaning solution. These mixtures do not etch Si, SiO_2, Si_3N_4 and Si_xN_y, thereby enabling the use of poly-Ge as a sacrificial substrate layer in polysilicon surface micromachining. Using the above-mentioned techniques, poly-Ge-based thermistors and Si_3N_4-membrane-based pressure sensors using poly-Ge sacrificial layers have been fabricated [Li et al., 1999]. In addition, poly-Ge microstructures, such as lateral resonant structures, have been fabricated on Si substrates containing CMOS structures with no process-related degradation in performance, thus showing the advantages of low deposition temperatures and compatible wet chemical etching techniques [Franke et al., 1999].

SiGe is an alloy of Si and Ge and has recently received attention for its usefulness in microelectronics; therefore, deposition technologies for SiGe thin films are readily available. While the requirements for SiGe-based electronic devices require single-crystal material, the requirements for MEMS are much less restrictive, allowing for the use of polycrystalline material in many applications. Polycrystalline SiGe (poly-SiGe) films retain many properties comparable to polysilicon but can be deposited at lower substrate temperatures. Deposition processes include LPCVD, atmospheric pressure chemical vapor deposition (APCVD) and RTCVD (rapid thermal CVD) using SiH_4 and GeH_4 as precursor gases. Deposition temperature range from 450°C for LPCVD [Franke et al., 2000] to 625°C for RTCVD [Sedky et al., 1998]. The LPCVD processes can be performed in horizontal furnace tubes similar in configuration and size to those used for the deposition of polyslicon films. In general, the deposition temperature is related to the concentration of Ge in the films, with higher Ge concentration resulting in lower deposition temperatures. Like polysilicon, poly-SiGe can be doped with B and P to modify its conductivity. In fact, it has been reported that as-deposited, *in situ*, B-doped poly-SiGe films have a resistivity of 1.8 mΩ-cm [Franke et al., 2000].

Poly-SiGe can be deposited on a number of sacrificial substrates, including SiO_2 [Sedky et al., 1998], PSG [Franke et al., 1999] and poly-Ge [Franke et al., 1999], which, as detailed in previous sections of this chapter, can also be deposited at relatively low processing temperatures. For films rich in Ge, a thin polysilicon seed layer is sometimes used on SiO_2 surfaces, as Ge does not readily nucleate on oxide surfaces. Because poly-SiGe is an alloy, variations in film stoichiometry can result in changes in physical properties. For instance, attack of poly-SiGe by H_2O_2, a main component in some Ge etchants, becomes problematic for Ge concentrations over 70%. As with most CVD thin films, residual stress is dependent on the substrate used and the deposition conditions; however, for *in situ* B-doped films, the as-deposited stresses are quite low at 10 MPa compressive [Franke et al., 2000].

In many respects, fabrication of devices made from poly-SiGe thin films follows processing methods used in polysilicon micromachining, as Si and Ge are quite compatible. The poly-SiGe/poly-Ge material system is particularly attractive for surface micromachining, as it is possible to use H_2O_2 as a release agent. It has been reported that in H_2O_2, poly-Ge etches at a rate of 0.4 µm/min, while poly-SiGe with Ge concentrations below 80% have no observable etch rate after 40 hr [Heck et al., 1999]. The ability to use H_2O_2 as a sacrificial etchant makes the poly-SiGe and poly-Ge combination perhaps the ideal material system for surface micromachining. To this end, several interesting devices have been fabricated from poly-SiGe. Due to the conformal nature of the poly-SiGe coating, poly-SiGe-based, high-aspect-ratio structural elements, such as gimbal/microactuator structures made using the Hexil process [Heck et al., 1999], can readily be fabricated. Capitalizing on the low substrate temperatures associated with the deposition of poly-SiGe and poly-Ge thin films, an integrated MEMS fabrication process on Si wafers has been demonstrated [Franke et al., 2000]. In this process, CMOS structures are first fabricated into standard Si wafers. Poly-SiGe thin-film mechanical structures are surface micromachined atop the CMOS devices using a poly-Ge sacrificial layer and H_2O_2 as an etchant. A significant advantage of this design lies in the fact that the MEMS structure is positioned directly above the CMOS structure, thus significantly reducing the parasitic capacitance and contact resistance characteristic of interconnects associated with the side-by-side integration schemes often used in integrated polysilicon MEMS. Use of H_2O_2 as the sacrificial etchant means that no special protective layers are required to protect the underlying CMOS layer during release. Clearly, the unique properties of the poly-SiGe/poly-Ge material system, used in conjunction with the Si/SiO_2 material system, enable fabrication of integrated MEMS that minimizes interconnect distances and potentially increases device performance.

15.7 Metals

Metals are used in many different capacities, ranging from hard etch masks and thin film conducting interconnects to structural elements in microsensors and microactuators. Metallic thin films can be deposited using a wide range of deposition techniques, the most common being evaporation, sputtering, CVD and electroplating. Such a wide range of deposition methods makes metal thin films one of the most versatile classes of materials used in MEMS devices. A complete review would constitute a chapter in itself; the following illustrative examples are included to give the reader an idea of how different metal thin films can be used.

Aluminum (Al) is probably the most widely used metal in micrfabricated devices. In MEMS, Al thin films can be used in conjunction with polymers such as polyimide because the films can be sputter-deposited at low temperatures. In most cases, Al is used as a structural layer; however, Al can be used as a sacrificial layer, as well. The polyimide/aluminum combination as structural and sacrificial materials, respectively, has also been demonstrated to be effective for surface micromachining [Schmidt et al., 1988; Mahadevan et al., 1990]. In this case, acid-based Al etchants can be used to dissolve the Al sacrificial layer. A unique feature of this material system is that polyimide is significantly more compliant than polysilicon and silicon nitride (e.g., its elastic modulus is nearly 50 times smaller). At the same time, polyimide can withstand large strains (up to 100% for some chemistries) before fracture. Finally, because both polyimide and Al can be processed at low temperatures (e.g., below 400°C), this material system can be used subsequent to the fabrication of ICs on the wafer. A drawback of polyimide is its viscoelastic properties (i.e., it creeps).

Tungsten (deposited by CVD) as a structural material and silicon dioxide as a sacrificial material have also been used for surface micromachining [Chen and MacDonald, 1991]. In this case, HF is used for removing the sacrificial oxide. In conjunction with high-aspect-ratio processes, nickel and copper are being used as structural layers with polyimide and other metals (e.g., chromium) as the sacrificial layers. The study of many of these material systems has been either limited or is just in the preliminary stages; as a result, their benefits are yet to be determined.

Metal thin films are among the most versatile MEMS materials, as alloys of certain metallic elements exhibit a behavior known as the shape-memory effect. The shape-memory effect relies on the reversible transformation from a ductile martensite phase to a stiff austenite phase upon the application of heat. The reversible nature of this phase change allows the shape-memory effect to be used as an actuation mechanism. Moreover, it has been found that high forces and strains can be generated from shape-memory thin films at reasonable power inputs, thus enabling shape memory actuation to be used in MEMS-based microfluidic devices such as microvalves and micropumps. Alloys of Ti and Ni, collectively known as TiNi, are among the most popular shape-memory alloys owing to their high actuation work densities (reported to be up to 50 MJ/m^3) and large bandwidth (up to 0.1 kHz) [Shih et al., 2001]. TiNi is also attractive because conventional sputtering techniques can be employed to deposit thin films of the alloy, as detailed in a recent report [Shih et al., 2001]. In this study, TiNi films deposited by two methods—co-sputtering elemental Ti and Ni targets and co-sputtering TiNi alloy and elemental Ti targets—were compared for use in microfabricated shape-memory actuators. In each case, the objective was to establish conditions so that films with the proper stoichiometry, and hence phase transition temperature, could be maintained. The sputtering tool was equipped with a substrate heater in order to deposit films on heated substrates as well as to anneal the films in vacuum after deposition. It was reported that co-sputtering from TiNi and Ti targets produced better films than co-sputtering from Ni and Ti targets, due to process variations related to roughening of the Ni target. The TiNi/Ti co-sputtering process has been successfully used as an actuation material in a silicon spring-based microvalve [Hahm et al., 2000].

Use of thin-film metal alloys in magnetic actuator systems is yet another example of the versatility of metallic materials in MEMS. From a physical perspective, magnetic actuation is fundamentally the same in the microscopic and macroscopic domains, with the main difference being that process constraints limit the design options of microscale devices. Magnetic actuation in microdevices generally requires the magnetic layers to be relatively thick (tens to hundreds of microns), so as to create structures that can be used to generate magnetic fields of sufficient strength to generate the desired actuation. To this end, magnetic materials are often deposited by thick-film methods such as electroplating. The thicknesses of these layers often exceeds what can feasibly be patterned by etching, so plating is often conducted in microfabricated molds usually made from X-ray-sensitive materials such as polymethylmethacrylate (PMMA). The PMMA mold thickness can exceed several hundred microns, so X-rays are used as the exposure source. In some cases, a thin-film seed layer is deposited by sputtering or other conventional means before the plating process begins. At the completion of the plating process, the mold is dissolved, freeing the metallic component. This process, commonly known as LIGA, has been used to produce high-aspect-ratio structures such as microgears from NiFe magnetic alloys [Leith and Schwartz, 1999]. LIGA is not restricted to the creation of magnetic actuator structures and, in fact, has been used to make such structures as Ni fuel atomizers [Rajan et al., 1999]. In this application, Ni was selected for its desirable chemical, wear and temperature properties, not its magnetic properties.

15.8 Silicon Carbide

Use of Si as a mechanical and electrical material has enabled the development of MEMS for a wide range of applications. Of course, use of MEMS is restricted by the physical properties of the material, which in the case of Si-based MEMS limits the devices to operating temperatures of about 200°C in low-wear and benign chemical environments. Therefore, alternate materials are necessary to extend the usefulness of MEMS to areas classified as "harsh environments." In a broad sense, harsh environments include all conditions where use of Si is prohibited by its electrical, mechanical and chemical properties. These would include high-temperature, high-radiation, high-wear and highly acidic and basic chemical environments. To be a direct replacement for Si in such applications, the material would have to be a chemically inert, extremely hard, temperature-insensitive, micromachinable semiconductor. These requirements pose significant fabrication challenges, as micromachining requires the use of chemical and mechanical processes to remove unwanted material. In general, a class of wide bandgap semiconductors that includes silicon carbide (SiC) and diamond embodies the electrical, mechanical and chemical properties required for many harsh environment applications, but until recently these materials found little usefulness in MEMS because the necessary micromachining processes did not exist. The following two sections review the development of SiC and diamond for MEMS applications.

SiC has long been recognized as a semiconductor with potential for use in high-temperature and high-power electronics. SiC is a material that is polymorphic, meaning that it exists in multiple crystalline structures, each sharing a common stoichiometry. SiC exists in three main polytypes: cubic, hexagonal and rhombehedral. The cubic polytype, called 3C-SiC, has an electronic bandgap of 2.3 eV, which is over twice that of Si. Numerous hexagonal and rhombehedral polytypes have been identified, the two most common being the 4H-SiC and 6H-SiC hexagonal polytypes. The electronic bandgap of 4H- and 6H-SiC is even higher than 3C-SiC, being 2.9 and 3.2 eV, respectively. SiC in general has a high thermal conductivity, ranging from 3.2 to 4.9 W/cm-K, and a high breakdown field (30×10^5 V/cm). SiC films can be doped to create n- and p-type material. The stiffness of SiC is quite large relative to Si, with measured Young's modulus values in the range of 300 to 700 GPa, which makes it very attractive for micromachined resonators and filters, as the resonant frequency increases with increasing modulus. SiC is not etched in any wet chemistries commonly used in Si micromachining. SiC can be etched in strong bases like KOH, but only at temperatures in excess of 600°C. SiC is a material that does not melt, but rather sublimes at temperatures in excess of 1800°C. Single-crystal 4H- and 6H-SiC wafers are commercially available, although they are smaller (3-in. diameter) and much more expensive than Si. With this list of properties, it is little wonder why SiC is being actively researched for MEMS applications.

SiC thin films can be grown or deposited using a number of different techniques. For high-quality single-crystal films, APCVD and LPCVD processes are most commonly employed. The high crystal quality is achieved by homoepitaxial growth of 4H- and 6H-SiC films on substrates of like crystal type. These processes usually employ dual precursors to supply Si and C, with the common sources being SiH_4 and C_3H_8. Typical epitaxial growth temperatures range from 1500 to 1700°C. Epitaxial films with p- or n-type conductivity can be grown using such dopants as Al and B for p-type films and N or P for n-type films. In fact, doping with N is so effective at modifying the conductivity that growth of undoped SiC is virtually impossible because the concentrations of residual N in these deposition systems can be quite high. At these temperatures, the crystal quality of the epilayers is sufficient for the fabrication of electronic device structures.

Both APCVD and LPCVD can be used to deposit the only known polytype to grow epitaxially on a non-SiC substrate, namely 3C-SiC on Si. Heteroepitaxy is possible because 3C-SiC and Si have similar lattice structures. The growth process involves two key steps. The first step, called *carbonization*, involves converting the near-surface region of the Si substrate to 3C-SiC by simply exposing it to a propane/hydrogen mixture at a substrate temperature of about 1300°C. The carbonized layer forms a crystalline template on which a 3C-SiC film is grown by adding silane to the hydrogen/propane mix. A 20% lattice mismatch between Si and 3C-SiC results in the formation of crystalline defects in the 3C-SiC film. The density is highest in the carbonization layer, but it decreases with increasing thickness, although not to a level comparable with epitaxial 6H- and 4H-SiC films. Regardless, the fact that 3C-SiC does grow on Si substrates enables the use of Si bulk micromachining techniques to fabrication of a host of SiC-based MEMS structures such as pressure sensors and resonant structures.

Polycrystalline SiC, hereafter referred to as poly-SiC, has proven to be a very versatile material for SiC MEMS. Unlike single-crystal versions of SiC, poly-SiC can be deposited on a variety of substrate types, including common surface micromachining materials such as polysilicon, SiO_2 and Si_3N_4. Moreover, poly-SiC can be deposited using a much wider set of processes than epitaxial films; LPCVD, APCVD, PECVD and reactive sputtering have all been used to deposit poly-SiC films. The deposition of poly-SiC requires much lower substrate temperatures than epitaxial films, ranging from roughly 500 to 1200°C. The microstructure of poly-SiC films is temperature and substrate dependent [Wu et al., 1999]. In general, grain size increases with increasing temperature. For amorphous substrates such as SiO_2 and Si_3N_4, poly-SiC films tend to be randomly oriented with equiaxed grains, with larger grains deposited on SiO_2 substrates. In contrast, for oriented substrates such as polysilicon, the texture of the poly-SiC film matches that of the substrate as a result of grain-to-grain epitaxy [Zorman et al., 1996]. This variation in microstructure suggests that device performance can be tailored by selecting the proper substrate and deposition conditions.

Direct bulk micromachining of SiC is very difficult, due to its outstanding chemical durability. Conventional wet chemical techniques are not effective; however, several electrochemical etch processes have been demonstrated. These techniques are selective to certain doping types, so dimensional control of the etched structures depends on the ability to form doped layers, which can only be formed by *in situ* or ion implantation processes, as solid source diffusion is not possible at reasonable processing temperatures. This constraint limits the geometrical complexity of fabricated devices. To fabricate thick (hundreds of microns), three-dimensional, high-aspect-ratio SiC structures, a molding technique has been developed [Rajan et al., 1999]. The molds are fabricated from Si substrates using deep reactive ion etching, a dry etch process that has revolutionized Si bulk micromachining. The micromachined Si molds are then filed with SiC using a combination of thin epitaxial and thick polycrystalline film CVD processes. The thin-film process, which produces 3C-SiC films with a featureless SiC/Si interface, is used to ensure that the molded structure has smooth outer surfaces. The mold-filling process coats all surfaces of the mold with a very thick SiC film. To remove the mold and free the SiC structure, the substrate is first mechanically polished to expose sections of the mold, then the substrate is immersed in a Si etchant to completely dissolve the mold. Because SiC is not attacked by Si etchants, the final SiC structure is released without the need of any special procedures. This process has been successful in the fabrication of solid SiC fuel atomizers, and a variant has been used to fabricate SiC structures in Si-based, micro, gas turbines [Lohner et al., 1999]. In both cases, the process capitalizes on the chemical inertness of SiC in conjunction with the reactivity of Si to create structures that could otherwise not be fabricated with existing technologies.

Although SiC cannot be etched using conventional wet etch techniques, thin SiC films can be patterned using conventional dry etching techniques. RIE processes using fluorinated compounds such as CHF_3 and SF_6 combined with O_2 and sometimes with an inert gas or H_2 are used. The high oxygen content in these plasmas generally prohibits the use of photoresist as a masking material; therefore, hard masks made of metals such as Al and Ni are often used. RIE processes are generally effective patterning techniques; however, a phenomenon called *micromasking*, which results in the formation of etch-field grass, can sometimes be a problem. Nonetheless, RIE-based SiC surface-micromachining processes using polysilicon and SiO_2 sacrificial layers have been developed [Fleischman et al., 1996; 1998]. These processes are effective means to fabricate single-layer SiC structures, but multilayer structures are very difficult to fabricate because the etch rates of the sacrificial layers are much higher than the SiC structural layers. The lack of a robust etch stop makes critical dimensional control in the thickness direction unreliable, thus making RIE-based SiC multilayer processes impractical.

To address the materials compatibility issues facing RIE-based SiC surface micromachining in the development of a multilayer process, a micromolding process for SiC patterning on sacrificial layer substrates has been developed [Yasseen et al., 2000]. In essence, the micromolding technique is the thin film analog to the molding technique presented earlier. The cross-sectional schematic shown in Figure 15.6 illustrates the steps to fabricate a SiC lateral resonant structure. The micromolding process utilizes polysilicon and SiO_2 films as sacrificial molds, Si_3N_4 as an electrical insulator, and SiO_2 as a sacrificial substrate. These films are deposited and patterned by conventional methods, thus leveraging the well-characterized and highly selective processes developed for polysilicon MEMS. Poly-SiC films are deposited into and onto the micromolds. Mechanical polishing with a diamond-based slurry is used to remove poly-SiC from atop the molds, then the appropriate etchant is used to dissolve the molds and sacrificial layers. An example of device structure fabricated using this method is shown in Figure 15.7. The micromolding method clearly utilizes the differences in chemical properties of the three materials in this system in a way that bypasses the difficulties associated with chemical etching of SiC.

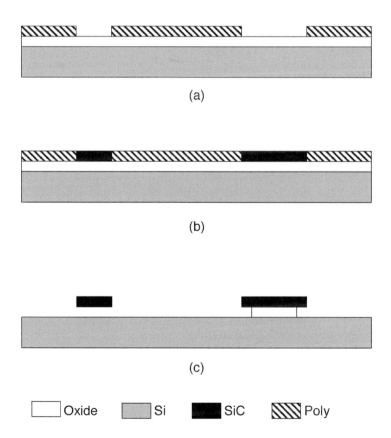

FIGURE 15.6 Cross-sectional schematics of the SiC micromolding process: (a) after micromold fabrication, (b) after SiC deposition and planarization, and (c) after mold and sacrificial layer release.

FIGURE 15.7 SEM micrograph of a SiC lateral resonant structure fabricated using the micromolding process.

15.9 Diamond

Along with SiC, diamond is a leading material for MEMS applications in harsh environments. It is commonly known as nature's hardest material, an ideal property for high-wear environments. Diamond has a very large electronic bandgap (5.5 eV) which is well suited for stable high-temperature operation. Diamond is a high-quality insulator with a dielectric constant of 5.5; however, it can be doped with B to create p-type conductivity. In general, diamond surfaces are chemically inert in the same environments as SiC. Diamond has a very high Young's modulus (1035 GPa), making it the ideal material for high-frequency micromachined resonators. Perhaps the only disadvantage with diamond from a materials properties perspective is that a stable oxide cannot be grown on the diamond surface. Thermal oxidation results in the formation of CO and CO_2, which, of course, are gaseous substances under standard conditions. This complicates the fabrication of diamond-based electronic devices as deposited insulating thin films must be used. Operation of diamond-based sensors at high temperatures requires the use of passivation coatings to protect the diamond structures from oxidation. These limitations, however, can be overcome and do not severely restrict the use of diamond films in harsh environment applications.

Unlike SiC, fabrication of diamond MEMS structures is restricted to polycrystalline and amorphous material. Although diamond epitaxy has been demonstrated, the epi films were grown on small, irregular, single-crystalline pieces, because single-crystalline diamond wafers are not yet available. 3C-SiC thin films have been used to deposit highly oriented diamond films on Si substrates. Polycrystalline diamond films can be deposited on Si and SiO_2 substrates, but the surfaces often must be seeded either by damaging the surface with diamond powders or by biasing the surface with a negative charge, a process called *bias enhanced nucleation*. In general, diamond nucleates much more readily on Si surfaces than on SiO_2 surfaces, and this fact can be exploited to pattern diamond films into microstructures, such as a micromachined atomic force microscope (AFM) cantilever probe, using a selective growth process in conjunction with SiO_2 molding masks [Shibata et al., 2000]. As mentioned previously, diamond can be made insulating or semiconducting, and it is relatively straightforward to produce both types in polycrystalline diamond. This capability enables the fabrication of "all diamond" microelectromechanical structures, thus eliminating the need for Si_3N_4 as an insulating layer.

Bulk micromachining of diamond is more difficult than SiC because electrochemical etching techniques have not been demonstrated. Using a strategy similar to that used in SiC, bulk micromachined diamond structures have been fabricated using bulk micromachined Si molds [Bjorkman et al., 1999]. The Si molds were fabricated using conventional micromachining techniques and filed with polycrystalline diamond deposited by HFCVD. The HFCVD process uses hydrogen as a carrier gas and methane as the carbon source. A hot tungsten wire is used to crack the methane into reactive species as well as to heat the substrate. The process was performed at a substrate temperature of 850 to 900°C and a pressure of 50 mtorr. The Si substrate was seeded with diamond particles suspended in an ethanol solution prior to deposition. After diamond deposition, the top surface of the diamond structure was polished using a hot iron plate. The material removal rate was reported to be around 2 μm/hr. After polishing, the Si mold was removed in a Si etchant, leaving behind the micromachined diamond structure. This process was used to produce all-diamond, high-aspect-ratio, capillary channels for microfluidic applications [Rangsten et al., 1999].

Surface micromachining of polycrystalline diamond thin films requires modifications of conventional micromachining practices to compensate for the nucleation and growth mechanisms of diamond thin films on sacrificial substrates. Early work in this area focused on developing thin-film patterning techniques. Conventional RIE methods are generally ineffective, so effort was focused on developing selective growth methods. One early method used selective seeding to form patterned templates for diamond nucleation. The selective seeding process was based on the lithographic patterning of photoresist mixed with diamond powders [Aslam and Schulz, 1995]. The diamond-loaded photoresist was deposited onto a Cr-coated Si wafer, exposed and then developed, leaving a patterned structure on the wafer surface. During the diamond deposition process, the photoresist rapidly evaporates, leaving behind the diamond seed particles in the desired structural shapes, which then serve as a template for diamond growth.

A second process has been developed for selective deposition directly on sacrificial substrate layers. This processes combines a conventional diamond seeding technique with photolithographic patterning and etching to fabricate micromachined diamond structures using SiO_2 sacrificial layers [Ramesham, 1999]. The process can be executed in one of two approaches. The first approach begins with the formation of a SiO_2 layer by thermal oxidation on a Si wafer. The wafer is then seeded with diamond particles, coated with photoresist, and photolithographically patterned to form a mask for SiO_2 etching. Unmasked regions of the seeded SiO_2 film are then partially etched in BOE to form a surface unfavorable for diamond growth. The photoresist is then removed and a diamond film is selectively deposited. The second approach begins with an oxidized Si wafer, which is coated with photoresist. The resist is photolithographically patterned and the wafer is then seeded, with the photoresist protecting select regions of the SiO_2 surface from the damage caused by the seeding process. The photoresist is removed and selective diamond deposition is performed. In each case, once the diamond film is patterned, the structures could be released using conventional means. These techniques have been used to fabricate cantilever beams and bridge structures.

A third method to surface-micromachine polycrystalline diamond films follows the conventional approach of film deposition, dry etching and release. The chemical inertness of diamond renders most conventional plasma chemistries useless for etching diamond films. Oxygen-based ion beam plasmas, however, can be used to etch diamond thin films [Yang et al., 1999]. The oxygen ion beam prohibits the use of photoresist masks, so hard masks made from metals such as Al are required. A simple ion-beam, etching-based, surface- micromachining process begins with the deposition of a Si_3N_4 film on a Si wafer and is followed by the deposition of a polysilicon sacrificial layer. The polysilicon layer is seeded with a diamond slurry, and a diamond film is deposited by HFCVD. To prepare the diamond film for etching, an Al masking film is deposited and patterned. The diamond films are then etched in the O_2 ion beam plasma, and the structures are released by etching the polysilicon with KOH. This process has been used to create lateral resonant structures; although the patterning process was successful, the devices were not operable because of a significant stress gradient in the film. With a greater understanding of the structure–property relationships of diamond thin films, such problems with surface micromachined structures should be solvable, thus enabling the successful fabrication of a new class of highly functional devices.

15.10 III–V Materials

Galium arsenide (GaAs), indium phosphide (InP) and other III–V compounds are attractive electronic materials for various types of sensors and optoelectronic devices. In general, III–V compounds have favorable piezoelectric and optoelectric properties, high piezoresistive constants and wide electronic bandgaps (relative to Si). In addition, III–V materials can be deposited as ternary and quaternary alloys that have lattice constants closely matched to the binary compounds from which they are derived (e.g., $Al_xGa_{1-x}As$ and GaAs), thus permitting the fabrication of a wide variety of heterostructures that facilitate device performance. Although the III–V class of materials is quite large, this section of the chapter will focus on GaAs and InP for MEMS applications.

Crystalline GaAs has a zinc-blend crystal structure. It has an electronic bandgap of 1.4 eV, enabling GaAs electronic devices to function at temperatures as high as 350°C [Hjort et al., 1994]. High-quality single-crystal wafers are commercially available, as are well-developed metallorganic chemical vapor deposition (MOCVD) and molecular beam epitaxy (MBE) growth processes for epitaxial layers of GaAs and its alloys. GaAs does not outperform Si in terms of mechanical properties; however, its stiffness and fracture toughness are still suitable for micromechanical devices. A favorable combination of mechanical and electrical properties makes GaAs attractive for certain MEMS applications.

Micromachining of GaAs is relatively straightforward, as many of its lattice-matched ternary and quaternary alloys have sufficiently varying chemical properties to allow their use as sacrificial layers. For example, the most common ternary alloy for GaAs is $Al_xGa_{1-x}As$. For values of $x \leq 0.5$, etchants containing mixtures of HF and H_2O etch $Al_xGa_{1-x}As$ without attacking GaAs. In contrast, etchants consisting of NH_4OH and H_2O_2 mixtures attack GaAs isotropically but do not etch $Al_xGa_{1-x}As$, thereby enabling the bulk micromachining of GaAs wafers with lattice-matched etch stops. An extensive review of III–V etch processes can be found in Hjort (1996). By taking advantage of the single-crystal heterostructures that can be formed on GaAs substrates, both surface micromachined and bulk micromachined devices can be fabricated from GaAs. The list of devices is widely varying and includes comb-drive lateral resonant structures [Hjort, 1996], pressure sensors [Fobelets et al., 1994; Dehe et al., 1995b], thermopile sensors [Dehe et al., 1995a] and Fabry–Perot detectors [Dehe et al., 1998].

Micromachining of InP closely resembles the techniques used for GaAs. Many of the properties of InP are similar to GaAs in terms of crystal structure, mechanical stiffness and hardness; however, the optical properties of InP make it particularly attractive for micro-optomechanical devices to be used in the 1.3- to 1.55-µm wavelengths [Seassal et al., 1996]. Like GaAs, single-crystal wafers of InP are readily available. Ternary and quaternary lattice-matched alloys of InP include InGaAs, InAlAs, InGaAsP and InGaAlAs compounds, and, like GaAs, some of these can be used as either etch stop and/or sacrificial layers, depending on the etch chemistry. For instance, InP structural layers deposited on $In_{0.53}Al_{0.47}As$ sacrificial layers can be released using $C_6H_8O_7:H_2O_2:H_2O$ etchants. At the same time, InP films and substrates can be etched in $HCl:H_2O$-based solutions with $In_{0.53}Ga_{0.47}As$ films as etch stops. A comprehensive list of wet chemical etchs for InP and related alloys is reviewed in Hjort (1996). Using InP-based micromachining techniques, multi-air-gap filters [Leclerq et al., 1998], bridge structures [Seassal et al., 1996] and torsional membranes [Dehe et al., 1998] have been fabricated from InP and its related alloys.

15.11 Piezoelectric Materials

Piezoelectric materials play an important role in MEMS technology, mainly for mechanical actuation but also to a lesser extent for sensing applications. In a piezoelectric material, mechanical stress polarizes the material which results in the production of an electric field. The effect also works in reverse; that is, an applied electric field acts to produce a mechanical strain. Many materials retain some sort of piezoelectric behavior, such as quartz, GaAs and ZnO, to name a few. Recent work in MEMS has focused on the development of the compound lead zirconate titanate, $Pb(Zr_xTi_{1-x})O_3$ (PZT). PZT is attractive because it has high piezoelectric constants that lead to high mechanical transduction.

PZT can be deposited by a wide variety of methods, including co-sputtering, CVD and sol-gel processing. Sol-gel processing has been receiving attention lately, due to being able to control the composition and the homogeneity of the deposited material over large surface areas. The sol-gel process uses liquid precursor containing Pb, Ti, Zr and O to create PZT solutions [Lee et al., 1996]. The solution is then deposited on the substrate using a spin-coating process. The substrates in this example consist of a Si wafer with a $Pt/Ti/SiO_2$ thin-film multilayer on its surface. The deposition process produces a PZT film in multilayer fashion, with each layer consisting of a spin-coated layer that is dried at 110°C for 5 min and then heat treated at 600°C for 20 min. After building up the PZT layer to the desired thickness, the multilayer was heated at 600°C for up to 6 hr. Prior to this anneal, a PbO top layer was deposited on the PZT surface. A Au/Cr electrode was then sputter-deposited on the surface of the piezoelectric stack. This process was used to fabricate a PZT-based force sensor. Like Si, PZT films can be patterned using dry etch techniques based on chlorine chemistries, such as Cl_2/CCl_4, as well as ion beam milling using inert gases such as Ar.

15.12 Conclusions

The early development of MEMS can be attributed to the recognition of silicon as a mechanical material. The rapid expansion of MEMS over the last decade is due in part to the inclusion of new structural materials that have expanded the functionality of microfabricated devices beyond what is achievable in silicon. As shown by the examples in this chapter, the materials science of MEMS is not only about the structural layers, but also about the associated sacrificial and masking layers and how these layers interact during the fabrication process in order to realize the final device. In simple terms, MEMS is about material systems; therefore, analysis of what makes MEMS devices work (or, in many cases, not work) relies on a thorough understanding of this fact.

References

Abe, T., and Reed, M.L. (1996) "Low Strain Sputtered Polysilicon for Micromechanical Structures," in *Proc. 9th Int. Workshop on Microelectromechanical Systems,* February, 11–15, San Diego, CA, pp. 258–262.

Adams, A.C. (1988) "Dielectric and Polysilicon Film Deposition," *VLSI Technology,* 2nd ed., McGraw-Hill, New York.

Alley, R.L., Cuan, C.J., Howe, R.T., and Komvopoulos, K. (1992) "The Effect of Release Etch Processing on Surface Microstructure Stiction," *Technical Digest—Solid-State Sensor and Actuator Workshop,* June, Hilton Head, SC, pp. 202–207.

Anderson, R., Muller, R.S., and Tobias, C.W. (1994) "Porous Polycrystalline Silicon: A New Material For MEMS," *J. MEMS* **3**, pp. 10–18.

Aslam, M., and Schulz, D. (1995) "Technology of Diamond Microelectromechanical Systems," in *Proc. 8th Int. Conf. Solid-State Sensors and Actuators,* June 25–29, Stockholm, Sweden, pp. 222–224.

Bjorkman, H., Rangsten, P., Hollman, P., and Hjort, K. (1999) "Diamond Replicas from Microstructured Silicon Masters," *Sensors and Actuators A* **73**, pp. 24–29.

Chen, L.Y., and MacDonald, N. (1991) "A Selective CVD Tungsten Process for Micromotors," in *Technical Digest—6th Int. Conf. on Solid-State Sensors and Actuators,* June, San Francisco, CA, pp. 739–742.

Danel, J.S., Michel, F., and Delapierre, G. (1990) "Micromachining of Quartz and Its Application to an Acceleration Sensor," *Sensors and Actuators A* **21-23**, pp. 971–977.

Dehe, A., Fricke, K., and Hartnagel, H.L. (1995a) "Infrared Thermopile Sensor Based on AlGaAs-GaAs Micromachining," *Sensors and Actuators A* **46-47**, pp. 432–436.

Dehe, A., Fricke, K., Mutamba, K., and Hartnagel, H.L. (1995b) "A Piezoresistive GaAs Pressure Sensor With GaAs/AlGaAs Membrane Technology," *J. Micromech. Microeng.,* 5, pp. 139–142.

Dehe, A., Peerlings, J., Pfeiffer, J., Riemenschneider, R., Vogt, A., Streubel, K., Kunzel, H., Meissner, P., and Hartnagel, H.L. (1998) "III–V Compound Semiconductor Micromachined Actuators for Long Resonator Tunable Fabry–Perot Detectors," *Sensors and Actuators A* **68**, pp. 365–371.

Ding, X., Ko, W.H., and Mansour, J. (1990) "Residual and Mechanical Properties of Boron-Doped p+ Silicon Films," *Sensors and Actuators A* **21-23**, pp. 866–871.

Fleischman, A.J., Roy, S., Zorman, C.A., and Mehregany, M. (1996) "Polycrystalline Silicon Carbide for Surface Micromachining," in *Proc. 9th Int. Workshop on Microelectromechanical Systems,* February 11–15, San Diego, CA, pp. 234–238.

Fleischman, A.J., Wei, X., Zorman, C.A., and Mehregany, M. (1998) "Surface Micromachining of Polycrystalline SiC Deposited on SiO_2 by APCVD," *Mater. Sci. Forum* **264-268**, pp. 885–888.

Fobelets, K., Vounckx, R., and Borghs, G. (1994) "A GaAs Pressure Sensor Based on Resonant Tunnelling Diodes," *J. Micromech. Microeng.* **4**, pp. 123–128.

Folkmer, B., Steiner, P., and Lang, W. (1995) "Silicon Nitride Membrane Sensors with Monocrystalline Transducers," *Sensors and Actuators A* **51**, pp. 71–75.

Franke, A., Bilic, D., Chang, D.T., Jones, P.T., King, T.J., Howe, R.T., and Johnson, C.G. (1999) "Post-CMOS Integration of Germanium Microstructures," in *Proc. 12th Int. Conf. on Microelectromechanical Systems,* January 17–21, Orlando, FL, pp. 630–637.

Franke, A.E., Jiao, Y., Wu, M.T., King, T.J., and Howe, R.T. (2000) "Post-CMOS Modular Integration of Poly-SiGe Microstructures Using Poly-Ge Sacraficial Layers," in *Technical Digest—Solid-State Sensor and Actuator Workshop,* June 4–8, Hilton Head, SC, pp. 18–21.

French, P.J., Sarro, P.M., Mallee, R., Fakkeldij, E.J.M., and Wolffenbuttel, R.F. (1997) "Optimization of a Low-Stress Silicon Nitride Process for Surface Micromachining Applications," *Sensors and Actuators A* **58**, pp. 149–157.

Funk, K., Emmerich, H., Schilp, A. Offenberg, M., Neul, R., and Larmer, F. (1999) "A Surface Micromachined Silicon Gyroscope Using a Thick Polysilicon Layer," in *Proc. 12th Int. Conf. on Microelectromechanical Systems,* January 17–21, Orlando, FL, pp. 57–60.

Gennissen, P., Bartek, M., French, P.J., and Sarro, P.M. (1997) "Bipolar-Compatible Epitaxial Poly for Smart Sensors: Stress Minimization and Applications," *Sensors and Actuators A* **62**, pp. 636–645.

Ghandhi, S.K. (1983) *VLSI Fabrication Principles—Silicon and Gallium Arsenide,* John Wiley & Sons, New York.

Greek, S., Ericson, F., Johansson, S., Furtsch, M., and Rump, A. (1999) "Mechanical Characterization of Thick Polysilicon Films: Young's Modulus and Fracture Strength Evaluated with Microstructures," *J. Micromech. Microeng.* **9**, pp. 245–251.

Guckel, H., Randazzo, T., and Burns, D.W. (1985) "A Simple Technique for the Determination of Mechanical Strain in Thin Films with Application to Polysilicon," *J. Appl. Phys.* **57**, pp. 1671–1675.

Guckel, H., Burns, D.W., Visser, C.C.G., Tilmans, H.A.C., and Deroo, D. (1988) "Fine-Grained Polysilicon Films with Built-In Tensile Strain," *IEEE Trans. Electron. Devices* **ED-35**, pp. 800–801.

Guckel, H., Sniegowski, J.J., Christianson, T.R., and Raissi, F. (1990) "The Applications of Fine-Grained Polysilicon to Mechanically Resonant Transducers," *Sensors and Actuators A* **21-23**, pp. 346–351.

Hahm, G., Kahn, H., Phillips, S.M., and Heuer, A.H. (2000) "Fully Microfabricated Silicon Spring Biased Shape Memory Actuated Microvalve," in *Technical Digest—Solid-State Sensor and Actuator Workshop,* June 4–8, Hilton Head Island, SC, pp. 230–233.

Heck, J.M., Keller, C.G., Franke, A.E., Muller, L., King, T.-J., and Howe, R.T. (1999) "High Aspect Ratio Polysilicon-Germanium Microstructures," in *Proc. 10th Int. Conf. on Solid-State Sensors and Actuators,* June 7–10, Sendai, Japan, pp. 328–334.

Hjort, K. (1996) "Sacrificial Etching of III–V Compounds for Micromechanical Devices," *J. Micromech. Microeng.* **6**, pp. 370–365.

Hjort, K., Soderkvist, J., and Schweitz, J.-A. (1994) "Galium Arsenide as a Mechanical Material," *J. Micromech. Microeng.* **4**, pp. 1–13.

Honer, K., and Kovacs, G.T.A. (2000) "Sputtered Silicon for Integrated MEMS Applications," in *Technical Digest—Solid-State Sensor and Actuator Workshop,* June 4–8, Hilton Head, SC, pp. 308–311.

Howe, R.T., and Muller, R.S. (1983) "Stress in Polysilicon and Amorphous Silicon Thin Films," *J. Appl. Phys.* **54**, pp. 4674–4675.

Kamins, T. (1998) *Polycrystalline Silicon for Integrated Circuits and Displays,* 2nd ed., Kluwer Academic, Berlin.

Kim, T.W., Gogoi, B., Goldman, K.G., McNeil, A.C., Rivette, N.J., Garling, S.E., and Koch, D.J. (1998) "Substrate and Annealing Influences on the Residual Stress of Polysilicon," in *Technical Digest—Solid-State Sensor and Actuator Workshop,* June 8–11, Hilton Head, SC, pp. 237–240.

Lang, W., Steiner, P., and Sandmaier, H. (1995) "Porous Silicon: A Novel Material For Microsystems," *Sensors and Actuators A* **51**, pp. 31–36.

Lange, P., Kirsten, M., Riethmuller, W., Wenk, B., Zwicker, G., Morante, J.R., Ericson, F., and Schweitz, J.A. (1996) "Thick Polycrystalline Silicon for Surface-Micromechanical Applications: Deposition, Structuring, and Mechanical Characterization," *Sensors and Actuators A* **54**, pp. 674–678.

Leclerq, J., Ribas, R.P., Karam, J.M., and Viktorovitch, P. (1998) "III–V Micromachined Devices for Microsystems," *Microelectron. J.* **29**, pp. 613–619.

Lee, C., Itoh, T., and Suga, T. (1996) "Micromachined Piezoelectric Force Sensors Based on PZT Thin Films," *IEEE Trans. Ultrasonics, Ferroelectrics, Frequency Control* **43**, pp. 553–559.

Leith, S.D., and Schwartz, D.T. (1999) "High-Rate Through-Mold Electrodeposition of Thick (>200 micron) NiFe MEMS Components with Uniform Composition," *J. MEMS* **8**, pp. 384–392.

Li, B., Xiong, B., Jiang, L., Zohar, Y., and Wong, M. (1999) "Germanium as a Versatile Material for Low-Temperature Micromachining," *J. MEMS* **8**, pp. 366–372.

Liu, R., Vasile, M.J., and Beebe, D.J. (1999) "The Fabrication of Nonplanar Spin-On Glass Microstructures," *J. MEMS* **8**, pp. 146–151.

Lohner, K., Chen, K.S., Ayon, A.A., and Spearing, M.S. (1999) "Microfabricated Silicon Carbide Microengine Structures," *Mater. Res. Soc. Symp. Proc.* **546**, pp. 85–90.

Mahadevan, R., Mehregany, M., and Gabriel, K.J. (1990) "Application of Electric Microactuators to Silicon Micromechanics," *Sensors and Actuators A* **21-23**, pp. 219–225.

Maseeh, F., and Senturia, D.D. (1990) "Plastic Deformation of Highly Doped Silicon," *Sensors and Actuators A* **21-22**, pp. 861–865.

Monk, D.J., Soane, D.S., and Howe, R.T. (1993) "Enhanced Removal of Sacrificial Layers for Silicon Surface Micromachining," in *Technical Digest—The 7th Int. Conf. on Solid-State Sensors and Actuators,* June, Yokohama, Japan, pp. 280–283.

Mulhern, G.T., Soane, D.S., and Howe, R.T. (1993) "Supercritical Carbon Dioxide Drying of Microstructures," in *Technical Digest—7th Int. Conf. on Solid-State Sensors and Actuators,* June, Yokohoma, Japan, pp. 296–299.

Pilskin, W.A. (1977) "Comparison of Properties of Dielectric Films Deposited by Various Methods," *J. Vacuum Sci. Technol.* **21**, pp. 1064–1081.

Rajan, N., Mehregany, M., Zorman, C.A., Stefanescu, S., and Kicher, T. (1999) "Fabrication and Testing of Micromachined Silicon Carbide and Nickel Fuel Atomizers for Gas Turbine Engines," *J. MEMS* **8**, pp. 251–257.

Ramesham, R. (1999) "Fabrication of Diamond Microstructures for Microelectromechanical Systems (MEMS) by a Surface Micromachining Process," *Thin Solid Films* **340**, pp. 1–6.

Rangsten, P., Bjorkman, H., and Hjort, K. (1999) "Microfluidic Components in Diamond," in *Proc. of the 10th Int. Conf. on Solid-State Sensors and Actuators,* June 7–10, Sendai, Japan, pp. 190–193.

Rosler. R.S. (1977) "Low Pressure CVD Production Processes for Poly, Nitride, and Oxide," *Solid-State Technol.* **20**, pp. 63–70.

Schmidt, M.A., Howe, R.T., Senturia, S.D., and Haritonidis, J.H. (1988) "Design and Calibration of a Microfabricated Floating-Element Shear-Stress Sensor," *Trans. Electron. Devices* **ED-35**, pp. 750–757.

Seassal, C., Leclercq, J.L., and Viktorovitch, P. (1996) "Fabrication of Inp-Based Freestanding Microstructures by Selective Surface Micromachining," *J. Micromech. Microeng.* **6**, pp. 261–265.

Sedky, S., Fiorini, P., Caymax, M., Loreti, S., Baert, K., Hermans, L., and Mertens, R. (1998) "Structural and Mechanical Properties of Polycrystalline Silicon Germanium for Micromachining Applications," *J. MEMS* **7**, pp. 365–372

Sekimoto, M., Yoshihara, H., and Ohkubo, T. (1982) "Silicon Nitride Single-Layer X-ray Mask," *J. Vacuum Sci. Technol.* **21**, pp. 1017–1021.

Shibata, T., Kitamoto, Y., Unno, K., and Makino, E. (2000) "Micromachining of Diamond Film for MEMS Applications," *J. MEMS* **9**, pp. 47–51.

Shih, C.L., Lai, B.K., Kahn, H., Phillips, S.M., and Heuer, A.H. (2001) "A Robust Co-Sputtering Fabrication Procedure for TiNi Shape Memory Alloys for MEMS," *J. MEMS* **10**, pp. 69–79.

Smith, C.S. (1954) "Piezoresistive Effect in Germanium and Silicon," *Phys. Rev.* **94**, pp. 1–10.

Walker, J.A., Gabriel, K.J., and Mehregany, M. (1991) "Mechanical Integrity of Polysilicon Films Exposed to Hydrofluoric Acid Solutions," *J. Electronic Mater.* **20**, pp. 665–670.

Wolfe, S., and Tauber, R. (1999) *Silicon Processing for the VLSI Era,* 2nd ed., Lattice Press, Sunset Beach, CA.

Wu, C.H., Zorman, C.A., and Mehregany, M. (1999) "Growth of Polycrystalline SIC Films on SiO$_2$ and Si$_3$N$_4$ By APCVD," *Thin Solid Films* **355-356**, pp. 179–183.

Yang, Y., Wang, X., Ren, C., Xie, J., Lu, P., and Wang, W. (1999) "Diamond Surface Micromachining Technology," *Diamond Related Mater.* **8**, pp. 1834–1837.

Yang, J., Kahn, H., He, A.-Q., Phillips, S.M., and Heuer, A.H. (2000) "A New Technique for Producing Large-Area as-Deposited Zero-Stress LPCVD Polysilicon Films: The Multipoly Process," *J. MEMS* **9**, pp. 485–494.

Yasseen, A., Cawley, J.D., and Mehregany, M. (1999) "Thick Glass Film Technology for Polysilicon Surface Micromachining," *J. MEMS* **8**, pp. 172–179.

Yasseen, A., Wu, C.H., Zorman, C.A., and Mehregany, M. (2000) "Fabrication and Testing of Surface Micromachined Polycrystalline SiC Micromotors," *Electron. Device Lett.* **21**, pp. 164–166.

Zhang, X., Zhang, T.Y., Wong, M., and Zohar, Y. (1998) "Rapid Thermal Annealing of Polysilicon Thin Films," *J. MEMS* **7**, pp. 356–364.

Zhang, X., Ghodssi, R., Chen, K.S., Ayon, A.A., and Spearing, S.M. (2000) "Residual Stress Characterization of Thick PECVD TEOS Film for Power MEMS Applications," in *Technical Digest—Solid-State Sensor and Actuator Workshop,* June 4–8, Hilton Head, SC, pp. 316–319.

Zorman, C.A., Roy, S., Wu, C.H., Fleischman, A.J., and Mehregany, M. (1996) "Characterization of Polycrystalline Silicon Carbide Films Grown by Atmospheric Pressure Chemical Vapor Deposition on Polycrystalline Silicon," *J. Mater. Res.* **13**, pp. 406–412.

For Further Information

A comprehensive review of polysilicon as a material for microelectronics and MEMS is presented in *Polycrystalline Silicon for Integrated Circuits and Displays*, 2nd ed., by Ted Kamins. The Materials Research Society holds an annual symposium on the materials science of MEMS at the Fall meetings. The proceedings from these symposia have been published as volumes 546B, 605B and 657B of the *Materials Research Society Symposium Proceedings*. Several regularly published journals contain contributed and review papers concerning materials aspects of MEMS, including: (1) *The Journal of Microelectromechanical Systems*, (2) *Journal of Micromachining and Microengineering*, (3) *Sensors and Actuators* and (4) *Sensors and Materials*. These journals are carried by most engineering and science libraries and may be accessible online.

16
MEMS Fabrication[1]

Marc J. Madou
Nanogen, Inc.

16.1 Wet Bulk Micromachining: Introduction **16**-2
16.2 Historical Note **16**-3
16.3 Silicon Crystallography **16**-6
 Miller Indices • Crystal Structure of Silicon • Geometric Relationships between Some Important Planes in the Silicon Lattice
16.4 Silicon as a Substrate and Structural Material **16**-16
 Silicon as Substrate • Silicon as a Structural Element in Mechanical Sensors
16.5 Wet Isotropic and Anisotropic Etching **16**-29
 Wet Isotropic and Anisotropic: Empirical Observations • Chemical Etching Models
16.6 Etching With Bias and/or Illumination of the Semiconductor **16**-64
 Electropolishing and Microporous Silicon
16.7 Etch-Stop Techniques **16**-72
 Introduction • Boron Etch Stop • Electrochemical Etch-Stop Technique • Photo-Assisted Electrochemical Etch Stop (for n-Type Silicon) • Photo-Induced Preferential Anodization, PIPA (for p-Type Silicon) • Etch Stop at Thin Films-Silicon on Insulator
16.8 Problems with Wet Bulk Micromachining **16**-82
 Introduction • Extensive Real Estate Consumption • Corner Compensation
16.9 Wet Bulk Micromachining Examples **16**-90
16.10 Surface Micromachining: Introduction **16**-95
16.11 Historical Note **16**-97
16.12 Mechanical Properties of Thin Films **16**-98
 Introduction • Adhesion • Stress in Thin Films • Strength of Thin Films
16.13 Surface Micromachining Processes **16**-117
 Basic Process Sequence • Fabrication Step Details • Control of Film Stress • Dimensional Uncertainties • Sealing Processes in Surface Micromachining • IC Compatibility
16.14 Poly-Si Surface Micromachining Modifications **16**-133
 Porous Poly-Si • Hinged Polysilicon • Thick Polysilicon • Milli-Scale Molded Polysilicon Structures

---

[1]This chapter was originally published in Madou, M.J. (1997) "Wet Bulk Micromachining" and "Surface Micromachining" (chaps. 4 and 5), in *Fundamentals of Microfabrication*, CRC Press, Boca Raton, FL.

16.15 Non-Poly-Si Surface Micromachining
Modifications **16**-140
Silicon on Insulator Surface Micromachining

16.16 Resists as Structural Elements and Molds
in Surface Micromachining **16**-146
Introduction • Polyimide Surface Structures • UV Depth
Lithography • Comparison of Bulk Micromachining with
Surface Micromachining

16.17 Materials Case Studies.. **16**-149
Introduction • Polysilicon Deposition and Material Structure
• Amorphous and Hydrogenated Amorphous Silicon • Silicon
Nitride • CVD Silicon Dioxides

16.18 Polysilicon Surface Micromachining Examples....... **16**-160

16.1 Wet Bulk Micromachining: Introduction

In wet bulk micromachining, features are sculpted in the bulk of materials such as silicon, quartz, SiC, GaAs, InP, Ge and glass by orientation-dependent (anisotropic) and/or by orientation-independent (isotropic) wet etchants. The technology employs pools as tools [Harris, 1976], instead of the plasmas studied in Madou (1997, chap. 2). A vast majority of wet bulk micromachining work is based on single crystal silicon. There has been some work on quartz, some on crystalline Ge and GaAs, and a minor amount on GaP, InP and SiC. Micromachining has grown into a large discipline, comprising several tool sets for fashioning microstructures from a variety of materials. These tools are used to fabricate microstructures either in parallel or serial processes. Madou (1997, Table 7.7) summarizes these tools. It is important to evaluate all the presented micromanufacturing methods before deciding on one specific machining method optimal for the application at hand—in other words, to zero-base the technological approach [Block, B., private communication]. The principle commercial Si micromachining tools used today are the well-established wet bulk micromachining and the more recently introduced surface micromachining (see Section 16.10). A typical structure fashioned in a bulk micromachining process is shown in Figure 16.1. This type of piezoresistive membrane structure, a likely base for a pressure sensor or an accelerometer, demonstrated that batch fabrication of miniature components does not need to be limited to integrated circuits (ICs). Despite all the emerging new micromachining options, Si wet bulk micromachining, being the best characterized micromachining tool, remains

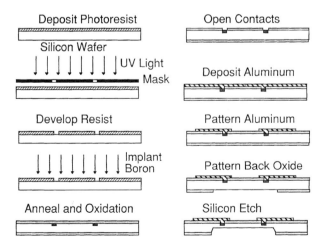

FIGURE 16.1 A wet bulk micromachining process is used to craft a membrane with piezoresistive elements. Silicon micromachining selectively thins the silicon wafer from a starting thickness of about 400 μm. A diaphragm having a typical thickness of 20 μm or less with precise lateral dimensions and vertical thickness control results.

most popular in industry. An emphasis in this chapter is on the wet etching process itself. Other machining steps typically used in conjunction with wet bulk micromachining, such as additive processes and bonding processes, are covered in Madou (1997, chaps. 3 and 8, respectively).

Wet bulk micromachining had its genesis in the Si IC industry, but further development will require the adaptation of many different processes and materials. To emphasize the need for micromachinists to look beyond Si as the ultimate substrate and/or building material, we have presented many examples of non-Si micromachinery throughout the book. The need to incorporate new materials and processes is especially urgent for progress in chemical sensors and micro-instrumentation which rely on non-IC materials and often are relatively large. The merging of bulk micromachining with other new fabrication tools such as surface micromachining and electroplating and the adaptation of new materials such as Ni and polyimides has fostered a powerful new, nontraditional precision engineering method. A truly multidisciplinary engineering education will be required to design miniature systems with the most appropriate building philosophy.

After a short historical note on wet bulk micromachining we begin this chapter with an introduction to the crystallography of single crystal Si and a listing of its properties, clarifying why Si is such an important sensor material. Some empirical data on wet etching are reviewed and different models for anisotropic and isotropic etching behavior follow. Then, etch stop techniques, which catapulted micromachining into an industrial manufacturing technique, are discussed. Subsequently, a discussion of problems associated with bulk micromachining such as IC incompatibility, extensive real-estate usage, and issues involving corner compensation is presented. Finally, examples of applications of wet bulk micromachining in mechanical and chemical sensors are given.

16.2 Historical Note

The earliest use of wet etching of a substrate, using a mask (wax) and etchants (acid-base), appears to be in the fifteenth century for **_decorating armor_** [Harris, 1976]. Engraving hand tools were not hard enough to work the armor and more powerful acid-base processes took over. By the early seventeenth century, etching to decorate arms and armor was a completely established process. Some pieces stemming from that period have been found where the chemical milling was accurate to within 0.5 mm. The masking in this traditional chemical milling was accomplished by cutting the maskant with a scribing tool and peeling the maskant off where etching was wanted. Harris (1976) describes in detail all of the improvements that, by the mid-1960s, made this type of chemical milling a valuable and reliable method of manufacturing. It is especially popular in the aerospace world. The method enables many parts to be produced more easily and cheaply than by other means and in many cases provides a means to design and produce parts and configurations not previously possible. Through the introduction of photosensitive masks by Niépce in 1822 [Madou, 1997, chap. 1], chemical milling in combination with lithography became a reality and a new level of tolerances came within reach. The more recent major applications of lithography-based chemical milling are the manufacture of printed circuit boards, started during the Second World War, and, by 1961, the fabrication of Si-integrated circuitry. Photochemical machining is also used for such precision parts as color television shadow masks, integrated circuit lead frames, light chopper and encoder discs, and decorative objects such as costume jewelry [Allen, 1986]. The geometry of a 'cut' produced when etching silicon integrated circuits is similar to the chemical-milling cut of the aerospace industry, but the many orders of magnitude difference in size and depth of the cut account for a major difference in achievable accuracy. Accordingly, the tolerances for fashioning integrated circuitry are many orders of magnitude smaller than in the chemical milling industry.

In this book we are concerned with lithography and chemical machining used in the IC industry and in microfabrication. A major difference between these two fields is in the aspect ratio (height-to-width ratio) of the features crafted. In the IC industry one deals with mostly very small, flat structures with aspect ratios of 1 to 2. In the microfabrication field, structures typically are somewhat larger and aspect ratios might be as high as 400.

Decorating armor. (From Harris, T. W., *Chemical Milling,* Clarendon Press, Oxford, 1976. With permission.)

Isotropic etching has been used in silicon semiconductor processing since its beginning in the early 1950s. Representative work from that period is the impressive series of papers by Robbins and Schwartz (1959, 1960) and Schwartz and Robbins (1961, 1976) on chemical isotropic etching, and Uhlir's paper on electrochemical isotropic etching [Uhlir, 1956]. The usual chemical isotropic etchant used for silicon was HF in combination with HNO_3 with or without acetic acid or water as diluent [Robbins and Schwartz, 1959; 1960; Schwartz and Robbins, 1961; 1976]. The early work on isotropic etching in an electrochemical cell (i.e., 'electropolishing') was carried out mostly in nonaqueous solutions, avoiding black or red deposits that formed on the silicon surface in aqueous solutions [Hallas, 1971]. Turner showed that if a critical current density is exceeded, silicon can be electropolished in aqueous HF solutions without the formation of any deposits [Turner, 1958].

In the mid-1960s, the Bell Telephone Laboratories started the work on anisotropic Si etching in mixtures of, at first, KOH, water and alcohol and later in KOH and water. This need for high aspect ratio cuts in silicon arose for making dielectrically isolated structures in integrated circuits such as for beam leads. Chemical and electrochemical anisotropic etching methods were pursued [Stoller and Wolff, 1966; Stoller, 1970; Forster and Singleton, 1966; Kenney, 1967; Lepselter, 1966; 1967; Waggener, 1970; Kragness and Waggener, 1973; Waggener et al., 1967a; 1967b; Bean and Runyan, 1977]. In the mid-1970s, a new surge of activity in anisotropic etching was associated with the work on V-groove and U-groove transistors [Rodgers et al., 1976; 1977; Ammar and Rodgers, 1980].

The first use of Si as a micromechanical element can be traced back to a discovery and an idea from the mid-1950s and early 1960s, respectively. The discovery was the large piezoresistance in Si and Ge by Smith in 1954 [Smith, 1954]. The idea stems from Pfann et al. in 1961 [Pfann, 1961], who proposed a diffusion technique for the fabrication of Si piezoresistive sensors for stress, strain and pressure. As early as 1962 Tufte et al. (1962), at Honeywell, followed up on this suggestion. By using a combination of a wet isotropic etch, dry etching and oxidation processes, Tufte et al. made the first thin Si piezoresistive diaphragms, of the type shown in Figure 16.1, for pressure sensors [Tufte et al., 1962]. Sensym/National Semiconductor (sold to Hawker Siddley in 1988) became the first to make stand-alone Si sensor products

	Some Private California MEMS Companies
1972	Foxboro ICT (called SenSym ICT since 1999)
1972	SenSym (called SenSym ICT since 1999)
1975	Endevco
1975	IBM Micromachining
1976	Cognition (sold to Rosemount in 1978)
1980	Irvine Sensors Corp.
1981	ChemIcon Inc.
1981	Microsensor Technology (sold to Tylan in 1986)
1982	ICSensors (sold to EG&G in 1994)
1982	Transensory Devices (sold to ICSensors in 1987)
1985	NovaSensor (sold to Lucas in 1990)
1986	Captor (sold to Dresser in 1991)
1987	Aura
1988	Nanostructures
1988	Redwood Microsystems
1988	TiNi Alloys
1989	Abaxis
1989	Advanced Recording Technologies
1991	Incyte Genomics
1991	Sentir
1992	Silicon Microstructures
1993	Affymetrix
1993	Fluid IC (dissolved in 1995)
1993	Nanogen
1993	Silicon Micromachines
1994	Berkeley Microsystems Incorporated (BMI)
1995	Aclara Biosciences
1995	Integrated Micromachines
1995	MicroScape
1996	Caliper
1996	Cepheid
1997	Microsensors
1997	Mycometrix
1998	Quantum Dot
1998	Zyomyx
1999	Symyx

(1972). By 1974, National Semiconductor described a broad line of Si pressure transducers, in the first complete silicon pressure transducer catalog [Editorial, 1974]. Other early commercial suppliers of micromachined pressure sensor products were Foxboro/ICT, Endevco, Kulite and Honeywell's Microswitch. Other micromachined structures began to be explored by the mid- to late-1970s: Texas Instruments produced a thermal print head (1977) [Editorial, 1977]; Hewlett Packard made thermally isolated diode detectors (1980) [O'Neill, 1976]; fiberoptic alignment structures were made at Western Electric [Boivin, 1974]; and IBM produced ink jet nozzle arrays (1977) [Bassous et al., 1977]. Many **Silicon Valley microsensor companies** played and continue to play a pivotal role in the development of the market for Si sensor products.

European and Japanese companies followed the U.S. lead more than a decade later; for example, Druck Ltd. in the U.K. started exploiting Greenwood's micromachined pressure sensor in the mid-1980s [Greenwood, 1984].

Petersen's 1982 paper, extolling the excellent mechanical properties of single crystalline silicon, helped galvanize academia to get involved in Si micromachining in a major way [Petersen, 1982]. Before that time most efforts had played out in industry and practical needs were driving the technology (market pull). The new generation of micromachined devices often constitutes gadgetry only, and the field is perceived by many as a technology looking for applications (technology push). It has been estimated that

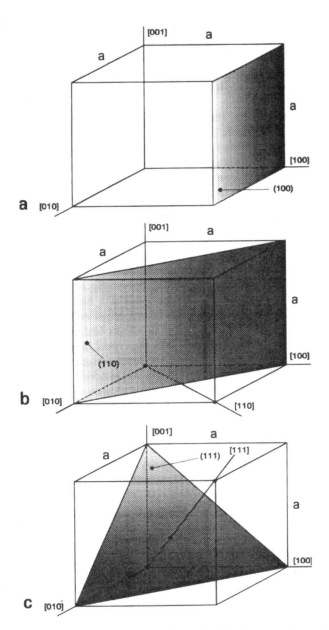

FIGURE 16.2 Miller indices in a cubic lattice: planes and axes. Shaded planes are: a (100), b (110), c (111).

today there are more than 10,000 scientists worldwide involved in Si sensor research and development [Middlehoek and Dauderstadt, 1994]. In order to justify the continued investment, it has become an absolute priority to understand the intended applications better and to be able to select a more specific micromachining tool set intelligently and to identify large market applications [Madou, 1997, chap. 10].

16.3 Silicon Crystallography

16.3.1 Miller Indices

The periodic arrangement of atoms in a crystal is called the lattice. The unit cell in a lattice is a segment representative of the entire lattice. For each unit cell, basis vectors (a_1, a_2 and a_3) can be defined such that if that unit cell is translated by integral multiples of these vectors, one arrives at a new unit cell identical to the original. A simple cubic-crystal unit cell for which $a_1 = a_2 = a_3$ and the axes angles are $\alpha = \beta = \gamma = 90°$ is shown in Figure 16.2. In this figure, the dimension 'a' is known as the lattice constant. To identify a plane or a direction, a set of integers h, k and l called the Miller indices is used. To determine the Miller indices of a plane, one takes the intercept of that plane with the axes and expresses these intercepts as multiples of the basis vectors a_1, a_2, a_3. The reciprocal of these three integers is taken, and, to obtain whole numbers, the three reciprocals are multiplied by the smallest common denominator. The resulting set of numbers is written down as (hkl). By taking the reciprocal of the intercepts, infinities (∞) are avoided in the plane identification. A direction in a lattice is expressed as a vector with components as multiples of the basis vectors. The rules for determining the Miller indices of an orientation are: translate the orientation to the origin of the unit cube and take the normalized coordinates of its other vertex. For example, the body diagonal in a cubic lattice as shown in Figure 16.2 is 1a, 1a and 1a or a diagonal along the [111] direction. Directions [100], [010] and [001] are all crystallographically equivalent and are jointly referred to as the family, form or group of <100> directions. A form, group or family of faces which bear like relationships to the crystallographic axes—for example, the planes (001), (100), (010), (00$\bar{1}$), ($\bar{1}$00) and (0$\bar{1}$0)—are all equivalent and they are marked as {100} planes. For illustration, in Figure 16.3, some of the planes of the {100} family of planes are shown.

16.3.2 Crystal Structure of Silicon

Crystalline silicon forms a covalently bonded structure, the diamond-cubic structure, which has the same atomic arrangement as carbon in diamond form and belongs to the more general zinc-blend classification [Kittel, 1976]. Silicon, with its four covalent bonds, coordinates itself tetrahedrally, and these tetrahedrons make up the diamond-cubic structure. This structure can also be represented as two interpenetrating face-centered cubic lattices, one displaced (1/4,1/4,1/4)a with respect to the other, as shown in Figure 16.4. The structure is face-centered cubic (fcc), but with two atoms in the unit cell. For such a cubic lattice, direction [hkl] is perpendicular to a plane with the three integers (hkl), simplifying further discussions about the crystal orientation, i.e., the Miller indices of a plane perpendicular to the [100] direction are (100). The lattice parameter 'a' for silicon is 5.4309 Å and silicon's diamond-cubic lattice is surprisingly wide open, with a packing density of 34%, compared to 74% for a regular face-centered cubic lattice. The {111} planes present the highest packing density and the atoms are oriented such that three bonds are below the plane. In addition to the diamond-cubic structure, silicon is known to have several stable high-pressure crystalline phases [Hu et al., 1986] and a stress-induced metastable phase with a wurtzite-like structure, referred to as diamond-hexagonal

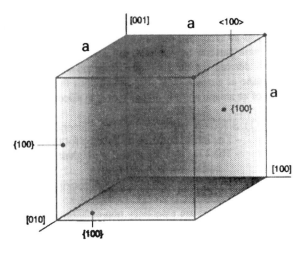

FIGURE 16.3 Miller indices for some of the planes of the {100} family of planes.

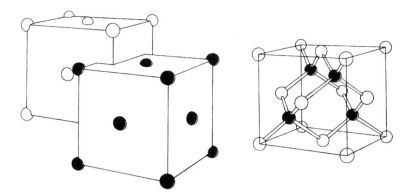

FIGURE 16.4 The diamond-type lattice can be constructed from two interpenetrating face-centered cubic unit cells. Si forms four covalent bonds, making tetrahedrons.

silicon. The latter has been observed after ion-implantation [Tan et al., 1981] and hot indentation [Eremenko and Nikitenko, 1972].

When ordering silicon wafers, the crystal orientation must be specified. The most common orientations used in the IC industry are the <100> and <111> orientation; in micromachining, <110> wafers are used quite often as well. The <110> wafers break or cleave much more cleanly than other orientations. In fact, it is the only major plane that can be cleaved with exactly perpendicular edges. The <111> wafers are used less, as they cannot be etched anisotropically except when using laser-assisted etching [Alavi et al., 1992]. On a <100> wafer, the <110> direction is often made evident by a flat segment, also called an orientation flat. The precision on the flat is about 3°. The position of the flat on (110)-oriented wafers varies from manufacturer to manufacturer, but often parallels a (111) direction. Flat areas help orientation determination, placement of slices in cassettes and fabrication equipment (large primary flat), and help identify orientation and conductivity type (smaller secondary flat; see Madou [1997, chap. 3, Si growth]). Primary and secondary flats on <111> and <100> silicon wafers are indicated in Figure 16.5.

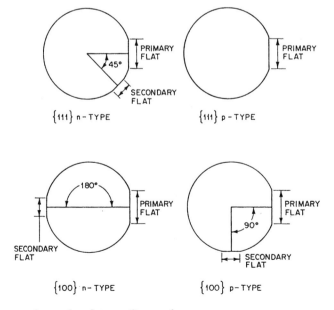

FIGURE 16.5 Primary and secondary flats on silicon wafers.

16.3.3 Geometric Relationships between Some Important Planes in the Silicon Lattice

To better appreciate the different three-dimensional shapes resulting from anisotropically etched single crystal Si (SCS) and to better understand the section further below on corner compensation, some of the more important geometric relationships between different planes within the Si lattice need further clarification. We will consider only silicon wafers with a (100) or a (110) as the surface planes. We will also accept, for now, that in anisotropic alkaline etchants the {111} planes, which have the highest atom-packing density, are nonetching compared to the other planes. As the {111} planes are essentially not attacked by the etchant, the sidewalls of an etched pit in SCS will ultimately be bounded by this type of plane, given that the etch time is long enough for features bounded by other planes to be etched away. The types of planes introduced initially depend on the geometry and the orientation of the mask features.

16.3.3.1 [100]-Oriented Silicon

In Figure 16.6, the unity cell of a silicon lattice is shown together with the correct orientation of a [100]-type wafer relative to this cell [Peeters, 1994]. It can be seen from this figure that intersections of the nonetching {111} planes with the {100} planes (e.g., the wafer surface) are mutually perpendicular and lying along the <110> orientations. Provided a mask opening (say, a rectangle or a square) is accurately aligned with the primary orientation flat, i.e., the [110] direction, only {111} planes will be introduced as sidewalls from the very beginning of the etch. Since the nonetching character of the {111} planes renders an exceptional degree of predictability to the recess features, this is the mask arrangement most often utilized in commercial applications. During etching, truncated pyramids (square mask) or truncated V-grooves (rectangular mask) deepen but do not widen (Figure 16.7). The edges in these structures are <110> directions, the ribs are <211> directions, the sidewalls are {111} planes and the bottom is a (100) plane parallel with the wafer surface. After prolonged etching, the {111} family of planes is exposed down to their common intersection and the (100) bottom plane disappears, creating a pyramidal pit (square mask) or a V-groove (rectangular mask) (Figure 16.7). As shown in Figure 16.7, no underetching of the

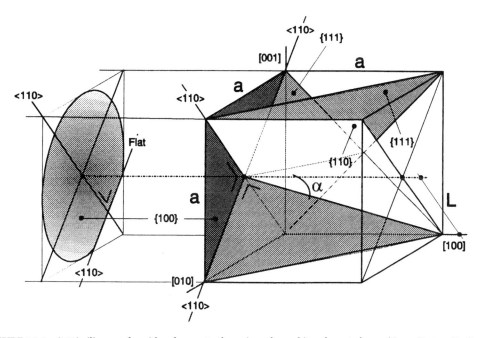

FIGURE 16.6 (100) silicon wafer with reference to the unity cube and its relevant planes. (From Peeters, E., *Process Development for 3D Silicon Microstructures with Application to Mechanical Sensor Design*, KUL, Belgium, 1994. With permission.)

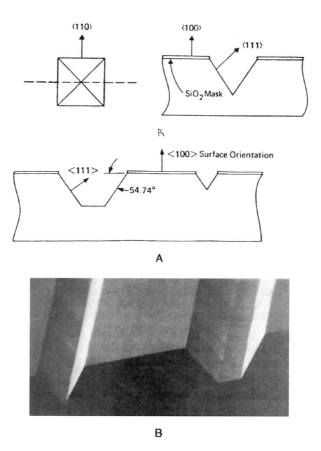

FIGURE 16.7 Anisotropically etched features in a (100) wafer with (A) Square mask (schematic) and (B) Rectangular mask (scanning electron microscope micrograph of resulting actual V- and U-grooves).

etch mask is observed, due to the perfect alignment of the concave oxide mask opening with the <110> direction. Misalignment still results in pyramidal pits, but the mask will be undercut. For a mask opening with arbitrary geometry and orientation (for example, a circle) and for sufficiently long etch times, the anisotropically etched recess in a {100} wafer is pyramidal with a base perfectly circumscribing the circular mask opening [Peeters, 1994]. Convex corners (>180°) in a mask opening will always be completely undercut by the etchant after sufficiently long etch times. This can be disadvantageous (for example, when attempting to create a mesa rather than a pit) or it can be advantageous for undercutting suspended cantilevers or bridges. In the section of this chapter on corner compensation, the issue of undercutting will be addressed in detail. The slope of the sidewalls in a cross section perpendicular to the wafer surface and to the wafer flat is determined by the angle α as in Figure 16.6 depicting the off-normal angle of the intersection of a (111) sidewall and a (110) cross-secting plane, and can be calculated from:

$$\tan \alpha = \frac{L}{a} \qquad (16.1)$$

with $L = a \times \frac{\sqrt{2}}{2}$ or $\alpha = \arctan \frac{\sqrt{2}}{2} = 35.26°$, or 54.74° for the complementary angle. The tolerance on this slope is determined by the alignment accuracy of the wafer surface with respect to the (100) plane. Wafer manufacturers typically specify this misalignment to 1° (0.5° in the best cases).

The width of the rectangular or square cavity bottom plane, W_0, in Figure 16.8, aligned with the <110> directions, is completely defined by the etch depth, z, the mask opening, W_m, and the above-calculated sidewall slope:

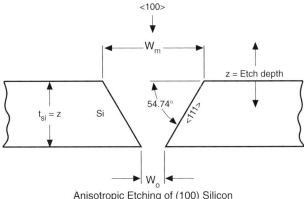

FIGURE 16.8 Relation of bottom cavity plane width with mask opening width.

$$W_0 = W_m - 2\cotan(54.74°)z$$

or

$$W_0 = W_m - \sqrt{2}\, z \tag{16.2}$$

The larger the opening in the mask, the deeper the point at which the {111} sidewalls of the pit intersect. The etch stop at the {111} sidewalls' intersection occurs when the depth is about 0.7 times the mask opening. If the oxide opening is wide enough, $W_m > 849$ μm (for a typical 6-in. wafer with thickness $t_{si} = z = 600$ μm), the {111} planes do not intersect within the wafer. The etched pit in this particular case extends all the way through the wafer, creating a ***small orifice*** or ***via***. If a high density of such vias through the Si is required, the wafer must be made very thin.

Corners in an anisotropically etched recess are defined by the intersection of crystallographic planes, and the resulting corner radius is essentially zero. This implies that the size of a silicon diaphragm is very well defined, but it also introduces a considerable stress concentration factor. The influence of the zero corner radius on the yield load of diaphragms can be studied with finite element analysis (FEA).

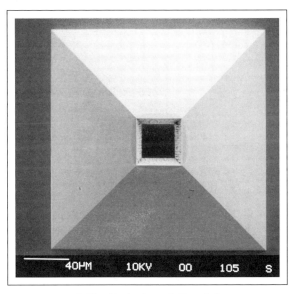

Orifice (A via through Si wafer).

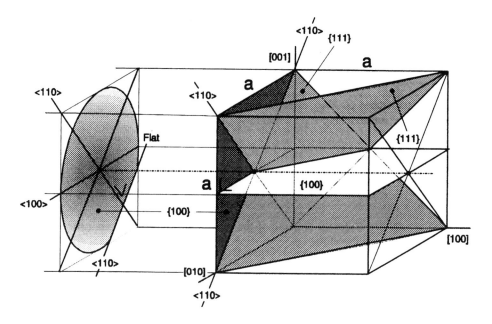

FIGURE 16.9 (100) silicon wafer with <100> mask-aligned features introduces vertical sidewalls. (From Peeters, E., *Process Development for 3D Silicon Microstructures with Application to Mechanical Sensor Design*, KUL, Belgium, 1994. With permission.)

One way to obtain vertical sidewalls instead of 54.7° sidewalls using a [100]-oriented Si wafer is illustrated in Figure 16.9. It can be seen in this figure that there are {100} planes perpendicular to the wafer surface and that their intersections with the wafer surface are <100> directions. These <100> directions enclose a 45° angle with the wafer flat (i.e., the <110> directions). By aligning the mask opening with these <100> orientations, {100} facets are initially introduced as sidewalls. The {110} planes etch faster than the {100} planes and are not introduced. As the bottom and sidewall planes are all from the same {100} group, lateral underetch equals the vertical etch rate and rectangular channels, bounded by slower etching {100} planes, result (Figure 16.10). Since the top of the etched channels is exposed to the etchant longer than the bottom, one might have expected the channels in Figure 16.10 to be wider at the top than at the bottom. With some minor corrections in Peeters' derivation [Peeters, 1994], we can use his explanation for why the sidewalls stay vertical. Assume the width of the mask opening to be W_m. At a given depth, z, into the wafer, the underlying Si is no longer masked by W_m, but rather by the intersection of the previously formed {100} facets with the bottom surface. The width of this new mask is larger than the lithography mask W_m by the amount the latter is being undercut. Let's call the new mask width W_2 the effective mask width at a depth z. The relation

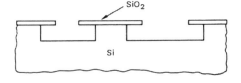

FIGURE 16.10 Vertical sidewalls in a (100) wafer.

between W_m and W_z is given by the lateral etch rate of a {100} facet and the time that facet was exposed to the etchant at depth z, i.e.:

$$W_z = W_m + 2R_{xy}\Delta t_z \qquad (16.3)$$

where R_{xy} is the lateral underetch rate (i.e., etch rate in the x-y plane) and Δt_z the etch time at depth z. The underetching, U_z, of the effective mask opening W_z is given by:

$$U_z = TR_{xy} - R_{xy}\Delta t_z \qquad (16.4)$$

where T is the total etch time so far. The width of the etched pit, W_{tot}, at depth z is further given by the sum of W_m and twice the underetching for that depth:

$$W_{tot} = W_z + 2U_z = W_m + 2TR_{xy} \qquad (16.5)$$

Or, since T can also be written as the measured total etch depth z divided by the vertical etch rate R_z, Eq. (16.5) can be rewritten as

$$\begin{aligned} W_{tot} &= W_m + 2z(R_{xy}/R_z) \quad \text{or since } R_{xy} = R_z, \\ W_{tot} &= W_m + 2z \end{aligned} \qquad (16.6)$$

The width of the etched recess is therefore equal to the photolithographic mask width plus twice the etch depth—independent of that etch depth, in other words; the walls remain vertical independent of the depth z.

For sufficiently long etch times, {111} facets take over eventually from the vertical {100} facets. These inward sloping {111} facets are first introduced at the corners of a rectangular mask and grow larger at the expense of the vertical sidewalls until the latter ultimately disappear altogether. Alignment of mask features with the <100> directions in order to obtain vertical sidewalls in {100} wafers, therefore, is not very useful for the fabrication of diaphragms. However, it can be very effective for anticipating the undercutting of convex corners on {100} wafers. This useful aspect will be revisited when discussing corner compensation.

16.3.3.2 [110]-Oriented Silicon

In Figure 16.11, we show a unit cell of Si properly aligned with the surface of a (110) Si wafer. This drawing will enable us to predict the shape of an anisotropically etched recess on the basis of elementary geometric crystallography. Whereas the intersections of the {111} planes with the (100) wafer surface are mutually perpendicular, here they enclose an angle γ in the (110) plane. Moreover, the intersections are not parallel (<110>) or perpendicular (<100>) to the main wafer flat (assumed to be <110> in this case), but rather enclose angles δ or $\delta + \gamma$. It follows that a mask opening that will not be undercut (i.e., oriented such that resulting feature sidewalls are exclusively made up by {111} planes) cannot be a rectangle aligned with the flat, but must be a parallelogram skewed by $\gamma - 90°$ and δ degrees off-axis. The angles γ and δ are calculated as follows [Peeters, 1994] (see Figures 16.11 and 16.12):

$$\tan\beta = \frac{\frac{1}{2}a\frac{\sqrt{2}}{2}}{\frac{a}{2}} = \frac{\sqrt{2}}{2} \qquad (16.7)$$

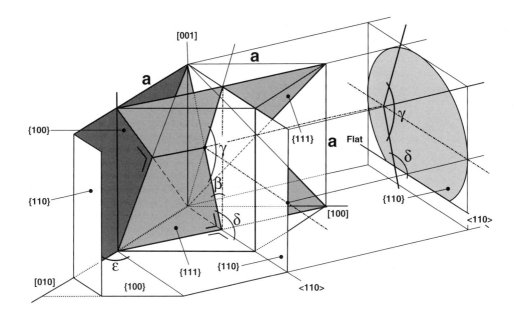

FIGURE 16.11 (110) silicon wafer with reference to the unity cube and its relevant planes. The wafer flat is in a <110> direction. (From Peeters, E., *Process Development for 3D Silicon Microstructures with Application to Mechanical Sensor Design*, KUL, Belgium, 1994. With permission.)

$$\gamma = 180° - 2\beta = 180° - 2\arctan\left(\frac{\sqrt{2}}{2}\right) = 109.47°$$

$$\delta = 90° + \beta = 90° + \arctan\left(\frac{\sqrt{2}}{2}\right) = 125.26° \qquad (16.8)$$

$$\varphi = 270° - \delta = 144.74°$$

From Figure 16.11, it can also be seen that the {111} planes are oriented perpendicular to the (110) wafer surface. This makes it possible to etch pits with vertical sidewalls (Figure 16.12). The bottom of the pit shown here is bounded by {110} and/or {100} planes, depending on the etch time. At short etch times, one mainly sees a flat {110} bottom. As the {110} planes are etching slightly faster than the {100} planes, the flat {110} bottom is getting smaller and smaller and a V-shaped bottom bounded by {100} planes eventually results. The angle ε as shown in Figure 16.12 equals 45°, being the angle enclosed by the intersections of a {100} and a {110} bottom plane. The general rule does apply that an arbitrary window opening is circumscribed by a parallelogram with the given orientation and skewness for sufficiently long etch times. Another difference between (100)- and (110)-oriented Si wafers is that on the (110) wafers it is possible to etch under microbridges crossing at a 90° angle a shallow V-groove (formed by (111) planes). In order to undercut a bridge on a (100) plane, the bridge cannot be perpendicular to the V-groove; it must be oriented slightly off normal [Elwenspoek et al., 1994].

16.3.3.3 Selection of [100]- or [110]-Oriented Silicon

In Table 16.1, we compare the main characteristics of etched features in [100]- and [110]-oriented wafers. This guide can help decide which orientation to use for a specific microfabrication application at hand.

From this table it is obvious that for membrane-based sensors, [100] wafers are preferred. The understanding of the geometric considerations with [110] wafers is important, though, if one wants to fully appreciate all the possible single crystal silicon (SCS) micromachined shapes, and it is especially helpful

MEMS Fabrication

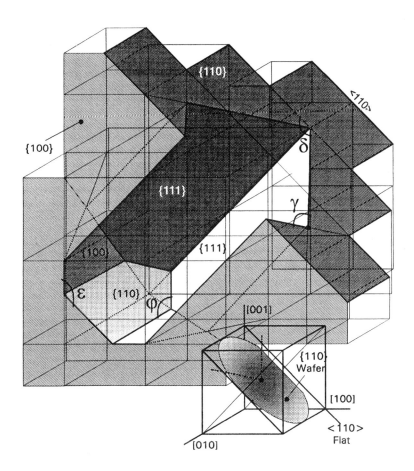

FIGURE 16.12 (110) silicon wafer with anisotropically etched recess inscribed in the Si lattice. $\gamma = 109.47°$; $\delta = 125.26°$; $\varphi = 144.74°$; and $\varepsilon = 45°$. (From Peeters, E., *Process Development for 3D Silicon Microstructures with Application to Mechanical Sensor Design*, KUL, Belgium, 1994. With permission.)

to understand corner compensation schemes (see below). Moreover, all processes for providing dielectric isolation require that the silicon be separated into discrete regions. To achieve a high component density with anisotropic etches on (100) wafers, the silicon must be made very thin because of the aspect ratio limitations due to the sloping walls (see above). With vertical sidewall etching in a (100) wafer, the etch mask is undercut in all directions to a distance approximately equal to the depth of the etching. Vertical etching in (110) surfaces relaxes the etching requirement dramatically and enables more densely packed

TABLE 16.1 Selection of Wafer Type

[100] Orientation	[110] Orientation
Inward sloping walls (54.74°)	Vertical {111} walls
The sloping walls cause a lot of lost real estate	Narrow trenches with high aspect ratio are possible
Flat bottom parallel to surface is ideal for membrane fabrication	Multifaceted cavity bottom ({110} and {100} planes) makes for a poor diaphragm
Bridges perpendicular to a V-groove bound by (111) planes cannot be underetched	Bridges perpendicular to a V-groove bound by (111) planes can be undercut
Shape and orientation of diaphragms convenient and simple to design	Shape and orientation of diaphragms awkward and more difficult to design
Diaphragm size, bounded by nonetching {111} planes, is relatively easy to control	Diaphragm size is difficult to control; the <100> edges are not defined by nonetching planes

structures such as beam leads or image sensors. Kendall describes and predicts a wide variety of applications for (110) wafers such as fabrication of trench capacitors, vertical multi-junction solar cells, diffraction gratings, infrared interference filters, large area cathodes and filters for bacteria [Kendall, 1975; 1979].

16.4 Silicon as a Substrate and Structural Material

16.4.1 Silicon as Substrate

For many mechanical sensor applications, single crystal Si, based on its intrinsic mechanical stability and the feasibility of integrating sensing and electronics on the same substrate, often presents an excellent substrate choice. For chemical sensors, on the other hand, Si, with few exceptions,[2] is merely the substrate and the choice is not always that straightforward.

In Table 16.2, we show a performance comparison of substrate materials in terms of cost, metallization ease and machinability. Both ceramic and glass substrates are difficult to machine, and plastic substrates are not readily amenable to metallization. Silicon has the highest material cost per unit area, but this cost often can be offset by the small feature sizes possible in a silicon implementation. Si with or without passivating layers, due to its extreme flatness, relative low cost and well-established coating procedures, often is the preferred substrate especially for thin films. A lot of thin-film deposition equipment is built to accommodate Si wafers and, as other substrates are harder to accommodate, this lends Si a convenience advantage. There is also a greater flexibility in design and manufacturing with silicon technology compared to other substrates. In addition, although much more expensive, the initial capital equipment investment is not product specific. Once a first product is on line, a next generation or new products will require changes in masks and process steps but not in the equipment itself.

Disadvantages of using Si usually are most pronounced with increasing device size and low production volumes and when electronics do not need or cannot be incorporated on the same Si substrate. The latter could be either for cost reasons (e.g., in the case of disposables such as glucose sensors) or for technological reasons (e.g., the devices will be immersed in conductive liquids or they must operate at temperatures above 150°C).

An overwhelming determining factor for substrate choice is the final package of the device. A chemical sensor on an insulating substrate almost always is easier to package than a piece of Si with conductive edges in need of insulation.

Sensor packaging is so important in sensors that as a rule sensor design should start from the package rather than from the sensor. In this context, an easier to package substrate has a huge advantage. The latter is the most important reason why recent chemical sensor development in industry has retrenched from a move toward integration on silicon in the 1970s and early 1980s to a hybrid thick film on ceramic approach in the late 1980s and early 1990s. In academic circles in the U.S. chemical sensor integration with electronics continued until the late 1980s; in Europe and Japan, such efforts are still going on [Madou, 1994].

TABLE 16.2 Performance Comparison of Substrate Materials

Substrate	Cost	Metallization	Machinability
Ceramic	Medium	Fair	Poor
Plastic	Low	Poor	Fair
Silicon	High	Good	Very good
Glass	Low	Good	Poor

---

[2]A notable exception is the ion-sensitive field effect transistor (ISFET) where the Si space charge is modulated by the presence of chemicals for which the ISFET chemical coating is sensitive [Madou, 1997, chap. 10].

16.4.2 Silicon as a Structural Element in Mechanical Sensors

16.4.2.1 Introduction

In mechanical sensors the active structural elements convert a mechanical external input signal (force, pressure, acceleration etc.) into an electrical signal output (voltage, current or frequency). The transfer functions in mechanical devices describing this conversion are mechanical, electro-mechanical and electrical.

In the mechanical conversion, a given external load is concentrated and maximized in the active member of the sensor. Structurally active members are typically high-aspect-ratio elements such as suspended beams or membranes. The electromechanical conversion is the transformation of the mechanical quantity into an electrical quantity such as capacitance, resistance, charge etc. Often the electrical signal needs further electrical conversion into an output voltage, frequency or current. For electrical conversion into an output voltage, a Wheatstone bridge may be used as in the case of a piezoresistive sensor, and a charge amplifier may be used in the case of a piezoelectric sensor. To optimize all three transfer functions, detailed electrical and mechanical modeling is required. One of the most important inputs required for the mechanical models are the experimentally determined independent elasticity constants or moduli. In what follows, we describe what makes Si such an important structural element in mechanical sensors and present its elasticity constants.

16.4.2.2 Important Characteristics of Mechanical Structural Elements: Stress-Strain Curve and Elasticity Constants

Yield, tensile strength, hardness and creep of a material all relate to the elasticity curve, i.e., the stress–strain diagram of the material as shown in Figure 16.13. For small strain values, Hooke's law applies, i.e., stress (force per unit area, N/m^2) and strain (displacement per unit length, dimensionless) are proportional and the stress–strain curve is linear, with a slope corresponding to the elastic modulus E (Young's modulus-N/m^2). This regime as in Figure 16.13 is marked as the elastic deformation regime. For isotropic media such as amorphous and polycrystalline materials, the applied axial force per unit area or tensile stress, σ_a, and the axial or tensile strain, ε_a, are thus related as:

$$\sigma_a = E\varepsilon_a \tag{16.9}$$

with ε_a given by the dimensionless ratio of $L_2 - L_1/L_1$, i.e., the ratio of the wire's elongation to its original length. The elastic modulus may be thought of as stiffness or a material's resistance to elastic deformation.

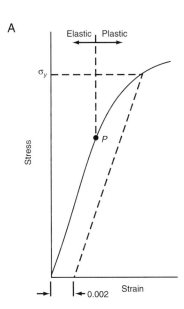

FIGURE 16.13 Typical stress-strain behavior for a material showing elastic and plastic deformations, the proportional limit P and the yield strength σ_y, as determined using the 0.002 strain offset method (see text).

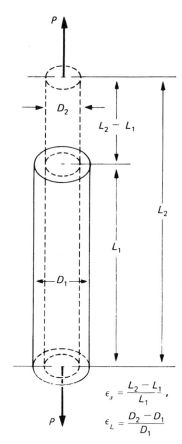

FIGURE 16.14 Metal wire under axial or normal stress; normal stress creates both elongation and lateral contraction.

The greater the modulus, the stiffer the material. A tensile stress usually also leads to a lateral strain or contraction (Poisson effect), ε_l, given by the dimensionless ratio of $D_2 - D_1/D_1$ ($\Delta D/D_1$), where D_1 is the original wire diameter and ΔD is the change in diameter under axial stress (see Figure 16.14). The Poisson ratio is the ratio of lateral over axial strain:

$$\nu = -\frac{\varepsilon_l}{\varepsilon_a} \tag{16.10}$$

The minus sign indicates a contraction of the material. For most materials, ν is a constant within the elastic range. Normally, some slight volume change does accompany the deformation, and, consequently, ν is smaller than 0.5. The magnitude of the Young's modulus ranges from 4.1×10^4 MPa (the N/m^2 unit is called the Pascal, Pa) for magnesium, to 40.7×10^4 MPa for tungsten and 144 GPa for Invar. With increasing temperature, the elastic modulus diminishes. The Poisson ratios for aluminum and cast steel are 0.34 and 0.28, respectively. The value of ν fluctuates for different materials over a relatively narrow range. Generally, it is on the order of 0.25 to 0.35. In extreme cases, values as low as 0.1 (certain types of concrete) and as high as 0.5 (rubber) occur. A value of 0.5 is the largest value possible. It is attained by materials during plastic flow and indicates a constant volume. For an elastic isotropic medium subjected to a triaxial state of stress, the resulting strain component in the x direction, ε_x, is given by the summation of elongation and contraction:

$$\varepsilon_x = \frac{1}{E}[\sigma_x - \nu(\sigma_y + \sigma_z)] \tag{16.11}$$

and so on for the y and z directions (three equations in total).

For an analysis of mechanical structures we must consider not only compressional and tensile strains but also shear strains. Whereas normal stresses create elongation plus lateral contraction with accompanying volume changes, shear stresses (e.g., by twisting a body) create shape changes without volume changes, i.e., shear strains. The one-dimensional shear strain, γ, is produced by the shear stress, τ (N/m^2). For small strains, Hooke's law may be applied again:

$$\gamma = \frac{\tau}{G} \tag{16.12}$$

where G is called the elastic shear modulus or the modulus of rigidity. For any three-dimensional state of shear stress, three equations of this type will hold. Isotropic bodies are characterized by two independent elastic constants only, since the shear modulus G, as can be shown [Chou and Pagano, 1967], relates the Young's modulus and the Poisson ratio as:

$$G = \frac{E}{2(1+\nu)} \tag{16.13}$$

Crystal materials, whose elastic properties are anisotropic, require more than two elastic constants, the number increasing with decreasing symmetry. Cubic crystals (bcc, fcc), for example, require 3 elastic constants, hexagonal crystals require 5, and materials without symmetry require 21 [Kittel, 1976; Chou and Pagano, 1967]. The relation between stresses and strains is more complex in this case and depends greatly on the spatial orientation of these quantities with respect to the crystallographic axes. Hooke's law in the most generic form is expressed in two formulas:

$$\sigma_{ij} = E_{ijkl} \cdot \varepsilon_{kl} \quad \text{and} \quad \varepsilon_{ij} = S_{ijkl} \cdot \sigma_{kl} \tag{16.14}$$

where σ_{ij} and σ_{kl} are stress tensors of rank 2 expressed in N/m^2; ε_{ij} and ε_{kl} are strain tensors of rank 2 and are dimensionless; E_{ijkl} is a stiffness coefficient tensor of rank 4 expressed in N/m^2; and S_{ijkl} is a compliance coefficient tensor of rank 4 expressed in m^2/N. The first expression is analogous to Eq. (16.9) and the second expression is the inverse, giving the strains in terms of stresses. The tensor representations in Eq. (16.14) can also be represented as two matrices:

$$\sigma_m = \sum_{n=1}^{6} E_{mn}\varepsilon_n \quad \text{and} \quad \varepsilon_m = \sum_{n=1}^{6} S_{mn}\sigma_n \tag{16.15}$$

Components of tensors E_{ijkl} and S_{ijkl} are substituted by elements of the matrices E_{mn} and S_{mn}, respectively. To convert the ij indices to m and the kl indices to n, the following scheme applies:

11 → 1, 22 → 2, 33 → 3, 23 and 32 → 4, 13 and 31 → 5, 12 and 21 → 6, E_{ijkl} → E_{mn} and S_{ijkl} → S_{mn} when m and n = 1,2,3; 2 S_{ijkl} → S_{mn} when m or n = 4,5,6; 4S_{ijkl} → S_{mn} when m and n = 4,5,6; σ_{ij} → σ_m when m = 1,2,3; and ε_{ij} → ε_m when m = 4,5,6

With these reduced indices there are thus six equations of the type:

$$\sigma_x = E_{11}\varepsilon_x + E_{12}\varepsilon_y + E_{13}\varepsilon_z + E_{14}\gamma_{yz} + E_{15}\gamma_{zx} + E_{16}\gamma_{xy} \tag{16.16}$$

and hence 36 moduli of elasticity or E_{mn} stiffness constants. There are also six equations of the type:

$$\varepsilon_x = S_{11}\sigma_x + S_{12}\sigma_y + S_{13}\sigma_z + S_{14}\tau_{yz} + S_{15}\tau_{zx} + S_{16}\tau_{xy} \tag{16.17}$$

defining 36 S_{mn} constants which are called the compliance constants; see also Madou (1997, Eq. (9.29)). It can be shown that the matrices E_{mn} and S_{mn}, each composed of 36 coefficients, are symmetrical; hence, a material without symmetry elements has 21 independent constants or moduli. Due to symmetry of crystals, several more of these may vanish until, for our isotropic medium, they number two only (E and v). The stiffness coefficient and compliance coefficient matrices for cubic-lattice crystals with the vector of stress oriented along the [100] axis are given as:

$$E_{mn} = \begin{bmatrix} E_{11} & E_{12} & E_{12} & 0 & 0 & 0 \\ E_{12} & E_{11} & E_{12} & 0 & 0 & 0 \\ E_{12} & E_{12} & E_{11} & 0 & 0 & 0 \\ 0 & 0 & 0 & E_{44} & 0 & 0 \\ 0 & 0 & 0 & 0 & E_{44} & 0 \\ 0 & 0 & 0 & 0 & 0 & E_{44} \end{bmatrix} \quad S_{mn} = \begin{bmatrix} S_{11} & S_{12} & S_{12} & 0 & 0 & 0 \\ S_{12} & S_{11} & S_{12} & 0 & 0 & 0 \\ S_{12} & S_{12} & S_{11} & 0 & 0 & 0 \\ 0 & 0 & 0 & S_{44} & 0 & 0 \\ 0 & 0 & 0 & 0 & S_{44} & 0 \\ 0 & 0 & 0 & 0 & 0 & S_{44} \end{bmatrix} \quad (16.18)$$

In cubic crystals, the three remaining independent elastic moduli are usually chosen as E_{11}, E_{12} and E_{44}. The S_{mn} values can be calculated simply from these E_{mn} values. Expressed in terms of the compliance constants, one can show that $1/S_{11} = E =$ Young's modulus, $-S_{12}/S_{11} = v =$ Poisson's ratio, and $1/S_{44} = G =$ shear modulus. In the case of an isotropic material, such as a metal wire, there is an additional relationship:

$$E_{44} = \frac{E_{11} - E_{12}}{2} \quad (16.19)$$

reducing the number of independent stiffnesses constants to two. The anisotropy coefficient α is defined as:

$$\alpha = \frac{2E_{44}}{E_{11} - E_{12}} \quad (16.20)$$

making $\alpha = 1$ for an isotropic crystal. For an anisotropic crystal, the degree of anisotropy is given by the deviation of α from 1. Single crystal silicon has moderately anisotropic elastic properties [Brantley, 1973; Nikanorov et al., 1972], with $\alpha = 1.57$. Brantley (1973) gives the non-zero stiffness components, referred to the [100] crystal orientation as: $E_{11} = E_{22} = E_{33} = 166 \times 10^9$ N/m^2, $E_{12} = E_{13} = E_{23} = 64 \times 10^9$ N/m^2 and $E_{44} = E_{55} = E_{66} = 80 \times 10^9$ N/m^2:

$$\begin{vmatrix} \sigma_x \\ \sigma_y \\ \sigma_z \\ \tau_{xy} \\ \tau_{xz} \\ \tau_{yz} \end{vmatrix} = \begin{vmatrix} 166(E_{11}) & 64(E_{12}) & 64(E_{12}) & 0 & 0 & 0 \\ 64(E_{12}) & 166(E_{11}) & 64(E_{12}) & 0 & 0 & 0 \\ 64(E_{12}) & 64(E_{12}) & 166(E_{11}) & 0 & 0 & 0 \\ 0 & 0 & 0 & 80(E_{44}) & 0 & 0 \\ 0 & 0 & 0 & 0 & 80(E_{44}) & 0 \\ 0 & 0 & 0 & 0 & 0 & 80(E_{44}) \end{vmatrix} \times \begin{vmatrix} \varepsilon_x \\ \varepsilon_y \\ \varepsilon_z \\ \gamma_{xy} \\ \gamma_{xz} \\ \gamma_{yz} \end{vmatrix} \quad (16.21)$$

with σ normal stress, τ shear stress, ε normal strain and γ shear strain. The values for E_{mn}, in Eq. (16.21), compare with a Young's modulus of 207 GPa for a low carbon steel. Variations on the values of the elastic constants on the order of 30%, depending on crystal orientation, must be considered; doping level (see below) and dislocation density have minor effects as well. From the stiffness coefficients the compliance coefficients of Si can be calculated as $S_{11} = 7.68 \times 10^{-12}$ m^2/N, $S_{12} = -2.14 \times 10^{-12}$ m^2/N, and $S_{44} = 12.6 \times 10^{-12}$ m^2/N [Khazan, 1994]. A graphical representation of elastic constants

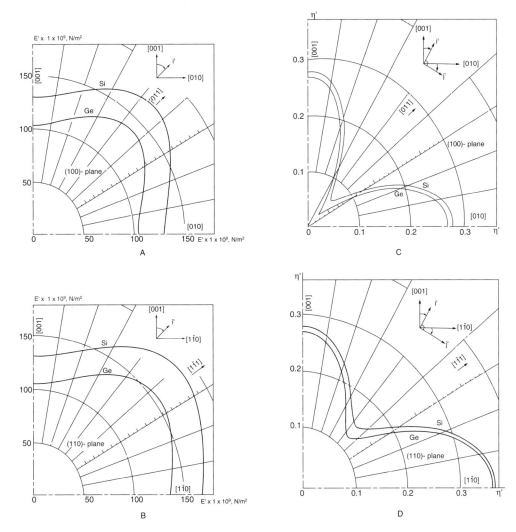

FIGURE 16.15 Elasticity constants for Si and Ge. (A) Young's modulus as a function of direction in the (100) plane. (B) Young's modulus as a function of direction in the (110) plane. (C) Poisson ratio as a function of direction in the (100) plane. (D) Poisson's ratio as a function of direction in the (110) plane. (From Worthman, J. J. and R. A. Evans, *J. Appl. Phys.*, 36, 153–156, 1965. With permission.)

for different crystallographic directions in Si and Ge is given in Worthman and Evans (1965) and is reproduced in Figure 16.15. Figure 16.15A to D displays E and ν for Ge and Si in planes (100) and (110) as functions of direction. Calculations show that E, G and ν are constant for any direction in the (111) plane. In other words, a plate lying in this plane can be considered as having isotropic elastic properties. A review of independent determinations of the Si stiffness coefficients, with their respective temperature coefficients, is given in Metzger and Kessler (1970). Some of the values from that review are reproduced in Table 16.3. Values for Young's modulus and the shear modulus of Si can also be found in Greenwood (1988) and are reproduced in Table 16.4 for the three technically important crystal orientations.

16.4.2.3 Residual Stress

Most properties, such as the Young's modulus, for lightly and highly doped silicon are identical. From Section 16.10 on, we will see that the Young's modulus for polycrystalline silicon is about 161 GPa.

TABLE 16.3 Stiffness Coefficients and Temperature Coefficient of Stiffness for Si [Metzger and Kessler, 1970]

Stiffness Coefficients (GPa = 10^9 N/m^2)	Temperature Coefficient of Stiffness ($\delta E/\delta T/E$)
E_{11} = 164.8 ± 0.16	-122×10^{-6}
E_{12} = 63.5 ± 0.3	-162×10^{-6}
E_{44} = 79.0 ± 0.06	-97×10^{-6}

TABLE 16.4 Derived Values for Young's Modulus and Shear Modulus for Si [Greenwood, 1988]

Miller Index for Orientation	Young's Modulus (E) (GPa)	Shear Modulus (G) (GPa)
[100]	129.5	79.0
[110]	168.0	61.7
[111]	186.5	57.5

Residual stress and associated stress gradients in highly boron doped single crystal silicon present some controversy. Highly boron-doped membranes, which are usually reported to be tensile, also have been reported compressive [Huff and Schmidt, 1992; Maseeh and Senturia, 1990]. From a simple atom-radius argument, one expects that a large number of substitutional boron atoms would create a net shrinkage of the lattice compared to pure silicon and that the residual stress would be tensile with a stress gradient corresponding to the doping gradient. That is, an etched cantilever would be expected to bend up out of the plane of the silicon wafer. Maseeh and Santuria (1990) believe that the appearance of compressive behavior in heavily boron-doped single crystal layers results from the use of an oxide etch mask. They suggest that plastic deformation of the p$^+$ silicon beneath the compressively stressed oxide can explain the observed behavior. Ding and Ko (1991), who also found compressive behavior for nitride-covered p$^+$ Si thin membranes, believe that the average stress in p$^+$ silicon is indeed tensile, but great care is required to establish this fact because the combination of heavy boron doping and a high-temperature drive-in under oxidizing conditions can create an apparent reversal of both the net stress (to compressive) and of the stress gradient (opposite to the doping gradient). A proposed explanation is that at the oxide-silicon interface, a thin compressively stressed layer is formed during the drive-in which is not removed in buffered HF. It can be removed by reoxidation and etching in HF, or by etching in KOH.

16.4.2.4 Yield, Tensile Strength, Hardness and Creep

As a material is deformed beyond its elasticity limit, yielding or plastic deformation—permanent, non-recoverable deformation—occurs. The point of yielding in Figure 16.13 is the point of initial departure from linearity of the stress–strain curve and is sometimes called the proportional limit indicated by a letter "P." The Young's modulus of mild steel is ±30,000,000 psi,[3] and its proportional limit (highest stress in the elastic range) is approximately 30,000 psi. Thus, the maximum elastic strain in mild steel is about 0.001 under a condition of uniaxial stress. This gives an idea as to the magnitude of the strains we are dealing with. A convention has been established wherein a straight line is constructed parallel to the elastic portion of the stress–strain curve at some specified strain offset, usually 0.002. The stress corresponding to the intersection of this line and the stress–strain curve as it bends over in the plastic region is defined as the yield strength, σ_y (see Figure 16.13). The magnitude of the yield strength of a material

---

[3] In the sensor area it is still mandatory to be versatile in the different unit systems especially with regards to expressing units for quantities such as pressure and stress. In this book we are using mostly Pascal, Pa (=N/m^2), but in industry it is still customary to use psi when dealing with metal properties, torr when dealing with vacuum systems, and dyne/cm^2 when dealing with surface tension.

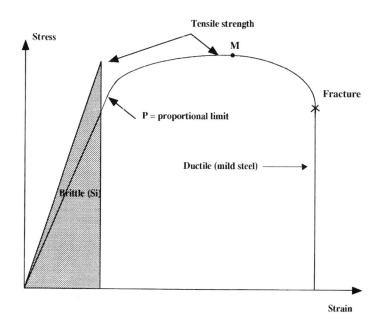

FIGURE 16.16 Stress–strain curve. Tensile strength of a metal is the stress at the maximum of this curve. Abrupt brittle fracture of a high modulus material with no plastic deformation region like Si is indicated as well.

is a measure of its resistance to plastic deformation. Yield strengths may range from 35 MPa (5000 psi) for a soft and weak aluminum to over 1400 MPa (200,000 psi) for high strength steels. The tensile strength is the stress at the maximum of the stress–strain curve (Figure 16.16). This corresponds to the maximum stress that can be sustained by a structure in tension; if the stress is applied and maintained, fracture will result. Both tensile strength and hardness are indicators of a metal's resistance to plastic deformation. Consequently, they are roughly proportional [Callister, 1985]. Material deformation occurring at elevated temperatures and static material stresses is termed creep. It is defined as a time-dependent and permanent deformation of materials when subjected to a constant load or stress.

Silicon exhibits no plastic deformation or creep below 800°C; therefore, Si sensors are inherently very insensitive to fatigue failure when subjected to high cyclic loads. Silicon sensors have actually been cycled in excess of 100 million cycles with no observed failures. This ability to survive a very large number of duty cycles is due to the fact that there is no energy absorbing or heat generating mechanism due to intergranular slip or movement of dislocations in silicon at room temperature. However, single crystal Si, as a brittle material, will yield catastrophically, when stress beyond the yield limit is applied, rather than deform plastically as metals do (see Figure 16.16). At room temperature, high modulus materials such as Si, SiO_2 and Si_3N_4 often exhibit linear-elastic behavior at lower strain and transition abruptly to brittle-fracture behavior at higher strain. Plastic deformation in metals is based on stress-induced dislocation generation in the grain boundaries and a subsequent dislocation migration that results in a macroscopic deformation from inter-grain shifts in the material. No grain boundaries exist in single crystal silicon (SCS), and plastic deformation can only occur through migration of the defects originally present in the lattice or of those that are generated at the surface. As the number of these defects is very low in SCS, the material can be considered a perfect elastic material at normal temperatures. Perfect elasticity implies proportionality between stress and strain (i.e., load and flexure) and the absence of irreversibilities or mechanical hysteresis. The absence of plastic behavior also accounts for the extremely low mechanical losses in SCS, which enable the fabrication of resonating structures that exhibit exceptionally high Q-factors. Values of up to 10^8 in vacuum have been reported. At elevated temperatures, and with metals and polymers at ordinary temperatures, complex behavior in the stress–strain curve can occur. Considerable plasticity can be induced in SCS at elevated temperatures (>800°C), when the mobility of defects in the lattice is substantially increased. Huff and Schmidt (1992) actually report a pressure switch exhibiting hysteresis based on buckling of plastically deformed silicon membranes. To eliminate plastic deformation of Si wafers, it is important that during high temperature steps the presence of films that could stress or even warp the wafer in an asymmetric way, typically oxides or nitrides, be avoided.

16.4.2.5 Piezoresistivity in Silicon

Piezoresistance is the fractional change in bulk resistivity induced by small mechanical stresses applied to a material. Most materials exhibit piezoresistivity, but the effect is particularly important in some semiconductors (more than an order of magnitude higher than that of metals). Monocrystalline silicon has a high piezoresistivity and, combined with its excellent mechanical and electronic properties, makes a superb material for the conversion of mechanical deformation into an electrical signal. Actually, the history of silicon-based sensors started with the discovery of the piezoresistance effect in Si and Ge more than four decades ago [Smith, 1954]. The two main classes of piezoresistive sensors are membrane-type structures (typically pressure and flow sensors) and cantilever beams (typically acceleration sensors) with in-diffused resistors (boron, arsenic or phosphorus) strategically placed in zones of maximum stress.

For a three-dimensional anisotropic crystal, the electrical field vector (E) is related to the current vector (i) by a 3-by-3 resistivity tensor [Khazan, 1994]. Experimentally the nine coefficients are always found to reduce to six, and the symmetric tensor is given by:

$$\begin{bmatrix} E_1 \\ E_2 \\ E_3 \end{bmatrix} = \begin{bmatrix} \rho_1 & \rho_6 & \rho_5 \\ \rho_6 & \rho_2 & \rho_4 \\ \rho_5 & \rho_4 & \rho_3 \end{bmatrix} \cdot \begin{bmatrix} i_1 \\ i_2 \\ i_3 \end{bmatrix} \qquad (16.22)$$

For the cubic Si lattice, with the axes aligned with the <100> axes, ρ_1, ρ_2 and ρ_3 define the dependence of the electric field on the current along the same direction (one of the <100> directions); ρ_4, ρ_5 and ρ_6 are cross-resistivities, relating the electric field to the current along a perpendicular direction.

The six resistivity components in Eq. (16.22) depend on the normal (σ) and shear (τ) stresses in the material as defined in the preceding section. Smith (1954) was the first to measure the resistivity coefficients π_{11}, π_{12} and π_{44} for Si at room temperature. Table 16.5 lists Smith's results [Smith, 1954]. The piezoresistance coefficients are largest for π_{11} in n-type silicon and π_{44} in p-type silicon, about -102.10^{-11} and 138.10^{-11} Pa^{-1}, respectively.

Resistance change can now be calculated as a function of the membrane or cantilever beam stress. The contribution to resistance changes from stresses that are longitudinal (σ_l) and transverse (σ_t) with respect to the current flow is given by:

$$\frac{\Delta R}{R} = \sigma_l \pi_l + \sigma_t \pi_t \qquad (16.23)$$

where

σ_l = Longitudinal stress component, i.e., stress component parallel to the direction of the current.
σ_t = Transversal stress component, i.e., the stress component perpendicular to the direction of the current.
π_l = Longitudinal piezoresistance coefficient.
π_t = Transversal piezoresistance coefficient.

The piezoresistance coefficients π_l and π_t for (100) silicon as a function of crystal orientation are reproduced from Kanda in Figure 16.17A (for p-type) and B (for n-type) [Kanda, 1982]. By maximizing the expression for the stress-induced resistance change in Eq. (16.23), one optimizes the achievable sensitivity in a piezoresistive silicon sensor.

TABLE 16.5 Resistivity and Piezoresistance at Room Temperature [Smith, 1954; Khazan, 1994]

	ρ (Ω cm)	π_{11}[a]	π_{12}[a]	π_{44}[a]
p-Si	7.8	+6.6	−1.1	+138.1
n-Si	11.7	−102.2	+53.4	−13.6

[a] Expressed in 10^{-12} cm^2 dyne^{-1} or 10^{-11} Pa^1.

The orientation of a membrane or beam is determined by its anisotropic fabrication. The surface of the silicon wafer is usually a (100) plane; the edges of the etched structures are intersections of (100) and (111) planes and are thus <110> directions. p-Type piezoresistors are most commonly used because the orientation of maximum piezoresistivity (<110>) happens to coincide with the edge orientation of a conventionally etched diaphragm and because the longitudinal coefficient is roughly equal in magnitude but opposite in sign as compared to the transverse coefficient (Figure 16.17A)[Peeters, 1994]. With the values in Table 16.5, π_l and π_t now can be calculated numerically for any orientation. The longitudinal piezoresistive coefficient in the <110> direction is $\pi_l = 1/2(\pi_{11} + \pi_{12} + \pi_{44})$. The corresponding transverse coefficient is $\pi_t = 1/2(\pi_{11} + \pi_{12} - \pi_{44})$. From Table 16.5 we know that, for p-type resistors, π_{44} is more important than the other two coefficients and Eq. (16.23) is approximated by:

$$\frac{\Delta R}{R} = \frac{\pi_{44}}{2}(\sigma_l - \sigma_t) \tag{16.24}$$

For n-type resistors, π_{44} can be neglected, and we obtain:

$$\frac{\Delta R}{R} = \frac{\pi_{11} + \pi_{12}}{2}(\sigma_l + \sigma_t) \tag{16.25}$$

Equations (16.24) and (16.25) are valid only for uniform stress fields or if the resistor dimensions are small compared with the membrane or beam size. When stresses vary over the resistors they have to be integrated, which is most conveniently done by computer simulation programs.

To convert the piezoresistive effect into a measurable electrical signal, a Wheatstone bridge is often used. A balanced Wheatstone bridge configuration is constructed as in Figure 16.18A by locating four p-piezoresistors midway along the edges of a square diaphragm as in Figure 16.18B (location of maximum stress). Two resistors are oriented so that they sense stress in the direction of their current axes and two are placed to sense stress perpendicular to their current flow. Two longitudinally stressed resistors (A) are balanced against two transversally stressed resistors (B); two of them increase in value and the other two decrease in value upon application of a stress. In this case, from Eq. (16.24),

$$\frac{\Delta R}{R} \approx 70 \cdot 10^{-11}(\sigma_l - \sigma_t) \tag{16.26}$$

with σ in Pa. For a realistic stress pattern where σ_l = 10 MPa and σ_t = 50 MPa, Eq. (16.26) gives us a $\Delta R/R \approx 2.8\%$ [Peeters, 1994].

By varying the diameter and thickness of the silicon diaphragms, piezoresistive sensors in the range of 0 to 200 MPa have been made. The bridge voltages are usually between 5 and 10 volts, and the sensitivity may vary from 10 mV/kPa for low pressure to 0.001 mV/kPa for high pressure sensors.

Peeters (1994) shows how a more sensitive device could be based on n-type resistors when all the n-resistors oriented along the <100> direction are subjected to an uniaxial stress pattern in the longitudinal axis, as shown in Figure 16.19. The overall maximum piezoresistivity coefficient (π_l in the <100> direction) is substantially higher for n-silicon than it is for p-type silicon in any direction (maximum π_t and π_l in the

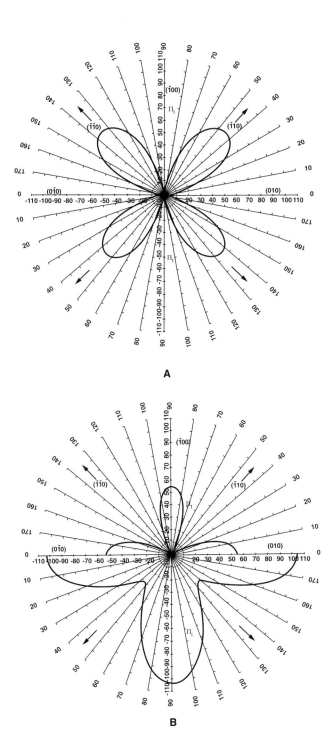

FIGURE 16.17 Piezoresistance coefficients π_l and π_t for (100) silicon. (A) For p-type in the (001) plane (10^{-12} cm^2/dyne). (B) For n-type in the (001) plane (10^{-12} cm^{-2}/dyne). (From Kanda, Y., *IEEE Trans. Electr. Dev.*, ED-29, 64–70, 1982. With permission.)

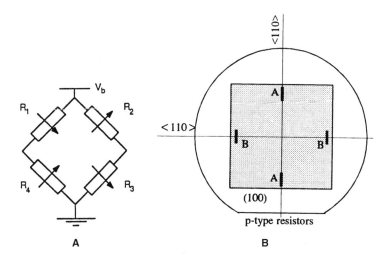

FIGURE 16.18 Measuring on a membrane with piezoresistors. (A) Wheatstone-bridge configuration of four in-diffused piezoresistors. The arrows indicate resistance changes when the membrane is bent downward. (B) Maximizing the piezoresistive effect with p-type resistors. The A resistors are stressed longitudinally and the B resistors are stressed transversally. (From Peeters, E., Ph.D. thesis, KUL, Belgium, 1994. With permission.)

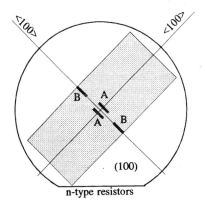

FIGURE 16.19 Higher pressure sensitivity by strategic placement of in-diffused piezoresistors proposed by Peeters (1994). The n-resistors are stressed longitudinally with the A resistors under tensile stress, and the B resistors under compressive stress. (From Peeters, E., Ph.D. thesis, KUL, Belgium, 1994. With permission.)

<100> direction, Figure 16.17B). Exploitation of these high piezoresistivity coefficients is less obvious, though, since the resistor orientation for maximum sensitivity (<100>) is rotated over 45° with respect to the <110> edges of an anisotropically etched diaphragm. Also evident from Figure 16.17B is that a transversally stressed resistor cannot be balanced against a longitudinally stressed resistor. Peeters has circumvented these two objections by a uniaxial, longitudinal stress pattern in the rectangular diaphragm represented in Figure 16.19. With a (100) substrate and a <100> orientation (45° to wafer flat) we obtain:

$$\frac{\Delta R}{R} \approx 53 \cdot 10^{-11} \cdot \sigma_t - 102 \cdot 10^{-11} \cdot \sigma_l \qquad (16.27)$$

with σ in Pa. Based on Eq. (16.27), with $\sigma_t = 10$ MPa and $\sigma_l = 50$ MPa, $\Delta R/R \approx -4.6\%$. In the proposed stress pattern it is important to minimize the transverse stress by making the device truly uniaxal as the

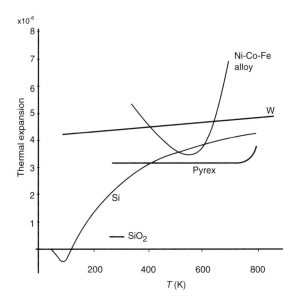

FIGURE 16.20 Thermal expansion coefficient vs. absolute temperature. (From Greenwood, J.C., *J. Phys. E: Sci. Instrum.*, 21 1114–1128, 1988. With permission.)

longitudinal and transverse stress components have opposite effects and can even cancel out. In practice, a pressure sensor with an estimated 65% gain in pressure sensitivity over the more traditional configurations could be made in the case of an 80% uniaxiality [Peeters, 1994].

The piezoresistive effect is often described in terms of the gauge factor, *G*, defined as:

$$G = \frac{1}{\varepsilon} \frac{\Delta R}{R} \qquad (16.28)$$

which is the relative resistance change divided by the applied strain. The gauge factor of a metal strain gauge is typically around 2, for single crystal Si it is 90, and for poly-crystalline Si it is about 30 (see also Section 16.10).

16.4.2.6 Thermal Properties of Silicon

In Figure 16.20, the expansion coefficient of Si, W, SiO_2, Ni-Co-Fe alloy and Pyrex® is plotted vs. absolute temperature. Single crystal silicon has a high thermal conductivity (comparable with metals such as steel and aluminum) and a low thermal expansion coefficient. Its thermal expansion coefficient is closely matched to Pyrex glass but exhibits considerable temperature dependence. A good match in thermal expansion coefficient between the device wafer (e.g., Si) and the support substrate (e.g., Pyrex) is required. A poor match introduces stress, which degrades the device performance. This makes it difficult to fabricate composite structures of Pyrex and Si that are stress-free over a wide range of temperatures. Drift in silicon sensors often stems from packaging. In this respect, several types of stress-relief, subassemblies for stress-free mounting of the active silicon parts play a major role; using silicon as the support for silicon sensors is highly desirable. The latter aspect is addressed in Madou (1997, chap. 8).

Although the Si band-gap is relatively narrow, by employing silicon on insulator (SOI) wafers, high temperature sensors can be fashioned. For the latter application, relatively highly doped Si, which is relatively linear in its temperature coefficient of resistance and sensitivity over a wide range, typically is employed.

When fabricating thermally isolated structures on Si, the large thermal conductivity of Si poses a considerable problem as the major heat leak occurs through the Si material. For thermally isolated structures, machining in glass or quartz with their lower thermal conductivity represents an important alternative.

16.5 Wet Isotropic and Anisotropic Etching

16.5.1 Wet Isotropic and Anisotropic: Empirical Observations

16.5.1.1 Introduction

Wet etching of Si is used mainly for cleaning, shaping, polishing and characterizing structural and compositional features [Uhlir, 1956]. Wet chemical etching provides a higher degree of selectivity than dry etching techniques. Wet etching often is also faster; compare a few microns to several tens of microns per minute for isotropic etchants and about 1 μm/min for anisotropic etchants vs. 0.1 μm/min in typical dry etching. More recently though, with ECR dry etching, rates of up to 6 μm/min were achieved (see Madou [1997, chap. 2]). Modification of wet etchant and/or temperature can alter the selectivity to silicon dopant concentration and type and, especially when using alkaline etchants, to crystallographic orientation. Etching proceeds by reactant transport to the surface (1), surface reaction (2), and reaction product transport away from the surface (3). If (1) or (3) is rate determining, etching is diffusion limited and may be increased by stirring. If (2) is the rate-determining step, etching is reaction rate limited and depends strongly on temperature, etching material and solution composition. Diffusion-limited processes have lower activation energies (of the order of a few Kcal/mol) than reaction-rate controlled processes and therefore are relatively insensitive to temperature variations. In general, one prefers reaction rate limitation as it is easier to reproduce a temperature setting than a stirring rate. The etching apparatus needs to have both a good temperature controller and a reliable stirring facility [Kaminsky, 1985; Stoller et al., 1970].

Isotropic etchants, also polishing etchants, etch in all crystallographic directions at the same rate; they usually are acidic, such as $HF/HNO_3/CH_3COOH$ (HNA), and lead to rounded isotropic features in single crystalline Si. They are used at room temperature or slightly above (<50°C). Historically they were the first Si etchants introduced [Robbins and Schwartz, 1959; 1960; 1961; 1976; Uhlir, 1956; Hallas, 1971; Turner, 1958; Kern, 1978; Klein and D'Stefan, 1962]. Later it was discovered that some alkaline chemicals will etch anisotropically, i.e., they etch away crystalline silicon at different rates depending on the orientation of the exposed crystal plane. Typically the pH stays above 12, while more elevated temperatures are used for these slower type etchants (>50°C). The latter type of etchants surged in importance in the late 1960s for the fabrication of dielectrically isolated structures in silicon [Stoller and Wolff, 1966; Stoller, 1970; Forster and Singleton, 1966; Kenney, 1967; Lepselter, 1966; 1967; Waggener, 1970; Kragness, 1973; Waggener et al., 1967a; 1967b; Bean and Runyan, 1977; Rodgers et al., 1977; Rodgers et al., 1976; Ammar and Rodgers, 1980; Schnable and Schmidt, 1976]. Isotropic etchants typically show diffusion limitation, while anisotropic etchants are reaction rate limited.

Preferential or selective etching (also structural etchants) usually are isotropic etchants that show some anisotropy [Kern and Deckert, 1978]. These etchants are used to produce a difference in etch rate between different materials or between compositional or structural variations of the same material on the same crystal plane. These type of etches often are the fastest and simplest techniques to delineate electrical junctions and to evaluate the structural perfection of a single crystal in terms of slip, lineage and stacking faults. The artifacts introduced by the defects etch into small pits of characteristic shape. Most of the etchants used for this purpose are acids with some oxidizing additives [Yang, 1984; Chu and Gavaler, 1965; Archer, 1982; Schimmel and Elkind, 1973; Secco d'Aragona, 1972].

16.5.1.2 Isotropic Etching

Usage of Isotropic Etchants

When etching silicon with aggressive acidic etchants, rounded isotropic patterns form. The method is widely used for:

1. Removal of work-damaged surfaces
2. Rounding of sharp anisotropically etched corners (to avoid stress concentration)
3. Removing of roughness after dry or anisotropic etching

4. Creating structures or planar surfaces in single-crystal slices (thinning)
5. Patterning single-crystal, polycrystalline, or amorphous films
6. Delineation of electrical junctions and defect evaluation (with preferential isotropic etchants)

For isotropic etching of silicon, the most commonly used etchants are mixtures of nitric acid (HNO_3) and hydrofluoric acids (HF). Water can be used as a diluent, but acetic acid (CH_3COOH) is preferred because it prevents the dissociation of the nitric acid better and so preserves the oxidizing power of HNO_3 which depends on the undissociated nitric acid species for a wide range of dilution [Robbins and Schwartz, 1960]. The etchant is called the HNA system; we will return to this etch system below.

Simplified Reaction Scheme
In acidic media, the Si etching process involves hole injection into the Si valence band by an oxidant, an electrical field, or photons. Nitric acid in the HNA system acts as an oxidant; other oxidants such as H_2O_2 and Br_2 also work [Tuck, 1975]. The holes attack the covalently bonded Si, oxidizing the material. Then follows a reaction of the oxidized Si fragments with OH^- and subsequent dissolution of the silicon oxidation products in HF. Consider the following reactions that describe these processes.

The holes are, in the absence of photons and an applied field, produced by HNO_3, together with water and trace impurities of HNO_2:

$$HNO_3 + H_2O + HNO_2 \rightarrow 2HNO_2 + 2OH^- + 2h^+ \quad \text{(Reaction 16.1)}$$

The holes in Reaction 16.1 are generated in an autocatalytic process; HNO_2 generated in the above reaction re-enters into the further reaction with HNO_3 to produce more holes. With a reaction of this type, one expects an induction period before the oxidation reaction takes off, until a steady-state concentration of HNO_2 has been reached. This has been observed at low HNO_3 concentrations [Tuck, 1975]. After hole injection, OH^- groups attach to the oxidized Si species to form SiO_2, liberating hydrogen in the process:

$$Si^{4+} + 4OH^- \rightarrow SiO_2 + H_2 \quad \text{(Reaction 16.2)}$$

Hydrofluoric acid (HF) dissolves the SiO_2 by forming the water-soluble H_2SiF_6. The overall reaction of HNA with Si looks like:

$$Si + HNO_3 + 6HF \rightarrow H_2SiF_6 + HNO_2 + H_2O + H_2 \text{ (bubbles)} \quad \text{(Reaction 16.3)}$$

The simplification in the above reaction scheme is that only holes are assumed. In the actual Si acidic corrosion reaction, both holes and electrons are involved. The question of hole and/or electron participation in Si corrosion will be considered after the introduction of the model for the Si/electrolyte interfacial energetics. We will learn from that model that the rate-determining step in acidic etching involves hole injection in the valence band, whereas in alkaline anisotropic etching it involves electron injection in the conduction band by surface states. The reactivity of a hole injected in the valence band is significantly greater than that of an electron injected in the conduction band. The observation of isotropy in acidic etchants and anisotropy in alkaline etchants centers on this difference in reactivity.

Iso-Etch Curves
By the early 1960s, the isotropic HNA silicon etch was well characterized. Schwartz and Robbins published a series of four very detailed papers on the topic between 1959 and 1976 [Robbins and Schwartz, 1959; 1960; 1961; 1976]. Most of the material presented below is based on their work.

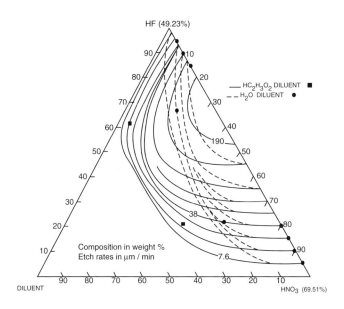

FIGURE 16.21 Iso-etch curves. From Robbins and Schwartz (1960), recalculated for one-sided Si etching and expressed in μm/min.

HNA etching results, represented in the form of iso-etch curves, for various weight percentages of the constituents are shown in Figure 16.21. For this work, normally available concentrated acids of 49.2 wt% HF and 69.5 wt% HNO_3 are used. Water as diluent is indicated by dash-line curves and acetic acid by solid-line curves. Also, as in Wong's representation [Wong, 1990], we have recalculated the curves from Schwartz et al to express the etch rate in μm/min and divided the authors' numbers by 2 as we are considering one-sided etching only. The highest etch rate is observed around a weight ratio HF-HNO_3 of 2:1 and is nearly 100 times faster than anisotropic etch rates. Adding a diluent slows down the etching. From these curves, the following characteristics of the HNA system can be summarized:

1. At high HF and low HNO_3 concentrations, the iso-etch curves describe lines of constant HNO_3 concentrations (parallel to the HF-diluent axis); consequently, the HNO_3 concentration controls the etch rate. Etching at those concentrations tends to be difficult to initiate and exhibits an uncertain induction period (see above). In addition, it results in relatively unstable silicon surfaces proceeding to slowly grow a layer of SiO_2 over a period of time. The etch is limited by the rate of oxidation, so that it tends to be orientation dependent and affected by dopant concentration, defects and catalysts (sodium nitrate often is used). In this regime the temperature influence is more pronounced, and activation energies for the etching reaction of 10 to 20 Kcal/mol have been measured.

2. At low HF and high HNO_3 concentrations, iso-etch curves are lines parallel to the nitric-diluent axis, i.e., they are at constant HF composition. In this case, the etch rate is controlled by the ability of HF to remove the SiO_2 as it is formed. Etches in this regime are isotropic and truly polishing, producing a bright surface with anisotropies of 1% or less (favoring the <110> direction) when used on <100> wafers [Wise et al., 1981]. An activation energy of 4 Kcal/mol is indicative of the diffusion limited character of the process; consequently, in this regime, temperature changes are less important.

3. In the region of maximal etch rate both reagents play an important role. The addition of acetic acid, as opposed to the addition of water, does not reduce the oxidizing power of the nitric acid until a fairly large amount of diluent has been added. Therefore, the rate contours remain parallel with lines of constant nitric acid over a considerable range of added diluent.

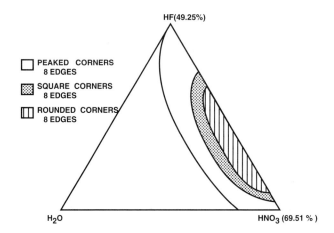

FIGURE 16.22 Topology of etched Si surfaces. (From Schwartz, B. and H. Robbins, *J. Electrochem. Soc.*, 123, 1903–1909, 1976. With permission.)

4. In the region around the HF vertex the surface reaction rate-controlled etch leads to rough, pitted Si surfaces and sharply peaked corners and edges. In moving towards the HNO_3 vertex, the diffusion-controlled reaction results in the development of rounded corners and edges and the rate of attack on (111) planes and (110) planes becomes identical in the polishing regime (anisotropy less than 1%; see point 2).

In Figure 16.22 we summarize how the topology of the Si surfaces depends strongly on the composition of the etch solution. Around the maximum etch rates the surfaces appear quite flat with rounded edges, and very slow etching solutions lead to rough surfaces [Schwartz and Robbins, 1976].

Arrhenius Plot for Isotropic Etching

The effect of temperature on the reaction rate in the HNA system was studied in detail by Schwartz and Robbins (1961). An Arrhenius plot for etching Si in 45% HNO_3, 20% HF and 35% $HC_2H_3O_2$, culled from their work, is shown in Figure 16.23. Increasing the temperature increases the reaction rate. The graph shows two straight-line segments, indicating a higher activation energy below 30°C and a lower one above this temperature. In the low temperature range, etching is preferential and the activation energy is associated with the oxidation reaction. At higher temperatures the etching leads to smooth surfaces and the activation energy is lower and associated with diffusion limited dissolution of the oxide [Schwartz and Robbins, 1961].

With isotropic etchants the etchant moves downward and outwards from an opening in the mask, undercuts the mask, and enlarges the etched pit while deepening it (Figure 16.24). The resulting isotropically etched features show more symmetry and rounding when agitation accompanies the etching (the process is diffusion limited). This agitation effect is illustrated in Figure 16.24. With agitation the etched feature approaches an ideal round cup; without agitation the etched feature resembles a rounded box [Petersen, 1982]. The flatness of the bottom of the rounded box generally is poor, since the flatness is defined by agitation.

Masking for Isotropic Silicon Etchants

Acidic etchants are very fast; for example, an etch rate for Si of up to 50 μ min^{-1} can be obtained with 66% HNO_3 and 34% HF (volumes of reagents in the normal concentrated form) [Kern, 1978; Kern and Deckert, 1978]. Isotropic etchants are so aggressive that the activation barriers associated with etching the different Si planes are not differentiated; all planes etch equally fast, making masking a real challenge.

Although SiO_2 has an appreciable etch rate of 300 to 800 Å/min in the HF:HNO_3 system, one likes to use thick layers of SiO_2 as a mask anyway, especially for shallow etching, as the oxide is so easy to form

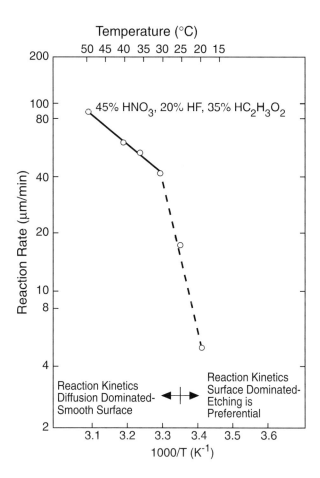

FIGURE 16.23 Etching Arrhenius plot. Temperature dependence of the etch rate of Si in HF:HNO$_3$:CH$_3$:COOH (1:4:3). (From Schwartz, B. and H. Robbins, *J. Electrochem. Soc.*, 108, 365–372, 1961. With permission.)

and pattern. A mask of nonetching Au or Si$_3$N$_4$ is needed for deeper etching. Photoresists do not stand up to strong oxidizing agents such as HNO$_3$, and neither does Al.

Silicon itself is soluble to a small extent in pure HF solutions; for a 48% HF, at 25°C, a rate of 0.3 Å/min was observed for n-type, 2-ohm cm (111)-Si. It was established that Si dissolution in HF is not due to oxidation by dissolved oxygen. Diluted HF etches Si at a higher rate because the reaction in aqueous solutions proceeds by oxidation of Si by OH$^-$ groups [Hu and Kerr, 1967]. A typically buffered HF (BHF) solution has been reported to etch Si at radiochemically measured rates of 0.23 to 0.45 Å/min, depending on doping type and dopant concentration [Hoffmeister, 1969].

By reducing the dopant concentration (n or p) to below 10^{17} atoms/cm^3 the etch rate of Si in HNA is reduced by ~150 [Muraoka et al., 1973]. The doping dependence of the etch rate provides yet another means of patterning a Si surface (see next section). A summary of masks that can be used in acidic etching is presented in Table 16.6.

Dopant Dependence of Silicon Isotropic Etchants

The isotropic etching process is fundamentally a charge-transfer mechanism. This explains the etch rate dependence on dopant type and concentration. Typical etch rates with an HNA system (1:3:8) for n- or p-type dopant concentrations above 10^{18} cm^3 are 1 to 3 µm/min. As presented in the preceding section, a reduction of the etch rate by 150 times is obtained in n- or p-type regions with a dopant concentration of 10^{17} cm^{-3} or smaller [Muraoka et al., 1973]. This presumably is due to the lower mobile carrier

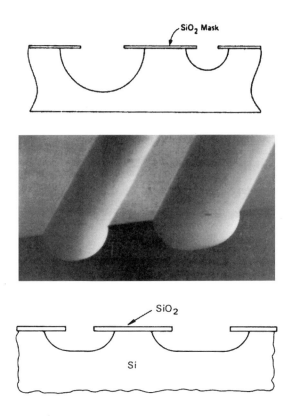

FIGURE 16.24 Isotropic etching of Si with (A) and without (B) etchant solution agitation.

TABLE 16.6 Masking Materials for Acidic Etchants[a]

	Etchants		
Masking	Piranha (4:1, $H_2O_2:H_2SO_4$)	Buffered HF (5:1 NH_4F:conc. HF)	HNA
Thermal SiO_2		0.1 µm/min	300–800 Å/min; limited etch time, thick layers often are used due to ease of patterning
CVD (450°C) SiO_2		0.48 µm/min	0.44 µm/min
Corning 7740 glass		0.063 µ/min	1.9 µ/min
Photoresist	Attacks most organic films	Good for short while	Resists do not stand up to strong oxidizing agents like HNO_3 and are not used
Undoped Si, polysilicon	Forms 30 Å of SiO_2	0.23 to 0.45 Å/min	Si 0.7 to 40 µm/min at room temperature; at a dopant concentration $<10^{17}$ cm^{-3} (n or p)
Black wax			Usable at room temperature
Au/Cr	Good	Good	Good
LPCVD Si_3N_4		1 Å/min	Etch rate is 10–100 Å/min; preferred masking material

[a] The many variables involved necessarily means that the given numbers are approximate only.

concentration available to contribute to the charge transfer mechanisms. In any event, heavily doped silicon substrates with high conductivity can be etched more readily than lightly doped materials. Dopant-dependent isotropic etching can also be exploited in an electrochemical set-up as described in the next section. Although doping does change the chemical etch rate, attempts to exploit these differences for industrial production have failed so far [Seidel, 1989]. This situation is different in electrochemical isotropic etching (see next section).

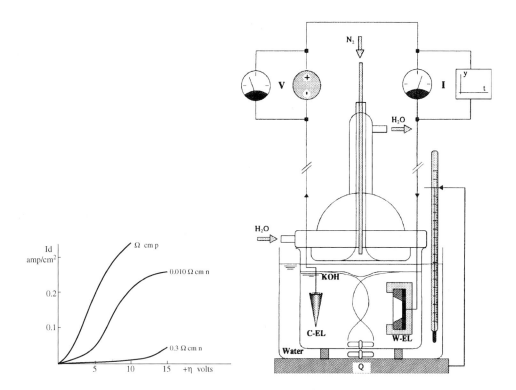

FIGURE 16.25 Electrochemical etching apparatus. W-EL: working electrode (Si), C-EL: counter electrode (e.g., Pt), Q = heat supplied. Inset: Current-voltage (I_d/η) curves in electrochemical etching of Si of various doping. Etch rate dependence on dopant concentration and dopant type for HF-anodic etching of silicon. (From van Dijk, H.J.A. and J. de Jonge, *J. Electrochem. Soc.*, 117, 553–554, 1970. With permission.)

Electrochemical Isotropic Silicon Etch-Etch Stop

Sometimes, a high temperature or extremely aggressive chemical etching process can be replaced by an electrochemical procedure utilizing a much milder solution, thus allowing a simple photoresist mask to be employed [Kern and Deckert, 1978]. In electrochemical acidic etching, with or without illumination of the corroding Si electrode, an electrical power supply is employed to drive the chemical reaction by supplying holes to the silicon surface (W-EL see Figure 16.25). A voltage is applied across the silicon wafer and a counter electrode (C-EL usually platinum) arranged in the same etching solution. Oxidation is promoted by a positive bias applied to the silicon causing an accumulation of holes in the silicon at the silicon/electrolyte interface. Under this condition, oxidation at the surface proceeds rapidly while the oxide is readily dissolved by the HF solution. No oxidant such as HNO_3 is needed to supply the holes; excess electron-hole pairs are created by the electrical field at the surface and/or by optical excitation, thereby increasing the etch rate. This technique proved successful in removing heavily doped layers, leaving behind the more lightly doped membranes in all possible dopant configurations: p on p^+, p on n^+, n on p^+, and n on n^+ [Theunissen et al., 1970; Meek, 1971]. This electrochemical etch-stop technique is demonstrated in the inset of Figure 16.25 [van Dijk et al., 1970]. A 5% HF solution is used, the electrolyte cell is kept in the dark at room temperature, and the distance between the Si anode and the Pt cathode in the electrochemical cell is 1 to 5 cm. Instead of using HF, one can substitute NH_4F (5 wt%) for the electrochemical etching as described by Shengliang et al. (1987). Shengliang reports a selectivity of n-silicon to n^+-silicon (0.001 Ωcm) of 300 with the latter etchant. In the inset in Figure 16.25 the current density vs. applied voltage across the anode and cathode during dissolution is plotted. The current density is related to the dissolution rate of silicon. It can be seen that p-type and heavily doped n-type materials can be dissolved at relatively low voltages, whereas n-type silicon with a lower doping level does not dissolve at the same low voltages. Experiments in this same set-up with homogeneously doped silicon wafers show that n-type silicon of about 3.10^{18} cm^{-3} (<0.01 Ωcm) completely dissolves in these etching conditions, whereas n-type silicon of donor concentrations lower than 2.10^{16} cm^{-3} (>0.3 Ωcm) barely dissolves. For p-type silicon, dissolution is initiated when the acceptor concentration is higher than 5.10^{15} cm^{-3} (<3 Ωcm) and the dissolution rate further increases with increasing acceptor concentration. Under specific circumstances, namely high HF concentrations and low etching currents, porous Si may form [Bomchil et al., 1986].

The acidic electrochemical technique has not been used much in micromachining and is primarily used to polish surfaces. Since the etching rate increases with current density, high spots on the surface are more rapidly etched and very smooth surfaces result. The method of isotropic electrochemical etching has some major advantages which could make it a more important micromachining tool in the future. The etched surfaces are very smooth (say with an average roughness, R_a, of 7 nm), the process is room temperature and IC compatible, simpler resists schemes can be used as the process is much milder than etching in HNA, and etching can be controlled simply by switching a voltage on or off. We will pick up the discussion of anodic polishing, photo etching, and formation of porous silicon in HF solutions after gathering more insight in various etching models.

Preferential Etching
A variety of additives to the HNA system, mainly oxidants, can be included to modify the etch rate, surface finish or isotropy, rendering the etching baths preferential. It is clear that the effect of these additives will only show up in the reaction-controlled regime. Only additives that change the viscosity of the solution could modify the etch rate in the diffusion-limited regime, thereby changing the diffusion coefficient of the reactants [Tuck, 1975; Bogenschutz et al., 1967]. We will not review the effect of these additives any further; refer to Table 16.7 and the cited literature for further study [Yang, 1984; Chu and Gavaler, 1965; Archer, 1982; Schimmel and Elkind, 1973; Secco d'Aragona, 1972].

Problems with Isotropic Etchants
There are several problems associated with isotropic etching of Si. First, it is difficult to mask with high precision using a desirable and simple mask such as SiO_2 (etch rate is 2 to 3% of the silicon etch rate). Second, the etch rate is very agitation sensitive in addition to being temperature sensitive. This makes it difficult to control lateral as well as vertical geometries. Electrochemical isotropic etching (see above) and the development of anisotropic etchants in the late 1960s (see below) overcame many of these problems.

A comprehensive review of isotropic etchants solutions can be found in Kern and Deckert (1978). These authors also give a review of different techniques practiced in chemical etching such as immersion etching, spray etching, electrolytic etching, gas-phase etching, and molten salt etching (fusion techniques). In Table 16.7 some isotropic and preferential etchants and their specific applications are listed.

16.5.1.3 Anisotropic Etching
Introduction
Anisotropic etchants shape, also 'machine,' desired structures in crystalline materials. When carried out properly, anisotropic etching results in geometric shapes bounded by perfectly defined crystallographic planes. Anisotropic wet etching techniques, dating back to the 1960s at the Bell Laboratories, were developed mainly by trial and error. Going over some experimental data, before embarking upon the models, seems fitting. Moreover, we must keep in mind that for higher index planes most models fail.

Figure 16.1 shows a cross-section of a typical shape formed using anisotropic etching. The thinned membranes with diffused resistors could be used for a piezoresistive pressure sensor or an accelerometer. In the usual application, the wafer is selectively thinned from a starting thickness of 300 to 500 μm to form a diaphragm having a final thickness of 10 to 20 μm with precisely controlled lateral dimensions and a thickness control of the order of 1 μm or better. A typical procedure involves the steps summarized in Table 16.8 [Elwenspoek et al., 1994].

TABLE 16.7 Isotropic and Preferential Defect Etchants and Their Specific Applications

Etchant	Application	Remark
HF: 8 vol%, HNO$_3$: 75 vol%, and CH$_3$COOH: 17 vol%	n- and p-type Si, all planes, general etching	Planar etch; e.g., 5 μm/min at 25°C
1 part 49% HF, 1 part of (1.5 M CrO$_3$) (by volume)	Delineation of defects on (111), (100), and (110) Si without agitation	Yang etch[a]
5 vol parts nitric acid (65%), 3 vol parts HF (48%), 3 vol parts acetic acid (96%), 0.06 parts bromine	Polishing etchant used to remove damage introduced during lapping	So-called CP4 etchant; Heidenreich, U.S. Patent 2619414
HF	SiO$_2$	Si etch rate for 48% HF at 25°C is 0.3 Å/min with n-type 2 Ω cm (111) Si
1HF, 3HNO$_3$, 10CH$_3$COOH (by volume)	Delineates defects in (111) Si; etches p$^+$ or n$^+$ and stops at p$^-$ or n$^-$	Dash etch; p- and n-Si at 1300 Å/min in the [100] direction and 46 Å/min in the [111] direction at 25°C[b]
1HF, 1(5 M CrO$_3$) (by volume)	Delineates defects in (111); needs agitation; does not reveal etch pits well on (100) well	Sirtl etch[c]
2HF, 1(0.15 M K$_2$Cr$_2$O$_7$) (by volume)	Yields circular (100) Si dislocation etch pits; agitation reduces etch time	Secco etch[d]
60 ml HF, 30 ml HNO$_3$, 30 ml (5 M CrO$_3$), 2 g Cu(NO$_3$)$_2$, 60 ml CH$_3$COOH, 60 ml H$_2$O	Delineates defects in (100) and (111) Si; requires agitation	Jenkins etch[e]
2HF, 1 (1 M CrO$_3$) (by volume)	Delineates defects in (100) Si without agitation; works well on resistivities 0.6–15.0 Ωcm n- and p-types)	Schimmel etch[f]
2HF, 1 (1 M CrO$_3$), 1.5 (H$_2$O) (by volume)	Works well on heavily doped (100) silicon	Modified Schimmel[f]
HF/KMnO$_4$/CH$_3$COOH	Epitaxial Si	
H$_3$PO$_4$	Si$_3$N$_4$	160–180°C
KOH + alcohols	Polysilicon	85°C
H$_3$PO$_4$/HNO$_3$/HC$_2$H$_3$O$_2$	Al	40–50°C
HNO$_3$/BHF/water	Si and polysilicon	0.1 μm min^{-1} for single crystal Si

[a] Yang, K.H., *J. Electrochem. Soc.*, 131, 1140 (1984).
[b] Dash, W.C., *J. Appl. Phys.*, 27, 1193 (1956).
[c] Sirtl, E. and Adler, A., *Z. Metallkd.*, 52, 529 (1961).
[d] Secco d'Aragona, F., *J. Electrochem. Soc.*, 119, 948 (1972).
[e] Jenkins, M.W., *J. Electrochem. Soc.*, 124, 757 (1977).
[f] Schimmel, D.G., *J. Electrochem. Soc.*, 126, 479 (1979).

TABLE 16.8 Summary of the Process Steps Required for Anisotropic Etching of a Membrane [Elwenspoek et al., 1994]

Process	Duration	Process Temperature (°C)
Oxidation	Variable (hours)	900–1200
Spinning at 5000 rpm	20–30 sec	Room temperature
Prebake	10 min	90
Exposure	20 sec	Room temperature
Develop	1 min	Room temperature
Post-bake	20 min	120
Stripping of oxide (BHF:1:7)	±10 min	Room temperature
Stripping resist (acetone)	10–30 sec	Room temperature
RCA1 (NH$_3$(25%) + H$_2$O + H$_2$O$_2$:1:5:1)	10 min	Boiling
RCA2 (HCl + H$_2$O + H$_2$O$_2$:1:6:1)	10 min	Boiling
HF-dip (2% HF)	10 sec	Room temperature
Anisotropic etch	From minutes up to one day	70–100

The development of anisotropic etchants solved the lateral dimension control lacking in isotropic etchants. Lateral mask geometries on planar photoengraved substrates can be controlled with an accuracy and reproducibility of 0.5 µm or better, and the anisotropic nature of the etchant allows this accuracy to be translated into control of the vertical etch profile. Different etch stop techniques, needed to control the membrane thickness, are available. The invention of these etch stop techniques truly made items as shown in Figure 16.1 manufacturable.

While anisotropic etchants solve the lateral control problem associated with deep etching, they are not without problems. They are slower, even in the fast etching <100> direction, with etch rates of 1 µm/min or less. That means that etching through a wafer is a time-consuming process: to etch through a 300-µm-thick wafer one needs 5 hours. They also must be run hot to achieve these etch rates (85 to 115°C), precluding many simple masking options. Like the isotropic etchants, their etch rates are temperature sensitive; however, they are not particularly agitation sensitive, considered to be a major advantage.

Anisotropic Etchants

A wide variety of etchants have been used for anisotropic etching of silicon, including alkaline aqueous solutions of KOH, NaOH, LiOH, CsOH, NH_4OH, and quaternary ammonium hydroxides, with the possible addition of alcohol. Alkaline organics such as ethylenediamine, choline (trimethyl-2-hydroxy-ethyl ammonium hydroxide) or hydrazine with additives such as pyrocathechol and pyrazine are employed as well. Etching of silicon occurs without the application of an external voltage and is dopant insensitive over several orders of magnitude, but in a curious contradiction to its suggested chemical nature, it has been shown to be bias dependent [Allongue et al., 1993; Palik et al., 1987]. This contradiction will be explained with the help of the chemical models presented below.

Alcohols such as propanol and isopropanol butanol typically slow the attack on Si [Linde and Austin, 1992; Price, 1973]. The role of pyrocathechol [Finne and Klein, 1967] is to speed up the etch rate through complexation of the reaction products. Additives such as pyrazine and quinone have been described as catalysts by some [Reisman et al., 1979]; but this is contested by other authors [Seidel et al., 1990]. The etch rate in anisotropic etching is reaction rate controlled and thus temperature dependent. The etch rate for all planes increases with temperature and the surface roughness decreases with increasing temperature, so etching at the higher temperatures gives the best results. In practice, etch temperatures of 80 to 85°C are used to avoid solvent evaporation and temperature gradients in the solution.

Arrhenius Plots for Anisotropic Etching

A typical set of Arrhenius plots for <100>, <110> and <111> silicon etching in an anisotropic etchant (EDP, or ethylene-diamine/pyrocathechol) is shown in Figure 16.26 [Seidel et al., 1990]. It is seen that the temperature dependence of the etch rate is quite large and is less dependent on orientation. The slope differs for the different planes, i.e., (111) > (100) > (110). Lower activation energies in Arrhenius plots correspond to higher etch rates. The anisotropy ratio (AR) derived from this figure is

$$AR = (hkl)_1 \text{ etch rate}/(hkl)_2 \text{ etch rate} \qquad (16.29)$$

The AR is approximately 1 for isotropic etchants and can be as high as 400/200/1 for (110)/(100)/(111) in 50 wt% KOH/H_2O at 85°C. Generally, the activation energies of the etch rates of EDP are smaller than those of KOH. The (111) planes always etch slowest but the sequence for (100) and (110) can be reversed (e.g., 50/200/8 in 55 vol% ethylenediamine ED/H_2O; also at 85°C). The (110) Si plane etches 8 times slower and the (111) 8 times faster in KOH/H_2O than in ED/H_2O, while the (100) etches at the same rate [Kendall and Guel, 1985]. Working with alcohols and other organic additives often changes the relative etching rate of the different Si planes. Along this line, Seidel et al. (1990a; 1990b) found that the decrease in etch rate by adding isopropyl alcohol to a KOH solution was 20% for <100>, but almost 90% for <110>. As a result of the much stronger decrease of the etch rate on a (110) surface, the etch ratio of (100):(110) is reversed.

MEMS Fabrication

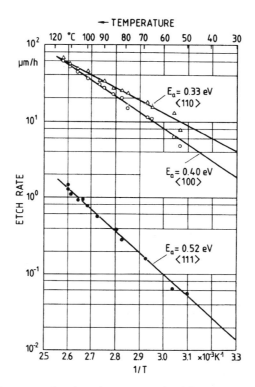

FIGURE 16.26 Vertical etch rates as a function of temperature for different crystal orientations: (100), (110) and (111). Etch solution is EDP (133 ml H$_2$O, 160 g pyrocatechol, 6 g pyrazine and 1 l ED). (From Seidel, H. et al., *J. Electrochem. Soc.*, 137, 3612–3626, 1990. With permission.)

Important Anisotropic Etchant Systems

In choosing an etchant, a variety of issues must be considered:

- Ease of handling
- Toxicity
- Etch rate
- Desired topology of the etched bottom surface
- IC-compatibility
- Etch stop
- Etch selectivity over other materials
- Mask material and thickness of the mask

The principal characteristics of four different anisotropic etchants are listed in Table 16.9. The most commonly used are KOH [Stoller and Wolff, 1966; Stoller, 1970; Forster and Singleton, 1966; Kenney, 1967; Lepselter, 1966; 1967; Waggener, 1970; Kragness and Waggener, 1973; Waggener et al., 1967a; 1967b; Bean and Ruyan, 1977; Rodgers et al., 1976; 1977; Ammar and Rodgers, 1980; Seidel et al., 1969; 1990; Kendall and de Guel, 1985; Lee, 1990; Noworolski et al., 1995; Waggener and Dalton, 1972; Weirauch, 1975; Clemens, 1973; Bean et al., 1974; Declercq et al., 1975] and ethylene-diamine/ pyrocatechol + water (EDP) [Finne and Klein, 1967; Reisman et al., 1979; Wu et al., 1986]; hydrazine-water rarely is used [Declercq et al., 1975; Mehregany and Senturia, 1988]. More recently, quaternary ammonium hydroxide solutions such as tetraethyl ammonium hydroxide (TEAH) have become more popular [Asano et al., 1976; Tabata et al., 1990]. Each has its advantages and problems. NaOH is not used much anymore [Pugacz-Muraszkiewicz and Hammond, 1977].

TABLE 16.9 Principal Characteristics of Four Different Anisotropic Etchants[a]

Etchant/Diluent/Additives/ Temperature	Etch Stop	Etch Rate (100) (μm/min)	Etch Rate Ratio (100)/(111)	Remarks	Mask (Etch Rate)
KOH/water, isopropyl alcohol additive, 85°C	$B > 10^{20}$ cm^{-3} reduces etch rate by 20	1.4	400 and 600 for (110)/(111)	IC incompatible, avoid eye contact, etches oxide fast, lots of H_2 bubbles	Photoresist (shallow etch at room temperature); Si_3N_4 (not attacked); SiO_2 (28 Å/min)
Ethylene diamine pyrocatechol (water), pyrazine additive, 115°C	$\geq 5 \times 10^{19}$ cm^{-3} reduces the etch rate by 50	1.25	35	Toxic, ages fast, O_2 must be excluded, few H_2 bubbles, silicates may precipitate	SiO_2 (2–5 Å/min); Si_3N_4 (1 Å/min); Ta, Au, Cr, Ag, Cu
Tetramethyl ammonium hydroxide (TMAH) (water), 90°C	$> 4 \times 10^{20}$ cm^{-3} reduces etch rate by 40	1	From 12.5 to 50	IC compatible, easy to handle, smooth surface finish, few studies	SiO_2 etch rate is 4 orders of magnitude lower than (100) Si LPCVD Si_3N_4
N_2H_4/(water), isopropyl alcohol, 115°C	$> 1.5 \times 10^{20}$ cm^{-3} practically stops the etch	3.0	10	Toxic and explosive, okay at 50% water	SiO_2 (<2 Å/min) and most metallic films; does not attack Al according to some authors [Wise, 1985]

[a] Given the many possible variables, the data in the table are only typical examples.

Hydrazine-water is explosive at high hydrazine concentrations (rocket fuel) and is a suspected carcinogen. Its use should be avoided for safety reasons. A 50% hydrazine/water solution is stable, though, and, according to Mehregany and Senturia (1988), excellent surface quality and sharply defined corners are obtained in Si. Also on the positive side, the etchant has a very low SiO_2 etch rate and will not attack most metal masks except for Al, Cu and Zn. According to Wise, on the other hand, Al does not etch in hydrazine either, but the etch produces rough Si surfaces [Wise, 1985].

Ethylenediamine in EDP reportedly causes allergic respiratory sensitization, and pyrocathechol is described as a toxic corrosive. The material is also optically dense, making end-point detection harder, and it ages quickly; if the etchant reacts with oxygen, the liquid turns to a red-brown color and it loses its good properties. If cooled down after etching, one gets precipitation of silicates in the solution. Sometimes one even gets precipitation during etching, spoiling the results. When preparing the solution, the last ingredient added should be the water, since water addition causes the oxygen sensitivity. All of the above make the etchant quite difficult to handle. But, a variety of masking materials can be used in conjunction with this etchant and it is less toxic than hydrazine. No sodium or potassium contamination occurs with this compound and the etch rate of SiO_2 is slow. The ratio of etch rates of Si and SiO_2 using EDP can be as large as 5000:1 (about 2 Å/min of SiO_2 compared to 1 μm/min of Si) which is much larger than the ratio in KOH a ratio of as high as 400:1 has been reported [Bean, 1978]. Importantly, the etch rate slows down at a lower boron concentration than with KOH. A typical fastest-to-slowest hierarchy of Si etch rates with EDP at 85°C according to Barth (1984) is (110) > (411) > (311) > (511) > (211) > (100) > (331) > (221) > (111).

The simple KOH water system is the most popular etchant. A KOH etch, in near saturated solutions (1:1 in water by weight) at 80°C, produces a uniform and bright surface. Nonuniformity of etch rate gets considerably worse above 80°C. Plenty of bubbles are seen emerging from the Si wafer while etching in KOH. The etching selectivity between Si and SiO_2 is not very good in KOH, as it etches SiO_2 too fast. KOH is also incompatible with the IC fabrication process and can cause blindness when it gets in contact with the eyes. The etch rate for low index planes is maximal at around 4 M (see Figure 16.27A [Peeters, 1994] and Lambrechts and Sansen [1992]). The surface roughness continuously decreases with increasing concentration as can be gleaned from Figure 16.27B. Since the difference in etch rates for different KOH concentrations is small, a highly concentrated KOH (e.g., 7 M) is preferred to obtain a smooth surface on low index planes.

Except at very high concentrations of KOH, the etched (100) plane becomes rougher the longer one etches. This is thought to be due to the development of hydrogen bubbles, which hinder the transport of fresh solution to the silicon surface [Ternz, 1988]. Average roughness, R_a, is influenced strongly by fluid agitation. Stirring can reduce the R_a values over an order of magnitude, probably caused by the more efficient removal of hydrogen bubbles from the etching surface when stirring [Gravesen, 1986]. The silicon etch rate as a function of KOH concentration is shown in Figure 16.27C [Seidel et al., 1990].

Herr and Baltes (1991) found that the high-index crystal planes exhibit the highest etch rates for 6 M KOH and that for lower concentrations the etch bottoms disintegrate into microfacets. In 6 M KOH, the etch-rate order is (311) > (144) > (411) > (133) > (211) > (122). These authors could not correlate the particular etch rate sequence with the measured activation energies. This is in contrast to lower activation energies corresponding to higher etching rates for low index planes as shown in Figure 16.26. Their results obtained on large open area structures differ significantly from previous ones obtained by underetching special mask patterns. The vertical etching rates obtained here are substantially higher than the underetching rates described elsewhere, and the etch rate sequence for different planes is also significantly different. These results suggest that crevice effects may plan an important role in anisotropic etching.

Besides KOH [Clark and Edell, 1987], other hydroxides have been used, including NaOH [Allongue et al., 1993; Pugacz-Muraszkiewicz and Hammond, 1977], CsOH [Clark et al., 1988] and NH_4OH [Schnakenberg et al., 1990]. A major disadvantage of KOH is the presence of alkali ions, which are detrimental to the fabrication of sensitive electronic parts. Work is under way to find anisotropic etchants

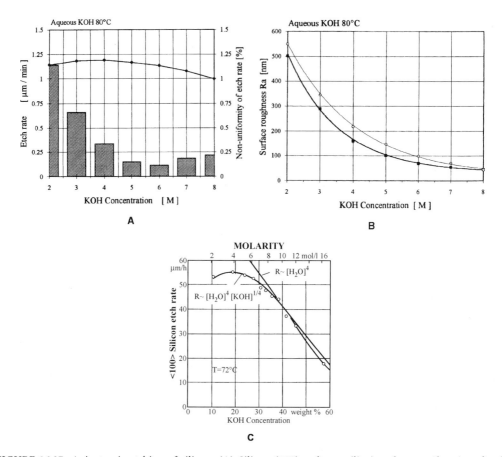

FIGURE 16.27 Anisotropic etching of silicon. (A) Silicon (100) etch rate (line) and nonuniformity of etch rate (column) in KOH at 80°C as a function of KOH concentration. The etch rate for all low index planes is maximal at around 4 M. (B) Silicon (100) surface roughness (Ra) in aqueous KOH at 80°C as a function of concentration for a 1-hour etch time (thin line) and for an etch depth of 60 µm (thick line). (C) Silicon (100) etch rate as a function of KOH concentration at a temperature of 72°C. (A and B from Peeters, E., Ph.D. thesis, 1996, KUL, Belgium. With permission. C from Seidel et al., *J. Electrochem. Soc.*, 137, 3612–3626, 1990. With permission.)

that are more compatible with CMOS processing and that are neither toxic nor harmful. Two examples are ammonium hydroxide-water (AHW) mixtures [Schnakenberg et al., 1990] and tetramethyl ammonium hydroxide-water (TMAHW) mixtures [Tabata et al., 1990; Schnakenberg et al., 1990]. TMAHW solutions do not decompose at temperatures below 130°C, a very important feature from the viewpoint of production. They are nontoxic and can be handled easily. TMAHW solutions also exhibit excellent selectivity to silicon oxide and silicon nitride. At a solution temperature of 90°C and 22 wt% TMAH, a maximum (100) silicon etch rate of 1.0 µm/min is observed, 1.4 µm/min for (110) planes (this is higher than those observed with EDP, AHW, hydrazine water and tetraethyl ammonium hydroxide [TEA], but slower than those observed for KOH and an anisotropy ratio, AR(100)/(111), of between 12.5 and 50 [Tabata et al., 1992]). From the viewpoint of fabricating various silicon sensors and actuators, a concentration above 22 wt% is preferable, since lower concentrations result in larger roughness on the etched surface. However, higher concentrations give a lower etch rate and lower etch ratio (100)/(111). Tabata (1995) also studied the etching characteristics of pH-controlled TMAHW. To obtain a low aluminum etching rate of 0.01 µm/min, pH values below 12 for 22 wt% TMAHW were required. At those pH values the Si(100) etching rate is 0.7 µm/min.

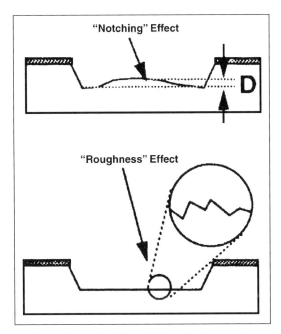

Macroscopic roughness (notching effect) and microscopic roughness.

Surface Roughness and Notching

Anisotropic etchants frequently leave too rough a surface behind, and a slight isotropic etch is used to 'touch-up'. A distinction must be made between **macroscopic** and **microscopic roughness**. Macroscopic roughness, also referred to as notching or pillowing, results when centers of exposed areas etch with a seemingly lower average speed compared with the borders of the areas, so that the corners between sidewalls and (100) ground planes are accentuated. Membranes or double-sided clamped beams (microbridges) therefore tend to be thinner close to the clamped edges than in the center of the structure. This difference can be as large as 1 to 2 μm, which is quite considerable if one is etching 10 to 20 μm thick structures. Notching increases linearly with etch depth but decreases with higher concentrations of KOH. The microscopic smoothness of originally mirror-like polished wafers can also be degraded into microscopic roughness. It is this type of short-range roughness we referred to in discussing Figure 16.27B above.

Masking for Anisotropic Etchants

Etching through a whole wafer (400 to 600 μm) takes several hours (a typical wet anisotropic etch rate being 1.1 μm/min), definitely not a fast process. When using KOH, SiO_2 cannot be used as a masking material for features requiring that long an exposure to the etchants. The SiO_2 etch rate as a function of KOH concentration at 60°C is shown in Figure 16.28. There is a distinct maximum at 35 wt% KOH of nearly 80 nm/hr. The shape of this curve will be explained further below on the basis of Seidel et al.'s model. Experiments have shown that even a 1.5-μm-thick oxide is not sufficient for the complete etching of a 380-μm-thick wafer (6 hours) because of pinholes in the oxide [Lambrechts and Sansen, 1992]. The etch rate of thermally grown SiO_2 in $KOH-H_2O$ somewhat varies and apparently depends not only on the quality of the oxide, but also on the etching container and the age of the etching solution, as well as other factors [Kendall, 1979]. The Si/SiO_2 selectivity ratio at 80°C in 7 M KOH is 30 ± 5. This ratio increases with decreasing temperature; reducing the temperature from 80 to 60°C increases the selectivity ratio from 30 to 95 in 7 M KOH [Kendall, 1975]. Thermal oxides are under strong compressive stress due to the fact that in the oxide layer one silicon atom takes nearly twice as much space as in single crystalline Si (see also Madou [1997, chap. 3]). This might have severe consequences; for example, if the oxide mask is stripped on one side of the wafer, the wafer will bend. Atmospheric pressure chemical vapor deposited (APCVD) SiO_2 tends to exhibit pinholes and etches much faster than thermal oxide.

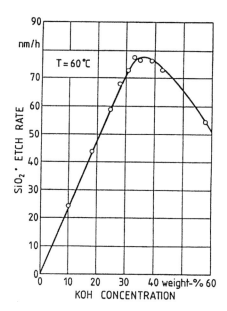

FIGURE 16.28 The SiO_2 etch rate in nm/hr as a function of KOH concentration at 60°C. (From Seidel, H. et al., *J. Electrochem. Soc.*, 137, 3612–3626, 1990. With permission.)

Annealing of APCVD oxide removes the pinholes but the etch rate in KOH remains greater by a factor of 2 to 3 than that of thermal oxide. Low pressure chemical vapor deposited (LPCVD) oxide is a mask material of comparable quality as thermal oxide. The etch rate of SiO_2 in EDP is smaller by two orders of magnitude than in KOH.

For prolonged KOH etching, a high density silicon nitride mask has to be deposited. A low pressure chemical vapor deposited (LPCVD) nitride generally serves better for this purpose than a less dense plasma deposited nitride [Puers, 1991]. With an etch rate of less than 0.1 nm/min, a 400-Å layer of LPCVD nitride suffices to mask against KOH etchant. The etch selectivity Si/Si_3N_4 was found to be better than 10^4 in 7 M KOH at 80°C. The nitride also acts as a good ion-diffusion barrier, protecting sensitive electronic parts. Nitride can easily be patterned with photoresist and etched in a CF_4/O_2-based plasma or, in a more severe process, in H_3PO_4 at 180°C (10 nm/min) [Buttgenbach, 1991]. Nitride films are typically under a tensile stress of about 1×10^9 Pa. If in the overall processing of the devices, nitride deposition does not pose a problem, KOH emerges as the preferential anisotropic wet etchant. For dopant dependent etching, EDP is the better etchant and generally better suited for deep etching since its oxide etch rate is negligible (<5 Å/min).

Oxide and nitride are masking for anisotropic etchants to varying degrees with both mask types being used. When these layers are used to terminate an etch in the [100] direction, a low etch rate of the mask layer allows overetching of silicon, to compensate for wafer thickness variations. A KOH solution etches SiO_2 at a relatively fast rate of 1.4 to 3 nm/min so that Si_3N_4 or Au/Cr must be used as a mask against KOH for deep and long etching.

Backside Protection

In many cases it is necessary to protect the backside of a wafer from an isotropic or anisotropic etchant. The backside is either mechanically or chemically protected. In the mechanical method the wafer is held in a holder, often made from Teflon. The wafer is fixed between Teflon-coated O-rings which are carefully aligned in order to avoid mechanical stress in the wafer. In the chemical method, waxes or other organic coatings are spun onto the back side of the wafer. Two wafers may be glued back-to-back for faster processing.

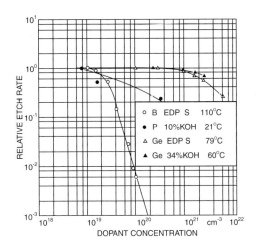

FIGURE 16.29 Relative etch rate for (100) Si in EDP and KOH solutions as a function of concentration of boron, phosphorus and germanium. (From Seidel, H. et al., *J. Electrochem. Soc.*, 137, 3626–3632, 1990. With permission.)

Etch Rate and Etch Stops

The Si etch rate, R, as a function of KOH concentration at 72°C, was shown in Figure 16.27C. The etch rate has a maximum at about 20% KOH. The best fit for this experimentally determined etch rate, for most KOH concentrations, is [Seidel et al., 1990a; 1990b]

$$R = k[H_2O]^4[KOH]^{\frac{1}{4}} \quad (16.30)$$

Any model of anisotropic etching will have to explain this peculiar dependency on the water and KOH concentration, as well as the fact that all anisotropic etchant systems of Table 16.9 exhibit drastically reduced etch rates for high boron concentrations in silicon ($\geq 5 \times 10^{19}$ cm^{-3} solid solubility limit). Other impurities (P, Ge) also reduce the etch rate, but at much higher concentrations (see Figure 16.29 [Seidel et al., 1990]). Boron typically is incorporated using ion implantation (thin layers) or liquid/solid source deposition (thick layers >1 µm). These doped layers are used as very effective etch stop layers (see below). Hydrazine or EDP, which displays a smaller (100)-to-(111) etch rate ratio (~35) than KOH, exhibits a stronger boron concentration dependency. The etch rate in KOH is reduced by a factor of 5 to 100 for a boron concentration larger than 10^{20} cm^{-3}. When etching in EDP, the factor climbs to 250 [Bogh, 1971]. With TMAHW solutions, the Si etch rate decreases to 0.01 µm/min for boron concentrations of about 4×10^{20} cm^{-3} [Steinsland et al., 1995]. The mechanism eludes us, but Seidel et al.'s model (see below) gives the most plausible explanation for now. Some of the different mechanisms to explain etch stop effects that have been suggested follow:

1. Several observations suggest that doping leads to a more readily oxidized Si surface. Highly boron- or phosphorus-doped silicon in aqueous KOH spontaneously can form a thin passivating oxide layer [Palik et al., 1982; 1985]. The boron-oxides and -hydroxides initially generated on the silicon surface are not soluble in KOH or EDP etchants [Petersen, 1982]. The substitutional boron creates local tensile stress in the silicon, increasing the bond strength so that a passivating oxide might be more readily formed at higher boron concentrations. Boron-doped silicon has a high defect density (slip planes), encouraging oxide growth.
2. Electrons produced during oxidation of silicon are needed in a subsequent reduction step (hydrogen evolution in Reaction 16.2). When the hole density passes 10^{19} cm^{-3} these electrons combine with holes instead, thus stopping the reduction process [Palik et al., 1982]. Seidel et al.'s model follows this explanation (see below).

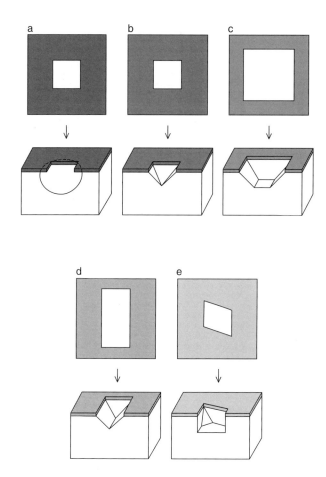

FIGURE 16.30 Isotropic and anisotropic etched features in <100> and <110> wafers. (a) isotropic etch; (b) to (e) anisotropic etch. (a) to (d): <100> oriented wafers and (e): <110> oriented wafer.

3. Silicon doped with boron is under tension as the smaller boron atoms enter the lattice substitutionally. The large local tensile stress at high boron concentration makes it energetically more favorable for the excess boron (above 5×10^{19} cm^{-3}) to enter interstitial sites. The strong B-Si bonds bind the lattice rigidly. With high enough doping the high binding energy can stop etching [Petersen, 1982]. This hypothesis is similar to item 1, except that no oxide formation is invoked.

In what follows we review the results of some typical anisotropic etching experiments.

16.5.1.4 Anisotropically Etched Structures

Examples
In Figure 16.30 we compare a wet isotropic etch (a) with examples of anisotropic etches (b to e). In the anisotropic etching examples, a square (b and c) and a rectangular pattern (d) are defined in an oxide mask with sides aligned along the <110> directions on a <100>-oriented silicon surface. The square openings are precisely aligned (within one or two degrees) with the <110> directions on the (100) wafer surface to obtain pits that conform exactly to the oxide mask rather than undercutting it. Most (100) silicon wafers have a main flat parallel to a <110> direction in the crystal, allowing for an easy alignment of the mask (see Figure 16.5). Etching with the square pattern results in a pit with well-defined {111} sidewalls (at angles of 54.74° to the surface) and a (100) bottom.

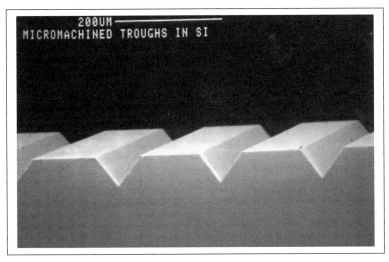

Long V-shaped grooves in a (100) Si wafer.

The dimensions of the hole at the bottom of the pit, as we saw above, are given by Eq. (16.2). The larger the square opening in the mask, the deeper the point where the {111} sidewalls of the pit intersect. If the oxide opening is wide enough, i.e., $W_m > \sqrt{2}\, z$ (with $z = 600$ μm for a typical 6-in. wafer, this means $W_m > 849$ μm), the {111} planes do not intersect within the wafer (see also earlier in this chapter). The etched pit in this particular case extends all the way through the wafer, creating a small square opening on the bottom surface. As shown in Figure 16.30 (b to d), no under-etching of the etch mask is observed due to the perfect alignment of the concave oxide mask opening with the <110> direction. In Figure 16.30a, an undercutting isotropic etch (acidic) is shown. Misalignment in the case of an anisotropic etch still results in pyramidical pits, but the mask will also be severely undercut. A rectangular pattern aligned along the <110> directions on a <100> wafer leads to **long V-shaped grooves** (see Figure 16.30d) or an open slit, depending on the width of the opening in the oxide mask.

Using a properly aligned mask on a <110> wafer, holes with four vertical walls ({111} planes) result (see Figure 16.30e and Figure 16.31A and C). Figure 16.31B shows that a slight mask misorientation

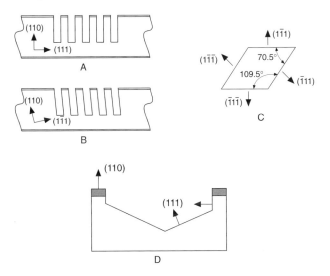

FIGURE 16.31 Anisotropic etching of <110> wafers. (A) Closely spaced grooves on correctly oriented (110) surface. (B) Closely spaced grooves on misoriented <110> wafer. (C) Orientations of the {111} planes looking down on a (110) wafer. (D) Shallow slanted (111) planes eventually form the bottom of the etched cavity.

leads to all skewed sidewalls. A U-groove based on a rectangular mask with the long sides along the <111> directions and anisotropically etched in (110) silicon has a complex shape delineated by six {111} planes, four vertical and two slanted (see Figure 16.31D). Before emergence of {111} planes, the U-groove is defined by four vertical (111) planes and a horizontal (110) bottom. Self-stopping occurs when the tilted end planes intersect at the bottom of the groove. It is easy to etch long, narrow U-grooves very deeply into a <110> silicon wafer. However, it is impossible to etch a short, narrow U-groove deeply into a slice of silicon [Kendall, 1979], because the narrow dimension of the groove is quickly limited by slow-etching {111} planes that subtend an angle of 35° to the surface and cause etch termination. At a groove of length $L = 1$ mm on the top surface, etching will stop when it reaches a depth of 0.289 mm, i.e., $D_{max} = L/2\sqrt{3}$. For very long grooves, the tilted end planes are too far apart to intersect in practical cases, making the end effects negligible compared to the remaining U-shaped part of the groove.

A laser can be used to melt or 'spoil' the shallow (111) surfaces, making it possible to etch deep vertical-walled holes through a (110) wafer as shown in Figure 16.32A [Schumacher et al., 1994; Barth et al., 1985; Seidel and Csepregi, 1988]. The technique is illustrated in Figure 16.32B. The absorbed energy of a Nd:YAG laser beam causes a local melting or evaporation zone enabling etchants to etch the shallow (111) planes in the line-of-sight of the laser. Etching proceeds until 'unspoiled' (111) planes are encountered. Some interesting resulting possibilities, including partially closed microchannels, are shown in Figure 16.32C [Schumacher et al., 1994; Alavi et al., 1991]. Note that with this method it is possible to use <111> wafers for micromachining. The light of the Nd:YAG laser is very well suited for this micromachining technique due to the 1.17-eV photon energy, just exceeding the band gap energy of Si. Details on this laser machining process can be found, for example, in Alavi et al. (1991; 1992).

Especially when machining surface structures by undercutting, the orientation of the wafer is of extreme importance. Consider, for example, the formation of a bridge in Figure 16.33A [Barth et al., 1985]. When using a (100) surface, a suspension bridge cannot form across the etched V-groove; two independent truncated V-grooves flanking a mesa structure result instead. To form a suspended bridge it must be oriented away from the <110> direction. This in contrast with a (110) wafer where a microbridge crossing a V-groove with a 90° angle will be undercut. Convex corners will be undercut by etchant, allowing formation of cantilevers as shown in Figure 16.33B. The diving board shown forms by undercutting starting at the convex corners.

To create vertical (100) faces, as shown in Figure 16.10, in general only KOH works (not EDP or TMAHW) and it has to happen in high-selectivity conditions (low temperature, low concentration: 25 wt% KOH, 60°C). Interestingly, high concentration KOH (45 wt%) at higher temperatures (80°C) produces a smooth sidewall, controllable and repeatable at an angle of 80°. EDP produces 45° angled planes and TMAHW usually makes a 30° angle [Palik et al., 1985].

Alignment Patterns
When alignment of a pattern is critical, pre-etch alignment targets become useful to delineate the planes of interest since the wafer flats often are aligned to ±1° only. In order to find the proper alignment for the mask, a test pattern of closely spaced lines can be etched (see Figure 16.34). The groove with the best vertical walls determines the proper final mask orientation. Along this line, Ciarlo (1987) made a set of lines 3 mm long and 8 µm wide, fanning out like spokes in a wagon wheel at angles 0.1° apart. This target was printed near the perimeter of the wafer and then etched 100 µm into the surface. Again, by evaluating the undercut in this target the correct crystal direction could be determined. Alignment with better than 0.05° accuracy was accomplished this way. Similarly, to obtain detailed experimental data on crystal orientation dependence of etch rates, Seidel et al. (1990) used a wagon-wheel or star-shaped mask (e.g., made from CVD-Si_3N_4; SiH_4 and NH_3 at 900°C), consisting of radially divergent segments with an angular separation of one degree. Yet finer 0.1° patterns were made around the principal crystal directions. The etch pattern emerging on a <100> oriented wafer covered with such a mask is shown in Figure 16.35A. The blossom-like figure is due to the total underetching of the passivation layer in the vicinity of the center of the wagon wheel,

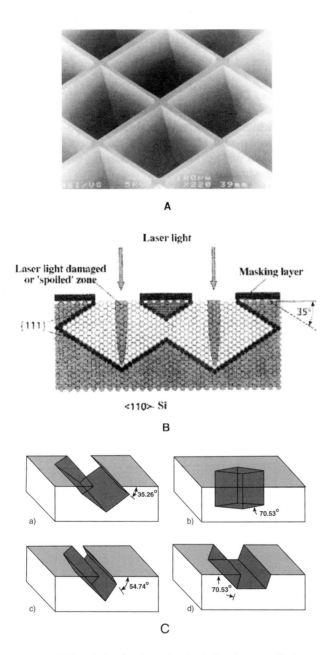

FIGURE 16.32 Laser/KOH machining. (A) Holes through a (110) Si wafer created by laser spoiling and subsequent KOH etching. The two sets of (111) planes making an angle of 70° are the vertical walls of the hole. The (111) planes making a 35° angle with the surface tend to limit the depth of the hole but laser spoiling enables one to etch all the way through the wafer (see also B [c]). (B) After laser spoiling, the line-of-sight (111) planes are spoiled and etching proceeds until unspoiled (111) planes are reached. (C) Some of the possible features rendered by laser spoiling. (From Schumacher, A. et al., *Technische Rundschau*, 86, 20–23, 1994. With permission.)

leaving an area of bare exposed Si. The radial extension of the bare Si area depends on the crystal orientation of the individual segments, leading to a different amount of total underetching. The observation of these blossom-like patterns was used for qualitative guidance of etching rates only. In order to establish quantitative numbers for the lateral etch rates, the width, w, of the overhanging passivation layer was measured with an optical line-width measurement system (see Figure 16.35B).

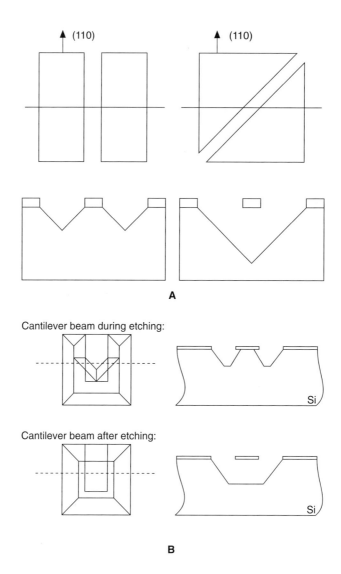

FIGURE 16.33 How to make (A) a suspension bridge from a (100) Si wafer and (B) a diving board from a (110) Si wafer [Barth et al., 1985].

Laser beam reflection was used to identify the crystal planes and ellipsometry was used to monitor the etching rate of the mask itself. Lateral etch rates determined in this way on <100>- and <110>- oriented wafers at 95°C in EDP (470 ml water, 11 ED, 176 g pyrocatechol) and at 78°C in a 50% KOH solution are shown in Figure 16.36. Etch rates shown are normal to the actual crystal surface and are conveniently described in a polar plot in which the distance from the origin to the polar plot surface (or curve in two dimensions) indicates the etch rate for that particular direction. Note the deep minima at the {111} planes. It can also be seen that in KOH the peak etch rates are more pronounced. A further difference is that with EDP the minimum at {111} planes is steeper than with KOH. For both EDP and KOH, the etch rate depends linearly on misalignment. All the above observations have important consequences for the interpretation of anisotropy of an etch (see below). The difference between KOH and EDP etching behavior around the {111} minima has the direct practical consequence that it is more important for etching in EDP to align the crystallographic direction more precisely than in KOH [Elwenspoek et al., 1994].

FIGURE 16.34 Test pattern of U-grooves in a <110> wafer to help in the alignment of the mask; final alignment is done with the groove that exhibits the most perfect long perpendicular walls.

When determining etch rates without using underetching masks but by using vertical etching of beveled silicon samples, results are quite different than when working with masked silicon [Herr and Baltes, 1991]. The etch rates on open areas of beveled structures are much larger than in underetching experiments with masked silicon, and different crystal planes develop. Herr et al. conclude that crevice effects may play an important role in anisotropic etching. Elwenspoek et al.'s model [Elwenspoek, 1993; Elwenspoek et al. 1994], analyzed below, is the only model which predicts such a crevice effect. He explains why, when etchants are in a small restricted crevice area and are not refreshed fast enough, etching rates slow down and increase anisotropy.

16.5.2 Chemical Etching Models

16.5.2.1 Introduction

Lots of conflicting data exist in the literature on the anisotropic etch rates of the different Si planes, especially for the higher index planes. This is not too surprising, given the multiple parameters influencing individual results: temperature, stirring, size of etching feature (i.e., crevice effect), KOH concentration, addition of alcohols and other organics, surface defects, complexing agents, surfactants, pH, cation influence etc. More rigorous experimentation and standardization will be needed, as well as better etching models, to better understand the influence of all these parameters on etch rates.

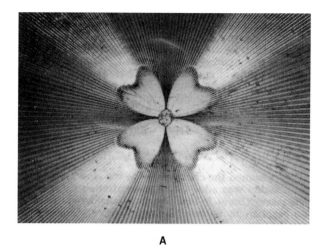

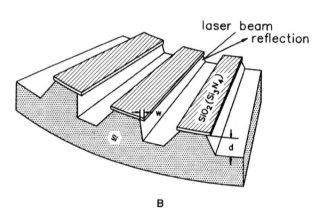

FIGURE 16.35 (A) Etch pattern emerging on a wagon wheel-masked, <100>-oriented Si wafer after etching in an EDP solution. (B) Schematic cross section of a silicon test chip covered with a wagon wheel-shaped masking pattern after etching. The measurement of 'w' is used to construct polar diagrams of lateral underetch rates as shown in Figure 16.36. (From Seidel et al., *J. Electrochem. Soc.*, 177, 3612–3626, 1990. With permission.)

Several chemical models explaining the anisotropy in etching rates for the different Si orientations have been proposed. Presently we will list all of the proposed models and compare the two most recent and most detailed models (the one by Seidel et al. [1990a; 1990b] and the one by Elwenspoek [1993] and Elwenspoek et al. [1994]). Different Si crystal properties have been correlated with the anisotropy in silicon etching.

1. It has been observed that the {111} Si planes present the highest density of atoms per cm^2 to the etchant and that the atoms are oriented such that ***three bonds are below the plane***. It is possible that these bonds become chemically shielded by surface bonded (OH) or oxygen, thereby slowing the etch rate.
2. It also has been suggested that etch rate correlates with available bond density, the surfaces with the highest bond density etching faster [Price, 1973]. The available bond densities in Si and other diamond structures follow the sequence 1:0.71:0.58 for the {100}:{110}:{111} surfaces. However, Kendall (1979) commented that bond density alone is an unlikely explanation because of the magnitude of etching anisotropy (e.g., a factor of 400), compared to the bond density variations of at most a factor of 2.

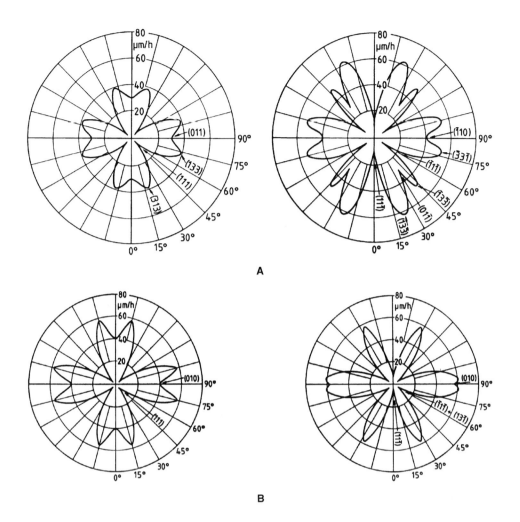

FIGURE 16.36 Lateral underetch rates as a function of orientation for (A) EDP (470 ml water, 1 l ED, 176 g pyrocatechol) at 95°C. (B) KOH (50% solution) at 78°C. Left, <100>- and right, <110>-oriented Si wafers. (From Seidel, H. et al., *J. Electrochem. Soc.*, 137, 3612–3626, 1990. With permission.)

On {111} planes three backbonds are below the plane.

3. Kendall (1979) explains the slow etching of {111} planes on the basis of their faster oxidation during etching; this does not happen on the other faces, due to greater distance of the atoms on planes other than (111). Since they oxidize faster, these planes may be better protected against etching. The oxidation rate in particular follows the sequence {111} > {110} > {100}, and the etch rate often follows the

reverse sequence (see also Madou [1997, chap. 3 on Si oxidation]). In the most used KOH-H$_2$O, however, the sequence is {110} > {100} > {111}.

4. In yet another model, it is assumed that the anisotropy is due to differences in activation energies and backbond geometries on different Si surfaces [Glembocki et al., 1985].
5. Seidel et al.'s model [Seidel et al., 1990a; 1990b] supports the previous explanation. They detail a process to explain anisotropy based on the difference in energy levels of backbond-associated surface states for different crystal orientations.
6. Finally, Elwenspoek (1993) and Elwenspoek et al. (1994) propose that it is the degree of atomic smoothness of the various surfaces that is responsible for the anisotropy of the etch rates. Basically, this group argues that the kinetics of smooth faces (the (111) plane is atomically flat) is controlled by a nucleation barrier that is absent on rough surfaces. The latter, therefore, would etch faster by orders of magnitude.

The reason why acidic media lead to isotropic etching and alkaline media to anisotropic etching was, until recently, not addressed in any of the models surveyed. In the following we will give our own model as well as Elwenspoek et al.'s to explain isotropic vs. anisotropic etching behavior.

It is our hope that the reading of this section will inspire more detailed electrochemistry work on Si electrodes. The refining of an etching model will be of invaluable help in writing more predictive Si etching software code.

16.5.2.2 Seidel et al.'s Model

Seidel et al.'s model is based on the fluctuating energy level model of the silicon/electrolyte interface and assumes the injection of electrons in the conduction band of Si during the etching process. Consider the situation of a piece of Si immersed in a solution without applied bias at open circuit. After immersion of the silicon crystal into the alkaline electrolyte a negative excess charge builds up on the surface due to the higher original Fermi level of the H$_2$O/OH$^-$ redox couple as compared to the Fermi level of the solid, i.e., the work function difference is equalized. This leads to a downward bending of the energy bands on the solid surface for both p- and n-type silicon (Figure 16.37A and B). The downward bending is more pronounced for p-type than for n-type due to the initially larger difference of the Fermi levels between the solid and the electrolyte.

Next, hydroxyl ions cause the Si surface to oxidize, consuming water and liberating hydrogen in the process. The detailed steps, based on suggestions by Palik et al. (1985), are:

$$\text{Si} + 2\text{OH}^- \rightarrow \text{Si(OH)}_2^{2+} + 2e^-$$
$$\text{Si(OH)}_2^{2+} + 2\text{OH}^- \rightarrow \text{Si(OH)}_4 + 2e^- \quad \text{(Reaction 16.4)}$$
$$\text{Si(OH)}_4 + 4e^- + 4\text{H}_2\text{O} \rightarrow \text{Si(OH)}_6^{2-} + 2\text{H}_2$$

Silicate species were observed by Raman spectroscopy [Palik et al., 1983].

The overall silicon oxidation reaction consumes four electrons first injected into the conduction band, where they stay near the surface due to the downward bending of the energy bands (see Figure 16.37). Evidence for injection of four electrons rather than a mixed hole and electron mechanism was first presented by Raley et al. (1984). The authors could explain the measured etch-rate dependence on hole concentration only by assuming that the proton or water reduction reaction is rate determining and that a four-electron injection mechanism with the conduction band is involved. These injected electrons are highly 'reducing' and react with water to form hydroxide ions and hydrogen:

$$4\text{H}_2\text{O} + 4e^- \rightarrow 4\text{H}_2\text{O}^- \quad \text{(Reaction 16.5)}$$

$$4\text{H}_2\text{O}^- \rightarrow 4\text{OH}^- + 4\text{H}^+ + 4e^- \rightarrow 4\text{OH}^- + 2\text{H}_2 \quad \text{(Reaction 16.6)}$$

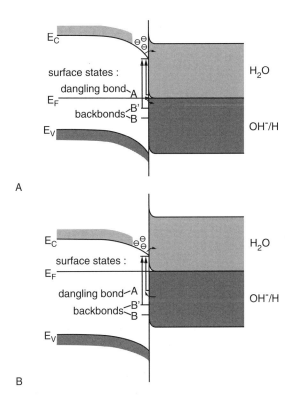

FIGURE 16.37 Band model of the silicon/electrolyte interface for moderately doped Si (electrolyte at pH > 12): (A) p-type Si and (B) n-type Si. We assume no applied bias and no illumination. The energy scale functions in respect to the saturated calomel electrode (SCE), an often-used reference in electrochemistry. Notice that p-type Si exhibits more band bending as its Fermi level is lower in the band gap. For simplicity, we show only one energetic position for surface states associated with dangling bonds and backbonds; in reality, there will be new surface states arising during reactions as the individual dangling bonds and backbonds are taking on different energies as new Si-OH bonds are introduced.

It is thought that the hydroxide ions in Reaction 16.6, generated directly at the silicon surface, react in the oxidation step. The hydroxide ions from the bulk of the solution may not play a major role, as they will be repelled by the negatively charged Si surface, whereas the hydroxide ions formed *in situ* do not need to overcome this repelling force. This would explain why the etch rates for an EDP solution with an OH^- concentration of 0.034 mol/l are nearly as large as those for KOH solutions with a hundred-fold higher OH^- concentration of 5 to 10 mol/l [Seidel et al., 1990]. The hydrogen formed in Reaction 16.6 can inhibit the reaction and surfactants may be added to displace the hydrogen (IBM, U.S. Patent 4,113,551, 1978). Additional support for the involvement of four water molecules (Reaction 16.5) comes from the experimentally observed correlation between the fourth power of the water concentration and the silicon etch rate for highly concentrated KOH solutions (Eq. (16.30) and Figure 16.27C). The weak dependence of the etching curve on the KOH concentration (~1/4 power) supports the assumption that the hydroxide ions involved in the oxidation reactions are mostly generated from water. A strong influence of water on the silicon etch rate was also observed for EDP solutions. In molar water concentrations of up to 60%, a large increase of the etch rate occurs [Finne and Klein, 1967]. The driving force for the overall Reaction 16.4 is given by the larger Si-O binding energy of 193 kcal/mol as compared to a Si-Si binding energy of only 78 kcal/mol. The role of cations, K^+, Na^+, Li^+, and even complicated cations such as $NH_2(CH_2)_2NH_3^+$ can probably be neglected [Seidel et al., 1990].

The four electrons in Reaction 16.5 are injected into the conduction band in two steps. In the case of {100} planes there are two ***dangling bonds*** per surface atom for the first two of the four hydroxide

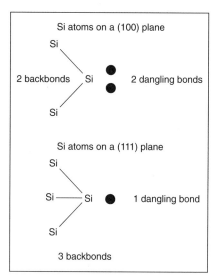

Dangling bonds.

ions to react with, injecting two electrons into the conduction band in the process. As a consequence of the strong electronegativity of the oxygen atoms, the two bonded hydroxide groups on the silicon atom reduce the strength of the two **silicon backbonds**. With two new hydroxide ions approaching, two more electrons (now stemming from the Si-Si backbonds) are injected into the conduction band and the silicon-hydroxide complex reacts with the two additional hydroxide ions. Seidel et al. (1990) claim that the step of activating the second two electrons from the backbonds into the conduction band is the rate-limiting step, with an associated thermal activation energy of about 0.6 eV for {100} planes. The electrons in the backbonds are associated with surface states within the bandgap (see Figure 16.37). The energy level of these surface states is assumed to be varying for different surface orientations, being lowest for {111} planes. The thermal activation of the backbonds corresponds to an excitation of the electrons out of these surface states into the conduction band. Since the energy for the backbond surface state level is the lowest within the bandgap for {111} planes, these planes will be hardest to etch. The {111} planes have only one dangling bond for a first hydroxide ion to react with. The second rate-limiting step involves breaking three lower energy backbonds. The lower energy of the backbond surface states for {111} Si atoms can be understood from the simple argument that their energy level is raised less by the electronegativity of a single binding hydroxide ion, compared to two in the case of the silicon atoms in {100} planes. The high etch rate generally observed on {110} surfaces is similarly explained by a high energy level of the backbond-associated surface states for these planes. Elwenspoek (1993) and Elwenspoek et al. (1994) do not accept this 'two vs. three backbonds' argument. They point out that the silicon atoms in the {110} planes also have three backbonds, and activation energy in these crystallographic directions should be comparable to that of {111} planes in contrast to experimental evidence. Seidel et al. would probably counter-argue here that the backbonds and the energy levels of the associated surface states is not necessarily the same for {111} and {110} planes, as that energy will also be influenced by the effect of the orientation of these bonds. Another argument in favor of the high etching rates of {110} planes is the easier penetrability of {110} surfaces for water molecules along channels in that plane.

The final step in the anisotropic etching is the removal of the reaction product $Si(OH)_4$ by diffusion. If the production of $Si(OH)_4$ is too fast, for solutions with a high water concentration, the $Si(OH)_4$ leads to the formation of a SiO_2-like complex before $Si(OH)_4$ can diffuse away. This might be observed experimentally as a white residue on the wafer surface [Wu et al., 1985]. The high pH values in anisotropic etching are required to obtain adequate solubility of the $Si(OH)_4$ reaction product and to remove the native oxide from the silicon surface. From silicate chemistry it is known that for pH values above 12 the $Si(OH)_4$ complex will undergo the following reaction by the detachment of two protons:

$$Si(OH)_4 \rightarrow SiO_2(OH)_2^{2-} + 2H^+ \qquad \text{(Reaction 16.7)}$$

$$2H^+ + 2OH^- \rightarrow 2H_2O \qquad \text{(Reaction 16.8)}$$

Pyrocathechol in an ethylenediamine etchant acts as complexing agent for reaction products such as $Si(OH)_4$, converting these products into more complex anions:

$$Si(OH)_4 + 2OH^- + 3C_6H_4(OH)_2 \rightarrow Si(C_6H_4O_2)_3^{2-} + 6H_2O \qquad \text{(Reaction 16.9)}$$

There is evidence by Abu-Zeid et al. (1985) of diffusion control contribution to the etch rate in EDP, probably because the hydroxide ion must diffuse through the layer of complex silicon reaction products (see Reaction 16.9). The same authors also found that the etch rate depends on the effective Si area being exposed and on its geometry (crevice effect). That is why the silicon wafer is placed in a holder and the solution is vigorously agitated in order to minimize the diffusion layer thickness. For KOH solutions, no effect of stirring on etching rate was noticed. Stirring here is mainly used to decrease the surface roughness, probably through removal of hydrogen bubbles.

The influence of alcohol on the KOH etching rate in Eq. (16.30) mainly is due to a change in the relative water concentration and its concomitant pH change; it does not participate in the reaction (this was confirmed by Raman studies by Palik et al. [1983]). The reversal of etch rates for {110} and {100} planes through alcohol addition to KOH/water etchants can be understood by assuming that the alcohol covers the silicon surface [Palik et al., 1983], thus canceling the 'channeling advantage' of the {110} planes. In the case of EDP, alcohol has no effect, as the water concentration can be freely adjusted without significantly influencing the pH value due to the incomplete dissociation of EDP.

For the etching of SiO_2 shown in Figure 16.28, Seidel et al. propose the following reaction:

$$SiO_2 + 2OH^- \rightarrow SiO_2(OH)_2^{2-} \qquad \text{(Reaction 16.10)}$$

At KOH concentrations up to 35% a linear correlation occurs between etch rate and KOH concentration. The SiO_2 etch rate in KOH solutions exceeds those in EDP by close to three orders of magnitude. For higher concentrations, the etch rate decreases with the square of the water concentration, indicating that water plays a role in this reaction. Seidel et al. speculate that at high pH values the silicon electrode is highly negatively charged (the point of zero charge of SiO_2 is 2.8), repelling the hydroxide ions while water takes over as reaction partner. An additional reason for the decrease is that the hydroxide concentration does not continue to increase with increasing KOH concentration for very concentrated solutions. The decrease of the Si/SiO_2 etch rate ratio with increasing temperature and pH value of the solution follows out of the larger activation energy of the SiO_2 etch rate (0.85 eV) and its linear correlation with the hydroxide concentration, whereas the silicon etch rate mainly depends on the water concentration.

The effect of water concentration and pH value on the etching process in the Seidel et al. model is summarized in Table 16.10 [Seidel, 1990]. For aqueous KOH solutions within a concentration range from 10 to 60% the following empirical formula for the calculation of the silicon etch rate R proved to be in close agreement with the experimental data:

$$R = k_0 [H_2O]^4 [KOH]^{\frac{1}{4}} e^{-\frac{E_a}{kT}} \qquad (16.31)$$

The values for the fitting parameters were $E_a = 0.595$ eV and $k_0 = 2480$ μm/hr (mol/l)$^{-4.25}$; for a (100) wafer $E_a = 0.6$ and $k_0 = 4500$ μm/hr (mol/l)$^{-4.25}$. For the SiO_2 etch, an activation energy of 0.85 eV was used.

TABLE 16.10 Effect of Water Concentration and pH Value on the Characteristics of Silicon Etching [Seidel, 1990]

	$-H_2O+$	$-pH+$
SiO_2 etch rate	No effect	$- \Leftrightarrow +$
Si etch rate	$- \Leftrightarrow +$	Little effect
Solubility	No effect	$- \Leftrightarrow +$
Si/SiO_2 ratio	$- \Leftrightarrow +$	$+ \Leftrightarrow -$
Diffusion effects	$- \Leftrightarrow +$	$+ \Leftrightarrow -$
Residue formation	$- \Leftrightarrow +$	$+ \Leftrightarrow -$
p^+ etch stop	$- \Leftrightarrow +$	$+ \Leftrightarrow -$
p, n etch stop	$- \Leftrightarrow +$	$+ \Leftrightarrow -$

In the section on etch stop techniques, we will see that the Seidel et al. model also nicely explains why all alkaline etchants exhibit a strong reduction in etch rate at high boron dopant concentration of the silicon; at high doping levels the conduction band electrons for the rate-determining reduction step are not confined to the surface anymore and the reaction basically stops.

The key points of the Seidel et al. model can be summarized as follows (see also Table 16.10) [Seidel et al., 1990]:

1. The rate-limiting step is the water reduction.
2. Hydroxide ions required for oxidation of the silicon are generated through reduction of water at the silicon surface. The hydroxide ions in the bulk do not contribute to the etching, since they are repelled from the negatively charged surface. This implies that the silicon etch rate will depend on the molar concentration of water and that cations will have little effect on the silicon etch rate.
3. The dissolution of silicon dioxide is assumed to be purely chemical with hydroxide ions. The SiO_2 etch rate depends on the pH of the bulk electrolyte.
4. For boron concentrations in excess of 3×10^{19} cm^{-3}, the silicon becomes degenerate, and the electrons are no longer confined to the surface. This prevents the formation of the hydroxide ions at the surface and thus causes the etching to stop.
5. Anodic biases will prevent the confinement of electrons near the surface as well and lead to etch stop as in the case of a p^+ material.

Points 4 and 5 will become more clear when we discuss the workings of etch-stop techniques. This model applies well for lower index planes (i.e., {nnn} with n < 2) where high etch rates always correspond to low activation energies. But, for higher index planes (i.e., {n11} and {1nn} with n = 2,3,4), Herr and Baltes (1991) found no correlation between activation energies and etch rates. For higher index planes, we must rely mainly on empirical data.

The Si etching reactions suggested by Seidel et al. are only the latest; in earlier proposed schemes, according to Ghandhi (1968) and Kern (1978), the silicon oxidation reaction steps suggested were injection of holes into the Si (raising the oxidation state of Si), hydroxylation of the oxidized Si species, complexation of the silicon reaction products and dissolution of the reaction products in the etchant solution. In this reaction scheme, etching solutions must provide a source of holes as well as hydroxide ions and they must contain a complexing agent with soluble reacted Si species in the etchant solution, e.g., pyrocatechol forming the soluble $Si(C_6H_4O_2)_3^{2-}$ species. This older model still seems to be guiding the current thinking of many micromachinists, although Seidel et al.'s energy level based model of the silicon/electrolyte interface proves more satisfying.

16.5.2.3 Elwenspoek et al. Model

Elwenspoek (1993) and Elwenspoek et al. (1994) introduced an alternative model for anisotropic etching of Si, a model built on theories derived from crystal growth. According to these authors, the Seidel et al.

model does not clearly explain the fast etching of {110} planes. Those planes, having three backbonds like the {111} planes, should etch equally slowly. The activation energy of the anisotropic etch rate depends on the etching system used; for example, etching in KOH is faster than in EDP, even when the pH of the solution is the same. Seidel et al. attribute this dependence to diffusion that plays a greater role in EDP than in KOH solutions. But Elwenspoek et al. point out that, at least for slow etching, the etch rate should not be diffusion controlled but governed by surface reactions. With surface reactions, diffusion should have a minor effect, analogous to growth at low pressure in an LPCVD reactor [Madou, 1997, chap. 3]. Another comment focuses on the lack of understanding why certain etchants etch isotropically and others etch anisotropically.

Elwenspoek et al. note the parallels in the process of etching and growing of crystals; slowly growing crystal planes also etch slowly! A key to understanding both processes, growing or dissolution (etching), pertains to the concept of the energy associated with the creation of a critical nucleus on a single crystalline smooth surface, i.e., the free energy associated with the creation of an island (growth) or a cavity (etching). Etching or growing of a material starts at active kink sites on steps. Kink sites are atoms with as many bonds to the crystal as to the liquid. Kinetics depend critically on the number of such kink sites. This aspect remained neglected in the discussion of etch rates of single crystals up to now.

The free energy change, ΔG, involved in creating an island or digging a cavity (of circular shape in an isotropic material) of radius r on or in an atomically smooth surface, is given by:

$$\Delta G = -N\Delta\mu + 2\pi r\gamma \tag{16.32}$$

where N is the number of atoms forming the island or the number of atoms removed from the cavity, $\Delta\mu$ is the chemical potential difference between silicon atoms in the solid state and in the solution, and γ is the step free energy. The step free energy in Eq. (16.32) will be different at different crystallographic surfaces. This can easily be understood from the following example. A perfectly flat {111} surface in the Si diamond lattice has no kink positions (three backbonds, one dangling bond per atom), while on the {001} face every atom has two backbonds and two dangling bonds, i.e., every position is a kink position. Consequently, creating an adatom-cavity pair on {111} surfaces costs energy: three bonds must be broken and only one is reformed. In the case of {001} faces, the picture is quite different. Creating an adatom-cavity pair now costs no energy because one has to break two bonds in order to remove an atom from the {001} face, but one gets them back by placing the atom back on the surface. The binding energy ΔE of an atom in a crystal slice with orientation (hkl) divided by kT (Boltzmann constant times absolute temperature) is known as the α factor of Jackson of that crystal face [Jackson, 1966], or:

$$\alpha = \frac{\Delta E}{kT} \tag{16.33}$$

At sufficiently low temperature, where entropy effects can be ignored, $kT\alpha$ is proportional to the step free energy γ and the number of adatom-cavity pairs is proportional to exp $(-\alpha)$. This number is very small on the {111} silicon faces at low temperature, but 1 on the {001} silicon faces at any temperature. The consequence for {111} and {001} planes is that, at sufficiently low temperatures, the first are atomically smooth and the latter are atomically rough. N in Eq. (16.32) can be further written out as:

$$N = \pi r^2 h \rho \tag{16.34}$$

where h is the height of the step, r is the diameter of the hole or island and ρ is the density (atoms per cm^3) of the solid material. The result is

$$\Delta G = -\pi r^2 h \rho \Delta\mu + 2\pi r\gamma \tag{16.35}$$

where $\Delta\mu$ is counted positive and γ is positive in any case. In Figure 16.38 we show a plot of ΔG vs. r. Eq. (16.35) exhibits a maximum at:

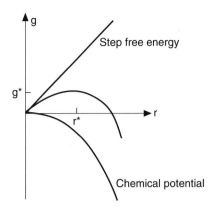

FIGURE 16.38 A plot of ΔG vs. r based on Eq. (16.35) exhibits a maximum.

$$r^* = \frac{\gamma}{h\rho\Delta\mu} \tag{16.36}$$

At r^* the free energy is

$$\Delta G^* = \Delta G(r^*) = \frac{\pi\gamma^2}{h\rho\Delta\mu} \tag{16.37}$$

Consequently, an island or an etch cavity of critical size exist on a smooth face. If by chance a cavity is dug into a crystal plane smaller than r^*, it will be filled rather than allowed to grow and an island that is too small will dissolve rather than continue to grow, since that is the easy way to decrease the free energy. With $r = r^*$, islands or cavities do not have any course of action, but in case of $r > r^*$ the islands or cavities can grow until the whole layer is filled or removed. In light of the above nucleation barrier theory, to remove atoms directly from flat crystal faces such as the {111} Si faces seems very difficult, since the created cavities increase the free energy of the system and filling of adjacent atoms is more probable than removal; in other words, a nucleation barrier has to be overcome. The growth and etch rates, R, of flat faces are proportional to:

$$R \sim \exp\left(-\frac{\Delta G^*}{kT}\right) \tag{16.38}$$

Since ΔG^* is proportional to γ^2, the activation energy is different for different crystallographic faces and both the etch rate and the activation energy are anisotropic. If $\Delta G^*/kT$ is large, the etch rate will be very small, as is the case for large step free energies and for small undersaturation (i.e., the 'chemical drive' or $\Delta\mu$ is small) (see Eq. (16.37)). Both $\Delta\mu$ and γ depend on the temperature and type of etchant and these parameters might provide clues to understanding the variation of etch rate, degree of anisotropy, temperature dependence etc., giving this model more bandwidth than the Seidel et al.'s model. According to Elwenspoek et al., the chemical reaction energy barrier and the transport in the liquid are isotropic and the most prominent anisotropy effect is due to the step free energy (absent on rough surfaces) rather than the surface free energy. The surface free energy and the step free energies are related, though, when comparing flat faces, those having a large surface free energy have a small step free energy, and vice versa. The most important difference in these two parameters is that the step-free energy is zero for a rough surface, whereas the surface energy remains finite.

Flat faces grow and etch with a rate proportional to ΔG^* which predicts that faces with a large free energy to form a step will grow and etch much slower than faces with smaller free energy. Elementary analysis indicates that the only smooth face of the diamond lattice is the (111) plane. There may be other flat faces, but with lower activation energies, due to reconstruction and/or adsorption, prominent candidates in this category are {100} and {110} planes. On the other hand, a rough crystal face grows and etches with a rate directly proportional to $\Delta\mu$. The temperature at which γ vanishes and a face transitions from smooth to rough is called the roughening transition temperature T_R [Elwenspoek and van der Weerden, 1987; Bennema, 1984]. Above T_R, the crystal is rough on the microscopic scale. Because the step free energy is equal to zero, new Si units may be added or removed freely to the surface without changing the number of steps. Rough crystal faces grow and dissolve with a rate proportional to $\Delta\mu$ and therefore proceed faster than flat surfaces. Imperfect crystals, e.g., surfaces with screw dislocations, etch even faster with R proportional to $\Delta\mu^2$.

For the state of a surface slightly above or below the roughening temperature, T_R, thermal equilibrium conditions apply. Etching, in most practical cases, is far from equilibrium and kinetic roughening might occur. Kinetic roughening [Bennema, 1984] occurs if the super- or undersaturation of the solution is so large that the thermally created islands or cavities are the size of the critical nucleus. One can show that if the super- or undersaturation is larger than $\Delta\mu_c$, given by:

$$\Delta\mu_c = \frac{\pi f_0 \gamma^2}{kT} \tag{16.39}$$

(f_0 being the area one atom occupies in a given crystal plane), the growth and etch mechanism changes from a nucleation barrier-controlled mechanism to a direct growth/etch mechanism. The growth rate and etch rate again become proportional to the chemical potential difference. It can thus be expected that if the undersaturation becomes high enough, even the {111} faces could etch isotropically, as they indeed do in acidic etchants. If the undersaturation becomes so large that $\Delta G^* \ll kT$, the nucleation barrier breaks down. Each single-atom cavity acts as a nucleus made in vast numbers by thermal fluctuations. The face in question etches with a rate comparable to the etch rate of a rough surface. This situation is called kinetic roughening. If all faces are kinetically rough, the etch rate becomes isotropic.

Isotropic etching requires conditions of kinetic roughening, because the etch rate is no longer dominated by a nucleation barrier but by transport processes in solution and the chemical reaction. To test this aspect of the model, Elwenspoek et al. show that there is a transition from isotropic to anisotropic etching if the undersaturation becomes too small. This can occur if one etches with an acidic etchant very long or if one etches through very small holes in a mask (crevice effect). In both cases, anisotropic behavior becomes evident as aging or limited transport of the solution causes the undersaturation to become very small. No proof is available to indicate that acidic etchants are much more undersaturated than the alkaline etchants. Still, the above explains some phenomena that the Seidel et al.'s model fails to address. Another nice confirmation of the Elwenspoek et al. model is in the effect of misalignment on etch rate. A misalignment of the mask close to smooth faces implies steps; there is no need for nucleation in order to etch. Since the density of steps is proportional to the angle of misalignment, the etch rate should be proportional to the misalignment angle, provided the distance between steps is not too large. Nucleation of new cavities becomes very probable. This has indeed been observed for the etch rate close to the <111> directions [Seidel et al., 1990].

Where the Elwenspoek et al.'s model becomes a bit murky is in the classification of which surfaces are smooth and which ones are rough. Elementary analysis classifies only the {111} planes as smooth at low temperatures. At this stage the model does not explain anything more than other models; every model has an explanation for the slower {111} etch rate. But these authors invoke the possibility of surface reconstruction and/or adsorption of surface species which, by decreasing the surface-free energy, could make faces such as {001} and {110} flat as well but with lower activation energies. They also take heart in the fact that CVD experiments often end up showing flat {110}, {100}, {331} and, strongest of all, {111} planes. Especially where the influence of the etchant is concerned a lot more convincing thermodynamic data to estimate $\Delta\mu$ and γ are needed.

16.5.2.4 Isotropic vs. Anisotropic Etching of Silicon

In contrast to alkaline etching, with an acidic etchant such as HF, holes are needed for etching Si. An n-type Si electrode immersed in HF in the dark will not etch due to lack of holes. The same electrode in an alkaline medium etches readily. A p-type electrode in an HF solution, where holes are available under the proper bias, will etch even in the dark. For HF etchants one might assume that the Ghandi and Kern model [Kern, 1978; Jackson, 1966], relying on the injection of holes, applies. In terms of the band model, this must mean that the silicon/electrolyte interface in acidic solutions exhibits quite different energetics from the alkaline case. It is not directly obvious why the energetics of the silicon/electrolyte interface would be pH dependent. To the contrary, since the flatband potential of most oxide semiconductors as well as most oxide-covered semiconductors (such as Si) and the Fermi level of the H_2O/OH^- redox couple in an aqueous solution are both expected to change by 59 mV for each pH unit change [Madou and Morrison, 1989], one would expect the energetics of the interface to be pH independent. Since the electronegativity in a Si-F bond is higher than for a Si-OH bond, one might even expect the backbond surface states to be raised higher in an HF medium, making an electron injection mechanism even more likely than in alkaline media. To clarify this contradiction we will analyze the band model of a Si electrode in an acidic medium in more detail. The band model shown in Figure 16.39 was constructed on the basis of a set of impedance measurements on an n-type Si electrode in a set of aqueous solutions at different pH values. From the impedance measurements Mott-Schottky plots were constructed to determine the pH dependency of the flat-band potential. From that, the position of the conduction band and valence band edges (E_{cs} and E_{vs}, respectively) of the Si electrode in an acidic medium at a fixed pH of 2.2 (the point of zero charge) was calculated at 0.74 eV vs. SCE (saturated calomel electrode as reference) for E_{cs}, and −0.36 eV vs. SCE for E_{vs} [Madou et al., 1980; 1981].

We have assumed in Figure 16.39 that the bands are bent upwards at open circuit (see below for justification), so that holes in the valence band are driven to the interface where they can react with Si atoms or with competing reducing agents from the electrolyte. Since we want to etch Si, we are only interested in the reactions where Si itself is consumed. Reactions of holes with reducing agents are of great importance in photoelectrochemical solar cells [Madou et al., 1981]. For n-type Si, holes can be (1) injected by oxidants from the solution (e.g., by adding nitric acid to the HF solution), (2) supplied at the electrolyte/semiconductor interface by shining light on a properly biased n-Si wafer, or (3) created by impact ionization, i.e., Zener breakdown, of a sufficiently high reverse biased n-Si electrode [Levy-Clement et al., 1992]. With a p-Si wafer, a small forward bias supplies all the holes needed for the oxidation of the lattice even without light, as the conduction happens via a hole mechanism. An important

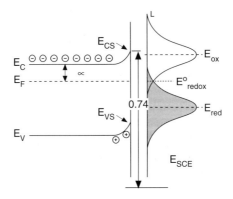

FIGURE 16.39 Band diagram for n-type Si in pH = 2.2 (no bias or illumination). Reference is the saturated calomel electrode (SCE). In Figure 16.37 no energy values were given; here we provide actual positions of the conduction band edge, E_{cs} = 0.74 eV vs. SCE, and the valence band edge, E_{vs} = 0.74 eV −1.1 eV = −0.36 eV vs. SCE (1.1 eV is the band gap of Si). These values were determined by means of Mott-Schottky plots [Madou et al., 1981]. The separation between the Fermi energy and the bottom of the conduction band is indicated by μ.

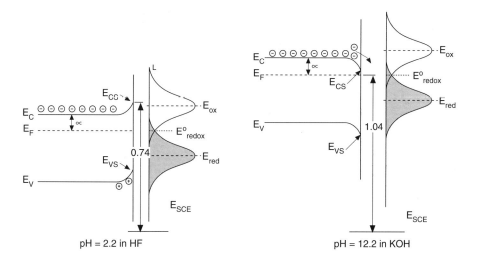

FIGURE 16.40 Band model comparison of the Si/electrolyte interface at low and high pH. Increasing the pH by 10 units shifts the redox-levels up by 600 mV, whereas the Si bands only move up by 300 mV. This leads to a different band-bending and a different reaction mechanism, i.e., electron injection in the alkaline media (anisotropic) and hole injection in acidic media (isotropic).

finding, explaining the different reaction paths in acidic and alkaline media comes from plotting the flat band potential as a function of pH. It was found that the band diagram of Si shifts with less than 59 mV per pH unit. Actually the shift is only about 30 mV per pH unit [Madou et al., 1981]. As shown in Figure 16.40, with increasing pH, the energy levels of the solution rise faster than the energy levels in the semiconductor. As a consequence it is more likely that electron injection takes place in alkaline media as the filled levels associated with the OH⁻ are closer to the conduction band, whereas in acidic media the filled levels of the redox system overlap better with the valence band, favoring a hole reaction. A lower position of the redox couple with respect to the conduction band edge, E_{cs}, in acidic media explains the upward bending of the bands as drawn in Figure 16.39. With isotropic etching in acidic media, the reaction starts with a hole in the valence band, equivalent to a broken Si-Si bond. In this case, the relative position of backbond related surface states in different crystal orientations are of no consequence and all planes etch at the same rate. A study of the interfacial energetics helps to understand why isotropic etching occurs in acidic media and anisotropic etching in alkaline media.

A few words of caution on our explanation for the reactivity difference between acidic and alkaline media are in order. Little is known about the width of the bell-shaped curves describing the redox levels in solution [Madou and Morrison, 1989]. Not knowing the surface concentrations of the reactive redox species involved in the etching reactions further hinders a better understanding of the surface energetics as the bell-shaped curves for oxidant and reductant will only be the same height (as they were drawn in Figures 16.39 and 16.40) if the concentration of oxidant and reductant are the same. Clearly, the above picture is oversimplified; several authors have found that the dissolution of Si in HF might involve both the conduction and valence band, a claim confirmed by photocurrent multiplication experiments [Matsumura and Morrison, 1983; Lewerenz et al., 1988]. These photomultiplication experiments showed that one or two holes generated by light in the Si valence band were sufficient to dislodge one Si [Brantley, 1973] unit, meaning that the rest of the charges were injected into the conduction band. Our contention here is only that the low pH dependence of the flat-band potential of a silicon electrode makes a conduction band mechanism more favorable with alkaline type etchants and a hole mechanism more favorable in acidic media.

Continued attempts at modeling the etch rates of all Si planes are under way. For example, Hesketh et al. (1993) attempted to model the etch rates of the different planes developing on silicon spheres in etching experiments with KOH and CsOH by calculating the surface free energy. The number of surface bonds per centimeter square on a Si plane is indicative of the surface free energy, which can be estimated by counting the bond density and multiplying by the bond energy. Using the unit cell dimension "a" of Si of 5.431 Å, and a silicon-to-silicon bond energy of 42.2 kcal/mol, the surface free energy, ΔG, can be related to, N_B, the bond density, by the following expression:

$$\Delta G = \frac{N_B}{2} \times 2.94 \times 10^{-19} \frac{J}{m^2} \qquad (16.40)$$

Although Hesketh et al. could not explain the etching differences observed between CsOH and KOH (these authors identified a cation effect on the etch rate!), a plot of the calculated surface free energy vs. orientation yielded minima for all low index planes such as {100}, {110} and {111}, as well as for the high index {522} planes. Fewer bonds per unit area on the low index planes produce a lower surface energy and lower etching rate. When Hesketh et al. added the in-plane bond density to the surface bond density, producing a total bond density, a correlation with the hierarchy of etch rates in CsOH and KOH was found, i.e., {311}, {522} > {100} > {111}. The surfaces with the higher bond density etched faster, suggesting that the etch rate might be a function of the number of electrons available at the surface. Hesketh et al. imply that their result falls in line with the Seidel et al.'s model, although it is unclear how the total bond density relates to surface state energies of backbonds. Moreover, Hesketh's model does not take into account the angles of the bonds, and in Elwenspoek's view, the surface free energy actually does not determine the anisotropy.

More research could focus on the modeling of Si etch rates. The semiconductor electrochemistry of corroding Si electrodes will be a major tool in further developments. Interested readers may consult Sundaram and Chang (1993) in an article on Si etching in hydrazine and Palik et al. (1985) on the etch-stop mechanism in heavily doped silicon; both explain in some detail the silicon/electrolyte energetics. A more generic treatise on semiconductor electrochemistry can be found in Morrison (1980).

16.6 Etching With Bias and/or Illumination of the Semiconductor

The isotropic and anisotropic etchants discussed so far require neither a bias nor illumination of the semiconductor. The etching in such cases proceeds at open circuit and the semiconductor is shielded from light. In a cyclic voltammogram, as shown in Figure 16.41, the operational potential is the rest potential, V_r, where anodic and cathodic currents are equal in magnitude and opposite in sign, resulting in the absence of flow of current in an external circuit. This does not mean that macroscopic changes do not occur at the electrode surface, since the anodic and cathodic currents may be part of different chemical reactions. Consider isotropic etching of Si in an HF/HNO$_3$ etchant at open circuit where the local anodic reaction is associated with corrosion of the semiconductor:

$$\text{Anode: } Si + 2H_2O + nh^+ \rightarrow SiO_2 + 4H^+ + (4-n)e^- \qquad \text{(Reaction 16.11)}$$

$$SiO_2 + 6HF \rightarrow H_2SiF_6 + 2H_2O \qquad \text{(Reaction 16.12)}$$

(the involvement of holes 'h$^+$' in the acidic reaction was discussed in the preceding section), while the local cathodic reaction could be associated with reduction of HNO$_3$:

$$\text{Cathode: } HNO_3 + 3H^+ \rightarrow NO + H_2O + 3H^+ \qquad \text{(Reaction (16.13)}$$

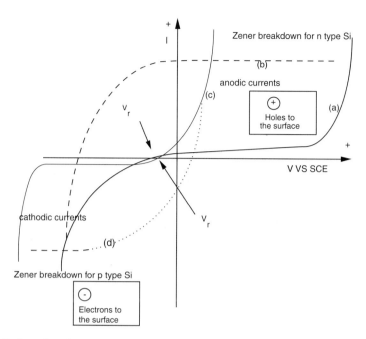

FIGURE 16.41 Basic cyclic voltammograms (I vs. V) for n- and p-type Si in an HF solution, in the absence of a hole injecting oxidant: (a) n-type Si without illumination; (b) n-type Si under illumination; (c) p-type Si without illumination; (d) p-type Si under illumination. If the reactions on the dark Si electrode determining the rest potential V_r for n- and p-type are the same, then V_r is expected to be the same as well. For the clarity of the figure we have chosen the V_rs different here; in practice, V_r for n- and p-type Si in HF are found to be identical.

We will now explore Si etching while illuminating and/or applying a bias to the silicon sample. To simplify the situation, we will first consider the case of an oxidant-free solution so that all the holes must come from within the semiconductor. Anodic dissolution of n-Si in an HF containing solution requires a supply of holes to the surface. For an n-type wafer under reverse bias, very few holes will show at the surface unless the high reverse (anodic) bias is sufficient to induce impact ionization or Zener breakdown (see Figure 16.41a). Alternatively, the interface can be illuminated, creating holes in the space charge region which the field pushes towards the semiconductor/electrolyte interface (Figure 16.41b). In the forward direction, electrons from the Si conduction band (majority carriers) reduce oxidizing species in the solution (e.g., reduction of protons to hydrogen). A p-type Si sample exhibits high anodic currents even without illumination at small anodic (forward) bias (Figure 16.41c). Here the current is carried by holes. A p-type electrode illuminated under reverse bias gives rise to a cathodic photocurrent (Figure 16.41d). At relatively low light intensities, the photocurrent plateaus for both n- and p-types (Figures 16.41b and 16.41d, respectively) depend linearly on light intensity. The photocurrent is cathodic for p-type Si (species are reduced by photoproduced electrons at the surface, e.g., hydrogen formation) and anodic for n-type Si (species are oxidized by photoproduced holes at the surface; either the lattice itself is consumed or reducing compounds in solution are). In Figure 16.42 we show the cyclic voltammograms of n-type and p-type Si in the presence of a hole-injecting oxidant. The most obvious effect is on the dark p-type Si electrode. The injection of holes in the valence band increases the cathodic dark current dramatically. The current level measured in this manner for varying oxidant concentration or different oxidants could be used to estimate the efficiency of different isotropic Si etchants; a pointer to the fact that semiconductor electrochemistry has been underutilized as a tool to study Si etching. When n-type Si is consumed under illumination, we experience photocorrosion (see Figure 16.41b). This photocorrosion phenomenon has been a major barrier to the long-term viability of photoelectrochemical cells [Madou et al., 1980]. In what follows, photocorrosion is put to use for electropolishing and formation of microporous and macroporous layers [Levy-Clement et al., 1992].

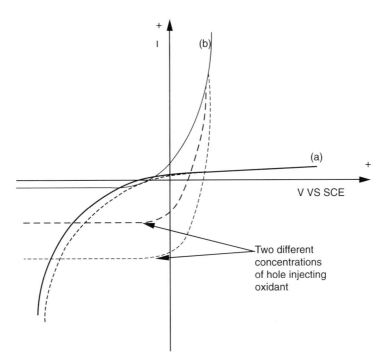

FIGURE 16.42 Basic cyclic voltammograms for n- and p-type Si in an HF solution and in the presence of a hole injecting oxidant, e.g., HNO_3: (a) n-type Si in the dark; (b) p-type Si in the dark. An increase of the cathodic dark current on the p-type electrodes is most obvious. The current level is proportional to the oxidant concentration.

16.6.1 Electropolishing and Microporous Silicon

16.6.1.1 Electropolishing

Photoelectrochemical etching (PEC-etching) involves photocorrosion in an electrolyte in which the semiconductor is generally chemically stable in the dark, i.e., no hole-injecting oxidants are present, as in the case of an HF solution (see Figure 16.41). For carrying out the experiments a set-up as shown in Figure 16.25 may be used with a provision to illuminate the semiconductor electrode. At high light intensities the anodic curves for n- and p-type Si are the same, except for a potential shift of a few hundred mV (see Figure 16.43). Because of this equivalence, several of the etching processes described apply for both forward biased p-type and n-type Si under illumination. The anodic curves in Figure 16.43 present two peaks, characterized by i_{CRIT} and i_{MAX}. At the first peak, i_{CRIT}, partial dissolution of Si in reactions such as:

$$Si + 2F^- + 2h^+ \rightarrow SiF_2 \qquad \text{(Reaction 16.14)}$$

$$SiF_2 + 2HF \rightarrow SiF_4 + H_2 \qquad \text{(Reaction 16.15)}$$

leads to the formation of porous Si while hydrogen formation occurs simultaneously. The porous Si typically forms when using low current densities in a highly concentrated HF solution—in other words, by limiting the oxidation of silicon due to a hole and OH^- deficiency. Above i_{CRIT}, the transition from the charge-supply-limited to the mass-transport-limited case, the porous film delaminates and bright electropolishing occurs at potentials positive of i_{MAX}. With dissolution of chemical reactants in the electrolyte rate limiting, HF is depleted at the electrode surface and a charge of holes builds up at the interface. Hills on the surface dissolve faster than depressions because the current density is higher on high spots. As a result, the surface becomes smoother, i.e., electropolishing takes place [Lehmann, 1993].

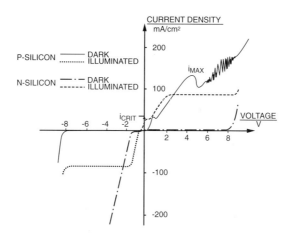

FIGURE 16.43 Cyclic voltammograms identifying porous Si formation regime and electropolishing regime. (From Levy-Clement, C. et al., *Electrochima Acta*, 37, 877–888, 1992. With permission.)

Electropolishing in this regime can be used to smooth silicon surfaces or to thin epitaxially grown silicon layers. The peak and oscillations in Figure 16.43 are explained as follows: at current densities exceeding i_{MAX} an oxide grows first on top of the silicon, leading to a decrease of the anodic photocurrent (explaining the i_{MAX} peak), until a steady state is reached in which dissolution of the oxide by HF through formation of a fluoride complex in solution (SiF_6^{2-}) equals the oxide growth rate. The oscillations observed in the anodic curve in Figure 16.43 can be explained by a nonlinear correlation between formation and dissolution of the oxide [Levy-Clement et al., 1992].

16.6.1.2 Porous Silicon

Introduction

The formation of porous Si was first discovered by Uhlir in 1956 [Uhlir, 1956]. His discovery is leading to all types of interesting new devices from quantum structures, permeable membranes, photoluminescent and electroluminescent devices to a basis for making thick SiO_2 and Si_3N_4 films [Bomchil et al., 1986; Smith and Collins, 1990]. Two types of pores exist: micropores and macropores. Their sizes can differ by three orders of magnitude and the underlying formation mechanism is quite different. Some important features of porous Si, as detailed later, are

- Pore sizes in a diameter range from 20 Å to 10 µm.
- Pores that follow crystallographic orientation [Chuang and Smith, 1988].
- Very high aspect ratio (~250) pores in Si maintained over several millimeters' distance [Lehmann, 1993].
- Porous Si is highly reactive, oxidizes and etches at very high rate.
- Porosity varies with the current density.

These important attributes contribute to the essential role porous Si plays in both micromachining [Barret et al., 1992] and the fashioning of quantum structures [Canham, 1990].

Microporous Silicon

Whether one is in the regime of electropolishing or porous silicon formation depends on both the anodic current density and the HF concentration. The surface morphology produced by the Si dissolution process critically depends on whether mass transport or hole supply is the rate-limiting step. Porous silicon formation is favored for high HF concentrations and low currents (weak light intensities for n-type Si), where the charge supply is limiting, while etching is favored for low HF concentration and high currents (strong light intensity for n-type Si), where mass transport is limiting. For current densities below i_{MAX} (Figure 16.43) holes are depleted at the surface and HF accumulates at the electrode/electrolyte interface. As a result, a dense network of fine holes forms [Unagami, 1980; Watanabe et al., 1975]. The formation of a porous silicon layer (PSL) in this regime has been explained as a self-adjusting electrochemical reaction due to hole depletion by a quantum confinement in the microporous structure [Lehmann and Gosele, 1991]. The structure of the pores in PSL can best be observed by transmission electron microscopy (TEM) and its thickness can be monitored with an IR microscope. The structure of the porous layer primarily depends on the doping level of the wafer and on the illumination during etching. The pore sizes decrease when etching occurs under illumination [Levy-Clement et al., 1992]. The pores formed in p-type Si show much smaller diameters than n-type ones for the same formation conditions. It is believed that the porous silicon layer (PSL) consists of silicon hydrides and oxides. The pore diameter typically ranges from 40 to 200 Å [Lehmann, 1996; Yamana et al., 1990; Arita and Sunohara, 1977]. This very reactive porous material etches or oxidizes rapidly. Heat treatment in an oxidizing atmosphere (1100°C in oxygen for 30 minutes is sufficient to make a 4-μm thick film) leads to oxidized porous silicon (OPS). The oxidation occurs throughout the whole porous volume and several micrometers-thick SiO_2 layers can be obtained in times that correspond to the growth of a few hundred nanometers on regular Si surfaces. Porous silicon is low density and remains single crystalline, providing a suitable substrate for epitaxial Si film growth. These properties have been used to obtain dielectric isolation in ICs and to make silicon-on-insulator wafers [Watanabe et al., 1975].

Porous Si can also be formed chemically through Reactions 16.11 and 16.13. In this case the difference between chemical polishing and porous Si formation conditions is more subtle. In the chemical polishing case, all reacting surface sites switch constantly from being local anode (Reaction 16.11) to being local cathode (Reaction 16.13), resulting in nonpreferential etching. If surface sites do not switch fast between being local anode and cathode, charges have time to migrate over the surface. In this case, the original local cathode site remains a cathode for a longer time, and the corresponding local anode site, somewhere else on the surface, also remains an anode in order to keep the overall reaction neutral. A preferential etching results at the localized anode sites, making the surface rough and causing a porous silicon to form [Jung et al., 1993]. Any inhomogeneity, e.g., some oxide or a kink site at the surface might increase such preferential etching [Jung et al., 1993]. Unlike PSLs fabricated by electrochemical means, the chemically etched porous Si film thickness is self limiting.

Besides its use for dielectric isolation and the fabrication of SOI wafers, porous silicon has been introduced in a wide variety of other applications: Luggin capillaries for electrochemical reference electrodes, high surface area gas sensors, humidity sensors, sacrificial layers in silicon micromachining, etc. Recently PSL was shown to exhibit photoluminescent and electroluminescent behavior; light-emitting porous silicon (LEPOS) was demonstrated. Visible light emission from regular Si is very weak due to its indirect bandgap. Pumping porous Si with a green light laser (argon) caused it to emit a red glow. If a LEPOS device could be integrated monolithically with other structures on silicon, a big step in micro-optics, photon data transmission, and processing would be achieved. To explain the blue shift of the absorption edge of LEPOS of about 0.5 eV compared to bulk silicon [Lehmann and Gosele, 1991] and room temperature photoluminescence [Canham, 1990], Searson et al. (1993) have proposed an energy-level diagram for porous silicon where the valence band is lowered with respect to bulk silicon to give a band-gap of about 1.8 eV. Not only may PSL formation, as seen above, be explained invoking quantum structures, but its remarkable optical properties may also be explained this way. Canham believes that the thin Si filaments may act as quantum wires. Significant quantum effects require structural sizes below 5 nm (see Madou [1997, chap. 7]) and the porous Si definitely can have structures of that size. By treatment of the porous Si with NH_3 at high temperatures, it is possible to make thick Si_3N_4 films. Even at 13 μm, these films show little evidence of stress in contrast to stoichiometric LPCVD nitride films [Smith and Collins, 1990].

Porous Si might represent a simple way of making the quantum structures of the future. Pore size of PSL can be influenced by both light intensity and current density. The quantum aspect adds significantly to the sudden big interest taken in porous Si.

Macroporous Silicon

In addition to micropores, well-defined macropores can also be made in Si by photo and/or bias etching in HF solutions. Macropores have sizes as large as 10 μm, visible with a scanning electron microscope (SEM) rather than TEM. The two types of pores often coexist, with micropores covering the walls of macropores. Sizes can differ by three orders of magnitude. This is not a matter of a broad fractal type distribution of pores, but the formation mechanism is quite different.

Electrochemical etching of macropores or macroholes has been reported for n-type silicon in 2.5 to 5% HF under high voltage (>10 V), low current density (10 mAcm^{-2}), under illumination, and in the dark [Levy-Clement et al., 1992]. In the latter case, Zener breakdown in silicon (electric field strength in excess of 3×10^5 V/cm) causes the hole formation. The macropores are formed only with lightly doped n-silicon at much higher anodic potentials than those used for micropore formation [Zhang, 1991]. By using a pore initiation pattern the macropores actually can be localized at any desired location. This dramatic effect is illustrated by comparing Figures 16.44A and B [Lehmann, 1993; 1996]. Pores orthogonal to the surface with depths up to a whole wafer thickness can be made and aspect ratios as large as 250 become possible. The formation mechanism in this case cannot be explained on the basis of depletion of holes due to quantum confinement in the fine porous structures, given that these macropores exhibit sizes well beyond 5 nm. As with microporous Si, the surface morphology produced by the dissolution process depends critically on whether mass transport or charge supply is the rate-limiting step. For pore

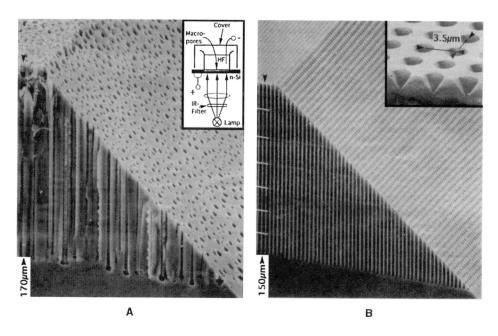

FIGURE 16.44 Macroporous Si; formation of random and localized macropores or macroholes. (A) Random: surface, cross section and a 45° bevel of an n-type sample (10^{15}/cm^3 phosphorus-doped) showing a random pattern of macropores. Pore initiation was enhanced by applying 10-V bias in the first minute of anodization followed by 149 min at 3 V. The current density was kept constant at 10 mA/cm^2 by adjusting the backside illumination. A 6% aqueous solution of HF was used as an electrolyte. The set-up used for anodization is sketched in the upper right corner. (B) Localized: surface, cross section and a 45° bevel of an n-type sample (10^{15}/cm^3 phosphorus-doped) showing a predetermined pattern of macropores (3 V, 350 min, 2.5% HF). Pore growth was induced by a regular pattern of pits produced by standard lithography and subsequent alkaline etching (inset upper right). To measure the depth dependence of the growth rate, the current density was kept periodically at 5 mA/cm^2 for 45 min and then reduced to 3.3 mA/cm^2 for 5 min. This reduction resulted in a periodic decrease of the pore diameter, as marked by white labels in the figure. (From Lehmann, V., *J. Electrochem. Soc.*, 140, 2836–2843, 1993. With permission.)

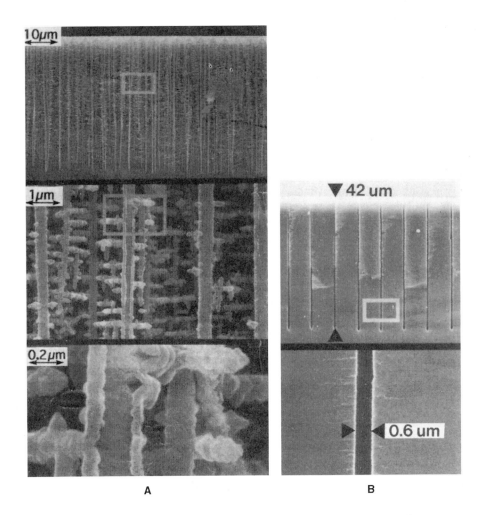

FIGURE 16.45 Comparison of macropores made with breakdown holes (A) and macropores made with light created holes (B). (A) An oxide replica of pores etched under weak backside illumination visualizes the branching of pores produced by generation of charge carriers due to electrical breakdown (5 V, 3% HF, room temperature, $10^{15}/cm^3$ phosphorus-doped). (From Lehmann, V., *J. Electrochem. Soc.*, 140, 2836–2844, 1993. With permission.) (B) Single pore associated with KOH pit. (From Lehmann, V. and H. Foll, *J. Electrochem. Soc.*, 137, 653–659, 1990. With permission.)

formation one must work again in a charge depletion mode. Macropore formation, as micropore formation, is a self-adjusting electrochemical mechanism. In this case, the limitation is due to the depletion of the holes in the pore walls in n-Si wafers causing them to passivate. Holes keep on being collected by the pore tip where they promote dissolution. No passivating layer is involved to protect the pore wall. The only decisive differences between pore tips and pore walls are their geometry and their location. Holes generated by light or Zener breakdown are collected at pore tips. Every depression or pit in the surface initiates pore growth because the electrical field at a curved pore bottom is much larger than that of a flat surface due to the effect of the radius of curvature. The latter leads to higher current and enhances local etching [Zhang, 1991]. Zener breakdown and illumination of n-Si lead to different types of pore geometry [Lehmann, 1993; Theunissen, 1972; Lehmann and Foll, 1990]. Branched pores with sharp tips form if holes are generated by breakdown (see Figure 16.45A) [Lehmann, 1993]. Unbranched pores with larger tip radii result from holes created by illumination (see Figure 16.45B) [Lehmann and Foll, 1990]. The latter difference can be understood as follows: the electric field strength is a function of bias, doping density and geometry. High doping level density or sharp pore tips will lower the required bias for breakdown, so macropores will tend to follow pores with the sharpest tips. Since every tip causes a new breakdown and hole generation, the position of the original pore tip becomes independent of the other pores, branching of the pores is possible, and fir-tree type pores can be observed. With illumination the pore radius may be larger as the breakdown field strength is not necessary to generate charge carriers, so the pores remain unbranched.

The Si anisotropy common with alkaline etchants surprisingly shows up here with an isotropic etchant such as HF. For example, with breakdown-supplied holes, <100>-directed macroholes with <110> branches form (see also Figure 16.45A) [Lehmann, 1993], leading to a complex network of caverns beneath the silicon surface. Pyramidal pore tips [Lehmann, 1993] also were observed when the current density was limited by the bias ($i < i_{CRIT}$). Isotropic pore pits form when the current is larger than the critical current density, i.e., isotropy in HF etching can be changed into anisotropy when the supply of holes is limited. We refer here to the Elwenspoek et al. model (see above) which predicts that in confined spaces etching will tend to be more anisotropic, even when using a normally isotropic etchant such as HF. It was also determined that macrohole formation depends on the wavelength of the light used. No hole formation occurs below about 800 nm. Depending on the wavelength, the shape of the hole can be manipulated as well [Lehmann and Foll, 1990]. For wavelengths above 867 nm, the depth profile of the holes changed from conical to cylindrical. The latter was interpreted in terms of the influence of the local minority carrier generation rate. Carriers generated deep in the bulk would promote the hole growth at the tips, whereas near-surface generation would lead to lateral growth.

Cahill et al. (1993) reported in 1993 the creation of 1- to 5-μm-size pores with pore spacings (center-to-center) from 200 to 1000 μm. Until this finding, the pores typically formed were spaced in the range of 4 to 30 μm center-to-center, while being 0.6 to 10 μm in diameter. Making highly isolated pores presents quite another challenge. In the previous work, the relative close spacing of the pores allowed the authors to conclude that the regions between the pores were almost totally depleted and that practically all carriers were collected by the pore tips. In such a case, neither the pore side walls nor the wafer surface etched as all holes were swept to the pore tips. Since the surface was not attacked by pore-forming holes, the quality of the pore initiation mask lost its relevance. In Cahill et al.'s case, on the other hand, a long-lived mask (>20 hours) needed to be developed to help prevent pore formation everywhere except at initiation pits. The mask eventually used is shown in Figure 16.46; it shows a SiC layer sandwiched between two layers of silicon nitride. The silicon nitride directly atop the silicon served to insulate the silicon carbide from the underlying substrate. As the silicon carbide proves very resistant to HF, loss of thickness did not show during the procedure. The top nitride served to protect the carbide during aniso-tropic pit formation. By lowering the bias to less than two volts with respect to a saturated calomel electrode (SCE) side-branching was avoided.

It seems very likely that this macropore formation phenomenon could extend to all types of n-type semiconductors. Some evidence with InP and GaAs supports this statement [Lehmann, 1993].

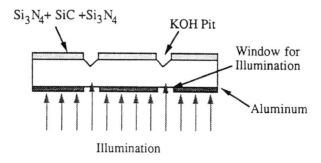

FIGURE 16.46 Long-lived macropore initiation mask.

16.7 Etch-Stop Techniques

16.7.1 Introduction

In many cases it is desirable to stop etching in silicon when a certain cavity depth or a certain membrane thickness is reached. Nonuniformity of etched devices due to nonuniformity of the silicon wafer thickness can be quite high. Taper of double-polished wafers, for example, can be as high as 40 μm [Lambrechts and Sansen, 1992]! Even with the best wafer quality the wafer taper is still around 2 μm. The taper and variation in etch depth lead to intolerable thickness variations for many applications. Etch-rate control typically requires monitoring and stabilization of:

- Etchant composition
- Etchant aging
  - Stabilization with N_2 sparging (especially with EDP and hydrazine)
  - Taking account of the total amount of material etched (loading effects)
- Etchant temperature
- Diffusion effects (constant stirring is required, especially for EDP)
  - Stirring also leads to a smoother surface through bubble removal
  - Trenching (also pillowing) and roughness decrease with increased stirring rate
- Light may affect the etch rate (especially with n-type Si)
- Surface preparation of the sample can have a big effect on etch rate (the native oxide retards etch start; a dip in dilute HF is recommended)

With good temperature, etchant concentration, and stirring control the variation in etch depth typically is 1% (see Figure 16.47) [Lambrechts and Sansen, 1992]. A good pretreatment of the surface to be etched is a standard RCA clean combined with a 5% HF dip to remove the native oxide immediately prior to etching in KOH.

In the early days of micromachining, one of the following techniques was used to etch a Si structure anisotropically to a predetermined thickness. In the simplest mode the etch time was monitored (Table 16.9 lists some etch rates for different etchants) or a bit more complex; the infrared transmittance was followed. For thin membranes the etch stop cannot be determined by a constant etch time method with sufficient precision. The spread in etch rates becomes critical if one etches membranes down to thicknesses of less than 20 μm; it is almost impossible to etch structures down to less than 10 μm with a timing technique. In the V-groove technique, V-grooves with precise openings (see Eq. (16.2)) were used such that the V-groove stopped etching at the exact moment a desired membrane thickness was reached (see Figure 16.48) [Samaun et al., 1973]. One can also design wider mask openings on the wafer's edge so that the wafer is etched through at those sites at the moment the membrane has reached the appropriate thickness. Although Nunn and Angell (1975) claimed that an accuracy of about 1 μm could be obtained using the V-groove method, none of the mentioned techniques are found to be production worthy. Nowadays the above methods are almost completely replaced by etch-stop techniques based on a change in etch rate dependant on doping level or chemical nature of a stopping layer. High resolution silicon micromachining relies on the availability of effective etch-stop layers. It is actually the existence of impurity-based etch stops in silicon that has allowed micromachining to become a high-yield production process.

16.7.2 Boron Etch Stop

The most widely used etch-stop technique is based on the fact that anisotropic etchants, especially EDP, do not attack boron-doped (p+) Si layers heavily. This effect was first noticed by Greenwood (1969). He assumed that the presence of a p-n junction was responsible.

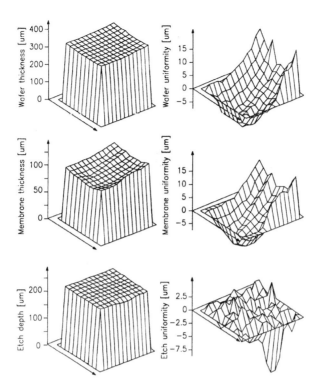

FIGURE 16.47 A map of the wafer thickness, the membrane thickness and the etch depth. (From Lambrechts, S.W. and W. Sansen, *Biosensors: Microelectrochemical Devices*, Institute of Physics Publishing, 1992. With permission.)

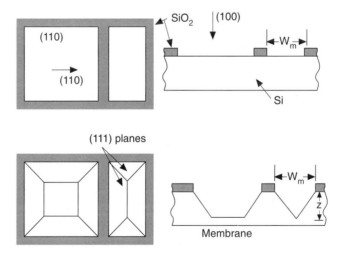

FIGURE 16.48 V-groove technique to monitor the thickness of a membrane. At the precise moment the V-groove is developed, the membrane has reached the desired thickness.

Bogh (1971) found that at an impurity concentration of about $7 \times 10^{19}/cm^3$ resulted in the etch rate of Si in EDP dropping sharply (see also Table 16.9), but without any requirement for a p-n junction. For KOH-based solutions, Price (1973) found a significant reduction in etch rate for boron concentrations above $5 \times 10^{18}\ cm^{-3}$. The model by Seidel et al., discussed above, provides an elegant explanation for the etch stop at high boron concentrations. At moderate dopant concentration, we saw that the electrons

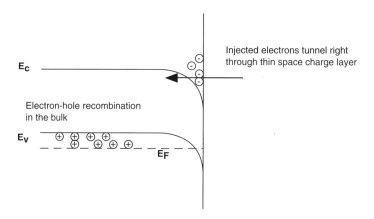

FIGURE 16.49 Si/electrolyte interface energetics at high doping level explaining etch-stop behavior.

injected into the conduction band stay localized near the semiconductor surface due to the downward bending of the bands (Figure 16.37). The electrons there have a small probability of recombining with holes deeper into the crystal even for p-type Si. This situation changes when the doping level in the silicon increases further. At a high dopant concentration, silicon degenerates and starts to behave like a metal. For a degenerate p-type semiconductor, the space charge thickness shrinks and the Fermi level drops into the valence band as indicated in Figure 16.49. The injected electrons shoot (tunnel) right through the thin surface charge layer into deeper regions of the crystal where they recombine with holes from the valence band. Consequently, these electrons are not available for the subsequent reaction with water molecules (Reaction 16.5), the reduction of which is necessary for providing new hydroxide ions in close proximity to the negatively charged silicon surface. These hydroxide ions are required for the dissolution of the silicon as $Si(OH)_4$. The remaining etch rate observed within the etch stop region is then determined by the number of electrons still available in the conduction band at the silicon surface. This number is assumed to be inversely proportional to the number of holes and thus the boron concentration. Experiments show that the decrease in etch rate is nearly independent of the crystallographic orientation and the etch rate is proportional to the inverse fourth power of the boron concentration in all alkaline etchants.

From the above it follows that a simple boron diffusion or implantation, introduced from the front of the wafer, can be used to create beams and diaphragms by etching from the back. The boron etch-stop technique is illustrated in Figure 16.50 for the fabrication of a micromembrane nozzle [Brodie and Muray, 1982]. The SiO_2 mesa in Figure 16.50b leads to the desired boron p^+ profile. The anisotropic etch from the back clears the lightly doped p-Si (Figure 16.50d). Layers of p^+ silicon having a thickness of 1 to 20 μm can be formed with this process. The boron etch stop constitutes a very good etch stop; it is not very critical when the operator takes the wafer out of the etchant. One important practical note is that the boron etch stop may become badly degraded in EDP solutions that were allowed to react with atmospheric oxygen. Since boron atoms are smaller than silicon, a highly doped, freely suspended membrane or diaphragm will be stretched; the boron-doped silicon is typically in tensile stress and the microstructures are flat and do not buckle. While doping with boron decreases the lattice constant, doping with germanium increases the lattice constant. A membrane doped with B and Ge still etches much slower than undoped silicon, and the stress in the layer is reduced. A stress-free, dislocation-free and slow etching layer (±10 nm/min) is obtained at doping levels of 10^{20} cm^{-3} boron and 10^{21} cm^{-3} germanium [Seidel et al., 1990; Heuberger, 1989]. One disadvantage with this etch-stop technique is that the extremely high boron concentrations are not compatible with standard CMOS or bipolar techniques, so they can only be used for microstructures without integrated electronics. Another limitation of this process is the fixed number and angles of (111) planes one can accommodate. The etch stop is less effective in KOH compared to EDP. Besides boron, other impurities have been tried for use in an etch stop in anisotropic

MEMS Fabrication

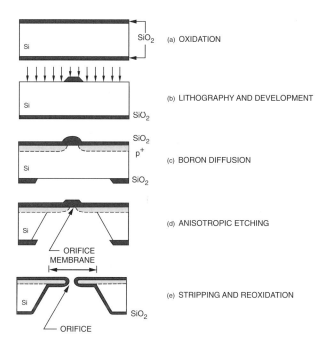

FIGURE 16.50 Illustration of the boron etch stop in the fabrication of a membrane nozzle. (From Brodie, I. and J. J. Muray, *The Physics of Microfabrication*, Plenum Press, New York, 1982. With permission.)

etchants. Doping Si with germanium has hardly any influence on the etch rate of either the KOH or EDP solutions. At a doping level as high as 5×10^{21} cm^{-3} the etch rate is barely reduced by a factor of two [Seidel et al., 1990].

By burying the highly doped boron layer under an epitaxial layer of lighter doped Si the problem of incompatibility with active circuitry can be avoided. A ±1% thickness uniformity is possible with modern epilayer deposition equipment (see, for example, *Semiconductor International*, July 1993, pp. 80–83). A widely used method of automatically measuring the epi thickness is with infrared (IR) instruments, especially Fourier transform infrared (FTIR) [Rehrig, 1990]; see also Epitaxy later in this chapter and in Madou (1997, chap. 5).

16.7.3 Electrochemical Etch-Stop Technique

For the fabrication of piezoresistive pressure sensors, the doping concentration of the piezo resistor must be kept smaller than 1×10^{19} cm^{-3} because the piezoresistive coefficients drop considerably above this value and reverse breakdown becomes an issue. Moreover, high boron levels compromise the quality of the crystal by introducing slip planes and tensile stress and prevent the incorporation of integrated electronics. As a result, a boron stop often cannot be used to produce well-controlled thin membranes unless, as suggested above, the highly doped boron layer is buried underneath a lighter doped Si epi layer. Alternatively, a second etch-stop method, an electrochemical technique, can be used. In this case, a lightly doped p-n junction is used as an etch stop by applying a bias between the wafer and a counter electrode in the etchant. This technique was first proposed by Waggener (1970). Other early work on electrochemical etch-stops with anisotropic etchants such as KOH and EDP was performed by Palik et al. (1982), Jackson et al. (1981), Faust and Palik (1983) and Kim and Wise (1983). In electrochemical anisotropic etching a p-n junction is made, for example, by the epitaxial growth of an n-type layer (phosphorus-doped, 10^{15} cm^{-3}) on a p-type substrate (boron-doped, 30 Ωcm). This p-n junction forms a large diode over the whole wafer. The wafer is usually mounted on an inert substrate, such as a sapphire plate, with an acid-resistant wax and is partly or wholly immersed in the solution. An ohmic contact to the n-type epilayer is connected to one pole of a voltage source and the other pole of the voltage source is connected

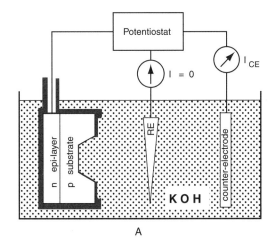

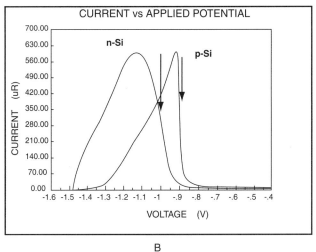

FIGURE 16.51 Electrochemical etch stop. (A) Electrochemical etching set-up with potentiostatic control (three-electrode system). Potentiostatic control, mainly used in research studies, enables better control of the potential as it is referenced now to a reference electrode such as a saturated calomel electrode (SCE). In industrial settings, electrochemical etching is often carried out in a simpler two-electrode system, i.e., a Pt counter electrode and Si working electrode [Kloeck et al., 1989]. (B) Cyclic voltammograms of n- and p-type silicon in an alkaline solution at 60°C. Flade potentials are indicated with an arrow.

via a current meter to a counterelectrode in the etching solution (see Figure 16.51A). In this arrangement the p-type substrate can be selectively etched away and the etching stops at the p-n junction, leaving a membrane with a thickness solely defined by the thickness of the epilayer. The incorporation of a third electrode (a reference electrode) in the three-terminal method depicted in Figure 16.51A allows for a more precise determination of the silicon potential with respect to the solution than a two-terminal set-up as we illustrated in Figure 16.25.

At the Flade potential in Figure 16.51B, the oxide growth rate equals the oxide etch rate; a further increase of the potential results in a steep fall of the current due to complete passivation of the silicon surface. At potentials positive of the Flade potential, all etching stops. At potentials below the Flade potential, the current increases as the potential becomes more positive. This can be explained by the formation of an oxide which etches faster than it forms, i.e., the silicon is etched away. Whereas electrochemists like to talk about the Flade potential, physicists like to discuss matters in terms of the flat-band potential. The flat-band potential is that applied potential at which there are no more fields within the semiconductor, i.e., the energy band diagram is flat throughout the semiconductor. The passivating SiO_2 layer is assumed to start growing as soon as the negative surface charge on the silicon electrode is cancelled by the externally applied positive bias, a bias corresponding to the flat band potential. At these potentials, the formation of $Si(OH)_x$ complexes does not lead to further dissolution of silicon, because two neighboring Si-OH HO-Si groups will react by splitting of water, leading to the formation of Si-O-Si bonds. As can be learned from Figure 16.51B, the value of the Flade potential depends on the dopant type. Consequently, if a wafer with both n and p regions exposed to the electrolyte is held at a certain potential in the passive range for n-type and the active range for p-type, the p-regions are etched away, whereas the n-regions are retained. In the case of the diode shown in Figure 16.51A, where at the start of the experiment only p-type Si is exposed to the electrolyte, one starts by applying a positive bias to the n-type epilayer (V_n). This reverse biases the diode, and only a reverse bias current can flow. The potential of the p-type layer in this regime is negative of the flat-band potential and active dissolution takes place. At the moment that the p-n junction is reached, a large anodic current can flow and the applied positive potential passivates the n-type epilayer. Etching continues on the areas where the wafer is thicker until the membrane is reached there, too. The thickness of the silicon membrane is thereby solely defined by the thickness of the epilayer; neither the etch uniformity nor the wafer taper will influence the result. A uniformity of better than 1% can be obtained on a 10-μm-thick membrane. The current vs. time curve can be used to monitor the etching process; at first, the current is relatively low, i.e., limited by the reverse bias current of the diode, then as the p-n junction is reached a larger anodic current can flow until all the p-type material is consumed and the current falls again to give a plateau. The plateau indicates we have reached the current associated with a passivated n-type Si electrode. The etch procedure can be stopped at the moment that the current plateau has been reached, and since the etch stop is thus basically one of anodic passivation one sometimes terms it an anodic oxidation etch stop. Registering an I vs. V curve as in Figure 16.51B will establish an upper limit on the applied voltage, V_n [McNeil et al., 1990] and such curves are used for *in situ* monitoring and controlling of the etch stop. A crude endpoint monitoring can be accomplished by the visual observation of cessation of hydrogen bubble formation accompanying Si etching. Palik et al. presented a detailed characterization of Si membranes made by electrochemical etch stop [Palik et al., 1988].

Hirata et al. (1988), using the same anodic oxidation etch stop in a hydrazine-water solution at 90 ± 5°C and a simple two-terminal electrochemical cell, obtained a pressure sensitivity variation of less than 20% from wafer to wafer (pressure sensitivity is inversely proportional to the square of membrane thickness). A great advantage of this etch-stop technique is that it works with Si with low doping levels of the order of 10^{16} cm^{-3}. Due to the low doping levels, it is possible to fabricate structures with a very low, or controllable, intrinsic stress. Moreover, active electronics and piezo membranes can be built into the Si without problems. A disadvantage is that the back of the wafer with the aluminum contact has to be sealed hermetically from the etchant solution which requires complex fixturing and manual wafer handling. The fabrication of a suitable etch holder is no trivial matter. The holder must (1) protect the epi-contact from the etchant, (2) provide a low-resistance ohmic contact to the epi and (3) must not introduce stress into wafers during etching [Peeters, 1994; Elwenspoek et al., 1994]. Stress introduced by etch holders easily leads to diaphragm or wafer fracture and etchant seepage through to the epi-side.

Using a four-electrode electrochemical cell, controlling the potentials of both the epitaxial layer and the silicon wafer, as shown in Figure 16.52, can further improve the thickness control of the resulting membrane by directly controlling the p/n bias voltage. The potential required to passivate n-type Si can be measured using the three-electrode system in Figure 16.51A, but this system does not take into account the diode leakage. If the reverse leakage is too large, the potential of the n-Si, V_n, will approach the potential of the p-Si, V_p. If there is a large amount of reverse diode leakage, the p-region may passivate prior to reaching the n-region and etching will cease. In the four-electrode configuration, the reverse leakage current is measured separately via a p-region contact, and the counter-electrode current may be monitored for end-point detection. The four-electrode approach allows etch stopping on lower

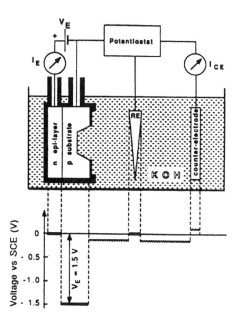

FIGURE 16.52 Four-electrode electrochemical etch-stop configuration. Voltage distribution with respect to the SCE reference electrode (RE) for the four-electrode case. The fourth electrode enables an external potential to be applied between the epitaxial layer and the substrate, thus maintaining the substrate at etching potentials. (From Kloeck et al., *IEEE Trans. Electron. Dev.*, 36, 663–669, 1989. With permission.)

quality epis (larger leakage current) and should also enable etch stopping of p-epi on n-substrates. Kloeck et al. (1989) demonstrated that using such an electrochemical etch-stop technique and with current monitoring, the sensitivity of pressure sensors fabricated on the same wafer could be controlled to within 4% standard deviation. These authors used a 40% KOH solution at 60 ± 0.5°C. Without the etch stop, the sensitivity from sensor to sensor on one wafer varied by a factor of 2.

The etching solution used in electrochemical etching can either be isotropic or anisotropic. Electrochemical etch stop in isotropic media was discussed above. In this case one uses HF/H_2O mixtures to etch the highly doped regions of p^+p, n^+n, n^+p, and p^+n systems [Theunissen et al., 1970; Meek, 1971; Theunissen, 1972; Wen and Weller, 1972; van Dijk, 1972]. The rate-determining step in etching with isotropic etchants does not involve reducing water with electrons from the conduction band as it does in anisotropic etchants and the etch stop mechanism, as we learned earlier is obviously different. In isotropic media, the etch stop is simply a consequence of the fact that higher conductivity leads to higher corrosion currents and the etch slows down on lower conductivity layer. A major advantage of the KOH electrochemical etch is that it retains all of the anisotropic characteristics of KOH without needing a heavily doped p^+ layer to stop the etch [Saro and van Herwaarden, 1986].

In Figure 16.53, we review the etch-stop techniques discussed so far: diffused boron etch stop, buried boron etch stop and electrochemical etch stop [Wise, 1985]. A comparison of the boron etch stop and the electrochemical etch stop reveals that the IC compatibility and the absence of built-in stresses, both due to the low dopant concentration, are the main assets of the electrochemical etch stop.

16.7.4 Photo-Assisted Electrochemical Etch Stop (for n-Type Silicon)

A variation on the electrochemical diode etch-stop technique is the photo-assisted electrochemical etch-stop method illustrated in Figure 16.54 [Mlcak et al., 1993]. An n-type silicon region on a wafer may be selectively etched in an HF solution by illuminating and applying a reverse bias across a p-n junction, driving the p-layer cathodic and the n-layer anodic. Etch rates up to 10 μm/min for the n-type material and a high resolution etch stop render this an attractive potential micromachining process. Advantages

MEMS Fabrication

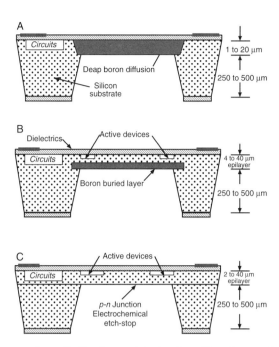

FIGURE 16.53 Typical cross sections of bulk micromachined wafers with various methods for etch-stop formation shown. (A) Diffused boron etch stop. (B) Boron etch stop as a buried layer. (C) Electrochemical etch stop. (From Wise, K.D., in *Silicon Micromachining and Its Application to High Performance Integrated Sensors,* Fung, C.D., Ed., Elsevier, New York, 1985. With permission.)

also include the use of lightly doped n-Si, bias- and illumination intensity-controlled etch rates, *in situ* process monitoring using the cell current and the ability to spatially control etching with optical masking or laser writing. Using this method [Mlcak et al., 1993] prepared stress-free cantilever beam test samples. They diffused boron into a 10^{15}-cm^{-3} (100) n-Si substrate through a patterned oxide mask, leaving exposed a small n-type region which defines two p-type cantilever beams (see Figure 16.54).

The boron diffusion resulted in a junction 3.3 μm underneath the surface. An ohmic contact on the backside of the wafer was used to apply a variable voltage across the p-n junction, and both p and n areas were exposed to the HF electrolyte. The exposed n-region was etched to a depth of 150 μm by shining light on the whole sample. The resulting n-type Si surface was at first found to be rough, as porous Si up to 5 μm in height forms readily in HF solutions. The Si surface could be made smoother by etching at higher bias (4.3 V vs. SCE) and higher light intensity (2 W/cm^2) to a finish with features of the order of 0.4 μm in height. Smoothing could also be accomplished by a 5-second dip in HNO_3:HF:H_3COOH or a 30-second dip in 25 wt%, 25°C, KOH. Yet another way of removing unwanted porous Si is a 1000°C wet oxidation to make oxidized porous Si (OPS) followed by an HF etch (see next section) [Yoshida et al., 1992].

16.7.5 Photo-Induced Preferential Anodization, PIPA (for p-Type Silicon)

Electrochemical etching requires the application of a metal electrode to apply the bias. The application of such a metal electrode often induces contamination and constitutes at least one extra process step, and extra fixturing is needed. With photo-induced preferential anodization (PIPA) it is not necessary to deposit metal electrodes. Here one relies on the illumination of a p-n junction to bias the p-type Si anodically, and the p-Si converts automatically into porous Si while the n-type Si acts as a cathode for the reaction. The principle of photo-biasing for etching purposes was known and patented by Shockley as far back as 1963 [Shockley, 1963]. In U.S. patent 3,096,262, he writes " … light can be used in place of electrical connections … for biasing of the sample …." This means a small isolated area of p-type

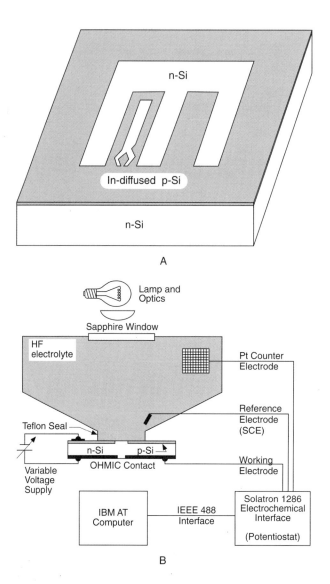

FIGURE 16.54 Photoelectrochemical etching. (A) Schematic of the photoeletrochemical etching experiment apparatus. (B) Schematic of the spatial geometry of the diffused p-Si layer into n-Si used to form cantilever beam structures. (From Mlcak, R. et al., Photo-Assisted Electromechanical Machining of Micromechanical Structures, presented at Micro Electromechanical Systems, Ft. Lauderdale, FL, 1993. With permission.)

material on an n-type body may be preferentially biased for removal of material beyond the junction by etching." The method was reinvented by Yoshida et al. (1992) and Peeters et al. (1993a; 1993b). The latter group called the method 'PHET' for photovoltaic electrochemical etch-stop technique, and the former group coined the PIPA acronym.

In PIPA, etch rates of up to 5 µm/min result in the formation of porous layers readily removed with Si etching solutions. An important advantage of the technique is that very small and isolated p-type islands can be anodized at the same time. The method also lends itself to fabricate three-dimensional structures using p-type Si as sacrificial layers [Yoshida et al., 1992]. A disadvantage of the technology is that one cannot control the process very well, as the current cannot be measured for endpoint detection. Application of PIPA to form a microbridge is shown in Figure 16.55. First, a buried p-Si layer doped to

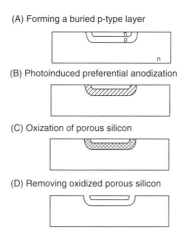

FIGURE 16.55 Photo-assisted electrochemical etch stop (for n-type Si). Fabrication process for a microbridge and SEM picture of Si structure before and after PIPA. (From Yoshida et al., Photo-Induced Preferential Anodization for Fabrication of Monocrystalline Micro-mechanical Structures, presented at Micro Electromechanical Systems, Travemunde, 1992. With permission.)

10^{18} cm^{-3} and an n-type layer doped to 10^{15} cm^{-3} are formed on an n-type substrate using epitaxy (Figure 16.55a). Then p-type is preferentially anodized in 10% HF solution under 30 mW/cm^2 light intensity for 180 minutes (Figure 16.55b), forming porous Si. The porous Si is then oxidized in wet oxygen at 1000°C (Figure 16.55c). Finally, the sacrificial layer of oxidized porous silicon is etched and removed with an HF solution (Figure 16.55d). The resulting surfaces of the n-type silicon are very smooth. It is interesting to consider making complicated three-dimensional structures by going immediately to the electropolishing regime instead. Yoshida et al. (1992) believe that porous silicon as a sacrificial intermediate is more suitable for fabricating complicated structures. The authors are probably referring to the fact that electropolishing is much more aggressive and could not be expected to lead to the same retention of the shape of the buried, sacrificial p-layers. Peeters et al. (1993) carry out their photovoltaic etching in KOH, thus skipping the porous Si stage of Yoshida et al. They, like Yosida et al., stress the fact that in one single etch step this technique can make a variety of complex shapes that would be impossible with electrochemical etching techniques. These authors found it necessary to coat the n-Si part of the wafer with Pt to get enough photovoltaic drive for the anodic dissolution; this metallization step makes the process more akin to the photoelectrochemical etching and some of the advantage of the photo-biasing process is being lost.

16.7.6 Etch Stop at Thin Films-Silicon on Insulator

Yet another distinct way (the fourth) to stop etching is by employing a change in composition of material. An example is an etch stopped at a Si_3N_4 diaphragm. Silicon nitride is very strong, hard and chemically inert and the stress in the film can be controlled by changing the Si/N ratio in the LPCVD deposition process. The stress turns from tensile in stoichiometric films to compressive in silicon-rich films (for details see later in this chapter). A great number of materials are not attacked by anisotropic etchants. Hence, a thin film of such a material can be used as an etch stop.

Another example is the SiO_2 layer in a silicon-on-insulator structure (SOI). A buried layer of SiO_2, sandwiched between two layers of crystalline silicon, forms an excellent etch stop because of the good selectivity of many etchants of Si over SiO_2. The oxide does not exhibit the good mechanical properties of silicon nitride and is consequently used rarely as a mechanical member in a microdevice. As with the photo-induced preferential anodization (PiPA), no metal contacts are needed with an SOI etch stop, greatly simplifying the process over an electrochemical etch stop technique.

16.8 Problems with Wet Bulk Micromachining

16.8.1 Introduction

Despite the introduction of better controllable etch-stop techniques, bulk micromachining remains a difficult industrial process to control. It is also not an applicable submicron technology, because wet chemistry is not able to etch reliably on that cale. For submicron structure definition dry etching is required (dry etching is also more environmentally safe). We will look now into some of the other problems associated with bulk micromachining, such as the extensive real estate consumption and difficulties in etching at convex corners, and detail the solutions that are being worked on to avoid, control or alleviate those problems.

16.8.2 Extensive Real Estate Consumption

16.8.2.1 Introduction

Bulk micromachining involves extensive real estate consumption. This quickly becomes a problem in making arrays of devices. Consider the diagram in Figure 16.56, illustrating two membranes created by etching through a <100> wafer from the backside until an etch stop, say a Si_3N_4 membrane, is reached. In creating two of these small membranes a large amount of Si real estate is wasted and the resulting device becomes quite fragile.

16.8.2.2 Real Estate Gain by Etching from the Front

One solution to limiting the amount of Si to be removed is to use thinner wafers, but this solution becomes impractical below 200 μm as such wafers break too often during handling. A more elegant solution is to etch from the front rather than from the back. Anisotropic etchants will undercut a masking material an amount dependent on the orientation of the wafer with respect to the mask. Such an etchant will etch any <100> silicon until a pyramidal pit is formed, as shown in Figure 16.57. These pits have sidewalls with a characteristic 54.7° angle with respect to the surface of a <100> silicon wafer, since the

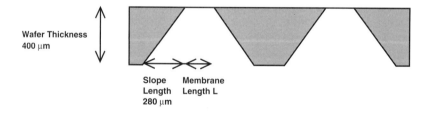

FIGURE 16.56 Two membranes formed in a <100>-oriented silicon wafer.

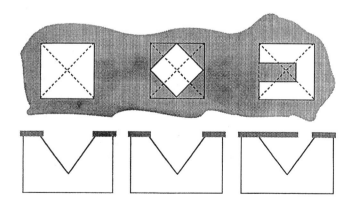

FIGURE 16.57 Three anisotropically etched pits etched from the front in a <100>-oriented silicon wafer.

MEMS Fabrication

delineated planes are {111} planes. This etch property makes it possible to form cantilever structures by etching from the front side, as the cantilevers will be undercut and eventually will be suspended over a pyramidal pit in the silicon. Once this pyramidal pit is completed, the etch rate of the {111} planes exposed is extremely slow and practically stops. Process sequences, which depend on achieving this type of a final structure, are therefore very uniform across a wafer and very controllable. The upper drawings in Figure 16.57 represent patterned holes in a masking material: a square, a diamond and a square with a protruding tab. The drawings immediately below represent the etched pit in the silicon produced by the anisotropic etchant. Note in the first drawing that the square mask produces a four-sided pyramidal pit. In the second drawing, a similar shape oriented at 45° produces an etched pit which is oriented parallel to the pit etched in the first drawing. In the second drawing the corners of the diamond are undercut by the etchant as it produces the final etch pit. The third drawing illustrates that any protruding member is eventually undercut by the anisotropic etchant, leaving a cantilever structure suspended over the etch pit.

16.8.2.3 Real Estate Gain by Using Silicon Fusion-Bonded Wafers

Using silicon fusion-bonded (SFB) wafers rather than conventional wafers also makes it possible to fabricate much smaller microsensors. The process is clarified in Figure 16.58 for the fabrication of a gauge pressure sensor [Bryzek et al., 1990]. The bottom, handle wafer has a standard thickness of 525 μm and is anisotropically etched with a square cavity pattern. Next, the etched handle wafer is fusion bonded to a top sensing wafer (see SFB and surface micromachining later in this chapter, and packaging in Madou [1997, chap. 8]). The sensor wafer consists of a p-type substrate with an n-type epilayer corresponding to the required thickness of the pressure-transducing membrane. The sensing wafer is thinned all the way to the epilayer by electrochemical etching and resistors are ion implanted. The handle wafer is ground and polished to the desired thickness. For gauge measurement, the anisotropically etched cavity is truncated by the polishing operation, exposing the backside of the diaphragm. For an absolute pressure sensor the cavity is left enclosed. With the same diaphragm dimensions and the same overall thickness of the chip, an SFB device is almost 50% smaller than a conventional machined device (see Figure 16.59).

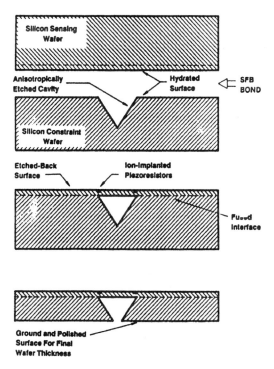

FIGURE 16.58 Fabrication process of an SFB-bonded gauge pressure sensor. (From Bryzek, J. et al., *Silicon Sensors and Microstructures*, Novasensor, Fremont, CA, 1990. With permission.)

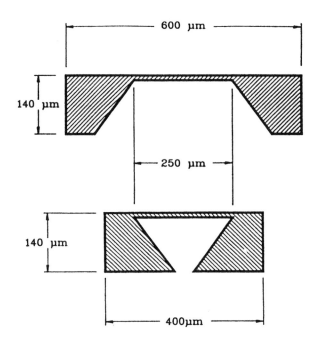

FIGURE 16.59 Comparison of conventional and SFB processes. The SFB process results in a chip which is at least 50% smaller than the conventional chip. (From Bryzek, J. et al., *Silicon Sensors and Microstructures*, Novasensor, Fremont, CA, 1990. With permission.)

16.8.3 Corner Compensation

16.8.3.1 Underetching

Underetching of a mask which contains no convex corners, i.e., corners turning outside in, in principle stems from mask misalignment and/or from a finite etching of the {111} planes. Peeters measured the widening of {111}-walled V-grooves in a (100) Si wafer after etching in 7 M KOH at 80 ± 1° over 24 hours as 9 ± 0.5 μm [Peeters, 1994]. The sidewall slopes of the V-groove are a well-defined 54.74°, and the actual etch rate R_{111} is related to the rate of V-groove widening R_v through:

$$R_{111} = \frac{1}{2}\sin(54.74°)R_v \quad \text{or} \quad R_{111} = 0.408 \cdot R_v \quad (16.41)$$

with R_{111} the etch rate in nm/min and R_v the groove widening, also in nm/min. The V-groove widening experiment then results in a R_{111} of 2.55 ± 0.15 nm/min. In practice, this etch rate implies a mask underetching of only 0.9 μm for an etch depth of 360 μm. For a 1-mm-long V-groove and a 1° misalignment angle, a total underetching of 18 μm is theoretically expected, with 95% due to misalignment and only 5% due to etching of the {111} sidewalls [Peeters, 1994]. The total underetching will almost always be determined by misalignment, rather than by etching of {111} walls.

Mask underetching with masks that do include convex corners is usually much larger than the underetching just described, as the etchant tends to circumscribe the mask opening with {111} walled cavities. This is usually called undercutting rather than underetching. It is advisable to avoid mask layouts with convex corners. Often mesa-type structures are essential, though, and in that case there are two possible ways to reduce the undercutting. One is by chemical additives, reducing the undercut at the expense of a reduced anisotropy ratio, and the other is by a special mask compensating the undercut at the expense of more lost real estate.

16.8.3.2 Undercutting

When etching rectangular convex corners, deformation of the edges occurs due to undercutting. This is an unwanted effect, especially in the fabrication of, say, acceleration sensors, where total symmetry and perfect 90° convex corners on the proof mass are mandatory for good device prediction and specification. The undercutting is a function of etch time and thus directly related to the desired etch depth. An undercut ratio is defined as the ratio of undercut to etch depth (δ/H).

Saturating KOH solutions with isopropanol (IPA) reduces the convex corner undercutting; unfortunately, this happens at the cost of the anisotropy of the etchant. This additive also often causes the formation of pyramidical or cone-shaped hillocks [Peeters, 1994; Gravesen, 1986]. Peeters claims that these hillocks are due to carbonate contamination of the etchant and he advises etching under inert atmosphere also and stock-piling all etchant ingredient under an inert nitrogen atmosphere [Peeters, 1994].

Undercutting can also be reduced or even prevented by so-called corner compensation structures which are added to the corners in the mask layout. Depending on the etching solution, different corner compensation schemes are used. Commonly used are square corner compensation (EDP or KOH) and rotated rectangle corner compensation methods (KOH). In Figure 16.60, these two compensation methods are illustrated. In the square corner compensation case, the square of SiO_2 in the mask, outlining the square proof mass feature for an accelerometer, is enhanced by adding an extra SiO_2 square to each corner (Figure 16.60a). Both the proof mass and the compensation squares are aligned with their sides parallel to the <110> direction. In this way, two concave corners are created at the convex corner to be protected. Thus, direct undercutting is prevented. The three 'sacrificial' convex corners at the protective square

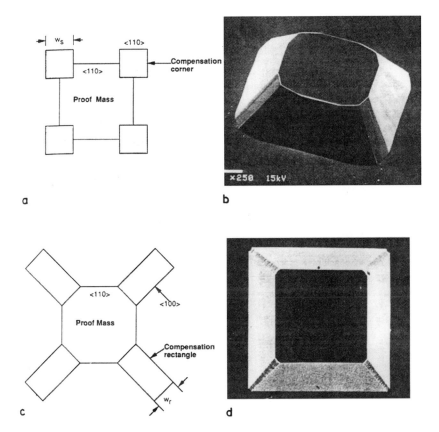

FIGURE 16.60 Formation of a proof mass by silicon bulk micromachining. (a,b) Square corner compensation method, using EDP as the etchant. (c,d) Rotated rectangle corner compensation method, using KOH as the etchant.

are undercut laterally by the fast etching planes during the etch process. The dimension of the compensation square, w_s, for a 500-μm-thick wafer (4 in.) is about 500 μm. The resulting mesa structure after EDP or KOH etching is shown in Figure 16.60b. In the rotated rectangle corner compensation method shown in Figure 16.60c, a properly scaled rectangle (w_r should be twice the thickness of the wafer) is added to each of the mask corners. The four sides of the mesa square are still aligned along the <110> direction, but the compensation rectangles are rotated (45°) with their longer sides along the <100> directions. Using KOH as an etchant reveals the mesa shown in Figure 16.60d. A proof mass is frequently dislodged by simultaneously etching from the front and the back. Corner compensation requires significant amount of space around the corners, making the design less compact, and the method is often only applicable for simple geometries.

Different groups around the world have been using different corner compensation schemes and they all claim to have optimized spatial requirements. For an introduction to corner compensation, refer to Puers and Sansen (1990), and Sandmaier et al. (1991) for KOH etching use <110>-oriented beams, <110>-oriented squares and <010>-oriented bands for corner compensation. They found that spatial requirements for compensation structures could be reduced dramatically by combining several of these compensation structures. The mask layouts for some of the different compensation schemes used by Sandmaier et al. are shown in Figure 16.61. To understand the choice and dimensioning of these compensation structures, as well as those in Figure 16.60, we will first look at the planes emerging at convex corners during KOH

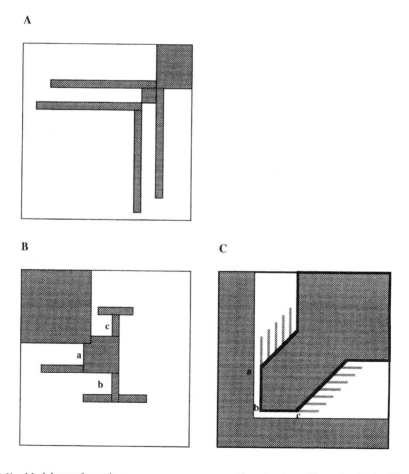

FIGURE 16.61 Mask layout for various convex corner compensation structures. (From Sandmaier, H. et al., Corner Compensation Techniques in Anisotropic Etching of (100) Silicon Using Aqueous KOH, presented at Transducers '91, San Francisco, CA, 1991. With permission.)

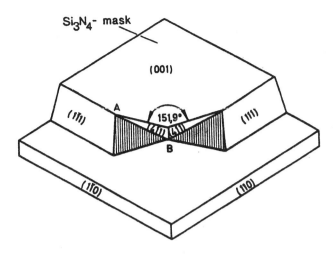

FIGURE 16.62 Planes occurring at convex corners during KOH etching. (From Mayer, G.K. et al., *J. Electrochem. Soc.*, 137, 3947–3951, 1990. With permission.)

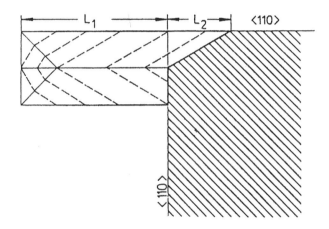

FIGURE 16.63 Dimensioning of the corner compensation structure with a <110>-oriented beam. (From Sandmaier, H. et al., Corner Compensation Techniques in Anisotropic Etching of (100) Silicon Using Aqueous KOH, presented at Transducers '91, San Francisco, CA, 1991. With permission.)

etching. Mayer et al. (1990) found that the undercutting of convex corners in pure KOH etch is determined exclusively by {411} planes. The {411} planes of the convex underetching corner, as shown in Figure 16.62, are not entirely laid free, however; rugged surfaces, where only fractions of the main planes can be detected, overlap the {411} planes under a diagonal line shown as AB in this figure. The ratio of {411} to {100} etching does not depend on temperature between 60 and 100°C. The value does decline with increasing KOH concentration from about 1.6 at 15% KOH to 1.3 at above 40% where the curve flattens out [Mayer et al., 1990]. Ideally one avoids rugged surfaces and searches for well-defined planes bounding the convex corner. In Figure 16.63 it is shown how a <110> beam is added to the convex corner to be etched. The fast etching {411} planes, starting at the two convex corners, are laterally underetching a <110>-oriented beam (broken lines in Figure 16.63). It is clear that the longer this <110>-oriented beam is, the longer the convex corner is protected from undercutting. It is essential that by the end of the etch the beam has disappeared to maintain a minimum of rugged surface at the convex edge. On the other hand, as is obvious from Figure 16.63, a complete disappearance of the beam leads to a beveling at the face of the convex corner. The dimensioning of the compensating <110> beam works then as follows: the length of the compensating beam is calculated primarily from the required etch depth (H) and the etch rate ratio $R\{411\}/R\{100\}$ ($\approx \delta/H$), at the concentration of the KOH solution used:

$$L = L_1 - L_2 = 2H\frac{R\{411\}}{R\{100\}} - \frac{B_{<110>}}{2\tan(30.9°)} \qquad (16.42)$$

with H the etch depth; $B_{<110>}$ the width of the <110>-oriented beam; tan (30.9°) the geometry factor [Mayer et al., 1990]. The factor 2 is the first term of this equation results as the etch rate of the {411} plane is determined normal to the plane and has to be converted to the <110> direction. The second term in Eq. (16.42) takes into account that the <110> beam needs to disappear completely by the time the convex corner is reached. The resulting beveling in Figure 16.63 can be reduced by further altering the compensation structures. This is done by decelerating the etch front, which largely determines the corner undercutting. One way to accomplish this is by creating more concave shapes right before the convex corner is reached. In Figure 16.61A, splitting of the compensation beam creates such concave corners, and by arranging two such double beams a more symmetrical final structure is achieved. By using these split beams, the bevelling at the corner is reduced by a factor of 1.4 to 2 and leads to bevel angles under 45°.

Corner compensation with <110>-oriented squares (as shown in Figure 16.60A) features considerably higher spatial requirements than the <110>-oriented beams. Since these squares are again undercut by {411} planes that are linked to the rugged surfaces described above, the squares do not easily lead to sharp {111}-defined corners. Dimensioning of the compensation square is done by using Eq. (16.42) again, where L_1 is half the side length of the square, and for $B_{<110>}$ the side length is used. All fast-etching planes have to reach the convex corner at the same time. As before, the spatial requirements of this compensation structure can be reduced if it is combined with <110>-oriented beams. Such a combination is shown in Figure 16.61B. The three convex corners of the compensation square are protected from undercutting by the added <110> beams. During the first etch step, the <110>-oriented side beams are undercut by the etchant. Only after the added beams have been etched does the square itself compensate the convex corner etching. The dimensioning of this combination structure is carried out in two steps. First the <110>-oriented square is selected with a size that is permitted by the geometry of the device to be etched. From these dimensions, the etch depth corresponding to this size is calculated from Eq. (16.42). For the remaining etch depth the <110>-oriented beams are dimensioned like any other <110>-oriented beam. If the side beam on corner b is selected about 30% longer than the other two side beams, the quality of the convex corner can be further improved. In this case, the corner is formed by the etch fronts starting at the corners a and b (Figure 16.61B).

A drawback to all the above proposed compensation schemes is the impossibility, due to rugged surfaces always accompanying {411} planes, of obtaining a clean corner in both the top and the bottom of a convex edge. Buser and de Rooij (1988) introduced a compensation scenario where a convex corner was formed by two {111} planes which were well defined all the way from the mask to the etch bottom. No rugged, undefined planes show in this case. The mask layout to create such an ideal convex corner has bands that are added to convex corners in the <100> direction (see Figure 16.61C and Figure 16.60c and d). These bands will be underetched by vertical {100} planes from both sides. With suitable dimensioning of such a band, a vertically oriented membrane results, thinning, and eventually freeing the convex edge shortly before the final intended etch depth is reached. In contrast to compensation structures undercut by {411} planes, posing problems with undefined rugged surfaces (see above), this compensation structure is mainly undercut by {100} planes. Over the temperature range of 50 to 100°C and KOH concentrations ranging from 25 to 50 wt%, no undefined surfaces could be detected in the case of structures undercut by {100} planes [Sandmaier et al., 1991]. The width of these <010>-oriented compensation beams, which determines the minimum dimension of the structures to etch,

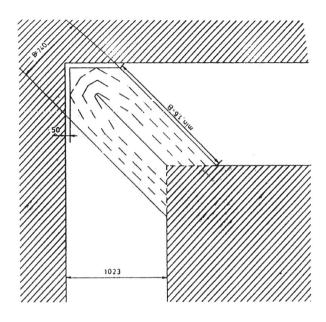

FIGURE 16.64 Beam structure open on one side. The beam is oriented in the <010> direction. Dimensions in microns. B is the width of the beam.

has to be twice the etching depth. These beams can either connect two opposite corners and protect both from undercutting simultaneously, or they can be added to the individual convex corners (open beam). With an open beam approach (see Figure 16.64) one has to be certain that the {100} planes reach the corner faster than the {411} planes. For that purpose the beams have to be wide enough to avoid complete underetching by {411} planes moving in from the front side, before they are completely underetched by {100} planes moving in from the side. For instance, in a 33% KOH etchant a ratio between beam length and width of at least 1.6 is required. To make these compensation structures smaller while at the same time maintaining {100} undercutting to define the final convex corner, Sandmaier et al. remarked that the shaping {100} planes do not need to be present at the beginning of the etching process. These authors implement delaying techniques by adding fan-like <110>-oriented side beams to a main <100>-oriented beam (see Figure 16.61C). As described above, these narrow beams are underetched by {411} planes and the rugged surfaces they entail until reaching the <100>-oriented beam. Then the {411} planes are decelerated in the concave corners between the side beams by the vertical {100} planes with slower etching characteristics. The length of the <110>-oriented side beams is calculated from:

$$L_{\langle 110 \rangle} = \left(H - \frac{B_{\langle 010 \rangle}}{2}\right)\frac{R\{411\}}{R\{100\}} \qquad (16.43)$$

with H being the etching depth at the deepest position of the device.

The width of the side beams does not influence the calculation of their required length. In order to avoid the rugged surfaces at the convex corner, the width of the side beams as well as the spaces between them should be kept as small as possible. For an etching depth of 500 µm, a beam width of 20 µm and a space width of 2 µm are optimal [Sandmaier et al., 1991].

Depending on the etchant, different planes are responsible for undercutting. From the above we learned that in pure KOH solutions undercutting, according to Mayer et al. (1990) and Sandmaier et al. (1991), mainly proceeds through {411} planes or {100} planes. That the {411} planes are the fastest undercutting planes was confirmed by Seidel (1986); at the wafer surface, the sectional line of a (411) and a (111) plane point in the <410> direction, forming an angle of 30.96° with the <110> direction, and it was in this direction that he found a maximum in the etch rate. In KOH and EDP etchants Bean (1978) identified the fast undercutting planes as {331} planes. Puers and Sansen (1990), for alkali/alcohol/water, identified the fast underetching planes as {331} planes, as well. Mayer et al. (1990), working with pure KOH, could not confirm the occurrence of such planes. Lee indicated that in hydrazine-water the fastest underetching planes are {211} planes [Lee, 1969]. Abu-Zeid (1984) reported that the main beveling planes are {212} planes in ethylene-diamine-water solution (no added pyrocatechol). Wv and Ko (1989) found the main beveling planes at undercut corners to be {212} planes whether using KOH, hydrazine, or EPW solutions are used. In view of our earlier remarks on the sensitivity of etching rates of higher index rates on a wide variety of parameters (temperature, concentration, etching size, stirring, cation effect, alcohol addition, complexing agent, etc.) these contradictory results are not too surprising. Along the same line, Wu and Ko (1989) and Peurs et al. (1990) have suggested triangles to compensate for underetching, but Mayer et al. (1997) found them to lead to rugged surfaces at the convex corner. Combining a chemical etchant with more limited undercutting (IPA in KOH) with Sandmaier's reduced compensation structure schemes could further decrease the required size of the compensation features while retaining an acceptable anisotropy.

Corner compensation for <110>-oriented Si was explored by Ciarlo [1987]. Ciarlo comments that both corner compensation and corner rounding can be minimized by etching from both surfaces so as to minimize the etch time required to achieve the desired features. This requires accurate front-to-back alignment and double-sided polished wafers.

Employing corner compensation offers access to completely new applications such as rectangular solids, orbiting V-grooves, truncated pyramids with low cross-sections on the wafer surface, bellow structures for decoupling mechanical stresses between micromechanical devices and their packaging, etc. [Sandmaier et al., 1991].

16.9 Wet Bulk Micromachining Examples

Example 1. Dissolved Wafer Process

Figure 16.65 illustrates the dissolved wafer process Cho is optimizing for the commercial production of low-cost inertial instruments [Cho, 1995]. This process, also in use by Draper Laboratory [Greiff, 1995; Weinberg et al., 1996] for the same application, involves a sandwich of a silicon sensor anodically bonded to a glass substrate. The preparation of the silicon part requires only two masks and three processing steps. A recess is KOH-etched into a p-type (100) silicon wafer (step 1 with mask 1), followed by a high-temperature boron diffusion (step 2 with no mask). In step one RIE may be used as well. Cho claims that by maintaining a high-temperature uniformity in the KOH etching bath (±0.1°C), the accuracy and absolute variation of the etch across the wafer, wafer-to-wafer, and lot-to-lot, can be maintained to <0.1 µm using pre-mixed, 45 wt% KOH. Cho is also using low-defect oxidation techniques (e.g., nitrogen annealing and dry oxidation) to form defect-free silicon surfaces. In the boron diffusion the key is optimizing the oxygen content. In general, the optimal flow of oxygen is on the order of 3 to 5% of the nitrogen flow, in which case the doping uniformity is on the order of ±0.2 µm. Varying the KOH etch depth and the shallow boron diffusion time, a wide variety of operating ranges and sensitivities for sensors can be obtained. Next, the silicon is patterned for a reactive ion etching (RIE) etch (step 3 with mask 2). Aspect ratios above 10 are accessible. Using some of the newest dry etching techniques depths in excess of 500 µm at rates above 4 µm/min (with a SF_6 chemistry) are now possible [Craven and Pandhumsporn, 1995]. The glass substrate (#7740 Corning glass) preparation involves etching a recess, depositing and in a one-mask step patterning a multi-metal system of Ti/Pt/Au. The electrostatic bonding of glass to silicon takes place at 335°C with a potential of 1000 V applied between the two parts (electrostatic or anodic bonding is explained in detail in Madou [1997, chap. 8]). Commercial bonders have alignment accuracies on the order of <1 µm. The lightly doped silicon is dissolved in an EDP solution at 95°C. The keys to uniform EDP etching are temperature uniformity

MEMS Fabrication

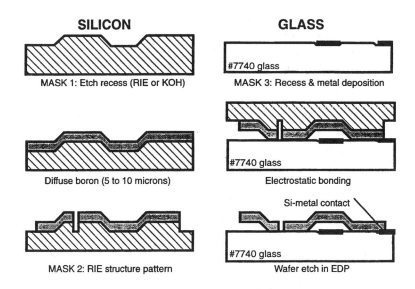

FIGURE 16.65 Dissolved wafer process. (From Greiff, P., SOI-Based Micromechanical Process, presented at Micromachining and Microfabrication Process Technology, Austin, TX, 1995. With permission.)

and suppression of etchant depletion through chemical aging or restricted flow (e.g., through bubbles). These effects can be minimized by techniques that optimize temperature control and reduce bubbling (e.g., proper wafer spacing, lower temperature, large bath). The structures are finally rinsed in DI water and a hot methanol bath.

Draper Laboratory, although obtaining excellent device results with the dissolved wafer process, is now exploring an SOI process as an alternative. The latter yields an all-silicon device while preserving many of the dissolved wafer process advantages (see also "SOI Surface Micromachining" later in this chapter).

Example 2. An Electrochemical Sensor Array Measuring pH, CO_2, and O_2 in a Dual Lumen Catheter

This sensor array developed by the author is shown in Figure 16.66, packaged and ready for *in vivo* monitoring of blood pH, CO_2 and O_2. The linear electrochemical array fits inside a 20-gauge catheter (750 μm in diameter) without taking up so much space as to distort the pressure signal monitored by a pressure sensor outside the catheter. A classical (macro) reference electrode, making contact with the blood through the saline drip, was used for the pH signal, while the CO_2 and O_2 had their own internal reference electrodes. The high impedance of the small electrochemical probes makes a close integration of the electronics mandatory; otherwise, the high impedance connector leads, in a typical hospital setting, act as antennas for the surrounding electronic noise. As can be seen from the computer-aided design (CAD) picture in Figure 16.67, the thickness of the sensor comes from two silicon pieces, the top piece containing electrochemical cells and the bottom piece containing the active electronics. Each wafer is 250 μm thick. The individual electrochemical cells are etched anisotropically into the top silicon wafer. The bottom piece is fabricated in a custom IC housing using standard IC processes. The process sequence to build a generic electrochemical cell in a 250-μm-thick Si wafer with one or more electrodes at the bottom of each well is illustrated in Figure 16.68. Electrochemical wells are etched from the front of the wafer and, after an oxidation step, access cavities for the metal electrodes are also etched from the back. The etching of the vias stops at the oxide-covered bottoms of the electrochemical wells (Figure 16.68A). Next, the wafers are oxidized a second time with the oxide thickness doubling everywhere except in the suspended window areas where no more Si can feed further growth (Figure 16.68B). The desired electrode metal is subsequently deposited from the back of the wafer into the access cavities and against the oxide window

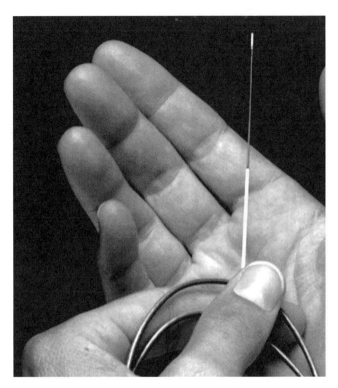

FIGURE 16.66 An *in vivo* pH, CO_2 and O_2 sensor based on a linear array of electrochemical cells.

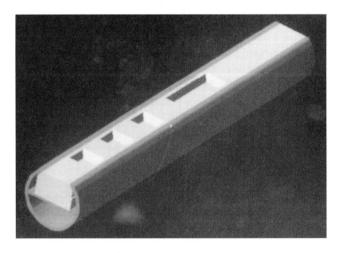

FIGURE 16.67 CAD of the electrochemical sensor array showing two pieces of Si (each 250 µm thick) on top of each other, mounted in a dual lumen catheter. The bottom part of the catheter is left open so pressure can be monitored and blood samples can be taken.

(Figure 16.68C). Finally, a timed oxide etch removes the sacrificial oxide window from above the underlying metal, while preserving the thicker oxide layer in the other areas on the chip (Figure 16.68D) [Joseph et al., 1989; Madou and Otagawa, 1989; Holland et al., 1988]. An SEM micrograph demonstrating step D in Figure 16.68 is shown in Figure 16.69. The electrodes in the electrochemical cells of the top wafer are further connected to the bottom wafer electronics by solder balls in the access vias of the top silicon wafer (see Figure 16.70). Separating the chemistry from the electronics in this way provides extra

MEMS Fabrication

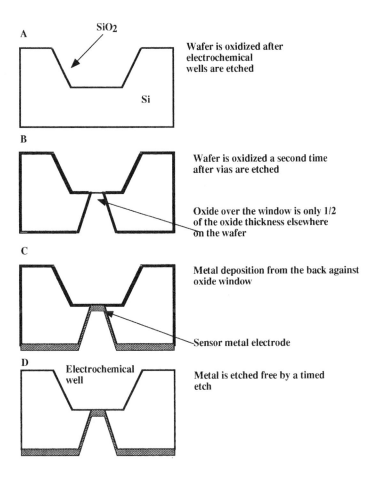

FIGURE 16.68 Fabrication sequence for a generic electrochemical cell in Si. Depth of the electrochemical well, number of electrodes and electrode materials can be varied.

FIGURE 16.69 SEM micrograph illustrating process step D from Figure 16.67. A 30 × 30-μm Pt electrode is shown at the bottom of an electrochemical well. This Pt electrode is further contacted to the electronics from the back.

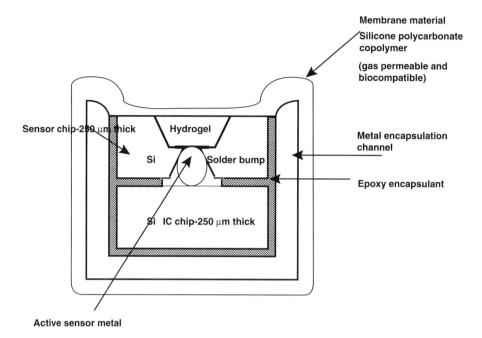

FIGURE 16.70 Schematic of the bonding scheme between sensor wafer and IC. The schematic is a cut-through of the catheter.

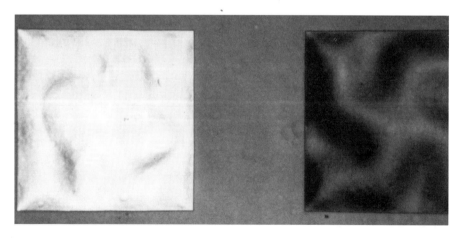

FIGURE 16.71 SEM micrograph of an Ag/AgCl (left) and an IrO_x (right) electrode at the bottom of an anisotropically etched well in Si. This electrochemical cell forms the basis for a Severinghaus CO_2 sensor.

protection for the electronics from the electrolyte as well as from the electronics, and chemical sensor manufacture can proceed independently. Depending on the type of sensor element, one or more electrodes are fabricated at the bottom of the electrochemical cells. For example, shown in Figure 16.71 is an almost completed (Severinghaus) CO_2 sensor with an Ag/AgCl electrode as the reference electrode (left in the figure) and an IrOx pH-sensitive electrode, both at the bottom of one electrochemical well. The metal electrodes are electrically isolated from each other by the SiO_2 passivation layer over the surface of the silicon wafer. To complete the CO_2 sensor, we silk-screen a hydrogel containing an electrolytic medium into the silicon sensor cavity and dip-coat the sensor into a silicone-polycarbonate rubber solution to form the gas-permeable membrane. For hydrogel inside the micromachined well, we use poly(2-hydroxyethyl)methacrylate (PHEMA) or polyvinylalcohol (PVA).

MEMS Fabrication

The concept of putting the sensor chemistries and the electronics on opposite sides of a substrate is a very important design feature we decided upon several years ago in view of the overwhelming problems encountered in building chemical sensors based on ISFETs or EGFETs [Madou and Morrison, 1989].

16.10 Surface Micromachining: Introduction

Bulk micromachining means that three-dimensional features are etched into the bulk of crystalline and noncrystalline materials. In contrast, surface micromachined features are built up, layer by layer, on the surface of a substrate (e.g., a single crystal silicon wafer). Dry etching defines the surface features in the x,y plane and wet etching releases them from the plane by undercutting. In surface micromachining, shapes in the x,y plane are unrestricted by the crystallography of the substrate. For illustration, in Figure 16.72 we compare an absolute pressure sensor based on poly-Si and made by surface micromachining with one made by bulk micromachining in single crystal Si.

The nature of the deposition processes involved determines the very flat surface micromachined features (Hal Jerman from EG&G's ICSensors called them 2.5 D features [Jerman, 1994]). Specifically, low-pressure chemical vapor deposited (LPCVD) polycrystalline silicon (poly-Si) films generally are only a few microns high (low z). In contrast with wet bulk micromachining only the wafer thickness limits the feature height. A low z may mean a drawback for some sensors. For example, it will be difficult to fashion a large inertial mass

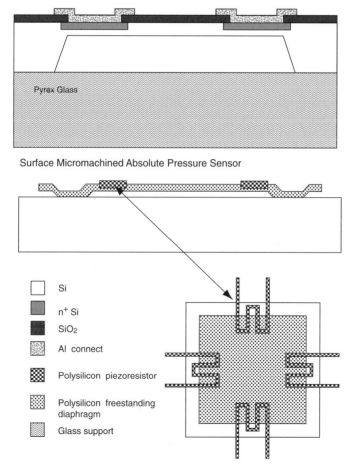

FIGURE 16.72 Comparison of bulk micromachined and surface micromachined absolute pressure sensors equipped with piezoresistive elements. (Top) Bulk micromachining in single crystal Si. (Bottom) Surface micromachining with poly-Si.

for an accelerometer from those thin poly-Si plates. Not only do many parameters in the LPCVD polysilicon process need to be controlled very precisely, subsequent high temperature annealing (say, at temperatures of about 580°C) is needed to transform the as-deposited amorphous silicon into polysilicon—the main structural material in surface micromachining. Even with the best possible process control, polysilicon has some material disadvantages over single crystal Si. For example, it generates a somewhat smaller yield strength (values between 2 and 10 times smaller have been reported) [Biebl and von Philipsborn, 1993; Greek et al., 1995] and has a lower piezoresistivity [Le Berre et al., 1996]. Moreover, since a grain diameter may constitute a significant fraction of the thickness of a mechanical member, the effective Young's modulus may exhibit significant variability from sensor to sensor [Howe, 1995]. An important positive attribute of poly-Si is that its material properties, although somewhat inferior to single crystal, are far superior compared to those of metal films, and, most of all, since they are isotropic, design is rendered dramatically simpler than with single crystal material. Dimensional uncertainties are of greater concern than material issues. Although absolute dimensional tolerances obtained with lithography techniques may be submicron, relative tolerances are poor, perhaps 1% on the length of a 100-µm-long feature. The situation becomes critical with yet smaller feature sizes [Madou, 1997, chap. 7, Figure 7.1]. Although the relatively coarse dimensional control in the microdomain is not specific to surface micromachining, there is no crystallography to rely on for improved dimensional control as in the case of wet bulk micromachining. Moreover, since the mechanical members in surface micromachining tend to be smaller, more post-fabrication adjustment of the features is required to achieve reproducible characteristics. Finally, the wet process for releasing structural elements from a substrate tends to cause sticking of suspended structures to the substrate, so-called stiction, introducing another disadvantage associated with surface micromachining.

Some of the mentioned problems associated with surface micromachining have recently been resolved by process modifications and/or alternative designs, and the technique has rapidly gained commercial interest, mainly because it is the most IC-compatible micromachining process developed to date. Moreover, in the last 5 to 10 years, processes such as silicon on insulator (SOI) [Diem et al., 1993], hinged poly-Si [Pister, 1992], Keller's molded milli-scale polysilicon [Keller and Ferrari, 1994], thick poly-Si (10 µm and beyond) [Lange et al., 1995], as well as LIGA[4] and LIGA-like processes, have further enriched the surface micromachining arsenal. Some preliminary remarks on each of these surface micromachining extensions follow.

Silicon crystalline features, anywhere between fractions of a micron to 100 µm high, can readily be obtained by surface micromachining of the epi-silicon or fusion-bonded silicon layer of SOI wafers [Noworolski et al., 1995]. Structural elements made from these single crystalline Si layers result in more reproducible and reliable sensors. SOI machining combines the best features of surface micromachining (i.e., IC compatibility and freedom in x,y shapes) with the best features of bulk micromachining (superior single crystal Si properties). Moreover, SOI surface micromachining frequently involves fewer process steps and offers better control over the thickness of crucial building blocks. Given the poor reproducibility of mechanical properties and generally poor electronic characteristics of polysilicon films, SOI machining may surpass the poly-Si technology for fabricating high performance devices.

The fabrication of poly-Si planar structures for subsequent vertical assembly by mechanical rotation around micromachined hinges dramatically increases the plethora of designs feasible with poly-Si [Pister, 1992]. Today, erecting these poly-Si structures with the probes of an electrical probe station or, occasionally, assembly by chance in the HF etch or DI water rinse [Chu et al., 1995] represents too complicated or unreliable postrelease assembly methods for commercial acceptance.

At the University of California at Berkeley, Keller introduced a combination of surface micromachining and LIGA-like molding processes [Keller and Ferrari, 1994] in the so-called HEXSIL[5] process, a technology enabling the fabrication of tall three-dimensional microstructures without postrelease assembly. Using CVD processes one can generally only deposit thin films (~2 to 5 µm) on flat surfaces. If however, these surfaces are the opposing faces of deep narrow trenches, the growing films will merge to form solid beams. Releasing

---
[4]LIGA is the acronym for the German Lithographie, Galvanoformung, Abformung (see chap. 17).
[5]HEXSIL is the acronym for HEXagonal honeycomb polySILicon.

of such polysilicon structures and the incorporation of electroplating steps expand the surface micromachining bandwidth in terms of choice of materials and accessible feature heights. In this fashion, high aspect ratio structures normally associated with LIGA can now also be made of CVD polysilicon.

Applying classical LPCVD to obtain poly-Si deposition is a slow process. For example, a layer of 10 μm typically requires a deposition time of 10 hours. Consequently, most micromachined structures are based on layer thicknesses in the 2- to 5-μm range. Based on dichlorosilane (SiH_2Cl_2) chemistry, Lange et al. (1995) developed a CVD process in a vertical epitaxy batch reactor with deposition rates as high as 0.55 μm/min at 1000°C. The process yields acceptable deposition times for thicknesses in the 10-μm range. The highly columnar poly-Si films were deposited on sacrificial SiO_2 layers and exhibit low internal tensile stress making them suitable for surface micromachining.

Finally, thick layers of polyimide and other new UV resists also receive a lot of attention as important new extensions of surface micromachining. Due to their transparency to exposing UV light, they can be transformed into tall surface structures with LIGA-like high aspect ratios. They may be both electroplated and molded.

In this part of the chapter, we first review thin film material properties in general, focusing on significant differences with bulk properties of the same material, followed by a review of the main surface micromachining processes. Next, we clarify the recent extensions of the surface micromachining technique listed above. Because of the complexity of the many parameters influencing thin film properties, we then present a set of case studies on the most commonly used thin film materials.

In the surface micromachining examples at the end of this chapter, we first look at a lateral resonator. Resonators have found an important industrial application in accelerometers introduced by Analog Devices. The other example concerns a micromotor: an electrostatic linear motor with a slider stepping between two poly-Si rails. Micromotors may lack practical use as of yet, but, just as the ion-sensitive field effect transistor (ISFET) galvanized the chemical sensor community in trying out new chemical sensing approaches, micromotors energized the micromachining research community to fervently explore miniaturization of a wide variety of mechanical sensors and actuators. Micromotors also brought about the christening of the micromachining field into 'Microelectromechanical Systems' or MEMS.

16.11 Historical Note

The first example of a surface micromachine for an electro-mechanical application consisted of an underetched metal cantilever beam for a resonant gate transistor made by Nathanson (1967). By 1970, a first suggestion for a magnetically actuated metallic micromotor emerged [Dutta, 1970]. Because of fatigue problems metals rarely are counted on as mechanical components. The surface micromachining method as we know it today was first demonstrated by Howe and Muller in the early 1980s and relied on polysilicon as the structural material [Howe and Muller, 1982]. These pioneers and Guckel [Guckel and Burns, 1985], an early contributor to the field, produced free-standing LPCVD polysilicon structures by removing the oxide layers on which the polysilicon features were formed. Howe's first device consisted of a resonator designed to measure the change of mass upon adsorption of chemicals from the surrounding air. A gas sensor does not necessarily represent a good application of a surface micromachined electrostatic structure since humidity and dust foul the thin air gap of such an unencapsulated microstructure in a minimal amount of time. Later mechanical structures, especially hermetically sealed mechanical devices, provided proof that the IC revolution could be extended to electromechanical systems [Howe, 1995]. In these structures, the height (z-direction) typically is limited to less than 10 μm, ergo the name surface micromachining.

The first survey of possible applications of poly-Si surface micromachining was presented by Gabriel et al. (1989). Microscale movable mechanical pin joints, springs, gears, cranks, sliders, sealed cavities and many other mechanical and optical components have been demonstrated in the laboratory [Muller, 1987; Fan et al., 1987]. In 1991, Analog Devices announced the first commercial product based on surface micromachining, namely, the ADXL-50, a 50-g accelerometer for activating air-bag deployment [Editor, 1991]. A new wave of major commercial applications for surface micromachining could be based on

Texas Instruments' Digital Micromirror Device™. This surface micromachined movable mirror is a digital light switch that precisely controls a light source for projection display and hard copy applications [Hornbeck, 1995]. The commercial acceptance of this application will likely determine the staying power of the surface micromachining option.

16.12 Mechanical Properties of Thin Films

16.12.1 Introduction

Thin films in surface micromachines must satisfy a large set of rigorous chemical, structural, mechanical and electrical requirements. Excellent adhesion, low residual stress, low pinhole density, good mechanical strength and chemical resistance all may be required simultaneously. For many microelectronic thin films, the material properties depend strongly on the details of the deposition process and the growth conditions. In addition, some properties may depend on post-deposition thermal processing, referred to as annealing. Furthermore, the details of thin-film nucleation and/or growth may depend on the specific substrate or on the specific surface orientation of the substrate. Although the properties of a bulk material might be well characterized, its thin-film form may have properties substantially different from those of the bulk. For example, thin films generally display smaller grain size than bulk materials. An overwhelming reason for the many differences stem from the properties of thin films, which exhibit a higher surface-to-volume ratio than large chunks of material and are strongly influenced by the surface properties.

For more details on deposition techniques in general, refer to Madou (1997, chap. 3); now we focus on the physical characteristics of the resulting thin deposits. Some terminology characterizing thin films and their deposition is introduced in Table 16.11.

Since thin films were not intended for load-bearing applications, their mechanical properties have largely been ignored. The last 10 years saw the development of a strong appreciation for understanding the mechanical properties as essential for improving the reliability and life-time in thin films, even in nonstructural applications [Vinci and Braveman, 1991]. Surface micromachining contributes heavily to this understanding. One of the most influential long-term contributions of surface micromachining might well lie in the elucidation of stress mechanism in thin films.

16.12.2 Adhesion

One cannot stress enough the importance of adhesion of various films to one another and to the substrate in overall IC performance and reliability. As mechanical pulling forces might be involved, adhesion is even more crucial in micromachining. If films lift from the substrate under a repetitive, applied mechanical force, the device will fail. Classical adhesion tests include the scotch tape test, abrasion method, scratching, deceleration (ultrasonic and ultracentrifuge techniques), bending, pulling etc. [Campbell, 1970]. Micromachined structures, because of their sensitivity to thin-film properties, enable some innovative new ways of *in situ* adhesion measurement. Figure 16.73 illustrates how a suspended membrane may be used for adhesion measurements. Figure 16.73A shows the membrane suspended but still adherent to the substrate. Figure 16.73B shows the membrane after it has been peeled from its substrate by an applied load (gas pressure). Figure 16.73C illustrates the accompanying P(ressure)-V(olume) cycle, in which the membrane is inflated, peeled and then deflated. The shaded portion of Figure 16.73C illustrates the P-V work creating the new surface, which equals the average work of adhesion for the film-substrate interface times the area peeled during the test [Senturia, 1987].

Cleanliness of a substrate is a *condition sine qua non* for good film adhesion. Roughness, providing more bonding surface area and mechanical interlocking further improve it. Adhesion also improves with increasing adsorption energy of the deposit and/or increasing number of nucleation sites in the early growth stage of the film. Sticking energies between film and substrate range from less than 10 kcal/mole in physisorption to more than 20 kcal/mole for chemisorption. The weakest form of adhesion involves van der Waals forces only (see also Table 16.11).

TABLE 16.11 Thin Film Terms Used in Characterizing Deposition

Term	Remark
Physisorbed film	Bond energy < 10 kcal/mole
Chemisorbed film	Bond energy > 20 kcal/mole
Nucleation	Adatoms forming stable clusters
Condensation	Initial formation of nuclei
Island formation	Nuclei grow in three dimensions, especially along the substrate surface
Coalescence	Nuclei contact each other and larger, rounded shapes form
Secondary nucleation	Areas between islands are filled in by secondary nucleation, resulting in a continuous film
Grain size of thin film	Generally smaller than for bulk materials and function of deposition and annealing conditions (higher T, larger grains)
Surface roughness	Lower at high temperatures except when crystallization starts; at low temperature the roughness is higher for thicker films; also oblique deposition and contamination increase roughness
Epitaxial and amorphous films	Very low surface roughness
Density	More porous deposits are less dense; density reveals a lot about the film structure
Crystallographic structure	Adatom mobility: amorphous, polycrystalline, single crystal or fiber texture, or preferred orientation

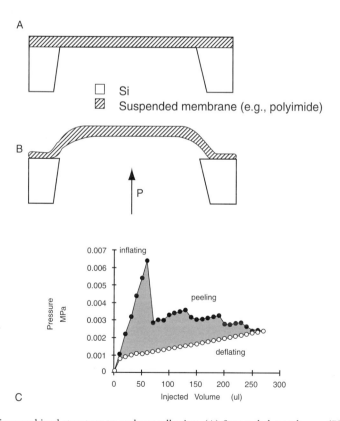

FIGURE 16.73 Micromachined structure to evaluate adhesion. (A) Suspended membrane. (B) Partially detached membrane-outward peel. (C) Pressure-volume curve during inflate-deflate cycle. (From Senturia, S., Can We Design Microrobotic Devices Without Knowing the Mechanical Properties of Materials?, presented at Micro Robots and Tekoperators Workshop, Hyannis, MA, 1987. With permission.)

It is highly advantageous to include a layer of oxide-forming elements between a metal and an oxide substrate. These adhesion layers, such as Cr, Ti, Al etc., provide good anchors for subsequent metallization. Intermediate film formation allowing a continuous transition from one lattice to the other results in the best adhesion. Adhesion also improves when formation of intermetallic metal alloys takes place.

16.12.3 Stress in Thin Films

16.12.3.1 Stress in Thin Films—Qualitative Description

Film cracking, delamination and void formation may all be linked to film stress. Nearly all films foster a state of residual stress, due to mismatch in the thermal expansion coefficient, nonuniform plastic deformation, lattice mismatch, substitutional or interstitial impurities and growth processes. Figure 16.74 lists stress-causing factors categorized as either intrinsic or extrinsic [Krulevitch, 1994]. The intrinsic stresses (also growth stresses) develop during the film nucleation. Extrinsic stresses are imposed by unintended external factors such as temperature gradients or sensor package-induced stresses. Thermal stresses, the most common type of extrinsic stresses, are well understood and often easy to calculate (see below). They arise either in a structure with inhomogeneous thermal expansion coefficients subjected to a uniform temperature change or in a homogeneous material exposed to a thermal gradient [Krulevitch, 1994]. Intrinsic stresses in thin films often are larger than thermal stresses. They usually are a consequence of the nonequilibrium nature of the thin-film deposition process. For example, in chemical vapor deposition, depositing atoms (adatoms) may at first occupy positions other than the lowest energy configuration. With too high a deposition rate and/or too low adatom surface mobility, these first adatoms may become pinned by newly arriving adatoms, resulting in the development of intrinsic stress. Other types of intrinsic stresses illustrated in Figure 16.74 include: transformation stresses occurring when part of a material under-

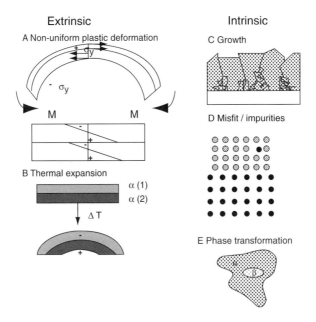

FIGURE 16.74 Examples of intrinsic and extrinsic residual stresses. (A) Nonuniform plastic deformation results in residual stresses upon unloading. M = bonding moment. (B) Thermal expansion mismatch between two materials bonded together. α (1) and α (2) are thermal expansion coefficients. (C) Growth stresses evolve during film deposition. (D) Misfit stresses due to mismatches in lattice parameters in an epitaxial film and stresses from substitutional or interstitial impurities. (E) Volume changes accompanying phase transformations cause residual stresses [Vinci and Braveman, 1991]. (After Krulevitch, P.A., Micromechanical Investigations of Silicon and Ni-Ti-Cu Thin Films, Ph.D. thesis, University of California, Berkeley, 1994.)

goes a volume change during a phase transformation, misfit stresses arising in epitaxial films due to lattice mismatch between film and substrate, and impurities either interstitial or substitutional which cause intrinsic residual stresses due to the local expansion or contraction associated with point defects. Intrinsic stress in a thin film does not suffice to result in delamination unless the film is quite thick. For example, to overcome a low adsorption energy of 0.2 eV, a relatively high stress of about 5×10^9 dyn cm^{-2} (10^7 dyn/cm^2 = 1 MPa) is required [Campbell, 1970]. High stress can result in buckling or cracking of films.

The stress developing in a film during the initial phases of a deposition may be compressive (i.e., the film tends to expand parallel to the surface), causing buckling and blistering or delamination in extreme cases (especially with thick films). Alternatively, thin films may be in tensile stress (i.e., the film tends to contract), which may lead to cracking if forces high enough to exceed the fracture limit of the film material are present. Subsequent rearrangement of the atoms, either during the remainder of the deposition or with additional thermal processing, can lead to further densification or expansion, decreasing remaining tensile or compressive stresses, respectively.

The mechanical response of thin-film structures is affected by the residual stress, even if the structures do not fail. For example, if the residual stress varies in the direction of film growth, the resulting built-in bending moment will warp released structures, such as cantilever beams. The presence of residual stress also alters the resonant frequency of thin-film, resonant microstructures (see Eq. (16.59) below) [Pratt et al., 1992]. Also, residual stress can lead to degradation of electrical characteristics and yield loss through defect generation, such as {311} defects [Krulevitch, 1994]. Another observation found the resistivity of stressed metallic films to be higher than that of their annealed counterparts. In a few cases residual stress has been used advantageously, such as in self-adjusting microstructures [Judy et al., 1991], and for altering the shape-set configuration in shape memory alloy films [Krulevitch, 1994].

In general, the stresses in films, by whatever means produced, are in the range of 10^8 to 5×10^{10} dyn/cm^{-2} and can be either tensile or compressive. For normal deposition temperatures (50 to a few hundred degrees C), the stress in metal films typically ranges from 10^8 to 10^{10} dyn cm^{-2} and is tensile, the refractory metals at the upper end, the soft ones (Cu, Ag, Au, Al) at the lower end. At low substrate temperatures, metal films tend to exhibit tensile stress. This often decreases in a linear fashion with increasing substrate temperature, finally going through zero or even becoming compressive. The changeover to compressive stress occurs at lower temperatures for lower melting point metals. The mobility of the adatoms is key to understanding the ranking for refractory and soft metals. A metal such as aluminum has a low melting temperature and a corresponding high diffusion rate even at room temperature, thus usually it is fairly stress free. By comparison, tungsten has a relatively high melting point and a low diffusion rate and tends to accumulate more stress when sputter-deposited. With dielectric films, stresses often are compressive and have slightly lower values than commonly noted in metals.

Tensile films result, for example, when a process by-product is present during deposition and later driven off as a gas. If the deposited atoms are not sufficiently mobile to fill in the holes left by these departing byproducts, the film will contract and go in tension. Nitrides deposited by plasma CVD usually are compressive due to the presence of hydrogen atoms in the lattice. By annealing, driving the hydrogen out, the films can turn highly tensile. Annealing also has a dramatic effect on most oxides. Oxides often are porous enough to absorb or give off a large amount of water. Full of water, they are compressive; devoid of water they are tensile. Thermal SiO_2 is compressive, though, even when dry. If atoms are jammed in place (such as with sputtering) the film tends to act compressively. The stress in a thin film also varies with depth. The RF power of a plasma-enhanced CVD (PECVD) deposition influences stress, e.g., a thin film may start out tensile, decrease as the power increases, and finally become compressive with further RF power increase. CVD equipment manufacturers concentrate on building stress-control capabilities into new equipment by controlling plasma frequency (see also Madou [1997, chap. 3]).

16.12.3.2 Stress in Thin Films on Thick Substrates—Quantitative Analysis

The total stress in a thin film typically is given by:

$$\sigma_{tot} = \sigma_{th} + \sigma_{int} + \sigma_{ext} \quad (16.44)$$

i.e., the sum of any intentional external applied stress (σ_{ext}), the thermal stress (σ_{th}, an unintended external stress) and different intrinsic components (σ_{int}). With constant stress through the film thickness, the stress components retain the form of

$$\begin{aligned}
\sigma_x &= \sigma_x(x, y) \\
\sigma_y &= \sigma_y(x, y) \\
\tau_{xy} &= \tau_{xy}(x, y) \\
\tau_{xz} &= \tau_{yz} = \sigma_z = 0
\end{aligned} \quad (16.45)$$

That is, the three nonvanishing stress components are functions of x and y alone. No stress occurs in the direction normal to the substrate (z). With x, y as principal axes, the shear stress τ_{xy} also vanishes [Chou and Pagano, 1967] and Eq. (16.45) reduces to the following strain–stress relationships:

$$\begin{aligned}
\varepsilon_x &= \frac{\sigma_x}{E} - \frac{\nu \sigma_y}{E} \\
\varepsilon_y &= \frac{\sigma_y}{E} - \frac{\nu \sigma_x}{E} \\
\sigma_z &= 0
\end{aligned} \quad (16.46)$$

In the isotropic case $\varepsilon = \varepsilon_x = \varepsilon_y$ so that $\sigma_x = \sigma_y = \sigma$, or:

$$\sigma = \left(\frac{E}{1 - \nu}\right)\varepsilon \quad (16.47)$$

where the Young's modulus of the film and the Poissons's ratio of the film act independently of orientation. The quantity $E/1 - \nu$ often is called the biaxial modulus. Uniaxial testing of thin films is difficult, prompting the use of the biaxial modulus, rather than Young's uniaxial modulus. Plane stress, as described here, presents a good approximation several thicknesses away from the edge of the film (say, three thicknesses from the edge).

Thermal Stress

Thermal stresses develop in thin films when high temperature deposition or annealing are involved, and usually are unavoidable due to mismatch of thermal expansion coefficients between film and substrate. The problem of a thin film under residual thermal stress can be modeled by considering a thought experiment involving a stress-free film at high temperature on a thick substrate. Imagine detaching the film from the high-temperature substrate and cooling the system to room temperature. Usually, the substrate dimensions undergo minor shrinkage in the plane while the film's dimensions may reduce significantly. In order to reapply the film to the substrate with complete coverage, the film needs stretching with a biaxial tensile load to a uniform radial strain ε, followed by perfect bondage to the rigid substrate and load removal. The film stress is assumed to be the same in the stretched and free-standing film as in the film bonded to the substrate, i.e., no relaxation occurs in the bonding process. To calculate the thermal residual stress from Eq. (16.47) the elastic moduli of the film must be known, as well as the volume change associated with the residual stress, i.e., the thermal strain, ε_{th}, resulting from the difference in the coefficients of thermal expansion between the film and the substrate.

Let us now consider whether, qualitatively, the above assumptions apply to the measurement of thin films on Si wafers. Such films typically measure 1 µm thick and are deposited on 4-in. wafers nominally

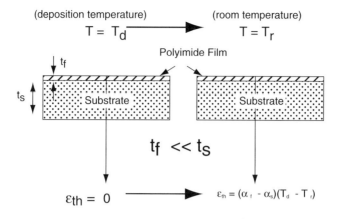

FIGURE 16.75 Thermal stress. Tension and compression are determined by the relative size of thermal expansion coefficients of film and substrate. Suppose a strain-free film at deposition temperature, T_d, is cooled to room temperature, T_r, on a substrate with a different coefficient of thermal expansion.

550 μm thick. In this case, the substrate measures nearly three orders of magnitude thicker than the film, and, because the bending stiffness is proportional to the thickness cubed, the substrate essentially is rigid relative to the film. The earlier assumptions clearly apply.

Figure 16.75 portrays a quantitative example where a polyimide film, strain-free at the deposition temperature (T_d) of 400°C, is cooled to room temperature T_r (25°C) on an Si substrate with a different coefficient of thermal expansion. The resulting strain is given by:

$$\varepsilon_{th} = \int (\alpha_f(T) - \alpha_s(T)) dt \qquad (16.48)$$

where α_f and α_s represent the coefficients of thermal expansion for the polyimide film and the Si substrate, respectively. The thermal strain can be of either sign, based on the relative values of α_f and α_s: positive is tensile, negative is compressive. Polyimide features a thermal expansion ($\alpha_f = 70 \times 10^{-6} {}^\circ C^{-1}$) larger than the thermal expansion coefficient of Si ($\alpha_s = 2.6 \times 10^{-6} {}^\circ C^{-1}$); hence, a tensile stress is expected. With SiO_2 grown or deposited on silicon at elevated temperatures, a compressive component [α_f ($0.35 \times 10^{-6} {}^\circ C^{-1}$) < α_s ($2.6 \times 10^{-6} {}^\circ C^{-1}$)] is expected. Assuming that the coefficients of thermal expansion are temperature independent, Eq. (16.48) simplifies to:

$$\varepsilon_{th} = (\alpha_f - \alpha_s)(T_d - T_r) \qquad (16.49)$$

The calculated thermal strain ε_{th} for polyimide on Si then measures 25×10^{-3} at room temperature. The biaxial modulus ($E/1 - \nu$), with $E = 3$ GPa and $\nu = 0.4$, equals 5 GPa, and the residual stress σ, from Eq. (16.47), is 125 MPa and tensile.

Intrinsic Stress

The intrinsic stress, σ_i, reflects the internal structure of a material and is less clearly understood than the thermal stress which it often dominates [Guckel et al., 1988]. Several phenomena may contribute to σ_i, making its analysis very complex. Intrinsic stress depends on thickness, deposition rate (locking in defects), deposition temperature, ambient pressure, method of film preparation, type of substrate used (lattice mismatch), incorporation of impurities during growth etc. Some semiquantitative descriptions of various intrinsic stress-causing factors follow:

- Doping ($\sigma_{int} > 0$ or $\sigma_{int} < 0$): When doping Si, the atomic or ionic radius of the dopant and the substitutional site determine the positive or negative intrinsic stress ($\sigma_{int} > 0$ or tensile, and $\sigma_{int} < 0$ or compressive). With boron-doped poly-Si, an atom small compared to Si, the film is expected to be tensile ($\sigma_{int} > 0$); with phosphorous doping, an atom large compared to Si, the film is expected to be compressive ($\sigma_{int} < 0$).
- Atomic peening ($\sigma_{int} < 0$): Ion bombardment by sputtered atoms and working gas densifies thin films, rendering them more compressive. Magnetron sputtered films at low working pressure (<1 Pa) and low temperature often exhibit compressive stress.
- Microvoids ($\sigma_{int} > 0$): Microvoids may arise when byproducts during deposition escape as gases and the lateral diffusion of atoms evolves too slowly to fill all the gaps, resulting in a tensile film.
- Gas entrapment ($\sigma_{int} < 0$): As an example we can cite the hydrogen trapped in Si_3N_4. Annealing removes the hydrogen, and a nitride film, compressive at first, may become tensile if the hydrogen content is sufficiently low.
- Shrinkage of polymers during cure ($\sigma_{int} > 0$): The shrinkage of polymers during curing may lead to severe tensile stress, as becomes clear in the case of polyimides. Special problems are associated with measuring the mechanical properties of polymers as they exhibit a time-dependent mechanical response (viscoelasticity), a potentially significant factor in the design of mechanical structures where polymers are subjected to sustained loads [Maseeh and Senturia, 1990].
- Grain boundaries ($\sigma_{int} = ?$): Based on intuition one expects that the interatomic spacing in grain boundaries differs depending on the amount of strain, thus contributing to the intrinsic stress. But the origin of, for example, the compressive stress in polysilicon and how it relates to the grain structure and interatomic spacing are not yet completely clear (see also below on coarse and fine-grated Si).

For further reading on thin-film stress, refer to Hoffman (1975; 1976).

16.12.3.3 Stress-Measuring Techniques

Introduction

A stressed thin film will bend a thin substrate by a measurable degree. A tensile stress will bend and render the surface concave; a compressive stress renders the surface convex. The most common methods for measuring the stress in a thin film are based on this substrate bending principle. The deformation of a thin substrate due to stress is measured either by observing the displacement of the center of a circular disk or by using a thin cantilevered beam as a substrate and calculating the radius of curvature of the beam and hence the stress from the deflection of the free end. More sophisticated local stress measurements use analytical tools such as X-ray [Wong, 1978], acoustics, Raman spectroscopy [Nishioka et al., 1985], infrared spectroscopy [Marco et al., 1991], and electron-diffraction techniques. Local stress does not necessarily mean the same as the stress measured by substrate bending techniques, since stress is defined microscopically, while deformations are induced mostly macroscopically. The relation between macroscopic forces and displacements and internal differential deformation, therefore, must be modeled carefully. Local stress measurements may also be made using *in situ* surface micromachined structures such as strain gauges made directly out of the film of interest itself [Mehregany et al., 1987]. The deflections of thin suspended and pressurized micromachined membranes may be measured by mechanical probe [Jaccodine and Schlegel, 1966], laser [Bromley et al., 1983], or microscope [Allen et al., 1987]. Intrinsic stress influences the frequency response of microstructures (see Eq. (16.59)) which can be measured by laser [Zhang et al., 1991], spectrum analyzer [Pratt et al., 1991], or stroboscope. Whereas residual stress can be determined from wafer curvature and microstructure deflection data, material structure of the film can be studied by X-ray diffraction and transmission electron microscopy (TEM). Krulevitch, among others, attempted to link the material structure of poly-Si and its residual strain [Krulevitch, 1994]. Following, we will review stress-measuring techniques, starting with the more traditional ones and subsequently clarifying the problems and opportunities in stress measuring with surface micromachined devices.

Disk Method

For all practical purposes only stresses in the x and y directions are of interest in determining overall thin-film stress, as a film under high stress can only expand or contract by bending the substrate and deforming it in a vertical direction. Vertical deformations will not induce stresses in a substrate because it freely moves in that direction. The latter condition enables us to obtain quite accurate stress values by measuring changes in bow or radius of curvature of a substrate. The residual stresses in thin films are large and sensitive optical or capacitive gauges may measure the associated substrate deflections.

The disk method, most commonly used, is based on a measurement of the deflection in the center of the disk substrate (say, a Si wafer) before and after processing. Since any change in wafer shape is directly attributable to the stress in the deposited film, it is relatively straightforward to calculate stress by measuring these changes. Stress in films using this method is found through the Stoney equation [Hoffman, 1976], relating film stress to substrate curvature:

$$\sigma = \frac{1}{R} \frac{E}{6(1-\nu)} \frac{T^2}{t} \qquad (16.50)$$

where R represents the measured radius of curvature of the bent substrate, $E/(1-\nu)$ the biaxial modulus of the substrate, T the thickness of the substrate, and t the thickness of the applied film [Singer, 1992]. The underlying assumptions include:

- The disc substrate is thin and has transversely isotropic elastic properties with respect to the film normal.
- The applied film thickness is much less than the substrate thickness.
- The film thickness is uniform.
- Temperature of the disk substrate/film system is uniform.
- Disc substrate/film system is mechanically free.
- Disc substrate without film has no bow.
- Stress is equi-biaxial and homogeneous over the entire substrate.
- Film stress is constant through the film thickness.

For most films on Si we assume that $t \leq T$; for example, t/T measures $\sim 10^{-3}$ for thin films on Si. The legitimacy of the uniform thickness, homogeneous, and equi-biaxial stress assumptions depend on the deposition process. Chemical vapor deposition (CVD) is such a widely used process as it produces relatively uniform films; however, sputter-deposited films can vary considerably over the substrate. In regard to the assumption of stress uniformity with film thickness, residual stress can vary considerably through the thickness of the film. Eq. (16.50) only gives an average film stress in such cases. In cases where thin films are deposited onto anisotropic single crystal substrates, the assumption of a substrate with transversely isotropic elastic properties with respect to the film normal is not always satisfied. Using single crystal silicon substrates possessing moderately anisotropic properties (Eq. (16.20)) (<100>- or <111>-oriented wafers) satisfies the transverse isotropy argument. Any curvature inherent in the substrate must be measured before film deposition and algebraically added to the final measured radius of curvature. To give an idea of the degree of curvature, 1 μm of thermal oxide may cause a 30-μm warp of a 4-in. silicon wafer, corresponding to a radius of curvature of 41.7 m.

Presently, five companies offer practical disk method-based instruments to measure stress on wafers: ADE Corp. (Newton, MA), GCA/Tropel (Fairport, NY), Ionic Systems (Salinas, CA), Scientific Measurements Systems, Inc. (San Jose, CA), and Tencor Instruments (Mountain View, CA) [Singer, 1992]. Figure 16.76A illustrates the sample output from Tencor's optical stress analysis system. Figure 16.76B represents the measuring principle of Ionic Systems' optical stress analyzer. None of the above tech-

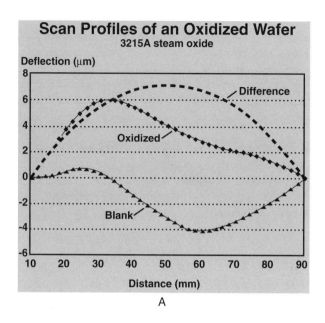

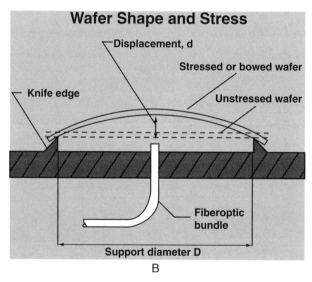

FIGURE 16.76 Curvature measurement for stress analysis. (A) Sample output from Tencor's FLX stress analysis instrument, showing how stress is derived from changes in wafer curvature. (B) The reflected light technique, used by Ionic Systems to measure wafer curvature. (From Singer, P., *Semicond. Int.*, 15, 54–58, 1992. With permission.)

niques satisfies the need for measuring stress in low modulus materials such as polyimide. For the latter applications the suspended membrane approach (see below) is more suited.

Uniaxial Measurements of Mechanical Properties of Thin Films
Many problems associated with handling thin films in stress test equipment may be bypassed by applying micromachining techniques. One simple example of problems encountered with thin films is the measurement of uniaxial tension to establish the Young's modulus. This method, effective for macroscopic samples, proves problematic for small samples. The test formula is illustrated in Figure 16.77A. The gauge length L in this figure represents the region we allow to elongate and the area A (=W × H) is the cross section of the specimen. A stress F/A is applied and measured with a load cell; the strain δL/L is measured

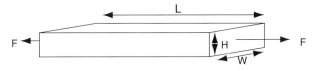

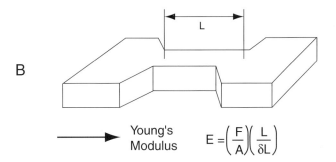

FIGURE 16.77 Measuring Young's modulus. (A) With a bar-shaped structure. (B) With a dog-bone, Instron specimen. (After Senturia and Howe, 1990.)

with an LVDT or another displacement transducer (a typical instrument used is the Instron 1123). The Young's modulus is then deduced from:

$$E = \left(\frac{F}{A}\right)\left(\frac{L}{\delta L}\right) \quad (16.51)$$

The obvious problem, for small or large samples, is how to grip onto the sample without changing A. Under elongation, A will indeed contract by $(w+H)\delta vL$. In general, making a dog-bone shaped structure (Instron specimen) solves that problem as shown in Figure 16.77B. Still, the grips introduced in an Instron sample can produce end effects and uncertainties in determining L. Making Instron specimens in thin films is even more of a challenge since the thin film needs to be removed from the surface, possibly changing the stress state, while the removal itself may modify the film.

As in the case of adhesion, some new techniques for testing stress in thin films, based on micromachining, are being explored. These microtechniques prove more advantageous than the whole wafer disc technique in that they are able to make local measurements.

The fabrication of micro-Instron specimens of thin polyimide samples is illustrated in Figure 16.78A. Polyimide is deposited on a p^+ Si membrane in multiple coats. Each coat is prebaked at 130°C for 15 min. After reaching the desired thickness the film is cured at 400°C in nitrogen for 1 hr (A). The polyimide is then covered with a 3000-Å layer of evaporated aluminum (B). The aluminum layer is patterned by wet etching (in phosphoric-acetic-nitric solution referred to as PAN etch) to the Instron specimen shape (C). Dry etching transfers the pattern to the polyimide (D). After removing the Al mask by wet etching, the p^+ support is removed by a wet isotropic etch (HNA) or a SF_6 plasma etch (E), and finally the side silicon is removed along four pre-etched scribe lines, releasing the residual stress (F). The remaining silicon acts as supports for the grips of the Instron [Meseeh-Tehrani, 1990]. The resulting structure can

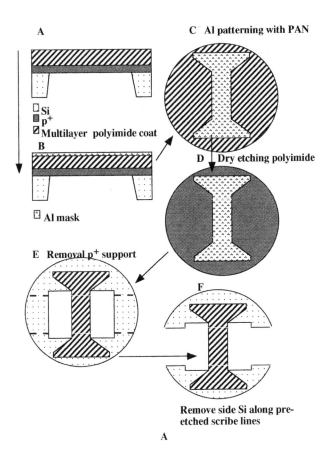

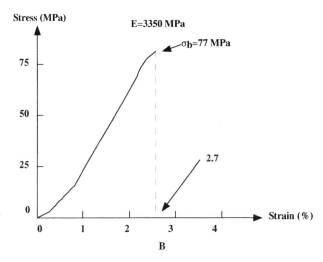

FIGURE 16.78 Uniaxial stress measurement. (A) Fabrication process of a dog bone sample for measurement of uniaxial strain. (B) Stress vs. strain for Du Pont's 2525 polyimide. (Courtesy of Dr. F. Maseeh-Tehrani.)

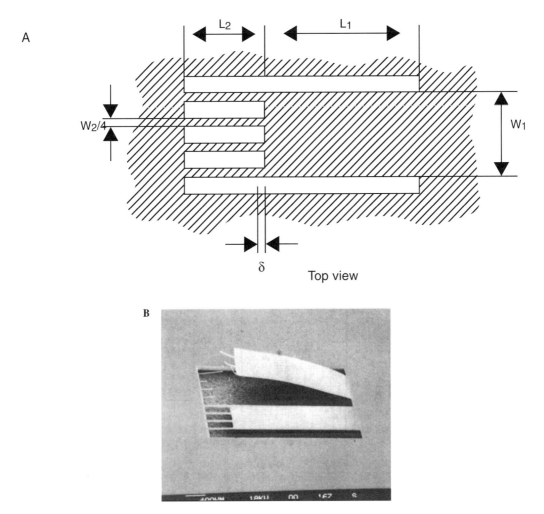

FIGURE 16.79 Ultimate strain. (A) Test structures for stress-to-modulus (strain) and ultimate stress measurements. (B) Two released structures, one of which has exceeded the ultimate strain of the film, resulting in fracture of the necks. (From Senturia, S., Can We Design Microbiotic Devices without Knowing the Mechanical Properties of the Materials?, presented at Micro Robots and Teleoperators Workshop, Hyannis, MA, 1987. With permission.)

be manipulated as any macrosample without the need for removal of the film from its substrate. This technique enables the gathering of stress/strain data for a variety of commercially available polyimides [Maseeh and Senturia, 1989]. A typical measurement result for Dupont's polyimide 2525, illustrated in Figure 16.78B, gives a break stress and strain of 77 MPa (σ_b) and 2.7% (ε_b), respectively, and 3350 MPa for the Young's modulus.

Figure 16.79A illustrates a micromachined test structure able to establish the strain and the ultimate stress of a thin film [Senturia, 1987]. A suspended rectangular polymer membrane is patterned into an asymmetric structure before removing the thin supporting Si. Once released, the wide suspended strip (width w_1) pulls on the thinner necks (total width w_2), resulting in a deflection δ from its original mask position toward the right to its final position after release. The residual tensile stress in the film drives the deformation δ as shown in Figure 16.79A. By varying the geometry, it is possible to create structures exhibiting small strain in the thinner sections as opposed to others that exceed the ultimate strain of the film. For structures where the strain is small enough to be modeled with linear elastic behavior, the deflection δ can be related to the strain as follows:

$$\varepsilon = \frac{\sigma}{E} = \frac{\delta\left(\frac{W_1}{L_1} + \frac{W_2}{L_2}\right)}{W_1 - W_2} \qquad (16.52)$$

where the geometries are defined as illustrated in Figure 16.79A. Figure 16.79B displays a photograph of two released structures: one with thicker necks, the other with necks so thin that they fractured upon release of the film. Based on the residual tensile strain of the film and the geometry of the structures that failed, the ultimate strain of the particular polyimide used was determined to be 4.5%.

Using similar micromachined tensile test structures [Biebl and von Philipsborn, 1993] measured the fracture strength of undoped and doped polysilicon and found 2.84 ± 0.09 GPa for undoped material and 2.11 ± 0.10 GPa in the case of phosphorous doping, 2.77 ± 0.08 GPa for boron doping and 2.70 ± 0.09 GPa for arsenic doping. No statistically significant differences were observed between samples released using concentrated HF or buffered HF. However, a 17% decrease of the fracture stress was observed for a 100% increase in etching time. These data contrast with Greek et al. (1995) *in situ* tensile strength test result of 768 MPa for an undoped poly-Si film. A mean tensile strength almost ten times less than that of single crystal Si (6 GPa) [Ericson and Schweitz, 1990]. We normally expect polycrystalline films to be stronger than single crystal films (see below under Strength of Thin Films). Greek et al. (1995) explain this discrepancy for poly-Si by pointing out that their polysilicon films have a very rough surface compared to single crystal material, containing many locations of stress concentration where a fracture crack can be initiated.

Biaxial Measurements of Mechanical Properties of Thin Films: Suspended Membrane Methods
We noted earlier that none of the disk stress-measuring techniques is suitable for measuring stress in low modulus tensile materials such as polyimides. Suspended membranes are very convenient for this purpose. The same micromachined test structure used for adhesion testing, sometimes called the blister test, as shown in Figure 16.73, can measure the tensile stress in low modulus materials. This type of test structure ensues from shaping a silicon diaphragm by conventional anisotropic etching, followed by applying the coating, and, finally, removing the supporting silicon from the back with an SF_6 plasma [Senturia, 1987]. By pressurizing one side of the membrane and measuring the deflection, one can extract both the residual stress and the biaxial modulus of the membrane. Pressure to the suspended film can be applied by a gas or by a point-load applicator [Vinci and Braveman, 1991]. The load-deflection curve at moderate deflections (strains less than 5%) answers to:

$$p = C_1 \frac{\sigma t d}{a^2} + C_2 \left(\frac{E}{1-\nu}\right) \frac{t d^3}{a^4} \qquad (16.53)$$

where p represents the pressure differential across the film; d, the center deflection; a, the initial radius; t, the membrane thickness; and σ, the initial film stress. In the simplified Cabrera model for circular membranes the constants C_1 and C_2 equal 4 respectively 8/3. For more rigorous solutions for both circular and rectangular membranes and references to other proposed models, refer to Maseeh-Tehrani [Maseeh-Tehrani, 1990]. The relation in Eq. (16.53) can simultaneously determine σ and the biaxial modulus $E/1 - \nu$; plotting pa^2/dt vs. $(d/a)^2$ should yield a straight line. The residual stress can be extracted from the intercept and the biaxial modulus from the slope of the least-squares-best-fit line [Senturia, 1987]. A typical result obtained via such measurements is represented in Figure 16.80. For the same Dupont polyimide 2525, measuring a Young's modulus of 3350 MPa in the uniaxial test (Figure 16.78B), the measurements give 5540 MPa for the biaxial modulus and 32 MPa for the residual stress. The residual stress-to-biaxial modulus ratio, also referred to as the residual biaxial strain, thus reaches 0.6%. The latter quantity must be compared to the ultimate strain when evaluating potential reliability problems associated with cracking of films. By loading the membranes to the elastic limit point, yield stress and strain can be determined as with the uniaxial test.

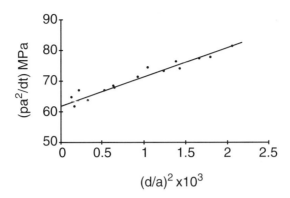

FIGURE 16.80 Load deflection data of a polyimide membrane (Du Pont 2525). From the intercept, a residual stress of 32 MPa was calculated and from the slope, a biaxial modulus of 5540 MPa. (From Senturia, S., Can We Design Microbiotic Devices without Knowing the Mechanical Properties of the Materials?, presented at Micro Robots and Teleoperators Workshop, Hyannis, MA, 1987. With permission.)

Poisson Ratio for Thin Films

The Poisson ratio for thin films presents us with more difficulties to measure than the Young's modulus as thin films tend to bend out of plane in response to in-plane shear. Maseeh and Senturia (1989) combine uniaxial and biaxial measurements to calculate the in-plane Poisson ratio of polyimides. For example, for the Dupont polyimide 2525, they determined 3350 MPa for E and 5540 MPa for the biaxial modulus ($E/1 - v$) leading to 0.41 ± 0.1 for the Poisson's ratio (v). The errors on both the biaxial and uniaxial measurements need to be reduced in order to develop more confidence in the extracted value of the Poisson's ratio. At present, the precision on the Poisson's ratio is limited to about 20%.

Other Surface Micromachined Structures to Gauge Intrinsic Stress

Various other surface micromachined structures have been used to measure mechanical properties of thin films. We will give a short review here, but the interested reader might want to consult the original references for more details.

Clamped-Clamped Beams. Several groups have used rows of clamped-clamped beams (bridges) with incrementally increasing lengths to determine the critical buckling load and hence deduce the residual compressive stress in polysilicon films (Figure 16.81A) [Sekimoto et al., 1982; Guckel et al., 1985]. The residual strain, $\varepsilon = \sigma/E$, is obtained from the critical length, L_c, at which buckling occurs (Euler's formula for elastic instability of struts):

$$\varepsilon = \frac{4\pi^2}{A} \frac{I}{L_c^2} \tag{16.54}$$

where A is the beam cross-sectional area and I the moment of inertia. As an example, with a maximum beam length of 500 μm and a film thickness of 1.0 μm, the buckling beam method can detect compressive stress as small as 0.5 MPa. This simple Euler approach does not take into considerations additional effects such as internal moments resulting from gradients in residual stress.

Ring Crossbar Structures. Tensile strain can be measured by a series of rings (Figure 16.81B) constrained to the substrate at two points on a diameter and spanned orthogonally by a clamped-clamped beam. After removal of the sacrificial layer, tensile strain in the ring places the spanning beam in compression; the critical buckling length of the beam can be related to the average strain [Guckel et al., 1988].

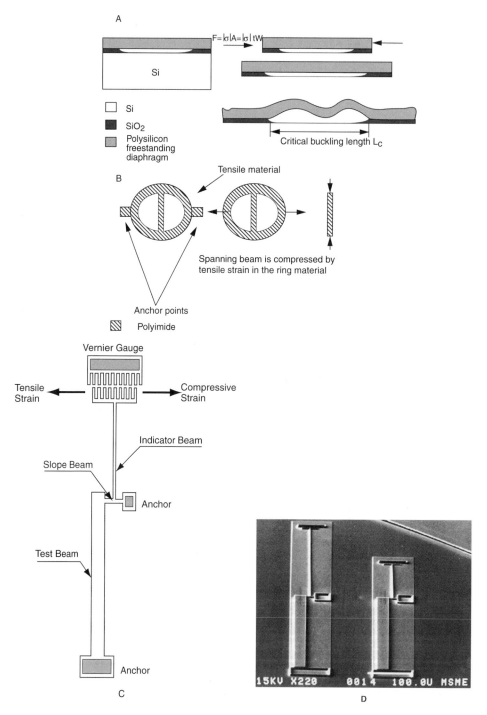

FIGURE 16.81 Some micromachined structures used for stress measurements. (A) Clamped-clamped beams: measuring the critical buckling length of clamped-clamped beams enables measurement of residual stress. (B) Crossbar rings: tensile stress can be measured by buckling induced in the crossbar of a ring structure. (C) A schematic of a strain gauge capable of measuring tensile or compressive stress. (D) SEM microphotograph of two strain gauges. (C and D from Lin, L., Selective Encapsulations of MEMS: Micro Channels, Needles, Resonators and Electromechanical Filters, Ph.D. thesis, University of California, Berkeley, 1993. With permission.)

Vernier Gauges. Both clamped-clamped beams and ring structures need to be implemented in entire arrays of structures. They do not allow easy integration with active microstructures due to space constraints. As opposed to proof structures one might use vernier gauges to measure the displacement of structures induced by residual strain [Lin, 1993]. The idea was first explored by Kim (1991), whose device consisted of two cantilever beams fixed at two opposite points. The end movement of the beams caused by the residual strain was measured by a vernier gauge. This method only requires one structure, but the best resolution for strain measurement reported is only 0.02% for 500-μm beams. Moreover, the vernier gauge device may indicate an erroneous strain when an out-of-plane strain gradient occurs [Lin, 1993]. Other types of direct strain measurement devices are the T- and H-shaped structures from Allen et al. (1987) and Mehregany et al. (1987) Optical measurement of the movement at the top of the T- or H-shape structures only becomes possible with very long beams (greater than 2.5 mm). They occupy large areas and their complexity requires finite element methods to analyze their output. The same is true for the strain magnification structure by Goosen et al. (1993). This structure measures strain by interconnecting two opposed beams such that the residual strain in the beams causes a third beam to rotate as a gauge needle. The rotation of the gauge needle quantifies the residual strain. A schematic of a micromachined strain gauge capable of measuring tensile or compressive residual stress, as shown in Figure 16.81C, was developed by Lin at University of California, Berkeley [Lin, 1993]. Figure 16.81D represents a scanning electron microscope (SEM) photograph of Lin's strain gauge. This gauge by far outranks the various *in situ* gauges explored. The strain gauge uses only one structure, can be fabricated *in situ* with active devices, determines tensile or compressive strain under optical microscopes, has a fine resolution of 0.001%, and resists to the out-of-plane strain gradient. When the device is released in the sacrificial etch step, the test beam (length L_t) expands or contracts, depending on the sign of the residual stress in the film, causing the compliant slope beam (length L_s) to deflect into an 's' shape. The indicator beam (length L_i), attached to the deforming beam at its point of inflection, rotates through an angle θ, and the deflection δ is read on the vernier scale. The residual strain is calculated as [Lin, 1993]:

$$\varepsilon_f = \frac{2L_s \delta}{3 L_i L_t C} \quad (16.55)$$

where C is a correction factor due to the presence of the indicator beam [Lin, 1993]. This equation was derived from simple beam theory relations and assumes that no out-of-plane motion occurs. The accuracy of the strain gauge is greatly improved because its output is independent of both the thickness of the deposited film and the cross section of the microstructure. Krulevitch used these devices to measure residual stress in *in situ* phosphorus-doped poly-Si films [Krulevitch, 1994], while Lin tested LPCVD silicon-rich silicon nitride films with it [Lin, 1993].

An improved micromachined indicator structure, inspired by Lin's work, was built by Ericson et al. (1995) By reading an integrated nonius scale in an SEM or an optical microscope, internal stress was measured with a resolution better than 0.5 MPa [Ericson et al., 1995; Benitez et al., 1995]. Both thick (10-μm) and thin (2-μm) poly-Si films were characterized this way.

Lateral Resonators. Biebl et al. (1995) extracted the Young's modulus of *in situ* phosphorus-doped polysilicon by measuring the mechanical response of poly-Si linear lateral comb-drive resonators (see Figure 16.88). The results reveal a value of 130 ± 5 GPa for the Young's modulus of highly phosphorus-doped films deposited at 610°C with a phosphine-to-silane mole ratio of 1.0×10^{-2} and annealing at 1050°C. For a deposition at 560°C with a phosphine to silane ratio of 1.6×10^{-3}, a Young's modulus of 147 ± 6 GPa was extracted.

Stress Nonuniformity Measurement by Cantilever Beams and Cantilever Spirals

The uniformity of stress through the depth of a film introduces an extremely important property to control. Variations in the magnitude and direction of the stress in the vertical direction can cause cantilevered

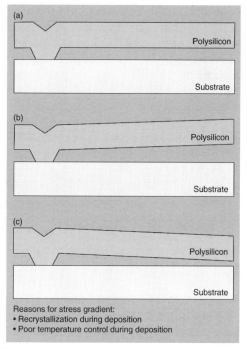

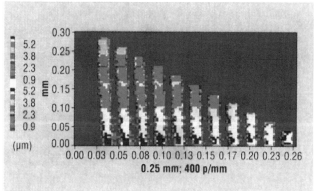

FIGURE 16.82 (A) Micro-cantilever deflection for measuring stress non-uniformity. (a) No gradient; (b) high tensile stress near the surface; and (c) lower tensile stress near the surface. (B) Topographical contour map of polysilicon cantilever array. (From Core, T.A., W.K. Tsang, and S.J. Sherman, *Solid State Technol.*, 36, 39–47, 1993. With permission.)

structures to curl toward or away from the substrate. Stress gradients present in the polysilicon film must thus be controlled to ensure predictable behavior of designed structures when released from the substrate. To determine the thickness variation in residual stress, noncontact surface profilometer measurements on an array of simple cantilever beams [Core et al., 1993; Chu et al., 1992] or cantilever spirals can be used [Fan et al., 1990].

Cantilever Beams. The deflections resulting from stress variation through the thickness of simple cantilever beams after their release from the substrate is shown in Figure 16.82A. The bending moment causing deflection of a cantilever beam follows out of pre-release residual stress and is given by (see Figure 16.82A):

$$M = \int_{-t/2}^{t/2} zb\sigma(z) \, dZ \qquad (16.56)$$

where σ(z) represents the residual stress in the film as a function of thickness and b stands for cantilever width. Assuming a linear strain gradient Γ (physical dimensions 1/length) such that $\sigma(z) = E\Gamma z$, Eq. (16.56) converts to:

$$\Gamma = \frac{M(12)}{Ebt^3} = \frac{M}{EI} \quad (16.57)$$

where the moment of inertia, I, for a rectangular cross section is given by $I = bt^3/12$. The measured deflection z, i.e., the vertical deflection of the cantilever's endpoint, from beam theory for a cantilever with an applied end moment, is given as:

$$z = \frac{ML^2}{2EI} = \frac{\Gamma L^2}{2} \quad (16.58)$$

Figure 16.82B represents topographical contour map of an array of polysilicon cantilevers. The cantilevers vary in length from 25 to 300 μm by 25-μm increments. Notice that the tip of the longest cantilever resides at a lower height (approximately 0.9 μm closer to the substrate) than the anchored support, indicating a downward bending moment [Core et al., 1993]. The gradients can be reduced or eliminated with a high-temperature anneal. With integrated electronics on the same chip, long high-temperature processing must be avoided. Therefore, stress gradients can limit the length of cantilevered structures used in surface-micromachined designs.

Cantilever Spirals. Residual stress gradients can also be measured by Fan's cantilever spiral as shown in Figure 16.83A [Fan et al., 1990]. Spirals anchored at the inside spring upwards, rotate, and contract with positive strain gradient (tending to curl a cantilever upwards), while spirals anchored at the outside deflect in a similar manner in response to a negative gradient. Theoretically, positive and negative gradients produce spirals with mirror symmetry [Fan et al., 1990]. The strain gradient can be determined from spiral structures by measuring the amount of lateral contraction, the change in height, or the amount of rotation. Krulevitch presented the computer code for the spiral simulation in his doctoral thesis [Krulevitch, 1994]. Figure 16.83B shows a simulated spiral with a bending moment of $\Gamma = \pm 3.0$ mm^{-1} after release.

Krulevitch compared all the above surface micromachined structures for stress and stress gradient measurements on poly-Si films. His comments are summarized in Table 16.12 [Krulevitch, 1994]. Krulevitch found that the fixed-fixed beam structures for determining compressive stress from the buckling criterion produced remarkably self-consistent and repeatable results. Wafer curvature stress profiling proved reliable for determining average stress and the true stress gradient as compared with micromachined spirals. Measurements of curled cantilevers could not be used much as the strain gradients mainly were negative for poly-Si, leading to cantilevers contacting the substrate. The strain gauge dial structures were useful over a rather limited strain-gradient range. With too large a strain gradient curling of the long beams overshadows expansion effects and makes the vernier indicator unreadable.

16.12.4 Strength of Thin Films

Due to the high activation energy for dislocation motion in silicon (2.2 eV), hardly any plastic flow occurs in single crystalline silicon for temperatures lower than 673°C. Grain boundaries in poly-Si block dislocation motion; hence, polysilicon films can be treated as an ideal brittle material at room temperature [Biebl et al., 1995]. High yield strengths often are obtained in thin films with values up to 200 times as large as those found in the corresponding bulk material [Campbell, 1970]. In this light, the earlier quoted fracture stresses of poly-Si, between two and ten times smaller than that of bulk single crystal, are

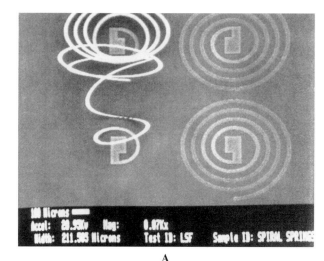
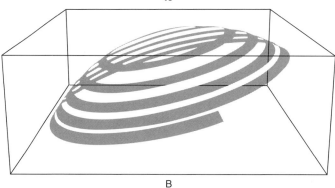

FIGURE 16.83 Cantilever spirals for stress gradient measurement. (A) SEM of micrographs of spirals from an as-deposited poly-Si. (Courtesy of Dr. L.S. Fan.) (B) Simulation of a thin-film micromachined spiral with $\Gamma = 3.0$ mm^{-1}. (From Krulevitch, P.A., Micromechanical Investigations of Silicon and Ni-Ti-Cu Thin Films, Ph.D. thesis, University of California, Berkeley, 1994. With permission.)

TABLE 16.12 Summary of Various Techniques for Measuring Residual Film Stress

Measurement Technique	Measurable Stress State	Remarks
Wafer curvature	Stress gradient, average stress	Average stress over entire wafer; provides true stress gradient; approx. 5 MPa resolution
Vernier strain gauges	Average stress	Local stress; small dynamic range; resolution = 2 MPa
Spiral cantilevers	Stress gradient	Local stress, provides equivalent linear gradient
Curling beam cantilevers	Large positive stress gradient	Local stress, provides equivalent linear gradient
Fixed-fixed beams	Average compressive stress	Local stress measurement

Source: Krulevitch, P.A., Micromechanical Investigations of Silicon and Ni-Ti-Cu Thin Films, Ph.D. thesis, University of California, Berkeley, 1994. With permission.

surprising. Greek et al. (1995) explain this deviation by pointing at the high surface roughness of poly-Si films compared to single crystal Si. They believe that a reduction in surface roughness would improve the tensile fracture strength considerably.

Indentation (hardness) testing is very common for bulk materials where the direct relationship between bulk hardness and yield strength is well known. It can be measured by pressing a hard, specially shaped point into the surface and observing indentation. This type of measurement is of little use for measuring thin films below 5×10^4 Å. Consequently, very little is known about the hardness of thin films. Recently, specialized instruments have been constructed (e.g., the Nanoindenter) in which load and displacement data are collected while the indentation is being introduced in a thin film. This eliminates the errors associated with later measurement of indentation size and provides continuous monitoring of load/displacement data similar to a standard tensile test. Load resolution may be 0.25 µN, displacement resolution 0.2 to 0.4 nm, and x-y sample position accuracy 0.5 µm. Empirical relations have correlated hardness with Young's modulus and with uniaxial strength of thin films. Hardness calculations must include both plastic and long-distance elastic deformation. If the indentation is deeper than 10% of the film, corrections for elastic hardness contribution of the substrate must also be included [Vinci and Braveman, 1991]. Mechanical properties such as hardness and modulus of elasticity can be determined on the micro- to picoscales using AFM [Bushan, 1996]. Bushan provides an excellent introduction to this field in *Handbook of Micro/Nanotribology* [Bushan, 1995].

16.13 Surface Micromachining Processes

16.13.1 Basic Process Sequence

A surface micromachining process sequence for the creation of a simple free-standing poly-Si bridge is illustrated in Figure 16.84 [Howe and Muller, 1983; Howe, 1985]. A sacrificial layer, also called a spacer layer or base, is deposited on a silicon substrate coated with a dielectric layer as the buffer/isolation layer (Figure 16.84A). Phosphosilicate glass (PSG) deposited by LPCVD stands out as the best material for the sacrificial layer because it etches even more rapidly in HF than SiO_2. In order to obtain a uniform etch rate, the PSG film must be densified by heating the wafer to 950–1100°C in a furnace or a rapid

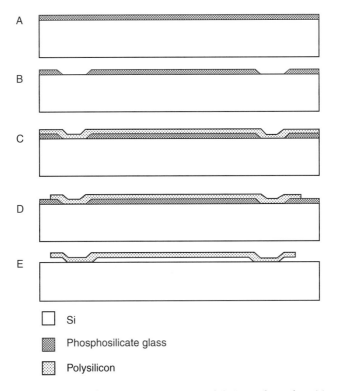

FIGURE 16.84 Basic surface micromachining process sequence. (A) Spacer layer deposition (the thin dielectric insulator layer is not shown). (B) Base patterning with mask 1. (C) Microstructure layer deposition. (D) Pattern microstructure with mask 2. (E) Selective etching of spacer layer.

thermal annealer (RTA) [Yun, 1992]. With a first mask the base is patterned as shown in Figure 16.84B. Windows are opened up in the sacrificial layer and a microstructural thin film, whether consisting of polysilicon, metal, alloy or a dielectric material, is conformally deposited over the patterned sacrificial layer (Figure 16.84C). Furnace annealing, in the case of polysilicon at 1050°C in nitrogen for one hour, reduces stress stemming from thermal expansion coefficient mismatch and nucleation and growth of the film. Rapid thermal annealing has been found effective for reducing stress in polysilicon as well [Yun, 1992]. With a second mask, the microstructure layer is patterned, usually by dry etching in a $CF_4 + O_2$ or a $CF_3Cl + Cl_2$ plasma (Figure 16.84D) [Adams, 1988]. Finally, selective wet etching of the sacrificial layer, say in 49% HF, leaves a free-standing micromechanical structure (Figure 16.84E). The surface micromachining technique is applicable to combinations of thin films and lateral dimensions where the sacrificial layer can be etched without significant etching or attack of the microstructure, the dielectric or the substrate.

16.13.2 Fabrication Step Details

16.13.2.1 Pattern Transfer to SiO$_2$ Buffer/Isolation Layer

A blanket n$^+$ diffusion of the Si substrate, defining a ground plane, often outlines the very first step in surface micromachining, followed by a passivation step of the substrate, for example, with 0.15-μm-thick LPCVD nitride on a 0.5-μm thermal oxide. Suppose the buffer/isolation passivation layer as illustrated in Figure 16.85 itself needs to be patterned, perhaps to make a metal contact pad onto the Si substrate. Then, the appropriate fabrication step is a pattern transfer to the thin isolation film as shown in Figure 16.85, illustrating a wet pattern transfer to a 1-μm-thick thermal SiO$_2$ film with a 1-μm resist layer. Typically, an isotropic etch such as buffered HF, e.g., BHF (5:1), which is 5 parts NH$_4$F and 1 part conc. HF, is used (unbuffered HF attacks the photoresist). This solution etches SiO$_2$ at a rate of 100 nm/min, and the creation of the opening to the underlying substrate takes about 10 min. The etch progress may be monitored optically (color change) or by observing the hydrophobic/hydrophilic behavior [Hermansson et al., 1991] of the etched layer. With a resist opening, L_m, the undercut typically measures the same thickness as the oxide thickness, t_{SiO_2}. In other words, the contact pad will have a size of $L_m + 2t_{SiO_2}$. The undercut worsens with loss of photoresist adhesion during etching. Using an adhesion promoter such

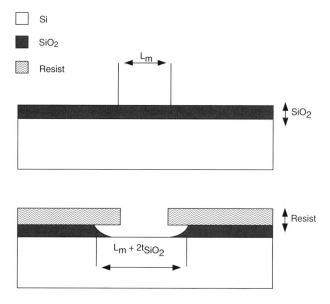

FIGURE 16.85 Wet etch pattern transfer to a thin thermal SiO$_2$ film for the fabrication of a contact pad to the Si substrate.

as HMDS (hexamethyldisilazane) prove useful in such case. A new bake after 5 min of etching is a good procedure to maintain the resist integrity. After the isotropic etch, the resist is stripped in a piranha etch bath. This strong oxidizer grows about 3 nm of oxide back in the cleared window. To remove the oxide resulting from the piranha, a dip in diluted BHF suffices. After cleaning and drying, the substrate is ready for contact metal and base material deposition. Applying a dry etch (say, a CF_4-H_2 plasma) to open up the window in the oxide would eliminate undercutting of the resist, but requires a longer set-up time. With an LPCVD low stress nitride on top of a thermal SiO_2, an often-used combination for etching the buffer/isolation layer is a dry etch (say, a SF_6 plasma) followed by a 5:1 BHF [Tang, 1990].

16.13.2.2 Base Layer (also Spacer or Sacrificial Layer) Deposition and Etching

A thin LPCVD phosphosilicate glass (PSG) layer (say, 2 μm thick) is a preferred base, spacer, or sacrificial layer material. Adding phosphorous to SiO_2 to produce PSG enhances the etch rate in HF [Monk et al., 1992; Tenney and Ghezzo, 1973]. Other advantages to using doped SiO_2 include its utility as a solid state diffusion dopant source to make subsequent polysilicon layers electrically conductive and helping to control window taper (see below). As deposited phosphosilicate displays a nonuniform etch rate in HF and must be densified, typically carried out in a furnace at 950°C for 30 min to 1 hr in a wet oxygen ambient. The etch rate in BHF can be used as the measure of the densification quality. The base window etching stops at the buffer isolation layer, often a Si_3N_4/SiO_2 layer which also forms the permanent passivation of the device. Windows in the base layer are used to make anchors onto the buffer/isolation layer for mechanical structures.

The edges of the etched windows in the base may need to be tapered to minimize coverage problems with subsequent structural layers, especially if these layers are deposited with a line-of-sight deposition technique. An edge taper is introduced through an optimization of the plasma etch conditions (see Madou [1997, chap. 2]), through the introduction of a gradient of the etch rate, or by reflow of the etched spacer. In Figure 16.86A we depict the reflow process of a PSG spacer after patterning in a dry etch. Viscous flow at higher temperature smoothes the edge taper. The ability of PSG to undergo viscous deformation at a given temperature primarily is a function of the phosphorous content in the glass; reflow profiles get progressively smoother the higher the phosphorous concentration—reflecting the corresponding

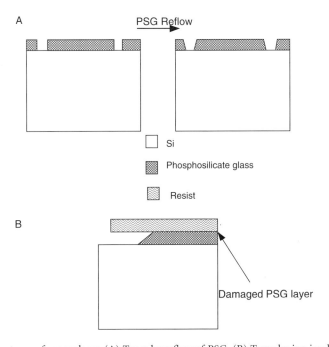

FIGURE 16.86 Edge taper of spacer layer. (A) Taper by reflow of PSG. (B) Taper by ion implantation of PSG.

enhancement in viscous flow [Levy and Nassau, 1986]. In Figure 16.86B, ion implantation of PSG has created a rapidly etching, damaged PSG layer [North et al., 1978; Goetzlich and Ryssel, 1981; White, 1980]. The steady-state taper is a function of the etch rate in PSG etchants (e.g., BHF).

16.13.2.3 Deposition of Structural Material

For the best step coverage of a structural material over the base window, chemical vapor deposition is preferred. If a physical deposition method must be used, sputtering is preferred over line-of-sight deposition techniques which lead to the poorest step coverage (see Madou [1997, chap. 3]). In the latter case, edge taper could be introduced advantageously.

The most widely used structural material in surface micromachining is polysilicon (poly-Si or simply poly). Polysilicon is deposited by low pressure (25 to 150 Pa) chemical vapor deposition in a furnace (a poly chamber) at about 600°C. The undoped material is usually deposited from pure silane, which thermally decomposes according to the reaction:

$$SiH_4 \rightarrow Si + 2H_2 \qquad \text{(Reaction 16.16)}$$

Typical process conditions may consist of a temperature of 605°C, a pressure of 550 mTorr (73 Pa), and a silane flow rate of 125 sccm. Under those conditions a normal deposition rate is 100 Å/min. To deposit a 1-μm film will take about 90 min. Sometimes the silane is diluted by 70 to 80% nitrogen. The silicon is deposited at temperatures ranging from 570 to 580°C for fine-grained poly-Si to 620 to 650°C for coarse-grained poly-Si. The characteristics of these two types of poly-Si materials will be compared in the case studies presented below. Furnace annealing of the poly-Si film at 1050°C in nitrogen for one hour is employed commonly to reduce stress stemming from thermal expansion coefficient mismatch and nucleation and growth of the polysilicon film.

To make parts of the microstructure conductive, dopants can be introduced in the poly-Si film by adding dopant gases to the silane gas stream, by drive-in from a solid dopant source or by ion implantation. When doping from the gas phase, the dopant can be readily controlled in the range of 10^{19} to $10^{21}/cm^3$. Polysilicon deposition rates, in the case of gas phase doping (*in situ* doping), may be significantly impacted. For example, decreases in poly-Si deposition rate by as much as a factor of 25 have been reported in phosphine and arsine doping. The effect is associated with the poisoning of reaction sites by phosphine and arsine [Adams, 1988]. The lower deposition rate of *in situ* phosphine doping can be mitigated by reducing the ratio of phosphine to silane flow by one third [Howe, 1995]. With the latter flow regime, deposition at 585°C (for a 2-μm-thick film), followed by 900°C rapid thermal annealing for 7 min, results in a polysilicon with low residual stress, negligible stress gradient and low resistivity [Howe, 1995; Biebl et al., 1995]. *In situ* boron doping, in contrast to arsine and phosphine doping, accelerates the polysilicon deposition rate through an enhancement of silane adsorption induced by the boron presence [Fresquet et al., 1995]. Film thickness uniformity for doped films typically is less than 1% and sheet resistance uniformity less than 2%. Alternatively, poly-Si may be doped from PSG films sandwiching the undoped poly-Si film. By annealing such a sandwich at 1050°C in N_2 for one hour, the polysilicon is symmetrically doped by diffusion of dopant from the top and the bottom layers of PSG. Symmetric doping results in a polysilicon film with a moderate compressive stress. The resulting uniform grain texture avoids gradients in the residual stress which would cause bending moments warping microstructures upon release. Finally, ion implantation of undoped polysilicon, followed by high temperature dopant drive-in, also leads to conductive polysilicon. This polysilicon has a moderate tensile stress, with a strain gradient that causes cantilevers to deflect toward the substrate [Core et al., 1993]. The poly-Si is now ready for patterning by RIE in, say, a CF_4-O_2 plasma.

Although the mechanical properties still are not well understood, microstructures based on poly-Si as a mechanical member have been commercialized [Editor, 1991; Core et al., 1993]. Other structural materials used in surface micromachining include single crystal Si (epi-Si or etched back, fusion bonded Si), SiO_2, silicon nitride, silicon oxynitride, polyimide, diamond, SiC, GaAs, tungsten, α-Si:H, Ni, W, Al etc. A few words about the merit of some of these materials as structural components follow.

Silicon nitride and silicon oxide also can be deposited by CVD methods but usually exhibit too much residual stress which hampers their use as mechanical components. However, CVD of mixed silicon oxynitride can produce substantially stress-free components.

Amorphous Si (α-Si) can be stress annealed at temperatures as low as 400°C [Chang et al., 1991]. This low-temperature anneal makes the material compatible with almost any active electronic component. Unfortunately, very little is known about the mechanical properties of amorphous Si.

Hydrogenated amorphous Si (α-Si:H), with its interesting electronic properties, is even less understood in terms of its mechanical properties. If the mechanical properties of hydrogenated amorphous Si were found as good as those of poly-Si, the material might make a better choice than poly-Si as a MEMS material given its better electronic characteristics.

Tungsten CVD deposition is IC compatible. The material has some unique mechanical properties (see Madou [1997, chap. 8, Table 8.4]). Moreover, the material can be applied selectively. Selective CVD tungsten has the unique property that tungsten will only nucleate on silicon or metal surfaces but does not deposit on dielectrics such as oxides and nitrides [Chen and MacDonald, 1991].

Metals and polyimides, because they are easily deformed, usually do not qualify as mechanical members but have been used, for example, in plastically deformable hinges [Suzuki et al., 1994; Hoffman et al., 1995]. Aluminum constitutes the mirror material in Texas Instruments' flexure-beam micromirror devices (FBMDs). The metal is used both for the L-shaped flexure hinges and the mirror itself [Lin, 1996]. Polycrystalline diamond films, deposited by CVD, are potential high-temperature, harsh environment MEMS candidates [Herb et al., 1990]. Problems include oxidation above 500°C for nonpassivated films, difficulty of making reliable ohmic contact to the material, and the reproducibility and surface roughness of the films [Obermeier, 1995]. Poly-SiC films have been deposited by an APCVD process on 4-in. polysilicon-coated, silicon wafers. A surface micromachining process using the underlying polysilicon film as the sacrificial layer was developed. Poly-SiC is projected for use as structural material for high temperatures and harsh environments, and to reduce friction and wear between moving components [Fleischman et al., 1996]. Surface micromachining of thin single crystalline Si layers in SOI and with polyimides is discussed separately below.

16.13.2.4 Selective Etching of Spacer Layer

Selective Etching

To create movable micromachines the microstructures must be freed from the spacer layers. The challenge in freeing microstructures by undercutting is evident from Figure 16.87. After patterning the poly-Si by

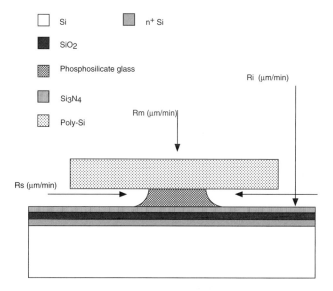

FIGURE 16.87 Selective etching of spacer layer.

RIE in, say, a SF_6 plasma, it is immersed in an HF solution to remove the underlying sacrificial layer, releasing the structure from the substrate. Commonly, a layer of sacrificial phosphosilicate glass, between 1 and 2000 μm long and 0.1 to 5 μm thick, is etched in concentrated, dilute or buffered HF. The spacer etch rate, R_s, should be faster than the attack on the microstructural element, R_m, and that of the insulator layer, R_i. For this type of complete undercutting, only wet etchants can be used. Etching narrow gaps and undercutting wide areas with BHF can take hours. To shorten the etch time, extra apertures in the microstructures sometimes are provided for additional access to the spacer layer. The etch rate of PSG, the most common spacer material, increases monotonically with dopant concentration, and thicker sacrificial layers etch faster than thinner layers [Monk et al., 1994].

The selectivity ratios for spacer layer, microstructure and buffer layer are not infinite [Lober and Howe, 1998], and in some instances even silicon substrate attack by BHF was observed under polysilicon/spacer regions [Fan et al., 1988; Mehregany et al., 1988]. Heavily phosphorous-doped polysilicon is especially prone to attack by BHF. Silicon nitride deposited by LPCVD etches much more slowly in HF than oxide films, making it a more desirable isolation film. When depositing this film with a silicon-rich composition, the etch rate is even slower (15 nm/min) [Tang, 1990]. Eaton et al. (1995) compared oxide and nitride etching in a 1:1 $HF:H_2O$ and in a 1:1 HF:HCl solution and concluded that the HCl-based etch yielded both faster oxide etch rates (617 nm/min vs. 330 nm/min) and slower nitride etch rates (2 nm/min vs. 3.6 nm/min), providing a much greater selectivity of the oxide to silicon nitride (310 vs. 91!). The same authors also studied the optimum composition of a sacrificial oxide for the fastest possible etching in their most selective 1:1 HF:HCl etch. The faster sacrificial layer etch limited the damage to nitride structural elements. Their results are summarized in Table 16.13. A densified CVD SiO_2 was used as a control, and a 5%/5% borophosphosilicate glass (BPSG) was found to etch the fastest.

Watanabe et al. (1995) using low pressure vapor HF, found high etch ratios of PSG and BPSG to thermal oxides of over 2000, with the BPSG etching slightly faster than the PSG. We will see further that low pressure vapor HF also leads to less stiction of structural elements to the substrate. In Table 16.14 we present etch rate and etch ratios for R_i and R_s in BHF (7:1) for a few selected materials.

Detailed studies on the etching mechanism of oxide spacer layers were undertaken by Monk et al. (1994a; 1994b) They found that the etching reaction shifts from kinetic controlled to diffusion controlled as the etch channel becomes longer. This affects mainly large-area structures, as diffusion limitations were observed only after approximately 200 μm of channel etching or 15 min in concentrated HF. Eaton and Smith (1996) developed a release etch model which is an extension of the work done by Monk et al. (1994a; 1994b) and Liu et al. (1993).

TABLE 16.13 Etch Rate in 1:1 HF:HCl of a Variety of Sacrificial Oxides

Thin Oxide	Lateral Etch Rate (Å/min)
CVD SiO_2 (densified at 1050°C for 30 min)	6170
Ion-implanted and densified CVD SiO_2 (P, $8 \times 10^{15}/cm^2$, 50 keV)	8330
Phosphosilicate (PSG)	11,330
5%/5% borophosphosilicate (BPSG)	41,670

Source: Adapted from Eaton, W.P. and Smith, J.H., A CMOS-compatible, Surface Micromachined Pressure Sensor for Aqueous Ultrasonic Application, presented at SPIE Smart Structure and Materials, 1995.

TABLE 16.14 Etching of Spacer Layer and Buffer Layer in BHF (7:1)

Property	Material		
	LPCVD Si_3N_4	LPCVD SiO_2	LPCVD 7% PSG
Etch rate	7–12 Å/min (R_i)	700 Å/min (R_s)	~10,000 Å/min (R_s)
Selectivity ratio	1	60–100	~800–1200

TABLE 16.15 Etchants-Spacer and Microstructural Layer

Etchant	Buffer/Isolation	Spacer	Microstructure
Buffered HF (5:1, NH$_4$F: conc. HF)[a]	LPCVD Si$_3$N$_4$/thermal SiO$_2$	PSG	Poly-Si
RIE using CHF$_3$ BHF (6:1)[b]	LPCVD Si$_3$N$_4$	LPCVD SiO$_2$	CVD Tungsten
KOH[c]	LPCVD Si$_3$N$_4$/thermal SiO$_2$	Poly-Si	Si$_3$N$_4$
Ferric chloride[d]	Thermal SiO$_2$	Cu	Polyimide
HF[e]	LPCVD Si$_3$N$_4$/thermal SiO$_2$	PSG	Polyimide
Phosphoric/acetic acid/nitric acid (PAN, or 5:8:1:1 Water:phosphoric:acetic:nitric)[f]	Thermal SiO$_2$	Al	PECVD Si$_3$N$_4$ Nickel
Ammonium iodide/iodine alcohol[g]	Thermal SiO$_2$	Au	Ti
Ethylene-diamine/pyrocathecol (EDP)	Thermal SiO$_2$	Poly-Si	SiO$_2$

(a) Howe and Muller (1983), Howe (1985), Guckel and Burns (1984); (b) Chen and MacDonald (1991); (c) Sugiyama et al. (1986; 1987); (d) Kim and Allen (1991); (e) Suzuki et al. (1994); (f) Chang et al. (1991), Scheeper et al. (1991); (g) Yamada and Kuriyama (1991)

Etching is followed by rinsing and drying. Extended rinsing causes a native oxide to form on the surface of the polysilicon structure. Such a passivation layer often is desirable and can be formed more easily by a short dip in 30% H_2O_2.

Etchant-Spacer-Microstructure Combinations

A wide variety of etchant, spacer, and structural material combinations have been used; a limited listing is presented in Table 16.15. One interesting case concerns poly-Si as the sacrificial layer. This was used, for example, in the fabrication of a vibration sensor at Nissan Motor Co [Nakamura et al., 1985]. In this case poly-Si is etched in KOH from underneath a nitride/polysilicon/nitride sandwich cantilever. Also, a solution of HNO_3 and BHF can be used to etch poly-Si, but it proves difficult to control. Using aqueous solutions of NR_4OH, where R is an alkyl group, provides a better etching solution for poly-Si, with greater selectivity with respect to silicon dioxide and phosphosilicate glass. The relatively slow etch rate enables better process control [Bassous and Liu, 1978] and the etchant does not contain alkali ions, making it more CMOS compatible. With tetramethylammonium hydroxide (TMAH) the etch rate of CVD poly-Si, deposited at 600°C from SiH_4, follows the rates of the (100) face of single crystal Si and is dopant dependent. The selectivity of Si/SiO$_2$ and Si/PSG, at temperatures below 45°C are measured to be about 1000. Hence, a layer of 500-Å PSG can be used as the etch mask for 10,000 Å of poly-Si.

16.13.2.5 Stiction

Stiction During Release

The use of sacrificial layers enables the creation of very intricate movable polysilicon surface structures. An important limitation of such polysilicon shapes is that large-area structures tend to deflect through stress gradients or surface tension induced by trapped liquids and attach to the substrate/isolation layer during the final rinsing and drying step, a stiction phenomenon that may be related to hydrogen bonding or residual contamination. Recently, great strides were made towards a better understanding and prevention of stiction.

The sacrificial layer removal with a buffered oxide etch followed by a long, thorough rinse in deionized water and drying under an infrared lamp typically represent the last steps in the surface micromachining sequence. As the wafer dries, the surface tension of the rinse water pulls the delicate microstructure to the substrate where a combination of forces, probably van der Waals forces and hydrogen bonding, keeps it firmly attached (see Figure 16.88) [Core et al., 1993]. Once the structure is attached to the substrate by stiction, the mechanical force needed to dislodge it usually is large enough to damage the microme-chanical structure [Lober and Howe, 1988; Guckel et al., 1987; Alley et al., 1988]. Basically, the same phenomena are thought to be involved in room temperature wafer bonding [Madou, 1997, chap. 8]. We will not further dwell upon the mechanics of the stiction process here, but the reader should refer to the theoretical and experimental analysis of the mechanical stability and adhesion of microstructures under capillary forces by Mastrangelo et al. (1993a; 1993b).

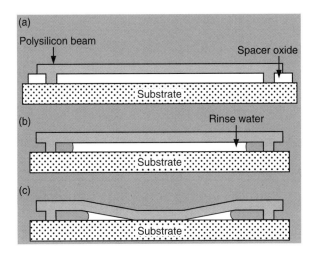

FIGURE 16.88 Stiction phenomenon in surface micromachining and the effect of surface tension on micromechanical structures. (a) Unreleased beam. (b) Released beam before drying. (c) Released beam pulled to the substrate by capillary forces as the wafer dries [conc].

Creating stand-off bumps on the underside of a poly-Si plate [Tang, 1990; Fan, 1989] or adding meniscus-shaping microstructures to the perimeter of the microstructure are mechanical means to help reduce sticking [Abe et al., 1995]. Fedder et al. (1992) used another mechanical approach to avoid stiction by temporarily stiffening the microstructures with polysilicon links. These very stiff structures are not affected by liquid surface tension forces and the links are severed afterward with a high current pulse once the potentially destructive processing is complete. Yet another mechanical approach to avoid stiction involves the use of sacrificial supporting polymer columns. A portion of the sacrificial layer is substituted by polymer spacer material, spun-on after partial etch of the oxide glass. After completion of the oxide etch, the polymer spacer prevents stiction during evaporative drying. Finally, an isotropic oxygen plasma etches the polymer to release the structure [Mastrangelo and Saloka, 1993].

Ideally, to ensure high yields, one should avoid contact between structural elements and the substrate during processing. In a liquid environment, however, this may become impossible due to the large surface tension effects. Consequently, most solutions to the stiction problem involve reducing the surface tension of the final rinse solution by physico-chemical means. Lober et al. (1988), for example, tried HF vapor and Guckel et al. (1989; 1990) used freeze-drying of water/methanol mixtures. Freezing and sublimating the rinse fluid in a low-pressure environment gives improved results by circumventing the liquid phase. Takeshima et al. (1991) used t-butyl alcohol freeze-drying. Since the freezing point of this alcohol lies at 25.6°C, it is possible to perform freeze-drying without special cooling equipment. More recently, attempts at supercritical drying resulted in high microstructure yields [Mulhern et al., 1993]. With this technique, the rinse fluid is displaced with a liquid that can be driven into a supercritical phase under high pressure. This supercritical phase does not exhibit surface tension. Typically, CO_2 under about 75 atm is used (see also Madou [1997, chap. 1 under Cleaning]).

Kozlowski et al. (1995) substituted HF in successive exchange steps by the monomer divinylbenzene to fabricate very thin (500 nm) micromachined polysilicon bridges and cantilevers. The monomer was polymerized under UV light at room temperature and was removed in an oxygen plasma. Analog Devices applied a proprietary technique involving only standard IC process technology in the fabrication of a micro-accelerometer to eliminate yield losses due to stiction [Core et al., 1993].

Stiction After Release, i.e., In-Use Stiction
Stiction remains a fundamental reliability issue due to contact with adjacent surfaces after release. Stiction free passivation that can survive the packaging temperature cycle is not known at present [Howe, 1995]. Attempted solutions are summarized below.

Adhesive energy may be minimized in a variety of ways, for example, by forming bumps on surfaces (see above) or roughening of opposite surface plates [Alley et al., 1993]. Also, self-assembled monolayer coatings have been shown to reduce surface adhesion and to be effective at friction reduction in bearings at the same time [Alley et al., 1992]. Making the silicon surface very hydrophobic or coating it with diamond-like carbon are other potential solutions for preventing postrelease stiction of polysilicon microstructures. Man et al. (1996) eliminate post-release adhesion in microstructures by using a thin conformal fluorocarbon film. The film eliminates the adhesion of polysilicon beams up to 230 μm long even after direct immersion in water. The film withstands temperatures as high as 400°C and wear tests show that the film remains effective after 10^8 contact cycles. Along the same line, ammonium fluoride-treated Si surfaces are thought to be superior to the HF-treated surfaces due to a more complete hydrogen termination, leading to a cleaner hydrophobic surface [Houston et al., 1995]. Gogoi and Mastrangelo (1995) introduced electromagnetic pulses for postprocessing release of stuck microstructures.

16.13.3 Control of Film Stress

After reviewing typical surface micromachining process sequences we are ready to investigate some of the mechanical properties of the fabricated mechanical members. Consider a lateral resonator as shown in Figure 16.89A. Electrostatic force is applied by a drive comb to a suspended shuttle. Its motion is detected capacitively by a sense comb. For many applications a tight control over the resonant frequency, f_0, is required. A simple analytical approximation for f_0 of this type of resonator can be deduced from Rayleigh's method [Howe, 1987]:

$$f_0 \approx \frac{1}{2\pi}\sqrt{\frac{4EtW^3}{ML^3} + \frac{24\sigma_r tW}{5ML}} \quad (16.59)$$

where E represents the Young's modulus of polysilicon; L, W and t are the length, width and thickness of the flexures; and M stands for the mass of the suspended shuttle (of the order of 10^{-9} kg, or less). For typical values ($L = 150$ μm and $W = t = 2$ μm) and a small tensile residual stress, the resonant frequency f_0 is between 10 and 100 kHz [Howe, 1995]. For typical values of L/W the stress term in Eq. (16.59) dominates the bending term. Any residual stress, σ_r, obviously will affect the resonant frequency. Consequently, stress and stress gradients represent critical stages for microstructural design. One of the many challenges of any surface-micromachining process is to control the intrinsic stresses in the deposited films. Several techniques can be used to control film stress. Some we detailed before, but we list them again for completeness:

- Large-grained poly-Si films, deposited around 625°C, have a columnar structure and are always compressive. Compressive stress can cause buckling in constrained structures. Annealing at high temperatures, between 900 and 1150°C, in nitrogen significantly reduces the compressive stress in as-deposited poly-Si [Guckel et al., 1985; Howe and Muller, 1983] and can eliminate stress gradients. No significant structural changes occur when annealing a columnar poly-Si film. The annealing process is not without danger in cases where active electronics are integrated on the same chip. Rapid thermal annealing might provide a solution (see IC Compatibility, below).
- Undoped poly-Si films are in an amorphous state when deposited at 580°C or lower. The stress and the structure of this low-temperature material depend on temperature and partial pressure of the silane. A low-temperature anneal leads to a fine-grained poly-Si with low tensile stress and very smooth surface texture [Guckel et al., 1989]. Tensile rather than compressive films are a necessity if lateral dimensions of clamped structures are not to be restricted by compressive buckling. Conducting regions are formed in this case by ion implantation. Fine-grained and large-grained poly-Si are compared in more detail further below.

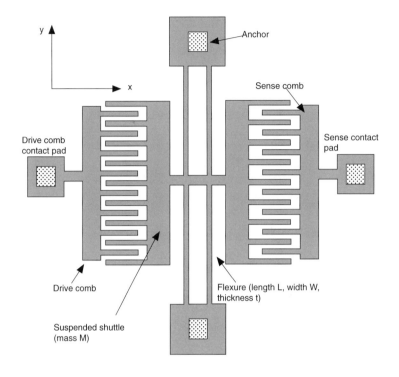
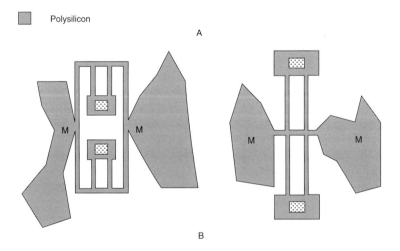

FIGURE 16.89 Layout of a lateral resonator with straight flexures. (A) Folded flexures (left) to release stress are compared with straight flexures (right). (B) M is shuttle mass.

- Phosphorus [Murarka and Retajczyk, 1983; Orpana and Korhonen, 1991; Lin et al., 1993], boron [Orpana and Korhonen, 1991; Choi and Hearn, 1984; Ding and Ko, 1991], arsenic [Orpana and Korhonen, 1991] and carbon [Hendriks et al., 1983] doping have all been shown to affect the state of residual stress in poly-Si films. In the case of single crystal Si, to compensate for strain induced by dopants, one can implant with atoms with the opposite atomic radius vs. silicon. Similar approaches would most likely be effective for poly-Si as well.
- Tang et al. (1990; 1989) developed a technique that sandwiches a poly-Si structural layer between a top and bottom layer of PSG and lets the high temperature anneal drive in the phosphorus symmetrically, producing low stress poly-Si with a negligible stress gradient.

- Another stress reduction method is to vary the materials composition, something readily done in CVD processes. An example of this method is the Si enrichment of Si_3N_4 which reduces the tensile stress [Sekimoto et al., 1982; Guckel et al., 1986].
- During plasma-assisted film deposition processes, one can influence stress dramatically. In a physical deposition process such as sputtering, stress control involves varying gas pressure and substrate bias. In plasma-enhanced chemical vapor deposition (PECVD) the RF power, through increased ion-bombardment, influences stress. In this way, the stress in a thin film starts out tensile, decreases as the power increases and finally becomes compressive with further RF power increase. PECVD equipment manufacturers also are working to build stress control capabilities into new equipment by controlling plasma frequency (see also Madou [1997, chap. 3]). In CVD, stress control involves all types of temperature treatment programs.
- A clever mechanical design might facilitate structural stress relief [Tang et al., 1989]. By folding the flexures in the lateral resonator in Figure 16.89B, and by the overall structural symmetry, the relaxation of residual polysilicon stress is possible without structural distortion. By folding the flexures, the resonant frequency, f_0, becomes independent of σ_r (see Eq. (16.59)). The springs in the resonator structure provide freedom of travel along the direction of the comb-finger motions (x) while restraining the structure from moving sideways (y), thus preventing the comb fingers from shorting out the drive electrodes. In this design, the spring constant along the y direction must be higher than along the x direction, i.e., $k_y \gg k_x$. The suspension should allow for the relief of the built-in stress of the polysilicon film and the axial stress induced by large vibrational amplitudes. The folded-beam suspensions meet both criteria. They enable large deflections in the x direction (perpendicular to the length of the beams) while providing stiffness in the y direction (along the length of the beams). Furthermore, the only anchor points (see Figure 16.89B) for the whole structure reside near the center, thus allowing the parallel beams to expand or contract in the y direction, relieving most of the built-in and induced stress [Tang, 1990]. Tang also modeled and built spiral and serpentine springs supporting torsional resonant plates. An advantage of the torsional resonant structures is that they are anchored only at the center, enabling radial relaxation of the built-in stress in the polysilicon film [Tang, 1990]. For some applications, the design approach with folded flexures is an attractive way to eliminate residual stress. However, a penalty for using flexures is increased susceptibility to out-of-plane warpage from residual stress gradients through the thickness of the polysilicon microstructure [Howe, 1995].
- Corrugated structural members, invented by Jerman for bulk micromachined sensors [Jerman, 1990], also reduce stress effectively. In the case of a single crystal Si membrane, stress may be reduced by a factor of 1000 to 10,000 [Spiering et al., 1991]. One of the applications of such corrugated structures is the decoupling of a mechanical sensor of its encapsulation, by reducing the influence of temperature changes and packaging stress [Spiering et al., 1991; Offereins et al., 1991]. Thermal stress alone can be reduced by a factor of 120 [Spiering et al., 1993]. Besides stress release, corrugated structures enable much larger deflections than do similar planar structures. This type of structural stress release was studied in some detail for single crystal Si [Zhang and Wise, 1994] (see also earlier in this chapter), polyimide [van Mullem et al. 1991] and LPCVD silicon nitride membranes [Scheeper et al., 1994]; but the quantitative influence of corrugated poly-Si structures still requires investigation. Figure 16.90 illustrates a fabrication sequence for a polyimide corrugated structure. The sacrificial Al in step 4 may be etched away by a mixture of phosphoric acidic: acetic acid: nitric acid (PAN, see Table 16.15).

16.13.4 Dimensional Uncertainties

The often-expressed concerns about run-to-run variability in material properties of polysilicon or other surface micromachined materials are somewhat misplaced, Howe points out [Howe, 1995]. He contrasts the relatively large dimensional uncertainties inherent to any lithography technique with poly-Si quality factors of up to 100,000 and long-term (>3 years) resonator frequency variation of less than 0.02 Hz.

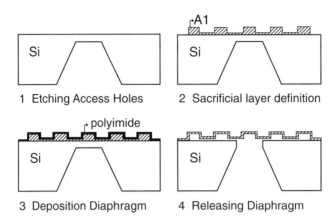

FIGURE 16.90 Schematic view of the fabrication process of a polyimide corrugated diaphragm. (From van Mullem, C.J. et al., Large Deflection Performance of Surface Micromachined Corrugated Diaphragms, presented at Transducers '91, San Francisco, CA, 1991. With permission.)

We follow his calculations here to prove the relative importance of the dimensional uncertainties. The shuttle mass M of a resonator as shown in Figure 16.89 is proportional to the thickness (t) of the polysilicon film, and neglecting the residual stress term, Eq. (16.59) reduces to:

$$f_0 \propto \left(\frac{W}{L}\right)^{\frac{3}{2}} \tag{16.60}$$

In case the residual stress term dominates in Eq. (16.59), the resonant frequency is expressed as:

$$f_0 \propto \left(\frac{W}{L}\right)^{\frac{1}{2}} \tag{16.61}$$

The width-to-length ratio is affected by systematic and random variations in the masking and etching of the microstructural polysilicon. For 2-μm-thick structural polysilicon, patterned by a wafer stepper and etched with a reactive-ion etcher, a reasonable estimate for the variation in linear dimension of etched features, Δ is about 0.2 μm (10% relative tolerance).

From Eq. (16.60) the variation Δ in lateral dimensions will result in an uncertainty δf_0 in the lateral frequency of:

$$\frac{\delta f_0}{f_0} \approx \frac{3}{2}\left(\frac{\Delta}{W}\right) \tag{16.62}$$

for a case where the residual stress can be ignored. With a nominal flexure width of $W = 2$ μm, the resulting uncertainty in resonant frequency is 15%. For the stress-dominated case, Eq. (16.61) indicates that the uncertainty is

$$\frac{\delta f_0}{f_0} \approx \frac{1}{2}\left(\frac{\Delta}{W}\right) \tag{16.63}$$

The same 2-μm-wide flexure would then lead to a 5% uncertainty in resonant frequency.

Interestingly, the stress-free case exhibits the most significant variation in the resonant frequency. In either case, resonant frequencies must be set by some postfabrication frequency trimming or other adjustment.

In Chapter 7 of Madou (1997) we draw further attention to the increasing loss of relative manufacturing tolerance with decreasing structure size (see Figure 7.1 in Madou [1997]).

16.13.5 Sealing Processes in Surface Micromachining

Sealing cavities to hermetically enclose sensor structures is a significant attribute of surface micromachining. Sealing cavities often embodies an integral part of the overall fabrication process and presents a desirable chip level, batch packaging technique. The resulting surface packages (microshells) are much smaller than typical bulk micromachined ones (see Madou [1997, chap. 8 on Packaging]).

16.13.6 IC Compatibility

Putting detection and signal conditioning circuits right next to the sensing element enhances the performance of the sensing system, especially when dealing with high impedance sensors. A key benefit of surface micromachining, besides small device size and single-sided wafer processing, is its compatibility with CMOS processing. IC compatibility implies simplicity and economy of manufacturing. In the examples at the end of this chapter we will discuss how Analog Devices used a mature 4-μm BICMOS process to integrate electronics with a surface micromachined accelerometer.

To develop an appreciation of integration issues involved in combining a CMOS line with surface micromachining, we highlight Yun's [Yun, 1992] comparison of CMOS circuitry and surface micromachining processes in Table 16.16A. Surface micromachining processes are similar to IC processes in several aspects. Both processes use similar materials, lithography and etching techniques. CMOS processes involve at least ten lithography steps where lateral small feature size plays an important role. Some processing steps, such as gate and contact patterning, are critical to the functionality and performance of the CMOS circuits. Furthermore, each processing step is strongly correlated with other steps. Change in any one of the processing steps will lead to modifications in a number of other steps in the process. In contrast, surface micromachining is relatively simple. It usually consists of two to six masks, and the feature sizes are much larger. The critical processing steps, such as structural poly-Si,

TABLE 16.16 Surface Micromachining and CMOS

A. Comparison of CMOS and Surface Micromachining		
	CMOS	Surface Micromachining
Common features	Silicon-based processes; same materials, same etching principles	
Process flow	Standard	Application specific
Vertical dimension	~1 μm	~1–5 μm
Lateral dimension	<1 μm	2–10 μm
Complexity (# masks)	>10	2–6

B. Critical Process Temperatures for Microstructures		
	Temperature (°C)	Material
LPCVD deposition	450	Low temperature oxide (LTO)/PSG
LPCVD deposition	610	Low stress poly-Si
LPCVD deposition	650	Doped poly-Si
LPCVD deposition	800	Nitride
Annealing	950	PSG densification
Annealing	1050	Poly-Si stress annealing

(After Yun, W., A Surface Micromachined Accelerometer with Integrated CMOS Detection Circuitry, Ph.D. Thesis, U.C. Berkeley, 1992.)

often are self-aligned which eliminates lithographic alignment. The CMOS process is mature, quite generic and fine-tuned, while surface micromachining strongly depends on the application and still needs maturing.

Table 16.16B presents the critical temperatures associated with the LPCVD deposition of a variety of frequently used materials in surface micromachining. Polysilicon is used for structural layers and thermal SiO_2, LPCVD SiO_2 and PSG are used as sacrificial layers; silicon nitride is used for passivation. The highest temperature process in Table 16.16B is 1050°C and is associated with the annealing step to release stress in the polysilicon layers. Doped polysilicon films deposited by LPCVD under conventional IC conditions usually are in a state of compression that can cause mechanically constrained structures such as bridges and diaphragms to buckle. The annealing step above about 1000°C promotes crystallite growth and reduces the strain. If one wants to build polysilicon microstructures after the CMOS active electronics have been implemented (a so-called post-CMOS procedure), one has to avoid temperatures above 950°C, as junction migration will take place at those temperatures. This is especially true with devices incorporating shallow junctions where migration might be a problem at temperatures as low as 800°C. The degradation of the aluminum metallization presents yet a bigger problem. Aluminum typically is used as the interconnect material in the conventional CMOS process. At temperatures of 400 to 450°C, the aluminum metallization will start suffering. Anneal temperatures (densification of the PSG and stress anneal of the poly-Si) only account for some of the concerns; in general, several compatibility issues must be considered: (1) deposition and anneal temperatures, (2) passivation during micromachining etching steps and (3) surface topography.

Yun (1992) compared three possible approaches to build integrated microdynamic systems: pre-, mixed and post-CMOS microstructural processes as shown in Figure 16.91. He concluded that building up the microstructures after implementation of the active electronics offers the best results.

In a post-CMOS process, the electronic circuitry is passivated to protect it from the subsequent micromachining processes. The standard IC processing may be performed at a regular IC foundry, while the surface micromachining occurs as an add-on in a specialized sensor fabrication facility. LPCVD silicon nitride (deposited at 800°C, see Table 16.16B) is stable in HF solutions and is the preferred passivation layer for the IC during the long release etching step. PECVD nitride can be deposited at around 320°C, but it displays relatively poor step coverage, while pin holes in the film allow HF to diffuse through and react with the oxide underneath. LPCVD nitride is conformably deposited. While it shows fewer pin holes, circuitry needs to be able to survive the 800°C deposition temperature.

Aluminum metallization must be replaced by another interconnect scheme in order to raise the post-CMOS temperature ceiling higher than 450°C. Tungsten, which is refractory, shows low resistivity and has a thermal expansion coefficient matching that of Si, is an obvious choice. One problem with tungsten metallization is that tungsten reacts with silicon at about 600°C to form WSi_2, implying the need for a diffusion barrier. The process sequence for the tungsten metallization developed at Berkeley is shown in Figure 16.92. A diffusion barrier consisting of $TiSi_2$ and TiN is used. The TiN film forms during a 30-sec sintering step to 600°C in N_2. Rapid thermal annealing with its reduced time at high temperatures (10 sec to 2 min) and high ramp rates (~150°C/sec) allows very precise process control as well as a dramatic reduction of thermal budgets, reducing duress for the active on-chip electronics. Titanium silicide is formed at the interface of titanium and silicon while titanium nitride forms simultaneously at the exposed surface of the titanium film. The $TiSi_2$/TiN forms a good diffusion barrier against the formation of WSi_2 and at the same time provides an adhesion and contact layer for the W metallization.

To avoid the junction migration in a post-CMOS process, rapid thermal annealing is used for both the PSG densification and polysilicon stress anneal: 950°C for 30 sec for the PSG densification and 1000°C for 60 sec for the stress anneal of the poly-Si. Alternatively, one could consider the use of fine-grained polysilicon which can yield a controlled tensile strain with low-temperature annealing [Guckel et al., 1988].

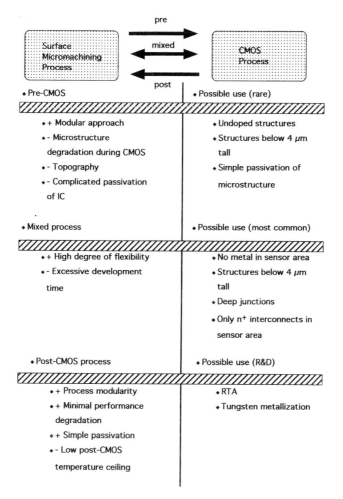

FIGURE 16.91 Comparison of various options for process integration. (Adapted from Yun, W., A Surface Micromachined Accelerometer with Integrated CMOS Detection Circuitry, Ph.D. thesis, University of California, Berkeley, 1992. With permission.)

Despite some advantages the post-CMOS process with tungsten metallization is not the preferred implementation. Hillock formation in the W lines during annealing and high contact resistance remain problems [Howe, 1995]. Moreover, the finely tuned CMOS fabrication sequence may also be affected by the heavily doped structural and sacrificial layers.

The mixed CMOS/micromachining approach implements a processing sequence which puts the processes in a sequence to minimize performance degradation for both electronic and mechanical components. According to Yun this requires significant modifications to the CMOS fabrication sequence. Nevertheless, Analog Devices relied on such an interleaved process sequence to build the first commercially available integrated micro-accelerometer (see Example 1). The modifications required on a standard BICMOS line were minimal: to facilitate integration of the IC and to surface micromachine the thickness of deposited microstructural films the line was limited to 1 to 4 μm. Relatively deep junctions permitted thermal processing for the sensor poly-Si anneal and interconnections to the sensor were made only via n^+ underpasses. No metallization is present in the sensor area. This industrial solution remains truer to the traditional IC process experience than the post-CMOS procedure. Howe recently detailed another example of such a mixed process [Howe, 1995]. In Howe's scenario, the micromachining sequence is inserted after the completion of the electronic

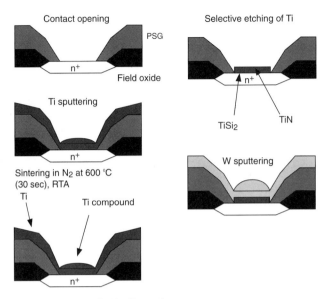

- Self-aligned process
- No peeling of W up to 1050 °C in Ar
- Thin layer of nitride forms during RTA in N$_2$

FIGURE 16.92 Tungsten metallization process in a modified CMOS process. (Adapted from Yun, W., A Surface Micromachined Accelerometer with Integrated CMOS Detection Circuitry, Ph.D. thesis, University of California, Berkeley, 1992. With permission.)

structures, but prior to contact etching or aluminum metallization. By limiting the polysilicon annealing to 7 min at 900°C, only minor dopant redistribution is expected. Contact and metallization lithography and etching become more complex now due to the severe topography of the poly-Si microstructural elements.

The pre-CMOS approach is to fabricate microstructural elements before any CMOS process steps. At first glance this seems like an attractive approach as no major modifications would be needed for process integration. However, due to the vertical dimensions of microstructures, step coverage is a problem for the interconnection between the sensor and the circuitry (the latest approach introduced by Howe faces the same dilemma). Passivation of the microstructure during the CMOS process can also become problematic. Furthermore, the fine-tuned CMOS fabrication sequence, such as gate oxidation, can be affected by the heavily doped structural layers. Consequently, this approach is only used for some special applications [Yun, 1992].

A unique pre-CMOS process was developed at Sandia National Labs [Smith et al., 1995]. In this approach, micromechanical devices are fabricated in a trench etched in a Si epilayer. After the mechanical components are complete, the trench is filled with oxide, planarized using chemical-mechanical polishing (CMP) (see Madou [1994, chap. 3]) and sealed with a nitride membrane. The flat wafer with the embedded micromechanical devices is then processed by means of conventional CMOS processing. Additional steps are added at the end of the CMOS process in order to expose and release the embedded micromechanical devices.

The SPIE "Smart Structures and Materials 1996" meeting in San Diego, CA, had two complete sessions dedicated to the crucial issue of integrating electronics with polysilicon surface micromachining [Varadan and McWhorter, 1996].

16.14 Poly-Si Surface Micromachining Modifications

16.14.1 Porous Poly-Si

In the first part of this chapter we discussed the transformation of single crystal Si into a porous material with porosity and pore sizes determined by the current density, type and concentration of the dopant and the hydrofluoric acid concentration. A transition from pore formation to electropolishing is reached by raising the current density and/or by lowering the hydrofluoric acid concentration [Memming and Schwandt, 1966; Zhang et al., 1989]. Porous silicon can also be formed under similar conditions from LPCVD poly-Si [Anderson et al., 1994]. In this case pores roughly follow the grain boundaries of the polysilicon. Figure 16.93 illustrates the masking and process sequence to prepare a wafer to make thin layers of porous Si between two insulating layers of low stress silicon nitride [Field, 1987]. The wafer, after the process steps outlined in Figure 16.93, is put in a Teflon test fixture, protecting the back from HF attack. An electrical contact is established on the back of the wafer and a potential is applied with respect to a Pt-wire counterelectrode immersed in the same HF solution. Electrolytes consisting of 5 to 49% HF (wt) and current densities from 0.1 to 50 A cm^{-2} are used. The advance of a pore-etching front, growing parallel to the wafer surface, may be monitored using a line-width measurement tool. The highest observed rate of porous-silicon formation is 15 µm min^{-1} (in 25%, wt HF). In the electro-polishing regime, at the highest currents, the etch rate is diffusion limited but the reaction is controlled by surface reaction kinetics in the porous Si growth regime.

By changing the conditions from pore formation to electropolishing and back to porous Si, an enclosed chamber may be formed with porous poly-Si walls (plugs) and 'floor and ceiling' silicon nitride layers. Sealing of the cavities by clogging the microporous poly-Si was attempted by room temperature oxidation in air and in a H_2O_2 solution. Leakage through the porous plug persisted after those room temperature oxidation treatments, but with a Ag deposition from a 400-mM AgNO$_3$ solution and subsequent atmospheric tarnishing (48 h, Ag$_2$S) the chambers appeared to be sealed, as determined from the lack of

1. Deposit CVD silicon nitride on silicon wafer.
2. Pattern and plasma-etch silicon nitride.

3. Deposit CVD polysilicon.
4. Deposit CVD silicon nitride.

2. Pattern and plasma-etch nitride and polysilicon.

FIGURE 16.93 Masking and process sequence preparing a wafer for laterally grown porous polysilicon. (From Field, L., *Low-Stress Nitrides for Use in Electronic Devices* (University of California, Berkeley, 1987), pp. 42 and 43. With permission.)

penetration by methanol. This technology might open up possibilities for filling cavities with liquids and gases under low-temperature conditions. The chamber provided with a porous plug might also make a suitable on-chip electrochemical reference electrolyte reservoir.

Hydrofluoric acid can penetrate thin layers of poly-Si either at foreign particle inclusion sites, or at other critical film defects such as grain boundaries (see above). This way the HF can etch underlying oxide layers, creating, for example, circular regions of free-standing poly-Si, so-called 'blisters.' The poly-Si permeability associated with blistering of poly-Si films has been applied successfully by Judy and Howe (1993a; 1993b) to produce thin-shelled hollow beam electrostatic resonators from thin poly-Si films deposited onto PSG. The possible advantage of using these hollowed structures is to obtain a yet higher resonator quality factor Q. The devices were made in such a way that the 0.3-μm-thick undoped poly-Si completely encased a PSG core. After annealing, the structures were placed in HF which penetrated the poly-Si shell and dissolved away the PSG, eliminating the need for etch windows. It was not possible to discern the actual pathways through the poly-Si using TEM.

Lebouitz et al. (1995) apply permeable polysilicon etch-access windows to increase the speed of creating microshells for packaging surface micromachined components. After etching the PSG through the many permeable Si windows, the shell is sealed with 0.8 μm of low stress LPCVD nitride.

16.14.2 Hinged Polysilicon

One way to achieve high vertical structures with surface micromachining is building large flat structures horizontally and then rotating them on a hinge to an upright position. Pister et al. [Burgett et al., 1992; Ross and Pister, 1994; Lin et al., 1995; Yeh et al., 1994; Lin et al., 1996] developed the poly-Si hinges shown in Figure 16.94A; on these hinges, long structural poly-Si features (1 mm and beyond) can be rotated out of the plane of a substrate. To make the hinged structures, a 2-μm-thick PSG layer (PSG-1) is deposited on the Si substrate as the sacrificial material, followed by the deposition of the first polysilicon layer (2-μm-thick poly-1). This structural layer of polysilicon is patterned by photolithography and dry etching to form the desired structural elements, including hinge pins to rotate them. Following the deposition and patterning of poly-1, another layer of sacrificial material (PSG-2) of 0.5-μm thickness is deposited. Contacts are made through both PSG layers to the Si substrate, and a second layer of polysilicon is deposited and patterned (poly-2), forming a staple to hold the first polysilicon layer hinge to the surface. The first and second layers of poly are separated everywhere by PSG-2 in order for the first polysilicon layer to freely rotate off the wafer surface when the PSG is removed in a sacrificial etch. After the sacrificial etch, the structures are rotated in their respective positions. This is accomplished in an electrical probe station by skillfully manipulating the movable parts with the probe needles. Once the components are in position, high friction in the hinges tends to keep them in the same position. To obtain more precise and stable control of position, additional hinges and supports are incorporated. To provide electrical contact to the vertical poly-Si structures one can rely on the mechanical contact in the hinges, or poly-Si beams (cables) can be attached from the vertical structure to the substrate.

Pister's research team made a wide variety of hinged microstructures, including hot wire anemometers, a box dynamometer to measure forces exerted by embryonic tissue, a parallel plate gripper [Burgett et al., 1992], a micro-windmill [Ross and Pister, 1994], a micro-optical bench for free-space integrated optics [Lin et al., 1995] and a standard CMOS single piezoresistive sensor to quantify rat single heart cell contractile forces [Lin et al., 1996]. One example from this group's efforts is illustrated in Figure 16.94B, showing an SEM photograph of an edge-emitting laser diode shining light onto a collimating micro-Fresnel lens [Lin et al., 1995]. The micro-Fresnel lens in the SEM photo is surface micromachined in the plane and erected on a polysilicon hinge. The lens has a diameter of 280 μm. Alignment plates at the front and the back sides of the laser are used for height adjustment of the laser spot so that the emitting spot falls exactly onto the optical axis of the micro-Fresnel lens. After assembly, the laser is electrically contacted by silver epoxy. Although this hardly outlines standard IC manufacturing practices, excellent collimating ability for the Fresnel lenses has been achieved. The eventual goal of this work is a micro-optical bench (MOB) in which microlenses, mirrors, gratings and other optical components are pre-aligned in the mask layout stage using computer-aided design. Additional fine adjustment would be achieved by on-chip micro-actuators and micropositioners such as rotational and translational stages.

Today, erecting these poly-Si structures with the probes of an electrical probe station or, occasionally, assembly by chance in the HF etch or deionized water rinse represent too complicated or too unreliable postrelease assembly methods for commercial acceptance.

Friction in poly-Si joints, as made by Pister, is high because friction is proportional to the surface area (s^2) and becomes dominant over inertial forces (s^3) in the microdomain (see Madou [1997, chap. 9]). Such joints are not suitable for microrobotic applications. Although attempts have been made to incorporate poly-Si hinges in such applications [Yeh et al., 1994], plastically deformable hinges make more sense for microrobot machinery involving rotation of rigid components. Noting that the external skeleton of insects incorporates hard cuticles connected by elastic hinges, Suzuki et al. (1994) fabricated rigid

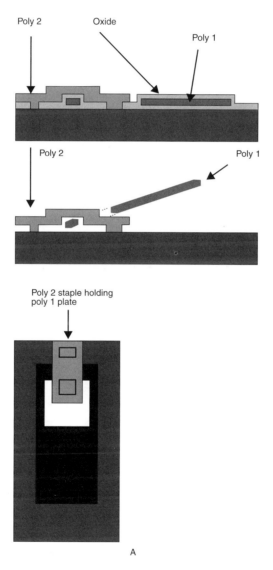

FIGURE 16.94 Microfabricated hinges. (A) Cross-section, side view and top view of a single-hinged plate before and after the sacrificial etch. (B) Schematic (top) and SEM micrograph of the self-aligned hybrid integration of an edge-emitting laser with a micro-Fresnsel lens. (From Lin, L.Y. et al., Micromachined Integrated Optics for Free-Space Interconnections, presented at MEMS '96, Amsterdam, 1995. With permission.)

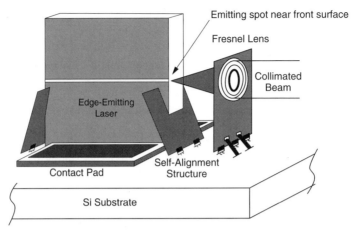

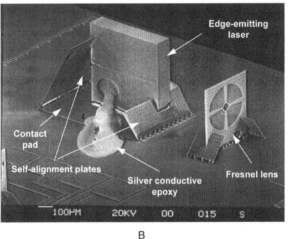

FIGURE 16.94 (Continued).

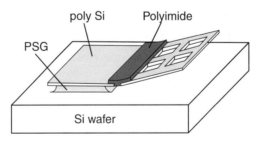

FIGURE 16.95 Flexible polyimide hinge and poly-Si plate (butterfly wing). (From Suzuki, K. et al., *J. Microelectromech. Syst.*, 3, 4–9, 1994. With permission.)

poly-Si plates (E = 140 GPa) connected by elastic polyimide hinges (E = 3 GPa) as shown in Figure 16.95 (see also section below on Polimide Surface Structures). Holes in the poly-Si plates shorten the PSG etch time compared to plates without holes. The plates without holes remain attached to the substrate while the ones with holes are completely freed. Using electrostatic actuators, a structure as shown in Figure 16.95 can be made to flap like the wing of a butterfly. By applying an AC voltage of 10 kHz, resonant vibration of such a flapping wing was observed [Suzuki et al., 1994].

More recently, Hoffman et al. (1995) demonstrated aluminum plastically deformable hinges on oxide movable thin plates. Oxide plates and Al hinges were etched free from a Si substrate by using XeF_2, a vapor phase etchant exhibiting excellent selectivity of Si over Al and oxide. According to the authors, this process, due to its excellent CMOS compatibility, might open the way to designing and fabricating sophisticated integrated CMOS-based sensors with rapid turnaround time (see also Madou [1997, chap. 2]).

16.14.3 Thick Polysilicon

Applying classical LPCVD to obtain poly-Si deposition is a slow process. For example, a layer of 10 µm typically requires a deposition time of 10 hr. Consequently, most micromachined structures are based on layer thicknesses in the 2- to 5-µm range. Basing their process on dichlorosilane (SiH_2Cl_2) chemistry, Lange et al. (1995) developed a CVD process in a vertical epitaxy batch reactor with deposition rates as high as 0.55 µm/min at 1000°C. The process yields acceptable deposition times for thicknesses in the 10-µm range (20 min). The highly columnar poly-Si films are deposited on sacrificial SiO_2 layers and exhibit low internal tensile stress making them suitable for surface micromachining. The surface roughness comprises about 3% of the thickness, which might preclude some applications.

Kahn et al. (1996) made mechanical property test structures from thick undoped and *in situ* B-doped polysilicon films. The elastic modulus of the B-doped polysilicon films was determined as 150 ± 30 GPa. The residual stress of as-deposited undoped thick polysilicon was determined as 200 ± 10 MPa.

16.14.4 Milli-Scale Molded Polysilicon Structures

The assembly of tall three-dimensional features in the described hinged polysilicon approach is complicated by the manual assembly of the fabricated microparts resembling building a miniature boat in a bottle. Keller at University of California, Berkeley, came up with an elegant alternative for building tall, high-aspect-ratio microstructures in a process that does not require postrelease assembly steps [Keller and Ferrari, 1994]. The technique involves deep dry etching of trenches in a Si substrate, deposition of sacrificial and structural materials in those trenches, and demolding of the deposited structural materials by etching away the sacrificial materials. CVD processes can typically only deposit thin films (~1 to 2 µm) on flat surfaces. If, however, these surfaces are the opposing faces of deep narrow trenches, the growing films will merge to form solid beams. In this fashion, high-aspect-ratio structures that would normally be associated with LIGA now also can be made of CVD polysilicon. The procedure is illustrated in Figure 16.96 [Keller and Ferrari, 1994; Keller and Howe, 1995]. The first step is to etch deep trenches into a silicon wafer. The depth of the trenches equals the height of the desired beams and is limited to about 100 µm with aspect ratios of about 10 (say, a 10-µm-diameter hole with a depth of 100 µm). For trench etching, Keller uses a Cl_2 plasma etch with the following approximate etching conditions: flow rates of 200 sccm for He and 180 sccm for Cl_2, a working pressure of 425 mTorr, a power setting of 400 W and an electrode gap of 0.8 cm. The etch rate for Si in this mode equals 1 µm/min. Thermal oxide and CVD oxide act as masks with 1 µm of oxide needed for each 20 µm of etch depth. Before the Cl_2 etch a short 7-sec SF_6 pre-etch removes any remaining native oxide in the mask openings. During the chlorine etch a white sidewall passivating layer must be controlled to maintain perfect vertical sidewalls. After every 30 min of plasma etching, the wafers are submerged in a silicon isotropic etch long enough to remove the residue [Keller and Howe, 1995]. Beyond 100 µm, severe undercutting occurs and the trench cross section becomes sufficiently ellipsoidal to prevent molded parts from being pulled out. Advances in dry cryogenic etching are continually improving attainable etch depths, trench profiles and minimum trench diameter. We can expect continuous improvements in the tolerances of this novel technique. After plasma etching, an additional 1 µm of silicon is removed by an isotropic wet etch to obtain a smoother trench wall surface. Alternatively, to smooth sidewalls and bottom of the trenches a thermal wet oxide is grown and etched away. The sacrificial oxide in step 2 is made by CVD phosphosilicate glass (PSG

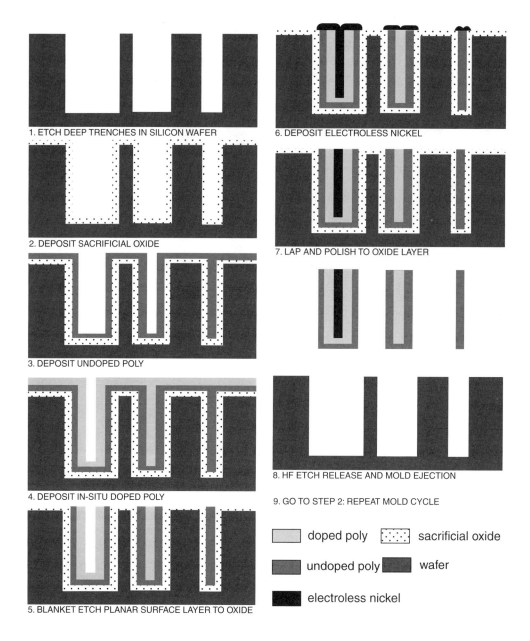

FIGURE 16.96 Schematic illustration of HEXSIL process. The mold wafer may be part of an infinite loop. (From Keller and Ferrari [1994]. Courtesy of Mr. C. Keller.)

at 450°C, 140 Å/min), CVD low-temperature oxide (LTO at 450°C) or CVD polysilicon (580°C, 65 Å/min). The latter is completely converted to SiO_2 by wet thermal oxidation at 1100°C. The PSG needs an additional reflowing and densifying anneal at 1000°C in nitrogen for 1 hr. This results in an etching rate of the sacrificial layer of ~20 µm/min in 49% HF. The mold shown in Figure 16.96 displays three different trench widths and can be used to build integrated micromachines incorporating doped and undoped poly-Si parts as well as metal parts. The remaining volume of the narrowest trench after oxide deposition is filled completely with the first deposition of undoped polysilicon (poly 1) in step 3. The undoped poly will constitute the insulating regions in the micromachine. Undoped CVD polysilicon was formed in this case at 580°C, with a 100 SCCM silane flow rate, and

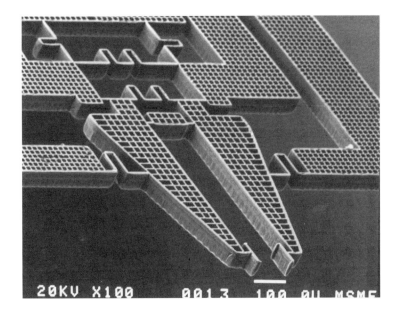

FIGURE 16.97 SEM micrograph of HEXSIL tweezers: 4 mm long, 2 mm wide and 80 μm tall. Lead wires for current supply are made from Ni-filled poly-Si beams; *in situ* phosphorus-doped polysilicon provides the resistor part for actuation. The width of the beam is 8 μm: 2 μm poly-Si, 4 μm Ni, and 2 μm poly-Si. (Courtesy of Mr. C. Keller.)

a 300-mTorr reactor pressure, resulting in a deposition rate of 0.39 μm/hr. The deposited film under these conditions is amorphous or very fine-grained. Since the narrowest trenches are completely filled in by the first deposition, they cannot accept material from later depositions. The trenches of intermediate width are lined with the first material and then completely filled in by the second deposition. In the case illustrated the second deposition (step 4) consists of *in situ* doped poly-Si and forms the resistive region in the micromachine under construction. To prevent diffusion of P from the doped poly deposited on top of the narrow undoped beams, a blanket etch in step 5 is used to remove the doped surface layer prior to the anneal of the doped poly. The third deposition, in step 6 of the example case, consists of electroless nickel plating on poly-Si surfaces but not on oxides surfaces and results in the conducting parts of the micromachine. By depositing structural layers in order of increasing conductivity, as done here, regions of different conductivity can be separated by regions of narrow trenches containing only nonconducting material. Lapping and polishing in step 7 with a 1-μm diamond abrasive in oil planarizes the top surface, readying it for HF etch release and mold ejection in step 8. Annealing of the polysilicon is required to relieve the stress before removing the parts from the wafer so they remain straight and flat. In step 8 the sacrificial oxide is dissolved in 49% HF. A surfactant such as Triton X100 is added to the etch solution to facilitate part ejection by reducing surface adhesion between the part and the mold. The parts are removed from the wafer, and the wafer may be returned to step 2 for another mold cycle. An example micromachine, resulting from the described process, is the thermally actuated tweezers shown in Figure 16.97. These HEXSIL tweezers measure 4 mm long, 2 mm wide and 80 μm tall. The thermal expansion beam to actuate the tweezers consists of the *in situ* doped poly-Si; the insulating parts are made from the undoped poly-Si material. Ni-filled poly-Si beams are used for the current supply leads. It is possible to combine the HEXSIL process with classical poly-Si micromachining, as illustrated in Figure 16.98, where HEXSIL forms a stiffening rib for a membrane filter fashioned by surface micromachining of a surface poly-Si layer. The surface poly-Si is deposited after HEXSIL. A critical need in HEXSIL technology is controlled mold ejection. Keller and Howe (1995) have experimented with HEXSIL-produced bimorphs, making the structure spring up after release.

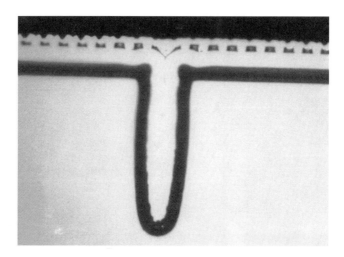

FIGURE 16.98 SEM micrograph of surface micromachined membrane filter with a stiffening rib (50 μm high). Original magnification 1000×. (Courtesy of Mr. C. Keller.)

16.15 Non-Poly-Si Surface Micromachining Modifications

16.15.1 Silicon on Insulator Surface Micromachining

16.15.1.1 Introduction

A major sensor use of silicon on insulator wafers is established in the production of high-temperature sensors. Compared to p-n junction isolation, which is limited to about 125°C, much higher temperature devices are possible based on the dielectric insulation of SOI. Recently, a wide variety of SOI surface micromachined structures have been explored, including pressure sensors, accelerometers, torsional micromirrors, light sources, optical choppers etc. [Diem et al., 1993; Noworolski et al., 1995]

Three major techniques currently are applied to produce *SOI* wafers (see also under Epitary in Madou [1997, chap. 3]): *SIMOX* (Separated by IMplanted OXygen), the Si fusion bonded (SFB) wafer technique and zone-melt recrystallized (ZMR) polysilicon. With SIMOX, standard Si wafers are implanted with oxygen ions and then annealed at high temperatures (1300°C). The oxygen and silicon combine to form a silicon oxide layer beneath the silicon surface. The oxide layer's thickness and depth are controlled by varying the energy and dose of the implant and the anneal temperature. In some cases, a CVD process deposits additional epitaxial silicon on the top silicon layer. Attempts have also been made to implant nitrogen in Si to create abrupt etch stops. At high enough energies the implanted nitrogen is buried 1/2 to 1 μm deep. At a high enough dose, the etching in that region stops. It is not necessary to implant the stoichiometric amount of nitrogen concentration; a dose lower by a factor of 2 to 3 suffices. After implantation, it is necessary to anneal the wafer because the implantation destroys the crystal structure at the surface of the wafer.

The bonded wafer process starts with an oxide layer grown (typically about 1 μm) on a standard Si wafer. That wafer is then bonded to another wafer, with the oxide sandwiched between. For the bonding no mechanical pressure or other forces are applied. The sandwich is annealed at 1100°C for 2 hr in a nitrogen ambient leading to a strong binding between the two wafers. One of the wafers is then ground to a thickness of a few microns using mechanical and CMP.

A third process for making SOI structures is to recrystallize polysilicon (e.g., with a laser, an electron-beam or a narrow strip heater) deposited on an oxidized silicon wafer. This process is called zone melting recrystallization (ZMR). This technique is used primarily for local recrystallization and has not yet been explored much in micromachining applications.

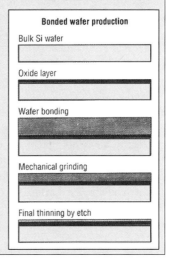

How SOI wafers are made.

Many different silicon-on-insulator materials have been developed over the years, but two are currently being used for IC production: SIMOX (Separated by IMplanted OXygen) and bonded wafers.

In the SIMOX process, a standard silicon wafer is implanted with oxygen ions, and then annealed at high temperatures; the oxygen and silicon combine to form a silicon oxide layer beneath the wafer surface. To minimize wafer damage, the oxygen is sometimes implanted in two or more passes, each followed by an anneal. The oxide layer's thickness and depth are controlled by varying the energy and dose of the implant and the anneal temperature. In some cases, a CVD process is used to deposit additional silicon on the top layer.

The bonded wafer process starts with an oxide layer of the desired thickness (typically 0.25 to 2 microns) being grown on a standard silicon wafer. That wafer is then bonded at high temperatures to another wafer, with the oxide sandwiched between. One of the wafers is then ground to a thickness of a few microns using a mechanical tool.

Because advanced devices require an even thinner layer, more silicon must be removed. In Hughes Danbury's AcuThin process, the wafer is etched with a confined plasma, between 3 and 30 mm wide, which is stepped across the wafer surface. A film thickness map is made for each wafer, and used to compute the dwell time for the plasma etcher at each stop. The process can be repeated for additional precision; Hughes Danbury offers silicon thicknesses of as little as 1000 to 3000 angstroms, with total thickness variation of 200 angstroms. IBM has also developed an etch-back process for bonded wafers. ■

From Dunn, P.N., *Solid State Technol.*, October, 32–35 (1993). With permission.

The crystalline perfection of conventional silicon wafers in SFB and ZMR is completely maintained in the SOI layer as the wafers do not suffer from implant-induced defects. By using plasma etching, wafers with a top Si layer thickness of as little as 1000 to 3000 Å with total thickness variations of less than 200 Å can be made [Dunn, 1993]. SOI layers of 2 μm thick are more standard [Abe and Matlock, 1990].

Kanda (1991) reviews different types of SOI wafers in terms of their micromachining and IC applications. Working with SOI wafers offers several advantages over bulk Si wafers: fewer process steps are needed for feature isolation, parasitic capacitance is reduced, and power consumption is lowered. In the IC industry these wafers are used for high-speed CMOS ICs, smart power ICs, three-dimensional ICs, and radiation-hardened devices [Kuhn and Rhee, 1973]. In micromachining, SOI wafers are employed to produce an etch stop in such mechanical devices as pressure and acceleration sensors and in high-temperature sensors, and ISFETs. Etched-back fusion-bonded Si wafers and SIMOX are already employed extensively to build micromachines. The two types of SOI wafers are commercially available. A tremendous amount of effort is spent in the IC industry on controlling the SOI thin Si layer thickness which will benefit any narrow tolerance micromachine. The silicon fusion bonded (SFB) method offers the more versatile MEMS approach due to the associated potential for thicker single crystal layers and the option of incorporating buried cavities, facilitating micromachine packaging. Sensors manufactured by means of SFB now are commercially available [Pourahmadi et al., 1992]. The SIMOX approach is less labor intensive and holds better membrane thickness control. An important expansion of the SOI technique is selective epitaxy. The latter enables a wide range of new mechanical structures (see Madou [1997, chap. 3]) and enables novel etch-stop methods [Gennissen et al., 1995], as well as electrical and/or thermal separation and independent optimization of active sensor and readout electronics [Bartek et al., 1995]. In all cases SOI machining involves dry anisotropic etching to etch a pattern into the Si layer on top of the insulator. These structures then become free by etching the sacrificial buried SiO_2 insulator layer, which displays a thickness with very high reproducibility (400 ± 5 nm) and uniformity (<±5 nm), especially in the case of SIMOX. Etched free cantilevers and membranes consist of single crystalline silicon with thicknesses ranging from microns and submicrons (SIMOX) up to hundreds of microns (SFB). Below we review three implementations of SOI techniques that may be crucial for future MEMS development.

16.15.1.2 Silicon Fusion Bonded Micromachining

Silicon fusion bonding enables the formation of thick single crystal layers with cavities built in. An example is shown in Figure 16.99 [Noworolski et al., 1995]. The device pictured involves two 4-in. <100> wafers: a handle wafer and a wafer used for the SOI surface. The p-type (3 to 7 Ωcm) handle wafer is thermally oxidized at 1100°C to obtain a 1-µm-thick oxide. Thermal oxidation enables thicker oxides than the ones formed in SIMOX by ion implantation and avoids the potential implantation damage in the working material. To make a buried cavity, the oxide is patterned and etched. To produce yet deeper cavities, the Si handle wafer may be etched as well (as in Figure 16.99). In the case shown, the top wafer consists of the same p-type substrate material as the handle wafer with a 2- to 30-µm-thick n-type epitaxial layer. The epitaxial layer determines the thickness of the final mechanical material. The epitaxial layer is fusion bonded to the cavity side of the handle wafer (2 hr at 1100°C). The top wafer is then partially thinned by grinding and polishing (Figure 16.99A). An insulator is deposited and patterned on the back side of the handle wafer to etch access holes to the insulator. After the insulator at the bottom of the etch hole is removed by a buffered oxide etch (BOE), aluminum is sputtered and sintered to make contact to the n-type epilayer for the electrochemical etch back of the remaining p-type material (Figure 16.99B). The final single crystal silicon thickness is uniform to within ±0.05 µm (std. dev.) and does not require a costly, high accuracy polish step.

Draper Laboratory is using SOI processes in the development of inertial sensors, gyros and accelerometers as an alternative to their current devices fabricated by the dissolved wafer process (see Example 1, Section 16.9). The main advantage is that the former consists of an all Si process rather than a Si/Pyrex sandwich [Greiff, 1995].

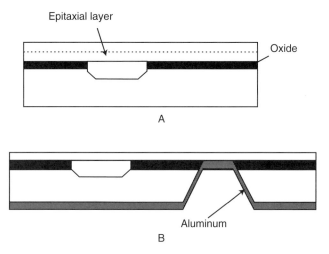

FIGURE 16.99 (A) A wafer sandwich after grind-and-polish step. (B) A wafer after electrochemical etch-back in KOH, buried oxide removal and aluminum deposition. (From Noworolski, J.M. et al., Fabrication of SOI Wafers with Buried Cavities Using Silicon Fusion Bending and Electrochemical Etchback, presented at Transducers '95, Stockholm, Sweden, 1995. With permission.)

16.15.1.3 SIMOX Surface Micromachining

Both capacitive and piezoresistive pressure sensors were microfabricated from SIMOX wafers [Diem et al., 1993]. Figure 16.100 illustrates the process sequence by Diem et al. (1993) for fabricating an absolute capacitive pressure sensor. The 0.2-µm silicon surface layer of the SIMOX wafer is thickened with doped epi-Si to 4 µm. An access hole is RIE etched in the Si layer, and vacuum cavity and electrode gap are obtained by etching the SiO_2 buried layer. Since the buried thick oxide layer exhibits a very high reproducibility and homogeneity over the whole wafer (0.4 µm ± 5 nm), the resulting vacuum cavity and electrode gap after etching also are very well controlled. The small gap results in relatively high capacitance values between the free membrane and bulk substrate (20 pf/mm^2). Diaphragm diameter, controlled by the SiO_2 etching is up to several hundreds micrometers (±2 µm). The etching hole is hermetically sealed under vacuum by plasma CVD deposition of nonstressed dielectric layer plugs.

With the above scheme Diem et al. realized an absolute pressure sensor with a size of less than 1.5 mm^2. The temperature dependence of a capacitive sensor is mainly due to the temperature coefficient of the offset capacitance. Therefore, a temperature compensation is needed for high accuracy sensors. A drastic reduction of the temperature dependence is obtained by a differential measurement, especially if the reference capacitor resembles the sensing capacitor. A reference capacitor was designed with the mem-

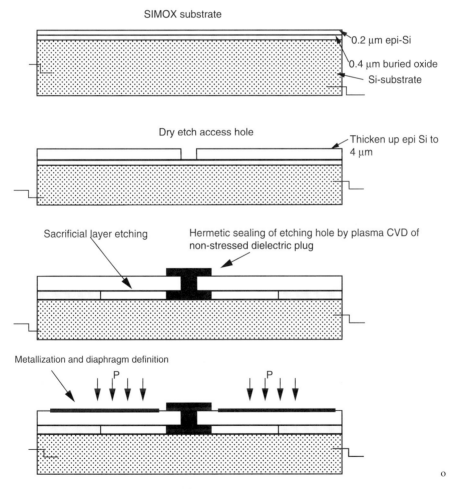

FIGURE 16.100 Process sequence of a SIMOX absolute capacitive pressure sensor by Diem et al. (From Diem, B. et al., SOI (SIMOX) as a Substrate for Surface Micromachining of Single Crystalline Silicon Sensors and Actuators, presented at Seventh Int. Conf. on Solid-State Sensors and Actuators, Yokohama, Japan, 1993. With permission.)

brane blocked by several plugs for pressure insensitivity. The localization and the number of plugs were modeled by finite element analysis (FEA) (ANSYS software was used) to get a deformation lower than 1% of the active sensor's deformation. Even without temperature calibration the high output of the differential signal resulted in an overall output error better than ±2% over the whole temperature range (−40°C to +125°C) compared to 10% for nondifferential measurements. The temperature coefficient of the sensitivity is about 100 ppm/°C which agrees with the theoretical variation of the Young modulus of silicon. A piezoresistive sensor could be achieved by implanted strain gauges in the membrane. Although SIMOX wafers are more expensive than regular wafers, they come with several process steps embedded and they make packaging easier.

16.15.1.4 Selective Epitaxy Surface Micromachining

In the discussion on epitaxy in Madou (1997, chap. 3) we drew attention to the potential of selective epitaxy for creating novel microstructures. The example in Figure 3.21 in Madou illustrates the selective deposition of epi-Si on a Si substrate through a SiO_2 window. The same figure also demonstrates the simultaneous deposition of poly-Si on SiO_2 and crystalline epi-Si on Si, creating the basis for a structure featuring an epi-Si anchor with poly-Si side arms.

Neudeck et al. (1990) and Schubert and Neudeck (1990) at Purdue and Gennissen et al. (1995) and Bartek et al. (1995) at Twente proved that selective epitaxy can also be applied for automatic etch stop on buried oxide islands. Figure 16.101 demonstrates how epitaxial lateral overgrowth (ELO) can bury oxide islands. After removal of the native oxides from the seed windows, epi is grown for 20 min at 950°C and at 60 torr using a $Si_2H_2Cl_2$-HCl-H_2 gas system. The epi growth front moves parallel to the wafer surface while growing in the lateral direction, leaving a smooth planar surface. During epi growth the HCl prevents poly nucleation on the nonsilicon areas. The epi quality is strongly dependent on the orientation of the seed holes in the oxide. Seed holes oriented in the <100> direction lead to the best epi material and surface quality. Selective epi's other big problem for fabrication remains sidewall defects [Bashir et al.]. The buried oxide islands stop the KOH etch of the substrate, enabling formation of beams and membranes as shown in Figure 16.101B. This technique might form the basis of many high performance microstructures. The Purdue and Twente groups also work on confined selective epitaxial growth (CSEG), a process pioneered by Schubert and Neudeck (1990). In this process a micromachined cavity is formed above a silicon substrate with a seed contact window to the silicon substrate and access windows for epi-Si (Figure 16.101C) [Bartek et al., 1995]. Low-stress, silicon-rich nitride layers act as structural layers to confine epitaxial growth; PSG is used as sacrificial material. This confined selective epitaxial growth technique allows electrical and/or thermal isolation separation, as well as independent optimization of active sensor and readout electronic areas.

16.15.1.5 SOI vs. Poly-Si Surface Micromachining

The power of poly-Si surface micromachining mainly lays in its CMOS compatibility. When deposited on an insulator, both poly-Si and single crystal layers enable higher operating temperatures (>200°C) than bulk micromachined sensors featuring p-n junction isolation only (130°C max) [Luder, 1986]. An additional benefit for SOI-based micromachining is IC compatibility combined with single crystal Si performance excellence. The maximum gauge factor (see Eq. (16.28)) of a poly-Si piezoresistor is about 30, roughly 15 times larger than that of a metal strain gauge but only one third of that of an indiffused resistor in single crystal Si [Obermeier and Kopystynski, 1992]. Higher piezoresistivity and fracture stress would seem to favor SOI for sensor manufacture. But there is an important counter argument: the piezoresistivity and fracture stress in poly-Si are isotropic, a major design simplification. Moreover, by laser recrystallization the gauge factor of poly-Si might increase to above 50 [Voronin et al., 1992], and by appropriate boron doping the temperature coefficient of resistance (TCR) can actually reach 0 vs. a TCR of, say, $1.7 \times 10^{-3} K^{-1}$ for single crystal p-type Si. Neither technical nor cost issues will be the deciding factor in determining which technology will become dominant in the next few years. Micromachining is very much a hostage to trends in the IC industry: promising technologies such as GaAs and micromachining do not necessarily take off in no small part because of the invested capital in some limited sets of standard silicon technologies. On this basis SOI surface micromachining is the favored candidate:

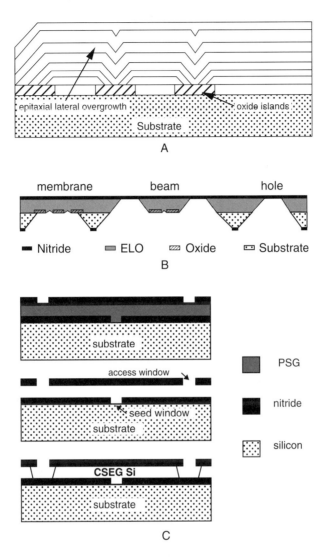

FIGURE 16.101 Micromachining with epi-Si. (A) Lateral overgrowth process of epi-Si (ELO, epitaxial lateral overgrowth) out of <100>-oriented holes in an oxide mask. (B) KOH etch stop on buried oxide islands or front side nitride. (C) Principle of confined selective epitaxial growth. (From papers presented at Transducers '95, Stockholm, Sweden, 1995. With permission.)

SOI extends silicon's technological relevance and experiences increasing investment from the IC industry, benefiting SOI micromachining [Kanda, 1991].

Based on the above we believe that SOI micromachining not only introduces an improved method of making many simple micromachines, but it also will probably become the favored approach of the IC industry. A summary of SOI advantages is listed below:

- IC industry use in all types of applications such as MOS, bipolar digital, bipolar linear, power devices, BICMOS, CCDs, heterojunction bipolar [Burggraaf, 1991] etc.
- Batch packaging through embedded cavities
- CMOS compatibility
- Substrate industrially available at lower and lower cost (about $200 today)
- Excellent mechanical properties of the single crystalline surface layer

- Freedom of shapes in the x-y dimensions and continually improving dry etching techniques, resulting in larger aspect ratios and higher features
- Freedom of choice of a very well-controlled range of thicknesses of epi surface layers
- SiO_2 buried layer as sacrificial and insulating layer and excellent etch stop
- Dramatic reduction of process steps as the SOI wafer comes with several 'embedded' process steps
- High temperature operation

16.16 Resists as Structural Elements and Molds in Surface Micromachining

16.16.1 Introduction

Polyimide and deep UV photoresists were covered already in Madou (1997, chap. 1 on lithography). We now reiterate some of the material covered there in the context of surface micromachining. Novel deep UV photoresists enable the molding of a wide variety of high aspect ratio microstructures in a wide variety of moldable materials or they are used directly as structural elements. LIGA (covered in Chapter 17), employing X-rays to pattern resists, is really just an extension of the same principles.

16.16.2 Polyimide Surface Structures

Polyimide surface structures, due to their transparency to exposing UV light, can be made very high and exhibit LIGA-like high aspect ratios. By using multiple coats of spun-on polyimide, thick suspended plates are possible. Moreover, composite polyimide plates can be made, depositing and patterning a metal film between polyimide coats. Polyimide surface microstructures are typically released from the substrate by selectively etching an aluminum sacrificial layer (see Figure 16.90), although Cu and PSG (e.g., in the butterfly wing in Figure 16.95) have been used as well (see Table 16.15).

An early result in this field was obtained at SRI International, where polyimide pillars (spacers) about 100 μm in height were used to separate a Si wafer, equipped with a field emitter array, from a display glass plate in a flat panel display [Bordie et al., 1990]. The flat panel display and an SEM picture of the pillars are shown in Figure 16.102. The Probimide 348 FC formulation of Ciba-Geigy was used. This viscous precursor formulation (48% by weight of a polyamic ester, a surfactant for wetting and a sensitizer) with a 3500-cs viscosity was applied to the Si substrate and formed into a film of a 125-μm thickness by spinning. A 30- to 40-min prebake at about 100°C removed the organic solvents from the precursor. The mask with the pillar pattern was then aligned to the wafer coated with the precursor and subjected to about 20 min of UV radiation. After driving off moisture by another baking operation the coating, still warm, was spray developed (QZ 3301 from Ciba-Geigy), revealing the desired spacer matrix. By baking the polyimide at 100°C in a high vacuum (10^{-9} Torr) the pillars shrunk to about 100 μm, while the polyimide became more dense and exhibited greater structural integrity.

More recently Frazier and Allen (1993) obtained a height-to-width aspect ratio of about 7 with polyimide structures. Ultraviolet was used to produce structures with heights in the range of 30 to 50 μm. At greater heights, the verticality of the sidewalls was relatively poor. Spun-on thickness in excess of 60 μm in a single coat was obtained for both Ciba-Geigy and Du Pont commercial UV-exposable, negative-tone polyimides. Using a G-line mask aligner, an exposure energy of 350 mJ/cm^2 was sufficient to develop a pattern with the Ciba-Geigy QZ 3301 developer. Allen and his team combined polyimide insert molds with electrodeposition to make a wide variety of metal structures [Ahn et al., 1993]. This polyimide application will be contrasted with LIGA in Chapter 17.

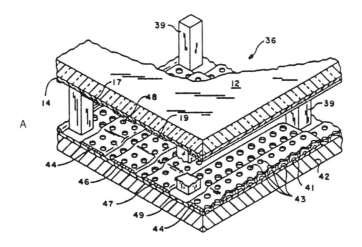

FIGURE 16.102 Polyimide structural elements. (A) Micromachined flat panel display. Number 39 represents one of the spacer pillars in the matrix of polyimide pillars (100 μm high). The spacer array separates the emitter plate from the front display plate. (B) SEM of the spacer matrix. (Courtesy of Dr. I. Brodie.) The height of the pillars is similar to what can be accomplished with LIGA. Since only a simple UV exposure was used, this polyimide is referred to as poor man's LIGA process, or pseudo-LIGA. (From Brodie, I. et al., U.S. Patent 4,923,421, 1990.)

16.16.3 UV Depth Lithography

Besides polyimides, research on novolak-type resists also is leading to higher three-dimensional features. Lochel et al. (1994; 1996a; 1996b) use novolak, positive tone resists of high viscosity (e.g., AZ 4000 series, Hoechst). They deposit in a multiple-coating process layers up to 200 μm thick in a specially designed spin coater incorporating a co-rotating cover. The subsequent UV lithography yields patterns with aspect ratios up to 10, steep edges (more than 88°) and a minimum feature size down to 3 μm. By combining this resist technology with sacrificial layers and electroplating, a wide variety of three-dimensional microstructures resulted.

Along the same line, researchers at IBM have started experimenting with Epon SU-8 (Shell Chemical), an epoxy-based, onium-sensitized, UV transparent negative photoresist used to produce high-aspect-ratio (>10:1) features as well as straight sidewalled images in thick film (>200 µm) using standard lithography [Acosta et al., 1995; LaBianca et al., 1995]. SU-8 imaged films were used as stencils to plate permalloy for magnetic motors [Acosta et al., 1995].

Patterns generated with these thick resist technologies should now be compared with LIGA-generated patterns, not only in terms of aspect ratio, where LIGA presumably still produces better results, but also in terms of sidewall roughness and sidewall run-out. Such a comparison will determine which surface machining technique to employ for the job at hand.

16.16.4 Comparison of Bulk Micromachining with Surface Micromachining

Surface and bulk micromachining have many processes in common. Both techniques rely heavily on photolithography; oxidation; diffusion and ion implantation; LPCVD and PECVD for oxide, nitride and oxynitride; plasma etching; use of polysilicon; metallizations with sputtered, evaporated and plated Al, Au, Ti, Pt, Cr and Ni. Where the techniques differ is in the use of anisotropic etchants, anodic and fusion bonding, (100) vs. (110) starting material, p^+ etch stops, double-sided processing and electrochemical etching in bulk micromachining, and the use of dry etching in patterning and isotropic etchants in release steps for surface micromachines. Combinations of substrate and surface micromachining also frequently appear. The use of polysilicon avoids many challenging processing difficulties associated with bulk micromachining and offers new degrees of freedom for the design of integrated sensors and actuators. The technology combined with sacrificial layers also allows the nearly indispensable further advantage of *in situ* assembly of the tiny mechanical structures, because the structures are preassembled as a consequence of the fabrication sequence. Another advantage focuses on thermal and electrical isolation of polysilicon elements. Polycrystalline piezoresistors can be deposited and patterned on membranes of other materials, e.g., on a SiO_2 dielectric. This configuration is particularly useful for high-temperature applications. The p-n junctions act as the only electrical insulation in the single crystal sensors, resulting in high leakage currents at high temperatures, whereas current leakage for the poly-Si/SiO_2 structure virtually does not exist. The limits of surface micromachining are quite striking. CVD silicon usually caps at layers no thicker than 1 to 2 µm because of residual stress in the films and the slow deposition process (thick poly-Si needs further investigation). A combination of a large variety of layers may produce complicated structures, but each layer is still limited in thickness. Also, the wet chemistry needed to remove the interleaved layers may require many hours of etching (except when using the porous Si option discussed above), and even then stiction often results.

The structures made from polycrystal silicon exhibit inferior electronic and slightly inferior mechanical properties compared to single crystal silicon. For example, poly-Si has a lower piezoresistive coefficient (resulting in a gauge factor of 30 vs. 90 for single crystal Si) and it has a somewhat lower mechanical fracture strength. Poly-Si also warps due to the difference of thermal expansion coefficient between polysilicon and single crystal silicon. Its mechanical properties strongly depend on processing procedures and parameters.

Table 16.17 extends a comparison of surface micromachining with polysilicon and wet bulk micromachining. The status depicted reflects the mid-1990s and only includes poly-Si surface micromachining. As discussed, SOI micromachining, thick poly-Si, hinged poly-Si, polyimide and millimeter-molded poly-Si structures have dramatically expanded the application bandwidth of surface micromachining. In Chapter 17 on LIGA we will see how X-ray lithography can further expand the z direction for new surface micromachined devices with unprecedented aspect ratios and extremely low surface roughness. In Table 16.18 we compare physical properties of single crystal Si with those of poly-Si.

Summarizing, although polysilicon can be an excellent mechanical material, it remains a poor electronic material. Reproducible mechanical characteristics are difficult and complex to consistently realize.

TABLE 16.17 A Comparison of Bulk Micromachining with Surface Micromachining

Bulk Micromachining	Surface Micromachining
Large features with substantial mass and thickness	Small features with low thickness and mass
Utilizes both sides of the wafer	Multiple deposition and etching required to build up structures
Vertical dimensions: one or more wafer thicknesses	Vertical dimensions are limited to the thickness of the deposited layers (~2 μm), leading to compliant suspended structures with the tendency to stick to the support
Generally involves laminating Si wafer to Si or glass	Surface micromachined device has its built-in support and is more cost effective
Piezoresistive or capacitive sensing	Capacitive and resonant sensing mechanisms
Wafers may be fragile near the end of the production	Cleanliness critical near end of process
Sawing, packaging, testing are difficult	Sawing, packaging, testing are difficult
Some mature products and producers	No mature products or producers
Not very compatible with IC technology	Natural but complicated integration with circuitry; integration is often required due to the tiny capacitive signals

Source: Adapted from Jerman, H., *Bulk Silicon Micromachining*, Banf, Canada, 1994.

TABLE 16.18 Comparison of Material Properties of Si Single Crystal with Crystalline Polysilicon

Material Property	Single Crystal Si	Poly-Si
Thermal conductivity (W/cm°K)	1.57	0.34
Thermal expansion (10^{-6}/°K)	2.33	2–2.8
Specific heat (cal/g°K)	0.169	0.169
Piezoresistive coefficients	n-Si ($\pi_{11} = -102.2$); p-Si ($\pi_{44} = +138.1$); e.g., gauge factor of 90	Gauge factor of 30 (>50 with laser recrystallization)
Density (cm^3)	2.32	2.32
Fracture strength (GPa)	6	0.8 to 2.84 (undoped poly-Si)
Dielectric constant	11.9	Sharp maxima of 4.2 and 3.4 eV at 295 and 365 nm, respectively
Residual stress	None	Varies
Temperature resistivity coefficient (TCR) (°K^{-1})	0.0017 (p-type)	0.0012 nonlinear, + or – through selective doping, increases with decreasing doping level, can be made 0!
Poisson ratio	0.262 max for (111)	0.23
Young's modulus ($10^{11} N/m^2$)	1.90 (111)	1.61
Resistivity at room temperature (ohm.cm)	Depends on doping	7.5×10^{-4} (always higher than for single crystal silicone)

Source: Based on Lin (1993), Adams (1988), and Heuberger (1989). (See also Madou [1997, chap. 8, Table 8.5].)

Fortunately, SOI surface micromachining and other newly emerging surface micromachining techniques can alleviate many of the problems [Petersen et al., 1991].

16.17 Materials Case Studies

16.17.1 Introduction

Thin-film properties prove not only difficult to measure but also to reproduce, given the many influencing parameters. Dielectric and polysilicon films can be deposited by evaporation, sputtering and molecular

TABLE 16.19 Comparison of Different Deposition Techniques

	Atmospheric Pressure CVD (APCVD)	Low Temperature LPCVD	Medium Temperature LPCVD	Plasma-Enhanced CVD (PECVD)
Temp (°C)	300–500	300–500	500–900	100–350
Materials	SiO_2, P-glass	SiO_2, P-glass, BP-glass	Poly-Si, SiO_2, P-glass, BP-glass, Si_3N_4, SiON	SiN, SiO_2, SiO_2, SiON
Uses	Passivation, insulation, spacer	Passivation, insulation spacer	Passivation, gate metal, structural element, spacer	Passivation, insulation, structural elements
Throughput	High	High	High	Low
Step coverage	Poor	Poor	Conformal	Poor
Particles	Many	Few	Few	Many
Film properties	Good	Good	Excellent	Poor

Note: P-glass = phosphorus-doped glass; BP-glass = borophosphosilicate glass.
Source: Adapted from Adams, A.C., in *VLSI Technology,* Sze, S.M., Ed., McGraw-Hill, New York, 1988, pp. 233–271.

TABLE 16.20 Approximate Mechanical Properties of Microelectronic Materials

	E(GPa)	ν	α(1/°C)	σ_0
Substrates				
Silicon	190	0.23	2.6×10^{-6}	—
Alumina	~415	—	8.7×10^{-6}	—
Silica	73	0.17	0.4×10^{-6}	—
Films				
Polysilicon	160	0.23	2.8×10^{-6}	Varies
Thermal SiO_2	70	0.20	0.35×10^{-6}	Compressive, e.g., 350 MPa
PECVD SiO_2	—	—	2.3×10^{-6}	—
LPCVD Si_3N_4	270	0.27	1.6×10^{-6}	Tensile
Aluminum	70	0.35	25×10^{-6} (high!)	Varies
Tungsten (W)	410 (stiff!)	0.28	4.3×10^{-6}	Varies
Polyimide	3.2	0.42	20–70×10^{-6} (very high)	Tensile

Note: E = Young's modulus, ν = Poisson ratio, α = coefficient of thermal expansion, σ_0 = residual stress.
Source: Based on lecture notes from S.D. Senturia and R.T. Howe.

beam techniques. In VLSI and surface micromachining, none of these techniques are as widely used as CVD techniques. The major problems associated with the former methods are defects caused by excessive wafer handling, low throughput, poor step coverage, and nonuniform depositions. From the comparison of CVD techniques in Table 16.19 we can conclude that LPCVD, at medium temperatures, prevails above all others. VLSI devices and integrated surface micromachines require low processing temperatures to prevent movement of shallow junctions uniform step coverage, few process-induced defects (mainly from particles generated during wafer handling and loading) and high wafer throughput to reduce cost. These requirements are best met by hot-wall, low-pressure depositions [Iscoff, 1991] (see also Madou [1997, chap. 3]). While depositing a material with LPCVD the following process parameters can be varied: deposition temperature, gas pressure, flow rate and deposition time.

Table 16.20 cites some approximate mechanical properties of microelectronic materials. The numbers for thin film materials must be approached as approximations; the various parameters affecting mechanical properties of thin films will become clear in the case studies below.

16.17.2 Polysilicon Deposition and Material Structure

16.17.2.1 Introduction

The IC industry applies polysilicon in applications ranging from simple resistors, gates for MOS transistors, thin-film transistors (TFT) (with amorphous hydrogenated silicon: α-Si:H), DRAM cell plates and trench fills, as well as in emitters in bipolar transistors and conductors for interconnects. For the last application, highly doped polysilicon is especially suited; it is easy to establish ohmic contact, it is light insensitive, corrosion resistant, and its rough surface promotes adhesion of subsequent layers. Doping elements such as arsenic, phosphorous or boron reduce the resistivity of the polysilicon. Polysilicon also has emerged as the central structural/mechanical material in surface micromachining, and a closer look at the influence of deposition methodology on its materials characteristics is warranted.

16.17.2.2 Undoped Poly-Si

The properties of low-pressure chemical vapor deposited (LPCVD) undoped polysilicon films are determined by the nucleation and growth of the silicon grains. LPCVD Si films, grown slightly below the crystallization temperature (about 600°C for LPCVD), initially form an amorphous solid that subsequently may crystallize during the deposition process [Krulevitch, 1994; Lietoila et al., 1982]. The CVD method results in amorphous films when the deposition temperatures are well below the melting temperature of Si (1410°C). The subsequent transition from amorphous to crystalline depends on atomic surface mobility and deposition rate. At low temperatures, surface mobility is low, and nucleation and growth are limited. Newly deposited atoms become trapped in random positions and, once buried, require a substantial amount of time to crystallize as solid state diffusion is significantly lower than surface mobility. That is why, for low temperature deposition, amorphous layers only start to crystallize after sufficient time at temperature in the reactor. Working at temperatures between 580 and 591.5°C, Guckel et al. (1990) produced mostly amorphous films. But Krulevitch, working at only slightly higher temperatures (605°C) and probably leaving the films longer in the LPCVD set-up, produced crystallized films. Upon crossing the transition temperature between amorphous and crystalline growth (see Madou [1997, chap. 3, Figure 3.18]), crystalline growth immediately initiates at the substrate due to the increased surface mobility which allows adatoms to find low energy, crystalline positions from the start of the deposition process. The deposition temperature at which the transition from amorphous to a crystalline structure occurs depends on many parameters, such as deposition rate, partial pressure of hydrogen, total pressure, presence of dopants and presence of impurities (O, N or C) [Adams, 1988]. In the crystalline regime, numerous nucleation sites form, resulting in a transition zone of a multitude of small grains at the film/substrate interface to columnar crystallites on top, as shown in the schematic of a 620 to 650°C columnar film in Figure 16.103. In this figure, a transition zone of small, randomly oriented grains is sketched near the SiO_2 layer. The rate of crystallization is faster here than the deposition rate. Columnar grains ranging between 0.03 and 0.3 µm in diameter form on top of the small grains [Adams, 1988]. The columnar coarse grain structure arises from a process of growth competition among the small grains,

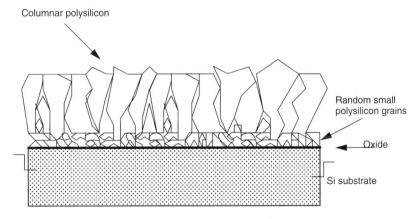

FIGURE 16.103 Schematic of compressive poly-Si formed at 620 to 650°C. The columnar coarse-grain structure arises from a process of grain growth competition among the small grains, during which those grains preferentially oriented for fast vertical growth survive at the expense of misoriented, slowly growing grain. (After Krulevitch, P.A., Micromechanical Investigations of Silicon and Ni-Ti-Cu Thin Films, Ph.D. thesis, University of California, Berkeley, 1994.)

during which those grains preferentially oriented for fast vertical growth survive at the expense of misoriented, slowly growing grains [van der Drift, 1967; Matson and Polysakov, 1977]. The lower the deposition temperature, the smaller the initial grain size will be. At 700°C, films also are columnar; however, the grains are cylindrical extending through the thickness of the entire film and there is no transition zone near the SiO_2 interface [Krulevitch, 1994].

Stress in poly-Si films was found to vary significantly with deposition temperature and silane pressure. Guckel et al. (1990) found that their mainly amorphous films, deposited at temperatures below 600°C, proved highly compressive with strain levels as high as −0.67%. At temperatures barely above 600°C, Krulevitch reports tensile films, whereas for yet higher temperatures ($\geq$620°C) the stress again turns compressive. While films deposited at temperatures greater than 630°C all turned out compressive, the magnitude of the compression decreased with increasing temperature. The stress gradient in the poly-Si films explains why compressive undoped and unannealed poly-Si beams tend to curl upward (positive stress gradient) when released from the substrate [Lober et al., 1988].

Using high resolution transmission electron microscopy, Guckel et al. (1990) and Krulevitch (1980) found a strong correlation between the material's microstructure and the exhibited stress. Guckel et al. found that in their mainly amorphous films, deposited at temperatures below 600°C, a region near the substrate interface crystallized during growth with grains between 100 and 4000 Å. Krulevitch found that tensile, low-temperature films (605°C) have Si grains dispersed throughout the film thickness. Krulevitch suggests that the compressive stress in the higher temperature compressive films ($\geq$620°C) relates to the competitive growth mechanism of the columnar grains. The same author concluded that thermal sources of stress are insignificant.

Importantly, Guckel et al. discovered that annealing in nitrogen or under vacuum converts the low-temperature films with compressive built-in strain (−0.007) to a tensile strain with controllable strain levels between 0 and + 0.003 (see Figure 16.104). During anneal no grain size increase was noticed (100 to 4000 Å) but a slight increase in surface roughness was measured. This type of poly-Si is referred to as fine-grain

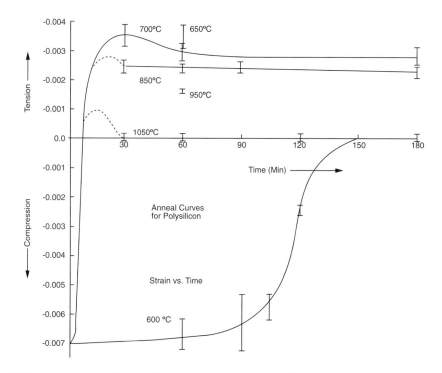

FIGURE 16.104 Anneal curves for poly-Si. Strain versus anneal time. Upper curves: low-temperature film. Lower curve: high-temperature film. (From Guckel, H. et al., *IEEE Trans. Electron. Dev.*, 35, 800–801, 1988. With permission.)

poly-Si (also Wisconsin poly-Si). Guckel et al. explain this strain field reversal as follows: as the amorphous region of the film crystallizes, it attempts to contract, but due to the substrate constrained newly crystallized region, a tensile stress results. Higher temperature films, during an anneal, also become less strained, but the strain remains compressive (see lower curve in Figure 16.104). Moreover, in this case grain size does increase and the surface turns considerably rougher. The latter is called coarse-grain poly-Si. Fine-grain poly-Si with its tensile strain is preferable; however, it cannot be doped to as low a resistivity as coarse-grain polysilicon. Hence, fine-grain polysilicon should be considered as a structural material rather than an electronic material.

Summarizing stress in poly-Si depends on the material's microstructure, with tension arising from the amorphous to crystalline transformation during deposition and compression from the competitive grain growth mechanism.

Polysilicon deposited at 600 to 650°C has a {110}-preferred orientation. At higher temperatures, the {100} orientation dominates. Dopants, impurities and temperature influence this preferred orientation [Adams, 1988]. Drosd and Washburn (1982) introduced a model explaining the experimental observation that regrowth of amorphized Si is faster for {100} surfaces, followed by {110} and {111} which are 2.3 and 20 times slower, respectively. Interestingly, the latter also pinpoints the order of fastest to slowest etching of the crystallographic planes in alkaline etchants. As discussed earlier in this chapter, Elwenspoek et al. (1994) used this observation of symmetry between etching and growing of Si planes as an important insight to develop a new theory explaining anisotropy in etching. In Table 16.21 we compare the discussed coarse- and fine-grain poly-Si forms.

The above picture might further be complicated by the controlling nature of the substrate. For example, depositing amorphous Si ($\alpha - Si$) at even lower temperatures of 480°C from disilane (Si_2H_6) and crystallizing it by subsequent annealing at 600°C demonstrated a large dependency of crystallite size on the underlying SiO_2 surface condition. Treating the surface with $HF:H_2O$ or $NH_4OH:H_2O_2:H_2O$ leads to poly-Si films with a large grain size, two or three times as large as without SiO_2 treatment, believed to be the consequence of nucleation rate suppression [Shimizu and Ishihara, 1995].

T. Abe and M.L. Reed made low strain polysilicon thin film by DC-magnetics sputtering and post-annealing. The films showed very small regional stress and very smooth texture. The deposition rate was 193 Å/min and the substrate was neither cooled nor heated. The average roughness was found to be comparable to the surface roughness of polished, bare silicon substrates [Abe and Reed, 1996].

16.17.2.3 Doped Poly-Si

To produce micromachines, doped poly-Si is used far more frequently than undoped poly-Si. Dopants decrease the resistivity to produce conductors and control stress. Polysilicon can be doped by diffusion, implantation or the addition of dopant gases during deposition (*in situ* doping). We only detail material properties of *in situ* doped poly-Si here.

TABLE 16.21 Comparison of Coarse-Grain and Fine-Grain Poly-Si

	Coarse-Grain Poly-Si	Fine-Grain Poly-Si
Temperature of deposition (°C)	620–650	570–591.5
Surface roughness	Rough, >50Å	Smooth, <15Å
Grain size	Undoped: 160–320 Å as deposited; *in situ* P-doped: 240–400 Å	Very small grains
As deposited strain	−0.007 (compressive)	−0.007 (compressive)
Effect of high-temperature anneal	Grains size increases, residual strain decreases but remains compressive, reduced bending moment	Grain size increases to 100 Å [Guckel et al., 1988]; others have found 700–900 Å; large variation in strain (see Figure 16.102) from compressive to tensile
Dry and wet etch rate	Higher for doped material; depends on dopant concentration	Higher for doped material; depends on dopant concentration
Texture	<110> as deposited, <311> *in situ* P-doped	No texture as deposited; depends on dopant concentration, <111> after 900–1000°C anneal [Harbeke et al., 1983]

Doping poly-Si films *in situ* reduces the number of processing steps required for producing doped micro-devices and also provides the potential for uniform doping through the film thickness. *In situ* doping of poly-Si with phosphorous is accomplished by maintaining a constant PH_3 to SiH_4 gas flow ratio of about 1 vol % in a hot-wall LPCVD set-up. At this ratio the phosphorus content in the film appears above the saturation limit and the excess dopant segregates at the grain boundaries [Adams, 1988]. *In situ* phosphorus-doped poly-Si undergoes the same amorphous to crystalline growth transformation observed in the undoped film, with the material's microstructure depending on deposition temperature as well as deposition pressure. The temperature of transformation is lower for the doped films than for the undoped poly-Si and occurs between 580 and 620°C [Mulder et al., 1990; Kinsbron et al., 1983]. Phosphorus doping thus enhances crystallization in amorphous silicon [Lietoila et al., 1982] and, due to passivation of the poly-Si surface by the phosphine gas, reduces the poly-Si deposition rate [Mulder et al., 1990]. Decreases in deposition rate by as much as a factor of 25 have been reported [Meyerson and Olbricht, 1984]. Slower deposition rates allow more time for adatoms to find crystalline sites, resulting in crystalline growth at lower temperatures. From Table 16.21 we read that the grain size of phosphorus-doped poly-Si tends to be larger (240 to 400 Å) than for the undoped material and that {311} planes show up as a texture facet in the doped material.

At lower deposition temperatures and higher pressures, the microstructure again consists of amorphous and crystalline regions, while at higher temperatures and lower pressures columnar films result and as deposited films exhibit compressive residual stress. The columnar films have a stress gradient that increases towards the film surface, as opposed to the gradient found in undoped columnar poly-Si. This gradient in stress most likely is due to nonuniform distribution of phosphorus throughout the film. Annealing at 950°C for 1 hour results in the same stress and stress gradient for initially columnar and initially amorphous/crystalline films (i.e., σ_f = −45 MPa and Γ = +0.2 mm^{-1}, respectively) [Krulevitch, 1994].

As with undoped poly-Si, phosphorus-doped poly-Si films with smooth surfaces (fine-grain) can be obtained by depositing *in situ* doped films in the amorphous state and then annealing [Harbeke et al., 1983; Hendriks and Mavero, 1991]. Phosphorus-doped poly-Si oxidizes faster than undoped poly-Si. The rate of oxidation is determined by the dopant concentration at the poly-Si surface [Adams, 1988].

Two drawbacks of *in situ* phosphorus doping are slower deposition rates [Kurokawa, 1982] and reduced film thickness uniformity [Meyerson and Olbricht, 1984], although uniformity can be improved by modifying the reactor geometry [Mulder et al., 1990]. As discussed before, the lower deposition rate of *in situ* phosphine doping can be mitigated by reducing the flow of phosphine/silane ratio by one third [Howe, 1995]. In contrast to *in situ* phosphine and arsine doping, which both decrease the deposition rate, diborane doping of poly-Si to make it p^+ accelerates the deposition rate [Adams, 1988].

The addition of oxygen to poly-Si increases the film's resistivity and the resulting coating, semi-insulating poly-Si (SIPOS) acts as a passivating coating for high voltage devices in the IC industry. The latter material does not seem to have emerged in surface micromachining yet.

The values for the fracture stress of boron-, arsenic- and phosphorus-doped polysilicon are 2.77 ± 0.08 Gpa, 2.70 ± 0.09 Gpa, and 2.11 ± 0.1 Gpa, respectively, compared with 2.84 ± 0.09 Gpa for undoped polysilicon. The lower value for phosphorus-doped material has been attributed to high surface roughness and with the large number of defects associated with extensive grain growth in highly phosphorus-doped films [Biebl and von Philipsborn, 1993].

The quest for low temperature poly-Si deposition makes the 320°C PECVD deposition method especially interesting. PECVD films, deposited in a 50-kHz parallel-plate diode reactor, can be doped *in situ* and crystallized by rapid thermal annealing (RTA: 1100°C, 100 seconds). It was shown that small-grained PECVD films annealed by RTA have good electrical properties and gauge factors between 20 and 30, i.e., similar to those reported for other alternative types of polycrystalline silicon [Compton, 1992].

16.17.3 Amorphous and Hydrogenated Amorphous Silicon

Amorphous silicon behaves quite different from either fine- or coarse-grain poly-Si. The amorphous material produces a high breakdown strength (7 to 9 MV/cm) oxide with low leakage currents (vs. a low breakdown voltage and large leakage currents for polycrystalline Si oxides). Amorphous polysilicon also attains a broad maximum in its dielectric function without the characteristic sharp structures near 295 and 365 nm (4.2 and 3.4 eV) of crystalline poly-Si. Approximate refractive index values at a wavelength of 600 nm are 4.1 for crystalline polysilicon and 4.5 for amorphous material [Adams, 1988]. As deposited, the material is under compression, but an anneal at temperatures as low as 400°C reduces the stress significantly, even leading to tensile behavior [Chang et al., 1991].

Hydrogenated amorphous silicon (α-Si:H) enables the fabrication of active semiconductor devices on foreign substrates at temperatures between 200 and 300°C. The technology, first applied primarily to the manufacturing of photovoltaic panels, now is quickly expanding into the field of large-area microelectronics such as active matrix liquid crystal displays (AMLCD). It is somehow surprising that micromachinists have not taken more advantage of this material either for powering surface micromachines or to implement electronics cheaply on non-Si substrates.

Spear and LeComber (1975) showed that, in contrast to α-Si, α-Si:H could be doped both n- and p-type. Singly bonded hydrogen, incorporated at the Si dangling bonds, reduces the electronic defect density from $\sim 10^{19}/cm^3$ to $\sim 10^{16}/cm^3$ (typical H concentrations are 5 to 10 atomic percent—several orders of magnitude higher than needed to passivate all the Si dangling bonds). The lower defect density results in a Fermi level that is free to move, unlike in ordinary amorphous Si, where it is pinned. Other interesting electronic properties are associated with α-Si:H—exposure of α-Si:H to light increases photoconductivity by four to six orders of magnitude, and its relatively high electron mobility (~ 1 cm^2/V sec^{-1}) enables fabrication of useful thin-film transistors. Only recently did Lee et al. (1995) note that hydrogenated amorphous silicon solar cells are an attractive means to realize an on-board power supply for integrated micromechanical systems. They point out that the absorption coefficient of α-Si:H is more than an order of magnitude larger than that of single crystal Si near the maximum solar photon energy region of 500 nm. Accordingly, the optimum thickness of the active layer in an α-Si:H solar cell can measure 1 µm, much smaller than that of single crystal Si solar cells. By interconnecting 100 individual solar cells in series, the measured open circuit potential reaches as high as 150 V under AM 1.5 conditions, a voltage high enough to drive on-board electrostatic actuators.

Hydrogenated amorphous silicon is manufactured by plasma-enhanced chemical vapor deposition from silane. Usually, planar RF-driven diode sources using SiH_4 or SiH_4/H_2 mixtures are used. Typical pressures of 75 mTorr and temperatures between 200 and 300°C allow silane decomposition with Si deposition as the dominant reaction. Decomposition occurs by electron impact ionization, producing many different neutral and ionic species [Crowley, 1992]. Deposition rates for usable device quality α-Si:H generally do not exceed ~ 2 to 5 Å/sec, due to the effects of temperature, pressure and discharge power. Table 16.22 gives state-of-the-art parameters for α-Si:H prepared by PECVD. Although its semiconducting properties are inferior to single crystal Si, the material is finding more and more applications. Some examples are TFT switches for picture elements in AMLCDs [Holbrook and McKibben, 1992], page-wide TFT-addressed document scanners, high-voltage TFTs capable of switching up to 500 V etc. [Bohm, 1988]. An excellent source for further information on amorphous silicon is the book *Plasma Deposition of Amorphous Silicon-Based Materials* [Bruno et al., 1995].

16.17.4 Silicon Nitride

Silicon nitride is a commonly used material in microcircuit and microsensor fabrication due to its many superior chemical, electrical, optical and mechanical properties. The material provides an extremely good barrier to the diffusion of water and to ions, particularly of Na^+. It also oxidizes slowly (about 30 times less than silicon) and has highly selective etch rates over SiO_2 and Si in many etchants. Some applications of silicon nitride are optical waveguides (nitride/oxide), encapsulant (diffusion barrier to water and ions), insulator (high dielectric strength), mechanical protection layer, etch mask, oxidation barrier and ion implant mask (density is 1.4 times that of SiO_2). Silicon nitride is also hard and can be used as, for example, a bearing material in micromotors [Pool, 1988].

TABLE 16.22 Typical Opto-Electronic Parameters Obtained for PECVD α-Si:H

	Symbol	Parameter
Undoped		
Hydrogen content		~10%
Dark conductivity at 300 K	σ_D	~10^{-10} $(\Omega\text{-cm})^{-1}$
Activation energy	E_σ	0.8–0.9 eV
Pre-exponent conductivity factor	σ_0	>10^3 $(\Omega\text{-cm})^{-1}$
Optical bank gap at 300 K	E_g	1.7–1.8 eV
Temperature variation of band gap	$E_g(T)$	2–4 × 10^{-4} eV/K
Density of states at the minimum	g_{min}	>10^{15}–10^{17} cm^3/eV
Density of states at the conduction band edge		~10^{15}/cm^3
ESR spin density	N_s	~10^{21}/cm^3–eV
Infrared spectra		2000/640 cm^{-1}
Photoluminescence peak at 77K		~1.25 eV
Extended state mobility		
Electrons	μ_n or μ_e	>10 cm^2/V–s
Holes	μ_p or μ_h	~1 cm^2/V–s
Drift mobility		
Electrons	μ_n or μ_e	~1 cm^2/V–s
Holes	μ_p or μ_h	~10^{-2} cm^2/V–s
Conduction band tail slope		25 meV
Valence band tail slope		40 meV
Hole diffusion length		~1 μm
Doped amorphous		
n-Type[a]	σ_D	10^{-2} $(\Omega\text{-cm})^{-1}$
	E_g	~0.2 eV
p-Type[b]	σ_D	10^{-3} $(\Omega\text{-cm})^{-1}$
	E_g	~0.3 eV
Doped microcrystalline		
n-Type[c]	σ_D	≥1 $(\Omega\text{-cm})^{-1}$
	E_g	≤0.05 eV
p-Type[d]	σ_D	≥1 $(\Omega\text{-cm})^{-1}$
	E_g	≤0.05 eV

[a] ~1% PH$_3$ added to gas phase.

[b] ~1% B$_2$H$_6$ added to gas phase.

[c] ~1% PH$_3$ added to dilute SiH$_4$/H$_2$, or 500 vppm PH$_3$ added to SiF$_4$/H$_2$ (8:1) gas mixtures. Relatively high powers are involved.

[d] ~1% B$_2$H$_6$ added to dilute SiH$_4$/H$_2$.

Source: Crowley, J.L., *Solid State Technol.*, February, 94–98, 1992. With permission.

Silicon nitride can be deposited by a wide variety of CVD techniques: APCVD, LPCVD and PECVD. Nitride often is deposited from SiH$_4$ or other Si containing gases and NH$_3$ in a reaction such as:

$$3SiCl_2H_2 + 4NH_3 \rightarrow Si_3N_4 + 6HCl + 6H_2 \qquad \text{(Reaction 16.17)}$$

In this CVD process, the stoichiometry of the resulting nitride can be moved toward a silicon-rich composition by providing excess silane or dichlorosilane compared to ammonia.

16.17.4.1 PECVD Nitride

Plasma-deposited silicon nitride, also plasma nitride or SiN, is used as the encapsulating material for the final passivation of devices. The plasma-deposited nitride provides excellent scratch protection, serves as a moisture barrier and prevents sodium diffusion. Because of the low deposition temperature, 300 to 350°C, the nitride can be deposited over the final device metallization. Plasma-deposited nitride

TABLE 16.23 Properties of Silicon Nitride[a]

Deposition	LPCVD	PECVD
Temperature (°C)	700–800	250–350
Density (g/cm^3)	2.9–3.2	2.4–2.8
Pinholes	No	Yes
Throughput	High	Low
Step coverage	Conformable	Poor
Particles	Few	Many
Film quality	Excellent	Poor
Dielectric constant	6–7	6–9
Resistivity (Ωcm)	10^{16}	10^6–10^{15}
Refractive index	2.01	1.8–2.5
Atom % H	4–8	20–25
Energy gap	5	4–5
Dielectric strength (10^6 V/cm)	10	5
Etch rate in conc. HF	200 Å/min	
Etch rate in BHF	5–10 Å/min	
Residual stress (10^9 dyn/cm^2)	1T	2 C–5T
Poisson ratio	0.27	
Young's modulus	270 GPa	
TCE	1.6×10^{-6}/°C	

Note: C = compressive; T = tensile.
[a] See Adams (1988), Sinha and Smith (1978) and Retajczyk and Sinha (1980).

and oxide both act as insulators between metallization levels, particularly useful when the bottom metal level is on aluminum or gold. The silicon nitride that results from PECVD in the gas mixture of Reaction 16.2 has two shortcomings: high hydrogen content (in the range of 20 to 30 atomic percent) and high stress. The high compressive stress (up to 5×10^9 dyn/cm^2) can cause wafer warping and voiding and cracking of underlying aluminum lines [Rosler, 1991]. The hydrogen in the nitride also leads to degraded MOSFET lifetimes. To avoid hydrogen incorporation, one may employ low or no hydrogen-containing source gases such as nitrogen instead of ammonia as the nitrogen source. Also, a reduced flow of SiH$_4$ results in less Si-H in the film. The hydrogen content and the amount of stress in the film are closely linked. Compressive stress, for example, changes toward tensile stress upon annealing to 490°C in proportion to the Si-H bond concentration. By adding N$_2$O to the nitride deposition chemistry an oxynitride forms with lower stress characteristics; however, oxynitrides are somewhat less effective as moisture and ion barriers than nitrides.

We discussed the effect of RF frequency on nitride stress, hydrogen content, density and the wet etch rate in detail in Madou (1997, chap. 3). We can conclude that low frequency (high energy bombardment) results in films with low compressive stress, lower etch rates and higher density. The higher the ion bombardment during PECVD, the higher the stress is. Stress also is affected by moisture exposure and temperature cycling (for a good review, see Wu and Rosler [1992]). We compare properties of silicon nitride formed by LPCVD and PECVD in Tables 16.23 and 16.24. Table 16.24 highlights typical process parameters.

16.17.4.2 LPCVD Nitride

In the IC industry, stoichiometric silicon nitride (Si$_3$N$_4$) is LPCVD deposited at 700 to 900°C and functions as an oxidation mask and as a gate dielectric in combination with thermally grown SiO$_2$. In micromachining, LPCVD nitride serves as an important mechanical membrane material and isolation/buffer layer. By increasing the Si content in silicon nitride the ensile film stress reduces (even to compressive), the film turns more transparent and the HF etch rate lowers. Such films result by increasing the dichlorosilane/ammonia ratio [Sakimoto et al., 1982].

Figure 16.105 illustrates the effect of gas flow ratio and deposition temperature on stress and the corresponding refractive index and HF etch rate [Sakimoto et al., 1982].

TABLE 16.24 Silicon Nitride PECVD Process Conditions

Flow (sccm)	SiH$_4$	190–270
	NH$_3$	1900
	N$_2$	1000
Temperature (°C)	T.C.	350 or 400
	Wafer	~330 or 380
Pressure (torr)		2.9
rf power (watt)	1100	
Deposition rate (Å/min)		1200–1700
Refractive index		2.0

Source: Wu, T.H.T. and R.S. Rosler, *Solid State Technol.*, May, 65–71, 1992. With permission.

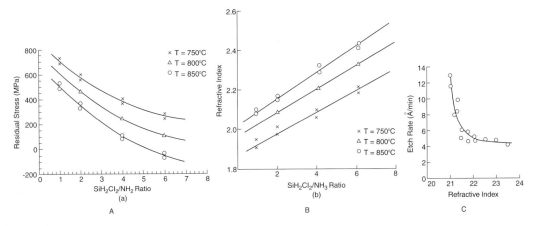

FIGURE 16.105 Silicon nitride LPCVD deposition parameters. (A) Effect of gas-flow ratio and deposition temperature on stress in nitride films. (B) The corresponding index of refraction. (C) The corresponding HF etch rate. (From Sakimoto, M. et al., *J. Vac. Sci. Technol.*, 21, 1017–1021, 1982. With permission.)

TABLE 16.25 Etching Behavior of LPCVD Si$_3$N$_4$

Etchant	Temperature (°C)	Etch Rate (Å/min)	Selectivity of Si$_3$N$_4$:SiO$_2$:Si
H$_3$PO$_4$	180	100	10:1:0.3
CF$_4$–4% O$_2$ plasma	—	250	3:2.5:17
BHF	25	5 to 10	1:200:±0
HF (40%)	25	200	1:>100:0.1

Silicon-rich or low-stress nitride emerges as an important micromechanical material. Low residual stress means that relatively thick films can be deposited and patterned without fracture. Low etch rate in HF means that films of silicon-rich nitride survive release etches better than stoichiometric silicon nitride. The etch characteristics of LPCVD SiN are summarized in Table 16.25. The properties of a LPCVD Si$_x$N$_y$ film, deposited by reaction of SiCl$_2$H$_2$ and NH$_3$ (5:1 by volume) at 850°C, were already summarized in Table 16.23.

16.17.5 CVD Silicon Dioxides

Silicon oxides, like other dielectrics, function as insulation between conducting layers, for diffusion and ion implementation masks, for diffusion from doped oxides, for capping doped oxides to prevent the loss of dopants, for gettering impurities and for passivation to protect devices from impurities, moisture and scratches. In micromachining, silicon oxides serve the same purposes but also act as sacrificial material. Phosphorus-doped glass (PSG), also called P-glass or phosphosilicate glass, and borophosphosilicate glasses (BPSG) soften and flow at lower temperatures enabling the smoothing of topography. They etch much faster than SiO_2, which benefits their application as sacrificial material. The deposition of thermal SiO_2 is covered in Madou (1997, chap. 3). Now we consider CVD techniques to deposit doped and undoped SiO_2.

16.17.5.1 CVD Undoped SiO_2

Several CVD methods will deposit SiO_2. CVD silicon dioxide can be deposited on the wafer out of the vapor phase from the reaction of silane (SiH_4) with oxygen at relative low temperatures (300 to 500°C) and low pressure. The silane/oxygen gas mixture has been applied at atmospheric pressure (AP) and plasma-enhanced conditions. The main advantage of silane/oxygen is the low deposition temperature; the main disadvantage is the poor step coverage.

In general, tetraethylorthosilicate (TEOS) brings about a better starting chemistry than the traditional silane-based CVD technologies. TEOS-based depositions lead to superior film quality in terms of step coverage and reflow properties. With an LPCVD reactor, the deposition temperature for TEOS is as high as 650 to 750°C, precluding its use over aluminum lines. In contrast, silicon dioxide is deposited by PECVD at 300 to 400°C from TEOS. CVD oxides in general feature porosity and low density. Low frequency (high energy) ion bombardment results in more compressive films, higher density and lower etch rates as well as better moisture resistance [Rosler, 1991]. Use of TEOS oxide, to replace SiH_4-based oxide in spin-on-glass (SOG) and photoresist planarization schemes, now has become commonplace for devices with small features.

Yet another promising silicon dioxide deposition technology is the subatmospheric pressure CVD (SACVD). Both undoped and borophosphosilicate glass have been deposited in this fashion. The process involves an ozone and TEOS reaction at a pressure of 600 Torr and at a temperature below 400°C. While offering the same good step coverage over submicron gaps, the films from the thermal reaction of ozone and TEOS exhibit relatively neutral stress and have a higher film density compared to low pressure processes. This increased density gives the oxide greater moisture resistance, lower wet etch rate and smaller thermal shrinkage. Compared to a 60-Torr process, the film density has increased from 2.09 to 2.15 g/cm^3; the wet etch rate decreased by more than 40%; and the thickness shrinkage changed from 12 to 4% after a 30-min anneal in dry N_2 at 1000°C [Lee et al., 1992].

16.17.5.2 CVD Phosphosilicate Glass Films

Adding a few percent of phosphine to the gas stream during deposition to obtain a lower melting point for the oxide (PH_3) results in a phosphosilicate glass. Numerous applications of this material exist, such as:

1. Interlevel dielectric to insulate metallization levels
2. Gettering and flow capabilities
3. Passivation overcoat to provide mechanical protection for the chip from its environment
4. Solid diffusion source to dope silicon with phosphorus
5. Fast etching sacrificial material in surface micromachining

Both wet and dry etching rates of PSG are faster than the undoped material and depend on the dopant concentration. Profiles over steps get progressively smoother with higher phosphorous concentrations, reflecting the corresponding enhancement in viscous flow [Levy and Nassau, 1986]. Increased ion bombardment (energy or density) in a PECVD PSG results in more stable phosphosilicate glass film with compressive as-deposited stress. Table 16.26 summarizes typical PSG process conditions [Wu et al., 1992].

The addition of boron to P-glass further lowers its softening temperature. Flow occurs at temperatures between 850 and 950°C, even with phosphorous concentration as low as 4 wt%. A BPSG doped at 4% boron and 6% phosphorus is normal. BPSG deposition conditions are shown in Table 16.27 [Bonifield et al., 1993].

TABLE 16.26 Typical PSG Process Conditions

Flow (sccm)	SiH$_4$	150–230
	N$_2$O	4500
	N$_2$	1500
	PH$_3$	150[a]
Temperature (°C)	T.C.	400
	Wafer	~380–390
Pressure (Torr)		2.2
RF power (watt)		1200
Deposition rate (Å/min)		4000–5000
Refractive index		1.46

[a] 10% PH$_3$ in N$_2$.
[b] 5% PH$_3$ in N$_2$.

Source: Wu, T.H.T. and R.S. Rosler, *Solid State Technol.*, May, 65–71, 1992. With permission.

TABLE 16.27 Typical BPSG Process Conditions

Parameter	Dimension	NSG	BPSG
Deposition temp.	°C	400	400
TEOS flow	g/min	0.33	0.66
O$_2$ flow	SLM	7.5	7.5
O$_3$/O$_2$	volume %	1	4.5
Carrier N$_2$ flow	SLM	18.0	18.0
B conc.	atomic %		4
P conc.	atomic %		6
Exhaust	mmH$_2$O	2.0	2.0
Growth rate	Å/min	1200	1800
Thickness	Å	1000	5800

Note: NSG = non-doped silicate glass; BPSG = borophosphosilicate glass.

From Bonifield, T., K. Hewes, B. Merritt, R. Robinson, S. Fisher, and D. Maisch, *Semicond. Int.*, July, 200–204, 1993. With permission.

16.18 Polysilicon Surface Micromachining Examples

Example 1. Analog Devices Accelerometer

Accelerometers based on lateral resonators represent the main application of surface micromachining today. To facilitate integration of their 50-g surface micromachined accelerometer (ADXL-50) with on-board electronics, Analog Devices opted for a mature 4-μm BICMOS process [Core et al., 1993]. Figure 16.106 presents a photograph of the finished capacitive sensor (center) with on-chip excitation, self-test and signal conditioning circuitry. The polysilicon-sensing element only occupies 5% of the total die area. The whole chip measures 500 × 625 μm and operates as an automotive airbag deployment sensor. The measurement accuracy is 5% over the ±50-g range. Deceleration in the axis of sensitivity exerts a force on the central mass that, in turn, displaces the interleaved capacitor plates, causing a fractional change in capacitance. In operation, the device has a force-balance electronic control loop to prevent the mass from actual movement. Straight flexures were used for the layout of the lateral resonator shuttle mass (see Figure 16.89B).

In the sensor design, n$^+$ underpasses connect the sensor area to the electronic circuitry, replacing the usual heat-sensitive aluminum connect lines. Most of the sensor processing is inserted into the BICMOS

FIGURE 16.106 Analog Devices' ADXL-50 accelerometer with a surface micromachined capacitive sensor (center), on-chip excitation, self-test, and signal-conditioning circuitry. (From Core, T.A. et al., *Solid State Technol.*, October, 39–47 (1993). With permission.)

process right after the borophosphosilicate glass planarization. After planarization, a designated sensor region or moat is cleared in the center of the die (Figure 16.107a). A thin oxide is then deposited to passivate the n^+ underpass connects, followed by a thin, low-pressure, vapor-deposited nitride to act as an etch stop (buffer layer) for the final poly-Si release etch (Figure 16.107b). The spacer or sacrificial oxide used is a 1.6-μm densified low-temperature oxide (LTO) deposited over the whole die (Figure 16.107c). In a first timed etch, small depressions that will form bumps or dimples on the underside of the polysilicon sensor are created in the LTO layer. These will limit stiction in case the sensor comes in contact with the substrate. A subsequent etch cuts anchors into the spacer layer to provide regions of electrical and mechanical contact (Figure 16.107c). The 2-μm-thick sensor poly-Si is then deposited, implanted, annealed and patterned (Figure 16.108a). The relatively deep junctions of the BICMOS process permit the polysilicon thermal anneal as well as brief dielectric densifications without resulting in degradation of the electronic functions. Next is the IC metallization which starts with the removal of the sacrificial spacer oxide from the circuit area along with the LPCVD nitride and LTO layer. A low-temperature oxide is deposited on the poly-Si sensor part and contact openings appear in the IC part of the die where platinum is deposited to form a platinum silicide (Figure 16.108b). The trimmable thin-film material, TiW barrier metal and Al/Cu interconnect metal are sputtered on and patterned in the IC area. The circuit area is then passivated in two separate deposition steps. First, plasma oxide is deposited and patterned (Figure 16.108c), followed by a plasma nitride to form a seal with the earlier deposited LPCVD nitride (Figure 16.109a). The nitride acts as an HF barrier in the subsequent long etch release. The plasma oxide left on the sensor acts as an etch stop for the removal of the plasma nitride (Figure 16.109a). Subsequently, the sensor area is prepared for the final release etch. The undensified dielectrics are removed from the sensor and the final protective resist mask is applied. The photoresist protects the circuit area from the long-term buffered oxide etch (Figure 16.109b). The final device cross section is shown in Figure 16.109c.

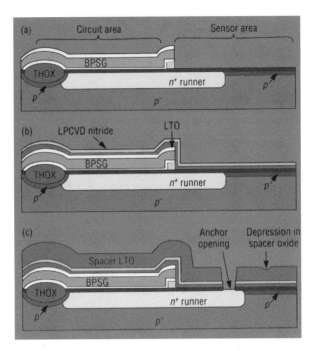

FIGURE 16.107 Preparation of IC chip for poly-Si. (a) Sensor area post-BPSG planatization and moat mask. (b) Blanket deposition of thin oxide and thin nitride layer. (c) Bumps and anchors made in LTO spacer layer. (From Core, T.A., W.K. Tsang, and S.J. Sherman, *Solid State Technol.*, 36, 39–47, 1993. With permission.)

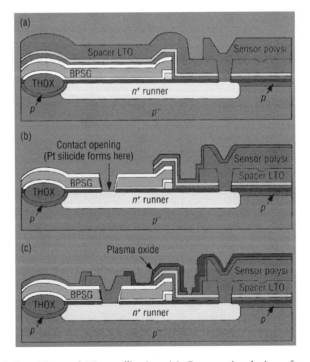

FIGURE 16.108 Poly-Si deposition and IC metallization. (a) Cross-sectional view after polysilicon deposition, implant, anneal, and patterning. (b) Sensor area after removal of dielectrics from circuit area, contact mask, and Pt silicide. (c) Metallization scheme and plasma oxide passivation and patterning. (From Core, T.A., W.K. Tsang, and S.J. Sherman, *Solid State Technol.*, 36, 39–47, 1993. With permission.)

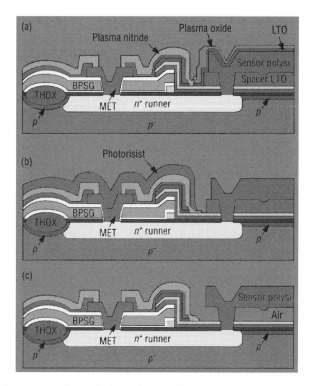

FIGURE 16.109 Pre-release preparation and release. (a) Post-plasma nitride passivation and patterning. (b) Photoresist protection of the IC. (c) Freestanding, released poly-Si beam. (From Core, T.A., W.K. Tsang, and S.J. Sherman, *Solid State Technol.*, 36, 39–47, 1993. With permission.)

Example 2. Polysilicon Stepping Slider

The principle of operation of the stepping slider is illustrated in Figure 16.110 [Akiyama, 1993; Akiyama and Shono, 1993]. Figure 16.110A shows a cross-sectional view of the polysilicon slider plate (length = 50 μm, width = 30 μm, height = 1.0 μm) and bushing (height = 1 μm) on an insulator film (Si_3N_4) as it sets one step. Figure 16.110B displays a schematic of the complete slider. At the rise of an applied voltage pulse on the slider rail, the polysilicon plate is pulled down. Since one end of the plate, supported by the bushing, cannot move, the other part is pulled down to come into contact with the surface of the insulator. The warp of the plate causes the bushing to shift. At the fall of the pulse, the distortion is released and the

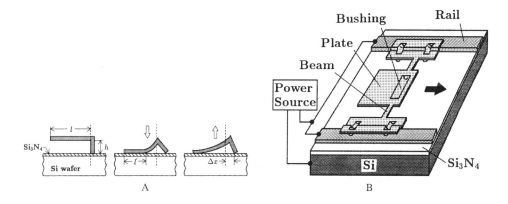

FIGURE 16.110 Principle of operation of stepping slider. (A) Cross-sectional view of the polysilicon plate and bushing on an insulator (Si_3N_4) on Si as the plate moves one step. (B) Schematic diagram of stepping slider.

plate snaps back to its original shape. If the size of the polysilicon step is Δx, the velocity of the polysilicon microstructure is given by:

$$v = \Delta x \cdot f \quad (16.64)$$

where f is the frequency of the pulse. The size of the step itself is given by:

$$\Delta x = \frac{h^2}{2(l - l')} \quad (16.65)$$

where l' stands for the length of the plate touching the insulator film.

Figure 16.111 illustrates the sequence of a stepping slider fabrication process. A layer of 0.3-μm silicon nitride is deposited at 900°C by the reaction of SiH_4 and NH_3 in N_2. Next, *in situ* phosphorus-doped polysilicon is deposited and patterned in a SF_6 plasma to form the sliding rail for the power supply of the slider (A) (mask 1). Thick sacrificial SiO_2 is deposited at 610°C by CVD of SiH_4 and O_2 in N_2. The thickness of the oxide (1 μm) determines the future bushing height. The oxide is patterned with a second mask exposing the silicon nitride in the area of the bushing. Fifty nm of CVD PSG is then deposited over the whole wafer at 610°C from the thermal reaction of SiH_4, PH_3 and O_2 in N_2 (B). In step C, 1.4-μm

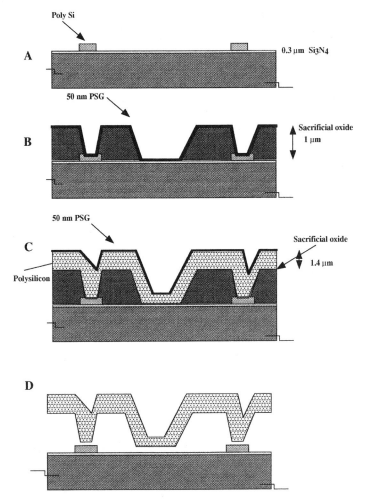

FIGURE 16.111 Process sequence for the fabrication of polysilicon stepper slider. (After Akiyama, T., *J. Microelectromech. Syst.*, 2, 106–110, 1993. With permission.)

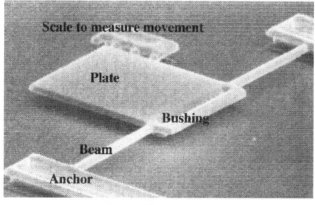

FIGURE 16.112 SEM micrograph of the stepping slider (top) and scanned image (bottom). (From Akiyama, T., *J. Microelectromech. Syst.*, 2, 106–110, 1993. With permission.)

polysilicon is deposited at 610°C from SiH$_4$, coated with PSG and patterned with mask 3 to shape the slider plate. The silicon wafer is then heated in N$_2$ at 1050°C for 60 min to release the residual strain in the polysilicon and at the same time to activate and diffuse phosphorus from the PSG into the polysilicon slider. The last step (D) in the fabrication process is the release of the polysilicon slider, by dipping the wafer into 50% HF to fully dissolve the sacrificial oxide. Finally, the wafer is rinsed in deionized water and IPA (isopropyl-alcohol) and dried in N$_2$. In the fabrication process, to avoid washing away the slider in the rinse, the slider is attached by polysilicon springs to the end of the rail. The springs are cut by the step motion of the slider and subsequently the sliders can move along the rail. The velocity at the peak voltage of 150 V is 30 μm/sec with a 1200-Hz pulse. Figure 16.112 shows an SEM photo of the resulting slider (top). Components of the slider are named on the scanned image (bottom).

References

Abe, T. and J. H. Matlock, "Wafer Bonding Technique for Silicon-on-Insulator Technology," *Solid State Technol.*, November, 39–40, 1990.

Abe, T. and M. L. Reed, "Low Strain Sputtered Polysilicon for Micromechanical Structures," The Ninth Annual International Workshop on Micro Electro Mechanical Systems, San Diego, CA, 1996, pp. 258–262.

Abe, T., W. C. Messner, and M. L. Reed, "Effective Methods to Prevent Stiction During Post–Release-Etch Processing," Proceedings. IEEE Micro Electro Mechanical Systems (MEMS '95), Amsterdam, Netherlands, 1995, pp. 94–9.

Abu-Zeid, M. M., "Corner Undercutting in Aniso-tropically Etched Isolation Contours," *J. Electrochem. Soc.*, 131, 2138–2142, 1984.

Abu-Zeid, M. M., D. L. Kendall, G. R. de Guel, and R. Galeazzi, "Abstract 275," JECS, Toronto, Canada, 1985, p. 400.

Acosta, R. E., C. Ahn, I. V. Babich, E. I. Cooper, J. M. Cotte, W. J. Horkans, C. Jahnes, S. Krongelb, K. T. Kwietniak, N. C. Labianca, E. J. M. O'Sullivan, A. T. Pomerene, D. L. Rath, L. T. Romankiw, and J. A. Tornello, "Integrated Variable Reluctance Magnetic Mini–Motor," *J. Electrochem. Soc.*, 95–2, 494–495, 1995.

Adams, A. C., "Dielectric and Polysilicon Film Deposition," in *VLSI Technology*, Sze, S. M., Ed., McGraw-Hill, New York, 1988, pp. 233–71.

Ahn, C. H., Y. J. Kim, and M. G. Allen, "A Planar Variable Reluctance Magnetic Micromotor with a Fully Integrated Stator and Wrapped Coils," Proceedings. IEEE Micro Electro Mechanical Systems (MEMS '93), Fort Lauderdale, FL, 1993, pp. 1–6.

Akiyama, T., "Controlled Stepwise Motion in Polysilicon Microstructures," *J. Microelectromech. Syst.*, 2, 106–110, 1993.

Akiyama, T. and K. Shono, "A New Step Motion of Polysilicon Microstructures," Proceedings IEEE Micro Electro Mechanical Systems (MEMS '93), Fort Lauderdale, FL, 1993, pp. 272–277.

Alavi, M., S. Buttgenbach, A. Schumacher, and H. J. Wagner, "Laser Machining of Silicon for Fabrication of New Microstructures," 6th International Conference on Solid-State Sensors and Actuators (Transducers '91), San Francisco, CA, 1991, pp. 512–515.

Alavi, M., S. Buttgenbach, A. Schumacher, and H. J. Wagner, "Fabrication of Microchannels by Laser Machining and Anisotropic Etching of Silicon," *Sensors and Actuators A*, 32, 299–302, 1992.

Allen, D. M., *The Principles and Practice of Photochemical Machining and Photoetching*, Adam Hilger, Bristol and Boston, 1986.

Allen, M. G., M. Mehregany, R. T. Howe, and S. D. Senturia, "Microfabricated Structures for the In Situ Measurement of Residual Stress, Young's Modulus and Ultimate Strain of Thin Films," *Appl. Phys. Lett.*, 51, 241–3, 1987.

Alley, R. L., G. J. Cuan, R. T. Howe, and K. Komvopoulos, "The Effect of Release Etch Processing on Surface Microstructure Stiction," Technical Digest of the 1988 Solid-State Sensor and Actuator Workshop., Hilton Head Island, SC, 1988, pp. 202–7.

Alley, R. L., R. T. Howe, and K. Komvopoulos, "The Effect of Release-Etch Processing on Surface Microstructure Stiction," Technical Digest of the 1992 Solid State Sensor and Actuator Workshop, Hilton Head Island, SC, 1992, pp. 202–7.

Alley, R. L., P. Mai, K. Komvopoulos, and R. T. Howe, "Surface Roughness Modifications of Interfacial Contacts in Polysilicon Microstructures," 7th International Conference on Solid-State Sensors and Actuators (Transducers '93), Yokohama, Japan, 1993, pp. 288–91.

Allongue, P., V. Costa-Kieling, and H. Gerischer, "Etching of Silicon in NaOH Solutions, Part I and II," *J. Electrochem. Soc.*, 140, 1009–1018 (Part I) and 1018–1026 (Part II), 1993.

Ammar, E. S. and T. J. Rodgers, "UMOS Transistors on (110) Silicon," *IEEE Trans. Electron Devices*, ED–27, 907–914, 1980.

Anderson, R. C., R. S. Muller, and C. W. Tobias, "Porous Polycrystalline Silicon: A New Material for MEMS," *J. Microelectromech. Syst.*, 3, 10–18, 1994.

Archer, V. D., "Methods for Defect Evaluation of Thin <100> Oriented Silicon in Epitaxial Layers Using a Wet Chemical Etch," *J. Electrochem. Soc.*, 129, 2074–2076, 1982.

Arita, Y. and Y. Sunohara, "Formation and Properties of Porous Silicon Film," *J. Electrochem. Soc.*, 124, 285–295, 1977.

Asano, M., T. Cho, and H. Muraoko, "Applications of Choline in Semiconductor Technology," *Electrochem. Soc. Extend. Abstr.*, 354, 76–2, 911–913, 1976.

Barret, S., F. Gaspard, R. Herino, M. Ligeon, F. Muller, and I. Ronga, "Porous Silicon as a Material in Micro-sensor Technology," *Sensors and Actuators A*, 33, 19–24, 1992.

Bartek, M., P. T. J. Gennissen, P. J. French, and R. F. Wolffenbuttel, "Confined Selective Epitaxial Growth: Potential for Smart Silicon Sensor Fabrication," 8th International Conference on Solid-State Sensors and Actuators (Transducers '95), Stockholm, Sweden, June 1995, pp. 91–94.

Barth, P., "Si in Biomedical Applications," *Micro-electronics—Photonics, Materials, Sensors and Technology*, 1984.

Barth, P. W., P. J. Shlichta, and J. B. Angel, "Deep Narrow Vertical-Walled Shafts in <110> Silicon," 3rd International Conference on Solid-State Sensors and Actuators, Philadelphia, PA, 1985, pp. 371–373.

Bashir, R. et al., "Characterization and Modeling of Sidewall Defects in Selective Epitaxial Growth of Silicon," *J. Vacuum Sci.*

Bassous, E. and C.-Y. Liu, *Polycrystalline Silicon Etching with Tetramethylammonium Hydroxide*, U.S. Patent 4,113,551, 1978.

Bassous, E., H. H. Taub, and L. Kuhn, "Ink Jet Printing Nozzle Arrays Etched in Silicon," *Appl. Phys. Lett.*, 31, 135–137, 1977.

Bean, K., "Anisotropic Etching of Silicon," *IEEE Trans. Electron Devices*, ED–25, 1185–1193, 1978.

Bean, K. E. and W. R. Runyan, "Dielectric Isolation: Comprehensive, Current and Future," *J. Electrochem. Soc.*, 124, 5C–12C, 1977.

Bean, K. E., R. L. Yeakley, and T. K. Powell, "Orientation Dependent Etching and Deposition of Silicon," *J. Electrochem. Soc.*, 121, 87C, 1974.

Benitez, M. A., J. Esteve, M. S. Benrakkad, J. R. Morante, J. Samitier, and J. Å. Schweitz, "Stress Profile Characterization and Test Structures Analysis of Single and Double Ion Implanted LPCVD Polycrystalline Silicon," 8th International Conference on Solid-State Sensors and Actuators (Transducers '95), Stockholm, Sweden, June 1995, pp. 88–91.

Bennema, P., "Spiral Growth and Surface Roughening: Developments since Burton, Cabrera, and Frank," *J. Cryst. Growth*, 69, 182–197, 1984.

Biebl, M. and H. von Philipsborn, "Fracture Strength of Doped and Undoped Panical Filters," Ph.D. thesis, University of California at Berkeley, 1993.

Biebl, M., G. Brandl, and R. T. Howe, "Young's Modulus of In Situ Phosphorous-Doped Polysilicon," 8th International Conference on Solid-State Sensors and Actuators (Transducers '95), Stockholm, Sweden, June 1995, pp. 80–3.

Biebl, M., G. T. Mulhern, and R. T. Howe, "In Situ Phosphorous-Doped Polysilicon for Integrated MEMS," 8th International Conference on Solid-State Sensors and Actuators (Transducers '95), Stockholm, Sweden, June 1995, pp. 198–201.

Block, B., *Zero-Base the Technological Approach*, private communication.

Bogenschutz, A. F., K.-H. Locherer, W. Mussinger, and W. Krusemark, "Chemical Etching of Semiconductors in HNO_3-HF-CH_3COOH," *J. Electrochem. Soc.*, 114, 970–973, 1967.

Bogh, A., "Ethylene Diamine-Pyrocatechol-Water Mixture Shows Etching Anomaly in Boron-Doped Silicon," *J. Electrochem. Soc.*, 118, 401–402, 1971.

Bohm, M., "Advances in Amorphous Silicon Based Thin Film Microelectronics," *Solid State Technol.*, September, 125–131, 1988.

Boivin, L. P., "Thin Film Laser–to Fiber Coupler," *Appl. Opt.*, 13, 391–395, 1974.

Bomchil, G., R. Herino, and K. Barla, "Formation and Oxidation of Porous Silicon for Silicon on Insulator Technologies," in *Energy Beam-Solid Interactions and Transient Thermal Processes*, Nguyen, V. T. and Cullis, A., Eds., Les Editions de Physique, Les Ulis, 1986, p. 463.

Bonifield, T., K. Hewes, B. Merritt, R. Robinson, S. Fisher, and D. Maisch, "Extended Run Evaluation of TEOS/Ozone BPSG Deposition," *Semicond. Int.*, July, 200–204, 1993.

Brantley, W. A., "Calculated Elastic Constants for Stress Problems Associated with Semiconductor Devices," *J. Appl. Phys.*, 44, 534–535, 1973.

Brodie, I. and J. J. Muray, *The Physics of Micro-fabrication*, Plenum Press, New York, 1982.

Brodie, I., H. R. Gurnick, C. E. Holland, and H. A. Moessner, *Method for Providing Polyimide Spacers in a Field Emission Panel Display,* U.S. Patent 4,923,421, 1990.

Bromley, E. I., J. N. Randall, D. C. Flanders, and R. W. Mountain, "A Technique for the Determination of Stress in Thin Films," *J. Vac. Sci. Technol.,* B1, 1364–6, 1983.

Bruno, G., P. Capezzuto, and A. Madan, Eds. *Plasma Deposition of Amorphous Silicon-Based Materials,* Academic Press, Boston, 1995.

Bryzek, J., K. Petersen, J. R. Mallon, L. Christel, and F. Pourahmadi, *Silicon Sensors and Microstructures,* Novasensor, Fremont, CA, 1990.

Burgett, S. R., K. S. Pister, and R. S. Fearing, "Three Dimensional Structures Made with Microfabricated Hinges," ASME 1992, Micromechanical Sensors, Actuators, and Systems, Anaheim, CA, 1992, pp. 1–11.

Burggraaf, P., "Epi's Leading Edge," *Semicond. Int.,* June, 67–71, 1991.

Buser, R. A. and N. F. de Rooij, "Monolithishes Kraftsensorfeld," *VDI-Berichte,* Nr. 677, 1988.

Bushan, B., "Nanotribology and Nanomechanics of MEMS Devices," Proceedings IEEE, Ninth Annual International Workshop on Micro Electro Mechanical Systems, San Diego, CA, 1996, pp. 91–98.

Bushan, B., Ed. *Handbook of Micro/Nanotribology,* CRC Press, Boca Raton, FL, 1995.

Buttgenbach, S., *Mikromechanik,* Teubner Studienbucher, Stuttgart, 1991.

Cahill, S. S., W. Chu, and K. Ikeda, "High Aspect Ratio Isolated Structures in Single Crystal Silicon," 7th International Conference on Solid-State Sensors and Actuators (Transducers '93), Yokohama, Japan, 1993, pp. 250–253.

Callister, D. W., *Materials Science and Engineering,* John Wiley & Sons, New York, 1985.

Campbell, D. S., "Mechanical Properties of Thin Films," in *Handbook of Thin Film Technology,* Maissel, L. I. and R. G., Eds., McGraw-Hill, New York, 1970.

Canham, L. T., "Silicon Quantum Wire Array Fabrication by Electrochemical and Chemical Dissolution of Wafers," *Appl. Phys. Lett.,* 57, 1046–1050, 1990.

Chang, S., W. Eaton, J. Fulmer, C. Gonzalez, B. Underwood, J. Wong, and R. L. Smith, "Micro-mechanical Structures in Amorphous Silicon," 6th International Conference on Solid-State Sensors and Actuators (Transducers '91), San Francisco, CA, 1991, pp. 751–4.

Chen, L.-Y. and N. C. MacDonald, "A Selective CVD Tungsten Process for Micromotors," 6th International Conference on Solid-State Sensors and Actuators (Transducers '91), San Francisco, CA, 1991, pp. 739–42.

Cho, S. T., "A Batch Dissolved Wafer Process for Low Cost Sensor Applications," Micromachining and Microfabrication Process Technology (Proceedings of the SPIE), Austin, TX, 1995, pp. 10–17.

Choi, M. S. and E. W. Hearn, "Stress Effects in Boron-Implanted Polysilicon Films," *J. Electrochem. Soc.,* 131, 2443–6, 1984.

Chou, P. C. and N. J. Pagano, *Elasticity: Tensor, Dyadic, and Engineering Approaches,* Dover Publications, Inc., New York, 1967.

Chu, P. B., P. R. Nelson, M. L. Tachiki, and K. S. J. Pister, "Dynamics of Polysilicon Parallel-Plate Electrostatic Actuators," 8th International Conference on Solid-State Sensors and Actuators (Transducers '95), Stockholm, Sweden, June 1995, pp. 356–359.

Chu, T. L. and J. R. Gavaler, "Dissolution of Silicon and Junction Delineation in Silicon by the CrO_3-HF-H_2O System," *Electrochim. Acta,* 10, 1141–1148, 1965.

Chu, W. H., M. Mehregany, X. Ning, and P. Pirouz, "Measurement of Residual Stress-Induced Bending Momemt of p+ Silicon Films," *Mat. Res. Soc. Symp.,* 239, 169, 1992.

Chuang, S. F. and R. L. Smith, "Preferred Crystallo-graphic Directions of Pore Propagation in Porous Silicon Layers," Technical Digest of the 1988 Solid-State Sensor and Actuator Workshop, Hilton Head Island, SC, 1988, pp. 151–153.

Ciarlo, D. R., "Corner Compensation Structures for (110) Oriented Silicon," Proceedings of the IEEE Micro Robots and Teleoperators Workshop, Hyannis, MA, 1987, pp. 6/1–4.

Clark, J. D. L., J. L. Lund, and D. J. Edell, "Cesium Hydroxide [CsOH]: A Useful Etchant for Micromachining Silicon," Technical Digest of the 1988 Solid-State Sensor and Actuator Workshop, Hilton Head Island, SC, 1988, pp. 5–8.

Clark, L. D. and D. J. Edell, "KOH:H_2O Etching of (110) Si, (111) Si, SiO_2, and Ta: an Experimental Study," Proceedings of the IEEE Micro Robots and Tele-operators Workshop, Hyannis, MA, 1987, pp. 5/1–6.

Clemens, D. P., "Anisotropic Etching of Silicon on Sapphire," 1973, p. 407.

Compton, R. D., "PECVD: A Versatile Technology," *Semicond. Int.,* July, 60–65, 1992.

Core, T. A., W. K. Tsang, and S. J. Sherman, "Fabrication Technology for an Integrated Surface-Micromachined Sensor," *Solid State Technol.,* 36, 39–47, 1993.

Craven, D., K. Yu, and T. Pandhumsoporn, "Etching Technology for "Through-The-Wafer" Silicon Etching," Micromachining and Microfabrication Process Technology (Proceedings of the SPIE), Austin, TX, 1995, pp. 259–263.

Crowley, J. L., "Plasma Enhanced CVD for Flat Panel Displays," *Solid State Technol.,* February, 94–98, 1992.

Declercq, M. J., J. P. DeMoor, and J. P. Lambert, "A Comparative Study of Three Anisotropic Etchants for Silicon," *Electrochem. Soc. Abstracts,* 75–2, 446, 1975.

Declercq, M. J., L. Gerzberg, and J. D. Meindl, "Optimization of the Hydrazine-Water Solution for Anisotropic Etching of Silicon in Integrated Circuit Technology," *J. Electrochem. Soc.,* 122, 545–552, 1975.

Diem, B., M. T. Delaye, F. Michel, S. Renard, and G. Delapierre, "SOI(SIMOX) as a Substrate for Surface Micromachining of Single Crystalline Silicon Sensors and Actuators," 7th International Conference on Solid-State Sensors and Actuators (Transducers '93), Yokohama, Japan, 1993, pp. 233–6.

Ding, X. and W. Ko, "Buckling Behavior of Boron-Doped P^+ Silicon Diaphragms," 6th International Conference on Solid-State Sensors and Actuators (Transducers '91), San Francisco, CA, 1991, pp. 201–204.

Drosd, R. and J. Washburn, "Some Observation on the Amorphous to Crystalline Transfomation in Silicon," *J. Appl. Phys.,* 53, 397–403, 1982.

Dunn, P. N., "SOI:Ready to Meet CMOS Challenge," *Solid State Technol.,* October, 32–5, 1993.

Dutta, B., "Integrated Micromotor Concepts," Int. Conf. on Microelectronic Circuits and Systems Theory, Sydney, Australia, 1970, pp. 36–37.

Eaton, W. P. and J. H. Smith, "A CMOS-compatible, Surface-Micromachined Pressure Sensor for Aqueous Ultrasonic Application," Smart Structures and Materials 1995. Smart Electronics (Proceedings of the SPIE), San Diego, CA, 1995, pp. 258–65.

Eaton, W. P. and J. H. Smith, "Release-Etch Modeling for Complex Surface Micromachined Structures," SPIE Proceedings, Micromachining and Microfabrication Process Technology II, Austin, TX, 1996, pp. 80–93.

Editor, "Analog Devices Combines Micromachining with BICMOS," *Semicond. Int.,* 14, 17, 1991.

Editorial, "Transducers, Pressure and Temperature, Catalog," National Semiconductor, Sunnyvale, CA, 1974.

Editorial, *Thermal Character Print Head,* Texas Instruments, Austin, 1977.

Elwenspoek, M., "On the Mechanism of Anisotropic Etching of Silicon," *J. Electrochem. Soc.,* 140, 2075–2080, 1993.

Elwenspoek, M. and J. P. van der Weerden, "Kinetic Roughening and Step Free Energy in the Solid-on-Solid Model and on Naphtalene Crystals," *J. Phys. A: Math. Gen.,* 20, 669–678, 1987.

Elwenspoek, M., H. Gardeniers, M. de Boer, and A. Prak, "Micromechanics," University of Twente, Report No. 122830, Twente, Netherlands, 1994.

Elwenspoek, M., U. Lindberg, H. Kok, and L. Smith, "Wet Chemical Etching Mechanism of Silicon," IEEE International Workshop on Micro Electro Mechanical Systems, MEMS '94, Oiso, Japan, 1994, pp. 223–8.

Eremenko, V. G. and V. I. Nikitenko, "Electron Microscope Investigation of the Microplastic Deformation Mechanisms by Indentation," *Phys. Stat. Sol. A,* 14, 317–330, 1972.

Ericson, F. and J. Å. Schweitz, "Micromechanical Fracture Strength of Silicon," *J. Appl. Phys.*, 68, 5840–4, 1990.

Ericson, F., S. Greek, J. Soderkvist, and J. Å. Schweitz, "High Sensitive Internal Film Stress Measurement by an Improved Micromachined Indicator Structure," 8th International Conference on Solid-State Sensors and Actuators (Transducers '95), Stockholm, Sweden, June 1995, pp. 84–7.

Fan, L. S., "Integrated Micromachinery: Moving Structures on Silicon Chips," Ph.D. thesis, University of California at Berkeley, 1989.

Fan, L.-S., Y.-C. Tai, and R. S. Muller, "Pin Joints, Gears, Springs, Cranks, and Other Novel Micromechanical Structures," 4th International Conference on Solid-State Sensors and Actuators (Transducers '87), Tokyo, Japan, 1987, pp. 849–52.

Fan, L. S., Y. C. Tai, and R. S. Mulller, "Integrated Movable Micromechanical Structures for Sensors and Actuators," *IEEE Trans. Electron Devices*, 35, 724–30, 1988.

Fan, L. S., R. S. Muller, W. Yun, J. Huang, and R. T. Howe, "Spiral Microstructures for the Measurement of Average Strain Gradients in Thin Films," Proceedings. IEEE Micro Electro Mechanical Systems, (MEMS '90), Napa Valley, CA, 1990, pp. 177–81.

Faust, J. W. and E. D. Palik, "Study of The Orientation Dependent Etching and Initial Anodization of Si in Aqueous KOH," *J. Electrochem. Soc.*, 130, 1413–1420, 1983.

Fedder, G. K., J. C. Chang, and R. T. Howe, "Thermal Assembly of Polysilicon Microactuators with Narrow–Gap Electrostatic Comb Drive," Technical Digest of the 1992 Solid State Sensor and Actuator Workshop, Hilton Head Island, SC, 1992, pp. 63–8.

Field, L., "Low-Stress Nitrides for Use in Electronic Devices," EECS/ERL Res. Summary, 42–43, University of California at Berkeley, 1987.

Finne, R. M. and D. L. Klein, "A Water-Amine-Complexing Agent System for Etching Silicon," *J. Electrochem. Soc.*, 114, 965–970, 1967.

Fleischman, A. J., S. Roy, C. A. Zorman, M. Mehregany, and L. Matus, G., "Polycrystaline Silicon Carbide for Surface Micromachining," The Ninth Annual International Workshop on Micro Electro Mechanical Systems, San Diego, CA, 1996, pp. 234–238.

Forster, J. H. and J. B. Singleton, "Beam–Lead Sealed Junction Integrated Circuits," *Bell Laboratories Record*, 44, 313–317, 1966.

Frazier, A. B. and M. G. Allen, "Metallic Microstructures Fabricated Using Photosensitive Polyimide Electro-plating Molds," *J. Microelectromech. Syst.*, 2, 87–94, 1993.

Fresquet, G., C. Azzaro, and J.-P. Couderc, "Analysis and Modeling of In Situ Boron-Doped Polysilicon Deposition by LP CVD," *J. Electrochem. Soc.*, 142, 538–47, 1995.

Gabriel, K., J. Jarvis, and W. Trimmer, "Small Machines, Large Opportunities: A Report on the Emerging Field of Microdynamics," National Science Foundation, 1989.

Gennissen, P. T. J., M. Bartke, P. J. French, P. M. Sarro, and R. F. Wolffenbuttel, "Automatic Etch Stop on Buried Oxide Using Epitaxial Lateral Overgrowth," 8th International Conference on Solid-State Sensors and Actuators (Transducers '95), Stockholm, Sweden, June 1995, pp. 75–8.

Ghandi, S. K., *The Theory and Practice of Micro-electronics*, John Wiley & Sons, New York, 1968.

Glembocki, O. J., R. E. Stahlbush, and M. Tomkiewicz, "Bias-Dependent Etching of Silicon in Aqueous KOH," *J. Electrochem. Soc.*, 132, 145–151, 1985.

Goetzlich, J. and H. Ryssel, "Tapered Windows in Silicon Dioxide, Silicon Nitride, and Polysilicon Layers by Ion Implantation," *J. Electrochem. Soc.*, 128, 617–9, 1981.

Gogoi, B. P. and C. H. Mastrangelo, "Post-Processing Release of Microstructures by Electromagnetic Pulses," 8th International Conference on Solid-State Sensors and Actuators (Transducers '95), Stockholm, Sweden, June 1995, pp. 214–7.

Goosen, J. F. L., B. P. van Drieenhuizen, P. J. French, and R. F. Wolfenbuttel, "Stress Measurement Structures for Micromachined Sensors," 7th International Conference on Solid-State Sensors and Actuators (Transducers '93), Yokohama, Japan, 1993, pp. 783–6.

Gravesen, P., "Silicon Sensors," Status report for the industrial engineering thesis, DTH Lyngby, Denmark, 1986.

Greek, S., F. Ericson, S. Johansson, and J.-Å. Schweitz, "In Situ Tensile Strength Measurement of Thick-Film and Thin-Film Micromachined Structures," 8th International Conference on Solid-State Sensors and Actuators (Transducers '95), Stockholm, Sweden, June 1995, pp. 56–9.

Greenwood, J. C., "Ethylene Diamine-Cathechol-Water Mixture Shows Preferential Etching of p-n Junction," *J. Electrochem. Soc.,* 116, 1325–1326, 1969.

Greenwood, J. C., "Etched Silicon Vibrating Sensor," *J. Phys. E, Sci. Instrum.,* 17, 650–652, 1984a.

Greenwood, J. C., "Silicon in Mechanical Sensors," *J. Phys. E, Sci. Instrum.,* 21, 1114–1128, 1988b.

Greiff, P., "SOI-based Micromechanical Process," Micromachining and Microfabrication Process Technology (Proceedings of the SPIE), Austin, TX, 1995, pp. 74–81.

Guckel, H. and D. W. Burns, "Planar Processed Polysilicon Sealed Cavities for Pressure Transducer Arrays," IEEE International Electron Devices Meeting. Technical Digest, IEDM '84, San Francisco, CA, 1984, pp. 223–5.

Guckel, H. and D. W. Burns, "A Technology for Integrated Transducers," International Conference on Solid-State Sensors and Actuators, Philadelphia, PA, 1985, pp. 90–2.

Guckel, H., T. Randazzo, and D. W. Burns, "A Simple Technique for the Determination of Mechanical Strain in Thin Films with Applications to Polysilicon," *J. Appl. Phys.,* 57, 1671–5, 1985.

Guckel, H., D. K. Showers, D. W. Burns, C. R. Rutigliano, and C. G. Nesler, "Deposition Techniques and Properties of Strain Compensated LP CVD Silicon Nitride," Technical Digest of the 1986 Solid State Sensor and Actuator Workshop, Hilton Head Island, SC, 1986.

Guckel, H., D. W. Burns, C. K. Nesler, and C. R. Rutigliano, "Fine Grained Polysilicon and Its Application to Planar Pressure Transducers," 4th International Conference on Solid-State Sensors and Actuators (Transducers '87), Tokyo, Japan, 1987, pp. 277–282.

Guckel, H., D. W. Burns, C. C. G. Visser, H. A. C. Tilmans, and D. Deroo, "Fine-grained Polysilicon Films with Built–in Tensile Strain," *IEEE Trans. Electron Devices,* 35, 800–1, 1988.

Guckel, H., D. W. Burns, H. A. C. Tilmans, C. C. G. Visser, D. W. DeRoo, T. R. Christenson, P. J. Klomberg, J. J. Sniegowski, and D. H. Jones, "Processing Conditions for Polysilicon Films with Tensile Strain for Large Aspect Ratio Microstructures," Technical Digest of the 1988 Solid State Sensor and Actuator Workshop, Hilton Head Island, SC, 1988, pp. 51–6.

Guckel, H., D. W. Burns, H. A. C. Tilmans, D. W. DeRoo, and C. R. Rutigliano, "Mechanical Properties of Fine Grained Polysilicon: The Repeatability Issue," Technical Digest of the 1988 Solid State Sensor and Actuator Workshop, Hilton Head Island, SC, 1988, pp. 96–9.

Guckel, H., J. J. Sniegowski, T. R. Christenson, S. Mohney, and T. F. Kelly, "Fabrication of Micromechanical Devices from Polysilicon Films with Smooth Surfaces," *Sensors and Actuators A,* 20, 117–21, 1989.

Guckel, H., J. J. Sniegowski, T. R. Christenson, and F. Raissi, "The Application of Fine-Grained, Tensile Polysilicon to Mechanically Resonant Transducers," *Sensors and Actuators A,* 21, 346–351, 1990.

Hallas, C. E., "Electropolishing Silicon," *Solid State Technol.,* 14, 30–32, 1971.

Harbeke, G., L. Krausbauer, E. F. Steigmeier, and A. E. Widmer, "LP CVD Polycrystalline Silicon: Growth and Physical Properties of In-Situ Phosphorous Doped and Undoped Films," *RCA Rev.,* 44, 287–313, 1983.

Harris, T. W., *Chemical Milling,* Clarendon Press, Oxford, 1976.

Hendriks, M. and C. Mavero, "Phosphorous Doped Polysilicon for Double Poly Structures. Part I. Morphology and Microstructure," *J. Electrochem. Soc.,* 138, 1466–1470, 1991.

Hendriks, M., R. Delhez, and S. Radelaar, "Carbon Doped Polycrystalline Silicon Layers," in *Studies in Inorganic Chemistry,* Elsevier, Amsterdam, 1983, p. 193.

Herb, J. A., M. G. Peters, S. C. Terry, and J. H. Jerman, "PECVD Diamond Films for Use in Silicon Microstructures," *Sensors and Actuators A,* 23, 982–7, 1990.

Hermansson, K., U. Lindberg, B. Hok, and G. Palmskog, "Wetting Properties of Silicon Surfaces," 6th International Conference on Solid-State Sensors and Actuators (Transducers '91), San Francisco, CA, 1991, pp. 193–6.

Herr, E. and H. Baltes, "KOH Etch Rates of High–Index Planes from Mechanically Prepared Silicon Crystals," 6th International Conference on Solid-State Sensors and Actuators (Transducers '91), San Francisco, CA, 1991, pp. 807–810.

Hesketh, P. J., C. Ju, S. Gowda, E. Zanoria, and S. Danyluk, "Surface Free Energy Model of Silicon Anisotropic Etching," *J. Electrochem. Soc.,* 140, 1080–1085, 1993.

Heuberger, A., *Mikromechanik,* Springer-Verlag, Heidelberg, 1989.

Hirata, M., K. Suzuki, and H. Tanigawa, "Silicon Diaphragm Pressure Sensors Fabricated by Anodic Oxidation Etch-Stop," *Sensors and Actuators A,* 13, 63–70, 1988.

Hoffman, E., B. Warneke, E. Kruglick, J. Weigold, and K. S. J. Pister, "3D Structures with Piezoresistive Sensors in Standard CMOS," Proceedings. IEEE Micro Electro Mechanical Systems (MEMS '95), Amsterdam, Netherlands, 1995, pp. 288–93.

Hoffman, R. W., "Stresses in Thin Films : The Relevance of Grain Boundaries and Impurities," *Thin Solid Films,* 34, 185–90, 1975.

Hoffman, R. W., "Mechanical Properties of Non-Metallic Thin Films," in *Physics of Nonmetallic Thin Films (NATO Advanced Study Institutes Series: Series B, Physics),* Dupuy, C. H. S. and Cachard, A., Eds., Plenum Press, New York, 1976, pp. 273–353.

Hoffmeister, W., "Determination of the Etch Rate of Silicon in Buffered HF Using a ^{31}Si Tracer Method," *Int. J. Appl. Radiation and Isotopes,* 2, 139, 1969.

Holbrook, D. S. and J. D. McKibben, "Microlithography for Large Area Flat Panel Display Substrates," *Solid State Technol.,* May, 166–172, 1992.

Holland, C. E., E. R. Westerberg, M. J. Madou, and T. Otagawa, *Etching Method for Producing an Electrochemical Cell in a Crystalline Substrate,* U.S. Patent 4,764,864, 1988.

Hornbeck, L. J., "Projection Displays and MEMS: Timely Convergence for a Bright Future," Micro-machining and Microfabrication Process Technology (Proceedings of the SPIE), Austin, TX, 1995, p. 2.

Houston, M. R., R. Maboudian, and R. Howe, "Ammonium Fluoride Anti-Stiction Treatment for Polysilicon Microstructures," 8th International Conference on Solid-State Sensors and Actuators (Transducers '95), Stockholm, Sweden, June 1995, pp. 210–3.

Howe, R. T., "Polycrystalline Silicon Microstructures," in *Micromachining and Micropackaging of Transducers,* Fung, C. D., Cheung, P. W., Ko, W. H., and Fleming, D. G., Eds., Elsevier, New York, 1985, pp. 169–87.

Howe, R. T., "Resonant Microsensors," 4th International Conference on Solid-State Sensors and Actuators (Transducers '87), Tokyo, Japan, 1987, pp. 843–8.

Howe, R. T., "Polysilicon Integrated Microsystems: Technologies and Applications," 8th International Conference on Solid-State Sensors and Actuators (Transducers '95), Stockholm, Sweden, June, 1995a, pp. 43–6.

Howe, R. T., "Recent Advances in Surface Micromachining," 13th Sensor Symposium. Technical Digest, Tokyo, Japan, 1995b, pp. 1–8.

Howe, R. T. and R. S. Muller, "Polycrystalline Silicon Micromechanical Beams," Spring Meeting of The Electrochemical Society, Montreal, Canada, 1982, pp. 184–5.

Howe, R. T. and R. S. Muller, "Polycrystalline Silicon Micromechanical Beams," *J. Electrochem. Soc.,* 130, 1420–1423, 1983a.

Howe, R. T. and R. S. Muller, "Stress in Polycrystalline and Amorphous Silicon Thin Films," *J. Appl. Phys.,* 54, 4674–5, 1983b.

Hu, J. Z., L. D. Merkle, C. S. Menoni, and I. L. Spain, "Crystal Data for High–Presure Phases of Silicon," *Phys. Rev. B,* 34, 4679–4684, 1986.

Hu, S. M. and D. R. Kerr, "Observation of Etching of n-type Silicon in Aqueous HF Solutions," *J. Electrochem. Soc.,* 114, 414, 1967.

Huff, M. A. and M. A. Schmidt, "Fabrication, Packaging and Testing of a Wafer Bonded Microvalve," Technical Digest of the 1992 Solid State Sensor and Actuator Workshop, Hilton Head Island, SC, 1992, pp. 194–197.

Iscoff, R., "Hotwall LP CVD Reactors: Considering the Choices," *Semicond. Int.,* June, 60–64, 1991.

Jaccodine, R. J. and W. A. Schlegel, "Measurements of Strains at Si-SiO$_2$ Interface," *J. Appl. Phys.*, 37, 2429–34, 1966.

Jackson, K. A., "A Review of the Fundamental Aspects of Crystal Growth," Crystal Growth, Boston, MA, 1966, pp. 17–24.

Jackson, T. N., M. A. Tischler, and K. D. Wise, "An Electrochemical P-N Junction Etch-Stop for the Formation of Silicon Microstructures," *IEEE Electron Device Lett.*, EDL–2, 44–45, 1981.

Jerman, H., "Bulk Silicon Micromachining," Hardcopies of viewgraphs, 1994, Banf, Canada.

Jerman, J. H., "The Fabrication and Use of Micro-machined Corrugated Silicon Diaphragms," *Sensors and Actuators A*, 23, 988–92, 1990.

Joseph, J., M. Madou, T. Otagawa, P. Hesketh, and A. Saaman, "Catheter-Based Micromachined Electrochemical Sensors," Catheter-Based Sensing and Imaging Technology (Proceedings of the SPIE), Los Angeles, CA, 1989, pp. 18–22.

Judy, M. W. and R. T. Howe, "Highly Compliant Lateral Suspensions Using Sidwall Beams," 7th International Conference on Solid-State Sensors and Actuators (Transducers '93), Yokohama, Japan, 1993a, pp. 54–57.

Judy, M. W. and R. T. Howe, "Hollow Beam Polysilicon Lateral Resonators," Proceedings IEEE Micro Electro Mechanical Systems (MEMS '93), Fort Lauderdale, FL, 1993b, pp. 265–271.

Judy, M. W., Y. H. Cho, R. T. Howe, and A. P. Pisano, "Self-Adjusting Microstructures (SAMS)," Proceedings. IEEE Micro Electro Mechanical Systems (MEMS '91), Nara, Japan, 1991, pp. 51–6.

Jung, K. H., S. Shih, and D. L. Kwong, "Developments in Luminescent Porous Si," *J. Electrochem. Soc.*, 140, 3046–3064, 1993.

Kahn, H., S. Stemmer, K. Nandakumar, A. H. Heuer, R. L. Mullen, R. Ballarini, and M. A. Huff, "Mechanical Properties of Thick, Surface Micromachined Polysilicon Films," Ninth Annual International Workshop on Micro Electro Mechanical Systems, San Diego, CA, 1996, pp. 343–348.

Kaminsky, G., "Micromachining of Silicon Mechanical Structures," *J. Vac. Sci. Technol.*, B3, 1015–1024, 1985.

Kanda, Y., "A Graphical Representation of the Piezoresistance Coefficients in Silicon," *IEEE Trans. Electron Devices*, ED–29, 64–70, 1982.

Kanda, Y., "What Kind of SOI Wafers are Suitable for What Type of Micromachining Purposes?" 6th International Conference on Solid-State Sensors and Actuators (Transducers '91), San Francisco, CA, 1991, pp. 452–5.

Keller, C. and M. Ferrari, "Milli-Scale Polysilicon Structures," Technical Digest of the 1994 Solid State Sensor and Actuator Workshop, Hilton Head Island, SC, 1994, pp. 132–137.

Keller, C. G. and R. T. Howe, "Hexsil Bimorphs for Vertical Actuation," 8th International Conference on Solid-State Sensors and Actuators (Transducers '95), Stockholm, Sweden, June 1995a, pp. 99–102.

Keller, C. G. and R. T. Howe, "Nickel-Filled Hexsil Thermally Actuated Tweezers," 8th International Conference on Solid-State Sensors and Actuators (Transducers '95), Stockholm, Sweden, June 1995b, pp. 376–379.

Kendall, D. L., "On Etching Very Narrow Grooves in Silicon," *Appl. Phys. Lett.*, 26, 195–198, 1975.

Kendall, D. L., "Vertical Etching of Silicon at Very High Aspect Ratios," *Ann. Rev. Mater. Sci.*, 9, 373–403, 1979.

Kendall, D. L. and G. R. de Guel, "Orientation of the Third Kind: The Coming of Age of (110) Silicon," in *Micromachining and Micropackaging of Transducers*, Fung, C. D., Ed., Elsevier, New York, 1985, pp. 107–124.

Kenney, D. M., *Methods of Isolating Chips of a Wafer of Semiconductor Material*, in U.S. Patent 3,332,137, 1967.

Kermani, A., K. E. Johnsgard, and W. Fred, "Single Wafer RT CVD of Polysilicon," *Solid State Technol.*, May, 71–73, 1991.

Kern, W. and C. A. Deckert, "Chemical Etching," in *Thin Film Processes*, Vossen, J. L. and Kern, W., Eds., Academic Press, Orlando, 1978.

Kern, W., "Chemical Etching of Silicon, Germanium, Gallium Arsenide, and Gallium Phosphide," *RCA Rev.*, 39, 278–308, 1978.

Khazan, A. D., *Transducers and Their Elements*, PTR Prentice Hall, Englewood Cliffs, NJ, 1994.

Kim, C. J., "Silicon Electromechanical Microgrippers: Design, Fabrication, and Testing," Ph.D. thesis, University of California, Berkeley, 1991.

Kim, S. C. and K. Wise, "Temperature Sensitivity in Silicon Piezoresistive Pressure Transducers," *IEEE Trans. Electron Devices*, ED–30, 802–810, 1983.

Kim, Y. W. and M. G. Allen, "Surface Micromachined Platforms Using Electroplated Sacrificial Layers," 6th International Conference on Solid-State Sensors and Actuators (Transducers '91), San Francisco, CA, 1991, pp. 651–4.

Kinsbron, E., M. Sternheim, and R. Knoell, "Crystallization of Amorphous Silicon Films during Low Pressure Chemical Vapor Deposition," *Appl. Phys. Lett.*, 42, 835–837, 1983.

Kittel, C., *Introduction to Solid State Physics*, John Wiley & Sons, New York, 1976.

Klein, D. L. and D. J. D'Stefan, "Controlled Etching of Silicon in the HF-HNO$_3$ System," *J. Electrochem. Soc.*, 109, 37–42, 1962.

Kloeck, B., S. D. Collins, N. F. de Rooij, and R. L. Smith, "Study of Electrochemical Etch-Stop for High-Precision Thickness Control of Silicon Membranes," *IEEE Trans. Electron Devices*, 36, 663–669, 1989.

Kozlowski, F., N. Lindmair, T. Scheiter, C. Hierold, and W. Lang, "A Novel Method to Avoid Sticking of Surface Micromachined Structures," 8th International Conference on Solid-State Sensors and Actuators (Transducers '95), Stockholm, Sweden, June 1995, pp. 220–3.

Kragness, R. C. and H. A. Waggener, *Precision Etching of Semiconductors*, U.S. Patent 3,765,969, 1973.

Krulevitch, P. A., "Micromechanical Investigations of Silicon and Ni–Ti–Cu Thin Films," Ph.D. thesis, University of California at Berkeley, 1994.

Kuhn, G. L. and C. J. Rhee, "Thin Silicon Film on Insulating Substrate," *J. Electrochem. Soc.*, 120, 1563–6, 1973.

Kurokawa, H., "P–Doped Polysilicon Film Growth Technology," *J. Electrochem. Soc.*, 129, 2620–4, 1982.

LaBianca, N. C., J. D. Gelorme, E. Cooper, E. O'Sullivan, and J. Shaw, "High Aspect Ratio Optical Resist Chemistry for MEMS Applications," JECS 188th Meeting, Chicago, IL, 1995, pp. 500–501.

Lambrechts, M. and W. Sansen, *Biosensors: Micro-electrochemical Devices*, Institute of Physics Publishing, Philadelphia, PA, 1992.

Lange, P., M. Kirsten, W. Riethmuller, B. Wenk, G. Zwicker, J. R. Morante, F. Ericson, and J. Å. Schweitz, "Thick Polycrystalline Silicon for Surface Micro-mechanical Applications: Deposition, Structuring and Mechanical Characterization," 8th International Conference on Solid-State Sensors and Actuators (Transducers '95), Stockholm, Sweden, June, 1995, pp. 202–5.

Learn, A. J. and D. W. Foster, "Deposition and Electrical Properties of In Situ Phosphorous-Doped Silicon Films Formed by Low–Pressure Chemical Vapor Deposition," *J. Appl. Phys.*, 61, 1898–1904, 1987.

Le Berre, M., P. Kleinmann, B. Semmache, D. Barbier, and P. Pinard, "Electrical and Piezoresin," V. K. and P. J. McWhorter, Eds. Smart Electronics and MEMS, Proceedings of the Smart Structures and Materials 1996 Meeting, Vol. 2722, SPIE, San Diego, CA, 1996.

Lebouitz, K. S., R. T. Howe, and A. P. Pisano, "Permeable Polysilicon Etch-Access Windows for Microshell Fabrication," 8th International Conference on Solid-State Sensors and Actuators (Transducers '95), Stockholm, Sweden, June 1995, pp. 224–227.

Lee, D. B., "Anisotropic Etching of Silicon," *J. Appl. Phys.*, 40, 4569–4574, 1969.

Lee, J. B., Z. Chen, M. G. Allen, A. Rohatgi, and R. Arya, "A Miniaturized High-Voltage Solar Cell Array as an Electrostatic MEMS Power Supply," *J. Microelectromech. Syst.*, 4, 102–108, 1995.

Lee, J. G., S. H. Choi, T. C. Ahn, C. G. Hong, P. Lee, K. Law, M. Galiano, P. Keswick, and B. Shin, "SA CVD: A New Approach for 16 Mb Dielectrics," *Semicond. Int.*, May, 115–120, 1992.

Lehmann, V., "The Physics of Macropore Formation in Low Doped n-Type Silicon," *J. Electrochem. Soc.*, 140, 2836–2843, 1993.

Lehmann, V., "Porous Silicon-A New Material for MEMS," The Ninth Annual International Workshop on Micro Electro Mechanical Systems, MEMS '96, San Diego, CA, 1996, pp. 1–6.

Lehmann, V. and H. Foll, "Formation Mechanism and Properties of Electrochemically Etched Trenches in n-Type Silicon," *J. Electrochem. Soc.*, 137, 653–659, 1990.

Lehmann, V. and U. Gosele, "Porous Silicon Formation: A Quantum Wire Effect," *Appl. Phys. Lett.,* 58, 856–858, 1991.

Lepselter, M. P., "Beam Lead Technology," *Bell. Sys. Tech. J.,* 45, 233–254, 1966.

Lepselter, M. P., *Integrated Circuit Device and Method,* U.S. Patent 3,335,338, 1967.

Levy, R. A. and K. Nassau, "Viscous Behavior of Phosphosilicate and Borophosphosilicate Galsses in VLSI Processing," *Solid State Technol.,* October, 123–30, 1986.

Levy-Clement, C., A. Lagoubi, R. Tenne, and M. Neumann-Spallart, "Photoelectrochemical Etching of Silicon," *Electrochim. Acta,* 37, 877–888, 1992.

Lewerenz, H. J., J. Stumper, and L. M. Peter, "Deconvolution of Charge Injection Steps in Quantum Yield Multiplication on Silicon," *Phys. Rev. Lett.,* 61, 1989–92, 1988.

Lietoila, A., A. Wakita, T. W. Sigmon, and J. F. Gibbons, "Epitaxial Regrowth of Intrinsic, ^{31}P–Doped and Compensated ($^{31P+11}$B-Doped) Amorphous Si," *J. Appl. Phys.,* 53, 4399–4405, 1982.

Lin, G., K. S. J. Pister, and K. P. Roos, "Standard CMOS Piezoresistive Sensor to Quantify Heart Cell Contractile Forces," The Ninth Annual International Workshop on Micro Electro Mechanical Systems, MEMS '96, San Diego, CA, 1996, pp. 150–155.

Lin, L., R. T. Howe, and A. P. Pisano, "A Novel In Situ Micro Strain Gauge," Proceedings IEEE Micro Electro Mechanical Systems, (MEMS '93), Fort Lauderdale, FL, 1993, pp. 201–206.

Lin, L., "Selective Encapsulations of MEMS: Micro Channels, Needles, Resonators and Electromechanical Filters," Ph.D. thesis, University of California at Berkeley, 1993.

Lin, L. Y., S. S. Lee, M. C. Wu, and K. S. J. Pister, "Micromachined Integrated Optics for Free-Space Interconnections," Proceedings IEEE Micro Electro Mechanical Systems, (MEMS '95), Amsterdam, Netherlands, 1995, pp. 77–82.

Lin, T.-H., "Flexure-Beam Micromirror Devices and Potential Expansion for Smart Micromachining," Proceedings SPIE, Smart Electronics and MEMS, San Diego, CA, 1996, pp. 20–29.

Linde, H. and L. Austin, "Wet Silicon Etching with Aqueous Amine Gallates," *J. Electrochem. Soc.,* 139, 1170–1174, 1992.

Liu, J., Y.-C. Tai, J. Lee, K.-C. Pong, Y. Zohar, and C.-H. Ho, "In Situ Monitoring and Universal Modeling of Sacrificial PSG Etching Using Hydrofluoric Acid," Proceedings of Micro Electro Mechanical Systems, IEEE, Fort Lauderdale, FL, 1993, pp. 71–76.

Lober, T. A. and R. T. Howe, "Surface Micromachining for Electrostatic Microactuator Fabrication," Technical Digest of the 1988 Solid State Sensor and Actuator Workshop, Hilton Head Island, SC, 1988, pp. 59–62.

Lober, T. A., J. Huang, M. A. Schmidt, and S. D. Senturia, "Characterization of the Mechanisms Producing Bending Moments in Polysilicon Micro-Cantilever Beams by Interferometric Deflection Measurements," Technical Digest of the 1988 Solid State Sensor and Actuator Workshop, Hilton Head Island, SC, 1988, pp. 92–95.

Lochel, B., A. Maciossek, H. J. Quenzer, and B. Wagner, "UV Depth Lithography and Galvanoforming for Micromachining," Electrochemical Microfabrication II, Miami Beach, FL, 1994, pp. 100–111.

Lochel, B., A. Maciossek, H. J. Quenzer, and B. Wagner, "Ultraviolet Depth Lithography and Galvanoforming for Micromachining," *J. Electrochem. Soc.,* 143, 237–244, 1996.

Loechel, B., M. Rothe, S. Fehlberg, G. Gruetzner, and G. Bleidiessel, "Influence of Resist Baking on the Pattern Quality of Thick Photoresists," SPIE–Micromachining and Microfabrication Process Technology II, Austin, TX, 1996, pp. 174–181.

Luder, E., "Polcrystalline Silicon-Based Sensors," *Sensors and Actuators A,* 10, 9–23, 1986.

Madou, M. J., "Compatibility and Incompatibility of Chemical Sensors and Analytical Equipment with Micromachining," Technical Digest of the 1994 Solid-State Sensor and Actuator Workshop, Hilton Head Island, SC, 1994, pp. 164–171.

Madou, M. J. and S. R. Morrison, *Chemical Sensing with Solid State Devices,* Academic Press, New York, 1989.

Madou, M. J. and T. Otagawa, *Microelectrochemical Sensor and Sensor Array,* U.S. Patent 4,874,500, 1989.

Madou, M. J., K. W. Frese, and S. R. Morrison, "Photoelectrochemical Corrosion of Semiconductors for Solar Cells," *SPIE*, 248, 88–95, 1980.

Madou, M. J., B. H. Loo, K. W. Frese, and S. R. Morrison, "Bulk and Surface Characterization of the Silicon Electrode," *Surf. Sci.*, 108, 135–152, 1981.

Man, P. F., B. P. Gogoi, and C. H. Mastrangelo, "Elimination of Post–Release Adhesion in Micro-structures Using Thin Conformal Fluorocarbon Films," Ninth Annual International Workshop on Micro Electro Mechanical Systems, MEMS '96, San Diego, CA, 1996, pp. 55–60.

Marco, S., J. Samitier, O. Ruiz, J. R. Morante, J. Esteve-Tinto, and J. Bausells, "Stress Measurements of SiO_2-Polycrystalline Silicon Structures for Micro-mechanical Devices by Means of Infrared Spectroscopy Technique," 6th International Conference on Solid-State Sensors and Actuators (Transducers '91), San Francisco, CA, 1991, pp. 209–12.

Maseeh, F. and S. D. Senturia, "Elastic Properties of Thin Polyimide Films," in *Polyimides: Materials, Chemistry and Characterization*, Feger, C., Khojasteh, M. M., and McGrath, J. E., Eds., Elesevier, Amsterdam, 1989, pp. 575–584.

Maseeh, F. and S. D. Senturia, "Plastic Deformation of Highly Doped Silicon," *Sensors and Actuators A*, 23, 861–865, 1990.

Maseeh, F. and S. D. Senturia, "Viscoelasticity and Creep Recovery of Polyimide Thin Films," Technical Digest of the 1990 Solid-State Sensor and Actuator Workshop, Hilton Head Island, SC, 1990, pp. 55–60.

Maseeh-Tehrani, F., "Characterization of Mechanical Properties of Microelectronic Thin Films," Ph.D. thesis, Massachusetts Institute of Technology, Cambridge, 1990.

Mastrangelo, C. H. and C. H. Hsu, "Mechanical Stability and Adhesion of Microstructures under Capillary Forces-Part I: Basic Theory," *J. Microelectromech. Syst.*, 2, 33–43, 1993a.

Mastrangelo, C. H. and C. H. Hsu, "Mechanical Stability and Adhesion of Microstructures under Capillary Forces-Part II: Experiments," *J. Microelectromech. Syst.*, 2, 44–55, 1993b.

Mastrangelo, C. H. and G. S. Saloka, "A Dry-Release Method Based on Polymer Columns for Microstructure Fabrication," Proceedings IEEE Micro Electro Mechanical Systems, (MEMS '93), Fort Lauderdale, FL, 1993, pp. 77–81.

Matson, E. A. and S. A. Polysakov, "On the Evolutionary Selection Principle in Relation to the Growth of Polycrystalline Silicon Films," *Phys. Sta. Sol. (a)*, 41, K93–K95, 1977.

Matsumura, M. and S. R. Morrison, "Photoanodic Properties of an n-Type Silicon Electrode in Aqueous Solution Containing Fluorides," *J. Electroanal. Chem.*, 144, 113–120, 1983.

Mayer, G. K., H. L. Offereins, H. Sandmeier, and K. Kuhl, "Fabrication of Non-Underetched Convex Corners in Ansiotropic Etching of (100)-Silicon in Aqueous KOH with Respect to Novel Micromechanic Elements," *J. Electrochem. Soc.*, 137, 3947–3951, 1990.

McNeil, V. M., S. S. Wang, K.-Y. Ng, and M. A. Schmidt, "An Investigation of the Electrochemical Etching of (100) Silicon in CsOH and KOH," Technical Digest of the 1990 Solid-State Sensor and Actuator Workshop, Hilton Head Island, SC, 1990, pp. 92–97.

Meek, R. L., "Electrochemically Thinned N/N+ Epitaxial Silicon-Method and Applications," *J. Electrochem. Soc.*, 118, 1240–1246, 1971.

Mehregany, M. and S. D. Senturia, "Anisotropic Etching of Silicon in Hydrazine," *Sensors and Actuators A*, 13, 375–390, 1988.

Mehregany, M., R. T. Howe, and S. D. Senturia, "Novel Microstructures for the In Situ Measurement of Mechanical Properties of Thin Films," *J. Appl. Phys.*, 62, 3579–84, 1987.

Mehregany, M., K. J. Gabriel, and W. S. N. Trimmer, "Integrated Fabrication of Polysilicon Mechanisms," *IEEE Trans. Electron Devices*, 35, 719–23, 1988.

Memming, R. and G. Schwandt, "Anodic Dissolution of Silicon in Hydrofluoric Acid Solutions," *Surf. Sci.*, 4, 109–24, 1966.

Metzger, H. and F. R. Kessler, "Der Debye-Sears Effect zur Bestimmung der Elastischen Konstanten von Silicium," *Z. Naturf.*, A25, 904–906, 1970.

Meyerson, B. S. and W. Olbricht, "Phosphorous-Doped Polycrystalline Silicon Via LPCVD I. Process Characterization," *J. Electrochem. Soc.*, 131, 2361–2365, 1984.

Middlehoek, S. and U. Dauderstadt, "Haben Mikrosensoren aus Silizium eine Zukunft?" *Technische Rundschau*, July, 102–105, 1994.

Mlcak, R., H. L. Tuller, P. Greiff, and J. Sohn, "Photo Assisted Electromechanical Machining of Micro-mechanical Structures," Proceedings IEEE Micro Electro Mechanical Systems (MEMS '93), Fort Lauderdale, FL, 1993, pp. 225–229.

Monk, D. J., D. S. Soane, and R. T. Howe, "LPCVD Silicon Dioxide Sacrificial Layer Etching for Surface Micromachining," Smart Materials Fabrication and Materials for Micro-Electro-Mechanical Systems, San Francisco, CA, 1992, pp. 303–310.

Monk, D. J., D. S. Soane, and R. T. Howe, "Hydrofluoric Acid Etching of Silicon Dioxide Sacrificial Layers. Part II. Modeling," *J. Electrochem. Soc.*, 141, 270–274, 1994a.

Monk, D. J., D. S. Soane, and R. T. Howe, "Hydrofluoric Acid Etching of Silicon Dioxide Sacrificial Layers. Part I. Experimental Observations," *J. Electrochem. Soc.*, 141, 264–269, 1994b.

Morrison, S. R., *Electrochemistry on Semiconductors and Oxidized Metal Electrodes*, Plenum Press, New York, 1980.

Mulder, J. G. M., P. Eppenga, M. Hendriks, and J. E. Tong, "An Industrial LP CVD Process for In Situ Phosphorus-Doped Polysilicon," *J. Electrochem. Soc.*, 137, 273–279, 1990.

Mulhern, G. T., D. S. Soane, and R. T. Howe, "Supercritical Carbon Dioxide Drying of Micro-structures," 7th International Conference on Solid-State Sensors and Actuators (Transducers '93), Yokohama, Japan, 1993, pp. 296–9.

Muller, R. S., "From IC's to Microstructures: Materials and Technologies," Proceedings of the IEEE Micro Robots and Teleoperators Workshop, Hyannis, MA, 1987, pp. 2/1–5.

Muraoka, H., T. Ohashi, and T. Sumitomo, "Controlled Preferential Etching Technology," Semiconductor Silicon 1973, Chicago, IL, 1973, pp. 327–338.

Murarka, S. P. and T. F. J. Retajczyk, "Effect of Phosphorous Doping on Stress in Silicon and Polycrystalline Silicon," *J. Appl. Phys.*, 54, 2069–2072, 1983.

Nakamura, M., S. Hoshino, and H. Muro, "Monolithic Sensor Device for Detecting Mechanical Vibration," Densi Tokyo, 24 (IEE Tokyo Section), 87–8, 1985.

Nathanson, H. C., W. E. Newell, R. A. Wickstrom, and J. R. Davis, "The Resonant Gate Transistor," *IEEE Trans. Electron Devices*, ED–14, 117–133, 1967.

Neudeck, G. W. and et. al., "Three Dimensional Devices Fabricated by Silicon Epitaxial Lateral Overgroth," *J. Electron. Mater.*, 19, 1111–1117, 1990.

Nikanorov, S. P., Y. A. Burenkov, and A. V. Stepanov, "Elastic Properties of Silicon," *Sov. Phys.-Solid State*, 13, 2516–2518, 1972.

Nishioka, T., Y. Shinoda, and Y. Ohmachi, "Raman Microprobe Analysis of Stress in Ge and GaAs/Ge on Silicon Dioxide-Coated Silicon Substrates," *J. Appl. Phys.*, 57, 276–81, 1985.

North, J. C., T. E. McGahan, D. W. Rice, and A. C. Adams, "Tapered Windows in Phosphorous-Doped Silicon Dioxide by Ion Implantation," *IEEE Trans. Electron Devices*, ED–25, 809–12, 1978.

Noworolski, J. M., E. Klaassen, J. Logan, K. Petersen, and N. Maluf, "Fabrication of SOI Wafers With Buried Cavities Using Silicon Fusion Bonding and Electro-chemical Etchback," 8th International Conference on Solid-State Sensors and Actuators (Transducers '95), Stockholm, Sweden, June 1995, pp. 71–4.

Nunn, T. and J. Angell, "An IC Absolute Pressure Transducer with Built–in Reference Chamber," Workshop on Indwelling Pressure Transducers and Systems, Cleveland, OH, 1975, pp. 133–136.

O'Neill, P., "A Monolithic Thermal Converter," *Hewlett-Packard J.*, 31, 12–13, 1980.

Obermeier, E., "High Temperature Microsensors Based on Polycrystalline Diamond Thin Films," 8th International Conference on Solid-State Sensors and Actuators (Transducers '95), Stockholm, Sweden, June 1995, pp. 178–81.

Obermeier, E. and P. Kopystynski, "Polysilion as a Material for Microsensor Applications," *Sensors and Actuators A*, 30, 149–155, 1992.

Offereins, H. L., H. Sandmaier, B. Folkmer, U. Steger, and W. Lang, "Stress Free Assembly Technique for a Silicon Based Pressure Sensor," 6th International Conference on Solid-State Sensors and Actuators (Transducers '91), San Francisco, CA, 1991, pp. 986–9.

Orpana, M. and A. O. Korhonen, "Control of Residual Stress in Polysilicon Thin Films by Heavy Doping in Surface Micromachining," 6th International Conference on Solid-State Sensors and Actuators (Transducers '91), San Francisco, CA, 1991, pp. 957–60.

Palik, E. D., J. W. Faust, H. F. Gray, and R. F. Green, "Study of the Etch-Stop Mechanism in Silicon," *J. Electrochem. Soc.*, 129, 2051–2059, 1982.

Palik, E. D., H. F. Gray, and P. B. Klein, "A Raman Study of Etching Silicon in Aqueous KOH," *J. Electrochem. Soc.*, 130, 956–959, 1983.

Palik, E. D., V. M. Bermudez, and O. J. Glembocki, "Ellipsometric Study of the Etch-Stop Mechanism in Heavily Doped Silicon," *J. Electrochem. Soc.*, 132, 135–141, 1985.

Palik, E. D., O. J. Glembocki, and J. I. Heard, "Study of Bias-Dependent Etching of Si in Aqueous KOH," *J. Electrochem. Soc.*, 134, 404–409, 1987.

Palik, E. D., O. J. Glembocki, and R. E. Stahlbush, "Fabrication and Characterization of Si Membranes," *J. Electrochem. Soc.*, 135, 3126–3134, 1988.

Peeters, E., "Process Development for 3D Silicon Microstructures, with Application to Mechanical Sensor Design," Ph.D. thesis, Catholic University of Louvain, Belgium, 1994.

Peeters, E., D. Lapadatu, W. Sansen, and B. Puers, "Developments in Etch-Stop Techniques," 4th European Workshop on Micromechanics (MME '93), Neuchatel, Switzerland, 1993a.

Peeters, E., D. Lapadatu, W. Sansen, and B. Puers, "PHET, an Electrodeless Photovoltaic Electrochemical Etch-Stop Technique," 7th International Conference on Solid-State Sensors and Actuators (Transducers '93), Yokohama, Japan, 1993b, pp. 254–257.

Petersen, K. E., "Silicon as a Mechanical Material," *Proc. IEEE,* 70, 420–457, 1982.

Petersen, K., D. Gee, F. Pourahmadi, R. Craddock, J. Brown, and L. Christel, "Surface Micromachined Structures Fabricated with Silicon Fusion Bonding," 6th International Conference on Solid-State Sensors and Actuators (Transducers '91), San Francisco, CA, 1991, pp. 397–399.

Pfann, W. G., "Improvement of Semiconducting Devices by Elastic Strain," *Solid State Electron.,* 3, 261–267, 1961.

Pister, K. S. J., "Hinged Polysilicon Structures with Integrated CMOS TFT's," Technical Digest of the 1992 Solid-State Sensor and Actuator Workshop, Hilton Head Island, SC, 1992, pp. 136–9.

Pool, R., "Microscopic Motor Is a First Step," *Res. News,* October, 379–380, 1988.

Pourahmadi, F., L. Christel, and K. Petersen, "Silicon Accelerometer with New Thermal Self–Test Mechanism," Technical Digest of the 1992 Solid State Sensor and Actuator Workshop, Hilton Head Island, SC, 1992, pp. 122–5.

Pratt, R. I., G. C. Johnson, R. T. Howe, and J. C. Chang, "Micromechanical Structures for Thin Film Characterization," 6th International Conference on Solid-State Sensors and Actuators (Transducers '91), San Francisco, CA, 1991, pp. 205–8.

Pratt, R. I., G. C. Johnson, R. T. Howe, and D. J. J. Nikkel, "Characterization of Thin Films Using Micromechanical Structures," *Mat. Res. Soc. Symp. Proc.,* 276, 197–202, 1992.

Price, J. B., "Anisotropic Etching of Silicon with KOH–H_2O-Isopropyl Alcohol," Semiconductor Silicon 1973, Chicago, IL, 1973, pp. 339–353.

Puers, B. and W. Sansen, "Compensation Structures for Convex Corner Micromachining in Silicon," *Sensors and Actuators A,* 23, 1036–1041, 1990.

Puers, R., "Mechanical Silicon Sensors at K.U. Leuven," Proceeding Themadag: SENSOREN, Rotterdam, Netherlands, 1991, pp. 1–8.

Pugacz-Muraszkiewicz, I. J. and B. R. Hammond, "Application of Silicates to the Detection of Flaws in Glassy Passivation Films Deposited on Silicon Substrates," *J. Vac. Sci. Technol.,* 14, 49–53, 1977.

Raley, N. F., F. Sugiyama, and T. Van Duzer, "(100) Silicon Etch-Rate Dependence on Boron Concentration in Ethylenediamine-Pyrocatechol-Water Solutions," *J. Electrochem. Soc.,* 131, 161–171, 1984.

Rehrig, D. L., "In Search of Precise Epi Thickness Measurements," *Semicond. Int.,* 13, 90–95, 1990.

Reisman, A., M. Berkenbilt, S. A. Chan, F. B. Kaufman, and D. C. Green, "The Controlled Etching of Silicon in Catalyzed Ethylene-Diamine-Pyrocatechol-Water Solutions," *J. Electrochem. Soc.,* 126, 1406–1414, 1979.

Retajczyk, T. F. J. and A. K. Sinha, "Elastic Stiffness and Thermal Expansion Coefficients of Various Refractory Silicides and Silicon Nitride Films," *Thin Solid Films,* 70, 241–247, 1980.

Robbins, H. R. and B. Schwartz, "Chemical Etching of Silicon-I. The System, HF, HNO_3 and H_2O," *J. Electrochem. Soc.,* 106, 505–508, 1959.

Robbins, H. and B. Schwartz, "Chemical Etching of Silicon-II. The System HF, HNO_3, H_2O, and $HC_2C_3O_2$," *J. Electrochem. Soc.,* 107, 108–111, 1960.

Rodgers, T. J., W. R. Hiltpold, J. W. Zimmer, G. Marr, and J. D. Trotter, "VMOS ROM," *IEEE J. Solid-State Circuits,* SC–11, 614–622, 1976.

Rodgers, T. J., W. R. Hiltpold, B. Frederick, J. J. Barnes, F. B. Jenné, and J. D. Trotter, "VMOS Memory Technology," *IEEE J. Solid-State Circuits,* SC–12, 515–523, 1977.

Rosler, R. S., "The Evolution of Commercial Plasma Enhanced CVD Systems," *Solid State Technol.,* June, 67–71, 1991.

Ross, M. and K. Pister, "Micro-Windmill for Optical Scanning and Flow Measurement," Eurosensors VIII (*Sensors and Actuators A*), Toulouse, France, 1994, pp. 576–9.

Sakimoto, M., H. Yoshihara, and T. Ohkubo, "Silicon Nitride Single-Layer X-Ray Mask," *J. Vac. Sci. Technol.,* 21, 1017–1021, 1982.

Samaun, S., K. D. Wise, and J. B. Angell, "An IC Piezo-resistive Pressure Sensor for Biomedical Instrumentation," *IEEE Trans. Biomed. Eng.,* 20, 101–109, 1973.

Sandmaier, H., H. Offereins, L., K. Kuhl, and W. Lang, "Corner Compensation Techniques in Anisotropic Etching of (100)-Silicon Using Aqueous KOH," 6th International Conference on Solid-State Sensors and Actuators (Transducers '91), San Francisco, CA, 1991, pp. 456–459.

Saro, P. M. and A. W. van Herwaarden, "Silicon Cantilever Beams Fabricated by Electrochemically Controlled Etching for Sensor Applications," *J. Electrochem. Soc.,* 133, 1722–1729, 1986.

Scheeper, P. R., W. Olthuis, and P. Bergveld, "Fabrication of a Subminiature Silicon Condenser Microphone Using the Sacrificial Layer Technique," 6th International Conference on Solid-State Sensors and Actuators (Transducers '91), San Francisco, CA, 1991, pp. 408–11.

Scheeper, P. R., W. Olthuis, and P. Bergveld, "The Design, Fabrication, and Testing of Corrugated Silicon Nitride Diaphragms," *J. Microelectromech. Syst.,* 3, 36–42, 1994.

Schimmel, D. G. and M. J. Elkind, "An Examination of the Chemical Staining of Silicon," *J. Electrochem. Soc.,* 125, 152–155, 1973.

Schnable, G. L. and P. F. Schmidt, "Applications of Electrochemistry to Fabrication of Semiconductor Devices," *J. Electrochem. Soc.,* 123, 310C–315C, 1976.

Schnakenberg, U., W. Benecke, and B. Lochel, "NH_4OH-Based Etchants for Silicon Micromachining," *Sensors and Actuators A,* 23, 1031–1035, 1990.

Schubert, P., J and G. W. Neudeck, "Confined Lateral Selective Epitaxial Growth of Silicon for Device Fabrication," *IEEE Electron Device Lett.,* 11, 181–183, 1990.

Schumacher, A., H.-J. Wagner, and M. Alavi, "Mit Laser und Kalilauge," *Tech. Rundsch.,* 86, 20–23, 1994.

Schwartz, B. and H. R. Robbins, "Chemical Etching of Silicon–III. A Temperature Study in the Acid System," *J. Electrochem. Soc.,* 108, 365–372, 1961.

Schwartz, B. and H. Robbins, "Chemical Etching of Silicon-IV. Etching Technology," *J. Electrochem. Soc.,* 123, 1903–1909, 1976.

Searson, P. C., S. M. Prokes, and O. J. Glembocki, "Luminescence at the Porous Silicon/Electrolyte Interface," *J. Electrochem. Soc.,* 140, 3327–3331, 1993.

Secco d'Aragona, F., "Dislocation Etch for (100) Planes in Silicon," *J. Electrochem. Soc.,* 119, 948–951, 1972.

Seidel, H., Ph.D. thesis, FU Berlin, Germany, 1986.

Seidel, H., "Nasschemische Tiefenatztechnik," in *Mikromechanik,* Heuberger, A., Ed., Springer, Heidelberg, 1989.

Seidel, H., "The Mechanism of Anisotropic Electro-chemical Silicon Etching in Alkaline Solutions," Technical Digest of the 1990 Solid-State Sensor and Actuator Workshop, Hilton Head Island, SC, 1990, pp. 86–91.

Seidel, H. and L. Csepregi, "Advanced Methods for the Micromachining of Silicon," 7th Sensor Symposium. Technical Digest, Tokyo, Japan, 1988, pp. 1–6.

Seidel, H., L. Csepregi, A. Heuberger, and H. Baumgartel, "Anisotropic Etching of Crystalline Silicon in Alkaline Solutions–Part I. Orientation Dependence and Behavior of Passivation Layers," *J. Electrochem. Soc.*, 137, 3612–3626, 1990a.

Seidel, H., L. Csepregi, A. Heuberger, and H. Baumgartel, "Anisotropic Etching of Crystalline Silicon in Alkaline Solutions-Part II. Influence of Dopants," *J. Electrochem. Soc.*, 137, 3626–3632, 1990b.

Senturia, S., "Can we Design Microrobotic Devices Without Knowing the Mechanical Properties of Materials?" Proceedings of the IEEE Micro Robots and Teleoperators Workshop, Hyannis, MA, 1987, pp. 3/1–5.

Senturia, S. D. and R. T. Howe, "Mechanical Properties and CAD," Lecture Notes, MIT, Boston, MA, 1990.

Shengliang, Z., Z. Zongmin, and L. Enke, "The NH_4F Electrochemical Etching Method of Silicon Diaphragm for Miniature Solid-State Pressure Transducer," 4th International Conference on Solid-State Sensors and Actuators (Transducers '87), Tokyo, Japan, 1987, pp. 130–133.

Shimizu, T. and S. Ishihara, "Effect of SiO_2 Surface Treatment on the Solid-Phase Crystallization of Amorphous Silicon Films," *J. Electrochem. Soc.*, 142, 298–302, 1995.

Shockley, W., *Method of making Thin Slices of Semiconductive Material*, U.S. Patent 3,096,262, 1963.

Singer, P., "Film Stress and How to Measure It," *Semicond. Int.*, 15, 54–8, 1992.

Sinha, A. K. and T. E. Smith, "Thermal Stresses and Cracking Resistance of Dielectric Films," *J. Appl. Phys.*, 49, 2423–2426, 1978.

Slater, T., *Vertical (100) Etching*, personal communication, October 1995.

Smith, C. S., "Piezoresistance Effect in Germanium and Silicon," *Phys. Rev.*, 94, 42–49, 1954.

Smith, J., S. Montague, J. Sniegowski, and P. McWhorter, "Embedded Micromechanical Devices for Monolithic Integration of MEMs with CMOS," IEEE International Electron Devices Meeting. Technical Digest, IEDM '95, Washington, D.C., 1995, pp. 609–12.

Smith, R. L. and S. D. Collins, "Thick Films of Silicon Nitride," *Sensors and Actuators A*, 23, 830–834, 1990.

Spear, W. E. and P. G. Le Comber, "Substitutional Doping of Amorphous Silicon," *Solid State Commun.*, 17, 1193–1196, 1975.

Spiering, V. L., S. Bouwstra, J. Burger, and M. Elwenspoek, "Membranes Fabricated with a Deep Single Corrugation for Package Stress Reduction and Residual Stress Relief," 4th European Workshop on Micromechanics (MME '93), Neuchatel, Switzerland, 1993, pp. 223–7.

Spiering, V. L., S. Bouwstra, R. M. E. J. Spiering, and M. Elwenspoek, "On-Chip Decoupling Zone for Package-Stress Reduction," 6th International Conference on Solid-State Sensors and Actuators (Transducers '91), San Francisco, CA, 1991, pp. 982–5.

Steinsland, E., M. Nese, A. Hanneborg, R. W. Bernstein, H. Sandmo, and G. Kittilsland, "Boron Etch-Stop in TMAH Solutions," 8th International Conference on Solid-State Sensors and Actuators (Transducers '95), Stockholm, Sweden, June 1995, pp. 190–193.

Stoller, A. I., "The Etching of Deep Vertical–Walled Patterns in Silicon," *RCA Rev.*, 31, 271–275, 1970.

Stoller, A. I. and N. E. Wolff, "Isolation Techniques for Integrated Circuits," Proc. Second International Symp. on Microelectronics, Munich, Germany, 1966.

Stoller, A. I., R. F. Speers, and S. Opresko, "A New Technique for Etch Thinning Silicon Wafers," *RCA Rev.*, 265–270, 1970.

Sugiyama, S., T. Suzuki, K. Kawahata, K. Shimaoka, M. Takigawa, and I. Igarashi, "Micro-Diaphragm Pressure Sensor," IEEE International Electron Devices Meeting. Technical Digest, IEDM '86, Los Angeles, CA, 1986, pp. 184–7.

Sugiyama, S., K. Kawakata, M. Abe, H. Funabashi, and I. Igarashi, "High-Resolution Silicon Pressure Imager with CMOS Processing Circuits," 4th International Conference on Solid-State Sensors and Actuators (Transducers '87), Tokyo, Japan, 1987, pp. 444–447.

Sundaram, K. B. and H.-W. Chang, "Electrochemical Etching of Silicon by Hydrazine," *J. Electrochem. Soc.,* 140, 1592–1597, 1993.

Suzuki, K., I. Shimoyama, and H. Miura, "Insect–Model Based Microrobot with Elastic Hinges," *J. Microelectromech. Syst.,* 3, 4–9, 1994.

Tabata, O., "pH-Controlled TMAH Etchants for Silicon Micromachining," 8th International Conference on Solid-State Sensors and Actuators (Transducers '95), Stockholm, Sweden, June 1995, pp. 83–86.

Tabata, O., R. Asahi, and S. Sugiyama, "Anisotropic Etching with Quarternary Ammonium hydroxide Solutions," 9th Sensor Symposium. Technical Digest, Tokyo, Japan, 1990, pp. 15–18.

Tabata, O., R. Asahi, H. Funabashi, K. Shimaoka, and S. Sugiyama, "Anisotropic Etching of Silicon in TMAH Solutions," *Sensors and Actuators A,* 34, 51–57, 1992.

Takeshima, N., K. J. Gabriel, M. Ozaki, J. Takahashi, H. Horiguchi, and H. Fujita, "Electrostatic Parallelogram Actuators," 6th International Conference on Solid-State Sensors and Actuators (Transducers '91), San Francisco, CA, 1991, pp. 63–6.

Tan, T. Y., H. Foll, and S. M. Hu, "On the Diamond-Cubic to Hexagonal Phase Transformation in Silicon," *Phil. Mag. A,* 44, 127–140, 1981.

Tang, W. C.–K., "Electrostatic Comb Drive for Resonant Sensor and Actuator Applications," Ph.D. thesis, University of California at Berkeley, 1990.

Tang, W. C., T. H. Nguyen, and R. T. Howe, "Laterally Driven Polysilicon Resonant Microstructures," *Sensors and Actuators A,* 20, 25–32, 1989.

Tang, W. C., T.-C. H. Nguyen, and R. T. Howe, "Laterally Driven Polysilicon Resonant Microstructures," Proceedings IEEE Micro Electro Mechanical Systems, (MEMS '89), Salt Lake City, UT, 1989, pp. 53–9.

Tenney, A. S. and M. Ghezzo, "Etch Rates of Doped Oxides in Solutions of Buffered HF," *J. Electrochem. Soc.,* 120, 1091–1095, 1973.

Ternez, L., Ph.D. thesis, Uppsala University, 1988.

Theunissen, M. J. J., "Etch Channel Formation During Anodic Dissolution of n-type Silicon in Aqueous Hydrofluoric Acid," *J. Electrochem. Soc.,* 119, 351–360, 1972.

Theunissen, M. J., J. A. Apples, and W. H. C. G. Verkuylen, "Applications of Preferential Electrochemical Etching of Silicon to Semiconductor Device Technology," *J. Electrochem. Soc.,* 117, 959–965, 1970.

Tuck, B., "Review-The Chemical Polishing of Semiconductors," *J. Mater. Sci.,* 10, 321–339, 1975.

Tufte, O. N., P. W. Chapman, and D. Long, " Silicon Diffused–element Piezoresistive Diaphragms," *J. Appl. Phys.,* 33, 3322, 1962.

Turner, D. R., "Electropolishing Silicon in Hydrofluoric Acid Solutions," *J. Electrochem. Soc.,* 105, 402–408, 1958.

Uhlir, A., "Electrolytic Shaping of Germanium and Silicon," *Bell Syst. Tech. J.,* 35, 333–347, 1956.

Unagami, T., "Formation Mechanism of Porous Silicon Layer by Anodization in HF Solution," *J. Electrochem. Soc.,* 127, 476–483, 1980.

van der Drift, A., "Evolutionary Selection, A Principle Governing Growth Orientation in Vapour–Deposited Layers," *Philips Res. Rep.,* 22, 267–288, 1967.

van Dijk, H. J. A., "Method of Manufacturing a Semiconductor Device and Semiconductor Device Manufactured by Said Method," U.S. Patent 3,640,807, 1972.

van Dijk, H. J. A. and J. de Jonge, "Preparation of Thin Silicon Crystals by Electrochemical Thinning of Epitaxially Grown Structures," *J. Electrochem. Soc.,* 117, 553–554, 1970.

van Mullem, C. J., K. J. Gabriel, and H. Fujita, "Large Deflection Performance of Surface Micromachined Corrugated Diaphragms," 6th International Conference on Solid-State Sensors and Actuators (Transducers '91), San Francisco, CA, 1991, pp. 1014–7.

Varadan, V. K. and P. J. McWhorter, Eds. *Smart Electronics and MEMS, Proceedings of the Smart Structures and Materials 1996 Meeting,* Vol. 2722, SPIE, San Diego, CA, 1996.

Vinci, R. P. and J. C. Braveman, "Mechanical Testing of Thin Films," 6th International Conference on Solid-State Sensors and Actuators (Transducers '91), San Francisco, CA, 1991, pp. 943–8.

Voronin, V. A., A. A. Druzhinin, I. I. Marjamora, V. G. Kostur, and J. M. Pankov, "Laser–Recrystallized Polysilicon Layers in Sensors," *Sensors and Actuators A,* 30, 143–147, 1992.

Waggener, H. A., "Electrochemically Controlled Thinning of Silicon," *Bell. Syst. Tech. J.,* 49, 473–475, 1970.

Waggener, H. A. and J. V. Dalton, "Control of Silicon Etch Rates in Hot Alkaline Solutions by Externally Applied Potentials," *J. Electrochem. Soc.,* 119, 236C, 1972.

Waggener, H. A., R. C. Krageness, and A. L. Tyler, "Two–way Etch," *Electronics,* 40, 274, 1967a.

Waggener, H. A., R. C. Kragness, and A. L. Tyler, "Anisotropic Etching for Forming Isolation Slots in Silicon Beam Leaded Integrated Circuits," IEEE International Electron Devices Meeting. Technical Digest, IEDM '67, Washington, D.C., 1967b, p. 68.

Watanabe, H., S. Ohnishi, I. Honma, H. Kitajima, H. Ono, R. J. Wilhelm, and A. J. L. Sophie, "Selective Etching of Phosphosilicate Glass with Low Pressure Vapor HF," *J. Electrochem. Soc.,* 142, 237–43, 1995.

Watanabe, Y., Y. Arita, T. Yokoyama, and Y. Igarashi, "Formation and Properties of Porous Silicon and Its Applications," *J. Electrochem. Soc.,* 122, 1351–1355, 1975.

Weinberg, M., J. Bernstein, J. Borenstein, J. Campbell, J. Cousens, B. Cunningham, R. Fields, P. Greiff, B. Hugh, L. Niles, and J. Sohn, "Micromachining Inertial Instruments," Micromachining and Micro-fabrication Process Technology II, Austin, TX, 1996, pp. 26–36.

Weirauch, D. F., "Correlation of the Anisotropic Etching of Single-Crystal Silicon Spheres and Wafers," J. Appl. Phys., 46, 1478–1483, 1975.

Wen, C. P. and K. P. Weller, "Preferential Electro-chemical Etching of p+ Silicon in an Aqueous $HF-H_2SO_4$ Electrolyte," *J. Electrochem. Soc.,* 119, 547–548, 1972.

White, L. K., "Bilayer Taper Etching of Field Oxides and Passivation Layers," *J. Electrochem. Soc.,* 127, 2687–93, 1980.

Wise, K. D., "Silicon Micromachining and Its Applications to High Performance Integrated Sensors," in *Micromachining and Micropackaging of Transducers,* Fung, C. D., Cheung, P. W., Ko, W. H. and Fleming, D. G., Eds., Elsevier, New York, 1985, pp. 3–18.

Wise, K. D., M. G. Robinson, and W. J. Hillegas, "Solid State Processes to Produce Hemispherical Components for Inertial Fusion Targets," *J. Vac. Sci. Technol.,* 18, 1179–1182, 1981.

Wong, A., "Silicon Micromachining," 1990, Viewgraphs, presented in Chicago, IL.

Wong, S. M., "Residual Stress Measurements on Chromium Films by X-ray Diffraction Using the sin2 Y Method," *Thin Solid Films,* 53, 65–71, 1978.

Worthman, J. J. and R. A. Evans, "Young's Modulus, Shear Modulus and Poisson's Ratio in Silicon and Germanium," *J. Appl. Phys.,* 36, 153–156, 1965.

Wu, T. H. T. and R. S. Rosler, "Stress in PSG and Nitride Films as Related to Film Properties and Annealing," *Solid State Technol.,* May, 65–71, 1992.

Wu, X. and W. A. Ko, "A Study on Compensating Corner Undercutting in Anisotropic Etching of (100) Silicon," 4th International Conference on Solid-State Sensors and Actuators (Transducers '87), Tokyo, Japan, 1987, pp. 126–129.

Wu, X.-P. and W. H. Ko, "Compensating Corner Undercutting in Anisotropic Etching of (100) Silicon," *Sensors and Actuators A,* 18, 207–215, 1989.

Wu, X. P., Q. H. Wu, and W. H. Ko, "A Study on Deep Etching of Silicon Using EPW," 3rd International Conference on Solid-State Sensors and Actuators (Transducers '85), Philadelphia, PA, 1985, pp. 291–4.

Wu, X. P., Q. H. Wu, and W. H. Ko, "A Study on Deep Etching of Silicon Using Ethylene-Diamine-Pyrocathechol-Water," *Sensors and Actuators A,* 9, 333–343, 1986.

Yamada, K. and T. Kuriyama, "A New Modal Mode Controlling Method for A Surface Format Surrounding Mass Accelerometer," 6th International Conference on Solid-State Sensors and Actuators (Transducers '91), San Francisco, CA, 1991, pp. 655–8.

Yamana, M., N. Kashiwazaki, A. Kinoshita, T. Nakano, M. Yamamoto, and W. C. Walton, "Porous Silicon Oxide Layer Formation by the Electrochemical Treatment of a Porous Silicon Layer," *J. Electrochem. Soc.,* 137, 2925–7, 1990.

Yang, K. H., "An Etch for Delineation of Defects in Silicon," *J. Electrochem. Soc.,* 131, 1140–1145, 1984.

Yeh, R., E. J. Kruglick, and K. S. J. Pister, "Towards an Articulated Silicon Microrobot," ASME 1994, Micromechanical Sensors, Actuators, and Systems, Chicago, IL, 1994, pp. 747–754.

Yoshida, T., T. Kudo, and K. Ikeda, "Photo-Induced Preferential Anodization for Fabrication of Monocrystalline Micromechanical Structures," Proceedings IEEE Micro Electro Mechanical Systems, (MEMS '92), Travemunde, Germany, 1992, pp. 56–61.

Yun, W., "A Surface Micromachined Accelerometer with Integrated CMOS Detection Circuitry," Ph.D. thesis, University of California at Berkeley, 1992.

Zhang, L. M., D. Uttamchandani, and B. Culshaw, "Measurement of the Mechanical Properties of Silicon Microresonators," *Sensors and Actuators A,* 29, 79–84, 1991.

Zhang, X. G., "Mechanism of Pore Formation on n-Type Silicon," *J. Electrochem. Soc.,* 138, 3750–3756, 1991.

Zhang, X. G., S. D. Collins, and R. L. Smith, "Porous Silicon Formation and Electropolishing of Silicon by Anodic Polarization in HF Solution," *J. Electrochem. Soc.,* 136, 1561–5, 1989.

Zhang, Y. and K. D. Wise, "Performance of Non-Planar Silicon Diaphragms under Large Deflections," *J. Microelectromech. Syst.,* 3, 59–68, 1994.

17
LIGA and Other Replication Techniques[1]

Marc J. Madou
Nanogen, Inc.

17.1 Introduction 17-1
17.2 LIGA Processes—Introduction
    to the Technology 17-2
    History • Synchrotron Orbital Radiation—General
    Applications • Synchrotron Radiation—Technical
    Aspects • Access to Technology • X-Ray Masks • LIGA
    Process Steps and Materials
17.3 LIGA Applications 17-42
    Introduction • Microfluidic Elements • Electronic
    Microconnectors and Packages • Micro-Optical Components •
    LIGA Accelerometer • Interlocking Gears • Electrostatic
    Actuators • LIGA Inductors • Electromagnetic Micromotor •
    Other Potential LIGA Applications
17.4 Technological Barriers and Competing
    Technologies 17-60
    Insertion Point of X-Ray Lithography • Competing
    Technologies

17.1 Introduction

LIGA is the German acronym for X-ray lithography (X-ray lithographie), electrodeposition (galvanoformung), and molding (abformtechnik). The process involves a thick layer of X-ray resist—from microns to centimeters—and high-energy X-ray radiation exposure and development to arrive at a three-dimensional resist structure. Subsequent electrodeposition fills the resist mold with a metal and, after resist removal, a free standing metal structure results (see Figure 17.1) [IMM, 1995]. The metal shape may be a final product or serve as a mold insert for precision plastic injection molding. Injection-molded plastic parts may in turn be final products or lost molds. The plastic mold retains the same shape, size and form as the original resist structure but is produced quickly and inexpensively as part of an infinite loop. The plastic lost mold may generate metal parts in a second electroforming process or ceramic parts in a slip casting process.

Micromachining techniques are reshaping manufacturing approaches for a wide variety of small parts; frequently, IC-based, batch microfabrication methods are being considered for their fabrication together with more traditional, serial machining methods. LIGA, as a 'handshake-technology' between IC and classical manufacturing technologies, has the potential of speeding up this process. The method borrows lithography from the IC industry and electroplating and molding from classical manufacturing. The

---

[1] This chapter was originally published in Madou, M.J. (1997) "LIGA" (chap. 6), in *Fundamentals of Microfabrication*, CRC Press, Boca Raton, FL.

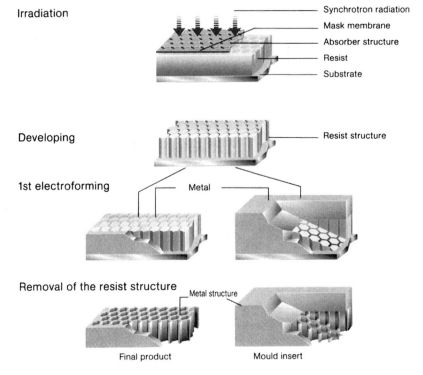

FIGURE 17.1A Basic LGA process steps X-ray deep-etch lithography and 1st electroforming. (From Lehr, H. and M. Schmidt, The LIGA Technique, commercial brocure, IMM Institut für Mikrotechnik GmbH, Mainz-Hechtscheim, 1995. With permission.)

capacity of LIGA for creating a wide variety of shapes from different materials makes it akin to classical machining with the added benefit of unprecedented aspect ratios and absolute tolerances. The LIGA bandwidth of possible sizes in three dimensions makes it potentially useful, not only for microstructure manufacture itself (micron and submicron dimensions), but also for the manufacture of microstructure packages (millimeter and centimeter dimensions) as well as for connectors from those packages (electrical, e.g., through-vias or physical, e.g., gas in- and outlets) to the 'macroworld.'

Given the cost of the LIGA equipment, various pseudo-LIGA processes are under development. They involve replication of molds created by alternate means such as deep, cryogenic dry etching and novel ultraviolet thick photoresists.

Some aspects of X-ray lithography for use in LIGA were explored in Madou (1997, chap. 1). After an historical introduction we will analyze all of the process steps depicted in Figure 17.1 in detail, starting with a description of the different applications and technical characteristics of synchrotron radiation, followed by an introduction to the crucial issues involved in making X-ray masks optimized for LIGA. Alternative process sequences, popular in LIGA micromanufacturing today, will be reviewed as well. After an introduction to the technology we will review current and projected LIGA applications and discuss key technological barriers to be overcome and consider competing 'LIGA-like' processes.

17.2 LIGA Processes—Introduction to the Technology

17.2.1 History

LIGA combines the sacrificial wax molding method, known since the time of the Egyptians, with X-ray lithography and electrodeposition. Combining electrodeposition and X-ray lithography was first carried

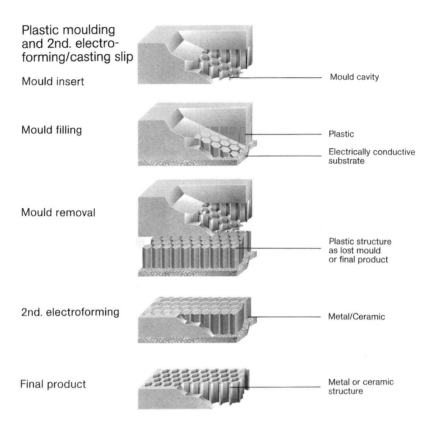

FIGURE 17.1B Plastic molding and 2nd electroforming/slip casting. (From Lehr, H. and M. Schmidt. The LGA Technique, commercial brochure, IMM Institut für Mikrotechnik GmbH, Mainz-Hechtsheim, 1995. With permission.)

out by Romankiw and co-workers at IBM in 1975 [Spiller et al., 1976]. These authors made high aspect ratio metal structures by plating gold in X-ray-defined resist patterns of up to 20 μm thick. They had, in other words, already invented 'LIG'; i.e., LIGA without the abformtech (molding) [Spiller et al., 1997]. This IBM work formed an extension to through-mask plating, also pioneered by Romankiw in 1969, geared toward the fabrication of thin-film magnetic recording heads [Romankiw et al., 1970]. The addition of molding to the lithography and plating process was realized by Ehrfeld et al. [Becker et al., 1982] at KfK in 1982. By adding molding, Ehrfeld and co-workers made the technology potentially low cost. These pioneers recognized the broader implications of LIGA as a new means of low-cost manufacturing of a wide variety of microparts previously impossible to batch fabricate [Becker et al., 1982]. In Germany, LIGA originally developed almost completely outside of the semiconductor industry. It was Guckel in the U.S. who repositioned the field in light of existing semiconductor capabilities and brought it closer to standard manufacturing processes.

The development of the LIGA process initiated by the Karlsruhe Nuclear Research Center (Kernforschungszentrum Karlsruhe, or KfK) was intended for the mass production of micron-sized nozzles for uranium-235 enrichment (see Figure 17.2) [Becker et al., 1982]. The German group used synchrotron radiation from a 2.5-GeV storage ring for the exposure of the poly(methylmethacrylate) (PMMA) resist.

Today LIGA is researched in many laboratories around the world. The technology has reinvigorated attempts at developing alternative micromolds for the large-scale production of precise micromachines. Mold inserts, depending on the dimensions of the microparts, the accuracy requirements and the fabrication costs are realized by e-beam writing, deep UV resists, excimer laser ablation, electrodischarge machining, laser cutting and X-ray lithography as involved in the LIGA technique.

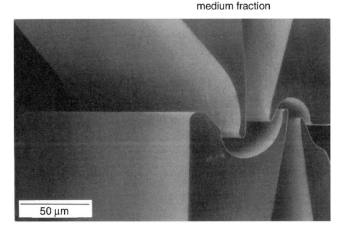

FIGURE 17.2 Scanning electron micrograph of a separation nozzle structure produced by electroforming with nickel using a micromolded PMMA template. This nozzle represents the first actual product ever made by LIGA. (From Hagmann, P. et al., Fabrication of Microstructures with Extreme Structural Heights by Reaction Injection Molding, presented at First Meeting of the European Polymer Federation, European Symp. on Polymeric Materials, Lyon, France, 1987. With permission.)

17.2.2 Synchrotron Orbital Radiation—General Applications

Lithography based on synchrotron radiation, also called synchrotron orbital radiation (SOR), is primarily being developed with the aim of adopting the technology as an industrial tool for the eventual large-scale manufacture of microelectronic circuits with characteristic dimensions in the submicron range [Waldo and Yanof, 1991; Hill, 1991]. Synchrotron radiation sources outshine all other sources of X-rays. They emit a much higher flux of usable collimated X-rays, thereby allowing shorter exposure times and larger throughputs. Pros and cons of X-ray radiation for lithography in IC manufacture, identified in Madou (1997, chap. 1), are summarized in Table 17.1.

Despite the many promising features of X-ray lithography, the technique lacks mainstream acceptance in the IC industry. Continued improvements in optical lithography outpace the industrial use of X-ray

TABLE 17.1 Pros and Cons of SOR X-Ray Lithography for IC Manufacture

Pros	Cons
Lithography process insensitive to resist thickness, exposure time, development time (large depth of focus)	Resist not very sensitive (not too important because of the intense light source)
Absence of backscattering results in insensitivity to substrate type, reflectivity and topography, pattern geometry and proximity, dust and contamination	Masks very difficult and expensive to make
High resolution < 0.2 µm	Very high start-up investment
High throughput	Not proven as a manufacturing system yet
	Radiation effects on SiO_2

lithography for IC applications. Its use for prototype development on a small scale will no doubt continue its course though. In 1991, experts projected that X-ray lithography could be in use by 1995 for 64 megabit DRAM manufacture, with critical dimensions (CDs) around 0.3 to 0.4 µm. With more certainty, they projected that the transition to X-rays would occur with the 0.2- to 0.3-µm CDs of 256 megabit DRAMs, expected to be in production by 1998 [Hill, 1991]. The first projected date has passed without materialization of the industrial use of X-rays; many are doubtful that the second prediction will be met [Stix, 1995].

In addition to the primary intended application—the production of the newest generation of DRAM chips—X-rays may also be used in the fabrication of microsensors and other microstructures, especially since the LIGA method enables the fabrication of high-aspect-ratio microstructures from a wide variety of materials. In LIGA, synchrotron radiation is used in the lithography step only. Other micromachining applications for SOR exist. Urisu and his colleagues, for example, explored the use of synchrotron radiation for radiation-excited chemical-vapor deposition and etching [Urisu and Kyuragi, 1987]. Micromachinists are well aware of the need to piggy-back X-ray lithography research and development efforts for the fabrication of micromachines onto major IC projects. The use of X-ray lithography for fabricating microdevices, other than integrated circuits, does not appear to present a large business opportunity yet. Not having a major IC product line associated with X-ray lithography makes it extra hard to justify the use of X-ray lithography for micromachining, especially since other, less expensive micromachining technologies have not yet opened up the type of mass markets one is used to in the IC world. The fact that the X-rays used in LIGA are shorter wavelength than in the IC application (2 to 10 Å vs. 20 to 50 Å) also puts micromachinists at a disadvantage. For example, the soft X-rays in the IC industry may eventually be generated from a much less expensive source, such as a transition radiation source [Goedtkindt et al., 1991]. Also, nontraditional IC materials are frequently employed in LIGA. Fabricating X-ray masks pose more difficulties than masks for IC applications. Rotation and slanting of the X-ray masks may be required to craft nonvertical walls. All these factors make exploring LIGA in the current economic climate a major challenge. However, given sufficient research and development money, large markets are almost certain to emerge over the next 10 years. These markets may be found in the manufacture of devices with stringent requirements imposed on resolution, aspect ratio, structural height and parallelism of structural walls. Optical applications seem particularly important product targets. The commercial exploitation of LIGA is being aggressively pursued by two German companies: Microparts[2] (formed in 1990) and IMM[3] (formed in 1991).

So far, the research community has primarily benefited from the availability of SOR photon sources. With its continuously tunable radiation across a very wide photon range, also highly polarized and directed into a narrow beam, SOR provides a powerful probe of atomic and molecular resonances. Other types of photon sources prove unsatisfactory for these applications in terms of intensity or energy spread. As can be concluded from Table 17.2, applications of SOR beyond lithography range from structural and chemical analysis to microscopy, angiography and even to the preparation of new materials.

---

[2] Dortmund, Griesbachstrasse 10, D-7500 Karlsruhe 21; tel: 0721/84990; fax: 0721/857865.
[3] Mainz-Hechtsheim, Carl-Zeiss-Str. 18-20, D-55129; tel: 06131/990-0; fax: 06131/990-205.

TABLE 17.2 SOR Applications

Application Area	Instruments/Technologies Needed
Structural analysis	
Atoms	Photoelectron spectrometers
Molecules	Absorption spectrometers
Very large molecules	Fluorescent spectrometers
Proteins	Diffraction cameras
Cells	Scanning electron microscope (to view topographical radiographs)
Polycrystals	Time-resolved X-ray diffractometers
Chemical analysis	
Trace	Photoelectron spectrometers
Surface	(Secondary ion) mass spectrometer
Bulk	Absorption/fluorescence spectrometers
	Vacuum systems
Microscopy	
Photoelectron	Photoemission microscopes
X-ray	X-ray microscopes SEM (for viewing)
	Vacuum systems
Micro/nanofabrication	
X-ray lithography	Steppers, mask making
Photochemical deposition of thin films	Vacuum systems
Etching	LIGA process
Medical diagnostics	
Radiography	X-ray cameras and equipment
Angiography and tomography	Computer-aided display
Photochemical reactions	
Preparation of novel materials	Vacuum systems
	Gas-handling equipment

Source: After Muray, J.J. and I. Brodie, Study on Synchroton Orbital Radiation (SOR) Technologies and Applications, SRI International 2019, 1991.

17.2.3 Synchrotron Radiation—Technical Aspects

Some important concepts associated with synchrotron radiation, such as the bending radius of the synchrotron magnet, magnetic field strength, beam current, critical wavelength and total radiated power, require introduction. Figure 17.3 presents a schematic of an X-ray exposure station. The cone of radiation in this figure is the electromagnetic radiation emitted by electrons due to the radial acceleration that keeps them in orbit in an electron synchrotron or storage ring. For high energy particle studies, this radiation, emitted tangential to the circular electron path, limits the maximum energy the electrons can attain. This so-called 'Bremsstrahlung' is a nuisance for studies of the composition of the atomic nucleus in which high energy particles are smashed into the nucleus. To minimize this problem, physicists desire ever bigger synchrotrons. For X-ray lithography applications, electrical engineers want to maximize the X-ray emission and build small synchrotrons instead (the radius of curvature for a compact, superconducting synchrotron, for example, is 2 m). The angular opening of the radiation cone in Figure 17.3 is determined by the electron energy, E, and is given by:

$$\theta \approx \frac{mc^2}{E_*} = \frac{0.5}{E(\text{GeV})}(\text{mard}) \qquad (17.1)$$

The X-ray light bundle with the cone opening, θ, describes a horizontal line on an intersecting substrate as the X-ray bundle is tangent to the circular electron path. In the vertical direction, the intensity of the

LIGA and Other Replication Techniques

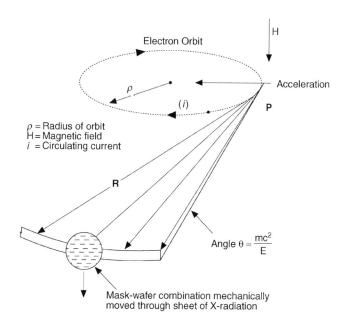

FIGURE 17.3 Schematic of an X-ray exposure station with a synchrotron radiation source. The X-ray radiation cone (opening, θ) is tangential to the electron's path, describing a line on an intersecting substrate.

beam exhibits a Gaussian distribution and the vertical exposed height on the intersecting substrate can be calculated knowing θ and R, the distance from the radiation point, P, to the substrate. With $E = 1$ GeV, $\theta = 1$ mrad and $R = 10$ m, the exposed area in the vertical direction measures about 0.5 cm. To expose a substrate homogeneously over a wider vertical range, the sample must be moved vertically through the irradiation band with a precision scanner (e.g., done at a speed of 10 mm/sec over a 100-mm scanning distance). Usually the substrate is stepped up and down repeatedly until the desired X-ray dose is obtained.

The electron energy E in Eq. (17.1) is given by:

$$E(\text{SeV}) = 0.29979\ B\ (\text{Tesla})\ \rho\ (\text{meters}) \quad (17.2)$$

with B the magnetic field and ρ the radius of the circular path of the electrons in the synchrotron.

The total radiated power can be calculated from the energy loss of the electrons per turn and is given by:

$$P(\text{kW}) = \frac{88.47 E^4 i}{\rho} \quad (17.3)$$

with i the beam current.

The spectral emission of the synchrotron electrons is a broad spectrum without characteristic peaks or line enhancements, and its distribution extends from the microwave region through the infrared, visible, ultraviolet and into the X-ray region. The critical wavelength, λ_c, is defined so that the total radiated power at lower wavelengths equals the radiated power at higher wavelengths, and is given by:

$$\lambda_c(\text{Å}) = \frac{5.59 \rho(\text{m})}{E^3(\text{GeV})} \quad (17.4)$$

TABLE 17.3 Access to Synchrotron Radiation Available at the Following Facilities in the U.S.

Facility	Institute	Contact (1997)
Advanced Photon Source	Argonne National Laboratory	Susan Barr, Derek Mancini
Cornell High Energy Synchrotron Source	Cornell University	Proposal Administrator
National Synchrotron Light Source	Brookhaven National Laboratory	Susan White-DePace
Standard Synchrotron Radiation Laboratory	Stanford University	Katherine Cantwell
SURF	National Institute of Standards and Technology	Robert Madden
Synchrotron Radiation Center	University of Wisconsin-Madison	Pamela Layton
Center for Advanced Microdevices (CAMD)	Louisiana State University	Volker Saile
Advanced Light Source (ALS)	Lawrence Berkeley Laboratory	Alfred Schlachter

Equation 17.3 shows that the total radiated power increases with the fourth power of the electron energy. From Eq. (17.4) we appreciate that the spectrum shifts toward shorter wavelengths with the third power of the electron energy.

The dose variation absorbed in the top vs. the bottom of an X-ray resist should be kept small so that the top layer does not deteriorate before the bottom layers are sufficiently exposed. Since the depth of penetration increases with decreasing wavelength, synchrotron radiation of very short wavelength is needed to pattern thick resist layers. In order to obtain good aspect ratios in LIGA structures, the critical wavelength ideally should be 2 Å. Bley et al. (1992) at KfK designed a new synchrotron optimized for LIGA. They proposed a magnetic flux density, B, of 1.6285 T; a nominal energy, E, of 2.3923 GeV; and a bending radius, ρ, of 4.9 m. With those parameters, Eq. (17.4) results in the desired λ_c of 2 Å and an opening angle of radiation, based on Eq. (17.1), of 0.2 mrad (in practice this angle will be closer to 0.3 mrad due to electron beam emittance).

The X-rays from the ring to the sample site are held in a high vacuum. The sample itself is either kept in air or in a He atmosphere. The inert atmosphere prevents corrosion of the exposure chamber, mask and sample by reactive oxygen species, and removal of heat is much faster than in air (the heat conductivity of He is high compared to air). In He, the X-ray intensity loss is also 500 times less than in air. A beryllium window separates the high vacuum from the inert atmosphere. For wavelengths shorter than 1 nm, Be is very transparent—i.e., an excellent X-ray window.

17.2.4 Access to Technology

Unfortunately, the construction cost for a typical synchrotron today totals over $30 million, restricting the access to LIGA dramatically. Obviously, one would like to find less expensive alternatives for generating intense X-rays. In Japan, companies such as Ishikawajima-Harima Heavy Industries (IHI) are building compact synchrotron X-ray sources (e.g., a 800-MeV synchrotron will be about 30 feet per side).

By the end of 1993, eight nonprivately owned synchrotrons were in use in the U.S. The first privately owned synchrotron was put into service in 1991 at IBM's Advanced Semiconductor Technology Center (ASTC) in East Fishkill, NY. Table 17.3 lists the eight U.S. synchrotron facilities.

Some of the facilities listed in Table 17.3 can perform micromachining work. For example, MCNC (Research Triangle Park, NC), in collaboration with the University of Wisconsin–Madison, announced its first multi-user LIGA process sponsored by ARPA in September 1993. CAMD at Louisiana State University has three beam lines on their synchrotron exclusively dedicated to micromachining work; ALS at Berkeley has one beam line available for micromachining.

Forschungszentrum Karlsruhe GmbH[4] also offers a multi-user LIGA service (LEMA, or LIGA-experiment for multiple applications).

---

[4] Forschungszentrum Karlsruhe GmbH; Postfach 3640; D-76021 Karlsruhe.

17.2.5 X-Ray Masks

17.2.5.1 Introduction

A short introduction to X-ray masks was presented in Madou (1997, chap. 1 on lithography). X-ray masks should withstand many exposures without distortion, be alignable with respect to the sample and be rugged. A possible X-ray mask architecture and its assembly with a substrate in an X-ray scanner are shown in Figure 17.4. The mask shown here has three major components: an absorber, a membrane or mask blank, and a frame. The absorber contains the information to be imaged onto the resist. It is made up of a material with a high atomic number (Z), often Au, patterned onto a membrane material with a low Z. The high-Z material absorbs X-rays, whereas the low-Z material transmits X-rays. The frame lends robustness to the membrane/absorber assembly so that the whole can be handled confidently.

The requirements for masks in LIGA differ substantially from those for the IC industry. A comparison is presented in Table 17.4 [Ehrfeld et al., 1986]. The main difference lies in the absorber thickness. In order to achieve a high contrast (>200), very thick absorbers (>10 µm vs. 1 µm) and highly transparent mask blanks (transparency > 80%) must be used because of the low resist sensitivity and the great depth of resist. Another difference focuses on the radiation stability of membrane and absorber. For conventional optical lithography, the supporting substrate is a relatively thick, near optically flat piece of glass or quartz highly transparent to optical wavelengths. It provides a highly stable (>10^6 µm) basis for the thin (0.1 µm) chrome absorber pattern. In contrast, the X-ray mask consists of a very thin membrane (2 to 4 µm) of

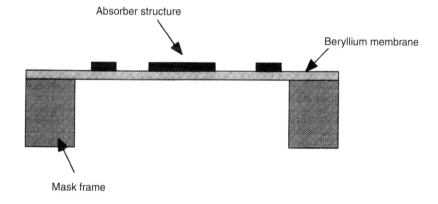

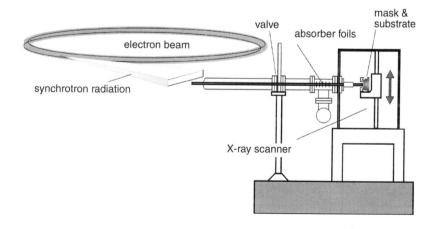

FIGURE 17.4 Schematic of a typical X-ray mask and mask and substrate assembly in an X-ray scanner. (The latter from IMM, 1995 Brochure. With permission.)

TABLE 17.4 Comparison of Masks in LIGA and the IC Industry
[Ehrfeld et al., 1986]

Attribute	Semiconductor Lithography	LIGA Process
Transparency	≥50%	≥80%
Absorber thickness	±1 μm	10 μm or higher
Field size	50 × 50 mm^2	100 × 100 mm^2
Radiation resistance	= 1	= 100
Surface roughness	<0.1 μm	<0.5 μm
Waviness	< ±1 μm	< ±1 μm
Dimensional stability	<0.05 μm	<0.1–0.3 μm
Residual membrane stress	~10^8 Pa	~10^8 Pa

low-Z material carrying a high-Z thick absorber pattern [Lawes, 1989]. A single exposure in LIGA results in an exposure dose a hundred times higher than in the IC case.

We will look into these different mask aspects separately before detailing a process with the potential of obviating the need for a separate X-ray mask by using a so-called transfer mask.

17.2.5.2 X-Ray Membrane (Mask Blank)

The low-Z membrane material in an X-ray mask must have a transparency for rays with a critical wavelength, λ_c, from 0.2 to 0.6 nm of at least 80% and should not induce scattering of those rays. To avoid pattern distortion, the residual stress, σ_r, in the membrane should be less than 10^8 dyn/cm^2. Mechanical stress in the absorber pattern can cause in-plane distortion of the supporting thin membrane, necessitating a high Young's modulus for the membrane material. Humidity or high deposited doses of X-ray might also distort the membrane. During one typical lithography step the masks may be exposed to 1 MJ/cm^2 of X-rays. Since most membranes must be very thin for optimum transparency, a compromise must be found between transparency, strength and form stability. Important X-ray membrane materials are listed in Table 17.5. The higher radiation dose in LIGA prevents the use of BN as well as compound mask blanks which incorporate a polyimide layer. Those mask blanks are perfectly appropriate for classical IC lithography work but will not do for LIGA work. Mask blanks of metals such as titanium (Ti) and beryllium (Be) were specifically developed for LIGA applications because of their radiation hardness [Ehrfeld et al., 1986; Schomburg et al., 1991]. In comparing titanium and beryllium membranes, beryllium can have a much greater membrane thickness, d, and still be adequately transparent. For example, a membrane transparency of 80%, essential for adequate exposure of a 500-μm-thick PMMA resist layer, can be achieved with a thin 2-μm titanium film, whereas with beryllium a thick 300-μm membrane must

TABLE 17.5 Comparison of Membrane Materials for X-Ray Masks

Material	X-Ray Transparency	Nontoxicity	Dimensional Stability	Remark
Si	0	++	0	Single crystal Si, well developed, rad hard, stacking faults cause scattering, material is brittle
SiN$_x$	0	++	0	Amorphous, well developed, rad hard if free of oxygen, resistant to breakage
SiC	+	++	++	Poly and amorphous, rad hard, some resistance to breakage
Diamond	+	++	++	Poly, research only, highest stiffness
BN	+	++	0	Not rad hard, i.e., not applicable for LIGA
Be	++	−	++	Research, especially suited for LIGA; even at 100 μm the transparency is good, 30 μm typical; difficult to electroplate; toxic material
Ti	−	++	0	Research, used for LIGA, not very transparent, films must not be more than 2 to 3 μm thick

TABLE 17.6 Comparison of Absorber Materials for X-Ray Masks

Material	Remark
Gold	Not the best stability (grain growth), low stress, electroplating only, defects repairable
Tungsten	Refractory and stable; special care is needed for stress control; dry etchable; repairable
Tantalum	Refractory and stable; special care is needed for stress control; dry etchable; repairable
Alloys	Easier stress control, greater thickness required to obtain 10 db

be used. The beryllium membrane can thus be used at a thickness permitting easier processing and handling. In addition, beryllium has a greater Young's modulus E than titanium (330 vs. 140 kN/mm^2) and, since it is the product of $E \times d$ which determines the amount of mask distortion, distortions due to absorber stress should be much smaller for beryllium blanks [Schomburg et al., 1991; Hein et al., 1992]. Beryllium comes forward as an excellent membrane material for LIGA because of its high transparency and excellent damage resistance. Such a mask should be good for up to 10,000 exposures and may cost $20 to 30 K ($10 to 15 K in quantity). Stoichiometric silicon nitride (Si$_3$N$_4$) used in X-ray mask membranes may contain numerous oxygen impurities absorbing X-rays and thus producing heat. This heat often suffices to prevent the use of nitride as a good LIGA mask. Single crystal silicon masks have been made (1 cm^2 and 0.4 µm thick and 10 cm^2 and 2.5 µm thick) by electrochemical etching techniques. For Si and Si$_3$N$_4$, the Young's modulus is quite low compared to CVD-grown diamond and SiC films, which are as much as three times higher. These higher stiffness materials are more desirable because the internal stresses of the absorbers, which can distort mask patterns, are less of an issue.

17.2.5.3 Absorber

The requirements on the absorber are high attenuation (>10 db), stability under radiation over extended periods of time, negligible distortion (stress < 10^8 dyn/cm^2), ease of patterning, repairable and low defect density. A listing of typical absorber materials is presented in Table 17.6. Gold is used most commonly. Several groups are looking at the viability of tungsten and other materials. In the IC industry an absorber thickness of 0.5 µm might suffice, whereas LIGA deals with thicker layers of resist requiring a thicker absorber to maintain the same resolution.

Figure 17.5 illustrates how X-rays, with a characteristic wavelength of 0.55 nm, are absorbed along their trajectory through a Kapton preabsorber filter, an X-ray mask and resist [Bley et al., 1991]. The low energy portion of the synchrotron radiation is absorbed mainly in the top portion of the resist layer, since absorption increases with increasing wavelength. The Kapton preabsorber filters out much of the low energy radiation in order to prevent overexposure of the top surface of the resist. The X-ray dose at which the resist gets damaged, D_{dm}, and the dose required for development of the resist, D_{dv}, as well as the 'threshold-dose' at which the resist starts dissolving in a developer, D_{th}, are all indicated in Figure 17.5. In the areas under the absorber pattern of the X-ray mask, the absorbed dose must stay below the 'threshold-dose', D_{th}; otherwise, the structures partly dissolve, resulting in poor feature definition. From Figure 17.5 we can deduce that the height of the gold absorbers must exceed 6 µm to reduce the absorbed radiation dose of the resist under the gold pattern to below the threshold dose, D_{th}. In Figure 17.6, the necessary thickness of the gold absorber patterns of an X-ray mask is plotted as a function of the thickness of the resist to be patterned; the Au must be thicker for thicker resist layers and for shorter characteristic wavelengths, λ_c, of the X-ray radiation. In order to pattern a 500-µm high structure with a λ_c of 0.225 nm, the gold absorber must be more than 11 µm high.

Exposure of yet more extreme photoresist thicknesses is possible if proper X-ray photon energies are used. At 3000 eV, the absorption length in PMMA roughly measures 100 µm, which enables the above-mentioned 500-µm exposure depth [Guckel et al., 1994]. Using 20,000-eV photons results in absorption lengths of 1 cm. PMMA structures up to 10 cm (!) thick have been exposed this way [Siddons and Johnson, 1994]. A high energy mask used by Guckel for these high energy exposures has an Au absorber 50 µm thick and a blank membrane of 400-µm-thick Si. Guckel obtained an absorption contrast of 400 when exposing a 1000-µm-thick PMMA sheet with this mask. An advantage of being able to use these

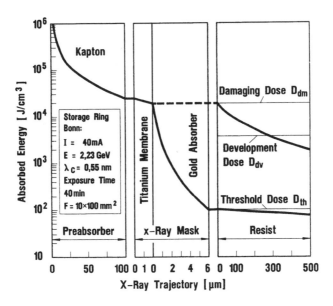

FIGURE 17.5 Absorbed energy along the X-ray trajectory including a 500-μm-thick PMMA specimen, X-ray mask and a Kapton preabsorber. (From Bley, P. et al., Application of the LIGA Process in Fabrication of Three-Dimensional Mechanical Microstructures, presented at 1991 Int. Micro Process Conf. With permission.)

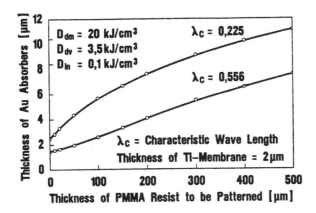

FIGURE 17.6 Necessary thickness of the gold absorbers of an X-ray mask. (From Bley, P. et al., Application of the LIGA Process in Fabrication of Three-Dimensional Mechanical Microstructures, presented at 1991 Int. Micro Process Conf. With permission.)

thick Si blank membranes is that larger resist areas can be exposed since one does not depend on a fragile membrane/absorber combination [Guckel, 1994].

17.2.5.4 Thick Absorber Fabrication

To make a mask with gold absorber structures of a height above 10 μm, one must first succeed in structuring a resist of that thickness. The height of the resist should in fact be a bit higher (say, 20%) than the absorber itself in order to accommodate the electrodeposited material fully in between the resist features. Currently, no means to structure a resist of that height with sufficient accuracy and perfect verticality of the walls exist, unless X-rays are used. Different procedures for producing X-ray masks with thicker absorber layers using a two-stage lithography process have been developed.

The KfK solution calls for first making an intermediate mask with photo or e-beam lithography. This intermediate mask starts with a 3-μm-thick resist layer, in which case the needed accuracy and steepness of the walls of printed features can be obtained through several means. After gold plating in between the resist features and stripping of the resist, this intermediate mask is used to write a pattern with X-rays in a 20-μm-thick resist. After electrodepositing and resist stripping, the latter will become the actual X-ray mask, i.e., the master mask.

Since hardly any accuracy is lost in the copying of the intermediate mask with X-rays to obtain the master mask, it is the intermediate mask quality that determines the ultimate quality of the LIGA-produced microstructures. The structuring of the resist in the intermediate mask is handled with optical techniques when the requirements on the LIGA structures are somewhat relaxed. The minimal lateral dimensions for optical lithography in a 3-μm-thick resist typically measure about 2.5 μm. Under optimum conditions, a wall angle of 88° is achievable. With electron-beam lithography, a minimum lateral dimension of 1 μm is feasible. The most accurate pattern transfer is achieved through reactive ion etching of a tri-level resist system. In this approach, a 3- to 4-μm-thick polyimide resist is first coated onto the titanium or beryllium membrane, followed by a coat of 10 to 15 nm titanium deposited with magnetron sputtering. The thin layer of titanium is an excellent etch mask for the polyimide; in an optimized oxygen plasma, the titanium etches 300 times slower than the polyimide. To structure the thin titanium layer itself, a 0.1-μm-thick optical resist is used. Since this top resist layer is so thin, excellent lateral tolerances result. The thin Ti layer is patterned with optical photo-lithography and etched in an argon plasma. After etching the thin titanium layer, exposing the polyimide locally, an oxygen plasma helps to structure the polyimide down to the titanium or beryllium membrane. Lateral dimensions of 0.3 μm can be obtained in this fashion. Patterning the top resist layer with an electron-beam increases the accuracy of the three-level resist method even further. Electrodeposition of gold on the titanium or beryllium membrane and stripping of the resist finishes the process of making the intermediate LIGA mask. To make a master mask, this intermediate mask is printed by X-ray radiation onto a PMMA resist coated master mask. The PMMA thickness corresponds to a bit more than the desired absorber thickness. Since the resist layer thickness is in the 10- to 20-μm range, a synchrotron X-ray wavelength of 10 Å is adequate for the making of the master mask. A further improvement in LIGA mask making is to fabricate intermediate and master mask on the same substrate, greatly reducing the risk for deviations in dimensions caused, for example, by temperature variations during printing [Becker et al., 1986]. The ultimate achievement would be to create a one-step process to make the master mask. Along this line of work, Hein et al. (1992) have started an investigation in the direct patterning of 10-μm-high resist layers with a 100-kV electron beam.

A completely different approach is suggested by Friedrich from the Institute of Manufacturing in Louisiana. He suggested the fabrication of an X-ray mask by traditional machining methods, such as micromilling, micro-EDM (electrodischarge machining) and lasers [Friedrich, 1994]. The advantages of this proposal are rapid turnaround (less than 1 day per mask), low cost and flexibility, as no intermediate steps interfere. Disadvantages are less dimensional edge acuity and nonsharp interior corners, as well as less absolute tolerance.

17.2.5.5 Alignment of the X-Ray Mask to the Substrate

The mask and resist-coated substrate must be properly registered to each other before they are put in an X-ray scanner. Alignment of an X-ray mask to the substrate is problematic since no visible light can pass through most X-ray membranes. To solve this problem, Schomburg et al. (1991) etched windows in their Ti X-ray membrane. Diamond membranes have a potential advantage here, as they are optically transparent and enable easy alignment for multiple irradiations without a need for etched holes.

Figure 17.7 illustrates an alternative, innovative X-ray alignment system involving capacitive pickup between conductive metal fingers on the mask and ridges on a small substrate area; Si, in this case U.S. Patent 4,607,213 (1986) and 4,654,581 (1987). When using multiple groups of ridges and fingers, two axis lateral and rotational alignment become possible.

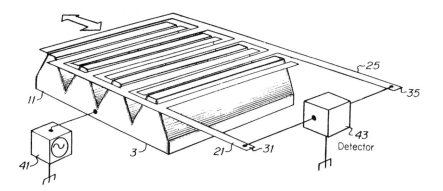

FIGURE 17.7 Mask alignment system in X-ray lithography. Conductive fingers on the mask and ridges on the Si are used for alignment. (From U.S. Patents 4,654,581, 1987 and 4,607,213, 1986.)

Another alternative may involve liquid nitrogen-cooled Si (Li) X-ray diodes as alignment detectors, eliminating the need for observation with visible light [Henck, 1984].

17.2.5.6 Transfer Mask for High Aspect Ratio Micro-Lithography

Recently Vladimirsky et al. (1995) developed a procedure to eliminate the need for an X-ray mask membrane. Unlike conventional masks, the so-called 'X-ray transfer mask' does not treat a mask as an independent unit. The technique is based on forming an absorber pattern directly on the resist surface. An example process sequence is shown in Figure 17.8. In this process, a transfer mask plating base is first prepared on the PMMA substrate plate by evaporating 70 Å of chromium (as adhesion layer) followed by 500 Å of gold using an electron beam evaporator. A 3-μm-thick layer of standard novolak-based AZ-type resist S1400-37 (Shipley Co.) is then applied over the plating base and exposed in contact mode through an optical mask using an ultraviolet exposure station. After development, the transfer mask is further completed by 3-μm gold electroplating on the exposed plating base. The remaining resist is removed by a blanket exposure and subsequent development, and the 500 Å of Au plating base is dissolved by a dip of 20 to 30 sec in a solution of KI (5%) and I (1.25%) in water; the Cr adhesion layer is removed by a standard chromium etch (from KTI). Fabrication of the transfer mask can thus be performed using standard lithography equipment available at almost any lithography shop. Depending on the resolution required, the X-ray transfer mask can be fabricated using known photon, electron-beam or X-ray lithography techniques. New lithography and post lithography techniques, such as *in situ* development, etching and deposition, can now be realized by using a transfer mask. The presence of the membrane in a conventional mask practically precludes use of *in situ* development and is not compatible with etching and deposition due to physical obstruction and material resistance considerations. The patterning of the PMMA resist with a transfer mask can be accomplished in multiple steps of exposure and development. An example of a cylindrical resonator made this way is shown in Figure 17.9. Each exposure/development step involved an exposure dose of about 8 to 12 J/cm^2. Subsequent 5-min development steps removed ~30 μm of PMMA. With seven steps, a self-supporting 1.5-mm-thick PMMA resist was patterned to a depth of more than 200 μm. The resist pattern shown in Figure 17.9 is 230 μm thick and exhibits a 2-μm gap between the inner cylinder and the pickup electrodes (aspect ratio is 100:1). The pattern was produced using soft (=10 Å) X-rays and a 3-μm-thick Au absorber only.

Vladimirsky et al. (1995) suggest that the forming of the transfer mask directly on the sample surface creates several additional new opportunities, besides the already mentioned *in situ* development, etching and deposition, these include exposure of samples with curved surfaces and dynamic deformation of a sample surface during the exposure (hemispherical structures would become possible this way). Elegant and cost-saving innovations like these could mainstream LIGA faster than currently believed.

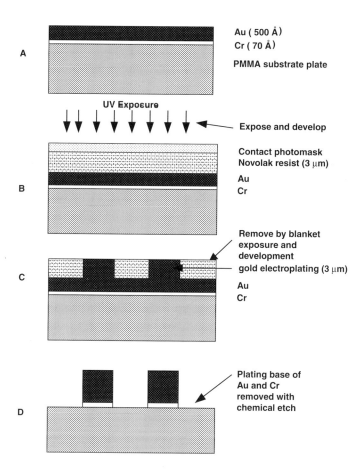

FIGURE 17.8 Sample transfer mask formation. (After Vladimirsky, Y. et al., Transfer Mask for High Aspect Ratio Micro-Lithography, presented at Microlithography '95, Santa Clara, CA, 1995.)

FIGURE 17.9 SEM micrograph of a cylindrical PMMA resonator made by the Transfer Mask method and multiple exposure/development steps. (From Vladimirsky, Y. et al., Transfer Mask for High Aspect Ratio Micro-Lithography, presented at Microlithography '95, Santa Clara, CA, 1995. With permission.) (Courtesy of Dr. V. Saile.)

17.2.6 LIGA Process Steps and Materials

17.2.6.1 Choice of Primary Substrate

In the LIGA process, the primary substrate, or base plate, must be a conductor or an insulator coated with a conductive top layer. A conductor is required for subsequent electrodeposition. Some examples of primary substrates that have been used successfully are austenite steel plate, Si wafers with a thin Ti or Ag/Cr top layer [Michel et al., 1993], and copper plated with gold, titanium or nickel [Becker et al., 1986]. Other metal substrates as well as metal-plated ceramic, plastic and glass plates have been employed [Rogner et al., 1992]. It is important that the plating base provide good adhesion for the resist. For that purpose, prior to applying the X-ray resist on copper or steel, the surface sometimes is mechanically roughened by microgrit blasting with corundum. Microgrit blasting may lead to an average roughness, R_a, of 0.5 µm, resulting in better physical anchoring of the microstructures to the substrate [Mohr et al., 1998]. In the case of a polished metal base, chemical preconditioning may be used to improve adhesion of the resist microstructures. During chemical preconditioning, a titanium layer, sputter-deposited onto the polished metal base plate (e.g., a Cu plate), is oxidized for a few minutes in a solution of 0.5 M NaOH and 0.2 M H_2O_2 at 65°C. The oxide produced typically measures 30 nm thick and exhibits a microrough surface instrumental to better securing resist to the base plate. The Ti adhesion layer may further be covered with a thin nickel seed layer (~150 Å) for electroless or electroplating of nickel in-between the resist microstructures. When using a highly polished Si surface, adhesion promoters need to be added to the resist resin (see also Madou [1997, chap. 1] and Resist Adhesion farther below). A substrate of special interest is a processed silicon wafer with integrated circuits. Integrating the LIGA process with IC circuitry on the same wafer might greatly improve sensor applications.

Thick resist plates can act as plastic substrates themselves. For example, using 20,000-eV rather than the more typical 3000-eV radiation [Guckel et al., 1994; Siddons and Johnson, 1994], exposed plates of PMMA up to 10 cm thick.

17.2.6.2 Resist Requirements

An X-ray resist ideally should have high sensitivity to X-rays, high resolution, resistance to dry and wet etching, thermal stability of greater than 140°C and a matrix or resin absorption of less than 0.35 μm^{-1} at the wavelength of interest [Madou, 1997, chap. 1; Moreau, 1988]. These requirements equal those for IC production with X-ray lithography [Lingnau et al., 1989]. To produce high-aspect-ratio microstructures with very tight lateral tolerances demands an additional set of requirements. The unexposed resist must be absolutely insoluble during development. This means that a high contrast (γ) is required. The resist must also exhibit very good adhesion to the substrate and be compatible with the electroforming process. The latter imposes a resist glass transition temperature (T_g) greater than the temperature of the electrolyte bath used to electrodeposit metals between the resist features remaining after development (say, at 60°C). To avoid mechanical damage to the microstructures induced by stress during development, the resist layers should exhibit low internal stresses [Mohr et al., 1989]. If the resist structure is the end product of the fabrication process, further specifications depend on the application itself, e.g., optical transparency and refractive index for optical components or large mechanical yield strength for load-bearing applications. For example, PMMA exhibits good optical properties in the visible and near-infrared range and lends itself to the making of all types of optical components [Gottert et al., 1991].

Due to excellent contrast and good process stability, known from electron-beam lithography, poly(methylmethacrylate) (PMMA) is the preferred resist for deep-etch synchrotron radiation lithography. Two major concerns with PMMA as a LIGA resist are a rather low lithographic sensitivity of about 2 J/cm^2 at a wavelength λ_c of 8.34 Å and a susceptibility to stress cracking. For example, even at shorter wavelengths, $\lambda_c = 5$ Å, over 90 min of irradiation are required to structure a 500-µm-thick resist layer with an average ring storage current of 40 mA and a power consumption of 2 MW at the 2.3-GeV ELSA synchrotron (Bonn, Germany) [Wollersheim et al., 1994]. The internal stress arising from the combination of a polymer and a metallic substrate can cause cracking in the microstructures during development, a phenomenon PMMA is especially prone to, as illustrated in the scanning electron microscope in Figure 17.10.

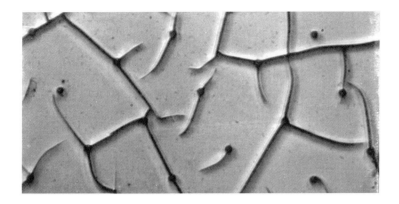

FIGURE 17.10 Cracking of PMMA resist. Method to test stress in thick resist layers. The onset of cracks in a pattern of holes with varying size (say, 1 to 4 µm) in a resist is shifted towards smaller hole diameter the lower the stress in the film. The SEM picture displays extensive cracking incurred during development of the image in a 5-µm-thick PMMA layer on an Au-covered Si wafer. The 5-µm-thick PMMA layer resulted from five separate spin coats. Annealing pushed the onset of cracking toward smaller holes until the right cycle was reached and no more cracks were visible [Madou and Murphy, 1995].

To make throughput for deep-etch lithography more acceptable to industry, several avenues to more sensitive X-ray resists have been pursued. For example, copolymers of PMMA were investigated: methylmethacrylate combined with methacrylates with longer ester side chains show sensitivity increases of up to 32% (with tertiary butylmethacrylate). Unfortunately, a deterioration in structure quality was observed [Mohr et al., 1988b]. Among the other possible approaches for making PMMA more X-ray sensitive we can count on the incorporation of X-ray absorbing high-atomic-number atoms or the use of chemically amplified photoresists [Madou, 1997, chap. 1]. X-ray resists explored more recently for LIGA applications are poly(lactides), e.g., poly(lactide-co-glycolide) (PLG), polymethacrylimide (PMI), polyoxymethylene (POM) and polyalkensulfone (PAS). Poly(lactide-co-glycolide) is a new positive resist developed by BASF AG, more sensitive to X-rays by a factor of 2 to 3 compared with PMMA. Its processing is less critical but it is not commercially available yet. From the comparison of different resists for deep X-ray lithography in Table 17.7, PLG emerges as the most promising LIGA resist. POM, a promising mechanical material, may also be suited for medical applications given its biocompatability. All of the resists shown in Table 17.7 exhibit significantly enhanced sensitivity compared to PMMA, and most exhibit a reduced stress corrosion [Wollersheim et al., 1994]. Negative X-ray resists have inherently higher sensitivities compared to positive X-ray resists, although their resolution is limited by swelling. Poly(glycidylmethacrylate-co-ethyl acrylate) (PGMA), a negative electron-beam resist (not shown in Table 17.7), has also been used in X-ray

TABLE 17.7 Properties of Resists for Deep X-Ray Lithography

	PMMA	POM	PAS	PMI	PLG
Sensitivity	−	+	++	0	0
Resolution	++	0	− −	+	++
Sidewall smoothness	++	− −	− −	+	++
Stress corrosion	−	++	+	− −	++
Adhesion on substrate	+	+	+	−	+

Note: PMMA = poly(methylmethacrylate), POM = polyoxymethylene, PAS = polyalkensulfone, PMI = polymethacrylimide, PLG = poly(lactide-co-glycolide). ++ = excellent, + = good, 0 = reasonable, − = bad, − − = very bad.
Source: After Ehrfeld, W., *LIGA at 1MM*, Baniff, Canada, 1994.

lithography. In general, resist materials sensitive to electron-beam exposure also display sensitivity to X-rays and function in the same fashion; materials positive in tone for electron-beam radiation typically are also positive in tone for X-ray radiation. A strong correlation exists between the resist sensitivities observed with these two radiation sources, suggesting that the reaction mechanisms might be similar for both types of irradiation (see Madou [1997, chap. 1]). IMM, in Germany, started developing a negative X-ray resist 20 times more sensitive than PMMA, but the exact chemistry has not been disclosed yet [IMM, 1995].

17.2.6.3 Resist Application

Multiple Spin Coats
Different methods to apply ultra-thick layers of PMMA have been studied. In the case of multilayer spin coating, high interfacial stresses between the layers can lead to extensive crack propagation upon developing the exposed resist. For example, in Figure 17.10 we present an SEM picture of a 5-μm-thick PMMA layer, deposited in five sequential spin coatings. Development resulted in the cracked riverbed mud appearance with the most intensive cracking propagating from the smallest resist features. The test pattern used to expose the resist consisted of arrays of holes ranging in size from 1 to 4 μm. We found that annealing the PMMA films shifted the cracking toward holes with smaller diameter compared to the unannealed film shown in Figure 17.10 [Madou and Murphy, 1995]. The described experiment might comprise a generic means of studying the stress in thick LIGA resist layers. It was found at CAMD that multiple spin coating can be used for up to 15-μm-thick resist layers and that with applying the appropriate annealing and developer (see below) no cracking results [private communication, 1995]. Further on we will learn that a prerequisite for low stress and small lateral tolerances in PMMA films is a high mean molecular weight. The spin-coated resist films in Figure 17.10 might not have a high enough molecular weight to lead to good enough selectivity between radiated and nonradiated PMMA during a long development process.

Commercial PMMA Sheets
High molecular weight PMMA is commercially available as prefabricated plate (e.g., GS 233; Rohm GmbH, Darmstadt, Germany) and several groups have employed free-standing or bonded PMMA resist sheets for producing LIGA structures [Guckel, 1994; Mohr et al., 1988]. After overcoming the initial problems encountered when attempting to glue PMMA foils to a metallic base plate with adhesives, this has become the preferred method [Mohr et al., 1988a]. Guckel used commercially available thick PMMA sheets (thickness > 3mm), XY-sized and solvent-bonded them to a substrate, and, after milling the sheet to the desired thickness, exposed the resist without cracking problems [Guckel, 1994].

Casting of PMMA
PMMA also can be purchased in the form of a casting resin, e.g., Plexit 60 (PMMA without added cross-linker) and Plexit 74 (PMMA with cross-linker added) from Rohm GmbH in Darmstadt, Germany. In a typical procedure, PMMA is *in situ* polymerized from a solution of 35 wt% PMMA of a mean molecular weight of anywhere from 100,000 g/mol up to 10^6 g/mol in methylmethacrylate (MMA). Polymerization at room temperature takes place with benzoyl peroxide (BPO) catalyst as the hardener (radical builder) and dimethylaniline (DMA) as the starter or initiator [Mohr et al., 1988a; 1988b]. The oxygen content in the resin, inhibiting polymerization, and gas bubbles, inducing mechanical defects, are reduced by degassing while mixing the components in a vacuum chamber at room temperature and at a pressure of 100 mbar for 2 to 3 min.

In a practical application, resin is then dispensed on a base plate provided with shims to define pattern and thickness and subsequently covered with a glass plate to avoid oxygen absorption. The principle of polymerization on a metal substrate is schematically represented in Figure 17.11. Due to the hardener, polymerization starts within a few minutes after mixing of the components and comes to an end within 5 min. The glass cover plate is coated with an adhesion preventing layer (e.g., Lusin L39, Firma Lange u. Seidel, Nurnberg). After polymerization the anti-adhesion material is removed by diamond milling and a highly polished surface results. *In situ* polymerization and commercial-cast PMMA sheets top the list of thick resist options in LIGA today.

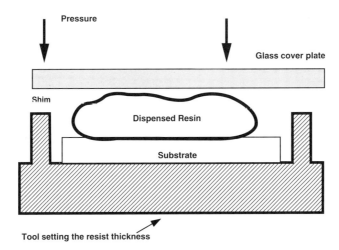

FIGURE 17.11 Principle of *in situ* polymerization of a thick resist layer on a metal substrate.

Plasma Polymerized PMMA

Guckel et al. explored plasma-polymerized PMMA to make conformal coatings more than 100 μm thick [Guckel et al., 1988]. As discussed in Madou (1997, chap. 3), plasma polymerization offers several advantages. Surprisingly, little work focuses on plasma polymerized X-ray resists.

17.2.6.4 Resist Adhesion

Adhesion promotion by mechanically or chemically modifying the primary substrate was introduced above under Choice of Primary Substrate. Smooth surfaces such as Si wafers with an average roughness, R_a, smaller than 20 nm pose additional adhesion challenges often solved by modifying the resist itself. To promote adhesion of resist to polished untreated surfaces, such as a metal-coated Si wafers, coupling agents must be used to chemically attach the resist to the substrate. An example of such a coupling agent is methacryloxypropyl trimethoxy silane (MEMO). With 1 wt% of MEMO added to the casting resin, excellent adhesion results. The adherence is brought about by a siloxane bond between the silane and the hydrolyzed oxide layer of the metal. As illustrated in Figure 17.12, the integration of this coupling agent in the polymer matrix is achieved via the double bond of the methacryl group of MEMO [Mohr et al., 1988b].

Hydroxyethyl methacrylate (HEMA) can improve PMMA adhesion to smooth surfaces but higher concentrations are needed to obtain the same adhesion improvement. Silanization of polished surfaces prior to PMMA casting, instead of adding adhesion promoters to the resin, did not seem to improve the PMMA adhesion. In the case of PMMA sheets, as mentioned before, one option is solvent bonding of the layers to a substrate. In another approach Galhotra et al. (1996) simply mechanically clamped the exposed and developed self-supporting PMMA sheet onto a 1.0-mm-thick Ni sheet for subsequent Ni plating. Rogers et al. (1996) have shown that cyanoacrylate can be used to bond PMMA resist sheets to a Ni substrate and that it can be lithographically patterned using the same process sequence used to pattern PMMA. For a 300-μm-thick PMMA sheet on a sputtered Ni coating on a silicon wafer a 10-μm-thick cyanoacrylate bonding layer was used. Such a thick cyanoacrylate layer caused some problems for subsequent uniform electrodeposition of metal. The dissolution rate of the cyanoacrylate is faster than the PMMA resist, resulting in metal posts with a wide profile at the base.

17.2.6.5 Stress-Induced Cracks in PMMA

The internal stress arising from the combination of a polymer on a metallic substrate can cause cracking in the microstructures during development. To reduce the number of stress-induced cracks (see Figure 17.10), both the PMMA resist and the development process must be optimized. Detailed measurements of the heat of reaction, the thermomechanical properties, the residual monomer content and the molecular weight distribution during polymerization and soft baking have shown the necessity to produce resist layers with

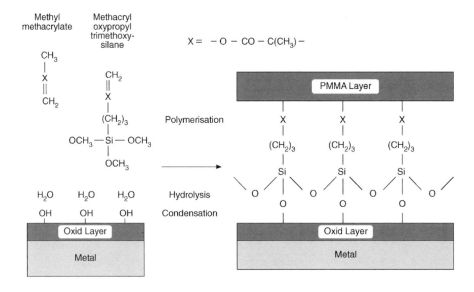

FIGURE 17.12 Schematic presentation of the adherence mechanism of methacryloxypropoyl trimethoxy silane (MEMO). (From Mohr, J. et al., *J. Vac. Sci. Technol.*, B6, 2264–2267, 1988. With permission.)

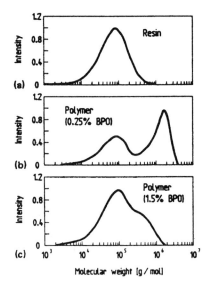

FIGURE 17.13 Molecular weight distribution of (a) the casting resin, (b) a resist layer polymerized at low hardener content, and (c) a resist layer polymerized at high hardener content determined by gel permeation chromatography. (From Mohr, J. et al., *J. Vac. Sci. Technol.*, B6, 2264–2267, 1988. With permission.)

a high molecular weight and with only a very small residual monomer content [Mohr et al., 1988a; 1988b]. Figure 17.13 compares the molecular weight distribution determined by gel permeation chromatography of a polymerized PMMA resist (two hardener concentrations were used) with the molecular weight distribution of the casting resin. The casting resin is unimodal, whereas the polymerized resist layer typically shows a bimodal distribution with peak molecular weights centered around 90,000 and 300,000 g/mol. The first low-molecular-weight peak belongs to the PMMA oligomer dissolved in the casting resin, and the second molecular weight peak results from the polymerization of the monomer. The molecular weight distribution is constant across the total resist thickness, except for the boundary layer at the base plate, where the average molecular weight can be significantly higher (~450,000 g/mol) [Mohr et al., 1989].

The amount of the high molecular weight portion in the polymerized resist depends on the concentration of the hardener. A low hardener content leads to a high molecular weight dominance and vice versa (see Figure 17.13) [Mohr et al., 1988b]. Since high molecular weight is required for low stress, a hardener concentration of less than 1% BPO (benzoyl peroxide) must be used. Ideally, for a low stress resist, the residual monomer content should be less than 0.5%. The residual monomer content decreases with increasing hardener content, and >1% BPO is needed to reduce the residual monomer content below 0.5%. The problem resulting from these opposite needs can be overcome by the addition of 1% of a cross-linking dimethacrylate (triethylene glycol dimethacrylate, TEDMA) to the resin. In such cross-linked PMMA, a smaller amount of BPO suffices to suppress the residual monomer content; crack-free PMMA can be obtained with 0.8% of BPO [Mohr et al., 1988a].

For solvent removal and to further minimize the defects caused by stress, the polymerized resin is cured at 110°C for one hour (soft bake). The measurement of the reaction enthalpy shows that post-polymerization reactions occur at room temperature and during heating to the glass transition temperature [Mohr et al., 1988b]. The rate of heating up to that temperature is 20°C/hr; after curing, the samples are cooled down from 110°C to room temperature at a very low rate of 5 to 10°C/hr [Mohr et al., 1988a; 1988b]. The soft-bake temperature is slightly below the glass transition temperature measured to be 115°C.

Another important factor reducing stress in thick PMMA resist layers is the optimization of the developer. Stress-induced cracking can be minimized with solvent mixtures whose dissolution parameters lie near the boundary of the PMMA solubility range, i.e., a nonaggressive solvent is preferred. This is discussed in more detail below, under Development. Small amounts of additives such as described above for reducing stress or to promote adhesion do not influence the mechanical stability of the microstructures or the sensitivity of the resist.

17.2.6.6 Optimum Wavelength and Deposited Dose

Optimum Wavelength

For a given polymer, the lateral dimension variation in a LIGA microstructure could, in principle, result from the combined influence of several mechanisms, such as Fresnel diffraction, the range of high energy photoelectrons generated by the X-rays, the finite divergence of synchrotron radiation and the time evolution of the resist profiles during the development process. The theoretical manufacturing precision obtainable by deep X-ray lithography was investigated by means of computer simulation of both the irradiation step and the development step by Becker et al. (1984) and by Munchmeyer (1984), and further tested experimentally and confirmed by Mohr et al. (1988a). The theoretical results demonstrate that the effect of Fresnel diffraction (edge diffraction), which increases as the wavelength increases, and the effect of secondary electrons in PMMA, which increases as the wavelength decreases, lead to minimal structural deviations when the characteristic wavelength ranges between 0.2 and 0.3 nm (assuming an ideal development process and no X-ray divergence). To fully utilize the accuracy potential of a 0.2- to 0.3-nm wavelength, the local divergence of the synchrotron radiation at the sample site should be less than 0.1 mrad. Under these conditions, the variation in critical lateral dimensions likely to occur between the ends of a 500-μm-high structure due to diffraction and secondary electrons are estimated to be 0.2 μm. The estimated Fresnel diffraction and secondary electron scattering effects are shown as a function of characteristic wavelength in Figure 17.14.

Using cross-linked PMMA, or linear PMMA with a unimodal and extremely high-molecular-weight distribution (peak molecular weight greater than 1,000,000 g/mol), the experimentally determined lateral tolerances on a test structure as shown in Figure 17.15 are 0.055 μm per 100 μm resist thickness, in good agreement with the 0.2 μm over 500 μm expected on a theoretical basis [Mohr et al., 1988a]. These results are obtained only when a resist/developer system with a ratio of the dissolution rates in the exposed and unexposed areas of approximately 1000 is used.

The use of resist layers, not cross-linked and displaying a relatively low bimodal molecular-weight distribution, as well as the application of excessively strong solvents such as used to develop thin PMMA resist

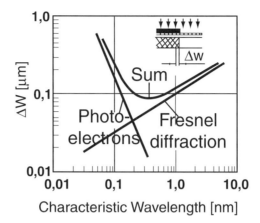

FIGURE 17.14 Fresnel diffraction and photoelectron generation as a function of characteristic wavelength, λ_c, and the resulting lateral dimension variation (ΔW). (After Menz, W.P., *Mikrosystemtechnik fur Ingenieure*, VCH Publishers, Germany, 1993.)

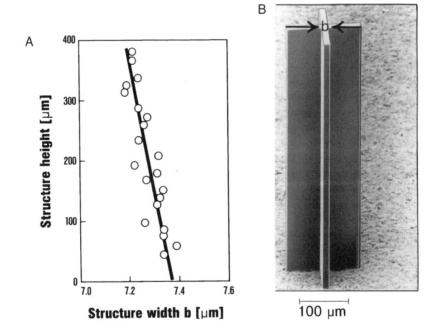

FIGURE 17.15 Structural tolerances. (A) SEM micrograph of a test structure to determine conical shape. (B) Structural dimensions as a function of structure height. The tolerances of the dimensions are within 0.2 μm over the total structure height of 400 μm [Mohr et al., 1988b]. (Courtesy of the Karlsruhe Nuclear Research Center.)

layers in the IC industry, lead to more pronounced conical shapes in the test structures of Figure 17.15. An illustration of the effect of molecular weight distribution on lateral geometric tolerances is that linear PMMA with a peak molecular weight below 300,000 g/mol shows structure tolerances of up to 0.15 μm/100 μm [Mohr et al., 1988a]. To obtain the best tolerances necessitates a PMMA with a very high molecular weight, also a prerequisite for low stress in the developed resist. Finally, if the synchrotron beam is not parallel to the absorber wall but at an angle greater than 50 mrad, greater coning angles may also result [Mohr et al., 1988a].

LIGA and Other Replication Techniques

Deposited Dose

As shown in Figure 17.16, depicting the average molecular weight of PMMA as a function of radiation dose, the X-ray irradiation of PMMA reduces the average molecular weight [Menz and Bley, 1993]. For one-component positive resists, this lowering of the average molecular weight (M_n^*; see Madou [1997, chap. 1, Eq. (1.7)]) causes the solubility of the resist in the developer to increase dramatically. The average molecular weight making dissolution possible is a sensitive function of the type of developer used and the development temperature. Interestingly, it can be observed from Figure 17.16 that above a certain dose (15 to 20 kJ/cm^3) the average molecular weight does not decrease any further. This can be understood by recalling the $G(s)$ and $G(x)$ values of a resist introduced in Madou (1997, chap. 1). These G values express the number of main chain scissions (in case of a positive resist), and cross-links (in case of a negative resist), respectively, per 100 eV of absorbed radiation energy. The average molecular weight decreases with radiation dose as long as $G(s)$ is larger than $G(x)$. When the G-values are equal the average molecular weight does not change any further. Our example in Figure 17.16 shows this to occur at a radiation dose between 15 and 20 kJ/cm^3.

The molecular weight distribution, measured after resist exposure, is unimodal with peak molecular weights ranging from 3000 to 18,000 g/mol, dependent on the dose deposited during irradiation (see Figure 17.16). The peak molecular weight increases nearly linearly with increasing resist depth, i.e., decrease of the absorbed dose [Mohr et al., 1989]. At the bottom of the resist layer, the absorbed dose must be higher than the development dose, D_{dv}, while at the top of the resist the absorbed dose must be lower than the damaging dose, D_{dm}. In Figure 17.5, where the absorption of X-rays along the path from source to sample was illustrated, the exposure time and the preabsorber were chosen so that the bottom of a 500-µm-thick PMMA layer received the necessary development dose D_{dv}, while the dose at the top of the layer stayed well below D_{dm}. Exposure of PMMA with longer wavelengths results in correspondingly longer exposure times and can lead to an overexposure of the top surface, where the lower energy radiation is mainly absorbed.

Menz and Bley (1993) describe the influence of the radiation dose on the quality of the resulting LIGA structures in a slightly different manner. Following their approach, Figure 17.17A illustrates a typical bimodal molecular weight distribution of PMMA before radiation, exhibiting an average molecular weight of 600,000. The gray region in this figure indicates the molecular weight region where PMMA readily dissolves, i.e., below the 20,000-g/mol level for the temperature and developer used. Since the fraction of

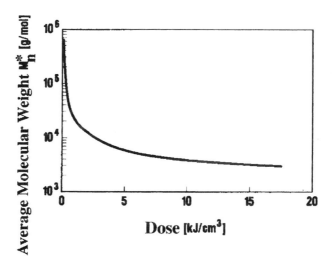

FIGURE 17.16 Average molecular weight M_n^* vs. X-ray radiation dose. (From Menz, W.P., *Mikrosystemtechnik fur Ingenieure*, VCH Publishers, Germany, 1993. With permission.)

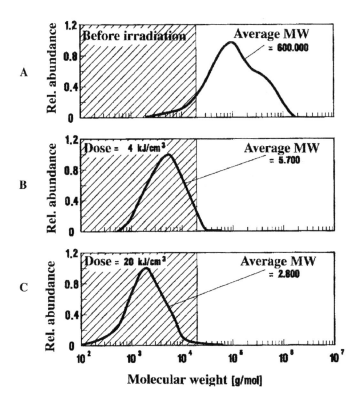

FIGURE 17.17 Molecular weight distribution of PMMA before (A) and after irradiation with 4 (B) to 20 kJ/cm^3 (C). The shaded areas indicate the domain in which PMMA is minimally 50% dissolved (at 38°C in the LIGA developer described in the text). (From Menz, W.P., *Mikrosystemtechnik fur Ingenieure*, VCH Publishers, Germany, 1993. With permission.)

PMMA with a 20,000 molecular weight is very small in nonirradiated PMMA, the developer hardly attacks the resist at all. After irradiation with a dose D_{dv} of 4 kJ/cm^3, the average molecular weight becomes low enough to dissolve almost all of the resist (Figure 17.17B). With a dose D_{dm} of 20 kJ/cm^3, all of the PMMA dissolves swiftly (Figure 17.17C). At a dose above D_{dm}, the microstructures are destroyed by the formation of bubbles. It follows that to dissolve PMMA completely and to make defect-free microstructures, the radiation dose for the specific type of PMMA used must lay between 4 and 20 kJ/cm^3. These two numbers also lock in a maximum value of 5 for the ratio of the radiation dose at the top and bottom of a PMMA structure. To make this ratio as small as possible, the soft portion of the synchrotron radiation spectrum is usually filtered out by a preabsorber, e.g., a 100-μm-thick polyimide foil (Kapton) in order to reduce differences in dose deposition in the resist.

X-ray exposure equipment developed by KfK has a vibration-free bedding; the exposure chamber is under thermostatic control (±0.2°C) and includes a precision scanner for the periodic movement of the sample through the irradiation plane. The polyimide window isolates the vacuum of the accelerator from the helium atmosphere (200 mbar) which serves as coolant for substrate and mask in the irradiation chamber [Becker et al., 1986]. IMM in collaboration with Jenoptik GmbH developed an X-ray scanner for deep lithography, enabling irradiation of up to 1000-μm-thick resist. A mask-to-resist registration within ±0.3 μm is claimed [Editorial, 1994c].

17.2.6.7 Development

X-ray radiation changes the polymer in the unmasked areas, and chemicals etch away the regions that have been exposed during development. To fully utilize the accuracy potential of synchrotron radiation lithography, it is essential that the resist/developer system has a ratio of dissolution rate in the exposed and unexposed areas of approximately 1000 (see above). The developer empirically arrived at by KfK consists of a mixture of 20 vol% tetrahydro-1,4-oxazine (an azine), 5 vol% 2-aminoethanol-1 (a primary amine), 60 vol% 2-(2-butoxyethoxy)ethanol (a glycolic ether) and 15 vol% water [Mohr et al., 1988a; 1989; Ghica and Glashauser, 1982]. This developer causes an infinitely small dissolution of unexposed, high-molecular-weight, cross-linked PMMA and achieves a sufficient dissolution rate in the exposed area. It also exhibits much less stress-induced cracking than developers conventionally used for thin PMMA resists. During development the developing agent flows toward the resist surface being developed, circulates and is filtered continuously and the temperature remains controlled at 35°C. To stop development, less concentrated developer solutions are applied to prevent the precipitation of already dissolved resist.

Systematic investigation of different organic solvents and mixtures of the above developer systems showed that solvents with a solubility parameter at the periphery of the solubility range of PMMA dissolve exposed PMMA slowly but selectively and without stress-induced cracking or swelling of unexposed areas. Solvents with a solubility parameter close to those of MMA show a much higher dissolution rate, but cause serious problems related to cracking and swelling. As we saw before, the application of excessively strong solvents also led to more pronounced conical shapes in the test structures shown in Figure 17.15. An improved developer found in the above systematic investigation is a mixture of tetrahydro-1,4-oxazine and 2-aminoethanol-1. Its sensitivity is 30% higher, but the process latitude is much narrower compared to the developer described above [Mohr et al., 1989].

At KfK, a dedicated machine was built for the development process, enabling the continuous and homogeneous transport of developing and rinsing agents into deep structure elements and the removal of the dissolved resist from these structures. Several substrates are arranged vertically on a rotor, with each structure surface facing to the outside. Three independent medium circuits are available for immersion and spraying processes [Becker et al, 1986].

After development the microstructures are rinsed with deionized water and dried in a vacuum. Alternatively, drying is done by spinning and blasting with dry nitrogen. At this stage the devices can be the final product, e.g., as micro-optical components, or they are used for subsequent metal deposition.

17.2.6.8 Electrodeposition/Electroless Deposition

Electrodeposition

Overview. In the electroforming of microdevices with LIGA (see Figure 17.1), a conductive substrate, carrying the resist structures, serves as the cathode. The metal layer growing on the substrate fills the gaps in the resist configuration, thus forming a complementary metal structure. The use of a solvent-containing development agent ensures a substrate surface completely free of grease and ready for plating. Most of the plating involves Ni. We will discuss Ni electroplating in detail and list some of the other materials that have been or could be plated.

The fabrication of metallic relief structures is a well-known art in the electroforming industry (see also Madou [1997, chaps. 3 and 7]). The technology is used, for example, to make the fabrication tools for records and videodiscs where structural details in the submicron range are transferred. Electrodeposition is a powerful tool for creating microstructures, not only in LIGA where resist molds are made by X-ray lithography, but also in combination with molds made by laser ablation, dry etching, Keller's polysilicon molding technique (see Chapter 16), bulk micromachining etc. Because of the extreme aspect ratio, several orders of magnitude larger than in the crafting of CDs, electroplating in LIGA poses new challenges. The near-ultimate in packing density, the extreme height-to-width aspect ratio and fidelity of reproduction of electrodeposition were first demonstrated in Romankiw's precursor work to LIGA in which gold was plated through polymeric masks generated by X-ray lithography. This author, using X-ray exposure, developed holes in PMMA resist features of 1-μm width and depths of up to 20 μm and electroplated gold in such features up to thicknesses of 8 μm (see also Madou [1997, chap. 7]; Spiller et al. [1976]). By 1975, the same author and his team had used the electroplating approach to create microwires 300 × 300 atoms in cross section [Romankiw, 1993].

Nickel Electrodeposition. The electrodeposition reaction of nickel on the cathode, i.e.,

$$Ni^{2+} + 2e^- \rightarrow Ni \qquad \text{(Reaction 17.1)}$$

competes with the hydrogen evolution reaction on the same electrode, i.e.,

$$2H^+ + 2e^- \rightarrow H_2 \qquad \text{(Reaction 17.2)}$$

The theoretical amount of Ni deposited from Reaction 17.1 can be calculated from Faraday's law (see Madou [1997, chap. 3, Eq. (3.29)]). Since some of the current goes into hydrogen evolution, the actual amount of Ni deposited usually is less than calculated from that equation. The electrodeposition yield, η, is then defined as the actual amount of deposited material, m_{act}, over the total calculated from Faraday's law, m_{max}, and can be calculated from [Romankiw, 1993]:

$$\eta[\%] = \frac{m_{act}}{m_{max}} \times 100 = \frac{m_{act}}{\frac{1}{z}iAt\frac{1}{F}M} \times 100 \qquad (17.5)$$

where

A = electrode surface
t = electrodeposition time
F = Faraday constant (96487 A/sec/mol)
i = current density
z = electrons involved in Reaction 17.1
M = atomic weight of Ni

The amount of hydrogen evolving and competing with Ni deposition depends on the pH, the temperature and the current density. Since one of the most important causes of defects in the metallic LIGA microstructures is the appearance of hydrogen bubbles, these three parameters need very precise control. Pollutants cause hydrogen bubbles to cling to the PMMA structures, resulting in pores in the nickel deposit, so the bath must be kept clean, e.g., by circulating through a membrane filter with 0.3-µm pore openings [Harsch, 1988]. Besides typical impurities, such as airborne dust or dissolved anode material, the main impurities are nickel hydroxide formed at increased pH values in the cathode vicinity and organic decomposition products from the wetting agent. The latter two can be avoided to some degree by monitoring and controlling the pH and by adsorption of the organic decomposition products on activated carbon.

Another cause of defects in electroplating is an incomplete wetting of the microchannels in the resist structure. The contact angle between PMMA and the plating electrolyte at 50°C lies between 70 and 80°. A wetting agent is thus indispensable for wetting the surface of the plastic structures in order for the electrolyte to penetrate into the microchannels. With a wetting agent, the contact angle between PMMA and the plating electrolyte can be reduced from 80 to 5° [Harsch et al., 1988]. In the electroforming of microdevices, a much higher concentration of wetting agent is necessary than in conventional electroplating. A dramatic illustration of this wetting effect can be seen in Figure 17.18A and B. In the set-up illustrated in Figure 17.18A, where only 2.5 ml/l of a wetting agent is added to the sulfamate nickel deposition solution, nickel posts with a diameter of 50 µm often fail to form. In the same experiment no posts with a diameter of 5 and 10 µm form at all. Increasing the wetting agent concentration to 10 ml/l results in the perfectly formed nickel posts with a diameter of 5 µm as shown in Figure 17.18B.

A microelectrode of the same size or less than the diffusion layer thickness, as shown in Figure 17.18B, could be expected to plate faster than a larger electrode because of the extra increment of current due to the nonlinear diffusion contribution [Madou, 1997, chap. 3, Eq. (3.53)]. On the other hand, as derived from Madou (1997, chap. 3, Eq. (3.55)), describing the current to a metal electrode recessed in a resist layer, we learned that the nonlinear diffusion contribution increases with decreasing radius r but decreases with increasing resist thickness, L. High-aspect-ratio features consequently will plate slower than low aspect ratio features. Moreover, the consumption of hydrogen ions in the high-aspect-ratio features causes

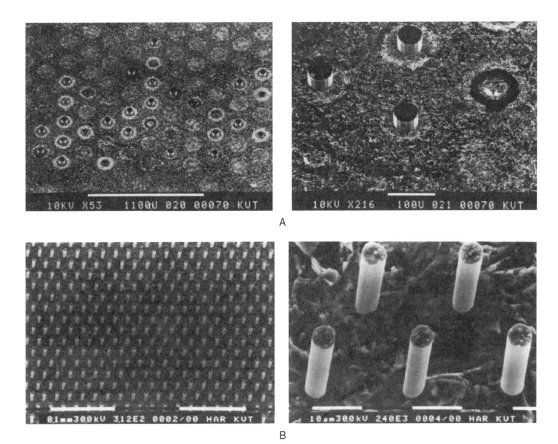

FIGURE 17.18 Electrodeposition of nickel posts of varying diameter. (A) nickel posts with diameter of 50 μm. Only 2.5 ml/l wetting agent added to the nickel sulfamate solution. Many 50-μm posts are missing and posts with a diameter of 10 or 5 μm do not even form. (B) Nickel posts with diameter of 5 μm. 10 ml/l wetting agent was added; all posts developed perfectly. (Courtesy of the Karlsruhe Nuclear Research Center.)

the pH to locally increase. As no intense agitation is possible in these crevices, an isolating layer of nickel hydroxide might form, preventing further metal deposition. This all contributes to making the deposition rate, important for an economical production, much smaller than the rates expected from the linear diffusion model of current density in large, low-aspect-ratio structures.

A nickel sulfamate bath composition, optimized for Ni electrodeposition, is given in Table 17.8. In addition to nickel sulfamate and boric acid as a buffer, a small quantity of an anion-active wetting agent

TABLE 17.8 Composition and Operating Conditions of Nickel Sulfamate Bath

Parameter	Value
Nickel metal (as sulfamate)	76–90 g/l
Boric acid	40 g/l
Wetting agent	10 ml/l
Current density	1–10 A/dm^2
Temperature	50–62°C
pH	3.5–4.0
Anodes	Sulfur depolarized

is added. Sulfur-depolarized nickel pellets are used as anode materials and are held in a titanium basket [Maner, 1988]. Table 17.8 also lists operational parameters. Nickel sulfamate baths produce low internal stress deposits without the need for additional agents avoiding a cause for more defects. The bath is operated at 50 to 62°C and at a pH value between 3.5 and 4.0. Metal deposition is carried out at current densities up to 10 A/dm^2. Growth rates vary from 12 (at 1 A/dm^2) to 120 µm/hr (at 10 A/dm^2).

Physical Properties of Electrodeposited Nickel. The hardness of the Ni deposits can be adjusted from 200 to 350 Vickers by varying the operating conditions. The hardness decreases with increasing current density. To reach a high hardness of 350 Vickers, the electroforming must proceed at a reduced current of 1 A/dm^2. Also, for low compressive stress of 20 N/mm^2 or less, a reduced current density must be used. Internal stress in the Ni deposits is not only influenced by current density but also by the layer thickness, pH, temperature and solution agitation. In the case of pulse plating (see below), pulse frequency has a distinct influence as well [Harsch et al., 1988]. From Figure 17.19 we can derive that for thin Ni deposits the stress is high and decreases very fast as a function of thickness. For thick Ni deposits (>30 µm), the stress as a function of thickness reaches a plateau. At a current density of 10 A/dm^2 these thick Ni films are under compressive stress; at 1 A/dm^2 they are under tensile stress; at 5 A/dm^2 the internal stress reduces to practically zero. Stirring of the plating solution reduces stress dramatically, indicating that mass transport to the cathode is an important factor in determining the ultimate internal stress. Consequently, since high-aspect-ratio features do not experience the same agitation of a stirred solution as bigger features do, this results in higher stress concentration in the smallest features of the electroplated structure. Since the stress is most severe in the thinnest Ni films, Harsch et al. (1988) undertook a separate study of internal stresses in 5-µm-thick Ni films. For three plating temperatures investigated (42°, 52° and 62°) 5-µm-thick films were found to exhibit minimal or no stress at a current density of 2 A/dm^2. At 62°C, the 5-µm-thick films show no stress at 2 A/dm^2 and remain at a low compressive stress value of about 20 N/mm^2 for the whole current range (1 to 10 A/dm^2). The internal stress at 2 A/dm^2 is minimal at a pH value between 3.5 and 4.5 of a sulfamate electrolyte (Table 17.8). Ni concentration (between 76 and 100 g/l) and wetting agent concentration do not seem to influence the internal stress. The higher the frequency of the pulse in pulse plating (see below), the smaller the internal stress.

The long-term mechanical stability of Ni LIGA structures was investigated by electromagnetic activation of Ni cantilever beams by Mohr et al. (1992). The number of stress cycles, N, necessary to destroy a mechanical structure depends on the stress amplitude, S, and is determined from fatigue curves or S-N curves (applied stress on the x-axis and number of cycles necessary to cause breakage on the y-axis). Experimental results show that for Ni cantilevers produced by LIGA the long-term stability reaches the range of comparable

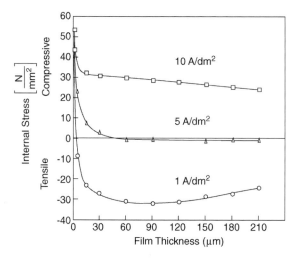

FIGURE 17.19 Influence of the nickel layer thickness on the internal stress. The electrolyte used is described in Table 17.5 (pH = 4; bath temperature = 52°C) [Harsch et al., 1998]. (Courtesy of the Karlsruhe Nuclear Research Center.)

literature data for bulk annealed and hardened nickel specimens. Usually, stress leads to crack initiation, which often starts at the surface of the structure. Since microstructures have a higher surface-to-volume ratio, one might have expected the S-N curves to differ from macroscopic structures, but so far this has not been observed. To the contrary, it seems that the smaller structures are more stable.

Pulse Plating. For the fabrication of microdevices with high deposition rates exceeding 120 μm/hr, a reduction of the internal stress can occur only by raising the temperature of the bath or by using alternative electrodeposition methods. Raising of the bath temperature is not an attractive option, but using alternative electrochemical deposition techniques deserves further exploration. For example, using pulsed (≥500 Hz) galvanic deposition instead of a DC method can be used to influence several important properties of the Ni deposit. Properties such as grain size, purity and porosity can be manipulated this way without the addition of organic additives [Harsch et al., 1988]. In pulse plating, the current pulse is characterized by three parameters: pulse current density, i_p; pulse duration, t_d; and pulse pause, t_p. These three independent variables determine the average current density, i_a, which is the important parameter influencing the deposit quality and is given by:

$$i_a = \frac{t_d}{t_d + t_p} * i_p \qquad (17.6)$$

Pulse plating leads to smaller metal grain size and smaller porosity due to a higher deposition potential. Because each pulse pause allows some time for Ni^{2+} replenishment at the cathode (Ni^{2+} enrichment) and for diffusion away of undesirable reaction products which might otherwise get entrapped in the Ni deposit, a cleaner Ni deposit results. The higher frequency of the pulse leads to the smaller internal stress in the resulting metal deposit [Harsch et al., 1988].

Pulse plating represents only one of many emerging electrochemical plating techniques considered for microfabrication. For more background on techniques such as laser-enhanced plating, jet plating, laser-enhanced jet plating and ultrasonically enhanced plating, refer to the review of Romankiw and Palumbo (1987) and references therein, as well as Madou (1997, chaps. 3 and 7).

Primary Metal Microstructure and Metal Mold Inserts. The backside of electrodeposited microdevices attaches to the primary substrate but can be removed from the substrate if necessary. In the latter case, the substrate may be treated chemically or electrochemically to induce poor adhesion. Ideally, excellent adhesion exists between substrate and resist and poor adhesion exists between the electroplated structure and plating base. Achieving these two contradictory demands is one of the main challenges in LIGA.

Slight differences in metal layer thickness cannot be avoided in the electroforming process. Finish-grinding of the metal samples with diamond paste is used to even out microroughness and slight variations in structural height. Finally, the remaining resist is stripped out from between the metal features (see also under PMMA Stripping) and a primary metal microstructure results.

If the metal part needs to function as a mold insert, it can be left on the primary plating base or metal can be plated several millimeters beyond the front faces of the resist structures to produce a monolithic micromold (see Figure 17.20). In the latter case, to avoid damage to the mold insert when separating it from the plating base, an intermediate layer sometimes is deposited on the base plate, ensuring adhesion of the resist structures while facilitating the separation of the electroformed mold insert from its plating base. In addition, it helps to prevent burrs from forming at the front face of the mold insert as a result of underplating of the resist structures. Underplating can occur because the resist does not adhere well to the substrate, allowing electrolyte solution to penetrate between the two, or the plating solution might attack the substrate/resist interface. Finally, microcracks at the interface might contribute to underplating. Burrs are easily eliminated then by dissolving the thin auxiliary metal layer with a selective etchant, removing the mold without the need for a mechanical load. In view of the observed underplating problems it is surprising that Galhotra et al. (1996) did obtain good plating results when simply mechanically clamping the exposed and developed PMMA sheet to an Ni plating box.

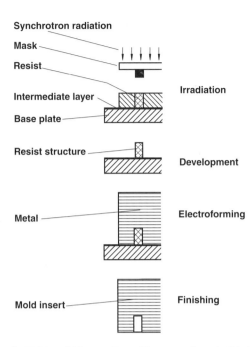

FIGURE 17.20 Fabrication of a LIGA mold insert. (From Hagmann, P. et al., Fabrication of Microstructures with Extreme Structural Heights by Reaction Injection Molding, presented at First Meeting of the European Polymer Federation, European Symp. on Polymeric Materials, Lyon, France, 1987. With permission.)

Plating Automation. An automated galvanoforming facility used by KfK is shown in Figure 17.21 [GmbH, 1994]. This set-up includes provisions for on-line measurement of each electrolyte constituent in flow-through cells and concentration corrections when tolerance limits are exceeded. A computer-controlled transport system moves the individual plating racks, holding the microdevice substrates, through the process stages whereby the substrates are degreased, rinsed, pickled, electroplated and dried and are then returned to a magazine, which can accommodate up to seven racks. The facility is designed as cleanroom equipment, since contamination of the microstructures must be prevented [Maner et al., 1988]. An instrument as described here can be bought, for example, from Reinhard Kissler (the μGLAV 750) [GmbH, 1994]. Although this type of automation is a must for the eventual commercialization of the LIGA technique, it should be recognized that at this explanatory stage over-automatization may be counterproductive.

Plating Issues
Two major sources of difficulty associated with plating of tall structures in photoresist molds are chemical and mechanical incompatibility. The chemical incompatibility means that the photoresist mask may be attacked by the plating solution, while the mechanical incompatibility is film stress in the plated layer which could cause the plated structure to lose adhesion to the substrate.

If the plating solution attacks the photoresist even slightly, considerable damage to the photoresist layer may have occurred by the time a 200-μm-thick structure has been created. The limiting thickness of a plated structure due to a chemical interaction is therefore dependent on both the photoresist chemistry and the plating bath itself. In general, the plating bath must have a pH in the range of 3 to around 9.5 (acidic to mildly alkaline), a fairly wide range that can accommodate many of the commercially available photoresists. A surprising number of plating baths do not fall into that range though. Baths, either very strongly acidic or more than mildly alkaline, tend to attack and destroy photoresist.

The plated structure must have extremely low stress to avoid cracking or peeling during the plating process. In addition, if the structure contains narrow features, the stress must be tensile as any compressive stress would result in buckling of the structure. The primary concern with regard to stresses is the incorporation of 'brightening agents' into the plating solution. These can be selenium, arsenic, thallium and

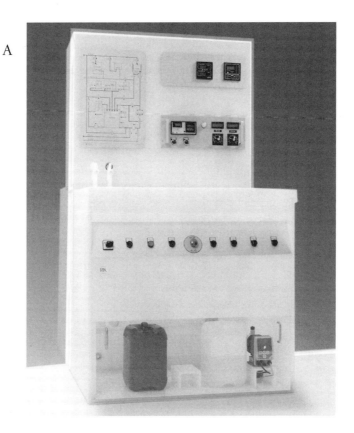

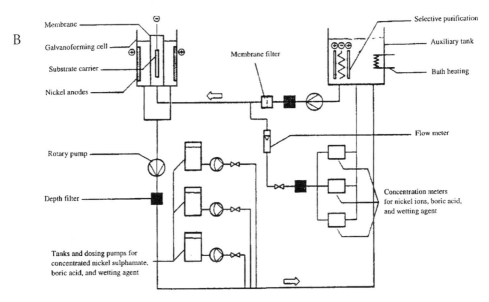

FIGURE 17.21 (A) The μGALV 750 comprises the galvanoforming cell with nickel anodes and substrate carrier and an auxiliary tank incorporating heating and purification auxiliary equipment. Three concentration meters measure nickel ions, boric acid and the concentration of the wetting agent. The concentrations are adjusted automatically by adding via metering pumps [GmbH, 1984]. (With permission from Reinhard Kissler, GmbH, 1994.) (B) Schematic drawing of a galvanoforming unit. (From Maner, A. et al., *Plating and Surf. Finishing*, 60–65, 1988. With permission.)

ammonium ions, as well as others. The brightening materials exhibit different atomic configurations and sizes compared to most materials which we would like to plate such as Cu, Au, Ni, FeNi, Pt, Ag, NiCo etc. This represents the source of a good deal of the stress intrinsic to the plating process. Ag and FeNi are notorious for their high stresses. The Ni-sulfamate bath, presently used extensively in LIGA, is used without brighteners, leading to Ni deposits with extremely low stress. Some other systems low in stress are the electroless plating baths (see section below). These auto-catalytic baths, such as Cu and Ni, coat indiscriminately, but have very low stress. They should be compatible with the LIGA process for certain applications, such as mold insert formation where significant overplating is required.

A third difficulty in finding plating baths compatible with LIGA is that most plating baths are not intended for use in the semiconductor industry. Information including normal deposition parameters, uniformity, stress, compatibility with various semiconductor processes and particulate contamination normally supplied by a vendor is not available. Since the majority of the semiconductor industry does not incorporate these kinds of "thin-film" processes, it may be difficult to determine whether a plating bath will be suitable simply from conversations with the vendor. In many instances, trial and error decide.

Electroless Metal Deposition
Electroless plating may be elected above electrodeposition due to the simplicity of the process since no special plating base is needed. This represents a major simplification for combining LIGA structures with active electronics where a plating base could short out the active electronics. Also, electroless Ni exhibits less stress than electrodeposited Ni, a fact of considerable importance in most mechanical structures. The major concern is the temperature of the electroless plating processes, which is often considerably higher than for electrochemical processes. Electroless deposition was reviewed in Madou (1997, chap. 3). Few studies of electroless plating in LIGA molds have been reported [Harsch et al., 1988]. We want to draw attention, though, to electroless plating applications with large market potential, i.e., the making of very high density read-write magnetic heads and vias and interconnects for three-dimensional ICs. It is interesing to note the similarity in traditional via plating efforts and making of magnetic read-write heads and the more recent LIGA work. Combining innovative plating techniques such as the ones pioneered by Romankiw (1984) and van der Putten and de Bakker (1993) with high-aspect-ratio LIGA molds will create many new opportunities for magnetic heads and vias and interconnects with unprecedented packing densities. Possibly, the progress made in LIGA microfabrication for these three-dimensional IC and magnetic head applications may pay off the most.

17.2.6.9 Summary of Electrochemical and Electroless Processes and Materials

Figure 17.22 features a list of choices for material deposition employing electrochemical and electroless techniques [Ehrfeld, 1994]. The first metal LIGA structures consisted of nickel, copper or gold electrodeposited from suitable electrolytes [Maner, 1988]. Nickel-cobalt and nickel-iron alloys were also experimented with. A nickel-cobalt electrolyte for deposition of the corresponding alloys has been developed especially for the generation of microstructures with increased hardness (400 Vickers at 30% cobalt) and elastic limit [Harsch et al., 1991]. The nickel-iron alloys permit tuning of magnetic and thermal properties of the crafted structures [Harsch et al., 1991; Thomes et al., 1992]. From Figure 17.22 it is obvious that many more materials could be combined with LIGA molds.

Beyond the deposition of a wide variety of materials in LIGA molds, additional shaping and manipulation of the electroless or electrodeposited metals during or after deposition will bring a new degree of freedom to micromachinists. Of special interest in this respect is the 'bevel plating' from van der Putten and de Bakker (1993). With 'bevel plating' van der Putten introduced anisotropy in the electroless Ni plating process resulting in beveled Ni microstructures (see Madou [1997, chap. 7, Figure 7.10]). Combining van der Putten's bevel plating with LIGA one can envision LIGA-produced contact leads with sharp contact points. In our own efforts we are attempting to make sharp Ni needles for use in scanning tunnelling microscopy work. A process to sharpen electrodeposited Ni posts is being developed by applying anodic potentials between an array of posts and a gate electrode, as shown in Madou (1997, chap. 1, Figure 1.8) [Madou, 1995; Akkaraju et al., 1996]. The anodic potentials force the posts to corrode in the electrolyte. Because of the concentric field

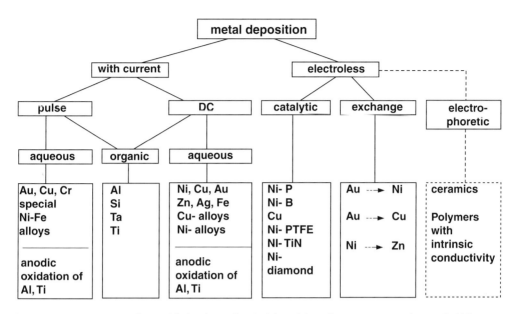

FIGURE 17.22 Processes and materials for electrochemical deposition of LIGA structures. (From Ehrfeld, W., *LIGA at IMM*, Banff, Canada, 1994. With permission.)

distribution between gate and posts, the corrosion works it way concentrically inward, leaving sharp tips. We believe that the only limitation in tip sharpness will be associated with the Ni crystal grain size which might be made very small by controlling the Ni deposition current density.

17.2.6.10 PMMA Stripping

After finish grinding an electroplated LIGA work piece, a primary metal shape results from removing the photoresist by ashing in an oxygen plasma or stripping in a solvent. In the case of cross-linked PMMA, the resist is exposed again to synchrotron radiation, guaranteeing sufficient solubility before being stripped.

17.2.6.11 Comparing Micromolds

LIGA PMMA features, as small as 0.1 µm, are replicated in the metal with almost no defects. When comparing mold inserts made by spark erosive cutting (see Madou [1997, chap. 7]) and by X-ray lithography combined with electroforming, the latter proves superior [Hagmann and Ehrfeld, 1988]. The electroformed structures have a superior surface quality with a surface roughness, R_{max}, of less than 0.02 µm [Hagmann et al., 1987]. A significant application of LIGA might thus be the fabrication of those metal molds that cannot be accomplished with other techniques because of the tight wall roughness tolerances, small size and high aspect ratios. Discussions with classical electroforming industries might prove useful in determining the application shortcomings of their technology.

Micromolds may also be formed by such techniques as precision electro discharge machines (EDM), excimer layer ablation and e-beam writing. In Table 17.9 LIGA molds are compared with laser machined and precision engineered molds [Weber et al., 1996].

From the table it is obvious that LIGA micromolds excel both in very low surface roughness and excellent accuracy.

17.2.6.12 Molding and Demolding

Overview

The previously described LIGA steps to produce primary PMMA structures or complementary primary metal mold inserts require intensive, slow and very costly labor. In order for LIGA to become an economically viable micromachining alternative, one needs to succeed in replicating microstructures

TABLE 17.9 Comparison of Micromolds

Parameters	LIGA	LASER	EDM
Aspect ratio	10–50	<10	Up to 100 (related to dimensions)
Surface roughness	<50 nm	100 nm	0.3–1 μm
Accuracy	<1 μm	A few microns	Some microns
Mask needed?	Yes	No	No
Maximum height	Some microns	A few 100 μm	Microns to millimeters

From Weber, L., W. Ehrfeld, H. Freimuth, M. Lacher, H. Lehr, and P. Pech, *SPIE, Macromachining and Microfabrication Process Technology II*, Austin, Texas, 1996, pp. 156–167. With permission.

successively, in a so-called infinite loop, without having to remake the primary structures. In the plastic molding process illustrated in Figure 17.1, a metal structure produced by electroforming serves as a mold insert used over and over without reverting back to X-ray exposure. For mass-produced plastic devices, one of the following molding techniques is suitable: reaction injection molding (RIM), thermoplastic injection molding and impression molding. As of the time of this writing, most of the plastic molding in LIGA has been executed in Germany with only some early efforts at CAMD in the U.S. [Galhotra et al., 1996; Rogers et al., 1996].

During reaction injection molding, the polymer components are mixed shortly before injection into the mold and the polymerization takes place in the mold itself (e.g., polyurethane and PMMA). Thermoplastic injection molding means heating the polymer above its glass transition temperature and introducing it in a more or less viscous form into the mold, where it hardens again by cooling—e.g., PVC (polyvinylchloride), PMMA and ABS (polyacrylnitrilbutadienstyrol). The pros and cons of these technologies are summarized in Table 17.10. Compression molding means heating the mold material above its glass transition temperature and patterning it in vacuum by impression of the molding tool.

The high-aspect-ratio, electrodeposited metal mold inserts generate new problems compared to the molding and demolding processes for the production of records and videodiscs. Even high-aspect-ratio metal structures can be molded and demolded with polymers quite easily, as long as a polymer with a small adhesive power and rubber-elastic properties, such as silicone rubber, is used. However, rubber-like plastics have low shape stability and would not be adequate. Shape-preserving polymers, on the other hand, after hardening, require a mold with extremely smooth inner surfaces to prevent form-locking between the mold and the hardened polymer. A mold release agent may be required for demolding. Using external mold release agents may prove difficult because of the small dimensions of the microstructures to be molded, as typically they are sprayed onto the mold. Consequently, internal mold release compounds

TABLE 17.10 A Comparison of Various Molding Techniques

	Injection Molding	Reaction Molding
Gas bubbles	Less of a problem	Reduced by evacuating the mold insert and the mold material; contributes to a high cost
Surface quality of mold insert	Less of an issue	Because of low viscosity demands on the surface quality of the mold, insert is very high
Chemicals	Easy to handle the molten plastic	Very reactive and explosive reagents
Reagent temperature (RT), form temperature (FT), and injection pressure (IP) of mold material	RT = 200°C, FT = 25°C, IP = 100 bar; temperature, time, and pressure need to be controlled very accurately	RT = 40°C FT = 70°C IP = 10–100 bar
Slow steps, cycle time	Injection time is long	Medium, with optimized release agent, 11.5 min cycle time
Viscosity of mold material	10^2–10^5 Pa/sec	0.1–1 Pa/sec

Source: Based on Menz, W. and P. Bley, *Mikrosystemtechnik for Ingenieure*, VCH Publishers, Germany, 1993.

LIGA and Other Replication Techniques

need to be mixed with the polymer, without significantly changing the polymer characteristics. Given the early stage of development of molding and demolding plastic LIGA shapes in this country, we have analyzed the German plastic replication work in some detail.

Reaction Injection Molding (RIM)

A laboratory-size reaction injection molding set-up, as shown in Figure 17.23, was used by the KfK group for their LIGA work [Hagmann and Ehrfeld, 1988]. The set-up consists of a container for mixing the various reactants, a vacuum chamber for evacuation of the mold cavity, the molding tool and a hydraulic clamping unit to open and close the vacuum chamber and the molding tool. After the vacuum chamber has been closed and evacuated, the tool is closed by the clamping unit. The mixed reagents are degassed in the materials container and, under a gas overpressure of up to 3 MPa, pushed through the opened inlet valve into the tool holding the evacuated insert mold. To compensate for shrinkage due to polymerization, an overpressure of up to 30 MPa can be applied during hardening of the casting resin. If the holding overpressure is too low, sunken spots appear in the plastic due to shrinkage of the polymer. For PMMA, the volume shrinkage is about 14%. If the mold material is not degassed and the mold cavity is not evacuated, bubbles develop in the molded piece, resulting in defects and possible partial filling of the mold. To harden the mold material and anneal material stress, the reaction injection molding machine can be operated at temperatures up to 150°C.

A variety of resins were tried to fabricate RIM products, including epoxy resins, silicone resins and acrylic resins. The most promising results are obtained with resins on a methylmethacrylate base to which an internal mold release agent is added in order to reduce adhesion of the molded piece to the walls of the metal mold. The mold insert in the evacuated tool is covered by means of an electrically conductive perforated plate, i.e., the gate plate (Figure 17.24A). For filling the mold, injection holes are positioned above large, free spaces in the metal structure and the low-viscosity reactants fill the smaller sections laterally. After hardening of the molded polymer, a form-locking connection between the produced part and the gate plate is established at the injection holes, permitting the demolding of the part from the insert (Figure 17.24B). No damage to the mold insert is observed after up to 100 mold–demold cycles, even at the level of a scanning electron microscope picture. The secondary plastic structures formed are exact replicas of the original PMMA structures obtained after X-ray irradiation and development. If the

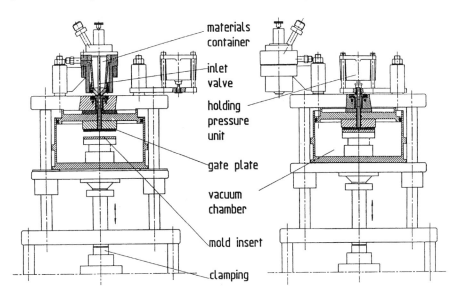

FIGURE 17.23 Schematic presentation of a vacuum molding set-up. With minor changes, this set-up can be used for reaction injection molding, thermoplastic injection molding and compression molding. (From Hagman, P.W., *J. Polymer Process. Soc.*, 4, 188–195, 1988. With permission.)

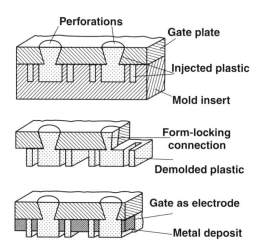

FIGURE 17.24 Molding process with a perforated gate plate. (A) Filling of the insert mold with plastic material. (B) Demolding. (C) Electrodeposition with the gate plate as plating base. (After Menz, W. and P. Bley, *Mikrosystemtechnik fur Ingenieure*, VCH Publishers, Germany, 1993.)

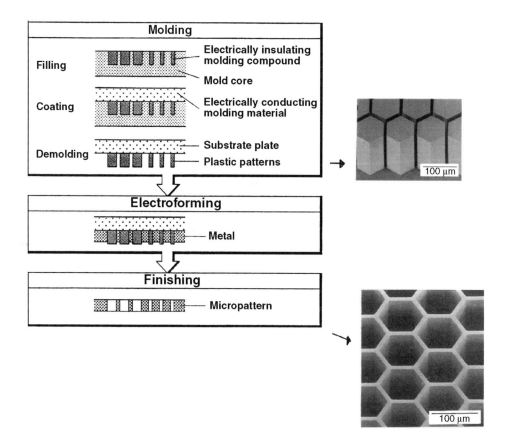

FIGURE 17.25 Ni honeycomb structure with cell diameter of 80 μm, wall thickness of 8 μm, and wall height of 70 μm. A structure like this could not be produced using a gate plate; a conductive plastic is used instead. The Ni plating occurs on the conductive plastic carrier pate, which is fused with the insulating plastic of the microstructure. (From Harmening, M. et al., *Makromol. Chem. Macromol. Symp.*, 50, 277–284, 1991. With permission.)

final product is plastic, the molding process can be simplified considerably by omitting the gate plate. The mold insert is then cast over with the molding polymer in a vacuum vessel, building a stable gate block over the mold insert after hardening. The polymeric gate block is used to demold the part from the insert.

The secondary plastic templates usually are not the final product but may be filled with a metal by electrodeposition as with the primary molds. In this case, the secondary plastic templates must be provided with an electrode or plating base. The gate plate can be used directly as the electrode for the deposition of metal in the secondary plastic templates (see Figure 17.24C). It is important that the mold insert is pressed tight against the gate plate so that no uniform isolating plastic film forms over the whole interface between the gate plate and the mold insert; otherwise, electrodeposition might be impossible. Safe sealing between mold insert and gate plate is achieved by use of soft-annealed aluminum gate plates into which the insert is pressed when closing the tool. It was found that the condition to be met for perfect deposition of metal does not require that the gate plate be entirely free of plastic [Hagmann and Ehrfeld, 1988]. It suffices that a number of electrical conducting points emerge from the plastic film. Transverse growth of the metal layer produced in electrodeposition allows a fault-free, continuous layer to be produced. The secondary metal microstructures are replicas of the primary metal structures and a comparison of, for example, a master separation nozzle mold insert with a secondary separation nozzle structure did not reveal any differences in quality. After the metal forming, the top surface must be polished and the plastic structures must be removed. Depending on the intended use of the secondary metal shape, one can leave it connected to the gate plate or the gate plate can be selectively dissolved or mechanically removed.

Gate plates can be applied only if the microstructures are interconnected by relatively large openings. The injection holes in the gate plate are produced by mechanical means with openings of about 1 mm in diameter and must be aligned with large openings in the mold insert. Many desirable microstructures cannot be built this way. For example, in Figure 17.25 we show a Ni honeycomb structure with a cell diameter of 80 µm, a wall thickness of 8 µm, and a height of 70 µm [Harmening et al., 1991]. The holes in a gate plate are too large to inject plastic into this structure and the interconnections in the honeycomb are too small to provide a good transverse movement of the polymerizing resin. Alternative approaches enabling the plating of any shape are discussed next.

Electrically Conductive Plastic Plates as Carriers and Cathodes. It is possible to use electrically conducting plastic as a starting layer for electrodeposition [Harmening et al., 1991]. In this process, illustrated in Figure 17.25, the mold insert is filled initially with the electrically insulating, e.g., unfilled, casting resin. The surplus molding material is scraped off the front of the mold insert followed by covering of the mold insert with the electrically conducting carrier plate. While the casting resin cures, the carrier plate is forced against the front of the mold insert. In the curing process, the electrically insulating material in the honeycomb structures fuses with the carrier plate. After demolding one obtains insulating plastic structures firmly connected with the electrically conducting carrier plate. The conductive plastic is used for the electrodeposition between the insulating plastic features and, after removal of the plastic with a solvent, a plate with metal microstructures results. To achieve the desired conductivity in the plastic carrier plates one adds metal powders (e.g., copper, silver or gold) or carbon powders of an average grain diameter of less than 3 µm to the cast resin. From the metallic powder additives, copper oxidizes too quickly, leading to a high resistivity composite material. Silver-filled resins, on the other hand, process easily and set into homogeneously conducting composites, i.e., free from bubbles and blisters. Carbon-filled resins only lead to homogeneous composite materials to a certain extent. At an Ag filler (Ag, Demetron type 6321-8000) content of 75 wt%, the PMMA attained a value of composite resistivity of 2.4×10^{-4} Ωcm. The PMMA composite must also contain an internal mold release agent to make the release of the metal shapes easy (see Demolding). The addition of a demolding agent decreases the glass transition temperature (T_g) and the temperature for the beginning of the softening process (T_s) with respect to the values for pure PMMA, but both properties are practically independent of filler concentration. The thermal and mechanical properties of the composites are dependent on the filler content. With increasing filler content, the composite becomes stiffer, as indicated by an increasing Young's modulus and a drop of the maximum loss factor, tan δ, at the glass transition temperature. The coefficient of thermal expansion decreases with increase in filler content. To ensure high dimensional stability, both a high Young's modulus and a low thermal expansion coefficient are beneficial. For uniform electrodeposition the composite material must also have a low, homogeneous resistivity. Metal deposition starts at the filler particles exposed at the surface. Those particles are connected to the bulk of the electrode by a network of conducting particles. Fault-free electroforming of patterns in the micrometer range requires a very small distance between the electrical points of contact, the starting points of electrodeposition, on the surface of the composite material. At a filler content of 75 wt%, the mean distance between the starting points of electrodeposition attained is 1.5 µm.

The micro-pattern illustrated in Figure 17.25 was made by molding electrically insulating micro-patterns of PMMA on the surface of substrate plates of PMMA filled with 75 wt% of silver. The pattern consists of a total of 74,400 single patterns on an area of 500 mm^2 and serves as an infrared high-pass filter [Harmening et al., 1991].

Injection Molding
Thermoplastic injection molding is the technique used to make compact discs involving features about 0.1 µm high and minimum lateral dimensions of 0.6 µm, i.e., an aspect ratio of 0.16. In LIGA, the aspect ratio easily can be 1000 times larger. Injection molding can work here as well, as long as the mold insert walls are kept at a uniform temperature above the glass transition temperature (T_g) of the plastic material, so that the injected molten plastic mass does not harden prematurely. The latter requires a rather slow filling of the mold insert with the plastic mass. To be able to fill features with aspect ratios higher than 10, the temperature of the tool holding the insertion mold should be higher than what is typically used in more conventional PMMA injection molding applications. One should stay well below the PMMA melting temperature of 240°C (say, 70°C less), as the microstructure might show temperature-induced defects if one works at a temperature closer to the melting temperature [Eicher et al., 1992]. For injection molding, an apparatus very similar to the one shown in Figure 17.23 can be used.

The mold insert must again have extremely smooth walls to be able to demold the structures. Walls with the required smoothness cannot be produced with a classical technique such as spark erosion. Because PMMA absorbs water in air (up to 0.3%), leading to poorer wall quality, the PMMA polymer flakes need to be dried for 4 to 6 hours at 70° before use in injection molding.

Compression Molding (also Relief Printing or Hot Embossing)
Compression molding or relief printing is the method of choice for incorporating LIGA structures on a processed Si wafer. Hot embossing takes place in a machine frame similar to that of a press. In this method, microstructures are generated by molding of a thermoplastic polymer (mold material) using a molding process as illustrated in Figure 17.26. In a first step, the mold material is applied by polymerization onto the processed wafer with an electrically isolating layer and overlayed with a metal plating base (Figure 17.26A). At the glass transition temperature, the plastic is in a viscoelastic condition suitable for the impression molding step. Above this temperature (about 160°C for PMMA) the mold material is patterned in vacuum by impression with the molding tool. In order to avoid damaging the electronic circuits by contact with the molding tool, the mold material is not completely displaced during molding. A thin, electrically insulating residual layer is left between the molding tool and the plating base on the processed wafer (Figure 17.26B). The residual layer is removed by reactive ion etching (RIE) in an oxygen plasma etch (Figure 17.26C), freeing the plating base for subsequent electrodeposition. The oxygen RIE process is as anisotropic as possible, so that the sidewalls of the structures do not deteriorate and the amount of material removed from the top of the plastic microstructures is small compared to the total height. After electrodeposition on the plating base the metal microstructures are laid bare by dissolution of the mold material (Figure 17.26D). Finally, using argon sputtering or chemical etching, the plating base in between the metal microstructures is removed so that the metal parts do not short-circuit (Figure 17.26E). In the case of chemical etching, a very fast etch should be employed to avoid etching of the plating base from underneath the electrodeposited structures [Michel et al., 1993; Eicher et al., 1992].

At KfK, hot embossing is mainly used for small series production, whereas injection molding is applied to mass production. Both techniques are applied to amorphous (PMMA, PC, PSU) and semicrystalline

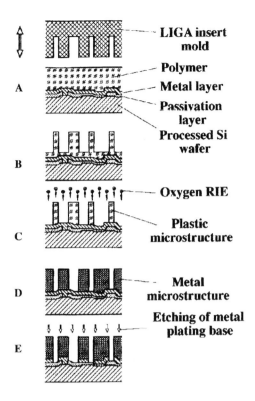

FIGURE 17.26 Fabrication of microstructures on a processed Si wafer. (A) Patterned isolation layer, conductive plating base and *in situ* polymerization of PMMA. (B) Impression molding using a LIGA primary insert mold. (C) Removal of remaining plastic from the plating base by a highly directional oxygen etch. (D) Electrodeposition of metal shape. (E) Removal of plating base from between the metal structures by argon sputtering. (After Menz, W. and P. Bley, *Mikrosystemtechnik fur Ingenieure*, VCH, Weinheim, Germany, 1993.)

thermoplastics (POM, PA, PVDF, PFA). Reaction injection molding is used for molding of thermoplastics (PMMA, PA), duroplastics, (PMMA, expoxide) or elastomers (silicones) [Ruprecht et al., 1995].

Demolding

The demolding step is carried out by means of a clamping unit at preset temperatures and rates (see Figure 17.23). Demolding is facilitated by an internal mold release agent such as PAT 665 (Wurtz GmbH, Germany), normally employed with polyester resins [Hagmann and Ehrfeld, 1988]. The optimum yield with this agent and Plexit M60 PMMA occurs at 3 to 6 wt%. The yield drops very fast below 3 wt% as adhesion between the molded piece and the mold insert becomes stronger. These adhesive forces are estimated by qualitatively determining the forces necessary to remove the plastic structures from the mold insert. An upper limit is 10 wt% where the MMA does not polymerize anymore. Above 5 wt%, the Young's modulus and, hence, the mechanical stability decrease, and above 6 wt%, pores start forming in the microstructures. The internal mold release agent also has a marked influence on the optimum demolding temperature. The demolding yield decreases quickly above 60°C for a 4 wt% PAT internal mold release agent. With 6 wt%, one can only obtain good yields at 20°C [Hagmann et al., 1987; Hagmann and Ehrfeld, 1988]. The molding process initially led to a production cycle time of 120 min. For a commercial process, much faster cycle times are needed, for instance, by optimizing the mold release agent. With that in mind, the KfK group started to work with a special salt of an organic acid leading to a 100% yield at a release agent content of 0.2 wt% only and a temperature of 40°C (at 0.05 wt% a 95% yield is still achieved). At 80°C, a 100% demolding yield was obtained and, significantly, a cycle time of 11.5 min was reached. During these experiments the mold was filled at 80°C and heated to 110°C within 7.5 min. As the curing occurs at 110°C, the material needs to be cooled down to 80°C for demolding. Moreover, the 0.2 wt% of the 'magic release agent' did not impact the Young's modulus and the glass transition temperature of Plexit M60 [Hagmann and Ehrfeld, 1988].

17.2.6.13 LIGA Extensions: Movable, Stepped, and Slanted Structures

Modifying the LIGA process enables movable microstructures and microstructures with different level heights, i.e., stepped and inclined sidewalls. These features often are essential for mechanical and optical devices [Mohr et al., 1992].

Movable Structures

As in surface micromachining (see Chapter 16), sacrificial layers make it possible to fabricate partially attached and freed metal structures in the primary mold process [Burbaum et al., 1991]. The ability to implement these features leads to assembled micromechanisms with submicron dimension accuracies, opening up many additional applications for LIGA, especially in the field of sensors and actuators. The sacrificial layer may be polyimide, silicon dioxide, polysilicon or some other metal [Guckel et al., 1991]. The sacrificial layer is patterned with photolithography and wet etching before polymerizing the resist layer over it. At KfK, a several-micron-thick titanium layer often acts as the sacrificial layer because it provides good adhesion of the polymer and it can be etched selectively against several other metals used in the process. If for exposure the X-ray mask is adjusted to the sacrificial layer, some parts of the microstructures will lie above the openings in the sacrificial layer, whereas other parts will be built up on it. These latter parts will be able to move after removal of the sacrificial layer [Mohr et al., 1990]. The fabrication of a movable LIGA structure is illustrated in Figure 17.27A.

Stepped Microstructures

In principle, stepped LIGA structures can be accomplished by means of X-ray lithography with a single mask with stepped absorber structures. In this manner, variable dose depositions can be achieved at the same resist heights. The variable dose results in different molecular weights and,

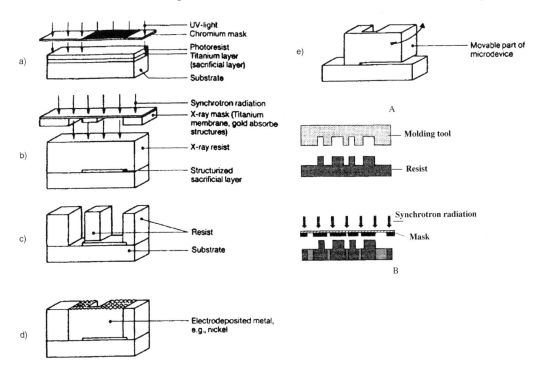

FIGURE 17.27 LIGA extensions. (A) Movable microstructures. (B) Stepped microstructures. (From Mohr et al. *Microsyst. Technol. 90*, Springer, Berlin, 1990. With permission.)

hence, in a different developing behavior. This technique unfortunately leads to rounded features and poor step-height control.

Alternatively, one can first relief print a PMMA layer, e.g., by using a Ni mold insert made from a first X-ray mask. Subsequently, the relief structure may be exposed to synchrotron radiation to further pattern the polymer layer through a precisely adjusted second X-ray mask. To carry out this process, a two-layer resist system needs to be developed consisting of a top PMMA layer which fulfills the requirement of the relief printing process, and a bottom layer which fulfills the requirements of the X-ray lithography [Harmening et al., 1992]. The bottom resist layer promotes high molecular weight and adhesion, while the top PMMA layer is of lower molecular weight and contains an internal mold-release agent. This process sequence combining plastic impression molding with X-ray lithography is illustrated in Figure 17.27B. The two-step resist then facilitates the fabrication of a mold insert by electroforming, which can be used for the molding of two-step plastic structures. Extremely large structural heights can be obtained from the additive nature of the individual microstructure levels. Alternatively, to make stepped structures one can also employ multiple masks and irradiation steps. The latter method only requires one blend of PMMA resist.

Slanted Microstructures

Changing the angle at which synchrotron radiation is incident upon the resist, usually 90°, enables the fabrication of microstructures with inclined sidewalls [Bley et al., 1991]. Slanted microstructures may be produced by a single oblique irradiation or by a swivel irradiation. One application of microstructures incorporating inclined sidewalls is the vertical coupling of light into waveguide structures using a 45° prism [Gottert et al., 1992]. Such optical devices must have a wall roughness of less than 50 nm. Due to the sharp decrease of the dose in the resist underneath the edge of the inclined absorber and because of the sharp decrease of the dissolution of the resist as a function of the molecular weight in the developer, usually no deviation of the inclination of the sidewall over the total height of the microstructure is found.

17.2.6.14 Alternative Materials in LIGA

Alternative X-ray resist materials and a variety of other metals for electroplating besides Ni were discussed above. Following we will discuss some alternative molding materials. Besides PMMA and POM, used in a commercially available form, semicrystalline polyvinylidenefluoride (PVDF), a piezoelectric material, has been used to make polymeric microstructures [Harmening et al., 1992]. The optimum molding temperature of the PVDF was found to be 180°C. The PVDF structures could be molded without using mold release agents. Fluorinated polymers such as PVDF will also enable higher temperature applications than PMMA. Polycarbonate (PC), well known for molding compact discs, is under development and should perform well [Rogner et al., 1992].

Other materials that have the ability to flow or can be sintered, such as glasses and ceramics, can be incorporated in the LIGA process as well. In the case of ceramics, for example, the plastic microstructures are used as disposable or lost molds. The molds get filled with a slurry in a slip-casting process. Before sintering at high temperatures, the plastic mold and the organic slurry components are removed completely in a burnout process at lower temperatures. First results have been obtained for zirconium oxide and aluminum oxide. The process is illustrated in Figure 17.1. An important application for ceramic microstructures is the fabrication of arrays of piezoceramic columns embedded in a plastic matrix. Since the performance of these actuators is linked to the height and width ratio of the individual ceramic columns, as well as the distance between the columns in the array, LIGA, with its tremendous capacity for tall, dense and high-aspect-ratio features is an ideal fabrication tool [Preu et al., 1991; Lubitz, 1989].

A very interesting method to make LIGA ceramic or glass structures in the future may be based on sol-gel technology. This technique involves relatively low temperatures and may enable LIGA products such as glass capillaries for gas chromatography (GC) or possibly even high-TC superconductor actuators. Sol-gel techniques should work well with LIGA-type molds if they are filled under a vacuum to eliminate trapped gas bubbles. In the sol-gel technique a solution is spun on a substrate which is then given an initial firing at around 200°C, driving off the solvent in the film. Subsequently, the

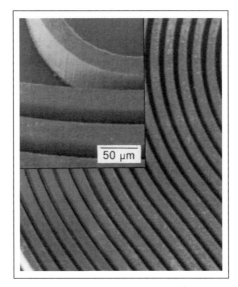

Copper coil. (Courtesy of IMM, Germany.)

substrate is given a high-temperature firing at 800 to 900°C to drive out the remaining solvents and crystallize the film. A major issue deals with the large shrinkage of the sol-gel films during the initial firing. For example, the maximum thickness of high-TC superconductor films currently achievable measures approximately 2 μm due to the high stress in the film caused during shrinkage. The latter is related to the ceramic yield or the amount of ceramic in the sol-gel compared to the amount of solvent. The sol-gel technique, in general, suits the production of thicker films well, as long as high ceramic yield sol-gels are used. As LIGA-style films tend to be thick, a reduction in the amount of solvents to allow a high ceramic yield after firing would be in order to make sol-gel compatible with LIGA. This would require some development. Also lead zirconate titanate (PZT) devices could be made with sol-gel technology and LIGA, most likely by applying the metal plated structure rather than the direct PMMA template for the creation of the device, as PZT must be processed in several elevated temperature steps. The PZT sol-gel contains no particles that would prevent flow into small channels in a plated mold. The sol-gel contains high-molecular-weight polymer chains that hold the constituent metal salts, which are further processed into the final ceramic film. Thus, the sol-gel could be formed into the mold in a process much like reaction-injection molding, with the mold being filled with the chemistry under vacuum.

17.3 LIGA Applications

17.3.1 Introduction

A plethora of potential applications of LIGA have been investigated. Examples include a capacitive accelerometer, a *copper coil* for an inductive position sensor, a *resonant mesh infrared band pass filter* with a cross-shaped dipole pattern of 18.5-μm slit length and 3-μm slit width and a distance between two crosses of 1.5 μm, electrical and optical interconnections, micron-sized nozzles, microfiltration membranes, spinneret plates for synthetic fiber production, microvalves, nozzles for fuel injection, ink-jet nozzles, microchannels and mixing elements. The economic and technical justification for applying LIGA for several of those applications requires further exploration. Nevertheless, these studies illustrate the potential of the method.

The following list of LIGA microstructures includes the more significant research devices realized as of this writing. The majority remain experimental and their performance cannot accurately be assessed yet, or they represent simple test structures designed to prove the feasibility of the process. Geometrical

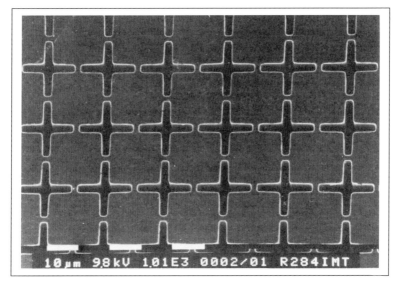

Resonant-mesh infrared band pass filter—3 μm slit width, 18.5 μm slit length. (Courtesy of IMM, Germany.)

specifications of LIGA structures depend on the design, the fabrication process and the material. Typical values of LIGA features incorporated in the listed applications are [Rogner et al., 1992]:

- Sructural height (typical): 20–500 μm
- Minimum dimension: 1–2 μm
- Smallest x, y surface detail: 0.25–0.5 μm
- Surface roughness: 0.03–0.05 μm (max. peak to valley)
- Maximum x, y dimension: 20 × 60 mm

The description of LIGA examples includes a discussion on the appropriateness of LIGA as a fabrication tool and complements the microfabrication applications discussed in Madou (1997, chap. 10).

17.3.2 Microfluidic Elements

17.3.2.1 Introduction

In fluid and gas dynamics a number of devices must have micrometer dimensions or have their efficiency and performance improved by miniaturization [Ehrfeld et al., 1987]. The impact of miniaturization in fluidics is discussed in detail in Madou (1997, chap. 9). Fluidic devices such as nozzles, valves, mixing chambers and pumps may at first glance appear good LIGA applications because both high aspect ratio and accuracy often drive the manufacturing choice. Moreover, the flexibility in the choice of materials and the wide dynamic range in size gives LIGA a significant advantage over bulk Si micromachining approaches. Surface micromachining in poly-Si does not suit microfluidics; the height of achievable structures is too low and the tendency of components to stick to each other would only be exacerbated when in contact with fluids. Fluidic devices (for example, fluidic logic devices) mostly made from photosensitive glass (see also Madou [1997, chap. 7]) reached their peak of popularity in the mid-1960s but lost out to electronic devices by the early 1970s. A new manufacturing technique will not revive fluidic logic; an approach proven noncompetitive. In the case of other fluidic devices, such as pneumatic sensors and analytical equipment incorporating fluidic elements, photosensitive glass again is an inexpensive alternative. LIGA might yet play an important role in those cases where tolerances are very tight, or extreme aspect ratios or very low wall roughness dominate. Examples include gas separation nozzles, spinneret nozzles and very sensitive flow meters.

17.3.2.2 Double Deflecting Gas Separation Nozzles for Uranium Enrichment

In separation nozzles, the optimum operating pressure is inversely proportional to the characteristic dimension of the flow; smaller nozzles make for an increased separation efficiency. The manufacture of tiny separation nozzles for separation of uranium isotopes at KfK [Becker et al., 1982; 1986] actually initiated the development work of LIGA (see Figure 17.2 for a picture of these micronozzles). In 1982, these nozzles were made by electroplating Ni in the primary resist structure; by 1985 they resulted from a combination of electroforming and plastic molding. The minimum slit width in the curved nozzle measures 3 µm, and the slit length is about 300 µm. Since a higher gas pressure results in a corresponding reduction of the compressor size and the piping system for a given throughput of a separation nozzle plant, the dimension reduction of the individual nozzles has a direct economic advantage [Ehrfeld et al., 1987].

The feasibility of the LIGA separation nozzles was demonstrated, but the parts were never implemented as government funding for the project stopped (the KfK work was carried out in Germany for the Brazilian government and the U.S. finally objected to the sale of the uranium separation system to Brazil). The accuracy and aspect ratio required for the reliable gas separation of uranium isotopes still make the gas separation nozzles one of the best demonstrations for the correct application of LIGA. No other technique can yet produce a mold insert with the required aspect ratios and tolerances (see also Spinneret Nozzles).

17.3.2.3 Spinneret Nozzles

Profiled capillaries (nozzles) in a *spinneret plate* for *spinning synthetic fibers* from a molten or dissolved polymer normally are produced by spark erosion (see Madou [1997, chap. 7]). This process establishes a practical lower limit of 20 to 50 µm for the minimum characteristic dimension. Smaller, more precise nozzles can be produced by LIGA [Maner et al., 1987].

Spinneret nozzles lend themselves to LIGA from the technical point of view. Compared to fabrication by spark erosion, the minimum characteristic dimensions with LIGA can be reduced by an order of

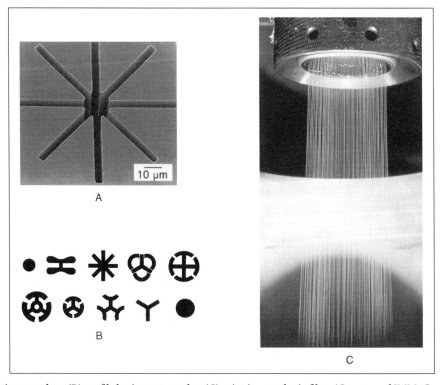

(A) Spinneret plate; (B) profiled spinneret nozzles; (C) spinning synthetic fiber. (Courtesy of IMM, Germany.)

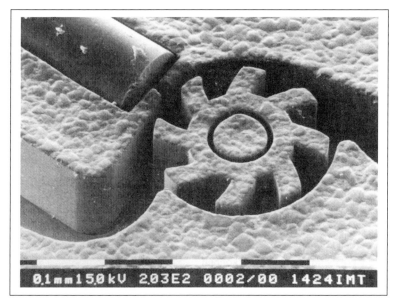

Turbine rotor with integrated optical fiber. (Courtesy KfK, Karlsruhe, Germany.)

magnitude, and a high capillary length with excellent surface finish can easily be obtained. Moreover, the LIGA process makes all the nozzles in parallel while the spark erosion operates as a serial technique. The market might focus on niche applications such as specialty, multilumen catheters. Also, medical use of atomizers for dispensing drugs is projected to increase dramatically in the coming years, this might cause a run for a wide variety of precise, inexpensive micronozzles.

17.3.2.4 LIGA Turbine

A small, released nickel **turbine rotor**, rotated by blowing gas through channels embedded in the surrounding structure, was produced using the LIGA technique. The turbine exemplifies one of the first movable LIGA structures. The axle, rotor and channels of the turbine are fabricated of plated nickel [Menz et al., 1991]. As the shaft and the gear are produced in one mask step, no alignment procedure of the finished components is necessary. Some data on the performance of the turbine were presented by Mohr et al. (1990; 1991). The maximum speed measured 2500 rotations per second, and a total of 85×10^6 rotations on one turbine have been recorded. To make this turbine into a volumetric flow sensor requires a very tight packaging scheme. The packaging must minimize gas flow around the turbine and channel it all through the turbine. Another requirement is for very hard materials to enhance life-time. An integrated optical fiber measures rotation speed in the microturbine flow sensor. Flow rates from 10 to 50 sccm min^{-1} have been detected this way [Himmelhaus et al., 1992].

More materials-related development work is needed to further mature this application. For example, the bonding of a top layer to enclose LIGA structures is an area that still needs a good solution. In principle, flow sensors embody a good LIGA opportunity since they require both accuracy and a high aspect ratio. Improved fabrication accuracy may make for more uniformity of flow sensors from and between large batches. The high aspect ratio combined with high manufacturing accuracy will ensure more sensitive flow sensors for special applications.

17.3.2.5 Microvalve

Microvalves represent one of the most eagerly pursued microdevices today, either as separate components or for integration in micromachined fluidic systems (pumps). No good solutions have been presented; valves are either too large or too expensive or they do not seal tightly enough.

Systems of *four LIGA microvalves* assembled from thin titanium membranes, an electroplated valve body in nickel or gold, and glass covers have been assembled [Schomburg et al., 1993]. The microvalves

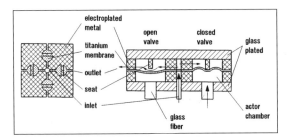

LIGA microvalves. Four valves connected to common inlet. (From Schomburg et al., *J. Micromech. Microeng.*, 4, 186–191, 1994. With permission.)

find a common support in the center of the four-valve system and can be actuated individually by an external pneumatic system. To close the flow channel, the 2.7-µm thin titanium membrane is pressed pneumatically onto a valve seat dividing each valve chamber into two parts. If the membrane is not actuated, it is deflected off the valve seat by the pumped medium. Besides the lithography and plating of the valve body, no other LIGA steps are involved and the valve mechanism is completely traditional [Schomburg et al., 1994]. One wonders why the manufacture of the current microvalves did not follow the traditional thermoplastic molding method as in the case of the injection molded polysulfone pump discussed below.

The LIGA manufacturing technique introduced here does not improve upon the technological shortcomings of these types of valves, i.e., the high cost and poor valve sealing. Several commercially available non-LIGA micromachined valves are compared in Madou (1997, chap. 10).

17.3.2.6 Micropump

For analysis and dosage of small amounts of liquids and gases, pumps capable of handling minute amounts of well-defined fluid quantities are needed. Many attempts to make such micropumps have been made but no clear solution yet stands out. Current micropumps are too expensive and, as we discussed above, do not have adequate valving systems. The following short history of LIGA pumps clearly illustrates that LIGA does not present the answer to every micromachining problem at hand.

The *first LIGA pump* for both water and air was a micromembrane pump with a case made of electroplated gold and incorporating a complex 1-mm-diameter valve consisting of a stiff titanium membrane working in tandem with a flexible polyimide membrane [Rapp et al., 1994]. The pump was manufactured by KfK using X-ray lithography to fabricate the gold pump case, photolithography for defining both the titanium and the polyimide membrane layers, and gluing on of a top glass plate. The LIGA pump was tested with an external pneumatic drive [Rapp et al., 1993; 1994]. The complexity of the first device led to the fabrication of a second *micro-pump* not involving any LIGA steps at all by the same KfK group. In this

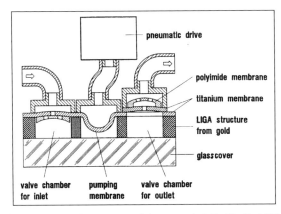

First LIGA pump. (From Rapp et al., *Sensors and Actuators A*, A40, 57–61, 1994. With permission.)

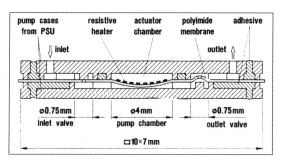

Alternative non-LIGA pump. (From Bustgens et al., Proceedings MEM '94, Oiso, Japan, 1994. With permission.)

case, the device was made by combining traditional thermoplastic molding with membrane techniques [Bustgens et al., 1994]. The pump case consisted of two pieces made by injection molding of polysulfone and had lateral dimensions of 7 by 10 mm. The depth and diameter of the pump chamber were 100 μm and 4 mm, respectively, and the chamber was covered by a 2.5-μm-thick polyimide membrane. The inlet and outlet valves were formed by orifices in the membrane positioned over a valve seat which is a part of the case. The membrane was driven in this case by an integrated thermopneumatic actuator.

The above short history of the LIGA pump demonstrates the fallacy of applying a new micromachining method to every micromachining issue without establishing a clear advantage beforehand. A new and expensive manufacturing technology should only be applied to a commercial need if it enables some new approaches (e.g., LIGA might provide much smoother valve seat surfaces to make a stronger actuator, etc.) or if it ultimately affords an old approach to be implemented more efficiently and at lower cost. In the case of the LIGA microvalves and micropumps, neither is the case and we are still struggling today to find a better, low-cost tightly closing microvalve. The micropump closest to market introduction is a bulk Si micromachined pump incorporating a piezoelectrically activated membrane developed by Debiotech (see Madou [1997, chap. 10]).

17.3.2.7 Fluidic Amplifiers

A LIGA *fluidic flip-flop*, also called a wall-attachment amplifier, is based on the Coanda effect. A jet emanating from a supply nozzle is attached to one of the two attachment walls. The jet can be switched from one position to the other by the temporary application of a small control pressure at the control port next to the attached jet. The jet reaching one of the two output ports builds up a pressure which

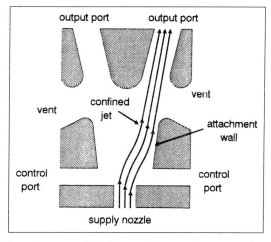

Fluidic flip-flop. Bistable wall attachment amplifier. (From Vollmer et al., *Sensors and Actuators A*, 43, 330–334, 1994. With permission.)

can be used, for example, to drive a microfluidic actuator. Using LIGA, bistable wall attachment amplifiers have been manufactured with heights up to 500 µm [Vollmer et al., 1994a; 1994b].

The LIGA work from the early and mid-1990s as applied to fluidic amplifiers lacks originality. More sophisticated fluidic components providing functions such as flip-flops, logic elements such as OR-NOR gates and AND gates, proportional amplifiers and fluid resistors already existed in the 1960s. The fluidic components made in those days still provide an excellent demonstration for the capabilities of photofabrication with photosensitive glass (see Madou [1997, chap. 7] and Humphrey and Tarumoto [1965]). Humphrey and colleagues made detailed technical and economic comparisons of the different manufacturing options for fluidic elements before embarking upon photofabrication as the manufacturing choice, a practice every micromachinist should adapt. In the 1980s, Si micromachining was applied to make the same fluidic logic devices, as well as fluidic valves and analytical equipment (for a review of the more recent work, see Elwenspoek et al. [1994]). Despite these attempts, neither LIGA nor any of the Si microfluidics produced today surpass the degree of maturity achieved with the photofabrication method. The 'fluidic' logic elements could achieve many of the functions of an electronic circuit and were more robust, simpler and better able to operate in hostile environments. They reacted much slower (milliseconds instead of nanoseconds) and were larger than their electronic counterparts. The two latter factors added up to the eventual demise of microfluidics in the 1960s.

In view of the above, the use of LIGA micromachining for fluidic logic is hardly justifiable. If fluidic logic would, for some unforeseen reason, become competitive again with an electronic approach, the earlier photofabrication machining approach would still be the better approach for most applications since cost, not accuracy, is the major concern.

As demonstrated in the fabrication of some of the fluidic devices discussed so far, zero-basing the validity of a new approach to the micromanufacture of a microdevice at hand should be the first step in any micromachine design. This requires a thorough knowledge of the intended application, suppression of the urge to apply a new manufacturing tool to anything small, and a literature search stretching beyond the last 5 years. The role of LIGA in analytical instrumentation, especially equipment involving optics, may still represent an opportunity as tolerances on wall roughness and accuracy may be very stringent. Even in these latter applications, advantages over established techniques are not always self-evident.

17.3.2.8 Membrane Filters

A key component in all of today's filtration membranes is a microporous structure with uniform and well-defined pore sizes and a high overall porosity. Most current microfiltration membranes are characterized by a relatively broad distribution with respect to the size and position of the pores. When using membranes made by the nuclear track technique, isoporous membranes result. However, these type of filters suffer from limited porosity. The process to make LIGA membrane filters, on the other hand, allows the production of isoporous structures with a very high porosity [Becker et al., 1986]. Also, much smaller capillaries can be realized than when applying spark erosion. Since the membrane thickness may be some hundred micrometers, a high mechanical stability can be realized in combination with an extremely high transparency [Ehrfeld et al., 1988].

The porous membrane application makes a lot of sense for LIGA. This application would be of great interest, especially if we could make large sheets of the porous material in a continuous process. In principle, large primary molds could be manufactured by rolling long sheets of PMMA, in a continuous process, through the X-ray exposure zone. Table 17.11 presents a summary of the expected benefits of LIGA microfilters (MicroParts).

A *transverse flow microfilter* (cross-flow) with porous membrane built up by X-ray lithography is described in a 1989 patent by Ehrfeld et al. (U.S. Patent 4,797, 211).

17.3.3 Electronic Microconnectors and Packages

The decrease in critical dimensions in ICs results in a corresponding demand for higher density electrical connections with subminiaturized dimensions. ***Electrical multipin microconnectors*** can be equipped with

TABLE 17.11 LIGA Microfilter Characteristics

Uniform pore size	Sharp separation boundary
Extremely high porosity	High filtration flow
Designable pore shape	Optimized filtration characteristics
Constant channel cross-section	Ideal surface filter
Integrated supporting structures	High strength
Wide range of materials	Suitable for corrosive and high temperature media

Source: From *The LIGA Technique*, courtesy of MicroParts.

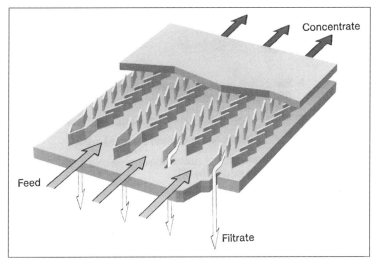

Transverse flow microfilter. (Courtesy of IMM, Germany.)

Electrical multipin microconnectors. (From Ehrfeld, W., Proceedings, Micro System Technologies '90, Berlin, 521–528, 1990. With permission.)

precise guiding structures which allow multiple nondestructive insertions and separations [Ehrfeld, 1990]. Applications of these arrays include detachable interconnectors of high quality ICs or miniaturized multipin sensor heads [Ehrfeld, 1990]. The typical mating force of such a LIGA multipin electrical connector is 2 and 15 cN/pin for the latching. The typical current capacity is 15 mA/pin and values up to 4 A before destruction have been observed.

Given that the required tolerances for microconnectors will become increasingly stringent, the electronic interconnect application does seem promising. Pseudo-LIGA techniques (for example, based on UV-sensitive polyimides) may prove more cost effective solutions for this application in the short run. But, further down the line, we may envision microconnectors being fabricated as part of the parallel batch IC process. LIGA with its wide dynamic range can cover the submicron dimensions of an IC as well as the millimeter dimensions of the connectors. Once the insertion point of X-ray lithography in the IC industry has become a fact, this 'integrated connector strategy' may become very important.

Most of the costs for producing an integrated chip relate to the package and introduction of the circuit into the package. Part of the cost differences follow from the vast difference between integrated circuit technology and packaging technology. In the future, we foresee packaging becoming more and more part of the IC process. The IC package, for example, could be replaced by a LIGA sealing wall around the edge of the chip and a seal of a metal diaphragm over the top. It may also be possible to plate completely over the top of the chip rather than sealing with a diaphragm. The leads to the chip would remain outside of the LIGA wall and the entire assembly could then be wire bonded to an inexpensive plastic package and still have the hermetic seal associated with more expensive packages. Additionally, this approach could be used for packaging of multi-chip modules. Other advantages would be built-in EMI (electromagnetic interference) protection and easy assembly of multi-chip modules, for instance, for stacking die vertically.

17.3.4 Micro-Optical Components

17.3.4.1 Introduction

Before LIGA, techniques suitable to micromanufacture micro-optical components such as prisms, lenses, beam splitters and optical couplings in a freely selectable configuration, material and structural size were hard to come by [Gottert et al., 1991]. LIGA enables the integrated fabrication of most of these components in a wide variety of materials and sizes. Micro-optics may develop into a significant LIGA application, especially for cases where cost can be overlooked. When very smooth vertical walls and high aspect ratio of the parts play a major role, LIGA may be the only available technique.

PMMA is transparent in the visible and near IR regions, making it a natural optical material. The simplest LIGA optical structure based on PMMA is a ***multimode waveguide*** of rectangular cross section consisting of a three-layer resist structure [Gottert et al., 1992]. To make such a planar waveguide, an epoxy phenolic resin substrate, insensitive to X-rays, is first coated with a cladding resist layer. A cladding material well-suited to X-ray patterning is a copolymer of PMMA (78%) and tetrafluoropropyl

Multimode waveguide. (From Gottert et al., Integrierte Optik und Mikrooptik mit Polymeren, Mainz, Germany, 1992. With permission).

LIGA and Other Replication Techniques 17-51

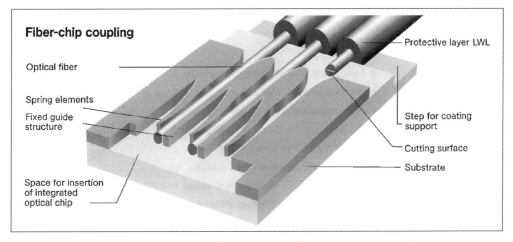

Coupling for monomode fibers. (Courtesy of IMM, Mainz, Germany.)

methacrylate (TFPMA) (22%) with a refractive index of 1.476. With this composition and a PMMA core layer (refractive index of 1.49), a numerical aperture of 0.2 is obtained. By varying the composition of the copolymer each refractive index between 1.49 and 1.425 can be obtained precisely, allowing the numerical aperture of the polymer waveguide to be fitted to the light-guiding fibers used to inject the light into the component. The planar cladding layer is milled down to a thickness corresponding to the thickness of a connecting optical fiber cladding. A foil of PMMA core material is then welded onto the first cladding layer and the core is milled down to the thickness of the same fiber core. Welding is performed slightly above the glass transition temperature and under pressure. The procedure causes only a small amount of interdiffusion of the two layers and a well-defined refractive index profile results; the variation of the refractive index between two successive layers takes place within an interface layer of less than 10 μm thickness. A cover foil consisting of the same copolymer is finally welded and milled to cladding thickness on the PMMA core foil. Bulk PMMA specimens show a mean attenuation of 0.2 dB/cm in the 600- to 1300-nm wavelength range. At absorption peaks of 900 and 1180 nm, this value rises to 0.4 and 3 dB/cm, respectively. To manufacture low-loss PMMA material for the near-infrared region, deuterated PMMA may be used as a core material. This causes the absorption peak at about 900 nm, caused by C-H resonances, to be shifted to longer wavelengths. Attenuation of the deuterated material in the spectral range from 600 to 1300 nm is only about 0.1 dB/cm. For the described three-layer resist structure, the attenuation is somewhat larger than for the bulk PMMA specimens due to the relatively large roughness of the side walls of the thus fabricated waveguides and by defects in the three-resist layer [Mohr et al., 1991].

Although LIGA is producing projected features rather than three-dimensional true shapes (such as, say, in a contact lens), this technique has brought micromachinists much closer to the possible realization of a totally integrated micro-optical bench. The lithographic alignment of the separate optical components; the high, extremely smooth, vertical walls; and the excellent optical properties of one of the key LIGA resists, PMMA, all combine to propel LIGA as an ideal technique for micro-optical applications within reach of commercialization [Menz, 1995]. Micro-optical benches were attempted via polysilicon surface micromachining [Lin et al., 1995] and bulk micromachining techniques [Wolffenbuttel and Kwa, 1991]. In the former method, the very thin optical components had to be lifted from the surface and secured upright mechanically in a postfabrication assembly step (see Chapter 16). In the latter method, the proposed structures consisted of complex multi components. We suggest that given the required tolerances neither of these alternative optical benches will ever be manufactured outside a research laboratory.

17.3.4.2 Fiberoptic Connectors

LIGA elements for *coupling monomode fibers* with integrated optical chips are under development. These prealignment arrays may utilize fixed nickel guiding structures in combination with leaf springs to ensure

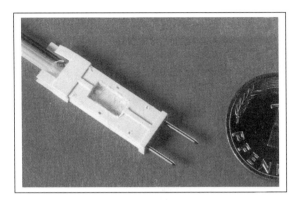

Pull-push LIGA connector. (Courtesy, IMM, Mainz, Germany.)

a precise alignment of the optical fibers relative to the optical chip with an accuracy in the submicron range. The thermal expansion coefficient of the substrate material is matched to the optical chip material—e.g., a glass substrate is used for coupling fibers to glass chips—and the use of spring elements simplifies the handling of the fibers [Rogner et al., 1991]. Alternatives involve the use of optical adhesives and the use of silicon V-groove arrays (see Chapter 16). Both technologies are more labor intensive and suffer from thermal expansion problems.

The use of microptical LIGA components is especially attractive for the coupling of multimode fibers; here the capability of exact lithographic positioning of mechanical mounting supports is used advantageously to position the multimode fibers very precisely with respect to the micro-optical components without need for any additional adjusting operations [Gottert et al., 1992a; 1992b]

A *pull-push LIGA* connector for single-mode fiber ribbons, incorporating a set of *precision microsprings*, coupling up to 12 fibers spaced at 250 µm, is close to commercial implementation and is one of the LIGA products with mass market appeal. To obtain good coupling efficiency, the fibers are positioned horizontally with a precision of 1 µm. LIGA is an excellent method to provide that precision. These connectors are under development at IMM [Editorial, 1994a]. In line with the micromanufacturing philosophy presented in this book—to optimize the use of each micromachining technique for its optimum cost/performance application—those parts of the mold insert which do not require such high accuracy are fabricated using other methods of precision machining such as EDM (electro discharge machining) (see Madou [1997, chap. 7]).

17.3.4.3 Fresnel Zone Plates

Self-supporting *Fresnel zone plates* have been made by LIGA [Munchmeyer and Ehrfeld, 1987]. One such lens was made of nickel with a distance between the rings of 74 µm on the inner side, decreasing to 10 µm on the external side. The width of the radial web was also 10 µm. A *Fresnel zone plate array in PMMA* with a 10-mm focal length and a zone plate diameter of 10 mm was made as well [Editorial, 1994b]. In a similar way, self-supporting gratings or infrared polarizers might be fabricated.

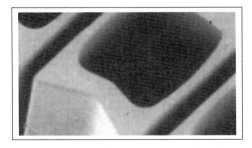

Precision microsprings. (Courtesy, IMM, Mainz, Germany.)

LIGA and Other Replication Techniques **17**-53

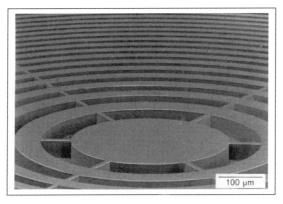

Nickel Fresnel lens. (Courtesy, IMM, Mainz, Germany.)

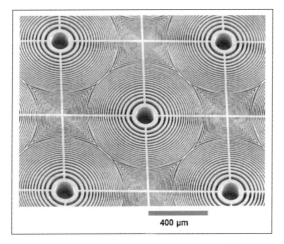

Fresnel zone plate array (PMMA). (Courtesy, IMM, Mainz, Germany.)

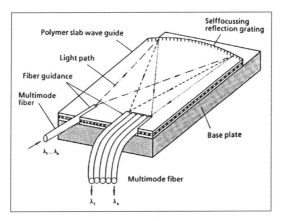

Spectrometer MUX/DMUX. (From Mohr et al., *Sensors and Actuators A*, 25-27, 571–575, 1991. With permission.)

17.3.4.4 Planar Grating Spectrograph-Integrated Optical Bench

Using the above described planar polymer waveguide in combination with a self-focusing reflection grating, a ***spectrometer*** 18 × 6.4 mm in size was developed [Anderer et al., 1988]. The polychromatic light, injected into the waveguide by an optical fiber, is dispersed at a curved grating coated with a sputtered gold layer. The spectrally divided portions of the light are focused on ten optical fibers, or, alternatively, on an array of diodes positioned at the focal line. The intensity at the focal line is about 25% of the incoupled intensity, and the spectral resolution is 0.2 nm/μm. As the fiber grooves are aligned exactly to the grating, no adjustment is necessary to make the component work. The grating is designed to work in a spectral range between 720 and 900 nm and is blazed to diffract in the second order. The grating pitch (groove spacing) is 3.4 μm, and the step of the groove is 0.5 μm (more recently 0.25 μm was realized), which is well within the accuracy of deep-etch X-ray lithography [Mohr et al., 1991]. Although the spectral resolution of this microspectrometer is relatively low, this type of grating can be used for optical analysis in the whole visible and near-IR-wavelength region (up to 1100 nm), e.g., for photometric analysis of gas mixtures, optical film thickness measurement and as a spectral detector in liquid chromatography [Staerk et al., 1996]. Another set of applications of the described microspectrometer is in the area of combining and separation of optical wavelengths. In this technology, signals of several different wavelengths are multiplexed on a single optical fiber. Multiplexer (MUX) and demultiplexer (DMUX) devices are required to combine and separate multichannel signals. Among the many proposed systems, the self-focusing concave reflection grating is the only one which can handle a large channel number (>5) with good performance and without the need of additional reflecting and focusing elements [Anderer et al., 1988].

For operation at wavelengths greater than 1100 nm, air must be used as a transmission medium instead of a PMMA-based waveguide. At NASA Ames, we have embarked upon the development of a gas analysis spectrometer including the capability of CO_2 detection at 4.2 μm. Instead of piping polychromatic light into the spectrometer on a multimode fiber, the intent here is to use a microfabricated heated filament integrated into the spectrometer. The microheater filament has already been realized by a combination of surface and bulk micromachining on an SOI substrate. Monochromatic light, at 4.2 μm, will be isolated from the broad-band filament radiation by reflecting the light off a self-focusing grating and then bouncing the monochromatic light from several LIGA micromirrors in an absorption cell-type arrangement. A 10-cm light path is required for a 1-ppm CO_2 sensitivity, and we found that such a path length is easily designed by arranging the mirrors on a 1×1-cm footprint. Crucial technical problems yet to be overcome are, first, the considerable self-diffraction of a small light beam at 4.2 μm in air spreads the light spot quickly beyond the individual mirror size, and, second, the light emanating from the microfilament, fabricated in the substrate plane, must be brought from the vertical into a horizontal plane so it can be reflected off the vertical self-focusing grating and micromirrors. Finally, the integrity of the micro-optical components (e.g., reflectivity) must be preserved when exposed to a gaseous uncontrolled environment [Madou, 1997, chap. 10; Desta et al., 1995].

17.3.4.5 Other Passive Optical Components

Other passive optical components that have been made using LIGA are Y-couplers [Rogner, 1992], cylindrical and prismatic lenses and beam splitters. The compact design of splitters and couplers with the light paths

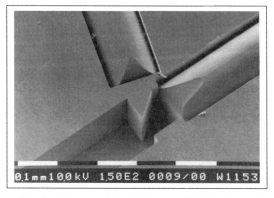

Fiber coupling (1×2 beam splitter). (Courtesy, KfK, Karlsruhe, Germany.)

LIGA and Other Replication Techniques

between two fiber ends in the range of the fiber diameter makes attenuation losses very small. For example, the excess loss of a *bidirectional fiber coupling element (1 × 2 beam splitter)* with 100/140 µm multimode fibers is less than 1 dB.

17.3.5 LIGA Accelerometer

The sacrificial layer technique was used to fabricate a LIGA *acceleration sensor* with a capacitive read-out. The width of the gap between stationary electrodes and a seismic mass (Ni in this case) changes when they are accelerated. The resulting change in capacity is what is measured. The advantage of a LIGA accelerometer over Si-technology is the material variety and therefore the choice of density of the seismic mass, as well as the choice of the Young's modulus of the bending beams. This flexibility could enable a

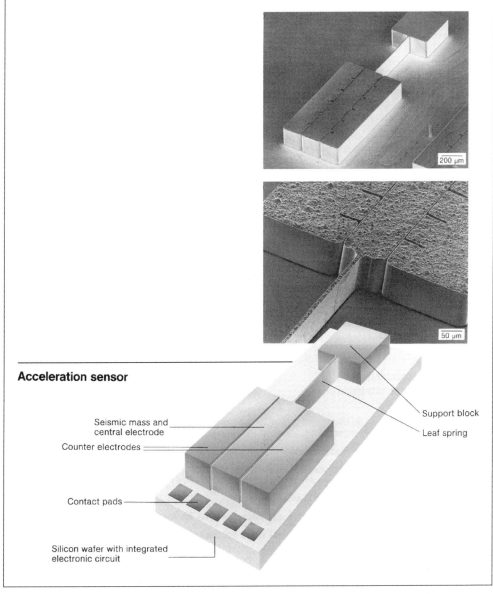

Acceleration sensor with capacitive read-out. (Courtesy, IMM, Mainz, Germany.)

family of accelerometers with a much wider g-range than bulk Si micromachined devices and definitely a wider range than poly-Si surface micromachined accelerometers which, due to the thin inertial mass, are limited in this respect. Also, the extreme aspect ratios attainable with LIGA bring about a better signal. The thickness of the sensor has no influence on the sensitivity, but it does affect the value of the absolute capacitance, which increases with greater thickness. A large sensor thickness also improves the stiffness of the cantilever which is advantageous to the cross-sensitivity. The width of the capacitor gap influences the sensitivity: the smaller the gap, the more sensitive the accelerometer. LIGA enables very narrow gaps and thick inertial plates. For example, the zero-point capacity given by a gap between the seismic mass and the electrodes of only 3 to 4 μm at a structural height of 100 μm is 0.7 pF. A sensor with this zero-point capacity was shown to have a sensitivity of 15% capacity change per g and a resonance frequency of 570 Hz. The temperature coefficient was found to be about 0.5% K^{-1}. By optimized sensor design, the zero-point capacity and the temperature behavior can further be improved. Extremely temperature-resistant accelerometers were achieved by designing the sensor partly with a positive and partly with a negative temperature coefficient. In this way, a LIGA accelerometer for 1 g with a measured temperature coefficient of offset (TCO) of 1.02×10^{-4} g/K in the temperature range −10 to +100°C was realized [Strohrmann et al., 1994]. For automotive applications a resonance frequency of more than 500 Hz is required, and for 1 g of acceleration the change in capacity should be larger than 20% [Burbaum et al., 1991]. In case the accelerometer is also electrostatically actuated, other benefits of using LIGA result as explained further below under electrostatic LIGA actuators.

It has been shown that LIGA may be compatible with substrates containing active electronics, and in principle integrated LIGA accelerometers could represent a good business opportunity. On the other hand, there is more accessible bulk and surface micromachining technology available, which is now feeding most of the industrial needs.

17.3.6 Interlocking Gears

A series of interlocking gears was fabricated in one process sequence without assembly. The gears are free to rotate and could be used in watches or in audio, video and measuring technologies. The precision of the LIGA process allows the realization of gears with minimized backlash and friction [Rogner et al., 1992]. At this point, it is not clear if the accuracy demands in those different application fields really do require LIGA or if LIGA-like processes could fulfill those needs.

17.3.7 Electrostatic Actuators

For the fabrication of actuator elements, where free lateral patterning is required, one usually resorts to surface micromachining. With polysilicon surface micromachining, the thickness of actuator elements is usually limited to heights of a few micrometers, which is particularly disadvantageous in the case of

Micro-optical switch with electrostatic comb-like structures.

microactuators. This can be understood from our discussions on scaling laws in Madou (1997, chap. 9). In practical realizations, linear electrostatic microactuators often have two *comb-like structures*. By applying a voltage the movable part of the comb moves closer to the fixed part. For a rectangular design, the force and the microactuator displacement is proportional to the ratio of structure height, T, vs. separation gap, d, and the number of fingers, N (see also Madou [1997, chap. 9]):

$$F = -\frac{\varepsilon_0 T V^2}{2d} N \qquad (17.7)$$

An efficient comb drive can be thus achieved by designing many high comb fingers with narrow gaps. In surface micromachining, the thickness, T, is limited to about 2 to 3 µm, whereas with LIGA several hundreds of microns are possible. Clearly, LIGA has the potential to produce better electrostatic actuators. Mohr et al. (1992) further showed that by making the comb fingers trapezoidal the maximum displacement of the interdigitated fingers can be further increased and smaller capacitor gaps are possible. For a more rigorous mathematical derivation of Eq. (17.7), we refer to the finite-element simulation of comb drives by Tang (1990). Muller and others have used LIGA electrostatic comb drives micromachined on top of a silicon wafer to fabricate micro-optical switches [Muller et al., 1993].

For better electrostatic actuators, LIGA will have to compete with surface micromachined structures on SOI wafers where the thickness of the epilayer can be much thicker than that of a poly-Si layer and where dry etching can also make good vertical walls and narrow gaps. But perhaps where ceramic actuators are involved, LIGA has a unique advantage.

17.3.8 LIGA Inductors

LIGA coils could find applications in inductive position or proximity sensing and all types of magnetic actuators. It is possible to achieve a high Q-factor as well as a high driving current as a result of the high aspect ratio and the high surface quality of the conductors. A very important application is to expand the work, initiated by Romankiw at IBM, to make finer and finer, batch-fabricated, thin-film magnetic recording heads (see Madou [1997, chap. 7, Figure 7.8]) [Romankiw, 1989]. Other applications of LIGA inductors are in RF oscillators or in inductive power transmission for rotating systems [Rogner et al., 1992]. Rogge et al. (1995) successfully demonstrated a fully batch-fabricated magnetic micro-actuator using a two-layer LIGA process. The contact forces at 1A for a permalloy relay made this way were well beyond those of miniaturized relays. Based on cost, only the most demanding applications such as the read-write heads might justify the use of LIGA; for most other applications LIGA-like methods will probably suffice.

17.3.9 Electromagnetic Micromotor

We have shown already how LIGA makes better electrostatic actuators than other Si micromachining techniques; the same is true for electromagnetic actuators. In Madou (1997, chap. 9 on scaling laws), it is stated that most Si micromachined motors today are producing negligible amounts of torque. In practical situations, what is needed are actuators in the millimeter range delivering torques of 10^{-6} to 10^{-7} Nm. Such motors can be fabricated with classical precision engineering or a combination of LIGA and precision engineering.

The performance with respect to torque and speed of a traditional miniature magnetic motor with a 1-mm diameter to 2-mm-long permanent magnetic rotor was demonstrated, in practice, to be incomparably better than the surface micromachined motors discussed in Chapter 16 [Goemans et al., 1993]. The small torque of the surface micromachined electrostatic motors is, to a large extent, due to the fact that they are so flat. The magnetic device made with conventional three-dimensional metal-working techniques has an expected maximum shaft torque of 10^{-6} to 10^{-7} Nm. The torque depends on the Maxwell shearing stress on the rotor surface integrated over the area of the latter; the longer magnetic rotor easily outstrips the flat electrostatic one with the same rotor diameter [Goemans et al., 1993]. In both electric and magnetic microactuators, force production is proportional to changes in stored energy. The amount of force that may be generated per unit substrate area is proportional to the height of the actuator. Large-aspect-ratio structures are therefore preferred.

With LIGA techniques **_large-aspect-ratio magnetic motors_** can be built [Ehrfeld, 1994]. Essentially, a soft magnetic rotor follows a rotating magnetic field produced by the currents in the stator coils. The motor manufacture is an example of how micromachining and precision engineering can be complementary techniques for producing individual parts, which have to be assembled afterward. Only the components with the smallest features are produced by means of microfabrication techniques, whereas the other parts are produced by traditional precision mechanical methods.

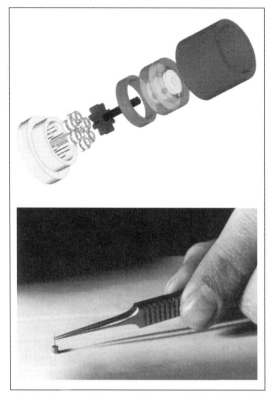

Large-aspect-ratio magnetic motor. (Courtesy IMM, Mainz, Germany.)

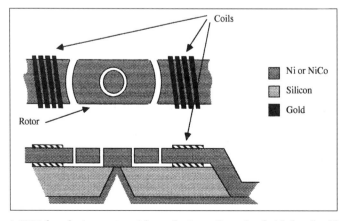

A LIGA-based micromotor with enveloping coils made of nickel and gold.

A magnetic LIGA drive was also demonstrated by the Wisconsin team [Guckel et al., 1991] and consisted of a plated nickel rotor which was free to rotate about a fixed shaft. Large plated poles (stators) came close to the rotor. The rotor was turned by spinning a magnet beneath the substrate. Further work by the same group involved **enveloping coils**, to convert current to magnetic field *in situ* rather than externally [Guckel et al., 1992]. It is necessary to plate two different materials—a high-permeability material such as nickel and a good conductor such as gold—to accomplish this.

The motor can also be used as electrostatic motor. By applying voltages to the stator arms, the rotor (which is grounded through the post about which it rotates) is electrostatically attracted and, by poling the voltages, can be made to turn. Design rules and tests of electrostatic LIGA micromotors were presented by Wallrabe et al. (1994). Minimum driving voltages needed were measured to be about 60 V; optimized design torques of the order of some μNm are expected. Cost of electromagnetic LIGA micromotors will often urge the investigation of LIGA-like technologies or hybrid approaches (LIGA combined with traditional machining) as more accessible and adequate machining alternatives.

17.3.10 Other Potential LIGA Applications

Besides the interconnect and packaging applications already mentioned, LIGA might have other applications in the microelectronics industry. Some of the circuit elements currently very expensive to fabricate on an integrated circuit could be replaced with LIGA circuit elements. The expense associated with large capacitors on an integrated circuit, for example, is due to the large amount of space they use. Shrinking capacitors to use less real estate is an area of great interest for research. A LIGA capacitor can be manufactured with tall, thin vertical plates to achieve a high capacitance without using a large amount of real estate.

An additional area of application would be the manufacture of supercapacitors, small devices with very large values of capacitance. These are used as discrete devices for a large range of applications, including the generation of extremely large current pulses (by trickle-charging the capacitor, then driving the capacitor plates together with explosive devices) for use in military equipment.

There are a number of methods for using LIGA as a means for creating much more precise and inexpensive optical components. Some possible applications, including a spectrograph for liquid and gas analysis, were mentioned already and some additional opportunities follow.

Presently, one of the largest areas for sensor research is in optical sensors. In general, an optical sensor is measuring a change in an optical property with the use of optical fibers. This change can be seen by measuring an amplitude change in light passing through or reflected off of some media or by a phase change resulting in interference patterns.

Different methods for connecting fiberoptic waveguides to sensor platforms, which can hold various chemistries or fluids or be inserted into a human body, still have not been satisfactorily developed. LIGA has a great deal of potential in this area, due to the ability to create both a fiberoptic connector and waveguide in one step. The waveguides can be of arbitrary shape, making it easy to include small cups to hold chemistries or to wrap extremely long waveguides into a small area. The number of ways which LIGA could be used for optical sensing is tremendous.

An additional item of note is that one of the limiting factors for the wide application of fiber optic sensors is the cost associated with connecting the sensor to a fiber with the required accuracy. This would not be a problem with LIGA-based sensor platforms. The optical applications of LIGA might have potential independent of developments in the IC industry as higher prices are often acceptable for precise optical components.

The use of LIGA as a method for creating on-chip waveguides would have a tremendous impact on the area of optical communication between chips. Presently, there is great interest in producing light-emitting diodes or semiconductor lasers on a silicon substrate by growing gallium arsenide (or other III-V semiconductors) directly on a silicon substrate and using this to create the light source. This source would then be used as a means of communication with the outside world, rather than electrical connections.

Coupling these light sources will require on-chip waveguides and a means for connecting the chips to optical fibers. LIGA is well suited to creating optical waveguides, as discussed above, and is an obvious and potentially inexpensive solution.

Micromanipulation is a field which is on the verge of explosive growth. The combination of techniques developed for use with integrated circuit lithography and probing completed microchips is also being applied to other areas of microelectronics—for instance, automated testing of wire bonds or failure analysis of integrated circuits.

Traditional micromanipulators are now quite large in size and movement compared to the devices they are to manipulate or probe. Not only can LIGA-based micromanipulator machines be built of extremely tiny LIGA parts, the actual microprobes can be manufactured by LIGA.

The ability to plate refractory materials such as rhenium, osmium and ruthenium would have use in the creation of incredibly precise two- and four-point probes for much higher accuracy in testing applications for integrated circuits. The relative precision with which a two- or four-point probe can test a substrate is related to how closely the probe points can be spaced. For two-point probes, the smallest spacing available is about 10 μm, and this can only be achieved by hand assembly and grinding operations by a skilled technician. With LIGA, the probe points could easily be spaced as close as 0.5 μm, perhaps as close as 0.2 μm. The utility of this for probing silicon wafers during the manufacturing process would be very high.

An additional area for application would be the creation of LIGA probe cards for testing of integrated circuits. These probe cards are difficult to manufacture and expensive, and periodic realignment of the probe points is necessary due to wear and tear. A cheap, disposable LIGA probe card would be of great use for probing integrated circuit test patterns, which are increasingly complicated and decreasing remarkably in size. Many major integrated circuit manufacturers are relocating electrical test patterns into the scribe lines of silicon wafers to save on real estate. This necessitates a major size reduction in the contact pads that the probe card must contact. In the area of extremely precise probe cards, LIGA would have a definite advantage.

17.4 Technological Barriers and Competing Technologies

17.4.1 Insertion Point of X-Ray Lithography

The original work demonstrating the LIGA process was performed by a group of scientists led by Dr. E. W. Becker at Karlsruhe. Subsequently, a group led by Prof. Guckel replicated most of the aspects of this process and demonstrated some new technologies as well. This achievement was accomplished in a rather short amount of time, approximately 18 months. This was due to the extensive literature available on the topic (the Karlsruhe group has been very prolific in publishing on the various aspects of LIGA processing), but it also shows that LIGA is a technology which, in principle, is ready to be taken up by industry. Wider commercial acceptance of micromachining will be linked to cost/performance advantages and availability of the technique. The latter may well be dependent on the insertion point of X-ray lithography in the production of the next generation of industrial ICs. As pointed out earlier in this chapter and in Madou (1997, chap. 1), the X-ray lithography insertion point keeps on being postponed, possibly keeping LIGA a tool for the research community only.

17.4.2 Competing Technologies

Deep dry etching (often cryogenic) and thick UV-sensitive resists such as polyimides, AZ-4000 and SU-8 (see Madou [1997, chap. 1] and chap. 16, this book) are contenders for several of the applications for which LIGA is being promoted. With respect to dry etching, higher and higher aspect ratios are being achieved; especially when using highly anisotropic etching conditions and cryogenic cooling, remarkable results are being achieved (see Madou [1997, chap. 2]). With respect to photosensitive polyimides, 100-μm high features can be accomplished readily, and minimum feature sizes of 2 μm are possible (see Madou [1997, chap. 1]). Both competing techniques are more accessible than LIGA and will continue to improve, taking more opportunities away from LIGA. Like LIGA, both alternative techniques can be coupled with plating, but neither technique can yet achieve the extreme low surface roughness and vertical walls LIGA is capable of. The lesson is that more comparative studies of the different micromachining technologies should be undertaken before one is being decided upon.

Other less obvious competing technologies for LIGA-like applications are ion-beam milling, laser ablation methods and even ultra-precision machining (see Madou [1997, chap. 7]). The latter three are serial processes and rather slow, but if we are considering making a mold then these technologies might be competitive.

References

Akkaraju, S., Y. M. Desta, B. Q. Li, and M. C. Murphy, "A LIGA-Based Family of Tips for Scanning Probe Applications," SPIE, Microlithography and Metrology in Micromachining II, Austin, TX, 1996, pp. 191–198.

Anderer, B., W. Ehrfeld, and D. Munchmeyer, "Development of a 10–Channel Wavelength Division Multiplexer/Demultiplexer Fabricated by an X-Ray Micromachining Process," *SPIE*, 1014, 17–24, 1988.

Becker, E. W., W. Ehrfeld, D. Munchmeyer, H. Betz, A. Heuberger, S. Pongratz, W. Glashauser, H. J. Michel, and V. R. Siemens, "Production of Separation Nozzle Systems for Uranium Enrichment by a Combination of X-Ray Lithography and Galvanoplastics," *Naturwissenschaften*, 69, 520–523, 1982.

Becker, E. W., W. Ehrfeld, and D. Munchmeyer, "Untersuchungen zur Abbildungsgenauigkeit der Rontgentiefenlitographie mit Synchrotonstrahlung," KfK, Report No. 3732, Karlsruhe, Germany, 1984.

Becker, E. W., W. Ehrfeld, P. Hagmann, A. Maner, and D. Munchmeyer, "Fabrication of Microstructures with High Aspect Ratios and Great Structural Heights by Synchrotron Radiation Lithography, Galvanoforming, and Plastic Molding (LIGA process)," *Microelectron. Eng.*, 4, 35–56, 1986.

Bley, P., J. Gottert, M. Harmening, M. Himmelhaus, W. Menz, J. Mohr, C. Muller, and U. Wallrabe, "The LIGA Process for the Fabrication of Micromechanical and Microoptical Components," in *Microsystem Technologies '91*, Krahn, R. and Reichl, H., Eds., VDE-Verlag, Berlin, 1991, pp. 302–314.

Bley, P., W. Menz, W. Bacher, K. Feit, M. Harmening, H. Hein, J. Mohr, W. K. Schomburg, and W. Stark, "Application of the LIGA Process in Fabrication of Three-Dimensional Mechanical Microstructures," 4th International Symposium on MicroProcess Conference, Kanazawa, Japan, 1991, pp. 384– 389.

Bley, P., D. Einfeld, W. Menz, and H. Schweickert, "A Dedicated Synchrotron Light Source for Micromechanics," EPAC92. Third European Particle Accelerator Conference, Berlin, Germany, 1992, pp. 1690–2.

Burbaum, C., J. Mohr, P. Bley, and W. Ehrfeld, "Fabrication of Capacitive Acceleration Sensors by the LIGA Technique," *Sensors and Actuators A*, 25, 559–563, 1991.

Bustgens, B., W. Bacher, W. Menz, and W. K. Schomburg, "Micropump Manufactured by Thermoplastic Molding," Proceedings. IEEE Micro Electro Mechanical Systems (MEMS '94), Oiso, Japan, January 1994, pp. 18–21.

Desta, Y. M., M. Murphy, M. Madou, and J. Hines, "Integrated Optical Bench for a CO_2 Gas Sensor," Microlithography and Metrology in Micromachining, (Proceedings of the SPIE), Austin, TX, 1995, pp. 172–177.

Editorial, "Fibre Ribbon Ferrule Insert Made by LIGA," Commercial brochure, 1994a.

Editorial, "Micro-Optics at IMM," Commercial brochure, 1994b.

Editorial, "X-Ray Scanner for Deep Lithography," Commercial brochure, 1994c.

Ehrfeld, W., "The LIGA Process for Microsystems," Proceedings. Micro System Technologies '90, Berlin, Germany, 1990, pp. 521–528.

Ehrfeld, W., "LIGA at IMM," Notes from handouts, 1994, Banff, Canada.

Ehrfeld, W., W. Glashauer, D. Munchmeyer, and W. Schelb, "Mask Making for Synchrotron Radiation Lithography," *Microelectron. Eng.*, 5, 463–470, 1986.

Ehrfeld, W., P. Bley, F. Gotz, P. Hagmann, A. Maner, J. Mohr, H. O. Moser, D. Munchmeyer, W. Schelb, D. Schmidt, and E. W. Becker, "Fabrication of Microstructures Using the LIGA Process," Proceedings of the IEEE Micro Robots and Teleoperators Workshop, Hyannis, MA, 1987, pp. 1–11.

Ehrfeld, W., R. Einhaus, D. Munchmeyer, and H. Strathmann, "Microfabrication of Membranes with Extreme Porosity and Uniform Pore Size," *J. Membr. Sci.*, 36, 67–77, 1988.

Eicher, J., R. P. Peters, and A. Rogner, VDI-Verlag, VDI-Verlag, Dusseldorf, Report No. VDI-Bericht 960, 1992.

Elwenspoek, M., T. S. J. Lammerink, R. Miyake, and J. H. J. Fluitman, "Micro Liquid Handling Systems," Proceedings of the Symposium, Onder the Loep Genomen, Koninghsh of Veldhoven, Netherlands, 1994.

Friedrich, C., "Complementary Micromachining Processes," Notes from handouts, 1994, Banff, Canada.

Galhotra, V., C. Marques, Y. Desta, K. Kelly, M. Despa, A. Pendse, and J. Collier, "Fabrication of LIGA Mold Inserts Using a Modified Procedure," SPIE, Micro-machining and Microfabrication Process Technology II, Austin, TX, 1996, pp. 168–173.

Ghica, V. and W. Glashauser, *Verfahren fur die Spannungsrissfreie Entwicklung von Bestrahlthen Polymethylmethacrylate-Schichten,* in Deutsche Offen-legungsschrift, Germany, Patent 3,039,110, 1982.

GmbH, R. K., "μGALV 750," Sales Brochure, RK Kissler, Daimlerstrasse 8, Speyer 67346, 1994.

Goedtkindt, P., J. M. Salome, X. Artru, P. Dhez, N. Maene, F. Poortmans, and L. Wartski, "X-Ray Lithography with a Transition Radiation Source," *Microelectron. Eng.*, 13, 327–330, 1991.

Goemans, P. A. F. M., E. M. H. Kamerbeeek, and P. L. A. J. Klijn, "Measurement of the Pull-out Torque of Synchronous Micromotors with PM Rotor," 6th International Conference on Electrical Machines and Drives, Oxford, 1993, pp. 4–8.

Gottert, J., J. Mohr, and C. Muller, "Mikrooptische Komponenten aus PMMA, Hergestellt Durch Roentgentiefenlithographie Werkstoffe der Mikrotechnik-Bais fur neue Producte," *VDI Berichte*, 249–263, 1991.

Gottert, J., J. Mohr, and C. Muller, "Coupling Elements for Multimode Fibers by the LIGA Process," Proceedings Micro System Technologies '92, Berlin, Germany, 1992a, pp. 297–307.

Gottert, J., J. Mohr, and C. Muller, "Examples and Potential Applications of LIGA Components in Microoptics," Integrierte Optik und Mikrooptik mit Polymeren, Mainz, Germany, 1992b.

Guckel, H., "Deep Lithography," Notes from handouts, 1994, Banff, Canada.

Guckel, H., J. Uglow, M. Lin, D. Denton, J. Tobin, K. Euch, and M. Juda, "Plasma Polymerization of Methyl Methacrylate: A Photoresist for 3D Applications," Technical Digest of the 1988 Solid-State Sensor and Actuator Workshop, Hilton Head Island, SC, 1988, pp. 9–12.

Guckel, H., K. J. Skrobis, T. R. Christenson, J. Klein, S. Han, B. Choi, and E. G. Lovell, "Fabrication of Assembled Micromechanical Components via Deep X-ray Lithography," Proceedings. IEEE Micro Electro Mechanical Systems (MEMS '91), Nara, Japan, 1991a, pp. 74–79.

Guckel, H., K. J. Skrobis, T. R. Christenson, J. Klein, S. Han, B. Choi, E. G. Lovell, and T. W. Chapma, "Fabrication and Testing of the Planar Magnetic Micromotor," *J. Micromech. Microeng.*, 4, 40–45, 1991b.

Guckel, H., T. R. Christenson, and K. Skrobis, "Metal Micromechanisms via Deep X-Ray Lithography, Electroplating and Assembly," *J. Micromech. Microeng.*, 2, 225–228, 1992.

Guckel, H., T. R. Christenson, T. Earles, J. Klein, J. D. Zook, T. Ohnstein, and M. Karnowski, "Laterally Driven Electromagnetic Actuators," Technical Digest of the 1994 Solid-State Sensor and Actuator Workshop, Hilton Head Island, SC, 1994, pp. 49–52.

Hagmann, P. and W. Ehrfeld, "Fabrication of Microstructures of Extreme Structural Heights by Reaction Injection Molding," *J. Polymer Process. Soc.*, 4, 188–195, 1988.

Hagmann, P., W. Ehrfeld, and H. Vollmer, "Fabrication of Microstructures with Extreme Structural Heights by Reaction Injection Molding," First Meeting of the European Polymer Federation, European Symp. on Polymeric Materials, Lyon, France, 1987, pp. 241–251.

Harmening, M., W. Bacher, and W. Menz, "Molding Plateable Micropatterns of Electrically Insulating and Electrically Conducting Poly(Methyl Methacrylate)s by the LIGA Technique," *Makromol. Chem. Macromol. Symp.*, 50, 277–284, 1991.

Harmening, M., W. Bacher, P. Bley, A. El-Kholi, H. Kalb, B. Kowanz, W. Menz, A. Michel, and J. Mohr, "Molding of Three-Dimensional Microstructures by the LIGA Process," Proceedings IEEE Micro Electro Mechanical Systems (MEMS '92), Travemunde, Germany, 1992, pp. 202–207.

Harsch, S., W. Ehrfeld, and A. Maner, "Untersuchungen zur Herstellung von Mikrostructuren grosser Strukturhohe durch Galvanoformung in Nickelsulfamatelek trolyten," KfK, Report No. 4455, Karlsruhe, Germany, 1988.

Harsch, S., D. Munchmeyer, and H. Reinecke, "A New Process for Electroforming Movable Microdevices," Proceedings of the 78th AESF Annual Technical Conference (SUR/FIN '91), Toronto, Canada, June, 1991.

Hein, H., P. Bley, J. Gottert, and U. Klein, "Elektro-nenstrahllithographie zur Herstellung von Rontgen masken fur das LIGA-Verfahren," *Feinw. Tech. Messtech,* 100, 387–389, 1992.

Henck, R., "Detecteurs au Silicium Pour Electrons et Rayons X, Principes de Fonctionnement, Fabrication et Performance," *J. Microsc. Spectrosc. Electron.*, 9, 131–133, 1984.

Hill, R., "Symposium on X-ray Lithography in Japan," 1991, National Academy of Sciences.

Himmelhaus, M., P. Bley, J. Mohr, and U. Wallrabe, "Integrated Measuring System for the Detection of the Revolution of LIGA Microturbines in View of a Volumetric Flow Sensors," *J. Micromech. Microeng.*, 2, 196–198, 1992.

Humphrey, E. F. and D. H. Tarumoto, Eds. *Fluidics,* Fluidic Amplifier Associates, Inc., Boston, MA, 1965.

IMM, "The LIGA Technique," Commercial brochure, IMM, 1995.

Lawes, R. A., "Sub-Micron Lithography Techniques," *Appl. Surf. Sci.*, 36, 485–499, 1989.

Lin, L. Y., S. S. Lee, M. C. Wu, and K. S. J. Pister, "Micromachined Integrated Optics for Free-Space Interconnections," Proceedings IEEE Micro Electro Mechanical Systems (MEMS '95), Amsterdam, Netherlands, 1995, pp. 77–82.

Lingnau, J., R. Dammel, and J. Theis, "Recent Trends in X-Ray Resists: Part 1," *Solid State Technol.*, 32, 105–112, 1989.

Lubitz, K., "Mikrostrukturierung von Piezokeramik," *VDI-Tagungsbericht,* 796, 1989.

Madou, M. J. and M. Murphy, "A Method for PMMA Stress Evaluation," unpublished results, 1995.

Maner, A., S. Harsch, and W. Ehrfeld, "Mass Production of Microstructures with Extreme Aspect Ratios by Electroforming," Proceedings of the 74th AESF Annual Technical Conference (SUR/FIN '87), Chicago, IL, July 1987, pp. 60–65.

Maner, A., S. Harsch, and W. Ehrfeld, "Mass Production of Microdevices with Extreme Aspect Ratios by Electro-forming," *Plating and Surface Finishing,* 60–65, 1988.

Menz, W., "LIGA and Related Technologies for Industrial Application," 8th International Conference on Solid-State Sensors and Actuators (Transducers '95), Stockholm, Sweden, June 1995, pp. 552–555.

Menz, W. and P. Bley, *Mikrosystemtechnik fur Ingenieure,* VCH Publishers, Weinheim, Germany, 1993.

Menz, W., W. Bacher, M. Harmening, and A. Michel, "The LIGA Technique—A Novel Concept for Microstructures and the Combination with Si-Technologies by Injection Molding," Proceedings IEEE Micro Electro Mechanical Systems (MEMS '91), Nara, Japan, 1991, pp. 69–73.

Michel, A., R. Ruprecht, and W. Bacher, "Abformung von Mikrostrukturen auf Prozessierten Wafern," KfK, Report No. 5171, 1993, Karlsruhe, Germany, 1993.

MicroParts, "The LIGA Technique," Commercial brochure.

Mohr, J. and M. Strohmann, "Examination of Long-term Stability of Metallic LIGA Microstructures by Electromagnetic Activation," *J. Micromech. Microeng.*, 2, 193–195, 1992.

Mohr, J., W. Ehrfeld, and D. Munchmeyer, "Analyse der Defectursachen und der Genauigkeit der Structuru-bertragung bei der Rontgentiefenlithographie mit Synchrotronstrahlung," KfK, Report No. 4414, Karlsruhe, Germany, 1988a.

Mohr, J., W. Ehrfeld, and D. Munchmeyer, "Requirements on Resist Layers in Deep-Etch Synchrotron Radiation Lithography," *J. Vac. Sci. Technol.*, B6, 2264–2267, 1988b.

Mohr, J., B. Anderer, and W. Ehrfeld, "Fabrication of a Planar Grating Spectrograph by Deep-etch Lithography with Synchrotron Radiation," *Sensors and Actuators A,* 25–27, 571–575, 1991.

Mohr, J., W. Ehrfeld, D. Munchmeyer, and A. Stutz, "Resist Technology for Deep-Etch Synchrotron Radiation Lithography," *Makromol. Chem. Macromol. Symp.*, 24, 231–251, 1989.

Mohr, J., Bley P., Strohrmann, M., and Wallrabe, U., "Microactuators Fabricated by the LIGA process," *J. Micromech. Microeng.*, 2, 234–241, 1992.

Mohr, J., C. Burbaum, P. Bley, W. Menz, and U. Wallrabe, "Movable Microstructures Manufactured by the LIGA Process as Basic Elements for Micro-systems," in *Microsystem Technologies 90*, Reichl, H., Ed., Springer, Berlin, 1990, pp. 529–537.

Mohr, J., P. Bley, C. Burbaum, W. Menz, and U. Wallrabe, "Fabrication of Microsensor and Microactuator Elements by the LIGA-Process," 6th International Conference on Solid-State Sensors and Actuators (Transducers '91), San Francisco, CA, 1991, pp. 607–609.

Mohr, J., W. Bacher, P. Bley, M. Strohmann, and U. Wallrabe, "The LIGA Process-A Tool for the Fabrication of Microstructures Used in Mechatronics," Proceedings 1st Japanese-French Congress of Mechatronics, Besancon, France, October 1992, pp. 1–7.

Moreau, W. M., *Semiconductor Lithography*, Plenum Press, New York, 1988.

Muller, A., J. Gottert, and J. Mohr, "LIGA Micro-structures on Top of Micromachined Silicon Wafers Used to Fabricate a Micro-Optical Switch," *J. Micromech. Microeng.*, 3, 158–160, 1993.

Munchmeyer, D., Ph.D. thesis, University of Karlsruhe, Germany, 1984.

Munchmeyer, D. and W. Ehrfeld, "Accuracy Limits and Potential Applications of the LIGA Technique in Integrated Optics," 4th International Symposium on Optical and Optoelectronic Applied Sciences and Engineering (Proceedings of the SPIE), The Hague, Netherlands, 1987, pp. 72–9.

Muray, J. J. and I. Brodie, "Study on Synchrotron Orbital Radiation (SOR) Technologies ans Applications," Report No. 2019, SRI International, Menlo Park, CA, 1991.

Preu, G., A. Wolff, D. Cramer, and U. Bast, "Micro-structuring of Piezoelectric Ceramic," Proceedings of the Second European Ceramic Society Conference (2nd ECerS '91), Augsburg, Germany, 1991, pp. 2005–2009.

Private Communication, Vladimirsky, O., *Spin Coating PMMA*, LSU, May 1995.

Rapp, R., P. Bley, W. Menz, and W. K. Schomburg, "Konzeption und Entwicklung einer Mikromembranpumpe in LIGA-Technik," KfK, Report No. 5251, Karlsruhe, Germany, 1993.

Rapp, R., W. K. Schomburg, D. Maas, J. Schulz, and W. Stark, "LIGA Micropump for Gases and Liquids," *Sensors and Actuators A*, 40, 57–61, 1994.

Rogge, B., J. Schulz, J. Mohr, A. Thommes, and W. Menz, "Fully Batch Fabricated Magnetic Microactuator," 8th International Conference on Solid-State Sensors and Actuators (Transducers '95), Stockholm, Sweden, June 1995, pp. 552–555.

Rogers, J., C. Marques, and K. Kelly, "Cyanoacrylate Bonding of Thick Resists for LIGA," SPIE Microlithography and Metrology in Micromachining II, Austin, TX, 1996, pp. 177–182.

Rogner, A., "Micromoulding of Passive Network Components," Proceedings Plastic Optical Fibres and Applications Conference, Paris, France, 1992, pp. 102–4.

Rogner, A., J. Eichner, D. Munchmeyer, R.-P. Peters, and J. Mohr, "The LIGA Technique-What Are the New Opportunities?" *J. Micromech. Microeng.*, 2, 133–140, 1992.

Rogner, A., W. Ehrfeld, D. Munchmeyer, P. Bley, C. Burbaum, and J. Mohr, "LIGA-Based Flexible Microstructures for Fiber-Chip Coupling," *J. Micromech. Microeng.*, 1, 167–170, 1991.

Romankiw, L. T., "Electrochemical Technology in Electronics Today and Its Future: A Review," *Oberflache-Surface*, 25, 238–247, 1984.

Romankiw, L. T., "Thin Film Inductive Heads; From One to Thirty One Turns," Proceedings of the Symposium on Magnetic Materials, Processes, and Devices, Hollywood, FL, 1989, pp. 39–53.

Romankiw, L. T., ""Think Small, One Day It May Be Worth A Billion," *Interface*, Summer, 17–57, 1993.

Romankiw, L. T. and T. A. Palumbo, "Electrodeposition in the Electronic Industry," Proceedings of the Symposium on Electrodeposition Technology, Theory and Practice, San Diego, CA, 1987, pp. 13–41.

Romankiw, L. T., I. M. Croll, and M. Hatzakis, "Batch-Fabricated Thin-Film Magnetic Recording Heads," *IEEE Trans. Magn.*, MAG-6, 597–601, 1970.

Ruprecht, R., W. Bacher, H. J. Hausselt, and V. Piotter, "Injection Molding of LIGA and LIGA-Similar Microstructures Using Filled and Unfilled Thermoplastics," Micromachining and Microfabrication Process Technology (Proceedings of the SPIE), Austin, TX, 1995, pp. 146–157.

Schomburg, W. K., H. J. Baving, and P. Bley, "Ti- and Be-X-Ray Masks with Alignement Windows for the LIGA Process," *Microelectron. Eng.*, 13, 323–326, 1991.

Schomburg, W. K., J. Fahrenberg, D. Maas, and R. Rapp, "Active Valves and Pumps for Microfluidics," *J. Micromech. Microeng.*, 3, 216–218, 1993.

Schomburg, W. K., J. Vollmer, B. Bustgens, J. Fahrenberg, H. Hein, and W. Menz, "Microfluidic Components in LIGA Technique," *J. Micromech. Microeng.*, 4, 186–191, 1994.

Siddons, D. P. and E. D. Johnson, "Precision Machining Using Hard X-Rays," *Synchrotron Radiation News*, 7, 16–18, 1994.

Spiller, E., R. Feder, J. Topalian, E. Castellani, L. Romankiw, and M. Heritage, "X-Ray Lithography for Bubble Devices," *Solid State Technol.*, April, 62–68, 1976.

Staerk, H., A. Wiessner, C. Muller, and J. Mohr, "Design Considerations and Performance of a Spectro-Streak Apparatus Applying a Planar LIGA Microspectrometer for Time-Resolved Ultrafast Fluorescence Spectrometry," *Rev. Sci. Instrum.*, 67, 2490–2495, 1996.

Stix, G., "Toward Point One," *Sci. Am.*, 272, 90–95, 1995.

Strohrmann, M., P. Bley, O. Fromhein, and J. Mohr, "Acceleration Sensor with Integrated Compensation of Temperature Effects Fabricated by the LIGA Process," *Sensors and Actuators A*, 41–42, 426–429, 1994.

Tang, W. C.-K., "Electrostatic Comb Drive for Resonant Sensor and Actuator Applications," Ph.D. thesis, University of California at Berkeley, 1990.

Thomes, A., W. Stark, H. Goller, and H. Liebscher, "Erste Ergebnisse zur Galvanoformung von LIGA Mikrostrukturen aus Eisen-Nickel Legierungen," Symp. Mikroelektrochemie, Friedrichsroda, 1992.

Urisu, T. and H. Kyuragi, "Synchrotron Radiation-Excited Chemical-Vapor Deposition and Etching," *J. Vac. Sci. Technol.*, B5, 1436–1440, 1987.

van der Putten, A. M. T. and J. W. G. de Bakker, "Anisotropic Deposition of Electroless Nickel-Bevel Plating," *J. Electrochem. Soc.*, 140, 2229–2235, 1993.

Vladimirsky, Y., O. Vladimirsky, V. Saile, K. Morris, and J. M. Klopf, "Transfer Mask for High Aspect Ratio Micro-Lithography," Microlithography '95. Proceedings of the Spie—The International Society for Optical Engineering, Santa Clara, CA, 1995, pp. 391–396.

Vollmer, J., H. Hein, W. Menz, and F. Walter, "Bistable Fluidic Elements in LIGA Technique for Flow Control in Fluidic Microactuators," *Sensors and Actuators A*, 43, 330–334, 1994a.

Vollmer, J., H. Hein, W. Menz, and F. Walter, "Bistable Fluidic Elements in LIGA-Technique as Microactuators," *Sensors and Actuators A*, 43, 330–4, 1994b.

Waldo, W. G. and A. W. Yanof, "0.25 Micron Imaging by SOR X-Ray Lithography," *Solid State Technol.*, 34, 29–31, 1991.

Wallrabe, U., P. Bley, B. Krevet, W. Menz, and J. Mohr, "Design Rules and Test of Electrostatic Micromotors Made by the LIGA Process," *J. Micromech. Microeng.*, 4, 40–45, 1994.

Weber, L., W. Ehrfeld, H. Freimuth, M. Lacher, H. Lehr, and B. Pech, "Micro-Molding—A Powerful Tool for the Large Scale Production of Precise Microstructures," SPIE, Micromachining and Microfabrication Process Technology II, Austin, TX, 1996, pp. 156–167.

Wolffenbuttel, R. F. and T. A. Kwa, "Integrated Monochromator Fabricated in Silicon Using Micromachining Techniques," 6th International Conference on Solid-State Sensors and Actuators (Transducers '91), San Francisco, CA, 1991, pp. 832–835.

Wollersheim, O., H. Zumaque, J. Hormes, J. Langen, P. Hoessel, L. Haussling, and G. Hoffman, "Radiation Chemistry of Poly(lactides) as New Polymer Resists for the LIGA Process," *J. Micromech. Microeng.*, 4, 84–93, 1994.

18
X-Ray-Based Fabrication

Todd Christenson
Sandia National Laboratories

18.1 Introduction **18**-1
18.2 DXRL Fundamentals... **18**-4
   X-Ray Mask Fabrication • Thick X-Ray Photoresist • DXRL Exposure (Direct LIGA Approach) • Development • PMMA Mechanical Properties
18.3 Mold Filling **18**-12
18.4 Material Characterization and Modification **18**-19
18.5 Planarization... **18**-23
18.6 Angled and Re-entrant Geometry **18**-24
18.7 Multilayer DXRL Processing **18**-25
18.8 Sacrificial Layers and Assembly................................... **18**-28
18.9 Applications **18**-30
   Precision Components • Microactuators • Magnetic Microactuator Applications • Other Applications
18.10 Conclusions **18**-40
   Acknowledgments **18**-41

18.1 Introduction

Originally conceived for the fabrication of smaller microelectronic features, X-ray lithography has also proven to possess attributes of great utility in micromechanical fabrication. In contrast to the many micromachining processes that have been developed from microelectronic processing, however, X-ray-based approaches may be largely carried out independent of a tightly controlled clean-room environment. The mode of X-ray-based microfabrication most commonly used places this type of processing in the additive category where a sacrificial mold is used to define the desired structural material. As a result, this technique lends itself to a very rich and ever-expanding material base including a variety of plastics, metals and glasses, as well as ceramics and composites. The idea of using X-rays to define molds extends from the 1970s, when its precedent involved the definition of high-density coils for magnetic recording read/write heads and high-density magnetic bubble memory overlays when ultimately the use of X-rays for very large-scale integration (VLSI) lithography was initially investigated [Romankiw et al., 1970; Romankiw, 1995; Spiller et. al., 1976; Spears and Smith, 1972]. The distinction from VLSI X-ray lithography is that the mold or photoresist thickness for micromachining interests is generally much greater than 50 μm and may be well over 1 mm. X-ray processing at these thicknesses has prompted the nomenclature *deep X-ray lithography* or DXRL-based microfabrication.

The primary utility of DXRL processing extends from its ability to precisely and accurately define a mold. Consequent component definition via mold filling is thus directly determined by mold acuity. Exceptional definition in this regard is possible with highly collimated X-rays that may be obtained via

TABLE 18.1 Synchrotron Radiation Facilities in the U.S. with Active DXRL-Devoted Beamlines

Storage Ring	Ring Parameters (Energy, Critical Wavelength, Current)	Site
Aladdin	0.8, 1 GeV, 22.7, 11.6 Å 260, 190 mA	Synchrotron Radiation Center (SRC), Stoughton, WI
CAMD (Center for Advanced Microstructure Devices)	1.3, 1.5 GeV, 7.4, 4.8 Å 400, 200 mA	Louisiana State University (LSU), Baton Rouge, LA
ALS (Advanced Light Source)	1.5, 1.9 GeV, 8.2, 4.1 Å 400 mA	Lawrence Berkeley Laboratory (LBL), Berkeley, CA
NSLS (National Synchrotron Light Source)	X-ray ring 2.584, 2.8 GeV, 2.2, 1.7 Å 300, 250 mA VUV ring 0.808 GeV, 20 Å 1000 mA	Brookhaven National Laboratory (BNL), Upton, NY
SPEAR [SPEAR3] (Stanford Positron Electron Accelerating Ring)	3.0 GeV [3.0 GeV], 2.6 [1.4] Å 100 mA [200 mA]	Stanford Synchrotron Radiation Laboraory (SSRL), Stanford, CA
APS (Advanced Photon Source)	7.0 GeV, 0.64 Å 100 mA	Argonne National Laboratory (ANL), Argonne, IL

synchrotron radiation from a storage ring. Such X-rays, in addition to possessing atomic rather than optical absorption character, thereby eliminating diffraction and standing wave effects, possess collimation on the order of 0.1 mrad. Combined with a developer that has high selectivity between unexposed and exposed photoresist, this exposure capability yields mold sidewall definition with less than 0.1 µm run-out for several hundred micron thickness. This radical form of lithography has resulted in a new type of foundry service to provide synchrotron radiation access. In the U.S., for example, these services may be obtained from the facilities listed in Table 18.1. Many more facilities are actively engaged in this activity throughout the world.

Two distinct X-ray-based microfabrication philosophies exist. The realization that relatively thick (>>100 µm) X-ray lithography was possible and could subsequently be applied to precision molding was first made by Ehrfeld for application to separation nozzle fabrication for uranium isotope enrichment [Becker et al., 1982; 1986; Ehrfeld et al., 1987; 1988a; 1988b]. The process was coined "LIGA," a German acronym for the three basic processing steps of deep X-ray *lithography,* mold filling by *galvanoformung* or electroforming and injection molding replication or *abformung*. The approach defined by the LIGA process aims to become mostly independent from a synchrotron radiation X-ray exposure step by defining a master metal mold insert used to replicate numerous further plastic molds via plastic injection molding. Recently, improved storage ring access and the ability to expose multiple samples simultaneously have resulted in another option of using DXRL to form all sacrificial molds. The typical DXRL process sequence is outlined in Figure 18.1. With either case, the ultimate result is the ability to batch fabricate prismatically shaped components with nearly arbitrary in-plane geometry at thicknesses of several hundred microns to millimeters while maintaining submicron dimensional control. This result translates to 100-ppm accuracy for millimeter and submillimeter dimensions. Precision, or repeatability, is obtained by the batch nature of this lithographically based process.

The novel implications of this result are many-fold. For precision-engineered componentry, an increase in resolution is possible over such conventional machining techniques as stamping or fine blanking, or

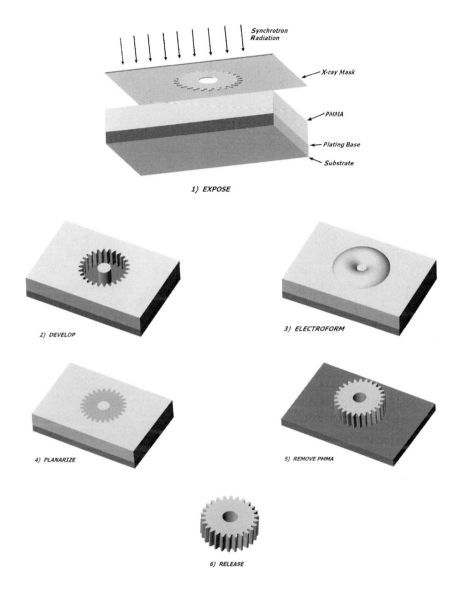

FIGURE 18.1 (Color figure follows p. 12-26.) DXRL-based (direct LIGA) microfabrication process.

electro-discharge machining (EDM), while the throughput associated with batch or parallel fabrication is acquired. Results of using DXRL-based molding for the fabrication of EDM electrodes have also shown that an arrayed type of tool may be used for batch plunge EDM of many precision parts in parallel. The conclusion is that DXRL-based processing is a tool appropriate for miniature precision piece-part component manufacture when material property requirements can be met through some additive deposition procedure or sequence of multiple molding steps. A fundamental issue for the fabrication of MEMS is the ability to batch fabricate complex three-dimensional mechanisms with appropriately scaled tolerances. The three-dimensional requirement is of particular concern for constructing microactuators where force output scales directly with the volume of stored energy as well as in micromachined sensors such as seismic sensors, for example, for the definition of inertial proof masses. Fabrication based on planar processing is immediately challenged in this regard and a so-called high-aspect-ratio (HAR) process is needed. Furthermore, the difficulty with process integration in realizing a microelectromechanical system

TABLE 18.2 DXRL-Based Processing Attributes

Type of structural geometry accommodated	Prismatic with arbitrary two-dimensional shape and sidewall angle
Structural thickness	Commonly 200 to 800 μm up to several millimeters (10 cm demonstrated)
Lateral run-out	<0.1 μm per 100 μm of vertical length
Minimum critical dimension	A few microns typical, function of photoresist stability
Surface roughness	10 to 20 nm RMS typical; as low as several nm
Thickness control	With conventional lapping: ± a few microns typical, as low as 0.5 μm across 4-in. diameter
Materials (electroformable metals)	Ni, Cu, Au, Ag, NiFe, NiCo, NiFeCo etc.
Materials (pressed powders, embossed, hot forged)	Alumina, PZT, ferrites, NdFeB, SmCo, variety of plastics and glasses

(MEMS) is many times a result of temperature effects. Most additive mold-filling processing used in DXRL-based processing takes place at less than 100°C and is therefore appropriate for postprocessed components. By appropriately accommodating for possible semiconductor X-ray damage, the possibility of integrating microelectronics and microsensors with scaled precision metal mechanisms becomes particularly attractive.

The flexibility of the X-ray-based approach is also becoming apparent in its application to optical MEMS or micro opto electro mechanical systems (MOEMS). Exposures at arbitrary angles with the substrate are possible, and multiple-angled exposures can be accommodated, as has recently been demonstrated in photonic band gap (PBG) structure fabrication that uses three 120°-rotated exposures at 35.26° to form a diamond lattice structure [Feirtag et al., 1997a; Cuisin et al., 2000]. Table 18.2 provides a summary of DXRL processing attributes. The fundamental fabrication issues involved in DXRL processing center on X-ray mask fabrication, thick photoresist application, deep X-ray exposure and selective development. Subsequent process issues pertaining to typical device interests include mold filling, planarization, multilayer processing, assembly and mechanism construction and device integration. Applications arising from this fabrication technique are summarized in component, actuator and sensor categories.

18.2 DXRL Fundamentals

18.2.1 X-Ray Mask Fabrication

The fidelity of the X-ray mask pattern largely determines the results one will obtain with DXRL-assisted processing. There are two components to an X-ray mask: A supporting substrate or membrane with sufficient X-ray transparency is used in conjunction with an absorber layer patterned upon it. The required X-ray transmission character for the supporting mask substrate material may be found by considering the X-ray absorption behavior of the X-ray photoresist plotted in Figure 18.2 for poly(methyl methacrylate) (PMMA), the most commonly used X-ray photoresist. This plot conveys the X-ray photon energies required to practically expose a given thickness of PMMA to a dose needed to make it susceptible to dissolution by a suitable solvent, the developer. For typical DXRL layer thicknesses of 100 μm to 1 mm, these energies range from 3.5 to 7 keV (or wavelength of from 3.5 to 1.7 Å). Possible mask substrates include Be, C and Si slices with thicknesses near 100 μm and diamond, Si, SiC, SiN and Ti membranes with thicknesses of 1 to 2 μm [Sekimoto et al., 1982; Visser et al., 1987; Guckel et al., 1989]. The transmission behavior for these mask substrates is plotted in Figure 18.3. What is practically desired is a mask substrate stable to large X-ray flux and with sufficient thickness or support to allow easy handling while maintain-ing the transmission required to readily achieve the required exposure dose, which, for PMMA, is near 3 kJ/cm^3. A corresponding source of X-ray radiation is also required. The nearly ideal light source for this task is found in the synchrotron radiation (SR) that is emitted by charged-particle

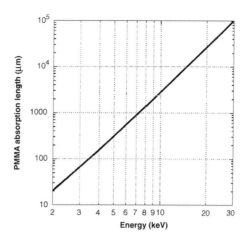

FIGURE 18.2 Absorption length of PMMA ($C_5H_8O_2$, density = 1.19 g/cc) as a function of X-ray photon energy.

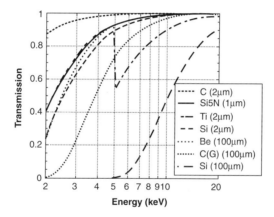

FIGURE 18.3 X-ray transmission of various X-ray mask substrates as a function of X-ray photon energy. The nonstoichiometric form of silicon nitride is a low-stress LPCVD (low-pressure chemical vapor deposition) form used in membrane fabrication. For the carbon mask at 100-µm thickness, graphite with a density of 1.65 g/cc was used.

storage rings. The properties of this light source that result from accelerating charged particles at relativistic energies include highly intense radiation over a broad spectral range from ultraviolet (UV) to X-ray wavelengths and collimation of the order of 0.1 mrad. The SR beam is emitted in a narrow horizontal stripe of light through a vacuum line (*beamline*) extending from the storage ring, requiring the mask and photoresist combination to be scanned vertically through the beam which additionally determines a local dose rate in the photoresist. The SR spectra are plotted in Figure 18.4 for U.S. synchrotron radiation facilities. Frequently used SR equations are listed in Table 18.3 [Margaritondo, 1988]. Low-energy (<2 keV) flux creates thermal problems generated by a locally peaked dose at the surface of the photoresist, requiring a filter to limit this dose to a value that avoids eventual photoresist damage found to occur in PMMA near a dose of 20 kJ/cm^3 [Ehrfeld et al., 1988b]. Filtering is achieved with slices of lower atomic number materials inserted in the SR beamline, including a window that vacuum-isolates the storage ring from the exposure beamline. Thus, a mask substrate material with high thermal conductance is desired to assist in heat dissipation due to absorbed radiation by the mask substrate and absorber. The data in Figures 18.3 and 18.4 suggest that Be and C in the form of graphite can be used as X-ray mask substrates

TABLE 18.3 Frequently Used Synchrotron Radiation Formulas

Wavelength/energy conversion	$\lambda(\text{Å}) = \dfrac{12398.5}{h\nu(\text{eV})}$
Bending magnet radius	$\rho(m) \cong \dfrac{3.336 E(\text{GeV})}{B(\text{T})}$
Critical wavelength	$\lambda_C(\text{Å}) \cong \dfrac{18.64}{B(\text{T}) E^2(\text{GeV})}$

Note: $h = 4.136(10)^{-15}$ eVs; ν = frequency; E = energy of circulating charge particle; B = bending magnet field.

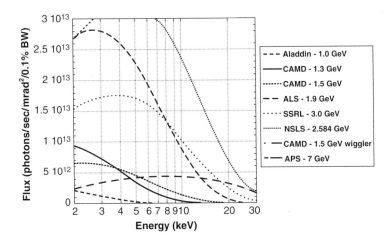

FIGURE 18.4 Synchrotron radiation spectra for U.S. storage rings listed in Table 18.1. The spectral curve for the APS source which has a peak flux near $1(10)^{14}$ photons/sec/mrad^2/0.1 %BW at 16 keV does not appear on the graph.

at thicknesses of 100 μm for all SR sources. The cost and hazards associated with Be processing make it less attractive, however, and graphite has proven to be a convenient mask material, particularly when its surface is treated in a manner that decreases roughness due to bulk porosity, thereby facilitating accurate lithography. Filling the porosity via a pyrolytic carbon coating, for example, and then subsequently polishing this surface layer may accomplish this.

Requirements for the X-ray absorber are set by the minimum required exposure contrast that may be defined as the exposure dose at the PMMA bottom surface in the mask transmission areas divided by the exposure dose delivered to the PMMA top surface under the absorber regions. Mask contrast is a function of the X-ray source, mask substrate and mask absorber and exposed photoresist thickness. The results reveal two categories of X-ray masks. One X-ray mask type is particularly well suited for low-energy exposures of PMMA up to several hundred micron thicknesses and requires a mask substrate with high transmission of flux near 4 keV such as membranes that are one to a few microns thick of strain-controlled silicon nitride, silicon carbide or silicon. Atomic number and density determine the figure of merit for an absorber material. Thus, materials such as platinum, tantalum, tungsten and gold are chosen, with gold being most prevalent as it is readily electroplated. The corresponding absorber for the low-energy mask is several microns of gold. For accurate dimensional control, this again leads to an additive technique in which a photoresist several micron thick is defined, usually via a UV optical microlithographic mask transfer, to provide for through-mask electroplating of gold. Vertical sidewall photoresist pattern transfer is essential to maintain accurate X-ray exposure definition, as an absorber sidewall taper leads to a sidewall taper that is X-ray exposed and developed and is the result of an insufficient absorber thickness. To achieve vertical photoresist patterns, a bi-layer deep UV (DUV) photoresist technique using spun-on PMMA has been implemented that takes advantage of the highly

X-Ray-Based Fabrication

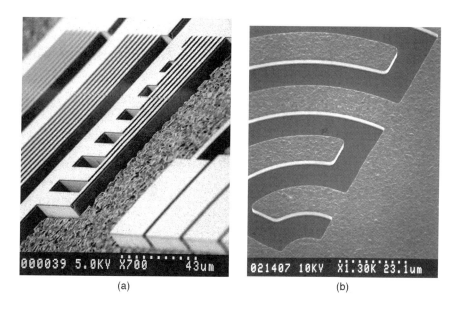

FIGURE 18.5 (a) PMMA 8-μm thick patterned with a conformal portable mask (CPM) or bi-layer process. A flood deep-UV source exposes the PMMA that was subsequently developed using G–G developing solution. (b) Resulting 6-μm-thick gold absorber pattern after PMMA stripping.

FIGURE 18.6 Frontside and backside view of 1-μm thick silicon nitride membrane X-ray mask 5 × 7 cm in area supported by a 4-in.-diameter silicon wafer with 8-μm-thick patterned and electroplated gold absorber layer.

selective developing system used with DXRL PMMA exposures [Lin, 1975; Guckel et al., 1990]. Use of a bi-layer or CPM (conformal portable mask) technique minimizes out-gassing during exposure that is of particular concern for membrane contact lithography, whereby mask–membrane separation may consequently arise. A result is shown in Figure 18.5a, which reveals the vertical sidewalls of an 8-μm thick PMMA layer. Subsequent gold electroplating and photoresist removal yields an absorber structure such as shown in Figure 18.5b. A number of recent commercial photoresist additions have also proven to be suitable for low-energy X-ray mask absorber definition, including Shipley SJR5740 and 220 developed for through-mask electroplating applications for magnetic read/write heads [Romankiw, 1995]. Thicker UV exposed photoresists have also been applied to through-mask plating for less critical applications [Allen, 1993; Loechel and Maciossek, 1995; Despont et al., 1997]. Figure 18.6 shows a typical membrane-type, low-energy X-ray mask.

Exposures of PMMA thicknesses of millimeters to over 1 cm entail substantially different X-ray masks. These exposures involve X-ray photon energies over 10 keV, and for sufficient contrast they require gold absorber thicknesses over 20 μm. Because of the difficulty in precisely patterning this thickness of photoresist to submicron tolerances with UV lithography, a two-step mask fabrication sequence is commonly employed. Consequently, a low-energy X-ray mask with 2 μm gold absorber is used to pattern

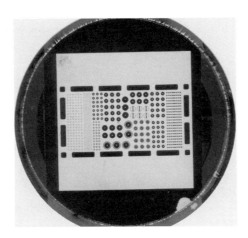

FIGURE 18.7 X-ray mask patterned on a 3-in.-diameter graphite support 100 μm thick with 20-μm-thick gold absorber pattern.

several tens of microns of PMMA directly upon the high-energy X-ray mask substrate. Necessary soft X-ray filtering may then be transferred to the X-ray mask substrate due to the increased amount of filtering of softer X-rays required to maintain low surface doses with higher energy storage-ring spectra. Thus, the high-energy mask substrate may be a relatively thick layer of low Z material, such as 100-μm thick silicon. Water cooling of the X-ray mask and exposure substrate also becomes a necessity with higher energy exposures due to an increased overall delivered power during exposure.

A mask manufacturability issue is evident which relates to the constraints of using a membrane-based, low-energy mask to achieve high-energy exposures. Figure 18.3 reveals that carbon as graphite at 100-μm thickness may serve both roles. The importance of graphite is apparent, as it possesses the qualities of being an inexpensive rugged mask substrate capable of being practically used for low- and high-energy DXRL exposures, having good thermal conductivity, and avoiding the hazards and expense of beryllium and other low-Z compounds and their associated processing. A carbon X-ray mask is depicted in Figure 18.7.

18.2.2 Thick X-Ray Photoresist

High-aspect-ratio processing, as with many MEMS technologies, is particularly sensitive to internal and process-induced strain, as well as adhesion issues. The control of these two areas substantially determines DXRL-based manufacturability. The major issue of process stability in this regard concerns realization of thick, stress-free photoresist application with exceptional adherance to a metallized substrate for electrolytic deposition or other substrate material to facilitate further processing. The difficulty that arises in most photoresist application methods is avoiding large internal tensile stress, which can lead to crazing during development and subsequent processing. A number of X-ray photoresist materials suitable for thick application have been examined [Ehrfeld et al., 1997]. Because it possesses relatively good mechanical stability and uniquely high resolution, PMMA has been the primary X-ray photoresist of use. The application procedure initially conceived uses direct polymerization with a casting resin [Mohr et al., 1988]. A crosslinking agent is typically added in the process to further increase the yield strength, thereby providing additional resistance to crazing. To improve adhesion to a metallized substrate, an adhesion-promoting chemistry or intermediate layer may be used [Khan Malek et al., 1998; DeCarlo et al., 1998]. The volume contraction of 20%, resulting during polymerization, leads to a tensile-stressed film which may be partially alleviated with annealing but cannot be completely removed due to thermal mismatch with typical substrate materials that have lower thermal expansion coefficients. The consequences of residual photoresist stress are component geometry design rule constraints and limits on processing procedures.

FIGURE 18.8 Bonded 1.5-mm-thick PMMA on 4-in.-diameter substrate.

An alternative approach makes use of precast, low-stress PMMA sheets [Guckel et al., 1996a]. In this method, a linear high-molecular-weight or crosslinked PMMA sheet is used to obtain PMMA plate cut-outs commensurate with a particular substrate geometry. Typical sheet thicknesses range from 500 µm to 1 cm. A bonding technique providing for attachment of these sheet cut-outs to a metallized substrate without a significant temperature variation or induced chemical or mechanical perturbation is then implemented. A well-proven technique uses a thin prespun layer of high-molecular-weight (>1 million) PMMA combined with the use of the monomer, methyl methacrylate (MMA), as the bonding solvent. A plate of PMMA is placed on the desired substrate with a spun PMMA layer typically 1 µm thick, and MMA is applied between the PMMA surfaces and wets the interface via capillary action. Because the diffusion of MMA through PMMA is extremely rapid even at room temperatures, curing may take place without a bake-out at elevated temperature. Figure 18.8 shows an example of the result, which is a highly repeatable procedure for obtaining thick low-stress PMMA films with arbitrary thickness. The bonded sheet may additionally be thinned via precision milling or flycutting. Use of a linear PMMA polymer allows dissolution of the PMMA subsequent to mold filling with a solvent such as methylene chloride.

The polymerization of organic vapors in a glow discharge is another method used to produce thin polymer films [Goodman, 1960; Biederman, 1987]. The use of MMA, in particular, has been studied for use as a plasma-deposited photoresist [Morita et al., 1980; Guckel et al., 1988]. The result is a highly conformal deposited film able to uniformly fill deep trenches with excellent adhesion to semiconductor materials. The plasma-polymerized MMA (PPMMA) film is highly crosslinked and has a stability dependence on reactor pressure, power and temperature. Practically, deposition rates as high as 140 Å/min may be achieved, and films as thick as 10 µm have been deposited. Increased power levels result in further decomposition of MMA and films less like PMMA.

Another interesting X-ray resist candidate is provided by the negative-working SU-8 epoxy-based photoresist more typically used for e-beam or thick near-UV lithography [Lee et al., 1995; Khan Malek, 1998]. The resist is based on the EPON® resin SU-8 from Shell Chemical which possesses excellent high-temperature stability with a glass transition temperature of more than 200°C when crosslinked [Shaw et al., 1997]. This property is due to a high epoxide group functionality (with an average near eight) that also leads to an exceptionally high sensitivity as a photoresist. Deep X-ray lithography results indicate a sensitivity near 52 J/cm^3 [Bogdanov and Peredkov, 2000] or roughly 40 times greater than PMMA. The resulting crosslinked epoxy also possesses a high optical refractive index near 1.65 but is difficult to chemically dissolve in this state.

18.2.3 DXRL Exposure (Direct LIGA Approach)

With the ability to expose and apply PMMA layers over 1 cm in thickness while most applications being pursued require 150 to 500 µm (10-cm exposures have been demonstrated [Siddons et al., 1994; Guckel, 1996]), the question of how to most beneficially use deep X-ray lithography arises. What becomes immediately apparent is that multiple sheets of PMMA may be stacked and exposed simultaneously

[Guckel, 1998; Guckel et al., 1999a; Siddons et al., 1997]. Accounting for the issues associated with the bonding or attachment of exposed PMMA sheets to a substrate accommodating development and mold filling, this procedure has significant throughput advantages. An examination of exposures obtained with the 7.5-Tesla wiggler insertion device [Manohara, 1998] at the Center for Advanced Microstructure Devices (CAMD), Louisiana State University, reveal these advantages. For a 1-cm-thick PMMA stack with a limited delivered top dose of 10 kJ/cm^3 and bottom dose of 3 kJ/cm^3, an exposure time slightly greater than 4 hr results for an exposure area of 12.5 cm × 5 cm. This may be compared to the time (depending on storage-ring beam current) of roughly 20 to 30 min to perform the same area exposure at 250-μm thickness using 1.5-GeV CAMD bending magnet radiation. The enhanced high-energy performance results in a throughput for a 250-μm PMMA thickness of over 200 substrates per day with a total exposure cost of less than \$0.25/cm^2, or \$15 per 4-in.-diameter equivalent substrate. The result reveals a potential approach for direct LIGA or using DXRL as the primary tool for precision mold fabrication.

18.2.4 Development

Although PMMA sensitivity, at roughly 500 mJ/cm^2, is much lower than other photoresists, the combination of high-intensity, well-collimated X-ray flux and the existence of a highly selective PMMA developing system facilitates exploitation of the high resolution of PMMA in a batch process for obtaining microminiature, ultra-precision molds. A developer composed of an aqueous mixture of solvents (80% diethylene glycol monobutyl ether, 20% morpholine, 5% ethanolamine, and 15% water), commonly referred to as the G–G developer after its patent holders, has been formulated to achieve very high contrast PMMA development [Ghica and Glashauser, 1983; Stutz, 1986]. The G–G developer is commonly used at a temperature of 35°C but may be also used successfully at lower temperatures down to 15°C with decreased development rates yet higher selectivity. The G–G developer solution also etches copper, which may require a protective layer if this material is desired as a plating base. Figure 18.9 shows development rate curves as a function of X-ray dose [McNamara, 1999]. Development rate temperature dependence has also been noted [Tan et al., 1998]. Additionally, the quality of PMMA definition has been found to be dependent on developer temperature and whether the PMMA is linear or crosslinked [Pantenburg et al., 1998]. Diffusion-limited development is another complicating factor, particularly for high-aspect-ratio features with critical dimensions of tens of microns and below. This issue has prompted the use of megasonic (~600 kHz to over 1 MHz) agitation of the developing solution which additionally alleviates redeposition of dissolved yet marginally high-molecular-weight photoresist, thereby decreasing sidewall particulates and roughness.

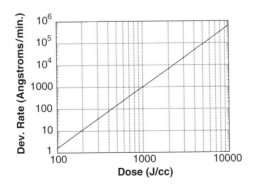

FIGURE 18.9 PMMA development rate as a function of X-ray exposure dose for ICI CQ-grade PMMA with G–G developer at 35°C.

Detailed discussions of PMMA photochemistry and pattern limits exist in several references [Ouano, 1978; Mohr, 1986; Münchmeyer and Ehrfeld, 1987; Pantenburg and Mohr, 1995; Feiertag et al., 1997b]. Modeling of fluorescence radiation has revealed that a dose of magnitude 500 J/cc may be deposited underneath absorber regions at areas adjacent to the mask absorber edge and at the PMMA–substrate interface due to emission of fluorescence radiation at the substrate (in this case, titanium) surface [Feiertag et al., 1997c]. This behavior may readily lead to PMMA undercutting during development and ultimately PMMA adhesion loss. The consequence is the possibility of over development and a resulting defect "foot" in the electroform. Electroplating a few microns of sacrificial metal before the structural material can resolve this situation in some circumstances, although it is not an issue for freestanding PMMA exposures. Further study of G–G developer behavior has shown that this alkaline solvent mixture induces a chemical reaction with exposed PMMA prior to dissolution and additionally that there is a strong development dependence on PMMA stereochemistry [Schmalz et al., 1996]. Another related issue concerns the origins and tailoring of PMMA sidewall roughness, which has direct implications for optical components and tribological behavior. Results from CAMD exposures have shown PMMA sidewall roughnesses as high as 25 nm and as low as 5 nm RMS with typical results near 12 nm. This result concurs with the typical resolution quoted for PMMA of roughly 50 to 100 Å [Van der Gaag and Sherer, 1990].

18.2.5 PMMA Mechanical Properties

The origin of design rules in the DXRL-based molding process rests on the mechanical limitations of the mold material, PMMA. A tensile curve for a high-molecular-weight linear PMMA material is shown in Figure 18.10 and was measured with a milliscale tensile pulling technique. A summary of properties for PMMA is listed in Table 18.4. Many of these parameters are also dependent on the molecular weight or degree of crosslinking as well as additives that are commonly used for enhanced UV absorption and stabilization. The large thermal coefficient of expansion leads to significant concerns of geometry distortion and strain-induced buckling. As an example, for a PMMA sheet bonded to a silicon substrate, a 10°C temperature change leads to a 0.11% strain in the PMMA. This situation typically manifests itself via buckling of PMMA plates defined by three built-in edge conditions at the substrate interface and side edges and a free edge condition at the top edge. The solution for this plate-buckling problem may be to generate design curves indicating mechanically limiting regions of geometry and thereby establish minimum

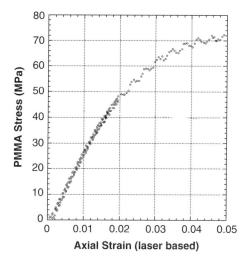

FIGURE 18.10 Tensile curve of PMMA (ICI CQ-grade) measured with load frame and laser extensometer system for a DXRL-patterned tensile coupon geometry with 750-μm gauge width and 150-μm thickness.

TABLE 18.4 Properties of PMMA[a]

E, bulk modulus	~3.0 GPa
G, shear modulus	~1.7 GPa
ν, Poisson's ratio	0.40
α, linear thermal coefficient of expansion	~7(10)$^{-5}$/°C
Water absorption (by wt%)	
(during immersion)	
24 hr	0.2%
7 days	0.5%
21 days	0.8%
48 days	1.1%
Glass transition temperature	105°C
σ_Y, tensile strength	~70 Mpa
δ, density	1.19 g/cc
κ, thermal conductivity	0.193 W/m·K
Heat capacity	1.42 J/g·K
n, refractive index at 365 nm, 1014 nm	1.514, 1.483
Abbe number	58.0
ε_r, dielectric constant	
60 Hz	3.5
1 kHz	3.0
1 MHz	2.6
30 GHz	2.57

[a] Nominal at 20°C.

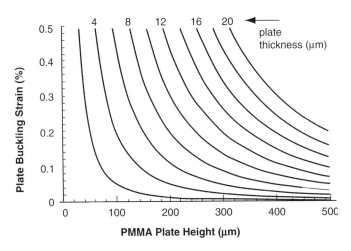

FIGURE 18.11 Buckling strain of PMMA plates typical of DXRL-defined vertical geometry with bottom side attached to a substrate and top side free as a function of plate height for various plate thicknesses (for plate length ≫ plate height).

PMMA design rules [Christenson, 1995]. One such graph is provided in Figure 18.11. The net effect is that this behavior constitutes the major restriction on aspect-ratio and process integration. This same effect, on the other hand, may be used to quantify residual PMMA strain.

18.3 Mold Filling

The material base that may be accessed largely defines application areas that may benefit from DXRL fabrication. PMMA mold forms produced by deep X-ray lithography are most commonly used to provide an electroforming mask, as depicted, for example, in Figure 18.12. In the case of electroforming, the DXRL mold substrate must be provided with a plating base suitable for an electrolyte of interest. An example plating base is a three-layer film of Ti/Cu/Ti or Ti/Au/Ti in which the lower titanium layer

FIGURE 18.12 DXRL-exposed and developed 200-μm-thick PMMA mold form.

provides adhesion to a substrate such as silicon or glass and the top titanium layer provides for photoresist (PMMA) adhesion while also protecting the electroplating seed layer, which in this case is copper or gold. The top titanium layer may be removed after PMMA development with a dilute (100:1) hydrofluoric acid etch. Additional lower adhesion layer materials include Cr, V and Nb.

A large number of elemental metals and alloys may be electroplated, but this capability alone does not lead to their use as engineering or structural materials or allow electrodeposition into high-aspect-ratio mold forms. Many electroplating procedures, for example, require high temperatures or electrolytes that will dissolve or react chemically with PMMA molds, but the main hindrance to extending the use of electroplated materials to microelectroforming is internal strain. Electrodeposit strain can impart delamination stresses onto adjacent PMMA mold material and eventually lead to adhesion loss of the deposit itself. Geometry and process control cannot be attained without minimizing deposit stress to an acceptable level. An integrated means to obtain local measurement of internal electrodeposit stress may be accomplished with patterned beam structures as depicted in Figure 18.13 [Guckel et al., 1992]. This approach employs arrays of clamped–clamped beams and ring-and-beam structures of various beam lengths that are sensitive to lateral buckling. The structures reveal internal strain levels by visual determination of a critical buckling beam length and are fabricated with the use of a lower sacrificial layer etched away to free the beam structure from the substrate. Considerable research directed to understand how stress arises during electrodeposition has been carried out [Marti, 1966; Weil, 1970; Harsch et al., 1988]. In the course of this research, certain electrolytes have proven to be particularly well suited for low-stress electroforming. The most notable of these is the sulfamate-based electroforming bath that has made intricate electroforming of nickel possible. This bath formulation, specified in Table 18.5, is used for most LIGA electroforming due to the relative ease and rapidity with which thick low-stress electroforms may be obtained. Example resultant electroformed nickel structures subsequent to PMMA dissolution are shown in Figure 18.14. Another material particularly convenient for electroforming is copper, for which the formulation listed in Table 18.5 is particularly well suited for high-aspect-ratio electroforming. An increase in the ratio of sulfuric acid to copper has been identified as increasing plating solution conductivity and throwing power, which is advantageous for high-aspect-ratio geometry. Complications arising from electroforming through varied geometry and component dimensions arise due to

TABLE 18.5 Common Electroforming Bath Compositions and Low Stress Plating Conditions

Nickel		Copper	
$Ni(NH_2SO_3)_2 \cdot 4H_2O$	440 g/l	$CuSO_4 \cdot 5H_2O$	68 g/l
Ni (as metal)	80 g/l	Cu (as metal)	17 g/l
Boric acid	48 g/l	Sulfuric acid	170 g/l
Wetting agent	0.4%/vol	Chloride ion	70 ppm
Temperature	50°C	Proprietary brightener, carrier and leveler additives	Per manufacturer
pH	3.8–4.0	Temperature	25°C
Current density	50 mA/cm^2	Current density	25 mA/cm^2
Anode	Sulfur-depolarized nickel in Ti basket with anode bag	Anode	Phosphorized copper (0.05% P) with anode bag

FIGURE 18.13 DXRL-patterned and electroformed nickel ring and beam structure used to measure tensile internal strain via mechanical buckling. The beam cross member, 2 µm wide and 18 µm thick, was originally straight after electroforming and has now buckled after release via a sacrificial etch as a result of tensile internal strain. Together with the outer and inner ring radius of 100 and 80 µm, respectively, the critical buckling geometry of this structure represents a 0.08% internal tensile strain.

locally varying current density as a result of current crowding and limited electrolyte conductivity, in addition to mass transport variation of the electrolyte to the electrodeposit surface [Mehdizadeh et al., 1993]. These issues may be partially alleviated with techniques such as pulse-plating and electrolyte agitation schemes, including jet or fountain plating and paddle cells [Andricacos et al., 1996]. What has been found is that extremely high-aspect-ratio cavities may be readily electroplated with high accuracy, as demonstrated by the structure in Figure 18.15. The electroform replication of the PMMA mold is also found to be essentially perfect to the subnanometer-length scale [Hall, 2000]. Other electroformable materials used with direct LIGA processing include gold, silver, indium, platinum, lead–tin and alloys of nickel, including nickel–iron and nickel–cobalt.

The ability to electroform ferromagnetic materials opens the possibility to implement a large class of magnetic devices. The nickel–iron alloy 78 Permalloy (78% Ni/ 22% Fe), for example, may be electroformed to 500-µm thicknesses, as shown in Figure 18.16. This soft (low coercivity) ferromagnetic material has near zero magnetostriction and a B–H behavior, as shown in Figure 18.17. Because microactuator

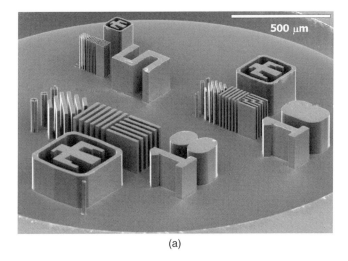

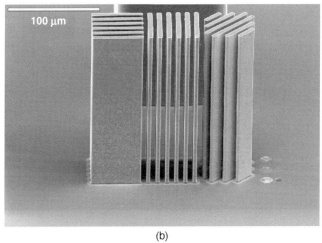

FIGURE 18.14 (a) Electroformed nickel test structures 200 μm tall; (b) close-up of 5-μm-wide nickel lines and spaces.

force is developed from the equivalent mechanical pressure generated in a magnetic field as defined by Eq. (18.1):

$$p_M = \frac{B^2}{2\mu_o} \ (\text{Pa}) \qquad (18.1)$$

where B is the magnetic flux density (T) and μ_o is the permeability of free space $(4\pi(10)^{-7}$ Hy/m), a need for soft ferromagnetic materials that can accommodate as large a B field as possible is evident. Recent work has demonstrated electroplated NiFeCo material with magnetic saturation flux density of 2.1 T [Osaka et al., 1998]. Such materials have great potential if they can be incorporated into DXRL processing.

Another area in magnetics requiring attention is identified by considering electromagnetic scaling issues for micromagnetic devices that show the advantage of using permanent magnets for efficient miniature electromagnetic transducers. Permanent magnets that posses the best qualities for microactuator applications are rare earth based and include the samarium cobalt (SmCo) and neodymium iron boron (NdFeB) families. These materials may be directly processed in a bonded form containing isotropic

FIGURE 18.15 Electroplated nickel in PMMA craze resulting from PMMA stress concentration test structure. The resulting nickel wall is 20 μm in height with a width of 0.12 μm.

FIGURE 18.16 Gear electroformed from 78/22 nickel/iron to a thickness of 500 μm. This material, in addition to possessing desirable soft ferromagnetic response, has been found to perform well as part of mechanisms where tribology is of concern due to its high hardness and yield strength.

rare-earth-based powder mixed with a binder material such as epoxy. The figure of merit of interest, maximum energy product, varies for these types of magnets from 6 to 10 MGOe (mega-gauss-oersted; 1 MGOe = $100/4\pi$ kJ/m^3) for bonded varieties to over 40 MGOe in fully dense anisotropic forms. Work with bonded rare-earth permanent magnet (REPM) material has demonstrated the direct molding of bonded REPM from DXRL PMMA mold forms [Christenson et al., 1999a]. The result is the batch fabrication of a multitude of prismatic REPM geometry with feature sizes as small as 5 μm and dimensional tolerances of 1 μm. The example of bonded permanent magnet geometry shown in Figure 18.18 was fabricated with the process outlined in Figure 18.19. An unmagnetized mixture of REPM powder and epoxy is applied to a substrate with a DXRL-defined PMMA mold while in a low-viscosity state by calendering and pressing. After curing in a press at pressures near 10 ksi, the substrate with pressed composite is planarized. The entire substrate of permanent magnets is then subjected to a magnetizing

X-Ray-Based Fabrication

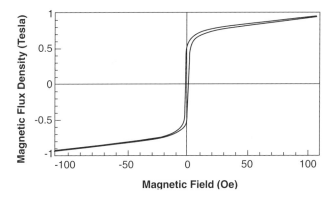

FIGURE 18.17 B–H response of an electroplated 78 Permalloy sample. The initial permeability is near 4000 and the magnetic saturation flux density is 1 T.

FIGURE 18.18 SEM photographs of bonded $Nd_2Fe_{14}B$ permanent magnets formed with DXRL-defined PMMA molds. Powder particle size ranges from 1 to 5 µm, and the binder is a methylene-chloride-resistant epoxy. These images show magnetized structures, thus mild distortion is present due to magnetic perturbations of the imaging electrons. Maximum energy products of 9 MGOe have been obtained with this process.

field of at least 35 kOe in the desired magnetization orientation and the magnets are subsequently released from the substrate by dissolving the PMMA and etching an underlying metal layer. A related result using a blended form of photoresist (epoxy-based SU-8) and Sm_2Co_{17} particle film uses direct UV lithography to pattern permanent magnets with 2.8-MGOe energy density [Dutoit et al., 1999].

Fabrication of REPM structures with a greater maximum energy product requires higher temperature processing, for which PMMA is not suited. An intermediate mold approach has therefore been examined. Direct slurry casting of alumina from PMMA mold forms has yielded replicated geometry with a minimum of distortion (<2 µm) by careful firing sequences [Garino et al., 1998]. Figure 18.20 depicts example glass/alumina composite structures. Other ceramic materials have been demonstrated by this technique, including ferrites and PZT [Ritzhaput-Kleissl et al., 1996; Pioter et al., 1997]. The resulting alumina mold may be used to perform high-temperature processing with powder metals. The complications of anisotropic REPM powder processing have prompted an approach that does not involve powders [Christenson et al., 1999b]. In a manner similar to hot embossing, a bulk anisotropic slab of REPM material is placed over the alumina mold that is decorated with a copper release layer. By vacuum hot pressing at 700°C and 2000 psi, strain rates of 22% for die-upset $Nd_2Fe_{14}B$ material are realized, thereby allowing filling of the alumina mold. Maximum energy products of 23 MGOe have been achieved in submillimeter permanent magnet shapes with this approach, with the test geometry results shown in Figure 18.21.

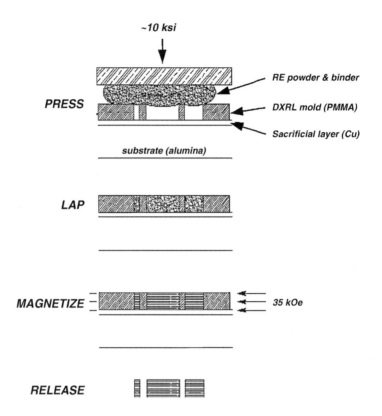

FIGURE 18.19 DXRL-based bonded REPM fabrication sequence.

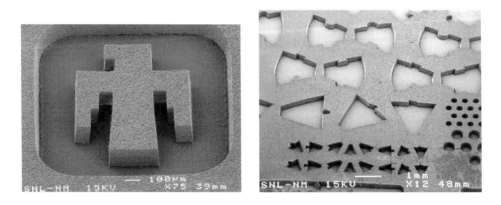

FIGURE 18.20 Alumina/glass structures created from DXRL-defined PMMA mold forms intended to be used for further high-temperature mold processing.

Further micromold filling techniques have been explored which involve approaches ranging from glass castings [Lee and Vasile, 1998] and embossing to flame spray [Christenson, 1998]. A promising batch microfabrication technique from the standpoint of dramatically widening the available material base is the use of DXRL-patterned electrodes for electro-discharge machining [Takahata et al., 1999]. The batch micro-EDM approach uses an electrode pattern defined with an array of components to simultaneously plunge cut a number of parts from a conductive material. Although currently not as accurate as electroforming, these latter techniques may suit applications requiring batch microfabrication of a particular material.

FIGURE 18.21 Hot-forged $Nd_2Fe_{14}B$ permanent magnet test bars extracted from intermediate DXRL-formed alumina mold.

18.4 Material Characterization and Modification

Material properties possessed by DXRL-fabricated components are unique in the sense that they depend on the processing parameters used to fabricate them. Thus, a means of *in situ* metrology is needed to generate data reflecting lot-to-lot and intrasubstrate variation. This methodology is based on the drop-in test die approach used in integrated circuit planar processing. For ultra-precise, high-aspect-ratio, DXRL-defined components, commensurate routine dimensional metrology remains a challenge. Mechanical property measurement techniques are readily available, however, and have revealed material property deficiencies to which MEMS are particularly sensitive. Post-process treatments show some promise for ameliorating these deficiencies.

Strain diagnostic structures were previously mentioned. This procedure readily provides localized stress data as a function of process conditions as with current density during electroplating, for example. A measure of tensile mechanical behavior has been achieved with a milliscale torsion tester comprising a tabletop servohydraulic load frame instrumented with a laser displacement measurement system [Christenson et al., 1998a]. This system is compatible with common DXRL structural dimensions and accommodates the tensile specimen geometry depicted in Figure 18.22. The specimen has a gauge section roughly 3000 µm long with a cross-section defined by a 762-µm width and a height set by the electrodeposition thickness which may be from 50 µm to 1 mm. Two pairs of extruding optical marker bars that are read by the laser displacement system define the gauge length. For these markers to negligibly interfere with the measurement and not rotate, a pair of spars 25 µm wide are used to attach the marker to the gauge section. The specimen shape may be used as a drop-in test structure on any mask definition to allow tensile property measurement on every processed substrate.

The tensile testing procedure has aided in building a mechanical property database. Particularly useful information that has been extracted includes sensitivity to plating current density and temperature cycle effects. Data for nickel are listed in Figure 18.23 and Table 18.6 and copper results are given in Figure 18.24 and Table 18.7. Nickel–iron deposits are found to have substantially higher tensile strength as has been previously noted [Safranek, 1986]. Figure 18.25 shows the tensile data for 78/22 Ni–Fe deposited from a sulfate/citrate bath at room temperature and a pH of 4.8. The high yield strength of the Ni–Fe deposit that is also common to other electrodeposited alloys renders these materials well suited for use in spring fabrication.

A mechanical property complication in electroformed deposits is apparent, however, from deposit cross-section micrographs and grain-orientation distribution analyses. For direct current nickel deposits, for example, the initial deposit is found to have a fine grain structure which evolves into a coarser lenticular grain structure oriented parallel to the deposition direction. Figure 18.26 shows examples of this structure. Indentation data reveals corresponding anisotropic and varying elastic and hardness properties throughout the electrodeposited structure [Buchheit et al., 1998]. Another result of the morphology variation is an internal strain gradient through the thickness of the deposit with usually an initially compressive deposit [Harsch et al., 1988]. Means to resolve this nonuniformity include the use of pulse plating and bath chemical additions.

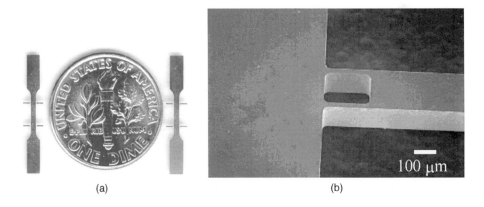

FIGURE 18.22 (a) DXRL-fabricated metal tensile specimen; (b) close-up of laser displacement tab attachment.

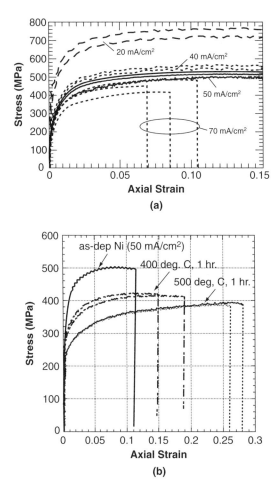

FIGURE 18.23 Tensile data for electrodeposited nickel from a sulfamate bath as a function of (a) current density and (b) anneal cycles.

TABLE 18.6 Electrodeposited Nickel Mechanical Property Current Density Dependence

Current Density/No. of Tests (mA/cm^2)	Elastic Modulus (Gpa)	0.2% Proof Stress (Mpa)	Maximum Stress (Mpa)
20/3	156 ± 9	441 ± 27	758 ± 28
40/3	155 ± 11	305 ± 12	562 ± 9
50/3	160 ± 20	277 ± 8	521 ± 19
70/4	131 ± 13	275 ± 18	460 ± 31

Note: see Figure 18.23a.

TABLE 18.7 Electrodeposited Copper Mechanical Properties

Anneal Conditions	Elastic Modulus (Gpa)	0.2% Proof Stress (Mpa)	Maximum Stress (Mpa)
As-deposited @ 30 mA/cm^2	110 ± 10	239 ± 5	377 ± 5
500°C, 1-hr anneal	76 ± 10	122 ± 5	327 ± 5

Note: See Figure 18.24.

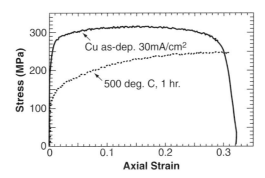

FIGURE 18.24 Tensile data for electrodeposited copper from a copper sulfate bath.

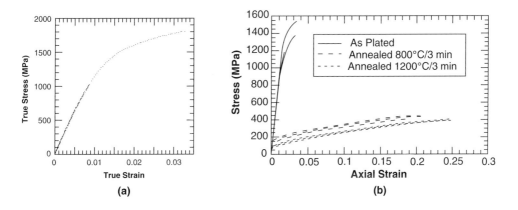

FIGURE 18.25 Tensile data for electrodeposited 78/22 nickel/iron from sulfate bath: (a) as-deposited, (b) temperature dependence.

Because scaled micromechanisms fabricated with DXRL-based processing contain components with much larger ratios of surface area to volume, surface behavior may substantially affect or dominate mechanism dynamics. In the limiting case, situations of metal component surfaces in contact over a period of time can lead to a static friction situation that prohibits mechanism operation. As a result of these surface sensitivities and material base limitations, surface treatments that improve tribology and

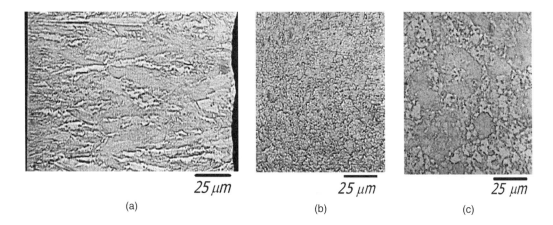

FIGURE 18.26 (a) Metallurgical cross-section photograph of electrodeposited nickel with deposition direction from left to right nucleating on an evaporated copper plating base; (b) nucleating surface; (c) top surface.

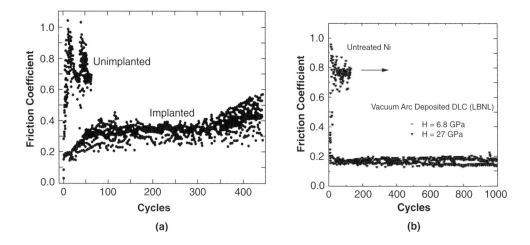

FIGURE 18.27 Friction coefficients of electroformed nickel and comparison with (a) Ti/C treatment and (b) diamond-like carbon deposition. Data were obtained in laboratory ambient with unidirectional sliding against a 1.6-mm-radius 440C steel ball with 9-gf load.

wear for existing electrodeposited material have been explored. An important requirement for these treatments is that they do not substantially alter dimensional tolerances and that they do not introduce additional adhesion or shedding problems. A method that has been very successful in this regard that has been previously used for iron and steels is a dual implantation of titanium and carbon [Myers et al., 1997]. The procedure involves implanting Ti and C to a dose of $2(10)^{17}/cm^2$ at respective energies of 180 and 45 keV. The result for electroplated nickel is found to be similar to iron and steel in that an amorphized region near the nickel surface is formed which possesses significantly enhanced mechanical properties. Table 18.8 shows a comparison summary of these properties. Furthermore, the implanted surface enhances tribological properties as Figure 18.27 explains. Fixturing to address the line-of-sight nature of the implant is needed to allow articulation of a substrate with attached DXRL components to yield an approximately even dose over all part topology while providing a batch wafer treatment.

Another approach to improving as-electroplated metal tribology has been explored with the use of diamond-like carbon coatings [Ager et al., 1998]. Experiments showed that carbon deposited via a pulsed vacuum-arc process has the advantage of providing partially conformal deposits. These films were deposited on nickel with thicknesses of 100 to 200 nm and possess hardnesses of over 30 GPa. The resulting coatings

TABLE 18.8 Surface Mechanical Properties of Implanted Electrodeposited Nickel [Myers et al., 1997]

	As-Deposited Ni (GPa)	Ti-C Implanted (GPa)
E (elastic modulus)	230	400
H (hardness)	6.3	13.8
σ_Y (yield strength)	1.4	4.8

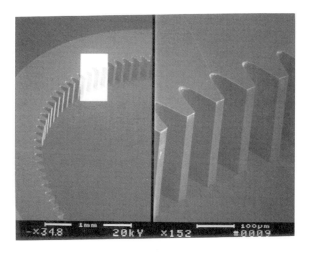

FIGURE 18.28 Copper internal gear that was planarized via the diamond nano-grinding process. Typical RMS surface roughness of less than 2 nm is possible.

provide enhancements in tribological performance similar to Ti/C implantation, as the test results in Figure 18.27 show.

18.5 Planarization

The demonstration of submicron in-plane tolerances for DXRL-fabricated components has resulted in similar requirements for thickness control for out-of-plane tolerances. Leveling during electroforming may be aided by bath composition, agitation and pulsed current techniques. Controlling electrodeposits several hundred microns thick to micron tolerances has proved impractical for the variety of component geometry of interest and in this case a concession is needed. An alternative approach is to let the electrodeposit form over the top of the photoresist mold and then planarize the plastic/metal geometry matrix to a desired thickness.

The planarization technique that has proven to be particularly appropriate for these materials is a diamond lapping procedure that has been referred to as "nano-grinding" [Christenson and Guckel, 1995; Gatzen et al., 1996]. Machining operations involving fly-cutting and milling are found to generate considerable impact loading and shear stress on microelectroformed parts. Standard lapping procedures involving loose alumina abrasive slurry on a glass plate, on the other hand, are found to preferentially remove the softer PMMA material, leaving a situation where the matrixed metal is prone to smearing, resulting in loss of edge definition. A locally rounded metal surface is also a consequence of the rolling abrasive. In the nano-grinding approach, a diamond slurry is applied to a composite soft metal plate into which it is embedded via a conditioning plate or ring. The fixed diamond-loaded surface is then used as a grinding wheel to cut the metal/PMMA composite layer on a substrate held with a precision vacuum-mounted fixture. The metal plate flatness may be maintained to 0.0001 in. over a 15-in. diameter by the proper use of diamond cutting rings, thus a lapped surface can be held to a thickness within 1 μm over a 3-in.-diameter substrate. Results are shown in Figure 18.28 for a copper part. Nearly any material

18.6 Angled and Re-entrant Geometry

A large amount of flexibility of X-ray mask/substrate orientation relative to the SR beam is available due to the highly collimated beam as well as the absence of reflection outside the grazing incidence angles. The first straightforward result is a prismatic angled exposure as shown in Figure 18.29. In this case, both mask and substrate are oriented at an angle relative to the SR beam and scanned across the X-ray mask area. X-ray mask contrast potentially becomes a problem in this case at the edge of the absorber definition, where a gradient in contrast exists across the subtended length defined by the absorber thickness. Thus, for critical applications, the X-ray mask absorber may require patterning at the desired structure angle. A reduced mask contrast, however, may be useful in intentionally generating tapered geometry such as that shown in Figure 18.30 [Guckel, 1996]. The PMMA flank is generated by a partial exposure underneath the X-ray mask absorber region. For adjacent exposed areas with relatively much higher

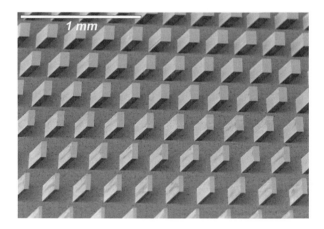

FIGURE 18.29 Copper structures fabricated from a 45° DXRL exposure in PMMA. Metal thickness is 200 μm normal to the substrate.

FIGURE 18.30 Tapered PMMA test structures realized with a low-contrast X-ray mask and NSLS exposure. PMMA thickness is 1.5 mm.

X-Ray-Based Fabrication

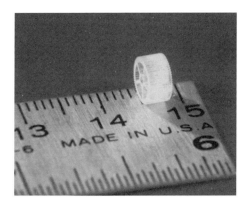

FIGURE 18.31 PMMA gear, 3 mm thick, patterned with high-energy NSLS exposure.

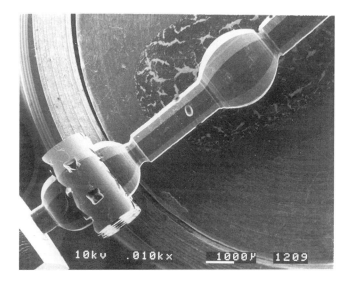

FIGURE 18.32 Re-entrant PMMA shape realized by rotation during X-ray exposure [Johnson, pers. comm.].

delivered dose, the flank profile is essentially constant. A transition to a parabolic flank dependence occurs in deeper regions of lower dose where diffusion-transport-limited development begins to contribute.

High-energy, ultra-deep X-ray exposures have yielded extremely thick photoresist exposures and structures [Siddons et al., 1997]. Results from the National Synchrotron Light Source (NSLS) depicted in Figure 18.31 reveal over 3-mm-thick structures. Extending this effort still further, using 35-keV X-rays has enabled a 110-mm-deep lithography result [Siddons et al., 1994]. The ability of DXRL to precisely define nearly perfectly vertical sidewall definition has additionally suggested continuously rotating exposures of an X-ray mask aperture and PMMA substrate combination about an axis normal to the X-ray beam. The result is directly patterned re-entrant or undercut PMMA geometry. Figure 18.32 shows an example. Using negative X-ray resist in this context also leads to additional novel microstructures [Ehrfeld and Schmidt, 1998].

18.7 Multilayer DXRL Processing

The use of multiple masks to form multilayer stepped prismatic structures makes possible such geometry as flow channels, stacked gears and gears with shafts, as well as integrated packaging structures. Several means have been explored to achieve a multilayer DXRL process. Direct plating has proven to be one viable approach with results depicted in Figure 18.33 [Massoud-Ansari et al., 1996]. In this technique, subsequent to planarization of the first layer, a second level of PMMA is bonded to the first composite metal/PMMA layer, X-ray exposed with an aligned X-ray mask, developed, electroformed and finally planarized. Consecutive layer geometry is typically restricted to be nonoverhanging with respect to previous geometry.

Another multilayer process has been demonstrated with the use of a batch diffusion bonding and release procedure [Christenson and Schmale, 1999]. The fabrication sequence is shown in cross section in Figure 18.34. Two planarized substrates decorated with DXRL patterned and electroformed layers and

FIGURE 18.33 Two-level nickel DXRL-formed test structures produced via direct electroforming.

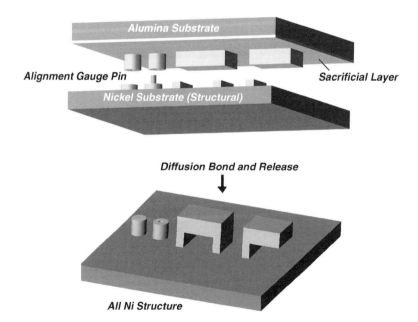

FIGURE 18.34 Batch wafer-scale diffusion bonding procedure for multilevel fabrication of DXRL-patterned geometry. This sequence may be directly repeated to integrate further levels.

FIGURE 18.35 Various DXRL-based nickel structures resulting from two-level wafer-scale diffusion bonding. (a) Small gear on large gear; (b) large gear on small gear; (c) hexagonal structured cantilever; (d) ring and beam tensile strain measurement structure.

with PMMA removed are aligned face-to-face with gauge pins that are press-fit into complementing alignment structures. The substrates are joined in a vacuum hot press, and the sacrificial layer on one substrate is etched entirely away, leaving one layer bonded to the other. Figure 18.35 reveals results for a test mask pattern. This procedure is compatible with arbitrary overlapping geometry and may be repeated indefinitely. Careful attention must be given to the temperature cycles that have the potential to degrade mechanical properties substantially. For this reason, a milliscale mechanical torsion tester suited for batch measurement of shear bond strength of high-aspect-ratio structures has been used to optimize diffusion bonding parameters [Christenson et al., 1998b]. Using this approach, a diffusion bond cycle of 450°C at 7 ksi for 4 hr with electrodeposited nickel was measured to yield a 130-MPa bond strength suitable for most applications. Alignment accuracy is measured with the vernier alignment structures shown in Figure 18.36 which indicate a less than 1-μm alignment error over a 3-in.-diameter substrate.

 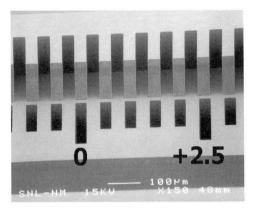

FIGURE 18.36 Vernier indicators of two-level diffusion bonding alignment that reveal an alignment accuracy of less than 1 µm. Each vernier mark indicates 0.5 µm of misalignment.

FIGURE 18.37 Assembled nickel gear on nickel shaft and 10× close-up showing 1-µm journal bearing clearance at 120-µm structural thickness.

18.8 Sacrificial Layers and Assembly

The multilayer fabrication schemes in the previous section aim to substantially alleviate the assembly burden. To date, however, most complex devices have required some form of micro-assembly. The reason for the reliance on assembly lies with the high-aspect-ratio nature of these types of components. Maintaining intercomponent tolerances of 1 µm or less is not feasible directly at very high aspect ratios, thus assembly may be used to allow biasing of 1 µm or less between two separately fabricated components which are then assembled to realize an acceptable precision fit. Figure 18.37 is an image of a resulting combination. This micro-assembly sequence realizes released components via a sacrificial layer etch that may also be used to partially release structures such as the one in Figure 18.38. The micro-assembly of DXRL-fabricated structures has been extended to more elaborate mechanisms via stacking (Figure 18.39) [Fischer and Guckel, 1998] and multistage element assembly (Figure 18.40) [Christenson, 1998]. The individual

FIGURE 18.38 Partially released nickel electrostatic comb drive.

FIGURE 18.39 Five-layer stacked 78 Permalloy linear actuator mechanism over 1 mm in height.

microcomponent assembly techniques that have been employed include the use of magnetic probes for ferromagnetic materials and liquid surface tension to pick up certain geometry, as well as parallel robotic schemes [Feddema and Christenson, 1999]. Additive wafer transfer schemes such as batch processes incorporating wafer-scale component transfer with wafer-scale bonding and sacrificial layer-release sequences ultimately have the most appeal.

FIGURE 18.40 Assembled two-layer gear train and rack mechanism fabricated from 78 Permalloy. The aluminum substrate was machined conventionally to accept press-fit tool steel gauge pins on which the keyed bushings and gears are assembled.

18.9 Applications

Two basic DXRL application areas will be discussed. One area includes the use of DXRL for the batch fabrication of precision-machined components requiring dimensional tolerances of 1 μm or less or those requiring a high aspect ratio with microscale critical dimensions. The other application area involves the use of the high-aspect-ratio nature of the process to generate increased actuation forces as well as increased inertial effects.

18.9.1 Precision Components

Devices in the precision component category range from mechanical elements such as springs that would require prohibitively difficult conventional machining approaches to precision alignment and optical structures. An example is shown in Figure 18.41, where a compact flexure with a 0.15-N/m lateral spring constant reveals the capability of realizing a power spring. The ability to pattern fine features within relatively large geometry allows accurate scaling of gears as depicted in Figure 18.42. Arbitrary gear teeth profiles are possible, thus any pressure angle for an involute geometry, for example, may be accommodated. These components have been applied to a miniature precision planetary gear box construction, as shown in Figure 18.43.

A large number of additional piece-part applications have centered around grid structures or arrayed apertures. The screen in Figure 18.44 illustrates a typical example. This rather large nickel component contains 50-μm wide spars with less than 0.25-μm deviation through its thickness of 150 ±1 μm. In this case, repeatability is of prime importance for which the lithographic dimensional extension afforded by DXRL-based processing is well suited. A related structure is desired for use in a "lobster-eye" optic for X-ray astronomy [Peele, 1999]. The requirement is for an array of square apertures that reflect X-ray flux at a grazing angle off of their orthogonal sidewalls. These reflections are directed into a central focus area if the axis of each aperture is radial to the center of a sphere. A desired material for this X-ray optic structure is nickel with a figure of merit that is strongly dependent on sidewall surface roughness. An acceptable surface roughness for the X-ray wavelengths of interest is calculated to be a few nanometers. Figure 18.45 depicts a test structure of the type needed for the optic element. Sidewall roughness for these structures is typically 10 nm and has been measured as low as 7 nm RMS. Methods to reduce this

FIGURE 18.41 Spiral spring with 3-μm-wide flexural element and 100-μm structural height fabricated from nickel.

FIGURE 18.42 Released nickel gears, 100 μm thick, with 7-μm involute gear tooth thickness.

value by roughly a factor of 5 are being investigated. The resulting screen would ultimately be deformed about a spherical mandrel.

Another device category requiring arrayed channels concerns heat exchangers. Reduced heat exchanger dimensions of 100 μm to 1 mm result in more effective heat transfer due to decreased thermal diffusion lengths. A DXRL-based fabrication method for cross-flow micro-heat exchangers has been realized [Harris et al., 2000]. The scheme uses a DXRL-patterned, two-layer nickel mold insert (Figure 18.46a) to hot emboss half of the heat exchanger panel (Figure 18.46b) [Despa et al., 1999]. Two panels are then placed face-to-face and bonded to realize the automobile-like radiator shown in Figure 18.46c. Estimated performance for this device realized in PMMA with an air pressure drop of 175 Pa and coolant pressure

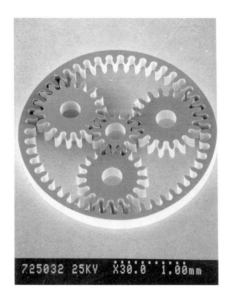

FIGURE 18.43 Planetary gear set fabricated from nickel for RMB Corp. by the Microelectronics Center of North Carolina (MCNC).

FIGURE 18.44 Precision screen part fabricated from 150-μm-thick nickel.

X-Ray-Based Fabrication

FIGURE 18.45 Nickel X-ray "lobster-optic" test structure 200 μm thick with 50-μm apertures and 3-μm sidewalls. The outer sidewalls were patterned at 1.5-μm thickness where partial development occurred.

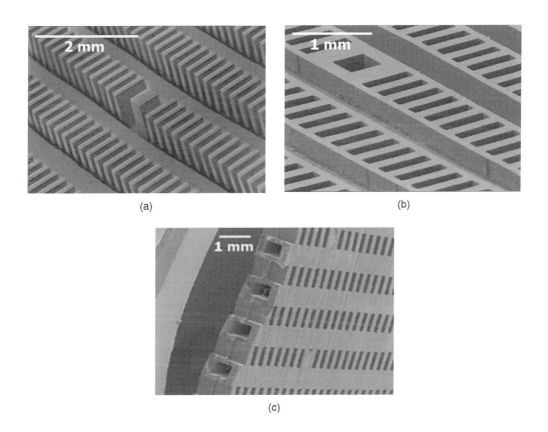

FIGURE 18.46 Cross-flow micro-heat exchanger realized by bonding two PMMA halves embossed from a nickel mold [Harris, 2000]. (a) Heat exchanger nickel mold insert; (b) one side of embossed PMMA heat exchanger; (c) assembled and bonded plastic heat exchanger showing cross section of coolant channels.

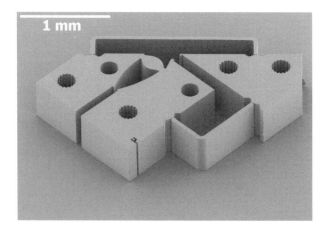

FIGURE 18.47 PMMA optical fiber alignment structure and focusing lens at 500-μm thickness. This structure, which provides for a minimum-sized spot to optically monitor a moving surface, is press-fit over 200-μm-diameter pins; subsequently, its two retaining breakable beams are detached, thereby providing a sprue function. The lens is a best-form aspheric designed for a 0.16-NA single-mode fiber at 632.8 nm with a center thickness of 200 μm.

drop of 5 kPa is a heat dissipation of 6.0 W/cm^2 or 33.3 W/cm^3, a substantial increase over larger, traditionally scaled devices.

The direct use of PMMA as a component material has also proven its utility for optical components. Figure 18.47 reveals an integrated fiber alignment guide and prismatic focusing optic structure fabricated with PMMA [Holswade, 1999]. Two support sprues hold two components together and are removed after pin press-fit mounting. The structure accommodates input and output fibers oriented at 90° to each other. The light from one fiber is focused into a line across a reflective encoded surface that reflects a signal back into the receiving fiber.

18.9.2 Microactuators

The ability to integrate magnetic material with batch DXRL-based microactuator fabrication has made possible high-pressure microactuators with low driving-point impedance. Both linear and rotary devices have profited in accord with a figure of merit for prismatic based actuators defined by the force generated in a working volume,

$$F_x = \rho_E h \frac{\partial A}{\partial x} \quad (18.2)$$

defined by a height h and area A tangent to the axis of motion in the x direction. The energy density multiplier, ρ_E, is limited for a magnetic actuator by the magnetic flux saturation density, B_{sat}, as was previously described, or for an electrostatic actuator by the electric breakdown field, E_{bkdn}, via:

$$\rho_{E_m}(\max) = \frac{B_{sat}^2}{2\mu_o} \quad (18.3)$$

$$\rho_{E_e}(\max) = \frac{1}{2}\varepsilon_o E_{bkdn}^2 \quad (18.4)$$

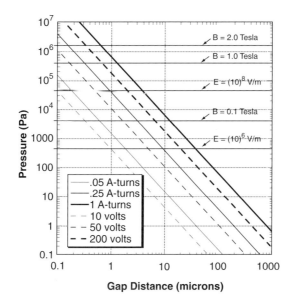

FIGURE 18.48 Plot of magnetic and electric pressure generated for various drive levels. Solid lines are magnetic with ni in ampere-turns and dashed lines are electric with V in volts as in Eqs. (18.5) and (18.6). Maximum pressures for limiting B and E fields are also indicated.

where $\mu_o = 4\pi(10)^{-7}$ Hy/m and $\varepsilon_o = 8.854(10)^{-12}$ F/m are the vacuum permeability and permittivity, respectively. In terms of driving electrical parameters, ni (ampere-turns) for magnetics and V (volts) for electrostatics, and working volume air gap d, the relationships for actuator pressure become:

$$\rho_{E_m} = \frac{1}{2}\mu_o \left(\frac{ni}{d}\right)^2 \tag{18.5}$$

$$\rho_{E_e} = \frac{1}{2}\varepsilon_o \left(\frac{V}{d}\right)^2 \tag{18.6}$$

The caveat in the magnetic case is that an efficient magnetic circuit is assumed with high permeability or soft ferromagnetic material. A plot of these equations in Figure 18.48 reveals the impedance–pressure trade-off for various working gaps.

Linear magnetic microactuators have been fabricated with three basic components. A movable plunger constrained by a folded spring flexure operates into a magnetic circuit portion that supports the transfer of flux to the working gap. Both components are fabricated from a soft ferromagnetic material such as 78/22 nickel/iron or 78 Permalloy. The third component, a coil, has been most effectively realized by conventional miniature coil winding techniques using a DXRL-defined mandrel that may be press-fit into the magnetic circuit. Figure 18.49 shows a typical construction for such a magnetic variable reluctance (VR) linear microactuator. Typical performance of this microactuator type at a structural thickness of 150 µm is a force output in excess of 1 mN, with a stroke of 450 µm for an input current of 40 mA and corresponding input power of 10 mW. The resonant characteristics include a resonant frequency near several hundred Hz and amplitude of ±200 µm at an input power of 200 µW with a quality factor in air near 200. With an inductance change of 1 µHy/µm, the possibility of using inductance sensing to control microactuator position has been exploited [Guckel et al., 1996b; 1998; Stiers and Guckel, 1999]. The device depicted in Figure 18.50 uses a sense coil as part of an oscillator circuit whose resonant frequency is compared to an input frequency representing the desired actuator position. A phase detector

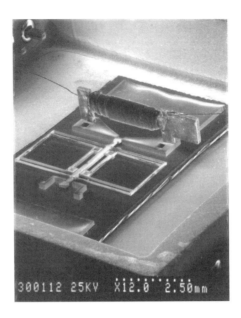

FIGURE 18.49 Packaged VR linear microactuator.

FIGURE 18.50 Linear magnetic microactuator with sensing for closed-loop control. All components are 78 Permalloy. Wound coil forms remain to be inserted into the serrated magnetic pad areas.

circuit that integrates the difference in drive frequency and sense frequency yields an effective phase-shift signal that constitutes a position error. Despite a significantly reduced electrical quality factor for microscale inductance-based oscillators, with the use of an integrating phase detection scheme a positional accuracy of 260 nm was realized with a settling time of roughly 100 ms.

Magnetic rotational microactuators have also been fabricated using a process similar to that of linear types, with the difference being the substitution of a rotor for the linear armature. Figure 18.51 shows a rotational magnetic micromotor integrated with silicon-rotor-position-sensing p–n junction photodetectors, demonstrating the ability to readily post-fabricate DXRL-based structures on semiconductor substrates with microelectronics [Guckel et al., 1993]. Maximum achieved rotor speeds for 140-μm

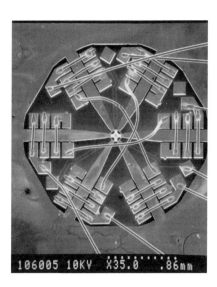

FIGURE 18.51 DXRL-based VR stepping micromotor with integrated photodiode position sensors.

diameter rotors are 150,000 rpm for a closed-slot stator design that minimizes torque ripple in this variable reluctance design. These types of micromotors have been operated continuously for over three months at over 10^{10} rotational cycles and reveal no change in operating characteristics, which for a journal bearing is quite acceptable. The low wear rate that is observed is suspected to be partially due to a vertical variable reluctance rotor suspension, a consequence of a thinner rotor than stator, that maintains rotor rotation off of the substrate. This category of micromotors with integrated wire-bonded coils is found to be rather inefficient, with output torques of the order of 10 nNm. A more efficient coil for rotational magnetic micromotors has been implemented with conventionally wound soft magnetic mandrels providing a large air gap flux and resulting in submillimeter-diameter, 200-µm-thick rotors capable of providing 0.2 µNm of torque [Christenson et al., 1996].

Much greater magnetic microactuator efficiency may be obtained by incorporating permanent magnets. The magnetization of permanent magnets is independent of scale which is an attribute making their use at the microscale much preferable to an electromagnet. An electromagnetic coil possesses a magnetic field production scaling dependence that is linear and brought about by current density constraints due to thermal and electromigration limitations. Coils utilizing superconductors share the same scaling relationship, which arises due to a similar critical current density limit. The use of rare-earth-based permanent magnets, in particular, is attractive due to their high magnetization and high demagnetization resistance making them amenable to a nearly arbitrary aspect ratio. The incorporation of DXRL-fabricated rare-earth permanent magnets has been demonstrated with a miniature brushless DC motor as depicted in Figure 18.52 which shows a low profile brushless DC motor with a 5-mm-diameter, four-pole, radially anisotropic bonded $Nd_2Fe_{14}B$ rotor [Christenson et al., 1999c]. The slotless four-pole, nine-slot configuration rotates at speeds of 15,000 rpm with 2-V, 5-mA, three-phase signals and generates a maximum torque near 10 µNm at 300-µm rotor thickness. The planar configuration of this motor is suitable for tasks that are present in low-profile, miniature hard-disk drives, for example.

18.9.3 Magnetic Microactuator Applications

Magnetic linear actuators fabricated via DXRL approaches have found many uses for optical devices. A spectrometer application, for example, uses a three-phase, magnetic, VR, linear stepping motor to stretch and compress a tunable infrared (IR) filter for miniaturized gas analyzers and multispectral IR systems [Ohnstein et al., 1995; 1996]. Figure 18.53 depicts an overview of the device that includes a

FIGURE 18.52 Brushless DC motor with 5-mm-diameter four-pole rotor fabricated with DXRL-formed bonded $Nd_2Fe_{14}B$ magnets. The rotor is mounted on a 320-mm-diameter shaft that extends into a 1.6-mm-diameter ball bearing residing under the rotor. The outer magnetic return path is 78 Permalloy with an 8-mm outer diameter. Nine coils with 120 windings per slot are arranged on a DXRL-defined PMMA coil form.

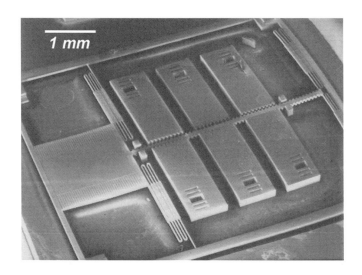

FIGURE 18.53 Overview of tunable IR filter with integrated three-phase magnetic VR stepping linear microactuator. The IR filter is seen on the left with a 2.5 × 2.5-mm area and is supported by a frame surrounding the device. Three coils are assembled into the stator forms located to the right [Ohnstein, 1996].

DXRL- fabricated wire grid IR filter that is also part of a distributed flexure attached to an armature of a linear stepping motor. All components are fabricated from electroplated 78 Permalloy. The IR filters, shown in detail in Figure 18.54, have an active area ranging from 1 mm × 1 mm to 2.5 mm × 2.5 mm, commensurate with an IR focal plane array and a wire spacing that may be deformed to yield cut-off between wavelengths of 8 and 32 µm. The variable filter response is shown in Figure 18.55. Total filter displacements of 1.7 mm were achieved with a stepping actuator, providing an average force of 2.4 mN/step at a drive current of 45 mA, which is more than sufficient for the filter spring constants at roughly 1 mN/mm.

A fiberoptic switching device has also been developed with a magnetic VR linear microactuator. A 1 × 2 single-mode fiber switch consisting of a two-phase VR linear spring constrained microactuator is depicted in Figure 18.56 [Guckel et al., 1999b]. Switch performance for a 125-µm-diameter fiber with 8-µm core

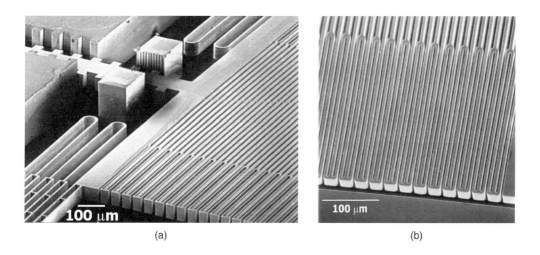

FIGURE 18.54 (a) IR wire filter shown connected to linear stepping microactuator armature; (b) close-up of Ni/Fe IR wire filter.

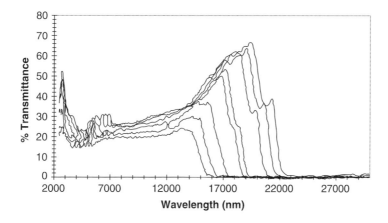

FIGURE 18.55 Transmission curves of the tunable IR filter.

includes a switching speed under 1 ms at a power dissipation of below 20 mW with as low as 0.5-dB insertion loss in air. Index matching fluids may also be used with this switch that may additionally incorporate permanent magnets for a latched state with zero power dissipation.

18.9.4 Other Applications

Deep X-ray lithography fabricated gears have been used to realize a miniature in-line gear pump [Deng et al., 1998]. A magnetic coupling consisting of a 78 Permalloy bar embedded in a 1.5-mm-diameter PMMA gear driven with an external motor and permanent magnet drives two counter-rotating gears in a PMMA cavity. Via a maximum calculated coupling torque of 12.4 µNm and a rotation speed of 3500 rpm with 200-µm-thick gears, pumping rates of 70 µl/min were achieved through 696-µm-diameter inlet and outlet ports. Such a pump is self-priming and in this case was able to pull a maximum of 45 mm of water.

By combining the high density of electroplated materials such as gold with high-aspect-ratio DXRL structures, two main issues for miniaturized inertial sensors are addressed. Figure 18.57 reveals an additional possibility of integrating mechanical damping to provide an acceleration-time response [Polosky et al., 1999]. In this device, a proof-mass with extended rack feature when accelerated is driven into a

FIGURE 18.56 Assembled single-mode 1 × 2 optical fiber switch.

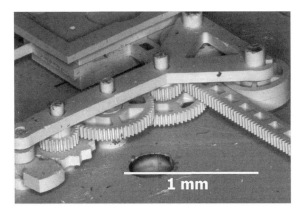

FIGURE 18.57 Assembled escapement mechanism used to provide damping for a spring-mass accelerometer structure via a rack-and-pinion coupling.

step-up gear train that drives an escapement and consequently provides damping. The resulting integrating accelerometer with total dimensions of 4 × 6 mm is capable of sensing 4- to 8-g accelerations.

18.10 Conclusions

Tools for the microfabrication of mechanical structures continue to be aided by advancing microelectronic concerns. Many applications of micromechanics in turn ultimately depend on its integration with microelectronics. A key aspect of micromachining is achieving precision at the submillimeter scale. Highly collimated X-rays enable commensurate lithographic precision at this scale while providing a unique high-aspect-ratio capability. The utility of this batch fabrication approach is additionally enhanced by the additive nature of mold-based processing that lends itself to accommodating a large material base. The result is a generalized prismatic-based microfabrication tool set able to realize many useful microelectromechanical components and ultimately devices through repeated additive batch microfabrication steps. Consequent microsystem integration and packaging remain a challenge that this technology is well positioned to address via its low-temperature character. Access to appropriate synchrotron radiation sources continues to improve as applications demanding high microscale precision and accuracy become more prevalent. The impending technological impact of these applications with entirely new characteristics looms large.

Acknowledgments

This work was performed at Sandia National Laboratories, a multiprogram laboratory operated by Sandia Corp., a Lockheed Martin company, for the U.S. Department of Energy under contract DE-AC04-84AL85000.

Defining Terms

Aspect ratio: Most commonly defined as the ratio of structure height to structure width; another use is as the sidewall slope expressed as a vertical to horizontal run-out ratio.

Beamline: A unidirectional vacuum line extending from a storage ring that captures a segment of synchrotron radiation and transfers it to an "end-station" where the radiation is used for X-ray exposure, for example.

Deep X-ray lithography (DXRL): The use of X-rays to pattern photoresist at large thicknesses (>100 µm) for substantially micromechanical uses; additionally results in particularly high-aspect-ratio microstructures.

LIGA: A German acronym representing three fundamental processing steps: *lithographie*, deep X-ray lithography (see DXRL); *galvanoformung*, electroforming (electroplating into a mold); *abformung*, injection molding replication.

Storage ring: A roughly circular structure where either highly relativistic electrons or protons may be stored in a circular path for many hours; the source of synchrotron radiation.

Synchrotron radiation: The radiation emitted by a charged particle traveling in a circular arc as a consequence of its radial acceleration. This radiation is highly intense and directional when electrons traveling at close to the speed of light are bent in a magnetic field.

X-ray mask: The structure that defines the pattern during X-ray lithography. This structure must accommodate atomic absorption effects at these wavelengths and thus requires a relatively thin low atomic number carrier (the equivalent of a glass plate in UV microlithography) as well as a relatively thick high atomic number absorber (the equivalent of chromium in UV microlithography).

References

Ager, J.W., III, Monteiro, O.R., Brown, I.G., Follstaedt, D.M., Knapp, J.A., Dugger, M.T., and Christenson, T.R. (1998) "Performance of Ultra Hard Carbon Wear Costings on Microgears Fabricated by LIGA," *Mater. Sci. MEMS Devices MRS Symp. Proc. Ser.* **546**.

Allen, M.G. (1993) "Polyimide Based Processes for the Fabrication of Thick Electroplated Microstructures," in *Proc. Transducers '93*, Yokohama, Japan, pp. 60–65.

Andricacos, P.C. et al. (1996) "Vertical Paddle Plating Cell," U.S. Patent 5,516,412.

Becker, E.W., Betz, H., Ehrfeld, W., Glashauser, W., Heuberger, A., Michel, H.J., Münchmeyer, D., Pongratz, S., and Siemens, R.V. (1982) "Production of Separation Nozzle Systems for Uranium Enrichment by a Combination of X-ray Lithography and Galvoplastics," *Naturwissenschaften* **69**, pp. 520–523.

Becker, E.W., Ehrfeld, W., Hagmann, P., Maner, A., and Münchmeyer, D. (1986) "Fabrication of Microstructures with High Aspect Ratios and Great Structural Heights by Synchrotron Radiation Lithography, Galvanoformung, and Plastic Molding (LIGA Process)," *Microelectronic Eng.* **4**, pp. 35–56.

Biederman, H. (1987) "Polymer Films Prepared by Plasma Polymerization and Their Potential Application," *Vacuum* **37**(3–4), pp. 367–373.

Bogdanov, A.L., and Peredkov, S.S. (2000) "Use of SU-8 Photoresist for Very High Aspect Ratio X-Ray Lithography," *Microelectronic Eng.* **53**, pp. 493–496.

Buchheit, T.E., Christenson, T.R., Schmale, D.T., and LaVan, D.A. (1998) "Understanding and Tailoring the Mechanical Properties of LIGA Fabricated Materials," *Mater. Sci. MEMS Devices MRS Symp. Proc. Ser.* **546**, pp. 121–126.

Christenson, T.R. (1995) "Micro-Electromagnetic Rotary Motors Realized by Deep X-Ray Lithography and Electroplating," Ph.D. thesis, University of Wisconsin, Madison, pp. 82–86.

Christenson, T.R. (1998) "Advances in LIGA-Based Post Mold Fabrication," *Proc. SPIE Micromachining Microfabrication Proc. Technol. IV* **3511**, pp. 192–203.

Christenson, T.R., and Guckel, H. (1995) "Deep X-Ray Lithography for Micromechanics," *Proc. SPIE Micromachining Microfabrication Proc. Technol.* **2639**, pp. 134–145.

Christenson, T.R., and Schmale, D.T. (1999) "A Batch Wafer Scale LIGA Assembly and Packaging Technique via Diffusion Bonding," *Proc. IEEE MEMS Workshop*, pp. 476–481.

Christenson, T.R., Guckel, H., and Klein, J. (1996) "A Variable Reluctance Stepping Microdynamometer," *Microsyst. Technol.* **2**(3), pp. 139–143.

Christenson, T.R., Buchheit, T.E., Schmale, D.T., and Bourcier, R.J. (1998a) "Mechanical and Metallographic Characterization of LIGA Fabricated Nickel and 80%Ni–20%Fe Permalloy," *Microelectromechanical Structures Mater. Res. Mater. Res. Proc.*, **518**.

Christenson, T.R., Buchheit, T.E., and Schmale, D.T. (1998b) "Torsion Testing of Diffusion Bonded LIGA Formed Nickel," *Mater. Sci. MEMS Devices MRS Symp. Proc. Ser.* **546**.

Christenson, T.R., Garino, T.J., and Venturini, E.L. (1999a) "Deep X-Ray Lithography Based Fabrication of Rare-Earth Based Permanent Magnets and Their Applications to Microactuators," *Electrochem. Soc. Proc.* **98-20**, pp. 312–323.

Christenson, T.R., Garino, T.J., and Venturini, E.L. (1999b) "Microfabrication of Fully Dense Rare-Earth Permanent Magnets via Deep X-Ray Lithography and Hot Forging," *HARMST Workshop*, Kisarazu, Japan, Book of Abstracts, pp. 82–83.

Christenson, T.R., Garino, T.J., Venturini, E.L., and Berry, D.M. (1999c) "Application of Deep X-Ray Lithography Fabricated Rare-Earth Permanent Magnets to Multipole Magnetic Microactuators," in *Digest of Technical Papers—10th Int. Conf. on Solid-State Sensors and Actuators—Transducers '99*, Sendai, Japan, pp. 98–101.

Cuisin, C., Chelnokov, A., Lourtioz, J.-M., Decanini, D., and Chen, Y. (2000) "Submicrometer Resolution Yablonovite Templates Fabricated by X-Ray Lithography," *Appl. Phys. Lett.* **77**(6), August 7.

DeCarlo, F., Song, J.J., and Mancini, D.C. (1998) "Enhanced Adhesion Buffer Layer for Deep X-Ray Lithography Using Hard X-Rays," *J. Vac. Sci. Technol. B* **16**(6), pp. 3539–3542.

Deng, K., Dewa, A.S., Ritter, D.C., Bonham, C., and Guckel, H. (1998) "Characterization of Gear Pumps Fabricated by LIGA," *Microsyst. Technol.* **4**(4), pp. 163–167.

Despa, M.S., Kelly, K.W., and Collier, J.R. (1999) "Injection Molding of Polymeric LIGA-HARMS," *Microsyst. Technol.* **6**(2), pp. 60–66.

Despont, M., Lorenz, H., Fahrni, N., Brugger, J., Renaud, P., and Vettiger, P. (1997) "High-Aspect Ratio, Ultrathick, Negative-Tone Near-UV Photoresist for MEMS Applications," *Proc. IEEE MEMS Workshop*, pp. 518–522.

Dutoit, B.M., Besse, P.-A., Blanchard, H., Guérin, L., and Popovic, R.S. (1999) "High Performance Micromachined Sm_2Co_{17} Polymer Bonded Magnets," *Sensors and Actuators A* **77**, pp. 178–182.

Ehrfeld, W., and Schmidt, A. (1998) "Recent Developments in Deep X-Ray Lithography," *J. Vacuum Sci. Technol. B* **16**(6), pp. 3526–3534.

Ehrfeld, W., Bley, P., Götz, F., Hagmann, P., Maner, A., Mohr, J., Moser, H.O., Münchmeyer, D., Schelb, W., Schmidt, D., and Becker, E.W. (1987) "Fabrication of Microstructures Using the LIGA Process," *Proc. of IEEE Micro Robots and Teleoperators Workshop*, pp. 1–11.

Ehrfeld, W., Götze, F., Münchmeyer, D., Schelb, W., and Schmidt, D. (1988a) "LIGA Process: Sensor Techniques via X-Ray Lithography," *Technical Digest, IEEE Solid-State Sensor and Actuator Workshop*, pp. 1–4.

Ehrfeld, W., Bley, P., Götz, F., Mohr, J., Münchmeyer, D., Schelb, W., Baving, H.J., and Beets, D. (1988b) "Progress in Deep-Etch Synchrotron Radiation Lithography," *J. Vacuum Sci. Technol. B* **6**(1), pp. 178–185.

Ehrfeld, W., Hessel, V., Lehr, H., Lowe, H., Schmidt, M., and Schenk, R. (1997) "Highly Sensitive Resist Material for Deep X-Ray Lithography," *Proc. of SPIE—Int. Society for Optical Engineering Conf.* **3049**, pp. 650–658.

Feddema, J.T., and Christenson, T.R. (1999) "Parallel Assembly of LIGA Components," in *Tutorial on Modeling and Control of Micro and Nano-Manipulation, IEEE Int. Conf. on Robotics and Automation*.

Feiertag, G., Ehrfeld, W., Freimuth, H., Kolle, H., Lehr, H., Schmidt, M., Sigalas, M.M., Soukoulis, C.M., Kiriakidis, G., Pedersen, T., Kuhl, J., and Koeneig, W. (1997a) "Fabrication of Photonic Crystals by Deep X-Ray Lithography," *Appl. Phys. Lett.* **71**(11), pp. 1441–1443.

Feiertag, G., Ehrfeld, W., Lehr, H., Schmidt, A., and Schmidt, M. (1997b) "Calculation and Experimental Determination of the Structure Transfer Accuracy in Deep X-Ray Lithography," *J. Micromech. Microeng.* **7**, pp. 323–331.

Feiertag, G., Ehrfeld, W., Lehr, H., Schmidt, A., and Schmidt, M. (1997c) "Accuracy of Structure Transfer in Deep X-Ray Lithography," *Microelectronic Eng.* **35**, pp. 557–560.

Fischer, K., and Guckel, H. (1998) "Long Throw Linear Magnetic Microactuators Stackable to One Millimeter of Structural Height," *Microsyst. Technol.* **4**(4).

Garino, T.J., Christenson, T.R., and Venturini, E.L. (1998) "Fabrication of MEMS Devices by Powder-Filling into DXRL-Formed Molds," *Mater. Sci. MEMS Devices MRS Symp. Proc. Ser.* **546**.

Gatzen, H.H., Maetzig, J.C., and Schwabe, M.K. (1996) "Precision Machining of Rigid Disk Head Sliders," *IEEE Trans. Magnetics* **32**(3), pp. 1843–1849.

Ghica, V., and Glashauser, W. (1983) "Method of Stress-Free Development of Irradiated Polymethylmethacrylate," U.S. Patent 4,393,129.

Goodman, J. (1960) "The Formation of Thin Polymer Films in the Gas Discharge," *J. Polym. Sci.* **44**(144), pp. 551–552.

Guckel, H. (1996) "LIGA and LIGA-Like Processing with High Energy Photons," *Microsyst. Technol.* **2**, pp. 153–156.

Guckel, H. (1998) "High Aspect-Ratio Micromachining Via Deep X-Ray Lithography," *Proc. IEEE* **86**(8), pp. 1586–1593.

Guckel, H., Uglow, J., Lin, M., Denton, D., Tobin, J., Euch, K., and Juda, M. (1988) "Plasma Polymerization of Methyl Methacrylate: A Photoresist for 3D Applications," *IEEE Solid-State Sensor and Actuator Workshop*, Hilton Head Island, SC, pp. 9–12.

Guckel, H., Burns, D.W., Christenson, T.R., and Tilmans, H.A.C. (1989) "Polysilicon X-Ray Masks," *Microelectronic Eng.* **9**, pp. 159–161.

Guckel, H., Christenson, T.R., Skrobis, K.J., Denton, D.D., Choi, B., Lovell, E.G., Lee, J.W., Bajikar, S.S., and Chapman, T.W. (1990) "Deep X-Ray and UV Lithographies for Micromechanics," *Technical Digest—IEEE Solid-State Sensor and Actuator Workshop*, pp. 118–122.

Guckel, H., Burns, D., Rutigliano, C., Lovell, E., and Choi, B. (1992) "Diagnostic Microstructures for the Measurement of Intrinsic Strain in Thin Films," *J. Micromech. Microeng.* **2**, pp. 86–95.

Guckel, H., Christenson, T.R., Skrobis, K.J., Klein, J., and Karnowsky, M. (1993) "Design and Testing of Planar Magnetic Micromotors Fabricated by Deep X-Ray Lithography and Electroplating," in *Digest of Technical Papers—7th Int. Conf. on Solid-State Sensors and Actuators—Transducers '93*, Yokohama, Japan, pp. 76–79.

Guckel, H., Christenson, T.R., and Skrobis, K. (1996a) "Formation of Microstructures using a Preformed Photoresist Sheet," U.S. Patent No. 5,576,147.

Guckel, H., Earles, T., Klein, J., Zook, J.D., and Ohnstein, T. (1996b) "Electromagnetic Linear Actuators with Inductive Position Sensing," *Sensors and Actuators A* **53**, pp. 386–391.

Guckel, H., Fischer, K., and Stiers, E. (1998) "Closed-Loop Controlled Large Throw, Magnetic Linear Microactuator with 1000 μm Structural Height," *Proc. of IEEE MEMS Workshop*, Heidelberg, Germany, pp. 414–418.

Guckel, H., Fischer, K., Stiers, E., Chaudhuri, B., McNamara, S., Ramotowski, M., Johnson, E.D., and Kirk, C. (1999a) "Direct, High Throughput LIGA for Commercial Applications: A Progress Report," *HARMST, '99 Workshop*, Book of Abstracts, Kisarazu, Japan, pp. 2–3.

Guckel, H., Fischer, K., Chaudhuri, B., Stiers, E., McNamara, S., and Martin, T. (1999b) "Single Mode Optical Fiber Switch," *HARMST '99 Workshop*, Book of Abstracts, Kisarazu, Japan, pp. 146–147.

Hall, A. (2000) Sandia National Laboratories, private communication.

Harris, C., Despa, M., and Kelly, K. (2000) "Design and Fabrication of Cross-Flow Micro Heat Exchanger," *IEEE J. Microelectromechanical Syst.* **9**(4), pp. 502–508.

Harsch, S., Ehrfeld, W., and Maner, A. (1988) "Untersuchungen zur Herstellung von Mikrostructuren grosser Strukturhöhe durch Galvanoformung in Nickel-sulfamatelektrolyten," Report No. 4455, Kernforschungszentrum Karlsruhe, Germany.

Holswade, S. (1999) Sandia National Laboratories, private communication.

Johnson, E. (1997) Brookhaven National Laboratory–NSLS, private communication.

Khan Malek, C. (1998) "Mask Prototyping for Ultra-Deep X-Ray Lithography: Preliminary Studies for Mask Blanks and High-Aspect-Ratio Absorber Patterns," *Proc. of SPIE Conf. on Materials and Device Characterization in Micromaching* **3512**, pp. 277–285.

Khan Malek, C.G., and Das, S.S. (1998) "Adhesion Promotion between Poly(methylmethacrylate) and Metallic Surfaces for LIGA Evaluated by Shear Stress Measurements," *J. Vacuum Sci. Technol. B* **16**(6), pp. 3543–3546.

Lee, C.-K.T., and Vasile, M.J. (1998) "A Direct Molding Technique to Fabricate Silica Micro-Optical Components," *Mater. Sci. MEMS Devices MRS Symp. Proc. Ser.* **546**.

Lee, K.Y., LaBianca, N., Rishton, S.A., Zolgharnain, S., Gelorme, J.D., Shaw, J., and Chang, T.H.-P. (1995) "Micromachining Applications of a High Resolution Ultrathick Photoresist," *J. Vacuum Sci. Technol. B* **13**(6), pp. 3012–3016.

Lin, B.J. (1975) "Deep UV Lithography," *J. Vacuum Sci. Technol.* **12**(6), pp. 1317–1320.

Loechel, B., and Maciossek, A. (1995) "Surface Micro Components Fabricated by UV Depth Lithography and Electroplating," *Proc. SPIE Micromachining Microfabrication Process Technol.* **2639**, pp. 174–184.

Manohara, H. (1998) "Development of Hard X-Ray Exposure Tool for Micromachining at CAMD," in *Proc. of 15th Int. Conf. on Application of Accelerators in Research and Industry*, Part 2, AIP Conf. Proc., J.L. Duggan and I.L. Morgan, Eds., American Institute of Physics, pp. 618–621.

Margaritondo, G. (1988) *Introduction to Synchrotron Radiation*, Oxford University Press, London.

Marti, J.L. (1966) "The Effect of Some Variables upon Internal Stress of Nickel as Deposited from Sulfamate Electrolytes," *Plating* **Jan.**, pp. 61–71.

Massoud-Ansari, S., Mangat, P.S., Klein, J., and Guckel, H. (1996) "A Multi-Level, LIGA-Like Process for Three-Dimensional Actuators," *Proc. of IEEE MEMS Workshop*, pp. 285–289.

McNamara, S. (1999) University of Wisconsin, private communication.

Mehdizadeh, S., Kukovic. J., Andricacos, P.C., Romankiw, L.T., and Chen, H.Y. (1993) "The Influence of Lithographic Patterning on Current Distribution in Electrochemical Microfabrication," *J. Electrochem. Soc.* **140**, pp. 3497.

Mohr, J. (1986) "Analyse der Defektursachen und der Genauigkeit der Strukturübertragung bei der Röntgentiefenlithographie mit Synchrotronstrahlung," Dissertation, Universität Karlsruhe.

Mohr, J., Ehrfeld, W., and Münchmeyer, D. (1988) "Requirements on Resist Layers in Deep-Etch Synchrotron Radiation Lithography," *J. Vacuum Sci. Technol. B* **6**(6), pp. 2264–2267.

Morita, S., Tamano, J., Hattori, S., and Ieda, M. (1980) "Plasma Polymerized Methyl-Methacrylate as an Electron-Beam Resist," *J. Appl. Phys.* **51**(7), pp. 3938–3941.

Münchmeyer, D., and Ehrfeld, W. (1987) "Accuracy Limits and Potential Applications of the LIGA Technique in Integrated Optics," in *Proc. SPIE 803—Micromachining of Elements with Optical and Other Submicrometer Dimensional and Surface Specifications*, M. Weck, Ed., Society of Photo Instrumentator Engineers, pp. 72–79.

Myers, S.M., Follstaedt, D.M., Knapp, J.A., and Christenson, T.R. (1997) "Hardening of Nickel Alloys by Implantation of Titanium and Carbon," *Mater. Res. Soc. Symp. Proc.* **444**, pp. 99–104.

Ohnstein, T.R. et al. (1995) "Tunable IR Filters Using Flexible Metallic Microstructures," *Proc. IEEE MEMS,* Amsterdam, The Netherlands, pp. 170–174.

Ohnstein, T.R., Zook, J.D., French, H.B., Guckel, H., Earles, T., Klein, J., and Mangat, P. (1996) "Tunable IR Filters with Integral Electromagnetic Actuators," *Technical Digest of 1996 Solid-State Sensor and Actuator Workshop,* Hilton Head Island, SC, pp. 196–199.

Osaka, T., Takai, M., Hayashi, K., and Sogawa, Y. (1998) "New Soft Magnetic CoNiFe Plated Films with High B_s = 2.0–2.1 T," *IEEE Trans. Mag.* **34**(4), pp. 1432–1434.

Ouano, A.C. (1978) "A Study on the Dissolution Rate of Irradiated Poly(Methyl Methacrylate)," *Polymer Eng. Sci.* **18**(4), pp. 306–313.

Pantenburg, F.J., and Mohr, J. (1995) " Influence of Secondary Effects on the Structure Quality in Deep X-Ray Lithography," *Nucl. Instrum. Methods Phys. Res. B, Beam Interact. Mater. At.* **B97**(1–4), pp. 551–556.

Pantenburg, F.J., Achenbach, S., and Mohr, J. (1998) "Influence of Developer Temperature and Resist Material on the Structure Quality in Deep X-Ray Lithography," *J. Vacuum Sci. Technol. B* **16**(6), pp. 3547–3551.

Peele, A. (1999) "Investigation of Etched Silicon Wafers for Lobster-Eye Optics" *Rev. Sci. Instrum.* **70**(2), pp. 1268–1273.

Pioter, V., Hanemann, T., Ruprecht, R., Thies, A., and Haußelt, J. (1997) "New Developments of Process Technologies for Microfabrication," *Proc. SPIE Micromachining and Microfabrication Process Technol. III* **3223**, pp. 91–99.

Polosky, M.A., Christenson, T.R., Jojola, A.J., and Plummer, D.W. (1999) "LIGA Fabricated Environmental Sensing Device," *HARMST '99 Workshop,* Book of Abstracts, Kisarazu, Japan, pp. 144–145.

Ritzhaupt-Kleissl, H.-J., Bauer, W., Günther, E., Laubersheimer, J., and Haußelt, J. (1996) "Development of Ceramic Microstructures," *Microsystem Technol.* **2**(3), pp. 130–134.

Romankiw, L.T. (1995) "Evolution of the Plating Through Lithographic Mask Technology," in *ECS Proc.,* Vol. 95-18, L.T. Romankiw and D. Herman, Jr., Eds., The Electrochemical Society, pp. 253–273.

Romankiw, L.T., Croll, I.M., and Hatzakis, M. (1970) "Batch-Fabricated Thin-Film Magnetic Recording Heads," *IEEE Trans. Mag.* **MAG-6**, pp. 597–601.

Safranek, W.H. (1986) *The Properties of Electrodeposited Metals and Alloys,* 2nd ed., AESF Society, Orlando, FL.

Schmalz, O., Hess, M., and Kosfeld, R. (1996) "Structural Changes in Poly(Methyl Methacrylate) during Deep-Etch X-Ray Synchrotron Radiation Lithography. III. Mode of Action of the Developer," *Die Angewandte Makromolekulare Chemie* **239**, pp. 93–106.

Sekimoto, M., Yoshihara, H., and Ohkubo, T. (1982) "Silicon Nitride Single-Layer X-Ray Mask," *J. Vacuum Sci. Technol.* **21**, pp. 1017–1021.

Shaw, J.M., Gelorme, J.D., LaBianca, N.C., Conley, W.E., and Holmes, S.J. (1997) "Negative Photoresists for Optical Lithography," *IBM J. Res. Develop.* **41**(1/2), pp. 81–94.

Siddons, D.P., Johnson, E.D., and Guckel, H. (1994) "Precision Machining with Hard X-Rays," *Synchrotron Radiation News* **7**(2), pp. 16–18.

Siddons, D.P., Johnson, E.D., Guckel, H., and Klein, J.L. (1997) "Method and Apparatus for Micromachining Using X-Rays," U.S. Patent 5, 679, 502.

Spears, D.L., and Smith, H.I. (1972) "High-Resolution Pattern Replication Using Soft X-Rays," *Electronic Lett.* **8**(4), pp. 102–104.

Spiller, E., Feder, R., Topalian, J., Castellani, E., Romankiw, L., and Heritage, M. (1976) "X-Ray Lithography for Bubble Devices," *Solid State Technol.* **April**, pp. 62–68.

Stiers, E., and Guckel, H. (1999) "A Closed-Loop Control Circuit for Magnetic Microactuators," *HARMST '99 Workshop,* Book of Abstracts, National Research Laboratory of Metrology, Kisarazu, Japan.

Stutz, A. (1986) "Untersuchungen Zum Entwicklungsverhalten eines Röntgenresist aus Vernetztem Polymethylmethacrylat," Diplomarbeit, Institut fur Kernverfahrenstechnik des Kernforschungszentrums Karlsruhe und der Universität Karlsruhe.

Takahata, K., Shibaike, N., and Guckel, H. (1999) "A Novel Micro Electro-Discharge Machining Method Using Electrodes Fabricated by the LIGA Process," *Technical Digest—IEEE MEMS '99*, pp. 238–243.

Tan, M.X., Bankert, M.A., and Griffiths, S.K. (1998) "PMMA Development Studies Using Various Synchrotron Sources and Exposure Conditions," *SPIE Conf. Materials and Device Characterization in Micromachining* **3512**, pp. 262–270.

Van der Gaag, B.P., and Sherer, A. (1990) "Microfabrication Below 10 nm," *Appl. Phys. Lett.* **56**, p. 481.

Visser, C.C.G., Uglow, J.E., Burns, D.W., Wells, G., Redaelli, R., Cerrina, F., and Guckel, H. (1987) "A New Silicon Nitride Mask Technology for Synchrotron Radiation X-Ray Lithography: First Results," *Microelectronic Eng.* **6**, pp. 299–304.

Weil, R. (1970) "The Origins of Stress in Electrodeposits," in *Plating*, AES Research Project 22, American Electroplating and Surface Finishing, pp. 1231–1237.

For Further Information

International workshops devoted to high-aspect-ratio microstructure technology (HARMST) were first held in 1995 with a particular emphasis on DXRL processing and LIGA technique. The meetings are held directly following the international Transducers conference every other year. HARMST '95 was held in Karlsruhe, Germany; HARMST '97, in Madison, WI; HARMST '99, in Kisarazu, Japan; HARMST '01, in Baden-Baden, Germany. A book of abstracts is published with the workshop and selected papers from the workshop are published in the *Microsystem Technologies* research journal published by Springer.

Information on VLSI X-ray lithography with some DXRL subject matter is covered by the international conference on Electron, Ion and Photon Beam Technology and Nanofabrication (EIPBN), with papers published in the *Journal of Vacuum Science and Technology B*.

Book chapters devoted to the topic of LIGA may be found in *Fundamentals of Microfabrication* (M. Madou, CRC Press, 1997) and *Handbook of Microlithography, Micromachining, and Microfabrication*, Vol. 2: *Micromachining and Microfabrication* (P. Rai-Choudhury, Ed., SPIE Press, 1997; see chapters "Plating Techniques" and "High Aspect Ratio Processing"). Volume 1 also contains a comprehensive review chapter on "X-Ray Lithography."

Many LIGA-related journal articles may be found in the IoP *Journal of Micromechanics and Microengineering: Structures, Devices, and Systems*.

Web sites where additional information on LIGA may be found include: http://www.fzk.de/imt/eimt.htm; http://www.imm.uni-mainz.de/; http://daytona.ca.sandia.gov/LIGA/; http://mems.engr.wisc.edu/liga.html.

19
Electrochemical Fabrication (EFAB™)

Adam L. Cohen
MEMGen Corporation

19.1 Introduction 19-1
19.2 Background... 19-2
    Solid Freeform Fabrication • SFF for Microfabrication
19.3 A New SFF Process 19-3
    Why Use Electrodeposition? • Selective Electrodeposition
19.4 Instant Masking... 19-5
    A Printing Plate for Metal • Instant Masking Performance
19.5 EFAB 19-8
    Multiple-Material Layers • Process Steps
19.6 Detailed Process Flow 19-11
    Mask Making • Substrate Preparation • EFAB Layer
    Cycle • Post-Processing
19.7 Microfabricated Structures... 19-15
19.8 Automated EFAB Process Tool................................... 19-16
19.9 Materials for EFAB.. 19-17
19.10 EFAB Performance 19-17
    Accuracy • Materials Properties
19.11 EFAB Compared with Other SFF Processes.............. 19-19
19.12 EFAB Compared with Other
    Microfabrication Processes... 19-19
19.13 EFAB Limitations and Shortcomings 19-20
19.14 EFAB Applications 19-21
19.15 The Future of EFAB... 19-22
    Acknowledgments 19-22
    Definitions 19-22

19.1 Introduction

Electrochemical fabrication (EFAB™) is an emerging micromachining technology invented at the University of Southern California and licensed to MEMGen Corporation [Cohen, 1998; 1999; Tseng et al., 1999]. It is based on multilayer electrodeposition of material using a new selective deposition technique called Instant Masking™ which provides a simpler, faster and readily automated alternative to through-mask electroplating. EFAB is based on the paradigm of solid freeform fabrication (SFF), rather than the semiconductor clean-room fabrication paradigm on which conventional micromachining is based. The technology is targeted at prototyping and volume production of functional microscale components, devices and systems.

19.2 Background

19.2.1 Solid Freeform Fabrication

Since the late 1980s, there has been a great deal of commercial and research activity in the field known as solid freeform fabrication, also known as rapid prototyping. Generally speaking, SFF is an approach to fabricating mechanical components that is additive vs. subtractive (such as with machining). In SFF, multiple layers of material, each representing a cross section of a desired three-dimensional structure, are deposited one at a time to form a laminated stack. By using hundreds or even thousands of such layers, extremely complex, freeform, three-dimensional shapes can be produced. SFF technologies in commercial use include stereolithography, three-dimensional printing, fused deposition modeling and laminated object manufacturing.

Solid freeform fabrication technologies are characterized by a number of significant features. First, they are extremely versatile, producing extremely complex shapes (including shapes having internal features that would be impossible to produce by subtractive methods). Second, they typically involve a fixed process with just a few simple process steps repeated again and again to form each layer, and so can easily be implemented in a single process tool and automated (indeed, SFF processes almost universally run unattended). Third, they are driven in a top-down fashion by a three-dimensional CAD model of the desired structure.

Until recently, SFF has been applied primarily to the fabrication of models and prototypes of macroscale mechanical parts (e.g., the automotive part shown in Figure 19.1), and to making tooling for volume production of these parts. However, there is now growing interest in applying the SFF paradigm to micromachining, and it is this approach that is taken by EFAB.

19.2.2 SFF for Microfabrication

At first glance, the idea of bridging the worlds of SFF and micromachining seems dubious, and, indeed, as may be seen from Table 19.1, conventional approaches to microfabrication are radically different in most ways from the approach used in SFF. Yet, if applying SFF principles to micromachining could be accomplished, it could offer a valuable alternative to conventional processes and help to make micromachining more versatile, rapid and cost-effective. SFF has the potential to offer microdevices virtually arbitrary three-dimensional geometry and thus a major increase in versatility and functionality; reduced lead times; fully automated, repeatable, unattended processing in a single, self-contained process tool;

FIGURE 19.1 A part fabricated using macroscale solid freeform fabrication; in this case, a full-size aluminum cylinder head casting for an internal combustion engine. (Courtesy of Soligen, Inc.)

TABLE 19.1 Conventional Microfabrication Compared with SFF

Characteristic	Conventional Microfabrication	SFF
Where device is manufactured	Clean room	Self-contained process tool in office or lab
Part complexity/number of structural layers	Up to 4 (to date)	Up to 1000s (highly complex three-dimensional shapes)
How device is manufactured	Mixture of manual and automated processing	Nearly total automation
Equipment required	Multiple process tools, each performing a different process	One process tool to build device (sometimes one other for post-processing)
CAD design level	Typically two dimensional (individual mask levels)	Three dimensional (solid model)
Steps required per layer	Many, usually different for each layer	Few, identical for each layer
Throughput	Fairly low (hours or days per layer)	High (minutes per layer)
Knowledge/skill level needed to design device	Rather high (commonly doctoral level)	Moderate (Bachelor's degree)
Knowledge/skill level needed to fabricate device	Rather high	Low (technician)
Capital expenditure required to produce devices	$5–100s of millions	$50–500,000

low-cost prototyping and manufacturing infrastructure with minimal need for a clean room; and easier design of devices.

19.3 A New SFF Process

While it is desirable to apply the principles of SFF to microscale manufacturing, the specific SFF technologies that have been developed for making macroscale models and prototypes do not readily lend themselves to this task. For example, most SFF processes cannot produce minimum features less than 75 to 100 μm wide, yet microscale devices may require minimum features in the range of 10 to 25 μm or smaller (e.g., for electrostatic actuators). Also, most SFF processes cannot produce layers thinner than about 75 to 150 μm, yet the overall height of many microdevices ranges from several microns to several hundred microns. Therefore, much thinner layers (e.g., a few microns) are required to define a multilayer three-dimensional geometry that minimizes dimensional distortions and "stairsteps" on curved or oblique surfaces.

While the costs of traditional SFF are acceptable on the macroscale as a means of making prototypes, for micromanufacturing one requires a method that is suitable for both prototyping *and* volume production, as there is no other means of fabricating large quantities of functional devices. This means that a viable microscale SFF process needs to both be economically scaleable and work with functional engineering materials. Unfortunately, virtually all SFF processes are not batch processes and generate a layer serially (i.e., one volume element at a time), so are too slow and costly for quantity production. Also, SFF processes typically are based on materials whose physical properties leave something to be desired in terms of their functionality. For example, they tend to suffer from low strength, low ductility and low temperature resistance, either due to the intrinsic properties of the material or the presence of pores.

A further problem with some conventional SFF processes for micromanufacturing is that they do not produce net-shape structures, but instead require that support structures—in effect, "scaffolding"—be fabricated along with the desired structure. When these supports are made of the same material as the desired structure they must be removed manually. This is difficult and quite costly if the structure is very small and is clearly not viable for large-volume production. Finally, some macroscale SFF processes produce structures that are shrunken or distorted with respect to the CAD design, usually due to phase changes of the material or thermal effects during processing.

Several teams have been working to develop microscale versions of one SFF process—stereolithography [e.g., Zhang et al., 1998]—which have succeeded admirably in generating interesting structures having small features and thin layers. However, most "microstereolithography" processes are (as on the macroscale) still serial[1] and require support structures for most geometries. Also, the materials used are still photopolymers, which do not offer the mechanical, electrical, thermal and optical properties desired for many microscale applications.

Given these issues with pre-existing SFF processes, it was determined necessary to develop a new SFF process suitable for microscale applications. The resulting new process is based on electrodeposition.

19.3.1 Why Use Electrodeposition?

Electrodeposition is a tremendously important industrial process used to finish metal parts (in the case of electroplating) and to manufacture metal parts by depositing metal over pre-formed mandrels (in the case of electroforming). Electrodeposition is carried out in an electrochemical cell comprising two electrodes—an anode and a cathode—and an electrodeposition bath. When depositing metals, the workpiece or mandrel serves as the cathode. When a current is applied, metal ions in solution migrate to the cathode where they become neutralized to form metal atoms which coat the cathode as a metallic deposit. In many plating processes, the anode also dissolves, producing metal ions that replenish those being deposited.

Compared with the additive processes on which most SFF processes are based, electrodeposition offers functional engineering materials such as metals and alloys that provide highly desirable properties. Moreover, electrodeposition allows material to be deposited on a very fine scale: essentially atom-by-atom (vs. particle-by-particle, drop-by-drop or sheet-by-sheet as in other SFF processes). This allows extremely thin, dense deposits to be produced with good material properties. Furthermore, with electrodeposition, material may be deposited over an entire layer (which may contain hundreds or thousands of devices) simultaneously, enabling batch processing. As a low-temperature process (i.e., 20 to 90°C), electrodeposition can achieve functional metal structures that exhibit low residual stress and no shrinkage. Finally, because more than one material can be readily deposited on a single layer, support structures formed from a different material than the structural material can be easily removed using an automated, batch process.

19.3.2 Selective Electrodeposition

A further key benefit of electrodeposition is that it can be used to define very small features. Some techniques for accomplishing selective electrodeposition are as follows.

19.3.2.1 Through-Mask Plating

Through-mask plating is the standard technique for high-precision selective electrodeposition. It is used in the manufacture of read/write heads for hard disk drives and serves as the basis of the LIGA process. The technique can be used to create features down to 1 μm or smaller. Initially it might appear that through-mask plating provides a suitable patterning method for an SFF-type process. Indeed, a patent [Guckel, 1993] describes a multilayer electrodeposition process that relies on this approach to define each layer. With through-mask plating, a metal substrate is cleaned, photoresist is applied and exposed through a photomask, the resist is developed and often descummed, the substrate is plated through apertures in the resist, and, finally, the resist is chemically stripped. The entire process involves a large number of distinct steps, the use of several different pieces of equipment, often located in a cleanroom, and the application of or immersion in multiple liquids. Though it is of course possible to automate all these steps, the result would be a process tool of high complexity and cost. Furthermore, total cycle time for

---

[1]Some work in this area uses an electronically addressable mask and illuminates the entire substrate simultaneously; however, there is a trade-off between resolution and substrate area.

through-mask plating is on the order of several hours, so repeating all these steps hundreds or thousands of times to fabricate a complex 3-D microstructure would seem impractical.

19.3.2.2 Other Conventional Approaches

Other approaches to selective electrodeposition are brush plating, in which an electrodeposition bath is applied locally to a substrate using a brush-type electrode, and laser-enhanced plating, in which the electrodeposition rate is locally accelerated by the rise in temperature due to absorbed laser radiation. These processes have difficulty producing very small, well-defined features and tend to be slow, as they are essentially serial.

19.3.2.3 Sharp Anode

Some work on fabricating microstructures using selective electrodeposition was performed at MIT in which an anode with a very sharp tip was moved in a three-dimensional path [Madden and Hunter, 1996]. Deposition was observed to occur preferentially in a localized volume due to the high current density near the tip, and spline-like structures such as springs were produced in metals such as nickel. As attractive as this process is in terms of its ability to fabricate structures directly from CAD design data, it would have difficulty generating arbitrary three-dimensional shapes (e.g., structures with overhangs), as it lacks a support material, and the features created are rather imprecisely defined. Also, even if the process were to incorporate multiple anodes, it would appear to be somewhat slow for mass production.

Given these limitations of known methods for selective electrodeposition, it was necessary to develop a new technique on which to base a viable SFF-based micromachining technology.

19.4 Instant Masking

This new technique goes by the name of Instant Masking™. It is similar to through-mask plating in that it uses photolithography to create a mask that defines the desired pattern of deposition and also allows material to be deposited simultaneously over an entire layer. Yet, unlike through-mask plating, Instant Masking: (1) allows the photolithographic processing to be performed completely *separate* from the device-building process; (2) allows all of the necessary masks to be produced as a batch, *prior* to part generation rather than *during* part generation; and (3) potentially allows the mask to be re-used.[2] By eliminating the need to repeat the photoresist processing inherent in through-mask plating over and over again for every layer, Instant Masking provides the enabling technology for a simple, low-cost, automated, self-contained SFF process tool that builds microdevices. Although the photolithographic steps required to generate micron-scale features may still need to be performed in a traditional clean room, the separation of mask fabrication from device fabrication allows for much-improved speed, efficiency and cost effectiveness. For example, the owner of the process tool does not need to spend millions of dollars to establish a clean room, as the Instant Mask™ can be purchased from an outside source, much like a printing plate, a photomask or an injection molding die is purchased now.

19.4.1 A Printing Plate for Metal

Instant Masking as a process is somewhat analogous to printing of ink on paper. With the latter process, a printing plate is fabricated which has a surface pattern that is the mirror image of the desired distribution of ink on paper. The plate is inked and pressed against the paper, then removed, resulting in a patterned deposit of ink. With Instant Masking, an Instant Mask is fabricated which has a surface pattern that is the mirror image of the desired distribution of metal on a substrate.[3] The Instant Mask is immersed in

---

[2] Compared with the patterned photoresist used in through-mask plating, which is chemically removed and disposed of after electrodeposition.

[3] The term *substrate* signifies either the actual substrate upon which a microstructure is fabricated or, equivalently, the previous layer of a multilayer structure which itself behaves as a substrate for the next layer.

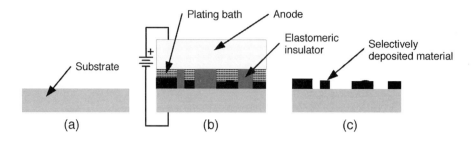

FIGURE 19.2 The Instant Masking process: (a) bare substrate; (b) during deposition, with the substrate and Instant Mask in intimate contact; (c) the resulting patterned deposit.

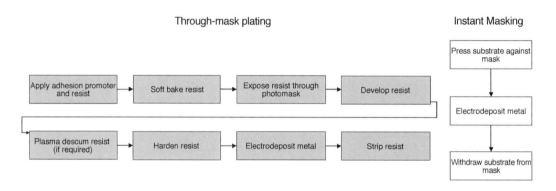

FIGURE 19.3 Instant Masking greatly reduces the number of steps required by conventional through-mask plating.

an electrodeposition bath and pressed against the substrate, current is applied and the mask is then removed to produce a micropatterned deposit of metal (Figure 19.2). Figure 19.3 compares the steps required for Instant Masking with those that may be required for through-mask plating, suggesting that simplification and improved speed are possible by bypassing the need for photoresist processing.

The Instant Mask is a two-component device, consisting of a layer of an insulating elastomeric mask material that is patterned with apertures and attached to an anode (Figure 19.2b). The elastomeric mask, having a typical hardness in the range of 5 to 300 Shore A, conforms to the topography of the substrate and prevents deposition of material other than within the region of the apertures. The anode serves its usual role in an electrodeposition cell but also provides mechanical support for the elastomer layer. Because the goal is to build nearly arbitrary three-dimensional structures, it is necessary to pattern nearly arbitrary two-dimensional cross sections, often with complex topology. Isolated "islands" of mask material are maintained in precise alignment by being attached to, and supported by, the anode.

19.4.2 Instant Masking Performance

The key performance issues with Instant Masking are (1) how selective the deposition process really is (i.e., is there any extraneous deposit, "flash," in the masked areas); (2) how small a feature can be patterned with sharply defined edges and without flash; and (3) how thick a deposit can be obtained (too thin a layer is not acceptable, as layers in the range of 2 to 10 µm would normally be desired for micromachining). Investigation of these issues was initially performed using the simple "printing press" apparatus shown schematically in Figure 19.4.

At the bottom of the apparatus, immersed in an electroplating bath, is an Instant Mask, with the patterned mask material facing upward. Located above the mask is a planarized metal substrate, mounted

Electrochemical Fabrication (EFAB™)

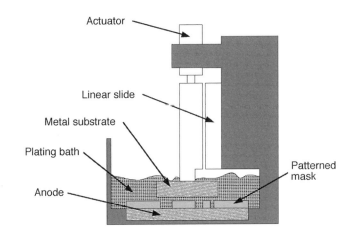

FIGURE 19.4 "Printing press" type of apparatus used to test Instant Masking.

FIGURE 19.5 An Instant Mask featuring a resolution test pattern. The largest bar is 200 μm wide and the smallest is 12 μm wide.

on a linear slide that enables it to be pressed against the mask. Figure 19.5 shows a typical Instant Mask, in this case, one having a resolution test pattern in which the smallest bar is 12 μm wide (along the short axis) and the largest is 200 μm wide. Experiments using this and other apparatus and masks have shown very good definition of the plated deposit, as long as the substrate is reasonably flat, smooth and parallel to the mask. For example, as shown in Figure 19.6, the edges of the deposit are sharp and show virtually no flash[4] on bars down to 25 μm in width, although 12-μm bars showed some definition problems.

Instant Masking involves conditions very different from those in standard electroplating baths, in which a large volume of bath is available. Indeed, a solid understanding of the electrochemical phenomena under these conditions is not yet available. An immediate concern one might have is deposit thickness, as during Instant Masking there is a very small volume (as little as tens of picoliters) of electrodeposition

---

[4]Traces of electrolyte remain on the substrate in this photograph.

FIGURE 19.6 A deposit of Cu on Ni produced with the mask of Figure 19.5.

bath within the electrochemical cell formed by the anode, cathode and patterned elastomer. This tiny volume is capable of supplying only enough ions to deposit a few nanometers of material. Fortunately, this problem can be solved if the Instant Mask anode is made from a metal that is electrochemically dissolved during deposition so as to provide "feedstock." Thus, Instant Mask anodes are normally made of the same metal that is to be deposited and normally erode several microns while depositing a layer. This erosion cannot be allowed to continue too long, as it can undercut and potentially delaminate the elastomer; however, to allow re-use of masks, it is believed that Instant Masks could be regenerated by plating metal back into the apertures to replace metal that has been lost (with the already-patterned elastomer serving as a mask).[5] Figure 19.7 is a micrograph of an Instant Mask used to create the thick (approximately 13 µm) deposit shown in Figure 19.8.

Other aspects of Instant Masking as a new method of electrodeposition still require further investigation. For example, chemical changes in the trapped volumes of the electrodeposition bath may occur during deposition. Also, because hydrogen is evolved when depositing materials with low current efficiency, deposition of these materials using Instant Masking may be problematic, as there is no escape route for these gasses. Difficulties with such materials have not been verified; however, efforts to date have focused on depositing metals such as copper having very high current efficiency. The strategy is generally to deposit the sacrificial material using Instant Masking and use conventional (i.e., open bath) plating methods to deposit the structural material, thus allowing for a considerable variety of structural materials without any of the constraints that Instant Masking might impose.

19.5 EFAB

Having demonstrated the ability of Instant Masking to selectively deposit material in a simple, fast and easily automated manner, a new SFF process based on it was developed called EFAB.

---

[5]This process has not yet been developed.

Electrochemical Fabrication (EFAB™)

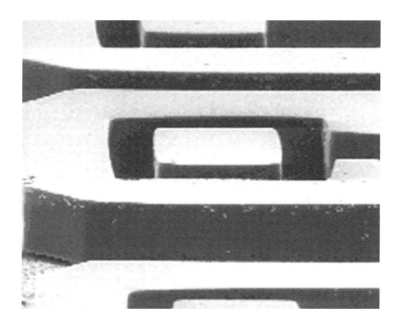

FIGURE 19.7 An Instant Mask used to pattern material for a single layer of a microstructure. The mask thickness (vertical dimension) is approximately 60 μm.

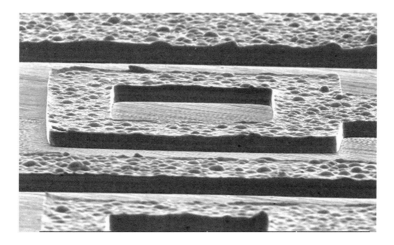

FIGURE 19.8 A thick (~13 μm) copper deposit produced with the Instant Mask of Figure 19.8.

19.5.1 Multiple-Material Layers

A completed layer produced with EFAB consists of both a structural and a sacrificial material, as illustrated in Figure 19.9. The sacrificial material in which EFAB-built devices are temporarily embedded serves the same purpose as sacrificial material in surface micromachining: mechanical support of structural material. This eliminates virtually all restrictions on microstructure geometry,[6] allowing the structural material on a layer to overhang—and even be disconnected from—that of the previous layer. Figure 19.10h shows a geometry that would suffer from a disconnected "island" on the fourth layer (right side) were there no

---

[6]The only restriction relates to the removal of sacrificial material: A hollow structure with no access to the interior cannot be fabricated, and certain geometries may include gaps, holes and slots that are difficult to clear of sacrificial material, especially if these features are long and narrow.

FIGURE 19.9 Plane view of a typical layer in EFAB, depicting structural material and surrounding sacrificial material.

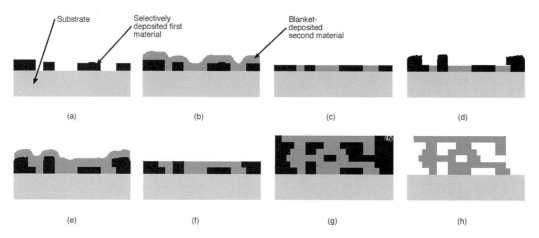

FIGURE 19.10 Steps in the EFAB process, as detailed in the text.

sacrificial material on the third layer. Such geometrical freedom also makes possible monolithically fabricated "assemblies" of discrete, interconnected parts (e.g., a gear train) without the need for actual assembly. The placement of the sacrificial material, and its ultimate removal in an automated batch process, is not geometry-specific (avoiding the need for explicit design and manual removal of support structures).

In EFAB, both the structural and the sacrificial materials are normally electrodeposited. For example, nickel can serve as the structural material, with copper as the sacrificial material. It is very important that both materials be conductive, as a conductive substrate is required for electrodeposition, and the goal is to allow layers to be deposited over one another regardless of the particular cross-sectional geometry of either layer.

19.5.2 Process Steps

Once Instant Masks representing each unique cross section of the desired microstructure and a suitable substrate have been prepared, the EFAB process can commence. This consists of the following three basic steps, repeated the same way for every layer: (1) selectively deposit a first material, (2) blanket deposit a second material and (3) planarize. The process flow is depicted in Figure 19.10 for a simple six-layer structure. In Figure 19.10a, the first material (here, the sacrificial material) has been selectively deposited onto a substrate using Instant Masking to yield a (partial) first layer. In Figure 19.10b, the second material (here, the structural material) has been blanket-deposited over this, contacting the substrate in those areas not covered with the first material. Then, as in Figure 19.10c, the entire two-material layer has been planarized to achieve precise layer thickness, smoothness and planarity. In Figure 19.10d, the first material has been selectively deposited for the second layer, while in Figure 19.10e the second material has been blanket-deposited over this. Figure 19.10f shows the second layer after planarization. After repeating these three steps for the remaining layers, the result is the embedded structure of Figure 19.10g. This is then etched to remove the sacrificial material, yielding the desired structure (Figure 19.10h).

19.6 Detailed Process Flow

19.6.1 Mask Making

All SFF processes require software to generate cross-sectional layer geometry from three-dimensional CAD geometry. With EFAB, one or more standard photomasks must be prepared which include the geometry of all the cross sections of the device. Specialized Microsoft Windows®-based software called Layerize™ has been developed which imports three-dimensional CAD geometry in a format that virtually all three-dimensional CAD systems can export (the SFF industry standard .STL file) and exports a two-dimensional file in a format accepted by commercial photomask pattern generators (e.g., a CIF or GDSII file). Layerize also takes as its inputs a set of rarely changed parameters located in an editable parameter file and a set of frequently changed parameters input by the user. It calculates the cross sections for each layer of the desired structure, formats them appropriately, and lays them out across the available photomask area. In a planned future version of Layerize, the geometries of cross sections will be compared and multiple layers will be assigned to a single mask, to reduce the total number of masks required.

For making prototype quantities of microdevices, the substrate is usually much smaller in area than the Instant Mask. Thus, the file that Layerize generates contains multiple smaller masks, or *submasks* (an example of which is shown in Figure 19.11), all of which are patterned on a common anode to form a single, easy-to-handle mask (Figure 19.12). One advantage of this approach is that all the submasks on the Instant Mask are patterned from the same photomask and are thus mutually aligned with the precision intrinsic to the photomask; this facilitates mask–substrate alignment later. Layerize also outputs a file that can control an automated EFAB process tool (e.g., instructing it which mask to use for each layer). Figure 19.13 illustrates the inputs and outputs of Layerize.

Once a photomask is produced (typically in two to four days), it is then used to fabricate the Instant Mask. One method of fabricating Instant Masks is based on micromolding. For example, the photomask can be used to expose a layer of thick photoresist deposited on a substrate, which when developed yields a two-level mold corresponding to the photomask geometry. A liquid elastomer can then be applied to the mold and a metal anode is then placed in contact with the elastomer and pressed to squeeze out

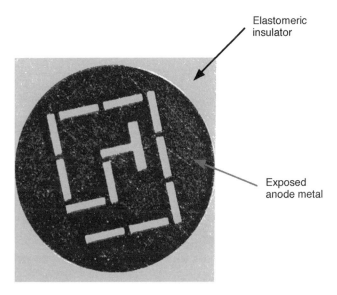

FIGURE 19.11 A single submask of an Instant Mask used to define a single cross section. The circle is approximately 1.7 mm in diameter.

FIGURE 19.12 An Instant Mask comprising multiple submasks.

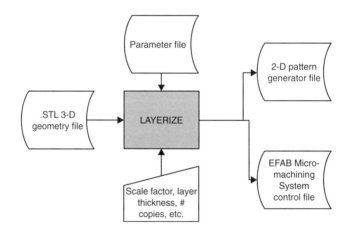

FIGURE 19.13 Inputs and outputs of Layerize software used for EFAB.

excess material. The elastomer can thereafter be hardened, after which it can be de-molded. Using this method, Instant Masks can be fabricated for all the required masks from a photomask within two days and then installed in the EFAB process tool.

19.6.2 Substrate Preparation

A conductive material to which the deposited materials are capable of adhering is needed as the substrate for the EFAB process. Typically, this is a metal plate, which is initially planarized before being installed in the EFAB process tool (it may alternatively be planarized online within the system), to provide a relatively flat, smooth starting surface for Instant Masking deposition.

19.6.3 EFAB Layer Cycle

The three steps performed on each layer in EFAB are described more fully as follows, with reference to Figure 19.14, which depicts the configuration and operation of the tabletop, manually operated EFAB process tool shown in Figure 19.15. The system consists of several subsystems: The first, a substrate fixture, includes a substrate onto which the layers are deposited, a linear slide to move the substrate up and down (to mate and unmate with the mask, to accommodate the addition of layers and to allow for motion during planarization) and a linear actuator. A second subsystem, for selective deposition, includes an Instant Mask (such as that shown in Figure 19.12) having multiple submasks, repeatable x- and y-travel stages, and a tank to contain the electrodeposition bath. A third subsystem, for blanket deposition, includes an anode and a tank containing an electrodeposition bath. Finally, a fourth subsystem is used for planarization—in this case, lapping (in the following, it is assumed that the substrate is already planarized before the first layer is created).

19.6.3.1 Selective Deposition

An Instant Mask is mounted within the tank of the selective deposition subsystem; the x- and y-coordinates of each submask are measured using a microscope and recorded. Then the substrate is immersed in an electrodeposition bath for the first material to be deposited. A substrate is mounted in the substrate fixture, which is then lowered on top of the selective deposition subsystem and aligned manually by mechanical means. The mask is positioned using the x- and y-stages to locate the submask for the current layer in the correct position below the substrate (Figure 19.14a), according to the recorded coordinates. The actuator then lowers the substrate onto the mask. Once the substrate and mask are mated, an electric current of a magnitude calculated to provide the desired current density is passed between the substrate and the anode of the mask for a time calculated to selectively deposit the desired thickness of the first material.

19.6.3.2 Blanket Deposition

The substrate fixture is placed on top of the blanket deposition subsystem (Figure 19.14b), and the substrate is lowered into an electrodeposition bath for the second material. An electric current is passed between the anode of the third subsystem and the substrate. The current is turned off after a time calculated to deposit the desired thickness of the second material over the first material (typically the same thickness as the first material).

19.6.3.3 Planarization

The substrate fixture is placed on top of a lapping plate (Figure 19.14c), and the two deposited materials are lapped until: (1) the second material no longer overlies the first, (2) the layer is reasonably flat and smooth and (3) the correct layer thickness is obtained. Electrodeposited materials are normally deposited with nonuniformities in thickness. By plating to a minimum thickness greater than that of the final layer thickness desired and then planarizing, such variations can easily be tolerated. Experience to date indicates that thickness nonuniformities are not excessive (e.g., Figure 19.8).

While it may be possible to selectively deposit both structural and sacrificial materials using Instant Masking, planarization is preferred for several reasons. For example, planarization (1) allows the use of blanket deposition of one material, thus requiring only half the masks and avoiding the need to align masks for one material with those for another; (2) avoids the need for either material to be deposited uniformly enough (e.g., due to pattern-dependent effects) to obtain a surface sufficiently planar and smooth to achieve good masking for the next layer; and (3) allows accuracy along the Z (vertical) axis to be precisely controlled.

Lapping and/or polishing may be performed using a variety of lapping plate and/or polishing pad materials and a variety of slurries, including those containing alumina, diamond and silicon carbide. Other lapping parameters include pressure and speed of the pad or plate. A means of endpoint detection must also be provided, so that lapping can be stopped when the layer is the proper thickness. Depending on the lapping materials and parameters used, surface roughnesses can be obtained on the order of 100 nm rms or smaller with very good planarity (e.g., <1 μm over several centimeters). Polishing can yield even

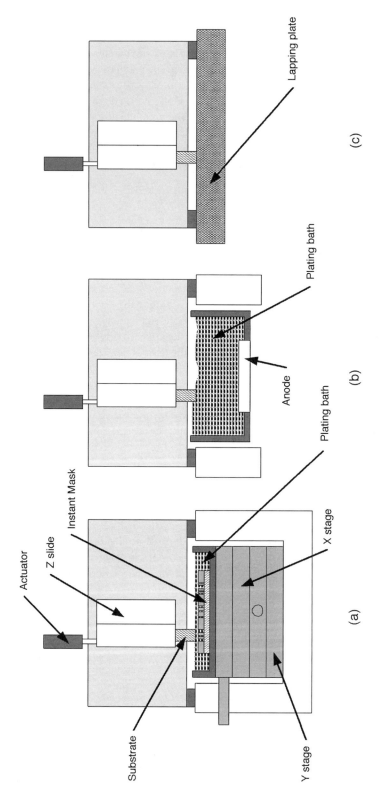

FIGURE 19.14 A schematic of a manually operated tabletop system used for EFAB, as described in the text.

Electrochemical Fabrication (EFAB^{TM})

FIGURE 19.15 The actual system shown in illustration in Figure 19.14.

FIGURE 19.16 Some five-layer nickel structures measuring ~40 μm high.

smoother surfaces, when required by some applications (e.g., optical mirrors or valve seats). Certain combinations of materials may exhibit differential planarization rates, causing one material to become recessed below the surface of the other, although no such effect has been measured, for example, with copper and nickel. By proper selection of planarization materials and parameters, these effects should be largely controllable.

19.6.4 Post-Processing

Once the above three steps have been repeated for all layers required, the sacrificial material is removed, typically by chemical etching. With copper as the sacrificial material, etching rates of several hundred microns per hour can be obtained. Material can be removed from fairly high aspect ratios; for example, copper has been etched from a volume with an exposed cross section of 25 × 300 μm to a depth of 3 mm.

19.7 Microfabricated Structures

As of this writing, EFAB has fabricated several fairly simple nickel structures such as the five-layer structure shown in Figure 19.16, whose overall height is ~40 μm. The 12-layer microchain shown in Figure 19.17 has links of ~290 μm. The horizontal links are free to move, while the vertical links are in this case attached to the substrate. However, by simply including a sacrificial layer beneath these, the entire chain

FIGURE 19.17 A 12-layer microchain fabricated in nickel using EFAB, without assembly. Overall height is ~100 μm and the width of a chain link is ~290 μm. All horizontal links are free to move.

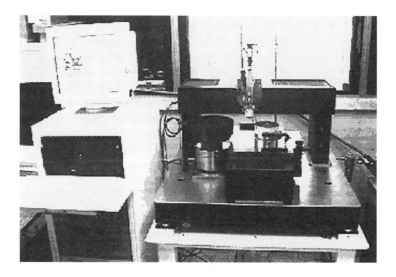

FIGURE 19.18 A prototype automated EFAB process tool, partly assembled.

could be released from the substrate. The entire chain is fabricated in a pre-assembled state, without the need for actual assembly. All structures were built using the tabletop EFAB process tool shown in Figure 19.15, located in a room that is anything but clean! Far more complex structures with many more layers are planned for the next-generation automated EFAB process tool.

19.8 Automated EFAB Process Tool

Figure 19.18 shows a partially assembled prototype of an automated process tool for EFAB, designated the EMS-Alpha (for EFAB Micromachining System). The system consists of two modules: (1) a cabinet housing a PC, motor drivers and other electronics and (2) a frame on which is mounted subsystems that perform the EFAB process (motorized motion stages, a planarization system etc.), as well as support subsystems such as electrodeposition bath pumps and filters. The footprint of the entire system when completed will be about 2.0 × 2.0 m. The system—and future commercial systems like it—could be installed in any reasonable lab space or shop floor and used to produce prototype quantities of microdevices—only compressed air and electricity vs. a clean room are required.

Whereas the manual tabletop system of Figure 19.15 has a throughput of about two completed, planarized, two-material, 8-µm-thick layers per day (in a single shift), the throughput of the EMS-Alpha is projected to be over 30 such layers per day (in three shifts). The system is designed for fully automated, unattended operation under the control of specialized software. The substrate is sized for prototyping purposes, measuring initially about 8 mm in diameter, though larger substrates can be accommodated. The system can handle several 100-µm Instant Masks, allowing structures with hundreds of unique cross sections to be fabricated. In summary, the EMS-Alpha, when completed, will be similar in operation to commercial solid freeform fabrication systems. As of this writing, the hardware and software are being debugged and operating parameters are being established.

19.9 Materials for EFAB

EFAB can be used to form microstructures from many materials that can be deposited by electrolytic or electroless deposition from aqueous solutions at temperatures ranging from 20 to 90°C. Such materials include many pure metals (e.g., nickel, copper, gold, silver, lead, zinc, iron, cobalt), a variety of metal alloys (e.g., nickel phosphorous, tin–lead solder), certain semiconductors (e.g., lead sulfide, bismuth telluride), composites (e.g., Ni-Teflon®, Ni-diamond) and even conductive polymers as well [Schlesinger and Paunovic, 2000]. A constraint is that there must be a depositable sacrificial material available that can be selectively removed (e.g., by etching) after all layers are formed. The material system that has received the greatest attention so far is nickel and copper, with copper as the sacrificial material.[7] Nickel has good strength and good temperature and corrosion resistance, while copper can be selectively etched with respect to nickel fairly easily. Other structural materials compatible with copper as the sacrificial material include nickel alloys, gold, platinum, iron, tin–lead solders and nickel composites.

19.10 EFAB Performance

Primary issues with any manufacturing process for mechanical structures are geometric accuracy and materials properties. The following issues are known as of this writing.

19.10.1 Accuracy

19.10.1.1 x–y Accuracy

Accuracy in the plane of the layers (defined as the x–y plane) is affected by multiple factors. The width of deposited metal features is determined by the width of the corresponding features in the Instant Mask during deposition. These in turn are the result of pattern transfer distortions with respect to the photomask associated with fabrication of the Instant Mask and possible mask distortion due to the mating pressure of mask and substrate. However, it is believed that, once characterized, both effects can be minimized using compensation algorithms built into Layerize software to yield a feature width accuracy of approximately ±1 µm. Mask–substrate misalignments can cause a shift in the relative position of layers, an effect that can be observed quite easily in Figure 19.19, where a five-layer, letter "E" shape[8] exhibits layers that are misregistered by as much as ±8 µm. Here, the problem was caused in part by a subsequently fixed defect in the manual EFAB system, but also by the nature of this apparatus: It uses mechanical, operator-dependent methods for registration. The EMS-Alpha, in contrast, uses precision, high repeatability

---

[7]This can be reversed, such that the nickel is etched and the copper is left behind.
[8]The bottom layer is still embedded in sacrificial material, as the etching was stopped prematurely for this structure.

FIGURE 19.19 A Ni microletter "E," illustrating some manufacturing artifacts. Overall height is ~40 μm.

motion stages to achieve an expected layer-to-layer alignment of about ±1 μm. The misalignment error will not accumulate from layer to layer, as all layers are aligned with respect to the substrate, rather than the previous layer.

Another alignment issue is *layer skew*, a problem caused by nonorthogonality between the z-travel of the substrate and the plane of the layers. Layer skew would cause a "vertical" wall formed by multiple layers that should be precisely 90° from the layer plane to lie at some other angle. By proper design and calibration of the system, this effect can be reduced to a few minutes of arc or better. A final x–y accuracy issue is *smearing*, an artifact of the planarization process which can be seen in Figures 19.16, 19.17 and 19.19 as a slight feathering of the edges of the layers in some locations. Smearing is the result of one material being deformed and pushed into another by shear forces associated with mechanical planarization processes. With nickel and copper, smearing typically causes very thin films of nickel to protrude 1 to 2 μm beyond the desired edge, yielding a ragged appearance. With process optimization, however, it is expected that smearing can be reduced to submicron levels, as is the case with read/write heads that are planarized using similar methods.

19.10.1.2 Z Accuracy

Dimensional accuracy along the z (layer stacking, or vertical) axis is determined largely by the ability to accurately control the planarization process. This further requires very accurate, high-resolution z-axis measurement and control. The methods developed for use with the manual EFAB system provide z accuracy in the range of ±1 to 2 μm per layer; however, improved methods developed for the EMS-Alpha should allow for a ±2 μm *overall* height accuracy. This nevertheless assumes that the z coordinates of all important features are specifically designed to be integer multiples of the layer thickness; if not, then because z-axis coordinates are quantized as with any multilayer process, additional inaccuracies will result.

Another potential source of inaccuracy along the z-axis is residual stress. It is not uncommon in SFF and conventional microfabrication processes to observe curling of cantilevers and other stress-sensitive geometries. This is often due to the significant residual stresses these processes may produce in deposited materials (e.g., due to phase change of materials, thermal effects etc.) With EFAB, residual stresses can be made quite low. Electrodeposited nickel, for example, can be fine-tuned so that its measured residual stress is 0 to 15 MPa (either compressive or tensile). Of course, this optimization is highly material and process-parameter dependent, and some electrodeposited materials may exhibit much higher stresses.

19.10.2 Materials Properties

As of this writing, materials properties of structures made with EFAB such as hardness and tensile strength have not been directly measured. However, it is expected that for a given deposition bath chemistry and set of process parameters, these would be similar to those measured for commercial electroformed metal parts, and perhaps even more similar to those measured for LIGA-produced structures. In general, electrodeposited materials, while different microstructurally from materials processed using mechanical and thermal methods, offer very useful engineering properties such as hardness, tensile strength and Young's modulus. In some cases, electrodeposited materials can even have superior properties (e.g., tensile and yield strength) to traditionally processed materials [Schlesinger and Paunovic, 2000].

Because EFAB, like all mainstream SFF, is a multilayer process, the issue of interlayer bonding is critical, especially as low tensile strength along the z-axis would compromise structural integrity and functionality in many applications. Ideally, the interlayer adhesion would be comparable to the bulk yield strength of the material. Preliminary experiments with nickel suggest that the bond strength is strongly dependent on the preparation of the substrate, and that bond strengths ranging from approximately 20% to a very large fraction of the bulk strength can be achieved.

A related matter is electrical contact resistance between layers. This is of particular importance in structures such as three-dimensional microinductors that might be made with EFAB, but which would require many interlayer junctions. Kelvin-type measurements performed with a four-point probe on stacked nickel layers having an 85×85-μm cross section indicated a per-layer contact resistance of less than 5 $\mu\Omega$. This low value is encouraging and is actually less than the bulk resistance of each nickel layer in the stack, therefore, it seems that complex microelectrical devices should indeed be feasible with EFAB.

19.11 EFAB Compared with Other SFF Processes

While EFAB is similar to, and inspired by, macroscale SFF processes, it is also different in some very significant ways. The fabrication of very small features and very thin layers is one obvious difference between EFAB and other SFF processes. Another is the emphasis on functional materials and the capability of depositing multiple materials (through successive masking operations on a given layer) to build sophisticated functional devices.

Still another difference is the batch fabrication capability needed for volume production, which leads to the use of fixed tooling (Instant Masks) in EFAB, rather than the "direct write" approach common to most SFF processes. The trade-off in general is between processes that are directly CAD driven but serial (e.g., stereolithography or e-beam lithography) and thus too slow and costly for volume production, and processes that use tooling (e.g., conventional semiconductor manufacturing using photomasks) but which can achieve high productivity and low cost when making devices in quantity. Because of the need for production volumes of microdevices rather than just prototypes, EFAB takes the latter approach, requiring the need for tooling. Nevertheless, to allow economical prototyping, EFAB is downward scalable, perhaps more easily so than conventional microfabrication and semiconductor fabrication. This is achieved by making use of smaller substrates than is practical with photoresist processing,[9] by combining multiple cross sections onto a single Instant Mask to reduce the cost of photomasks and by automation that minimizes substrate handling.

19.12 EFAB Compared with Other Microfabrication Processes

Table 19.2 compares the major microfabrication processes, as well as microstereolithography, with EFAB. While EFAB offers certain potential advantages over these other processes in some areas, there are clearly some applications for which it is not the best choice among alternative methods.

---

[9] Substrates as small as 1.7-mm diameter have been used.

TABLE 19.2 EFAB Compared with Other Micromachining Processes

Characteristic	EFAB	Bulk Micromachining	Surface Micromachining	LIGA	Microstereolithography
True three-dimensional geometry (100s–1000s of levels)	Y	N	N	N	Y
No cleanroom needed for device fabrication	Y	N	N	N	Y
Very high-aspect-ratio devices without a synchrotron	Y	N	N	N	Y
Automated processing/few process variables	Y	N	N	N	Y
IC-compatible processing	Y	Y	Difficult	Y[a]	N/A[b]
Batch fabrication (entire layer simultaneously)	Y	Y	Y	Y	N[c]
Device design at the three-dimensional CAD level, with automatic two-dimensional layout	Y	N	N	N/A	Y
Identical process performed on each layer	Y	N/A	Maybe	N/A	Y
Produces net-shape structures	Y	Y	Y	Y	N
Avoids the need for any tooling (direct-write)	N	N	N	N	Y
Minimum features <20 μm	N	(Y at present)	Y	Y	Y
Vertical wall smoothness	±1 μm	<20 nm[d]	N/A	30–50 nm[e]	Unknown (±1 μm?)
Dielectric structural materials available	N	(N/A for now)[f]	Y	Y	Y (limited)
Exotic structural materials available (e.g., piezoelectrics)	N[g]	N	Y	Y	N
Vertical build rate (after photolithography, where applicable)	~10 μm/hr	N/A	<<1 μm/hr	20–100 μm/hr[h]	Area dependent

[a] Exposure of thick resist to high-energy X-rays can, however, cause radiation damage to an underlying IC.
[b] The process is compatible in principle, but because the structure produced is an insulating polymer, IC integration is probably not very useful.
[c] At least one version uses a dynamic mask to expose the layer simultaneously.
[d] Vargo et al. (2000).
[e] http://greenmfg.me.berkeley.edu/LIGA/right.htm.
[f] Yes, if EFAB is supplemented by other processing, and with future extensions.
[g] Not directly, but this can be done by hybridizing EFAB with conventional deposition processes.
[h] Madou (1997).

19.13 EFAB Limitations and Shortcomings

Regarded as an SFF process, EFAB has certain limitations and shortcomings. The first is throughput. Electrodeposition is considerably faster (typically 25 to 100 μm/hr) than vacuum deposition processes used, for example, in surface micromachining, and the overall productivity (e.g., number of layers per day) of EFAB promises to be tens of times greater than that of surface micromachining. However, by macroscale SFF standards, the vertical build rate of EFAB is fairly low. For example, three-dimensional printing, a fast commercial SFF technology, can deposit centimeters of material per day, while EFAB can only deposit several hundred microns per day. On the other hand, there is no fundamental limit to the horizontal extents of the layer which EFAB produces. Thus, EFAB could simultaneously produce thousands or potentially millions of microdevices in a single process tool. When combined with the fact that microdevices tend to be only tens to hundreds of microns tall, the cost per device can still be very reasonable. Nevertheless, because of the build-rate limitations, EFAB is not expected to be economical in most cases for production of small macroscale devices (e.g., taller than 5 to 10 mm)[10] unless there is no alternative fabrication technology that can provide the required geometrical complexity or materials properties. Another limitation already noted is that EFAB requires tooling. There is, therefore, a tooling cost and a delay of several days following completion of the design before fabrication can begin. Finally, because of the use of sacrificial material, geometries are not entirely unrestricted: the need to remove the sacrificial material imposes some constraints on the design, though constraints are far fewer than if sacrificial material were not employed.

In the context of micromachining, the major liability of EFAB as an electrodeposition-based process, especially compared with surface micromachining, is probably the incorporation of dielectric materials into fabricated structures in an arbitrary fashion. Certainly, EFAB structures can be deposited onto a substrate that comprises dielectric materials, air can be used as a dielectric when the design permits, and structures can be backfilled with dielectric. But, the ultimate goal of allowing the designer the flexibility to place dielectric material anywhere within a structure is more challenging. Several approaches to extending EFAB in this direction are under consideration, and it is believed this limitation will be overcome with further development.

When compared with surface micromachining again, EFAB is somewhat more restricted in the choice of materials, as it is possible to vacuum-deposit (using sputtering, chemical vapor deposition, evaporation etc.) an extremely wide variety of materials. However, it is believed that EFAB can be combined with conventional deposition processes in a hybrid fashion so as to expand the range of materials that can be integrated. Finally, compared with alternative microfabrication processes, EFAB's minimum x–y features are—at 20×20 μm—some 4 to 10 times larger, at least at present. Because EFAB can deposit multiple thick layers, device footprints can be larger than usual without loss of stiffness, thus obviating the need for smaller features in many cases. Small gaps between structures can be achieved, however, when necessary by exploiting the multilayer nature of EFAB: Structures can be fabricated on different layers in a pre-aligned fashion and then deformed in a massively parallel manner to render them co-planar. Also, future improvements are expected to reduce as-fabricated minimum features to 5 to 10 μm.

19.14 EFAB Applications

Broadly speaking, it is believed that with further development EFAB will be capable of fabricating a wide variety of microscale parts, micromachines, microelectromechanical systems (MEMS) and microoptoelectromechanical systems (MOEMS). The market for microscale devices in many industries (e.g., medical, automotive, electronics, aerospace) is rapidly growing. Some general areas of application include:

- Three-dimensional microscale mechanical devices (e.g., springs, screws, helical gears, drive chains)
- Monolithically fabricated multicomponent micromachines (e.g., gearboxes, motors, valves, pumps, solenoids)
- Microelectromagnetic devices requiring three-dimensional coils (e.g., on-chip high-Q and high-inductance inductors, motors, solenoids, sensors)
- Metal tooling for mass production of three-dimensional microcomponents in polymers and ceramics

Some specific application areas being investigated for EFAB include micro-power generators, switches for optical networking, microsurgical instruments and implantable medical devices (e.g., stents), microrelays, electronics packaging and thermal management devices.

---

[10] Unless perhaps if the x–y extents are very small, allowing many devices to be made simultaneously.

19.15 The Future of EFAB

As a fabrication technology for the microscale, major developments expected for EFAB in the next few years include:

- Commercial EFAB machines for prototyping and volume production
- Monolithic, on-chip integration of EFAB-produced structures with integrated circuits
- The ability to process additional materials, especially dielectrics

Acknowledgments

The author wishes to express his gratitude to the Defense Advanced Research Projects Agency (DARPA), Defense Sciences Office and Microsystems Technology Office, for primary funding of EFAB development under the Mesoscale Machines and MEMS Programs, and to the University of Southern California/ Information Sciences Institute for additional funding. The author also wishes to acknowledge all the people who have been involved to date in the development of EFAB, most particularly Gang Zhang, Uri Frodis, Fan-Gang Tseng and John Evans.

Definitions

EFAB: A multilevel three-dimensional microfabrication process based on electrochemical deposition and the use of Instant Masking to define the geometry of the fabricated device on each layer.

Instant Masking: A micropatterning method involving a conformable patterned mask, an electrically active bath, and a source of current. When used for electrodeposition, the mask is mated with a substrate in the presence of an electrodeposition bath and current is applied to deposit material onto the substrate in regions not contacted by the mask.

Instant Mask: A mask used for Instant Masking consisting generally of a conformable, patterned insulating material laminated to a rigid support structure.

Layerize: Special-purpose software developed for use with EFAB that imports a three-dimensional geometry file produced by standard three-dimensional CAD systems and exports a two-dimensional geometry file in a format acceptable to standard photomask pattern generators. The photomask produced is then used to fabricate an Instant Mask.

Solid freeform fabrication (SFF): A group of fabrication processes that produce structures having complex, nearly freeform, three-dimensional shapes through additive means. Typically, SFF methods deposit and stack together multiple layers of material one by one and are driven by a three-dimensional CAD model of the desired structure.

Submask: A portion of an Instant Mask used to pattern a single cross section of a device.

References

Cohen, A. L., Zhang, G., Tseng, F. G., Mansfeld, F., Frodis, U., and Will, P. M. (1998) "EFAB: Batch Production of Functional, Fully-Dense Metal Parts with Micron-Scale Features," in *Proc. of the Solid Freeform Fabrication Symposium 1998,* D. L. Bourell et al., Eds., pp. 161–168, August 10–12, Austin, TX.

Cohen, A. L., Zhang, G., Tseng, F. G., Frodis, U., Mansfeld, F., and Will, P. M. (1999) "EFAB: Rapid, Low-Cost Desktop Micromachining of High Aspect Ratio True 3-D MEMS," in *Technical Digest—12th IEEE Int. Conf. on Microelectromechanical Systems,* pp. 244–251, January 17–21, Orlando, FL.

Guckel, H. (1993). "Formation of Microstructures by Multiple Level Deep X-Ray Lithography with Sacrificial Metal Layers," U.S. Patent 5,190,637.

Madou, M. (1997). *Fundamentals of Microfabrication,* CRC Press, Boca Raton, FL.

Madden, J., and Hunter, I. (1996) "Three-Dimensional Microfabrication by Localized Electrochemical Deposition," *J. MEMS* **5**(1), pp. 24–32.

Schlesinger, M., and Paunovic, M., Eds. (2000) *Modern Electroplating,* John Wiley & Sons, New York.

Tseng, F. G., Zhang, G., Frodis, U., Cohen, A. L., Mansfeld, F., and Will, P. M. (1999) "EFAB: High Aspect Ratio, Arbitrary 3-D Metal Microstructures Using a Low-Cost Automated Batch Process," in *Proc. Micro-Electro-Mechanical Systems (MEMS), 1999 ASME International Mechanical Engineering Congress and Exposition,* A. P. Lee et al., Eds., pp. 55–60, November 14–19, Nashville, TN.

Vargo, S. E., Wu, C., and Turner, T. (2000) "Reactive Ion Etching of Smooth, Vertical Walls in Silicon," *NASA Tech. Briefs* **24**(10).

Zhang, X., Jiang, X. N., and Sun, C. (1998), "Micro-Stereolithography for MEMS," Micro-Electro-Mechanical Systems: MEMS, DSC-Vol. 66, in *Proc. of the 1998 ASME Int. Mechanical Engineering Congress and Exposition,* X. Zhang et al., Eds., pp. 3–9, November 15–20, Anaheim, CA.

For Further Information

The University of California/Information Sciences Institute has a Web site for EFAB at www.isi.edu/efab. The site includes links to some SFF companies that can provide more information on SFF. Readers may also visit MEMGen Corporation's Web site at www.memgen.com.

EFAB, Instant Masking, Instant Mask and Layerize are trademarks of MEMGen Corporation. Windows is a registered trademark of Microsoft Corporation.

20
Fabrication and Characterization of Single-Crystal Silicon Carbide MEMS

Robert S. Okojie
NASA Glenn Research Center

20.1 Introduction 20-1
20.2 Photoelectrochemical Fabrication Principles of 6H-SiC 20-3
20.3 Characterization of 6H-SiC Gauge Factor 20-8
   Resistor–Diaphragm Modeling • Temperature Effect on Gauge Factor • Temperature Effect on Resistance
20.4 High-Temperature Metallization 20-17
   General Experimental and Characterization Procedure • Characterization of Ti/TiN/Pt Metallization • Ti/TaSi$_2$/Pt Scheme
20.5 Sensor Characteristics 20-26
20.6 Summary 20-29
   Acknowledgments 20-30

20.1 Introduction

For the purpose of precision instrumentation to better enable accurate measurements in high-temperature environments (>500°C), there is a growing need for sensing and electronic devices capable of operating reliably for a reasonable length of time in such a harsh environment. Typical applications for sensors that function at high temperature include automotive, aeropropulsion (both commercial and military), process control in materials engineering, and a host of others. Temperatures in these can go as high as 500°C or greater. However, most existing electronic components are limited to temperatures lower than 200°C, primarily due to the thermal limitations imposed by the conventional materials used in their manufacture (most notably silicon). Robust device architecture based on silicon-on-insulator (SOI) technology can extend device operation to near 400°C, either for brief period of time or with water-cooling-assisted packaging. However, at 500°C the thermomechanical deformation of silicon becomes the ultimate factor limiting high-temperature silicon microelectromechanical (MEMS) devices [Huff et al., 1991]. Therefore, to meet the increasing need for higher temperature instrumentation, new and innovative devices from materials more robust than silicon are being developed by various groups.

Technological advancement in the growth of wide band-gap semiconductor crystals such as silicon carbide (SiC) has made it possible to extend the operation of solid-state devices and MEMS beyond 500°C. Silicon carbide has long been viewed as a potentially useful semiconductor material for high-temperature

TABLE 20.1 Comparison of Properties of α-6H-SiC and Silicon

Properties	Silicon	6H-SiC
Bandgap (eV)	1.12	~3
Melting point (°C)	1420	>1800
Breakdown voltage ($\times 10^6$ V cm^{-1})	.3	2.5
Young's modulus of elasticity (GPa)	165	488
Thermal conductivity (W(cm-C)$^{-1}$)	1.5	5
Electron saturation velocity ($\times 10^7$ cm s^{-1})	1	2
Maximum operating temperature (°C)	300	1240

applications. Its excellent electrical characteristics—wide band-gap, high-breakdown electric field and low intrinsic carrier concentration—make it a superior candidate for high-temperature electronic applications [Pearson et al., 1957]. Table 20.1 shows a comparison of relevant electrical and mechanical properties between 6H-SiC and silicon. The fact that SiC exhibits excellent thermal and mechanical properties at high temperature combined with its fairly large piezoresistive coefficients makes it well suited for use in the fabrication of high-temperature electromechanical sensors. On the basis of the above discussion, there are growing efforts by different research groups to develop sensors based on SiC [Okojie et al., 1996; Berg et al., 1998; Mehregany et al., 1998]. SiC crystals appear in various crystal structures called *polytypes*. The polytypes most frequently available for use are the cubic-SiC (also referred to as 3C-SiC or beta-SiC) and two of the hexagonal polytypes, 4H-SiC and 6H-SiC. The main physical differences between the 4H- and the 6H-SiC are the stacking sequence of the silicon–carbon atomic bi-layers, the number of atoms per unit crystal cell, and the lattice constants. The hexagonal crystals are grown in large boules and sliced into wafers, with the current largest commercially available size of 76 mm in diameter. Homo-epitaxial growth by chemical vapor deposition (CVD) can be performed on the single-crystal substrate to obtain epilayers of thickness and doping level (nitrogen doping for n-type and aluminum or boron for p-type) as desired for various devices. In the case of 3C-SiC, there is no viable technology in existence for growing it as a single-crystal boule of acceptable quality. Therefore, hetero-epitaxial CVD was adopted to grow it on silicon substrate. However, the large lattice mismatch that exists at the 3C-SiC/Si heterojunction causes numerous crystal defects in the 3C-SiC hetero-epilayer that greatly degrades the performance of electronic devices fabricated in it. The existence of such defects in 3C-SiC and the fact that it has the lowest band gap (2.3 eV vs. 2.9 eV for 6H- and 3.2 eV for 4H-SiC) diminish its attractiveness for broader sensor applications, especially those that require significant high temperature electronics integration.

The discussions in this chapter will focus on the fundamental technological challenges that exist in the implementation of MEMS-based SiC pressure sensors. The SiC pressure sensor is chosen as representative of other SiC-based MEMS sensors due to its identification as a crucially needed technology and the fact that it is one of the most mature with the potential for commercialization. Furthermore, an application environment (turbine engines) already exists where it can be readily inserted. Appreciable advancements in its development have been made recently that bring it significantly closer to viable commercialization. This chapter is delineated into five sections to enable readers to better understand the fundamental challenges of this emerging technology. The basic fabrication principles of single-crystal silicon carbide are presented in Section 20.2, in which SiC device fabrication by photoelectrochemical etching (PECE) and conventional electrochemical etching (ECE) are discussed. Important etching parameters leading to the fabrication of resistor and diaphragm structures for sensors will be discussed in this section. Relevant SiC fabrication processes using more recently developed deep reactive ion etching are discussed in Chapter 21. In strain-sensor technology, the output of the device is a function of the gauge factor (GF) and temperature coefficient of resistance (TCR). Because the GF is a critical parameter used to determine the sensitivity to mechanical stimulus, Section 20.3 will discuss characterization of the SiC gauge factor and its temperature dependency, including the TCR. Metallization that is stable at the desired high operating temperature is recognized as critical for a successful implementation of high temperature sensor. Two specific approaches to realize thermally stable ohmic contacts on SiC sensor material are presented in Section 20.4, along with the deleterious effects of oxygen on the contacts. Section 20.5 discusses the process integration of various components leading to the realization of a 6H-SiC pressure sensor. Finally, Section 20.5 will discuss sensor testing and performance characteristics.

20.2 Photoelectrochemical Fabrication Principles of 6H-SiC

The difference between PECE and ECE is basically the photo-induced electron-hole pair (EHP) generation that is required for the former. The ECE process is a conventional anodization process in which high voltage or current is used. In both cases, the sample substrate acts as the anode electrode. The PECE process is adopted where anisotropic etching is critical. For example, in the PECE of resistors, it is important to minimize isotropic etching in order to maintain uniform resistor geometry with minimum undercut. Nonuniform geometry of the resistors typically leads to imbalance in the Wheatstone bridge circuit in which they are configured. The effect will be to induce high null output values, away from the ideal zero output voltage if the bridge is balanced.

Unlike silicon technology, processes associated with fabrication of structures in single-crystal SiC are limited generally to ECE and reactive ion etching. This arises from the nearly inert surface chemistry of SiC, which makes conventional wet chemical etching impossible at room temperature. Carrabba et al. (1989) used PECE process on SiC, but the visible light source used in his study resulted in very slow etch rates. Shor et al. (1992) utilized ultraviolet (uv) laser radiation and achieved very high rates of SiC etching, though the morphology of the etched surface was very rough. In this section, emphasis will be on electrochemical etching methods of SiC with the awareness that rapid progress has been made in deep reactive ion etching of SiC [Beheim and Salupo, 2000]. The process of using deep reactive ion etching for fabricating MEMS structures in SiC is described in detail in Chapter 21 of this book. The fabrication of piezoresistor-based sensors in single-crystal SiC requires at least one intentionally doped epilayer, usually n-type, grown on a p-type substrate. However, in PECE or ECE processes, it is often necessary to have two epilayers grown on an n-type substrate as shown on Figure 20.1. The bottom epilayer is p-type, followed by a top epilayer of n-type. The resistor sensing elements are fabricated in the n-type layer. The p-type epilayer primarily functions as an etch-stop during PECE of the n-type resistors. On the backside n-type substrate, the ECE process is applied with high voltage or current that enables high etch rates to form a cavity as deep as 200 μm.

While the fabrication principles of both photo and dark electrochemical etching are to be discussed in further detail, this discussion briefly describes the method used to fabricate both the pressure sensor and the beams that are utilized for characterizing the piezoresistor properties of single crystal SiC. The fabrication of the piezoresistor for use as a strain sensor requires a thorough investigation of the etching characteristics of the n^+-type SiC epilayer. The preference for the highly doped epilayer is due to the associated low resistivity and lesser sensitivity to temperature variations relative to lower doped epilayers. However, this preference leads to a trade-off in lower strain sensitivity due to reduction in gauge factor. In order to fabricate a SiC diaphragm transducer, a fully controllable etching process for thinning and forming of deep cavities in the backside must be developed. For diaphragm fabrication, the desirable etch

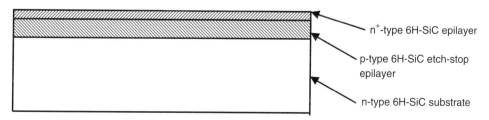

FIGURE 20.1 Typical SiC substrate with two epilayers for use in photoelectrochemical etching (PECE).

rates should be in the range of 1 μm/min. It has been shown [Okojie, 1996] that by utilizing electrochemical etching in dilute HF, cavities with depth greater than 100 μm in both p- and n-type 6H-SiC can be fabricated. In many cases of using dilute HF during ECE, a porous layer is formed in the process of the anodization. The texture of this layer depends on the concentration of HF in water and the voltage or current density applied. At low HF concentration, the anodization process leaves a soft, highly porous layer, while anodization at higher HF concentrations produces a hard layer of low porosity. Because these porous layers have large crystal surface areas, enhanced thermal oxidation occurs within it relative to the bulk. Subsequent stripping in conventional oxide etchant (such as buffered hydrofluoric acid, BHF) can easily remove all the porous material to reveal the etched cavity. In n-type SiC, ECE is possible only when the positive anode voltage (with the SiC as the anode electrode) is sufficiently high to cause the semiconductor space charge layer at the SiC/electrolyte interface to break down. Under this condition, electrons are injected from the SiC/HF interface into the bulk of the semiconductor, resulting in dissolution reactions at the surface. The voltage required for the breakdown depends on the doping level of the semiconductor and the concentration of the electrolyte. Two ECE experiments were conducted with n-type SiC: (1) anodization in a 0.625% HF in a two-electrode cell (SiC-anode with platinum counter-electrode) and subsequent thermal oxidation to remove the porous layer, and (2) anodization in a 0.625% HF in a three-electrode configuration (SiC-anode, platinum counter-electrode and reference standard calomel electrode) and oxidation to remove the porous layer. Both experiments demonstrated rapid etching of n-type SiC. In the first, the average etch rate was 0.3 μm/min, while in the second an etch rate of 0.8 μm/min was obtained. At average current densities ($J_{av} \sim 100$ mA/cm^2) used in both experiments higher etch rates should have been obtained if etching was the only process occurring. The actual etch rates obtained from the experiments suggested that other parallel processes were occurring (e.g., water decomposition) which used up part of the anodization current.

Materials must possess certain properties to be used as a mask in electrochemical etching. In dilute HF the mask should be inert in the concentrations used, and no electrolytic reaction should occur during anodization that might interfere with the etching process. Two masking materials were investigated. Platinum, which is very resistant to HF, generated gas bubbles at the high anodic potentials used, resulting in nonuniform etching. Silicon nitride did not form bubbles because it is an insulator, however, it is not completely resistant to HF. Because the process of deep etching to form a diaphragm takes a long time (200 min to etch a 200-μm-deep cavity at 1 μm/min), a certain minimum thickness of nitride is required to survive the dilute HF electrolyte during the anodization process in order to prevent the complete stripping of the nitride mask and consequent undesired etching of the entire SiC surface. The investigations briefly described above led to the following important experimental conclusions [Okojie, 1996]: At a fixed voltage of 12 V:

1. Etching in 2.5% HF occurs at higher rates than in 0.625% HF.
2. Etching in 2.5% HF produced deep etch pits and/or hillocks on the surface of SiC, while no etch pits or hillocks were observed on surfaces etched in 0.625% HF. The pits are attributed to the enhanced electrochemical etching around the micropipes in the SiC substrate.
3. The surface of the etched region was smoother and the depth more uniform when etched in 0.625% HF.

In an attempt to further optimize the etching process for forming n-type SiC diaphragms, an additional series of experiments was conducted with different etching parameters and concentrations of HF. Hillocks ranging in size from 10 to 50 μm formed randomly on the diaphragm surface. The presence of uncontrollable discontinuities of this size results in nonuniform stress distributions that adversely affect the operation of a sensor. Although the cause of the hillock formation was not conclusively determined, there was evidence that the formations were associated with bubbles generated during the high-rate etching. The existence of micropipes in the SiC wafers, which could lead to high current-density concentrations in localized areas, could result in selectively high etch rates around the defect sites, increasing the possibility of etch pit formation.

Although the etching potential of highly doped n-type SiC using the ECE process is much higher than that of p-type SiC, it is possible to stop the etching process at the n–p junction if the ohmic contact used for anodization potential control is made only to the n^+-SiC epilayer (refer to Figure 20.1). The positive anodic potential on the n^+-SiC layer will cause the junction to be reverse biased, thereby preventing the flow of current through the underlying p-SiC epilayer. In order for an etch-stop to be effective, the breakdown voltage of the n–p junction must be higher than the etching potential of the n^+-SiC epilayer. Also, the p-type layer should have a doping level significantly lower than that of the n^+-SiC epilayer to minimize the possibility of tunneling current. Details of this conductivity selective process can be found in Shor et al. (1993).

The process of utilizing PECE of SiC for the purpose of fabricating well-defined resistor structures is described below. As shown in Figure 20.1, the starting wafer is an n-type 6H-SiC substrate upon which a 5-μm-thick, lightly doped ($3 \times 10^{18} cm^{-3}$), p-type epilayer is grown by CVD, followed by a 2-μm-thick, n^+-type 6H-SiC epilayer. An ohmic contact metallization, preferably nickel, is deposited and patterned into a circular shape on the top n^+-SiC epilayer to enable anodization potential control during the PECE process. Platinum is sputtered on to the top of the wafer covering the patterned ohmic contact metal and the entire n^+-SiC epilayer. The platinum in direct contact with the epilayer is then patterned into the shape of serpentine resistor elements. This platinum acts as etch mask so that the serpentine resistor patterns can be transferred onto the n^+-SiC during the PECE process. Contact electrode wire is bonded on the section of the platinum mask that is in direct contact with the nickel ohmic contact. A thin layer of black wax is then applied over areas of the sample that will not be etched, including the contact electrode wires. The wafer is then immersed in the dilute HF electrolyte with the side to be etched facing up. This face is then exposed to a UV light source with the anodization potential set at 1.7 V_{SCE} (SCE stands for standard calomel electrode, which is the reference electrode against which the anode voltage is measured). Under this condition, the exposed sections of the n^+-SiC epilayer are photoelectrochemically etched while the sections under the serpentine-shaped platinum etch mask are unetched. After the anodization process, the wax is stripped in acetone while the platinum mask and nickel ohmic contacts are stripped in *aqua regia* and a 50:50 nitric/hydrochloric acid mixture, respectively. After stripping, the resistor patterns transferred to the n^+-epilayer are revealed. The current vs. time curves of a typical PECE are shown in Figure 20.2. For the epilayer thickness used, the photocurrent rises to a first maximum in the first 5 min, then drops rapidly due to the blocking action of the bubbles during the release of gaseous products. A second maximum

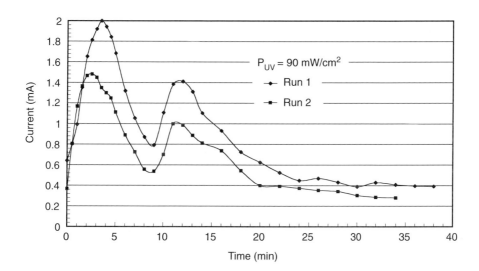

FIGURE 20.2 Current density vs. time during photoelectrochemical etching of n-type 6H-SiC in dilute HF electrolyte. Tow anodization I-t characteristics indicate the reproducibility of the process as long as ohmic contact is present.

appears in the curve after 12 min when the bubbles deflate, after which time the current density gradually decreased. The initial rise of the current is due to the rapid generation of electron–hole pairs (EHPs) during the initial stages of the etching process. These carriers are available near the surface of the SiC to participate in the chemical reaction process. A combination of various mechanisms leads to the gradual decay of the current as observed in Figure 20.2. As the etching progresses, the reacting species will need to diffuse through the porous SiC layer being formed to get to the bulk SiC surface just as the products of the reaction need to diffuse out. Therefore, as the porous SiC thickness increases, the mass transport process is slowed down, resulting in a reduced rate of reaction. The porous SiC will also shade the bulk SiC from the UV intensity, resulting in a drop in a the EHPs needed to keep the reaction going. Once the etching reaches the underlying p-type epilayer, the abrupt change in epilayer conductivity from n^+-SiC to that of the underlying p-type SiC causes the etching to stop. Because most of the electric field is confined within the n^+-SiC and that the applied 1.7-V_{SCE} anode potential causes the n–p junction to be reverse biased, only an insignificant leakage current will flow through the p-epilayer and be transported to participate in the etching process at the electrolyte interface. The current-time curves are repeatable between two etching runs, as can also be seen in Figure 20.2. It should be noted that the etch rates of the n^+-SiC epilayer are dependent on the UV intensity and the doping level of the epilayer. By increasing the UV intensity, which causes more EHPs to be generated, the n^+-SiC epilayer can be etched more rapidly and more selectivity between n- and p-type because fewer carriers are present at the p-epilayer to participate in the etching. SiC etch rates greater than 20 nm/min have been achieved. The wafer is thermally oxidized and then dipped in 49% HF to remove the porous layer that formed during anodization. After the process of selective etching, the n–p junction is sometimes not well isolated electrically, which leads to unsteady outputs during device operation. To prevent this from occurring, the etching time of the n^+ piezoresistor epilayer is increased so that the PECE process partially occurs in the p-epilayer. Poor electrical isolation can also be avoided if the junction isolation is verified immediately after PECE before the mask is stripped. Usually, a second thermal oxidation is carried out to ensure that the n–p junctions of all the devices are well isolated. A metal contact via is opened in the oxide by a conventional photolithographic process and BHF wet etching to expose sections of the resistors. Subsequently, ohmic contact metallization is deposited and patterned over the via to form the electrical contacts. This is followed by the deposition and patterning of the diffusion barrier to form the final device shown in the cross section of Figure 20.3a. The top view drawing of the sensor is shown in Figure 20.3b.

Another process that can be utilized to pattern n^+-SiC epilayers is the ECE. As stated earlier, when the small positive anodic potential is applied through the ohmic contact on the n^+-SiC, current cannot flow between the n^+-SiC epilayer and the underlying p-SiC substrate. Therefore, the p-type SiC epilayer acts as an etch-stop. However, experiments with ECE on the n^+-SiC epilayer demonstrated that by applying a reverse bias on the n–p junction, it is possible to achieve an etch-stop at the p-type SiC buried epilayer when the n^+-SiC top epilayer is being etched in the dark. It is critical that the junction leakage current be significantly low for this approach to work. However, etching selectivity may not be as high as in the PECE, as leakage current will increase for the voltage needed to perform a high etching rate with dark current.

After the platinum mask is removed, the surface of the underlying resistor is sometimes pitted. The pitting is the result of the pinholes in the platinum, which allow electrolyte to seep through the mask and etch the otherwise protected n^+-SiC surface. Therefore, it is important to ensure that the etch mask used for the pattern transfer is free of pinholes. Undercutting of the piezoresistors during etching may also occur as a result of the small dark current between the edge of the etching n^+-SiC and the electrolyte under bias of the anodic potential. Although this potential is made small enough to avoid reverse bias current flow between the n-type epilayer and the underlying p-type epilayer, lateral current conduction between the expanding side of the n-type SiC epilayer and the electrolyte allows undercutting etching to proceed during the PECE process. The above problems typically lead to undesirable nonuniformity in the resistor patterns. Using n-type epilayers with lower doping reduces dark etching at anodization potentials of about 1.7 V_{SCE} can minimize these effects. This is due to the fact that at that anodic potential fewer carriers from the dark current are injected into the electrolyte relative to the EHPs generated by the UV light source. However, this approach means the gauge factor and the temperature effects of the piezoresistor will change, as gauge factor is dependent on doping level. By substituting platinum with

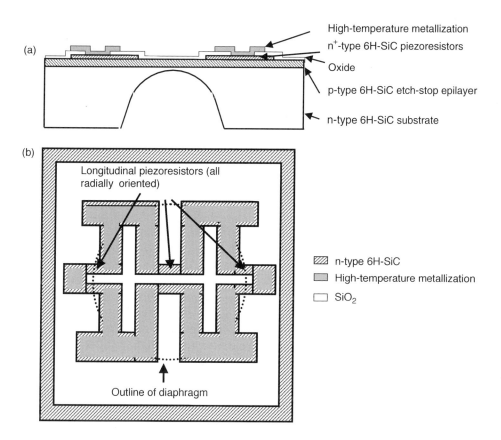

FIGURE 20.3 (a) Cross-section view of 6H-SiC after PECS of the top n^+-epilayer and ECE of the backside cavity. Notice the curvature in the cavity, which is characteristic of the ECE process. (b) Top view of patterned piezoresistors in n-type 6H-SiC.

another appropriate mask, such as polyimide or polyimide on silicon nitride, anodization reaction will only occur at the appropriate areas. The polyimide can prevent the pinhole formation while the silicon nitride can minimize the effect of undercutting. Polyimide is highly conformal and therefore will plug the pinholes. Silicon nitride is not conductive and is therefore electrically and chemical inactive during PECE. One effective method used to neutralize pinholes in platinum is by double-layer deposition. After the deposition of the first platinum layer, the film is sputter-etched, and subsequently a second platinum layer is deposited. This significantly reduces pinhole formation and pitting associated with the platinum etch mask. In many cases, the p-type SiC layer is not a fully effective etch-stop. This effect was observed in p-type SiC with low doping levels ($N_a \sim 10^{18} cm^{-3}$). Apparently, in lightly doped material, the electric field in the space-charge region is not high enough to prevent all the photogenerated carriers from reaching the surface to cause etching. In addition, the UV light incident on the n–p junction causes higher leakage currents across the junction than higher doped p-type SiC. Although the anodic voltage is applied only through the ohmic contact on the top n-type SiC epilayer, the light-induced current through the junction leads to etching of the p-type SiC. To avoid etching of the p-type SiC epilayer, the reference voltage (V_{SCE}) must be reduced to a level that curtails the drifting of photocarriers assisted by electric field when the p-epilayer is eventually exposed to the electrolyte. This fabrication procedure described above can be adopted to produce resistors in n-type epilayers with any doping level.

The characteristics of a diaphragm-based pressure sensor device are determined by the piezoresistors and by the dimensions of the diaphragm. Two key dimensions that characterize any circular diaphragm are thickness and radius. Because the radius is generally a fixed value determined by the pressure range and package specification, thickness is left as the main controlling variable. Therefore, the process of etching of the diaphragm to achieve a desired thickness and shape is of primary importance in diaphragm-based pressure-sensor technology. Until recently, the application of SiC diaphragms as pressure-sensing devices was almost nonexistent. Fortunately, SiC microfabrication technology has advanced significantly over the last decade, largely due to the ability to perform high-energy plasma etching of various types of structures. Some structures have been selectively etched in SiC using reactive ion etching (RIE), but the etch rates reported were too low for practical use in the wafer thinning and shaping [Palmour et al., 1987] needed for pressure-sensor fabrication. Rapid progress has been made in the area of deep reactive ion etching (DRIE) with etch rates up to 1 µm/min already reported.

20.3 Characterization of 6H-SiC Gauge Factor

Generally, the design of devices that sense physical phenomena and provide electrical readouts calls for the interpolation of two or more kinds of mathematical relationships. In the design of high-temperature pressure sensors, the equations that explain the physical phenomena (i.e., pressure and diaphragm deflection) are interpolated with the electrical equations that express the resulting output voltage. The equations that model deflecting diaphragms are classified in two main categories. One category models maximum diaphragm deflections less than the diaphragm thickness (linear case), while the other supports diaphragms with maximum deflections greater than the diaphragm thickness (nonlinear case). A fundamental reason for having separate expressions has to do with the problem of linearity of the membrane deflection. If the maximum deflection of the membrane is less than its thickness, as occurs in applications of short-range pressure measurement, there is generally a reasonable degree of linearity of diaphragm deflection in response to applied pressure. For wide-band pressure measurements, however, the deflection of the diaphragm in response to applied pressure is no longer linear. For the device to be used continuously over a long period of time, the membrane must be capable of repeatedly deflecting under applied pressure with precision and little hysterisis. To achieve this, the membrane must retain its elastic property after it is subjected to maximum applied pressures. To that effect, there is need to choose materials with an appreciable linear region on the stress-strain curve. For the diaphragm to retain its elastic integrity, the stress induced by pressure must not exceed the yield or fracture point. In essence, the maximum operating stress should be at a point below the yield/fracture stress limit. If the operating stress reaches the elastic or fracture limit, there is very strong likelihood the diaphragm would lose its elasticity, becoming permanently deformed, and/or fracture.

The choice of the resistance of the piezoresistors represents a compromise between several conflicting requirements. The pressure sensitivity of a bridge containing four active piezoresistors is directly proportional to the sheet resistance and the supply voltage. The temperature stability, on the other hand, is inversely proportional to the sheet resistance and the supply voltage. The requirement for good temperature stability conflicts with the requirements for pressure sensitivity.

The piezoresistive effect is associated with changes in resistance as a result of an applied mechanical stimulus. Based on this effect, strain gauges were developed. When an object is mechanically strained, the strain sensor changes its resistance. The resistance can be measured by incorporating the gauge within a Wheatstone bridge configuration and the resultant output voltage is related to the applied force. The piezoresistive effect is rather small in metals, but in semiconductors it is much more pronounced. The explanation can be found in Eq. (20.1) for the gauge factor:

$$GF = \frac{dR}{R\varepsilon} = 1 + 2\nu + \frac{d\rho}{\rho\varepsilon} \tag{20.1}$$

where GF = gauge factor; R = electrical resistance (Ω); ρ = electrical resistivity (Ω-cm) of the gauge material, ν = Poisson ratio (a geometrical effect), and ε = applied strain. The change in the resistance with strain is dependent on two terms: One is associated with the geometrical piezoresistive effect and the second originates from the strain dependency of the resistivity, as shown in Eq. (20.1). In metals, this latter term is zero, while it is significantly nonzero in semiconductors. In semiconductors, the contribution of the geometrical term to the gauge factor is of the same order as in metals. Therefore, the gauge factor in semiconductors is related to the large strain sensitivity of the resistivity. This can be explained on the basis of the energy-band structure of semiconductors. In n-type semiconductors that have multiple valleys in the conduction band, such as 6H-SiC, the piezoresistive effect is associated with a change in the relative energy positions of the multivalley minima under applied stress. This effect causes the electrons to transfer between the valleys, causing a net change in the mobility, which in turn has a dominant effect on the strain (stress) dependency of the resistivity [Herring and Vogt, 1956].

20.3.1 Resistor–Diaphragm Modeling

A resistor on a circular diaphragm arranged tangentially, with current flowing parallel to the resistor, will experience a longitudinal stress induced by the tangential strain component. It will also experience a transverse (radial) strain component (strain perpendicular to resistor length), which usually inserts a negative piezoresistance coefficient. On the other hand, as depicted in Figure 20.3b, the radially oriented resistor, with current flowing parallel along its length, will be dominated by the stress induced by the longitudinal strain component. The transverse effect is introduced via the tangential stress, with its corresponding negative piezoresistance coefficient. The above has been extensively verified and utilized in silicon. The output of the sensor is strongly affected by how the resistors are oriented. Therefore, the resistor geometry and orientation should be such that only one strain component exists while the other is suppressed.

When the maximum deflection, w, of a clamped circular plate is less than its thickness, the equation that describes it is expressed as [Timoshenko and Woinowsky-Krieger, 1959]:

$$w = 0.89 \frac{Pa^4}{64D} \tag{20.2}$$

where P is the applied pressure (Pa), a is the radius of the diaphragm (μm), and D is the flexural rigidity (N-m) of the membrane material. D is expressed as:

$$D = \frac{Et^3}{12(1-v^2)} \tag{20.3}$$

where E is the Young's modulus (Pa) and t = membrane thickness (m). The total stress on the membrane associated with such small deflection at the clamped edge is expressed as:

$$(\sigma_r)_r = \frac{3}{4} \frac{Pa^2}{t^2} \tag{20.4}$$

The choice of a large or small deflection equation, as stated before, is basically dictated by device application. A circular diaphragm can be easily mounted, and, in the case of materials with high elastic moduli, high pressures can be applied on diaphragms with reasonable diameter to thickness ratios. According to the theory of plates and shells by Timoshenko and Woinowsky-Krieger (1959), the radial and tangential stresses (σ_r and σ_t, respectively) at any point on the front side of a circular diaphragm with edges fixed can be related to the applied pressure, P, on the front of the diaphragm; its thickness, t; radius, a; and the distance, r, from the center of the point of interest as:

$$\sigma_r = -\frac{3}{4}\frac{Pa^2}{t^2}\left[(v+3)\frac{r^2}{a^2}-(v+1)\right] \tag{20.5}$$

and

$$\sigma_t = -\frac{3}{4}\frac{Pa^2}{t^2}\left[(1+3v)\frac{r^2}{a^2}-(v+1)\right] \tag{20.6}$$

The same set of equations can be used to define stresses on the back of the diaphragm, when pressure is applied there, but with negative sign. A simple analysis of the radial and tangential stress distribution on the front side of the diaphragm shows that both stresses change (as a result of applied pressure from the front side), at a certain distance from the center of the diaphragm, from compressive to tensile. When pressure is applied to the front side of the diaphragm, any piezoresistor on the back or on the front side of the diaphragm will be subjected to parallel and perpendicular stresses, depending on its location. Therefore, the functional relationship between the fractional change in electrical resistance ($\Delta R/R$) of the piezoresistor and the perpendicular and parallel stress components is given by:

$$\frac{\Delta R}{R} = \pi_t \sigma_\| + \pi_r \sigma_\perp \quad (20.7)$$

where π_t and π_r are parallel and perpendicular piezoresistive coefficients, respectively, and $\sigma_\|$ and $\sigma_\perp$ are parallel and perpendicular stress components, respectively. For tangentially and radially oriented piezoresistors on a circular diaphragm whose radius is large as compared with the resistor dimensions, this fractional change in resistance can be expressed as follows:
For tangential resistors:

$$\left(\frac{\Delta R}{R}\right)_t = \pi_t \sigma_t + \pi_r \sigma_r \quad (20.8)$$

For radial resistors:

$$\left(\frac{\Delta R}{R}\right)_r = \pi_t \sigma_r + \pi_r \sigma_t \quad (20.9)$$

In the case of semiconductors with hexagonal crystal structure (such as 6H-SiC) the problem is much more complicated in terms of resolving the piezoresistive constants in different directions. Earlier attempts by Russian scientists [Rapatskaya et al., 1968; Azimov et al., 1974; Guk et al., 1974a; 1974b] to characterize the piezoresistance of 6H-SiC as a function of crystallographic orientation yielded significant differences in the obtained values. Possible reasons for this discrepancy include the differences and imperfections of the Lely (1955) platelets, the only crystal available at that time, and also the variable quality of the metal ohmic contacts. Recent advances in SiC technology have led to the mass production of more reproducible, better quality SiC wafers and ohmic contacts, making it possible to obtain more reliable results.

In order to measure the gauge factor of 6H-SiC, resistors were etched in 6H-SiC epilayers (p- and n-type) by PECE means described in Section 20.2. The configuration of the resistors was such that two legs were transversely oriented while the other two were longitudinally oriented. The substrate was then diced into rectangular chips, each containing a pair of Wheatstone bridge circuits, one of which is shown in Figure 20.4a. The equivalent circuit of the Wheatstone bridge is shown in Figure 20.4b. It depicts one of each transverse and longitudinal piezoresistors arranged alternately. While all the resistors have current running parallel to their length, the transverse resistors, R_1 and R_4 will experience a perpendicular strain (strain perpendicular to current flow) while the longitudinal resistors, R_2 and R_3 will experience a parallel strain (strain parallel to current flow). The purpose of using two Wheatstone bridges on a single chip is to simultaneously compare the measured responses from both tensile and compressive states of the bridges. The rectangular chip is then attached to a machined metal diaphragm made of Incoloy^{TM}, as shown in the cross-section schematic of Figure 20.5. One Wheatstone bridge is at the edge of the metal diaphragm while the other one is at the push rod. Pressure was applied to the diaphragm through a back port (not shown), which caused the boss of the metal diaphragm to deflect the beam. One set of piezoresistors, closer to the center of the metal diaphragm, was placed in tension, while the other, closer to the periphery of the diaphragm, was in compression. The ΔR of each arm of the bridge was measured in order to obtain

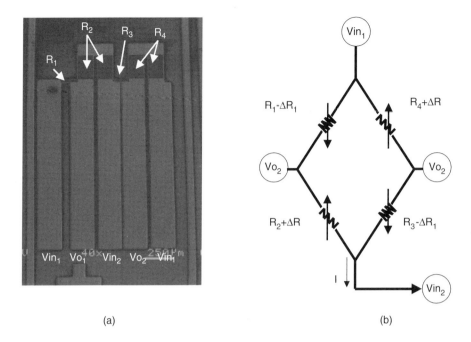

FIGURE 20.4 (a) Top view SEM picture of 6H-SiC beam with resistors, and (b) equivalent Wheatstone bridge configuration. R_1 and R_4 are the transverse piezoresistors, while R_2 and R_3 are the longitudinal piezoresistors.

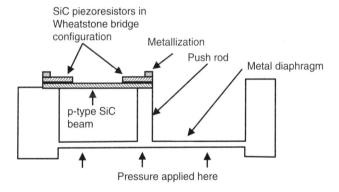

FIGURE 20.5 Integrated beam transducer used for gauge factor characterization.

the longitudinal and transverse gauge factors. All the beams were uniaxially deformed perpendicular to within 4° of the 6H-SiC crystal's basal plane, which is in the (0001) direction. Because the beam is integrated with the metal diaphragm, it is necessary to calculate the strain and stresses on the surface of the beam and to analyze their distribution across the beam. The problem is solved as a superposition of two systems, namely: (1) one with an edge-fixed diaphragm and (2) one with a beam, one edge of which is fixed (clamped) and the second guided. Deflection, w_p, at the center resulting from uniform loading (pressure) on the front of the metal diaphragm is expressed as:

$$w_p = \frac{Pa^4}{64D_m} \tag{20.10}$$

where D_m is the flexural rigidity (N-m) of the metal membrane material and is expressed as:

$$D = \frac{E_m t_m^3}{12(1-v_m^2)} \qquad (20.11)$$

where E_m represents the Young' modulus (Pa) of the diaphragm, t_m (m) is the metal diaphragm thickness and v_m is the metal Poisson constant. Deflection of the metal diaphragm, w_f, at the center resulting from a concentrated load at the center is

$$w_f = \frac{Fa^2}{16\pi D_m} \qquad (20.12)$$

where F (N) is the concentrated load or the contact force on the boss section of the metal diaphragm. The net deflection resulting from combined loading, when the concentrated load acts in the opposite direction of the applied pressure, is determined by:

$$w_{net} = w_p - w_f = \frac{a^2}{16 D_m}\left(\frac{Pa^2}{4} - \frac{F}{\pi}\right) \qquad (20.13)$$

Applying the Castigliano (1966) method solves the problem of deflection of a beam fixed (clamped) on one end and "guided" on the other (guided means that the slope at the guided point is always zero, but slight deflection is allowed and may change):

$$w_b = \frac{Fl^3}{12 E_b I_b} \qquad (20.14)$$

where w_b = beam deflection (m); l = length of the beam (m); E_b = modulus of elasticity of the beam (Pa); I_b = moment of inertia of the beam, expressed as:

$$I_b = \frac{bh^3}{12} \qquad (20.15)$$

where b = beam width (m) and h = beam thickness (m). During loading, the deflection of the beam will be equal to that of the diaphragm; therefore, Eqs. (20.13) and (20.14) can be set equal to solve for the contact force, F (which is contact force between the diaphragm boss and the beam). Because the radius of the diaphragm and the length of the beam are for all intents and purposes equal, the equation for F simplifies to:

$$F = \frac{3}{4}\left(\frac{Pa^2 \pi E_b bh^3}{4a E_m t_m^3 \pi + 3 E_b bh^3}\right) \qquad (20.16)$$

However, the maximum strains (ε_{max}) occur at either end of the beam and have opposite signs:

$$\varepsilon_{max} = \frac{Fat}{4 E_b I_b} \qquad (20.17)$$

The maximum stress, σ_{max}, at the edge of the beam can be calculated from Hook's Law:

$$\sigma_{max} = \varepsilon_{max} E_b \qquad (20.18)$$

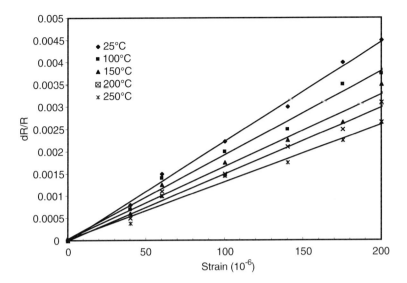

FIGURE 20.6 Normalized changes in resistivity vs. applied strain in longitudinal n-type 6H-SiC gauges with two doping levels.

Based on the dimensions of the bossed metal diaphragm and beam (diaphragm thickness, $t_m = 0.2$ mm; Young's modulus, $E_d = 207$ GPa; diaphragm radius, $a = 4$ mm; beam thickness, $h = 305$ μm; beam length, $l = 4$ mm; beam width, $b = 1.52$ mm; beam Young's modulus, $E_b = 488$ GPa), the strain in the SiC beam calculated from the dimensions of the metal diaphragm was approximately 1 nanostrain/Pa (~7 microstrains/psi). With the above modeling, it becomes possible to calculate the gauge factor of the 6H-SiC material, as there is a direct relationship between the applied strain and gauge factor, as indicated in Eq. (20.1).

The gauge factor and TCR of n-type 6H-SiC were analyzed in the basal (0001) plane. The characteristics of 6H-SiC piezoresistors were established individually in a four-arm Wheatstone bridge. The load on the beam was applied through pressure exerted on the metallic diaphragm, previously described, and transmitted by the boss acting as a push rod to the beam. Patterned strain gauges were fabricated in homoepitaxially grown 2-μm n-type epilayers on p-type 6H-SiC substrates with two n-type doping levels, namely $2 \times 10^{17} \text{cm}^{-3}$ and $3 \times 10^{18} \text{cm}^{-3}$. In hexagonal crystals, because the piezoresistance tensor is isotropic in the basal (0001) plane, the gauge may be rotated about the c-axis without affecting the piezoresistivity [Davis et al., 1988]. Longitudinal and transverse gauges were measured, yielding results corresponding to the piezoresistive coefficients π_{11} and π_{12}, respectively. Figure 20.6 shows normalized change in the resistance vs. applied strain in longitudinal n-type 6H-SiC gauges for the two n-type doping levels. The measured gauge factors were approximately –25 for the lower doping level and –20 for the higher doping level. Measurements conducted on the transverse piezoresistors ($N_d = 3 \times 10^{18} \text{cm}^{-3}$) yielded a gauge factor of approximately 11. The longitudinal and the transverse piezoresistance coefficients, π_{11} and π_{12}, can be calculated from the relationship [Rapatskaya et al., 1968]:

$$\frac{dR}{R} = \frac{dV}{V} = \pi_{11}\sigma \qquad (20.19)$$

where dV represents the bridge output of the Wheatstone bridge circuit, while V is the input voltage. By using the stress obtained from Eq. (20.18), the longitudinal gauge factor was found to be about -5.12×10^{-12} cm²/dyne for the SiC gauge with the lower doping level and -4.3×10^{-12} cm²/dyne for the SiC gauge with the higher doping level. One initial conclusion is that an increase in doping level results in a decrease of the gauge factor. Because the piezoresistive effect is an energy band transport phenomenon, factors such as impurities, donor ionization energy, mobility, defect density and overall quality of the crystal may influence it substantially. This behavior is similar to the one observed in silicon.

20.3.2 Temperature Effect on Gauge Factor

The piezoresistive properties of a 6H-SiC beam transducer were measured between 25 and 250°C. Measurements were carried out on a beam consisting of longitudinal and transverse n-type 6H-SiC (doping level $N_d = 2 \times 10^{19} cm^{-3}$) beams fabricated as described previously. Figure 20.7a shows the relative change in resistance of the longitudinal piezoresistors as a function of strain at different temperatures. At all temperatures a linear relationship is observed between $\Delta R/R$ and strain, but the strain sensitivity decreases with increasing temperature. Figure 20.7b shows the longitudinal gauge factor as a function of temperature. At 250°C the gauge factor is approximately 60% of its room-temperature resistance for this doping level. The bridge output as a function of pressure and temperature is shown in Figure 20.7c. The decrease in output due to increasing temperature is explained in terms of intravalley carrier transport, in which the external applied heat energy leads carriers to acquire more energy that enables more of them to be transported and occupy other energy minima. Therefore, when strain is applied under heat, only a few electrons can be transported to the energies yet occupied. As a result of the fewer electrons available for intravalley exchange, the piezoresistance decreases. The bridge gauge factor decreases linearly with temperature as seen in Figure 20.7b. The relative change in resistance vs. strain of the transverse piezoresistors is shown in Figure 20.8, from which the transverse gauge factor can also be calculated. In order to check the reproducibility of the measurements, another beam transducer structure was built and the bridge output as a function of pressure was measured. For comparison, Figure 20.9 shows the results obtained on both beams. In both cases, the dependence between the bridge output and the applied pressure is linear; however, one of the beams exhibited a slightly lower sensitivity. This was probably a result of either a geometrical factor (i.e., the metal diaphragm in both cases did not have exactly the same dimensions) and/or the mounting procedure. Random temperature variations made it difficult to measure the gauge factor of an individual bridge element, especially the transverse resistors, which exhibited very small changes in resistance with pressure.

20.3.3 Temperature Effect on Resistance

In addition to the gauge factor, another important consideration in the selection of a resistor for high- or low-temperature applications is the way its electrical resistance changes with the temperature. Resistance variation with temperature is usually expressed as a temperature coefficient of resistance (TCR), defined as:

$$\text{TCR} = \frac{1}{R_o} \frac{R_f - R_o}{T_f - T_o} \qquad (20.20)$$

where R_o = resistance at room or reference temperature (Ω), R_f = resistance at operating temperature, T_o = room or reference temperature (usually 25°C) and T_f = operating temperature. The TCR may be positive or negative and is usually given either in ppm/°C or in %/°C. Practically, the TCR can be influenced by resistor structure as well as by processing conditions such as uniformity of the resistivity across the wafer.

To evaluate the TCR behavior of SiC, the resistances of four individual gauges ($N_d = 2 \times 10^{19} cm^{-3}$) in a transducer were measured and plotted as a function of temperature (Figure 20.10). All measured resistances in this sample decreased with temperature up to 250°C, due to increasing ionization of the donors in the heavily doped SiC. In contrast, the initial resistance measurements carried out with the lower doped n-type 6H-SiC ($1.8 \times 10^{17} cm^{-3}$) decreased with temperature in the range from −60 to 25°C. Above 25°C, the resistance increased. Using Eq. (20.20), the average TCR value for the range 25 to 625°C was found to be 0.56%/°C. In the $3 \times 10^{18} cm^{-3}$ doped samples the resistance was observed to decrease

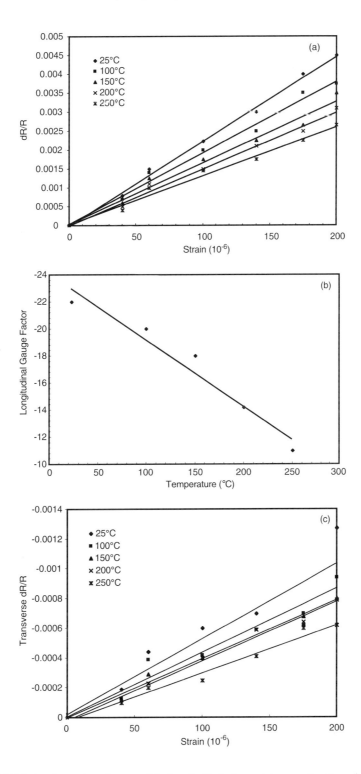

FIGURE 20.7 (a) Relative change in resistance of the longitudinal piezoresistors as a function of strain at different temperatures (n^+-6H-SiC, $N_d = 2 \times 10^{19}$ cm^{-3}). (b) Longitudinal gauge factor as a function of temperature. At 250°C the gauge factor is approximately 60% of its room-temperature value (n^+-6H-SiC, 2-μm-thick epilayer, $N_d = 2 \times 10^{19}$ cm^{-3}). (c) Net bridge output as a function of pressure at five different temperatures ($N_d = 2 \times 10^{19}$ cm^{-3}). Bridge input voltage is 5 V.

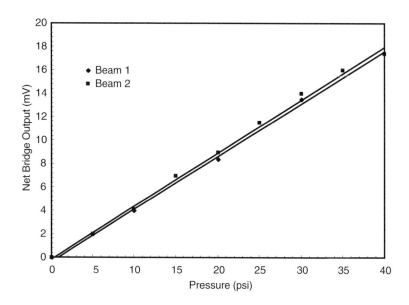

FIGURE 20.8 Relative change in resistance of the transverse piezoresistors as a function of strain at different temperatures ($N_d = 2 \times 10^{19}$ cm^{-3})

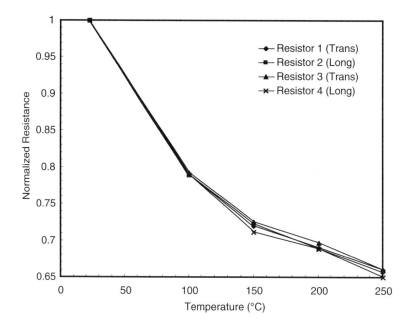

FIGURE 20.9 Net bridge output as function of pressure of two different beam sensors. In both cases the dependence is linear ($N_d = 2 \times 10^{19}$ cm^{-3}).

up to 100°C and then increase. The average TCR value for this sample in the range 100 to 625°C was found to be 0.28%/°C. The decrease in resistance with increase of temperature below a certain temperature limit, which typically lies between 0 and 25°C, is associated with the increasing ionization of dopant impurities. In this temperature range the semiconductor resistance is primarily controlled by carrier ionization. Once most dopant impurities have become ionized, carrier phonon-related lattice scattering

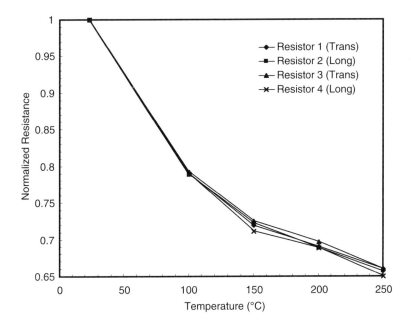

FIGURE 20.10 Change in normalized resistance of four individual gauges in a transducer as a function of temperature. All measured resistances decrease as the temperature increases (n^+-6H-SiC, $N_d = 2 \times 10^{19}$ cm^{-3}).

increases with the temperature to increase the resistance. This observed behavior is consistent with well-known semiconductor carrier statistics and carrier transport physics. In highly doped n-type SiC, the impurity ionization is completed at higher temperature due to the large number of impurities and its wide bandgap.

20.4 High-Temperature Metallization

There is growing demand for devices to operate in high-temperature environments beyond the capability of silicon and gallium arsenide technology. Most existing semiconductor electronic devices are limited to ambient temperatures below 200°C, primarily due to the degradation of the intrinsic properties of the associated materials (silicon and gallium arsenide). Silicon-on-insulator (SOI) technology was developed to extend device operation to about 300°C. However, long-term SOI reliability beyond this temperature remains unproven. SiC-based technology appears to be the most mature, wide-band-gap semiconductor material with the proven capability to function at temperatures above 500°C [Jurgens, 1982; Palmour et al., 1991]. However, the contact metallization of SiC typically undergoes severe degradation beyond this temperature due to enhanced thermochemical reactions and microstructural changes. The causative factors of contact failures include interdiffusion between layers, oxidation and compositional and microstructural changes. These mechanisms are potential device killers by way of contact failure. Liu et al. (1996) and Papanicolaou et al. (1998) have demonstrated stable ohmic contacts at 650°C for up to 3000 hours and 850°C for a short duration in vacuum. Vacuum aging is, however, not representative of the environmental condition in which SiC pressure-sensor devices are expected to operate.

In order to build any high-temperature electronic device, it is essential to fabricate ohmic contacts and diffusion barriers capable of withstanding the device operational temperatures. It is necessary to identify metals and alloys that form acceptable ohmic contact to 6H-SiC. For the ohmic contacts on n-type SiC, the metals include titanium and its alloys of nickel–titanium and titanium–tungsten. It has been shown [Zeller et al., 1987] that titanium contacts deposited on 3C-SiC withstand 20 hours at 650°C. Nickel is also known to be a good ohmic contact to n-type SiC but exhibits severe adhesion problems.

Nickel–titanium alloys may combine the favorable electrical properties of nickel with high reactivity and adhesion of titanium. Titanium–tungsten alloys exhibit good diffusion barrier properties and may also be ohmic.

Several research groups have also demonstrated the effectiveness of TiN as a diffusion barrier. This section will start with the examination of the use of TiN as a diffusion barrier in a multilayer metallization scheme. The disadvantages associated with this scheme will be discussed. A recently investigated alternative scheme that uses a $TaSi_2$ diffusion barrier will then be presented; this technique shows promising performance characteristics able to support stable device operation at 600°C in air.

20.4.1 General Experimental and Characterization Procedure

Several (0001)-oriented, Si-face, 6H-SiC substrates (tilt angle 3.5°) having n-type 1-µm-thick epilayers of different doping levels were used. The wafers were initially cleaned by modified RCA method and dipped in 49% HF for 5 sec, after which they were rinsed with deionized water and blow-dried with nitrogen. They were then thermally oxidized in dry oxygen ambient at 1150°C for 4 hr to yield an oxide thickness of about 60 nm. This oxide was then stripped in 49% HF, and the samples were rinsed and dried again. A second thermal oxidation for 5 hr at 1150°C was performed, yielding a cleaner thermal oxide layer. Photoresist was applied and patterned into circular patterns. Circular contact holes were then etched through the oxide with BHF to expose circular sections of the epilayer surface. The contact holes consisted of 12 rows, each row being made up of four circular contact holes of the same diameter, d, with 225-µm equidistant separations between their centers, as shown in Figure 20.11. The diameter of the contacts ranged from 6 to 28 µm. After stripping the photoresist, the samples were RCA cleaned again, but with no HF dip to ensure that a very clean epilayer surface with a monolayer of oxide is formed. The samples were immediately transferred into the sputtering chamber for metal depositions. A 300°C dehydration process in vacuum for 20 min followed to remove any water trapped within micropipes in the wafers.

The I–V characteristic measurement of epilayer contacts was conducted by probing two adjacent contacts from the same row. The contact resistivity was measured using the modified four-point method of Kuphal (1981) expressed as:

$$r_{cs} = \frac{A}{I_{AD}} \left[V_{AB} - V_{BC} \frac{\ln\left(\left(3\frac{s}{d}\right) - \left(\frac{1}{2}\right)\right)}{2\ln 2} \right] [\Omega\text{-cm}^2] \quad (20.21)$$

where V_{AB} is the voltage measured across contacts A and B in Figure 20.11 while V_{BC} measures the voltage between contacts B and C (epilayer resistance); A = contact area (µm); s = distance between adjacent contacts (µm); and d = diameter of the contact (µm). Current, I_{AD}, was passed through the circular contact pad, A, as shown in Figure 20.11, and out through pad D. Unless otherwise specified, the applied current was set at 1 mA. Measurements were made with four probes coming in contact with the Shockley

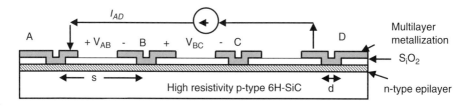

FIGURE 20.11 Cross-sectional view of multilayer metallization contact on n-type 6H-SiC epilayer for contact resistivity measurements.

pads that extend over the oxide. Therefore, the spreading and contact resistance, R_s and R_c, respectively, can be calculated by:

$$R_s + R_c = \frac{V_{AB} - V_{BC}}{I_{AD}} \quad (20.22)$$

The resistance of the probes in these measurements was negligible. Because the main parameter of interest in these measurements was the overall change in resistance under the contact, R_c and R_s were lumped together in determining the contact resistivity. As a result of lumping the spreading and contact resistances, the result obtained could be considered to be the high end of possible values of the specific contact resistivity.

20.4.2 Characterization of Ti/TiN/Pt Metallization

Several (0001)-oriented, highly resistive, Si-face, p-type 6H-SiC substrates each having n-type epilayers (1 μm thick) of different doping levels ranging between $3.3 \times 10^{17} cm^{-3}$ and $1.9 \times 10^{19} cm^{-3}$ were purchased from Cree Research, Inc. The wafers were initially cleaned by modified RCA method and dipped in 49% HF for 5 sec, followed by rinsing and blow-drying. An *ex situ* dehydration process followed this at 200°C in nitrogen ambient for 20 min to desorb water trapped within the micropipes. Depositions of Ti (50 nm)/TiN (50 nm)/Pt (100 nm) were made on the samples by sputtering without breaking vacuum. Titanium nitride was obtained by reactive sputtering of titanium in 20% nitrogen/argon ambient. The top platinum layer was etched in light *aqua regia* to form rectangular and circular probe pads that overlap the field oxide. The exposed TiN/Ti on the field oxide was selectively etched in 1:1 EDTA:H_2O_2 to electrically isolate it. The pads offered total coverage of the contact regions and facilitated broad area probe contact during testing.

In the as-deposited state the titanium contact on the n-type epilayer was ohmic for the sample with the highest doping level ($1.9 \times 10^{19} cm^{-3}$). The contact resistance using Eq. (20.21) was found to be approximately $1 \times 10^{-5} cm^{-2}$. In order to obtain ohmic contact to n-type 6H-SiC with lower doping levels (3.3×10^{17}–$10^{18} cm^{-3}$), high-temperature annealing was required.

The experimental results of the Ti/TiN/Pt ohmic contact are summarized in Table 20.2. The I–V characteristics of the as-deposited metallization on all samples were rectifying, except for the highest doped sample ($1.9 \times 10^{19} cm^{-3}$). After 30 to 60 sec of rapid thermal anneal at 1000°C in argon ambient, ohmic contact was achieved on all samples except for the lightest doped, which remained rectifying after 3.5 min of annealing.

The average barrier height before annealing was obtained from the forward I–V characteristic curve using the thermionic emission model:

$$J = J_s [e^{(qV/nkT)} - 1] \quad (20.23a)$$

TABLE 20.2 Summary Results of Electrical Characteristics of Ti/TiN/Pt Metallization on n-Type 6H-SiC Epilayers

Sample No.	Conc. (cm^{-3})	As-Deposited	Annealed	Total time (min)	SBH_{as-dep} (eV)	r_{cs} ($10^{-4} \Omega cm^2$)
A	3.3×10^{17}	Rectifying	Rectifying	3.5	0.84	—
B	1.4×10^{18}	Rectifying	Ohmic	.5	0.82	3.42
C	1.5×10^{18}	Rectifying	Ohmic	1	0.74	2.5
D	1.7×10^{18}	Rectifying	Ohmic	0.5	0.82	2.1
E	2.7×10^{18}	Rectifying	Ohmic	0.5	0.80	1.5
F	1.9×10^{19}	Ohmic	Ohmic	0.5	—	.15

where J is the forward current density (A/cm^2), V is the applied voltage, q is the electronic charge, k is the Boltzmann constant, T is the temperature (K) and n is the ideality factor that models the deviation from the theoretical ideal I–V characteristic depending on the integrity of the metal/epilayer interface. The saturation current density, J_s, is expressed as:

$$J_s = A^* T^2 e^{(-q\phi_B/kT)} \qquad (20.23b)$$

where A^* is the effective Richardson constant (Acm^{-2}K^{-2}) and ϕ_B (V) is the Schottky barrier height (SBH) between the metals in intimate contact with the 6H-SiC epilayer. The ideality factors before and after annealing ranged from 1 to 1.05 and J_s in the range of 9.44 × 10^{-8} to 4.4 × 10^{-3} (Acm^{-2}). In this work, the effective Richardson constant was estimated by:

$$A^* = 120(m_e^*/m_o)\,[\text{Acm}^{-2}\text{K}^{-2}] \qquad (20.23c)$$

where m_e^*/m_o is the ratio of the effective electron mass to the electronic rest mass. For a value of 0.45 m_o, A^* was calculated to be 54 Acm^{-2}K^{-2}. The obtained average SBH values ranged from 0.54 to 0.84 eV.

The ohmic contact obtained after annealing is believed to be due to the barrier-lowering effect caused by the change at the metal–SiC interface during annealing. The annealing is believed to cause the formation of low-work-function titanium carbide at the epilayer surface and changes in the density of surface states. The Auger electron spectroscopy (AES) depth profile of Figure 20.12a indicated distinct boundaries of the as-deposited metals on the SiC epilayer. However, in Figure 20.12b, intermixing and zone reactions after annealing are evident and a new layer consisting mainly of Pt, Ti, Si and C atoms is observed in direct contact with the epilayer. The synchronous tracking of Ti and C atoms at a constant ratio of almost 1:2 (discounting the primary Ti–N signal that was difficult to distinguish) strongly suggested the formation of a TiC species. Several groups had previously confirmed the formation of TiC$_{1-x}$ and Ti$_5$Si$_3$ species for wafers annealed in the range of 500 to 1200°C. The decrease in the SBH could be associated with the low work function of TiC (3.35 eV) compared to that of titanium (4.1 eV) [Smithells].

The contact resistivity plotted against the impurity concentration is shown in Figure 20.13, which exhibited the characteristic exponential dependence of contact resistivity on impurity concentration. In order to estimate the specific contact resistivity, r_{cs}, the spreading resistance, R_s, and the contact resistance, R_c, were decoupled. TiC was assigned R_s on the assumption that it was the new layer in contact with the 6H-SiC epilayer. Therefore, R_s was evaluated with respect to its thickness by using the method of Cox and Strack (1967):

$$R_s = \rho_{\text{TiC}}\frac{t}{\pi(d/2)^2} \qquad (20.24)$$

where d is the contact diameter (μm), ρ_{TiC} is the resistivity (μΩ-cm) of the assumed TiC layer and t is the thickness (~100 nm). Substituting for R_s with Eq. (20.24) in Eq. (20.22), we have:

$$\rho_{\text{TiC}}\frac{t}{\pi(d/2)^2} + R_c = \frac{1}{I_{AD}}[V_{AB} - V_{BC}] \qquad (20.25)$$

For a TiC resistivity of 200 μΩ-cm [Toth, 1971], the contact resistance, R_c, and the specific contact resistivity, ρ_c, was then evaluated. The values of obtained were such that $R_s \ll R_c$.

A comparison of the Figure 20.12 Auger depth profile of the annealed samples to the as-deposited samples reveals the reactions at the TiN/Pt and 6H-SiC/Ti interfaces. The oxygen content (at. 17%) between the former interface was an artifact of the deposition system. The degree of its effect on the

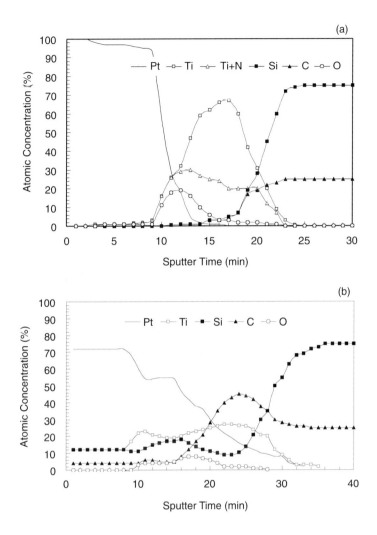

FIGURE 20.12 (a) Auger electron spectroscopy (AES) depth profile of as-deposited Ti/TiN/Pt metallization on n-type 6H-SiC. (b) AES of Ti/TiN/Pt after rapid thermal anneal at 1000°C for 30 sec in argon atmosphere. Synchronous 2:1 ratio tracking between titanium and carbon suggested the formation of TiC_{1-x} on the epilayer surface.

electrical characteristics was not known, but the 6H-SiC–Ti interface was relatively free of oxygen contamination. The surface of the top platinum layer exhibited a faint brown coloration after annealing, indicating the appearance of titanium species, which was confirmed by Auger surface spectral analysis.

The average contact resistivity ranging from 1.5×10^{-5} to 3.42×10^{-4} $\Omega\text{-cm}^2$ and a Schottky barrier height between 0.8 and 0.84 eV were in reasonable agreement with previously published results [Waldrop and Grant, 1993]. Interlayer delamination, sometimes attendant in the preprocessed titanium layer, was not observed. Auger electron spectroscopy revealed the out-diffusion of titanium–silicon species. The continued presence of the TiN layer after annealing at 1000°C suggested its partial survivability, but it did not offer a full barrier against platinum diffusion. In order for this multilayer high-temperature metallization to be applicable, the oxygen contamination in the metal must be kept to less than 3 at. % oxygen. High levels of oxygen contamination at the interfaces of Pt–TiN and/or TiN–Ti could be the result of full or partial decomposition of the titanium nitride layer at high temperature and replacement by a layer of titanium oxide due to the high affinity of titanium toward oxygen. Formation of titanium oxide results in two deleterious effects: (1) it greatly reduces the effectiveness of the diffusion barrier and (2) it forms a dielectric layer, leading to rectification and failure of the ohmic contact. Another destructive

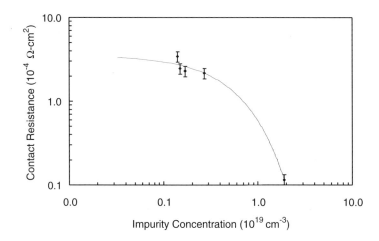

FIGURE 20.13 Bulk contact resistance as function of doping level.

effect is penetration of oxygen through the outer platinum layer. Inspection with scanning electron microscopy (SEM) indicated that the deposited platinum layer contained a high density of pinholes. At high temperature, oxygen would diffuse through the pinholes, thereby degrading the titanium nitride diffusion barrier. Oxygen also reacts with the titanium beneath the barrier, forming a solid titanium oxide layer with undesirable rectifying properties that extends to the 6H-SiC surface. For the metallization scheme to work effectively in air, the issue of oxygen contamination must be resolved. There is a need to investigate other schemes that would be more durable in air.

20.4.3 Ti/TaSi$_2$/Pt Scheme

This section discusses an ongoing effort to develop high reliability contact metallization schemes capable of supporting SiC sensors and electronics operating within a temperature range of 500 to 600°C in air. Oxygen contamination generally poses a big problem for the Ti/TiN/Pt scheme, but the risk is reduced if the device with such metallization is hermetically sealed. However, the diffusion barrier integrity of the TiN layer against platinum diffusion to the semiconductor interface is also undesirable. A robust high-temperature metallization scheme must have at a minimum the following attributes: (1) Ohmic contact with reasonably low contact resistance relative to the bulk epilayer, (2) long-term contact stability in the harsh environment, (3) compatibility with SiC large-scale integrated electronics fabrication technology, (4) good wirebond strength and (5) compatibility with high-temperature interconnect and packaging technology.

In order to meet the criteria mentioned above, it was necessary to identify metallization schemes that would both form an ohmic contact on n-type SiC and at the same time offer an excellent diffusion barrier against oxygen penetration as well as preventing migration of any top-layer metallization such as platinum toward the contact SiC interface. In addition, such a scheme should have a top surface that is essentially wire-bondable. In developing this new scheme, thermodynamic and thermochemical issues were taken into consideration with the recognition that at 600°C the activation energies of several metals are enough to promote reactions or intermixing between metals. Metal layers with low mutual diffusivities were identified in order to keep intermixing at a minimum. In the case where they mix, however, the alloys formed must be thermodynamically, mechanically and electrically stable. The scheme must maintain excellent diffusion barrier characteristics. By combining the ability of titanium to form ohmic contact on n-type SiC, the diffusion barrier characteristics of TaSi$_2$ and the relative stability of the interface of the two layers, a new scheme was developed with a result that proved far superior to the Ti/TiN scheme.

A sequential deposition of Ti (100 nm)/TaSi$_2$ (200 nm)/Pt (300 nm) multilayer contacts was performed in a three-gun, UHV/load–lock, sputtering system. Details of the deposition parameters are shown in

TABLE 20.3 Process Parameters for Deposition of Ti/TaSi$_2$/Pt

Layer	Thickness(nm)		Power (W)	Deposition Conditions Gas Flow (sccm)	Time (min)	Deposition Method
Ti	100	6 mTorr	200 R.F.	50 Ar	16.5	Sputtering
TaSi$_2$	200	6 mTorr	100 R.F.	50 Ar	33.3	Sputtering
Pt	300	9 mTorr	75 D.C.	50 Ar	6.3	Sputtering

Table 20.3. A 2-µm aluminum layer was deposited by e-beam evaporation and used as an etch mask during reactive ion etching to pattern the multiplayer metallization into Shockley probe pads over the contacts. Following the RIE, the etch mask was selectively removed with an aluminum etchant to expose the underlying platinum layer. The specific contact resistance, ρ_{cs}, was calculated using the modified four-point probe measurement method described earlier.

The specific contact resistances of the sample sets as a function of time at 500 and 600°C in air atmosphere are shown in Figures 20.14a and 20.14b, respectively. The samples treated at 500°C, shown in Figure 20.14a, exhibited higher contact resistance values initially which remained high for the first 40 hr. The contact resistance dropped after then and practically remained constant for the entire period between 70 and 600 hr. The results of the samples tested at 600°C in air for 150 hr are shown in Figure 20.14b. The specific contact resistance also increased after an initial 30-min anneal at 600°C in forming gas, but subsequent heat treatments in air, however, saw a nearly exponential decrease in the contact resistance values that appeared to taper off after 100 hr.

There is an obvious difference in the contact resistance values between the two sample sets in the first 40 hr, as depicted in Figure 20.14. This observed differences could perhaps be attributed to one or a combination of three things: (1) oxygen contamination of samples treated at 500°C in air, (2) the probable existence of surface states and (3) incomplete reaction product formation at the SiC interface after the initial 30-min anneal at 600°C in forming gas, which might be accelerated by the subsequent heat treating at 600°C. Because the heat treatments in air between the sample sets are different, variations in activation energies may cause differences in product formation, thereby leading to variations in electrical characteristics. After the contrast in results for times less than 100 hr between both sample sets, the average specific contact resistance for both sets leveled off to values around $2-3 \times 10^{-4}$ Ω-cm^2.

To begin understanding the active mechanisms, we examined the relationship between the electrical and thermochemical characteristics within the context of diffusion barrier formation and interfacial reactions. The Auger profiles after annealing at 600°C for 30 min in forming gas and after 50 hr at 500°C in air are shown in Figures 20.15a and 20.15b, respectively. Generally, the two figures are similar in terms of phenomenological changes taking place within the layers. Figure 20.15a shows the unidirectional migration of silicon preferentially into platinum and toward the surface. This is of significant importance as it forms the basis for the diffusion barrier characteristics of the metallization. This migration creates a silicon-depleted zone inside the metallization, as depicted in Figure 20.15b. The AES profiles show a build-up of silicon, at a nearly consistent platinum-to-silicon ratio of 2:1 within the platinum layer. Because we anticipate stable titanium silicide and titanium carbide as the reaction products between titanium and SiC, no new source of silicon exists that will migrate to the surface. Although no strong indications of titanium carbide and titanium silicide signals were observed at the epilayer interface, the extension of the contact boundary a few nanometers into the epilayer strongly suggests an underlying physical and chemical reaction.

The representative I–V characteristics after various conditions for both sets of samples are shown in Figure 20.16. The I–V curve shown in Figure 20.16a is that of the as-deposited condition, which was the same for both sample sets. The observed weak rectification could be related to low-level oxidation issues previously discussed in this chapter. The entire sample set exhibited linear I–V characteristics after annealing at 600°C in H$_2$ (5%)/N$_2$ forming gas for 30 min, as shown in Figure 20.16b. The slight curvature observed was attributed to the ongoing reaction at the SiC interface. Figure 20.16c shows that

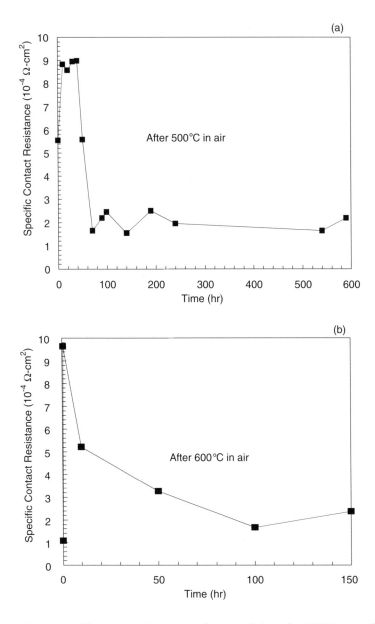

FIGURE 20.14 (a) Average specific contact resistance as a function of time after 500°C in air. The high contact resistance in the first few hours may be attributed to conditions stated in the text. (b) Average specific contact resistance as a function of time after 600°C in air. The high contact resistance in the first few hours may be attributed to condition sated in the text.

the representative I–V characteristic of the sample set after 630 hr at 500°C in air remained practically unchanged relative to Figure 20.16b. The observed slight change in the I–V slope was an artifact of the oxidation process occurring on the Shockley pads and necessitated scratching the surface with the probe tips to get good electrical contact between the pad and the probe tip. For the sample set treated at 600°C in air, the slope of the I–V characteristic after 150 hr, shown in Figure 20.16d, compares very well with that of Figure 20.16b and c, given that they were all from the same wafer (net epilayer doping level, $N_d = 2 \times 10^{19} cm^{-3}$). This result is an indication of the improved thermal stability of the new SiC interface.

An understanding of these chemical reactions can be drawn from works by Bellina and Zeller (1987) and Chamberlain (1980). The temperatures in which we performed the heat treatments were in a range

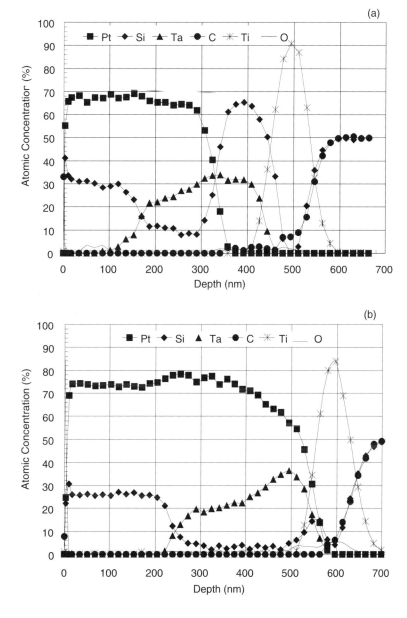

FIGURE 20.15 (a) Auger depth profile of 6H-SiC/Ti/TaSi$_2$/Pt after heat treatments at 600°C anneal in H$_2$ (5%) /N$_2$ for 30 min. (b) Auger depth profile of 6H-SiC/Ti/TaSi$_2$/Pt after heat treatments at 500°C in air for 50 hr.

consistent with the work of Chamberlain, who identified the formation of titanium carbide and silicide at similar temperatures. Applying Chamberlain's parabolic reaction rate relation gives an approximation of the minimum thickness of reaction products:

$$x^2 = x_o + 2K(t - t_o) \, [\text{cm}^2] \quad (20.26a)$$

$$K = K_o \exp\left(\frac{-262.7 \text{ KJmol}^{-1}}{RT}\right) [\text{cm}^2 \text{s}^{-1}] \quad (20.26b)$$

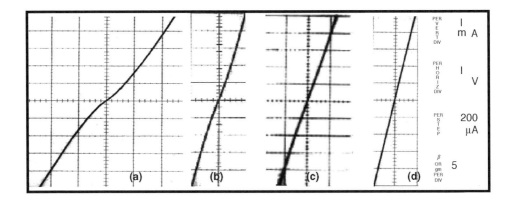

FIGURE 20.16 I–V characteristics of samples measured at different stages: (a) as-deposited, (b) 600°C annealed in H_2 (5%)/N_2 forming gas for 30 min, (c) sample treated at 500°C in air for 630 hr and (d) sample treated at 600°C in air for 150 hr. The current and voltage scales are the same for all I–V plots.

where k_o is the reaction rate constant ($cm^2 s^{-1}$); K is the temperature dependence of the rate constant; x_o and t_o are initial distance and time constants, respectively, which in our case were set to zero; and R is the universal gas constant (8.314 J-$mol^{-1}K^{-1}$). Without correction for pressure, a layer of no less than 2 nm of new products would have formed at the epilayer interface after 30 min of anneal at 600°C. While in-depth analysis is ongoing, thermodynamic and reaction-limited kinetics are tentatively proposed to be acting in sequence, such that the decomposition of $TaSi_2$ in the presence of platinum proceeds with a final reaction:

$$4Pt + TaSi_2 \rightarrow Ta + 2Pt_2Si \qquad (20.27)$$

If this reaction proceeds to the right, the limiting reactant would be silicon, as it is fully consumed by platinum. This proposition is made more likely given that the heat of formation of Pt_2Si ($-\Delta H_f \sim 10$ kcal mol^{-1}) is less than that of $TaSi_2$ ($-\Delta H_f \sim 28$ kcal mol^{-1}) [Andrews and Phillips, 1975].

It is important to examine the oxygen concentrations in Figures 20.15a and b. Both parts of the figure show very little migration of oxygen into the metallization. This implies a much more stable contact structure than the Ti/TiN/Pt approach [Okojie et al., 1999], which was affected by significant oxidation of the metals in the contact. Various groups have extensively studied the oxidation kinetics of the silicides. Murarka (1988) concluded that the oxidation mechanism for most silicides essentially have the same heats of formation as normal oxides. Lie et al. (1984) and Razouk et al. (1982) confirmed that the oxidation kinetics of tantalum disilicide has a parabolic rate. This implies that it would take an appreciable length of time for oxygen to diffuse through the entire contact.

20.5 Sensor Characteristics

On the backside of the wafer, a circular cavity mask was aligned to each set of piezoresistors that form a Wheatstone bridge network. Electrochemical etching as previously described was used to fabricate the circular diaphragm cavities. The irregularity of the sidewalls and base of the cavity indicates that a consistent control of the electrochemical etching process is required. This batch of sensors had a diaphragm thickness of about 50 µm and chip area of 1.48 mm^2. The above process was followed by a 20-hr wet oxidation at 1150°C to ensure complete p–n junction isolation and passivation of the active elements.

Contact holes were etched through the oxide using BHF to expose sections of the resistor elements. This was followed by *in vacuo* sputter deposition of the Ti/$TaSi_2$/Pt high-temperature metallization and patterning of the metallization by the method described earlier. The wafer was then diced into chips and individually mounted on specially designed pressure-sensor headers. Gold wires were bonded from the sensor to the header pins to facilitate external electrical connection.

TABLE 20.4 Performance Characteristics of 6H-SiC Pressure Sensor[a]

Temperature (°C)	Characteristics	Sensor 5	Sensor 6	Sensor 10
23	Full-scale output (mV)	41.7	39.51	66.42
	Linearity (%)	0.604	0.42	0.90
100	Full-scale output (mV)	37.5	34.47	57.23
	Linearity (%)	0.40	0.8	0.005
	TCGF (%/°C)	−0.13	−0.17	−0.20
	TCR(%/°C)	−0.24	−0.23	−0.23
200	Full-scale output (mV)	32.2	27.92	47.5
	Linearity (%)	0.93	0.86	0.12
	TCGF (%/°C)	−0.13	−0.17	−0.16
	TCR(%/°C)	−0.17	−0.17	−0.17
300	Full-scale output (mV)	28.2	24.56	36.54
	Linearity (%)	0.39	0.86	0.93
	TCGF (%/°C)	−0.12	−0.14	−0.16
	TCR(%/°C)	−0.12	−0.12	−0.12
400	Full-scale output (mV)	21.77	19.61	28.33
	Linearity (%)	0.95	0.30	0.41
	TCGF (%/°C)	−0.13	−0.13	−0.15
	TCR(%/°C)	0.04	0.04	0.05
500	Full-scale output (mV)	18.11	15.73	25.04
	Linearity (%)	0.57	0.20	0.12
	TCGF (%/°C)	−0.12	−0.13	−0.13
	TCR(%/°C)	0.07	0.07	0.06
600	Full-scale output (mV)	18.69	10.66	23.00
	Linearity (%)	0.65	0.94	0.16
	TCGF (%/°C)	−0.10	−0.13	−0.11
	TCR(%/°C)	0.07	0.07	0.08
23	Full-scale output (mV)	41.5	39.46	66.34

[a] Epilayer doping level $N_d = 2 \times 10^{19}$ cm^{-3}.

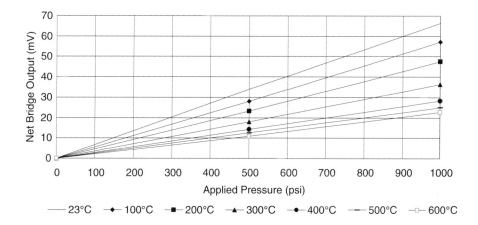

FIGURE 20.17 Net bridge output voltage of 6H-SiC pressure sensor as function of pressure at various temperature regimes.

The performance characteristics of the three sensors shown in Table 20.4 are representative of multiple batch fabrication and diaphragm thickness. The room-temperature net output voltage as a function of applied pressure at various temperatures is shown in Figure 20.17 for sensor 10. With a bridge input of 5 V, the full-scale output (FSO) was 66.42 mV at room temperature for an applied pressure of 1000 psi, indicating a sensitivity of 0.013 mV/V/psi. Very low hysterisis of 0.7% FSO and nonlinearity of −0.9% FSO were obtained. The 600°C output of 25.04 mV indicated a 62% output drop from the room-temperature value.

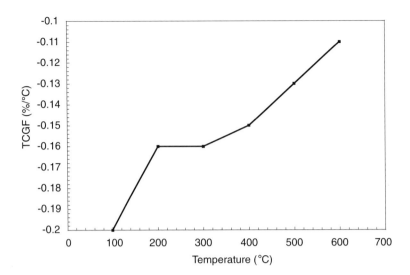

FIGURE 20.18 Temperature coefficient of gauge factor of 6H-SiC (calculated over 100°C increments) as function of temperature (epilayer doping level, $N_d = 2 \times 10^{19}$ cm^{-3}).

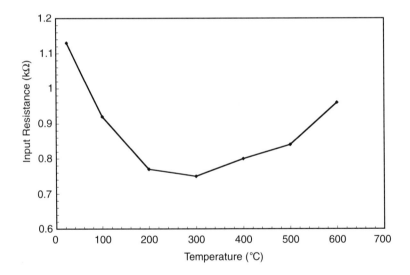

FIGURE 20.19 Bridge resistance of 6H-SiC piezoresistive pressure sensor as function of temperature.

The characterization of the gauge factor, described in Section 20.3, showed a linear drop in gauge factor with increased temperature. The output was observed to decrease as temperature increased, but became gradually insensitive to temperature as temperature approached 600°C. Keyes (1960) had previously predicted this behavior in silicon. The temperature coefficient of gauge factor (TCGF), a measure of the output sensitivity to temperature, is defined here as:

$$\text{TCGF} = \frac{1}{V_{(T_O)}} \frac{V_{(T)} - V_{(T_O)}}{T - T_O} 100 \ [\%/°C] \tag{20.28}$$

where $V_{(T_O)}$ and $V_{(T)}$ are the full scale outputs at room temperature and final temperature. The TCGF (calculated over 100°C increments) shown in Figure 20.18 indicated an initial pronounced sensitivity that approached smaller (less negative) values as the temperature increased. The TCGF response is expected to

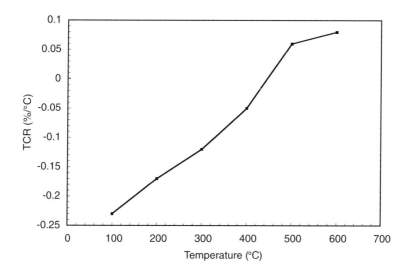

FIGURE 20.20 Temperature coefficient of resistance of 6H-SiC (calculated over 100°C increments) as function of temperature (epilayer doping level, $N_d = 2 \times 10^{19} \text{cm}^{-3}$).

be lower in magnitude for doping levels greater than $2 \times 10^{19} \text{cm}^{-3}$. The effect of temperature on the resistance is shown in Figure 20.19. It indicates a gradual decrease from a room-temperature bridge resistance value of 1.13 kΩ to about 750 Ω at 300°C due to carrier ionization. The upward swing of the resistance is associated with the growing dominance of lattice scattering mechanism [Streetman, 1990]. From this result, the temperature coefficient of resistance from Eq. (20.20) was calculated over 100°C increments and is shown in Figure 20.20. The negative TCR characteristic, relative to the room-temperature resistance, was consistent with the n-type 6H-SiC epilayer of this doping level ($2 \times 10^{19} \text{cm}^{-3}$). For more heavily doped crystals, the negative TCR will extend to higher temperatures, thereby allowing for a less complex compensation scheme.

20.6 Summary

It is noteworthy that the development of SiC sensor technology is fundamentally motivated by the need to perform instrumentation in extreme environments of temperature, vibration and harsh chemical media. Accurate measurement in such environment therefore requires a new generation of sensors that can survive in such environments. In that respect, this chapter has presented critical technology issues that have been investigated in recent years as part of the effort toward realizing a new generation of microelectromechanical systems (MEMS) in single-crystal silicon carbide. Three important areas were addressed, namely, electrochemical etching methodologies for resistors and diaphragm structures, piezoresistance characterization in terms of strain and pressure, and high-temperature metallization to support sensor and electronic operation at high-temperatures. Photoconductive selectivity as a method to fabricate structures on either n-type or p-type SiC was demonstrated. This principle was then utilized to fabricate piezoresistors in n-type epilayers by applying photoelectrochemical etching. The p-type epilayer beneath the n-type epilayer served as an etch-stop. The newly developed process for resistor fabrication can be applied with minor adjustments to fabricate resistors with any n-type doping level. This work also demonstrated preferential oxidation of porous SiC to facilitate good pattern definition and fast removal of residues. The thinning and cavity etching to form the diaphragm were carried out by applying dark etching on the back side of the n-SiC wafer. The resulting cavities are relatively free of etch pits and hillocks. The average etch rates were found to be 0.6 to 0.8 μm/min. Thermally stable Ti/TaSi$_2$/Pt ohmic contact was demonstrated on n-type SiC, which is predominantly the conductivity of choice for piezoresistive SiC sensors. On the basis of these efforts, a first-generation, batch-microfabricated,

6H-SiC diaphragm-based piezoresistive pressure sensor was produced. Efforts are underway by various groups to overcome existing technological challenges that act as barriers against global commercialization. Issues such as temperature compensation, packaging and overall cost reduction are currently being pursued.

Acknowledgments

The bulk of this work was funded by NASA Glenn Research Center (then NASA Lewis Research Center) under a Phase II SBIR contract NAS3-27011 awarded to Kulite Semiconductor Products, Leonia, NJ, and monitored by Dr. Lawrence G. Matus. My appreciation goes to Dr. Anthony D. Kurtz, the chairman of Kulite, for thrusting upon me the challenge to implement the objectives of this project, the bulk of which culminated in my doctoral thesis. My gratitude goes to Dr. William N. Carr for giving me the opportunity to be his graduate student. At Israel Institute of Technology, Technion, many thanks go to Drs. Ben Z. Weiss and Ilana Grimberg for their technical support. Also at Kulite, the full support of Alex N. Ned is greatly appreciated. I am grateful to the entire staff of department 200 with special reference to Gary Provost (formerly at Kulite) for metallization/dielectric depositions and other equipment support, to Mahesh Patel for support in testing the sensors and to Scott Goodman for material support in testing the sensors. Here at NASA Glenn Research Center, where some more recent advanced work in $Ti/TaSi_2/Pt$ metallization has been performed, I thank the various efforts of the technician cadre and Dr. Phillip G. Neudeck for his critical review of this chapter.

References

Andrews, J.M., and Phillips, J.C. (1975) *Phys. Rev. Lett.* **35**, p. 56.
Azimov, S.A., Mirzabaev, M.M., Reifman, M.B., Uribaev, O.U., Khairullaev, Sh., and Shashkov, Yu. M. (1974) *Soviet Phys. Semicond.* **8**(11), pp. 1427–1428.
Beheim, G., and Salupo, C.S. (2000) "Deep RIE Process for Silicon Carbide Power Electronics and MEMS," in *Proc. of the MRS 2000 Spring Meeting*, May 24–28, San Francisco, CA, MRS Proc. 622, paper T8.9.
Bellina, J.J., Jr., and Zeller, M.V. (1987) "Novel Refractory Semiconductors," in *Mater. Res. Soc. Symp Proc. 97*, eds. D. Emin, T.L. Aselage, and C. Wood, p. 265, Pittsburgh, PA.
Berg, J. von, Ziermann, R., Reichert, W., Obermeier, E., Eickhoff, M., Krötz, G., Thoma, U., Cavalloni, C., and Nendza, J.P. (1998) "Measurement of the Cylinder Pressure in Combustion Engines with a Piezoresistive β-SiC-on-SOI Pressure Sensor," *Technical Proc. 4th Int. High Temperature Electron. Conf.*, pp. 245–249.
Carrabba, M.M., Li, J., Hachey, J.P., Rauh, R.D., and Wang Y. (1989) *Electrochem. Soc. Extended Abstr.* **89–1**, pp. 727.
Castigliano, A. (1966) *The Theory of Equilibrium of Elastic Systems and Its Applications*, Dover, New York.
Chamberlain, M.B. (1980) *Thin Solid Films* **72**, pp. 305–311.
Cox, R.H., and Strack, H. (1967) "Ohmic Contacts for GaAs Devices," in *Solid-State Electronics*, Vol. 10, Pergamon Press, Elmsford, NY.
Davis, R.F., Sitar, Z., Williams, B.E., Kong, H.S., Kim, H.J., Palmour, J.W., Edmond, J.A., Ryu, J., Glass, J.T., and Carter, C.H., Jr. (1988) *Mater. Sci. Eng.* **B1**, pp. 77–104.
Herring, C., and Vogt, E. (1956) *Phys. Rev.* **101**, p. 944.
Huff, M.A., Nikolich, D., and Schmidt, M.A. (1991) "A Threshold Pressure Switch Utilizing Plastic Deformation of Silicon," in *Transducers '91, Int. Conf. on Solid State and Actuators, Digest of Technical Papers*, p. 177.
Guk, G.N., Usol'tseva, N.Ya., Shadrin, V.S., and Mundus-Tabakaev, A. F. (1974a) "Piezoresistance of α-SiC under Hydrostatic Pressures," *Soviet Phys. Semicond.* **8**(3), pp. 406–407.
Guk, G.N., Lyubimskii, V.M., Gofman, E.P., Zinov'ev, V.B., and Chalyi, E.A. (1974b) "Temperature Dependence of the Piezoresistance π_{11} of n-Type SiC(6H)," *Sov. Phys. Semicond.* **9**, p. 104.

Jurgens, R.F. (1982) *IEEE Trans. Ind. Electron.* **1E-29**(2), pp. 107–111.
Keyes, R.W. (1960) *Solid State Physics*, Vol. 11, Academic Press, New York.
Kuphal, E. (1981) "Low Resistance Ohmic Contacts to n-Type and p-InP," *Solid State Electron.* **24**, pp. 69–78.
Lely, J.A. (1955) *Bericht Deutsche Keram. Gesel.* **32**, p. 229.
Lie, L.N., Tiller, W.A., and Saraswat, K.C. (1984) "Thermal Oxidation of Silicides," *J. Appl. Phys.* **56**(7), pp. 2127–2132.
Liu, S., Reinhardt, K., Severt, C., Scofield, J., Ramalingam, M., and Tunstall, C., Sr. (1996) "Long-Term Thermal Stability of Ni/Cr/W Ohmic Contacts on N-Type SiC," in *Proc. 3rd Int. High Temp. Electron. Conf.*, pp. VII (9–13).
Mehregany, M., Zorman, C.A., Rajan, N., and Wu, C.H. (1998) "Silicon Carbide MEMS for Harsh Environments," *Proc. IEEE*, **86**(8), pp. 1594–1609.
Murarka, S.P. (1980) "Refractory Silicides for Integrated Circuits," *J. Vacuum Sci. Technol.* **17**(4), pp. 775–792.
Murarka, S.P. (1988) *J. Vacuum Sci. Technol.* **17**(4), pp. 775–792.
Okojie, R.S. (1996) "Characterization and Fabrication of α(6H)-SiC as a Piezoresistive Pressure Sensor for High Temperature Applications," Ph.D. thesis, New Jersey Institute of Technology, Newark.
Okojie, R.S., Ned, A.A., Kurtz, A.D., and Carr, W.N. (1996) "α(6H)-SiC Pressure Sensors for High Temperature Applications," in *Proc., 9th Int. Workshop on Micro Electro Mechanical Systems*, pp. 146–149.
Okojie, R.S., Ned, A.A., and Kurtz, A.D. (1997) "Operation of Alpha 6H-SiC Pressure Sensor at 500°C," *Solid-State Sensors and Actuators, 1997. Transducers '97* **2**, pp. 1407–1409.
Okojie, R.S., Ned, A.A., Kurtz, A.D., and Carr, W.N. (1999) "Electrical Characterization of Annealed Ti/TiN/Pt Contacts on N-Type 6H-SiC Epilayer," *IEEE Trans. Electron. Devices* **46**(2), pp. 269–274.
Palmour, J.W., Davis, R.F., Astell-Burt, P., and Blackborow, P. (1987) "Science and Technology of Microfabrication," in *Proc. Mat. Res. Soc.*, eds. R.E. Howard, E.L. Hu, S. Namba, and S.W. Pang, p. 185.
Palmour, J.W., Kong, H.S., Waltz, D.G., Edmond, J.A., and Carter, C.H., Jr. (1991) "6H-Silicon Carbide Transistors for High Temperature Operation," in *Proc. 1st Int. High Temp. Electron. Conf.*, pp. 229–236.
Papanicolaou, N.A., Edwards, A.E., Rao, M.V., Wickenden, A.E., Koleske, D.D., Henry, R.L., and Anderson. W.T. (1998) "A High Temperature Vacuum Annealing Method for Forming Ohmic Contacts on GaN and SiC," in *Proc. 4th Int. High Temp. Electron. Conf.*, pp. 122–127.
Pearson, G.L., Read, W.T., and Feldman, W.L. (1957) *Acta Met.* **5**, p. 181.
Rapatskaya, I.V., Rudashevskii, G.E., Kasaganova, M.G., Iglitsin, M.I., Reifman, M.B., and Fedotova, E.F. (1968) "Piezoresistance Coefficients of n-Type α-SiC," *Sov. Phys. Solid State* **9**(12), pp. 2833–2835.
Razouk, R.R., Thomas, M.E., and Pressaco, S.L., (1982) "Oxidation of Tantalum Disilicide/Polycrystalline Silicon Structures in Dry O_2," *J. Appl. Phys.* **53**(7), pp. 5342–5344.
Shor, J.S., Zhang, X.G., and Osgood, R.M. (1992) *J. Electrochem. Soc.* **139**, pp. 12–13.
Shor, J.S., Okojie, R.S., and Kurtz, A.D. (1993) *Institute of Physics Conference Series*, No. 137, Chap. 6, pp. 523–526.
Smithells, C.J., ed. *Metals Reference Book*, 5th ed., Butterworth & Co., London.
Streetman, B.E. (1990) *Solid State Electronic Devices*, 3rd ed., Prentice-Hall, Englewood Cliffs, NJ.
Timoshenko, S., and Woinowsky-Krieger, S. (1959) *Theory of Plates and Shells*, 2nd ed., McGraw-Hill, New York.
Ting, C.Y., and Wittmer, M. (1982) "The Use of Titanium-Based Contact Barrier Layers in Silicon Technology," *Thin Solid Films* **96**, pp. 327–345.
Toth, L.E. (1971) *Transistion Metals, Carbides and Nitrides*, Academic Press, New York.
Waldrop, J.R., and Grant, R.W. (1993) "Schottky Barrier Height and Interface Chemistry of Annealed Metal Contacts to Alpha 6H-SiC: Crystal Face Dependence," *Appl. Phys. Lett.* **62**(21), pp. 2685–2687.
Zeller, M.V., Bellina, J., Saha, N., Filar, J., Hargraeves, J., and Will, H. (1987) *Mat. Res. Soc. Symp. Proc.* **97**, pp. 283–288.

21
Deep Reactive Ion Etching for Bulk Micromachining of Silicon Carbide

Glenn M. Beheim
NASA Glenn Research Center

21.1 Introduction 21-1
21.2 Fundamentals of High-Density Plasma Etching 21-2
21.3 Fundamentals of SiC Etching Using
 Fluorine Plasmas 21-3
21.4 Applications of SiC DRIE: Review 21-5
21.5 Applications of SiC DRIE: Experimental Results 21-5
21.6 Applications of SiC DRIE: Fabrication
 of a Bulk Micromachined SiC Pressure Sensor 21-9
21.7 Summary 21-11

21.1 Introduction

It is often desired to insert microsensors and other microelectromechanical systems (MEMS) into harsh (e.g., hot or corrosive) environments. Silicon carbide (SiC) offers considerable promise for such applications, because SiC can be used to fabricate both high-temperature electronics and extremely durable microstructures. One of the attractive characteristics of SiC is the compatibility of its process technologies with those of silicon, which allows for the co-fabrication of SiC and silicon MEMS. However, a very important difference in the processing of these semiconductors arises from the chemical inertness of SiC, a characteristic that makes it attractive for use in corrosive environments but also makes it very difficult to micromachine.

Realization of the full potential of SiC MEMS will require the development of a set of micromachining tools for SiC comparable to the tool set available for silicon. Micromachining methods are generally classified as bulk, in which the wafer is etched, or surface, in which deposited surface layers are patterned. Surface micromachining methods for deposited SiC layers have been developed to a high level [Mehregany et al., 1998]. Silicon carbide can be readily etched to the required depths of just several microns using reactive ion etching (RIE) processes [Yih et al., 1997]. Further work remains to be done, however, in developing RIE processes with greater selectivity for SiC. Current RIE processes lack the selectivity needed to etch a SiC layer entirely through while minimally modifying an underlying silicon or silicon dioxide layer. This limitation has motivated the development of a micromolding method in which SiC is deposited into molds formed by RIE of silicon or silicon dioxide [Yasseen et al., 1999].

The emphasis here is bulk micromachining of SiC, for the fabrication of SiC microstructures with vertical dimensions from approximately 10 μm to several hundred microns. Three methods for bulk micromachining of SiC have been developed: electrochemical etching, micromolding and deep reactive ion etching (DRIE). Each of these has an important role to play in SiC MEMS fabrication.

Electrochemical etching of SiC can provide the high rates (>1 μm/min) needed for economical deep etching [Shor and Kurtz, 1994]. In addition, electrochemical etching provides a high degree of selectivity to other materials, and it allows for the use of p–n junction etch-stops for precise depth control. Etch directionality, however, is poor, resulting in significant undercutting of the mask. Highly directional (or anisotropic) etch processes, which provide vertical etch sidewalls, are needed to accurately transfer the mask pattern deeply into the wafer.

Bulk micromolding methods have been developed to circumvent the limitations of available deep-etch processes for SiC [Lohner et al., 1999]. Here, silicon molds are fabricated using highly developed silicon DRIE processes. The molds are filled with polycrystalline SiC using chemical vapor deposition (CVD) and then removed by wet etching. Thick, finely featured SiC microcomponents have been fabricated using this approach. The molded polycrystalline SiC, however, does not have the excellent mechanical and electronic properties of single-crystal SiC. In particular, deep etching of single-crystal SiC wafers is required if high-quality SiC electronics are to be integrated with thick SiC microstructures.

Conventional parallel-plate RIE is not well suited for deep etching (i.e., etch depths greater than about 10 μm) of SiC because it provides low etch rates, high rates of mask erosion and relatively poor directionality (although superior to that which can be achieved using electrochemical etching). The use of advanced high-density-plasma (HDP) reactors can alleviate all these difficulties. High-density plasma has enabled the DRIE process for silicon, and its usefulness for deep etching of SiC is becoming well established.

21.2 Fundamentals of High-Density Plasma Etching

The advantages of high-density plasma for DRIE are primarily the results of a more highly reactive plasma, which provides for increased etch rates, and a lower operating pressure, which improves directionality by minimizing the scattering of ions. There are several types of HDP systems, the most common being electron cyclotron resonance (ECR) and inductively coupled plasma (ICP). Inductively coupled plasma has come to dominate the HDP market [Bhardwaj and Ashraf, 1995], in large part due to lower complexity and cost (ICP uses RF frequencies, while ECR uses microwaves to generate the plasma). The focus here will be on etching with ICP, although ECR etching shares many of the same characteristics.

Inductively coupled plasma has several important advantages relative to conventional RIE. First, ICP can have a plasma density (ions per unit volume) which is two orders of magnitude higher than that of conventional RIE. This produces a much greater flux of ions and reactive species to the substrate surface. Also, ICP etching is performed at lower pressures, which helps minimize bowing of the etch sidewalls caused by scattering of ions. Lower pressures also facilitate the transport of etchants and etch products into and out of narrow trenches. In addition, the low operating pressure helps to provide smoother surfaces, as sputtered mask materials are less likely to be scattered back onto the etched surface, which can cause micromasking and the formation of "grass" or other etch residues. Another important advantage of ICP is the independent control it provides for the plasma density and the energy with which ions bombard the substrate. This allows for the adjustment of the chemical and mechanical components of the etch process to give a satisfactory trade-off between etch rate and the erosion rate of the mask. The ratio of the substrate and mask etch rates, the mask selectivity, is an important parameter for deep etching as it determines the maximum depth that can be etched with a given mask thickness.

A schematic of an inductively coupled plasma etching system is shown in Figure 21.1. Typically, a fluorinated gas, such as SF_6, is supplied to the reactor at a controlled rate. A high throughput pumping system maintains the system at a low pressure and enables a high gas flow to rapidly replenish depleted etchants. A plasma is generated by supplying RF power to a coil wrapped around the ceramic walls of the chamber. Power supplied to the coil produces a time-varying axial magnetic field inside the chamber. This induces an azimuthal electric field, which accelerates the electrons to high energies. The circumferential electric field helps to confine the electrons to the plasma, which results in the much higher plasma

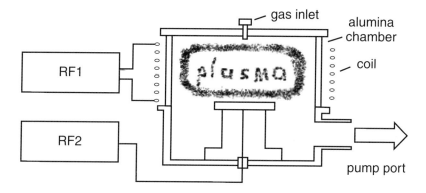

FIGURE 21.1 Schematic of an inductively coupled plasma (ICP) etching system. The RF generators are labeled RF1 and RF2.

densities characteristic of ICP. Because confinement of the electrons increases the probability that an electron will undergo an ionizing collision with a gas molecule before it leaves the plasma, the ionization rate is increased. The plasma can therefore be sustained at lower pressures (as low as 1 mTorr), which increases the mean path length the electrons travel between collisions. This enables the acceleration of the electrons to higher energies, which increases the likelihood that a collision with a gas molecule will break the molecule apart. An ICP reactor, therefore, is very effective at dissociating a relatively inert fluorinated feed gas, such as SF_6, to produce high concentrations of atomic fluorine and other highly reactive radicals.

At the same time the substrate is subjected to a high flux of reactive radicals, it can be bombarded with energetic ions. A second generator supplies RF power to an electrode, onto which the wafer is clamped. Upon application of the AC voltage, electrons and ions are alternately attracted. The less massive electrons are considerably more mobile so the electrode acquires a negative charge, which gives it a time-averaged negative potential with respect to the plasma. This negative bias potential acts to repel electrons, preventing further net accumulation of charge. The potential gradient between the plasma and substrate electrode causes ions drifting out of the plasma to be accelerated across the dark space (or sheath) between the plasma and substrate. Typically, the ions are accelerated to energies from several tens of electronvolts to several hundred electronvolts. The damage which these ions cause to the substrate can dramatically increase the etch rate—for example, by creating highly reactive dangling bonds. The ions strike the substrate at normal incidence, provided they do not undergo scattering collisions with gas molecules in the sheath. Inductively coupled plasma enables highly anisotropic etching, as the low operating pressures of ICP ensure that ions strike only the horizontal wafer surfaces and are not scattered to the etch sidewalls.

21.3 Fundamentals of SiC Etching Using Fluorine Plasmas

The use of fluorine-based plasma etch chemistries is most easily implemented, as, unlike chemistries based on other halogens such as chlorine, it allows the use of nontoxic feed gases such as SF_6. The plasma dissociates the relatively inert fluorinated gas (e.g., SF_6) to produce highly reactive radicals such as atomic fluorine. For DRIE of silicon, a fluorine etch chemistry is almost always used, in part because of convenience, but primarily because fluorine provides the high rates needed for economical deep etching. Fluorine plasmas are widely used in SiC etching, also. Limited experimentation with chlorine plasma etching of SiC has yielded relatively low etch rates [Wang et al., 1998].

The reactivity of SiC in a fluorine plasma contrasts strongly with that of silicon. Exposure to atomic fluorine produces rapid isotropic etching of silicon, as shown in Figure 21.2. Silicon reacts spontaneously with atomic fluorine, producing volatile etch products (e.g., SiF_4 and SiF_2) which rapidly desorb from

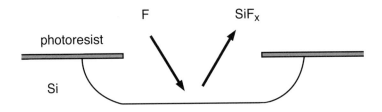

FIGURE 21.2 Etching of silicon in a fluorine plasma, using a photoresist etch mask.

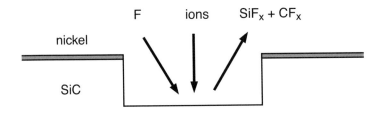

FIGURE 21.3 Etching of SiC in a fluorine plasma, using a nickel etch mask.

the surface. Etching proceeds laterally as well as vertically, which results in an isotropic profile. In contrast, the etch rate of SiC when exposed to atomic fluorine is extremely low, unless energy is supplied to the surface. In reactive ion etching, this energy is provided by ion bombardment, which results in an anisotropic etch profile, as shown in Figure 21.3. Energetic ions damage the SiC lattice (e.g., breaking Si–C bonds) which promotes reactions with atomic fluorine to produce volatile SiF_x and CF_x species. If the ions are well collimated, only the horizontal surfaces are etched, producing the desired anisotropy.

In fluorine-based DRIE of silicon, the required anisotropic profile is produced by modifying the process to cause the formation of a passivating layer on the etch sidewall. Most commonly, the reactor is programmed to switch back and forth between etching of silicon and deposition of an etch-resistant fluorocarbon polymer. Low energy ion bombardment is used during the etch step to remove the polymer from the horizontal surfaces, while leaving it intact on the etch sidewalls. Proper balancing of the etch and passivation steps produces a vertical sidewall, which at high magnification in a scanning electron microscope has a scalloped appearance.

In DRIE of SiC, the low reactivity of the substrate makes the etch profile inherently anisotropic. However, the low reactivity of SiC creates a number of other problems, including low etch rate, low selectivity to the mask and a tendency for the formation of residues on the etched surfaces. The net result is that DRIE of SiC is decidedly more difficult than DRIE of silicon. It is the high reactivity of silicon with fluorine that enables the high rate, photoresist etch masks and tunable sidewall slope of the DRIE process for silicon. Because the silicon DRIE process uses low ion energies, it provides a high selectivity with respect to photoresist, the most economical etch mask. Also, the silicon DRIE process provides excellent control of the etch profile, as the sidewall can be adjusted from outward to inward sloping by simply increasing the duration of the etch step relative to that of the passivation step. With inherently anisotropic etch processes, such as DRIE of SiC, it may prove more difficult to obtain the same high aspect ratios (etch depth divided by minimum feature size) that can be realized in silicon.

The high ion energies needed to etch SiC at acceptable rates make it impractical to use photoresist as an etch mask, as the etch rate of photoresist under these conditions is not much different from that of SiC. Nickel and indium–tin–oxide (ITO) are widely used because these materials do not form a volatile reaction product with fluorine and so are etched only by sputtering.

The formation of residues on the etched surface is a significant problem in SiC etching. The highly energetic ions cause significant erosion of the etch mask, and sputtered mask materials that deposit onto the SiC surface act as micromasks. Because the etch process is highly anisotropic, such micromasks are not undercut and pillar-like features and other residues can result. High pressures increase the likelihood of residue formation, as sputtered mask materials are more likely to be scattered back onto the SiC surface. Also, the use of aluminum masks, rather than nickel or ITO, has been found to promote the formation of etch residues.

21.4 Applications of SiC DRIE: Review

Throughput is a key economic factor, and much work has focused on the development of high-rate etch processes for SiC. Various types of HDP reactors have been used to demonstrate SiC etch rates greater than 1 µm/min, which compares fairly well with the 2-µm/min rates typical of silicon DRIE. In SiC, however, the attainment of such a high etch rate necessitates highly energetic ion bombardment, which dramatically increases the sputter erosion of the etch mask. This can severely limit the etch depths that can be obtained using practically feasible mask thicknesses. In addition, high-rate etching often does not provide the smooth etched surfaces needed, due to the increased tendency for micromasking.

Several groups have demonstrated deep reactive ion etching of SiC. In one case, a parallel-plate RIE was used to etch a via through an 80-µm-thick substrate at a rate of 0.23 µm/min [Sheridan et al., 1999]. The feed gas was NF_3, and the chamber pressure was relatively high, 225 mTorr, which provided for a high concentration of reactive species. The high operating pressure also resulted in relatively low ion energies, which minimized sputtering of the mask. The selectivity to the mask was only 25, and thick electroplated nickel–alloy masks were used for deep etching of SiC. Anisotropy was quite good for the roughly 100-µm-diameter via, but scattering of ions will limit the aspect ratios that can be achieved at such a high pressure.

Etching of 4H-SiC was reported using a helicon HDP reactor [Chabert et al., 2000]. The feed gas was SF_6 and 25% O_2. It was stated that the addition of O_2 did not affect the etch rate but prevented the deposition of sulfur compounds on the chamber walls. A 50-µm-thick nickel sheet was used as a shadow mask to etch a via through a 330-µm-thick 4H-SiC substrate. The wafer was etched through in approximately 6 hr, giving an etch rate of 0.9 µm/min. The selectivity with respect to the nickel mask was 40:1. Etch rates as high as 1.35 µm/min were obtained, but rough surfaces were produced at these high rates.

Using an ICP reactor with SF_6 and 33% O_2, holes were etched to a depth of 97 µm in 4H-SiC, at a rate of 0.32 µm/min [Cho et al., 2000]. A cross-sectional scanning electron micrograph showed that the bottom of the etched hole was relatively clean. The low operating pressure of 2 mTorr may have been the key to providing clean etched surfaces despite the use of an aluminum mask, which often results in micromasking. Selectivity with respect to the mask was reported to be greater than 50.

Using an ICP reactor with SF_6, very smooth etched surfaces were obtained for an etch depth of 45 µm, and a via was etched through a 100-µm-thick 6H-SiC wafer [Beheim and Salupo, 2000]. At a rate of 0.3 µm/min, the selectivity with respect to a nickel mask was 80. Since the presentation of these results, this laboratory has made substantial progress in SiC DRIE, as will be described in the next section.

21.5 Applications of SiC DRIE: Experimental Results

This section will present experimental results for DRIE of SiC using an inductively coupled plasma reactor (STS Multiplex ICP). The reactor has an automated load-lock and a 1000-L/sec turbopump with a dry backing pump. Helium is supplied to the backside of the electrostatically clamped wafer to provide effective heat transfer to a water-cooled chuck. The SiC substrates, which range in size from 10 mm^2 to 50-mm diameter, are attached to 100-mm silicon carrier wafers using a thin layer of photoresist. The use of a sacrificial carrier wafer (which etches at approximately 2 µm/min) helps to minimize roughness caused by the sputtering of etch-resistant materials onto the SiC surface.

Liftoff of electron-beam evaporated films is a convenient means of fabricating both nickel and ITO etch masks. Liftoff of ITO masks as thick as 3.5 µm was readily accomplished, while film stresses limited the evaporated nickel masks to thicknesses no greater than 2500 Å. For extremely deep etching, nickel masks with thicknesses up to 15 µm were fabricated by selective electroplating.

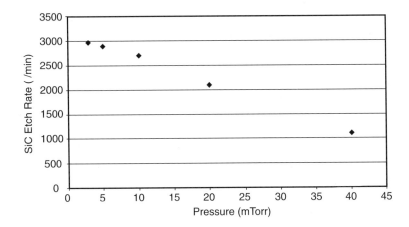

FIGURE 21.4 Etch rate of n-type 4H-SiC as a function of pressure for 800-W coil power, 75-W platen power and 100% SF_6.

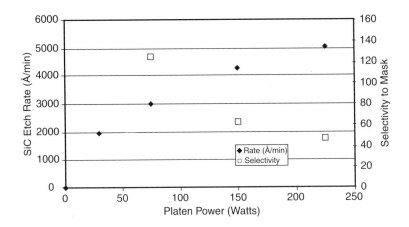

FIGURE 21.5 Etch rate of n-type 4H-SiC and selectivity to a nickel mask as functions of platen power for 800-W coil power, 5-mTorr pressure, and 100% SF_6.

A series of experiments was performed to determine the effects of various process parameters on the SiC etch rate and selectivity to the mask. During these experiments, the coil power was 800 W and the feed gas was 100% SF_6. The gas flow was maintained as high as practical, given the constraints of the pumping system and the mass flow controller. The 10-mm^2 n-type 4H-SiC substrates were masked using evaporated nickel.

Figure 21.4 shows the etch rate of 4H-SiC as a function of pressure for a fixed platen power of 75 W. The highest etch rate of 0.30 μm/min was obtained at a pressure of 5 mTorr. The etch rates for 4H-SiC are not significantly different from those reported for the 6H polytype [Beheim and Salupo, 2000]. The dopant type also has been found to have little effect on the etch rate.

Figure 21.5 shows the SiC etch rate and the selectivity to the nickel mask (ratio of the SiC and Ni etch rates) as functions of the power supplied to the substrate electrode or platen. The platen power determines the kinetic energy of the ions which bombard the substrate. Selectivity was found to decrease with increasing ion energies, due to increased sputtering of the mask. For a platen power of 75 W, the etch rate was 0.30 μm/min and the selectivity to the nickel mask was 125.

An investigation was performed to determine the effects of the various parameters on the surface morphology of deeply etched features. It was found that 5-mTorr pressure and 75-W platen power, in conjunction with 800-W coil power and 55-sccm SF_6, provide a deep etch that is satisfactory for many applications.

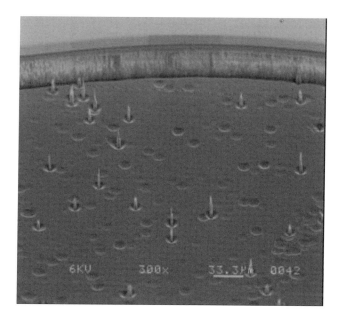

FIGURE 21.6 6H-SiC (n-type) etched to a depth of 45 µm using the baseline 100% SF_6 process. The SiC substrate was cleaned in hot sulfuric acid prior to etching. The electroplated nickel etch mask has not been stripped.

Relative to these baseline process parameters, higher platen powers give lower selectivity to the mask and higher pressures promote residue formation, while lower pressures cause increased trenching at the base of the sidewall.

The morphology of the etched surface was found to be strongly influenced by the cleanliness of the surface at the start of the etch process. A number of cleaning procedures (solvents, hot sulfuric acid etch, oxidation followed by etch in hydrofluoric acid) were tried, but none was found satisfactory. Figure 21.6 shows results typical of those obtained by cleaning a 6H-SiC substrate using standard methods prior to loading it in the reactor. After etching to a depth of 45 µm, the etched surface is covered with dimples and spike-like residues.

An *in situ* plasma cleaning procedure was found to be effective at providing smooth surfaces for etch depths up to roughly 75 µm. Figure 21.7 shows the result when the SiC surface was sputter cleaned in argon immediately prior to a 45-µm-deep etch using the baseline process. A disadvantage of the argon sputter clean is the significant mask erosion that it causes. The 10-min sputter etch in argon was found to remove 1 µm of ITO, or 2500 Å of nickel, while removing only 800 Å of SiC. An oxygen plasma cleaning process, with moderately energetic ion bombardment (75W platen power), was found as effective as the argon sputter cleaning, while causing much less mask erosion. In the case of ITO masks, however, the oxygen plasma pretreatment was found to leave etch-resistant residues on the SiC surface, unless the openings in the mask were quite wide (>100 µm).

When the SiC surface is plasma cleaned prior to deep etching, the etched surface is initially smooth, but with increasing etch depth, first dimples and then rough residues gradually appear. For etch depths less than roughly 75 µm the surfaces are generally smooth, but for greater depths the etched surfaces become increasingly rough. Figure 21.8 shows a 160-µm-deep etch of a p-type 6H-SiC substrate, which was oxygen-plasma cleaned prior to the start of the baseline etch process. Auger analysis of the rough residues showed small quantities of aluminum. The residues are apparently the result of micromasking caused by aluminum sputtered from the vacuum chamber. The rough residues are found only if dimples are also present. The surface of a dimple may react more readily with aluminum than the untextured SiC surface, thereby trapping aluminum to produce a micromask. If no dimples are present, aluminum may remain mobile on the surface until it is sputtered off, thereby causing no roughening of the surface. Experimentally, residual hydrocarbons and water were found to greatly increase the density of dimples,

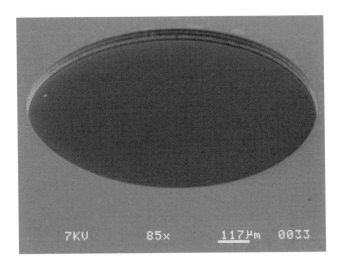

FIGURE 21.7 A 1-mm-diameter well etched to a depth of 45 μm in n-type 6H-SiC using the baseline 100% SF$_6$ etch process, after an *in situ* sputter cleaning in argon. The ITO etch mask has been stripped.

FIGURE 21.8 A 1-mm-diameter well etched to a depth of 160 μm in p-type 6H-SiC using the baseline 100% SF$_6$ etch process, after an *in situ* oxygen plasma clean using energetic ion bombardment. The ITO etch mask has been stripped.

in the absence of *in situ* plasma cleaning prior to etching. As determined by X-ray photon spectroscopy (XPS), the etched SiC surface was found to be covered with a thin (<10 Å) fluorocarbon layer, which mediates the etch reaction. Hydrogen is known to reduce the thickness and passivating properties of such a fluorocarbon layer on an etched silicon surface, so it is feasible that hydrogen-bearing contaminants may cause the local enhancement of etch rate which results in the dimples observed here. The texture-inducing contaminants are apparently found in the background vacuum of the chamber, as the dimples gradually appear with continued etching, despite an initial *in situ* cleaning of the surface.

Smooth surfaces, even for extremely deep etching, were obtained by diluting the SF$_6$ etchant gas with argon to cause continuous sputter cleaning of the surface throughout the etch process. In a series of

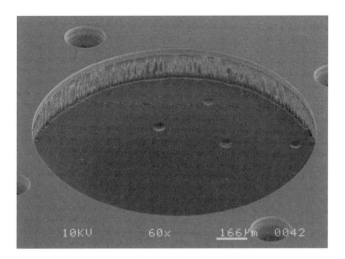

FIGURE 21.9 A 1-mm-diameter well etched to a depth of 245 μm in p-type 6H-SiC using the SF_6 + 85% Ar etch process, after an *in situ* sputter cleaning using argon. The electroplated nickel mask has not been stripped.

experiments, the argon concentration was gradually increased until smooth etched surfaces were obtained at 85% argon and 15% SF_6. Other parameters were unchanged from the baseline process (5-mTorr pressure, 800-W coil and 75-W platen powers). Dilution with 85% argon reduces the SiC etch rate from 0.30 to 0.22 μm/min. The selectivity to nickel is reduced from 125 to 55. The selectivity to ITO is only 20, insufficient for deep etching. For DRIE using the Ar-diluted SF_6 process, thick nickel masks were fabricated by selective electroplating. Figure 21.9 shows the smooth surface that resulted after etching p-type 6H-SiC to a depth of 245 μm, using SF_6 and 85% argon. The mask in this case was 6-μm-thick electroplated nickel.

A key attribute of the silicon DRIE process is its ability to accurately transfer finely detailed features deep into the wafer, as quantified by the aspect ratio, or etch depth divided by lateral feature size. Typically, DRIE of silicon can provide aspect ratios of 30 or greater. Deep RIE processes suitable for the fabrication of high-aspect-ratio SiC microstructures are still in the very early stages of development. Figure 21.10 shows trenches with moderately high aspect ratios which were fabricated by DRIE of n-type 6H-SiC. The etch depth was 60 μm, and the trenches which form the letters shown were 10 μm wide, for an aspect ratio of 6. The electroplated nickel mask (initially 6 μm thick) was stripped prior to SEM. The wafer was etched using the standard 100% SF_6 etch for 200 min, after a 10-min sputter clean in argon. In developing a high-aspect-ratio DRIE process for SiC, the primary objective will be to control the deposition of passivating polymer on the etch sidewall in order to minimize the sidewall roughness apparent in Figure 21.10. This can be accomplished by using a switched process, as for DRIE of silicon, or by modifying the SiC DRIE process to provide for simultaneous etching and polymer deposition.

21.6 Applications of SiC DRIE: Fabrication of a Bulk Micromachined SiC Pressure Sensor

This section will describe a practical application for SiC DRIE—specifically, a piezoresistive SiC pressure sensor, which can be used at temperatures as high as 500°C. Silicon piezoresistive pressure transducers were some of the first MEMS sensors and now represent a large fraction of the MEMS devices manufactured. Silicon pressure sensors with precisely controlled dimensions can be inexpensively fabricated using highly anisotropic wet etch processes. At moderate temperatures, silicon is a nearly ideal elastic material; however, at about 450°C, silicon begins to deform plastically. Silicon carbide is a superior material for high-temperature pressure sensors, because SiC maintains its excellent mechanical properties

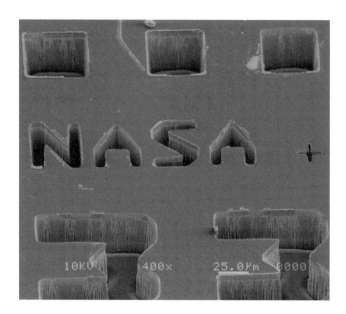

FIGURE 21.10 6H-SiC (n-type) etched to a depth of 60 μm using the 100% SF_6 baseline process, after an *in situ* sputter cleaning using argon. The electroplated nickel mask has been stripped. Visible on the tops of the sidewall are fragments of the polymer layer which had built up on the edge of the mask.

to very high temperatures (>1000°C). A production-worthy DRIE process for SiC is a key to the successful commercialization of high-temperature SiC pressure sensors, as DRIE is the only bulk-micromachining method for SiC that provides the required dimensional control.

The packaged SiC pressure sensor chip is shown in Figure 21.11 [Ned et al., 2001]. A SiC diaphragm is micromachined by using DRIE to etch a 1-mm-diameter well to a depth of about 60 μm in the backside of a 120-μm-thick 6H-SiC wafer. On the front side of the diaphragm, four SiC strain gauges are fabricated in an n-type epitaxial layer. An underlying p-type epitaxial layer provides electrical isolation between the strain gauges. The strain gauges are defined using a photoelectrochemical etch process which automatically stops when the p-type layer is reached. The strain gauges are configured in a Wheatstone bridge, with two gauges positioned over the edge of the diaphragm while the other two gauges are positioned in the center of the diaphragm. The gauges on the edge of the diaphragm have the greatest sensitivity to pressure, while the sensitivity of the gauges in the center of the diaphragm is smaller and opposite in sign. The output of the Wheatstone bridge responds to the difference in the resistances of the edge-mounted and center-mounted gauges. The temperature-induced resistance changes, therefore, are compensated, as they are largely the same for all four gauges. A three-layer metallization system, consisting of layers of titanium, tantalum silicide and platinum, is used to electrically contact the SiC piezoresistors.

The highly anisotropic DRIE process for SiC produces diaphragms that are uniform in thickness and have precisely positioned edges. This maximizes the sensitivity and provides a high uniformity of response from one sensor to the next, as the sensitivity is strongly influenced by the positions of the strain gauges in relation to the edge of the diaphragm. For a 60-μm etch depth, the 1-mm-diameter diaphragm was determined to have a thickness that was uniform to better than ±1 μm.

The SiC pressure sensor of Figure 21.11 was tested in the compressor discharge of a gas turbine engine. The sensor performed as expected during this hour-long test, in which it was exposed to gas temperatures as high as 520°C. Current efforts are focused on reducing the cross-sensitivity to temperature: first by refining the design and fabrication processes to more accurately balance the bridge, and, second, by incorporating an on-chip temperature sensor to enable the compensation of temperature-induced changes in gauge factor.

FIGURE 21.11 (Color figure follows p. 12-26.) Packaged SiC pressure-sensor chip. Through the semitransparent SiC can be seen the edge of the well that has been etched in the backside of the wafer to form the circular diaphragm. The metal-covered, n-type SiC that connects the strain gauges is highly visible, while the n-type SiC strain gauges are more faintly visible. There is a U-shaped strain gauge over the edge of the diaphragm at top and bottom, and there are two vertically oriented linear gauges in the center of the diaphragm.

21.7 Summary

Deep RIE of SiC is a key to enabling the development of a wide range of SiC MEMS for use in harsh environments. High-density plasma makes DRIE of SiC possible, as it provides the required high chemical reactivity, low operating pressure and independent control of ion energy. Deep RIE of silicon carbide is quite different from DRIE of silicon, due to the chemical inertness of SiC. The need for energetic ion bombardment to produce an appreciable etch rate causes RIE of SiC to be inherently anisotropic. Energetic ion bombardment necessitates the use of a nonreactive etch mask such as nickel or ITO. The attainable etch depth is typically limited by sputter erosion of the etch mask. Often, higher ion energies can be used to increase the SiC etch rate, but this can produce an unacceptable reduction in the selectivity with respect to the mask. The etch depths for which smooth residue-free surfaces are obtained can be limited by background contaminants in the reaction chamber. Argon can be introduced to the chamber to provide continuous sputter cleaning of the surface during the etch process, but this has detrimental effects on etch rate and mask selectivity. At present, SiC DRIE processes have been developed which are well suited for the fabrication of low-aspect-ratio microstructures such as pressure sensors. However, high-aspect-ratio DRIE processes are still in the early stages of development.

Defining Terms

Anisotropy: Degree of directionality of an etch process. In reactive ion etching, an ideal anisotropic process provides negligible etching in the lateral direction.

Aspect ratio: Etch depth divided by lateral feature size. Highly anisotropic etch processes are needed to fabricate microstructures with high aspect ratios.

Deep reactive ion etching (DRIE): Highly anisotropic plasma etching to depths greater than about 10 µm.

High-density plasma (HDP): A reactor that employs some means of electron confinement to produce a plasma density (ions per unit volume) which is typically two orders of magnitude greater than that of a conventional parallel plate RIE.

Inductively coupled plasma (ICP): A type of high-density plasma reactor in which power is coupled to the plasma using a coil. The time-varying magnetic field induced by the coil helps to confine electrons to the plasma, which increases the plasma density.

Mask selectivity: The substrate etch rate divided by the etch rate of the mask.

Micromasking: The formation of residues due to the masking effect of materials sputtered from the etch mask onto the etched surface. In some reactors, sputtering of the substrate electrode can also cause micromasking.

References

Beheim, G., and Salupo, C.S. (2000) "Deep RIE Process for Silicon Carbide Power Electronics and MEMS," in *Proc. of the MRS 2000 Spring Meeting*, May 24–28, San Francisco, CA, MRS Proc. 622, paper T8.9.

Bhardwaj, J.K., and Ashraf, H. (1995) "Advanced Silicon Etching Using High Density Plasmas," in *Micromachining and Microfabrication Process Technology*, ed. K.W. Markus, *Proc. SPIE* **2639**, pp. 224–233.

Chabert, P., Proust, N., Perrin, J., and Boswell, R.W. (2000) "High Rate Etching of 4H-SiC Using a SF_6/O_2 Helicon Plasma," *Appl. Phys. Lett.* **76**, pp. 2310–2312.

Cho, H., Leerungnawarat, P., Hays, D.C., Pearton, S.J., Chu, S.N.G., Strong, R.M., Zetterling, C.-M., Ostling, M., and Ren, F. (2000) "Ultradeep, Low-Damage Dry Etching of SiC," *Appl. Phys. Lett.* **76**, pp. 739–741.

Lohner, K.A., Chen, K.-S., Ayon, A.A., and Spearing, S.M. (1999) "Microfabricated Silicon Carbide Microengine Structures," in *Proc. of the MRS 1998 Fall Meeting—Symposium AA, Materials Science of MEMS*, December 1–2, 1998, Boston, MRS Proc. 546, pp. 85–90.

Mehregany, M., Zorman, C.A., Rajan, N., and Wu, C.H. (1998) "Silicon Carbide MEMS for Harsh Environments," *Proc. IEEE* **86**, pp. 1594–1610.

Ned, A.A., Kurtz, A.D., Masheeb, F., and Beheim, G. (2001) "Leadless SiC Pressure Sensors for High Temperature Applications," in *Proc. of ISA 2001, Instrument Society of America Annual Conf.*, September 10–13, Houston, TX.

Sheridan, D.C., Casady, J.B., Ellis, C.E., Siergiej, R.R., Cressler, J.D., Strong, R.M., Urban, W.M., Valek, W.F., Seiler, C.F., and Buhay, H. (1999) "Demonstration of Deep (80 µm) RIE Etching of SiC for MEMS and MMIC Applications," in *Proc. Int. Conf. on Silicon Carbide and Related Materials 1999*, October 10–15, Research Triangle Park, NC, pp. 1053–1056.

Shor, J.S., and Kurtz, A.D. (1994) "Photoelectrochemical Etching of 6H-SiC," *J. Electrochem Soc.* **141**, pp. 778–781.

Wang, J.J., Lambers, E.S., Pearton, S.J., Ostling, M., Zetterling, C.-M., Grow, J.M., Ren, F., and Shul, R.J. (1998) "ICP Etching of SiC," *Solid-State Electron.* **42**, pp. 2283–2288.

Yasseen, A.A., Zorman, C.A., and Mehregany, M. (1999) "Surface Micromachining of Polycrystalline SiC Films Using Microfabricated Molds of SiO_2 and Polysilicon," *J. MEMS* **8**, pp. 237–242.

Yih, P.H., Saxena, V., and Steckl, A.J. (1997) "A Review of SiC Reactive Ion Etching in Fluorinated Plasmas," *Phys. Stat. Sol. (b)* **202**, pp. 605–642.

22
Microfabricated Chemical Sensors for Aerospace Applications

Gary W. Hunter
NASA Glenn Research Center

Chung-Chiun Liu
Case Western Reserve University

Darby B. Makel
Makel Engineering, Inc.

22.1 Introduction 22-1
22.2 Aerospace Applications.. 22-3
   Leak Detection • Fire Safety Monitoring • Engine Emission Monitoring
22.3 Sensor Fabrication Technologies................................. 22-6
   Microfabrication and Micromachining Technology • Nanomaterials • SiC-Based High-Temperature Electronics
22.4 Chemical Sensor Development.................................... 22-8
   Si-Based Hydrogen Sensor Technology • Nanocrystalline Tin Oxide Thin Films for NO_x and CO Detection • Electrochemical Cell Oxygen Detection • SiC-Based Hydrogen and Hydrocarbon Detection • NASICON-Based CO_2 Detection
22.5 Future Directions, Sensor Arrays and Commercialization 22-17
   High-Selectivity Gas Sensors Based on Ceramic Membranes • Leak-Detection Array • High-Temperature Electronic Nose
22.6 Commercial Applications... 22-21
22.7 Summary 22-21
   Acknowledgments.. 22-22

22.1 Introduction

The advent of microelectromechanical systems (MEMS) technology is important in the development and use of chemical sensor technology for a range of applications, especially those that include operation in harsh environments or effect safety. As will be discussed in this chapter, chemical microsensors can provide unique information that can significantly improve safety and reliability while decreasing costs of a system or process. Such information can also be used to improve a system's performance and reduce its effect on the environment. Chemical sensor data also can complement data derived from physical measurements such as temperature, pressure, heat flux etc., further improving overall knowledge of a system and expanding its capabilities.

However, the application of even traditional macrosized chemical sensor technology can be problematic. Chemical sensors often need to be specifically designed (or tailored) to operate in a given environment. It is often the case that a chemical sensor that meets the needs of one application will not function adequately in another application. The more demanding the environment and specialized the requirement, the greater the need to modify existing chemical sensor technologies to meet these requirements or, as necessary, develop new chemical sensor technologies. Four common parameters are typically cited as relevant in determining whether a chemical sensor can meet the needs of an application: *sensitivity*, *selectivity*, *response time* and *stability*. Sensitivity refers to the ability of a sensor to detect the desired chemical species in the range of interest. Selectivity refers to the ability of the sensor to detect the species of interest in the presence of interfering gases, which also can produce a sensor response. Response time refers to the time it takes for the sensor to provide a meaningful signal (often defined as 90% of the steady-state signal) when the chemical environment is changed. Finally, stability refers to the degree to which the sensor baseline and response to a given environment change over time. Simply stated, for micro- and macrochemical sensing systems, one needs a sensor that will accurately determine the species of interest in a given environment with a response large and rapid enough to be of use in the application and whose response does not significantly drift over its designed operational lifetime. Depending on the application, it is often difficult to find a suitable sensor material that provides the required sensitivity, selectivity, response time and stability.

The microfabrication of chemical sensors involves much more than just making a macrosized sensor smaller. The processing used to produce a sensor material as a macroscopic bulk pellet can change considerably when it is desired to fabricate the material as part of a miniaturized system. For example, a chemical sensor in the form of a macroscopic bulk pellet can be fabricated from the powder of a starting material, pressed into a pellet containing lead wires and then sintered at high temperatures to form the resulting sensor. However, using the same starting material, pressing the material onto a substrate to form a smaller or even microscopic sensor is often not a viable option. Rather, a thin or thick film of the sensor material must be deposited onto a substrate, which itself, at a minimum, must support the sensor and allow for connections to the outside world. The underlying substrate with sensor film may not survive the sintering that typically is done with the pellet sensor material.

Further, given the surface-sensitive nature of many chemical sensors, the effects of miniaturization can be dramatic and include significant changes in sensor sensitivity and response time. This is in part due to the fact that the sensor film is often produced by techniques such as sputtering which may result in different material properties than those of bulk materials. The resulting surface-to-volume ratio of a thin film is larger than that of a bulk material: Surface effects that may affect only a small percentage of a sensor in the bulk form may occur within a significantly larger percentage volume of a thin-film sensor. This can strongly affect the response of the sensor. For example, oxidation may occur on the surface of a sensor exposed to high temperatures. In a bulk material, this oxidation may only be a small percentage of the sensor's volume, while in a thin-film material the same oxidation thickness may account for a sizable percentage of the sensor's volume. If the sensor's detection mechanism relies on bulk conduction, this oxidation could significantly affect the sensor response by changing the nature of the volume of the sensor. In addition, stresses in sensor thin films that degrade sensor response or catastrophically damage the sensor structure may be less of a factor in bulk materials [e.g., Hughes et al., 1987; Hunter et al., 1998]. Therefore, new technical challenges often must be overcome as sensor technology is miniaturized.

This chapter will discuss the development and application of microfabricated chemical sensor technology for aerospace (aeronautics and space) applications. These applications are particularly challenging, as often they include operation in harsh environments or have requirements that have not been previously emphasized by commercial suppliers. Section 22.2 discusses the chemical sensing needs of three important aerospace applications: leak detection, emission monitoring and fire detection. For each application, the use and advantages of MEMS-based approaches will be discussed. Each application has vastly different problems associated with the measurement of chemical species. Nonetheless, the development of a common base technology can address the measurement needs of a number of applications. Section 22.3 explores three base technologies being used to develop chemical sensors for these applications: microfabrication techniques as applied to chemical sensors, nanomaterials and high-temperature silicon carbide (SiC)-based electronics. Section 22.4 discusses the use of these base technologies to produce chemical sensors for a range of species: hydrogen (H_2), nitrogen oxides (NO_x), oxygen (O_2), hydrocarbons (C_xH_y), carbon monoxide (CO) and carbon dioxide (CO_2). The following two sections discuss the integration, application and future direction of chemical sensor technology in aerospace applications and provide a summary.

22.2 Aerospace Applications

Aerospace applications vary considerably in their requirements for chemical sensor technology. In general, there are three envisioned uses of chemical sensor technology in aerospace vehicle applications: (1) system development and ground testing and where the sensor provides information on the state of a system that does not fly; (2) vehicle health monitoring (VHM), which involves the long-term monitoring of a system in operation to determine the health of the vehicle system (e.g., whether or not the engine is losing performance, increasing emissions or developing a leak in the fuel system); (3) active control of the vehicle in a feedback mode where information from a sensor is used to change a system parameter in real time (e.g., fuel flow to the engine changing due to sensor measurements of emissions). A sensor could be used in all three applications: qualifying the system on the ground, monitoring its health in flight and using this information in real time to change the operation parameters. These applications have a strong impact on vehicle safety, performance and reliability. The design of each individual sensor depends on the requirements of each application. As will be discussed below, very different sensor technology is necessary depending on temperature range, ambient gas, interfering gases etc. For example, the sensor technology used for ground testing for leaks in an air ambient is very different from that used for leak testing in an inert (no oxygen) ambient. This section illustrates the needs of three specific aerospace applications and how MEMS-based technology can address these needs by examining a ground-based/VHM application (leak detection), an in-flight VHM application (fire detection) and an application (emission monitoring) that can be included in all three aerospace application areas: ground testing, VHM and active control.

22.2.1 Leak Detection

Launch vehicle safety requires the rapid and accurate detection of fuel leaks. One major application is the detection of low concentrations of H_2 associated with the launch and operation of the Space Shuttle. Hydrogen leaks can lead to hazardous situations and, unless their locations can be rapidly identified, explosive situations, delays in vehicle launches and significant costs. In the summer of 1990, hydrogen leaks on the Space Shuttle while it was on the launch pad temporarily grounded the fleet until the leak source was identified and repaired. The standard leak-detection system was a mass spectrometer connected to an array of sampling tubes placed throughout the region of interest. Although able to detect hydrogen in a variety of ambient environments, the mass spectrometer leak-detection approach suffered delay times, as gas had to be transported from the fuel pipe through the sampling tubes to the remotely located mass spectrometer. Pinpointing the exact location of the leak was also problematic, as often the fuel pipes were covered with insulation. A leak emitted from the pipe could travel in the insulation and exit the insulation in a location different from the leak source. Thus, the mass spectrometer leak-detection approach was unable to adequately isolate the location of the hydrogen leaks, leading to a delay in Shuttle launches. Problems with hydrogen leak detection on the Shuttle continue. As recently as 1999, the launch of STS-93 was delayed for 2 days due to an ambiguous signal in the mass spectrometer leak-detection system. The Shuttle hydrogen leak-detection system must be improved in order to allow improved safety and reduced operational costs.

No commercial hydrogen sensors existed at the time of the 1990 leaks that operated satisfactorily in this and other space-related applications, primarily due to the conditions in which the sensor must operate. The hydrogen sensor must be able to detect hydrogen from low concentrations (ppm range) through the lower explosive limit (LEL) which is 4% in air. The sensor must be able to survive and preferably function, even during exposure to 100% hydrogen, without damage or change in calibration. Further, the sensor may be exposed to gases emerging from cryogenic sources which could cool the sensor and cause undesired changes in sensor output. Thus, sensor temperature measurement and control are necessary. Operation in inert environments is also necessary, as the sensor may have to operate in areas purged with helium. The ability to operate in a vacuum is preferred to allow leak detection even in space. Commercially available sensors, which often required oxygen (air) to operate or depended upon moisture [Hunter, 1992a], did not meet the needs of this application (especially operation in inert environments), thus development of new types of hydrogen sensors was initiated [Hunter, 1992a].

NASA has been active in efforts to improve fuel leak-detection capabilities for propulsion systems. The long-term objective has been an automated detection system using MEMS-based, point-contact hydrogen sensors. The application of MEMS-based sensor technology allows the fabrication of sensors with minimal size, weight and power consumption to beneficially decrease vehicle weight and power requirements. A number of these sensors placed in the region of interest could allow multiplexing of the signals and "visualization" of the magnitude and location of the hydrogen leak. Potential locations for sensors include fuel lines, inside insulation and throughout chambers inside the vehicle. The ability to launch a vehicle on schedule depends on a leak-detection system that can monitor key regions in the vehicle and automatically locate a leak in a limited time which can reduce costs and manpower associated with the detection and location of leaks as well as improve overall system reliability.

Hydrogen is not the only fuel that may be used in launch vehicle applications. Other potential fuels considered for launch vehicles include propane, methane, ethanol and hydrazine. Leak-detection sensors for these fuels will have to meet many of the same requirements as the hydrogen sensor: sensitive detection of the fuel in possibly inert or cryogenic environments, sensor survival in high concentrations of the fuel and minimal size, weight and power consumption. Methane, which is a naturally occurring by-product of animal wastes, already exists in small quantities in the environment. This increases the difficulty of its measurement in some applications. Further, the detection mechanism that may readily allow the detection of hydrogen may not work for the detection of methane or ethanol [Chen et al., 1996]. Hydrazine is toxic, which adds further importance to leak-detection systems: The gas must be detected not only for reasons of explosion prevention but also for personnel health safety reasons. Just as hydrogen sensor technology in 1990 did not meet the needs of launch vehicle safety applications, suitable fuel leak-detection systems still do not exist for launch vehicles that use alternate fuels to hydrogen.

22.2.2 Fire Safety Monitoring

The detection of fires onboard commercial aircraft is extremely important to avoid catastrophic situations. Although dependable fire-detection equipment currently exists within the cabin, detection of fire within the cargo hold has been less reliable [Grosshandler, 1997]. In aircraft where cargo-hold fire detection is required, the fire-detection equipment used in many commercial aircraft relies on the detection of smoke. These smoke detectors either are optically based or depend on ionization of particles. Although highly developed, these sensors are subject to false alarms, with estimated false alarm rates varying from 10:1 to 500:1. Because some cargo areas are inaccessible to the flight crew during flight, visual confirmation is not possible. The presence of false alarms decreases the confidence of pilots in these systems and may potentially cause accidents if pilots react to reported fires that may not exist. These false alarms may be caused by a number of sources, including changes in humidity, condensation on the fire detector surface and contamination from animals, plants or other contents of the cargo bay.

A second, independent method of fire detection to complement the conventional smoke-detection techniques, such as the measurement of chemical species indicative of a fire, will help reduce false alarms and improve aircraft safety. Although many chemical species are indicative of a fire, two species of particular interest are CO and CO_2 [Grosshandler, 1995]. Some requirements for these chemical sensors differ from those of leak detection and emission monitoring. The sensor must withstand temperatures ranging from −30 to 50°C, pressures from 18.6 to 104 kPa and relative humidities from 0 to 95%. Different types of fires produce different chemical signatures [Grosshandler, 1995]; the sensor must be able to detect the presence of a real fire and must not be affected by the presence of gases produced by contents of the cargo hold. The response must be quick and reliable and able to provide relevant information to the pilot.

The use of MEMS-based chemical sensing technology for fire detection would be similar to that for the leak-detection application above: a number of sensors placed throughout a region that give an indication of the chemical signature of a fire and its location. Similar to the leak-detection application, a smaller system is preferred. In larger sensor systems, fewer sensors can be placed in a cargo bay, and the likelihood of them being dislodged is greater because they stick out into the chamber. The ideal is to have a multifunctional sensor array that provides all the information necessary to determine the presence of the fire while being easily integrated with existing fire-detection systems and cargo bay.

22.2.3 Engine Emission Monitoring

One of the NASA agency goals is to significantly reduce emissions of future aircraft. Control of emissions from aircraft engines is an important component in the development of the next generation of these engines. A reduction of engine emissions can be achieved, in principle, if the emissions of the engine are monitored in real time and that information is fed back and used to modify the combustion process of the engine. This active combustion control also depends on the development of actuators that control the combustion process as well as a control system to interpret the sensor data and appropriately adjust the engine parameters. The ability to monitor the type and quantity of emissions being generated by an engine is also important in determining the operational status of the engine. For example, by monitoring the emission output of the engine and correlating changes in the emissions produced with previously documented changes in engine performance, long-term degradation of engine components might be monitored. If such an emission-monitoring system were developed, an initial use would be measurement of emissions during the developmental stage of the engine.

Very few sensors available commercially are able to measure the components of engine emissions *in situ*. The harsh conditions, such as high temperatures, inherent near reaction chambers of the engine render most sensors inoperable. A notable exception to this limitation in sensor technology is the commercially available oxygen sensor currently used in automobile engines [Logothetis, 1991]. This sensor, which is based on the properties of zirconium dioxide (ZrO_2), has been instrumental in decreasing automotive engine emissions. However, comparable sensors for other components of the gas stream do not exist. Highly sensitive monitoring of emissions of nitrogen oxides, hydrogen and hydrocarbons is not currently possible *in situ* with point-contact sensors placed near the engine. Even the traditional ZrO_2-based sensor has sensitivity limits as well as size, weight and power consumption requirements that discourage its use in some applications.

The development of a new class of sensors for monitoring of emissions from aircraft engines is necessary. The difficulties inherent with larger sensors placed in the engine stream include: (1) added weight to the engine; (2) reduced ability to pinpoint the chemical signature of the stream at a given location if the sensors vary in position due to their size; (3) greater turbulence experienced by the array in the engine flow stream and more interference with the flow of gases in the engine; (4) size of the projectile if the sensor array is dislodged from the measuring site and emitted into the engine. Thus, an emission sensor array should be as small as possible. This implies the use of MEMS-based technology.

The environment in which a sensor must operate in emission-monitoring applications varies drastically from the environment of launch vehicle propellant leak applications. The sensor must operate at high temperatures and detect low concentrations of the gases to be measured. Although the measurement of NO_x is important in these applications, the measurement of other gases present in the emission stream such as C_xH_y, CO, CO_2 and O_2 is also of interest. Table 22.1 provides a partial listing of gases of interest in an emissions stream identified by the Glennan Microsystems Initiative (see For Further Information section). The measurement ranges depend on the chemical species and the engine, but generally the detection of NO_x and C_xH_y may be necessary at sensitivities down to 5 to 10 ppm, CO down to 50 ppm, O_2 from 0 to near 25% and CO_2 from 2 to 14%. Such a wide range of detection requirements demands multiple types of sensor technology, each of which should have stable or at least predictable performance for extended periods of time. Given the limited number of appropriate sensors for this application, development of new sensor technology is also necessary for this application.

TABLE 22.1 Partial Listing of Gases of Interest in an Emissions Stream as Identified by the Glennan Microsystems Initiative

Chemical Species	Concentration Range
CO	50–2000 ppm
CO_2	2–14%
NO_x	10–2000 ppm
O_2	0–25%
C_xH_y	5–5000 ppm

22.3 Sensor Fabrication Technologies

In order to meet the chemical sensor needs of aerospace applications described in Section 22.2, the development of a new generation of sensor technologies is necessary. Active development of chemical sensor technology to meet these needs is taking place at NASA Glenn Research Center (NASA GRC) and Case Western Reserve University (CWRU) based on recent progress in three technology areas: (1) micromachining and microfabrication technology to fabricate miniaturized sensors, (2) the use of nanocrystalline materials to develop sensors with improved stability combined with higher sensitivity and (3) the emergence of silicon carbide (SiC) semiconductor technology to provide electronic devices and sensors operable at high temperatures. This section will give a brief overview of these technologies, while Section 22.4 discusses chemical sensor technology development and application.

22.3.1 Microfabrication and Micromachining Technology

Microfabrication and micromachining technology is derived from advances in the semiconductor industry [Madou, 1997]. A significant number of silicon-based microfabrication processes were developed for the integrated circuit (IC) industry. Of the various processing techniques, lithographic reduction, thin-film metallization, photoresist patterning and chemical etching have found extensive use in chemical sensor applications. These processes allow the fabrication of microscopic sensor structures. The ability to mass-fabricate many of these sensors on a single wafer (batch processing) using currently available semiconductor processing techniques significantly decreases the fabrication costs per sensor. A large number of sensors are fabricated simultaneously with the same processing, resulting in high reproducibility of response between the sensors. The smaller sensor size also enables better sensor performance, especially in low concentration measurements, where the small surface area of the sensor allows a smaller number of molecules to have a larger relative effect on the sensor output.

However, these processes produce mainly two-dimensional planar structures, which have limited application. By combining these processes with micromachining technology, three-dimensional structures can be formed that have a wider range of application to chemical-sensing technology. Micromachining technology is generally defined as the means to produce three-dimensional micromechanical structures using both bulk and surface micromachining techniques. The techniques used in micromachining fabrication include chemical anisotropic and dry etching, the sacrificial layer method and LIGA (*lithographie, galvanoformung, abformung*). Chemical anisotropic etching is an etching procedure that depends on the crystalline orientation of the substrate. For silicon etching, potassium hydroxide (KOH) and tetramethyl ammonium hydroxide (TMAH) solutions are most commonly used as etching agents. Dry-etching processes include ion milling, plasma etching, reactive ion etching and reactive ion beam etching. These dry-etching processes are not dictated or limited by the crystalline structure. LIGA techniques have been used to produce high-aspect-ratio multistructures [Madou, 1997].

The sacrificial layer method employs a deposited underlayer that can be chemically removed. This technique can be used to make a chamber electrode structure to protect the integrity of the sensor element. For many applications, temperature control is necessary. Incorporation of a microfabricated heating element and a temperature detector allows feedback control of the operating temperature. In these microstructures, a small thermal mass is desirable in order to minimize heat loss and heat energy consumption. This is accomplished by micromachining techniques that allow the selective removal of the underlying silicon substrate, producing a suspended diaphragm structure. This diaphragm structure underneath the sensing element, combined with a heater and temperature detector integrated as part of the sensor, is useful as a sensor platform that has small thermal mass that quickly reaches thermal equilibrium with less heat loss and energy consumption. The sacrificial layer method can also be used to make a chamber structure surrounding the sensor and protecting the integrity of the sensing element.

Microelectromechanical systems processes allow the mass production of sensors tailored for a given applications which have minimal size, weight and power consumption. This sensor processing can be done using silicon (Si) either as a substrate on which a structure is built or as a semiconductor that is part of an electrical circuit. Alternative materials to Si may be used as substrates as long as they are compatible with IC processing; however, micromachining of materials other than Si is less developed, and different processes are often necessary to produce the same structures in, for example, ceramics.

22.3.2 Nanomaterials

Nanocrystalline materials are materials whose grain size is less than 100 nm. Nanocrystalline materials have several inherent advantages over conventionally fabricated materials including increased stability and sensitivity at high temperature. The improved stability is due to the fact that the mechanisms involved in the material grain growth are different for nanocrystalline materials than for materials with micro- or macrosized grains. This means that these materials are more stable than conventional materials. Drift in the properties of gas-sensitive materials such as tin oxide (SnO_2) with long-term heating due to grain boundary annealing have been previously noted [e.g., Jin et al., 1998; Hunter, 1992a and references therein]. This drift causes changes in the sensor output with time and reduces sensor sensitivity. The use of nanocrystalline materials such as nanocrystalline SnO_2 is being investigated to stabilize the sensor materials for long-term operation. Further, as the detection mechanism of materials such as SnO_2 depends on the number of grain boundaries [Hunter, 1992a], nanocrystalline materials have a higher density of grains and thus a higher sensitivity. However, while nanomaterials may provide advantages as sensor materials, their use in microsensor platforms depends on the capability to process them in a manner compatible with the processing used for the other components of the sensor structure.

22.3.3 SiC-Based High-Temperature Electronics

Silicon carbide-based semiconductor electronic devices are currently being developed for use in conditions under which conventional semiconductors cannot adequately perform. Due to its wide band gap and low intrinsic carrier concentration, SiC operates as a semiconductor at significantly higher temperatures than possible with silicon (Si) semiconductor technology. SiC is now available commercially, and processing difficulties associated with fabrication of SiC devices are being addressed [Neudeck, 1995].

Silicon carbide occurs in many different crystal structures (polytypes) with each crystal structure having its own unique properties. In many device applications, the exceptionally high breakdown field (>5 times that of Si), wide bandgap energy (>2 times that of Si) and high thermal conductivity (>3 times that of Si) of SiC could lead to substantial performance gains. Due to its other material properties, such as superior mechanical toughness, SiC is an excellent material for use in a wide range of harsh environments.

The capability of SiC to function in extreme conditions is expected to enable significant improvements to a wide range of applications and systems, including: (1) greatly improved high-voltage switching for energy savings in public electric power distribution; (2) more powerful microwave electronics for radar and communications; (3) electronics, sensors and controls for cleaner burning, more fuel-efficient jet aircraft and automobile engines. A potentially major application of SiC as a semiconductor is in a gas-sensing structure. One set of approaches used to produce SiC-based gas sensors is discussed in Section 22.4.4.

22.4 Chemical Sensor Development

As discussed in the previous sections, the needs of aerospace applications are driving the development of a new generation of chemical microsensor technology. While the sensor design and sensing approach depend strongly on the application, the application needs nevertheless have some common factors:

1. Optimally, the detection of gas should take place *in situ*, thus, it is strongly preferred that the sensor operate in the environment of the application.
2. The sensor must have minimal size, weight and power consumption to allow placement of multiple sensors in a number of locations without, for example, significantly increasing the weight or power consumption of the vehicle.
3. Ideally, the sensors should be readily multiplexed to allow a number of sensors to be placed in a given region and the results of the measurement fed back to monitoring hardware or software. This will allow the accurate monitoring of a region and, by coordinating the signals from an array of sensors determine, for example, the location of a leak or fire.

This section discusses the work at NASA GRC and CWRU using microfabrication technology, nanomaterials and SiC semiconductor technology to develop H_2, C_xH_y, NO_x, CO, O_2 and CO_2 sensors. Three different sensor platforms are used: a Schottky diode, a resistor and an electrochemical cell. A brief description is given of each sensor configuration and its current stage of development. The Si-based hydrogen sensor is at a relatively mature stage of development while the state of development of the other sensors ranges from the proof of concept level to prototype stage. Integration of the sensor technology with hardware and software as well as system demonstrations are also taking place in specific applications with Makel Engineering, Inc. (MEI).

22.4.1 Si-Based Hydrogen Sensor Technology

The needs of hydrogen detection for the aerospace leak detection applications discussed above require that the sensor work in a wide concentration range: from the ppm range to 100% hydrogen in inert environments. The types of sensors that work in these environments are limited, and finding one sensor to completely meet the needs of this application is difficult. For example, metal films that change resistance upon exposure to hydrogen have a response proportional to the partial pressure of H_2: $(P_{H_2})^{1/2}$. This dependence is due to the sensor detection mechanism: migration of hydrogen into the bulk of the metal changing the bulk conductance of the metal [Hunter 1992b]. This results in low sensitivity at low hydrogen concentrations but a continued response over a wide range of hydrogen concentrations. In contrast, Schottky diodes, composed of a metal in contact with a semiconductor (metal-semiconductor, MS) or a metal in contact with a very thin oxide on a semiconductor (metal-oxide semiconductor, MOS), have a very different detection mechanism. For hydrogen detection, the metal is hydrogen sensitive (e.g., palladium, Pd). For Pd–SiO_2–Si Schottky diodes, hydrogen dissociates on the Pd surface and diffuses to the Pd–SiO_2 interface, affecting the electronic properties of the MOS system and resulting in an exponential response of the diode current to hydrogen [Lundstrom, 1989].

This exponential response has higher sensitivity at low concentrations and decreasing sensitivity at higher concentrations as the sensor saturates. Thus, by combining both a resistive sensor and a Schottky diode, sensitive detection of hydrogen throughout the range of interest can be accomplished. Further, temperature control is necessary for both sensor types for an accurate reading. MEMS-based technology enables fabrication in a limited area of both sensors (resistor and Schottky diode) with temperature control while having minimal size, weight and power consumption.

NASA GRC and CWRU have developed palladium alloy Schottky diodes and resistors for hydrogen sensors on Si substrates. Palladium alloys are more resilient than pure Pd to damage caused by exposure to high concentrations of hydrogen [Hughes et al., 1987]. Although the sensor signal is affected by the presence of oxygen, the sensor does not require oxygen to function. The NASA GRC/CWRU design uses a palladium–chrome (PdCr) alloy due to its ability to withstand exposure to 100% hydrogen

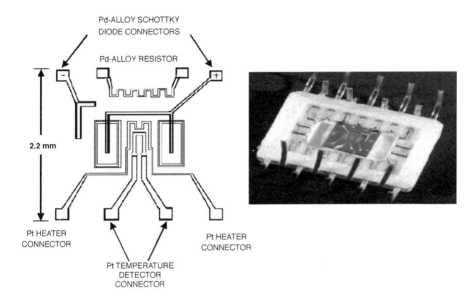

FIGURE 22.1 Schematic diagram and picture of the silicon-based hydrogen sensor. The Pd alloy Schottky diode (rectangular regions) resides symmetrically on either side of a Pt heater and temperature detector. The Pd alloy Schottky diode is used for low-concentration measurements, while the resistor is included for high-concentration measurements.

[Hunter et al., 1998]. The sensor structure is shown in Figure 22.1 and includes a PdCr Schottky diode, a PdCr resistor, a temperature detector and a heater all incorporated in the same chip. The sensor dimensions are approximately 2.2 mm on a side. The combination of the Schottky diode and resistor sensing elements results in a sensor with a broad detection range; the Schottky diode provides sensitive detection of low concentrations of hydrogen up to near 1%, while the resistor provides sensitivity from 1% up to 100% hydrogen. However, if the sensor is only exposed to low concentrations of hydrogen, then a different alloy, palladium silver (PdAg), can be used in the Schottky diode structure for higher sensitivity.

The MEMS-based hydrogen sensors using both PdCr and PdAg alloys have been selected for use and/or demonstrated in several applications. These sensors have been used on the assembly line of Ford Motor Company (see Section 22.6) and on the Space Shuttle. They have also been selected for use on the safety system of the experimental NASA X-33 vehicle and the water purification system for the International Space Station. Supporting hardware and software have also been included with the sensor to provide signal conditioning and control of the sensor temperature and to record sensor data [Hunter et al., 1998; 2000]. Each application has had different requirements; the entire system from sensors to supporting hardware and software has been tailored for each application. It should be noted that the successful operation of these MEMS-based sensors depends on this supporting technology and associated packaging.

This complete hydrogen-detection system (two sensors on a chip with supporting electronics) flew on the STS-95 mission of the Space Shuttle (launched in October 1998) and again on STS-96 (launched in May 1999) [Hunter et al., 1999a; 2000]. The hydrogen-detection system was installed in the aft main engine compartment of the Shuttle and used to monitor the hydrogen concentration in that region. Currently, a mass spectrometer monitors the hydrogen concentration in the aft compartment before launch, while after launch "grab" bottles are used. The inside of these "grab" bottles is initially at vacuum. During flight, the "grab" bottles are pyrotechnically opened for a brief period of time and the gas in the aft compartment is captured in the bottle. Several of these bottles are opened at different times during the takeoff and after the Shuttle returns to earth and their contents are analyzed to determine the time profile of the gases in the aft chamber. However, this information is only available after the flight and cannot be used to monitor the Shuttle condition in real time.

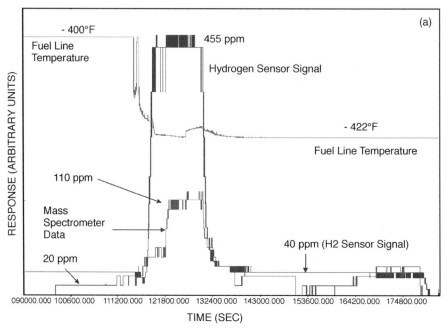

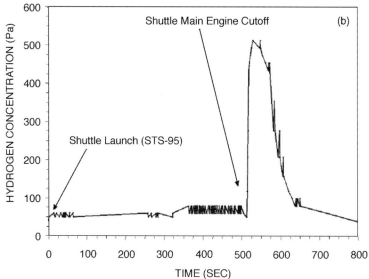

FIGURE 22.2 The response of a MEMS-based PdCr hydrogen sensor in the aft compartment of the Space Shuttle: (a) Compared to the response of a mass spectrometer during loading of liquid hydrogen while on the launchpad; the time scale, in seconds, refers to a clock associated with the Shuttle monitoring system. (b) Recorded data in flight during the launch of the Shuttle; the time scale, in seconds, refers to time after launch.

Data from the STS-95 mission are shown in Figure 22.2. The response of the hydrogen sensors was compared to that of the mass spectrometer and the information obtained by the "grab" bottles. During ground monitoring, the hydrogen sensor (Schottky diode) response was compared with that of the mass spectrometer. The sensor and the mass spectrometer intakes were located at different points in the aft compartment. The results are shown in Figure 22.2a, where the hydrogen sensor signal, the mass spectrometer hydrogen signal and the hydrogen fuel line temperature are shown as a function of time.

The decrease in the fuel-line temperature corresponds to the onset of fuel flow to the Shuttle hydrogen tanks. As this fueling begins, a small increase in hydrogen concentration is observed with both the hydrogen sensor and mass spectrometer. After the initial signal, the hydrogen concentration is seen to drop by both the sensor and the mass spectrometer. Overall, the hydrogen sensor response is seen to generally parallel that of the mass spectrometer but with a larger signal and quicker response time (perhaps due to the relative location of the each measuring device with respect to the hydrogen source). These results are for just one sensor monitoring the chamber; a number of these MEMS-based sensors placed around the chamber could be used to "visualize" the location of the leak.

The hydrogen sensor response during the launch phase of flight is shown in Figure 22.2b. No significant sensor response is seen until the cutoff of the Shuttle main engine. Near this time, a spike in the hydrogen concentration is observed that decreases with time back to baseline levels. These results are qualitatively consistent with the leakage of very small concentrations of unburned fuel from the engines into the aft compartment after engine cut-off. These observations also qualitatively agree with those derived by analyzing the contents of the grab bottles. Moreover, the advantage of this microsensor approach is that the hydrogen monitoring of the compartment is continuous and, in principle, could be used for real-time health monitoring of the vehicle in flight.

22.4.2 Nanocrystalline Tin Oxide Thin Films for NO_x and CO Detection

The detection of NO_x and CO can be accomplished using nanocrystalline SnO_2 materials deposited on a microfabricated and/or micromachined substrate [Hunter et al., 2000]. The substrate can be either Si or ceramic. Figure 22.3 shows the basic structure of the sensor design with a Si substrate. It consists of two components, Si and glass, which are fabricated separately and then bonded together. The microfabrication process allows the sensor to be small in size with low heat loss and minimal energy consumption. Energy consumption is further reduced by depositing the sensor components (temperature detector, heater and gas-sensing element) over a diaphragm region. This minimizes the thermal mass of the sensing area, thereby decreasing power consumption for heating and the time to reach thermal equilibrium. The substrate utilized for the Si-based design is a 0.15-mm-thick glass. On one side of the glass are the heater and temperature detector, while the sputtered platinum interdigitated fingers reside on the other side. The width of the fingers and the gap between them is 30 µm each. The overall sensor dimensions are approximately 300 µm on a side with a height of 250 µm.

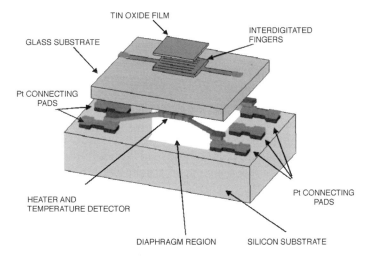

FIGURE 22.3 Structure of a SnO_2-based sensor on a Si substrate. The sensor has Si and glass components and includes a temperature detector and heater.

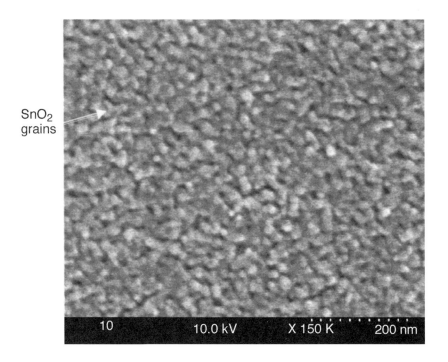

FIGURE 22.4 Nanocrystalline SnO_2 after annealing at 700°C for 1 hr. Each SnO_2 grain is on the order of 10 nm.

The SnO_2 is placed across the interdigitated fingers using a sol–gel technique [Jin et al., 1998]. The gel containing the nanocrystalline SnO_2 particles is spun on the substrate and patterned across the interdigitated finger region using conventional photolithography. This film is subsequently fired, yielding a nanocrystalline SnO_2 film whose conductivity is measured across the fingers. An example of a fired nanocrystalline SnO_2 film is shown in Figure 22.4. The size of each grain of SnO_2 is on the order of 10 nm, with the porosity between the grains being of roughly equal dimensions. Changes in the conductivity of SnO_2 across the interdigitated electrodes are measured and correlated to CO and NO_x concentration. The detection of NO_2 has been demonstrated down to the 5-ppm level at 360°C, with the highest level of sensitivity in the lower ppm with a very stable response [Hunter et al., 1999a]. Similar results with the detection of CO have been obtained [Jin et al., 1998].

Two major components of this development work are to optimize the SnO_2 response to a given gas and to stabilize the SnO_2 for long-term, high-temperature operation. Doping the SnO_2 can improve both the selectivity of the sensor to a specific gas [Jin et al., 1998] as well as sensor stability. For example, the inclusion of Pt has been shown to improve sensitivity to CO, while the inclusion of SiO_2 has been shown to significantly decrease the grain growth of the SnO_2 [Hunter et al., 2000]. Another component of the work is to fabricate the nanocrystalline SnO_2 films on alternate substrates for different applications. For example, for fire detection applications a Si substrate is generally preferred. For emission-sensing applications, deposition on a ceramic substrate, as shown for the sensors in Figure 22.5, is often advantageous. While the configuration of interdigitated fingers, temperature detector and heater are easily transferable to ceramics, micromachining and substrate bonding for high-temperature applications are more problematic.

A tin oxide-based NO_x sensor has been tested in a ground-based power generation turbine engine. The sensor was fabricated on an alumina substrate as shown in Figure 22.5 and placed in a region where the exhaust flow stream from the engine is extracted and passed over the sensor. Supporting electronics were placed in a cooler region near the sensor. The temperature of the gas flowing over the sensor is near 350°C, and the sensor temperature is maintained at that temperature using the built-in temperature detector and heater. The response of the sensor was compared to an industry-standard, continuous emission monitoring (CEM) system which was simultaneously measuring the NO_x production. As seen

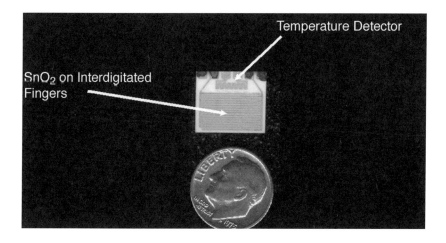

FIGURE 22.5 Tin oxide-based sensors fabricated on ceramic substrates for high-temperature operation. Shown are the interdigitated fingers on which SnO_2 is deposited and the temperature detector. The heater is on the backside of the ceramic.

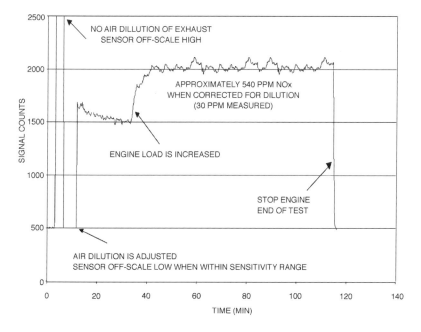

FIGURE 22.6 The response of a SnO_2-based NO_x sensor in a ground-based turbine engine emission test. The emission stream is diluted with air by 18:1 near the beginning of the test to avoid saturation of the sensor. The sensor output is display as signal counts processed by the supporting electronics.

in Figure 22.6, the sensor is initially saturated by the high level of NO_x. By diluting the flow into the sensor region with air by a factor of 18 (i.e., 18 parts air to 1 part exhaust), the sensor response is brought back into detection range. The engine load is then increased, which correspondingly increases the NO_x production. This is reflected by the both response of the microfabricated sensor (Figure 22.6) as well as the readings of the CEM. The two readings show good agreement: The sensor measured 540 ppm NO_x (after correction for the dilution) compared to the CEM measurement of 593 ppm. These results support the viability of replacing instrument-rack-sized CEM systems with MEMS-based chemical sensor technology.

22.4.3 Electrochemical Cell Oxygen Detection

A microfabricated O_2 sensor has applications both in fuel leak detection and, as demonstrated in the automotive emissions control example in Section 22.2.3, emissions control. Commercially available O_2 sensors are typically electrochemical cells using ZrO_2 as a solid electrolyte and platinum as the anode and cathode. The anode is exposed to a reference gas, usually air, while the cathode is exposed to the gas to be detected. Zirconium dioxide becomes an ionic conductor of $O^=$ at temperatures of 600°C and above. This property of ZrO_2 to ionically conduct O_2 means that the electrochemical potential of the cell can be used to measure the ambient oxygen concentration at high temperatures [Logothetis, 1991]. However, the operation of these commercially available sensors in this potentiometric mode limits the range of oxygen detection. This sensor is very sensitive near 0% O_2 but has limited sensitivity at higher O_2 concentrations. Further, the current manufacturing procedure of this sensor, using sintered ZrO_2, is relatively costly and results in a complete sensor package with a power consumption on the order of several watts.

The objective of the research described below is to develop a ZrO_2 solid electrolyte O_2 sensor using microfabrication and micromachining techniques. As noted in previous sections, the presence of O_2 often affects the response of H_2, C_xH_y and NO_x sensors. An accurate measurement of the O_2 concentration would help to better quantify the response of other sensors in environments where the O_2 concentration is varying. Thus, the combination of an O_2 sensor with other microfabricated gas sensors is envisioned to optimize the ability to monitor emissions.

A schematic cross section of the microfabricated O_2 sensor design is shown in Figure 22.7. As discussed in Section 22.4.2, CO and NO_x detection, microfabricating the sensor components onto a micromachined diaphragm region allows the sensor to be small in size and to have decreased energy consumption and reduced time to reach thermal equilibrium. When operated in the amperometric (current measuring) mode, the current of this cell is a linear function of the surrounding O_2 concentration. This linear response to oxygen concentration significantly increases the O_2 sensor detection range over O_2 sensors operated in the potentiometric mode. A chamber structure with a well-defined orifice can be micromachined to cover the sensing area. This orifice provides a pathway to control oxygen diffusion which is important in amperometric measurements. A filter may also be placed over the sensor to minimize flow and provide protection. As discussed in the introduction to this chapter, fabrication of a MEMS-based version of a macrodevice is not just a matter of decreasing the dimensions of the macrodevice. As discussed in Ward (1999), the fabrication of a microfabricated version of the oxygen sensor is a significant technical challenge with obstacles not encountered in the fabrication of the macrosized sensor. These challenges include adhesion between the various layers to each other and the substrate, pinhole electrical leaks between the layers and diffusion effects caused by heating the sensor to high temperatures. Testing of a complete O_2 sensor has been accomplished and further improvements on the design are planned.

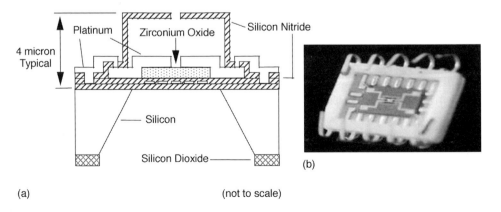

FIGURE 22.7 The structure of a microfabricated amperometric oxygen sensor and a packaged sensor without a protective orifice.

22.4.4 SiC-Based Hydrogen and Hydrocarbon Detection

Hydrogen and hydrocarbon sensors are being developed using SiC semiconductors in order to meet the needs of leak detection and emission monitoring applications. Like the Si-based system discussed in Section 22.4.1, the SiC-based Schottky diode does not need oxygen to operate. In contrast to the Si-based Schottky diode sensors, which are limited by the properties of the Si semiconductor to lower temperature operation (significantly below 350°C), SiC can operate at high enough temperatures (at least 600°C) as a semiconductor to allow the detection of hydrocarbons or NO_x. This opens the possibility of using SiC-based gas sensors in engine emission-sensing applications. The development of high-temperature SiC-based gas sensors for use in harsh environments at GRC/CWRU has focused on the development of a stable gas-sensitive Schottky diode. One main advantage of a Schottky diode sensing structure is its high sensitivity, which is useful in emission sensing applications where the concentrations to be measured are low (see Table 22.1).

The detection mechanism of SiC-based Schottky diodes is similar to that for Si-based diodes discussed in Section 22.4.1. For hydrogen, this involves the dissociation of the hydrogen on the surface of a catalytic metal. The atomic hydrogen migrates to the interface of the metal and the insulator, or the metal and the semiconductor, forming a dipole layer. Given a fixed forward-bias voltage, this dipole layer affects the barrier height of the diode, resulting in an exponential change of current flow with the change in barrier height. The magnitude of this effect can be correlated with the amount of hydrogen and other gas species (especially oxygen) present in the surrounding ambient. The detection of gases such as hydrocarbons is made possible if the sensor is operated at a temperature high enough to dissociate hydrocarbons and produce atomic hydrogen. The resulting atomic hydrogen affects the sensor output in the same way as molecular hydrogen [Baranzahi et al., 1995; Hunter et al., 1995].

The major advantage of SiC as a semiconductor for Schottky diode gas sensors is shown in Figure 22.8, which shows the zero-bias capacitive response of a Pd/SiC Schottky diode to one hydrocarbon, 360 ppm propylene, in nitrogen at a range of temperatures [Chen et al., 1996 and references therein]. The sensor temperature is increased from 100 to 400°C in steps of 100°C and the sensor response is observed. At a

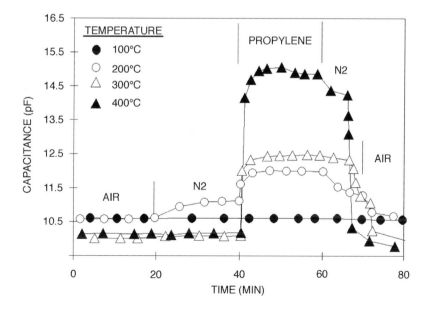

FIGURE 22.8 The temperature dependence of the zero-bias capacitance of a Pd/SiC Schottky diode to various gas mixtures. The sensor response to propylene is seen to require a temperature above a minimum temperature (~200°C) and to be strongly temperature dependent.

given temperature, the sensor is exposed to air for 20 minutes, N_2 for 20 minutes, 360 ppm of propylene in N_2 for 20 minutes, N_2 for 10 minutes and then 10 minutes of air. There are two points to note in the sensor behavior as shown in Figure 22.8. First, the baseline capacitance in air does not change between 100 and 200°C, but decreases to a slightly lower value as the temperature is raised to 300°C and above, suggesting the possibility that temperature-dependent chemical reactions that affect the diode electronic properties may have an onset at temperatures above 200°C. Second, Figure 22.8 clearly shows that the magnitude of sensor response to 360 ppm propylene depends strongly on the operating temperature. A sensor operating temperature of 100°C is too low for propylene to dissociate on the Pd surface, so the device does not respond at all. The three other curves for 200, 300 and 400°C show that elevating the temperature increases the response of the sensor to propylene. The presence of propylene can be detected at any of these higher temperatures, with 200°C being the minimum operating temperature determined in this study. Because the standard long-term operating temperature of Si is usually below 200°C, these results demonstrate the significant advantages of using SiC rather than Si in gas-sensing applications.

Several SiC-based Schottky diode structures have been investigated. Two MS structures investigated are Pd on SiC (Pd/SiC) and PdCr on SiC (PdCr/SiC). Very sensitive detection, on the order of 100 ppm of hydrogen and hydrocarbons in both inert and oxygen containing environments, has been demonstrated. Although the Pd/SiC sensor response is affected by prolonged high-temperature heating at 400°C, the PdCr/SiC structure has been shown to be much more stable with comparable sensitivity [Hunter et al., 1997]. The difference in behavior appears to be linked to reduced migration of Si into the PdCr metal from the SiC substrate and less formation of SiO_2 on the surface of the metal. Metal contacts to SiC for durable high-temperature operation, either as chemical sensors or as electrical device connections (see elsewhere in this book), are areas of continued investigation.

The third structure involves the incorporation of chemically reactive oxides into the SiC-based Schottky diode structure [Hunter et al., 1997; 1999a]. A wide variety of materials (e.g., metal oxides such as SnO_2) are sensitive to C_xH_y and NO_x at high temperatures. These materials could be incorporated as a sensitive component in MOS structures. Unlike silicon, SiC-based devices can be operated at high enough temperatures for these materials to be reactive to gases such as C_xH_y and NO_x. This results in a different type of gas sensitive structure: a metal reactive-oxide semiconductor structure (MROS). The potential advantages of this type of SiC-based structure include: (1) increased sensor sensitivity, as the diode responds to gas reactions due to both the catalytic metal as well as the reactive oxide; (2) improved sensor stability, as the gas reactive oxide can act as a barrier layer between the metal and SiC, potentially stabilizing the structure of the sensor against degradation at high temperature; and (3) the ability to vary sensor selectivity by varying the reactive oxide element. This MROS approach has been demonstrated using a SnO_2 film resulting in a $Pd/SnO_2/SiC$ structure. This structure has been shown to have improved stability and significantly different responses than those of the Pd/SiC structure. Further, the geometry and processing conditions by which the SnO_2 is deposited affects the sensor response. Multiple types of oxide layers are envisioned to produce different selectivities for the SiC-based Schottky diodes which would improve the ability to selectively determine the concentration of a specific gas (see Section 22.5.4).

Development of SiC-based sensor packaging to allow sensor operation at temperatures greater than 300°C is also an active research area. For example, temperatures above 300 to 400°C will likely be necessary to optimally measure gases such as propylene and methane. Unlike the SnO_2 sensors in Section 22.4.2 which can be directly deposited onto a MEMS-based structure, the SiC-based sensors are semiconductor chips processed separately and then packaged. The type of packaging depends on the application. For leak-detection applications in a room-temperature ambient, one difficulty with operating sensing elements functioning at higher temperatures than ambient is that a considerable amount of power (order of several watts) may be necessary to properly heat the sensing element. Such power demands to achieve optimal sensing temperatures may limit sensor use in some applications. For these applications, one approach towards minimizing the heating element power demand is to mount the sensor in a MEMS-based package that thermally isolates the sensors and decreases the thermal mass. An early prototype SiC-based packaged sensor design is discussed in Hunter et al. (1999b). For emissions applications, the objective is to package the sensor, with connections to the outside world, for operation in high-temperature engine environments. Thus, issues such as stable electrical contacts to SiC and the integrity of electrical insulation at high temperatures are of importance. Micromachining processes for SiC are being developed to enable formation of, for example, diaphragm structures that decrease the thermal mass and power consumption of the sensor in the same manner as is done with Si structures. These issues are discussed in more detail elsewhere in this book (see Chapter 21).

22.4.5 NASICON-Based CO_2 Detection

The detection of CO_2, like the detection of oxygen, can be based on the use of a solid electrolyte; however, the solid electrolyte CO_2 sensor is NASICON (sodium super ionic conductor) rather than zirconia. NASICON is an ionic conductor composed of $Na_3Zr_2Si_2PO_{12}$ that is useful for CO_2 detection. A sol–gel technique will be used to fabricate the NASICON. The sensor structure will be similar to that of Figure 22.7: a microfabricated, miniaturized sensor structure that can be incorporated into a package with other sensors such as the CO sensor. The ability to detect CO_2 over a range of concentrations has been demonstrated [Hunter et al., 1999a]. However, the largest challenges to application of this technology involves incorporation of NASICON onto the MEMS-based structure: film adherence, material compatibility and regeneration of the electrolyte are technical challenges associated with this approach.

22.5 Future Directions, Sensor Arrays and Commercialization

The long-term goals of NASA include significantly improving safety and decreasing the cost of space travel, significantly decreasing the amount of emissions produced by aeronautic engines, and improving the safety of commercial airline travel. In order to reach these goals, the development of "smart" vehicle technology is envisioned: a vehicle that can monitor itself and adapt to changing conditions. This monitoring, maintenance and management must take place throughout the operational lifetime of the vehicle to ensure high safety, integrity and reliability with decreased maintenance and costs. In order for self-monitoring to be effective, both physical and chemical parameters must be measurable throughout the vehicle, including the propulsion system. Thus, implementation of sensor technology in a variety of locations including harsh environments, vacuum or turbulent conditions will be necessary. Large, single sensors measuring a single parameter using a single detection mechanism will no longer be a viable technology approach. Rather, MEMS-based microsensor arrays which can measure several parameters of interest simultaneously and are integrated on a single device will be necessary to meet the objective of a "smart" vehicle.

One of the factors behind the reluctance to apply sensor technology throughout a vehicle is the reliability of the sensor technologies. Widespread use of sensors has been hampered, for example, by the fact that sensors can be damaged, require calibration, or produce unreliable results as they drift from calibration. Thus, integrated hardware and software that can monitor and condition the sensor, detect problems, perform self-calibration as necessary and provide this information to the user are also required: The accompanying electronics are what make a MEMS-based sensor system "smart." Further, these complete "smart" systems must meet the system requirements of the vehicle on which it is installed. As seen below for space-based systems, this may include military-grade components or being able to use the power available on the vehicle.

The sections that follow discuss work ongoing toward the development of MEMS-based chemical microsensor arrays. This discussion includes high-temperature membranes selective to individual gases as well as examples of arrays in development with supporting electronics. The particular examples given are a leak-detection array and a high-temperature electronic nose. The most advanced of these activities is the leak-detection array work: some of the same techniques being developed for that system can also be applied to other array systems. Finally, the non-aerospace, commercial applications of this technology are then discussed.

22.5.1 High-Selectivity Gas Sensors Based on Ceramic Membranes

Chemical sensors often lack selectivity and display similar responses to several different gaseous species [Hunter et al., 2000]. The selectivity of these sensors can be improved through the use of a gas selective membrane over the sensor to exclude other interfering gaseous species from reaching the sensing element. A selective membrane with a narrow distribution of pore sizes can filter gas molecules according to molecular size, thereby offering size selectivity to species of interest. Because the gas molecules of interest are on the order of several angstroms, the pore size would also have to be only several angstroms in diameter.

Zeolites appear particularly well suited to this filtering application. Zeolites are crystalline ceramic materials that contain open channels (or pores) as a natural part of their structure. Zeolites are composed primarily of aluminum, silicon and oxygen. The elements combine to form SiO_4 and AlO_4 tetrahedra which arrange in three-dimensional networks yielding uniform channels through the ceramic crystal. While these channels are identical in size or diameter within each crystallite of a zeolite, the diameters of these channels can vary significantly (from 3 to 12 Å), depending mainly on the ratio of aluminum to silicon. The effective diameters of the crystalline channels can be modified by the size of ions within the channels. Different ions may be incorporated into the crystalline channels by ion exchange. Further, as a ceramic material, zeolites are capable of withstanding the high temperatures and harsh environments that may be encountered in gas measuring applications.

Composite membranes containing zeolites have been prepared using various processing techniques to achieve desired selectivity, including sol-gel and electrochemical vapor deposition methods [Hunter et al., 2000]. Figure 22.9 shows the grain structure of one of the prototype films. The gas-selective membranes are being integrated with electrochemical sensors to further enhance the selectivity to gas molecules of interest.

22.5.2 Leak-Detection Array

The development of miniaturized chemical sensor microsystems using the hydrogen and oxygen sensor elements combined in one package is currently underway. A prototype MEMS-based leak-detection sensor system has integrated both hydrogen and oxygen sensors into a single substrate and realized

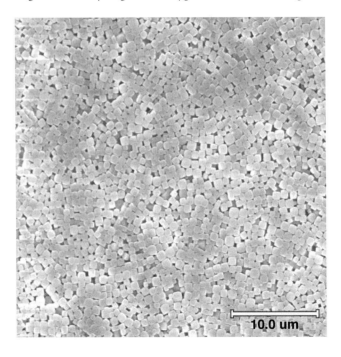

FIGURE 22.9 A scanning electron micrograph of a prototype of a selectively permeable zeolite thin-film membrane.

Microfabricated Chemical Sensors for Aerospace Applications

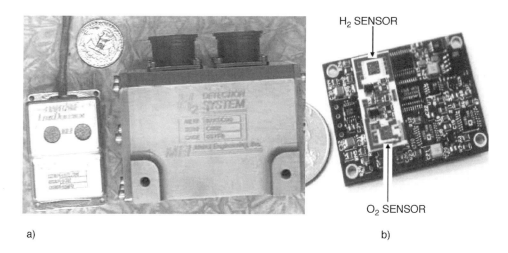

FIGURE 22.10 (a) (right) Shuttle system hardware (H_2 sensor with electronics); (left) prototype miniature smart sensor with hydrogen and oxygen sensors and control electronics with protective cover. (b) Electronics of miniaturized systems showing the H_2 and O_2 sensors.

improvements in the hardware supporting the sensor array. The size of the electronics is determined, in part, by the aerospace application. For space-based applications, electronics must often meet military specifications, be radiation hard, or function on vehicle voltage supplies not optimized for small-scale electronics (e.g., 28 V rather than 5 V). The smallest, most powerful electronics may not meet these specifications, or a bulky transformer may have to be added to the system to accommodate power restrictions.

Some results of these restrictions in one case are shown in the right-hand side of Figure 22.10a for the flight hardware that flew on the Shuttle. While the sensor size is 2 mm on each side and wafer thin, the associated packaging and hardware necessary for a system qualified for manned space flight are orders of magnitude larger. Nevertheless, significant improvements are possible. Figure 22.10b shows a prototype of an existing system: two sensors, hydrogen and oxygen, are integrated on the same board as the "smart" electronics. The smart sensor package contains all required analog and digital electronics, including CPU, memory and communications devices. The total volume of the package is less than 5 cm^3. The supporting electronics have been chosen for flight readiness while other components are still in the process of being qualified.

The operation of the prototype sensor unit with combined hydrogen and oxygen sensors is shown in Figure 22.11. The responses to changes in gas concentrations of the two hydrogen-sensing elements (a hydrogen-sensitive resistor and Schottky diode) as well as an oxygen sensor are shown. In a nitrogen ambient, all three sensors are first exposed to 21% O_2 and the concentration of hydrogen is varied from 0% to a maximum of 1.5% and back to 0%. Then, the oxygen background is changed from 21 to 4% O_2 and the hydrogen concentration stepwise increased and then decreased through the same hydrogen concentration range (0 to 1.5%). The hydrogen-sensitive resistor provides higher sensitivity at higher hydrogen concentrations, while the Schottky diode provides a larger response at lower hydrogen concentrations. The O_2 concentration is measured independently. The hydrogen sensor responses are seen to be somewhat affected by changing the O_2 concentration. However, unlike a number of other sensing techniques, the detection mechanisms of the sensors do not require O_2 for operation [Hunter et al., 1998 and references therein].

The combined responses from both the H_2 and O_2 sensors yield a powerful tool for leak detection. Because the H_2 sensor response weakly depends on the O_2 concentration, the O_2 signal can be used to more accurately determine the H_2 concentration. More importantly, because the LEL of the combustible gas depends, in part, on the corresponding concentration of the oxidizer, knowing the relative ratio of the gases is necessary to determine if an explosive condition exists. Using the sensor system shown in Figure 22.11, information is provided on both the concentration of the combustible gas (H_2) and the

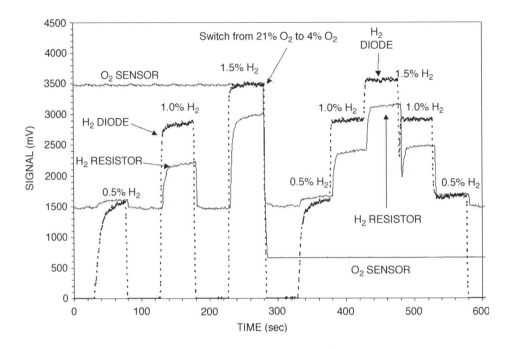

FIGURE 22.11 The response of a hydrogen- (two sensors) and oxygen-detection system with associated electronics to multiple concentrations of hydrogen in a changing oxygen ambient. The responses of all three sensors to varying hydrogen concentrations at two different oxygen concentrations are shown.

oxidizer (O_2) simultaneously, thus a more accurate determination of the hazardous conditions can be determined. This work demonstrates the beneficial use of multiple sensors with different detection mechanisms yielding improved information about the chemical constituents of an environment. It is envisioned that this multiple microsensor approach, combined with MEMS-based fabrication of the component parts, can allow the placement of sensors in a number of locations to better determine the safety of the region of interest.

22.5.3 High-Temperature Electronic Nose

Integration of a number of the individual high temperature gas sensors discussed above into a single platform will enable the formation of a sensor array. This array, composed of a range of sensors, will allow for the detection of a number of gases on a single chip operable in high-temperature environments such as an engine. The formation of an array of the high-temperature-capable sensors discussed in this chapter could detect H_2, C_xH_y, NO_x, CO, O_2 and CO_2. Development of a such a microfabricated gas sensor array operable at high temperatures and high flow rates would be a dramatic step toward realizing the goal of monitoring and control of emissions produced by an aeronautic engine. Such a gas sensor array would, in effect, be a high-temperature electronic nose and be able to detect a variety of gases of interest. Several of these arrays could be placed around the exit of the engine exhaust to monitor the emissions produced by the engine. The signals produced by this nose could be analyzed to determine the constituents of the emission stream, and this information then could be used to control those emissions. Microfabrication of these sensors is necessary: a conventional bulky system would add weight to the aircraft and impede the flow of gases leaving the engine exhaust. The component parts and packaging of this sensor array for a high-temperature environment would be fundamentally different than those used in the leak-detection example discussed in the previous section. The concept of an electronic nose has been in existence for a number of years [Gardner and Bartlett, 1994].

Commercial electronic noses for near-room-temperature applications currently exist and there are a number of efforts to develop other electronic noses. However, these electronic noses often depend significantly on the use of polymers and other lower temperature materials to detect the gases of interest. These polymers are generally unstable above 400°C and thus would not be appropriate for use in harsh engine environments. Thus, a separate development is necessary for a high-temperature electronic nose.

The development of such a high-temperature electronic nose has begun by using the high-temperature sensors discussed in this chapter. In addition to the higher temperature limit of the sensors that constitute this nose, another major difference in approach compared to conventional noses is that there are three very different sensor types that constitute the high-temperature electronic nose: resistors, electrochemical cells and Schottky diodes. As discussed earlier in this chapter, each sensor type relies on different mechanisms to detect gases so that each sensor type gives fundamentally different responses to the gas environment. For example, in general, electrochemical cells have logarithmic responses to different gas concentrations while Schottky diodes have exponential responses; resistors may have a sensitivity to many gases, while electrochemical cells can be operated to detect only gases that have certain binding energies; the time response of a Schottky diode is largely determined by migration of a chemical species into a film, while the time response of resistors such as SnO_2 is related to surface effects. Thus, each sensor type provides very different types of information on the environment. This is in contrast to a conventional array of sensors that generally consist of elements of the same type (e.g., SnO_2 resistors doped differently for different selectivities). Each sensor in this conventional system provides information available through the SnO_2 resistors (reactions occurring on the surface of the sensor film) but do not provide information determined by electrochemical cells or Schottky diodes. It is envisioned that the high-temperature electronic nose will sense responses from this array of very different high-temperature MEMS-based sensors (resistors, diodes and electrochemical cells) and integrate this information with neural net processing to allow a more accurate determination of the chemical constituents of harsh, high-temperature environments.

22.6 Commercial Applications

The gas sensors being developed at NASA GRC and CWRU are meant for aerospace applications but they can also be used in a variety of commercial applications, as well. For example, an early design of the Pd alloy hydrogen sensors, using PdAg, was adapted from space applications for use in an automotive application. GenCorp Aerojet Corporation, in conjunction with NASA Marshall Space Flight Center, developed hardware and software to monitor and control the NASA GRC/CWRU sensors to enable the sensors to monitor and display the condition of the tank of a natural gas vehicle. Several of these systems have been purchased for use on the Ford Motor Company assembly line to test for leaks in the valves and fittings of natural gas vehicles. This complete system received a 1995 R&D 100 Award as one of the 100 most significant inventions of that year.

Likewise, the high-temperature C_xH_y, NO_x and O_2 sensors are being developed for aeronautic applications but can also be applied to commercial applications. For example, the conditions in an aeronautic engine are similar to those of an automotive engine. Thus, sensors that work in aeronautic engine applications may be operable in automotive engine applications. The hydrogen sensors that can detect leaks in propulsion systems can be used to detect hydrogen evolution in fuel cells. Other possible applications include combustion process monitoring, catalytic reactor monitoring, alarms for high-temperature pressure vessels and piping, chemical plant processing, polymer production and volatile organics detection.

22.7 Summary

The needs of aerospace applications require the development of chemical gas sensors with capabilities beyond those of conventional commercial gas sensors. These requirements include operation in harsh environments, high sensitivity and minimal size, weight and power consumption. MEMS-based sensor technology is being developed to address these requirements using microfabrication and micromachining technology as well as SiC semiconductor technology. The combination of these technologies allows for the fabrication of a wide variety of sensor designs with responses and properties that can be tailored to the given application. Integration of these sensors with supporting packaging technology, hardware and software is necessary for their application. Sensors designed for aerospace applications also have significant commercial applications. Although each application is different and the sensor must be tailored for that environment, the base technology being developed for aerospace applications can have a significant impact on a broad range of fields.

Acknowledgments

The authors would like to acknowledge the contributions of Dr. Jih-Fen Lei, Dr. Philip G. Neudeck, Mr. Gustave Fralick, Dr. Daniel L. P. Ng and Mr. Paul Raitano of NASA GRC; Dr. Liang-Yu Chen and Dr. Valarie Thomas of the Ohio Aerospace Institute; Mr. Q. H. Wu and Mr. Benjamin Ward of the CWRU Electronics Design Center; Mr. William Rauch and Dr. M. Liu of the Georgia Institute of Technology and Mr. Greg Hall and Mr. José Perotti of the NASA Kennedy Space Center; also, the technical assistance of Mr. Drago Androjna and Mr. Carl Salupo of Akima/NASA GRC.

Defining Terms

Ambient: The gas composition dominant in the surrounding environment (air, pure nitrogen, pure helium, etc.).

Detection mechanism: The chemical and/or physical reaction by which a sensor responds to a given chemical species.

Interfering gases: Gases that can cause a competing response in a sensor and can thus mask or interfere in the response of the sensor to a given chemical species. For example, many sensors that respond to hydrogen can also respond to carbon monoxide; thus, carbon monoxide is an interfering gas in the measurement of hydrogen.

LIGA: *Lithographie, galvanoformung, abformung* (i.e., lithography, plating, molding), a German acronym signifying a combination of deep-etch X-ray lithography, electroplating and molding.

Lithographic reduction: The reduction in size of patterns, such as those used for microfabricated structures, onto a photolithographic mask. These patterns are eventually transferred from a mask to a thin film in the formation of a sensor structure.

Lower explosive limit (LEL): The lowest concentration at which a flammable gas becomes explosive. This limit depends on the flammable gas and the corresponding amount of oxidant. For hydrogen in air, this limit is a hydrogen concentration of 4%.

Micromachining technology: The fabrication of three-dimensional miniature structures using processing techniques such as etching and LIGA.

Photoresist patterning: The use of a polymer film, photoresist, to create miniature features after exposure to ultraviolet radiation and removal by a developer.

Response time: The time it takes for a sensor to respond to the environment. Because some sensors never completely stabilize and reach a stable maximum value, often a value equal to 90% of the steady-state value is cited.

Sacrificial layer: A thin layer of material added to a multilayer structure for the purpose of later being removed (or sacrificed) to open up that region of that structure.

Selectivity: The ability of a sensor to respond to only one chemical species.

Sensitivity: The amount of change in output of a sensor from baseline to a given chemical species.

Sol–gel: The formation of a compound through the use of organic compounds and water which, upon drying, form a gel that can be applied to a substrate. Higher temperature processing removes the organics and leaves the chemically sensitive material.

Stability: The reproducibility and repeatability of a sensor signal and baseline over time.

Thin Film: Typically, a film whose thickness is less than 1 μm.

Vehicle health monitoring (VHM): The capability to efficiently perform checkout, testing and monitoring of vehicles, subsystems and components before, during and after operation. This includes the ability to perform timely status determination, diagnostics and prognostics. The use of both sensors to determine the vehicle health and software to interpret the data is necessary.

References

Baranzahi, A., Spetz, A. L., Glavmo, M., Nytomt, J., and Lundstrom, I. (1995) "Fast Responding High Temperature Sensors For Combustion Control," in *Proc. 8th Int. Conf. Solid-State Sensors and Actuators and Eurosensors IX,* Stockholm, Sweden, pp. 741–744.

Chen, L.-Y., Hunter, G. W., Neudeck, P. G., Knight, D., Liu, C. C., and Wu, Q. H. (1996) "SiC-Based Gas Sensors," in *Proc. Third Int. Symp. on Ceramic Sensors,* 190th Meeting of the Electrochemical Society, Inc., October 6–11, San Antonio, TX, pp. 92–105.

Gardner, J. W., and Bartlett, P. N. (1994) "A Brief History of Electronic Noses," *Sensors and Actuators B* **18**(1–3), pp. 211–220.

Grosshandler, W. L. (1995) "A Review of Measurements and Candidate Signatures for Early Fire Detection," NISTIR 555, National Institute of Standards and Technology, Gaithersburg, MD.

Grosshandler, W. L., ed. (1997) *Nuisance Alarms in Aircraft Cargo Areas and Critical Telecommunications Systems, Proc. Third NIST Fire Detector Workshop,* December, Gaithersburg, MD, NISTIR 6146.

Hughes, R. C., Schubert, W. K., Zipperian, T. E., Rodriguez, J. L., and Plut, T. A. (1987) "Thin Film Palladium and Silver Alloys and Layers for Metal-Insulator-Semiconductor Sensors," *J. Appl. Phys.* **62**, pp. 1074–1038.

Hunter, G. W. (1992a) "A Survey and Analysis of Commercially Available Hydrogen Sensors," NASA Technical Memorandum 105878.

Hunter, G. W. (1992b) "A Survey and Analysis of Experimental Hydrogen Sensors," NASA Technical Memorandum 106300.

Hunter, G. W., Neudeck, P. G., Chen, L. Y., and Knight, D. (1995) "Silicon Carbide-Based Hydrogen and Hydrocarbon Gas Detection," paper presented at the 31st AIAA Joint Propulsion Conference, July 10–12, San Diego, CA, AIAA Paper 95-2647.

Hunter, G. W., Neudeck, P. G., Chen, L.-Y., Knight, D., Liu, C. C., and Wu, Q. H. (1997) "SiC-Based Schottky Diode Gas Sensors," in *Proc. Int. Conf. on SiC and Related Materials,* September 1–7, Stockholm, Sweden.

Hunter G. W. et al. (1998) "A Hazardous Gas Detection System for Aerospace and Commercial Applications," paper presented at 31st Joint Propulsion Conference, July 13–15, Cleveland, OH, AIAA Paper 98-3614.

Hunter, G. W., Neudeck, P. G., Chen, L.-Y., Liu, C. C., Wu, Q.H., Sawayda, S., Jin, Z., Hammond, J. D., Makel, D. B., Liu, M., Rauch, W. A., and Hall, G. (1999a) "Chemical Sensors for Aeronautic and Space Applications III," a short course presented at Sensors Expo '99, September, Cleveland, OH, NASA Technical Memorandum 1999-209450.

Hunter, G. W., Neudeck, P. G., Chen, L.-Y., Makel, D. B., Liu, C. C., and Wu, Q. H. (1999b) "SiC-Based Gas Sensor Development," in *Proc. Int. Conf. Silicon Carbide and Related Materials,* October 10–15, Research Triangle Park, MD, pp. 1439–1442.

Hunter, G. W., Neudeck, P. G., Fralick, G., Liu, C. C., Wu, Q. H., Sawayda, S., Jin, Z., Makel, D. B., Liu, M., Rauch, W. A., and Hall, G. (2000) "Chemical Microsensors for Aerospace Applications," in *Microfabricated Systems and MEMS V, Proc. Int. Symp.,* Hesketh, P. J. et al., eds., 198th Meeting of the Electrochemical Society, October 22–27, Phoenix, AZ, pp. 126–141.

Jin, Z. H., Zhou, H. J., Jin, Z. L., Savinell, R. F., and Liu, C. C. (1998) "Application of Nanocrystalline Porous Tin Oxide Thin Film for CO Sensing," *Sensors and Actuators B* **52**, pp. 188–194.

Logothetis, E. M. (1991) "Automotive Oxygen Sensors," in *Chemical Sensor Technology,* Vol. 3, ed. N. Yamazoe, Kodansha, Ltd., pp. 89–104.

Lundstrom, I. (1989) "Physics with Catalytic Metal Gate Chemical Sensors," *CRC Crit. Rev. Solid State Mater. Sci.* **15**, pp. 201–278.

Madou, M. (1997) *Fundamentals of Microfabrication,* CRC Press, Boca Raton, FL.
Neudeck, P. G. (1995) "Progress in Silicon Carbide Semiconductor Electronics Technology," *J. Electron. Mater.* **24**, pp. 283–288.
Ward, B. (1999) "Production Methods for an Amperometric Oxygen Sensor Incorporating Silicon Based Microfabrication and Micromachining Techniques," Masters thesis, Case Western Reserve University, Cleveland, OH.

For Further Information

For more information on the development of high-temperature sensor materials, see http://cism.ohio-state.edu/mse/cism/home_page.html for activities of the Center for Industrial Sensors and Measurements, Ohio State University. More information on the Glennan Microsystem Initiative, a consortium of NASA, industry and Ohio State universities to address microsystems for Harsh environments, can be found at http//www.glennan.org. For further information on semiconductor device technology, see Sze, S.M. (1981) *Physics of Semiconductor Devices,* 2nd ed., John Wiley & Sons, New York.

23
Packaging of Harsh-Environment MEMS Devices

Liang-Yu Chen
NASA Glenn Research Center

Jih-Fen Lei
NASA Glenn Research Center

23.1 Introduction 23-1
23.2 Material Requirements for Packaging Harsh-Environment MEMS 23-2
  Substrates • Metallization/Electrical Interconnection System • Die-Attach • Hermetic Sealing
23.3 High-Temperature Electrical Interconnection System 23-4
  Thick-Film Metallization • Thick-Film-Based Wirebond • Conductive Die-Attach
23.4 Thermomechanical Properties of Die-Attach 23-10
  Governing Equations and Material Properties • Thermomechanical Simulation of Die-Attach
23.5 Discussion 23-19
  Innovative Materials • Innovative Structures • Innovative Processes
  Acknowledgments 23-21

23.1 Introduction

Microelectromechanical system (MEMS) devices, as they are defined, are both electrical and mechanical devices. Via microlevel mechanical operation, MEMS devices, as sensors, transform mechanical, chemical, optical, magnetic and other nonelectrical parameters to electrical/electronic signals; as actuators, MEMS devices transform electrical/electronic signal to nonelectrical/electronic operations. Therefore, MEMS devices very often interact with the environment electrically, magnetically, optically, chemically and mechanically. In order to support these nonconventional device operations (i.e., the device mechanical operation and the nonelectrical interactions between the device and their environments), new packaging capabilities beyond those provided by conventional integrated circuit (IC) packaging technology are required [Madou, 1997]. A chemically inert, optically dark and electromagnetically "quiet" environment for packaging conventional ICs, provided by hermetic sealing and electromagnetic screening, is no longer suitable for packaging most MEMS devices. Because MEMS devices have very specific requirements for their immediate packaging environment, it is expected that the design of MEMS packaging will be very device dependent. This is in contradiction to the conventional IC packaging practice in which a universal package design can accommodate many different ICs. Compared to conventional IC packaging, the most distinct issue of MEMS packaging is to meet the requirements imposed by the mechanical operability and reliability of MEMS devices.

NASA is interested in using harsh-environment-operable MEMS and electronic devices [Neudeck et al., 2000; McCluskey and Pecht, 1999; Kirschman, 1999; Willander and Hartnagel, 1997] to characterize *in situ* combustion environments of aerospace engines and the atmosphere of inner solar planets such as Venus. The operation environment of a high-temperature pressure sensor, one of the most wanted sensors by NASA and the aerospace industry for diagnosis and control of a new generation of aerospace engines, specifies the general requirements for packaging these harsh-environment MEMS: This pressure sensor must operate in an engine combustion chamber and be completely exposed to the combustion environment in order to measure *in situ* combustion chamber pressure in real time. The specifications of the high-temperature pressure sensor and the standard of *in situ* operation environment have all been determined by the Propulsion Instrumentation Working Group [PIWG, 2001]. The sensor operates at temperatures up to 500°C in a gas ambience composed of chemically reactive species such as oxygen in air, hydrocarbon/hydrogen in fuel, and catalytically poisoning species such as NO_x and SO_x in combustion products. The sensor operation environment is summarized as high temperature, high dynamic pressure and chemically corrosive. This is indeed a typical harsh environment compared to the standard operation conditions for most advanced (commercial) sensors/electronics, so the packaging materials and basic components—including substrate, metallization material(s), electrical interconnections (such as wirebond) and die-attach—must be able to withstand an environment that is 500°C, corrosive and especially oxidizing/reducing and with high dynamic pressure. As discussed in Chapter 20, one of the suggested high-temperature pressure sensors is a semiconductor piezoresistive device. The sensing mechanism of this device depends on the mechanical deformation of semiconductor resistors residing on a diaphragm fabricated by micromachining. Therefore, this device is very sensitive to external forces applied to the device. The major source of undesired external force is the thermal mechanical stress from the die-attach structure due to mismatches of coefficients of thermal expansion (CTE) of the die material (such as SiC), the substrate material and the die-attaching material. The thermal stress of the die-attach must be suppressed in order to achieve precise and reliable device operation because the thermally induced stress may generate unwanted device response to the changes in thermal environment, and in the extreme case thermal stress can cause permanent mechanical damage to the die-attach.

A high-temperature pressure sensor is needed for a planned space mission to determine the atmospheric pressure profile of Venus, where the temperature is up to 500°C and the gas ambient is chemically corrosive. Normal conditions of the operating environment of these pressure sensors include high temperatures (500°C), chemical corrosion and high dynamic pressure, under which most conventional electronic packaging materials can no longer operate. Therefore, development of revolutionary MEMS devices operable in harsh environments presents significant challenges to the device packaging field. In the next section, general material requirements for the major components of harsh-environment MEMS packaging are analyzed by adopting the basic structure of conventional IC packaging, in which the die is attached to a substrate using a die-attaching material layer and the die is electrically interconnected to the package by wirebond.

23.2 Material Requirements for Packaging Harsh-Environment MEMS

A summary of the basic requirements of the major materials needed for packaging harsh environment MEMS may help to establish the general guidelines for material selections.

23.2.1 Substrates

The basic function of the packaging substrate is to provide a framework for attaching the device die, metallization for electrical interconnection (such as wirebond), mounting the leads connecting the chip to the external environment, building nonelectrical signal paths and mechanical and possibly also electromagnetical shielding. Plastic/polymer-type materials are not suitable for 500°C operation because of melting and depolymerization at temperatures above 350°C [McCluskey et al., 1997; Pecht et al., 1999].

Most metal and alloy materials suffer severely from corrosion, especially, from oxidation at temperatures approaching 500°C in air. So the remaining material system suitable for substrate is ceramics, which meets the basic requirements for a substrate: excellent and stable chemical and electrical properties in a harsh environment, especially at high temperatures and corrosive gas ambience. After selecting the substrate material, a metallization scheme (both materials and processing) and associated sealing materials matching the substrate material must be identified or developed simultaneously. In order to reduce the thermal stress of the die-attach, the CTE of the substrate material must match that of the device material (such as SiC). The properties of substrate surface and the interfaces formed at high temperature with other packaging materials, such as the die-attach material, also become very important. At high temperatures, the surfaces of some ceramics (nitrides and carbides) gradually react with gas ambience, such as oxygen and water vapor, and therefore would lead to changes in properties such as surface resistivity and surface adhesiveness. Concerns such as these regarding the surface properties of "well-known" ceramic materials in high-temperature corrosive gas ambience may lead to valuable results from research into packaging materials for high-temperature MEMS.

23.2.2 Metallization/Electrical Interconnection System

Most metals and alloys, including some noble metals, oxidize at temperatures approaching 500°C in air. So the metals and alloys commonly used in conventional IC packaging such as Cu, Al and Au/Ni coated Kovar are excluded from packaging applications at 500°C and above, unless a perfectly hermetically sealed inert/vacuum condition is achievable. Intermetallic phases form at the interfaces of different metals such as Al and Cu at temperatures above 200°C. The intermetallic phase very often may reduce mechanical strength of an interconnection system, so achieving material consistency of an interconnection system becomes extremely important to avoid thermal mechanical failure at high temperatures.

Because of their chemical stability and good electrical conductivity, precious metals are naturally considered for applications in substrate metallization and electrical interconnection. Some precious metals (e.g., platinum and palladium) react with atomic hydrogen to form H-rich alloys at elevated temperatures [Lewis, 1967]. Though this is desirable for gas-sensing devices [Hunter et al., 1999] it is not for electrical interconnection applications, because the phase transition may cause significant changes in the physical and electrical properties. Gold (Au) is widely used for both substrate metallization and wirebond in packaging and hybridizing conventional high-frequency, high-reliability ICs. Besides the high conductivities and superior chemical stability at high temperatures, Au also has a low Young's modulus and a narrow elastic region. If Au is used as a die-attaching material, these properties of Au are helpful to reduce thermal stress generated at the Au–die and the Au–substrate interfaces due to CTE mismatches. These features of Au are especially desirable for applications in packaging 500°C-operable MEMS because a wide range of operating temperature is of concern. It has been reported that Au thin film/wire with small grain size suffers from electromigration of Au atoms at grain boundaries at high temperature under extreme current density ($\sim 10^6$ A/cm^2) operation [Goetz and Dawson, 1996]. Surface modification and coating have been suggested for the Au conductor to withstand high-temperature and high-current-density operation [Goetz and Dawson, 1996]. The electrical migration effect is proportional to J^{-2} (J is the current density); therefore, the obtainable lifetime of a Au conductor operated at high temperature but low current density may still be substantial even without the surface modification as demonstrated in the next section. Developing a low-cost, highly conductive and thermally and chemically stable conductor with excellent thermal–mechanical properties for electrical interconnection is a valuable goal for packaging harsh-environment (high-temperature) devices [Grzybowski and Gericke, 1999; Harman, 1999].

23.2.3 Die-Attach

The basic function of die-attach is to provide mechanical, electrical and thermal support to the die. Most die-attaching (adhesive) materials used for packaging conventional ICs for operation at temperatures below 250°C are not suitable for 500°C operation. The basic failure mechanisms are depolymerization of epoxy-type materials, oxidizing and melting of eutectic-type materials and softening and melting of glass-type materials. Generally, we expect the die-attach materials to be electrically and thermally conductive and physically and chemically stable at high temperatures and to permit a low-temperature attaching process. For applications in packaging MEMS devices, the thermal mechanical properties of the die-attach material such as CTE, Young's modulus, fatigue/creep properties and their temperature dependences are very much of concern, as the die-attaching material is the material in intimate contact, both thermally and mechanically, with the die and the substrate. The material properties, especially the CTE, are expected to match those of the die and the substrate materials. In case the CTE of the die is not completely consistent with that of the substrate, the die-attaching material is ideally expected to thermomechanically compensate for any possible CTE mismatch between the die and the substrate materials by absorbing the thermal strain. Another major concern regarding die-attaching materials for packaging high-temperature MEMS devices is the long-term chemical and mechanical stability of the interfaces formed with the die (or the metallization of the die) and substrate (or the metallization of the substrate) materials at high temperatures. If the die-attach is expected to be electrically conductive, then the electronic properties of its interface with the die (or the metallization materials on the die) would also become critically important. Combining all these thermal mechanical, chemical and electrical requirements for die-attaching material, it is apparent that a material system for die attach with a suitable process is critical to success in packaging high-temperature MEMS device.

The thermomechanical stress in the die-attach due to the differences in properties of the materials of the die-attach structure (particularly CTE mismatch) may cause degradation and failure of packaged devices. The extreme case of die-attach failure caused by thermal stress is cracking of the die or the attaching material. Besides this catastrophic failure due to fatigue and creep of the die or the attaching material(s), thermal stress in die-attach can also cause the high-temperature MEMS device an undesired thermal response, irreproducibility of device operation, and output signal drift. Therefore, thermal mechanical failure due to CTE mismatches of the materials involved in the die-attach is expected to be a common and important thermal mechanical issue that must be addressed, especially for high-temperature MEMS devices.

23.2.4 Hermetic Sealing

The basic purpose of hermetic sealing is to create and maintain a stable and sometimes inert ambience for the packaged device. This simple function at low (room) temperature is difficult to achieve at temperatures approaching 500°C because, first, at such high temperatures most soft or flexible sealing materials such as plastic/polymer-based materials can no longer operate; second, sealing very often applies between different materials. The CTE mismatches of these materials make hermetic sealing difficult over a wide temperature range, especially under thermal cycling condition. Third, high temperatures activate and significantly promote thermal processes such as diffusion and degassing at material surfaces [Palmer, 1999], thus it becomes difficult to maintain an ambience in a small enclosure by sealing. Therefore, it is expected that creative sealing concepts are necessary to meet the requirements for packaging many high-temperature MEMS devices.

Harsh-environment MEMS packaging offers new challenges to the device packaging field. Discussions on the basic requirements of properties of the materials necessary for packaging harsh-environment MEMS are followed by addressing an electrical interconnection system (both electrically and mechanically) that is based on a ceramic substrate and thick-film metallization, with a compatible conductive die-attach scheme for chip-level packaging of low-power, harsh-environment SiC MEMS devices. The thermal mechanical optimization of Au thick-film material-based die-attach is discussed using nonlinear finite element analysis (FEA) results.

23.3 High-Temperature Electrical Interconnection System

As discussed in the last section, ceramic materials are naturally selected as packaging substrate materials because of their superior high-temperature chemical and electrical stabilities. Aluminum oxide (Al_2O_3) is a low-cost substrate widely used for conventional IC packaging. Another advantage of using Al_2O_3 is that many thick-film materials have been developed for metallization of Al_2O_3. In comparison to Al_2O_3, aluminum nitride (AlN) has a higher thermal conductivity and a low CTE that is very close to that of SiC. These features of AlN make it be useful in packaging high-temperature and high-power devices [Martin and Bloom, 1999]. If the basic framework of conventional IC packaging is adopted, the next step after selecting the substrate is to identify metallization material(s). In this section we review a Au thick-film-based 500°C operable electrical interconnection system for chip-level, low-power, harsh-environment MEMS packaging [Chen et al., 2000a; 2000c].

Thick-film metallization materials are usually composed of fine metal (such as gold) powder, inorganic binder (such as metal oxides) and organic vehicle. The screenprinting technique is usually used for thick-film coating for thickness control and patterning. During the initial drying process (at 100 to 150°C), the organic vehicle evaporates and the paste becomes a semisolid phase mixture of metal powder and binder. In the final curing process (~850°C recommended for most thick-film products), the inorganic binder molecules migrate to the metal/substrate (e.g., Au–ceramic) interface and form reactive binding chains. Au thin wires can be bonded directly to Au thick-film metallization pads using commercialized wirebond equipment to provide electrical interconnection in the packaging. Some new thick-film materials may be applicable to various ceramic substrates such as alumina (Al_2O_3) [Keusseyan et al., 1996] and aluminum nitride [Chitale et al., 1994; Shaikh, 1994; Keusseyan et al., 1996]. Compared with direct thin-film metallization on ceramic substrate, thick-film metallization offers low cost, simple process, low resistance, and better adhesion provided by the reactive binders at metal–substrate interface. Both the electrical and mechanical properties of Au thick-film materials for applications in hybrid-packaging conventional ICs have been extensively validated at $T < 150°C$. In order to evaluate Au thick-film materials for 500°C application, both electrical and mechanical tests at 500°C are necessary. Test results of Au thick-film-material-based electrical interconnections (Au thick-film printed wires and thick-film-metallization-based Au wirebond) and a conductive die-attach scheme using Au thick-film as die-attaching material for operation up to 500°C [Chen et al., 2000c] are reviewed as follows.

23.3.1 Thick-Film Metallization

A Au thick-film-printed wire circuit, as shown in Figure 23.1, was screenprinted on a ceramic substrate (AlN or 96% Al_2O_3) and cured at 850°C in air using the recommended curing process, forming printed wire and metallization pads for wirebond on ceramic substrate. The circuit was electrically and thermally tested at 500°C in air for a total of ~1500 hr by four-probe resistance measurement. The electrical resistance of the thick-film wire/circuit was first measured at room temperature; afterward, the temperature

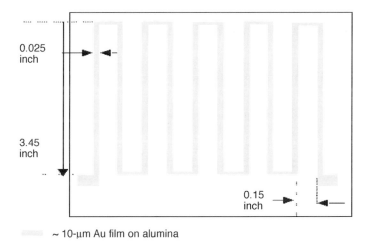

FIGURE 23.1 Schematic diagram of Au thick-film printed wire for high-temperature tests.

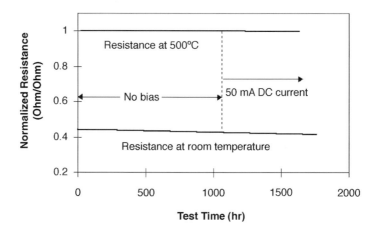

FIGURE 23.2 Normalized resistance of thick-film wire at various temperatures with and without dc bias.

was ramped up to 500°C for ~1000 hr without electrical current flow, and the resistance of the wire was measured periodically during this period of time. The resistance slightly fluctuated within ±0.1% during the 1000 hours test, as shown in Figure 23.2. After testing for 1000 hr without electrical bias, the circuit was biased with 50 mA dc current and the resistance was continuously monitored by four-probe resistance measurement. The resistance fluctuated slightly within ±0.1% for 500 hr with electrical bias. This very small change in resistance is acceptable for almost all envisioned high-temperature device packaging application.

As discussed earlier, Au thick-film materials for various substrates have been systematically validated both electrically and mechanically for conventional IC packaging; however, in order to be reliably applicable at high temperatures, these Au thick-film material systems must be mechanically evaluated at elevated temperatures in addition to the electrical validation discussed above. The tensile strength of Au thick-film metallization on a 96% Al_2O_3 substrate has been tested at room temperature after extended storage at 500°C [Salmon et al., 1998]. In order to examine the mechanical strength and the thermal dynamic stability of the binding system of Au thick films at high temperatures, the shear strength of selected Au thick-film metallization on 96% Al_2O_3 substrate was tested at temperatures up to 500°C [Chen et al., 2001]. The shear strength (breaking point) at 500°C reduced by a factor of ~0.80 with respect to that at 350°C, while the shear strength (breaking point) at 350°C was close to that at room temperature. The shear strengths of selected Au thick films designed for AlN were not as high as those for Al_2O_3 but were sufficient for application in microsystem packaging operable at 500°C.

23.3.2 Thick-Film-Based Wirebond

After the qualifications of Au thick-film metallization, the thick-film-based wirebond needs to be evaluated for high-temperature operation. The electrical test circuit including Au thick-film printed wires/pads and multiple thin gold wires bonded to the thick-film pads is illustrated in Figure 23.3. The geometry of the printed thick-film conductor was designed so that the electrical resistance of the test circuit was dominated by the resistance of thin (0.0254-mm diameter) bonded gold wires. The thick-film conductive wires/pads were processed according to the standard drying and curing processes [DuPont, 1999] suggested by the material manufacturer. The thin gold wires were bonded to the thick-film pads on the substrate by a thermal-compression wirebonding technique.

The electrical resistance of a thick-film-metallization-based wirebond test circuit (Figure 23.3) includes those of thick-film conductive wire/pads, bonded thin Au wires, and the interfaces of the wirebonds. The resistances of 22 units (44 bonds) in series were measured at room temperature and 500°C vs. accumulated

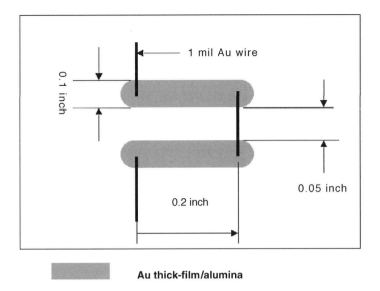

FIGURE 23.3 Schematic diagram of one unit of thick-film-metallization-based wirebond test circuit; 0.0254-mm (0.001 in.) Au wires were bonded to thick-film pads using the thermal-compression technique.

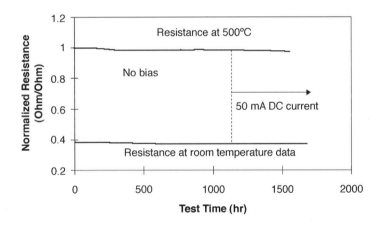

FIGURE 23.4 Normalized resistance of thick-film-metallization-based wirebond at various temperatures with and without dc bias.

testing time at 500°C. The resistance was first measured at room temperature, after which the temperature was ramped to 500°C and the resistance was monitored in air without electrical bias for 670 hr. As shown in Figure 23.4, the temperature was then lowered back to room temperature and the resistance was recorded again. After this thermal cycle, the circuit resistance was continuously monitored for a total of 1200 hr at 500°C in air without electrical bias (current flow), followed by another 500 hr at 500°C with 50-mA (dc) current. The resistances under all these conditions were desirably low (less than 0.5 Ω per unit) and decreased slowly and slightly at an average rate of 2.7% over the 1500-hr testing period. The rate of resistance decrease under the dc bias is close to that without electrical bias.

An identical wirebond sample was electrically tested in a dynamic thermal environment. The same wirebond circuit was tested in thermal cycles between room temperature and 500°C with an initial temperature rate of 32°C/min for 123 cycles and a dwell time of 5 min at 500°C, then at the temperature rate of 53°C/min (higher than thermal shock rate) for an additional 100 cycles with 50-mA dc current.

The maximum change in electrical resistance, during the thermal cycle test, was 1.5% at room temperature and 2.6% at 500°C [Chen et al., 2001]. The electrical stability of this Au thick-film-based wirebond system should meet most interconnection needs for high-temperature, low-power applications.

23.3.3 Conductive Die-Attach

23.3.3.1 SiC Test Die

The fabrication of a high-temperature SiC Schottky diode used to test the conductive die-attach scheme was previously reported in detail [Chen et al., 2000b]. An N-type (nitrogen, resistivity less than 0.03 Ω-cm), Si-terminated 4H-SiC wafer was used to fabricate test Schottky diode. The backside wafer (unpolished side) was first coated with a nickel (Ni) thin film by electron beam evaporation. The SiC wafer was then annealed at 950°C in argon in a tube furnace for 5 min, forming an ohmic contact on the backside of the SiC wafer. The device structure on the front side of the SiC wafer was fabricated by electron-beam evaporation of thin titanium (Ti) and thin Au films on cleaned SiC wafer and patterned with the liftoff technique.

23.3.3.2 Conductive Die-Attach

After dicing, the 1-mm × 1-mm SiC diode chips were attached to a ceramic substrate (either AlN or 96% alumina) using selected Au thick-film materials, as shown in Figure 23.5. An optimized thick-film die-attaching process for SiC device results in a low-resistance, conductive die-attach that is very often required for packaging devices requiring backside electrical contact.

The SiC test die with a Ni contact on the back was attached to a ceramic substrate using an optimized two-step Au thick-film processing [Chen et al., 2000b]. A thick-film layer was first screenprinted on the substrate and cured at 850°C using the standard process. The SiC die was then attached to the cured thick-film pattern with a minimal amount of subsequent thick film. A slower drying process (120 to 150°C) was critical to keeping the thick-film bonding layer uniform and the die parallel to the substrate after the curing process. Following the drying, the attached die was processed at a lower final curing temperature (600°C).

This optimized Au thick-film die-attach process allows sufficient diffusion of inorganic binders toward the thick film–substrate interface, resulting in a good strength of binding to the ceramic substrate. Meanwhile, it prevents the attached semiconductor chip from being exposed to temperatures above the ultimate operation temperature of SiC devices (600°C) during the die-attach process. The second advantage of this die-attach process is that the distribution of the thick film (after the final curing) between the chip and the substrate can be better controlled because only a minimal amount of thick-film material is necessary to attach the chip to a cured thick-film pad. Therefore, many potential problems caused by nonuniform thick-film distribution at the chip–substrate interface might be avoided.

A 0.001-in.-diameter Au wire was bonded onto the top of the Au thin-film metallization area covered with a Au thick-film overlayer by the thermal-compression bonding technique. Thick-film material was also used to reinforce the top Au thin film for better wire bonding. The Au thin-film metallization area

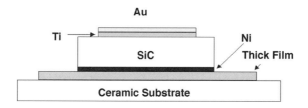

FIGURE 23.5 Schematic diagram of as-fabricated SiC device and die-attach structure.

was coated with thick film on the top then dried at 150°C for 10 min. The thick film on the device top was cured during the final die-attach process (at 600°C).

23.3.3.3 Electrical Test

The attached SiC test diode (Figure 23.5) was characterized by current–voltage (I–V) measurements at both room temperature and 500°C for various heating times at 500°C. A minimum dynamic resistance (dV/dI) under forward bias which is reduced from the I–V curve was used to estimate the upper limit of resistance of the die-attach structure (both interfaces and materials) and to monitor the resistance stability of the die-attach, as shown in Figure 23.6. This dynamic resistance includes the forward dynamic resistance of the Au–Ti–SiC interface, SiC wafer bulk resistance, the die-attach materials/interfaces resistance, bonded wire resistance and the test leads resistance in series. The resistance contributed from test leads and bonded wire were measured independently and subtracted. The attached device was first characterized by I–V measurements at room temperature. The device exhibited rectifying behavior, and the minimum dynamic resistance after subtracting the test-leads/bond-wire resistance (which also applies to all of the following discussion) measured under forward bias was ~2.6 Ω, as shown in Figure 23.7.

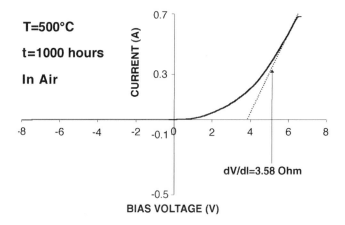

FIGURE 23.6 I–V curve of an attached SiC test diode characterized at 500°C after being tested for 1000 hr in 500°C oxidizing air.

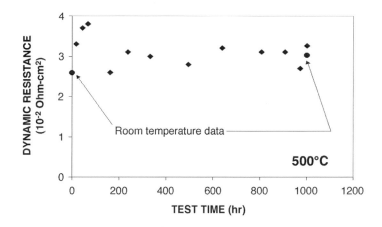

FIGURE 23.7 Minimum specific dynamic resistance (normalized to device area) calculated from I–V data vs. heating time at 500°C. This resistance includes resistances contributed from the Au(Ti)–SiC rectifying interface, SiC wafer and die-attach materials/interfaces. Resistances of the bonded wire and the test leads have been subtracted.

The temperature was then ramped up to 500°C (in air) and the diode was *in situ* characterized periodically by I–V measurement for ~1000 hr. During the first 70 hr at 500°C, the minimum dynamic resistance under forward bias increased slightly from 3.3 to 3.8 Ω. After that, the minimum dynamic resistance decreased slightly and remained at an average of 3.1 Ω. The diode was then cooled down to room temperature and characterized again. The minimum forward dynamic resistance measured at room temperature was 3.3 Ω. It is worth noting that the I–V curve of the device changed somewhat with time during heat treatment at 500°C. However, the minimum dynamic resistance of the attached diode remained comparatively low over the entire duration of the test and the entire temperature range, indicating a low and relatively stable die-attach resistance.

23.4 Thermomechanical Properties of Die-Attach

By adopting a conventional die-attach structure, in which the die backside is attached to a substrate using a thin attaching material layer, the die-attach using Au thick-film material discussed in the last section (illustrated in Figure 23.5) is a typical example of this structure. As discussed previously, the most important issue that concerns the thermal mechanical reliability of a packaged MEMS is the die-attach thermal stress and the guidelines for material selection, structure design and optimization of the attaching process to minimize the overall die-attach thermal stress effects on the mechanical operation of the device.

We start with a list of static equations governing the thermomechanical behavior of the die-attach structure to analyze the material properties and other factors determining the thermal mechanical properties of the structure. Following that, optimization of the die-attach thermal mechanical reliability is discussed using the simulation results of nonlinear finite element analysis [Lin, 2001].

23.4.1 Governing Equations and Material Properties

Assuming that the temperature distribution in the die-attach is static ($\frac{\partial T}{\partial t} = 0$) and uniform ($\frac{\partial T}{\partial x} = \frac{\partial T}{\partial y} = \frac{\partial T}{\partial z} = 0$), and the external force on the die-attach structure is zero, the general thermal mechanical governing equations for the die-attach system are

$$\frac{\partial^2 u}{\partial x^2} + \frac{\partial^2 v}{\partial x \partial y} + \frac{\partial^2 w}{\partial x \partial z} + (1 - 2v)\nabla^2 u = 0 \tag{23.1}$$

$$\frac{\partial^2 v}{\partial y^2} + \frac{\partial^2 u}{\partial x \partial y} + \frac{\partial^2 w}{\partial y \partial z} + (1 - 2v)\nabla^2 v = 0 \tag{23.2}$$

$$\frac{\partial^2 w}{\partial z^2} + \frac{\partial^2 u}{\partial x \partial z} + \frac{\partial^2 v}{\partial y \partial z} + (1 - 2v)\nabla^2 w = 0 \tag{23.3}$$

where $\nabla^2 = \frac{\partial^2}{\partial x^2} + \frac{\partial^2}{\partial y^2} + \frac{\partial^2}{\partial z^2}$; u, v and w are the displacements in the x, y and z directions, respectively; v is Poisson's ratio of the material. The normal stress distribution are determined by material properties, the temperature and the displacements [Lau et al., 1998]:

$$\sigma_x = \frac{\lambda}{v}\left[(1 - v)\frac{\partial u}{\partial x} + \left(\frac{\partial v}{\partial y} + \frac{\partial w}{\partial z}\right)\right] - \beta(T - T_o) \tag{23.4}$$

$$\sigma_y = \frac{\lambda}{v}\left[(1 - v)\frac{\partial v}{\partial y} + \left(\frac{\partial w}{\partial z} + \frac{\partial u}{\partial x}\right)\right] - \beta(T - T_o) \tag{23.5}$$

$$\sigma_z = \frac{\lambda}{v}\left[(1 - v)\frac{\partial w}{\partial z} + \left(\frac{\partial u}{\partial x} + \frac{\partial v}{\partial y}\right)\right] - \beta(T - T_o) \tag{23.6}$$

Shear stress are reduced from u, v and w as:

$$\tau_{xy} = \frac{E}{2+2\nu}\left(\frac{\partial u}{\partial y} + \frac{\partial v}{\partial x}\right) \quad (23.7)$$

$$\tau_{yz} = \frac{E}{2+2\nu}\left(\frac{\partial v}{\partial z} + \frac{\partial w}{\partial y}\right) \quad (23.8)$$

$$\tau_{zx} = \frac{E}{2+2\nu}\left(\frac{\partial w}{\partial x} + \frac{\partial u}{\partial z}\right) \quad (23.9)$$

where $\lambda = \frac{\nu E}{(1+\nu)(1-2\nu)}$, and $\beta = \frac{\alpha E}{1-2\nu}$. E is Young's modulus and α is the coefficient of linear thermal expansion (CTE). Material properties α, E and ν generally vary from the die to the attaching layer and the substrate. If there is no residual stress at a certain temperature, such as the die-attaching temperature, then the mismatch of CTEs of the die-attach is the major source of thermal mechanical stress. This would become more apparent if the governing equations, Eqs. (23.1) to (23.3), are listed for the die, the attaching layer and the substrate separately. There would be an external force term in these equations because of the mechanical interactions at the boundaries—the interface between the die and the attaching layer and the interface between the attaching layer and the substrate. These external force terms are generated by the mismatch of material CTEs or the residual stresses. In case there is no residual stress and the CTEs match with each other, these external force terms vanish and solutions for the equations would be trivial.

In the temperature range for conventional IC packaging, all these material properties can be approximated as constants. However, in the operation temperature range from room temperature to 500°C, the temperature dependences of these material properties must be quantitatively considered for a precise solution for these equations.

The boundary conditions at the interface between the die and the attaching layer and the interface between the attaching layer and the substrate are largely determined by the bonding quality and the thermal mechanical stability of the die/attaching layer interface and the attaching layer/substrate interface. The simplest model of boundary conditions is ideal bonding: The continuity of all displacement components at interfaces in the entire temperature range $u|_{S_s+} = u|_{S_s-}$, $v|_{S_s+} = v|_{S_s-}$, $w|_{S_s+} = w|_{S_s-}$, $u|_{S_a+} = u|_{S_a-}$, $v|_{S_a+} = v|_{S_a-}$ and $w|_{S_a+} = w|_{S_a-}$. S_s is the interface between the die and the attaching material, and S_a is the interface between the attaching material and the substrate, respectively.

The effects of thermal mechanical stress on the performance of a material/component are usually evaluated by either Von-Mises stress, $\sigma_{VM} = 1/2 S_{ij} S_{ij}$ (the deviation stress tensor, $S_{ij} = \sigma_{ij} - 1/3\sigma_{\alpha\alpha}\delta_{ij}$, where $\delta_{ij} = 1$ for $i = j$, and $\delta_{ij} = 0$ for $i \neq j$), or the maximum principle stress σ_{MP} [Giacomo, 1996]. Assume:

1. The substrate size is much larger than that of die and the material is amorphous/polycrystalline (isotropic).
2. One of the basal planes of the chip crystal is parallel to the die-attach interface.
3. Boundary conditions at the interfaces are determined by ideal bonding; all displacement components are continuous at both the interface between the die and the attaching layer and the interface between the attaching layer and the substrate.
4. The die-attach structure is in a uniform temperature field.

Then, the maximum thermal stress distribution, either Von-Mises stress or maximum principle stress (MPS), is basically a function of the thickness of the die (θ_d); the thickness of the attaching material layer (θ_a); the properties E, α and ν of the die material (E_d, α_d and ν_d); the properties of the attaching material (E_a, α_a and ν_a); the properties of the substrate material (E_s, α_s and ν_s); and the temperature deviation from the thermal stress relaxing temperature, $T - T_R$, where the T_R is thermal stress relaxing temperature:

$$\sigma = \sigma(x, y, z, \theta_d, \theta_a, \theta_s, \vec{M}_d, \vec{M}_a, \vec{M}_s, (T - T_R)) \quad (23.10)$$

where $\vec{M}_{d,a,s}$ are material property parameters of the die, the attaching material and the substrate, respectively. For 500°C operable packaging, because the operation temperature range is, relatively, much wider compared to that for conventional ICs, the temperature dependencies of these material properties and their nonlinearity have to be considered for a precise assessment of the stress/strain distribution. Usually σ is symmetric about the vertical die central axis if the package dimensions are much larger than those of the die. For example, if the die is in a square shape, σ has approximately lateral fourfold (90°) symmetry about the vertical die central axis. In case the die surface is off the crystal basal plane, then σ depends on the angle off the basal plan, and the horizontal symmetry of σ reduces from 90° to 180°. Ideally, if the CTEs of the die, attaching materials and substrate are consistent with each other in the entire operation temperature range and stress is relaxed as an initial condition, then at any temperature:

$$\sigma = \sigma(x, y, z, \theta_d, \theta_a, \vec{M}_d = \vec{M}_a = \vec{M}_s, (T - T_R)) = 0 \quad (23.11)$$

At the relaxing temperature, T_R, the global thermal stress is largely relaxed: $\sigma = \sigma(x, y, z, \theta_d, \theta_a, \vec{M}_d, \vec{M}_a, \vec{M}_s, (T = T_R)) = 0$. Sometime the stress distribution at relaxing temperature may not be completely relaxed but reaches a minimum; this minimum stress depends on the die-attaching material, the physical and chemical processes of die-attaching and the thermal experience of the die-attach.

Consider thick-film material-based die-attach as an example. T_R is basically the final curing temperature if the cooling process after the final curing is so rapid that thermal stress does not have a chance to relax through sufficient diffusion at lower temperatures. However, T_R can be lower than the final curing temperature if the cooling process is so slow that a relaxing configuration is reached at lower temperatures. Therefore, T_R is initially determined by a die-attach processing temperature but it may change with the thermal history (temperature vs. time) after the attachment. The dependence of T_R on the thermal history is not desired because this may cause a reproducibility problem for device response. The basic physical/chemical attaching process certainly has significant impact on T_R. For some die-attach materials and attaching process, such as phase-transition material, T_R may not be the die-attach processing temperature.

In addition to the direct impact from material properties of the attaching layer, the thermomechanical properties of the interfaces between the die, the attaching layer and the substrate are also important to the thermal stress–strain configuration of the die-attach structure. The interfacial shear elastic/plastic properties, the interfacial fatigue/creep behavior, and their temperature dependencies are determined by the chemical/physical interactions between the materials composing the interface. If the materials chemically react at the interface, an interphase may form and dominate the interfacial thermomechanical properties. The thermal mechanical properties of these interphases can certainly be critical to the overall thermomechanical properties of the die-attach structure.

In order to improve the thermal mechanical reliability of the packaged device the thermomechanical properties of the die-attach need to be optimized either locally or globally through material selection, structure design and process control. The guideline for local optimization is to minimize the maximum local thermal stress where the device mechanically operates to optimize mechanical operation with the least thermal stress effects. The guideline for global optimization is to minimize the weighted global stress. The parameters that can be adjusted to thermomechanically optimize the die-attach structure are $\vec{M}_a$, $\vec{M}_s$, θ_a and T_R after determination of die material and die size.

The thermal mechanical governing Eqs. (23.1) to (23.3) are so complicated that closed nontrivial (analytical) solutions of these equations are very difficult to obtain even for simple boundary geometry. The nonlinear temperature dependences of the material properties make it almost impossible to solve the equations with closed analytical solutions; however, numerical computing methods such as finite element analysis provide a powerful tool for simulation and optimization of the thermomechanical properties of a die-attach structure through material selection, structure design and processing control. Numerical methods make it possible to calculate the thermomechanical configuration of the packaging components/materials in a wide temperature range; however, it is still not trivial to optimize a die-attach structure for operation in a wide temperature range using FEA simulation because of the following reasons:

1. The numerical calculations involved in FEA sometimes can be difficult, especially when the boundary geometry of the MEMS devices is complicated and the dimensions of the mechanical structures are much smaller than the die-attach size [Rudd, 2000].
2. Currently, thermomechanical and fatigue/creep properties of many electronic and packaging materials in such a wide and high temperature range are often not completely available.
3. If the die-attaching process is based on phase-transition phenomena, then detailed changes in the thermomechanical properties of the attaching material, before and after the phase transition, may not be completely known.
4. Quantitative modeling of interfaces between the die, attaching layer, and substrate can be difficult.

Lin and Zou used a Si stress-monitoring chip to measure surface stress–strain of a die-attach [Lin et al., 1997a; Zou et al., 1999]. This unique *in situ* stress–strain monitoring method can be modified by using a high-temperature-operable SiC stress-monitoring chip to optimize the die-attach for high-temperature MEMS packaging.

23.4.2 Thermomechanical Simulation of Die-Attach

For high-temperature MEMS packaging, the thermomechanical reliability of a die-attach structure needs to be addressed at two levels: (1) mechanical damages of a die-attach structure resulted from thermal mechanical stress, and (2) thermal stress–strain effects on the mechanical operation of the device. Failures at both levels are rooted in thermomechanical stress in the die-attach. In the remaining part of this section FEA simulation results [Lin, 2001] of the Au thick-film-based SiC die-attach reviewed in the last section are analyzed as an example of die-attach optimization for high-temperature MEMS packaging.

23.4.2.1 High-Temperature Material Properties

The basic thermal and mechanical properties of SiC, Au, AlN and 96% Al_2O_3 and their temperature dependences used for FEA simulation are listed in Table 23.1 [Lin, 2001]. The temperature dependence of Young's modulus of AlN in a wide temperature range has not been reported, so it is extrapolated as a constant from the data at room temperature. The Poisson's ratios of 4H single-crystal SiC and AlN are not available; they were estimated according to those of other carbides and nitrides. The thermal and mechanical properties of 4H-SiC were assumed to be isotropic. Because the yield strength of gold thick-film material has not been published either, the simulation was conducted in a range of values from 650 to 3000 psi. However, computations using the yield strength at the low end very often diverged so the calculation was difficult.

23.4.2.2 Stress Distribution and Die Size Effects

Figure 23.8 shows Von-Mises stress and maximum principle stress contours of a quarter of a SiC–Au–AlN die-attach structure at room temperature assuming the relaxing temperature is 600°C. Horizontally, at the interface with the Au thick-film layer, the stress in the die basically increases with the distance from the center of the die-attach interface; the stress reaches a maximum at the area close to the die edges, especially at the corner area. This is understandable. The source of thermal stress of the die-attach structure is CTE mismatch at bonding interfaces. At the area far from the neutral point (larger distance from neutral point, DNP) the trends of relative displacement with respect to Au thick-film, driven by the CTE mismatch, would be larger. It can be predicted with these results that the maximum thermal stress would increase tremendously with the increase of die size or the attaching area. Vertically, the stress attenuates rapidly with the distance from the interface. At the die center region the stress near the die surface attenuates by a factor of 1/80 with respect to the stress at the interface. This picture of the thermal stress distribution indicates that flip-chip bonding would not be recommended for high-temperature MEMS

TABLE 23.1 Basic Material Properties of SiC, Au, 96% Alumina and AlN Used for FEA Simulation for Die-Attach Structure

Temp (deg C)	CIE (×E–6/C)	E (×E6 psi)	
AlN Material Properties			
−15	3.14	50.00	0.25
20	3.90	50.00	0.25
105	5.36	50.00	0.25
205	6.51	50.00	0.25
305	7.25	50.00	0.25
405	7.76	50.00	0.25
505	8.25	50.00	0.25
605	8.72	50.00	0.25
705	9.09	50.00	0.25
Au Material Properties			
−15	14.04	11.09	0.44
20	14.24	10.99	0.44
105	14.71	10.83	0.44
205	15.23	10.59	0.44
305	15.75	10.28	0.44
405	16.29	9.92	0.44
505	16.89	9.51	0.44
605	17.58	9.03	0.44
705	18.38	8.50	0.44
96% Alumina Properties			
−15	5.34	44.00	0.21
20	6.20	44.00	0.21
105	7.83	43.58	0.21
205	8.48	43.06	0.21
305	8.89	42.51	0.21
405	9.28	41.93	0.21
505	9.65	41.34	0.21
605	10.00	40.74	0.21
705	10.33	40.13	0.21
SiC Material Properties			
−15	2.93	66.72	0.3
20	3.35	66.72	0.3
105	3.97	66.42	0.3
205	4.23	66.08	0.3
305	4.46	65.74	0.3
405	4.68	65.39	0.3
505	4.89	65.05	0.3
605	5.10	64.71	0.3
705	5.29	64.36	0.3

devices if the CTE mismatch is not well controlled, and a significantly thicker die (or a stress buffer layer of the same material as that of the chip) can reduce the thermal stress at the die surface region.

Figure 23.8c shows equivalent plastic strain (EPS) in the Au thick-film layer. Because of the same physical mechanism as that for the SiC die, the highest EPS in the Au layer is located in an area close to the corner where the DNP is larger. This indicates again that a smaller die size or attaching area may better satisfy die-attach thermomechanical reliability requirements.

23.4.2.3 Effects of Substrate Material

In comparison with the AlN substrate, a 96% Al_2O_3 substrate has relatively higher CTE ($\sim 6.2 \times 10^{-6}$/°C compared with $\sim 3.9 \times 10^{-6}$/°C of AlN). In order to evaluate the substrate material effects on the thermomechanical stress of the die-attach structure, FEA simulation was used to compare the maximum stress in both the SiC die and the gold thick-film attaching layer using both substrates [Lin, 2001]. The results indicate that using AlN substrate would result in an improvement of maximum Von-Mises stress in the SiC die by a factor of 0.29, an improvement of Von-Mises stress in the substrate by a factor of

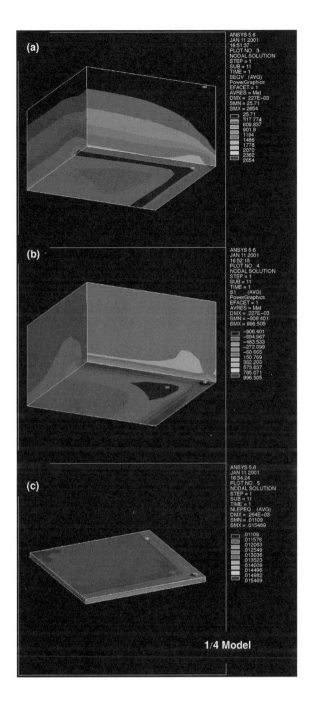

FIGURE 23.8 (Color figure follows p. **12**-26.) Thermal stress and strain distribution in SiC die and Au thick-film layer. (a) Von-Mises stress distribution in SiC die; (b) principle stress distribution in SiC die; (c) equivalent plastic strain in Au thick-film layer.

0.33, and an improvement of MPS in the Au thick-film layer by a factor of 0.42 assuming the yield strength of Au thick-film is 3000 psi. This improvement of thermal stress–strain corresponds to an improvement of fatigue lifetime by a factor of 4.3 to 9.0 (assuming that the power law exponent in the Coffin–Mason model, C, is −0.4 and −0.6, respectively). So, in terms of the thermomechanical reliability of the die-attach, AlN is suggested for packaging SiC high-temperature MEMS devices when compared to 96% Al_2O_3 owing to the fact that the CTE of AlN is closer to that of SiC.

SiC is another candidate substrate material for packaging SiC devices, but currently the cost of single-crystal SiC material is much higher compared with other ceramics suitable for packaging. The CTEs of α- and β-polycrystalline SiC are very close to that of single-crystal SiC so thermomechanically they are ideal substrates for packaging large SiC die for high-temperature operation. However, both the dielectric constants and the dissipation factors of these materials are relatively high [Johnson, 1999]. In order to be used as packaging substrates, the surface properties of polycrystalline SiC must be modified, but these materials might still be suitable only for low-frequency application because of the high dissipation factors.

23.4.2.4 Effects of Thickness of the Attaching Layer

The Au attaching layer contributes to the thermomechanical properties of the die-attach structure in two ways:

1. *Direct interface effect.* The Au layer forms interfaces with both the SiC die and the substrate, so the material properties and the configurations of the Au layer directly influence the thermal stress distributions in both the die and the substrate.
2. *Coupling effects.* As an interlayer between the die and the substrate, the Au layer couples the die with the substrate mechanically.

When the interlayer is very thick, the substrate and die are decoupled from each other, thus the second effect vanishes and the first effect dominates; if the thickness of the interlayer → 0, then the die and the substrate are directly coupled with each other, the influence of Au material properties vanishes, and the second effect dominates.

Figure 23.9 shows maximum Von-Mises stress in the SiC die and the substrates vs. the thickness of the Au attaching layer, θ_a. The maximum stress in the SiC die decreases by a factor of 0.75 with respect to the change in θ_a from 20 to 50 μm for an AlN substrate, while the maximum Von-Mises stress in the SiC die decreases by a factor of 0.5 with respect to the change in θ_a from 20 to 50 μm for the alumina substrate. The increase of Au layer thickness also significantly reduces the stress in the alumina substrate, as shown in Figure 23.9b. Figure 23.9c shows the maximum equivalent plastic strain in the Au layer vs. θ_a. The increase of θ_a from 20 to 50 μm significantly reduces the EPS in the Au layer in both AlN and Al_2O_3 cases. The maximum stress in SiC die and the substrate and EPS in Au attaching layer were calculated against the yield strength of Au thick-film.

23.4.2.5 Effects of Relaxing Temperature

In addition to the dependence of thermal stress in the die-attach on material properties of the die, the attaching layer and the substrate, the thermomechanical stress in the die-attach structure also depends on the temperature deviation from the relaxing temperature, T_R, at which the structure is largely relaxed. Therefore, the relaxing temperature of the die-attach structure is another important factor determining thermomechanical configuration (stress and strain) of a die-attach structure at a certain temperature. Generally, the more the temperature deviates from the relaxing temperature, the higher the thermomechanical stress that exists in the die-attach structure. For the type of die-attaching using diffusion-based bonding (e.g., thick-film-material-based die-attaching), the relaxing temperature is likely to be close to the processing (curing) temperature. For phase-transition-phenomena-based die-attach processes, the residual thermal stress could exist at the processing/attaching temperature because of the changes in material properties during the phase transition. In this case, the die-attach structure reaches a minimum stress configuration at certain temperature but the residual stress may not be completely relaxed.

In order to assess the relaxing temperature effects on the thermal stress of die-attach structure, the stress distribution of the die-attach at room temperature is simulated by FEA assuming that the structure is relaxed at various temperatures (from 300 to 600°C). Figure 23.10 shows the maximum Von-Mises stress in the die and EPS in the Au thick-film layer vs. the relaxing temperature. If the relaxing temperature could be lowered from 600 to 300°C, the maximum Von-Mises stress in the die could be reduced by a factor of 0.8, and the maximum EPS in the Au thick-film layer could be reduced by a factor of 0.5. The fatigue lifetime corresponding to the stress reduction is improved by a factor of 3 (assuming the exponent

Packaging of Harsh-Environment MEMS Devices

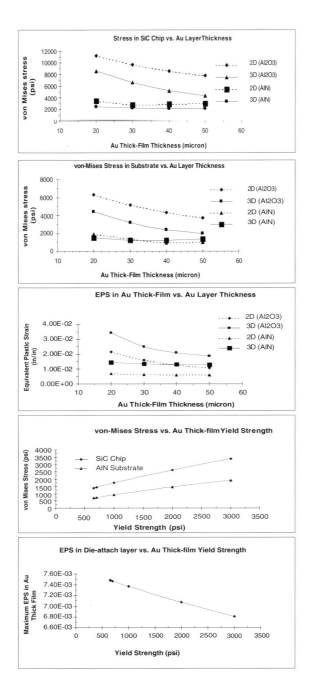

FIGURE 23.9 Dependence of stress–strain in SiC die, substrate and Au attaching layer on the thickness and the yield strength of Au thick-film layer. (From Lin, 2001. With permission.)

in the Coffin-Mason model of fatigue, C, is −0.6). Ideally, the thermal mechanical property of the die-attach structure is optimized if the relaxing temperature could be set at the middle of the operation temperature range, but physically this is not always realistic.

Mathematically, optimization of the thermomechanical property of a die-attach is a complicated multi-parameter problem. So, even if the numerical computing is possible, both the skill and the amount of calculations required for the optimization can be considerable.

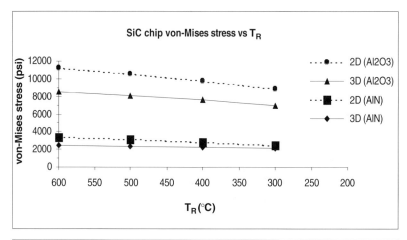
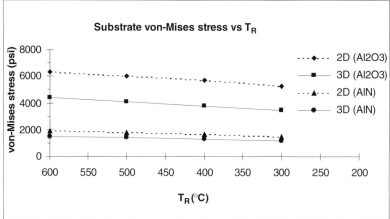
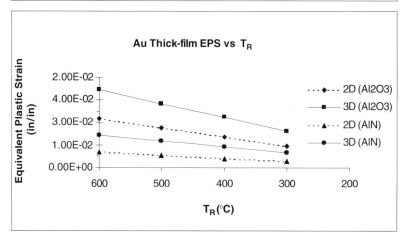

FIGURE 23.10 Relaxing temperature dependence of maximum stress and strain in SiC chip, substrate and Au attaching layer. (From Lin, 2001. With permission.)

23.5 Discussion

The combination of versatile functions of MEMS devices and survivability/operability in harsh environments leads to a new generation of microdevices, harsh-environment MEMS, with revolutionary capabilities for aerospace and civil applications. Packaging these revolutionary devices, however, has generated new and challenging research in the device-packaging field. In order to meet these challenges, innovative packaging materials with superior physical/chemical properties suitable for harsh-environment operation, innovative packaging structures and designs to meet the requirements of microstructures and micromechanical operations and innovative packaging processes to utilize these innovative packaging materials to fabricate innovative packaging structures are expected.

23.5.1 Innovative Materials

An ideal die-attaching material suitable for use in a wide temperature range is an illustration of the need for innovative materials to package harsh-environment MEMS devices. In addition to the superior chemical and electrical stabilities at high temperatures and in a corrosive environment, the die-attaching materials must possess good thermal and electrical conductivity, unique features of metallic material; meanwhile, the CTE of such a material should match those of the die and the ceramic substrate. Ideally, if the CTE of the substrate slightly mismatches that of the die, the die-attach material should be able to compensate the CTE mismatch between the die and the substrate. A material with low Young's modulus and narrow elastic region certainly would help to absorb thermal strain, thus reducing the stresses in the die and the substrate. However, this would reduce the lifetime of the die-attach material layer in a dynamic thermal environment because of the accumulated permanent strain. Comparing the extraordinary material requirements listed above with the features of carbon nano tubes (CNT) we discover that CNT is an ideal die-attach interlayer that matches all these requirements: CNTs have been reported to have superior electrical and thermal conductivities [Saito et al., 1998; Hone et al., 2000]. The longitudinal mechanical strength (both elastic modulus and breaking point) of CNTs is very high in comparison with steel [Wong et al., 1997; Hernandez et al., 1998; Poncharal et al., 1999; Salvetat et al., 1999; Yu et al., 2000], but the shear modulus of single-walled CNTs is low [Salvetat, 1999]. The graphitic bond between neighboring in-wall carbon atoms of each CNT makes the interaction between the neighboring tube walls weak [Girifalco et al., 2000]. Combining all these features of CNTs we can visualize an innovative die-attaching interlayer material. If CNT can be vertically grown on the C face of a SiC (the C face of a SiC is usually used for die-attach because the Si face is favored for fabrication of most electronic devices) wafer, it may provide an ideal interlayer with excellent thermal and electrical conductivities, superior mechanical strength and very low lateral Young's modulus and thus possess the unique capability to manage the CTE mismatch between the die and the substrate. A schematic diagram using CNT as die-attach interlayer that is expected to decouple the die mechanically from the substrate is illustrated in Figure 23.11.

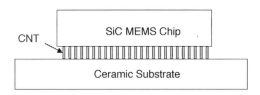

FIGURE 23.11 Schematic diagram of die-attach structure using carbon nano tube as an interlayer with capability to match the CTEs of both SiC chip and the substrate.

23.5.2 Innovative Structures

As discussed in previous sections, suppressing thermal stress in the die-attach in a wide temperature range is a common thermomechanical concern for high-temperature MEMS packaging. A bellow structure was invented (for conventional MEMS packaging) to absorb thermal strain and thus release thermal stress in the area where the device operates mechanically [Garcia and Sniegowski, 1995]. The bellow structure can be fabricated between the mechanical operation part and the chip base so the mechanical operation part is mechanically decoupled laterally from the rest of the device chip. Or, the bellow structure can be fabricated between the device chip and the packaging substrate so the entire chip is mechanically decoupled from the packaging environment. A similar structure can also be used for harsh-environment MEMS packaging. Various thermal stress suppression methods for packaging conventional MEMS were summarized by Madou (1997). Some of these methods may be modified for applications in packaging high-temperature MEMS.

As discussed in Section 23.4.2.2, the thermomechanical stress in the SiC die of the die-attach structure discussed in the last section attenuates rapidly with the vertical distance to the interface composed of SiC and Au with different CTEs. Horizontally, the maximum thermal stress increases rapidly with the die size or die-attaching area. These results suggest a simple but effective thermal stress suppression method, side die-attach, which attaches the die through one of the die sides to the packaging base rather than the bottom of the die. This die-attach scheme allows a much larger distance from the mechanical operation area to the die-attach interface and reduces the die-attach area. The side die-attach structure is expected to be especially effective to release thermomechanical stress–strain in the direction vertical to the die-attach interface because the chip has no direct restraint in this direction.

23.5.3 Innovative Processes

In order to accommodate the microsize mechanical features of MEMS devices and support their microlevel mechanical operation, the immediate device packaging environment may have to meet micromechanical requirements in both alignment and assembly. An innovative, organic-solution-based, self-assembly method was used to integrate microsized GaAs light emission diodes (LED) onto a micromachined Si substrate [Yeh and Smith, 1994]. GaAs LEDs are suspended in ethanol above a micromachined Si wafer. Macrovibration enables the LEDs to randomly walk in the solution. When a LED hits a vacancy on the Si wafer it fits in and gains the lowest potential there because of surface–ethanol–surface interactions. These interactions are so strong that perturbation from vibration would not free trapped LEDs. This is a good example of how innovative packaging processes at various levels beyond conventional IC packaging technology are introduced to package MEMS devices.

Madou (1997) summarized oxidation bonding, fusion bonding, field-assisted thermal bonding and modified field-assisted thermal bonding techniques developed for Si MEMS packaging. Low-pressure and low-temperature hermetic wafer bonding using microwave heating was introduced for Si wafer packaging [Budraa et al., 1999]. Some of these wafer-bonding methods may be modified for applications in high-temperature MEMS wafer materials such as SiC.

A microcavity is a common microstructure necessary to support the operation of many MEMS devices such as an absolute pressure sensor or a microresonator [Ikeda et al., 1990; Hanneborg and Øhlckers, 1990]. These microcavities are often included in the device fabrication instead of being assigned later to conventional packaging that typically is macrolevel processing. This reflects the trend that micromachining/fabrication processes initially used only for device fabrication now are also used to package MEMS devices. An extreme example of using microfabrication to package MEMS devices would be wireless "bug sensors" for operations *in vivo* and other harsh environments. The packaged device must be miniaturized; therefore, both the device and the package have to be microfabricated and micro-assembled. These MEMS devices become "self-packaged" because their fabrication and device packaging processes are completely merged [Santos, 1999]. These "self-packaged" MEMS devices may have the unique advantage of material consistency for operation at high temperatures.

Acknowledgments

One of the authors (L.C.) is very grateful to Dr. Gary W. Hunter for introducing the author to the challenging and vivid research field of high-temperature device packaging. Dr. Lawrence G. Matus is thanked for proofreading the manuscript. Discussions with Drs. Philip G. Neudeck and Robert S. Okojie drawing upon their device perspectives were very helpful to writing this material. The FEA modeling of the die-attach was conducted by Dr. Shun-Tien (Ted) Lin at United Technologies Research Center with support of NASA Electronic Parts and Packaging Program through NASA Glenn Research Center. The authors would like to thank Dr. Lin for his very helpful input on the FEA modeling results. The harsh-environment microsystem packaging work at NASA Glenn Research Center is currently supported by the NASA Glennan Microsystems Initiative (GMI) and NASA Electronic Parts and Packaging (NEPP) program.

References

Budraa, K. N., Jakson, H. W., William, T. P., and Mai, J. D. (1999) "Low Pressure and Low Temperature Hermetic Wafer Bonding Using Microwave Heating," in *IEEE Electro Mechanical Systems Technical Digest*, IEEE Catalog No. 99CH36291C.

Chen, L.-Y., Hunter, G. W., and Neudeck, P. G. (2000a) "Thin and Thick Film Materials Based Interconnection Technology for 500°C Operation, in *Trans. of First Int. AVS Conf. on Microelectronics and Interfaces*, February 7–11, Santa Clara, CA.

Chen, L.-Y., Hunter, G. W., and Neudeck, P. G. (2000b) "Silicon Carbide Die Attach Scheme for 500°C Operation," in *MRS 2000 Spring Meeting Proc.—Wide-Bandgap Electronic Devices (Symp. T)*, April 10–14, San Francisco, CA.

Chen, L.-Y., and Neudeck, P. G. (2000c) "Thick and Thin Film Materials Based Chip Level Packaging for High Temperature SiC Sensors and Devices," in *Proc. 5th Int. High Temperature Electronics Conference (HiTEC)*, June 11–16, Albuquerque, NM.

Chen, L.-Y., Okojie, R. S., Neudeck, P. G., Hunter, G. W., and Lin, S.-T. (2001) "Material System for Packaging 500°C MicroSystem," in *Symp. N: Microelectronic, Optoelectronic, and MEMS Packaging*, April 16–20, San Francisco, CA.

Chitale, S. M., Huang, C., and Sten, S. J. (1994) "ELS Thick Film Materials for AlN," *Advancing Microelectron.* **21**(1) pp. 22–23.

DuPont Electronic Materials (1999) *DuPont Processing and Performance Data of Thick Film Materials*, DuPont Electronic Materials, Research Triangle Park, NC.

Garcia, E., and Sniegowski, J. (1995) "Surface Micromachined Microengine," *Sensors Actuators A*, **A48**.

Giacomo, G. D. (1996) *Reliability of Electronic Packages and Semiconductor Devices*, McGraw-Hill, New York.

Girifalco, L. A., Hodak, M., and Lee, R. S. (2000) "Carbon Nanotubes, Buckyballs, Ropes, and a Universal Graphitic Potential," *Phys. Rev. B* **62**, p. 19.

Goetz, G. G., and Dawson, W. M. (1996) "Chromium Oxide Protection of High Temperature Conductors and Contacts," in *Transactions*, Vol. 2, Third Int. High Temperature Electronics Conf. (HiTEC), June 9–14, Albuquerque, NM.

Grzybowski, R. R., and Gericke, M. (1999) "Electronics Packaging and Testing Fixturing for the 500°C Environment," in *High-Temperature Electronics*, ed. R. Kirschman, IEEE Press, Piscataway, NJ.

Hannegorg, A., and Øhlckers, P. (1990) "A Capacitive Silicon Pressure Sensor with Low TCO and High Long-Term Stability," *Sensors and Actuators A*, **A21**, pp. 151–156.

Harman, G. G. (1999) "Metallurgical Bonding System for High-Temperature Electronics," in *High-Temperature Electronics*, ed. R. Kirschman, IEEE Press, Piscataway, NJ.

Hernandez, E., Goze, C., Bernier, P., and Rubio, A. (1998). "Elastic Properties of C and B C N Composite Nanotubes," *Phys. Rev. Lett.* **80**, p. 20.

Hone, J., Batlogg, Z. B., Johnson, A. T., and Fisher, J. E. (2000) "Quantized Phonon Spectrum of Single-Wall Carbon Namotubes," *Science*, **289**, p. 8. (See news release at the Web site http://www.eurekalert.org/releases/up-iac083000.html.)

Hunter, G. W., Neudeck, P. G., Fralick, G. C. et al. (1999) "SiC-Based Gas Sensors Development," in *Proc. Int. Conf. on SiC and Related Materials,* October 10–15, Raleigh, NC.

Ikeda, K., Kuwayama, H., Kobayashi, T., Watanabe, T., Nishikawa, T., Yoshida, T., and Harada, K. (1990) "Silicon Pressure Sensor Integrates Resonant Strain Gauge on Diaphragm," *Sensors and Actuators A* **21**, pp. 146–150.

Johnson, R. W. (1999) "Hybrid Materials, Assembly, and Packaging," in *High-Temperature Electronics,* ed. R. Kirschman, IEEE Press, Piscataway, NJ.

Keusseyan, R. L., Parr, R., Speck, B. S., Crunpton, J. C., Chaplinsky, J. T., Roach, C. J., Valena, K., and Horne, G. S. (1996) *New Gold Thick Film Compositions for Fine Line Printing on Various Substrate Surfaces,* ISHM Symp.

Kirschman, R., ed. (1999) *High-Temperature Electronics,* IEEE Press, Piscataway, NJ.

Lau, J., Wong, C. P., Prince, J. L., and Nakayama, W., (1998) *Electronic Packaging—Design, Materials, Process, and Reliability,* McGraw-Hill, New York.

Lewis, F. A. (1967) *The Palladium-Hydrogen System,* Academic Press, New York.

Lin, S. T. (2001) *Packaging Reliability of High Temperature SiC Devices,* NASA Contract Report.

Lin S. T., Benoit, J. T., Grzybowski, R. R., Zou, Y., Suhling, J. C., and Jaeger, R. C. (1997a) *High Temperature Die-Attach Effects in Die Stresses,* HiTeC '97 Proc., Albuquerque, NM.

Lin, S. T., Han, B., Suhling, J. C., Johnson, R. W., and Evans, J. L. (1997b) *Finite Element and Moire Interferometry Study of Chip Capacitor Reliability,* InterPack, Kohala, HI.

MacKay, C. A. (1991) "Bonding Amalgam and Method Making," U.S. Patent 5,053,195.

Madou, M. (1997) *Fundamentals of Microfabrication,* CRC Press, Boca Raton, FL.

Martin, T., and Bloom, T. (1999) "High Temperature Aluminum Nitride Packaging," in *High-Temperature Electronics,* ed. R. Kirschman, IEEE Press, Piscataway, NJ.

McCluskey, P., and Pecht, M. (1999) "Pushing the Limit: The Rise of High Temperature Electronics," in *High-Temperature Electronics,* ed. R. Kirschman, IEEE Press, Piscataway, NJ.

McCluskey, F. P., Grzybowski, R., and Podlesak, T. (1997) *High Temperature Electronics,* CRC Press, Boca Raton, FL.

Neudeck, P. G., Beheim, G. M., and Salupo, C. (2000) "600°C Logic Gates Using Silicon Carbide JFETs," 2000 Government Microcircuit Applications Conf. March 20–23, Anaheim, CA. See earlier review article: Davis, R. F., Kelner, G., Shur M., Palmour, J. W., and Edmond, J. A. (1991) "Thin Film Deposition and Microelectronic and Optoelectronic Device Fabrication and Characterization in Monocrystalline Alpha and Beta Silicon Carbide," Special Issue on Large Bandgap Electronic Materials and Components, *Proc. IEEE* **79**, p. 5.

Palmer, D. W. (1999) "High-Temperature Electronics Packaging," in *High-Temperature Electronics,* ed. R. Kirschman, IEEE Press, Piscataway, NJ.

Pecht, M. G., Agarwal, R., McCluskey, P., Dishongh, T., Javadpour, S., and Mahajan, R. (1999) *Electronic Packaging—Materials and Their Properties,* CRC Press, Boca Raton, FL.

Poncharal, P., Wang, Z. L., Ugarte, D., and de Heer, W. A. (1999) "Electrostatic Deflections and Electromechanical Resonances of Carbon Nanotubes," *Science* **283**, p. 5.

Propulsion Instrumentation Working Group (2001) http://www.oai.org/PIWG/table/table2.html.

Rudd, R. E. (2000) "Coupling of Length Scales in MEMS Modeling: the Atomic Limit of Finite Elements," in *Design, Test, Integration, and Packaging of MEMS/MOEMS,* eds. B. Courtois et al., May 9–11, Paris, France.

Saito, R., Dresselhaus, G., and Dresselhaus, M. S. (1998) *Physical Properties of Carbon Nanotubes,* Imperial College Press, London.

Salmon, J. S., Johnson, R., and Palmer, M. (1998) "Thick Film Hybrid Packaging Techniques for 500°C Operation," in *Trans. Fourth Int. High Temperature Electronics Conf.* (HiTEC), June 15–19, Albuquerque, NM.

Salvetat, J.-P., Briggs, G. A. D., Bonard, J.-M., Bacsa, R. R., Kulik, A. J., Stöckli, T., Burnham, N. A., and Forró, L. (1999) "Elastic and Shear Moduli of Single-Walled Carbon Nanotubes Ropes," *Phys. Rev. Lett.* **82**, p. 5.

Santos, H. J. D.-L. (1999) *Introduction to Microelectromechanical (MEM) Microwave Systems,* Artech House, Boston, MA.

Shaikh, A. (1994) "Thick-Film Pastes for AlN Substrates," *Advancing Microelectron.* **21**(1), pp. 18–221.

Willander, M., and Hartnagel, H. L. (1997) *High Temperature Electronics,* Chapman & Hall, London.

Wong, E. W., Sheehan, P. E., and Lieber, C. M. (1997) "Nanobeam Mechanics: Elasticity, Strength, and Toughness of Nanorods and Nanotubes," *Science* **277**, p. 26.

Yeh, H. J., and Smith J. S. (1994) "Fluidic Self-Assembly of Microstructures and Its Application to the Integration of GaAs on Si," in *IEEE Int. Workshop on Microelectromechanical Systems,* Oiso, Japan.

Yu, M.-F., Lourie, O., Dyer, M. L., Moloni, K., Kelly, T. F., and Ruoff R. S. (2000) "Strength and Breaking Mechanism of Multiwalled Carbon Nanotubes under Tensile Load," *Science* **287**, p. 28.

Zou, Y., Suhling, J. C., Jaeger, R. C., Lin, S. T., Benoit, J. T., and Grzybowski, R. R. (1999). "Die Surface Stress Variation during Thermal Cycling and Thermal Aging Reliability Tests," in *Proc. of IEEE Electronic Components and Technology Conference,* San Diego, CA.

Application of MEMS

24 Inertial Sensors *Paul L. Bergstrom, Gary G. Li* 24-1
Introduction • Applications of Inertial Sensors • Basic Acceleration Concepts
• Accelerometer Design Parameters • Micromachining Technologies for Inertial
Sensing • Micromachining Technology Manufacturing Issues • System Issues for Inertial
Sensors • Rotational Gyroscopes: Specific Design Considerations • Concluding Remarks

25 Micromachined Pressure Sensors *Jae-Sung Park, Chester Wilson
and Yogesh B. Gianchandani* 25-1
Device Structure and Performance Measures • Piezoresistive Pressure Sensors
• Capacitive Pressure Sensors • Other Approaches • Conclusions

26 Sensors and Actuators for Turbulent Flows *Lennart Löfdahl,
Mohamed Gad-el-Hak* 26-1
Introduction • MEMS Fabrication • Turbulent Flows • Sensors
for Turbulence Measurements and Control • Microactuators
for Flow Control • Microturbomachines • Concluding Remarks

27 Surface-Micromachined Mechanisms *Andrew D. Oliver, David W. Plummer* 27-1
Introduction • Material Properties and Geometric Considerations
• Machine Design • Applications • Failure Mechanisms in MEMS

28 Microrobotics *Thorbjörn Ebefors, Göran Stemme* 28-1
Introduction • What is Microrobotics? • Where To Use Microrobots?
• How To Make Microrobots? • Microrobotic Devices • Multirobot System
(Microfactories and Desktop Factories) • Conclusion and Discussion

29 Microscale Vacuum Pumps *E. Phillip Muntz, Stephen E. Vargo* 29-1
Introduction • Fundamentals • Pump Scaling • Alternative Pump
Technologies • Conclusions

30 Microdroplet Generators *Fan-Gang Tseng* 30-1
Introduction • Operation Principles of Microdroplet Generators • Physical
and Design Issues • Fabrication of Microdroplet Generators • Characterization
of Droplet Generation • Applications • Concluding Remarks

31 Micro Heat Pipes and Micro Heat Spreaders *G. P. "Bud" Peterson* 31-1
Introduction • Individual Micro Heat Pipes • Arrays of Micro Heat Pipes
• Flat-Plate Micro Heat Spreaders • New Designs • Summary
and Conclusions

32 **Microchannel Heat Sinks** *Yitshak Zohar* 32-1
Introduction • Fundamentals of Convective Heat Transfer in Microducts • Single-Phase Convective Heat Transfer in Microducts • Two-Phase Convective Heat Transfer in Microducts • Summary

33 **Flow Control** *Mohamed Gad-el-Hak* 33-1
Introduction • Unifying Principles • The Taming of the Shrew • Control of Turbulence • Suction • Coherent Structures • Scales • Sensor Requirements • Reactive Flow Control • Magnetohydrodynamic Control • Chaos Control • Soft Computing • Use of MEMS for Reactive Control • Parting Remarks

24
Inertial Sensors

Paul L. Bergstrom
Michigan Technological University

Gary G. Li
OMM Inc.

24.1	Introduction	24-1
24.2	Applications of Inertial Sensors	24-2
24.3	Basic Acceleration Concepts	24-4
24.4	Accelerometer Design Parameters	24-5
	Converting Force to Displacement: The Seismic Mass and Elastic Spring • Device Damping: The Dashpot • Device Damping: Impact Dynamics • Mechanical to Electrical Transduction: The Sensing Method	
24.5	Micromachining Technologies for Inertial Sensing	24-13
24.6	Micromachining Technology Manufacturing Issues	24-15
	Stiction • Material Stability • High-Aspect-Ratio Etching • Inertial Sensor Packaging	
24.7	System Issues for Inertial Sensors	24-17
	System Partitioning: One-Chip or Multi-Chip • Sensor Integration Approaches • System Methodologies: Open- or Closed-Loop Control • System Example: Motorola Two-Chip x-Axis Accelerometer System	
24.8	Rotational Gyroscopes: Specific Design Considerations	24-22
	System Considerations: Quadrature Error and Coupled Sensitivity • System Examples: Vibratory Ring Gyroscopes and Coupled Mass Gyroscopes	
24.9	Concluding Remarks	24-24

24.1 Introduction

Inertial sensors are designed to convert, or transduce, a physical phenomenon into a measurable signal. The physical phenomenon is an inertial force. Often this force is transduced into a linearly scaled voltage output with a given sensitivity. The methodologies used for macroscopic inertial sensors can and have been utilized for micromachined sensors in many applications. It is worth considering further what motivations have led to the introduction of micromachined inertial sensors. As will be demonstrated in this chapter, differences in accelerometer and gyroscopic application requirements do impact the choice of micromachining technology, transducer design and the system architecture. The definition of the system requirements often delineates micromachining technology options very clearly, although most sensing mechanisms and micromachining technologies have been applied to inertial sensors. First, design parameters will be reviewed for accelerometers. Accelerometers are the most straightforward inertial sensor system and will demonstrate the major physical mechanisms and micromachining technologies implemented to sense inertial displacement. Next, gyroscopic sensors, or sensors for rotational inertia, will be reviewed and will demonstrate how the technology can impact system and sensor, or transducer, design. Inertial device

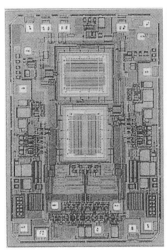 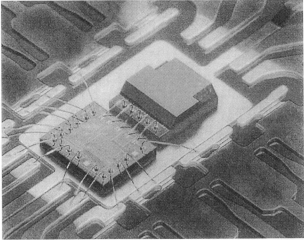

FIGURE 24.1 (Color figure follows p. 12.26.) Examples of two high-volume accelerometer products. On the left is Analog Devices, Inc. ADXL250 two-axis lateral monolithically integrated accelerometer. On the right is a Motorola, Inc. wafer-scale packaged accelerometer and control chips mounted on a lead frame prior to plastic injection molding. (Photos courtesy of Analog Devices, Inc. and Motorola, Inc.)

parameters are tightly coupled with lessons from Part I of this handbook, applying material from Chapters 2 to 5 by Trimmer and Stroud, Sharpe, Gad-el-Hak, and Kirby et al., in particular. A thorough background in micromachining technologies that have been implemented in micromachined inertial devices can be found in Part II of this handbook. Systems issues of inertial sensing tie in with the control theory discussions found in Chapters 12 and 13 by Goodwine and Bewley, respectively.

24.2 Applications of Inertial Sensors

Three primary areas often motivate micromachined device applications: packaged volume or size, system cost and sometimes performance. Often these three drivers cannot be met in a single technology choice. Packaged volume or overall system size is often an easy goal for micromachined inertial sensors compared to the macroscopic counterparts. Micromachining technologies are capable of reducing the sensor element and electronics board components to the scale of one integrated or two co-packaged chips, potentially as small as an 8-pin dual inline package (DIP) or surface mount (SMT) plastic package. Figure 24.1 shows two such examples: (left) an integrated accelerometer technology produced by Analog Devices, Inc., and (right) a co-packaged accelerometer produced by Motorola, Inc.

Reducing the system cost is also a goal easily achieved by micromachined inertial sensors as compared to macroscopic systems. Due to their relation to the microelectronics industry, micromachined sensors can be batch fabricated, sharing process costs over large volumes of sensors, reducing the overall process constraints significantly. While many individual processes may be significantly more expensive than the macroscopic counterparts, the impact is greatly reduced because the benefits of scale can be applied.

Meeting a targeted device performance with sufficient profitability requires improvements in production costs per unit. One factor in this cost improvement is die area utilization. Maximizing device sensitivity per unit area minimizes die area. Current sensor device requirements for occupant safety systems have allowed the initial incorporation of surface micromachined technologies. Surface micromachining technologies use the successive deposition of sacrificial and structural layers to produce an anchored yet freestanding device typically made of polycrystalline silicon with a structural thickness of less than 3 μm [Ristic et al., 1992]. These technologies have been successful for current design requirements but are being replaced by technologies that demonstrate application flexibility and improved die area utilization.

TABLE 24.1 Inertial Sensor Applications in the Automobile

	Range	Application Comments
Applications	±1 g	Antilock braking systems (ABS), traction control systems (TCS), virtual reality (VR)
	±2 g	Vertical body motion
	±50 g	Front-airbag deployment, wheel motion
	±100–250 g	Side (B-pillar) airbag deployment
	±100–250°/s	Roll and yaw rate for safety and stability control
Resolution	<0.1%	Full-scale, all applications
Linearity	<1%	Full-scale, all applications
Output noise	<0.005–0.05 %FS/$\sqrt{Hz}$	Full-scale signal, all applications
Offset drift	<1 g/s	Accelerometer
	<0.1°/s/s	Gyroscope
Temperature range	–40 to 85°C	Operational conditions
	–55 to 125°C	Storage conditions
Cross-axis sensitivity	<1 to 3%	Application dependent
Frequency response	dc to 1–5 kHz	Airbag deployment
	dc to 10–100 Hz	Gyroscope and 1- to 2-g accelerometers
Shock survivability	>500 g	Powered all axes
	>1500 g	Unpowered all axes

Source: Adapted from MacDonald, N.C. (1990) *Sensors and Actuators A*, **21**, 303–307.

The system area least often demonstrating improvement due to micromachined sensors as compared to the macroscopic counterparts is system performance. The trade-off between system size and cost and performance is often directly coupled, although notable exceptions exist to that rule. Performance improvements have been the primary technology drivers for micromachined inertial sensors. Micromachined technologies have succeeded in applications driven by size and cost with modest performance requirements. One such application is the automobile.

The automobile is significantly driven by system cost and currently motivates most micromachined inertial sensor technology development. Automotive inertial sensor applications are listed in Table 24.1. Accelerometers are among the largest volume micromachined products used in airbag control systems in automobile occupant-restraining systems. Accelerometers designed for airbag systems typically require inertial sensitivities of 20 to 100 g full-scale for front-impact airbags and 100 to 250 g full-scale for side-impact airbags. (One g represents the acceleration due to Earth's gravity.) Single-axis inertial sensors are also used in vehicle dynamics for active suspension systems with typical required inertial sensitivities ranging from 0.5 to 10 g. Future occupant safety systems will likely require more sensors to tailor a system response to the conditions of the crash event. Crash variables can include impact location, occupant position and weight, use of seat belts and crash severity. A future occupant safety system may use many multi-axis transducers distributed around the automobile to determine whether an airbag should be deployed and at what rate.

Vehicle dynamics and occupant safety systems are increasing in complexity and capability. Used in active suspensions, traction control and rollover safety systems, low-g accelerometers, yaw rate sensors and tilt gyros are employed and tied into engine, steering and antilock braking systems to return control of the vehicle to the driver in an out-of-control situation. Combined with front- and/or side-impact airbag systems, the system will determine the severity of the event and deploy seat belt pretensioners, side bags, head bags, window bags and interaction airbags between occupants as necessary. Versions of these systems are being introduced on more and more vehicles today. The merger of many of these systems in the future will provide greater system capability and complexity.

Gyro technologies, encompassing pitch, roll and yaw sensing, require significantly higher sensitivities than the analogous accelerometer. Such devices typically exhibit milli-g resolution in order to produce stable measurements of rotational inertia with less than one degree per minute drift. Design considerations for such devices encompass the same micromachining technologies but add significant device and system complexity to achieve stable and reliable results.

Outside the automotive marketplace there are many applications for sub-g inertial sensor products including virtual reality systems, intelligent toys, industrial motion control, hard drive head protection systems, video camera image stabilization systems, shipping damage detectors, robotic warehouse operations, GPS receivers and inertial navigation systems.

24.3 Basic Acceleration Concepts

Inertial sensing is dependent on the reference frame and what is being measured. Practically speaking, two primary applications define the required reference frames and measurements for most devices: the linear accelerometer and the rotational gyroscope. The accelerometer is generally defined in a Cartesian reference frame and measures the kinematic force due to a linear acceleration, as shown in Figure 24.2. The rotational gyroscope can be defined in a cylindrical or Cartesian reference frame and measures the angular velocity of a rotation about its primary axis due to the coupled Coriolis force on a rotating or vibrating body.

Linear acceleration, **a**, can be defined as:

$$\mathbf{a} = \frac{d^2\mathbf{r}}{dt^2} = \frac{d\mathbf{v}}{dt} \tag{24.1}$$

where **r** denotes linear displacement in meters (m) and **v** is the linear velocity in meters per second (m/s). The equation is written in vector notation to indicate that, while in most systems only one axis of motion (where the scalar, x, would replace **r**) is allowed, off-axis interactions need to be considered in complex system design.

Rotational gyroscopes measure angular velocity, ω, which is defined as:

$$\omega = \frac{d\theta}{dt} \tag{24.2}$$

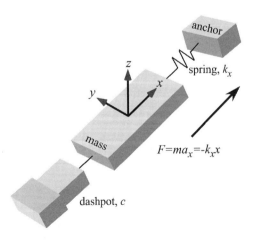

FIGURE 24.2 Cartesian reference frame for linear accelerometers. The figure shows an x-axis device for reference, including anchor, spring, seismic mass and dashpot.

where θ is the angular displacement in radians, and the angular velocity, ω, is measured in radians per second (rad/s). Angular acceleration, α, may also be measured and is defined as:

$$\alpha = \frac{d^2\theta}{dt^2} = \frac{d\omega}{dt} \qquad (24.3)$$

To measure the angular velocity of a moving body, the Coriolis effect is most often utilized in micromachined *vibratory* gyroscopic devices, although rotating micromachined gyroscopes have been demonstrated [Shearwood et al., 1999]. If a proof mass is driven into oscillation in one axis, rotation of the gyroscope reference frame will displace the oscillating mass into a second orthogonal axis. This coupling is the Coriolis effect and is given by:

$$\mathbf{F}_c = 2m\mathbf{v} \times \Omega \qquad (24.4)$$

which is a force produced when a vibratory mass, *m*, moving at a velocity, **v**, is placed in a rotation, Ω. This Coriolis-induced force is orthogonal to the vibratory motion as defined by the vector cross product.

24.4 Accelerometer Design Parameters

The successful implementation of micromachining technology to produce an inertial sensor requires more than the micromachining technology. The trade-offs between transducer and circuitry define the system approach and often demonstrate the complexity of the machined systems these micromechanical systems intend to replace. All micromachining technologies exhibit limitations that must be accounted for in the design of the overall system. The design of the transducer and the design of the overall system are interwoven. The transducer design and technology capability must first be addressed. The integration of the transducer function at the system level also defines the partitioning and complexity of the technology.

The manufacturability of a transducer structure is of paramount importance and should be vigorously considered during the design stage. In theory, one may design a sensor structure as sensitive as desired, but if the structure cannot be manufactured in a robust manner the effort is wasted. For example, a too-large proof mass or a too-soft spring may dramatically increase the probability of stiction, resulting in yield loss. In this regard, the design must follow a certain guideline of design rules (design rules will differ from technology to technology) in order to improve the yield of manufacturing.

Two primary axes of inertial sensitivity are produced for accelerometers: in-plane, often called *x*-axis or *x*-lateral accelerometers, and normal to the wafer surface, or *z*-axis accelerometers. The choice of axis is primarily driven by the application. Front-airbag systems require in-plane sensing, or *x*-axis sensing, while side- and satellite-airbag sensors are often mounted vertically, calling for *z*-axis sensing. Gyroscopic sensors to measure angular rate require a minimum of two orthonormal axes of motion in order to suitably measure the small Coriolis force exerted on a resonating proof mass during rotation. Rollover gyros typically resonate in plane and sense normal to the surface. Axes of sensitivity for gyroscopic sensors are shown in Figure 24.3.

Any inertial sensor design requires a seismic mass (also called a proof mass), an elastic spring, a dashpot and a method to measure the displacement of the seismic mass. The mass is used to generate an inertial force due to an acceleration or deceleration event, and the elastic spring to mechanically support the proof mass and to restore the mass to its neutral position after the acceleration is removed. The dashpot is usually the volume of air, or controlled ambient, captured inside the cavity and is designed to control the motion of the seismic mass and to obtain favorable frequency-response characteristics. The sense methodology converts the mechanical displacement to an electrical output.

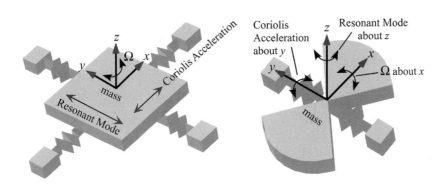

FIGURE 24.3 Reference frames for rotational gyroscopes based on the Coriolis effect showing axes of sensing for yaw and roll applications.

24.4.1 Converting Force to Displacement: The Seismic Mass and Elastic Spring

The application of an inertial load exerts a force on the proof mass, which is translated into a displacement by the elastic spring. The simple force equation for a static load is shown in Eq. (24.5), where m is the proof mass (kg), a is the static or steady-state inertial acceleration (m/s^2), k is the elastic spring constant (N/m), and x is the displacement of the mass with respect to the reference frame generated by the inertial load. In the ideal case, the proof-mass displacement is well controlled and in one axis only. The ratio of the proof mass to spring constant defines the sensitivity of the system to inertial loads:

$$\text{sensitivity} = \frac{\text{output (displacement)}}{\text{input (acceleration)}} = \frac{x}{a} = \frac{m}{k} \qquad (24.5)$$

The accelerometer design needs to precisely control the displacement of the seismic mass. The proof mass should be sufficiently massive and rigid to act in a well-behaved manner. In general, the design of a proof mass should provide at least an order of magnitude greater stiffness than the elastic spring stiffness in the axis of sensitivity. For a lateral sensor, the design of a sufficiently massive proof mass benefits by thicker structures. For a given transducer area, the mass increases linearly with thickness, and the stiffness out of plane increases by the third power.

In most applications, sufficient proof-mass stiffness is not difficult to achieve; however, design considerations may be impacted by other system requirements. For example, in lateral capacitive structures the incorporation of interdigitated sense fingers attached to the periphery of the proof-mass can complicate the proof-mass behavior. These sense fingers are often only a few times stiffer than the elastic springs and introduce a nonideal behavior to the sensitivity for large inertial loads. Minimizing this impact is often a significant design trade-off between device area and nominal capacitance for a device. Again, the thickness of the sense fingers improves the stiffness of a beam design and increases nominal capacitance per finger.

The elastic spring is required to provide kinematic displacement of the proof mass in the axis of sensitivity to produce a suitable sense signal while being sufficiently rigid in other axes to eliminate cross-axis sensitivity. The spring constant of a beam is governed by its geometry and material properties. In general, the spring constant of a beam of uniform material is defined by the proportion of its length to its cross-sectional area. Micromachined springs are often folded, or bent around a radius, to improve the performance of the overall device. Bent beams are often used to relieve residual intrinsic stress in the micromachined material and stabilize device parameters across wafers and production lots. Folded beams perform this function and reduce the device topology for the overall structure. Folded-beam elastic spring designs may also improve the sensitivity of the inertial sensor to package stress by reducing the spacing between anchored points of the springs around the periphery of the sensor structure.

For a lateral accelerometer, increasing the out-of-plane stiffness proves to be one of the major challenges for many micromachining technologies. Drop-shock immunity and cross-axis sensitivity are strongly impacted by the stiffness out of plane. Increasing the thickness of the beam increases the out-of-plane stiffness by the third power while the spring constant in the lateral axis of sensitivity only increases linearly. The aspect ratio, or ratio of the height to width of the spring, impacts the relative difference in spring constants in and out of plane. Because the spring constant in the lateral axis of sensitivity is defined to be the chosen geometry of the device, simply increasing its geometrical length or reducing the beam width can compensate the increase in spring constant of a beam. While this reduces the out-of-plane stiffness somewhat, it has a considerably smaller linear impact compared to the cubic increase in beam stiffness due to the increase in thickness.

For a rotational gyroscopic sensor, two axes of motion must be adequately controlled for the device to function properly. Drop-shock immunity and cross-axis sensitivity impact gyros, as well. Producing devices with sufficient stiffness off-axis to minimize the introduction of error and noise signals in a gyro is a major undertaking. Simply increasing the thickness of the micromachined layer impacts both axes of motion in a gyro, greatly increasing the complexity of the design effort.

24.4.2 Device Damping: The Dashpot

While the proof mass and elastic spring can be designed for a static condition, the design of the dashpot should provide an optimal dynamic damping coefficient through squeeze-film damping by the proper choice of the sensor geometry and packaging pressures. The extent of this effect is defined by the aspect ratio of the space between the plates and the ambient pressure. A large area-to-gap ratio results in significantly higher squeeze numbers, resulting in greater damping as described by Starr (1990).

Micromachined inertial sensor devices are often operated in an isolated environment filled with nitrogen or other types of gas such that the gas functions as a working fluid and dissipates energy. A gas film between two closely spaced parallel plates oscillating in normal relative motion generates a force, due to compression and internal friction, which opposes the motion of the plates. The damping due to such a force, related to energy loss in the system, is referred to as squeeze-film damping. In other cases, two closely spaced parallel plates oscillate in a direction parallel to each other, and the damping generated by a gas film in this situation is referred to as shear damping. Under the small-motion assumption, the flow-induced force is linearly proportional to the displacement and velocity of the moving plates. The coefficient of the velocity is the damping coefficient, c:

$$m\frac{d^2\Delta x}{dt^2} + c\frac{d\Delta x}{dt} + k_x \Delta x = -ma_x \tag{24.6}$$

A single-axis dynamic model of an accelerometer is shown in Eq. (24.6). The solution to the differential equation demonstrates the fundamental resonant mode for the primary axis of motion shown in Eq. (24.6). More generally, inertial sensors have multiple on- and off-axis resonant modes to be controlled, but the resonance in the axis of sensitivity is the primary effect with the greatest impact on device operation. The magnitude of the resonance peak is determined by the magnitude of the damping coefficient, c. This is the major contributing factor in the stability of the system and demonstrates the trade-off between spring constant and seismic mass. Typically, the design intent is to push the resonant modes far beyond the typical frequencies of interest for operation. Too large a seismic mass or too low a device spring constant drops the magnitude of the fundamental resonant frequency and reduces the system margin for operation.

For a given inertial sensor design, the magnitude of the damping is impacted by the aspect ratio of the spaces surrounding the seismic mass, the packaged pressure and other internal material losses in the system. Figure 24.4 shows how packaged pressure impacts the magnitude of the resonant peak for a surface micromachined lateral polysilicon accelerometer. By reducing the magnitude of the resonance

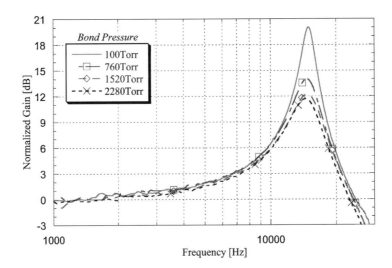

FIGURE 24.4 Normalized mechanical response over frequency for a 3-μm-thick lateral accelerometer capped at various pressures.

peak using squeeze-film damping at higher pressures, the motion of the seismic mass is brought under better control. In this case, for a given aspect ratio, increasing the pressure increases the damping coefficient, c, by approximately a factor of two. Alternatively, increasing the aspect ratio has a significant impact on the damping coefficient, c, for a given ambient pressure. In the case of high-pressure packaging, improving the aspect ratio of a design can minimize technology complexity by reducing the required packaging pressure to manageable and controllable values.

While the unit "N/m" (Newton per meter) for spring constant is well understood, the unit "kg/s" (kilogram per second) for damping coefficient is somewhat abstract, and it is often difficult to grasp its magnitude. Consequently, for a given mass-spring-dashpot oscillator, a nondimensional damping ratio, ξ, is often used in practice. An under-damped system corresponds to $\xi < 1$, a critically damped system corresponds to $\xi = 1$, and an over-damped system corresponds to $\xi > 1$. To correlate the dimensional damping, c, with the nondimensional damping, ξ, consider the following governing equation of motion for a forced simple harmonic oscillator:

$$m\ddot{x} + c\dot{x} + kx = f_0 \sin \omega t \quad (24.7)$$

where m is mass, c is damping coefficient, and k is spring constant. We then introduce a nondimensional damping, ξ, such that:

$$\xi = \frac{c}{2}\sqrt{\frac{m}{k}} \equiv \frac{c}{2m\omega_n} \quad (24.8)$$

where $\omega_n = \sqrt{k/m}$ is the system's natural frequency. With ξ and ω_n defined, Eq. (24.7) can be written into a standard form of:

$$\ddot{x} + 2\xi\omega_n \dot{x} + \omega_n^2 x = \frac{f_0}{m} \sin \omega t \quad (24.9)$$

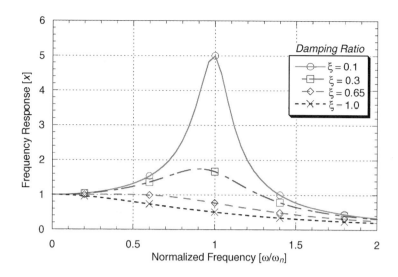

FIGURE 24.5 The frequency response $\bar{x}$ vs. normalized frequency ratio ω/ω_n.

The steady-state solution to Eq. (24.9) is therefore:

$$x(t) = \frac{1}{\sqrt{\left[1-\left(\frac{\omega}{\omega_n}\right)^2\right]^2 + \left[2\xi\frac{\omega}{\omega_n}\right]^2}} \times \frac{f_0}{k}\sin(\omega t - \phi) \equiv \bar{x}\sin(\omega t - \phi) \qquad (24.10)$$

where $\bar{x}$ is the amplitude of the response, and ϕ is the phase difference. The damping ratio, ξ, has a large influence on the amplitude in the frequency region near resonance. For $\frac{\omega}{\omega_n} \ll 1$, the amplitude of response is simply the static displacement f_0/k. Near resonance, the amplitude of the response may be amplified greatly for small values of ξ. Although the critical damping condition of $\xi = 1$ would eliminate the vibration altogether, the maximum flatness of the system response can be achieved at a damping of $\xi = 0.65$. This can be clearly seen in Figure 24.5.

Another terminology related to damping is the quality factor, Q. In forced vibration, Q is a measure of the sharpness of resonance. It is defined as a ratio of the system's natural frequency ω_n over a bandwidth $\omega_2 - \omega_1$, where ω_1 and ω_2 are two frequencies at which the response, $\bar{x}$, is 0.707 of its peak value. Manipulating Eq. (24.10), it can be shown that, for small damping,

$$Q = \frac{\omega_n}{\omega_2 - \omega_1} = \frac{1}{\sqrt{1+2\xi} - \sqrt{1-2\xi}} \cong \frac{1}{2\xi} \qquad (24.11)$$

Therefore, at low damping, the quality factor Q is approximately inversely proportional to damping ξ. A quality factor of 5 means a damping ratio of 0.1.

To eliminate resonance in a transducer, a damping of $\xi > 0.65$ is required [Blech, 1983]. Both ambient pressure and the transducer structure influence the damping. To illustrate this, consider a sensing structure modeled for a 3-μm polysilicon technology. The simulated damping at 1 atm is about 0.11. If a thicker sensing structure were used, the damping would increase approximately in a quadratic manner with thickness, t. At $t = 10$ μm, the damping ratio is 1.1, a factor of 10 increase. If the spring stiffness were kept at constant while increasing the thickness, t, the pace of damping increase would go up faster. At $t = 10$ μm, the damping is more than 2 and becomes over-damped.

If damping is too low in a micromachined lateral accelerometer, the severe degree of resonance of the accelerometer, upon an impact of external force, may produce a large signal that overloads the control circuitry, resulting in system failure. High damping (near critical) is generally desired for accelerometers. As for yaw-rate gyroscopic sensors, on the other hand, low damping is required in order to achieve sufficient sensitivity of the system under a given driving force and for certain types of applications. Therefore, in designing a MEMS device, the consideration of damping must be taken into account at the earliest stage.

In general, the capping pressure for a micromachined system is below or much below the atmospheric pressure. As pressure decreases, the mean free path of the gas molecules (nitrogen for example) increases. When the mean free path is comparable to the air gap between two plates, one may no longer be able to treat the gas as a continuum. Therefore, an effective viscosity coefficient is introduced such that governing equations of motion for fluid at relatively high pressures can still be used to treat fluid motion at low pressures where the mean free path is comparable or even larger than the air gap of the plates.

Based on earlier lubrication theory [Burgdorfer, 1959; Blech, 1983], Veijola et al. (1995, 1997, 1998) and Andrews et al. (1993, 1995) conducted extensive studies on squeeze-film damping. They showed that the viscous damping effect of the air film dominates at low frequencies or squeeze numbers, but the flow-induced spring becomes more prominent at high frequencies or squeeze numbers. For systems where only squeeze-film damping is present and for small plate oscillations, experimental studies [Andrews et al., 1993; Veijola and Ryhanen, 1995] for large, square, parallel plates are in general agreement with theory. In particular, the study carried out by Corman et al. (1997) is especially important for systems exposed to very low ambient pressures. For bulk micromachined resonator structures in silicon, they presented a squeeze-film damping comparison between theory and experiments for pressures as low as 0.1 mbar. The agreement between theory and experiment is considered to be reasonable. By employing the technique of Green's function solution, Darling et al. (1998) also conducted a theoretical study on squeeze-film damping for different boundary conditions. The resulting models are computationally compact and thus applicable for dynamic system simulation purposes. Additional studies by van Kampen and Wolffenbuttel (1998), Reuther et al. (1996) and Pan et al. (1998) also dealt with squeeze-film damping with an emphasis on numerical aspects of simulations. From these studies, one may get a comprehensive understanding of the micromachined systems and their dynamic behavior and gain insight on how to fine-tune the design parameters to achieve higher sensitivity and better overall performance.

It is also worth mentioning the experimental work by Kim et al. (1999) and Gudeman et al. (1998) on microelectromechanical systems (MEMS) damping. Kim and coworkers investigated the squeeze-film damping for a variety of perforated structures by varying the size and number of perforations. They found from finite element analysis that the model underestimated the squeeze-film damping by as much as 66% of the experimental values. Using a doubly supported MEMS ribbon of a grating light valve device, Gudeman and coworkers were able to characterize the damping by introducing a concept of "damping time." A simple linear relationship is found between the damping time of the ribbons and the gas viscosity when corrected for rarefaction effects.

All the above-mentioned literature is on squeeze-film damping for parallel plates oscillating in the normal direction. There are relatively few studies on shear damping in lateral accelerometers. A study by Cho et al. (1994) investigated viscous damping for laterally oscillating microstructures. It was found that Stokes-type fluid motion models viscous damping more accurately than Couette-type flow field. The theoretical damping was also compared with experimental data, and a discrepancy of about 20% still remains between theoretically estimated (from Stokes-type model) and measured damping.

24.4.3 Device Damping: Impact Dynamics

Micromachined devices may demonstrate weaknesses not typically found in macroscopic sensor systems that occur during the system assembly of the micromachined inertial sensor with other electronic components. A common test to determine how robust a micromachined sensor is to system assembly is the drop test. When a packaged micromachined system is dropped from a tabletop to a hard-surface floor, both the package and the microstructure undergo sudden changes in their respective velocities. Assuming that there is no energy loss during the flight of the drop, both the package and the

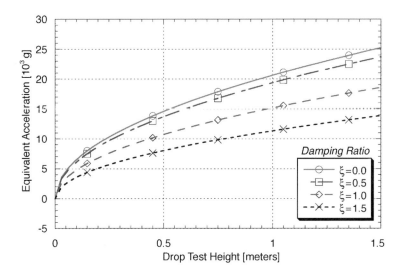

FIGURE 24.6 Calculation of the equivalent g load on an accelerometer due to an inelastic drop shock for several damping ratios. The device conditions are $m = 0.6$ μg, $k = 5$ N/m, $r = 0$.

microstructure would have a downward velocity of $v = \sqrt{2gh}$ immediately prior to impact. After impact, the package may be stuck to the ground or may bounce back with a smaller velocity. The motion of the package can be calculated by ignoring the influence of the microstructure, which is at least five orders of magnitude smaller than the mass of the package itself [Li and Shemansky, 2000]. The motion of the microstructure as a result of the impact is governed by a second order ordinary differential equation of a standard form, as in Eq. (24.9) [Meirovitch, 1975]. Equation (24.12) gives the maximum displacement:

$$z_{max} = \sqrt{\frac{2mgh}{k}} d_0(\xi, r), \qquad (24.12)$$

where $d_0(\xi, r)$ is a unit-less scaling function only of the damping ratio, ξ, and the elasticity of the collision as defined by a restitution coefficient, r, with $0 \leq r \leq 1$ [Li and Shemansky, 2000]. Knowing the seismic-mass travel as a result of a mechanical drop, the equivalent g load on the structure can be determined in a similar fashion and is graphically shown in Figure 24.6 for a specific device application.

At a 1 m inelastic ($r = 0$) drop, the impact experienced by the accelerometer is similar to a situation where it is subjected to an acceleration greater than 20,000 g. At a damping ratio of 1.5 (over-damped), the equivalent acceleration is 14,000 g. These values of acceleration would be doubled if the package impact with the floor is elastic ($r = 1$). Nevertheless, it is clear that the drop-induced acceleration is large, much larger than one normally expects.

Analysis on a lumped mass and spring model helps to provide a picture regarding the magnitude of proof-mass travel and the large g-force induced by a drop. A MEMS structure, however, has distributed properties such as mass and stiffness. Therefore, it is sometimes necessary to carry the analysis one step further to take the flexibility of the micromachined structure into account. This is especially important for those lateral sensors and actuators of comb-type designs where the conductive movable component is large in lateral dimensions and is susceptible to bending. The maximum possible displacement in a spring-supported structure is the sum of the travel as a lumped mass and the bending caused by dropping.

A typical lateral accelerometer is basically comprised of an array of sensing and actuating electrodes, a central spine plate and multiple supporting springs. In such a design, both the movable and stationary sensing electrodes can be modeled as cantilevers such that when subjected to an acceleration, the fixed end follows the motion of the attachment and the free end deflects. As for the central spine plate, it can be modeled either as a hinged–hinged beam or a clamped–clamped beam, depending on the position of supporting springs.

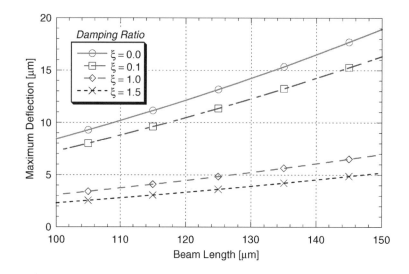

FIGURE 24.7 Calculation of the maximum beam displacement, z_{max}, for a 2-μm-thick cantilever beam with different damping ratios and lengths. (Adapted from Li, G.X., and Shemansky, F.A. (2000) *Sensors and Actuators A*, **85**, 280–286.)

The maximum displacement, z_{max}, for a 2-μm-thick polysilicon cantilever is graphically shown in Figure 24.7, where $h = 1.2$ m, $E = 161,000$ MPa, $I = 4/3$ μm^4, and $r = 0.5$. Assuming there is no damping, a 100-μm cantilever would bend approximately 8.5 μm near its free end, and the bending would be 19 μm for a beam 150 μm long. When damping is included, the beam deflection becomes smaller as indicated by the dashed and dotted curves in Figure 24.7. The damping ratio for a nominal lateral accelerometer of 2-μm-thick polysilicon and 1.5-μm finger gap is approximately 0.1. Therefore, depending on finger dimension, the drop-induced bending could be very excessive and cause structural damage.

24.4.4 Mechanical to Electrical Transduction: The Sensing Method

Many viable approaches have been implemented to produce inertial sensors. The capability of any technology needs to be tempered by the cost and market focus required by the application. Many sensing methodologies have been successfully demonstrated for inertial sensors. Piezoresistive, resonant frequency modulation, capacitive, floating-gate field-effect transistor (FET) sensing, strain FET sensing and tunneling-based sensing will be discussed briefly below. Other variations continue to be demonstrated. A study of the materials used in micromachining can be found in Part II of this handbook. Detailed explanations of the electronic properties of materials utilized for sensing methodologies can be found in references such as *Semiconductor Physics: An Introduction* [Seeger, 1985] and others.

Piezoresistive sensing has been successfully demonstrated in bulk micromachined and single-crystal inertial sensors [Roylance and Angell, 1979; Partridge et al., 1998]. This sense method has been utilized for many years in pressure-sensor structures quite successfully, demonstrating very sensitive devices with marketable costs [Yoshii et al., 1997; Andersson et al., 1999]. This technique is sensitive to temperature variations but can be compensated electronically. Yoshii et al. (1997) and Ding et al. (1999) have demonstrated monolithic integration of circuitry with piezoresistive elements. Thermal sensitivity, junction noise and junction leakage are piezoresistive sensing issues that require compensation for highly sensitive systems.

Resonant frequency shifts in a structure caused by inertial forces has been applied to some of the most sensitive and highest performing inertial sensor products on the market today. Resonant-beam, tuning-fork-style inertial sensors are in production for many high-end applications. Zook et al. (1999) and others have demonstrated resonant systems providing a greater than 100-g full-scale range with milli-g resolution.

These techniques are sensitive to temperature and generally require sensitive and complex control circuitry to keep the transducers resonating at a controlled magnitude.

Of all the alternatives, capacitive sensing has the broadest application to current inertial sensor products. There are several reasons why capacitive sensing is enjoying broad interest. The methodology is inherently temperature insensitive, and appropriate precautions in transducer and circuit design result in nearly zero temperature coefficients of offset and sensitivity [Lee and Wise, 1981]. Scaling the method to suit different sensing ranges can be implemented by scaling the device capacitances to provide larger output signals for small inertial loads. The technique can be implemented in a wide variety of micromechanical process techniques ranging from bulk micromachining to surface micromachining [Hermann et al., 1995; Delapierre, 1999]. All axes of motion can be sensed capacitively. Complementary metal oxide semiconductor (CMOS) control circuitry is especially well suited to measure capacitances, leading to broader application of this technique being implemented with switched capacitor sense circuitry.

Kniffin et al. (1998) and others have demonstrated floating-gate FET structures to measure inertial forces. The technique allows a direct voltage transduction from the inertial force via a floating-gate FET structure. The technique is complicated, however, by the sensitivity of the air gap in such a structure to variations in work function and the difficulty in providing a stable bias condition for the FET device. Variations in packaged environment can result in quite large offsets in the response, which has made this methodology difficult to implement industrially.

The use of FET-based strain sensors has not been broadly studied for micromachined inertial sensors. Strain measurement in packaging development for large-scale CMOS circuits has been studied extensively [Jaeger et al., 1997] and has been extended to inertial sensing, providing direct voltage sensitivity to inertial forces as demonstrated by Haronian (1999). Concerns remain regarding the manufacturability and stability of a FET device at the base of a micromachined strain-gauge beam. However, this technique holds promise for direct integration of inertial sensing devices with CMOS.

Devices based on electron tunneling can provide extreme sensitivity to displacement. Because tunneling current is so strongly dependent on the space between the cathode and anode in the system, closed-loop operation is the most common configuration considered. Recent efforts to produce tunneling-based inertial sensors can be found in Rockstad et al. (1995), Wang et al. (1996) and Yeh and Najafi (1997). Significant challenges remain for this methodology. Drift and issues related to the long-term stability of the tunneling tip, as well as low-frequency noise sources remain [Grade et al., 1996]. As tunneling-based technologies expand in application, solutions will be found to mitigate the current limitations of the methodology.

24.5 Micromachining Technologies for Inertial Sensing

Coupled to the physical principle used to sense the inertial displacement is the micromachining technology implemented to produce the transducer device. Far more comprehensive details regarding these technological developments are described in Section II of this handbook. Bulk silicon micromachined technologies were first implemented for inertial sensors; however, polysilicon-based surface micromachined technologies dominate the current marketplace for micromachined inertial sensors. The trend for inertial sensors is to higher aspect ratio "surface" micromachined technologies.

Surface micromachined capacitive inertial sensors were broadly demonstrated commercially as a result of the collaboration between Analog Devices, Inc., and the University of California, Berkeley, in the introduction of the Analog Devices, Inc. iMEMS™ bipolar complementary metal oxide semiconductor (BiCMOS) integrated surface-micromachined accelerometer process technology, as shown in Figure 24.1a, which embeds a 2-μm micromechanical polysilicon layer into a BiCMOS process flow [Chau et al., 1996]. Sandia National Laboratories [Smith et al., 1995], Motorola Sensor Products Division [Ristic et al., 1992] and Siemens [Hierold et al., 1996] have all demonstrated industrial surface-micromachined inertial sensor technologies. Limitations to surface-micromachining are primarily related to its limitations in producing high-aspect-ratio structures that have demonstrated benefits for sensitivity, mechanical damping properties and insensitivity to off-axis motion.

Epitaxially deposited polysilicon eliminates the aspect-ratio limitations of standard low-pressure chemical vapor deposition (LPCVD) polysilicon deposition typically used in surface micromachining. This technology, sometimes referred to as "epipoly" technology, also allows the monolithic integration of CMOS or BiCMOS circuitry with higher aspect ratio capacitive transducers, typically on 10- to 12-μm-thick epitaxial layers [Kirsten et al., 1995; Offenberg et al., 1995; Geiger et al., 1999]. This technology has very few limitations. Epitaxial deposition of silicon is cost competitive for micromachining to thicknesses of 50 μm. These high-aspect-ratio transducer structures are relatively insensitive to out-of-plane motion and provide suitable mechanical damping at reasonable packaging pressures. This material has been demonstrated to have desirable film properties with nearly immeasurable intrinsic stress and a high deposition rate [Gennissen and French, 1996].

With a reasonably flexible interconnect scheme, "epipoly" technologies have demonstrated monolithic integration with CMOS and BiCMOS circuitry as well as device thicknesses ranging from 8 μm to over 50 μm. The technology is not without challenges, however. Co-deposited epitaxial silicon and polycrystalline silicon have differing deposition rates complicating production. High-temperature polycrystalline films typical of epitaxial deposition also suffer from severe surface roughness and very large semiconical crystalline grains. Solutions to many of these issues have been documented by various sources [Kirsten et al., 1995; Gennissen and French, 1996].

Direct wafer bond (DWB) technology has long demonstrated the successful incorporation of thick capacitively or piezoresistively sensed inertial sensor structures. Recent advances in this technique have demonstrated improved device interconnect through the use of a silicon-on-insulator (SOI) handle wafer with defined interconnect [Ishihara et al., 1999]. Piezoresistive elements have also been incorporated on the sidewalls of very-high-aspect-ratio DWB structures to provide transducers with both piezoresistive and capacitive sensing mechanisms [Partridge et al., 1998]. DWB transducer technologies provide great process and device flexibility. Very-high-aspect-ratio structures are possible, approaching bulk-wafer thicknesses if necessary, providing excellent out-of-plane insensitivity and mechanical damping properties. Monolithic integration with CMOS is also possible. The technology requires significant process capability to successfully produce DWB structures at high yield.

As silicon-on-insulator microelectronic device technologies gain popularity for high-performance mainstream CMOS process technologies, the substrate material required for micromachining becomes cost competitive with alternative transducer technology approaches, making SOI more appealing for inertial sensing applications. SOI technology, as a descendant of DWB technology, provides technological flexibility with desirable device properties, including the out-of-plane insensitivity and high damping associated with high-aspect-ratio structures. SOI technology provides the advantage of single-crystal silicon sensor structures with very well-behaved mechanical properties and extraordinary flexibility for device thickness, as with DWB technologies. Thicknesses can range from submicron to hundreds of microns for structural layers. Unlike DWB, SOI technologies often lack the flexibility of pre-bond processing of the handle and active wafers to form microcavities or buried contact layers that are often implemented in DWB technologies. Another technology hurdle has been the choice of methodology to minimize parasitics to the handle wafer. Even so, SOI has demonstrated a significant increase in its popularity as a micromachining substrate [Delapierre, 1999; Lemkin et al., 1999; Park et al., 1999; McNie et al., 1999; Noworolski and Judy, 1999; Lehto, 1999; Usenko and Carr, 1999]. While technological hurdles still need to be overcome for broad industrialization of SOI MEMS devices, the technology holds great promise for a broad technological platform with few limitations. Lemkin et al. (1999) demonstrated the monolithic integration of SOI inertial sensors with CMOS.

MacDonald and Zhang (1993) at Cornell University developed a process known as SCREAM (single-crystal reactive etching and metallization) which produces SOI-like high-aspect-ratio single-crystal silicon transducers using a two-stage dry-etching technique on a bulk silicon substrate. This technology had been limited by the difficulty in electrically isolating the transducer structure from the surrounding substrate. Sridhar et al. (1999) demonstrated that this technology can now produce fully isolated high-aspect-ratio transducers in bulk silicon substrates. With fully isolated structures, this technology can produce 20:1-aspect-ratio devices for thicknesses to 50 μm and can be monolithically integrated with circuitry. Xie et al. (2000) have also demonstrated a related two-stage release methodology on capacitive inertial devices formed in a CMOS integrated technology. This technology shows promise for full integration of high-aspect-ratio lateral inertial structures with CMOS. Also, Haronian (1999) demonstrated an integrated FET readout for an inertial mass released using the SCREAM process.

The development of metal micromachined structures by electroforming has demonstrated 300-μm-thick nickel structures with submicron gaps formed using LIGA (*Lithographie, Galvanoformung, Abformung*) techniques [Ehrfeld et al., 1987]. LIGA-*like* process techniques using reasonably high-resolution thick ultraviolet photoresist processes have resulted in inertial sensor development in nickel, permalloy and gold post-CMOS micromachining [Putty and Najafi, 1994; Wycisk et al., 1999]. Very thick structures are possible using these techniques with effective "buried" contacts to the underlying circuitry in the substrate. As a single-layer process addition, the technology adds minimal cost to the overall sensing system. High-aspect-ratio structures are possible. However, the material properties of electroformed materials are difficult to stabilize and can be prone to creep.

Traditionally a bulk micromachined technology, the application of deep anisotropic etching of (110)-oriented silicon wafers has produced novel inertial sensors with very high aspect ratios. Aspect ratios up to 200:1 have been demonstrated using this technique, although the practical application of the technology may limit the maximum aspect ratio to below 100:1 [Hölke and Henderson, 1999]. This technology allows an elegant solution that provides extremely high aspect ratios compared to anisotropic dry-etching techniques. The technology is somewhat limited in its application flexibility, as the deep trenches are crystallographically defined by the intersection of the (111) planes with the surface of the wafer and must be arranged in parallelograms in the etch mask. Circular and truly orthogonal structures are not easily configured for this technology.

24.6 Micromachining Technology Manufacturing Issues

The manufacturability of a transducer structure should be considered as important as the performance of the device. In theory, a sensor structure may be designed to be as sensitive as desired, but if the structure cannot be manufactured in a robust manner the effort is futile. Issues such as release or in-use stiction, stability of material properties and control of critical processes in the manufacture of inertial sensors should be investigated and understood. The impact of high-aspect-ratio technologies creates new challenges in controlling and maintaining processes.

24.6.1 Stiction

Stiction is a term used in micromachining to describe two conditions: release and in-use stiction. Release stiction is the irreversible latching of some part of the moveable structure in the device caused during the release etch and drying processes. In-use stiction is the irreversible latching of the moveable structure during device operation. All high-aspect-ratio technologies require a step to release the moveable structure from the supporting substrate or sidewalls at some point in the process flow. This process is not always, but most often is, a wet etch of a dielectric layer, often using a solution containing hydrofluoric acid and water. Release stiction typically occurs during the drying step following a wet solution process as the surface tension forces in the liquid draw the micromachined structure into intimate contact with adjacent surfaces. The close contact and typically hydrated surfaces result in van der Waals attraction along the smooth parallel surfaces, bonding the layers to each other [Mastrangelo and Hsu, 1993]. Too large a proof mass or too soft a spring may dramatically increase the probability of stiction, resulting in yield loss.

Many techniques are employed to reduce or eliminate release stiction. Supercritical CO_2 drying processes avoid surface tension forces completely and often result in very good stiction yields. This technique has been difficult to implement in industrial process conditions. Surface modifications, often based on fluorinated polymer coatings, have been used to reduce surface tension forces on the micromachined structure during release and drying with some success. As hydrophobic materials, these monolayer coatings require significant surface treatment and have not found broad industrial utilization yet. Other techniques have been employed with some success, all at the cost of additional process complexity and structural compromises. However, the problem with stiction yield loss is increased with aspect ratio because the surface tension forces act over a larger area. High-aspect-ratio structures must be designed with care to minimize the complications from release stiction.

In-use stiction issues are also increased with the aspect ratio of a device. For capacitive accelerometers, the proof mass closing in on an actuated electrode can cause electrostatic latching if the electrostatic force becomes larger than the elastic spring's restoring force. This condition is called pull-in or electrostatic latching and is design dependent. High-aspect-ratio designs result in more capacitive coupling force for a given device topography and can be more prone to latching. Many devices are designed with over-travel stops to reduce the risk from this compromising situation.

24.6.2 Material Stability

While providing design performance and off-axis stiffness, high-aspect-ratio devices remain sensitive to the stability of material properties. This is particularly important for polycrystalline silicon devices, as the deposition process can result in variations in the average intrinsic stress of a film as well as generate stress gradients throughout the sensor layer. However, all associated materials result in significant impacts on the device performance and repeatability. Stability and uniformity of backside film stacks, plasma-assisted deposited dielectric films and even the proximity of metallizations in the front-end process can impact the uniform and controlled behavior of a device.

24.6.3 High-Aspect-Ratio Etching

High-aspect-ratio silicon based structures require high aspect ratio etching. In the case of (110) silicon technologies, this etch is a wet anisotropic etchant and uses crystallographic planes in the device material to control the profile of the trenches formed. Control of such processing requires accurate alignment of the etch mask to the crystallographic planes in order to successfully control the aspect ratio to a designed parameter [Hölke and Henderson, 1999]. The technique is also sensitive to impurities in the crystal. In most high-aspect-ratio technologies, however, a deep dry reactive ion etch (DRIE) of the trenches forms the structure. Deep trench etching has been implemented using various techniques, but an etch pioneered by Bosch [Laerme et al., 1999] has demonstrated a clear predominance in the field of alternatives. Control of the etch properties and profile is the challenge for high-aspect-ratio technologies. Many potential process conditions can degrade the etch profile or complicate the uniformity of the process for across-wafer and wafer-to-wafer variations in the process. These variations in profile and width strongly impact the design parameters such as spring constant and damping, etc.

24.6.4 Inertial Sensor Packaging

Package interactions are just as critical as device technology choices and often contribute significant performance shifts from package to package [Li et al., 1998]. Micromachined inertial sensors, while robust on a microscale, are fragile at the assembly scale, are easily damaged, and often require two levels of packaging: (1) wafer level packaging, which is usually hermetic to provide damping control and to protect the MEMS devices from the subsequent assembly operations; and (2) conventional electronic packaging of die-bonding, wire-bonding and molding to provide a housing for handling, mounting and board-level interconnection. The package must fulfill several basic functions: (1) provide electrical connections and isolation, (2) dissipate heat through thermal conduction, and (3) provide mechanical support and isolate stress. An industrially relevant packaging process must be stable, robust and easily automated and must take testability into account.

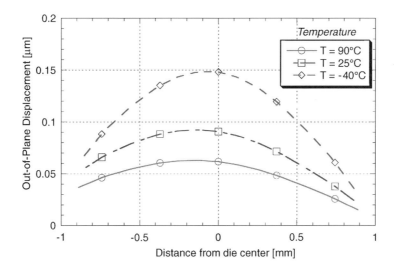

FIGURE 24.8 Die deformation due to a chip-packaging scheme, demonstrating the impact of packaging technologies on the industrialization of microelectromechanical systems. (Adapted from Li, G.X. et al. (1998) in *Proc. 3rd Int. Symp. on Electronic Package Technology* (ISEPT '98), pp. 553–562.)

Wafer-level packaging techniques include silicon-to-glass anodic bonding [Dokmeci et al., 1997], thermocompression bonding using glass frit [Audet et al., 1997] or eutectic [Wolffenbuttel and Wise, 1994; Cheng et al., 2000], direct wafer bonding [Huff et al., 1991] and monolithic capping technologies [Burns et al., 1995] utilizing one or more wafers. These techniques allow the transducer device to be sealed at the wafer level to protect the movable components from damage during assembly. The wafer level package also provides the sensor with a controlled ambient to preserve the damping characteristics of the proof mass. Figure 24.1b shows a Motorola accelerometer die with a wafer-level cap in silicon to the right of the co-packaged CMOS control integrated circuit.

The unique challenge of sensor packaging is that in addition to providing a mounting foundation to a PC board, stresses induced by mismatch in the thermal expansion coefficients of the materials used to fabricate the package and the external thermal loading of the package must be controlled and kept at a level low enough to avoid impact to the sensor or to control circuitry performance. An example of this challenge is illustrated in Figure 24.8, where external package-induced stresses on an accelerometer die produced a 0.15-μm curvature out of plane for the die from center to edge, resulting in a device offset that would not be present for the die in wafer form. For capacitive devices capable of resolving displacements at the nanometer or subnanometer scale, excessive curvature due to stress on the die at a late point in the assembly can be catastrophic. In general, different MEMS devices have different stress-tolerance levels. Therefore, each MEMS package must be uniquely designed and evaluated to meet special requirements [Dickerson and Ward, 1997; Tang et al., 1997].

24.7 System Issues for Inertial Sensors

The partitioning of the transduction and controlled output of an inertial sensing system has led to many variations on what an inertial sensor technology should look like. The coupling between the micromachined structure and the microsystem, including the method of control and the choice of interface electronics, is very close. A change in an aspect of one strongly impacts the requirements for the other. The motivations and potential for integration of the integrated circuit with the transducer and the impact of the control circuit architecture on the overall system reveal the many system design trade-offs required to produce a complete inertial sensing system.

24.7.1 System Partitioning: One-Chip or Multi-Chip

The sense methodologies and transducer technologies discussed above have all been demonstrated as both multicomponent and monolithically integrated systems. One-chip vs. package-level multichip integration of transducer and circuitry has generated many passionate discussions regarding the viability of each approach and motivations to choose one over the other. System cost and system capability are the two primary motivations for the choice to monolithically integrate or to co-package components and have been used as justification to pursue both types of integration.

Silicon die area utilization, process complexity, wafer-scale testing requirements and packaging costs all significantly contribute to the overall production costs for any sensor system. Some multicomponent systems have mitigated process complexity by isolating transducer and integrated circuit processes and co-packaging the unique system pieces. System requirements have been met by this multichip methodology. Front-end silicon process complexity and cost have been traded for back-end testing and packaging costs, which make up a large fraction of the overall product costs. There is no clear application boundary for multichip system partitioning compared to single-chip integration. Industrialized sensor products have demonstrated two-chip solutions implemented successfully for even the most complex and demanding system requirements in military and aerospace applications [Delapierre, 1999].

Monolithically integrated transducer technologies have advantages in approaching the fundamental sensing performance limitations for many applications by mitigating inter-chip parasitics associated with multichip integration. Trade-offs in system costs are moving toward the incorporation of additional front-end complexity to design for testability and mitigate back-end testing and packaging costs. An additional benefit to monolithic integration of transducer and circuitry is the minimization of silicon die area. Die area may become *the* driving motivation to develop monolithic integration for future sensor applications. The drive to smaller outline package surface mount technologies also requires a smaller silicon footprint in the package, again driving toward integration.

There are significant reasons to consider monolithic integration of transducer technologies with circuitry for system performance and system cost. Each application should be reviewed carefully to provide system requirements with the least expensive high-performance technology. With future applications demanding greater performance in an increasingly small package, considering integratable transducer technologies for future technology applications is prudent. In monolithically integrated technology, the method of integration defines the manufacturability, process cost, testability and ability to integrate the technology. This decision includes both the integration method and the choice of integrated circuit technology and should fully address the system requirements.

24.7.2 Sensor Integration Approaches

Three methodologies classify the integration of a micromachined element into an integrated circuit process: transducer first, transducer middle or interleaved and transducer last integrations. All have demonstrated monolithically integrated inertial sensor technologies. All have noted strengths and weaknesses. The *transducer first* integration incorporates the transducer element processing prior to the integrated circuit fabrication. A notable example of a transducer first integration is the buried transducer process, now know as Sandia's integrated microelectromechanical systems technology, or IMEMS, demonstrated by Smith et al. (1995) at Sandia National Laboratories. The transducer is formed in a recessed region in the field of the wafer and sealed in a stack of dielectric films prior to the beginning of a CMOS process flow. This dielectrically sealed moat is planarized using chemical mechanical polishing (CMP) prior to the start of the CMOS flow with electrical interconnections formed as part of the CMOS process, with the transducer released as a final step in the process flow. A cross-sectional diagram is shown in Figure 24.9. The benefit of this approach is that the transducer is largely decoupled from the remainder of the integrated circuit process and can be transported between integrated circuit processes. This allows the use of standard integrated circuit process steps, with minimal impact to the integrated circuit processing steps and only limited impact on thermal budget, allowing optimization of the transducer without impact on the standard integrated circuit process. The challenge of this versatile approach

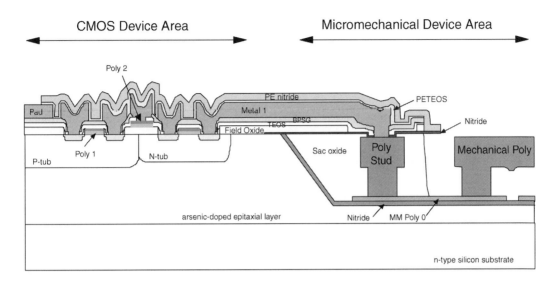

FIGURE 24.9 Cross-sectional diagram of the IMEMS process developed at Sandia National Laboratories demonstrating the transducer formed in a recessed moat and sealed prior to the high-density CMOS process commencing. (Photograph courtesy of Sandia National Laboratories.)

lies in the fact that all of the integrated circuit processing steps are still ahead, increasing the risk of contamination.

The *transducer middle* integration merges the process that created the transducer element into the standard integrated circuit process by reusing and minimally adjusting and inserting processing steps for the formation of the transducer. A well-known example of this technique is the Analog Devices iMEMS™ process [Chau et al., 1996]. This technology merges a 2-µm-thick micromechanical polysilicon layer for the transducer with a high-density mixed-signal BiCMOS process technology, interleaving the transducer process with the integrated circuit flow. An example of this technology is shown in Figure 24.1a. The benefit of this approach lies in the reuse of existing integrated circuit process steps used in the creation of the transducer, which has the potential to minimize the number of additional process steps required to implement this approach. The risk for this approach lies in its specialization. Because the transducer is merged into an integrated circuit process, a unique process flow is generated for the specific system. A merged process will impact the integrated circuit device parameters as well as the transducer parameters. The reuse of existing process steps complicates the optimization of the transducer as well as the integrated circuit.

The *transducer last* integration merges the process of the transducer with the standard integrated circuit process by inserting the transducer formation at the end of the integrated circuit processing steps. This integration can take on many forms, including the approach taken by Xie et al. (2000) which uses layers and structures formed during the CMOS process to define post-processed regions for subsequent deep reactive ion etching of high-aspect-ratio structures, or that of Franke et al. (2000) or of Honer and Kovacs (2000) using low-temperature post-processing of structural layers to integrate micromechanical structures with circuitry. The benefit of this approach lies in the reuse of a minimal set of existing integrated circuit back-end processing steps used in the creation of the transducer, thereby having the potential to minimize the number of additional process steps required to implement this approach. Because the transducer is formed after the completion of all the standard integrated circuit device formation steps, the impact on integrated circuit device parametrics is defined primarily by exposure of the circuitry to the maximum temperature of film deposition or the plasma processing required to form the transducer. In addition, there is some measure of transportability of the transducer process between existing integrated circuit platforms. The difficulty with this approach lies in the potential complexity of the integration approach and its impact on the isolation and interconnect aspects of the transducer. Because the integrated circuit processing has been completed before formation of the transducer, the patterning of the transducer around the existing interconnect scheme becomes difficult, and the release etch in the absence of an adequate etch stop can be a real challenge. Solutions to that challenge have been proposed through dry release [Xie et al., 2000] or CMOS-benign release in hydrogen peroxide [Franke et al., 2000]. While transportability between different integrated circuit platforms is possible, a degree of optimization is required, which will be specific to the platform used.

The introduction of monolithically integrated inertial sensor technologies requires clever system design to understand the impact of the device behavior in the system. Improvements in test methodology and the potential to isolate transducer and integrated circuit behavior is required. In particular, suitable calibration of self-test forces and the decoupling of drive and sense responses are even more important for the active transducers required for rotational inertia sensing.

24.7.3 System Methodologies: Open- or Closed-Loop Control

Sensing methodologies utilized in inertial sensing generally fall under two categories: open-loop or closed-loop control architectures. The chapters by Goodwine and Bewley consider these concepts in greater detail. Suffice it to say that there are motivations to pursue both open-loop and closed-loop force-feedback control for inertial sensing systems. Open-loop control systems measure changes in the sense signal, whether it be a change in piezoresistance or capacitance or other change, as a result of the inertial load displacing the seismic mass from its zero state position. These signals are typically amplified, compensated, filtered, buffered and output as control variables either as analog voltages or digital control signals to the larger system. Open-loop control schemes tend to be relatively immune to small production variations in the transducer element, are inherently stable systems relying on no feedback signals, provide ratiometric output signals and, possibly most importantly, are often smaller in die area than their closed-loop counterparts.

Closed-loop control schemes rely on feedback to control the position of the seismic mass via a force feedback, or force rebalancing, at its rest position. The force feedback required is proportional to the magnitude of the inertial load. The potential for this methodology is great. This force feedback defines the sensitivity of the system and acts as an electrostatic spring force, which is added in Eq. (24.5). It also contributes to Eq. (24.6), impacting the dynamic behavior of the system and modifying the damping conditions. The most prevalent system configuration utilizing force rebalancing is in capacitive inertial devices. A notable example of an inertial sensing system utilizing force rebalancing is the Analog Devices ADXL50, which is a lateral 50-g accelerometer product as described in Goodenough (1991) and Sherman et al. (1992). Closed-loop force feedback systems have the potential for very high sensitivity and have been implemented in gyroscopic systems due to the minute forces and displacements generated by the Coriolis effect [Putty and Najafi, 1994; Greiff and Boxenhorn, 1995].

24.7.4 System Example: Motorola Two-Chip x-Axis Accelerometer System

Motorola's 40-g x-lateral accelerometer can be used as an example of a high-volume product that will demonstrate the design trade-offs and broad classes of issues required to produce inertial sensors. The packaged system, as shown in perspective in Figure 24.1b on a lead frame prior to final assembly, is comprised of a two-chip copackaged system in a dual in-line plastic package. The sensor die, capped with a hermetically sealed silicon cap, is wire-bonded to the adjacent integrated circuit control die. Low-stress adhesive and coating materials are used to minimize mechanical coupling of the sensor die to the metal lead frame and the plastic injection molding materials and to minimize system offsets.

The transducer, a portion of which is shown in a perspective-view scanning electron microscopy (SEM) image in Figure 24.10, is produced using surface micromachining in polysilicon to form a suspended polysilicon, lateral, double-sided capacitive structure. The central proof-mass plate is configured to produce an over-damped z-axis response at a given capping pressure. The x-axis damping is defined by the aspect ratio between the suspended sense electrodes and the fixed capacitor plates forming the lateral

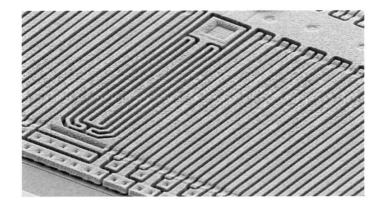

FIGURE 24.10 SEM perspective view of a portion of an *x*-axis capacitive accelerometer in polysilicon. (Photograph courtesy of Motorola, Inc.)

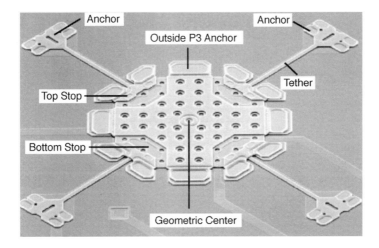

FIGURE 24.11 SEM perspective view of a *z*-axis capacitive accelerometer in a three-level polysilicon process. (Photograph courtesy of Motorola, Inc.)

capacitive structure. The folded-beam spring design, the anchor of which is at the top center of Figure 24.10, minimizes coupling to external mechanical strains on the chip, controls the *x*-axis spring constant to the designed parameter, maximizes the *y*-axis spring constants to minimize off-axis coupling and provides the electrical connection to the center electrode for the lateral capacitive measurement. The *z*-axis motion in this device is constrained by a series of motion limiters, limiting the out-of-plane motion to a designed tolerance. The double-sided capacitor structure is formed by the definition of a plurality of adjacent left and right electrodes to the electrodes formed on the proof mass in a configuration often described as a capacitive "comb" structure. This comb structure, through the multiplication of the small capacitive coupling in each left, center and right set of electrodes provides sufficient capacitive coupling to suitably sense the inertial deflection of the proof mass center electrode with respect to the left and right differential electrodes.

This device demonstrates a clear system partitioning. Minimizing process complexity, the two-chip approach allows the potential maximization of process capability for both sensor and control circuitry for the given application. Alternatively, fabrication process and assembly costs can be minimized for the performance required by the application by isolating the micromechanical and circuit elements. This technique has been implemented in many other accelerometer and gyroscopic systems over the past decade. Figure 24.11 shows a Motorola 40-*g* *z*-axis accelerometer implemented in a two-chip system configuration similar to the *x*-lateral accelerometer as described in Ristic et al. (1992). Other inertial sensor systems have been demonstrated as two-chip approaches with vastly different system requirements [Greiff et al., 1991; Yazde and Najafi, 1997; Ayazi and Najafi, 1998; Hiller et al., 1998; Delapierre, 1999].

Two-chip methodologies do have some limitations. Inter-chip parasitic capacitances pre-load the control circuitry with a nonsensitive capacitance, requiring the control system to discriminate small capacitance changes in a total capacitance several times as large as the capacitance of the accelerometer device alone. This inter-chip parasitic capacitance can be on the order of two to five times the transducer nominal capacitance, requiring clever circuit techniques to minimize the impact of the parasitic coupling. Other capacitive sensor systems have implemented single-chip integration to eliminate the inter-chip parasitic coupling at the expense of increased process complexity.

24.8 Rotational Gyroscopes: Specific Design Considerations

Unlike an accelerometer, the transducer of a yaw-rate sensor needs to be driven into oscillation in order to function as a sensor. This requirement comes from the coupling of vibratory motion by the Coriolis effect to produce a positional shift sufficient for sensing. This requirement adds both transducer and circuit complexity to the system. Upon a rotation of the transducer about its sense axis, a Coriolis force is generated in the presence of a rotational velocity of the reference frame, which in turn drives the transducer structure orthogonally, as given in Eq. (24.4). The scalar governing equation of motion for a gyroscopic device with a resonating mass in the *y*-axis, rotated about the *z*-axis is given by:

$$\frac{d^2 x}{dt^2} + 2\xi\omega_n \frac{dx}{dt} + \omega_n^2 x = 2\Omega_z \frac{dy}{dt} \qquad (24.13)$$

where Ω_z is the rate of rotation and *y* is linear velocity of the structure due to the drive. One may make the analogy that a yaw-rate sensor is acting just like an accelerometer if the Coriolis term ($2\,\Omega_z\,dy/dt$) is considered as an acceleration. According to a typical automotive spec where the full range of angular velocity is 100 deg/s, then an equivalent acceleration, *a*, is given by:

$$a = 3.5 \frac{dy}{dt} \qquad (24.14)$$

In general, the driving frequency is near resonance and the vibration amplitude of the transducer structure is about 1 µm. Assuming a natural frequency of 10 kHz, the resulting Coriolis acceleration of Eq. (24.14) is

$$a = 0.022\ mg \qquad (24.15)$$

demonstrating that this Coriolis-force-induced acceleration is very small.

For most applications, a single-axis angular rotation measurement is required. Such a single axis rate sensor can be built by sensing induced displacement from an oscillating rotor or from a linearly oscillating structure. Although these two types of rate sensor designs appear to be very different, the operation principles are the same. In both cases, when the reference frame, or device substrate, experiences a rotation along the input axis, the oscillating mass, either translational or rotary, in a direction perpendicular to the input axis (referred to as the drive axis) would induce a Coriolis force or torque in a direction perpendicular to both the input axis and the drive axis. With the amplitude of the drive oscillation fixed and controlled, the amplitude of the sensing oscillation is proportional to the rate of rotation of the mounting foundation.

Because the coupling of the Coriolis effect is orthogonal to the vibratory motion in a micromachined gyroscopic device, two degrees of mechanical freedom are required. One degree of freedom is utilized for the excitation of the vibratory motion, and the second degree of freedom orthogonal to the first is required for sensing. This requirement couples tightly into the technology choice for gyroscopic inertial sensors, as the axis of sensitivity defines which mechanical degrees of freedom are required to sense it. For example, a very thick high-aspect-ratio technology such as is possible with direct-wafer-bonded structures might not be the most suitable technology for a device required to move out of the plane of the wafer. However, as with accelerometers, most technologies and sensing methodologies have been applied to vibratory gyroscopes with new combinations of methodologies always under consideration.

Putty and Najafi (1994) provide a discussion of the varieties of vibrating gyroscopes, including vibrating prismatic beams [Greiff et al., 1991], tuning fork designs [Voss et al., 1997; Hiller et al., 1998], coupled accelerometers [Lutz et al., 1997; Kobayashi et al., 1999; Park et al., 1999] and vibrating shells [Putty and Najafi, 1994; McNie et al., 1999]. Multiple-axis systems have also been demonstrated [Juneau et al., 1997; Fujita et al., 1997]. In all cases, the vibrating structure is displaced orthogonal to the direction of the vibrating motion. This configuration can lead to system errors related to the transducer structure and the electronics. The primary error related to the transducer is called quadrature error.

24.8.1 System Considerations: Quadrature Error and Coupled Sensitivity

Most of the earlier gyro sensor designs involve a single proof mass for both driving and sensing. The proof mass is supported by a set of multiple slender beam linkages, usually made of the same semiconductor materials as the proof mass, to allow for movement in two mutually perpendicular axes. A major drawback in a single proof-mass design is the cross-axis coupling between the drive axis and the sense axis, a phenomenon commonly referred to as quadrature error. This coupling can be attributed to defects or small nonorthogonalities in the mechanical structure. Because the sense displacement is a minute fraction of the typical drive displacement, small structural defects can generate large quadrature errors in the system. Quadrature is compensated by enhanced structural design, as will be demonstrated in the examples below, as well as through the generation of quadrature-canceling force feedback of position in the control electronics [Geen, 1998]. Fortunately, the quadrature error coupled from the drive vibration is 90 degrees out of phase with respect to Coriolis-induced vibration and can be phase-discriminated to a large degree in the control circuitry at the expense of additional control circuit complexity. However, the continued increasing complexity in the structural design in modern micromachined gyroscopes indicates that quadrature error cancellation cannot be completely resolved by the control circuit.

The sensitivity requirements for gyroscopes far exceed those for most accelerometer systems, both in terms of the transducer design and the circuit complexity. In vibrational gyroscopic systems, structural sensitivity and absolute stability in the control electronics are required to accurately measure rotational rate. Because the magnitude of the driven vibration is directly proportional to the magnitude of the Coriolis-induced output displacement, the structure and electronics are designed to maximize the coupling and stability of the magnitude. Structurally, the driven oscillations can have large displacements, on the order of microns or tens of microns in some cases. The devices are also commonly operated in a near-vacuum environment to minimize the impact of mechanical damping on the structure to maximize the resonant response of the system. Electronically, precise control of the driven vibration amplitude is paramount. Phase discriminating circuitry such as phase-locked loop (PLL) control is used to drive the device displacement at or near resonance to maximize the displacement while precise amplitude control is maintained. Phase discrimination and synchronous phase demodulators are also required to sense the Coriolis force displacement and cancel quadrature effects [Geen, 1998; Kobayashi et al., 1999]. As with accelerometer systems, the sense circuitry can be operated in open-loop or closed-loop force-feedback configurations to sense displacement with the system trade-offs discussed in the preceding section.

24.8.2 System Examples: Vibratory Ring Gyroscopes and Coupled Mass Gyroscopes

As an example, a study by Ayazi and Najafi (1998) presented a detailed analysis of the design and scaling limits of vibrating ring gyroscopes and their implementation using a combined bulk and surface micromachining technology. A high-aspect-ratio p++/polysilicon trench and refill fabrication technology was used to realize the 30- to 40-µm-thick polysilicon ring structure with 0.9-µm ring-to-electrode gap spacing. The theoretical analysis of the ring gyroscope shows that several orders of magnitude improvement in performance can be achieved through materials development and design. By taking advantage of the high-quality factor of polysilicon, submicron ring-to-electrode gap spacing, the high-aspect-ratio polysilicon ring structure produced using deep dry etching, and the all-silicon feature of this technology, tactical-grade vibrating ring gyroscopes with a random walk as small as 0.05 deg/$\sqrt{Hz}$ were realized. Ayazi et al. (2000) enhanced this technology with an even higher aspect ratio 60-µm-thick ring. This device demonstrated a random walk of 0.04 deg/$\sqrt{Hz}$ with a theoretical Brownian noise floor of 0.01 deg/$\sqrt{Hz}$.

As an alternative to single proof-mass gyroscope designs, a design concept involving two coupled oscillating masses has emerged, one mass for driving and one mass for sensing. One of the first such designs was documented by Hsu et al. (1999), who have used an outer ring as the drive mass and an inner disk as the sense mass. The driving mass is actuated by a set of rotary comb structures and is oscillating about the z-axis (or the vertical axis). The sensing disk is anchored to the substrate in such a way that the stiffness about the z-axis is significantly greater than the stiffness about the other axes. The outer ring and the inner disk are connected by a set of flexible beams or linkages. When there is no input of angular rotation, the oscillation of the drive mass about the z-axis has virtually no impact on the motion of the sense disk. When the device experiences a rotation about either the x- or y-axis, the Coriolis-force-induced torque drives the inner disk into a rock motion about the y- or x-axis. Electrode pads underneath the disk measure the variation of capacitance, which is proportional to the input angular rate. Another advantage of this two-mass design is that the dual proof-mass structure permits the ring and the disk to be excited independently so that each can be dynamically compensated for manufacturing nonuniformity.

Several other vibrational gyroscopes have been demonstrated that also involving two mutually perpendicular oscillating masses [Kobayashi et al., 1999; Park et al., 1999]. In these designs, the drive mass is forced to oscillate along the y-axis by comb actuators and the sense mass is forced to oscillate along the x-axis by a Coriolis force. The magnitude of this Coriolis force is proportional to the input angular rate along the z-axis. The angular rate of rotation is measured by detecting changes in capacitance with interdigitated comb structures attached to the sense mass. The linkage between the drive and sense masses is designed in such a way that the Coriolis force is transferred from the drive mass to the sense mass in an efficient way and yet the feedback from the motion of the sense mass to the drive mass is kept to a minimum by frequency matching.

24.9 Concluding Remarks

Inertial sensors, both accelerometers and rotational gyroscopes, have seen broad commercial and industrial application of micromachining technologies driving system cost, size and performance. Many of the early automotive applications, driven initially by cost and package size with modest performance requirements, resulted in a niche-product field in which micromachined accelerometers could successfully compete with macroscopic competitors. As the technology field has matured and broadened, increasing performance expectations of what could now be considered *traditional* micromachined product areas (such as front-airbag crash sensing and future utilization of distributed sensor systems in automobiles for such systems as stability control, ride control, future-generation occupant safety systems, rollover and a cadre of potential new applications) will push the technology frontiers much further. Not only will micromachined inertial sensors continue to produce the low-cost, small-size parts as they have done so successfully, but they will also have to demonstrate significant system performance gains while maintaining modest costs and packages in order to continue to flourish.

References

Andersson, G.I., Hedenstierna, N., Svensson, P., and Pettersson, H. (1999) "A Novel Bulk Gyroscope," in *Technical Digest IEEE Int. Conf. on Solid-State Sensors and Actuators (Transducers '99)*, pp. 902–905, Institute of Electrical and Electronics Engineers, New York.

Andrews, M.K., and Harris, P.D. (1995) "Damping and Gas Viscosity Measurements Using a Microstructure," *Sensors and Actuators A* **49**, pp. 103–108.

Andrews, M.K., Harris, I., and Turner, G. (1993) "A Comparison of Squeeze Film Theory with Measurements on a Microstructure," *Sensors and Actuators A* **36**, pp. 79–87.

Audet, S.A., Edenfeld, K.M., and Bergstrom, P.L. (1997) "Motorola Wafer-Level Packaging for Integrated Sensors," *Micromachine Devices* **2**, pp. 1–3.

Ayazi, F., and Najafi, K. (1998) "Design and Fabrication of High-Performance Polysilicon Vibrating Ring Gyroscope," in *Technical Digest IEEE Int. Micro Electro Mechanical Systems Workshop (MEMS '98)*, pp. 621–626, Institute of Electrical and Electronics Engineers, New York.

Ayazi, F., Chen, H.H., Kocer, F., He, G., and Najafi, K. (2000) "A High Aspect-Ratio Polysilicon Vibrating Ring Gyroscope," *Technical Digest, Solid-State Sensor and Actuator Workshop*, Hilton Head, SC, pp. 289–292, Transducers Research Foundation, Cleveland, OH.

Blech, J.J. (1983) "On Isothermal Squeeze Films," *J. Lubrication Technol.* **105**, pp. 615–620.

Burgdorfer, A. (1959) "The Influence of the Molecular Mean Free Path on the Performance of Hydrodynamic Gas Lubricated Bearings," *J. Basic Eng.* **March**, pp. 94–100.

Burns, D.W., Zook, J.D., Horning, R.D., Herb, W.R., and Guckel, H. (1995) "Sealed-Cavity Resonant Microbeam Pressure Sensor," *Sensors and Actuators A* **48**, pp. 179–186.

Chau, K.H.-L., Lewis, S.R., Zhao, Y., Howe, R.T., Bart, S.F., and Marcheselli, R.G. (1996) "An Integrated Force-Balanced Capacitive Accelerometer for Low-g Applications," *Sensors and Actuators A* **54**, pp. 472–476.

Cheng, Y.T., Lin, L., and Najafi, K. (2000) "Localized Silicon Fusion and Eutectic Bonding for MEMS Fabrication and Packaging," *J. MEMS* **9**, pp. 3–8.

Cho, Y.-H., Pisano, A.P., and Howe, R.T. (1994) "Viscous Damping Model for Laterally Oscillating Microstructures," *J. MEMS* **3**, pp. 81–87.

Corman, T., Enoksson, P., and Stemme, G. (1997) "Gas Damping of Electrostatically Excited Resonators," *Sensors and Actuators A* **61**, pp. 249–255.

Darling, R.B., Hivick, C., and Xu, J. (1998) "Compact Analytical Modeling of Squeeze Film Damping with Arbitrary Venting Conditions using a Green's Function Approach," *Sensors and Actuators A* **70**, pp. 32–41.

Delapierre, G. (1999) "MEMS and Microsensors: From Laboratory to Industry," in *Technical Digest IEEE Int. Conf. on Solid-State Sensors and Actuators (Transducers '99)*, pp. 6–11, Institute of Electrical and Electronics Engineers, New York.

Dickerson, T., and Ward, M. (1997) "Low Deformation and Stress Packaging of Micro-Machined Devices," *IEE Colloq. on Assembly and Connections in Microsystems*, pp. 7/1–7/3, IEE, London.

Ding, X., Czarnocki, W., Schuster, J.P., and Roeckner, B. (1999) "DSP-Based CMOS Monolithic Pressure Sensor for High Volume Manufacturing," in *Technical Digest IEEE Int. Conf. on Solid-State Sensors and Actuators (Transducers '99)*, pp. 362–365, Institute of Electrical and Electronics Engineers, New York.

Dokmeci, M.R., von Arx, J.A., and Najafi, K. (1997) "Accelerated Testing of Anodically Bonded Glass-Silicon Packages in Salt Water," in *Technical Digest Int. Conf. on Solid-State Sensors and Actuators (Transducers '97)*, pp. 283–286, Institute of Electrical and Electronics Engineers, New York.

Ehrfeld, W., Bley, B., Götz, F., Hagmann, P., Maner, A., Mohr, J., Moser, H.O., Münchmeyer, D., Schelb, W., Schmidt, D., and Becker, E.W. (1987) "Fabrication of Microstructures Using the LIGA Process," in

Franke, A.E., Jiao, Y., Wu, M.T., King, T.-J., and Howe, R.T. (2000) "Post-CMOS Modular Integration of Poly-SiGe Microstructures," in *Technical Digest, Solid-State Sensor and Actuator Workshop*, Hilton Head, SC, pp. 18–21, Transducers Research Foundation, Cleveland, OH.

Fujita, T., Mizuno, T., Kenny, R., Maenaka, K., and Maeda, M. (1997) "Two-Dimensional Micromachined Gyroscope," in *Technical Digest IEEE Int. Conf. on Solid-State Sensors and Actuators (Transducers '97)*, pp. 887–890, Institute of Electrical and Electronics Engineers, New York.

Geen, J. (1998) "A Path to Low Cost Gyroscopy," in *Technical Digest, Solid-State Sensor and Actuator Workshop*, Hilton Head, SC, pp. 51–54, Transducers Research Foundation, Cleveland, OH.

Geiger, W., Merz, J., Fischer, T., Folkmer, B., Sandmaier, H., and Lang, W. (1999) "The Silicon Angular Rate Sensor System MARS–RR," in *Technical Digest IEEE Int. Conf. on Solid-State Sensors and Actuators (Transducers '99)*, pp. 1578–1581, Institute of Electrical and Electronics Engineers, New York.

Gennissen, P.T.J., and French, P.J. (1996) "Applications of Bipolar Compatible Epitaxial Polysilicon," *Proc. SPIE* **2882**, pp. 59–65.

Gianchandani, Y.B., and Najafi, K. (1992) "A Bulk Silicon Dissolved Wafer Process for Microelectromechanical Devices," *J. MEMS* **1**, pp. 77–85.

Goodenough, F. (1991) "Airbags Boom When IC Accelerometer Sees 50 g," *Electro. Design* **39** (15), 1991.

Grade, J., Barzilai, A., Reynolds, J.K., Liu, C.H., Partridge, A., Kenny, T.W., VanZandt, T.R., Miller, L.M., and Podosek, J.A. (1996) "Progress in Tunnel Sensors," in *Technical Digest, 1996 Solid State Sensor and Actuator Workshop*, Hilton Head, SC, pp. 72–75, Transducers Research Foundation, Cleveland, OH.

Greiff, P., and Boxenhorn, B. (1995) "Micromechanical Gyroscopic Transducer with Improved Drive and Sense Capabilities," U.S. Patent 5,408,877.

Greiff, P., Boxenhorn, B., King, T., and Niles, L. (1991) "Silicon Monolithic Micromechanical Gyroscope," in *Technical Digest IEEE Int. Conf. on Solid-State Sensors and Actuators (Transducers '91)*, pp. 966–968, Institute of Electrical and Electronics Engineers, New York.

Gudeman, C.S., Staker, B., and Daneman, M. (1998) "Squeeze Film Damping of Doubly Supported Ribbons in Noble Gas Atmospheres [MEMS Light Valve]," in *Technical Digest, 1998 Solid-State Sensor and Actuator Workshop*, Hilton Head, SC, pp. 288–291, Transducers Research Foundation, Cleveland, OH.

Haronian, D. (1999) "Direct Integration (DI) of Solid State Stress Sensors with Single Crystal Micro-Electro-Mechanical Systems for Integrated Displacement Sensing," in *Technical Digest IEEE Workshop on Micro Electro Mechanical Systems (MEMS '99)*, pp. 88–93, Institute of Electrical and Electronics Engineers, New York.

Hermann, J., Bourgeois, C., Porret, F., and Kloeck, B. (1995) "Capacitive Silicon Differential Pressure Sensor," in *Technical Digest IEEE Int. Conf. on Solid-State Sensors and Actuators (Transducers '95)*, pp. 620–623, Institute of Electrical and Electronics Engineers, New York.

Hierold, C., Hildebrant, A., Näher, U., Scheiter, T., Mensching, B., Steger, M., and Tielert, R. (1996) "A Pure CMOS Micromachined Integrated Accelerometer," in *Technical Digest IEEE Workshop on Micro Electro Mechanical Systems (MEMS '96)*, pp. 174–179, Institute of Electrical and Electronics Engineers, New York.

Hiller, K., Wiemer, M., Billep, D., Huhnerfurst, A., Geßner, T., Pyrko, B., Breng, U., Zimmermann, S., and Gutmann, W. (1998) "A New Bulk Micromachined Gyroscope with Vibration Enhancement Coupled Resonators," in *Technical Digest 6th Int. Conf. on Micro Electro, Opto, Mechanical Systems and Components (Micro Systems Technologies '98)*, pp. 115–120, VDE Verlag, Berlin.

Hölke, A., and Henderson, H.T. (1999) "Ultra-Deep Anisotropic Etching of (110) Silicon," *J. Micromech. Microeng.* **9**, pp. 51–57.

Honer, K.A., and Kovacs, G.T.A. (2000) "Sputtered Silicon for Integrated MEMS Applications," in *Technical Digest, Solid-State Sensor and Actuator Workshop*, Hilton Head, SC, pp. 308–311, Transducer Research Foundation, Cleveland, OH.

Hsu, Y., Reeds, J.W., and Saunders, C.H. (1999) "Multi-Element Micro-Gyro," U.S. Patent 5,955,668.

Huff, M.A., Nikolich, A.D., and Schmidt, M.A. (1991) "A Threshold Pressure Switch Utilizing Plastic Deformation of Silicon," in *Technical Digest IEEE Int. Conf. on Solid-State Sensors and Actuators (Transducers '91)*, pp. 177–180, Institute of Electrical and Electronics Engineers, New York.

Ishihara, K., Yung, C.-F., Ayón, A.A., and Schmidt, M.A. (1999) "An Inertial Sensor Technology using DRIE and Wafer Bonding with Enhanced Interconnect Capability," in *Technical Digest IEEE Int. Conf. on Solid-State Sensors and Actuators (Transducers '99)*, pp. 254–257, Institute of Electrical and Electronics Engineers, New York.

Jaeger, R.C., Suhling, J.C., Liechti, K.M., and Liu, S. (1997) "Advances in Stress Test Chips," in *Technical Digest ASME 1997 Int. Mechanical Engineering Congress and Exposition: Applications of Experimental Mechanics to Electronic Packaging*, pp. 1–5, American Society of Mechanical Engineers, New York.

Juneau, T., Pisano, A.P., and Smith, J.H. (1997) "Dual Axis Operation of a Micromachined Rate Gyroscope," in *Technical Digest, IEEE Int. Conf. on Solid-State Sensors and Actuators (Transducers '97)*, pp. 883–886, Institute of Electrical and Electronics Engineers, New York.

Kaiser, T.J., and Allen, M.G. (2000) "A Micromachined Pendulous Oscillating Gyroscopic Accelerometer," in *Technical Digest, Solid-State Sensor and Actuator Workshop*, Hilton Head, SC, pp. 85–88, Transducers Research Foundation, Cleveland, OH.

Kim, E.S., Cho, Y.H., and Kim, M.U. (1999) "Effect of Holes and Edges on the Squeeze Film Damping of Perforated Micromechanical Structures," in *Technical Digest IEEE Workshop on Micro Electro Mechanical Systems (MEMS '99)*, pp. 296–301, Institute of Electrical and Electronics Engineers, New York.

Kirsten, M., Wenk, B., Ericson, F., Schweitz, J.Å., Riethmüller, W., and Lange, P. (1995) "Deposition of Thick Doped Polysilicon Films with Low Stress in an Epitaxial Reactor for Surface Micromachining Applications," *Thin Solid Films* **259**, pp. 181–187.

Kniffin, M.L., Wiegele, T.G., Masquelier, M.P., Fu, H., and Whitfield, J.D. (1998) "Modeling and Characterization of an Integrated FET Accelerometer," in *Technical Digest IEE Int. Conf. on Modeling and Simulation of Microsystems, Semiconductors, Sensors and Actuators*, pp. 546–551, Computational Publications, Cambridge, MA.

Kobayashi, S., Hara, T., Oguchi, T., Asaji, Y., Yaji, K., and Ohwada, K. (1999) "Double-Frame Silicon Gyroscope Packaged under Low Pressure by Wafer Bonding," in *Technical Digest IEEE Int. Conf. on Solid-State Sensors and Actuators (Transducers '99)*, pp. 910–913, Institute of Electrical and Electronics Engineers, New York.

Laerme, F., Schilp, A., Funk, K., and Offenberg, M. (1999) "Bosch Deep Silicon Etching: Improving Uniformity and Etch Rate for Advanced MEMS Applications," in *Technical Digest IEEE Int. Con. on Micro Electro Mechanical Systems (MEMS '99)*, pp. 211–216, Institute of Electrical and Electronics Engineers, New York.

Lee, K.W., and Wise, K.D. (1981) "Accurate simulation of high-performance silicon pressure sensors," in *Technical Digest IEEE Int. Electron Devices Meeting (IEDM '81)*, pp. 471–474, Institute of Electrical and Electronics Engineers, New York.

Lehto, A. (1999) "SOI Microsensors and MEMS," in *Proc. Silicon-on-Insulator Technology and Devices IX*, ECS Proc. **99–3**, pp. 11–24.

Lemkin, M., Juneau, T.N., Clark, W.A., Roessig, T.A., and Brosnihan, T.J. (1999) "A Low-Noise Digital Accelerometer Using Integrated SOI–MEMS Technology," in *Technical Digest IEEE Int. Conf. on Solid-State Sensors and Actuators (Transducers '99)*, pp. 1294–1297, Institute of Electrical and Electronics Engineers, New York.

Li, G.X., and Shemansky, F.A. (2000) "Drop Test and Analysis on Micro-Machined Structures," *Sensors and Actuators A* **85**, pp. 280–286.

Li, G.X., Zhang, Z.L., and Shemansky, F.A. (1997) "Transient and Impact Dynamics of a Micro-Accelerometer," in *Proc. ASME Int. Intersociety Electronic and Photonic Packaging Conf. (INTERpack '97)*, pp. 439–446, American Society of Mechanical Engineers, New York.

Li, G.X., Bergstrom, P.L., Ger, M.-L., Foerstner, J., Schmiesing, J.E., Shemansky, F.A., Mahadevan, D., and Shah, M.K. (1998) "Low Stress Packaging of a Micro-Machined Accelerometer," in *Proc. 3rd Int. Symp. on Electronic Package Technology (ISEPT '98)*, pp. 553–562.

Lotters, J.C., Olthuis, W., Veltink, P.H., and Bergveld, P. (1999) "A Sensitive Differential Capacitance to Voltage Converter for Sensor Applications," *IEEE Trans. Instrum. Meas.* **48**, pp. 89–96.

Lutz, M., Golderer, W., Gerstenmeier, J., Marek, J., Maihöfer, B., Mahler, S., Münzel, H., and Bischof, U. (1997) "A Precision Yaw Rate Sensor in Silicon Micromachining," in *Technical Digest IEEE Int. Conf. on Solid-State Sensors and Actuators (Transducers '97)*, pp. 847–850, Institute of Electrical and Electronics Engineers, New York.

MacDonald, G.A. (1990) "A Review of Low Cost Accelerometers for Vehicle Dynamics," *Sensors and Actuators A* **21**, pp. 303–307.

MacDonald, N.C., and Zhang, Z.L. (1993) "RIE Process for Fabricating Submicron, Silicon Electromechanical Structures," U.S. Patent 5,198,390.

Mastrangelo, C.H., and Hsu, C.H. (1993) "Mechanical Stability and Adhesion of Microstructures Under Capillary Forces. II. Experiments," *J. MEMS* **2**, pp. 44–55.

Matsumoto, Y., and Esashi, M. (1993) "Integrated Silicon Capacitive Accelerometer with PLL Servo Technique," *Sensors and Actuators A* **39**, pp. 209–217.

Matsumoto, Y., Iwakiri, M., Tanaka, H., Ishida, M., and Nakamura, T. (1995) "A Capacitive Accelerometer Using SDB–SOI Structure," in *Technical Digest IEEE Int. Conf. on Solid-State Sensors and Actuators (Transducers '95)*, **2**, pp. 550–553, Institute of Electrical and Electronics Engineers, New York.

McNie, E., Burdess, J.S., Harris, A.J., Hedley, J. and Young, M. (1999) "High Aspect Ratio Ring Gyroscopes Fabricated in [100] Silicon on Insulator (SOI) Material," in *Technical Digest IEEE Int. Conf. on Solid-State Sensors and Actuators (Transducers '99)*, pp. 1590–1593, Institute of Electrical and Electronics Engineers, New York.

Meirovitch, L. (1975) *Elements of Vibration Analysis*, McGraw-Hill, New York.

Noworolski, J.M., and Judy, M. (1999) "VHARM: Sub-Micrometer Electrostatic MEMS," in *Technical Digest IEEE Int. Conf. on Solid-State Sensors and Actuators (Transducers '99)*, pp. 1482–1485, Institute of Electrical and Electronics Engineers, New York.

Offenberg, M., Lärmer, F., Elsner, B., Münzel, H., and Riethmüller, W. (1995) "Novel Process for a Monolithic Integrated Accelerometer," in *Technical Digest IEEE Int. Conf. on Solid-State Sensors and Actuators (Transducers '95)*, pp. 589–592, Institute of Electrical and Electronics Engineers, New York.

Pan, F., Kubby, J., Peeters, E., Tran, A.T., and Mukherjee, S. (1998) "Squeeze Film Damping on the Dynamic Response of a MEMS Torsion Mirror," *J. Micromech. Microeng.* **8**, pp. 200–208.

Park, Y., Jeong, H.S., An, S., Shin, S.H., and Lee, C.W. (1999) "Lateral Gyroscope Suspended by Two Gimbals Through High Aspect Ratio ICP Etching," in *Technical Digest IEEE Int. Conf. on Solid-State Sensors and Actuators (Transducers '99)*, pp. 972–975, Institute of Electrical and Electronics Engineers, New York.

Partridge, A., Reynolds, J.K., Chui, B.W., Chow, E.M., Fitzgerald, A.M., Zhang, L., Cooper, S.R., Kenny, T.W., and Maluf, N.I. (1998) "A High Performance Planar Piezoresistive Accelerometer," in *Technical Digest, Solid-State Sensor and Actuator Workshop*, Hilton Head, SC, pp. 59–64, Transducers Research Foundation, Cleveland, OH.

Putty, M., and Najafi, K. (1994) "A Micromachined Vibrating Ring Gyroscope," in *Technical Digest, Solid-State Sensor and Actuator Workshop*, Hilton Head, SC, pp. 213–220, Transducers Research Foundation, Cleveland, OH.

Reuther, H.M., Weinmann, M., Fischer, M., von Munch, W., and Assmus, F. (1996) "Modeling Electrostatically Deflectable Microstructures and Air Damping Effects," *Sensors Materials* **8**, pp. 251–269.

Ristic, Lj., Gutteridge, R., Dunn, B., Mietus, D., and Bennett P. (1992) "Surface Micromachined Polysilicon Accelerometer," in *Technical Digest IEEE Solid-State Sensor and Actuator Workshop*, Hilton Head, SC, pp. 118–121, Institute of Electrical and Electronics Engineers, New York.

Rockstad, H.K., Reynolds, J.K., Tang, T.K., Kenny, T.W., Kaiser, W.J., and Gabrielson, T.B. (1995) "A Miniature, High-Sensitivity, Electron Tunneling Accelerometer," in *Technical Digest IEEE Int. Conf. on Solid-State Sensors and Actuators (Transducer '95)*, **2**, pp. 675–678, Institute of Electrical and Electronics Engineers, New York.

Roylance, L.M., and Angell, J.B. (1979) "A Batch-Fabricated Silicon Accelerometer," *IEEE Trans. Electron. Devices* **26**, pp. 1911–1917.

Seeger, K. (1985) *Semiconductor Physics: An Introduction*, Springer-Verlag, New York.

Shearwood, C., Ho, K.Y., and Gong, H.Q. (1999) "Testing of a Micro-Rotating Gyroscope," in *Technical Digest IEEE Int. Conf. on Solid-State Sensors and Actuators (Transducers '99)*, pp. 984–987, Institute of Electrical and Electronics Engineers, New York.

Sherman, S.J., Tsang, W.K., Core, T.A., Payne, R.S., Quinn, D.E., Chau, K.H.-L., Farash, J.A., and Baum, S.K. (1992) "A Low Cost Monolithic Accelerometer; Product/Technology Update," in *Technical Digest IEEE Int. Electron. Devices Meeting*, pp. 501–504, Institute of Electrical and Electronics Engineers, New York.

Smith, J., Montague, S., and Sniegowski, J. (1995) "Material and Processing Issues for the Monolithic Integration of Microelectronics with Surface-Micromachined Polysilicon Sensors and Actuators," *SPIE* **2639**, pp. 64–73.

Sparks, D.R., Zarabadi, S.R., Johnson, J.D., Jiang, Q., Chia, M., Larsen, O., Higdon, W., and Castillo-Borelley, P. (1997) "A CMOS Integrated Surface Micromachined Angular Rate Sensor: Its Automotive Applications," in *Technical Digest IEEE Int. Conf. on Solid-State Sensors and Actuators (Transducers '97)*, pp. 851–854, Institute of Electrical and Electronics Engineers, New York.

Sridhar, U., Lau, C.H., Miao, Y.B., Tan, K.S., Foo, P.D., Liu, L.J., Sooriakumar, K., Loh, Y.H., John, P., Lay Har, A.T., Austin, A., Lai, C.C., and Bergstrom, J. (1999) "Single Crystal Silicon Microstructures Using Trench Isolation," in *Technical Digest IEEE Int. Conf. on Solid-State Sensors and Actuators (Transducers '99)*, pp. 258–261, Institute of Electrical and Electronics Engineers, New York.

Starr, J.B. (1990) "Squeeze-Film Damping in Solid-State Accelerometers," in *Technical Digest IEEE Solid-State Sensor and Actuator Workshop*, Hilton Head, SC, pp. 44–47, Institute of Electrical and Electronics Engineers, New York.

Tang, T.K., Gutierrez, R.C., Stell, C.B., Vorperian, V., Arakaki, G.A., Rice, J.T., Li, W.J., Chakraborty, I., Shcheglov, K., Wilcox, J.Z., and Kaiser, W.J. (1997) "A Packaged Silicon MEMS Vibratory Gyroscope for Microspacecraft," in *Technical Digest IEEE Int. Workshop on Micro Electro Mechanical Systems (MEMS '97)*, Nagoya, Japan, pp. 500–505, Institute of Electrical and Electronics Engineers, New York.

Usenko, Y., and Carr, W.N. (1999) "SOI Technology for MEMS Applications," in *Proc. Silicon-on-Insulator Technology and Devices IX, ECS Proc.* **99–3**, pp. 347–352.

van Kampen, R.P., and Wolffenbuttel, R.F. (1998) "Modeling the Mechanical Behavior of Bulk-Micro-Machined Silicon Accelerometers," *Sensors and Actuators A* **4**, pp. 137–150.

Veijola, T., and Ryhanen, T. (1995) "Model of Capacitive Micromechanical Accelerometer Including Effect of Squeezed Gas Film," in *Technical Digest 1995 IEEE Symp. on Circuits and Systems*, pp. 664–667, Institute of Electrical and Electronics Engineers, New York.

Veijola, T., Kuisma, H., and Lahdenperä, J. (1997) "Model for Gas Film Damping in a Silicon Accelerometer," in *Technical Digest IEEE Int. Conf. on Solid-State Sensors and Actuators (Transducers '97)*, **2**, pp. 1097–1100, Institute of Electrical and Electronics Engineers, New York.

Veijola, T., Kuisma, H., and Lahdenperä, J. (1998) "The Influence of Gas–Surface Interaction on Gas–Film Damping in a Silicon Accelerometer," *Sensors Actuators A* **66**, pp. 83–92.

Voss, R., Bauer, K., Ficker, W., Gleissner, T., Kupke, W., Rose, M., Sassen, S., Schalk, J., Seidel, H., and Stenzel, E. (1997) "Silicon Angular Rate Sensor for Automotive Applications with Piezoelectric Drive and Piezoresistive Read-Out," in *Technical Digest IEEE Int. Conf. on Solid-State Sensors and Actuators (Transducers '97)*, pp. 879–882, Institute of Electrical and Electronics Engineers, New York.

Wang, J., McClelland, B., Zavracky, P.M., Hartley, F., and Dolgin, B. (1996) "Design, Fabrication and Measurement of a Tunneling Tip Accelerometer," in *Technical Digest, 1996 Solid-State Sensor and Actuator Workshop,* Hilton Head, SC, pp. 68–71, Transducers Research Foundation, Cleveland, OH.

Wenk, B., Ramos-Martos, J., Fehrenbach, M., Lange, P., Offenberg, M., and Riethmüller, W. (1995) "Thick Polysilicon Based Surface Micromachined Capacitive Accelerometer with Force Feedback Operation," *Proc. SPIE* **2642**, pp. 84–94.

Wolffenbuttel, R.F., and Wise, K.D. (1994) "Low-Temperature Silicon Wafer-to-Wafer Bonding using Gold at Eutectic Temperature," *Sensors and Actuators A* **43**, pp. 223–229.

Wycisk, M., Tönnesen, T., Binder, J., Michaelis, S., and Timme, H.-J. (1999) "Low Cost Post-CMOS Integration of Electroplated Microstructures for Inertial Sensing," in *Technical Digest IEEE Int. Conf. on Solid-State Sensors and Actuators (Transducers '99),* pp. 1424–1427, Institute of Electrical and Electronics Engineers, New York.

Xie, H., Erdmann, L., Zhu, X., Gabriel, K.J., and Fedder, G.K. (2000) "Post-CMOS Processing for High-Aspect-Ratio Integrated Silicon Microstructures," in *Technical Digest, 2000 Solid State Sensor and Actuator Workshop,* Hilton Head, SC, pp. 77–80, Transducers Research Foundation, Cleveland, OH.

Yazde, N., and Najafi, K. (1997) "An All-Silicon Single-Wafer Fabrication Technology for Precision Microaccelerometers," in *Technical Digest IEEE Int. Conf. on Solid-State Sensors and Actuators (Transducers '97),* pp. 1181–1184, New York.

Yeh, C., and Najafi, K. (1997) "Micromachined Tunneling Accelerometer with a Low-Voltage CMOS Interface Circuit," in *Technical Digest IEEE Int. Conf. on Solid-State Sensors and Actuators (Transducers '97),* pp. 1213–1216, Institute of Electrical and Electronics Engineers, New York.

Yoshii, Y., Nakajo, A., Abe, H., Ninomiya, K., Miyashita, H., Sakurai, N., Kosuge, M., and Hao, S. (1997) "1 Chip Integrated Software Calibrated CMOS Pressure Sensor with MCU, A/D Converter, D/A Converter, Digital Communication Port, Signal Conditioning Circuit and Temperature Sensor," in *Technical Digest IEEE Int. Conf. on Solid-State Sensors and Actuators (Transducers '97),* pp. 1485–1488, Institute of Electrical and Electronics Engineers, New York.

Zook, J.D., Herb, W.R., Bassett, C.J., Stark, T., Schoess, J.N., and Wilson, M.L. (1999) "Fiber-Optic Vibration Sensor Based on Frequency Modulation of Light-Excited Oscillators," in *Technical Digest IEEE Int. Conf. on Solid-State Sensors and Actuators (Transducers '99),* pp. 1306–1309, Institute of Electrical and Electronics Engineers, New York.

25
Micromachined Pressure Sensors

Jae-Sung Park
University of Wisconsin–Madison

Chester Wilson
University of Wisconsin–Madison

Yogesh B. Gianchandani
University of Wisconsin–Madison

25.1 Device Structure and Performance Measures **25**-2
25.2 Piezoresistive Pressure Sensors **25**-4
25.3 Capacitive Pressure Sensors **25**-7
   Design Variations • Circuit Integration and Device Compensation
25.4 Other Approaches **25**-14
   Resonant Beam Pick-Off • Servo-Controlled Pressure Sensors • Tunneling Pick-Off Pressure Sensors • Optical Pick-Off Pressure Sensors • Pirani Sensors
25.5 Conclusions **25**-19

Pressure sensors represent one of the greatest successes of micromachining technology. They have benefited from developments in this field for about four decades, in which time both commercially available and research-oriented devices have been developed for a variety of automotive, biomedical and industrial applications. Automotive applications include pressure sensors for engine manifolds, fuel lines, exhaust gases, tires, seats and other uses. Biomedical applications that have been proposed or developed include implantable devices for measuring ocular, cranial or bowel pressure and devices built into catheters that can assist in procedures such as angioplasty. Many industrial applications related to monitoring manufacturing processes exist. In the semiconductor sector, for example, process steps such as plasma etching or deposition and chemical vapor deposition are very sensitive to operating pressures.

In the long history of the use of micromachining technology for pressure sensors, device designs have evolved as the technology has progressed, allowing pressure sensors to serve as a technology demonstration vehicle in some sense [Wise, 1994]. A number of sensing approaches that offer different relative merits have evolved, and there has been a steady march toward improving performance parameters such as sensitivity, resolution and dynamic range. Although multiple options exist, silicon has been a popular choice for the structural material of micromachined pressure sensors partly because its material properties are adequate and partly because significant manufacturing capacity and know-how can be borrowed from the integrated circuit industry. The primary focus in this chapter is on schemes that use silicon as the structural material. The chapter is divided into four sections. The first section introduces structural and performance concepts that are common to a number of micromachined pressure sensors. The second and third sections focus in some detail on piezoresistive and capacitive pick-off schemes for detecting pressure. These two schemes form the basis of the vast majority of micromachined pressure sensors available commercially and studied by the microelectromechanical systems (MEMS) research community. Fabrication, packaging and calibration issues related to these devices are also addressed in these sections. The fourth section concludes with a survey of alternative approaches that may be suitable for selected applications. It includes a few schemes that have been explored with nonmicromachined apparatus but may be suitable for miniaturization in the future.

25.1 Device Structure and Performance Measures

The essential feature of most micromachined pressure sensors is an edge-supported diaphragm that deflects in response to a transverse pressure differential across it. This deformation is typically detected by measuring the stresses that it produces in the diaphragm or by measuring the displacement of the diaphragm. An example of the former approach is the piezoresistive pick-off, in which resistors are formed at specific locations of the diaphragm to measure the stress; an example of the latter approach is the capacitive pick-off, in which an electrode is located on a substrate some distance below the diaphragm to capacitively measure its displacement. The choice of silicon as a structural material is amenable to both approaches because it has a relatively large piezoresistive coefficient and because it can serve as an electrode for a capacitor as well.

The deflection of a diaphragm and the stresses associated with it can be calculated analytically in many cases. It is generally worthwhile to make some simplifying assumptions regarding the dimensions and boundary conditions. A commonly made assumption is that, although the edges of the diaphragm are rigidly affixed (built-in) to the support around its perimeter, they are approximated as being simply supported. It is a reasonable approximation if the thickness of the diaphragm, h, is much smaller than its radius, a. This condition prevents transverse displacement of the neutral surface at the perimeter, while allowing rotational and longitudinal displacement. Mathematically, it permits the second derivative of the deflection to be zero at the edge of the diaphragm. Under this assumption, the stress on the lower surface of a circular diaphragm can be expressed in polar coordinates by [Timoshenko and Woinowsky-Krieger, 1959; Samaun et al., 1973; Middleoek and Audet, 1994]:

$$\sigma_r = \frac{3 \cdot \Delta P}{8h^2}[a^2(1+\nu) - r^2(3+\nu)] \tag{25.1}$$

$$\sigma_t = \frac{3 \cdot \Delta P}{8h^2}[a^2(1+\nu) - r^2(1+3\nu)] \tag{25.2}$$

where Eq. (25.1) denotes the radial component and Eq. (25.2) the tangential component, a and h are the radius and thickness of the diaphragm, r is the radial coordinate, ΔP is the pressure applied to the upper surface of the diaphragm, and ν is Poisson's ratio (Figure 25.1). In the (100) plane of silicon, Poisson's ratio shows fourfold symmetry and varies from 0.066 in the (011) direction to 0.28 in the (001) direction [Evans and Evans, 1965; Madou, 1997]. These equations indicate that both components of stress vary from the same tensile maximum at the center of the diaphragm to different compressive maxima at its periphery. Both components are zero at separate values of r between zero and a. In general, piezoresistors located at the points of highest compressive and tensile stress will provide the largest responses. The deflection of a circular diaphragm under the stated assumptions is given by:

$$d = \frac{3 \cdot \Delta P (1-\nu^2)(a^2-r^2)^2}{16Eh^3} \tag{25.3}$$

where E is the Young's modulus of the structural material. Like Poisson's ratio, the Young's modulus for silicon shows fourfold symmetry in the (100) plane, varying from 168 GPa in the (011) direction to 129.5 GPa

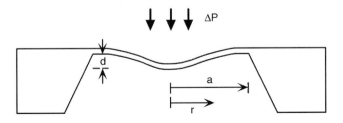

FIGURE 25.1 Deflection of a diaphragm under applied pressure.

in the (100) direction [Greenwood, 1988; Madou, 1997]. When polycrystalline silicon (polysilicon) is used as the structural material, the composite effect of grains of varying size and crystalline orientation can cause substantial variations. It is important to note that additional variations in mechanical properties may arise from crystal defects caused by doping and other disruptions of the lattice. Equation (25.3) indicates that the maximum deflection of a diaphragm is at its center, which comes as no surprise. More importantly, it is dependent on the radius to the third power and on the thickness to the fourth power, making it extremely sensitive to inadvertent variations in these dimensions. This can be of some consequence in controlling the sensitivity of capacitive pressure sensors.

It should be noted that the analysis presented here assumes that the residual stress in the diaphragm is negligibly small. Although mathematically convenient, this is often not the case. In reality, a tensile stress of 5 to 50 MPa is not uncommon. This may significantly reduce the sensitivity of certain designs, particularly if the diaphragm is very thin.

Pressure sensors are typically divided into three categories: absolute, gauge and differential (relative) pressure sensors. Absolute pressure sensors provide an output referenced to vacuum and often accomplish this by vacuum sealing a cavity underneath the diaphragm. The output of a gauge pressure sensor is referenced to atmospheric pressure. A differential pressure sensor compares the pressure at two input ports, which typically transfer the pressure to different sides of the diaphragm.

A number of different units are used to denote pressure, which can lead to some confusion when comparing performance ratings. One atmosphere of pressure is equivalent to 14.696 pounds per square inch (psi), 101.33 kPa, 1.0133 bar (or centimeters of H_2O at 4°C), or 760 Torr (or millimeters of Hg at 0°C).

The performance criteria of primary interest in pressure sensors are sensitivity, dynamic range, full-scale output (FSO), linearity and the temperature coefficients of sensitivity and offset. These characteristics depend on the device geometry, the mechanical and thermal properties of the structural and packaging materials, and the selected sensing scheme. Sensitivity is defined as a normalized signal change per unit pressure change to reference signal:

$$S = \frac{1}{\theta}\frac{\partial \theta}{\partial P} \qquad (25.4)$$

where θ is output signal and $\Delta\theta$ is the change in it due to the applied pressure ΔP. Dynamic range is the pressure range over which the sensor can provide a meaningful output. It may be limited by the saturation of the transduced output signal such as the piezoresistance or capacitance. It also may be limited by yield and failure of the pressure diaphragm. The full-scale output of a pressure sensor is simply the algebraic difference of the end points of the output. Linearity refers to the proximity of the device response to a specified straight line. It is the maximum separation between the output and the line, expressed as a percentage of full-scale output. Generally, capacitive pressure sensors provide highly nonlinear outputs, and piezoresistive pressure sensors provide fairly linear output.

The temperature sensitivity of a pressure sensor is an important performance metric. The definition of temperature coefficient of sensitivity (TCS) is

$$TCS = \frac{1}{S}\frac{\partial S}{\partial T} \qquad (25.5)$$

where S is sensitivity. Another important parameter is the temperature coefficient of offset (TCO). The offset of a pressure sensor is the value of the output signal at a reference pressure, such as when $\Delta P = 0$. Consequently, the TCO is

$$TCO = \frac{1}{\theta_0}\frac{\partial \theta_0}{\partial T} \qquad (25.6)$$

where θ_0 is offset, and T is temperature. Thermal stresses caused by differences in expansion coefficients between the diaphragm and the substrate or packaging materials are some of the many possible contributors to these temperature coefficients.

25.2 Piezoresistive Pressure Sensors

The majority of commercially available micromachined pressure sensors are bulk micromachined piezoresistive devices. These devices are etched from single-crystal silicon wafers, which have relatively well-controlled mechanical properties. The diaphragm can be formed by etching the back of a (100)-oriented Si wafer with an anisotropic wet etchant such as potassium hydroxide (KOH). An electrochemical etch-stop, dopant-selective etch-stop, or a layer of buried oxide can be used to terminate the etch and control the thickness of the unetched diaphragm, which is supported at its perimeter by a portion of the wafer that was not exposed to the etchant and remains at full thickness (Figure 25.1). The piezoresistors are fashioned by selectively doping portions of the diaphragm to form junction-isolated resistors. Although this form of isolation can permit significant leakage current at elevated temperatures and the resistors present sheet resistance per unit length that depends on the local bias across the isolation diode, it permits the designer to exploit the substantial piezoresistive coefficient of silicon and locate the resistors at the points of maximum stress on the diaphragm.

Surface micromachined piezoresistive pressure sensors have also been reported. In Sugiyama et al. (1991), the structural material for the diaphragm was silicon nitride. Polycrystalline silicon (polysilicon) was used both as a sacrificial material and to form the piezoresistors. This fabrication approach permits small devices with high packing density to be fabricated. However, the maximum deflection of the diaphragm is limited to the thickness of the sacrificial layer and can constrain the dynamic range much like the capacitive device to be discussed in the following section. In Guckel et al. (1986), polysilicon was used to form both the diaphragm and the piezoresistors.

In an anisotropic material such as single-crystal silicon, resistivity is defined by a tensor which relates the three-directional components of the electric field to the three-directional components of current flow. In general, the tensor has nine elements expressed in a 3×3 matrix, but they reduce to six independent values from symmetry considerations:

$$\begin{bmatrix} \varepsilon_1 \\ \varepsilon_2 \\ \varepsilon_3 \end{bmatrix} = \begin{bmatrix} \rho_1 & \rho_6 & \rho_5 \\ \rho_6 & \rho_2 & \rho_4 \\ \rho_5 & \rho_4 & \rho_3 \end{bmatrix} = \begin{bmatrix} j_1 \\ j_2 \\ j_3 \end{bmatrix} \quad (25.7)$$

where ε_i and j_i represent electric field and current density components, and ρ_i represents resistivity components. Following the treatment in Kloeck and Rooij (1994) and Middleoek and Audet (1994), if the Cartesian axes are aligned to the (100) axes in a cubic crystal structure such as silicon, ρ_1, ρ_2 and ρ_3 will be equal, as they all represent resistance along the (100) axes and are denoted by ρ. The remaining components of the resistivity matrix, which represent cross-axis resistivities, will be zero because unstressed silicon is electrically isotropic. When stress is applied to silicon, the components in the resistivity matrix change. The change in each of the six independent components, $\Delta\rho_i$, will be related to all the stress components. The stress can always be decomposed into three normal components (σ_i) and three shear components (τ_i), as shown in Figure 25.2. The change in the six components of the resistivity matrix (expressed as a fraction of the unstressed resistivity ρ) can then be related to the six stress components by a 36-element tensor. However, due to symmetry conditions, this tensor is populated by only three non-zero components, as shown:

$$\begin{bmatrix} \rho_1 \\ \rho_2 \\ \rho_3 \\ \rho_4 \\ \rho_5 \\ \rho_6 \end{bmatrix} = \begin{bmatrix} \rho \\ \rho \\ \rho \\ 0 \\ 0 \\ 0 \end{bmatrix} + \begin{bmatrix} \Delta\rho_1 \\ \Delta\rho_2 \\ \Delta\rho_3 \\ \Delta\rho_4 \\ \Delta\rho_5 \\ \Delta\rho_6 \end{bmatrix}; \quad \frac{1}{\rho}\begin{bmatrix} \Delta\rho_1 \\ \Delta\rho_2 \\ \Delta\rho_3 \\ \Delta\rho_4 \\ \Delta\rho_5 \\ \Delta\rho_6 \end{bmatrix} = \begin{bmatrix} \pi_{11} & \pi_{12} & \pi_{12} & 0 & 0 & 0 \\ \pi_{12} & \pi_{11} & \pi_{12} & 0 & 0 & 0 \\ \pi_{12} & \pi_{12} & \pi_{11} & 0 & 0 & 0 \\ 0 & 0 & 0 & \pi_{44} & 0 & 0 \\ 0 & 0 & 0 & 0 & \pi_{44} & 0 \\ 0 & 0 & 0 & 0 & 0 & \pi_{44} \end{bmatrix} \begin{bmatrix} \sigma_1 \\ \sigma_2 \\ \sigma_3 \\ \tau_1 \\ \tau_2 \\ \tau_3 \end{bmatrix} \quad (25.8)$$

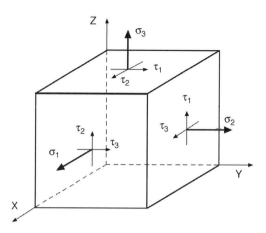

FIGURE 25.2 Definition of normal and shear stresses.

where the π_{ij} coefficients, which have units of Pa^{-1}, may be either positive or negative and are sensitive to doping type, doping level and operating temperature. It is evident that π_{11} relates the resistivity in any direction to stress in the same direction, whereas π_{12} and π_{44} are cross-terms.

Equation (25.8) was derived in the context of a coordinate system aligned to the (100) axes and is not always convenient to apply. A preferred representation is to express the fractional change in an arbitrarily oriented diffused resistor by:

$$\frac{\Delta R}{R} = \pi_l \sigma_l + \pi_t \sigma_t \tag{25.9}$$

where π_l and σ_l are the piezoresistive coefficient and stress parallel to the direction of current flow in the resistor (i.e., parallel to its length), and π_t and σ_t are the values in the transverse direction. The piezoresistive coefficients referenced to the direction of the resistor may be obtained from those referenced to the (100) axes in Eq. (25.8) by using a transformation of the coordinate system. It can then be stated that:

$$\pi_l = \pi_{11} + 2(\pi_{44} + \pi_{12} - \pi_{11})(l_1^2 m_1^2 + l_1^2 n_1^2 + n_1^2 m_1^2) \tag{25.10}$$

$$\pi_t = \pi_{12} - (\pi_{44} + \pi_{12} - \pi_{11})(l_1^2 l_2^2 + m_1^2 m_2^2 + n_1^2 n_2^2) \tag{25.11}$$

where l_1, m_1 and n_1 are the direction cosines (with respect to the crystallographic axes) of a unit-length vector which is parallel to the current flow in the resistor, whereas l_2, m_2 and n_2 are those for a unit-length vector perpendicular to the resistor. Thus, $l_i^2 + m_i^2 + n_i^2 = 1$. As an example, for the (111) direction, in which projections to all three crystallographic axes are equal, $l_i^2 = m_i^2 = n_i^2 = 1/3$.

A sample set of piezoresistive coefficients for Si are listed in Table 25.1. It is evident that π_{44} dominates for p-type Si, with a value that is more than 20 times larger than the other coefficients. By using the dominant coefficient and neglecting the smaller ones, Eqs. (25.10) and (25.11) can be further simplified. It should be noted, however, that the piezoresistive coefficient can vary significantly with doping level and operating temperature of the resistor. A convenient way in which to represent the changes is to normalize the piezoresistive coefficient to a value obtained at room temperature for weakly doped silicon [Kanda, 1982]:

$$\pi(N,T) = P(N,T) \pi_{ref.} \tag{25.12}$$

Figure 25.3 shows the variation of parameter P for p-type and n-type Si.

TABLE 25.1 Sample of Room-Temperature Piezoresistive Coefficients in Si in 10^{-11} Pa^{-1}

Resistivity	π_{11}	π_{12}	π_{44}
7.8 Ω-cm, p-type	6.6	−1.1	138.1
11.7 Ω-cm, n-type	−102.2	53.4	−13.6

Source: From Smith, C.S. (1954) *Physical Rev.*, **94**, 42–49.

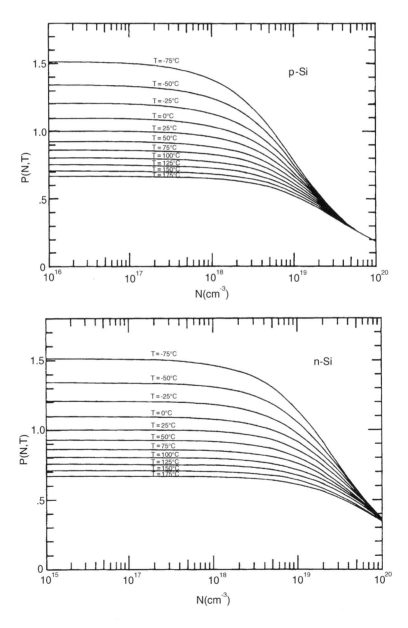

FIGURE 25.3 Variation of piezoresistive coefficient for n-type and p-type Si. (From Kanda, Y. (1982) *IEEE Trans. Electron. Devices*, **29**, 64–70. With permission.)

Figure 25.4 plots the longitudinal and transverse piezoresistive coefficients for resistor orientations on the surface of a (100) silicon wafer. Note that each figure is split into two halves, showing π_l and π_t simultaneously for p-type Si in one case and n-type Si in the other case. Each curve would be reflected in the horizontal axis if drawn individually. Also note that for p-type Si, both π_l and π_t peak along (110), whereas for n-type Si they peak along (100). Since anisotropic wet etchants make trenches aligned to (110) on these wafer surfaces, p-type piezoresistors, which can be conveniently aligned parallel or perpendicular to the etched pits, are favored.

Consider two p-type resistors aligned to the (110) axes and near the perimeter of a circular diaphragm on a silicon wafer. Assume that one resistor is parallel to the radius of the diaphragm, whereas the other is perpendicular to it. Using the equations presented above, it can be shown that as pressure is applied, the fractional change in these resistors is equal and opposite:

$$\left(\frac{\Delta R}{R}\right)_{ra} = -\left(\frac{\Delta R}{R}\right)_{ta} = -\Delta P \cdot \frac{3\pi_{44} a^2 (1-\nu)}{8h^2} \quad (25.13)$$

where the subscripts denote radial and tangentially oriented resistors. The complementary change in these resistors is well suited to a bridge-type arrangement for readout, as shown in Figure 25.5. (The bridge-type readout arrangement is suitable for square diaphragms as well.) The output voltage in this case is given by:

$$\Delta V_0 = V_s \frac{\Delta R}{R} \quad (25.14)$$

Since the output is proportional to the supply voltage V_s, the output voltage of piezoresistive pressure sensors is generally presented as a fraction of the supply voltage per unit change in pressure. Thus, a typical sensitivity is on the order of 100 ppm/Torr. The maximum fractional change in a piezoresistor is on the order of 1 to 2%. It is evident from Eqs. (25.13) and (25.14) that the temperature coefficient of sensitivity, which is the fractional change in the sensitivity per unit change in temperature, is primarily governed by the temperature coefficient of π_{44}.

A valuable feature of the resistor bridge is the relatively low impedance that it presents. This permits the remainder of the sensing circuit to be located at some distance from the diaphragm, without deleterious effects from parasitic capacitance that may be incorporated. As will be discussed in the next section, this stands in contrast to the output from capacitive pick-off pressure sensors, for which the high output impedance creates significant challenges.

The resistors present a scaling limitation for the pressure sensors. As the length of a resistor is decreased, the resistance decreases and the power consumption rises, which is not favorable. As the width is decreased, minute variations that may occur because of nonideal lithography or other processing limitations will have a more significant impact on the resistance. These issues constrain how small a resistor can be made. Then, as the size of the diaphragm is reduced, the resistors will span a larger area between its perimeter and the center. Since the maximum stresses occur at these locations, a resistor that extends between them will be subject to stress averaging, and the sensitivity of the readout will be compromised. In addition, if the nominal values of the resistors vary, the two legs of the bridge become unbalanced, and the circuit presents a non-zero signal even when the diaphragm is undeflected. This offset varies with temperature and cannot be easily compensated because it is in general unsystematic.

25.3 Capacitive Pressure Sensors

Capacitive pressure sensors were developed in the late 1970s and early 1980s. As indicated in Section 25.1, in these devices the flexible diaphragm serves as one electrode of a capacitor, whereas the other electrode is located on a substrate beneath it. As the diaphragm deflects in response to applied pressure, the average gap between the electrodes changes, leading to a change in the capacitance. The concept is illustrated in Figure 25.6.

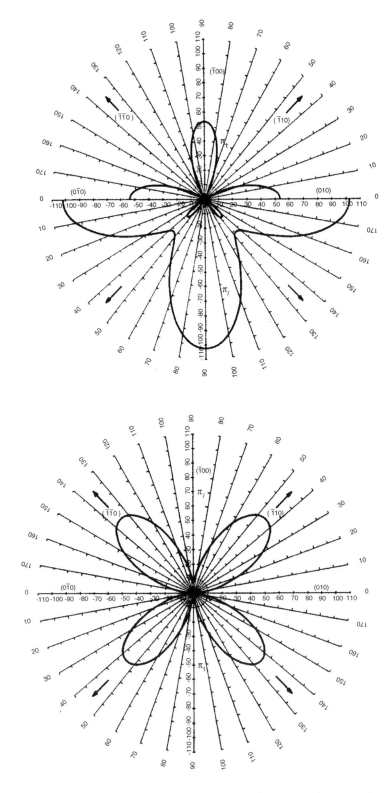

FIGURE 25.4 Longitudinal and transverse piezoresistive coefficients for p-type and n-type resistors on the surface of a (100) Si wafer. (From Kanda, Y. (1982) *IEEE Trans. Electron. Devices*, **29**, 64–70. With permission.)

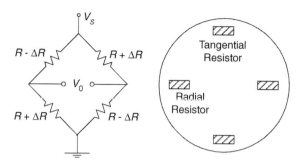

FIGURE 25.5 Schematic representation of Wheatstone bridge configuration (left) and placement of radial and tangential strain sensors on a circular diaphragm (right).

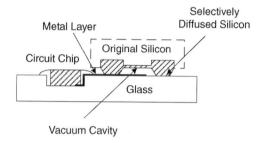

FIGURE 25.6 Schematic of a pressure sensor fabricated on a glass substrate using a dopant-selective etch stop. (From Chau, H., and Wise, K. (1988) *IEEE Trans. Electron. Devices,* **35**(12), 2355–2362. With permission.)

Although many fabrication schemes can be conceived, an attractive bulk micromachining approach is described in Chau and Wise (1988). Areas of a silicon wafer were first recessed, leaving plateaus that served as anchors for the diaphragm. Boron diffusion was then used to define the regions that would eventually form the structure. The top surface of the silicon wafer was then anodically bonded to a glass wafer that had been inlaid with a thin film of metal that served as the stationary electrode and provided lead transfer to circuitry. The undoped portions of silicon were finally dissolved in a dopant selective etchant such as ethylene diamine pyrocatechol (EDP). In order to maintain a low profile and small size, the interface circuit was hybrid-packaged into a recess in the same glass substrate, creating a transducer that was small enough to be located within a cardiovascular catheter of 0.5-mm outer diameter. Using a 2-μm-thick, 560×280-μm^2 diaphragm with a 2-μm capacitor gap and a circuit chip 350 μm wide, 1.4 mm long and 100 μm thick, pressure resolution of <2 mmHg was achieved [Ji et al., 1992].

The capacitive sensing scheme circumvents some of the limitations of piezoresistive sensing. For example, because the resistors do not have to be fabricated on the diaphragm, scaling down the device dimensions is easier because concerns about stress averaging and resistor tolerance are eliminated. In addition, the largest contributor to the temperature coefficient of sensitivity—the variation of π_{44}—is also eliminated. The full-scale output swing can be 100% or more, in comparison to about 2% for piezoresistive sensing. There is virtually no power consumption in the sense element as the dc current component is zero. However, capacitive sensing presents other limitations: The capacitance changes nonlinearly with diaphragm displacement and applied pressure. Even though the fractional change in the sense capacitance may be large, the absolute change is small and considerable caution must be exercised in designing the sense circuit. The output impedance of the device is large, which also affects the interface circuit design, and the parasitic capacitance between the interface circuit and the device output can have a significant negative impact on the readout, which means that the circuit must be placed in close proximity to the device in a hybrid or monolithic implementation. An additional concern is related to lead transfer and packaging. In the case of absolute pressure sensors the cavity beneath the diaphragm must be sealed in

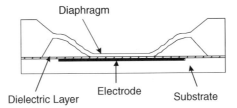

FIGURE 25.7 A touch mode capacitive pressure sensor.

vacuum. Transferring the signal at the counter electrode out of the cavity in a manner that retains the hermetic seal can present a substantial manufacturing challenge.

The capacitance between two parallel electrodes can be expressed as:

$$C = \frac{\varepsilon_0 \varepsilon_r A}{d} \qquad (25.15)$$

where ε_0, ε_r, A and d are permittivity of free space (8.854×10^{-14} F/cm), relative dielectric constant of material between the plates (which is unity for air), effective electrode area and gap between the plates, respectively. Since the gap is not uniform when the diaphragm deflects, finite element analysis is commonly used to compute the response of a capacitive pressure sensor. However, it can be shown that for deflections that are small compared to the thickness of the diaphragm, the sensitivity, which is defined as the fractional change in the capacitance per unit change in pressure, has the following proportionality [Chau and Wise, 1987]:

$$S_{cap} = \frac{\Delta C}{C \cdot \Delta P} \alpha \frac{1-\nu^2}{E} \frac{a^4}{h^3 g} \qquad (25.16)$$

where g is the nominal gap between the diaphragm and the electrode, and the remaining variables represent the same quantities as for Eq. (25.3). Comparing to Eq. (25.13), which provides the fractional resistance change in a piezoresistive device, it is evident that capacitive devices have an increased dependence on a/h, the ratio of the radius to the thickness of the diaphragm. In addition, they are dependent on g, the capacitive gap. The sensitivity of capacitive pressure sensors is generally in the range of 1000 ppm/Torr, which is about 10× greater than piezoresistive pressure sensors. This is an important advantage, but it carries a penalty for linearity and dynamic range.

The dynamic range of a capacitive pressure sensor is conventionally perceived to be limited by the "full-scale" deflection, which is defined to exist at the point that $\Delta C = C$. However, in reality the capacitance keeps rising as the pressure increases even beyond the threshold at which the diaphragm touches the substrate beneath it. As the pressure increases, the contact area grows, extending the useful operating range of the device (Figure 25.7). As long as the electrode is electrically isolated from the diaphragm, such as by an intervening thin film of dielectric material, the sensor can continue to be used. This mode of operation is called "touch mode" [Cho et al., 1992; Ko et al., 1996]. Since most of the diaphragm may in fact be in contact with the substrate in this mode the output capacitance is relatively large. For example, a 4-μm-thick p^{++} silicon diaphragm of 1500×447-μm^2 area and a nominal capacitive gap of 10.4 μm was experimentally shown to have a dynamic range of 120 psi in touch mode, whereas the conventional range was only 50 psi [Ko et al., 1996]. The concept of touch-mode operation exploits the fact that in a capacitive pressure sensor the substrate provides a natural over-pressure stop that can delay or prevent rupture of the diaphragm. This is a feature lacking in most piezoresistive pressure sensors.

One cause of temperature coefficients in capacitive pressure sensors is expansion mismatch between the substrate and diaphragm, which can be minimized by the proper choice of materials and fabrication sequence. For example, when a glass substrate is used, it is important to select one that is expansion-matched to the silicon. Hoya™ SD-2 is designed to be very closely matched to silicon, and #7740 Pyrex™

glass provides acceptable performance for a lower price. Naturally, the best results can be expected when the substrate material is the same as the diaphragm material. Anodic bonding of the silicon structure to a silicon-substrate wafer that was coated with 2.5 to 5 μm of glass has been shown to produce devices with low TCO [Hanneborg and Ohlckers, 1990]. The typical TCO of silicon pressure sensors that use anodically bonded glass substrates is <100 ppm/K. Since the sensitivity is in the range of 1000 ppm/Torr, the TCO can be stated as 0.1 Torr/K.

For absolute pressure sensors, the primary source of temperature sensitivity can be expansion of gas entrapped in the sealed reference cavity. This gas may come from an imperfect vacuum ambient during the sealing. If the gas is unreactive, this component can be estimated by the ideal gas equation. Another source of entrapped gas is out-diffusion from the cavity walls after sealing, which is more difficult to quantify [Henmi et al., 1994]. To some extent, this can be blocked by using a thin-film metal coating as a diffusion barrier [Cheng et al., 2001]. Various sealing methods have been developed based on the principle of creating a cavity by etching a subsurface sacrificial layer though a narrow access hole and then sealing off the access by depositing a thin film at a low pressure [Murakami, 1989]. The reactive sealing method reported by Guckel et al. (1987, 1990) provides some of the lowest sealed-cavity pressures. In this approach, chemical-vapor-deposited polysilicon is used as the sealant. A subsequent thermal anneal causes the trapped gas to react with the interior walls of the cavity, leaving a high vacuum.

Although appropriate for surface micromachined cavities, the thermal budget of the reactive sealing process that uses polysilicon is too high for devices that use glass substrates. A similar effect can be achieved with metal films called nonevaporable getters (NEG) which are suitable for use with glass wafers. In Henmi et al. (1994), the NEG used was a Ni/Cr ribbon covered with a mixture of porous Ti and Zr–V–Fe alloy. To achieve the best results it was initially heated to 300°C to desorb gas, following which it was sealed within the reference cavity and finally heated to 400°C to be activated. The final cavity pressure achieved was reported to be $<10^{-5}$ Torr. It should be noted that gas entrapped in the sealed reference cavity changes pressure as the diaphragm is deflected and the cavity volume changed. This can lead to erroneous readings and a loss of sensitivity [Ji et al., 1992].

For capacitive pressure sensors, lead transfer to the sense electrode located within the sealed cavity has been a persistent challenge. For the type of device shown in Figure 25.6, in which the anodic bond must form a hermetic seal over the metal lead patterned on the glass substrate, the metal layer thickness may not exceed 50 nm [Lee and Wise, 1982]. In addition, a thin-film dielectric must be used to separate the silicon anchor from the lead. Epoxy is sometimes applied at the end of the fabrication sequence to strengthen the seal over the lead. However, this is not a favored solution, and alternatives have been developed. In one approach, a hole is etched or drilled in the substrate or in a rigid portion of the microstructure next to the flexible diaphragm, the lead is transferred through it, and then it is sealed with epoxy or metals [Wang and Esashi, 1997; Giachino et al., 1981]. The approach used by Wang and Esashi (1997) is shown in Figure 25.11. In another approach, subsurface polysilicon leads are used with the help of chemical–mechanical polishing to achieve a planar bonding surface that allows a hermetic seal [Chavan and Wise, 1997].

25.3.1 Design Variations

A number of variations of the conventional choices for structures and materials have been reported. These include ultra-thin dielectric diaphragms, bossed or corrugated diaphragms, double-walled diaphragms with embedded rigid electrodes and structures with sense electrodes located external to the sealed cavity.

Dielectric material compatible with silicon processing can be useful alternatives as structural materials for pressure sensors. One promising option is stress-compensated silicon nitride because it is chemically and mechanically stable. One concern about a larger and thinner diaphragm is high stress developed under applied pressure, which may cause device failure. However, the yield strength of silicon nitride is twice that of silicon. High sensitivities were reached by using a large and thin silicon nitride diaphragm of 2-mm diameter and 0.3-μm thickness [Zhang and Wise, 1994b]. Silicon nitride is under high tensile residual stress after low-pressure chemical vapor deposition (LPCVD), which reduces the diaphragm deflection. To compensate for this residual stress, a silicon dioxide layer, which creates compressive stress, is put between two silicon nitride layers (60 nm/180 nm/60 nm). To compensate for the nonlinear behavior due to large deflections, a 3-μm-thick boss (p^{++} Si) is located in the center of the diaphragm. Its diameter is 60% that of the diaphragm; a greater percentage lowers sensitivity, and a smaller percentage adversely affects linearity. The sensitivity has been shown to be 10,000 ppm/Pa (5 fF/mTorr), which is 10× greater than the typical value for capacitive devices. The minimum resolution pressure was 0.1 mTorr. The TCO and TCS were, respectively, 910 ppm/K and −2900 ppm/K. The dynamic range was >130 Pa.

One issue of great importance to very small and very sensitive pressure sensors is noise. Many sources of noise, both electrical and mechanical, may contribute to the output signal. In general, these include Brownian motion from gas surrounding the diaphragm, noise from the sense capacitor (or resistor), electrostatic pressure variation from the sense circuit and electrical noise in the sense circuit itself [Chau and Wise, 1987].

Introducing corrugation into a pressure-sensor diaphragm allows longer linear travel and a larger dynamic range in capacitive pressure sensors. Corrugations can be manufactured with wet or dry etching. Residual stress in the diaphragm can result in bending when corrugations are present. The use of a boss in the center was shown to considerably improve this but causes a deflection in the opposite direction [Ding, 1992; Zhang and Wise, 1994a]. These papers presented development of bossed and corrugated structures manufactured under tensile, neutral and compressive stress. Both papers develop a square root dependence of deflection vs. pressure for unbossed, corrugated pressure sensors.

A differential capacitive pressure sensor consisting of a double-layer diaphragm with an embedded electrode was reported by Wang et al. (2000). One capacitor was formed between the upper diaphragm and middle electrode, and another between the electrode and lower diaphragm. The diaphragms were mechanically connected to each other and moved in the same direction, which was dependent on the pressure difference between the upper and lower surfaces. The advantage of this pressure sensor is that the cavity is sealed from the outside but still can generate a signal based on pressure difference. The diaphragm size was 150 × 150 μm^2, the thickness of the diaphragm was 2 μm and the capacitor gaps were nominally 1 μm.

Most masking steps in pressure sensors are consumed on electrical lead transfer from the inside of the vacuum cavity to the outside as shown in the previous devices. A pressure sensor that eliminates the problem of the sealed lead transfer by locating the pick-off capacitance outside the sealed cavity is illustrated in Figure 25.8 [Park and Gianchandani, 2000]. A skirt-shaped electrode extends outward from the periphery of the vacuum-sealed cavity, acting as the electrode that deflects under pressure. The stationary electrode is metal patterned on the substrate below this skirt. As the external pressure increases,

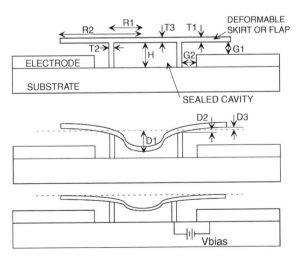

FIGURE 25.8 Electrostatic attraction between the electrode and skirt opposes the deflection due to external pressure.

the center of the diaphragm deflects downwards, and the periphery of the skirt rises, reducing the pick-off capacitance. This deflection continues monotonically as the external pressure increases beyond the value at which the center diaphragm touches the substrate, so this device can be operated in touch mode for expanded dynamic range. To fabricate this a silicon wafer was first dry etched to the desired height of the cavity and then selectively diffused with boron to define the radius of the pressure sensor. The depth of the boron diffusion determined the eventual thickness of the structural layer. The silicon wafer was then flipped over and anodically bonded to a glass wafer that had been inlaid with a Ti/Pt metal that serves as the interconnect and provides the bond pads. The undoped Si was finally dissolved in EDP, leaving the pressure sensor on the glass substrate.

A single-crystal silicon pressure diaphragm using epitaxial deposition was reported by Mastrangelo (1995). The advantage of this kind of sensor is that it does not require bonding on glass as most p^{++} silicon pressure sensors do, and the diaphragm stress condition and material properties are predictable, as single-crystal silicon provides more consistent performance than polysilicon.

25.3.2 Circuit Integration and Device Compensation

An interface circuit for capacitive pressure sensors which offers high sensitivity and immunity from parasitic capacitance is illustrated in Figure 25.9. This is basically a switched-capacitor charge integrator which utilizes a two-phase clock to compare the variable sense capacitor (C_X) to a reference capacitor (C_R) [Park and Wise, 1990]. A reset pulse initially nulls the output. Subsequently, when the clock switches, a charge proportional to the clock amplitude and the difference between the reference and variable capacitor is forced into the feedback capacitor C_F, producing an output voltage given by:

$$V_{out} = \frac{V_p(C_X - C_R)}{C_F} \quad (25.17)$$

Since this scheme measures the difference between two capacitors, it is unaffected by parasitic capacitance values that are common to both.

Capacitive pressure sensors offer numerous advantages to piezoresistive sensors, but they are highly nonlinear. Piezoresistive devices are more linear but are limited to about 0.5% of full scale (8 bits) when uncompensated. One of the first efforts at compensation of pressure sensors was by Crary et al. (1990). The compensation was provided with an off-chip computer. Two possible strategies would be to use polynomial fitting or full lookup tables. The authors used polynomial fitting, using stepwise regression. They characterized a piezoresistive polysilicon full-bridge membrane device over 153 fixed points, and temperature and pressure compensated the device. Before compensation, endpoint-to-endpoint nonlinearity was 29% of full scale. After compensation, the device was improved a quoted 4.4 bits.

One of the most highly integrated devices as of this writing was reported by Chavan and Wise (2000). The developed capacitive pressure sensor offered 15-bit resolution. It was unique in that the on-chip circuitry could interface with five different transducers, as it had a 3-bit multiplexer–transducer select register. It included programmable gain and offset circuitry to implement the circuitry into a variety of ranges and scales and had a buffered dc analog output. The system offered a calibration and operation

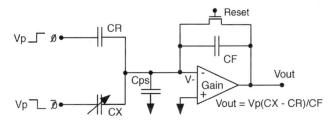

FIGURE 25.9 A switched capacitor charge integrator. (From Park, Y.E., and Wise, K.D., in *Microsensors*, R.S. Muller et al., Eds. (1990) IEEE Press, New York, pp. 329–333. With permission.)

mode and had external debugging pins. The finished array had five different transducers, covering a 300-Torr range. Due to the diameter dependence of capacitive sensors, the membrane diameter only varied 1000 to 1100 μm. The membrane was 3.7/0.3-μm Si–SiO$_2$ with a boss center. The fabrication was a 20-mask process, 15 for the on-device p-well circuitry, three for the transducers and two for the glass processing. The circuitry was fabricated in recessed cavities 2.5 μm deep, and the transducers in 8-μm-deep cavities, allowing the bonding anchors to be the highest points on the wafer. Then, a 2-μm-thick level of polysilicon was deposited for electrical feedthroughs, followed by chemical–mechanical polishing, to make the device flat for bonding. The sensor was vacuum sealed, resulting in a 9.8-μm deflection at atmosphere, leaving a 0.5-μm gap and a resulting 8- to 10-pF capacitance. Ti/Pt/Au electrodes were used to getter the device. The sensor was measured with a three-stage programmable switched capacitor circuit (polysilicon reference capacitors). The front end was a differential charge integrator with a folded cascode amplifier. The operation was circuitry controlled so that a global sensor is used first, then the chip decides which transducer is used next to achieve maximum resolution. The clocking scheme used correlated double sampling, to lower the 1/f noise and amplifier offset.

25.4 Other Approaches

Several pressure sensor concepts that do not rely on capacitive and piezoresistive sensing have been developed. The motivation to pursue alternative approaches is typically driven by a unique application or by stringent requirements in a certain performance category. Resonant beam devices offer particularly high resolution and a pseudo-digital output. Closed-loop servo controlled devices offer a wider operating range. Tunneling devices offer high sensitivity, optical devices are ideal for harsh environments, and Pirani devices are well suited for vacuum applications.

25.4.1 Resonant Beam Pick-Off

Frequency measurement can be considerably more robust and accurate than the measurement of a resistor or capacitor. Resonating devices, therefore, will be more affected by the mechanical qualities than electrical qualities of the device. A common rule of thumb is that resonating devices can achieve ten times greater measurement accuracies than the more conventional capacitive and piezoresistive methods. In addition, because the output is naturally quantized, interfacing to a digital system can be easier.

Resonant devices make use of the fact that imparted strain on a resonating beam will change its effective spring constant, hence the resonating frequency. The typical resonating beam pressure sensor will incorporate end-clamped beams to detect the strain in a diaphragm which can be related to deformation as in piezoresistive sensing. The resonating beam may be driven by electrostatic, magnetic or even optical stimulus. The resonant frequency can be measured by these methods or by implanted piezoresistors on the beam itself.

One of the first commercialized devices was able to achieve accuracies better than 0.01% of the full-scale range [Ikeda et al., 1990]. This device utilized two resonating beams, one in the center of the diaphragm and one at the perimeter to achieve a differential signal (Figure 25.10). The beams were vacuum encapsulated through a fairly complex fabrication process using four levels of epitaxial growth. The cavity was evacuated to a pressure of 1 mTorr, which is allowed by the low pressures used in epitaxial growth processing. This allowed high Q factors of above 50,000 to be achieved for the beam. Temperature sensitivity of the device was limited by the thermal expansion of silicon, with a coefficient of –40 ppm/K.

The beam was excited to resonance by driving an ac current through the resonating beam, and applying a dc magnetic field. The current was controlled by active feedback circuitry to keep the oscillation amplitude constant. The smaller resonance amplitude helps minimize second-order nonlinearities in the resonance, in particular hysteresis of the frequency response. The high accuracy that was a result of this design addressed the aerospace market, where the sensor has been commercialized.

Another resonating beam pressure sensor used a thin single-crystal silicon membrane [Peterson et al., 1991]. This was high-temperature fusion-bonded to a second wafer, which formed the square diaphragm

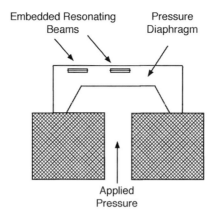

FIGURE 25.10 Schematic of a resonating beam pressure sensor.

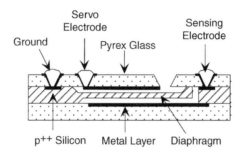

FIGURE 25.11 The structure of the electrostatic servo-controlled pressure sensor. (From Wang, Y., and Esashi, M. (1997) in *IEEE Int. Conf. on Solid-State Sensors and Actuators (Transducers '97)*, June, pp. 1457–1460. With permission)

of the pressure sensor. A single resonant beam was fabricated by etching two slots through the membrane. This beam was excited to resonance electrostatically by a voltage on the order of a few millivolts.

Burns et al. (1994) developed a unique resonant beam sensor that utilized a modulated laser beam as the driving mechanism of the resonator. This structure was developed with a single resonator on the membrane, however, a reference resonator can be located off the diaphragm to provide a differential output. The beam could also be driven electrostatically and was optimized to reduce secondary vibration modes. The cited sensitivity was 3880 counts per second per psi with a dynamic range of 0 to 10 psi. The temperature stability was 0.01 ppm per degree K, and the sensor operated with 12-V off-device circuitry. Compensation circuitry improved device linearity. The final noise variance was 0.2 to 0.4 Hz. One design strategy of the device was to select particular ratios of microbeam length to diaphragm diameter to optimize frequency sensitivity. Then the pressure sensor range was attained by varying diaphragm thickness, which allowed the same masks to be used for different models.

25.4.2 Servo-Controlled Pressure Sensors

When a measured signal from a pressure change is used for feedback control to restore the signal to its reference, the pressure sensor is called a *closed-loop pressure sensor*. This concept has been widely used in micromachined inertial sensors to increase sensitivity, linearity and dynamic range and can be beneficial to pressure sensors, as well. To implement this concept, an actuator is necessary to drive the pressure sensor. The most common choice for this purpose is an electrostatic actuator, because it is simple and easy to design and fabricate, although the actuation force is small.

In the structure shown in Figure 25.11, the Si diaphragm is suspended between two glass wafers on which the sensing and feedback electrodes are located [Wang and Esashi, 1997]. A perforation in one glass wafer provides access to applied pressure. Lead transfer for the lower electrode is through the silicon wafer—this provides access to all the electrodes from the upper surface of the device. This device was fabricated by bulk silicon micromachining and required two glass wafers and one Si wafer. For 10-Torr (1.33-kPa) applied pressure, 70 V was applied to the electrostatic actuator.

A surface micromachined device that provided low-voltage operation was presented by Gogoi and Mastrangelo (1999). To accomplish this, the electrode area was designed to be 100× larger than the pressure diaphragm area to reduce the restoring voltage. The device has a 20-μm × 20-μm diaphragm, and 250-μm × 250-μm footprint area. The diaphragm as well as the feedback electrodes were made from polysilicon. The fabrication process required 15 masks, and the device was self-encapsulated, which can greatly simplify packaging. For 100-kPa applied pressure, the actuation voltage was less than 12 V.

To avoid structural complications and high operating voltage over a wide pressure range, another closed-loop pressure sensor has been devised [Park and Gianchandani, 2001]. In the proposed structure in Figure 25.8, the flexible diaphragm that deforms in response to applied pressure is not directly used for sensing. Instead, it is attached to a skirt-like external electrode which deflects *upward* as the diaphragm deflects *downward*. This movement is detected capacitively by a counter electrode located under it. For closed-loop operation, this setup permits deflection of the skirt to be balanced by a voltage bias on the sense electrode without an extra structural layer on the top of the diaphragm. Moreover, the concentric layout of the sealed cavity and the skirt permits the electrode to naturally occupy a larger area than the diaphragm, as preferred for electrostatic feedback. A point of distinction from other implementations is that in this feedback scheme the skirt, and not necessarily the diaphragm, is restored to its reference position. Low operating voltages are possible because there is no need to completely balance the pressure force, and the electrode is much more flexible than the diaphragm.

The closed-loop operation of the pressure sensor was demonstrated by varying the chamber pressure and setting the bias voltage to provide the capacitance that was measured at zero gauge pressure [Park and Gianchandani, 2001]. The applied voltage was varied from 31.2 to 73.2 V as the pressure was varied from 20 to 100 kPa, providing an average sensitivity of 0.516 V/kPa. Over this range, the response deviates from linearity by ≤3.22% of the full-scale output. The sensitivity can be electrically tuned by applying a bias to reduce the gap between the electrodes.

25.4.3 Tunneling Pick-Off Pressure Sensors

Certain applications require a particularly high sensitivity to pressure. High-sensitivity capacitive sensors require the fabrication of very thin and large-area pressure diaphragms in order to provide measurable deflection. However, tunneling current established between an electrode and the diaphragm can be used to measure nanometer scale deflections. This is because the tunneling current is an exponential function of the separation distance between the tunneling tip and counter electrode [Wickramasinghe, 1989]. One key advantage of this device is its ability to scale dimensions. A comparably sensitive device using traditional capacitive sensors would be quite large. In addition, fabrication yield is likely to be improved, as it would not require the fabrication of very thin and wide membranes.

Tunneling-based pressure sensors reported by Yeh and Najafi (1994) used two levels of silicon and one level of glass. The first silicon layer is fabricated as a cantilever beam with a tunneling tip, and the second silicon layer is fabricated as a thin pressure diaphragm. The glass substrate has a patterned metal electrode and is used to deflect the cantilever beam down using electrostatic forces. Two modes of operation are available. In the first mode, the diaphragm deflects with applied external pressure, and the tunneling current changes at a constant applied voltage. In the second closed-loop mode, the tunneling current is kept constant by applying an appropriate deflection voltage between the beam and deflection electrode. An increase of 10 V in either the deflection voltage or the tunneling voltage produces two orders of magnitude change in the tunneling current. Based on the beam deflection as a function of pressure, signal/pressure sensitivity was found to be 0.21 nA/mTorr.

Due to the exponential tunneling current, the dynamic range becomes limited to 1–10 mTorr in the open-loop mode. The range is significantly improved in closed-loop mode. The range limitation now

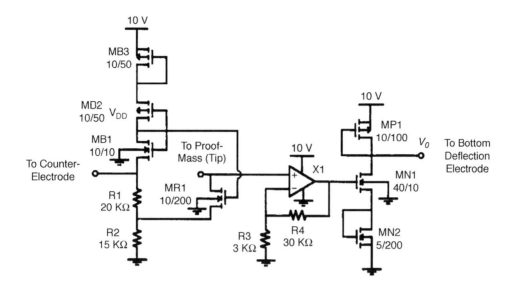

FIGURE 25.12 Circuitry for tunneling-based sensor. (From Yeh, C. and Najafi, K. (1998) *J. MEMS*, March, 6–9. With permission.)

becomes the mechanical displacement limits of the device. Noise is a key issue in this device, as the measured quantity is in nanoamps, but for the same reason power consumption is uniquely small. Tunneling points can wear with time, so the long-term performance can drift. Tunneling electrodes, however, do not necessarily have to be tips and may be manufactured as flat electrodes.

Tunneling-based pressure sensors require special circuitry for their operation. Several circuits to operate these devices were presented by Yeh and Najafi (1998). One of the circuits tested is shown in Figure 25.12. Transistor MR1, a long NMOS, is biased in the linear region. As the tunneling current goes through MR1 its voltage increases roughly linearly. This is amplified 10× with the following noninverting amplifier and is inverted and buffered by the final stage. The final stage output controls the deflection electrode, which regulates the tunneling distance by deflecting a proof mass on a membrane. The final stage is a high-impedance output to prevent over-current should the proof mass and the deflection electrode contact. When MR1 is replaced with a diode, the current is exponential with the voltage across the diode, as the tunneling current is. This provides a deflection voltage that is linear with the gap distance. The final 10-V circuit is suitable for portable battery operation.

25.4.4 Optical Pick-Off Pressure Sensors

Optical devices offer particular advantages in speed and remote sensing and may be utilized more in the future for specific niche applications. As the sensors can be addressed and read with light, they are potentially useful for hazardous environments. In areas that are flammable, remote or corrosive, it may not be practical to run electrical cabling.

A unique device was developed to measure surface pressure fluctuations. Measuring high-speed turbulent behavior on aeronautical surfaces requires remote sensing and good spatial and temporal resolution [Miller et al., 1997]. The objective for this application was to couple pressure fluctuations and mechanical vibrations. An array of pressure cavity devices was constructed that acts as Fabry–Perot interferometric etalons, in conjunction with an external near-infrared laser. The optical nature of the device allows no electrical connections to disturb the pressure fluctuations, as the device is interrogated with a laser externally. The device was fabricated with 0.55-mm-thick silicon-on-insulator (SOI) wafers, with a 55-μm-thick upper layer, and a 0.4-μm-thick buried SiO_2 layer. The etalon cavities were formed by wet-etching the front surface down to the oxide layer using KOH. The buried oxide ensured an even depth and a smooth reflective trench surface. A second SOI wafer was then bonded to it face to face. The silicon on the second wafer was then etched, forming a 55-µm-thick diaphragm. A half-wave coating of SiO_2 was deposited on the diaphragm. The formed etalons were 1000-µm square, with an internal pressure of 0.8 atm. The device was interrogated with a 1.5-µm wavelength laser, and interference patterns were observed with a video camera corresponding to quarter-wavelength deformations.

Various device concepts using optical sensing have been proposed by Greywall (1997). The devices are located at the end of an optical fiber and include a rigid substrate, followed by an evacuated or fixed-pressure air gap, followed by a flexible membrane. The device acts as a Fabry–Perot interferometer. The author defines a parameter, β, which is proportional to the square of the radius and inversely proportional to membrane stress and thickness and to the air/vacuum gap. The larger the β, the larger the midpoint displacement and therefore the larger the slope of reflectivity vs. pressure. As a consequence, larger β improves sensitivity but limits the range of operation. In this device concept, the optical intensity is the same for several values of pressure in some cases, as the gap can cycle through several multiples of quarter wavelengths. Luminous intensity can be a function of stray light leaking into the system, surface roughness of the device and the index of refraction of the medium being measured.

An in-line Fabry–Perot device was developed for use in pressure sensing of oil wells during the drilling process [Maron, 2000]. This sensor is categorized as an intrinsic Fabry–Perot interferometer, in that the fiberoptic waveguide has two mirrors internal to the fiber. The mirrors are partially transparent and act as an interferometer to light directed down the fiber. These sensors were packaged in a metal case, and they were able to measure peak pressures in the 15,000- to 30,000-Torr range, with resolution on the order of 300 Torr. These devices are in contrast to external Fabry–Perot interferometers, such as Greywall's, where at the end of the fiber is a flexible membrane that acts as the interferometer. The device reported by Kao and Taylor (1996) is a hybrid between the two, where a fiber with internal mirrors is epoxied to a pressure diaphragm and encapsulated in a cylindrical housing. The fiber is epoxied to the membrane, pulled under tension and connected to the other end of the sensor housing, which is sealed at a constant pressure. As the external pressure increases, the diaphragm deforms, and the optical length between the two mirrors changes accordingly. The fiber is interrogated with a laser, and Fabry–Perot interferometry is used to determine the optical distance. Modeling was done for the mechanics of the diaphragm, and the device was tested and found to be able to measure pressure from 1 to 100 Torr with a resolution of 0.4 Torr.

There are a few limited applications where the pressures are so high as to essentially preclude mechanical sensors. One application would be in a diamond anvil, where high-pressure physics experiments are conducted and pressures can reach the megabar range. These pressures are usually performed in very small volumes, and it is impractical to utilize any feedthroughs, so remote sensing is a must. A common sensor for this is a ruby crystal [Trzeciakowski et al., 1992]. The crystal is illuminated with a laser, and the very narrow fluorescent atomic excitation lines due to the chromium impurity shift to lower energies with pressure. The resolution is around 1 to 2 kbar, which is good for the Mbar range, but inadequate for lower ranges. Measuring pressure from the wavelength of atomic excitations offers advantages over intensity measurements, as roughness and reflection coefficients are no longer an issue. Spectral lines shift with pressure in many materials, but the effect is limited to fairly low (<77 K) temperatures due to phonon interaction. Intra-impurity transitions (as in ruby, due to the chrome) are not very sensitive to pressure.

Trzeciakowski (1992) constructed multiple quantum-well (MQW) structures using 10 or more consecutive GaAs/AlGaAs layers of 10-nm thickness which confine electrons and holes as excitons. The spectral lines shift in pressure similar to the energy gap in quantum-well-based materials. GaAs and AlGaAs emission can be excited with Ar, Kr and HeNe gas lasers, so the material is placed in the pressure chamber and externally stimulated with an off-the-shelf laser. Optical attenuation through several layers induce a temperature gradient from the absorbed light energy causing broadening of the lines. Better results were found in wider GaAs wells, with higher aluminum concentration in the AlGaAs regions. The heterostructure sample showed about 20 times the sensitivity of a ruby crystal with an accuracy cited at 0.02 kBar over a 40-kbar range. The temperature coefficient was about 40 bar/K.

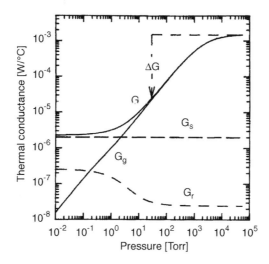

FIGURE 25.13 Thermal conductance as a function of pressure. (From Chou, B.C.S. and Shie, J.S. (1997) in *IEEE Int. Conf. on Solid-State Sensors and Actuators (Transducers '97)*, June, pp. 1465–1468. With permission.)

25.4.5 Pirani Sensors

Pirani sensors are vacuum sensors that utilize a temperature-sensitive resistor as a heating and sensing element to measure heat conduction across an air gap. This concept is widely used in conventional sensors, and micromachined versions have been reported [Chou and Shie, 1997]. Figure 25.13 illustrates thermal conductance as a function of pressure. The total thermal conductance, G, is a traditional S curve. G_s is the solid heat transfer-through leads and various contact points. G_r is radiative thermal conduction, and G_g is the gas heat conduction.

Gas heat conduction dominates radiative heat conduction above 1 Torr and is essentially linear with pressure up to a saturation point. Since constants such as solid conduction can be neglected, the thermal conduction is linear over some range of pressure. This range can be increased by minimizing the air gap across which the thermal conductivity is measured. Chao and Wise (2000) built a floating membrane 50×50 μm^2 with plasma-enhanced chemical vapor deposition (PECVD) low-stress silicon nitride, followed by a 100-nm platinum film acting as a 200-Ω temperature-sensitive resistor. This was fabricated on top of a 0.3-μm-thick polysilicon layer, which after being etched away exposed the air gap to the substrate. This was constructed as an element in a bridge-like circuit to measure temperature-dependent resistivity as a function of pressure. An identical platinum resistor located on the substrate served as a reference. The smaller air gap allowed an operating region from 10^{-6} to 10^4 Torr, extending the upper limit into above-atmospheric-pressure regimes.

25.5 Conclusions

In this chapter, we have attempted to provide a perspective of the breadth and depth of developments in pressure sensors, which represent one of the richest applications of micromachining technology to date. Although most of the commercially available options use piezoresistive sense schemes, capacitive sensing has been a favorite research theme for a number of years, and other options have also emerged. Novel structures have been proposed to address special needs such as on-chip encapsulation and feedback control. Highly sensitive devices have been developed using thin-film dielectric diaphragms and by using tunneling current for sensing. Optical sensing appears to be particularly promising for applications in harsh environments. There is little doubt that this standard-bearer for the MEMS field will continue to evolve with exciting new structures and technologies in the future.

References

Burns, D.W., Zook, J.D., Horning, R.D., Herb, W.R., and Guckel, H.G. (1994) "A Digital Pressure Sensor Based on Resonant Microbeams," in *Solid-State Sensor and Actuator Workshop,* pp. 221–224.

Chau, H., and Wise, K. (1987) "Scaling Limits in Batch-Fabricated Silicon Pressure Sensors," *IEEE Trans. Electron. Devices* **34**, 85–88.

Chau, H., and Wise, K. (1988) "An Ultra-Miniature Solid-State Pressure Sensor for Cardiovascular Catheter," *IEEE Trans. Electron. Devices* **35**(12), 2355–2362.

Chavan, A.V., and Wise, K.D. (1997) "A Batch-Processed Vacuum-Sealed Capacitive Pressure Sensor," in *IEEE Int. Conf. on Solid-State Sensors and Actuators (Transducers '97),* pp. 1449–1451.

Chavan, A.V., and Wise, K.D. (2000) "A Monolithic Fully Integrated Vacuum Sealed CMOS Pressure Sensor," in *Proc. IEEE Int. Conf. on Microelectromechanical Systems,* pp. 341–346.

Cheng, Y.T., Hsu, W.T., Lin, L., Nguyen, C.T., and Najafi, K. (2001) "Vacuum Packaging Technology Using Localized Aluminum/Silicon-to-Glass Bonding," in *Proc. IEEE Int. Conf. on Microelectromechanical Systems,* pp. 18–21.

Cho, S.T., Najafi, K., Lowman, C.L., and Wise, K.D. (1992) "An Ultrasensitive Silicon Pressure-Sensor-Based Microflow Sensor," *IEEE Electron. Devices* **ED-39**, 825–835.

Chou, B.C.S., and Shie, J.S. (1997) "An Innovative Pirani Pressure Sensor," in *IEEE Int. Conf. on Solid-State Sensors and Actuators (Transducers '97),* pp. 1465–1468.

Crary, S., Baer, W.G., Cowles, J.C., and Wise, K.D. (1990) "Digital Compensation of High Performance Silicon Pressure Transducers," *Sensors and Actuators A* **21**, 70–72.

Ding, X. (1997) "Behavior and Application of Silicon Diaphragms with a Boss and Corrugations," in *Int. Conf. on Solid-State Sensors and Actuators (Transducers)* pp. 166–169.

Evans, J.J., and Evans, R.A. (1965) *J. Appl. Phys.* **36**, 153–156.

Giachino, J.M. et al. U.S. Patents 4,261,086 (1981); 4,386,453 (1983); 4,586,109 (1986).

Gogoi, B., and Mastrangelo, C.H. (1999) "A Low Voltage Force-Balanced Pressure Sensor with Hermetically Sealed Servomechanism," in *Proc. IEEE Int. Conf. on Microelectromechanical Systems,* pp. 493–498.

Greenwood, J.C. (1988) "Silicon in Mechanical Sensors," *J. Phys. E, Sci. Instrum.* **21**, pp. 1114–1128.

Greywall, D.S. (1997) "Micromechanical Light Modulators, Pressure Gauges, and Thermometers Attached to Optical Fibers," *J. Micromech. Microeng.* **7**(4), 343–352.

Guckel, H., Burns, D.W., and Rutigliano, C.T. (1986, 1990) "Design and Construction Techniques for Planar Polysilicon Pressure Transducers with Piezoresistive Read-Out," in *Solid State Sensor and Actuator Workshop* and *Microsensors,* R.S. Muller et al., Eds., IEEE Press, New York, pp. 350–351.

Guckel, H., Sniegowski, J.J., and Christenson, T.R. (1990) "Construction and Performance Characteristics of Polysilicon Resonating Beam Force Transducers," in *Integrated Micro-Motions: Micromachining, Control, and Applications: A Collection of Contributions Based on Lectures Presented at the Third Toyota Conference,* F. Harashima, Ed., Elsevier, Amsterdam, pp. 393–404.

Guckel, H., Burns, D.W., Rutigliano, C.R., Showers, D.K., and Uglow, J. (1987) "Fine Grained Polysilicon and Its Application to Planar Pressure Transducers," *Int. Conf. on Solid-State Sensors and Actuators (Transducers),* pp. 277–282.

Hanneborg, A., and Ohlckers, P. (1990) "A Capacitive Silicon Pressure Sensor with Low TCO and High Long-Term Stability," *Sensors and Actuators A* **21-23**, 151–154.

Henmi, H., Shoji, S., Shoji, Y., Yoshmi, K., and Esashi, M. (1994) "Vacuum Packaging for Microsensors by Glass-Silicon Anodic Bonding," *Sensors and Actuators A* **43**, 243–248.

Ikeda, K., Kuwayama, H., Kobayashi, T., Watanabe, T., Nishikawa, T., Yoshida, T., and Harada, K. (1990) "Silicon Pressure Sensor Integrates Resonant Strain Gauge on Diaphragm," *Sensors and Actuators A* **21**, 146–150.

Ji, J., Cho, S.T., Zhang, Y., and Najafi, K. (1992) "An Ultraminiature CMOS Pressure Sensor for a Multiplexed Cardiovascular Catheter," *IEEE Trans. Electron. Devices* **39**(10), 2260–2267.

Kanda, Y. (1982) "A Graphical Representation of the Piezoresistive Coefficients in Si," *IEEE Trans. Electron. Devices* **29**, 64–70.

Kao, T.W., and Taylor, H.F. (1996) "High-Sensitivity Intrinsic Fiber-Optic Fabry–Perot Pressure Sensor," *Optics Lett.* **21**(8), 615–618.

Kloeck, B., and de Rooij, N.F. (1994) "Mechanical Sensors," in *Semiconductor Sensors,* S.M. Sze, Ed., John Wiley, New York.

Ko, W.H., Wang, Q., and Wang, Y. (1996) "Touch Mode Pressure Sensors for Industrial Applications," in *Solid-State Sensor and Actuator Workshop,* pp. 244–248.

Lee, Y.S., and Wise, K.D. (1982) "A Batch-Fabricated Silicon Capacitive Pressure Transducer with Low Temperature Sensitivity," *IEEE Trans. Electron. Devices* **29**(1), 42–48.

Madou, M. (1997) *Fundamentals of Microfabrication,* CRC Press, Boca Raton, FL, pp. 460–464.

Maron, R.J. (2000) "High Sensitivity Fiber Optic Pressure Sensor for Use in Harsh Environments," U.S. Patent 6,016,702.

Mastrangelo, C.H. (1995) "Method for Producing a Silicon-on-Insulator Capacitive Surface Micromachined Absolute Pressure Sensor," U.S. Patent 5,470,797.

Middleoek, S., and Audet, S.A. (1994) *Silicon Sensors,* Delft University of Technology, The Netherlands, 1994.

Miller, M.F., Allen, M.G., Arkilic, E., Breuer, K.S., and Schmidt, M.A. (1997) "Fabry–Perot Pressure Sensor Arrays for Imaging Surface Pressure Distributions," in *IEEE Int. Conf. on Solid-State Sensors and Actuators (Transducers '97),* pp. 1469–1472.

Murakami (1989)"Pressure Transducer and Method for Fabricating Same," U.S. Patent 4,838,088.

Park, J.-S., and Gianchandani, Y.B. (2000) "A Capacitive Absolute-Pressure Sensor with External Pick-Off Electrodes," *J. Micromech. Microeng.* **10**(4), 528–533.

Park, J.-S., and Gianchandani, Y.B. (2001) "A Servo-Controlled Capacitive Pressure Sensor with a Three-Mask Fabrication Sequence," in *Int. Conf. on Solid-State Sensors and Actuators (Transducers),* pp. 506–509.

Park, Y.E., and Wise, K.D. (1983, 1990) "An MOS Switched Capacitor Readout Amplifier for Capacitive Pressure Sensors," in *Record of the IEEE Custom IC Conference,* and *Microsensors,* R.S. Muller et al., Eds., IEEE Press, New York, pp. 329–333.

Petersen, K., Pourahmadi, F., Brown, J., Parsons, P., Skinner, M., and Tudor, J. (1991) "Resonant Beam Pressure Sensor Fabricated with Silicon Fusion Bonding," in *Solid-State Sensor and Actuator Workshop,* pp. 664–667.

Samaun, Wise, K.D., and Angell, J.B. (1973) "An IC Piezoresistive Pressure Sensor for Biomedical Instrumentation," *IEEE Trans. Biomed. Eng.* **20**, 101–109.

Smith, C.S. (1954) "Piezoresistance Effect in Germanium and Silicon," *Physical Rev.* **94**, 42–49.

Sugiyama, S., Shimaoka, K., and Tabata, O. (1991) "Surface Micromachines Micro-Diaphragm Pressure Sensors," in *Int. Conf. on Solid-State Sensors and Actuators (Transducers),* pp. 188–191.

Timoshenko, S., and Woinowsky-Krieger, S. (1959) *Theory of Plates and Shells,* McGraw-Hill, New York.

Trzeciakowski, W., Perlin, P., Teisseyre, H., Mendonca, C. A., Micovic, M., Ciepielewski, P., and Kaminska, E. (1992) "Optical Pressure Sensors Based on Semiconductor Quantum Wells," *Sensors and Actuators A* **32**, 632–638.

Wang, Y., and Esashi, M. (1997) "A Novel Electrostatic Servo Capacitive Vacuum Sensor," in *IEEE Int. Conf. on Solid-State Sensors and Actuators (Transducers '97),* pp. 1457–1460.

Wang, C.C., Gogoi, B.P., Monk, D.J., and Mastrangelo, C.H. (2000) "Contamination Insensitive Differential Capacitive Pressure Sensors," in *Proc. IEEE Int. Conf. on Microelectromechanical Systems,* pp. 551–555.

Wickramasinghe, H.K. (1989) "Scanned-Probe Microscopes," *Sci. Am.* **October**, 98–105.

Wise, K.D. (1994) "On Metrics for MEMS," *MEMS Newslett.* **1**(1).

Yeh, C., and Najafi, K. (1994) "Bulk Silicon Tunneling Based Pressure Sensors," *Solid-State Sensor and Actuator Workshop,* pp. 201–204.

Yeh, C., and Najafi, K. (1998) "CMOS Interface Circuitry for a Low Voltage Micromachined Tunneling Accelerometer," *J. MEMS* **March**, **7**(1), 6–9.

Zhang, Y., and Wise, K.D. (1994a) "Performance of Non-Planar Silicon Diaphragms Under Large Deflections," *J. MEMS* **June**, **3**(2), 59–68.

Zhang, Y., and Wise, K.D. (1994b) "An Ultra-Sensitive Capacitive Pressure Sensor with a Bossed Dielectric Diaphragm," in *Solid-State Sensor and Actuator Workshop,* pp. 205–208.

26
Sensors and Actuators for Turbulent Flows

Lennart Löfdahl
Chalmers University of Technology

Mohamed Gad-el-Hak
University of Notre Dame

26.1 Introduction 26-1
26.2 MEMS Fabrication... 26-3
  Background • Microfabrication
26.3 Turbulent Flows 26-8
  Introductory Remarks • Definition of Turbulence • Methods for Analyzing Turbulence • Scales • Sensor Requirements
26.4 Sensors for Turbulence Measurements and Control 26-14
  Introductory Remarks • Background • Velocity Sensors • Wall-Shear-Stress Sensors • Pressure Sensors
26.5 Microactuators for Flow Control................................ 26-53
  Background on Three-Dimensional Structures • The Bi-Layer Effect • Electrostatic and Magnetic External Forces • Mechanical Folding • One-Layer Structures
26.6 Microturbomachines 26-63
  Micropumps • Microturbines • Microbearings
26.7 Concluding Remarks 26-68
  Acknowledgments 26-69

26.1 Introduction

During the last decade, microelectromechanical systems (MEMS) devices have had major impacts on both industrial and medical applications. Examples of the former group are accelerometers for automobile airbags, scanning electron microscope tips to image single atoms, micro-heat-exchangers for electronic cooling and micromirrors used for light-beam steering. Micropumping is useful for ink-jet printing and cooling of electronic equipment. Existing and prospective medical applications include reactors for separating biological cells, controlled delivery and measurement of minute amount of medication, pressure sensors for catheter tips and development of artificial pancreas. The concept of MEMS includes a variety of devices, structures and systems, and this chapter is limited to focus only on MEMS devices that generally can be categorized as microsensors and microactuators. Particularly, attention is directed toward microdevices used for turbulent-flow diagnosis and control.

MEMS are created using specialized techniques derived and developed from integrated (IC) technology in a process often called *micromachining*. A vast variety of sensors and actuators and several associated technologies for their fabrication exist today. However, three main technologies are usually distinguished when discussing micromachining: bulk micromachining, surface micromachining and micromolding. Bulk micromachining involves the different techniques that use simple, single-crystal, silicon wafer as structural material. Using anisotropic silicon etching and wafer bonding, three-dimensional structures such as pressure sensors, accelerometers, flow sensors, micropumps and various resonators have been fabricated. This branch has been under development for more than 20 years and may now be considered a well-established technology.

In the second group, surface micromachining, the silicon substrate is used as support material, and thin films such as polysilicon, silicon dioxide and silicon nitride provide sensing elements and electrical interconnections as well as structural, mask and sacrificial layers. The basis of surface micromachining is sacrificial etching where freestanding, thin-film structures (polysilicon) are free-etched by lateral underlying sacrificial layer (silicon dioxide). Surface micromachining is very simple but powerful, and, despite the two-dimensional nature of this technique, various complex structures such as pressure sensors, micromotors and actuators have been fabricated.

The third group, micromolding (the LIGA technique), is more similar to conventional machining in concept. A metal mold is formed using lithographic techniques, which allow fine feature resolution. Typically, tall structures with submicrometer resolution are formed. Products created by micromolding techniques include thermally actuated microrelays, micromotors and magnetic actuators. In a rapidly growing field such as MEMS, numerous surveys on fabrication have been published and here we give only a few: Petersen (1982), Linder et al. (1992), Brysek et al. (1994), Diem et al. (1995) and Tien (1997). The books by Madou (1997) and Kovacs (1998) are valuable references, and the entire second part of this handbook focuses on MEMS design and fabrication. As mentioned, the emphasis in this chapter is on MEMS applications for turbulence measurements and flow control.

For more than 100 years, turbulence has been a challenge for scientists and engineers. Unfortunately, no simple solution to the "closure problem" of turbulence exists, so for the foreseeable future turbulence models will continue to play a crucial role in all engineering calculations. The modern development of turbulence models is basically directed toward applications to high-Reynolds-number flows ($Re > 10^6$). This development will be a joint effort between direct numerical simulations of the governing equations and advanced experiments. However, an "implicit closure problem" is inherent in the experiments, as an increase in Reynolds number will automatically generate smaller length scales and shorter time scales, which both in turn require small and fast sensors for a correct resolution of the flowfield. MEMS offer a solution to this problem, as sensors with length and time scales on the order of the relevant Kolmogorov microscales can now be fabricated. Additionally, these sensors are produced with high accuracy at a relatively low cost per unit. For instance, MEMS pressure sensors can be used to determine fluctuating pressures beneath a turbulent boundary layer with a spatial resolution that is about one order of magnitude finer than what can be achieved with conventional transducers.

MEMS sensors can be closely spaced together on one chip, and such multisensor arrays are of significant interest when measuring correlations of fluctuating pressure and velocity and, in particular, for their applications in aeroacoustics. Moreover, the low cost and energy consumption per unit device will play a key role when attempting to cover a large macroscopic surface with sensors to study coherent structures. More elaborate discussion on turbulence and the closure problem can be found in textbooks in the field [e.g., Tennekes and Lumley, 1972; Hinze, 1975] or in surveys on turbulence modeling [e.g., Speziale, 1991; Speziale et al., 1991; Hallbäck, 1993]. The role and importance of fluctuating pressures and velocities in aeroacoustics is well covered in the books by Goldstein (1976), Dowling and Ffowcs-Williams (1983) and Blake (1986).

Reactive flow control is another application where microdevices may play a crucial future role. MEMS sensors and actuators provide opportunities for targeting the small-scale coherent structures in macroscopic turbulent shear flows. Detecting and modulating these structures may be essential for a successful control of turbulent flows to achieve drag reduction, separation delay and lift enhancement. To cover areas of significant spatial extension, many devices are required that utilize small-scale, low-cost and low-energy-use components. In this context, the miniaturization, low-cost fabrication and low energy consumption of microsensors and microactuators are of utmost interest and promise a quantum leap in control-system performance. Combined with modern computer technologies, MEMS yield the essential matching of the length and time scales of the phenomena to be controlled. These issues and other aspects of flow control are well summarized in a number of reviews on reactive flow control [e.g., Wilkinson, 1990;

Gad-el-Hak, 1994; Moin and Bewley, 1994], and in the books by Gad-el-Hak et al. (1998) and Gad-el-Hak (2000). The topic is also detailed in Chapter 33 of this handbook.

In this chapter, we focus on specific applications of MEMS in fluid dynamics, namely to measure turbulence and to reactively control fluid flows in general and turbulent flows in particular. To place these applications in perspective, we start by giving a brief description of MEMS fabrication; the next section is devoted to a brief general introduction to turbulence, discussion on tools necessary for the analysis of turbulent flows and some fundamental findings made in turbulence. We pay specific attention to small scales, which are of significant interest both in turbulence measurements and in reactive flow control, and we discuss the spatial- and temporal-resolution requirements. MEMS sensors for velocities, wall-shear stress and pressure measurements are then discussed.

As compared to conventional technologies, an extremely small measuring volume can be achieved using MEMS-based velocity sensors. Most commonly, the velocity sensors are designed as hot-wires with the sensitive part made of polysilicon, but other principles are also available and these are elucidated and discussed.

A significant parameter for control purposes is the fluctuating wall-shear stress, as it determines the individual processes transferring momentum to the wall. MEMS offer a unique possibility for direct as well as indirect measurements of this local flow quantity. Different design principles of conventional and MEMS-based wall-shear-stress sensors are discussed together with methods for calibrating those sensors.

The discussion of pressure sensors is focused on measurements of the fluctuating pressure field beneath turbulent boundary layers. Some basic design criteria are given for MEMS pressure sensors and advantages and drawbacks are elucidated. Significant quantities such as rms values, correlations and advection velocities of pressure events obtained with MEMS sensors, yielding spatial resolution of 5 to 10 viscous units, are compared to conventional measurements.

In Section 26.5, we address a real challenge and a necessity for reactive flow-control, MEMS-based flow actuators. Our interest is focused on three-dimensional structures, and we discuss actuators working with the bilayer effect as a principle. Electrostatic and magnetic actuators operating through external forces are also discussed together with actuators operating with mechanical folding. We also summarize the one-layer structure technology and discuss the out-of-plane rotation technology that has been made possible with this method. In connection with the actuator section, we discuss in Section 26.6 MEMS-fabricated devices such as pumps and turbines. Finally, we present our concluding remarks and reflections on the future possibilities that can be achieved in turbulence measurements and flow control by using MEMS technology.

26.2 MEMS Fabrication

26.2.1 Background

MEMS can be considered as a logical step in the silicon revolution, which took off when silicon microelectronics revolutionized the semiconductor and computer industries with the manufacturing of integrated circuits. An additional dimension is now being added by micromachines, as they allow the integrated circuit to break the confines of the electronic world and interact with the environment through sensors and actuators. This is a giant step, and it can be foreseen that MEMS will have in the near future the same impact on the society and economy as the IC has had since the early 1960s. The key element for the success of MEMS will be, as pointed out by Tien (1997), "the integration of electronics with mechanical components to create high-functionality, high-performance, low-cost, integrated microsystems." In other words, the material silicon and the MEMS fabrication processes are crucial to usher in a new era of micromachines.

Silicon is a well-characterized material. It is strong, being essentially similar to steel in the modulus of elasticity, stronger than stainless steel in yield strength and exceeding aluminum in strength-to-weight ratio. Silicon has high thermal conductivity and a low bulk expansion coefficient, and its electronic properties are well defined and sensitive to stress, strain, temperature and other environmental factors. In addition, the lack of hysteresis and the property of being communicative with electronic circuitry make silicon an almost perfect material for fabricating microsensors and microactuators for a broad variety of applications.

In MEMS fabrication, silicon can be chemically etched into various shapes, and associated thin-film materials such as polysilicon, silicon nitride and aluminum can be micromachined in batches into a vast variety of mechanical shapes and configurations. Several technologies are available for MEMS fabrication, but three main technologies are usually distinguished: bulk micromachining, surface micromachining and micromolding. An important characteristic of all micromachining techniques is that they can be complemented by standard IC batch-processing techniques such as ion implantation, photolithography, diffusion epitaxy and thin-film deposition. In this section, we will give a background of the three main technologies from a user viewpoint. Readers who are interested in more comprehensive information on fabrication are referred to more elaborate work in the field [e.g., Petersen, 1982; O'Connor, 1992; Bryzek et al., 1994; Tien, 1997; Madou, 1997; Kovacs, 1998]. Part II of this handbook also focuses on MEMS design and fabrication.

26.2.2 Microfabrication

26.2.2.1 Bulk Micromachining

Bulk micromachining is the oldest technology for making MEMS. The technique has been used to fabricate sensors for about 20 years. The mechanical structures are created within the confines of a silicon wafer by selectively removing parts of the wafer material, and this is done by using orientation-dependent etching of single-crystal silicon substrate. Etch-stopping techniques and masking films are crucial in the bulk micromachining process. The etching can be either isotropic or anisotropic, or a combination of both. In isotropic etching, the etch rate is identical in all directions, while in anisotropic etching the etch rate depends on the crystallographic orientation of the wafer. Two commonly used etchants are ethylene diamine pyrocatechol (EDP) and an aqueous solution of potassium hydroxide (KOH). It is important to be able to stop the etching process at a precise location, and to do so etch-stopping techniques have been developed. One such method is based on the fact that heavily doped regions etch more slowly than undoped regions; hence, by doping a portion of the material, the etch process can be made selective. Another technique for etch stopping is electrochemical in nature and is based on the fact that etching stops upon encountering a region of different polarity in a biased p–n junction.

A good illustration of the various steps in the bulk microfabrication process is the following example. Tien (1997) has summarized the processing steps necessary for micromachining a hole and a diaphragm in a wafer, and this is shown in Figure 26.1. Silicon nitride is used as an etch mask, as it is not etched

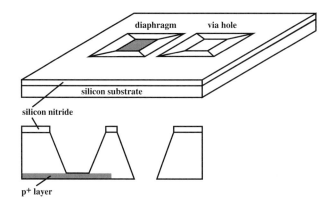

FIGURE 26.1 Bulk micromachined structures, diaphragm and via hole in a silicon substrate. Depositioned silicon nitride is the mask for the wet etch, and the doped silicon layer serves as an etch stop for the diaphragm formation. (From Tien, N.C. (1997) *Microscale Thermophys. Eng.*, **1**, 275–292. Reproduced by permission of Taylor & Francis, Inc.)

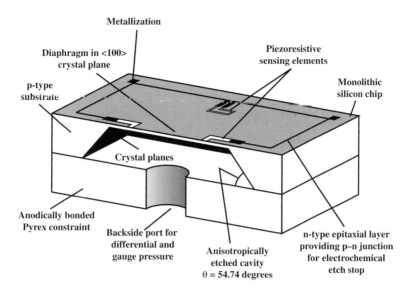

FIGURE 26.2 A bulk micromachined pressure sensor shown in cross section. The sensor contains a thin silicon diaphragm formed by etching the silicon wafer with alkaline hydroxide. The diaphragm deflection depends on pressure and is sensed by boron-doped piezoresistors. (From Bryzek, J. et al. (1994) *IEEE Spectrum*, **31**, 20–31. © 1994 IEEE. With permission.)

by either EDP or KOH. To stop the etch process at a specific location, and thereby form the diaphragm, a region heavily doped with boron is used. Holes and diaphragms, as shown in Figure 26.1, constitute the basis for many mechanical devices, such as, for example, pressure transducers which today are commercially available for measurements in the range of 60 Pa to 68 MPa.

The fabrication of a pressure transducer is straightforward, as has been summarized by Bryzek et al. (1994). As illustrated in Figure 26.2, the process starts with a silicon substrate that is polished on both sides. Boron-doped piezoresistors and both p^+ and n^+ enhancement regions are introduced by means of diffusion and ion implantation. Piezoresistors are the sensitive elements in pressure and acceleration sensors, as their resistance varies with stress and temperature, the latter being the unwanted part of the signal if the objective is to measure force. A thin layer of deposited aluminum or other metal creates the ohmic contacts and connects the piezoresistors into a Wheatstone bridge. Finally, the device side of the wafer is protected and the back is patterned to allow formation of an anisotropically etched diaphragm. After stripping and cleaning, the wafer is anodically bonded to Pyrex® and finally diced.

Bulk micromachining is the most mature of the micromachining technologies and constitutes the base for many microdevices such as silicon pressure sensors and silicon accelerometers. The fabrication process is straightforward and does not require much elaborate equipment, but the technique is afflicted with some severe, limiting drawbacks. Because the geometries of the structures are restricted by the aspect ratio inherent in the fabrication method, the devices tend to be larger than those made with other micromachining technologies. As a consequence of this, expensive silicon "real state" is wasted. Another drawback is the use of alkaline etchants, which unfortunately are not compatible with IC manufacturing. However, strategies to circumvent these drawbacks are available and details on such methods can be found in Bryzek et al. (1994) and Tien (1997).

26.2.2.2 Surface Micromachining

In contrast to bulk micromachining techniques, surface micromachining does not penetrate the wafer. Instead, thin-film materials are selectively added to and removed from the substrate during the processing. Polysilicon is used as the mechanical material, with sacrificial material such as silicon dioxide sandwiched between layers of polysilicon. Both materials are commonly deposited using low-pressure chemical vapor

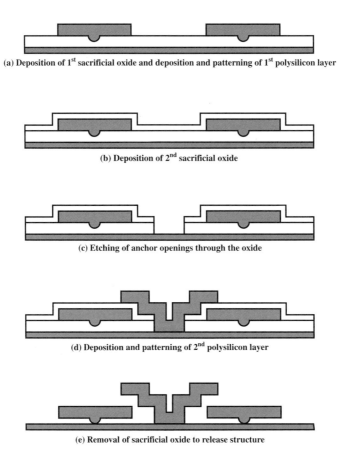

FIGURE 26.3 Polysilicon surface micromachining process for the fabrication of a slider with a central rail. (From Tien, N.C. (1997) *Microscale Thermophys. Eng.*, **1**, 275–292. Reproduced by permission of Taylor & Francis, Inc.)

deposition (LPCVD). Both wet and dry etching are essential, and the sacrificial layers constitute the basis of surface micromachining.

To illustrate the processes necessary for surface micromachining, a simplified fabrication of a polysilicon slider with a central rail has been summarized by Tien (1997), and the basic steps are illustrated in Figure 26.3. Two layers of structural polysilicon and sacrificial oxide are needed for this design; Figure 26.3a illustrates the first sacrificial oxide layer and how the deposition and patterning of the first polysilicon layer have been completed. Figures 26.3b and c show the deposition of the second sacrificial oxide layer together with the free etching of the anchor openings through the oxide. The next step is the deposition and patterning of the second polysilicon layer, followed by removal of the sacrificial oxide used to release the structure. More details including used etchants, sacrificial layers and other "tricks" used in the fabrication process can be found in Tien (1997).

An essential advantage of surface micromachining is that there is no constraint on the miniaturization of the devices fabricated other than those raised by limitations in the lithography technology. Another important benefit is that structurally complex mechanical systems including freestanding or moveable parts can be created by stacking multiple layers of material. In addition, surface micromachining offers a high degree of compatibility with IC processing, an important trait assuming that the future success of MEMS will be linked to the integration of electronics with mechanical systems. The main drawback of surface micromachining is that it is a thin-film technology that creates essentially two-dimensional structures. However, this has been circumvented by creative designs [e.g., Pister et al., 1992; Tien et al., 1996a; 1996b]. We shall return to the issue of three-dimensional structures in the microactuators section.

26.2.2.3 Micromolding

Although the micromolding technique is more similar to conventional machining in concept, it should be discussed in connection with micromachining because it is capable of producing minute devices using advanced IC lithography. In this group, the LIGA process is the most common method and was introduced in the late 1980s by Ehrfeld et al. (1988). (LIGA is a German acronym for Lithographie, Galvanoformoung, Abformung; in English, lithography, electroforming and molding). The method basically relies on forming a metal mold using lithographic techniques. To form the mold, a thick layer of photoresist placed on top of a conductive substrate is exposed and developed using X-ray lithography. As illustrated in Figure 26.4, the metal is then electroplated from the substrate through the openings in the photoresist. After removing the photoresist, the metal mold can be used for pouring low-viscosity polymers such as polyimide, polymethyl metacrylathe and other plastic resins. After curing, the mold is removed, leaving behind microreplicas of the original pattern. Products created by LIGA are three dimensional and include, for example, thermally actuated microrelays, micromotors, magnetic actuators, micro-optics and microconnectors, as well as a host of micromechanical components such as joints, springs, bearings and gears.

An extension of the LIGA process, the SLIGA technique, gains another degree of design freedom by combining LIGA with the use of sacrificial layers. Keller and Howe (1995) have presented the HEXSIL, which also includes a sacrificial layer and creates polysilicon components. A drawback of the micromolding method is that the assembly of small parts and the integration of electronics with mechanical devices can be real challenges. Additionally, the X-ray equipment necessary for the fabrication is quite expensive.

To conclude this section, it is worth mentioning that much of what is known about the design of mechanical structures scales down to the microstructure level very nicely; however, the same cannot be said for the properties of materials moving from the bulk to the thin-film regimes. For instance, residual stresses within thin films can produce unwanted tension or compression within the microstructure.

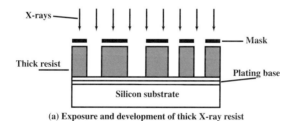

(a) Exposure and development of thick X-ray resist

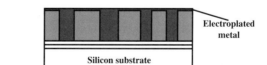

(b) Electroplating of metal (i.e., nickel) through the resist

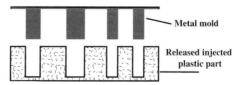

(c) Metal mold is released from the substrate and used to fabricate plastic parts

FIGURE 26.4 The basic LIGA process. (From Tien, N.C. (1997) *Microscale Thermophys. Eng.,* **1**, 275–292. Reproduced by permission of Taylor & Francis, Inc.)

Microdefects can be ignored for thicknesses greater than 10 μm but become important in the 1-μm range, which is typical for surface micromachining. Finally, microfriction, surface tension and van der Waals forces can create undesired stiction or adhesion [e.g., Israelachvili, 1991].

26.3 Turbulent Flows

26.3.1 Introductory Remarks

For more than 100 years, turbulence has been a fascinating challenge to scientists in fluid mechanics. It is very easy to observe a turbulent flow and to form a picture of its nature by looking at, for instance, the plume of a smoke stack. Such visualization shows clearly that the turbulent flowfield contains numerous eddies of various size, orientation and intensity. The largest eddies have a spatial extension of approximately the same size as the width of the flowfield, while the smallest eddies are of the size where viscous effects become dominant and energy is transferred from kinetic into internal. To qualitatively analyze turbulent flowfields, the eddies are conveniently described by length, time and velocity scales.

In this section, we provide a general discussion on the classification of small and large length scales and their importance in analyzing and modeling turbulent flows. We find that the width of the wave-number spectrum is proportional to the Reynolds number in such a way that high Re generates smaller scales. Because turbulent flows are high-Reynolds-number flows, it is clear that a knowledge of scales, and in particular the small scales, is essential for the analysis and modeling of turbulence. MEMS offer through miniaturization of sensors and actuators unique opportunities to resolve, as well as to target for control, the smallest scales even at high Re and thus provide this essential information. The scale discussion here constitutes a cornerstone for the following sections, where we consider the use of MEMS sensors and actuators for measuring and controlling turbulence. For the benefit of readers new to turbulence, we start with a very gentle introduction to the subject, leading to simple ways for estimating typical scales and sensor requirements for a particular flow field.

26.3.2 Definition of Turbulence

During the century in which turbulence has been formally studied many different definitions have been contemplated. The first attempt to define turbulence was made in the late nineteenth century by Osborne Reynolds who simply stated that turbulence was a "sinuous motion." Later, more comprehensive and detailed definitions have been given, and it is noteworthy that each definition commonly has been associated with the current fashion of approaching the closure problem of turbulence. Hence, the definition by G. I. Taylor in the 1930s had clear links to the statistical treatments of turbulence by Peter Bradshaw in the 1960s for hot-wire measurements and by Marcel Lesieur in the late 1980s for large-eddy and direct numerical simulations.

The most pragmatic definition is probably the one given by Tennekes and Lumley (1972), who provide not quite a definition but instead seven characteristics of turbulence. The authors stated that turbulence is irregular, or random, and this makes a deterministic approach impossible so in the analysis one must rely on statistical methods. Diffusivity is another crucial feature of turbulence, as it causes rapid mixing and increased rates of momentum, heat and mass transfer. Turbulent flows occur always at high Reynolds numbers which implies that they are always associated with small scales and complex interactions between the viscous and the nonlinear inertia terms in the equations of motion. All turbulent flows are three dimensional and are characterized by high levels of vorticity fluctuations. The viscous shear stresses perform deformation work, which increases the internal energy at the expense of the kinetic energy, meaning that all turbulent flows are strongly dissipative. If no energy is supplied, turbulent flows eventually decay. Under ordinary circumstances turbulence is a continuum phenomenon, so turbulent flows obey the continuum hypothesis, and the governing equations of fluid mechanics are applicable instantaneously. Even the smallest scales of a turbulent field are under normal conditions much larger than any molecular length scale. Finally, turbulent flows are flows, and because all turbulent flows are unique no general solution to problems associated with turbulence is in sight. In spite of the latter statement, turbulent flows have many characteristics in common, and this fact is exploited in the following section dealing with methods of analysis.

26.3.3 Methods for Analyzing Turbulence

Turbulence is one of the unsolved problems in classical physics, and it is still almost impossible to make accurate quantitative predictions for turbulent flows without relying heavily on empirical data. This is basically due to the fact that no methodology exists for obtaining stochastic solutions to the nonlinear partial differential equations describing the instantaneous three-dimensional flow. Moreover, statistical studies of the equations of motion always lead to a situation where there are more unknowns than equations, the closure problem of turbulence. This can easily be derived and is shown in most textbooks in the field [e.g., Tennekes and Lumley, 1972; Hinze, 1975; Pope, 2000]. Excluding direct numerical simulations of the governing equations, which thus far have been used only for simple geometries and low Reynolds numbers, it can be stated that all computations, both scientific and engineering, of turbulent flows will even in the foreseeable future require experiment, modeling and analysis.

One powerful tool in the study of turbulent flows is dimensional analysis because it may be possible under certain conditions to argue that the structure of turbulence depends only on a few independent variables. Then, dimensional analysis dictates the relation between the dependent and independent variables, and the solution is known except for a numerical coefficient. An example where dimensional analysis has been successful is in the derivation of the region called the "inertial subrange" in the turbulence kinetic energy spectrum. Here the slope obeys the so-called −5/3 law.

Because turbulent flows are characterized by high Reynolds numbers, it is reasonable to expect that a description of turbulence should behave properly as the Reynolds number approaches infinity. This method of analysis is called *asymptotic invariance* and has been successfully used in the development of the theory for turbulent boundary layers. In analyzing turbulence, the concept of local invariance or "self-preservation" is often invoked. This tool is powerful when the turbulent flow can be characterized as if it was controlled mainly by its immediate environment, and this situation occurs typically in the far downstream region of a wake, jet or free-shear layer. There, the time and length scales vary only slowly in the downstream direction, and if the turbulence time scales are sufficiently small it can be assumed that the flow has sufficient time to adjust to its gradually changing environment. The turbulence then is dynamically similar everywhere provided the average quantities are nondimensionalized with local length and time scales.

More details on the physics of turbulence can be found in classical textbooks in the field [Tennekes and Lumley, 1972; Hinze, 1975; Monin and Yaglom, 1975; Townsend, 1976]. There are also many good modern books available on the subject [McComb, 1990; Lesieur, 1991; Holmes et al., 1996; Pope, 2000]. The important point here is that almost all methods for analyzing turbulence are heuristic and are not derived from first principles; therefore, detailed measurements of flow quantities will continue to be an essential component of the arsenal of attacks on the turbulence problem. In this context, MEMS-based sensors have widened the horizon of experiments and can be used for measuring turbulence reliably and inexpensively at high Reynolds numbers.

26.3.4 Scales

As mentioned, turbulence is a high-Reynolds-number phenomenon that is characterized by the existence of numerous length and time scales. The spatial extension of the length scales is bounded from above by the dimensions of the flowfield and from below by the diffusive and dissipative action of the molecular viscosity. If we limit our interest to shear flows, which are basically characterized by two large length scales, one in the streamwise direction (the convective or longitudinal length scale) and the other perpendicular to the flow direction (the diffusive or lateral length scale), we obtain a more well-defined problem.

Moreover, at sufficiently high Reynolds numbers the boundary layer approximation applies and it is assumed that there is a wide separation between the lateral and the longitudinal length scales. This leads to some attractive simplifications in the equations of motion—for instance, the elliptical Navier-Stokes equations are transferred to the parabolic boundary-layer equations [e.g., Hinze, 1975]. So in this approximation, the lateral scale is approximately equal to the extension of the flow perpendicular to the flow direction (the boundary-layer thickness), and the largest eddies typically have this spatial extension.

The large eddies are most energetic and play a crucial role both in the transport of momentum and contaminants. A constant energy supply is needed to maintain the turbulence, and this energy is extracted from the mean flow into the largest most energetic eddies. The lateral length scale is also the relevant scale for analyzing this energy transfer. However, there is an energy destruction in the flow due to the action of the viscous forces (the dissipation), and for the analysis of this process other smaller length scales are needed.

As the eddy size decreases, viscosity becomes a more significant parameter, as one property of viscosity is its effectiveness in smoothing out velocity gradients. The viscous and the nonlinear terms in the momentum equation counteract each other in the generation of small-scale fluctuations. While the inertial terms try to produce smaller and smaller eddies, the viscous terms check this process and prevent the generation of infinitely small scales by dissipating the small-scale energy into heat. In the early 1940s, the universal equilibrium theory was developed by Kolmogorov (1941a; 1941b). One cornerstone of this theory is that the small-scale motions are statistically independent of the relatively slower large-scale turbulence. An implication of this is that the turbulence at the small scales depends only on two parameters—namely, the rate at which energy is supplied by the large-scale motion and the kinematic viscosity. In addition, it is assumed in the equilibrium theory that the rate of energy supply to the turbulence should be equal to the rate of dissipation. Hence, in the analysis of turbulence at small scales, the dissipation rate per unit mass ε is a relevant parameter together with the kinematic viscosity v. Kolmogorov (1941a; 1941b) used simple dimensional arguments to derive length, time and velocity scales relevant for the small-scale motion, given respectively by:

$$\eta = \left(\frac{v^3}{\varepsilon}\right)^{1/4} \tag{26.1}$$

$$\tau = \left(\frac{v}{\varepsilon}\right)^{1/2} \tag{26.2}$$

$$\upsilon = (v\varepsilon)^{1/4} \tag{26.3}$$

These scales are accordingly called the *Kolmogorov microscales*, or sometimes the inner scales of the flow. As they are obtained through a physical argument, these scales are the smallest scales that can exist in a turbulent flow and they are relevant for both free-shear and wall-bounded flows.

In boundary layers, the shear-layer thickness provides a measure of the largest eddies in the flow. The smallest scale in wall-bounded flows is the viscous wall unit, which will be shown below to be of the same order as the Kolmogorov length scale. Viscous forces dominate over inertia in the near-wall region, and the characteristic scales there are obtained from the magnitude of the mean vorticity in the region and its viscous diffusion away from the wall. Thus, the viscous time scale t_v is given by the inverse of the mean wall vorticity:

$$t_v = \left[\left.\frac{\partial \overline{U}}{\partial y}\right|_w\right]^{-1} \tag{26.4}$$

where $\overline{U}$ is the mean streamwise velocity. The viscous length-scale a ℓ_v is determined by the characteristic distance by which the (spanwise) vorticity is diffused from the wall, and is thus given by:

$$\ell_v = \sqrt{v t_v} = \sqrt{\frac{v}{\partial \overline{U}/\partial y|_w}} \quad (26.5)$$

where v is the kinematic viscosity. The wall velocity-scale (so-called friction velocity, u_τ) follows directly from the above time and length scales:

$$u_\tau = \frac{\ell_v}{t_\tau} = \sqrt{v \frac{\partial U}{\partial y}\bigg|_w} = \sqrt{\frac{\tau_w}{\rho}} \quad (26.6)$$

where τ_w is the mean shear stress at the wall, and ρ is the fluid density. A wall unit implies scaling with the viscous scales, and the usual ( )$^+$ notation is used; for example, $y^+ = y/\ell_v = y u_\tau/v$. In the wall region, the characteristic length for the large eddies is y itself, while the Kolmogorov scale is related to the distance from the wall y as follows:

$$\eta^+ = \frac{\eta u_\tau}{v} \approx (\kappa y^+)^{1/4} \quad (26.7)$$

where κ is the von Karman constant (≈ 0.41). As y^+ changes in the range of 1 to 5 (the extent of the viscous sublayer), η changes from 0.8 to 1.2 wall units.

We now have access to scales for the largest and smallest eddies of a turbulent flow. To continue our analysis of the cascade energy process, it is necessary to find a connection between these diverse scales. One way of obtaining such a relation is to use the fact that at equilibrium the amount of energy dissipating at high wavenumbers must equal the amount of energy drained from the mean flow into the energetic large-scale eddies at low wavenumbers. In the inertial region of the turbulence kinetic energy spectrum, the flow is almost independent of viscosity and, because the same amount of energy dissipated at the high wavenumbers must pass this "inviscid" region, an inviscid relation for the total dissipation may be obtained by the following argument. The amount of kinetic energy per unit mass of an eddy with a wavenumber in the inertial sublayer is proportional to the square of a characteristic velocity for such an eddy, u^2. The rate of transfer of energy is assumed to be proportional to the reciprocal of one eddy turnover time, u/ℓ, where ℓ is a characteristic length of the inertial sublayer. Hence, the rate of energy that is supplied to the small-scale eddies via this particular wavenumber is of order u^3/ℓ, and this amount of energy must be equal to the energy dissipated at the highest wavenumber, expressed as:

$$\varepsilon \approx \frac{u^3}{\ell} \quad (26.8)$$

Note that this is an inviscid estimate of the dissipation, as it is based on large-scale dynamics and does not either involve or contain viscosity. More comprehensive discussion of this issue can be found in Taylor (1935) and Tennekes and Lumley (1972). From an experimental perspective, this is a very important expression as it offers one way of estimating the Kolmogorov microscales from quantities measured in a much lower wavenumber range.

Because the Kolmogorov length and time scales are the smallest scales occurring in turbulent motion, a central question will be how small these scales can be without violating the continuum hypothesis. By looking at the governing equations, it can be concluded that high dissipation rates are usually associated with large velocities, and this situation is more likely to occur in gases than in liquids so it would be sufficient to show that for gas flows the smallest turbulence scales are normally much larger than the molecular scales of motion. The relevant molecular length scale is the mean free path, Λ, and the ratio between this length and the Kolmogorov length scale, η, is the microstructure Knudsen number and can be expressed as [see Corrsin, 1959]:

$$Kn = \frac{\Lambda}{\eta} \approx \frac{Ma^{1/4}}{Re} \qquad (26.9)$$

where the turbulence Reynolds number Re and the turbulence Mach number Ma are used as independent variables. It is obvious that a turbulent flow will interfere with the molecular motion only at a high Mach number and low Reynolds number, a very unusual situation occurring only in certain gaseous nebulae. (Note that in microduct flows and the like, the Re is usually too small for turbulence to even exist, so the issue of turbulence Knudsen number is moot in those circumstances even if rarefaction effects become strong.) Thus, under normal conditions the turbulence Knudsen number falls in the group of continuum flows. Noteworthy, however, is that measurements using extremely thin hot-wires, small MEMS sensors or flows within narrow MEMS channels can generate values in the slip-flow regime and even beyond, and this implies, for instance, that the no-slip condition may be questioned.

26.3.5 Sensor Requirements

The ultimate goal of all measurements in turbulent flows is to resolve both the largest and smallest eddies that occur in the flow. At the lower wavenumbers, the largest and most energetic eddies occur, and normally there are no problems associated with resolving these eddies. Basically, this is a question of having access to computers with sufficiently large memory for storing the amount of data acquired from a large number of distributed probes, each collecting data for a time period long enough to reduce the statistical error to a prescribed level. However, at the other end of the spectrum, both the spatial and the temporal resolutions are crucial, putting severe limitations on the sensors to be used. It is possible to obtain a relation between the small and large scales of the flow by substituting the inviscid estimate of the total dissipation rate, Eq. (26.8), into the expressions for the Kolmogorov microscales, Eqs. (26.1) to (26.3); thus,

$$\frac{\eta}{\ell} \approx \left(\frac{u\ell}{v}\right)^{3/4} = Re^{3/4} \qquad (26.10)$$

$$\frac{\tau u}{\ell} \approx \left(\frac{u\ell}{v}\right)^{-1/2} = Re^{-1/2} \qquad (26.11)$$

$$\frac{\upsilon}{u} \approx \left(\frac{u\ell}{v}\right)^{-1/4} = Re^{-1/4} \qquad (26.12)$$

where Re is the Reynolds number based on the speed of the energy containing eddies u and their characteristic length ℓ. Because turbulence is a high-Reynolds-number phenomenon, these relations show that the small length, time and velocity scales are much less than those of the larger eddies, and that the separation in scales widens considerably as the Reynolds number increases. Moreover, this also implies that the assumptions made on the statistical independence and the dynamical equilibrium state of the small structures will be most relevant at high Reynolds numbers. Another interesting conclusion that can be drawn from the above relations is that if two turbulent flow fields have the same spatial extension (i.e., same large scale) but different Reynolds numbers, there would be an obvious difference in the small-scale structure in the two flows. The low-Reynolds-number flow would have a relatively coarse small-scale structure, while the high-Reynolds-number flow would have much finer small eddies.

To spatially resolve the smallest eddies, sensors approximately the same size as the Kolmogorov length scale for the particular flow under consideration are needed. This implies that as the Reynolds number increases smaller sensors are required. For instance, in the self-preserving region of a plane-cylinder wake at a modest Reynolds number, based on the cylinder diameter of 1840, the value of η varies in the range of 0.5 to 0.8 mm [Aronson and Löfdahl, 1994]. For this case, conventional hot-wires can be used for turbulence measurements. However, an increase in the Reynolds number by a factor of ten will yield Kolmogorov scales in the micrometer range and call for either extremely small conventional hot-wires or MEMS-based sensors. Another illustrating example of the Reynolds number effect on the requirement of small sensors is a simple two-dimensional, flat-plate boundary layer. At a momentum thickness Reynolds number of $Re_\theta = 4000$, the Kolmogorov length scale is typically of the order of 50 μm, and in order to resolve these scales it is necessary to have access to sensors that have a characteristic active measuring length of the same spatial extension.

Severe errors will be introduced in the measurements by using sensors that are too large, as such a sensor will integrate the fluctuations due to the small eddies over its spatial extension, and the energy content of these eddies will be interpreted by the sensor as an average "cooling." When measuring fluctuating quantities, this implies that these eddies are counted as part of the mean flow and their energy is "lost." The result will be a lower value of the turbulence parameter, and this will wrongly be interpreted as a measured attenuation of the turbulence [Ligrani and Bradshaw, 1987]. However, because turbulence measurements deal with statistical values of fluctuating quantities, it may be possible to loosen the spatial constraint of having a sensor of the same size as η, to allow sensor dimensions slightly larger than the Kolmogorov scale—say, on the of order of η.

For boundary layers, the wall unit has been used to estimate the smallest necessary size of a sensor for accurately resolving the smallest eddies. For example, Keith et al. (1992) state that 10 wall units or less is a relevant sensor dimension for resolving small-scale pressure fluctuations. Measurements of fluctuating velocity gradients, essential for estimating the total dissipation rate in turbulent flows, are another challenging task. Gad-el-Hak and Bandyopadhyay (1994) argue that turbulence measurements with probe lengths greater than the viscous sublayer thickness (about 5 wall units) are unreliable, particularly near the surface. Many studies have been conducted on the spacing between sensors necessary to optimize the formed velocity gradients [see Aronson et al., 1997, and references therein]. A general conclusion from both experiments and direct numerical simulations is that a sensor spacing of 3 to 5 Kolmogorov lengths is recommended. When designing arrays for correlation measurements the spacing between the coherent structures will be the determining factor. For example, when studying the low-speed streaks in a turbulent boundary layer, several sensors must be situated along a lateral distance of 100 wall units, the average spanwise spacing between streaks. All this requires quite small sensors, and many attempts have been made to meet these conditions with conventional sensor designs. However, in spite of the fact that conventional sensors such as hot-wires have been fabricated in the micrometer-size range (for their diameter but not their length), they are usually handmade, difficult to handle and too fragile, and it is here that MEMS technology has really opened a door for new applications.

It is clear from the above that the spatial and temporal resolutions for any probe to be used to resolve high-Reynolds-number turbulent flows are extremely tight. For example, both the Kolmogorov scale and the viscous length scale change from a few microns at the typical field Reynolds number—based on the momentum thickness—of 10^6 to a couple of hundred microns at the typical laboratory Reynolds number of 10^3. MEMS sensors for pressure, velocity, temperature and shear stress are at least one order of magnitude smaller than conventional sensors [Löfdahl et al., 1994a; 1994b; Ho and Tai, 1998]. Their small size improves both the spatial and temporal resolutions of the measurements, typically a few microns and few microseconds, respectively. For example, a micro-hot-wire (called a *hot-point*) has very small thermal inertia, and the diaphragm of a micro-pressure-transducer has correspondingly fast dynamic response. Moreover, the extreme miniaturization and low energy consumption of microsensors make them ideal for monitoring the flow state without appreciably affecting it. Also, literally hundreds of microsensors can be fabricated on the same silicon chip at a reasonable cost, making them well suited for distributed measurements and control. The UCLA/Caltech team [Ho and Tai, 1996; 1998, and references therein] has been very effective in developing many MEMS-based sensors and actuators for turbulence diagnosis and control.

In the next section we focus our attention on sensors used for measurements in turbulent flows. Specifically, we discuss sensors for velocity, pressure and wall-shear stress, quantities that so far have been difficult to measure and for which the introduction of MEMS has created a completely new perspective.

26.4 Sensors for Turbulence Measurements and Control

26.4.1 Introductory Remarks

By definition, a transducer is a device actuated by power from one system that supplies power, usually in another form, to a second system. Hence, in an electromechanical transducer one connection to the environment typically conducts electrical energy and another conducts mechanical energy. Microphones, loudspeakers, strain gauges and electric motors are examples of electromechanical transducers, which in turn may be categorized into sensors and actuators. A sensor is a device that responds to physical stimulus such as velocity, pressure and temperature and transmits a resulting impulse for either measurement or control purposes. The output of a sensor may depend on more than one variable. Ideally, the sensor monitors the state of a system without affecting it, while an actuator operates on the system the other way around: It imposes a state without regard to the system load.

Typically, a sensor converts the physical parameter to be measured into an electrical signal, which is subsequently analyzed and interpreted. The studied physical parameters are usually classified into different groups, such as chemical, mechanical and thermal signals, and the transducer or sensor "helps" the electronic to "see," "hear," "smell," "taste" or "touch." In many applications, a sensor can be divided into a sensing part and a converting part. For instance, for a piezoresistive pressure sensor, the sensing part is the deflecting diaphragm and the converting part is the piezoresistor, which converts the deflection of the diaphragm into an electrical signal. It is generally more difficult to control a system than to monitor it, and in the last decade it has been more challenging for scientists and engineers to design and build microsensors than microactuators. As a consequence, progress in sensors lags behind that in actuators.

26.4.2 Background

A typical MEMS sensor is well below 1 mm in size, and at least one order of magnitude smaller than traditional sensors used to measure instantaneous flow quantities such as velocity, pressure and temperature. The small spatial extension implies that both the inertial mass and the thermal capacity are reduced which make these sensors suited for measurements of flow quantities in high-Reynolds-number turbulent flows where both high-frequency response and fine spatial resolution are essential. For instance, pressure and velocity sensors with, respectively, diaphragm side length and wire length of less than 100 μm are in use today. MEMS sensors are not handmade but are produced by photolithographic methods, implying that each unit is fabricated to extremely low tolerance and at low cost. (Normally the fabrication of a prototype sensor, the first batch, is very costly, but once the fabrication principle has been outlined the unit cost drops dramatically.) The latter trait makes it possible to use a large number of sensors to cover large areas/volumes of the flowfield. This in turn makes it feasible to study coherent structures and to effectively execute reactive control for turbulent shear flows. An additional important advantage of microfabrication is that it enables packing of sensors in arrays on the same silicon chip. A major difficulty, however, is that the leads of each sensor have to be connected to an external signal-conditioning instrument; also, the handling of numerous signal paths is tedious and they occupy a large portion of the precious surface area of the chip. For example, Ho and Tai (1998) state that their array of wall-shear-stress sensors containing 85 elements occupies 1% of the area, while the leads take about 50% of the surface. However, current research is attempting to solve this problem.

In this section, we look at MEMS-based sensors for use in turbulence measurements and reactive flow-control applications. In particular, we review the state of the art of microsensors used to measure the instantaneous velocity, wall-shear stress and pressure, which we deem as quantities of primary importance in turbulence diagnosis and control. For each group, we give a general background, design criteria and calibration procedure and provide examples of measurements conducted with MEMS-based sensors. When possible, we compare the results to conventional measurements. Each subsection concludes with brief remarks on future prospects.

26.4.3 Velocity Sensors

26.4.3.1 Background

Turbulence is one of the unsolved problems of classical physics, and it is almost impossible to make predictions for turbulent flows without relying heavily on empirical data. Because turbulence obeys the continuum hypothesis, the governing equations are known and an analysis of these equations shows that mean velocities, higher order moments of fluctuating velocities and products of gradients of fluctuating velocities are needed for future development of turbulence models. To this end, thermal anemometers have been the most significant tool for measuring these quantities. The introduction of MEMS has extended the range of applicability of the thermal anemometer and has provided incentives for conducting new measurements in high-Reynolds-number flows. This progress is basically achieved by the increased spatial and temporal resolutions that are feasible through the miniaturization and formation of sensor arrays.

According to King (1914), the first experiments using thermal anemometers were conducted already in 1902, but otherwise the work of King (1915; 1916) on the design of hot-wires and on the theory of heat convection from cylinders is considered the starting point for the era of thermal anemometry research. Early experiments were usually limited to measurements of mean velocities. In the late 1920s, however, the emphasis shifted toward measurements of fluctuating velocities [e.g., Dryden and Kuethe, 1930]. Since then, numerous papers in the field describe measurements that have been conducted using thermal anemometry with hot-wires or hot-films as sensing elements. Many researchers have made significant contributions to thermal anemometry and should be given credit in any complete review of the subject; however, here we focus only on the advantages gained from MEMS for improving measurements with current thermal anemometry. For a literature review of conventional hot-wires, the reader is referred to the survey papers by Comte-Bellot (1976), Freymuth (1983; 1992) and Fingerson and Freymuth (1977) and to the books by Hinze (1975), Lomas (1986) and Perry (1982).

In this section, we recall the principle of thermal anemometry for velocity measurements and summarize the characteristics of hot-wire sensors operated in a constant-temperature circuit. Because the governing equations of thermal anemometry are the same whether the sensor is a conventional hot-wire or a MEMS-based probe, we discuss these equations for one, two and three sensors and remark on their applicability for MEMS. We provide an overview of current MEMS-based sensors used for velocity measurements and discuss the results of experiments conducted with these probes. Finally, the section ends with the future outlook for MEMS velocity sensors.

26.4.3.2 Thermal-Sensor Principle

For our purpose, we may consider the thermal anemometer as a device used for measuring significant quantities for turbulence diagnosis and for reactive flow control. The sensors used are small, conducting elements which are heated by an electric current and cooled by the flow. All modes of heat transfer are present but forced convection is usually the dominate mode. From the temperature, or rather resistance, attained by the sensor, it is possible to gain the desired information on the instantaneous velocity vector. In order to thoroughly investigate a turbulent flowfield, usually more than one sensor is needed and multisensor arrangements are commonly used in measurements forming the base for development of turbulence models.

The generally small sensor size yields a good spatial resolution and frequency response [Freymuth, 1977], making thermal anemometry especially suited for studying flow details in turbulent flows. A simple thermal anemometer is shown schematically in Figure 26.5. The minute resistor mounted between two

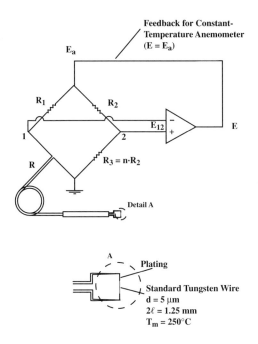

FIGURE 26.5 Basic elements of a hot-wire anemometer. (From Fingerson, L.M., and Freymuth, P. (1977) in *Fluid Mechanics Measurements,* 2nd ed., R.J. Goldstein, Ed., Taylor & Francis, Washington, D.C., pp. 115–173. Reproduced by permission of Routledge, Inc.)

prongs constitutes the sensing part and forms one arm in a Wheatstone bridge. For moderate temperature changes, the hot-wire resistance changes linearly with its temperature:

$$R = R_r[1 + \alpha(T_m - T_r)] \qquad (26.13)$$

where R_r is the resistance at the reference temperature T_r, T_m is the mean sensor temperature along its length and α is the temperature coefficient of resistance. The latter value is critical, because if the hot-wire sensor did not vary in resistance with temperature there would be no signal from a thermal anemometer. The ambient fluid temperature T_a is often used as reference temperature T_r, and the value of α depends on the reference temperature used.

The Wheatstone bridge arrangement shown in Figure 26.5 is designed so that resistance R_1 is large compared to that of the sensor. Then, current I through the hot-wire is nearly constant. This implies that any increase in heat transfer from the hot-wire to its surrounding will cause the sensor to cool, and this in turn will decrease the hot-wire resistance R (a decrease in the voltage E_{12} and a decrease in the amplifier output E). A decrease in heat transfer between sensor and fluid will have the opposite effect and create an increase in E. Without the feedback amplifier, the principle scheme shown in Figure 26.5 is basically an uncompensated, constant-current, hot-wire anemometer, and this kind of system was dominate in the infancy of the thermal anemometer era. Since then, advances have taken place in hot-wire fabrication, electronic control circuits and data acquisition. The result is that the constant-current operation of thermal sensors has been largely replaced by the constant-temperature operation, which offers much better stability and frequency response through high-gain feedback amplifiers. Stability criteria and techniques for checking the frequency response are now well understood for constant-temperature systems, and the introduction of digital techniques have significantly expanded the capabilities for analyzing the resulting data. Today, the nonlinear output is no longer a limitation; correlations, power spectra and amplitude probability distributions are all readily obtainable.

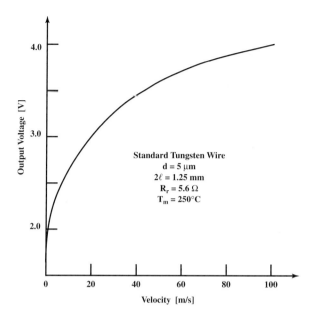

FIGURE 26.6 Typical calibration curve for a hot-wire anemometer. (From Fingerson, L.M., and Freymuth, P. (1977) in *Fluid Mechanics Measurements,* 2nd ed., R.J. Goldstein, Ed., Taylor & Francis, Washington, D.C., pp. 115–173. Reproduced by permission of Routledge, Inc.)

26.4.3.3 Calibration

A calibration curve is usually obtained by measuring a number of distinct velocities in a nearly inviscid flow—for example, in the freestream outside a boundary layer. A Pitot tube is used as standard for the velocity, and at the same time the output voltage of the anemometer is recorded. A typical calibration curve for a single hot-wire is shown in Figure 26.6. As is evident, this curve shows a nonlinear output with the velocity and a sensitivity decrease as the velocity increases. However, the sensitivity as a percentage of reading stays nearly constant which makes the hot-wire anemometer useful over a wide range of velocities. Under normal conditions the hot-wire is calibrated in a well-defined flow. Today the nonlinearity of the sensor characteristics is not a major constraint and many different transfer functions and cooling laws are available [e.g., Perry, 1982].

Once the calibration curve is obtained, good measurements in the unknown environment can be made directly provided that the fluid temperature, composition and density are the same as those during the calibration. In addition, it is assumed that the turbulence intensity is small (say, less than 15% of the mean velocity) and that the flow is incompressible. Once these conditions are fulfilled, the single hot-wire can be used to measure the streamwise mean and fluctuating velocities. Today, most modern laboratories use fast analog-to-digital converters (ADCs), implying that all the above functions, including the linearization, are performed with high accuracy employing digital computers.

26.4.3.4 Hot-Wire Sensors

In the design of hot-wire sensors one has to deal with many compromises. For example, the length and diameter are optimizations between several conflicting criteria. Concerning the hot-wire length, a short sensor is desired to maximize the spatial resolution and minimize the aerodynamic stress. However, a long sensor is desired to minimize conduction losses to the support, provide more uniform temperature distribution and reduce prong interference. Regarding the sensor diameter, a small diameter is desired to eliminate output noise due to separated flow around the sensor, to maximize the frequency response of the wire due to thermal inertia, to create higher heat transfer coefficient, to maximize the spatial resolution and to improve the signal-to-noise ratio at high frequencies. On the other hand, the solidity of the wire is essential, and a large diameter is desirable to increase the probe strength and to reduce contamination effects due to particles in the fluid.

For research work, typical values of the traditional wire diameter are in the range of 2.5 to 5 μm, and the wire length-to-diameter ratios are usually in the range of $100 < \ell/d < 600$. The choice of wire material is also a compromise. Tungsten, platinum and platinum–iridium are common hot-wire materials. Tungsten is desirable because of its high temperature coefficient of resistance and high strength. Platinum is available in small diameters, has good temperature coefficients of resistance and does not oxidize. Platinum–iridium is a wire that does not oxidize and has good strength, but unfortunately it has a low temperature coefficient of resistance. The article by Fingerson and Freymuth (1977) contains an elaborate section on wire materials and summarizes the properties of the most commonly used materials. An analytical relation between the flow velocity and anemometer output is desirable when physically interpreting the heat-transfer process in complex flow situations not covered by the calibration and when correcting for various sources of error and approximations. For a constant-temperature anemometer, the heat transfer from the sensor to its environment, Q, can be expressed in terms of the anemometer output voltage E as follows:

$$Q = \frac{E^2 R}{(R + R_1)^2} = \phi + K \tag{26.14}$$

where R represents the resistance of the sensor only. The rate of heat transfer, Q, is equal to the electrical power input to the sensor and consists of two parts: convective heat transfer, ϕ, between the heated portion of the sensor and the flowing fluid and conductive heat transfer, K, between the heated portion of the sensor and its support prongs. Our aim is now to establish a relation between the heat transfer, or electrical power input, and the flow velocity. Bremhorst and Gilmore (1976) concluded that the static- and dynamic-calibration coefficients agree to within 3% for the velocity range of 3 to 32 m/s. Based on this, and neglecting radiation, we can for the stationary case focus our interest on ϕ, which can be expressed as:

$$\phi = Nu \times 2\pi \ell \kappa_f (T_m - T_a) \tag{26.15}$$

where Nu is the Nusselt number, $\equiv h_c d / \kappa_f$, where d is the sensor diameter, h_c is the coefficient of convective heat transfer and κ_f is the fluid thermal conductivity at $T_f = (T_m + T_a)/2$, where T_m is the mean cylinder temperature, T_a is the ambient fluid temperature and 2ℓ is the length of the sensitive portion of the wire.

The main problem now is to find a representative expression of Nusselt number in terms of fluid and sensor parameters. General expressions of this quantity can be found in Hinze (1975), Perry (1982) and Fingerson and Freymuth (1977). Fortunately, most applications permit a reduction of the parameters needed, and a commonly used simplified expression is

$$Nu = A + B \, Re^{0.5} \tag{26.16}$$

where A and B are constants determined through the calibration, and Re is the Reynolds number based on the wire diameter and the speed being measured. This is a very useful expression for ordinary measurements, and its only drawback is that it does not represent the Nu number over a wide range of velocities. More elaborate expressions have, however, been proposed by, for example, Collis and Williams (1959):

$$Nu = (A + B \, Re^n)\left(1 + \frac{\alpha_T}{2}\right)^{0.17} \tag{26.17}$$

where α_T is the overheat ratio, and also by Kramers (1946):

$$Nu = 0.42 \, Pr^{0.26} + 0.57 \, Pr^{0.53} \, Re^{0.50} \tag{26.18}$$

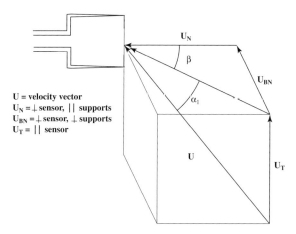

FIGURE 26.7 Velocity components relative to a hot-wire sensor. (From Fingerson, L.M., and Freymuth, P. (1977) in *Fluid Mechanics Measurements,* 2nd ed., R.J. Goldstein, Ed., Taylor & Francis, Washington, D.C., pp. 115–173. Reproduced by permission of Routledge, Inc.)

where Pr is the Prandtl number. The latter expression covers the ranges $0.71 < Pr < 525$ and $2 < Nu < 20$ and can be used even when the fluid is water.

A discussion on parameters neglected in the general expression can be found in the references cited above. Noteworthy is the fact that the ambient temperature, wire aspect ratio and conduction to wire supports have a large influence on the Nusselt number. Additionally, the Knudsen number (defined in the turbulence section) has been shown to be crucial in measurements using fine wires and small MEMS devices. In the next section, we consider the important issue of the angular dependence of the heat transfer, and the equations necessary for measurement of one, two and three velocity components.

26.4.3.5 Mean and Fluctuating Velocities

Figure 26.7 illustrates the instantaneous velocity components sensed by a hot-wire element. The simplest analytical expression for the angular sensitivity for an infinitely long wire is the so-called "cosine law":

$$U_{eff} = U \cos \alpha_1 \qquad (26.19)$$

where U_{eff} is the effective cooling velocity past the sensor. This equation essentially states that the velocity component along the wire ($U \sin \alpha_1$) has no cooling effect, and that the sensor is rotationally symmetrical in both construction and response. However, the real hot-wire has a finite length and the velocity component parallel to the sensor must be considered. Therefore, Champagne (1965) suggested an extended expression for the effective velocity:

$$U_{eff} = U \sqrt{\cos^2 \alpha_1 + k_T^2 \sin^2 \alpha_1} \qquad (26.20)$$

where k_T is an empirically determined factor which varies as a function of both α_1 and the velocity. Champagne (1965) found that k_T decreases nearly linearly with ℓ/d from a value of $k_T = 0.2$ at $\ell/d = 200$, to zero at $\ell/d = 600$–800. Clearly, a large-aspect-ratio sensor decreases the influence of the velocity component along the wire. Comte-Bellot et al. (1971) have shown that aerodynamic effects from both support needles and probe body must be considered, and Jörgensen (1971) has suggested a cooling law to take these effects into account:

$$U_{eff} = \sqrt{U_N^2 + k_T^2 U_T^2 + k_N^2 U_{BN}^2} \qquad (26.21)$$

where U_{BN} is the velocity perpendicular to both sensor and supporting prongs. The value of k_N can range from 1.0 to 1.2, depending on the design of probe support and needles [Drubka et al., 1977]. It is important to note that neither k_T nor k_N can be considered as constants for all angles and velocities. The cooling laws are basically methods for calibrating a hot-wire. More elaborate computer-oriented cooling-law relations have been presented. For example, Lueptov et al. (1988) have suggested the so-called "lookup" table, where a direct transfer of voltages into velocities and angles of attack is conducted.

It is common practice to use a single hot-wire probe perpendicular to the flow direction to measure the time-averaged velocity $\overline{U}$ and corresponding fluctuating velocity mean-square $\overline{u^2}$. In Figure 26.7, the wire is oriented with the prongs in the streamwise direction so that $U_N = U_1$, $U_T = U_2$, and $U_{BN} = U_3$, where U_1, U_2 and U_3 are the orthogonal components of the velocity vector U. The effective instantaneous cooling velocity can then be expressed as:

$$U_{eff} = \sqrt{U_1^2 + k_T^2 U_2^2 + k_N^2 U_3^2} \tag{26.22}$$

If the mean flow is in the U_1 direction and if Reynolds decomposition is introduced, then the mean velocities $\overline{U}_2 = \overline{U}_3 = 0$, and the effective cooling velocity may be expressed as:

$$U_{eff} = \sqrt{(\overline{U}_1 + u_1)^2 + k_T^2 u_2^2 + k_N^2 u_3^2} \tag{26.23}$$

Because k_T is small and $k_N \approx 1$, we can write:

$$U_{eff} = \sqrt{(\overline{U}_1 + u_1)^2 + u_3^2} \tag{26.24}$$

If it is further assumed that u_3 is small, then:

$$\overline{U}_{eff} = \overline{U}_1 = \overline{U} \tag{26.25}$$

$$\sqrt{\overline{u_1^2}} = \sqrt{\overline{u^2}} \tag{26.26}$$

Two velocity components can be measured with a cross hot-wire or by rotating a single probe and taking several sequential measurements [Fujita and Kovasznay, 1968]. Three velocity components can be measured with a triple hot-wire or by rotating a cross hot-wire. Figure 26.8 shows a cross wire with two

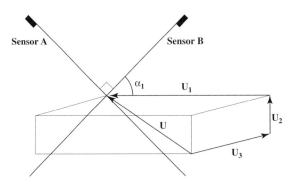

FIGURE 26.8 Configuration of a cross hot-wire sensor. (From Fingerson, L.M., and Freymuth, P. (1977) in *Fluid Mechanics Measurements,* 2nd ed., R.J. Goldstein, Ed., Taylor & Francis, Washington, D.C., pp. 115–173. Reproduced by permission of Routledge, Inc.)

sensors A and B located so that they are sensitive to U_2. The effective cooling velocities of the two sensors can be expressed as:

$$U_{A,eff}^2 = (U_1\cos\alpha_1 - U_2\sin\alpha_1)^2 + k_T^2(U_1\cos\alpha_1 + U_2\sin\alpha_1)^2 + k_N^2 U_3^2 \qquad (26.27)$$

$$U_{B,eff}^2 = (U_1\sin\alpha_1 + U_2\cos\alpha_1)^2 + k_T^2(U_1\cos\alpha_1 - U_2\sin\alpha_1)^2 + k_N^2 U_3^2 \qquad (26.28)$$

If the coordinates are selected so that the mean velocity $\overline{u}_3 = 0$, and if it is assumed once again that the sensors are sufficiently long, then $k_T \to 0$, $k_N \to 1$, and Eqs. (26.27) and (26.28) are reduced to:

$$U_{A,eff}^2 = (U_1\cos\alpha_1 - U_2\sin\alpha_1)^2 + u_3^2 \qquad (26.29)$$

$$U_{B,eff}^2 = (U_1\sin\alpha_1 + U_2\cos\alpha_1)^2 + u_3^2 \qquad (26.30)$$

Assuming that u_3 is small and that the sensors are oriented so that $\alpha_1 = 45°$, the streamwise and normal velocities are then expressed, respectively, as:

$$U_1 = \frac{1}{\sqrt{2}}(U_{A,eff} + U_{B,eff}) \qquad (26.31)$$

$$U_2 = \frac{1}{\sqrt{2}}(U_{A,eff} - U_{B,eff}) \qquad (26.32)$$

It is a common practice to orient the wire in the mean flow direction so that $\overline{U}_2 = 0$. In this case, the expressions for the mean velocity and Reynolds stresses are

$$\overline{U} = \frac{1}{\sqrt{2}}(U_{A,eff} + U_{B,eff}) \qquad (26.33)$$

$$\overline{u_1^2} = \frac{1}{2}(u_{A,eff} + u_{B,eff})^2 \qquad (26.34)$$

$$\overline{u_2^2} = \frac{1}{2}(u_{A,eff} - u_{B,eff})^2 \qquad (26.35)$$

$$\overline{u_1 u_2} = \frac{1}{2}\overline{(u_{A,eff} + u_{B,eff})(u_{A,eff} - u_{B,eff})} \qquad (26.36)$$

Equations (26.34) to (26.36) are the fundamental relations used in most cross hot-wire measurements. As mentioned earlier, more information about the flowfield can be obtained if all three velocity components are measured. This can, for instance, be done as shown schematically in Figure 26.9 by adding a third sensor C whose axis is at an angle α_3 to U_1 and in the (U_1, U_3) plane. The effective cooling velocity of the third wire will then be

$$U_{C,eff}^2 = (U_1\sin\alpha_3 + U_3\cos\alpha_3)^2 + k_T^2(U_1\cos\alpha_3 - U_3\sin\alpha_3)^2 + k_N^2 U_2^2 \qquad (26.37)$$

Together with the equations for the cross hot-wire we have three equations and three unknowns, so the instantaneous velocity vector can be determined. Once the instantaneous components are available, all turbulence parameters can be calculated. Again, the equations can be simplified if the sensors are sufficiently

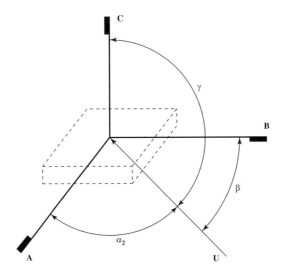

FIGURE 26.9 Direction sensitivity of a triple hot-wire sensor. (From Fingerson, L.M., and Freymuth, P. (1977) in *Fluid Mechanics Measurements,* 2nd ed., R.J. Goldstein, Ed., Taylor & Francis, Washington, D.C., pp. 115–173. Reproduced by permission of Routledge, Inc.)

long so that $k_T \to 0$ and $k_N \to 1$, and if the sensors are oriented so that $\overline{U_2} = \overline{U_3} = 0$, and $\alpha_1 = \alpha_2 = 45°$. The simplified equations are given below and were used by Fabris (1978) to make three-component velocity measurements:

$$U_{A,eff}^2 = U^2(\sin^2\alpha_2 + k^2\cos^2\alpha_2) \tag{26.38}$$

$$U_{B,eff}^2 = U^2(\sin^2\beta + k^2\cos^2\beta) \tag{26.39}$$

$$U_{C,eff}^2 = U^2(\sin^2\gamma + k^2\cos^2\gamma) \tag{26.40}$$

$$U^2 = \frac{U_A^2 + U_B^2 + U_C^2}{2 + k^2} \tag{26.41}$$

All the equations developed in this section for velocity measurements using one, two and three hot-wires are used in the same way when making velocity measurements using MEMS-based sensors. Few variants of the thermal principle for velocity measurements are used in microsensors, and these are discussed in the next section.

26.4.3.6 MEMS Velocity Sensors

Because turbulence is a high-Reynolds-number phenomenon with small length and time scales, it is generally preferred that sensors used to measure instantaneous quantities have a small physical size. Although it is true that conventional hot-wires with sensor diameter in the micrometer range have been reported

Sensors and Actuators for Turbulent Flows

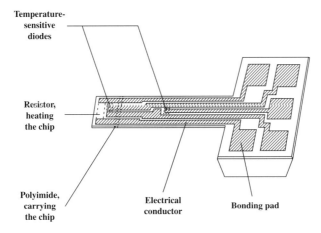

FIGURE 26.10 Schematic of a single-wire MEMS sensor consisting of three parts: the base plate 1.5 × 1.0 × 0.3 mm, the silicon beam 1.6 × 0.4 × 0.03 mm, and the chip 0.4 × 0.3 × 0.03 mm. (From Löfdahl, L. et al. (1992) *Exp. Fluids*, **12**, 391–393. © 1992 Springer-Verlag. With permission.)

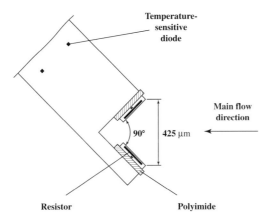

FIGURE 26.11 Double chip sensor for the determination of fluctuating velocity correlations. (From Löfdahl, L. et al. (1992) *Exp. Fluids*, **12**, 391–393. © 1992 Springer-Verlag. With permission.)

in the literature, these wires are usually very fragile, expensive and handmade. Conventional velocity sensors are difficult to use and not feasible to employ in large numbers. MEMS sensors are not afflicted with these drawbacks, and as is clear from earlier discussion are well suited to being fabricated in dense arrays at a moderate cost.

The first MEMS-based sensor for the determination of mean velocities and turbulence intensities was presented by Löfdahl et al. (1992). A schematic drawing of this sensor is depicted in Figure 26.10. Different overheat ratios and length-to-width ratios of the sensor were used: 4/3, 5/1 and 10/1. The principle of operation is very similar to that used in conventional hot-wires, but some deviations are important to note. Instead of using a bridge balance, the operation relies on a voltage difference which is formed between two temperature-sensitive p–n junction diodes. The "hot" diode monitors the temperature of the chip, which is electrically heated by an integrated resistor and cooled by the flow. The "cold" diode, which is positioned on the beam as shown in Figure 26.10, adjusts the set overheat ratio relative to the ambient air temperature. Thus, the power dissipated in the heated resistor is a measure of the instantaneous flow velocity.

Both single and double chip sensors were fabricated by Löfdahl et al. (1992), and Figure 26.11 shows the perpendicular arrangement of two heated sensors for measuring two velocity components. The performance

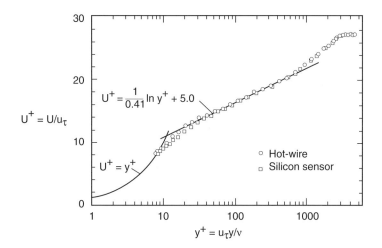

FIGURE 26.12 Typical mean velocity profile measured with a MEMS sensor and a conventional single hot-wire. Here, $U/u_\tau = f(u_\tau y/\nu)$, where u_τ is the friction velocity. (From Löfdahl, L. et al. (1992) *Exp. Fluids*, **12**, 391–393. © 1992 Springer-Verlag. With permission.)

of this micro-velocity-sensor was tested in a two-dimensional, flat-plate boundary layer at a Reynolds number based on distance from the leading edge of 4.2×10^6. Figure 16.12 compares the mean velocity profile in a turbulent boundary layer as measured with a conventional and MEMS-based sensor. The microsensor measured mean velocities with the same accuracy as a corresponding conventional hot-wire. Moreover, it was also demonstrated that the microsensor had spatial and temporal resolutions that made it suited for turbulence measurements. Figure 26.13 shows the streamwise turbulence intensity and the Reynolds stress as measured by a conventional X-wire and by two silicon microsensors. The MEMS-based sensors operated with good resolution even when the temperature of the heated part was reduced considerably. A clear drawback of this micro-hot-film is the proximity of the heated part of the sensor to the surface of the chip, rendering the probe insensitive to changes in flow direction. This makes the silicon sensor unsuited for use in three-dimensional flows where the primary flow direction is not known *a priori*.

Jiang et al. (1994) presented a MEMS-based velocity sensor with the hot-wire freestanding in space without any nearby structures, so that cooling velocities can be determined in the same way as with a conventional hot-wire. Their sensor is shown in Figure 26.14 and has a polysilicon hot-element that is greatly reduced in size, typically about 0.5 μm thick, 1 μm wide, and 10–160 μm long. The dynamic performance and sensitivity of this sensor have been tested. A heating time of 2 μs and a cooling time of 8 μs for the 30-μm-long sensor in constant-current mode have been achieved. For constant-temperature operation, a time constant of 0.5 μs for the 10-μm-long sensor has been recorded. The corresponding cut-off frequency is 1.4 MHz. The calibration curves of a 20-μm-long micro-hot-wire at two different angles are shown in Figure 26.15. The average sensitivity was found to be 20 mV/m/s at an input current of 0.5 mA. No turbulence measurements have been reported using this sensor. Noteworthy is that the silicon hot-wires have a trapezoidal cross section, which might cause severe uncontrolled errors in turbulence measurements.

A severe drawback of commercially available triple hot-wire probes is the large measuring volume, typically a sphere with a diameter of 3 mm. This is far too large to be acceptable for turbulence measurements at realistic Reynolds numbers. Ebefors et al. (1998) have presented a MEMS-based triple-hot-wire sensor, shown schematically in Figure 26.16. The x- and y-hot-wires are located in the wafer plane while the z-wire is rotated out of the plane using a radial polyimide joint. The silicon chip size is $3.5 \times 3.0 \times 0.5$ mm^3, and the three wires are each $500 \times 5 \times 2$ μm^3. The sensor is based on the thermal anemometer principle, and the polyimide microjoint technique is used to create a well-controlled,

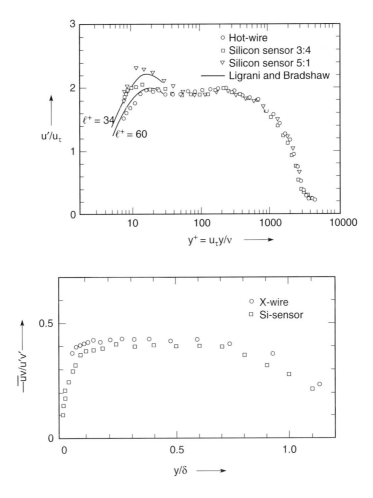

FIGURE 26.13 Streamwise velocity fluctuation (top plot), and turbulent shear stress (bottom plot) measured with MEMS sensors and conventional hot-wires. (From Löfdahl, L. et al. (1992) *Exp. Fluids*, **12**, 391–393. © 1992 Springer-Verlag. With permission.)

out-of-plane rotation of the third wire. The time constant of the hot-wire resistance change caused by heating without any flow was measured, and the cooling and heating time constants were found to be 120 and 330 µs, respectively. By operating the sensor at a constant temperature, these response times can be improved. Only some very preliminary velocity measurements have been conducted with this sensor in channel flow.

26.4.3.7 Outlook for Velocity Sensors

MEMS velocity sensors are superior to conventional hot-wires in the sense that they can be reduced in size to the order of microns. This is of course an important advantage, but for velocity measurements it is presumably the MEMS properties of accurate fabrication at a low unit cost that are even more relevant. For engineering purposes this means that velocity sensors, which may not necessarily be extremely small, can be used for controlling and optimizing system performance. Due to the low price per unit, numerous control points can be sensed, which is of utmost importance, for instance, in controlling room ventilation.

In turbulence research, small sensors are required for correct interpretation of the flow phenomena; here, the most significant advantage of the MEMS sensors is that they can be spaced close together and fabricated in arrays for studies of coherent structures in the flow. In particular, further development of devices for measuring the instantaneous, three-dimensional velocity using, for example, the triple-hot-wire presented

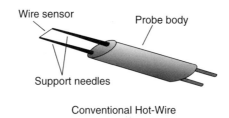
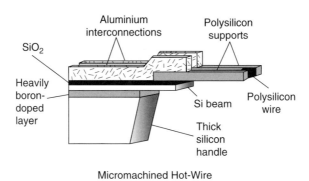

FIGURE 26.14 Structure of hot-wire probes: (a) conventional hot-wire, (b) micromachined hot-wire. (From Jiang, F. et al. (1994) in *Solid-State Sensor and Actuator Workshop,* Hilton Head, SC. © 1994 IEEE. With permission.)

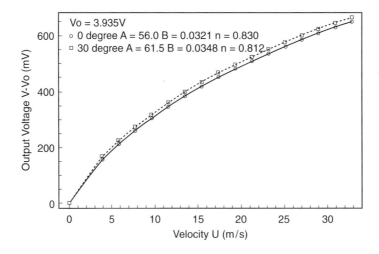

FIGURE 26.15 Calibration curves of a 20-μm-long, MEMS-based hot-wire sensor at two different angles. (From Jiang, F. et al. (1994) in *Solid-State Sensor and Actuator Workshop,* Hilton Head, SC. © 1994 IEEE. With permission.)

by Ebefors et al. (1998) would be of great interest. Further miniaturization of this sensor and positioning of each sensor in closely spaced arrays would offer the possibility of measuring flow events, both in the direction normal to the surface and in the spanwise direction. An integration of such sensor arrays with arrays of pressure transducers would be beneficial in instantaneous mapping of the flowfield for studies of coherent structures. Combined with proper actuators, MEMS-based velocity sensors can be used for effective reactive control of difficult-to-tame turbulent shear flows.

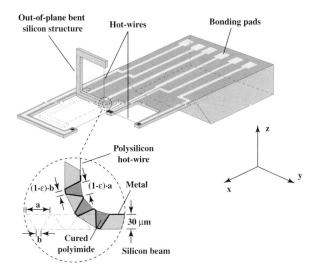

FIGURE 26.16 MEMS triple hot-wire. (From Ebefors, T. et al. (1998) *J. Micromech. Microeng.*, **8**, 188–194. With permission.)

26.4.4 Wall-Shear-Stress Sensors

26.4.4.1 Background

From scientific and engineering perspectives, the wall-shear stress is an essential quantity to compute, measure or infer in a wall-bounded turbulent flow. Time-averaged values of this quantity are indicative of the global state of the flow along a surface and can be used to determine body-averaged properties such as skin-friction drag. The time-resolved part of the wall-shear stress is a measure of the unsteady structures in the flow which are responsible for the individual momentum transfer events and is an indicator of the coherent portion of the turbulence activities. Spatially distributed values of the instantaneous wall-shear stress can be used in a feedforward or feedback control loop to effect beneficial changes in the boundary layer. The majority of wall-shear-stress studies conducted rely on the premise that the mean velocity-gradient and the heat-transfer rate near or at the wall are both proportional to the wall-shear stress. The former relation was established already by the early 1920s by Stanton et al. (1920), and the latter a few years later by Leveque (1928).

During the last few decades, numerous experiments have been conducted where measurements of wall-shear stress have been the kernel. The success of those endeavors depends basically on the complexity of the flow, the geometry of the solid boundaries and the limitations of the measuring device used. Some facts have been established, such as the wall-shear-stress distribution along flat plates and simple bodies of revolution, but in general our knowledge of wall-shear stress and in particular its fluctuating component is limited. An explanation of this is that it is both difficult and cumbersome to measure wall-shear stress. In general it is a parameter of small magnitude, and some typical values can be of interest to keep in mind for the following discussion. A submarine cruising at 30 km/hr has an estimated value of the shear stress of about 40 Pa; an aircraft flying at 420 km/hr, 2 Pa; and a car moving at 100 km/hr, 1 Pa. Such small forces per unit area put heavy demands on the resolution of the measuring devices used. Those estimates are approximate and were obtained by Munson et al. (1990), who treated the body as a collection of parts; for instance, the drag of an aircraft was approximated by adding the drag contributions produced by wings, fuselage, tail, etc.

Many different methods for the determination of wall-shear stress have been developed, and the required spatial and temporal resolutions are specific for each application and environment. In laminar flows, the sensors must be capable of measuring the time-averaged shear stress, while in turbulent flows both the mean shear and its fluctuating component are of interest. An attempt to classify the techniques available for wall-shear-stress measurements was made by Haritonidis (1989), who divided the technologies into direct

TABLE 26.1 Operation and Detection Principle of Silicon Micromachined Wall-Shear-Stress Sensors

Author	Operation	Detection Principle	Active Sensor Area (mm^2)
Oudheusden and Huijsing (1988)	Thermal	Thermopile	4×3
Schmidt et al. (1988)	Floating element	Capacitive	0.5×0.5
Ng et al. (1991)	Floating element	Piezoresistive	0.12×0.14
Liu et al. (1995)	Thermal	Anemometer	0.2×0.2
Pan et al. (1995)	Floating element	Capacitive	$\approx 0.1 \times 0.1$
Padmanabhan (1997)	Floating element	Optical	0.12×0.12
Kälvesten (1994)	Thermal	Anemometer	$\approx 0.1 \times 0.1$

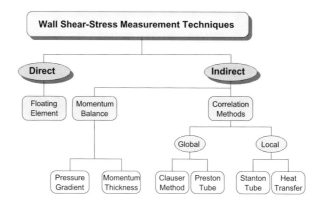

FIGURE 26.17 Classification of wall-shear-stress measurements techniques. (Adapted from Haritonidis, 1989.)

or indirect methods, depending upon whether the method measures the wall-shear stress directly or infers it from other measured properties. Figure 26.17 shows a bird's eye view of Haritonidis' classification. Because the number of different methods is enormous, it is out of the scope of this chapter to review each technique but a trend common to all measurements and methods can be highlighted. Since the mid-1950s, the evolution of the probes used has been directed toward utilizing smaller and smaller sensors in order to improve the accuracy, flexibility and resolution. Several review papers describe shear-stress sensors and discuss in detail the merits and drawbacks of the methods used in a vast variety of flow situations. To cite a few, Winter (1977) gives a comprehensive review of available conventional methods and a good discussion of measurements in turbulent flows. Haritonidis (1989) summarizes conventional methods and introduces the first micromachined wall-shear-stress sensor. Hakkinen (1991) lists the merits and drawbacks of conventional techniques. More recently, Hanratty and Campbell (1996) discuss the relevant experimental issues associated with the use of various wall-shear-stress sensors.

A major problem faced in wall-shear-stress measurements is that conventional fabrication of sensors and balances allows only a certain degree of miniaturization, as these devices are more or less handmade. Spatial and temporal resolutions better than what conventional probes can provide are needed in typical turbulence studies and reactive flow control. MEMS has the potential of circumventing the conventional sensor limitations, as microfabrication offers a high degree of miniaturization and associated increased resolution together with the fact that the sensors are not handmade, implying that each unit is fabricated to extremely low tolerance at reasonable cost. Table 26.1 lists the operation and detection principles as well as the sensor area of some recently developed MEMS-based shear-stress probes.

A glance at the classification scheme of Figure 26.17 shows that the floating-element principle and the heat-transfer method (or more correctly thermal-element method) are both well suited for MEMS fabrication. Figure 26.18 shows a schematic of the principle of operation of the two methods. The thermal-element method relies on measuring the amount of energy necessary for keeping a wall-mounted, electrically heated resistor at constant temperature, despite a time-dependent, convective heat transfer to

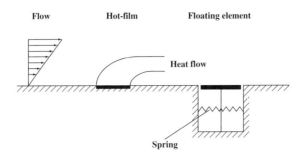

FIGURE 26.18 Wall-shear-stress sensors of the hot-film and floating-element type.

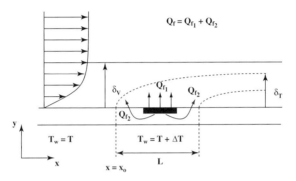

FIGURE 26.19 An illustration of a surface mounted hot-film.

the flow. A high freestream velocity yields steep velocity and temperature gradients at the heated surface and thereby a good cooling, which requires more energy, while the opposite is true for a low freestream velocity. In the floating-element principle, a flush-mounted wall-element, moveable in the plane of the wall-shear stress, is displaced laterally by the tangential viscous force and this movement can be measured using resistive, capacitive or optical detection principles.

In the following eight sections, we describe the principles of the thermal sensor and the floating element, as both strategies are particularly suited for measurements of fluctuating wall-shear stress as well as for MEMS fabrication. In the description of the thermal-sensor method, we briefly recall the classical equations, and based on these we highlight possible uncertainties and sources of errors in measuring instantaneous wall-shear stress. For the floating-element method, we discuss the general principle and pitfalls associated with steady and time-dependent measurements. For both principles we review current MEMS-based sensors. Because calibration is a crucial part of wall-shear-stress measurements, particularly in the determination of fluctuating quantities, we summarize some common calibration methods and discuss possibilities for dynamic calibration. Finally, we provide an outlook for the use of MEMS-based wall-shear-stress sensors for future turbulence studies and flow control.

26.4.4.2 Thermal-Sensor Principle

The thermal or hot-film sensor benefits from the fact that the heat transfer from a sufficiently small heated surface depends only on the flow characteristics in the viscous region of the boundary layer. A schematic drawing of a thermal sensor is depicted in Figure 26.19. The hot-film gauge consists of a thin metallic film positioned into a substrate. Usually the gauge forms one part of a Wheatstone bridge, and an electric current is passed through the film in order to maintain it at a constant temperature as heat is continuously being transferred from the film to the moving fluid. The ohmic heating in the device q_W is transferred both to the fluid and to the surrounding substrate. This can be expressed as:

$$q_W = q_T + q_C \tag{26.42}$$

where q_T represents the heat transferred to the fluid directly from the heated surface and indirectly through the heated portion of the substrate, and q_C represents the heat lost to the substrate. If the flow is steady and laminar and if the streamwise pressure gradient can be neglected, the resulting heat-transfer rate to the fluid q_T has been shown to be related to the wall-shear stress according to the classical relation:

$$\tau_w \propto q_T^3 \qquad (26.43)$$

This expression has been derived by many researchers [e.g., Ludwieg, 1950; Liepmann and Skinner, 1954; Bellhouse and Schultz, 1966]. However, this simple relation breaks down in unsteady and turbulent flows as has been pointed out by Bellhouse and Schultz (1966). In a turbulent environment, most assumptions made in the derivation are violated and the heat conduction to the substrate changes instantaneously. In spite of this, we will use an extended version of Equation (26.43) to illustrate the assumptions made and to pinpoint possible pitfalls in using hot-film gauges for fluctuating wall-shear-stress measurements. In the following argument, we will follow a path outlined by Bellhouse and Schultz (1966) and later by Menendez and Ramaprian (1985), who derived the extended version for boundary-layer flows subjected to a periodic freestream velocity of the form:

$$U_a(t) = U_o(1 + \varepsilon \sin \omega t) \qquad (26.44)$$

where ω is the angular frequency of the oscillation, U_o is the mean freestream velocity, and ε is the relative amplitude of the oscillation. The derivation below is strongly abbreviated, and the reader is referred to Menendez and Ramaprian (1985) for a more elaborate version. The problem to be considered is shown in Figure 26.19. A thermal boundary layer develops within a turbulent boundary layer over a heated film having an effective streamwise length L. The thermal boundary layer is produced by a sudden jump in the surface temperature, from a constant value equal to the ambient temperature T_a to the higher constant value T_w at the location $x = x_o$. It is assumed that the thermal boundary layer is totally embedded in the viscous region, so effects of turbulent diffusion can be neglected. A relation between the local wall-shear stress τ_w and the heat-transfer rate from the wall to the fluid q_T is derivable from the first law of thermodynamics. For a two-dimensional flow, neglecting dissipation and compressibility effects and applying the usual boundary-layer approximation, the energy equation reads:

$$\frac{\partial T}{\partial t} + U \frac{\partial T}{\partial x} + V \frac{\partial T}{\partial y} = \frac{\kappa_f}{\rho c_p} \frac{\partial^2 T}{\partial y^2} \qquad (26.45)$$

where $T(x, y, t)$ is the desired temperature distribution; t is time; U and V are the velocity components along the coordinates x and y, respectively; and ρ, κ_f and c_p are, respectively, the fluid density, thermal conductivity and specific heat. Using the equation of continuity and integrating Eq. (26.45) across the thermal boundary layer up to a point $y = y_e$ beyond the edge of the thermal boundary layer, we get:

$$\frac{\partial}{\partial t}\int_0^{y_e}(T-T_0)dy + \frac{\partial}{\partial x}\int_0^{y_e}U(T-T_e)dy = -\frac{1}{\rho c_p}q_W \qquad (26.46)$$

where T_e is the temperature at the edge of the thermal boundary layer. It is now assumed that within the viscous region where the thermal boundary layer resides, inertial effects can be neglected. The velocity profile can be approximated by:

$$U = \frac{\tau_w}{\mu}y + \frac{1}{2\mu}\left(\frac{\partial p}{\partial x}\right)y^2 \qquad (26.47)$$

where p is the pressure, and μ is the dynamic viscosity of the fluid. It is further assumed that the temperature distribution in the thermal boundary layer is self-similar at any instant of time and that the thermal boundary layer thickness δ_T can be expressed as follows:

$$\frac{T - T_e}{T_w - T_e} = f(\xi, t) \tag{26.48}$$

$$\frac{\delta_T(x, t)}{\delta_T(x_o + L, t)} = \left[\frac{x - x_o}{L}\right]^n \tag{26.49}$$

where $\xi \equiv y/\delta_T$ is the similarity coordinate, and n is an unknown exponent which can in general depend on time. After some algebra and assuming n is independent of time, an expression for the local wall-shear stress at $x = x_o + L$ is obtained as a function of the heat-transfer rate from the wall to the fluid:

$$\tau_w = (A_1 q_W^3) + A_2 \frac{\cos \omega t}{q_W} + A_3 \frac{\partial q_W}{\partial t} \tag{26.50}$$

where

$$A_1 \equiv \frac{\mu L}{(1 - n) a \rho c_p \kappa^2 (\Delta T_o)^3} \tag{26.51}$$

$$A_2 \equiv \frac{b \kappa (\Delta T_o) a \rho \varepsilon \omega U_o}{2a} \tag{26.52}$$

$$A_3 \equiv \frac{\mu L c}{(1 + n) a \kappa (\Delta T_o)} \tag{26.53}$$

In these expressions, the "shape parameters," a, b and c, are basically functions of the streamwise coordinate x. According to Menendez and Ramaprian (1985), Eq. (26.50) can be regarded as the basic relationship between the instantaneous value of the wall-shear stress and the wall heat flux. By taking the ensemble average of Eq. (26.50) over a large number of cycles and assuming that the turbulent fluctuations are small, a linearized expression for the periodic wall-shear stress is obtained.

To summarize, the classical expression, Eq. (26.43), gives a relation between the wall-shear stress and the heat transfer from the wall. This expression assumes steady, laminar, zero-pressure-gradient flow and is not valid in a turbulent environment. Menendez and Ramaprian (1985) have derived an extended version of Eq. (26.43) valid for a periodically fluctuating freestream velocity, Eq. (26.50). The latter relation contains, however, some assumptions that are questionable for turbulent flows. For instance, in the thermal boundary layer it is assumed that the temperature distribution is self-similar and that the local thickness varies linearly. These assumptions are relevant for a streamwise velocity oscillation and a weak fluctuation, but certainly not in a turbulent flow that is strongly unstable in all directions. Following a brief look at available calibration methods, we discuss the spatial and temporal resolutions of wall-mounted hot-films together with some MEMS-based sensors for wall-shear-stress determination.

26.4.4.3 Calibration

Several methods and formulas are in use for calibrating hot-film shear probes operated in the constant-temperature mode, and the choice of method depends on flow conditions and sensor used. In this section, two static calibration methods are discussed. Both are based on the theoretical analysis leading to the relation between rate of heat transfer and wall-shear stress, as discussed in the previous section. The challenge is, of course, to be able to use the shear-stress sensor in a turbulent environment.

If a laminar flow facility is used to calibrate the wall-shear-stress sensor, then Eq. (26.43) can be rewritten more conveniently in time-averaged form:

$$\overline{\tau_w}^{1/3} = A\overline{e^2} + B \qquad (26.54)$$

where $\overline{\tau_w}$ is the desired mean wall-shear stress, $\overline{e^2}$ is the square of the mean output voltage, and A and B are calibration constants. The term B represents the heat loss to the substrate in a quiescent surrounding, and this procedure is similar to a conventional calibration of a hot-wire [King, 1916]. However, a laminar flow is often difficult to realize in the desired range of turbulence wall-shear stress, and it is more practical to calibrate without moving the sensor between calibration site and measurement site. In that case, the calibration is made in a high-turbulence environment and the high-order moments of the voltage must also be considered.

Ramaprian and Tu (1983) proposed an improved calibration method, and the instantaneous version of Eq. (26.54), can be rewritten and time-averaged to give:

$$\overline{\tau_w} = \overline{(Ae^2 + B)^3} = A^3\overline{e^6} + 3A^2B\overline{e^4} + 3AB^2\overline{e^2} + B^3 \qquad (26.55)$$

This can be rewritten as:

$$\overline{\tau_w} = C_6\overline{e^6} + C_4\overline{e^4} + C_2\overline{e^2} + C_0 \qquad (26.56)$$

where the Cs are the new calibration constants. The high-order moments of the voltage can easily be calculated with a computer, but care must be taken that they are fully converged. The relation between e and τ_w can also be represented by a full polynomial function, so if the frequency response of the sensor is sufficiently flat, it is possible to write:

$$\overline{\tau_w} = C_0 + C_1\overline{e} + C_2\overline{e^2} + \cdots + C_M\overline{e^N} \qquad (26.57)$$

where M is the number of calibration points and N is the order of the polynomial above. A system of linear equations is obtained where the calibration coefficients can be computed by a numerical least-square method. The mean wall-shear stress on the left-hand side of Eq. (26.57) can be measured with, for example, a Preston tube using the method of Patel (1965). The second calibration technique described here is called "stochastic" calibration by Breuer (1995), who also demonstrated that the validity of the calibration polynomial may extend well beyond the original calibration range, although this requires careful determination of the higher order statistics as well as a thorough understanding of the sensor response function.

The experiments of Bremhorst and Gilmore (1976) showed that the static and dynamic calibration coefficients for hot-wires agree to within a standard error of 3% for the velocity range of 3 to 32 m/s. Thus, they recommend the continued use of static calibration for dynamic measurements. As pointed out in the previous section, this is not true for a wall-mounted hot-film. One solution to this problem may be to calibrate the hot-film in pulsatile laminar flow with a periodic freestream velocity $U_a(t)$, and make use of the Menendez and Ramaprian's formula, Eq. (26.50):

$$\overline{\tau_w} = (A\overline{e^2} + B)^3 + \frac{c_1}{A\overline{e^2} + B}\frac{dU_a}{dt} + c_2 A\frac{d\overline{e^2}}{dt}$$

to obtain the additional calibration constants c_1 and c_2. This step is necessary in order to characterize the gauge dynamic response at relatively high frequencies. The constants A and B are obtained from a steady-state calibration.

One difficulty in performing a dynamic calibration is generating a known sinusoidal wall-shear-stress input. Bellhouse and Rasmussen (1968) and Bellhouse and Schultz (1966) achieved this in two different ways. One method is to mount the hot-film on a plate which can be oscillated at various known frequencies and amplitudes. The main drawback to this arrangement is the limited amplitudes and frequencies that can be achieved when attempting to vibrate a relatively heavy structure. An alternative strategy is to generate the shear-stress variations by superimposing a monochromatic sound field of different frequencies on a steady, laminar flowfield. A hot-wire close to the wall can be used as a reference.

26.4.4.4 Spatial Resolution

Equation (26.50) is an extended version of Eq. (26.43), and by assuming a steady flow and zero-pressure-gradient in the streamwise direction it reduces to the classical formula. The latter expression can be used to estimate the effective streamwise length of a hot-film:

$$\tau_w = -A_1 q_W^3 \qquad (26.58)$$

By assuming a stepwise temperature variation and introducing the average heat flux $\overline{q_W}$ over the heated area which is assumed to have a streamwise length L, the desired relation reads:

$$\overline{q_W} = 0.807 \kappa_f \Delta T_w Pr^{1/3} \left(\frac{\tau_w}{L\nu\mu}\right)^{1/3} \qquad (26.59)$$

It is convenient to rewrite the above expression in a nondimensional form, and in doing so we keep L explicit:

$$\overline{Nu} \equiv \frac{h_c L}{\kappa_f} = 0.807 (Pr L^{+2})^{1/3} \qquad (26.60)$$

where $\overline{Nu}$ is the Nusselt number averaged over the heated area, L^+ is the streamwise length of the heated area normalized with the viscous length-scale ν/u_τ, and h_c is the convective heat-transfer coefficient. This equation has been derived for flows with pressure gradient by Brown (1967). The dimensionless sensor length L^+ is a crucial parameter when examining the assumptions made.

The lower limit on L^+ is imposed by the boundary-layer approximation, as there is an abrupt change of temperature close to the leading and trailing edge zones of the heated strip where the neglected diffusive terms in Eq. (26.45) become significant. Tardu et al. (1991) have conducted a numerical simulation of the heat transfer from a hot-film, and found a peak of the local heat transfer at the leading and trailing edges. They conclude that if the hot-film is too narrow, the heat transfer would be completely dominated by these edge effects. Ling (1963) has studied the same problem in a numerical investigation and concludes that the diffusion in the streamwise direction can be neglected if the Péclet number is larger than 5000. (The Péclet number here is defined as the ratio of heat transported by convection and by molecular diffusion, $Pe = Pr L^{+2}$ [see Brodkey, 1967]). Pedley (1972) concludes that, provided $0.5 \, Pe^{-0.5} < x/L < (1 - 0.7 \, Pe^{-0.5})$, there exists a central part of the hot-film where the boundary-layer solution predicts the heat transfer within 5%. Figure 26.20 shows the relation proposed by Pedley (1972), and it can be seen that the heat transfer is correctly described over a large part of the hot-film area, but as the Péclet number decreases, the influence from the diffusive terms must be considered. For Péclet numbers larger than 40, the heat transfer is correctly described by more than 80% of the heated area.

The upper limit of L^+ is crucial, as there the hot-film thermal boundary layer may not be entirely submerged in the viscous sublayer. Equation (26.58) has been included in Figure 26.20, and it can be seen that Péclet numbers larger than 40 correspond approximately to $\overline{\tau_w}^{1/3} > 0.95$. A relatively simple calculation of the upper limit of L^+ can then be made by assuming that the viscous sublayer is about five viscous units, and this yields an upper limit of L^+ which for air can be estimated to be approximately 47.

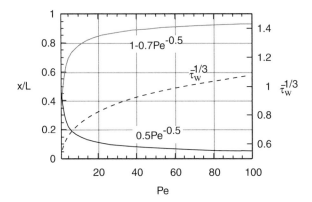

FIGURE 26.20 Pedley's (1972) relation showing the region of validity of the boundary-layer solution as a function of Péclet number. In the area in between the two continuous curves, the boundary-layer approximation predicts the heat transfer accurately to within 5% of the correct value. (From Pedley, T.J. (1972) *J. Fluids Eng.*, **55**, pp. 329–357. With permission.)

In summary, then, it can be concluded that the streamwise extent of the hot-film cannot be too small; otherwise, the boundary-layer approximations made are not applicable. On the other hand, a sensor that is too large will cause the thermal boundary layer to grow beyond the viscous region. Additionally, the spatial resolution will be adversely affected if the sensor is too large, as the smallest eddies imposed by the flow structures above the wall will then be integrated along the sensor length.

26.4.4.5 Temporal Resolution

The temporal resolution of the thermal probe is affected by the different time constants of the hot-film and the substrate. The hot-film usually has a much shorter time constant than the substrate. The higher the percentage of the total heat that leaks into the substrate, the lower is the sensitivity of the device to shear-stress fluctuations, and this changes the sensor characteristics sufficiently to invalidate the static calibration made. An example of this phenomenon is given by Haritonidis (1989), who showed that a hot-film sensor in a fluctuating wall-shear-stress environment will respond quickly to the instantaneous shear stress, while the substrate due to its much larger thermal inertia will react slowly. Haritonidis (1989) also showed that the ratio of the fluctuating sensitivity S_f to the average sensitivity S_a can be related to the ratio of the effective lengths under dynamic and static conditions:

$$\frac{S_f}{S_a} = \left(\frac{L_f}{L_a}\right)^{2/3} < 1 \qquad (26.61)$$

where L_a is the average effective length during static calibration, and L_f is the effective length during dynamic calibration. Due to this, the hot-film becomes less sensitive to shear-stress fluctuations at higher frequencies and the static calibration in a laminar flow will not give a correct result. These length scales can be considerably larger than the probe true extent. For example, Brown (1967) reported that the effective length scale from a static calibration was about twice the physical length.

Both the substrate material and the amount of heat lost to the substrate are crucial when determining the temporal or frequency response of the hot-film sensor. At low frequencies, the thermal waves through the substrate and into the fluid are quasi-static, which means that the fluctuating sensitivity of the hot-film is determined by the first derivative of the static calibration curve. In this range, which is of the order of a couple of cycles per second, the heat transfer through the substrate responds without time lag to wall-shear-stress fluctuations. Basically, this frequency range does not cause any major problems.

For the high-frequency range at the other end of the spectrum, it is possible to estimate the substrate role by considering the propagation of heat waves through a semi-infinite solid slab subjected to periodic temperature fluctuations at one end. This has been reported by Blackwelder (1981) who compared the

Sensors and Actuators for Turbulent Flows

wavelength of the heat wave to the hot-film length. He showed that the amplitude of the thermal wave would attenuate to a fraction of a percent over a distance equal to its wavelength. A relevant quantity to consider in this context turned out to be the ratio of the wavelength to the length of the substrate, as this is an indication of the extent to which the substrate will partly absorb heat from the heated surface and partly return it to the flow. Haritonidis (1989) computed this ratio for a number of fluids and films and the main conclusion drawn is that at high frequencies the substrate would not participate in the heat-transfer process. However, this conclusion should be viewed with some caution, as the frequencies studied were the highest that could be expected in a wall flow.

The most difficult problem occurs for frequencies in the intermediate range, resulting in a clear substrate influence and an associated deviation from the static calibration. Hanratty and Campbell (1996) showed that damping by the thermal boundary layer for pipe flow turbulence is important when:

$$[L^{+2}(5.65 \times 10^{-2})^3 Pr]^{1/2} \leq 1 \qquad (26.62)$$

For $Pr = 0.72$, this requires the dimensionless length in the streamwise direction L^+ to be less than 90. For turbulence applications, this is most disturbing, as it is in this frequency range where the most energetic eddies are situated. The primary conclusion to be reached from this discussion is that all wall-shear-stress measurements in turbulent flows require dynamical calibration of the hot-film sensor.

26.4.4.6 MEMS Thermal Sensors

Kälvesten (1996) and Kälvesten et al. (1996b) have developed a MEMS-based, flush-mounted, wall-shear-stress sensor that relies on the same principle of operation as the micro-velocity-sensor presented by Löfdahl et al. (1992). The shear sensor is based on the cooling of a thermally insulated, electrically heated part of a chip. As depicted in Figure 26.21, the heated portion of the chip is relatively small, $300 \times 60 \times 30\ \mu m^3$,

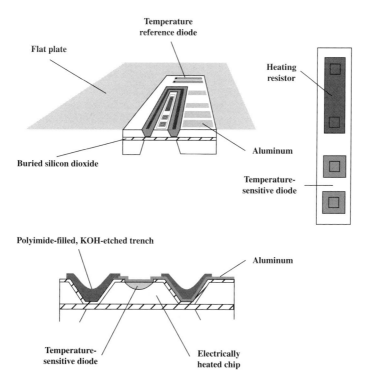

FIGURE 26.21 Flush-mounted wall-shear-stress sensor: (a) schematic cross section, (b) top view. (From Kälvesten, E. (1996) *Pressure and Wall Shear Stress Sensors for Turbulence Measurements*, Royal Institute of Technology, TRITA-ILA-9601, Stockholm, Sweden. © 1996 IEEE. With permission.)

TABLE 26.2 Some Calculated and Measured Characteristics of the MEMS-Based Wall-Shear-Stress Sensors Fabricated by Kälvesten et al. (1994)

	Theory		Measurements	
	300×60	1200×600	300×60	1200×600
Heated chip top area, $A = w \times \ell$ (µm²)	300×60	1200×600	300×60	1200×600
Heated chip thickness, h (µm)	30	30	30	30
Thermal conduction conductance, G_c (µW/K)	372	510	426	532
Thermal convection conductance, G_f (µW/K):				
Perpendicular configuration at 50 m/s	6,7	207	21.0	231
Parallel configuration at 50 m/s	5.6	192	17.3	160
Thermal capacity (µJ/K)	1.2	37	—	—
Thermal time constant at zero flow (ms)	3	72	6	55
Time response (electronic feedback) (µs)	—	—	25	—

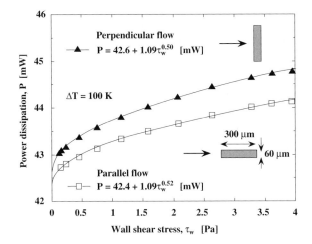

FIGURE 26.22 Total steady-state power dissipation calibration as a function of the wall-shear stress for two different probe orientations. (From Kälvesten, E. (1996) *Pressure and Wall Shear Stress Sensors for Turbulence Measurements*, Royal Institute of Technology, TRITA-ILA-9601, Stockholm, Sweden. © 1996 IEEE. With permission.)

and is thermally insulated by polyimide-filled, KOH-etched trenches. The rectangular top area, with a ratio of side length to side length of 5:1, yields a directional sensitivity for the measurements of the two perpendicular in-plane components of the fluctuating wall-shear stress. Due to the etch properties of KOH, the 30-µm-deep, thermally insulating trenches have sloped walls with a bottom and top width of about 30 µm. The sensitive part of the chip is electrically heated by a polysilicon piezoresistor and its temperature is measured by an integrated diode. For the ambient temperature, a reference diode is integrated on the substrate chip, far away from the heated portion of the chip. (Note that for backup two hot diodes and two cold diodes are fabricated on the same chip.)

Kälvesten (1996) performed a static wall-shear-stress calibration in the boundary layer of a flat plate. A Pitot tube and a Clauser plot were used to determine the time-averaged wall-shear stress. The power consumed to maintain the hot part of the sensor at a constant temperature was measured, and Figure 26.22 shows the data for two different probe orientations. For a step-wise increase of electrical power, the response time was about 6 ms, which is double the calculated value. This response was considerably shortened to 25 µs when the sensor was operated in a constant-temperature mode using feedback electronics. Table 26.2 lists some calculated and measured characteristics of the Kälvesten MEMS-based wall-shear-stress sensor.

Jiang et al. (1994; 1996) have developed an array of wall-shear-stress sensors based on the thermal principle. The primary objective of their experiment was to map and control the low-speed streaks in

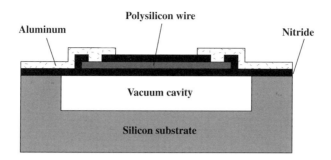

FIGURE 26.23 Cross-sectional structure of a single shear-stress sensor. (From Jiang, F. et al. (1997) in *Proc. IEEE MEMS Workshop (MEMS '97)*, pp. 465–470. © 1997 IEEE. With permission.)

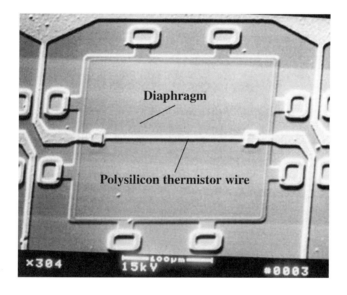

FIGURE 26.24 SEM photo of a single wall-shear-stress sensor. (From Jiang, F. et al. (1997) in *Proc. IEEE MEMS Workshop (MEMS '97)*, pp. 465–470. © 1997 IEEE. With permission.)

the wall region of a turbulent channel flow. Therefore, to properly capture the streaks, each sensor was made smaller than a typical streak width. For a Reynolds number based on the channel half-width and centerline velocity of 10^4, the streaks are estimated to be about 1 mm in width, so each sensor was designed to have a length less than 300 µm. Figure 26.23 shows a schematic of one of Jiang et al.'s sensors. It consists of a diaphragm with a thickness of 1.2 µm and a side length of 200 µm. The polysilicon resistor wire is located on the diaphragm and is 3 µm wide and 150 µm long. Below the diaphragm there is a 2-µm-deep vacuum cavity so that the device will have a minimal heat conduction loss to the substrate. When the wire is heated electrically, heat is transferred to the flow by heat convection, resulting in an electrically measurable power change that is a function of the wall-shear stress. Figure 26.24 shows a photograph of a portion of the 2.85 × 1.00-cm^2 streak-imaging chip, containing just one probe. The sensors were calibrated in a fully developed channel flow with known average wall-shear-stress values. Figure 26.25 depicts the calibration results for ten sensors in a row. The output of these sensors is sensitive to the fluid temperature, and the measured data must be compensated for this effect. Measurements of the fluctuating wall-shear stress using the sensor of Jiang et al. (1997) have been reported by Österlund (1999) and Lindgren et al. (2000).

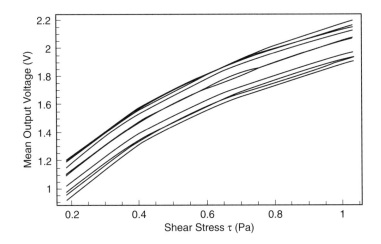

FIGURE 26.25 Calibration curves of 10 different sensors in an array. (From Jiang, F. et al. (1997) in *Proc. IEEE MEMS Workshop (MEMS '97)*, pp. 465–470. © 1997 IEEE. With permission.)

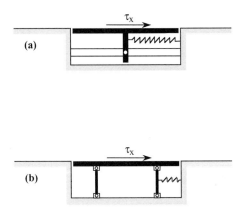

FIGURE 26.26 Schematic drawing of a floating-element device, with (a) pivoted support, and (b) parallel linkage support. (From Haritonidis, J.H. (1989) in *Advances in Fluid Mechanics Measurements*, M. Gad-el-Hak, Ed., Springer-Verlag, Berlin, pp. 229–261. © 1989 Springer-Verlag. With permission.)

26.4.4.7 Floating-Element Sensors

The floating-element technique is a direct method for sensing skin friction, which means a direct measurement of the tangential force exerted by the fluid on a specific portion of the wall. The advantage of this method is that the wall-shear stress is determined without having to make any assumptions either about the flowfield above the device or about the transfer function between the wall-shear stress and the measured quantity. The sensing wall-element is connected to a balance which determines the magnitude of the applied force. Basically, two arrangements are distinguished to accomplish this: either direct measurement of the distance the wall-element is moved by the wall-shear stress, so-called displacement balance, or measurement of the force required to maintain the wall element at its original position when actuated on by the wall-shear stress, the null balance.

The principle of a floating-element balance is shown in Figure 26.26. In spite of the fact that the force measurement is simple, the floating-element principle is afflicted with some severe drawbacks which strongly limit its use, as has been summarized by Winter (1977). It is difficult to choose the relevant size

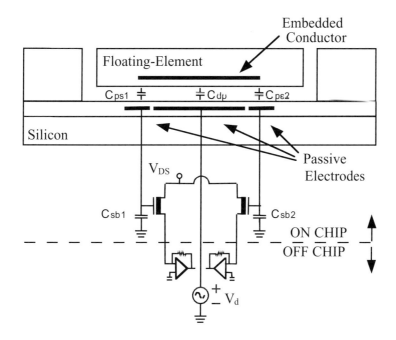

FIGURE 26.27 Schematic of the MEMS floating-element balance of Schmidt et al. (From Schmidt et al. (1988) *IEEE Trans. Electron. Devices* **35**, pp. 750–757. © 1988 IEEE. With permission.)

of the wall element, in particular, when measuring small forces and in turbulence applications. Misalignments and the gaps around the element, especially when measuring small forces, are constant sources of uncertainty and error. Effects of pressure gradients, heat transfer and suction/blowing cause large uncertainties in the measurements, as well. If the measurements are conducted in a moving frame of reference, effects of gravity, acceleration and large transients can also severely influence the results. Haritonidis (1989) discussed the mounting of floating-element balances and errors associated with the gaps and misalignments. In addition, floating-element balances fabricated with conventional techniques have in general poor frequency response and are not suited for measurements of fluctuating wall-shear stress. To summarize, the idea of direct force measurements by a floating-element balance is excellent in principle, but all the drawbacks taken together make them difficult and cumbersome to work with in practice. It was not until the introduction of microfabrication in the late 1980s that floating-element force sensors experienced revitalized interest, particularly for turbulence studies and reactive flow control.

26.4.4.8 MEMS Floating-Element Sensors

Schmidt et al. (1988) were the first to present a MEMS-based floating-element balance for operation in low-speed turbulent boundary layers. A schematic of their sensor is shown in Figure 26.27. A differential capacitive sensing scheme was used to detect the floating-element movements. The area of the floating element used was $500 \times 500 \ \mu m^2$, and it was suspended by four tethers which acted both as supports and restoring springs. The floating element had a thickness of 30 µm, and was suspended 3 µm above the silicon substrate on which it was fabricated. The gap on either side of the tethers and between the element and surrounding surface was 10 µm, while the element top was flush with the surrounding surface within 1 µm. The element and its tethers were made of polyimide, and the sensor was designed to have a bandwidth of 20 kHz. A static calibration of this force gauge indicated linear characteristics, and the sensor was able to measure a shear stress as low as 1 Pa. However, the sensor showed sensitivity to electromagnetic interference due to the high-impedance capacitance used, and drift problems attributed to water-vapor absorption by the polyimide were observed. No measurements of fluctuating wall-shear stress were made, as the signal amplitude available from the device itself was too low in spite of the fact that the first-stage amplification was fabricated directly on the chip.

Since the introduction of Schmidt et al.'s sensor, other floating-element sensors based on transduction, capacitive and piezoresistive principles have been developed [Ng et al., 1991; Goldberg et al., 1994; Pan et al., 1995]. Ng et al.'s sensor was small and had a floating element of size 120×40 μm^2. It operated on a transduction scheme and was basically designed for polymer-extrusion applications so it operated in the shear stress range of 1 to 100 kPa. Goldberg et al.'s sensor has a larger floating element size, 500×500 μm^2. It has the same application and the same principle of operation as Ng et al.'s balance. Neither of these two sensors is of interest in turbulence and flow-control applications, as their sensitivities are far too low. The capacitive floating-element sensor of Pan et al. (1995) is a force-rebalance device designed for wind-tunnel measurements and is fabricated using a surface micromachining process. Unfortunately, this particular fabrication technique can lead to nonplanar floating-element structures. The sensor has only been tested in laminar flow, and no dynamic response of this device has been reported.

Padmanabhan (1997) presented a floating-element wall-shear-stress sensor based on optical detection of instantaneous element displacement. The probe is designed specifically for turbulent boundary layer research and has a measured resolution of 0.003 Pa and a dynamic response of 10 kHz. A schematic illustrating the sensing principle is shown in Figure 26.28. The sensor is comprised of a floating element suspended by four support tethers. The element moves in the plane of the chip under the action of wall-shear stress.

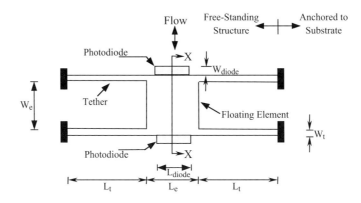

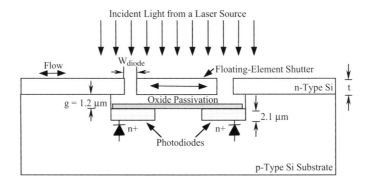

FIGURE 26.28 Schematic of the floating-element balance of Padmanabhan (1997). (From Padmanabhan, A. (1997) Silicon Micromachined Sensors and Sensor Arrays for Shear-Stres Measurements in Aerodynamic Flows, Ph.D. thesis, Department of Aeronautics and Astronautics, Massachusetts Institute of Technology, Cambridge, MA. © 1997 IEEE. With permission.)

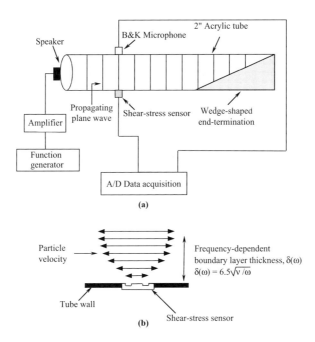

FIGURE 26.29 An acoustic plane-wave tube for the calibration of a floating-element shear sensor: (a) overall experiment setup and (b) probe region.

Two photodiodes are placed symmetrically underneath the floating element, at the leading and trailing edges, and a displacement of the element causes a "shuttering" of the photodiodes. Under uniform illumination from above, the differential current from the photodiodes is directly proportional to the magnitude and sign of shear stress. Analytical expressions were used to predict the static and dynamic response of the sensor, and based on the analysis two different floating-element sizes were fabricated, $120 \times 120 \times 7$ μm^3 and $500 \times 500 \times 7$ μm^3. The device has been calibrated statically in a laminar flow over a stress range of four orders of magnitude, 0.003–10 Pa. The gauge response was linear over the entire range of wall-shear stress. The sensor also showed good repeatability and minimal drift.

A unique feature of the shear sensor described above is that its dynamic response has been experimentally determined to 10 kHz. Padmanabhan (1997) described how oscillating wall-shear stress of a known magnitude and frequency can be generated using an acoustic plane-wave tube. A schematic of the calibration experiment is shown in Figure 26.28. The set-up is comprised of an acrylic tube with a speaker/compression driver at one end and a wedge-shaped termination at the other end. The latter is designed to minimize reflections of sound waves from the tube end and thereby set up a purely traveling wave in the tube. A signal generator and an amplifier drive the speaker to radiate sound at different intensities and frequencies. At some distance downstream, the waves become a plane and at this location a condenser microphone, which measures the fluctuating pressure, and the shear-stress sensor are mounted. The flowfield inside the plane–wave tube is very similar to a classical fluid dynamics problem—the Stokes second problem. The only difference is that instead of an oscillating wall with a semi-infinite stationary fluid, the plane–wave tube has a stationary wall and oscillating fluid particles far away from the wall. Solutions to the Stokes problem can be found in many textbooks [Brodkey, 1967; Sherman, 1990; White, 1991]. Padmanabhan (1997) converted the boundary conditions and derived corresponding analytical expression for the plane–wave tube. The analytical solution of the fluctuating wall-shear stress was compared to the measured output of the shear-stress sensor as a function of frequency and a transfer function of the sensor was determined. As expected, the measured shear stress showed a square-root dependence on frequency.

26.4.4.9 Outlook for Shear-Stress Sensors

Because coherent structures play a significant role in the dynamics of turbulent shear flows, the ability to control these structures will have important technological benefits such as drag reduction, transition control, mixing enhancement and separation delay. In particular, the instantaneous wall-shear stress is of interest for reactive control of wall-bounded flows to accomplish any of those goals. An anticipated scenario to realize this vision would be to cover a fairly large portion of a surface (for instance, parts of an aircraft wing or fuselage) with sensors and actuators. Spanwise arrays of actuators would be coupled with arrays of wall-shear-stress sensors to provide a locally controlled region. The basic idea is that sensors upstream of the actuators detect the passing coherent structures, and sensors downstream of the actuators provide a performance measure of the control. Fast, small and inexpensive wall-shear-stress sensors like the microfabricated thermal or floating-element sensors discussed in this section would be a necessity in accomplishing this kind of futuristic control system. Control theory, control algorithms and the use of microsensors and microactuators for reactive flow control are among the topics discussed in several chapters of this handbook.

As mentioned in the introduction, the fluctuating wall-shear stress is an indicator of the turbulence activity of the flow. In the ongoing development of existing turbulence models for application in high-Reynolds-number flows, absolute values of the fluctuating wall-shear stress are of significant interest. To accomplish this kind of measurement, reliable methods for conducting dynamical calibration of wall-shear-stress sensors are needed and must be developed. The calibration method of Padmanabhan (1997) is of interest and other strategies are likely to develop in the future. Another intriguing possibility that would be challenging is to design a microsensor where the dynamic effects of the gauge could be controlled in such a way that only a static calibration of the sensor would be adequate.

26.4.5 Pressure Sensors

26.4.5.1 Background

In turbulence modeling, flow control and aeroacoustics, the fluctuating wall pressure beneath a wall-bounded flow is a crucial parameter. By measuring this random quantity much information can be gleaned about the boundary layer itself without disturbing the interior of the flow. The fluctuating wall-pressure is coupled via a complex interaction to gradients of both mean-shear and velocity fluctuations as described by the transport equations for Reynolds stresses [Tennekes and Lumley, 1972; Hinze, 1975; Pope, 2000].

The characteristics of the fluctuating wall-pressure field beneath a turbulent boundary layer have been extensively studied in both experimental and theoretical investigations, and reviews of earlier work may be found in Blake (1986), Eckelmann (1990) and Keith et al. (1992). However, from the experimental perspective, knowledge of pressure fluctuations is far from being as comprehensive as that of velocity fluctuations, as there is a lack of a generally applicable pressure-measuring instrument that can be used in the same wide variety of circumstances as the hot-wire anemometer. In spite of this, some general facts have been established for wall-pressure fluctuations—for example, the order of magnitude of the root-mean-square (rms) value, the general shape of the power spectra and the space–time correlation characteristics [Harrison, 1958; Willmarth and Wooldridge, 1962; Bull, 1967; Bull and Thomas, 1976; Schewe, 1983; Blake, 1986; Lauchle and Daniels, 1984; Farabee and Casarella, 1991].

A clear shortcoming in many of the experiments designed to measure pressure fluctuations is the quality of the data. In the low-frequency range, the data may be contaminated by facility-related noise, while in the high-frequency range the spatial resolution of the transducers limits the accuracy. The former difficulty is usually circumvented by noise-cancellation techniques [Lauchle and Daniels, 1984] or by using a free-flight glider as an experimental platform. This problem may not be considered as a major obstacle today. At the other end of the spectrum, the spatial-resolution problem is more difficult to handle. The main criticism raised is that in many experiments the size of the pressure transducer used has been far too large in relation to the thickness of the boundary layer in context, let alone in relation to the characteristic small scale. The ultimate solution is of course to use small sensors, but testing in tunnels containing fluids with high viscosity—to increase the viscous length scale in the flow—has been conducted. Using highly viscous fluids creates its own set of problems. Very specialized facilities and instrumentation are needed when oil or glycerin, for example, are used as the working fluid. An oil tunnel is expensive to build and to operate. Moreover, the Reynolds numbers achieved are generally low, and it is not clear how to extrapolate the results to higher Reynolds number flows [Gad-el-Hak and Bandyopadhyay, 1994].

Pinhole microphones have also been utilized in an attempt to improve the sensor's spatial resolution. Unfortunately, results from this type of arrangement are in general questionable, and to this end the use of pinhole microphones must be considered an open question. Bull and Thomas (1976) concluded that the use of pinhole sensors in air may lead to severe errors in the measured spectra, while Farabee (1986), Gedney and Leehey (1989) and Farabee and Casarella (1991) all claim that the pinhole sensors are most effective for wall-pressure measurements.

Based on the above arguments, it seems then that the only realistic solution to improve measurements of fluctuating wall pressure is to use small sensors. For this reason microfabrication offers a unique opportunity for reducing the diaphragm size by at least one order of magnitude. MEMS also provides an opportunity to fabricate inexpensive, dense arrays of pressure sensors for correlation measurements and studies of coherent structures in turbulent boundary layers. In this section, we describe the basic principles used for MEMS-based pressure sensors/transducers/microphones. We look specifically at the design of pressure sensors utilizing the piezoresistive principle, which is particularly suited for measurements of wall-pressure fluctuations in turbulent flows. The section also contains a summary of measurements conducted with MEMS pressure sensors and when possible a comparison to conventional data. Finally, we provide an outlook for the use of MEMS-based pressure sensors for turbulence measurements and flow control.

26.4.5.2 Pressure-Sensor Principles

Many different methods have been advanced for the detection of pressure fluctuations. The principles available are based on detecting the vibrating motion of a diaphragm using piezoelectric, piezoresistive and capacitive techniques. These principles were already known at the beginning of the twentieth century, but the introduction of photolithographic fabrication methods during the last two decades has provided strong impetus as this technology offers sensors fabricated to extremely low tolerance, increased resolution due to a high degree of miniaturization and low unit-cost. This method of fabrication is also compatible with other IC techniques so electronic circuitry such as preamplifiers can be integrated close together with the sensor, an important factor for improving pressure sensor performance. Here we will give a short background of the above-mentioned principles for pressure sensor operations. In the description we use the word *microphone* for a device whereby sound waves are caused to generate an electric current for the purpose of transmitting or recording sound.

A piezoelectric sensor consists of a thin diaphragm, which is either fabricated in a piezoelectric material or mechanically connected to a cantilever beam consisting of two layers of piezoelectric material with opposite polarization. A vertical movement of the diaphragm causes a stress in the piezoelectric material and generates an electric output voltage. Royer et al. (1983) presented the first MEMS-based piezoelectric sensor, shown in Figure 26.30. This sensor consists of a 30-μm-thick silicon diaphragm with a diameter

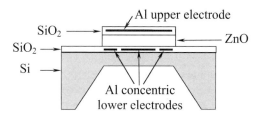

FIGURE 26.30 Cross-sectional view of a piezoelectric silicon microphone. (From Royer, M. et al. (1983) *Sensors and Actuators A,* **44,** 1–11. © 1983 IEEE. With permission.)

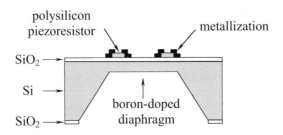

FIGURE 26.31 Cross-sectional view of a piezoresistive silicon microphone. (From Schellin, R. and Hess, G. (1992) *Sensors and Actuators A*, **32**, 555–559. © 1992 IEEE. With permission.)

of 3 mm. On top of the diaphragm a layer of 3- to 5-μm ZnO was deposited, sandwiched between two SiO_2 layers that contained the upper and lower aluminum electrodes. The sensor can be provided with an integrated preamplifier and is basically used for microphone applications. A sensitivity of 50 to 250 μV/Pa and a frequency response in the range of 10 Hz to 10 kHz (flat within 5 dB) were recorded. Other researchers have presented similar piezoelectric silicon microphones [Kim et al., 1991; Kuhnel, 1991; Schellin and Hess, 1992; Schellin et al., 1995], and the sensitivities of these microphones were in the range of 0.025 to 1 mV/Pa. However, applications of the piezoelectric microphone to turbulence measurements are strongly limited because of its high noise level, which has been found to be in the range of 50 to 72 dB(A)SPL. These noise levels are most commonly measured using an A-weighted filter, in dBs relative to 2×10^{-5} Pa, which is the lowest sound level detectable by the human ear. The A-weighted filter corrects for the frequency characteristics of the human ear and provides a measure of the audibility of the noise.

A piezoresistive sensor consists of a diaphragm which is usually provided with four piezoresistors in a Wheatstone bridge configuration. One common arrangement is to locate two of the gauges in the middle and two at the edge of the diaphragm. When the diaphragm deflects, the strains at the middle and at the edge of the diaphragm would have opposite signs, causing an opposite effect on the piezoresistive gauges. The most important advantage of this detection principle is its low output-impedance and its high sensitivity. The main drawback is that the piezoresistive material is sensitive to both stress and temperature. Unfortunately, this gives the piezoresistive sensor a strong temperature dependency. Schellin and Hess (1992) presented the first MEMS-fabricated piezoresistive sensor, which is shown in Figure 26.31. This particular sensor was used as a microphone and had a diaphragm made of 1-μm-thick, highly boron-doped silicon with an area of 1 mm^2. The diaphragm was equipped with a 250-nm-thick, p-type polysilicon resistors, which were isolated from the diaphragm by a 60-nm silicon dioxide layer. Using a bridge supply voltage of 6 V, this transducer showed a sensitivity of 25 μV/Pa and a frequency response in the range of 100 Hz to 5 kHz (+3 dB). However, the sensitivity was lower than expected by a factor of 10, which was explained by the initial static stress in the highly boron-doped silicon diaphragm. To improve the sensitivity of the piezoresistive sensor, various diaphragm materials have been explored such as polysilicon and silicon nitride [Guckel, 1987; Sugiyama et al., 1993].

Most MEMS-fabricated pressure sensors are based on the capacitive detection principle, and a vast majority of these sensors are used as microphones. Figure 26.32 shows a cross-sectional view of such a condenser microphone together with the associated electrical circuit. The latter must be included in the discussion, as the preamplifier is a vital part of determining the sensitivity of the capacitive probe. Basically, a condenser microphone consists of a backchamber (with a pressure-equalizing hole), a backplate (with acoustic holes), a spacer and a diaphragm covering the air gap created by the spacer located on the backplate. The condenser C_m and a dc-voltage source V_b constitute the sensing part of the microphone. Fluctuations in the flow pressure field above the diaphragm cause it to deflect which in turn changes the capacitance C_m. These changes are amplified in the preamplifier H_o which acts as an impedance converter with a bias resistor R_b and an input capacitance C_i. In this figure, C_p is a parasitic capacitance which is of interest when determining the microphone attenuation. In discussing the sensitivity

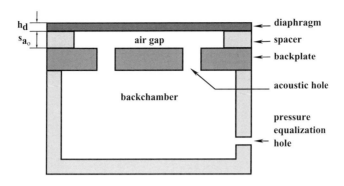

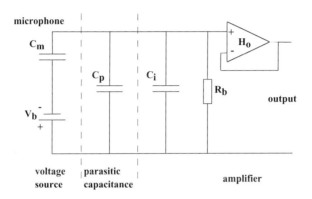

FIGURE 26.32 Cross-sectional view of a condenser microphone. The microphone is connected to an external dc bias voltage source and loaded by a parasitic capacitance, bias resistor and a preamplifier with an input capacitance. (From Löfdahl, L. et al. (1996) *J. Fluids Eng.*, **118**, 457–463. With permission.)

of a capacitive sensor, the open-circuit sensitivity is a relevant quantity and is considered to consist of two components, namely the mechanical sensitivity S_m and the electrical sensitivity S_e. The total sensitivity is a weighted value of both. The former sensitivity S_m is defined as the increase of the diaphragm deflection dw resulting from an increase in the pressure dp acting on the microphone:

$$S_m = \frac{dw}{dp} \qquad (26.63)$$

From Figure 26.32 we obtain the relation $dw = -ds_{a_o}$, where s_{a_o} is the thickness of the air gap between the diaphragm and the backplate. The electrical sensitivity of the microphone is given by the change in the voltage across the air gap dV resulting from a change in the air-gap thickness ds_{a_o}. Thus,

$$S_e = \frac{dV}{ds_{a_o}} \qquad (26.64)$$

The quasi-static, open-circuit sensitivity S_{open} of a condenser microphone may be defined as $S_{\text{open}} = -S_m S_e$. The output signal e_m can then be expressed as

$$e_m = S_m S_e p \qquad (26.65)$$

where p is the fluctuating pressure. For details of this derivation, see Scheeper et al. (1994).

TABLE 26.3 Summary of Silicon Micromachined Microphones

Author	Transducer Principle	Diaphragm Area (mm²)	Upper Frequency (kHz)	Sensitivity (mV/Pa)	Equivalent Noise Lavel (dB(A))
Royer et al. (1983)	Piezoelectric	3 × 3	40	0.25	66
Hohm (1985)	Capacitive	0.8 × 0.8	>20	4.3	54
Muller (1987)	Piezoelectric	3 × 3	7.8	0.05	72
Franz (1988)	Piezoelectric	0.8 × 0.9	45	0.025	68
Sprenkels (1988)	Capacitive	3 × 3	>10	7.5	25
Sprenkels et al. (1989)	Capacitive	3 × 3	>10	25	—
Hohm and Hess (1989)	Capacitive	0.8 × 0.8	2	1	—
Voorthuysen et al. (1989)	Capacitive	2.45 × 2.45	15	19	60
Murphy et al. (1989)	Capacitive	3 × 3	>15	4–8	30
Bergqvist and Rudolf (1991)	Capacitive	2 × 2	4	13	31.5
Schellin and Hess (1992)	Piezoresistive	1 × 1	10	0.025 (at 6V)	—
Kuhnel (1991)	FET	0.85 × 1.3	>20	5	62
Kuhnel and Hess (1992)	Capacitive	0.8 × 0.8	16	0.4–10	<25
Scheeper et al. (1992)	Capacitive	1.5 × 1.5	14	2	35
Bourouina et al. (1992)	Capacitive	1 × 1	2.5	3.5	35
Ried et al. (1993)	Piezoelectric	2.5 × 2.5	18	0.92	57
Kälvesten (1994)	Piezoresistive	0.1 × 0.1	>25	0.0009 (at 10V)	90
Scheeper et al. (1994)	Capacitive	2 × 2	14	5	30
Bergqvist (1994)	Capacitive	2 × 2	>17	11	30
Schellin et al. (1995)	Piezoelectric	1 × 1	10	0.1	60
Kovacs and Stoffel (1995)	Capacitive	0.5 × 0.5	20	0.065	58
Kronast et al. (1995)	FET	0.5 × 0.5	25	0.01	69

Hohm (1985) presented the first electret microphone based on MEMS technology. The backplate, 1-cm × 1-cm silicon, was provided with one circular acoustic hole with a diameter of 1 mm. A 2-μm-thick SiO_2 layer was used as an electret and was charged to about 350 V. The diaphragm was a metallized 13-μm-thick foil with a diameter of 8 mm. Later, polymer foil diaphragms were used in condenser microphones by Sprenkels (1988) and Murphy et al. (1989). In these microphones, the fabrication was made more compatible with standard thin-film technology. Bergqvist and Rudolf (1991) showed that MEMS-fabricated microphones can achieve a high sensitivity. For example, microphones with a 2-cm × 2-cm diaphragm had an open-circuit sensitivity in the range of 1.4 to 13 mV/Pa. In these microphones, diaphragms with thickness of 5 to 8 μm were fabricated using an anisotropic etching in a KOH solution and by applying an electrochemical etch stop. More details on the design and performance of the capacitive pressure sensors can be found in Scheeper et al. (1994).

An overview of the most significant dimensions, measured sensitivity, noise level and high-frequency response of first-generation MEMS pressure sensors is provided in Table 26.3. Inspecting the data, the following remarks can be made. The piezoelectric sensors seem to have the highest noise levels. Although simple in design, they have fairly low sensitivity and relatively large spatial extension of their diaphragms. The piezoresistive sensors seem to be most flexible, as the major advantage of locating piezoresistive gauges on a diaphragm is the high sensitivity accomplished with this arrangement. The drawback is the high temperature sensitivity, although this problem can be circumvented by a Wheatstone bridge arrangement of the gauges. The capacitive sensors are most common, as these sensors hold the largest commercial interest and are mainly designed to operate in conventional microphone applications such as music and human communications with pressure amplitudes as low as 20 μPa. The capacitive sensors have relatively large spatial extension and are commonly equipped with on-chip electronics for signal amplification. This implies a reduction of the noise levels but unfortunately also yields a very complex fabrication process.

The main advantages of the capacitive sensors are their high sensitivity to pressure and low sensitivity to temperature. Their primary disadvantages are the relatively large diaphragm areas and the decreased sensitivity for high frequencies due to the air-streaming resistance of the narrow air gap inherent in the principle.

26.4.5.3 Requirements for Turbulent Flows

There is no simple way to calculate the required pressure range of a sensor for turbulence and flow-control applications. Tennekes and Lumley (1972) estimate the fluctuating pressure to be a weighted integral of the Reynolds stresses, so its length scales should in general be larger than those of the velocity fluctuations. Moreover, it is plausible to assume that the order of magnitude of the fluctuating pressure would not be less than the Reynolds stresses, giving a good hint as to the intensity of the fluctuating pressure. This intensity, of course, depends strongly on the particular flow under consideration, but for a typical flat-plate boundary layer at, say, $Re_\theta = 4000$, this implies that the fluctuating pressure rms would be of the order of 10 Pa. Higher Reynolds numbers yield higher magnitudes of the fluctuating wall pressure. As compared to other applications, where the fluctuating pressure has a major interest, as, for example, in combustion processes, the pressure magnitudes in incompressible turbulent flows are extremely small. There is also a spatial constraint in the sensor design. For the same boundary layer considered here, we have a Kolmogorov length scale of about 50 µm, requiring a diaphragm size in the range of 100 to 300 µm. The required temporal resolution of the pressure sensor is probably the easiest to estimate as we have access to good turbulence kinetic energy spectra and these show that the energy content in the flow above 10 kHz is almost negligible. Based on these physical arguments, a frame for the design of a pressure transducer for turbulence applications can be established; namely, the sensor should have a pressure sensitivity of ±10 Pa, a diaphragm size of 100 µm, and flat frequency characteristics in the range of 10 Hz to 10 kHz. Of course, the signal-to-noise ratio must be sufficiently high so that an ordinary data acquisition can be made.

A critical scrutiny of the different principles for designing a pressure sensor shows that the most suitable principle for turbulence applications is the piezoresistive, as the required spatial and temporal resolutions can easily be achieved, a simple fabrication in MEMS is possible, and the temperature drift can be controlled. In the next section, we outline the design procedure of such a piezoresistive pressure transducer for turbulence and flow control applications.

26.4.5.4 Piezoresistive Pressure Sensors

Micromachined piezoresistive pressure sensors were introduced on the market already in the late 1950s by companies such as Kulite, Honeywell and MicroSystems. These early transducers were all very simple devices and typically the piezoresistive gauges were glued by hand onto the diaphragm. As the technology developed, more advanced fabrication methods were introduced, and today pressure transducers are all fabricated using batch-processing technologies and no hand assembly is required. A large variety of different pressure transducers are available commercially and the range of applications is very wide, spanning from relatively slow but stable tools for meteorological observations to sensors used to optimize rapid combustion processes. However, an examination of these commercial sensors with turbulence and control applications in mind will show that very few, almost none, can yet be applied for these specific purposes. If the sensors have either a frequency response or sensitivity that is acceptable, then the sensing area is far too large and the necessary spatial resolution is not fulfilled. Or, vice versa, if the diaphragm size is acceptable—which is very rare—then either the frequency response or the sensitivity is far too low for the fulfillment of, respectively, the required temporal resolution or an acceptable signal-to-noise ratio. Moreover, a literature review will confirm this statement, and the sensors of Table 26.3 are the ones that are fabricated using MEMS technology and fall in the range of interest for turbulence and flow-control applications.

Based on the earlier mentioned criteria, it may be concluded that a piezoresistive sensor for turbulence applications should be fabricated with a diaphragm size of $100 \times 100 \times 0.4$ µm. In order to give the pressure sensor the necessary sensitivity, the diaphragm must be very thin, and the air gap behind the diaphragm

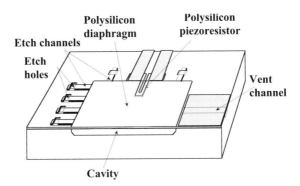

FIGURE 26.33 Isometric view of a pressure sensor/microphone for turbulence applications. The diaphragm side-length is 100 μm and its thickness is 0.4 μm. (From Löfdahl, L. et al. (1996) *J. Fluids Eng.*, **118**, 457–463. With permission.)

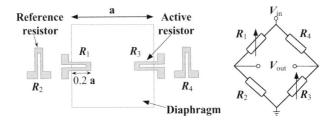

FIGURE 26.34 Top view of the piezoresistor layout with the electrical Wheatstone bridge configuration. (From Löfdahl, L. et al. (1996) *J. Fluids Eng.*, **118**, 457–463. With permission.)

(the cavity) must be relatively deep (say, 4 μm) to reduce the air stiffening effect. For equalization of the static pressure between the cavity and the ambient, a narrow vent channel with a length of 3 mm and a cross section of 5 × 0.1 μm must be included in the design. An isometric view of an acoustic sensor built to those specifications by Kälvesten (1994) is shown in Figure 26.33.

When a diaphragm is deflected by a pressure load, stress is induced in the diaphragm and this should be detected by piezoresistive strain gauges located at appropriate positions on the diaphragm. Noting that the geometrical piezoresistive effect in semiconductors can be neglected, the general relation between the relative change of resistance and the strain components in the longitudinal (streamwise), $\langle \varepsilon_L \rangle$, and transverse (spanwise), $\langle \varepsilon_T \rangle$, directions can, according to Guckel (1991), be expressed as:

$$\frac{\Delta R}{R} = G_{par} \langle \varepsilon_L \rangle + G_{per} \langle \varepsilon_T \rangle \tag{26.66}$$

where G_{par} and G_{per} are the longitudinal and transverse gauge factors. The pressure sensors fabricated by Kälvesten (1994) use two active strain gauges positioned at the points of maximum stress of the diaphragm (i.e., at the middle of the edges of the diaphragm), and the other two are integrated beside the diaphragm to serve as passive reference gauges (see Figure 26.34). This arrangement makes it possible to connect them in a half-active Wheatstone bridge for high pressure and low temperature sensitivity.

The expected theoretical acoustical pressure sensitivity, S, for small pressure amplitudes has been derived by Kälvesten (1994) and reads:

$$S = \frac{1}{V_{in}} \frac{\partial V_{out}}{\partial p} \approx \frac{0.077 G_{par}(1-\mu^2)}{\left(\frac{Et^2}{a^2} + \frac{3(1+\mu)E\varepsilon_o}{4\pi} + \frac{\alpha^{-1}(1-\mu^2)a^2}{16\pi^2 td} \right)} \tag{26.67}$$

TABLE 26.4 Some Calculated and Measured Characteristics of the MEMS-Based Microphone Fabricated by Kälvesten (1994)

	Theory		Measurements	
Diaphragm side length, a (μm)	100	300	100	300
Diaphragm thickness, t (μm)	0.4	0.4	0.4	0.4
Lower frequency limit (Hz)	2×10^{-3}	5×10^{-6}	10^a	10^a
Upper frequency limit (Hz)	1310	894	25^a	25^a
Acoustical sensitivity at 10 V (μV/Pa)	1.0	0.2	0.9	0.3
Static sensitivity at 10 V (μV/Pa)	1.7	2.2	1.2	1.4
Noise (A-weighted) (dB(A))	—	—	≈90	≈90

[a] Limits set by the calibration measurement setup.

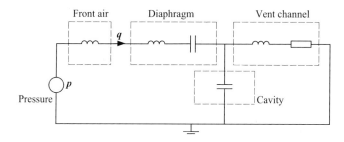

FIGURE 26.35 An electrical analogy describing the equivalent acoustical circuit of the pressure sensor/microphone. (From Löfdahl, L. et al. (1996) *J. Fluids Eng.*, 118, 457–463. With permission.)

where V_{in} and V_{out} are, respectively, the bridge supply and output voltages; p is the pressure acting on the diaphragm; α is the isothermal compressibility coefficient of air; d is the cavity depth; and G_{par} is the longitudinal gauge factor of the polysilicon piezoresistor. For the polysilicon diaphragm, μ is the Poisson's ratio, E is the Young's modulus, ε_o is the built-in strain, a is the side length, and t is the thickness. Assuming a linear diaphragm deflection, this equation can be used to derive the static pressure sensitivity of the sensor. In doing so it is necessary to assume a vacuum-sealed cavity by setting the third term of the denominator (air-gap compressive forces) to zero; see Rossi (1988). Table 26.4 compares the measured and calculated characteristics of the microphone developed by Kälvesten (1994).

A piezoresistive sensor is a combination of mechanical (the small and thin diaphragm), acoustical (the air-gap cavity, the air in the front of the diaphragm and the narrow vent channel) and electrical (polysilicon piezoresistors) elements. In order to optimize the dimensions of the sensor, an electrical analogy model is very useful for the task; see Rossi (1988). The important advantage of this analogy—from which the microphone frequency response and the diaphragm center deflection can be derived—is that it allows a good overview of even very complex systems. The analog model relies on the fact that there is a relationship between the geometry of the mechano-acoustical elements and their equivalent electrical impedance. Hence, pressure p (N/m^2) is equivalent to voltage; flow rate q (m^3/s), to current; and acoustic mass M (kg/m^4), to inductance. Damping in the system, which is due to the viscosity of the gas, is described by resistances R (N s/m^5) and capacitors C (m^5/N). These quantities then describe the diaphragm flexibility and the gas compressibility. However, a limitation of this "equivalent circuit method" is that analytical expressions for the equivalent electrical impedance are available only for simple mechanical and acoustical structures. One possibility for the derivation of the equivalent impedance is demonstrated in Kälvesten (1994) and Kälvesten et al. (1995; 1996a), where an energy method is used for the calculation of the energy and power contributions of the different elements. These expressions are then identified as equivalent electrical impedance. Figure 26.35 shows a simplified circuit diagram of their microphone.

The piezoresistive pressure probe fabricated by Kälvesten (1994) has perhaps the smallest diaphragm built specifically for turbulence diagnosis. Its spatial resolution is adequate to measure the small scales, and the small diaphragm has a correspondingly high resonance frequency which leads to a wide frequency bandwidth. The main drawbacks of this small sensor are its relatively low acoustical sensitivity and high equivalent noise level. It should be noted, however, that for turbulence applications, the pressure is of the order of 1 Pa, while pressure amplitudes as low as 20 µPa are encountered in human communication and music applications.

The frequency characteristics of Kälvesten's pressure sensor/microphone were determined by using a Bruel & Kjaer, Type 4226, closed coupler with a built-in reference microphone at a sound pressure level of 114 dB. All measurements were performed for a bridge supply voltage of 10 V and show acoustical sensitivities of about 0.9 and 0.3 µV/Pa for diaphragm side lengths of, respectively, 100 and 300 µm. The deviations from the theoretical estimates for the acoustical sensitivities can be explained by the approximations made in the sensitivity calculation and the uncertainty of the level of the built-in stresses in the polysilicon diaphragm. As can be seen from Table 26.4, and as is further elucidated in Kälvesten et al. (1995; 1996a), the required frequency range for turbulence applications of 10 Hz < f < 10 kHz is safely covered by the theoretical range of about 1 Hz < f < 1 MHz. The noise levels of the sensors are nearly the same for the two sensor versions, corresponding to an acoustic level of about 90 dB(A). More details on this sensor can be found in Kälvesten (1994) and Löfdahl et al. (1996).

26.4.5.5 MEMS Pressure Sensors

Genuine MEMS sensors for turbulence measurements were not introduced until the 1990s, so very few reliable measurements for studying wall-pressure fluctuations beneath turbulent boundary layers have been conducted. Otherwise, many MEMS devices have been launched, but fluid dynamics verification of these devices has been made at a surprisingly low level, typically in some undefined channel or pipe flow.

Schellin et al. (1995) presented a subminiature microphone based on the piezoresistive effect in polysilicon using only one chip. This sensor was classified as an acoustical sensor fabricated with a complementary metal oxide semiconductor (CMOS) process and standard microfabrication technology. The diaphragm area was 1 mm^2 and the sensitivity was 0.025 mV/Pa at 6 V, so for turbulence application this sensor is a bit on the large side as well as too insensitive. The frequency response was determined to be flat from 100 Hz to 5 kHz. Unfortunately, no fluid dynamics measurements have been reported with this sensor.

Kälvesten (1994), Kälvesten et al. (1995; 1996a) and Löfdahl et al. (1996) have designed, fabricated and used a silicon-based pressure transducer for studies focused on the high-frequency portion of the wall-pressure spectrum in a two-dimensional, flat-plate boundary layer. The momentum thickness Reynolds number in their study was Re_θ = 5072. A large value of the ratio between the boundary layer thickness δ and the diaphragm side-length d was used. The side lengths of the diaphragm were 100 µm (d^+ = 7.2) and 300 µm (d^+ = 21.6). This gives a ratio of the boundary-layer thickness to the diaphragm side length of the order of 240 and a resolution of eddies with wavenumbers less than 10 viscous units. Power spectra were measured for the frequency range 13 Hz < f < 13 kHz, and these were scaled in outer and inner variables. A clear overlap region between the mid- and high-frequency parts of the spectrum was found, and in this region the slope was estimated to be ω^{-1}. For the high-frequency region, the slope was proportional to ω^{-5}. The normalized rms pressure fluctuations was shown to depend strongly on the dimensionless diaphragm size with an increase connected to the resolution of the high-frequency region as shown in Figure 26.36. Classical data in the field are also plotted in the same figure for comparison.

Correlation measurements in both the longitudinal and transverse directions were performed by Löfdahl et al. (1996). Longitudinal space–time correlations, including the high-frequency range, indicated an advection velocity of the order of half the freestream velocity as can be seen in Figure 26.37. In this figure, U_c is the advection velocity, U_∞ is the freestream velocity, x_1 is the streamwise distance between two probes, and δ^* is the local displacement thickness. The advection velocities computed by Löfdahl et al. (1996) are consistently lower than those obtained by Willmarth and Wooldridge (1962) and Bull (1967). A broad-band filtering of the longitudinal correlation confirmed that the high-frequency part of the

Sensors and Actuators for Turbulent Flows

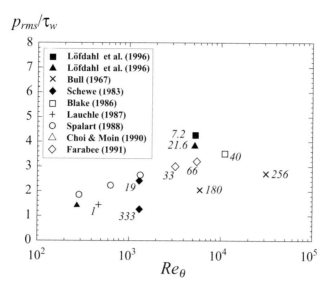

FIGURE 26.36 Dependence of normalized rms pressure fluctuations on Reynolds number; uncertainty is less than 10%. The numbers next to selected data points indicate the diaphragm side length in wall units. (From Löfdahl, L. et al. (1996) *J. Fluids Eng.*, **118**, 457–463. With permission.)

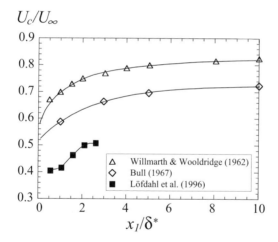

FIGURE 26.37 Local advection velocity of the pressure producing eddies for various frequency ranges; uncertainty is less than 10%. (From Löfdahl, L. et al. (1996) *J. Fluids Eng.*, **118**, 457–463. With permission.)

spectrum is associated with the smaller eddies from the inner part of the boundary layer, resulting in a reduction of the correlation, as shown in Figure 26.38.

Miller et al. (1997) designed, fabricated and tested a unique micromachined sensor array for making optical measurements of surface-pressure distributions. Each sensor element consists of a Fabry–Perot etalon fabricated from single-crystal silicon. Illuminating the sensor array with a near-infrared, tunable, diode laser and detecting the reflected light intensity with an infrared camera allow remote, simultaneous reading of the reflected signals from the entire sensor array. The results obtained demonstrate the basic feasibility of optically measuring surface pressure using micromachined Fabry–Perot pressure sensor arrays. Future work in this context will concentrate on demonstrating temporally resolved pressure measurements. This technology is interesting, as it offers the possibility of putting the arrays on curved surfaces.

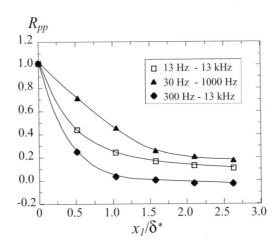

FIGURE 26.38 Longitudinal spatial correlation of the pressure fluctuations in three frequency ranges: 30 Hz < f < 1000 Hz; 300 Hz < f < 13 kHz; and 13 Hz < f < 13 kHz. Momentum thickness Reynolds number is Re = 5072. (From Löfdahl, L. et al. (1996) *J. Fluids Eng.*, **118**, 457–463. With permission.)

26.4.5.6 Outlook for Pressure Sensors

MEMS pressure transducers have certainly opened many doors to formulating relationships between wall pressure and turbulence. In this section, we will focus on some issues that should be possible to sort out in the near future by further miniaturization and array arrangements of pressure transducers. The Reynolds number effect on the rms values of the wall pressure is one key issue. The data shown in Figure 26.36 seem to suggest a trend that is also depicted schematically in the figure; that is, they come from a family of d^+. In a review of Reynolds number effects, Gad-el-Hak and Bandyopadhyay (1994) concluded that meaningful trends in the inner layer can be extracted only when the sensors are of the order of the viscous sublayer thickness. At all Reynolds numbers in the figure, the asymptotic values are reached by sensors approaching the size $d^+ = 5$, and the slope of the family of d^+ lines drops as d^+ values are increased, indicating a decreasing sensitivity of sensors to Reynolds number effects. It should be noted that the data shown in Figure 26.36 come from various sources with varying degrees of background noise, differences in data acquisition and corrections applied and different inherent errors in the sensors and instrumentation. Hence, the uncertainties in each dataset are different and probably not known accurately. There are also gaps in the data, so a useful contribution would be a systematic variation of the sensor dimension and Reynolds number so the data of Figure 26.36 could be completed.

Using arrays of wall-pressure sensors will offer the possibilities of more extensive studies of space-time correlations of pressure fluctuations. The primary contribution of the small transducers is that the eddies of the flowfield can be resolved with a spatial resolution of about five viscous units (i.e., the thickness of the viscous sublayer). Based on the measured correlations it is possible to examine the advection velocities of the different wavenumbers. It has earlier been shown [e.g., Bull, 1967] that the advection velocities of the turbulent eddies based on broad- and narrow-band frequency analysis approach 80% and 60%, respectively, of the freestream velocity. This can be questioned, as in all earlier investigations sensors that were too large have been used. In other words, the ratio between the boundary layer thickness and the characteristic dimension of the sensor has been too low. With MEMS sensors, about ten times larger values of this ratio can be achieved as compared with conventional sensors.

Having access to well-designed pressure-sensor arrays, the advection velocities for the different wavenumbers can be determined. Such data also offer the possibility of examining whether or not the structures are produced by pressure sources inside the boundary layer with a wide range of advection velocities. The so-called "two family" concept suggested by Bull (1967) can then be scrutinized. It was claimed by Bull (1967) that one family of high-wavenumber components was associated with the turbulent motion in the constant-stress layer. Those components were longitudinally coherent for times proportional to the time needed for structures to be advected distances equal to their wavelengths and laterally coherent over distances proportional to their own wavelengths. The other wavenumber family was associated with large-scale eddy motion in the boundary layer and loses coherence as a group independent of the wavelength. To this end, it has not been possible to question the Bull's hypothesis mainly because of the lack of small pressure transducers and associated arrays.

The concepts discussed above in relation to MEMS-based pressure sensors are all key issues and significant to aeroacoustics, energy transport in vibrating structures interacting with turbulent flows, turbulence modeling of the PVC terms and last but certainly not least to reactive control of the flow.

26.5 Microactuators for Flow Control

26.5.1 Background on Three-Dimensional Structures

It has recently been suggested by Busch-Vishniac (1998) that it is generally more difficult to control a system than to monitor it. For reactive control, the state of the system at a point or points in space–time must be monitored then accordingly made to change by a suitable actuation process. In the last decade it has been more challenging for scientists and engineers to design and build microsensors than microactuators; as a consequence, the progress in sensors has been left behind that for actuators. In this section, the terse coverage of microactuators relative to that of microsensors is a reflection of the relative progress in the two fields.

To reactively control a flowfield to reduce the drag, for example, it is necessary to develop specific actuators. Most conveniently the actuators should be flush-mounted on the body and aimed to influence near-wall phenomena, which in turbulent flows are dominated by small scales. To actuate these scales, it is required that the spatial extent of the actuator be of the same order as the small scales. As mentioned, the small scales decrease in spatial extension as the Reynolds number increases so sensor/actuator miniaturization is crucial for most field applications of the targeted control idea. In addition, the actuators are commonly intended to operate in a feedback loop with sensors on an extremely short time scale, implying that the operation of small actuators must additionally be very fast. Other constraints of control actuators are that they should yield minimal pressure drag especially when not in use and that they should be fabricated in large numbers at low cost so large portions of drag-inducing surfaces may be covered. Additionally, the actuator must be able to apply a significant disturbance to the flow while expending little energy when in operation. Finally, the sensor/actuator system must be able to withstand the typically harsh field environment. Taken together, it can be seen that the design of actuators for reactive flow control constitutes a real challenge.

Numerous devices for actuating a variety of flowfields have been advanced during the last few years. This has mostly been motivated by the quest to achieve efficient reactive control of turbulent flows. Wiltse and Glezer (1993) originated an interesting path when they used piezoelectric-driven cantilevers as flow-control actuators to beneficially affect free-shear layers. Following the idea of Wiltse and Glezer (1993), Jacobson and Reynolds (1995) designed an actuator that has the piezoelectric cantilever mounted flush with a boundary layer surface. The cantilever was oriented in the streamwise direction and fixed to the wall at its upstream end and could flex out into the flow and down into a cavity in the wall. The under-cantilever cavity was filled with fluid that was connected to the flow through the gaps on the sides and at the tip of the cantilever. Hence, when the cantilever flexed into the cavity, it forced fluid out of the cavity into the external flow, and when the cantilever flexed out of the cavity, fluid was drawn into the cavity from the external flow. Neither Wiltse and Glezer (1993) nor Jacobson and Reynolds (1995) used MEMS technology to fabricate their actuators, but the main conclusion drawn in Jacobson and Reynolds (1995) was that future actuators must be MEMS-based in order to match the relevant scales of turbulent flows.

The major challenge in designing MEMS-based actuators is that a motion perpendicular to the plane of the surface is needed to influence selective flow structures. It is implicitly complex to accomplish this, as micromachining and MEMS devices are basically two-dimensional structures. Some kind of quasi-three-dimensional structures have been fabricated using bulk and surface micromachining (see, for example, the passive cantilever structures developed by Bandyopadhyay, 1995). While these techniques usually have good resolution in the surface plane, in the direction normal to this plane the resolution is less impressive. Genuine three-dimensional structures can be achieved by using the so-called out-of-plane rotation of microstructures. A major advantage of this method is that it utilizes conventional high-resolution surface lithography and thus simultaneously provides access to the third dimension. A number of techniques based on out-of-plane rotation or folding of silicon microstructures have been advanced, and in this section we present the characteristics of some of these three-dimensional methods which may form a base for future development of MEMS actuators for reactive flow control. Actuating mechanisms based on the bilayer effect, electrostatic or electromagnetic forces, mechanical folding and one-layer structures are presented and discussed. The section will conclude with a brief look at some recently developed microturbomachines whose primary applications may be in fields other than flow control.

26.5.2 The BiLayer Effect

Out-of-plane rotation of microstructures can be achieved by controlled stress engineering in sandwiched structures, the bilayer effect method. The volume change between the two different materials used is the key to controlling the rotation, and several methods are available for this. A significant advantage of this technique is that the microstructure is self-assembled, so no external manipulators are needed to raise the structure out of plane. However, because the bilayer effect is based on the volume change of different materials, the rotation is limited to a fairly large radius of curvature which for certain applications is certainly a severe drawback.

It has been more than 70 years since Timoshenko (1925) presented the basic theory for bending sandwiched structures that consist of two materials with different coefficients of thermal expansion, the bimetal effect. By heating a bimetal cantilever, controlled bending can be achieved and this has been studied by, for example, Riethmuller and Benecke (1988) and Ataka et al. (1993), who used different standard IC materials. Commonly aluminum, chromium or gold, all having relatively high coefficients of thermal expansion, are used in combinations with silicon, polysilicon, silicon dioxide or silicon nitride, which all have low thermal expansion coefficients. As mentioned, the drawback of the bilayer method is that the bending radii are always large, and very thin and fragile structures are needed to reduce the radius of curvature. This pitfall reduces the impact of this particular technology.

Figure 26.39 illustrates a surface-micromachined actuator achieved by utilizing the large residual stress between two different polyimide films. The out-of-plane curling can be controlled either with thermal actuation using an integrated heater and the bimetal effect or by electrostatic actuation [Elwenspoek et al., 1992; Lin et al., 1995]. The electrostatic actuation requires unfortunately relatively high driving voltage. To circumvent this, Yasuda et al. (1997) have proposed using a serial connection of several bending actuators as depicted in Figure 26.40.

Smela et al. (1995) have shown that when a conjugated or conducting polymer is changed from insulating to metallic by applying a small voltage, the volume of the polymer is changed. A bilayer structure consisting of the polymer and a gold layer will curl, and this can be used as a hinge. Large reversible bending angles with small radii of curvature can be obtained with this technique. However, the price to pay is that the tuning requires an electrolyte which normally is not compatible with sensor applications. The polymer material is also more unstable than conventional integrated circuit materials and has a large potential for degradation.

Volume changes that result in bending can be achieved by using controlled heating to obtain a phase change in a material, so-called shape memory alloys (SMAs); see Otsuka (1995). An integrated heater is used to obtain the phase transformation of the SMA. Additionally, volume changes can also be obtained by applying a voltage over a piezoelectric material deposited on a thin cantilever. When such material expands, the cantilever bends and it is possible to obtain a voltage-controlled, out-of-plane rotation. Unfortunately, both the SMA and piezoelectric techniques give relatively large bending radii.

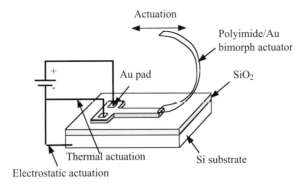
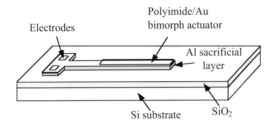
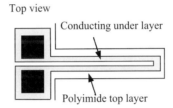

FIGURE 26.39 Surface micromachined three-dimensional structures using a bimorph structure consisting of two different polyimides with different thermal expansion coefficients. Control of the out-of-plane rotation can be obtained either with integrated gold heater or by electrostatic forces. (From Lin, G. et al. (1995) *in Technical Digest, Eighth Int. Conf. on Solid-State Sensors and Actuators (Transducers '95 and Eurosensors 9)*, June 25–29, Stockholm, Sweden, pp. 416–419. © 1995 IEEE. With permission.)

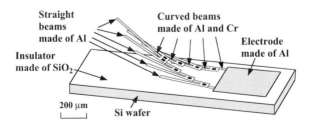

FIGURE 26.40 The difference between residual tensile stresses in the Al–Cr films initially curls the beam upwards. Electrostatic forces obtained through an applied voltage between the serially connected beams and the Si wafer are used to control the out-of-plane rotation. (From Yasuda, T. et al. (1997) in *Proc. IEEE Tenth Int. Workshop MEMS (MEMS '97)*, January 26–30, Nagoya, Japan. © 1997 IEEE. With permission.)

26.5.3 Electrostatic and Magnetic External Forces

The most straightforward way of bending a structure is simply to apply an active force on the structure. Different forces can be used to repel or attract freestanding structures so that they bend out of the wafer plane. Both electrostatic and magnetic forces have been utilized. Electrostatic actuators are easy to fabricate, but their operation is limited by small force and displacement outputs. The electromagnetic actuators are, on the other hand, more robust and can be used when larger forces and displacements are required—for example, for separation control or drag reduction. MEMS-based electrostatic actuators have been used to control aerodynamic instabilities. Huang et al. (1996) used an electrostatic actuator to control screech in high-speed jets. With the design of a 70-µm peak-to-peak displacement at the resonant frequency of 5 kHz, it has been shown that disturbances in the jet can be suppressed. Figure 26.41 illustrates the external-force principle using a permanent magnet to achieve an out-of-plane rotation [Liu et al., 1995]. Miller et al. (1996) used a microelectromagnetic flap with a 30-turn copper coil and a layer of permalloy on a 4-mm × 4-mm silicon plate to obtain a 2-mm tip motion of the actuator. These devices have been shown to be effective for changing the location of flow separation [Liu et al., 1995] and to achieve drag reduction [Tsao et al., 1994; Ho et al., 1997].

To summarize, it may be stated that the electrostatic actuators seem simple to fabricate, however, their operations are limited by small force and displacement outputs. The electromagnetic actuators are more robust and can be used when larger forces and displacements are required (e.g., for separation control or drag reduction). Promising results were reported by Yang et al. (1997), who used thermopneumatic

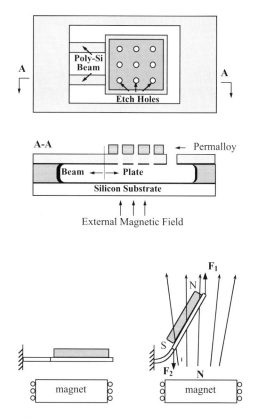

FIGURE 26.41 Three-dimensional actuation achieved by applying external magnetic forces. (From Liu, C. et al. (1995) in *Technical Digest, Eighth Int. Conf. on Solid-State Sensors and Actuators* (*Transducers '95 and Eurosensors 9*), June 25–29, Stockholm, Sweden, pp. 328–331. © 1995 IEEE. With permission.)

Sensors and Actuators for Turbulent Flows

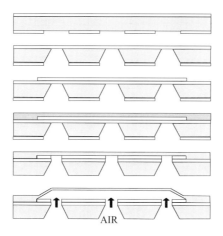

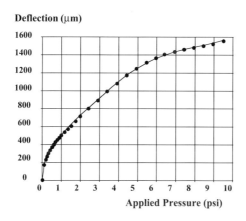

FIGURE 26.42 Steps involved in fabricating a microballoon actuator (top figure), and applied pressure vs. deflection (bottom plot) for a 2.3 × 6.6-mm balloon actuator. (From Yang, X. et al. (1997) in *Proc. IEEE Tenth Int. Workshop MEMS (MEMS '97)*, January 26–30, Nagoya, Japan. © 1997 IEEE. With permission.)

actuators or microballoons. Typically, these actuators have a heater in a cavity covered by a diaphragm, and when the fluid inside the cavity is heated a significant pressure rise is achieved. This causes a deflection of the diaphragm and for silicon rubber diaphragms more than 1-mm displacement has been measured by Yang et al. (1997). Figure 26.42 depicts the fabrication steps and the deflection achieved by Yang et al.'s microballoon actuator.

26.5.4 Mechanical Folding

A simple method to achieve out-of-plane bending of a microstructure is to use an external manipulator and fold a flexible or elastic joint/hinge, as shown in Figure 26.43. Unfortunately, this technique relies on the use of an external manipulator and is accordingly not batch compatible. However, to solve this problem various methods with integrated actuators have been presented by Fukuta et al. (1995; 1997) and Garcia (1998). The main drawback of these methods seems to be that they are quite complex and fragile. Elastic joints in polyimide were developed by Suzuki et al. (1992) as part of an attempt to fabricate a microrobot mimicking insects. Figure 26.44 shows their three-dimensional structure with the elastic polyimide joint and the surface micromachining steps involved in fabricating it. This structure is manually assembled into its final position just like paper folding, and bending angles up to 70° can be achieved.

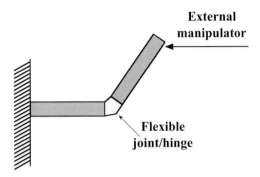

FIGURE 26.43 A three-dimensional microstructure undergoing a mechanical folding caused by external manipulator. (From Ebefors, T. et al. (1997) *Eurosensors XI*, September 21–24, Warsaw, Poland. © 1997 IEEE. With permission.)

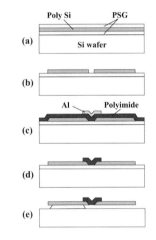

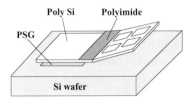

FIGURE 26.44 Steps involved in fabricating a three-dimensional structure that uses mechanical folding of an elastic polyimide joint (top figure), and isometric view of the resulting microactuator (bottom figure). (From Suzuki, K. et al. (1992) in *Proc. IEEE Fifth Int. Workshop MEMS (MEMS '92)*, February 4–7, Travemunde, Germany, pp. 190–195. © 1992 IEEE. With permission.)

An approach similar to the elastic polyimide-joint was employed to fabricate three-dimensional piezoresistive sensors using post-process etching of standard CMOS wafers [Hoffman et al., 1995; Kruglick et al., 1998]. External manipulators were used to rotate the structures out of plane with a flexible aluminum hinge, as shown in Figure 26.45. The advantage here is that three-dimensional actuators can readily be integrated on the same chip with the electronics. Moreover, the electrical interconnection to the folded structure is automatically obtained by the aluminum hinge; however, the main drawback is the non-self-assembling fabrication involved which requires interlocking braces. Furthermore, the resulting

Sensors and Actuators for Turbulent Flows

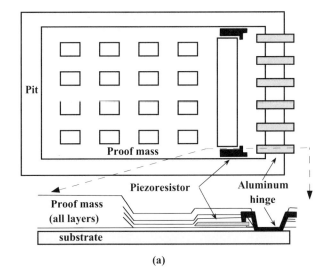

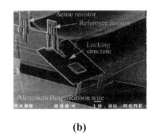
(b)

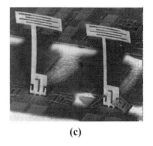
(c)

FIGURE 26.45 (a) Overview of a three-dimensional accelerometer design based on aluminum hinges with piezoresistive detection. The aluminum hinges are formed by front-bulk silicon etching of standard CMOS-wafers. (b) Close-up of the aluminum hinge with an interlocking brace. (c) SEM-photo of raised test structures. (Parts (a) and (b) from Kruglick, E.J.J. et al. (1998) in *Proc. IEEE Eleventh Int. Workshop MEMS (MEMS '98)*, January 25–29, Heidelberg, Germany, pp. 631–636; part (c) from Hoffman, E. et al. (1995) in *Proc. IEEE Eighth Int. Workshop MEMS (MEMS '95)*, January 29–February 2, Amsterdam, The Netherlands, pp. 288–293. © 1995 IEEE. With permission.)

structure is usually extremely fragile. A micromachined polysilicon hinge has been presented by Pister et al. (1990). This hinge is fabricated using surface micromachining with two layers of pholysilicon and posphosilicate glass as the sacrificial layer, as shown in Figure 26.46. A variety of applications using the polysilicon hinge have been presented—for example, micro-optical devices such as mirrors and lenses [Lin et al., 1995; Wu et al., 1995; Gunawan et al., 1995; Solgaard et al., 1995; Tien et al., 1996a; Fan et al., 1997]. However, a drawback of these surface-made microstructures is that they cannot stand by themselves, so they require some kind of interlocking arrangement to stay bent out of the wafer which complicates the assembly process. The structure usually consists of many stacked layers resulting in rugged surfaces. To be useful for three-dimensional applications, the difficulties in achieving electrical interconnections to the assembled structure have to be resolved.

Fukuta et al. (1995) have presented a reshaping technique using Joule's heating to effect thermal plastic deformation for permanent three-dimensional polysilicon structures. Figure 26.47 illustrates the experimental procedure for the three-dimensional structure realization. To achieve self-assembling structures without the drawback of using an external micromanipulator, a technique using an integrated scratch-drive actuator for folding was presented by Akiyama et al. (1997). Figure 26.48 shows the principle of

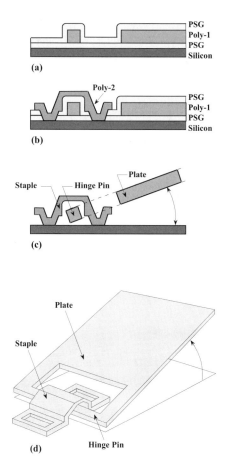

FIGURE 26.46 (a)–(c) Cross sections of the hinge during fabrication. (d) A perspective view of a hinged plate after release by external manipulator. (From Pister, K.S.J. et al. (1992) *Sensors and Actuators A*, **33**, 249–256. © 1992 IEEE. With permission.)

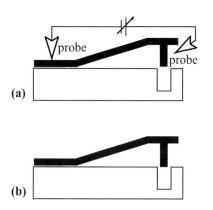

FIGURE 26.47 An external manipulator is used to fold and supply the structure with a reshaping voltage. The Joule's heating raises the temperature of the arm to cause an annealing effect which results in stress release and plastic deformation. After removal of the voltage, the arm cools down and the structure retains its three-dimensional shape. (From Fukuta, Y. et al. (1995) in *Technical Digest, Int. Conf. on Solid-State Sensors and Actuators (Transducers '95)*, June 25–29, Stockholm, Sweden, pp. 174–177. With permission.)

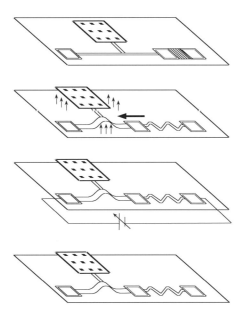

FIGURE 26.48 Illustration of a three-dimensional self-assembled polysilicon structure based on scratch-drive actuators. (From Akiyama, T. et al. (1997) *J. MEMS*, **6**, 10–17. © 1997 IEEE. With permission.)

the self-assembling process which produces permanent three-dimensional structure by using integrated scratch-drive actuator and reshaping technology. Fukuta et al. (1995) also discuss the self-assembling of three-dimensional structures with reshaping technology.

26.5.5 One-Layer Structures

All the techniques described in the previous subsection require external manipulators to raise the microstructures. To overcome this drawback, a self-assembling technique for fabrication of three-dimensional structures was presented by Green et al. (1996). The method is based on the surface tension force of a molten solder. The main drawback is that nonstandard IC material is used, meaning that very accurate process control is required. In addition, it seems difficult to achieve electrical connections to the out-of-plane rotated structures. To circumvent the lack of IC compatibility, Syms (1998) proposed using glass as the meltable material to achieve a more controlled assembly; however, the glass requires a high temperature to melt which makes it impossible to integrate metal interconnections and electronics together with the three-dimensional structure. Figure 26.49 shows the preparation of a three-dimensional structure based on surface tension forces. The structure after the out-of-plane rotation is shown in the bottom of this figure.

Another self-assembling microjoint technique was presented by Ebefors et al. (1998). In this technique, polyimide is used to create controlled out-of-plane rotations yielding small silicon structures with detailed features. The basic principle of the polyimide joint is shown in Figure 26.50. Polyimide in the V-grooves shrinks when cured in such a way that the absolute contraction length of the polyimide is larger at the top of the groove than at the bottom. This creates a rotation of the material which implies a bending of the freestanding structure out of the wafer plane. Larger bending angles can be achieved by connecting several V-grooves in a row. Figure 26.51 shows a schematic view of a polyimide joint with three V-grooves. The static, irreversible bending angle is obtained during the thermal curing process when cross-links between the molecules are effected and various solvents in the polyimide are out-gassed, yielding a reduction in the polyimide volume. The shrinkage depends on the curing temperature, which makes it possible to control the final bending angle rather well. A reversible bending angle can be obtained as a result of the thermal expansion of the cured polyimide. A metal conductor is used as a resistive heater to produce local power

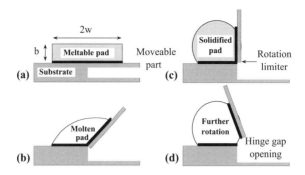

FIGURE 26.49 Production of a three-dimensional structure with Si mechanical parts based on surface tension forces (top figure), and SEM photo of the resulting three-dimensional structure (bottom figure). (From Syms, R.R.A. (1998) *Sensors and Actuators A,* **65**, 238–243. © 1998 IEEE. With permission.)

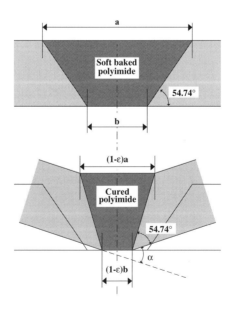

FIGURE 26.50 Principle of the polyimide V-groove joint. The polyimide in the V-groove shrinks when the polyimide is cured. The absolute lateral-contraction length of the polyimide is larger at the top of the V-groove than at the bottom, resulting in a rotation that bends the free-standing structure out of the wafer plane. (From Ebefors, T. et al. (1997) *Eurosensors XI,* September 21–24, Warsaw, Poland. © 1997 IEEE. With permission.)

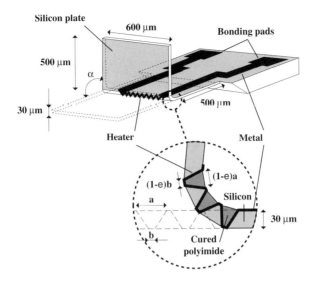

FIGURE 26.51 Isometric view of a three-dimensional structure based on a polyimide joint with three V-grooves. The metal lead-wires through the V-grooves are used to realize electrical connections to the out-of-plane rotated structure. Exploiting the thermal expansion of the polyimide, the metal can also be used as a heater in the V-grooves to obtain reversible movements. (From Ebefors, T. et al. (1997) in *Transducers '97*, June 16–19, Chicago, IL. With permission.)

dissipation in the joint. A temperature increase results in an expansion of the polyimide and a dynamic change in the bending angle.

26.6 Microturbomachines

26.6.1 Micropumps

In the several studies of microfabricated pumps, some have used nonmechanical effects. The so-called Knudsen pump uses the thermal-creep effect to move rarefied gases from one chamber to another. Ion drag is used in electrohydrodynamic pumps [Bart et al., 1990; Richter et al., 1991; Fuhr et al., 1992]. These rely on the electrical properties of the fluid and are thus not suited for many applications. Valveless pumping by ultrasound has been proposed by Moroney et al. (1991), but it produces very little pressure difference. Mechanical pumps based on conventional centrifugal or axial turbomachinery will not work at micromachine scales where the Reynolds numbers are typically small, on the order of 1 or less. Centrifugal forces are negligible and, furthermore, the Kutta condition through which lift is normally generated is invalid when inertial forces are vanishingly small. In general there are three ways in which mechanical micropumps can work:

1. *Positive-displacement pumps:* These are mechanical pumps with a membrane or diaphragm actuated in a reciprocating mode and with unidirectional inlet and outlet valves. They work on the same physical principle as their larger cousins. Micropumps with piezoelectric actuators have been fabricated [Van Lintel et al., 1988; Esashi et al., 1989; Smits, 1990]. Other actuators such as thermopneumatic, electrostatic, electromagnetic or bimetallic can be used [Pister et al., 1990; Döring et al., 1992; Gabriel et al., 1992]. These exceedingly minute positive-displacement pumps require even smaller valves, seals and mechanisms, a not-too-trivial micromanufacturing challenge. In addition, there are long-term problems associated with wear or clogging and consequent leaking around valves. The pumping capacity of these pumps is also limited by the small displacement and frequency involved. Gear pumps are a different kind of positive-displacement device.

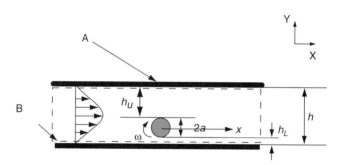

FIGURE 26.52 Schematic of micropump developed by Sen et al. (1996). (From Sen, M., Wajerski, D., and Gad-el-Hak, M. (1996) *J. Fluids Eng.* **118**, pp. 624–627. With permission.)

2. *Continuous, parallel-axis rotary pumps:* A screw-type, three-dimensional device for low Reynolds numbers was proposed by Taylor (1972) for propulsion purposes and is shown in his seminal film. It has an axis of rotation parallel to the flow direction, implying that the powering motor must be submerged in the flow, the flow turned through an angle, or that complicated gearing would be needed.

3. *Continuous, transverse-axis rotary pumps:* This class of machines has been developed by Sen et al. (1996). They have shown that a rotating body, asymmetrically placed within a duct, will produce a net flow due to viscous action. The axis of rotation can be perpendicular to the flow direction, and the cylinder can thus be easily powered from outside a duct. A related viscous-flow pump was designed by Odell and Kovasznay (1971) for a water channel with density stratification; however, their design operates at a much higher Reynolds number and is too complicated for microfabrication.

As evidenced from the third item above, it is possible to generate axial fluid motion in open channels through the rotation of a cylinder in a viscous fluid medium. Odell and Kovasznay (1971) studied a pump based on this principle at high Reynolds numbers. Sen et al. (1996) carried out an experimental study of a different version of such a pump. The novel viscous pump, shown schematically in Figure 26.52, consists simply of a transverse-axis cylindrical rotor eccentrically placed in a channel, so that the differential viscous resistance between the small and large gaps causes a net flow along the duct. The Reynolds numbers involved in Sen et al.'s work were low ($0.01 < Re \equiv 2\omega a^2/\nu < 10$, where ω is the radian velocity of the rotor, and a is its radius), typical of microscale devices, but were achieved using a macroscale rotor and a very viscous fluid. The bulk velocities obtained were as high as 10% of the surface speed of the rotating cylinder. Sen et al. (1996) have also tried cylinders with square and rectangular cross sections, but the circular cylinder delivered the best pumping performance.

A finite-element solution for low-Reynolds-number, uniform flow past a rotating cylinder near an impermeable plane boundary has already been obtained by Liang and Liou (1995). However, detailed two-dimensional Navier–Stokes simulations of the pump described above have been carried out by Sharatchandra et al. (1997), who extended the operating range of Re beyond 100. The effects of varying the channel height H and the rotor eccentricity ε have been studied. It was demonstrated that an optimum plate spacing exists and that the induced flow increases monotonically with eccentricity, the maximum flow rate being achieved with the rotor in contact with a channel wall. Both the experimental results of Sen et al. (1996) and the two-dimensional numerical simulations of Sharatchandra et al. (1997) have verified that, at $Re < 10$, the pump characteristics are linear and therefore kinematically reversible. Sharatchandra et al. (1997; 1998a; 1998b) also investigated the effects of slip flow on the pump performance as well as the thermal aspects of the viscous device. Wall slip does reduce the traction at the rotor surface and thus lowers the performance of the pump somewhat. However, the slip effects appear to be significant only for Knudsen numbers greater than 0.1, which is encouraging from the point of view of microscale applications.

In an actual implementation of the micropump, several practical obstacles need to be considered. Among these are the larger stiction and seal design associated with rotational motion of microscale devices. Both the rotor and the channel have a finite (in fact, rather small) width. DeCourtye et al. (1998) numerically investigated the viscous micropump performance as the width of the channel W becomes exceedingly small. The bulk flow generated by the pump decreased as a result of the additional resistance to the flow caused by the side walls. However, effective pumping was still observed with extremely narrow channels. Finally, Shartchandra et al. (1998b) used a genetic algorithm to determine the optimum wall shape to maximize the micropump performance. Their genetic algorithm uncovered shapes that were nonintuitive but yielded vastly superior pump performance.

Though most of the micropump discussion above is of flow in the steady state, it should be possible to give the eccentric cylinder a finite number of turns or even a portion of a turn to displace a prescribed minute volume of fluid. Numerical computations will easily show the order of magnitude of the volume discharged and the errors induced by acceleration at the beginning of the rotation and deceleration at the end. Such system can be used for microdosage delivery in medical applications.

26.6.2 Microturbines

DeCourtye et al. (1998) have described the possible utilization of the inverse micropump device as a turbine. The most interesting application of such a microturbine would be as a microsensor for measuring exceedingly small flow rates on the order of nanoliters (i.e., microflow metering for medical and other applications).

The viscous pump described operates best at low Reynolds numbers and should therefore be kinematically reversible in the creeping-flow regime. A microturbine based on the same principle should, therefore, lead to a net torque in the presence of a prescribed bulk velocity. The results of three-dimensional numerical simulations of the envisioned microturbine are summarized in this section. The Reynolds number for the turbine problem is defined in terms of the bulk velocity, as the rotor surface speed is unknown in this case:

$$Re = \frac{\bar{U}(2a)}{\nu} \tag{26.69}$$

where $\bar{U}$ is the prescribed bulk velocity in the channel, a is the rotor radius and ν is the kinematic viscosity of the fluid.

Figure 26.53 shows the dimensionless rotor speed as a function of the bulk velocity for two dimensionless channel widths, $W = \infty$ and $W = 0.6$. In these simulations, the dimensionless channel depth

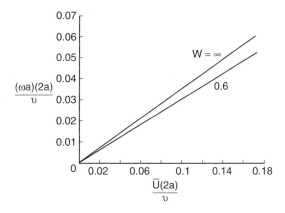

FIGURE 26.53 Turbine rotation as a function of the bulk velocity in the channel. (From DeCourtye, D. et al. (1998) *Int. J. Comp. Fluid Dyn.*, **10**, 13–25. With permission by Taylor & Francis Ltd.)

is $H = 2.5$ and the rotor eccentricity is $\varepsilon/\varepsilon_{max} = 0.9$. The relation is linear as was the case for the pump problem. The slope of the lines is 0.37 for the two-dimensional turbine and 0.33 for the narrow channel with $W = 0.6$. This means that the induced rotor speeds are, respectively, 0.37 and 0.33 of the bulk velocity in the channel. (The rotor speed can never, of course, exceed the fluid velocity even if there is no load on the turbine. Without load, the integral of the viscous shear stress over the entire surface area of the rotor is exactly zero, and the turbine achieves its highest albeit finite rpm.) For the pump, the corresponding numbers were 11.11 for the two-dimensional case and 100 for the three-dimensional case. Although it appears that the side walls have a greater influence on the pump performance, it should be noted that in the turbine case a vastly higher pressure drop is required in the three-dimensional duct to yield the same bulk velocity as that in the two-dimensional duct (dimensionless pressure drop of $\Delta p^* \equiv \Delta p(2a)^2/\rho \nu^2 = -29$ vs. $\Delta p^* = -1.5$).

The turbine characteristics are defined by the relation between the shaft speed and the applied load. A turbine load results in a moment on the shaft, which at steady state balances the torque due to viscous stresses. At a fixed bulk velocity, the rotor speed is determined for different loads on the turbine. Again, the turbine characteristics are linear in the Stokes (creeping) flow regime, but the side walls have a weaker, though still adverse, effect on the device performance as compared to the pump case. For a given bulk velocity, the rotor speed drops linearly as the external load on the turbine increases. At large enough loads, the rotor will not spin, and maximum rotation is achieved when the turbine is subjected to zero load.

At present, it is difficult to measure flow rates on the order of 10^{-12} m^3/s (1 nl/s). One possible way is to directly collect the effluent over time. This is useful for calibration but is not practical for online flow measurement. Another way is to use heat transfer from a wire or film to determine the local flow rate as in a thermal anemometer. Heat transfer from slowly moving fluids is mainly by conduction so that temperature gradients can be large. This is undesirable for biological and other fluids easily damaged by heat. The viscous mechanism that has been proposed and verified for pumping may be turned around and used for measuring. As demonstrated in this section, a freely rotating cylinder eccentrically placed in a duct will rotate at a rate proportional to the flow rate due to a turbine effect. In fact, other geometries such as a freely rotating sphere in a cylindrical tube should also behave similarly. The calibration constant, which depends on system parameters such as geometry and bearing friction, should be determined computationally to ascertain the practical viability of such a microflow meter. Geometries that are simplest to fabricate should be explored and studied in detail.

26.6.3 Microbearings

Many of the micromachines use rotating shafts and other moving parts that carry a load and need fluid bearings for support, most of them operating with air or water as the lubricating fluid. The fluid mechanics of these bearings is very different compared to that of their larger cousins. Their study falls in the area of microfluid mechanics, an emerging discipline which has been greatly stimulated by its applications to micromachines and which is the subject of this chapter.

Macroscale journal bearings develop their load-bearing capacity from large pressure differences which are a consequence of the presence of a viscous fluid, an eccentricity between the shaft and its housing, a large surface speed of the shaft and a small clearance-to-diameter ratio. Several closed-form solutions of the no-slip flow in a macrobearing have been developed. Wannier (1950) used modified Cartesian coordinates to find an exact solution to the biharmonic equation governing two-dimensional journal bearings in the no-slip, creeping flow regime. Kamal (1966) and Ashino and Yoshida (1975) worked in bipolar coordinates; they assumed a general form for the stream function with several constants determined using the boundary conditions. Though all these methods work if there is no slip, they cannot be readily adapted to slip flow. The basic reason is that the flow pattern changes if there is slip at the walls and the assumed form of the solution is no longer valid.

Microbearings are different in the following aspects: (1) being so small, they are difficult to manufacture with a clearance much smaller than the diameter of the shaft; (2) because of the small shaft size, its surface speed, at normal rotational speeds, is also small (the microturbomachines being developed currently at MIT operate at shaft rotational speeds on the order of 1 million rpm and are therefore operating at different flow regime from that considered here); and (3) air bearings in particular may be small enough for noncontinuum effects to become important. For these reasons the hydrodynamics of lubrication is very different at microscales. The lubrication approximation that is normally used is no longer directly applicable and other effects come into play. From an analytical point of view there are three consequences of the above: Fluid inertia is negligible, slip flow may be important for air and other gases, and relative shaft clearance need not be small.

In a recent study, Maureau et al. (1997) analyzed microbearings represented as an eccentric cylinder rotating in a stationary housing. The flow Reynolds number is assumed small, the clearance between shaft and housing is not small relative to the overall bearing dimensions, and there is slip at the walls due to nonequilibrium effects. The two-dimensional governing equations are written in terms of the streamfunction in bipolar coordinates. Following the method of Jeffery (1920), Maureau et al. (1997) succeeded in obtaining an exact infinite-series solution of the Navier–Stokes equations for the specified geometry and flow conditions. In contrast to macrobearings and due to the large clearance, flow in a microbearing is characterized by the possibility of a recirculation zone which strongly affects the velocity and pressure fields. For high values of the eccentricity and low slip factors the flow develops a recirculation region, as shown in the streamlines plot in Figure 26.54.

From the infinite-series solution the frictional torque and the load-bearing capacity can be determined. The results show that both are similarly affected by the eccentricity and the slip factor: They increase with the former and decrease with the latter. For a given load, there is a corresponding eccentricity which generates

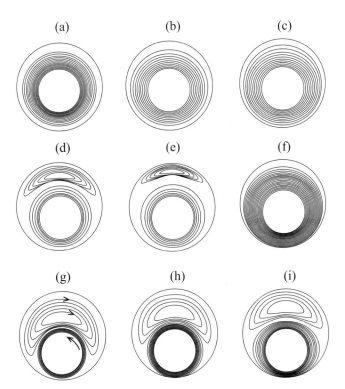

FIGURE 26.54 Effect of slip factor and eccentricity on the microbearing streamlines. From top to bottom, eccentricity changes as $\varepsilon = 0.2, 0.5, 0.8$. From left to right, slip factor changes as $S \equiv [(2 - \sigma_v)/\sigma]$, $Kn = 0, 0.1, 0.5$. (From Maureau, J. et al. (1997) *J. Micromech. Microeng.*, **7**, 55–64. With permission.)

a force sufficient to separate the shaft from the housing (i.e., sufficient to prevent solid-to-solid contact). As the load changes, the rotational center of the shaft shifts a distance necessary for the forces to balance.

It is interesting to note that for a weight that is vertically downwards, the equilibrium displacement of the center of the shaft is in the horizontal direction. This can lead to complicated rotor dynamics governed by mechanical inertia, viscous damping and pressure forces. A study of this dynamics may of interest. Real microbearings have finite shaft lengths, and end walls, and other three-dimensional effects influence the bearing characteristics. Numerical simulations of the three-dimensional problem can readily be carried out and may also be of interest to the designers of microbearings. Other potential research includes determination of a criterion for onset of cavitation in liquid bearings. From the results of these studies, information related to load, rotational speed and geometry can be generated that would be useful for the designer.

Finally, Piekos et al. (1997) have used full Navier–Stokes computations to study the stability of ultra-high-speed gas microbearings. They conclude that it is possible, despite significant design constraints, to attain stability for specific bearings to be used with the MIT microturbomachines [Epstein and Senturia, 1997; Epstein et al., 1997], which incidentally operate at much higher Reynolds numbers (and rpm) than the micropumps/microturbines/microbearings considered thus far. According to Piekos et al. (1997), high-speed bearings are more robust than low-speed ones due to their reduced running eccentricities and the large loads required to maintain them.

26.7 Concluding Remarks

In a presentation to the 1959 annual meeting of the American Physical Society, Richard Feynman anticipated the extension of electronic miniaturization to mechanical devices. That vision is now a reality. Microelectromechanical systems, a fledgling field that took off just this decade, are already finding numerous applications in a variety of industrial and medical fields. This chapter focused on MEMS-based sensors and actuators especially as used for the diagnosis and control of turbulent flows. The miniaturization of sensors leads to improved spatial and temporal resolutions for measuring useful turbulence quantities at high Reynolds numbers. The availability of inexpensive, low-energy-usage microsensors/microactuators that can be packed densely on a single chip promises a quantum leap in the performance of reactive flow control systems. Such control is now in the realm of the possible for future vehicles and other industrial devices. In a turbulent flow, an increase in Reynolds number will automatically generate smaller length scales and shorter time scales, which both in turn require small and fast sensors for correct resolution of the flow field. MEMS offer a solution to this problem, as sensors with length and time scales of the order of the relevant Kolmogorov microscales can now be fabricated. Additionally, these sensors are produced with high accuracy at a relatively low cost per unit. For instance, a MEMS pressure sensor can be used to determine fluctuating pressures beneath a turbulent boundary layer with a spatial resolution that is about one order-of-magnitude finer than what can be achieved with conventional microphones.

In this chapter, we have reviewed the state-of-the-art of microsensors used to measure the instantaneous velocity, wall-shear stress and pressure, which we deem as quantities of primary importance in turbulence diagnosis. For each group, we gave a general background, design criteria and calibration procedure; provided examples of measurements conducted with MEMS-based sensors; and, when possible, compared the results to conventional measurements. Microsensors can be fabricated at low unit costs and can be spaced close together in dense arrays. These traits are particularly useful for studies of coherent structures in wall-bounded turbulent flows.

Reactive flow control is another application for which microdevices may play a crucial future role. MEMS sensors and actuators provide opportunities for targeting the small-scale coherent structures in macroscopic turbulent shear flows. Detecting and modulating these structures may be essential for successful control of wall-bounded turbulent flows to achieve drag reduction, separation delay and lift enhancement. To cover areas of significant spatial extension, many devices are needed that utilize small-scale, low-cost and low-energy-use components. In this context, the miniaturization, low-cost fabrication and low-energy consumption of microsensors and microactuators are of utmost interest and promise a quantum leap in control system performance. Combined with modern computer technologies, MEMS yield the essential matching of the length and time scales of the phenomena to be controlled.

Numerous actuators have been developed during the past few years. In this chapter, we reviewed the state-of-the-art of microactuators based on the bilayer effect, electrostatic or electromagnetic forces, mechanical folding and one-layer structures. We have also briefly described recently advanced ideas for viscous micropumps and microturbines. Future challenges include achieving significant actuation perpendicular to the plane of what is basically a two-dimensional chip, further reducing unit cost and energy expenditure of microactuators, and designing microdevices that are capable of withstanding the harsh field environment of, for example, an aircraft. These are not easy tasks, but the payoff if, for example, the drag of air, water, or land vehicles could be reduced by a few percentage points, would come as fuel savings in the billions of dollars as well as tremendous benefits to the environment.

Microelectromechanical systems have witnessed phenomenal advances in a mere 10-year period. The 1960s and 1970s were arguably the decades of the transistor, 10 years after its discovery, and it is likely that the first years of the third millennium will be the MEMS decades. Medical and industrial breakthroughs are inevitable with every advance in MEMS technology, and the future worldwide market for micromachines is bound to be in the tens of billions of dollars.

Acknowledgments

The authors would like to acknowledge the valuable help of Dr. Andrey Bakchinov and Mr. Peter Johansson for preparing the figures. Our thanks are extended to Professors Haim Bau, Ali Beskok, Kenneth Breuer, Chih-Ming Ho, Stuart Jacobson and George Karniadakis, who all shared with us several of their reports and papers. Our sincere appreciation to Professor Mihir Sen for sharing his ideas regarding shear-sensor calibration and microturbomachines.

References

Akiyama, T., Collard, D., and Fujita, H. (1997) "Scratch Drive Actuator with Mechanical Links for Self-Assembly of Three-Dimensional MEMS," *J. MEMS* **6**, pp. 10–17.

Aronson, D., and Löfdahl, L. (1994) "The Plane Wake of a Cylinder: An Estimate of the Pressure Strain Rate Tensor," *Phys. Fluids* **6**, pp. 2716–2721.

Aronson, D., Johansson, A.V., and Löfdahl, L. (1997) "A Shear-Free Turbulent Boundary Layer-Experiments and Modeling," *J. Fluid Mech.* **338**, pp. 363–385.

Ashino, I., and Yoshida, K. (1975) "Slow Motion between Eccentric Rotating Cylinders," *Bull. JSME* **18** (117), pp. 280–285.

Ataka, M., Omodaka, A., and Fujita, H. (1993) "A Biomimetic Micro System—A Ciliary Motion System," in *Proc. Seventh Int. Conf. on Solid-State Sensors and Actuators (Transducers '93)*, pp. 38–41, Yokohama, Japan.

Bandyopadhyay, P.R. (1995) "Microfabricated Silicon Surfaces for Turbulence Diagnostics and Control," in *Proc. Int. Symp. on Active Control of Sound and Vibration (Active '95)*, S. Sommerfeldt and H. Hamada, Eds., pp. 1327–1338, July 6–8, Newport Beach, CA.

Bart, S.F., Tavrow, L.S., Mehregany, M., and Lang, J.H. (1990) "Microfabricated Electrohydrodynamic Pumps," *Sensors and Actuators A* **21–23**, pp. 193–197.

Bellhouse, B.J., and Rasmussen, C.G. (1968) "Low-Frequency Characteristics of Hot-Film Anemometers," *DISA Inform.* **6**, pp. 3–10.

Bellhouse, B.J., and Schultz, L. (1966) "Determination of Mean and Dynamic Skin Friction Separation in Low-Speed Flow with a Thin-Film Heated Element," *J. Fluid Mech.* **24**, pp. 379–400.

Bergqvist, J. (1994) "Modelling and Micromachining of Capacitive Microphones," Ph.D. thesis, Uppsala University, Uppsala, Sweden.

Bergqvist, J., and Rudolf, F. (1991) "A New Condenser Microphone in Silicon," *Sensors and Actuators A* **21**, pp. 123–125.

Blackwelder, R.F. (1981) "Hot-Wire and Hot-Film Anemometers," in *Methods of Experimental Physics—Fluid Dynamics*, Vol. 18A, R.J. Emrich, Ed., pp. 259–314, Springer-Verlag, Berlin.

Blake, W.K. (1986) *Mechanics of Flow Induced Sound and Vibrations*, Academic Press, New York.

Bourouina, T., Spirkovitch, S., Baillieu, F., and Vauge, C. (1992) "A New Condenser Microphone with a p^+ Silicon Membrane," *Sensors and Actuators A* **31**, pp. 149–152.

Bremhorst, K., and Gilmore, D.B. (1976) "Comparison of Dynamic and Static Hot-Wire Anemometer Calibrations for Velocity Perturbation Measurements," *J. Physics E: Sci. Instrum.* **9**, pp. 1097–1100.

Breuer, K.S. (1995) "Stochastic Calibration of Sensors in Turbulent Flow Fields," *Exp. Fluids* **19**, pp. 138–141.

Brodkey, R.S. (1967) *The Phenomena of Fluid Motions*, Addison-Wesley, New York.

Brown, G.L. (1967) "Theory and Application of Heated Films for Skin Friction Measurements," in *Proc. Heat Transfer and Fluid Mechanics Institute*, pp. 361–381.

Bryzek, J., Peterson, K., and McCulley, W. (1994) "Micromachines on the March," *IEEE Spectrum* **31**, pp. 20–31.

Bull, M.K. (1967) "Wall-Pressure Fluctuations Associated with Subsonic Turbulent Boundary Layer Flow," *J. Fluid Mech.* **28**, pp. 719–754.

Bull, M.K., and Thomas, A.S.W. (1976) "High Frequency Wall-Pressure Fluctuations in Turbulent Boundary Layers," *Phys. Fluids* **19**, pp. 597–599.

Busch-Vishniac, I.J. (1998) "Trends in Electromechanical Transduction," *Physics Today* **51**, pp. 28–34.

Champange, F.H. (1965) "Turbulence Measurements with Inclined Hot-Wires," Boeing Scientific Research Laboratories, Flight Science Laboratory Report No.103, Boeing Company, Seattle, WA.

Collis, D.C., and Williams, M.J. (1959) "Two-Dimensional Convection from Heated Wires at Low Reynolds Number," *J. Fluid Mech.* **6**, pp. 357–384.

Comte-Bellot, G. (1976) "Hot-Wire Anemometry," *Annu. Rev. Fluid Mech.* **8**, pp. 209–231.

Comte-Bellot, G., Strohl, A., and Alcarez, E. (1971) "On Aerodynamic Disturbances Caused by a Single Hot-Wire Probe," *J. Appl. Mech.* **93**, pp. 767–774.

Corrsin, S. (1959) "Outline of Some Topics in Homogenous Turbulent Flow," *J. Geophys. Res.* **64**, pp. 2134–2150.

DeCourtye, D., Sen, M., and Gad-el-Hak, M. (1998) "Analysis of Viscous Micropumps and Microturbines," *Int. J. Comp. Fluid Dyn.* **10**, pp. 13–25.

Diem, B., Rey, P., Renard, S., Viollet, P., Bosson, M., Bono, H., Michel, F., Delaye, M., and Delapierre, G. (1995) "SOI 'SIMOX': From Bulk to Surface Micromachining, a New Age for Silicon Sensors and Actuators," *Sensors and Actuators A* **46–47**, pp. 8–16.

Döring, C., Grauer, T., Marek, J., Mettner, M.S., Trah, H.-P., and Willmann, M. (1992) "Micromachined Thermoelectrically Driven Cantilever Structures for Fluid Jet Deflection," in *Proc. IEEE MEMS '92*, pp. 12–18, Feb. 4–7, Travemunde, Germany.

Dowling, A.P., and Ffowcs-Williams, J.E. (1983) *Sound and Sources of Sound*, Wiley, New York.

Drubka, R.E., Tan-Atichat, J., and Nagib, H.M. (1977) "On Temperature and Yaw Dependence of Hot-Wires," IIT Fluid and Heat Transfer Report No. R77-1, Illinois Institute of Technology, Chicago.

Dryden, H.L., and Kuethe, A.M. (1930) "The Measurement of Fluctuations of Air Speed by Hot-Wire Anemometer," NACA Technical Report No. NACA-TR-320, Washington, D.C.

Ebefors, T., Kälvesten, E., and Stemme, G. (1997a) "Dynamic Actuation of Polyimide V-Grooves Joints by Electrical Heating," in *Eurosensors XI*, Sept. 21–24, Warsaw, Poland.

Ebefors, T., Kälvesten, E., Vieider, C., and Stemme, G. (1997b) "New Robust Small Radius Joints Based on Thermal Shrinkage of Polyimide in V-grooves," in *Proc. Ninth Int. Conf. on Solid-State Sensors and Actuators (Transducers '97)*, June 16–19, Chicago, IL.

Ebefors, T., Kälvesten, E., and Stemme, G. (1998) "New Small Radius Joints Based on Thermal Shrinkage of Polyimide in V-Grooves for Robust Self-Assembly 3-D Microstructures," *J. Micromech. Microeng.* **8**, pp. 188–194.

Eckelmann, H. (1989) "A Review of Knowledge on Pressure Fluctuations," in *Near-Wall Turbulence*, S.J. Kline and N.H. Afgan, Eds., pp. 328–347, Hemisphere, New York.

Ehrfeld, W., Götz, F., Münchmeyer, D., Schleb, W., and Schmidt, D. (1988) "LIGA Process: Sensor Construction Techniques Via X-Ray Lithography," in *Technical Digest, Solid-State Sensor and Actuator Workshop,* pp. 1–4, June 6–9, Hilton Head, SC.

Elwenspoek, M., Smith, L., and Hök, B. (1992) "Active Joints for Microrobot Limbs," *J. Micromech. Microeng.* **2**, pp. 221–223.

Epstein, A.H., and Senturia, S.D. (1997) "Macro Power from Micro Machinery," *Science* **276**, p. 1211.

Epstein, A.H., Senturia, S.D., Al-Midani, O., Anathasuresh, G., Ayon, A., Breuer, K., Chen, K.-S., Ehrich, F.F., Esteve, E., Frechette, L., Gauba, G., Ghodssi, R., Groshenry, C., Jacobson, S.A., Kerrebrock, J.L., Lang, J.H., Lin, C.-C., London, A., Lopata, J., Mehra, A., Mur Miranda, J.O., Nagle, S., Orr, D.J., Piekos, E., Schmidt, M.A., Shirley, G., Spearing, S.M., Tan, C.S., Tzeng, Y.-S., and Waitz, I.A. (1997) "Micro-Heat Engines, Gas Turbines, and Rocket Engines-The MIT Microengine Project," AIAA Paper No. 97-1773, AIAA, Reston, VA.

Esashi, M., Shoji, S., and Nakano, A. (1989) "Normally Closed Microvalve Fabricated on a Silicon Wafer," *Sensors and Actuators A* **20**, pp. 163–169.

Fabris, G. (1978) "Probe and Method for Simultaneous Measurement of 'True' Instantaneous Temperature and Three Velocity Components in Turbulent Flow," *Rev. Sci. Instrum.* **49**, pp. 654–664.

Fan, L., Wu, M.C., Choquette, K.D., and Crawford, M.H. (1997) "Self-Assembled Microactuated XYZ Stages for Optical Scanning and Alignment," in *Proc. Ninth Int. Conf. on Solid-State Sensors and Actuators (Tranducers '97),* pp. 319–322, June 16–19, Chicago, IL.

Farabee, T.M. (1986) "An Experimental Investigations of Wall Pressure Fluctuations Beneath Non-Equilibrium Turbulent Flows," David Taylor Naval Ship Research and Development Center, DTNSRDC Technical Report No. 86/047, Bethesda, MD.

Farabee, T.M., and Casarella, M. (1991) "Spectral Features of Wall Pressure Fluctuations Beneath Turbulent Boundary Layers," *Phys. Fluids* **3**, pp. 2410–2419.

Fingerson, L.M., and Freymuth, P. (1977) "Thermal Anemometers," in *Fluid Mechanics Measurements,* second ed., R.J. Goldstein, Ed., pp. 115–173, Taylor & Francis, Washington, D.C.

Franz, J. (1988) "Aufbau, Funktionsweise und technische Realisierung eines piezoelectrischen Siliziumsensors für akustische Grössen," VDI-Berichte no. 677, pp. 299–302, Germany.

Freymuth, P. (1977) "Frequency Response and Electronic Testing for Constant-Temperature Hot-Wire Anemometers," *J. Physics E: Sci. Instrum.* **10**, pp. 705–710.

Freymuth, P. (1983) "History of Thermal Anemometry," in *Handbook of Fluids in Motion,* M.P. Cheremisinoff, Ed., Ann Arbor Publishers, Ann Arbor, MI.

Freymuth, P. (1992) "A Bibliography of Thermal Anemometry," TSI Inc., St. Paul, MN.

Fuhr, G., Hagedorn, R., Muller, T., Benecke, W., and Wagner, B. (1992) "Microfabricated Electrohydrodynamic (EHD) Pumps for Liquids of Higher Conductivity," *J. MEMS* **1**, pp. 141–145.

Fuijta, H., and Kovasznay, L.S.G. (1968) "Measurement of Reynolds Stress by a Single Rotated Hot-Wire Anemometer," *Rev. Sci. Instrum.* **39**, pp. 1351–1355.

Fukuta, Y., Akiyama, T., and Fujita, H. (1995) "A Reshaping Technology with Joule Heat for Three-Dimensional Silicon Structures," in *Proc. Eighth Int. Conf. on Solid-State Sensors and Actuators (Tranducers '95),* pp. 174–177, June 25–29, Stockholm, Sweden.

Fukuta, Y., Collard, D., Akiyama, T., Yang, E.H., and Fujita, H. (1997) "Microactuated Self-Assembling of 3-D Polysilicon Structures with Reshaping Technology," in *Proc. IEEE Tenth Int. Workshop on MEMS (MEMS '97),* pp. 477–481, January 26–30, Nagoya, Japan.

Gabriel, K.J., Tabata, O., Shimaoka, K., Sugiyama, S., and Fujita, H. (1992) "Surface-Normal Electrostatic/Pneumatic Actuator," in *Proc. IEEE MEMS '92,* pp. 128–131, Feb, 4–7, Travemunde, Germany.

Gad-el-Hak, M. (1994) "Interactive Control of Turbulent Boundary Layers: A Futuristic Overview," *AIAA J.* **32**, pp. 1753–1765.

Gad-el-Hak, M. (1998) "Frontiers of Flow Control," in *Flow Control: Fundamentals and Practices,* M. Gad-el-Hak, A. Pollard, and J.-P. Bonnet, Eds., Lecture Notes in Physics, Vol. 53, pp. 109–153, Springer-Verlag, Berlin.

Gad-el-Hak, M. (1999) "The Fluid Mechanics of Microdevices—The Freeman Scholar Lecture," *J. Fluids Eng.* **121**, pp. 5–33.

Gad-el-Hak, M. (2000) *Flow Control: Passive, Active, and Reactive Flow Management,* Cambridge University Press, London.

Gad-el-Hak, M. and Bandyopadhyay, P.R. (1994) "Reynolds Number Effect in Wall Bounded Flows," *Appl. Mech. Rev.* **47**, pp. 307–365.

Gad-el-Hak, M., Pollard, A., and Bonnet, J.-P. (1998) *Flow Control: Fundamentals and Practices,* Springer-Verlag, Berlin.

Garcia, E.J. (1998) "Micro-Flex Mirror and Instability Actuation Technique," in *Proc. IEEE Eleventh Int. Workshop on MEMS (MEMS '98),* pp. 470–475, Jan. 25–29, Heidelberg, Germany.

Gedney, C.J., and Leehey, P. (1989) "Wall Pressure Fluctuations During Transition on a Flat Plate," in *Proc. Symp. on Flow Induced Noise Due to Laminar-Turbulent Transition Process,* ASME NCA-Vol. 5, American Society of Mechanical Engineers, New York.

Goldberg, H.D., Breuer, K.S., and Schmidt, M.A. (1994) "A Silicon Wafer-Bonding Technology for Microfabricated Shear-Stress Sensors with Backside Contacts," in *Technical Digest, Solid-State Sensor and Actuator Workshop,* pp. 111–115, Hilton Head, SC.

Goldstein, M.E. (1976) *Aeroacoustics,* McGraw-Hill, New York.

Goldstein, R.J. (1996) *Fluid Mechanics Measurements,* second ed., Taylor & Francis, Washington, D.C.

Green, P.W., Syms, R.R.A., and Yeatman, E.M. (1996) "Demonstration of Three-Dimensional Microstructure Self-Assembly," *J. MEMS* **4**, pp. 170–176.

Guckel, H. (1987) "Fine Grained Films and Its Application to Planar Pressure Transducers," in *Proc. Fourth Int. Conf. on Solid-State Sensors and Actuators (Transducers '87),* pp. 277–282, June 2–5, Tokyo, Japan.

Guckel, H. (1991) "Surface Micromachined Pressure Transducers," *Sensors Actuators A* **28**, pp. 133–146.

Gunawan, D.S., Lin, L.Y., Lee, S.S., and Pister, K.S.J. (1995) "Micromachined Free-Space Integrated Micro-Optics," *Sensors and Actuators A* **50**, pp. 127–134.

Hakkinen, R.J. (1991) "Survey of Skin Friction Measurements Techniques," AIAA Minisymposium, Dayton, OH.

Hallbäck, M. (1993) "Development of Reynolds Stress Closures of Homogenous Turbulence through Physical and Numerical Experiments," Ph.D. thesis, Department of Mechanics, Royal Institute of Technology, Stockholm, Sweden.

Hanratty, T.J., and Campbell, J.A. (1996) "Measurement of Wall-Shear Stress," in *Fluid Mechanics Measurements,* second ed., R. Goldstein, Ed., pp. 575–640, Taylor & Francis, Washington, D.C.

Harrison, M. (1958) "Pressure Fluctuations on the Wall Adjacent to Turbulent Boundary Layers," David Taylor Naval Ship Research and Development Center, DTNSRDC Tech. Rep. No. 1260, Bethesda, MD.

Haritonidis, J.H. (1989) "The Measurements of Wall-Shear Stress," in *Advances in Fluid Mechanics Measurements,* M. Gad-el-Hak, Ed., pp. 229–261, Springer-Verlag, Berlin.

Hinze, J.O. (1975) *Turbulence,* second ed., McGraw-Hill, New York.

Ho, C.-M., and Tai, Y.-C. (1996) "Review: MEMS and Its Applications for Flow Control," *J. Fluids Eng.* **118**, pp. 437–447.

Ho, C.-M., and Tai, Y.-C. (1998) "Micro-Electro-Mechanical Systems (MEMS) and Fluid Flows," *Annu. Rev. Fluid Mech.* **30**, pp. 579–612.

Ho, C.-M., Tung, S., Lee, G.B., Tai, Y.-C., Jiang, F., and Tsao, T. (1997) "MEMS-A Technology for Advancements in Aerospace Engineering," AIAA Paper No. 97-0545, AIAA, Reston, VA.

Hoffman, E., Warneke, B., Kruglick, E., Weigold, J., and Pister, K.S.J. (1995) "3D Structures with Pizoresistive Sensors in Standard CMOS," in *Proc. IEEE Eighth Int. Workshop on MEMS (MEMS '95),* pp. 288–293, Jan. 29–Feb. 2, Amsterdam, The Netherlands.

Hohm, D. (1985) "Subminiatur-Silizium-Kondensatormikrofon, Fortschritte der Akustik," in *DAGA '85,* pp. 847–850.

Hohm, D., and Hess, G. (1989) "A Subminiature Condenser Microphone with Silicon-Nitride Membrane and Silicon Backplate," *J. Acoustical Soc. of Am.* **85**, pp. 476–480.

Holmes, P., Lumley, J., and Berkooz, G. (1996) *Turbulence, Coherent Structures, Dynamical Systems and Symmetry,* Cambridge University Press, London.

Huang, J.-B., Tung, S., Ho, C.-M., Liu, C., and Tai, Y.-C. (1996) "Improved Micro Thermal Shear-Stress Sensor," *IEEE Trans. Instrum. Measure.* **45**(2).

Israelachvili, J.N. (1991) *Intermolecular and Surface Forces,* second ed., Academic Press, New York.

Jacobson, S.A., and Reynolds, W.C. (1995) "An Experimental Investigation Towards the Active Control of Turbulent Boundary Layers," Department of Mechanical Engineering Report No. TF-64, Stanford University, Stanford, CA.

Jeffery, G.B. (1920) "Plane Stress and Plane Strain in Bipolar Co-ordinates," *Phil. Trans. Roy. Soc. A* **221**, pp. 265–289.

Jiang, F., Tai, Y.-C., Ho, C.-M., and Li, W.J. (1994) "A Micromachined Polysilicon Hot-Wire Anemometer," in *Solid-State Sensor and Actuator Workshop,* pp. 264–267, Hilton Head, SC.

Jiang, F., Tai, Y.-C., Gupta, B., Goodman, R., Tung, S., Huang, J., and Ho, C.-M. (1996) "A Surface Micromachined Shear-Stress Imager," in *Proc. IEEE Ninth Int. Workshop on MEMS,* pp. 110–115, February, San Diego, CA.

Jiang, F. Tai, Y.-C., Walsh, K., Tsao, T., Lee, G.B., and Ho, C.-H. (1997) "A Flexible MEMS Technology and Its First Application to Shear Stress Skin," in *Proc. IEEE MEMS Workshop (MEMS '97),* pp. 465–470.

Jörgensen, F.E. (1971) "Directional Sensitivity of Wire and Fiber Film Probes," *DISA Inf.* **11**, pp. 31–37.

Kälvesten, E. (1994) "Piezoresistive Silicon Microphones for Turbulence Measurements," Royal Institute of Technology, TRITA–IL-9402, Stockholm, Sweden.

Kälvesten, E. (1996) "Pressure and Wall Shear Stress Sensors for Turbulence Measurements," Royal Institute of Technology, TRITA-ILA-9601, Stockholm, Sweden.

Kälvesten, E., Löfdahl, L., and Stemme, G. (1995) "Small Piezoresistive Silicon Microphones Specially Designed for the Characterization of Turbulent Gas Flows," *Sensors and Actuators A* **46–47**, pp. 151–155.

Kälvesten, E., Löfdahl, L., and Stemme, G. (1996a) "An Analytical Characterization of a Piezoresistive Square Diaphragm Silicon Microphone," *Sensors Mater.* **8**, pp. 113–136.

Kälvesten, E., Vieider, C., Löfdahl, L., and Stemme, G. (1996b) "An Integrated Pressure-Velocity Sensor for Correlation Measurements in Turbulent Gas Flows," *Sensors and Actuators A* **52**, pp. 51–58.

Kamal, M.M. (1966) "Separation in the Flow Between Eccentric Rotating Cylinders," *Trans. ASME Series D* **88**, pp. 717–724.

Keith, W., Hurdis, D., and Abraham, B. (1992) "A Comparison of Turbulent Boundary Layer Wall-Pressure Spectra," *J. Fluids Eng.* **114**, pp. 338–347.

Keller, C.G., and Howe, R.T. (1995) "Nickel-Filled Hexsil Thermally Actuated Tweezers," in *Eighth Int. Conf. on Solid-State Sensors and Actuators and Eurosensors-9,* Digest of Technical Papers, Vol. 2, pp. 376–379, Stockholm, Sweden.

Kim, E.S., Kim, J.R., and Muller, R.S. (1991) "Improved IC-Compatible Piezoelectric Microphones and CMOS Process," *Proc. Sixth Int. Conf. on Solid-State Sensors and Actuators (Transducers '91),* June 24–28, San Fransisco, CA.

King, L.V. (1914) "On the Convection of Heat from Small Cylinders in a Stream of Fluid: Determination of the Convection of Small Platinum Wires with Applications to Hot-Wire Anemometry," *Phil. Trans. Roy. Soc. London* **214**, pp. 373–432.

King, L.V. (1915) "On the Precition Measurement of Air Velocity by Means of the Linear Hot-Wire Anemometer," *Phil. Mag. and J. Sci. Sixth Series* **29**, pp. 556–577.

King, L.V. (1916) "The Linear Hot-Wire Anemometer and Its Applications in Technical Physics," *J. Franklin Inst.* **181**, pp.1–25.

Kolmogorov, A.N. (1941a) "Local Structure of Turbulence in Incompressible Fluid at Very High Reynolds Numbers," *Dokl. Akad. Nauk. SSSR* **30**, pp. 299–303.

Kolmogorov, A.N. (1941b) "Energy Dissipation in Locally Isotropic Turbulence," *Dokl. Akad. Nauk. SSSR* **32**, pp. 19–21.

Kovacs, A., and Stoffel, A. (1995) "Integrated Condenser Microphone with Polysilicon Electrodes," in *MME '95, Micromechanics Europe,* pp. 132–135, Sept. 3–5, Copenhagen, Denmark.

Kovacs, G.T.A. (1998) *Micromachined Transducers Sourcebook*, McGraw-Hill, New York.

Kramers, H. (1946) "Heat Transfer from Spheres to Flowing Media," *Physica* **12**, pp. 61–80.

Kronast, W., Muller, B., and Stoffel, A. (1995) "Miniaturized Single-Chip Silicon Membrane Microphone with Integrated Field-Effect Transistor," in *MME '95, Micromechanics Europe*, pp. 136–139, Sept. 3–5, Copenhagen, Denmark.

Kruglick, E.J.J., Warneke, B.A., and Pister, K.S.J. (1998) "CMOS 3-Axis Accelerometers with Integrated Amplifier," in *Proc. IEEE Eleventh Int. Workshop on MEMS (MEMS '98)*, pp. 631–636, Jan. 25–29, Heidelberg, Germany.

Kuhnel, W. (1991) "Silicon Condenser Microphone with Integrated Field-Effect Transistor," *Sensors and Actuators A* **25–27**, pp. 521–525.

Kuhnel, W., and Hess, G. (1992) "Micromachined Subminiature Condenser Microphones in Silicon," *Sensors and Actuators A* **32**, pp. 560–564.

Lauchle, G.C. and Daniels, M.A. (1984) "Wall-Pressure Fluctuations in Turbulent Pipe Flow," *Phys. Fluids* **30**, pp. 3019–3024.

Lesieur, M. (1991) *Turbulence in Fluids*, second ed., Kluwer, Dordrecht.

Leveque, M.A. (1928) "Transmission de Chaleur par Convection," *Ann. Mines.* **13**, p. 283.

Liang, W.J. and Liou, J.A. (1995) "Flow Around a Rotating Cylinder Near a Plane Boundary," *J. Chinese Inst. Eng.* **18**, pp. 35–50.

Liepman H.W., and Skinner, G.T. (1954) "Shearing-Stress Measurements by Use of a Heated Element," NACA Technical Note No. 3268, Washington, D.C.

Ligrani, P.M., and Bradshaw, P. (1987) "Spatial Resolution and Measurements of Turbulence in the Viscous Sublayer Using Subminiature Hot-Wire Probes," *Exp. Fluids* **5**, pp. 407–417.

Lin, G., Kim, C.J., Konishi, S., and Fujita, H. (1995) "Design, Fabrication and Testing of a C-Shaped Actuator," in *Technical Digest, Eighth Int. Conf. on Solid-State Sensors and Actuators (Transducers '95 and Eurosensors 9)*, pp. 416–419, June 25–29, Stockholm, Sweden.

Linder, C., Paratte, L., Gretillat, M., Jaecklin, V., and de Rooij, N. (1992) "Surface Micromachining," *J. Micromech. Microeng.* **2**, pp. 122–132.

Lindgren, B., Österlund, J., and Johansson, A.V. (2000) "Flow Structures in High Reynolds Number Turbulent Boundary Layers," in *Proc. Eighth European Turbulence Conf., Adv. in Turbulence* **8**, pp. 399–402, June 27–30, Barcelona, Spain.

Ling, S.C. (1963) "Heat Transfer from a Small Isothermal Spanwise Strip on an Insulated Boundary," *J. Heat Transfer* **C85**, pp. 230–236.

Liu, C., Tsao, T., Tai, Y.-C., Liu, W., Will, P., and Ho, C.-M. (1995) "A Micromachined Permalloy Magnetic Actuator Array for Micro Robotic Assembly System," in *Proc. Eighth Int. Conf. on Solid-State Sensors and Actuators (Transducers '95 and Eurosensors 9)*, pp. 328–331, June 25–29, Stockholm, Sweden.

Löfdahl, L., Stemme, G., and Johansson, B. (1992) "Silicon Based Flow Sensors for Mean Velocity and Turbulence Measurements," *Exp. Fluids* **12**, pp. 391–393.

Löfdahl, L., Kälvesten, E., Hadzianagnostakis, T., and Stemme, G. (1994a) "An Integrated Silicon Based Pressure-Shear Stress Sensor for Measurements in Turbulence Flows," in *Symp. on Application of Micro-Fabrication to Fluid Mechanics*, ASME Winter Annual Meeting, pp. 245–253, Nov. 17–22, Atlanta, GA.

Löfdahl, L., Kälvesten, E., and Stemme, G. (1994b) "Small Silicon Based Pressure Transducers for Measurements in Turbulent Boundary Layers," *Exp. Fluids* **17**, pp. 24–31.

Löfdahl, L., Kälvesten, E., and Stemme, G. (1996) "Small Silicon Pressure Transducers for Space-Time Correlation Measurements in a Flat Plate Boundary Layer," *J. Fluids Eng.* **118**, pp. 457–463.

Lomas, C.G. (1986) *Fundamentals of Hot-Wire Anemometry*, Cambridge University Press, London.

Ludwieg, H. (1950) "Instrument for Measuring the Wall Shearing Stress of Turbulent Boundary Layers," NACA Technical Memorandum No. 1284, Washington, D.C.

Lueptov, R.M., Breuer, K.S., and Haritonidis, J.H. (1988) "Computer-Aided Calibration of X-Probes Using Look-Up Table," *Exp. Fluids* **6**, pp. 115–188.

Madou, M. (1997) *Fundamentals of Microfabrication,* CRC Press, Boca Raton, FL.

Maureau, J., Sharatchandra, M.C., Sen, M., and Gad-el-Hak, M. (1997) "Flow and Load Characteristics of Microbearings with Slip," *J. Micromech. Microeng.* **7**, pp. 55–64.

McComb, W.D. (1990) *The Physics of Fluid Turbulence,* Clarendon Press, Oxford.

Menendez, A.N. and Ramaprian, B.R. (1985) "The Use of Flush-Mounted Hot-Film Gauges to Measure Skin Friction in Unsteady Boundary Layers," *J. Fluid Mech.* **161**, pp. 139–159.

Miller, M.F., Allen, M.G., Arkilic, E.B., Breuer, K.S., and Schmidt, M.A. (1997) "Fabry–Perot Pressure Sensor Arrays for Imaging Surface Pressure Distributions," in *Proc. Int. Conf. on Solid-State Sensors and Actuators (Transducers '97),* pp. 1469–1472, June 16–19, Chicago, IL.

Miller, R., Burr, G., Tai, Y.-C., Psaltis, D., Ho, C.M., and Katti, R.R. (1996) "Electromagnetic MEMS Scanning Mirrors for Holographic Data Storage," in *Technical Digest, Solid-State Sensor and Actuator Workshop,* pp. 183–186, Hilton Head, SC, Catalog No. 96 TRF-0001.

Moin, P., and Bewley, T. (1994) "Feedback Control of Turbulence," *Appl. Mech. Rev.* **47** (6, part 2), pp. S3–S13.

Monin, A.S., and Yaglom, A.M. (1975) *Statistical Fluid Mechanics: Mechanics of Turbulence,* Vols. 1 and 2, MIT Press, Cambridge, MA.

Moroney, R.M., White, R.M., and Howe, R.T. (1991) "Ultrasonically Induced Microtransport," in *Proc. IEEE MEMS '91,* Nara, Japan, pp. 277–282, Institute of Electrical and Electronics Engineers, New York.

Muller, R.S. (1987) "Strategies for Sensor Research," in *Proc. Fourth Int. Conf. on Solid-State Sensors and Actuators (Transducers '87),* pp. 107–111, June 2–5, Tokyo, Japan.

Munson, B.H., Young, D.F., and Okiishi, T.H. (1990) *Fundamentals of Fluid Mechanics,* Wiley, New York.

Murphy, P., Hubschi, K., DeRooij, N., and Racine, C. (1989) "Subminiature Silicon Integrated Electret Microphone," *IEEE Trans. Electron Insul.* **24**, pp. 495–498.

Ng, K., Shajii, K., and Schmidt, M. (1991) "A Liquid Shear-Stress Sensor Fabricated Using Wafer Bonding Technology," in *Proc. Sixth Int. Conf. on Solid-State Sensors and Actuators (Transducers '91),* pp. 931–934, June 24–27, San Francisco, CA.

O'Connor, L. (1992) "MEMS: Micromechanical Systems," *Mech. Eng.* **114**, pp. 40–47.

Odell, G.M., and Kovasznay, L.S.G. (1971) "A New Type of Water Channel with Density Stratification," *J. Fluid Mech.* **50**, pp. 535–543.

Österlund, J. (1999) "Experimental Studies of Zero Pressure-Gradient Turbulent Boundary Layer Flow," Ph.D. thesis, Department of Mechanics, Royal Institute of Technology, Stockholm, Sweden.

Otsuka, K. (1995) "Fundamentals of Shape Memory Alloys in View of Intelligent Materials," in *Proc. Int. Symp. on Microsystems, Intelligent Materials and Robots,* pp. 225–230, Sept. 27–29, Sendai, Japan.

Oudheusden, B., and Huijsing, J. (1988) "Integrated Flow Friction Sensor," *Sensors and Actuators A* **15**, pp. 135–144.

Padmanabhan, A. (1997) "Silicon Micromachined Sensors and Sensor Arrays for Shear-Stress Measurements in Aerodynamic Flows," Ph.D. thesis, Department of Aeronautics and Astronautics, Massachusetts Institute of Technology, Cambridge, MA.

Pan, T., Hyman, D., Mehregany, M., Reshotko, E., and Willis, B. (1995) "Calibration of Microfabricated Shear Stress Sensors," in *Proc. Eighth Int. Conf. on Solid-State Sensors and Actuators (Transducers '95),* pp. 443–446, June 25–29, Stockholm, Sweden.

Patel, V.C. (1965) "Calibration of the Preston Tube and Limitations on Its Use in Pressure Gradients," *J. Fluid Mech.* **23**, pp. 185–208.

Pedley, T.J. (1972) "On the Forced Heat Transfer from a Hot-Film Embedded in the Wall in Two-Dimensional Unsteady Flow," *J. Fluid Mech.* **55**, pp. 329–357.

Perry, A.E. (1982) *Hot-Wire Anemometry,* Clarendon Press, Oxford.

Petersen, K.E. (1982) "Silicon as a Mechanical Material," *Proc. IEEE 70,* pp. 420–457, Institute of Electrical and Electronics Engineers, New York.

Piekos, E.S., Orr, D.J., Jacobson, S.A., Ehrich, F.F., and Breuer, K.S. (1997) "Design and Analysis of Microfabricated High Speed Gas Journal Bearings," AIAA Paper No. 97-1966, AIAA, Reston, VA.

Pister, K.S.J., Fearing, R.S., and Howe, R.T. (1990) "A Planar Air Levitated Electrostatic Actuator System," IEEE Paper No. CH2832-4/90/0000-0067, Institute of Electrical and Electronics Engineers, New York.

Pister, K.S.J., Judy, S.R., Burgett, S.R., and Fearing, R.S. (1992) "Micro-Fabricated Hinges," *Sensors and Actuators A* **33**, pp. 249–256.

Pope, S.B. (2000) *Turbulent Flows,* Cambridge University Press, London.

Ramaprian, B.R., and Tu, S.W. (1983) "Calibration of a Heat Flux Gage for Skin Friction Measurement," *J. Fluids Eng.* **105**, pp. 455–457.

Richter, A., Plettner, A., Hofmann, K.A., and Sandmaier, H. (1991) "A Micromachined Electrohydrodynamic (EHD) Pump," *Sensors and Actuators A* **29**, pp. 159–168.

Ried, R., Kim, E., Hong, D., and Muller, R. (1993) "Pizoelectric Microphone with On-Chip CMOS Circuits," *J. MEMS* **2**, pp. 111–120.

Riethmuller, W. and Benecke, W. (1988) "Thermally Exited Silicon Microactuators," *IEEE Trans. Electron. Devices* **35**, pp. 758–763.

Robinson, S.K. (1991) "Coherent Motions in the Turbulent Boundary Layer," *Annu. Rev. Fluid Mech.* **23**, pp. 601–639.

Rossi, M. (1988) *Acoustics and Electroacoustics,* Artech House, Boston, MA.

Royer, M., Holmen, P., Wurm, M., Aadland, P., and Glenn, M. (1983) "ZnO on Si Integrated Acoustic Sensor," *Sensors and Actuators A* **4**, pp. 357–362.

Scheeper, P., van der Donk, A., Olthuis, W., and Bergveld, P. (1992) "Fabrication of Silicon Condenser Microphones Using Silicon Wafer Technology," *J. MEMS* **1**, pp. 147–154.

Scheeper, P., van der Donk, A., Olthuis, W., and Bergveld, P. (1994) "A Review of Silicon Microphones," *Sensors and Actuators A* **44**, pp. 1–11.

Schellin, R., and Hess, G. (1992) "A Silicon Microphone Based on Piezoresistive Polysilicon Strain Gauges," *Sensors and Actuators A* **32** pp. 555–559.

Schellin, R., Pedersen, M., Olthuis, W., Bergveld, P., and Hess, G. (1995) "A Monolithically Integrated Silicone Microphone with Piezoelectric Polymer Layers," in *MME '95, Micromechanics Europe,* pp. 217–220, Sept. 3–5, Copenhagen, Denmark.

Schewe, G. (1983) "On the Structure and Resolution of Wall-Pressure Fluctuations Associated with Turbulent Boundary-Layer Flow," *J. Fluid Mech.* **134**, pp. 311–328.

Schmidt, M., Howe, R., Senturia, S., and Haritonidis, J. (1988) "Design and Calibration of a Microfabricated Floating-Element Shear-Stress Sensor," *IEEE Trans. Electron. Devices* **35**, pp. 750–757.

Sen, M., Wajerski, D., and Gad-el-Hak, M. (1996) "A Novel Pump for MEMS Applications," *J. Fluids Eng.* **118**, pp. 624–627.

Sessler, G. (1991) "Acoustic Sensors," *Sensors and Actuators A* **25–27**, pp. 323–330.

Sharatchandra, M.C., Sen, M., and Gad-el-Hak, M. (1997) "Navier-Stokes Simulations of a Novel Viscous Pump," *J. Fluids Eng.* **119**, pp. 372–382.

Sharatchandra, M.C., Sen, M., and Gad-el-Hak, M. (1998a) "Thermal Aspects of a Novel Micropumping Device," *J. Heat Transfer* **120**, pp. 99–107.

Sharatchandra, M.C., Sen, M., and Gad-el-Hak, M. (1998b) "A New Approach to Constrained Shape Optimization Using Genetic Algorithms," *AIAA J.* **36**, pp. 51–61.

Sherman, F.S. (1990) *Viscous Flow,* McGraw-Hill, New York.

Smela, E., Inganäs, O., and Lundström, I. (1995) "Controlled Folding of Micrometer-Size Structures," *Science* **268**, pp. 221–223.

Smits, J.G. (1990) "Piezoelectric Micropump with Three Valves Working Peristaltically," *Sensors and Actuators A* **21–23**, pp. 203–206.

Solgaard, O., Daneman, M., Tien, N.C., Friedberger, A., Muller, R.S., and Lau, K.Y. (1995) "Optoelectronic Packing Using Silicon Surface-Micromachined Alignment Mirrors," *IEEE Photon. Tech. Lett.* **7**, pp. 41–43.

Speziale, C.G. (1991) "Analytical Methods for the Development of Reynolds Stress Closures in Turbulence," *Annu. Rev. Fluid Mech.* **23**, pp. 107–157.

Speziale, C.G., Sarkar, S., and Gatski, T.B. (1991) "Modelling the Pressure Strain Correlation of Turbulence: An Invariant Dynamical System Approach," *J. Fluid Mech.* **227**, pp. 245–272.

Sprenkels, A.J. (1988) "A Silicon Subminiature Electret Capacitive Microphone," Ph.D. thesis, University of Utrecht, Utrecht, The Netherlands.

Sprenkels, A.J., Groothengel, R.A., Verloop, A.J., and Bergveld, P. (1989) "Development of an Electret Microphone in Silicon," *Sensors and Actuators A* **17**, pp. 509–512.

Stanton, T.E., Marshall, D., and Bryant, C.N. (1920) "On the Conditions at the Boundary of a Fluid in Turbulent Motion," *Proc. Roy. Soc. London A* **97**, pp. 413–434.

Sugiyama, S., Shimaoka, K., and Tabata, O. (1993) "Surface Micromachined Microdiaphragm Pressure Sensor," *Sensors and Materials* **4**, pp. 238–243.

Suzuki, K., Shimoyama, I., Miuara, H., and Ezura, Y. (1992) "Creation of an Insect-Based Microrobot with an External Skeleton and Elastic Joints," in *Proc. IEEE Fifth Int. Workshop MEMS (MEMS '92)*, pp. 190–195, Feb. 4–7, Travemunde, Germany.

Syms, R.R.A. (1998) "Demonstration of Three-Dimensional Microstructure Self-Assembly," *Sensors and Actuators A* **65**, pp. 238–243.

Tardu, S., Pham, C.T., and Binder, G. (1991) "Effects of Longitudinal Diffusion in the Fluid and Heat Conduction to the Substrate on Response of Wall Hot-Film Gauges," in *Advances in Turbulence*, Vol 3, A.V. Johansson and P.H. Alfredsson, Eds., Springer-Verlag, Berlin.

Taylor, G.I. (1935) "Statistical Theory of Turbulence," *Proc. Roy. Soc. London A* **151**, pp. 421–478.

Taylor, G. (1972) "Low-Reynolds-Number Flows," in *Experiments in Fluid Mechanics*, pp. 47–54, MIT Press, Cambridge, MA.

Tennekes, H., and Lumley, J.L. (1972) *A First Course in Turbulence*, MIT Press, Cambridge, MA.

Tien, N.C. (1997) "Silicon Micromachined Thermal Sensors and Actuators," *Microscale Thermophys. Eng.* **1**, pp. 275–292.

Tien, N.C., Kiang, M.H., Daneman, M.J., Solgaard, O., Lau, K.Y., and Muller, R.S. (1996a) "Actuation of Polysilicon Surface-Micromachined Mirrors," in *SPIE Proc. Miniaturized System with Micro-Optical and Micromechanics*, Vol. 2687, pp. 53–59, Jan. 30–31, San Jose, CA.

Tien, N.C., Solgaard, O., Kiang, M.H., Daneman, M.J., Lau, K.Y., and Muller, R.S. (1996b) "Surface Micromachined Mirrors for Laser Beam Positioning," *Sensors and Actuators A* **52**, pp. 76–80.

Timoshenko, S. (1925) "Analysis of Bi-Metal Thermostats," *J. Optical Soc. Am.* **11**, pp. 233–255.

Townsend, A.A. (1976) *The Structure of Turbulent Shear Flows*, second ed., Cambridge University Press, London.

Tsao, T., Liu, C., Tai, Y.-C., and Ho, C.-M. (1994) "Micromachined Magnetic Actuator for Active Flow Control," in *ASME FED-197*, pp. 31–38, American Society of Mechanical Engineers, New York.

Van Lintel, H.T.G., Van de Pol, F.C.M., and Bouwstra, S. (1988) "A Piezoelectric Micropump Based on Micromachining of Silicon," *Sensors and Actuators* **15**, pp. 153–167.

Voorthuysen, J.A., Bergveld, P., and Sprenkels, A.J. (1989) "Semiconductor-Based Electret Sensors for Sound and Pressure," *IEEE Trans. Electr. Insul.* **24**, pp. 267–276.

Wannier, G.H. (1950) "A Contribution to the Hydrodynamics of Lubrication," *Q. Appl. Math.* **8**, pp. 1–32.

White, F.M. (1991) *Viscous Fluid Flow*, second ed., McGraw-Hill, New York.

Wilkinson, S.P. (1990) "Interactive Wall Turbulence Control," in *Viscous Drag Reduction in Boundary Layers*, D.M. Bushnell and J.N. Hefner, Eds., pp. 479–509, AIAA, Washington, D.C.

Willmarth, W.W., and Wooldridge, C.E. (1962) "Measurements of the Fluctuating Pressure at the Wall Beneath a Thick Turbulent Boundary Layer," *J. Fluid Mech.* **14**, pp. 187–210.

Wiltse, J.M., and Glezer, A. (1993) "Manipulation of Free Shear Flows Using Piezoelectric Actuators," *J. Fluid Mech.* **249**, pp. 261–285.

Winter, K. (1977) "An Outline of the Techniques Available for the Measurement of Skin Friction in Turbulent Boundary Layers," *Prog. Aero. Sci.* **18**, pp. 1–55.

Wu, M.C., Lin, L.Y., Lee, S.S., and Pister, K.S.J. (1995) "Micromachined Free-Space Integrated Micro-Optics," *Sensors and Actuators A* **50**, pp.127–134.

Yang, X, Grosjean, C., Tai, Y.-C., and Ho, C.-M., (1997) "A MEMS Thermopneumatic Silicone Membrane Valve," in *Proc. IEEE Tenth Int. Workshop MEMS (MEMS '97)*, Jan. 26–30, Nagoya, Japan.

Yasuda, T., Shimonoyama, I., and Mirura, H. (1997) "CMOS Drivable Electrostatic Microactuator with Large Deflection," in *Proc. IEEE Tenth Int. Workshop MEMS (MEMS '97)*, pp. 90–95, Jan. 26–30, Nagoya, Japan.

27
Surface-Micromachined Mechanisms

Andrew D. Oliver
Sandia National Laboratories

David W. Plummer
Sandia National Laboratories

27.1 Introduction 27-1
27.2 Material Properties and Geometric Considerations..... 27-2
    Stress and Strain • Young's Modulus • Poisson's Ratio • Contact Stresses • Stress in Films and Stress Gradients • Wear • Stiction
27.3 Machine Design 27-7
    Compliance Elements—Columns, Beams and Flexures • Columns • Beams • Stress Concentration • Cantilever Beam Springs • Fixed Beam Springs • Flexures • Springs in Combinations • Buckling
27.4 Applications 27-18
    Microengine • Countermeshing Gear Discriminator • Micro-Flex Mirror
27.5 Failure Mechanisms in MEMS 27-27
    Vertical Play and Mechanical Interference in Out-of-Plane Structures • Electrical Shorts • Lithographic Variations • Methods of Increasing the Reliability of Mechanisms
    Acknowledgments 27-33

27.1 Introduction

Surface-micromachining technologies have offered the following advantages to mechanism users and designers: smaller machines, different physical effects that dominate at the microscale, and reduced assembly costs. The advantages of smaller machines are sometimes very important. For example, small machines are important in aviation and space applications where a decrease in size and weight corresponds to an increased range or a reduction in the amount of fuel required for a given mission. The advantages of the different physical effects that dominate at the microscale are less obvious. Mechanisms that operate at high frequencies can greatly benefit (or suffer) from the reduced influence of inertia (typical of surface-micromachined devices) if the aim is to start and stop quickly. Smaller scales also mean surface-micromachined mechanisms are more resistant to shock and vibration than macrosized mechanisms because the component strength decreases as the square of the dimensions while the mass and inertia decrease as the cube of the dimensions. Another difference is that forces, such as van der Waals forces and electrostatic attraction, are much more important at the microscale than at the macroscale. Reduced assembly costs are another advantage of surface-micromachined mechanisms. Surface micromachining in most cases allows the creation of machines that are assembled at the same time as their constituent components. Instead of using skilled workers to assemble intricate mechanisms by hand or investing in complicated machinery, the assembly is done as a batch process during the integrated-circuit-derived fabrication process. Preassembly does impose certain limitations on the designer, such as the inability to build devices with as-fabricated stored mechanical energy. Instead, structures such as springs must have energy mechanically added to them. Preassembly also requires structures that operate out of the plane of the substrate be erected prior to use.

Perhaps the greatest disadvantage of surface-micromachined mechanisms is the limit on the number and type of layers available to a designer. Usually only one structural material with a restricted number of layers (usually two or three) is available to the designer. The thickness of the layers is restricted by the limited deposition rate of equipment and stress in the thin films. The uninterrupted span of a structure in the plane of the substrate is limited by the speed of the release etch. Having only one structural material also leads to tribology issues. The friction and wear created by like materials rubbing on each other are one of the greatest problems in surface-micromachined mechanisms.

The intent of this chapter is to provide theoretical and practical information to designers of surface-micromachined mechanisms. The design of surface-micromachined mechanisms has much in common with the design of macrosized mechanisms but there are also many differences. For example, the function of four bar linkages is similar for both macro- and microsized but the joints and connections between the linkages differ at the microscale. Rotating joints are commonly used in macrosized mechanisms, and flexible joints are more common in surface-micromachined mechanisms. The design of surface-micromachined mechanisms requires a good understanding of the fabrication process as well as an appreciation of the unique advantages of designing in the microdomain.

This chapter covers some of the limitations inherent in the design and performance of surface-micromachined mechanisms. It also covers several advantages. Some of the basics of the mechanics of materials, with an emphasis on those concepts important in the design of simple surface-micromachines, are also described. The chapter concludes by describing some failure modes in microelectromechanical systems (MEMS) mechanisms.

27.2 Material Properties and Geometric Considerations

Loads and forces acting on a body or structure will cause it to undergo a change in shape. Most engineering materials will return to their original shape after the load is removed, providing that the load is not too large. If the body returns to its original shape after the load is removed, it has experienced elastic deformation. If the load exceeds the elastic limit, the body will not completely return to its original shape after the load is removed. If the body retains some permanent change in shape, it has experienced plastic or nonelastic deformation.

Most mechanisms are designed so that the expected loads cause only elastic deformation. In fact, plastic deformation is usually deemed to be mechanism failure. The designer must be able to quantify the threshold between elastic and plastic deformation. The equations typically used in design are based on certain assumptions about the material and its geometric configuration. The theory of elasticity contains expressions for relating deformation and load that do not require simplifying assumptions, but those are usually too mathematically complex for designers to use.

One assumption made in the derivation of simple design equations is that the material is isotropic. An isotropic material is one whose elastic properties are the same in all directions. The designer is cautioned about blindly accepting the assumption of isotropy. Single-crystal silicon [Worthman and Evans, 1965], individual crystals of polysilicon [Madou, 1997] and electroformed materials in the LIGA process [Sharpe et al., 1997] are known to show nonisotropic behavior.

A second common assumption is homogeneity. A homogeneous body is one that has the same properties throughout its entire extent. Actually, no material is homogeneous. Materials consist of an aggregate of very small crystals spread out through the body. When a cross section of the body contains a large number of crystals, the effects of forces and loads are spread out and shared. In this case, the assumption of homogeneity is fulfilled. Again, the designer is cautioned about accepting this assumption. With microfabrication techniques, mechanism dimensions can approach the size scale of the crystals. In those cases, the critical cross section may contain only a few crystals. The effect of the load cannot be averaged and nonhomogeneity occurs. With the cautions described above, the equations to be presented are the best starting point for mechanism designers.

27.2.1 Stress and Strain

When a body supports a load, the material is under stress. Stress is a measure of the intensity of the body's reaction to the load. Stress is a field quantity. Like most fields, the solution for stress can be obtained by the use of cutting planes and exposed area. Stress is measured as the force per unit of exposed cross-sectional area. In SI units, stress is measure in Pascals (Newtons per square meter).

In many instances, the distribution of stress across the area is ignored so that only the average stress is considered. This is especially true in the cases of axial loads along slender bodies that cause only lengthening or shortening. The average stress, σ, is equal to the load divided by the cross-sectional area:

$$\sigma = \frac{F}{A} \tag{27.1}$$

Stress that tends to lengthen a body is called *tension*; stress that tends to shorten a body is called *compression*.

The deformation, δ, of a body under load is dependent on the size and shape of the body. The amount of deformation normalized by the dimensions of the body is called *strain*, ε. Strain is represented mathematically as:

$$\varepsilon = \frac{\delta}{L} \tag{27.2}$$

27.2.2 Young's Modulus

In a body undergoing elastic deformation in the normal direction, the ratio of stress to strain is known as the Young's modulus, E, which is also called the modulus of elasticity:

$$E = \frac{\sigma}{\varepsilon} \tag{27.3}$$

If the body is undergoing shear, the corresponding proportionality constant is G, the modulus of rigidity or the shear modulus of elasticity. Both of these proportionality constants have units of Pascals (Pa). These material properties describe the stiffness of the material. A compliant material, such as rubber, has a small Young's modulus, 0.5 GPa; a hard material, such as stainless steel (18–8), has a large Young's modulus, 190 GPa; and a single-crystal diamond has a Young's modulus of 1035 GPa.

27.2.3 Poisson's Ratio

Poisson's ratio, υ, describes the ratio of the transverse strain to the axial strain when a body is subjected to an axial load:

$$\upsilon = -\frac{\varepsilon_{transverse}}{\varepsilon_{axial}} \tag{27.4}$$

This can be visualized by thinking about pieces of rubber. If a rubber band is stretched (a positive axial strain), then the rubber band becomes narrower, which is a negative transverse strain. Conversely, if a pencil eraser is compressed, then it becomes wider. In most cases, υ has a value between 0.2 and 0.5. A value of 0.5 corresponds to a conservation of volume.

In situations where the shear modulus is not known, it can be estimated using E and υ in the following relationship:

$$E = 2G(1 + \upsilon) \tag{27.5}$$

if the materials are assumed to be isotropic.

27.2.4 Contact Stresses

When two bodies are pressed together, the point or line of contact expands to area contact. As the surfaces move together, their progress is interrupted when an asperity from one body contacts an asperity from the other body. These asperities attempt to carry the entire contact force, but quickly yield and plastically deform. This allows the bodies to move closer together until more asperities touch. This process of contact and yield continues until there is enough total contact area to support the force at the yield strength of the material. These regions of contact, called *a-spots*, typically form a circular pattern known as the Hertzian stress circle.

If the contacting surfaces have a radius of curvature, the radius of the resulting Hertzian stress circle can be calculated using the equation originally developed by Hertz:

$$a = \sqrt[3]{\frac{3F}{4} \times \frac{(1-v_1^2)/E_1 + (1-v_2^2)/E_2}{(1/r_1 + 1/r_2)}} \tag{27.6}$$

where F is the force of contact, v is Poisson's ratio, E is Young's modulus and r is the radii of curvature for the two contacting surfaces [Johnson, 1985]. For the case of a curved surface contacting a plane, the radius of curvature of the plane is equal to infinity. It is a good design practice to ensure that the apparent contact area (determined by the nominal dimensions) is larger than the Hertzian contact area. A more detailed description can be found in Roark and Young (1989).

27.2.5 Stress in Films and Stress Gradients

One of the important nonidealities in designing surface-micromachined mechanisms is stress in the thin films. Stress has different sources and different effects depending on the application and the geometries involved. One of the sources of stress in thin films is the difference in the thermal expansion coefficients between the film and the substrate, especially if the film was deposited at an elevated temperature. Other sources include the doping of semiconductor films if the dopant atoms are a different size than the host atoms and the method of film deposition. An example of stress in thin films is thermally grown silicon dioxide (SiO_2) that is in compression because the silicon dioxide molecules are larger than the lattice spacing of the host silicon atoms.

There are also many effects of stresses in thin films. Excessive stress can result in cracking or delamination of thin films. Stress in films can also be relieved as strain or deformed microstructures. For example, a compressive stress will buckle a clamped–clamped beam if the stress is above a certain level. This will be discussed later in the section on buckling. Another effect of stress is that it will change the frequency of resonant devices. For a lateral resonator with four cantilever beam springs, Madou (1997) discusses the effect of residual stress on resonant frequency. A drawing of the device is shown in Figure 27.1 and its resonant frequency is given by:

$$f_0 \approx \frac{1}{2\pi}\sqrt{\frac{4Etb^3}{mL^3} + \frac{24\sigma_r tb}{5mL}} \tag{27.7}$$

where m is the mass of the resonator plate, E is the Young's modulus, t is the flexure thickness, L is the flexure length, b is the width and σ_r is the residual stress in the flexures. The assumption in this equation is that the mass of the springs is negligible compared to the mass of the resonator plate.

Example

Find the resonant frequency of the device in Figure 27.1. Then recalculate the resonant frequency assuming that the residual stress increases by 30%. Assume the device is constructed of polycrystalline silicon with a Young's modulus of 155 GPa and a density of 2.33×10^3 kg/m^3. The support beams are

Surface-Micromachined Mechanisms

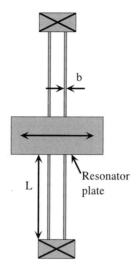

FIGURE 27.1 Lateral resonator. The "×" at each end of the device depicts an anchor and the mass is at the center with an arrow depicting the direction of motion.

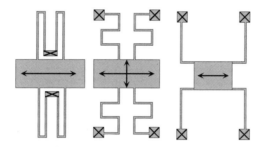

FIGURE 27.2 Three flexure designs. From left to right they are folded flexures, meandering flexures and crab-leg flexures.

2.25 μm thick, 2 μm wide and 200 μm long. The shuttle is 2.25 μm thick, 120 μm wide and 250 μm long. The residual stress is 50 MPa.

The first step is to calculate the mass of the shuttle:

$$\text{mass} = (2.25 \times 10^{-6}\,\text{m})(1.2 \times 10^{-4}\,\text{m})(2.5 \times 10^{-4}\,\text{m}) \times 2.33 \times 10^{3}\,\text{kg/m}^{3} = 160 \times 10^{-12}\,\text{kg}$$

The resonant frequency is equal to:

$$f_0 = \frac{1}{2\pi}\sqrt{\frac{4 \times 155\,\text{GPa} \times 2.25\,\mu\text{m} \times (2\,\mu\text{m})^3}{160 \times 10^{-12}\,\text{kg} \times (200\,\mu\text{m})^3} + \frac{24 \times 50\,\text{MPa} \times 2.25\,\mu\text{m} \times 2\,\mu\text{m}}{5 \times 160 \times 10^{-12}\,\text{kg} \times 200\,\mu\text{m}}} = 33\,\text{KHz}$$

If the residual stress increases by 30% (not a huge change even for a production facility), then the resonant frequency shifts to 37 kHz. Notice that designers need to design their machines to be resistant to changes in the fabrication process.

One method of relieving stresses in thin-film mechanical elements is the use of folded flexures. Folded flexures relieve axial stress because each flexure is free to expand or contract in the axial direction. This eliminates most of the stress caused by the fabrication process and by thermal expansion mismatches between the flexures and the substrate. Crab-leg flexures and meandering flexures also relieve some of the residual stress. All three flexure types are shown in Figure 27.2.

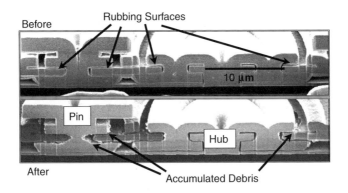

FIGURE 27.3 A pin joint and hub joint before and after accelerated wear. (Courtesy of D. Tanner, Sandia National Laboratories.)

Another nonideality of MEMS is stress gradients in the films. Stress gradients are variations in the stress of the film through the thickness of the film and cause cantilevers to either curl away from or toward the substrate. This can be partially compensated for by folding the flexures. Stress gradients will also warp thin plates or gears into dome shapes and cause spiral-shaped cantilever beams to deform [Fan et al., 1990].

27.2.6 Wear

One of the limitations of many types of mechanisms, including surface-micromachined mechanisms, is wear. This is not surprising as surface-micromachined mechanisms can have touching, rubbing or impacting surfaces and both surfaces are typically constructed of the same material. This is different than conventional machines, which commonly use dissimilar materials to reduce wear. Also, surface-micromachines typically run without liquid lubrication. A great deal of research into the wear of polysilicon surface micromachines has been done by Tanner (2000) and Dugger et al. (1999). They report that the wear of polysilicon surface micromachines has similarities to the wear of macroscopic mechanisms. Friction and wear have been correlated with the normal forces on a surface akin to the coefficient of friction concept. The greater the force, the higher the wear rate. Also, the wear rate was found to increase at operating frequencies above the resonant frequency. It has been theorized that wear is caused by adhesion and stick-slip behavior between two moving surfaces. The accumulation of wear debris that results in three-body wear (the two surfaces and the wear debris) has also been observed in surface-micromachined mechanisms. Figure 27.3 shows the effects of extreme, accelerated wear on a surface-micromachined pin joint and a hub. Another source of information on the tribology of surface-micromachined devices is Chapter 16 of the *Handbook of Micro/Nano Tribology*, 2nd ed., by Bhushan (1999).

27.2.7 Stiction

Potentially, one of the most destructive influences on the performance of MEMS mechanisms is stiction, which is the unintended adhesion of surface-micromachined parts to each other or to the substrate. It is a combination of various effects (including capillary forces, van der Waals forces, electrostatic attraction and hydrogen bonding) that conspire to pull flat, compliant MEMS structures together and to keep them stuck together. Contributing factors to stiction include very flat (nanometer-scale roughness) surfaces, large surface areas, compliant structures, electrostatic attraction, humidity and hydrophilic surfaces. There are two major categories of stiction: stiction due to the release process and in-use stiction. The main cause of stiction in the release process is capillary forces of the liquids in the release process that attempt to pull the parts together. Release-stiction can be countered with various hydrophobic coatings or sublimation drying processes. The mechanism designer can combat release-stiction by designing structures that are stiff in the out-of-plane direction. This is very important because it increases the amount of force that must be expended to pull the structure down to the substrate. Another way the mechanism designer can help the release process is by reducing the surface area of the parts—for example, by using a spoked gear instead of a solid gear. Another way of reducing the surface area is to use dimples, which are minimum feature-sized protrusions on the bottom of a surface-micromachined part. Dimples greatly reduce the surface area in contact with the substrate and also increase the out-of-plane stiffness.

The prevalence of in-use stiction can be reduced by many of the same ways as release stiction. Hydrophobic coatings, reduced surface area, dimples and increasing the stiffness of the parts all help to reduce stiction. These are very important tools to reduce stiction and should be utilized even if the release process is relatively robust. A cause of stiction not found in the release process is electrostatic attraction. This usually involves insulating materials, such as silicon nitride, that can trap a charge. The trapped charge causes an electrostatic force whose magnitude can shift in an unpredictable manner. The best method of avoiding this problem is to use a conductive ground plane to cover any exposed insulating layers and to connect every object to the ground. If this is not possible (for instance, with a rotating gear), the hub at a minimum should be tied to the ground. Packaging is another area where the mechanism designer can combat stiction. A high-humidity environment increases the adhesive energy between parts. If water vapor in the package condenses on the surface micromachine, the water will cause capillary adhesive forces similar to those found in the release process. Therefore, a package with a dewpoint well below the minimum storage or operating temperature of the mechanism is essential.

27.3 Machine Design

The fundamental difference between structures and machines is that machines have intended degrees of freedom—usually one. Micromachines can be further divided into two classes by the element used to provide the motion. Elastic-mode machines use compliance elements, such as springs and flexures, to allow motion. Rigid-body-mode (or gross-motion) machines permit motion through joints and bearings. Gross-motion machines allow parts to accumulate motion. As an example of gross motion, a gear spinning on a hub accumulates angular displacement. Elastic-mode machines force the machine parts to oscillate about a fixed point or axis.

27.3.1 Compliance Elements—Columns, Beams and Flexures

Compliance elements or springs allow machine parts to move. Springs come in many forms, but they are primarily realized in microsystems as beams or flexures. Elastic-mode machines are usually focused on resonating devices. In some cases, the spring element also serves as the resonating mass. In those cases, the spring must be evaluated using the beam equations from continuum mechanics. If the spring supports another element whose mass is significantly larger, then the spring can be treated as a lumped element described by a parameter called the *spring constant* or *spring rate*. The spring rate is the ratio of applied force to deflection:

$$k = \frac{F}{\delta} \qquad (27.8)$$

where k is the spring constant, F is the applied force and δ is the resulting deflection.

27.3.2 Columns

A column is a structural member that supports pure tension or compression. Columns are loaded along their longitudinal axis and are generally stiff in comparison to beams and flexures. Columns are not frequently used by themselves but more often as part of more complex structures and, consequently, are important to understand. If loaded in compression beyond a critical value, columns fail by buckling.

FIGURE 27.4 Illustration showing the elongation of a column. L is the original length of the column, and δ is the change in the length of the column.

Buckling is a large lateral deformation caused by an axial load. Columns that buckle but do not yield can provide a mechanism for out-of-plane movement [Garcia, 1998]. The force P on a column creates an elongation δ, shown in Figure 27.4, and whose relationship is derived from Eqs. (27.1) to (27.3):

$$\frac{P}{A} = E\frac{\delta}{L} \tag{27.9}$$

where E is Young's modulus, A is the cross-sectional area and L is the length of the column. The resulting spring rate for columns is

$$k = \frac{EA}{L} \tag{27.10}$$

27.3.3 Beams

Beams are structural members characterized by their ability to support bending and they can provide structures with a linear degree of freedom. Moments and transverse loads applied to beams deform them into curves, as opposed to axial loads that only make them longer or shorter. As the loaded beam takes on a curved shape, one side of the beam must grow longer while the opposite side grows shorter. The outside surface of the beam is in tension and the inside surface of the beam is in compression. A plane through the geometric center of the beam, called the *neutral axis*, does not change in length at all and is unstressed as a result of bending.

For simple cases, the maximum stress in a beam occurs at the surface and is independent of the material composing the beam. The magnitude of the stress of a beam bent into a curved shape by either a moment or a transverse load is

$$\sigma = \frac{Mc}{I} \tag{27.11}$$

where M is the moment causing the curvature, c is the distance from the neutral axis to the surface of the beam (usually half the thickness) and I is the area moment of inertia of the beam's cross section.

Moment of inertia is a measure of the dispersion of area about the centroid. For those familiar with statistics, the centroid is directly analogous to the mean, and the moment of inertia is directly analogous to the standard deviation. In fact, moment of inertia is calculated similarly to standard deviation for a continuous distribution. There are different types of moment of inertia. What is commonly called "moment of inertia" is calculated when the axis of rotation is in the same plane as the cross section. The "polar moment of inertia" is calculated for cases when the axis of rotation is normal to the cross section. Expressions for moment of inertia for common geometric shapes can be found in most mathematics and mechanical engineering handbooks, as well as mechanics of materials textbooks. The moments of inertia for the two common shapes encountered in surface-micromachined mechanisms are illustrated in Figure 27.5. The lines through the cross sections in Figure 27.5 show the axis of rotation for the curving beam. The moment of inertia for a rectangular beam with a rotation about the y-axis

Surface-Micromachined Mechanisms

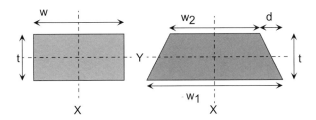

FIGURE 27.5 Moments of inertia for two common beam shapes.

of rotation is

$$I = \frac{wt^3}{12} \tag{27.12}$$

and for rotation about the x-axis:

$$I = \frac{tw^3}{12} \tag{27.13}$$

where w and t are the two dimension shown in Figure 27.5. For a trapezoidal beam about the y-axis the moment of inertia is

$$I = \frac{t^3(w_1^2 + 4w_1 w_2 + w_2^2)}{36(w_1 + w_2)} \tag{27.14}$$

and the moment of inertia about the x-axis is

$$I = \frac{t}{36(w_1 + w_2)} [w_1^4 + w_2^4 + 2w_1 w_2(w_2^2 + w_1^2) - d(w_1^3 + 3w_1^2 w_2 - 3w_1 w_2^2 - w_2^3) \\ + d^2(w_1^2 + 4w_1 w_2 + w_2^2)] \tag{27.15}$$

where the dimensions w_1, w_2, d and t are shown in Figure 27.5.

Example

At a given cross section in a beam, the moment acting on it is 2 nN-m. The beam is rectangular with a thickness of 2 μm and a width of 30 μm. What is the maximum stress in the beam?

The stress caused by bending in the beam is $\sigma = Mc/I$, where c is the distance from the neutral axis to the surface of interest. Because we want the maximum stress, c is half the thickness. The moment of inertia must be calculated from the expression for rectangular cross sections:

$$I = \frac{30 \text{ μm} \times (2 \text{ μm})^3}{12} = 2 \times 10^{-23} \text{ m}^4$$

The resulting maximum stress at the cross section is

$$\sigma = \frac{2 \text{ nN-m} \times 1 \text{ μm}}{2 \times 10^{-23} \text{ m}^4} = 10^8 \text{ N/m}^2 = 100 \text{ MPa}$$

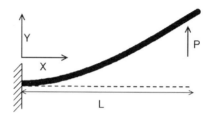

FIGURE 27.6 Cantilever beam spring. The deformation of the beam is due to the load P and can be calculated using Eqs. (27.19) and (27.20).

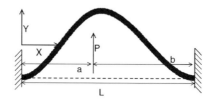

FIGURE 27.7 Deflection of a fixed-beam spring due to a point load. The applied load is P, the load's location is shown by a and b on the x-axis, and the spring deflects in the direction of the y-axis.

Note that in this example it was not necessary to know the length of the beam or the location of the applied loads.

Beams can be categorized by their supports. A beam held rigid on one end and free to move on the other is called a *cantilever beam*. A beam held rigid at both ends is called a *fixed beam*. In general, cantilever beams have more flexibility than fixed beams, so cantilevers will have lower natural frequencies and larger deflections. A diving board at a swimming pool is a classic cantilever beam. A shelf on a bookcase (if only the ends of the shelf are attached to the bookcase) is an example of a fixed beam. Drawings of a cantilever beam spring and a fixed beam spring are shown in Figures 27.6 and 27.7, respectively.

The equations for deflection of cantilever beams and fixed beams are relatively straightforward and can be applied accurately in most cases. In situations of complex geometry or intricate loading or if more precision is required, alternate methods, such as finite element analysis, should be employed.

27.3.4 Stress Concentration

Stress equations are developed for homogeneous bodies with mathematically simple geometries. In real devices, the geometry is usually complex and this can lead to regions of highly localized stresses not predicted by the simple equations. Stress concentrations are significant for ductile materials only when the loads are repeated. For the case of static loading, stress concentration is important only for materials that are both brittle and relatively homogeneous. These amplifications of stress must be accounted for in design.

Stress concentration is an empirical contrivance for design engineers. Stress concentration factors better predict the stress than the nominal stress equations. More importantly, information about stress concentration can be used as design guidance to avoid performance complications or failures. Because changes in geometry can occur in an infinite variety of ways, it is not possible to find an empirical stress concentration that applies for every case. Thus, engineering judgement with considerable estimation is essential.

As a general rule, practical approaches to accommodate micromachine fabrication are adverse for managing stress. Etch release holes serve as stress concentrations in plates. Sharp corners in Manhattan (right-angle) geometries also create stress concentrations. The stress concentration factor is used to adjust the stress calculated from the standard equations, such as Eqs. (27.1) and (27.11), which were discussed

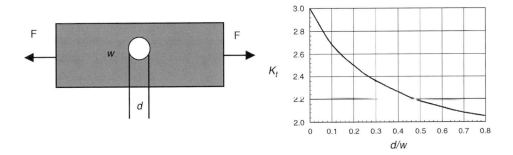

FIGURE 27.8 Stress concentrations for a flat plate loaded axially with a circular hole.

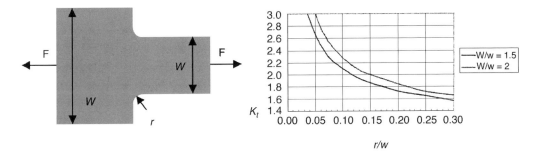

FIGURE 27.9 Stress concentrations for a flat plate loaded axially with two different widths and fillet radius r.

earlier, where K_t is the stress concentration factor:

$$\sigma_{CONCENTRATION} = K_t \sigma_{NOMINAL} \qquad (27.16)$$

For flat plates loaded axially with a circular hole in them, the nominal stress is given by:

$$\sigma_{NOM} = \frac{F}{t(w-d)} \qquad (27.17)$$

where w is the width of the plate and d is the diameter of the hole. The stress concentration K_t is shown in Figure 27.8 along with a drawing of the geometry. For square etch release holes, refer to Figure 27.8 and Eq. (27.17), with the length of the sides of the square opening being equal to the diameter of the circle.

For flat plates with two different widths the nominal stress is given by Eq. (27.18). A drawing of the geometry and a graph of the stress concentration factor are shown in Figure 27.9. Note that the stress concentration factor depends on the radius of the fillet that is denoted as r. For surface-micromachines designed with a Manhattan geometry, r is nominally zero but the fabrication process will create small non-zero radii in the corners. The process engineer should be able to provide a reasonable estimate for r.

$$\sigma_{NOM} = \frac{F}{wt} \qquad (27.18)$$

27.3.5 Cantilever Beam Springs

The deformation y at any arbitrary point x along a cantilever beam, illustrated in Figure 27.6, can be calculated by:

$$y = \frac{P}{6EI}(3Lx^2 - x^3) \qquad (27.19)$$

where L is the length of the beam, E is Young's modulus of the beam material and P is the applied load. The maximum deflection occurs at the end of the beam and can be calculated by:

$$y_{MAX} = \frac{PL^3}{3EI} \quad (27.20)$$

The maximum bending stress in a cantilever beam occurs at the attachment or fixed end where the internal moment is PL. For a rectangular cross section of thickness t and width w, the maximum stress can be calculated with Eq. (27.11) to be

$$\sigma_{MAX} = \frac{6PL}{bh^2} \quad (27.21)$$

The fundamental resonant frequency in Hertz of a simple cantilever beam is

$$f_1 = \frac{3.52}{2\pi} \sqrt{\frac{EI}{\rho A L^4}} \quad (27.22)$$

where E is Young's modulus, A is the cross-sectional area and ρ is the density (mass per unit volume).

Example

A cantilever beam fabricated in polycrystalline silicon with a rectangular cross section is pulled down to an electrode on a substrate by electrostatic attraction. The beam is 50 μm wide, 6 μm thick and 200 μm long. The gap between the bottom of the beam and the electrode is 2 μm. What is the required electrostatic force? If the beam were suddenly released, at what frequency would it vibrate?

The equation for cantilever beam deflection can be solved for force:

$$P = \frac{3EIy}{L^3} \quad (27.23)$$

A reasonable value for the Young's modulus for polycrystalline silicon is 155 GPa or 0.155 N/μm². The area moment of inertia for a rectangular cross section is $I = wt^3/12 = (50\ \mu m)(6\ \mu m)^3/12 = 900\ \mu m^4$. So,

$$P = \frac{3(0.155\ \text{N}/\mu\text{m}^2)(900\ \mu\text{m}^4)(2\ \mu\text{m})}{(200\ \mu\text{m})^3} = 105\ \mu\text{N}$$

A reasonable value for the density of solid polycrystalline silicon is 2.33×10^{-15} kg/μm³. The cross-sectional area of the beam is $A = wt = (50\ \mu m)(6\ \mu m) = 300\ \mu m^2$:

$$f = \frac{3.52}{2\pi} \sqrt{\frac{(0.155\ \text{N}/\mu\text{m}^2)(900\ \mu\text{m}^4)(10^6\ \mu\text{m/m})}{[2.33(10^{-15})\ \text{kg}/\mu\text{m}^3](50\ \mu\text{m} \times 6\ \mu\text{m})(200\ \mu\text{m})^4}} = 198\ \text{kHz}$$

27.3.6 Fixed Beam Springs

The deflection at any point $0 < x < a$ along a fixed beam, shown in Figure 27.7, can be calculated using the following equation:

$$y = \frac{Pb^2x^2[3aL - (3a+b)x]}{6L^3EI} \quad (27.24)$$

Surface-Micromachined Mechanisms

The equation above is based on a point-applied load somewhere along the length of the beam. The deflection caused by a distributed load is somewhat different. Deflection can be estimated by assuming the entire load is applied at the center with some error. An exact solution can be found in texts on mechanics of materials such as Timoshenko (1958).

The fundamental frequency for a fixed beam is

$$f_1 = \frac{22.4}{2\pi} \sqrt{\frac{EI}{\rho A L^4}} \qquad (27.25)$$

The form of this equation is similar Eq. (27.22) for cantilever beams. Note that fixing the other end of the beam increases the fundamental frequency by more than a factor of six.

Example

Assuming the same parameters as the cantilever beam example, except that it is fixed on both ends, calculate the force applied at the center necessary to pull it down to the substrate and the resulting frequency of vibration when released.

For this case, $x = a = b = L/2$, so the equation for deflection reduces to:

$$y_{MAX} = \frac{PL^3}{192EI} \qquad (27.26)$$

and solving for load:

$$P = \frac{192EIy}{L^3} = \frac{192(0.155 \text{ N}/\mu\text{m}^2)(900 \text{ }\mu\text{m}^4)(2 \text{ }\mu\text{m})}{(200 \text{ }\mu\text{m})^3} = 6700 \text{ }\mu\text{N}$$

$$f = \frac{22.4}{2\pi} \sqrt{\frac{(0.155 \text{ N}/\mu\text{m}^2)(900 \text{ }\mu\text{m}^4)(10^6 \text{ }\mu\text{m/m})}{[2.33(10^{-15}) \text{ kg}/\mu\text{m}^3](6 \text{ }\mu\text{m} \times 50 \text{ }\mu\text{m})(200 \text{ }\mu\text{m})^4}} = 1.26 \text{ MHz}$$

27.3.7 Flexures

Flexures are typically used when a rotational degree of freedom is desired. As opposed to beams where the applied moments cause curvature, torque applied to flexures causes twisting. The ends of the flexure will rotate relative to one another through some angle θ. The resultant twisting creates shear stress in the element. Strictly speaking, the only cross section for which the simple torsion equation applies is circular. Circular cross sections do not appear frequently in micromachine design; rather, rectangular cross sections are the norm. The simple equation can be adapted for use with little problem.

The shear stress in a circular bar arising from an applied torque is

$$\tau = \frac{Tc}{J} \qquad (27.27)$$

where τ is the shear stress, T is the applied torque, c is the distance from the central axis to the point where the stress is desired and J is the polar area moment of inertia.

The equation above must be modified to account for rectangular cross sections violating a basic assumption used in deriving the equation above. More precise analysis suggests the following equation:

$$\tau = \frac{9T}{2wt^2} \qquad (27.28)$$

where w is the width (larger dimension) of the flexure and t is the thickness (smaller dimension). The deflection created by the applied torque is

$$\theta = \frac{TL}{JG} \qquad (27.29)$$

where L is the length of the flexure, G is the shear modulus and J is the polar moment of inertia. The polar moment of inertia is defined as:

$$J = I_x + I_Y \qquad (27.30)$$

Example

A 150-μm square plate made of polysilicon is supported by 6-μm-wide torsional flexures forming an axis of rotation through the plate's midpoint. The plate is 4 μm thick and the flexures are 20 μm long. If one edge is pulled down to the substrate 3 μm from the bottom of the plate, what is the resulting shear stress in the flexure? When the force is released, at what frequency will the system vibrate?

Assuming that the plate does not significantly deform, the flexures are required to rotate through an angle of 2.3° for the plate to touch the substrate. This angle can be estimated by:

$$\theta = \tan^{-1}\left(\frac{3\,\mu m}{(150\,\mu m/2)}\right) = 2.3° = 0.040 \text{ rad} \qquad (27.31)$$

The reader is challenged to understand why the arctan approach is an approximation instead of an exact expression.

A reasonable value for the modulus of rigidity is 0.0772 N/μm². The polar area moment of inertia for a rectangular section is

$$J = \frac{wt}{12}(w^2 + t^2) = \frac{(6\,\mu m)(4\,\mu m)}{12}[(6\,\mu m)^2 + (4\,\mu m)^2] = 104\,\mu m^4 \qquad (27.32)$$

The torque required to rotate one flexure through the angle θ is

$$T = \frac{JG\theta}{L} = \frac{(104\,\mu m^4)(0.0772\,N/\mu m^2)(0.04\,\text{rad})}{20\,\mu m} = 1.6 \times 10^{-2}\,\mu Nm \qquad (27.33)$$

This is the torque per flexure. The total torque on the plate is twice that value because there are two flexures. If it could be applied at the edge, the value of the force needed to create this much rotation is

$$F = 2\frac{T}{r} = 2 \times \frac{1.6 \times 10^{-2}\,\mu N\text{-}m}{(150\,\mu m/2)} = 4.3 \times 10^{-4}\,N \qquad (27.34)$$

The resulting shear stress in each flexure is

$$\tau = \frac{9T}{2wt^2} = \frac{9 \times 1.6 \times 10^{-2}\,N \cdot \mu m}{2(6\,\mu m)(4\,\mu m)^2} = 7.5 \times 10^{-4}\,N/\mu m^2 = 0.75\,\text{GPa} \qquad (27.35)$$

In a dynamics context, the system can be a modeled as a rigid body vibrating about a spring. The differential equation of motion for a simple mass–spring system and the associated natural frequency are

$$i\ddot{\varphi} + K\varphi = 0 \qquad (27.36)$$

$$f_n = \frac{1}{2\pi}\sqrt{\frac{K}{i}} \qquad (27.37)$$

where i is the mass moment of inertia (different than area moment of inertia used to calculate stress) and K is the torsional spring rate. The spring rate of the flexures is

$$K = \frac{T}{\theta} = \frac{JG}{L} = \frac{(104\ \mu m^4)(0.0772\ N/\mu m^2)}{20\ \mu m} = 0.40\ \mu Nm/rad$$

The mass of the plate is

$$m = \rho w l t = [2.33 \times 10^{-18}\ kg/\mu m^3](150\ \mu m)^2(4\ \mu m) = 2.1 \times 10^{-13}\ kg$$

The mass moment of inertia for a thin rectangular plate vibrating about its center is

$$i = \frac{1}{12}mw^2 = \frac{[2.1 \times 10^{-13}\ kg](150\ \mu m)^2}{12} = 3.93 \times 10^{-10}\ kg \cdot \mu m^2 \qquad (27.38)$$

The natural frequency is

$$f_n = \frac{1}{2\pi}\sqrt{\frac{(0.40\ N \cdot \mu m)(10^6)\ \mu m/m}{3.93 \times 10^{-10}\ kg \cdot \mu m^2}} = 5.1\ MHz$$

27.3.8 Springs in Combinations

Springs are energy storage elements analogous to electrical capacitors, and the laws for calculating equivalent spring rates are the same as the laws for calculating equivalent capacitance. Springs are in parallel if they undergo the same deflection; springs are in series if they support the same loading. The equivalent spring rate for springs in parallel is

$$k_{eq} = k_1 + k_2 + k_3 + \cdots \qquad (27.39)$$

where springs in parallel experience the same deflection δ.

The equivalent spring rate for spring in series is

$$\frac{1}{k_{eq}} = \frac{1}{k_1} + \frac{1}{k_2} + \frac{1}{k_3} + \cdots \qquad (27.40)$$

where springs in series can be identified because each spring experiences the same force. Springs in parallel and series are shown in Figures 27.10 and 27.11, respectively.

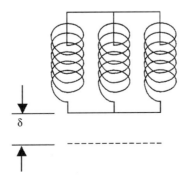

FIGURE 27.10 Springs in parallel.

FIGURE 27.11 Springs in series.

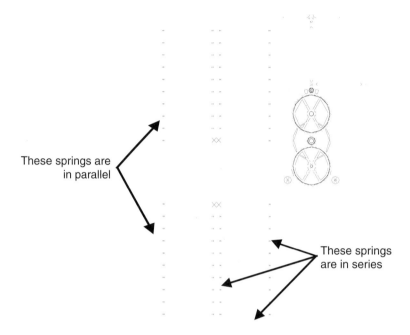

FIGURE 27.12 Surface-micromachined oscillating mechanism supported by springs in series and parallel. (The original drawing was courtesy of G. Vernon and M. Polosky at Sandia National Laboratories.)

To tailor the performance or to conserve real estate, it is sometimes desirable to use springs in combinations. Figure 27.12 shows an oscillating structure supported by a complex suspension system. The suspension is comprised of a parallel combination of four spring systems and each spring system is a series combination of three individual beam and column elements. In the case shown in Figure 27.12, the U-shaped springs are all attached to the substrate and to the resonator shuttle. When the shuttle moves, each of the four U-shaped springs deflects the same amount—the motion of the shuttle. There are four different paths from the shuttle to the ground, which is another indication that the springs are in parallel.

Closer inspection of a single U-shaped spring reveals three beams in series. Because the beams form a single path between the shuttle and ground, they must carry the same load. A free-body diagram shows that the load from one beam is transmitted fully to the next until the load to transmitted to ground.

Example

Calculate the equivalent spring of the U-shaped spring shown in Figure 27.13.

The spring system consists of three springs in series: two cantilever beam springs connected by one column. It is reasonable to treat spring 2 simply as a column even though it supports reaction moments from springs 1 and 3 because none of the resulting bend in spring 2 contributes to the motion of the shuttle. The springs are clearly in series, as the load from one spring is transmitted fully to the next and each spring is allowed to deflect whatever amount is required to support the load.

For springs in series, the equivalent spring rate is given by:

$$\frac{1}{k_{eq}} = \frac{1}{k_1} + \frac{1}{k_2} + \frac{1}{k_3} \tag{27.40}$$

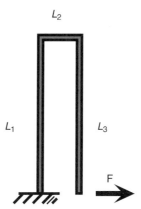

FIGURE 27.13 U-shaped spring comprised of three beam springs connected in series and parallel.

For MEMS, it is reasonable to assume that all the springs are fabricated of the same material. Springs 1 and 3 are cantilever beams and their rates, respectively, are $k_1 = \frac{3EI_1}{L_1^3}$ and $k_3 = \frac{3EI_3}{L_3^3}$. The rate for the column, spring 2, is

$$k_2 = \frac{EA_2}{L_2} \tag{27.41}$$

The equivalent rate for the three-spring combination is

$$\frac{1}{k_{eq}} = \frac{1}{k_1} + \frac{1}{k_2} + \frac{1}{k_3} = \frac{L_1^3}{3EI_1} + \frac{L_2}{EA_2} + \frac{L_3^3}{3EI_3} \tag{27.42}$$

27.3.9 Buckling

When a short column is loaded in compression, the average compressive stress is calculated by simply dividing the load by the cross-sectional area as in Eq. (27.1). However, when the column is long and slender, the situation is complicated by the possibility of static instability, also known as *lateral buckling*.

Buckling occurs in columns loaded in compression beyond their critical value. When loaded below the buckling threshold, a column under a compressive load will get shorter while remaining essentially straight. After the buckling threshold is exceeded, the column deflects normally to the axis of the column, and the stress increases rapidly. In addition to the direct compression, the beam experiences bending stress created by the couple between the offset beam and the line of action of the force. Recall that bending creates compressive stress on the inside surface of the beam. The direct and bending compressive stresses add to and cause failure by rupture.

The phenomenon of buckling is much different from the phenomenon of bending. A beam will begin to bend as soon as any moment (bending load) is present. A column will not exhibit any lateral deflection until the critical or buckling load is reached. Additional loading above the critical load causes large increases in lateral deflection. Because of the tensile properties of polysilicon, it is possible to design a column to buckle without exceeding the yield strength. If yield does not occur, the column will return to its original straight position after the load is removed. If designed appropriately, a buckling column can be used as an out-of-plane flexure.

Buckling is a complex nonlinear problem and predicting the resulting shape is beyond the scope of this text. It is, however, covered in several references including Timoshenko and Gere (1961), Brush and Almroth (1975), Hutchinson and Koiter (1970) and Fang and Wickert (1994). But, the equation for predicting the onset of buckling is relatively straightforward. For a column with one end fixed and the other free to move in any direction (i.e., a cantilever), the critical load to cause buckling is

$$F_{CR} = \frac{\pi^2 EI}{4L^2} \tag{27.43}$$

It is important to remember that the critical load for buckling cannot exceed the maximum force supported by the material. In other words, if the load needed to cause buckling is larger than the load needed to exceed the compressive strength, the column will fail by rupture before buckling.

27.4 Applications

The applications section of this chapter uses the material that has been covered thus far. It describes some surface-micromachined mechanisms and discusses them with regard to concepts such as spring constant and critical buckling load. The chapter concludes with a discussion of some design rules and lessons learned in the design of surface-micromachined mechanisms.

27.4.1 Microengine

One important element of many polysilicon mechanism designs is the "microengine." This device, described by Garcia and Sniegowski (1995) and shown in Figures 27.14 to 27.16, uses a comb drive connected to a pinion gear by a slider-crank mechanism with a second comb drive to move the pinion past the top and bottom dead center. Comb drives are electrostatic devices that use fringe fields to produce a uniform force over a large displacement. Two comb drives are necessary because the torque on a pinion produced by a single actuator has a dependence on angle and is given by the following equation:

$$Torque = F_0 r |\sin(\theta)| \tag{27.44}$$

where F_0 is the force of the comb drive, θ describes the angle between hub of the output gear and the linkage, and r is the distance between the hub and the linkage. The torque produced by the "Y" actuator has a similar dependence on angle. A second reason for two comb drives is that the inertia of the pinion gear is not great enough to rotate the gear past the top and bottom dead center. A drawing of a microengine is shown in Figure 27.15. One important feature of this device is that the conversion from linear motion to rotary motion requires the beams between the actuator and the driven gear to bend. The bending is permitted by a polysilicon linkage that is 40 µm in length and 1.5 µm in width with a thickness of 2.5 µm.

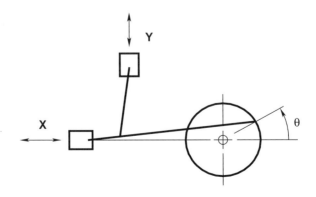

FIGURE 27.14 Mechanical representation of a microengine.

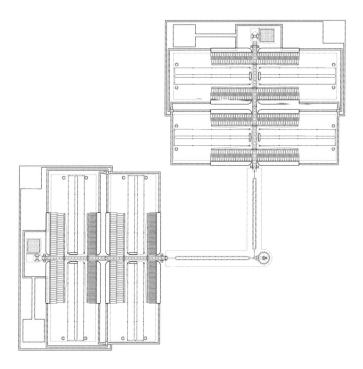

FIGURE 27.15 Drawing of a microengine. The actuator measures 2.2 mm × 2.2 mm and produces approximately 55 pN-m of torque.

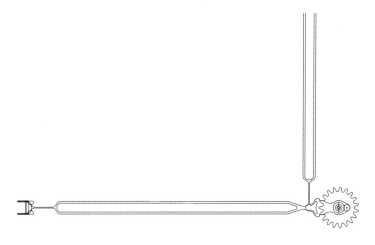

FIGURE 27.16 Detail of Figure 27.15 showing the thin linkages that connect the comb drives to the gear.

Example

The comb drive labeled "X" in Figure 27.14 has an actuator arm that must bend 17 μm in the lateral direction as the gear rotates from 0 to 90°. The gear is connected to the comb drive via a 50-μm-wide beam that is 500 μm long in series with a thin flexible link that is 1.5 μm wide and 45 μm long. The thin link is connected to the comb drive actuator and both beams have a Young's modulus of 155 GPa. Given that both beams are 2.5 μm thick, approximately how much force does it take to rotate the gear from $\theta = 0$ to $\theta = 90°$ if friction and surface forces are neglected? A drawing of this linkage is shown in Figure 27.16.

Assume that each segment is a cantilever beam spring that has one end fixed while the other end is undergoing a small deflection. For each beam:

$$y_{max} = \frac{PL^3}{3EI} \tag{27.45}$$

and

$$k = \frac{P}{y_{MAX}} = \frac{3EI}{L^3} \tag{27.46}$$

Recall that for rectangular cross sections,

$$I = \frac{tw^3}{12} \tag{27.13}$$

Using substitution, the equivalent spring constant for the long beam is

$$k = \frac{Etw^3}{4L^3} = \frac{155 \text{ GPa} \times (2.5 \text{ μm})(50 \text{ μm})^3}{4 \times (500 \text{ μm})^3} = 97 \text{ N/m}$$

and for the short beam,

$$k = \frac{Etw^3}{4L^3} = \frac{155 \text{ GPa} \times (2.5 \text{ μm})(1.5 \text{ μm})^3}{4 \times (45 \text{ μm})^3} = 3.6 \text{ N/m}$$

By Eq. (27.40), the equivalent spring rate is

$$\frac{1}{k_{eq}} = \frac{1}{k_1} + \frac{1}{k_2} = \frac{1}{97 \text{ N/m}} + \frac{1}{3.6 \text{ N/m}} = 3.5 \text{ N/m}$$

We can employ some simple trigonometry and calculus to determine the needed deflection of the flexure. From the section on cantilever beam springs,

$$y = \frac{P}{6EI}(3Lx^2 - x^3) \tag{27.19}$$

Because the majority of the bending occurs in the thin flexible link the desired slope of the flexible link at its end is

$$\frac{dy}{dx} = \frac{17 \text{ μm}}{545 \text{ μm}} = \frac{P}{6EI}(6Lx - 3x^2) \tag{27.47}$$

For the case of the small flexible link, I is equal to:

$$I = \frac{tw^3}{12} = \frac{2.5 \text{ μm} \times (1.5 \text{ μm})^3}{12} = 7 \times 10^{-25} \text{ m}^4$$

Because $x = L = 45$ μm, P can be calculated as:

$$P = \frac{17 \text{ μm}}{545 \text{ μm}} \times 6 \times 155 \text{ GPa} \times 7.0 \times 10^{-25} \text{ m}^4 \times \frac{1}{(6 \times (45 \text{ μm})^2 - 3 \times (45 \text{ μm})^2)} = 3.4 \text{ μN}$$

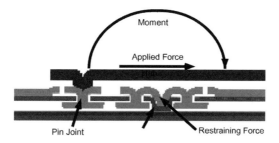

FIGURE 27.17 Cross section of the hub and pin joint. This joint connects the comb drive actuators to the gear and the gear to the substrate. The arrows show the applied forces and the moment that results from them. The hub is on the center of the drawing; the pin joint is to the left of it and is attached to the top layer of polysilicon.

FIGURE 27.18 The countermeshing gear discriminator. The two code wheels are the large gears with five spokes in the center of the drawing; the counter-rotation pawls are connected to the comb drives and the long beams in the upper right and lower left portion of the photograph. (Drawing courtesy of M. Polosky, Sandia National Laboratories.)

A second important aspect of this device is the pin joint that connects the actuators to the gear. This joint rotates with respect to the gear and allows the actuators to apply force to the gear at any angle. The linkage does not mechanically interfere with the hub because it is constructed of a different layer of polysilicon and passes over the hub. Flanges on the hub restrict the amount of vertical play. However, because the linkage is above the plane of the gear, there may be rotational moments that will attempt to push or pull the gear out of plane. A cross section of a pin joint and a rotating hub are shown in Figure 27.17.

27.4.2 Countermeshing Gear Discriminator

One example of a surface-micromachined mechanism that incorporates comb drives and many other components is the countermeshing gear discriminator that was invented by Polosky et al. (1998). This device has two large wheels with coded gear teeth that are driven by microengines. Counter-rotation pawls restrain each wheel so that it can rotate counterclockwise and is prevented from rotating clockwise. The wheels have three levels of teeth that are designed so that the teeth will interfere if the wheels are rotated in the incorrect sequence. Only the correct sequence of drive signals will allow the wheels to rotate and open an optical shutter. If mechanical interference occurs, the mechanism is immobilized in the counterclockwise direction by the interfering gear teeth and by the counter-rotation pawls in the clockwise direction. A drawing of the device and a closeup of the teeth are shown in Figures 27.18 and 27.19, respectively. The wheels have three levels of intermeshing gear teeth that will allow only one

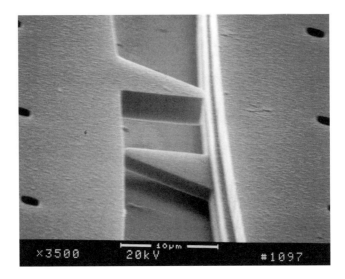

FIGURE 27.19 Photograph of teeth in the countermeshing gear discriminator. The gear tooth on the left is on the top level of polysilicon and the gear tooth on the right is on the bottom layer of teeth. If the gears do not tilt or warp, the teeth should pass without interfering with each other. (Photograph courtesy of M. Polosky, Sandia National Laboratories.)

sequence of rotations out of the more than the 16 million that are possible. Because the gear teeth on one level are not intended to interfere with gear teeth on another level and because the actuators must remain meshed with the code wheels, the vertical displacement of the code wheels must be restricted. This was accomplished with dimples on the underside of the coded wheels that limited the vertical displacement to 0.5 μm. Warpage of the large 1.9-mm-diameter coded wheels is reduced by adding ribs with an additional layer of polysilicon.

The large coded wheels are prevented from rotating backward by the counter-rotation pawls. These devices must be compliant in one direction and capable of preventing rotation in the other direction. The next example discusses counter-rotation pawls.

Example

Figure 27.20 shows a counter-rotation pawl. The spring is 180 μm long from the anchor to a stop (labeled as l) and 20 μm long from the stop to the end of the flexible portion (denoted as a), with a width of 2 μm and a thickness of 3 μm. The Young's modulus is 155 GPa. Assume that the tooth on the free end of the beam does not affect the stiffness and that the width of the stop is negligible. Find the spring constant of the pawl if the gear is rotated in the counterclockwise direction. Comment on the spring constant if the gear is rotated in the clockwise direction.

In the counterclockwise direction, the spring constant k is

$$k = \frac{Ew^3 t}{4L^3} = 0.12 \text{ N/m}$$

using a length $l + a$ of 200 μm. In the clockwise direction, it is tempting to redefine the spring length as 20 μm. If you do so, you will obtain a result of 116 N/m. Unfortunately, this is an oversimplification because the beam will deform around the stop. The exact equation is in Timoshenko's *Strength of Materials* [Timoshenko, 1958] and in Eq. (27.48):

$$k = \frac{1}{\left(\frac{a^3}{3EI} + \frac{a^2 l}{4EI}\right)} \tag{27.48}$$

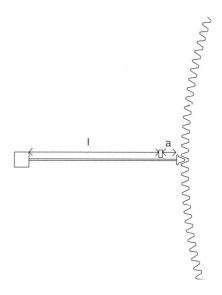

FIGURE 27.20 Example of a simple counter-rotation pawl. The stop is assumed to be a point load with a width of 0.

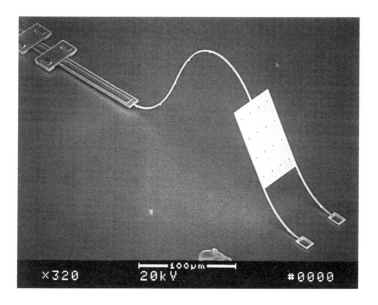

FIGURE 27.21 A flexible pop-up mirror that operates via buckling. (Photograph courtesy of E. Garcia, Sandia National Laboratories, a further description of this device can be found in Garcia, 1998.)

This equation results in a spring constant of 15 N/m, which is still very stiff but not as stiff as the results of the oversimplified calculation.

27.4.3 Micro-Flex Mirror

The next example is a device that deforms out of plane [Garcia, 1998]. It can be built in any surface micromachining process that has at least one level of released structural material. This device can be configured as a mirror, an optical shutter, a valve or any structure that requires a plate or beam to move out of the plane of fabrication. A scanning electron microscope (SEM) photograph of the device is shown in Figure 27.21. The mirror consists of a long flexible beam connected to a plate that in turn is connected to two anchors via two additional flexible beams. The device operates by buckling. When a force is placed on the long flexible beam in the direction of the anchors, the structure is placed under a compressive force. When the force exceeds the critical value described by Eq. (27.43), the structure buckles. Because the long flexible beam is larger in the direction parallel to the plane of the substrate than it is in the direction away from the substrate, it preferentially buckles out of the plane of the substrate. Also, because the plate and the two anchor beams are wider than the long flexible beam, the majority of the bending occurs in the long flexible beam and not in the plate or the anchor beams. The next example discusses the buckling criteria.

Example

Determine how much axial force is needed for a micromachined polysilicon mirror to buckle given the following dimensions. The main beam is 300 μm long, 4 μm wide and 1 μm thick with a Young's modulus of 155 GPa. Assume that the beam can be simplified to be a cantilever. Neglect the buckling of the anchors and the mirror.

From the mechanics of materials section we know that the minimum force to buckle the beam is

$$F_{cr} = \frac{\pi^2 EI}{4L^2} \qquad (27.49)$$

The moment of inertia for this situation is

$$I = \frac{wt^3}{12} = 3.3 \times 10^{-25} \, m^4$$

Substituting this into the equation for force and using 155 GPa as the value of Young's modulus we have the following:

$$F_{cr} = \frac{\pi^2 \times 155 \times 10^9 \, N/m^2 \times 3.3 \times 10^{-25} \, m^4}{4 \times (300 \times 10^{-6} \, m)^2} = 1.4 \, \mu N$$

As the example shows it is necessary to have a great deal of force in order to buckle the flexible mirror. In the original paper by Garcia (1998), a transmission, shown in Figure 27.22, was used to increase the force on the mirror. The transmission traded displacement for force to ensure that the mirror buckled. The amount of deflection after buckling is, unfortunately, a nonlinear dynamics problem and beyond the scope of this book. For those readers interested in the subject, good references include the work of Timoshenko and Gere (1961), Brush and Almroth (1975) and Hutchinson and Koiter (1970). Fang and Wickert (1994) have examined the subject for micromachined beams. One interesting point about buckling is that it is impossible to determine if the beam will initially buckle toward or away from the substrate. If the beam buckles away from the substrate, it will continue to deflect away from the substrate. If it initially buckles toward the substrate, it will contact the substrate and further compression of the beam will result in the structure buckling away from the substrate.

Example

For the mechanism shown in Figure 27.22, how much mechanical force is made available by the transmission if the microengine output gear has a diameter of 50 μm and a torque of 50 pN-m, the large spoked wheel has a diameter of 1600 μm, and the pin joint that links the spoked wheel to the mirror is 100 μm from the hub? Find the radial force at the pin joint and determine the force on the mirror when the pin joint, hub and mirror are perfectly aligned.

The torque available at the pin joint is

$$torque = 50 \, pNm \times \frac{1600 \, \mu m}{50 \, \mu m} = 1600 \, pNm$$

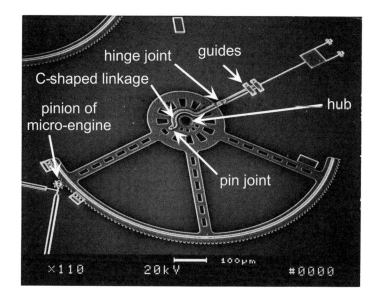

FIGURE 27.22 Photograph showing a flexible mirror in its initial state and a large gear with spokes. The spoked wheel acts as a transmission and is used to gain enough mechanical advantage to buckle the beam. Two important features are the hinge joint and guides which convert the off-axis motion of the "c"-shaped linkage to motion that is aligned with the mirror. (Photograph courtesy of E. Garcia, Sandia National Laboratories.)

and the radial force at the pin joint is

$$F_{radial} = \frac{1600 \text{ pNm}}{100 \text{ μm}} = 16 \text{ μN}$$

At the mirror, the force is given by:

$$F_{mirror} = \frac{F_{radial}}{\sin(\theta)} \tag{27.50}$$

where θ is the angle between a line connecting the pin joint and the mirror and a line connecting the pin joint and the hub. Because $\theta = 0$ when all the points are collinear, the force applied on the mirror approaches infinity for angles approaching zero degrees. This example is taken after Garcia (1998), who included a more detailed derivation that accounts for the rotation of a "c"-shaped linkage.

A different type of rigid-body-mode machine is a vertical axis hinge. These hinges are used in surface micromachining to create three-dimensional structures out of a two-dimensional surface micromachining process. In the surface micromachining process, hinges are constructed by stapling one layer of polysilicon over another layer of polysilicon with a sacrificial layer between them. A cross section of a simple hinge is illustrated in Figure 27.23. Photographs of more complex hinges are shown in Figures 27.24 and 27.25. Hinges have some advantages over flexible joints. One advantage is that no stress is transmitted to the hinged part so greater angles of rotation are possible. Also, the performance of hinged structures is not influenced by the thickness of the material, and deformation of the machine is not required. The limitations of hinged structures are that the sacrificial layers must be thick enough for the hinge pin to rotate and at least two released layers of material are required.

Hinges in the surface micromachining process are not always trouble free. There is a problem with friction in hinges because the sliding surfaces are neither round nor lubricated. There is also a substantial

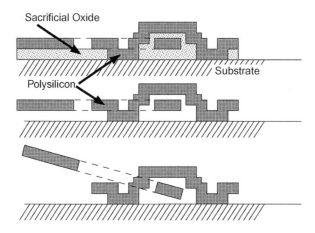

FIGURE 27.23 Cross section of polysilicon mirror hinge. The top figure shows the device before release, the middle drawing is after release, and the bottom depicts the mirror actuated.

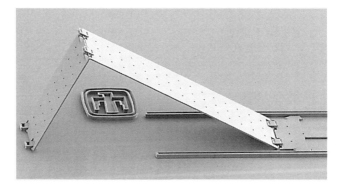

FIGURE 27.24 Hinged polysilicon micromirror fabricated in the Sandia National Laboratories SUMMiT™ process. (Photograph courtesy of Sandia National Laboratories.)

FIGURE 27.25 Detail of polysilicon hinge. (Photograph courtesy of M. Polosky and A. Jojola, Sandia National Laboratories.)

amount of backlash built into the system. A third problem relates to the ability of the hinge to pivot. Often there is not a large rotational moment to rotate the mirror out of plane when planar hinged structures are connected to planar-surface-micromachined actuators. This means that the designer must take great care to ensure that the hinge always rotates in the correct direction.

27.5 Failure Mechanisms in MEMS

One of the most effective ways we can learn is to learn from our own mistakes. This can be a memorable experience but in the field of MEMS it can be a very expensive and inefficient one. One reason is that the time between design completion and testing is usually measured in months and the price per fabrication run is many thousands of dollars. As of the year 2000, there were only rudimentary modeling and simulation tools available for surface-micromachined mechanisms. This section seeks to share learning obtained at the expense of others by describing some mechanical failures in surface-micromachined mechanisms. The hope is that the reader will gain a deeper appreciation for the complexities of surface-micromachined mechanism design and learn about some of the pitfalls.

27.5.1 Vertical Play and Mechanical Interference in Out-of-Plane Structures

Surface-micromachined parts typically have a thickness that is very small in relationship to their width or breadth. In the out-of-plane direction, the thickness is limited to a few micrometers due to the limited deposition rates of low-pressure chemical vapor deposition (LPCVD) systems and the stresses in the deposited films. In the plane of the substrate, structures can be millimeters across. These factors typically lead to surface-micromachined structures that have a very small aspect ratio as well as stiffness issues in the out-of-plane direction due to the limited thickness of the parts. The result is that designers of surface-micromachines need to design structures in three dimensions and account for potential movements out of the plane of the substrate. One instance of a potential problem is when gears fabricated in the same structural layer of polysilicon fail to mesh because one or both of the gears are tilted. Another is when structures moving above or below another structure mechanically interfere with each other when it was the intended for them to clear each other without touching. Both of these instances will be examined separately.

An example of the out-of-plane movement of gears is illustrated in Figure 27.26. In this instance, the driven gear in the top of the figure has been wedged underneath the large load gear at the bottom of the photograph. The way to prevent this situation is to understand the forces that create the out-of-plane motion

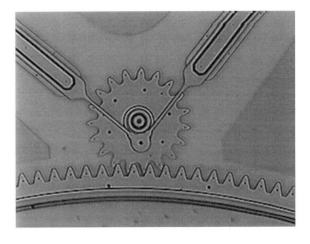

FIGURE 27.26 The gear teeth of the small gear are wedged underneath the teeth of the large diameter gear. In this case, gear misalignment is about 2.5 μm in the vertical direction.

and to reduce or restrain them. One way of reducing the relative vertical motion of meshed gears is to increase the ratio of the radius of the hub to the radius of the gear. Mathematically, the maximum displacement of the outside edge of a gear from its position parallel to the plane of the substrate is

$$Y_{max} = R \times Y/X \qquad (27.51)$$

where the radius of the hub is X, the vertical play in the hub is Y, the radius of the gear is R and the maximum vertical displacement is Y_{max}. The increase in hub diameter, while it can increase the amount of adhesive forces in the hub, does not increase the amount of frictional forces in the rotating or sliding structure.

The vertical displacement of a gear can also be limited by contact with the substrate, and one way of limiting the vertical displacement of the gear is to place dimples on the underside of the gear. Dimples are available in several commercially available surface micromachining processes including the MUMPS process offered by Cronos Integrated Microsystems and the SUMMiT^{TM} process from Sandia National Laboratories. The dimples are generally spaced well apart from each other, depending on the stiffness of the gear and the prevalence of stiction. They typically occupy less than 1% of the surface area.

Another method of restraining the vertical motion of surface-micromachined parts involves the use of clips above or below the gear or other moving structure as shown in Figure 27.27. Generally, a maximum of three or four clips are used, which is enough to restrain the moving structure while limiting the increase in the amount of frictional forces on the gear. Dimples on the bottom of the clips can further reduce the amount of play. Of course, all of these techniques can be combined to reduce vertical play.

Murphy's law for surface-micromachined mechanisms causes a completely different type of reliability problem. It can be stated as, "Whatever can be disconnected will become disconnected." The following example illustrates the point: A pin and socket joint was the connection between an actuator and an optical shutter. A cross section of the joint is shown in Figure 27.28. Unfortunately, the side walls of the socket are slightly sloped and the arm is not restrained in the vertical direction. In Figure 27.29, the force between a pin and the optical shutter was relieved not by the horizontal movement of the shutter but by the vertical movement of the pin, which slid up the sloped side walls of the shutter and out of the socket. The proper way to design this joint would be to use an additional layer of polysilicon below the

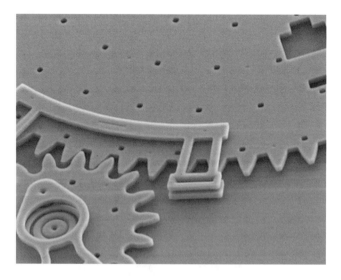

FIGURE 27.27 SEM photograph of clip-over gear. The purpose of the clip is to limit the vertical displacement of the gear and to help ensure that the gear remains meshed. The line in the center of the clip is a dimple that further reduces the amount of vertical play in the gear. (Photograph courtesy of the Intelligent Micromachine Department at Sandia National Laboratories.)

FIGURE 27.28 A cross section of a pin and socket joint.

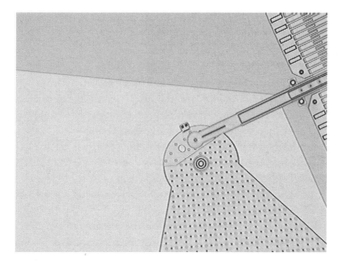

FIGURE 27.29 The pin and socket joint that connects the actuator to the shutter has failed. The pin, which should have been restrained in the vertical direction, was originally located near the center of the shutter.

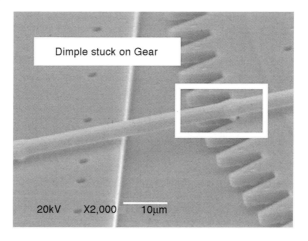

FIGURE 27.30 SEM image of polysilicon beam with dimples that is caught on surface-micromachined gear. The main problem with the beam is that it was designed to be compliant in the vertical direction. A secondary problem was that the dimple was small enough to be caught in the gear teeth.

shutter to restrain the linkage in the vertical direction. An example of this type of connection is shown in Figure 27.17. An alternative approach would be to attach the linkage to the shutter and make the linkage compliant to compensate for the rotation of the shutter.

Unintended vertical motion of surface-micromachined parts can be exacerbated by structures that are soft in the vertical direction. Figure 27.30 shows a beam with a dimple that is caught on a gear. Note that there should be no interaction between the two elements, as the dimple should clear the gear by 0.5 μm. Unfortunately, in this case neither the gear nor the spring is adequately grounded, which may have caused some attractive force between the two. Stiction may also have contributed to the proximity of the two parts. The spring constant of this beam in the vertical direction is much too soft at 1.5×10^{-3} N/m. The solution to the problem is to make the surface-micromachined part stiff in the vertical direction. As noted earlier in Eq. (27.46):

$$k = \frac{Ew^3 t}{4L^3} \tag{27.46}$$

for a rectangular cantilever beam. Therefore, to increase the stiffness in the vertical direction the thickness of the structure should be increased and the length should be decreased. If the spring constant is increased, it takes more energy to pull the structure down to the substrate. Another, but less effective, option is to increase the width. A completely different but complementary approach would be to change the size of the dimple so that it is larger than the space between the gear teeth.

Another source of vertical motion in surface-micromachined parts is electrostatic forces. Unintended electrostatic forces in surface micromachines have two basic causes. One is fixed charge buildup in dielectric layers. Charges in silicon nitride and silicon dioxide, especially silicon nitride, are not mobile and tend to attract other structures. This can be a great problem in accelerometers or gyroscopes because they are designed to be sensitive to the vertical displacement of the proof mass. Unfortunately, techniques from the integrated circuit industry that seek to avoid trapped charges are not typically applicable to MEMS, aside from careful processing, which seeks to avoid the incorporation of sodium in the dielectrics. Currently, the best way to avoid the influences of trapped charge is to avoid them by minimizing exposure of the dielectrics to radiation and energetic electrons. The other method is to shield the moveable mechanism parts from the charge buildup in the dielectric by using a conductive ground plane between the movable structure and the dielectrics.

Another instance when unintended electrostatic forces are important is when features and structures in a design are left electrically floating. In this situation, as in the dielectric charging case, the floating potential can eventually cause mobile structures to be attracted. The cure for this situation is to fix all potentials to a known level. For dielectrics, a grounded conductor should shield them from other structures and layers. For rotating gears or other structures that are not electrically connected to the substrate, the hubs or mechanical restraints should be connected to the ground. If the restraints and the moving component momentarily touch, the moving part can discharge through the support to the substrate. Electrostatic attraction of floating components can be a maddening problem, as the potentials on the various conductors and dielectrics can seemingly shift randomly and cause structures to be unexpectedly attracted to each other.

27.5.2 Electrical Shorts

One problem that is unique to surface-micromachines made of conductive materials is the problem of shorting between different parts of the machine. One common situation is when the fingers of a comb drive touch. Another situation is when the dimple of a moving structure such as a gear or a shutter comes in contact with an underlying conductive trace. An example is shown in Figure 27.31. The problem is that bare polysilicon structures (and silicon–germanium) are neither good conductors nor good insulators even when coated by a thin anti-stiction coating or native oxides. The native oxide coatings and anti-stiction coatings are prone to breakdown under the high voltages commonly used to power electrostatic actuators. Because insulating dielectrics are not normally used between layers, the designer must resort to minimizing the chances for contact between conductors at different potentials. One method of doing this is to make structures, especially comb drives, stiff in all directions but the direction of intended motion. Another is to use stops or other mechanical restraints to prevent contact between conductors at different potentials.

Surface-Micromachined Mechanisms

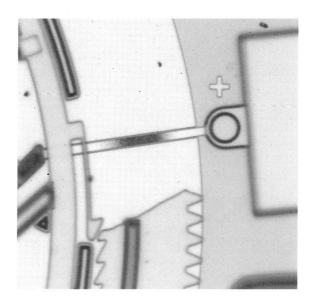

FIGURE 27.31 (Color figure follows p. 12.26.) The blackened trace near the "+" sign was caused by a dimple on a grounded gear touching a high potential signal line. (Photograph courtesy of S. Barnes, Sandia National Laboratories.)

27.5.3 Lithographic Variations

One problem common to all types of mechanism design is tolerances in the manufacturing process. This is an especially grave problem in surface-micromachined mechanisms because the manufacturing tolerances approach the size of the parts. For example, assume a cantilever beam spring with a width of 2 μm and a variation in the width of 0.2 μm. Cantilever beam springs have a spring constant that has been given in Eq. (27.46):

$$k = \frac{Ew^3 t}{4L^3} \qquad (27.46)$$

Due to the cubic relation between w and k, a small change in w will result in a large change in k. For example, if the linewidth changes by 1%, the resulting spring constant changes by about 3%:

$$\Delta k = \left(\frac{\partial k}{\partial w}\right)\Delta w = \frac{3Ew^2}{4L^3}\Delta w \qquad (27.52)$$

$$\frac{\Delta k}{k} = 3\frac{\Delta w}{w} \qquad (27.53)$$

There are a few ways of minimizing the effects of features that vary in size. The first is not to design minimum-sized features. The linewidth is not reduced as a fixed percentage of the feature size but instead all features are reduced by the same amount. For example, a 0.2-μm reduction in the size of a 5-μm-wide beam spring is only a 4% reduction in the width of the spring and results in approximately a 12% change in the spring constant, while a 1-μm-wide spring reduced by the same amount changes the linewidth by 20% and the spring constant by 49%.

The second method of minimizing the variation of mechanism performance due to linewidth is to borrow techniques from integrated analog circuit layout techniques. A good introduction to this topic can be found in the text by Johns and Martin (1997). Probably the most applicable technique they discuss

FIGURE 27.32 Dummy columns next to a column on a micro-flex mirror. The structures ensure that the micro-flex mirror is matched to other micro-flex mirrors on the same substrate and is exposed to similar processing conditions. Note that the dummy columns do not extend the length of the center column to avoid mechanically interfering with the structure attached to the center column as it moves from right to left.

FIGURE 27.33 Gears with spokes that were designed to reduce stiction. (Photograph courtesy of M. Polosky, Sandia National Laboratories.)

is to build dummy features adjacent to the intended mechanical features. Dummy features make processing boundary conditions (for example, in reactive ion etching or chemical mechanical polishing) similar for all parts of the mechanism and reduce the variation in feature size between mechanism parts. An example of this technique is shown in Figure 27.32 where dummy columns are placed adjacent to the column on the micro-flex mirror. This will help ensure that the column on the micro-flex mirror is etched to a reproducible width and will be matched to other columns elsewhere on the chip. Care must be taken, however, so that the dummy parts do not interfere with the mechanism parts.

27.5.4 Methods of Increasing the Reliability of Mechanisms

There are several ways to increase the reliability of surface-micromachined mechanisms. One common problem is adhesion of the surface-micromachined mechanism to the substrate or stiction. Stiction can occur during the release process or afterward. One method of avoiding stiction is to make mechanisms stiff in the vertical direction. A second method of stiction avoidance is to minimize the overlapping surface area between flat parts, such as using dimples. Dimples reduce the contact area by more than a factor of a 1000 if a 2-μm × 2-μm dimple is used every 75 μm. A second method of reducing stiction is to use large cutouts in gears and other structures. In most cases, a surface-micromachined gear does not need the stiffness provided by the material between the teeth and the hub. A gear with spokes, shown in Figure 27.33, performs as well as one without spokes and has a reduced probability of stiction.

Another way of increasing the reliability of surface-micromachined mechanisms is to know the limitation of the technology. This is important as few practitioners in the field of MEMS can change a fabrication process without a compelling reason. Therefore, it is imperative that designers operate within the available fabrication and packaging processes and know the limitations of the fabrication process. Designers must be able to predict problems that might occur during the fabrication process and ensure that their designs will still work in spite of the problems. For example, a release process with an abundance of conductive particles is not compatible with designs that employ electrostatic comb drives because the particles may become lodged between the comb fingers and short them together. Thermal actuators over a nonconductive substrate would be a better alternative. Also, a nonhermetic packaging process would not be compatible with devices that are susceptible to humidity-induced stiction. A third example is that a release process that is prone to stiction is more compatible with designs that have limited surface area and are stiff in the vertical direction.

A final method of improving the reliability of surface-micromachined mechanisms is to design with regard to the packaging of the MEMS. While the package protects the surface-micromachined mechanism from particles, humidity and handling, it can also generate problems such as the outgassing of water vapor from the die-attach adhesive. Many types of electronic packages are not designed for surface-micromachined mechanisms both in terms of hermeticity and in interfacing the mechanism with the next assembly. Many times a MEMS device is designed without regard to the final package or the packaging process; however, better systems typically result if the packaging is contemplated at the beginning of the design process instead of at the end.

Acknowledgments

The authors would like to thank their colleagues in the Electromechanical Engineering and Intelligent Micromachines Departments at Sandia National Laboratories for their efforts. Many of the devices and concepts reported here are due to their efforts. Special thanks are due to Jim Allen, Steve Barnes, Jennifer Brecheisen, Chris Dyck, Ernie Garcia, Adam Harris, Larry Hostetler, Andy Jojola, Alice Kilgo, Bob Kipp, Sita Mani, Bill Miller, Marc Polosky, Rosemarie Renn, Danelle Tanner and Ed Vernon for their contributions to this effort. The authors are with Sandia National Laboratories. Sandia National Laboratories is a multiprogram laboratory operated by the Sandia Corporation, a Lockheed Martin Company, for the U.S. Department of Energy under contract DE-AC04-94AL8500.

Defining Terms

Beam: A structural member that can bend.
Buckling: Lateral deformations of a column caused by an axial load.
Column: A structural member that supports pure tension or compression.
Dimple: Small protusion on the bottom of a structure that acts to reduce the area in contact with the structure below.
Isotropic: The same in all directions.
Microengine: A surface-micromachined electric motor invented by Garcia and Sniegowski at Sandia National Laboratories [Garcia and Sniegowski, 1995].
Moment of inertia: A measure of the dispersion of area about the centroid. It is calculated when the axis of rotation is in the same plane as the cross section. In rectangular coordinates for simple situations it is given by:

$$I_x = \int_A y^2 dA \tag{27.54}$$

$$I_y = \int_A x^2 dA \tag{27.55}$$

Poisson's ratio: The ratio of transverse to axial strain when a body is subjected to an axial load. Its symbol is υ.
Polar moment of inertia: The moment of inertia for objects whose axis of rotation is normal to the cross section.

Stiction: Adhesion or sticking of smooth surfaces together by one or more of the following: capillary forces, van der Waals forces, electrostatic forces and hydrogen bonding.

Stress: The intensity of a body's reaction to a load.

Stress gradients: Variations in the stress of the film through the thickness of the film. They cause released cantilevers to either curl away from or toward the substrate and can warp other released structures.

Strain: The deformation per unit length.

Tribology: The study of friction, lubrication and wear.

Young's modulus: Also known as the modulus of elasticity, it is the ratio of stress to strain in a body undergoing elastic deformation in the normal direction. It is normally denoted by E.

References

Bhushan, B., Ed., (1999) *Handbook of Micro/Nano Tribology*, 2nd ed., CRC Press, Boca Raton, FL.

Brush, D.O., and Almroth, B.O. (1975) *Buckling of Bars, Plates, and Shells*, McGraw-Hill, New York.

Chironis, N.P., and Sclater, N., Eds. (1996) *Mechanisms and Mechanical Devices Sourcebook*, 2nd ed., McGraw-Hill, New York.

Dugger, M.T., Poulter, G.A., and Ohlhausen, J.A. (1999) "Surface Passivation for Reduced Friction and Wear in Surface Micromachined Devices," in *Proc. Fall MRS Symp.*, December, Boston, MA.

Fan, L.S. et al. (1990) "Spiral Microstructures for the Measurement of Average Strain Gradients in Thin Films," in *Proc. 1990 IEEE Conf. on Micro Electro Mechanical Systems (MEMS '90)*, pp. 177–181.

Fang, W., and Wickert, J.A. (1994) "Post-Buckling of Micromachined Beams," *J. Micromech. Microeng.* **4**(3), pp. 116–122.

Garcia, E.J. (1998) "Micro-Flex Mirror and Instability Actuation Technique," in *Proc. 1998 IEEE Conf. on Micro Electro Mechanical Systems (MEMS '98)*, pp. 470–474.

Garcia, E.J., and Sniegowski, J.J. (1995) "Surface Micromachined Microengine as the Driver for Micro-mechanical Gears," in *Digest Int. Conf. on Solid-State Sensors and Actuators*, June, Stockholm, Sweden, pp. 365–368.

Hutchinson, J.W., and Koiter, W.T., (1970) "Postbuckling Theory," *Appl. Mech. Rev.* **23**, pp. 1353–1366.

Johns, D.A., and Martin, K. (1997) *Analog Integrated Circuit Design*, John Wiley & Sons, New York.

Johnson, K.L. (1985) *Contact Mechanics*, Cambridge University Press, Cambridge, U.K.

Kovacs, G.T. (1998) *Micromachined Transducers Sourcebook*, WCB/McGraw-Hill, Boston.

Madou, M. (1997) *Fundamentals of Microfabrication*, CRC Press, Boca Raton, FL.

Polosky, M.A., Garcia, E.J., and Allen, J.J. (1998) "Surface Micromachined Counter-Meshing Gears Discrimination Device," in *SPIE 5th Annu. Int. Symp. on Smart Structures and Materials*, San Diego, CA.

Roark, R., and Young, W. (1989) *Formulas for Stress and Strain*, McGraw-Hill, New York.

Sharpe, W.N., Jr., LaVan, D.A., and Edwards, R.L. (1997) "Mechanical Properties of LIGA-Deposited Nickel for MEMS Transducers," in *Proc. 1997 Int. Conf. on Sensors and Actuators*, Chicago, IL, pp. 607–610.

Shigley, J.E., and Mischke, C.R. (1989) *Mechanical Engineering Design*, 5th ed., McGraw-Hill, New York.

Tanner, D.M. (Invited Keynote) (2000) *Proc. 22nd Int. Conf. on Microelectronics*, Nis, Yugoslavia, pp. 97–104.

Timoshenko, S.P. (1958) *Strength of Materials*, Van Nostrand, New York.

Timoshenko, S.P., and Gere, S.P. (1961) *Theory of Elastic Stability*, McGraw-Hill, New York.

Worthman, J.J., and Evans, R.A. (1965) "Young's Modulus, Shear Modulus and Poisson's Ratio in Silicon and Germanium," *J. Appl. Phys.* **36**, pp. 153–156.

For Further Reading

Among the several excellent sources of information on surface-micromachining and surface-micromachined devices are the recommended books by Madou (1997) and Kovacs (1998). Many conferences—for example, IEEE Micro Electro Mechanical Systems Conferences, biannual conferences on Solid State Sensors and Actuators (Transducers)—are mainly focused on device fabrication are also good sources for some good device and mechanism papers. SPIE and ASME also sponsor MEMS conferences or workshops. Some of the heavily read journals in the field of MEMS include *Sensors and Actuators* and the *Journal of Microelectromechanical Systems* (*J. MEMS*). Of the two, *J. MEMS* is the better source for surface-micromachined mechanism papers. For information on designing and building surface micromachines, both Cronos and Sandia provide information about their processes. The internet address for Cronos is http://www.memsrus.com/; for Sandia, the address is http://www.mdl.sandia.gov/Micromachine/. Good references on mechanical engineering and mechanism design include *Mechanical Engineering Design* [Shigley and Mischke, 1989] and the *Mechanisms and Mechanical Devices Sourcebook* [Chironis and Sclater, 1996].

28
Microrobotics

Thorbjörn Ebefors
Royal Institute of Technology

Göran Stemme
Royal Institute of Technology

28.1 Introduction 28-1
  MEMS as the Motivation for Robot Miniaturization
28.2 What is Microrobotics? 28-2
  Task-Specific Definition of Microrobots • Size- and
  Fabrication-Technology-Based Definitions of Microrobots •
  Mobility- and Functional-Based Definition of Microrobots
28.3 Where To Use Microrobots? 28-6
  Applications for MEMS-Based Microrobots • Microassembly
28.4 How To Make Microrobots? 28-10
  Arrayed Actuator Principles for Microrobotic
  Applications • Microrobotic Actuators and Scaling
  Phenomena • Design of Locomotive Microrobot Devices
  Based on Arrayed Actuators
28.5 Microrobotic Devices 28-14
  Microgrippers and Other Microtools • Microconveyers •
  Walking MEMS Microrobots
28.6 Multirobot System (Microfactories
  and Desktop Factories) 28-32
  Microrobot powering • Microrobot communication
28.7 Conclusion and Discussion 28-35

28.1 Introduction

28.1.1 MEMS as the Motivation for Robot Miniaturization

The microelectromechanical systems (MEMS) field has traditionally been dominated by silicon micromachining. In the early days, efforts were concentrated on fabricating various silicon structures and relatively simple components and devices were then developed. For describing this kind of microelectromechanical *structures* the acronym MEMs is used. A growing interest in manufacturing technologies other than the integrated circuit (IC)-inspired silicon wafer and batch MEMs fabrication is evident in the microsystem field today. This interest in alternative technologies has surfaced with the desire to use new MEMs materials, that enable a greater degree of geometrical freedom than materials that rely on planar photolithography as a means to define the structure. One such new MEMs material is plastic, which can be used to produce low-cost, disposable microdevices through microreplication. The micromachining field has also matured and grown from a technology used to produce simple devices to a technology used for manufacturing complex miniaturized systems which has shifted the acronym from representing *structures* to microelectromechanical *systems*. Microsystems encompass microoptical systems (microoptoelectromechanical systems, MOEMS), microfluidics (micro-total analysis systems, μ-TAS) etc. These systems contain micromechanical components including moveable mirrors and lenses, sensors, light sources, pumps and valves and passive components such as optical and fluidic waveguides, as well as electrical components and power sources of various types.

With the growing focus on systems perspective—for example, the integration of various functions such as sensors, actuators and processing capabilities into miniaturized systems (which may also be fabricated by different fabrication technologies in the MEMS field)—new problems arise. For example, how does one assemble these very small devices to form larger systems? As early as 1959 Professor Feynman addressed "the problem of manipulating and controlling things on small scale" in his famous APS talk [Feynman, 1960]. Feynman's solution to microassembling problems was the use of micromachines (or microrobotic devices as they will be named in this chapter) which would be used to build or assemble other micromachines (microsystems) consisting of different microdevices. This approach can be regarded as a hybrid technology or *serial* (one-at-a-time) fabrication/assembly, as opposed to the monolithic approach, where all integration is done on one single silicon chip, preferably using wafer-level assembly. Wafer-level assembly performed on several wafers at the same time (so-called batch fabrication) is a *parallel* (several-at-a-time) manufacturing process originally developed for the IC industry but now also commonly used for MEMS fabrication. General and good reviews of the different assembly approaches for MEMS have been presented [Cohn et al., 1998; *MSTnews*, 2000]. The different aspects of MEMS assembly will be further discussed in Section 28.3.2.

All systems, no matter what kind of components they consist of, must be assembled in some way. The assembly of complete microsystems can be done either *monolithically* (i.e., several systems simultaneously assembled at wafer level) or in a *hybrid* fashion (i.e., by serial assembly of several individual microcomponents fabricated by different technologies and on different wafers). In both cases *microrobotic* devices could simplify the assembly process (i.e., microrobots for making MEMS); however, these microrobots must be highly miniaturized. A common way to obtain miniaturization is by using MEMS technologies (i.e., MEMS for making microrobots). The benefits of scaling down all the subsystems of a robot to the same scale as the control systems by integrating motors, sensors, logic and power supplies onto a single piece of silicon have been discussed by Flynn et al. (1989). Enormous advantages can be obtained in the form of mass producability, lower costs and fewer connector problems encountered when interconnecting these discrete subsystems in comparison to the macro- (or miniature) robot concepts.

The hybrid microassembling approach, allowing microdevices to be fabricated by various techniques and in various materials to be assembled into more complex systems, is one of the driving motives for robotic miniaturization, in general, and MEMS-based microrobotics, in particular. Of course, miniaturized robots are also very attractive for assembling all kinds of miniaturized components (e.g., watch components). Besides the use of microrobotic devices for assembling purposes, other application fields such as medical technology may benefit greatly from the use of miniaturized robot devices. One such example is the steerable catheters used for minimal invasive surgery (MIS) shown in Figure 28.1. Such microrobotic catheters have been fabricated using MEMS technologies [Haga et al., 1998; Park and Esashi, 1999].

28.2 What is Microrobotics?

Today we can see a growing worldwide interest in microrobotic devices and their potential, including micromanipulation tools and microconveyers and/or microrobots as locomotive mechanisms [Ebefors, 2000]. The term *microrobotics* covers several different types of small robot devices and systems. The term mainly refers to the complete robot system—that is, what type of work the microrobot(s) perform, how to accomplish those tasks and other system-oriented aspects. Often, the term *microrobotics* also includes more fundamental buildings blocks, such as steerable links, microgrippers, conveyer systems and locomotive robots, as well as the technologies used to fabricate these devices and the control algorithms used to carry out the robot's task. The term *microrobotic* can be compared with the MEMS (or MEMs) term in that the former often encompasses not only the system itself, but also fabrication and material issues, as well as simple devices used as building blocks for the system. Although the microrobotics field currently is an area of intensive research, no clear definition of the term *microrobot* exists. As stated by other authors reviewing the microrobotics field [Dario et al., 1992; Fatikow and Rembold, 1997], many "micro" terms such as *micromechatronics, micromechanism, micromachines* and *microrobots* are used synonymously to

FIGURE 28.1 Example of a microrobotic application. In minimal invasive surgery (MIS), MEMS-based microrobotic devices such as steerable catheters, microgrippers and other microtools are expected to have a deep impact on endoscopic applications in the near future. (Illustration published with permission from Surgical Vision, Ltd., U.K. Courtesy of Graham Street.)

indicate a wide range of devices whose function is related to the "fuzzy" concept of operating at a "small" scale; however, "small" scale is a relative term so a more clear definition is needed.

28.2.1 Task-Specific Definition of Microrobots

The basic definition of a microrobot parallels the features attributed to a robot in the macroworld; i.e., a microrobot should be able to move, apply forces, manipulate objects, etc. in the same way as a "macrorobot." The obvious difference between a macrorobot and a microrobot is the size of the robot. Thus, one definition of a microrobot is "a device having dimensions smaller than classical watch-making parts (i.e., μm to mm) and having the ability to move, apply forces and manipulate objects in a workspace with dimensions in the micrometer or submicrometer range" [Johansson, 1995]. However, in many cases it is important that the robot can move over much larger distances. The above-stated task-specific definition is quite wide and includes several types of very small robots as well as stationary micromanipulation systems, which are a few decimeters in size but can carry out very precise manipulation (in the micron or even nanometer range) [Fatikow and Rembold, 1997]. Because the fabrication technologies used to fabricate these devices play an important role, another more precise subdivision is desirable in order to help identify the practical capabilities of the different technologies.

28.2.2 Size- and Fabrication-Technology-Based Definitions of Microrobots

Following the classification scheme of microrobots used by Dario et al. (1992) and Fatikow and Rembold (1997), one can categorize the microrobot into many different groups. In the definition made by Dario et al., the microrobots were separated into three different subcategories, each characterized by the fabrication technology used to obtain the robot and the size of the device, as illustrated in Table 28.1.

Some good examples of "miniature robots" are those that competed in the annual "Micro Robot Maze Contest 1992–99" at the International Symposium on Micromachine and Human Science (MHS) in Nagoya, Japan [Dario et al., 1998; Ishihara, 1998]. MEMS-based microrobots will be further described in the rest of this chapter.

While MEMS-based microrobots consist of billions of atoms and are fabricated using photolithography techniques and various etching methods, nanorobots will most likely be built by assembling individual

Table 28.1 Classification of Microrobots According to Size and Fabrication Technology

Robot Class	Size and Fabrication Technology
Miniature robots or minirobots:	Having a size on the order of a few cubic centimeters and fabricated by assembling conventional miniature components as well as some micromachines (such as MEMS-based microsensors)
MEMS-based microrobots (or microrobots[a])	Defined as a sort of "modified chip" fabricated by silicon MEMS-based technologies (such as batch-compatible bulk or surface micromachining or by micromolding and/or replication method) having features in the micrometer range
Nanorobots	Operating at a scale similar to the biological cell (on the order of a few hundred nanometers) and fabricated by nonstandard mechanical methods such as protein engineering

[a] To distinguish a MEMS-based microrobot with *micro*meter-sized components from the whole class of microrobots (including mini-, micro- and nanorobots), several more or less confusing notations have been proposed. In this publication, the term "MEMS-based microrobot" is introduced and used. The term "MEMS-based microrobot" differs from the notation originally used by Dario et al., but the content is the same.

Source: Adapted from Dario et al., 1992.

atoms or molecules one at a time. Instead of using conventional mechanical principles, an approach based on chemical self-assembly will most likely be the technology used to fabricate nanorobots. Examples of these nanorobots (nanomachines or molecular machines) and some proposed fabrication methods used to achieve the different components, such as gears, bearings and harmonic drives, needed for the realization of a nanorobot are found in the textbook of Drextler (1992). In 1999, the first reports of a successful molecular motor were presented [Kelly et al., 1999]. That nanomotor, consisting of a 78-atom molecular paddle wheel, was able to rotate one third of a complete revolution in approximately 3 to 4 hours. Researchers in the field of theoretical physics pushing for further miniaturization have introduced the concept of quantum machines and quantum computers, which will consist of stretched parts of a single atom [Wilson, 1997]. If we stretch the term *microrobotic* a bit, devices such as the atomic force microscope (AFM) or scanning tunneling microscope (STM) could also be categorized as microrobotic tools. The sharp probe tip of these microscope devices, which most often are fabricated using MEMS technologies, have been used to manipulate single atoms [Eigler and Schweizer, 1990]. To extend this further, nanorobots and other nanoelectromechanical systems (NEMS) may be fabricated using sharp silicon tips and other microrobotic tools (i.e., MEMS for nanorobots and NEMS).

28.2.3 Mobility- and Functional-Based Definition of Microrobots

Besides classification by task or size, microrobots can also be classified by their mobility and functionality [Hayashi, 1991; Dario et al., 1992]. Many robots, including the three different classes of microrobots mentioned in Table 28.1, usually consist of sensors and actuators, a control unit and an energy source. Depending on the arrangement of these components, one can classify microrobots according to the following criteria:

- Locomotive and positioning possibility (yes or no)
- Manipulation possibility (yes or no)
- Control type (wireless or wires)
- Autonomy (energy source onboard or not onboard)

Figure 28.2 illustrates 15 different possible microrobot configurations by combining the four criteria. As depicted in Figure 28.2, the classification is dependent on the following microrobot components: the control (CU), the power source (PS), the actuators necessary for moving the robot platform (i.e., robot drive for locomotion and positioning; AP), and the actuators necessary for operation (i.e., manipulation

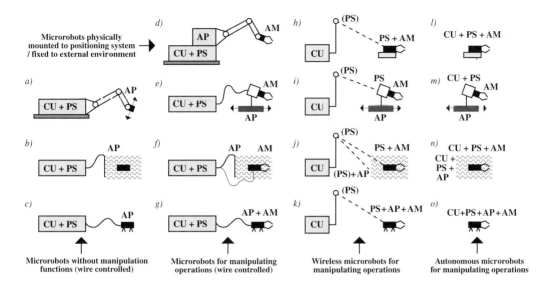

FIGURE 28.2 Classification of microrobots by functionality (modification of earlier presented classification schemes [Hayashi, 1991; Dario et al., 1992; Fatikow and Rembold, 1997]). CU indicates the control unit; PS, the power source or power supply; AP, the actuators for positioning; AM, the actuators for manipulation.

using robot arms and hands; AM). Besides the different actuation functions, sensory functions are also needed—for example, tactile sensors for microgrippers or charge-coupled device (CCD) cameras for endoscope applications (compare Figures 28.2d and a).

The ultimate goal is to create a fully autonomous, wireless mobile microrobot equipped with suitable microtools according to Figure 28.2o. Because this is a very difficult task, a good start is to investigate the possibility of making silicon microrobot platforms that are steered and powered through wires, like the one in Figure 28.2c, and to study their locomotion capability.

The majority of MEMS-based microrobotic devices developed so far could be categorized as moveable links—microcatheters [Haga et al., 1998; Park and Esashi, 1999], according to Figure 28.2a, or microgrippers [Kim et al., 1992; Keller and Howe, 1995; 1997; Greitmann, 1996; Keller, 1998a; 1998b; Ok et al., 1999], as those in Figure 28.2d, or the microgrippers [Jager, 2000a, 2000b] shown in Figure 28.2e. Among the research publications covering locomotive microrobots, most publications have addressed microconveyance systems (Figure 28.2b) [Kim et al., 1990; Pister et al., 1990; Ataka et al., 1993a; 1993b; Konishi and Fujita, 1993; 1994; Goosen and Wolffenbuttel, 1995; Liu et al., 1995; Böhringer et al., 1996; 1997; Nakazawa et al., 1997; Suh et al., 1997; 1999; Hirata et al., 1998; Kladitis et al., 1999; Nakazawa et al., 1999; Ruffieux and Rooij, 1999; Smela et al., 1999]; see Table 28.3 for more details. Robots using external sources for locomotion could be used (compare Figures 28.2b, f, j, n). According to Fatiokow and Rembold (1997), several researchers are working on methods to navigate micromechanisms through human blood vessels; however, these microrobots are difficult to control. Examples of partially autonomous systems (compare Figure 28.2j) are the concept for so-called "smart pills." Centimeter-sized pills for sensing temperature and/or pH inside the body have been presented [Zhou, 1989; Uchiyama, 1995] as well as pills equipped with video cameras [Carts-Powell, 2000]. The pill is swallowed and transported to the part of the body where one wants to measure or record a video sequence. The information of the measured parameter or the signals from the camera are then transmitted (telemetrically) out of the body. More sophisticated approaches involving actuators (AM) for drug delivery of various kinds have also been proposed [Uchiyama, 1995; Fatikow and Rembold, 1997]. The position of the pill inside the body is located by an X-ray monitor or ultrasound. As soon as the pill reaches an infected area, a drug encapsulated in the pill can be released by the actuators onboard. External communication could be realized through radio signals.

Several important results have been presented regarding walking microrobots (see Figures 28.2c,g,k) fabricated by MEMS technologies and batch manufacturing. Different approaches for surface-micromachined robots [Yeh et al., 1996; 2000; Kladitis, 1999] and for a piezoelectric dry-reactive-ion-etched microrobot should be mentioned. A suitable low-power Application Specific Integrated Circuit (ASIC) for robot control has been successfully tested and is planned to be integrated on a walking microrobot [Ruffieux and Rooij, 1999]. The large European Esprit project *MINIMAN* (1997) has the goal of developing moveable microrobotic platforms with integrated tools with six degrees of freedom for applications such as microassembly within a scanning electron microscope (SEM) and involves different MEMS research groups from several universities and companies across Europe. One of the MINIMAN robots, as well as other MEMS-based microrobots, will be further described in Section 28.5.3. Further, miniature robot systems with MEMS/MST (MicroSystem Technology) components have been developed [Breguet, 1996; Breguet and Renaud, 1996]. Several research publications on "gnat" minirobots and actuator technologies for MEMS microrobots [Flynn et al., 1989; 1992] were reported by U.S. researchers in the early 1990s, and also several groups in Japan are currently developing miniaturized robots based on MEMS devices [Takeda, 2001]. In Japan, an extensive 10-year program on "micromachine technology," supported by the Ministry of International Trade and Industry (MITI), started in 1991. One of the goals of this project is to create micro- and miniature robots for microfactory, medical technology and maintenance applications. Several microrobotic devices including locomotive robots, and microconveyers have been produced within this program. Miniature robot devices or vehicles [Teshigahara et al., 1995] for locomotive tasks, containing several MEMS components, have been presented. Even though great efforts have been made on robot miniaturization using MEMS technologies, no experimental results on MEMS batch-fabricated microrobots suitable for autonomous walking (i.e., robust enough to be able to carry its own power source or to be powered by telemetric means) have been presented yet (as of year 2000). The first batch-fabricated MEMS-based microrobot platform able to walk was presented in 1999 [Ebefors et al., 1999] (see Figures 28.17 and 28.23). However, this robot was powered through wires and not equipped with manipulation actuators. Besides the walking microrobotic devices, several reports on flying [Arai et al., 1995; Mainz, 1999; Miki and Shimoyama, 1999; 2000] and swimming [Fukuda et al., 1994; 1995] robots have been published. Micromotors and gear boxes made using LIGA technology (a high-precision, lithographically defined plating technology) are used to build small flying microhelicopters, which are commercially available from Institute of Microtechnology in Mainz, Germany, as rather expensive demonstration objects [Mainz, 1999]. Besides the pure mechanical microrobots, hybrid systems consisting of electromechanical components and living organisms such as cockroaches have also been reported [Shimoyama, 1995a].

28.3 Where To Use Microrobots?

28.3.1 Applications for MEMS-Based Microrobots

Already in the late 1950s, Feynman (1961) foresaw the utility and possibility for miniaturization of machines and robots. According to Feynman, the concepts of *microfactory* and *microsurgery* (based on tiny microdevices inserted in a patient's blood vessels) were very interesting possibilities for small machines.

Microrobotic research has gone from being only theoretical ideas in the 1960s and 1970s to actual building blocks when complex systems and "micromachines" on a chip started to emerge in the late 1980s. These building blocks took the form of surface-micromachined micromotors made of polysilicon fabricated on a silicon chip [Muller, 1990]. In the late 1980s and early 1990s, more concrete suggestions on how one should realize MEMS-based microrobotic devices using such micromotors and potential applications were published [Flynn et al., 1989; Trimmer and Jebens, 1989]. Now, 10 years later one finds research publications on microrobotic devices for practical use in various fields. One of these application fields is medical technology [Dario et al., 1997; Haga et al., 1998; Tendick et al., 1998; Park and Esashi, 1999]. In surgery, the use of steerable catheters and endoscopes, as illustrated in Figure 28.1, is very attractive and they may be further miniaturized by MEMS-based microrobotic devices, allowing advanced computer-assisted surgery (CAS) and even surgery over the Internet. For such applications, smaller and more flexible active endoscopes that can react to instructions in real time are needed in order to assist the surgeon. They may enter into the blood vessels and enter various cavities (anglioplasty) by remote control, where they carry out complex measurements and manipulations (gripping, cutting, applying tourniquets, making incisions, suction and rinsing operations etc.) To meet these demands, an intelligent endoscope must have a microprocessor, several sensors and actuators, a light source and possibly an integrated image processing unit. These microrobotic devices will revolutionize classical surgery, but their realization is still a problem because of friction, poor navigability, biocompatibility issues, etc.; they are also not small enough yet [Fatikow and Rembold, 1997].

Other interesting areas for microrobotic devices are production (microassembly [Fatikow and Rembold, 1996; Cohn et al.,1998] and microfactory [Ishikawa, 1997; Kawahara et al., 1997]), metrology (automated testing of microelectronic chips, surface characterization, etc.) [Li et al., 2000], inspection and maintenance [Suzumori et al., 1999; Takeda, 2001], biology (capturing, sorting and combining cells [Ok, 1999]), bio-engineering [Takeuchi and Shimoyama, 1999] and microoptics (positioning of microoptical chips, microlenses and prisms) [Frank, 1998]. Many of these applications require automated handling and assembly of small parts with accuracy in the submicron range. Different approaches for microassembly, among them microrobotic-assisted assembly, are described in the following section.

28.3.2 Microassembly

Two different aspects of MEMS microrobotics and microassembly should be addressed. The first aspect is the use of MEMs for making microrobots using the silicon monolithic chip fabrication technology (several robots at a time; see Table 28.1). Because many microrobotic devices need three-dimensional (i.e., out-of-plane) arms and legs, the folding of these component is an essential assembly issue and constitutes one side of the assembling technology (i.e., MEMs for making microrobots). The second aspect is the use of microrobots for assembling more complex microsystems (i.e., microrobots for making MEMS). The use of microrobots for assembly of microcomponents and devices (i.e., fundamental building blocks) into complete microsystems is classified as hybrid manufacturing (one system at a time). Following is an introduction to and background of the two different assembling approaches.

The MEMS fabrication technology, especially surface-micromachining, allows batch-fabricated mechanisms to be preassembled *in situ*, and all structures on the entire wafer are released simultaneously (parallel assembly) by etching of sacrificial layers. This, together with the ability to integrate electronics on the chip, is a very attractive feature. Micromechanical systems of impressive complexity such as linear motors, rotating gears and linkages between these components have been made with alignment tolerances in the micrometer range by the use of parallel fabrication methods [Rodgers et al., 1999]. In the MOEMS and microrobotic fields, several devices make use of both parallel and serial assembling. To fold micromirrors out of the wafer plane to achieve, for example, microoptical benches, several serially addressed motors fabricated by standard parallel processing (i.e., lithography, thin-film depositions and etching) can be used to achieve the assembling of each individual micromirror. These mirrors are controlled externally by electrical signals connected to the integrated micromotors [Rodgers et al., 1999]. Figure 28.3f shows a complex free-space microoptical bench (FSMOB) that integrates several different optical elements [Lin et al., 1996]. To implement free-space optics monolithically, tall out-of-plane optical elements (moveable mirrors lenses etc.) with optical axes parallel to the substrate are needed. A common way to realize this kind of system is by using the surface-micromachined microhinge technology [Pister et al., 1992], which enables out-of-plane rotated optical elements to be fabricated using planar processes as shown in Figures 28.3a to e. The planar elements are then assembled into three-dimensional structures after fabrication either by built in-motors for self-assembly or by the use of microrobotic devices and external manipulation. Also integrated in the FSMOB system, shown in Figure 28.3, is a self-alignment technique that incorporates hybridly mounted active optoelectronic devices such as semiconductor lasers and vertical cavity surface-emitting lasers [Lin et al., 1996]. One key advantage of FSMOB is that the optics can be "prealigned" during the design stage by CAD.

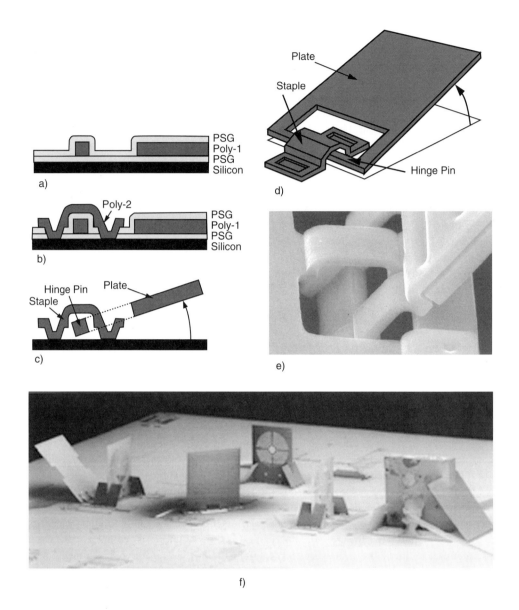

FIGURE 28.3 (a)–(c) Cross sections of a microhinge during fabrication. (d) A perspective view of a hinged plate after release by external manipulators. (e) SEM photograph of a polysilicon hinge. (f) Example of a complex microoptical system based on out-of-plane rotated optical elements using the surface-micromachined microhinge and hybrid mounting of a laser [Lin et al., 1996]. (Photograph of the optical bench taken by David Scharf and published with permission. Courtesy of M. C. Wu, UCLA and K. Pister, BSAC-Berekley.)

For some applications with very complex out-of-plane folding, the serial self-assembly technology using integrated micromotors is not feasible; therefore, quasi-manual or semi-automated assembly using external manipulators is required. To assemble several folded microstructures into more complex out-of-plane devices, MEMS-based microrobotic tools are used mainly because of their precise and small dimensions. Different MEMS technologies can be used to achieve out-of-plane rotated structures [Ebefors, 2000]. Besides MOEMS applications, out-of-plane rotated microstructures are also essential for the realization of most MEMS-based microrobotic devices. However, for most microrobot application the robustness of surface-micromachined legs and arms that are folded out of the wafer are limited.

Therefore, bulk micromachining is an attractive technology to fabricate more robust legs and arms for various microrobots.

Besides by use of surface-micromachining, impressive and complex microsystems can also be achieved using bulk micromachining where different kinds of wafers can be stacked on top of each other by bonding [Tong and Gösle, 1999]. However, by using bulk micromachining the size is generally one order of magnitude larger than for surface micromachined devices. Several packaging schemes have been used to assemble complete microsystems on the wafer level (i.e., parallel batch fabrication) [Corman, 1999]. However, for many applications the use of lithography, etching and bonding alone cannot fulfill the requirements on the system level. Monolithic integration of electronic and micromechanical functions is not always the most suitable. By combining microelectronics, micromechanics, microfluidics and/or microoptics into a single fabrication process one faces the risk that one may have to compromise the performance of all subsystems.

A disadvantage for all integration of electronic and mechanical components is that the electronics (complementary metal oxide semiconductor [CMOS], bipolar CMOS [BiCMOS], etc.) and MEMS fabrication processes often are not compatible, due to thermal budgets, wafer size, etc. Wafer bonding (flip chip, fusion bonding, etc.) and the monolithic integration approaches of electronics and MEMS on the same chip are then not realistic assembling alternatives. In those cases, the serial assembly (i.e., "pick-and-place") can be the only assembly approach that allows integration of electrical and mechanical components. Such micro "pick-and-place" systems can be achieved by microrobotic devices in the form of microtweezers and microgrippers [Kim et al., 1992; Keller and Howe, 1995; 1997; Greitmann, 1996; Keller, 1998a; 1998b; Ok et al., 1999] and will be further described in Section 28.5.1. To extend this serial assembling approach further, other microrobotic components such as conveyers and miniaturized robots (preferably fabricated using MEMS technologies) are also required; see Section 28.6 regarding the concepts on microfactories and desktop micromanipulation stations.

Besides the monolithic assembling on the wafer level and the serial "pick-and-place" assembly, parallel approaches for hybrid assembling are also possible. Pister et al. [see Tahhan et al., 1999] presented a concept where folded boxes using three-dimensional microstructures based on aluminum hinges were used to obtain wafer-sized pallets for automated microparts assembly and inspection, as illustrated in Figure 28.4. The microparts (e.g., small diced MEMS or electronic chips) are supposed to be randomly transported on the pallet, by an external vibration field or by integrated microconveyers (see Section 28.5.2). The boxes with three walls, connected with microlocks, have one opening to capture the conveyed microparts. The microparts are sorted by their size or geometrical dimensions and kept in place by the walls of the boxes for handling, testing or assembling tasks. In the proposed concept, the fixtures should have integrated sensors and actuators by which the clamps actively close the box when the parts enter the fixture. For this concept, large fixture arrays and electrostatic or optical sensors integrated into the fixture cell should trigger the clamping function. Each cell operates autonomously and no global control should then be necessary.

Parallel sorting and assembling can also be obtained by other MEMS-based microconveyer strategies. Autonomous distributed micromachines (ADMs) are composed of several microcells and are smart enough to decide their behavior by themselves (i.e., by the use of integrated sensors) [Konishi and Fujita, 1995]. This concept has been validated for conveyance systems using a computer simulation model. Theories on programmable vector fields for advanced control of microconveyance systems have been presented [Böhringer et al., 1997]. Recently, these algorithms, which do not require sensing or feedback, were experimentally tested using polyimide thermal bimorph ciliary microactuator arrays with integrated CMOS electronics [Suh et al., 1999].

Besides the assembling of complete microsystems, other types of assembling are also required in MEMS. One such example, which already has been mentioned, is the need for three-dimesnsional MEMS structures for which one wants to create structures or sensor elements with feature sizes so small (micron or submicron) that they require photolithography in all three dimensions. Because normal IC-inspired lithography is planar, one commonly used method is to fabricate all structures requiring high-resolution patterns in the plane and then in a final process step-fold these structures out-of-plane using various

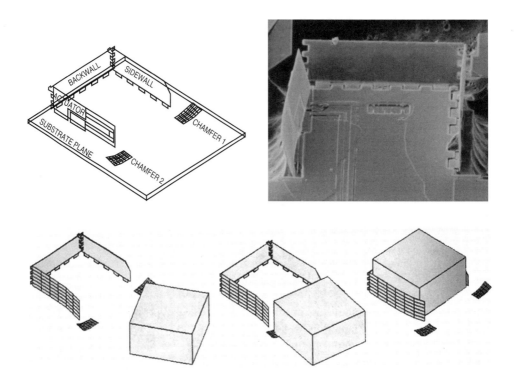

FIGURE 28.4 Illustrations and SEM micrograph of a folded fixture cell for automated microparts assembly and inspection. From: http://www.ee.washington.edu/research/mems/. Illustration published with permission and courtesy of K. Böhringer, University of Washington.)

types of microhinges or microjoints [Ebefors et al., 1998b]. The out-of-plane microstructure assembling can be obtained in the same way as the system level assembling by: (1) serial self-assembling (built-in actuation by micromotors); (2) parallel batch self-assembly (using built-in mechanisms such as surface tension forces [Green et al., 1995; Syms, 1998; 1999] or volume shrinkage effects in polymers [Ebefors, 1998a]) for the assembling; or (3) the use of external manipulators (not self-assembly). Self-assembling of three-dimensional microstructures have been obtained using a variety of techniques [Ebefors, 2000]. All of these techniques are of great interest, as manual assembly is very time, labor and cost intensive. However, since self-assembly is not feasible in all applications, quasi-manual or semi-automatic assembling using microrobotic tools is one way of reducing costs and time. Complex three-dimesnsional structures, such as the clock tower model shown in Figure 28.5, have been assembled using MEMS-based microrobotic tools connected to motorized micromanipulators, as illustrated in Figure 28.9 [Hui et al., 2000]. For most MEMS-based microrobotic devices robust out-of-plane working structures are essential (e.g., arms and legs fabricated by bulk micromachining). To be useful as microrobots, these structures also need to produce large strokes and forces when actuated. The next section describes how this can be realized.

28.4 How To Make Microrobots

During the design of a MEMS-based microrobotic device, trade-offs among several parameters such as range of motion, strength, speed (actuation frequency), power consumption, control accuracy, system reliability, robustness, force generation and load capacity must be taken into consideration. These parameters strongly depend on the actuation principle. In the following section, the physical parameters influencing the performance, as well as different aspects, of actuators are described.

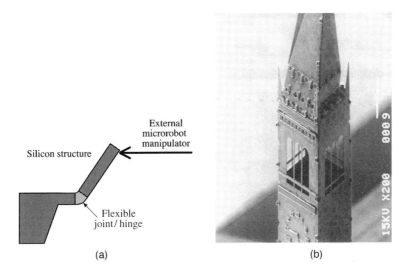

FIGURE 28.5 (a) Illustration of an out-of-plane folded silicon microstructure using an external microrobot tool for the assembly manipulation. The three-dimensional silicon structure which is structured in-plane could be fabricated both with surface- and bulk-micromachining techniques. (b) Example of a complex three-dimensional structure assembled by a MEMS-robotic tool. The three-dimensional model shows the University of California, Berkeley, Companile clock tower having a total height of 1.8 mm above the wafer surface. The microassembly was made manually by the use of the microtool shown in Figure 28.9. (Published with permission and courtesy of Hui Elliot, BSAC-Berkeley, CA.)

28.4.1 Arrayed Actuator Principles for Microrobotic Applications

For most microactuators, the force that can be generated and the load the actuator withstands are limited. This constitutes a fundamental drawback for all miniaturization in the robotic field. However, by using MEMS microfabrication it is easy to fabricate large arrays of actuators where the actuators are working together and thereby increasing the total force and load capability. This approach, which takes advantage of lithographic fabrication, also has the potential for integrating electronics on the robot chip to carry out more sophisticated tasks (e.g., autonomous microrobots). The concept of array configuration of microactuators, where the cooperative work of many coordinated simple actuators generates interaction with the macro world, was introduced by Fujita and Gabriel in 1991 [Fujita and Gabriel, 1991]. They called their technique *distributed micromotion systems* (DMMS). The driving schemes for the actuators in the array can be in either a synchronous mode (all actuators are switched on and off at the same time) or an asynchronous mode (different actuators are switched on and off at different times). Most often, the asynchronous mode is favorable as it is more effective and smoother movements can be achieved. Intelligent control of the actuation schemes for the actuators can be achieved by integration of sensors in the system [Suh et al., 1997]. Based on information from these sensors (i.e., object weighing and/or positioning), advanced control of each actuator can be obtained for improved functionality of the micromotion system.

28.4.2 Microrobotic Actuators and Scaling Phenomena

Several theoretical reports on the scale effects and the applicability of various actuation principles for microrobotic devices have been published [Trimmer and Jebens, 1989; Dario et al., 1992; Shimoyama, 1995b; Fearing, 1998; Thornell, 1998]. As stated in the previous section, most microactuators generate relatively small forces and strokes, but these could somewhat be increased by array configurations. However, there are large variations between how different actuation principles scale to microscale; some principles scale more favorably than others. Some actuators (for example, piezoelectric- and electrostatic-based actuators) have the advantage of low power consumption and can be driven at high speeds (kHz regime and above).

TABLE 28.2 Comparison of a Selection of Microactuators for Microrobotic Applications

Actuator Type	Volume (10^{-9} m^3)	Speed (s^{-1} or rad/s*)	Force (N) / Torque (Nm)**	Stroke (m)	Power Density (W/m^3)	Power Consumption (W)	Ref.
Linear electrostatic	400	5000	10^{-7}	6×10^{-6}	200	NA	[Kim et al., 1992]
Rotational electrostatic	$\pi/4 \times 0.5^2 \times 3$	40*	$2 \times 10^{-7**}$		900	NA	[Nakamura et al., 1995]
Rotational piezoelectric	$\pi/4 \times 1.5^2 \times 0.5$	30*	$2 \times 10^{-11**}$		0.7	NA	[Udayakumar et al., 1991]
Rotational piezoelectric	$\pi/4 \times 4.5^2 \times 4.5$	1.1*	$3.75 \times 10^{-3**}$		90×10^3	2.5% efficiency	[Bexell and Johansson, 1999; Johansson, 2000]
Linear magnetic	$0.4 \times 0.4 \times 0.5$	1000	2.9×10^{-6}	10^{-4}	3000	NA	[Liu et al., 1994]
Scratch drive actuator (SDA)	$0.07 \times 0.05 \times 0.5$	50	6×10^{-5}	160×10^{-9}	300	NA	[Akiyama and Fujita, 1995]
Rotational magnetic	$2 \times 3.7 \times 0.5$	150*	10^{-6**}		3×10^3	0.002% efficiency	[Teshigahara et al., 1995]
Rotational magnetic	$10 \times 2.5 \times ??$	20*	$350 \times 10^{-**}$		3×10^4	8% efficiency	[Stefanni et al., 1996]
EAP (Ppy–Au bimorph)	$1.91 \times 0.04 \times 0.00008$	0.2		1.25×10^{-3}	1.4×10^4	0.2% efficiency	[Smela et al., 1999]
Thermal polysilicon heatuator	Approx. $0.27 \times 0.02 \times 0.002$	2	$1/96 \times 30 \times 10^{-3}$	3.75×10^{-6}	2×10^4	NA	[Kladitis et al., 1999]
Thermal bimorph polyimide	Approx. $0.4 \times 0.4 \times 0.01$	1–60	69×10^{-6} N/mm^2	2.6–9×10^{-6}	$<10^4$	$\approx 16.7 \times 10^{-3}$	[Suh et al., 1997]
PGV-joint actuator	$0.75 \times 0.6 \times 0.03$	3–300	10^{-3}	10–150×10^{-6}	10^5	200 mW; 0.001% efficiency	[Ebefors, 2000]

* A rotational actuator that has speed measured in rad/s compared to a linear actuator (s^{-1}).
** It is impossible to separate the torque = force × stroke into force and stroke for a rotational actuator; therefore, only the torque is given.

However, they generally show relatively low force capacity and small strokes, while others such as magnetic and thermal actuation principles have the potential to exert large forces and displacements when the driving speed is the limiting factor. The thermal and magnetic actuator principles rely on significant currents and power levels which may require forced cooling. During the design of a microrobotic device, the trade-offs among range of motion, strength, speed (actuation frequency), power consumption, control accuracy, system reliability, robustness, load capacity, etc. must be taken into consideration. In an extensive actuator review focusing on microrobotics [Fearing, 1998], the general conclusion was that actuators generating large strokes and high forces are best suited for microrobotics applications. The speed criteria are of less importance as long as the actuation speed is in the range of a couple of hertz and above. Table 28.2 provides some data on stroke and force generation capability as well as power densities and efficiencies for a small selection of actuators suitable for microrobotic applications.

28.4.3 Design of Locomotive Microrobot Devices Based on Arrayed Actuators

Living organisms very often offer good models for designing microrobotic systems [Ataka et al., 1993; Zill and Seyfarth, 1996]. Mimicry of a six-legged insect's gait [Zill, 1996] has been proposed for the design of multilegged robots implemented using microfabrication techniques [Kladitis, 1999]. The first

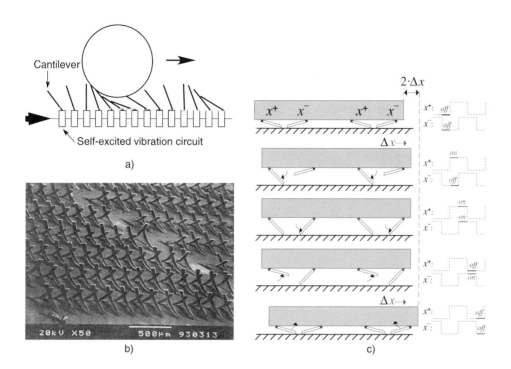

FIGURE 28.6 The ciliary motion principle used in most contact-based microrobotic systems. (a) Conveyance system used in nature to convey object. The cantilevers (ciliary hair) are actuated by a certain phase shift. (b) SEM photograph of the first microconveyer system based on the ciliary motion principle using bimorph polyimide legs. (Illustrations published with permission and courtesy of H. Fujita, University of Tokyo, Japan.) (Note: A video of the CMS microconveyance systems is available at http://www.s3.kth.se/mst/research/dissertations/thorbjornedoc/multimedia.html.) (c) A two-phase CMS for a walking microrobot platform when implemented using MEMS technology for out-of-plane working three-dimensional actuators. By using two rows of legs it is possible to steer in the forward–backward, left–right and up–down directions. More details of the robot are given in Section 28.5.3.1. (Note: A video animation on the dynamic, asynchronous operation of arrayed out-of-plane erected legs for micro robotics is available at http://www.s3.kth.se/mst/research/dissertations/thorbjornedoc/multimedia.html, Figure 47.)

proposed [Benecke, 1988] and realized [Ataka et al., 1993] MEMS contact conveyance transportation system was based on the ciliary motion principle adopted from nature. The principle for a *ciliary motion system* (CMS) is illustrated in Figure 28.6. The CMS principle relies on an asynchronous driving technique which requires at least two spatially separated groups of actuators turned on and off at different times, alternately holding and driving the device. Higher speeds and smoother motions can be achieved with such asynchronous driving than with synchronous driving.

For locomotive microrobotic applications such as conveyance systems and walking robots it is essential to have actuators that can both generate forces to lift an object or the robot itself out from the plane (i.e., to avoid surface sticking) and also generate forces that cause in-plane movements. Two fundamental principles exist, the first being contact-free (CF) systems. Here, different force fields, such as electrostatic, magnetic or pneumatic forces, are used to create a cushion to separate the object from the surface, as illustrated in Figure 28.7. To drive the device in the in-plane direction, these force fields need to have a direction dependence to force the device forward (i.e., directed air streams for a pneumatic system [Hirata et al., 1998] or the Meissner effect for levitation and orthogonal Lorentz force for driving a magnetic actuator system [Kim et al., 1990]).

The second fundamental principle is contact (C) systems, where a structure is in contact with the moving object (e.g., legs for walking), as illustrated in Figure 28.6c). To avoid surface stiction, these structures must create out-of-plane movements. A review of different techniques available to create such

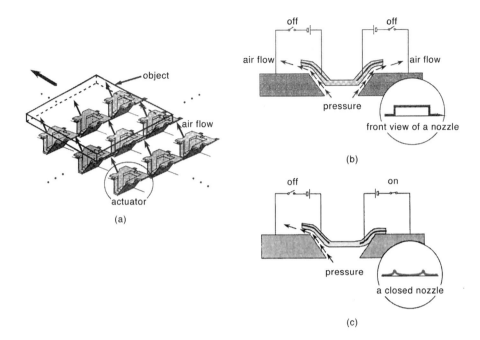

FIGURE 28.7 Electrostatic-controlled pneumatic actuators for a one-dimensional contact-free conveyance system. (a) Concept for the arrayed pneumatic conveyer (i.e., contact-free operation). (b) and (c) Mechanism for flow control by electrostatic actuation of nozzle when (b) the electrostatic nozzle is in the normal situation (off) and (c) when voltage is applied to one electrode (on). (Illustration published with permission and courtesy of H. Fujita, University of Tokyo.)

three-dimensional actuators have been presented [Ebefors, 2000]. Most of these techniques could be arranged in array configurations for distributed micromotion systems (DMMS).

For out-of-plane actuators using an external force field it is difficult to control each individual actuator in a large array of folded structures. Therefore, a synchronous jumping mode is used to convey the objects or move the device itself. This jumping mode involves quick actuation of all the actuators simultaneously, which forces the object to jump. When the object lands on the actuators (located in their off position), the object has moved a small distance and the actuators can be actuated again to move (walk or convey) further [Liu et al., 1995].

A critical aspect of large distributed micromotion systems based on arrayed actuators and distributed (or collective) actuation is the problem associated with the need for the very high yield of the actuators [Ruffieux and Rooij, 1999]. Just one nonworking actuator could destroy the entire locomotion principle. Therefore, special attention must be paid to achieve high redundancy by parallel designs wherever possible. These aspects will be further addressed in Section 28.5.3.

28.5 Microrobotic Devices

As pointed out in Section 28.2, a microrobotic device can be either a simple catheter with a steerable joint (see Figure 28.2a), or a complex autonomous walking robot equipped with various microtools as in Figure 28.2o). Between these two extremes one finds microgrippers and microtools of various kinds, as well as microconveyers and walking microrobot platforms. Each of these three microrobotic devices will be presented more in detail in the following discussion. Section 28.6 describes more complex microrobotic systems where both microtools and actuators for locomotion are integrated to form so-called microfactories or desk-to-manipulation stations. Also, multirobot systems and communication between microrobots in such multirobot systems will be discussed.

28.5.1 Microgrippers and Other Microtools

The first presented microrobotic device was based on in-plane electrostatic actuation [Kim et al., 1992]. This microgripper had two relatively thin gripping arms (i.e., thin-film deposited polysilicon) as shown in Figure 28.8. Other microgrippers based on quasi-three-dimensional structures with high-aspect-ratios fabrication techniques (i.e., beams perpendicular rather than parallel to the surface) have also been presented [Keller and Howe, 1995; 1997] (see Figure 28.9). These kinds of grippers, so-called over-hanging tools, are formed by etching away the substrate under the gripper.

One critical parameter for the in-plane technique is how to achieve actuators with large displacement and force generation capabilities. Thermal actuators are known for their ability to generate high forces. A thermal actuator made from a single material would be easy to fabricate, but the displacement due to thermal expansion of a simple beam, for example, is quite small. This is a general drawback for in-plane

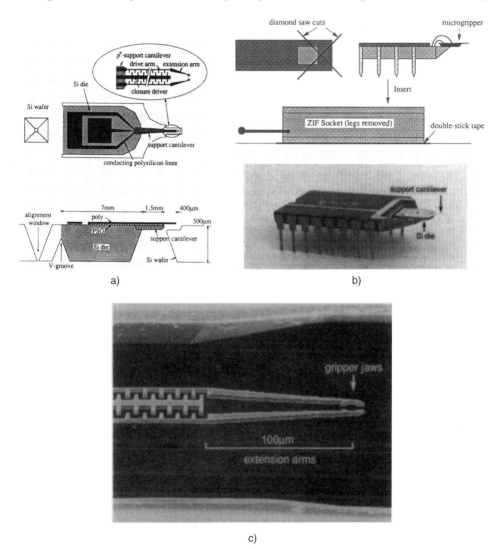

FIGURE 28.8 (a) Schematics of the microgripper unit (top and cross-sectional view). (b) Schematic showing gripper packaging and electrical access and a photograph of the packaged gripper. (c) SEM photograph showing a close-up of the gripper jaws and the comb-drive structures and extension arms. (Illustration published with permission and courtesy of C.-J. Kim, UCLA.)

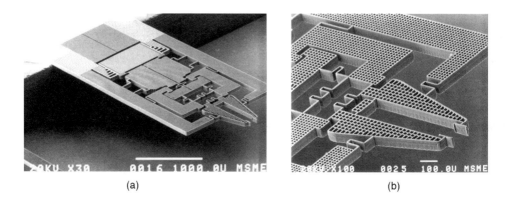

FIGURE 28.9 Photograph of fabricated HEXSIL tweezers. (a) Overview of the overhanging microtweezers with a compliant linkage system. (b) Close-up of the 80-µm-tall HEXSIL tweezers. The tip displacement between the closed and open position is typically 40 µm with a time constant <0.5 s for a typical actuation power of 0.5 A at 6 V. (Published with permission from MEMS Precision Instruments, Berkeley, CA; courtesy of C. Keller.)

actuators that occurs independently of the fabrication technique used. However, by using mechanical leverage, large displacements can be obtained, as was demonstrated by Keller and Howe (1995). They used a replication and micromolding technique, named HEXSIL, to fabricate thermally actuated microtweezers made from nickel and later in polysilicon [Keller and Howe, 1997]. In the HEXSIL process [Keller, 1998a] the mold is formed by deep trench etching in the silicon substrate. A sacrificial layer of oxide is deposited in the silicon mold which is then filled with deposited polysilicon. Then the polysilicon structure is released from the mold by sacrificial etching of the oxide. Afterwards, the mold can be reused by a new oxide and polysilicon deposition process. One advantage of this process is the ability to make thick (100 µm or greater) polysilicon structures (i.e., quasi-three-dimesnsional structures) on which electronics can be integrated. Figure 28.9b shows a close-up of the leverage design for the HEXSIL microtweezer; a large beam is resistively heated by the application of current, and subsequently expansion causes other beams in the link system to rotate and open the tweezer tips. When cooled, the contraction of the thermal element closes the tweezers.

Leverage and linkage systems (sometimes combined with gears for force transfer) are useful techniques for obtaining large displacements or forces that can be used for thermal actuation as well as electrostatic comb-drive actuators [Rodgers et al., 1999]. Several publications on design optimization schemes for various leverage techniques applied to thermal actuators (so-called compliant microstructures) have been presented [Jonsmann et al., 1999]. Another way to achieve a leverage effect is to use clever geometrical designs for single material expansion. One such method is the polyimide-filled V-groove (PVG) joint technology shown in Figure 28.10. The PVG joint technique has also been used for microconveyers and walking microrobots, as will be described in Sections 28.5.2 and 28.5.3. The purpose of the PVG joint microgripper in Figure 28.10 is easy integration with a walking microrobot platform.

Several publications on LIGA-based microgrippers have been presented. The reason for using LIGA is to get quasi-three-dimesnsional structures (i.e., thick structures) similar to the HEXIL tweezers in Figure 28.9. The LIGA process has also been used to produce single material (unimorph) in-plane thermal actuators for micropositioning applications. Guckel et al. (1992) presented an asymmetric LIGA structure with one "cold" and one "hot" side to generate large displacements (tenths of a millimeter) with relatively low power consumption, as illustrated in Figure 28.20a). More recently, this approach was used by Bright et al. [Comotis and Bright, 1996] for surface micromachined polysilicon thermal actuators. With this in-plane actuator they have successfully fabricated over-hanging microgrippers.

As an alternative to single-material expansion actuators, bimorph structures could also be used for out-of-plane acting gripping arms [Greitmann and Buser, 1996]. A bimorph microgripper for automated handling of microparts is shown in Figure 28.11. This device consists of two gripping arm chips assembled

Microrobotics   **28**-17

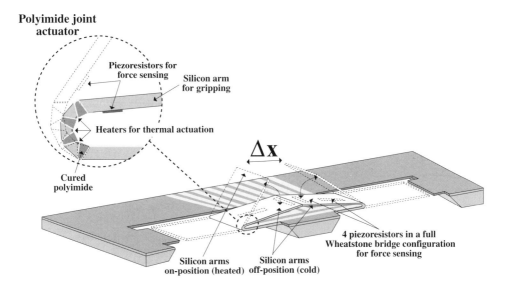

FIGURE 28.10 Concept of a microgripper fabricated by polyimide V-groove joints [Ebefors, 2000]. The polyimide in the V-grooves expands due to heating and the gripping arms are opened. Self-assembling out-of-plane rotation of the arms as well as the leverage effect for single material expansion are accomplished by the well-controlled geometrical shape of the V-groove. Polysilicon resistors are used both as resistive heaters and as strain gauges for force sensing.

FIGURE 28.11 Photograph of a microgripper based on bimorph thermal actuation and piezoresistive tactile sensing. (From Greitmann, G., and Buser, R. (1996) *Sensors and Actuators A*, **53**(1–4), 410–415. With permission.)

together. Each gripper arm has integrated heating resistors for actuation of the bimorph and tactile piezoresistors for force sensing.

Several different approaches to obtain three-dimensional microgrippers working out-of-plane like the ones in Figures 28.10 and 28.11 exist. One commonly used approach is use of the surface-micromachined polysilicon microhinge technology shown in Figure 28.3a–e. Such microhinges have been used both for microgrippers and for articulated microrobot components [Pister et al., 1992; Yeh et al., 1996].

One major drawback of these microgrippers (mainly based on thermal or electrostatic actuation) is found in biological applications. Microgrippers based on thermal, magnetic or high-voltage electric actuation could easily kill or destroy biological and living samples. The pneumatic microgripper presented by Kim et al. [Ok et al., 1999] avoids such problems. An alternative to heating grippers (such as the ones shown in Figures 28.10 and 28.11, which require relatively high heating temperatures) is the use of shape memory alloy (SMA) actuators. Microgrippers based on SMA often require lower temperatures than thermally actuated bimorph or unimorph grippers. SMA-based three-dimensional microgrippers have been used to grip (clip) an insect nerve for recording the nerve activity of various insects [Takeuchi and Shimoyama, 1999].

In the biotechnology filed—for example, the growing area of genomics and proteomics—microtools for manipulation of single cells are of major importance. In particular, massive parallel single-cell manipulation and characterization by the use of microrobotic tools are very attractive. In this type of application, the microgrippers usually must operate in aqueous media. Most of the microgrippers presented so far in this review cannot operate in water because of electrical short-circuiting, etc. One possible solution is to use conductive polymers. Such conductive polymers, which undergo volume changes during oxidation and reduction, are often referred to as electroactive polymers (EAPs) or micromuscles. These kinds of micromuscles have been used as joint material for microrobotic arms for single-cell manipulation devices [Smela et al., 1995; Jager, 2000a,b]. Figure 28.12 describes the fabrication of a microrobot arm based on a polypyrole (PPy) conductive polymer. During electrochemical doping of PPy, volume changes take place which can be used to achieve movement of micrometer-size actuators. The actuator joints consist of a PPy and gold bimorph structure, and the rigid parts between the joints consist of benzocyclobutene (BCB). The conjugated polymer is grown electrochemically on the gold electrode, and the electrochemical doping reactions take place in a water solution of a suitable salt. The voltages required to drive the motion are in the range of a few volts.

One of the many experiments conducted with the various robot arms fabricated with the PPy micromuscles is shown in Figure 28.13. The drawback of microrobotic devices based on the conductive polymer hinge (or "micromuscles") is that they cannot operate in dry media.

28.5.2 Microconveyers

Recently, a variety of MEMS concepts for realization of locomotive microrobotic systems in the form of microconveyers have been presented [Riethmüller and Benecke, 1989; Kim et al., 1990; Pister et al., 1990; Ataka et al., 1993a,b; Konishi and Fujita, 1993; 1994; Goosen and Wolffenbuttel, 1995; Liu et al., 1995; Böhringer et al., 1996; 1997; Nakazawa et al., 1997; 1999; Suh et al., 1997; 1999; Hirata et al., 1998; Kladitis et al., 1999; Ruffieux and Rooij, 1999; 2000; Smela et al., 1999; Ebefors et al., 2000]. The characteristics for some of these devices are summarized in Table 28.3, where the microconveyers are classified in two groups: contact-free or contact systems, depending on whether the conveyer is in contact with the moving object or not, and synchronous or asynchronous, depending on how the actuators are driven. Examples of both contact and contact-free microconveyance systems were shown in Figures 28.6 and 28.7, respectively.

Contact-free systems have been realized using pneumatic, electrostatic or electromagnetic forces to create a cushion on which the mover levitates. Magnetic levitation can be achieved by using permanent magnets, electromagnets or diamagnetic bodies (i.e., a superconductor). The main advantage of the contact-free systems is low friction. The drawback of these systems is their high sensitivity to the cushion thickness (i.e., load dependent), while the cushion thickness can also be quite difficult to control. Also, this kind of conveyance system often has low load capacity.

Systems where the actuators are in contact with the moving object have been realized based on arrays of moveable legs erected from the silicon wafer surface. The legs have been actuated by using different principles such as thermal, electrostatic and magnetic actuation. Both synchronous driving [Liu et al., 1995] and the more complex, but also more effective, asynchronous driving modes have been used.

The magnetic [Nakazawa et al., 1997; 1999] and pneumatic [Hirata et al., 1998] actuation principles for contact-free conveyer systems have the disadvantage that they require a specially designed magnet

Microrobotics

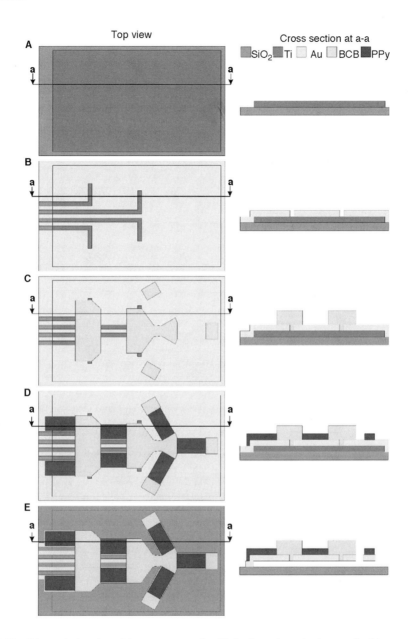

FIGURE 28.12 Schematic drawing of the process steps for fabricating microrobotic arms (in this case, an arm with three fingers arranged in a 120° configuration) based on hinges (micromuscle joints) consisting of PPy(DBS)/Au bimorph structures. (a) Deposition and patterning of a sacrificial Ti layer. (b) Deposition of a structural Au layer and etching of the isolating slits. (c) Patterning of BCB rigid elements. (d) Electrodeposition of PPy (conductive polymer). (e) Etching of the final robot and electrode structure and removal of the sacrificial layer. Each microactuator is 100 μm × 50 μm. The total length of the robot is 670 μm, and the width at the base is either 170 or 240 μm (depending on the wire width). (Published with permission and courtesy of E. Jager, LiTH-IFM, Sweden.)

mover or slider which limits its usefulness. With a contact system based on thermal actuators it is possible to move objects of various kinds (nonmagnetic, nonconducting, unpatterned, unstructured, etc.); however, the increased temperature of the leg in contact with the conveyed object may be a limitation in some applications. The contact-free techniques have been developed mainly to meet the necessary criteria for a cleanroom environment, where a contact between the conveyer and the object may generate particles that would then serve to restrict its applicability for conveyance in clean rooms.

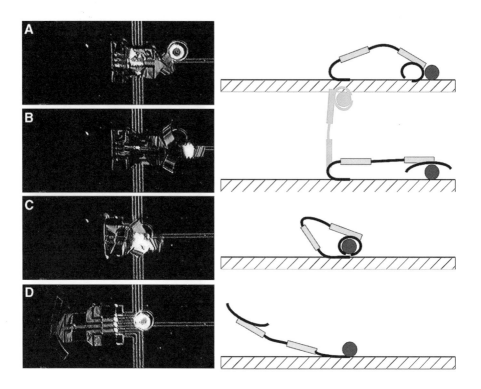

FIGURE 28.13 (a)–(d) Sequence of pictures (left) showing the grabbing and lifting of a 100-μm glass bead and schematic drawings (right) of the motion. In this case, the arm has three fingers, placed 120° from each other (see Figure 28.12). The pictures do not illustrate the fact that the bead is actually lifted from the surface before it is placed at the base of the robot arm. This is illustrated in the second sketch on the right where the lifting stage is shown in gray. (Published with permission and courtesy of E. Jager, LiTH-IFM, Sweden.) (Note: Video sequences showing the full movement of various robot arms for the different experiments are available at http://www.sciencemag.org/feature/data/1050465.shl and http://www.ifm.liu.se/Applphys/ConjPolym/research/micromuscles/CPG_micromuscles.html.)

A microconveyer structure based on very robust polyimide V-groove actuators has been realized [Ebefors et al., 2000]. This conveyer is shown in Figure 28.14. In contrast to most of the previously presented microconveyance systems, the PVG joint approach has the advantage of producing robust actuators with high load capacity. Another attractive feature of this approach is the built-in self-assembly by which one avoids time-consuming manual erection of the conveyer legs out of plane. Some of the conveyers listed in Table 28.3 require special movers (e.g., magnets or sliders with accurate dimensions). The PVG joint conveyer solution is more flexible because one can move flat objects of almost any material and shape. The large actuator displacement results in a fast system which is less sensitive to the surface roughness of the moving object. By using individually controlled heaters in each actuator, an efficient asynchronous driving mode has been realized, which also allows a parallel design giving relatively high redundancy for actuator failure. The first experiments with the conveyer showed good performance, and one of the highest reported load capacities for MEMS-based microconveyers was obtained. The maximum load successfully conveyed on the structure had a weight of 3500 mg and was placed on a 115-mg silicon mover, as shown in Figure 28.14d. Conveyance velocities up to 12 mm/s have been measured. Both forward–backward and simple rotational conveyance movements have been demonstrated. The principle for rotating an object by a two-row conveyer is shown in Figure 28.14a. The lifetime of the PVG joints actuator exceeds 2×10^8 load cycles and so far no device has broken due to fatigue.

TABLE 28.3 Overview of Some Microconveyance Systems

Principle[a]	Maximum Velocity	Moved Object/Load Capacity	Length per Stroke/Frequency	No. of Actuators/Size	Ref.
CF: pneumatic air bearing (for low-friction levitation) + electrostatic force for driving	Slow	Flat Si pieces/ <1.8 mg	100–500 µm/ max at 1–2 Hz	Not specified	[Pister et al., 1990]
CF: magnetic levitation (Meissner effect) + magnetic Lorentz force for driving	7.1 mm/s[b]	Nd–Fe–B magnet slider/ 8–17 mg	Not specified	Not specified	[Kim et al., 1990]
C: array of thermobimorph polyimide legs[c] (cantilevers); electrical heating (asyn)	0.027–0.5 mm/s	Flat Si piece/ 2.4 mg	$\Delta x = 80$ µm ($f < f_c$; 33 mW)/ $f_c = 10$ Hz	8 × 2 × 16 legs/ 500 µm/total area: 5 × 5 mm^2	[Ataka et al., 1993]
CF: array of pneumatic valves; electrostatically actuated	Not specified	Flat Si piece/ 0.7 mg	Not specified/ $f = 1$ Hz (pressure)	9 × 7 valves/100 × 200 µm^2/total area: 2 × 3 mm^2	[Konishi and Fujita, 1994]
C: array of magnetic in-plane flap actuators; external magnet for actuation (syn)	2.6 mm/s[d]	Flat Si pieces/ <222 mg	$\Delta x = 500$ µm/ $f_c = 40$ Hz	4 × 7 × 8 flaps/ 1400 µm/total area: 10 × 10 mm^2	[Liu et al., 1995]
C: array of torsional 5 µm high; Si-tips; electrostatic actuation (asyn)	Slow	Flat glass piece/ ≈1 mg	$\Delta x = 5$ µm/f_c high kHz-range	15,000 tips 180 × 240 mm^2/total area: 1000 mm^2	[Böhringer et al., 1996; 1997]
C: array of thermobimorph polyimide legs;[c] thermal electrostatic actuation (asyn)	0.2 mm/s	Silicon chips/ 250 µN/mm^2	$\Delta x = 20$ µm/f_c not specified	8 × 8 × 4 legs/ 430 µm/total area: 10 × 10 mm^2	[Böhringer et al., 1997; Suh et al., 1997]
CF: Pneumatic (air jets)	35 mm/s[e] (for flat objects)	Sliders of Si/ <60 mg	Not specified	2 × 10 slits/ 50 µm/ total area: 20 × 30 mm^2	[Hirata et al., 1998]
C: array of erected[f] Si-legs; thermal actuation (asyn)	0.00755 mm/s	Piece of plastic film/3.06 mg	$\Delta x = 3.75$ µm ($f<f_c$; 175 mW)/ $f_c = 3$ Hz	96 legs/270 µm/ total area: 10 × 10 mm^2	[Kladitis et al., 1999]
CF: array of planar electromagnets	28 mm/s[g] (unloaded)	Flat magnet + external load/ <1200 mg[g]	Not specified	≈40 × 40 coils/ 1 × 1 mm^2/total area: 40 × 40 mm^2	[Nakazawa et al., 1997; 1999]
C: array of non-erected Si-legs; piezoelectric or thermal actuation (asyn)	Not specified	Not specified	$\Delta x = 10$ µm/($f = 1$ Hz, 20 mW) $f_c \approx 30$ Hz (thermal)/f_c high kHz-range (piezoelectric)	125 triangular cells (legs) 400 µm (300 µm) long on a hexagonal chip approx. 18 mm^2	[Ruffieux and Rooij, 1999]
C: array of erected[c] Si-legs; thermal actuation of polyimide joints (asyn)	12 mm/s[h]	Flat Si pieces + external load/ 3500 mg	$\Delta x = 170$ µm[h] ($f<f_c$; 175 mW)/ $f_c = 3$ Hz[i]	2 × 6 legs/500 µm/ total area: 15 × 5 mm^2	[Ebefors et al., 2000]

[a] C = contact; CF = contact-free; asyn = asynchronous; syn = synchronous.
[b] The superconductor requires low temperature (77 K).
[c] Self-assembled erection of the legs.
[d] Estimated cycletime ≈25 ms (faster excitation results in uncontrolled jumping motion) and 0.5 mm movements on 8 cycles [Liu, 1995].
[e] For flat sliders. The velocity depends on the surface of the moving slider (critical tolerances of the slider dimensions).
[f] Manual assembly of the erected leg.
[g] Depends on the magnet and surface treatment.
[h] Possible to improve with longer legs and more V-grooves in the joint.
[i] Possible to improve. Thinner legs with smaller polyimide mass to heat would increase the cut-off frequency, f_c => larger displacements at higher frequencies.

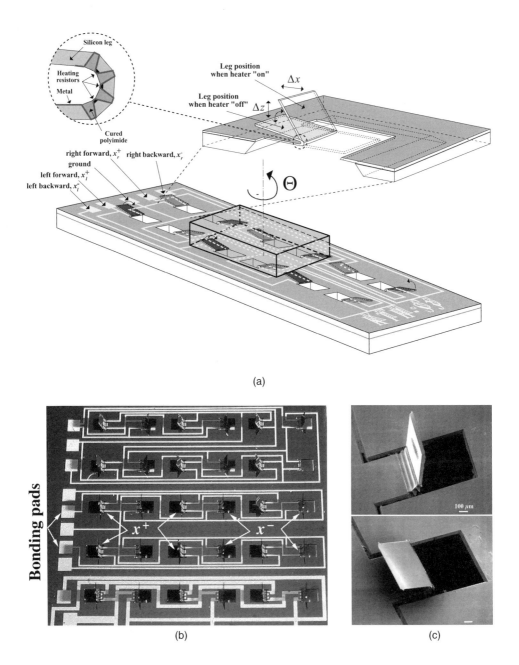

FIGURE 28.14 (a) Principle for the rotational movements on a test conveyer using robust PVG joints. The left and right side can be actuated separately like a caterpillar. Each leg has a size of 500 × 600 × 30 μm. (b) Photograph showing different (undiced) structures used to demonstrate the function of the polyimide-joint-based microconveyer. One conveyer consists of two rows of legs (12 silicon legs in total). Two sets of legs (six each of x^+ and x^-) are indicated in the photograph (compare Figure 28.6c). For the conveyer with five bonding pads, the right and left rows of legs can be controlled separately for possible rotational conveyance. (c) SEM photographs showing Si legs with a length of 500 μm. (d) The microconveyer during a load test. The 2-g weight shown in the photograph is equivalent to 350 mg on each leg or 16,000 times the weight of the legs. (Note: Videos of various experiments involving this microconveyer are available at http://www.s3.kth.se/mst/research/gallery/conveyer.html/ or http://www.iop.org/Journals/jm.)

(d)

FIGURE 28.14 Continued.

The most sophisticated microrobotic device fabricated to date is the two-dimensional microconveyer system with integrated CMOS electronics for control which have been fabricated by Suh et al. (1999). The theories on programmable vector fields for advanced control of microconveyance systems presented by Böhringer et al. (1997) were tested on this conveyer. Several different versions of these conveyers have been fabricated throughout the years [Suh et al., 1997; 1999]. All versions are based on polyimide thermal bimorph ciliary microactuator arrays, as shown in Figure 28.15.

28.5.3 Walking MEMS Microrobots

In principle, most of the microconveyer structures described in the previous section could be turned upside down to realize locomotive microrobot platforms. For contact systems, that means that the device will have legs for walking or jumping. The contact-free systems relying on levitation forces will "float" over the surface rather than walk. Such systems seem more difficult to realize than the contact-operating robots. The focus for the rest of this section is, therefore, on contact microrobot systems for walking.

Although it seems straightforward to turn a microconveyer upside down, most of the existing conveyers do not have enough load capacity to carry their own weight. Further, there are problems on how to supply the robot with the required power. As illustrated in Figure 28.16, power supply through wires may influence the robot operation range, and the stiffness of the wires may degrade the controllability too much. On the other hand, telemetric or other means of wireless power transmission require complex electronics on the robot. Because many actuators proposed for microrobotics require high power consumption, the limited amount of power that can be transmitted through wireless transmission is a big limitation for potential autonomous robot applications. To avoid the need for interconnecting wires, designs based on solar cells have been proposed, as have low-power-consuming piezoelectric actuators [Ruffieux and Rooij, 1999; Ruffieux, 2000], electrostatic comb-drives [Yeh et al., 1996] or inch-worm actuators [Yeh and Pister, 2000] suited for such wireless robots. For wire-powered robots, a limited amount of wires is preferable, which implies that simple leg actuation schemes (i.e., on–off actuation as used for the walking robot platform in Figure 28.6c) are required if complex onboard steering electronics are to be avoided. Several proposals for making totally MEMS-based microrobots include the possibility of locomotion (e.g., walking) powered with or without wires.

28.5.3.1 Examples of Walking Microrobots Designs

Several different principles used to actuate the different legs on a walking microrobot have been proposed, most of them mimicking principles used in nature. Some of the most feasible principles for walking MEMS microrobot platforms are listed below:

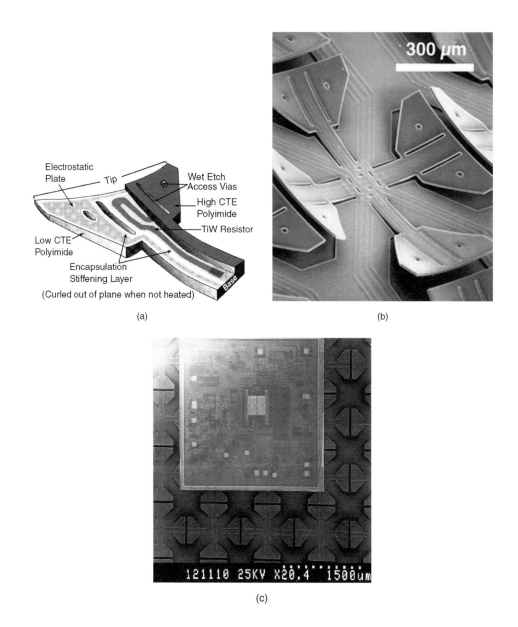

FIGURE 28.15 (a) Principal of operation of an organic microactuator (bimorph polyimide legs) using thermal and electrostatic forces for actuation. Half of the upper polyimide and silicon nitride encapsulation/stiffening layer have been removed along the cilium's axis of symmetry to show the feature details. (b) SEM photograph of a polyimide cilia motion pixel. Four actuators in a common center configuration make up a motion pixel. Each cilium is 430 μm long and bends up to 120 μm out of the plane. (c) Micro cilia device moving an ADXL50 accelerometer chip. (d) The CMS-principle for two-dimensional conveyance (compare Figure 28.6). The state of the four actuators (north, east, west, south) is encoded with small letters (e.g., n) for down, and capital letters (e.g., S) for up. (e)–(h) Images (video frames) of a 3 × 3-mm^2 IC chip rotating from a rotation demonstration. (Published with permission and courtesy of Suh, Böhringer and Kovacs, Stanford University.) (Note: Videos and animations of this microconveyance systems are available at http://www.ee.washington.edu/research/mems/Projects/Video/.)

Microrobotics

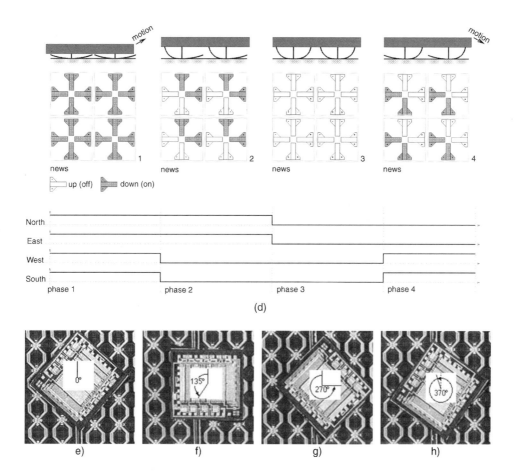

FIGURE 28.15 Continued

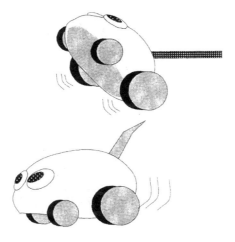

FIGURE 28.16 Influence of energy cable on microrobots. (Illustration published with permission and courtesy of T. Fukuda, Nagoya University, Japan.)

FIGURE 28.17 (Color figure follows p. 12.26.) Photograph of the microrobot platform, used for walking, during a load test. The load of 2500 mg is equivalent to maximum 625 mg/leg (or more than 30 times the weight of the robot itself). The power supply is maintained through three 30-µm-thin and 5- to 10-cm-long bonding wires of gold. The robot walks using the asynchronous ciliary motion principle described in Figure 28.6(c). The legs are actuated using the polyimide V-groove joint technology described in Figure 28.10. (Photograph by P. Westergård and published with permission.) (Note: Videos of various experiments performed using the microrobot shown here are available at http://www.s3.kth.se/mst/research/gallery/microrobot_video.html.)

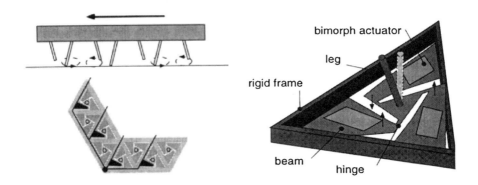

FIGURE 28.18 The rotational leg walking principle. A few hundred cells (shown in the right picture) are arranged into a hexagonal array, inside a triangular grid frame that provides stiffness and room for interconnection. The simplest form of gait requires two phases so that half the actuators are in contact with the ground, where friction transmits their motion to the device, while the other half is preparing for the next step (compare to the CMS technique depicted in Figure 28.6). The bottom left figure illustrates the serial interconnections of the actuator resulting in six independent groups of beams (From Ruffieux, D. et al., in *Proc. IEEE 13th Int. Conf. on Micro Electromechanical Systems (MEMS '2000)*, pp. 662–667, 2000. With permission.)

- *Ciliary motion* was used by Ebefors et al. (1999) for eight-legged robots, as illustrated in Figures 28.6 and 28.17.
- *Elliptical leg movements* adopted from the animal kingdom have been proposed by Ruffieux et al. [Ruffieux and Rooij, 1999; Ruffieux, 2000] and are illustrated in Figure 28.18. Similar to this concept is the principle used by Simu and Johansson (2001). Their microrobot concept consists of both a walking and a microtool unit for highly flexible micromanipulation. Instead of using thin-film piezolayers and silicon legs as did Ruffieux et al., Johansson and Simu have made robot

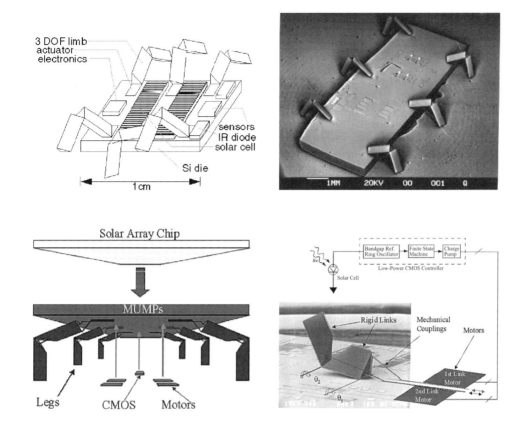

FIGURE 28.19 First version of the microrobot prototype based on surface micromachined microhinges for joints; each leg has three degrees of freedom and is comprised of two 1.2-mm-long, rigid polysilicon links and electrostatic step motors for movement (not included in the SEM-photo). (Bottom) Concept for a new design based on a solar-powered silicon microrobot. Various components can be made separately and then assembled. Surface micromachined hinges are used for folding the legs (see Figures 28.3d and e). Each leg has two links, and each link will be actuated by an inchworm motor [Yeh and Pister, 2000]. (Published with permission and courtesy of K. Pister, BSAC-Berkeley, CA.)

legs in a solid, multilayer, piezoceramic material mounted on a glass body [Simu and Johansson, 1999] (see Figure 28.22).

- Gait mimicry of *six-legged* insects (similar to that of a crab [Zill and Seyfarth, 1996]) has been proposed by several research groups—for example, the six-legged microrobot by Yeh et al. [Yeh et al., 1996; Yeh and Pister, 2000] and the multilegged microrobot prototype by Kladitis et al. [Kladitis, 1999]. These concepts are illustrated in Figures 28.19 and 28.20.
- *Inch-worm* robots [Thornell, 1998] or *slip-and-stick* robots [Breguet and Renaud, 1996] that mimick an inch-worm or caterpillar are attractive microrobot walking principles, as these techniques take advantage of the frictional forces rather than trying to avoid them. The scratch-drive actuator principle [Mita et al., 1999] works according to this principle by firmly attaching to the surface only half of the robot or actuator body, then extending the spine (or middle part) of the body before anchoring the other half, releasing the first grip, shortening the spine, swapping the grip, and so on in a repeatable forward motion.
- Vibration fields and resonating thin-film silicon legs with polyimide joints have been used by Shimoyama et al. [Yasuda et al., 1993], as illustrated in Figure 28.21.

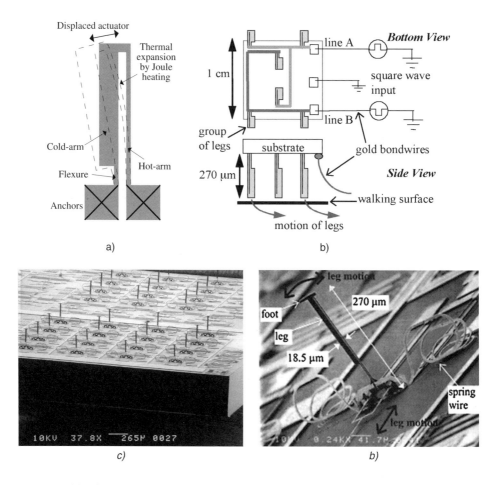

FIGURE 28.20 (a) Schematic illustration of a single-material asymmetric thermal actuator for in-plane actuation (also called "heatuator"). The actuator structure was originally made of nickel in a LIGA process, although surface-micromachined polysilicon structures have also been used. (b) An out-of-plane folded heatuator for microrobotic applications. (c) Photograph of the microrobotic device in a conveyance mode (belly up). (d) Close-up of one of the heatuator legs erected out of plane. (Part (a) from Guckel, H. et al. (1992) in *Technical Digest, Solid-State Sensor and Actuator Workshop*, pp. 73–75. With permission. Part (b) from Kladitis, P. et al. (1999) in *Proc. IEEE 12th Int. Conf. on Micro Electro Mechanical Systems (MEMS '99)*, pp. 570–575. With permission. Photographs courtesy of P. Kladitis and V. Bright, MEMS Group at the University of Colorado at Boulder.)

Because friction in many cases poses a restriction on microrobots due to their small size, solutions that take advantage of this effect (e.g., inch-worm robots) rather than trying to avoid it are the best suited for microgait. This trend can also be seen in the evolution of micromotors for microrobotic and other applications. Currently, many researcher try to avoid bearings or sliding contacts in their motors and instead make use of friction in wobble and inch-worm motors [Yeh and Pister, 2000] or actuators with flexible joints [Suzuki et al., 1992; Ebefors et al., 1999] without the drawback of wearing out.

The main problem associated with the fabrication of silicon robots is to achieve enough strength in the moveable legs and in the rotating joints. Most efforts to realize micromachined robots utilize surface micromachining techniques which results in relatively thin and fragile legs. Pister et al. [Yeh et al., 1996] have proposed the use of surface micromachined microhinges for joints, poly-Si beams for linkage to the triangular polysilicon legs and linear electrostatic stepper motor actuation for the realization of a microrobot, as illustrated in Figure 28.19.

A similar approach for walking microrobots was used by Kladitis et al. (1999). They also used the microhinge technique to fold the leg out of plane, but instead of area-consuming comb drives for

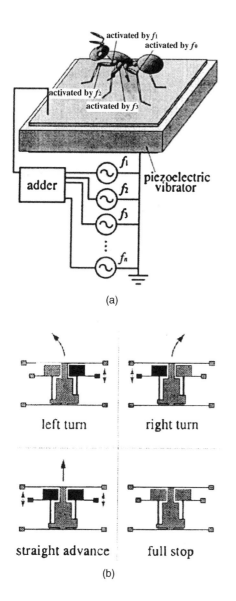
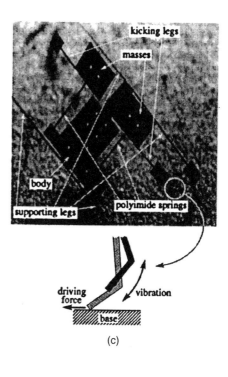

FIGURE 28.21 (a) The principle for a selective power supply through a vibrating energy field [Yasuda et al., 1993]. The microrobot has several resonant actuators with mutually exclusive resonance frequencies. The power and control signals to the robot are obtained via the vibrating table. (b) The four possible kinds of actions. (c) Photograph of the 1.5 × 0.7-mm² surface micromachined microrobot. The legs have different spring constants and masses, resulting in differnt resonance frequencies. Polyimide is used for the soft springs and the polyimide joints are used for the erected leg. (From Yasuda et al. (1993) *Technical Digest, 7th Int. Conf. on Solid-State Sensors and Actuators* (Transducers '93), pp. 42–45, June 7–10, Yokohama, Japan. © 1993 IEEE/ASME. With permission.)

actuation they used the thermal "heatuator" principle integrated in the leg, as illustrated in Figure 28.20. To further improve the robustness and load capacity of their robot, they used 96 legs arranged in six groups instead of using a six-legged robot. However, that robot structure could "only" withstand a load four times the dead weight of the robot. That was not enough to obtain locomotive walking but was high enough for conveyance applications.

For both microrobots described above, which were based on surface micromachined polysilicon microhinges, the polysilicon legs were manually erected out of plane. The use of the microhinge technique may cause problems because of wear after long-term actuation. Shimoyama et al. [Suzuki et al., 1992; Miura et al., 1995] introduced the concept of creating insect-like microrobots with exoskeletons made from surface-micromachined polysilicon plates and rigid polyimide joints that have low friction and therefore lower the risk for wearing out. These microrobots were powered externally by a vibrating field (i.e., no cables were needed) as illustrated in Figure 28.21. By cleverly designed robot legs having different masses and spring constants and thus different mechanical resonance frequencies, the leg to be actuated can be selected by applying a certain external vibrating (resonance) frequency. The robot can then be "shaken" in a controlled way to walk forward and left or right, as illustrated in Figure 28.21b. The need for a vibrating table to achieve the locomotion strongly limits the applications for this robot. Also, the thin legs cause the robot to stick easily to the vibrating table by surface forces, which means that the surface on which the robot walks must be insulating.

In a large European microrobot project, named MINIMAN, different miniaturized microrobots have been developed and more are under development. By the year 2000, a few demonstrators of minirobots and also a microrobot, having five degrees of freedom micromanipulation capability, had been fabricated. This prototype microrobot is illustrated in Figure 28.22. Milling with a high-precision CNC (computer numerical controlled) machine tool in a green ceramic body has been used to make a stator unit where the drive elements are monolithically integrated with the base plate. The robot is made out of two almost identical piezoelectric stator units (shown in Figure 28.22d) which are put together back to back. One stator unit consists of six individually controlled legs. The lower stator, the micropositioning unit, should move laterally on a flat surface while the upper stator, the micromanipulating unit, will turn a ball equipped with a tool arbitrarily. Figure 28.22c illustrates the driving mechanism of the six-legged piezo-ceramic positioning and manipulating units to gain rotation and linear motion. Each three-axial element (leg) in the stator unit has four electrically separated parts able to produce the different motions needed for control of the robot. The elements are divided into two sets (different shading for sets I and II) phase-shifted 180°. Therefore, three elements are always in friction contact with the counter surface, producing a conveying (or walking) motion of the ball (or the robot itself).

The mechanical walking mechanism with solid multilayer piezoactuators is designed to give high handling speed, force and precision. A future version will include integration of the power drivers with the control circuitry and soldering the unit to a flexible printed circuit connecting the actuator elements in the stator unit. The electronic chip would then be placed between the two stator units.

Figure 28.22e shows the experimental setup for testing the first prototype of the micromanipulating stator unit. These stators in combination with prototypes of ICs intended for integration have been used to evaluate the performance. This setup will also be combined with a larger positioning platform, MINIMAN III (i.e., a walking minirobot platform), to obtain flexible microrobotic devices for micromanipulation.

28.5.3.2 Redundancy Criteria for Walking Microrobots

As was already pointed out in Section 28.4.3 one critical aspect when designing large distributed micro-motion systems based on arrayed actuators and distributed (or collective) actuation is the problem associated with the need for a very high yield of the actuators [Ruffieux and Rooij, 1999]. One nonworking actuator can undo the entire walking principle. Therefore, special attention has to be paid to achieve high redundancy by parallel designs wherever possible. The conveyer legs of the robot prototype design presented in Figure 28.22 is fabricated by stacking several layers of thick-film piezoelectric layers on top of each other with metal electrodes in between. Because one layer is approximately 50 μm thick, a piezostructure for a 1-mm-long leg consists of several layers and electrodes, which require very high yield in the fabrication. The metal electrodes are connected in parallel by the metal on the edges of the piezolegs to obtain higher redundancy for failure. Besides the need for parallel designs, the interconnecting wires must be kept at a minimum for walking microrobots (see Figure 28.16). For the piezorobot prototype shown in Figure 28.22e, the stator unit is soldered to a flexible printed circuit board as it will be in the complete microrobot. The only difference is that the conductors do not lead to a surface-mounted connector, but instead to surface mounted integrated circuits. By using such a configuration, the number

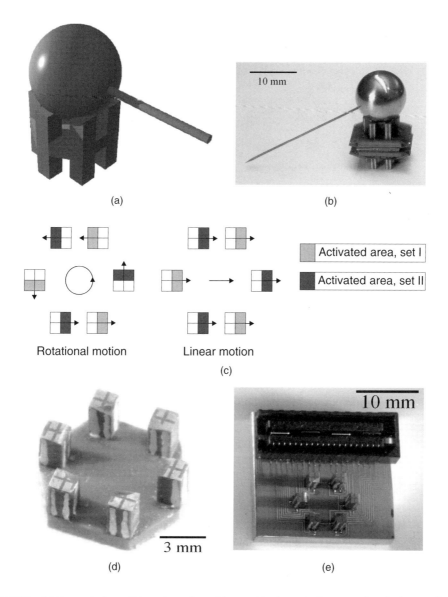

FIGURE 28.22 (a) Concept of a walking microrobot with a positioning unit (lower part) and micromanipulating unit (upper part) for turning the ball equipped with a tool. (b) A first prototype of the robot; this version is unable to walk. (c) Driving mechanism for the stator unit. (d) A photograph showing a six-legged piezoelement stator. (e) Experimental setup for the micropositioning unit. Note the amount of interconnection wires for this prototype. (Courtesy of U. Simu, Uppsala University, Sweden.)

of cables required to steer the robot will be minimized. In the current prototype version, the amount of interconnecting wires is quite large, as can be seen in Figure 28.22e.

For the microrobots shown in Figures 28.17 and 28.23, all the legs are connected in parallel so all of the actuators or legs are not needed to have perfect functionality. If one leg is broken, all the others are still functional, as the power supply line is connected in parallel (see Figure 28.23) instead of the serial connection, which is the most often used design for distributed microsystems. However, it is important that the nonfunctional (or broken) legs do not limit the walking function by blocking the action of the other legs. To further increase redundancy, each heating resistor in the PVG joint (see Figures 28.10 and 28.14) is divided into separate parts, allowing functional heating even if some part of the resistors (or the via contacts between the resistor and the metal) is not working.

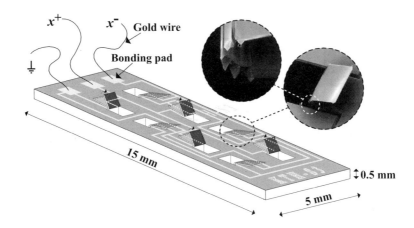

FIGURE 28.23 An upside-down view of the PVG joint-based microrobot platform used to study walking microrobots. The robot consists of two sets of legs (four of each x^+ and x^-). The metal interconnection is done in a parallel way to obtain the highest possible redundancy against failure. With three bonding pads the robot can walk forward and backward as well as lift up and down. By driving the legs on the left and right side at different speeds or stroke lengths, like a caterpillar (requires five wires), the robot can make left–right turns. The SEM photographs show the silicon leg and a close-up of a five-V-groove polyimide joint [Ebefors et al., 1999].

28.5.3.3 Steering of Microrobots

As shown in Figure 28.23, the simplest way to steer a CMS-based walking microrobot is by controlling the phase of the two rows of actuators on the robot. By driving one side forward and the other side backwards (or by not actuating it at all) the robot makes turns imitating a caterpillar. Other approaches to steer CMS microrobots to the left and right include:

- *Power*: Longer stroke lengths (changes in the x direction, or Δx) on one side can be produced by increasing the power, but this would also produce higher steps (changes in the y direction, or Δy), so the robot would walk with a stoop.
- *Frequency*: Increasing the number of steps on one side would produce smaller steps (both Δx and Δy), so again the robot would walk with a stoop.
- *Combination of frequency and power*: Equal stroke lengths, but faster on one side, would allow "smooth" robot movements.

Another approach to obtain steerable walking robots is the use of large two-dimensional arrays with several rows of legs oriented in both the x and y directions. For these systems, based on the principle described in Figure 28.15, onboard electronics are probably required for minimizing the amount of wires required for the steering signals. Monolithic integration of CMOS electronics has been shown to work for conveyer systems [Suh et al., 1999], but the integration is not an easy task. Hybrid integration of electronics on a separate chip, as is planned for the microrobot shown in Figure 28.22, is not an easy task either, when many interconnect wires are needed.

28.6 Multirobot System (Microfactories and Desktop Factories)

In the famous 1959 American Physical Society presentation by Feynman (1961), describing a new and exciting field of physics (e.g., miniaturization of systems, or "*there's plenty of room at the bottom*"), the use of small machines to build smaller machines or systems was anticipated as a way to obtain miniaturization. However, this has not yet been realized in the MEMS field. Instead, we use larger and larger machines to achieve miniaturization and study physics at a very small scale (e.g., large vacuum chambers to evaporate the very thin films used in MEMS or the huge accelerator facilities used to study the smallest particles in the atom). However, at the research level, interest in such microfactories, where small machines build even smaller machines, has been growing since the mid-1990s.

Often MEMS is seen as the next step in the silicon system revolution (e.g., miniaturization) in that silicon integration breaks the confines of the electrical world and allows the ability to interact with the environment by means of sensors and actuators. Most of the commercially successful processes employed in MEMS technology have been adopted from the fabrication scheme of integrated circuits. This kind of processing equipment is usually very large and expensive but has the very attractive feature of batch fabrication, which allows production of low-cost devices in large quantities. However, as already described in this chapter, manufacturing concepts other than the IC-inspired batch silicon microfabrication are being developed today. All IC-based micromachining techniques require large volumes to be cost effective, as the equipment is large and expensive. Emerging alternative fabrication technologies allow for more flexible geometry (e.g., true three-dimensional structures) not feasible by planar IC process and integration of materials with properties other than those of semiconductors and thin metal films [Thornell and Johansson, 1998; Thornell, 1999]. These technologies utilize miniaturized versions of traditional tools typical for local processing (such as drilling, milling, etc. [Masuzawa, 1998]) rather than the parallel (batch) approach used in the IC industry. The main challenge in adopting these processes for commercial application is the slow processing time associated with high-precision machining. One potential solution to this problem might be to use microfactories, where a great number of microrobotic devices (conveyers and robots) are working in parallel to support the microtools with material [Hirai et al., 1993; Aoyama et al., 1995; Ishihara et al., 1995]. This kind of multirobot system, illustrated in Figure 28.24, introduces special requirements for the microrobots. These requirements include a wireless power supply (i.e., onboard power sources or telemetrically supplied power) and interrobot communication.

During the last decade, efforts have been made to develop various multi-(micro)robot system concepts, especially in Japan [Ishikawa and Kitahara, 1997; Kawahara et al., 1997], where the term *microfactories* is used, but also in Europe and U.S., where the most frequently used term is *desktop station fabrication*. The idea behind the microfactory approach is to allow production of small devices. Often, other factors such as saving energy, space and resources are also considered important, so the manufacturing process should also be cost effective for low-volume microsystems. It is still impossible to predict if these techniques will ever be competitive with standard parallel-batch MEMS processing, but two important advantages have already been demonstrated: flexibility and geometrical freedom.

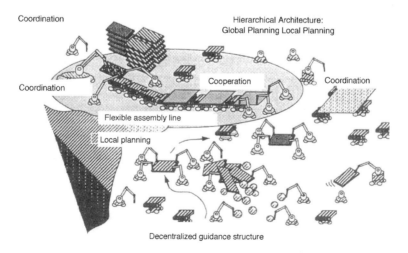

FIGURE 28.24 Concept for flexible decentralized multi-(micro)robot system. (Translated from Fatikow, S., *Mikroroboter and Mikromontage*, B. G. Tenbner, Stuttgart-Leipzig (in German). With permission.)

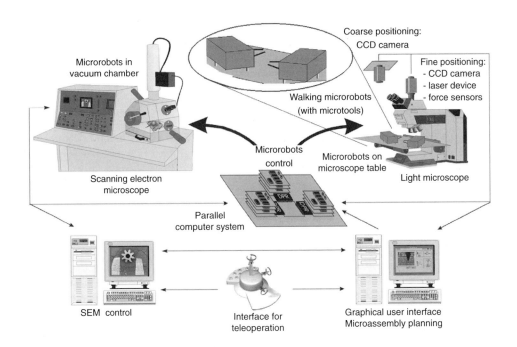

FIGURE 28.25 Concept of the flexible microrobot-based microassembly station (FMMS). (Published with permission and courtesy of S. Fatikow, University of Kassel, Germany.)

Similar to the microfactory concept is the "automated desktop station using micromanipulation robots" proposed by Fatikow et al. [Fatikow and Rembold, 1996; 1997; Fatikow, 2000], as illustrated in Figure 28.25. Here, the focus is on assembly of different types of microdevices, fabricated by batch MEMS techniques, to build various microsystems rather than fabricate subcomponents in the systems.

28.6.1 Microrobot Powering

The ultimate goal for many microrobotics researchers is to achieve true autonomous microrobots like the ones shown in Figure 28.2o for work in multirobot systems such as microfactories. Some of the locomotive microrobot platforms presented in this chapter can be used as first steps toward that goal, but much research remains. Currently, there are no energy sources available to power a fully autonomous microrobot. Some robot actuators, such as the polyimide V-groove joint actuator shown in Figure 28.17 and other thermal actuators, have the drawback of high power consumption. Further research concerning the reduction of power is needed. This is a general problem: Robust and efficient microactuators usually have such high energy consumption that an onboard power supply cannot provide enough electric power over an extended period of time. Using cables to supply the energy to a microrobot obstructs the action (especially in multirobot systems) of the microrobot due to the stiffness of the cables; therefore, the robot has to be supplied with energy without using cables. Several research groups are working on developing microenergy sources—for example, various microgenerators, pyroelectric elements, solar batteries, fuel cells and other power-MEMS concepts [Lin et al., 1999]. An overview of thin-film batteries that may be suitable for microrobot applications is given by Kovacs (1998). Low-weight batteries have been used for a flying microrobot [Pornsin-Sirirak et al., 2000]. Even though the lifetime of such batteries and dc–dc converter systems is short, it demonstrates the possibilities for battery-powered microrobots. Solar cells made of a-Si–H junction for microrobot applications have also been proposed [Ruffieux and Rooij, 1999; Ruffieux, 2000]. By connecting several a-Si–H junctions a voltage of about 10 V is expected to be generated.

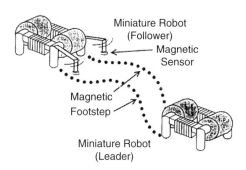 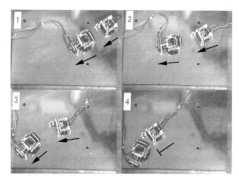

FIGURE 28.26 Communication between miniature robots. The following robot follows the leader by a magnetic trace. (Published with permission and courtesy of H. Aoyama, University of Electro-Communications, Tokyo.)

28.6.2 Microrobot Communication

Besides the problems associated with the power sources, more research is required on the wireless transmission of steering signals to the microrobot and other means of communication between microrobots. Wireless microrobots and other microdevices can be achieved by using acoustic [Suzumori et al., 1999], optical [Ishihara and Fukuda, 1998], electromagnetic [Hierold et al., 1998] or thermal transmission of the control signal to the robot.

Common to all multirobot system concepts is the big challenge of how to obtain autonomous communication between different microrobotic devices. Normally, various telemetric solutions similar to those described in the previous section for wireless power transfer are also proposed for communication, but other approaches seem to be feasible as well. In nature, the ant creates an odor trail to guide fellow ants. Shimoyama (1995a,b) has been working on hybrid systems consisting of mechanical parts and living organisms. By using "living sensors" in the form of insect antennae, Shimoyama and co-workers were able to steer and control a mobile robot by pheromone stimulation [Kuwana et al., 1995]. Aoyama et al. [Aoyama et al., 1997] presented inch-sized robots who leave behind a magnetic footprint that can be detected by other robots for communication and motion guidance, as illustrated in Figure 28.26. These footprints fade out with time, allowing continuous and advanced motion involving several miniature robots.

28.7 Conclusion and Discussion

Even though the microrobotic field and especially MEMS-based microrobotics are areas of constant expansion and we see a growing worldwide interest in the concept and possible applications, it still remains to be seen if these devices will come to practical and commercial use in applications such as hybrid assembling and manufacturing of microsystems. In this review, two fundamentally different aspects of microrobotics have been addressed: (1) MEMs for making microrobots, and (2) microrobots for making MEMS. The MEMs fabrication technology is an attractive approach for obtaining miniaturized robots. By scaling down the remaining robotic subsystems to the same scale as the control systems by integrating actuators, sensors, logic and power supplies onto a single piece of silicon, enormous advantages can be obtained in terms of mass producability, lower costs and fewer connector problems encountered when interconnecting these discrete subsystems. The other aspect of MEMS and microrobotics is the application of these microrobots. Among several applications for microrobots, their use in medical applications (e.g., minimal invasive surgery) and biotechnology (e.g., cell-sorting) should be mentioned as the most promising for future commercial success of MEMS-based microrobotic devices. Further, the use of various microrobotic devices and robot systems for assembling complex microsystems and perhaps also for more effective fabrication by hybrid technologies is an attractive approach for making more cost-effective devices and systems when low-volume or prototyping devices and systems are needed (i.e., microrobots for making MEMS). The micromanipulation tools, microconveyers and microrobots for locomotive mechanisms described in this chapter should be seen as the first steps in the long chain of developments needed to achieve autonomous microrobots for work in multirobotic systems such as microfactories for fabrication of other microsystems. However, much research remains in this field and it could be worth remembering the words of the great physicist Feynman when he "invented" the field of MEMS back in 1959:

> *What are the possibilities of small but movable machines? They may or they may not be useful, but they surely would be fun to make.*

The devices described in this chapter illustrate the basic building blocks and some of the progress made in the young research field of MEMS-based microrobotics during the 1990s. Therefore, let us have fun, feel the wonder of microrobotics when "small movable machines" are fabricated and when new frontiers for science and research are explored. Beside the element of fun, there are tremendous opportunities for microrobotic applications, thus a large potential for industrial innovation and competition.

Defining Terms

Ciliary motion system (CMS): A transportation system based on the ciliary motion principle adopted from nature. The CMS principle relies on an asynchronous driving technique that requires at least two spatially separated groups of actuators that are turned on and off at different times, alternately holding and driving the device. Higher speeds and smoother motions can be achieved with such asynchronous driving than with synchronous driving.

Distributed micromotion systems (DMMS): The concept of an array configuration of microactuators, where the cooperative work of many coordinated simple actuators generates interaction with the macroworld.

Hybrid assembly: Assembly of several individual microcomponents fabricated by different technologies and on different wafers. If microrobotic devices are used for the assembly (microrobots for making MEMS), extremely small components can be assembled.

MEMS-based microrobotics: A sort of "modified chip" fabricated by silicon MEMS-based technologies (such as batch-compatible bulk- or surface-micromachining or by micromolding and/or replication method) having features in the micrometer range.

Microfactory: A complete automatic factory containing conveyers, robot arms and other machine tools at a very small level. A synonym for microfactory is desktop factory.

Microrobotics: A general term for devices having dimensions smaller than classical watch-making parts (i.e., µm to mm) with the ability to move, apply forces and manipulate objects in a workspace with dimensions in the micrometer or submicrometer range. Included in the term are both system aspects and more fundamental building-block aspects of a robot system. Examples of these fundamental building blocks in the microrobotic field are microgrippers, microconveyers and locomotive robots (arms for pick-and-place and/or legs for walking).

Monolithic assembly: All integration is done on one single (silicon) chip preferably using wafer-level assembly. This technology is preferable when fabricating microrobotic devices that rely on actuator arrays (MEMS for making microrobots).

Parallel (batch) manufacturing: Fabrication performed on several wafers at the same time (several-devices-at-a-time process) originally developed for the IC industry but now also commonly used for MEMS fabrication.

Serial manufacturing: One-device-at-a-time fabrication/assembly contrary to the monolithic approach where all integration is done on one single silicon chip using wafer-level assembly.

References

Akiyama, T., and H. Fujita (1995) "A Quantiative Analysis of Scratch Drive Actuator Using Buckling Motion," in *Proc. of IEEE 8th Int. Workshop on Micro Electro Mechanical Systems (MEMS '95)*, pp. 310–315, Jan. 29–Feb. 2, Amsterdam, The Netherlands.

Aoyama, H. et al. (1995) "Desktop Flexible Manufacturing System by Movable Miniature Robots—Miniature Robots with Micro Tool and Sensor," in *IEEE Int. Conf. on Robotics and Automation*, pp. 660–665, May 21–27, Nagoya, Aichi, Japan.

Aoyama, H. et al. (1997) "Pursuit Control of Micro-Robot Based Magentic Footstep," in *Proc. Int. Conf. on Micromechatronics for Information and Precision Equipment (MIPE '97)*, pp. 256–260, July 20–23, Tokyo, Japan.

Arai, K.I. et al. (1995) "Magnetic Small Flying Machines," in *Tech. Digest Transducers '95 and Eurosensors IX*, pp. 316–319, June 25–29, Stockholm, Sweden.

Ataka, M. et al. (1993a) "A Biometric Micro System—A Ciliary Motion System," in *Tech. Digest, 7th Int. Conf. Solid-State Sensors and Actuators (Transducers '93)*, pp. 38–41, June 7–10, Yokohama, Japan.

Ataka, M. et al. (1993b) "Fabrication and Operation of Polyimide Bimorph Actuators for a Ciliary Motion System," *IEEE J. MEMS* **2**(4) pp. 146–150.

Benecke, W.R. (1988) "Thermally Excited Silicon Microactuators," *IEEE Trans. Electron. Devices* **35**(6), pp. 758–763.

Bexell, M., and Johansson, S. (1999) "Fabrication and Evaluation of a Piezoelectric Miniature Motor," *Sensors and Actuators A* **75**, pp. 8–16.

Böhringer, K.F. et al. (1996) "Single-Crystal Silicon Actuator Arrays for Micro Manipulation Tasks," in *Proc. of IEEE 9th Int. Workshop on Micro Electro Mechanical Systems (MEMS '96)*, pp. 7–12, Feb. 11–15, 1996, San Diego, CA, USA.

Böhringer, K.F. et al. (1997) "Computational Methods for Design and Control of MEMS Micromanipulator Arrays," *IEEE Comput. Sci. Eng.* Jan.–March, pp. 17–29.

Breguet, J.-M. (1996) "Stick and Slip Actuators and Parallel Architectures Dedicated to Microrobotics," in *SPIE Conf. on Microrobotics, Components and Applications, SPIE Photonics East*, pp. 13–24, Nov., Boston, MA.

Breguet, J.-M., and Renaud, P. (1996) "A 4 Degrees-of-Freedoms Microrobot with Nanometer Resolution," *Robotics* **14**, pp. 199–203.

Carts-Powell, Y. (2000) "Tiny Camera in a Pill Extends Limits of Endoscopy," *SPIE: OE-Reports* Aug. (2000), (available on Internet at: http://www.spie.org).

Cohn, M.B. et al. (1998) "Microassembly Technologies for MEMS," in *SPIE Conf. on Microfluidic Devices and Systems*, pp. 2–16, Sept. 21–22, Santa Clara, CA.

Comotis, J., and Bright, V. (1996) "Surface Micromachined Polysilicon Thermal Actuator Arrays and Applications," in *Solid-State Sensor and Actuator Workshop, (Hilton Head '96)*, pp. 174–177, June 2–6, Hilton Head, SC.

Corman, T. (1999) "Vacuum-Sealed and Gas-Filled Micromachined Devices," Ph.D. thesis, Instrumentation Laboratory, Department of Signals, Sensors and Systems (S3), Royal Institute of Technology (KTH), Stockholm, Sweden (ISBN 91-7170-482-5, available for downloading at http://www.s3.kth.se/mst/research/dissertations/thierrycdoc.html).

Dario, P. et al. (1992) "Review—Microactuators for Microrobots: A Critical Survey," *J. Micromech. Microeng.* **2**(3), pp. 141–157.

Dario, P. et al. (1997) "A Micro Robotic System for Colonoscopy," in *Int. Conf. on Robotics and Automation (ICRA 97)*, pp. 1567–1572, April 20–25, Albuquerque, NM.

Dario, P. et al. (1998) "Contest Winner—A Mobile Microrobot Actuated by a New Electromagnetic Wobbler Micromotor," *IEEE/ASME Trans. Mechatronics* **3**(1).

Drextler, E. (1992) *Nanosystems—Molecular Machinery, Manufacturing, and Computation,* John Wiley & Sons, New York.

Ebefors, T. (2000) "Polyimide V-Groove Joints for Three-Dimensional Silicon Transducers Exemplified Through a 3D Turbulent Gas Flow Sensor and Micro-Robotic Devices," Ph.D. thesis, Instrumentation Laboratory, Department of Signals, Sensors and Systems (S3), Royal Institute of Technology (KTH), Stockholm, Sweden (ISBN 91–7170–586–6, available for downloading at http://www.s3.kth.se/mst/research/dissertations/thorbjornedoc.html).

Ebefors, T. et al. (1998a) "New Small Radius Joints Based on Thermal Shrinkage of Polyimide in V-grooves for Robust Self-Assembly 3-D Microstructures," *J. Micromech. Microeng.*, **8**(3), pp. 188–194.

Ebefors, T. et al. (1998b) "Three Dimensional Silicon Triple-Hot-Wire Anemometer Based on Polyimide Joints," in *IEEE 11th Int. Workshop on Micro Electro Mechanical Systems (MEMS '98)*, pp. 93–98, Jan. 25–29, Heidelberg, Germany.

Ebefors, T. et al. (1999) "A Walking Silicon Micro-Robot," in *10th Int. Conf. on Solid-State Sensors and Actuators (Transducers '99)*, pp. 1202–1205, June 7–10, Sendai, Japan.

Ebefors, T. et al. (2000) "A Robust Micro Conveyer Realized by Arrayed Polyimide Joint Actuators," *IOP J. Micromech. Microeng.* **10**(3), pp. 337–349.

Eigler, D.M., and Schweizer, E.K. (1990) "Positioning Single Atoms with a Scanning Tunneling Microscope," *Nature* **344**, pp. 524–526.

Fatikow, S. (2000) *Mikroroboter und Mikromontage*, B.G. Teubner, Stuttgart-Leipzig.

Fatikow, S., and Rembold, U. (1996) "An Automated Microrobot-Based Desktop Station for Micro Assembly and Handling of Micro-Objects," in *IEEE Conf. on Emerging Technologies and Factory Automation (EFTA '96)*, pp. 586 –592, Nov. 18–21, Kauai, HI.

Fatikow, S., and Rembold, U. (1997) *Microsystem Technology and Microrobotics*, Springer-Verlag, Berlin.

Fearing, R.S. (1998) "Powering 3-Dimensional Microrobots: Power Density Limitations," in *IEEE Int. Conf. on Robotics and Automation*, tutorial on "Micro Mectronics and Micro Robotics," May.

Feynman, R.P. (1961) "There's Plenty of Room at the Bottom," Presented at the annual meeting of the American Physical Society (APS) at the California Institute of Technology, December 29, 1959; published by Reinhold Publishing Corp. in "Miniaturization," edited by Horance D. Gilbert (reprinted by S. Senturia in *IEEE/ASME J. MEMS*, **1**(1), pp. 60–66).

Flynn, A.M. et al. (1989) "The World's Largest One Cubic Inch Robot," in *Proc. of IEEE 2nd Int. Workshop on Micro Electro Mechanical Systems (MEMS '89)*, pp. 98–101, Feb. 20–22, Salt Lake City, UT.

Flynn, A. et al. (1992) "Piezoelectric Micromotors for Microrobots," *IEEE J. MEMS* **1**(1), pp. 44–51.

Frank, T. (1998) "Two-Axis Electrodynamic Micropositioning Devices," *J. Micromech. Microeng.* **8**, pp. 114–118.

Fujita, H., and Gabriel, K.J. (1991) "New Opportunities for Micro Actuators," in *Technical Digest of the 6th Int. Conf. on Solid-State Sensors and Actuators (Transducers '91)*, pp. 14–20, June 24–27, San Francisco, CA.

Fukuda, T. et al. (1994) "Mechanism and Swimming Experiment of Micro Mobile Robot in Water," in *Proc. IEEE 7th Int. Workshop on Micro Electro Mechanical Systems (MEMS '94)*, pp. 273–278, Jan. 25–28, Oiso, Japan.

Fukuda, T. et al. (1995) "Steering Mechanism and Swimming Experiment of Micro Mobile Robot in Water," in *Proc. IEEE 8th Int. Workshop on Micro Electro Mechanical Systems (MEMS '95)*, pp. 300–305, Jan. 29–Feb. 2, Amsterdam, The Netherlands.

Goosen, J., and Wolffenbuttel, R. (1995) "Object Positioning Using a Surface Micromachined Distributed System," in *Technical Digest of the 8th Int. Conf. on Solid-State Sensors and Actuators (Transducers '95 and Eurosensors IX)*, pp. 396–399, June 25–29, Stockholm, Sweden.

Green, P.W. et al. (1995) "Demonstration of Three-Dimensional Microstructure Self-Assembly," *J. MEMS* **4**(4), pp. 170–176.

Greitmann, G., and Buser, R. (1996) "Tactile Microgripper for Automated Handling of Microparts," *Sensors and Actuators A* **53**(1–4), pp. 410–415.

Guckel, H. et al. (1992) "Termo-Magnetic Metal Flexure Actuators," in *Technical Digest of the Solid-State Sensor and Actuator Workshop (Hilton Head '92)*, pp. 73–75, June 13–16, Hilton Head, SC.

Haga, Y. et al. (1998) "Small Diameter Active Catheter Using Shape Memory Alloy," in *IEEE 11th Int. Workshop on Micro Electro Mechanical Systems (MEMS '98)*, pp. 419–424, Jan. 25–29, Heidelberg, Germany.

Hayashi, T. (1991) "Micro Mechanism," *J. Robotics Mechatronics* **3**, pp. 2–7.

Hierold, C. et al. (1998) "Implantable Low Power Integrated Pressure Sensor System for Minimal Invasive Telemetric Patient Monitoring," in *Proc. IEEE 11th Int. Workshop on Micro Electro Mechanical Systems (MEMS '98)*, pp. 568–573, Jan. 25–29, Heidelberg, Germany.

Hirai, S. et al. (1993) "Cooperative Task Execution Technology for Multiple Micro Robot Systems," in *Proc. IATP Workshop on Micromachine Technologies and Systems*, pp. 32–37, Tokyo, Japan.

Hirata, T. et al. (1998) "A Novel Pneumatic Actuator System Realized by Microelectrodischarge Machining," in *Proc. IEEE 11th Int. Workshop on Micro Electro Mechanical Systems (MEMS '98)*, pp. 160–165, Jan. 25–29, 1998, Heidelberg, Germany.

Hui, E. et al. (2000) "Single-Step Assembly of Complex 3-D Microstructures," in Proc. *IEEE 13th Int. Conf. on Micro Electro Mechanical Systems (MEMS '2000)*, pp. 602–607, Jan. 23–27, Miyazaki, Japan.

Ishihara, H. (1998) "International Micro Robot Maze Contest Information," available at http://www.mein.nagoya-u.ac.jp/maze/index.html.

Ishihara, H., and Fukuda, T. (1998) "Optical Power Supply System for Micromechanical System," in *SPIE Conf. on Microrobotics and Micromanipulation*, pp. 140–145, Nov. 4–5, 1998, Boston, Massachusetts, USA.

Ishihara, H. et al. (1995) "Approach to Distributed Micro Robatic System—Development of Micro Line Trace Robot and Autonomous," in *Proc. IEEE Int. Conf. on Robotics and Automation*, pp. 375–380, May 21–27, 1995, Nagoya, Japan.

Ishikawa, Y. and T. Kitahara (1997) "Present and Future of Micromechatronics," in *Int. Symp. on Micromechanics and Human Science (MHS '97)*, pp. 13–20, Oct. 5–8, 1997, Nagoya, Japan.

Jager, E.W.H. et al. (2000a) "Microrobots for Micrometer-Size Objects in Aqueous Media: Potential Tools for Single Cell Manipulation," *Science* **288**, pp. 2335–2338.

Jager, E.W.H. et al. (2000b) "Microfabricating Conjugated Polymer Actuators," *Science* **290**, pp. 1540–1545.

Johansson, S. (1995) "Micromanipulation for Micro- and Nanomanufacturing," in *INRIA/IEEE Symp. on Emerging Technologies and Factory Automation (ETFA '95)*, pp. 3–8, Oct. 10–13, Paris, France.

Jonsmann, J. et al. (1999) "Multi Degrees of Freedom Electro-Thermal Microactuators," in *Proc. IEEE 10th Int. Conf. on Solid-State Sensors and Actuators (Transducers '99)*, pp. 1373–1375, June 7–10, Sendai, Japan.

Kawahara, N. et al. (1997) "Microfactories: New Applications of Micromachine Technology to the Manufacture of Small Products," *Res. J. Microsystem Technol.* **3**(2), pp. 37–41.

Keller, C. (1998a) "Microfabricated High Aspect Ratio Silicon Flexures: HEXSIL, RIE, and KOH Etched Design & Fabrication" (http://www.memspi.com/book.html).

Keller, C. (1998b) "Microfabricated Silicon High Aspect Ratio Flexures for In-Plane Motion," Ph.D. Thesis, Dept. of Materials Science and Mineral Engineering, University of California at Berkeley.

Keller, C.G., and Howe, R.T. (1995) "Nickel-Filled Hexsil Thermally Actuated Tweezers," in *Technical. Digest of the Transducers '95 and Eurosensors IX*, pp. 376–379, June 25–29, Stockholm, Sweden.

Keller, C.G., and Howe, R.T. (1997) "HEXSIL Tweezers for Teleoperated Micro-assembly," in *Proc. IEEE 10th Int. Workshop on Micro Electro Mechanical Systems (MEMS '97)*, pp. 72–77, Jan. 26–30, 1997, Nagoya, Japan.

Kelly, T.R. et al. (1999) "Unidirectional Rotary Motion in a Molecular System," *Nature* **401**(6749), pp. 150–152.

Kim, C.-J. et al. (1992) "Silicon-Processed Overhanging Microgripper," *IEEE/ASME J. MEMS* **1**(1), pp. 31–36.

Kim, Y.-K. et al. (1990) "Fabrication and Testing of a Micro Superconducting Actuator Using the Meissner Effect," in *Proc. IEEE 3rd Int. Workshop on Micro Electro Mechanical Systems (MEMS '90)*, pp. 61–66, Feb. 11–14, Napa Valley, CA.

Kladitis, P. et al. (1999) "Prototype Microrobots for Micro Positioning in a Manufacturing Process and Micro Unmanned Vehicles," in *Proc. IEEE 12th Int. Conf. on Micro Electro Mechanical Systems (MEMS '99)*, pp. 570–575, Jan. 17–21, Oralndo, FL.

Konishi, S., and Fujita, H. (1993) "A Conveyance System Using Air Flow Based on the Concept of Distributed Micro Motion Systems," in *Technical Digest of the 7th Int. Conf. on Solid-State Sensors and Actuators (Transducers '93)*, pp. 28–31, June 7–10, Yokohama, Japan.

Konishi, S., and Fujita, H. (1994) "A Conveyance System Using Air Flow Based on the Concept of Distributed Micro Motion Systems," *IEEE J. MEMS* **3**(2), pp. 54–58.

Konishi, S., and Fujita, H. (1995) "System Design for Cooperative Control of Arrayed Microactuators," in *Proc. IEEE 8th Int. Workshop on Micro Electro Mechanical Systems (MEMS '95)*, pp. 322–327, Jan. 29–Feb. 2, Amsterdam, the Netherlands.

Kovacs, G.T.A. (1998) "Thin-Film Batteries," in *Micromachined Transducers Sourcebook*, pp. 745–750, McGraw-Hill, New York.

Kuwana, Y. et al. (1995) "Steering Control of a Mobile Robot Using Insect Antennae," in *IEEE/RSJ Int. Conf. on Intelligent Robots and Systems 95: Human Robot Interaction and Cooperative Robots*, pp. 530–535.

Li, M.-H. et al. (2000) "High Performance Scanning Thermal Probe Using a Low Temperature Polyimide-Based Micromachining Process," in *Proc. IEEE 13th Int. Conf. on Micro Electro Mechanical Systems (MEMS '2000)*, pp. 763–768, Jan. 23–27, Miyazaki, Japan.

Lin, C.-C. et al. (1999) "Fabrication and Characterization of a Micro Turbine/Bearing Rig," in *Proc. IEEE 12th Int. Conf. on Micro Electro Mechanical System (MEMS '99)*, pp. 529–533, Jan. 17–21, 1999, Orlando, FL.

Lin, L.Y. et al. (1996) "Realization of Novel Monolithic Free-Space Optical Disk Pickup Heads by Surface Micromachining," *Optics Lett.* **21**(2), pp. 155–157.

Liu, C. et al. (1994) "Surface Micromachined Magnetic Actuators," in *Proc. IEEE 7th Int. Conf. on Micro Electro Mechanical Systems (MEMS '94)*, pp. 57–62, Jan. 25–28, Oiso, Japan.

Liu, C. et al. (1995) "A Micromachined Permalloy Magnetic Actuator Array for Micro Robotics Assembly System," in *Technical Digest of the 8th Int. Conf. on Solid-State Sensors and Actuators (Transducers '95 and Eurosensors IX)*, pp. 328–331, June 25–29, Stockholm, Sweden.

Mainz (Institute of Technology) (1999) "Micro-motors: The World's Tiniest Helicopter" (http://www.imm-mainz.de/english/developm/products/hubi.html).

Masuzawa, T. (1998) "Micromachining by Machine Tools," in *Micromechanical Systems—Principles and Technology*, Vol. 6, pp. 63–82, T. Fukuda and W. Menz, Eds., Elsevier Science, Amsterdam.

Miki, N., and Shimoyama, I. (1999) "Flight Performance of Micro-Wings Rotating in an Alternating Magnetic Field," in *Proc. IEEE 12th Int. Conf. on Micro Electro Mechanical Systems (MEMS '99)*, pp. 153–158, Jan. 17–21, Oralndo, FL.

Miki, N., and Shimoyama, I. (2000) "A Micro-Flight Mechanism with Rotational Wings," in *Proc. IEEE 13th Int. Conf. on Micro Electro Mechanical Systems (MEMS '2000)*, pp. 158–163, Jan. 23–27, Miyazaki, Japan.

MINIMAN (1997) MINIMAN—Miniaturised Robot for Micro MANipulation," final report (part II), Second Phase Project Proposal, Open Scheme LTR Project ESPRIT, Project No. 23915.

Mita, Y. et al. (1999) "An Inverted Scratch-Drive-Actuators Array for Large Area Actuation of External Objects," in *Technical Digest of Transducers '99*, pp. 1196–1197, June 7–10, Sendai, Japan.

Miura, H. et al. (1995) "Insect-Model-Based Microrobot," in *Technical Digest of the 8th Int. Conf. on Solid-State Sensors and Actuators (Transducers '95)*, pp. 392–395., June 16–19, Stockholm, Sweden.

MSTnews (2000) Special volume dedicated to packaging and modular microsystems (No. 1/00).

Muller, R.S. (1990) "Microdynamics," *Sensors and Actuators A* **21–23**, pp. 1–8.

Nakamura, K. et al. (1995) "Evaluation of the Micro Wobbler Motor Fabricated by Concentric Buildup Process," in *Proc. IEEE 8th Int. Workshop on Micro Electro Mechanical Systems (MEMS '95)*, pp. 374–379, Jan. 29–Feb. 2, Amsterdam, The Netherlands.

Nakazawa, H. et al. (1997) "The Two-Dimensional Micro Conveyer," in *Technical Digest of the 9th Int. Conf. Solid-State Sensors and Actuators (Transducers '97)*, pp. 33–36., June 16–19, Chicago, IL.

Nakazawa, H. et al. (1999) "Electromagnetic Micro-Parts Conveyer with Coil-Diod Modules," in *Technical Digest of the 10th Int. Conf. on Solid-State Sensors and Actuators (Transducers '99)*, pp. 1192–1195, June 7–10, Sendai, Japan.

Ok, J. et al. (1999) "Pneumatically Driven Microcage for Micro-Objects in Biological Liquid," in *Proc. IEEE 12th Int. Conf. on Micro Electro Mechanical System (MEMS '99)*, pp. 459–463, Jan. 17–21, 1999, Orlando, FL.

Park, K.-T., and Esashi, M. (1999) "A Multilink Active Catheter with Polyimide-Based Integrated CMOS Interface Circuits," *IEEE J. MEMS* **8**(4), pp. 349–357.

Pister, K. et al. (1990) "A Planar Air Levitated Electrostatic Actuator System," in *Proc. IEEE 3rd Int. Workshop on Micro Electro Mechanical Systems (MEMS '90)*, pp. 67–71, Feb. 11–14, Napa Valley, CA.

Pister, K. et al. (1992) "Micro-Fabricated Hinges," *Sensors and Actuators A* **33**, pp. 249–256.

Pornsin-Sirirak, N., et al. (2000) "MEMS Wing Technology for a Battery-Powered Ornithopter," in *Proc. IEEE 13th Int. Conf. on Micro Electro Mechanical Systems (MEMS '2000)*, pp. 799–804, Jan. 23–27, Miyazaki, Japan.

Riethmüller, W., and Benecke, W. (1989) "Application of Silicon-Microactuators Based on Bimorph Structures," in *Proc. IEEE 2nd Int. Workshop on Micro Electro Mechanical Systems (MEMS '89)*, pp. 116–120., Feb. 20–22, Salt Lake City, UT.

Rodgers, S. et al. (1999) "Intricate Mechanisms-on-a-Chip Enabled by 5-Level Surface Micromachining," in *Proc. IEEE 10th Int. Conf. on Solid-State Sensors and Actuators (Transducers '99)*, pp. 990–993, June 7–10, Sendai, Japan.

Ruffieux, D., and Rooij, N.F.d. (1999) "A 3-DOF Bimorph Actuator Array Capable of Locomotion," in *13th European Conf. on Solid-State Transducers (Eurosensors XIII)*, pp. 725–728, Sept. 12–15, The Hauge, The Netherlands.

Ruffieux, D. et al. (2000) "An AlN Piezoelectric Microacctuator Array," in *Proc. IEEE 13th Int. Conf. on Micro Electro Mechanical Systems (MEMS '2000)*, pp. 662–667, Jan. 23–27, 2000, Miyazaki, Japan.

Shimoyama, I. (1995a) "Hybrid System of Mechanical Parts and Living Organisms for Microrobots," in *Proc. IEEE 6th Int. Symp. on Micro Machine and Human Science (MHS '95)*, p. 55, Oct. 4–6, Nagoya, Japan.

Shimoyama, I. (1995b) "Scaling in Microrobots," *IEEE '95*, pp. 208–211.

Simu, U., and Johansson, S. (1999) "Multilayered Piezoceramic Microactuators Formed by Milling in the Green State," in *SPIE Conf. on Devices and Process Technologies for MEMS and Microelectronics,* **Vol. 3892**, *part of SPIE's Int. Symp. on Microelectronics and MEMS (MICRO/MEMS '99)*, 27–29 Oct. Royal Pines Resort, Gold Coast, Queensland, Australia.

Simu, U., and Johansson, S. (2001) "A Monolithic Piezoelectric Miniature Robot with 5 DOF," in 11th International Conf. Solid-State Sensors and Actuators (Transducers 2001, Eurosensors XV), pp. 690–693, June 10–14, Munich, Germany.

Smela, E. et al. (1995) "Controlled Folding of Micrometer-Size Structures," *Science* **268**(June), pp. 1735–1738.

Smela, E. et al. (1999) "Electrochemically Driven Polypyrrole Bilayers for Moving and Positioning Bulk Micromachined Silicon," *J. MEMS* **8**(4), pp. 373–383.

Stefanni, C. et al. (1996) "A Mobile Micro-Robot Driven by a New Type of Electromagnetic Micromotor," in *Proc. IEEE 7th Int. Symp. on Micro Machine and Human Science (MHS '96)*, Oct., Nagoya, Japan.

Suh, J. et al. (1997) "Organic Thermal and Electrostatic Ciliary Microactuator Array for Object Manipulation," *Sensors and Actuators A* **58**, pp. 51–60.

Suh, J. et al. (1999) "CMOS Integrated Ciliary Actuator Array as a General-Purpose Micromanipulation Tool for Small Objects," *IEEE/ASME J. MEMS* **8**(4), pp. 483–496.

Suzuki, K. et al. (1992) "Creation of an Insect-based Microrobot with an External Skeleton and Elastic Joints," in *Proc. IEEE 5th Int. Workshop on Micro Electro Mechanical Systems (MEMS '92)*, pp. 190–195, Feb. 4–7, Travemünde, Germany.

Suzumori, K. et al. (1999) "Micro Inspection Robot for 1-in Pipes," *IEEE/ASME Trans. on Mechatronics* **4**(3), pp. 286–292.

Syms, R.R.A. (1998) "Rotational Self-Assembly of Complex Microstructures by the Surface Tension of Glass," *Sensors and Actuators A* **65**(2,3), pp. 238–243.

Syms, R.R.A. (1999) "Surface Tension Powered Self-Assembly of 3-D Micro-Optomechanical Structures," *J. MEMS* **8**(4), pp. 448–455.

Tahhan, I.N. et al. (1999) "MEMS Fixtures for Handling and Assembly of Microparts," in *SPIE Micromachined Devices and Components V*, pp. 129–139, Santa Clara, CA.

Takeda, M. (2001) "Applications of MEMS to Industrial Inspection," in *Proc. IEEE 14th Int. Conf. on Micro Electro Mechanical Systems (MEMS '2001)*, pp. 182–191, Jan. 21–25, Interlaken, Switzerland.

Takeuchi, S., and Shimoyama, I. (1999) "Three Dimensional SMA Microelectrodes with Clipping Structure for Insect Neural Recording," in *Proc. 12th IEEE Int. Conf. on Micro Electro Mechanical System (MEMS '99)*, pp. 464–469, Jan. 17–21, Orlando, FL.

Tendick, F. et al. (1998) "MIS Application of Micromechatronics in Minimally Invasive Surgery," *IEEE/ASME Trans. on Mechatronics* **3**(1).

Teshigahara, A. et al. (1995) "Performance of a 7-mm Microfabricated Car," *IEEE/ASME J. MEMS* **4**(2), pp. 76–80.

Thornell, G. (1998) "Lilliputian Reflections," in *Proc. Micro Structure Workshop (MSW '98)*, pp. 24.1–24.6, March 24–25, Uppsala, Sweden.

Thornell, G. (1999) "Minuscular Sculpturing," Ph.D. thesis No. 456, Acta Univertsitatis Upsaliensis, Faculty of Science and Technology, Uppsala University, Uppsala, Sweden (ISBN 91–554–4480–6).

Thornell, G., and Johansson, S. (1998) "Microprocessing at the Fingertips," *J. Micromechan. Microeng.* **8**, pp. 251–262.

Tong, Q.-Y., and Gösle, U. (1999) *Semiconductor Wafer Bonding—Science and Technology*, John Wiley & Sons, New York.

Trimmer, W., and Jebens, R. (1989) "Actuators for Micro Robots," in *IEEE Int. Conf. on Robotics and Automation (MEMS '89)*, pp. 1547–1552, May 14–19.

Uchiyama, A. (1995) "Endoradiosonde Needs Micro Machine Technology," in *Proc. IEEE 6th Int. Symp. on Micro Machine and Human Science (MHS '95)*, pp. 31–37.

Udayakumar, K.R. et al. (1991) "Ferroelectric Thin-Film Ultrasonic Micromotors," in *Proc. IEEE 4th Int. Workshop on Micro Electro Mechanical Systems (MEMS '91)*, pp. 109–113, Jan. 30–Feb. 2, Nara, Japan.

Wilson, J. (1997) "Shrinking Micromachines," *Popular Mechanics* **Nov.** (available at http://popularmechanics.com/popmech/sci/9711STROP.html).

Yasuda, T. et al. (1993) "Microrobot Actuated by a Vibration Field," in *Technical Digest, 7th Int. Conf. on Solid-State Sensors and Actuators (Transducers '93)*, pp. 42–45, June 7–10, Yokohama, Japan.

Yeh, R., and Pister, K. (2000) "Design of Low-Power Articulated Microrobots," in *Proc. Int. Conf. on Robotics and Automation Workshop on Mobile Micro-Robots*, pp. 21–28, April 23–28, San Francisco, CA.

Yeh, R. et al. (1996) "Surface-Micromachined Components for Articulated Microrobots," *J. MEMS* **5**(1) pp. 10–17.

Zhou, G.-X. (1989) "Swallowable or Implantable Body Temperature Telemeter-Body Temperature Radio Pill," in *Proc. IEEE 15th Annual Northeast Bioengineering Conf.*, pp. 165–166.

Zill, S.N., and Seyfarth, E.-A. (1996) "Exoskeltal Sensors for Walking," *Sci. Am.* **July**, pp. 70–74.

Further Information

A good introduction to MEMS-based microrobotics is presented in *Microsystem Technology and Microrobotics* (Fatikows and Rembold, Springer-Verlag, Berlin, 1997) and *Mikroroboter und Mikromontage* (in German); Fatikow, B.G. Teubner, Stuttgart-Leipzig, 2000). A good review on actuators suitable for microrobots as well as different aspects of the definition of microrobots is found in the paper, "Review—Microactuators for Microrobots: A Critical Survey," by Dario et al. and published in IOP *Journal of Micromechanics and Microengineering (JMM)*, Vol. 2, pp. 141–157, 1992. Proceedings from the "Microrobotics and Micromanipulation" conference have been published annually by SPIE (the International Society for Optical Engineering) since 1995. These proceedings document the latest development in the field of microrobotics each year.

29
Microscale Vacuum Pumps

E. Phillip Muntz
University of Southern California

Stephen E. Vargo
SiWave, Inc.

29.1 Introduction **29**-1
29.2 Fundamentals **29**-3
  Basic Principles • Conventional Types of Vacuum Pumps • Pumping Speed and Pressure Ratio • Definitions for Vacuum and Scale
29.3 Pump Scaling **29**-5
  Positive-Displacement Pumps • Kinetic Pumps • Capture Pumps • Pump-Down and Ultimate Pressures for MEMS Vacuum Systems • Operating Pressures and $\dot{N}$ Requirements in MEMS Instruments • Summary of Scaling Results
29.4 Alternative Pump Technologies **29**-16
  Outline of Thermal Transpiration Pumping • Accommodation Pumping
29.5 Conclusions **29**-25
  Acknowledgments **29**-25

29.1 Introduction

Numerous potential applications for meso- and microscale sampling instruments are based on mass spectrometry [Nathanson et al., 1995; Ferran and Boumsellek, 1996; Orient et al., 1997; Piltingsrud, 1997; Wiberg et al., 2000; White et al., 1998; Freidhoff et al., 1999; Short et al., 1999] and gas chromatography [Terry et al., 1979]. Other miniaturized instruments utilizing electron optics [Chang et al., 1990; Park et al., 1997; Callas, 1999] will require both high-vacuum and repeated solid-sample transfers from higher pressure environments. The mushrooming interest in chemical laboratories on chips will likely evolve toward some manifestations requiring vacuum capabilities. At present, there are no microscale or mesoscale vacuum pumps to pair with the embryonic instruments and laboratories that are being developed. Certainly, small vacuum pumps will not always be necessary. Some of the new devices are attractive because of low quantities of waste and rapidity of analysis, not directly because they are small, energy efficient, or portable. However, for other applications involving portability and/or autonomous operations, appropriately small vacuum pumps with suitably low power requirements will be necessary. This chapter addresses the question of how to approach providing microscale and mesoscale vacuum pumping capabilities consistent with the volume and energy requirements of meso- and microscale instruments and processes in need of similarly sized vacuum pumps. It does not review existing microscale pumping devices because none are available with attractive performance characteristics (a review of the attempts has recently been presented by Vargo, 2000; see also NASA/JPL, 1999).

In the macroscale world, vacuum pumps are not very efficient machines, ranging in thermal efficiencies from very small fractions of one percent to a few percent. They generally do not scale advantageously to

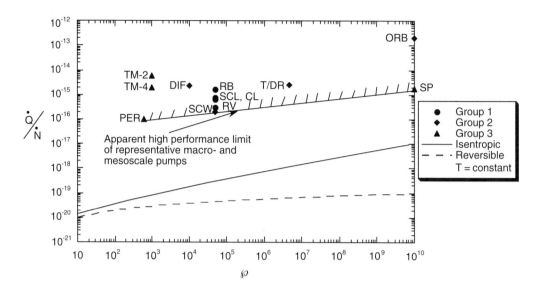

FIGURE 29.1 Representative performance ($\dot{Q}/\dot{N}$) of selected macro- and mesoscale vacuum pumps (from Table 29.1) as a function of pressure ratio ($\wp$).

small sizes, as is discussed in the section on pump scaling. Because of the continuing significant efforts to miniaturize instruments and chemical processes, it is a reasonable hypothesis that there is not much attraction in using oversized, power-intensive vacuum pumps to permit them to operate. This is probably generally true even for situations where pump size and power are not critical issues. More specifically, because microscale-generated vacuums are unavailable today, serious limits are imposed on the potential microscale applications of many high-performance analytical instruments and chemical processes where portability or autonomous operations are necessary.

To illustrate this point, the performance characteristics of several types of macroscale and mesoscale vacuum pumps are presented in Figure 29.1 and Table 29.1. Pumping performance is measured by $\dot{Q}/\dot{N}$, which is derived from the power required ($\dot{Q}, W$) and the pump's upflow in molecules per unit time ($\dot{N}$, #/s). Representative vacuum pumping tasks are indicated by the inlet pressure p_I and the pressure ratio $\wp$ through which $\dot{N}$ is being pumped. Also shown in the figure is the reversible, constant-temperature compression power required per molecule of upflow ($\dot{Q}/\dot{N}$) as a function of $\wp$. The adiabatic reversible (isentropic) compression power that would be required is also shown in Figure 29.1. The constant temperature comparison is most appropriate for vacuum systems. Note the 3- to 5-decade gap in the ideal $\dot{Q}/\dot{N}$ and the actual values for the macroscale vacuum pumps. For low pump inlet pressures, the gases are very dilute and high volumetric flows are required to pump a given $\dot{N}$. The result is relatively large machines with significant size and frictional overheads. Unfortunately, scaling the pumps to smaller sizes generally increases the overhead relative to the upflow $\dot{N}$.

In addition to the power requirement, vacuum pumps tend to have large volumes, so that another indicator of relevance to miniaturized pumps is a pump's volume (V_P) per unit upflow ($V_P/\dot{N}$). The two measures $\dot{Q}/\dot{N}$ and $V_P/\dot{N}$ will be used throughout the following discussions for evaluating different approaches to the production of microscale vacuums.

The bias adopted in this chapter is to address only the production of appropriate vacuums where throughput or continuous gas sampling or, alternatively, multiple sample insertions are required. In some cases, so-called capture pumps (sputter ion pumps, getter pumps) may provide a convenient high- and ultra-high-vacuum pumping capacity. However, because of the finite capacity of these pumps before regeneration, they may or may not be suitable for long-duration studies. Because such trade-offs are very situation dependent, only the sputter ion and orbitron ion capture pumps are considered in the present

TABLE 29.1 Data with Sources for Conventional Vacuum Pumps that Might be Considered for Miniaturization

Type	p_I, p_E (mbar)	S_P (l/s)	$\dot{Q}$ (W)	$\dot{Q}/\dot{N}$ (W/ #/s)	$\wp(—)$	Comments and Sources
Group 1: Macroscale Positive Displacement						
Roots blower (RB)	2E-2, 1E3	2.8	2100	1.6E-15	5E4	5 stage (Lafferty, 1998, p. 161)
Claw (CL)	2E-2, 1E3	6.1	2100	7.5E-16	5E4	4 stage (Lafferty, 1998, p. 165)
Screw (SCW)	2E-2, 1E3	3.6	500	3E-16	5E4	(Lafferty, 1998, p. 167)
Scroll (SCL)	2E-2, 1E3	1.4	500	7.7E-16	5E4	(Lafferty, 1998, p. 168)
Rotary vane (RV)	2E-2, 1E3	0.7	250	6.2E-16	5E4	Catalog
Group 2: Macroscale Kinetic and Ion						
Drag (DR)	2E-2, 1E3	36	3300	1.9E-16	5E4	Molecular/regenerative (Lafferty, 1998, p. 253)
Diffusion (DIF)	1E-5, 1E-1	2.5E4	1.4E3	2.5E-15	1E4	Zyrianka catalog
Turbo/drag (T/DR)	1E-5, 4.5E1	30	20	2.5E-15	4.5E6	Alcatel catalog (30H+30)
Orbitron (ORB)	1E-7, 1E3	1700	750	2E-13	1E10	(Denison, 1967)
Group 3: Mesoscale Pumps						
Turbomolecular (TM-4)	1E-5, 1E-2	4	2	2E-15	1E3	Experimental, $f_P \cup 1.7$ E3, 4-cm dia., Creare web site
Peristaltic (PER)	1.6, 1E3	3.3E-3	20	1E-16	6E2	(Piltingsrud, 1996)
Sputter ion (SP)	1E-7, 1E3	2.2	1.1E-2	1.7E-15	1E10	Based on Suetsugu (1993), 1.5-cm dia., 3.1-cm length
Turbomolecular (TM-2)	1E-5, 1E-2	1.5	2	5E-15	1E3	Muntz scaling of Creare pump to 2-cm dia., using Bernhardt (1983)

study. Both of these pump "active" (N_2, O_2, etc.) and inert (noble and hydrocarbon) gases, whereas other nonevaporable and evaporable getters only pump the active gases efficiently [Lafferty, 1998].

29.2 Fundamentals

29.2.1 Basic Principles

Several basic relationships that are derived from the kinetic theory of gases [*cf.* Bird, 1994; Lafferty, 1998; Cercignani, 2000] are important to the discussion of both macroscale and microscale vacuum pumps. The conductance or volume flow in a channel under free molecule or collisionless flow conditions can be written as:

$$C_L = C_A \alpha \tag{29.1}$$

where C_L is the channel volume flow in one direction for a channel of length L. The conductance of the upstream aperture is C_A and α is the probability that a molecule, having crossed the aperture into the channel, will travel through the channel to its end (this includes those that pass through without hitting a channel wall and those that have one or more wall collisions). Employing the kinetic theory expression for the number of molecules striking a surface per unit time per unit area ($n_g \overline{C}'/4$) gives

$$C_A = (\overline{C}'/4)A_A = \{(8kT_g/\pi m)^{1/2}/4\}A_A \tag{29.2}$$

where: A_A is the aperture's area; $\overline{C}' = \{8kT_g/\pi n\}^{1/2}$ is the mean thermal speed of the gas molecules of mass m; k is Boltzmann's constant; and T_g and n_g are the gas temperature and number density, respectively. The probability, α, can be determined from the length and shape of the channel and the rules governing the reflection at the channel's walls [*c.f.* Lafferty, 1998; Cercignani, 2000].

Several terms associated with wall reflection will be used. *Diffuse reflection* of molecules occurs when the angle of reflection from the wall is independent of the angle of incidence, with any reflected direction in the gas space equally probable per unit of projected surface area in that direction. The reflection is said to be *specular* if the angle of incidence equals the angle of reflection and both the incident and reflected velocities lie in the same plane and have equal magnitude.

The condition for effectively collisionless flow (no significant influence of intermolecular collisions) is reached when the mean free path (λ) of the molecules between collisions in the gas is significantly larger than a representative lateral dimension (l) of the flow channel. Usually this is expressed by the Knudsen number (Kn), such that:

$$Kn_l = \lambda/l \geq 10 \tag{29.3}$$

The mean free path λ can have, for our purposes, the elementary kinetic theory form,

$$\lambda = 1/(\sqrt{2}\Omega n_g) \tag{29.4}$$

with Ω being the temperature-dependent, hard-sphere, total-collision cross section of a gas molecule (for a hard-sphere gas of diameter d, $\Omega = \pi d^2$). As an example, the mean free path for air at 1 atm and 300 K is $\lambda \approx 0.06$ µm.

Expressions for conductance analogous to Eq. (29.1) can be obtained for transitional ($10 > Kn_l > 0.01$) flows and continuum ($Kn_l \leq 10^{-3}$) viscous flows [Cercignani, 2000, and references therein]. For our discussion, the major interest is in collisionless and early transitional flow ($K_n \geq 0.1$).

The performance of a vacuum pump is conventionally expressed as its pumping speed or volume of upflow (S_p) measured in terms of the volume flow of low-pressure gas from the chamber that is being pumped (see Lafferty, 1998, for details on the size and shape of the chamber). Following the recipe of Eq. (29.1), the pumping speed can be written as:

$$S_P = C_{AP}\alpha_P \tag{29.5}$$

Once a molecule has entered the pump's aperture of conductance C_{AP}, it will be "pumped" with probability α_P. Clearly, $(1 - \alpha_P)$ is the probability that the molecule will return or be backscattered to the low-pressure chamber. A pump's upflow in this chapter will generally be described in terms of the molecular upflow in molecules per unit time ($\dot{N}$, #/s). For a chamber pressure p_l and temperature T_l, the number density n_l is given by the ideal gas equation of state, $p_l = n_l k T_l$, and the molecular upflow is

$$\dot{N} = S_P n_l \tag{29.6}$$

29.2.2 Conventional Types of Vacuum Pumps

The several types of available vacuum pumps have been classified [c.f. Lafferty, 1998] into convenient groupings, from which potential candidates for microelectromechanical systems (MEMS) vacuum pumps can be culled. The groupings are

- *Positive-displacement* (vane, piston, scroll, Roots, claw, screw, diaphragm)
- *Kinetic* (vapor jet or diffusion, turbomolecular, molecular drag, regenerative drag)
- *Capture* (getter, sputter ion, orbitron ion, cryopump)

In systems requiring pressures $<10^{-3}$ mbar, the positive-displacement, molecular and regenerative drag pumps are used as "backing" or "fore" pumps for turbomolecular or diffusion pumps. In their operating pressure range ($<10^{-4}$ mbar), the capture pumps require no backing pump but have a more or less limited storage capacity before requiring "regeneration." Also, some means to pump initially to about 10^{-4} mbar from the local atmospheric pressure is necessary. There is insufficient space here to discuss the details of these pump types, and the interested reader is referred to J.M. Lafferty's excellent book [Lafferty, 1998]. The historical roots of this reference are also of interest [Dushman, 1949; Dushman and Lafferty, 1962]. Perhaps all that is necessary to remember at present is that the positive-displacement pumps mechanically trap gas in a volume at a low pressure, the volume decreases, and the trapped gas is rejected at a higher pressure. The kinetic pumps continuously add momentum to the pumped gas so that it can overcome adverse pressure gradients and be "pumped." The storage pumps trap gas or ions on and in a nanoscale lattice or, in the case of cryocondensation pumps, simply condense the gas. In either case, the storage pumps have a finite capacity before the stored gas has to be removed (pump regeneration) or fresh adsorption material supplied.

29.2.3 Pumping Speed and Pressure Ratio

For all pumps, except for the ion pumps, there is a trade-off between upflow S_P and the pressure ratio, $\wp$, that is being maintained by the pump. It has been found convenient to identify a pump's performance by two limiting characteristics [e.g., Bernhardt, 1983]: (1) the maximum upflow, $S_{P,MAX}$, which is achieved when the pressure ratio $\wp = 1$; and (2) the maximum pressure ratio $\wp_{MAX}$ which is obtained for $S_P = 0$. In many cases, a simple expression relating pumping speed (S_P) and pressure ratio ($\wp$) to $S_{P,MAX}$ and $\wp_{MAX}$ describes the trade-off between speed and pressure ratio:

$$S_P/S_{P,MAX} = \frac{(1 - \wp/\wp_{MAX})}{(1 - 1/\wp_{MAX})} \tag{29.7}$$

This relationship is not strictly correct because in many pumps the conductances that result in backflow losses relative to the upflow change dramatically as pressure increases. For the critical lower pressure ranges (10^{-1} mbar in macroscale pumps but significantly higher in microscale pumps), Eq. (29.7) is a reasonable expression for the trade-off between speed and pressure ratio. Equation (29.7) is convenient because $\wp_{MAX}$ and $S_{P,MAX}$ are identifiable and measurable quantities that can then be generalized by Eq. (29.7).

29.2.4 Definitions for Vacuum and Scale

The terms *vacuum* and *MEMS* have both flexible and strict definitions that depend on the beholders. For our discussion, the following categories of vacuum in reduced-scale devices will be used. The pressure range from 10^{-2} to 10^3 mbar will be defined as *low* or *roughing vacuum*. The range from 10^{-2} to 10^{-7} mbar will be refered to as *high vacuum*, and for pressure below 10^{-7} mbar, *ultra-high vacuum*.

For the foreseeable future most small-scale vacuum systems are unlikely to fall within the strict definition of MEMS devices (maximum component dimension <100 µm). Typically, device dimensions somewhat larger than 1 cm are anticipated. They will be fabricated using MEMS techniques but the total construct will be better termed *mesoscale*. Device scale lengths 10 cm and greater indicate macroscale devices.

29.3 Pump Scaling

In this section, the sensitivities to size reduction of the performances of several generic, conventional vacuum pump configurations are discussed. Positive-displacement, turbomolecular (also molecular drag kinetic pumps), sputter and orbitron ion capture pumps are the major focus. Other possibilities, such as diffusion kinetic pumps, diaphragm positive-displacement pumps, getter capture pumps, cryocondensation and cryosorption pumps do not appear to be attractive for MEMS applications. This is due to vaporization and condensation of a separate working fluid (diffusion pumps), large backflow due to valve leaks relative to upflow (diaphragm pumps), low saturation gas loadings (getter capture pumps), an inability to pump the noble gases (getter capture pumps), and the difficulty in providing energy-efficient cryogenic temperatures for MEMS-scale cryocondensation or cryosorption pumps.

29.3.1 Positive-Displacement Pumps

Consider a generic positive-displacement pump that traps a volume V_T of low-pressure gas with a frequency of f_T trappings per unit time. In order to derive a phenomenological expression for pumping speed several inefficiencies must be taken into account: backflow due to clearances, which is particularly important for dry pumps and the volumetric efficiency of the pump's cycle, including both time-dependent inlet conductance effects and dead volume fractions. The generic positive-displacement pump is illustrated in Figure 29.2. Without going into detail, the pumping speed (S_P) for an intake number density n_I and an exhaust number density n_E (or $\wp$ corresponding to the pressures p_I and p_E, as the process gas temperature is assumed to be constant in the important low-pressure pumping range) can be derived:

$$S_p = (1 - \wp \wp_G^{-1})(1 - e^{-(C_{LI}/V_{T,I} f_T)\beta_1}) V_{T,I} f_T - (\wp - 1) C_{LB} \beta_2 \tag{29.8}$$

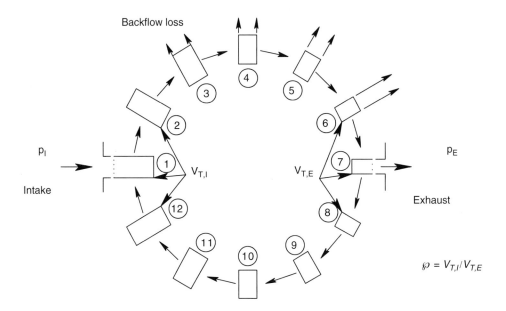

① → ② Intake from p_I to $V_{T,I}$ at pressure p_E/p_G, partial filling of $V_{T,I}$ due to limited inlet time. $V_{T,I}$ closed at $\leq p_I$.

② → ③ Volume decreases and backflow loss begins.

④ → ⑥ Volume continues to decrease to $V_{T,E}$, backflow increases, pressure in $V_{T,E} > p_E$.

⑦   Exhaust of excess pressure from $V_{T,E}$ to p_E.

⑧ $V_{T,E}$ closed with pressure p_E.

⑦ → ⑫ Volume expands to $V_{T,I}$, pressure drops to $p_E/\wp_G$.

①   Cycle repeats.

FIGURE 29.2 Generic positive-displacement vacuum pump.

where $\wp$ is the pressure ratio $p_E/p_I \equiv n_E/n_I$ (assuming $T_I = T_E$); C_{LI} is the pump's inlet conductance; $V_{T,I}$ is the trapping volume at the inlet; C_{LB} is the conductance of the backflow channels between exhaust and inlet pressures; $\wp_G = V_{T,I}/V_{T,E}$ is the geometric trapped-volume ratio between inlet and exit; β_1 is the fraction of the trapping cycle during which the inlet aperture is exposed; and β_2 is the fraction of the cycle during which the backflow channels are exposed to the pressure ratio $\wp$.

The pumping speed expression of Eq. (29.8) applies most directly to a single compression stage. The backflow conductance is assumed to be approximately constant because the flow is in the "collisionless" flow regime that exists in the first few stages of a typical dry pumping system. The inlet conductance for a dry microscale system will be in the collisionless flow regime at low pressure (say 10^{-2} mbar). The inlet conductance per unit area can increase significantly (amount depends on geometry) for transitional inlet pressures [Sone and Itakura, 1990; Lafferty, 1998; Sharipov and Seleznev, 1998]. The performance of macroscale (inlet apertures of several centimeters and larger) positive-displacement pumps at a given inlet pressure will thus benefit from the increased inlet conductance per unit area compared to their reduced scale counterparts in the important low-pressure range of 10^{-3} to 10^{-1} mbar.

The term $(1 - \wp \wp_G^{-1})$ in Eq. (29.8) represents an inefficiency due to a finite dead volume in the exhaust portion of the cycle. The effect of incomplete trapped volume filling during the open time of the inlet aperture is represented by $(1 - e^{-(C_{LI}/V_{T,I}f_T)\beta_1})$. The ideal (no inefficiencies) pumping speed is $V_{T,I}f_T$. The backflow inefficiency is $(\wp - 1)C_{LB}\beta_2$. The dimensions of all groupings are volume per unit time. As in all pumps, a maximum upflow ($S_{P,MAX}$) can be found by assuming $\wp = 1$ in Eq. (29.8). Similarly, the maximum pressure ratio ($\wp_{MAX}$) can be obtained from Eq. (29.8) by setting $S_P = 0$. With some manipulations, Eq. (29.8) can be rewritten as:

$$S_P = V_{T,I}f_T \wp_G^{-1} \{1 + (C_{LB}\beta_2/V_{T,I}f_T)\wp_G - \exp(-C_{LI}\beta_1/V_{T,I}f_T)\}(\wp_{MAX} - \wp) \qquad (29.9)$$

The relationship between $\wp$, $\wp_{MAX}$, S_P and $S_{P,MAX}$ can be found by setting $\wp = 1$ in Eq. (29.9). Substituting back into Eq. (29.9) gives the same expression as in Eq. (29.7).

For the purposes of the present discussions, the upflow of molecules per unit time ($\dot{N} = S_P n_I$) is a useful measure of pumping speed. The form of Eq. (29.9) has been checked by fitting it successfully to observed pumping curves (S_P vs. p_I) for several positive-displacement pumps (using data in Lafferty, 1998) between inlet pressures of 10^{-2} and 10^0 mbar. This is done by following Eq. (29.9) and replotting the experimental results using S_P and $(\wp_{MAX} - \wp)$ as the two variables. The variation of inlet conductance with Kn, $C_{LI}(Kn)$, is important in matching Eq. (29.9) to the observed pumping performances.

At low inlet pressures, the energy use of positive-displacement pumps is dominated by friction losses due to the relative motion of their mechanical components. Taking the two possibilities of sliding and viscous friction as limiting cases, the frictional energy losses can be represented as:

$$\dot{Q}_{sf} = \bar{u}\mu_{sf}A_s\tilde{F}_N \qquad (29.10a)$$

for sliding friction, and for viscous friction:

$$\dot{Q}_\mu = \bar{u}\mu A_s(\bar{u}/h) \qquad (29.10b)$$

Here, $\dot{Q}_{sf}$ and $\dot{Q}_\mu$ are the powers required to overcome sliding friction and viscous friction, respectively. The coefficients are, respectively, μ_{sf} and μ; A_s is the effective area involved; $\tilde{F}_N$ is the normal force per unit area; and $\bar{u}$ is a representative relative speed of the two surfaces that are in contact for sliding friction or separated by a distance h for the viscous case. The contribution to viscous friction in clearance channels exposed to the process gas at low pumping pressures is usually not important, but there may be significant viscous contributions from bearings or lubricated sleeves. An estimate of the power use per unit of upflow can be obtained by combining Eqs. (29.9) and (29.10) to give $(\dot{Q}/\dot{N})$, with units of power per molecule per second or energy per molecule.

Consider the geometric scaling to smaller sizes of a "reference system" macroscopic positive-displacement vacuum pump. The scaling is described by a scale factor s_i applied to all linear dimensions. Inevitable manufacturing difficulties when s_i is very small are put aside for the moment. As a result of the geometric scaling, the operating frequency $f_{T,i}$ needs to be specified. It is convenient to set $f_{T,i} = s_i^{-1} f_{T,R}$, which keeps the component speeds constant between the reference and scaled versions. This may not be possible in many cases; another more or less arbitrary condition would be to have $f_{T,i} = f_{T,R}$. Using the scaling factor and Eq. (29.9), an expression for the scaled pump upflow, $\dot{N}_i$, can be written as a function of s_i and values of the pump's important characteristics at the reference or $s_i = 1$ scale (e.g., $V_{T,I,i} = s_i^3 V_{T,I,R} s_i^{-1} f_{T,R}$). For the case of $f_{T,i} = s_i^{-1} f_{T,R}$,

$$\dot{N}_i = n_I s_i^3 V_{T,I,R} s_i^{-1} f_{T,R} \wp_G^{-1} [1 + (s_i^2 C_{LB,R} \beta_2 / s_i^3 V_{T,I,R} s_i^{-1} f_{T,R}) \wp_G$$
$$-\exp\{-(s_i^2 C_{LI,R} \beta_1 / s_i^3 V_{T,I,R} s_i^{-1} f_{T,R})\}](\wp_{MAX} - \wp) \quad (29.11)$$

Similarly, the scaled version of Eq. (29.10b) becomes:

$$\dot{Q}_{\mu,i} = \bar{u}^2 \mu s_i^2 A_{S,R} / s_i h_R \quad (29.12)$$

Forming the ratios $(\dot{Q}_{\mu,i}/\dot{Q}_{\mu,R})/(\dot{N}_i/\dot{N}_R) \equiv (\dot{Q}_{\mu,i}/\dot{N}_i)/(\dot{Q}_{\mu,R}/\dot{N}_R)$ eliminates many of the reference system characteristics and highlights the scaling. In this case, for the viscous energy dissipation per molecule of upflow in the scaled system compared to the same quantity in the reference system, a scaling relationship is obtained:

$$(\dot{Q}_{\mu,i}/\dot{N}_i)/(\dot{Q}_{\mu,R}/\dot{N}_R) = s_i^{-1} \quad (29.13)$$

A summary of the results of this type of scaling analysis applied to positive-displacements pumps is presented in Table 29.2, along with all the other types of pumps that are considered below. In Table 29.2 it is convenient to summarize the performance using $\wp_{MAX}$ ($S_P = 0$) and $S_{P,MAX}$ or $\dot{N}_{MAX}$ ($\wp = 1$) in order to eliminate $\wp$ appearing explicitly as a variable in the scaling expressions through the dependency of $\dot{N}$ on the pressure ratio (refer to Eq. (29.11)).

29.3.2 Kinetic Pumps

Because of their sensitivity to orientation and their potential for contamination, diffusion pumps do not appear to be suitable for MEMS-scale vacuum pumps, except possibly for situations permitting fixed installations. The other major kinetic pumps, turbomolecular and molecular drag, require high-rotational-speed components but are dry and in macroscale versions can be independent of orientation, at least in time-independent situations. More or less arbitrarily, only the turbomolecular and molecular drag pumps are discussed in this chapter.

A simplified model of turbomolecular pumping has been developed by Bernhardt (1983). Following this description, the maximum pumping speed ($\wp = 1$) of a turbomolecular pumping stage (rotating blade row and a stator row) can be written as:

$$S_{P,MAX} = A_I(\bar{C}'/4)(v_c/\bar{C}')/[(1/qd_f) + (v_c/\bar{C}')] \quad (29.14)$$

In Eq. (29.14): A_I is the inlet area to the rotating blade row; v_c is an average tangential speed of the blades; q is the trapping probability of the rotating blade row for incoming molecules, and besides blade geometry it is a function of $(v_c/\bar{C}')$. The term, d_f, accounts for a reduction in transparency due to blade thickness. It is assumed in Eq. (29.14) that the blades are at an angle of 45° to the rotational plane of the blades. As an instructive convenience, write $v_r = (v_c/\bar{C}')$; v_r is similar to but not identical with the Mach number.

TABLE 29.2a Effect of Scaling on Performance

Type	$\dfrac{\wp_{MAX,i}}{\wp_{MAX,R}}$	$\dfrac{S_{P,MAX,i}}{S_{P,MAX,R}}$	$\dfrac{(\dot{Q}_{sf}/\dot{N}_{MAX})_i}{(\dot{Q}_{sf}/\dot{N}_{MAX})_R}$	$\dfrac{(\dot{Q}_\mu/\dot{N}_{MAX})_i}{(\dot{Q}_\mu/\dot{N}_{MAX})_R}$	$\dfrac{(V_P/\dot{N}_{MAX})_i}{(V_P/\dot{N}_{MAX})_R}$
		$f_{T,i} = s_i^{-1} f_{T,R}$	$(\bar{u}_i = \bar{u}_R)$		
Positive displacement	1	s_i^2	1	s_i^{-1}	s_i
Turbomolecular	1	s_i^2	1	s_i^{-1}	s_i
		$f_{T,i} = f_{T,R}$	$(\bar{u}_i = s_i \bar{u}_R)$		
Positive displacement	$\mathcal{O}[1]$ to $\mathcal{O}[s_i]$	$\mathcal{O}[s_i^3]$	$>\mathcal{O}[1]$	$>\mathcal{O}[1]$	$>\mathcal{O}[1]$
Turbomolecular	$(\wp_{MAX,R})^{(s_i-1)}$	$\mathcal{O}[s_i^4]$	$>\mathcal{O}[s_i^{-1}]$	$>\mathcal{O}[s_i^{-1}]$	$>\mathcal{O}[s_i^{-1}]$
	$\dfrac{\wp_{MAX,i}}{\wp_{MAX,R}}$	$\dfrac{S_{P,MAX,i}}{S_{P,MAX,R}}$	$(\dot{Q}/\dot{N}_{MAX})_i/(\dot{Q}/\dot{N}_{MAX})_R$		$\dfrac{(V_P/\dot{N}_{MAX})_i}{(V_P/\dot{N}_{MAX})_R}$
Sputter ion (inactive and active gases)	1	s_i^2	$V_{D,i}/V_{D,R} \approx 1$		s_i
Orbitron ion active gases	1	s_i^2		1	s_i
Obitron ion inactive gases (ions)	1	s_i		1	s_i^2

Notes: For the case of $f_{T,i} = f_{T,R}$ the expressions that result are sensitive to the particular importance of backflow and inlet losses in each case. Estimates have been made, using typical values for the losses, of the order of magnitude of these expressions in order to simplify the presentation. The detailed expressions appear below in Table 29.2b. The scaling of $V_{P,i}/V_{P,R}$ is assumed to be as s_i^3 when obtaining the $(V_P/\dot{N}_{MAX})_i/(V_P/\dot{N}_{MAX})_R$ scaling.

The maximum pressure ratio ($S_P = 0$) can be written as [Bernhardt, 1983]:

$$\wp_{MAX} = e^{\xi v_r} \tag{29.15}$$

where ξ is a constant that depends on blade geometry. Dividing Eq. (29.14) by $A_I(\bar{C}'/4)$ gives the pumping probability:

$$\alpha_{P,MAX} = v_r/[(1/qd_f) + v_r] \tag{29.16}$$

The $\wp_{MAX}$ for turbomolecular stages is generally large ($>\mathcal{O}[10^5]$) for gases other than He and H$_2$ due to the exponential expression of Eq. (29.15). During operation in a multistage pump, the stages can be employed at pressure ratios of $\wp \ll \wp_{MAX}$. Thus, from Eq. (29.7) (which also applies to turbomolecular drag stages), the pumping speed approaches S_{PMAX}.

The scaling characteristics of turbomolecular pumps can be derived using Eqs. (29.14) and (29.15). It is assumed that the pump blades remain similar during the scaling. The tangential speed (v_c) will be written as $2\pi R f_p$, where R is a characteristic radius of the blade row and f_p is the rotational frequency in rps. From Eqs. (29.14) and (29.15):

$$(S_{P,MAX})_i = s_i^2 A_{I,R}(\bar{C}'/4)\{[2\pi s_i R_R f_{P,i}/\bar{C}']/[(1/q_i d_{f,i}) + (2\pi s_i R_R f_{P,i}/\bar{C}')]\}$$
$$\wp_{MAX,i} = \exp(\xi 2\pi s_i R_R f_{Pi}/\bar{C}') \tag{29.17}$$

TABLE 29.2b Detailed Expressions for Size Scaling with $f_{T,i} = f_{T,R}$

Type	$\dfrac{\wp_{MAX,i}}{\wp_{MAX,R}}$	$\dfrac{S_{P,MAX,i}}{S_{P,MAX,R}}$	$\dfrac{(\dot{Q}_{sf}/\dot{N}_{MAX})_i}{(\dot{Q}_{sf}/\dot{N}_{MAX})_R}$	$\dfrac{(\dot{Q}_\mu/\dot{N}_{MAX})_i}{(\dot{Q}_\mu/\dot{N}_{MAX})_R}$
Positive displacement	$\left\{\dfrac{1-\exp(-s_i^{-1}K_{I,R})+s_i^{-1}K_{B,R}}{1-\exp(-K_{I,R})+K_{B,R}}\right\} \times \left\{\dfrac{1-\exp(-K_{I,R})+\wp_G K_{B,R}}{1-\exp(-s_i^{-1}K_{I,R})+s_i^{-1}\wp_G K_{B,R}}\right\}$	$s_i^3\left[\dfrac{1-\exp(-s_i^{-1}K_{I,R})}{1-\exp(-K_{I,R})}\right]$	$\left[\dfrac{1-\exp(-K_{I,R})}{1-\exp(-s_i^{-1}K_{I,R})}\right]$	$\left[\dfrac{1-\exp(-K_{I,R})}{1-\exp(-s_i^{-1}K_{I,R})}\right]$
Turbomolecular	$(\wp_{MAX,R})^{(s_i-1)}$	$s_i^3\left[\dfrac{(q_R d_{f,R})^{-1}+(v_{c,R}/\overline{C}')}{(q_R d_{f,R})^{-1}+s_i(v_{c,R}/\overline{C}')}\right]$	$\left[\dfrac{(q_R d_{f,R})^{-1}+(s_i v_{c,R}/\overline{C}')}{(q_R d_{f,R})^{-1}+(v_{c,R}/\overline{C}')}\right]$	$\left[\dfrac{(q_R d_{f,R})^{-1}+(s_i v_{c,R}/\overline{C}')}{(q_R d_{f,R})^{-1}+(v_{c,R}/\overline{C}')}\right]$

Notes: The $K_{I,R}$ and $K_{B,R}$ are associated with the inlet losses and backflow losses, respectively (larger $K_{I,R}$ leads to less inlet, losses, but the larger $K_{B,R}$ the more serious the backflow loss). $K_{I,R} = C_{I,LR}\beta_I/V_{T,LR}f_{T,LR}$, $K_{B,R} = C_{LB,R}\beta_2/V_{T,LR}f_{T,LR}$. The symbols are defined in the section on positive displacement pumps.

For example, the case of $f_{P,i} = f_{P,R}$, $d_{f,i} = d_{f,R}$, gives

$$\frac{(S_{P,MAX})_i}{(S_{P,MAX})_R} = s_i^3[(1/q_R d_{f,R}) + (v_{c,R}/\overline{C}')] / [(1/q_i d_{f,R}) + (s_i v_{c,R}/\overline{C}')] \tag{29.18}$$

$$\wp_{MAX,i} = (\wp_{MAX,R})^{s_i-1} \tag{29.19}$$

For the case of $f_{P,i} = s_i^{-1} f_{P,R}$ (constant v_c)

$$\frac{(S_{P,MAX})_i}{(S_{P,MAX})_R} = s_i^2, \quad \wp_{MAX,i} = \wp_{MAX,R} \tag{29.20}$$

A complete set of scaling results is presented in Table 29.2.

29.3.3 Capture Pumps

29.3.3.1 Sputter Ion Pumps

The sputter ion pump (SIP) is an option for high-vacuum MEMS pumping. The application of simple scaling approaches to these pumps is difficult; however, centimeter-scale pumps are already available. The SIP has the basic configuration illustrated in Figure 29.3. A cold cathode discharge (Penning discharge), self-maintained by a several-thousand-volt potential difference and an externally imposed magnetic field that restricts the loss of discharge electrons, causes ions created in the discharge by electron-neutral collisions to bombard the cathodes. The energetic ions (with energy some fraction of the driving potential) both sputter cathode material (usually Ti) and imbed themselves in the cathodes. Sputtered material deposits on the anode and portions of the opposing cathodes. The freshly deposited material acts as a continuously refreshed adsorption pumping surface for "active" gases (most things other than the noble

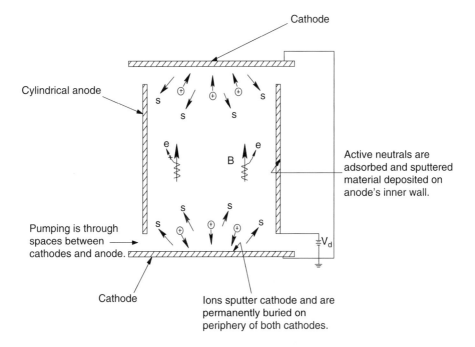

FIGURE 29.3 Sputter ion pump schematic.

gases and hydrocarbons). Small (down to an 8-mm-diameter anode cylinder and a cathode separation of 3.6 cm) pumps have been studied theoretically by Suetsugu (1993). His results compare reasonably well to experimental results for the particular case of a 1.5-cm-diameter anode. For a discharge voltage of 3000 V, a magnetic field strength of 0.3 T and a 0.8-cm-diameter anode, a pumping speed of slightly greater than 0.5 l/s is predicted at 10^{-8} Torr. For these conditions, the discharge is operating in the high-magnetic-field (HMF) mode, which results in a maximum pumping speed. The pumping speed increases slowly as pressure increases.

SIPs have a finite but relatively long life and may be useful when ultra-high vacuums are required in small-scale systems. Their scaling to true MEMS sizes is uncertain because they require several thousand volts to operate reasonably effectively (ion impact energies approaching 1000 V are required for efficient sputtering and rare gas ion burial). The description of the SIP operation developed by Suetsugu (1993) can be used to provide a scaling expression (that needs to be employed cautiously). The power used by the discharge can be obtained knowing the applied potential difference, V_D, and the ion current, I_{ion}. Suetsugu (1993) gives for the pumping speed:

$$S_P = (K_G q \eta I_{ion})/(3.3 \times 10^{19}) e p_I \quad (29.21)$$

where p_I is the pressure in Torr, η is the sputtering coefficient of cathode material due to the impact of energetic ions, and q is the sticking coefficient for active gases on the sputtered material. The charge on an electron is e, and K_G is a nondimensional geometric parameter derived from the electrode configuration that remains constant with geometric scaling. An expression for the power required per pumped molecule becomes:

$$(\dot{Q}/\dot{N})_i = (V_D I_{ion})_i / [\{(K_G q \eta I_{ion})/((3.3 \times 10^{19}) e p_I)\}_i \{10^{-3} n_I\}] \quad (29.22)$$

and

$$(\dot{Q}/\dot{N})_i/(\dot{Q}/\dot{N})_R = V_{D,i}/V_{D,R} \quad (29.23)$$

This assumes $q_i = q_R$, $\eta_i = \eta_R$.

The scaling expression for pumping speed becomes:

$$S_{P,i}/S_{P,R} \equiv S_{P,MAX,i}/S_{P,MAX,R} = (I_{ion})_i/(I_{ion})_R \quad (29.24)$$

where the $S_{P,MAX}$ is employed to be consistent with the previous usage for other pumps, although the Eq. (29.7) relationship does not really apply in this case.

The ion current is obtained by an iterative numerical solution involving the number density of trapped electrons [Suetsugu, 1993]. The scaling expression for $\dot{Q}/\dot{N}$ in Eq. (29.23) is particularly simple because the ion currents cancel. The scaling represented by Eq. (29.23) appears reasonably valid providing η remains relatively constant, which implies that the ion energy should be relatively constant. Ion burial in the cathodes, which is the mechanism by which SIPs pump rare gases, is not discussed in detail here but can be considered within the framework of Suetsugu's analysis. Scaling results are summarized in Table 29.2.

29.3.3.2 Orbitron Ion Pump

The orbitron ion pump [Douglas, 1965; Denison, 1967; Bills, 1967] was developed based on an electrostatic electron trap best known for application in ion pressure gauges. A sketch is presented in Figure 29.4. Injected electrons orbit an anode; the triode version illustrated in Figure 29.4 [Denison, 1967; Bills, 1967] has an independent sublimator that provides a continuous active getter (Ti) coating of the ion collector. The getter permits active gas pumping as well as permanent burial of rare gas and other ions that are

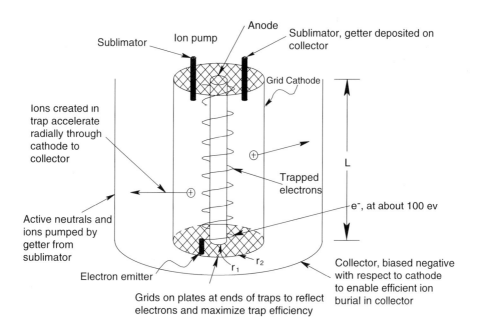

FIGURE 29.4 Orbitron pump schematic.

accelerated out of the trap through the cathode mesh by the radial electric fields. Initial work on reducing the size of an orbitron to MEMS scales has recently been reported by Wilcox (1999).

For a given geometry and potential difference between the anode and cathode mesh, the cylindrical capacitor represented by the trap geometry has a limiting maximum net negative charge of orbiting electrons. The corresponding ionization rate in the trap can be written as:

$$\dot{N}_{MAXions} = \frac{2X\pi\varepsilon_0 V_D^{3/2} \Omega_I L n_g}{(em_e)^{1/2} \ln(r_2/r_1)^{3/2}} \qquad (29.25)$$

where X is the fraction of the maximum charge that permits stable electron orbits (less than 0.5), V_D is the applied potential difference, m_e is the electron mass, ε_0 is the permittivity of free space, Ω_I is the electron-neutral ionization cross section, L is the trap's length, and r_2 and r_1 are the radii of the trap's cathode and anode, respectively (Figure 29.4).

The orbitron's noble gas pumping speed and the trap's volume can be scaled based on the expression for ionization rate in Eq. (29.25):

$$\frac{(S_{P,MAXions})_i}{(S_{P,MAXions})_R} = s_i \frac{\Omega_i}{\Omega_R} \frac{(V_D^{3/2})_i}{(V_D^{3/2})_R}; \qquad \frac{(V_P/\dot{N}_{MAXions})_i}{(V_P/\dot{N}_{MAXions})_R} = s_i^2 \frac{\Omega_{I,i}(V_D)_i^{3/2}}{\Omega_{I,R}(V_D)_R^{3/2}} \qquad (29.26)$$

Note that this is favorable scaling.

The sublimator's scaling, assuming the temperature of the sublimating getter is constant, can be written for neutrals and ions as:

$$\frac{(\dot{Q}_s/\dot{N}_{MAXneut})_i}{(\dot{Q}_s/\dot{N}_{MAXneut})_R} = 1; \qquad \frac{(\dot{Q}_{ion}/\dot{N}_{MAXions})_i}{(\dot{Q}_{ion}/\dot{N}_{MAXions})_R} = 1 \qquad (29.27)$$

Complete scaling results are presented in Table 29.2.

29.3.4 Pump-Down and Ultimate Pressures for MEMS Vacuum Systems

What is the consequence of size-scaling a vacuum system? Consider an elementary system made up of a pump and a vacuum chamber of volume V_c and surface area A_{sc}. The pump is modeled by writing the pumping speed as:

$$S_{P,i} = \{A_{I,P}(\overline{C}'/4)\alpha_P\}_i \qquad (29.28)$$

where $A_{I,P}$ is the area of the pump's inlet aperture from the chamber and α_P is the probability that once through the aperture a molecule will be pumped. For a geometrically similar size change $S_{P,i} = s_i^2(\alpha_{P,i}/\alpha_{P,R})S_{P,R}$, and the pumping speed per unit surface area of the vacuum chamber is

$$(S_P/A_{sc})_i/(S_P/A_{sc})_R = \alpha_{P,i}/\alpha_{P,R} \qquad (29.29)$$

Assuming equal outgassing rates per unit area for the reference and scaled systems, the ultimate system pressure will depend only on the pumping probability ratio, $\alpha_{P,i}/\alpha_{P,R}$. The pump-down time for the system can be measured using the ratio $S_P/V_c = \tau_P$:

$$\tau_{P,i}/\tau_{P,R} = s_i^{-1}(\alpha_{P,i}/\alpha_{P,R}) \qquad (29.30)$$

For geometric scaling and the same outgassing rates, a MEMS vacuum system will have a significantly shorter pump-down time assuming $\alpha_{P,i}/\alpha_{P,R}$ can be kept near one.

In practice MEMS-scale vacuum systems are likely to have pump apertures relatively much larger than their macroscopic counterparts. The economic deterrent to having large aperture pumps that exist at macroscales does not apply at the MEMS scale. At MEMS scales, it appears that technical issues associated with pump construction will favor making the pumps as large as possible. Consequently, relatively large pump apertures with areas about the same as the cross section of the pumped volume are anticipated.

29.3.5 Operating Pressures and $\dot{N}$ Requirements in MEMS Instruments

The selection of vacuum pumps for MEMS instruments and processes will depend on operating pressure and $\dot{N}$ requirements. Because these can be determined reliably only when the task, instrument and detector or a particular process has been specified, it is virtually impossible to discuss significant, general size-scaling tendencies. For example, there has been speculation [Young, 1999] that a MEMS mass spectrometer sampling instrument might operate at upper pressures specified by keeping the Knudsen number based on, say, the quadrupole length constant compared to similar macroscale instruments. This can typically lead to tolerable upper operating pressures for microscale instruments of 10^{-3} to 10^{-2} mbar, depending on the scaling factor (see also Ferran and Boumsellek, 1996). On the other hand, the default response of many mass spectroscopists is 10^{-5} mTorr, independent of scaling. Vargo (2000) based $\dot{N}$ requirements for a miniaturized sampling mass spectrometer on the goal of replacing the entire volume of gas in the instrument (30 cm^3 in Vargo's case) every second at an operating pressure of 10^{-4} mTorr, giving $\dot{N} = 1.4 \times 10^{14}$ molecules/s. A point to remember is that for a constant Kn system the equilibrium quantity of adsorbed gas in the system increases compared to unadsorbed gas as the s_i decreases [Muntz, 1999].

A careful consideration of the operating pressure and $\dot{N}$ requirements for a particular situation is obviously important, but impossible within the confines of the present discussion. It is clear that because of the difficulty in supplying volume- and power-compatible microscale vacuum systems, it will be important for overall system design to define operating conditions that are based on real needs.

29.3.6 Summary of Scaling Results

The scaling analyses outlined above have been applied to several pump types, with the results appearing in Table 29.2. The operating frequencies were selected to give two extremes: maintaining a constant average speed $\bar{u}_i = \bar{u}_R$ (tangential for rotating devices or linear for reciprocating) by using the frequency scaling, $f_{T,i} = s_i^{-1} f_{T,R}$; or maintaining a constant frequency, $f_{T,i} = f_{T,R}$, resulting in $\bar{u}_i = s_i \bar{u}_R$. Two alternative types of frictional drag, sliding and viscous, have been included, again as extremes of the likely possibilities.

The scaling expressions for the case $\bar{u}_i = \bar{u}_R$ are simple when normalized by their respective reference scale values. For the second case, where $\bar{u}_i = s_i \bar{u}_R$, the expressions are more complex and the results depend on the relative magnitude of the quantities $K_{l,R}$, $K_{B,R}$, $q_R d_{f,R}$, etc. (see Table 29.2b). For the cases involving more complex expressions, order of magnitude estimates of the scaling based on typical pump characteristics have been included in Table 29.2a.

The sputter ion pumps have been included assuming that permanent magnets provide the required field strengths. Note that the mesoscale SIP performance presented in Figure 29.1 and Table 29.1 is for HMF operation.

In order to put the scaling results appearing in Table 29.2 in perspective it is important to remember that they are for geometrically accurate scale reductions. It is assumed that the relative dimensional accuracy of the components is the same in the reduced scale realization as in the reference macroscopic pumps. This is a very idealized assumption. The dimensional accuracy that can currently be attained in micromechanical parts as a function of size is illustrated in Figure 29.5, which is derived in part from Madou (1997). It is clear from Figure 29.5 that the scaling results of Table 29.2 may be very optimistic if true MEMS-scale pumps (component sizes 100 µm or less) are required. On the other hand, the scalings do represent the best scaled performances that could be expected and as such are a useful guide. Note from Figure 29.5 that the smallest fractional tolerances can be achieved by precision machining techniques for approximately 1-cm size components. As a result, mesoscale pumps may be possible from a tolerance (although perhaps not economic) perspective using precision machining techniques.

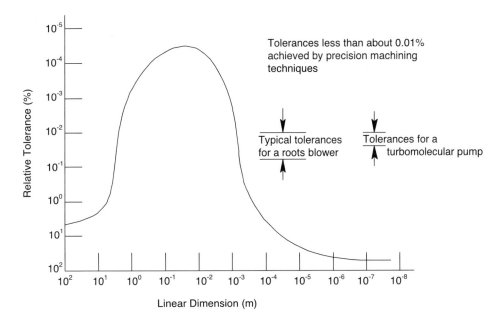

FIGURE 29.5 Dimensional accuracy of manufactured components as a function of size. (Adapted from Madou, 1997.)

Several points should be noted in the Table 29.2 results. For the case of $\bar{u}_i = \bar{u}_R$, the ideal scalings to small sizes are reasonable (remember s_i will range between 10^{-1} and 10^{-4}). In the case of viscous friction losses, the energy use per molecule of upflow becomes large at small scales (increases as s_i^{-1}) while the pump volume per unit upflow decreases as s_i decreases. For the case of positive-displacement pumps and $\bar{u}_i = s_i \bar{u}_R$, the energy use per molecule of upflow scales satisfactorily, but upflow scales as s_i^3 so that the volume scaling is of $\mathcal{O}[1]$ rather than the s_i^{-1} for the $\bar{u}_i = \bar{u}_R$ case. The $\wp_{MAX}$ scaling to small scales for $\bar{u}_i = s_i \bar{u}_R$ is a disaster for turbomolecular pumps. Also, the upflow, energy and pump volume all scale badly for the turbomolecular pump with $\bar{u}_i = s_i \bar{u}_R$. These scalings are all a result of the trapping coefficient $q \sim s_i$ for low peripheral speeds. Although not explicitly included, molecular drag pumps can be expected to scale similarly to the turbomolecular pump. For positive-displacement pumps, the pressure ratio scales well if there are no losses but can scale as badly as s_i depending on specific pump characteristics. For the cases where the pressure ratio scales badly, more pump stages would be required for a given task, leading to larger pumps as indicated by the > symbol in the energy use and pump volume per unit upflow columns.

The sputter ion pump scales well to smaller sizes. The major concern here for microscales will be the fundamental requirement for relatively high voltages. Also, thermal control will be difficult as it is complicated by the need for high field strengths (0.5–1 T) using co-located permanent magnets.

The orbitron ion pump scales well to smaller sizes but unfortunately, as seen in Figure 29.1 and Table 29.1, begins with a very poor performance as measured by $\dot{Q}/\dot{N}$.

Generally speaking, for the positive-displacement and turbomolecular pumps, the idealized geometric scaling results presented in Table 29.2 demonstrate that there is a mixed bag of possibilities, ranging from decreased performance to maintaining performance, with a few cases showing improvement on macro-scale performance by going to smaller scales. From a vacuum pump perspective, with ideal scaling there is little to no advantage based on performance to go to small scales, except for the ion pumps.

The actual performance of small-scale pumps is likely to be significantly poorer than the idealized scaling results shown in Table 29.2. For instance, it has proven to be very difficult to attain the high rotational speeds necessary to satisfy the $\bar{u}_i = \bar{u}_R$ requirements in MEMS-scale devices; on the other hand, recent progress in air-bearing technology [Fréchette et al., 2000] has been reported for mesoscale gas turbine wheels. Mesoscale sputter ion pumps have been operated, and the investigation of orbitron scaling is just beginning. Whether either can be scaled to true MEMS sizes is unclear, but they may be the only alternative for achieving high vacuum with MEMS pumps.

Keeping the preceding comments in mind, it is useful to re-visit the macroscale vacuum pump performances reviewed in Figure 29.1. Consider a typical energy requirement from Figure 29.1 of 3×10^{-15} W/molecule of upflow for macroscale systems, and assume that this can be maintained at mesoscales to pump through a pressure ratio of 10^6 (10^{-3} mbar to 1 bar). A typical upflow, assuming a 3-cm^3 volume at 10^{-3} mbar is changed every second, is 1.1×10^{14} molecules/s and the required energy is 0.33 W. This is somewhat high but perhaps tolerable for a mesoscale system. However, with the expected degradation of the performance of complex macroscale pumps at meso- and microscales, it is clear that it is important to search for alternative, unconventional pumping technologies that will be buildable and operate efficiently at small scales.

29.4 Alternative Pump Technologies

The previous section on scaling indicates that searching for appropriate alternative technologies as a basis for MEMS vacuum pumps is necessary. There has been some effort in this regard during the past decade. In the fall of 1993, Muntz, Pham-Van-Diep and Shiflett hypothesized that the rarefied gas dynamic phenomenon of thermal transpiration might be particularly well suited for MEMS-scale vacuum pumps. Thermal transpiration is the application of a more general phenomenon, thermal creep, that can be used to provide a pumping action in flow channels for Knudsen numbers ranging from very large to about 0.05. The observation resulted in a publication [Pham-Van-Diep et al., 1995] and has very recently led to the construction of a prototype micromechanical pump stage by Vargo (2000). A MEMS thermal transpiration pump has also been proposed by Young (1999). There is one fundamental problem with thermal transpiration or thermal creep pumps—for reasons explained later, they are staged devices that require part of each stage to have a minimum size corresponding to a dimension greater than about 0.2 molecular mean free paths (λ) in the pumped gas. At 1 mTorr (1.32×10^{-3} mbar) $\lambda \approx 0.05$ m in air, resulting in required passages no smaller than about 1 cm. Thus, at low inlet pressures the pumps can be unacceptably large for MEMS applications.

An interestingly different version of a thermal creep pump has been suggested by Sone and his co-workers [Sone et al., 1996; Aoki et al., 2000], although it also has a low-pressure use limit similar to the one mentioned above.

Another alternative, the accommodation pump, which is superficially similar to thermal transpiration pumps but based on a different physical phenomenon, has been investigated [Hobson and Pye, 1972]. It can in principle be used to provide pumping at arbitrarily low pressures without minimum size restrictions.

29.4.1 Outline of Thermal Transpiration Pumping

Two containers, one with a gas at temperature T_L and one at T_H are separated by a thin diaphragm of area A_i having single or multiple apertures, each of which has an area A_a and a size $\sqrt{A_a} \ll \lambda_L$ or λ_H (refer to Figure 29.6). The number of molecules hitting a surface per unit time per unit area in a gas is $n\bar{C}'/4$. In Figure 29.6 this means that there are $(n_L \bar{C}'_L/4) A_a$ molecules passing from cold to hot through an aperture. Similarly, there are $(n_H \bar{C}'_H/4) A_a$ molecules per unit time passing from hot into the cold chamber.

Assume m is the same for the molecules in both chambers, and assume that the inlet and outlet are adjusted so that $p_L = p_H$. Under these circumstances, the net number flow of molecules from cold to hot is, with the help of the equation of state:

$$\dot{N}_{MAX} = A_a (2\pi m k)^{-1/2} p_L [T_H^{1/2} - T_L^{1/2}]/(T_L T_H)^{1/2} \tag{29.31}$$

If, on the other hand, there is no net flow:

$$(p_H/p_L)_{\dot{N}=0} = \wp_{MAX} = (T_H/T_L)^{1/2} \tag{29.32}$$

For p_H between p_L and $p_L(T_H/T_L)^{1/2}$ there will be both a pressure increase and a net flow, which are the necessary conditions for a pump! This effect is known as *thermal transpiration*. If there are Γ apertures

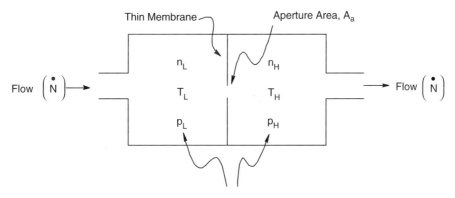

FIGURE 29.6 Elementary single stage of a thermal transpiration compressor.

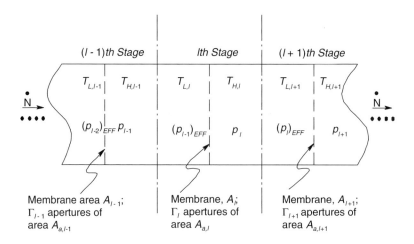

FIGURE 29.7 Cascade of thermal transpiration stages to form a Knudsen Compressor (the pressure difference, $(p_\ell)_{EFF} - p_\ell$, drives the flow through the connector, as illustrated in Figure 29.8).

the total upflow of molecules is

$$\dot{N} = \Gamma A_a (2\pi m k)^{-1/2} [p_L/T_L^{1/2} - p_H/T_H^{1/2}] \qquad (29.33)$$

If the hot gas at T_H is allowed to cool and is sent to another stage as indicated in Figure 29.7, for the condition that $\lambda \ll \sqrt{A_j}$ where A_j is the stage area (but also $\lambda \gg \sqrt{A_{aj}}$), the pressure $p_{L,j+1} = P_{H,j}$. Thus, a cascade of stages with net upflow is a pump with no net temperature increase over the pump cascade. This cascade of thermal transpiration pump stages has no moving parts and no close tolerances between shrouds and impellers, so it is an ideal candidate for a MEMS pump. However, the compressor stages sketched in Figure 29.7 have a serious problem. The thin films containing the apertures are not practical in most applications because of heat-transfer considerations. A more practical stage is shown in Figure 29.8, where the thin membrane has been replaced by a bundle of capillary tubes. The temperature and pressure profiles in the stage are also illustrated in Figure 29.8 along with nomenclature that will be used below.

The stage configuration in Figure 29.8 can be put in a cascade (Figure 29.7) to form a pump as proposed by Pham-Van-Diep et al. (1995) and called a Knudsen Compressor (after original work by Knudsen, 1910a; 1910b). Several investigations implementing thermal transpiration in macroscale pumps have been made over the years [Turner, 1966], but with no result beyond initial analysis and laboratory experimental studies. For the Knudsen Compressor, a performance analysis was presented for the case of collisionless flow in the capillaries and continuum flow in the connectors [Pham-Van-Diep et al., 1995]. Energy requirements per molecule of upflow ($\dot{Q}/\dot{N}$) were estimated. Recently, Muntz and several collaborators [Muntz et al., 1998; Vargo, 2000] have extended the analysis to situations where the flow in both the capillaries and connector section can be in the transitional flow regime ($10 > Kn > 0.05$). An initial look at the minimization of cascade energy consumption and volume has been conducted [Muntz et al., 1998; Vargo, 2000] by searching for minimums in $\dot{Q}/\dot{N}$ and $V_p/\dot{N}$. Based on these studies, a preprototype micromechanical Knudsen Compressor stage has been constructed and tested [Vargo and Muntz, 1996; 1998; Vargo et al., 2000].

For transitional flow in a capillary tube with a longitudinal temperature gradient imposed on the tube's wall, flow is driven from the cold to hot ends, as illustrated in Figure 29.9. The increased pressure at the hot end drives the return flow. For small temperature differences and thus small pressure increases, the Boltzmann equation can be linearized and results for the flow through a cylindrical capillary driven by small wall temperature gradients obtained [Sone and Itakura, 1990; Loyalka and Hamoodi, 1990;

Microscale Vacuum Pumps

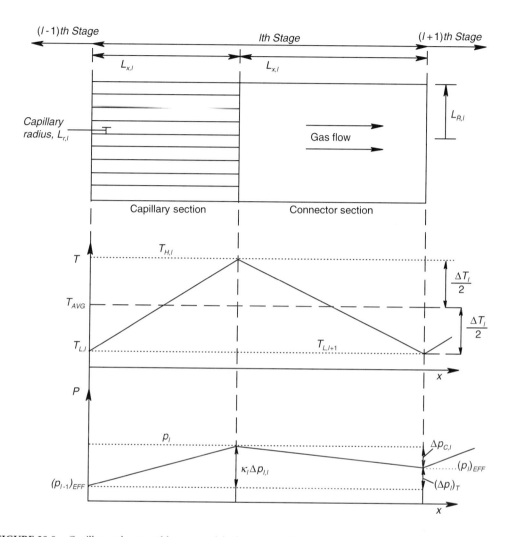

FIGURE 29.8 Capillary tube assembly as a model of a practical Knudsen Compressor.

Loyalka and Hickey, 1991]. The definition of the flow coefficients (Q_T and Q_P) is implied in the following expression [Sone and Itakura, 1990]:

$$\dot{M} = p_{AVG}(2(k/m)T_{AVG})^{-1/2} A \left[\frac{L_r}{T_{AVG}} \frac{dT}{dx} Q_T - \frac{L_r}{p_{AVG}} \frac{dp}{dx} Q_P \right] \quad (29.34)$$

Here, A is the cross-sectional area of the capillary, L_r is its radius, Q_P is the backflow due to the pressure increase, and Q_T is the thermally driven upflow near the walls. The sum determines the net mass flow through the tube from cold to hot. For large Kn ($= \lambda/L_r$), the $\dot{M} = 0$ pressure increase provided by a tube is $p_H/p_C = Q_T/Q_P = (T_H/T_C)^{1/2}$, identical to the aperture case discussed earlier. In the large Kn case, the backflow and upflow completely mingle over the entire tube cross section. For transitional Kns the upflow is confined near the wall in a layer on the order of λ thick, as illustrated in Figure 29.9.

The values of Q_T and Q_P vary markedly throughout the transitional flow regime, as shown in Figure 29.10. The details of their functional variations as well as the ratio Q_T/Q_P are important to any pump using

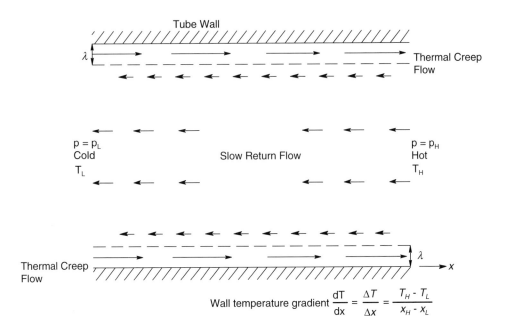

FIGURE 29.9 Transitional flow configuration in a capillary tube.

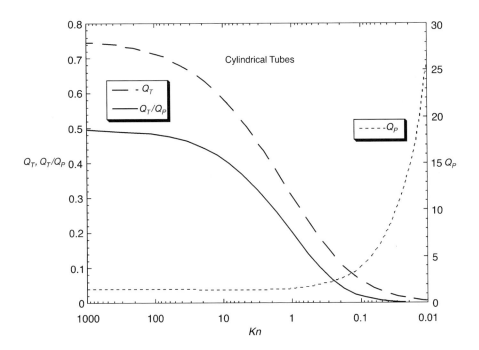

FIGURE 29.10 Transitional flow coefficients as a function of Knudsen number ($Kn = \lambda/L_r$) for cylindrical capillaries. (Adapted from Vargo, 2000.)

thermal transpiration. Their roles are best illustrated by the expression for pumping speed and pressure ratio of the jth stage of a Knudsen Compressor [Muntz et al., 1998]:

$$S_{P,j} = \pi^{1/2}(1-\kappa_j)\left|\frac{\Delta T}{T_{AVG}}\right|\left[\frac{Q_T}{Q_P} - \frac{Q_{T,C}}{Q_{P,C}}\right]_j \left[\frac{L_x/L_r}{FQ_P} + \frac{L_X/L_R}{F_C(A_c/A)Q_P}\right]_j^{-1}(C_A)_j \qquad (29.35)$$

In Eq. (29.35), the κ_j is a parameter that sets the fraction of the $\dot{M} = 0$ pressure rise that is realized in a finite upflow situation. The pressure ratio for the stage, assuming $|\Delta T/T_{AVG}| \gg 1$ [Muntz et al., 1998], is

$$\wp_j = 1 + \kappa_j\left\{\left|\frac{\Delta T}{T_{AVG}}\right|\left[\frac{Q_T}{Q_P} - \frac{Q_{T,C}}{Q_{P,C}}\right]_j\right\} \qquad (29.36)$$

In Eqs. (29.35) and (29.36), $|\Delta T|$ is the temperature change across both the capillary section and the connector section; F is the fraction of the capillary section's area, A_j, that corresponds to the open area of the capillary tubes; and F_C is the fraction of the connector section's area, $A_{C,j}$, that is open (usually $F_{C,j} = 1$). The dimensions $L_{x,j}$, $L_{X,j}$, $L_{r,j}$, $L_{R,j}$ are defined in Figure 29.8.

From Eq. (29.36), if $\kappa_j = 0$ then $\wp_j = 1$. The corresponding maximum upflow is obtained by substituting $\kappa_j = 0$ into Eq. (29.35) to give $(S_{PMAX})_j$. The result is

$$S_{P,j}/(S_{P,MAX})_j = (1 - \kappa_j) \qquad (29.37)$$

which can be combined with the expression for $\wp_{j,MAX}$ (using $\kappa_j = 1$ in Eq. (29.35)) to give the familiar relationship for a vacuum pumping system, Eq. (29.7), between $S_{P,j}$, $S_{P,MAX,j}$, $\wp_j$ and $\wp_{MAX,j}$.

The expression for pumping speed in Eq. (29.35) can also be used to compute the pumping probability $\alpha_{P,j}$, where from Eq. (29.5) $\alpha_{P,j} = S_{P,j}/C_{A,j}$. Using Eqs. (29.35) and (29.36) and remembering that $\dot{N}_j = S_{P,j}n_{I,j}$, where $n_{I,j}$ is the number density at the inlet to the stage, cascade performance calculations can be accomplished. The approach is presented in detail in Muntz et al. (1998). The results of several cascade simulations are given by Vargo (2000). As shown by Muntz, the majority of the energy required is the thermal conduction loss across the capillary tube bundle. The physical realization of the capillary section of a Knudsen Compressor is frequently not a bundle of capillary tubes, but rather a porous membrane. Nevertheless, the theoretical behavior based on "equivalent" capillary tubes appears to be consistent with the results obtained experimentally [c.f. Vargo, 2000, and the discussions presented therein].

Selected results of micro/meso and macroscale pump cascade simulations reported by Vargo (2000) are plotted in Figure 29.11 and presented in Table 29.3. It is important to note that all the Knudsen Compressor cascade results are from simulations; a micromechanical cascade has yet to be constructed, although several mesoscale stages have been operated in series [Vargo and Muntz, 1996]. In his thesis, Vargo (2000) describes the construction of a laboratory prototype micromechanical Knudsen Compressor stage and the experimental results obtained from it. A scheme for constructing a cascade is also presented, along with predicted energy requirements, but this has yet to be built and tested. The Knudsen Compressor performance presented in Figure 29.11 is based on a significant body of preliminary work but has not yet been demonstrated experimentally. The results are for varying initial pressures p_I and pumping to 1 atm (the detailed conditions are presented in Table 29.3. In MEMS devices, pressures below 10 mTorr extract a significant energy penalty, tending to preclude Knudsen Compressor applications to lower pressures at MEMS scale. This restriction, however, depends on the $\dot{N}$ that is required. The energy penalty originates from the difficulty in effectively reaching continuum flow in the connector section for a reasonable size. Note that the macroscale simulations presented in Figure 29.11 give efficient pumping to significantly lower p_I (higher $\wp$).

TABLE 29.3 Results of Simulations of Micro-, Meso- and Macroscale Knudsen Compressors for Several Pumping Tasks

p_I (mTorr)	p_E (mTorr)	$\dot{Q}/\dot{N}$ W/(#/s)	$V_P/\dot{N}$ cm^3/(#/s)	$\wp$
		Micro-/Mesoscale[a]		
0.1	7.6E5	1.4E-12	7.8E-12	7.6E6
1.0	7.6E5	9E-15	5.2E-14	7.6E5
10	7.6E5	9E-17	5.2E-16	7.6E4
10	5.25E3	1.4E-17	1.6E-16	5.25E2
50	7.6E5	2.9E-17	1.6E-16	1.5E4
		Macroscale[b]		
0.1	7.6E5	9.1E-16	1.3E-12	7.6E6
1.0	7.6E5	2.8E-17	4.1E-14	7.6E5
10	7.6E5	1.5E-17	2.1E-14	7.6E4

[a] From Vargo (2000, Tables 3, 4 and 9).
[b] From Vargo (2000, Tables 3).

Notes: Micro-/mescoscale cascade: $L_r = 500$ μm, $L_R/L_r = 5$ from p_I to 50 mTorr, then constant Kn_j. Macroscale cascade: $L_r = 5$ mm, $L_R/L_r = 20$ from p_I to 10 mTorr, then constant Kn_j.

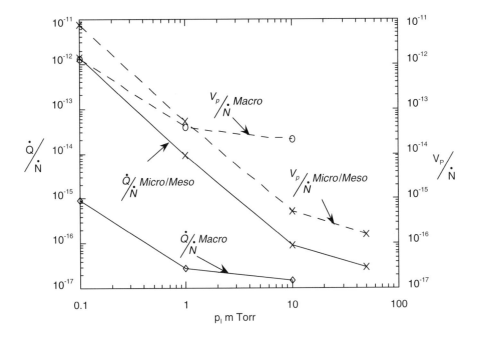

FIGURE 29.11 Simulation results for several macro- and microscale pumping tasks from p_I to $p_E = 1$ atm using Knudsen Compressors (as presented in Table 29.3).

The size scaling of Knudsen Compressors at low inlet pressures is dominated by the ratio L_R/L_r, which can be achieved in the early stages of a particular design. The ratio L_R/L_r determines the Knudsen number ratio, which in turn determines the Q_T/Q_P ratio (Figure 29.10). By referring to Eq. (29.36), the effects on the stage $\wp_j$ are immediately clear. In the limit of $L_R/L_r \to 1$, $\wp = 1$. For $L_R/L_r \gg 1$, $Q_{T,C}/Q_{P,C} \ll 1$ and $\wp_j$ reaches a maximum value that depends only on κ_j and $|\Delta T/T_{AVG}|$.

Minimizing the energy requirement of a Knudsen Compressor cascade depends on a number of important parameters describing the individual stage configurations. The results reported by Vargo (2000)

Microscale Vacuum Pumps

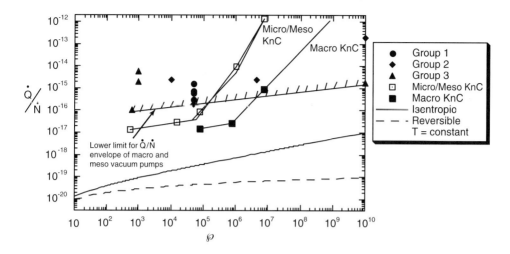

FIGURE 29.12 Comparison of macro- and microscale Knudsen Compressor performances from simulations [Vargo, 2000] to representative available macro- and mescoscale vacuum pumps.

are only a beginning attempt at optimizing Knudsen Compressor performance (there are extended discussions of optimization issues in Muntz et al., 1998). The predicted Knudsen Compressor performance presented in Figure 29.11 is quite promising. Generally, as illustrated in Figure 29.12, the energy requirement for the micro-/mesoscale version is at least an order of magnitude better than macroscale, positive-displacement mechanical backing pumps. It also appears very competitive with the macroscale molecular drag kinetic backing pumps used in conjunction with turbomolecular pumps.

Another MEMS thermal transpiration compressor has been proposed by Young (1999). Its configuration is different from the Knudsen Compressor, relying on temperature gradients established in the gas rather than along a wall. No experimental or theoretical analysis of the flow or pressure-rise characteristics of this configuration has been reported. There are theoretical reasons to believe that the flow effects induced by thermal gradients in a gas are significantly smaller than the effects supported by wall thermal gradients [Cercignani, 2000].

29.4.2 Accommodation Pumping

In 1970, Hobson introduced the idea of high-vacuum pumping by employing a characteristic of surface scattering that had been observed in molecular beam experiments [*cf.* the review by Smith and Saltzburg, 1964]. For some surfaces that give quasi-specular reflection under controlled conditions, when the beam temperature differs from the surface temperature, the quasi-specular scattering lobe moves toward the surface normal for cold beams striking hot surfaces. For hot beams striking cold surfaces, the lobe moves away from the normal. Hobson's initial study was followed in short order by several investigations adopting the same approach [Hobson and Pye, 1972; Hobson and Salzman, 2000; Doetsch and Ryce, 1972; Baker et al., 1973]. Other investigators, relying on the same phenomenon, have used different geometrical arrangements [Tracy, 1974; Hemmerich, 1988]. Recently Hudson and Bartel (1999) analyzed a sampling of previous experiments with the Direct Simulation Monte Carlo technique. To date, the most successful pumping arrangement, which is sketched in Figure 29.13, has been that of Hobson.

Following Hobson's lead it is easy to write expressions for $\wp_{MAX,j} = p_3/p_1$ and $S_{P,MAX,j}$ for the stage in Figure 29.13. The accommodation pump works best under collisionless flow conditions, so assume $\lambda \geq 10d$, where d is the tube diameter. Assuming steady-state conditions, conservation of molecule flow equations for each volume can be solved simultaneously for zero net upflow to give:

$$\wp_{MAX,j} = \frac{\alpha_{23}}{\alpha_{32}} \cdot \frac{\alpha_{12}}{\alpha_{21}} = \frac{\alpha_{12}}{\alpha_{21}} \qquad (29.38)$$

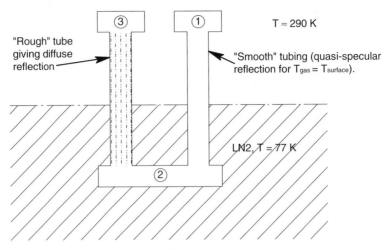

"Smooth" tube is normal Pyrex® tubing.

"Rough" tube is leached Pyrex® to provide suppression of specular reflection for the temperature differences of the experiment.

FIGURE 29.13 Schematic representation of Hobson's accommodation pump configuration. (Adapted from Hobson, 1970).

Here, α_{kl} is the probability for a molecule having entered a tube from the kth chamber reaching the lth chamber. In this case, because the tube joining chambers 2 and 3 is assumed always to reflect diffusely, $\alpha_{23} = \alpha_{32}$.

The same molecule flow equations can be solved along with Eq. (29.38) to provide maximum upflow or pumping speed ($\wp_j = 1$):

$$S_{P,MAX,j} = \frac{\bar{C}'_H A_j}{4}\{(\wp_{MAX,j} - 1)\alpha_{2,1}/[1 + (\alpha_{2,1})/(8L_r/3L_x)]\} \qquad (29.39)$$

From the earlier discussions, it is reasonable to anticipate that $\alpha_{2,1}$ will have a value that approximately corresponds to diffuse reflection, which for a long tube is $(8L_r/3L_x)$. A simple approximate expression for maximum pumping speed becomes:

$$S_{P,MAX,j} = (\bar{C}'_H/4)A_j(4L_r/3L_x)(\wp_{MAX,j} - 1) \qquad (29.40)$$

If $\wp_{MAX,j}$ is known from experiments, $S_{P,MAX,j}$ can be found from Eq. (29.40).

Scaling accommodation pumping is not difficult because, unlike the Knudsen Compressor, the flow is collisionless everywhere. If the minimum radius of the two channels in a stage is such that $L_r \geq 100$ nm, there is little reason to believe that $\wp_{MAX,j}$ or the transmission probabilities will vary with scale. For geometric scaling:

$$\frac{S_{P,MAX,j}}{S_{P,MAX,R}} = \hat{s}_i^2 \qquad (29.41)$$

No analysis or measurements of the energy requirements of accommodation pumping are available, but the temperature differences required to obtain an effect are quite large (the temperature ratio in various experiments ranges from 3 to 4). Careful thermal management would be required at MEMS scales. Additionally, the stage $\wp_{MAX,j}$ values are quite low (Hobson's are typically around 1.15). For an upflow of $S_{P,MAX,j}/2$, to pump through a pressure ratio of 10^2 would require about 125 stages. For comparison, a Knudsen Compressor operating between the same temperatures would require 10 stages. Of course, as outlined earlier, the Knudsen Compressor could not pump effectively at very low pressures, whereas accommodation pumps can.

Using the miniature Creare turbomolecular pump scaled down with constant $f_{P,i}$ to a 2-cm diameter, the authors have estimated an air pumping probability of about 0.05. A Hobson-type accommodation pump with 2-cm-diameter, $L_r/L_x = 0.05$ Pyrex® tubes and operating between room and LN_2 temperatures would have, from Eq. (29.40), a pumping probability of about 0.01.

Scaling Hobson's pump to MEMS scales would permit operation with molecular flow up to pressures much greater than 10^{-2} mbar, providing the surface reflection characteristics were not negatively affected by adsorbed gases. This is a serious uncertainty, as all of the available experiments have been reported for pressures between 1 and several orders of magnitude less than 10^{-2} mbar.

29.5 Conclusions

The performances of a representative sampling of macroscale vacuum pumps have been reviewed based on parameters relevant to micro- and mesoscale vacuum requirements. If the macroscale performances of these generally complex devices could be maintained at meso- and microscales, they might be barely compatible with the energy and volume requirements of meso- and microsampling instruments. A general scaling analysis of macroscale pumps considered as possible candidates for scaling (although going to microscales with these scaled pumps is currently beyond the state of the manufacturing art) is presented and provides little to no incentive for simply trying to scale existing pumps, with the possible exception of ion pumps. When the difficulty in developing MEMS manufacturing techniques with adequate tolerances is considered, the possibility of maintaining the macroscale performance at microscales is remote. As a consequence, two alternative pump technologies are examined. The first, the Knudsen Compressor, is based on thermal creep flow or thermal transpiration and appears attractive as a result of both analysis and initial experiments. The second, also a thermal creep pump, is under study but requires further analysis before its microscale suitability can be established. Both, however, have a characteristic that prevents them from operating efficiently when pumping high vacuums (<10^{-3} to 10^{-2} mbar).

The second possibility, accommodation pumping, although superficially similar to the Knudsen Compressor, is based on surface molecular reflection characteristics and will pump to ultra-high vacuums. Further analysis and experiments are required to be able to provide performance estimates of accommodation pumping in meso- and microscale contexts.

The alternative technology pumps have no moving parts; no liquids are required for seals, pumping action or lubrication. These are characteristics shared with the ion pumps. They all have the capability of providing mesoscale pumping and the potential for truly MEMS-scale pumping. To realize these alternative technology pumps, significant and focused research and development efforts will be required.

Acknowledgments

The authors thank Andrew Jamison for preparing the figures and Tariq El-Atrache and Tricia Harte for patiently working on the manuscript through many revisions. Dr. Andrew Ketsdever, Professor Geoff Shiflett, and Mr. Dean Wiberg have provided helpful comments. Discussions with Dr. David Salzman over several years have been both entertaining and informative.

Defining Terms

Capture pumps: Vacuum pumps where pumping action relies on sequestering pumped gas in a solid matrix until it can be removed in a separate, off-line operation.

Conductance: Measure of the capability of a vacuum system component (tube, channel) for handling gas flows; has units of gas volume per unit time.

Knudsen number: Indication of the degree of rarefaction of flow in a vacuum system component; for $Kn > 10$, molecules collide predominantly with the walls, and, for $Kn < 10^{-3}$, intermolecular collisions dominate.

Microscale, mesoscale and macroscale: In reference to a device, these terms as used in this chapter correspond to the following typical dimensions:

- *Microscale*: Smallest components <100 µm, device <1 cm
- *Mesoscale*: Smallest component 100 µm to 1 mm, device 1 cm to 10 cm
- *Macroscale*: Smallest components >1 mm, device >10 cm

Pumping speed: Measure of the pumping ability of a vacuum pump in terms of volume per unit time of inlet pressure gas that can be pumped.

Regeneration: Off-line removal of sequestered gas from a capture pump.

Vacuum: For the purposes of this chapter, the pressure ranges corresponding to various descriptions attached to vacuums are

- *Low or roughing*: 10^3 to 10^{-2} mbar
- *High*: 10^{-2} to 10^{-7} mbar
- *Ultra-high*: $<10^{-7}$ mbar

References

Aoki, K., Sone, Y., Takata, S., Takahashi, K., and Bird, G. A. (2000) "One-Way Flow of a Rarefied Gas Induced in a Circular Pipe with a Periodic Temperature Distribution," in *Proc. 22nd Rarefied Gas Dynamics Symp.*, G. A. Bird, T. J. Bartel, and M. A. Gallis, Eds., July 9–14, Sydney, Australia.

Baker, B. G., Hobson, J. P., and Pye, A. W. (1973) "Further Measurements of Physical Factors Influencing Accommodation Pumps," *J. Vacuum Sci. Technol.* **10**(1), pp. 241–245.

Baum, H. (1957) *Vakuum-Technik,* **7**, pp. 154–159.

Bernhardt, K. H. (1983) "Calculation of the Pumping Speed of Turbomolecular Vacuum Pumps by Means of Simple Mechanical Data," *J. Vacuum Sci. Technol. A* **1**(2), pp. 136–139.

Bills, D. G. (1967) "Electrostatic Getter-Ion-Pump Design," *J. Vacuum Sci. Technol.* **4**(4), pp. 149–155.

Bird, G. A. (1994) *Molecular Gas Dynamics and the Direct Simulation of Gas Flows,* Clarendon Press, Oxford.

Callas, J. L. (1999) "Vacuum System Requirements for a Miniature Scanning Electron Microscope," in *NASA/JPL Miniature Vacuum Pumps Workshop,* Glendale, CA, JPL website http://cissr.jpl.nasa.gov/pumpsworkshop/.

Cercignani, C. (2000) *Rarefied Gas Dynamics,* Cambridge University Press, Cambridge, U.K.

Chang, T. H. P., Kern, D. P., and Muray L. P. (1990) "Microminiaturization of Electron Optical Systems," *J. Vacuum Sci. Technol. B* **8**, pp. 1698–1705.

Denison, D. R. (1967) "Performance of a New Electrostatic Getter-Ion Pump," *J. Vacuum Sci. Technol.* **4**(4), pp. 156–162.

Doetsch, I. H., and Ryce, S. A. (1972) "Separation of Gas Mixtures by Accommodation Pumping," *Can. J. Chem.* **50**(7), pp. 957–960.

Douglas, R. A. (1965) "Orbitron Vacuum Pump," *Rev. Sci. Instrum.* **36**(1), pp. 1–6.

Dushman, S. (1949) *Scientific Foundations of Vacuum Technique,* Wiley, New York.

Dushman, S., and Lafferty, J. M. (1962) *Scientific Foundations of Vacuum Technique,* 2nd ed., Wiley, New York.

Ferran, R. J., and Boumsellek, S. (1996) "High-Pressure Effects in Miniature Arrays of Quadrupole Analyzers for Residual Gas Analysis from 10^{-9} to 10^{-2} Torr," *J. Vacuum Sci. Technol.* **14**, pp. 1258–1265.

Fréchette, L. G., Stuart, A. J., Breuer, K. S., Ehrich, F. F., Ghodssi, R., Kanna, R., Wang, C.W., Zhong, X., Schmidt, M. A., and Epstein, A. H. (2000) "Demonstration of a Microfabricated High-Speed Turbine Supported on Gas Bearings," in *Solid-State Sensor and Actuator Workshop,* pp. 43–47, Hilton Head Island.

Freidhoff, C. B., Young, R. M., Sriram, S. S., Braggins, T. T., O'Keefe, T. W., Adam, J. D., Nathanson, H. C., Symms, R. R. A., Tate, T. J., Ahmad, M. M., Taylor, S., and Tunstall, J. (1999) "Chemical Sensing Using Nonoptical Microelectromechanical Systems," *J. Vacuum Sci. Technol. A* **17**, pp. 2300–2307.

Hemmerich, J. L. (1988) "Primary Vacuum Pumps for the Fusion Reactor Fuel Cycle," *J. Vacuum Sci. Technol. A* **6**(1), pp. 144–153.

Hobson, J. P. (1970) "Accommodation Pumping—A New Principle for Low Pressure," *J. Vacuum Sci. Technol.* **7**(2), pp. 301–357.

Hobson, J. P., and Pye, A. W. (1972) "Physical Factors Influencing Accommodation Pumps," *J. Vacuum Sci. Technol.* **9**(1), pp. 252–256.

Hobson, J. P., and Salzman, D. B. (2000) "Review of Pumping by Thermomolecular Pressure," *J. Vacuum Sci. Technol., A* **18**(4), pp. 1758–1765.

Hopfinger, E. J., and Altman, M. (1969) "A Study of Thermal Transpiration for the Development of a New Type of Gas Pump," *J. Eng. Power, Trans. ASME* **91**, pp. 207–215.

Hudson, M. L., and Bartel T. J. (1999) "DSMC Simulation of Thermal Transpiration Pumps," in *Rarefied Gas Dynamics*, Vol. 1, R. Brun et al., Eds., pp. 719–726, Cépaduès Éditions, Toulouse, France.

Knudsen, M. (1910a) "Eine Revision der Gleichgewichtsbedingung der Gase. Thermische Molekularströmung," *Annalen der Physik* **31**, pp. 205–229.

Knudsen, M. (1910b) "Thermischer Molekulardruck der Gase in Röhre," *Annalen der Physik* **33**, pp. 1435–1448.

Lafferty, J. M. (1998) *Foundations of Vacuum Science and Technology,* John Wiley & Sons, New York.

Loyalka, S. K., and Hamoodi, S. A. (1990) "Poiseuille Flow of a Rarefied Gas in a Cylindrical Tube: Solution of Linearized Boltzmann Equation," *Phys. Fluids A* **2**(11), pp. 2061–2065.

Loyalka, S. K., and Hickey, K. A. (1991) "Kinetic Theory of Thermal Transpiration and the Mechanocaloric Effect: Planar Flow of a Rigid Sphere Gas with Arbitrary Accommodation at the Surface," *J. Vacuum Sci. Technol. A* **9**(1), pp. 158–163.

Madou, M. (1997) *Fundamentals of Microfabrication,* CRC Press, Boca Raton, FL.

Muntz, E. P. (1999) "Surface Dominated Rarefied Flows and the Potential of Surface Nanomanipulations," in *Rarefied Gas Dynamics*, R. Brun, R. Campargue, and J. C. Lengrand, Eds., pp. 3–15, Cépaduès Édition, Toulouse, France.

Muntz, E. P., Sone, Y., Aoki, K., and Vargo, S. (1998) "Performance Analysis and Optimization Considerations for a Knudsen Compressor in Transitional Flow," U.S.C. Aerospace and Mechanical Engineering Report No. 98001, University of Southern California, Los Angeles.

NASA/JPL (1999) *NASA/JPL Miniature Vacuum Pumps Workshop,* http://cissr.jpl.nasa.gov/.

Nathanson, H. C., Liberman, I., and Freidhoff, C. (1995) In *Proc. IEEE 8th Annu. Int. Workshop on Micro Electro Mechanical Systems (MEMS '95)*, pp. 72–76, IEEE, Amsterdam, The Netherlands.

Orient, O. J., Chutjian, A., and Garkanian, V. (1997) "Miniature, High-Resolution, Quadrupole Mass-Spectrometer Array," *Rev. Sci. Instrum.* **68**, pp. 1393–1397.

Orner, P. A., and Lammers, G. B. (1970) "The Application of Thermal Transpiration to a Gaseous Pump," *J. Basic Eng., Trans. ASME* **92**, pp. 294–302.

Park, J.-Y., et al. (1997) "Fabrication of Electron-Beam Microcolumn Aligned by Scanning Tunneling Microscope," *J. Vacuum Sci. Technol. A* **15**, pp. 1499–1502.

Pham-Van-Diep, G., Keeley, P., Muntz, E. P., and Weaver, D. P. (1995), "A Micromechanical Knudsen Compressor," in *Proc. 19th Int. Symp. on Rarefied Gas Dynamics*, J. Harvey and G. Lord, Eds., pp. 715–721, Oxford University Press, London.

Piltingsrud, H. V. (1994) "Miniature Cryosporption Vacuum Pump for Portable Instruments," *J. Vacuum Sci. Technol. A* **12**, pp. 235–240.

Piltingsrud, H. V. (1996) "Miniature Peristaltic Vacuum Pump for Use in Portable Instruments," *J. Vacuum Sci. Technol. A* **14**, pp. 2610–2617.

Piltingsrud, H. V. (1997) "A Field Deployable Gas Chromatograph/Mass Spectrometer for Industrial Hygiene Applications," *Am. Ind. Hygiene Assoc. J.* **58**, pp. 564–577.

Sharipov, F., and Seleznev, V. (1998) "Data on Internal Rarefied Gas Flows," *J. Phys. Chem. Ref. Data* **27**, pp. 657–706.

Short, R. T., Fries, D. P., Toler, S. K., Lembke, C. E., and Byrne, R. H. (1999) "Development of an Underwater Mass-Spectrometry System for *In Situ* Chemical Analysis," *Meas. Sci. Technol.* **10**, pp. 1195–1201.

Smith, J. N., and Saltzburg, H. (1964), "Recent Studies of Molecular Beam Scattering from Continuously Deposited Gold Films," in *Proc. 4th Int. Symp. on Rarefied Gas Dynamics*, Vol. 2, J. H. de Leeuw, Ed., pp. 491–504, Academic Press, New York.

Sone, Y., and Itakura, E. (1990) "Analysis of Poiseuille and Thermal Transpiration Flows for Arbitrary Knudsen Numbers by a Modified Knudsen Number Expansion Method and Their Database," *J. Vacuum Soc. Jpn.* **33**, pp. 92–94.

Sone Y., and Yamamoto, K. (1968) "Flow of a Rarefied Gas through a Circular Pipe," *Phys. Fluids*, **11**, pp. 1672–1678.

Sone, Y., Waniguchi, Y., and Aoki, K. (1996) "One-Way Flow of a Rarefied Gas Induced in a Channel with a Periodic Temperature Distribution," *Phys. Fluids* **8**, pp. 2227–2235.

Suetsugu, Y. (1993) "Numerical Calculation of an Ion Pump's Pumping Speed," *Vacuum* **46**(2), pp. 105–111.

Terry, S. C., Jerman, J. H., and Angell J. B. (1979) "A Gas Chromatographic Air Analyzer Fabricated on a Silicon Wafer," *IEEE Trans. Electron. Devices* **ED-26**, pp. 1880–1886.

Tracy, D. H. (1974) "Thermomolecular Pumping Effect," *J. Phys. E: Sci. Instrum.* **7**, pp. 533–563.

Turner, D. J. (1966) "A Mathematical Analysis of a Thermal Transpiration Vacuum Pump," *Vacuum* **16**, pp. 413–419.

Vargo, S. E. (2000) "The Development of the MEMS Knudsen Compressor as a Low Power Vacuum Pump for Portable and *In Situ* Instruments," Ph.D. thesis, University of Southern California, Los Angeles.

Vargo, S. E., and Muntz, E.P. (1996), "An Evaluation of a Multi-Stage Micromechanical Knudsen Compressor and Vacuum Pump," in *Proc. 20th Int. Symp. on Rarefied Gas Dynamics*, C. Shen, Ed., pp. 995–1000, Peking University Press, Beijing, China.

Vargo, S. E., and Muntz, E. P. (1998) In *Proc. 21st Int. Symp. on Rarefied Gas Dynamics*, R. Brun, R. Campargue, and J. C. Lengrand, Eds., pp. 711–718, Cépaduès Éditions, Marseille, France.

Vargo, S. E., Muntz,, E. P., Shiflett, G. R., and Tang, W. C. (1999) "The Knudsen Compressor as a Micro and Macroscale Vacuum Pump without Moving Parts or Fluids," *J. Vacuum Sci. Technol. A* **17**, pp. 2308–2313.

White, A. J., Blamire, M. G., Corlette, C. A., Griffiths, B. W., Martin D. M., Spencer S. B., and Mullock, S. J. (1998) "Development of a Portable Time-of-Flight Membrane Inlet Mass Spectrometer for Environmental Analysis," *Rev. Sci. Instrum.* **69**, pp. 565–571.

Wiberg, D., Scheglar, K., White, V., Orient, O., and Chutjian, A. (2000) "Toward a Micro Gas Chromatograph/ Mass Spectrometer (GC/MS) System," presented at the *3rd Annu. Int. Conf. Integrated Nano/Microtechnology Space Applications*, Houston, TX.

Wilcox, J. Z., Feldman, J., George, T., Wilcox, M., and Sherer A. (1999) "Miniaturization of High Vacuum Pumps: Ring-Anode Orbitron," presented at *AVS 46th Int. Symp.*, Oct. 25–29, Seattle, WA.

Young, R. M. (1999) "Analysis of a Micromachine Based Vacuum Pump on a Chip Actuated by the Thermal Transpiration Effect," *J. Vacuum Sci. Technol. B* **17**, pp. 280–287.

Further Information

A concise up-to-date review of the mathematical background of internal rarefied gas flow is presented in the book, *Rarefied Gas Dynamics*, by Carlo Cercignani [Cercignani, 2000]. Vacuum terminology and important considerations for the description of vacuum system performance and pump types are presented in the excellent publication by J. M. Lafferty, *Foundations of Vacuum Science and Technology* [Lafferty, 1998]. A detailed presentation of work on rarefied transitional flows is documented by the proceedings of the biannual International Symposia on Rarefied Gas Dynamics (various publishers over the last 45 years, all listed in Cercignani, 2000). This reference source is generally only useful for those with the time and inclination for academic study of undigested research, although there are several review papers in each publication.

30
Microdroplet Generators

Fan-Gang Tseng
National Tsing Hua University

30.1 Introduction 30-1
30.2 Operation Principles of Microdroplet Generators 30-2
Pneumatic Actuation • Piezoelectric Actuation •
Thermal-Bubble Actuation • Thermal-Buckling
Actuation • Acoustic-Wave Actuation • Electrostatic
Actuation • Inertial Actuation
30.3 Physical and Design Issues 30-7
Frequency Response • Thermal/Hydraulic Cross-Talk and
Overfill • Satellite Droplets • Puddle Formation • Material
Issues
30.4 Fabrication of Microdroplet Generators 30-15
Multiple Pieces • Monolithic Fabrication
30.5 Characterization of Droplet Generation 30-17
Droplet Trajectory • Ejection Direction • Ejection
Sequence/velocity and Droplet Volume • Flow Field
Visualization
30.6 Applications 30-20
Inkjet Printing • Biomedical and Chemical Sample
Handling • Fuel Injection and Mixing Control • Direct
Writing and Packaging • Optical Component Fabrication
and Integration • Solid Freeforming • Manufacturing
Process • Integrated Circuit Cooling
30.7 Concluding Remarks 30-26

30.1 Introduction

Microdroplet generators are becoming an important research area in microelectromechanical systems (MEMS), not only because of the valuable marketing device—inkjet printhead—but also because of the many other emerging applications for precise or micro-amount fluidic control. There has been a long history of development of microdroplet generators ever since the initial inception by Sweet (1964, 1971), who used piezo actuation, and by Hewlett-Packard and Cannon [Nielsen et al., 1985] in the late 1970s, who used thermal bubble actuation. Tremendous research activities regarding inkjet applications have been devoted to this exciting field. Emerging applications in the biomedical, fuel-injection, chemical, pharmaceutical, electronic fabrication, microoptical device, integrated circuit cooling, and solid freeform fields have fueled these research activities. Thus, many new operation principles, designs, fabrication processes and materials related to microdroplet generation have been explored and developed recently, supported by micromachining technology.

In this chapter, microdroplet generators are defined as droplet generators generating microsized droplets in a controllable manner; that is, droplet size and number can be accurately controlled and counted. Thus, atomizer, traditional fuel-injector or similar droplet-generation devices that do not offer such control are not discussed here.

Microdroplet generators usually employ mechanical actuation to generate high pressure to overcome liquid surface tension and viscous force for droplet ejection. Depending on the droplet size, the applied pressure is usually greater than several atmospheres. The operation principles, structure/process designs and materials often play key roles in the performance of droplet generators.

The applications of microdroplet generators, in addition to the well-known application of inkjet printing, cover a wide spectrum of fields, including direct writing, fuel injection, solid freeform, solar cell fabrication, light emitting polymer display (LEPD) fabrication, packaging, microoptical components, particle sorting, microdosage, plasma spraying, drug screening/delivery/dosage, micropropulsion, integrated circuit cooling and chemical deposition. Many of these applications may become key technologies for integrated microsystems in the near future.

This chapter provides the reader with an overview of the operation principles, physical properties, design issues, fabrication process and issues, characterization methods and applications of microdroplet generators.

30.2 Operation Principles of Microdroplet Generators

Many attempts have been made to generate controllable microdroplets [Buehner et al., 1977; Twardeck, 1977; Carmichael, 1977; Ashley et al., 1977; Bugdayci et al., 1983; Darling et al., 1984; Lee et al., 1984; Myers and Tamulis, 1984; Nielsen, 1985; Bhaskar and Aden, 1985; Allen et al., 1985; Krause et al., 1995; Chen and Wise, 1995; Tseng et al., 1996; Hirata et al., 1996; Zhu et al., 1996]. Most of these methods have employed the principle of creating pressure differences, either by lowering the outer pressure or increasing the inner pressure of a nozzle, to push or pull liquid out of the nozzle to form droplets. Typical examples are pneumatic, piezoelectric, thermal bubble, thermal buckling, focused acoustic-wave and electrostatic actuations. The basic principles of those droplet generators are introduced in the following sections. An ejection method by acceleration is also included in the last section, in which inertial force is employed for droplet generation.

30.2.1 Pneumatic Actuation

The spray nozzle is one of the most commonly used devices for generating droplets nowadays for airbrush or sprayer applications. Two types of spray nozzles are shown in Figure 30.1. Figure 30.1a shows that the air brush generates lower pressure at the outer edge of the capillary tube by blowing air across the tube

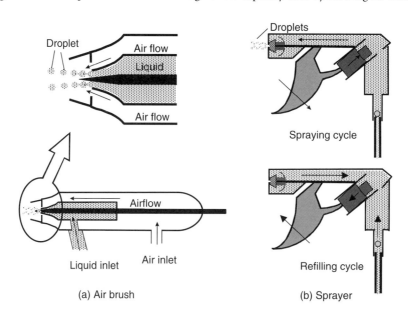

FIGURE 30.1 Operation principle of airbrush and sprayer. (After Tseng, 1998.)

Microdroplet Generators 30-3

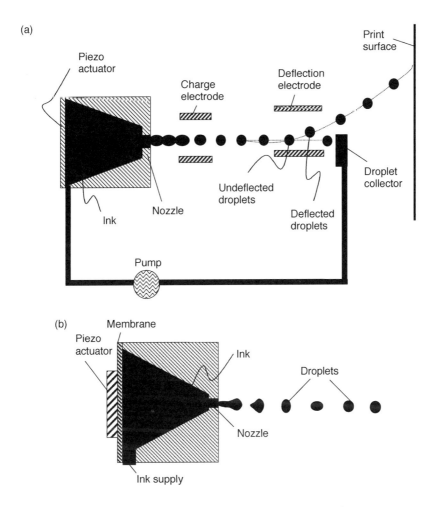

FIGURE 30.2 (a) Operation principle of a piezoelectric-actuated continuous droplet generator. (After Buehner et al., 1977.) (b) Operation principle of a piezoelectric-actuated droplet-on-demand droplet generator. (After Lee et al., 1984.)

end, which forces liquid to move out of the tube and form droplets. The sprayer shown in Figure 30.1b employs high pressure to push liquid through a small nozzle to form droplets. Typical sizes of the droplets generated by the spray nozzle range from tens to hundreds of microns in diameter. The device can be fabricated in microsize using micromachining technology, however, it is difficult to control each nozzle separately in an array format.

30.2.2 Piezoelectric Actuation

The droplet ejection by piezoelectric actuation was invented by Sweet in 1964. Based on the piezoelectric technology, there are two types of piezoelectric devices. One is called the *continuous inkjet* [Buehner et al., 1977; Twardeck, 1977; Carmichael, 1977; Ashley et al., 1977]. Figure 30.2a shows the operation principle of this type of inkjet. Conductive ink is forced out of the nozzles by pressure. The jet breaks up continuously into droplets with random sizes and spacing. Uniformity of the size and spacing of the droplets can be controlled by applying an ultrasonic wave with fixed frequency to the ink through a piezo-electric transducer. The continuously generated droplets pass through a charge plate, and only the desired ones are charged by the electric field and deflected to printout, while the nondesired ones are collected in a gutter and recycled. One piezoelectric transducer can support multiple nozzles, so the nozzle spacing can be as small as desired for high-resolution arrays, however, the complexity of the droplet charging and collecting system is a major obstacle to practical use of this device.

The other device, called droplet-on-demand inkjet, utilizes a piezoelectric tube or disc for droplet ejection only when printing a spot is desired [Bugdayci et al., 1983; Darling et al., 1984; Lee et al., 1984]. Figure 30.2b shows a typical drop-on-demand drop generator. The operational principle is based on the generation of an acoustic wave in a fluid-filled chamber by a piezoelectric transducer through the application of a voltage pulse. The acoustic wave interacts with the free meniscus surface at the nozzle to eject a single drop. The major advantage of the drop-on-demand method is that a complex system for droplet deflection and collection is not required. However, the main drawback is that the size of the piezoelectric transducer tube or disc, on the order of submillimeters to several millimeters, is too large for high-resolution applications. It has been reported that the typical frequency for stable operation of a piezoelectric inkjet would be tens of kilohertz [Chen et al., 1999].

30.2.3 Thermal-Bubble Actuation

The thermal bubble jet was developed by Hewlett-Packard in the U.S. and by Canon in Japan in the early 1980s [Nielsen, 1985; Bhaskar and Aden, 1985; Allen et al., 1985]. There are also many other designs reported in the literature [e.g., Krause et al., 1995; Chen and Wise, 1995; Tseng et al., 1996; Tseng, 1998]. Figure 30.3 shows the cross section of a thermal bubble jet. Liquid in the chamber is heated by a pulse current applied to the heater under the chamber. The temperature of the liquid covering the surface of the heater rises to around the liquid critical point in microseconds, and then a bubble grows on the surface of the heater, which serves as a pump. The bubble pump pushes liquid out of the nozzle to form a droplet. After the droplet is ejected, the heating pulse is turned off and the bubble starts to collapse. Liquid refills the chamber by surface tension on the free surface of the meniscus for recovery to the original position. The second pulse starts again to generate another droplet. The energy consumption for ejecting each droplet is around 0.04 mJ for the H-P ThinkJet printhead. Because bubbles can deform freely, the chamber size of the thermal bubble jet would be smaller than that of other actuation means, which is important for high-resolution applications. The resolution reported in the literature ranges from 150 to

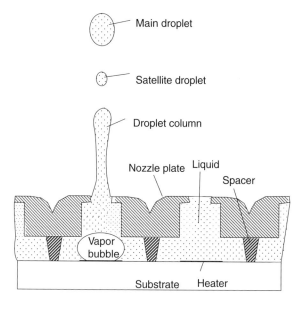

FIGURE 30.3 Operation principle of a thermal-bubble-actuated droplet generator. (After Allen et al., 1985.)

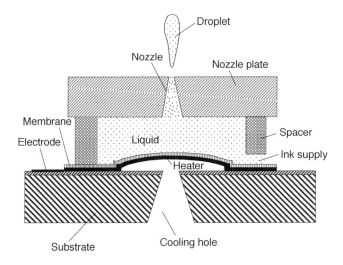

FIGURE 30.4 Operation principle of a thermal-buckling-actuated droplet generator. (After Hirata et al., 1996.)

600 dpi [Krause et al., 1995] and 1016 dpi [Chen and Wise, 1995]. The typical operational frequency for the contemporary thermal bubble jets is from several to tens of kilohertz.

30.2.4 Thermal-Buckling Actuation

Hirata et al. (1996) employed a buckling diaphragm for droplet generation. Figure 30.4 shows the basic operation principle. A composite circular membrane, consisting of silicon dioxide and nickel layers, is fixed on the border with a small gap between it and the substrate. A heater is placed at the center of the composite membrane and electrically isolated from it. Pulsed current is sent to the heater and then the membrane is heated for several microseconds. When the thermally induced stress is greater than the critical stress, the diaphragm buckles up abruptly and ejects a droplet out of the nozzle. The power required to generate a droplet at a speed of 10 m/s is around 0.1 mJ for a 300-μm-diameter diaphragm. The power consumption and the device size of buckling membrane inkjet are much larger than those of the thermal bubble jet. The reported frequency response of membrane buckling jet ranges from 1.8 to 5 kHz, depending on the desired droplet velocity.

30.2.5 Acoustic-Wave Actuation

Figure 30.5 shows a lensless liquid ejector using a constructive interference of acoustic waves to generate droplets [Zhu et al., 1996]. A PZT thin-film actuator with the help of an on-chip Fresnel lens was employed to generate and focus acoustic waves on the air–liquid interface for droplet formation. The actuation comes from excitation of a piezoelectric film under a burst of radio-frequency (RF) signal. The device does not require a nozzle to define droplets, thus reducing the troublesome clogging problem occurring in most of the droplet generators employing nozzles. Droplet size can also be controlled by acoustic waves with specific frequencies, however, due to vigorous agitation of the acoustic wave in the liquid, it is difficult to maintain a quiet interface for reliable and repeatable droplet generation. As a result, a "nozzle area" is still desired to maintain the interface at a stable level. The applied RF ranges from 100 to 400 MHz and the burst period is 100 μs. The power consumption for one droplet is around 1 mJ, which is high compared to other principles. The droplet size ranges from 20 to 100 μm, depending on the RF. The reported size of the device is 1×1 mm^2, which is much larger than other droplet generators mentioned in the previous sections.

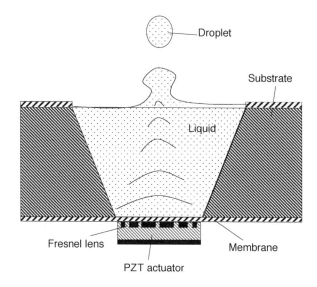

FIGURE 30.5 Operation principle of an acoustic-wave-actuated droplet generator. (After Zhu et al., 1996.)

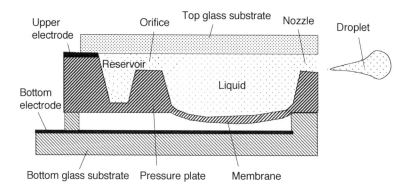

FIGURE 30.6 Operation principle of an electrostatic-actuated droplet generator. (After Kamisuki et al., 1998.)

30.2.6 Electrostatic Actuation

The electrostatically driven inkjet printhead was first introduced by Seiko–Epson Corp. [Kamisuki et al., 1998, 2000] for commercial printing purposes. As shown in Figure 30.6, the actuation beings by application of a dc voltage between the electrode plate and pressure plate to deflect the pressure plate for ink filling. When the voltage turns off, the pressure plate reflects back to push the droplet out of the nozzle. This device was developed for use in electric calculators, as it offers low power consumption of less than 0.525 mW/nozzle. The driving voltage for a SEAJet™ is 26.5 V and the driving frequency can be up to 18 kHz with uniform ink ejection. A device with 128 nozzles/chip with 360-dpi pitch resolution has also been demonstrated to have high printing quality (for bar coding), high-speed printing, low power consumption, a long lifetime under heavy-duty usage (lifetime more than 4 billion ejections) and low acoustic noise. However, the fabrication comprises complex bonding processes among three different micromachined pieces. Besides, the pressure plate requires a very precise etching process to control the accuracy and uniformity of the thickness. Due to the deformation limitation of solid materials and alignment accuracy required in the bonding process, the nozzle pitch may not easily be reduced further for higher resolution applications.

Microdroplet Generators 30-7

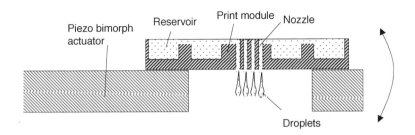

FIGURE 30.7 Operation principle of an inertia-actuated droplet generator. (After Gruhler et al., 1999.)

30.2.7 Inertial Actuation

Inertial droplet actuators apply high acceleration to the nozzle chip for droplet ejection. Such an apparatus is shown in Figure 30.7 [Gruhler et al., 1999]. The print module consists of large reservoirs on the top plate, which connects to the nozzles on the bottom plate. The print module is mounted on a long cantilever beam with a piezo–bimorph actuator for acceleration generation. It requires 500 µs to generate 1 nl of droplets from 100-µm-diameter nozzles. Twenty-four liquid droplets of different solution types were demonstrated to be ejected simultaneously from the nozzles in a 500-µm-pitch. The smallest droplet claimed to be generated is 100 pl from 50-µm-diameter nozzles. This principle provides a gentle ejection process for bioreagent applications. However, the ejection of smaller droplets may encounter strong surface tension and flow drag forces at the microscale that are much larger than the droplet inertial. Also, the droplets cannot be selectively and individually ejected from the desired nozzles which limits its applications.

30.3 Physical and Design Issues

The sequence of droplet generation involves wide ranges of physical issues and design concerns, including microfluid flow, heat transfer, wave propagation, surface properties, material properties and structure strength. The following sections discuss frequency response, thermal cross-talk, hydraulic cross-talk, overfill, satellite droplets, puddle formation and material issues commonly seen in microdroplet generators.

30.3.1 Frequency Response

The frequency response of droplet generators is an important measure for assessing the performance of a device. The reported typical frequency response for thermal bubble jets, piezo jets, thermal buckling jets, acoustic wave jets and electrostatic and inertial jets ranges from a kilohertz to tens of kilohertz. Piezo and acoustic wave jets typically have higher frequency responses than the others. Recently, the novel design by Tseng (1998) improved the frequency response of thermal bubble jets by about three times (35 kHz), which is comparable to the speed of piezo jets. Higher speed devices (in the range of hundreds of kilohertz) are currently under development by major inkjet printer makers and further breakthroughs will follow in the next few years.

The three important time constants related to the frequency response of microdroplet generators are actuation, droplet ejection and liquid refilling, as shown in Figure 30.8.

The typical time constant from heating to bubble formation in a thermal bubble jet is around 5 to 10 µs for a chamber size ranging from 20 to 100 µm [Tseng, 1998]. Thermal buckling and electrostatic microdroplet generators have actuation times of tens to hundreds of microseconds (estimated from Hirata et al., 1996), due to the large actuation plate required for sufficient displacement for droplet formation. The inertial type required even longer actuation time (estimated from Gruhler et al., 1999), typically hundreds to thousands of microseconds, due to the large cantilever structure necessary for generating large droplets with sufficient inertial force to overcome liquid surface tension and viscosity. The actuation

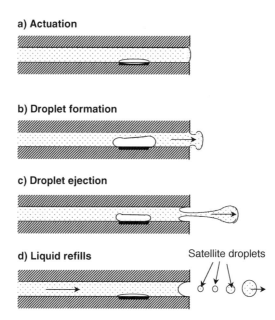

FIGURE 30.8 The sequence of droplet generation.

time for piezo or acoustic wave types of generators may be shorter [Darling et al., 1984; Zhu et al., 1996], around one microsecond to tens of microseconds.

After application of actuation pressure to the liquid, the droplet starts to eject. The ejection sequence usually takes a couple to hundreds of microseconds for a droplet volume of 1 pl to 1 nl [Tseng, 1998], which does not vary much for different operation principles.

The liquid refills back automatically by surface tension force after the droplet ejection. Liquid refilling time can vary by three orders of magnitude (e.g., from less than 10 μs to over 1 s), depending on the length and geometry of the refilling path. In most of the commercial inkjet printhead designs, a chamber neck [Nielsen, 1985], elongated chamber channel [Neilsen, 1985] or physical valve [Karz et al., 1994] is used to prevent hydraulic cross-talk and maintain a high pressure in the firing chamber. However, those designs, if not arranged properly, may greatly increase the refilling time, causing a reduction in the frequency response. As a result, how to prevent the hydraulic cross-talk without sacrificing the device speed becomes an important issue in the design of super-high-speed and high-resolution droplet generators. Tseng et al. (1998a; 1998b; 1998c) introduced the concept of a virtual chamber neck to speed up the refilling process of thermal bubble jets and to suppress cross-talk. The virtual neck consists of vapor bubbles that provide sealing pressure while the droplet is ejecting and opens up to reduce flow resistance while liquid is refilling, thus increasing the frequency response. The concept is shown in Figure 30.9.

For simulation of droplet actuation and the formation sequence, much work [Curry and Portig, 1977; Lee, 1977; Levanoni, 1977; Pimbley and Lee, 1977; Fromm, 1984; Bogy and Talke, 1984; Asai et al., 1987; 1988; Asai, 1989; 1991; 1992; Mirfakhraee, 1989; Chen et al., 1997a; 1997b; 1988a; 1988b; 1999; Rembe et al., 2000] has been conducted on the bubble formation sequence, droplet generation, and aerodynamics of droplets traveling in the atmosphere for thermal bubble as well as piezo jets. Interested readers can refer to those works for details.

30.3.2 Thermal/Hydraulic Cross-Talk and Overfill

When nozzle pitch becomes small, two types of cross-talk—hydraulic and thermal (for thermal bubble jet)—become significant in multiple-nozzle droplet generators. Hydraulic cross-talk relates to the transportation of a pressure wave from the firing chamber to the neighboring chambers, as shown in Figure 30.10. The vibration of the meniscus of the neighboring chambers may result in poor droplet volume control

Microdroplet Generators

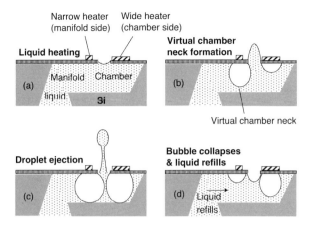

FIGURE 30.9 Operation principle of virtual chamber neck. (After Tseng et al., 1998c.)

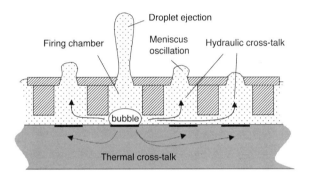

FIGURE 30.10 Cross-talk among neighboring microchambers.

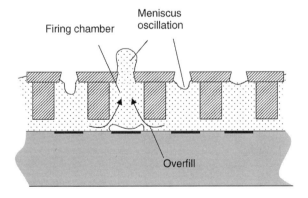

FIGURE 30.11 Overfill phenomenon.

or, even worse, unexpected droplet ejection. Thermal cross-talk, which only appears in the thermal bubble jet, is the phenomenon of thermal energy transportation from the firing chamber to neighboring chambers, resulting in poor droplet volume control. After droplet ejection, the refilling process of liquid sometimes causes meniscus oscillation, posing another issue of overfill. Overfill, similar to cross-talk, increases the waiting time for the next droplet ejection and even causes undesired droplet ejection. The phenomenon of overfill is illustrated in Figure 30.11.

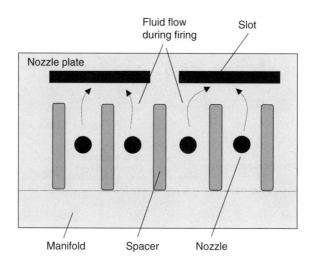

FIGURE 30.12 Method of applying parallel compliant (reservoir) to overcome cross-talk.

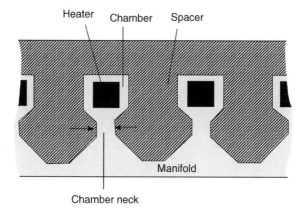

FIGURE 30.13 The cross-talk overcome by employing a chamber neck.

These issues stem from the fact that there is not enough flow compliance among nozzles. One of the solutions developed by IBM [Nielsen, 1985] is to increase the channel length for each chamber. However, the increased serial compliance of the lengthened channel between the reservoir and chamber increases the flow resistance and inertia significantly, increasing the liquid refilling time.

HP tried to solve this problem by using either a parallel compliant (reservoir) or a chamber neck. Figure 30.12 shows a slot used as a reservoir beside the nozzles to store energy while the bubble is exploding and to release energy while the bubble is collapsing [Nielsen, 1985]. Figure 30.13 shows the second approach by narrowing the inlet of the chamber to form a chamber neck [Buskirk et al., 1988].

The effects from the aforementioned methods were simulated by Buskirk et al. (1988). Figure 30.14 shows the simulation results of the meniscus position of the firing chamber and neighboring chamber for various chamber designs. In the first figure, without any design of flow compliance, the meniscuses on both the firing chamber and the neighboring chamber show huge fluctuations. Doubling the channel length (series compliance) damped down some of the fluctuation, but not completely. Only the chamber neck design almost completely damped down the meniscus fluctuation; however, it has the slowest rising response among the three methods.

In contrast to fixed chamber neck design, Xerox [Karz et al., 1994] has used a flexible plate as a valve to address the cross-talk issue. However, this design may suffer from low frequency response and material

Microdroplet Generators 30-11

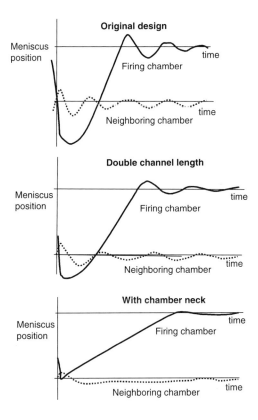

FIGURE 30.14 Meniscus oscillation in various chamber structures. (Data from Buskirk et al., 1988.)

reliability problems. Tseng et al. (2000) used a novel design of a virtual chamber neck, employing the bubble as a virtual valve to reduce the cross-talk problem while maintaining the high frequency response of the droplet generator, as shown in Figure 30.9. Their work demonstrated that the bubble has faster response and more reliable operation performance in the microscale than do solid valves.

In addition to hydraulic cross-talk, thermal cross-talk, in Tseng's work, was also reduced by placing the heater on the chamber top of a thin film with low thermal conduction, instead of leaving it on the thick substrate through which heat can be conducted to the neighbors [Tseng et al., 1998c; 2001a]. It is clear that as the nozzle pitch becomes closer, the cross-talk problem becomes more severe in the operation of very-high-resolution and high-speed droplet generators.

30.3.3 Satellite Droplets

Satellite droplets result from the breakdown of the long ejected liquid column by the interaction among surface tension, air drag force and inertial force. The velocity mismatch along the liquid column, resulting from the variation of actuation velocity, promotes the breakdown. The droplet ejection sequence captured from a commercial inkjet reveals the detailed steps of satellite droplet formation, as shown in Figure 30.15 [Tseng et al., 1998c]. When applied to printing, the quality is degraded from the occurrence of satellite drops, as revealed in Figure 30.16. Satellite droplets also reduce the accuracy for precise liquid dispensing control.

A literature survey reveals that there have been many attempts to predict the droplet formation. Asai et al. (1987; 1988; Asai, 1989; 1991; 1992) conducted both numerical simulation and experimental measurements to obtain the temporal variation of droplet length at a thermal bubble jet. However, the formation process of satellite droplets is not included. In the drop-on-demand inkjet, Fromm (1984)

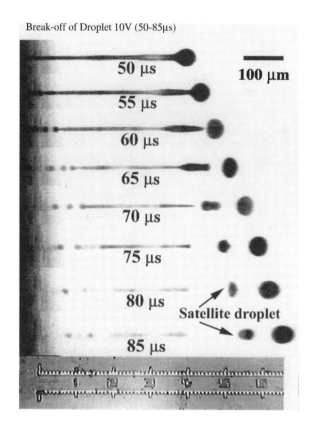

FIGURE 30.15 Droplet ejection sequence of HP 51626A printhead. (After Tseng et al., 1998c.)

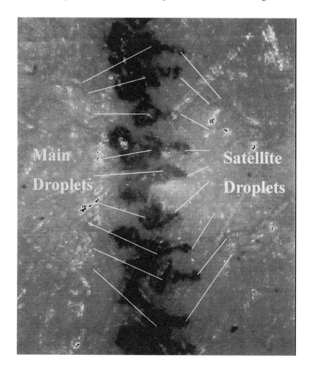

FIGURE 30.16 A printed vertical line smeared by satellite droplets. (After Tseng et al., 1998c.)

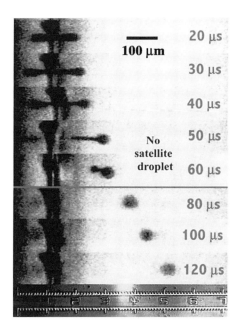

FIGURE 30.17 Droplet ejection without satellite droplets. (After Tseng et al., 1998a.)

and Chen et al. (1997a; 1997b; 1998a; 1998b; 1999) solved the Navier–Stokes equation to predict the droplet formation, and showed the process of satellite droplet formation. Pimbley and Lee (1977) and Chen et al. (1997b) demonstrated the evolution of the droplet as well as satellite droplet formation by flow visualization, however, the detailed method for eliminating the separation of satellite droplet from the main droplet was not discussed.

Various efforts have been made to eliminate satellite droplets from the commercial product. In piezo droplet generators, triangular waves were used to eliminate satellite drops [Chen et al., 1999]. For thermal bubble jets, Tseng et al. (1998a; 1998c; 2000) proposed a novel method employing the bubble as a trimmer to cut off the long droplet tail to eliminate satellite droplets, as shown in Figure 30.17. The tail of the liquid column, shortened by the bubble trimmer in Tseng's work, is drawn back by surface tension into the main droplet, thus eliminating satellite drops.

30.3.4 Puddle Formation

A liquid puddle forms when liquid is pushed to flow outward and accumulates on the outside surface of the nozzle. Puddles impose a great blocking force on droplet ejection, causing distortion or even disruption of droplet ejection. One of the major causes of puddle formation is the hydrophilic nozzle surface. When it comes into contact with working fluid from the chamber, the liquid accumulates on the outer surface of the nozzle. It was observed by (Tseng, 1998) that the puddle appears after several continuous operations; that is, the chamber surface does not acquire a puddle until getting wet after several runs. If the operation of droplet ejection stops, the puddle is drawn back into the chamber by surface tension. However, once the chamber surface becomes wet, puddle formation always occurs when operation starts again. Figure 30.18 shows the formation of a puddle during the running of a microinjector [Tseng, 1998]. Notice that the droplet ejection position is away from the nozzle, due to distortion by the liquid puddle.

One way to eliminate puddle formation is to coat the chamber outer surface with a nonwetting material to prevent the wetting process of the working fluid, the inner surface of the chamber must remain hydrophilic

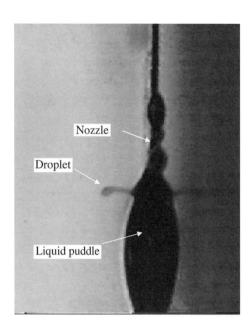

FIGURE 30.18 Liquid puddle formation outside the micro nozzles. (After Tseng, 1998.)

for liquid refill. Even with the coating, however, there is still no guarantee that the puddle will not form. More research is under way to fully understand the mechanism of the puddle formation process.

30.3.5 Material Issues

Material issues, including stress, erosion, durability and compatibility, are very complex issues in the design of microdroplet generators. Material compatibility, stress and durability problems are commonly discussed from the processing perspective. Material compatibility issues result from the processing temperature, processing environment (oxidation, reactive gas, etc.), etching method used and adhesion ability; stress issues are usually a result of processing temperature as well as doping conditions; material durability issues either are material intrinsic properties or are due to the mechanical forces induced during the process (i.e., fluid flow force, surface tension force, vacuum forces or handling force). Lots of care needs to be taken in the process flow design to eliminate the material issues, such as compensating material stress during or after the fabrication process, conducting the high-temperature process earlier than the low-temperature material, finishing aggressive wet etching before metal film deposition or using low-temperature bonding material and processes to protect integrated circuit and microdevices.

From the operation aspect, durability, stress and erosion issues are the major concerns. Due to the cycling nature of the droplet generation process, the materials chosen for actuation face challenges not only from stress but also from fatigue. HP reported that possible reasons for failure of the heater passivation materials are cavitation and thermal stress [Bhaskar and Aden, 1985]. Silicon, low-stress silicon nitride, silicon carbide, silicon dioxide and some metals are usually used to overcome the aforementioned problems. In addition to proper material selection, reducing sharp corners in the design is also an important key to preventing the material from cracking by eliminating stress concentration points. Moreover, the erosion of structure materials from the working fluid is another serious issue. Lee et al. (1999) reported the erosion of spacer material in a commercial inkjet head when diesel fuel was used as the working fluid. In contrast, materials, including silicon and silicon nitride, used by Tseng et al. [Tseng et al., 1998c; Tseng, 1998] and Lee et al. (1999b) in the microinjector are free of this problem and can be applied to a wide variety of fluids, including solvents and chemicals. Selecting materials wisely, arranging them correctly in process and properly designing the materials in the microstructures are the three important concepts in reducing material issues.

30.4 Fabrication of Microdroplet Generators

The common structures in most of the microdroplet generators include a manifold for liquid storage, microchannels for liquid transportation, microchambers for holding liquid, nozzles for droplet size and direction definition and actuation mechanisms for droplet generation. Occasionally, droplet generators may not have nozzles but instead use energy focusing as a means to generate droplets locally, such as acoustic wave droplet generators [Zhu et al., 1996]. Before micromachining processes became popular, most of the fabrication processes of microdroplet generators stemmed from the same concept: Nozzle plates, fluid-handling plates and actuation plates are manufactured separately and then integrated into one final device. However, as the nozzle resolution becomes finer, bonding processes pose severe alignment, yield, material and integrated-circuit compatibility problems. Also, interconnection lines may not have enough space to pull out individually from each chamber when nozzle resolution is higher than 600 dpi. As a result, monolithic ways to fabricate high-resolution integrated circuit droplet generators have become very important. In the following sections, examples of different fabrication means are introduced.

30.4.1 Multiple Pieces

Figure 30.19 shows the traditional fabrication of microdroplet generators by the bonding of fabricated structure pieces [Tseng, 1999]. Actuation plates are fabricated separately from the nozzle plates. In the thermal bubble jet, heaters are usually sputtered or evaporated and then patterned with an integrated circuit on the bottom plate, while piezo, thermal buckling, electrostatic and inertial actuators consist of

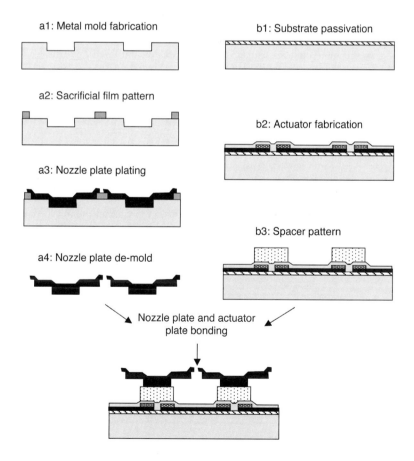

FIGURE 30.19 Conventional fabrication process flow of microdroplet generators.

more complex structures, such as piezo disc, thin plate structure, or cantilever beam. Parallel to the actuator fabrication, nozzles are fabricated by electroforming [Ta et al., 1988], molding or laser drilling [Keefe et al., 1997]. The combination of those processed pieces is assembled either by using polymer spacer material as intermediate layers [Siewell et al., 1985; Askeland et al., 1988; Hirata et al., 1996; Keefe et al., 1997] or by directly adhering several pieces through anodic bonding [Kamisuki et al., 1998; 2000], fusion bonding [Gruhler et al., 1999], eutectic bonding or low-temperature chemical bonding. However, most of the bonding means are chip-level not wafer-level processes and face the similar challenges of alignment, bonding quality and material/process compatibility. As the nozzle resolution becomes higher than 600 dpi, alignment accuracy of 4 μm (10% of the nozzle pitch) is difficult to achieve. Higher alignment accuracy significantly increases the fabrication cost, especially for the chip-level process. Bonding quality is another important issue corresponding to the fabrication yield of large-array and high-resolution devices. The bonding materials (mostly polymers) chosen may not be suitable for the application environments and working fluids. Also, the bonding process, involving heat, pressure, high voltage or chemical situations, restricts integrated circuit integration with the droplet generators which is essential for large-array and high-resolution applications.

30.4.2 Monolithic Fabrication

In order to address the aforementioned issues, monolithic processes utilizing micromachining technology have been widely employed since the early 1990s. Two major methods have been introduced: One is the combination of bulk and surface micromachining and the other is the use of bulk micromachining and deep-ultraviolet (UV) lithography associated with electroforming (or UV lithography only).

For example, Tseng, (1998) combined surface and bulk micromachining to fabricate a microdroplet generator array with potential nozzle resolution up to 1200 dpi (printing resolution can be 2400 dpi or higher). In this design, double bulk-micromachining processes were utilized to fabricate a fluid-handling system that included a manifold, microchannels and microchambers. Surface micromachining was used for heater, interconnection line and nozzle fabrication. The whole process was finished on (100)-crystal-orientation silicon wafers. Figures 30.20 and 30.21 show the three-dimensional structure of the microinjectors and the monolithic fabrication process, respectively. The ejection of 0.9-pl droplets has also been demonstrated by Tseng et al. (2001b) for high-resolution microinjectors. The structure materials used in the microinjector are silicon silicon nitride and silicon oxide, durable in high temperature and suitable for various liquids (even some harsh chemicals). Integrated circuits can be easily integrated with this device on the same silicon substrate.

The second case can be found in Lee et al.'s (1999a) work. In this design, the multi-exposure and signal development (MESD) lithography method was used to define microchannel and microchamber structures

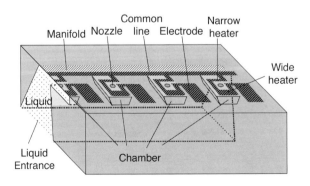

FIGURE 30.20 Schematic three-dimensional structure view of microinjectors. (After Tseng et al., 1998c.)

Microdroplet Generators

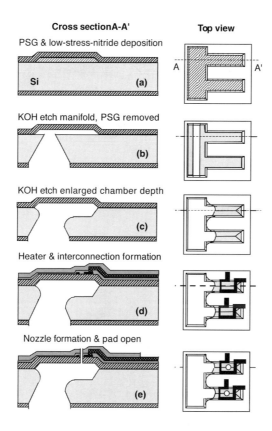

FIGURE 30.21 Fabrication process flow of monolithic micro injectors. (After Tseng et al., 1998c.)

(using a photoresist as the sacrificial layer), and the physical structures were constructed by electroformed metal. The manifold was manufactured from the wafer backside by electrochemical methods [Lee et al., 1995]. This device demonstrated compatibility with the very-high-resolution array and integrated circuit process. Another method, using a photoresist as the sacrificial layer and polyimide as the structure layer, was introduced by Chen et al. (1998b) for applications compatible with high-resolution and integrated circuit applications.

30.5 Characterization of Droplet Generation

Droplet trajectory, volume, ejection direction and ejection sequence/velocity are four important quantitative measures for assessing the ejection quality of microdroplet generators. The following sections briefly introduce the basic methods for the testing of droplet generation.

30.5.1 Droplet Trajectory

Droplet trajectory visualized by utilizing a flashing light on the ejection stream, as shown in Figure 30.22, was introduced by Tseng et al. (1998a). The white dots in Figure 30.23 indicate the droplet stream. The visualized droplet trajectory follows an exponential curve, very different from the parabolic one expected for normal-size objects with similar initial horizontal velocity. The droplet trajectory was also estimated by Tseng et al. (1998a) by solving a set of ordinary differential equations from the force balance, in both horizontal and vertical direction, of a single droplet flying through air.

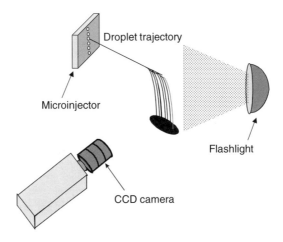

FIGURE 30.22 Experimental setup for droplet stream visualization.

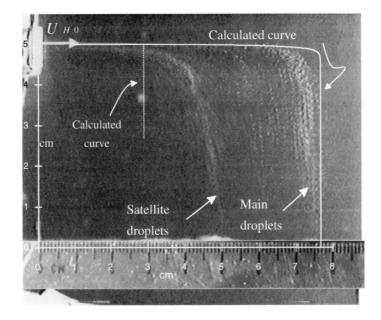

FIGURE 30.23 Visualized droplet stream and estimated trajectory of a microdroplet generator. (After Tseng et al., 1998a.)

From the analysis, the vertical position Y and horizontal position X of the droplet can be expressed by:

$$Y = U_{v\infty}\left[t - \frac{U_{v\infty}}{g}\left(1 - e^{\frac{-6\pi\mu r_0}{m}t}\right)\right] \quad (30.1)$$

$$X = \frac{U_{H0}m}{6\pi\mu r_0}\left(1 - e^{\frac{-6\pi\mu r_0}{m}t}\right) \quad (30.2)$$

where g is the acceleration due to gravity, t is the time, m is the mass, r_0 is the radius of the droplet, μ is the viscosity of air, $U_{v\infty} = \frac{mg}{6\pi\mu r_0}$ is the droplet terminal velocity, and U_{H0} is the initial horizontal velocity.

The trajectory is drawn in Figure 30.23 with the experimental result and fits the visualized trajectory well except at the end, suggesting an interaction among droplets. From this simple analysis, the maximum flying distance of a droplet with a known diameter can be estimated as:

$$X_{max} = \frac{U_{H0} m}{6\pi \mu r_0} = \frac{2\rho_{liquid}}{9\mu_{air}}(U_{H0} r_0^2), \text{ when } t \sim \infty \qquad (30.3)$$

Here, the maximum distance is proportional to the droplet velocity and droplet radius to the second power. For droplets of varying sizes and the same initial velocity, the maximum flying distance is reduced quickly for the smaller ones. To obtain 1-mm flying distance, a droplet with 10-m/s initial velocity must have a minimum radius of 2.7 μm. From the above estimation, droplet size should be maintained beyond a certain value to ensure enough flying distance for printing. Printing with very fine droplets (diameter smaller than a couple of micrometers) requires either increasing the initial velocity of the droplets or printing in a special vacuum environment to overcome the resistant force from air drag.

30.5.2 Ejection Direction

Droplet direction can be decided by the visualized trajectory. Many parameters, including nozzle shape, roughness, aspect ratio and wetting property, as well as actuation direction and chamber design, affect droplet direction. In general, symmetric structure design and accurate alignment can help control droplet direction.

30.5.3 Ejection Sequence/velocity and Droplet Volume

To characterize the detailed droplet ejection sequence, a visualization system [Chen et al., 1997b; Tseng et al., 1998c], as shown in Figure 30.24, has been widely used. In this system, an LED was placed under the droplet generator to back-illuminate the droplet stream. Two signals, synchronized with adjustable time delay, were sent to a microinjector and an LED, respectively. Droplets were ejected from a droplet generator continuously, and the droplet images were frozen by the flashlight from the LED at specified time delays, as shown in Figure 30.15. Droplet volume can be determined from the images by assuming the droplet is axisymmetric or by weighing certain numbers of droplets. Droplet velocity can be estimated by measuring the flying distance difference of the droplet fronts in two successive images.

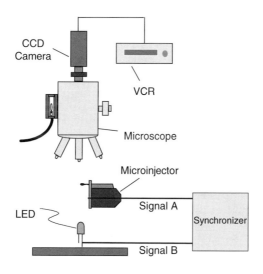

FIGURE 30.24 Experimental setup for droplet ejection sequence visualization.

30.5.4 Flow Field Visualization

To better understand flow properties such as cross-talk, actuation sequence, liquid refill and droplet formation inside microdroplet generators, flowfield visualization is one of the most direct and effective ways. Flow visualization at a small scale has some difficulties that do not occur at a large scale, such as limited viewing angles, hard-to-apply light sheet reflection from the particles trapped on the wall, short response time and small spatial scale. Meinhart and Zhang (2000) adopted a micrometer-resolution particle image velocimetry system to measure instantaneous velocity fields in an electrostatically actuated inkjet head. In the setup, 700-nm-diameter fluorescent particles were introduced for flow tracing. The spatial as well as temporal resolutions of the image velocimetry were found to be 5 to 10 µm and 2 to 5 µs, respectively. The four primary phases of the injection operation, including infusion, inversion, ejection and relaxation, were clearly captured and quantitatively analyzed.

30.6 Applications

More than a hundred applications have been explored employing microdroplet generators. This section provides a summary of some of the applications.

30.6.1 Inkjet Printing

Inkjet printing involves arranging small droplets on a printing medium to form texts, figures or images and is the most well-known application. The smaller and cleaner the droplets are, the sharper the printing is, however, smaller droplets cover a smaller printing area and thus increase the printing time. Therefore, in printing applications, high-speed microdroplet generation with stable and clean microsized droplets is desired for fast, high-quality printing. The printing media can be paper, textile, skin, cans or other surfaces that can adsorb or absorb printing solutions. Inkjet printing generated a more than $10 billion worldwide market in 2000 and continues to grow.

30.6.2 Biomedical and Chemical Sample Handling

The application of microdroplet generators to biomedical sample handling is an emerging field and is drawing much attention. Much research effort has been focused on droplet volume control, droplet size miniaturization, compatibility issues, variety of samples and high-throughput parallel methods.

Luginbuhl et al. (1999), Miliotis et al. (2000) and Wang et al. (1999) developed piezo- and pneumatic-type droplet injectors, for mass spectrometry. Figure 30.25 shows the design of the injectors, which are utilized to generate submicron- to micron-sized bioreagent droplets for sample separation and analysis in a mass spectrometer, as shown in Figure 30.26. Lugnbuhl et al. (1999) employed silicon bulk micromachining to fabricate a silicon nozzle plate and Pyrex® glass actuation plate, while Wang et al. (1999)

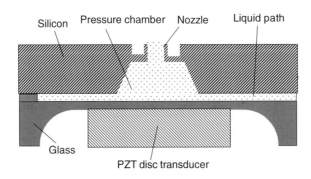

FIGURE 30.25 The injector design for mass spectrometry. (After Luginbuhl et al., 1999).

Microdroplet Generators 30-21

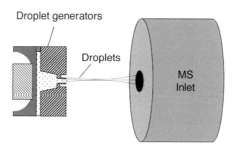

FIGURE 30.26 Operation principle of mass spectrometry using microdroplet generators.

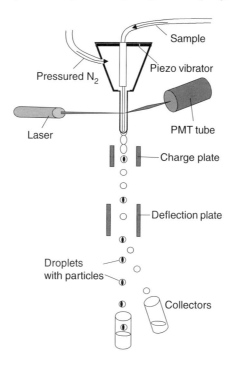

FIGURE 30.27 Particle sorting using droplet generators. (After Asano et al., 1995.)

employed the combination of surface and bulk micromachining to fabricate a droplet generator. These injectors are part of the lab-on-a-chip system for incorporating a microchip with a macro-instrument.

Microdroplet generators were also used by Koide et al. (2000), Nilsson et al. (2000), Goldmann and Gonzalez (2000) and Szita et al. (2000), for the accurate dispensing of biological solutions. Piezo- and thermal-type injectors were used in this research for protein, peptide, enzyme or DNA dispensing. In those applications, the operation principles of the employed devices are similar to those for inject printing. A single biological droplet can be precisely dispensed and deposited onto a desired medium, and the dispensing of droplet arrays can also be carried out. The arrayed bioreagents can be further bioprocessed for high-throughput analysis.

Continuous-jet-type droplet generators were reported by Asano et al. (1995) to effectively focus and sort particles by electrostatic force. The experimental setup is shown in Figure 30.27. A syringe pump pressurizes the sample fluid containing the particles, which pass through a nitrogen sheath flow for focusing. The sample is then ejected from a piezoelectric-transducer-disturbed nozzle to form a droplet. The droplets containing the desired particles are charged at the breakup point and deflected into collectors. The reported separation probability for 5-, 10-, and 15-μm particle can be as high as 99%. However, the inner jet diameter limits the particle size for separation. Other than solid particle separation, this method can potentially also be applied to cell sorting for biomedical applications.

In addition to biomedical reagent handling, microdroplet generators have been widely used in chemical handling. For example, Shah et al. (1999) used an inkjet to print catalyst patterns for electroless metal deposition. In their system, a Pt solution was employed and ejected by a commercial inkjet printhead as a seed layer for Cu electroless plating. The lines produced by this method were reported to be 100 µm wide and 0.2 to 2 µm high.

30.6.3 Fuel Injection and Mixing Control

Microdroplet generators can be used for fuel injection to dispense controllable and uniform droplets, important for mixing and combustion applications. Combustion efficiency depends on the mixing rate of the reactants. The reactants in a shear flow are first entrained by large vortical structures (Brown and Roshko, 1974) and then mixed by fine-scale eddies. The entrainment can be greatly enhanced by controlling the evolution of large-scale vortices either actively (Ho and Huang, 1982) or passively (Ho and Gutmark, 1987). The effectiveness of controlling large-scale vortical structures by increasing the combustion efficiency has also been experimentally demonstrated (Shadow et al., 1987). Although much work has been done on improving the mixing efficiency in combustion chambers, a significant challenge in combustion research is to improve the small-scale mixing and to reduce the evaporation time of the liquid fuel.

Traditional injectors with a nozzle diameter of tens to hundreds of microns can neither supply uniform microdroplets for reducing evaporation time and fine-scale mixing nor eject droplets which can be controlled individually to modulate vortex structure [Lee et al., 1999b]. To overcome those limitations, Tseng et al. (1996) proposed a microdroplet injector array fabricated using micromachining technologies for fuel injection. The droplets ejected from microinjectors are uniform and the size can be one to tens of microns in diameter, which is close to the microscale of low-turbulence eddies. The fine-scale mixing can be carried out by the reaction of the low-turbulence eddies directly with the microdroplets. The evaporation time is also greatly reduced by increasing the evaporation surface from the reduced and unformed droplet size. In addition, appropriate selection of microinjectors distributed around the nozzle of a dump combustor (Figure 30.28) provides spatial coherent perturbations to control the large vortices. Two types of coherent structures (i.e., spanwise and streamwise vortices) can be influenced by imposing subharmonics of the most unstable instability frequency of the air jet. Control of the spanwise vortices can be accomplished by applying temporal amplitude modulation on the injection. If the ejecting phases of the

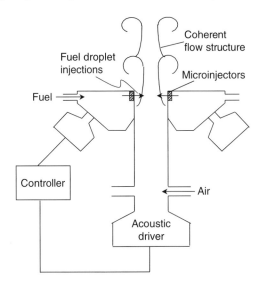

FIGURE 30.28 Control of mixing and fuel injection by microdroplet generators. (After Tseng et al., 1996.)

microdroplets along the azimuthal direction are the same, the mode zero instability (Brown and Roshko, 1974) is enhanced. When a certain defined phase lag is imposed on these microinjectors, higher mode instability (Brown and Roshko, 1974) waves are generated which are usually beneficial for mass transfer enhancement. Because about a thousand injectors are placed around the nozzle, the spatial modulation in the azimuthal direction can perturb the streamwise vortices. The interaction of streamwise and spanwise vortices by microinjectors brings forth fine-scale mixing.

30.6.4 Direct Writing and Packaging

Microdroplet generators offer an alternative to the lithography process for electronics and optoelectronics manufacturing. This approach has the advantages of precise volume control of dispensed materials, data-driven flexibility, low cost, high speed and low environmental impact, as mentioned by Hayes and Cox (1998). Materials used with this process include adhesives for component bonding, filled polymer systems for direct resistor writing and oxide deposition and solder for solder-bumping of flip-chip ball grid arrays (BGAs), printed circuit boards (PCBs), and chip-scale packages (CSPs) [Teng et al., 1988; Hays et al., 1999]. In these printing applications, the temperature should be elevated to 100 ~ 200°C, and the viscosity of fluids should be around 40 cps; in some cases, an inert process environment, such as nitrogen flow, is required to prevent oxidation of the materials.

Direct writing by inkjet printing can eliminate the fabrication difficulty inherent in photolithography or screen-printing processes for solar-cell metallization and light-emitting polymer (LEP) deposition of light-emitting polymer displays (LEPDs). In solar-cell metallization, metallo-organic decomposition (MOD) silver ink is used to inkjet print directly onto solar-cell surfaces for avoidance of p–n junction degradation under the traditional screenprinting method requiring 600 to 800°C for firing process. Inkjet printing also allows formation of a uniform line of film on rough solar-cell surfaces [Tang and Vest, 1988a; 1988b; Somberg, 1990] which is not easily achieved by traditional photolithography.

Organic light-emitting devices that require deposition of many organic layers to perform full-color operation face similar problems. Due to the solubility of those organic layers in many solvents and aqueous solutions, conventional methods such as photolithography, screen printing and evaporation, which require a wet patterning process, are not suitable [Hebner et al., 1998; Shimoda et al., 1999; Kobayashi et al., 2000]. Thus, direct writing of organic materials by inkjet printing has become a promising solution to provide a safe, patternable process without a wet etching procedure. However, due to pinholes appearing on the patterned materials, high-quality polymer devices may not be easily inkjet printed. Yang's group proposed a hybrid way of combining an inkjet-printed layer with another uniform spin-coated polymer layer to overcome the problem [Bharathan and Young, 1998]. In such a system, the uniform layer serves as a buffer to seal the pinholes and the inkjet-printed layer contains the desired patterns [Bharathan and Young, 1998].

30.6.5 Optical Component Fabrication and Integration

Integrated microoptics has become a revolutionized concept in the optics field, as it provides the advantages of low cost, miniaturization, improved spatial resolution and time response and reduction of the assembly process on optical systems, which is not possible by traditional means. As a result, fabricating and integrating miniaturized optical components with performance similar or even better than traditional components are critical issues in integrated microoptics systems. Standard bulk or surface micromachining provides various ways to fabricate active/passive micromirrors, wave-guides and Fresnel lenses, but fabrication of refractive lenses with curved surfaces is not easy. Compared to photolithography, which utilizes patterned and melted photoresist columns as lenses, the inkjet printing method provides more flexibility as far as the process, material choices and system integration. Cox et al. [Cox et al., 1994; Hays and Cox, 1998] employed inkjet printing technology to eject heated polymer material to fabricate a microlenslet array. The shape of the lens was controlled by the viscosity of the droplets at the impact point, the substrate wetting condition and the cooling/curing rate of the droplets [Hays et al., 1998]. A 70- to 150-µm-diameter lens with a density greater than 15,000/cm^2 has been successfully fabricated and

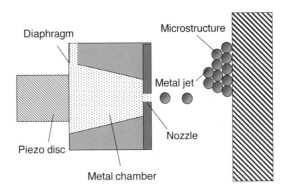

FIGURE 30.29 Operation principle of three-dimensional structure fabrication by microdroplet generator. (After Yamaguchi et al., 2000.)

has focal lengths between 50 and 150 μm. Besides the lens, wave-guides have also been demonstrated by Cox et al. (1994) using inkjet technology.

Because the optical components can be selectively deposited onto the desired region with varying properties, integration of those components with fabricated integrated circuit or other devices is possible and efficient.

30.6.6 Solid Freeforming

Two-dimensional patterns and three-dimensional solid structures can be generated by microdroplet generators. Orme and Huang (1993) and Marusak (1993) reported the application of molten metal drops for solid freeform fabrication. Evans' group demonstrated the application of continuous and drop-on-demand inkjets for ceramic printing to fabricate three-dimensional structures as well as functionally graded materials [Mott and Evans, 1999; Blasdell and Evans, 2000; Yamaguchi et al. 2000] used a metal jet to print functional three-dimensional microstructures, and Figure 30.29 shows the operation principle. Employing multijets for structure and sacrificial material deposition to print an overhanging structure was also proposed by Yamaguchi et al. (2000), while Fuller and Jacobson (2000) used laminated poly(methyl methacrylate) (PMMA) film as the supporting material for ejection of metal cantilever beams. The fabrication principle is shown in Figure 30.30.

30.6.7 Manufacturing Process

Droplet generators also provide novel material processing. For example, submicron ceramic particles can be plasma sprayed, as introduced by Blazdell and Juroda, for surface coating [Blazdell and Kuroda, 2000]. The operation principle is shown in Figure 30.31. A continuous-jet printer was used for droplet formation from ceramic solution. The produced ceramic stream was delivered into the hottest part of the plasma jet and then sprayed onto the working piece. The splats produced by the plasma spray are similar in morphology to those produced using conventional plasma spraying of a coarse powder but are significantly smaller in size, which may provide unique characteristics such as extension of solid-state solubility, refinement of grain size, formation of metastable phases and a high concentration of point defects [Blazdell and Kuroda, 2000].

30.6.8 Integrated Circuit Cooling

Conventionally, blowing fans and fins are widely used to cool integrated-circuit chips, especially for central processing unit (CPUs). Recently, as the heating power has increased greatly with increasing CPU size, more advanced methods, such as heat pipes, CPL and impinging air jets, have been introduced for quick heat removal. However, no matter how the designs improve, the limitation of heat removal ability

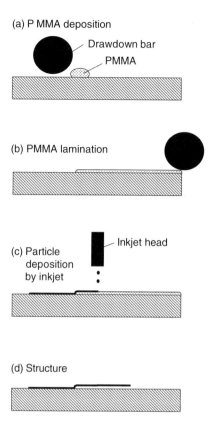

FIGURE 30.30 Fabrication of three-dimensional structures by droplet ejection and polymer lamination. (After Fuller and Jacobson, 2000.)

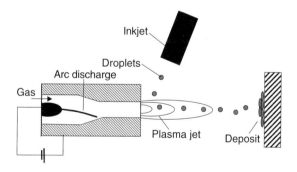

FIGURE 30.31 Operation principle of plasma spraying by microdroplet generators. (After Blazdell and Kuroda, 2000.)

for those devices is on the order of tens of watts per square centimeter. In addition, being able to detect hot spots and selectively remove the heat only from hot regions to preserve energy is highly desired but not easy to perform by traditional ways. As a result, the concept of transporting latent heat through the droplet evaporation process holds promise. This method can, in principle, remove three to four orders more heat than conventional methods. Also, the cooling spot can be selected and monitored through the integrated micro-temperature sensor and integrated circuit array. The conceptual design by Tseng (2001), as shown in Figure 30.32, used a two-dimensional array of microinjectors to selectively deposit liquid droplets onto the chip surface. The applied droplet frequency and numbers can be adjusted to maintain a dry chip surface with constant temperature. The estimated maximum heat removed by this device is

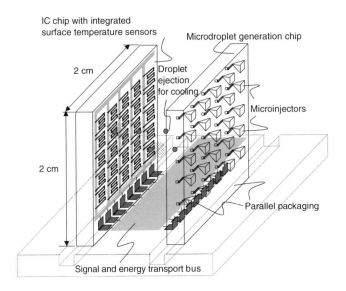

FIGURE 30.32 Conceptual design of integrated circuit cooling by microdroplet injections.

around 300,000 W/cm^2, more than 1000 times greater than by conventional means. Temperature sensors as well as a control circuit can also be fabricated on the same chip to form a self-contained smart system.

30.7 Concluding Remarks

The droplet generator is an important fluid-handling device for precise liquid-dosing control. MEMS technology makes microsized droplet generators possible and popular for many applications. Various methods of droplet generation, including piezoelectric, thermal bubble, thermal buckling, focused acoustic wave, electrostatic and inertial actuation, have been employed. Compared to other methods, the thermal bubble approach has greater actuation deformation, simpler design/fabrication and less limitation on the chamber volume, but it has the drawbacks of being temperature sensitivity and influenced by liquid properties. The piezo-type jet has the advantage of high frequency response, controllable droplet size, and no satellite drops, but it has the limitation of finite actuation deformation, thus limiting miniaturization of the chamber volume. The electrostatic and thermal buckling jets have size limitations similar to the piezo type. Despite the electrostatic generator having the benefit of low power consumption, both have limited frequency response due to size limitations. The acoustic-wave droplet generator, on the other hand, is not mature and stable enough for commercial applications, while the inertial actuation method offers limited miniaturization due to its operation principle. More types of microdroplet generators are under development and may one day replace the ones we have been using for decades.

Physical properties, design issues and manufacturing aspects are important concerns in the design and fabrication of microdroplet generators. The associated issues, including frequency response, cross-talk, satellite droplets, puddle formation, material selection and integration, require great care at the various design and fabrication levels. MEMS technology provides some of the key solutions to those practical issues.

Many aspects of the applications have made microdroplet generators important and exciting ever since their inception. Inkjet printing is the traditional application and has generated a significant amount of revenue in the printing market. Moreover, hundreds of applications are yet emerging, including bioreagent handling, fine chemical handling, drug delivery, direct writing, solid freeform, integrated circuit cooling and fuel injection, which show promising results and many potential markets. Many exciting applications of microdroplet generators are yet to be discovered.

References

Allen, R.R., Meyer, J.D., and Knight, W.R., (1985) "Thermodynamics and Hydrodynamics of Thermal Ink Jets," *Hewlett-Packard J.* **May**, pp. 21–27.

Asai, A. (1989), "Application of the Nucleation Theory to the Design of Bubble Jet Printers," *Jpn. J. Appl. Phys.* **28**, pp. 909–915.

Asai., A. (1991), "Bubble Dynamics in Boiling Under High Heat Flux Pulse Heating," *ASME J. Heat Transfer* **113**, pp. 973–979.

Asai, A. (1992), "Three-Dimensional Calculation of Bubble Growth and Drop Ejection in a Bubble Jet Printer," *Trans. ASME* **114**, pp. 638–641.

Asai, A., Hara, T., and Endo, I. (1987) "One Dimensional Model of Bubble Growth and Liquid Flow in Bubble Jet Printers," *Jpn. J. Appl. Phys.* **26**(10) pp. 1794–1801.

Asai, A., Hirasawa, S., and Endo, I. (1988) "Bubble Generation Mechanism in the Bubble Jet Recording Process," *J. Imaging Technics* **14**, pp. 120–124.

Asano, K., Funayama, Y., Yatsuzuka, K., and Higashiyama, Y. (1995) "Spherical Particle Sorting by Using Droplet Deflection Technology," *J. Electrostatics* **35**, pp. 3–12.

Ashley, C.T., Edds, K.E., and Elbert, D.L. (1977) "Development and Characterization of Ink for an Electrostatic Ink Jet Printer," *IBM J. Res. Develop.* **21**(1), pp. 69–74.

Askeland, R.-A., Childers W.-D., and Sperry, W.-R. (1988) "The Second-Generation Thermal InkJet Structure," *Hewlett-Packard J.* **Aug.**, pp. 28–31.

Bharathan, J., and Yang, Y. (1998) "Polymer Electroluminescent Devices Processed by Inkjet Printing. I. Polymer Light-Emitting Logo," *Appl. Phys. Lett.* **72**(21), pp. 2660–2662.

Bhaskar, E.V., and Aden, J.S. (1985) "Development of the Thin-film Structure for the Thinkjet Printhead," *Hewlett-Packard J.* **May**, pp. 27–33.

Blasdell, P.F., and Evans, J.R.G. (2000) "Application of a Continuous Ink Jet Printer to Solid Freeforming of Ceramics," *J. Mater. Proc. Technol.* **99**, pp. 94–102.

Blazdell, P., and Kuroda, S. (2000) "Plasma Spraying of Submicron Ceramic Suspensions Using a Continuous Ink Jet Printer," *Surface Coatings Technol.* **123**, pp. 239–246.

Bogy, D.B., and Talke, F.E. (1984) "Experimental and Theoretical Study of Wave Propagation Phenomena in Drop-on Demand Ink Jet Devices," *IBM J. Res. Develop.* **28**(3), pp. 314–321.

Brown, G.L., and Roshko, A. (1974) "On Density Effects and Large Structure in Turbulent Mixing Layers," *J. Fluid Mech.* **64**, pp. 775–816.

Buehner, W.L., Hill, J.D., Williams, T. H., and Woods, J.W. (1977) "Application of Ink Jet Technology to a Word Processing Output Printer," *IBM J. Res. Develop.* **21**(1), pp. 2–9.

Bugdayci, N., Bogy. D.B., and Talke, F.E. (1983) "Axisymmetric Motion of Radially Polarized Piezoelectric Cylinders Used in Ink jet Printing," *IBM J. Res. Develop.* **27**(2), pp. 171–180.

Buskirk, W.-A., Hackleman, D.-E., Hall, S.-T., Hanarek, P.-H., Low, R.-N., Trueba, K.-E., and Van de Poll, R.-R. (1988) "Development of a High-Resolution Thermal Inkjet Printhead," *Hewlett-Packard J.* **Oct**, pp. 55–61.

Carmichael, J.M. (1977) "Controlling Print Height in an Ink Jet Printer," *IBM J. Res. Develop.* **21**(1), pp. 52–55.

Chen, J.-K., and Wise, K.D. (1995) "A High-Resolution Silicon Monolithic Nozzle Array for Inkjet Printing," in *Proc. 8th Int. Conf. on Solid-State Sensors and Actuators and Eurosensors IX,* June 25–29, Stockholm, Sweden, pp. 321–324.

Chen, J.-K., Juan, W., Kubby, J., and Hseih, B.-C. (1998b) "A Monolithic Polyimide Nozzle Aray for Inkjet Printing," *Technical Digest IEEE Solid-State Sensor and Actuator Workshop,* Hilton Head Island, SC.

Chen, P.-H., Chen, W.-C., and Chang, S.-H. (1997a) "Bubble Growth and Ink Ejection Process of a Thermal Ink Jet Printhead," *Int. J. Mech. Sci.* **39**, pp. 687–695.

Chen, P.-H., Chen, W.-C., and Chang, S.-H. (1997b) "Visualization of Drop Ejection Process of a Thermal Bubble Ink Jet Printhead," in *Proc. 1st Pacific Symp. Flow Visualization and Image Processing,* Honolulu, HI, pp. 132–137.

Chen, P.-H., Chen, W.-C., Ding, P.-P., and Chang, S.-H. (1998a) "Droplet Formation of a Thermal Sideshooter Inkjet Printhead," *Int. J. Heat and Fluid Flow,* **19**, pp. 382–390.

Chen, P.-H., Peng, H.-Y., Liu, H.-Y., Chang, S.-L., Wu, T.-I., and Cheng C.-H. (1999) "Pressure Response and Droplet Ejection of a Piezoelectric Inkjet Printhead," *Int. J. Mech. Sci.* **41**, pp. 235–248.

Cox, W.R., Chen, T., Hayes, D.J., MacFarlane, D.L., Narayan, V., and Jatum, J.A., (1994) "Microjet Fabrication of Microlens Arrays," *IEEE Photonics Technol. Lett.* **6**(9), pp. 1112–1114.

Curry, S.A., and Portig, H. (1977) "Scale Model of an Ink Jet," *IBM J. Res. Develop.* **21**(1), pp. 10–20.

Darling, R.H., Lee, C.-H., and Kuhn, L. (1984) "Multiple-Nozzle Ink Jet Printing Experiment," *IBM J. Res. Develop.* **28**(3), pp. 300–306.

Fromm, J.E. (1984) "Numerical Calculation of the Fluid Dynamics of Drop-On-Demand Jets," *IBM J. Res. Develop.* **28**(3), pp. 322–333.

Fuller, S., and Jacobson, J. (2000) "Ink Jet Fabricated Nanoparticle MEMS," in *Proc. IEEE MEMS '2000,* Miyazaki, Japan, pp. 138–141.

Goldmann, T., and Gonzalez, J.S. (2000) "DNA Printing: Utilization of a Standard Inkjet Printer for the Transfer of Nucleic Acids to Solid Supports," *J. Biochem. Biophys. Methods* **42**, pp. 105–110.

Gruhler, H., Hey, N., Muller, M., Bekesi, S., Freygang, M., Sandmaier, H., and Zengerle, R. (1999), "Topspot—A New Method for the Fabrication of Biochips," *Proc. IEEE MEMS '99,* Orlando, FL, pp. 7413–417.

Hayes, D.J., and Cox, W.R. (1998) "Micro Jet Printing of Polymers for Electronics Manufacturing," in *Adhesive Joining and Coating Technology in Electronics Manufacturing, Proc. 3rd Int. Conf.,* pp. 168–173.

Hayes, D.J., Grove, M.E., and Cox, W.R. (1999) "Development and Application by Ink-Jet Printing of Advanced Packaging Materials," in *Proc. Int. Symp. on Advanced Packaging Materials,* pp. 88–93.

Hebner, T.R., Wu, C.C., Marcy, D., Lu, M.H., and Sturm, J.C. (1998) "Ink-Jet Printing of Doped Polymer for Organic Light Emitting Devices," *Applied Physics Letters,* **72**(5), pp. 519–521.

Hirata, S., Ishii, Y., Matoba, H., and Inui, T. (1996) "An Ink-Jet Head Using Diaphragm Microactuator," in *Proc. 9th IEEE Micro Electro Mechanical Systems Workshop,* San Diego, CA, pp. 418–423.

Ho, C.M., and Gutmark, E. (1987) "Vortex Induction and Mass Entrainment in a Small Aspect Ratio Elliptic Jet," *J. Fluid Mech.* **179**, pp. 383–405.

Ho, C.M., and Huang, L.S. (1982) "Subharmonics and Vortex Merging in Mixing Layers," *J. Fluid Mech.* **119**, pp. 443–473.

Kamisuki, S., Hagata, T., Tezuka, C., Nose, Y., Fuji, M., and Atobe, M. (1998) "A Low Power, Small, Electrostatically Driven Commercial Inkjet Head," in *Proc. IEEE MEMS '98,* Heidelberg, Germany, pp. 63–68.

Kamisuki, S., Fuji, M., Takekoshi, T., Tezuka, C., and Atobe, M. (2000) "A High Resolution, Electrostatically Driven Commercial Inkjet Head," in *Proc. IEEE MEMS '2000,* Miyazaki, Japan, pp. 793–798.

Karz, R.S., O'Neill, J.F., and Daneile, J.J. (1994) "Ink Jet Printhead with Ink Flow Direction Valves," U.S. Patent 5,278,585.

Keefe, B.-J., Ho, M.-F., Courian, K.-J., Steinfield, S.-W., Childers, W.-D., Tappon, E.-R., Trueba, K.-E., Chapman, T.-I., Knight, W.-R., Moritz, J.-G. (1997) "Inkjet Printhead Architecture for High Speed and High Resolution Printing," U.S. Patent 5,648,805.

Kobayashi, H. et al. (2000) "A Novel RGB Multicolor Light-Emitting Polymer Display," *Synthetic Metals* **111–112**, pp. 125–128.

Koide, A., Sasaki, Y., Yoshimura, Y., Miyake R., and Terayama, T. (2000) "Micromachined Dispenser with High Flow Rate and High Resolution," in *Proc. IEEE MEMS '2000,* Miyazaki, Japan, pp. 424–428.

Krause, P., Obermeier, E., and Wehl, W. (1995) "Backshooter—A New Smart Micromachined Single-Chip Inkjet Printhead," in *Proc. 8th Int. Conf. on Solid-State Sensors and Actuators and Eurosensors IX,* June 25–29, Stockholm, Sweden, pp. 325–328.

Lee, F.C., Mills, R.N., and Talke, F.E. (1984) "The Application of Drop-on-Demand Ink Jet Technology to Color Printing," *IBM J. Res. Develop.* **28**(3), pp. 307–313.

Lee, H.C. (1977) "Boundary Layer Around a Liquid Jet," *IBM J. Res. Develop.* **21**(1), pp. 48–51.

Lee, J.-D., Lee H.-D., Lee H.-J., Yoon, J.-B., Han, K.-H., Kim, J.-K., Kim, C.-K., and Han, C.-H. (1995) "A Monolithic Thermal Inkjet Printhead Utilizing Electrochemical Etching and Two-Step Electroplating Techniques," *IEEE IEOM 95–60L*, pp. 23.3.1–23.3.4.

Lee, J.-D., Yoon, J.-B., Kim, J.-K., Chung H.-J., Lee C.-S., Lee H.-D., Lee H.-J., Kim, C.-K., and Han, C.-H. (1999a) "A Thermal Inkjet Printhead with a Monolithically Fabricated Nozzle Plate and Self-Aligned Ink Feed Hole," *J. MEMS* **8**(3) pp. 229–236.

Lee, Y.-K., Yi, U., Tseng, F.-G., Kim, C.J., and Ho, C.-M. (1999b) "Diesel Fuel Injection by a Thermal Microinjector," in *Proc. MEMS, ASME IMECE '99*, Nashville, TN, pp. 419–425.

Levanoni, M. (1977) "Study of Fluid Flow Through Scaled-Up Ink Jet Nozzles," *IBM J. Res. Develop.* **21**(1), pp. 56–68.

Luginbuhl, P. et al. (1999) "Micromachined Injector for DNA Spectrometry," in *Proc. IEEE Transducers '99*, Sendai, Japan, pp. 1130–1133.

Marusak, R.E. (1993) "Picoliter Solder Droplet Dispensing," in *Proc. Solid Freeform Fabrication Symp.*, Austin, TX, pp. 81–87.

Meinhart, C.-D., and Zhang H. (2000) "The Flow Structure Inside a Microfabricated Inkjet Printhead," *J. MEMS* **9**(1), pp. 67–75.

Miliotis, T., Eficsson, D., Marko-Varga, G., Ekstrom, S., Nilsson, J., and Laurell, T. (2000) "Interfacing Protein and Peptide Separation to Maldi-tof MS Using Microdispensing and On-Target Enrichment Strategies," in *Proc. µTAS '2000*, Enschede, The Netherlands, pp. 387–391.

Mirfakhraee, A. (1989) "Growth and Collapse of Vapor Bubbles in Ink-Jet Printers," Ph.D. thesis, University of California, Berkeley.

Mott, M., and Evans, J.R.G. (1999) "Zirconia/Alumina Functionally Graded Material Made by Ceramic Ink Jet Printing," *Mater. Sci. Eng.* **A271**, pp. 344–352.

Myers, R.A., and Tamulis, J.C. (1984) "Introduction to Topical Issue on Non-Impact Printing Technologies," *IBM J. Res. Develop.* **28**(3), pp. 234–240.

Nielsen, N.J. (1985) "History of ThinkJet Printhead Development," *Hewlett-Packard J.* **May**, pp. 4–10.

Nilsson, J., Bergkvist, J., and Laurell, T. (2000) "Optimization of the Droplet Formation in a Piezo-Electric Flow-Through Microdispenser," in *Proc. µTAS '2000*, Enschede, The Netherlands, pp. 75–78.

Orme, M., and Huang, C. (1993) "Thermal Design Parameters Critical to The Development of Solid Freefrom Fabrication of Structural Materials with Controlled Nanoliter Droplets," in *Proc. Solid Freeform Fabrication Symp.*, Austin, TX, pp. 88–94.

Pimbley, W.T., and Lee, H.C. (1977) "Satellite Droplet Formation in a Liquid Jet," *IBM J. Res. Develop.* **21**(1), pp. 21–30.

Rembe, C., Siesche, S.A.D., and Hofer, E.P. (2000) "Thermal Ink Jet Dynamics: Modeling, Simulation, and Testing," *Microelectron. Reliability* **40**, pp. 525–532.

Shadow, K.C., Gutmark, E., Wilson, K.J., Parr, D.M., Mahan, V.A., and Ferrell, G.B. (1987) "Effect of Shear-Flow Dynamics in Combustion Processes," *Combustion Sci. Technol.* **54**, pp. 103–116.

Shah, P., Kevrekides, Y., and Benziger, J. (1999), "Ink-Jet Printing of Catalyst Patterns for Electroless Metal Deposition," *Langmuir* **15**, pp. 1584–1587.

Shimoda, T., Kimura M., Seki S., Kobayashi, H., Kanbe, S., and Miyahita, S. (1999) "Technology for Active Matrix Light Emitting Polymer Displays," in *Proc. IEEE IEDM '99*, pp. 107–110.

Siewell, G.-L., Boucher, W.R., and McClelland, P.H. (1985) "The ThinkJet Orifice Plate: A Part with Many Functions," *Hewlett-Packard J.* **May**, pp. 33–37.

Somberg, H. (1990) "Inkjet Printing for Metallization on Very Thin Solar Cells," in *Conf. Record of the 21st IEEE in Photovoltaic Specialists Conf.*, pp. 666–667.

Sweet, R.G. (1964) "High Frequency Recording with Electrostatically Deflected Ink Jets," *Stanford Electronics Laboratories Technical Report No. 1722-1*, Stanford University, Stanford, CA.

Sweet, R.G. (1971) "Fluid Droplet Recorder," U.S. Patent 3,576,275.

Szita, N., Sutter, R., Dual, J., and Buser, R. (2000) "A Fast and Low-Volume Pipettor with Integrated Sensors for High Precision," in *Proc. IEEE MEMS '2000*, Miyazaki, Japan, pp. 409–413.

Ta, C.-C., Chan, L.-W., Wield, P.-J., and Nevarez R. (1988). "Mechanical Design of Color Graphics Printer," *Hewlett-Packard J.* **Aug.**, pp. 21–27.

Teng, K.F., and Vest, R.W. (1988a) "Metallization of Solar Cells with Ink Jet Printing and Silver Metallo-Organic Inks," *IEEE Trans. Components, Hybrids, and Manuf. Technol.* **11**(3), pp. 291–297.

Teng, K.F., and Vest R.W. (1988b) "Application of Ink Jet Technology on Photovoltaic Metallization," *IEEE Electron. Device Lett.* **9**(11).

Teng, K.F., Azadpour, M.A., and Yang, H.Y. (1988) "Rapid Prototyping of Multichip Packages Using Computer-Controlled, Ink Jet Direct-Write," in *Proc. 38th Electronics Components Conf.*, pp. 168–173.

Tseng, F.-G. (1998) "A Micro Droplet Generation System," Ph.D. thesis, University of California, Los Angeles.

Tseng, F.-G. (1999) Presentation material.

Tseng, F.-G. (2001) "Droplet Impinging Micro Cooling Arrays," proposal for NSC project.

Tseng, F.-G., Linder, C., Kim, C.-J., and Ho, C.-M (1996) "Control of Mixing with Micro Injectors for Combustion Application," in *Proc. Micro-Electro-Mechanical Systems* (MEMS '96), DSC-Vol.59, Atlanta, GA, pp. 183–187.

Tseng, F.-G., Kim, C.-J., and Ho, C.-M (1998a) "A Microinjector Free of Satellite Drops and Characterization of the Ejected Droplets," in *Proc. MEMS, ASME IMECE '98*, Anaheim, CA, pp. 89–95.

Tseng, F.-G., Shih, C., Kim, C.-J., and Ho, C.-M (1998b) "Characterization of Droplet Injection Process of a Microinjector," in *Abstract Book of 13th U.S. Congress of Applied Mechanics*, University of Florida, Gainesville, p. TB3.

Tseng, F.-G., Kim, C.-J., and Ho, C.-M (1998c) "A Novel Microinjector with Virtual Chamber Neck," in *Technical Digest of the 11th IEEE Int. Workshop on Micro Electro Mechanical Systems*, Heidelberg, Germany, pp. 57–62.

Tseng, F.-G., Kim, C.-J., and Ho, C.-M. (2000) "Apparatus and Method for Using Bubble as Virtual Valve in Micro Injector to Eject Fluid," U.S. and international patent (pending).

Tseng, F.-G., Kim, C.-J., and Ho, C.-M. (2001a) "A Monolithic, High Frequency Response, High-Resolution Microinjector Array Ejecting Sub Picoliter Droplets without Satellite Drops. Part I. Concepts, Designs and Molding," submitted to *J. MEMS*.

Tseng, F.-G., Kim, C.-J., and Ho, C.-M. (2001b) "A Monolithic, High Frequency Response, High-Resolution Microinjector array Ejecting Sub Picoliter Droplets without Satellite Drops. Part II. Fabrication, Characterization and Performance Comparison," submitted to *J. MEMS*.

Twardeck, T.G. (1977) "Effect of Parameter Variations on Drop Placement in an Electrostatic Ink Jet Printer," *IBM J. Res. Develop.* **21**(1), pp. 31–36.

Wang, X.Q., Desai, A., Tai, Y.C., Licklider, L., and Lee, T.D. (1999) "Polymer-Based Electrospray Chips for Mass Spectrometry," in *Proc. IEEE MEMS '99*, Orlando, FL, pp. 523–528.

Yamaguchi, K., Sakai, K., Ymanaka, T., and Hirayama, T. (2000) "Generation of Three-Dimensional Micro Structure Using Metal Jet," *Precision Eng.* **24**, pp. 2–8.

Zhu, X., Tran, E., Wang, W., Kim, E.S., and Lee, S.Y. (1996) "Micromachined Acoustic-Wave Liquid Ejector," in *Proc. Solid-State Sensor and Actuator Workshop*, June 2–6, Hilton Head, SC, pp. 280–282.

31
Micro Heat Pipes and Micro Heat Spreaders

G. P. "Bud" Peterson
Rensselaer Polytechnic Institute

31.1 Introduction 31-1
 Capillary Limitation • Viscous Limitation • Sonic
 Limitation • Entrainment Limitation • Boiling
 Limitation • Heat Pipe Thermal Resistance
31.2 Individual Micro Heat Pipes 31-8
 Modeling Micro Heat Pipe Performance • Testing of
 Individual Micro Heat Pipes
31.3 Arrays of Micro Heat Pipes 31-12
 Modeling of Heat Pipe Arrays • Testing of Arrays of Micro
 Heat Pipes • Fabrication of Arrays of Micro Heat Pipes
31.4 Flat-Plate Micro Heat Spreaders 31-15
 Modeling of Micro Heat Spreaders • Testing of Micro Heat
 Spreaders • Fabrication of Micro Heat Spreaders
31.5 New Designs 31-18
31.6 Summary and Conclusions 31-22

31.1 Introduction

As described by Peterson (1994), a heat pipe operates on a closed two-phase cycle in which heat added to the evaporator region causes the working fluid to vaporize and move to the cooler condenser region, where the vapor condenses, giving up its latent heat of vaporization. In traditional heat pipes, the capillary forces existing in a wicking structure pump the liquid back to the evaporator. While the concept of utilizing a wicking structure as part of a device capable of transferring large quantities of heat with a minimal temperature drop was first introduced by Gaugler (1944), it was not until much more recently that the concept of combining phase-change heat transfer and microscale fabrication techniques (i.e., microelectromechanical systems, or MEMS, devices for the dissipation and removal of heat) was first proposed by Cotter (1984). This initial introduction envisioned a series of very small "micro" heat pipes incorporated as an integral part of semiconductor devices. While no experimental results or prototypes designs were presented, the term *micro heat pipe* was first defined as one "so small that the mean curvature of the liquid–vapor interface is necessarily comparable in magnitude to the reciprocal of the hydraulic radius of the total flow channel" [Babin et al., 1990]. Early proposed applications of these devices included the removal of heat from laser diodes (Mrácek, 1988) and other small localized heat-generating devices [Peterson, 1988a; 1988b]; thermal control of photovoltaic cells [Peterson, 1987a; 1987b]; removal or dissipation of heat from the leading edge of hypersonic aircraft [Camarda et al., 1997]; applications involving the nonsurgical treatment of cancerous tissue through either hyper- or hypothermia [Anon., 1989; Fletcher and Peterson, 1993]; and space applications in which heat pipes are embedded in silicon radiator panels to dissipate the large amounts of waste heat generated [Badran et al., 1993].

While not all of these applications have been implemented, micro heat pipes ranging in size from 1 mm in diameter and 60 mm in length to 30 μm in diameter and 10 mm in length have been analyzed, modeled and fabricated, and the larger of these are currently commonplace in commercially available products, such as laptop computers or high-precision equipment where precise temperature control is essential. More recently, this work has been expanded to include micro heat spreaders fabricated in silicon or in new metallized polymeric materials, which can be used to produce highly conductive, flexible heat spreaders capable of dissipating extremely high heat fluxes over large areas, thereby reducing the source heat flux by several orders of magnitude.

Since the initial introduction of the micro heat pipe concept, the study of microscale heat transfer has grown enormously and has encompassed not only phase-change heat transfer, but also the entire field of heat transfer, fluid flow and, particularly for a large number of fundamental studies, thin film behavior, as described elsewhere in this book. Microscale fluid behavior and heat transfer at the microscale, along with the variations between the behavior of bulk thermophysical properties and those that exist at the micro- or nanoscale levels are all areas of considerable interest. While the division between micro- and macroscale phase-change behavior is virtually indistinguishable, in applications involving phase-change heat transfer devices, such as micro heat pipes and micro heat spreaders, it can best be described by applying the dimensionless expression developed by Babin and Peterson (1990) and described later in this chapter. This expression relates the capillary radius of the interface and the hydraulic radius of the passage and provides a good indicator of when the forces particular to the microscale begin to dominate.

A number of previous reviews have summarized the literature published prior to 1996 [Peterson and Ortega, 1990; Peterson, 1992; Peterson et al., 1996]; however, significant advances have been made over the past several years, particularly in the development of a better understanding of the thin-film behavior that governs the operation of these devices. The following review begins with a very brief overview of the early work in this area and then looks at advances made in individual micro heat pipes, arrays of micro heat pipes and more recent investigations of flat-plate microscale heat spreaders.

For heat pipes operating in steady state, there are a number of fundamental mechanisms that limit the maximum heat transfer. These have been summarized and described in a concise format (which will be summarized here) by Marto and Peterson (1988) and include the capillary wicking limit, viscous limit, sonic limit, entrainment and boiling limits. The first two of these deal with the pressure drops occurring in the liquid and vapor phases. The sonic limit results from pressure-gradient-induced vapor velocities that may result in choked vapor flow, while the entrainment limit focuses on the entrainment of liquid droplets in the vapor stream, which inhibits the return of the liquid to the evaporator and ultimately leads to dryout. Unlike these limits, which depend upon the axial transport, the boiling limit is reached when the heat flux applied in the evaporator portion is high enough that nucleate boiling occurs in the evaporator wick, creating vapor bubbles that partially block the return of fluid.

While a description of the transient operation and start-up dynamics of these devices is beyond the scope of this work, it is appropriate to include a brief description of the methods for determining the steady-state limitations. For additional information on the theory and fundamental phenomena that cause each of these limitations, readers are referred to Tien et al. (1975), Chi (1976), Dunn and Reay (1982) and Peterson (1994).

31.1.1 Capillary Limitation

The operation and performance of heat pipes are dependent on many factors, including the shape, working fluid and wick structure. The primary mechanism by which these devices operate results from the difference in the capillary pressure across the liquid–vapor interfaces in the evaporator and condenser. To operate properly, this pressure difference must exceed the sum of all the pressure losses throughout the liquid and vapor flow paths. This relationship can be expressed as:

$$\Delta P_c \geq \Delta P_+ + \Delta P_- + \Delta P_l + \Delta P_v \tag{31.1}$$

where

ΔP_c = net capillary pressure difference.
ΔP_+ = normal hydrostatic pressure drop.
ΔP_- = axial hydrostatic pressure drop.
ΔP_l = viscous pressure drop occurring in the liquid phase.
ΔP_v = viscous pressure drop occurring in the vapor phase.

As long as this condition is met, liquid is returned to the evaporator. For situations where the summation of the viscous pressure losses, ΔP_l and ΔP_v, and the hydrostatic pressure losses, ΔP_+ and ΔP_-, is greater than the capillary pressure difference between the evaporator and condenser, the wicking structure becomes starved of liquid and dries out. This condition, referred to as the capillary wicking limitation, varies according to the wicking structure, working fluid, evaporator heat flux and operating temperature.

31.1.1.1 Capillary Pressure

The capillary pressure difference at a liquid–vapor interface, ΔP_c, is defined by the LaPlace–Young equation, which for most heat pipe applications reduces to:

$$\Delta P_{c,m} = \left(\frac{2\sigma}{r_{c,e}}\right) - \left(\frac{2\sigma}{r_{c,c}}\right) \tag{31.2}$$

where $r_{c,e}$ and $r_{c,c}$ represent the radii of curvature in the evaporator and condenser regions, respectively.

During normal heat pipe operation, the vaporization occurring in the evaporator causes the liquid meniscus to recede into the wick, reducing the local capillary radius, $r_{c,e}$, while condensation in the condenser results in increases in the local capillary radius, $r_{c,c}$. It is this difference in the two radii of curvature that "pumps" liquid from the condenser to the evaporator. During steady-state operation, it is generally assumed that the capillary radius in the condenser, $r_{c,c}$, approaches infinity, so that the maximum capillary pressure for a heat pipe operating at steady state can be expressed as a function of only the capillary radius of the evaporator wick,

$$\Delta P_{c,m} = \left(\frac{2\sigma}{r_{c,e}}\right) \tag{31.3}$$

Values for the effective capillary radius, $r_{c,e}$, can be found theoretically for simple geometries [Chi, 1976] or experimentally for other more complicated structures.

31.1.1.2 Hydrostatic Pressure Drops

The normal and axial hydrostatic pressure drops, ΔP_+ and ΔP_-, are the result of the local gravitational body force. The normal and axial hydrostatic pressure drops can be expressed as:

$$\Delta P_+ + \rho_l g d_v \cos \psi \tag{31.4}$$

and

$$\Delta P_- = \rho_l g L \sin \psi \tag{31.5}$$

where ρ_l is the density of the liquid, g is the gravitational acceleration, d_v is the diameter of the vapor portion of the pipe, Ψ is the angle the heat pipe makes with respect to the horizontal, and L is the length of the heat pipe.

In a gravitational environment, the axial hydrostatic pressure term may either assist or hinder the capillary pumping process depending upon whether the tilt of the heat pipe promotes or hinders the flow of liquid back to the evaporator (i.e., the evaporator lies either below or above the condenser). In a zero-g environment or for cases where the surface tension forces dominate, such as micro heat pipes, both of these terms can be neglected.

31.1.1.3 Liquid Pressure Drop

As the liquid returns from the condenser to the evaporator, it experiences a viscous pressure drop, ΔP_l, which can be written in terms of the frictional drag,

$$\frac{dP_l}{dx} = -\frac{2\tau_l}{(r_{h,l})} \tag{31.6}$$

where τ_l is the frictional shear stress at the liquid–solid interface and $r_{h,l}$ is the hydraulic radius, defined as twice the cross-sectional area divided by the wetted perimeter.

This pressure gradient is a function of the Reynolds number, Re_l, and drag coefficient, f_l, defined as:

$$Re_l = \frac{2(r_{h,l})\rho_l V_l}{\mu_l} \tag{31.7}$$

and

$$f_l = \frac{2\tau_l}{\rho_l V_l^2} \tag{31.8}$$

respectively, where V_l is the local liquid velocity which is related to the local heat flow,

$$V_l = \frac{q}{\varepsilon A_w \rho_l \lambda} \tag{31.9}$$

A_w is the wick cross-sectional area, ε is the wick porosity, and λ is the latent heat of vaporization.

Combining these expressions yields an expression for the pressure gradient in terms of the Reynolds number, drag coefficient and the thermophysical properties:

$$\frac{dP_l}{dx} = \left(\frac{(f_l Re_l)\mu_l}{2\varepsilon A_w (r_{h,l})^2 \lambda \rho_l}\right) q \tag{31.10}$$

which in turn can be written as a function of the permeability, K, as:

$$\frac{dP_l}{dx} = \left(\frac{\mu_l}{K A_w \lambda \rho_l}\right) q \tag{31.11}$$

where the permeability expressed as:

$$K = \frac{2\varepsilon (r_{h,l})^2}{f_l Re_l} \tag{31.12}$$

For steady-state operation with constant heat addition and removal, Eq. (31.11) can be integrated over the length of the heat pipe to yield:

$$\Delta P_l = \left(\frac{\mu_l}{K A_w \lambda \rho_l}\right) L_{\text{eff}}\, q \tag{31.13}$$

where L_{eff} is the effective heat pipe length defined as:

$$L_{\text{eff}} = 0.5 L_e + L_a + 0.5 L_c \tag{31.14}$$

31.1.1.4 Vapor Pressure Drop

The methods for calculating the vapor pressure drop in heat pipes is similar to that used for the liquid pressure drop described above but is complicated by the mass addition and removal in the evaporator and condenser, respectively, and by the compressibility of the vapor phase. As a result, accurate computation of the total pressure drop requires that the dynamic pressure be included. In-depth discussions of the methodologies for determining the overall vapor pressure drop have been presented previously by Chi (1976), Dunn and Reay (1982) and Peterson (1994). The resulting expression is similar to that developed for the liquid:

$$\Delta P_v = \left(\frac{C(f_v Re_v) \mu_v}{2(r_{h,v})^2 A_v \rho_v \lambda} \right) L_{eff} \, q \tag{31.15}$$

where $(r_{h,v})$ is the hydraulic radius of the vapor space and C is a constant that depends on the Mach number.

Unlike the liquid flow, which is driven by the capillary pressure difference and hence is always laminar, the vapor flow is driven by the temperature gradient and for high heat-flux applications may result in turbulent flow conditions. As a result, it is necessary to determine the vapor flow regime as a function of the heat flux by evaluating the local axial Reynolds number, defined as:

$$Re_v = \frac{2(r_{h,v}) q}{A_v \mu_v \lambda} \tag{31.16}$$

Due to compressibility effects, it is also necessary to determine if the flow is compressible. This is accomplished by evaluating the local Mach number, defined as:

$$Ma_v = \left(\frac{q}{A_v \rho_v \lambda (R_v T_v \gamma_v)} \right)^{1/2} \tag{31.17}$$

where R_v is the gas constant, T_v is the vapor temperature, and γ_v is the ratio of specific heats, which is equal to 1.67, 1.4 or 1.33 for monatomic, diatomic and polyatomic vapors, respectively (Chi, 1976).

Previous investigations summarized by Kraus and Bar-Cohen (1983) have demonstrated that the following combinations of these conditions can be used with reasonable accuracy:

$$\begin{aligned} Re_v &< 2300, \quad Ma_v < 0.2 \\ (f_v Re_v) &= 16 \\ C &= 1.00 \end{aligned} \tag{31.18}$$

$$\begin{aligned} Re_v &< 2300, \quad Ma_v > 0.2 \\ (f_v Re_v) &= 16 \\ C &= \left[1 + \left(\frac{\gamma_v - 1}{2} \right) Ma_v^2 \right]^{-1/2} \end{aligned} \tag{31.19}$$

$$\begin{aligned} Re_v &> 2300, \quad M_v < 0.2 \\ (f_v Re_v) &= 0.038 \left(\frac{2(r_{h,v}) q}{A_v \mu_v \lambda} \right)^{3/4} \\ C &= 1.00 \end{aligned} \tag{31.20}$$

$$Re_v > 2300, \quad Ma_v > 0.2$$

$$(f_v Re_v) = 0.038\left(\frac{2(r_{h,v})q}{A_v \mu_v \lambda}\right)^{3/4} \quad (31.21)$$

$$C = \left[1 + \left(\frac{\gamma_v - 1}{2}\right)Ma_v^2\right]^{-1/2}$$

The solution procedure is to first assume laminar, incompressible flow and then to compute the Reynolds and Mach numbers. Once these values have been found, the initial assumptions of laminar, incompressible flow can be evaluated and the appropriate modifications made.

31.1.2 Viscous Limitation

At very low operating temperatures, the vapor pressure difference between the closed end of the evaporator (the high-pressure region) and the closed end of the condenser (the low-pressure region) may be extremely small. Because of this small pressure difference, the viscous forces within the vapor region may prove to be dominant and hence limit the heat pipe operation. Dunn and Reay (1982) discuss this limit in more detail and suggest the criterion:

$$\frac{\Delta P_v}{P_v} < 0.1 \quad (31.22)$$

for determining when this limit might be of a concern. For steady-state operation, or applications in the moderate operating temperature range, the viscous limitation will normally not be important.

31.1.3 Sonic Limitation

The sonic limitation in heat pipes is the result of vapor velocity variations along the length of the heat pipe due to the axial variation of the vaporization and condensation. Much like the effect of decreased outlet pressure in a converging–diverging nozzle, decreased condenser temperature results in a decrease in the evaporator temperature up to but not beyond that point where choked flow occurs in the evaporator, causing the sonic limit to be reached. Any further decreases in the condenser temperature do not reduce the evaporator temperature or the maximum heat transfer capability, due to the existence of choked flow.

The sonic limitation in heat pipes can be determined as:

$$q_{s,m} = A_v \rho_v \lambda \left(\frac{\gamma_v R_v T_v}{2(\gamma_v + 1)}\right)^{1/2} \quad (31.23)$$

where T_v is the mean vapor temperature within the heat pipe [Chi, 1976].

31.1.4 Entrainment Limitation

In an operating heat pipe, the liquid and vapor typically flow in opposite directions, resulting in a shear stress at the interface. At very high heat fluxes, liquid droplets may be picked up or entrained in the vapor flow. This entrainment results in dryout of the evaporator wick due to excess liquid accumulation in the condenser. The Weber number, We, which represents the ratio of the viscous shear force to the force resulting from the liquid surface tension, can be used to determine at what point this entrainment is likely to occur:

$$We = \frac{2(r_{h,w})\rho_v V_v^2}{\sigma} \quad (31.24)$$

To prevent the entrainment of liquid droplets in the vapor flow, the Weber number must therefore be less than one, which implies that the maximum transport capacity based on the entrainment limitation may be determined as:

$$q_{e,m} = A_v \lambda \left(\frac{\sigma \rho_v}{2(r_{h,w})} \right)^{1/2} \quad (31.25)$$

where $(r_{h,w})$ is the hydraulic radius of the wick structure (Dunn and Reay, 1983).

31.1.5 Boiling Limitation

As mentioned, all of the limits discussed previously depend upon the axial heat transfer. The boiling limit, however, depends upon the evaporator heat flux and occurs when the nucleate boiling in the evaporator wick creates vapor bubbles that partially block the return of fluid. The presence of vapor bubbles in the wick requires both the formation of bubbles and also the subsequent growth of these bubbles. Chi (1976) has developed an expression for the boiling limit, which can be written as

$$q_{b,m} = \left(\frac{2\pi L_{eff} k_{eff} T_v}{\lambda \rho_v \ln(r_i/r_v)} \right) \left(\frac{2\sigma}{r_n} - \Delta P_{c,m} \right) \quad (31.26)$$

where k_{eff} is the effective thermal conductivity of the liquid–wick combination, r_i is the inner radius of the heat pipe wall, and r_n is the nucleation site radius [Dunn and Reay, 1982].

31.1.6 Heat Pipe Thermal Resistance

Once the maximum transport capacity is known, it is often useful to determine the temperature drop between the evaporator and condenser. The overall thermal resistance for a heat pipe is comprised of nine resistances of significantly different orders of magnitude, arranged in a series/parallel combination. These resistances can be summarized as follows:

R_{pe} = radial resistance of the pipe wall at the evaporator.
R_{we} = resistance of the liquid–wick combination at the evaporator.
R_{ie} = resistance of the liquid–vapor interface at the evaporator.
R_{va} = resistance of the adiabatic vapor section.
R_{pa} = axial resistance of the pipe wall.
R_{wa} = axial resistance of the liquid–wick combination.
R_{ic} = resistance of the liquid–vapor interface at the condenser.
R_{wc} = resistance of the liquid–wick combination at the condenser.
R_{pc} = radial resistance of the pipe wall at the condenser.

Previous investigations have indicated that typically the resistance of the vapor space, the axial resistances of the pipe wall and liquid–wick combinations, can all be neglected. In addition, the liquid–vapor interface resistances and the axial vapor resistance can, in most situations, be assumed to be negligible. This leaves only the pipe wall radial resistances and the liquid–wick resistances at both the evaporator and condenser.

As presented by Peterson (1994), the radial resistances at the pipe wall can be computed from Fourier's law as:

$$R_{pe} = \frac{\delta}{k_p A_e} \quad (31.27)$$

for flat plates, where δ is the plate thickness and A_e is the evaporator area, or

$$R_{pe} = \frac{\ln(d_o/d_i)}{2\pi L_e k_p} \qquad (31.28)$$

for cylindrical pipes, where L_e is the evaporator length. An expression for the equivalent thermal resistance of the liquid–wick combination in circular pipes is

$$R_{we} = \frac{\ln(d_o/d_i)}{2\pi L_e k_{eff}} \qquad (31.29)$$

where k_{eff} is the effective conductivity of the liquid–wick combination.

Combining these individual resistances allows the overall thermal resistance to be determined, which when combined with the maximum heat transport found previously will yield an estimation of the overall temperature drop.

31.2 Individual Micro Heat Pipes

The earliest embodiments of micro heat pipes typically consisted of a long thin tube with one or more small noncircular channels that utilized the sharp-angled corner regions as liquid arteries. While initially quite novel in size (see Figure 31.1), it was soon apparent that devices with characteristic diameters of approximately 1 mm functioned in nearly the same manner as larger, more conventional liquid artery heat pipes. Heat applied to one end of the heat pipe vaporizes the liquid in that region and forces it to move to the cooler end, where it condenses and gives up the latent heat of vaporization. This vaporization and condensation process causes the liquid–vapor interface in the liquid arteries to change continually along the pipe, as illustrated in Figure 31.2 and results in a capillary pressure difference between the evaporator and condenser regions. This capillary pressure difference promotes the flow of the working fluid from the condenser back to the evaporator through the triangular-shaped corner regions. These corner regions serve as liquid arteries, thus no wicking structure is required [Peterson, 1990; 1994]. The following sections present a summary of the analytical and experimental investigations conducted on individual micro heat pipes, arrays of micro heat pipes, flat-plate microscale heat spreaders, and the latest advances in the development of highly conductive, flexible, phase-change heat spreaders.

31.2.1 Modeling Micro Heat Pipe Performance

The first steady-state analytical models of individual micro heat pipes utilized the traditional pressure-balance approach developed for use in more conventional heat pipes and described earlier in this chapter. These models provided a mechanism by which the steady-state and transient performance characteristics of micro heat pipes could be determined and indicated that, while the operation was similar to that observed in larger more conventional heat pipes, the relative importance of many of the parameters is quite different. Perhaps the most significant difference was the relative sensitivity of the micro heat pipes

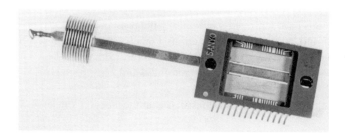

FIGURE 31.1 Micro-heat-pipe-cooled ceramic chip carrier. (From Peterson, G.P. (1994) *An Introduction to Heat Pipes: Modeling, Testing and Applications,* John Wiley & Sons, New York. With permission.)

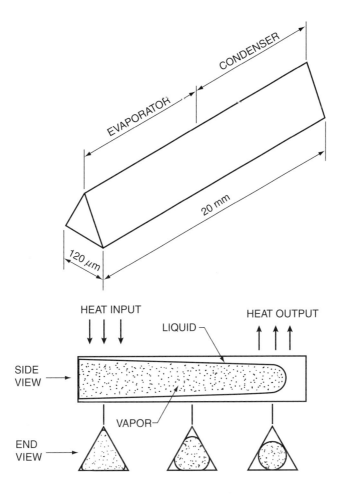

FIGURE 31.2 Micro heat pipe operation. (From Peterson, G.P. et al. (1996) in *Microscale Energy Transport* (C.L. Tien et al., Eds.), Taylor & Francis, Washington, D.C. Reproduced by permission of Routledge, Inc.)

to the amount of working fluid present. These early steady-state models later led to the development of both transient numerical models and three-dimensional numerical models of the temperature and pressure distribution within individual micro heat pipes [Peterson, 1992; 1994; Peterson et al., 1996].

31.2.1.1 Steady-State Modeling

The first steady-state model specifically designed for use in the modeling of micro heat pipes was developed by Cotter (1984). Starting with the momentum equation and assuming uniform cross-sectional area and no-slip conditions at the boundaries, this expression was solved for both the liquid and vapor pressure differentials and then combined with the continuity expression. The result was a first-order ordinary differential equation, which related the radius of curvature of the liquid–vapor interface to the axial position along the pipe. Building upon this model, Peterson (1988a) and Babin et al. (1990) developed a steady-state model for a trapezoidal micro heat pipe, using the conventional steady-state modeling techniques outlined by Chi (1976) and described earlier in this chapter. The resulting model demonstrated that the capillary pumping pressure governed the maximum heat transport capacity of these devices.

A comparison of the performance limitations resulting from the models presented by Cotter (1984) and by Babin et al. (1990) indicated significant differences in the capillary limit predicted by the two models. These differences have been analyzed and found to be the result of specific assumptions made in the initial formulation of the models [Peterson, 1992].

A comparative analysis of these two early models was performed by Gerner et al. (1992), who indicated that the most important contributions of Babin et al. (1990) were the inclusion of the gravitational body force and the recognition of the significance of the vapor pressure losses. In addition, the assumption that the pressure gradient in the liquid flow passages was similar to that occurring in Hagen–Poiseuille flow was questioned, and a new scaling argument for the liquid pressure drop was presented. In this development, it was assumed that the average film thickness was approximately one fourth the hydraulic radius, resulting in a modified expression for the capillary limitation.

A significant contribution made by Gerner et al. (1992) was the recognition that the capillary limit may never actually be reached due to Kelvin–Helmholtz-type instabilities occurring at the liquid–vapor interface. Using stability analysis criteria for countercurrent flow in tubes developed by Tien et al. (1979) and minimizing the resulting equations, the wavelength was found to be approximately 1 cm for atmospheric water and methanol. Because this length was long with respect to the characteristic wavelength, it was assumed that gravity was the dominant stabilizing mechanism. The decision as to whether to use the traditional capillary limit proposed by Babin et al. (1990) or the interfacial instability limit proposed by Gerner et al. (1992) should be governed by evaluating the shape and physical dimensions of the specific micro heat pipe being considered.

31.2.1.2 Transient Modeling

As heat pipes diminish in size, the transient nature becomes of increasing interest. The ability to respond to rapid changes in heat flux coupled with the need to maintain constant evaporator temperature in modern high-powered electronics necessitates a complete understanding of the temporal behavior of these devices. The first reported transient investigation of micro heat pipes was conducted by Wu and Peterson (1991). This initial analysis utilized the relationship developed by Collier (1981) and was used later by Colwell and Chang (1984) to determine the free-molecular-flow mass flux of evaporation. The most interesting result from this model was the observation that reverse liquid flow occurred during the startup of micro heat pipes. As explained in the original reference [Wu et al., 1991], this reverse liquid flow is the result of an imbalance in the total pressure drop and occurs because the evaporation rate does not provide an adequate change in the liquid–vapor interfacial curvature to compensate for the pressure drop. As a result, the increased pressure in the evaporator causes the meniscus to recede into the corner regions, forcing liquid out of the evaporator and into the condenser. During startup, the pressure of both the liquid and vapor is higher in the evaporator and gradually decreases with position, promoting flow away from the evaporator. Once the heat input reaches full load, the reverse liquid flow disappears and the liquid mass flow rate into the evaporator gradually increases until a steady-state condition is reached. At this time, the change in the liquid mass flow rate is equal to the change in the vapor mass flow rate for any given section [Wu and Peterson, 1991].

Several, more detailed transient models have been proposed. Badran et al. (1993) developed a conjugate model to account for the transport of heat within the heat pipe and conduction within the heat pipe case. This model indicated that the specific thermal conductivity of micro heat pipes (effective thermal conductivity divided by the density) could be as high as 200 times that of copper and 100 times that of Gr/Cu composites. Longtin et al. (1994) developed a one-dimensional, steady-state model that indicated that the maximum heat transport capacity varied with respect to the cube of the hydraulic diameter, and Khrustalev and Faghri (1994) presented a detailed mathematical model of the heat- and mass-transfer processes in micro heat pipes which described the distribution of the liquid and the thermal characteristics as a function of the liquid charge. The liquid flow in the triangular-shaped corners of a micro heat pipe with a polygonal cross section was considered by accounting for the variation of the curvature of the free liquid surface and the interfacial shear stresses due to the liquid–vapor interaction. A comparison of the predicted results with the experimental data obtained by Wu and Peterson (1991) and Wu et al. (1991) indicated the importance of the liquid charge, the contact angle and the shear stresses at the liquid–vapor interface in predicting the maximum heat-transfer capacity and thermal resistance of these devices.

Ma et al. (1996) developed a closed mathematical model of the liquid friction factor for flow occurring in triangular grooves. This model, which built upon the earlier work of Ma et al. (1994), considered the interfacial shear stresses due to liquid–vapor frictional interactions for countercurrent flow. Using a coordinate transformation and the Nachtsheim–Swigert iteration scheme, the importance of the liquid–vapor interactions on the operational characteristics of micro heat pipes and other small phase-change devices was demonstrated. The solution resulted in a method by which the velocity distribution for countercurrent liquid–vapor flow could be determined and allowed the governing liquid flow equations to be solved for cases where the liquid surface is strongly influenced by the vapor flow direction and velocity. The results of the analysis were verified using an experimental test facility constructed with channel angles of 20, 40 and 60 degrees. The experimental and predicted results were compared and found to be in good agreement [Ma and Peterson 1996a; 1996b; Peterson and Ma, 1996a].

31.2.2 Testing of Individual Micro Heat Pipes

As fabrication capabilities have developed, experimental investigations on individual micro heat pipes have been conducted on progressively smaller and smaller devices, beginning with early investigations on what now appear to be relatively large micro heat pipes, approximately 3 mm in diameter, and progressing to micro heat pipes in the 30-µm-diameter range. These investigations have included both steady-state and transient investigations.

31.2.2.1 Steady-State Experimental Investigations

In the earliest experimental tests of this type reported in the open literature by Babin et al. (1990), several micro heat pipes approximately 1 mm in external diameter were evaluated. The primary purpose of this investigation was to determine the accuracy of the previously described steady-state modeling techniques, to verify the micro heat pipe concept, and to determine the maximum heat-transport capacity. The fabrication techniques used to produce these test articles were developed by Itoh Research and Development Company, Osaka, Japan [Itoh, 1988]. As reported previously, four test articles were evaluated, two each from silver and copper. Two of these test pipes were charged with distilled, deionized water and the other two were used in an uncharged condition to determine the effect of the vaporization–condensation process on the overall thermal conductivity of these devices. Steady-state tests were conducted over a range of tilt angles to determine the effect of the gravitational body force on the operational characteristics. An electrical resistance heater supplied the heat into the evaporator. Heat rejection was achieved through the use of a constant-temperature ethyl–glycol solution, which flowed over the condenser portion of the heat pipe. The axial temperature profile was continuously monitored by five thermocouples bonded to the outer surface of the heat pipe using a thermally conductive epoxy. Three thermocouples were located on the evaporator: one on the condenser and one on the outer surface of the adiabatic section. Throughout the tests, the heat input was systematically increased and the temperature of the coolant bath adjusted to maintain a constant adiabatic wall temperature [Babin et al., 1990].

The results of this experiment have been utilized as a basis for comparison with a large number of heat pipe models. As previously reported [Peterson et al., 1996], the steady-state model of Babin et al. (1990) overpredicted the experimentally determined heat-transport capacity at operating temperatures below 40°C and underpredicted it at operating temperatures above 60°C. These experimental results represented the first successful operation of a "micro" heat pipe that utilized the principles outlined in the original concept of Cotter (1984) and as such paved the way for numerous other investigations and applications.

31.2.2.2 Transient Experimental Investigations

While the model developed by Babin et al. (1990) was shown to predict the steady-state performance limitations and operational characteristics of the trapezoidal heat pipe reasonably well for operating temperatures between 40 and 60°C, little was known about the transient behavior of these devices. As a result, Wu et al. (1991) undertook an experimental investigation of the transient characteristics of these devices. This experimental investigation again utilized micro heat pipe test articles developed by Itoh (1988); however, this particular test pipe was designed to fit securely under a ceramic chip carrier and had small fins at the condenser end of the heat pipe for removal of heat by free or forced convection, as shown in Figure 31.1. Startup and transient tests were conducted in which the transient response characteristics of the heat pipe as a function of incremental power increases, tilt angle, and mean operating temperature were measured.

Itoh and Polásek (1990a; 1990b) presented the results of an extensive experimental investigation on a series of micro heat pipes ranging in size and shape from 1 to 3 mm in diameter and 30 to 150 mm in length that utilized both cross-sectional configurations, similar to those presented previously, and a conventional internal wicking structure (Polásek, 1990; Fejfar et al., 1990). The unique aspect of this particular investigation was the use of neutron radiography to determine the distribution of the working fluid within the heat pipes [Itoh and Polásek, 1990a; 1990b; Ikeda, 1990]. Using this technique, the amount and distribution of the working fluid and noncondensable gases were observed during real-time operation along with the boiling and/or re-flux flow behavior. The results of these tests indicated several important results [Peterson, 1992]:

- As is the case for conventional heat pipes, the maximum heat-transport capacity is principally dependent upon the mean adiabatic vapor temperature.
- Micro heat pipes with smooth inner surfaces were found to be more sensitive to overheating than those with grooved capillary systems.
- The wall thickness of the individual micro heat pipes had a greater effect on the thermal performance than did the casing material.
- The maximum transport capacity of heat pipes utilizing axial channels for return of the liquid to the evaporator was found to be superior to that of those utilizing a formal wicking structure.

31.3 Arrays of Micro Heat Pipes

31.3.1 Modeling of Heat Pipe Arrays

The initial conceptualization of micro heat pipes by Cotter (1984) envisioned fabricating micro heat pipes directly into semiconductor devices as shown schematically in Figure 31.3. While many of the previously discussed models can be used to predict the performance limitations and operational characteristics of individual micro heat pipes, it is not clear from the models or analyses how the incorporation of an array of these devices might affect the temperature distribution or the resulting thermal performance. Mallik et al. (1991) developed a three-dimensional numerical model capable of predicting the thermal performance of an array of parallel micro heat pipes constructed as an integral part of semiconductor chips, similar to that illustrated in Figure 31.4. In order to determine the potential advantages of this concept, several different thermal loading configurations were modeled, and reduction in maximum surface temperature, the mean chip temperature and the maximum temperature gradient across the chip was determined [Peterson, 1994].

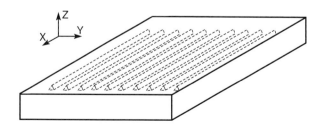

FIGURE 31.3 Array of micro heat pipes fabricated as an integral part of a silicon wafer.

FIGURE 31.4 Silicon wafer into which an array of micro heat pipes has been fabricated.

Although the previous investigations of Babin et al. (1990), Wu and Peterson (1991) and Wu et al. (1991) indicated that an effective thermal conductivity greater than 10 times that of silicon could be achieved, additional analyses were conducted to determine the effect of variations in this value. Steady-state analyses were performed using a heat pipe array comprised of 19 parallel heat pipes. Using an effective thermal conductivity ratio of 5, the maximum and mean surface temperatures were 37.69 and 4.91°C, respectively. With an effective thermal conductivity ratio of 10, the maximum and mean surface temperatures were 35.20 and 4.21°C, respectively. Using an effective thermal conductivity ratio of 15, the maximum and mean surface temperatures were 32.67 and 3.64°C, respectively [Peterson, 1994]. These results illustrate how the incorporation of an array of micro heat pipes can reduce the maximum wafer temperature, reduce the temperature gradient across the wafers and eliminate localized hot spots. In addition, this work highlighted the significance of incorporating these devices into semiconductor chips, particularly those constructed in materials with thermal conductivities significantly less than that of silicon, such as gallium arsenide.

This work was further extended to determine transient response characteristics of an array of micro heat pipes fabricated into silicon wafers as a substitute for polycrystalline diamond or other highly thermally conductive heat spreader materials [Mallik and Peterson, 1991; Mallik et al., 1992]. The resulting transient, three-dimensional, numerical model was capable of predicting the time-dependent temperature distribution occurring within the wafer when given the physical parameters of the wafer and the locations of the heat sources and sinks and indicated that significant reductions in the maximum localized wafer temperatures and thermal gradients across the wafer could be obtained through the incorporation of an array of micro heat pipes. Utilizing heat sinks located on the edges of the chip perpendicular to the axis of the heat pipes and a cross-sectional area porosity of 1.85%, reductions in the maximum chip temperature of up to 40% were predicted.

31.3.2 Testing of Arrays of Micro Heat Pipes

Peterson et al. (1991) fabricated, charged and tested micro heat pipe arrays incorporated as an integral part of semiconductor wafers. These tests represented the first successful operation of these devices reported in the open literature. In this investigation, several silicon wafers were fabricated with distributed heat sources on one side and an array of micro heat pipes on the other, as illustrated in Figure 31.4. Since that time, a number of experimental investigations have been conducted to verify the micro heat pipe array concept and determine the potential advantages of constructing an array of micro heat pipes as an integral part of semiconductor devices [Peterson et al., 1993; Peterson, 1994]. The arrays tested have typically been fabricated in silicon and have ranged in size from parallel rectangular channels, 30 μm wide, 80 μm deep and 19.75 mm long, machined into a silicon wafer 20 mm square and 0.378 mm thick with an interchannel spacing of 500 μm, to etched arrays of triangular channels, 120 μm wide and 80 μm

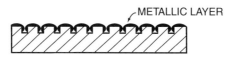

FIGURE 31.5 Vapor deposition process for fabricating micro heat pipes.

deep machined into 20-mm-square silicon wafers 0.5 mm thick [Peterson et al., 1993]. In addition, arrays of micro heat pipes fabricated using a vapor deposition process first proposed by Peterson (1990) and illustrated in Figure 31.5 were tested by Mallik et al. (1995).

In this work, wafers with arrays of 34 and 66 micro heat pipes were evaluated using an infrared thermal imaging system in conjunction with a VHS video recorder. These arrays occupied 0.75% and 1.45% of the wafer cross-sectional area, respectively. The wafers with micro heat pipe arrays demonstrated a 30 to 45% reduction in the thermal time constant when compared to that obtained for plain silicon wafers, which led to a significant reduction in the maximum wafer temperature. The experimental results were then used to validate the transient numerical model described previously [Peterson and Mallik, 1995].

31.3.3 Fabrication of Arrays of Micro Heat Pipes

Considerable information is available on the methods used to fabricate micro heat pipes with hydraulic diameters on the order of 20 to 150 μm into silicon or gallium arsenide wafers. These early investigations included the use of conventional techniques such as the machining of small channels [Peterson, 1988b; Peterson et al., 1991], the use of directionally dependent etching processes to create rectangular- or triangular-shaped channels [Peterson, 1988b; Gerner, 1990; Mallik et al., 1991; Gerner et al., 1992] or other more elaborate techniques that utilize the multisource vapor deposition process illustrated in Figure 31.5 [Mallik et al., 1991; Weichold et al., 1993] to create an array of long, narrow channels of triangular cross section lined with a thin layer of copper. Peterson (1994) has summarized these. The earliest fabricated arrays were machined into a silicon wafer 2 cm square and 0.378 mm thick, with an interchannel spacing of 500 μm. Somewhat later, Adkins et al. (1994) reported on a different fabrication process used for an array of heat pipes with a segmented vapor space. Peterson (1988b), Gerner (1990), Peterson et al. (1993), Ramadas et al. (1993) and Gerner et al. (1994) have described other processes. All of these techniques are similar in nature and typically utilize conventional photolithography masking techniques, coupled with an orientation-dependent etching technique.

Perhaps the most important aspects of these devices are the shape and relative areas of the liquid and vapor passages. A number of investigations have been directed at the optimization of these grooves. These include investigations by Ha and Peterson (1994) that analytically evaluated the axial dryout of the evaporating thin liquid film, one by Ha and Peterson (1996) that evaluated the interline heat transfer and others that examined other important aspects of the problem. [Ha and Peterson, 1998a; 1998b; Peterson and Ha, 1998; Ma and Peterson, 1998]. These studies and others have shown both individual and arrays of micro heat pipes to be extremely sensitive to flooding [Peterson, 1992], and for this reason several different charging methods have been developed and described in detail [Duncan and Peterson, 1995]. These vary from those that are similar to the methods utilized on larger more conventional heat pipes to a method in which the working fluid is added and then the wafer is heated to above the critical temperature of the working fluid so that the working fluid is in the supercritical state and exists entirely as a vapor. The array is then sealed and allowed to cool to below the critical temperature, allowing the vapor to cool and condense. Because, when in the critical state, the working fluid is uniformly distributed throughout the individual micro heat pipes, the exact charge can be carefully controlled and calculated.

31.4 Flat-Plate Micro Heat Spreaders

While arrays of micro heat pipes have the ability to significantly improve the effective thermal conductivity of silicon wafers and other conventional heat spreaders, they are of limited value in that they only provide heat transfer along the axial direction of the individual heat pipes. To overcome this problem, flat-plate heat spreaders, capable of distributing heat over a large two-dimensional surface have been proposed by Peterson (1992; 1994). In this application, a wicking structure is fabricated in silicon multichip module substrates to promote distribution of the fluid and vaporization of the working fluid (Figure 31.6). This wick structure is the key element in these devices and several methods for wick manufacture have been considered [Peterson et al. 1996].

In the most comprehensive investigation of these devices to date, a flat-plate micro heat pipe similar to that described by Peterson et al. (1996) was fabricated in silicon multichip module substrates 5 mm × 5 mm square [Benson et al., 1996a; 1996b]. These devices, which are illustrated in Figure 31.6, utilized two separate silicon wafers. On one of the two wafers, the wick pattern was fabricated leaving a small region around the perimeter of the wafer unpatterned to allow the package to be hermetically sealed. The other silicon wafer was etched in such a manner that a shallow well was formed that corresponded to the wick area. The two pieces were then wafer-bonded together along the seal ring. Upon completion

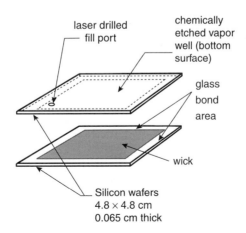

FIGURE 31.6 Flat-plate micro heat spreader. (From Benson, D.A. et al. (1996a) *Proc. IEEE Multichip Module Conf.*, Santa Clara, CA. © 1996 IEEE. With permission.)

of the fabrication, the flat-plate micro heat pipe was filled through a small laser-drilled port located in one corner of the wafer. Because the entire wicking area was interconnected, the volume of the liquid required to charge was of sufficient volume that conventional charging techniques could be utilized [Benson et al., 1996b].

31.4.1 Modeling of Micro Heat Spreaders

Analytical investigations of the performance of these micro heat spreaders or flat-plate heat pipes have been underway for some time; Benson et al. (1996a; 1996b) and Peterson (1996) have summarized the results. These investigations have demonstrated that these devices can provide an effective mechanism for distributing the thermal load in semiconductor devices and reducing the localized hot spots resulting from active chip sites [Peterson, 1996]. The models indicate that the performance of these devices is excellent. In addition, because these devices can be made from silicon, Kovar or a wide variety of other materials, an excellent match between the coefficient of thermal expansion (CTE) can be achieved, while keeping the material and fabrication costs very low. A number of different wicking structures have been considered. Among these are wicks fabricated using a silicon dicing saw (Figure 31.7), wicks fabricated using conventional anisotropic etching techniques (Figure 31.8) and wicks fabricated using a deep plasma etching technique (Figure 31.9). Recent modeling has focused on the development of optimized wicking structures that could be fabricated directly into the wafer and provide maximum capillary pumping while optimizing the thin-film region of the meniscus in order to maximize the heat flux [Wayner et al., 1976; Peterson and Ma, 1996b; 1999].

FIGURE 31.7 Wick pattern prepared with bidirectional saw cuts on a silicon wafer. (From Benson, D.A. et al. (1996b) *Advances in Design, Materials and Processes for Thermal Spreaders and Heat Sinks Workshop,* April 19–21, Vail, CO. © 1996 IEEE. With permission.)

FIGURE 31.8 Chemically etched orthogonal, triangular groove wick. (From Mallik, A.K., and Peterson, G.P. (1991) in *3rd ASME–JSME Thermal Engineering Joint Conf. Proc.,* Vol. 2, March 17–22, Reno, NV, pp. 394–401. With permission.)

TABLE 31.1 Thermal Conductivity, Coefficient of Thermal Expansion, Cost Estimates and Scaling Trends of Current and Potential Substrate Materials

Materials	Thermal Conductivity (W/cm-K)	Coefficient of Thermal Expansion (10^{-6}/K)	Cost of Substrate ($/in^2)	Scaling with Area Cost Trend
Alumina	0.25	6.7	0.09	6" limit
FR-4	Depends on copper	13.0	0.07	Constant to 36"
AlN	1.00–2.00	4.1	0.35	6" limit
Silicon	1.48	4.7	1.00	6–10" limit
Heat pipe in silicon	8.00 → 20.00 (?)	4.7	3.00	6–10" limit
Al	2.37	41.8	0.0009	Scales as area
Cu	3.98	28.7	0.0015	Scales as area
Diamond	10.00–20.00	1.0–1.5	1000.00	Scales as area^2
Kovar	0.13	5.0	0.027	Scales as area
Heat pipe in Kovar	>8.00	5.0	0.10	Scales as area
AlSiC	2.00 (at 70%)	7.0 (?)	1.00	Casting size limited

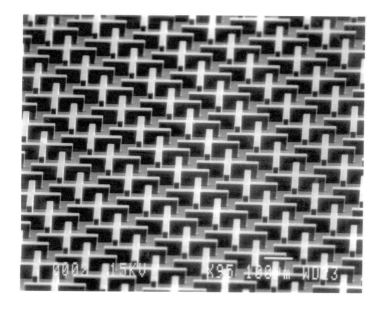

FIGURE 31.9 Wick pattern on silicon prepared by a photomask and deep plasma etch technique; 25-μm-wide × 50-μm-deep wafer. (From Benson, D.A. et al. (1996b) *Advances in Design, Materials and Processes for Thermal Spreaders and Heat Sinks Workshop,* April 19–21, Vail, CO. © 1996 IEEE. With permission.)

The results of these optimization efforts have demonstrated that these microscale flat-plate heat spreaders allow the heat to be dissipated in any direction across the wafer surface, thereby vastly improving performance. The resulting effective thermal conductivities can approach and perhaps exceed those of diamond coatings of equivalent thicknesses. Table 31.1 [Benson et al., 1998] illustrates the relative comparison of these flat-plate heat pipes and other types of materials traditionally utilized in the electronics industry for heat spreading. In this comparison, it is important to note that the ideal heat spreader would have the thermal conductivity of diamond, a coefficient of thermal expansion of silicon, and a cost comparable to aluminum. As shown, flat-plate heat pipes fabricated in either silicon or Kovar compare very favorably with diamond in terms of thermal conductivity and have a coefficient of thermal expansion relatively close to silicon (or exactly, in the case of silicon), and a projected cost that is quite low. Based upon this comparison, it would appear that these flat-plate heat pipes have tremendous commercial potential.

31.4.2 Testing of Micro Heat Spreaders

As described by Benson et al. (1998), a number of different flat-plate micro heat pipe test articles have been evaluated using an infrared camera to determine the spatially resolved temperature distribution. Using this information and a technique initially described by Peterson (1993) for arrays of micro heat pipes, the effective thermal conductivities of charged and uncharged flat-plate micro heat pipes and a series of micro heat spreaders were evaluated experimentally. The results indicated that an effective thermal conductivity between 10 and 20 W/cm-K was possible over a fairly broad temperature range. These values of thermal conductivity approach those of polycrystalline diamond substrates, or are more than five times that of a solid silicon substrate even at elevated temperatures (50°C) and power levels (15 W/cm^2). The cost of such advanced silicon substrates is estimated at $.60/cm^2 (see Table 31.1). Any other inexpensive material with a CTE close to that of the chip may also be a potential option for the heat pipe case material. For example, many alloys in the Fe/Ni/Co family have CTEs closely matching those of semiconductor materials [Benson et al., 1996].

As noted by Peterson (1992), several aspects of the technology remain to be examined before flat-plate micro heat spreaders can come into widespread use, but it is clear from the results of these early experimental tests that spreaders such as the ones discussed here, fabricated as integral parts of silicon chips, present a feasible alternative cooling scheme that merits serious consideration for a number of heat-transfer applications.

31.4.3 Fabrication of Micro Heat Spreaders

The fabrication of these micro heat spreaders is basically just an extension of the methods used by several early investigations to fabricate individual micro heat pipes with hydraulic diameters on the order of 20 to 150 μm. As discussed previously, a number of different wicking structures have been utilized. These wicking structures have been Kovar, silicon or gallium arsenide and include the use of conventional techniques such as machining, directionally dependent etching and deep plasma etching multisource vapor deposition processes. Charging of these devices is somewhat easier than for the individual arrays of micro heat pipes and, while these devices are still sensitive to undercharge, they can accommodate an overcharge much more readily.

31.5 New Designs

In addition to the designs described above, several new designs are currently being developed and evaluated for use in conventional electronic applications and for advanced spacecraft applications. The function of these designs is to provide lightweight, flexible flat-plate heat pipes capable of collecting heat from high heat-flux sources and transporting it to large surface areas where it can be dissipated. In electronic applications, this may entail the collection of heat from a microprocessor and transport of it to a conventional heat spreader or to a more readily available heat sink, such as the screen of a laptop computer. In advanced spacecraft applications, these devices may be used to fabricate highly flexible radiator fin structures for use on long-term spacecraft missions.

To date, several new designs have been proposed. The first of these consists of a flexible, micro heat pipe array, fabricated by sintering an array of aluminum wires between two thin aluminum sheets as shown in Figure 31.10. In this design, the sharp corner regions formed by the junction of the plate and the wires act as the liquid arteries. When made of aluminum with ammonia or acetone as the working fluid, these devices become excellent candidates for use as flexible radiator panels for long-term spacecraft missions and can have a thermal conductivity that greatly exceeds the conductivity of an equivalent thickness of any known material.

A numerical model, combining both conduction and radiation effects, has been developed to predict the heat-transfer performance and temperature distribution of these types of radiator fins in a simulated space environment [Wang et al., 2000]. Three different configurations were analyzed and experimentally

TABLE 31.2 Configurations of Microheat Pipes

	Prototype		
	No. 1	No. 2	No. 3
Material	Aluminum	Aluminum	Aluminum
Working fluid	Acetone	Acetone	Acetone
Total dimension (mm)	152 × 152.4	152 × 152.4	152 × 152.4
Thickness of sheet (mm)	0.40	0.40	0.40
Diameter of wire (mm)	0.50	0.80	0.50
Number of wires	43	43	95

Source: From Wang, Y. et al. (2000) Paper No. AIAA-2000-0969, 38th Aerospaces Sciences Meeting, January 10–13, Reno, NV. With permission.

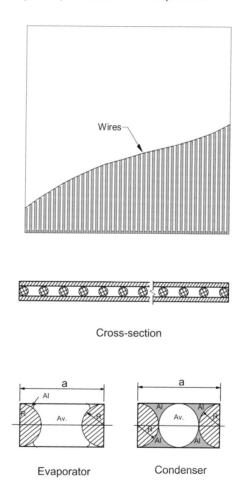

FIGURE 31.10 Flexible wire bonded heat pipe. (From Wang, Y. et al. (2000) Paper No. AIAA-2000-0696, 38th Aerospace Sciences Meeting, January 10–13, Reno, NV. With permission.)

evaluated, and the results were compared. Each of the three configurations was modeled both with and without a working fluid charge in order to determine the reduction in the maximum temperature, mean temperature and temperature gradient on the radiator surface. Table 31.2 lists the physical specifications of the three micro heat pipe arrays fabricated. Acetone was used as the working fluid in both the modeling effort and also in the actual experimental tests.

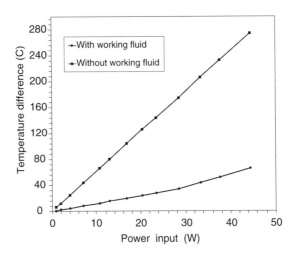

FIGURE 31.11 Temperature difference of micro heat pipe arrays with or without working fluid. (From Wang, Y. et al. (2000) Paper No. AIAA-2000-0696, 38th Aerospace Sciences Meeting, January 10–13, Reno, NV. With permission.)

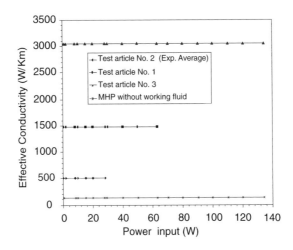

FIGURE 31.12 Effective thermal conductivity of micro heat pipe arrays. (From Wang, Y. et al. (2000) Paper No. AIAA-2000-0696, 38th Aerospace Sciences Meeting, January 10–13, Reno, NV. With permission.)

The results of the preliminary tests conducted on these configurations are shown in Figure 31.11. As indicated, the heat transport was proportional to the temperature difference between the evaporator and condenser; i.e., the effective thermal conductivity of the micro heat pipe array was constant with respect to the temperature. From the temperature difference and heat transport obtained as shown in Figure 31.11 the effective conductivity was obtained. As illustrated in Figure 31.12, the effective thermal conductivities of micro heat pipe arrays No.1, 2 and 3 were 1446.2, 521.3 and 3023.1 W/Km, respectively. For the micro heat pipe arrays without any working fluid, the effective conductivities in the x-direction were 126.3, 113.0 and 136.2 W/Km, respectively. Comparison of the predicted and experimental results indicated that these flexible radiators, with the arrays of micro heat pipes, have an effective thermal conductivity between 15 and 20 times that of the uncharged version. This results in a more uniform temperature distribution, which could significantly improve the overall radiation effectiveness, reduce the overall size, and meet or exceed the baseline design requirements for long-term manned missions to Mars.

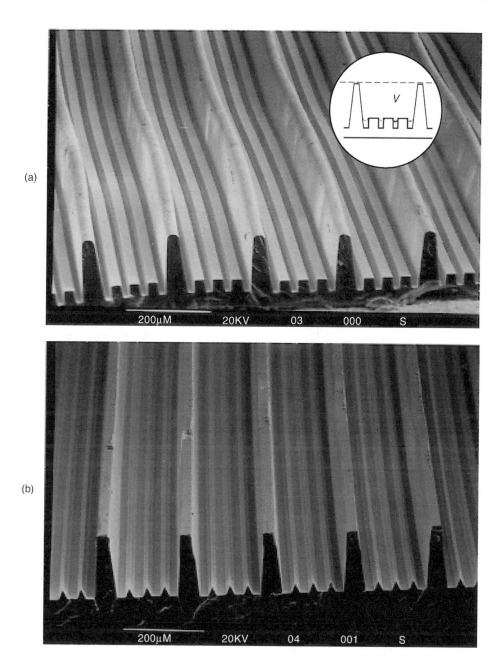

FIGURE 31.13 Flexible polymer micro heat pipe. (a) Rectangular grooves; (b) trapezoidal grooves.

The second design currently being considered consists of an array of flexible micro heat pipes fabricated in a polymer material, as illustrated in Figure 31.13a. This material is extruded in such a fashion that it has a series of large rectangular grooves that serve as the actual heat pipes, each approximately 200 µm wide. Within each of these micro heat pipes is a series of smaller grooves that serve as the liquid arteries (see inset). As shown in Figure 31.13a, these grooves can be rectangular in nature or, as shown in Figure 31.13b, they can be trapezoidal. In both cases, the material is polypropylene, and the internal dimension of the individual heat pipes is approximately 200 µm. The smaller grooves within each of the individual heat pipes are designed to transport the fluid from the evaporator to the condenser. While only preliminary experimental test data are available, this design appears to hold great promise for both spacecraft radiator applications and also for flexible heat spreaders for use in Earth-based electronic applications.

31.6 Summary and Conclusions

It is clear from the preceding review that the concept of using microscale heat pipes and/or flat-plate micro heat spreaders is feasible, practical and cost effective. A number of different concepts and sizes have been shown to be acceptable from both an experimental and theoretical perspective, and a number of these devices are already in widespread use. Steady-state and transient models have been developed and verified experimentally and are capable of predicting the operational limits and performance characteristics of micro heat pipes with diameters less than 100 microns with a high degree of reliability. These models are currently being expanded for use in both individual heat pipes and also with arrays of heat pipes constructed as an integral part of semiconductor devices.

In addition to the analytical work, experimental evaluation has indicated that these devices can be effective in dissipating and transporting heat from localized heat sources and are presently being used in a number of commercial applications. Arrays of micro heat pipes on the order of 35 microns have been successfully fabricated, charged and tested, and incorporated as an integral part of semiconductor devices. Extensive testing has indicated that these heat pipes can provide an effective method for dissipating localized heat fluxes, eliminating localized hot spots, reducing the maximum wafer temperatures and thereby improving the wafer reliability.

Finally, several new designs have been and continue to be developed that have applications ranging from spacecraft radiator applications to land-based electronics applications. These new designs incorporate optimized wicking structures and clever new fabrication schemes, along with materials not previously utilized for heat pipe applications.

Nomenclature

A = area (m^2)
C = constant (defined in text)
d = diameter (m)
f = drag coefficient (dimensionless)
k = thermal conductivity (W/m-K)
K = wick permeability (m^2)
L = length (m)
Ma = Mach number (dimensionless)
P = pressure (N/m^2)
q = heat flow rate (W)
R = thermal resistance (K/W) or universal gas constant (J/kg-K)
Re = Reynolds number (dimensionless)
r = radius (m)
T = temperature (K)
V = velocity (m/s)
w = groove width (m) or wire spacing (m)
We = Weber number (dimensionless)

Greek Symbols

λ = latent heat of vaporization (J/kg)
μ = dynamic viscosity (kg/m-s)
ρ = density (kg/m^3)

σ = surface tension (N/m)
τ = shear stress (N/m^2)
Ψ = angle of inclination (degrees or radians)

Subscripts

- b = boiling
- c = capillary, capillary limitation, condenser
- e = entrainment, evaporator section
- eff = effective
- f = fin
- h = hydraulic
- i = inner
- l = liquid
- m = maximum
- o = outer
- p = pipe
- s = sonic
- v = vapor
- w = wire spacing, wick
- − = axial hydrostatic pressure
- + = normal hydrostatic pressure

References

Adkins, D.R., Shen, D.S., Palmer, D.W., and Tuck, M.R., (1994) "Silicon Heat Pipes for Cooling Electronics," in *Proc. 1st Annu. Spacecraft Thermal Control Symp.*, November 16–18, Albuquerque, NM.

Anon. (1989) "Application of Micro Heat Pipes in Hyperthermia," Annual Report of the Itoh Research and Development Laboratory, Osaka, Japan.

Babin, B.R., and Peterson, G.P. (1990) "Experimental Investigation of a Flexible Bellows Heat Pipe for Cooling Discrete Heat Sources," *ASME J. Heat Transfer* **112**(3), pp. 602–607.

Babin, B.R., Peterson, G.P., and Wu, D. (1990) "Steady-State Modeling and Testing of a Micro Heat Pipe," *ASME J. Heat Transfer* **112**(3), pp. 595–601.

Badran, B., Gerner, F.M., Ramadas, P., Henderson, H.T., and Baker, K.W. (1993) "Liquid Metal Micro Heat Pipes," 29th National Heat Transfer Conference, Atlanta, CA, *HTD*, **236**, pp. 71–85.

Benson, D.A., Mitchell, R.T., Tuck, M.R., Adkins, D.R., and Palmer, D.W. (1996a) "Micro-Machined Heat Pipes in Silicon MCM Substrates," in *Proc. IEEE Multichip Module Conf.*, Santa Clara, CA.

Benson, D.A., Adkins, D.R., Peterson, G.P., Mitchell, R.T., Tuck, M.R., and Palmer, D.W. (1996b) "Turning Silicon Substrates into Diamond: Micromachining Heat Pipes," in *Advances in Design, Materials and Processes for Thermal Spreaders and Heat Sinks Workshop*, April 19–21, Vail, CO.

Benson, D.A., Adkins, D.R., Mitchell, R.T., Tuck, M.R., Palmer, D.W., and Peterson, G.P. (1998) "Ultra High Capacity Micro Machined Heat Spreaders," *Microscale Thermophys. Eng.* **2**(1), pp. 21–29.

Camarda, C.J., Rummler, D.R., and Peterson, G.P. (1997) "Multi Heat Pipe Panels," *NASA Technical Briefs*, LAR-14150.

Chi, S.W. (1976) *Heat Pipe Theory and Practice*, McGraw-Hill, New York.

Collier, J.G. (1981) *Convective Boiling and Condensation*, McGraw-Hill, New York.

Colwell, G.T., and Chang, W.S. (1984) "Measurements of the Transient Behavior of a Capillary Structure under Heavy Thermal Loading," *Int. J. Heat Mass Transfer* **27**(4), pp. 541–551.

Cotter, T.P. (1984) "Principles and Prospects of Micro Heat Pipes," in *Proc. 5th Int. Heat Pipe Conf.*, Tsukuba, Japan, pp. 328–335.

Duncan, A.B., and Peterson, G.P. (1995) "Charge Optimization of Triangular Shaped Micro Heat Pipes," *AIAA J. Thermophys. Heat Transfer* **9**(2), pp. 365–367.

Dunn, P.D., and Reay, D.A. (1982) *Heat Pipes,* 3rd ed., Pergamon Press, New York.

Fejfar, K., Polásek, F., and Stulc, P. (1990) "Tests of Micro Heat Pipes," Annual Report of the SVÚSS, Prague.

Fletcher, L.S., and Peterson, G.P. (1993) "A Micro Heat Pipe Catheter for Local Tumor Hyperthermia," U.S. Patent 5,190,539.

Gaugler, R.S. (1944) "Heat Transfer Devices," U.S. Patent 2,350,348.

Gerner, F.M. (1990) "Micro Heat Pipes," AFSOR Final Report No. S-210-10MG-066, Wright-Patterson AFB, Dayton, OH.

Gerner, F.M., Longtin, J.P., Ramadas, P., Henderson, T.H., and Chang, W.S. (1992) "Flow and Heat Transfer Limitations in Micro Heat Pipes," in *28th National Heat Transfer Conf.,* August 9–12, San Diego, CA.

Gerner, F.M., Badran, B., Henderson, H.T., and Ramadas, P. (1994) "Silicon-Water Micro Heat Pipes," *Thermal Sci. Eng.* **2**(1), pp. 90–97.

Ha, J.M., and Peterson, G.P. (1994) "Analytical Prediction of the Axial Dryout of an Evaporating Liquid Film in Triangular Micro Channels," *ASME J. Heat Transfer* **116**(2), pp. 498–503.

Ha, J.M., and Peterson, G.P. (1996) "The Interline Heat Transfer of Evaporating Thin Films along a Micro Grooved Surface," *ASME J. Heat Transfer* **118**(4), pp. 747–755.

Ha, J.M., and Peterson, G.P. (1998a) "Capillary Performance of Evaporating Flow in Micro Grooves: An Analytical Approach for Very Small Tilt Angles," *ASME J. Heat Transfer* **120**(2), pp. 452–457.

Ha, J.M., and Peterson, G.P. (1998b) "The Maximum Heat Transport Capacity of Micro Heat Pipes," *ASME J. Heat Transfer* **120**(4), pp. 1064–1071.

Ikeda Y. (1990) "Neutron Radiography Tests of Itoh's Micro Heat Pipes," private communication, Nagoya University to F. Polásek.

Itoh, A. (1988) *Micro Heat Pipes,* Prospectus of the Itoh R and D Laboratory, Osaka, Japan.

Itoh, A., and Polásek, F. (1990a) "Development and Application of Micro Heat Pipes," in *Proc. 7th Int. Heat Pipe Conf.,* May 21–25, Minsk, USSR.

Itoh, A., and Polásek, F. (1990b) "Micro Heat Pipes and Their Application in Industry," in *Proc. Czechoslovak–Japanese Symp. on Heat Pipes,* Rícany, Czechoslovakia.

Kendall, D.L. (1979) "Vertical Etching of Silicon at Very High Aspect Ratios," *Ann. Rev. Mater. Sci.,* **7**, pp. 373–403.

Khrustalev, D., and Faghri, A. (1994) "Thermal Analysis of a Micro Heat Pipe" *ASME J. Heat Transfer* **116**(1), 189–198.

Kraus, A.D., and BarCohen, A. (1983) *Thermal Analysis and Control of Electronic Equipment,* McGraw-Hill, New York.

Longtin, J.P., Badran, B., and Gerner, F.M. (1994) "A One-Dimensional Model of a Micro Heat Pipe during Steady-State Operation," *ASME J. Heat Transfer* **116**, pp. 709–715.

Ma, H.B., and Peterson, G.P. (1996a) "Experimental Investigation of the Maximum Heat Transport in Triangular Grooves," *ASME J. Heat Transfer* **118**(4), pp. 740–746.

Ma, H.B., and Peterson, G.P. (1996b) "Temperature Variation and Heat Transfer in Triangular Grooves with an Evaporating Film," *AIAA J. Thermophys. Heat Transfer* **11**(1), pp. 90–98.

Ma, H.B., and Peterson, G.P. (1998) "Disjoining Pressure Effect on the Wetting Characteristics in a Capillary Tube," *Microscale Thermophys. Eng.* **2**(4), pp. 283–297.

Ma, H.B., Peterson, G.P., and Lu, X.J. (1994) "The Influence of the Vapor-Liquid Interactions on the Liquid Pressure Drop in Triangular Microgrooves," *Int. J. Heat Mass Transfer* **37**(15), pp. 2211–2219.

Ma, H.B., Peterson, G.P., and Peng, X.F. (1996) "Experimental Investigation of Countercurrent Liquid–Vapor Interactions and Its Effect on the Friction Factor," *Exp. Thermal Fluid Sci.* **12**(1), pp. 25–32.

Mallik, A.K., and Peterson, G.P. (1991) "On the Use of Micro Heat Pipes as an Integral Part of Semiconductors," in *3rd ASME-JSME Thermal Engineering Joint Conf. Proc.,* Vol. 2, March 17–22, Reno, NV, pp. 394–401.

Mallik, A.K., and Peterson, G.P. (1995) "Steady-State Investigation of Vapor Deposited Micro Heat Pipe Arrays," *ASME J. Electronic Packaging* **117**(1), pp. 75–81.

Mallik, A.K., Peterson, G.P., and Weichold, W. (1991) "Construction Processes for Vapor Deposited Micro Heat Pipes," in *10th Symp. on Electronic Materials Processing and Characteristics,* June 3–4, Richardson, TX.

Mallik, A.K., Peterson, G.P., and Weichold, M.H. (1992) "On the Use of Micro Heat Pipes as an Integral Part of Semiconductor Devices," *ASME J. Electronic Packaging* **114**(4), pp. 436–442.

Mallik, A.K., Peterson, G.P., and Weichold, M.H. (1995) "Fabrication of Vapor Deposited Micro Heat Pipes Arrays as an Integral Part of Semiconductor Devices," *ASME J. MEMS* **4**(3), pp. 119–131.

Marto, P.J., and Peterson, G.P. (1988) "Application of Heat Pipes to Electronics Cooling," in *Advances in Thermal Modeling of Electronic Components and Systems* (A. Bar-Cohen and A.D. Kraus, Eds.), pp. 283–336. Hemisphere Publishing, New York.

Mrácek, P. (1988) "Application of Micro Heat Pipes to laser Diode Cooling," Annual Report of the VÚMS, Prague, Czechoslovakia.

Peterson, G.P. (1987a) "Analysis of a Heat Pipe Thermal Switch," in *Proc. 6th Int. Heat Pipe Conf.,* Vol. 1, May 25–28, Grenoble, France, pp. 177–183.

Peterson, G.P. (1987b) "Heat Removal Key to Shrinking Avionics," *Aerospace Am.* **8**(10), pp. 20–22.

Peterson, G.P. (1988a) "Investigation of Miniature Heat Pipes," Final Report, Wright Patterson AFB, Contract No. F33615-86-C-2733, Task 9.

Peterson, G.P. (1988b) "Heat Pipes in the Thermal Control of Electronic Components," invited paper, *Proc. 3rd Int. Heat Pipe Symp.,* September 12–14, Tsukuba, Japan, pp. 2–12.

Peterson, G.P. (1990) "Analytical and Experimental Investigation of Micro Heat Pipes," in *Proc. 7th Int. Heat Pipe Conf.,* Paper No. A-4, May 21–25, Minsk, USSR.

Peterson, G.P. (1992) "An Overview of Micro Heat Pipe Research," *Appl. Mechanics Rev.* **45**(5), pp. 175–189.

Peterson, G.P. (1993) "Operation and Applications of Microscopic Scale Heat Pipes," in *Encyclopedia of Science and Technology,* Vol. 20, pp. 197–200. McGraw-Hill, New York.

Peterson, G.P. (1994) *An Introduction to Heat Pipes: Modeling, Testing and Applications,* John Wiley & Sons, New York.

Peterson, G.P. (1996) "Modeling, Fabrication and Testing of Micro Heat Pipes: An Update," *Appl. Mechanics Rev.* **49**(10), pp. 175–183.

Peterson, G.P., and Ha, J.M. (1998) "Capillary Performance of Evaporating Flow in Micro Grooves: Approximate Analytical Approach and Experimental Investigation," *ASME J. Heat Transfer* **120**(3), pp. 743–751.

Peterson, G.P., and Ma, H.B. (1996a) "Analysis of Countercurrent Liquid-Vapor Interactions and the Effect on the Liquid Friction Factor," *Exp. Thermal Fluid Sci.* **12**(1), pp. 13–24.

Peterson, G.P., and Ma, H.B. (1996b) "Theoretical Analysis of the Maximum Heat Transport in Triangular Grooves: A Study of Idealized Micro Heat Pipes," *ASME J. Heat Transfer* **118**(4), pp.734–739.

Peterson, G.P., and Ma, H.B. (1999) "Temperature Response and Heat Transfer in a Micro Heat Pipe," *ASME J. Heat Transfer* **121**(2), pp. 438–445.

Peterson, G.P., and Mallik, A.K. (1995) "Transient Response Characteristics of Vapor Deposited Micro Heat Pipe Arrays," *ASME J. Electronic Packaging* **117**(1), pp. 82–87.

Peterson, G.P., and Ortega, A. (1990) "Thermal Control of Electronic Equipment and Devices," in *Advances in Heat Transfer,* Vol. 20 (J. P. Hartnett and T. F. Irvine, Eds.), pp. 181–314. Pergamon Press, New York.

Peterson, G.P., Duncan, A.B., Ahmed, A.K., Mallik, A.K., and Weichold, M.H. (1991) "Experimental Investigation of Micro Heat Pipes in Silicon Devices," in *1991 ASME Winter Ann. Meeting,* ASME Vol. DSC-32, December 1–6, Atlanta, GA, pp. 341–348.

Peterson, G.P., Duncan, A.B., and Weichold, M.H. (1993) "Experimental Investigation of Micro Heat Pipes Fabricated in Silicon Wafers," *ASME J. Heat Transfer* **115**(3), pp. 751–756.

Peterson, G.P., Swanson, L.W., and Gerner, F.M. (1996) "Micro Heat Pipes," in *Microscale Energy Transport* (C. L. Tien, A. Majumdar, and F. M. Gerner, Eds.), Taylor & Francis, Washington, D.C.

Polásek, F. (1990) "Testing and Application of Itoh's Micro Heat Pipes," Annual Report of the SVÚSS, Prague, Czechoslovakia.

Ramadas, P., Badran, B., Gerner, F.M., Henderson, T.H., and Baker, K.W. (1993) "Liquid Metal Micro Heat Pipes Incorporated in Waste-Heat Radiator Panels," in *Tenth Symp. on Space Power and Propulsion*, January 10–14, Albuquerque, NM.

Tien, C.L., Chung, K.S., and Lui, C.P. (1979) "Flooding in Two-Phase Countercurrent Flows," EPRI NP-1283, U.S. Department of Energy, Washington, D.C.

Wang, Y., Ma, H.B., and Peterson, G.P. (2000) "Investigation of the Temperature Distributions on Radiator Fins with Micro Heat Pipes," Paper No. AIAA-2000-0969, 38th Aerospace Sciences Meeting, January 10–13, Reno, NV.

Wayner, Jr., P.C., Kao, Y.K., and LaCroix, L.V. (1976) "The Interline Heat-Transfer Coefficient of an Evaporating Wetting Film," *Int. J. Heat Mass Transfer* **19**(3), pp. 487–492.

Weichold, M.H., Peterson, G.P., and Mallik, A. (1993) *Vapor Deposited Micro Heat Pipes*, U.S. Patent 5,179,043.

Wu, D., and Peterson, G.P. (1991) "Investigation of the Transient Characteristics of a Micro Heat Pipe," *AIAA J. Thermophys. Heat Transfer* **5**(2), pp. 129–134.

Wu, D., Peterson, G.P., and Chang, W.S. (1991) "Transient Experimental Investigation of Micro Heat Pipes," *AIAA J. Thermophys. Heat Transfer* **5**(4), pp. 539–545.

32
Microchannel Heat Sinks

Yitshak Zohar
Hong Kong University of Science and Technology

32.1 Introduction 32-1
32.2 Fundamentals of Convective Heat Transfer in Microducts 32-2
   Modes of Heat Transfer • The Continuum Hypothesis • Thermodynamic Concepts • General Laws • Particular Laws • Governing Equations • Size Effects
32.3 Single-Phase Convective Heat Transfer in Microducts 32-8
   Flow Structure • Entrance Length • Governing Equations • Fully Developed Gas Flow Forced Convection • Fully Developed Liquid Flow Forced Convection
32.4 Two-Phase Convective Heat Transfer in Microducts 32-18
   Boiling Curves • Critical Heat Flux • Flow Patterns • Bubble Dynamics • Modeling of Forced Convection Boiling
32.5 Summary 32-27
   Acknowledgments 32-28

32.1 Introduction

The last decade has witnessed impressive progress in micromachining technology enabling the fabrication of micron-sized mechanical devices, which have become more prevalent in both commercial applications and scientific research. These micromachines have had a major impact on many disciplines, including biology, chemistry, medicine, optics, aerospace and mechanical and electrical engineering. This emerging field not only provides miniature transducers for sensing and actuation in a domain that we could not examine in the past but also allows us to venture into research areas in which the surface effects dominate most of the physical phenomena [Ho and Tai, 1998]. Fundamental heat-transfer problems posed by the development and processing of advanced integrated circuits (ICs) and microelectromechanical systems (MEMS) are becoming a major consideration in the design and application of such systems. The demands on heat-removal and temperature-control functions in modern devices that have highly transient thermal loads require an approach providing high cooling rates and uniform temperature distributions.

As the field of microfluidics and micro heat transfer continues to grow, it becomes increasingly important to understand the mechanisms and fundamental differences involved with heat transfer in single- and two-phase flow in microducts. The idea of fabricating microchannel heat sinks is not new. As early as two decades ago, Tuckerman and Pease (1981) pioneered the use of microchannels for the cooling of planar integrated circuits. They demonstrated that by flowing water through small cooling channels etched in a silicon substrate, heat-transfer rates of about 10^5 W/m^{2}K could be achieved. This rate is about two orders of magnitude higher than that in the state-of-the-art commercial technologies for cooling arrays of ICs. Subsequently, similar experiments were conducted by, among others, Goldberg (1984), Wu and Little (1984), Mahalingam (1985), and Nayak et al. (1987), who passed either liquids or gases in channels ranging in cross-sectional size from 100 μm to 1000 μm. The heat-transfer performance of flows in microchannels has been theoretically analyzed by a number of researchers [e.g., Keyes, 1984; Samalam, 1989; Weisberg et al., 1992]. These works include a number of simplifications and approximations; the major ones are that the heat-transfer coefficient along the channel walls is uniform, the heat transport occurs through the vertical fins, the fluid temperature at each cross section is uniform, and the gas flow is incompressible. In reality, however, the heat-transfer coefficient is a function of the local velocity profile and varies appreciably along the walls. A significant amount of heat is likely to be transported through the channel top or bottom. The fluid temperature field is not one but is at least two dimensional. Most of these early studies dealt with single-phase flow, either liquid or gas, through microchannels, as it proved to be complicated enough. However, it is clear that utilizing the latent heat associated with phase change can dramatically enhance the performance of microchannel heat sinks.

The area of two-phase forced convection heat transfer in microchannels is relatively young and most of the work has been carried out within the last decade [Stanley et al., 1995]. By far, the majority of the reported research work in this area has been empirical in nature. Peng and Wang (1993) investigated the flow boiling through microchannels with a cross section of 0.6 mm × 0.7 mm. They reported that no partial nucleate boiling existed, and that the velocity and liquid subcooling had no obvious effect on the flow nucleate boiling. Moreover, no bubbles were observed throughout the investigation. Peng et al. (1995) argued that the flow boiling was initiated at once and that immediately fully developed nucleate boiling took place. The liquid species and concentration were all found to affect the boiling heat-transfer coefficients [Peng et al., 1996]. Bowers and Mudawar (1994) investigated the pressure drop and critical heat flux in mini- and microchannels. They found that the thermal resistance in a microchannel was lower; however, in the nucleation regime, the pressure drop along the microchannel rose drastically with the increased heat flux. They modeled the pressure drop using a homogeneous equilibrium model to account for acceleration and friction effects. Ravigururajan (1998) studied the impact of channel geometry on two-phase flow in microchannels. The diamond cross section resulted in a lower heat-transfer coefficient but higher critical heat flux compared to the triangular cross section. Only recently have efforts to derive analytical models from basic principles rather than empirical correlations been reported. Peles and Haber (2000) calculated the steady-state, one-dimensional, evaporating, two-phase flow in a triangular microchannel. Although the physical model has been simplified to render the mathematical model tractable, it is the first serious attempt to calculate such a complex flow field.

In this article, the discussion is limited to microchannel heat sinks under steady-state operation. The supplied heat is removed by either single- or two-phase flow forced through microducts, while the single phase can be either water or vapor. Unsteady phenomena in the operation of microchannel heat sinks are naturally of great interest and should be covered in a separate article, as the aim here is to highlight size effects on heat and mass transport in microchannels.

32.2 Fundamentals of Convective Heat Transfer in Microducts

In the science of thermodynamics, which deals with energy in its various forms and with its transformation from one form to another, two particularly important transient forms are defined: work and heat. These energies are termed transient since, by definition, they exist only when there is an exchange of energy between two systems or between a system and its surrounding. When such an exchange takes place without the transfer of mass from the system and not by means of a temperature difference, the energy is said to have been transferred through the performance of work. If the exchange of energy between the systems is the result of a temperature difference, the exchange is said to have been accomplished via the transfer of heat. The existence of a temperature difference is the distinguishing feature of the energy exchange form known as heat transfer. Microchannel heat sinks, of which a typical schematic is shown in Figure 32.1, are a class of devices that can be applied for the removal of thermal energy from very small areas.

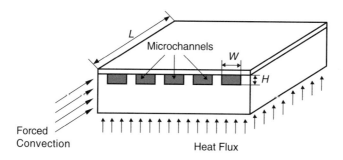

FIGURE 32.1 A schematic of a typical microchannel heat sink.

32.2.1 Modes of Heat Transfer

The mechanism by which heat is transferred in an energy conversion system is complex; however, three basic and distinct modes of heat transfer have been classified: conduction, convection and radiation [Sadik and Yaman, 1995]. Convection is the heat-transfer mechanism, which occurs in a fluid by the mixing of one portion of the fluid with another portion due to gross movements of the mass of fluid. Although the actual process of energy transfer from one fluid particle or molecule to another is still heat conduction, the energy may be transported from one point in space to another by the displacement of the fluid itself. An analysis of convective heat transfer is, therefore, more involved than that of heat transfer by conduction alone, as the motion of the fluid must be studied simultaneously with the energy transfer process. The fluid motion may be caused by external mechanical means (e.g., pump), in which case the process is called forced convection. If the fluid motion is caused by density differences created by the temperature differences existing in the fluid mass, the process is termed free or natural convection. The important heat transfers in liquid–vapor phase-change processes (i.e., boiling and condensing) are also classified as convective mechanisms, as fluid motion is still involved, with the additional complication of a latent heat exchange. Hence, heat transfer in microchannel heat sinks belongs to this class of forced convection heat transfer with or without phase change [Stephan, 1992].

32.2.2 The Continuum Hypothesis

In an analysis of convective heat transfer in a fluid, the motion of the fluid must be studied simultaneously with the heat transport process. In its most fundamental form, the description of the motion of a fluid involves a study of the behavior of all the discrete particles (e.g., molecules) that make up the fluid. The most fundamental approach in analyzing convective heat transfer, therefore, would be to apply the laws of mechanics and thermodynamics to each individual particle, or a statistical group of particles, subsequent to some initial conditions. Such an approach, kinetic theory or statistical mechanics, would give an insight into the details of the energy transfer processes; however, it is not practical for most scientific problems and engineering applications.

In most applications, the primary interest lies not in the molecular behavior of the fluid, but rather in the average or macroscopic effects of many molecules. It is these macroscopic effects that we ordinarily perceive and measure. In the study of convective heat transfer, therefore, the fluid is treated as an infinitely divisible substance, a continuum, while the molecular structure is neglected. The continuum model is valid as long as the size and the mean free path of the molecules are small enough compared with other dimensions existing in the medium such that a statistical average is meaningful.

However, the continuum assumption breaks down whenever the mean free path of the molecules becomes the same order of magnitude as the smallest significant dimension of the problem. In gas flows, the deviation of the state of the fluid from continuum is represented by the Knudsen number, defined as $Kn \equiv \lambda/L$. The mean free path, λ, is the average distance traveled by the molecules between successive collisions, and L is the characteristic length scale of the flow. The appropriate flow and heat-transfer models depend on the range of the Knudsen number, and a classification of the different gas flow regimes is as follows [Schaaf and Chambre, 1961]:

$Kn < 10^{-3}$ continuum flow
$10^{-3} < Kn < 10^{-1}$ slip flow
$10^{-1} < Kn < 10^{+1}$ transition flow
$10^{+1} < Kn$ free molecular flow

In the slip-flow regime, the continuum flow model is still valid for the calculation of the flow properties away from solid boundaries. However, the boundary conditions have to be modified to account for the incomplete interaction between the gas molecules and the solid boundaries. Under normal conditions, Kn is less than 0.1 for most gas flows in microchannel heat sinks with a characteristic length scale on the order of 1 μm. Therefore, only the regime of slip flow, but neither transition nor free molecular flow, will be discussed. The continuum assumption is of course valid for liquid flows in microchannel heat sinks.

32.2.3 Thermodynamic Concepts

The most convenient framework within which heat-transfer problems can be studied is the system, which is a quantity of matter, not necessarily constant, contained within a boundary. The boundary can be physical, partly physical and partly imaginary, or wholly imaginary. The physical laws to be discussed are always stated in terms of a system. A control volume is any specific region in space across the boundaries of which mass, momentum and energy may flow and within which mass, momentum and energy storage may take place and on which external forces may act. The complete definition of a system or a control volume must include, implicitly at least, the definition of a coordinate system, as the system may be moving or stationary. The characteristic of a system of interest is its state, which is a condition of the system described by its properties. A property of a system can be defined as any quantity that depends on the state of the system and is independent of the path (i.e., previous history) by which the system arrived at the given state. If all the properties of a system remain unchanged, the system is said to be in an equilibrium state.

A change in one or more properties of a system necessarily means that a change in the state of the system has occurred. The path of the succession of states through which the system passes is called the process. When a system in a given initial state goes through a number of different changes of state or processes and finally returns to its initial state, the system has undergone a cycle. The properties describe the state of a system only when it is in equilibrium. If no heat transfer takes place between any two systems when they are placed in contact with each other, they are said to be in thermal equilibrium. Any two systems are said to have the same temperature if they are in thermal equilibrium with each other. Two systems that are not in thermal equilibrium have different temperatures, and heat transfer may take place from one system to the other. Therefore, temperature is a property that measures the thermal level of a system.

When a substance exists as part liquid and part vapor at a saturation state, its quality is defined as the ratio of the mass of vapor to the total mass. The quality χ may be considered a property ranging between 0 and 1. Quality has meaning only when the substance is in a saturated state (i.e., at saturated pressure and temperature). The amount of energy that must be transferred in the form of heat to a substance held at constant pressure so that a phase change occurs is called the latent heat. It is the change in enthalpy, which is a property of the substance at the saturated conditions, of the two phases. The heat of vaporization, boiling, is the heat required to completely vaporize a unit mass of saturated liquid.

32.2.4 General Laws

The general laws when referring to an open system (e.g., microchannel heat sink) can be written in either an integral or a differential form. The law of conservation of mass simply states that in the absence of any mass–energy conversion the mass of the system remains constant. Thus, in the absence of a source or sink, $Q = 0$, the rate of change of mass in the control volume (CV) is equal to the mass flux through the control surface (CS). Newton's second law of motion states that the net force, F, acting on a system in an inertial coordinate system is equal to the time rate of change of the total linear momentum of the system. Similarly, the law of conservation of energy for a control volume states that the rate of change of the total energy E of the system is equal to the sum of the time rate of change of the energy within the control volume and the energy flux through the control surface.

The first law of thermodynamics, which is a particular statement of conservation of energy, states that the rate of change in the total energy of a system undergoing a process is equal to the difference between the rate of heat transfer to the system and the rate of work done by the system. The second law of thermodynamics leads to the introduction of entropy, S, as a property of the system. It states that the rate of change in the entropy of the system is either equal to or larger than the rate of heat transfer to the system divided by the system temperature during the heat-transfer process. Even in cases where entropy calculations are not of interest, the second law of thermodynamic is sill important as it is equivalent to stating that heat cannot pass spontaneously from a lower to a higher temperature system.

32.2.5 Particular Laws

Fourier's law of heat conduction, based on the continuum concept, states that the heat flux due to conduction in a given direction (i.e., the heat-transfer rate per unit area) within a medium (solid, liquid or gas) is proportional to temperature gradient in the same direction, namely:

$$\mathbf{q}'' = -k\nabla T \tag{32.1}$$

where $\mathbf{q}''$ is the heat flux vector, k is the thermal conductivity, and T is the temperature.

Newton's law of cooling states that the heat flux from a solid surface to the ambient fluid by convection, q'', is proportional to the temperature difference between the solid surface temperature, T_w, and the fluid free-stream temperature, T_∞, as follows:

$$q'' = h(T_w - T_\infty) \tag{32.2}$$

where h is the heat transfer coefficient.

32.2.6 Governing Equations

The integral form of the conservation laws is useful for the analysis of the gross behavior of the flowfield. However, detailed point-by-point knowledge of the flowfield can be obtained only from the equations of fluid motion in differential form. Microchannel heat sinks typically incorporate arrays of elongated microchannels varying in cross-sectional shape; therefore, it is most convenient to use the governing equations derived either in a rectangular or cylindrical coordinate system. The governing equations for forced convection heat transfer in differential form include conservation of mass, momentum and energy as follows:

$$\frac{D\rho}{Dt} + \rho(\nabla \cdot \mathbf{U}) = 0 \tag{32.3}$$

$$\rho\frac{D\mathbf{U}}{Dt} = -\nabla P + \rho\mathbf{B} + \mu\nabla^2\mathbf{U} + (\mu + \eta)\nabla(\nabla \cdot \mathbf{U}) \tag{32.4}$$

$$\rho c_p \frac{DT}{Dt} = k\nabla^2 T + \frac{DP}{Dt} + \phi + \theta \tag{32.5}$$

In this set of equations, ρ is the density; P is the thermodynamic pressure; $\mathbf{B}$ is the body force (e.g., gravity); μ and η are the shear and the bulk viscosity coefficients, respectively; c_p is the specific heat; θ is the heat source or sink; and ϕ is the viscous dissipation given by:

$$\phi = 2\mu\left[\left(\frac{\partial u}{\partial x}\right)^2 + \left(\frac{\partial v}{\partial y}\right)^2 + \left(\frac{\partial w}{\partial z}\right)^2 + \frac{1}{2}\left(\frac{\partial u}{\partial y} + \frac{\partial v}{\partial x}\right)^2 + \frac{1}{2}\left(\frac{\partial v}{\partial z} + \frac{\partial w}{\partial y}\right)^2 + \frac{1}{2}\left(\frac{\partial u}{\partial z} + \frac{\partial w}{\partial x}\right)^2\right] \quad (32.6)$$

where u, v and w are the three components of the velocity vector $\mathbf{U}$ in a rectangular coordinate system (x, y, z). The state of a simple compressible pure substance, or a mixture of gases, is defined by two independent properties. From experimental observations, it has been established that the behavior of gases at low density is closely given by the ideal-gas equation of state:

$$P = \rho RT \quad (32.7)$$

where R is the specific gas constant. At very low density, all gases and vapors approach ideal-gas behavior; however, the behavior may deviate substantially from that at higher densities. Nevertheless, due to its simplicity, the ideal gas equation of state has been widely used in thermodynamic calculations.

32.2.7 Size Effects

Length scale is a fundamental quantity that dictates the type of forces or mechanisms governing physical phenomena. Body forces are scaled to the third power of the length scale. Surface forces depend on the first or the second power of the characteristic length. This difference in slopes means that a body force must intersect a surface force as a function of the length scale. Empirical observations in biological studies and MEMS show that 1 mm is approximately the order of the demarcation scale [Ho and Tai, 1998]. The characteristic scale of microsystems is smaller than 1 mm; therefore, body forces such as gravity can be neglected in most cases, even in liquid flows, in comparison with surface forces. The large surface-to-volume ratio is another inherent characteristic of microsystems. This ratio is typically inversely proportional to the smaller length scale of the device cross section and is about 1 µm in surface-micromachined devices. The large surface-to-volume ratio in microdevices accentuates the role of surface effects.

32.2.7.1 Noncontinuum Mechanics

The characteristic length scale of a microchannel (i.e., the hydraulic diameter) is typically on the order of a few micrometers. When gas is the working fluid, the mean free path is about 10 to 100 nm, resulting in a Knudsen number of about 0.05. Thus, the flow is considered to be in the slip regime, $0.001 < Kn < 0.1$, where deviations from the state of continuum are relatively small. Consequently, the flow is still governed by Eqs. (32.3) to (32.5), derived and based on the continuum assumption. The rarefaction effect is modeled through Maxwell's velocity-slip and Smoluchowski's temperature-jump boundary conditions [Beskok and Karniadakis, 1994]:

$$U_s - U_w = \frac{2 - \sigma_U}{\sigma_U} \lambda \frac{\partial U}{\partial n}\bigg|_w \quad (32.8a)$$

$$T_j - T_w = \frac{2 - \sigma_T}{\sigma_T} \frac{2\gamma}{\gamma + 1} \frac{k}{\mu c_p} \lambda \frac{\partial T}{\partial n}\bigg|_w \quad (32.8b)$$

U_w and T_w are the wall velocity and temperature, respectively; U_s and T_j are the gas flow velocity and temperature at the boundary; n is the direction normal to the solid boundary; $\gamma = c_p/c_v$ is the ratio of specific heats; and σ_U and σ_T are the momentum and energy accommodation coefficients, respectively, which model the momentum and energy exchange of the gas molecules impinging on the solid boundary. Experiments with gases over various surfaces show that both coefficients are approximately 1.0. This essentially means a diffuse reflection boundary condition, where the impinging molecules are reflected at a random angle uncorrelated with the incident angle.

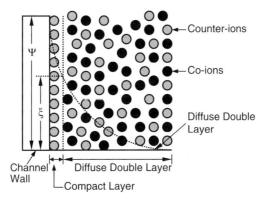

FIGURE 32.2 A schematic representation of an electric double layer (EDL) at the channel wall.

32.2.7.2 Electric Double Layer

Most solid surfaces are likely to carry electrostatic charge (i.e., an electric surface potential) due to broken bonds and surface charge traps. When a liquid containing a small amount of ions is forced through a microchannel under hydrostatic pressure, the solid-surface charge will attract the counter-ions in the liquid to establish an electric field. The arrangement of the electrostatic charges on the solid surface and the balancing charges in the liquid is called the electric double layer (EDL), as illustrated in Figure 32.2. Counter-ions are strongly attracted to the surface and form a compact layer, about 0.5 nm thick, of immobile counter-ions at the solid–liquid interface due to the surface electric potential. Outside this layer, the ions are affected less by the electric field and are mobile. The distribution of the counter-ions away from the interface decays exponentially within the diffuse double layer, with a characteristic length inversely proportional to the square root of the ion concentration in the liquid. The thickness of the diffuse EDL ranges from a few up to several hundreds of nanometers, depending on the electric potential of the solid surface, the bulk ionic concentration and other properties of the liquid. Consequently, EDL effects can be neglected in macrochannel flow. In microchannels, however, the EDL thickness is often comparable to the characteristic size of the channel, and its effect on the fluid flow and heat transfer may not be negligible.

Consider a liquid between two parallel plates, separated by a distance H, containing positive and negative ions in contact with a planar positively charged surface. The surface bears a uniform electrostatic potential ψ_0, which decreases with the distance from the surface. The electrostatic potential, ψ, at any point near the surface is approximately governed by the Debye–Huckle linear approximation [Mohiuddin Mala et al., 1997]:

$$\frac{d^2\psi}{dy^2} = \frac{2n_0\zeta^2 e^2}{\varepsilon\varepsilon_0 k_b T}\psi \qquad (32.9)$$

where ε is the dielectric constant of the medium, and ε_0 is the permittivity of vacuum; ζ is the valence of negative and positive ions; e is the electron charge; k_b is the Boltzmann constant; and n_0 is the ionic concentration. The characteristic thickness of the EDL is the Debye length given by $k_d^{-1} = (\varepsilon\varepsilon_0 k_b T/2n_0\zeta^2 e^2)^{1/2}$. For the boundary conditions when $\psi = 0$ at the midpoint, $y = 0$, and $\psi = \xi$ on both walls, $y = \pm H/2$, the solution is

$$\psi = \frac{\xi}{\sinh(k_d H/2)}|\sinh(k_d y)| \qquad (32.10)$$

where ξ is the electric potential at the boundary between the diffuse double layer and the compact layer.

32.2.7.3 Polar Mechanics

In classical nonpolar mechanics, the mechanical action of one part of a body on another is assumed to be equivalent to a force distribution only. However, in polar mechanics, the mechanical action is assumed to be equivalent to not only a force but also a moment distribution. Thus, the state of stress at a point in nonpolar mechanics is defined by a symmetric second-order tensor, which has six independent components. On the other hand, in polar mechanics, the state of stress is determined by a stress tensor and a couple-stress tensor. The most important effect of couple stresses is to introduce a size-dependent effect that is not predicted by the classical nonpolar theories [Stokes, 1984].

In micropolar fluids, rigid particles contained in a small volume can rotate about the center of the volume element described by the microrotation vector. This local rotation of the particles is in addition to the usual rigid body motion of the entire volume element. In micropolar fluid theory, the laws of classical continuum mechanics are augmented with additional equations that account for conservation of microinertia moments. Physically, micropolar fluids represent fluids consisting of rigid, randomly oriented particles suspended in a viscous medium, where the deformation of the particles is ignored. The modified momentum, angular momentum and energy equations are

$$\rho \frac{D\mathbf{U}}{Dt} = \nabla \cdot \tau + \rho \mathbf{f} \tag{32.11}$$

$$\rho I \frac{D\mathbf{\Omega}}{Dt} = \nabla \cdot \sigma + \rho \mathbf{g} + \tau_x \tag{32.12}$$

$$\rho c_p \frac{DT}{Dt} = k \nabla^2 T + \tau : (\nabla \mathbf{U}) + \sigma : (\nabla \mathbf{\Omega}) - \tau_x \cdot \mathbf{\Omega} \tag{32.13}$$

where $\mathbf{\Omega}$ is the microrotation vector and I is the associated micro-inertia coefficient; $\mathbf{f}$ and $\mathbf{g}$ are the body and couple force vectors, respectively, per unit mass; τ and σ are the stress and couple-stress tensors; $\tau : (\nabla \mathbf{U})$ is the dyadic notation for $\tau_{ji} U_{i,j}$, the scalar product of τ and $\nabla \mathbf{U}$. If $\sigma = 0$ and $\mathbf{g} = \mathbf{\Omega} = 0$, then the stress tensor τ reduces to the classical symmetric stress tensor, and the governing equations reduce to the classical model [Lukaszewicz, 1999].

32.3 Single-Phase Convective Heat Transfer in Microducts

Flows completely bounded by solid surfaces are called internal flows and include flows through ducts, pipes, nozzles, diffusers, etc. External flows are flows over bodies in an unbounded fluid. Flows over a plate, a cylinder or a sphere are examples of external flows, and they are not within the scope of this article. Only internal flows, in either liquid or gas phase, in microducts will be discussed, with an emphasis on size effects, which may potentially lead to different behavior in comparison with similar flows in macroducts.

32.3.1 Flow Structure

Viscous flow regimes are classified as laminar or turbulent on the basis of flow structure. In the laminar regime, flow structure is characterized by smooth motion in laminae, or layers. The flow in the turbulent regime is characterized by random, three-dimensional motions of fluid particles superimposed on the mean motion. These turbulent fluctuations enhance the convective heat transfer dramatically. However, turbulent flow occurs in practice only as long as the Reynolds number, $Re = \rho U_m D_h / \mu$, is greater than a critical value, Re_{cr}. The critical Reynolds number depends on the duct inlet conditions, surface roughness, vibrations imposed on the duct walls and the geometry of the duct cross section. Values of Re_{cr} for various duct cross-section shapes have been tabulated elsewhere [Bhatti and Shah, 1987]. In practical applications,

Microchannel Heat Sinks

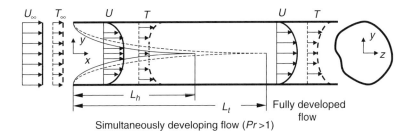

FIGURE 32.3 A schematic of hydrodynamically and thermally developing flow, followed by hydrodynamically and thermally fully developed flow.

though, the critical Reynolds number is estimated to be

$$Re_{cr} = \frac{\rho U_m D_h}{\mu} \cong 2300 \quad (32.14)$$

where U_m is the mean flow velocity, and $D_h = 4A/S$ is the hydraulic diameter, with A and S being the cross-section area and the wetted perimeter, respectively. Microchannels are typically larger than 1000 μm in length with a hydraulic diameter of about 10 μm. The mean velocity for gas flow under a pressure drop of about 0.5 MPa is less than 100 m/s, and the corresponding Reynolds number is less than 100. The Reynolds number for liquid flow will be even smaller due to the much higher viscous forces. Thus, in most applications, the flow in microchannels is expected to be laminar. Turbulent flow may develop in short channels with large hydraulic diameter under high-pressure drop and therefore will not be discussed here.

32.3.2 Entrance Length

When a viscous fluid flows in a duct, a velocity boundary layer develops along the inside surfaces of the duct. The boundary layer fills the entire duct gradually, as sketched in Figure 32.3. The region where the velocity profile is developing is called the hydrodynamics entrance region, and its extent is the hydrodynamic entrance length. An estimate of the magnitude of the hydrodynamic entrance length, L_h, in laminar flow in a duct is given by Shah and Bhatti (1987):

$$\frac{L_h}{D_h} = 0.056 Re \quad (32.15)$$

The region beyond the entrance region is referred to as the hydrodynamically fully developed region. In this region, the boundary layer completely fills the duct and the velocity profile becomes invariant with the axial coordinate.

If the walls of the duct are heated (or cooled), a thermal boundary layer will also develop along the inner surfaces of the duct, shown in Figure 32.3. At a certain location downstream from the inlet, the flow becomes fully developed thermally. The thermal entrance length, L_t, is then the duct length required for the developing flow to reach fully developed condition. The thermal entrance length for laminar flow in ducts varies with the Reynolds number, Prandtl number ($Pr = \mu c_p/k$) and the type of the boundary condition imposed on the duct wall. It is approximately given by:

$$\frac{L_t}{D_h} \cong 0.05 Re\, Pr \quad (32.16)$$

More accurate discussion on thermal entrance length in ducts under various laminar flow conditions can be found elsewhere [e.g., Shah and Bhatti, 1987].

In most practical applications of microchannels, the Reynolds number is less than 100 while the Prandtl number is on the order of 1. Thus, both the hydrodynamic and thermal entrance lengths are less than 5 times the hydraulic diameter. Because the length of microchannels is typically two orders of magnitude larger than the hydraulic diameter, both entrance lengths are less than 5% of the microchannel length and can be neglected.

32.3.3 Governing Equations

Representing the flow in rectangular ducts as flow between two parallel plates, the two-dimensional governing equations can be simplified as follows (Sadik and Yaman, 1995):

Continuity:

$$\frac{\partial(\rho u)}{\partial x} + \frac{\partial(\rho v)}{\partial y} = 0 \tag{32.17}$$

x-momentum:

$$\frac{\partial(\rho uu)}{\partial x} + \frac{\partial(\rho vu)}{\partial y} = -\frac{\partial P}{\partial x} + \mu\left(\frac{\partial^2 u}{\partial x^2} + \frac{\partial^2 u}{\partial y^2}\right) + \frac{\mu}{3}\frac{\partial}{\partial x}\left(\frac{\partial u}{\partial x} + \frac{\partial v}{\partial y}\right) \tag{32.18}$$

y-momentum:

$$\frac{\partial(\rho uv)}{\partial x} + \frac{\partial(\rho vv)}{\partial y} = -\frac{\partial P}{\partial y} + \mu\left(\frac{\partial^2 v}{\partial x^2} + \frac{\partial^2 v}{\partial y^2}\right) + \frac{\mu}{3}\frac{\partial}{\partial y}\left(\frac{\partial u}{\partial x} + \frac{\partial v}{\partial y}\right) \tag{32.19}$$

Energy:

$$u\frac{\partial T}{\partial x} + v\frac{\partial T}{\partial y} = \frac{k}{\rho c_p}\left(\frac{\partial^2 T}{\partial x^2} + \frac{\partial^2 T}{\partial y^2}\right) + \frac{2\mu}{\rho c_p}\left[\left(\frac{\partial u}{\partial x}\right)^2 + \left(\frac{\partial v}{\partial y}\right)^2 + \frac{1}{2}\left(\frac{\partial u}{\partial y} + \frac{\partial v}{\partial x}\right)^2\right] \tag{32.20}$$

32.3.4 Fully Developed Gas Flow Forced Convection

Analytical solution of Eqs. (32.17) to (32.20) is not available. Some solutions can be obtained upon further simplification of the mathematical model. Indeed, incompressible gas flows in macroducts with different cross sections subjected to a variety of boundary conditions are available [Shah and Bhatti, 1987]. However, the important features of gas flow in microducts are mainly due to rarefaction and compressibility effects. Two more effects due to acceleration and nonparabolic velocity profile were found to be of second order compared to the compressibility effect (van den Berg et al., 1993). The simplest system for demonstration of the rarefaction and compressibility effects is the two-dimensional flow between parallel plates separated by a distance H, with L being the channel length ($L/H \gg 1$). If $MaKn \ll 1$, all streamwise derivatives can be ignored except the pressure gradient, which is the driving force. The Mach number, $Ma = U/a$, is the ratio between the fluid speed and the speed of sound a. In such a case, the momentum equation reduces to:

$$-\frac{dP}{dx} + \mu\frac{d^2 u}{dy^2} = 0 \tag{32.21}$$

with the symmetry condition at the channel centerline, $y = 0$, and the slip boundary conditions at the walls, $y = \pm H/2$, as follows:

$$\frac{du}{dy} = 0 \quad @ \quad y = 0 \tag{32.22}$$

$$u = -\lambda \frac{du}{dy}\bigg|_{y=H/2} \quad @ \quad y = \pm H/2 \tag{32.23}$$

Integration of Eq. (32.21) twice with respect to y, assuming $P = P(x)$, yields the following velocity profile [Arkilic et al., 1997]:

$$u(y) = -\frac{H^2}{8\mu}\frac{dP}{dx}\left[1 - \left(\frac{y}{H/2}\right)^2 + 4Kn(x)\right] \tag{32.24}$$

where $Kn(x) = \lambda(x)/H$. The streamwise pressure distribution, $P(x)$, calculated based on the same model is given by:

$$\frac{P(x)}{P_o} = -6Kn_o + \sqrt{\left(6Kn_o + \frac{P_i}{P_o}\right)^2 - \left[\left(\frac{P_i^2}{P_o^2} - 1\right) + 12Kn_o\left(\frac{P_i}{P_o} - 1\right)\right]\left(\frac{x}{L}\right)} \tag{32.25}$$

where P_i is the inlet pressure, P_o the outlet pressure, and Kn_o is the outlet Knudsen number. It is difficult to verify experimentally the cross-stream velocity distribution, $u(y)$, within a microchannel. However, detailed pressure measurements have been reported [Liu et al., 1993; Pong et al., 1994]. A picture of a microchannel integrated with pressure sensors for such experiments is shown in Figure 32.4a. Indeed, the calculated pressure distributions based on Eq. (32.25) were found to be in a close agreement with the measured values as shown in Figure 32.4b [Li et al., 2000]. Furthermore, the mass flow rate, Q_m, as a function of the inlet and outlet conditions is obtained by integrating the velocity profile with respect to x and y as follows:

$$Q_m = \frac{H^3 W P_o^2}{24\mu RTL}\left[\left(\frac{P_i}{P_o}\right)^2 - 1 + 12Kn_o\left(\frac{P_i}{P_o} - 1\right)\right] \tag{32.26}$$

where W is the width of the channel. This simple equation was found to yield accurate results for three different working gasses: nitrogen, helium and argon, with ambient temperatures ranging from 20 to 60°C, as demonstrated in Figure 32.5 [Jiang et al., 1999a].

The microchannel flow temperature distribution and heat flux depend on the boundary conditions, and extensive analytical work has been conducted (Harley et al., 1995; Beskok et al., 1996). However, closed-formed analytical solutions in general are still not available. Numerical simulations of Eqs. (32.17) to (32.20) were carried out for constant wall temperature and constant heat flux boundary conditions by Kavehpour et al. (1997), and the results are summarized in Figure 32.6. The heat transfer rate from the wall to the gas flow decreases while the entrance length increases due to the rarefaction effect (i.e., increasing Knudsen number). This may not be a universal result, however, as the slip flow conditions include two competing effects (Zohar et al., 1994). The velocity slip at the wall increases the flow rate, thus enhancing the cooling efficiency. On the other hand, the temperature jump at the boundary acts as a barrier to the flow of heat to the gas, thus reducing the cooling efficiency. The net result of these effects depends on the specific material properties and specific geometry of the system.

A microchannel integrated with suspended temperature sensors was constructed, Figure 32.7a, for an initial attempt to experimentally assess the slip flow effects on heat transfer in microchannels

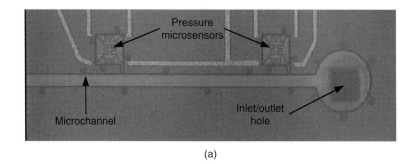

FIGURE 32.4 Slip flow effect on a microchannel flow: (a) microchannel, 40 μm wide, integrated with pressure microsensors; (b) a comparison between calculated (dash lines) and measured (symbols) streamwise pressure distributions. (Reprinted from Li, X. et al. (2000) *Sensors and Actuators A*, **83**, 277–283. With permission from Elsevier Science.)

[Jiang et al., 1999a]. The resulting temperature distributions along the microchannel are shown in Figure 32.7b for different wall temperatures and pressure drops. In all cases, the temperature along the channel is almost uniform and equal to the wall temperature, and no cooling effect has been observed. Indeed, on one hand, the slip flow effects are small, but, on the other hand, the sensitivity of the experimental system is not sufficient. Thus, experiments with higher resolution and greater sensitivity are required to accurately verify the weak slip flow effects on the temperature and the heat-transfer coefficient predicted by theoretical analyses and numerical simulations.

32.3.5 Fully Developed Liquid Flow Forced Convection

Liquid flow is considered to be incompressible even in microducts, as the distance between the molecules is much smaller than the characteristic scale of the flow. Hence, no rarefaction effect is encountered, and the classical model in Eq. (32.21) should be valid. Again, in such a case, extensive data are readily available [Shah and Bhatti, 1987]. However, two unique features of liquid flow in microducts, polarity and EDL, could affect the flow behavior.

Microchannel Heat Sinks

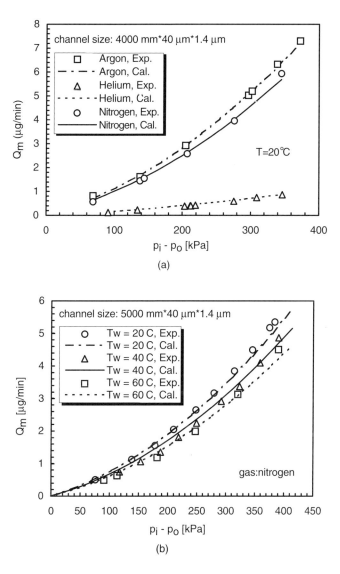

FIGURE 32.5 Slip flow effect on microchannel mass flow rate as a function of the total pressure drop for various working gases (a) and wall temperatures (b). (From Jiang, L. et al. (1999) *J. Micromech. Microeng.*, **9**, 422–428. With permission.)

The characteristic length scale of the electric double layer is inversely proportional to the square root of the ion concentration in the liquid. For example, in pure water the scale is about 1 μm, while in 1 mole of NaCl solution the EDL length scale is only 0.3 nm. Thus, in microducts, liquid flow with low ionic concentration and the associated heat transfer can be affected by the presence of the EDL. The x-momentum and energy equations for a two-dimensional duct flow can be reduced to (Mohiuddin Mala et al., 1997):

$$\mu \frac{d^2 u}{dy^2} - \frac{dP}{dx} - \varepsilon \varepsilon_0 \frac{E_s}{L} \frac{d^2 \psi}{dy^2} = 0 \qquad (32.27)$$

$$\rho c_p \left(u \frac{\partial T}{\partial x} \right) = k \left(\frac{\partial^2 T}{\partial y^2} + \frac{\partial^2 T}{\partial x^2} \right) + \mu \left(\frac{\partial u}{\partial y} \right)^2 \qquad (32.28)$$

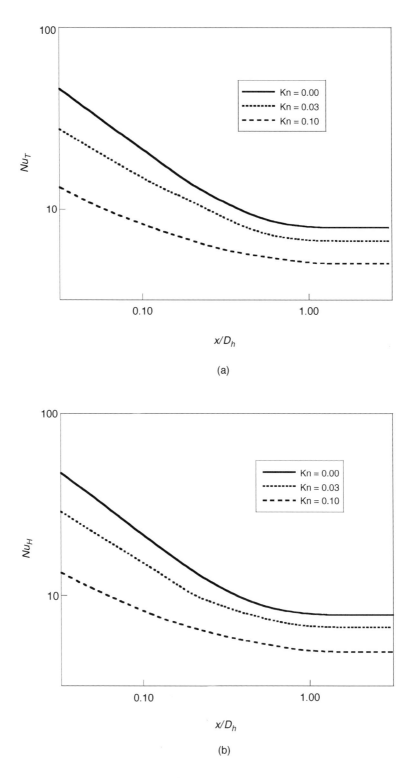

FIGURE 32.6 Numerical simulations of the effect of the inlet Knudsen number, Kn_i, on the Nusselt number, Nu, along a microchannel for uniform wall temperature (Nu_T) (a) and heat flux (Nu_H) (b) boundary conditions. (From Kavehpour, H.P. et al. (1997) *Numerical Heat Transfer A*, **32**, 677–696. Reproduced by permission of Taylor & Francis, Inc.)

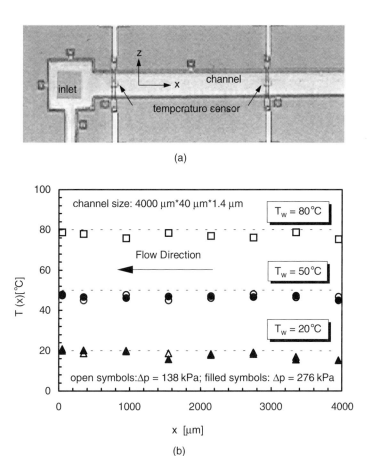

FIGURE 32.7 Slip flow effect on microchannel flow: (a) microchannel integrated with suspended temperature sensors; (b) measured streamwise temperature distributions for different ambient temperature and pressure drop. (From Jiang, L. et al. (1999) *J. Micromech. Microeng.*, **9**, 422–428. With permission.)

where E_s is the steaming potential and L is the duct length. Equation (32.27) was solved analytically, and Eq. (32.28) was solved numerically for constant wall temperature boundary condition for a given inlet liquid temperature. The results showed that both the temperature gradient at the wall and the difference between the wall and the bulk temperature decrease with downstream distance. The value of the temperature gradient decreases much faster, resulting in a decreasing Nusselt number, $Nu = hD_h/k$, along the channel, as plotted in Figure 32.8. However, with no double layer effects (i.e., $\xi = 0$) a higher heat-transfer rate (higher Nu) is obtained. The EDL results in a reduced flow velocity (higher apparent viscosity), thus decreasing the heat-transfer rate.

In order to evaluate micropolar effects on microchannel heat transfer, Jacobi (1989) considered the steady, fully developed, laminar flow in a cylindrical microtube with uniform heat flux, for which the energy equation is given by:

$$\rho c_p \left(u \frac{\partial T}{\partial x} \right) = \frac{k}{r} \left(\frac{\partial T}{\partial r} + r \frac{\partial^2 T}{\partial r^2} \right) \quad (32.29)$$

where r is the radial coordinate. Both the velocity and temperature radial distributions were analytically estimated. Based on the temperature field, the heat-transfer rate was calculated and the results are shown in Figure 32.9 for different values of Γ, a length scale that depends on the viscosity coefficients of the micropolar fluid. The Nusselt number is smaller than the classical value of $Nu = 4.3636$ by as much as

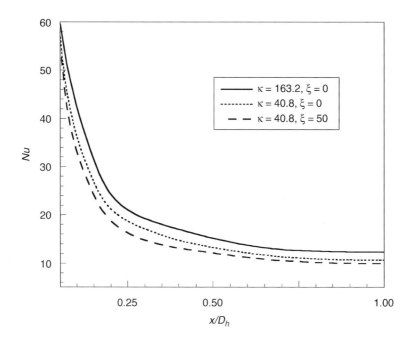

FIGURE 32.8 Electric double layer effect on the variation of the local Nusselt number, *Nu*, along the channel length. (Reprinted from Mohiddin Mala, G. et al. (1997) *Int. J. Heat Mass Transfer*, **40**, 3079–3088. With permission from Elsevier Science.)

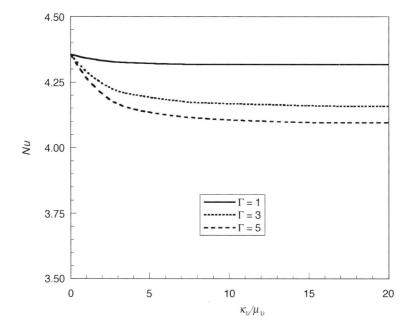

FIGURE 32.9 Micopolar fluid effect on the Nusselt number, *Nu*, as a function of the viscosity ratio, k_v/μ_v, for different values of Γ. (From Jacobi, A.M. (1989) *J. Heat Transfer*, **111**, 1083–1085. With permission.)

7% for this micropolar flow. Although the micropolar fluid theory has been applied to many situations, however, the drawback to these analyses is still the unknown viscosity coefficients.

Clearly, the EDL and micropolar fluid effects on liquid forced convection in microducts are indirect; namely, the velocity is modified due to these effects and, as a consequence, the heat-transfer rate is affected.

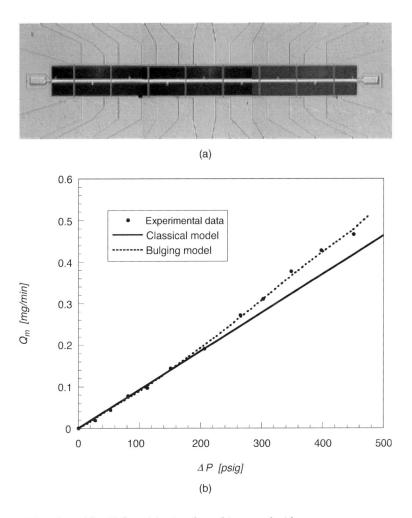

FIGURE 32.10 Microchannel liquid flow: (a) microchannel integrated with temperature sensors on the channel roof; (b) a comparison between liquid flow rate measurements as a function of the pressure drop and theoretical calculations based on classical and bulging models. (From Wu, P. et al. (1998) in *Proc. 11th Int. Workshop on Micro Electro Mechanical Systems* (*MEMS'98*), pp. 87–92. ©1998/2000 IEEE. With permission.)

Thus, it is important to first verify the hydrodynamic effects. Indeed, it has been suggested in a few reports that theoretical calculations based on the classical model did not agree with experimental measurements of liquid flow properties in microchannels [Pfahler et al., 1990; Peng et al., 1994; Peng and Peterson, 1996]. An experimental study of water flow in a microchannel with a cross-section area of 600 μm × 30 μm was carried out specifically to evaluate micropolar effects by Papautsky et al. (1998). They concluded that micropolar fluid theory provides a better approximation to the experimental data than the classical theory. However, a close examination of the results shows that the difference between the results of the two theories is smaller than the difference between the experimental data and the predictions of either theory.

In carefully conducted experiments of water flow through a suspended microchannel (Figure 32.10a) with a cross-section area of 20 μm × 2 μm and under a pressure drop of up to 500 psi, none of these effects has been observed [Wu et al., 1998]. The slight mismatch between theory and experiment was found to be a result of the bulging effect of the channel roof under the high pressure. The deformation of the channel roof can be measured accurately. Once the corrected cross-section area has adequately been accounted for in the calculations, the classical theory results agree well with the experimental measurements as evident in Figure 32.10b. However, more research work is required to verify these observations, as these discrepancies may have to do more with experimental errors rather than true size effects.

32.4 Two-Phase Convective Heat Transfer in Microducts

Micro heat sinks have been constructed as micro heat exchangers for cooling of thermal microsystems, which were developed and investigated either experimentally or theoretically. It is a common finding that the cooling rates in such microchannel heat exchangers should increase significantly due to a decrease in the convective resistance to heat transport caused by a drastic reduction in the thickness of the thermal boundary layers. The potentially high heat-dissipation capacity of such a micro heat sink is based on the large heat-transfer surface-to-volume ratio of the microchannel heat exchanger. In order to increase the heat flux from a microchannel with single-phase flow while maintaining practical limits on surface temperature, it is necessary to increase the heat-transfer coefficient by either increasing the flow rate or decreasing the hydraulic diameter. Both are accompanied by a large increase in the pressure drop. However, forced-convection flow with phase change can achieve a very high heat-removal rate for a constant flow rate, while maintaining a relatively constant surface temperature determined by the saturation properties of the cooling fluid. The advantage of using two-phase over single-phase micro heat sinks is clear. Single-phase heat sinks compensate for high heat flux by a large streamwise increase in both coolant and heat sink temperature. Two-phase heat sinks, in contrast, utilize latent heat exchange, which maintains streamwise uniformity both in the coolant and the heat sink temperature at a level set by the coolant saturation temperature. Therefore, it is expected that two-phase heat transfer may lead to significantly more efficient heat transfer, and a two-phase micro heat exchanger would be the most promising approach for cooling in microsystems [Stanley et al., 1995].

Heat transfer during boiling of a liquid in free convection is essentially determined by the difference between the heating-surface and boiling temperatures, the properties of the liquid, and the properties of the heating surface. Thus, the heat-transfer coefficient can be represented by a simple empirical correlation of the form $h \propto q^m$. During boiling in forced convection, however, the flow velocities of the vapor and liquid phases and the phase distribution play additional roles. Consequently, the mass flow rate and the quality are additional limiting quantities, giving rise to a correlation of the form $h \propto q^m Q_m^n f(\chi)$. Forced convection boiling is complex not only due to the co-existence of two separate phases having different properties but also, especially, due to the existence of a highly convoluted vapor–liquid interface resulting in a variety of flow patterns. Typical patterns that have experimentally been observed in macroducts, such as bubble, slug, churn, annular and drop flow, are sketched in Figure 32.11. Accordingly, flow pattern maps have been suggested in which the duct orientation on heat-transfer boiling is significant due to gravity effects [Stephan, 1992].

32.4.1 Boiling Curves

Forced convection boiling is attractive as it ensures low device temperature for high power dissipation or, alternatively, it allows higher power dissipation for a given device temperature. Measurements of either the inner wall or the fluid bulk temperature distributions along a microduct under forced convection boiling are not available yet, due to the difficulty in integrating sensors at the desired locations.

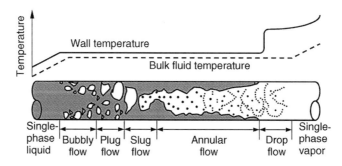

FIGURE 32.11 Wall and mean fluid temperature, flow patterns, and the accompanying heat-transfer ranges in a typical heated duct. (From Stephan, K., *Heat Transfer in Condensation and Boiling*, Springer-Verlag, Berlin, 1992. With permission.)

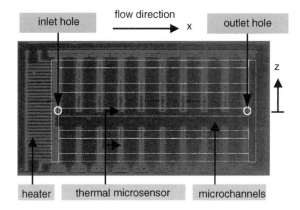

FIGURE 32.12 Photograph of a microchannel heat sink showing the localized heater, the buried microchannel array and the temperature microsensor array. (From Jiang, L. et al. (1999) *J. MEMS*, **8**, 358–365. © 1999/2000 IEEE. With permission.)

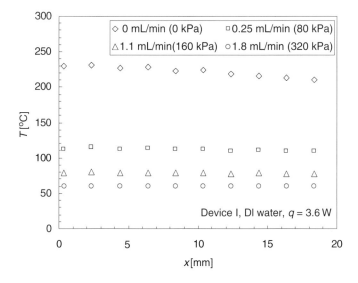

FIGURE 32.13 Flow rate effect on the temperature distribution along the microchannel heat sink centerline. (From Jiang, L. et al. (1999) *J. MEMS*, **8**, 358–365. © 1999/2000 IEEE. With permission.)

However, measurements of the surface temperature of a microchannel heat sink device have been reported [Jiang et al., 1999b]. A picture of the integrated microsystem, consisting of an array of microducts, a local micro heater and an array of temperature microsensors, is shown in Figure 32.12. The 35 diamond-shaped microducts, each with a hydraulic diameter of about 40 µm, are buried between two bonded silicon wafers. A significant reduction of the device temperature is demonstrated in Figure 32.13. Initially, the device temperature and its temperature gradient, for a given power dissipation (3.6 W), is high. The maximum temperature of about 230°C is measured close to the heater. The device temperature drops sharply to about 115°C even for the low flow rate of 0.25 mL/min (average liquid velocity of about 6.7 cm/s within each duct). Increasing the water flow rate leads to further reduction of the device temperature to a level below the saturation temperature of about 100°C. This is expected, as a higher flow rate results in a higher heat-transfer rate. Consequently, the device internal energy (i.e., the device temperature) decreases. Furthermore, the temperature distribution becomes more uniform as well, which suggests that the local heat-transfer rate is highly nonuniform. It should be emphasized, though, that the flow is in

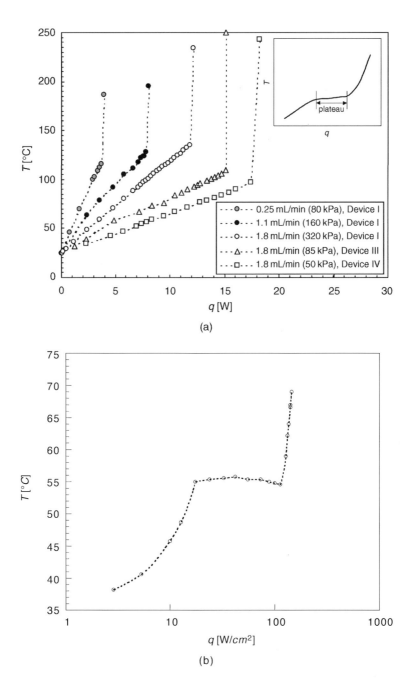

FIGURE 32.14 Boiling curves of device temperature as a function of the input power for microchannels with: (a) water and $D_h = 40\,\mu m$ or $80\,\mu m$ [Jiang et al., 1999b], and (b) R-113 and $D_h = 510\,\mu m$ [Bowers and Mudawar, 1994]. (Reprinted with permission from Elsevier Science.)

single-liquid phase for the high-flow-rate case and in two-phase for the low-flow-rate case, as indicated by the exit fluid quality. Hence, the heat-transfer mechanism changes character as the flow rate varies.

The measured spanwise temperature distributions were found to be uniform, similar to the streamwise temperature distributions plotted in Figure 32.13. Thus, the average temperature along the device centerline can characterize the device temperature. In order to obtain a complete boiling curve, the device temperature was recorded as the input power increased by small increment while maintaining the inlet water flow rate constant at room temperature (22°C). This experiment was repeated several times for different devices with varying flow rate [Jiang et al., 1999b], and the results are summarized in Figure 32.14a. In all curves, the device average temperature increases monotonically, almost linearly, with the power level. At a certain input power, known as critical heat flux (CHF), the temperature increases sharply. The exit flow changes from single-liquid phase, quality zero, through two-phase flow of liquid–vapor, to a single vapor phase, quality one, under CHF conditions. These boiling curves are in contrast to the previously reported data of Bowers and Mudawar (1994), plotted in Figure 32.14b for a microchannel 510 μm in diameter. The typical boiling plateau illustrated at the inset of Figure 32.14a has not been observed under all tested conditions. The plateau in the boiling curve is due to the saturated nucleate boiling, where bubbles continuously form, grow and detach such that the temperature is kept uniform and constant although the heat dissipation is increasing until the CHF condition is approached. The curves in Figure 32.14a suggest that the saturated nucleate boiling does not develop in such microducts due to size effect, which could be verified by flow visualization of the boiling pattern.

32.4.2 Critical Heat Flux

The critical heat flux is the most important factor used to determine the upper limit of the heat sink cooling ability. When the CHF condition is approached, a sudden dryout takes place at the heat-transfer surface. This is accompanied by a drastic reduction of the heat-transfer coefficient and a sharp rise in surface temperature. The exit flow quality is one, as the entire liquid passing through the heat sink changes phase into vapor. Therefore, it is reasonable that the critical heat flux, q_{CHF}, increases linearly with the flow rate (Figure 32.15a), as most of the input power is converted into latent heat at about the saturation temperature (Jiang et al., 1999b). An important parameter associated with the CHF condition is the corresponding device temperature. The dependence of the average temperature under the CHF condition, T_{CHF}, on the water flow rate, Q_v, is shown in Figure 32.15b for three different devices. For the large heat sinks, $D_h = 80$ μm, the CHF temperature depends neither on the flow rate nor on the number of channels. Furthermore, T_{CHF} is slightly higher than the saturation temperature of water under atmospheric pressure, 100°C. The higher CHF temperature may be due to the higher pressure, larger than 1 atm, throughout the microducts. However, for the small heat sink, $D_h = 40$ μm, the CHF temperature increases almost linearly with the water flow rate, which cannot be attributed to the higher pressure. The difference between the two heat sinks is puzzling, and more experiments with wider flow rate ranges are required to confirm this observation.

Bowers and Mudawar (1994) reported similar dependence of the critical heat flux on liquid flow rate in a study comparing the performance of mini- and microchannel heat sinks. The data exhibited a lack of subcooling effect on the CHF for both heat sinks and all operating conditions. This was attributed to fluid reaching the saturation temperature within a short distance into the heated section of the channel. However, they did notice a distinct separation between mini- and microchannel curves, which was explained as a result of the large difference in L/D ratio (L and D being the channel length and diameter, respectively): 3.94 for the minichannel and 19.6 for the microchannel. Consequently, they proposed the following CHF correlation:

$$\frac{q_{mp}}{Gh_{fg}} = 0.16 We^{-0.19} \left(\frac{L}{D}\right)^{-0.54} \quad (32.30)$$

where q_{mp} is the CHF based upon the heated channel inside area, G is the mass velocity, and h_{fg} is the latent heat of evaporation. $We = G^2 L/\beta \rho$ is the Weber number, where β and ρ are the liquid surface tension and density, respectively. The authors argued that the small diameter of the channels resulted in an increased frequency and effectiveness of droplet impact on the channel wall. This could have increased the heat-transfer coefficient and enhanced the CHF compared to droplet flow regions in larger tubes. The small overall size of the heat sinks seemed to contribute to delaying CHF by conducting heat away from the downstream region undergoing partial or total dryout to the boiling region of the channel. Thus, a higher heat-transfer rate is required to trigger CHF conditions along the entire microchannel rather than just at the downstream region.

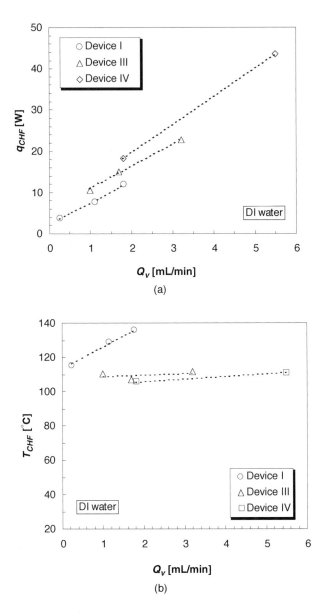

FIGURE 32.15 Flow rate effect on the critical heat flux (a) and the corresponding device temperature (b). (From Jiang, L. et al. (1999) *J. MEMS*, **8**, 358–365. © 1999/2000 IEEE. With permission.)

32.4.3 Flow Patterns

Two-phase flow patterns in ducts are the result of the detailed heat transfer between the solid boundary and the working fluid. The flow patterns are important, as the temperature distributions in both the solid boundary and the fluid flow are directly determined by these patterns. Mudawar and Bowers (1999) suggested that low- and high-velocity flows are characterized by drastically different flow patterns as well as unique CHF trigger mechanisms. Whereas the low flow exhibits a succession of bubbly, slug and annular flow, the high flow is characterized by a bubbly flow near the wall with a liquid core. Unfortunately, limited results of flow patterns have been reported thus far, so it is not clear whether this distinction is valid for microchannel heat sinks.

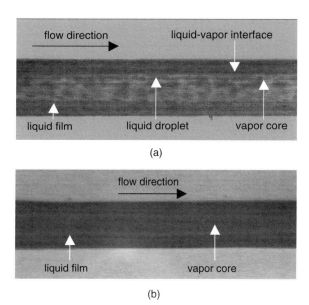

FIGURE 32.16 Pictures of flow patterns during forced convection boiling: (a) unstable annular flow with liquid droplets in the vapor core ($q/q_{CHF} = 0.6$), and (b) stable annular flow ($q/q_{CHF} = 0.8$) (channel width, 50 μm; 35 channels in the microdevice). (From Jiang, L. et al. (2001) *J. MEMS*, **10**, pp. 80–87. © 2000 IEEE. With permission.)

An integrated microsystem, similar to the one shown in Figure 32.12, has been fabricated to study the forced convection boiling flow patterns [Jiang et al., 2000]. The triangular-shaped grooves etched in the silicon wafer were covered by a bonding glass wafer, rather than a silicon wafer, in order to facilitate flow visualizations. In microducts, body forces such as gravity are negligible with respect to surface forces (i.e., surface tension or capillary forces). Consequently, the microduct orientation has little effect on forced convection boiling, and no difference between the flow patterns in horizontal and vertical microducts could be detected experimentally. Furthermore, the boiling modes identified in these microducts are different from the classical patterns sketched in Figure 32.11. At moderate power levels, an annular flow mode with liquid droplets within the vapor core could be observed, as shown in Figure 32.16a, while the vapor–liquid interface in the channel appears to be wavy. This mode should be regarded as an unstable, transition stage, as it was not always detected. Moreover, when it did appear, it was short lived. An annular flow mode, shown in Figure 32.16b, was observed to be a stable pattern for a wide range of input power levels, $0.6 < q/q_{CHF} < 0.9$. A thin liquid film coated each channel wall, and an interface between the liquid film and the vapor core was clearly distinguishable. No liquid droplets existed within the vapor core, indicating that the vapor-core temperature was higher than the liquid saturation temperature.

Evaporation at the liquid film–vapor core interface dominated the heat transfer from the channel wall to the fluid in the annular flow mode. Because the heat is conducted through the liquid film to the interface, the temperature at the wall has to increase to allow a higher heat-transfer rate enforced by the increased input power. The temperature would increase linearly with the input power if the film thickness stays constant. However, the film thickness decreased with increased power, due to the evaporation process, resulting in higher quality of the two-phase exit flow. Thus, the input power is converted into: (1) latent heat required for evaporation at the liquid–vapor interface due to the phase change, and (2) internal energy of the liquid film manifested by the increased liquid and wall temperature. The combination of the two mechanisms resulted in a monotonic temperature increase with decreasing slope as the input power increased. It is not clear whether the annular flow is a general pattern in microchannels due to size effect, or if it is unique only to triangular channel cross sections due to the strong capillary forces at the sharp corners (similar to micro heat pipes).

32.4.4 Bubble Dynamics

Boiling is a phase-change process in which vapor bubbles are formed either on a heated surface or in a superheated liquid layer adjacent to the heated surface. It differs from evaporation at predetermined vapor–liquid interfaces because it also involves creation of these interfaces at discrete sites on the heated surface. Nucleate boiling is a very efficient mode of heat transfer, and it is used in various energy-conversion systems. The number density of sites that become active increases as wall heat flux or wall superheat increases. Clearly, addition of new nucleation sites influences the heat-transfer rate from the solid surface to the working fluid. Knowledge of the nucleation site density as a function of wall superheat is, therefore, needed to develop a credible model for prediction of the heat flux. Several other parameters also affect the site density, including the surface finish, surface wettability and material thermophysical properties. After inception, a bubble continues to grow until forces causing it to detach from the surface exceed those pushing the bubble against the wall. Bubble dynamics, which plays an important role in determining the heat-transfer rate, includes the processes of bubble growth, bubble departure and bubble release frequency [Dhir, 1998].

Jiang et al. (2000) reported that the first experimentally observed mode of phase change, local nucleation boiling, was detected in the microchannel heat sink at an input power level as low as $q/q_{CHF} \cong 0.5$. The working fluid was water and the corresponding device temperature was about 70°C. Bubbles could be seen forming at specific locations along the channel walls at a few active nucleation sites. Bubble generation, growth and explosion at a fairly high frequency inside the microducts were recorded; a mature bubble is shown in Figure 32.17. However, there were very little, if any, active nucleation sites along the channel walls. Furthermore, most of the nucleation sites became inactive after one or two runs, suggesting that they may have been residues of the fabrication process. Therefore, no attempt was made to characterize the bubble release frequency. At slightly higher input power level, $0.5 < q/q_{CHF} < 0.6$, large bubbles were generated at the inlet/outlet common passages, which connect the microchannel array to the device common inlet/outlet. The boiling activity at these larger passages, shown in Figure 32.18, became more intense with increasing input power. Furthermore, the upstream bubbles were forced through the microducts as shown in Figure 32.19. The bubbles typically grew to a size larger than the microduct cross section. Therefore, upon departure from their nucleation sites, these bubbles blocked the duct entrances, as pictured in Figure 32.19a, until the upstream pressure was high enough to force them into the microduct. In some cases, the bubbles traveled slowly along the channel as slug flow as shown in Figure 32.19b. In most instances, however, the bubbles were ejected at high speed through the microduct and could not be detected until they reappeared at the channel exit, as shown in Figure 32.19c. Further increase of the input power level, $q/q_{CHF} > 0.7$, resulted in the annular flow pattern, and the nucleation sites on the duct walls could no longer be observed. The corresponding device temperature was about 90°C. It seems very likely that suppression of the nucleation sites within the microduct was the result of the activity of the

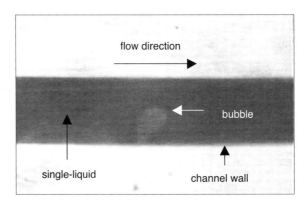

FIGURE 32.17 Active nucleation site within the microchannel exhibiting bubble formation, growth and explosion. (From Jiang, L. et al. (2001) *J. MEMS*, **10**, pp. 80–87. © 2001 IEEE. With permission.)

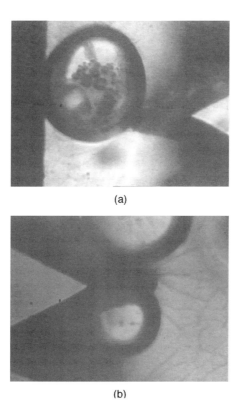

FIGURE 32.18 Bubble formation and growth at inlet (a) and outlet (b) common passages of a microchannel array (q/q_{CHF} = 0.5; 35 channels, each 50 μm in width). (From Jiang, L. et al. (2001) *J. MEMS*, **10**, pp. 80–87. © 2001 IEEE. With permission.)

upstream bubbles as they passed through the ducts rather than a genuine size effect. Similar bubble activity was reported by Peles et al. (1999), who conducted experiments with an almost identical microchannel heat sink.

It is reasonable to expect the bubble dynamics after inception, unlike the nucleation site density, to be affected by the channel size. However, it is clear that in channels with a hydraulic diameter as small as 25 μm, bubble growth and departure have been observed. Thus, the lack of partial nucleate boiling of subcooled liquid flowing through microchannels cannot be attributed to a direct fundamental size effect suppressing bubble dynamics (i.e., bubbles cannot grow and detach due to the small size of the channel). However, it is very plausible that the absence of partial nucleate boiling is an indirect size effect. Namely, another boiling mode such as annular flow becomes dominant due to small channel size (i.e., strong capillary forces), and as a result the bubble dynamics mechanisms are suppressed.

32.4.5 Modeling of Forced Convection Boiling

Phase change from liquid to gas within a microchannel presents a formidable challenge for physical and mathematical modeling. It is not surprising, therefore, that very little work has been reported on this subject. One of the first attempts to address this problem is the derivation of Peles et al. (2000), which was based on fundamental principles rather than empirical formulations. The idealized pattern of the flow in a heated microduct is depicted in Figure 32.20. In this model, the microchannel entrance flow is in single-liquid phase and the exit flow in single-vapor phase. The two phases are separated by a meniscus at a location determined by the heat flux. Such a flow is characterized by a number of specific properties due to the existence of the interfacial surface, which is infinitely thin with a jump in pressure and velocity while the temperature is continuous. Within the single-liquid or vapor phase, heat transfer

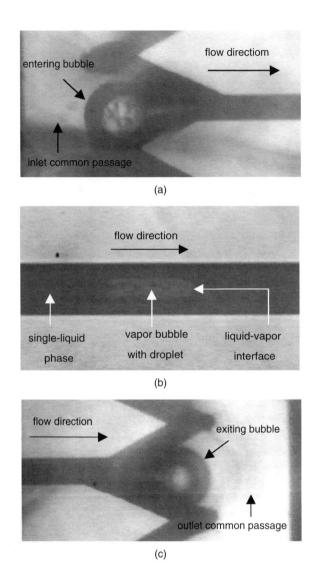

FIGURE 32.19 Sequence of pictures of a bubble entering into (a), traveling through (b), and exiting out of (c) a microchannel 50 μm in width (q/q_{CHF} = 0.5). (From Jiang, L. et al. (2001) *J. MEMS*, **10**, pp. 80–87. © 2001 IEEE. With permission.)

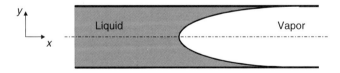

FIGURE 32.20 Physical model of forced convection boiling in a microchannel showing the single-liquid and single-vapor phases separated by an interface. (Reprinted from Peles, Y.P. et al. (2000) *Int. J. Multiphase Flow*, **26**, 1063–1093. With permission from Elsevier Science.)

from the wall to the fluid is accompanied by a streamwise increase of the liquid or vapor temperature and velocity. At the liquid–vapor interface, heat flux causes the liquid to move downstream and evaporate.

The mathematical model includes, on top of the standard equations of conservation and state for each phase, conditions corresponding to the interface surface. For stationary capillary flow, these conditions can be expressed by the equations of continuity of mass, thermal flux across the interface and the balance of all forces acting on the interface. For a capillary with evaporative meniscus, the governing equations take the following form [Peles et al., 2000]:

$$\sum_{b=1}^{2} \rho^{(b)} \mathbf{U}^{(b)} n_i^{(b)} = 0 \qquad (32.31)$$

$$\sum_{b=1}^{2} \left(c_p^{(b)} \rho^{(b)} \mathbf{U}^{(b)} T^{(b)} + k^{(b)} \frac{\partial T^{(b)}}{\partial x_i} \right) n_i^{(b)} = 0 \qquad (32.32)$$

$$\sum_{b=1}^{2} (P^{(b)} + \rho^{(b)} u_i^{(b)} u_j^{(b)}) n_i^{(b)} = (\sigma_{ij}^{(2)} - \sigma_{ij}^{(1)}) n_j + \beta \left(\frac{1}{r_1} + \frac{1}{r_2} \right) n_i^{(2)} + \frac{\partial \beta}{\partial x_i} \qquad (32.33)$$

where n_i and n_j correspond to the normal and tangent directions, respectively, and σ_{ij} is the tensor of viscous tension. The superscript b represents either vapor ($b = 1$) or liquid ($b = 2$). When the interface surface is expressed by a function $x = f(y, z)$, the general radii of curvature, r_i, are found from the equation:

$$A_2 r_i^2 + A_1 r_i + A_0 = 0 \qquad (32.34)$$

The coefficients A_i depend on the shape of the interface. This model, though simplified a great deal, clearly demonstrates the complexity of the theoretical models, which will have to be developed in order to obtain meaningful results. Peles et al. further assumed a quasi-one-dimensional flow and derived a set of equations for the average parameters in order to solve the system of equations. A comparison between the calculations and measurements will not be very useful at this stage, as the physical model, on which the mathematical model is based, has not been observed experimentally in microducts. However, the reported flow patterns could be the result of the specific triangular cross section used in the experiment [Peles et al., 1999; Jiang et al., 2000]. It may well be that with a different cross-section shape (e.g., square or triangular) the flow pattern depicted in Figure 32.20 could develop as a stable phase-change mode.

32.5 Summary

Single-phase forced convection heat transfer in macrosystems has been investigated and well documented. The theoretical calculations and the empirical correlations should be applied to microsystems as well, unless a clear size effect has been identified to require modification of these results. Most known size effects—velocity slip, electric double layer and micropolar fluid—affect the thermal performance indirectly by modifying the velocity field. The only size effect directly related to the heat-transfer mechanism is the temperature jump boundary condition for gas flow, which is the result of incomplete energy exchange between the impinging molecules and the solid boundary due to the size of the channel. Therefore, research in this area should first concentrate on the size effect on the velocity field. Theoretical calculations of size effects on the velocity and temperature distributions are available to some extent. With the advent of micromachining technology, it is possible to fabricate microchannels integrated with microsensors to collect precise experimental data, which can lead to sharp conclusions. Very few studies have been reported to date, and more work is required. Although solid experimental verification is still lacking, initial results show that most size effects are quite small and oftentimes are within the measurement experimental errors.

Two-phase convective heat transfer in microchannels appears to be a technology that can satisfy the demand for the dissipation of high fluxes associated with electronic and laser devices. However, in this area, both experimental and theoretical research related to phase change in microchannels is still very limited. Bubble dynamics during boiling is a complex phenomenon, and size effects can be very significant, much more so than in single-phase forced convection. It is, therefore, vital to establish a credible set of experimental data to provide guidance for theoretical modeling. Initial results show that classical bubble dynamics could still be observed in microchannels, but not classical flow patterns related to bubble activity. Again, this might be an indirect effect, as other phase-change modes become more dominant in microchannels. Clearly, local as well as global measurements are required to supply adequate information in order to understand the heat-transfer mechanisms in microchannels. The integration of microsensors (temperature, pressure, capacitance, etc.) in microchannel systems and flowfield visualizations (flow patterns, bubble dynamics, phase-change evolution) are becoming key components in the march to understand microscale forced convection heat transfer.

Acknowledgments

The author would like to thank Dr. Linan Jiang for her help with the assembling of the material and the critical review of the manuscript.

Defining Terms

Annular flow: A flow pattern in which liquid flow coats the solid boundaries while vapor flows in the central core of duct.
Boiling: A phase-change process in which vapor bubbles are formed either on a heated surface or in a superheated liquid layer adjacent to the heated surface; involves the creation of vapor–liquid interfaces at discrete sites on the heated surface.
Boiling curves: Curves of the heating surface temperature as a function of the heat-transfer rate.
Bubble dynamics: The processes of bubble formation at specific locations, growth and departure at certain frequencies.
Capillary force: The force balance between adhesion (attraction between solid and liquid molecules) and cohesion (mutual attraction of liquid molecules) forces at a liquid–gas–solid interface.
Convective heat transfer (convection): The process of energy transport by the combined effect of heat conduction and the movement of fluid.
Critical heat flux: A critical value beyond which an incremental increase of the heat flux results in a drastic increase of the heating surface temperature.
Entrance length: The entry section of a duct where either the velocity or temperature profile is still developing.
Forced convection: Convective heat transfer where the fluid motion involved in the process is induced by external means (e.g., pump or fan).
Fully developed flow: The duct section in which either the temperature or velocity profile is invariant with the axial coordinate along the duct.
Latent heat: The amount of energy that must be transferred in the form of heat to a substance held at constant pressure in order for a phase change to occur.
Mean free path: The average distance a molecule travels before it collides with another molecule.
Microchannel heat sink: A device comprising a microchannel array to remove heat utilizing forced convection heat transfer with or without boiling.
Nucleation sites: Vapor/gas trapped in imperfections such as cavities and scratches on a heated surface, which serve as nuclei for bubbles.

References

Arkilic, E.B., Schmidt, M.A., and Breuer, K.S. (1997) "Gaseous Slip Flow in Long Microchannels," *J. MEMS* **6**, pp. 167–178.
Beskok, A., and Karniadakis, G.E. (1994) "Simulation of Heat and Momentum Transfer in Complex Microgeometries," *J. Thermophys. Heat Transfer* **8**, pp. 647–655.

Beskok, A., Karniadakis, G.E., and Trimmer, W. (1996) "Rarefaction and Compressibility Effects in Gas Microflows," *J. Fluids Eng.* **118**, pp. 448–456.

Bhatti, M.S., and Shah, R.K. (1987) "Turbulent and Transition Flow Convective Heat Transfer in Ducts," in *Handbook of Single Phase Convective Heat Transfer* (S. Kakac, R.K. Shah, and W. Aung Eds.), Chapter 4, John Wiley & Sons, New York.

Bowers, M.B., and Mudawar, I. (1994) "High Flux Boiling in Low Flow Rate, Low Pressure Drop Mini-Channel and Micro-Channel Heat Sinks," *Int. J. Heat Mass Transfer* **37**, pp. 321–332.

Dhir, V.K. (1998) "Boiling Heat Transfer," *Annu. Rev. Fluid Mech.* **30**, pp. 365–401.

Goldberg, N. (1984) "Narrow Channel Forced Air Heat Sink," *IEEE Trans. Components, Hybrids, Manu. Technol.* **CHMT-7**, pp. 154–159.

Harley, J.C., Huang, Y., Bau, H.H., and Zemel, J.N. (1995) "Gas Flow in Micro-Channels," *J. Fluid Mech.* **284**, pp. 257–274.

Ho, C.M., and Tai, Y.C. (1998) "Micro-Electro-Mechanical-Systems (MEMS) and Fluid Flows," *Annu. Rev. Fluid Mech.* **30**, pp. 570–612.

Jacobi, A.M. (1989) "Flow and Heat Transfer in Microchannels Using a Microcontinuum Approach," *J. Heat Transfer* **111**, pp. 1083–1085.

Jiang, L., Wang, Y., Wong, M., and Zohar, Y. (1999a) "Fabrication and Characterization of a Microsystem for Microscale Heat Transfer Study," *J. Micromech. Microeng.* **9**, pp. 422–428.

Jiang, L., Wong, M., and Zohar, Y. (1999b) "Phase Change in Micro-Channel Heat Sinks with Integrated Temperature Sensors," *J. MEMS* **8**, pp. 358–365.

Jiang, L., Wong, M., and Zohar, Y. (2001) "Forced Convection Boiling in a Microchannel Heat Sink," *J. MEMS* **10**, pp. 80–87.

Kavehpour, H.P., Faghri, M., and Asako, Y. (1997) "Effects of Compressibility and Rarefaction on Gaseous Flows in Microchannels," *Numerical Heat Transfer A* **32**, pp. 677–696.

Keyes, R.W. (1984) "Heat Transfer in Forced Convection Through Fins," *IEEE Trans. Electron Devices*, **ED-31**, pp. 1218–1221.

Li, X., Lee, W.Y., Wong, M., and Zohar, Y. (2000) "Gas Flow in Constriction Microdevices," *Sensors and Actuators A* **83**, pp. 277–283.

Liu, J.Q., Tai Y.C., Pong, K.C., and Ho, C.M. (1993) "Micromachined Channel/Pressure Sensor Systems for Micro Flow Studies," in *Proc. 7th Int. Conf. on Solid-State Sensors and Actuators* (*Transducers '93*), pp. 995–997.

Lukaszewicz, G. (1999) *Micropolar Fluids: Theory and Applications,* Birkhauser, Boston.

Mahalingam, M. (1985) "Thermal Management in Semiconductor Device Packaging," *Proc. IEEE* **73**, pp. 1396–1404.

Mohiuddin Mala, G., Li, D., and Dale, J.D. (1997) "Heat Transfer and Fluid Flow in Microchannels," *Int. J. Heat Mass Transfer* **40**, pp. 3079–3088.

Mudawar, I., and Bowers, M.B. (1999) "Ultra-High Critical Heat Flux (CHF) for Subcooled Water Flow Boiling. I. CHF Data and Parametric Effects for Small Diameter Tubes," *Int. J. Heat Mass Transfer* **42**, pp. 1405–1428.

Nayak, D., Hwang, L., Turlik, I., and Reisman, A. (1987) "A High-Performance Thermal Module for Computer Packaging," *J. Electronic Mater.* **16**, pp. 357–364.

Papautsky, I., Brazzle, J., Ameel, T.A., and Frazier, A.B. (1998) "Microchannel Fluid Behavior Using Micropolar Fluid Theory," *Proc. 11th Int. Workshop on Micro Electro Mechanical Systems* (*MEMS '98*), pp. 544–549.

Peles, Y.P., and Haber, S. (2000) "A Steady State, One-Dimensional Model for Boiling Two-Phase Flow in Triangular Micro-Channel," *Int. J. Multiphase Flow* **26**, pp. 1095–1115.

Peles, Y.P., Yarin, L.P., and Hetsroni, G. (1999) "Evaporating Two-Phase Flow Mechanism in Microchannels," in *Proc. SPIE Int. Soc. Opt. Eng.* **3680**, pp. 226–236.

Peles, Y.P., Yarin, L.P., and Hetsroni, G. (2000) "Thermodynamic Characteristics of Two-Phase Flow in a Heated Capillary," *Int. J. Multiphase Flow* **26**, pp. 1063–1093.

Peng, X.F., and Peterson, G.P. (1996) "Convective Heat Transfer and Flow Friction for Water Flow in Microchannel Structures," *Int. J. Heat Mass Transfer* **39**, pp. 2599–2608.

Peng, X.F., and Wang, B.X. (1993) "Forced Convection and Flow Boiling Heat Transfer for Liquid Flowing Through Microchannels," *Int. J. Heat Mass Transfer* **36**, pp. 3421–3427.

Peng, X.F., Peterson, G.P., and Wang, B.X. (1994) "Heat Transfer Characteristics of Water Flowing Through Microchannels," *Exp. Heat Transfer* **7**, pp. 265–283.

Peng, X.F., Wang, B.X., Peterson, G.P., and Ma, H.B. (1995) "Experimental Investigation of Heat Transfer in Flat Plates with Rectangular Microchannels," *Int. J. Heat Mass Transfer* **38**, pp. 127–137.

Peng, X.F., Peterson, G.P., and Wang, B.X. (1996) "Flow Boiling of Binary Mixtures in Microchanneled Plates," *Int. J. Heat Mass Transfer* **39**, pp. 1257–1264.

Pfahler, J., Harley, J.C., Bau, H.H., and Jemel, J.N. (1990) "Liquid Transport in Micron and Submicron Channels," *Sensors and Actuators A* **21-23**, pp. 431–434.

Pong, K.C., Ho, C.M., Liu, J., and Tai, Y.C. (1994) "Non-Linear Pressure Distribution in Uniform Micro-Channels," *ASME FED* **197**, pp. 51–56.

Ravigururajan, T.S. (1998) "Impact of Channel Geometry on Two-Phase Flow Heat Transfer Characteristics of Refrigerants in Microchannel Heat Exchangers," *J. Heat Transfer,* **120**, pp. 485–491.

Sadik, K., and Yaman, Y. (1995) *Convective Heat Transfer,* 2nd ed., CRC Press, Boca Raton, FL.

Samalam, V.K. (1989) "Convective Heat Transfer in Microchannels," *J. Electronic Mater.* **18**, pp. 611–617.

Schaaf, S.A., and Chambre, P.L. (1961) *Flow of Rarefied Gases,* Princeton University Press, Princeton, NJ.

Shah, R.K., and Bhatti, M.S. (1987) "Laminar Convective Heat Transfer in Ducts," in *Handbook of Single Phase Convective Heat Transfer* (S. Kakac, R.K. Shah, and W. Aung, Eds.), Chapter 3, John Wiley & Sons, New York.

Stanley, R.S., Ameel, T.A., and Warrington, R.O. (1995) "Convective Flow Boiling in Microgeometries: A Review and Application," in *Proc. Convective Flow Boiling,* an international conference held at the Banff Center for Conferences, Banff, Alberta, Canada, April 30–May 5, pp. 305–310.

Stephan, K. (1992) *Heat Transfer in Condensation and Boiling,* Springer-Verlag, Berlin.

Stokes, V.K. (1984) *Theories of Fluids with Microstructure,* Springer-Verlag, Berlin.

Tuckerman, D.B., and Pease, R.F.W. (1981) "High-Performance Heat Sinking for VLSI," *IEEE Electron Device Lett.* **2**, pp. 126–129.

van den Berg, H.R. ten Seldam, C.A., and van der Gulik, P.S. (1993) "Compressible Laminar Flow in a Capillary," *J. Fluid Mech.* **246**, pp. 1–20.

Weisberg, A., Bau, H. H., and Zemel, J. N. (1992) "Analysis of Microchannels for Integrated Cooling," *Int. J. Heat Mass Transfer* **35**, pp. 2465–2474.

Wu, P., and Little, W.A. (1984) "Measurement of the Heat Transfer Characteristics of Gas Flow in Fine Channel Heat Exchangers Used for Microminiature Refrigerators," *Cryogenics* **24**, pp. 415–420.

Wu, S., Mai, J., Zohar, Y., Tai, Y.C., and Ho, C.M. (1998) "A Suspended Microchannel with Integrated Temperature Sensors for High-Pressure Flow Studies," in *Proc. 11th Int. Workshop on Micro Electro Mechanical Systems (MEMS '98),* pp. 87–92.

Zohar, Y., Chu, W.K.H., Hsu, C.T., and Wong, M. (1994) "Slip Flow Effects on Heat Transfer in Micro-Channels," *Bull. Am. Phys. Soc.* **39**, p. 1908.

33
Flow Control

Mohamed Gad-el-Hak
University of Notre Dame

33.1 Introduction 33-1
33.2 Unifying Principles 33-2
   Control Goals and Their Interrelation • Classification Schemes • Free-Shear and Wall-Bounded Flows • Regimes of Reynolds and Mach Numbers • Convective and Absolute Instabilities
33.3 The Taming of the Shrew... 33-9
33.4 Control of Turbulence.. 33-9
33.5 Suction 33-10
33.6 Coherent Structures.. 33-11
33.7 Scales... 33-15
33.8 Sensor Requirements 33-17
33.9 Reactive Flow Control 33-19
   Introductory Remarks • Targeted Flow Control • Reactive Feedback Control • Required Characteristics
33.10 Magnetohydrodynamic Control................................ 33-26
   Introductory Remarks • The von Kármán Equation • Vorticity Flux at the Wall • EMHD Tiles for Reactive Control
33.11 Chaos Control 33-30
   Nonlinear Dynamical Systems Theory • Chaos Control
33.12 Soft Computing.. 33-35
   Neural Networks for Flow Control
33.13 Use of MEMS for Reactive Control 33-39
33.14 Parting Remarks 33-41

I would like to shock you by stating that I believe that with today's technology we can easily—I say easily—construct motors one fortieth of this size on each dimension. That's sixty-four thousand times smaller than the size of McLellan's motor [400 microns on a side]. And in fact, with our present technology, we can make thousands of these motors at a time, all separately controllable. Why do you want to make them? I told you there's going to be lots of laughter, but just for fun, I'll suggest how to do it—it's very easy.

(From the talk "Infinitesmal Machinery," delivered by Richard P. Feynman at the Jet Propulsion Laboratory, Pasadena, CA, February 23, 1983.)

33.1 Introduction

The subject of flow control—particularly, reactive flow control—is broadly introduced in this chapter, leaving some of the details to other chapters which will deal with the issues of control theory, distributed and optimal control, soft computing tools such as neural networks, genetic algorithms and fuzzy logic, diagnostics and control of turbulent flows and futuristic control specifically targeting the coherent structures in turbulent flows. Distributed flow control can greatly benefit from the availability of inexpensive microsensors and microactuators, hence the relevance of this topic to *The MEMS Handbook*. In reactive control of turbulent flows, large arrays of minute sensors and actuators form feedback or feedforward control loops whose performance can be enhanced by employing optimized control algorithms and soft computing tools, thus the inclusion in this book of chapters on these specialized topics.

The ability to manipulate a flowfield actively or passively to effect a desired change is of immense technological importance, and this undoubtedly accounts for the fact that the subject is more hotly pursued by scientists and engineers than any other topic in fluid mechanics. The potential benefits of realizing efficient flow-control systems range from saving billions of dollars in annual fuel costs for land, air and sea vehicles to achieving economically and environmentally more competitive industrial processes involving fluid flows. Methods of control to effect transition delay, separation postponement, lift enhancement, drag reduction, turbulence augmentation and noise suppression are considered. Prandtl (1904) pioneered the modern use of flow control in his epoch-making presentation to the Third International Congress of Mathematicians held at Heidelberg, Germany. In just eight pages, Prandtl introduced the boundary-layer theory, explained the mechanics of steady separation, opened the way for understanding the motion of real fluids and described several experiments in which the boundary layer was controlled. He used active control of the boundary layer to show the great influence such a control exerted on the flow pattern. Specifically, Prandtl used suction to delay boundary-layer separation from the surface of a cylinder.

Notwithstanding Prandtl's success, aircraft designers in the three decades following his convincing demonstration were accepting lift and drag of airfoils as predestined characteristics with which no man could or should tamper [Lachmann, 1961]. This predicament changed mostly due to the German research in boundary-layer control pursued vigorously shortly before and during World War II. In the two decades following the war, extensive research on laminar flow control, where the boundary layer formed along the external surfaces of an aircraft is kept in the low-drag laminar state, was conducted in Europe and the U.S., culminating in the successful flight test program of the X–21 where suction was used to delay transition on a swept wing up to a chord Reynolds number of 4.7×10^7. The oil crisis of the early 1970s brought renewed interest in novel methods of flow control to reduce skin-friction drag even in turbulent boundary layers. In the 1990s, the need to reduce the emissions of greenhouse gases and to construct supermaneuverable fighter planes, faster/quieter underwater vehicles and hypersonic transport aircraft (e.g., the U.S. National Aerospace Plane) provides new challenges for researchers in the field of flow control.

The books by Gad-el-Hak et al. (1998) and Gad-el-Hak (2000) provide up-to-date overviews of the subject of flow control. In this chapter, following a description of the unifying principles of flow control, we focus on the concept of targeted control in which distributed arrays of microsensors and microactuators, connected in open or closed control loops, are used to target the coherent structures in turbulent flows in order to effect beneficial flow changes such as drag reduction, lift enhancement, noise suppression, etc.

33.2 Unifying Principles

A particular control strategy is chosen based on the kind of flow and the control goal to be achieved. Flow control goals are strongly, often adversely, interrelated, and therein lies the challenge of making tough compromises. There are several different ways for classifying control strategies to achieve a desired effect. The presence or lack of walls, Reynolds and Mach numbers, and the character of the flow instabilities are all important considerations for the type of control to be applied. All these seemingly disparate issues are what place the field of flow control in a unified framework. They will each be discussed in turn in the following.

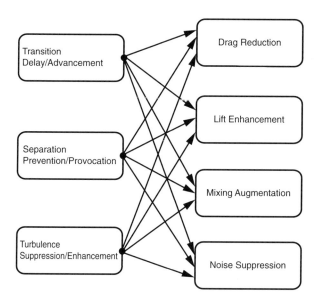

FIGURE 33.1 Engineering goals and corresponding flow changes. (From Gad-el-Hak, M. (2000) *Flow Control: Passive, Active, and Reactive Flow Management*. Reprinted with permission of Cambridge University Press, New York.)

33.2.1 Control Goals and Their Interrelation

What does the engineer want to achieve when attempting to manipulate a particular flowfield? Typically, engineers aim at reducing the drag; enhancing the lift; augmenting the mixing of mass, momentum or energy; suppressing the flow-induced noise or a combination thereof. To achieve any of these useful end results, for either free-shear or wall-bounded flows, transition from laminar to turbulent flow may have to be either delayed or advanced, flow separation may have to be either prevented or provoked and, finally, turbulence levels may have to be either suppressed or enhanced. All of those engineering goals and the corresponding flow changes intended to effect them are depicted in Figure 33.1. None of them is particularly difficult if taken in isolation, but the challenge is to achieve a goal using a simple device that is inexpensive to build as well as to operate and, most importantly, has minimum side effects. For this latter hurdle, the interrelation between control goals must be elaborated, and this is what is attempted below.

Consider the technologically very important boundary layers. An external wall-bounded flow, such as that developing on the exterior surfaces of an aircraft or a submarine, can be manipulated to achieve transition delay, separation postponement, lift increase, skin-friction and pressure[1] drag reduction, turbulence augmentation, heat-transfer enhancement or noise suppression. These objectives are not necessarily mutually exclusive. The schematic in Figure 33.2 is a partial representation of the interrelation between one control goal and another. To focus the discussion further, consider the flow developing on a lifting surface such as an aircraft wing. If the boundary layer becomes turbulent, its resistance to separation is enhanced and more lift could be obtained at increased incidence. On the other hand, the skin-friction drag for a laminar boundary layer can be as much as an order of magnitude less than that for a turbulent one. If transition is delayed, lower skin friction as well as lower flow-induced noise are achieved. However, the laminar boundary layer can support only a very small adverse pressure gradient without separation, and subsequent loss of lift and increase in form drag occur. Once the laminar boundary layer separates, a free-shear layer forms, and for moderate Reynolds numbers transition to turbulence takes place. Increased entrainment of high-speed fluid due to the turbulent mixing may result

---

[1]Pressure drag includes contributions from flow separation, displacement effects, induced drag, wave drag and, for time-dependent motion of a body through a fluid, virtual mass.

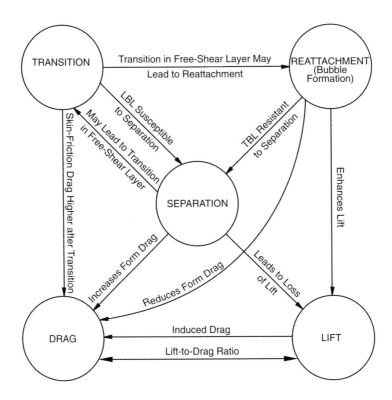

FIGURE 33.2 Interrelations among flow control goals. (From Gad-el-Hak, M. (2000) *Flow Control: Passive, Active, and Reactive Flow Management*. Reprinted with permission of Cambridge University Press, New York.)

in reattachment of the separated region and formation of a laminar separation bubble. At higher incidence, the bubble breaks down, either separating completely or forming a longer bubble. In either case, the form drag increases and the slope of the lift curve decreases. The ultimate goal of all this is to improve the performance of the airfoil by increasing the lift-to-drag ratio. However, induced drag is caused by the lift generated on a lifting surface with a finite span. Moreover, more lift is generated at higher incidence, but form drag also increases at these angles.

All of the above point to potential conflicts as one tries to achieve a particular control goal only to adversely affect another goal. An ideal method of control that is simple, inexpensive to build and operate and does not have any trade-offs does not exist, and the skilled engineer has to make continuous compromises to achieve a particular design goal.

33.2.2 Classification Schemes

Various classification schemes exist for flow control methods. One is to consider whether the technique is applied at the wall or away from it. Surface parameters that can influence the flow include roughness, shape, curvature, rigid-wall motion, compliance, temperature and porosity. Heating and cooling of the surface can influence the flow via the resulting viscosity and density gradients. Mass transfer can take place through a porous wall or a wall with slots. Suction and injection of primary fluid can have significant effects on the flowfield, influencing particularly the shape of the velocity profile near the wall and thus the boundary layer susceptibility to transition and separation. Different additives, such as polymers, surfactants, microbubbles, droplets, particles, dust or fibers, can also be injected through the surface in water or air wall-bounded flows. Control devices located away from the surface can also be beneficial. Large-eddy breakup devices (LEBUs; also called outer-layer devices, or OLDs), acoustic waves bombarding a shear layer from outside, additives introduced in the middle of a shear layer, manipulation of freestream

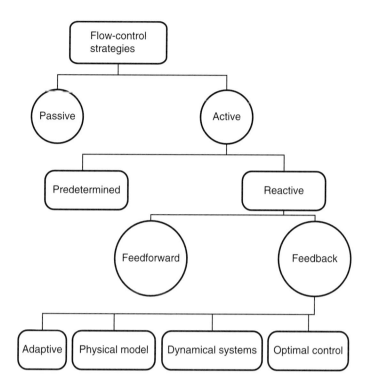

FIGURE 33.3 Classification of flow control strategies. (From Gad-el-Hak, M. (2000) *Flow Control: Passive, Active, and Reactive Flow Management*. Reprinted with permission of Cambridge University Press, New York.)

turbulence levels and spectra, gust and magneto- and electrohydrodynamic body forces are examples of flow control strategies applied away from the wall.

A second scheme for classifying flow control methods considers energy expenditure and the control loop involved. As shown in Figure 33.3, a control device can be passive, requiring no auxiliary power and no control loop, or active, requiring energy expenditure. As for the action of passive devices, some prefer to use the term *flow management* rather than *flow control* [Fiedler and Fernholz, 1990], reserving the latter terminology for dynamic processes. Active control requires a control loop and is further divided into predetermined or reactive. Predetermined control includes the application of steady or unsteady energy input without regard to the particular state of the flow. The control loop in this case is open, as shown in Figure 33.4a, and no sensors are required. Because no sensed information is being fed forward, this open control loop is not a feedforward one. This subtle point is often confused in the literature, blurring predetermined control with reactive, feedforward control. Reactive control is a special class of active control where the control input is continuously adjusted based on measurements of some kind. The control loop in this case can either be an open, feedforward one (Figure 33.4b) or a closed, feedback loop (Figure 33.4c). Classical control theory deals, for the most part, with reactive control.

The distinction between feedforward and feedback is particularly important when dealing with the control of flow structures that convect over stationary sensors and actuators. In feedforward control, the measured variable and the controlled variable differ. For example, the pressure or velocity can be sensed at an upstream location, and the resulting signal is used together with an appropriate control law to trigger an actuator which in turn influences the velocity at a downstream position. Feedback control, on the other hand, requires that the controlled variable be measured, fed back and compared with a reference input. Reactive feedback control is further classified into four categories: adaptive, physical model based, dynamical systems based and optimal control [Moin and Bewley, 1994].

Yet another classification scheme considers whether the control technique directly modifies the shape of the instantaneous/mean velocity profile or selectively influence the small dissipative eddies. An inspection

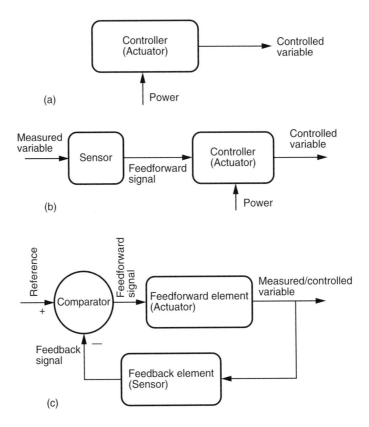

FIGURE 33.4 Different control loops for active flow control: (a) predetermined, open-loop control; (b) reactive, feedforward, open-loop control; (c) reactive, feedback, closed-loop control. (From Gad-el-Hak, M. (2000) *Flow Control: Passive, Active, and Reactive Flow Management*. Reprinted with permission of Cambridge University Press, New York.)

of the Navier–Stokes equations written at the surface [Gad-el-Hak, 2000] indicates that the spanwise and streamwise vorticity fluxes at the wall can be changed, either instantaneously or in the mean, via wall motion/compliance, suction/injection, streamwise or spanwise pressure gradient, normal viscosity gradient or a suitable streamwise or spanwise body force. These vorticity fluxes determine the fullness of the corresponding velocity profiles. For example, suction (or downward wall motion), a favorable pressure gradient or lower wall viscosity results in vorticity flux away from the wall, making the surface a source of spanwise and streamwise vorticity. The corresponding fuller velocity profiles have negative cur-vature at the wall and are more resistant to transition and to separation but are associated with higher skin-friction drag. Conversely, an inflectional velocity profile can be produced by injection (or upward wall motion), an adverse pressure gradient or higher wall viscosity. Such a profile is more susceptible to transition and to separation and is associated with lower, even negative, skin friction. Note that many techniques are available to effect a wall-viscosity gradient—for example, surface heating/cooling, film boiling, cavitation, sublimation, chemical reaction, wall injection of lower/higher viscosity fluid and the presence of a shear thinning/thickening additive.

Flow control devices can target certain scales of motion rather than globally changing the velocity profile. Polymers, riblets and LEBUs, for example, appear to selectively damp only the small dissipative eddies in turbulent wall-bounded flows. These eddies are responsible for the (instantaneous) inflectional profile and the secondary instability in the buffer zone, and their suppression leads to increased scales, a delay in the reduction of the (mean) velocity-profile slope and consequent thickening of the wall region. In the buffer zone, the scales of the dissipative and energy-containing eddies are roughly the same; hence, the energy-containing eddies will also be suppressed, resulting in reduced Reynolds stress production, momentum transport and skin friction.

33.2.3 Free-Shear and Wall-Bounded Flows

Free-shear flows, such as jets, wakes and mixing layers, are characterized by inflectional mean-velocity profiles and are therefore susceptible to inviscid instabilities. Viscosity is only a damping influence in this case, and the prime instability mechanism is vortical induction. Control goals for such flows include transition delay/advancement, mixing enhancement and noise suppression. External and internal wall-bounded flows, such as boundary layers and channel flows, can also have inflectional velocity profiles but, in the absence of an adverse pressure gradient and similar effects, are characterized by noninflectional profiles, and viscous instabilities are then to be considered. This kind of viscosity-dominated, wall-bounded flow is intrinsically stable and therefore is generally more difficult to control. Free-shear flows and separated boundary layers, on the other hand, are intrinsically unstable and lend themselves more readily to manipulation.

Free-shear flows originate from some kind of surface upstream, be it a nozzle, a moving body or a splitter plate, and flow control devices can therefore be placed on the corresponding walls, albeit far from the fully developed regions. Examples of such control include changing the geometry of a jet exit from circular to elliptic [Gutmark and Ho, 1986], using periodic suction/injection in the lee side of a blunt body to affect its wake [Williams and Amato, 1989] and vibrating the splitter plate of a mixing layer [Fiedler et al., 1988]. These and other techniques are extensively reviewed by Fiedler and Fernholz (1990), who offer a comprehensive list of appropriate references, and more recently by Gutmark et al. (1995), Viswanath (1995) and Gutmark and Grinstein (1999).

33.2.4 Regimes of Reynolds and Mach Numbers

The Reynolds number (Re) is the ratio of inertial to viscous forces and, absent centrifugal, gravitational, electromagnetic and other unusual effects, Re determines whether the flow is laminar or turbulent. It is defined as $Re \equiv v_o L/\nu$, where v_o and L are, respectively, suitable velocity and length scales, and ν is the kinematic viscosity. For low-Reynolds-number flows, instabilities are suppressed by viscous effects and the flow is laminar, as can be found in systems with large fluid viscosity, small length scale or small velocity. The large-scale motion of the highly viscous volcanic molten rock and air/water flow in capillaries and microdevices are examples of laminar flows. Turbulent flows seem to be the rule rather than the exception, occurring in or around most important fluid systems such as airborne and waterborne vessels, gas and oil pipelines, material processing plants and the human cardiovascular and pulmonary systems.

Because of the nature of their instabilities, free-shear flows undergo transition at extremely low Reynolds numbers as compared to wall-bounded flows. Many techniques are available to delay laminar-to-turbulence transition for both kinds of flows, but none would do that to indefinitely high Reynolds numbers. Therefore, for Reynolds numbers beyond a reasonable limit, one should not attempt to prevent transition but rather deal with the ensuing turbulence. Of course early transition to turbulence can be advantageous in some circumstances—for example, to achieve separation delay, enhanced mixing or augmented heat transfer. The task of advancing transition is generally simpler than trying to delay it. Numerous books and articles specifically address the control of laminar-to-turbulence transition [e.g., Gad-el-Hak, 2000, and references therein]. For now, we briefly discuss transition control for various regimes of Reynolds and Mach numbers.

Three Reynolds number regimes can be identified for the purpose of reducing skin friction in wall-bounded flows. First, if the flow is laminar, typically at Reynolds numbers based on a distance from the leading edge $<10^6$, then methods of reducing the laminar shear stress are sought. These are usually velocity-profile modifiers—for example, adverse-pressure gradient, injection, cooling (in water) and heating (in air)—that reduce the fullness of the profile at the increased risk of premature transition and separation. Second, in the range of Reynolds numbers from 1×10^6 to 4×10^7, active and passive methods to delay transition as far back as possible are sought. These techniques can result in substantial savings and are broadly classified into two categories: stability modifiers and wave cancellation. The skin-friction coefficient in the laminar flat plate can be as much as an order of magnitude less than that in the turbulent case. Note, however, that all the stability modifiers, such as a favorable pressure gradient, suction or heating (in liquids), result in an increase in the skin friction over the unmodified Blasius layer. The object is, of course, to keep this penalty below the potential saving; i.e., the net drag will be above that of the flat-plate laminar boundary layer but presumably well below the viscous drag in the flat-plate turbulent flow. Third, for $Re > 4 \times 10^7$, transition to turbulence cannot be delayed with any known practical method without incurring a penalty that exceeds the saving. The task is then to reduce the skin-friction coefficient in a turbulent boundary layer. Relaminarization [Narasimha and Sreenivasan, 1979] is an option, although achieving a net saving here is problematic at present.

The Mach number (Ma) is the ratio of a characteristic flow velocity to the local speed of sound, $Ma \equiv v_o/a_o$. It determines whether the flow is incompressible ($Ma < 0.3$) or compressible ($Ma > 0.3$). The latter regime is further divided into subsonic ($Ma < 1$), transonic ($0.8 < Ma < 1.2$), supersonic ($Ma > 1$) and hypersonic ($Ma > 5$). Each of those flow regimes lends itself to different optimum methods of control to achieve a given goal. Take laminar-to-turbulence transition control as an illustration [Bushnell, 1994]. During transition, the field of initial disturbances is internalized via a process termed *receptivity*, and the disturbances are subsequently amplified by various linear and nonlinear mechanisms. Assuming that bypass mechanisms, such as roughness or high levels of freestream turbulence, are identified and circumvented, delaying transition then is reduced to controlling the variety of possible linear modes: Tollmien–Schlichting modes, Mack modes, cross-flow instabilities and Görtler instabilities. Tollmien–Schlichting instabilities dominate the transition process for two-dimensional boundary layers having $Ma < 4$ and are damped by increasing the Mach number, by wall cooling (in gases) and by the presence of a favorable pressure gradient. Contrast this to the Mack modes that dominate for two-dimensional hypersonic flows. Mack instabilities are also damped by increasing the Mach number and by the presence of a favorable pressure gradient but are destabilized by wall cooling. Cross-flow and Görtler instabilities are caused by, respectively, the development of inflectional cross-flow velocity profile and the presence of concave streamline curvature. Both of these instabilities are potentially harmful across the speed range but are largely unaffected by Mach number and wall cooling. The cross-flow modes are enhanced by a favorable pressure gradient, while the Görtler instabilities are insensitive. Suction suppresses, to different degrees, all the linear modes discussed here.

33.2.5 Convective and Absolute Instabilities

In addition to grouping the different kinds of hydrodynamic instabilities as inviscid or viscous, one could also classify them as convective or absolute based on the linear response of the system to an initial localized impulse [Huerre and Monkewitz, 1990]. A flow is convectively unstable if, at any fixed location, this response eventually decays in time (in other words, if all growing disturbances convect downstream from their source). Convective instabilities occur when there is no mechanism for upstream disturbance propagation, as, for example, in the case of rigid-wall boundary layers. If the disturbance is removed, then the perturbation propagates downstream and the flow relaxes to an undisturbed state. Suppression of convective instabilities is particularly effective when applied near the point where the perturbations originate.

If any of the growing disturbances has zero group velocity, the flow is absolutely unstable. This means that the local system response to an initial impulse grows in time. Absolute instabilities occur when a mechanism exists for upstream disturbance propagation, as, for example, separated flow over a backward-facing step where the flow recirculation provides such a mechanism. In this case, some of the growing disturbances can travel back upstream and continually disrupt the flow even after the initial disturbance is neutralized. Therefore, absolute instabilities are generally more dangerous and more difficult to control; nothing short of complete suppression will work. In some flows (for example, two-dimensional blunt-body wakes), certain regions are absolutely unstable while others are convectively unstable. The upstream addition of acoustic or electric feedback can change a convectively unstable flow to an absolutely unstable one, and self-excited flow oscillations can thus be generated. In any case, identifying the character of flow instability facilitates its effective control (i.e., suppressing or amplifying the perturbation as needed).

33.3 The Taming of the Shrew

For the rest of this chapter, we focus on reactive flow control specifically targeting the coherent structures in turbulent flows. By comparison with laminar flow control or separation prevention, the control of turbulent flow remains a very challenging problem. Flow instabilities magnify quickly near critical flow regimes; therefore, delaying transition or separation is a relatively easier task. In contrast, classical control strategies are often ineffective for fully turbulent flows. Newer ideas for turbulent flow control to achieve, for example, skin-friction drag reduction focus on the direct onslaught on coherent structures. Spurred by the recent developments in chaos control, microfabrication and soft computing tools, reactive control of turbulent flows, where sensors detect oncoming coherent structures and actuators attempt to favorably modulate those quasi-periodic events, is now in the realm of the possible for future practical devices.

Considering the extreme complexity of the turbulence problem in general and the unattainability of first-principles analytical solutions in particular, it is not surprising that controlling a turbulent flow remains a challenging task, mired in empiricism and unfulfilled promises and aspirations. Brute force suppression, or taming, of turbulence via active, energy-consuming control strategies is always possible, but the penalty for doing so often exceeds any potential benefits. The artifice is to achieve a desired effect with minimum energy expenditure. This is, of course, easier said than done. Indeed, suppressing turbulence is as arduous as *The Taming of the Shrew*.

33.4 Control of Turbulence

Numerous methods of flow control have already been successfully implemented in practical engineering devices. Delaying laminar-to-turbulence transition to reasonable Reynolds numbers and preventing separation can readily be accomplished using a myriad of passive and predetermined active control strategies. Such classical techniques have been reviewed by, among others, Bushnell (1983; 1994), Wilkinson et al. (1988), Bushnell and McGinley (1989), Gad-el-Hak (1989; 2000), Bushnell and Hefner (1990), Fiedler and Fernholz (1990), Gad-el-Hak and Bushnell (1991a; 1991b), Barnwell and Hussaini (1992), Viswanath (1995) and Joslin et al. (1996). Yet, very few of the classical strategies are effective in controlling free-shear or wall-bounded turbulent flows. Serious limitations exist for some familiar control techniques when applied to certain turbulent flow situations. For example, in attempting to reduce the skin-friction drag of a body having a turbulent boundary layer using global suction, the *penalty* associated with the control device often exceeds the *savings* derived from its use. What is needed is a way to reduce this penalty to achieve a more efficient control.

Flow control is most effective when applied near the transition or separation points, in other words, near the critical flow regimes where flow instabilities magnify quickly. Therefore, delaying or advancing laminar-to-turbulence transition and preventing or provoking separation are relatively easier tasks to accomplish. Reducing the skin-friction drag in a nonseparating turbulent boundary layer, where the mean flow is quite stable, is a more challenging problem. Yet, even a modest reduction in the fluid resistance to the motion of, for example, the worldwide commercial airplane fleet is translated into fuel savings estimated to be in the billions of dollars. Newer ideas for turbulent flow control focus on the direct onslaught on coherent structures via reactive control strategies that utilize large arrays of microsensors and microactuators.

The primary objective of this chapter is to advance possible scenarios by which viable control strategies of turbulent flows might be realized. As will be argued in the following sections, future systems for control of turbulent flows, in general, and turbulent boundary layers, in particular, could greatly benefit from the merging of the science of chaos control, the technology of microfabrication and the newest computational tools collectively termed *soft computing*. Control of chaotic, nonlinear dynamical systems has been demonstrated theoretically as well as experimentally, even for multi-degree-of-freedom systems. Microfabrication is an emerging technology that has the potential for producing inexpensive, programmable sensor/actuator chips that have dimensions on the order of a few microns. Soft computing tools include neural networks, fuzzy logic and genetic algorithms and are now more advanced as well as more widely used as compared to just few years ago. These tools could be very useful in constructing effective adaptive controllers.

Such futuristic systems are envisaged as consisting of a large number of intelligent, interactive, microfabricated wall sensors and actuators arranged in a checkerboard pattern and targeted toward specific organized structures that occur quasi-randomly within a turbulent flow. Sensors detect oncoming coherent structures, and adaptive controllers process the sensor information and provide control signals to the actuators, which in turn attempt to favorably modulate the quasi-periodic events. A finite number of wall sensors perceive only partial information about the entire flowfield above. However, a low-dimensional dynamical model of the near-wall region used in a Kalman filter can make the most of the partial information from the sensors. Conceptually, all of this is not too difficult, but in practice the complexity of such a control system is daunting and much research and development work still remain.

The following discussion is organized into ten sections. A particular example of a classical control system—suction—is described in the following section. This will serve as a prelude to introducing the selective suction concept. The different hierarchy of coherent structures that dominate a turbulent boundary layer and that constitute the primary target for direct onslaught are then briefly recalled. The characteristic lengths and sensor requirements of turbulent flows are then discussed in the two subsequent sections. This is followed by a description of reactive flow control and the selective suction concept. The number, size, frequency and energy consumption of the sensor/actuator units required to tame the turbulence on a full-scale air or water vehicle are estimated in that same section. This is followed by an introduction to the topic of magnetohydrodynamics and a reactive flow control scheme using electromagnetic body forces. The emerging areas of chaos control and soft computing, particularly as they relate to reactive control strategies, are then briefly discussed in the two subsequent sections. This is followed by a discussion of the specific use of microelectromechanical systems (MEMS) devices for reactive flow control. Finally, brief concluding remarks are given in the last section.

33.5 Suction

To set the stage for introducing the concept of targeted or selective control, I shall first discuss in this section global control as applied to wall-bounded flows. A viscous fluid that is initially irrotational will acquire vorticity when an obstacle is passed through the fluid. This vorticity controls the nature and structure of the boundary-layer flow in the vicinity of the obstacle. For an incompressible, wall-bounded flow, the flux of spanwise or streamwise vorticity at the wall (and, hence, whether the surface is a sink or a source of vorticity) is affected by the wall motion (e.g., in the case of a compliant coating), transpiration (suction or injection), streamwise or spanwise pressure gradient, wall curvature, normal viscosity gradient near the wall (caused by, for example, heating or cooling of the wall or introduction of a shear-thinning/shear-thickening additive into the boundary layer) and body forces (such as electromagnetic ones in a conducting fluid). These alterations separately or collectively control the shape of the instantaneous as well as the mean velocity profiles, which in turn determine the skin friction at the wall, the boundary layer ability to resist transition and separation and the intensity of turbulence and its structure.

For illustration purposes, I shall focus on global wall suction as a generic control tool. The arguments presented here and in subsequent sections are equally valid for other global control techniques, such as geometry modification (body shaping), surface heating or cooling, electromagnetic control, etc. Transpiration provides a good example of a single control technique that is used to achieve a variety of goals. Suction leads to a fuller velocity profile (vorticity flux away from the wall) and can, therefore, be employed to delay laminar-to-turbulence transition, postpone separation, achieve an asymptotic turbulent boundary layer (i.e., one having constant momentum thickness) or relaminarize an already turbulent flow. Unfortunately, global suction cannot be used to reduce the skin-friction drag in a turbulent boundary layer. The amount of suction required to inhibit boundary-layer growth is too large to effect a net drag reduction. This is a good illustration of a situation where the penalty associated with a control device might exceed the saving derived from its use.

Small amounts of fluid withdrawn from the near-wall region of a boundary layer change the curvature of the velocity profile at the wall and can dramatically alter the stability characteristics of the flow. Concurrently, suction inhibits the growth of the boundary layer, so that the critical Reynolds number based on thickness may never be reached. Although laminar flow can be maintained to extremely high Reynolds numbers provided that enough fluid is sucked away, the goal is to accomplish transition delay with the minimum suction flow rate. This will reduce not only the power necessary to drive the suction pump, but also the momentum loss due to the additional freestream fluid entrained into the boundary layer as a result of withdrawing fluid from the wall. That momentum loss is, of course, manifested as an increase in the skin-friction drag.

The case of uniform suction from a flat plate at zero incidence is an exact solution of the Navier–Stokes equation. The asymptotic velocity profile in the viscous region is exponential and has a negative curvature at the wall. The displacement thickness has the constant value $\delta^* = \nu/|v_w|$, where ν is the kinematic viscosity, and $|v_w|$ is the absolute value of the normal velocity at the wall. In this case, the familiar von Kármán integral equation reads $C_f = 2C_q$. Bussmann and Münz (1942) computed the critical Reynolds number for the asymptotic suction profile to be $Re_{\delta^*} \equiv U_\infty \delta^*/\nu = 70{,}000$. From the value of δ^* given above, the flow is stable to all small disturbances if $C_q \equiv |v_w|/U_\infty > 1.4 \times 10^{-5}$. The amplification rate of unstable disturbances for the asymptotic profile is an order of magnitude less than that for the Blasius boundary layer [Pretsch, 1942]. This treatment ignores the development distance from the leading edge needed to reach the asymptotic state. When this is included into the computation, a higher $C_q = 1.18 \times 10^{-4}$ is required to ensure stability [Iglisch, 1944; Ulrich, 1944].

In a turbulent wall-bounded flow, the results of Eléna (1975; 1984) and Antonia et al. (1988) indicate that suction causes an appreciable stabilization of the low-speed streaks in the near-wall region. The maximum turbulence level at $y^+ \approx 13$ drops from 15 to 12% as C_q varies from 0 to 0.003. More dramatically, the tangential Reynolds stress near the wall drops by a factor of 2 for the same variation of C_q. The dissipation length scale near the wall increases by 40% and the integral length scale by 25% with the suction.

The suction rate necessary for establishing an asymptotic turbulent boundary layer independent of streamwise coordinate (i.e., $d\delta_\theta/dx = 0$) is much lower than the rate required for relaminarization ($C_q \approx 0.01$) but is still not low enough to yield a net drag reduction. For a Reynolds number based on distance from the leading edge, $Re_x = \mathcal{O}[10^6]$, Favre et al. (1966), Rotta (1970) and Verollet et al. (1972), among others, report an asymptotic suction coefficient of $C_q \approx 0.003$. For a zero-pressure-gradient boundary layer on a flat plate, the corresponding skin-friction coefficient is $C_f = 2C_q = 0.006$, indicating higher skin friction than if no suction was applied. To achieve a net skin-friction reduction with suction, the process must be further optimized. One way to accomplish this is to target the suction toward particular organized structures within the boundary layer and not to use it globally as in classical control schemes. This point will be revisited later, but the coherent structures to be targeted and the length scales to be expected are first detailed in the following two sections.

33.6 Coherent Structures

The discussion above indicates that achieving a particular control goal is always possible. The challenge is reaching that goal with a penalty that can be tolerated. Suction, for example, would lead to a net drag reduction, if only we could reduce the suction coefficient necessary for establishing an asymptotic turbulent boundary layer to below one half of the unperturbed skin-friction coefficient. A more efficient way of using suction, or any other global control method, is to target particular coherent structures within the turbulent boundary layer. Before discussing this selective control idea, I shall briefly describe in this and the following section the different hierarchy of organized structures in a wall-bounded flow and the expected scales of motion.

The classical view that turbulence is essentially a stochastic phenomenon having a randomly fluctuating velocity field superimposed on a well-defined mean has been changed in the last few decades by the realization that the transport properties of all turbulent shear flows are dominated by quasi-periodic, large-scale vortex motions [Laufer, 1975; Cantwell, 1981; Fiedler, 1988; Robinson, 1991]. Despite the extensive research work in this area, no generally accepted definition of what is meant by coherent motion has emerged. In physics, coherence stands for well-defined phase relationship. For our purpose, we adopt the rather restrictive definition given by Hussain (1986): "A coherent structure is a connected turbulent fluid mass with instantaneously phase-correlated vorticity over its spatial extent." In other words, underlying the random, three-dimensional vorticity that characterizes turbulence, there is a component of large-scale vorticity which is instantaneously coherent over the spatial extent of an organized structure. The apparent randomness of the flowfield is, for the most part, due to the random size and strength of the different types of organized structures comprising that field.

In a wall-bounded flow, a multiplicity of coherent structures have been identified mostly through flow visualization experiments, although some important early discoveries have been made using correlation measurements [Townsend, 1961; 1970; Bakewell and Lumley, 1967]. Although the literature on this topic is vast, no research-community-wide consensus has been reached, particularly on the issues of the origin of and interaction between the different structures, regeneration mechanisms and Reynolds number effects. What follow are somewhat biased remarks addressing those issues, gathered mostly via low-Reynolds-number experiments. The interested reader is referred to the book edited by Panton (1997) and the large number of review articles available [e.g., Kovasznay, 1970; Laufer, 1975; Willmarth, 1975a; 1975b; Saffman, 1978; Cantwell, 1981; Fiedler, 1986; 1988; Blackwelder, 1988; 1998; Robinson, 1991; Delville et al., 1998]. The paper by Robinson (1991) in particular summarizes many of the different, sometimes contradictory, conceptual models offered thus far by different research groups. Those models are aimed ultimately at explaining how the turbulence maintains itself and range from the speculative to the rigorous, but none, unfortunately, is self-contained and complete. Furthermore, the structure research dwells largely on the kinematics of organized motion and little attention is given to the dynamics of the regeneration process.

In a boundary layer, the turbulence production process is dominated by three kinds of quasi-periodic—or, depending on one's viewpoint, quasi-random—eddies: (1) the large outer structures, (2) the intermediate Falco eddies, and (3) the near-wall events. The large, three-dimensional structures scale with the boundary-layer thickness, δ, and extend across the entire layer [Kovasznay et al., 1970; Blackwelder and Kovasznay, 1972]. These eddies control the dynamics of the boundary layer in the outer region, such as entrainment, turbulence production, etc. They appear randomly in space and time, and seem to be, at least for moderate Reynolds numbers, the residue of the transitional Emmons spots [Zilberman et al., 1977; Gad-el-Hak et al., 1981; Riley and Gad-el-Hak, 1985].

The Falco eddies are also highly coherent and three dimensional. Falco (1974; 1977) named them typical eddies because they appear in wakes, jets, Emmons spots, grid-generated turbulence and boundary layers in zero, favorable and adverse pressure gradients. They have an intermediate scale of about 100 ν/u_τ (100 wall units; u_τ is the friction velocity, and ν/u_τ is the viscous length scale). The Falco eddies appear to be an important link between the large structures and the near-wall events.

The third kind of eddies exist in the near-wall region ($0 < y < 100\ \nu/u_\tau$), where the Reynolds stress is produced in a very intermittent fashion. Half of the total production of turbulence kinetic energy ($-\overline{uv}\ \partial \overline{U}/\partial y$) takes place near the wall in the first 5% of the boundary layer at typical laboratory Reynolds numbers (a smaller fraction at higher Reynolds numbers), and the dominant sequence of eddy motions found there is termed the *bursting phenomenon*. This dynamically significant process, identified during the 1960s by researchers at Stanford University [Kline and Runstadler, 1959; Runstadler et al., 1963; Kline et al., 1967; Kim et al., 1971; Offen and Kline, 1974; 1975], was reviewed by Willmarth (1975a), Blackwelder (1978), Robinson (1991) and, most recently, Panton (1997) and Blackwelder (1998).

To focus the discussion on the bursting process and its possible relationships to other organized motions within the boundary layer, we refer to the schematic in Figure 33.5. Qualitatively, the process, according to at least one school of thought, begins with elongated, counter-rotating, streamwise vortices having diameters of approximately 40 wall units or 40 ν/u_τ. The estimate for the diameter of the vortex is obtained from the conditionally averaged spanwise velocity profiles reported by Blackwelder and Eckelmann (1979). There is a distinction, however, between vorticity distribution and a vortex [Saffman and Baker, 1979; Robinson et al., 1989; Robinson, 1991], and the visualization results of Smith and Schwartz (1983) may indicate a much smaller diameter. In any case, referring to Figure 33.6, the counter-rotating

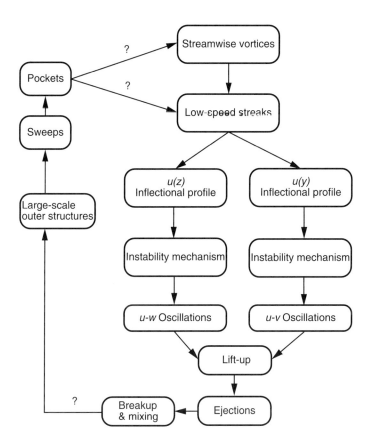

FIGURE 33.5 Proposed sequence of the bursting process. The arrows indicate the sequential events, and the "?" indicates relationships with less supporting evidence. (Adapted from Blackwelder, 1998; from Gad-el-Hak, M. (2000) *Flow Control: Passive, Active, and Reactive Flow Management*. Reprinted with permission of Cambridge University Press, New York.)

vortices exist in a strong shear and induce low- and high-speed regions between them. Those low-speed streaks were first visualized by Francis Hama at the University of Maryland [see Corrsin, 1957], although Hama's contribution is frequently overlooked in favor of the subsequent and more thorough studies conducted at Stanford University and cited above. The vortices and the accompanying eddy structures occur randomly in space and time. However, their appearance is sufficiently regular that an average spanwise wavelength of approximately 80 to 100 v/u_τ has been identified by Kline et al. (1967) and others.

It might be instructive at this point to emphasize that the distribution of streak spacing is very broad. The standard of deviation is 30 to 40% of the more commonly quoted mean spacing between low-speed streaks of 100 wall units. Both the mean and standard deviation are roughly independent of Reynolds number in the rather limited range of reported measurements (Re_θ = 300–6500; see Smith and Metzler, 1983; Kim et al., 1987). Butler and Farrell (1993) have shown that the mean streak spacing of 100 v/u_τ is consistent with the notion that this is an optimal configuration for extracting "the most energy over an appropriate eddy turnover time." In their work, the streak spacing remains 100 wall units at Reynolds numbers based on friction velocity and channel half-width of a^+ = 180–360.

Kim et al. (1971) observed that the low-speed regions grow downstream, lift up and develop (instantaneous) inflectional $U(y)$ profiles.[2] At approximately the same time, the interface between the low- and high-speed fluid begins to oscillate, apparently signaling the onset of a secondary instability. The low-speed

---

[2] According to Swearingen and Blackwelder (1984), inflectional $U(z)$ profiles are just as likely to be found in the near-wall region and can also be the cause of subsequent bursting events (see Figure 33.5).

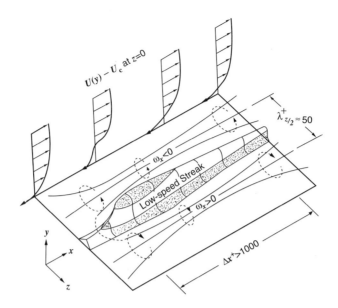

FIGURE 33.6 Model of near-wall structure in turbulent wall-bounded flows. (Adapted from Blackwelder, 1978; from Gad-el-Hak, M. (2000) *Flow Control: Passive, Active, and Reactive Flow Management*. Reprinted with permission of Cambridge University Press, New York.)

region lifts up away from the wall as the oscillation amplitude increases, and then the flow rapidly breaks up into a completely chaotic motion. The streak oscillations commence at $y^+ \approx 10$, and the abrupt breakup takes place in the buffer layer, although the ejected fluid reaches all the way to the logarithmic region. Because the breakup process occurs on a very short time scale, Kline et al. (1967) called it a burst.

Virtually all of the net production of turbulence kinetic energy in the near-wall region occurs during these bursts. Corino and Brodkey (1969) showed that the low-speed regions are quite narrow (i.e., $z = 20\,\nu/u_\tau$) and may also have significant shear in the spanwise direction. They also indicated that the ejection phase of the bursting process is followed by a large-scale motion of upstream fluid that emanates from the outer region and cleanses (sweeps) the wall region of the previously ejected fluid. The sweep phase is, of course, required by the continuity equation and appears to scale with the outer-flow variables. The sweep event seems to stabilize the bursting site, in effect preparing it for a new cycle.

Considerably more has been learned about the bursting process during the last two decades. For example, Falco (1980; 1983; 1991) has shown that when a typical eddy, which may be formed in part by ejected wall-layer fluid, moves over the wall it induces a high uv sweep (positive u and negative v). The wall region is continuously bombarded by pockets of high-speed fluid originating in the logarithmic and possibly the outer layers of the flow. These pockets appear to scale—at least in the limited Reynolds number range where they have been observed, $Re_\theta = \mathcal{O}[1000]$—with wall variables and tend to promote and/or enhance the inflectional velocity profiles by increasing the instantaneous shear leading to a more rapidly growing instability. The relation between the pockets and the sweep events is not clear, but it seems that the former forms the highly irregular interface between the latter and the wall-region fluid. More recently, Klewicki et al. (1994) conducted four-element, hot-wire probe measurements in a low-Reynolds-number canonical boundary layer to clarify the roles of velocity–spanwise vorticity field interactions regarding the near-wall turbulent stress production and transport.

Other significant experiments were conducted by Tiederman and his students [Donohue et al., 1972; Reischman and Tiederman, 1975; Oldaker and Tiederman, 1977; Tiederman et al., 1985], Smith and his colleagues [Smith and Metzler, 1982; 1983; Smith and Schwartz, 1983] and the present author and his collaborators. The first group conducted extensive studies of the near-wall region, particularly the viscous sublayer, of channels with Newtonian as well as drag-reducing non-Newtonian fluids. Smith's group, using a unique, two-camera, high-speed video system, was the first to indicate a symbiotic relationship between the occurrence of low-speed streaks and the formation of vortex loops in the near-wall region. Gad-el-Hak and Hussain (1986) and Gad-el-Hak and Blackwelder (1987a) have introduced methods by which the bursting events and large-eddy structures are artificially generated in a boundary layer. Their experiments greatly facilitate the study of the uniquely controlled simulated coherent structures via phase-locked measurements.

Blackwelder and Haritonidis (1983) have shown convincingly that the frequency of occurrence of the bursting events scales with the viscous parameters consistent with the usual boundary-layer scaling arguments. An excellent review of the dynamics of turbulent boundary layers has been provided by Sreenivasan (1989). More information about coherent structures in high-Reynolds-number boundary layers is given by Gad-el-Hak and Bandyopadhyay (1994). The book edited by Panton (1997) emphasizes the self-sustaining mechanisms of wall turbulence.

33.7 Scales

In this and the following section, we develop the equations needed to compute the characteristic lengths and sensor requirements of turbulent flows, in general, and wall-bounded flows, in particular. Turbulence is a high-Reynolds-number phenomenon which is characterized by the existence of numerous length and time scales [Tennekes and Lumley, 1972]. The spatial extension of the length scales is bounded from above by the dimensions of the flowfield and from below by the diffusive and dissipative action of the molecular viscosity.

If we limit our interest to shear flows, which are basically characterized by two large length scales, one in the streamwise direction (the convective or longitudinal length scale) and the other perpendicular to the flow direction (the diffusive or lateral length scale), we obtain a more well-defined problem. Moreover, at sufficiently high Reynolds numbers the boundary-layer approximation applies and it is assumed that there is a wide separation between the lateral and longitudinal length scales. This leads to some attractive simplifications in the equations of motion—for instance, the elliptical Navier–Stokes equations are transferred to the parabolic boundary-layer equations [e.g., Hinze, 1975]. So, in this approximation, the lateral scale is approximately equal to the extension of the flow perpendicular to the flow direction (e.g., the boundary-layer thickness), and the largest eddies typically have this spatial extension. These eddies are most energetic and play a crucial role both in the transport of momentum and contaminants. A constant energy supply is needed to maintain the turbulence, and this energy is extracted from the mean flow into the largest most energetic eddies. The lateral length scale is also the relevant scale for analyzing this energy transfer. However, there is an energy destruction in the flow due to the action of the viscous forces (the dissipation), and for the analysis of this process other, smaller length scales are needed.

As the eddy size decreases, viscosity becomes a more significant parameter, as one property of viscosity is its effectiveness in smoothing out velocity gradients. The viscous and the nonlinear terms in the momentum equation counteract each other in the generation of small-scale fluctuations. While the inertial terms try to produce smaller and smaller eddies, the viscous terms check this process and prevent the generation of infinitely small scales by dissipating the small-scale energy into heat. In the early 1940s, the universal equilibrium theory was developed by Kolmogorov (1941a; 1941b). One cornerstone of this theory is that the small-scale motions are statistically independent of the relatively slower large-scale turbulence. An implication of this is that the turbulence at the small scales depends only on two parameters—namely, the rate at which energy is supplied by the large-scale motion and the kinematic viscosity. In addition, it is assumed in the equilibrium theory that the rate of energy supply to the turbulence should be equal to the rate of dissipation. Hence, in the analysis of turbulence at small scales, the dissipation rate per unit mass ε is a relevant parameter together with the kinematic viscosity ν. Kolmogorov (1941a) used simple dimensional arguments to derive length, time and velocity scales relevant to the small-scale motion given, respectively, by:

$$\eta = \left(\frac{\nu^3}{\varepsilon}\right)^{\frac{1}{4}} \tag{33.1}$$

$$\tau = \left(\frac{\nu}{\varepsilon}\right)^{\frac{1}{2}} \tag{33.2}$$

$$\upsilon = (\nu\varepsilon)^{\frac{1}{4}} \tag{33.3}$$

These scales are accordingly called the Kolmogorov microscales, or sometimes the inner scales of the flow. As they are obtained through a physical argument, these scales are the smallest scales that can exist in a turbulent flow and they are relevant for both free-shear and wall-bounded flows.

In boundary layers, the shear-layer thickness provides a measure of the largest eddies in the flow. The smallest scale in wall-bounded flows is the viscous wall unit, which will be shown below to be of the same order as the Kolmogorov length scale. Viscous forces dominate over inertia in the near-wall region, and the characteristic scales there are obtained from the magnitude of the mean vorticity in the region and its viscous diffusion away from the wall. Thus, the viscous time scale, t_v, is given by the inverse of the mean wall vorticity:

$$t_v = \left[\left.\frac{\partial \overline{U}}{\partial y}\right|_w\right]^{-1} \tag{33.4}$$

where $\overline{U}$ is the mean streamwise velocity. The viscous length-scale, ℓ_v, is determined by the characteristic distance by which the (spanwise) vorticity is diffused from the wall, and is thus given by:

$$\ell_v = \sqrt{\nu t_v} = \sqrt{\frac{\nu}{\left.\frac{\partial \overline{U}}{\partial y}\right|_w}} \tag{33.5}$$

where ν is the kinematic viscosity. The wall velocity scale (so-called friction velocity, u_τ) follows directly from the above time and length scales:

$$u_\tau = \frac{\ell_v}{t_v} = \sqrt{\left.\nu\frac{\partial \overline{U}}{\partial y}\right|_w} = \sqrt{\frac{\tau_w}{\rho}} \tag{33.6}$$

where τ_w is the mean shear stress at the wall, and ρ is the fluid density. A wall unit implies scaling with the viscous scales, and the usual $(\cdots)^+$ notation is used; for example, $y^+ = y/\ell_v = yu_\tau/\nu$. In the wall region, the characteristic length for the large eddies is y itself, while the Kolmogorov scale is related to the distance from the wall y as follows:

$$\eta^+ \equiv \frac{\eta u_\tau}{\nu} \approx (\kappa y^+)^{\frac{1}{4}} \tag{33.7}$$

where κ is the von Kármán constant (≈ 0.41). As y^+ changes in the range of 1 to 5 (the extent of the viscous sublayer), η changes from 0.8 to 1.2 wall units.

We now have access to scales for the largest and smallest eddies of a turbulent flow. To continue our analysis of the cascade energy process, it is necessary to find a connection between these diverse scales.

One way of obtaining such a relation is to use the fact that at equilibrium the amount of energy dissipating at high wavenumbers must equal the amount of energy drained from the mean flow into the energetic large-scale eddies at low wavenumbers. In the inertial region of the turbulence kinetic energy spectrum, the flow is almost independent of viscosity, and because the same amount of energy dissipated at the high wavenumbers must pass this 'inviscid' region an inviscid relation for the total dissipation may be obtained by the following argument. The amount of kinetic energy per unit mass of an eddy with a wavenumber in the inertial sublayer is proportional to the square of a characteristic velocity for such an eddy, u^2. The rate of transfer of energy is assumed to be proportional to the reciprocal of one eddy turnover time, u/ℓ, where ℓ is a characteristic length of the inertial sublayer. Hence, the rate of energy that is supplied to the small-scale eddies via this particular wavenumber is of order of u^3/ℓ, and this amount of energy must be equal to the energy dissipated at the highest wavenumber, expressed as:

$$\varepsilon \approx \frac{u^3}{\ell} \tag{33.8}$$

Note that this is an inviscid estimate of the dissipation, as it is based on large-scale dynamics and does not either involve or contain viscosity. More comprehensive discussion of this issue can be found in Taylor (1935) and Tennekes and Lumley (1972). From an experimental perspective, this is a very important expression, as it offers one way of estimating the Kolmogorov microscales from quantities measured in a much lower wavenumber range.

Because the Kolmogorov length and time scales are the smallest scales occurring in turbulent motion, a central question will be how small these scales can be without violating the continuum hypothesis. By looking at the governing equations, it can be concluded that high dissipation rates are usually associated with large velocities, and this situation is more likely to occur in gases than in liquids, so it would be sufficient to show that for gas flows the smallest turbulence scales are normally much large than the molecular scales of motion. The relevant molecular length scale is the mean free path, $\mathcal{L}$, and the ratio between this length and the Kolmogorov length scale, η, is the microstructure Knudsen number and can be expressed as (see Corrsin, 1959):

$$Kn = \frac{\mathcal{L}}{\eta} \approx \frac{Ma^{\frac{1}{4}}}{Re} \tag{33.9}$$

where the turbulence Reynolds number, Re, and the turbulence Mach number, Ma, are used as independent variables. It is obvious that a turbulent flow will interfere with the molecular motion only at a high Mach number and low Reynolds number, and this is a very unusual situation occurring only in certain gaseous nebulae[3]. Thus, under normal conditions the turbulence Knudsen number falls in the group of continuum flows. However, measurements using extremely thin hot-wires, small MEMS sensors or flows within narrow MEMS channels can generate values in the slip-flow regime and even beyond; this implies that, for instance, the no-slip condition may be questioned, as thoroughly discussed in Part I of this book.

33.8 Sensor Requirements

It is the ultimate goal of all measurements in turbulent flows to resolve both the largest and smallest eddies that occur in the flow. At the lower wavenumbers, the largest and most energetic eddies occur, and normally no problems are associated with resolving these eddies. Basically, this is a question of having access to computers with sufficiently large memory for storing the amount of data that may have to be acquired from a large number of distributed probes, each collecting data for a time period long enough

---

[3]Note that in microduct flows and the like, the Re is usually too small for turbulence to even exist. So the issue of a turbulence Knudsen number is moot in those circumstances even if rarefaction effects become strong.

to reduce the statistical error to a prescribed level. However, at the other end of the spectrum, both the spatial and the temporal resolutions are crucial, and this puts severe limitations on the sensors to be used. It is possible to obtain a relation between the small and large scales of the flow by substituting the inviscid estimate of the total dissipation rate, Eq. (33.8), into the expressions for the Kolmogorov microscales, Eqs. (33.1) to (33.3). Thus,

$$\frac{\eta}{\ell} \approx \left(\frac{u\ell}{\nu}\right)^{-\frac{3}{4}} = Re^{-\frac{3}{4}} \qquad (33.10)$$

$$\frac{\tau u}{\ell} \approx \left(\frac{u\ell}{\nu}\right)^{-\frac{1}{2}} = Re^{-\frac{1}{2}} \qquad (33.11)$$

$$\frac{\upsilon}{u} \approx \left(\frac{u\ell}{\nu}\right)^{-\frac{1}{4}} = Re^{-\frac{1}{4}} \qquad (33.12)$$

where Re is the Reynolds number based on the speed of the energy containing eddies u and their characteristic length ℓ. Because turbulence is a high-Reynolds-number phenomenon, these relations show that the small length, time and velocity scales are much less than those of the larger eddies and that the separation in scales widens considerably as the Reynolds number increases. Moreover, this also implies that the assumptions made on the statistical independence and the dynamical equilibrium state of the small structures will be most relevant at high Reynolds numbers. Another interesting conclusion that can be drawn from the above relations is that if two turbulent flowfields have the same spatial extension (i.e., same large-scale) but different Reynolds numbers, there would be an obvious difference in the small-scale structure in the two flows. The low-Re flow would have a relatively coarse small-scale structure, while the high-Re flow would have much finer small eddies.

To spatially resolve the smallest eddies, we need sensors that are of approximately the same size as the Kolmogorov length scale for the particular flow under consideration. This implies that as the Reynolds number increases smaller sensors are required. For instance, in the self-preserving region of a plane-cylinder wake at a modest Reynolds number, based on the cylinder diameter of 1840, the value of η varies in the range of 0.5 to 0.8 mm [Aronson and Löfdahl, 1994]. For this case, conventional hot-wires can be used for turbulence measurements. However, an increase in the Reynolds number by a factor of 10 will yield Kolmogorov scales in the micrometer range and call for either extremely small conventional hot-wires or MEMS-based sensors. Another illustrating example of the Reynolds number effect on the requirement of small sensors is a simple two-dimensional, flat-plate boundary layer. At a momentum thickness Reynolds number of $Re_\theta = 4000$, the Kolmogorov length scale is typically of the order of 50 μm; in order to resolve these scales, it is necessary to have access to sensors that have a characteristic active measuring length of the same spatial extension.

Severe errors will be introduced in the measurements by using sensors that are too large, as such sensors will integrate the fluctuations due to the small eddies over their spatial extension, and the energy content of these eddies will be interpreted by the sensors as an average "cooling." When measuring fluctuating quantities, this implies that these eddies are counted as part of the mean flow and their energy is "lost." The result will be a lower value of the turbulence parameter, and this will wrongly be interpreted as a measured attenuation of the turbulence [e.g., Ligrani and Bradshaw, 1987]. However, because turbulence measurements deal with statistical values of fluctuating quantities, it may be possible to loosen the spatial constraint of having a sensor of the same size as η, to allow sensor dimensions which are slightly larger than the Kolmogorov scale, say on the order of η.

For boundary layers, the wall unit has been used to estimate the smallest necessary size of a sensor for accurately resolving the smallest eddies. For instance Keith et al. (1992) state that 10 wall units or less is a relevant sensor dimension for resolving small-scale pressure fluctuations. Measurements of fluctuating velocity gradients, essential for estimating the total dissipation rate in turbulent flows, are another challenging task. Gad-el-Hak and Bandyopadhyay (1994) argue that turbulence measurements with probe lengths greater than the viscous sublayer thickness (about 5 wall units) are unreliable, particularly near the surface. Many studies have been conducted on the spacing between sensors necessary to optimize the formed velocity gradients [Aronson et al., 1997, and references therein]. A general conclusion from both experiments and direct numerical simulations is that a sensor spacing of 3 to 5 Kolmogorov lengths is recommended. When designing arrays for correlation measurements or for targeted control, the spacing between the coherent structures will be the determining factor. For example, when targeting the low-speed streaks in a turbulent boundary layer, several sensors must be situated along a lateral distance of 100 wall units, the average spanwise spacing between streaks. All this requires quite small sensors, and many attempts have been made to meet these conditions with conventional sensor designs. However, in spite of the fact that conventional sensors such as hot-wires have been fabricated in the micrometer-size range (for their diameter but not their length), they are usually handmade, difficult to handle and too fragile, and here the MEMS technology has really opened a door for new applications.

It is clear from the above that the spatial and temporal resolutions for any probe to be used to resolve high-Reynolds-number turbulent flows are extremely tight. For example, both the Kolmogorov scale and the viscous length scale change from a few microns at the typical field Reynolds number—based on the momentum thickness—of 10^6 to a couple of hundred microns at the typical laboratory Reynolds number of 10^3. MEMS sensors for pressure, velocity, temperature and shear stress are at least one order of magnitude smaller than conventional sensors [Ho and Tai, 1996; 1998; Löfdahl et al., 1996; Löfdahl and Gad-el-Hak, 1999]. Their small size improves both the spatial and temporal resolutions of the measurements, typically a few microns and a few microseconds, respectively. For example, a micro-hot-wire (called hot-point) has very small thermal inertia and the diaphragm of a micro-pressure-transducer has correspondingly fast dynamic response. Moreover, the microsensors' extreme miniaturization and low energy consumption make them ideal for monitoring the flow state without appreciably affecting it. Finally, literally hundreds of microsensors can be fabricated on the same silicon chip at a reasonable cost, making them well suited for distributed measurements and control. The UCLA/Caltech team [e.g., Ho and Tai, 1996; 1998, and references therein] has been very effective in developing many MEMS-based sensors and actuators for turbulence diagnosis and control.

33.9 Reactive Flow Control

33.9.1 Introductory Remarks

Targeted control implies sensing and reacting to particular quasi-periodic structures in a turbulent flow. For a boundary layer, the wall seems to be the logical place for such reactive control, because of the relative ease of placing something in there, the sensitivity of the flow in general to surface perturbations and the proximity and therefore accessibility to the dynamically all-important, near-wall coherent events. According to Wilkinson (1990), very few actual experiments have used embedded wall sensors to initiate a surface actuator response [Alshamani et al., 1982; Wilkinson and Balasubramanian, 1985; Nosenchuck and Lynch, 1985; Breuer et al., 1989]. This 10-year-old assessment is quickly changing, however, with the introduction of microfabrication technology that has the potential for producing small, inexpensive, programmable sensor/actuator chips. Witness the more recent reactive control attempts by Kwong and Dowling (1993), Reynolds (1993), Jacobs et al. (1993), Jacobson and Reynolds (1993a; 1993b; 1994; 1995; 1998), Fan et al. (1993), James et al. (1994) and Keefe (1996). Fan et al. and Jacobson and Reynolds even consider the use of self-learning neural networks for increased computational speeds and efficiency. Recent reviews of reactive flow control include those by Gad-el-Hak (1994; 1996), Lumley (1996), McMichael (1996), Mehregany et al. (1996), Ho and Tai (1996) and Bushnell (1998).

Numerous methods of flow control have already been successfully implemented in practical engineering devices. Yet, limitations exist for some familiar control techniques when applied to specific situations. For example, in attempting to reduce the drag or enhance the lift of a body having a turbulent boundary layer using global suction, global heating/cooling or global application of electromagnetic body forces, the actuator's energy expenditure often exceeds the saving derived from the predetermined active control strategy. What is needed is a way to reduce this penalty to achieve a more efficient control. Reactive control geared specifically toward manipulating the coherent structures in turbulent shear flows, though considerably more complicated than passive control or even predetermined active control, has the potential to do just that. As will be argued in this and the following sections, future systems for control of turbulent flows, in general, and turbulent boundary layers, in particular, could greatly benefit from the merging of the science of chaos control, the technology of microfabrication and the newest computational tools collectively termed soft computing. Such systems are envisaged as consisting of a large number of intelligent, communicative wall sensors and actuators arranged in a checkerboard pattern and targeted toward controlling certain quasi-periodic, dynamically significant coherent structures present in the near-wall region.

33.9.2 Targeted Flow Control

Successful techniques to reduce the skin friction in a turbulent flow, such as polymers, particles or riblets, appear to act indirectly through local interaction with discrete turbulent structures, particularly small-scale eddies, within the flow. Common characteristics of all these passive methods are increased losses in the near-wall region, thickening of the buffer layer and lowered production of Reynolds shear stress [Bandyopadhyay, 1986]. Active control strategies that act directly on the mean flow, such as suction or lowering of near-wall viscosity, also lead to inhibition of Reynolds stress. However, skin friction is increased when any of these velocity-profile modifiers is applied globally.

Could these seemingly inefficient techniques (e.g., global suction) be used more sparingly and optimized to reduce their associated penalty? It appears that the more successful drag-reducing methods (e.g., polymers) act selectively on particular scales of motion and are thought to be associated with stabilization of the secondary instabilities. It is also clear that energy is wasted when suction or heating/cooling is used to suppress the turbulence throughout the boundary layer when the main interest is to affect a near-wall phenomenon. One ponders what would become of wall turbulence if specific coherent structures are targeted by the operator through a reactive control scheme for modification. The myriad organized structures present in all shear flows are instantaneously identifiable, quasi-periodic motions [Cantwell, 1981; Robinson, 1991]. Bursting events in wall-bounded flows, for example, are both intermittent and random in space as well as time. The random aspects of these events reduce the effectiveness of a predetermined active control strategy. If such structures are nonintrusively detected and altered, on the other hand, net performance gain might be achieved. It seems clear, however, that temporal phasing as well as spatial selectivity would be required to achieve proper control targeted toward random events.

A nonreactive version of the above idea is the selective suction technique which combines suction to achieve an asymptotic turbulent boundary layer and longitudinal riblets to fix the location of low-speed streaks. Although far from indicating net drag reduction, the available results are encouraging and further optimization is needed. When implemented via an array of reactive control loops, the selective suction method is potentially capable of a skin-friction reduction that approaches 60%.

The genesis of the selective suction concept can be found in the papers by Gad-el-Hak and Blackwelder (1987b; 1989) and the patent by Blackwelder and Gad-el-Hak (1990). These researchers suggest that one possible means of optimizing the suction rate is identifying where a low-speed streak is located and applying a small amount of suction under it. Assuming that the production of turbulence kinetic energy is due to the instability of an inflectional $U(y)$ velocity profile, one needs to remove only enough fluid so that the inflectional nature of the profile is alleviated. An alternative technique that could conceivably reduce the Reynolds stress is to inject fluid selectively under the high-speed regions. The immediate effect of normal injection would be to decrease the viscous shear at the wall, resulting in less drag. In addition, the velocity profiles in the spanwise direction, $U(z)$, would have a smaller shear, $\partial U/\partial z$, because the suction/injection would create a more uniform flow. Swearingen and Blackwelder (1984) and Blackwelder and Swearingen (1990) have found that inflectional $U(z)$ profiles occur as often as inflection points are

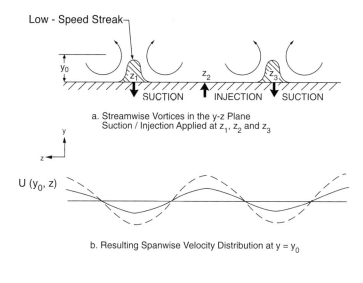
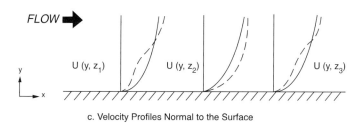

FIGURE 33.7 Effects of suction/injection on velocity profiles. Broken lines, reference profiles; solid lines, profiles with transpiration applied. (a) Streamwise vortices in the y–z plane, suction/injection applied at z_1, z_2 and z_3. (b) Resulting spanwise velocity distribution at $y = y_o$. (c) Velocity profiles normal to the surface. (From Gad-el-Hak, M. (2000) *Flow Control: Passive, Active, and Reactive Flow Management*. Reprinted with permission of Cambridge University Press, New York.)

observed in $U(y)$ profiles, thus suction under the low-speed streaks and/or injection under the high-speed regions would decrease this shear and hence the resulting instability.

The combination of selective suction and injection is sketched in Figure 33.7. In Figure 33.7a, the vortices are idealized by a periodic distribution in the spanwise direction. The instantaneous velocity profiles without transpiration at constant y and z locations are shown by the dashed lines in Figures 33.7b and c, respectively. Clearly, the $U(y_o, z)$ profile is inflectional, having two inflection points per wavelength. At z_1 and z_3, an inflectional $U(y)$ profile is also evident. The same profiles with suction at z_1 and z_3 and injection at z_2 are shown by the solid lines. In all cases, the shear associated with the inflection points would have been reduced. Because the inflectional profiles are all inviscidly unstable with growth rates proportional to the shear, the resulting instabilities would be weakened by the suction/injection process.

The feasibility of the selective suction as a drag-reducing concept has been demonstrated by Gad-el-Hak and Blackwelder (1989) and is shown in Figure 33.8. Low-speed streaks were artificially generated in a laminar boundary layer using three spanwise suction holes as per the method proposed by Gad-el-Hak and Hussain (1986), and a hot-film probe was used to record the near-wall signature of the streaks. An open, feedforward control loop with a phase lag was used to activate a predetermined suction from a longitudinal slot located in between the spanwise holes and the downstream hot-film probe. An equivalent suction coefficient of $C_q = 0.0006$ was sufficient to eliminate the artificial events and prevent bursting. This rate is five times smaller than the asymptotic suction coefficient for a corresponding turbulent

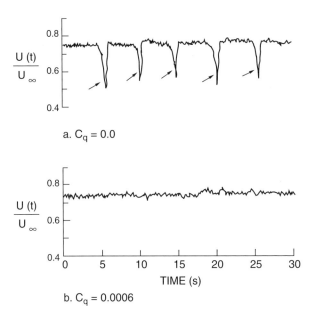

FIGURE 33.8 Effects of suction from a streamwise slot on five artificially induced burstlike events in a laminar boundary layer: $C_q = 0$, (b) $C_q = 0.0006$. (From Gad-el-Hak, M., and Blackwelder, R.F. (1989) *AIAA J.*, 27, 308–314. With permission.)

boundary layer. If this result is sustained in a naturally developing turbulent boundary layer, a skin-friction reduction of close to 60% would be attained.

Gad-el-Hak and Blackwelder (1989) propose to combine suction with nonplanar surface modifications. Minute longitudinal roughness elements, if properly spaced in the spanwise direction, greatly reduce the spatial randomness of the low-speed streaks [Johansen and Smith, 1986]. By withdrawing the streaks forming near the peaks of the roughness elements, less suction should be required to achieve an asymptotic boundary layer. Experiments by Wilkinson and Lazos (1987) and Wilkinson (1988) combine suction/blowing with thin-element riblets. Although no net drag reduction has yet been attained in these experiments, the results indicate some advantage of combining suction with riblets as proposed by Gad-el-Hak and Blackwelder (1987b; 1989).

The recent numerical experiments of Choi et al. (1994) also validate the concept of targeting suction/injection to specific near-wall events in a turbulent channel flow. Based on complete interior flow information and using the rather simple, heuristic control law proposed earlier by Gad-el-Hak and Blackwelder (1987b), Choi et al.'s direct numerical simulations indicate a 20% net drag reduction accompanied by significant suppression of the near-wall structures and the Reynolds stress throughout the entire wall-bounded flow. When only wall information was used, a drag reduction of 6% was observed; a rather disappointing result considering that sensing and actuation took place at every grid point along the computational wall. In a practical implementation of this technique, even fewer wall sensors would perhaps be available, measuring only a small subset of the accessible information and thus requiring even more sophisticated control algorithms to achieve the same degree of success. Low-dimensional models of the near-wall flow and soft computing tools can help in constructing more effective control algorithms. Both of these topics will be discussed later.

Time sequences of the numerical flowfield of Choi et al. (1994) indicate the presence of two distinct drag-reducing mechanisms when selective suction/injection is used: (1) deterring the sweep motion, without modifying the primary streamwise vortices above the wall, and consequently moving the high-shear regions from the surface to the interior of the channel, thus directly reducing the skin friction; and (2) changing the evolution of the wall vorticity layer by stabilizing and preventing lifting of the near-wall spanwise vorticity, thus suppressing a potential source of new streamwise vortices above the surface and interrupting a very important regeneration mechanism of turbulence.

Three modern developments have relevance to the issue at hand. First, the recently demonstrated ability to revert a chaotic system to a periodic one may provide optimal nonlinear control strategies for further reduction in the amount of suction (or the energy expenditure of any other active wall-modulation technique) needed to attain a given degree of flow stabilization. This is important because, as can be seen from the Kármán integral momentum equation, net drag reduction achieved in a turbulent boundary layer increases as the suction coefficient decreases. Second, selectively removing the randomly occurring low-speed streaks, for example, would ultimately require reactive control. In that case, an event is targeted, sensed and subsequently modulated. Microfabrication technology provides opportunities for practical implementation of the required large array of inexpensive, programmable sensor/actuator chips. Third, newly introduced soft computing tools include neural networks, fuzzy logic and genetic algorithms and are now more advanced as well as more widely used compared to just a few years ago. These tools could be very useful in constructing effective adaptive controllers.

33.9.3 Reactive Feedback Control

As was schematically depicted in Figure 33.3, a control device can be passive, requiring no auxiliary power, or active, requiring energy expenditure. Active control is further divided into predetermined or reactive. Predetermined control includes the application of steady or unsteady energy input without regard to the particular state of the flow. The control loop in this case is open, as was shown in Figure 33.4a, and no sensors are required. Because no sensed information is being fed forward, this open control loop is not a feedforward one. Reactive control is a special class of active control where the control input is continuously adjusted based on measurements of some kind. The control loop in this case can either be an open, feedforward one (Figure 33.4b) or a closed, feedback loop (Figure 33.4c).

Moin and Bewley (1994) categorize reactive feedback control strategies by examining the extent to which they are based on the governing flow equations. Four categories are discerned: adaptive, physical model-based, dynamical systems-based and optimal control (Figure 33.3). Note that, except for adaptive control, the other three categories of reactive feedback control can also be used in the feedforward mode or the combined feedforward–feedback mode. Also, in a convective environment such as that for a boundary layer, a controller would perhaps combine feedforward and feedback information and may include elements from each of the four classifications. Each of the four categories is briefly described below.

Adaptive schemes attempt to develop models and controllers via some learning algorithm without regard to the details of the flow physics. System identification is performed independently of the flow dynamics or the Navier–Stokes equations that govern the dynamics. An adaptive controller tries to optimize a specified performance index by providing a control signal to an actuator. In order to update its parameters, the controller thus requires feedback information relating to the effects of its control. The most recent innovation in adaptive flow control schemes involves the use of neural networks, which relate the sensor outputs to the actuator inputs through functions with variable coefficients and nonlinear, sigmoid saturation functions. The coefficients are updated using the so-called back-propagation algorithm, and complex control laws can be represented with a sufficient number of terms. Hand tuning is required, however, to achieve good convergence properties. The nonlinear adaptive technique has been used with different degrees of success by Fan et al. (1993) and Jacobson and Reynolds (1993b; 1995; 1998) to control, respectively, the transition process in laminar boundary layers and the bursting events in turbulent boundary layers.

Heuristic physical arguments can instead be used to establish effective control laws. That approach obviously will work only in situations in which the dominant physics are well understood. An example of this strategy is the active cancellation scheme, used by Gad-el-Hak and Blackwelder (1989) in a physical experiment and by Choi et al. (1994) in a numerical experiment, to reduce the drag by mitigating the effect of near-wall vortices. As mentioned earlier, the idea is to oppose the near-wall motion of the fluid, caused by the streamwise vortices, with an opposing wall control, thus lifting the high-shear region away from the surface and interrupting the turbulence regeneration mechanism.

Nonlinear dynamical systems theory allows turbulence to be decomposed into a small number of representative modes whose dynamics are examined to determine the best control law. The task is to stabilize the attractors of a low-dimensional approximation of a turbulent chaotic system. The best known strategy is the Ott–Grebogi–Yorke (OGY) method which, when applied to simpler, small number of degrees of freedom systems, achieves stabilization with minute expenditure of energy. This and other chaos control strategies, especially as applied to the more complex turbulent flows, will be discussed later.

Finally, optimal control theory applied directly to the Navier–Stokes equations can, in principle, be used to minimize a cost function in the space of the control. This strategy provides perhaps the most rigorous theoretical framework for flow control. As compared to other reactive control strategies, optimal control applied to the full Navier–Stokes equations is also the most computer-time intensive. In this method, feedback control laws are derived systematically for the most efficient distribution of control effort to achieve a desired goal. Abergel and Temam (1990) developed such an optimal control theory for suppressing turbulence in a numerically simulated, two-dimensional Navier–Stokes flow, but their method requires impractical full flowfield information. Choi et al. (1993) developed a more practical, wall-information-only, suboptimal control strategy which they applied to the one-dimensional stochastic Burgers equation. Later application of the suboptimal control theory to a numerically simulated turbulent channel flow has been reported by Moin and Bewley (1994) and Bewley et al. (1997; 1998). The recent book edited by Sritharan (1998) provides eight articles that focus on the mathematical aspects of optimal control of viscous flows.

33.9.4 Required Characteristics

The randomness of the bursting events necessitates temporal phasing as well as spatial selectivity to effect selective control. Practical applications of methods targeted at controlling a particular turbulent structure to achieve a prescribed goal would therefore require implementing a large number of surface sensors/actuators together with appropriate control algorithms. That strategy for controlling wall-bounded turbulent flows has been advocated by, among others and in chronological order, Gad-el-Hak and Blackwelder (1987b; 1989), Lumley (1991; 1996), Choi et al. (1992), Reynolds (1993), Jacobson and Reynolds (1993a; 1993b; 1994; 1995; 1998), Gad-el-Hak (1993; 1994; 1996; 1998; 2000), Moin and Bewley (1994), McMichael (1996), Mehregany et al. (1996), Blackwelder (1998), Delville et al. (1998) and Perrier (1998).

It is instructive to estimate some representative characteristics of the required array of sensors/actuators. Consider a typical commercial aircraft cruising at a speed of $U_\infty = 300$ m/s and at an altitude of 10 km. The density and kinematic viscosity of air and the unit Reynolds number in this case are, respectively, $\rho = 0.4$ kg/m^3, $\nu = 3 \times 10^{-5}$ m^2/s, and $Re = 10^7$/m. Assume further that the portion of fuselage to be controlled has turbulent boundary-layer characteristics identical to those for a zero-pressure-gradient flat plate at a distance of 1 m from the leading edge. In this case, the skin-friction coefficient[4] and the friction velocity are, respectively, $C_f = 0.003$ and $u_\tau = 11.62$ m/s. At this location, one viscous wall unit is only $\nu/u_\tau = 2.6$ μm. In order for the surface array of sensors/actuators to be hydraulically smooth, it should not protrude beyond the viscous sublayer, or $5\,\nu/u_\tau = 13$ μm.

Wall-speed streaks are the most visible, reliable and detectable indicators of the preburst turbulence production process. The detection criterion is simply low velocity near the wall, and the actuator response should be to accelerate (or to remove) the low-speed region before it breaks down. Local wall motion, tangential injection, suction, heating or electromagnetic body force, all triggered on sensed wall-pressure or wall-shear stress, could be used to cause local acceleration of near-wall fluid.

---

[4] Note that the skin friction decreases as the distance from the leading edge increases. It is also strongly affected by such things as the externally imposed pressure gradient. Therefore, the estimates provided in here are for illustration purposes only.

The recent numerical experiments of Berkooz et al. (1993) indicate that effective control of a bursting pair of rolls may be achieved by using the equivalent of two wall-mounted shear sensors. If the goal is to stabilize or to eliminate *all* low-speed streaks in the boundary layer, a reasonable estimate for the spanwise and streamwise distances between individual elements of a checkerboard array is, respectively, 100 and 1000 wall units,[5] or 260 µm and 2600 µm, for our particular example. A reasonable size for each element is probably one tenth of the spanwise separation, or 26 µm. A 1-m × 1-m portion of the surface would have to be covered with about $n = 1.5$ million elements. This is a colossal number, but the density of sensors/actuators could be considerably reduced if we moderate our goal of targeting every single bursting event (and also if less conservative assumptions are made).

It is well known that not every low-speed streak leads to a burst. On the average, a particular sensor would detect an incipient bursting event every wall-unit interval of $P^+ = P u_\tau^2 / \nu = 250$, or $P = 56$ µs. The corresponding dimensionless and dimensional frequencies are $f^+ = 0.004$ and $f = 18$ kHz, respectively. At different distances from the leading edge and in the presence of a nonzero pressure gradient, the sensors/actuators array would have different characteristics, but the corresponding numbers would still be in the same ballpark as estimated in here.

As a second example, consider an underwater vehicle moving at a speed of $U_\infty = 10$ m/s. Despite the relatively low speed, the unit Reynolds number is still the same as estimated above for the air case, $Re = 10^7$/m, due to the much lower kinematic viscosity of water. At 1 m from the leading edge of an imaginary flat plate towed in water at the same speed, the friction velocity is only $u_\tau = 0.39$ m/s, but the wall unit is still the same as in the aircraft example, $\nu / u_\tau = 2.6$ µm. The density of the required sensors/actuators array is the same as computed for the aircraft example, $n = 1.5 \times 10^6$ elements/m^2. The anticipated average frequency of sensing a bursting event is, however, much lower at $f = 600$ Hz.

Similar calculations have been recently made by Gad-el-Hak (1993; 1994; 1998; 2000), Reynolds (1993) and Wadsworth et al. (1993). Their results agree closely with the estimates made here for typical field requirements. In either the airplane or the submarine case, the actuator's response need not be too large. As will be shown later, wall displacement on the order of 10 wall units (26 µm in both examples), suction coefficient of about 0.0006 or surface cooling and heating on the order of 40°C and 2°C (in the first and second example, respectively) should be sufficient to stabilize the turbulent flow.

As computed in the two examples above, both the required size for a sensor/actuator element and the average frequency at which an element would be activated are within the presently known capabilities of microfabrication technology. The number of elements needed per unit area is, however, alarmingly large. The unit cost of manufacturing a programmable sensor/actuator element would have to come down dramatically, perhaps matching the unit cost of a conventional transistor,[6] before the idea advocated in here would become practical.

An additional consideration besides the size, amplitude and frequency response is the energy consumed by each sensor/actuator element. Total energy consumption by the entire control system obviously has to be low enough to achieve net savings. Consider the following calculations for the aircraft example. One meter from the leading edge, the skin-friction drag to be reduced is approximately 54 N/m^2. Engine power needed to overcome this retarding force per unit area is 16 kW/m^2, or 10^4 µW/sensor. If a 60% drag-reduction is achieved,[7] this energy consumption is reduced to 4320 µW/sensor. This number will increase by the amount of energy consumption of a sensor/actuator unit, but (it is hoped) not back to

---

[5] These are equal to, respectively, the average spanwise wavelength between two adjacentt streaks and the average streamwise extent for a typical low-speed region. One can argue that those estimates are too conservative: Once a region is relaminarized, it would perhaps stay as such for quite a while as the flow convects downstream. The next row of sensors/actuators may therefore be relegated to a downstream location well beyond 1000 wall units. Relatively simple physical or numerical experiments could settle this issue.

[6] The transistor was invented in 1947. In the mid-1960s, a single transistor sold for around $70. In 1997, Intel's Pentium II processor (microchip) contained 7.5×10^6 transistors and cost around $500, which is less than $0.00007 per transistor!

[7] A not-too-farfetched goal according to the selective suction results presented earlier.

the uncontrolled levels. The voltage across a sensor is typically in the range of $V = 0.1–1$ V, and its resistance is in the range of $R = 0.1–1$ MΩ. This means a power consumption by a typical sensor in the range of $P = V^2/R = 0.1–10$ µW, well below the anticipated power savings due to reduced drag.

For a single actuator in the form of a spring-loaded diaphragm with a spring constant of $k = 100$ N/m and oscillating up and down at the bursting frequency of $f = 18$ kHz with an amplitude of $y = 26$ µm, the power consumption is $P = (1/2)ky^2 f = 600$ µW/actuator. If suction is used instead, $C_q = 0.0006$ and, assuming a pressure difference of $\Delta p = 10^4$ N/m^2 across the suction holes/slots, the corresponding power consumption for a single actuator is $P = C_q U_\infty \Delta p/n = 1200$ µW/actuator. It is clear then that when the power penalty for the sensor/actuator is added to the lower level drag, a net saving is still achievable. The corresponding actuator power penalties for the submarine example are even smaller ($P = 20$ µW/actuator for the wall motion actuator and $P = 40$ µW/actuator for the suction actuator), and larger savings are therefore possible.

33.10 Magnetohydrodynamic Control

33.10.1 Introductory Remarks

Magnetohydrodynamics (MHD) is the science underlying the interaction of an electrically conducting fluid with a magnetic field. Several decades ago a number of researchers from around the world were vigorously attempting to exploit the electric current and thus electric power that result when a fluid conductor moves in the presence of a magnetic field. Thus, the fluid internal energy (or enthalpy) is directly converted into electrical energy, eliminating the traditional intermediate mechanical step. The usual turbine and generator in conventional power plants are therefore combined in a single unit, in essence an electromagnetic turbine with no moving parts. The inverse device is an electromagnetic pump where, in the presence of a magnetic field, a conducting fluid is caused to move when electric current passes through it. An excellent primer for the topic of MHD can be found in the book by Shercliff (1965).

Examples of conducting fluids (in descending order of electrical conductivities) include liquid metals, plasmas, molten glass and sea water. The electrical conductivity of mercury is $\sigma \approx 10^6$ mhos/m, and that for seawater is $\sigma \approx 4$ mhos/m. Possible useful applications include MHD power generation, propulsion and liquid metals used as coolants for magnetically confined fusion reactors. Though fictitious, the Soviet submarine in Tom Clancy's bestseller *The Hunt for Red October* was endowed with an extremely quiet MHD propulsion system. With the collapse of communism, the interest in this field has shifted somewhat to more peaceful uses, such as, for example, flow control or, in other words, the ability to manipulate a flowfield in order to achieve a beneficial goal such as drag reduction, lift enhancement and mixing augmentation. Here the so-called Lorentz body forces that result when a conducting fluid moves in the presence of a magnetic field are exploited to effect desired changes in the flowfield—for example, to suppress turbulence.

Of particular interest to this chapter is the development of efficient reactive control strategies that employ the Lorentz forces to enhance the performance of sea vessels. Methods to delay transition, prevent separation and reduce skin-friction drag in turbulent boundary layers are sought. Control strategies targeted toward certain coherent structures in a turbulent flow are particularly sought. We start here by developing the von Kármán momentum integral equation and the instantaneous equations of motion at the wall, both in the presence of Lorentz force. This should prove useful to make quick estimates of system behavior under different operating conditions. Results are usually not as accurate as solving the differential conservation equations themselves but are more accurate than dimensional analysis. This is followed by an outline of a suggested reactive control strategy that exploits electromagnetic body forces.

33.10.2 The von Kármán Equation

The Lorentz force (body force per unit volume) is given by the cross-product of the current density vector $\vec{j}$ and the magnetic flux density vector $\vec{B}$:

$$\vec{F} = \vec{j} \times \vec{B} \tag{33.13}$$

where the current is the sum of the applied and induced contributions:

$$\vec{j} = \sigma(\vec{E} + \vec{u} \times \vec{B}) \tag{33.14}$$

where σ is the electrical conductivity of the fluid, $\vec{E}$ is the applied electric field vector, and $\vec{u}$ is the fluid velocity vector. Thus, the Lorentz force is, in general, given by:

$$\vec{F} = \sigma(\vec{E} \times \vec{B}) + \sigma(\vec{u} \times \vec{B}) \times \vec{B} \tag{33.15}$$

Due to the rather low electrical conductivity of seawater, the induced Lorentz force is typically very small in such an application, and the only way to effect a significant body force is to cross the magnetic field with an applied electric field. For such application, therefore, the term electromagnetic forcing (EMHD) is commonly used in place of MHD.

Assume a two-dimensional boundary-layer flow of a conducting fluid. Further assume that the magnetic and electric fields are generated using alternating electrodes and magnets parallel to the streamwise direction (see Figure 33.9). Basically, both fields are in the plane (y, z). The resulting Lorentz force has components in the streamwise (x), normal (y) and spanwise (z) directions given by, respectively:

$$F_x = \sigma(E_y B_z - E_z B_y) - \sigma u(B_z^2 + B_y^2) \tag{33.16}$$

$$F_y = 0 - \sigma v B_z^2 \tag{33.17}$$

$$F_z = 0 - \sigma v B_z B_y \tag{33.18}$$

where u and v are, respectively, the streamwise and normal velocity components. Each of the first terms on the right-hand sides of Eqs. (33.16) to (33.18) is due to the applied electric field. The second term is

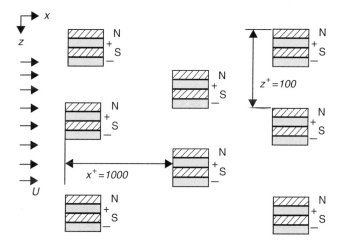

FIGURE 33.9 EMHD tiles for reactive control of turbulent boundary layers. (From Gad-el-Hak, M. (2000) *Flow Control: Passive, Active, and Reactive Flow Management*. Reprinted with permission of Cambridge University Press, New York.)

the induced Lorentz force and is negligible for low-conductivity fluids such as seawater. Therefore, for such an application and particular arrangement of electrodes/magnets, the electromagnetic body force is predominately in the streamwise direction, $F_x \approx \sigma(E_y B_z - E_z B_y)$.

We now write the continuity and streamwise momentum equations. For an incompressible, two-dimensional flow and neglecting the gravitational body force:

$$\frac{\partial u}{\partial x} + \frac{\partial v}{\partial y} = 0 \qquad (33.19)$$

$$\rho \frac{\partial u}{\partial t} + \rho u \frac{\partial u}{\partial x} + \rho v \frac{\partial u}{\partial y} = -\frac{\partial p}{\partial x} + \frac{\partial \tau}{\partial y} + \sigma(E_y B_z - E_z B_y) - \sigma u(B_z^2 + B_y^2) \qquad (33.20)$$

where τ is the viscous shear stress. The pressure term can be written in terms of the velocity outside the boundary layer U_o which in general is a function of x and t. Note that the total drag should be the sum of the usual viscous and form drags plus (or minus) the Lorentz force. In Eq. (33.20), the induced Lorentz force is always negative (for positive u). The applied force can be positive or negative depending on the direction of the applied electric field, $\vec{E}$, relative to the magnetic field, $\vec{B}$. In any case, when the Lorentz force is positive, it constitutes a thrust (negative drag).

At a given x location, the streamwise momentum equation can readily be integrated in y from the wall to the edge of the boundary layer. Invoking the continuity equation and using the usual definitions for the skin friction coefficient (C_f), and the displacement and momentum thicknesses (δ^* and δ_θ), the resulting integral equation reads:

$$C_f \equiv \frac{2\tau_w}{\rho U_o^2} = \frac{2}{U_o^2} \frac{\partial(U_o \delta^*)}{\partial t} + 2\frac{\partial \delta^*}{\partial x} - 2\frac{v_w}{U_o} + 2\delta_\theta\left(2 + \frac{\delta^*}{\delta_\theta}\right)\frac{1}{U_o}\frac{\partial U_o}{\partial x}$$

$$+ \frac{2\sigma}{\rho U_o^2}\int_0^\infty [E_y B_z - E_z B_y]\,dy - \frac{2\sigma}{\rho U_o^2}\int_0^\infty u[B_z^2 + B_y^2]\,dy \qquad (33.21)$$

where v_w is the injection (or suction) velocity normal to the wall. Note that upward and downward wall motions are equivalent to wall injection and suction. If σ is not constant, it has to be included in the two integrals on the right-hand side of Eq. (33.21).

It is clear that the following effects contribute to an *increase* in skin friction: temporal acceleration of freestream, growth of momentum thickness, wall suction (or downward wall motion), favorable pressure gradient and positive streamwise Lorentz force. When such force points opposite of the flow direction, the skin friction decreases and the velocity profile tends to become more inflectional. But, once again remember to add (or subtract) the drag (or thrust) due to the applied as well as induced Lorentz forces.

Note that Eq. (33.21) is valid for steady or unsteady flows, for Newtonian or non-Newtonian fluids and for laminar or turbulent flows (mean quantities are used in the latter case). The two integrals in Eq. (33.21) can be computed once the electric and magnetic fields are known and an approximate velocity profile is assumed. For seawater, the last integral in Eq. (33.21) is neglected.

33.10.3 Vorticity Flux at the Wall

Another very useful equation is the differential momentum equation written at the wall ($y = 0$). Here, we assume a Newtonian fluid but allow viscosity to vary spatially and the flow to be three dimensional. For a non-moving wall,[8] the instantaneous streamwise, normal and spanwise momentum equations read,

---

[8]The wall is not moving in the streamwise and spanwise directions. Normal wall motions are equivalent to suction/injection and hence are allowed in this formulation.

respectively:

$$\rho v_w \frac{\partial u}{\partial y}\bigg|_{y=0} + \frac{\partial p}{\partial x}\bigg|_{y=0} - \frac{\partial \mu}{\partial y}\bigg|_{y=0}\frac{\partial u}{\partial y}\bigg|_{y=0} - \sigma[E_y B_z - E_z B_y]_{y=0} = \mu \frac{\partial^2 u}{\partial y^2}\bigg|_{y=0} \quad (33.22)$$

$$0 + \frac{\partial p}{\partial y}\bigg|_{y=0} - 0 - F_y\big|_{y=0} = \mu \frac{\partial^2 v}{\partial y^2}\bigg|_{y=0} \quad (33.23)$$

$$\rho v_w \frac{\partial w}{\partial y}\bigg|_{y=0} + \frac{\partial p}{\partial z}\bigg|_{y=0} - \frac{\partial \mu}{\partial y}\bigg|_{y=0}\frac{\partial w}{\partial y}\bigg|_{y=0} - F_z\big|_{y=0} = \mu \frac{\partial^2 w}{\partial y^2}\bigg|_{y=0} \quad (33.24)$$

where $F_y|_{y=0}$ and $F_z|_{y=0}$ are, respectively, the wall values of the Lorentz force in the normal and spanwise directions. In Eq. (33.22), the streamwise Lorentz force at the wall, $F_x|_{y=0}$, is due only to the applied electric field (the induced force is zero at the wall as u vanishes there for nonmoving walls). The expression for F_x in Eq. (33.22) is for an array of alternating electrodes and magnets parallel to the streamwise direction,[9] but the streamwise Lorentz force can readily be computed for any other configuration of magnets and electrodes. If the electric conductivity varies spatially, the value of σ at the wall should be used in Eq. (33.22).

The right-hand side of Eq. (33.22) is the (negative of) wall flux of spanwise vorticity, while that of Eq. (33.24) is the (negative of) wall flux of streamwise vorticity. Much like transpiration, pressure-gradient or viscosity variations, the wall value of the Lorentz force determines the sign and intensity of wall vorticity flux. Streamwise force contributes to spanwise vorticity flux, and spanwise force contributes to streamwise vorticity flux. Take, for example, a positive streamwise Lorentz force. This is a negative term on the left-hand side of Eq. (33.22), which makes the curvature of the streamwise velocity profile at the wall more negative (i.e., instantaneously fuller velocity profile). The wall is then a source of spanwise vorticity, and the flow is more resistant to transition and separation. In a turbulent flow, a positive streamwise Lorentz force leads to suppression of normal and tangential Reynolds stresses.

33.10.4 EMHD Tiles for Reactive Control

In this section, I shall outline a reactive control strategy that exploits the Lorentz forces to modulate the flow of an electrically conducting fluid such as seawater. The idea here is to target the low-speed streaks in the near-wall region of a turbulent boundary layer. The electric field is applied only when and where it is needed; hence, the power consumed by the reactive control system is kept far below that consumed by a predetermined active control system. According to recent numerical and experimental results, brute force application of steady or time-dependent Lorentz force—to reduce drag, for example—do not achieve the breakeven point, the reason being the high energy expenditure by predetermined control systems (see, for example, the two meeting proceedings edited by Gerbeth, 1997, and Meng, 1998).

The top view in Figure 33.9 depicts the proposed distribution of EMHD tiles to achieve targeted control in a turbulent wall-bounded flow. Each tile consists of two streamwise strips constituting the north and south poles of a permanent magnet and another two strips constituting positive and negative electrodes.[10] The tiles are staggered in a checkerboard configuration and are separated by 100 wall units in the spanwise direction (z) and 1000 wall units in the streamwise direction (x). The length and width of each tile are about 50 wall units.[11]

---

[9] Note that, for such an array, the applied part of both $F_y|_{y=0}$ and $F_z|_{y=0}$ is identically zero, but the induced part is nonzero if $v_w \neq 0$. In seawater applications, the induced Lorentz force is negligible regardless of any suction/injection or normal wall motion.

[10] The cathode and anode are interchangeable depending on the desired direction of the Lorentz force.

[11] The length and width of each magnet pole (or electrode) are thus 50 and 12.5 wall units, respectively.

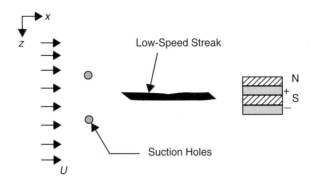

FIGURE 33.10 Control of a single artificial low-speed streak. (From Gad-el-Hak, M. (2000) *Flow Control: Passive, Active, and Reactive Flow Management.* Reprinted with permission of Cambridge University Press, New York.)

A shear-stress sensor having a spatial resolution of 2 to 5 wall units is placed just upstream and along the centerline of each tile. Each sensor and corresponding actuator is connected via a closed feedback control loop. The tile is activated to give a positive or negative streamwise Lorentz force when a low- or high-speed region is detected. The control law used can be based on simple physical arguments or a more complex self-learning neural network. Other possible albeit more sophisticated control laws can be based on nonlinear dynamical systems theory or optimal/suboptimal control theory.

For a simpler physical or numerical experiment that does not require sensors and complex closed-loop control, a single low-speed streak is artificially generated in a laminar boundary layer using two suction holes as depicted in the (x–z) view in Figure 33.10. The two holes are separated in the spanwise direction by 100 wall units, and each is about 0.5-wall unit in diameter. Such a method was successfully used by Gad-el-Hak and Hussain (1986) to generate artificial low-speed streaks as well as simulated bursting events in a laminar environment. In the proposed experiment, high-speed regions and counter-rotating streamwise vortices are also generated, and, if desired, one can readily target any of these simulated events for elimination. The necessary actuation would, of course, depend on the kind of coherent structure to be targeted.

A single EMHD tile is placed about 50 wall units directly downstream of the suction holes and is activated with a suitable time delay after a suction pulse is applied, in essence using a simple, predetermined, open-loop control. A single electric pulse[12] triggers a positive Lorentz force to eliminate the artificially generated low-speed streak. The experiment would of course be repeated several times, and the actuator location, phase lag, strength and duration could be optimized to achieve the desired goal.

The streamwise Lorentz force necessary to eliminate the resulting artificial low-speed streak could be computed from the measurements, thus the energy expenditure necessary to eliminate all the low-speed streaks in a real turbulent boundary layer could readily be estimated.

33.11 Chaos Control

In the theory of dynamical systems, the so-called butterfly effect denotes sensitive dependence of nonlinear differential equations on initial conditions, with phase–space solutions initially very close together separating exponentially. The solution of nonlinear dynamical systems of three or more degrees of freedom may be in the form of a strange attractor whose intrinsic structure contains a well-defined mechanism to produce a chaotic behavior without requiring random forcing. Chaotic behavior is complex and aperiodic and, though deterministic, appears to be random.

---

[12] Note that in this simulated environment the electric pulse would have the same polarity if only low-speed streaks are to be obliterated.

A question arises naturally: Just as small disturbances can radically grow within a deterministic system to yield rich, unpredictable behavior, can minute adjustments to a system parameter be used to reverse the process and control (i.e., regularize) the behavior of a chaotic system? Recently, that question was answered in the affirmative theoretically as well as experimentally, at least for system orbits that reside on low-dimensional strange attractors (see the review by Lindner and Ditto, 1995). Before describing such strategies for controlling chaotic systems, we first summarize the recent attempts to construct a low-dimensional dynamical systems representation of turbulent boundary layers. Such construction is a necessary first step to be able to use chaos control strategies for turbulent flows. Additionally, as argued by Lumley (1996), a low-dimensional dynamical model of the near-wall region used in a Kalman filter [Banks, 1986; Petersen and Savkin, 1999] can make the most of the partial information assembled from a finite number of wall sensors. Such a filter minimizes in a least-square sense the errors caused by incomplete information and thus globally optimizes the performance of the control system.

33.11.1 Nonlinear Dynamical Systems Theory

Boundary-layer turbulence is described by a set of nonlinear partial differential equations and is characterized by an infinite number of degrees of freedom. This makes it rather difficult to model the turbulence using a dynamical systems approximation. The notion that a complex, infinite-dimensional flow can be decomposed into several low-dimensional subunits is, however, a natural consequence of the realization that quasi-periodic coherent structures dominate the dynamics of seemingly random turbulent shear flows. This implies that low-dimensional, localized dynamics can exist in formally infinite-dimensional extended systems—such as open turbulent flows. Reducing the flow physics to finite-dimensional dynamical systems enables a study of its behavior through an examination of the fixed points and the topology of their stable and unstable manifolds. From the dynamical systems theory viewpoint, the meandering of low-speed streaks is interpreted as hovering of the flow state near an unstable fixed point in the low-dimensional state space. An intermittent event that produces high wall stress—a burst—is interpreted as a jump along a heteroclinic cycle to different unstable fixed point that occurs when the state has wandered too far from the first unstable fixed point. Delaying this jump by holding the system near the first fixed point should lead to lower momentum transport in the wall region and, therefore, to lower skin-friction drag. Reactive control means sensing the current local state and through appropriate manipulation keeping the state close to a given unstable fixed point, thereby preventing further production of turbulence. Reducing the bursting frequency by, say, 50%, may lead to a comparable reduction in skin-friction drag. For a jet, relaminarization may lead to a quiet flow and very significant noise reduction.

In one significant attempt, the proper orthogonal, or Karhunen-Loève, decomposition method has been used to extract a low-dimensional dynamical system from experimental data of the wall region [Aubry et al. 1988; Aubry, 1990]. Aubry et al. (1988) expanded the instantaneous velocity field of a turbulent boundary layer using experimentally determined eigenfunctions in the form of streamwise rolls. They expanded the Navier–Stokes equations using these optimally chosen, divergence-free, orthogonal functions, applied a Galerkin projection, and then truncated the infinite-dimensional representation to obtain a ten-dimensional set of ordinary differential equations. These equations represent the dynamical behavior of the rolls and are shown to exhibit a chaotic regime as well as an intermittency due to a burst-like phenomenon. However, Aubry et al.'s ten-mode dynamical system displays a regular intermittency, in contrast both to that in actual turbulence as well as to the chaotic intermittency encountered by Pomeau and Manneville (1980) in which event durations are distributed stochastically. Nevertheless, the major conclusion of Aubry et al.'s study is that the bursts appear to be produced autonomously by the wall region even without turbulence but are triggered by turbulent pressure signals from the outer layer. More recently, Berkooz et al. (1991) generalized the class of wall-layer models developed by Aubry et al. (1988) to permit uncoupled evolution of streamwise and cross-stream disturbances. Berkooz et al.'s results suggest that the intermittent events observed in Aubry et al.'s representation do not arise solely because of the effective closure assumption incorporated, but rather are rooted deeper in the dynamical phenomena of the wall region. The book by Holmes et al. (1996) details the Cornell University research group attempts at describing turbulence as a low-dimensional dynamical system.

In addition to the reductionist viewpoint exemplified by the work of Aubry et al. (1988) and Berkooz et al. (1991), attempts have been made to determine directly the dimension of the attractors underlying specific turbulent flows. Again, the central issue here is whether or not turbulent solutions to the infinite-dimensional Navier–Stokes equations can be asymptotically described by a finite number of degrees of freedom. Grappin and Léorat (1991) computed the Lyapunov exponents and the attractor dimensions of two- and three-dimensional periodic turbulent flows without shear. They found that the number of degrees of freedom contained in the large scales establishes an upper bound for the dimension of the attractor. Deane and Sirovich (1991) and Sirovich and Deane (1991) numerically determined the number of dimensions needed to specify chaotic Rayleigh-Bénard convection over a moderate range of Rayleigh numbers, Ra. They suggested that the intrinsic attractor dimension is $\mathcal{O}[Ra^{2/3}]$.

The corresponding dimension in wall-bounded flows appears to be dauntingly high. Keefe et al. (1992) determined the dimension of the attractor underlying turbulent Poiseuille flows with spatially periodic boundary conditions. Using a coarse-grained numerical simulation, they computed a lower bound on the Lyapunov dimension of the attractor to be approximately 352 at a pressure-gradient Reynolds number of 3200. Keefe et al. (1992) argue that the attractor dimension in fully resolved turbulence is unlikely to be much larger than 780. This suggests that periodic turbulent shear flows are deterministic chaos and that a strange attractor does underlie solutions to the Navier–Stokes equations. Temporal unpredictability in the turbulent Poiseuille flow is thus due to the exponential spreading property of such attractors. Although finite, the computed dimension invalidates the notion that the global turbulence can be attributed to the interaction of a *few* degrees of freedom. Moreover, in a physical channel or boundary layer, the flow is not periodic and is open. The attractor dimension in such case is not known but is believed to be even higher than the estimate provided by Keefe et al. for the periodic (quasi-closed) flow.

In contrast to closed, absolutely unstable flows, such as Taylor–Couette systems, where the number of degrees of freedom can be small, local measurements in open, convectively unstable flows, such as boundary layers, do not express the global dynamics, and the attractor dimension in that case may inevitably be too large to be determined experimentally. According to the estimate provided by Keefe et al. (1992), the colossal data required (about 10^D, where D is the attractor dimension) for measuring the dimension simply exceed current computer capabilities. Turbulence near transition or near a wall is an exception to that bleak picture. In those special cases, a relatively small number of modes are excited and the resulting simple turbulence can therefore be described by a dynamical system of a reasonable number of degrees of freedom.

33.11.2 Chaos Control

There is another question of greater relevance here. Given a dynamical system in the chaotic regime, is it possible to stabilize its behavior through some kind of active control? While other alternatives have been devised [e.g., Fowler, 1989; Hübler and Lüscher, 1989; Huberman, 1990; Huberman and Lumer, 1990], the recent method proposed by workers at the University of Maryland [Ott et al., 1990a; 1990b; Shinbrot et al., 1990; 1992a; 1992b; 1992c; 1998; Romeiras et al., 1992] promises to be a significant breakthrough. Comprehensive reviews and bibliographies of the emerging field of chaos control can be found in the articles by Shinbrot et al. (1993), Shinbrot (1993; 1995; 1998) and Lindner and Ditto (1995).

Ott et al. (1990a) demonstrated, through numerical experiments with the Henon map, that it is possible to stabilize a chaotic motion about any pre-chosen, unstable orbit through the use of relatively small perturbations. The procedure consists of applying minute time-dependent perturbations to one of the system parameters to control the chaotic system around one of its many unstable periodic orbits. In this context, targeting refers to the process whereby an arbitrary initial condition on a chaotic attractor is steered toward a prescribed point (target) on this attractor. The goal is to reach the target as quickly as possible using a sequence of small perturbations (Kostelich et al., 1993a).

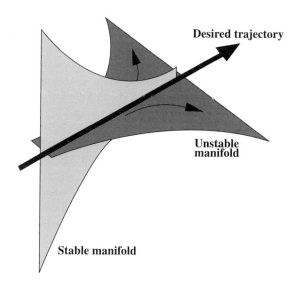

FIGURE 33.11 The OGY method for controlling chaos. (From Gad-el-Hak, M. (2000) *Flow Control: Passive, Active, and Reactive Flow Management*. Reprinted with permission of Cambridge University Press, New York.)

The success of the Ott–Grebogi–Yorke's (OGY) strategy for controlling chaos hinges on the fact that beneath the apparent unpredictability of a chaotic system lies an intricate but highly ordered structure. Left to its own recourse, such a system continually shifts from one periodic pattern to another, creating the appearance of randomness. An appropriately controlled system, on the other hand, is locked into one particular type of repeating motion. With such reactive control the dynamical system becomes one with a stable behavior.

The OGY-method can be simply illustrated by the schematic in Figure 33.11. The state of the system is represented as the intersection of a stable manifold and an unstable one. The control is applied intermittently whenever the system departs from the stable manifold by a prescribed tolerance; otherwise, the control is shut off. The control attempts to put the system back onto the stable manifold so that the state converges toward the desired trajectory. Unmodeled dynamics cause noise in the system and a tendency for the state to wander off in the unstable direction. The intermittent control prevents that and the desired trajectory is achieved. This efficient control is not unlike trying to balance a ball in the center of a horse saddle [Moin and Bewley, 1994]. There is one stable direction (front/back) and one unstable direction (left/right). The restless horse is the unmodeled dynamics, intermittently causing the ball to move in the wrong direction. The OGY control needs to be applied, in the most direct manner possible, only when the ball wanders off in the left/right direction.

The OGY method has been successfully applied in a relatively simple experiment in which reverse chaos was obtained in a parametrically driven, gravitationally buckled, amorphous magnetoelastic ribbon [Ditto et al., 1990; Ditto and Pecora, 1993]. Garfinkel et al. (1992) applied the same control strategy to stabilize drug-induced cardiac arrhythmias in sections of a rabbit ventricle. Other extensions, improvements and applications of the OGY strategy include higher dimensional targeting [Auerbach et al., 1992; Kostelich et al., 1993b]; controlling chaotic scattering in Hamiltonian (i.e., nondissipative, area conservative) systems [Lai et al., 1993a; 1993b]; synchronization of identical chaotic systems that govern communication, neural or biological processes [Lai and Grebogi, 1993]; use of chaos to transmit information [Hayes et al., 1994a; 1994b]; control of transient chaos [Lai et al., 1994] and taming spatio-temporal chaos using a sparse array of controllers [Chen et al., 1993; Qin et al., 1994; Auerbach, 1994].

In a more complex system, such as a turbulent boundary layer, there exist numerous interdependent modes and many stable as well as unstable manifolds (directions). The flow can then be modeled as coherent structures plus a parameterized turbulent background. The proper orthogonal decomposition

(POD) is used to model the coherent part because POD guarantees the minimum number of degrees of freedom for a given model accuracy. Factors that make turbulence control a challenging task are the potentially quite large perturbations caused by the unmodeled dynamics of the flow, the nonstationary nature of the desired dynamics and the complexity of the saddle shape describing the dynamics of the different modes. Nevertheless, the OGY control strategy has several advantages that are of special interest in the control of turbulence: (1) the mathematical model for the dynamical system need not be known, (2) only *small* changes in the control parameter are required, and (3) noise can be tolerated (with appropriate penalty).

Recently, Keefe (1993a; 1993b) made a useful comparison between two nonlinear control strategies as applied to fluid problems: the OGY feedback method described above and the model-based control strategy originated by Hübler [e.g., Hübler and Lüscher, 1989; Lüscher and Hübler, 1989], the H method. Both novel control methods are essentially generalizations of the classical perturbation cancellation technique: Apply a prescribed forcing to subtract the undesired dynamics, and impose the desired one. The OGY strategy exploits the sensitivity of chaotic systems to stabilize existing periodic orbits and steady states. Some feedback is needed to steer the trajectories toward the chosen fixed point, but the required control signal is minuscule. In contrast, Hübler's scheme does not explicitly make use of the system sensitivity. It produces general control response (periodic or aperiodic) and requires little or no feedback, but its control inputs are generally large. The OGY strategy exploits the nonlinearity of a dynamical system; indeed, the presence of a strange attractor and the extreme sensitivity of the dynamical system to initial conditions are essential to the success of this method. In contrast, the H method works equally for both linear and nonlinear systems.

Keefe (1993a) first examined numerically the two schemes as applied to fully developed and transitional solutions of the Ginzburg–Landau equation, an evolution equation that governs the initially weakly nonlinear stages of transition in several flows and that possesses both transitional and fully chaotic solutions. The Ginzburg–Landau equation has solutions that display either absolute or convective instabilities and is thus a reasonable model for both closed and open flows. Keefe's main conclusion is that control of nonlinear systems is best obtained by making maximum possible use of the underlying natural dynamics. If the goal dynamics is an unstable nonlinear solution of the equation and the flow is nearby at the instant control is applied, both methods perform reliably and at low-energy cost in reaching and maintaining this goal. Predictably, the performance of both control strategies degrades due to noise and the spatially discrete nature of realistic forcing.

Subsequently, Keefe (1993b) extended the numerical experiment in an attempt to reduce the drag in a channel flow with spatially periodic boundary conditions. The OGY method reduces skin friction to 60–80% of the uncontrolled value at a mass-flux Reynolds number of 4408. The H method fails to achieve any drag reduction when starting from a fully turbulent initial condition but shows potential for suppressing or retarding laminar-to-turbulence transition. Keefe (1993a) suggests that the H strategy might be more appropriate for boundary-layer control, while the OGY method might best be used for channel flows.

It is also relevant to note here the work of Bau and his colleagues at the University of Pennsylvania [Singer et al., 1991; Wang et al., 1992] who devised a feedback control to stabilize (relaminarize) the naturally occurring chaotic oscillations of a toroidal thermal convection loop heated from below and cooled from above. Based on a simple mathematical model for the thermosyphon, Bau and his colleagues constructed a reactive control system that was used to alter significantly the flow characteristics inside the convection loop. Their linear control strategy, perhaps a special version of the OGY chaos control method, consists simply of sensing the deviation of fluid temperatures from desired values at a number of locations inside the thermosyphon loop and then altering the wall heating either to suppress or to enhance such deviations. Wang et al. (1992) also suggested extending their theoretical and experimental method to more complex situations such as those involving Bénard convection [Tang and Bau, 1993a; 1993b]. Hu and Bau (1994) used a similar feedback control strategy to demonstrate that the critical Reynolds number for the loss of stability of planar Poiseuille flow can be significantly increased or decreased.

Other attempts to use low-dimensional dynamical systems representation for flow control include the work of Berkooz et al. (1993), Corke et al. (1994), and Coller et al. (1994a; 1994b). Berkooz et al. (1993)

applied techniques of modern control theory to estimate the phase–space location of dynamical models of the wall-layer coherent structures and used these estimates to control the model dynamics. Because discrete wall sensors provide incomplete knowledge of phase–space location, Berkooz et al. maintain that a nonlinear observer, which incorporates past information and the equations of motion into the estimation procedure, is required. Using an extended Kalman filter, they achieved effective control of a bursting pair of rolls with the equivalent of two wall-mounted shear sensors. Corke et al. (1994) used a low-dimensional dynamical system based on the proper orthogonal decomposition to guide control experiments for an axisymmetric jet. By sensing the downstream velocity and actuating an array of miniature speakers located at the lip of the jet, their feedback control succeeded in converting the near-field instabilities from spatial–convective to temporal–global. Coller et al. (1994a; 1994b) developed a feedback control strategy for strongly nonlinear dynamical systems, such as turbulent flows, subject to small random perturbations that kick the system intermittently from one saddle point to another along heteroclinic cycles. In essence, their approach is to use local, weakly nonlinear feedback control to keep a solution near a saddle point as long as possible, but then to let the natural, global nonlinear dynamics run its course when bursting (in a low-dimensional model) does occur. Though conceptually related to the OGY strategy, Coller et al.'s method does not actually stabilize the state but merely holds the system near the desired point longer than it would otherwise stay.

Shinbrot and Ottino (1993a; 1993b) offer yet another strategy presumably most suited for controlling coherent structures in area-preserving turbulent flows. Their geometric method exploits the premise that the dynamical mechanisms that produce the organized structures can be remarkably simple. By repeated stretching and folding of "horseshoes" present in chaotic systems, Shinbrot and Ottino have demonstrated numerically as well as experimentally the ability to create, destroy and manipulate coherent structures in chaotic fluid systems. The key idea to create such structures is to intentionally place folds of horseshoes near low-order periodic points. In a dissipative dynamical system, volumes contract in state space and the co-location of a fold with a periodic point leads to an isolated region that contracts asymptotically to a point. Provided that the folding is done properly, it counteracts stretching. Shinbrot and Ottino (1993a) applied the technique to three prototypical problems: a one-dimensional chaotic map, a two-dimensional one and a chaotically advected fluid. Shinbrot (1995; 1998) and Shinbrot et al. (1998) provide recent reviews of the stretching/folding as well as other chaos control strategies.

33.12 Soft Computing

The term *soft computing* was coined by Lotfi Zadeh of the University of California, Berkeley, to describe several ingenious modes of computations that exploit tolerance for imprecision and uncertainty in complex systems to achieve tractability, robustness and low cost [Yager and Zadeh, 1992; Bouchon-Meunier et al., 1995a; 1995b; Jang et al., 1997]. The principle of complexity provides the impetus for soft computing: as the complexity of a system increases, the ability to predict its response diminishes until a threshold is reached beyond which precision and relevance become almost mutually exclusive [Noor and Jorgensen, 1996]. In other words, precision and certainty carry a cost. By employing modes of reasoning—probabilistic reasoning—that are approximate rather than exact, soft computing can help in searching for globally optimal design or achieving effectual control while taking into account system uncertainties and risks.

Soft computing refers to a domain of computational intelligence that loosely lies in between purely numerical (hard) computing and purely symbolic computations. Alternatively, one can think about symbolic computations as a form of artificial intelligence lying in between biological intelligence and computational intelligence (soft computing). The schematic in Figure 33.12 illustrates the general idea. Artificial intelligence relies on symbolic information-processing techniques and uses logic as representation and inference mechanisms. It attempts to approach the high level of human cognition. In contrast, soft computing is based on modeling low-level cognitive processes and strongly emphasizes modeling of uncertainty as well as learning. Computational intelligence mimics the ability of the human brain to employ modes of reasoning that are approximate. Soft computing provides a machinery for the numeric

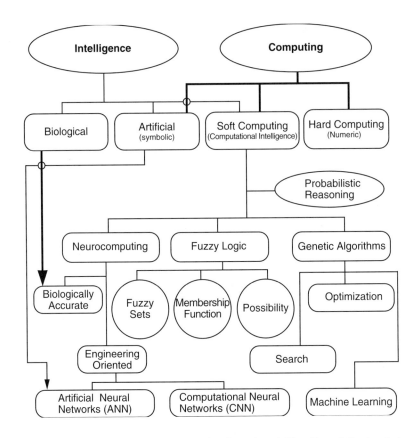

FIGURE 33.12 Tools for soft computing. (From Gad-el-Hak, M. (2000) *Flow Control: Passive, Active, and Reactive Flow Management*. Reprinted with permission of Cambridge University Press, New York.)

representation of the types of constructs developed in the symbolic artificial intelligence. The boundaries between these paradigms are, of course, *fuzzy*.

The principal constituents of soft computing are neurocomputing, fuzzy logic and genetic algorithms, as depicted in Figure 33.12. These elements, together with probabilistic reasoning, can be combined in hybrid arrangements resulting in better systems in terms of parallelism, fault tolerance, adaptivity and uncertainty management. To my knowledge, only neurocomputing has been employed for fluid flow control, but the other tools of soft computing may be just as useful to construct powerful controllers and have in fact been used as such in other fields such as large-scale subway controllers and video cameras. A brief description of these three constituents follows.

Neurocomputing is inspired by the neurons of the human brain and how they work. Neural networks are information-processing devices that can learn by adapting synaptic weights to changes in the surrounding environment; can handle imprecise, fuzzy, noisy and probabilistic information; and can generalize from known tasks (examples) to unknown ones. Actual engineering oriented hardware are termed artificial neural networks (ANN) while algorithms are called computational neural networks (CNN). The nonlinear, highly parallel networks can perform any of the following tasks: classification, pattern matching, optimization, control and noise removal. As modeling and optimization tools, neural networks are particularly useful when good analytic models are either unknown or extremely complex.

An artificial neural network consists of a large number of highly interconnected processing elements— essentially equations known as "transfer functions"—that are analogous to human neurons and are tied together with weighted connections that are analogous to human synapses. A processing unit takes weighted signals from other units, possibly combines them, and gives a numeric result. The behavior of neural networks—how they map input data—is influenced primarily by the transfer functions of the processing elements, how the transfer functions are interconnected, and the weights of those interconnections. Learning typically occurs by example, through exposure to a set of input–output data, where the training algorithm adjusts the connection weights (synapses). These connection weights store the knowledge necessary to solve specific problems. As an example, it is now possible to use neural networks to sense (smell) odors in many different applications [Ouellette, 1999]. The electronic noses (e-noses) are on the verge of finding commercial applications in medical diagnostics, environmental monitoring and the processing and quality control of foods. Neural networks as used in fluid flow control will be covered in the following subsection.

Fuzzy logic was introduced by Lotfi Zadeh in 1965 as a mathematical tool to deal with uncertainty and imprecision. The book edited by Yager and Zadeh (1992) is an excellent primer to the field. For computing and reasoning, general concepts (such as size) are implemented into a computer algorithm by using mostly words (such as small, medium or large). Fuzzy logic, therefore, provides a unique methodology for computing with words. Its rationalism is based on three mathematical concepts: fuzzy sets, membership function and possibility. As dictated by a membership function, fuzzy sets allow a gradual transition from belonging to not belonging to a set. The concept of possibility provides a mechanism for interpreting factual statements involving fuzzy sets. Three processes are involved in solving a practical problem using fuzzy logic: fuzzification, analysis and defuzzification. Given a complex, unsolvable problem in real space, those three steps involve enlarging the space and searching for a solution in the new superset, then specializing this solution to the original real constraints.

Genetic algorithms are search algorithms based loosely on the mechanics of natural selection and natural genetics. They combine survival of the fittest among string structures with structured yet randomized information exchange and are used for search, optimization and machine learning. For control, genetic algorithms aim at achieving minimum cost function and maximum performance measure while satisfying the problem constraints. The books by Goldberg (1989), Davis (1991) and Holland (1992) provide gentle introductions to the field.

In the Darwinian principle of natural selection, the fittest members of a species are favored to produce offspring. Even biologists cannot help but being awed by the complexity of life observed to evolve in the relatively short time suggested by the fossil records. A living being is an amalgam of characteristics determined by the (typically tens of thousands) genes in its chromosomes. Each gene may have several forms or alternatives called *alleles* which produce differences in the set of characteristics associated with that gene. The chromosomes are therefore the organic devices through which the structure of a creature is encoded, and this living being is created partly through the process of decoding those chromosomes. Genes transmit hereditary characters and form specific parts of a self-perpetuated deoxyribonucleic acid (DNA) in a cell nucleus. Natural selection is the link between the chromosomes and the performance of their decoded structures. Simply put, the process of natural selection causes those chromosomes that encode successful structures to reproduce more often than those that do not.

In an attempt to solve difficult problems, John H. Holland of the University of Michigan introduced in the early 1970s the manmade version of the procedure of natural evolution. The candidate solutions to a problem are ranked by the genetic algorithm according to how well they satisfy a certain criterion, and the fittest members are the most favored to combine amongst themselves to form the next generation of the members of the species. Fitter members presumably produce even fitter offspring and therefore better solutions to the problem at hand. Solutions are represented by binary strings, and each trial solution is coded as a vector called a *chromosome*. The elements of a chromosome are described as *genes*, and its varying values at specific positions are called *alleles*. Good solutions are selected for reproduction based on a fitness function using genetic recombination operators such as crossover and mutation. The main advantage of genetic algorithms is their global parallelism, in which the search efforts in many regions of the search area are simultaneously allocated.

Genetic algorithms have been used for the control of different dynamical systems, such as, for example, optimization of robot trajectories, but to my knowledge and at the time of writing this chapter reactive control of turbulent flows has yet to benefit from this powerful soft computing tool. In particular, when a finite number of sensors is used to gather information about the state of the flow, a genetic algorithm

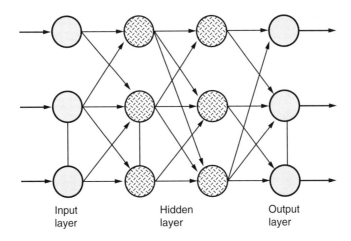

FIGURE 33.13 Elements of a neural network. (From Gad-el-Hak, M. (2000) *Flow Control: Passive, Active, and Reactive Flow Management*. Reprinted with permission of Cambridge University Press, New York.)

perhaps combined with a neural network can adapt and learn to use current information to eliminate the uncertainty created by insufficient *a priori* information.

33.12.1 Neural Networks for Flow Control

Biologically inspired neural networks are finding increased applications in many fields of science and technology. Modeling of complex dynamical systems, adaptive noise canceling in telephones and modems, bomb sniffers, mortgage-risk evaluators, sonar classifiers, and word recognizers are but a few of the existing uses of neural nets. The book by Nelson and Illingworth (1991) provides a lucid introduction to the field, and the review article by Antsaklis (1993) focuses on the use of neural nets for the control of complex dynamical systems. For flow control applications, neural networks offer the possibility of adaptive controllers that are simpler and potentially less sensitive to parameter variations as compared to conventional controllers. Moreover, if a colossal number of sensors and actuators is to be used, the massively parallel computational power of neural nets will surely be needed for real-time control.

The basic elements of a neural network are schematically shown in Figure 33.13. Several inputs are connected to the nodes (neurons or processing elements) that form the input layer. There are one or more hidden layers, followed by an output layer. Note that the number of connections is higher than the total number of nodes. Both numbers are chosen based on the particular application and can be arbitrarily large for complex tasks. Simply put, the multitask—albeit simple—job of each processing element is to evaluate each of the input signals to that particular element, calculate the weighted sum of the combined inputs, compare that total to some threshold level and finally determine what the output should be. The various weights are the adaptive coefficients which vary dynamically as the network learns to perform its assigned task; some inputs are more important than others. The threshold, or transfer, function is generally nonlinear, the most common one being the continuous sigmoid, or S-shaped, curve, which approaches a minimum and maximum value at the asymptotes. If the sum of the weighted inputs is larger than the threshold value, the neuron generates a signal; otherwise, no signal is fired. Neural networks can operate in feedforward or feedback mode.[13] Complex systems, for which dynamical equations may not be known or may be too difficult to solve, can be modeled using neural nets.

---

[13] Note that this terminology refers to the direction of information through the network. When a neural net is used as a controller, the overall control loop is, however, a feedback, closed loop: The self-learning network dynamically updates its various parameters by comparing its output to a desired output, thus requiring feedback information relating to the effect of its control.

For flow control, neural networks provide convenient, fast, nonlinear, adaptive algorithms to relate sensor outputs to actuator inputs via variable-coefficient functions and nonlinear, sigmoid saturation functions. With no prior knowledge of the pertinent dynamics, a self-learning neural network develops a model for the dynamics through observations of the applied control and sensed measurements. The network is by nature nonlinear and can therefore better handle nonlinear dynamical systems, a difficult task when classical (linear or weakly nonlinear) control strategies are attempted. The feedforward type of neural network acts as a nonlinear filter forming an output from a set of input data. The output can then be compared to some desired output, and the difference (error) is typically used in a back-propagation algorithm that updates the network parameters.

The number of researchers using neural networks to control fluid flows is growing rapidly. Here, we provide only a small sample. Using a pre-trained neural network, Fan et al. (1993) conducted a conceptual reactive flow control experiment to delay laminar-to-turbulence transition. Numerical simulations of their flow control system demonstrate almost complete cancellation of single and multiple artificial wave disturbances. Their controller also successfully attenuated a natural disturbance signal with developing wave packets from an actual wind-tunnel experiment.

Jacobson and Reynolds (1993b; 1995; 1998) used neural networks to minimize the boundary velocity gradient of three model flows: the one-dimensional stochastic Burgers equation, a two-dimensional computational model of the near-wall region of a turbulent boundary layer and a real-time turbulent flow with a spanwise array of wall actuators together with upstream and downstream wall sensors. For all three problems, the neural network successfully learned about the flow and developed into proficient controllers. For the laboratory experiments, however, Jacobson and Reynolds (1995) report that the neural network training time was much longer and the performance was no better than a simpler *ad hoc* controller that they developed. Jacobson and Reynolds emphasize that alternative neural net configurations and convergence algorithms may, however, greatly improve the network performance.

Using the angle of attack and angular velocity as inputs, Faller et al. (1994) trained a neural network to model the measured unsteady surface pressure over a pitching airfoil (see also Schreck et al., 1995). Following training and using the instantaneous angle of attack and pitch rate as the only inputs, their network was able to accurately predict the surface pressure topology as well as the time-dependent aerodynamic forces and moments. The model was then used to develop a neural network controller for wing-motion actuator signals which in turn provided direct control of the lift-to-drag ratio across a wide range of time-dependent motion histories.

As a final example, Kawthar-Ali and Acharya (1996) developed a neural network controller for use in suppressing the dynamic stall vortex that periodically develops in the leading edge of a pitching airfoil. Based on the current state of the unsteady pressure field, their control system specified the optimum amount of leading-edge suction to achieve complete vortex suppression.

33.13 Use of MEMS for Reactive Control

Current usage for microelectromechanical systems (MEMS) includes accelerometers for airbags and guidance systems, pressure sensors for engine air intake and blood analysis, rate gyroscopes for antilock brakes, microrelays and microswitches for semiconductor automatic test equipment and microgrippers for surgical procedures [Angell et al., 1983; Gabriel et al., 1988; 1992; O'Connor, 1992; Gravesen et al., 1993; Bryzek et al., 1994; Gabriel, 1995; Ashley, 1996; Ho and Tai, 1996; 1998; Hogan, 1996; Ouellette, 1996; Paula, 1996; Robinson et al., 1996a; 1996b; Madou, 1997; Tien, 1997; Amato, 1998; Busch-Vishniac, 1998; Kovacs, 1998; Knight, 1999; Epstein, 2000; O'Connor and Hutchinson, 2000]. Considerable work is under way to include other applications, one example being the micro steam engine described by Lipkin (1993), Garcia and Sniegowski (1993; 1995) and Sniegowski and Garcia (1996). A second example is the 3-cm × 1.5-cm digital light processor, which contains 0.5 to 2 million individually addressable micromirrors, each typically measuring 16 µm on a side. Texas Instruments, Inc., is currently producing such a device with a resolution of 2000 × 1000 pixels for high-definition televisions and other display equipments. The company maintains that when mass produced such a device would cost on the order of $100 (i.e., less than $0.0001 per actuator).

MEMS would be ideal for the reactive flow control concept advocated in this chapter. Methods of flow control targeted toward specific coherent structures involve nonintrusive detection and subsequent modulation of events that occur randomly in space and time. To achieve proper targeted control of these quasi-periodic vortical events, temporal phasing as well as spatial selectivity are required. Practical implementation of such an idea necessitates the use of a large number of intelligent, communicative wall sensors and actuators arranged in a checkerboard pattern. In this chapter, we have provided estimates for the number, characteristics and energy consumption of such elements required to modulate the turbulent boundary layer which develops along a typical commercial aircraft or nuclear submarine. An upper-bound number to achieve total turbulence suppression is about 1 million sensors/actuators per square meter of the surface, although as argued earlier the actual number needed to achieve effective control could perhaps be one to two orders of magnitude below that.

The sensors would be expected to measure the amplitude, location and phase or frequency of the signals impressed upon the wall by incipient bursting events. Instantaneous wall-pressure or wall-shear stress can be sensed, for example. The normal or in-plane motion of a minute membrane is proportional to the respective point force of primary interest. For measuring wall pressure, microphone-like devices respond to the motion of a vibrating surface membrane or an internal elastomer. Several types are available, including variable-capacitance (condenser or electret), ultrasonic, optical (e.g., optical-fiber and diode-laser) and piezoelectric devices [e.g., Löfdahl et al., 1993; 1994; Löfdahl and Gad-el-Hak, 1999]. A potentially useful technique for our purposes has been tried at MIT [Warkentin et al., 1987; Young et al., 1988; Haritonidis et al., 1990a; 1990b]. An array of extremely small (0.2 mm in diameter), laser-powered microphones (termed *picophones*) was machined in silicon using integrated circuit fabrication techniques and was used for field measurement of the instantaneous surface pressure in a turbulent boundary layer. The wall-shear stress, though smaller and therefore more difficult to measure than pressure, provides a more reliable signature of the near-wall events.

Actuators are expected to produce a desired change in the targeted coherent structures. The local acceleration action needed to stabilize an incipient bursting event can be in the form of adaptive wall, transpiration, wall heat transfer or electromagnetic body force. Traveling surface waves can be used to modify a locally convecting pressure gradient such that the wall motion follows that of the coherent event causing the pressure change. Surface motion in the form of a Gaussian hill with height $y^+ = \mathcal{O}[10]$ should be sufficient to suppress typical incipient bursts [Lumley, 1991; Carlson and Lumley, 1996]. Such time-dependent alteration in wall geometry can be generated by driving a flexible skin using an array of piezoelectric devices (dilate or contract, depending on the polarity of current passing through them), electromagnetic actuators, magnetoelastic ribbons (made of nonlinear materials that change their stiffness in the presence of varying magnetic fields) or Terfenol-d rods (a novel metal composite developed at Grumman Corporation that changes its length when subjected to a magnetic field). Note should also be made of other exotic materials that can be used for actuation. For example, electrorheological fluids [Halsey and Martin, 1993] instantly solidify when exposed to an electric field and may thus be useful for the present application. Recently constructed microactuators specifically designed for flow control include those by Wiltse and Glezer (1993), James et al. (1994), Jacobson and Reynolds (1995), Vargo and Muntz (1996) and Keefe (1996).

Suction/injection at many discrete points can be achieved by simply connecting a large number of minute streamwise slots, arranged in a checkerboard pattern, to a low-pressure/high-pressure reservoir located underneath the working surface. The transpiration through each individual slot is turned on and off using a corresponding number of independently controlled microvalves. Alternatively, positive-displacement or rotary micropumps [e.g., Sen et al., 1996; Sharatchandra et al., 1997] can be used for blowing or sucking fluid through small holes/slits. Based on the results of Gad-el-Hak and Blackwelder (1989), equivalent suction coefficients of about 0.0006 should be sufficient to stabilize the near-wall region. Assuming that the skin-friction coefficient in the uncontrolled boundary layer is $C_f = 0.003$ and assuming further that the suction used is sufficient to establish an asymptotic boundary layer ($d\delta_\theta/dx = 0$, where δ_θ is the momentum thickness), the skin friction in the reactively controlled case is then $C_f = 0 + 2C_q = 0.0012$, or 40% of the original value. The net benefit will, of course, be reduced by the energy expenditure of the suction pump (or micropumps) as well as the array of microsensors and microvalves.

Finally, if the bursting events are to be eliminated by lowering the near-wall viscosity, direct electric-resistance heating can be used in liquid flows, and thermoelectric devices based on the Peltier effect can be used for cooling in the case of gaseous boundary layers. The absolute viscosity of water at 20°C decreases by approximately 2% for each 1°C rise in temperature, while for room-temperature air μ decreases by approximately 0.2% for each 1°C drop in temperature. The streamwise momentum equation written at the wall can be used to show that a suction coefficient of 0.0006 has approximately the same effect on the wall curvature of the instantaneous velocity profile as a surface heating of 2°C in water or a surface cooling of 40°C in air [Liepmann and Nosenchuck, 1982; Liepmann et al., 1982].

Sensors and actuators of the types discussed in this section can be combined on individual electronic chips using microfabrication technology. The chips can be interconnected in a communication network controlled by a massively parallel computer or a self-learning neural network, perhaps each sensor/actuator unit communicating only with its immediate neighbors. In other words, it may not be necessary for one sensor/actuator to exchange signals with another far-away unit. Factors to be considered in an eventual field application of chips produced using microfabrication processes include sensitivity of sensors, sufficiency and frequency response of actuators' action, fabrication of large arrays at affordable prices, survivability in the hostile field environment and energy required to power the sensors/actuators. As argued by Gad-el-Hak (1994; 1996; 2000), sensor/actuator chips currently produced are small enough for typical field application and can be programmed to provide a sufficiently large/fast action in response to a certain sensor output (see also Jacobson and Reynolds, 1995). Current prototypes are, however, still quite expensive as well as delicate, but so was the transistor when first introduced! It is hoped that the unit price of future sensor/actuator elements would follow the same dramatic trends witnessed in the cases of the simple transistor and the even much more complex integrated circuit. The price anticipated by Texas Instruments for an array of 1 million mirrors hints that the technology is well on its way to mass-producing phenomenally inexpensive microsensors and microactuators. Additionally, current automotive applications are a rigorous proving ground for MEMS: under-the-hood sensors can already withstand harsh conditions such as intense heat, shock, continual vibration, corrosive gases and electromagnetic fields.

33.14 Parting Remarks

In this chapter, I have emphasized the frontiers of the field of flow control, reviewing the important advances that took place during the past few years and providing a blueprint for future progress. The future of flow control is in the taming of turbulence by targeting its coherent structures: *reactive control*. Recent developments in chaos control, microfabrication and soft computing tools are making it more feasible to perform reactive control of turbulent flows to achieve drag reduction, lift enhancement, mixing augmentation and noise suppression. Field applications, however, have to await further progress in those three modern areas.

The outlook for reactive control is quite optimistic. Soft computing tools and nonlinear dynamical systems theory are developing at a fast pace. MEMS technology is improving even faster. The ability of Texas Instruments to produce an array of one million individually addressable mirrors for around 0.01 ¢ per actuator foretells spectacular advances in the near future. Existing automotive applications of MEMS have already proven the ability of such devices to withstand the harsh environment under the hood. For the first time, targeted control of turbulent flows is now in the realm of the possible for future practical devices. What is needed now is a focused, well-funded research and development program to make it all come together for field application of reactive flow control systems.

In parting, it may be worth recalling that a mere 10% reduction in the total drag of an aircraft translates into a saving of $1 billion in annual fuel cost (at 1999 prices) for the commercial fleet of aircraft in the

U.S. alone. Contrast this benefit to the annual worldwide expenditure of perhaps a few million dollars for all basic research in the broad field of flow control. Taming turbulence, though arduous, will pay for itself in gold. Reactive control, as difficult as it seems, is neither impossible nor pie in the sky. Beside, lofty goals require strenuous efforts. Easy solutions to difficult problems are likely to be wrong, as implied by the witty words of the famed journalist Henry Louis Mencken (1880–1956): "*There is always an easy solution to every human problem—neat, plausible and wrong.*"

References

Abergel, F., and Temam, R. (1990) "On Some Control Problems in Fluid Mechanics," *Theor. Comput. Fluid Dyn.* **1**, pp. 303–325.

Alshamani, K.M.M., Livesey, J.L., and Edwards, F.J. (1982) "Excitation of the Wall Region by Sound in Fully Developed Channel Flow," *AIAA J.* **20**, pp. 334–339.

Amato, I. (1998) "Formenting a Revolution, in Miniature," *Science* **282**(5388), pp. 402–405.

Angell, J.B., Terry, S.C., and Barth, P.W. (1983) "Silicon Micromechanical Devices," *Faraday Transactions I* **68**, pp. 744–748.

Antonia, R.A., Fulachier, L., Krishnamoorthy, L.V., Benabid, T., and Anselmet, F. (1988) "Influence of Wall Suction on the Organized Motion in a Turbulent Boundary Layer," *J. Fluid Mech.* **190**, pp. 217–240.

Antsaklis, P.J. (1993) "Control Theory Approach," in *Mathematical Approaches to Neural Networks* (J.G. Taylor, Ed.), pp. 1–23. Elsevier, Amsterdam.

Aronson, D., and Löfdahl, L. (1994) "The Plane Wake of a Cylinder: An Estimate of the Pressure Strain Rate Tensor," *Phys. Fluids* **6**, pp. 2716–2721.

Aronson, D., Johansson, A.V., and Löfdahl, L. (1997) "A Shear-Free Turbulent Boundary Layer—Experiments and Modeling," *J. Fluid Mech.* **338**, pp. 363–385.

Ashley, S. (1996) "Getting a Microgrip in the Operating Room," *Mechanical Engineering* **118**, pp. 91–93.

Aubry, N. (1990) "Use of Experimental Data for an Efficient Description of Turbulent Flows," *Appl. Mech. Rev.* **43**, pp. S240–S245.

Aubry, N., Holmes, P., Lumley, J.L., and Stone, E. (1988) "The Dynamics of Coherent Structures in the Wall Region of a Turbulent Boundary Layer," *J. Fluid Mech.* **192**, pp. 115–173.

Auerbach, D. (1994) "Controlling Extended Systems of Chaotic Elements," *Phys. Rev. Lett.* **72**, pp. 1184–1187.

Auerbach, D., Grebogi, C., Ott, E., and Yorke, J.A. (1992) "Controlling Chaos in High Dimensional Systems," *Phys. Rev. Lett.* **69**, pp. 3479–3482.

Bakewell, H.P., and Lumley, J.L. (1967) "Viscous Sublayer and Adjacent Wall Region in Turbulent Pipe Flow," *Phys. Fluids* **10**, pp. 1880–1889.

Bandyopadhyay, P.R. (1986) "Review—Mean Flow in Turbulent Boundary Layers Disturbed to Alter Skin Friction," *J. Fluids Eng.* **108**, pp. 127–140.

Banks, S.P. (1986) *Control Systems Engineering*, Prentice-Hall, Englewood Cliffs, NJ.

Barnwell, R.W., and Hussaini, M.Y., Eds. (1992) *Natural Laminar Flow and Laminar Flow Control*, Springer-Verlag, New York.

Berkooz, G., Holmes, P., and Lumley, J.L. (1991) "Intermittent Dynamics in Simple Models of the Turbulent Boundary Layer," *J. Fluid Mech.* **230**, pp. 75–95.

Berkooz, G., Fisher, M., and Psiaki, M. (1993) "Estimation and Control of Models of the Turbulent Wall Layer," *Bull. Am. Phys. Soc.* **38**, p. 2197.

Bewley, T.R., Moin, P., and Temam, R. (1997) "Optimal and Robust Approaches for Linear and Nonlinear Regulation Problems in Fluid Mechanics," AIAA Paper No. 97-1872, AIAA, Reston, VA.

Bewley, T.R., Temam, R., and Ziane, M. (1998) "A General Framework for Robust Control in Fluid Mechanics," Center for Turbulence Research No. CTR-Manuscript-169, Stanford University, Stanford, CA.

Blackwelder, R.F. (1978) "The Bursting Process in Turbulent Boundary Layers," in *Workshop on Coherent Structure of Turbulent Boundary Layers* (C.R. Smith and D.E. Abbott, Eds.), pp. 211–227. Lehigh University, Bethlehem, PA.

Blackwelder, R.F. (1988) "Coherent Structures Associated with Turbulent Transport," in *Transport Phenomena in Turbulent Flows* (M. Hirata and N. Kasagi, Eds.), pp. 69–88. Hemisphere, New York.

Blackwelder, R.F. (1998) "Some Notes on Drag Reduction in the Near-Wall Region," in *Flow Control: Fundamentals and Practices* (M. Gad-el-Hak, A. Pollard, and J.-P. Bonnet, Eds.), pp. 155–198. Springer-Verlag, Berlin.

Blackwelder, R.F., and Eckelmann, H. (1979) "Streamwise Vortices Associated with the Bursting Phenomenon," *J. Fluid Mech.* **94**, pp. 577–594.

Blackwelder, R.F., and Gad-el-Hak, M. (1990) "Method and Apparatus for Reducing Turbulent Skin Friction," U.S. Patent 4,932,612.

Blackwelder, R.F., and Haritonidis, J.H. (1983) "Scaling of the Bursting Frequency in Turbulent Boundary Layers," *J. Fluid Mech.* **132**, pp. 87–103.

Blackwelder, R.F., and Kovasznay, L.S.G. (1972) "Time-Scales and Correlations in a Turbulent Boundary Layer," *Phys. Fluids* **15**, pp. 1545–1554.

Blackwelder, R.F., and Swearingen, J.D. (1990) "The Role of Inflectional Velocity Profiles in Wall Bounded Flows," in *Near-Wall Turbulence: 1988 Zoran Zaric Memorial Conference* (S.J. Kline and N.H. Afgan, Eds.), pp. 268–288, Hemisphere, New York.

Bouchon-Meunier, B., Yager, R.R., and Zadeh, L.A., Eds. (1995a) *Fuzzy Logic and Soft Computing*, World Scientific, Singapore.

Bouchon-Meunier, B., Yager, R.R., and Zadeh, L.A., Eds. (1995b) *Advances in Intelligent Computing—IPMU'94*, Lecture Notes in Computer Science, Vol. 945, Springer-Verlag, Berlin.

Breuer, K.S., Haritonidis, J.H., and Landahl, M.T. (1989) "The Control of Transient Disturbances in a Flat Plate Boundary Layer through Active Wall Motion," *Phys. Fluids A* **1**, pp. 574–582.

Bryzek, J., Peterson, K., and McCulley, W. (1994) "Micromachines on the March," *IEEE Spectrum* **31**, May, pp. 20–31.

Busch-Vishniac, I.J. (1998) "Trends in Electromechanical Transduction," *Physics Today* **51**, July, pp. 28–34.

Bushnell, D.M. (1983) "Turbulent Drag Reduction for External Flows," AIAA Paper No. 83-0227, AIAA, New York.

Bushnell, D.M. (1994) "Viscous Drag Reduction in Aeronautics," in *Proc. 19th Congress of the Int. Council of the Aeronautical Sciences*, Vol. 1, pp. XXXIII–L VI, Paper No. ICAS-94-0.1, AIAA, Washington, D.C.

Bushnell, D.M. (1998) "Frontiers of the 'Responsibly Imaginable' in Aeronautics," Dryden Lecture, AIAA Paper No. 98-0001, AIAA, Reston, VA.

Bushnell, D.M., and Hefner, J.N., Eds. (1990) *Viscous Drag Reduction in Boundary Layers*, AIAA, Washington, D.C.

Bushnell, D.M., and McGinley, C.B. (1989) "Turbulence Control in Wall Flows," *Annu. Rev. Fluid Mech.* **21**, pp. 1–20.

Bussmann, K., and Münz, H. (1942) "Die Stabilität der laminaren Reibungsschicht mit Absaugung," *Jahrb. Dtsch. Luftfahrtforschung* **1**, pp. 36–39.

Butler, K.M., and Farrell, B.F. (1993) "Optimal Perturbations and Streak Spacing in Wall-Bounded Shear Flow," *Phys. Fluids A* **5**, pp. 774–777.

Cantwell, B.J. (1981) "Organized Motion in Turbulent Flow," *Ann. Rev. Fluid Mech.* **13**, pp. 457–515.

Carlson, H.A., and Lumley, J.L. (1996) "Flow Over an Obstacle Emerging from the Wall of a Channel," *AIAA J.* **34**, pp. 924–931.

Chen, C.-C., Wolf, E.E., and Chang, H.-C. (1993) "Low-Dimensional Spatiotemporal Thermal Dynamics on Nonuniform Catalytic Surfaces," *J. Phys. Chemistry* **97**, pp. 1055–1064.

Choi, H., Moin, P., and Kim, J. (1992) "Turbulent Drag Reduction: Studies of Feedback Control and Flow Over Riblets," Department of Mechanical Engineering Report No. TF-55, Stanford University, Stanford, CA.

Choi, H., Temam, R., Moin, P., and Kim, J. (1993) "Feedback Control for Unsteady Flow and Its Application to the Stochastic Burgers Equation," *J. Fluid Mech.* **253**, pp. 509–543.

Choi, H., Moin, P., and Kim, J. (1994) "Active Turbulence Control for Drag Reduction in Wall-Bounded Flows," *J. Fluid Mech.* **262**, pp. 75–110.

Coller, B.D., Holmes, P., and Lumley, J.L. (1994a) "Control of Bursting in Boundary Layer Models," *Appl. Mech. Rev.* **47**(6, part 2), pp. S139–S143.

Coller, B.D., Holmes, P., and Lumley, J.L. (1994b) "Control of Noisy Heteroclinic Cycles," *Physica D* **72**, pp. 135–160.

Corino, E.R., and Brodkey, R.S. (1969) "A Visual Investigation of the Wall Region in Turbulent Flow," *J. Fluid Mech.* **37**, pp. 1–30.

Corke, T.C., Glauser, M.N., and Berkooz, G. (1994) "Utilizing Low-Dimensional Dynamical Systems Models to Guide Control Experiments," *Appl. Mech. Rev.* **47**(6, part 2), pp. S132–S138.

Corrsin, S. (1957) "Some Current Problems in Turbulent Shear Flow," in *Symp. on Naval Hydrodynamics* (F.S. Sherman, Ed.), pp. 373–400. National Academy of Sciences/National Research Council Publication No. 515, Washington, D.C.

Corrsin, S. (1959) "Outline of Some Topics in Homogenous Turbulent Flow," *J. Geophys. Research* **64**, pp. 2134–2150.

Davis, L., Ed. (1991) *Handbook of Genetic Algorithms*, Van Nostrand-Reinhold, New York.

Deane, A.E., and Sirovich, L. (1991) "A Computational Study of Rayleigh–Bénard Convection. Part 1. Rayleigh-Number Scaling," *J. Fluid Mech.* **222**, pp. 231–250.

Delville, J., Cordier L., and Bonnet, J.-P. (1998) "Large-Scale-Structure Identification and Control in Turbulent Shear Flows," in *Flow Control: Fundamentals and Practices* (M. Gad-el-Hak, A. Pollard, and J.-P. Bonnet, Eds.), pp. 199–273. Springer-Verlag, Berlin.

Ditto, W.L., and Pecora, L.M. (1993) "Mastering Chaos," *Sci. Am.* **269**, August, pp. 78–84.

Ditto, W.L., Rauseo, S.N., and Spano, M.L. (1990) "Experimental Control of Chaos," *Phys. Rev. Lett.* **65**, pp. 3211–3214.

Donohue, G.L., Tiederman, W.G., and Reischman, M.M. (1972) "Flow Visualization of the Near-Wall Region in a Drag-Reducing Channel Flow," *J. Fluid Mech.* **56**, pp. 559–575.

Eléna, M. (1975) "Etude des Champs Dynamiques et Thermiques d'un Ecoulement Turbulent en Conduit avec Aspiration á la Paroi," Thése de Doctorat des Sciences, Université d'Aix-Marseille, Marseille, France.

Eléna, M. (1984) "Suction Effects on Turbulence Statistics in a Heated Pipe Flow," *Phys. Fluids* **27**, pp. 861–866.

Epstein, A.H. (2000) "The Inevitability of Small," *Aerospace America* **38**, March, pp. 30–37.

Falco, R.E. (1974) "Some Comments on Turbulent Boundary Layer Structure Inferred from the Movements of a Passive Contaminant," AIAA Paper No. 74-99, AIAA, New York.

Falco, R.E. (1977) "Coherent Motions in the Outer Region of Turbulent Boundary Layers," *Phys. Fluids* **20**(10, part II), pp. S124–S132.

Falco, R.E. (1980) "The Production of Turbulence Near a Wall," AIAA Paper No. 80-1356, AIAA, New York.

Falco, R.E. (1983) "New Results, a Review and Synthesis of the Mechanism of Turbulence Production in Boundary Layers and Its Modification," AIAA Paper No. 83-0377, AIAA, New York.

Falco, R.E. (1991) "A Coherent Structure Model of the Turbulent Boundary Layer and Its Ability to Predict Reynolds Number Dependence," *Phil. Trans. R. Soc. London A* **336**, pp. 103–129.

Faller, W.E., Schreck, S.J., and Luttges, M.W. (1994) "Real-Time Prediction and Control of Three-Dimensional Unsteady Separated Flow Fields Using Neural Networks," AIAA Paper No. 94-0532, AIAA, Washington, D.C.

Fan, X., Hofmann, L., and Herbert, T. (1993) "Active Flow Control with Neural Networks," AIAA Paper No. 93-3273, AIAA, Washington, D.C.

Favre, A., Dumas, R., Verollet, E., and Coantic, M. (1966) "Couche Limite Turbulente sur Paroi Poreuse avec Aspiration," *J. Mécanique* **5**, pp. 3–28.

Fiedler, H.E. (1986) "Coherent Structures," in *Advances in Turbulence* (G. Comte-Bellot and J. Mathieu, Eds.), pp. 320–336. Springer-Verlag, Berlin.

Fiedler, H.E. (1988) "Coherent Structures in Turbulent Flows," *Prog. Aero. Sci.* **25**, pp. 231–269.

Fiedler, H.E., and Fernholz, H.-H. (1990) "On Management and Control of Turbulent Shear Flows," *Prog. Aero. Sci.* **27**, pp. 305–387.

Fiedler, H.E., Glezer, A., and Wygnanski, I. (1988) "Control of Plane Mixing Layer: Some Novel Experiments," in *Current Trends in Turbulence Research* (H. Branover, M. Mond, and Y. Unger, Eds.), pp. 30–64. AIAA, Washington, D.C.

Fowler, T.B. (1989) "Application of Stochastic Control Techniques to Chaotic Nonlinear Systems," *IEEE Trans. Autom. Control* **34**, pp. 201–205.

Gabriel, K.J. (1995) "Engineering Microscopic Machines," *Sci. Am.* **260**, September, pp. 150–153.

Gabriel, K.J., Jarvis, J., and Trimmer, W., Eds. (1988) *Small Machines, Large Opportunities: A Report on the Emerging Field of Microdynamics,* National Science Foundation, AT&T Bell Laboratories, Murray Hill, NJ.

Gabriel, K.J., Tabata, O., Shimaoka, K., Sugiyama, S., and Fujita, H. (1992) "Surface-Normal Electrostatic/Pneumatic Actuator," in *Proc. IEEE Micro Electro Mechanical Systems (MEMS'92)*, pp. 128–131, Feb. 4–7, Travemünde, Germany.

Gad-el-Hak, M. (1989) "Flow Control," *Appl. Mech. Rev.* **42**, pp. 261–293.

Gad-el-Hak, M. (1993) "Innovative Control of Turbulent Flows," AIAA Paper No. 93-3268, AIAA, Washington, D.C.

Gad-el-Hak, M. (1994) "Interactive Control of Turbulent Boundary Layers: A Futuristic Overview," *AIAA J.* **32**, pp. 1753–1765.

Gad-el-Hak, M. (1996) "Modern Developments in Flow Control," *Appl. Mech. Rev.* **49**, pp. 365–379.

Gad-el-Hak, M. (1998) "Frontiers of Flow Control," in *Flow Control: Fundamentals and Practices* (M. Gad-el-Hak, A. Pollard, and J.-P. Bonnet, Eds.), pp. 109–153. Springer-Verlag, Berlin.

Gad-el-Hak, M. (2000) *Flow Control: Passive, Active, and Reactive Flow Management,* Cambridge University Press, London.

Gad-el-Hak, M., and Bandyopadhyay, P.R. (1994) "Reynolds Number Effects in Wall-Bounded Flows," *Appl. Mech. Rev.* **47**, pp. 307–365.

Gad-el-Hak, M., and Blackwelder, R.F. (1987a) "Simulation of Large-Eddy Structures in a Turbulent Boundary Layer," *AIAA J.* **25**, pp. 1207–1215.

Gad-el-Hak, M., and Blackwelder, R.F. (1987b) "A Drag Reduction Method for Turbulent Boundary Layers," AIAA Paper No. 87-0358, AIAA, New York.

Gad-el-Hak, M., and Blackwelder, R.F. (1989) "Selective Suction for Controlling Bursting Events in a Boundary Layer," *AIAA J.* **27**, pp. 308–314.

Gad-el-Hak, M., and Bushnell, D.M. (1991a) "Status and Outlook of Flow Separation Control," AIAA Paper No. 91-0037, AIAA, New York.

Gad-el-Hak, M., and Bushnell, D.M. (1991b) "Separation Control: Review," *J. Fluids Eng.* **113**, pp. 5–30.

Gad-el-Hak, M., and Hussain, A.K.M.F. (1986) "Coherent Structures in a Turbulent Boundary Layer. Part 1. Generation of 'Artificial' Bursts," *Phys. Fluids* **29**, pp. 2124–2139.

Gad-el-Hak, M., Blackwelder, R.F., and Riley, J.J. (1981) "On the Growth of Turbulent Regions in Laminar Boundary Layers," *J. Fluid Mech.* **110**, pp. 73–95.

Gad-el-Hak, M., Pollard, A., and Bonnet, J.-P. (1998) *Flow Control: Fundamentals and Practices,* Springer-Verlag, Berlin.

Garcia, E.J., and Sniegowski, J.J. (1993) "The Design and Modelling of a Comb-Drive-Based Microengine for Mechanism Drive Applications," in *Proc. 7th Int. Conf. on Solid-State Sensors and Actuators (Transducers '93)*, pp. 763–766, June 7–10, Yokohama, Japan.

Garcia, E.J., and Sniegowski, J.J. (1995) "Surface Micromachined Microengine," *Sensors and Actuators A* **48**, pp. 203–214.

Garfinkel, A., Spano, M.L., Ditto, W.L., and Weiss, J.N. (1992) "Controlling Cardiac Chaos," *Science* **257**, pp. 1230–1235.

Gerbeth, G., Ed. (1997) *Proceedings of the International Workshop on Electromagnetic Boundary Layer Control (EBLC) for Saltwater Flows,* July 7–8, Forschungszentrum Rossendorf, Dresden, Germany.

Goldberg, D.E. (1989) *Genetic Algorithms in Search, Optimization, and Machine Learning,* Addison-Wesley, Reading, MA.

Grappin, R., and Léorat, J. (1991) "Lyapunov Exponents and the Dimension of Periodic Incompressible Navier–Stokes Flows: Numerical Measurements," *J. Fluid Mech.* **222**, pp. 61–94.

Gravesen, P., Branebjerg, J., and Jensen, O.S. (1993) "Microfluidics—A Review," *J. Micromech. Microeng.* **3**, pp. 168–182.

Gutmark, E.J., and Grinstein, F.F. (1999) "Flow Control with Noncircular Jets," *Annu. Rev. Fluid Mech.* **31**, pp. 239–272.

Gutmark, E.J., and Ho, C.-M. (1986) "Visualization of a Forced Elliptical Jet," *AIAA J.* **24**, pp. 684–685.

Gutmark, E.J., Schadow, K.C., and Yu, K.H. (1995) "Mixing Enhancement in Supersonic Free Shear Flows," *Annu. Rev. Fluid Mech.* **27**, pp. 375–417.

Halsey, T.C., and Martin, J.E. (1993) "Electrorheological Fluids," *Sci. Am.* **269**, October, pp. 58–64.

Haritonidis, J.H., Senturia, S.D., Warkentin, D.J., and Mehregany, M. (1990a) "Optical Micropressure Transducer," U.S. Patent 4,926,696.

Haritonidis, J.H., Senturia, S.D., Warkentin, D.J., and Mehregany, M. (1990b) "Pressure Transducer Apparatus," U.S. Patent 4,942,767.

Hayes, S., Grebogi, C., and Ott, E. (1994a) "Communicating with Chaos," *Phys. Rev. Lett.* **70**, pp. 3031–3040.

Hayes, S., Grebogi, C., Ott, E., and Mark, A. (1994b) "Experimental Control of Chaos for Communication," *Phys. Rev. Lett.* **73**, pp. 1781–1784.

Hinze, J.O. (1975) *Turbulence,* 2nd ed., McGraw-Hill, New York.

Ho, C.-M., and Tai, Y.-C. (1996) "Review: MEMS and Its Applications for Flow Control," *J. Fluids Eng.* **118**, pp. 437–447.

Ho, C.-M., and Tai, Y.-C. (1998) "Micro-Electro-Mechanical Systems (MEMS) and Fluid Flows," *Annu. Rev. Fluid Mech.* **30**, pp. 579–612.

Hogan, H. (1996) "Invasion of the Micromachines," *New Scientist* **29**, June, pp. 28–33.

Holland, J.H. (1992) *Adaptation in Natural and Artificial Systems,* MIT Press, Cambridge, MA.

Holmes, P., Lumley, J.L., and Berkooz, G. (1996) *Turbulence, Coherent Structures, Dynamical Systems and Symmetry,* Cambridge University Press, Cambridge, U.K.

Hu, H.H., and Bau, H.H. (1994) "Feedback Control To Delay or Advance Linear Loss of Stability in Planar Poiseuille Flow," in *Proc. Roy. Soc. London A* **447**, pp. 299–312.

Huberman, B. (1990) "The Control of Chaos," in *Proc. Workshop on Applications of Chaos,* Dec. 4–7, San Francisco, CA.

Huberman, B.A., and Lumer, E. (1990) "Dynamics of Adaptive Systems," *IEEE Trans. Circuits Syst.* **37**, pp. 547–550.

Hübler, A., and Lüscher, E. (1989) "Resonant Stimulation and Control of Nonlinear Oscillators," *Naturwissenschaften* **76**, pp. 67–69.

Huerre, P., and Monkewitz, P.A. (1990) "Local and Global Instabilities in Spatially Developing Flows," *Annu. Rev. Fluid Mech.* **22**, pp. 473–537.

Hussain, A.K.M.F. (1986) "Coherent Structures and Turbulence," *J. Fluid Mech.* **173**, pp. 303–356.

Iglisch, R. (1944) "Exakte Berechnung der laminaren Reibungsschicht an der längsangeströmten ebenen Platte mit homogener Absaugung," *Schriften Dtsch. Akad. Luftfahrtforschung B* **8**, pp. 1–51.

Jacobs, J., James, R., Ratliff, C., and Glazer, A. (1993) "Turbulent Jets Induced by Surface Actuators," AIAA Paper No. 93-3243, AIAA, Washington, D.C.

Jacobson, S.A., and Reynolds, W.C. (1993a) "Active Control of Boundary Layer Wall Shear Stress Using Self-Learning Neural Networks," AIAA Paper No. 93-3272, AIAA, Washington, D.C.

Jacobson, S.A., and Reynolds W.C. (1993b) "Active Boundary Layer Control Using Flush-Mounted Surface Actuators," *Bull. Am. Phys. Soc.* **38**, p. 2197.

Jacobson, S.A., and Reynolds, W.C. (1994) "Active Control of Transition and Drag in Boundary Layers," *Bull. Am. Phy. Soc.* **39**, p. 1894.

Jacobson, S.A., and Reynolds, W.C. (1995) "An Experimental Investigation Towards the Active Control of Turbulent Boundary Layers," Department of Mechanical Engineering Report No. TF-64, Stanford University, Stanford, CA.

Jacobson, S.A., and Reynolds, W.C. (1998) "Active Control of Streamwise Vortices and Streaks in Boundary Layers," *J. Fluid Mech.* **360**, pp. 179–211.

James, R.D., Jacobs, J.W., and Glezer, A. (1994) "Experimental Investigation of a Turbulent Jet Produced by an Oscillating Surface Actuator," *Appl. Mech. Rev.* **47**(6, part 2), pp. S127–S1131.

Jang, J.-S.R., Sun, C.-T., and Mizutani, E. (1997) *Neuro-Fuzzy and Soft Computing*, Prentice-Hall, Upper Saddle River, NJ.

Johansen, J.B., and Smith, C.R. (1986) "The Effects of Cylindrical Surface Modifications on Turbulent Boundary Layers," *AIAA J.* **24**, pp. 1081–1087.

Joslin, R.D., Erlebacher, G., and Hussaini, M.Y. (1996) "Active Control of Instabilities in Laminar Boundary Layers—Overview and Concept Validation," *J. Fluids Eng.* **118**, pp. 494–497.

Kawthar-Ali, M.H., and Acharya, M. (1996) "Artificial Neural Networks for Suppression of the Dynamic-Stall Vortex over Pitching Airfoils," AIAA Paper No. 96-0540, AIAA, Washington, D.C.

Keefe, L.R. (1993a) "Two Nonlinear Control Schemes Contrasted in a Hydrodynamic Model," *Phys. Fluids A* **5**, pp. 931–947.

Keefe, L.R. (1993b) "Drag Reduction in Channel Flow Using Nonlinear Control," AIAA Paper No. 93-3279, AIAA, Washington, D.C.

Keefe, L.R. (1996) "A MEMS-Based Normal Vorticity Actuator for Near-Wall Modification of Turbulent Shear Flows," in *Proc. Workshop on Flow Control: Fundamentals and Practices* (J.-P. Bonnet, M. Gad-el-Hak, and A. Pollard, Eds.), pp. 1–21, July 1–5, Institut d'Etudes Scientifiques des Cargèse, Corsica, France.

Keefe, L.R., Moin, P., and Kim, J. (1992) "The Dimension of Attractors Underlying Periodic Turbulent Poiseuille Flow," *J. Fluid Mech.* **242**, pp. 1–29.

Keith, W.L., Hurdis, D.A., and Abraham, B.M. (1992) "A Comparison of Turbulent Boundary Layer Wall-Pressure Spectra," *J. Fluids Eng.* **114**, pp. 338–347.

Kim, H.T., Kline, S.J., and Reynolds, W.C. (1971) "The Production of Turbulence Near a Smooth Wall in a Turbulent Boundary Layer," *J. Fluid Mech.* **50**, pp. 133–160.

Kim, J., Moin, P., and Moser, R.D. (1987) "Turbulence Statistics in Fully Developed Channel Flow at Low Reynolds Number," *J. Fluid Mech.* **177**, pp. 133–166.

Klewicki, J.C., Murray, J.A., and Falco, R.E. (1994) "Vortical Motion Contributions to Stress Transport in Turbulent Boundary Layers," *Phys. Fluids* **6**, pp. 277–286.

Kline, S.J., and Runstadler, P.W. (1959) "Some Preliminary Results of Visual Studies of the Flow Model of the Wall Layers of the Turbulent Boundary Layer," *J. Appl. Mech.* **26**, pp. 166–170.

Kline, S.J., Reynolds, W.C., Schraub, F.A., and Runstadler, P.W. (1967) "The Structure of Turbulent Boundary Layers," *J. Fluid Mech.* **30**, pp. 741–773.

Knight, J. (1999) "Dust Mite's Dilemma," *New Scientist* **162**(2180), May 29, pp. 40–43.

Kolmogorov, A.N. (1941a) "The Local Structure of Turbulence in Incompressible Viscous Fluid for Very Large Reynolds Number," *Dokl. Akad. Nauk SSSR* **30**, pp. 301–305. (Reprinted in *Proc. R. Soc. Lond. A* **434**, pp. 9–13, 1991.)

Kolmogorov, A.N. (1941b) "On Degeneration of Isotropic Turbulence in an Incompressible Viscous Liquid," *Dokl. Akad. Nauk SSSR* **31**, pp. 538–540.

Kostelich, E.J., Grebogi, C., Ott, E., and Yorke, J.A. (1993a) "Targeting from Time Series," *Bul. Am. Phys. Soc.* **38**, p. 2194.

Kostelich, E.J., Grebogi, C., Ott, E., and Yorke, J.A. (1993b) "Higher-Dimensional Targeting," *Phys. Rev. E* **47**, pp. 305–310.

Kovacs, G.T.A. (1998) *Micromachined Transducers Sourcebook*, McGraw-Hill, New York.

Kovasznay, L.S.G. (1970) "The Turbulent Boundary Layer," *Annu. Rev. Fluid Mech.* **2**, pp. 95–112.

Kovasznay, L.S.G., Kibens, V., and Blackwelder, R.F. (1970) "Large-Scale Motion in the Intermittent Region of a Turbulent Boundary Layer," *J. Fluid Mech.* **41**, pp. 283–325.

Kwong, A., and Dowling, A. (1993) "Active Boundary Layer Control in Diffusers," AIAA Paper No. 93-3255, Washington, D.C.

Lachmann, G.V., Ed. (1961) *Boundary Layer and Flow Control*, Vols. 1 and 2, Pergamon Press, Oxford, Great Britain.

Lai, Y.-C., and Grebogi, C. (1993) "Synchronization of Chaotic Trajectories Using Control," *Phys. Rev. E* **47**, pp. 2357–2360.

Lai, Y.-C., Deng, M., and Grebogi, C. (1993a) "Controlling Hamiltonian Chaos," *Phys. Rev. E* **47**, pp. 86–92.

Lai, Y.-C., Tél, T., and Grebogi, C. (1993b) "Stabilizing Chaotic-Scattering Trajectories Using Control," *Phys. Rev. E* **48**, pp. 709–717.

Lai, Y.-C., Grebogi, C., and Tél, T. (1994) "Controlling Transient Chaos in Dynamical Systems," in *Towards the Harnessing of Chaos* (M. Yamaguchi, Ed.), Elsevier, Amsterdam.

Laufer, J. (1975) "New Trends in Experimental Turbulence Research," *Annu. Rev. Fluid Mech.* **7**, pp. 307–326.

Liepmann, H.W., and Nosenchuck, D.M. (1982) "Active Control of Laminar–Turbulent Transition," *J. Fluid Mech.* **118**, pp. 201–204.

Liepmann, H.W., Brown, G.L., and Nosenchuck, D.M. (1982) "Control of Laminar Instability Waves Using a New Technique," *J. Fluid Mech.* **118**, pp. 187–200.

Ligrani, P.M., and Bradshaw, P. (1987) "Spatial Resolution and Measurements of Turbulence in the Viscous Sublayer Using Subminiature Hot-Wire Probes," *Exp. Fluids* **5**, pp. 407–417.

Lindner, J.F., and Ditto, W.L. (1995) "Removal, Suppression and Control of Chaos by Nonlinear Design," *Appl. Mech. Rev.* **48**, pp. 795–808.

Lipkin, R. (1993) "Micro Steam Engine Makes Forceful Debut," *Science News* **144**, September, p. 197.

Löfdahl, L., and Gad-el-Hak, M. (1999) "MEMS Applications in Turbulence and Flow Control," *Prog. Aero. Sci.* **35**, pp. 101–203.

Löfdahl, L., Glavmo, M., Johansson, B., and Stemme, G. (1993) "A Silicon Transducer for the Determination of Wall-Pressure Fluctuations in Turbulent Boundary Layers," *Appl. Sci. Res.* **51**, pp. 203–207.

Löfdahl, L., Kälvesten, E., and Stemme, G. (1994) "Small Silicon Based Pressure Transducers for Measurements in Turbulent Boundary Layers," *Exp. Fluids* **17**, pp. 24–31.

Löfdahl, L., Kälvesten, E., and Stemme, G. (1996) "Small Silicon Pressure Transducers for Space-Time Correlation Measurements in a Flat Plate Boundary Layer," *J. Fluids Eng.* **118**, pp. 457–463.

Lumley, J.L. (1991) "Control of the Wall Region of a Turbulent Boundary Layer," in *Turbulence: Structure and Control* (J.M. McMichael, Ed.), pp. 61–62, April 1–3, Ohio State University, Columbus.

Lumley, J.L. (1996) "Control of Turbulence," AIAA Paper No. 96-0001, AIAA, Washington, D.C.

Lüscher, E., and Hübler, A. (1989) "Resonant Stimulation of Complex Systems," *Helv. Phys. Acta* **62**, pp. 544–551.

Madou, M. (1997) *Fundamentals of Microfabrication*, CRC Press, Boca Raton, FL.

McMichael, J.M. (1996) "Progress and Prospects for Active Flow Control Using Microfabricated Electromechanical Systems (MEMS)," AIAA Paper No. 96-0306, AIAA, Washington, D.C.

Mehregany, M., DeAnna, R.G., and Reshotko, E. (1996) "Microelectromechanical Systems for Aerodynamics Applications," AIAA Paper No. 96-0421, AIAA, Washington, D.C.

Meng, J.C.S., Ed. (1998) *Proc. Int. Symp. on Sea Water Drag Reduction*, July 22–23, Naval Undersea Warfare Center, Newport, RI.

Moin, P., and Bewley, T. (1994) "Feedback Control of Turbulence," *Appl. Mech. Rev.* **47**(6, part 2), pp. S3–S13.

Narasimha, R., and Sreenivasan, K.R. (1979) "Relaminarization of Fluid Flows," in *Advances in Applied Mechanics*, Vol. 19 (C.-S. Yih, Ed.), pp. 221–309. Academic Press, New York.

Nelson, M.M., and Illingworth, W.T. (1991) *A Practical Guide to Neural Nets*, Addison-Wesley, Reading, MA.

Noor, A., and Jorgensen, C.C. (1996) "A Hard Look at Soft Computing," *Aerospace Am.* **34**, September, pp. 34–39.

Nosenchuck, D.M., and Lynch, M.K. (1985) "The Control of Low-Speed Streak Bursting in Turbulent Spots," AIAA Paper No. 85-0535, AIAA, New York.

O'Connor, L. (1992) "MEMS: Micromechanical Systems," *Mech. Eng.* **114**, February, pp. 40–47.

O'Connor, L., and Hutchinson, H. (2000) "Skyscrapers in a Microworld," *Mech. Engi.* **122**, March, pp. 64–67.

Offen, G.R., and Kline, S.J. (1974) "Combined Dye-Streak and Hydrogen-Bubble Visual Observations of a Turbulent Boundary Layer," *J. Fluid Mech.* **62**, pp. 223–239.

Offen, G.R., and Kline, S.J. (1975) "A Proposed Model of the Bursting Process in Turbulent Boundary Layers," *J. Fluid Mech.* **70**, pp. 209–228.

Oldaker, D.K., and Tiederman, W.G. (1977) "Spatial Structure of the Viscous Sublayer in Drag-Reducing Channel Flows," *Phys. Fluids* **20**(10, part II), pp. S133–144.

Ott, E., Grebogi, C., and Yorke, J.A. (1990a) "Controlling Chaos," *Phys. Rev. Lett.* **64**, pp. 1196–1199.

Ott, E., Grebogi, C., and Yorke, J.A. (1990b) "Controlling Chaotic Dynamical Systems," in *Chaos: Soviet–American Perspectives on Nonlinear Science* (D.K. Campbell, Ed.), pp. 153–172. American Institute of Physics, New York.

Ouellette, J. (1996) "MEMS: Mega Promise for Micro Devices," *Mech. Eng.* **118**, October, pp. 64–68.

Ouellette, J. (1999) "Electronic Noses Sniff Out New Markets," *Ind. Phys.* **5**(1), pp. 26–29.

Panton, R.L., Ed. (1997) *Self-Sustaining Mechanisms of Wall Turbulence*, Computational Mechanics Publications, Southampton, U.K.

Paula, G. (1996) "MEMS Sensors Branch Out," *Aerospace Am.* **34**, September, pp. 26–32.

Perrier, P. (1998) "Multiscale Active Flow Control," in *Flow Control: Fundamentals and Practices* (M. Gad-el-Hak, A. Pollard, and J.-P. Bonnet, Eds.), pp. 275–334. Springer-Verlag, Berlin.

Petersen, I.R., and Savkin, A.V. (1999) *Robust Kalman Filtering for Signals and Systems with Large Uncertainties*, Birkhäuser, Boston, MA.

Pomeau, Y., and Manneville, P. (1980) "Intermittent Transition to Turbulence in Dissipative Dynamical Systems," *Commun. Math. Phys.* **74**, pp. 189–197.

Prandtl, L. (1904) "Über Flüssigkeitsbewegung bei sehr kleiner Reibung," in *Proc. Third Int. Math. Cong.*, pp. 484–491, Heidelberg, Germany.

Pretsch, J. (1942) "Umschlagbeginn und Absaugung," *Jahrb. Dtsch. Luftfahrtforschung* **1**, pp. 54–71.

Qin, F., Wolf, E.E., and Chang, H.-C. (1994) "Controlling Spatiotemporal Patterns on a Catalytic Wafer," *Phys. Rev. Lett.* **72**, pp. 1459–1462.

Reischman, M.M., and Tiederman, W.G. (1975) "Laser-Doppler Anemometer Measurements in Drag-Reducing Channel Flows," *J. Fluid Mech.* **70**, pp. 369–392.

Reynolds, W.C. (1993) "Sensors, Actuators, and Strategies for Turbulent Shear-Flow Control," presented at *AIAA Third Flow Control Conference*, July 6–9, Orlando, FL.

Riley, J.J., and Gad-el-Hak, M. (1985) "The Dynamics of Turbulent Spots," in *Frontiers in Fluid Mechanics* (S.H. Davis and J.L. Lumley, Eds.), pp. 123–155. Springer-Verlag, Berlin.

Robinson, E.Y., Helvajian, H., and Jansen, S.W. (1996a) "Small and Smaller: The World of MNT," *Aerospace Am.* **34**, September, pp. 26–32.

Robinson, E.Y., Helvajian, H., and Jansen, S.W. (1996b) "Big Benefits from Tiny Technologies," *Aerospace Am.* **34**, October, pp. 38–43.

Robinson, S.K. (1991) "Coherent Motions in the Turbulent Boundary Layer," *Annu. Rev. Fluid Mech.* **23**, pp. 601–639.

Robinson, S.K., Kline, S.J., and Spalart, P.R. (1989) "A Review of Quasi-Coherent Structures in a Numerically Simulated Turbulent Boundary Layer," NASA Technical Memorandum No. TM-102191, Washington, D.C.

Romeiras, F.J., Grebogi, C., Ott, E., and Dayawansa, W.P. (1992) "Controlling Chaotic Dynamical Systems," *Physica D* **58**, pp. 165–192.

Rotta, J.C. (1970) "Control of Turbulent Boundary Layers by Uniform Injection and Suction of Fluid," in *Proc. Seventh Cong. of the Int. Council of the Aeronautical Sciences,* Paper No. 70-10, ICAS, Rome, Italy.

Runstadler, P.G., Kline, S.J., and Reynolds, W.C. (1963) "An Experimental Investigation of Flow Structure of the Turbulent Boundary Layer," Department of Mechanical Engineering Report No. MD-8, Stanford University, Stanford, CA.

Saffman, P.G. (1978) "Problems and Progress in the Theory of Turbulence," in *Structure and Mechanisms of Turbulence II* (H. Fiedler, Ed.), pp. 273–306. Springer-Verlag, Berlin.

Saffman, P.G., and Baker, G.R. (1979) "Vortex Interactions," *Annu. Rev. Fluid Mech.* **11**, pp. 95–122.

Schreck, S.J., Faller, W.E., and Luttges, M.W. (1995) "Neural Network Prediction of Three-Dimensional Unsteady Separated Flow Fields," *J. Aircraft* **32**, pp. 178–185.

Sen, M., Wajerski, D., and Gad-el-Hak, M. (1996) "A Novel Pump for MEMS Applications," *J. Fluids Eng.* **118**, pp. 624–627.

Sharatchandra, M.C., Sen, M., and Gad-el-Hak, M. (1997) "Navier–Stokes Simulations of a Novel Viscous Pump," *J. Fluids Eng.* **119**, pp. 372–382.

Shercliff, J.A. (1965) *A Textbook of Magnetohydrodynamics*, Pergamon Press, Oxford.

Shinbrot, T. (1993) "Chaos: Unpredictable Yet Controllable?" *Nonlinear Science Today* **3**, pp. 1–8.

Shinbrot, T. (1995) "Progress in the Control of Chaos," *Adv. Phys.* **44**, pp. 73–111.

Shinbrot, T. (1998) "Chaos, Coherence and Control," in *Flow Control: Fundamentals and Practices* (M. Gad-el-Hak, A. Pollard, and J.-P. Bonnet, Eds.), pp. 501–527. Springer-Verlag, Berlin.

Shinbrot, T., and Ottino, J.M. (1993a) "Geometric Method To Create Coherent Structures in Chaotic Flows," *Phys. Rev. Lett.* **71**, pp. 843–846.

Shinbrot, T., and Ottino, J.M. (1993b) "Using Horseshoes To Create Coherent Structures in Chaotic Fluid Flows," *Bull. Am. Phys. Soc.* **38**, p. 2194.

Shinbrot, T., Ott, E., Grebogi, C., and Yorke, J.A. (1990) "Using Chaos To Direct Trajectories to Targets," *Phys. Rev. Lett.* **65**, pp. 3215–3218.

Shinbrot, T., Ditto, W., Grebogi, C., Ott, E., Spano, M., and Yorke, J.A. (1992a) "Using the Sensitive Dependence of Chaos (the "Butterfly Effect") To Direct Trajectories in an Experimental Chaotic System," *Phys. Rev. Lett.* **68**, pp. 2863–2866.

Shinbrot, T., Grebogi, C., Ott, E., and Yorke, J.A. (1992b) "Using Chaos To Target Stationary States of Flows," *Phys. Lett. A* **169**, pp. 349–354.

Shinbrot, T., Ott, E., Grebogi, C., and Yorke, J.A. (1992c) "Using Chaos To Direct Orbits to Targets in Systems Describable by a One-Dimensional Map," *Phys. Rev. A* **45**, pp. 4165–4168.

Shinbrot, T., Grebogi, C., Ott, E., and Yorke, J.A. (1993) "Using Small Perturbations To Control Chaos," *Nature* **363**, pp. 411–417.

Shinbrot, T., Bresler, L., and Ottino, J.M. (1998) "Manipulation of Isolated Structures in Experimental Chaotic Fluid Flows," *Exp. Thermal Fluid Sci.* **16**, pp. 76–83.

Singer, J., Wang, Y.-Z., and Bau, H.H. (1991) "Controlling a Chaotic System," *Phys. Rev. Lett.* **66**, pp. 1123–1125.

Sirovich, L., and Deane, A.E. (1991) "A Computational Study of Rayleigh-Bénard Convection. Part 2. Dimension Considerations," *J. Fluid Mech.* **222**, pp. 251–265.

Smith, C.R., and Metzler, S.P. (1982) "A Visual Study of the Characteristics, Formation, and Regeneration of Turbulent Boundary Layer Streaks," in *Developments in Theoretical and Applied Mechanics*, Vol. XI (T.J. Chung and G.R. Karr, Eds.), pp. 533–543. University of Alabama, Huntsville.

Smith, C.R., and Metzler, S.P. (1983) "The Characteristics of Low-Speed Streaks in the Near-Wall Region of a Turbulent Boundary Layer," *J. Fluid Mech.* **129**, pp. 27–54.

Smith, C.R., and Schwartz, S.P. (1983) "Observation of Streamwise Rotation in the Near-Wall Region of a Turbulent Boundary Layer," *Phys. Fluids* **26**, pp. 641–652.

Sniegowski, J.J., and Garcia, E.J. (1996) "Surface Micromachined Gear Trains Driven by an On-Chip Electrostatic Microengine," *IEEE Electron Device Lett.* **17**, July, p. 366.

Sreenivasan, K.R. (1989) "The Turbulent Boundary Layer," in *Frontiers in Experimental Fluid Mechanics* (M. Gad-el-Hak, Ed.), pp. 159–29. Springer-Verlag, New York.

Sritharan, S.S., Ed. (1998) *Optimal Control of Viscous Flow*, SIAM, Philadelphia, PA.

Swearingen, J.D., and Blackwelder, R.F. (1984) "Instantaneous Streamwise Velocity Gradients in the Wall Region," *Bull. Am. Phys. Soc.* **29**, p. 1528.

Tang, J., and Bau, H.H. (1993a) "Stabilization of the No-Motion State in Rayleigh–Bénard Convection through the Use of Feedback Control," *Phys. Rev. Lett.* **70**, pp. 1795–1798.

Tang, J., and Bau, H.H. (1993b) "Feedback Control Stabilization of the No-Motion State of a Fluid Confined in a Horizontal Porous Layer Heated from Below," *J. Fluid Mech.* **257**, pp. 485–505.

Taylor, G.I. (1935) "Statistical Theory of Turbulence," *Proc. Roy. Soc. London. A* **151**, pp. 421–478.

Tennekes, H., and Lumley, J.L. (1972) *A First Course in Turbulence*, MIT Press, Cambridge, MA.

Tiederman, W.G., Luchik, T.S., and Bogard, D.G. (1985) "Wall-Layer Structure and Drag Reduction," *J. Fluid Mech.* **156**, pp. 419–437.

Tien, N.C. (1997) "Silicon Micromachined Thermal Sensors and Actuators," *Microscale Thermophys. Eng.* **1**, pp. 275–292.

Townsend, A.A. (1961) "Equilibrium Layers and Wall Turbulence," *J. Fluid Mech.* **11**, pp. 97–120.

Townsend, A.A. (1970) "Entrainment and the Structure of Turbulent Flow," *J. Fluid Mech.* **41**, pp. 13–46.

Ulrich, A. (1944) "Theoretische Untersuchungen über die Widerstandsersparnis durch Laminarhaltung mit Absaugung," *Schriften Dtsch. Akad. Luftfahrtforschung B* **8**, p. 53.

Vargo, S.E., and Muntz, E.P. (1996) "A Simple Micromechanical Compressor and Vacuum Pump for Flow Control and Other Distributed Applications," AIAA Paper No. 96-0310, AIAA, Washington, D.C.

Verollet, E., Fulachier, L., Dumas, R., and Favre, A. (1972) "Turbulent Boundary Layer with Suction and Heating to the Wall," in *Heat and Mass Transfer in Boundary Layers,* Vol. 1 (N. Afgan, Z. Zaric, and P. Anastasijevec, Eds.), pp. 157–168. Pergamon Presss, Oxford.

Viswanath, P.R. (1995) "Flow Management Techniques for Base and Afterbody Drag Reduction," *Prog. Aero. Sci.* **32**, pp. 79–129.

Wadsworth, D.C., Muntz, E.P., Blackwelder, R.F., and Shiflett, G.R. (1993) "Transient Energy Release Pressure Driven Microactuators for Control of Wall-Bounded Turbulent Flows," AIAA Paper No. 93-3271, AIAA, Washington, D.C.

Wang, Y., Singer, J., and Bau, H.H. (1992) "Controlling Chaos in a Thermal Convection Loop," *J. Fluid Mech.* **237**, pp. 479–498.

Warkentin, D.J., Haritonidis, J.H., Mehregany, M., and Senturia, S.D. (1987) "A Micromachined Microphone with Optical Interference Readout," in *Proc. Fourth Int. Conf. on Solid-State Sensors and Actuators (Transducers '87),* June, Tokyo, Japan.

Wilkinson, S.P. (1988) "Direct Drag Measurements on Thin-Element Riblets with Suction and Blowing," AIAA Paper No. 88-3670-CP, AIAA, Washington, D.C.

Wilkinson, S.P. (1990) "Interactive Wall Turbulence Control," in *Viscous Drag Reduction in Boundary Layers* (D.M. Bushnell and J.N. Hefner, Eds.), pp. 479–509, AIAA, Washington, D.C.

Wilkinson, S.P., and Balasubramanian, R. (1985) "Turbulent Burst Control through Phase-Locked Surface Depressions," AIAA Paper No. 85-0536, AIAA, New York.

Wilkinson, S.P., and Lazos, B.S. (1987) "Direct Drag and Hot-Wire Measurements on Thin-Element Riblet Arrays," in *Turbulence Management and Relaminarization* (H.W. Liepmann and R. Narasimha, Eds.), pp. 121–131. Springer-Verlag, New York.

Wilkinson, S.P., Anders, J.B., Lazos, B.S., and Bushnell, D.M. (1988) "Turbulent Drag Reduction Research at NASA Langley: Progress and Plans," *Int. J. Heat Fluid Flow* **9**, pp. 266–277.

Williams, D.R., and Amato, C.W. (1989) "Unsteady Pulsing of Cylinder Wakes," in *Frontiers in Experimental Fluid Mechanics* (M. Gad-el-Hak, Ed.), pp. 337–364. Springer-Verlag, New York.

Willmarth, W.W. (1975a) "Structure of Turbulence in Boundary Layers," *Adv. Appl. Mech.* **15**, pp. 159–254.

Willmarth, W.W. (1975b) "Pressure Fluctuations Beneath Turbulent Boundary Layers," *Annu. Rev. Fluid Mech.* **7**, pp. 13–37.

Wiltse, J.M., and Glezer, A. (1993) "Manipulation of Free Shear Flows Using Piezoelectric Actuators," *J. Fluid Mech.* **249**, pp. 261–285.

Yager, R.R., and Zadeh, L.A., Eds. (1992) *An Introduction to Fuzzy Logic Applications in Intelligent Systems,* Kluwer Academic, Boston, MA.

Young, A.M., Goldsberry, J.E., Haritonidis, J.H., Smith, R.I., and Senturia, S.D. (1988) "A Twin-Interferometer Fiber-Optic Readout for Diaphragm Pressure Transducers," in *Proc. IEEE Solid-State Sensor and Actuator Workshop,* June 6–9, Hilton Head, SC.

Zilberman, M., Wygnanski, I., and Kaplan, R.E. (1977) "Transitional Boundary Layer Spot in a Fully Turbulent Environment," *Phys. Fluids* **20**(10, part II), pp. S258–S271.

IV

The Future

34 Reactive Control for Skin-Friction Reduction *Haecheon Choi*..................................... 34-1
Introduction • Near-Wall Streamwise Vortices • Opposition Control—Control
Based on Physical Intuition • Reactive Controls Based on Sensing at the Wall
• Active Wall Motion—Numerical Studies • Concluding Remarks

35 Towards MEMS Autonomous Control of Free-Shear Flows *Ahmed Naguib*............... 35-1
Introduction • Free-Shear Flows: A MEMS Control Perspective • Shear-Layer
MEMS Control System Components and Issues • Control of the Roll Moment
on a Delta Wing • Control of Supersonic Jet Screech • Control of Separation
over Low-Reynolds-Number Wings • Reflections on the Future

36 Fabrication Technologies for Nanoelectromechanical Systems
Gary H. Bernstein, Holly V. Goodson, Gregory L. Snider... 36-1
Introduction • NEMS-Compatible Processing Techniques • Fabrication of Nanomachines:
The Interface with Biology • Summary

34
Reactive Control for Skin-Friction Reduction

Haecheon Choi
Seoul National University

34.1 Introduction 34-1
34.2 Near-Wall Streamwise Vortices 34-2
34.3 Opposition Control—Control Based
      on Physical Intuition 34-2
34.4 Reactive Controls Based on Sensing at the Wall 34-4
      Neural Network • Suboptimal Control • Vorticity Flux Control
34.5 Active Wall Motion—Numerical Studies 34-10
      Streak Control • Near-Wall Streamwise-Vortex Control
34.6 Concluding Remarks 34-12
      Acknowledgments 34-13

34.1 Introduction

Recently, interest in how to control thermal and fluid phenomena has increased significantly due to an immense need of control technologies aimed at drag reduction, noise reduction and mixing enhancement. In the case of large-scale vehicles, such as an airplane or oil tanker, the skin-friction drag is about 70 to 90% of the total drag. Therefore, reducing turbulent fluctuations to reduce skin-friction is one of the most important problems in fluid mechanics. On the other hand, a great benefit can be obtained by increasing turbulence. For example, turbulence should be increased in order to mix air and fuel efficiently or to decrease form drag of bluff bodies such as a golf ball.

In spite of its importance, research on flow control has been limited because most flows occurring in industrial machinery and transportation vehicles are turbulent. Before the 1960s, turbulent flow was considered to be a completely irregular motion, but now we know that coherent structures exist in turbulent flow, which play an important role in momentum transport [Cantwell, 1981; Robinson, 1991]. One of the most important coherent structures in wall-bounded flows may be the near-wall streamwise vortex, which is responsible for the turbulence production and skin-friction increase near the wall [Kline et al., 1967; Kim et al., 1971; Kravchenko et al., 1993; Choi et al., 1994]. Therefore, attempts to control turbulent flows for engineering applications have been focused on the manipulation of these coherent structures.

In the past (before the 1990s), most turbulence-control strategies for wall-bounded flows were based on passive approaches such as riblets or LEBUs (large-eddy breakup devices). Such devices play a passive role in that there is no feedback (or feedforward) loop to sense or manipulate flow structures; also, no energy input is required. So far, it is known that the riblet is one of the most successful passive devices for skin-friction reduction, which produces maximum 8% drag reduction [Walsh, 1982]. However, there is a limit to reducing skin-friction drag by passive devices alone.

Since 1990, there have been a few studies on the active feedback or feedforward (i.e., reactive) control of dynamically significant coherent structures to achieve skin-friction reduction using direct numerical simulation. The basic strategy of the most successful controls is to manipulate the coherent structures, especially the near-wall streamwise vortices. For example, in Choi et al. (1994), the blowing or suction velocity at the wall was given as exactly opposite to the wall-normal component of the velocity at $y^+ = yu_\tau/\nu = 10$, which is associated with the wall-normal velocity induced by the near-wall streamwise vortices, resulting in more than 25% drag reduction. Here, y is the wall-normal distance from the wall, u_τ is the wall-shear velocity, and ν is the kinematic viscosity. However, sensing at $y^+ = 10$ is not an easy task, so controls based on sensing at the wall have been suggested in the literature [Lee et al., 1997; 1998; Koumoutsakos, 1999]. On the other hand, blowing or suction from a hole or slot is difficult to do in real life; for example, at high altitude, icing blocks the hole such that the actuation mechanism completely fails. Thus, wall movement instead of blowing and suction has also been suggested in the literature [Carlson and Lumley, 1996; Endo et al., 2000; Kang and Choi, 2000]. In spite of all these successful results, real implementation of the reactive-control algorithm is still a very difficult task, mainly because the size of near-wall streamwise vortices is so small (submillimeter) and their occurrence is irregular in space and time. Therefore, detecting the vortices is very difficult, as is controlling them. Fortunately, the recent development of micromachining technology makes it possible to fabricate mechanical parts of micron size [Ho and Tai, 1996; 1998]. Such technology provides us with microsensors and microactuators that can be applied to reactive flow control.

Quite a few review papers on flow control exist in the literature. Therefore, we do not intend to cover all the different aspects of flow control here but rather will limit our discussion to reactive flow controls based on sensing of coherent structures for skin-friction reduction. For those who are interested in the general aspects of flow control, refer to Bushnell and McGinley (1989), Moin and Bewley (1994), Gad-el-Hak (1994, 1996) and Ho and Tai (1996, 1998).

In this chapter, we will first talk about the near-wall streamwise vortices in a boundary layer and then review some successful reactive control strategies for skin-friction reduction using blowing and suction or active wall motion based on sensing of coherent structures. Finally, some issues on practical implementation will be given in the last section.

34.2 Near-Wall Streamwise Vortices

The most well-known structure in the turbulent boundary layer is the streaky structure as shown in Figure 34.1a. Here, the streaky structure denotes the alternating low- and high-speed fluid pattern in the spanwise direction. Kline et al. (1967) showed through flow visualization that turbulent production in the boundary layer is closely associated with lift up, oscillation and breakup of the streaky structures. Corino and Brodkey (1969) reported that turbulent production occurs locally when high-speed fluids approach the wall (called *sweep*). This streaky structure is closely related to the near-wall streamwise vortices that produce sweep and ejection and thus generate turbulence production. Recently, Choi et al. (1994) and Kravchenko et al. (1993) showed that high skin friction on the wall has high correlation with the near-wall streamwise vortices, and the sweep event by them significantly increases the skin friction nearby (Figure 34.1b).

As a successful passive device, the riblets produce maximum 8% drag reduction, in which viscous drag is reduced by restricting the location of streamwise vortices above the wetted surface such that only a limited area of the riblets is exposed to downwash of the high-speed fluid induced by them (Choi et al., 1993a). Therefore, it is a natural consequence to search for a reactive control method that is efficient and more effective in controlling the near-wall streamwise vortices to achieve greater skin-friction reduction than the riblets are.

34.3 Opposition Control—Control Based on Physical Intuition

Gad-el-Hak and Blackwelder (1989) experimentally investigated the feasibility of removing some or all of the turbulence-producing eddy structures in a turbulent boundary layer using selective suction with knowledge of the spatial location of the eddy structure. They showed that the selective suction eliminates the low-speed streak and prevents bursting at much lower suction strength than that required from the conventional suction method, implying that selective suction may be a good control strategy for skin-friction reduction.

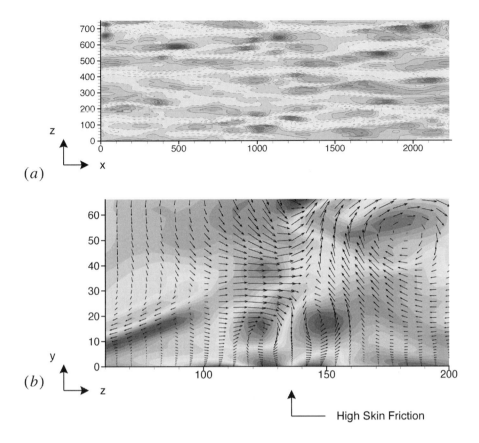

FIGURE 34.1 (a) Contours of the skin-friction fluctuations at the wall. Negative contours are dashed. (b) Contours of the streamwise vorticity and cross-flow vector (v, w) in a (y, z) plane. Contours in gray denote high magnitudes of the streamwise vorticity.

Choi et al. (1994) proposed two active control strategies for the purpose of drag reduction using direct numerical simulation of turbulent channel flow: control with the wall-normal velocity (blowing and suction) at the wall (v control; Figure 34.2a) and control with the spanwise velocity at the wall (w control; Figure 34.2b). The control algorithm was that the control-input velocity at the wall is negatively proportional to the instantaneous velocity at a location near the wall. For instance, in the case of v control, the blowing and suction velocity at the wall was exactly opposite to the wall-normal component of the velocity at a prescribed y location, y_d; i.e., $v_w = -v|_{y=y_d}$ (Figure 34.2a).

Figure 34.3a shows the time histories of the pressure gradients that were required to drive a fixed-mass flow rate for the unmanipulated, fully developed channel flow and for manipulated (v control) channel flows. Substantial skin-friction reduction was obtained ($\approx 25\%$ on each wall) with $y_d^+ \approx 10$. For other y_d^+ locations, either the drag was substantially increased ($y_d^+ \approx 26$) or the reduction was small ($y_d^+ \approx 5$). Noting that the streamwise vortices lead to strong spanwise velocity as well as wall-normal velocity, the out-of-phase boundary condition was applied to the spanwise velocity at the surface; i.e., $w_w = -w|_{y=y_d}$ (Figure 34.2b). Several sensor locations ranging from $y_d^+ \approx 5$ to 26 were tested, and the best result was obtained with $y_d^+ \approx 10$, yielding about 30% drag reduction (Figure 34.3b), which is slightly better than the optimum v control. With $y_d^+ > 20$, the drag was increased.

The velocity, pressure and vorticity fluctuations as well as the Reynolds shear stress were significantly reduced throughout the channel (see Figure 34.4 for the change in strength of the streamwise vorticity). Instantaneous flowfields showed that streaky structures below $y^+ \approx 5$ were clearly diminished by the active control, and the physical spacing of the streaky structure above $y^+ \approx 5$ was increased in the

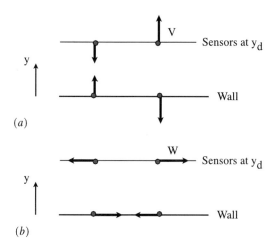

FIGURE 34.2 Schematic diagram of controls used by Choi et al. (1994): (a) v control; (b) w control. (Reprinted from Choi, H. et al. (1994) *J. Fluid Mech.* **262**, p. 75. With permission of Cambridge University Press.)

manipulated channels. The active blowing and suction or spanwise velocity at the wall significantly affected turbulence statistics above the wall.

A schematic diagram of the drag-reduction mechanism is shown in Figure 34.5. The high shear-rate regions on the wall are moved to the interior of the channel ($y^+ \approx 5$) by the control schemes. The sweep motion due to strong streamwise vortices is directly deterred by active controls. Note that, in the case of w control, a wall-normal velocity is induced very near the wall by the imposed spanwise velocity distribution at the wall; from the continuity equation at the wall, $\partial v/\partial y = -\partial w/\partial z \neq 0$, leading to higher values of v near the wall.

34.4 Reactive Controls Based on Sensing at the Wall

Even though Choi et al. (1994) showed that turbulent skin friction could be reduced by suppressing the dynamically significant coherent structures present in the wall region (i.e., near-wall streamwise vortices), practical implementation of their control scheme is nearly impossible at this time because one has to measure spatial distribution of the wall-normal velocity at $y^+ = 10$. They also considered control of turbulent boundary layer by placing sensors only at the wall for practical implementation; using the Taylor series expansion, the wall-normal velocity at $y^+ = 10$ was approximated as $v|_{y^+=10} \sim -\frac{\partial}{\partial z}\frac{\partial w}{\partial y}|_w$. However, a control based on this variable yielded only about 6% drag reduction, which is smaller than the maximum drag reduction by the riblets.

Motivated by the drag reduction techniques discussed in Choi et al. (1994), two successful active control methods based on wall measurement for skin-friction reduction have been presented. One uses the neural network [Lee et al., 1997] and the other uses the suboptimal control theory [Lee et al., 1998]. Both approaches applied blowing and suction to turbulent channel flow. Very recently, Koumoutsakos (1999) presented a feedback-control result in a turbulent channel flow, where the sensing variable is the vorticity flux at the wall and the actuation is blowing and suction at the wall. In this section, we summarize these three studies.

34.4.1 Neural Network

The objective of the work reported in Lee et al. (1997) was to seek wall actuations to obtain skin-friction reduction, in the form of blowing and suction at the wall, based on sensing of the wall shear stress. This requires knowledge of how the wall shear stress responds to wall actuation. Because of the complexity of turbulent flow, however, it is not possible to find such a correlation in a closed form or to approximate

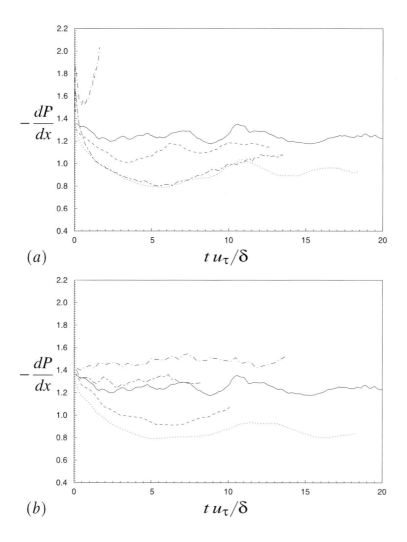

FIGURE 34.3 Time history of the pressure gradient required to drive a fixed-mass flow rate in the cases of (a) v control and (b) w control. Note: ———, unmanipulated channel; ––––, manipulated channel with sensors at $y_d^+ \approx 5$; ············ $y_d^+ \approx 10$; —·—, $y_d^+ \approx 20$; and –··—··, $y_d^+ \approx 26$. (Reprinted from Choi, H. et al. (1994) *J. Fluid Mech.*, **262**, 75. With permission of Cambridge University Press.)

it in a simple form. Thus, they used a neural network to approximate the correlation that predicts the optimal wall actuation to achieve minimum skin-friction drag.

A standard two-layer feedforward network with hyperbolic tangent hidden units and a linear output unit were used. The functional form of the final neural network was

$$v_w(x, z, t) = W_a \tanh\left(\sum_{j=-(N-1)/2}^{(N-1)/2} W_j \left.\frac{\partial w}{\partial y}\right|_w (w, z + j\Delta z, t) - W_b \right) - W_c \tag{34.1}$$

where W denotes the weights, N is the total number of input weights, and Δz is the grid spacing in the spanwise direction. Seven neighboring points ($N = 7$), including the point of interest, in the spanwise direction (corresponding to approximately 90 wall units) were found to provide enough information to adequately train and control the near-wall structures responsible for high skin friction. The network was trained to minimize the sum of a weighted-squared error that is proportional to the square of the

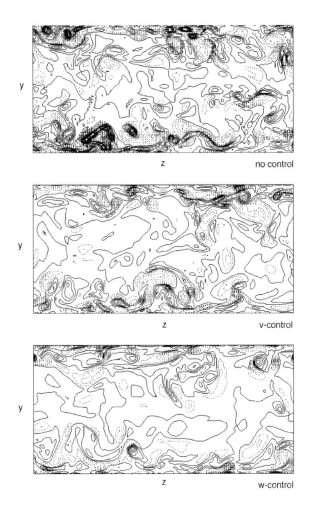

FIGURE 34.4 Contours of the streamwise vorticity in a cross-flow plane. The contour levels are the same for the cases with and without control, and negative contours are dotted. (Reprinted from Choi, H. et al. (1994) *J. Fluid Mech.* **262**, p. 75. With permission of Cambridge University Press.)

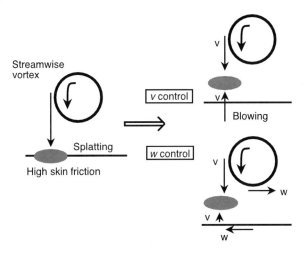

FIGURE 34.5 Schematic diagram of the drag-reduction mechanism by active controls. (Reprinted from Choi, H. et al. (1994) *J. Fluid Mech.*, **262**, 75. With permission of Cambridge University Press.)

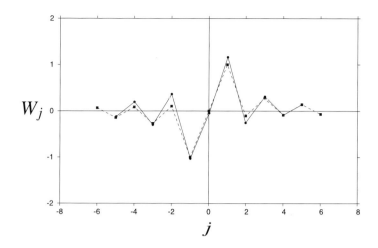

FIGURE 34.6 Weight distribution: ——, from neural network; - - - -, suboptimal control.

difference between the desired control input and the network control input. In their off-line training and control, the wall-normal velocity at $y^+ = 10$ was considered as a desired control input. After training was completed, the weight distribution in the spanwise direction was obtained, which is shown in Figure 34.6. Applying this control scheme to a turbulent channel flow at low Reynolds number resulted in about 20% drag reduction. The computed flow fields were examined to determine the mechanism by which the drag reduction was achieved. The most salient feature of the controlled case was that the strength of the near-wall streamwise vortices was substantially reduced. This result further substantiates the notion that a successful suppression of the near-wall streamwise vortices leads to a significant reduction in drag.

Lee et al. (1997) further considered a neural network with continuous on-line training, from which a very simple control scheme was obtained:

$$\hat{v}_w = C \frac{ik_z}{|k_z|} \frac{\partial \hat{w}}{\partial y}\bigg|_w \tag{34.2}$$

Here, the hat (^) denotes the Fourier component, k_z is the spanwise wavenumber, and C is a positive scale factor that determines the root-mean-square (rms) value of the actuation. This control worked equally well as Eq. (34.1).

34.4.2 Suboptimal Control

A systematic approach based on an optimal control theory was proposed by Abergel and Temam (1990). Choi et al. (1993b) proposed a "suboptimal" control procedure, in which the iterations required for global optimal control were avoided by seeking an optimal condition over a short time period. Lee et al. (1998) derived a simple control scheme by minimizing cost functions that are related to the near-wall streamwise vortices. They demonstrated that a wise choice of the cost function coupled with a variation of the formulation can lead to a more practical control law. They showed how to choose a cost function and how to minimize it to yield simple feedback control laws that require quantities measurable only at the wall. One of the laws requires spatial information on the wall pressure over the entire wall and the other requires information on one component of the wall shear stress. They then derive more practical control schemes that require only local wall pressure or local wall shear-stress information and show that the schemes work equally well. Here, we will explain the suboptimal control approach based on sensing of the wall shear stress.

In Lee et al. (1998), a cost function was chosen based on observation of the successful controls by Choi et al. (1994). The blowing and suction by Choi et al. (1994), equal and opposite to the wall-normal

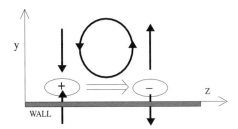

FIGURE 34.7 Effect of active blowing and suction on the presure and spanwise velocity gradient near the wall. (Reprinted from Lee, C. et al. (1998) *J. Fluid Mech.*, 358, 245. With permission of Cambridge University Press.)

velocity component at $y^+ = 10$, effectively suppress a streamwise vortex by counteracting up-and-down motion induced by the vortex. This blowing and suction action creates locally high pressure in the near-wall region marked + and low pressure in the region marked − in Figure 34.7. A crucial aspect of their analysis is the observation that this blowing and suction action *increases* the pressure gradient in the spanwise direction under the streamwise vortex near the wall and subsequently increases the spanwise velocity gradient $\partial w/\partial y$ near the wall. The above argument suggests that we should seek the blowing and suction velocity that increases $\partial w/\partial y$ near the wall for a short time period (i.e., in the suboptimal sense), in order to achieve a drag reduction similar to that achieved by Choi et al. (1994). The cost function $J(v_w)$ to be minimized is then:

$$J(v_w) = \frac{l}{2A\Delta t}\int_S \int_t^{t+\Delta t} v_w^2 \, dt \, dS - \frac{1}{2A\Delta t}\int_S \int_t^{t+\Delta t} \left(\frac{\partial w}{\partial y}\right)_w^2 dt \, dS \qquad (34.3)$$

where the integrations are over the wall, S, in space and over a short duration in time, Δt, which typically corresponds to the time step used in the numerical computation; l is the relative price of the control, as the first term on the right-hand side represents the cost of the actuation v_w. Note that there is a minus sign in front of the second term, as we want to maximize $\partial w/\partial y$ at the wall. It should be noted that the spanwise velocity gradient at the wall will *eventually* be reduced when the strength of the near-wall streamwise vortices is reduced through successful control. Here, blowing and suction that increase the spanwise-velocity gradient for the next step are sought as a suboptimal control. After some mathematical procedure, the blowing and suction velocity of minimizing the cost function is obtained as:

$$\hat{v}_w = C\frac{ik_z}{k}\frac{\partial \hat{w}}{\partial y}\bigg|_w, \qquad (34.4)$$

where the hat denotes the Fourier coefficient; $k^2 = k_x^2 + k_z^2$, k_x and k_z are the streamwise and spanwise wavenumbers, respectively; and C is a positive scale factor that determines the cost of the actuation. The above equation indicates that the optimum wall actuation should be proportional to the spanwise derivative of the spanwise shear at the wall, ($\hat{v}_w \sim \frac{\partial}{\partial z}\frac{\partial w}{\partial y}\big|_w$), with the high wavenumber components reduced by $1/k$. Note that the scale factor, C, in Eq. (34.4) is arbitrary: In the simulations, the rms value of v_w was taken to be equal to that of the wall-normal velocity at $y^+ = 10$, which gives the same rms value of wall actuations as that of Choi *et al.* (1994). Applying Eq. (34.4) to turbulent channel flow produced about 22% drag reduction. Turbulence modification was also very similar to those shown in Choi et al. (1994) and Lee et al. (1997).

Note that the control law from the neural network (Lee et al. 1997) shown in Eq. (34.2) produced almost the same distribution of blowing and suction as Eq. (34.4), because the near-wall structures have a relatively slow variation in the streamwise direction (the $k_x = 0$ component is dominant). This blowing and suction distribution is also very similar to the one based on $y^+ = 10$ data.

The control scheme presented above, however, is still impractical to implement, because it is expressed in terms of the Fourier components (i.e., in wavenumber space), which require information over the entire spatial domain. Therefore, the inverse transform of ik_z/k was sought numerically so that the convolution integral could be used to express the control law in physical space. For example, the discrete representation of the control law, Eq. (34.4), then became:

$$v_w(x, z, t) = C \sum_j W_j \frac{\partial w}{\partial y}\bigg|_w (x, z + j\Delta z, t) \tag{34.5}$$

The weights, W_j, decayed rapidly with distance from the point of interest, suggesting that the optimum actuation can be obtained by a local weighted average of $\partial w/\partial y|_w$. The result obtained using an 11-point average yielded about the same drag reduction as that obtained from full integration using Eq. (34.4). It is remarkable that localized information can produce such a significant drag reduction. The weight distribution, W_j, was very similar to the one found by applying the neural network to the same turbulent flow (see Figure 34.6).

34.4.3 Vorticity Flux Control

For an incompressible viscous flow over a stationary wall, the vorticity fluxes at the wall are directly proportional to the pressure gradients:

$$v\frac{\partial \omega_x}{\partial y}\bigg|_w = \frac{1}{\rho}\frac{\partial p}{\partial z}\bigg|_w, \quad -v\frac{\partial \omega_z}{\partial y}\bigg|_w = \frac{1}{\rho}\frac{\partial p}{\partial x}\bigg|_w \tag{34.6}$$

where p is the pressure, and ω_x and ω_z are the streamwise and spanwise vorticity components. The flux of the wall-normal vorticity, ω_y, can be determined from the kinematic condition ($\nabla \cdot \omega = 0$).

A feedback-control vorticity-flux algorithm was presented and applied to a turbulent channel flow in Koumoutsakos (1999), where the wall pressure was sensed and its gradient (the wall vorticity flux) was calculated. Unsteady blowing and suction at the wall were the control inputs and the strength was calculated explicitly by formulating Lighthill's mechanism of vorticity generation at a no-slip wall. He considered a series of vorticity flux (or, equivalently, pressure gradient) sensors on the wall at locations (x_i, z_i), $i = 1, 2, 3, \ldots, M$, and determined the blowing and suction strengths necessary to achieve a desired vorticity flux profile at the wall at a time instant, k, by solving the linear set of:

$$Bu_k + X_{k-1} = D_k, \tag{34.7}$$

where D_k is an $M \times 1$ vector of the *desired* vorticity flux at the sensor locations, X_{k-1} is an $M \times 1$ vector of the *measured* vorticity flux at the sensor locations, and u_k is an $N \times 1$ vector of blowing and suction strengths at the actuator locations. B is an $M \times N$ matrix whose elements B_{ji} are a function of the relative locations of the sensors (j index) and actuators (i index) (see Koumoutsakos, 1999, for more details). The unknown blowing and suction strengths were determined by solving Eq. (34.7).

An additional equation that determines how the desired and measured vorticity flux components are related was given as:

$$\begin{pmatrix} v\dfrac{\partial \omega_x}{\partial y} \\ v\dfrac{\partial \omega_z}{\partial y} \end{pmatrix}_{control} = \begin{pmatrix} a & b \\ c & d \end{pmatrix} \begin{pmatrix} v\dfrac{\partial \omega_x}{\partial y} \\ v\dfrac{\partial \omega_z}{\partial y} \end{pmatrix}_{measured} \tag{34.8}$$

where the coefficients, a, b, c, d, can be chosen *a priori* and may be constant or spatially varying. The simulations presented in Koumoutsakos (1999) have been conducted with the set of parameters $a = b = c = 0$ and $d = \pm 1$, equivalent to considering the *in-* and *out-of-phase* control, respectively, of the *spanwise* vorticity flux. Here, the in-phase control implies enhancement of the wall vorticity flux, whereas the out-of-phase control implies cancellation of the induced vorticity flux. He presented the results for the case of out-of-phase control.

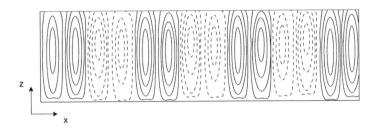

FIGURE 34.8 Highly spanwise correlated pattern of the blowing and suction as seen in Koumoutsakos (1999): ——, blowing; ----, suction. (Reprinted from Koumoutsakos, p. (1999) *Phys. Fluids* **11**, p. 248. With permission of American Institute of Physics.)

For the out-of-phase control, the streaks were eliminated and highly spanwise correlated patterns were established for the spanwise vorticity flux, shear stresses and actuation strengths (see, for example, Figure 34.8). These spanwise-correlated structures and further flow visualizations suggested the formation of unsteady spanwise vortical rollers in the inner layer of the wall. These spanwise vortical rollers resulted in the formation of positive and negative shear stresses at the wall. The elimination of streaks and disruption of the near-wall processes by establishment of the particular vortical rollers resulted in skin-friction reduction on the order of 40%. The regularity in the resulting actuation strength as shown in Figure 34.8 suggests that it may be possible to devise an open-loop control law, but further study in that direction was not conducted.

34.5 Active Wall Motion—Numerical Studies

Although the control schemes presented in the previous section (i.e., sensing at the wall) are a significant improvement over that of Choi et al. (1994), which requires velocity information above the wall, other issues must be resolved before these control schemes can be implemented in real situations (e.g., time delay between sensing and actuation). Other approaches, such as active wall motion using deformable actuators, may be more practical. So far, three studies involving active wall motion have been carried out using numerical simulation. A brief overview of these studies is given in this section.

34.5.1 Streak Control

Carlson and Lumley (1996) simulated turbulent flow in a minimal channel with time-dependent wall geometries in order to investigate the effect of wall movement on skin friction. The control device consisted of *one* actuator (Gaussian shape and about 12 wall units in height) on one of the channel walls:

$$h(x, z, t) = \varepsilon(t)\exp\left\{-\sigma^{-2}\left[(x-x_o)^2 + \left(\frac{z-z_o}{\mu}\right)^2\right]\right\} \quad (34.9)$$

where h is the time-varying (due to $\varepsilon(t)$) height of the actuator; (x_o, z_o) is the center location of the actuator; $\sigma = 0.18\partial$; $\mu = 1\partial$ or 2∂; and ∂ is the channel half-width. The control device located underneath either a high-speed streak or a low-speed streak (note that the minimal channel investigated in Carlson and Lumley has only one vortical structure on one side of the channel, thus only one pair of high- and low-speed streaks exists on the controlled wall).

Raising the actuator underneath a low-speed streak increased drag, but raising it underneath a high-speed streak reduced drag. They claimed that fast-moving fluid in the high-speed region is lifted by the actuator away from the wall, allowing the adjacent low-speed region to expand, thereby lowering the average wall shear stress. Conversely, raising an actuator underneath a low-speed streak allows the adjacent high-speed region to expand, which increases the skin friction (Figure 34.9). However, they could not present the amount of time-averaged drag reduction due to wall motion because only a short time control was conducted in their study.

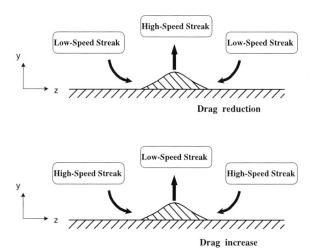

FIGURE 34.9 Schematic diagram of the mechanism for drag reduction or increase due to wall motion as seen in Carlson and Lumley (1996). (Reprinted from Carlson, A., and Lumley, J.L. (1996) *J. Fluid Mech.* **329**, p. 341. With permission of Cambridge University Press.)

Later, Endo et al. (2000) designed a realistic array of deformable actuators elongated in the streamwise direction. Each actuator moved up and down and induced the wall-normal velocity at the wall. The wall velocity of the actuator was determined by:

$$v_w(x, z, t) = v_m(t) f(x) \exp\left[-\frac{(z-z_o)^2}{\sigma_z^2}\right] \sin\left[\frac{2\pi(z-z_o)}{m_z}\right] \quad (34.10)$$

$$v_m(t) = \begin{cases} \alpha \tanh\left(\frac{\partial}{\partial z}\frac{\partial u}{\partial y}\frac{1}{\beta}\right) - \gamma y_m & \text{if } \frac{\partial}{\partial z}\frac{\partial w}{\partial y} < 0, \\ -\gamma y_m & \text{otherwise} \end{cases} \quad (34.11)$$

where $\alpha, \beta, \gamma, \sigma_z, m_z$ and y_m are the control parameters; $f(x)$ is a hyperbolic-tangent function (see Endo et al., 2000); and v_m is the wall velocity at the center of the actuator (x_o, z_o). Each sensor measured the spanwise distribution of the streamwise wall shear rate, $\frac{\partial}{\partial z}\frac{\partial u}{\partial y}\big|$ (see Eq. (34.11)), located in front of the actuator.

With the control scheme above, Endo et al. (2000) obtained about 10% drag reduction. Because measurement of the streamwise shear rate was a concern in their approach, the main mechanism of the drag reduction is believed to attenuate and stabilize streak meandering and thus to hamper the regeneration cycle of the near-wall turbulence [Hamilton et al., 1995].

34.5.2 Near-Wall Streamwise-Vortex Control

Kang and Choi (2000) conducted numerical simulations to investigate a possibility of reducing skin-friction drag in a turbulent channel flow with active wall motions. The wall was locally (and continuously) deformed according to the successful control strategy developed from the suboptimal control [Lee et al., 1998] and thus the induced velocity at the wall by wall motion was represented as:

$$\hat{v}_w = C\frac{ik_z}{k}\frac{\partial \hat{w}}{\partial y}\bigg|_w \quad (34.12)$$

which is the same as Eq. (34.4). Because the Fourier transformation is not realistic (one needs information on an entire wall), a more realistic wall-motion control strategy was devised by applying the 11-point weighted average of $\partial w/\partial y|_w$ in physical space (see Eq. (34.5)).

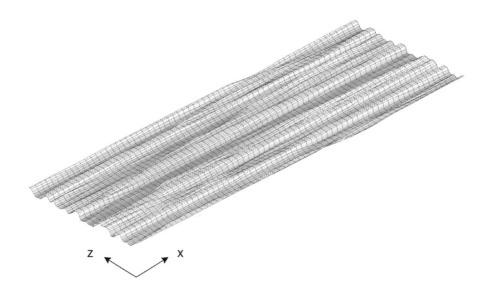

FIGURE 34.10 Instantaneous wall shape. (Reprinted from Kang, S., and Choi, H. (2000) *Phys. Fluids*, **12**, 3301. With permission of American Institute of Physics.)

As a result, about 13 to 15% drag reduction was obtained due to active wall motions. Active wall motions significantly weakened the strength of the streamwise vortices near the wall. An instantaneous wall shape during control is shown in Figure 34.10. The rms amplitude of wall deformation was about 3 wall units. The wall shapes were elongated in the streamwise direction and resembled riblets in appearance (a similar wall shape was also observed by Endo et al., 2000). The average peak-to-peak distance of the grooved wall shape was around 100 wall units, with which riblets hardly reduce the skin-friction drag; therefore, the mechanism of the drag reduction is essentially different from that of riblets. That is, the active wall motion directly suppresses the near-wall streamwise vortices by the induced blowing and suction, whereas passive riblets restrict the location of the streamwise vortices above the wetted surface such that only a limited area of the riblets is exposed to the downwash motion that the vortices induce.

34.6 Concluding Remarks

There have been many experimental studies on passive or open-loop active control in a turbulent boundary layer. Most work has focused on understanding the flow response to external disturbances, such as uniform or periodic blowing and suction, or surface roughness. This work did not seriously consider drag reduction performance because measuring the skin friction in a perturbed turbulent boundary layer is not easy. To the best of the author's knowledge, no study has shown an experimental evidence of drag reduction using an active open-loop control strategy with a constraint of zero-net mass flow rate. Only a very few preliminary experimental studies so far have considered reactive (feedback or feedforward) control for skin-friction reduction in a turbulent or laminar boundary layer. Jacobson and Reynolds (1998) implemented a spring-board-type actuator in a laminar boundary layer, which produced a quasi-steady pair of counter-rotating streamwise vortices. They showed that transition was delayed by 40 displacement thicknesses, and the mean and spanwise variation of wall shear stress was substantially reduced downstream of a single transverse array of actuators. Rathnasingham and Breuer (1997) constructed a linear feedforward control algorithm and applied it to a turbulent boundary layer for skin-friction reduction. It was shown that using three wall-based sensors and a single actuator, the control produced a maximum of 31% reduction in the rms streamwise velocity fluctuations. However, they did not show the detailed skin-friction variation due to the measurement difficulty.

In this chapter, we have shown that the skin friction can be reduced by manipulating the near-wall streamwise vortices or streaks using either blowing and suction or wall motion. For the past 10 years, quite a few different successful control strategies have been presented using numerical simulations, and they have provided information about what to manipulate for skin-friction reduction. The maximum reduction of skin-friction drag obtained so far is relaminarization at low Reynolds number (see Choi et al., 1994, for v and w control; Bewley et al., 2000, for optimal control). The control methods based on wall sensing using neural network or suboptimal control produced a maximum of 20% drag reduction. However, in a practical implementation, sensing and controlling the vortices are difficult tasks. One of the main reasons is that the near-wall structures appear randomly in space and in time. Thus, one has to place the sensors and actuators over the entire surface in order to obtain a substantial decrease of drag. Another reason is that the typical diameter of the near-wall streamwise vortices is about 30 wall units [Kim et al., 1987], which is about 100 μm in laboratory experiments. In order to detect and control these vortices, the size of sensors and actuators should be comparable to that of the vortices.

Practical implementation of reactive control algorithms developed from numerical studies has been hampered by difficulties in manufacturing microsensors and actuators using traditional machining tools. However, recent development of micromachining technology now makes it possible to fabricate micron-size mechanical parts, thus microsensors and microactuators can be mass produced. So far, microelectro-mechanical system (MEMS) technology has been successfully applied to active open-loop control of flow over bluff bodies [e.g., Ho and Tai, 1998, and the references therein]. Also, microsensors and microactuators for boundary-layer flow have been developed using MEMS technology (Ho and Tai, 1998). Spatial distribution of the skin friction has been measured using microsensors, and changes in boundary-layer flow due to microactuators have been thoroughly investigated. Researchers are now implementing the reactive control algorithms in laboratory experiments using MEMS technology and are expecting a significant amount of skin-friction reduction comparable to that obtained from numerical studies.

In most numerical studies mentioned in the previous sections, sensors and actuators are collocated and distributed all over the computational domain. In reality, sensors and actuators cannot be collocated and actuators should be located downstream of the sensors. Therefore, how to efficiently distribute the sensors and actuators is an important issue. The second issue is how to develop a new control method that is well suited for an experimental approach (note that most control methods investigated so far started from numerical studies). The third issue is the Reynolds number effect. Relaminarization of turbulent flow due to control, as has been observed in numerical studies, might happen because of a low Reynolds numbers. At low Reynolds numbers, the dynamics of boundary-layer flow is mostly governed by near-wall phenomena, whereas it is not only affected by the near-wall behavior but also by the fluid motion in the buffer or outer layer. Therefore, various control algorithms may have to be developed depending on the Reynolds number.

Acknowledgments

This work has been supported by the Creative Research Initiatives of the Korean Ministry of Science and Technology.

References

Abergel, F., and Temam, R. (1990) "On Some Control Problems in Fluid Mechanics," *Theor. Comp. Fluid Dynamics* **1**, p. 303.

Bewley, T.R., Moin, P., and Temam, R. (2001) "DNS-Based Predictive Control of Turbulence: An Optimal Benchmark for Feedback Algorithms," to be published in *J. Fluid Mech.*

Bushnell, D.M., and McGinley, C.B. (1989) "Turbulence Control in Wall Flows," *Ann. Rev. Fluid Mech.* **21**, p. 1.

Cantwell, B.J. (1981) "Organized Motion in Turbulent Flow," *Ann. Rev. Fluid Mech.* **13**, p. 457.

Carlson, A., and Lumley, J.L. (1996) "Active Control in the Turbulent Wall Layer of a Minimal Flow Unit," *J. Fluid Mech.* **329**, p. 341.

Choi, H., Moin, P., and Kim, J. (1993a) "Direct Numerical Simulation of Turbulent Flow Over Riblets," *J. Fluid Mech.* **255**, p. 503.

Choi, H., Temam, R., Moin, P., and Kim, J. (1993b) "Feedback Control for Unsteady Flow and Its Application to the Stochastic Burgers Equation," *J. Fluid Mech.* **245**, p. 509.

Choi, H., Moin, P., and Kim, J. (1994) "Active Turbulence Control for Drag Reduction in Wall-Bounded Flows," *J. Fluid Mech.* **262**, p. 75.

Corino, E.R., and Brodkey, R.S. (1969) "A Visual Investigation of the Wall Region in Turbulent Flow," *J. Fluid Mech.* **37**, p. 1.

Endo, T., Kasagi, N., and Suzuki, Y. (2000) "Feedback Control of Wall Turbulence with Wall Deformation," *Int. J. Heat Fluid Flow* **21**, p. 568.

Gad-el-Hak, M. (1994) "Interactive Control of Turbulent Boundary Layers: A Futuristic Overview," *AIAA J.* **32**, p. 1753.

Gad-el-Hak, M. (1996) "Modern Developments in Flow Control," *Appl. Mech. Rev.* **49**, p. 365.

Gad-el-Hak, M., and Blackwelder, R.F. (1989) "Selective Suction for Controlling Bursting Events in a Boundary Layer," *AIAA J.* **27**, p. 308.

Hamilton, J.M., Kim, J., and Waleffe, F. (1995) "Regeneration Mechanisms of Near-Wall Turbulence Structures," *J. Fluid Mech.* **287**, p. 317.

Ho, C.-H., and Tai, Y.-C. (1996) "Review: MEMS and Its Applications for Flow Control," *J. Fluids Eng.* **118**, p. 437.

Ho, C.-M., and Tai, Y.-C. (1998) "Micro-Electro-Mechanical Systems (MEMS) and Fluid Flows," *Ann. Rev. Fluid Mech.* **30**, p. 579.

Jacobson, S.A., and Reynolds, W.C. (1998) "Active Control of Streamwise Vortices and Streaks in Boundary Layers," *J. Fluid Mech.* **360**, p. 179.

Kang, S., and Choi, H. (2000) "Active Wall Motions for Skin-Friction Drag Reduction," *Phys. Fluids* **12**, p. 3301.

Kim, H.T., Kline, S.T., and Reynolds, W.C. (1971) "The Production of Turbulence Near a Smooth Wall in a Turbulent Boundary Layer," *J. Fluid Mech.* **50**, p. 133.

Kim, J., Moin, P., and Moser, R. (1987) "Turbulence Statistics in Fully Developed Channel Flow at Low Reynolds Number," *J. Fluid Mech.* **177**, p. 133.

Kline, S.J., Reynolds, W.C., Schraub, F.A., and Runstadler, P.W. (1967) "The Structure of Turbulent Boundary Layers," *J. Fluid Mech.* **30**, p. 741.

Koumoutsakos, P. (1999) "Vorticity Flux Control for a Turbulent Channel Flow," *Phys. Fluids* **11**, p. 248.

Kravchenko, A.G., Choi, H., and Moin, P. (1993) "On the Relation of Near-Wall Streamwise Vortices to Wall Skin Friction in Turbulent Boundary Layer," *Phys. Fluids A* **5**, p. 3307.

Lee, C., Kim, J., Babcock, D., and Goodman, R. (1997) "Application of Neural Networks to Turbulence Control for Drag Reduction," *Phys. Fluids* **9**, p. 1740.

Lee, C., Kim, J., and Choi, H. (1998) "Suboptimal Control of Turbulent Channel Flow for Drag Reduction," *J. Fluid Mech.* **358**, p. 245.

Moin, P., and Bewley, T. R. (1994) "Feedback Control of Turbulence," *Appl. Mech. Rev.* **47**, p. S3.

Rathnasingham, R., and Breuer, K. S. (1997) "System Identification and Control of a Turbulent Boundary Layer," *Phys. Fluids* **9**, p. 1867.

Robinson, S.K. (1991) "Coherent Motions in the Turbulent Boundary Layer," *Ann. Rev. Fluid Mech.* **23**, p. 601.

Walsh, M.J. (1982) "Turbulent Boundary Layer Drag Reduction Using Riblets," AIAA Paper No. 82-0169, AIAA, Washington, D.C.

35
Towards MEMS Autonomous Control of Free-Shear Flows

Ahmed Naguib
Michigan State University

35.1 Introduction 35-1
35.2 Free-Shear Flows: A MEMS Control Perspective 35-2
35.3 Shear-Layer MEMS Control System
 Components and Issues.. 35-3
 Sensors • Actuators • Closing the Loop: The Control Law
35.4 Control of the Roll Moment on a Delta Wing............. 35-9
 Sensing • Actuation • Flow Control • System Integration
35.5 Control of Supersonic Jet Screech 35-19
 Sensing • Actuation • Flow Control • System Integration
35.6 Control of Separation over
 Low-Reynolds-Number Wings..................................... 35-30
 Sensing • Flow Control
35.7 Reflections on the Future 35-33

35.1 Introduction

Interest in application of microelectromechanical systems (MEMS) technology to flow control and diagnostics started around the early 1990s. During this relatively short time period, a handful attempts have been aimed at using the new technology to develop and implement reactive control of various flow phenomena. The ultimate goal of these attempts has been to capitalize on the unique ability of MEMS to integrate sensors, actuators, driving circuitry and control hardware in order to attain *autonomous* active flow management. To date, however, there has been no demonstration of a fully functioning MEMS flow control system, whereby the multitude of information gathered from a distributed MEMS sensor array is successfully processed in real time using on-chip electronics to produce an effective response by a distributed MEMS actuator array.

Notwithstanding the inability of research efforts in the 1990s to realize autonomous MEMS control systems, the lessons learned are invaluable in understanding the strengths and limitations of MEMS in flow applications. In this chapter, the research work aimed at MEMS-based autonomous control of free-shear flows over the past decade is reviewed. The main intent is to use the outcome of these efforts as a telescope to peek through and project a vision of future MEMS systems for free-shear flow control. The presentation of the material is organized as follows: First, important classifications of free-shear flows are introduced in order to facilitate subsequent discussions, and, second, a fundamental analysis concerning the usability of MEMS in different categories of free-shear flows is provided. This is followed by an outline of autonomous flow control system components, with a focus on MEMS in free-shear flow applications, and related issues. Finally, the bulk of the chapter is used to review prominent research efforts in the 1990s, leading to a vision of future systems.

35.2 Free-Shear Flows: A MEMS Control Perspective

Free-shear flows refer to the class of flows that develop without the influence imposed by direct contact with solid boundaries. However, in the absence of thermal gradients, nonuniform body-force fields or similar effects, the vorticity in free-shear flows is actually acquired through contact with a solid boundary at one point in the history of development of the flow. The "free-shear state" of the flow is attained when it separates from the solid wall, carrying with it whatever vorticity is contained in the boundary layer at the point of separation. The mean velocity profile of the shear layer at the point of separation is inflectional and hence is inviscidly unstable; that is, it is extremely sensitive to small perturbations if excited at the appropriate frequencies. This point is of fundamental importance to MEMS-based control.

Because MEMS devices are micron in scale, they are only capable of delivering proportionally small energies when used as actuators. Therefore, if it were not for the high sensitivity of free-shear flows to disturbances at the point of separation, there would be no point in attempting to use MEMS actuators for shear-layer control. Moreover, it should be emphasized that active control of free-shear flows using MEMS as a disturbance source must be applied at, or extremely close to, the point of separation. The same statement can be extended to the more powerful conventional actuators, such as glow-discharge, large-scale piezoelectric devices, large-scale flaps, etc., if high-gain or efficient control is desired.

From a control point of view, it is useful to classify free-shear flows according to whether the separation line is stationary or moving. Stationary separation line (SSL) flows include jets, single- or two-stream shear layers and backward-facing step flows. In these flows, separation takes place at the sharply defined trailing edge of a solid boundary at the "origin" of the free-shear flow. On the other hand, moving separation line (MSL) flows include dynamic separation over pitching airfoils and wings, periodic flows through compressors and forward-facing step flows. The separation point in the latter, although steady on average, jitters around due to the lack of a sharp definition of the geometry at separation (such as the nozzle lip for the shear layer surrounding a jet).

Whether the free-shear flow to be controlled is of the SSL or MSL type is extremely important, not only from the point of view of control feasibility and ease but also concerning the need for MEMS vs. conventional technology. In particular, in SSL flows, actuators (MEMS or conventional) can be located directly at the known point of separation where they would be effective. As the operating conditions change (for example, through a change in the speed of a jet) the same set of actuators can be used to affect the desired control. In contrast, in an MSL flow, the instantaneous location of the separation line has to be known and only actuators located along this line should be used for control.

To fix the ideas further, consider controlling dynamic separation over a pitching two-dimensional airfoil. As the airfoil is pitched from, say, zero to a sufficiently large angle of attack, a separated shear layer is formed. The separation line of this shear layer moves with increasing angle of attack. In order to control this separating flow using MEMS or other conventional actuators, it is necessary to track the location of the separation point in real time in order to activate only those actuators located along the separation line. This would require distributed, or array, measurements on the surface of the airfoil. Furthermore, because the separation-line-locating algorithm is likely to employ spatial derivatives of the surface measurements, the surface sensors must have high spatial resolution and be packed densely. Therefore, in MSL problems, it seems inevitable to use MEMS sensor arrays if autonomous control is to be successful. Note that this statement is believed to be true regardless of whether MEMS or conventional technology is used for actuation.

When considering MEMS control systems, another useful classification is that associated with the characteristic size of the flow. With the advent of MEMS, it is now common to observe flows confined to domains that are no larger than a few hundred microns in characteristic size. Such flows may be found in different microdevices including pumps, channels, nozzles, turbines, etc. Therefore, it is important to distinguish between microscale (MIS) flows and their macro counterpart (MAS). In particular, there should be no ambiguity about whether or not only MEMS, or perhaps even NEMS (nanoelectromechanical systems), are the only feasible method of control within a microdevice.

A good example of where free-shear flows may be encountered in microdevices is the MIT microengine project [Epstein et al., 1997]. In the intricate, complex devices engineered in this project, shear layers exist in the microflows over the stator and rotor of the compressor and turbine and in the sudden expansion leading to the combustion chamber. Whether or not those shear layers and their susceptibility to excitation within the confines of the microdevice mimic those of macroscale shear layers is yet to be established. However, as indicated above, if control is to be exercised, the limitation imposed by the scale of the device dictates that any actuators and or sensors have to be as small, if not smaller, than the device itself.

It should be noted here that active control of MIS flows is an area that is yet to be explored. This, in part, is because of the fairly short time since interest in such flows started. Therefore, further discussion of the subject would be appropriately left to a future date when sufficient literature is available.

35.3 Shear-Layer MEMS Control System Components and Issues

To facilitate subsequent discussion of the research work pertaining to MEMS-based control of free-shear flows, the different components of the control system will be discussed and analyzed here. Of particular importance are the issues specific to utilization of MEMS technology to realize the control system components. Figure 35.1 displays a general functional block diagram for feedback control systems in SSL and MSL flows. As seen from Figure 35.1, for both types of flows the information obtained from surface-mounted sensor arrays is fed to a flowfield estimator in order to predict the state of the flowfield being controlled. If the desired flowfield, and deviations from it, can be defined in terms of a signature measurable at the surface, the flow estimator may be bypassed all together. Any difference between the measured and desired flow states is used to drive surface-mounted actuators in a manner that would force the flow towards the desired state. In the case of the MSL flow, the current location of the separation line must also be identified and fed to the controller in order to operate the actuator set located nearest or at the position of separation (see Figure 35.1, bottom).

35.3.1 Sensors

When attempting to use MEMS sensors for implementing autonomous control of free-shear flows, several issues should be considered.

35.3.1.1 Sensor Types

From a practical point of view, it is very difficult, if not impossible, to use sensors that are embedded inside the flow to achieve the information feedback necessary for implementing closed-loop control. This is primarily due to the inability to use a sufficiently large number of sensors to provide the needed information without blocking and/or significantly altering the flow. Therefore, the deployment of the MEMS sensor arrays should be restricted to the surface. Flows where no heat transfer occurs limit the measurements to primarily those of the surface stresses: wall shear (τ_w), or tangential component, and wall pressure (p_w), or normal component. The former has one component in the streamwise (x) and the other in the spanwise (z) direction. For a comprehensive review of MEMS surface shear and pressure sensors, the reader is referred to Chapters 25 and 26 in this handbook.

In addition to surface stresses, near-wall measurements of flow velocity may be achieved nonintrusively via optical means. An example of such a system is currently being developed by Gharib et al. (1999), who are utilizing miniature diode lasers integrated with optics in a small package to develop a mini-LDA (laser Doppler anemometer) for measurements of the wall shear and near-wall velocity. With the continued miniaturization of components, it is not difficult to envision an array of MEMS-based LDA sensors deployed over surfaces of aero- and hydrodynamic devices. Of course, with LDA sensors, flow seeding is necessary; hence, such optical techniques may only be practical for flows where natural contamination is present or when practical means for seeding the flow locally in the vicinity of the measurement volume can be devised.

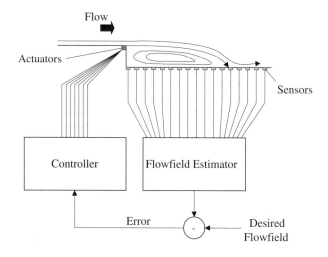

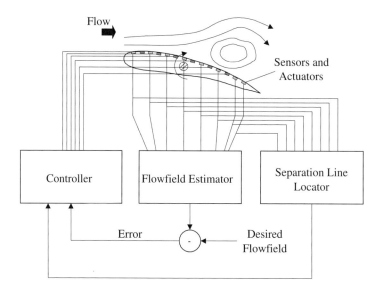

FIGURE 35.1 Conceptual block diagram of autonomous control systems for SSL (top) and MSL (bottom) flows.

It should be added here that, whereas for MSL flows surface sensors should cover the surface surrounding the flow to be controlled, in SSL flows, sensors would typically be placed at the trailing edge. However, in certain instances, placement of the sensors downstream of the trailing edge may be feasible. Examples include sudden expansion and backward-facing step flows (Figure 35.1, top) and jet flows, where a sting may be extended at the center of the jet for sensor mounting.

35.3.1.2 Sensor Characteristics

Properly designed and fabricated MEMS sensors should possess very high spatial and temporal resolution because of their extremely small size. Therefore, when considering the *response* of MEMS sensors for surface measurements, only the sensitivity and signal-to-noise ratio (SNR) are of concern. The need for high sensitivity and SNR is particularly important in detecting separation, as the wall shear values are near zero in the vicinity of the separation line ($\tau_w = 0$ at separation for steady flows). Also, hydrodynamic surface pressure fluctuations are typically small for low-speed flows and require microphone-like sensitivities for measurements.

The concern regarding MEMS sensor sensitivity does not include indirect measurements of surface shear, such as that conducted using thermal anemometry, which can be achieved with higher sensitivity using MEMS sensors as compared to conventional ones [e.g., Liu et al., 1994; Cain et al., 2000]. On the other hand, both direct measurements of the surface shear (using a floating element) and pressure measurements (through a deflecting diaphragm) rely on the force produced by those stresses acting on the area of the sensor. Because typical MEMS sensor dimensions are less than 500 μm on the side, the resulting force is extremely small. Because of this, currently no MEMS pressure sensor is known to have a sensitivity comparable to that of 1/8-in. capacitive microphones [Naguib et al., 1999a]. Also, notwithstanding the creativity involved in developing a number of floating-element designs and detection schemes [Padmanabhan et al., 1996; Reshotko et al., 1996], the signal-to-noise ratio of such sensors remains significantly below that achievable with thermal sensors. It should be pointed out here, however, that because of the direction-reversing nature of separated flows, thermal shear sensors are generally not too useful in MSL flows. Direct or other direction-sensitive sensors are required for conducting appropriate measurements in such flows.

35.3.1.3 Nature of the Measurements

Fundamentally, the information inferred from measurements of the wall-shear stress and near-wall velocity differs from that obtained from the surface pressure. The latter is known to be a "global" quantity that is influenced by both near-surface and remote flow structures. On the other hand, measurements of the shear stress and flow velocity provide information concerning flow structures that are in "direct contact" with the sensors.

Because of the global nature of pressure measurements, pressure sensors are probably the most suitable type of sensors for conducting measurements at the trailing edge in SSL flows. More specifically, in most, if not all, instances involving SSL flows, the control objective is aimed at controlling the flow downstream of the trailing edge, and local measurements of the wall shear and velocity would not be of great use in predicting the flow structure downstream of the trailing edge. A possible exception is when hydrodynamic feedback mechanisms are present, such as the case in a backward-facing step, where some structural influences downstream are naturally fed back to the trailing edge. In such flows, however, the instability of the shear layer may be *absolute* rather than *convective* [Huerre and Monkewitz, 1990]. Fiedler and Fernholz (1990) suggested that absolutely unstable flows are less susceptive to local periodic excitation such as that discussed here. Control provisions aimed at blocking the hydrodynamic feedback loop seem to be more effective in such cases.

Another important issue concerning the nature of surface pressure measurements is that they inevitably contain contributions from hydrodynamic as well as acoustic sources. The latter could be either a consequence of the flow itself (such as jet noise) or environmental, emanating from other surrounding sources that are not related to the flowfield. Typically, when the sound is produced by the flow, knowledge of the general characteristics of the acoustic field (direction of propagation, special symmetries, etc.) enables separation of the hydrodynamic and acoustic contributions to pressure measurements. In general, attention must be paid to ensure that the appropriate component of the pressure measurements is extracted.

35.3.1.4 Robustness and Packaging

Perhaps the primary challenge for the use of MEMS in practical systems is whether the minute, fairly fragile devices can withstand the operating environment. One solution is to package the devices in isolation from their environment. This solution has been adopted, for example, in the commercially available MEMS accelerometers from Analog Devices, Inc. Such a solution, although possibly feasible for the mini-LDA systems discussed above, is in general not useful for flow applications. Most wall-shear and -pressure sensors and all actuators must interact directly with the flow. Therefore, the ability of the minute devices to withstand harsh, high-temperature, chemically reacting environments is of concern. Also, the possibility for mechanical failure during routine operation and maintenance must be accounted for.

35.3.1.5 Ability to Integrate with Actuators and Electronics

Although many types of MEMS sensors have been fabricated and characterized for use in flow applications (see Chapters 24 through 26 for more details), most of these sensors have been designed for isolated operation individually or in arrays. If autonomous control is to be achieved using the unique capability of MEMS for integration of components, the sensor fabrication technology must be compatible with that of the actuators and the circuitry. More specifically, if a particular fabrication sequence is successful in constructing a pressure sensor with certain desired characteristics, the same sequence may be unusable when the sensors are to be integrated with actuators on the same chip. Examples of the few integrated MEMS systems that were developed for autonomous control include the fully integrated system used by Tsao et al. (1997) for controlling the drag force in turbulent boundary layers and the actuator-sensor systems from Huang et al. (1996) aimed at controlling supersonic jet screech (discussed in detail later in this chapter).

Tsao et al. (1997) utilized a CMOS-compatible technology to fabricate three magnetic flap-type actuators integrated with 18 wall-shear stress sensors and control electronics. In contrast, Huang et al. (1996; 1998) developed hot-wire and pressure sensor arrays integrated with resonant electrostatic actuators. Typically two actuators were integrated with three to four sensors in the immediate vicinity of the actuators.

35.3.2 Actuators

35.3.2.1 Actuator Types and Receptivity

When thinking of ways to excite a shear layer, there seems to be a multitude of possibilities that differ in their nature of excitation (mechanical, fluidic, acoustic, thermal, etc.; see Chapters 26 and 27), relative orientation to the flow (e.g., tangential vs. lateral actuation), domain of influence (local vs. global) and specific positioning with respect to the shear layer. Given the wide range of actuator types, it is rather confusing to choose the most appropriate type of actuation for a particular flow-control application. Because of the micron-level disturbances introduced by MEMS actuators, it is, however, obvious that the actuation scheme providing "the most bang for the money" should be utilized—that is, the type of actuation that is most efficient or to which the flow has the largest *receptivity*. The ambiguity in picking and choosing from the list of actuators is for the most part due to our lack of understanding of the receptivity of flows to the different types of actuation.

When selecting a MEMS (or conventional, for that matter) actuator for flow control, one is faced with a few fundamental questions: What type of actuator will achieve the control objective with minimum input energy? That is, what type of actuator produces the largest receptivity? Is it a mechanical, fluidic, acoustic or other type of actuator? If mechanical, should it oscillate in the normal, streamwise or spanwise direction? What actuator amplitude is needed in order to generate a certain flow-velocity disturbance magnitude? In the vicinity of the point of separation, is there an optimal location for the chosen type of actuator? All of these questions are very fundamental to not only the design of the actuation system, but also the assessment of the feasibility of using MEMS actuators to accomplish the control goal.

35.3.2.2 Forcing Parameters

Perhaps one of the most limiting factors on the applicability of MEMS actuators in flow control is their micron-size forcing amplitude. Typical mechanical MEMS actuators are capable of delivering oscillation amplitudes ranging from a few microns to tens of microns. In shear flows evolving from the separation of a thin laminar boundary layer, such small amplitudes can be comparable to the momentum thickness of the separating boundary layer, resulting in significant flow disturbance. This is particularly true in SSL flows, where the actuator location can be maintained in the immediate vicinity of the separation point.

To appreciate the susceptibility of shear layers to very low-level actuation, it is instructive to consider an order of magnitude analysis. For example, consider a situation where it is desired to attenuate, or eliminate, naturally existing two-dimensional disturbances in a laminar incompressible single-stream shear layer. One approach to attaining this goal is through "cancellation" of the disturbance during its initial linear growth phase, where the streamwise development of the instability velocity amplitude, or rms, is given by:

$$u'(x) = u'_o e^{-\alpha_i(x-x_o)} \tag{35.1}$$

where, u' is an integral measure of the instability amplitude at streamwise location x from the point of separation of the shear layer (trailing edge in this example), u'_o is the *initial* instability amplitude, x_o is the "virtual origin" of the shear layer, and α_i is the imaginary wavenumber component, or spatial growth rate, of the instability wave.

To cancel the flow instability, a periodic disturbance at the same frequency but 180° out of phase may be introduced at some location $x = x^*$ near the point of separation of the shear layer. The amplitude of the disturbance introduced by the control should be of the same order of magnitude as that of the natural flow instability at the location of the actuator. This may be estimated from Eq. (35.1) as follows:

$$u'(x^*) = u'_o e^{-\alpha_i(x^*-x_o)} \tag{35.2}$$

The location of the actuator should be less than one to two wavelengths (λ) of the flow instability downstream of the trailing edge to be within the linear region. This allows superposition of the control and natural instabilities, leading to cancellation of the latter. At the end of the linear range (approximately 2λ), the natural instability amplitude may be calculated from Eq. (35.1) as:

$$u'(2\lambda) = u'_o e^{-\alpha_i(2\lambda-x_o)} \tag{35.3}$$

Dividing Eq. (35.2) by Eq. (35.3), one may obtain an estimate for the required actuator-induced disturbance amplitude in terms of the flow instability amplitude at the end of the linear growth zone:

$$u'(x^*) = u'(2\lambda)e^{-\alpha_i(x^*-2\lambda)} \tag{35.4}$$

The instability amplitude at the end of the linear region may be estimated from typical amplitude saturation levels (about 10% of the freestream velocity, U). Also, if the natural instability corresponds to the most unstable mode, its frequency would be given by [e.g., Ho and Huerre, 1984]:

$$St = f\theta/U = 0.016 \tag{35.5}$$

where St is the Strouhal number, and θ is the local shear layer thickness. For the most unstable mode, the instability convection speed is equal to $U/2$ and $-\alpha_i\theta = 0.1$ [e.g., Ho and Huerre, 1984]. Using this information and writing the frequency in terms of the wavelength and convection velocity, Eq. (35.5) reduces to:

$$-\alpha_i\lambda \approx (2\times 0.16)^{-1} \tag{35.6}$$

for the most unstable mode.

Now, for the sake of the argument, assuming the actuator to be located at the shear-layer separation point ($x^* = 0$), estimating $u'(2\lambda)$ as $0.1U$, and substituting from Eq. (35.6) into Eq. (35.4), one obtains:

$$u'(x^*) = 0.1U e^{-\frac{2}{2\times 0.16}} \approx 10^{-4}U \tag{35.7}$$

That is, the required actuator–generated disturbance velocity amplitude should be about 0.01% of the free-stream velocity. Also, the actual forcing amplitude is not that given by Eq. (35.7) but rather one that is related to it through an amplitude receptivity coefficient. That is,

$$u'_{act.} = u'(x^*) \times R \tag{35.8}$$

where R is the receptivity coefficient, and $u'_{act.}$ is the actuation amplitude. Thus, if R is a number of order one, the required actuator velocity amplitude for exciting flow structures of comparable strength to the natural coherent structures in a high-speed shear layer (say, $U = 100$ m/s) is equal to 1 cm/s (i.e., of the order of a few cm/s). If the actuator can oscillate at a frequency of 10 kHz (easily achievable with MEMS), the corresponding actuation amplitude is only about a few microns!

In MSL flows, the ability of the sensor array and associated search algorithm to locate the instantaneous separation location of the shear layer, and the actuator-to-actuator spacing, may not permit flow excitation as close to the separation point as desired. Thus, in this case, even for extremely thin laminar shear layers, stronger actuators may be needed to compensate for the possible suboptimal actuation location. MEMS actuators that are capable of delivering hundreds of microns up to order of 1 mm excitation amplitude have been devised. These include the work of Miller et al. (1996) and Yang et al. (1997).

The more powerful, large-amplitude actuators, although providing much needed "muscle" to the miniature devices, nullify one of the main advantages of MEMS technology: the ability to fabricate actuators that can oscillate mechanically at frequencies of hundreds of kHz. Traditionally, high-frequency (few to tens of kHz) excitation of flows was only possible via acoustic means. With the ability of MEMS to fabricate devices with "micro inertia," it is now possible to excite high-speed flows using mechanical devices [e.g., see Naguib et al., 1997].

To estimate the order of magnitude of the required excitation frequency, the frequency of the linearly most unstable mode in two-dimensional shear layers may be used. This frequency may be estimated from Eq. (35.5). Using such an estimate, it is straightforward to show that the most unstable mode frequency for typical high-speed MAS (macroscale) shear layers, such as those found in transonic and supersonic jets, is of the order of tens to hundreds of kHz. On the other hand, if one considers a microscale (MIS) shear layer, with a momentum thickness in the range of 1 to 10 μm and a modest speed of 1 m/s, the corresponding most unstable frequency is in the range of 1.6 to 16 kHz. Of course, this estimate assumes that the MIS shear layer instability characteristics are similar to those of the MAS shear layer, an assumption that awaits verification.

A characteristic inherent to high-frequency MEMS actuators with oscillation amplitudes of tens of microns is that they tend to be of the resonant type. Moreover, the Q factor of such "large-amplitude" microresonators tends to be large. Therefore, these high-frequency actuators are typically useful only at or very close to the resonant frequency of the actuator. This is a limiting factor not only from the perspective of shear layer control under different flow conditions, but also when multimode forcing is desired. For instance, Corke and Kusek (1993) have shown that resonant subharmonic forcing of the shear layer surrounding an axisymmetric jet leads to a substantial enhancement in the growth of the shear layer. To implement this forcing technique it was necessary to force the flow at two different frequencies simultaneously using an array of miniature speakers. Although the modal shapes of the forcing employed by Corke and Kusek (1993) can be easily implemented using a MEMS actuator array distributed around the jet exit, only one frequency of forcing can be targeted with a single actuator design as discussed above. However, a possible remedy is one where the MEMS capability of fabricating densely packed structures is used to develop an interleaved array of two different actuators with two different resonant frequencies.

35.3.2.3 Robustness, Packaging and Ability to Integrate with Sensors and Electronics

Similar to sensors, robustness of the MEMS actuators is essential for them to be useful in practice. In fact, actuators tend to be more vulnerable to adverse effects of the flow environment than sensors are, as they typically protrude further into the flow, exposing themselves to higher flow velocities, temperatures, forces, etc. Additionally, as discussed earlier, the fabrication processes of the actuators must be compatible

Towards MEMS Autonomous Control of Free-Shear Flows

with those of the sensors and circuitry if they are to be packaged together into an autonomous control system.

35.3.3 Closing the Loop: The Control Law

Perhaps one of the most challenging aspects of realizing MEMS autonomous systems in practice is one that is not related to the microfabrication technology itself. Given an integrated array of MEMS sensors and actuators that meets the characteristics described above, a fundamental question arises: How should the information from the sensors be processed to decide on where, when and how much actuation should be exercised to maintain a desired flow state? That is, what is the appropriate control law?

Of course, to arrive at such a control law, one needs to know the response of the flow to the range of possible actuation. This, however, is far from a straightforward task given the nonlinearity of the system being controlled by that governed by the Navier–Stokes equations. Moin and Bewley (1994) provide a good summary of various approaches that have been attempted to develop control laws for flow applications. Additional discussion on this topic can be found in this handbook in Chapter 34. Detailed discussion of the topic is not part of this chapter and it is only mentioned in passing here to underline its significance for implementation of autonomous MEMS control systems.

35.4 Control of the Roll Moment on a Delta Wing

In this section, a MEMS system aimed at realizing autonomous control of the roll moment acting on a delta wing is discussed. The material is based on the work of fluid dynamics researchers from the University of California Los Angeles (UCLA) in collaboration with microfabrication investigators from Caltech. The premise of the control pursued by the UCLA/Caltech group is based on the dominating influence of the suction-side vortices of a delta wing on the lift force. In particular, when a delta wing is placed at a high angle of attack, the shear layer separating around the side edge of the wing rolls up into a persistent vortical structure. A pair of these vortices (one from each side of the wing) is known to be responsible for generating about 40% of the lift force. Plan- and end-view flow visualization images of such vortices, captured by the UCLA group, are given in Figure 35.2. Because the vortical structures evolve from a separating shear layer, their characteristics (e.g., location above the wing and strength) may be manipulated indirectly through alteration of the shear layer at or near the point of separation. Such a manipulation may be used to change the characteristics of only one of the vortices, thus breaking the symmetry of the flow structure and leading to a net rolling moment.

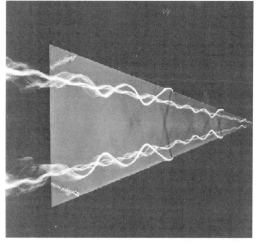

FIGURE 35.2 End (left) and plan (right) visualization of the flow around a delta wing at high angle of attack.

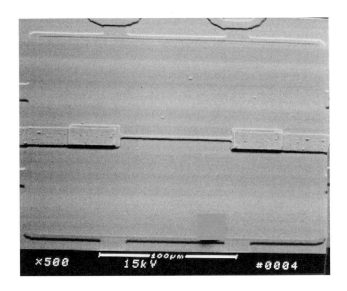

FIGURE 35.3 SEM image of the UCLA/Caltech shear stress sensor.

The shear layer separating from the edge of the delta wing is thin (order of 1 mm for the UCLA/Caltech work) and very sensitive to minute changes in the geometry. Therefore, as discussed earlier in this chapter, the use of microactuators to alter the shear-layer, and ultimately the vortical-structure, characteristics has good potential for success. Furthermore, when the edge of the wing is rounded, rather than sharp, the specific separation point location will vary with the distance from the wing apex, the flow velocity and the position of the wing relative to the flow. Therefore, a distributed sensor/actuator array is needed to cover the area around the edge of the delta wing for detection of the separation line and actuation in the immediate vicinity of it.

35.4.1 Sensing

To detect the location of the separation line around the edge of the delta wing, the UCLA/Caltech group utilized an array of MEMS hot-wire shear sensors. The sensors, which are described in detail by Liu et al. (1994), consisted of 2-μm wide × 80-μm long polysilicon resistors that were micromachined on top of an evacuated cavity (an SEM view of one of the sensors is provided in Figure 35.3). The vacuum cavity provided thermal isolation against heat conduction to the substrate in order to maximize sensor cooling by the flow. The resulting sensitivity was about 15 mV/Pa, and the frequency response of the sensors was 10 kHz. For more comprehensive coverage of this and other MEMS hot-wire sensors, the reader is referred to Chapter 26.

Because of directional ambiguity of hot-wire measurements and the three-dimensionality of the separation line, it was not possible to identify the location of separation from the instantaneous shear-stress values measured by the MEMS sensors. Instead, Lee et al. (1996) defined the location of the separation line as that separating the pressure- and suction-side flows in the vicinity of the edge of the wing. The distinction between the pressure and suction sides was based on the *rms* level of the wall-shear signal. This was possible, as the unsteady separating flow on the suction side produced a highly fluctuating wall-shear signature in comparison to the more steady attached flow on the pressure side.

A typical variation in the *rms* value of the wall-shear sensor is shown as a function of the position around the leading edge of the wing in Figure 35.4. Note that the position around the edge is expressed in terms of the angle from the bottom side of the edge, as demonstrated by the insert in Figure 35.4. It should also be pointed out that, because the *rms* is a time-integrated quantity, the detection criterion was primarily useful in identifying the average location of separation. In a more dynamic situation, where,

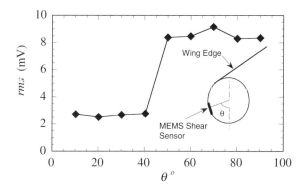

FIGURE 35.4 Distribution of surface-shear stress *rms* values around the edge of the delta wing.

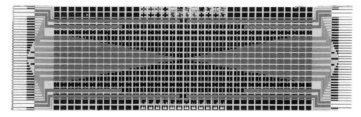

FIGURE 35.5 UCLA/Caltech flexible shear-stress sensor array skin.

for example, the wing is undergoing a pitching motion, different criteria or sensor types should be used to track the instantaneous location of separation.

To map the separation position along the edge of the wing, Ho et al. (1998) utilized 64 hot-wire sensors which were integrated during the fabrication process on a flexible shear stress skin. This 80-µm-thick skin, shown in Figure 35.5, covered a 1- × 3-cm^2 area and was wrapped around the curved leading edge of the delta wing. Using the *rms* criterion discussed in the previous paragraph, the separation line was identified for different flow speeds. A plot demonstrating the results is given in Figure 35.6, where the angle at which separation occurs (see insert in Figure 35.4 for definition of the angle) is plotted as a function of the distance from the wing apex. As seen from the figure, the separation line is curved, demonstrating the three-dimensional nature of the separation. Additionally, the separation point seems to move closer to the pressure side of the wing with increasing flow velocity.

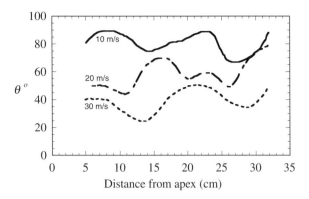

FIGURE 35.6 Separation line along the edge of the delta wing for different flow velocities.

FIGURE 35.7 SEM view of passive-type flap actuators.

35.4.2 Actuation

Two different types of actuators were fabricated for use in the MEMS rolling-moment control system: (1) magnetic flap actuators, and (2) bubble actuators. The electromagnetic driving mechanism of the former was selected because of its ability to provide larger forces and displacements than the more common electrostatic type. Passive as well as active versions of the magnetic actuator were conceived. In the passive design, a 1- × 1-mm^2 ploysilicon flap (Figure 35.7) was supported by two flexible straight beams on a substrate. A 5-μm-thick permalloy (80% Ni and 20% Fe) was electroplated on the surface of the 1.8-μm-thick flap. The permalloy layer caused the flap to align itself with the magnetic field lines of a permanent magnet. Hence, it was possible to move the flap up or down by rotating a magnet embedded inside the edge of the wing, as seen in Figure 35.8. The actuator is described in more detail in Liu et al. (1995).

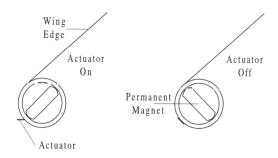

FIGURE 35.8 Permanent magnet actuation mechanism inside delta wing.

FIGURE 35.9 SEM view of active-type flap actuator.

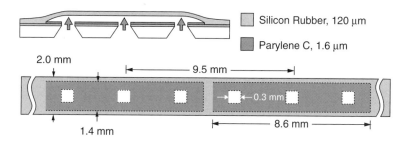

FIGURE 35.10 A schematic of bubble actuator details.

A photograph of the active flap actuator is shown in Figure 35.9. The construction of this device is generally the same as the passive one, except for a copper coil that is deposited on the silicon nitride flap. A time-varying current can be passed through the coil to modulate the flap motion around an average position (determined by the permalloy layer electroplated on the flap and the magnetic field imposed by an external permanent magnet). The actuator response was characterized by Tsao et al. (1994). The results demonstrated the ability of the actuator to produce tip displacements of more than 100 μm at frequencies of more than 1 kHz.

To develop more robust actuators that are usable not only in wind tunnels but also in practice, Grosjean et al. (1998) fabricated "balloon" or bubble actuators. The basic principle of the actuators is based on inflating flush-mounted, flexible silicon membranes using pressurized gas. As seen in Figure 35.10, the

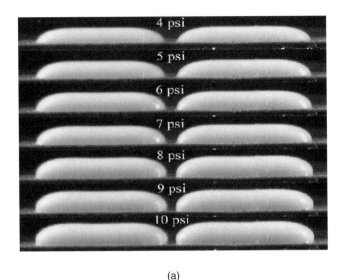

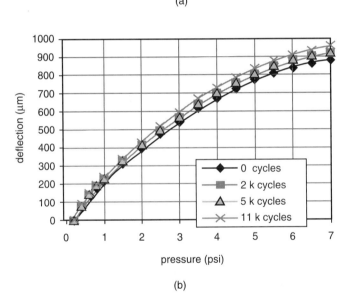

FIGURE 35.11 Bubble actuators (a) and their characteristics (b).

gas can be supplied through ports that are integrated under the membrane during the microfabrication process. When inflated, the bubbles can extend to heights close to 1 mm. Figure 35.11 demonstrates bubble inflation with increasing pressure.

35.4.3 Flow Control

To test the ability of the actuators to produce a significant change in the roll moment on a delta wing, a 56.5° swept-angle model was used. The model, which was 30 cm long and 1.47 cm thick, was placed inside a 91-cm^2 test section of an open return wind tunnel. For different angles of attack and flow speeds, the moment and forces on the wing were measured using a six-component force/moment transducer.

Figures 35.12a–d display measurements of the change in the rolling moment as a function of the location of the actuator around the leading edge of the wing. Each of the four plots in Figure 35.12

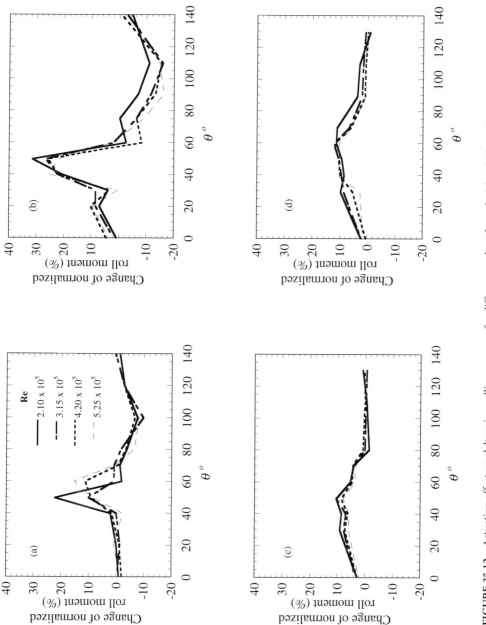

FIGURE 35.12 Actuation effect on delta wing rolling moment for different angles of attack: (a) 20°; (b) 25°; (c) 30°; (d) 35°.

represents data acquired at a different angle of attack: α = 20°, 25°, 30° and 35°, for plots a through d, respectively. Also, different lines represent different Reynolds numbers.

The results shown in Figure 35.12 were obtained for the magnetic flap actuators; however, similar results were also produced using bubble actuators [Ho et al., 1998]. In all cases, the actuators were deflected by 2 mm only on one side of the wing. The results demonstrate a significant change in the rolling moment (up to 40% for α = 25°) for all angles of attack. The largest positive roll-moment (rolling towards the actuation side) change is observed around an actuator location approximately 50° from the pressure side of the wing. The location of maximum influence was found to be slightly upstream of the separation line (identified earlier using the MEMS sensor array; see Figure 35.6) as anticipated. Another important feature in Figure 35.12 is the apparent collapse of the different lines, which suggests that the actuation impact is affected very little, if any, by changing Reynolds number.

In addition to producing a net positive roll moment, the miniature actuators are also capable of producing net negative moment at the lower angles of attack as implied by the negative peak depicted in Figures 35.12a and b. Although the largest negative roll value (found for an actuator location approximately 100° from the pressure side) is not as strong as the largest positive moment, it can be used to augment the total roll moment control authority in a particular direction by using a two-sided actuation scheme. That is, if actuators at the appropriate locations on both edges of the delta wing were operated simultaneously to produce an upward motion on one side and a downward one on the other, the two effects would superimpose to produce a larger net rolling moment. This type of moment augmentation was verified by Lee et al. (1996), as seen from the plot in Figure 35.13. An interesting aspect of the results shown in Figure 35.13 is that the net rolling moment produced by the two-sided actuation is equal to the sum of the moments produced by actuation on either side individually. This seems to suggest that the manipulation of the vortices is independent of each one.

From a physical point of view, the production of a net rolling moment by the upward motion of the actuator is caused by a resulting shift in the location of the nearby vortex core. This was demonstrated clearly in the flow visualization images from Ho et al. (1998). Two of these images may be seen in Figure 35.14. In each image, a light-sheet cut was used to make the cores of the two edge vortices visible. The top image corresponds to the flow without actuation, whereas the bottom one was captured when the right-side actuators (of the bubble type in this case) were activated. A clear outward shift of approximately one vortex diameter is observed for the right-side vortex core. A corresponding shift in the surface-pressure signature of the vortex presumably produces the observed net rolling moment.

The amount of vortex core shift, and resulting net moment, appears to be proportional to the actuation stroke. This can be seen from the results in Figure 35.15, where the change in rolling moment due to actuation on one side is displayed for two different actuator displacements: 1 and 2 mm. As observed from the data, the peak moment produced by the 2-mm actuator is about five times larger than that due

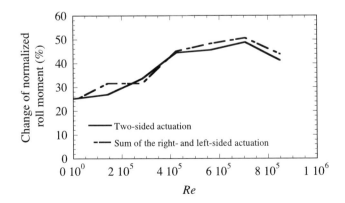

FIGURE 35.13 Roll-moment augmentation using two-sided actuation.

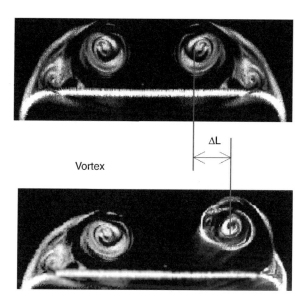

FIGURE 35.14 Flow visualization of the vortex structure on the suction side of the delta wing without (top) and with (bottom) actuation.

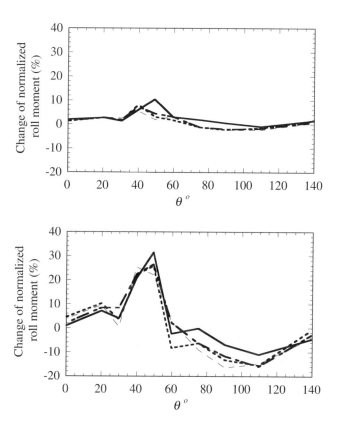

FIGURE 35.15 Effect of actuation amplitude on roll moment: 1 mm (top) and 2 mm (bottom).

to the 1-mm actuator. It is unclear, however, if the increase in the moment is a linear or nonlinear function of the actuator size, or stroke. In either case, the ability to adjust the amount of moment produced through the actuation scheme is essential if proportional control of the rolling moment is desired.

Finally, Lee et al. (1996) also demonstrated that the utility of the delta wing control system is not limited to the roll component of the moment vector. The ability to control the pitch and yaw moments was demonstrated as well.

35.4.4 System Integration

Although sensing and actuation components of the MEMS roll-moment control system were fabricated and demonstrated successfully by the UCLA/Caltech group, demonstration of a fully integrated system in operation has yet to take place—that is, demonstration of a real-time control action whereby the separation location is detected by the hot-wire sensor array and the output is fed to the appropriate algorithms to engage the actuators along the separation line, all being done autonomously and for dynamic separation-line conditions. Such a demonstration is perhaps more challenging from a fluid mechanics perspective than from the technology, or MEMS, point of view.

So far, the work conducted by the UCLA/Caltech group has shown the ability to use MEMS to fabricate the sensors and actuators necessary to significantly modify the rolling moment on a delta wing (at least at the scale of the test model and under controlled wind-tunnel conditions). This includes the compatibility of the fabrication processes of the sensor and actuators so that they can be integrated in arrays on the same substrate. However, from a fluid mechanics point of view, the use of hot-wires to detect the location of dynamic three-dimensional separation is almost impossible. This is particularly true if real-time control is to be implemented where the separation position is to be located instantaneously rather than on average.

Finally, it should be added here that the UCLA/Caltech group has recently collaborated with AeroVironment to develop an instrumented research UAV (unmanned aerial vehicle) for testing of advanced MEMS control concepts for maneuvering of aircraft. The vehicle, called "GRYPHON," is shown in the Figure 35.16 photograph.

FIGURE 35.16 GRYPHON: UCLA/Caltech MEMS flight research unmanned aerial vehicle.

35.5 Control of Supersonic Jet Screech

Reminiscent of several of the early efforts aimed at developing MEMS sensor/actuator systems for flow control, this study involved the collaboration of fluid dynamics researchers from the Illinois Institute of Technology (IIT) and microfabrication experts from the Center for Integrated Sensors and Circuits, Solid-State Electronics Laboratory at University of Michigan (UM). The selection of supersonic screech noise to test the scope of a MEMS control system was motivated by the desire to push, and therefore, identify the limits of MEMS technology, rather than the eventual implementation of such a system in practice. That is, the jet screech environment provided a harsh high-speed flow surroundings within which the devices not only had to operate but also produce a significant disturbance of the flow.

To explain further, it is useful to describe the general characteristics of jet screech. When a supersonic jet is operating at an off-design Mach number (i.e., the jet is under- or over-expanded), a set of shock cells forms. These cells interact with the axisymmetric or helical flow structures that originate from the instability of the shear layer surrounding the jet at its exit to produce acoustic noise. The noise is primarily radiated in the upstream direction, where it excites the jet shear layer regenerating the flow structure. Thus, a self-sustained feedback cycle (first recognized by Powel, 1953) is established during screech.

The basic principle of the proposed IIT/UM screech-control strategy is that of mode cancellation. Because the existence of axisymmetric or helical shear layer modes is a necessary link in the screech-generation chain of events, their elimination, or weakening, is anticipated to result in a corresponding cancellation, or attenuation, of screech noise. Such an elimination is to be achieved by the independent excitation of a flow structure of the same mode shape as (but out of phase with) that existing in the shear layer using MEMS actuators distributed around the nozzle lip. Because the MEMS forcing is achieved at the jet lip where the shear layer stability growth is likely to be linear, the natural and forced disturbances will superpose destructively to weaken the shear layer structure. Also, the placement of the actuators in the *immediate* vicinity of the jet lip, or point of separation of the shear layer, renders the flow susceptible to the minute disturbances produced by the MEMS actuators.

The spatially fixed location of the shear-layer separation point in the jet flow alleviates the need for using sensor arrays to detect and locate the point of separation. Nevertheless, sensors are needed to provide reference information concerning the mode shape and phase of the existing shear-layer structure. In this manner, the appropriate forcing phase of the actuators can be selected and adjusted with varying flow conditions.

35.5.1 Sensing

Two types of sensors were developed by the IIT/UM group to provide the necessary measurements: hot-wires and microphones. The former sensors were intended for direct measurements of the disturbance velocity in the shear layer near the nozzle lip, while the latter ones were aimed at measuring the feedback acoustic waves that generate the disturbances during screech, as explained above. A close-up view of one of the MEMS hot-wires overhanging a glass substrate is shown in Figure 35.17. Briefly, the hot-wire is made from p^{++} polysilicon with typical dimensions of 12-μm thickness, 4-μm width and 100-μm length. The corresponding resistance is in the range of 200 to 800 Ω. The glass substrate under the wire provides a good thermal-insulating base. Fabrication details are provided in Huang et al. (1996).

A sample calibration of one of the hot-wires, shown in Figure 35.18, was performed when the wire was operated at an overheat ratio of 1.05. Three different calibrations separated by time intervals of 1 hour and 10 days are included in the figure. The calibration results display a change of about 2 V of the hot-wire output for a velocity range of up to 35 m/s. A conventional hot-wire at even larger overheats of more than 1.5 would exhibit a voltage output range of less than 1 V. Hence, Figure 35.18 demonstrates the high sensitivity of MEMS (polysilicon) hot-wires. Furthermore, considering the calibration results from the three different trials, very little change in the sensor characteristics is observed. Given the long time interval between some of the calibrations, it is seen that the MEMS sensors are generally very stable. Further characterization of the sensors may be found in Naguib et al. (1999b).

FIGURE 35.17 SEM view of the IIT/UM hot-wire sensor.

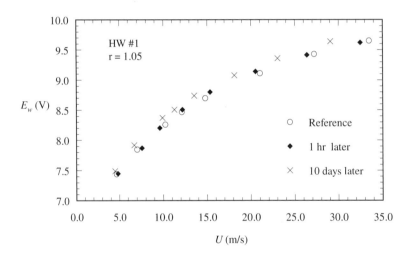

FIGURE 35.18 Static calibration of one of the IIT/UM hot-wire sensors.

On the other hand, the MEMS acoustic sensor, shown using a SEM view in Figure 35.19, consists of a stress-compensated plasma-enhanced chemical vapor deposition (PECVD) silicon nitride/oxide, 0.4-μm-thick diaphragm together with four monocrystalline ion-implanted p^{++} silicon piezoesistors. The coefficient, π_{44}, for this type of piezoresistor is about four times larger than that based on p-type polysilicon, thus leading to a higher transducer sensitivity. The piezoresistors are arranged in a full Wheatstone-bridge configuration for detection of the diaphragm deflection. For fabrication details, the reader is referred to Huang et al. (1998).

Figure 35.20 provides the frequency characteristics of a 710- × 710-μm^2 MEMS sensor obtained by Naguib et al. (1999a). For reference, the sensitivities of commercial silicon Kulite sensor and 1/8-in. B&K microphone are also provided in the figure. The open and closed circles represent calibration results for the same sensor obtained 2 months apart. The results demonstrate that very good agreement is found between the open and closed circles. This agreement indicates that the MEMS transducers are stable and reliable. For frequencies below 1 kHz, the sensitivity of the MEMS sensors seems to be almost an order of magnitude larger than the sensitivity of the commercial Kulite sensor. At higher frequencies, however, the MEMS sensitivity is attenuated.

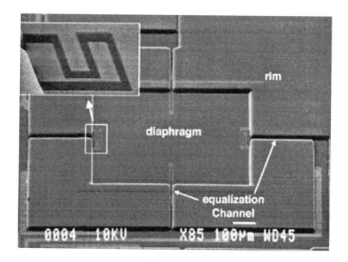

FIGURE 35.19 SEM view of IIT/UM acoustic sensor.

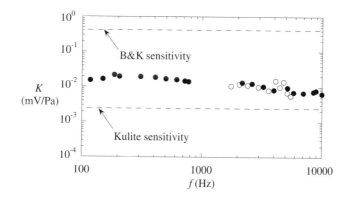

FIGURE 35.20 Frequency response of one of the IIT/UM acoustic sensors.

35.5.2 Actuation

The most common form of excitation of high-speed jets and shear layers has been via internal and external acoustic sources [e.g., Moore, 1977]. The popularity of acoustic forcing has been primarily due to ease of implementation and the ability to generate the high-frequency disturbances required for effective forcing of high-speed shear flows. However, the receptivity of the jet to acoustic excitation requires matching of the wavenumbers of the acoustic and instability waves [Tam, 1986]. Although such a condition is achievable in supersonic jets, at low Mach numbers the instability wavelength is much smaller than the acoustic wavelength. In this case, conversion of acoustic energy into instability waves is dependent on the "efficiency" of the unsteady Kutta condition at the nozzle lip.

Historically, studies utilizing mechanical actuators have been limited to low flow speeds due to the inability of the mechanical actuators, with their large inertia, to operate at frequencies higher than tens to hundreds of hertz. Therefore, the emergence of MEMS carried with it the potential to break this barrier through the fabrication of mechanical actuators with "microinertia" that can oscillate at high frequency. Ultimately, this may lead to the realization of more efficient actuators that can act directly on the flow without the need for a Kutta-like condition. For these reasons, micromechanical actuators were selected for the screech control problem.

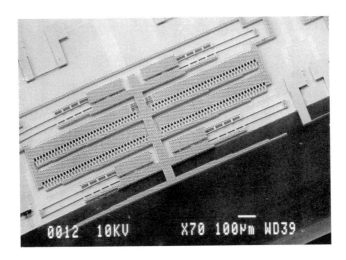

FIGURE 35.21 SEM view of IIT/UM electrostatic actuator.

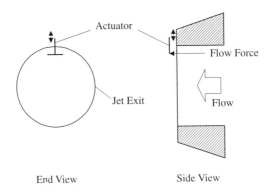

FIGURE 35.22 A schematic demonstrating the direction of flow forces acting on the MEMS actuator.

Figure 35.21 shows a SEM photograph of one of the electrostatic actuators used in the jet study. The device consists of a free-floating section that is mounted to a glass substrate underneath using elastic, folded beams. At the lower end of the floating section is a roughly 6-μm × 30-μm × 1300-μm T-shaped actuator head. The actuator head is made to oscillate at an amplitude up to 70 μm and a resonant frequency that can be tailored by adjusting the mass and stiffness of the actuator structure. Also, notice the porous actuator body, which results in a reduced dynamic load on the actuator head. For more details on the construction and manufacturing of the actuators, see Huang et al. (1996).

Two actuator designs were developed with resonant frequencies of 5 and 14 kHz. These values were selected to match the frequency of different screech modes in the High-Speed Jet Facility (HSJF) at IIT. The actuators were driven using a combination of 20-V dc and 40-V peak-to-peak sinusoidal voltage at the actuator resonant frequency.

The relative placement of one of the T-actuators with respect to the jet is illustrated in Figure 35.22. As depicted in the figure, the rigidity of the actuator with respect to out-of-plane deflection caused by flow forces is primarily dependent on the thickness of the device. For the device shown in Figure 35.21, the thickness was limited to a value of less than 12 μm because of the wet-etching methodology employed in the fabrication sequence. An alternate fabrication approach, which utilized deep RIE (reactive ion etching) etching, was later utilized to fabricate thicker, more rigid actuators. Using this approach, actuators with a thickness of 50 μm were obtained. A SEM view of one of the thick actuators is provided in Figure 35.23.

FIGURE 35.23 SEM view of a thick (50-μm) actuator.

35.5.3 Flow Control

The ability of the microactuators to disturb the shear layer surrounding the jet was tested for flow speeds of 70, 140 and 210 m/s. The corresponding Mach (M_j) and Reynolds numbers (Re; based on jet diameter, D) were 0.2, 0.4 and 0.6 and 118533, 237067 and 355600, respectively. The corresponding, linearly most unstable frequency of the jet shear layer was estimated to be 16, 44 and 91 kHz, in order of increasing Mach number. Because the largest effect on the flow is achieved when forcing as close as possible to the most unstable frequency, it is seen that the actuator frequency of 14 kHz was substantially lower than desired for the two largest jet speeds.

Using hot-wire measurements on the shear layer center line, velocity spectra of the natural and forced jet flows at the different Mach numbers were obtained. Those results are shown in Figure 35.24. The different lines in each of the plots represent spectra obtained at different streamwise (x) locations. For a Mach number of 0.2, the spectra obtained in the natural jet seem to be wideband, except for two fairly small peaks at approximately 7 and 13 kHz. When MEMS forcing is applied, a very strong peak is observed at the forcing frequency (Figure 35.24, top right). The magnitude of the peak seems to rise initially with downstream distance and then fall. In addition to the peak at the forcing frequency, a second strong peak at 28 kHz is observed at all streamwise locations. A third peak is also depicted at a frequency of 21 kHz at the second and third x locations. The existence of multiple peaks in the spectrum at frequencies which are multiples of the forcing frequency and its subharmonic suggests that the MEMS-introduced disturbance has a large enough amplitude to experience nonlinear effects of the flow.

The disturbance spectra for the natural jet at Mach number of 0.4 are shown on the left side of the middle part of Figure 35.24. The corresponding spectra for the forced jet are shown on the opposite side. As seen from Figure 35.24, the natural jet spectra possess a fairly large and broad peak at a frequency of about 13 kHz. Because the most unstable frequency of the jet shear layer at this Mach number is expected to be around 44 kHz, the 13-kHz peak does not seem to correspond to the natural mode of the shear layer or its subharmonic. When operating the MEMS actuator, a clear peak is depicted in the spectrum at the forcing frequency. The peak magnitude initially magnifies to reach a magnitude of more than an order of magnitude larger than the fairly strong peak depicted in the natural spectrum at 13 kHz. The MEMS-induced peak is also significantly sharper than the natural peak, presumably because of the more organized nature of forced modes.

Finally, the spectra obtained at a Mach number of 0.6 are displayed in the bottom of Figure 35.24. Similar to the results for 0.4 Mach number, a broad, fairly strong peak is depicted in the natural jet spectra. The frequency of this peak appears to be about 16 kHz at $x/D = 0.1$, which is considerably lower than the estimated most unstable frequency of 91 kHz. The peak frequency value decreases with increasing x.

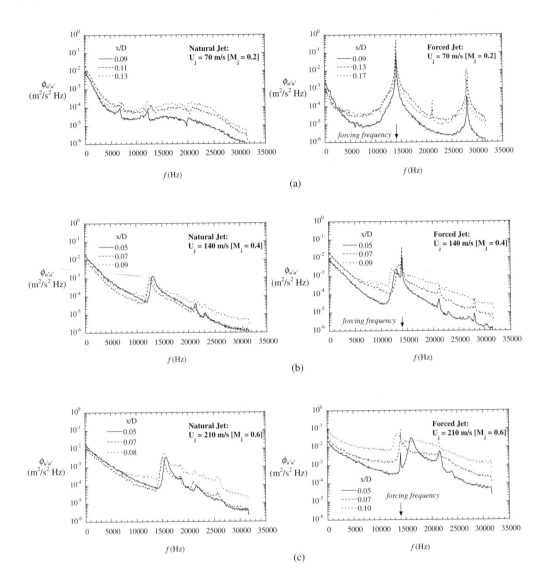

FIGURE 35.24 Natural (left) and forced (right) spectra in the IIT high-speed jet for a Mach number of: (a) 0.2; (b) 0.4; (c) 0.6.

This decrease in the peak frequency with x may be symptomatic of probe feedback effects [Hussein and Zaman, 1978]. In spite of this strong peak, when forced using the MEMS actuator a clearly observable peak at the forcing frequency is depicted in the spectrum of the forced shear layer. Unlike the results for the forced jet at Mach numbers of 0.2 and 0.4, the forced jet spectrum for Mach number of 0.6 does not contain spectral peaks at higher harmonics of the forcing frequency.

To evaluate the level of the disturbance introduced into the shear layer by the MEMS actuator, the energy content of the spectral peak at the forcing frequency was calculated from the phase-averaged spectra (to avoid inclusion of background turbulence energy). The forced disturbance energy ($<u_{rms,f}>$) dependence on the streamwise location is shown for all three Mach numbers in Figure 35.25. The disturbance energy is normalized by the jet velocity (U_j), and the streamwise coordinate is normalized by the jet diameter. Inspection of Figure 35.25 shows that for both Mach numbers of 0.2 and 0.4 no region of linear growth is detectable. For these two Mach numbers, $<u_{rms,f}>$ only increases slightly before

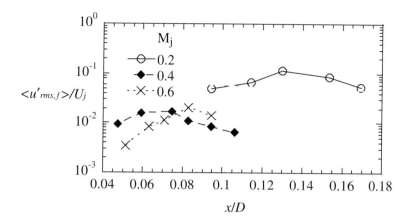

FIGURE 35.25 Streamwise development of the MEMS-induced disturbance *rms*.

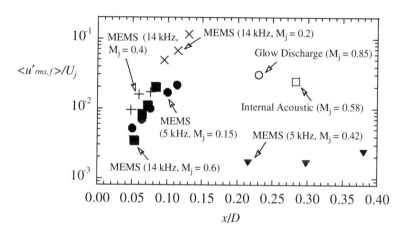

FIGURE 35.26 MEMS-induced disturbance *rms* compared to "macroscale" forcing schemes.

reaching a peak followed by a gradual decrease in value, a process reminiscent of nonlinear amplitude saturation.

On the other hand, the disturbance energy corresponding to a Mach number of 0.6 appears to experience linear growth over the first four streamwise positions before saturating. It may be noticed also that for only 0.6 Mach number the disturbance *rms* level is appreciably lower than 1% of the jet velocity at the first streamwise location. In the work of Drubka et al. (1989), the fundamental and subharmonic modes in an acoustically excited, incompressible, axisymmetric jet were seen to saturate when their *rms* value exceeded 1 to 2% of the jet velocity. Therefore, it seems that the IIT/UM MEMS actuator is capable of providing an excitation to the shear layer that is sufficient not only to disturb the flow but also to produce nonlinear forcing levels.

The magnitude of the MEMS forcing may also be appreciated further by comparison to other types of "macroscale" forcing. Thus, the disturbance *rms* value produced by internal acoustic (Lepicovsky et al., 1985) and glow discharge (Corke and Cavalieri, 1996) forcing is compared to the corresponding *rms* values produced by MEMS forcing in Figure 35.26. The results for MEMS forcing contained in the figure are those from Naguib et al. (1997) using the high-frequency MEMS actuators as well as those from the earlier study by Alnajjar et al. (1997), using the same type of MEMS actuators but at a forcing frequency of 5 kHz.

As seen from Figure 35.26, for most cases of MEMS forcing, the MEMS-generated disturbance grows to a level that is similar to that produced by glow discharge and acoustic forcing. For Mach number of 0.2, the forcing frequency of 14 kHz is almost equal to the most unstable frequency of the shear layer and the resulting flow disturbance is at a level which is significantly higher than even that produced by the "macroscale" forcing methods. On the other hand, at the 0.42 Mach number, the 5-kHz MEMS actuator excites the flow at a frequency which is almost an order of magnitude lower than the most amplified frequency of the shear layer. The small amplification rate associated with a disturbance at a frequency which is significantly smaller than the natural frequency of the flow is believed to be responsible for the resulting small disturbance level at the Mach number of 0.42 when forcing with the 5-kHz actuator.

The ability of the MEMS devices to excite the flow in par with other macro forcing devices is believed to be due to the ability to position the MEMS extremely close to the point of high receptivity at the nozzle lip where the flow is sensitive to minute disturbances. To investigate this matter further, the radial position of the MEMS actuator with respect to the nozzle lip (y_{off}) was varied systematically. For all actuator positions, the flow was maintained at 70 m/s, while the actuator was traversed in the range from about 50 μm (outside the flow) to −150 μm (inside the flow) relative to the nozzle lip. The boundary layer at the exit of the jet at 70 m/s was laminar and had a momentum thickness of about 72 μm. Therefore, at its innermost location, the actuator penetrated into the flow a distance which was less than 20% of the boundary layer thickness, and hence, based on a Blasius profile, it was exposed to a velocity less than one fifth of the jet speed.

The energy content of the spectral peak at the forcing frequency was calculated from the disturbance spectra obtained at $x/D = 0.1$ for different radial actuator locations. The results are displayed in Figure 35.27. For reference, a dimension indicating the momentum thickness of the boundary layer emerging at the exit of the jet (θ_o) is included in the figure. As observed from the figure, the largest disturbance energy is produced when the actuator is closest to the nozzle lip ($y_{\text{off}} = 0$) and into the flow. If the actuator is placed a distance as small as 75 μm (less than the size of a human hair) off the position corresponding to maximum response, an order of magnitude reduction in disturbance *rms* value is observed.

The reduction in actuator effectiveness as it is traversed radially outward is presumably because of the increased separation between the actuator and the shear layer. However, the reasons for observing the same trend when locating the actuator further into the flow are not equally clear. In fact, recently, Alnajjar et al. (2000), using piezoelectric actuators, found the response of the same jet flow used for MEMS testing to be fairly insensitive to the radial location of the actuator inside the shear layer, with the largest response found near the center of the shear layer. This suggests that the reduced actuation impact seen in Figure 35.27

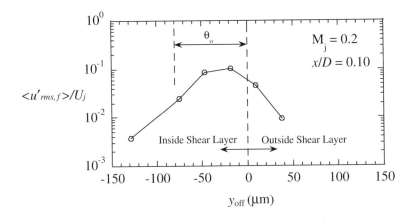

FIGURE 35.27 Effect of MEMS actuator radial position on the generated shear layer disturbance level.

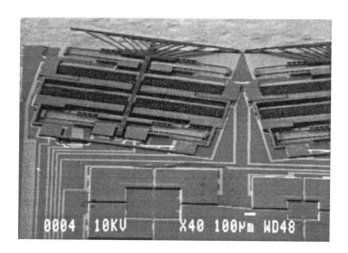

FIGURE 35.28 Two acoustic sensors integrated with two actuators.

as the actuator is positioned further into the flow is most likely due to a change in the behavior of the actuator itself. In particular, it is not unreasonable to expect that, as the actuator head is exposed to larger flow speeds for positions further into the flow, it may experience larger out-of-plan deflections. These deflections would cause the overlap area between the stationary and moving parts of the comb drives to become smaller, leading to a smaller actuation force and amplitude.

35.5.4 System Integration

The fabrication sequence of the sensors for the screech-control system was developed carefully to ensure its compatibility with that of the actuators. As discussed at the beginning of the chapter, the compatibility of the fabrication sequence of the individual components is necessary if one is to capitalize on the MEMS technology advantage of integrating component arrays to form full autonomous systems. For instance, to machine the MEMS acoustic sensors shown in Figure 35.19 integrated with the actuators on the same chip, a total of nine masks were required [Huang et al., 1998]. In contrast, fabrication of a similar sensor in isolation would probably require no more than three to four masks. SEM views of the electrostatic actuators integrated with hot wires and acoustic sensors are given in Figures 35.21 and 35.28, respectively.

To make external input–output connections to the sensors and actuators the glass substrate supporting the MEMS devices was fixed to a sector-shaped printed-circuit board using epoxy. Close-up and full views of the board may be seen in Figure 35.29. The full sensor/actuator array was then assembled by mounting the 16 sector boards on a special jet faceplate. Each of the individual boards was supported on a radial miniature traversing system to enable precise placement of the actuators at the location of maximum response (as dictated by the results in Figure 35.27). Figure 35.30 shows a schematic diagram of the jet faceplate, sector PC-boards and miniature traverses. An image of the full MEMS array as mounted on the jet can be seen in Figure 35.31.

The distributed 16-actuator array provided the capability of exciting flow structures with azimuthal modes up to a mode number (m) of eight. This was verified using single hot-wire measurements at different locations around the perimeter of the jet. At each measurement location the wire was located approximately at the center of the shear layer. The amplitude and phase information of the measured disturbance was obtained from the cross-spectrum between the hot wire data and the driving signal of one of the actuators. The results are demonstrated in Figure 35.32. As seen from the amplitude data, a fair amount of scatter is observed around an average value over the full circumference. The observed

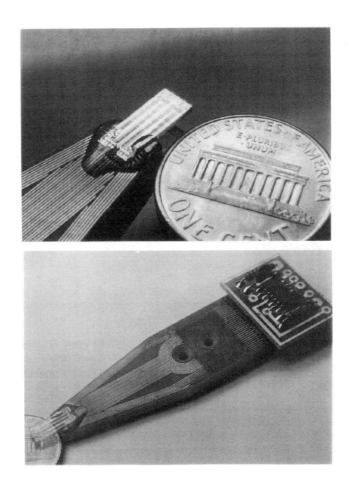

FIGURE 35.29 Packaging of the IIT/UM sensor-actuator system.

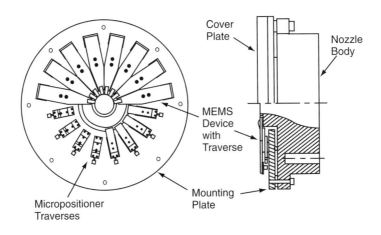

FIGURE 35.30 MEMS PC boards mounting provisions on the jet faceplate.

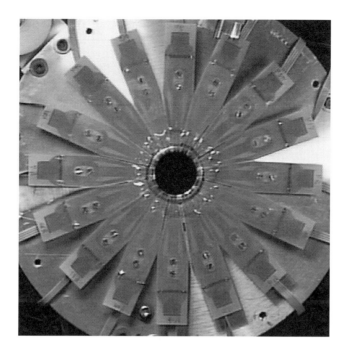

FIGURE 35.31 Photograph of the MEMS sensor/actuator array mounted on the jet.

scatter is believed to be due to imprecise positioning of the hot wire at the center of the shear layer and variation in the amplitude of the different actuators. On the other hand, the phase results (bottom plot in Figure 35.32) indicate that the disturbance phase measured in the shear layer agrees with the excited flow modes.

The nonuniformity of the MEMS actuator amplitudes can be easily remedied by adjustment of the individual actuator driving voltages to provide uniform forcing. A more significant problem encountered in forcing with the full array was the slight deviations in resonant frequency of the individual devices. For the most part, this deviation did not exceed ±2% of the average resonant frequency for a particular batch. This may not seem to be a substantial deviation; however, MEMS resonant devices tend to have very small damping ratios (high Q factor) which lead to very narrow resonant peaks. As a result, the devices cannot be operated at frequencies that deviate more than 0.5 to 1% from the resonance frequency while maintaining more than 20% of their resonance amplitude.

With the different actuators operating at slightly different frequencies it was not possible to sustain a specific phase relationship amongst the different actuators for extended periods of time. Therefore, the data shown in Figure 35.32 were acquired using a transient forcing scheme whereby the actuators were used to force a particular azimuthal mode only for as long as the largest phase deviation did not exceed an acceptable tolerance. The best remedy to the problem, with the current MEMS technology, appears to be the fabrication of a large batch of devices, so only those with matching resonant frequencies are used.

In summary, the IIT/UM work has resulted in the realization of integrated actuators and a sensors system capable of autonomous control of shear layers in high-speed jets. The system was not actually used to control screech, as the actuators operated successfully up to a Mach number of 0.8 but not in the highly unsteady-pressure environment encountered during screech. Nevertheless, the ability of the less-than-hair-width actuators to operate up to such a high speed while producing significant disturbance into the flow (even at frequencies less than 1/10 of the most unstable frequency) was quite an impressive demonstration of the potential of the technology.

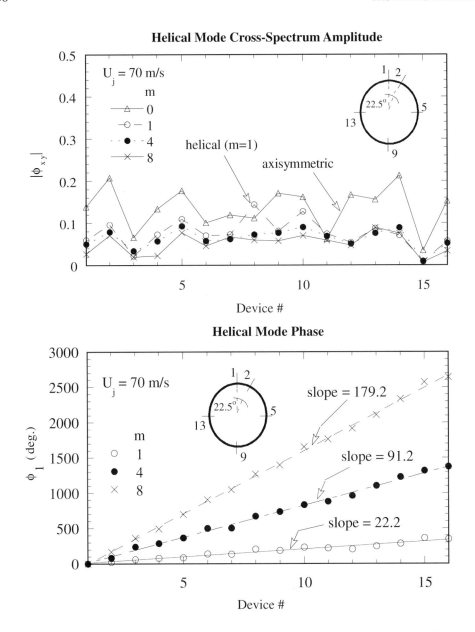

FIGURE 35.32 Azimuthal amplitude (top) and phase (bottom) distribution of the MEMS-induced disturbance.

35.6 Control of Separation over Low-Reynolds-Number Wings

Recently, researchers from the University of Florida have proposed a MEMS system for controlling separation at low Reynolds numbers. The primary motivation of the proposed system was to enhance the lift-to-drag ratio in micro-air-vehicle (MAV) flight. Because of their small size (a few centimeters' characteristic size) and low speed, MAVs experience low-Reynolds-number flow phenomena during flight. One of these is an unsteady laminar separation that occurs near the leading edge of the wing and affects the aerodynamic efficiency of the wing adversely.

Figure 35.33 displays a schematic of the proposed control-system components and test-model geometry. The main idea is based on the deployment of integrated MEMS sensors and actuators near the

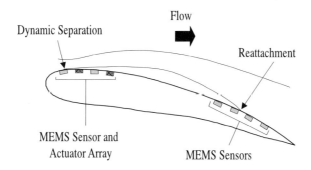

FIGURE 35.33 Control system components for University of Florida low-Reynolds-number wing-control project.

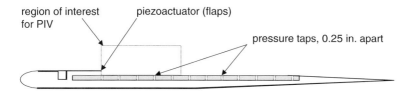

FIGURE 35.34 Test model for separation control experiments of University of Florida.

leading edge of an airfoil, or wing section. Additional sensor arrays are to be used near the trailing edge of the wing. The leading-edge sensors are intended for detection of the separation location in order to activate those actuators closest to that location for efficient control, as discussed in previous sections. On the other hand, the trailing edge sensors are to be utilized to sense the location of flow reattachment. In this manner, it would be possible to adapt the magnitude and location of actuation in response to changes in the flow in order to, for instance, maintain the flow attached at a particular location on the wing. The ultimate benefit of such a control system is the manipulation of the aerodynamic forces on the wing for increased efficiency as well as maneuverability without the use of cumbersome mechanical systems.

In actual implementation, the University of Florida group adopted a *hybrid* approach, whereby conventional-scale piezoelectric devices were used for actuation and MEMS sensors were used for measurements. Additionally, it appears that because of the difficulty in detecting the instantaneous separation location, as discussed in the delta wing control problem, a small step in the surface of the wing was introduced near the leading edge at the actuation location. Thus, the location of separation was fixed and there was no need for using leading-edge sensors for initial testing of the controllability of the flow. The flow control test model is shown in Figure 35.34.

35.6.1 Sensing

To measure the unsteady wall shear stress, platinum surface hot-wire sensors were microfabricated. The devices consisted of a 0.15-µm-thick × 4-µm-wide × 200-µm-long platinum wire deposited on top of a 0.15 silicon nitride membrane. Beneath the membrane is a 10-µm-deep vacuum cavity with a diameter of 200 µm. Similar to the UCLA/Caltech sensor, the evacuated cavity was incorporated in the sensor design to maximize the thermal insulation to cooling effects other than that due to the flow. As a result, the sensor exhibited a static sensitivity as high as 11 mV/Pa when operating at an overheat ratio (operating resistance/cold resistance) of up to 2.0. The sensor details can be seen in the SEM image in Figure 35.35. For a detailed characterization of the static and dynamic response of the sensor, the reader is referred to Chandrasekaran (2000) and Cain (2000).

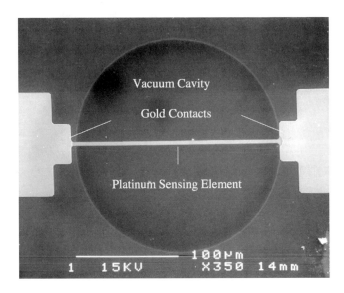

FIGURE 35.35 (Color figure follows p. 12.26.) SEM view of University of Florida MEMS wall-shear sensor.

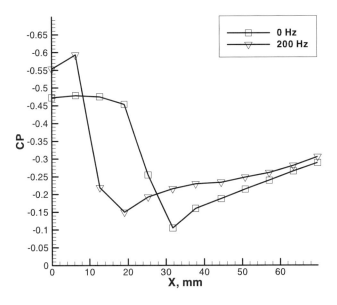

FIGURE 35.36 Static pressure distribution with and without control.

35.6.2 Flow Control

Static surface pressure measurements and particle image velocimetery (PIV) images were used by Fuentes et al. (2000) to characterize the response of the reattaching flow to forcing with the piezoelectric actuators. The 51-mm-wide × 16-m-long flap-type actuators (see Figure 35.34) were operated at their resonance frequency of 200 Hz. The resulting static pressure (plotted as a coefficient of pressure, CP) distribution downstream of the 1.4-mm-high step is given in Figure 35.36. Similar results without forcing are also provided in the figure for comparison. As seen from the figure, the minimum negative peak of CP, corresponding to the location of reattachment, shifts upstream with excitation. The extent of the shift is fairly significant, mounting to about 30% or so of the uncontrolled reattachment length.

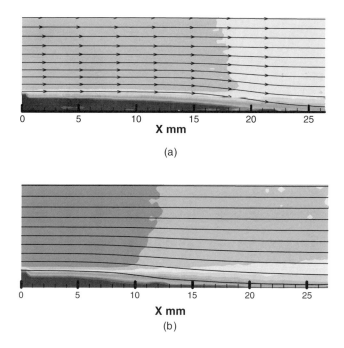

FIGURE 35.37 (Color figure follows p. **12.26**.) Streamlines of the flow over the step without (a) and with (b) actuation.

The reduction in the reattachment length with forcing can also be seen in the streamline plots obtained from PIV measurements (see Figure 35.37). However, the real benefit of the PIV data was revealing the nature of the flow structure associated with actuation by capturing images that were phase-locked to different points of the forcing cycle. Those results are provided in Figure 35.38 for an approximately full cycle of the forcing. A convecting vortex structure is clearly seen in the sequence of streamline plots (a) through (d). The observed vortex structures were periodic when an actuation amplitude of about 22 µm was used. For substantially smaller forcing amplitude, the generated vortices were found to be aperiodic.

Similar to the UCLA/Caltech and IIT/UM efforts, the University of Florida work has demonstrated the ability to alter the flow significantly through low-level forcing. Additionally, high-sensitivity MEMS sensors were developed and tested. However, for all three efforts there remains to be seen a demonstration of a fully autonomous system in operation.

35.7 Reflections on the Future

When it comes to the potential for using MEMS for flow control, it is not difficult to find contradictory views within the fluid dynamics community. This is not too surprising given the number of challenges facing the implementation and use of this fairly young technology. Challenges aside, however, one needs to recognize that there are certain capabilities that can only be achieved with MEMS technology. Examples include tens-of-kHz distributed mechanical actuators, sensor arrays that are capable of resolving the spatiotemporal character of the flow structure in high-Reynolds-number flows, integration of actuators, sensors and electronics, etc.—that is, the kind of capabilities that seem to be needed if we are to have any hope of controlling such a difficult system as that governed by the Navier–Stokes equations. Therefore, it is significantly much more constructive for us to identify the challenges facing the use of MEMS and strive to find solutions to these challenges, rather than simply dismiss the technology (along with its potential benefits). In this section, some of the leading challenges facing the attainment of

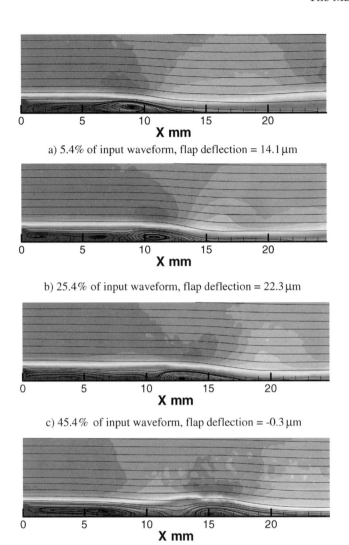

a) 5.4% of input waveform, flap deflection = 14.1 μm

b) 25.4% of input waveform, flap deflection = 22.3 μm

c) 45.4% of input waveform, flap deflection = −0.3 μm

d) 65.4% of input waveform, flap deflection = −22.5 μm

FIGURE 35.38 (Color figure follows p. 12.26.) Phase-averaged streamline plots at different phases of the forcing cycle.

autonomous MEMS control systems for shear layer control are highlighted. These are accompanied by the author's perspective on the hope of overcoming these challenges.

One of the main concerns regarding the implementation of MEMS devices is regarding their *robustness*, particularly if they have to be operated in harsh, high-temperature environments. For the most part, this concern stems from the micron size of the MEMS devices which renders them vulnerable to large external forces. However, it is important to remember that as one shrinks the size of a structure, the flow forces acting on it decrease along with its ability to sustain such forces. That is, to a certain extent, the microscale devices may be as strong as, if not stronger than, their larger scale equivalent (at least if they are designed well). That is probably why the actuators from Naguib et al. (1997) operated properly while immersed in a Mach-0.8 shear layer, and the actuators and sensors of Huang et al. (2000) successfully completed a test flight while attached on the outside of an F-15 fighter jet. Furthermore, as new microfabrication techniques are devised for more resilient, chemically inert, harder materials other than silicon, it will be possible to construct microdevices for harsh, high-temperature, chemically reacting environments. Some of the current notable efforts in this area are those concerned with micromachining of silicon carbide and diamond.

The robustness question is probably more critical from a practical point of view; that is, whereas a MEMS array of surface stress sensors deployed over an airplane wing may survive during flight, the array may easily be crushed by a person during routine maintenance. However, it appears that such issues should, and could, be addressed at the design stage where, for instance, the sensor array may be designed to be normally hidden away but to deploy during flight only. Additionally, the inherent array-fabricating ability of MEMS could be used to increase system robustness through redundancy. If a few sensors break, other on-chip sensors may be used instead. If the number of malfunctioning sensors becomes unacceptable, the entire chip could be replaced with a new one. The economics of replacing MEMS system modules will likely be justified, as it seems natural that MEMS will eventually follow in the path of the integrated circuit chip with its low-cost, bulk-fabrication technologies.

Beyond robustness, there is a need to develop innovative approaches to enhance the *signal-to-noise ratio* of MEMS sensors. As discussed before, when the size of sensors is shrunk, often, but not always, their sensitivity decreases proportionally. Because, for the most part, traditional transduction approaches have been used with the smaller sensors, the overall signal-to-noise ratio cannot be maintained at desired levels. Hence, there is a need to identify *ultra-sensitive* transduction methods. An example of such methods is the intra-grain, poly-diamond piezoresistive technology developed recently by Salhi and Aslam (1998). This technology promises the ability to integrate inexpensive poly-diamond piezoresitive gauges with a gauge factor of up to 4000 (20 times more sensitive than the best silicon sensors) into microsensors.

Finally, when it comes to actuation, one of the most challenging issues that need to be addressed is the sufficiency of MEMS actuation amplitudes. Notwithstanding the successful demonstrations of the IIT/UM and UCLA/Caltech groups discussed earlier in this chapter, in practice boundary layers tend to be significantly thicker and more turbulent at separation than those encountered in the cited experiments. Therefore, it is most likely that the use of MEMS actuators will be confined to controlled experiments in the laboratory (where they may be used, for example, for proof-of-concept experiments) and flows in microdevices. For large-scale flows, successful autonomous control systems will most probably be hybrid, consolidating macroactuators with MEMS sensors arrays, as was done, for instance, in the University of Florida work discussed earlier. This will require developing clever techniques for integrating the fabrication processes of MEMS with those of large-scale devices in order to capitalize on the full advantage of MEMS.

References

Alnajjar, E., Naguib, A.M., Nagib, H.M., and Christophorou, C. (1997) "Receptivity of High-Speed Jets to Excitation Using an Array of MEMS-Based Mechanical Actuators," in *Proc. ASME Fluids Engineering Division Summer Meeting*, Paper No. FEDSM97-3224, June 22–26, Vancouver, B.C.

Alnajjar, E., Naguib, A., and Nagib, H. (2000) "Receptivity of an Axi-Symmetric Jet to Mechanical Excitation Using a Piezoelectric Actuator," in *AIAA Fluids 2000*, AIAA Paper No. 2000-2557, June 19–22, Denver, CO.

Cain, A., Chandrasekaran, V., Nishida, T., and Sheplak, M. (2000) "Development of a Wafer-Bonded, Silicon-Nitride, Membrane Thermal Shear-Stress Sensor with Platinum Sensing Element," in *Solid-State Sensor and Actuator Workshop*, June 4–8, Hilton Head Island, SC.

Chandrasekaran, V., Cain, A., Nishida, T., and Sheplak, M. (2000) "Dynamic Calibration Technique for Thermal Shear Stress Sensors with Variable Mean Flow," in *38th Aerospace Sciences Meeting and Exhibit*, AIAA Paper No. 2000-0508, Jan. 10–13, Reno, NV.

Corke, T.C., and Cavalieri, D. (1996) "Mode Excitation in a Jet at Mach 0.85," *Bull. Am. Phys. Soc.* **41**(9), p. 1700.

Corke, T.C., and Kusek, S.M. (1993) "Resonance in Axisymmetric Jets with Controlled Helical-Mode Input," *J. Fluid Mech.* **249**, pp. 307–336.

Drubka, R.E., Reisenthel, P., and Nagib, H.M. (1989) "The Dynamics of Low Initial Disturbance Turbulent Jets," *Phys. Fluids A* **1**(10), pp. 1723–1735.

Epstein, A.H., Senturia, S.D., Al-Midani, O., Anathasuresh, G., Ayon, A., Breuer, K., Chen, K.-S., Ehrich, F.E., Esteve, E., Frechette, L., Gauba, G., Ghodssi, R., Groshenry, C., Jacobson, S., Kerrebrok, J.L., Lang, J.H., Lin, C.-C., London, A., Lopata, J., Mehra, A., Mur Miranda, J.O., Nagle, S., Orr, D.J., Piekos, E., Schmidt, M.A., Shirley, G., Spearing, S.M., Tan, C.S., Tzeng, Y.-S., and Waitz, I.A. (1997) "Micro-Heat Engines, Gas Turbines, and Rocket Engines—The MIT Microengine Project," in *AIAA Fluid Mechanics Summer Meeting,* AIAA Paper No. 97-1773, June 29–July 2, Snowmass, Co.

Fiedler, H.E., and Fernholz, H.-H. (1990) "On Management and Control of Turbulent Shear Flows," *Prog. Aero. Sci.* **27**, pp. 305–387.

Fuentes, C., He, X., Carroll, B., Lian, Y., and Shyy, W. (2000) "Low Reynolds Number Flows Around an Airfoil with a Movable Flap. Part 1. Experiments," in *AIAA Fluids 2000 and Exhibit,* AIAA Paper No. 2000-2239, June 19–22, Denver, CO.

Gharib, M., Modarress, D., Fourguette, D., and Taugwalder, F. (1999) "Design, Fabrication and Integration of Mini-LDA and Mini-Surface Stress Sensors for High Reynolds Number Boundary Layer Studies," in *ONR Turbulence and Wakes Program Review,* Sept. 9–10, Palo Alto, CA.

Grosjean, C., Lee, G., Hong, W., Tai, Y.C., and Ho, C.M. (1998) "Micro Balloon Actuators for Aerodynamic Control," in *Proc. 11th Annual Int. Workshop on Micro Electro Mechanical Systems* (*MEMS '98*), pp. 166–171, Jan. 25–29, Heidelberg, Germany.

Ho, C.M., and Huerre, P. (1984) "Perturbed Free Shear Layers," *Ann. Rev. Fluid Mech.* **16**, pp. 365–424.

Ho, C.M., Huang, P.H., Lew, J., Mai, J., Lee, G.B., and Tai, Y.C. (1998) "MEMS: An Intelligent System Capable of Sensing-Computing-Actuating," in *Proc. 4th Int. Conf. on Intelligent Materials,* pp. 300–303.

Huang, A., Ho, C.M., Jiang, F., and Tai, Y.C. (2000) "MEMS Transducers for Aerodynamics—A Paradigm Shift," in *AIAA 38th Aerospace Sciences Meeting and Exhibit,* AIAA Paper No. 00-0249, Jan. 10–13, Reno, NV.

Huang, C.C., Papp, J., Najafi, K., and Nagib, H.M. (1996) "A Microactuator System for the Study and Control of Screech in High-Speed Jets," in *Proc. of Micro Electro Mechanical Systems Workshop* (*MEMS '96*), San Diego, CA.

Huang, C.C., Najafi, K., Alnajjar, E., Christophorou, C., Naguib, A., and Nagib, H.M. (1998) "Operation and Testing of Electrostatic Microactuators and Micromachined Sound Detectors for Active Control of High Speed Flows," in *Proc. of the IEEE/ASME Micro Electro Mechanical Systems Workshop,* Heidelberg, Germany.

Huerre, P., and Monkewitz, P.A. (1990) "Local and Global Instabilities in Spatially Developing Flows," *Annu. Rev. Fluid Mech.* **22**, pp. 473–537.

Hussain, A.K.M.F., and Zaman, K.B.M.Q. (1978) "The Free Shear Layer Tone Phenomenon and Probe Interference," *J. Fluid Mech.* **87**(2), pp. 349–384.

Lee, G.B., Ho, C.M., Jiang, F., Liu, C., Tsao, T., Tai, Y.C., and Scheuer, F. (1996) "Control of Roll Moment by MEMS," in *ASME MEMS*.

Lepicovsky, J., Ahuja, K.K., and Burrin, R.H. (1985) "Tone Excited Jets. Part III. Flow Measurements," *J. Sound Vibration* **102**(1), pp. 71–91.

Liu, C., Tai, Y.C., Huang, J.B., and Ho, C.M. (1994) "Surface-Micromachined Thermal Shear Stress Sensor," in *Application of Microfabrication to Fluid Mechanics,* Vol. 197, ASME, pp. 9–16.

Liu, C., Tsao, T., Tai, Y.C., Leu, J., Ho, C.M., Tang, W.L., and Miu, D. (1995) "Out-of-Plane Permanent Magnetic Actuators for Delta Wing Control," in *Proc. IEEE Micro Electro Mechanical Systems,* pp. 7–12, Amsterdam, Netherlands.

Miller, R., Burr, G., Tai, Y.C., Psaltis, D., Ho, C.M., and Katti, R. (1996) "Electromagnetic MEMS Scanning Mirrors for Holographic Data Storage," in *Technical Digest, Solid-State Sensor and Actuator Workshop,* pp. 183–186.

Moin, P., and Bewley, T. (1994) "Feedback Control of Turbulence," *Appl. Mech. Rev.* **47**(6, part 2), pp. S3–S13.

Moore, C.J. (1977) "The Role of Shear-Layer Instability Waves in Jet Exhaust Noise," *J. Fluid Mech.* **80**(2), pp. 321–367.

Naguib, A., Christophorou, C., Alnajjar, E., and Nagib, H. (1997) "Arrays of MEMS-Based Actuators for Control of Supersonic Jet Screech," in *AIAA Summer Fluid Mechanics Meeting*, AIAA Paper No. 97-1963, June 29–July 2, Snowmass, CO.

Naguib, A., Soupos, E., Nagib, H., Huang, C., and Najafi, K. (1999a) "A Piezoresistive MEMS Sensor for Acoustic Noise Measurements," in *5th AIAA/CEAS Aeroacoustics Conf.*, AIAA Paper No. 99-1992, May 10–12, Bellevue, WA.

Naguib, A., Benson, D., Nagib, H., Huang, C., and Najafi, K. (1999b) "Assessment of New MEMS-Based Hot Wires," in *Proc. 3rd ASME/JSME Joint Fluids Engineering Conf.*, July 18–22, San Francisco, CA.

Padmanabhan, A., Goldberg, H.D., Breuer, K.S., and Schmidt, M.A. (1996) "A Wafer-Bonded Floating-Element Shear-Stress Microsensor with Optical Position Sensing by Photodiodes," *J. MEMS* **5**(4), pp. 307–315.

Powel, A. (1953) "On the Mechanism of Choked Jet Noise," *Proc. Phys. Soc. London* **66**(12, no. 408B), pp. 1039–1056.

Reshotko, E., Pan, T., Hyman, D., and Mehregany, M. (1996) "Characterization of Microfabricated Shear Stress Sensors," in *Eighth Beer-Sheva International Seminar on MHD Flows and Turbulence*, Jerusalem, Israel.

Salhi, S., and Aslam, D.M. (1998) "Ultra-High Sensitivity Intra-Grain Poly-Diamond Piezoresistors," *Sensors and Actuators A* **71**(3), pp. 193–197.

Tam, C.K.W. (1986) "Excitation of Instability Waves by Sound—A Physical Interpretation," *J. Sound Vibration* **105**(1), pp. 169–172.

Tsao, T., Liu, C., Tai, Y.C., and Ho, C. M. (1994) "Micromachined Magnetic Actuator for Active Fluid Control," in *Application of Microfabrication to Fluid Mechanics*, Vol. 197, ASME.

Tsao, F., Jiang, R., Miller, A., Tai, Y.C., Gupta, B., Goodman, R., Tung, S., and Ho, C.M. (1997) "An Integrated MEMS System for Turbulent Boundary Layer Control," *Technical Digest* (*Transducers '97*) **1**, pp. 315–318.

Yang, X., Tai, Y.C., and Ho, C.M. (1997) "Micro Bellow Actuators," in *Technical Digest* (*Transducers '97*) **1**, pp. 45–48.

36
Fabrication Technologies for Nanoelectromechanical Systems

Gary H. Bernstein
University of Notre Dame

Holly V. Goodson
University of Notre Dame

Gregory L. Snider
University of Notre Dame

36.1 Introduction 36-1
36.2 NEMS-Compatible Processing Techniques 36-3
 Electron Beam Lithography • X-Ray Lithography • Other Parallel Nanoprinting Techniques • Achieving Atomic Resolution
36.3 Fabrication of Nanomachines: The Interface with Biology 36-16
 Inspiration from Biology • Practical Fabrication of Biological Nanotechnology
36.4 Summary 36-20
 Acknowledgments 36-20

36.1 Introduction

As discussed in previous chapters of this volume, microelectromechanical systems (MEMS) are typically constructed on the micrometer scale, with some thin layers being perhaps in the nanometer range. As has already been demonstrated by microelectronic circuits, the lateral dimensions of MEMS are being pushed into the nanometer range as well. This advance has been dubbed "nanoelectromechanical systems," or NEMS. The ultimate utility of nanomachining (that is to say, the application of capable robots on the molecular scale to solving a range of problems) is limitless. Such a regime will likely be attainable only by the "bottom-up" approach in which atoms are individually manipulated to construct macromolecules or molecular machines. Properties of pure molecules, such as heat conduction, electrical conduction (low power dissipation), speed of performance and strength, without the limits of boundaries to other molecules and resulting materials defects, vastly exceed those of bulk materials.

Drexler wrote about molecular machinery that could be modeled after the ultimate existing nanoelectromechanical system, the biological cell [Drexler, 1981]. He discussed analogs within the cell for such mechanical devices as cables, solenoids, drive shafts, bearings, etc. Proteins exhibit a remarkable range of functionality, and compared with current MEMS technology, are extremely small. Reasoning that proteins are ideal models upon which to design nanomachines, Drexler envisioned the development of machinery that would allow us to artificially produce such nanoscale mechanical components as those listed above. Imagining complex machinery operating at the molecular scale gives rise to images of nanorobots repairing our organs [Langreth, 1993], smart materials that intelligently conform to our bodies by adjusting a vast number of rigid nanoplates and incorporating robots that scour our clothing for debris and keep them clean [Forrest, 1995] and ones that build yet more complex nanomachinary. The list of possibilities is endless.

However, at this time, such intriguing visions still reside in the realm of science fiction. If the creation of complex nanosystems of any kind is to be accomplished, it will likely happen soonest by manipulation of biological systems to perform feats of engineering that exploit their existing programming, or by chemical synthesis of ever more complex molecules in a beaker. Although the field of molecular electronics is progressing rapidly, molecules are currently combined with relatively large and cumbersome connections in order to study their properties [Reed, 1999], and to our knowledge, no molecular mechanical system has yet been synthesized from its basic atomic constituents. The future is long, though, and progress is unpredictable and often rapid. The advent of scanning probe microscopes, including scanning tunneling and atomic force microscopes, was a turning point in the advancement of nanoscience and technology, allowing us to manipulate single atoms [Binnig and Rohrer, 1987]. Even this great advancement is trivial compared with the requirements of assembling thousands of atoms in complex shapes, powering them, observing them, and controlling them. The problems posed in reaching this capability are staggering.

For the purposes of this chapter, we will define NEMS as the extension of current MEMS technology. However, this chapter includes a discussion of atomic-scale lithography and concludes with a discussion of the synthesis of nanoscale systems by purely biological and molecular synthetic means. Although these latter techniques are in their infancy, it is likely that in the nearer term, using more conventional fabrication techniques, we will be able to exploit new functionality possible only on the nanoscale. These might include frictional, thermal or viscosity properties of materials; interaction with electromagnetic radiation at very small wavelengths; or ultrahigh-frequency resonances. These regimes will be approachable by the more conventional "top-down" approach of defining small areas by high-resolution lithography and pattern transfer. In this chapter, we will discuss mainly top-down approaches, except for that of atomic holography, and scanning probes based on scanning tunneling or atomic force microscopes, which can be used in either approach. In top-down patterning, several techniques for achieving nanometer resolution of arbitrary shapes have been developed. These include mainly electron beam lithography (EBL), ion beam lithography (IBL) and X-ray lithography (XRL). Here we will discuss electron beam lithography and X-ray lithography, as due to expense, complexity and somewhat lower resolution than EBL, IBL is not as commonly used. We will also discuss two related techniques for achieving a parallel process for high throughput and small features, namely microcontact printing and nanoimprint lithography.

Using these processing techniques, it is certainly possible to machine silicon and related materials in the nanometer regime. Namatsu et al. (1995) used EBL and reactive ion etching (RIE), coupled with etching in KOH/propanol solutions, to create 100-nm vertical walls with a thickness of 6 to 9 nm. Using nanomachined tips (fabricated in a manner similar to contamination resist lithography, discussed in this chapter), Irmer et al. (1998) used a scanning probe "ploughing" technique to modify thin Al layers, resulting in 100×100 nm^2 Josephson junctions. Others have sought to exploit very high frequency resonances in ultrasmall beams. Erbe et al. (1998) used electromechanical resonators with dimensions of about 150 nm to control physically adjustable tunneling contacts at frequencies up to 73 MHz at room temperature. In a comprehensive review of nanoelectromechanical systems, Roukes (2000) discusses the limits of mechanical resonators made of various semiconducting materials. He reports that at the 10- to 100-nm scale, SiC resonators should oscillate in the 10-GHz range, and at molecular scales nanodevices should exhibit resonant frequencies in the THz range.

Finally, we recognize that biological systems are the ultimate nanomachines, with internal operation of a single cell being more complicated than any human-made factory. In each cell, tens of thousands of different kinds of proteins and enzymes go about their complex tasks in a highly elaborate interplay of activity. We conclude our chapter with a discussion of how some cellular activities are being harnessed as future nanomachines.

36.2 NEMS-Compatible Processing Techniques

36.2.1 Electron Beam Lithography

Electron beam lithography is a technique in which the energy imparted by a directed stream of electrons modifies a film on the surface of a substrate. In general, EBL is utilized for direct writing of patterns on semiconductor substrates, as well as defining mask patterns for optical and X-ray lithographies. In direct-write mode, a key issue in manufacturing is throughput, a characteristic for which EBL has had difficulty competing with other technologies such as optical projection lithography. Variations of EBL include exposures by Gaussian distributed beams, shaped beams in which the pattern of energy deposition is shaped by apertures, and a promising technique for manufacturing called "scattering with angular limitation in projection electron beam lithography," or SCALPEL [Berger et al., 1991]. SCALPEL works by the imaging on a substrate of a pattern of mask features through which electrons are either transmitted without scattering or are scattered at obstructions on the mask. An aperture in the optical column passes preferentially those electrons that are not scattered so the pattern of unscattered electrons is printed on the wafer. Although shaped beams offer improved throughput compared with Gaussian beams, SCALPEL is superior in this regard and, among the electron beam technologies, offers the most promise for manufacturing. For an excellent discussion of electron beam systems, see McCord and Rook (1997). In this chapter we are concerned primarily with resolution, as opposed to other aspects of more interest to mass production of patterns. Although these issues cannot be ignored for a technology to be ultimately viable, we discuss those issues that limit our ability to create nanomachines at the limits of physical possibility.

It is fortuitous that the reliance of nano-EBL on very narrow electron beams is the same as that required of high-resolution scanning electron microscopy (SEM), which now is a mature field. It is common for researchers to modify SEMs [Bazán and Bernstein, 1993],[1] to control the position of the beam, and, in conjunction with a beam blanker that modulates the beam current, obtain very high resolution patterning. Most commercial EBL systems are optimized for such issues as throughput and placement accuracy and make it less of a priority to achieve resolution in the nanometer range.

36.2.1.1 Resolution Limits

It is useful to understand the dependence of the ultimate size of patterning by EBL on all of the various steps involved. These include creation of the narrowest and brightest beam, choice of electron beam resist, forward- and backscattering of electrons in the resist and substrate, beam–resist interactions, resist development and, finally, pattern transfer to the substrate. Each of these issues will be addressed in turn.

36.2.1.1.1 Beam Size

The formation of a narrow, bright beam is critical to the formation of the smallest possible patterns. In all scanning electron optical systems, the column demagnifies an image of the electron source and projects this demagnified image onto the substrate. Electron lenses use magnetic fields created by current in coils to produce an effect of focusing much like that of optical lenses. The central function of the column is to demagnify the first crossover point at the source to a small, intense spot in the plane of the wafer. Because the diameter of the beam at the target is some fraction of the diameter of the source, starting with a small, bright source is of central importance. The five types of electron emitters commonly used in electron beam systems can be broadly classified as either thermionic emitters or field emitters. In order of decreasing source size, these are (a) tungsten hairpin filament, or thermionic; (b) lanthanum hexaboride, LaB_6, and the related CeB_6; (c) Schottky emitter; (d) thermal field emitter; and (e) cold cathode field emitter.

---

[1] EBL systems for SEMs are available commercially from Raith USA, Inc., 6 Beech Road, Islip, NY 11751-4907; and J.C. Nabity Lithography Systems, P.O. Box 5354, Bozeman, MT 59717.

TABLE 36.1 Properties of Various Cathodes Used for Electron Beam Lithography

Type	Vacuum (Pa)	Work Function (eV)	Cathode Temp. (K)	ΔE (eV)	Source Diameter or Tip Radius (μm)	Emission Current Density (A/cm^2)	Brightness (A/cm^2/Sr)	Probe Diameter (nm)	Lifetime (hours)
W	10^{-3}	4.6	2700	2	50	2	10^5	4	50
LaB$_6$	10^{-5}	2.7	1800	1	10	40	10^6	3	2000
ZrO/W Schottky	10^{-7}	2.8	1800	0.8	1	500	10^{10}	2	Several thousand
CCFE	10^{-8}	4.6	300	0.3	0.1	10^5	10^8	1	Several thousand

Table 36.1 lists the important properties of each emitter, except for the thermal field emitter. Note that these are representative values only and are highly dependent on manufacturers' specifications, beam energy, column design, etc. Tungsten hairpin filaments are historically the most common and easiest sources to use. These emitters operate by passing current through the filament to raise its temperature to the vicinity of 2700 K, so that electrons can overcome the work function and enter the electron optical column. They require the least stringent vacuum conditions (approximately 10^{-3} Pa), so the overall systems tend to be less expensive. Most importantly, the effective diameter of the emission spot at the crossover below the filament is about 50 μm, resulting in a typical spot size of about 3 to 5 nm. Also important is the brightness of the source, expressed in units of Amperes/cm^2/steradian. As spot size is reduced through changes in column conditions, a large fraction of electrons is lost by scattering outside of the optical path. In a typical tungsten filament system, 100 μA of current from the source translates to perhaps 10 pA at the sample in obtaining the smallest spot size. This makes imaging at the highest resolution more difficult, so in addition to increased throughput at a given spot size a practical benefit of increased brightness is the ability to form a crisp image during focusing, thereby forming the narrowest possible spot. This translates to better spot size control and ultimately higher resolution patterning.

Besides suffering from a large size at the source for tungsten filaments, the thermal energy spread of the beam is about 4.5 eV, which can contribute to an increase in the beam diameter. Although this is a small fraction of the accelerating voltage (typically 30 to 100 kV), it is enough to create noticeable chromatic aberrations, so that focusing of the electrons at a single spot is hampered by the differing focal lengths for electrons with different energies. Because the minimum spot size for each energy is achieved at different focal lengths, the overall minimum spot at any one distance (i.e., at the surface of the substrate) is enlarged.

Closely related to the tungsten hairpin filament is the LaB$_6$, lanthanum hexaboride emitter, and related CeB$_6$, cerium hexaboride. These materials are often referred to as lab-6 and cebix, respectively. This emitter is a sharpened piece of single crystal that is heated either by passing current directly through it or heating it by a separate heater, to about 1800 K. The tip is machined for some optimum shape and produces an effective source diameter of typically 10 μm. Due to its lower work function of 2.7 eV and attending lower energy spread of about 1 eV, it is brighter and produces less chromatic aberration. CeB$_6$ offers a slightly lower work function and greater robustness but is not as commonly used. In order to protect the tip from "poisoning," a better vacuum, typically 10^{-5} Pa, is required of a LaB$_6$ system. In most cases, given the quality of vacuum system, tungsten hairpin and LaB$_6$ sources can be used interchangeably. One important advantage is an operating lifetime for LaB$_6$ of up to 2000 hours vs. less than 100 hours for thermionic tungsten.

A significant improvement in resolution and brightness over the previous sources is provided by the Schottky emitter. This source consists of a ZrO-coated W tip sharpened to about 0.5 μm. The reduced work function (from 4.6 eV in W to 2.8 eV) and enhanced electric field, which lowers the barrier (Schottky effect), allow the electrons to be extracted at a temperature of about 1800 K. The advantages of this source are high stability and brightness combined with much higher resolution (smaller spot size) than the thermionic sources. Spot sizes of about 3 nm or better are achievable in these systems. Schottky emitters are becoming the most common in advanced EBL systems for their combination of beam stability, long lifetime (a few thousand hours), high brightness and small spot size.

The highest resolution beams for scanning electron microscopy are produced using field emission sources. These consist of a very sharp, single-crystal tungsten tip ($r = 0.01$ to 0.1 μm) operated under a high electric field. One variation, called thermal field emission (TFE), is to heat the tip; another, called cold cathode field emission (CCFE), is to run the tip at room temperature. The TFE emitter has not commonly been used since the advent of the Schottky emitter. In the CCFE case, a high extraction field renders the potential barrier so thin (about 10 nm) that electrons can tunnel through even at relatively low temperatures. Because of this, thermal energy is reduced to about 0.2 eV, and chromatic aberrations are significantly reduced. Because of the reduced area of emission and reduced chromatic aberrations, probe sizes as small as about 0.5 to 1 nm can be produced. CCFE sources are not commonly used in EBL applications because the beam is inherently unstable due to adsorption of gas layers that affect the emission properties. Heating the tip, as in the case of TFE cathodes, solves this problem. In CCFE SEMs, this instability is dealt with by feeding back a current signal instantaneously to the video monitor to produce a clean image with little or no evidence of current fluctuations. However, for purposes of beam writing, this can lead to fluctuations in instantaneous dose at the wafer and therefore to poor linewidth control. In spite of this, Dial et al. (1998) have reported excellent lithographic results using a CCFE SEM by building feedback into their exposure rates. Their system has demonstrated a regular array of dots with diameters of 10 nm on a pitch of 25 nm in polymethylmethacrylate (PMMA) at 30 keV.

36.2.1.1.2 ·Electron Scattering

Formation of a high-quality beam is only the first step in pattern formation. For very narrow beams, a larger contribution to pattern size is scattering of electrons in the resist and in the substrate. When an electron beam impinges on a resist layer, the electrons scatter through both elastic and inelastic processes. Elastic scattering results in backscattering of the electrons at energies close to the primary beam and in negligible energy transfer to the resist and substrate. Inelastic scattering leads to spreading of the primary beam and generation of low-energy secondary electrons that, in turn, expose the resist. In PMMA, a positive resist, the mechanism of exposure is the scission of bonds in the high-molecular-weight polymer to create regions of lower molecular weight. The lower molecular weight is more soluble in suitable developer solutions, so the resist is removed where exposed. In negative resist, energy deposition results in cross-linking of the polymer, and the resist is rendered less soluble and so remains behind after development and removal of unexposed resist.

The spread of the primary beam in the resist layer can be described by:

$$b = 625 \frac{Z}{E_0} t^{3/2} \left(\frac{\rho}{A}\right)^{1/2} \text{ cm} \qquad (36.1)$$

where b is the spread of the beam at a distance t into the resist; Z, A and ρ are the atomic number, atomic weight (g/mol) and density (g/cm^3), respectively; and E_0 is the beam energy (keV). For PMMA (($C_5O_2H_8)_n$), $Z = 3.6$, $A = 6.7$, and ρ is 1.2. For a delta function beam (i.e., an ideal beam with zero width) of energy 40 keV, the beam spreads out by 7.5 nm after 100 nm. This leads to a slight undercut profile that helps in lifting off metal patterns, so it is not necessarily undesirable. However, it does suggest that thin resists are preferable for achieving the highest resolution and pattern density. In thicker resists, it is often the case that the width of a lifted-off metal line is wider than the opening at the top surface, as metal can be deposited over the entire width of the line at the wafer surface; therefore, it can sometimes be deceiving to assume that the width of achievable resolution is that observed at the resist surface.

The Bethe retardation law,

$$\frac{dE}{dx} = -7.85 \times 10^4 \left(\frac{Z\rho}{AE_m}\right) \ln\left(\frac{1.66E}{J}\right) \frac{\text{keV}}{\text{cm}} \qquad (36.2)$$

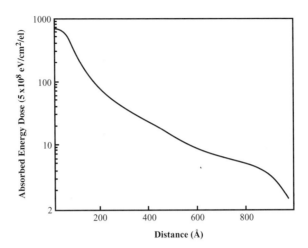

FIGURE 36.1 Absorbed energy due to exposure by a 2-nm beam of 100-keV electrons in 100 nm of PMMA. Energy deposition is primarily due to secondary electrons, and minimum feature size is limited to approximately 10 nm. (Adapted from Joy, 1983.)

(where dE/dX is the energy loss of the beam per unit distance in the solid; E_m is the mean electron energy along the path; Z, A and ρ are as defined above; and J is the mean ionization potential (keV)), tells us that low-energy electrons are more efficient than high-energy electrons at transferring energy to the resist and are in fact responsible for breaking bonds in PMMA. Also, although secondary electrons are generated at energies up to a significant fraction of the beam energy, those at lower energies are again much more effective at exposing resist.

Using the Monte Carlo approach, Joy (1983) has investigated the minimum possible features achievable in PMMA. Because low-energy secondary electrons lose energy more readily than high-energy electrons, they are the dominant agents for transferring energy to the resist. Figure 36.1 shows the absorbed energy due to exposure by a 2-nm beam of 100 keV electrons in 100 nm of PMMA. For finite contrast of the developer/resist system, the plateau of energy absorption is developed uniformly, and a spatial resolution of approximately 10 nm results. Smaller lines can result with high-contrast developers and very careful processing to take advantage of slight differences in absorbed energy within the plateau region. Chen and Ahmed (1993a) demonstrated linewidths approaching 5 nm with ultrasonic developers and reactive ion etching for pattern transfer. These narrow linewidths are not necessarily developed all the way through the resist, which is necessary to allow pattern transfer to the substrate by liftoff. Chen and Ahmed demonstrated lifted-off lines that were slightly less than 10 nm [Chen and Ahmed, 1993b], in good agreement with Joy's results.

The interaction volume of primary electrons in a substrate can be described by the Kanaya–Okayama (K–O) range [Goldstein, 1981], given by:

$$R_{KO} = 0.0276 \frac{A E_0^{1.67}}{Z^{0.889} \rho} \, \mu m \qquad (36.3)$$

where A, E_0, Z and ρ are as defined above. Electrons that are backscattered from the substrate are usually modeled as being Gaussian distributed with a characteristic width of about half of the K–O range. The effect of backscattered electrons is to expose the resist at some (usually) undesirable background level that depends on the surrounding patterns. This phenomenon is referred to as the "proximity effect" and much effort has gone into techniques for modifying pattern files to compensate exposure doses. The effect of backscattered electrons on the final resolution of an isolated pattern is less pronounced but can adversely affect the minimum resolution obtainable in dense patterns. Figure 36.2 shows the effect on

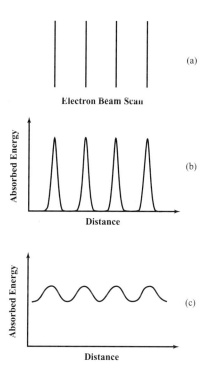

FIGURE 36.2 Absorbed energy in resist due to proximity effect. (a) Line scans, (b) energy deposition without proximity effect, and (c) total energy deposition including proximity effect. As the space between the line scans absorbs more energy, it is more difficult for the developer to resolve the patterns, and linewidths increase. (Adapted from Huang et al., 1993.)

absorbed energy of exposing lines closer together. Note that the spaces between the lines absorb more energy, and the difference in effective exposure dose between the intended patterns and spaces is reduced. This makes the job of developing only the lines a more difficult one for the developer to accomplish, as explained in more detail below.

A common technique for reducing the effects of backscattered electrons is to expose patterns on thin membranes (typically silicon nitride) formed by patterning the backside of a silicon wafer and etching through to expose the membrane over a usable area. Although not very practical for making electronic devices, this technique has been used to demonstrate very high resolution lithography and to allow direct imaging of the results in a transmission electron microscope.

36.2.1.1.3 Conventional Resists and Developers

The quality of a developer for a particular resist is defined in terms of the sensitivity and contrast. A typical dose/exposure plot showing normalized thickness of the resist after development for positive resist is shown in Figure 36.3. (A curve for negative resist is zero at low doses and maximum at high doses.) The threshold dose for positive resist can be defined as that which results in clearing of the resist in the exposed areas. It is generally desirable for the threshold dose to be as small as possible for purposes of throughput, but if it is too small, then statistical effects due to the number of electrons required to expose a single pixel can become important. In general, resists that exhibit highest resolution are also the slowest (least sensitive).

The contrast of a resist (using a particular developer under a particular set of conditions) is defined as:

$$\gamma = \log\left(\frac{D_i}{D_f}\right)^{-1} \qquad (36.4)$$

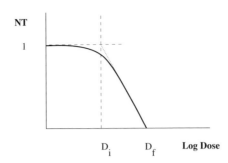

FIGURE 36.3 Dose/exposure plot of normalized positive resist thickness after development.

where (for positive resist) D_i is the dose at the steepest part of the contrast curve extrapolated to full thickness, and D_f is that dose extrapolated to zero thickness. The role of contrast is central to obtaining the finest pattern dimensions. As discussed above, each exposure results in some spatially varying energy deposition. Contrast (for positive resist) is the ability of the developer to selectively remove resist that has absorbed more than a critical value and ignore that resist that has absorbed less. For an infinite contrast, the contrast curve would be vertical; resist above a sharp value of absorbed energy would be perfectly developed and below that value would be totally unaffected. In this case, we could arbitrarily approach the peak in Figure 36.1 and obtain patterns equal to the beam width. In addition, we could make the lines arbitrarily close together, as the lines could be perfectly selectively removed. In reality, the slope of a real dose/exposure curve is finite, so a range of absorbed energy contours is removed, and it is difficult to achieve much less than the 10 nm discussed by Joy.

Typical unexposed, high-resolution PMMA has a molecular weight of about 10^6 Da, which is reduced to a few thousand Da after exposure [Harris, 1973]. The role of the developer is to selectively remove the lower molecular mass polymer components that result from chain scission of larger ones. The two most common developers of PMMA are methyl isobutyl ketone (MIBK) mixed with IPA in a ratio of 1:3, and 2-ethoxyethanol (Cellosolve) mixed with MeOH in a ratio of 3:7. The γ values provided by these mixtures are about −6 to −9 [Bernstein et al., 1992]. Bernstein et al. reported that the addition of about 1% methyl ethyl ketone (MEK) increases the contrast of these developers to greater than −12. This is ascribed to enhanced removal of higher molecular weight components that are not sufficiently broken at lower doses, without significantly affecting the solubility of the low-molecular-weight components.

As mentioned above, Chen and Ahmed (1993a) found that developing with ultrasonic agitation offers benefits to resolution and using this technique demonstrated 5- to 7-nm lines. They assert that, for exposures smaller than 10 nm, the resulting exposed PMMA fragments are trapped in a potential well created by the adjoining unexposed molecules. By exposing lines wider than 10 nm, the developer is able to dissolve the exposed resist but cannot remove them in the narrower lines. In the case of narrower lines, ultrasonic agitation can provide additional energy by which the exposed fragments can escape this potential well in the presence of the developer. (They used the Cellosolve solution mentioned above.) Although lift-off was not demonstrated, the patterns were transferred to a Si substrate by reactive ion etching. Even smaller features have been demonstrated by allowing regions of exposure by secondary electrons to overlap. Using this technique, Cumming et al. (1996) have created metal lines by lift-off as small as 3 nm. These are perhaps the smallest features ever created by exposure of PMMA.

Finally, upon exposure to very high doses (i.e., greater than about 10^{-3} C/cm^2), PMMA will cross-link to act as a negative resist. Tada and Kanayama (1995) report that with careful control, highly dosed PMMA developed in acetone can yield features as small as 10 nm. As the dose is increased, resistance to dry etching increases as well, to about twice that of unexposed PMMA. Using an electron cyclotron resonance (ECR) etcher, they created Si pillars with a width of 10 nm and a height of 95 nm.

The process of nanomachining can benefit from the use of high-resolution negative resists as well. One of the highest resolution (conventional) negative resists is Shipley's SAL 601-ER7, a member of the

Shipley ANR series of resists. It has been demonstrated to provide features as small as 50 nm [Bernstein et al., 1988]. This resist is one example of a chemically amplified resist, consisting of three components: an acid generator, cross-linker and resin matrix. Exposure by the electron beam generates an acid that, upon heating, catalyzes the cross-linking of the linker and the resin matrix, leading to very high sensitivity. Typical threshold exposure doses are about an order of magnitude less than for PMMA. Other negative resists that exhibit nearly similar resolution include IBM APEX-E and Shipley UVIIHS.

Because many deep ultraviolet optical resists are also sensitive to electron beams, a plethora of choices exists for resists that maximize sensitivity, resolution, dry-etch resistance or other properties. These include PBS (Mead), RE-4200N (Hitachi), APEX-E and KRS (IBM), UVIIHS and UV-5 (Shipley), AZ-PF514 and AZ-PN114 (Hoechst Celanese) and ZEP-520 (Nippon Zeon). Several of these are capable of achieving sub-100-nm resolution.

36.2.1.1.4 Unconventional Resists

Many other alternative resist schemes for defining patterns with an electron beam have been investigated. One common technique is the condensation of carbon compounds from the vapors present in electron beam vacuum systems. This is referred to as "contamination resist" lithography because it utilizes the same mechanism that often causes contamination and degradation of imaging quality in electron microscopes. As early as 1964, Broers (1964), using a 10-nm electron beam, obtained a resolution of 50 nm after ion milling a contamination resist mask. It is generally accepted that the mechanism of deposition is the cracking by secondary electrons of adsorbed hydrocarbons at the sample surface. As with PMMA, the resolution of this technique benefits by the use of membranes for supporting the patterns. Contamination resist cannot be used in a lift-off mode, but it does perform reasonably well as an etch mask. Broers (1995) later demonstrated 5-nm Au–Pd lines created by deposition on a membrane followed by ion milling.

In an effort to merge with molecular electronics, self-assembled monolayers (SAMs) have more recently been investigated as electron beam resists. Because these layers are only a few nanometers thick, it is thought that the resolution achieved in this system will ultimately be higher than with PMMA. Also, very low energies can be used to expose thin layers, which can be beneficial to backscattering issues and also allow the use of exposure by low-energy scanning probe tips. SAMs consist of an ordered layer of single molecules adsorbed to a surface. Examples include *n*-octadecyltrichlorosilane (OTS), *n*-octadecanethiolate (ODT), [Lercel et al., 1993], octadecylsiloxanes (ODS) [Lercel et al., 1996], and dodecanethiol (DDT). Their value lies in the presence of alkyl groups at the outer surface which give them hydrophobic properties and resistance to wet etching. Exposure by an electron beam (or ultraviolet radiation) removes the alkyl group, rendering them hydrophilic so exposed areas can be etched. They are, therefore, positive tone in nature. Using ODS SAMs, Lercel et al. exposed individual dots with diameters of approximately 6 nm. These are roughly the size of individual molecules that could be anchored for NEMS systems.

Another technique employing an electron beam for lithography is that of direct sublimation of material. This is often referred to as simply "drilling" through a layer of inorganic material. This is quite literally a nanomachining process. Materials that are amenable to drilling include Al_2O_3, LiF, NaCl, CaF_2, MgO and AlF_3. Mochel et al. (1983) has demonstrated 1-nm holes drilled in Al_2O_3, which is perhaps the highest resolution ever demonstrated with a high-energy beam of electrons. In a related technique, Hiroshima et al. (1995) created lines on a 15-nm pitch by exposure of SiO_2. Development was performed in a buffered HF solution. Such a technique might have applications in nanomachining.

36.2.2 X-Ray Lithography

X-ray lithography (XRL) is basically an extension of optical proximity printing using much higher energy X-ray photons, with wavelengths in the range of about 0.1 to 10 nm. It was first described by Spears and Smith (1972). For MEMS technology, it has been extended to the LIGA process for creating microscale machined pieces such as gears and other mechanical parts. (This is described in detail in Chapter 17 of this text.) Because the use of X-rays is so common for micromachining applications, it is worthwhile to

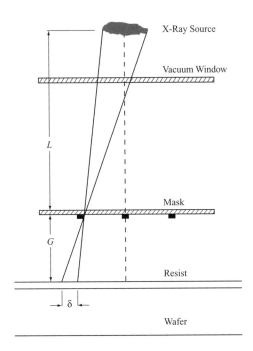

FIGURE 36.4 Schematic diagram of basic X-ray lithography exposure configuration. δ is the amount of penumbral blurring due to a finite-sized source.

investigate the limits of resolution for nanomachined systems. An excellent review of XRL is provided by Cerrina and Rai-Choudhury (1997).

Figure 36.4 shows schematically a basic X-ray exposure configuration. The main issue in achieving workable X-ray systems in manufacturing is the development of suitable X-ray sources, masks and alignment techniques. The problems involved in developing bright, collimated X-ray sources are, in general, shared with the LIGA community but are even more pronounced for nanolithography. For example, the geometrical considerations of a finite-sized source, such as penumbral blurring and runout produced by an X-ray tube or plasma source, would be more severe (as a fraction of their total size) for nanometer-scale features. Because synchrotron sources produce nearly parallel beams, these would be most advantageous for printing ultrasmall patterns. In the late 1980s, IBM installed a minisynchrotron source [Andrews and Wilson, 1989] capable of servicing more than a dozen steppers. The cost of such an X-ray source today would be on the order of about $20M [Cerrina, 2000] and might compete economically with advanced optical steppers.

For proximity printing in general, resolution is limited by diffraction effects as the minimum linewidth is approximately:

$$W = k\sqrt{G\lambda} \qquad (36.5)$$

where k is a process-dependent parameter in the range of 0.6 to 1 [Cerrina, 1996], G is the gap, and λ is the wavelength. The k parameter depends on the processing conditions, including type of mask, resist and other conditions. With the use of phase-shift mask technology, 40-nm features at 20-μm gaps have been achieved [Chen et al., 1997].

With conventional masks, for a λ of 1 Å and a relatively narrow gap of 5 μm, we can expect to achieve feature sizes of a few tens of nanometers. This is well within the useful range of nanomachining. For even smaller gaps, better results can be achieved. However, there are other potential limitations to the maximum resolution. Feldman and Sun (1992) and Chlebek et al. (1987) have discussed several potential contributions to degradation of resolution, shown in Figure 36.5. First, photoelectrons and Auger

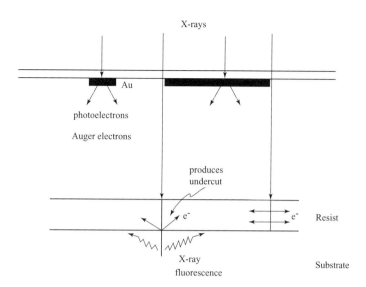

FIGURE 36.5 Contributions by scattered energy deposition in resist for X-ray lithography, including photoelectrons and Auger electrons from the mask and photoelectrons from the resist and substrate.

electrons that are energetic enough to expose the resist are released from the mask. Second, photons that pass along the edge of the absorber create photoelectrons in the resist that expose regions at the edge and under the absorber shadow. Third, strong photoabsorption at the substrate creates additional photoelectrons that can lead to undercutting of the resist. According to Feldman and Sun, the high-energy photoelectrons do not contribute significantly to the limits to resolution, and according to Chlebek et al. neither do fluorescence photons.

It appears, then, that the limits to X-ray lithography resolution in PMMA are nearly identical to those determined by Joy (1983) as discussed above, with the source of energy, either X-ray photons or electrons, not being a significant factor. In fact, Flanders (1980) performed experiments demonstrating very high resolution XRL of PMMA. The masks used in this experiment were nanomachined (partially employing angled shadow evaporation) to have alternating, vertical, 100-nm-thick bands of 17.5-nm-wide tungsten and carbon layers. As seen vertically by the X-ray source, this resulted in a mask of alternating regions of differing transmission values. In fact, successful contact printing of this pattern demonstrated that resolution of XRL in PMMA is comparable to that of high-resolution EBL. Using a similar technique, Early et al. (1990) demonstrated 30-nm lines independent of wavelengths ranging from 0.83 to 4.5 nm. The limit to useful printing scales in practical XRL systems will depend on other factors, such as patterning accuracy and mask distortions. Even including all of these effects, XRL will likely be an important technology for future NEMS applications.

36.2.3 Other Parallel Nanoprinting Techniques

Two other printing methods deserve mention due to their simplicity, potential high-resolution capability and potentially high throughput, which will be necessary for a commercially feasible technology. These are "microcontact printing," developed by Whitesides, and "nanoimprint lithography," developed by Chou. Both of these techniques function by physical contact between a patterned stamp and a substrate.

In the first case, the stamp is a pliable, elastomeric material, namely polydimethylsiloxane (PDMS), that is coated with a SAM (typically one of those mentioned above) and then pressed onto a substrate, resulting in printing of a monolayer of SAM molecules. The stamps are formed by first preparing a master pattern etched into a hard substrate made typically of glass, resists, polymers or silicon. The master is made using standard photolithography and etching techniques. The elastomer is poured onto the master, cured, peeled off and then treated with an appropriate SAM "ink." Finally, the stamp is pressed directly onto the target material by hand or with some other simple pressing system. Using this technique, remarkably high-quality patterns and processing applications are possible. St. John and Craighead (1995) have demonstrated the use of OTS as ink, followed by plasma etching to demonstrate 5-μm patterns in silicon. Kumar and Whitesides (1993) used a variety of thiol-based SAMs to pattern Au layers that were used for further processing of substrates. They demonstrated features as small as 200 nm using hexadecanethiolate as the ink, and wet etching of underlying gold layers using KCN solutions. Additional techniques demonstrated were selective area plating of nickel and selective area condensation of water vapor in hydrophilic regions for use as optical diffraction gratings.

The second stamping technique utilizes a rigid stamp in contact with a viscoelastic polymer, typically PMMA. The stamp is pressed into a resist layer that is heated above the glass transition temperature, T_g. Polymers such as PMMA are mixtures of amorphous and crystalline morphologies. T_g is an important parameter because it is the temperature at which the polymer chains begin to slip in their amorphous arrangement, and the material can flow easily when subjected to pressure. (Note that relaxation of the crystalline portion of the arrangement is referred to as "melting," which is not important to this discussion.) T_g for bulk, heterotactic PMMA is approximately 105°C. According to Kleideiter et al. (1999), T_g in general can vary depending on the thickness of the film and the substrate upon which it is cast. It is important for a successful resist used in nanoimprinting to have a low glass-transition temperature, low viscosity, low sticking forces to the stamp, high stability and high dry-etch resistance. Typical imprint temperatures are about 90°C greater than T_g.

Chou et al. (1997) have demonstrated 10-nm holes on a 40-nm pitch in PMMA. In order to ensure a clean surface after imprinting, the residual PMMA at the bottom of the patterns is etched away using an oxygen plasma. These patterns are then suitable for lift-off of metal patterns. By additional etching of the stamp features, 6-nm metal dots were lifted off. Schulz et al. (2000) have investigated alternative resists including thermoplastic polymers based on aromatic polymethacrylates and thermoset polymers based on multifunctional allylesters. These exhibit overall lower T_g (as low as 39°C in the case of the allylesters before imprinting), improved dry-etch resistance and imprint performance comparable to PMMA.

36.2.4 Achieving Atomic Resolution

The above lithographic techniques are not capable of achieving atomic resolution, but proximal, or scanning, probes offer a path for scaling fabrication down to atomic dimensions. The two basic types of scanned probes used for lithography include atomic force microscopes (AFMs) and scanning tunneling microscopes (STMs). With its proven atomic resolution, STM has been used to pattern substrates with the highest resolution reported to date, approximately 0.5 nm. AFM patterning has yet to attain this level of resolution, but the introduction of carbon nanotube tips promises to increase the resolution toward 5 nm.

36.2.4.1 STM Patterning

The most successful technique demonstrated for STM patterning has been the passivation and selective depassivation of silicon. In this method, pioneered by Lyding et al. (1997), a silicon surface is prepared and hydrogen passivated in an ultrahigh-vacuum (UHV) environment. Surface preparation is, of course, crucial to this process. The sample must be carefully cleaned before inserting it into the UHV chamber, which must achieve pressures on the order of 10^{-8} Pa to preserve the cleanliness of the surface. The sample is flash-heated to a temperature of 1200 to 1300°C for 1 minute, while maintaining 4×10^{-8}-Pa pressure, to drive off any surface contamination and oxides. At a temperature of 650°C, hydrogen is introduced into the chamber to bring the pressure to ~10^{-4} Pa. The hydrogen molecules are cracked using a tungsten filament, and hydrogen atoms attach to the dangling bonds on the silicon surface. The result for a (100) Si surface is a 2 × 1 reconstructed surface passivated by hydrogen.

To perform lithography, an STM tip is brought close to the surface (still in UHV), and a tip voltage of approximately −5 V relative to the sample is applied. Electrons streaming from the tip to the substrate transfer energy to the hydrogen atoms and break bonds between the hydrogen and silicon. Approximately 10^6 electrons are needed to remove each hydrogen atom. As the tip is scanned across the surface, lines

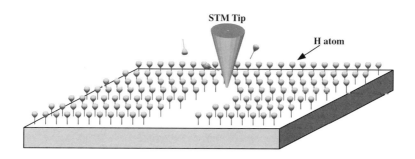

FIGURE 36.6 Depassivation of a silicon surface using an STM tip. Hydrogen atoms are knocked off the surface by electrons tunneling from tip to substrate.

of hydrogen atoms are removed, leaving depassivated Si atoms, as shown in Figure 36.6. Lyding et al. (1997) have demonstrated that, with proper control of the depassivation process, individual hydrogen atoms can be removed, yielding true atomic resolution.

The depassivated silicon atoms are far more chemically active than when bonded to hydrogen, and this difference can be used to transfer the pattern into the substrate by selective chemistry. Perhaps the simplest method is to dose the surface with O_2 molecules at 10^{-4} Pa. These molecules will react with the depassivated silicon to form a thin SiO_2 layer but leave the passivated silicon unchanged. While the oxide layers obtained by dosing with O_2 at low pressures are sub-monolayer, thicker (~4-nm) layers have been formed with oxygen at higher pressures, although with higher defect levels [Shen et al., 1995]. Thin nitride layers can be formed by similar treatments with NH_3. While these layers are very thin, layers of similar thickness have been used to transfer patterns into silicon [Snow et al., 1996].

The formation of nanoscale metal lines by STM is also being pursued. All future applications of nanofabrication, both electronic and nanomechanical, will require very small lines of metal to connect circuits and subsystems or to connect the nanosystem to the outside world. Work in this area using STM has been attempted but with limited success. Some of the techniques attempted include the field evaporation of gold from a gold tip, which yielded acceptable metal dots but gave only discontinuous lines [Mamin et al., 1990]. STM-assisted deposition of metals from organometallic precursors has been attempted, but the resulting lines are contaminated with impurities [Rubel et al., 1994]. Selective metal deposition has also been investigated. The most successful method uses thermal chemical vapor deposition (CVD) of metal on selectively depassivated substrates. Deposition can be performed at a temperature of ~400°C, which is low enough to preserve the hydrogen passivation. Proper choice of a precursor produces deposition on the depassivated areas, while the passivated areas are unaffected. Using trimethylphosphine-Au, a selective deposition was achieved on depassivated areas of Si.

While STM has demonstrated atomic-level patterning on silicon substrates, many challenges remain before it can be incorporated into the fabrication of NEMS and nanomachines. First, the patterns produced are very fragile. It will be a significant challenge to a pattern-transfer process to retain this resolution while producing the three-dimensional structures often needed for NEMS. There is also the issue of throughput, which applies to all scanned probe fabrication. Because the probe must form each feature sequentially, complex patterns will take a very long time to write. Work by Quate [Wilder et al., 1999] on multi-tip systems offers the possibility of high-throughput lithography systems. Scanned probe lithography can also be used to form the masks for other high-resolution, but more parallel, fabrication techniques such as the nanoimprint lithography discussed above.

36.2.4.2 AFM Patterning

Atomic force microscopes can also be used for ultra-high-resolution lithography. Approaches investigated range from the use of the AFM tip as scriber to locally activated oxidation of silicon or metals. AFM lithography in general has a much lower resolution than that of STM lithography. While STMs achieve

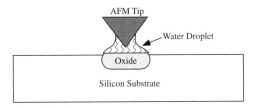

FIGURE 36.7 Anodic oxidation of a silicon surface using an AFM tip. A water droplet forms between the tip and substrate. The substrate is oxidized when a voltage is applied between tip and substrate.

atomic resolution by tunneling electrons out of only a tiny area of an already very sharp tip, AFM resolution is limited by the shape and size of the full tip. Indeed, a trade-off must be made between resolution and tip durability in many applications, especially those that require a mechanical modification of the surface by the tip. In scribing lithography, the tip is used to scratch the substrate surface. The substrate is typically covered with a soft material, such as PMMA, which can be displaced by the tip without excessive wear of the tip. A limitation of this technique is that the soft film is stressed and distorted as material is moved, and the displaced material has to go somewhere. Because this displaced material can intrude on nearby features, the packing density is limited. The resolution is strongly influenced by the size and shape of the tip. Work has been done using electron-beam-grown "super-tips," which are formed from contamination, as discussed previously, but these ultra-sharp tips are more prone to breakage and wear-induced degradation.

Anodic oxidation induced by an AFM tip is a technique that, while having a resolution lower than that of STM, produces patterns that are more durable. To perform the anodic oxidation of a silicon substrate, the substrate is first cleaned using standard silicon cleaning techniques and finishing with an HF dip and water rinse, leaving the surface hydrogen passivated. The sample is then placed in a controlled humidity atmosphere (humidity = 20–80%) at ambient pressure, and an AFM tip is brought to the surface. A meniscus of water forms between the tip and sample, as shown in Figure 36.7, and the surface oxidizes when a voltage of −2 to −10 V is applied to the tip. If no voltage is applied, the surface does not oxidize, so the surface can be imaged independently of the oxidation. The resulting oxide is approximately 1.5 nm thick [Snow et al., 2000]. While this is a very thin oxide, it is sufficiently thick for transfer into the underlying substrate. Both wet- and dry-etching techniques have been used. Wet etching is attractive because its lack of a physical sputtering component allows a very high selectivity with the proper choice of etchant. Snow et al. (1996) used a 1-nm oxide mask to etch through the 0.3-µm-thick top silicon layer of an SOI wafer using 70% hydrazine. Wet etching has limitations due to the isotropic or crystallographic nature of the etch which can degrade the resolution of the pattern transfer. Dry etching can give very anisotropic etch profiles, but the physical component of the etch reduces the selectivity of the etch relative to the mask, a serious concern when the mask is only 1 to 1.5 nm thick. For this reason, dry etching is best carried out in high-density plasma systems, such as electron cyclotron resonance (ECR) or inductively coupled plasma (ICP), where the physical component is separately controlled and can be kept low. ECR etching has been investigated as a pattern-transfer mechanism [Snow et al., 1995]. Silicon structures 10 nm wide and 30 nm deep were etched using a Cl_2/O_2 plasma. Oxidation of metallic films has also been demonstrated and may be useful for the fabrication of fine metallic lines and the electronic devices associated with a nanomechanical device. Ti, Al and Nb have been selectively oxidized in this fashion [Snow and Campbell, 1995; Schmidt et al., 1998; Matsumoto, 1999], and the technique has been used to fabricate a room-temperature, single-electron transistor [Matsumoto, 2000].

The resolution of anodic oxidation is largely limited by the size of the AFM tip. Using conventional tips, features as small as 10 nm have been demonstrated. The development of carbon nanotube tips opens the way for dramatic improvement in the resolution, and features as small as 5 nm may be possible. As with all scanning probe techniques, AFM patterning is a serial technique and is therefore slow, but as with STM lithography multi-tip systems may improve the throughput. AFM lithography has some advantages over STM. Because it is an ambient pressure technique, the experiment setup is much simpler and less expensive.

36.2.4.3 Atomic Holography

It is also possible to directly control the placement of individual atoms as they deposit on a surface using a variation of holography. Holography is familiar as a technique using optical waves to create three dimensional images by utilizing both the amplitude and phase information carried by the light. However, holography can be applied to any wave, including matter waves [Sone et al., 1999]. In matter waves, the "image" produced is made up of atoms and, if a substrate is placed in the image plane, the atoms will deposit on the surface. In this way, it is possible to create very small patterns. While optical holography uses a photographic plate to manipulate the optical wavefront, in atomic holography a nanofabricated membrane is used to manipulate the atomic wavefront. The design of this hologram is critical to the formation of the desired image and requires intensive calculations. Using the pattern that is to be created on the substrate, one must calculate back to find what wavefront must leave the membrane. The hologram then becomes a series of holes etched into the surface of the membrane which allows portions of the matter front to pass through in the required areas. The interference of the atomic wave from each hole as it propagates toward the surface produces the desired image. The membrane is typically 100-nm-thick silicon nitride, and the holes etched through the membrane have a typical dimension of 200 nm.

The key to atomic holography is to produce a mono-energetic beam of atoms (one wavelength), analogous to the coherent light used in optical holography. This is done by producing a beam of ionized atoms in a dc discharge, extracting the atoms and then choosing a particular species with a mass selector. The atoms are then passed through a Zeeman Slower, then caught in an atom trap, a magneto-optical trap with four laser beams, as shown in Figure 36.8. The cooled atoms are removed from the trap by the transfer laser and fall vertically under the gravitational force. The atoms are passed through the etched membrane hologram, where the beam is diffracted to form a reconstructed image at the substrate. In initial demonstrations, this substrate is a microchannel-plate (MCP) electron multiplier that detects the impact of atoms and produces a corresponding image on a fluorescent screen. This allows an image to be directly viewed and recorded by computer. The spatial resolution of the image is therefore limited by the resolution of the MCP, which is typically a few tens of microns. The actual resolution of the pattern is limited by the wavelength of the atomic beam, which is typically on the nanometer scale, in contrast to optical beams, for which the wavelength is typically more than 200 nm.

A challenge to be faced by atomic holography is to deposit useful materials by this technique. Initial demonstrations have used atoms of limited utility, such as neon and sodium [Behringer et al., 1996;

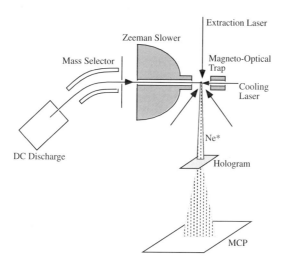

FIGURE 36.8 Schematic of an atomic holography system. (Adapted from Sone et al., 1999.)

Matsui and Fujita, 1998]. These atoms were chosen due to properties that make them attractive for atomic beam manipulations rather than for their usefulness as materials. In order for this technique to become widely applied, depositions of materials such as aluminum or copper must be demonstrated.

36.3 Fabrication of Nanomachines: The Interface with Biology

The preceding pages have described the current state and predicted future of MEMS manufacturing based on real and projected advances in electronics, materials science and chemistry. As discussed, it is expected that many of the most important advances will come as a result of cross-fertilizations between these fields. However, there is a growing consensus that the greatest advances, indeed the future of nanotechnology itself, lie in integration of biological tools and systems into nanotechnological design, function and manufacturing.

The basis for this statement is the obvious fact that nature has performed feats of nanotechnological engineering that we are only beginning to describe or understand. Modern approaches to biological research have revealed molecular machines that are remarkable in their beauty, surprising in their efficiency and sometimes astounding in their self-assembled, three-dimensional complexity. On the one hand are "simple" machines such as the bacterial flagellar motor, which powers the movement of organisms such as *Escherichia coli* through viscous media (Figure 36.9; see DeRosier, 1998, for an electron micrograph). The striking beauty of this rotating assembly (complete with drive shaft, bushing and universal joint) is all the more remarkable when one realizes that it is only 50 nm in diameter, is capable of rotating at 300 revolutions per second and produces a torque of 550 pN-nm [DeRosier, 1998]. Another example is bacteriophage T4, a virus (thankfully specific for bacteria) that bears a

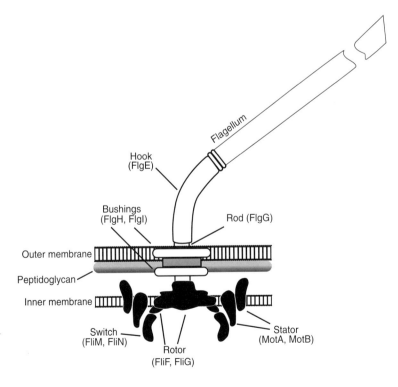

FIGURE 36.9 Representation of the bacterial flagellar motor. Gene names for proteins comprising individual components are given in parentheses. Protons pass through MotA to drive rotation of the propeller (the "flagellum"). Figure is drawn approximately to scale (width of motor = 50 nm). (Adapted from Berg, 1995 and Mathews, 2000.)

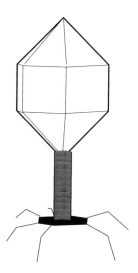

FIGURE 36.10 Representation of bacteriophage T4 (a bacterial virus). The phage attaches to the bacterium with specific receptors on its "legs" and injects the DNA stored in its "head" through the tailpiece and into the bacterium to initiate the process of infection. The head is approximately 85 to 115 nm.

remarkable resemblance to a lunar lander (Figure 36.10). For electron micrographs, see, for example, Nelson and Cox (2000) or http://www.asmusa.org/division/m/foto/T4Mic.html. This nanosyringe (only 85 nm in diameter) lands on and attaches specifically to the surface of a susceptible bacterium and then uses its tail projection to inject its DNA into the bacterium. Once the DNA is inside the cell, it essentially reprograms the bacterium, converting it into a factory for making more viruses. Though these examples exist on the nanoscale, the existence of our brains and bodies demonstrates that biological nanotechnology can be integrated and coordinated into three-dimensional structures more than 7 orders of magnitude larger.

36.3.1 Inspiration from Biology

Producing an engineered version of a system as complex as the human brain is obviously still in the realm of science fiction, but nanotechnological engineers in the present are finding inspiration in a number of biological sources. The most obvious sources are existing biological molecules. Biological molecules are finding application both as functional components and as structural components. Some molecules have natural functions that engineers hope to harness directly. These include the molecular motors such as the flagellar motor described above [DeRosier, 1998; Mehta et al., 1999; Vale and Milligan, 2000] (see also below) as well as electric switches or ion pumps like the voltage-gated ion channels found in nerve cells. More complicated systems, such as the photosynthetic apparatus, are also key targets [Gust and Moore, 1994]. One example of proteins that are naturally useful in a structural way is bacterial "S proteins." These pore-forming proteins found on the extracellular surface of bacteria can be extracted and then experimentally reassembled (on a variety of surfaces) into matrices with specific characteristics as variable as the number of different species of bacteria [Sleytr et al., 1997; Sleytr and Beveridge, 1999]. One could imagine that naturally assembling filamentous proteins such as cytoskeletal proteins could be utilized in similar ways.

Some of the most exciting developments may come from application of the natural characteristics of biological molecules in completely novel ways. For example, DNA has been used as a "rope" of defined length that can be positioned in specific places by virtue of its sequence-specific "sticky" cohesive ends and then used as a scaffold on which silver wires are deposited [Braun et al., 1998]. The self-organizing capacity of these sequence-specific "sticky" ends has also been utilized to assemble simple mixtures of cross-linked DNA units into well-organized two-dimensional "DNA crystals" with repeating surface features of various widths [Winfree et al., 1998]. Other DNA molecules have been used to construct more complicated three-dimensional structures, including cubes and truncated octahedrons [Seeman, 1999]. The fact that specific positions in these DNA molecules can be derivatized in a large variety of ways suggests that DNA may become an important structural tool for the design and manufacture of molecular electronic and other nanotechnological devices [Winfree et al., 1998].

Though biological molecules are useful, it is the harnessing of biological ideas, such as self-assembly and evolution by "natural selection," which may in the end have the greatest impact. Two obvious problems for nanotechnology are (1) how does one design a molecule with the activity desired, and (2) how does one assemble it? As noted above, nature has solved both of these problems to create "machines" of daunting complexity and remarkable ability. It has been argued that there is no reaction for which Nature has not developed an enzyme. This array of well-honed activities has been achieved by random mutation followed by reproductive "selection" of the more efficient variants. In design-based engineering, we are limited by our imagination and our ability to model. With mutation/selection-based engineering, one is limited only by the number of variants that can be tested and the efficiency with which improved variants can be selected. The difficult part of applying this idea is figuring out how to identify and isolate or reproduce the molecules with desired characteristics. Combinatorial chemists are actively using this approach to identify new drugs [Floyd et al., 1999], while biologists have used it to develop entirely new enzymatic activities, either *in vivo* (bioremediation) [Chen et al., 1999] or *in vitro* [Wilson and Szostak, 1999]. The breadth of practical future applications for "directed evolution" is difficult to predict, but it is clear that the idea is powerful.

The problem of three-dimensional self-assembly is closely related to that of structure and design. Indeed, natural machines have been selected by evolution to self-assemble (or to assemble in the assisting presence of other natural machines called "chaperones"; see Nelson, 2000, for an introduction to this process). For example, an enzyme is a linear chain of amino acids which folds into a complex and reproducible three-dimensional structure. This precise array of amino acid side chains endows it with specific affinities for other molecules and a specific chemical activity. To properly place these side chains artificially in three-dimensional space would be difficult, if not impossible, with existing technology. On the other hand, many proteins will spontaneously fold into the correct three-dimensional structure in seconds. These proteins can them assemble into larger machines of remarkable complexity and function. One beautiful example is the ribosome, a biological machine that functions (according to the analogy of Peter Moore) rather like a programmable loom [Ban et al., 2000; Nissen et al., 2000]. The ribosome "reads" the sequence of a given messenger RNA molecule and strings together amino acids according to this mRNA program, thus synthesizing the prescribed protein. The ribosome is indeed a programmable, positional, assembly-fabrication device existing on the nanometer scale, thus demonstrating that such nanofabrication devices are possible [Merkle, 1999].

The problem of how to predict three-dimensional structure from the linear sequence of amino acids is one of the remaining "Holy Grails" of molecular biology. If one could figure out how to predict three-dimensional structure from a sequence, it seems likely that one could figure out how to design a sequence necessary to produce a particular structure (see DeGrado et al. (1999) and Koehl and Levitt (1999a; 1999b) for discussions of current attempts to achieve this aim). However, there is the very real problem of whether this thermodynamically designed protein fold could be achieved in a kinetically realistic amount of time. One possibility is to model new proteins on old (evolutionarily selected) folds [Koehl and Levitt, 1999a; 1999b]. It also seems likely that engineers could use *in vitro* evolution to optimize the self-assembly (folding ability) of these designed protein machines. If single proteins could be designed *de novo*, why not entire complexes? If (when) this aim of *de novo* protein design is achieved, the machines of nanotechnology could very well start assembling themselves.

36.3.2 Practical Fabrication of Biological Nanotechnology

In order to begin to understand how devices based on biological molecules might be fabricated now, it is necessary to review some general techniques for the production of biological molecules. Many of the DNA-based applications described above require small DNA molecules (<100 basepairs), and small DNA molecules are relatively easy to make in large amounts either chemically (available commercially; 0.1 μmol of a 20 mer oligonucleotide can be made for less than $1/base, 10 μmol for less than $20/base by Sigma Genosys, for example) or by biological replication. One advantage of the synthetic approach is that the nucleotides at specific sites can be chemically derivatized in a wide variety of ways, limited only by the compatibility of the derivatization with the chemical process of DNA synthesis. Another advantage of chemical synthesis is that the engineer defines the sequence. For pieces of DNA larger than 100 bases (and up to chromosome size), biological manufacture is often preferable. Biological manufacture of DNA is performed by replicating an existing molecule, and it is done either *in vitro* or *in vivo*. *In vitro* replication is performed by putting the starting DNA and the necessary purified components and enzymes into a test tube, and letting the enzymes do their job. The polymerase chain reaction (PCR) is one particularly useful enzymatic approach for producing large numbers of replicas of a DNA fragment in the range of 0.1 to 10 kilobasepairs. In order to replicate a desired piece of DNA *in vivo*, it is ligated into a plasmid (a different piece of DNA that can replicate in bacteria independently of the chromosome), the plasmid is grown in bacteria, and then the plasmid (containing the DNA of interest) is purified out of the bacteria. The desired DNA can then be cut out of the plasmid using sequence-specific restriction enzymes and purified away from the rest of the plasmid DNA. In this process, the amount of DNA produced is limited only by the quantity of bacteria grown (indeed possible on an industrial scale).

Proteins also can be made either chemically or biologically, but most useful proteins are too large to be easily made in reasonable amounts chemically. Therefore, they are made biologically by inserting a piece of DNA encoding the desired protein into a cell type suitable for making the protein. Bacteria are usually the most productive hosts for protein production, but many human proteins cannot be grown in bacteria either because they are not folded properly into their three-dimensional shape or because they are not modified properly (proteins can require phosphorylations, sugar additions, etc.). Yeast cells, insect cells and human cells are (in that order) more likely to make active protein but are also less productive and (considerably) more expensive.

After the problem of how to express active protein is solved, the protein must be purified away from other cellular proteins. This problem can be made considerably easier by engineering into the DNA (and therefore the protein) an "affinity tag." This tag is a specific amino acid sequence that enables the engineered protein to stick specifically to a particular affinity matrix (several types exist), allowing the tagged protein to be retained on the matrix and the vast majority of undesired proteins to wash away. In addition, a variety of tags or modifications can be chemically added to the protein after it is purified. A tag of one form or another is often necessary to allow positioning of the protein in the desired place and in the desired orientation; it also may be necessary to harness the protein to other mechanical devices (for example, in the case of a molecular motor). For more details on the processes of working with DNA or protein, see Ausubel et al. (2000) and Coligan et al. (2000).

To understand these issues in more practical terms, it is useful to describe the approaches used by two engineers in trying to harness the work of the molecular motor enzyme, F1 ATPase. The F1 ATPase is a rotary motor present in essentially every living cell. Its natural purpose is to synthesize the energy carrier molecule adenosine triphosphate (ATP) by harnessing the flow of protons down an electrochemical gradient. During ATP synthesis, the central subunit rotates in a clockwise direction. However, if given a supply of ATP, the motor will utilize the ATP to drive rotation in a counter-clockwise direction. Montemagno and Bachand (1999) have produced and purified a tailored form of this molecule that they have attached specifically and in a specific orientation to a substrate. They attached the central subunit specifically to a latex bead and used it to drive the rotation of the latex bead.

In order to do this, Montemagno and Bachand first isolated the pieces of DNA encoding the subunits of the ATP synthase using standard molecular biology procedures. They spliced these pieces of DNA into a bacterial plasmid (see above) engineered so that production of the ATPase proteins could be turned on or off in response to an externally applied chemical signal (the sugar analog IPTG). They then tailored the DNA encoding the ATPase subunits to provide solutions to the other problems. For example, they modified the DNA of one of the subunits at the base of the ATPase by fusing it to a His tag, which is simply a string of 6 to 10 histidines that allows any protein (or complex of proteins) to bind specifically to the element nickel. This His tag allowed the ATPase complex to be purified on a column of nickel resin.

It also allowed the ATPase complex to bind to a nickel substrate with a defined orientation. As a final bit of genetic engineering, Montemagno and Bachand changed to cysteine one of the amino acids at the tip of the central rotating subunit (the "driveshaft"). Cysteine can be chemically cross-linked to a variety of different molecules, and this engineered cysteine allowed them to cross-link the driveshaft to biotin. Because biotin binds streptavidin specifically and tightly, the biotin-linked driveshaft bound specifically to a streptavidin-coated bead. This bead was therefore coupled to the rotating driveshaft, enabling the motor to be harnessed to move the bead.

Though as yet these experiments are just a demonstration of principle, it is easy to imagine replacing the latex bead with some form of molecular drivetrain. Thus far, it is true that the integration of biology into nanotechnology has yet to live up to expectations. However, with the breathtaking advances in biotechnology and the enormous breadth of biochemical biodiversity, the expectations are great indeed.

36.4 Summary

This chapter has attempted to provide an overview of technologies from established to nascent that might be used to create nanostructures suitable for future nanoelectromechanical systems. There exist many visions of what is possible in the near and further future for ultrasmall mechanical devices, so it is difficult to tell at this time which of the various fabrication technologies available today might become the most useful in the future. It is likely that some combination of many technologies will find applications. In order to address this, we discussed the most common nanofabrication technologies, such as electron beam and X-ray lithographies, and what limits their resolution in achieving the smallest possible pattern dimensions. We also discussed an alternative method, microcontact printing, for achieving high throughput at very small dimensions. We showed that these techniques were capable of achieving 10-nm resolution in a straightforward manner. Using exotic techniques, much smaller features are attainable.

In order to attain true atomic resolution, scanning probe techniques have proven to be viable technologies. We discussed the use of scanning tunneling microscopy and atomic force microscopy in defining Angstrom-scale patterns and pointed out a method of achieving high throughput with this technique, as well. One distinctive technique discussed is that of atomic holography which might someday be used to build up NEMS devices by direct deposition of the constituent atoms.

We concluded this chapter with a discussion of how biological systems qualify as NEMS and how future technologists might exploit existing cellular structures to create artificial nanomachines. We discussed specific cellular functions that are similar in function to potential nanomachines, including flagellar motors, voltage switches, ion pumps, cytoskeletal proteins for structural support and even DNA for use as a patterning template. We discussed how biologists succeed in customizing and fabricating batch quantities of these potential NEMS structural elements.

Acknowledgments

GHB is grateful to F. Cerrina for helpful discussions. This work was supported in part by ONR/DARPA, and the Indiana 21st Century Research and Technology Fund.

References

Andrews, D.E., and Wilson, M.N. (1989) "High-Energy Lithography Illumination by Oxford's Synchrotron: A Compact Superconducting Synchrotron X-Ray Source," *J. Vacuum Sci. Technol. B* **7**, pp. 1696–1701.

Ausubel, F., Brent, M., Kingston, R.E., Moore, D.D., Seidman, J.G., and Struhl, K., Eds. (2000) *Current Protocols in Molecular Biology,* John Wiley & Sons, New York.

Ban, N., Nissen, P., Hansen, J., Moore, P.B., and Steitz, T.A. (2000) "The Complete Atomic Structure of the Large Ribosomal Subunit at 2.4 Å Resolution," *Science* **289**, pp. 905–920.

Bazán, G., and Bernstein, G.H. (1993) "Electron Beam Lithography Over Very Large Scan Fields," *J. Vacuum Sci. Technol. A* **11**, pp. 1745–1752.

Behringer, R.E., Natarajan, V., and Timp, G. (1996) "Laser Focused Atomic Deposition: A New Lithography Tool," *Appl. Phys. Lett.* **68**, pp. 1034–1036.

Berger, S.D., Gibson, J.M., Camarda, R.M., Farrow, R.C., Huggins, H.A., Kraus, J.S., and Liddle, J.A. (1991) "Projection Electron-Beam Lithography: A New Approach," *J. Vacuum Sci. Technol. B* **9**, pp. 2996–2999.

Bernstein, G.H., Liu, W.P., Khawaja, Y.N., Kozicki, M.N., Ferry, D.K., and Blum, L. (1988) "High-Resolution Electron Beam Lithography with Negative Organic and Inorganic Resists," *J. Vacuum Sci. Technol. B* **6**, pp. 2298–2302.

Bernstein, G.H., Hill, D.A., and Liu, W.P. (1992) "New High-Contrast Developers for Poly(Methyl Methacrylate) Resist," *J. Appl. Phys.* **71**, pp. 4066–4075.

Binnig, G., and Rohrer, H. (1987) "Scanning Tunneling Microscopy—From Birth to Adolescence," *Rev. Mod. Phys.* **59**, pp. 615–625.

Braun, E., Eichen, Y., Sivan, U., and Ben-Yoseph, G. (1998) "DNA-Templated Assembly and Electrode Attachment of a Conducting Silver Wire," *Nature* **391**, pp. 775–778.

Broers, A.N. (1964) "Micromachining by Sputtering Through a Mask of Contamination Laid Down by an Electron Beam," in *Proc. 1st Int. Conf. on Electron and Ion Beam Science and Technology* (R. Bakish, Ed.), pp. 191–204, Wiley, New York.

Broers, A.N. (1995) "Fabrication Limits of Electron Beam Lithography, and of UV, X-Ray, and Ion-Beam Lithographies," *Phil. Trans. R. Soc. London A* **353**, pp. 291–311.

Cerrina, F. (1996) "The Limits of Patterning in X-Ray Lithography," *Mat. Res. Soc. Symp. Proc.* **380**, pp. 173–177.

Cerrina, F. (2000) Private communication.

Cerrina, F., and Rai-Choudhury, P., Eds. (1997) "Handbook of Microlithography, Micromachining, and Microfabrication." Vol. 1: *Microlithography,* SPIE Optical Engineering Press, Bellingham, WA.

Chen, W., and Ahmed, H. (1993a) "Fabrication of 5–7 nm Wide Etched Lines in Silicon Using 100 KeV Electron-Beam Lithography and Polymethylmethacrylate Resist," *Appl. Phys. Lett.* **62**, pp. 1499–1501.

Chen, W., and Ahmed, H. (1993b) "Fabrication of Sub-10 nm Structures by Lift-Off and by Etching After Electron-Beam Exposure of Poly(methylmethacrylate) Resist on Solid Substrates," *J. Vacuum Sci. Technol. B* **11**, pp. 2519–2523.

Chen, W., Bruhlmann, F., Richins, R.D., and Mulchandani, A. (1999) "Engineering of Improved Microbes and Enzymes for Bioremediation," *Curr. Opin. Biotechnol.* **10**, pp. 137–141.

Chen, Z.G., Leonard, Q.T., Khan, M., Cerrina, F., and Seeger, D.E. (1997) "X-Ray Phase-Mask: Nanostructures," *Proc. SPIE, Emerging Lithographic Technol.* **3048**, pp. 183–192.

Chlebek, J., Betz, H., Heuberger, A., and Huber, H.-L. (1987) "The Influence of Photoelectrons and Fluorescence Radiation on Resolution in X-Ray Lithography," *Microelectr. Eng.* **6**, pp. 221–226.

Chou, S.Y., Krauss, P.R., Wei-Zhang, Lingjie-Guo, Lei-Zhuang (1997) "Sub-10 nm Imprint Lithography and Applications" *J. Vac. Sci. Technol. B* **15**, pp. 2897–2904.

Coligan, J.E., Dunn, B.M., Ploegh, H.L., Speiche, Da.W., and Wingfield, P.T., Eds. (2000) *Current Protocols in Protein Science,* John Wiley & Sons, New York.

Cumming, D.R.S., Thoms, S., Beaumont, S.P., and Weaver, J.M.R. (1996) "Fabrication of 3 nm Wires Using 100 keV Electron Beam Lithography and Poly(methyl methacrylate) Resist," *Appl. Phys. Lett.* **68**, pp. 322–324.

DeGrado, W.F., Summa, C.M., Pavone, V., Nastri, F., and Lombardi, A. (1999) "De Novo Design and Structural Characterization of Proteins and Metalloproteins," *Ann. Rev. Biochem.* **68**, pp. 779–819.

DeRosier, D.J. (1998) "The Turn of the Screw: The Bacterial Flagellar Motor," *Cell* **93**, pp. 17–20.

Dial, O., Cheng, C.C., and Scherer, A. (1998) "Fabrication of High-Density Nanostructures by Electron Beam Lithography," *J. Vacuum Sci. Technol. B* **16**, pp. 3887–3890.

Drexler, K.E. (1981) "Molecular Engineering: An Approach to the Development of General Capabilities for Molecular Manipulation," *Proc. Natl. Acad. Sci. USA* **78**, pp. 5275–5278.

Early, K., Schattenburg, M.L., and Smith, H.L. (1990) "Absence of Resolution Degradation in X-Ray Lithography for λ from 4.5 nm to 0.83 nm," *Microelec. Eng.* **11**, pp. 317–321.

Erbe, A., Blick, R.H., Kriele, A., and Kotthaus, J.P. (1998) "A Mechanically Flexible Tunneling Contact Operating at Radio Frequencies," *Appl. Phys. Lett.* **73**, pp. 3751–3753.

Feldman, M., and Sun, J. (1992) "Resolution Limits in X-Ray Lithography," *J. Vacuum Sci. Technol. B* **10**, pp. 3173–3176.

Flanders, D.C. (1980) "Replication of 175-Å Lines and Spaces in Polymethylmethacrylate Using X-Ray Lithography," *Appl. Phys. Lett.* **36**, pp. 93–96.

Floyd, C.D., Leblanc, C., and Whittaker, M. (1999) "Combinatorial Chemistry as a Tool For Drug Discovery," *Prog. Med. Chem.* **36**, pp. 91–168.

Forest, D.R. (1995) "The Future Impact of Molecular Nanotechnology on Textile Technology and on the Textile Industry," in *Proc. of Discover Expo '95: Industrial Fabric & Equipment Exposition,* Charlotte, NC.

Goldstein, J.I., Newbury, D.E., Echlin, P., Joy, D.C., Fiori, C., and Lifshin, E. (1981) *Scanning Electron Microscopy and X-Ray Microanalysis,* Plenum Press, New York.

Gust, D., and Moore, T. (1994) "Photosynthesis Mimics as Molecular Electronic Devices," *IEEE Eng. Med. Biol.* **13**, pp. 58–66.

Harris, R.A. (1973) "Polymethyl Methacrylate as an Electron Sensitive Resist," *J. Electrochem. Soc. Solid-State Sci. Technol.* **120**, pp. 270–274.

Hiroshima, H., Okayama, S., Ogura, M., Komuro, M., Nakazawa, H., Nakagawa, Y., Ohi, K., and Tanaka, K. (1995) "Nanobeam Process System: An Ultrahigh Vacuum Electron Beam Lithography System with 3 nm Probe Size," *J. Vacuum Sci. Technol. B* **13**, pp. 2514–2517.

Huang, X., Bernstein, G.H., Bazán, G., and Hill, D.A. (1993) "Spatial Density of Lines Exposed in Poly (methylmethacrylate) by Electron Beam Lithography," *J. Vacuum Sci. Technol. A* **11**, pp. 1739–1744.

Irmer, B., Blick, R.H., Simmel, F., Godel, W., Lorenz, H., and Kotthaus, J.P. (1998) "Josephson Junctions Defined by a Nanoplough," *Appl. Phys. Lett.* **73**, pp. 2051–2053.

Joy, D.C. (1983) "The Spatial Resolution Limit of Electron Lithography," *Microelec. Eng.* **1**, pp. 103–119.

Kleideiter, G., Prucker, O., Bock, H., and Frank, C.W. (1999) "Polymer Thin Film Properties as a Function of Temperature and Pressure," *Macromolec. Symp.* **145**, pp. 95–102.

Koehl, P., and Levitt, M. (1999a) "De Novo Protein Design. I. In Search of Stability and Specificity," *J. Molec. Biol.* **293**, pp. 1161–1181.

Koehl, P., and Levitt, M. (1999b) "De Novo Protein Design. II. Plasticity in Sequence Space," *J. Molec. Biol.* **293**, pp. 1183–1193.

Kumar, A., and Whitesides, G.M. (1993) "Features of Gold Having Micrometer to Centimeter Dimensions Can Be Formed Through a Combination of Stamping with an Elastomeric Stamp and an Alkanethiol "Ink" Followed by Chemical Etching," *Appl. Phys. Lett.* **63**, pp. 2002–2004.

Langreth, R. (1993) "Molecular Marvels," *Pop. Sci.* pp. 91–94.

Lercel, M.J., Tiberio, R.C., Chapman, P.F., Craighead, H.G., Sheen, C.W., Parikh, A., and Allara, D.L. (1993) "Self-Assembled Monolayer Electron-Beam Resists on GaAs and Sio2," *J. Vacuum Sci. Technol. B* **11**, pp. 2823–2828.

Lercel, M.J., Craighead, H.G., Parikh, A.N., Seshadri, K., and Allara, D.L. (1996) "Sub-10 nm Lithography with Self-Assembled Monolayers," *Appl. Phys. Lett.* **68**, pp. 1504–1506.

Lyding, J.W. (1997) "UHV STM Nanofabricatrion: Progress, Technology Spin-Offs, and Challenges," *Proc. IEEE* **85**, pp. 589–600.

Mamin, H.J., Guethner, P.H., and Rugar, D. (1990) "Atomic Emission from a Gold Scanning-Tunneling-Microscope Tip," *Phys. Rev. Lett.* **65**, pp. 2418–2421.

Matsui, S., and Fujita, J.I. (1998) "Material-Wave Nanotechnology: Nanofabrication Using a de Broglie Wave," *J. Vacuum Sci. Technol. B* **16**, pp. 2439–2443.

Matsumoto, K. (1999) "Room Temperature Operated Single Electron Transistor Made by a Scanning Tunnelling Microscopy/Atomic Force Microscopy Nano-oxidation Process," *Int. J. Electr.* **86**, pp. 641–662.

Matsumoto, K. (2000) "Room-Temperature Single Electron Devices by Scanning Probe Process," *Int. J. High Speed Electr. Sys.* **10**, pp. 83–91.

McCord, M.A., and Rook, M.J. (1997) "Electron Beam Lithography," in *Handbook of Microlithography, Micromachining, and Microfabrication*, Vol. 1: *Microlithography* (P. Rai-Choudhury, Ed.), pp. 139–250. SPIE Press, Bellingham, WA.

Mehta, A.D., Rief, M., Spudich, J.A., Smith, D.A., and Simmons, R.M. (1999) "Single-Molecule Biomechanics with Optical Methods," *Science* **283**, pp. 1689–1695.

Merkle, R.C. (1999) "Biotechnology as a Route to Nanotechnology," *Trends Biotechnol.* **17**, pp. 271–274.

Mochel, M.E., Humphreys, C.J., Mochel, J.M., and Eades, J.A. (1983) "Cutting of 20 Å Holes and Lines in Metal-Aluminas," in *Proc. 41st Annual Meeting of the Electron Microscopy Society of America*, pp. 100–101.

Montemagno, C., and Bachand, G. (1999) "Constructing Nanomechanical Devices Powered by Biomolecular Motors," *Nanotechnology* **10**, pp. 225–231.

Namatsu, H., Nagase, M., Kurihara, K., Iwadate, K., Furuta, T., and Murase, K. (1995) "Fabrication of Sub-10-nm Silicon Lines with Minimum Fluctuation," *J. Vacuum Sci. Technol. B* **13**, pp. 1473–1476.

Nelson, D.L., and Cox, M.M. (2000) *Lehninger Principles of Biochemistry*, Worth, New York.

Nissen, P., Hansen, J., Ban, N., Moore, P.B., and Steitz, T.A. (2000) "The Structural Basis of Ribosome Activity in Peptide Bond Synthesis," *Science* **289**, pp. 920–930.

Reed, M.A. (1999) "Molecular-Scale Electronics," *Proc. IEEE* **87**, pp. 652–658.

Roukes, M.L. (2000) "Nanoelectromechanical Systems," in *Proc. Solid-State Sensor and Actuator Workshop*, Hilton Head Island, SC, pp. 367–376.

Rubel, S., Trochet, M., Ehrichs, E.E., Smith, W.F., and de-Lozanne, A.L. (1994) "Nanofabrication and Rapid Imaging with a Scanning Tunneling Microscope," *J. Vacuum Sci. Technol. B* **12**, pp. 1894–1897.

Schmidt, T., Martel, R., Sandstrom, R.L., and Avouris, P. (1998) "Current-Induced Local Oxidation of Metal Films: Mechanism and Quantum-Size Effects," *Appl. Phys. Lett.* **73**, pp. 2173–2175.

Schulz, H., Scheer, H.-C., Hoffmann, T., Sotomayor Torres, C.M., Pfeiffer, K., Bleidiessel, G., Grutzner, G., Cardinaud, Ch., Gaboriau, F., Peignon, M.-C., Ahopelto, J., and Heidari, B. (2000) "New Polymer Materials for Nanoimprinting," *J. Vacuum Sci. Technol. B* **18**, pp. 1861–1865.

Seeman, N.C. (1999) "DNA Engineering and Its Application to Nanotechnology," *Trends Biotechnol.* **17**, pp. 437–443.

Shen, T.C., Wang, C., Lyding, J.W., and Tucker, J.R. (1995) "Nanoscale Oxide Patterns on Si(100) Surfaces," *Appl. Phys. Lett.* **66**, pp. 976–978.

Sleytr, U.B., and Beveridge, T.J. (1999) "Bacterial S-Layers," *Trends Microbiol.* **7**, pp. 253–260.

Sleytr, U.B., Bayley, H., Sara, M., Breitwieser, A., Kupcu, S., Mader, C., Weigert, S., Unger, F.M., Messner, P., Jahn-Schmid, B., Schuster, B., Pum, D., Douglas, K., Clark, N.A., Moore, J.T., Winningham, T. A., Levy, S., Frithsen, I., Pankovc, J., Beale, P., Gillis, H.P., Choutov, D.A., and Martin, K.P. (1997) "Applications of S-Layers," *FEMS Microbiol. Rev.* **20**, pp. 151–175.

Snow, E.S., and Campbell, P.M. (1995) "AFM Fabrication of Sub-10-nanometer Metal-Oxide Devices with In Situ Control of Electrical Properties," *Science* **270**, pp. 1639–1641.

Snow, E.S., Juan, W.H., Pang, S.W., and Campbell, P.M., (1995) "Si Nanostructures Fabricated by Anodic Oxidation with an Atomic Force Microscope and Etching with an Electron Cyclotron Resonance Source," *Appl. Phys. Lett.* **66**, pp. 1729–1731.

Snow, E.S., Campbell, P.M., and McMarr, P.J., (1996) "AFM-Based Fabrication of Free-Standing Si Nanostructures," *Nanotechnology* **7**, pp. 434–437.

Snow, E.S., Jernigan, G.G., and Campbell, P.M. (2000) "The Kinetics and Mechanism of Scanned Probe Oxidation of Si," *Appl. Phys. Lett.* **76**, pp. 1782–1784.

Sone, J., Fujita, J., Ochiai, Y., Manako, S., Matsui, S., Nomura, E., Baba, T., Kawaura, H., Sakamoto, T., Chen, C.D., Nakamura, Y., and Tsai, J.S. (1999) "Nanofabrication Toward Sub-10 nm and Its Application to Novel Nanodevices," *Nanotechnology* **10**, pp. 135–141.

Spears, D., and Smith, H.I. (1972) "High-Resolution Pattern Replication Using Soft X-Rays," *Electr. Lett.* **8**, pp. 102–104.

St. John, P.M., and Craighead, H.G. (1995) "Microcontact Printing and Pattern Transfer Using Trichlorosilanes on Oxide Substrates," *Appl. Phys. Lett.* **68**, pp. 1022–1024.

Tada, T., and Kanayama, T. (1995) "Fabrication of Silicon Nanostructures with a Poly(methylmethacrylate) Single-Layer Process," *J. Vacuum Sci. Technol. B* **13**, pp. 2801–2804.

Vale, R.D., and Milligan, R.A. (2000) "The Way Things Move: Looking Under the Hood of Molecular Motor Proteins," *Science* **288**, pp. 88–95.

Wilder, K., Soh, H.T., Atalar, A., and Quate, C.F. (1999) "Nanometer-Scale Patterning and Individual Current-Controlled Lithography Using Multiple Scanning Probes," *Rev. Sci. Inst.* **70**, pp. 2822–2827.

Wilson, D.S., and Szostak, J.W. (1999) "In Vitro Selection of Functional Nucleic Acids," *Annu. Rev. Biochem.* **68**, pp. 611–647.

Winfree, E., Liu, F., Wenzler, L.A., and Seeman, N.C. (1998) "Design and Self-Assembly of Two-Dimensional DNA Crystals," *Nature* **394**, pp. 539–544.

Index

A

Absolute instabilities, **33**–8
Absolute pressure sensors, **25**–11
Absorber, **17**–11 to 13
Acceleration concepts, **24**–4 to 5
Acceleration sensor, **17**–55
Accelerometers. *See also* Inertial sensors
   design, **24**–5 to 13, **24**–20 to 22
   lateral, **24**–7, **24**–9
   LIGA technique, **17**–55 to 56
   polysilicon surface micromachining, **16**–160 to 162
Accommodation pumping, **29**–23 to 25
Accuracy, EFAB performance, **19**–17 to 19
Acidic electrochemical technique, **16**–36
Acidic vs. alkaline reactivity difference, **16**–63
Acoustic-wave actuation, **30**–5 to 6
Active wall motion studies, **34**–10 to 12
Actuators
   acoustic-wave actuation, **30**–5 to 6
   bubble, **35**–12 to 16
   delta wing, roll control, **35**–12 to 14
   effectiveness reduction, **35**–26 to 27
   electrostatic actuation, **17**–56 to 57, **30**–6
   free-shear flows, **35**–6 to 8
   inertial actuation, **30**–7
   LIGA technique, **17**–56 to 7
   magnetic actuation principles, **28**–18 to 19
   magnetic microactuator applications, **18**–37 to 39
   microactuators, **18**–34 to 39
   microrobotics principles, **28**–11 to 14
   nonuniformity actuator amplitudes, **35**–29
   piezo actuation, **30**–1
   pneumatic actuation, **30**–2 to 3
   reactive control, **33**–40 to 41
   receptivity, **35**–6
   supersonic jet screech, **35**–21 to 22
   thermal-bubble actuation, **30**–4
   thermal-buckling actuation, **30**–5
   turbulent flows, **26**–1 to 69, **26**–53 to 63
   types, **35**–6
Adaptive control algorithms, **13**–38
Adaptive neurocontrol, **14**–9 to 12
Adhesion, resist, **17**–19
Adiabatic reversible (isentropic) compression, **29**–2
Adjoint fields, **13**–27 to 33, **13**–33 to 34
Aerospace applications, **22**–3 to 6
AFM patterning, **36**–13 to 14
AFMs. *See* Atomic force microscopes (AFMs)
AHDLs. *See* Analog hardware description languages (AHDLs)
Alcohol, **16**–38, **16**–57, **16**–124
Algorithms. *See also* Models and modeling
   adaptive control, **13**–38
   convergence, **13**–36
   genetic, **14**–13 to 18
   grid velocity, **5**–12 to 13
   time-stepping, **5**–19 to 20
Alignment patterns, **16**–48 to 51
Alkaline vs. acid reactivity difference, **16**–63
Alternative inference systems, **14**–30 to 31
Amorphous silicon, **16**–154 to 155
Amplifiers, fluidic, **17**–47 to 48
Analog devices accelerometer, **16**–160 to 162
Analog hardware description languages (AHDLs), **5**–6
Anemometry, thermal, **26**–15
Anisotropic crystal, **16**–24
Anisotropic etching
   Arrhenius plot, **16**–38 to 39
   fundamentals, **16**–36 to 38
   single-crystal silicon, **15**–2
   structures, **16**–46 to 51
   types, **16**–38
   vs. isotropic, **16**–62 to 64
ANN. *See* Artificial neural networks (ANN)
Anodic oxidation, **36**–14
ANSYS, **5**–7
Applications of MEMS
   Feynman, Richard, **1**–3
   flow control, **33**–1 to 42

generators, microdroplet, **30**–1 to 26
heat sinks, microchannel, **32**–1 to 28
inertial sensors, **24**–1 to 24
pipes and spreaders, **31**–1 to 23
pressure sensors, micromachined, **25**–1 to 19
robotics, **28**–1 to 36
spreaders and pipes, **31**–1 to 23
surface-machined mechanisms, **27**–1 to 34
turbulent flow sensors and actuators, **26**–1 to 69
vacuum pumps, microscale, **29**–1 to 36
Arrays
    distributed architectures, flow control, **13**–14 to 16
    electrochemical sensor array, **16**–91 to 95
    heat pipes, **31**–12 to 15
    Routh array, **12**–6
Arrhenius plot, **16**–32, **16**–38
Artificial neural networks (ANN)
    feedforward, **14**–3 to 4
    fundamentals, **14**–2 to 3, **33**–36
    heat exchanger application, **14**–6 to 12
    implementation issues, **14**–5 to 6
    neurocontrol, **14**–6
    training, **14**–4 to 5
A-spots, **27**–4
Assembly of microrobotics, **28**–7 to 10
Atomic force microscopes (AFMs), **36**–12
Atomic-level patterning, **36**–13
Atomic peening, **16**–104
Atomic resolution, **36**–12 to 16
ATPase complex, **36**–20
Automated EFAB process tool, **19**–16 to 17
Automotive applications, **24**–3 to 4
Avogadro's number, **8**–4
Axial pressure gradient, **6**–31
Axisymmetric hypersonic blunt body flow, **7**–18 to 23

B

Backside protection, **16**–44
Balloon actuators. *See* Bubble actuators
Base layer deposition and etching, **16**–119 to 120
Batch diffusion and release procedure, **18**–26
Beams
    electron beam lithography, **36**–3 to 9
    fixed ends test, **3**–10 to 11
    machine design, **27**–7, **27**–8 to 10
    size, **36**–3 to 5
Beams, cantilever, **16**–113 to 115
Bearing clearance oxide, **15**–12
Bearings. *See also* Journal bearings
    constraints on, **9**–11 to 12
    journal type, **9**–14 to 21
    minimum etchable clearance, **9**–12
    thrust type, **9**–12 to 14
    turbulent flows, **26**–66 to 68
Beer's law, **10**–34
Bénard convection, **33**–34
Bending vs. buckling, **27**–17
Bend tests, **3**–9 to 12
Bernoulli's equation, compressibility, **4**–10
Beskok's slip-boundary conditions, **7**–12, **7**–24, **8**–11 to 12
Bessel function, **6**–23
Bethe retardation, **36**–5
Biaxial measurements, **16**–110 to 111
Bibliographic references. *See also* Glossaries; Nomenclature
    actuators, **26**–69 to 77
    autonomous control, **35**–35 to 37
    background and fundamentals, MEMS, **1**–4 to 5
    bubble/drop transport, **11**–12 to 13
    bulk micromachining, **21**–12
    Burnett simulations, **7**–34 to 36
    chemical sensors, **22**–23 to 24
    control theory, **12**–32 to 34
    deep reactive ion etching, **21**–12
    device scaling, **2**–9
    distributed architecture, **13**–44 to 45
    droplet generators, micro-, **30**–27 to 30
    electrochemical fabrication, **19**–22 to 23
    fabrication, **16**–165 to 183, **18**–41 to 46, **19**–22 to 23, **36**–20 to 24
    flow control, **33**–42 to 51
    flow physics, **4**–34 to 38
    free-shear flow, **35**–35 to 37
    generators, microdroplet, **30**–27 to 30
    heat sinks, **32**–28 to 30
    inertial sensors, **24**–25 to 30
    integrated simulation, **5**–23 to 25
    LIGA technique, **17**–61 to 65
    liquid flows, **6**–35 to 38
    lubrication, **9**–26 to 27
    materials, MEMS, **15**–23 to 26
    mechanical properties of MEMS materials, **3**–25 to 33
    MEMS fabrication, **16**–165 to 183
    microchannel heat sinks, **32**–28 to 30
    microdroplet generators, **30**–27 to 30
    microfluidic simulation, molecular-based, **8**–24 to 28
    microrobotics, **28**–37 to 42
    microscale vacuum pumps, **29**–26 to 28
    model-based flow control, **13**–44 to 45
    nanoelectromechanical systems, **36**–20 to 24
    packaging, **23**–21 to 23
    pipes and spreaders, **31**–23 to 26
    pressure sensors, **25**–20 to 21
    reactive control, skin-friction reduction, **34**–13 to 14
    replication techniques, **17**–61 to 65
    robotics, micro-, **28**–37 to 42
    sensors, **24**–25 to 30, **25**–20 to 21, **26**–69 to 77
    single-crystal silicon carbide MEMS, **20**–30 to 31
    skin-friction reduction, **34**–13 to 14
    soft computing, **14**–33 to 36
    spreaders and pipes, **31**–23 to 26
    surface-micromachined mechanisms, **27**–34
    thin liquid films, **10**–44 to 49
    turbulent flows, **26**–69 to 77
    vacuum pumps, **29**–26 to 28
    X-ray based fabrication, **18**–41 to 46
Bidirectional fiber coupling element, **17**–55
Bilayer effect, turbulent flows, **26**–54 to 55
Bimorph microgripper, **28**–16
Biological fluids, **6**–11

Biology interface, **36**–16 to 20
Biomedical sample handling, **30**–20 to 22
Biot numbers, **10**–30
Blunt body flow, two-dimensional hypersonic, **7**–15 to 18
Bode plot, **12**–11 to 13
Boiling
    bubble dynamics, **32**–24 to 25
    curves, **32**–18 to 21
    limitation, **31**–7
Boltzmann methods
    Burnett equations, **7**–3 to 4, **7**–7 to 11
    continuum model, **4**–7
    electric double layers, **6**–20, **32**–7
    equation research, **8**–17 to 19
    flowfield modeling, **4**–4
    hybrid with continuum simulation, **8**–19
    lattice, **8**–19 to 20
    metallization, **20**–20
    molecular-based models, **4**–18, **4**–21 to 23
    software integration, **5**–16
    vacuum pumps, **29**–3
Boron etch stop, **16**–72 to 75
Bouguer's law, **10**–34
Boundary conditions
    Burnett simulations, **7**–11 to 12
    die-attach, **23**–11
    disturbances, **13**–34 to 35
    flow physics, **4**–11 to 17
    kinematic, **10**–10
    slip, **7**–12, **7**–18, **7**–27, **8**–3
Boundary layers
    coherent structures, **33**–12 to 15
    control goals and interrelation, **33**–3
    laminar, **13**–25
    near-wall streamwise vortices, **34**–2
    nonlinear dynamical systems theory, **33**–31
    sensor requirements, **26**–13
    theory, **33**–2
Bremsstrahlung effect, **17**–6
Bretherton analysis, **11**–4 to 8, **11**–12
Bubble actuators, **35**–12 to 16
Bubble/drop transport
    Bretherton problem, **11**–4 to 8
    electrokinetic flow, **11**–8 to 12
    fundamentals, **11**–1 to 4
    future directions, **11**–12
Bubble dynamics, **6**–4, **32**–24 to 25
Bubble jet design, **30**–4, **30**–9
Buckling, **3**–15, **27**–17 to 18
Buckling diaphragm, **30**–5
Bulk micromachining. *See also* Surface micromachining; Wet bulk micromachining
    microfabrication, **26**–4 to 5
    single-crystal silicon, **15**–2
    vs. surface micromachining, **16**–148 to 149
    wet bulk problems, **16**–82 to 90
Burnett simulations
    axisymmetric hypersonic blunt body flow, **7**–18 to 23
    direct simulation Monte Carlo methods, **8**–11
    fundamentals, **7**–1 to 5, **7**–33
    governing equations, **7**–7 to 11
    history, **7**–5 to 7
    hypersonic shock structure, **7**–15
    linearized stability analysis, **7**–12 to 14
    molecular-based models, **4**–22 to 23
    NACA 0012 airfoil, **7**–23
    nomenclature, **7**–33 to 34
    numerical simulations, **7**–15 to 33
    subsonic flow in microchannel, **7**–24 to 27
    supersonic flow in microchannel, **7**–27 to 33
    two-dimensional hypersonic blunt body flow, **7**–15 to 18
    wall-boundary conditions, **7**–11 to 12
Bursting process, **33**–14, **33**–41, **34**–2
Butterfly effect, **33**–30

C

C4. *See* Controlled collapse chip connection (C4)
Calibration, **26**–17, **26**–31 to 33
Cantilever beams, spirals, and springs, **3**–11, **16**–113 to 115, **27**–11 to 12
Capacitive pressure sensors, **25**–7 to 14
Capillaries
    electrokinetic, **6**–31
    limitations, **31**–2 to 6
    stationary flow, **32**–26
    thin liquid films capillarity, **10**–40 to 42
Capture pumps, **29**–4, **29**–11 to 13
Carriers and cathodes, **17**–37 to 38
Cartesian properties
    Burnett equations, **7**–7, **7**–12
    inertial sensors, **24**–4
    piezoresistive pressure sensors, **25**–4
Castigliano method, **20**–12
CCFE. *See* Cold cathode field emission (CCFE)
Cell oxygen detection, **22**–14
Centralized approach, massive arrays, **13**–14 to 16
Ceramic membranes, **22**–18
Chain of integrators, **12**–26
Chaos control, **33**–30 to 35
Chapman-Enskog theory
    Burnett simulations, **7**–4, **7**–8 to 11
    molecular-based models, **4**–22
    molecular magnitudes, **8**–5
Chemical etching
    Elwenspoek model, **16**–58 to 61
    fundamentals, **16**–51 to 54
    isotropic vs. anisotropic etching, **16**–62 to 64
    Seidel model, **16**–54 to 58
Chemical sample handling, **30**–20 to 22
Chemical sensors
    applications, **22**–3 to 6, **22**–21
    cell oxygen detection, **22**–14
    ceramic membranes, **22**–18
    commercialization, **22**–17 to 21
    development, **22**–8 to 17
    engine emission monitoring, **22**–5 to 6
    fire safety monitoring, **22**–4 to 5
    fundamentals, **22**–1 to 3, **22**–21 to 22
    future trends, **22**–17 to 21

high-temp electronics, **22**–7, **22**–20 to 21
hydrogen and hydrocarbon detection, **22**–15 to 17
hydrogen technology, **22**–8 to 11
leak detection, **22**–3 to 4, **22**–18 to 20
nanocrystalline tin oxide thin films, **22**–11 to 13
nanomaterials, **22**–7
NASICON-based CO_2 detection, **22**–17
sensor fabrication technologies, **22**–6 to 8
Chemical vapor deposit (CVD), **16**–158 to 160
CHF. *See* Critical heat flux (CHF)
Chips, system partitioning, **24**–18, **24**–20 to 22
Choi numerical flowfield, **33**–22 to 23
Ciliary motion system (CMS), **28**–13, **28**–26
Circuit integration, pressure sensors, **25**–13 to 14
Circuit-microfluidic device simulation, **5**–15 to 20
Circuit simulator, **5**–8 to 9
Circular diaphragms, **20**–9, **25**–7
Clamped-clamped beams, **16**–111
Clancy, Tom, **33**–26
Classification schemes, flow control, **33**–4 to 6
Clauser plot, **26**–36
Closed loop control, **24**–20
CMOS circuitry, **16**–129 to 132
CMS. *See* Ciliary motion system (CMS)
CNN. *See* Computational neural networks (CNN)
Cockroaches, **28**–6
CODECS, **5**–2 to 3
Coffin-Mason model, **23**–17
Coherent structures, **33**–11 to 15
Cold cathode discharge, **29**–11
Cold cathode field emission (CCFE), **36**–5
Columns, **27**–7 to 8
Comb drives, **27**–18
Compatible processing techniques (NEMS), **36**–3 to 16
Compensator reduction, **13**–19 to 21
Complex geometries, **9**–9 to 10
Compliance elements, **27**–7
Compressibility, **4**–9 to 11, **5**–4, **6**–4, **6**–17
Compression molding, **17**–38 to 39
Computational neural networks (CNN), **33**–36
Condensation, **10**–36 to 44
Connectors and packages, **17**–48 to 50
Contact-free systems, **28**–13, **28**–18 to 19, **28**–23
Contact stresses, **27**–4
Contact systems, **28**–13 to 14
Continuous inkjet droplet generators, **30**–3, **30**–21
Continuous rotary pumps, **26**–64
Continuum equation models
    applicability of hypothesis, **9**–3 to 4
    coupling with direct simulation Monte Carlo methods, **8**–17
    flow physics, **4**–6 to 8
    hydrodynamics pressure driven flow, **6**–5 to 6
Control, flow. *See* Flow control
Control laws, **34**–7, **35**–9
Controlled collapse chip connection (C4), **10**–5 to 6
Control loop, **33**–5
Control theory
    classical linear control, **12**–2 to 17
    feedback linearization, **12**–24 to 28
    frequency response design methods, **12**–11 to 15
    fundamentals, **12**–1 to 2
    hybrid systems, **12**–30
    lead-lag compensation, **12**–15 to 17
    linear quadratic regulator, **12**–19 to 21
    Lyapunov stability theory, **12**–29 to 30
    mathematical preliminaries, **12**–2 to 5
    modern control, **12**–17 to 24
    nonlinear control, **12**–24 to 30
    PID control, **12**–7 to 8
    pole placement, **12**–18 to 19
    robust control, **12**–21 to 24
    root locus design method, **12**–8 to 11
    system analysis and design, **12**–5 to 17
Convection heat transfer
    continuum hypothesis, **32**–3 to 4
    electric double layer, **32**–7
    flow structure, **32**–8 to 8 to 9
    fundamentals, **32**–2
    general laws, **32**–4 to 5
    governing equations, **32**–5 to 6
    heat transfer mechanism, **32**–3
    noncontinuum mechanics, **32**–6
    particular laws, **32**–5
    polar mechanics, **32**–8
    single-phase, **32**–8 to 17
    size effects, **32**–6 to 8
    thermodynamic concepts, **32**–4
Convective instabilities, **33**–8
Conveyors, micro, **28**–18 to 23
Convolution kernels, **13**–16 to 18, **13**–38
Copper coil, **17**–42
Coriolis effect, **24**–4 to 5, **24**–22 to 24
Corner compensation, **16**–84 to 90
Corrugated structural materials, **16**–127
Cost functional, **13**–28 to 31, **13**–36
Couette properties, **4**–26, **9**–5 to 7
Coulomb forces, **4**–32, **11**–11
Counter ions, **32**–7
Countermeshing gear discriminator, **27**–21 to 23
Couple Device and Circuit Simulator (CODECS). *See* CODECS
Coupled mass gyroscopes, **24**–24
Couple sensitivity, **24**–23
Coupling monomode fibers, **17**–51 to 52
Crab-leg flexures, **27**–5
Creare turbomolecular pump, **29**–25
Creep
    silicon, **16**–22 to 24
    tests, **3**–16
    thermal, **4**–14, **7**–12, **8**–3, **29**–25
Crisp vs. fuzzy sets, **14**–24 to 27
Critical heat flux (CHF), **32**–21 to 22
Cross-flow instabilities, **33**–8
Crosstalk, **30**–8 to 11
Cross-wafer uniformity, **9**–22
Crystallography, silicon
    crystal structure, **16**–7 to 8
    geometric relationships, **16**–9 to 16
    Miller indices, **16**–6 to 7

[100]-oriented silicon, **16**–9 to 13, **16**–14 to 16
[110]-oriented silicon, **16**–13 to 16
Cube-square law, **9**–2
"Curse of dimensionality," **13**–28
CVD. *See* Chemical vapor deposit (CVD)

D

Damped natural frequency, **12**–9
Damping, **9**–3, **24**–7 to 10
Damping ratio, **12**–9
Dangling bonds, **16**–55
Darcy friction factor, **6**–7 to 8
Darwinian principle, **33**–37
Dashpot, **24**–7 to 10
Debye-Huckle equations
  electrical double layers, **6**–21
  electric double layer, **32**–7
  electro-osmotic flow, **6**–23 to 26
Decentralized approach, massive arrays, **13**–16
Decomposition, **13**–37
Deep reactive ion etching (DRIE)
  applications, **21**–5
  experimental results, **21**–5 to 9
  fluorine plasmas, **21**–3 to 5
  fundamentals, **21**–1 to 2, **21**–11
  high-aspect-ratio etching, **24**–16
  plasma etching, **21**–2 to 3
  pressure sensor, **21**–9 to 11
  restrictive aspects, **9**–11
  SiC etching, **21**–3 to 5
  uniformity, **9**–23
Delta wing, roll control, **35**–9 to 18
Demolding, **17**–39 to 40. *See also* Molding
Depassivated silicon atoms, **36**–13
Deposited dose, LIGA technique, **17**–23 to 24
Deposition
  base layer, **16**–119 to 120
  conditions, **15**–4
  EFAB layer cycle, **19**–13
  electrodeposition, **19**–4 to 5
  structural material, **16**–120 to 121
Design, microrobotics, **28**–10 to 14
Design and fabrication. *See also* Fabrication
  channel sensors, **22**–1 to 23
  deep reactive ion etching, **21**–1 to 23
  electrochemical (EFAB), **19**–1 to 22
  materials, **15**–1 to 23
  packaging, **23**–1 to 21
  silicon carbide (SiC), **20**–1 to 30, **21**–1 to 23
  X-ray based, **18**–1 to 41
Desktop factories, **28**–32 to 35
Developers, NEMS, **36**–7 to 9
Device damping, **24**–7 to 12
Devices, microrobotics, **28**–12 to 32
Diamond materials, **15**–20 to 21
Diamond membranes, **17**–13
Diaphragms
  anchors, **25**–9
  circular, **20**–9, **25**–7

deflection, **25**–2
double-layer, **25**–12
metal, **20**–11
piezoresistive strain gauges, **25**–48
temperature coefficients, **25**–10
Die-attach
  conductive, **23**–8 to 10
  electrical test, **23**–9 to 10
  governing equations, **23**–10 to 13
  material properties, **23**–10 to 13
  packaging material requirements, **23**–3 to 4
  relaxing temperature, **23**–16 to 18
  SiC test die, **23**–8
  simulation, **23**–13 to 18
  stress distribution, **23**–13 to 14
  substrate material effects, **23**–14 to 16
  thermal mechanical properties, **23**–10 to 18
  thickness, **23**–16
Diffuse reflection, **29**–4
Dimension measurements, **3**–5 to 6
Directed evolution, **36**–18
Direct simulation Monte Carlo (DSMC). *See also* Monte Carlo simulations
  Burnett equations, **7**–3 to 5, **7**–20 to 23, **7**–28 to 33
  limitations, **8**–9 to 11
  MEMS flows, **5**–4
  overview, **8**–7 to 9
Direct simulation Monte Carlo (DSMC), **4**–23
Direct wafer bond (DWB) technology, **24**–14
Direct writing, **30**–23
Disk method measurement, **16**–105 to 106
Dispersion force, **10**–18
Displacement measurements, **3**–6
Dissolved wafer process, **16**–90 to 95
Distributed architectures, flow control
  adaptive approach, **13**–38
  compensator reduction, **13**–19 to 21
  decentralization, **13**–14 to 16
  decomposition, **13**–37
  engineering trade-offs, **13**–39 to 40
  fundamentals, **13**–2 to 3
  future trends, **13**–41 to 43
  general framework synthesis, **13**–36 to 37
  global stabilization, **13**–37
  ideal control targets, **13**–38 to 39
  implementation, **13**–39 to 40
  language usage, **13**–41
  linear control of nonlinear systems, **13**–21 to 25
  linearization, **13**–3 to 6
  linear stabilization, **13**–6 to 14
  localization, **13**–16 to 19
  massive arrays, **13**–14 to 16
  nonlinear optimization, **13**–26 to 34
  robustification, **13**–34 to 36
  spatially developing flows, **13**–25
Distributed micromotion systems (DMMS), **28**–11, **28**–14
Disturbance spectra, **35**–23
DMMS. *See* Distributed micromotion systems (DMMS)
DNA, **36**–17 to 19
Dopant dependence, **16**–33 to 35

Doped polysilicon, **16**–153 to 154
Doping, **15**–6, **16**–104
Double deflecting gas nozzles, **17**–44
Double-layer diaphragm, **25**–12
Drag, **33**–3 to 4, **33**–41 to 42
DRIE. *See* Deep reactive ion etching (DRIE)
Droplet generation. *See* Microdroplet generators
Droplet-on-demand inkjet, **30**–4
Drop transport. *See* Bubble/drop transport
Dry etching, **17**–60, **36**–14
Dry friction, **4**–31
DSMC. *See* Direct simulation Monte Carlo (DSMC)
Dual lumen catheter, **16**–91 to 95
DWB technology. *See* Direct wafer bond (DWB) technology

E

Eckert number, **4**–14
ECR. *See* Electron cyclotron resonance (ECR)
Eddies
   breakup devices, **33**–4 to 5
   Falco, **33**–12
   scales, **26**–32, **33**–15 to 16
EDL. *See* Electrical double layers (EDL)
EFAB. *See* Electrochemical fabrication (EFAB)
Ejection of droplets, **30**–19
Elastic spring and seismic mass, **24**–5 to 7
Electret microphones, **26**–46
Electrical analogy and microfluidic networks, **6**–32
Electrical double layers (EDL)
   gas flows, **8**–20 to 21
   liquid flows, **6**–19 to 21, **32**–12 to 17
Electrical interconnection system, **23**–4 to 10
Electrically conductive plastic plates, **17**–37 to 38
Electrical multipin microconnectors, **17**–48 to 50
Electrical shorts, **27**–30 to 31
Electrical transduction, **24**–12 to 13
Electrochemical etch-etch stop, **16**–35 to 36
Electrochemical fabrication (EFAB). *See also* Design and fabrication; Fabrication
   applications, **19**–21
   automated process tool, **19**–16 to 19
   comparison of processes, **19**–19 to 20
   deposition, **19**–4, **19**–13
   electrodeposition, **19**–4
   fundamentals, **19**–1 to 3, **19**–8 to 10
   future trends, **19**–22
   layer cycle, **19**–13
   limitations and shortcomings, **19**–20 to 21
   mask processes, **19**–4 to 8, **19**–11 to 12
   materials, **19**–17, **19**–19
   microfabricated structures, **19**–15 to 16
   multiple-material layers, **19**–9 to 10
   performance, **19**–17 to 19
   printing plate, **19**–5 to 6
   process flow, **19**–10, **19**–11 to 15
   rapid prototyping, **19**–2
   SFF, **19**–2 to 5
   sharp anode, **19**–5
   solid freeform fabrication (SFF), **19**–2

   substrate preparation, **19**–12
   x-y accuracy, **19**–17 to 18
   z accuracy, **19**–18
Electrochemical sensor array, **16**–91 to 95
Electrodeposition, **17**–25 to 33, **19**–4 to 5
Electrokinetic effects. *See also* Kinetic effects
   applications, inadvertent, **6**–32
   background, **6**–18 to 19
   Bretherton analysis, **11**–12
   bubble transport, **8**–12
   electrical analogy and microfluidic networks, **6**–31
   electrical double layers, **6**–19 to 21
   electro-osmotic flow, **6**–21 to 28, **6**–30 to 31, **8**–22
   electrophoresis, **6**–24 to 28
   flowrate and pressure, **6**–30 to 31
   gas flows, **8**–20 to 21
   liquids and solids interface, **6**–3
   microchips, **6**–27 to 30
Electroless deposition, **17**–32 to 33
Electromagnetic forces, **4**–32
Electromagnetic micromotors, **17**–57 to 59
Electron beam lithography, **36**–2, **36**–3 to 9
Electron cyclotron resonance (ECR), **36**–8
Electronic microconnectors and packages, **17**–48 to 50
Electronic noses, **22**–20 to 21
Electronics, free-shear flows, **35**–6, **35**–8
Electron micrographs Web site, **36**–17
Electron scattering, **36**–5 to 7
Electro-osmotic flow
   gas flows, **8**–22
   liquid flows in microchannels, **6**–21 to 28, **6**–30 to 31
Electrophoresis, **6**–24 to 28
Electroplating. *See* Electrodeposition
Electropolishing, **16**–66 to 67
Electrostatic actuation, **17**–56 to 57, **30**–6
Electrostatically driven inkjet printhead, **30**–6
Electrostatic forces, **16**–125, **26**–56 to 57, **27**–30
Elliptical leg movements, **28**–26
Elwenspoek model, **16**–58 to 61
Embossing, hot, **17**–38 to 39
EMS-Alpha, **19**–16 to 17
Endo control scheme, **34**–11
Energy expenditure, **33**–5
Engine emission monitoring, **22**–5 to 6
Engines, **27**–18 to 21
Entrainment limitation, **31**–6
Entrance length, **6**–8, **32**–9 to 10
Enveloping coils, **17**–59
Environments, harsh, **23**–1 to 20. *See* Chemical sensors; Packaging
Epitaxy, **16**–144, **24**–14, **25**–13
Epon SU-8, **16**–148
Equations. *See* Algorithms; Models and modeling; *specific equation or formula*
Equation solution, **5**–8 to 9
Etchable clearance, **9**–12
Etching
   anisotropic, **16**–38
   base layer, **16**–119 to 120
   chemical, **16**–54 to 64

Index

deep, **9**–23, **17**–17, **17**–60
dry, **17**–60, **36**–14
high-aspect-ratio, **24**–16
history, **16**–3 to 6
6H-SiC, **20**–4 to 5
minimum clearance, **9**–12
rates, **16**–45 to 46, **16**–55, **16**–57
reactive ion, **15**–3
real estate gain, **16**–82 to 83
selective, **16**–121 to 123
semiconductor bias and/or illumination, **16**–64 to 71
spacer layer, **16**–121 to 123
stops, **16**–45 to 46
systems, **16**–39 to 42
underetching, **16**–84
wet isotropic, **16**–29 to 36
Etch stop
boron, **16**–72 to 75
electrochemical, **16**–35 to 36, **16**–75 to 78
fundamentals, **16**–72
photo-assisted electrochemical, **16**–78 to 79
photo-induced preferential anodization (PIPA), **16**–79 to 81
single-crystal silicon, **15**–3
thin-films, **16**–81
Ethylenediamine, **16**–41
Euler equations
Burnett equations, **7**–5, **7**–8
continuum model, **4**–8 to 9
flowfield modeling, **4**–5
flow-structure interactions, **5**–12
molecular-based models, **4**–22
Eu's evolution equations, **7**–5
Evaporation
bubble dynamics, **32**–24
flow patterns, **32**–23
thin liquid films, **10**–36 to 44
Extended hydrodynamic equations, **7**–3
Extensions, LIGA, **17**–40 to 41

F

Fabrication. *See also* Design and fabrication
anisotropic etching, **16**–29 to 64
biological nanotechnology, **36**–18 to 20
bulk micromachining, **16**–82 to 90
cross-wafer uniformity, **9**–22
deep etch uniformity, **9**–23
electrochemical (EFAB), **19**–1 to 22
etch-stop techniques, **16**–72 to 81
fundamentals, **9**–22, **16**–2 to 3
future trends, **36**–1 to 20
history, **16**–3 to 6, **16**–97 to 98
material properties, **9**–23 to 24
materials case studies, **16**–149 to 160
microdroplet generators, **30**–15 to 17
nanomachines, **36**–16 to 20
non-poly-Si modifications, **16**–140 to 146
pipes, heat, **31**–14 to 15
polysilicon, **16**–133 to 139, **16**–160 to 165
processes, surface micromachining, **16**–117 to 133
resists, **16**–146 to 149
semiconductor bias/illumination, **16**–64 to 71
silicon crystallography, **16**–6 to 16
solid freeform fabrication (SFF), **19**–1 to 4, **19**–19 to 20
spreaders, heat, **31**–18
steps, **16**–118 to 125
substrate and structural material, silicon, **16**–16 to 28
surface micromachining, **16**–95 to 97
thin films, mechanical properties, **16**–98 to 117
wet bulk micromachining, **16**–90 to 95
wet isotropic etching, **16**–29 to 64
X-ray based, **18**–1 to 41
Fabry-Perot devices, **25**–18, **26**–51
Failure mechanisms, **27**–27 to 33
Faraday's constant and laws, **6**–20
Fatigue tests, **3**–16
Feedback control, **34**–9
Feedback kernels, **13**–16
Feedback linearization, **12**–24 to 28
Feedback loop stabilization, **14**–9
Feedforward ANN, **14**–3 to 4
Feedforward network, **34**–5
FET-based strain sensors, **24**–13
Feynman, Richard P., **1**–1, **1**–3, **4**–33, **4**–34, **28**–2, **28**–6, **28**–32, **28**–36, **33**–1
Fiberoptic connectors, **17**–51 to 52
Fiberoptic switching device, **18**–38 to 39
Film stress, **16**–125 to 127, **27**–4 to 6
Filter, resonant mesh infrared band pass, **17**–42
Fire safety monitoring, **22**–4 to 5
Fixed beam springs, **27**–12 to 13
Fixed chamber neck design, **30**–10
Fixed ends test, beams, **3**–10 to 11
Fixtures, **27**–7
Flade potential, **16**–76 to 77
Flat-plate heat spreaders. *See* Spreaders, heat
Flexible plate design, **30**–10 to 11
Flexures, **27**–5, **27**–13 to 15
Flip-flop, fluidic, **17**–47
Floating-element sensors, **26**–38 to 41
Flow control
absolute instabilities, **33**–8
chaos control, **33**–30 to 35
classification schemes, **33**–4 to 6
coherent structures, **33**–11 to 15
convective instabilities, **33**–8
delta wing, roll control, **35**–14 to 18
devices, **33**–6
EMDH tiles, **33**–29 to 30
feedback control, **33**–23 to 24
free-shear, **33**–7
fundamentals, **33**–1 to 2, **33**–41 to 42
goals and interrelation, **33**–3 to 4
low-Reynolds-number wings, **35**–32 to 33
Mach number, **33**–7 to 8
magnetohydrodynamic control, **33**–26 to 30
MEMS usage, **33**–39 to 41
neural networks, **33**–38 to 39
nonlinear dynamical systems theory, **33**–31 to 32

principles, **33**–2 to 8
reactive, **33**–19 to 26, **33**–29 to 30, **33**–39 to 41
required characteristics, **33**–24 to 26
Reynolds number, **33**–7 to 8
scales, **33**–15 to 17
sensor requirements, **33**–17 to 19
separation control, **35**–32 to 33
soft computing, **33**–35 to 39
suction, **33**–10 to 11
supersonic jet screech, **35**–23 to 27
targeted, **33**–20 to 23
turbulence, **33**–9 to 10
von Kármán equation, **33**–27 to 28
vorticity flux, **33**–28 to 29
wall-bounded flows, **33**–7
Flow control, distributed architectures
adaptive approach, **13**–38
compensator reduction, **13**–19 to 21
decentralization, **13**–14 to 16
decomposition, **13**–37
engineering trade-offs, **13**–39 to 40
fundamentals, **13**–2 to 3
future trends, **13**–41 to 43
general framework synthesis, **13**–36 to 37
global stabilization, **13**–37
ideal control targets, **13**–38 to 39
implementation, **13**–39 to 40
language usage, **13**–41
linear control of nonlinear systems, **13**–21 to 25
linearization, **13**–3 to 6
linear stabilization, **13**–6 to 14
localization, **13**–16 to 19
massive arrays, **13**–14 to 16
nonlinear optimization, **13**–26 to 34
robustification, **13**–34 to 36
spatially developing flows, **13**–25
Flow field modeling, **4**–3 to 6
Flow field visualization, **30**–20
Flow pattern maps, **32**–18
Flowrate and pressure, electro-osmotic flow, **6**–30 to 31
Flows. *See also* Gas flows; Liquid flows; Turbulent flows
boundary conditions, **4**–11 to 17
compressibility, **4**–9 to 11
fluid modeling, **4**–3 to 6
fundamentals, **4**–1 to 3, **4**–33 to 34
liquid flows, **4**–24 to 30
molecular-based models, **4**–18 to 23
spatial development, **13**–25
structure, **5**–12, **32**–8 to 9
surface phenomena, **4**–30 to 33
two-phase convection patterns, **32**–22 to 23
Flow temperature distribution, **32**–11 to 12
Fluctuating velocities, **26**–19 to 22
Fluidic amplifiers, **17**–47 to 48
Fluid modeling, **4**–3 to 6
Fluid simulator, **5**–9 to 11
Fluorine plasmas, **21**–3 to 5
Fluxes
critical heat, **32**–11, **32**–21 to 22
momentum, **10**–37

vorticity, **33**–28 to 29, **34**–9 to 10
Flux-vector splitting method, Steger-Warming, **7**–15
Fomblin-Zdol lubricant, **9**–25
Force measurements, **3**–6
Force to displacement conversion, **24**–6 to 7
Forcing parameters, free-shear flows, **35**–6 to 8
Fourier's equations, laws, and transforms
continuum hydrodynamics, **6**–5
heat sinks, **32**–5
linearization, **13**–4
nonspatially invariant systems, **13**–20
space compensator reduction, **13**–19 to 20
three-dimensional systems, **13**–16 to 17
Fracture tests, **3**–16
Fracture toughness, **3**–3
Free-molecule flow, **4**–16, **4**–22
Free-shear flows
actuators, **35**–6 to 8, **35**–12 to 14, **35**–21 to 22
control law, **35**–9
delta wing, roll control, **35**–9 to 18
electronics, **35**–6, **35**–8
flow control, **33**–7, **35**–14 to 18, **35**–23 to 27, **35**–32 to 33
forcing parameters, **35**–6 to 8
fundamentals, **35**–1
future trends, **35**–33 to 35
low-Reynolds-number wings, **35**–30 to 33
measurements, **35**–5
MEMS control perspective, **35**–2 to 3
packaging, **35**–5, **35**–8
robustness, **35**–5, **35**–8
roll control, **35**–9 to 18
sensors, **35**–3 to 6, **35**–8, **35**–10 to 12, **35**–19 to 20, **35**–31 to 32
separation control, **35**–30 to 33
shear layer components and issues, **35**–3 to 9
supersonic jet screech, **35**–19 to 30
system integration, **35**–18, **35**–27 to 30
Free-space microoptical bench (FSMOB), **28**–7
Freeze-drying, **16**–124
Frequency response, **30**–7 to 8
Frequency response design methods, **12**–11 to 15
Fresnel lens and zone plates, **16**–134, **17**–52 to 53
Friction factor, **6**–7 to 8, **6**–12 to 13
Front-to-back mask alignment, **9**–22
FSMOB. *See* Free-space microoptical bench (FSMOB)
Fuel injection, **30**–22 to 23
Full state feedback linearization, **12**–26 to 28, **12**–29
Full-system simulation, **5**–2 to 3
Fusion-bonded wafers, **16**–83 to 84
Future trends
bubble/drop transport, **11**–12
flow control, distributed architectures, **13**–41 to 43
free-shear flows, **35**–1 to 35
nanoelectromechanical systems, **36**–1 to 20
skin-friction reduction, **34**–1 to 13
Fuzzy logic and control
alternative inference systems, **14**–30 to 31
fundamentals, **14**–18, **14**–24 to 28, **33**–37
fuzzy vs. crisp sets, **14**–24 to 27
implementation example, **14**–18 to 24
operations on fuzzy sets, **14**–27 to 28

G

Gait mimicry, **28**–27
Galerkin projection
    flow-structure interactions, **5**–12
    fluid simulator, **5**–9 to 10
    nonlinear dynamical systems theory, **33**–31
    structural simulator, **5**–14
Galium arsenide (GaAs), **15**–21 to 22
Gas entrapment, **16**–104
Gases vs. liquids and solids, **4**–24
Gas flows. *See also* Flows; Liquid flows; Turbulent flows
    Boltzmann equation research, **8**–17 to 19
    compressibility, **5**–4
    coupling with continuum equations, **8**–17
    dense gas flows, **8**–20 to 23
    direct simulation Monte Carlo method, **8**–7 to 17
    electric double layer, **8**–20 to 21
    electrokinetic effects, **8**–20 to 21
    electro-osmotic flow, **8**–22
    flow past spheres in pipes, **8**–16
    fundamentals, **8**–2 to 4, **8**–23 to 24
    hybrid Boltzmann/continuum simulation, **8**–19
    information preservation, **8**–16
    lattice Boltzmann methods, **8**–19 to 20
    liquid gas flows, **8**–20 to 23
    molecular dynamics method, **8**–22
    molecular magnitudes, **8**–4 to 7
    moving boundaries, **8**–16 to 17
    parallel algorithms, **8**–17
    separated rarefied gas flows, **8**–15 to 16
    single-phase convection, **32**–10 to 12
    treatment of surfaces, **8**–23
Gate plates, **17**–37
Gauge factor, single-crystal silicon carbide, **20**–8 to 17
Gauss' laws, **6**–26
Generalized hydrodynamic equations, **7**–3
Generators. *See* Microdroplet generators
Genetic algorithms, **14**–13 to 18, **33**–37
Germanium-based materials, **15**–14 to 15
Ginzburg-Landau equation, **33**–34
Glossaries. *See also* Bibliographic references; Nomenclature
    chemical sensors, **22**–22 to 23
    control theory, **12**–31 to 32
    deep reactive ion etching, **21**–11 to 12
    electrochemical fabrication, **19**–22
    fuzzy logic and control, **14**–32 to 33
    microchannel heat sinks, **32**–28 to 30
    microrobotics, **28**–36
    surface-micromachined mechanisms, **27**–33 to 34
    vacuum pumps, microscale, **29**–25 to 26
    X-ray based fabrication, **18**–41
Gold absorber structures, **17**–12 to 13, **17**–14
Görtler instabilities, **33**–8
Gouy-Chapman diffuse layer, **6**–19
Gradients, **13**–32 to 33, **27**–6
Grain boundaries, **16**–104
Grain-boundary growth, **15**–5
Gravitational effects, **2**–5 to 6, **4**–30, **10**–15 to 18
Grid velocity algorithm, **5**–12 to 13
Grippers, **28**–15 to 18
Gulliver's Travels, **1**–1, **1**–4, **4**–1
Gyroscopes, **24**–7, **24**–22 to 24. *See also* Inertial sensors

H

Hamiltonian system, **33**–33
Hamker constant, **10**–19
$H\infty$ approach, **13**–7 to 10
Hardness, silicon, **16**–22 to 24
Heat exchanger application
    adaptive neurocontrol, **14**–9 to 12
    feedback loop stabilization, **14**–9
    fundamentals, **14**–6
    genetic algorithms, **14**–17
    optimal control, **14**–12
    steady state, **14**–6 to 7
    thermal neurocontrol, **14**–7 to 12
Heat sinks. *See* Microchannel heat sinks
Heatuator principle, **28**–29
Hellman-Feyman theorem, **4**–32
Helmholtz-Smoluchowski equation, **6**–25 to 26
HEMA. *See* Hydroxyethyl methacrylate (HEMA)
Henon map, **33**–33
Hermetic sealing, **23**–4
Hertzian equation, **27**–4
Heterogeneous substrates, **10**–25 to 27
Heuristic physical arguments, **33**–23
HEXIL process, **16**–96, **16**–139, **28**–16
High-aspect-ratio electrodeposited metal mold inserts, **17**–34
High-aspect-ratio etching, **24**–16
High aspect ratio micro-lithography, **17**–14 to 15
High-definition televisions, **33**–39
High-temp electronics, chemical sensors, **22**–7, **22**–20 to 21
Hinges
    polysilicon, **16**–134 to 137, **26**–59
    vertical axis, **27**–25 to 27
HNA etching system, **16**–29 to 33
Hobson's pump, **29**–23 to 25
Holography, atomic, **36**–15 to 16
Homogeneity, **27**–2
Homogeneous substrates, **10**–20 to 25
Hooke's law, **16**–19
Horseshoes, **33**–35
Hot embossing, **17**–38 to 39
Hot-wire sensors, **26**–17 to 19
6H-SiC gauge factor, **15**–17, **20**–8 to 17
The Hunt for Red October, **33**–26
Hybrid meshes, **5**–10
Hybrid systems in control theory, **12**–30
Hydraulic cross-talk, **30**–11
Hydraulic diameter, **6**–6 to 7
Hydrazine water, **16**–41
Hydrocarbon detection, **22**–15 to 17
Hydrodynamic effects, **32**–16 to 17
Hydrodynamic surface pressure fluctuation, **35**–4
Hydrofluoric acid, **16**–134
Hydrogenated amorphous silicon, **16**–154 to 155
Hydrogen detection, **22**–15 to 17

Hydrogen sensors, **22**–8 to 11
Hydrogen technology, **22**–8 to 11
Hydrostatic pressure drops, **31**–3
Hydrostatic system operation, **9**–20 to 21
Hydroxides, **16**–41
Hydroxyethyl methacrylate (HEMA), **17**–19
Hyperbolic tangent hidden units, **34**–5
Hypersonic flow calculations
    axisymmetric, blunt body flow, **7**–18 to 23
    shock structure, **7**–15
    two-dimensional, blunt body, **7**–15 to 18

I

IC compatibility, **16**–129 to 132
III-V compounds, **15**–21 to 22
Impact dynamics, **24**–10 to 12
Implementation of distributed control techniques, **13**–39 to 40
Impurity drag, **15**–5
Inch-worm robots, **28**–27
Indentation tests, **3**–15
Inductively coupled plasma, **21**–2
Inductors, **17**–57
Inertial droplet actuators, **30**–7
Inertial sensors. *See also* Sensors
    acceleration concepts, **24**–4 to 5
    accelerometer design, **24**–5 to 13
    applications, **24**–2 to 4
    chips, **24**–18
    converting force to displacement, **24**–6 to 7
    coupled sensitivity, **24**–23
    dashpot, **24**–7 to 10
    device damping, **24**–7 to 12
    etching, **24**–16
    example, **24**–20 to 22, **24**–24
    fundamentals, **24**–1 to 2, **24**–24
    impact dynamics, **24**–10 to 12
    integration approaches, **24**–18 to 20
    loop control, **24**–20
    manufacturing issues, **24**–15 to 17
    material stability, **24**–16
    micromachining technologies, **24**–13 to 15
    packaging, **24**–16 to 17
    partitioning, **24**–18
    quadrature error, **24**–23
    rotational gyroscopes, **24**–22 to 24
    sensing method, **24**–12 to 13
    stiction, **24**–15 to 16
    system issues, **24**–17 to 22
    transduction, **24**–12 to 13
Infrared band pass filter, resonant mesh, **17**–42
Injection, and suction, **33**–21 to 22, **33**–40
Injection molding, **17**–34, **17**–38
Inkjet printing, **30**–20. *See also* Microdroplet generators
In-plane bending, **3**–11
Insertion point, LIGA technique, **17**–60
Instant masking, **19**–5 to 8
Insulator surface micromachining, **16**–140 to 146
Integrated circuit cooling, **30**–24 to 26

Integrated simulation of flows
    circuit-microfluidic device simulation, **5**–15 to 20
    circuit simulator, **5**–8 to 9
    comparison of circuit, fluid, and solid, **5**–15
    coupled circuit-device simulation, **5**–7 to 8
    coupled-domain problems, **5**–4 to 5
    demonstrations, **5**–20 to 22
    equation solution, **5**–8 to 9
    flow-structure interactions, **5**–12
    fluid simulator, **5**–9 to 11
    full-system simulation, **5**–2 to 3
    fundamentals, **5**–2 to 6, **5**–22 to 23
    grid velocity algorithm, **5**–12 to 13
    linearization, **5**–8
    MEMS flow complexity, **5**–3 to 4
    microfluidic system, **5**–20 to 21
    Nektar, **5**–9 to 11, **5**–21
    overview, **5**–1 to 2
    prototype problem, **5**–4 to 6
    simulator overview, **5**–8 to 15
    SPICE3, **5**–8 to 9, **5**–21
    structural simulator, **5**–13 to 15
    time discretization, **5**–8
IntelliCAD, **5**–3
Interfacial conditions, **10**–36 to 39
Interferometers, Fabry-Perot, **25**–18
Interlocking gears, **17**–56
Internal stress, **17**–19 to 21
Intrinsic stress, **16**–103 to 104
In-use stiction, **16**–124 to 125
Invariant systems, **13**–20 to 21
Inverted pendulum system, **12**–18 to 19, **12**–20 to 21
Ion beam lithography, **36**–2
Ion implantation, **15**–6
Iso-etch curves, **16**–30 to 32
Isothermal films, **10**–15 to 29
Isotropic etching. *See also* Wet isotropic etching
    history, **16**–4
    single-crystal silicon, **15**–2
    vs. anisotropic etching, **16**–62 to 64

J

Jet screech, supersonic, **35**–19 to 30
Joule heating, **6**–32, **26**–59
Joule-Thomson refrigerators, **6**–10 to 11
Journal bearings. *See also* Bearings
    advanced designs, **9**–17 to 19
    hydrodynamic operation, **9**–14
    hydrostatic operation, **9**–20 to 21
    macroscale, **26**–66
    side pressurization, **9**–19 to 20
    stability, **9**–16 to 17
    static bearing behavior, **9**–15 to 16

K

Kalman filter
    chaos control, **33**–31, **33**–35
    turbulence control, **33**–10

Kanaya-Okayama (K-O) range, **36**–6
Kang and Choi drag reduction, **34**–11 to 12
Kapton preabsorber, **17**–11
Karhunen-Loève decomposition method, **33**–31
von Kármán equation
    magnetohydrodynamic control, **33**–26 to 28
    scaling, **26**–11, **33**–16
    suction, **33**–11
    targeted flow control, **33**–23
Kelvin-Hemholtz-type instabilities, **31**–10
Kernels, **13**–16 to 18, **13**–38
Kern model, **16**–62
Kinetic effects
    boundary conditions, **10**–10
    energy, **33**–14
    pumps, **29**–3, **29**–4, **29**–8 to 11
    theory of dilute gases, **4**–19
    viscosity, **6**–4, **26**–11, **33**–15 to 16
Knudsen Compressor, **29**–18 to 23, **29**–25
Knudsen number
    boundary conditions, **4**–13 to 16
    Burnett equations, **7**–1 to 7, **7**–13 to 14, **7**–27
    continuum hypothesis, **9**–3, **32**–3 to 4
    flowfield modeling, **4**–5 to 6
    fluid simulators, **5**–10
    gas flows, **4**–34, **8**–2 to 7, **32**–11
    liquid flows, **4**–25, **4**–27
    MEMS flows, **5**–4
    molecular-based models, **4**–21 to 23
    scales, **26**–12, **33**–17
    vacuum pumps, **29**–14, **29**–16
Knudsen pump, **4**–13
KOH water system
    anisotropic etching, **16**–41 to 45
    etch stop techniques, **16**–72 to 73
    Seidel's model, **16**–55 to 57
    silicon carbide, **15**–17
    undercutting, **16**–85 to 86
Kolmogorov length scale, **26**–10 to 13, **26**–47, **33**–15 to 19
Koumoutsakos drag reduction, **34**–9 to 10
Kronecker delta, **4**–7
Kulite sensor, **35**–20
Kutta condition, **26**–63, **35**–21

L

Lagrangian forms, **5**–12, **13**–4
Laminar boundary conditions, **13**–25
Langmuir's slip-boundary conditions, **7**–12, **7**–24 to 25
Language usage, **13**–41
LaPlace-Young equation, **31**–3
Laplacian equations
    electro-osmotic flows, **6**–27
    flow-structure interactions, **5**–12
    $H\infty$ approach, **13**–9
    linear control, **12**–2 to 5
Large aspect ratio magnetic motors, **17**–58
Lateral resonators, **16**–113
Lattice Boltzmann methods, **8**–19 to 20
Launch vehicle applications, **22**–3 to 4

Lead-lag compensation, **12**–15 to 17
Leak detection, **22**–3 to 4, **22**–18 to 20
Learning rate, artificial neural networks, **14**–5
Length scales, **32**–6
Lennard-Jones potential
    constant surface tension, **10**–19
    liquid flows, **4**–26, **4**–29
    molecular dynamics method, **8**–22
    surface phenomena, **4**–32 to 33
Leverage and linkage systems, **28**–16
LIGA technique. *See also* X-ray lithography
    accelerometers, **17**–55 to 56
    actuators, electrostatic, **17**–56 to 7
    adhesion, resist, **17**–19
    alternative materials, **17**–41 to 42
    amplifiers, fluidic, **17**–47 to 48
    applications, **17**–42 to 60
    carriers and cathodes, **17**–37 to 38
    comparing micromolds, **17**–33
    competing technologies, **17**–60 to 61
    compression molding, **17**–38 to 39
    connectors and packages, **17**–48 to 50
    demolding, **17**–39 to 40
    deposited dose, **17**–23 to 24
    development, **17**–24 to 25
    double deflecting gas nozzles, **17**–44
    electrically conductive plastic plates, **17**–37 to 38
    electrodeposition, **17**–25 to 33
    electroless deposition, **17**–32 to 33
    electromagnetic micromotors, **17**–57 to 59
    electronic microconnectors and packages, **17**–48 to 50
    electrostatic actuators, **17**–56 to 57
    extensions, **17**–40 to 41
    fiberoptic connectors, **17**–51 to 52
    fluidic amplifiers, **17**–47 to 48
    Fresnel zone plates, **17**–52 to 53
    fundamentals, **17**–1 to 2, **17**–42 to 43, **18**–2, **22**–6
    history, **17**–2 to 4
    inductors, **17**–57
    injection molding, **17**–38
    insertion point, **17**–60
    interlocking gears, **17**–56
    membrane filters, **17**–48
    metal microstructures, **17**–29, **24**–15
    microconnectors and packages, electronic, **17**–48 to 50
    microgrippers, **28**–16
    micromolding, **17**–33, **26**–7 to 8
    micro-optical components, **17**–50 to 55
    micropump, **17**–23 to 24
    microvalves, **17**–45 to 46
    mircofluidic elements, **17**–43 to 48
    molding, **17**–29, **17**–33 to 40, **18**–13 to 14
    motors, electromagnetic micro, **17**–57 to 59
    movable structures, **17**–40
    multiple spin coats, **17**–18
    nickel electrodeposition, **17**–25 to 29
    nozzles, **17**–44 to 45
    optical components, micro, **17**–50 to 55
    optimum wavelength, **17**–21 to 22
    packages, electronic microconnectors and, **17**–48 to 50

patterns, **16**–148
plastic plates, electrically conductive, **17**–37 to 38
plating automation, **17**–30 to 32
PMMA, **17**–18 to 21, **17**–33
primary substrate choice, **17**–16
probe cards, **17**–60
processes, **17**–16 to 42
pull-push connectors, **17**–52
pulse plating, **17**–29
pump, micro, **17**–23 to 24
reaction injection molding (RIM), **17**–35 to 38
resist requirements and applications, **17**–16 to 19
slanted microstructures, **17**–41
spectrometers, **17**–53 to 54
spinneret nozzles, **17**–44 to 45
stepped microstructures, **17**–40 to 41
stress-induced cracks, **17**–19 to 21
stripping PMMA, **17**–33
synchrotron orbital radiation, **17**–4 to 8
technology access, **17**–8
technology barriers, **17**–60 to 61
turbine rotors, **17**–45
uranium enrichment, **17**–44
valves, micro, **17**–45 to 46
X-ray masks, **17**–9 to 15
Lighthill's mechanism, **34**–9
Linear control. *See* Control theory
Linearization
    Burnett equations, **7**–12 to 14
    feedback linearization, **12**–24 to 28
    flow control modeling, **13**–3 to 6
    full state feedback linearization, **12**–26 to 28, **12**–29
    perturbation field, **13**–30
    simulation overview, **5**–8
    stabilization, **13**–6 to 14
Linear magnetic microactuators, **18**–35
Linear output units, **34**–5
Linear quadratic regulator (LQR), **12**–19 to 21
Linguistic variables, **14**–28
Liouville equation, **4**–19 to 21, **8**–5
Liquid flows. *See also* Flows; Gas flows; Turbulent flows
    capacitive effects, **6**–18
    electric double layer, **32**–12
    electrokinetics background, **6**–18 to 32
    entrance length development, **6**–8
    experimental studies, **6**–10 to 18
    flow physics, **4**–24 to 30
    fundamentals, **6**–1 to 10, **6**–32 to 33
    granularity, **5**–4
    hydraulic diameter, **6**–6 to 7
    measurements of velocity, **6**–15 to 17
    microchannel uniqueness, **6**–3 to 4
    nomenclature, **6**–33 to 35
    noncircular channels, **6**–9 to 10
    nonlinear channels, **6**–17
    polarity, **32**–12
    pressure driven flow, **6**–5 to 6
    round capillaries, **6**–7 to 8
    single-phase convection, **32**–12 to 17
    transition to turbulent, **6**–8 to 9

Liquid refilling time, **30**–8
Liquids
    compressibility, **6**–18
    pressure drop, **31**–4
    vs. gases and solids, **4**–24
Lithography
    applications, **17**–5
    deep-etch, **17**–17
    electron beam, **36**–2, **36**–3 to 9
    high aspect ratio micro-lithography, **17**–14 to 15
    ion beam, **36**–2
    MEMS failure mechanisms, **27**–31 to 32
    nanoimprint lithography, **36**–11 to 12
    synchrontron radiation, **17**–4 to 8
    UV depth, **16**–147 to 148
    variations, **27**–31 to 32
    X-ray (XRL), **36**–2, **36**–9 to 11
Locomotive devices, **28**–12 to 14, **28**–13
Log plot (scaling theory), **2**–2 to 4
London force, **10**–18
Long-wave theory approach, **10**–8
Loop control, **24**–20
Loopshaping, **12**–23
Loop transfer recovery (LTR), **13**–6
Lorentz forces, **6**–26, **28**–13, **33**–27 to 30
Low-pressure chemical vapor deposit (LPCVD), **16**–119
    base layer, **16**–119
    germanium based materials, **15**–14
    materials case studies, **16**–150 to 153
    nitride, **16**–157 to 158
    polysilicon, **15**–7 to 9, **16**–95 to 97, **16**–137
    surface micromachining process, **16**–117, **26**–5 to 6
    vertical play, **27**–27
Low-speed regions, **33**–14
Low-z membrane material, **17**–10
LPCVD. *See* Low-pressure chemical vapor deposit (LPCVD)
LQR. *See* Linear quadratic regulator (LQR)
LTR. *See* Loop transfer recovery (LTR)
Lubrication
    constraints, **9**–11 to 12
    continuum hypothesis, **9**–3 to 4
    Couette-flow damping, **9**–5 to 7
    cross-wafer uniformity, **9**–22
    cube-square law, **9**–2 to 3
    deep etch uniformity, **9**–23
    device aspect ratio, **9**–11 to 12
    fabrication issues, **9**–22 to 24
    fundamentals, **9**–1 to 2, **9**–25 to 26
    governing equations, **9**–4 to 5, **9**–7 to 9
    hydrodynamic operation, **9**–14
    hydrostatic operation, **9**–20 to 21
    journal bearings, **9**–14 to 21
    material properties, **9**–23 to 24
    minimum etchable clearance, **9**–12
    molecular law, **9**–6 to 7
    reduced-order models, **9**–9 to 10
    rotating devices, **9**–10
    scaling issues, **9**–2 to 4
    side pressurization, **9**–19 to 20

silicon, **9**–25
squeeze-film damping, **9**–7 to 10
stability of journal bearings, **9**–16 to 17
static journal bearing behavior, **9**–15 to 16
stiction, **9**–25
surface roughness, **9**–4
theory, **24**–10
thrust bearings, **9**–12 to 14
tribology and wear, **9**–25
vent holes, **9**–9
Lyapunov stability theory, **12**–20, **12**–29 to 30

M

Machine design, **27**–7 to 18
Mach number
 Burnett equations, **7**–28
 compressibility, **4**–9, **4**–10 to 11
 flow control, **33**–7 to 8, **33**–17, **35**–33
 flowfield modeling, **4**–4, **4**–6
 gas flows, **4**–34
 vapor pressure drop, **31**–5
Mack modes, **33**–8
Macroporous silicon, **16**–69 to 71
Macroscale (MAS) flows, **35**–2
Macroscopic roughness, **16**–43
Magnetic actuation principles, **28**–18 to 19
Magnetic flap actuators, **35**–12 to 16
Magnetic forces, **26**–56 to 57
Magnetic microactuator applications, **18**–37 to 39
Magnetic rotational microactuators, **18**–36 to 37
Magnetohydrodynamics (MHD)
 EMD tiles, **33**–29 to 30
 fundamentals, **33**–26
 von Kármán equation, **33**–27 to 28
 propulsion system, **33**–26
 vorticity flux, **33**–28 to 29
Manhattan geometries, **27**–10
Manipulation, micro, **17**–60
Marangoni effect, **10**–29, **10**–31, **11**–7
MAS flows. *See* Macroscale (MAS) flows
Masking processes
 anisotropic silicon etchants, **16**–43 to 44
 electrochemical fabrication (EFAB), **19**–4 to 8, **19**–11 to 12
 front-to-back alignment, **9**–22
 isotropic silicon etchants, **16**–32 to 33
 LIGA technique, **18**–4 to 8, **18**–9 to 15
 material requirements, **20**–4
 underetching, **16**–84
 X-ray based fabrication, **18**–4 to 8
Massive arrays, designing for, **13**–14 to 16
Material case studies, surface micromachining
 amorphous silicon, **16**–154 to 155
 CVD phosphosilicate glass films, **16**–159 to 160
 CVD silicon dioxides, **16**–158 to 160
 CVD undoped SiO_2, **16**–159
 doped polysilicon, **16**–153 to 154
 fundamentals, **16**–149 to 150
 hydrogenated amorphous silicon, **16**–154 to 155
 LPCVD nitride, **16**–157 to 158
 PECVD nitride, **16**–156 to 157
 polysilicon, **16**–150 to 154
 silicon nitride, **16**–155 to 158
 undoped polysilicon, **16**–151 to 153
Materials
 diamond, **15**–20 to 21
 fundamentals, **15**–1, **15**–23
 germanium-based materials, **15**–14 to 15
 III-V compounds, **15**–21 to 22
 metals, **15**–16 to 17
 piezoelectric materials, **15**–22 to 23
 polysilicon, **15**–3 to 9
 properties, **9**–23 to 24, **27**–2 to 7
 silicon carbide, **15**–17 to 20
 silicon dioxide, **15**–9 to 11
 silicon nitride, **15**–11 to 14
 single-crystal silicon, **15**–2 to 3
 stability, **24**–16
Mathematical preliminaries in control theory, **12**–2 to 5
Matlab software, **12**–19, **14**–22. *See also* Software
Maximum overshoot, **12**–10
Maxwell equation applications, **4**–22, **5**–10
Maxwell-Smoluchowski slip-boundary conditions, **7**–12, **7**–18, **7**–27
Meandering flexures, **27**–5
Mean velocities, **26**–19 to 22
Measurements
 biaxial, **16**–110 to 111
 boiling curves, **32**–18
 cantilever beams and spirals, **16**–113 to 115
 clamped-clamped beams, **16**–111
 disk method, **16**–105 to 106
 free-shear flows, **35**–5
 lateral resonators, **16**–113
 micro-PIV, **6**–15 to 16
 moment of inertia, **27**–8
 nonuniformity of stress, **16**–113 to 115
 Poisson ratio, **16**–111
 ring crossbar structures, **16**–111 to 112
 spanwise temperature distributions, **32**–20
 surface temperature, **32**–19
 suspended membrane methods, **16**–110 to 111
 transverse piezoresistors, **20**–13
 turbulence sensors, **26**–14 to 53
 uniaxial, **16**–106 to 110
 vernier gauges, **16**–113
 wall-shear-stress, **26**–28
Mechanical/electrical transduction, **24**–12 to 13
Mechanical folding, **26**–57 to 61
Mechanical interference, **27**–27 to 30
Mechanical properties, MEMS materials. *See* MEMS materials, mechanical properties
MEDUSA, **5**–7
Membrane filters, **17**–48
Membrane tests, **3**–14
MEMCAD, **5**–3
MEMS materials, mechanical properties, **3**–7 to 9
 fundamentals, **3**–1 to 3
 initial design values, **3**–24 to 25

measurements, **3**–5 to 6
mechanical properties, **3**–17 to 23
specimen preparation, **3**–4 to 5
test methods, **3**–3 to 16
MEMS (microelectromechanical systems)
  applications, **1**–3
  basics, **1**–1 to 4
  bubble/drop transport, **11**–1 to 12
  Burnett simulations, **7**–1 to 33
  control perspectives, **35**–2 to 3
  control theory fundamentals, **12**–1 to 30
  distributed architecture flow control, **13**–1 to 43
  fabrication, **26**–3 to 8
  flow physics, **4**–1 to 34
  integrated simulation, **5**–1 to 23
  lubrication, **9**–1 to 26
  mechanical properties, **3**–1 to 25
  molecular-based microfluidic simulation, **8**–1 to 24
  reactive control, **33**–39 to 41
  scaling, **2**–1 to 9
  soft computing in control, **14**–1 to 32
  thin liquid films, physics of, **10**–1 to 44
Mencken, Henry Louis, **33**–42
Menendez and Ramaprian formula, **26**–30 to 32
Mesoscale sputter ion pumps, **29**–16. *See also* Sputter ion pumps
Metal diaphragms, **20**–11
Metallic relief structures, **17**–25
Metallization
  electrical interconnection system, **23**–3
  single-crystal silicon carbide, **20**–17 to 26
  thick-film, **23**–5 to 6
Metal microstructures, **17**–29
Metals, **15**–16 to 17
MHD. *See* Magnetohydrodynamics (MHD)
Microactuators, **18**–34 to 39, **26**–53 to 63
Microbearings, **26**–66 to 68
Microchannel heat sinks, **32**–2 to 8
  boiling curves, **32**–18 to 21
  bubble dynamics, **32**–24 to 25
  continuum hypothesis, **32**–3 to 4
  convective heat transfer, **32**–2 to 17
  critical heat flux, **32**–21 to 22
  electric double layer, **32**–7
  entrance length, **32**–9 to 10
  flow patterns, **32**–22 to 23
  flow structure, **32**–8 to 9
  fundamentals, **32**–1 to 2, **32**–4 to 5, **32**–18, **32**–27 to 28
  gas flow, **32**–10 to 12
  general laws, **32**–4 to 5
  governing equations, **32**–5 to 6, **32**–10
  liquid flow, **32**–12 to 17
  modeling, **32**–25 to 27
  modes of heat transfers, **32**–3
  noncontinuum mechanics, **32**–6
  particular laws, **32**–5
  ploar mechanics, **32**–8
  size effects, **32**–6 to 8
  thermodynamic concepts, **32**–4
  two-phase convection, **32**–18 to 27

Microchannels, liquid flows
  advantages, **6**–1
  applications, **6**–1 to 2
  capacitive effects, **6**–18
  continuum hydrodynamics, **6**–5 to 6
  electrical analogy, **6**–32
  electrical double layers, **16**–19 to 21
  electrokinetics, **6**–17 to 32
  electro-osmotic flow, **6**–21 to 24, **6**–25 to 31
  electrophoresis, **6**–24 to 28
  entrance length development, **6**–8
  experimental studies, **6**–10 to 18
  fundamentals, **6**–1 to 10, **6**–32 to 33
  hydraulic diameter, **6**–6 to 7
  measured behavior, **6**–13 to 15
  microchip, electrokinetic, **6**–27 to 30
  microfluidic networks, **6**–32
  noncircular channels, **6**–9 to 10
  nonlinear channels, **6**–17 to 18
  practical considerations, **6**–32
  pressure driven flow, **6**–5 to 6
  round capillaries, **6**–7 to 8
  transition to turbulent flow, **6**–8 to 9
  unique aspects, **6**–3 to 4
  velocity measurements, **6**–15 to 17
Microchips, **1**–1, **6**–28 to 30
Microconnectors and packages, electronic, **17**–48, **17**–48 to 50
Microcontact printing, **36**–11 to 12
Microconveyors, **28**–18 to 23
Microdroplet generators
  acoustic-wave actuation, **30**–5 to 6
  applications, **30**–20 to 26
  biomedical sample handling, **30**–20 to 22
  characterization of droplet generation, **30**–17 to 20
  chemical sample handling, **30**–20 to 22
  crosstalk, **30**–8 to 11
  direct writing, **30**–23
  droplet trajectory, **30**–17 to 19
  ejection of droplets, **30**–19
  electrostatic actuation, **30**–6
  fabrication, **30**–15 to 17
  flow field visualization, **30**–20
  frequency response, **30**–7 to 8
  fuel injection, **30**–22 to 23
  fundamentals, **30**–1 to 2, **30**–26
  inertial actuation, **30**–7
  inkjet printing, **30**–20
  integrated circuit cooling, **30**–24 to 26
  manufacturing process, **30**–24
  material issues, **30**–14
  mixing control, **30**–22 to 23
  monolithic fabrication, **30**–16 to 17
  multiple pieces, **30**–15 to 16
  operation principles, **30**–2 to 7
  optical components, **30**–23 to 24
  overfill phenomenon, **30**–8 to 11
  packaging, **30**–23
  physical and design issues, **30**–7 to 14
  piezoelectric actuation, **30**–3 to 4

pneumatic actuation, **30**–2 to 3
puddle formation, **30**–13 to 14
sample handling, **30**–20 to 22
satellite droplets, **30**–11 to 13
sequence, **30**–19
solid freeforming, **30**–24
thermal-bubble actuation, **30**–4
thermal-buckling actuation, **30**–5
thermal/hydraulic crosstalk and overfill, **30**–8 to 11
velocity, **30**–19
volume, **30**–19
Microengines, **27**–18 to 21
Microfabricated structures, **19**–15 to 16
Microfactories, **28**–32 to 35
Micro-flex mirror, **27**–23 to 27, **27**–32
Microfluids, **5**–20 to 21, **6**–32
Microgrippers, **28**–15 to 18
Microjoint technique, self-assembling, **26**–61
Microliquid dosing system, **5**–20
Micromachining technologies. *See also* Surface micromachining
  bulk micromachining, **16**–82 to 90
  bulk vs. surface micromachining, **16**–148 to 149
  corner compensation, **16**–84 to 90
  dissolved wafer process, **16**–90 to 91
  dual lumen catheter, **16**–91 to 95
  electrochemical sensor array, **16**–91 to 95
  examples, **16**–90 to 95
  front etching, **16**–82 to 83
  fundamentals, **16**–2 to 3, **16**–82
  inertial sensing, **24**–13 to 15
  insulator surface micromachining, **16**–140 to 146
  manufacturing issues, **24**–15 to 17
  microfabrication, **26**–4 to 5
  polysilicon surface micromachining, **16**–160 to 162
  processes, surface micromachining, **16**–117 to 133
  real estate consumption, **16**–82 to 84
  silicon fusion-bonded wafers, **16**–83 to 84
  silicon-on-insulator (SOI) wafers, **16**–142 to 146
  single-crystal silicon, **15**–2
  surface micromachining process, **16**–95 to 97, **16**–117, **26**–5 to 6
  undercutting, **16**–85 to 90
  underetching, **16**–84
  vs. surface micromachining, **16**–148 to 149
  wet bulk micromachining, **16**–90 to 95
  wet bulk problems, **16**–82 to 90
Micromanipulation, **17**–60
Micromolds, **17**–33, **26**–7 to 8
Micro-optical components, **17**–50 to 55
Microoptoelectromechanical systems (MOEMS), **28**–1, **28**–7
Microphones
  electret, **26**–46
  pinhole, **26**–43
  subminiature, **26**–50
Micro-PIV measurements, **6**–15 to 16
Microporous silicon, **16**–67 to 71
Micropumps, **17**–23 to 24, **26**–63 to 65. *See also* Vacuum pumps, microscale

Microrobotics
  actuator principles, **28**–11 to 14
  applications, **28**–6 to 10
  assembly, **28**–7 to 10
  communication, **28**–35
  conveyors, **28**–18 to 23
  design, **28**–10 to 14
  desktop factories, **28**–32 to 35
  devices, **28**–12 to 32
  examples, **28**–23 to 30
  fundamentals, **28**–1 to 6, **28**–35 to 36
  grippers, **28**–15 to 18
  locomotive devices, **28**–12 to 14
  microassembly, **28**–7 to 10
  microfactories, **28**–32 to 35
  multirobot systems, **28**–32 to 35
  powering, **28**–34
  redundancy criteria, **28**–30 to 32
  scaling phenomena, **28**–11 to 12
  steering, **28**–32
  tools, **28**–15 to 18
  walking microrobots, **28**–23 to 32
Microscale (MIS) flows, **35**–2 to 3
Microscale vacuum pumps. *See* Vacuum pumps, microscale
Microscopic roughness, **16**–43
Microturbomachines, **26**–63 to 68
Microvalves, **17**–45 to 46
Microvoids, **16**–104
Miller indices, **16**–6 to 7
Milliscale molded polysilicon structures, **16**–137 to 140
MIMO full state feedback linearization, **12**–29
Mini-LAD (laser Doppler anemometer), **35**–3
MINIMAN/MINIMAN III, **28**–6, **28**–30
Minimum etchable clearance, **9**–12
Mircofluidic elements, **17**–43 to 48
MIS flows. *See* Microscale (MIS) flows
Mixing control, **30**–22 to 23
Models and modeling. *See also* Algorithms
  compact models for devices, **5**–17 to 20
  continuum equation model, **4**–6 to 8, **8**–17
  flow control for distributed architectures, **13**–1 to 43
  flowfields, **4**–3 to 6
  flow sensor, **5**–17 to 18
  fluid modeling, **4**–3 to 6
  lumped element for devices, **5**–17 to 20
  molecular-based models, **4**–18 to 23
  nonlinear macromodeling approach, **5**–5
  piezoelectric transducers, **5**–17
  pipes, heat, **31**–8 to 11, **31**–12 to 13
  reduced-order for complex geometries, **9**–9 to 10
  resistor-diaphragm, **20**–9 to 13
  spreaders, heat, **31**–16 to 17
  steady-state, **31**–9 to 10
  transient, **31**–10 to 11
  two-phase convection, **32**–25 to 27
Modern control theory, **12**–17 to 24, **13**–6 to 14
MOEMS. *See* Microoptoelectromechanical systems (MOEMS)
Molded polysilicon structures, **16**–137 to 140
Molding, **17**–33 to 40, **26**–7 to 8. *See also* Demolding

Mold inserts, metal, **17**–29
Molecular-based microfluidic simulation
    Boltzmann equation research, **8**–17 to 19
    coupling with continuum equations, **8**–17
    dense gas flows, **8**–20 to 23
    direct simulation Monte Carlo (DSMC) method, **8**–7 to 17
    electric double layer, **8**–20 to 21
    electrokinetic effects, **8**–20 to 21
    electro-osmotic flow, **8**–22
    flow past spheres in pipes, **8**–16
    fundamentals, **8**–1 to 2, **8**–23 to 24
    gas flows, **8**–2 to 23
    hybrid Boltzmann/continuum simulation, **8**–19
    information preservation, **8**–16
    lattice Boltzmann methods, **8**–19 to 20
    liquid gas flows, **8**–20 to 23
    molecular dynamics method, **8**–22
    molecular magnitudes, **8**–4 to 7
    moving boundaries, **8**–16 to 17
    parallel algorithms, **8**–17
    separated rarefied gas flows, **8**–15 to 16
    treatment of surfaces, **8**–23
Molecular effects in liquids, **6**–4
Molecular flow limits, **9**–6 to 7
Molecular magnitudes, **8**–4 to 7
Molecular weight, **17**–21, **17**–23
Moment of inertia, **27**–8
Monolithically integrated transducer technologies, **24**–18 to 20
Monolithic fabrication, **30**–16 to 17
Monte Carlo simulations
    coupling with continuum equations, **8**–17
    electron scattering, **36**–6
    flow past spheres in pipes, **8**–16
    gas flows, **8**–11 to 16
    Hobson's pump, **29**–23
    information preservation, **8**–16
    limitations and error sources, **8**–9 to 11
    molecular-based modeling, **4**–23
    moving boundaries, **8**–16 to 17
    overview, **8**–7 to 9
    parallel algorithms, **8**–17
    recent advances, **8**–16 to 17
    separated rarefied gas flows, **8**–15 to 16
    surface phenomena, **4**–31
Moody chart, **6**–9
Motors, electromagnetic micro, **17**–57 to 59
Movable structures, **17**–40
Moving separation line (MSL), **35**–2
MQW structures. *See* Multiple quantum-well (MQW) structures
MSL. *See* Moving separation line (MSL)
Multimode waveguide, **17**–50
Multipin microconnectors, electrical, **17**–48 to 50
Multiple-material layers, EFAB, **19**–9 to 10
Multiple quantum-well (MQW) structures, **25**–18
Multiple spin coats, **17**–18
Multipoly process, **15**–7
Multirobot systems, **28**–32 to 35
Murphy's law, **13**–34 to 36, **27**–28

N

NACA 0012 airfoil, **7**–23
Nanocrystalline tin oxide thin films, **22**–7, **22**–11 to 13
Nanoelectromechanical systems (NEMS)
    AFM patterning, **36**–13 to 14
    atomic resolution, **36**–12 to 16
    beam size, **36**–3 to 5
    biology interface, **36**–16 to 20
    compatible processing techniques, **36**–3 to 16
    conventional resists, **36**–7 to 9
    developers, **36**–7 to 9
    electron beam lithography, **36**–3 to 9
    electron scattering, **36**–5 to 7
    free-shear flows, **35**–3
    fundamentals, **36**–1 to 2, **36**–20
    holography, atomic, **36**–15 to 16
    microcontact printing, **36**–11 to 12
    nanoimprint lithography, **36**–11 to 12
    resolution limits, **36**–3 to 9
    STM patterning, **36**–12 to 13
    unconventional resists, **36**–9
    X-ray lithography, **36**–2, **36**–9 to 11
Nanoimprint lithography, **36**–11 to 12
Nanomaterials, **22**–7
Nanomotor, **28**–4
Nanoscale metal lines, **36**–13
NASICON-based CO_2 detection, **22**–17
Navier-Stokes equation
    boundary conditions, **4**–12, **4**–14 to 15
    Burnett equations, **7**–1 to 7, **7**–13 to 17, **7**–27
    classification schemes, **33**–5 to 6
    continuum model, **4**–7 to 8, **9**–3
    control law, **35**–9
    Couette damping, **9**–5 to 7
    coupled circuit/microfluidic device simulator, **5**–6
    flowfield modeling, **4**–3, **4**–5 to 6
    flow physics, **4**–2
    flow-structure interactions, **5**–12
    fluid simulation, **5**–9 to 10
    gas flows, **8**–2
    linearization, **13**–4, **13**–30
    liquid flows, **4**–25 to 30
    MEMS flows, **5**–4
    microbearings, **26**–68
    molecular-based models, **4**–19, **4**–22 to 23
    nonlinear optimization, **13**–26 to 34
    reactive feedback control, **33**–23 to 24
    satellite droplets, **30**–13
    software integration, **5**–16
    squeeze-film damping, **9**–8
    well-posedness, **13**–35
Near-wall streamwise vortices, **34**–2, **34**–11 to 12
Nektar, **5**–9 to 11, **5**–14 to 21
NEMS. *See* Nanoelectromechanical systems (NEMS)
Nernst-Einstein equation, **6**–20
Neumann condition, **11**–8
Neural networks, **33**–36, **33**–38 to 39, **34**–4 to 7, **34**–13
Neurocomputing, **33**–36
Neurocontrol. *See* Artificial neural networks (ANN)

Newtonian fluids, **6**–5, **6**–14, **33**–28
Newton's laws, **10**–30, **32**–5
Nickel electrodeposition, **17**–25 to 29
Nitride, **16**–156 to 157, **16**–157 to 158
Noise, pressure sensor sensitivity, **25**–12
Nomenclature. *See also* Bibliographic references; Glossaries
    Burnett simulations, **7**–33 to 34
    liquid flows in microchannels, **6**–32 to 33
    pipes and spreaders, heat, **7**–22 to 23
Noncircular channels, flow in, **6**–9 to 10
Nonlinear channels, flow in, **6**–17
Nonlinear control systems, **13**–21 to 25. *See also* Control theory
Nonlinear dynamical systems theory, **33**–24, **33**–31 to 32
Nonlinear macromodeling approach, **5**–5
Nonlinear optimization, **13**–26 to 34
Non-normal systems, modern control design, **13**–11 to 13
Nonphysical assumption relaxation, **13**–16 to 19
Non-poly-Si modifications. *See also* Polysilicon
    fundamentals, **16**–140 to 142
    insulator surface micromachining, **16**–140 to 146
    selective epitaxy, **16**–144
    silicon fusion-bonding, **16**–142
    SIMOX wafers, **16**–143 to 144
    SOI vs. poly-Si, **16**–144 to 146
Nonspatially invariant systems, **13**–20 to 21
Normalization, artificial neural networks, **14**–5
Noses, electronic, **22**–20 to 21
No-slip boundary conditions, **4**–12, **4**–33, **8**–2
Notching, surface, **16**–43
Nozzles, **17**–44 to 45
N requirements, **29**–14
Nucleation sites, **32**–24
Numerical simulations. *See* Models and modeling
Nusselt number
    hot wire sensors, **26**–18 to 19
    liquid flows, **32**–15
    spatial resolutions, **26**–33
Nyquist plot, **12**–11, **12**–13 to 15

O

On-chip waveguides, LIGA applications, **17**–59
One-layer structures, **26**–61 to 63
Open loop control, **24**–20
Operating pressures, vacuum pumps, **29**–14
Opposition control, **34**–2 to 4
Optical components, **17**–50 to 55, **30**–23 to 24
Optical Doppler tomography, **1**–16
Optical pick-off pressure sensors, **25**–17 to 18
Optical sensors, **17**–59, **25**–18
Optimal control theory, **33**–24
Optimum wavelength, **17**–21 to 22
Orbitron iron pump, **29**–12 to 13, **29**–16
[100]-orientation, **16**–9 to 13
[110]-orientation, **16**–13 to 16
Orr-Sommerfeld/Squire formulation, **13**–4
Ott-Grebogi-Yorke's (OGY) method, **33**–33 to 35
Out-of-phase control streaks, **34**–10
Out-of-plane bending test, **3**–9 to 10

Out-of-plane structures, **27**–27 to 30
Overfill phenomenon, **30**–8 to 11

P

Packaging
    die-attach, **23**–3 to 4, **23**–8 to 18
    electronic microconnectors and, **17**–48 to 50
    free-shear flows, **35**–5, **35**–8
    fundamentals, **23**–1 to 2, **23**–19 to 20
    hermetic sealing, **23**–4
    high-temp electrical interconnection system, **23**–4 to 10
    inertial sensors, **24**–16 to 17
    material requirements, **23**–2 to 4
    mechanism reliability, **27**–33
    metallization/electrical interconnection system, **23**–3
    microdroplet generators, **30**–23
    sensors, **16**–16, **24**–17
    substrates, **23**–2 to 3
    thick film metallization, **23**–5 to 6
    wafer level, techniques, **24**–17
    wirebond, **23**–6 to 8
Parallel-axis rotary pumps, continuous, **26**–64
Parallel fabrication/assembly, **28**–2, **28**–9
Parameters, forcing, **35**–6 to 8
Partitioning, system, **24**–18
Passive optical components, **17**–54 to 55
Pattern transfer, **16**–118 to 119
PDMS. *See* Polydimethylsiloxane (PDMS)
Peak time, **12**–10
PEC-etching. *See* Photoelectrochemical etching (PEC-etching)
Péclet number, **26**–33
PECVD. *See* Plasma-enhanced chemical vapor deposit (PECVD)
Penning discharge, **29**–11
Perturbation field, **13**–30
Photo-assisted electrochemical etch stop, **16**–78 to 79
Photoelectrochemcial fabrication principles, **20**–3 to 8
Photoelectrochemical etching (PEC-etching), **16**–66 to 67
Photo-induced preferential anodization (PIPA) etch stop, **16**–79 to 81
Photolithography, **3**–16
Photoresist, **17**–30, **18**–5 to 8, **18**–8 to 9
Photosensitive resists, **17**–60
Physical intuition based control, **34**–2 to 4
Physical-space compensator reduction, **13**–20
Physical-space kernels, **13**–16
Physics
    fluid flow, **4**–1 to 34
    impact on MEMS, **1**–4
    thin liquid films, **10**–1 to 44
PI control. *See* Proportional-Integral (PI) control
Piezo actuation, **30**–1
Piezoelectric properties
    actuation, **30**–3 to 4
    materials, **15**–22 to 23
    transducers, **5**–17
Piezoresistant properties
    sensors, **25**–4 to 7, **25**–12, **26**–47 to 50

silicon, **16**–24 to 28
silicon nitride, **15**–12
Pinhole microphones, **26**–43
PIPA. *See* Photo-induced preferential anodization (PIPA) etch stop
Pipes, heat
　arrays, **31**–12 to 15
　boiling limitations, **31**–7
　capillary limitation, **31**–2 to 6
　designs, new, **31**–18 to 21
　entrainment limitation, **31**–6 to 7
　fundamentals, **31**–1 to 8, **31**–8, **31**–22
　hydrostatic pressure drops, **31**–3
　liquid pressure drop, **31**–4
　modeling, **31**–8 to 11, **31**–12 to 13
　sonic limitation, **31**–6
　steady-state modeling, **31**–9 to 10, **31**–11
　testing, **31**–11 to 12, **31**–13 to 14
　thermal resistance, **31**–7 to 8
　transient modeling, **31**–10 to 11, **31**–12
　vapor pressure drop, **31**–5 to 6
　viscous limitation, **31**–6
Pirani sensors, **25**–19
Pister's studies, **16**–134 to 135
Pitot tube, **26**–36
Planar grating spectrograph, **17**–53 to 54
Planarization, **19**–13
Plasma cleaning procedure, **21**–7 to 8
Plasma-enhanced chemical vapor deposit (PECVD)
　film stress, **16**–101
　materials case studies, **16**–154
　nitride, **16**–156 to 157
　pirani sensors, **25**–19
　polysilicon, **15**–10
Plasma etching, **21**–2 to 3
Plasma polymerized PMMA, **17**–19`
Plastic plates, electrically conductive, **17**–37 to 38
Plating
　automation, **17**–30 to 32
　issues, **17**–30 to 32
　through-mask, **19**–4
Plug flow profile, **6**–30
PMMA. *See* Poly(methyl methacrylate) (PMMA)
Pneumatic actuation, **28**–18 to 19, **30**–2 to 3
Point-wise velocity measurements, **6**–16
Poiseuille laws and applications, **6**–8, **33**–34
Poisson ratio
　bubble transport, **11**–8 to 9
　contact stresses, **27**–4
　coupled domain problems, **5**–5
　diaphragm deflection, **25**–2
　electrical double layers, **6**–20, **8**–20
　film stress, **16**–102
　material properties, **27**–3
　mechanical properties, **3**–20
　membrane tests, **3**–14
　resistor-diaphragm modeling, **20**–12
　silicon material, **16**–18 to 20
　thin films, **16**–111
Polar liquid flows, **4**–25

Pole placement, **12**–18 to 19
Polydimethylsiloxane (PDMS), **36**–11
Polyimide surface structures, **16**–146 to 147
Polymer shrinkage, **16**–104
Poly(methyl methacrylate) (PMMA)
　beam size, **36**–5
　cross-linked, **17**–21
　development, **18**–10 to 11, **36**–8
　electron scattering, **36**–6
　Fresnel zone plates, **17**–52
　large-grained films, **16**–125
　LIGA technique, **17**–18 to 21, **17**–33, **18**–9 to 10
　linear, **17**–21
　mechanical properties, **18**–11 to 12
　metals, **15**–17
　microdroplet generators, **30**–24
　microoptical components, **17**–50 to 51
　mold filling, **18**–12 to 17
　photo resist, **18**–8 to 9
　resists, **17**–16, **36**–8
　undoped films, **16**–125
　X-ray mask fabrication, **18**–4 to 8
Polysilicon. *See also* Non-poly-Si modifications
　analog devices accelerometer, **16**–160 to 162
　doped, **16**–153 to 154
　electrical properties, **15**–7
　epitaxially deposited, **24**–14
　grains, **15**–4 to 6
　growth, **15**–5 to 6
　hinged, **16**–134 to 137, **26**–59
　material properties, **15**–3 to 9, **16**–148 to 149
　materials case studies, **16**–150 to 154
　MEMS popularity, **3**–19, **3**–24
　molded structures, **16**–137 to 140
　porous, **16**–133 to 134, **16**–144 to 146
　resistivity, **15**–6
　stepping slider, **16**–163 to 165
　structural material deposition, **16**–120 to 121
　surface micromachining examples, **16**–160 to 165
　thermal conductivity, **15**–7
　thick, **16**–137
　undoped, **16**–151 to 153
Porous membrane application, **17**–48
Porous silicon, **15**–8 to 9, **16**–67 to 71
Positive-displacement pumps, **26**–63, **29**–4, **29**–6 to 8
Powering microrobotics, **28**–34
Prandtl number
　boundary conditions, **4**–13
　compressibility, **4**–10
　flow control, **33**–2
　flow structure, **32**–9
　hot wire sensors, **26**–19
　thermal effects, **10**–30
Precision microsprings, **17**–52
Pressure, capillary, **31**–3 to 6
Pressure drag, **33**–3 to 4
Pressure driven flow, **6**–5 to 6, **11**–4 to 8
Pressure ratio, vacuum pumps, **29**–5
Pressure sensors
　absolute, **25**–11

capacitive, **25**–7 to 14
circuit integration, **25**–13 to 14
design variations, **25**–11 to 13
device compensation, **25**–13 to 14
device structure, **25**–2 to 3
fundamentals, **25**–1, **25**–19, **26**–42 to 43
measurements, **35**–5
MEMS, **26**–50 to 52
noise, **25**–12
optical pick-off, **25**–17 to 18
outlook for, **26**–52 to 53
performance measures, **25**–2 to 3
piezoelectric, **26**–47 to 50
piezoresistant, **25**–4 to 7
pirani, **25**–19
principles, **26**–43 to 47
resonant beam pick-off approach, **25**–14 to 15
servo-controlled, **25**–15 to 16
SiC DRIE applications, **21**–9 to 11
tunneling pick-off, **25**–16 to 17
turbulent flow requirements, **26**–47
Principle of natural selection, **33**–37
Principle of the argument, **12**–14
Probe cards, LIGA technique, **17**–60
Proof mass, **24**–5 to 7
Proportional-integral (PI) control, **13**–6
Proximity effect, **36**–6
PSPICE, **5**–7
Puddle formation, **30**–13 to 14
Pull-push connectors, **17**–52
Pulse plating, **17**–29
Pumping speed, **29**–5
Pumps, micro, **17**–23 to 24, **26**–63 to 65. *See also* Vacuum pumps, microscale
Pump scaling, **29**–5 to 16

Q

Q-factor, **9**–1, **35**–8
Quadrature error, **24**–23

R

Rapid prototyping process, **15**–12
Rarefaction effects, **5**–4, **24**–10
Rathnasingham and Breuer drag reduction, **34**–12 to 13
Rayleigh numbers, **10**–29, **33**–32
Rayleigh-Taylor instability, **10**–17, **10**–33, **10**–37
Reaction injection molding (RIM), **17**–35 to 38
Reactive flow control
 feedback control, **33**–23 to 24
 fundamentals, **33**–19 to 20
 future trends, **34**–1 to 13
 MEMS sensors and actuators, **26**–2
 required characteristics, **33**–24 to 26
 targeted, **33**–20 to 23
Reactive ion etching, **15**–3
Real estate consumption, wet bulk micromachining, **16**–82 to 84
Reduced-order models, **5**–3, **9**–9 to 10

Redundancy criteria, **28**–30 to 32
Reflections, **29**–4
Refractory materials, **17**–60
Relaminarization, **33**–8, **34**–13
Reliability of mechanisms, **27**–32 to 33
Relief printing, **17**–38 to 39
Replication techniques. *See* LIGA technique
Reservoir effect, **10**–41
Resistor-diaphragm modeling, **20**–9 to 13
Resists
 adhesion, **17**–19
 bulk vs. surface micromachining, **16**–148 to 149
 fundamentals, **16**–146
 layers, **17**–21
 LIGA technique, **17**–16 to 19
 NEMS, **36**–7 to 9
 photosensitive, **17**–60
 polyimide surface structures, **16**–146 to 147
 UV depth lithography, **16**–147 to 148
 UV sensitive, **17**–60
Resolution limits, NEMS, **36**–3 to 9
Resonant beam pick-off approach, **25**–14 to 15
Resonant frequency shifts, **24**–12
Resonant mesh infrared band pass filter, **17**–42
Resonant structure tests, **3**–12 to 14
Resonating thin-film silicon legs, **28**–27
Resonators, lateral, **16**–113
Reversed reservoir effect, **10**–41
Reynolds number
 boundary conditions, **4**–14 to 15
 coherent structures, **33**–12, **33**–14
 continuum model, **4**–8
 electro-osmosis, **6**–22, **6**–26
 flow control, regimes, **33**–7 to 8
 flowfield modeling, **4**–4 to 5
 flow structure, **32**–8 to 10
 friction factor, **6**–7 to 8, **6**–13
 gas flows, **8**–2
 liquid pressure drop, **31**–4
 low-Reynolds-number wings, **35**–30 to 33
 lubrication, **9**–4
 MEMS flows, **5**–3 to 4
 microbearings, **26**–67
 micropumps, **26**–63
 neural networks, **34**–7
 scales, **26**–12, **33**–15, **33**–17
 sensor requirements, **33**–18
 skin-friction reduction, **34**–13
 suction, **33**–2
 turbulent flow, **6**–8 to 9, **26**–8 to 9, **26**–47
 vapor pressure drop, **31**–5
 velocities, **26**–21 to 24
 wavenumber pairs, **13**–13
Riblet, **34**–1 to 2, **34**–4
Riccati equations, **12**–20, **13**–20 to 21
RIM. *See* Reaction injection molding (RIM)
Ring crossbar structures, **16**–111 to 112
Rise time, **12**–10
Robot arm, PID control, **12**–7 to 8
Robotics. *See* Microrobotics

Robustification, **13**–34 to 36
Robustness
  control theory, **12**–21 to 24
  free-shear flows, **35**–5, **35**–8
  future trends, **35**–35
  genetic algorithms, **14**–13
Roll control, **35**–9 to 18
Root locus design and analysis, **12**–8 to 11, **13**–13
Rotary pumps, continuous parallel, **26**–64
Rotating devices lubrication, **9**–10
Rotating discs, thin liquid films, **10**–27 to 29, **10**–42 to 44
Rotational gyroscopes, **24**–22 to 24. *See also* Inertial sensors
Roughness, surface, **16**–43
Round capillaries, liquid flows, **6**–7 to 8
Round robin tests, **3**–16
Routh array, **12**–6

S

Sacrificial layer, **16**–119 to 120, **16**–123, **22**–6
Sample handling, **30**–20 to 22
SAMs. *See* Self-assembled monolayers (SAMs)
Satellite droplets, **30**–11 to 13
Scales, turbulent flows, **26**–9 to 12, **33**–15 to 17
Scaling
  accommodation pumping, **29**–24
  bubbles, pressure drop, **11**–3
  continuum hypothesis, **9**–3 to 4
  cube-square law, **9**–2 to 3
  fundamentals, **2**–1 to 2
  log plot, **2**–2 to 4
  mechanical systems, **2**–4 to 9
  microrobotics, **28**–11 to 12
  surface roughness, **9**–4
  vacuum pumps, **29**–15 to 16, **29**–24
Scanning tunneling microscopes (STMs), **36**–12 to 13
Schmidt sensor, **26**–40
Schottky emitter, **36**–4 to 5
Schrödinger equation, **4**–32
SCREAM. *See* Single-crystal reactive etching and metallization (SCREAM)
Screech, supersonic jets, **35**–19 to 30
SEAJet, **30**–6
Sealing processes, **16**–129
Seawater applications, Lorentz forces, **33**–29
Seidel model, **16**–54 to 58
Seismic mass and elastic spring, **24**–5 to 7
Self-assembled monolayers (SAMs), **36**–9 to 12
Self-assembling microjoint technique, **26**–61
Semiconductors, etching
  electropolishing, **16**–66 to 67
  fundamentals, **16**–64 to 65
  microporous silicon, **16**–67 to 71
Sensing at wall, **34**–4 to 10
Sensing method, transduction, **24**–12 to 13
Sensor arrays, electrochemical, **16**–91 to 95
Sensor fabrication technologies, **22**–6 to 8
Sensors. *See also* Inertial sensors
  absolute, **25**–11
  acceleration, **17**–55
  channel, **22**–1 to 23
  characteristics, **35**–4 to 5
  chemical, **22**–1 to 23
  deep reactive ion etching (DRIE), **21**–9 to 11
  delta wing, roll control, **35**–10 to 12
  FET-based, **24**–13
  floating-element, **26**–38 to 41
  flow control, **33**–17 to 19
  flow model, **5**–17 to 18
  free-shear flows, **35**–3 to 6, **35**–8
  hot-wire, **26**–17 to 19
  hydrogen, **22**–8 to 11
  LIGA applications, **17**–59
  low-Reynolds-number wings, **35**–31 to 32
  optical, **17**–59
  packaging, **16**–16, **24**–17
  piezoresistive pressure, **26**–47 to 50
  pressure sensors, **25**–11, **26**–42 to 53
  reactive control, **33**–40 to 41
  Schmidt, **26**–40
  separation control, **35**–31 to 32
  shear-stress, **26**–42
  single-crystal silicon carbide characteristics, **20**–26 to 29
  spacing, **26**–2
  supersonic jet screech, **35**–19 to 20
  thermal-type, **26**–15 to 17
  turbulent flow, and actuators, **26**–1 to 69
  types, **35**–3 to 4
  UCLA/Caltech group, **35**–18, **35**–31
  velocity, **26**–15 to 27
  wall-shear-stress type, **26**–27 to 42
  yaw-rate, **24**–22
Separation control, **35**–30 to 33
Serial fabrication/assembly, **28**–2, **28**–9
Servo-controlled pressure sensors, **25**–15 to 16
Settling time, **12**–10
SFF. *See* Solid freeform fabrication (SFF)
Sharp anode, deposition, **19**–5
Shear layer components and issues, **35**–3 to 9
Shear plane, **6**–19
Shear-stress sensors, **26**–42
Shock structure, hypersonic, **7**–15
SiC. *See* Silicon carbide (SiC)
Signal-to-noise ratio, **35**–5
Silicon
  amorphous type, **16**–154 to 155
  backbonds, **16**–56
  crystal structure, **16**–7 to 8
  CVD silicon dioxides, **16**–158 to 160
  CVD undoped SiO_2, **16**–159
  geometric relationships, **16**–9 to 16
  hydrogenated amorphous type, **16**–154 to 155
  insulator surface micromachining, **16**–140 to 146
  isotropic vs. anisotropic etching, **16**–62 to 64
  macroporous, **16**–69 to 71
  [100]-orientation, **16**–9 to 13, **16**–14 to 16
  [110]-orientation, **16**–13 to 16
  piezoresistivity, **16**–24 to 28
  porous, **16**–67 to 71
  structural element, **16**–17 to 28

substrate material, **16**–16
tribology of, **9**–25
Silicon carbide (SiC)
  deep reactive ion etching, **21**–5 to 12
  fluorine plasmas, **21**–3 to 5
  fundamentals, **21**–1 to 2
  material properties, **15**–17 to 20
Silicon crystallography
  crystal structure, **16**–7 to 8
  geometric relationships, **16**–9 to 16
  Miller indices, **16**–6 to 7
  [100]-oriented silicon, **16**–9 to 13, **16**–14 to 16
  [110]-oriented silicon, **16**–13 to 16
Silicon dioxide materials, **15**–9 to 11
Silicon fusion-bonded wafers, **16**–83 to 84, **16**–140 to 146
Silicon nitride, **16**–155 to 158
Silicon nitride materials, **15**–11 to 14
Silicon-on-insulator (SOI) device technologies, **24**–14
Silicon-on-insulator (SOI) wafers
  epitaxy surfaces, **16**–144
  fundamentals, **16**–140 to 142
  micromachining, **16**–142 to 146
  SIMOX surface, **16**–143 to 144
  vs. poly-Si surfaces, **16**–144 to 146
SIMOX, **16**–140 to 146
Simulation Program with Integrate Circuit Emphasis (SPICE). *See* SPICE/SPICE3
Simulators and simulations
  circuit, fluid and solid differences, **5**–15
  circuit type, **5**–8 to 9
  decomposition, **13**–37
  equation solution, **5**–8 to 9
  flow-structure interactions, **5**–12
  fluid type, **5**–9 to 11
  full-system simulation, **5**–2 to 3
  grid velocity algorithm, **5**–12 to 13
  linearization, **5**–8
  molecular dynamics method, **4**–18, **8**–22 to 23
  Monte Carlo, **8**–7 to 17
  Nektar, **5**–9 to 11, **5**–21
  SPICE 3, **5**–8 to 9, **5**–21
  structural simulator, **5**–13 to 15
  time discretization, **5**–8
Single-crystal reactive etching and metallization (SCREAM), **24**–14
Single-crystal silicon carbide
  fundamentals, **20**–1 to 3, **20**–29 to 30
  6H-SiC gauge factor, **20**–8 to 17
  metallization, **20**–17 to 26
  photoelectrochemcial fabrication principles, **20**–3 to 8
  resistor-diaphragm modeling, **20**–9 to 13
  sensor characteristics, **20**–26 to 29
  temperature effects, **20**–14 to 26
  Ti/TaSi$_2$/Pt metallization, **20**–22 to 26
  Ti/TiN/Pt metallization, **20**–19 to 22
Single-crystal silicon materials, **15**–2 to 3
Single-phase convective heat transfer
  entrance length, **32**–9 to 10
  flow structure, **32**–8 to 9
  fundamentals, **32**–8

gas flow, **32**–10 to 12
governing equations, **32**–10
liquid flow, **32**–12 to 17
vs. two phase convection, **32**–18
SISO feedback linearization, **12**–24 to 28
6H-SiC gauge factor, **15**–17, **20**–8 to 17
Size effects, **32**–27
Skin-friction reduction
  active wall motion studies, **34**–10 to 12
  fundamentals, **34**–1 to 2, **34**–12 to 13
  history, **33**–2
  near-wall streamwise vortices, **34**–2, **34**–11 to 12
  neural network, **34**–4 to 7
  opposition control, **34**–2 to 4
  passive devices, **34**–1
  riblets, **34**–1 to 2, **34**–4
  sensing at wall, **34**–4 to 10
  streak control, **34**–10 to 11
  suboptimal control, **34**–7 to 9
  suction, **33**–10 to 11
  targeted flow control, **33**–20
  vorticity flux control, **34**–9 to 10
Slanted microstructures, **17**–41
Slider, stepping, **16**–163 to 165
Sliding-mode fuzzy logic controllers (SMFLCs), **14**–31
Slip-boundary conditions, **7**–12, **7**–18, **7**–27, **8**–3
Slip-flow regime, **4**–16, **4**–34, **4**–126, **8**–2, **32**–4
SMFLCs. *See* Sliding-mode fuzzy logic controllers (SMFLCs)
S-N curve, **3**–3
Soft computing
  artificial neural networks, **14**–2 to 12
  flow control, **33**–35 to 39
  fundamentals, **14**–1 to 2
  fuzzy logic and fuzzy control, **14**–18 to 32
  genetic algorithms, **14**–13 to 18
  targeted flow control, **33**–23
  turbulence, **33**–9
Software, **5**–16 to 17. *See also* Matlab software
Sol-gel technology, **17**–41 to 42
Solid freeform fabrication (SFF), **19**–1 to 4, **19**–19 to 20. *See also* Electrochemical fabrication (EFAB)
Solid freeforming, **30**–24
SOLIDIS, **5**–3
Solid simulators comparison, **5**–15
Solids vs. liquids and gases, **4**–24
Solvent removal, **17**–21
Sonic limitation, **31**–6
SOR. *See* Synchrotron orbital radiation (SOR)
Spacer layer, **16**–119 to 123
Spatial resolution, **26**–33 to 34
Spectral emission, synchrotron electrons, **17**–7
Spectrometers, **17**–53 to 54
Specular reflection, **29**–4
SPICE/SPICE3, **5**–2 to 3, **5**–8 to 9, **5**–15 to 22
Spin-coating applications, **10**–28, **10**–42 to 43
Spinneret nozzles, **17**–44 to 45
Spinodal dewetting, **10**–20
Spin-on-glass, **15**–11
Spiral groove bearings, **9**–14

Spirals, cantilever, **16**–113 to 115
Spreaders, heat
    arrays, **31**–12 to 15
    boiling limitations, **31**–7
    capillary limitation, **31**–2 to 6
    designs, new, **31**–18 to 21
    entrainment limitation, **31**–6 to 7
    fabrication, **31**–18
    fundamentals, **31**–1 to 8, **31**–15 to 16, **31**–22
    hydrostatic pressure drops, **31**–3
    liquid pressure drop, **31**–4
    modeling, **31**–16 to 17
    sonic limitation, **31**–6
    testing, **31**–18
    thermal resistance, **31**–7 to 8
    vapor pressure drop, **31**–5 to 6
    viscous limitation, **31**–6
Spring combinations, **27**–15 to 17
Spring constant, **27**–7
Sputter ion pumps, **29**–11 to 12, **29**–16
Squeeze-film damping, **9**–7 to 10
SSL. *See* Stationary separation line (SSL)
Stability
    analysis of Burnett equations, **7**–12 to 14
    global, **13**–37
    inertial sensor material, **24**–16
Stand-off bumps, **16**–124
Stationary capillary flow, **32**–26
Stationary separation line (SSL), **35**–2
Steady state, **14**–6 to 7, **31**–9 to 11
Steering microrobots, **28**–32
Steger-Warming flux-vector splitting method, **7**–15
Stepped microstructures, **17**–40 to 41
Stepping slider, **16**–163 to 165
Stern layer, **6**–19
Stiction
    fabrication steps, **16**–123 to 125
    manufacturing issues, **24**–15 to 16
    surface-micromachined mechanisms, **27**–6 to 7
    surface phenomena, **4**–30 to 31
    tribology and wear, **9**–25
STMs. *See* Scanning tunneling microscopes (STMs)
Stokes theories and laws
    continuum hydrodynamics, **6**–5
    continuum model, **4**–7 to 8
    polar mechanics, **32**–8
Strain-induced growth, **15**–5
Strain measurements, **3**–6
Streaks
    control, **34**–10 to 11
    low-speed, **33**–24, **34**–2
    spacing, **33**–14
    suction effects, **34**–2
    targeted flow control, **33**–21 to 22
    wall-speed, **33**–24
Strength of thin films, **16**–115 to 117
Stress
    biaxial measurements, **16**–110 to 111
    cantilever beams and spirals, **16**–113 to 115
    clamped-clamped beams, **16**–111

    concentration, **27**–10 to 11
    disk method measurement, **16**–105 to 106
    gradients, **27**–4 to 6
    induced cracks, PMMA, **17**–19 to 21
    lateral resonators, **16**–113
    measuring techniques, **16**–104 to 115
    mechanical structural elements, **16**–17 to 21
    nonuniformity measurements, **16**–113 to 115
    Poisson ratio, **16**–111
    residual, **16**–21 to 22
    ring crossbar structures, **16**–111 to 112
    strain, **27**–3
    thin films, **16**–100 to 115
    uniaxial measurements, **16**–106 to 110
    vernier gauges, **16**–113
Strouhal number, **35**–7
Structural materials
    corrugated, **16**–127
    deposition, **16**–120 to 121
    silicon, **16**–17 to 28
Structural simulator, **5**–13 to 15
Structures
    anisotropically etched, **16**–46 to 51
    coherent, **33**–11 to 15
    milli-scale molded polysilicon, **16**–137 to 140
    polyimide surface, **16**–146 to 147
Subminiature microphones, **26**–50
Suboptimal control
    skin-friction reduction, **34**–7 to 9, **34**–13
    theory, **33**–24
Subsonic flow in microchannels, **7**–24 to 27
Substrates
    EFAB process, **19**–12
    heterogeneous, **10**–25 to 27
    homogeneous, **10**–20 to 25
    LIGA technique, **17**–16
    packaging, **23**–2 to 3
    reactive radicals, **21**–3
    silicon, **16**–16
    temperature coefficients, **25**–10
    thick, **10**–35 to 36, **16**–101 to 104
    X-ray mask alignment, **17**–13 to 14
Suction, **33**–2, **33**–10 to 11, **33**–20 to 21, **33**–40
Supercritical CO_2 drying processes, **24**–15 to 16
Supersonic flow in microchannels, **7**–27 to 33
Supersonic jet screech, **35**–19 to 30
Surface interactions, **8**–23
Surface micromachining. *See also* Bulk micromachining;
    Wet bulk micromachining
    adhesion, **16**–98 to 100
    applications, **27**–18 to 27
    beams, **27**–7, **27**–8 to 10
    biaxial measurements, **16**–110 to 111
    buckling, **27**–17 to 18
    cantilever beam springs, **27**–11 to 12
    clamped-clamped beams, **16**–111
    columns, **27**–7 to 8
    compliance elements, **27**–7
    contact stresses, **27**–4
    countermeshing gear discriminator, **27**–21 to 23
    dimensional uncertainties, **16**–127 to 129

disk method measurement, **16**–105 to 106
electrical shorts, **27**–30 to 31
engines, **27**–18 to 21
failure mechanisms, **27**–27 to 33
film stress, **16**–125 to 127, **27**–4 to 6
fixed beam springs, **27**–12 to 13
fixtures, **27**–7
flexures, **27**–13 to 15
fundamentals, **16**–95 to 97, **26**–5 to 6, **27**–1 to 2
geometric considerations, **27**–2 to 7
history, **16**–97 to 98
IC compatibility, **16**–129 to 132
lateral resonators, **16**–113
lithographic variations, **27**–31 to 32
machine design, **27**–7 to 18
material case studies, **16**–149 to 160
material properties, **27**–2 to 7
mechanical interference, **27**–27 to 30
microengines, **27**–18 to 21
micro-flex mirror, **27**–23 to 27, **27**–32
non-poly-Si modifications, **16**–140 to 146
nonuniformity measurements, **16**–113 to 115
out-of-plane structures, **27**–27 to 30
Poisson ratio, **16**–111, **27**–3
polysilicon, **16**–133 to 134, **16**–160 to 165
processes, **16**–117 to 133
reliability of mechanisms, **27**–32 to 33
resists, **16**–146 to 149
ring crossbar structures, **16**–111 to 112
sealing processes, **16**–129
silicon fusion-bonded wafers, **16**–140 to 146
spring combinations, **27**–15 to 17
stiction, **27**–6 to 7
stress, **16**–100 to 115, **27**–3 to 6, **27**–10 to 11
studies, **16**–149 to 160
suspended membrane measurement methods, **16**–110 to 111
thin film mechanical properties, **16**–98 to 117
uniaxial measurements, **16**–106 to 110
vernier gauges, **16**–113
vertical play, **27**–27 to 30
vs. bulk micromachining, **16**–148 to 149
wear, **27**–6
Young's modulus, **27**–3
Surface parameters, **33**–4
Surface phenomena, **4**–30 to 33
Surface tension and gravity, **2**–8 to 9, **10**–15 to 20, **10**–29 to 35
Suspended membrane measurement methods, **16**–110 to 111
Sutherland's law, **7**–16
Swift, Jonathan, **1**–1, **1**–4, **4**–1
Synchrotron orbital radiation (SOR), **17**–4 to 8
System integration, free-shear flows, **35**–18, **35**–27 to 30
System partitioning, **24**–21

T

Takagi-Sugeno fuzzy logic controllers (TSFLCs), **14**–31
Taylor-Couette systems, **33**–32

Taylor series, **4**–13, **4**–15
Temperature effects, **20**–14 to 26
Temperature-jump-boundary condition, **4**–15
Temporal resolution, **26**–34 to 35
Tensile films, **16**–101
Tensile strength, silicon, **16**–22 to 24
Tensile tests, **3**–6, **3**–7 to 9, **18**–19
Tests methods, MEMS materials, **3**–3 to 16
Texas Instruments, Inc., **33**–39
TFE. *See* Thermal field emission (TFE)
Theory of dilute gases, **4**–19
Thermal anemometry, **26**–15
Thermal-bubble actuation, **30**–4
Thermal-buckling actuation, **30**–5
Thermal creep, **4**–14, **7**–12, **8**–3, **29**–25
Thermal field emission (TFE), **36**–5
Thermal/hydraulic crosstalk and overfill, **30**–8 to 11
Thermal neurocontrol, **14**–7 to 12
Thermal properties, silicon, **16**–28
Thermal resistance, **31**–7 to 8
Thermal-sensors, **26**–15 to 17, **26**–29 to 31, **26**–35 to 38
Thermal stress, **16**–102 to 103
Thermal transpiration pumping, **29**–17 to 23
Thermocapillarity, **10**–29 to 35, **10**–40 to 42
Thermodymamic concepts, **32**–4 to 5
Thermoplastic injection molding, **17**–38
Thick-films, **23**–5 to 6, **23**–6 to 8
Thick polysilicon, **16**–137
Thick substrate, thin liquid films, **10**–35 to 36, **16**–101 to 104
Thin films
   adhesion, **16**–98 to 100
   capillarity, **10**–40 to 42
   condensation, **10**–36 to 44
   etch stop, **16**–81
   evaporation, **10**–36 to 44
   evolution equation for solid surfaces, **10**–9 to 15
   fundamentals, **10**–1 to 9, **10**–44, **16**–98
   heterogeneous substrates, **10**–25 to 27
   homogeneous substrates, **10**–20 to 25
   interfacial conditions, **10**–36 to 39
   isothermal films, **10**–15 to 29
   Poisson ratio, **16**–11
   rotating discs, **10**–27 to 29, **10**–42 to 44
   strength, **16**–115 to 117
   stress, **16**–101, **16**–110 to 115
   surface tension and gravity, **10**–15 to 20, **10**–29 to 35
   thermal effects, **10**–29 to 36
   thermocapillarity, **10**–29 to 35, **10**–40 to 42
   thick substrate, **10**–35 to 36
   vapor recoil, **10**–40 to 42
   van der Waals forces, **10**–18 to 20
Thompson and Trojan MD simulations, **4**–26 to 30
Three-dimensional self-assembly, **36**–18
Three-dimensional structures, **26**–53 to 54
Through-mask plating, **19**–4
Thrust bearings, **9**–12 to 14
Time discretization, **5**–8
Time-stepping algorithms, **5**–19 to 20
Ti/TaSi$_2$/Pt metallization, **20**–22 to 26

Ti/TiN/Pt metallization, **20**–19 to 22
Tollmien-Schlichting modes, **33**–8
Tools, microrobotics, **28**–15 to 18
Top-down patterning, **36**–2
Training, artificial neural networks, **14**–4 to 5, **14**–6
Transducer integration technologies, **24**–18 to 20
Transduction, mechanical to electrical, **24**–12 to 13
Transient energy growth, **13**–11, **13**–13
Transistors, **1**–1, **33**–25
Transitional flow regime, **4**–16, **4**–23, **4**–34, **8**–11, **9**–3
Transverse-axis rotary pumps, continuous, **26**–64
Transverse flow microfilter, **17**–48
Tribology, **9**–25, **18**–22
Trimmer, W., **1**–4
Trimmer brackets, **2**–7
TSFLCs. *See* Takagi-Sugeno fuzzy logic controllers (TSFLCs)
Tungsten filament, **15**–16, **36**–4
Tunneling pick-off pressure sensors, **25**–16 to 17
Turbine rotors, **17**–45
Turbomachines, **26**–63 to 68
Turbomolecular pumps, **29**–9, **29**–25
Turbulence
 drag reduction, **33**–41 to 42
 flow control, **33**–9 to 10
 linear feedback control, **13**–22
 sensors, measurements, **26**–14 to 53
 soft computing, **33**–9
Turbulent flows. *See also* Flows; Gas flows; Liquid flows
 actuators, **26**–53 to 63
 analyzing, **26**–9
 bearings, **26**–66 to 68
 bilayer effect, **26**–54 to 55
 bulk micromachining, **26**–4 to 5
 calibration, **26**–17, **26**–31 to 33
 control, **33**–9 to 10
 electrostatic forces, **26**–56 to 57
 floating-element sensors, **26**–38 to 41
 fluctuating velocities, **26**–19 to 22
 fundamentals, **26**–1 to 3, **26**–8 to 9, **26**–14 to 15, **26**–68 to 69
 hot-wire sensors, **26**–17 to 19
 magnetic forces, **26**–56 to 57
 mean velocities, **26**–19 to 22
 measurements and control, **26**–14 to 53
 mechanical folding, **26**–57 to 61
 MEMS fabrication, **26**–3 to 8
 microactuators, **26**–53 to 63
 microbearings, **26**–66 to 68
 micromolding, **26**–7 to 8
 micropumps, **26**–63 to 65
 microturbomachines, **26**–63 to 68
 molding, **26**–7 to 8
 one-layer structures, **26**–61 to 63
 piezoresistive pressure sensors, **26**–47 to 50
 pressure sensors, **26**–42 to 53
 pumps, **26**–63 to 65
 requirements, **26**–47
 scales, **26**–9 to 12
 sensors, **26**–12 to 53
 shear-stress sensors, **26**–42

 spatial resolution, **26**–33 to 34
 surface micromachining, **26**–5 to 6
 temporal resolution, **26**–34 to 35
 thermal-sensors, **26**–15 to 17, **26**–29 to 31, **26**–35 to 38
 three-dimensional structures, **26**–53 to 54
 turbomachines, **26**–63 to 68
 velocity sensors, **26**–15 to 27
 wall-shear-stress sensors, **26**–27 to 42
Two-dimensional hypersonic blunt body flow, **7**–15 to 18
Two-layer feedforward network, **34**–5
Two-phase convection
 boiling curves, **32**–18 to 21
 bubble dynamics, **32**–24 to 25
 critical heat flux, **32**–21 to 22
 flow patterns, **32**–22 to 23
 fundamentals, **32**–2, **32**–18
 modeling, **32**–25 to 27
 vs. single phase convection, **32**–18

U

UCLA/Caltech sensor, **35**–18, **35**–31
Undercutting, **16**–85 to 90
Underetching, **16**–84
Undoped polysilicon, **16**–151 to 153
Uniaxial measurements, **16**–106 to 110
Unification, **13**–36 to 37
Uranium enrichment, **17**–44
UV depth lithography, **16**–147 to 148
UV sensitive resists, **17**–60

V

Vacuum aging, **20**–17
Vacuum pumps, microscale. *See also* Pumps, micro
 accommodation pumping, **29**–23 to 25
 alternative technologies, **29**–16 to 25
 capture pumps, **29**–11 to 13
 fundamentals, **29**–1 to 5, **29**–25
 kinetic pumps, **29**–8 to 11
 N requirements, **29**–14
 operating pressures, **29**–14
 orbitron iron pump, **29**–12 to 13
 positive-displacement pumps, **29**–6 to 8
 pressure ratio, **29**–5
 pumping speed, **29**–5
 pump scaling, **29**–5 to 16
 scaling results, **29**–15 to 16
 sputter ion pumps, **29**–11 to 12
 thermal transpiration pumping, **29**–17 to 23
 types, **29**–4 to 5
Valves, micro, **17**–45 to 46
Vapor pressure drop, **31**–5 to 6
Vapor recoil, **10**–37 to 38, **10**–40 to 42
Velocity
 centerline, **8**–13
 droplet generation, **30**–19
 freestream, **26**–29
 point-wise measurements, **6**–16
 profiles, **33**–6

sensors, **26**–15 to 27
Velocity measurements, point-wise, **6**–16
Vent holes, **9**–9
Vernier gauges, **16**–113, **18**–27
Vertical axis hinges, **27**–25 to 27
Vertical play, **27**–27 to 30
V-groove technique, **16**–72, **26**–61
Vibration fields, **28**–27
Vibratory ring gyroscopes, **24**–24
Viscosity, **6**–4, **6**–14
Viscous limitation, **31**–6
Volume, microdroplet generators, **30**–19
Von-Mises stress, **23**–11 to 16
Vortices
  magnetohydrodynamic control, **33**–28 to 29
  near-wall streamwise, **34**–2, **34**–11 to 12
  vorticity flux control, **34**–9 to 10

W

van der Waals forces, **4**–32 to 33, **10**–18 to 20
Wafers, **16**–83 to 84, **24**–17, **28**–2
Walking microrobots, **28**–23 to 32
Wall-attachment amplifiers, **17**–47
Wall-bounded flows, **33**–7, **33**–12
Walls, sensing at, **34**–4 to 10
Wall-shear-stress sensors, **26**–27 to 42
Wall unit, **33**–18 to 19
Waveguides, **17**–50, **17**–59
Wavenumber pairs, **13**–4, **13**–11 to 14, **13**–19
Wear, surface-micromachined mechanisms, **27**–6
Weber number, **31**–6 to 7, **32**–21
Well-posedness, **13**–35
Wet bulk micromachining. *See also* Bulk micromachining; Surface micromachining
  corner compensation, **16**–84 to 90
  dissolved wafer process, **16**–90 to 91
  dual lumen catheter, **16**–91 to 95
  electrochemical sensor array, **16**–91 to 95
  examples, **16**–90 to 95
  front etching, **16**–82 to 83
  fundamentals, **16**–2 to 3, **16**–82
  real estate consumption, **16**–82 to 84
  silicon fusion-bonded wafers, **16**–83 to 84
  undercutting, **16**–85 to 90
  underetching, **16**–84
Wet isotropic etching
  Arrhenius plot, **16**–32
  dopant dependence, **16**–33 to 35
  electrochemical etch-etch stop, **16**–35 to 36
  etchants usage, **16**–29 to 30
  fundamentals, **16**–29
  iso-etch curves, **16**–30 to 32
  masking, **16**–32 to 33
  preferential, **16**–36
  problems, **16**–36
  reaction scheme, **16**–30
Wetting liquid, **11**–2 to 3
Wheatstone-bridge
  6H-Sic, **20**–3
  integrated simulation approach, **5**–21
  resistor-diaphragm modeling, **20**–13
  sensor characteristics, **20**–26
  silicon piezoresistivity, **16**–25
  thermal sensor principle, **26**–16, **26**–29
Wirebond, **23**–6 to 8

X

X-ray based fabrication
  angled exposure, **18**–24 to 25
  applications, **18**–30 to 40
  direct LIGA approach, **18**–9 to 10
  DXRL processing, **18**–4 to 12, **18**–25 to 28
  fundamentals, **18**–1 to 4, **18**–40 to 41
  LIGA technique, **17**–9 to 15, **18**–9 to 10
  masks, **17**–9 to 15, **18**–4 to 8
  material properties, **18**–19 to 23
  microactuators, **18**–34 to 39
  mold filling, **18**–12 to 19
  photoresist applications, **18**–8 to 9
  planarization, **18**–23
  PMMA mechanical properties, **18**–11 to 12
  precision components, **18**–30 to 34
  reentrant geometry, **18**–24 to 25
  sacrificial layers, **18**–28 to 30
X-ray imaging techniques, **6**–16
X-ray lithography, **36**–2, **36**–9 to 11. *See also* LIGA technique

Y

Yaw-rate sensors, **24**–22
Yielding point, **16**–22 to 24
Young's modulus
  cantilever beam springs, **27**–12
  columns, **27**–8
  comb drives, **27**–19
  contact stresses, **27**–4
  coupled domain problems, **5**–5
  diaphragm deflection, **20**–9, **25**–2
  film stress, **16**–102, **16**–125
  indentation tests, **3**–14
  lateral resonators, **16**–113
  LIGA accelerometer, **17**–55
  material properties, **3**–3 to 4, **23**–19, **27**–3
  membrane tests, **3**–14
  piezoresistive pressure sensors, **26**–49
  resistor-diaphragm modeling, **20**–12
  resonant structure tests, **3**–12 to 13
  silicon material, **16**–18 to 22
  structural simulator, **5**–14
  suspended membrane methods, **16**–110
  X-ray membrane, **17**–10 to 11

Z

Zeeman Slower, **36**–15
Zeolites, **22**–18
Zero dynamics, **12**–28

Zeta potential
  bubble transport, **11**–9 to 11
  electrical double layers, **6**–19
  electro-osmosis, **6**–23
  pressure-driven flow components, **6**–32